SHOWCASE EXAMPLES

These special examples stretch across the entire page and have a tan background that makes them ___. *Examples* introduce key topics and provide "how-to" instruction by walking step-by-step through the problem-solving process. With this format, the left and middle "columns" are like the instructor's voice offering an explanation (left column) and then summarizing information (middle column) during a classroom lecture.

EXAMPLE 1 **How to Solve a Linear Equation Using the Addition and Multiplication Properties of Equality**

Solve the equation $2z - 3 = 9$.

Step-by-Step Solution

Step 1: Isolate the term containing the variable.

$$2z - 3 = 9$$

Apply the Addition Property of Equality and add 3 to each side of the equation:
$$2z - 3 + 3 = 9 + 3$$
$$2z = 12$$

Step 2: Get the coefficient of the variable to be 1.

Apply the Multiplication Property of Equality and divide both sides by 2 $\left(\text{this is the same as multiplying both sides by } \dfrac{1}{2}\right)$:

$$\frac{2z}{2} = \frac{12}{2}$$
$$z = 6$$

Step 3: Check Confirm that $z = 6$ is the solution of the equation.

Substitute $z = 6$ into the original equation:

$$2z - 3 = 9$$
$$2(6) - 3 \stackrel{?}{=} 9$$
$$12 - 3 \stackrel{?}{=} 9$$
$$9 = 9 \quad \text{True}$$

Because $z = 6$ satisfies the equation, the solution of the equation is 6, or the solution set is $\{6\}$.

As you can see, *Showcase Examples* provide three "columns" of information where the left column describes a step...

...the middle column provides a brief annotation, as needed, to explain the step...

...and the right column presents the algebra.

Prep for Exams with the CHAPTER TEST PREP VIDEO CD

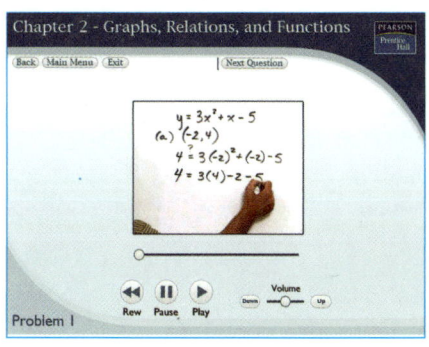

This video CD found at the back of your text contains worked-out solutions to every exercise in each Chapter Test in the text. To make the most of valuable study time when preparing for exams, follow these three steps:

1. Take the Chapter Test found at the end of the textbook chapter.

2. Check your answers in the back of the text.

3. Use the Chapter Test Prep Video CD to review every step of the worked-out solution to the specific questions you need to review.

Additional Resources to Help You Succeed

Student Study Pack

A single, easy-to-use package, available bundled with your textbook or by itself, for purchase through your bookstore. This package contains the following resources to help you succeed:

Student Solutions Manual

- Solutions to the odd-numbered section exercises
- Solutions to the Quick Check exercises
- Solutions to the Preparing for This Section, Putting the Concepts Together (mid-chapter review), Chapter Review, Chapter Test, and Cumulative Review exercises

Prentice Hall Math Tutor Center

- Staffed by qualified math instructors who provide students with tutoring on examples and odd-numbered exercises from the textbook. Tutoring is available via toll-free telephone, toll-free fax, email, or the Internet.

CD Lecture Series

- Perfect for review of a section or a specific topic, these mini-lectures cover the key concepts from each section of the text in approximately 10-15 minutes.
- Includes fully worked-out solutions to exercises marked with a CD video icon.

Online Homework and Tutorial Resources

MyMathLab

MyMathLab is a series of text specific, easily customizable, online courses for Prentice Hall textbooks in mathematics and statistics. MyMathLab is powered by CourseCompass™—Pearson Education's online teaching and learning environment—and by MathXL®—our online homework, tutorial, and assessment system. MyMathLab gives instructors the tools they need to deliver all or a portion of their course online, whether students are in a lab setting or working from home. MyMathLab provides a rich and flexible set of course materials, featuring free-response exercises that are algorithmically generated for unlimited practice and mastery. Students can also use online tools, such as video lectures, animations, and a multimedia textbook, to independently improve their understanding and performance. MyMathLab is available to qualified adopters. For more information, visit our Web site at www.mymathlab.com or contact your Prentice Hall sales representative. (MyMathLab must be set up and assigned by your instructor.)

MathXL® www.mathxl.com

MathXL is a powerful online homework, tutorial, and assessment system that accompanies the text. With MathXL, instructors can create, edit, and assign online homework and tests using algorithmically generated exercises correlated to your textbook. All student work is tracked in MathXL's online gradebook. Students can take chapter tests in MathXL and receive personalized study plans based on their test results. The study plan diagnoses weaknesses and links students directly to tutorial exercises for the objectives they need to study and retest. Students can also access supplemental animations and video clips directly from selected exercises. MathXL is available to qualified adopters. For more information, visit our Web site at www.mathxl.com or contact your Prentice Hall sales representative for a product demonstration. (MathXL must be set up and assigned by your instructor.)

Elementary & Intermediate Algebra

Michael Sullivan, III
Joliet Junior College

Katherine R. Struve
Columbus State Community College

Janet Mazzarella
Southwestern Community College

PEARSON

Prentice Hall

Upper Saddle River, New Jersey 07458

Library of Congress Cataloging-in-Publication Data

Sullivan, Michael.
   Intermediate algebra/Michael Sullivan, III, Katherine R. Struve, Janet Mazzarella
     p. cm.
   Includes indexes.
   ISBN 0-13-191505-3
   1. Algebra—Textbooks.   I. Title.

   CIP data available.

Executive Editor: Paul Murphy
Editor in Chief: Christine Hoag
Executive Project Manager: Ann Heath
Production Editor: Lynn Savino Wendel
Senior Managing Editor: Linda Mihatov Behrens
Executive Managing Editor: Kathleen Schiaparelli
Media Project Manager: Audra J. Walsh
Media Production Editor: Raegan Keida
Managing Editor, Digital Supplements: Nicole M. Jackson
Manufacturing Buyer: Maura Zaldivar
Manufacturing Manager: Alexis Heydt-Long
Director of Marketing: Patrice Jones
Senior Marketing Manager: Kate Valentine
Marketing Assistant: Jennifer de Leeuwerk
Development Editor: Don Gecewicz
Editor in Chief, Development: Carol Trueheart
Editorial Assistant: Abigail Rethore
Project Manager/Class Testing: Dawn Nuttall
Art Director: Jonathan Boylan
Interior Designers: Wanda España, Mary Siener
Cover Designer: Wanda España
Art Editor: Thomas Benfatti
Creative Director: Juan R. López
Director of Creative Services: Paul Belfanti
Director, Image Resource Center: Melinda Reo
Manager, Rights and Permissions: Zina Arabia
Manager, Visual Research: Beth Brenzel
Manager, Cover Visual Research & Permissions: Karen Sanatar
Image Permission Coordinator: Richard Rodrigues
Photo Researcher: Teri Stratford
Cover Photo: © Celia Pearson/Pearson Photography
Art Studios: Precision Graphics, Laserwords
Compositor: Interactive Composition Corporation

 © 2007 Pearson Education, Inc.
Pearson Prentice Hall
Pearson Education, Inc.
Upper Saddle River, New Jersey 07458

Pearson Prentice Hall™ is a trademark of Pearson Education, Inc.
Printed in the United States of America

10 9 8 7 6 5 4 3 2 1

ISBN 0-13-191505-3

Pearson Education LTD., *London*
Pearson Education Australia PTY, Limited, *Sydney*
Pearson Education Singapore, Pte. Ltd
Pearson Education North Asia Ltd, *Hong Kong*
Pearson Education Canada, Ltd., *Toronto*
Pearson Educación de Mexico, S.A. de C.V.
Pearson Education—Japan, *Tokyo*
Pearson Education Malaysia, Pte. Ltd

About the Authors

With training in mathematics, statistics, and economics, Michael Sullivan, III has a varied teaching background that includes 15 years of instruction in both high school and college-level mathematics. He is currently a full-time professor of mathematics at Joliet Junior College. Michael has numerous textbooks in publication, including an Introductory Statistics series, and a Precalculus series, which he writes with his father, Michael Sullivan.

Michael believes that his experiences writing texts for college-level math and statistics courses give him a unique perspective as to where students are headed once they leave the developmental mathematics tract. This experience is reflected in the philosophy and presentation of his developmental text series. When not in the classroom or writing, Michael enjoys spending time with his three children, Michael, Kevin, and Marissa, and playing golf. Now that his two sons are getting older, he has the opportunity to do both at the same time!

Kathy Struve has been a classroom teacher for nearly 25 years, first at the high school level, and, for the past 13 years, at Columbus State Community College. Kathy emphasizes classroom diversity: diversity of age, learning styles, and previous learning success. She is aware of the challenges of teaching mathematics at a large, urban community college, where students have varied mathematics backgrounds, and may enter college with a high level of mathematics anxiety.

Kathy served as Lead Instructor of the Developmental Algebra sequence at Columbus State where she developed curriculum and provided leadership to adjunct faculty in implementing graphing calculator technology in the classroom. She has authored classroom activities at the Elementary Algebra, Intermediate Algebra, and College Algebra levels. In her spare time Kathy enjoys biking, hiking, traveling, and reading British detective mysteries.

Born and raised in San Diego county, Janet Mazzarella spent her career teaching in culturally and economically diverse High Schools before taking a position at Southwestern Community College, 15 years ago. Janet has taught a wide range of mathematics courses from arithmetic through calculus for math/science/engineering majors and has training in mathematics, education, engineering and accounting.

Janet has worked to incorporate technology into the curriculum by participating in the development of Interactive Math and Math Pro. At Southwestern Community College, she helped develop the self-paced developmental mathematics program and spent 2 years serving as its director. In addition, she has been the Chair of the Mathematics Department, the faculty union president, and the faculty coordinator for Intermediate Algebra. In the past, free time consisted of racing motorcycles off-road in the Baja 500 and rock climbing, but recently she has given up the adrenaline rush of these activities for the thrill of traveling in Europe.

Contents

Preface   viii

CHAPTER 1 | **Operations on Real Numbers and Algebraic Expressions   1**

1.1 Success in Mathematics   2
1.2 The Number Systems and the Real Number Line   8
1.3 Adding, Subtracting, Multiplying, and Dividing Integers   18
1.4 Adding, Subtracting, Multiplying, and Dividing Rational Numbers
Expressed as Fractions and Decimals   28
Putting the Concepts Together (Sections 1.2–1.4)   40
1.5 Properties of Real Numbers   41
1.6 Exponents and the Order of Operations   49
1.7 Simplifying Algebraic Expressions   57
Chapter 1 Activity: The Math Game   65
Chapter 1 Review   66
Chapter 1 Test   73

CHAPTER 2 | **Equations and Inequalities in One Variable   74**

2.1 Linear Equations: The Addition and Multiplication Properties of Equality   75
2.2 Linear Equations: Using the Properties Together   84
2.3 Solving Linear Equations Involving Fractions and Decimals;
Classifying Equations 93
2.4 Evaluating Formulas and Solving Formulas for a Variable   103
Putting the Concepts Together (Sections 2.1–2.4)   116
2.5 Introduction to Problem Solving: Direct Translation Problems   117
2.6 Problem Solving: Direct Translation Problems Involving Percent   131
2.7 Problem Solving: Geometry and Uniform Motion   139
2.8 Solving Linear Inequalities in One Variable   150
Chapter 2 Activity: Pass to the Right   164
Chapter 2 Review   165
Chapter 2 Test   172

CHAPTER 3 | **Introduction to Graphing and Equations of Lines   174**

3.1 The Rectangular Coordinate System and Equations in Two Variables   175
3.2 Graphing Equations in Two Variables   191
3.3 Slope   206
3.4 Slope-Intercept Form of a Line   217
3.5 Point-Slope Form of a Line   227
3.6 Parallel and Perpendicular Lines   235
Putting the Concepts Together (Sections 3.1–3.6)   244
3.7 Variation   245
3.8 Linear Inequalities in Two Variables   252
Chapter 3 Activity: Graphing Practice   261
Chapter 3 Review   262
Chapter 3 Test   270
Cumulative Review Chapters 1–3   271

CHAPTER 4 | **Systems of Linear Equations and Inequalities   272**

4.1 Solving Systems of Linear Equations by Graphing   273
4.2 Solving Systems of Linear Equations Using Substitution   285
4.3 Solving Systems of Linear Equations Using Elimination   296
Putting the Concepts Together (Sections 4.1–4.3)   306
4.4 Solving Direct Translation, Geometry, and Uniform Motion Problems Using Systems of
Linear Equations   307
4.5 Solving Mixture Problems Using Systems of Linear Equations   314
4.6 Systems of Linear Inequalities   326
Chapter 4 Activity: Find the Numbers   334
Chapter 4 Review   334
Chapter 4 Test   341

CHAPTER 5 | **Exponents and Polynomials   343**

5.1 Adding and Subtracting Polynomials   344
5.2 Multiplying Monomials: The Product and Power Rules   354

5.3 Multiplying Polynomials   360
5.4 Dividing Monomials: The Quotient Rule and Integer Exponents   371
   Putting the Concepts Together (Sections 5.1–5.4)   383
5.5 Dividing Polynomials   383
5.6 Applying Exponent Rules: Scientific Notation   391
   Chapter 5 Activity: What Is the Question?   398
   Chapter 5 Review   400
   Chapter 5 Test   404
   Cumulative Review Chapters 1–5   405

CHAPTER 6 | **Factoring Polynomials   407**

6.1 Greatest Common Factor and Factoring by Grouping   408
6.2 Factoring Trinomials of the Form $x^2 + bx + c$ 418
6.3 Factoring Trinomials of the Form $ax^2 + bx + c, a \neq 1$ 427
6.4 Factoring Special Products   436
6.5 Summary of Factoring Techniques   444
   Putting the Concepts Together (Sections 6.1–6.5)   450
6.6 Solving Polynomial Equations by Factoring   451
6.7 Modeling and Solving Problems with Quadratic Equations   460
   Chapter 6 Activity: Which One Does Not Belong?   468
   Chapter 6 Review   469
   Chapter 6 Test   473

Getting Ready for Intermediate Algebra: A Review of Chapters 1–6   474

CHAPTER 7 | **Rational Expressions and Equations   476**

7.1 Simplifying Rational Expressions   477
7.2 Multiplying and Dividing Rational Expressions   485
7.3 Adding and Subtracting Rational Expressions with a Common Denominator   494
7.4 Finding the Least Common Denominator and Forming Equivalent
   Rational Expressions   501
7.5 Adding and Subtracting Rational Expressions with Unlike Denominators   509
7.6 Complex Rational Expressions   520
   Putting the Concepts Together (Sections 7.1–7.6)   528
7.7 Rational Equations   529
7.8 Models Involving Rational Equations   541
   Chapter 7 Activity: Correct the Quiz   555
   Chapter 7 Review   556
   Chapter 7 Test   561

Getting Ready for Intermediate Algebra: A Review of Chapters 1–7   563

CHAPTER 8 | **Graphs, Relations, and Functions   565**

8.1 Graphs of Equations   566
8.2 Relations   577
   Putting the Concepts Together (Sections 8.1–8.2)   584
8.3 An Introduction to Functions   584
8.4 Functions and Their Graphs   597
8.5 Linear Functions   607
8.6 Compound Inequalities   621
8.7 Absolute Value Equations and Inequalities   632
   Chapter 8 Activity: Shifting Discovery   644
   Chapter 8 Review   644
   Chapter 8 Test   656

CHAPTER 9 | **Radicals and Rational Exponents   659**

9.1 Square Roots   660
9.2 nth Roots and Rational Exponents   666
9.3 Simplify Expressions Using the Laws of Exponents   674
9.4 Simplifying Radical Expressions   680
9.5 Adding, Subtracting, and Multiplying Radical Expressions   692
9.6 Rationalizing Radical Expressions   699
   Putting the Concepts Together (Sections 9.1–9.6)   705
9.7 Functions Involving Radicals   705
9.8 Radical Equations and Their Applications   712
9.9 The Complex Number System   722

Chapter 9 Activity: Which One Does Not Belong? 733
Chapter 9 Review 734
Chapter 9 Test 714
Cumulative Review Chapters 1–9 741

CHAPTER 10   Quadratic Equations and Functions   743

10.1 Solving Quadratic Equations by Completing the Square 744
10.2 Solving Quadratic Equations by the Quadratic Formula 757
10.3 Solving Equations Quadratic in Form 772
   Putting the Concepts Together (Sections 10.1–10.3) 780
10.4 Graphing Quadratic Functions Using Transformations 781
10.5 Graphing Quadratic Functions Using Properties 795
10.6 Quadratic Inequalities 811
10.7 Rational Inequalities 821
   Chapter 10 Activity: Presidential Decision Making 827
   Chapter 10 Review 828
   Chapter 10 Test 835

CHAPTER 11   Exponential and Logarithmic Functions   837

11.1 Composite Functions and Inverse Functions 838
11.2 Exponential Functions 852
11.3 Logarithmic Functions 870
   Putting the Concepts Together (Sections 11.1–11.3) 883
11.4 Properties of Logarithms 884
11.5 Exponential and Logarithmic Equations 894
   Chapter 11 Activity: Correct the Quiz 903
   Chapter 11 Review 904
   Chapter 11 Test 909
   Cumulative Review Chapters 1–11 910

CHAPTER 12   Conics   912

12.1 Distance and Midpoint Formulas 913
12.2 Circles 920
12.3 Parabolas 927
12.4 Ellipses 938
12.5 Hyperbolas 949
   Putting the Concepts Together (Sections 12.1–12.5) 957
12.6 Nonlinear Systems of Equations 958
   Chapter 12 Activity: How Do You Know That . . . ? 966
   Chapter 12 Review 966
   Chapter 12 Test 972

CHAPTER 13   Sequences, Series, and the Binomial Theorem   973

13.1 Sequences 974
13.2 Arithmetic Sequences 982
13.3 Geometric Sequences and Series 990
   Putting the Concepts Together (Sections 13.1–13.3) 1001
13.4 The Binomial Theorem 1002
   Chapter 13 Activity: Pass to the Right 1008
   Chapter 13 Review 1008
   Chapter 13 Test 1012
   Cumulative Review Chapters 1–13 1013

Appendix A: Fractions, Decimals, and Percents A1

Appendix B: Synthetic Division B1

Appendix C: Geometry Review C1

Appendix D: The Library of Functions D1

Appendix E: More on Systems of Linear Equations E1

Appendix F: Table of Square Roots F1

Answers to Quick Check Exercises QC-1

Answers to Selected Exercises AN-1

Applications Index AP-1

Subject Index I-1

Photo Credits P-1

A Word about Textbook Design and Student Success

As students and instructors have related in Prentice Hall focus groups and market research surveys, developmental math textbooks should not look "cluttered" or "busy." A busy design can distract a student from what is most important in the text. It can also heighten math anxiety.

As a result of this research, the design of this text is understated and focused on the most important pedagogical elements. Students and instructors helped us to identify the primary elements of this text, which are central to student success. They include:

- Exercise Sets
- Examples and *Quick Check* exercises (practice problems)
- Rules, Property, and Definition boxes
- Study Aids: *In Words, Work Smart,* and *Work Smart: Study Skills*

As you will notice, these primary features are the most prominent elements in the design. We have made every attempt to ensure that these components are the features to which the eye is drawn. The remaining features, the secondary elements, blend into the "fabric" or "grain" of the overall design.

Our thanks go to all of the students and instructors who helped us develop the design of this text. Their feedback proved invaluable in helping us to make the right decisions. We are confident the design of this text will be both practical and engaging as it serves its educational and learning purposes.

Sincerely,

Paul Murphy

Executive Editor
Developmental Mathematics
Prentice Hall

Kate Valentine

Senior Marketing Manager
Developmental Mathematics
Prentice Hall

Preface

The Elementary & Intermediate Algebra course serves a diverse group of students. Some of them have never been exposed to algebra, while others have been introduced to the material but have not yet grasped all the concepts. Still other students have succeeded in the course some time ago and need a refresher. Not only do the backgrounds of students vary with regard to their mathematical abilities but students' motivation, study skills, and reading skills also range considerably.

This diversity makes teaching Elementary & Intermediate Algebra challenging. It is imperative that a new text recognize the diversity of the classroom and address the array of needs of the many students.

Elementary & Intermediate Algebra introduces students to the logic and precision of mathematics. We expect our students to leave the course with an appreciation of this precision as well as the power of mathematics. Our students have to understand that the concepts we teach in the course form the basis for future mathematics courses. Once they have a conceptual understanding of algebra, students recognize that the material is not a series of unconnected topics. Instead, they see a story in which each new chapter builds on concepts learned in previous chapters.

To reinforce this idea, we remind the students of a helpful fact—mathematics is about taking a problem and reducing it to another problem that we have already seen. Taking a problem and reducing it to parts that are easier to solve helps students to see the forest for the trees (and, to carry the metaphor further, prevents them from feeling that they are lost in the woods).

In short, to address the many needs of today's Elementary & Intermediate Algebra students, we established the following as our goals when we began to write this text:

- Present a variety of study aids and tips so the student quickly comes to view the text as a useful and reliable tool that can increase their success in the course.

- Offer comprehensive exercise sets that build students' skills, show various intriguing applications of mathematics, begin to build up mathematical thinking, and reinforce mathematical concepts.

- Provide students with ample opportunity to see the connections among the various topics learned in the course.

- Provide the student with a strong conceptual foundation in mathematics through a clear, comprehensive presentation of topics and a special emphasis on functions.

Developing an Effective Text for Use In and Out of the Classroom

Given the hectic lives led by most students, coupled with the anxiety and trepidation with which they approach this course, an outstanding developmental mathematics text must provide pedagogical support that makes the text valuable to students as they study and do assignments. Pedagogy must be presented within a framework that teaches students how to study math; pedagogical devices must also address what students see as the "mystery" of mathematics—and solve that mystery.

To encourage students and to clarify the material, we developed a set of pedagogical features that help students develop good study skills, garner an understanding of the connections between topics, and work smarter in the process. The pedagogy used in this text is based upon the more than 55 years of classroom teaching experience that the authors bring to this text.

Examples are often the determining factor in how valuable a textbook is to a student. Students look to Examples to provide them with guidance and instruction when they need it most—the times when they are away from the instructor and the classroom. We have developed several Example formats in an attempt to provide superior guidance and instruction for the students. The formats include:

Innovative *Sullivan/Struve Examples*

The innovative *Sullivan/Struve Example* has a two-column format in which annotations are provided to the **left** of the algebra, rather than the right, as is the practice in most texts. Because we read from **left-to-right,** placing the annotation on the left will make more sense to the student. It becomes clear that the annotation describes what we are about to do instead of what was just done. The annotations may be thought of as the teacher's voice offering clarification immediately before writing the solution on the board. Consider the following:

EXAMPLE 1   Multiplying Rational Numbers (Fractions)

Find the product: $\dfrac{2}{9} \cdot \left(-\dfrac{15}{19} \right)$

Solution

We begin by rewriting the rational number $-\dfrac{15}{19}$ as $\dfrac{-15}{19}$. Then we multiply the numerators and multiply the denominators.

$$\frac{2}{9} \cdot \left(\frac{-15}{19} \right) = \frac{2 \cdot (-15)}{9 \cdot 19}$$

Write the numerator and the denominator as the product of prime factors:
$$= \frac{2 \cdot 3 \cdot (-5)}{3 \cdot 3 \cdot 19}$$

Showcase Examples

Showcase Examples are used strategically to introduce key topics or important problem-solving techniques. These examples provide "how-to" instruction by offering a guided, step-by-step approach to solving a problem. Students can then immediately see how each of the steps is employed. We remind students that the *Showcase Example* is meant to provide "how-to" instruction by including the words "how to" in the example title.

The *Showcase Example* has a three-column format in which the left column describes a step, the middle column provides a brief annotation, as needed, to explain the step, and the right column presents the algebra. With this format, the left and middle columns can be thought of as the instructor's voice offering an explanation (left) and then summarizing information (middle) during a lecture.

EXAMPLE 1   How to Solve a Linear Equation Using the Addition and Multiplication Properties of Equality

Solve the equation $2z - 3 = 9$.

Step-by-Step Solution

Step 1: Isolate the term containing the variable.

$$2z - 3 = 9$$

Apply the Addition Property of Equality and add 3 to each side of the equation:
$$2z - 3 + 3 = 9 + 3$$
$$2z = 12$$

Step 2: Get the coefficient of the variable to be 1.

Apply the Multiplication Property of Equality and divide both sides by 2 $\left(\text{this is the same as multiplying both sides by } \dfrac{1}{2} \right)$:

$$\frac{2z}{2} = \frac{12}{2}$$
$$z = 6$$

Quick Check Exercises

Placed at the conclusion of every Example, the *Quick Check* exercises provide students with an opportunity for immediate reinforcement. By working the problems that mirror the example just presented, students get instant feedback and gain confidence in their understanding of the concept. All the answers to *Quick Check* exercises are provided in the back of the text. We think that the *Quick Check* exercises will make the text more accessible and encourage students to read, consult, and use the text regularly.

Study Skills and Student Success

We have included study skills and student success as regular themes throughout this text starting with *Section 1.1: Success in Mathematics*. In addition to this dedicated section that covers many of the basics that are essential to success in any math course, we have included several recurring study aids that appear in the margin. These features were designed to anticipate the student's needs and to provide immediate help—as if the teacher were looking over his or her shoulder. These margin features include *In Words; Work Smart;* and *Work Smart: Study Skills*.

> *Section 1.1: Success in Mathematics* focuses the student on the basics of study skills, including what to do during the first week of the semester; what to do before, during, and after class; how to use the text effectively; and how to prepare for an exam.

> *In Words* helps to address the difficulty that students have in reading mathematically precise definitions and theorems by explaining them in plain English.

> *Work Smart* provides "tricks of the trade" hints, tips, reminders, and alerts. It also identifies some common errors to avoid and helps students work more efficiently.

> *Work Smart: Study Skills* reminds students of study skills that will help them to succeed at various points in the course. Attention to these practices will help them to become better, more proficient, learners.

Test Preparation and Student Success

The Chapter Tests in this text and the companion Chapter Test Prep Video CD have been designed to help students make the most of their valuable study time.

> ### Chapter Test
> In preparation for their classroom test, students should take the practice test to make sure they understand the key topics in the chapter. The exercises in the Chapter Tests have been crafted to reflect the level and types of exercises a student is likely to see on a classroom test.

> ### Chapter Test Prep Video CD
> Packaged with each new copy of the text, the Chapter Test Prep Video CD provides students with help at the critical juncture when they are studying for a test. The CD Video presents step-by-step solutions to the exact exercises found in each of the book's Chapter Tests. Easy video navigation allows students to access instantly the worked-out solutions to the exercises they want to study or review.

Superior Exercise Sets: Paired with Purpose

Students learn algebra by doing algebra. The superior end-of-section exercise sets in this text provide students with ample practice of both procedures and concepts. The exercises are paired and present problem types with every possible derivative.

The exercises also present a gradual increase in difficulty level. The early, basic exercises keep the student's focus on as few "levels of understanding" as possible. The later or higher-numbered exercises are "multi-task" (or Mixed Practice) exercises where students are required to utilize multiple skills, concepts, or problem-solving techniques.

Throughout the textbook, the exercise sets are grouped into seven categories—some of which appear only as needed:

1. **Concepts and Vocabulary** exercises are fill-in-the-blank, true/false, and open-ended questions that test a student's understanding of the vocabulary and concepts

presented within the section. We have found that students do not succeed if they do not become familiar with the basic vocabulary of mathematics.

2. **Building Skills** exercises are drill problems that develop the student's understanding of the procedures and skills in working with the methods presented in the section. Often these exercises can be linked back to a single example in the section.

3. **Mixed Practice** exercises are also drill problems, but they offer a comprehensive assessment of the skills learned in the section by asking problems that relate to more than one concept or objective.

4. **Applying the Concepts** exercises are problems that allow students to see the relevancy of the material learned within the section. Problems in this category are either situational problems that use material learned in the section to solve "real-world" problems or they are problems that ask a series of questions to enhance a student's conceptual understanding of the mathematics presented in the section.

5. **Extending the Concepts** exercises can be thought of as problems that go beyond the basics. Within this block of problems an instructor will find a variety of problems to sharpen students' critical-thinking skills.

6. Starting with Chapter 10, we provide **Synthesis Review** exercises to help students grasp the "big picture" of algebra—once they have a sufficient conceptual foundation to build upon from their work in Chapters 1 through 9. Synthesis review exercises ask students to perform a single operation (adding, solving, and so on) on several objects (polynomials, rational expressions, and so on). The student is then asked to discuss the similarities and differences in performing the same operation on the different objects.

7. Finally, we also include coverage of the **graphing calculator** beginning with Chapter 8. Instructors' philosophies about the use of graphing devices vary considerably. Because instructors disagree about the value of this tool, we have made an effort to make graphing technology entirely optional. When appropriate, technology exercises are included at the close of a section's exercise set.

How It All Fits Together: The Big Picture

Another important role of the pedagogy in this text is to help students see and understand the connection between the mathematical topics being presented. Several section-opening and margin features help to reinforce connections:

The Big Picture: Putting It Together (Chapter Opener)
This feature is based on how we start each chapter in the classroom—with a quick sketch of what we plan to cover. Before tackling a chapter, we tie concepts and techniques together by summarizing material covered previously and then relate these ideas to material we are about to discuss. It is important for students to understand that content truly builds from one chapter to the next. We find that students need to be reminded that the familiar operations of addition, subtraction, multiplication, and division are being applied to different or more complex objects.

Preparing for This Section
As part of this building process, we think it is important to remind students of specific material that they will need from earlier in the course to be successful within a given section. The *Preparing for . . .* feature that begins each section not only provides a list of prerequisite skills that a student should understand before tackling the content of a new section but also offers a short quiz to test students' preparedness. Answers to the quiz are provided as a footnote on the same page, and a cross-reference to the material in the text is provided so that the student can remediate when necessary. We believe the *Preparing for . . .* problems will be especially useful for students who place into the Intermediate portion of the course.

Putting the Concepts Together (Mid-Chapter Review)
Each chapter has a group of exercises at the appropriate point in the chapter, entitled *Putting the Concepts Together*. These exercises serve as a review—synthesizing material introduced up to that point in the chapter. The exercises in

these mid-chapter reviews are carefully chosen to assist students in seeing the "big picture."

Cumulative Review

Learning algebra is a building process and building involves considerable reinforcement. The cumulative review exercises at the end of each odd-numbered chapter, starting with Chapter 3, help students to reinforce and solidify their knowledge by revisiting concepts and using them in context. This way, studying for the final exam should be fairly easy.

Getting Ready for Intermediate Algebra

The transition from the Elementary Algebra portion of the course to the Intermediate Algebra portion of the course can be difficult for students. As instructors, we want our students to be aware of any deficiencies they may have entering when starting Intermediate Algebra. We have written two transition quizzes for students, one appears after Chapter 6 and the other is after Chapter 7. The answers to all problems on the quizzes are located in the back of the text. Plus, a cross reference to material presented earlier in the course is provided so that students can review important concepts, if necessary.

A Strong Foundation Through a Functions Approach

The approach that we take in Intermediate Algebra portion of the course is that the function is the overriding theme of the text. The reason for this stress on functions is twofold. First, Intermediate Algebra is not a terminal course but rather a gateway to the future, and functions form the basis for much study in mathematics. The introduction of functions helps make the "jump" from Intermediate Algebra to College Algebra less severe because students feel more comfortable with functions and function notation. Second, today's students like to learn in context so that they can see the relevancy of the material. The function provides a great way to present the usefulness of the material we are teaching.

In Closing

When we started writing this textbook, we discussed what improvements we could make in organization and in a textbook's approach to fundamental concepts. We also wanted to enhance staples such as examples and problems, and we "re-engineered" selected pedagogical features to make them truly useful. After writing and rewriting, and reading many thoughtful reviews from instructors, we focused on the following features of this Elementary & Intermediate Algebra text to help students get a better sense of the "big picture" and be more successful in this course:

- The **innovative *Sullivan/Struve Examples*** and ***Showcase Examples*** provide students with superior guidance and instruction when they need it most—when they are away from the instructor and the classroom.

- The ***Quick Check*** exercises provide students with immediate reinforcement and instant feedback to determine their understanding of the concepts presented in the examples.

- We developed each of the margin features such as ***In Words, Work Smart,*** and ***Work Smart: Study Skills*** with the goals of improving study skills, making the textbook easier to navigate, and increasing student success.

- ***Exercise Sets: Paired with Purpose***—The exercise sets are structured to assess student understanding of vocabulary, concepts, drill, problem solving, and applications. The exercise sets are graded in difficulty level to build confidence and to enhance students' mathematical thinking.

- ***Putting the Concepts Together*** and ***Synthesis Review*** help students see the big picture and provide a structure for learning each new concept and skill in the course.

- The organization of this text provides a distinct transition from skill-based Elementary Algebra to the functions-based Intermediate Algebra approach employed by this author team. In doing so, the Sullivan/Struve/Mazzarella text prepares students for the quantitative courses they will take after Elementary & Intermediate Algebra.

Instructor and Student Resources

The following resources are available to help instructors and students use this text more effectively.

Instructor Resources

Annotated Instructor's Edition (0-13-191506-1)

Instructor Solutions Manual (0-13-191519-3)

Instructor's Resource Manual with Tests (0-13-191509-6)

CD Lecture Series—Lab Pack (0-13-134605-9)

TestGen (0-13-191514-2)

- Enables instructors to build, edit, print, and administer tests.
- Features a computerized bank of questions developed to cover all text objectives.
- Available on dual-platform Windows/Macintosh CD-ROM.

MyMathLab Instructor Version (0-13-147898-2)

MyMathLab is a series of text specific, easily customizable, online courses for Prentice Hall textbooks in mathematics and statistics. MyMathLab is powered by Course Compass™—Pearson Education's online teaching and learning environment—and by MathXL®—our online homework, tutorial, and assessment system.

MathXL® Instructor Version (0-13-147895-8) www.mathxl.com

MathXL is a powerful online homework, tutorial, and assessment system that accompanies the text. With MathXL, instructors can create, edit, and assign online homework and tests using algorithmically generated exercises correlated to your textbook.

Student Resources

Student Solutions Manual (0-13-191507-X)

- Solutions to the odd-numbered section exercises.
- Solutions to the Quick Check exercises.
- Solutions to the Preparing for This Section, Putting the Concepts Together (mid-chapter review), Chapter Review, Chapter Test, and Cumulative Review exercises.

Prentice Hall Math Tutor Center (0-13-191508-8)

- Staffed by qualified math instructors who provide students with tutoring on examples and odd-numbered exercises from the textbook.
- Tutoring is available via toll-free telephone, toll-free fax, e-mail, or the Internet.
- White board technology allows tutors and students to see problems worked while they "talk" in real time over the Internet during tutoring sessions.

Elementary & Intermediate Algebra Student Study Pack (0-13-228683-1)

The Student Study Pack includes:
- CD Lecture Series
- Student Solutions Manual
- Prentice Hall Math Tutor Center access code

Chapter Test Prep Video CD—Standalone (0-13-227112-5)

- Includes fully worked-out solutions to every problem from each Chapter Test in the text.

MathXL® Tutorial on CD (0-13-191513-4)

- Provides algorithmically generated practice exercises that correlate to exercises at the end of sections.
- Every exercise is accompanied by an example and a guided solution; selected exercises include a video clip.
- The software recognizes student errors and provides feedback. It can also generate printed summaries of students' progress.

Interact Math® Tutorial Web Site www.interactmath.com

Get practice and tutorial help online! This interactive tutorial Web site provides algorithmically generated practice exercises that correlate directly to the exercises in your textbook.

Acknowledgments

Textbooks are written by authors but evolve through the efforts of many people. We would like to extend our thanks to the following individuals for their important contributions to the project. From Prentice Hall: Paul Murphy, who saw the vision of this text from its inception and made it happen; Kate Valentine and Patrice Jones for their innovative marketing ideas; Ann Heath for her dedication, enthusiasm, and attention to detail (quite honestly, Ann was the cement of the project); Chris Hoag for her support and encouragement; Dawn Nuttall for her perseverance, publishing acumen, and attention to detail with the class testing effort; Lynn Savino Wendel for her attention to detail throughout production; Jonathan Boylan and the design team for the attractive and functional design; Thomas Benfatti for his attentive eye in overseeing the creation of literally thousands of pieces of art; Maura Zaldivar for coordinating the scheduling of this project with the compositor and the printer; Linda Behrens for her watchful eye and management over countless production details; and finally, the Prentice Hall sales team, for their confidence and support of our books. Thanks also to Don Gecewicz for his ability to manage and maintain a single voice as well as his talent to make us think about each sentence.

We would like to offer special thanks to a number of instructors who helped us with this project, including: Cindy Trimble and Ondine Parker for help with solutions and to Kevin Bodden and Randy Gallaher for their attention to detail in preparing the solutions manuals; Darren Wiberg for lending his teaching style and caring to the CD Lecture Series; Kimberly Neuburger for her participation in the Chapter Prep Test Video CD; and Andreana Grimaldo and Denise Robichaud for their creativity with Chapter Activities. We would also like to thank Rafiq Ladhani, Jon Stockdale, and Fred Landwehr for their dedication to accuracy in checking the art, examples, and answers; and Sarah Streett and Jenny Crawford for attention to detail and consistency in accuracy checking revised text pages and answers.

We offer many thanks to all the instructors from across the country who participated in reviewer conferences and focus groups, and reviewed and/or class-tested some aspect of the manuscript. Their insights and ideas form the backbone of this text. Hundreds of instructors contributed their time, energy, and ideas to help us shape this text. We will attempt to thank them all here. We apologize for any omissions. *Note:* At the time this book went to press, more class tests were being secured. Our thanks also go to those instructors who tested the manuscript with their students after the printing deadline.

Class Testers

Marwan Abu-Sawwa, *Florida Community College—Jacksonville*

Mary Lou Baker, *Columbia State Community College*

Donna Beatty, *Ventura College*

Becky Bradshaw, *Lake Superior College*

Tim Britt, *Jackson State Community College*

Beverly Broomell, *SUNY Suffolk*

Hien Bui, *Hillsborough Community College—Dale Mabry*

Elena Catoiu, *Joliet Junior College*

John Close, *Salt Lake Community College*

Shirley Davis, *South Plains College*

Erica Egizio, *Joliet Junior College*

Sanford Geraci, *Broward Community College*

Susan Grody, *Broward Community College*

Pete Herrera, *Southwestern College*

Becky Hubiak, *Tidewater Community College—Virginia Beach*

Sally Jackman, *Richland College*

Nancy Johnson, *Broward Community College*

Mike Kirby, *Tidewater Community College—Virginia Beach*

Carla Kulinsky, *Salt Lake Community College*

Lynn Marecek, *Santa Ana College*

Janet Mazzarella, *Southwestern College*

Michael McComas, *Marshall University*

Judy Meckley, *Joliet Junior College*

Ron Moore, *Florida Community College—Jacksonville*

Hossein Navid-Tabrizi, *Houston Community College*

Charlotte Newsom, *Tidewater Community College—Virginia Beach*

Charles Odion, *Houston Community College*

Eugenia Peterson, *Daley College*

Elise Price, *Tarrant County Community College*

RB Pruitt, *South Plains College*

William Radulovich, *Florida Community College—Jacksonville*

Pavlov Rameau, *Miami Dade Community College—Wolfson*

Nancy Ressler, *Oakton Community College*

George Rhys, *College of the Canyons*

Togba Sapolucia, *Houston Community College*

Gisela Spieler-Persad, *Rio Hondo Community College*

Patrick Stevens, *Joliet Junior College*

Jennifer Strehler, *Oakton Community College*

Katalin Szucs, *East Carolina University*

Jo Tucker, *Tarrant County Community College*

Richard Watkins, *Tidewater Community College*

Reviewers

Darla Aguilar, *Pima State University*

Grant Alexander, *Joliet Junior College*

Philip Anderson, *South Plains College*

Mary Lou Baker, *Columbia State Community College*

Bill Bales, *Rogers State*

Tony Barcellos, *American River College*

John Beachy, *Northern Illinois University*

David Bell, *Florida Community College—Jacksonville*

Sandy Berry, *Hinds Community College*

Lori Braselton, *Georgia Southern University*

Beverly Broomell, *Suffolk Community College*

Joanne Brunner, *Joliet Junior College*

Connie Buller, *Metropolitan Community College*

Annette Burden, *Youngstown State University*

James Butterbach, *Joliet Junior College*

Marc Campbell, *Daytona Beach Community College*

Elena Catoiu, *Joliet Junior College*

Nancy Chell, *Anne Arundel Community College*

John Close, *Salt Lake Community College*

Bobbi Cook, *Indian River Community College*

Carlos Corona, *San Antonio College*

Faye Dang, *Joliet Junior College*

Vivian Dennis-Monzingo, *Eastfield College*

Alvio Dominguez, *Miami Dade Community College—Wolfson*

Karen Driskell, *South Plains College*

Brenda Dugas, *McNeese State University*

Doug Dunbar, *Okaloosa-Walton Junior College*

Laura Dyer, *Southwestern Illinois State University*

Bill Echols, *Houston Community College—Northwest*

Erica Egizio, *Joliet Junior College*

Jason Eltrevoog, *Joliet Junior College*

Nancy Eschen, *Florida Community College—Jacksonville*

Mike Everett, *Santa Ana College*

Phil Everett, *Ohio State University*

Scott Fallstrom, *Shoreline Community College*

Betsy Farber, *Bucks County Community College*

Fitzroy Farqharson, *Valencia Community College—West*

Dorothy French, *Community College of Philadelphia*

Donna Gerken, *Miami Dade Community College—Kendall*

Adrienne Goldstein, *Miami Dade Community College—Kendall*

Marion Graziano, *Montgomery County Community College*

Susan Grody, *Broward Community College*

Tom Grogan, *Cincinnati State University*

Barbara Grover, *Salt Lake Community College*

Shawna Haider, *Salt Lake Community College*

Margaret Harris, *Milwaukee Area Technical College*

Teresa Hasenauer, *Indian River Community College*

Mary Henderson, *Okaloosa-Walton Junior College*

Celeste Hernandez, *Richland College*

Bob Hervey, *Hillsborough Community College—Dale Mabry*

Teresa Hodge, *Broward Community College*

Sandee House, *Georgia Perimeter College*

Becky Hubiak, *Tidewater Community College—Virginia Beach*

John Jarvis, *Utah Valley State College*

Steven Kahn, *Anne Arundel Community College*

Linda Kass, *Bergen Community College*

Donna Katula, *Joliet Junior College*

Mohammed Kazemi, *University of North Carolina—Charlotte*

Doreen Kelly, *Mesa Community College*

Mike Kirby, *Tidewater Community College—Virginia Beach*

Keith Kuchar, *College of Dupage*

Carla Kulinsky, *Salt Lake Community College*

Julie Labbiento, *Leigh Carbon Community College*

Kathy Lavelle, *Westchester Community College*

Deanna Li, *North Seattle Community College*

Brian Macon, *Valencia Community College—West*

Jim Matovina, *Community College of Southern Nevada*

Jean McArthur, *Joliet Junior College*

Mikal McDowell, *Cedar Valley College*

Lee McEwen, *Ohio State University*

Angela McNulty, *Joliet Junior College*

Debbie McQueen, *Fullerton College*

Judy Meckley, *Joliet Junior College*

Lynette Meslinsky, *Erie Community College—City Campus*

Kausha Miller, *Lexington Community College*

Chris Mizell, *Okaloosa Walton Junior College*

Jim Moore, *Madison Area Technical College*

Ronald Moore, *Florida Community College—Jacksonville*

Elizabeth Morrison, *Valencia Community College—West*

Roya Namavar, *Rogers State*

Hossein Navid-Tabrizi, *Houston Community College*

Carol Nessmith, *Georgia Southern University*

Kim Neuburger, *Portland Community College*

Larry Newberry, *Glendale Community College*

Elsie Newman, *Owens Community College*

Charlotte Newsome, *Tidewater Community College*

Charles Odion, *Houston Community College*

Viann Olson, *Rochester Community and Technical College*

Linda Padilla, *Joliet Junior College*

Carol Perry, *Marshall Community and Technical College*

Faith Peters, *Miami Dade Community College—Wolfson*

Philip Pina, *Florida Atlantic University*

Carol Poos, *Southwestern Illinois University*

William Radulovich, *Florida Community College—Jacksonville*

David Ray, *University of Tennessee—Martin*

Michael Reynolds, *Valencia Community College—West*

George Rhys, *College of the Canyons*

Jorge Romero, *Hillsborough Community College—Dale Mabry*

David Ruffato, *Joliet Junior College*

Carol Rychly, *Augusta State University*

David Santos, *Community College of Philadelphia*

Doug Smith, *Tarrant Community College*

Gisela Spieler-Persad, *Rio Hondo Community College*

Raju Sriram, *Okaloosa-Walton Junior College*

Patrick Stevens, *Joliet Junior College*

Bryan Stewart, *Tarrant Community College*

Elizabeth Suco, *Miami Dade Community College—Wolfson*

Katalin Szucs, *East Carolina University*

KD Taylor, *Utah Valley State College*

Suzanne Topp, *Salt Lake Community College*

Suzanne Trabucco, *Nassau Community College*

Jo Tucker, *Tarrant Community College*

Bob Tuskey, *Joliet Junior College*

Mary Vachon, *San Joaquin Delta College*

Carol Walker, *Hinds Community College*

Kim Ward, *Eastern Connecticut State University*

Natalie Weaver, *Daytona Beach Community College*

Darren Wiberg, *Utah Valley State College*

Rachel Wieland, *Bergen Community College*

Christine Wilson, *Western Virginia University*

Brad Wind, *Miami Dade Community College—North*

Roberta Yellott, *McNeese State University*

Steve Zuro, *Joliet Junior College*

Additional Acknowledgments

We also would like to extend thanks to our colleagues at Joliet Junior College and Columbus State Community College, who provided encouragement, support, and the teaching environment where the ideas and teaching philosophies in this text were developed.

Michael Sullivan, III

Katherine R. Struve

Janet Mazzarella

CHAPTER

1

Operations on Real Numbers and Algebraic Expressions

Did you know that the arrangement of seeds in a sunflower can be described using a sequence of numbers called the Fibonacci sequence? See Problem 121 in Section 1.3.

OUTLINE

1.1 Success in Mathematics
1.2 The Number Systems and the Real Number Line
1.3 Adding, Subtracting, Multiplying, and Dividing Integers
1.4 Adding, Subtracting, Multiplying, and Dividing Rational Numbers Expressed as Fractions and Decimals

   Putting the Concepts Together (Sections 1.2–1.4)

1.5 Properties of Real Numbers
1.6 Exponents and the Order of Operations
1.7 Simplifying Algebraic Expressions

   Chapter 1 Activity: The Math Game
   Chapter 1 Review
   Chapter 1 Test

The Big Picture: Putting It Together

Welcome to Elementary Algebra! This course is taken by a diverse group of individuals. Some of you may never have taken an algebra course, while others may have taken algebra at some time in the past. In any case, we have written this text with both groups in mind.

The first chapter of the text serves as a review of the topics that you may have learned in an arithmetic course. The material is presented with an eye on the future, which is algebra. This means that we will slowly build our discussion so that the shift from arithmetic to algebra is painless. Take care to study the methods used in this section because these same methods will be used again in later chapters.

1.1 Success in Mathematics

OBJECTIVES

1. What to Do the First Week of the Semester
2. What to Do Before, During, and After Class
3. How to Use the Text Effectively
4. How to Prepare for an Exam

In Words
Doing crunches doesn't solve the problem of running a race, but they are truly effective at conditioning your body.

Let's start by having a frank discussion about the "big picture" goals of the course and how this book can help you to be successful at mathematics. The first "big picture" goal of the class is to develop algebraic skills and gain an appreciation for the power of algebra and mathematics. But there is also a second "big picture" goal. By studying mathematics, we develop our sense of logic and exercise the part of our brain that deals with logical thinking. The examples and problems that appear throughout the text are like the crunches that we do in a gym to exercise our body. The goal of running or walking is to get from point A to point B, so doing fifty crunches on a mat does not accomplish this goal, but crunches do make our upper bodies, backs, and heart stronger when we need to run or walk.

Logical thinking can assist us in solving difficult everyday problems, so solving algebra problems "builds the muscles" in the part of our brain that performs logical thinking. So, when you are studying algebra, and getting frustrated with the amount of work that needs to be done, and you say, "My brain hurts," remember the phrase that we all use in the gym, "No pain, no gain."

Another phrase to keep in mind is, "Success Breeds Success." Mathematics is everywhere. You already are successful at doing some everyday mathematics. With practice, you can take your initial successes and become even more successful. Have you ever done any of the following everyday activities?

- Compare the price per ounce of different sizes of jars of peanut butter or jam.
- Leave a tip at a restaurant.
- Figure out how many calories your bowl of breakfast cereal gives you.
- Take an opinion survey along with many other people.
- Measure the distances between cities as you plan your summer vacation.
- Order the appropriate number of gallons of paint to cover the walls of a room that you are renovating.
- Buy a car and take out a car loan with interest.
- Double a cookie recipe.
- Change American dollars for Canadian dollars.
- Fill up a basketball or soccer ball with air (balls are spheres, after all).
- Coach a Little League team (scores, statistics, catching, and throwing all involve math).
- Check the percentages of saturated and unsaturated fats in a chocolate bar.

We just listed twelve of the many everyday mathematical activities, and you may do five or ten in a single day! The everyday mathematics that you already know is the foundation for your success in this course.

① What to Do the First Week of the Semester

You have enrolled in an Elementary & Intermediate Algebra course. The first week of the semester gives you the opportunity to prepare for a successful course. Here are the things that you should do:

1. **Pick a good seat.** As you enter the classroom for the first time, choose a seat that gives you a good view of the room. Sit close enough to the front so that you can easily see the board and hear the professor.

2. **Read the syllabus to learn about your instructor and the course.** Be sure to take note of your instructor's name, office location, e-mail address, telephone number, and office hours. Also, pay attention to any additional help that can be found on campus such as tutoring centers, videos in the library, software, on-line tutorials, and so on. Make sure that you fully understand all of the instructor's policies for

the class. This includes the policy on absences, missed exams or quizzes, and homework. Ask questions.

3. **Learn the names of some of your classmates and exchange contact information.** One of the best ways to learn math is through group study sessions. Try to create time each week to study with your classmates. Knowing how to get in contact with classmates is also useful if you ever miss class because you can obtain the assignment for the day.

4. **Budget your time.** Most students have a tendency to "bite off more than they can chew." To help with time management, consider the following general rule for studying mathematics: You should plan on studying *at least* two hours outside of class for each hour in class. So, if you enrolled in a four-hour math class, you should set aside at least eight hours each week to study for the course. If this is not your only course, you will have to set aside time for other courses as well. Consider your work schedule and personal life when creating your budget as well.

Work Smart: Study Skills
Plan on studying two hours outside of class for each hour in class every week.

② What to Do Before, During, and After Class

Now that the semester is underway, we present the following ideas for what to do before, during, and after each class meeting. While these suggestions may sound overwhelming, we guarantee that by following them, you will be successful in mathematics (and other courses). Also, you will find that studying for exams becomes much easier by following this plan.

Work Smart: Study Skills
Take a few minutes to plan out the next academic term.

Before Class Begins

1. Make sure you are mentally prepared for class. This means that your mind should be alert and ready to concentrate for the entire class period. (Invest in a cup of coffee and eat lots of protein for breakfast!)

2. Read the section or sections that will be covered in the upcoming class meeting.

3. Based upon your reading, prepare a list of questions. Jot them down. In many cases, your questions will be answered through the lecture. You can then ask any questions that are not answered completely.

During Class

1. Arrive early enough to prepare your mind and material for the lecture.

2. Stay alert. Do not doze off or daydream during class. It will be very difficult to understand the lecture when you "return to class."

3. Take thorough notes. It is normal not to get certain topics the first time that you hear them through the lecture. However, this does not mean that you throw your hands up in despair. Rather, continue to write your class notes.

4. You can ask questions when appropriate. Do not be afraid to ask questions. In fact, instructors love when students ask questions, for two reasons. First, we know as teachers that if one student has a question, there are many more in class with the same question. Second, by asking questions, you are teaching the teacher what topics cause difficulty.

Work Smart: Study Skills
Be sure to ask questions during class.

After Class

1. Reread (and possibly rewrite) your class notes. In our experience as students, we were amazed how often confusion that existed during class went away after studying our in-class notes later when we had more time to absorb the material.

2. Reread the section. This is an especially important step. Once you have heard the lecture, the section will make more sense and you will understand much more.

Work Smart: Study Skills
The reason for homework is to build your skill and confidence. Don't skip assignments.

3. Do your homework. **Homework is not optional.** There is an old Chinese proverb that says,

> I hear … and I forget
>
> I see … and I remember
>
> I do … and I understand

This proverb applies to any situation in life in which you want to succeed. Would a pianist expect to be the best if she didn't practice? The only way you are going to learn algebra is by doing algebra. Remember: Success breeds success.

4. And don't forget, when you get a problem wrong, try to figure out why you got the problem wrong. If you can't discover your error, be sure to ask for help.

5. If you have questions, visit your professor during office hours. You can also ask someone in your study group or go to the tutoring center on campus, if available.

Math Courses: No Brain Freezes Here!

Learning algebra is a building process. Learning is the art of making connections between thousands of neurons (specialized cells) in the brain. Memory is the ability to reactivate these neural networks—it is a conversation among neurons.

Math isn't a mystery. You already know some math. But you do have to practice what you know and expand your knowledge. Why? The brain contains thousands of neurons. Through repeated practice, a special coating forms that allows the signals to travel faster and reduces interference. The cells "fire" more quickly and connections are made faster and with less effort. Practice forms the pathways that allow us to retrieve concepts and facts at test time. Remember those crunches, which are a way of making your body more robust and nimble—learning does the same to your brain.

Have We Mentioned Asking Questions?

To move information from short-term memory to long-term memory, we need to think about the information, comprehend its meaning, and ask questions about it.

③ How to Use the Text Effectively

When we sat down to write this text, we knew based upon experience from teaching our own students that students typically do not read their mathematics text. Rather than saying to you, "Ah, but our book is different—it can be read!," we decided to accept how students study math.

Students usually go through the following steps:

1. Attend the lecture and watch the instructor do some problems on the board. Perhaps work some problems in class.

2. Go home and work on the homework assignment.

3. After each problem, check the answer in the back of the text. If right, move on, but if wrong, go back and see where the solution went wrong.

4. Maybe, the mistake can be identified, but if not, go to the class notes or try to find a similar example in the text. With a little luck, a similar example can be found and the student can determine where the solution went wrong in the problem.

Work Smart: Study Skills
Learn what the different features of this book are designed to do. Decide which ones you may need the most.

5. If not, mark the problem and ask about it in the next class meeting, which leads us back to step 1.

So with this model in mind, we started to develop this text so that there is more than one way to extract the information you need from it.

All of the features have been included in the text to help you succeed. We list each feature in the order they appear and briefly explain its purpose and how it can be used to help you succeed in this course:

Preparing for This Section: Warming Up

Immediately after the title of the section, each section (after Section 1.1) begins with a short "readiness quiz." The readiness quiz asks questions about material that was presented earlier in the course that is needed for the upcoming section. You should take the readiness quiz to be sure that you understand the material that the new section will be based on. Answers to the readiness quiz appear as footnotes on the page of the quiz. Check your answers. If you get a problem wrong, or don't know how to do a problem, go back to the section listed and review the material.

Objectives: A "Road Map" through the Course

To the left of the readiness quiz, we present a list of objectives to be covered in the section. If you follow the objectives, you will get a good idea of the section's "big picture"—the important concepts, techniques, and procedures. Once you have completed your homework for the section, you should be able to answer "yes" to the statement, "You should be able to ..." for each objective.

The objectives are numbered. (See the numbered headline at the beginning of this section.) When we begin discussing a particular objective, the number appears in the left-hand column of the text.

Examples: Where to Look for Information

You look to examples to provide you with guidance and instruction when you need it most—when you are away from the instructor and the classroom. With this in mind, we have developed two example formats.

Step-by-Step Examples have a three-column format where the left column describes a step, the middle column provides a brief explanation of the step, and the right column presents the algebra. With this format, the left and middle columns can be thought of as your instructor's voice during a lecture. *Step-by-Step Examples* are used to introduce key topics or important problem-solving strategies. They are meant to provide easy-to-understand, practical instructions by including the words "how to" in the examples' headline.

Annotated Examples have a two-column format in which explanations are provided to the left of the algebra. Because we read from left to right, placing the explanation on the left clearly describes what we are about to do in the order that we will do it. Again, the annotations can be thought of as your instructor's voice right before he or she writes the solution on the board.

Quick Check: Practice for the Examples

After a concept has been presented, we will provide between 1 and 4 (occasionally up to 8 for some concepts) problems to solve. The answers to these problems are provided in the back of the book, so that you can quickly check your answers. This feature allows you to get immediate feedback about whether or not you understand the concept presented. The Quick Checks usually follow each example in the text. This feature came from our belief that students use the examples as a template for solving problems, so we decided to place the Quick Checks just after an example. If you can do the Quick Check problems, you will be prepared for the end-of-section exercises, which are a comprehensive review of the material in the section.

In Words: Math in Everyday Language

Have you ever been given a math definition in class and said, "What in the world does that mean?" As teachers, we have heard that from our students. So we added the "In Words" feature, which takes definitions that are in their mathematical form and restates them in everyday language. These boxes will help you to understand the language of mathematics better.

Work Smart

These are "tricks of the trade" that can be used to help you solve problems. They also show alternative approaches to solving problems. Yes, there is more than one way to solve a math problem!

Work Smart—Study Skills

Working smart also means studying smart. We provide tips throughout the text to help you understand the study skills required for success in this and other mathematics courses.

Chapter Review

The chapter review is arranged section by section. For each section, we list key concepts, key terms, and objectives. For each objective, we provide the examples from the text, along with page references, that illustrate the objective. Also, for each objective, we list the problems in the review exercises that test your understanding. If you get a problem wrong, use this feature to help you to identify how to work the problem.

Chapter Test

We have included a chapter test. Once you think that you are prepared for the exam, take the chapter test. If you do well on the chapter test, chances are you will do well on your in-class exam. Be sure to take the chapter test under the conditions that you will face in class. If you are unsure how to solve a problem in the chapter test, watch the Chapter Test Prep Video CD, which is a video of an instructor solving each problem in the chapter test.

Cumulative Review: Reinforcing Your Knowledge

As we mentioned, learning algebra is a building process. Building involves a lot of reinforcement. To do so, we provide cumulative reviews at the end of every odd-numbered chapter starting with Chapter 3. Do these cumulative reviews after each chapter test, so that you are always refreshing your memory—making those neurons do their calisthenics. This way, studying for the final exam should be fairly easy.

④ How to Prepare for an Exam

The following steps are time-tested suggestions to help you prepare for an exam.

Step 1: Revisit your homework and the chapter review problems: Beginning about one week before your exam, start to redo your homework assignments. If you don't understand a topic, be sure to seek out help. You should also work the problems given in the chapter review. The problems are keyed to the objectives in the course. If you get a problem wrong, identify the objective and examples that illustrate the objective. Then review this material and try the problem in the chapter review again. If you get the problem wrong again, seek out help.

Step 2: Test yourself: A day or two before the exam, take the chapter test under test conditions. Be sure to check your answers. If you got any problems wrong, determine why you got them wrong and remedy the situation. Don't forget about the Chapter Test Prep Video CD, which is a video of an instructor solving each problem in the chapter test.

Work Smart: Study Skills
Do not "cram" for an exam by pulling an "all nighter."

Step 3: Follow these rules as you train: Be sure to arrive early at the location of the exam. Prepare your mind for the exam. Also, be sure that you are well rested. Don't try to pull "all-nighters." If you need to study all night long for an exam, then your time management is poor and you should rethink how you are using your time or whether you have enough time set aside for the course.

Step 4: **Relax, and work on the exam thoughtfully:** While taking the exam, be sure to read the instructions. Show all your work, and be neat so that your instructor can follow your work and find your solution. Also, taking an exam is not a race, so there is no reason to turn your exam in early. If you finish early, go over each problem and check your answers.

1.1 Exercises

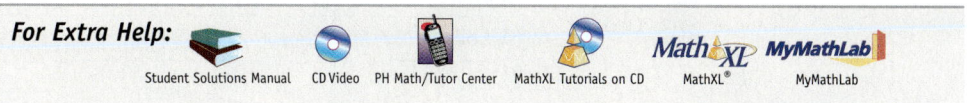

For Extra Help: Student Solutions Manual   CD Video   PH Math/Tutor Center   MathXL Tutorials on CD   MathXL®   MyMathLab

1. Why do you want to be successful in mathematics? Are your goals positive or negative? If you stated your goal negatively ("Just get me out of this course!"), can you restate it positively?

2. Name three activities in your daily life that involve the use of math (for instance, scoring a game of poker, bridge, or pinochle, operating your computer, or reading a credit-card bill).

3. What is your instructor's name?

4. What are your instructor's office hours? Where is your instructor's office?

5. Does your instructor have an e-mail address? If so, what is it?

6. Does your class have a Web site? Do you know how to access it? What information is located on the Web site?

7. Are there tutors available for this course? If so, where are they located? When are they available?

8. Name two other students in your class. What is their contact information? When can you meet with them to study?

9. List some of the things that you should do before class begins.

10. List some of the things that you should do during class.

11. List some of the things that you should do after class.

12. What is the point of the Chinese proverb on page 4?

13. What is the "readiness quiz"? How should it be used?

14. Name three features that appear in the margins. What is the purpose of each of them?

15. How should the chapter review material be used?

16. How should the chapter test be used?

17. How should the cumulative review be used?

18. List the four steps that should be followed when preparing for an exam.

19. Use the chart below to help manage your time. Be sure to fill in time allocated to various activities in your life including school, work and leisure.

	Monday	Tuesday	Wednesday	Thursday	Friday	Saturday	Sunday
7 A.M.							
8 A.M.							
9 A.M.							
10 A.M.							
11 A.M.							
Noon							
1 P.M.							
2 P.M.							
3 P.M.							
4 P.M.							
5 P.M.							
6 P.M.							
7 P.M.							
8 P.M.							
9 P.M.							

20. What is the Chapter Test Prep Video CD?

1.2 The Number Systems and the Real Number Line

OBJECTIVES

1. Classify Numbers
2. Plot Points on the Real Number Line
3. Use Inequalities to Order Real Numbers
4. Compute the Absolute Value of a Real Number

Work Smart

The use of the word "real" to describe numbers leads us to question "Are there 'nonreal' numbers?" The answer is "yes." We use the word "imaginary" to describe nonreal numbers. Imaginary does not mean that these numbers are made up, however. Imaginary numbers have many interesting applications in areas such as biology and the development of high-definition antennas. We will discuss imaginary numbers in Section 9.9. For most of the text, we concentrate on real numbers.

Preparing for the Number Systems and the Real Number Line

Before getting started, take the following readiness quiz. If you get a problem wrong, go to the section cited and review the material.

1. Write $\dfrac{5}{8}$ as a decimal. [Appendix, Section A.2, pp. A9–A10]

2. Write $\dfrac{9}{11}$ as a decimal. [Appendix, Section A.2, pp. A9–A10]

The goal of this section is to discuss the *real number system.* We use the real numbers every day in our lives, so it is an idea that you are already familiar with. In short, real numbers are numbers that we use to count or measure things: For instance, there might be 25 students in your class. Your car might get 18.4 miles per gallon. You might have a $130 debt.

As we proceed through the course, we will be dealing with various types of numbers. The kinds of numbers that we deal with are organized in *sets*. A **set** is a collection of objects. For example, we can identify the students enrolled in Elementary Algebra at your college as a set. The collection of numbers 0, 1, 2, 3, 4, 5, 6, 7, 8, and 9 may also be identified as a set. If we let A represent this set of numbers, then we can write

$$A = \{0, 1, 2, 3, 4, 5, 6, 7, 8, 9\}$$

In this notation, braces { } are used to enclose the objects, or **elements,** in the set. When a set has no elements in it, we say that the set is an **empty set.** Empty sets are denoted by the symbol $\varnothing$ or { }.

Preparing for...Answers **1.** 0.625
2. $0.8181\ldots$ or $0.\overline{81}$

EXAMPLE 1   Writing a Set

Write the set that represents the vowels.

Solution

The vowels are a, e, i, o, and u. If we let V represent this set, then

$$V = \{a, e, i, o, u\}$$

QUICK ✓

1. Write the set that represents the first 4 positive, odd numbers.

2. Write the set that represents the states that begin with the letter A.

3. Write the set that represents the states that begin with the letter Z.

① Classify Numbers

We will develop the real number system by looking at the history of numbers. The first types of numbers that humans worked with are called the *natural numbers* or *counting numbers*.

> **DEFINITION**
>
> The **natural numbers, or counting numbers,** are the numbers in the set $\{1, 2, 3, \ldots\}$. The three dots, called *ellipsis,* indicate that the pattern continues indefinitely.

Figure 1
The natural numbers.

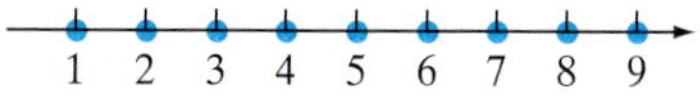

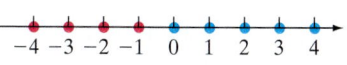

As their name implies, the counting numbers are often used to count things. For example, we can count the number of cars that arrive at a Wendy's drive-thru between 12 noon and 1:00 P.M. We can represent the counting numbers graphically using a number line. See Figure 1. The arrow on the right is used to indicate the direction in which the numbers increase.

Since we do not count the number of cars waiting in the drive-thru by saying, "zero, one, two, three …," zero is not a natural, or counting, number. When we add the number 0 to the set of counting numbers, we get the set of *whole numbers.*

> **DEFINITION**
>
> The **whole numbers** are the numbers in the set $\{0, 1, 2, 3, \ldots\}$.

Figure 2
The whole numbers.

Figure 2 represents the whole numbers on the number line. Notice that the set of natural numbers, $\{1, 2, 3, \ldots\}$ is included in the set of whole numbers.

By expanding the numbers to the left of zero on the number line, we have the set called *integers.*

> **DEFINITION**
>
> The **integers** are the numbers in the set $\{\ldots, -3, -2, -1, 0, 1, 2, 3, \ldots\}$.

Figure 3
The integers.

Figure 3 represents the integers on the number line. Notice that the whole numbers and natural numbers are included in the set of integers.

Integers are useful in many situations. For example, we could not discuss temperatures above 0°F (positive counting numbers) and temperatures below 0°F (negative counting numbers) without integers. A debt of 300 dollars can be represented as an integer as -300 dollars.

Work Smart: Study Skills
Do you need a refresher on fractions, decimals, or percents? If so, go to Appendix A and review your skills.

How can you represent a part of a whole? For example, how do you represent a part of last night's leftover pizza or part of a dollar? To address this problem, we enlarge our number system to include *rational numbers*.

> **DEFINITION**
>
> A **rational number** is a number that can be expressed as a fraction (or quotient) of two integers. That is, a rational number is a number that can be written in the form $\frac{p}{q}$, where p and q are integers. However, q cannot equal zero.

Work Smart
Remember, all integers are also rational numbers. For example $42 = \frac{42}{1}$.

Examples of rational numbers are $\frac{2}{5}, \frac{5}{2}, \frac{0}{8}, -\frac{7}{9}$, and $\frac{31}{4}$. Because $\frac{p}{1} = p$ for any integer p, it follows that all integers are also rational numbers. For example, 7 is an integer, but it is also a rational number because it can be written as $\frac{7}{1}$. We illustrate this idea below.

> Here, $\frac{7}{1}$ is written as a rational number . . .

$$\frac{7}{1} = 7$$

> . . . but over here, it is written as the integer 7, which is also a natural number

In addition to representing rational numbers as fractions, we can also represent rational numbers in decimal form as either a repeating decimal or a terminating decimal. Table 1 shows various rational numbers in fraction form and decimal form.

Table 1		
Fraction Form of Rational Number	**Decimal Form of Rational Number**	**Terminating or Repeating Decimal**
$\frac{7}{2}$	3.5	Terminating
$\frac{5}{11}$	$0.454545\ldots = 0.\overline{45}$	Repeating
$-\frac{3}{8}$	-0.375	Terminating
$-\frac{23}{21}$	$-1.095238095238\ldots = -1.\overline{095238}$	Repeating

For example, the repeating decimal $0.\overline{3}$ and the terminating decimal 0.27 are rational numbers because they can be converted to fractions (see Section A.2). The repeating decimal $0.\overline{3}$ is equivalent to the fraction $\frac{1}{3}$. The terminating decimal 0.27 is equivalent to the fraction $\frac{27}{100}$.

Decimals that neither terminate nor repeat are called irrational numbers. This means that irrational numbers cannot be written as the quotient of two integers. An example of an irrational number is the symbol π, whose value is approximately 3.141592. Another example of an irrational number is the symbol $\sqrt{2}$, whose value is approximately 1.41421.

> **DEFINITION**
>
> An **irrational number** is a number that cannot be written as the quotient of two integers. That is, it cannot be expressed as a terminating decimal or a repeating decimal.

Now that we have defined the set of rational numbers and the set of irrational numbers, we are ready for a formal definition of the set of *real numbers*.

> **DEFINITION**
> The set of rational numbers combined with the set of irrational numbers is called the set of **real numbers.**

Figure 4 shows the relationship among the various types of numbers. Notice the oval that represents the whole numbers surrounds the oval that represents the natural numbers. This means the whole numbers include all the natural numbers.

Figure 4

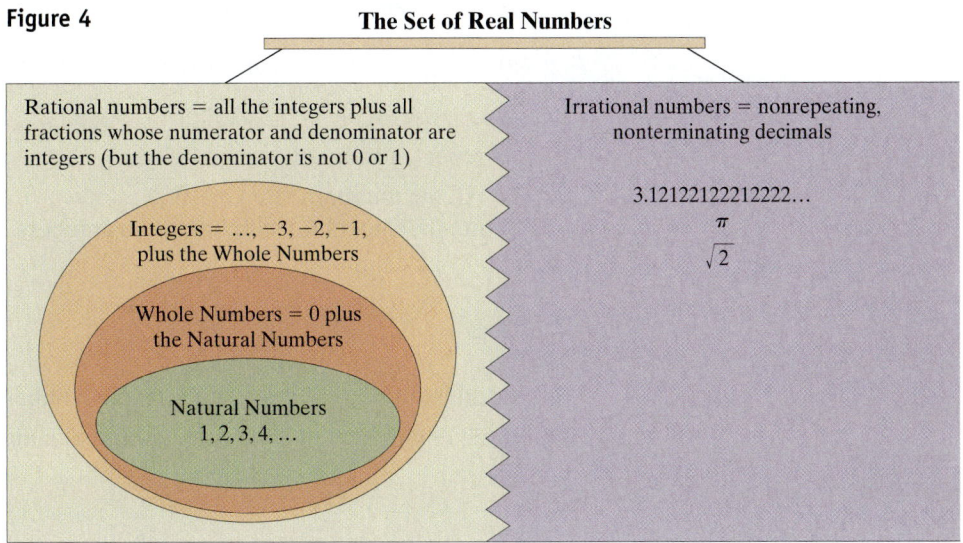

The Set of Real Numbers

Rational numbers = all the integers plus all fractions whose numerator and denominator are integers (but the denominator is not 0 or 1)

Integers = ..., −3, −2, −1, plus the Whole Numbers

Whole Numbers = 0 plus the Natural Numbers

Natural Numbers 1, 2, 3, 4, ...

Irrational numbers = nonrepeating, nonterminating decimals

3.12122122212222...
π
$\sqrt{2}$

Summary of the Set of Real Numbers		
Classification	**Elements of the Set**	**Description**
Natural numbers (also called counting numbers)	$\{1, 2, 3, \dots\}$	
Whole numbers	$\{0, 1, 2, 3, \dots\}$	Natural numbers and 0
Integers	$\{\dots, -3, -2, -1, 0, 1, 2, 3, \dots\}$	Positive and negative natural numbers and 0
Rational numbers	A number that can be written in the form $\frac{p}{q}$ where p and q are integers. However, q cannot equal zero. Examples: $3, 7, -11, -\frac{4}{5}, 7.\overline{16}, 4.125$	Any number that can be written as the quotient of two integers with the denominator not equal to 0 Terminating decimals or nonterminating, repeating decimals
Irrational numbers	Numbers that **cannot** be written as the quotient of two integers with a nonzero denominator Examples: $1.1121231234\dots, 3.14159\dots, 1.8975314253648607\dots$	Nonrepeating, nonterminating decimals
Real numbers	Set of rational numbers combined with the set of irrational numbers.	Rational numbers (any number that can be written as the quotient of two integers with the denominator not equal to 0; terminating decimals or nonterminating, repeating decimals) and Irrational numbers (nonrepeating, nonterminating decimals.)

EXAMPLE 2 **Classifying Numbers in a Set**

List the numbers in the set

$$\left\{9, -\frac{2}{7}, -4, 0, -4.010010001\ldots, 3.\overline{632}, 18.3737\ldots\right\}$$

that are

(a) Natural numbers   **(b)** Whole numbers   **(c)** Integers

(d) Rational numbers   **(e)** Irrational numbers   **(f)** Real numbers

Solution

(a) 9 is the only natural number

(b) 0 and 9 are the whole numbers

(c) 9, −4, and 0 are the integers

(d) $9, -\frac{2}{7}, -4, 0, 3.\overline{632}$, and $18.3737\ldots$ are the rational numbers.

(e) $-4.010010001\ldots$ is the only irrational number because the decimal does not repeat, nor does it terminate.

(f) All the numbers listed are real numbers. Real numbers consist of rational numbers together with irrational numbers.

QUICK ✔ *List the numbers in the set* $\left\{\frac{11}{5}, -5, 12, 2.\overline{76}, 0, \pi, \frac{18}{4}\right\}$ *that are*

4. Natural numbers   **5.** Whole numbers   **6.** Integers

7. Rational numbers   **8.** Irrational numbers   **9.** Real numbers

② Plot Points on the Real Number Line

In Words
We can think of the real number line as the graph of the set of all real numbers.

Look back at Figure 3. Notice that there are gaps between the integers plotted on the number line. These gaps are filled in with the real numbers that are not integers.

To construct a **real number line,** pick a point on a line somewhere in the center, and label it O. This point, called the **origin,** corresponds to the real number 0. See Figure 5.

Figure 5
The real number line.

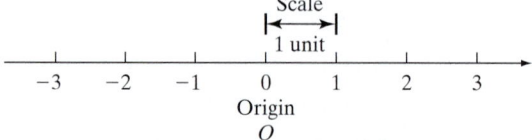

The point 1 unit to the right of O corresponds to the real number 1. The distance between 0 and 1 determines the **scale** of the number line. For example, the point associated with the number 2 is twice as far from O as 1 is. Notice that an arrowhead on the right end of the line indicates the direction in which the numbers increase. Points to the left of the origin correspond to the real numbers −1, −2, and so on.

> **DEFINITION**
>
> The real number associated with a point P is called the **coordinate** of P.

EXAMPLE 3   Plotting Points on the Real Number Line

On the real number line, label the points with coordinates $0, 6, -2, 2.5, -\frac{1}{2}$.

Solution

We draw a real number line and then plot the points. See Figure 6. Notice that 2.5 is midway between 2 and 3. Also notice that $-\frac{1}{2}$ is midway between -1 and 0.

Figure 6

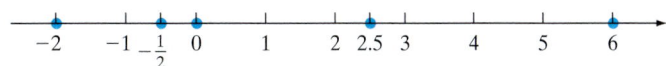

QUICK ✓

10. On the real number line, label the points with coordinates $0, 3, -2, \frac{1}{2}$, and 3.5.

The real number line consists of three classes (or categories) of real numbers, as shown in Figure 7.

Figure 7

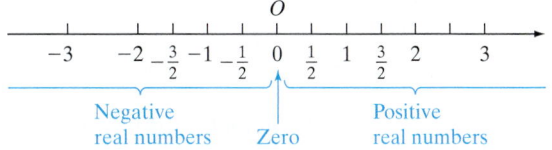

- The **negative real numbers** are the coordinates of points to the left of the origin, O.
- The real number **zero** is the coordinate of the origin denoted O.
- The **positive real numbers** are the coordinates of points to the right of the origin, O.

The **sign** of a number refers to whether the number is a positive real number or a negative real number. For example, the sign of -4 is negative while the sign of 100 is positive.

③ Use Inequalities to Order Real Numbers

An important property of the real number line follows from the fact that given two numbers (points) a and b, either a is to the left of b, denoted $a < b$, a is the same as b, denoted $a = b$, or a is to the right of b, denoted $a > b$. See Figure 8.

If a is either less than or equal to b, we write $a \leq b$. Similarly, $a \geq b$ means that a is either greater than or equal to b. Collectively, the symbols $<, >, \leq$, and $\geq$ are called **inequality symbols.** The "arrowhead" in an inequality always points to the smaller number. For $3 < 5$, the "arrowhead" points to the 3.

Note that $a < b$ and $b > a$ mean the same thing. For example, it does not matter whether we write $2 < 3$ (2 is to the left of 3) or $3 > 2$ (3 is to the right of 2).

Figure 8

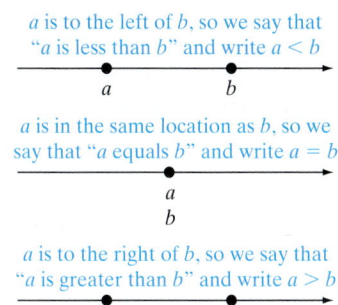

a is to the left of b, so we say that "a is less than b" and write $a < b$

a is in the same location as b, so we say that "a equals b" and write $a = b$

a is to the right of b, so we say that "a is greater than b" and write $a > b$

EXAMPLE 4   Using Inequality Symbols

(a) We know that having $3 is less than having $7 or having 3 apples is fewer than having 7 apples. Using the real number line, we say $3 < 7$ because the coordinate 3 lies to the left of the coordinate 7 on the real number line.

(b) Being $2 in debt is not as bad as being $5 in debt, so $-2 > -5$. Using the real number line, $-2 > -5$ because the coordinate -2 lies to the right of the coordinate -5 on the real number line.

(c) $2.7 > \dfrac{5}{2}$ because $\dfrac{5}{2} = 2.5$ and $2.7 > 2.5$.

(d) $\dfrac{5}{6} > \dfrac{4}{5}$ because $\dfrac{5}{6} = \dfrac{25}{30}$ and $\dfrac{4}{5} = \dfrac{24}{30}$. Having 25 parts out of 30 is more than having 24 parts out of 30. We could also write $\dfrac{5}{6} = 0.8\overline{3}$ and $\dfrac{4}{5} = 0.8$. Because $0.8\overline{3}$ is greater than 0.8, the coordinate $\dfrac{5}{6}$ lies to the right of the coordinate $\dfrac{4}{5}$ on the real number line.

Work Smart
Write fractions with a common denominator or change fractions to decimals to compare the location of the numbers on the number line.

QUICK ✓ *Replace the question mark by* $<$, $>$, *or* $=$, *whichever is correct.*

11. $2 ? 9$

12. $-5 ? -3$

13. $\dfrac{4}{5} ? \dfrac{1}{2}$

14. $\dfrac{4}{7} ? 0.5$

15. $\dfrac{4}{3} ? \dfrac{20}{15}$

If you look carefully at the results of Example 4, you should notice that the direction of the inequality symbol always points to the smaller number.

Based upon the discussion so far, we conclude that

$a > 0$	is equivalent to	a is positive
$a < 0$	is equivalent to	a is negative

We sometimes read $a > 0$ by saying that "a is positive." If $a \geq 0$, then either $a > 0$ or $a = 0$, and we may read this as "a is nonnegative" or "a is greater than or equal to zero."

④ Compute the Absolute Value of a Real Number

The real number line can be used to describe the concept of *absolute value*.

In Words
Think of absolute value as the number of units you must count to get from 0 to a number. The absolute value of a number can never be negative because it represents a distance.

DEFINITION

The **absolute value** of a number a, written $|a|$, is the distance from 0 to a on the real number line.

For example, because the distance from 0 to 3 on the real number line is 3, the absolute value of 3, $|3|$, is 3. Because the distance from 0 to -3 on the real number line is 3, the absolute value of -3, $|-3|$, is 3. See Figure 9.

Figure 9

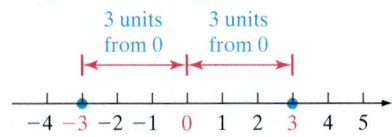

EXAMPLE 5 **Computing Absolute Value**

Evaluate each of the following:

   (a) $|-7|$ **(b)** $|-1.5|$ **(c)** $|6|$ **(d)** $|0|$

Solution

   (a) $|-7| = 7$ because the distance from 0 to -7 on the real number line is 7.

   (b) $|-1.5| = 1.5$ because the distance from 0 to -1.5 on the real number line is 1.5.

   (c) $|6| = 6$ because the distance from 0 to 6 on the real number line is 6.

   (d) $|0| = 0$ because the distance from 0 to 0 on the real number line is 0.   ▬

QUICK ✓ *Evaluate each expression.*

16. $|-15|$ **17.** $\left|-\dfrac{3}{4}\right|$

1.2 Exercises

For Extra Help: Student Solutions Manual   CD Video   PH Math/Tutor Center   MathXL Tutorials on CD   MathXL®   MyMathLab

Concepts and Vocabulary

In Problems 1–3, fill in the blanks.

  1. Real numbers that can be represented with a terminating decimal are called _____ numbers.

  2. Real numbers that cannot be represented as a ratio of *integers* are called _____ numbers.

  3. The distance from zero to a point on the number line is called the _____ _____ of the number.

In Problems 4–6, answer True or False to each statement.

  4. When comparing two real numbers, the number that is farther to the right on the number line is always the greater of the two.

  5. Every integer is a rational number.

  6. 0 is neither positive nor negative.

  7. Investigate the history of the number zero. Where and when was zero first used?

  8. Write a definition of "irrational number" in your own words. Describe the characteristics to look for when placing numbers into this set.

Building Skills

In Problems 9–14, write each set.

  9. *A* is the set of *whole* numbers less than 5.

  10. *B* is the set of *natural* numbers less than 25.

  11. *D* is the set of *natural* numbers less than 5.

  12. *C* is the set of *integers* between -6 and 4, not including -6 or 4.

  13. *E* is the set of even *natural* numbers between 4 and 15, not including 4 or 15.

  14. *F* is the set of odd *natural* numbers less than 1.

In Problems 15–20, use the set $\left\{-4, 3, \dfrac{-13}{2}, 0, 2.303003000\ldots\right\}$. *List all the elements that are…*

15. natural numbers   **16.** whole numbers

17. integers   **18.** rational numbers

19. irrational numbers   **20.** real numbers

In Problems 21–26, use the set $\left\{-4.2, 3.\overline{5}, \pi, \dfrac{5}{5}\right\}$. *List all the elements that are…*

21. real numbers   **22.** rational numbers

23. irrational numbers   **24.** integers

25. whole numbers   **26.** natural numbers

In Problems 27 and 28, plot the points in each set on a real number line.

27. $\left\{0, \dfrac{3}{3}, -1.5, -2, \dfrac{4}{3}\right\}$ **28.** $\left\{\dfrac{3}{4}, \dfrac{0}{2}, -\dfrac{5}{4}, -0.5, 1.5\right\}$

In Problems 29–36, evaluate each expression.

29. $|-12|$ **30.** $|-8|$ **31.** $|4|$ **32.** $|7|$

33. $\left|-\dfrac{3}{8}\right|$ **34.** $\left|-\dfrac{13}{9}\right|$ **35.** $|-2.1|$ **36.** $|-3.2|$

In Problems 37–44, determine whether the statement is True or False.

37. $-2 > -3$ **38.** $0 < -5$ **39.** $-6 \le -6$ **40.** $-3 \ge -5$

41. $\dfrac{3}{2} = 1.5$ **42.** $4.7 = 4.\overline{7}$ **43.** $\pi = 3.14$ **44.** $\dfrac{1}{3} = 0.33$

In Problems 45–52, replace the ? with the correct symbol: >, <, =.

45. $-1 \, ? \, 0$ **46.** $-8 \, ? \, -8.5$ **47.** $\dfrac{5}{8} \, ? \, \dfrac{6}{11}$ **48.** $\dfrac{5}{12} \, ? \, \dfrac{2}{3}$

49. $\dfrac{6}{13} \, ? \, 0.46$ **50.** $\dfrac{5}{11} \, ? \, 0.\overline{45}$ **51.** $|-7| \, ? \, |7|$ **52.** $\left|-\dfrac{3}{4}\right| \, ? \, \dfrac{3}{5}$

Mixed Practice

In Problems 53 and 54, (a) plot the points on the real number, (b) write the numbers in ascending order, (c) list the numbers that are (i) integers, (ii) rational numbers.

53. $\left\{\dfrac{3}{5}, -1, -\dfrac{1}{2}, 1, 3.5, |-7|, -4.5\right\}$ **54.** $\left\{8, -2, |-4|, -1.5, -\dfrac{4}{3}, 0, -\dfrac{15}{3}\right\}$

Applying the Concepts

In Problems 55–62, place a ✓ in the box if the given number belongs to the set.

		Natural	Whole	Integers	Rational	Irrational	Real
55.	-100						
56.	0						
57.	-10.5						
58.	π						
59.	$\dfrac{75}{25}$						
60.	4						
61.	$7.56556555\ldots$						
62.	$6.\overline{45}$						

In Problems 63–72, determine whether the statement is True or False.

63. Every *whole* number is also an *integer*.

64. Every decimal number is a *rational* number.

65. There are numbers which are both *rational* and *irrational*.

66. 0 is a positive number.

67. Every *natural* number is also a *whole* number.

68. Every *integer* is also a *real* number.

69. Every terminating decimal is a *rational* number.

70. Some numbers in the form $\dfrac{p}{q}$, $q \neq 0$ are *integers*.

71. 0 is a nonnegative *integer*.

72. -1 is a nonpositive *integer*.

In Problems 73–78, give the name of the set or elements of the set that matches the following description:

73. nonterminating and nonrepeating decimals

74. nonnegative integers

75. the set of rational numbers combined with the set of irrational numbers

76. terminating or repeating decimals

77. numbers which are both nonnegative and nonpositive

78. numbers which are both negative and positive

Extending the Concepts

If every element of set A is also an element of set B, we say A is a **subset** *of B and we write $A \subseteq B$. In Problems 79–82, use this definition and following sets to answer True or False to each statement.*

$$X = \{a, b, c, d, e\} \quad Y = \{c, e\} \quad Z = \{c, e, f\}$$

79. $Y \subseteq X$ **80.** $Z \subseteq Y$

81. $Y \subseteq Z$ **82.** $Z \subseteq X$

The **intersection** *of two sets is the set that contains the elements common to both A and B and is written $A \cap B$. The* **union** *of two sets is the set of all elements that are in either A or B and is written $A \cup B$. In Problems 83–86, write the elements of each set, using sets A, B, and C below.*

$$A = \{7, 8, 9, 10, 11, 12\} \quad B = \{10, 11, 12, 13, 14, 15\} \quad C = \{11, 12, 13, 14, 15\}$$

83. $A \cup B$ **84.** $A \cap B$ **85.** $B \cup C$

86. $B \cap C$ **87.** $A \cap C$ **88.** $A \cup C$

89. If $A = \{\text{even integers}\}$ and $B = \{\text{whole numbers less than } 11\}$, find $A \cap B$.

90. If $X = \{48, 49, 50, \ldots\}$ and $Y = \{60, 62, 64, \ldots 80\}$, find $X \cap Y$.

91. When writing subsets, it is important to be orderly when creating the list. Think of a pattern and then answer the following:

  (a) List all possible subsets of set Z where $Z = \{1, 2, 3, 4\}$.
  (b) How many subsets did you find?

92. Use the set $M = \{a, b, c\}$ to answer the following:

  (a) List all possible subsets of M.
  (b) How many subsets did you find?
  (c) Determine a rule for finding the number of subsets of a set that has n elements.

1.3 Adding, Subtracting, Multiplying, and Dividing Integers

OBJECTIVES

1. Add Integers
2. Determine the Additive Inverse of a Number
3. Subtract Integers
4. Multiply Integers
5. Divide Integers

Preparing for Adding, Subtracting, Multiplying, and Dividing Integers

Before getting started, take the following readiness quiz. If you get a problem wrong, go back to the section cited and review the material.

1. Write $\dfrac{16}{36}$ as a fraction in lowest terms.   [Appendix, Section A.1, pp. A5–A6]

In this section, we perform addition, subtraction, multiplication, and division, called **operations,** on integers. The symbols used in algebra for addition, subtraction, multiplication, and division are $+$, $-$, $\cdot$, and $/$, respectively. The words used to describe the results of these four operations are **sum, difference, product,** and **quotient.** Table 2 summarizes these ideas.

Table 2		
Operation	**Symbols**	**Words**
Addition	$a + b$	Sum: a plus b
Subtraction	$a - b$	Difference: a minus b
Multiplication	$a \cdot b, \ (a) \cdot b, \ a \cdot (b), \ (a) \cdot (b),$ $ab, \ (a)b, \ a(b), \ (a)(b)$	Product: a times b
Division	a/b or $\dfrac{a}{b}$	Quotient: a divided by b

In algebra, we avoid using the multiplication sign $\times$ used in arithmetic. Instead, when two expressions are placed next to each other without an operation symbol, as in ab, or in parentheses, as in $(a)(b)$, we know that the expressions are to be multiplied.

A **mixed number** is a whole number followed by a fraction. We do not use mixed numbers in algebra. When mixed numbers are used, addition is understood. For example,

$$3\frac{2}{5} \text{ means } 3 + \frac{2}{5}.$$

When you see a mixed number, rewrite it as a fraction. Do you remember how to do this? To write $3\frac{2}{5}$ as a fraction, we multiply the whole number 3 by the denominator 5 and obtain 15. Then add this result to the numerator 2 and obtain 17. This result is the numerator of the fraction. The denominator remains 5. So

$$3\frac{2}{5} = \frac{17}{5}$$

In algebra, using mixed numbers becomes confusing because we interpret the lack of an operation symbol between two terms to mean multiplication. To avoid confusion, write $3\frac{2}{5}$ as 3.4 or as $\dfrac{17}{5}$.

Work Smart
Do not use mixed numbers in algebra.

1 Add Integers

Adding Integers with the Same Sign Using a Number Line

We will use the real number line to discover a pattern for adding integers. When we add a positive integer, we move to the right on the number line, and when we add a negative integer, we move to the left on the number line.

Remember, the *sign* of a number refers to whether the number is positive or negative. For example, the sign of 4 is positive, while the sign of -12 is negative. We will first consider adding integers with the same sign.

Preparing for...Answer **1.** $\dfrac{4}{9}$

EXAMPLE 1 **Adding Two Positive Integers Using a Number Line**

Find the sum: $5 + 3$

Solution

We begin at 5 on the number line and move 3 spaces to the right, so $5 + 3 = 8$.
See Figure 10.

Figure 10

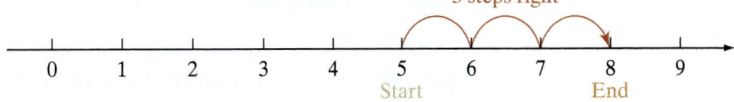

EXAMPLE 2 **Adding Two Negative Integers Using a Number Line**

Find the sum: $-7 + (-4)$

Solution

We begin at -7 on the number line and move 4 spaces to the left, so $-7 + (-4) = -11$.
See Figure 11.

Figure 11

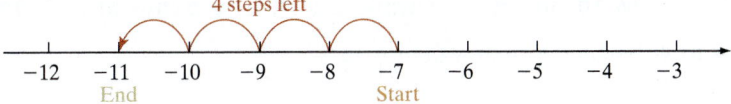

QUICK ✔ *Use a number line to find each sum.*

1. $8 + 19$ **2.** $-3 + (-5)$ **3.** $8 + 6$
4. $-3 + (-4)$ **5.** $-5 + (-12)$

Adding Integers with Different Signs Using a Number Line

We now consider the sum of two integers with different signs.

EXAMPLE 3 **Adding Integers with Different Signs Using a Number Line**

Find the sum: $-5 + 3$

Solution

We begin at -5 and then move 3 units to the right. From Figure 12 we see that
$-5 + 3 = -2$.

Figure 12

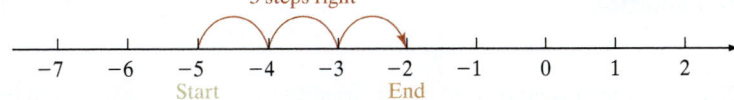

EXAMPLE 4 **Adding Integers with Different Signs Using a Number Line**

Find the sum: $7 + (-4)$

Solution

We begin at 7 and move 4 spaces to the left. We see that $7 + (-4) = 3$. See Figure 13.

Figure 13

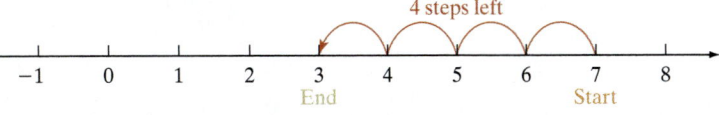

QUICK ✓ *Use a number line to find each sum.*

6. $-1 + 4$ **7.** $3 + (-4)$ **8.** $-8 + 10$ **9.** $-8 + 4$

10. $-17 + (-3)$ **11.** $-12 + 6$ **12.** $15 + (-5)$

Adding Integers Using Absolute Value

Did you discover a pattern for adding integers from Examples 1–4? When we add integers with the same sign (both positive or both negative), we add the absolute values of the integers being added and attach the common sign. When we add integers with different signs (one positive and one negative), we subtract the smaller absolute value from the larger absolute value and attach the sign of the integer having the larger absolute value.

EXAMPLE 5 **How to Add Real Numbers with the Same Sign Using Absolute Value**

Find $-16 + (-24)$ using absolute value.

Step-by-Step Solution

Step 1: Add the absolute values of the two integers.	We have $\lvert-16\rvert = 16$ and $\lvert-24\rvert = 24$. So $$16 + 24 = 40$$
Step 2: Attach the common sign, either positive or negative.	Both integers are negative in the original problem, so $$-16 + (-24) = -40$$

EXAMPLE 6 **How to Add Real Numbers with Different Signs Using Absolute Value**

Find $-31 + 16$ using absolute value.

Step-by-Step Solution

Step 1: Subtract the smaller absolute value from the larger absolute value.	We have $\lvert-31\rvert = 31$ and $\lvert16\rvert = 16$. The smaller absolute value is 16, so we compute $$31 - 16 = 15$$
Step 2: Attach the sign of the integer with the larger absolute value.	The larger absolute value is 31, which was a negative number in the original problem. So, the sum is negative. Therefore, $$-31 + 16 = -15$$

Work Smart

When we add integers with the same sign, we add the absolute values of the integers and attach the common sign. When we add integers with different signs, we subtract the smaller absolute value from the larger absolute value and attach the sign of the integer having the larger absolute value.

> **SUMMARY: Steps to Add Two Nonzero Integers Using Absolute Value**
>
> To add integers with the same sign (both positive or both negative),
>
> **Step 1:** Add the absolute values of the two integers.
>
> **Step 2:** Attach the common sign, either positive or negative.
>
> To add integers with different signs (one positive and one negative),
>
> **Step 1:** Subtract the smaller absolute value from the larger absolute value.
>
> **Step 2:** Attach the sign of the integer with the larger absolute value.

QUICK ✓ *Use absolute value to find each sum.*

13. $-11 + 7$ **14.** $5 + (-8)$ **15.** $-8 + (-16)$ **16.** $-94 + 38$

② Determine the Additive Inverse of a Number

What is $3 + (-3)$? What is $10 + (-10)$? What is $-143 + 143$? Of course, the answer to all three of these questions is 0! In fact, these results are true in general.

> **ADDITIVE INVERSE PROPERTY**
>
> For any real number a other than 0, there is a real number $-a$, called the **additive inverse,** or **opposite,** of a, having the following property.
>
> $$a + (-a) = -a + a = 0$$

Work Smart

The additive inverse of a, $-a$, is sometimes called the *negative* of a. Be careful when using this term because it suggests that the opposite is a negative number, which may not be true! For example, the additive inverse of -11 is 11, a positive number.

Any two numbers whose sum is zero are additive inverses, or opposites.

EXAMPLE 7 **Finding an Additive Inverse**

   (a) The additive inverse of 9 is -9 because $9 + (-9) = 0$.
   (b) The additive inverse of -12 is $-(-12) = 12$ because $-12 + 12 = 0$.

Notice from Example 7(b) that $-(-a) = a$ for any real number a.

QUICK ✓ *Determine the additive inverse of the given real number.*

17. 7   **18.** $\dfrac{3}{7}$ **19.** -21 **20.** $-\dfrac{8}{5}$ **21.** -5.75

③ Subtract Integers

Now that we understand addition and have the ability to determine the additive inverse of a number, we can proceed to subtract integers. Good news! There is not a new set of rules to follow for subtracting integers. We simply rewrite the subtraction problem as an addition problem and use the addition rules that were presented earlier.

From arithmetic, we write $10 - 6 = 4$. Using the additive inverse, we can write $10 - 6 = 4$ as $10 + (-6) = 4$.

DEFINITION

The **difference** $a - b$, read "a minus b" or a less b," is defined as

$$a - b = a + (-b)$$

In words, to subtract b from a, add the "opposite" of b to a.

So, subtracting b from a is the same as adding the additive inverse (opposite) of b to a.

EXAMPLE 8   **How to Subtract Integers**

Compute the difference: $-18 - (-40)$

Step-by-Step Solution

Step 1: Change the subtraction problem to an equivalent addition problem.	$-18 - (-40) = -18 + 40$
Step 2: Find the sum.	$= 22$

In Words
The problem in Example 8 is read "negative eighteen minus negative 40" not "negative 18 minus minus 40."

SUMMARY: Steps to Subtract Nonzero Integers

1. Change the subtraction problem to an equivalent addition problem using $a - b = a + (-b)$.

2. Find the sum.

QUICK ✓ *Find the value of each expression.*

22. $59 - (-21)$ **23.** $-32 - 146$ **24.** 17 minus 35

25. -382 subtracted from -2954

For the remainder of this text, the direction **evaluate** will mean to find the numerical value of an expression. When we evaluate a numerical expression in which there is both addition and subtraction, change all subtraction to addition. Then add left to right.

EXAMPLE 9   **Evaluating an Expression Containing Three Integers**

Evaluate the expression: $10 - 18 + 25$

Solution

$$10 - 18 + 25 = 10 + (-18) + 25$$

Add from left to right: $= -8 + 25$

$$= 17$$

Work Smart
When adding and subtracting more than two numbers, add in order from left to right.

EXAMPLE 10   **Evaluating an Expression Containing Four Integers**

Evaluate the expression: $162 - (-46) + 80 - 274$

Solution

$$162 - (-46) + 80 - 274 = 162 + 46 + 80 + (-274)$$

Add from left to right: $= 208 + 80 + (-274)$

$$= 288 + (-274)$$

$$= 14$$

QUICK ✓   *Evaluate each expression.*

26. $8 - 13 + 5 - 21$

27. $-27 - 49 + 18$

28. $3 - (-14) - 8 + 3$

29. $-825 + 375 - (-735) + 265$

---

④ **Multiply Integers**

A quick review of the language of multiplication is needed: In the multiplication statement $9 \cdot 2 = 18$, the numbers 9 and 2 are called **factors** and the number 18 is the **product.**

$$\underset{\text{factor}}{\underline{9}} \quad \cdot \quad \underset{\text{factor}}{\underline{2}} \quad = \quad \underset{\text{product}}{\underline{18}}$$

When we first learned how to multiply natural numbers, we were told that we can think of multiplication as repeated addition. For example, $3 \cdot 4$ is equivalent to adding 4 three times. That is,

$$3 \cdot 4 = \underset{\substack{\text{Add } 4 \\ \text{three times}}}{\underline{4 + 4 + 4}} = 12$$

Notice that the product of two positive factors produces a positive product. We knew that from arithmetic! But what is the product $3 \cdot (-4)$?

$$3 \cdot (-4) = \underset{\substack{\text{Add } -4 \\ \text{three times}}}{\underline{-4 + (-4) + (-4)}} = -12$$

We conclude that the product of two real numbers with *different signs* is negative.

What about the product of two negative numbers? Consider the pattern shown below.

$$-4 \cdot 3 = -12$$
$$-4 \cdot 2 = -8$$
$$-4 \cdot 1 = -4$$
$$-4 \cdot 0 = 0$$

Notice that each time the second factor decreases by 1, the product increases by four. Assuming this pattern continues, we would have

$$-4 \cdot -1 = 4$$
$$-4 \cdot -2 = 8$$

The pattern suggests that the product of two negative numbers is a positive number.

In Words
The sign of the product is *positive* if the signs of the two factors are the *same* and *negative* if the signs of the two factors are *different.*

> **RULES OF SIGNS FOR MULTIPLYING TWO INTEGERS**
>
> **1.** If we multiply two positive integers, the product is positive.
> **2.** If we multiply one positive integer and one negative integer, the product is negative.
> **3.** If we multiply two negative integers, the product is positive.

EXAMPLE 11   **Multiplying Integers**

(a) $2(-4) = -8$

(b) $-6(5) = -30$

(c) $(-7)(-8) = 56$

(d) $-25(18) = -450$

QUICK ✓ *Find the product.*

30. $-3(7)$ **31.** $13(-4)$ **32.** $5 \cdot 16$ **33.** $-9(-12)$ **34.** $(-13)(-25)$

Find the Product of Several Integers

When we find the product of more than two integers, we multiply in order, left to right.

EXAMPLE 12 **Multiplying Three Integers**

Find the product: $3 \cdot (-4) \cdot 7$

Solution

$$\text{Multiply left to right:} \quad 3 \cdot (-4) \cdot 7 = -12 \cdot 7$$
$$= -84$$

EXAMPLE 13 **Multiplying Four Integers**

Find the product: $-8 \cdot (-1) \cdot 4 \cdot (-5)$

Solution

$$\text{Multiply left to right:} \quad -8 \cdot (-1) \cdot 4 \cdot (-5) = 8 \cdot 4 \cdot (-5)$$
$$= 32 \cdot (-5)$$
$$= -160$$

Work Smart

If we multiply an *even* number of negative factors the product is *positive*.

If we multiply an *odd* number of negative factors the product is *negative*.

QUICK ✓ *Find the product.*

35. $-3 \cdot 9 \cdot (-4)$ **36.** $(-3) \cdot (-4) \cdot (-5) \cdot (-6)$

⑤ Divide Integers

When we divide, the numerator is called the **dividend,** the denominator is called the **divisor,** and the answer is called the **quotient.**

$$\text{dividend} \rightarrow \frac{28}{7} = 4 \leftarrow \text{quotient} \quad \text{or} \quad 28 \div 7 = 4$$
$$\text{divisor} \rightarrow \qquad \qquad \qquad \text{dividend} \quad \text{divisor} \quad \text{quotient}$$

Two nonzero numbers are *reciprocals* if their product is 1. Reciprocal is another word for the term *multiplicative inverse*.

In Words

Any two numbers whose product is 1 are called multiplicative inverses or reciprocals of each other.

> **MULTIPLICATIVE INVERSE (RECIPROCAL) PROPERTY**
>
> For each *nonzero* real number a, there is a real number $\frac{1}{a}$, called the **multiplicative inverse** or **reciprocal** of a, having the following property:
>
> $$a \cdot \frac{1}{a} = \frac{1}{a} \cdot a = 1 \quad a \neq 0$$

EXAMPLE 14 **Finding the Multiplicative Inverse (or Reciprocal) of an Integer**

(a) The multiplicative inverse or reciprocal of 5 is $\frac{1}{5}$.

(b) The multiplicative inverse or reciprocal of -8 is $-\frac{1}{8}$.

QUICK ✓ *Find the reciprocal of each integer.*

37. 6   **38.** −2

We use the idea behind the multiplicative inverse to define division of real numbers.

DEFINITION

If b is a nonzero real number, the **quotient** $\dfrac{a}{b}$, read as "a divided by b" or "the **ratio** of a to b," is defined as

$$\frac{a}{b} = a \cdot \frac{1}{b} \qquad \text{if } b \neq 0$$

For example, $\dfrac{40}{8} = 40 \cdot \dfrac{1}{8}$ and $\dfrac{12}{7} = 12 \cdot \dfrac{1}{7}$. Because division can be represented as multiplication, the same rules of signs that apply to multiplication also apply to division.

In Words
These are the same rules as before . . . a positive divided by a positive is positive; positive divided by negative is negative, and so on.

RULES OF SIGNS FOR DIVIDING TWO INTEGERS

1. If we divide two positive integers, the quotient is positive. That is, $\dfrac{+a}{+b} = \dfrac{a}{b}$.

2. If we divide one positive integer and one negative integer, the quotient is negative. That is, $\dfrac{-a}{b} = \dfrac{a}{-b} = -\dfrac{a}{b}$.

3. If we divide two negative integers, the quotient is positive. That is, $\dfrac{-a}{-b} = \dfrac{a}{b}$.

Finding the quotient of two integers is identical to writing a fraction in lowest terms.

EXAMPLE 15 **Finding the Quotient of Two Integers**

Find each quotient:

    (a) $\dfrac{-90}{20}$ **(b)** $200 \div (-5)$

Solution

(a)
$$\frac{-90}{20} = \frac{-9 \cdot 10}{2 \cdot 10}$$

Divide out the common factor:
$$= \frac{-9 \cdot \cancel{10}}{2 \cdot \cancel{10}}$$

$$= \frac{-9}{2}$$

$$= -\frac{9}{2}$$

(b)
$$200 \div (-5) = \frac{200}{-5}$$

$$= \frac{40 \cdot 5}{-1 \cdot 5}$$

Divide out the common factor:
$$= \frac{40 \cdot \cancel{5}}{-1 \cdot \cancel{5}}$$

$$= -40$$

QUICK ✓ *Find the quotient.*

39. $\dfrac{20}{-4}$　　　　**40.** $\dfrac{707}{-101}$　　　　**41.** $-63 \div -7$　　　　**42.** $\dfrac{-54}{4}$

1.3 Exercises

For Extra Help:
Student Solutions Manual　CD Video　PH Math/Tutor Center　MathXL Tutorials on CD　MathXL®　MyMathLab

Concepts and Vocabulary

In Problems 1–3, fill in the blanks.

1. The answer to an addition problem is called the _____.

2. The answer to a subtraction problem is called the _____.

3. The answer to a multiplication problem is called the _____.

In Problems 4–6, answer True or False to each statement.

4. When adding two negative numbers, we subtract their absolute values.

5. The subtraction problem $-3 - 10$ is equivalent to $3 + (-10)$.

6. The quotient of two negative numbers is negative.

7. Write a sentence or two that justifies the fact that the product of two negative integers is positive. You may use an example.

8. Explain how a subtraction problem can be written as an equivalent addition problem. Explain how $42 \div 4$ can be written as a multiplication problem.

Building Skills

In Problems 9–24, find the sum.

9. $8 + 7$　　　　**10.** $6 + 4$　　　　**11.** $-5 + 9$　　　　**12.** $-4 + 12$

13. $9 + (-5)$　　　**14.** $13 + (-7)$　　　**15.** $-11 + (-8)$　　　**16.** $-13 + (-5)$

17. $-16 + 37$　　　**18.** $-32 + 49$　　　**19.** $-119 + (-209)$　　**20.** $-145 + (-68)$

21. $-14 + 21 + (-18)$　　　　　　**22.** $(-13) + 37 + (-22)$

23. $74 + (-13) + (-23) + 5$　　　　**24.** $-34 + 46 + (-12) + 72$

In Problems 25–28, determine the additive inverse of each real number.

25. -325　　　　**26.** -34　　　　**27.** 125　　　　**28.** 7

In Problems 29–44, find the difference.

29. $23 - 12$　　　**30.** $35 - 23$　　　**31.** $9 - 17$　　　**32.** $12 - 19$

33. $-20 - 8$　　　**34.** $-15 - 9$　　　**35.** $13 - (-41)$　　　**36.** $14 - (-18)$

37. $-36 - (-36)$　　**38.** $-15 - (-15)$　　**39.** $0 - 41$　　　**40.** $0 - 18$

41. $-93 - (-62)$　　**42.** $46 - (-25)$　　**43.** $86 - (-86)$　　**44.** $49 - (-49)$

In Problems 45–60, find the product.

45. $5 \cdot 8$　　　　**46.** $7 \cdot 9$　　　　**47.** $8(-7)$　　　　**48.** $9(-7)$

49. $(0)(-21)$　　　**50.** $-21 \cdot 0$　　　**51.** $(-48)(-3)$　　　**52.** $(-22)(-5)$

53. $(-42)3$　　　**54.** $(-128)7$　　　**55.** $-5 \cdot 6 \cdot 3$　　　**56.** $-6 \cdot 4 \cdot 8$

57. $-10(3)(-7)$　　　　　　**58.** $-8(2)(-9)$

59. $(-2)(4)(-1)(3)(5)$　　　　**60.** $(-3)(-4)(6)(-1)$

In Problems 61–66, find the multiplicative inverse (or reciprocal) of each number.

61. 8 **62.** 10 **63.** −4

64. −3 **65.** 1 **66.** 2

In Problems 67–78, divide.

67. $10 \div 2$ **68.** $36 \div 9$ **69.** $\dfrac{-56}{-8}$ **70.** $\dfrac{-63}{-7}$

71. $\dfrac{-45}{3}$ **72.** $\dfrac{-144}{6}$ **73.** $\dfrac{35}{10}$ **74.** $\dfrac{20}{16}$

75. $\dfrac{60}{-42}$ **76.** $\dfrac{120}{-66}$ **77.** $\dfrac{-105}{-12}$ **78.** $\dfrac{-80}{-12}$

Mixed Practice

In Problems 79–94, evaluate the expression.

79. $-4 \cdot 18$ **80.** $7 \cdot (-15)$ **81.** $-16 - (-76)$ **82.** $87 - 19$

83. $-9 \cdot (-19)$ **84.** $7 \cdot 209$ **85.** $\dfrac{120}{-8}$ **86.** $\dfrac{-156}{-26}$

87. $-98 + 56$ **88.** $103 + (-66)$ **89.** $\dfrac{75}{|-20|}$ **90.** $\dfrac{|-42|}{12}$

91. $|-14| + |-26|$ **92.** $|-10| + (-62)$

93. $|-389| - 627$ **94.** $|-193| - (-20)$

In Problems 95–102, write each expression using mathematical symbols. Then evaluate the expression.

95. the sum of 28 and −21 **96.** the sum of 32 and −64

97. −21 minus 47 **98.** −16 minus −85

99. −12 multiplied by 18 **100.** 32 multiplied by −8

101. −36 divided by −108 **102.** −40 divided by 100

Applying the Concepts

In Problems 103–108, write the positive or negative number for each of the following.

103. Stock The price of IBM stock fell by 3.25 points.

104. Temperature The current temperature in Juneau, Alaska is 14° below zero.

105. Chargers Football The Chargers lost 6 yards on the play.

106. Profit Mark's BMW dealership showed a profit of $125,000 this quarter.

107. Checking Account Leila's checking account is now overdrawn by $48.

108. Census The number of people in Brian's home town grew by 12,368.

In Problems 109–116, write each problem as a numerical expression and evaluate.

109. Hiking Loren and Richard went on a hiking trip. They walked 8 miles to the base of Snow Creek Falls where they set up camp. They went another 3 miles to see the Falls and then returned to their campsite. How many miles did they walk that day?

110. Football The St. Louis Rams took over possession of the ball on their own 15 yard line. The following plays occurred: QB sack, loss of 7 yards; Marshall Faulk ran for 14 yards; Isaac Bruce caught a pass for 26 yards. What yard-marker is the ball on now?

111. Bank Balance When Martha balanced her checkbook, she had $563 in the account. Then the following transactions occurred: She wrote a check to Home Depot for $46; deposited $233; wrote a check to Vons for $63; and wrote a check to Petco for $32. What is Martha's new balance?

112. Checkbook Balance Josie began the month with $399 in her bank account. She deposited her paycheck of $839. She paid her $69 telephone bill, the electric bill for $78, and rent of $739. How much does she have left for spending money?

113. Warehouse Inventory The warehouse began the month with 725 cases of soda. During the month the following occurred: 120 cases were shipped out, 590 cases were shipped out, and a delivery to the warehouse of 310 cases. Does the warehouse have enough stock on hand to fill an order for 450 cases of soda? What is the difference between what they have and what has been requested?

114. Altitude of an Airplane A pilot leveled off his airplane at 35,000 feet at the beginning of the flight. The following adjustments were made during the trip: gained 4290 feet, dropped 10,400 feet and then dropped 2605 feet. At what altitude is the plane currently flying?

115. Distance An airplane flying at 25,350 feet is directly over a submarine that is 375 feet below sea level. What is the distance between a person in the plane and a person in the submarine?

116. Elevation The highest point in California is Mt. Whitney at an elevation of 14,495 feet and the lowest point is in Death Valley at 280 feet below sea level. What is the maximum difference in elevation between two people in California?

Extending the Concepts

117. Find two integers whose sum is -8 and whose product is 15.

118. Find two integers whose sum is 2 and whose product is -24.

119. Find two integers whose sum is -10 and whose product is -24.

120. Find two integers whose sum is -18 and whose product is 45.

121. The Fibonacci Sequence The Fibonacci sequence is a famous sequence of numbers that were discovered by Leonardo Fibonacci of Pisa. The numbers in the sequence are 1, 1, 2, 3, 5, 8, 13, 21, 34, 55, ... where each term after the second term is the sum of the two preceding terms.

(a) Form fractions of consecutive terms in the sequence. Find the decimal approximation to $\frac{1}{1}, \frac{2}{1}, \frac{3}{2}, \frac{5}{3}$, and so on.

(b) What number does the ratio get close to? This number is called the **golden ratio** and has application in many different areas.

(c) Research Fibonacci numbers and cite three different applications.

1.4 Adding, Subtracting, Multiplying, and Dividing Rational Numbers Expressed as Fractions and Decimals

OBJECTIVES

1. Multiply Rational Numbers Expressed as Fractions
2. Divide Rational Numbers Expressed as Fractions
3. Add and Subtract Rational Numbers Expressed as Fractions
4. Add, Subtract, Multiply, and Divide Rational Numbers Expressed as Decimals

Preparing for *Adding, Subtracting, Multiplying, and Dividing Rational Numbers Expressed as Fractions and Decimals*

Before getting started, take the following readiness quiz. If you get a problem wrong, go to the section cited and review the material.

1. Find the least common denominator of $\frac{5}{12}$ and $\frac{3}{16}$. [Appendix, Section A.1, pp. A4–A5]

2. Rewrite $\frac{4}{5}$ as an equivalent fraction with a denominator of 30. [Appendix, Section A.1, pp. A3–A5]

Preparing for...Answers **1.** LCD = 48

2. $\frac{4}{5} = \frac{24}{30}$

Now that we are comfortable with operations on integers, we expand our skill set so that we can perform operations on rational numbers. Remember, rational numbers can be expressed either as fractions or as a terminating decimal or a nonterminating, repeating decimal. We begin with operations on rational numbers expressed as fractions and end the section with the operations on rational numbers in decimal form.

Work Smart

Remember, any integer can be written as a fraction with 1 in the denominator. For example, $7 = \frac{7}{1}$.

All the skills we learn in this section apply to integers as well.

Finding the quotient of two integers is identical to writing a fraction in lowest terms. So, the direction to "find the quotient $\frac{-90}{20}$" is the same as "write $\frac{-90}{20}$ in lowest terms."

For example, the rational number $\frac{-90}{20}$ is written in lowest terms as $-\frac{9}{2}$.

QUICK ✓ *Write each rational number in lowest terms.*

1. $\dfrac{-4}{14}$ 　　　 **2.** $-\dfrac{18}{30}$ 　　　 **3.** $\dfrac{24}{-4}$

---

(1) Multiply Rational Numbers Expressed as Fractions

The following property tells us how to multiply two rational numbers when they are in fractional form:

In Words

When finding the product of two or more fractions, multiply the numerators together. Then multiply the denominators together. Write the fraction in lowest terms, if necessary.

MULTIPLYING FRACTIONS

$$\frac{a}{b} \cdot \frac{c}{d} = \frac{a \cdot c}{b \cdot d}, \text{ where } b \text{ and } d \neq 0$$

The rules of signs that apply to integers also apply to rational numbers: The product of two positive rational numbers is positive; the product of a positive rational number and a negative rational number is negative, and so on.

EXAMPLE 1　**Multiplying Rational Numbers (Fractions)**

Find the product: $\dfrac{2}{9} \cdot \left(-\dfrac{15}{19}\right)$

Solution

We begin by rewriting the rational number $-\dfrac{15}{19}$ as $\dfrac{-15}{19}$. Then we multiply the numerators and multiply the denominators.

$$\frac{2}{9} \cdot \left(\frac{-15}{19}\right) = \frac{2 \cdot (-15)}{9 \cdot 19}$$

Write the numerator and the denominator as the product of prime factors:
$$= \frac{2 \cdot 3 \cdot (-5)}{3 \cdot 3 \cdot 19}$$

Divide out common factors:
$$= \frac{2 \cdot \cancel{3} \cdot (-5)}{\cancel{3} \cdot 3 \cdot 19}$$

$$= \frac{2 \cdot (-5)}{3 \cdot 19}$$

Perform the multiplication:
$$= -\frac{10}{57}$$

QUICK ✓ *Find each product, and write in lowest terms if possible.*

4. $\dfrac{3}{4} \cdot \dfrac{9}{8}$ 　　 **5.** $\dfrac{-5}{7} \cdot \dfrac{56}{15}$ 　　 **6.** $\dfrac{12}{45} \cdot \left(-\dfrac{18}{20}\right)$ 　　 **7.** $-\dfrac{25}{75} \cdot \left(-\dfrac{9}{4}\right)$ 　　 **8.** $\dfrac{7}{3} \cdot \dfrac{1}{14} \cdot \left(-\dfrac{9}{11}\right)$

② Divide Rational Numbers Expressed as Fractions

To divide rational numbers, we must know how to determine the reciprocal of a rational number. We introduced the term *reciprocal* in Section 1.3 by saying that two numbers are reciprocals, or multiplicative inverses, if the product of the numbers is 1. This definition applies to any non-zero real number. So, for example,

$\dfrac{3}{2}$ and $\dfrac{2}{3}$ are reciprocals because $\dfrac{3}{2} \cdot \dfrac{2}{3} = 1$ 9 and $\dfrac{1}{9}$ are reciprocals because $9 \cdot \dfrac{1}{9} = 1$

$$-\dfrac{4}{7} \text{ and } -\dfrac{7}{4} \text{ are reciprocals because } -\dfrac{4}{7} \cdot \left(-\dfrac{7}{4}\right) = 1$$

QUICK ✓ *Find the reciprocal of each number.*

9. 12 **10.** $\dfrac{7}{5}$ **11.** $-\dfrac{1}{4}$ **12.** $-\dfrac{31}{20}$

We divide rational numbers by rewriting the division as an equivalent multiplication problem, according to the following.

> **DIVIDING RATIONAL NUMBERS EXPRESSED AS FRACTIONS**
> $$\dfrac{a}{b} \div \dfrac{c}{d} = \dfrac{a}{b} \cdot \dfrac{d}{c} = \dfrac{a \cdot d}{b \cdot c}, \quad \text{where } b, c, d \neq 0$$

EXAMPLE 2 **How to Divide Rational Numbers (Fractions)**

Find the quotient: $\dfrac{3}{10} \div \dfrac{12}{25}$

Step-by-Step Solution

Step 1: Write the equivalent multiplication problem.

$$\dfrac{3}{10} \div \dfrac{12}{25} = \dfrac{3}{10} \cdot \dfrac{25}{12}$$

Step 2: Write the product in factored form and divide out common factors.

$$= \dfrac{3 \cdot 5 \cdot 5}{5 \cdot 2 \cdot 4 \cdot 3}$$
$$= \dfrac{\cancel{3} \cdot \cancel{5} \cdot 5}{\cancel{5} \cdot 2 \cdot 4 \cdot \cancel{3}}$$
$$= \dfrac{5}{2 \cdot 4}$$

Step 3: Multiply the remaining factors.

$$= \dfrac{5}{8}$$

So $\dfrac{3}{10} \div \dfrac{12}{25} = \dfrac{5}{8}$.

QUICK ✓ *Find the quotient.*

13. $\dfrac{5}{7} \div \dfrac{7}{10}$ **14.** $-\dfrac{9}{12} \div \dfrac{14}{7}$ **15.** $\dfrac{8}{35} \div \left(\dfrac{-1}{10}\right)$ **16.** $-\dfrac{18}{63} \div \left(-\dfrac{54}{35}\right)$

③ Add and Subtract Rational Numbers Expressed as Fractions

Consider Figure 14, which shows a circle divided into 8 equal parts. We have two regions shaded, each of which represents the fraction $\frac{1}{8}$. Together, the two shaded regions make up $\frac{1}{4}$ of the circle. This implies that

Figure 14

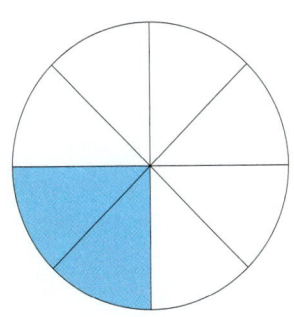

$$\frac{1}{8} + \frac{1}{8} = \frac{1+1}{8}$$
$$= \frac{2}{8}$$
$$= \frac{1}{4}$$

Or, suppose Bobby has \$0.25 and his grandma gives him \$0.50. How much does he now have? Of course, \$0.25 + \$0.50 = \$0.75 or $\frac{3}{4}$ of a dollar. Because $0.25 = \frac{1}{4}$ and $0.50 = \frac{1}{2}$, we can determine Bobby's good fortune using fractions:

$$\frac{1}{4} + \frac{1}{2} = \frac{1}{4} + \frac{2}{4}$$
$$= \frac{3}{4}$$

Based on these results, we might conclude that to add fractions with the same denominators, we add the numerators and write the result over the common denominator. This conclusion is correct. Also, because we can write any subtraction problem as an equivalent addition problem, we have the following methods for adding and subtracting rational numbers.

In Words
When adding or subtracting fractions with a common denominator, you add or subtract the numerators and retain the denominator.

> **ADDING OR SUBTRACTING RATIONAL NUMBERS (FRACTIONS) WITH THE SAME DENOMINATOR**
>
> $$\frac{a}{c} + \frac{b}{c} = \frac{a+b}{c} \quad \text{where } c \neq 0 \qquad \frac{a}{c} - \frac{b}{c} = \frac{a-b}{c} = \frac{a+(-b)}{c} \quad \text{where } c \neq 0$$

EXAMPLE 3   **Adding Rational Numbers (Fractions) with the Same Denominator**

Find the sum and write in lowest terms, if necessary: $-\frac{1}{8} + \frac{3}{8}$

Solution

$$-\frac{1}{8} + \frac{3}{8} = \frac{-1}{8} + \frac{3}{8}$$

Write the numerators as a sum over the common denominator:
$$= \frac{-1+3}{8}$$

Add the numerators:
$$= \frac{2}{8}$$

Factor 8:
$$= \frac{1 \cdot 2}{4 \cdot 2}$$

Divide out the 2s:
$$= \frac{1 \cdot \cancel{2}}{4 \cdot \cancel{2}}$$

$$= \frac{1}{4}$$

So $-\frac{1}{8} + \frac{3}{8} = \frac{1}{4}$.

EXAMPLE 4 Subtracting Rational Numbers (Fractions) with the Same Denominator

Find the difference and write in lowest terms, if necessary: $\dfrac{9}{16} - \dfrac{3}{16}$

Solution

$$\frac{9}{16} - \frac{3}{16} = \frac{9 - 3}{16}$$

Rewrite as an addition problem: $\quad = \dfrac{9 + (-3)}{16}$

Add the numerators: $\quad = \dfrac{6}{16}$

Factor 6; Factor 16: $\quad = \dfrac{3 \cdot 2}{8 \cdot 2}$

Divide out the 2s: $\quad = \dfrac{3 \cdot \cancel{2}}{8 \cdot \cancel{2}}$

$$= \frac{3}{8}$$

So $\dfrac{9}{16} - \dfrac{3}{16} = \dfrac{3}{8}$.

Work Smart: Study Skills
Notice the title of Example 5: "How to Add Rational Numbers with Unlike Denominators." This three-column example provides a guided, step-by-step approach to solving a problem so you can immediately see each of the steps. Cover the third column and try to work the example yourself. Then look at the entries in the third column and check your solution. Was it correct?

QUICK ✓ *Find the sum or difference, and write in lowest terms if necessary.*

17. $-\dfrac{9}{10} - \dfrac{3}{10}$ **18.** $\dfrac{8}{11} + \dfrac{2}{11}$ **19.** $-\dfrac{18}{35} + \dfrac{3}{35}$ **20.** $\dfrac{19}{63} - \dfrac{10}{63}$

Adding and Subtracting Rational Numbers (Fractions) with Unlike Denominators

Work Smart: Study Skills
Do you need a refresher on finding the LCD? If so, go to the Appendix on pages A4–A5 and review your skills.

How do we add rational numbers with different denominators? We must find the least common denominator of the two rational numbers. The **least common denominator (LCD)** is the smallest number that each denominator has as a common multiple.

EXAMPLE 5 How to Add Rational Numbers (Fractions) with Unlike Denominators

Find the sum: $\dfrac{5}{6} + \dfrac{3}{8}$

Step-by-Step Solution

Step 1: Find the least common denominator of the denominators.	Write each denominator as the product of prime factors, arranging like factors vertically: Find the product of each of the prime factors the greatest number of times it appears in any factorization:	$6 = 2 \quad\ \cdot 3$ $8 = 2 \cdot 2 \cdot 2$ $\text{LCD} = 2 \cdot 2 \cdot 2 \cdot 3$ $\quad\quad = 24$
Step 2: Write each rational number with the denominator found in Step 1.	Use $1 = \dfrac{4}{4}$ to change the denominator 6 to 24, and use $1 = \dfrac{3}{3}$ to change the denominator 8 to 24:	$\dfrac{5}{6} + \dfrac{3}{8} = \dfrac{5}{6} \cdot \dfrac{4}{4} + \dfrac{3}{8} \cdot \dfrac{3}{3}$ $= \dfrac{20}{24} + \dfrac{9}{24}$

Step 3: Add the numerators and write the result over the common denominator.	$= \dfrac{20 + 9}{24}$
	$= \dfrac{29}{24}$

Step 4: Write in lowest terms.	In this case, the rational number is already in lowest terms.

$$\text{Thus, } \frac{5}{6} + \frac{3}{8} = \frac{29}{24}.$$

EXAMPLE 6   How to Subtract Rational Numbers (Fractions) with Unlike Denominators

Find the difference: $-\dfrac{9}{14} - \dfrac{1}{6}$

Step-by-Step Solution

Step 1: Find the least common denominator of the denominators.	Write each denominator as the product of prime factors, aligning like factors vertically: Find the product of each of the prime factors the greatest number of times it appears in any factorization:	$14 = 2 \cdot 7$ $6 = 2 \quad \cdot 3$ $\text{LCD} = 2 \cdot 7 \cdot 3$ $= 42$
Step 2: Write each rational number with the denominator found in Step 1.	Use $1 = \dfrac{3}{3}$ to change the denominator 14 to 42, and use $1 = \dfrac{7}{7}$ to change the denominator 6 to 42:	$-\dfrac{9}{14} - \dfrac{1}{6} = \dfrac{-9}{14} \cdot \dfrac{3}{3} - \dfrac{1}{6} \cdot \dfrac{7}{7}$ $= \dfrac{-27}{42} - \dfrac{7}{42}$
Step 3: Subtract the numerators and write the result over the common denominator.		$= \dfrac{-27 - 7}{42}$ $= \dfrac{-34}{42}$
Step 4: Write in lowest terms.	Factor -34 and 42: Write in lowest terms by dividing out like factors:	$\dfrac{-34}{42} = \dfrac{2 \cdot (-17)}{2 \cdot 21}$ $= \dfrac{\cancel{2} \cdot (-17)}{\cancel{2} \cdot 21}$ $= -\dfrac{17}{21}$

$$\text{We see that } -\frac{9}{14} - \frac{1}{6} = -\frac{17}{21}.$$

SUMMARY: Steps to Add or Subtract Rational Numbers (Fractions) with Unlike Denominators

Step 1: Find the LCD of the rational numbers.

Step 2: Write each rational number with the LCD.

Step 3: Add or subtract the numerators and write the result over the common denominator.

Step 4: Write the result in lowest terms, if possible.

QUICK ✓ *Find each sum or difference, and write in lowest terms if necessary.*

21. $\dfrac{5}{12} - \dfrac{5}{18}$ **22.** $\dfrac{3}{14} + \dfrac{10}{21}$ **23.** $-\dfrac{23}{6} + \dfrac{7}{12}$ **24.** $\dfrac{3}{5} + \left(-\dfrac{4}{11}\right)$

Remember, the direction "evaluate" means to find the value of the expression, so the direction "evaluate $4 - \dfrac{2}{3}$" means to find the difference between 4 and $\dfrac{2}{3}$.

EXAMPLE 7 **Evaluating an Expression Containing Rational Numbers**

Evaluate and write in lowest terms if necessary: $4 - \dfrac{2}{3}$

Solution

The key is to remember that $4 = \dfrac{4}{1}$.

$$4 - \frac{2}{3} = \frac{4}{1} - \frac{2}{3}$$

Rewrite each fraction with LCD = 3: $= \dfrac{4}{1} \cdot \dfrac{3}{3} - \dfrac{2}{3}$

$$= \frac{12}{3} - \frac{2}{3}$$

$$= \frac{10}{3}$$

QUICK ✓ *Evaluate and write in lowest terms, if necessary.*

25. $-2 + \dfrac{7}{16}$ **26.** $6 + \left(\dfrac{-9}{4}\right)$

④ Add, Subtract, Multiply, and Divide Rational Numbers Expressed as Decimals

Adding and Subtracting Decimals

To add or subtract decimals, we arrange the numbers in a column with the decimals aligned. Then add or subtract the digits in the like place values, and place the decimal point in the answer directly below the decimal point in the problem.

EXAMPLE 8 **Adding and Subtracting Decimals That Are the Same Sign**

Evaluate each expression:

(a) $2.93 + 7.2 + 3.026$ **(b)** $76.4 - 4.95$

Solution

(a) Add zeros as placeholders:

$$\begin{array}{r} 2.930 \\ 7.200 \\ +3.026 \\ \hline 13.156 \end{array}$$

Work Smart

A whole number has an implied decimal point. For example,

$$74 = 74.000$$

(b) Add a zero as a placeholder:

$$\begin{array}{r} {\scriptstyle 5\ 13\ 10} \\ 76.4\cancel{0} \\ -4.95 \\ \hline 71.45 \end{array}$$

EXAMPLE 9 Adding and Subtracting Decimals That Are Different Signs

Evaluate each expression:

(a) $100.32 - (-32.015)$ (b) $-23.03 + 18.49$

Solution

(a) Remember, $a - (-b) = a + b$, so we can write $100.32 - (-32.015)$ as $100.32 + 32.015$.

$$
\begin{array}{r}
100.320 \\
+32.015 \\
\hline
132.335
\end{array}
$$

(b) To add real numbers with different signs (one positive and one negative), we subtract the smaller absolute value from the larger absolute value and attach the sign of the larger absolute value. Because $|-23.03| = 23.03$ and $|18.49| = 18.49$, we compute $23.03 - 18.49$ and attach a negative sign to the difference.

$$
\begin{array}{r}
23.03 \\
-18.49 \\
\hline
4.54
\end{array}
$$

So $-23.03 + 18.49 = -4.54$.

QUICK ✓ *Find the sum or difference.*

27. $9.67 + 11.344$ **28.** $81.96 - 17.39$ **29.** $14.95 + 7.118 + 0.3$

30. $345.67 - 8.0912$ **31.** $-180.782 + 100.3 + 9.07$ **32.** $-74.28 - 14.832$

Multiplying Decimals

The rules for multiplying decimals come from the rules for multiplying rational numbers written as fractions. For example,

$$
\underbrace{-0.7}_{\substack{1 \text{ decimal} \\ \text{place}}} \times \underbrace{0.03}_{\substack{2 \text{ decimal} \\ \text{places}}} = \frac{-7}{10} \times \frac{3}{100} = \frac{-21}{1000} = \underbrace{-0.021}_{\substack{3 \text{ decimal} \\ \text{places}}}
$$

Notice that there are three digits to the right of the decimal point in the answer. This is equal to the sum of the number of digits to the right of the decimal point in the factors.

Notice also that the rules of signs that we learned in Section 1.3 apply to decimals as well. We use this result to demonstrate how to multiply decimals.

EXAMPLE 10 Multiplying Decimals

Find the product:

(a) 3.43×2.6 (b) -3.17×0.02

Solution

(a)

$$
\begin{array}{r}
3.43 \quad \text{(two digits to the right of the decimal point)} \\
\times \ 2.6 \quad \text{(one digit to the right of the decimal point)} \\
\hline
2058 \\
686 \quad\quad \\
\hline
8.918 \quad \text{(three digits to the right of the decimal point)}
\end{array}
$$

(b)

$$
\begin{array}{r}
-3.17 \\
\times 0.02 \\
\hline
-0.0634
\end{array}
$$

four digits to the right of the decimal point

two digits to the right of the decimal point

two digits to the right of the decimal point

We summarize the procedure for multiplying decimals.

Work Smart

The number of digits to the right of the decimal point in the product is the *sum* of the number of digits to the right of the decimal point in the factors.

> **SUMMARY: Steps to Multiply Decimals**
>
> **Step 1:** Multiply the factors as if they were whole numbers.
>
> **Step 2:** Count the total number of digits to the right of the decimal point in the factors.
>
> **Step 3:** Place the decimal point in the product so that it contains the *sum* of the number of digits to the right of the decimal point in the factors.

QUICK ✔️ *Find the product.*

33. 23.9×0.2 **34.** 9.1×7.24 **35.** -3.45×0.03

36. $257 \times (-3.5)$ **37.** $-0.03 \times (-0.45)$ **38.** 9.9×0.002

Dividing Decimals

First, we review vocabulary. In the division problem $2\overline{)7.94}$ where the quotient is 3.97, the number 2 is called the *divisor,* 7.94 is called the *dividend,* and 3.97 is the *quotient.* You see from this example that when we divide by a whole number, we place the decimal point in the quotient directly above the decimal point in the dividend. In algebra, we typically write this division problem as $\dfrac{7.94}{2} = 3.97$.

To divide decimals, we want the divisor to be a whole number, so we multiply the dividend and the divisor by a power of 10 that will make the divisor a whole number. Then divide as though we were working with whole numbers. The decimal point in the quotient lies directly above the decimal point in the dividend.

EXAMPLE 11 Dividing Decimals

(a) Divide: $\dfrac{22.26}{15.9}$ **(b)** Divide: $\dfrac{0.03724}{-0.38}$

Solution

(a) Since the divisor 15.9 is fifteen and nine-tenths, we multiply $\dfrac{22.26}{15.9}$ by $\dfrac{10}{10}$ to make the divisor a whole number. $\dfrac{22.26}{15.9} \cdot \dfrac{10}{10} = \dfrac{222.6}{159}$. Now we divide.

$$
\begin{array}{r}
1.4 \\
159\overline{)222.6} \\
\underline{159} \\
63\ 6 \\
\underline{63\ 6} \\
0
\end{array}
$$

So $\dfrac{22.26}{15.9} = 1.4$.

(b) Because we have a positive number divided by a negative number, the quotient will be negative. Now, perform the division $\dfrac{0.03724}{0.38}$. Once we obtain this quotient, we will make it negative. The divisor 0.38 is thirty-eight hundredths, so we multiply $\dfrac{0.03724}{0.38}$ by $\dfrac{100}{100}$ to make the divisor a whole number. $\dfrac{0.03724}{0.38} \cdot \dfrac{100}{100} = \dfrac{3.724}{38}$. Now we divide.

$$
\begin{array}{r}
0.098 \\
38\overline{)3.724} \\
\underline{3\ 42} \\
304 \\
\underline{304} \\
0
\end{array}
$$

Therefore, $\dfrac{0.03724}{-0.38} = -0.098$.

We generalize the process for dividing decimals.

In Words
To divide decimals, change the divisor to a whole number and divide.

Steps to Divide Decimals

Step 1: Multiply the dividend and divisor by a power of ten that will make the divisor a whole number.

Step 2: Divide as though working with whole numbers.

Step 3: Place the decimal point in the quotient above the decimal point in the dividend.

QUICK ✓ *Find the quotient.*

39. $\dfrac{18.25}{73}$ **40.** $\dfrac{1.0032}{0.12}$ **41.** $\dfrac{-4.2958}{45.7}$ **42.** $\dfrac{0.1515}{-5.05}$

1.4 Exercises

For Extra Help:

Student Solutions Manual   CD Video   PH Math/Tutor Center   MathXL Tutorials on CD   MathXL®   MyMathLab

Concepts and Vocabulary

In Problems 1–3, fill in the blank.

1. Two numbers are _____, or _____ _____, if the product of the numbers is 1.

2. The _____ _____ _____ is the smallest number that each denominator has as a common multiple.

3. $\dfrac{a}{c} - \dfrac{b}{c} = \dfrac{-}{}, c \neq 0$

In Problems 4–6, answer True or False to each statement.

4. To add two rational expressions with the same denominator, we add the numerators and add the denominators.

5. $\dfrac{2}{3} \cdot \dfrac{5}{3} = \dfrac{10}{3}$

6. The reciprocal of $\dfrac{2}{3}$ is $-\dfrac{2}{3}$.

✎ **7.** What does it mean for a rational number to be written in lowest terms?

✎ **8.** Write an example that illustrates why $\frac{1}{2} + \frac{1}{2} = 1$.

Building Skills

In Problems 9–20, write each rational number in lowest terms.

9. $\frac{14}{21}$ **10.** $\frac{9}{15}$ **11.** $\frac{38}{-18}$ **12.** $\frac{81}{-36}$

13. $-\frac{22}{44}$ **14.** $-\frac{24}{27}$ **15.** $\frac{32}{40}$ **16.** $\frac{49}{63}$

17. $\frac{-15}{25}$ **18.** $\frac{-34}{51}$ **19.** $\frac{150}{225}$ **20.** $\frac{144}{156}$

In Problems 21–32, find the product, and write in lowest terms, if necessary.

21. $\frac{6}{5} \cdot \frac{2}{5}$ **22.** $\frac{7}{8} \cdot \frac{10}{21}$ **23.** $\frac{5}{-2} \cdot 10$ **24.** $\frac{3}{-7} \cdot 63$

25. $-\frac{3}{2} \cdot \frac{4}{9}$ **26.** $-\frac{5}{2} \cdot \frac{16}{25}$ **27.** $-\frac{22}{3} \cdot \left(-\frac{12}{11}\right)$ **28.** $-\frac{60}{75} \cdot \left(-\frac{25}{36}\right)$

29. $5 \cdot \frac{31}{15}$ **30.** $9 \cdot \frac{5}{18}$ **31.** $\frac{3}{4} \cdot \frac{8}{11}$ **32.** $\frac{4}{7} \cdot \frac{9}{16}$

In Problems 33–36, find the reciprocal of each number.

33. $\frac{3}{5}$ **34.** $\frac{9}{4}$ **35.** -5 **36.** -8

In Problems 37–48, find the quotient, and write in lowest terms, if necessary.

37. $\frac{4}{9} \div \frac{8}{15}$ **38.** $\frac{1}{2} \div \frac{3}{6}$ **39.** $-\frac{1}{3} \div 3$ **40.** $-\frac{1}{4} \div 4$

41. $\frac{5}{6} \div \left(-\frac{5}{4}\right)$ **42.** $\frac{4}{3} \div \left(-\frac{9}{10}\right)$ **43.** $\frac{36}{28} \div \frac{22}{14}$ **44.** $\frac{44}{63} \div \frac{11}{21}$

45. $-8 \div \left(\frac{2}{3}\right)$ **46.** $-3 \div \frac{7}{9}$ **47.** $-8 \div \left(-\frac{1}{4}\right)$ **48.** $-3 \div \left(-\frac{1}{6}\right)$

In Problem 49–72, find the sum or difference and write in lowest terms, if necessary.

49. $\frac{3}{4} + \frac{3}{4}$ **50.** $\frac{6}{11} + \frac{16}{11}$ **51.** $\frac{9}{8} - \frac{5}{8}$

52. $\frac{12}{5} - \frac{2}{5}$ **53.** $\frac{6}{7} - \left(-\frac{8}{7}\right)$ **54.** $\frac{2}{3} - \left(-\frac{7}{3}\right)$

55. $-\frac{5}{3} + 2$ **56.** $-\frac{7}{8} + 4$ **57.** $6 - \frac{7}{2}$

58. $3 - \frac{5}{3}$ **59.** $-\frac{4}{3} + \frac{1}{4}$ **60.** $-\frac{2}{5} + \left(-\frac{2}{3}\right)$

61. $\frac{7}{5} + \left(-\frac{23}{20}\right)$ **62.** $\frac{3}{4} + \left(-\frac{3}{8}\right)$ **63.** $\frac{7}{15} - \left(-\frac{4}{3}\right)$

64. $\frac{8}{3} - \left(-\frac{29}{9}\right)$ **65.** $\frac{8}{15} - \frac{7}{10}$ **66.** $\frac{17}{6} - \frac{13}{9}$

67. $-\frac{33}{10} - \left(-\frac{33}{8}\right)$ **68.** $-\frac{29}{6} - \left(-\frac{29}{20}\right)$ **69.** $\frac{19}{12} - \left(-\frac{41}{18}\right)$

70. $\frac{13}{12} - \frac{35}{16}$ **71.** $-\frac{2}{3} + \left(-\frac{5}{9}\right) + \frac{5}{6}$ **72.** $-\frac{1}{2} + \frac{3}{8} + \left(-\frac{3}{4}\right)$

In Problems 73–92, perform the indicated operation(s).

73. $-10.5 + 4$ **74.** $-13.2 + 7$ **75.** $-(-3.5) + 4.9$

76. $-(-32.9) + 10.3$ **77.** $39.1 - (-16.82)$ **78.** $29.23 - (-12.98)$

79. $-5.21 - (-6.7)$ **80.** $-4.94 - (-3.87)$ **81.** $45 - 2.45$

82. $32 - 5.68$ **83.** 4.3×5.8 **84.** 3.1×10.9

85. 0.075×120 **86.** 0.065×340 **87.** $\dfrac{136.08}{5.6}$

88. $\dfrac{332.59}{7.9}$ **89.** $\dfrac{25.48}{0.052}$ **90.** $\dfrac{48}{0.03}$

91. $-1.25 - (-0.6) + 1.6$ **92.** $-5.82 - (-2.9) + (-2.74)$

Mixed Practice

In Problems 93–124, evaluate and write in lowest terms, if necessary.

93. $-\dfrac{5}{6} + \dfrac{7}{15}$ **94.** $-\dfrac{8}{9} + \left(-\dfrac{16}{21}\right)$ **95.** $-\dfrac{10}{21} \cdot \dfrac{14}{5}$

96. $\dfrac{24}{5} \cdot \left(-\dfrac{35}{4}\right)$ **97.** $\dfrac{3}{8} \div \left(-\dfrac{9}{16}\right)$ **98.** $\dfrac{-12}{7} \div \dfrac{4}{-21}$

99. $-\dfrac{5}{12} + \dfrac{2}{12}$ **100.** $-\dfrac{4}{9} + \dfrac{1}{9}$ **101.** $-\dfrac{2}{7} - \dfrac{17}{5}$

102. $-\dfrac{3}{4} - \dfrac{1}{5}$ **103.** $-8.7 - (-10.3)$ **104.** $-4.63 - (-12.9)$

105. $\dfrac{1}{12} + \left(-\dfrac{5}{28}\right)$ **106.** $\dfrac{3}{16} + \left(-\dfrac{7}{40}\right)$ **107.** -12.03×4.2

108. $34.2 \times (-8.43)$ **109.** $36 \cdot \left(-\dfrac{4}{9}\right)$ **110.** $-\dfrac{8}{3} \cdot 15$

111. $-27 \div \dfrac{9}{5}$ **112.** $-24 \div \dfrac{8}{7}$ **113.** $3.62 - 10.2$

114. $4.75 - 6.2$ **115.** $\dfrac{-145.518}{18.42}$ **116.** $\dfrac{-297.078}{22.17}$

117. $\dfrac{12}{7} - \dfrac{17}{14} - \dfrac{48}{21}$ **118.** $\dfrac{9}{4} - \dfrac{21}{6} - \dfrac{11}{8}$

119. $54.2 - 18.78 - (-2.5) + 20.47$ **120.** $90.3 - 100.9 - (-34.26) + 32.95$

121. $400 \times 25.8 \times 0.003$ **122.** $500 \times 12.4 \times 0.02$

123. $-\dfrac{11}{12} - \left(-\dfrac{1}{6}\right) + \dfrac{7}{8}$ **124.** $\dfrac{8}{15} - \left(-\dfrac{7}{9}\right) + \dfrac{2}{3}$

Applying the Concepts

125. Watching TV If Rachel spends $\dfrac{1}{8}$ of her life watching TV, how many hours of TV does she watch in one week?

126. Halloween Candy Henry decided to make $\dfrac{2}{3}$ oz. bags of candy for treats at Halloween. If he bought 16 oz. of candy, how many bags will he have to give away?

127. Biology Class Susan's biology class begins with 36 students. If $\dfrac{2}{3}$ will finish the course and $\dfrac{3}{4}$ of those get a passing grade, how many students will pass Susan's biology class this term?

128. Pizza Time Joyce and Ramie bought a pizza. Joyce ate $\dfrac{2}{5}$ of the pizza and Ramie ate $\dfrac{1}{9}$ of what was left. What fraction of the pizza remains uneaten?

129. Overdrawn Maria's checking account was overdrawn by \$43.29. The bank charged overdraft fees amounting to \$25.50. What is the balance in Maria's checking account now?

130. Temperature The temperature in Yellowstone National Park fell from 25.6° above zero to 13.7° below zero. How much did the temperature drop?

131. Stock Prices The price per share of Intel stock has been up and down lately. On Monday it rose 2.75; on Tuesday it rose 0.87; on Wednesday it dropped 1.12; on Thursday it rose 0.52; and on Friday it fell 0.62. What was the net change in Intel's stock price per share for the week?

132. Bank Balance Henry started the month with $43.68 in his checking account. During the month the following transactions occurred: He deposited his paycheck of $929.30; and he wrote checks for rent $650, phone $33.49, credit card $229.50, cable service $75.50, and groceries $159.30. How much does he have in his account now?

Extending the Concepts

Problems 133–136 use the following definition.

If P and Q are two points on a real number line with coordinates a and b, respectively, the **distance between P and Q**, denoted by $d(P, Q)$, is

$$d(P, Q) = |b - a|$$

133. Find the distance between the points P and Q on the real number line if $P = -9.7$ and $Q = 3.5$.

134. Find the distance between the points P and Q on the real number line if $P = -12.5$ and $Q = 2.6$.

135. Find the distance between the points P and Q on the real number line if $P = -\dfrac{13}{3}$ and $Q = \dfrac{7}{5}$.

136. Find the distance between the points P and Q on the real number line if $P = -\dfrac{5}{6}$ and $Q = 4$.

PUTTING THE CONCEPTS TOGETHER (Sections 1.2–1.4)

These problems cover important concepts from Sections 1.2 to 1.4. We designed these problems so that you can review the chapter so far and show your mastery of the concepts. Take time to work these problems before proceeding with the next section. The answers to these problems are located at the back of the text starting on page AN-1.

1. Use the set $\left\{-12, -\dfrac{14}{7}, -1.25, 0, \sqrt{2}, 3, 11.2\right\}$. List all of the elements that are:

(a) integers   **(b)** rational numbers

(c) irrational numbers   **(d)** real numbers

2. Replace the ? with the correct symbol $>, <, =:$ $\dfrac{1}{8}$? 0.5

In Problems 3–23, perform the indicated operation. Reduce to lowest terms if necessary.

3. $17 + (-28)$ **4.** $-23 + (-42)$ **5.** $18 - 45$

6. $3 - (-24)$ **7.** $-18 - (-12.5)$ **8.** $(-5)(2)$

9. $25(-4)$ **10.** $(-8)(-9)$ **11.** $\dfrac{-35}{7}$

12. $\dfrac{-32}{-2}$ **13.** $27 \div -3$ **14.** $-\dfrac{4}{5} - \dfrac{11}{5}$

15. $7 - \dfrac{4}{5}$ **16.** $\dfrac{7}{12} + \dfrac{5}{18}$ **17.** $-\dfrac{5}{12} - \dfrac{1}{18}$

18. $\dfrac{6}{25} \cdot 15 \cdot \dfrac{1}{2}$ **19.** $\dfrac{2}{7} \div (-8)$ **20.** $\dfrac{0}{-8}$

21. $3.56 - (-7.2)$ **22.** $18.946 - 11.3$ **23.** $62.488 \div 42.8$

24. $(7.94)(2.8)$

1.5 Properties of Real Numbers

OBJECTIVES

1. Understand and Use the Identity Properties of Addition and Multiplication
2. Understand and Use the Commutative Properties of Addition and Multiplication
3. Understand and Use the Associative Properties of Addition and Multiplication
4. Understand the Multiplication and Division Properties of 0

Preparing for Properties of Real Numbers
Before getting started, take the following readiness quiz. If you get a problem wrong, go to the section cited and review the material.

1. Find the sum: $12 + 3 + (-12)$ [Section 1.3, pp. 18–21]

2. Find the product: $\dfrac{3}{4} \cdot 11 \cdot \dfrac{4}{3}$ [Section 1.4, p. 29]

This section is dedicated to presenting properties of real numbers. A property in mathematics is a rule that is always true. These properties will be used throughout this course and future math courses, so it is extremely important that you understand these properties and know how to use them.

① Understand and Use the Identity Properties of Addition and Multiplication

The real number 0 has an interesting property. It turns out that 0 is the only number that when added to any real number a results in the same real number a.

In Words
The word "identity" comes from a Latin word that means "the same" or "alike." The Identity Property of Addition means adding zero to a real number keeps the number the same.

> **IDENTITY PROPERTY OF ADDITION**
> For any real number a,
> $$0 + a = a + 0 = a$$
> That is, the sum of any number and 0 is that number. We call 0 the **additive identity.**

The real number 1 also has an interesting property. Recall that the expression $5 \cdot 3$ is equivalent to adding 5 three times. That is, $5 \cdot 3 = 5 + 5 + 5$. Therefore, $5 \cdot 1$ means to add 5 once, so, $5 \cdot 1 = 5$. This result is true in general.

> **MULTIPLICATIVE IDENTITY**
> For any real number a,
> $$a \cdot 1 = 1 \cdot a = a$$
> That is, the product of any number and 1 is that number. We call 1 the **multiplicative identity.**

Preparing for...Answers **1.** 3 **2.** 11

We also use the multiplicative identity throughout our math careers to create new expressions that are equivalent to previous expressions. For example, the expressions

$$\frac{4}{5} \quad \text{and} \quad \frac{4}{5} \cdot \frac{3}{3}$$

are equivalent because $\frac{3}{3} = 1$.

Conversion

A practical use of the multiplicative identity is conversion. **Conversion** is changing the units of measure (such as inches or pounds) from one measure to a different measure. For example, we might change a length from inches to feet or a weight from ounces to pounds.

EXAMPLE 1 Converting from Inches to Feet

Janice measures her family room and finds that its length is 184 inches long. How many feet long is Janice's family room? (*Note:* 12 inches = 1 foot)

Solution

The trick in doing conversions is to make sure that the units of measure that you are trying to remove get divided out and the new unit of measure remains. In this problem, we want the inches to divide out with feet remaining. Because 12 inches equals 1 foot, multiplying 184 inches by $\dfrac{1 \text{ foot}}{12 \text{ inches}}$ is like multiplying by one.

$$184 \text{ inches} = 184 \text{ inches} \cdot \overbrace{\frac{1 \text{ foot}}{12 \text{ inches}}}^{\text{Multiplying by 1}}$$

The inches "divide out":
$$= \frac{184}{12} \text{feet}$$

$184 = 2 \cdot 2 \cdot 2 \cdot 23; \ 12 = 2 \cdot 2 \cdot 3$:
$$= \frac{2 \cdot 2 \cdot 2 \cdot 23}{2 \cdot 2 \cdot 3} \text{feet}$$

Divide out common factors:
$$= \frac{\cancel{2} \cdot \cancel{2} \cdot 2 \cdot 23}{\cancel{2} \cdot \cancel{2} \cdot 3} \text{feet}$$

$$= \frac{46}{3} \text{feet}$$

Work Smart

```
      15
   3)46
     -3
     ──
      16
     -15
     ───
       1
```

So 184 inches equals $\dfrac{46}{3}$ feet. Because 46 divided by 3 is 15 with a remainder of 1 (Why? See the Work Smart), $\dfrac{46}{3}$ feet is equivalent to $15\dfrac{1}{3}$ feet. Since $\dfrac{1}{3} \text{foot} \cdot \dfrac{12 \text{ inches}}{1 \text{ foot}} = 4$ inches, we have that $\dfrac{46}{3}$ feet is equivalent to 15 feet, 4 inches. ∎

QUICK ✓ *Convert each measurement to the indicated unit of measurement.*

1. 96 inches = ? feet [1 foot = 12 inches]

2. 500 minutes = ? hours [60 minutes = 1 hour]

3. 88 ounces = ? pounds [16 ounces = 1 pound]

(2) ## Understand and Use the Commutative Properties of Addition and Multiplication

We illustrate another property of real numbers in the next example.

EXAMPLE 2 Illustrating the Commutative Properties

(a) $4 + 7 = 11$ and
$7 + 4 = 11$ so
$4 + 7 = 7 + 4$

(b) $3 \cdot 8 = 24$ and
$8 \cdot 3 = 24$ so
$3 \cdot 8 = 8 \cdot 3$ ∎

The results of Example 2 are true in general.

COMMUTATIVE PROPERTIES OF ADDITION AND MULTIPLICATION

If a and b are real numbers, then

$$a + b = b + a \quad \text{and} \quad a \cdot b = b \cdot a$$

In Words
The **Commutative Property** of real numbers states that the order in which we add or multiply real numbers does not affect the final result.

We add real numbers from left to right. We multiply real numbers from left to right. The **Commutative Property** allows us to write $3 + 5$ as $5 + 3$ or $3 \cdot 5$ as $5 \cdot 3$ without affecting the value of the expression. Why is this important? As the next example illustrates, by rearranging addition or multiplication problems, some expressions become a lot easier to evaluate.

EXAMPLE 3 Using the Commutative Property of Addition

Evaluate the expression: $18 + 3 + (-18)$

Solution

$$18 + 3 + (-18) = 18 + (-18) + 3$$

Add $18 + (-18)$: $= 0 + 3$

Add $3 + 0$: $= 3$

It is also true here that if we add in order, left to right, we also obtain the sum $18 + 3 + (-18) = 21 + (-18) = 3$. Notice that rearranging the numbers made the problem easier! ∎

Does subtraction obey the Commutative Property? In other words, does $3 - 14 = 14 - 3$? Because $3 - 14 = 3 + (-14) = -11$ while $14 - 3 = 14 + (-3) = 11$, we see that **subtraction is not commutative.**

QUICK ✓ *Use the Commutative Property of Addition to find the sum of the real numbers.*

4. $(-8) + 22 + 8$ **5.** $\dfrac{8}{15} + \dfrac{3}{20} + \left(-\dfrac{8}{15}\right)$ **6.** $2.1 + 11.98 + (-2.1)$

EXAMPLE 4 Using the Commutative Property of Multiplication

Find each product.

(a) $-27 \cdot 7 \cdot \left(-\dfrac{2}{9}\right)$ **(b)** $100 \cdot 307.5 \cdot 0.01$

Solution

(a)
$$-27 \cdot 7 \cdot \left(-\frac{2}{9}\right) = -27 \cdot \left(-\frac{2}{9}\right) \cdot 7$$
$$= -\overset{3}{27} \cdot \left(-\frac{2}{\underset{}{9}}\right) \cdot 7$$
$$= -3 \cdot (-2) \cdot \overset{1}{7}$$
$$= 6 \cdot 7$$
$$= 42$$

(b)
$$100 \cdot 307.5 \cdot 0.01 = 100 \cdot 0.01 \cdot 307.5$$
$$= 1 \cdot 307.5$$
$$= 307.5$$

QUICK ✓ *Find the product of the real numbers.*

7. $-8 \cdot (-13) \cdot \left(-\frac{3}{4}\right)$ **8.** $\frac{5}{22} \cdot \frac{18}{331} \cdot \left(-\frac{44}{5}\right)$ **9.** $100,000 \cdot 349 \cdot 0.00001$

Work Smart

Neither division nor subtraction is commutative.

Does division obey the Commutative Property? That is, does $a \div b = b \div a$? To see the answer, we ask does $8 \div 2 = 2 \div 8$? Because $8 \div 2 = \frac{8}{2} = 4$, but $2 \div 8 = \frac{2}{8} = \frac{1}{4}$, we conclude that **division is not commutative.**

③ **Understand and Use the Associative Properties of Addition and Multiplication**

Sometimes **grouping symbols,** such as parentheses ( ), brackets [ ], or braces { } are used to indicate that the operation within the grouping symbols is to be performed first. For example, $3 \cdot (8 + 3)$ indicates that we should first add 8 and 3 and then multiply this sum by 3.

Earlier we mentioned that addition is to be performed from left to right. We also stated that multiplication should be performed from left to right. But does the order in which we add (or multiply) three or more numbers matter? Let's see.

EXAMPLE 5 **Illustrating the Associate Properties**

(a) $2 + (8 + 6) = 2 + 14 = 16$ and $(2 + 8) + 6 = 10 + 6 = 16$

so
$$2 + (8 + 6) = (2 + 8) + 6$$

(b) $-4 \cdot (9 \cdot 2) = -4 \cdot (18) = -72$ and $(-4 \cdot 9) \cdot 2 = -36 \cdot 2 = -72$

so
$$-4 \cdot (9 \cdot 2) = (-4 \cdot 9) \cdot 2$$

Examples 4(a) and (b) illustrate the **Associative Properties of Addition and Multiplication.**

ASSOCIATIVE PROPERTIES OF ADDITION AND MULTIPLICATION

If $a, b,$ and c are real numbers, then

$$a + (b + c) = (a + b) + c = a + b + c$$
$$a \cdot (b \cdot c) = (a \cdot b) \cdot c = a \cdot b \cdot c$$

EXAMPLE 6   **Using the Associative Property of Addition**

Use the Associative Property to evaluate $23 + 453 + (-453)$.

Solution

There are three numbers to be added. However, we see that the second and third numbers are additive inverses, so we use the Associative Property of Addition and insert parentheses around $453 + (-453)$ and perform this operation first.

$$23 + 453 + (-453) = 23 + (453 + (-453))$$
$$= 23 + 0$$
$$= 23$$

EXAMPLE 7   **Using the Associative Property of Multiplication**

Use the Associative Property to evaluate $-\dfrac{3}{11} \cdot \dfrac{9}{4} \cdot \dfrac{8}{3}$.

Solution

There are three factors. If we look carefully at the problem, we notice that the second and third factors have common factors that can be divided out. So, we use the Associative Property of Multiplication and insert parentheses around $\dfrac{9}{4} \cdot \dfrac{8}{3}$ and perform this operation first.

$$-\frac{3}{11} \cdot \frac{9}{4} \cdot \frac{8}{3} = -\frac{3}{11} \cdot \left(\frac{9}{4} \cdot \frac{8}{3}\right)$$
$$= -\frac{3}{11} \cdot \frac{\overset{3}{\cancel{9}}}{\underset{1}{\cancel{4}}} \cdot \frac{\overset{2}{\cancel{8}}}{\underset{1}{\cancel{3}}}$$
$$= -\frac{3}{11} \cdot (3 \cdot 2)$$
$$= -\frac{3}{11} \cdot 6$$
$$= -\frac{18}{11}$$

QUICK ✓   *Use the Associative Property to evaluate each expression.*

10. $14 + 101 + (-101)$

11. $14 \cdot \dfrac{1}{5} \cdot 5$

12. $-34.2 + 12.6 + (-2.6)$

13. $\dfrac{19}{2} \cdot \dfrac{4}{38} \cdot \dfrac{50}{13}$

---

④ **Understand the Multiplication and Division Properties of 0**

We know that when we multiply any real number a by 1, the product is a. Now let's look at multiplication by zero.

> **MULTIPLICATION PROPERTY OF ZERO**
>
> For any real number a, the product of a and 0 is always 0; that is,
> $$a \cdot 0 = 0 \cdot a = 0$$

We now introduce some division properties of the number 0.

> **DIVISION PROPERTIES OF ZERO**
>
> For any nonzero real number a,
>
> **1.** The quotient of 0 and a is 0. That is, $\dfrac{0}{a} = 0$.
>
> **2.** The quotient of a and 0 is **undefined.** That is, $\dfrac{a}{0}$ is undefined.

Work Smart
Division by zero is not allowed. That is, 0 cannot be used as a divisor.

To see why these statements are true, consider the following. When we divide, we can check the quotient by multiplication. For example, $\dfrac{12}{4} = 3$ because $4 \cdot 3 = 12$. In the same way, $\dfrac{0}{4} = 0$ because $4 \cdot 0 = 0$. However, what is the value of $\dfrac{12}{0}$? To determine this quotient, we should be able to determine a real number such that $0 \cdot \square = 12$. But since the product of 0 and every real number is 0, there is no replacement value for $\square$.

EXAMPLE 8   Using Zero as a Divisor and a Dividend

Find the quotient:

(a) $\dfrac{23}{0}$ (b) $\dfrac{0}{17}$

Solution

(a) $\dfrac{23}{0}$ is undefined because 0 is the divisor.

(b) $\dfrac{0}{17} = 0$ because 0 is the dividend.

QUICK ✓ *Tell if the quotient is zero or undefined.*

14. $\dfrac{0}{22}$ **15.** $\dfrac{-11}{0}$ **16.** $-\dfrac{0}{5}$ **17.** $\dfrac{5678}{0}$

We conclude this section with a summary of the properties of addition, multiplication, and division.

Work Smart
The Commutative Property changes **order** and the Associative Property changes **grouping.**

> **SUMMARY: Properties of Addition**
>
> **Identity Property of Addition**
> For any real number a, $0 + a = a + 0 = a$.
>
> **Commutative Property of Addition**
> If a and b are real numbers, then $a + b = b + a$.
>
> **Additive Inverse Property**
> For any real number a, $a + (-a) = -a + a = 0$.
>
> **Associative Property of Addition**
> If a, b, and c are real numbers, then $a + (b + c) = (a + b) + c$.

> ### SUMMARY: Properties of Multiplication and Division
>
> **Commutative Property of Multiplication**
> If a and b are real numbers, then $a \cdot b = b \cdot a$.
>
> **Multiplication Property of Zero**
> For any real number a, the product of a and 0 is always 0; that is, $a \cdot 0 = 0 \cdot a = 0$.
>
> **Multiplicative Identity**
> $a \cdot 1 = 1 \cdot a = a$ for any real number a.
>
> **Associative Property of Multiplication**
> If a, b, and c are real numbers, then $a \cdot (b \cdot c) = (a \cdot b) \cdot c$.
>
> **Multiplicative Inverse Property**
>
> $$a \cdot \frac{1}{a} = \frac{1}{a} \cdot a = 1 \quad \text{provided that } a \neq 0$$
>
> **Division Properties of Zero**
> For any nonzero number a,
>
> **1.** The quotient of 0 and a is 0. That is, $\dfrac{0}{a} = 0$.
>
> **2.** The quotient of a and 0 is undefined. That is, $\dfrac{a}{0}$ is undefined.

1.5 Exercises

For Extra Help: Student Solutions Manual · CD Video · PH Math/Tutor Center · MathXL Tutorials on CD · MathXL® · MyMathLab

Concepts and Vocabulary

In Problems 1–3, fill in the blank.

1. The _____ _____ of _____ states that for any real number a, $0 + a = a + 0 = a$.

2. Because $a \cdot 1 = 1 \cdot a = a$ for any real number a, we call 1 the _____ _____.

3. The quotient of a and 0 is _____.

In Problems 4–6, answer True or False to each statement.

4. Division is commutative.

5. Addition is associative.

6. The Commutative Property of Addition says that changing the order of an addition problem will not change the sum.

7. How is the Identity Property of Addition related to the Additive Inverse Property? How is the Multiplicative Identity Property related to the Multiplicative Inverse Property?

8. In your own words, explain the Associative Property of Addition. Explain the key elements to look for when identifying this property and when you would use it to evaluate an expression.

Skill Building

In Problems 9–20, state the property of real numbers that is being illustrated.

9. $16 + (-16) = 0$

10. $4 \cdot 63 \cdot \dfrac{1}{4} = 4 \cdot \dfrac{1}{4} \cdot 63$

11. $\dfrac{3}{4}$ is equivalent to $\dfrac{3}{4} \cdot \dfrac{5}{5}$

12. $4 + 5 + (-4)$ is equivalent to $4 + (-4) + 5$

13. $12 \cdot \dfrac{1}{12} = 1$

14. $-236 + 236 = 0$

15. $34.2 + (-34.2) = 0$

16. $(4 \cdot 5) \cdot 7 = 4 \cdot (5 \cdot 7)$

17. $\dfrac{2}{3} \cdot \left(-\dfrac{12}{43}\right) \cdot \dfrac{3}{2} = \dfrac{2}{3} \cdot \dfrac{3}{2} \cdot \left(-\dfrac{12}{43}\right)$

18. $\dfrac{5}{12} \cdot \dfrac{12}{5} = 1$

19. $5.23 + 4.98 + (-5.23) = 5.23 + (-5.23) + 4.98$

20. $16.4 \cdot 0 = 0$

In Problems 21–30, convert each measurement to the indicated unit of measurement. Use the following conversions:

1 foot = 12 inches	3 feet = 1 yard	1 gallon = 4 quarts
100 centimeters = 1 meter	16 ounces = 1 pound	

21. 13 feet to inches

22. 130 feet to yards

23. 4500 centimeters to meters

24. 5900 centimeters to meters

25. 42 quarts to gallons

26. 58 quarts to gallons

27. 180 ounces to pounds

28. 120 ounces to pounds

29. 16,200 seconds to hours

30. 22,500 seconds to hours

Mixed Practice

In Problems 31–50, evaluate each expression by using the properties of real numbers.

31. $54 + 29 + (-54)$

32. $46 + 59 + (-46)$

33. $\dfrac{9}{5} \cdot \dfrac{5}{9} \cdot 18$

34. $\dfrac{4}{9} \cdot \dfrac{9}{4} \cdot 28$

35. $-25 \cdot 13 \cdot \dfrac{1}{5}$

36. $36 \cdot (-12) \cdot \dfrac{1}{6}$

37. $347 + 456 + (-456)$

38. $593 + 306 + (-306)$

39. $\dfrac{9}{2} \cdot \left(-\dfrac{10}{3}\right) \cdot 6$

40. $\dfrac{13}{2} \cdot \dfrac{8}{39} \cdot \dfrac{39}{4}$

41. $\dfrac{7}{0}$

42. $\dfrac{0}{100}$

43. $100(-34)(0.01)$

44. $4000(0.5)(0.001)$

45. $569.003 \cdot 0$

46. $104 \cdot \dfrac{1}{104}$

47. $\dfrac{45}{3902} + \left(-\dfrac{45}{3902}\right)$

48. $30 \cdot \dfrac{4}{4}$

49. $-\dfrac{5}{44} \cdot \dfrac{80}{3} \cdot \dfrac{11}{5}$

50. $\dfrac{7}{48} \cdot \left(-\dfrac{21}{4}\right) \cdot \dfrac{12}{7}$

Applying the Concepts

51. Balancing the Checkbook Alberto's checking account balance at the start of the month was $321.03. During the month, he wrote checks for $32.84, $85.03, and $120.56. He also deposited a check for $120.56. What is Alberto's balance at the end of the month?

52. Stock Price Before the opening bell on Monday, a certain stock was priced at $32.04. On Monday the stock was up $0.54, on Tuesday it was down $0.32, and on Wednesday it was down $0.54. What was the closing price of the stock on Wednesday?

53. In your own words, explain why 0 does not have a multiplicative inverse.

54. Why does $2(4 \cdot 5)$ not equal $(2 \cdot 4) \cdot (2 \cdot 5)$?

55. Why does $\dfrac{0}{4} = 0$? Why is $\dfrac{4}{0}$ undefined?

Extending the Concepts

In Problems 56–59, insert parentheses to make the statement true.

56. $-3 - 4 - 10 = 3$

57. $-6 - 4 + 10 = -20$

58. $-15 + 10 - 4 - 8 = -1$

59. $25 - 6 - 10 - 1 = 28$

60. Convert 30 miles per hour to feet per second. (*Note:* 1 mile = 5280 feet)

61. Convert 40 miles per hour to feet per second. (*Note:* 1 mile = 5280 feet)

1.6 Exponents and the Order of Operations

OBJECTIVES

1. Evaluate Exponential Expressions
2. Apply the Rules for Order of Operations

Preparing for Exponents and the Order of Operations

Before getting started, take this readiness quiz. If you get a problem wrong, go back to the section cited and review the material.

1. Find the sum: $9 + (-19)$ [Section 1.3, pp. 18–21]

2. Find the difference: $28 - (-7)$ [Section 1.3, pp. 21–23]

3. Find the product: $-7 \cdot \dfrac{8}{3} \cdot 36$ [Section 1.4, p. 29]

4. Find the quotient: $\dfrac{100}{-15}$ [Section 1.3, pp. 24–26]

① Evaluate Exponential Expressions

Suppose that we wanted to multiply 2 eight times. We would write this as

$$2 \cdot 2 \cdot 2 \cdot 2 \cdot 2 \cdot 2 \cdot 2 \cdot 2$$

That's a lot of writing! To reduce the amount of writing that is needed to represent this repeated multiplication, we introduce some new notation. Using this new notation, called **exponential notation,** we would write $2 \cdot 2 \cdot 2 \cdot 2 \cdot 2 \cdot 2 \cdot 2 \cdot 2$ as 2^8. The number 2 is called the **base** and the number 8 is called the **exponent.**

> **EXPONENTIAL NOTATION**
>
> If n is a natural number and a is a real number, then
>
> $$a^n = \underbrace{a \cdot a \cdot a \cdot \cdots \cdot a}_{n \text{ factors}}$$
>
> where a is called the **base** and the natural number n is called the **exponent** or **power.** The exponent tells the number of times the base is used as a factor.

An expression written in the form a^n is said to be in **exponential form.** The expression 6^2 is read "six squared," 8^3 is read "eight cubed," and the expression 11^4 is read "eleven to the fourth power." In general, we read a^n as "a to the nth power."

EXAMPLE 1   Writing a Numerical Expression in Exponential Form

Write each expression in exponential form.

    (a) $5 \cdot 5 \cdot 5$ **(b)** $(-4) \cdot (-4) \cdot (-4) \cdot (-4) \cdot (-4) \cdot (-4)$

Solution

    (a) The expression $5 \cdot 5 \cdot 5$ contains three factors of 5, so $5 \cdot 5 \cdot 5 = 5^3$.

    (b) The expression $(-4) \cdot (-4) \cdot (-4) \cdot (-4) \cdot (-4) \cdot (-4)$ contains 6 factors of -4, so $(-4) \cdot (-4) \cdot (-4) \cdot (-4) \cdot (-4) \cdot (-4) = (-4)^6$.

QUICK ✓ *Write each expression in exponential form.*

1. $11 \cdot 11 \cdot 11 \cdot 11 \cdot 11$ **2.** $7 \cdot 7 \cdot 7 \cdot 7 \cdot 7 \cdot 7 \cdot 7 \cdot 7$ **3.** $(-2) \cdot (-2) \cdot (-2)$

Preparing for...Answers **1.** -10 **2.** 35
3. -672 **4.** $-\dfrac{20}{3}$

To evaluate an exponential expression, we write the expression in **expanded form.** For example, 2^8 in expanded form would be $2 \cdot 2 \cdot 2 \cdot 2 \cdot 2 \cdot 2 \cdot 2 \cdot 2$. We then perform the multiplication.

EXAMPLE 2 **Evaluating an Exponential Expression**

Evaluate each exponential expression:

(a) 6^4 　　　　　(b) $\left(\dfrac{5}{3}\right)^5$

Solution

(a) $6^4 = 6 \cdot 6 \cdot 6 \cdot 6$

$\qquad = 1296$

(b) $\left(\dfrac{5}{3}\right)^5 = \left(\dfrac{5}{3}\right)\left(\dfrac{5}{3}\right)\left(\dfrac{5}{3}\right)\left(\dfrac{5}{3}\right)\left(\dfrac{5}{3}\right)$

$\qquad = \dfrac{5 \cdot 5 \cdot 5 \cdot 5 \cdot 5}{3 \cdot 3 \cdot 3 \cdot 3 \cdot 3}$

$\qquad = \dfrac{3125}{243}$

EXAMPLE 3 **Evaluating an Exponential Expression—Odd Exponent**

Evaluate each exponential expression:

(a) $(-5)^3$ 　　　　　(b) -5^3

Solution

(a) $(-5)^3 = (-5) \cdot (-5) \cdot (-5)$

$\qquad = -125$

(b) $-5^3 = -(5 \cdot 5 \cdot 5)$

$\qquad = -125$

EXAMPLE 4 **Evaluating an Exponential Expression—Even Exponent**

Evaluate each exponential expression:

(a) $(-5)^4$ 　　　　　(b) -5^4

Work Smart

There is a difference between finding $(-5)^4$ and -5^4.

The parentheses around (-5) tell us to use four factors of -5. However, in the expression -5^4, we use 5 as a factor four times and then multiply the result by -1. We could also read -5^4 as "Take the opposite of the quantity 5^4."

Solution

(a) $(-5)^4 = (-5) \cdot (-5) \cdot (-5) \cdot (-5)$

$\qquad = 625$

(b) $-5^4 = -(5 \cdot 5 \cdot 5 \cdot 5)$

$\qquad = -625$

QUICK ✓ *Evaluate each exponential expression.*

4. 2^4 　　　　　**5.** $(-8)^2$ 　　　　　**6.** $\left(-\dfrac{1}{6}\right)^3$

7. $(0.9)^2$ 　　　　　**8.** -2^4 　　　　　**9.** $(-2)^4$

---

(2) **Apply the Rules for Order of Operations**

Suppose you wish to evaluate the mathematical expression $3 \cdot 5 + 4$. Do you multiply first, then add to get $15 + 4 = 19$ *or* add first, then multiply to get $3 \cdot 9 = 27$?

Because $3 \cdot 5$ is equivalent to $5 + 5 + 5$, we have

In Words

Multiply first, then add.

$$3 \cdot 5 + 4 = 5 + 5 + 5 + 4$$

$$= 19$$

Based on this, **whenever the two operations of addition and multiplication appear in the same expression, the multiplication operation is always performed first, followed by the addition operation.**

Because any division problem can be written as a multiplication problem, we perform division before addition as well. Also, because any subtraction problem can be written as an addition problem, we perform multiplication and division before addition and subtraction.

EXAMPLE 5   **Finding the Value of an Expression Containing Multiplication and Addition**

Evaluate each expression:

(a) $11 + 2 \cdot (-6)$

(b) $7 + 12 \div 3 \cdot 5$

Solution

(a)   Multiply first: $11 + 2 \cdot (-6) = 11 + (-12)$
   Add: $= -1$

(b)   Multiply/divide left to right: $7 + 12 \div 3 \cdot 5 = 7 + 4 \cdot 5$
   Multiply: $= 7 + 20$
   $= 27$

QUICK ✓ *Evaluate each expression.*

10. $1 + 7 \cdot 2$

11. $-11 \cdot 3 + 2$

12. $18 + 3 \div \left(-\dfrac{1}{2}\right)$

13. $9 \cdot 4 - 5$

14. $\dfrac{15}{2} \div (-5) - \dfrac{3}{2}$

Parentheses

If we want to add two numbers first and then multiply, we use parentheses and write $(3 + 5) \cdot 4$. In other words, **the computation in any expression in parentheses is always done first.**

EXAMPLE 6   **Finding the Value of an Expression Containing Parentheses**

Evaluate each expression:

(a) $(5 + 3) \cdot 2$

(b) $\left(\dfrac{3}{2} - \dfrac{5}{2}\right)\left(\dfrac{7}{3} + \dfrac{2}{3}\right)$

Solution

(a) $(5 + 3) \cdot 2 = 8 \cdot 2$
   $= 16$

(b) $\left(\dfrac{3}{2} - \dfrac{5}{2}\right)\left(\dfrac{7}{3} + \dfrac{2}{3}\right) = \left(-\dfrac{2}{2}\right)\left(\dfrac{9}{3}\right)$
   $= (-1)(3)$
   $= -3$

QUICK ✓ *Evaluate each expression.*

15. $(8 \cdot 2) + 3$

16. $(2 - 9) \cdot (5 + 4)$

17. $\left(\dfrac{6}{7} + \dfrac{8}{7}\right) \cdot \left(\dfrac{11}{8} + \dfrac{5}{8}\right)$

Work Smart

The division bar (also called the *vinculum*) acts like parentheses. The word "vinculum" means "a link." So the division bar "ties things together."

The Division Bar

Another type of problem involves dividing numerical expressions that contain a division bar. When we divide numerical expressions in this form, we agree to treat the terms above and below the division bar as if they were in parentheses. To evaluate the numerical expression $\dfrac{3+5}{9+7}$, we see that

$$\frac{3+5}{9+7} = \frac{(3+5)}{(9+7)} = \frac{8}{16} = \frac{8 \cdot 1}{8 \cdot 2} = \frac{1}{2}$$

EXAMPLE 7 Finding the Value of an Expression That Contains a Division Bar

Evaluate each expression:

(a) $\dfrac{7 \cdot 3}{3 + 9 \cdot 2}$

(b) $\dfrac{1 + 7 \div \frac{1}{5}}{-6 \cdot 2 + 8}$

Solution

(a) Multiply: $\dfrac{7 \cdot 3}{3 + 9 \cdot 2} = \dfrac{21}{3 + 18}$

Add: $= \dfrac{21}{21}$

$= 1$

(b) Write division as multiplication: $\dfrac{1 + 7 \div \frac{1}{5}}{-6 \cdot 2 + 8} = \dfrac{1 + 7 \cdot 5}{-6 \cdot 2 + 8}$

Multiply: $= \dfrac{1 + 35}{-12 + 8}$

Add: $= \dfrac{36}{-4}$

$= \dfrac{9 \cdot 4}{-1 \cdot 4}$

$= -9$

QUICK ✓ Evaluate each expression.

18. $\dfrac{2 + 5 \cdot 6}{-3 \cdot 8 - 4}$

19. $\dfrac{(12 + 14) \cdot 2}{13 \cdot 2 + 13 \cdot 5}$

20. $\dfrac{4 + 3 \div \frac{1}{7}}{2 \cdot 9 - 3}$

In Words

When there is more than one set of grouping symbols in an expression, evaluate the innermost grouping symbol first, and work outward.

Embedded Grouping Symbols

We now introduce more grouping symbols. Grouping symbols include parentheses ( ), brackets [ ], and braces { }, and absolute value symbols, | |. They are used to group numbers and mathematical expressions together so that operations within the grouping symbols are performed first. **When multiple pairs of grouping symbols exist and are nested inside one another, we evaluate the information in the innermost grouping symbol first and work our way outward.**

EXAMPLE 8 Finding the Value of an Expression Containing Grouping Symbols

Evaluate each expression:

(a) $2 \cdot [3 \cdot (6 + 3) - 7]$

(b) $\left[4 + \left(\dfrac{2}{3} \cdot (-9) \right) \right] \cdot 3$

Solution

(a) Perform the operation in parentheses first: $2 \cdot [3 \cdot (6 + 3) - 7] = 2 \cdot [3 \cdot 9 - 7]$

$$= 2 \cdot [27 - 7]$$

Perform the operations in brackets, multiply first:

$$= 2 \cdot [20]$$

$$= 40$$

(b) Perform the operation in parentheses first: $\left[4 + \left(\dfrac{2}{3} \cdot (-9) \right) \right] \cdot 3 = [4 + (-6)] \cdot 3$

Perform the operation in brackets:

$$= -2 \cdot 3$$

$$= -6$$

QUICK ✓ *Evaluate each expression.*

21. $4 \cdot [2 \cdot (3 + 7) - 15]$ **22.** $2 \cdot \{4 \cdot [26 - (9 + 7)] - 15\} - 10$

In Words
Evaluate exponents before multiplication.

When do we evaluate exponents in the order of operations? Consider the expression $2 \cdot 4^3$. Do we multiply first and then evaluate the exponent to obtain $2 \cdot 4^3 = 8^3 = 512$, or do we evaluate the exponent first and then multiply to obtain $2 \cdot 4^3 = 2 \cdot 64 = 128$? Because $2 \cdot 4^3 = 2 \cdot 4 \cdot 4 \cdot 4 = 128$, we **evaluate exponents before multiplication.**

EXAMPLE 9 **Finding the Value of an Expression Containing Exponents**

Evaluate each of the following:

(a) $2 + 7(-4)^2$ **(b)** $\dfrac{2 \cdot 3^2 + 4}{3(2 - 6)}$

Solution

(a) Evaluate the exponent: $2 + 7(-4)^2 = 2 + 7 \cdot 16$

Multiply: $= 2 + 112$

Add: $= 114$

(b) Evaluate the exponent: $\dfrac{2 \cdot 3^2 + 4}{3(2 - 6)} = \dfrac{2 \cdot 9 + 4}{3(-4)}$

Find products: $= \dfrac{18 + 4}{-12}$

Add terms in numerator: $= \dfrac{22}{-12}$

Write in lowest terms: $= \dfrac{2 \cdot 11}{2 \cdot -6}$

$$= -\dfrac{11}{6}$$

QUICK ✓ *Evaluate each of the following:*

23. $\dfrac{7 - 5^2}{2}$ **24.** $3(7 - 3)^2$ **25.** $\dfrac{(-3)^2 + 7(1 - 3)}{3 \cdot 2 + 5}$ **26.** $2 + 5 \cdot 3^2 - \dfrac{3}{2} \cdot 2^2$

We now summarize the rules for performing operations on mathematical expressions.

> **ORDER OF OPERATIONS**
>
> Perform all operations within *grouping symbols* first. When an expression has nested grouping symbols, begin within the innermost pair of grouping symbols and work outward.
>
> **Step 1:** Evaluate expressions containing exponents.
>
> **Step 2:** Perform *multiplication and division* in the order they occur, working from *left to right*.
>
> **Step 3:** Perform *addition and subtraction* in the order they occur, working from *left to right*.

EXAMPLE 10 **How to Evaluate an Expression Using Order of Operations**

Evaluate: $18 + 7(2^3 - 26) + 5^2$

Step-by-Step Solution

Step 1: We evaluate the expression in the parentheses first. In the parentheses, evaluate the expression containing the exponent first.

$$18 + 7(2^3 - 26) + 5^2 = 18 + 7(8 - 26) + 25$$

Evaluate $8 - 26$ in the parentheses: $\quad = 18 + 7(-18) + 25$

Step 2: Perform multiplication and division in the order they occur, working from left to right.

Multiply $7(-18)$: $\quad = 18 - 126 + 25$

Step 3: Perform addition and subtraction in the order in which they occur, working from left to right.

$$= -108 + 25$$
$$= -83$$

EXAMPLE 11 **Evaluating a Numerical Expression Using Order of Operations**

Evaluate: $\left(\dfrac{2^3 - 6}{10 - 2 \cdot 3}\right)^2$

Solution

Evaluate exponential expression inside the parentheses:
$$\left(\dfrac{2^3 - 6}{10 - 2 \cdot 3}\right)^2 = \left(\dfrac{8 - 6}{10 - 2 \cdot 3}\right)^2$$

Multiply inside the parentheses:
$$= \left(\dfrac{8 - 6}{10 - 6}\right)^2$$

Add/subtract inside the parentheses:
$$= \left(\dfrac{2}{4}\right)^2$$

$$= \left(\dfrac{1 \cdot 2}{2 \cdot 2}\right)^2$$

Divide out common factors:
$$= \left(\dfrac{1}{2}\right)^2$$

Evaluate exponential expression:
$$= \dfrac{1}{4}$$

QUICK ✓ *Evaluate each expression.*

27. $\dfrac{(4-10)^2}{2^3-5}$　　**28.** $-3[(-4)^2-5(8-6)]^2$　　**29.** $\dfrac{(2.9+7.1)^2}{5^2-15}$

1.6 Exercises

For Extra Help:　　　　
Student Solutions Manual　　CD Video　PH Math/Tutor Center　MathXL Tutorials on CD　MathXL®　MyMathLab

Concepts and Vocabulary

In Problems 1–3, fill in the blanks.

1. In the expression 3^5, 3 is called the _____.

2. In the expression 3^5, 5 is called the _____.

3. ( ) and [ ] are called _____ _____.

In Problems 4–6, answer True or False to each statement.

4. When evaluating an expression with more than one operation, always multiply before dividing.

5. The square of 4 is 2.

6. To evaluate -3^3, multiply $-3 \cdot -3 \cdot -3$

7. What does it mean to square a number? What does it mean to cube a number? In general, what does it mean to raise a real number to the nth power?

8. Explain the difference between -3^2 and $(-3)^2$. Identify the distinguishing characteristics between the two problems, and explain how you correctly identify which number to use as the base.

Building Skills

In Problems 9–12, write in exponential form.

9. $5 \cdot 5$　　　　**10.** $4 \cdot 4 \cdot 4 \cdot 4 \cdot 4$　　**11.** $\dfrac{3}{5} \cdot \dfrac{3}{5} \cdot \dfrac{3}{5}$　　**12.** $(-2)(-2)(-2)$

In Problems 13–32, evaluate each exponential expression.

13. 8^2　　　　**14.** 4^3　　　　**15.** $(-8)^2$　　　　**16.** $(-4)^3$

17. 10^3　　　**18.** 2^4　　　**19.** $\left(\dfrac{3}{4}\right)^3$　　　**20.** $\left(\dfrac{5}{2}\right)^4$

21. $(1.5)^2$　　**22.** $(0.04)^2$　　**23.** -3^2　　　**24.** -5^4

25. -1^{20}　　**26.** $(-1)^{19}$　　**27.** 0^4　　　**28.** 1^6

29. $\left(-\dfrac{1}{2}\right)^6$　　**30.** $\left(-\dfrac{3}{2}\right)^5$　　**31.** $\left(-\dfrac{1}{3}\right)^3$　　**32.** $\left(-\dfrac{3}{4}\right)^2$

In Problems 33–64, evaluate each expression.

33. $2+3\cdot4$　　**34.** $12+8\cdot3$　　**35.** $-5\cdot3+12$　　**36.** $-3\cdot12+9$

37. $100\div2\cdot50$　　**38.** $50\div5\cdot4$　　**39.** $156-3\cdot2+10$　　**40.** $86-4\cdot3+6$

41. $(2+3)\cdot4$　　**42.** $(7-5)\cdot\dfrac{5}{2}$　　**43.** $8\div4\cdot2$　　**44.** $4\div7\cdot21$

45. $\dfrac{4+2}{2+8}$　　**46.** $\dfrac{5+3}{3+15}$　　**47.** $\dfrac{14-6}{6-14}$　　**48.** $\dfrac{15-7}{7-15}$

49. $13-[3+(-8)4]$　　　　　**50.** $12-[7+(-6)3]$

51. $(-8.75-1.25)\div(-2)$　　　**52.** $(-11.8-15.2)\div(-2)$

53. $4-2^3$　　**54.** $10-4^2$　　**55.** $15+4\cdot5^2$　　**56.** $10+3\cdot2^4$

57. $-2^3 + 3^2 \div (2^2 - 1)$

58. $-5^2 + 3^2 \div (3^2 + 9)$

59. $12 \div 6(-2)^3 - \left(\dfrac{1}{2}\right)^2$

60. $42 \div 21 \cdot (-3)^3 - \left(\dfrac{1}{2}\right)^2$

61. $-2 \cdot [5 \cdot (9 - 3) - 3 \cdot 6]$

62. $3 \cdot [6 \cdot (5 - 2) - 2 \cdot 5]$

63. $\left(\dfrac{4}{3} + \dfrac{5}{6}\right)\left(\dfrac{2}{5} - \dfrac{9}{10}\right)$

64. $\left(\dfrac{3}{4} + \dfrac{1}{2}\right)\left(\dfrac{2}{3} - \dfrac{1}{2}\right)$

Mixed Practice

In Problems 65–88, evaluate each expression.

65. $-2.5 + 4.5 \div 1.5$

66. $7.2 - 10.4 \div 5.2$

67. $4 + 2 \cdot (6 - 2)$

68. $3 + 6 \cdot (9 - 5)$

69. $\dfrac{12 - 16 \div 4 + (-24)}{16 \cdot 2 - 4 \cdot 0}$

70. $\dfrac{6 + 15 \div 3 + 16}{6 + 10 \cdot 0}$

71. $\dfrac{1}{2} + \dfrac{3}{4} \cdot \left[-2 \cdot \left(\dfrac{1}{4} + \dfrac{5}{12}\right) + \dfrac{5}{3}\right]$

72. $\dfrac{2}{5} + \dfrac{4}{11} \cdot \left[-3 \cdot \left(\dfrac{1}{5} + \dfrac{7}{20}\right) + \dfrac{3}{10}\right]$

73. $\dfrac{5^2 - 10}{3^2 + 6}$

74. $\dfrac{12(2)^3}{4^2 + 4 \cdot 5}$

75. $\left|6 \cdot (5 - 3^2)\right|$

76. $-6 \cdot (2 + |2 \cdot 3 - 4^2|)$

77. $\dfrac{81}{8} + \dfrac{13}{4} \div \dfrac{1}{2}$

78. $\dfrac{5}{12} \div \dfrac{1}{3} - \dfrac{7}{2}$

79. $\dfrac{-7}{20} + \dfrac{3}{8} \div \dfrac{1}{2}$

80. $-\dfrac{4}{5} + \dfrac{3}{10} \div \dfrac{2}{9}$

81. $\dfrac{21 - 3^2}{1 + 3}$

82. $\dfrac{5 + 3^2}{2 + 5}$

83. $\dfrac{3}{4} \cdot \left[\dfrac{5}{4} \div \left(\dfrac{3}{8} - \dfrac{1}{8}\right) - 3\right]$

84. $\left[\dfrac{9}{10} \div \left(\dfrac{2}{5} + \dfrac{1}{5}\right) + \dfrac{7}{2}\right] \cdot \dfrac{1}{10}$

85. $\left(\dfrac{4}{3}\right)^3 - \left(\dfrac{1}{2}\right)^2 \cdot \left(\dfrac{8}{3}\right) + 2 \div 3$

86. $\dfrac{1}{18} \cdot \dfrac{46}{5} - \left(\dfrac{2}{3}\right)^2$

87. $\dfrac{5^2 - 3^3}{|4 - 4^2|}$

88. $\dfrac{3 \cdot 2^3 - 2^2 \cdot 12}{3 + 3^2}$

Applying the Concepts

In Problems 89–92, express each number as the product of prime factors. Write the answer in exponential form.

89. 72

90. 675

91. 48

92. 200

In Problems 93–98, insert grouping symbols so that the expression has the desired value.

93. $4 \cdot 3 + 6 \cdot 2$ results in 36

94. $4 \cdot 7 - 4^2$ results in -36

95. $4 + 3 \cdot 4 + 2$ results in 42

96. $6 - 4 + 3 - 1$ results in 0

97. $6 - 4 + 3 - 1$ results in 4

98. $4 + 3 \cdot 2 - 1 \cdot 6$ results in 42

99. Cost of a TV The total amount paid for a flat screen television that costs \$479, plus state tax of 7.5% is found by evaluating the expression $479 + 0.075(479)$. Evaluate this expression rounded to the nearest cent.

100. Manufacturing Cost Evaluate the expression $3000 + 6(100) - \dfrac{100^2}{1000}$ to find the weekly production cost of manufacturing 100 calculators.

 △**101. Surface Area** The surface area of a right circular cylinder whose radius is 6 inches and height is 10 inches is given approximately by $2 \cdot 3.1416 \cdot 6^2 + 2 \cdot 3.1416 \cdot 6 \cdot 10$. Evaluate this expression rounded to two decimal places.

△ **102. Volume of a Cone** The volume of a cone whose radius is 3 centimeters and whose height is 12 centimeters is given approximately by $\frac{1}{3} \cdot 3.1416 \cdot 3^2 \cdot 12$.

Evaluate this expression rounded to two decimal places.

103. Investing If \$1000 is invested at 3% annual interest and remains untouched for 2 years, the amount of money that is in the account after 2 years is given by the expression $1000(1 + 0.03)^2$. Evaluate this expression, rounded to the nearest cent.

104. Investing If \$5000 is invested at 4.5% annual interest and remains untouched for 5 years, the amount of money that is in the account after 5 years is given by the expression $5000(1 + 0.045)^5$. Evaluate this expression, rounded to the nearest cent.

Extending the Concepts

The Angle Addition Postulate from geometry states that the measure of an angle is equal to the sum of the measures of its parts. Refer to the figure. Use the Angle Addition Postulate to answer Problems 105 and 106.

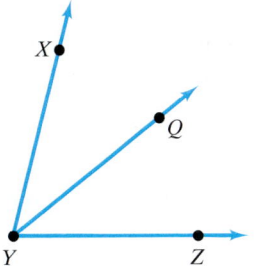

△ **105.** If $\angle XYQ = 46.5°$ and $\angle QYZ = 69.25°$, find the measure of $\angle XYZ$.

△ **106.** If $\angle QYZ = 18°$ and $\angle XYZ = 57°$, find the measure of $\angle XYQ$.

1.7 Simplifying Algebraic Expressions

OBJECTIVES

1. Evaluate Algebraic Expressions
2. Identify Like Terms and Unlike Terms
3. Use the Distributive Property
4. Simplify Algebraic Expressions by Combining Like Terms

Preparing for Simplifying Algebraic Expressions
Before getting started, take this readiness quiz. If you get a problem wrong, go back to the section cited and review the material.

1. Find the sum: $-3 + 8$ [Section 1.3, pp. 18–21]

2. Find the difference: $-7 - 8$ [Section 1.3, pp. 21–23]

3. Find the product: $-\frac{4}{3}(27)$ [Section 1.4, p. 29]

What is algebra? The word "algebra" is derived from the Arabic word, *al-jabr,* which means "restoration." Today algebra means more. According to the American Heritage Dictionary, **algebra** is a branch of mathematics in which symbols, usually letters of the alphabet, represent numbers or members of a set. These symbols are used to represent quantities and to express general relationships that hold for all members of the set. In this course, the set of numbers referred to in the definition is the set of real numbers.

1 Evaluate Algebraic Expressions

In arithmetic, we work with numbers. As stated in the definition, in algebra, we use letters such as $x, y, a, b,$ and c to represent numbers.

> **DEFINITION**
> When a letter represents any number from a set of numbers, it is called a **variable.**

The set of numbers that we use in this textbook is the set of real numbers.

> **DEFINITION**
> A **constant** is either a fixed number, such as 5, or a letter or symbol that represents a fixed number.

For example, in Einstein's Theory of Relativity, $E = mc^2$, E and m are variables that represent total energy and mass, respectively, while c is a constant that represents the speed of light (299,792,458 meters per second).

> **DEFINITION**
>
> An **algebraic expression** is any combination of variables, constants, grouping symbols, and mathematical operations such as addition, subtraction, multiplication, division, and exponents.

Some examples of algebraic expressions are

$$x - 5, \quad \frac{1}{2}x, \quad 2x - 7, \quad x^2 + 3 \quad \text{and} \quad \frac{x-1}{x+1}$$

Recall that a variable represents a number from a set of numbers. One of the procedures we perform on algebraic expressions is *evaluating an algebraic expression.*

> **DEFINITION**
>
> To **evaluate an algebraic expression,** substitute the numerical value for each variable into the expression and simplify the result.

EXAMPLE 1   Evaluating an Algebraic Expression

Evaluate each expression for the given value of the variable.

(a) $2x + 5$ for $x = 8$ (b) $a^2 - 2a + 4$ for $a = -3$

Solution

(a) We substitute 8 for x in the expression $2x + 5$:

$$2(8) + 5 = 16 + 5$$
$$= 21$$

(b) We substitute -3 for a in $a^2 - 2a + 4$:

$$(-3)^2 - 2(-3) + 4 = 9 + 6 + 4$$
$$= 19$$

EXAMPLE 2   An Algebraic Expression for Revenue

The expression $4.50x + 2.50y$ represents the total amount of money, in dollars, received at a school play where x represents the number of adult tickets sold and y represents the number of student tickets sold. Evaluate $4.50x + 2.50y$ for $x = 50$ and $y = 82$. Interpret the result.

Solution

We substitute 50 for x and 82 for y in the expression $4.50x + 2.50y$.

$$4.50(50) + 2.50(82) = 225 + 205 = 430$$

So \$430 was collected by selling 50 adult tickets and 82 student tickets.

QUICK ✓ *Evaluate each expression for the given value of the variable.*

1. $-3k + 5$ for $k = 4$ **2.** $\frac{5}{4}t - 6$ for $t = 12$ **3.** $-2y^2 - y + 8$ for $y = -2$

4. The Amadeus Coffee Shop wishes to blend two types of coffee to create a breakfast blend. They will mix x pounds of a mild coffee that sells for \$7.50 per pound with y pounds of a robust coffee that sells for \$10.00 per pound. An algebraic expression that represents the value of the breakfast blend, in dollars, is $7.50x + 10y$. Evaluate this expression for $x = 8$ and $y = 16$.

2   **Identify Like Terms and Unlike Terms**

Many algebraic expressions consist of the sum or difference of terms.

> **DEFINITION**
>
> A **term** is a constant or the product or quotient of a constant and one or more variables raised to powers.

EXAMPLE 3   **Identifying the Terms in an Algebraic Expression**

Identify the terms in the following algebraic expressions.

(a) $4a^3 + 5b^2 - 8c + 12$ 　　　　　　　　 (b) $\dfrac{x}{4} - 7y + 8z$

Solution

(a) The algebraic expression $4a^3 + 5b^2 - 8c + 12$ can be written as

$$4a^3 + 5b^2 + (-8c) + 12$$

so the terms are $4a^3$, $5b^2$, $-8c$, and 12. There are a total of 4 terms in the algebraic expression.

(b) The algebraic expression $\dfrac{x}{4} - 7y + 8z$ has three terms: $\dfrac{x}{4}$, $-7y$, and $8z$. ▬

QUICK ✓

Identify the terms in each algebraic expression.

5. $5x^2 + 3xy$ 　　　　　　　　 **6.** $9ab - 3bc + 5ac - ac^2$

7. $\dfrac{2mn}{5} - \dfrac{3n}{7}$ 　　　　　　　　 **8.** $\dfrac{m^2}{3} - 8$

> **DEFINITION**
>
> The **coefficient** of a term is the numerical factor of the term.

For example, the coefficient of $7x$ is 7; the coefficient of $-2x^2y$ is -2. For terms that have no number as a factor, such as mn, the coefficient is 1. The coefficient of $-y$ is -1 since $-y = -1 \cdot y$. If a term consists of just a constant, the coefficient is the number itself. We say the coefficient of 14 is 14.

EXAMPLE 4   **Determining the Coefficient of a Term**

Determine the coefficient of each term:

(a) $\dfrac{1}{2}xy^2$ 　　 (b) $-\dfrac{t}{12}$

Solution

(a) The coefficient of $\dfrac{1}{2}xy^2$ is $\dfrac{1}{2}$.

(b) The coefficient of $-\dfrac{t}{12}$ is $-\dfrac{1}{12}$ because $-\dfrac{t}{12}$ can be written as $-\dfrac{1}{12} \cdot t$. ▬

EXAMPLE 5 **Determining the Coefficient of a Term**

Determine the coefficient of each term:

(a) ab^3 **(b)** 12

Solution

(a) The coefficient of ab^3 is 1 because ab^3 can be written as $1 \cdot ab^3$.

(b) The coefficient of 12 is 12 because the coefficient of a constant is the number itself.

QUICK ✓ *Determine the coefficient of each term.*

9. $2z^2$ **10.** xy **11.** $-b$ **12.** 5 **13.** $-\dfrac{2}{3}z$ **14.** $\dfrac{x}{6}$

Sometimes we have to rewrite algebraic expressions by combining like terms.

Work Smart

Like terms can have different coefficients, but they cannot have different variables or different exponents on those variables.

DEFINITION

Terms that have the same variable factor(s) with the same exponent(s) are called **like terms.**

For example, $3x^2$ and $-7x^2$ are like terms because the variable x is raised to the second power, but $3x^2$ and $-7x^3$ are not like terms because the variable x is raised to the second power in $3x^2$ but to the third power in the term $-7x^3$. Constant terms are like terms: -9 and 3 are like terms.

EXAMPLE 6 **Classifying Terms as Like or Unlike**

Classify the following pairs of terms as *like* or *unlike:*

(a) $2p^3$ and $-5p^3$ **(b)** $7kr$ and $\dfrac{1}{4}k^2r$ **(c)** 5 and 8

Solution

(a) $2p^3$ and $-5p^3$ are *like* terms. They have the same variable p raised to the same power, 3.

(b) $7kr$ and $\dfrac{1}{4}k^2r$ are *unlike* terms. Although they both have the variables k and r, the variable k is raised to the first power in $7kr$ and to the second power in $\dfrac{1}{4}k^2r$.

(c) 5 and 8 are *like* terms. They are both constants.

QUICK ✓ *Tell if the terms are like or unlike.*

15. $-\dfrac{2}{3}p^2; \dfrac{4}{5}p^2$ **16.** $\dfrac{m}{6}; 4m$ **17.** $3a^2b; -2ab^2$ **18.** $8a; 11$

③ Use the Distributive Property

Now let's illustrate another property of real numbers. The Distributive Property of real numbers will be used throughout this course and in future courses.

> **THE DISTRIBUTIVE PROPERTY**
>
> If a, b, and c are real numbers, then
>
> $$a \cdot (b + c) = a \cdot b + a \cdot c$$
> $$(a + b) \cdot c = a \cdot c + b \cdot c$$
>
> That is, multiply each of the terms inside the parentheses by the factor on the outside.

Because $b - c = b + (-c)$, it is also true that $a(b - c) = a \cdot b - a \cdot c$. One of the purposes of the Distributive Property is to remove parentheses from an algebraic expression.

EXAMPLE 7   Using the Distributive Property to Remove Parentheses

Use the Distributive Property to remove the parentheses.

(a) $3(x + 5)$ (b) $-\dfrac{1}{3}(6x - 12)$

Solution

(a) To use the Distributive Property, we multiply each term in the parentheses by 3:

$$3(x + 5) = 3 \cdot x + 3 \cdot 5$$
$$= 3x + 15$$

Work Smart

The long name for the Distributive Property is The Distributive Property of Multiplication over Addition. This name helps to remind us that we do not distribute across multiplication. For example,

$$6x(5xy) \neq 6x \cdot 5x \cdot 6x \cdot y$$

(b) Multiply each term in the parentheses by $-\dfrac{1}{3}$:

$$-\frac{1}{3}(6x - 12) = -\frac{1}{3} \cdot 6x - \left(-\frac{1}{3}\right) \cdot 12$$
$$= -2x + 4$$

QUICK ✓ *Use the Distributive Property to remove the parentheses.*

19. $6(x + 2)$ **20.** $-5(x + 2)$ **21.** $-2(k - 7)$ **22.** $(8x + 12)\dfrac{3}{4}$

④ Simplify Algebraic Expressions by Combining Like Terms

An algebraic expression that contains the sum or difference of like terms may be simplified using the Distributive Property "in reverse." When we use the Distributive Property to add coefficients of like terms we say that we are **combining like terms.**

EXAMPLE 8   Using the Distributive Property to Combine Like Terms

Combine like terms:

(a) $2x + 7x$ (b) $x^2 - 5x^2$

Solution

(a) $2x + 7x = (2 + 7)x$
$$= 9x$$

(b) $x^2 - 5x^2 = (1 - 5)x^2$
$$= -4x^2$$

Look carefully at the results of Example 8. Notice that when we combine like terms, we add the coefficients of the like terms and keep the variables and exponents the same.

QUICK ✓ *Combine like terms.*

23. $3x - 8x$

24. $-5x^2 + x^2$

25. $-7x - x + 6 - 3$

26. $4x - 12x - 3 + 17$

Sometimes we must rearrange the terms in an algebraic expression using the Commutative Property of Addition.

EXAMPLE 9   **Combining Like Terms Using the Commutative Property**

Combine like terms: $4x + 5y + 12x - 7y$

Solution

Rearrange terms: $4x + 5y + 12x - 7y = 4x + 12x + 5y - 7y$

Use the Distributive Property "in reverse": $= (4 + 12)x + (5 - 7)y$

Combine like terms: $= 16x + (-2y)$

Write the answer in simplest form: $= 16x - 2y$

QUICK ✓ *Combine like terms.*

27. $3a + 2b - 5a + 7b - 4$

28. $(5ac + 2b) + (7ac - 5a) + (-b)$

29. $5ab^2 + 7a^2b + 3ab^2 - 8a^2b$

30. $\dfrac{4}{3}rs - \dfrac{3}{2}r^2 + \dfrac{2}{3}rs - 5$

Often, we need to first remove parentheses by using the Distributive Property before we can combine like terms. Recall that the rules for order of operations of real numbers place multiplication before addition or subtraction. In this section, the direction **simplify** will mean to remove all parentheses and combine like terms.

EXAMPLE 10   **Combining Like Terms Using the Distributive Property**

Simplify the algebraic expression: $3 - 4(2x + 3) + 5x - 1$

Solution

Use the Distributive Property
to remove parentheses: $3 - 4(2x + 3) + 5x - 1 = 3 - 8x - 12 + 5x - 1$

Rearrange terms using the Commutative Property of Addition: $= -8x + 5x + 3 - 12 - 1$

Combine like terms: $= -3x - 10$

Work Smart

Remember that multiplication comes before subtraction. In the first step of Example 10, do not compute $3 - 4$ first to obtain $-1(2x + 3)$.

QUICK ✓ *Simplify each expression.*

31. $3x + 2(x - 1) - 7x + 1$

32. $m + 2n - 3(m + 2n) - (7 - 3n)$

33. $2(a - 4b) - (a + 4b) + b$

34. $\dfrac{1}{2}(6x + 4) - \dfrac{1}{3}(12 - 9x)$

We summarize the steps for simplifying an algebraic expression below.

> **SUMMARY: Simplifying an Algebraic Expression**
>
> **Step 1:** Remove any parentheses using the Distributive Property.
>
> **Step 2:** Combine any like terms.

1.7 Exercises

For Extra Help:
Student Solutions Manual   CD Video   PH Math/Tutor Center   MathXL Tutorials on CD   MathXL®   MyMathLab

Concepts and Vocabulary

In Problems 1–3, fill in the blanks.

1. The _____ Property is used to remove parentheses in an expression.

2. _____ _____ are terms having the same variable with the same exponents.

3. The coefficient of the expression $-xy^2$ is _____.

In Problems 4–6, answer True or False to each statement.

4. The coefficient of $\frac{x}{8}$ is 1.

5. The terms $-3x^2$ and $9x^2$ are unlike terms.

6. The algebraic expression $\frac{3x + 5y}{4}$ contains 2 terms.

7. Explain why the sum $2x^2 + 4x^2$ is *not* equivalent to $6x^4$. What is the correct answer? Choose any value for x and then compare the results for each of these expressions.

8. Use $x = 4$ and $y = 5$ to answer parts (a), (b), and (c).
   (a) Evaluate $x^2 + y^2$.
   (b) Evaluate $(x + y)^2$.
   (c) Are the results the same? Explain your response.

Building Skills

In Problems 9–12, for each expression, identify the terms and then name the coefficient of each term.

9. $2x^3 + \dfrac{x^2}{4} - x + 6$

10. $3m^4 - m^3n^2 + \dfrac{5n}{7} - 1$

11. $z^2 + \dfrac{2y}{3}$

12. $t^3 - \dfrac{t}{4}$

In Problems 13–26, evaluate each expression using the given value of the variables.

13. $2x + 5$ for $x = 4$

14. $3x + 7$ for $x = 2$

15. $x^2 + 3x - 1$ for $x = 3$

16. $n^2 - 4n + 3$ for $n = 2$

17. $4 - k^2$ for $k = -5$

18. $-2p^2 + 5p + 1$ for $p = -3$

19. $\dfrac{5x}{y} + y^2$ for $x = 8, y = 10$

20. $m^2 - \dfrac{3n}{m}$ for $m = 2, n = 4$

21. $\dfrac{9x - 5y}{x + y}$ for $x = 3, y = 2$

22. $\dfrac{3y + 2z}{y - z}$ for $y = 4, z = -3$

23. $(x + 3y)^2$ for $x = 3, y = 4$

24. $(a - 2b)^2$ for $a = 1, b = -2$

25. $x^2 + 9y^2$ for $x = 3, y = 4$

26. $a^2 - 4b^2$ for $a = 1, b = -2$

In Problems 27–34, determine if the terms are like or unlike.

27. $8x$ and 8 **28.** $11p$ and 11 **29.** 54 and -21 **30.** -13 and 38

31. $12b$ and $-b$ **32.** $6a^2$ and $-3a^2$ **33.** r^2s and rs^2 **34.** x^2y^3 and y^2x^3

In Problems 35–42, use the Distributive Property to remove the parentheses.

35. $3(m + 2n)$ **36.** $3(4s + 2t)$ **37.** $(3n^2 + 2n - 1)6$ **38.** $(6a^4 - 4a^2 + 2)3$

39. $-(x - y)$ **40.** $-5(k - n)$ **41.** $(8x - 6y)(-0.5)$ **42.** $(16a + 12b)(-0.4)$

In Problems 43–68, simplify each expression by combining like terms, if possible.

43. $5x - 2x$ **44.** $14k - 11k$

45. $4z - 6z + 8z$ **46.** $9m - 8m + 2m$

47. $2m + 3n + 8m + 7n$ **48.** $x + 2y + 5x + 7y$

49. $0.3x^7 + x^7 + 0.9x^7$ **50.** $1.7n^4 - n^2 + 2.1n^4$

51. $-3y^6 + 13y^6$ **52.** $-7p^5 + 2p^5$

53. $-(6w + 12y - 13z)$ **54.** $-(-6m + 9n - 8p)$

55. $5(k + 3) - 8k$ **56.** $3(7 - z) - z$

57. $7n - (3n + 8)$ **58.** $18m - (6 + 9m)$

59. $-6(n - 3) + 2(n + 1)$ **60.** $-9(7r - 6) + 9(10r + 3)$

61. $\dfrac{2}{3}x + \dfrac{1}{6}x$ **62.** $\dfrac{3}{5}y + \dfrac{7}{10}y$

63. $\dfrac{1}{2}(8x + 5) - \dfrac{2}{3}(6x + 12)$ **64.** $\dfrac{1}{5}(60 - 15x) + \dfrac{3}{4}(12 - 24x)$

65. $2(0.5x + 9) - 3(1.5x + 8)$ **66.** $3(0.2x + 6) - 5(1.6x + 1)$

67. $3.2(x + 1.6) + 1.4(2x - 3.7)$ **68.** $1.8(x + 2.5) + 1.1(3x - 2.8)$

Mixed Practice

In Problems 69–80, (a) evaluate the expression for the given value(s) of the variable(s) before combining like terms, (b) simplify the expression by combining like terms and then evaluate the expression for the given value(s) of the variable(s). Compare your results.

69. $5x + 3x; x = 4$ **70.** $8y + 2y; y = -3$

71. $-2a^2 + 5a^2; a = -3$ **72.** $4b^2 - 7b^2; b = 5$

73. $4z - 3(z + 2); z = 6$ **74.** $8p - 3(p - 4); p = 3$

75. $5y^2 + 6y - 2y^2 + 5y - 3; y = -2$ **76.** $3x^2 + 8x - x^2 - 6x; x = 5$

77. $\dfrac{1}{2}(4x - 2) - \dfrac{2}{3}(3x + 9); x = 3$ **78.** $\dfrac{1}{5}(5x - 10) - \dfrac{1}{6}(6x + 12); x = -2$

79. $3a + 4b - 7a + 3(a - 2b); a = 2, b = 5$

80. $-4x - y + 2(x - 3y); x = 3, y = -2$

Applying the Concepts

In Problems 81–86, evaluate each expression using the given value of the variables.

81. $\dfrac{1}{2}h(b + B); h = 4, b = 5, B = 17$ **82.** $\dfrac{1}{2}h(b + B); h = 9, b = 3, B = 12$

83. $\dfrac{a - b}{c - d}; a = 6, b = 3, c = -4, d = -2$ **84.** $\dfrac{a - b}{c - d}; a = -5, b = -2, c = 7, d = 1$

85. $b^2 - 4ac; a = 7, b = 8, c = 1$ **86.** $b^2 - 4ac; a = 2, b = 5, c = 3$

87. Renting a Truck The cost of renting a truck from Hamilton Auto Rental is $59.95 per day plus $0.15 per mile. The expression $59.95 + 0.15m$ represents the cost of renting a truck for one day and driving it m miles. Evaluate $59.95 + 0.15m$ for $m = 125$.

88. Renting a Car The cost of renting a compact car for one day from CMH Auto is $29.95 plus $0.17 per mile. The expression $29.95 + 0.17m$ represents the total daily cost. Evaluate the expression $29.95 + 0.17m$ for $m = 245$.

89. **Ticket Sales** The Center for Science and Industry sells adult tickets for $12 and children's tickets for $7. The expression $12a + 7c$ represents the total revenue from selling a adult tickets and c children's tickets. Evaluate the algebraic expression $12a + 7c$ for $a = 156$ and $c = 421$.

90. **Ticket Sales** A community college theatre group sold tickets to a recent production. Student tickets cost $5 and nonstudent tickets cost $8. The algebraic expression $5s + 8n$ represents the total revenue from selling s student tickets and n nonstudent tickets. Evaluate $5s + 8n$ for $s = 76$ and $n = 63$.

△ 91. **Rectangle** The width of a rectangle is w yards and the length of the rectangle is $(3w - 4)$ yards. The perimeter of the rectangle is given by the algebraic expression $2w + 2(3w - 4)$.

    (a) Simplify the algebraic expression $2w + 2(3w - 4)$.
    (b) Determine the perimeter of a rectangle whose width w is 5 yards.

△ 92. **Rectangle** The length of a rectangle is l meters and the width of the rectangle is $(l - 11)$ meters. The perimeter of the rectangle is given by the algebraic expression $2l + 2(l - 11)$.

    (a) Simplify the expression $2l + 2(l - 11)$.
    (b) Determine the perimeter of a rectangle whose length l is 15 meters.

93. **Finance** Novella invested some money in two investment funds. She placed s dollars in stocks that yield 5.5% annual interest and b dollars in bonds that yield 3.25% annual interest. Evaluate the expression $0.055s + 0.0325b$ for $s = \$2950$ and $b = \$2050$. Round your answer to the nearest penny.

94. **Finance** Jonathan received an inheritance from his grandparents. He invested x dollars in a Certificate of Deposit that pays 2.95% and y dollars in an off-shore oil drilling venture that is expected to pay 12.8%. Evaluate the algebraic expression $0.0295x + 0.128y$ for $x = \$2500$ and $y = \$1000$.

Extending the Concepts

95. Explain how you can simplify this algebraic expression (cleverly!!) using the Distributive Property – in reverse! $2.75(-3x^2 + 7x - 3) - 1.75(-3x^2 + 7x - 3)$.

96. Simplify by using the Distributive Property in reverse:
$11.23(7.695x + 81.34) + 8.77(7.695x + 81.34)$.

CHAPTER 1 ACTIVITY: THE MATH GAME

Focus: Performing order of operations and simplifying expressions

Time: 10 minutes

Group size: 2–4

- The instructor will announce when the groups may begin solving the problems below.
- When your group has completed all of the problems, ask the instructor to check the answers. The instructor will tell you how many answers are correct, but not which ones.
- The first group to complete all of the problems correctly will win a prize, as determined by the instructor.

1. Evaluate: $-8 \div 2^2 \cdot 6 + (-2)^3$

2. Evaluate: $\dfrac{6(-3) + 4^2}{25 + 4(-9 + 4)}$

3. Evaluate: $x^3 - x^2$ for $x = -3$

4. Evaluate: $\dfrac{(x + 2y)^2}{xy}$ for $x = 1, y = -2$

5. Simplify: $-2(4x + 3) - (5x - 1)$

6. Simplify: $\dfrac{3}{4}(8x^2 + 16) - 2x^2 + 3x$

CHAPTER 1 REVIEW

Section 1.2 The Number Systems and the Real Number Line

KEY CONCEPTS

- $a < b$ means a is to the left of b on the real number line
- $a = b$ means a and b are in the same position on the real number line
- $a > b$ means a is to the right of b on the real number line
- $|a|$ is the distance from 0 to a on the real number line

KEY TERMS

Set	Real number line
Elements	Origin
Empty set	Scale
Natural numbers	Coordinate
Counting numbers	Negative real numbers
Whole numbers	Zero
Integers	Positive real numbers
Rational number	Sign
Irrational number	Inequality symbols
Real numbers	Absolute value

YOU SHOULD BE ABLE TO . . .	EXAMPLE	REVIEW EXERCISES
1 Classify numbers (p. 9)	Example 2	1–10, 27, 28
2 Plot points on the real number line (p. 12)	Example 3	11, 12
3 Use inequalities to order real numbers (p. 13)	Example 4	13–16, 21–26
4 Compute the absolute value of a real number (p. 14)	Example 5	17–20, 25, 26

In Problems 1–4, write each set.

1. A is the set of whole numbers less than 7.

2. B is the set of natural numbers less than or equal to 3.

3. C is the set of integers greater than -3 and less than or equal to 5.

4. D is the set of integers greater than or equal to -2 and less than 4.

In Problems 5–10, use the set $\left\{ -6, -3.25, 0, 5.030030003\ldots, \dfrac{9}{3}, 11, \dfrac{5}{7} \right\}$. *List all the elements that are*

5. natural numbers

6. whole numbers

7. integers

8. rational numbers

9. irrational numbers

10. real numbers

11. Plot the points $\left\{ -3, -\dfrac{4}{3}, 0, \dfrac{4}{2}, 3.5 \right\}$ on a real number line.

12. Plot the points $\left\{ -4, -\dfrac{5}{2}, \dfrac{6}{2}, 5.5 \right\}$ on a real number line.

In Problems 13–16, determine whether the statement is True or False.

13. $-3 > -1$ **14.** $5 \leq 5$ **15.** $-5 \leq -3$ **16.** $\dfrac{1}{2} = 0.5$

In Problems 17–20, evaluate each expression.

17. $-\left| \dfrac{1}{2} \right|$ **18.** $|-7|$ **19.** $-|-6|$ **20.** $-|-8.2|$

In Problems 21–26, replace the ? with the correct symbol: $>, <, =.$

21. $\dfrac{1}{4}$? 0.25 **22.** -6 ? 0 **23.** 0.83 ? $\dfrac{3}{4}$ **24.** -2 ? -10

25. $|-4|$? $|-3|$ **26.** $\dfrac{4}{5}$? $\left| -\dfrac{5}{6} \right|$

27. Explain the difference between a rational number and an irrational number. Be sure that your explanation includes a discussion of terminating decimals and nonterminating decimals.

28. What do we call the set of positive integers?

Section 1.3   Adding, Subtracting, Multiplying, and Dividing Integers

KEY CONCEPTS	KEY TERMS
• **Rules of Signs for Multiplying Two Integers** 1. If we multiply two positive integers, the product is positive. 2. If we multiply one positive integer and one negative integer, the product is negative. 3. If we multiply two negative integers, the product is positive. • **Rules of Signs for Dividing Two Integers** 1. If we divide two positive integers, the quotient is positive. That is, $$\frac{+a}{+b} = \frac{a}{b}$$ 2. If we divide one positive integer and one negative integer, the quotient is negative. That is, $$\frac{-a}{b} = \frac{a}{-b} = -\frac{a}{b}$$ 3. If we divide two negative integers, the quotient is positive. That is, $$\frac{-a}{-b} = \frac{a}{b}$$	Operations Sum Difference Product Quotient Mixed number Additive Inverse Opposite Evaluate Factors Dividend Divisor Multiplicative Inverse Reciprocal Ratio Golden ratio

YOU SHOULD BE ABLE TO. . .	EXAMPLE	REVIEW EXERCISES
① Add integers (p. 18)	Examples 1 through 6	29–36, 41–42, 59–60, 63, 69–70
② Determine the additive inverse of a number (p. 21)	Example 7	57–58
③ Subtract integers (p. 21)	Examples 8 through 10	37–42, 61, 62, 64, 69–71
④ Multiply integers (p. 23)	Examples 11 through 13	43–48, 65–66, 72
⑤ Divide integers (p. 24)	Example 15	49–56, 67–68

In Problems 29–56, perform the indicated operation.

29. $-2 + 9$ **30.** $6 + (-10)$ **31.** $-23 + (-11)$

32. $-120 + 25$ **33.** $-|-2 + 6|$ **34.** $-|-15| + |-62|$

35. $-110 + 50 + (-18) + 25$ **36.** $-28 + (-35) + (-52)$

37. $-10 - 12$ **38.** $18 - 25$ **39.** $-11 - (-32)$

40. $0 - (-67)$ **41.** $34 - 18 + 10$ **42.** $-49 - 8 + 21$

43. $-6(-2)$ **44.** $4(-10)$ **45.** $13(-86)$

46. -19×423 **47.** $(11)(13)(-5)$ **48.** $(-53)(-21)(-10)$

49. $\dfrac{-20}{-4}$ **50.** $\dfrac{60}{-5}$ **51.** $\dfrac{|-55|}{11}$

52. $-\left|\dfrac{-100}{4}\right|$ **53.** $\dfrac{120}{-15}$ **54.** $\dfrac{64}{-20}$

55. $\dfrac{-180}{54}$ **56.** $\dfrac{-450}{105}$

In Problems 57 and 58, determine the additive inverse of each number.

57. 13   **58.** -45

In Problems 59–68, write the expression using mathematical symbols, and then evaluate the expression.

59. -43 plus 101   **60.** 45 plus -28

61. -10 minus -116 **62.** 74 minus 56

63. the sum of 13 and -8 **64.** the difference between -60 and -10

65. -21 multiplied by -3 **66.** 54 multiplied by -18

67. -34 divided by -2 **68.** -49 divided by 14

69. Football Clinton Portis had three possessions of the football within the first few minutes of the game. On his first possession he gained 20 yards, on his second possession he lost 6 yards, and on his third possession he gained 12 yards. What was his total yardage?

70. Temperature On a winter day in Detroit, Michigan, the temperature was 10°F in the morning. The temperature rose 12°F in the afternoon, and then fell 25°F by midnight. What was the temperature at midnight in Detroit?

71. Temperature One day in Bismarck, North Dakota, the high temperature was 6°F above zero and the low temperature was 18°F below zero. What was the difference between the high and the low temperature on that day in Bismarck?

72. Test Score Ms. Rosen awards 5 points for each correct multiple choice question and awards 8 points for each correct free response question. On one of Ms. Rosen's tests, Sarah got 11 multiple choice questions correct and 4 free response questions correct. What was Sarah's test score?

Section 1.4	Adding, Subtracting, Multiplying, and Dividing Rational Numbers Expressed as Fractions and Decimals

KEY CONCEPTS	KEY TERM
• **Multiplying Fractions**	Least common denominator

• **Multiplying Fractions**

$$\frac{a}{b} \cdot \frac{c}{d} = \frac{a \cdot c}{b \cdot d}, \text{ where } b \text{ and } d \neq 0$$

• **Dividing Fractions**

$$\frac{a}{b} \div \frac{c}{d} = \frac{a}{b} \cdot \frac{d}{c} = \frac{a \cdot d}{b \cdot c}, \text{ where } b, c, d \neq 0$$

• **Adding or Subtracting Fractions with the Same Denominator**

$$\frac{a}{c} + \frac{b}{c} = \frac{a + b}{c}, \text{ where } c \neq 0$$

$$\frac{a}{c} - \frac{b}{c} = \frac{a - b}{c} = \frac{a + (-b)}{c}, \text{ where } c \neq 0$$

• **Adding or Subtracting Fractions with the Unlike Denominators**

Step 1: Find the LCD of the fractions.

Step 2: Find equivalent fractions with the LCD by multiplying by a factor of 1.

Step 3: Add or subtract the numerators and write the result over the common denominator.

Step 4: Simplify the result.

YOU SHOULD BE ABLE TO . . .	EXAMPLE	REVIEW EXERCISES
1 Multiply rational numbers expressed as fractions (p. 29)	Example 1	77–80, 110
2 Divide rational numbers expressed as fractions (p. 30)	Example 2	81–84
3 Add and subtract rational numbers expressed as fractions (p. 31)	Examples 3 through 7	85–96, 111
4 Add, subtract, multiply, and divide rational numbers expressed as decimals (p. 34)	Examples 8 through 11	97–109, 112

In Problems 73–76, write each rational number in lowest terms.

73. $\dfrac{32}{64}$ **74.** $-\dfrac{27}{81}$ **75.** $\dfrac{-100}{150}$ **76.** $\dfrac{35}{-25}$

In Problems 77–96, perform the indicated operation. Write in lowest terms if necessary.

77. $\dfrac{2}{3} \cdot \dfrac{15}{8}$ **78.** $-\dfrac{3}{8} \cdot \dfrac{10}{21}$ **79.** $\dfrac{5}{8} \cdot \left(-\dfrac{2}{25}\right)$ **80.** $5 \cdot \left(-\dfrac{3}{10}\right)$

81. $\dfrac{24}{17} \div \dfrac{18}{3}$ **82.** $-\dfrac{5}{12} \div \dfrac{10}{16}$ **83.** $-\dfrac{27}{10} \div 9$ **84.** $20 \div \left(-\dfrac{5}{8}\right)$

85. $\dfrac{2}{9} + \dfrac{1}{9}$ **86.** $-\dfrac{6}{5} + \dfrac{4}{5}$ **87.** $\dfrac{5}{7} - \dfrac{2}{7}$ **88.** $\dfrac{7}{5} - \left(-\dfrac{8}{5}\right)$

89. $\dfrac{3}{10} + \dfrac{1}{20}$ **90.** $\dfrac{5}{12} + \dfrac{4}{9}$ **91.** $-\dfrac{7}{35} - \dfrac{2}{49}$ **92.** $\dfrac{5}{6} - \left(-\dfrac{1}{4}\right)$

93. $-2 - \left(-\dfrac{5}{12}\right)$ **94.** $-5 + \dfrac{9}{4}$

95. $-\dfrac{1}{10} + \left(-\dfrac{2}{5}\right) + \dfrac{1}{2}$ **96.** $-\dfrac{5}{6} - \dfrac{1}{4} + \dfrac{3}{24}$

In Problems 97–108, perform the indicated operation.

97. $30.3 + 18.2$ **98.** $-43.02 + 18.36$ **99.** $201.37 - 118.39$

100. $-35.1 - 18.64$ **101.** $(-0.04)(-2.01)$ **102.** $(87.3)(-2.98)$

103. $\dfrac{69.92}{3.8}$ **104.** $-\dfrac{1.08318}{0.042}$ **105.** $12.5 - 18.6 + 8.4$

106. $-13.5 + 10.8 - 20.2$ **107.** $12.9 \times 1.4 \times (-0.3)$ **108.** $2.4 \times 6.1 \times (-0.05)$

109. Checking Account Lee had a balance of \$256.75 in her checking account. Lee wrote a check for \$175.68 on Wednesday and wrote a check for \$180.00 on Thursday. What is her checking account balance now? Is Lee's account overdrawn?

110. Super Bowl Party Jarred had 36 friends at his Super Bowl XXXVIII party. Two-thirds of his friends wanted the Carolina Panthers to win. How many of Jarred's friends wanted the Panthers to win the Super Bowl?

111. Ribbon Cutting Tara has a piece of ribbon that is 15 inches long. If she cuts off a $3\dfrac{1}{2}$ inch piece of the ribbon, what is the length of the piece that remains?

112. Buying Clothes While shopping at her favorite store, Sierra bought 5 sweaters. If the sweaters cost \$35 each and sales tax is 6.75% of the net price (net price = price × quantity), how much did Sierra spend on the clothes?

Section 1.5	Properties of Real Numbers	
KEY CONCEPTS		**KEY TERMS**

- **Identity Property of Addition**
  For any real number a, $0 + a = a + 0 = a$.
- **Commutative Property of Addition**
  If a and b are real numbers, then $a + b = b + a$.
- **Additive Inverse Property**
  For any real number a, $a + (-a) = -a + a = 0$.
- **Associative Property of Addition**
  If a, b, and c are real numbers, then $a + (b + c) = (a + b) + c$.
- **Commutative Property of Multiplication**
  If a and b are real numbers, then $a \cdot b = b \cdot a$.
- **Multiplication Property of Zero**
  For any real number a, the product of a and 0 is always 0; that is, $a \cdot 0 = 0 \cdot a = 0$.
- **Multiplicative Identity**
  $a \cdot 1 = 1 \cdot a = a$ for any real number a.
- **Associative Property of Multiplication**
  If a, b, and c are real numbers, then $a \cdot (b \cdot c) = (a \cdot b) \cdot c$.
- **Multiplicative Inverse Property**

  $a \cdot \dfrac{1}{a} = \dfrac{1}{a} \cdot a = 1$ provided that $a \neq 0$

- **Division Properties of Zero**
  For any nonzero number a,

  1. The quotient of 0 and a is 0. That is, $\dfrac{0}{a} = 0$.

  2. The quotient of a and 0 is undefined. That is, $\dfrac{a}{0}$ is undefined.

Additive Identity
Multiplicative Identity
Conversion
Commutative Property
Grouping symbols
Associative Property
Undefined

YOU SHOULD BE ABLE TO . . .	EXAMPLE	REVIEW EXERCISES
(1) Understand and use the Identity Properties of Addition and Multiplication (p. 41)	Example 1	114, 115, 118–120, 122, 125–130, 138–140
(2) Understand and use the Commutative Properties of Addition and Multiplication (p. 43)	Examples 2 through 4	116, 117, 121, 125–128, 131–132, 135–136, 141–142
(3) Understand and use the Associative Properties of Addition and Multiplication (p. 44)	Examples 5 through 7	113, 124
(4) Understand the Multiplication and Division Properties of 0 (p. 45)	Example 8	123, 133–134, 137

In Problems 113–124, state the property of real numbers that is being illustrated.

113. $(5 \cdot 12) \cdot 10 = 5 \cdot (12 \cdot 10)$

114. $20 \cdot \dfrac{1}{20} = 1$

115. $\dfrac{8}{3} \cdot \dfrac{3}{8} = 1$

116. $\dfrac{5}{3} \cdot \left(-\dfrac{18}{61}\right) \cdot \dfrac{3}{5} = \dfrac{5}{3} \cdot \dfrac{3}{5} \cdot \left(-\dfrac{18}{61}\right)$

117. $9 \cdot 73 \cdot \dfrac{1}{9} = 9 \cdot \dfrac{1}{9} \cdot 73$

118. $23.9 + (-23.9) = 0$

119. $36 + 0 = 36$

120. $-49 + 0 = -49$

121. $23 + 5 + (-23)$ is equivalent to $23 + (-23) + 5$

122. $\dfrac{7}{8}$ is equivalent to $\dfrac{7}{8} \cdot \dfrac{3}{3}$ **123.** $14 \cdot 0 = 0$

124. $-5.3 + (5.3 + 2.8) = (-5.3 + 5.3) + 2.8$

In Problems 125–142, evaluate each expression, if possible, by using the properties of real numbers.

125. $144 + 29 + (-144)$ **126.** $76 + 99 + (-76)$

127. $\dfrac{19}{3} \cdot 18 \cdot \dfrac{3}{19}$ **128.** $\dfrac{14}{9} \cdot 121 \cdot \dfrac{9}{14}$

129. $3.4 + 42.56 + (-42.56)$ **130.** $5.3 + 3.6 + (-3.6)$

131. $\dfrac{9}{7} \cdot \left(-\dfrac{11}{3}\right) \cdot 7$ **132.** $\dfrac{13}{5} \cdot \dfrac{18}{39} \cdot 5$ **133.** $\dfrac{7}{0}$

134. $\dfrac{0}{100}$ **135.** $1000(-334)(0.001)$ **136.** $400(0.5)(0.01)$

137. $43{,}569{,}003 \cdot 0$ **138.** $154 \cdot \dfrac{1}{154}$ **139.** $\dfrac{3445}{302} + \left(-\dfrac{3445}{302}\right)$

140. $130 \cdot \dfrac{42}{42}$ **141.** $-\dfrac{7}{48} \cdot \dfrac{20}{3} \cdot \dfrac{12}{7}$ **142.** $\dfrac{9}{8} \cdot \left(-\dfrac{25}{13}\right) \cdot \dfrac{48}{9}$

Section 1.6   Exponents and the Order of Operations

KEY CONCEPTS

- **Exponential Notation**

  If n is a natural number and a is a real number, then

  $$a^n = \underbrace{a \cdot a \cdot a \cdot \ldots \cdot a}_{n \text{ factors}}$$

  where a is called the base and the natural number n is called the exponent or power.

- **Rules for Order of Operations**

  Perform all operations within *grouping symbols* first. When an expression has multiple grouping symbols, begin with the innermost pair of grouping symbols and work outward.

  Step 1: Evaluate expressions containing *exponents*.

  Step 2: Perform *multiplication and division* in the order in which they occur, working from *left to right*.

  Step 3: Perform *addition and subtraction* in the order in which they occur, working from *left to right*.

KEY TERMS

Exponential notation
Base
Exponent
Power
Exponential form
Expanded form

YOU SHOULD BE ABLE TO . . .	EXAMPLE	REVIEW EXERCISES
① Evaluate exponential expressions (p. 49)	Examples 1 through 4	143–152
② Apply the rules for order of operations (p. 50)	Examples 5 through 11	153–160

In Problems 143–146, write in exponential form.

143. $3 \cdot 3 \cdot 3 \cdot 3$ **144.** $\dfrac{2}{3} \cdot \dfrac{2}{3} \cdot \dfrac{2}{3}$ **145.** $(-4)(-4)$ **146.** $(-3)(-3)(-3)$

In Problems 147–152, evaluate each expression.

147. 5^3 **148.** 2^5 **149.** $(-3)^4$

150. $(-4)^3$ **151.** -3^4 **152.** $\left(\dfrac{1}{2}\right)^6$

In Problems 153–160, evaluate each expression.

153. $-2 + 16 \div 4 \cdot 2 - 10$ **154.** $-4 + 3[2^3 + 4(2 - 10)]$

155. $(12 - 7)^3 + (19 - 10)^2$

156. $5 - (-12 \div 2 \cdot 3) + (-3)^2$

157. $\dfrac{2 \cdot (4 + 8)}{3 + 3^2}$

158. $\dfrac{3 \cdot (5 + 2^2)}{2 \cdot 3^3}$

159. $\dfrac{6 \cdot [12 - 3 \cdot (5 - 2)]}{5 \cdot [21 - 2 \cdot (4 + 5)]}$

160. $\dfrac{4 \cdot [3 + 2 \cdot (8 - 6)]}{5 \cdot [14 - 2 \cdot (2 + 3)]}$

Section 1.7   Simplifying Algebraic Expressions

KEY CONCEPT	KEY TERMS
• **Distributive Property** If a, b, and c are real numbers, then $a \cdot (b + c) = a \cdot b + a \cdot c$ and $(a + b) \cdot c = a \cdot c + b \cdot c$	Algebra Variable Constant Algebraic expression Evaluate an algebraic expression Term Coefficient Like terms Combining like terms Simplify

YOU SHOULD BE ABLE TO . . .	EXAMPLE	REVIEW EXERCISES
① Evaluate algebraic expressions (p. 57)	Examples 1 and 2	161–164, 179
② Identify like terms and unlike terms (p. 59)	Examples 3 through 6	165–170
③ Use the Distributive Property (p. 61)	Example 7	174–178
④ Simplify algebraic expressions by combining like terms (p. 61)	Examples 8 through 10	171–178

In Problems 161–164, evaluate each expression using the given values of the variables.

161. $x^2 - y^2$ for $x = 5, y = -2$

162. $x^2 - 3y^2$ for $x = 3, y = -3$

163. $(x + 2y)^3$ for $x = -1, y = -4$

164. $\dfrac{a - b}{x - y}$ for $a = 5, b = -10, x = -3, y = 2$

In Problems 165 and 166, for each expression, identify the terms and then name the coefficient of each term.

165. $3x^2 - x + 6$

166. $2x^2y^3 - \dfrac{y}{5}$

In Problems 167–170, determine if the terms are like or unlike.

167. $4xy^2, -6xy^2$

168. $-3x, 4x^2$

169. $-6y, -6$

170. $-10, 4$

In Problems 171–178, simplify each algebraic expression.

171. $4x - 6x - x$

172. $6x - 10 - 10x - 5$

173. $0.2x^4 + 0.3x^3 - 4.3x^4$

174. $-3(x^4 - 2x^2 - 4)$

175. $20 - (x + 2)$

176. $-6(2x + 5) + 4(4x + 3)$

177. $5 - (3x - 1) + 2(6x - 5)$

178. $\dfrac{1}{6}(12x + 18) - \dfrac{2}{5}(5x + 10)$

179. Moving Van The cost of renting a moving van for one day is $19.95 plus $0.25 per mile. The expression $19.95 + 0.25m$ represents the total cost of renting the truck for one day and driving m miles. Evaluate the expression $19.95 + 0.25m$ for $m = 315$.

CHAPTER 1 TEST

Remember to use your Chapter Test Prep Video CD to see fully worked-out solutions to any of these problems you would like to review.

In Problems 1–9, perform the indicated operation. Write in lowest terms if necessary.

1. $\dfrac{4}{15} - \left(-\dfrac{2}{30}\right)$

2. $\dfrac{21}{4} \cdot \dfrac{3}{7}$

3. $-16 \div \dfrac{3}{20}$

4. $14 - 110 - (-15) + (-21)$

5. $-14.5 + 2.34$

6. $(-4)(-1)(-5)$

7. $16 \div 0$

8. -6 subtracted from -20

9. -110 divided by -2

10. Use the set $\left\{-2, -\dfrac{1}{2}, 0, 2.5, 6\right\}$. List all of the elements that are:

   (a) natural numbers   **(b)** whole numbers   **(c)** integers
   (d) rational numbers   **(e)** irrational numbers   **(f)** real numbers

In Problems 11 and 12, replace the ? with the correct symbol $>$, $<$, or $=$.

11. $-|-14| \; ? \; -12$

12. $\left|-\dfrac{2}{5}\right| \; ? \; 0.4$

In Problems 13–15, evaluate each expression.

13. $-16 \div 2^2 \cdot 4 + (-3)^2$

14. $\dfrac{4(-9) - 3^2}{25 + 4(-6 - 1)}$

15. $8 - 10[6^2 - 5(2 + 3)]$

16. Evaluate $(x - 2y)^3$ for $x = -1$ and $y = 3$.

In Problems 17 and 18, simplify each algebraic expression.

17. $-6(2x + 5) - (4x - 2)$

18. $\dfrac{1}{2}(4x^2 + 8) - 6x^2 + 5x$

19. **Bank Account** Latoya started with $675.15 in her bank account. She wrote a check for $175.50, withdrew $78.00 in cash, and made a deposit of $110.20. How much money does Latoya have in her bank account now?

20. **Perimeter** The length of a rectangle is 5 feet more than its width. The algebraic expression $2(x + 5) + 2x$ represents the perimeter of the rectangle. Simplify the expression $2(x + 5) + 2x$.

2 Equations and Inequalities in One Variable

You plan to remodel your bathroom and you've chosen 1 foot-by-1 foot ceramic tiles to cover the floor. The bathroom is 7 feet 6 inches long and 8 feet 2 inches wide. How many tiles do you need to cover the floor of your bathroom? If each tile costs $6, how much will it cost to tile your floor? Suppose that the store selling the tile offers a discount of 10% on orders over $350. Does your order qualify for the discount?

You can solve everyday problems such as these using mathematical formulas and models. See Problem 79 in Section 2.4.

OUTLINE

2.1 Linear Equations: The Addition and Multiplication Properties of Equality

2.2 Linear Equations: Using the Properties Together

2.3 Solving Linear Equations Involving Fractions and Decimals; Classifying Equations

2.4 Evaluating Formulas and Solving Formulas for a Variable

Putting the Concepts Together
(SECTIONS 2.1–2.4)

2.5 Introduction to Problem Solving: Direct Translation Problems

2.6 Problem Solving: Direct Translation Problems Involving Percent

2.7 Problem Solving: Geometry and Uniform Motion

2.8 Solving Linear Inequalities in One Variable

Chapter 2 Activity: Pass to the Right
Chapter 2 Review
Chapter 2 Test

The Big Picture: Putting It Together

In Chapter 1, we reviewed arithmetic skills that will be needed throughout the course. We also introduced algebraic expressions and discussed how to simplify and evaluate algebraic expressions.

In this chapter we dive into the discussion of algebra. The word "algebra" is derived from the Arabic word, *al-jabr*. The word *al-jabr* means "restoration." This is a reference to the fact that, if a number is added to one side of an equation, then it must also be added to the other side in order to "restore" the equality. While algebra now means a whole lot more than "restoration," we will concentrate on the "restoration" part of algebra in this chapter.

2.1 Linear Equations: The Addition and Multiplication Properties of Equality

OBJECTIVES

1. Determine If a Number Is a Solution of an Equation
2. Use the Addition Property of Equality to Solve Linear Equations
3. Use the Multiplication Property of Equality to Solve Linear Equations

Preparing for Linear Equations: The Addition and Multiplication Properties of Equality

Before getting started, take this readiness quiz. If you get a problem wrong, go back to the section cited and review the material.

1. Determine the additive inverse of 3.   [Section 1.3, p. 21]

2. Determine the multiplicative inverse of $-\dfrac{4}{3}$.   [Section 1.3, pp. 24–25]

3. Evaluate: $\dfrac{2}{3}\left(\dfrac{3}{2}\right)$ [Section 1.4, p. 29]

4. Use the Distributive Property to simplify $-4(2x + 3)$.   [Section 1.7, p. 61]

---

1   Determine If a Number Is a Solution of an Equation

We begin with a definition.

> **DEFINITION**
>
> A **linear equation in one variable** is an equation that can be written in the form $ax + b = c$, where a, b, and c are real numbers and a does not equal 0.

Examples of linear equations (in the variable x) are

$$x - 5 = 8 \qquad \frac{1}{2}x - 7 = \frac{3}{2}x + 8 \qquad 0.2(x + 5) - 1.5 = 4.25 - (x + 3)$$

The algebraic expressions in the equation are called the **sides** of the equation. For example, in the equation $\frac{1}{2}x - 7 = \frac{3}{2}x + 8$, the algebraic expression $\frac{1}{2}x - 7$ is the **left side** of the equation and the algebraic expression $\frac{3}{2}x + 8$ is the **right side** of the equation.

An equation may be true or false. For example, the equation $x - 5 = 8$ is true if the variable x is replaced by the number 13, but the equation $x - 5 = 8$ is false if the variable x is replaced by the number 2.

Because replacing x by 13 in the equation $x - 5 = 8$ results in a true statement, we say that 13 is a *solution* of the equation $x - 5 = 8$. We also say that $x = 13$ *satisfies* the equation.

> **DEFINITION**
>
> The **solution** of a linear equation is the value or values of the variable that make the equation a true statement. The set of all solutions of an equation is called the **solution set.** We sometimes say that the solution **satisfies** the equation.

We use set notation to indicate the solution set of an equation. For example, because $x = 13$ satisfies the equation $x - 5 = 8$, we say that the solution set is $\{13\}$.

To determine whether a number satisfies an equation we replace the variable with the number and find out whether the left side of the equation equals the right side of the equation. If it does, we have a true statement and the replacement value is a solution of the equation.

Preparing for...Answers   **1.** -3 **2.** $-\dfrac{3}{4}$

3. 1   **4.** $-8x - 12$

EXAMPLE 1   Determine Whether a Number Is a Solution of an Equation

Determine if the given value of the variable is a solution of the equation $4x + 7 = 19$.

(a) $x = -2$ **(b)** $x = 3$

Solution

(a)
$$4x + 7 = 19$$

Replace x with -2: $\quad 4(-2) + 7 \stackrel{?}{=} 19$

Simplify: $\quad -8 + 7 \stackrel{?}{=} 19$

$$-1 = 19 \quad \text{False}$$

Since the left side does not equal the right side when we replace x by -2, $x = -2$ is *not* a solution of the equation.

(b)
$$4x + 7 = 19$$

Replace x with 3: $\quad 4(3) + 7 \stackrel{?}{=} 19$

Simplify: $\quad 12 + 7 \stackrel{?}{=} 19$

$$19 = 19 \quad \text{True}$$

Since the left side equals the right side when we replace x by 3, $x = 3$ is a solution of the equation.

QUICK ✓ *Determine if the given number is a solution of the equation.*

1. $a - 4 = -7; a = -3$

2. $\dfrac{1}{2} + x = 10; x = \dfrac{21}{2}$

3. $3x - (x + 4) = 8; x = 6$

4. $-9b + 3 + 7b = -3b + 8; b = -3$

② Use the Addition Property of Equality to Solve Linear Equations

We solve linear equations by writing a series of steps that result in the equation

$$x = a\ number$$

One method for solving equations algebraically requires that a series of *equivalent equations* be developed from the original equation until a solution results.

> **DEFINITION**
>
> Two or more equations that have precisely the same solutions are called **equivalent equations.**

We form equivalent equations using mathematical properties that transform the original equation into a new equation that has the same solution. The first property is called the *Addition Property of Equality*.

In Words
The Addition Property of Equality says that whatever you add to one side of the equation, you must also add to the other side.

> **ADDITION PROPERTY OF EQUALITY**
>
> The **Addition Property of Equality** states that for real numbers a, b, and c,
>
> $$\text{if } a = b, \text{ then } a + c = b + c$$

Also, because $a - b$ is equivalent to $a + (-b)$, the Addition Property of Equality can be used to add a real number to each side of the equation, or to subtract a real number from each side of the equation.

Because the goal in solving a linear equation is to get the variable by itself with a coefficient of 1, we say that we want to **isolate the variable.**

EXAMPLE 2 **How to Use the Addition Property of Equality to Solve a Linear Equation**

Solve the linear equation $x - 6 = 11$.

Step-by-Step Solution

Since the coefficient of the variable x in this equation is already 1, we just need to "get x by itself."

Step 1: Isolate the variable x on the left side of the equation.		$x - 6 = 11$
	Add 6 to each side of the equation:	$x - 6 + 6 = 11 + 6$
Step 2: Simplify the left and right sides of the equation.	Apply the Additive Inverse Property, $a + (-a) = 0$:	$x + 0 = 17$
	Apply the Additive Identity Property, $a + 0 = a$:	$x = 17$
Step 3: Check Verify that $x = 17$ is the solution.		$x - 6 = 11$
	Replace 17 for x in the original equation to see if a true statement results:	$17 - 6 \overset{?}{=} 11$
		$11 = 11$ True

Because $x = 17$ satisfies the original equation, the solution is 17, or the solution set is $\{17\}$.

EXAMPLE 3 **Using the Addition Property of Equality to Solve a Linear Equation**

Solve the linear equation $x + \dfrac{5}{2} = \dfrac{1}{4}$.

Solution

$$x + \frac{5}{2} = \frac{1}{4}$$

Subtract $\dfrac{5}{2}$ from each side of the equation:
$$x + \frac{5}{2} - \frac{5}{2} = \frac{1}{4} - \frac{5}{2}$$

$a + (-a) = 0$; LCD = 4:
$$x = \frac{1}{4} - \frac{5}{2} \cdot \frac{2}{2}$$

$$x = \frac{1}{4} - \frac{10}{4}$$

$$x = -\frac{9}{4}$$

Check Verify that $x = -\dfrac{9}{4}$ is the solution.

$$x + \frac{5}{2} = \frac{1}{4}$$

$$-\frac{9}{4} + \frac{5}{2} \overset{?}{=} \frac{1}{4}$$

$$-\frac{9}{4} + \frac{10}{4} \overset{?}{=} \frac{1}{4}$$

$$\frac{1}{4} = \frac{1}{4} \text{True}$$

Because $x = -\dfrac{9}{4}$ satisfies the original equation, the solution is $-\dfrac{9}{4}$, or the solution set is $\left\{-\dfrac{9}{4}\right\}$.

QUICK ✓ *Solve each equation using the Addition Property of Equality.*

5. $x - 11 = 21$

6. $y + 7 = 21$

7. $-8 + a = 4$

8. $12 + c = -3$

9. $z - \dfrac{2}{3} = \dfrac{5}{3}$

10. $p + \dfrac{5}{4} = \dfrac{1}{4}$

11. $w - \dfrac{1}{4} = \dfrac{3}{8}$

12. $\dfrac{5}{4} + x = \dfrac{1}{6}$

EXAMPLE 4 **How Much Is the CD Set?**

The total cost for purchasing a two-CD set including a sales tax of $1.56 was $27.56. To find the price p of the CD set before tax, solve the equation $p + 1.56 = 27.56$ for p.

Solution

$$p + 1.56 = 27.56$$

Subtract 1.56 from each side of the equation: $\quad p + 1.56 - 1.56 = 27.56 - 1.56$

Apply the Additive Inverse Property, $a + (-a) = 0$: $\quad p + 0 = 26$

Apply the Additive Identity Property, $a + 0 = a$: $\quad p = 26$

Since $p = 26$, the two-CD set cost $26 before tax.

QUICK ✓

13. The total cost for a new car, including tax, title, and dealer preparation charges of $1472.25 is $13,927.25. To find the price p of the car before the extra charges, solve the equation $p + 1472.25 = 13,927.25$ for p.

---

③ **Use the Multiplication Property of Equality to Solve Linear Equations**

A second property that allows us to create an equivalent equation is called the *Multiplication Property of Equality*.

In Words

The Multiplication Property of Equality says that when you multiply one side of an equation by a nonzero quantity, you must also multiply the other side by the same nonzero quantity.

> **MULTIPLICATION PROPERTY OF EQUALITY**
>
> The **Multiplication Property of Equality** states that for real numbers a, b, and c, where c does not equal 0,
>
> $$\text{if } a = b, \quad \text{then} \quad ac = bc$$

Let's see how to use the Multiplication Property of Equality to solve an equation in the next example.

| **EXAMPLE 5** | **How to Solve a Linear Equation Using the Multiplication Property of Equality** |

Solve the equation $5x = 30$.

Step-by-Step Solution

Step 1: Get the coefficient of the variable x to be 1. $\qquad\qquad\qquad\qquad\qquad\qquad\qquad 5x = 30$

Apply the Multiplication Property of Equality
and multiply each side of the equation by $\frac{1}{5}$: $\quad \frac{1}{5}(5x) = \frac{1}{5}(30)$

Step 2: Simplify the left and right sides of the equation.

Regroup factors using the Associative Property of Multiplication: $\quad \left(\frac{1}{5}\cdot 5\right)x = \frac{1}{5}(30)$

Apply the Multiplicative Inverse Property, $a\cdot\frac{1}{a} = 1$: $\quad 1\cdot x = 6$

Apply the Multiplicative Identity, $1\cdot a = a$: $\quad x = 6$

Step 3: Check Verify the solution. $\qquad\qquad\qquad\qquad\qquad\qquad\qquad\qquad 5x = 30$

Replace $x = 6$ in the original equation: $\quad 5(6) \overset{?}{=} 30$

$$30 = 30 \quad \text{True}$$

Because $x = 6$ satisfies the original equation, the solution is 6, or the solution set is $\{6\}$.

Work Smart
We multiply when the coefficient of the variable is a fraction, and we divide when the coefficient is an integer.

In Step 1 of Example 5, we multiplied both sides of the equation by $\frac{1}{5}$ to get the coefficient of x to equal 1. Instead of multiplying by $\frac{1}{5}$, we could also divide both sides of the equation by 5 because dividing by 5 is the same as multiplying by the reciprocal of 5, $\frac{1}{5}$.

$$5x = 30$$

Divide each side of the equation by 5: $\quad \dfrac{5x}{5} = \dfrac{30}{5}$

Simplify: $\quad x = 6$

Which approach do you prefer?

| **EXAMPLE 6** | **Solving a Linear Equation Using the Multiplication Property of Equality** |

Solve the equation $-4n = 18$.

Solution

$$-4n = 18$$

Multiply each side of the equation by $-\frac{1}{4}$: $\quad -\dfrac{1}{4}(-4n) = -\dfrac{1}{4}\cdot 18$

Regroup: $\quad \left(-\dfrac{1}{4}\cdot -4\right)n = -\dfrac{18}{4}$

Simplify: $\quad n = -\dfrac{9}{2}$

Check Verify the solution by replacing $n = -\dfrac{9}{2}$ in the original equation to see if a true statement results.

$$-4n = 18$$

$$-4\left(-\frac{9}{2}\right) \stackrel{?}{=} 18$$

Divide out the common factor: $\overset{-2}{\cancel{-4}}\left(-\dfrac{9}{\underset{1}{\cancel{2}}}\right) \stackrel{?}{=} 18$

$$18 = 18$$

Because $n = -\dfrac{9}{2}$ satisfies the original equation, the solution is $-\dfrac{9}{2}$, or the solution set is $\left\{-\dfrac{9}{2}\right\}$.

QUICK ✓ *Solve each equation using the Multiplication Property of Equality.*

14. $8p = 16$ **15.** $-7n = 14$ **16.** $6z = 15$ **17.** $-12b = 28$

EXAMPLE 7   Solving a Linear Equation with a Fraction as a Coefficient

Solve the equation $12 = \dfrac{2}{3}x$.

Solution

Work Smart
When the variable is on the right side of an equation, we isolate the variable in the same way we did when it was on the left side of the equation.

$$12 = \frac{2}{3}x$$

Multiply both sides of the equation by $\dfrac{3}{2}$, the reciprocal of $\dfrac{2}{3}$: $\quad \dfrac{3}{2}(12) = \dfrac{3}{2}\left(\dfrac{2}{3}x\right)$

Use the Associative Property of Multiplication: $\quad \dfrac{3}{2}(12) = \left(\dfrac{3}{2}\cdot\dfrac{2}{3}\right)x$

Apply the Multiplicative Inverse Property, $a\cdot\dfrac{1}{a} = 1$: $\quad 18 = (1)x$

Apply the Multiplicative Identity Property, $1\cdot a = a$: $\quad 18 = x$

Check Verify the solution by replacing $x = 18$ in the original equation to see if a true statement results.

$$12 = \frac{2}{3}x$$

$$12 \stackrel{?}{=} \frac{2}{3}(18)$$

$$12 = 12 \quad \text{True}$$

Because $x = 18$ satisfies the original equation, the solution is 18, or the solution set is $\{18\}$.

QUICK ✓ *Solve each equation using the Multiplication Property of Equality.*

18. $\dfrac{4}{3}n = 12$ **19.** $\dfrac{7}{3}k = -21$ **20.** $15 = -\dfrac{9}{2}z$

EXAMPLE 8   **Solving a Linear Equation with Fractions**

Solve the equation $\dfrac{4}{5} = -\dfrac{2}{15}p$.

Solution

$$\frac{4}{5} = -\frac{2}{15}p$$

Multiply both sides by $-\dfrac{15}{2}$: $-\dfrac{15}{2} \cdot \dfrac{4}{5} = -\dfrac{15}{2} \cdot \left(-\dfrac{2}{15}p\right)$

Simplify: $-\dfrac{\overset{3}{\cancel{15}}}{\underset{1}{\cancel{2}}} \cdot \dfrac{\overset{2}{\cancel{4}}}{\underset{1}{\cancel{5}}} = \left(-\dfrac{15}{2} \cdot -\dfrac{2}{15}\right)p$

$$-6 = p$$

Check   Let $p = -6$ in the original equation to see if a true statement results.

$$\frac{4}{5} = -\frac{2}{15}p$$

Let $p = -6$: $\dfrac{4}{5} \overset{?}{=} -\dfrac{2}{15}(-6)$

Simplify: $\dfrac{4}{5} \overset{?}{=} -\dfrac{2}{\underset{5}{\cancel{15}}}(-\overset{2}{\cancel{6}})$

$$\frac{4}{5} = \frac{4}{5} \quad \text{True}$$

The solution is -6, or the solution set is $\{-6\}$.   ▬

Working with fractional coefficients can be tricky. The three equations $\dfrac{-x}{3} = 7$, $-\dfrac{1}{3}x = 7$, and $\dfrac{x}{-3} = 7$ are all equivalent. Do you see why? In the case of $\dfrac{-x}{3} = 7$, we can write $\dfrac{-x}{3}$ as $\dfrac{-1x}{3}$, which is equivalent to $-\dfrac{1}{3}x$. In the case of $\dfrac{x}{-3} = 7$, we can write $\dfrac{x}{-3}$ as $\dfrac{1x}{-3}$, which is equivalent to $-\dfrac{1}{3}x$.

Work Smart

Whenever an equation is in the form $ax = b$, if a is an integer, either multiply by the reciprocal of a, or divide by a.

If a is a noninteger rational number, multiply by the reciprocal of a.

$2x = 7$ $5x = -\dfrac{15}{2}$ $-3x = 72$ $\dfrac{4}{5}x = -16$

$\dfrac{1}{2}(2x) = \dfrac{1}{2} \cdot 7$ $\dfrac{1}{5}(5x) = \dfrac{1}{5}\left(-\dfrac{15}{2}\right)$ $\dfrac{-3x}{-3} = \dfrac{72}{-3}$ $\dfrac{5}{4}\left(\dfrac{4}{5}x\right) = \dfrac{5}{4}(-16)$

$x = \dfrac{7}{2}$ $x = -\dfrac{3}{2}$ $x = -24$ $x = \dfrac{5}{\underset{1}{\cancel{4}}} \cdot \dfrac{\overset{-4}{\cancel{-16}}}{1}$

$x = -20$

QUICK ✓ *Solve each equation using the Multiplication Property of Equality.*

21. $\dfrac{3}{8}b = \dfrac{9}{4}$ **22.** $-\dfrac{4}{9} = \dfrac{-t}{6}$ **23.** $\dfrac{1}{4} = -\dfrac{7}{10}m$

2.1 Exercises

For Extra Help:

Student Solutions Manual   CD Video   PH Math/Tutor Center   MathXL Tutorials on CD   MathXL®   MyMathLab

Concepts and Vocabulary

In Problems 1–3, fill in the blanks.

1. The _____ _____ _____ _____ says that whatever number you add to one side of an equation you must also add to the other side.

2. Two or more equations that have precisely the same solution are called _____ _____.

3. To solve the equation $p + 9 = -11$, we _____ _____ (from/to) each side of the equation.

In Problems 4–6, answer True or False to each statement.

4. The equation $3x^2 - 5x = 9$ is a linear equation.

5. To find the solution to the equation $\dfrac{4}{5}y = 16$, we subtract $\dfrac{4}{5}$ from each side of the equation.

6. The solution to $-\dfrac{3}{11}b = 12$ is $b = -44$.

7. Explain what is meant by finding the *solution* to an equation.

8. Consider the equation $3z = \dfrac{15}{4}$. You could either multiply both sides of the equation by $\dfrac{1}{3}$ or you could divide both sides by 3. Explain which operation you would choose and why you think it is easier.

9. Explain in your own words the difference between an algebraic expression and an equation. Write an equation involving the algebraic expression $x - 10$. Then solve the equation.

10. A classmate suggests that to solve the equation $4x = 2$, you must divide by 4 and the result will be $x = 2$. Explain why this reasoning is or is not correct.

Building Skills

In Problems 11–18, determine if the given value is a solution to the equation. Answer Yes or No.

11. $3x - 1 = 5; x = 2$ **12.** $4t + 2 = 16; t = 3$

13. $4 - (m + 2) = 3(2m - 1); m = 1$ **14.** $3(x + 1) - x = 5x - 9; x = -3$

15. $8k - 2 = 4; k = \dfrac{3}{4}$ **16.** $-15 = 3x - 16; x = \dfrac{1}{3}$

17. $r + 1.6 = 2r + 1; r = 0.6$ **18.** $3s - 6 = 6s - 3.4; s = -1.2$

In Problems 19–34, solve the equation using the Addition Property of Equality. Be sure to check your solution.

19. $x - 9 = 11$ **20.** $y - 8 = 2$ **21.** $x + 4 = -8$ **22.** $r + 3 = -1$

23. $12 = n - 7$ **24.** $13 = u - 6$ **25.** $-8 = x + 5$ **26.** $-2 = y + 13$

27. $x - \dfrac{2}{3} = \dfrac{4}{3}$ **28.** $x - \dfrac{1}{8} = \dfrac{3}{8}$ **29.** $z + \dfrac{1}{2} = \dfrac{3}{4}$ **30.** $n + \dfrac{3}{5} = \dfrac{7}{10}$

31. $\dfrac{5}{12} = x - \dfrac{3}{8}$ **32.** $\dfrac{3}{8} = y - \dfrac{1}{6}$ **33.** $w + 3.5 = -2.6$ **34.** $z + 4.9 = -2.6$

In Problems 35–56, solve the equation using the Multiplication Property of Equality. Be sure to check your solution.

35. $5c = 25$ **36.** $8b = 48$ **37.** $-7n = 28$ **38.** $-8s = 40$

39. $4k = 14$ **40.** $4z = 30$ **41.** $-6w = 15$ **42.** $-8p = 20$

43. $\dfrac{5}{3}a = 35$ **44.** $\dfrac{4}{3}b = 16$ **45.** $-\dfrac{3}{11}p = -33$ **46.** $-\dfrac{6}{5}n = -36$

47. $\dfrac{6}{5} = 2x$ **48.** $\dfrac{9}{2} = 3b$ 🔘 **49.** $5y = -\dfrac{5}{3}$ **50.** $4r = -\dfrac{12}{5}$

51. $\dfrac{1}{2}m = \dfrac{9}{2}$ **52.** $\dfrac{1}{4}w = \dfrac{7}{2}$ **53.** $-\dfrac{3}{8}t = \dfrac{1}{6}$ **54.** $\dfrac{3}{10}q = -\dfrac{1}{6}$

55. $\dfrac{5}{24} = \dfrac{-y}{8}$ **56.** $\dfrac{11}{36} = -\dfrac{t}{9}$

Mixed Practice

In Problems 57–84, solve the equation. Be sure to check your solution.

57. $n - 4 = -2$ **58.** $m - 6 = -9$ **59.** $b + 12 = 9$

60. $c + 4 = 1$ **61.** $2 = 3x$ **62.** $9 = 5y$

63. $-4q = 24$ **64.** $-6m = 54$ **65.** $-39 = x - 58$

66. $-637 = c - 142$ **67.** $-18 = -301 + x$ **68.** $-46 = -51 + q$

69. $\dfrac{x}{5} = -10$ **70.** $\dfrac{z}{3} = -12$ **71.** $m - 56.3 = -15.2$

72. $p - 26.4 = -471.3$ **73.** $-40 = -6c$ **74.** $-45 = 12x$

75. $14 = -\dfrac{7}{2}c$ **76.** $12 = -\dfrac{3}{2}n$ **77.** $\dfrac{3}{4} = -\dfrac{x}{16}$

78. $\dfrac{5}{9} = -\dfrac{h}{36}$ **79.** $x - \dfrac{5}{16} = \dfrac{3}{16}$ **80.** $w - \dfrac{7}{20} = \dfrac{9}{20}$

81. $-\dfrac{3}{16} = -\dfrac{3}{8} + z$ **82.** $\dfrac{5}{2} = -\dfrac{13}{4} + y$

83. $\dfrac{5}{6} = -\dfrac{2}{3}z$ **84.** $-\dfrac{4}{9} = \dfrac{8}{3}b$

Applying the Concepts

85. New Car The total cost for a new car, including tax, title, and dealer preparation charges of \$1562.35, is \$20,062.15. To find the price of the car without the extra charges, solve the equation $y + 1562.35 = 20{,}062.15$, where y represents the price of the car without the extra charges.

86. New Kayak The total cost for a new kayak is \$862.92, including sales tax of \$63.92. To find the cost of the kayak without tax, solve the equation $k + 63.92 = 862.92$, where k represents the cost of the kayak.

87. Discount The cost of a sleeping bag has been discounted by \$17, so that the sale price of the bag is \$51. Find the original price of the sleeping bag, p, by solving the equation $p - 17 = 51$.

88. Discount The cost of a computer has been discounted by \$239, so that the sale price is \$1230. Find the original price of the computer, c, by solving the equation $c - 239 = 1230$.

89. Eating Out Rebecca purchased several "Happy Meals" at McDonald's for her child's play group. Each Happy Meal costs $4 and she spent a total of $48 on the food. To determine the number of Happy Meals, h, she purchased, solve the equation $4h = 48$.

90. Paperback Books Anne bought 3 paperback books to read on the flight to Europe. She paid $36 for the books (without sales tax). To find the price, p, of each book solve the equation $3p = 36$.

91. Interest Suppose you have a credit card debt of $3,000. Last month, the bank charged you $45 interest on the debt. The solution to the equation $45 = \dfrac{3000}{12} \cdot r$ represents the annual interest rate, r, on the credit card. Find the annual interest rate on the credit card.

92. Interest Suppose you have a credit card debt of $4,000. Last month, the bank charged you $40 interest on the debt. The solution to the equation $40 = \dfrac{4000}{12} \cdot r$ represents the annual interest rate, r, on the credit card. Find the annual interest rate on the credit card.

Extending the Concepts

93. Let λ represent some real number. Solve the equation $x + \lambda = 48$ for x.

94. Let β represent some real number. Solve the equation $x - \beta = 25$ for x.

95. Let θ represent some real number except 0. Solve the equation $14 = \theta x$ for x.

96. Let ψ represent some real number except 0. Solve the equation $\dfrac{2}{5} = \psi x$ for x.

97. Find the value of λ in the equation $x + \lambda = \dfrac{16}{3}$ so that the solution is $-\dfrac{2}{9}$.

98. Find the value of β in the equation $x - \beta = -13.6$ so that the solution is -4.79.

99. Find the value of θ in the equation $-\dfrac{3}{4} = \theta x$ so that the solution is $\dfrac{7}{8}$.

100. Find the value of ψ in the equation $-0.31 = \psi x$ so that the solution is 2.98.

2.2 Linear Equations: Using the Properties Together

OBJECTIVES

1. Apply the Addition and the Multiplication Properties of Equality to Solve Linear Equations
2. Combine Like Terms and Apply the Distributive Property to Solve Linear Equations
3. Solve a Linear Equation with the Variable on Both Sides of the Equation
4. Use Linear Equations to Solve Problems

Preparing for Linear Equations: Using the Properties Together
Before getting started, take this readiness quiz. If you get a problem wrong, go back to the section cited and review the material.

1. Simplify by combining like terms: $6 - (4 + 3x) + 8$ [Section 1.7, pp. 61–63]
2. Evaluate the expression $2(3x + 4) - 5$ for $x = -1$.   [Section 1.7, pp. 57–58]

1 Apply the Addition and the Multiplication Properties of Equality to Solve Linear Equations

In the last section, we solved equations such as $x + 3 = 7$ and $\dfrac{1}{2}z = 8$, which required that we use either the Addition Property or the Multiplication Property of Equality, but not both. We now consider equations such as $2x - 3 = 7$ and $\dfrac{1}{2}z + 6 = 18$, which require using both the Addition and Multiplication Properties of Equality to solve for the variable. For example, the equation $2x + 3 = 7$ can be read as "Two *times* a number x *plus* three equals seven." We must undo both the multiplication and addition to find the solution of the equation.

EXAMPLE 1 **How to Solve a Linear Equation Using the Addition and Multiplication Properties of Equality**

Solve the equation $2z - 3 = 9$.

Step-by-Step Solution

Step 1: Isolate the term containing the variable.

$$2z - 3 = 9$$

Apply the Addition Property of Equality and add 3 to each side of the equation:

$$2z - 3 + 3 = 9 + 3$$
$$2z = 12$$

Step 2: Get the coefficient of the variable to be 1.

Apply the Multiplication Property of Equality and divide both sides by 2 (this is the same as multiplying both sides by $\frac{1}{2}$):

$$\frac{2z}{2} = \frac{12}{2}$$
$$z = 6$$

Step 3: Check Confirm that $z = 6$ is the solution of the equation.

Substitute $z = 6$ into the original equation:

$$2z - 3 = 9$$
$$2(6) - 3 \overset{?}{=} 9$$
$$12 - 3 \overset{?}{=} 9$$
$$9 = 9 \quad \text{True}$$

Because $z = 6$ satisfies the equation, the solution of the equation is 6, or the solution set is $\{6\}$.

EXAMPLE 2 **Solving a Linear Equation Using the Addition and Multiplication Properties of Equality**

Solve the equation $\dfrac{3}{2}p + 3 = 12$.

Solution

$$\frac{3}{2}p + 3 = 12$$

Subtract 3 from both sides of the equation:

$$\frac{3}{2}p + 3 - 3 = 12 - 3$$

Simplify:

$$\frac{3}{2}p = 9$$

Multiply both sides of the equation by $\dfrac{2}{3}$:

$$\frac{2}{3}\left(\frac{3}{2}p\right) = \frac{2}{3}(9)$$

Simplify:

$$p = 6$$

Check Let $p = 6$ in the original equation to verify the solution.

$$\frac{3}{2}p + 3 = 12$$

Let $p = 6$: $\quad \frac{3}{2}(6) + 3 \stackrel{?}{=} 12$

Simplify: $\quad\quad 9 + 3 \stackrel{?}{=} 12$

$$12 = 12 \quad \text{True}$$

Because $p = 6$ satisfies the equation, the solution is $p = 6$, or the solution set is $\{6\}$.

QUICK ✓ *Solve each equation.*

1. $5x - 4 = 11$

2. $8 - 5r = -2$

3. $\frac{2}{3}k - 4 = 8$

4. $-\frac{3}{2}n + 2 = -\frac{1}{4}$

---

② ## Combine Like Terms and Apply the Distributive Property to Solve Linear Equations

Often, we must combine like terms before we can use the Addition or Multiplication Properties of Equality.

| **EXAMPLE 3** | **Combining Like Terms to Solve a Linear Equation** |

Solve the equation $2x - 6 + 3x = 14$.

Solution

$$2x - 6 + 3x = 14$$

Combine like terms: $\quad 5x - 6 = 14$

Add 6 to each side of the equation: $\quad 5x - 6 + 6 = 14 + 6$

$$5x = 20$$

Divide both sides by 5: $\quad \frac{5x}{5} = \frac{20}{5}$

$$x = 4$$

Check Verify that $x = 4$ is the solution.

$$2x - 6 + 3x = 14$$

Substitute 4 for x in the original equation: $\quad 2(4) - 6 + 3(4) \stackrel{?}{=} 14$

$$8 - 6 + 12 \stackrel{?}{=} 14$$

$$14 = 14$$

Since $x = 4$ results in a true statement, the solution of the equation is 4, or the solution set is $\{4\}$.

QUICK ✓ *Solve the equation.*

5. $7b - 3b + 3 = 11$

6. $-3a + 4 + 4a = 13 - 27$

7. $6c - 2 + 2c = 18$

8. $-12 = 5x - 3x + 4$

When an equation contains parentheses, we use the Distributive Property to eliminate parentheses before we use the Addition or Multiplication Properties of Equality.

EXAMPLE 4   **Solve a Linear Equation Using the Distributive Property**

Solve the equation $4(2x + 3) - 7 = -11$.

Solution

$$4(2x + 3) - 7 = -11$$

Apply the Distributive Property to remove parentheses: $8x + 12 - 7 = -11$

Combine like terms: $8x + 5 = -11$

Subtract 5 from each side of the equation: $8x + 5 - 5 = -11 - 5$

$$8x = -16$$

Divide both sides by 8: $\dfrac{8x}{8} = \dfrac{-16}{8}$

$$x = -2$$

Check   Verify that $x = -2$ is the solution.

$$4(2x + 3) - 7 = -11$$

Substitute -2 for x in the original equation: $4[2(-2) + 3] - 7 \overset{?}{=} -11$

$$4(-4 + 3) - 7 \overset{?}{=} -11$$

$$4(-1) - 7 \overset{?}{=} -11$$

$$-4 - 7 \overset{?}{=} -11$$

$$-11 = -11 \quad \text{True}$$

Because $x = -2$ results in a true statement, the solution of the equation is -2, or the solution set is $\{-2\}$. ∎

QUICK ✓   *Solve the equation.*

9. $2(y + 5) - 3 = 11$

10. $\dfrac{1}{2}(4 - 6x) + 5 = 3$

11. $4 - (6 - x) = 11$

12. $8 + \dfrac{2}{3}(2n - 9) = 10$

③   Solve a Linear Equation with the Variable on Both Sides of the Equation

When solving a linear equation, our goal is to get the terms that contain the variable on one side of the equation and the constants on the other side.

EXAMPLE 5   **Solve a Linear Equation with a Variable on Both Sides of the Equation**

Solve the equation $9y - 5 = 5y + 9$.

Solution

$$9y - 5 = 5y + 9$$

Subtract 5y from each side of the equation: $9y - 5 - 5y = 5y + 9 - 5y$

$$4y - 5 = 9$$

Add 5 to each side of the equation: $4y - 5 + 5 = 9 + 5$

$$4y = 14$$

Divide both sides by 4: $\dfrac{4y}{4} = \dfrac{14}{4}$

Simplify: $y = \dfrac{7}{2}$

Work Smart

When an equation is one in which the variable appears on both sides of the equation, work with the variable expression first. Once you have the expression containing the variable isolated, then add or subtract the constant to get it on the other side.

Check Confirm that $y = \dfrac{7}{2}$ is the solution.

$$9y - 5 = 5y + 9$$

Substitute $\dfrac{7}{2}$ for y in the original equation: $\quad 9\left(\dfrac{7}{2}\right) - 5 \stackrel{?}{=} 5\left(\dfrac{7}{2}\right) + 9$

$$\dfrac{63}{2} - 5 \stackrel{?}{=} \dfrac{35}{2} + 9$$

$$\dfrac{63}{2} - \dfrac{10}{2} \stackrel{?}{=} \dfrac{35}{2} + \dfrac{18}{2}$$

$$\dfrac{53}{2} = \dfrac{53}{2} \quad \text{True}$$

Because $y = \dfrac{7}{2}$ results in a true statement, the solution of the equation is $\dfrac{7}{2}$, or the solution set is $\left\{\dfrac{7}{2}\right\}$.

QUICK  *Solve the equation.*

13. $3x + 4 = 5x - 8$

14. $10m + 3 = 6m - 11$

We now summarize the steps that you should follow to solve an equation in one variable. Not all steps may be necessary to solve every equation, but you should use this summary as a guide.

SUMMARY: Steps for Solving an Equation in One Variable

Step 1: Remove any parentheses using the Distributive Property.

Step 2: Combine like terms on each side of the equation.

Step 3: Use the Addition Property of Equality to get the terms with the variable on one side of the equation and the constants on the other side.

Step 4: Use the Multiplication Property of Equality to get the coefficient of the variable term to be 1.

Step 5: Check the solution to verify that it satisfies the original equation.

EXAMPLE 6 | **Solving a Linear Equation in One Variable Using the Steps**

Solve the equation $2(z - 4) + 3z = 4 - (z + 2)$.

Step-by-Step Solution

Step 1: Remove any parentheses using the Distributive Property.		$2(z - 4) + 3z = 4 - (z + 2)$
		$2z - 8 + 3z = 4 - z - 2$
Step 2: Combine like terms on each side of the equation.		$5z - 8 = 2 - z$
Step 3: Use the Addition Property of Equality to get the terms with the variable on one side of the equation and the constants on the other side.	Add z to both sides of the equation:	$5z - 8 + z = 2 - z + z$
	Simplify:	$6z - 8 = 2$
	Add 8 to both sides of the equation:	$6z - 8 + 8 = 2 + 8$
	Simplify:	$6z = 10$

Step 4: Use the Multiplication Property of Equality to get the coefficient of the variable term to be 1.

Divide both sides of the equation by 6: $\dfrac{6z}{6} = \dfrac{10}{6}$

Simplify: $z = \dfrac{5}{3}$

Step 5: Check the solution to verify that it satisfies the original equation.

We leave the check to you.

The solution to the equation is $z = \dfrac{5}{3}$, or the solution set is $\left\{\dfrac{5}{3}\right\}$.

QUICK ✓ *Solve the equation.*

15. $-9x + 3(2x - 3) = -10 - 2x$ **16.** $3 - 4(p + 5) = 5(p + 2) - 12$

---

④ Use Linear Equations to Solve Problems

We continue to solve problems that are modeled by algebraic equations.

EXAMPLE 7 **How Much Does Alejandro Make in an Hour?**

Alejandro works as an engineer for a hospital. His contract calls for him to earn double time on all hours worked in excess of 40 hours for any given week. One week, Alejandro worked 46 hours and earned \$1326 before taxes. To determine Alejandro's hourly wage, w, solve the equation $40w + 6(2w) = 1326$.

Solution

$$40w + 6(2w) = 1326$$
Simplify: $40w + 12w = 1326$
Combine like terms: $52w = 1326$
Divide both sides of the equation by 52: $\dfrac{52w}{52} = \dfrac{1326}{52}$
Simplify: $w = 25.5$

Alejandro makes \$25.50 per hour.

QUICK ✓

17. Marcella works at a clothing store. Whenever she works more than 40 hours in a week, she gets paid twice her regular hourly wage of \$10 per hour. One week, Marcella earned \$640 before taxes. To determine how many hours, h, Marcella worked, solve the equation $400 + 20(h - 40) = 640$ for h.

2.2 Exercises

Concepts and Vocabulary

In Problems 1–3, fill in the blanks.

1. To solve the equation $2x - 8 = 41$, the first step is to _____ _____ to both sides of the equation.

2. To solve the equation $3 - 2(7x + 1) + 8x = 12$, first use the _____ Property to remove the parentheses.

3. To solve the equation $3x = -x - 19$, the first step is to _____ _____ to each side of the equation.

In Problems 4–6, answer True or False to each statement.

4. We use the Distributive Property to simplify $3(4x)$.

5. When solving the equation $z + 6z = 35$, we first combine z and $6z$ and obtain $7z = 35$.

6. It is not necessary to check solutions to equations.

7. Explain the difference between $6x - 2(x + 1)$ and $6x - 2(x + 1) = 6$. In general, what is the difference between an algebraic expression and an algebraic equation?

8. In your own words, explain the Addition and Multiplication Properties of Equality. In the Multiplication Property of Equality, why do you think we cannot multiply both sides of the equation by 0?

9. A classmate begins to solve the equation $7x + 3 - 2x = 9x - 5$ by adding $2x$ in the following manner:

$$7x + 3 - 2x = 9x - 5$$
$$\underline{+2x \qquad +2x}$$

Will this lead to the correct solution? Why or why not? Write an explanation telling the steps you would use to solve this equation.

10. A *corollary* is a rule or theorem that is closely related to a previous rule. Write a corollary to the Addition Property of Equality and title it the Subtraction Property of Equality. Write a corollary to the Multiplication Property of Equality, called the Division Property of Equality. What restrictions would you place on the Division Property of Equality? Why do you think these properties were not included in the text?

Building Skills

In Problems 11–46, solve the equation. Check your solution.

11. $3x + 4 = 7$

12. $5t + 1 = 11$

13. $2y - 1 = -5$

14. $6z - 2 = -8$

15. $-3p + 1 = 10$

16. $-4x + 3 = 15$

17. $8y + 3 = 15$

18. $6z - 7 = 3$

19. $5 - 2z = 11$

20. $1 - 3k = 4$

21. $\dfrac{2}{3}x + 1 = 9$

22. $\dfrac{5}{4}a + 3 = 13$

23. $\dfrac{7}{2}y - 1 = 13$

24. $\dfrac{1}{5}p - 3 = 2$

25. $3x - 7 + 2x = -17$

26. $5r + 2 - 3r = -14$

27. $2k - 7k - 8 = 17$
28. $2b + 5 - 8b = 23$
29. $2(x + 1) = -14$
30. $3(t - 4) = -18$
31. $-3(2 + r) = 9$
32. $-5(6 + z) = -20$
33. $2x + 9 = x + 1$
34. $7z + 13 = 6z + 8$
35. $2t - 6 = 3 - t$
36. $3 + 8x = 21 - x$
37. $14 - 2n = -4n + 7$
38. $6 - 12m = -3m + 3$
39. $-3(5 - 3k) = 6k + 6$
40. $-4(10 - 7x) = 3x + 10$
41. $2(2x + 3) = 3(x - 4)$
42. $3(5 + x) = 2(2x + 11)$
43. $-3(2y + 3) - 1 = -4(y + 6) + 2y$
44. $-5(b + 2) + 3b = -2(1 + 5b) + 6$
45. $9(6 + a) + 33a = 10a$
46. $5(12 - 3w) + 25w = 2w$

Mixed Practice

In Problems 47–62, solve the equation. Check your solution.

47. $-5x + 11 = 1$
48. $-6n + 14 = -10$
49. $4m + 5 = 2$
50. $7x + 1 = -9$
51. $-2(3n - 2) = 2$
52. $-5(2n - 3) = 10$
53. $\frac{1}{4} = \frac{3}{8} - 2x$
54. $\frac{6}{7} = -\frac{5}{14} + 3x$
55. $2y + 36 = 6 + 6y$
56. $7a - 26 = 13a + 2$
57. $\frac{1}{2}(-4k + 28) = 6 + 14k$
58. $\frac{2}{3}(9a - 12) = -6a - 11$
59. $-\frac{5}{2}(x + 6) + \frac{3}{2}x = -8$
60. $\frac{4}{3}a - \left(\frac{7}{3}a + 6\right) = -15$
61. $8[4 - 6(x - 1)] + 5[(2x + 3) - 5] = 18x - 338$
62. $3[10 - 4(x - 3)] + 2[(3x + 6) - 2] = 2x + 360$

Applying the Concepts

63. Burger King A Burger King chicken salad contains 4 more grams of fat than a McDonald's Crispy Chicken California Cobb salad. Find the number of grams of fat in each salad if there are 50 grams of fat in the two salads by solving the equation $x + (x + 4) = 50$, where x represents the number of fat grams in the McDonald's Crispy Chicken California Cobb salad and $x + 4$ represents the number of fat grams in the Burger King chicken salad. (*Source: Newsweek,* May 16, 2003)

64. Taco Bell A Taco Bell Express taco salad contains 3.5 fewer fat grams than a Wendy's mandarin chicken salad. Find the number of grams of fat in the Wendy's mandarin chicken salad if there are 65.5 grams of fat in the two salads by solving the equation $x + (x - 3.5) = 65.5$, where x represents the number of grams of fat in a Wendy's salad and $x - 3.5$ represents the number of fat grams in the Taco Bell salad. (*Source: Newsweek,* May 16, 2003)

△ **65. Dog Run** The length of a rectangular dog run is two feet more than twice the width, w, and the perimeter of the dog run is 30 feet. Solve the equation $2w + 2(2w + 2) = 30$ to find the width, w, of the dog run. Then find the length, $2w + 2$, of the dog run.

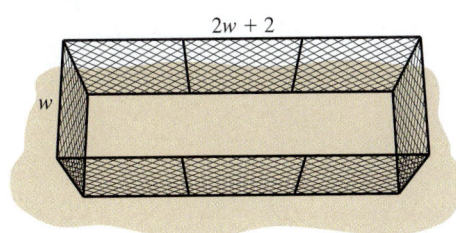

△ **66. Garden** The width of a rectangular garden is one yard more than one-half the length, L. The perimeter of the garden is 26 yards. Find the length, L, of the garden by solving the equation $2L + 2\left(\dfrac{1}{2}L + 1\right) = 26$.

67. Overtime Pay Jennifer worked 44 hours last week, including 4 hours of overtime, and earned \$368. She is paid at a rate of 1.5 times her regular hourly rate for overtime hours. Solve the equation $40x + 4(1.5x) = 368$ to find her regular hourly pay rate, x.

68. Overtime Pay Juan worked a total of 50 hours last week and earned \$498.75. He earned 1.5 times his regular hourly rate for 6 hours, and double his hourly rate for 4 holiday hours. Solve the equation $40x + 6(1.5x) + 4(2x) = \$498.75$, where x is Juan's regular hourly rate.

△ **69. Hanging Wallpaper** Becky purchased a remnant of 42 feet of a wallpaper border to hang in a rectangular bedroom. She knows that one wall of the bedroom is 5 feet longer than the other wall. Let x represent the length of the shorter wall. Assuming the length of the shorter wall is 8 feet, does Becky have enough wallpaper to hang the border? Use the equation $2x + 2(x + 5) = 42$ to answer the question.

△ **70. Perimeter of a Triangle** The perimeter of a triangle is 210 inches. If the sides are made up of 3 consecutive even integers, find the lengths of each of the 3 sides by solving the equation $x + (x + 2) + (x + 4) = 210$, where x represents the length of the shortest side.

Extending the Concepts

In Problems 71–74, use a calculator to solve the equation and round your answer to the indicated place.

71. $3(36.7 - 4.3x) - 10 = 4(10 - 2.5x) - 8(3.5 - 4.1x)$ to the nearest hundredth

72. $12(2.3 - 1.5x) - 6 = -3(18.4 - 3.5x) - 6.1(4x + 3)$ to the nearest tenth

73. $3.5\{4 - [6 - (2x + 3)] + 5\} = -18.4$ to the nearest tenth

74. $9\{3 - [4(2.3z - 1)] + 6.5\} = -406.3$ to the nearest hundredth

In Problems 75–78, determine the value of d to make the statement true.

75. In the equation $3d + 2x = 12$, the solution is -4.

76. In the equation $5d - 2x = -2$, the solution is -6.

77. In the equation $\dfrac{2}{3}x - d = 1$, the solution is $-\dfrac{3}{8}$.

78. In the equation $\dfrac{2}{5}x + 3d = 0$, the solution is $\dfrac{15}{8}$.

2.3 Solving Linear Equations Involving Fractions and Decimals; Classifying Equations

OBJECTIVES

1. Use the Least Common Denominator to Solve a Linear Equation Containing Fractions
2. Solve a Linear Equation Containing Decimals
3. Classify Linear Equations as Identity, Conditional, or Contradiction
4. Use Linear Equations to Solve Problems

Work Smart

A review of finding the least common denominator (LCD) can be found on pages A4–A5 in the Appendix, Section A.1.

Work Smart

Although removing fractions is not required to solve an equation, it frequently makes the arithmetic easier.

Preparing for Solving Linear Equations Involving Fractions and Decimals; Classifying Equations

Before getting started, take this readiness quiz. If you get a problem wrong, go back to the section cited and review the material.

1. Find the LCD of $\dfrac{3}{5}$ and $\dfrac{3}{4}$.   [Appendix, Section A.1, pp. A4–A5]

2. Find the LCD of $\dfrac{3}{8}$ and $-\dfrac{7}{12}$.   [Appendix, Section A.1, pp. A4–A5]

1 Use the Least Common Denominator to Solve a Linear Equation Containing Fractions

Sometimes equations are easier to solve if they do not contain fractions or decimals. To solve a linear equation containing fractions, we can multiply each side of the equation by the least common denominator (LCD) to clear the equation of fractions. Recall that the LCD is the smallest number that each denominator has as a common multiple. The Multiplication Property of Equality allows us to multiply both sides of the equation by the LCD.

EXAMPLE 1   **How to Solve a Linear Equation That Contains Fractions**

Solve the linear equation: $\dfrac{1}{2}x + \dfrac{2}{3}x = \dfrac{14}{3}$

Step-by-Step Solution

Before we follow the steps given in Section 2.2 on page 88, we will eliminate the fractions by multiplying both sides of the equation by the LCD. The LCD of 2 and 3 is 6, so we multiply both sides of the equation by 6.

$$6\left(\dfrac{1}{2}x + \dfrac{2}{3}x\right) = 6\left(\dfrac{14}{3}\right)$$

Now we can follow Steps 1–5 from the summary in Section 2.2 to solve the equation.

Step 1: Apply the Distributive Property to remove parentheses.	$6\left(\dfrac{1}{2}x + \dfrac{2}{3}x\right) = 6\left(\dfrac{14}{3}\right)$
	Use the Distributive Property: $6\left(\dfrac{1}{2}x\right) + 6\left(\dfrac{2}{3}x\right) = 6\left(\dfrac{14}{3}\right)$
	$3x + 4x = 28$
Step 2: Combine like terms.	$7x = 28$
Step 3: Use the Addition Property of Equality to get the terms with the variable on one side of the equation and the constants on the other side.	This step is not necessary because the variable is already on one side of the equation in this problem.

Step 4: Get the coefficient of the variable to be 1.

Divide both sides of the equation by 7:

$$\frac{7x}{7} = \frac{28}{7}$$

$$x = 4$$

Step 5: Confirm that $x = 4$ is the solution of the equation.

$$\frac{1}{2}x + \frac{2}{3}x = \frac{14}{3}$$

Substitute 4 for x in the original equation:

$$\frac{1}{2}(4) + \frac{2}{3}(4) \overset{?}{=} \frac{14}{3}$$

$$2 + \frac{8}{3} \overset{?}{=} \frac{14}{3}$$

$$\frac{6}{3} + \frac{8}{3} \overset{?}{=} \frac{14}{3}$$

$$\frac{14}{3} = \frac{14}{3} \quad \text{True}$$

The solution of the equation is 4, or the solution set is $\{4\}$.

QUICK ✓ *Solve each equation by multiplying by the LCD.*

1. $\dfrac{2x}{5} - \dfrac{x}{4} = \dfrac{3}{2}$

2. $\dfrac{5}{6}x + \dfrac{1}{9} = -\dfrac{1}{6}x - \dfrac{1}{6}$

EXAMPLE 2 Solve a Linear Equation Using the LCD

Solve the linear equation: $\dfrac{7n + 5}{8} = 2 + \dfrac{3n + 15}{10}$

Solution

Because the equation contains fractions, we multiply both sides of the equation by the LCD. Because the LCD of 8 and 10 is 40, we multiply both sides of the equation by 40.

$$\frac{7n + 5}{8} = 2 + \frac{3n + 15}{10}$$

Multiply both sides by the LCD, 40:
$$40\left(\frac{7n + 5}{8}\right) = 40\left(2 + \frac{3n + 15}{10}\right)$$

Use the Distributive Property to multiply all terms of the equation by the LCD:
$$40\left(\frac{7n + 5}{8}\right) = 40(2) + 40\left(\frac{3n + 15}{10}\right)$$

Divide out common factors:
$$\overset{5}{40}\left(\frac{7n + 5}{\underset{1}{8}}\right) = 40(2) + \overset{4}{40}\left(\frac{3n + 15}{\underset{1}{10}}\right)$$

$$5(7n + 5) = 40(2) + 4(3n + 15)$$

Use Distributive Property to remove parentheses:
$$35n + 25 = 80 + 12n + 60$$

Combine like terms:
$$35n + 25 = 140 + 12n$$

Isolate n:
$$35n + 25 - 12n = 140 + 12n - 12n$$

$$23n + 25 = 140$$

$$23n + 25 - 25 = 140 - 25$$

$$23n = 115$$

Divide both sides by 23:
$$\frac{23n}{23} = \frac{115}{23}$$

$$n = 5$$

Does $n = 5$ satisfy the original equation? Let's see.

Work Smart

When clearing fractions from an equation, first count the number of terms in the equation. You must have that many products when you apply the Distributive Property. So in Example 2, remember to multiply the "2" by 40 also.

Check

$$\frac{7n + 5}{8} = 2 + \frac{3n + 15}{10}$$

$$\frac{7(5) + 5}{8} \stackrel{?}{=} 2 + \frac{3(5) + 15}{10}$$

$$\frac{35 + 5}{8} \stackrel{?}{=} 2 + \frac{15 + 15}{10}$$

$$\frac{40}{8} \stackrel{?}{=} 2 + \frac{30}{10}$$

$$5 \stackrel{?}{=} 2 + 3$$

$$5 = 5 \quad \text{True}$$

The solution $n = 5$ checks, so the solution is 5, or the solution set is $\{5\}$. ▬

QUICK ✔ *Solve each equation by multiplying by the LCD.*

3. $\dfrac{a}{3} - \dfrac{1}{3} = -5$

4. $\dfrac{3x - 3}{4} - 1 = \dfrac{3}{5}x$

---

② Solve a Linear Equation Containing Decimals

When decimals occur in linear equations, we can clear the decimal using the same techniques that we used to clear the equation of fractions. That is, we multiply both sides of the equation by a power of 10 so that the decimal is cleared. For example, $0.8x$ is equivalent to $\dfrac{8}{10}x$, so we multiply by 10 to clear the decimal from the equation. Because 0.34 is equivalent to $\dfrac{34}{100}$, we multiply by 100 to clear the decimal from the equation.

EXAMPLE 3   Solving a Linear Equation with a Decimal Coefficient

Solve the equation: $0.3x - 4 = 11$

Solution

We want to clear the decimal from the equation. We can do this by multiplying both sides of the equation by 10. Do you see why? $0.3 = \dfrac{3}{10}$, so multiplying by the LCD = 10 will clear the decimal.

Work Smart
The equation $0.3x - 4 = 11$ is equivalent to

$$\frac{3}{10}x - 4 = 11$$

and may be solved by multiplying both sides of the equation by 10 to clear the fraction.

$$0.3x - 4 = 11$$

Multiply both sides of the equation by 10: $\quad 10 \cdot (0.3x - 4) = 10 \cdot 11$

Distribute: $\quad 10 \cdot 0.3x - 10 \cdot 4 = 110$

$$3x - 40 = 110$$

Add 40 to both sides of the equation: $\quad 3x = 150$

Divide both sides of the equation by 3: $\quad x = 50$

Check We verify our solution by replacing $x = 50$ in the original equation to see if a true statement results.

$$0.3x - 4 = 11$$
$$0.3(50) - 4 \overset{?}{=} 11$$
$$15 - 4 \overset{?}{=} 11$$
$$15 = 15 \quad \text{True}$$

The solution of the equation $0.3x - 4 = 11$ is 50, or the solution set is $\{50\}$.　■

It's not a requirement to clear fractions or decimals before solving an equation. Here is another way to solve $0.3x - 4 = 11$.

$$0.3x - 4 = 11$$

Add 4 to each side: $\quad 0.3x - 4 + 4 = 11 + 4$

$$0.3x = 15$$

Divide each side by 0.3: $\quad \dfrac{0.3x}{0.3} = \dfrac{15}{0.3}$

$$x = 50$$

Which method do you prefer?

QUICK ☑ *Solve each equation by clearing the decimal.*

5. $0.2z = 20$　　　　　　　　**6.** $0.15p - 2.5 = 5$

EXAMPLE 4　**Solving a Linear Equation Containing Decimals**

Solve the equation: $p + 0.08p = 129.6$

Solution

We can clear the decimals by multiplying both sides of the equation by 100. Do you see why? $0.08 = \dfrac{8}{100}$ and $129.6 = \dfrac{1296}{10}$, so multiplying by the LCD = 100 will clear the decimals. However, before we multiply both sides of the equation by 100, we will first combine like terms on the left-hand side of the equation.

$$p + 0.08p = 129.6$$

$p = 1 \cdot p$: $\quad 1p + 0.08p = 129.6$

Combine like terms: $\quad 1.08p = 129.6$

Multiply both sides of the equation by 100: $\quad 100 \cdot 1.08p = 100 \cdot 129.6$

Simplify: $\quad 108p = 12{,}960$

Divide both sides of the equation by 108: $\quad \dfrac{108p}{108} = \dfrac{12{,}960}{108}$

Simplify: $\quad p = 120$

We leave the check to you. The solution of the equation $p + 0.08p = 129.6$ is 120, or the solution set is $\{120\}$.　■

QUICK ☑ *Solve each equation by first clearing the decimal.*

7. $p + 0.05p = 52.5$　　　　　　**8.** $c - 0.25c = 120$

EXAMPLE 5   **Solving a Linear Equation That Contains Decimals**

Solve the equation: $0.05x + 0.08(10{,}000 - x) = 680$

Solution

Before we clear the decimals, we will distribute the 0.08.

$$0.05x + 0.08(10{,}000 - x) = 680$$

$$0.05x + 800 - 0.08x = 680$$

Combine like terms: $-0.03x + 800 = 680$

Subtract 800 from both sides: $-0.03x = -120$

Multiply both sides of the equation by 100 to eliminate the decimal: $100(-0.03x) = 100(-120)$

$$-3x = -12{,}000$$

Divide both sides by -3: $x = 4000$

We leave the check to you. The solution to the equation $0.05x + 0.08(10{,}000 - x) = 680$ is 4000, or the solution set is $\{4000\}$.

QUICK ✓   *Solve each equation.*

9. $0.36y - 0.5 = 0.16y + 0.3$

10. $0.12x + 0.05(5000 - x) = 460$

---

(3) **Classify Linear Equations as Identity, Conditional, or Contradiction**

All of the linear equations we have solved so far have had a single solution. This one value of the variable made the equation a true statement, while all other values of the variable make the equation false. We give these types of equations a special name.

> **DEFINITION**
>
> A **conditional equation** is an equation that is true for some values of the variable and false for other values of the variable.

For example, the equation

$$x + 5 = 11$$

is a conditional equation because it is true when $x = 6$ and false for every other real number x. All the equations that we have studied to this point have been conditional equations.

There are equations that are false for all values of the variable.

> **DEFINITION**
>
> A **contradiction** is an equation that is false for every replacement value of the variable.

For example, the equation

$$2x + 3 = 7 + 2x$$

is a contradiction because it is false for any replacement value for x. Contradictions are identified through the process of creating equivalent equations. For example, if we subtract $2x$ from both sides of $2x + 3 = 7 + 2x$, we obtain $3 = 7$, which is clearly false. **Contradictions have no solution, so the solution set is the empty set, written as $\{\ \}$ or $\varnothing$.**

Work Smart
Do not write the empty set as $\{\varnothing\}$.

EXAMPLE 6 Solving a Linear Equation That Is a Contradiction

Solve the equation: $3y - (5y + 4) = 12y - 7(2y - 1)$

Solution

$$3y - (5y + 4) = 12y - 7(2y - 1)$$

Use the Distributive Property to remove parentheses:
$$3y - 5y - 4 = 12y - 14y + 7$$

Combine like terms:
$$-2y - 4 = -2y + 7$$

Isolate y:
$$-2y + 2y - 4 = -2y + 2y + 7$$

$$-4 = 7$$

Work Smart

The term "contradiction" describes the type of equation. The solution set is $\varnothing$.

The last statement states that $-4 = 7$. This is a false statement, so the equation is a contradiction. The solution set is $\varnothing$ or $\{\ \}$.

Some equations are true for all real numbers for which the equation is defined.

> **DEFINITION**
>
> An **identity** is an equation that is satisfied for all values of the variable for which both sides of the equation are defined.

An example of an identity is the equation

$$2(x - 5) + 1 = 5x - (9 + 3x)$$

This equation is an identity because any real number x makes the equation a true statement. Just as with contradictions, identities are recognized through the process of creating equivalent equations. For example, if we use the Distributive Property and combine like terms in the equation $2(x - 5) + 1 = 5x - (9 + 3x)$, we obtain $2x - 9 = 2x - 9$, which is true for all replacement values for x. **The solution set of linear identities is the set of all real numbers.**

EXAMPLE 7 Solving a Linear Equation That Is an Identity

Solve the equation: $2(x + 5) = 4x - (2x - 10)$

Solution

$$2(x + 5) = 4x - (2x - 10)$$

Use the Distributive Property to remove parentheses:
$$2x + 10 = 4x - 2x + 10$$

Combine like terms:
$$2x + 10 = 2x + 10$$

Isolate x:
$$2x - 2x + 10 = 2x - 2x + 10$$

$$10 = 10$$

The last statement states that $10 = 10$. This is a true statement for all real numbers x. The solution set is the set of all real numbers.

QUICK ✓ *Solve the equation and state the solution set.*

11. $3(x + 4) = 4 + 3x + 18$

12. $\dfrac{1}{3}(6x - 9) - 1 = 6x - [4x - (-4)]$

13. $-5 - (9x + 8) + 23 = 7 + x - (10x - 3)$

14. $\dfrac{3}{2}x - 8 = x + 7 + \dfrac{1}{2}x$

SUMMARY:

- A *conditional* equation is true for some values of the variable and false for others.
- A *contradiction* is false for all values of the variable.
- An *identity* is true for all of the permitted values of the variable.

EXAMPLE 8   Classifying a Linear Equation

Solve the equation $2(2 + a) - 12a = -a + 5 - 9a$. Tell if the equation is a contradiction, an identity, or a conditional equation.

Solution

$$2(2 + a) - 12a = -a + 5 - 9a$$

Use the Distributive Property to remove parentheses: $4 + 2a - 12a = -a + 5 - 9a$

Combine like terms: $4 - 10a = 5 - 10a$

Isolate a: $4 - 10a + 10a = 5 - 10a + 10a$

$$4 = 5$$

The last statement states that $4 = 5$. This is a false statement, so the equation is a contradiction. The solution set is $\varnothing$ or $\{\ \}$.

QUICK ✓   *Solve the equation and tell if each equation is a contradiction, an identity, or a conditional equation.*

15. $2(x - 7) + 8 = 6x - (4x + 2) - 4$ **16.** $\dfrac{4(7 - x)}{3} = x$

17. $\dfrac{1}{2}(4x - 6) = 6\left(\dfrac{1}{3}x - \dfrac{1}{2}\right) + 4$ **18.** $4(5x - 4) + 1 = -2 + 20x$

---

④ Use Linear Equations to Solve Problems

We continue to solve problems that are modeled by algebraic equations.

EXAMPLE 9   Solving a Problem from Finance

You have $5000 to invest, and your financial advisor advises you to put part of the money in a Certificate of Deposit (CD) that earns 2% simple interest compounded annually, and the rest in bonds that earn 4% simple interest compounded annually. To determine the amount you should invest in the CD to earn $170 interest at the end of one year, solve the equation $0.02x + 0.04(5000 - x) = 170$, where x is the amount of money invested in CDs.

Solution

We want to solve the equation for x, the amount of money invested in CDs.

$$0.02x + 0.04(5000 - x) = 170$$

Use the Distributive Property to remove parentheses: $0.02x + 200 - 0.04x = 170$

Combine like terms: $-0.02x + 200 = 170$

Isolate x: $-0.02x + 200 - 200 = 170 - 200$

$$-0.02x = -30$$

Multiply both sides by 100 to eliminate the decimal: $-2x = -3000$

Divide both sides by -2: $x = 1500$

Since $x = 1500$, you must invest \$1500 in Certificates of Deposit to earn \$170 in interest at the end of one year.

QUICK ✓ *Solve the equation for the unknown quantity.*

19. Janet Majors invested part of her lottery winnings in a savings account that pays 4% annual interest, and \$250 more than that in a mutual fund that pays 6% annual interest. Her total interest was \$65. To determine the amount she invested in the savings account, solve the equation $0.04x + 0.06(x + 250) = 65$, where x represents the amount invested in the savings account.

2.3 Exercises

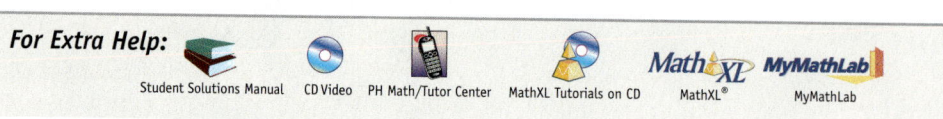

For Extra Help: Student Solutions Manual   CD Video   PH Math/Tutor Center   MathXL Tutorials on CD   MathXL®   MyMathLab

Concepts and Vocabulary

In Problems 1–3, fill in the blanks.

1. A(n) _____ _____ is an equation that is true for some values of the variable and false for other values of the variable.

2. To clear fractions from an equation, we multiply each side of the equation by the _____ _____ _____.

3. To clear decimals from the equation $0.25x + 5 = 7 - 0.3x$, we multiply both sides of the equation by _____.

In Problems 4–6, answer True or False to each statement.

4. A linear equation can have one solution, no solution, or infinitely many solutions.

5. An equation that is satisfied for every choice of the variable for which both sides of the equation are defined is called a contradiction.

6. To solve the equation $\frac{1}{10}x + 8 = \frac{1}{5}x - 3$, we multiply both sides of the equation by 5 to clear fractions.

7. A student solved the equation $3(x + 8) = \frac{1}{2}(6x + 4)$ and wrote the answer $24 = 2$. The instructor did not give full credit for this answer. Explain the student's error and determine the correct solution.

8. When solving the equation $\frac{2}{3}x - 5 = \frac{1}{2}x$, a student decided to multiply both sides by the LCD and wrote $6 \cdot \frac{2}{3}x - 5 = \frac{1}{2}x \cdot 6$. This resulted in the next line $4x - 5 = 3x$. Is this correct? Explain the steps necessary to finish by this technique and then suggest another list of steps that would arrive at the correct solution.

Building Skills

In Problems 9–44, solve the equation. Check your solution.

9. $\dfrac{2k - 1}{4} = 2$

10. $\dfrac{3a + 2}{5} = -1$

11. $\dfrac{3x + 2}{4} = \dfrac{x}{2}$

12. $\dfrac{2x - 3}{5} = \dfrac{3x}{10}$

13. $\dfrac{1}{5}x + \dfrac{3}{2} = \dfrac{3}{10}$

14. $\dfrac{3}{2}n - \dfrac{4}{11} = \dfrac{91}{22}$

15. $\dfrac{-2x}{3} + 1 = \dfrac{5}{9}$

16. $\dfrac{3m}{8} - 1 = \dfrac{5}{6}$

17. $0.4w = 12$

18. $0.3z = 6$

19. $-1.3c = 5.2$

20. $-1.7q = -8.5$

21. $1.05p = 52.5$

22. $1.06z = 31.8$

23. $p + 1.5p = 12$

24. $2.5a + a = 7$

25. $p + 0.05p = 157.5$

26. $p + 0.04p = 260$

27. $\dfrac{a}{4} - \dfrac{a}{3} = -\dfrac{1}{2}$

28. $\dfrac{3}{2}b - \dfrac{4}{5}b = \dfrac{28}{5}$

29. $\dfrac{5}{4}(2a - 10) = -\dfrac{3}{2}a$

30. $\dfrac{2}{3}(6 - x) = \dfrac{5x}{6}$

31. $\dfrac{y}{10} + 3 = \dfrac{y}{4} + 6$

32. $\dfrac{p}{8} - 1 = \dfrac{7p}{6} + 2$

33. $\dfrac{4x - 9}{3} + \dfrac{x}{6} = \dfrac{x}{2} - 2$

34. $\dfrac{3x + 2}{4} - \dfrac{x}{12} = \dfrac{x}{3} - 1$

35. $0.3x + 2.3 = 0.2x + 1.1$

36. $0.7y - 4.6 = 0.4y - 2.2$

37. $0.65x + 0.3x = x - 3$

38. $0.5n - 0.35n = 2.5n + 9.4$

39. $3 + 1.5(z + 2) = 3.5z - 4$

40. $5 - 0.2(m - 2) = 3.6m + 1.6$

41. $0.02(2c - 24) = -0.4(c - 1)$

42. $0.3(6a - 4) = -0.10(2a - 8)$

43. $0.15x + 0.10(250 - x) = 28.75$

44. $0.03t + 0.025(1000 - t) = 27.25$

In Problems 45–56, solve each equation. State whether the equation is a contradiction, an identity, or a conditional equation.

45. $4z - 3(z + 1) = 2(z - 3) - z$

46. $4(y - 2) = 5y - (y + 1)$

47. $6q - (q - 3) = 2q + 3(q + 1)$

48. $-3x + 2 + 5x = 2(x + 1)$

49. $9a - 5(a + 1) = 2(a - 3)$

50. $7b + 2(b - 4) = 8b - (3b + 2)$

51. $\dfrac{4x - 9}{6} - \dfrac{x}{2} = \dfrac{x}{6} + 3$

52. $\dfrac{2m + 1}{4} - \dfrac{m}{6} = \dfrac{m}{3} - 1$

53. $\dfrac{5z + 1}{5} = \dfrac{2z - 3}{2}$

54. $\dfrac{2y - 7}{4} = \dfrac{3y - 13}{6}$

55. $\dfrac{q}{3} + \dfrac{4}{5} = \dfrac{5q + 12}{15}$

56. $\dfrac{2x}{3} + \dfrac{x + 3}{12} = \dfrac{3x + 1}{4}$

Mixed Practice

In Problems 57–82, solve each equation.

57. $-3(2n + 4) = 10n$

58. $-15(z - 3) = 25z$

59. $-2x + 5x = 4(x + 2) - (x + 8)$

60. $5m - 3(m + 1) = 2(m + 1) - 5$

61. $-6(x - 2) + 8x = -x + 10 - 3x$

62. $3 - (x + 10) = 3x + 7$

63. $\dfrac{3}{4}x = \dfrac{1}{2}x - 5$

64. $\dfrac{1}{3}x = 2 + \dfrac{5}{6}x$

65. $\dfrac{1}{2}x + 2 = \dfrac{4x + 1}{4}$

66. $\dfrac{x}{2} + 4 = \dfrac{x + 7}{3}$

67. $0.3p + 2 = 0.1(p + 5) + 0.2(p + 1)$

68. $1.6z - 4 = 2(z - 1) - 0.4z$

69. $-0.7x = 1.4$

70. $0.2a = -6$

71. $\dfrac{3(2y - 1)}{5} = 2y - 3$

72. $\dfrac{4(2n + 1)}{3} = 2n - 6$

73. $0.6x - 0.2(x - 4) = 0.4(x - 2)$

74. $0.3x - 1 = 0.5(x + 2) - 0.2x$

75. $\dfrac{3x - 2}{4} = \dfrac{5x - 1}{6}$

76. $\dfrac{x - 1}{4} = \dfrac{x - 4}{6}$

77. $0.3x + 2.6x = 5.7 - 1.8 + 2.8x$

78. $0.3(z - 10) - 0.5z = -6$

79. $\dfrac{3}{2}x - 6 = \dfrac{2(x - 9)}{3} + \dfrac{1}{6}x$

80. $\dfrac{2x - 3}{4} + 5 = \dfrac{3(x + 3)}{4} - \dfrac{x}{2} + 2$

81. $\dfrac{2}{3}\left[4 - \left(\dfrac{x}{2} + 6\right) - 2x\right] + 3 = \dfrac{5x}{6}$

82. $\dfrac{1}{2}\left[3 - \left(\dfrac{2x}{3} - 1\right) + 3x\right] = \dfrac{-4x + 1}{3} + 1$

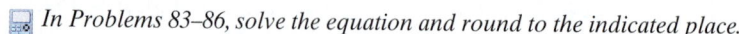 *In Problems 83–86, solve the equation and round to the indicated place.*

83. $2.8x + 13.754 = 4 - 2.95x$ to the nearest hundredth

84. $-4.88x - 5.7 = 2(-3.41x) + 1.2$ to the nearest whole number

85. $x - \{1.5x - 2[x - 3.1(x + 10)]\} = 0$ to the nearest tenth

86. $-3x - 2\{4 + 3[x - (1 + x)]\} = 12$ to the nearest hundredth

Applying the Concepts

87. Sales Tax The price of a pair of jeans including sales tax of 6% is $53. To find the price p of the jeans, solve the equation $1.06p = 53$.

88. Gardening Bob Adams rented a rototiller for x hours at a cost of $7.50 per hour. Bob paid a bill of $37.50. Find the number of hours he rented the tiller by solving the equation $7.50x = 37.50$.

89. Purchasing a Car The total cost (including 6% sales tax) for the purchase of an automobile was $19,080. To determine the cost of the auto before the sales tax was added, solve the equation $x + 0.06x = 19,080$, where x represents the cost of the car before taxes.

90. Purchasing a Kayak The total cost, including 5.5% sales tax, for the purchase of a kayak was $1266. Find the price of the kayak, k, before sales tax, by solving the equation $k + 0.055k = 1266$.

91. Hourly Pay Bob recently received a 4% pay increase. His hourly wage is now $8.84. To determine his hourly wage, w, before the 4% pay raise, solve the equation $w + 0.04w = 8.84$ for w.

92. Hourly Pay A union representing airline employees recently agreed to a 6% cut in hourly wage in order to help the airline avoid filing for bankruptcy. A baggage handler will now earn $26.32 per hour. To find the baggage handler's hourly wage, w, before the pay cut solve the equation $w - 0.06w = 26.32$.

93. CD Player Tamara purchased a CD player at a "25% off" sale for $71.25. To determine the original price, p, of the CD player, solve the equation $p - 0.25p = 71.25$.

94. Team Sweatshirt Your favorite college has logo sweatshirts on sale for $33.60. The sweatshirt has been marked down by 30%. To find the original price of the sweatshirt, x, solve the equation $x - 0.30x = 33.60$.

95. Piggy Bank Celeste saves dimes and quarters in a piggy bank. She opened the bank and discovered that she had $7.05 and the number of dimes was 3 more than twice the number of quarters. To find the number of quarters, q, solve the equation $0.25q + 0.10(2q + 3) = 7.05$.

96. Clean Car Pablo cleaned out his car and found nickels and quarters in the car seats. He found $4.25 in change and noticed that the number of quarters was 5 less than twice the number of nickels. Solve the equation $0.05n + 0.25(2n - 5) = 4.25$ to find n, the number of nickels Pablo found.

97. Comparing Perimeters The two rectangles shown below have the same perimeter. Solve the equation $2x + 2(x + 3) = 2\left(\dfrac{1}{2}x\right) + 2(x + 6)$ for x, the width of the first rectangle.

△ **98. Comparing Perimeters** The square and the rectangle shown have the same perimeter. Solve the equation $4x = 2\left(\frac{1}{2}x\right) + 2(x + 5)$ for x, the length of the side of the square.

x

$\frac{1}{2}x$

$x + 5$

99. Paying Your Taxes You are single and just determined that you paid \$1442.50 in federal income taxes for 2004. The solution to the equation $1442.50 = 0.15(x - 7150) + 715$ represents your adjusted gross income in 2004. Determine your adjusted gross income in 2004. (*Source:* Internal Revenue Service)

100. Paying Your Taxes You are married and just determined that you paid \$12,225 in federal income taxes in 2004. The solution to the equation $12,225 = 0.25(x - 58,100) + 8000$ represents the adjusted gross income of you and your spouse in 2004. Determine the adjusted gross income of you and your spouse in 2004. (*Source:* Internal Revenue Service)

Extending the Concepts

101. Make up a linear equation that has one solution. Make up a linear equation that has no solution. Make up a linear equation that is an identity. Comment on the differences and similarities in making up each equation.

2.4 Evaluating Formulas and Solving Formulas for a Variable

OBJECTIVES

1 Evaluate a Formula
2 Solve a Formula for a Variable

Preparing for Evaluating Formulas and Solving Formulas for a Variable
Before getting started, take this readiness quiz. If you get a problem wrong, go back to the section cited and review the material.

1. Evaluate the expression $2L + 2W$ for $L = 7$ and $W = 5$.   [Section 1.7, pp. 57–58]
2. Round the expression 0.5873 to hundredths place.   [Section A.2, p. A9]

1 Evaluate a Formula

In this section, we use *formulas* to solve mathematical problems.

> **DEFINITION**
> A mathematical **formula** is an equation that describes how two or more variables are related.

For example, a formula for the area of a rectangle is *area* = *length · width*, or $A = l \cdot w$. You use mathematical formulas every day, many times without even realizing it. For instance, if you plan to paint your apartment, you must find the surface area of the walls to compute the amount of paint you will purchase. To determine the number of gallons of gasoline you can afford to put in your car, you may also use a formula, as illustrated in the next example.

EXAMPLE 1 **Evaluating a Formula**

The number of gallons of gasoline that you put in your car is found by using the formula $\frac{C}{p} = n$, where C is the total cost, p is the price per gallon, and n is the number of gallons. How many gallons of gasoline can you purchase for $10.00 if gas costs $2.50 per gallon?

Solution

$$\frac{C}{p} = n$$

Replace C with cost, $10, and p with price per gallon, $2.50: $\quad \dfrac{\$10.00}{\$2.50} = n$

Evaluate the numerical expression: $\quad 4 = n$

You can purchase 4 gallons for $10.

QUICK ✔ *Evaluate each formula for the unknown quantity.*

1. Your best friend, who is on spring break in Europe, e-mailed you that the temperature in Paris today is 15° Celsius. Use the formula $F = \dfrac{9}{5}C + 32$, where F is degrees Fahrenheit and C is degrees Celsius, to find the approximate temperature in degrees Fahrenheit.

2. The size of a dress purchased in Europe is different from one purchased in the U.S. The equation $c = a + 30$ gives the European (Continental) dress size c in terms of the size in the U.S., a. Find the Continental dress size that corresponds to a size 10 dress in the United States.

EXAMPLE 2 **Evaluating a Formula Containing a Percent**

The formula $S = P - 0.25P$ gives the sale price S of an item which originally cost P dollars and that was reduced by 25%. Find the sale price of a pair of jeans that originally cost $40.00.

Solution

$$S = P - 0.25P$$

Replace P with the price of the jeans, $40: $\quad S = 40 - 0.25(40)$

Evaluate the numerical expression: $\quad S = 40 - 10$

$$S = 30$$

The sale price of the jeans is $30.

QUICK ✔ *Evaluate each formula for the unknown quantity.*

3. The formula $E = 250 + 0.05S$ is a formula for the earnings, E, of a salesman who receives $250 per week plus 5% commission on all weekly sales, S. Find the earnings of a salesman who had weekly sales of $1250.

4. The formula $N = p + 0.06p$ models the new population, N, of a town whose current population, p, is given, if the town expects population growth of 6% next year. Find the new population of a town whose current population is 5600 persons.

The Simple Interest Formula

Interest is money paid for the use of money. The total amount borrowed is called the **principal.** The principal can be in the form of a loan (an individual borrows from the bank) or a deposit (the bank borrows from the individual). The **rate of interest,**

expressed as a percent, is the amount charged for the use of the principal for a given period of time, usually on yearly basis.

> **SIMPLE INTEREST FORMULA**
>
> If an amount of money, P, called the **principal** is invested for a period of t years at an annual interest rate r, expressed as a decimal, the amount of interest I earned is
>
> $$I = Prt$$
>
> Interest earned according to this formula is called **simple interest.**

EXAMPLE 3   Evaluating Using the Simple Interest Formula

Janice invested \$500 in a two-year Certificate of Deposit (CD) paying 4% simple interest. Find the amount of interest Janice will earn in 6 months. Also find the total amount of money Janice will have at the end of 6 months.

Solution

Work Smart

Be sure to express t in years when using the simple interest formula.

We see that the amount of money Janice invested, P is \$500. The interest rate r is $4\% = 0.04$. Because 6 months is $\frac{1}{2}$ a year, we have that $t = \frac{1}{2}$.

$$I = Prt$$

Replace $P = \$500$, $r = 0.04$, and $t = \frac{1}{2}$: $I = 500 \cdot 0.04 \cdot \frac{1}{2}$

Evaluate the numerical expression: $I = 10$

Janice earned \$10 on her investment. At the end of 6 months, Janet will have $\$500 + \$10 = \$510$.

QUICK ✓   *Find the unknown quantity.*

5. Bill invested his \$2500 Virginia Lottery winnings in an 8-month Certificate of Deposit that earns 3% simple interest. Find the amount of interest Bill's investment will earn at the end of 8 months. Also find the total amount of money bill will have at the end of 8 months.

Geometry Formulas

Let's review a few common terms from geometry.

> **DEFINITIONS**
>
> The **perimeter** is the sum of the lengths of all the sides of a figure.
>
> The **area** is the amount of space enclosed by a two-dimensional figure measured in units squared.
>
> The **surface area** of a solid is the sum of the areas of the surfaces of a three-dimensional figure.
>
> The **volume** is the amount of space occupied by a three-dimensional figure measured in units cubed.
>
> The **radius** r of a circle is the line segment that extends from the center of the circle to any point on the circle.
>
> The **diameter** of a circle is any line segment that extends from one point on the circle through the center to a second point on the circle. The diameter is two times the length of the radius, $d = 2r$.
>
> In circles, we use the term **circumference** to mean the perimeter.

Formulas from geometry are useful in solving many types of problems. We list some of these formulas in Table 1.

Table 1		
Plane Figures		**Formulas**
Square		**Area:** $A = s^2$ **Perimeter:** $P = 4s$
Rectangle		**Area:** $A = lw$ **Perimeter:** $P = 2l + 2w$
Triangle		**Area:** $A = \dfrac{1}{2}bh$ **Perimeter:** $P = a + b + c$
Trapezoid		**Area:** $A = \dfrac{1}{2}h(B + b)$ **Perimeter:** $P = a + b + c + B$
Parallelogram		**Area:** $A = bh$ **Perimeter:** $P = 2a + 2b$
Circle		**Area:** $A = \pi r^2$ **Circumference:** $C = 2\pi r = \pi d$
Solids		**Formulas**
Cube		**Volume:** $V = s^3$ **Surface Area:** $S = 6s^2$
Rectangular Solid		**Volume:** $V = lwh$ **Surface Area:** $S = 2lw + 2lh + 2wh$
Sphere		**Volume:** $V = \dfrac{4}{3}\pi r^3$ **Surface Area:** $S = 4\pi r^2$
Right Circular Cylinder		**Volume:** $V = \pi r^2 h$ **Surface Area:** $S = 2\pi r^2 + 2\pi rh$
Cone		**Volume:** $V = \dfrac{1}{3}\pi r^2 h$

EXAMPLE 4 Evaluating a Formula for the Perimeter of a Rectangle

The number of yards of fencing that must be purchased to fence a rectangular garden is given by $P = 2l + 2w$, where P is the perimeter of the garden, l is the length of the garden and w is its width. Find the number of yards of fencing that must be purchased to enclose a garden that is 10.5 yards long and 4.25 yards wide as illustrated in Figure 1.

Figure 1

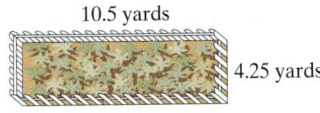

10.5 yards

4.25 yards

Solution

$$P = 2l + 2w$$

Replace l with 10.5 and w with 4.25: $P = 2(10.5) + 2(4.25)$

Evaluate the numerical expression: $P = 21 + 8.5$

$$P = 29.5$$

We find that 29.5 yards of fencing must be purchased.

QUICK ✓ *Evaluate the formula for the unknown quantity.*

6. Find the area of a trapezoid whose height is 4.5 inches, $B = 9$ inches, and $b = 7$ inches. Use $A = \dfrac{1}{2}h(B + b)$.

Sometimes we need to use more than one geometry formula in order to solve a problem.

EXAMPLE 5 Finding the Area of a Lawn and the Cost of Sod

A circular swimming pool whose diameter is 30 feet is to be installed in a rectangular yard that is 100 feet by 60 feet. Once the pool is installed, grass is to be installed on the remaining land. See Figure 2.

(a) Determine the area of land that is to receive grass.

(b) If sod costs $0.25 per square foot installed, what will be the cost of the lawn?

Figure 2

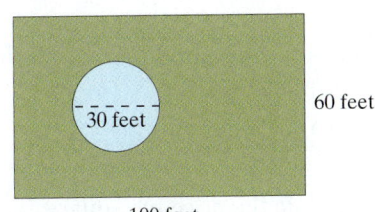

30 feet

60 feet

100 feet

Solution

(a) The area of a rectangle is $A = lw$ and the area of a circle is $A = \pi r^2$. We find the area of the remaining lawn by subtracting the area of the circle from the area of the lawn.

$$\text{Area of remaining lawn} = \text{Area of rectangle} - \text{Area of circle}$$
$$= lw - \pi r^2$$

The length l of the lawn is 100 feet and the width is 60 feet. The radius of the circle is one-half its diameter, so the radius is 15 feet. Substituting into the formulas, we obtain

$$\text{Area of remaining lawn} = (100 \text{ feet})(60 \text{ feet}) - \pi(15 \text{ feet})^2$$
$$= 6000 \text{ feet}^2 - 225\pi \text{ feet}^2$$

$\pi \approx 3.14159$: $\approx 5293 \text{ feet}^2$

There is approximately 5293 square feet of land that needs sod.

In Words

The word "per" means for each one. For example, the sod costs $0.25 for each square foot installed. When you see the word "per," think fraction. So $0.25 per square foot is represented as

$$\dfrac{\$0.25}{1 \text{ square foot}}$$

(b) We determine the cost of sod by multiplying the square footage by the cost per square foot.

$$\text{Cost for sod} = 5293 \text{ square feet} \cdot \dfrac{\$0.25}{1 \text{ square foot}}$$
$$= \$1323.25$$

The sod will cost $1323.25.

It is worth noting in Example 5(b) that the units "square feet" divided out, so that the answer is measured in dollars. Because the answer is in the appropriate units, we have more confidence that our answer is correct.

QUICK ✓ *Solve for the unknown quantity.*

7. A circular water feature whose diameter is 4 feet is to be installed in a rectangular garden that is 20 feet by 10 feet. Once the water feature is installed, grass is to be installed on the remaining land.

   (a) Determine the area, to the nearest square foot, of land that is to receive grass.

   (b) If sod costs $0.25 per square foot installed, what will be the cost of the lawn?

Geometry formulas can even be used to make us savvy consumers.

EXAMPLE 6 Determining the Better Buy

At Mamma da Vinci's Pizza Parlor, a 16″ pizza costs $14.99 and a 12″ pizza costs $9.99. Which is the "better" buy?

Solution

The "better" buy will be the pizza that costs less per square inch. Our plan is: (1) find the area of each pizza and then (2) find the price per square inch of each pizza. We must consider, too, that a 16″ pizza has a *diameter* of 16 inches so that the radius is 8″ and a 12″ pizza has a *diameter* of 12 inches so that the radius is 6″.

1. Find the area of each pizza.

Since pizzas are circular, we compute the area of the pizza using the formula for the area of a circle.

	Area of the 16″ pizza:	Area of the 12″ pizza:
	$A = \pi r^2$	$A = \pi r^2$
Replace r with its given value:	$A = \pi \cdot 8^2$	$A = \pi \cdot 6^2$
Evaluate the numerical expression:	$A \approx 201.06$ sq in.	$A \approx 113.10$ sq in.

2. Find the price per square inch of each pizza.

The price per square inch can be found by dividing the price of the pizza by the number of square inches of area.

price per square inch of large pizza: $\dfrac{\$14.99}{201.06} = \$0.075$

price per square inch of medium pizza: $\dfrac{\$9.99}{113.10} = \$0.088$

The price per square inch of the large pizza is about 7.5¢, while the price per square inch of the medium pizza is about 8.8¢. The large pizza is the "better" buy. ■

QUICK ✓ *Evaluate each formula for the unknown quantity.*

8. A homeowner plans to construct a circular brick pad for his barbeque grill. The diameter of the brick pad is to be 6 feet. Find, to the nearest hundredth of a square foot, the area of the barbeque pad.

9. An extra large (18″) pizza at Dante's Pizza costs $16.99 and a small (9″) pizza costs $8.99. Which is the better buy?

② Solve a Formula for a Variable

The expression "solve for a variable" means to isolate the variable with a coefficient of 1 on one side of the equation and all other variables and constants, if any, on the other side by forming equivalent equations. For example, in the formula for the area of a rectangle, $A = lw$, the formula is solved for A because A is by itself with a coefficient of 1 on one side of the equation while all other variables are on the other side.

The steps that we follow when solving formulas for a certain variable are identical to those that we followed when solving equations.

Solve for x:	$15 = \dfrac{3}{2}x$		**Solve for h:**	$A = \dfrac{1}{2}bh$
Multiply both sides of the equation by 2 to clear the fraction.	$2(15) = 2\left(\dfrac{3}{2}x\right)$		Multiply both sides of the equation by 2 to clear the fraction.	$2(A) = 2\left(\dfrac{1}{2}bh\right)$
	$30 = 3x$			$2A = bh$
Divide by 3 to isolate the variable x.	$\dfrac{30}{3} = \dfrac{3x}{3}$		Divide by b to isolate the variable h.	$\dfrac{2A}{b} = \dfrac{bh}{b}$
	$10 = x$			$\dfrac{2A}{b} = h$

EXAMPLE 7 How to Solve a Formula for a Variable

Solve the formula $P = 2l + 2w$ for w. *Note:* $P = 2l + 2w$ is the formula for the perimeter of a rectangle.

Step-by-Step Solution

Step 1: Isolate the term containing the variable.

Subtract $2l$ from both sides of the equation to isolate the term $2w$:

$$P = 2l + 2w$$
$$P - 2l = 2l + 2w - 2l$$
$$P - 2l = 2w$$

Step 2: Get the coefficient of the variable to be 1.

Divide both sides of the equation by 2:

$$\frac{P - 2l}{2} = \frac{2w}{2}$$
$$\frac{P - 2l}{2} = w$$

Work Smart

The Symmetric Property states that if $a = b$, then $b = a$.

We can write $\dfrac{P - 2l}{2} = w$ as $w = \dfrac{P - 2l}{2}$ using the *Symmetric Property*. We have solved the equation $P = 2l + 2w$ for w because w is isolated with a coefficient of 1 on one side of the equation and all other terms are on the other side of the equation.

QUICK ✓ *Solve for the specified variable.*

10. To convert from degrees Celsius to degrees Fahrenheit, we use the formula $F = \dfrac{9}{5}C + 32$. Solve this formula for C.

11. The formula $A = 2\pi rh + 2\pi r^2$ represents the surface area A of a right circular cylinder whose radius is r and height is h. Solve the formula for h.

This next example illustrates a skill that will be extremely important when we study graphing lines in Chapter 3.

EXAMPLE 8 Solving for a Variable in a Formula

Solve the equation $3x + 2y = 6$ for y.

Solution

We wish to solve for y. That is, we want to get y isolated on one side of the equation and the variable x and constants on the other side. We begin by isolating the term containing y on the left side of the equation.

$$3x + 2y = 6$$

Subtract 3x from both sides of the equation
to isolate the term containing y: $\quad 3x + 2y - 3x = 6 - 3x$

Simplify both sides of the equation: $\quad 2y = 6 - 3x$

Divide both sides by 2: $\quad \dfrac{2y}{2} = \dfrac{6 - 3x}{2}$

Simplify: $\quad y = \dfrac{6 - 3x}{2}$

So when we solve the equation $3x + 2y = 6$ for y, we obtain the equation $y = \dfrac{6 - 3x}{2}$. This equation may also be written in the form $y = 3 - \dfrac{3}{2}x$ or $y = -\dfrac{3}{2}x + 3$ by dividing 2 into each term in the numerator. Do you see why these three equations are equivalent?

QUICK ✓ *Solve each equation for the indicated variable.*

12. $x + 2y = 7$; for y

13. $5x - 3y = 15$; for y

14. $\dfrac{3}{4}a + 2b = 7$; for b

15. $3rs + \dfrac{1}{2}t = 12$ for t.

EXAMPLE 9 **Solving a Formula for a Specified Variable and Evaluating the Formula**

The amount of profit, P, earned by a manufacturer is given by the formula $P = R - C$, where R represents the manufacturer's revenue and C represents the manufacturer's costs.

(a) Solve the equation for R, the manufacturer's revenue.

(b) Find the amount of revenue if a manufacturer has a \$12,500 profit and \$6000 in costs.

Solution

(a) We wish to solve the equation for R, so we must isolate R on one side of the equation.

$$P = R - C$$

Add C to both sides of the equation: $\quad P + C = R - C + C$

Simplify: $\quad P + C = R$

Thus, $R = P + C$.

(b) Evaluate $R = P + C$ if $P = \$12{,}500$ and $C = \$6000$.

$$R = P + C$$

Replace P with 12,500 and C with 6000: $\quad R = 12{,}500 + 6000$

Evaluate: $\quad R = 18{,}500$

The manufacturer has revenue of \$18,500 when it has \$12,500 profit and \$6000 in costs.

QUICK ✓ *Solve each equation for the indicated variable and evaluate.*

16. (a) Solve the formula $d = rt$ for t, where d represents distance, r represents an average speed, and t represents time.

   (b) Suppose that a family drives from Columbus, Ohio, to Raleigh, North Carolina, a distance of 550 miles. Their average speed for the trip is 60 mph. How long will it take the family to reach their destination?

17. (a) Solve the simple interest formula $I = Prt$ for t.

   (b) Find the number of years that $1000 must be invested at 7% annual interest to earn $35 interest. **Hint:** Convert the percent to a decimal.

2.4 Exercises

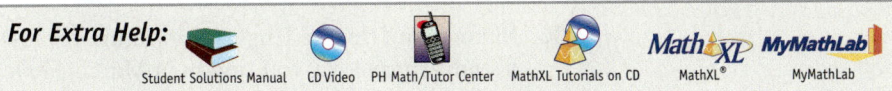

For Extra Help: Student Solutions Manual   CD Video   PH Math/Tutor Center   MathXL Tutorials on CD   MathXL®   MyMathLab

Concepts and Vocabulary

In Problems 1–3, fill in the blanks.

1. A mathematical _____ is a statement that describes how two or more variables are related.

2. To solve the formula $\dfrac{a + b + 2c}{4} = G$ for a, we first _____ both sides of the equation by _____.

3. To solve the formula $P = 2l + 2w$ for w, we first _____ $2l$ (to/from) both sides of the equation.

In Problems 4–6, answer True or False to each statement.

4. Solving $d = rt$ for t results in $t = \dfrac{d}{r}$.

5. The perimeter of a rectangle is found by multiplying length and width.

6. Solving $x - y = 6$ for y results in $y = -x + 6$.

7. A student solved the equation $x + 2y = 6$ for y and obtained the result $y = \dfrac{-x + 6}{2}$. Another student solved the same equation for y and obtained the result $y = -\dfrac{1}{2}x + 3$. Are both solutions correct? Explain why or why not.

8. Make up an example of a linear equation whose coefficients are integers. Make up a similar example where the coefficients are constants (letters). Solve both and then explain which steps in the solution are alike and which are different.

Building Skills

In Problems 9–28, substitute the given values into the formula and then evaluate to find the unknown quantity. Label units in the answer. If the answer is not exact, round your answer to the nearest hundredth.

9. Height of a Dam The formula $f = 3.281m$ converts a length in meters m to feet f. The world's highest dams, the Rogun and the Nurek, are both in Tajikistan. The Rogun is 335 meters high and the Nurek is 300 meters high. Convert these heights to feet.

10. Length of a Bridge The formula $m = 0.3048f$ converts length in feet f to length in meters m. The George Washington Bridge in New York City is 3500 feet long. How long is the George Washington Bridge in meters?

11. Buying a CD Player The formula $S = P - 0.20P$ gives the sale price S of an item which originally cost P dollars that was reduced by 20%. Find the sale price of a CD player that originally cost $24.00.

12. Buying a Computer The formula $S = P - 0.35P$ gives the sale price S of an item which originally cost P dollars that was reduced by 35%. Find the sale price of a computer that originally cost $950.00.

13. Salesperson's Earnings The formula $E = 500 + 0.15S$ is a formula for the earnings, E, of a salesperson who receives $500 per week plus 15% commission on all sales, S. Find the earnings of a salesperson who had weekly sales of $1000.

14. Salesperson's Earnings The formula $E = 750 + 0.07S$ is a formula for the earnings, E, of a salesperson who receives $750 per week plus 7% commission on all sales, S. Find the earnings of a salesperson who had weekly sales of $1200.

15. Planning a Trip A businessperson is planning a trip from Tokyo to San Diego, California, in March. She learns that the average high temperature in San Diego in March is 68°. Use the formula $C = \dfrac{5}{9}(F - 32)$ to convert 68° Fahrenheit to degrees Celsius.

16. Planning a Trip An English Literature professor plans to take twenty students on a study-abroad trip to London in March. He learns that the average daytime temperature in London in March is 10 degrees Celsius. Use the formula $F = \dfrac{9}{5}C + 32$ to convert 10 degrees Celsius to degrees Fahrenheit.

17. Lottery Earnings Therese invested her $200 West Virginia Lottery winnings in a 6-month Certificate of Deposit (CD) that earns 3% simple interest. Use the formula $I = Prt$ to find the amount of interest Therese's investment will earn.

18. Investing an Inheritance Christopher invested part of his $5000 inheritance from his grandmother in a 9-month Certificate of Deposit that earns 4% simple interest. Use the formula $I = Prt$ to find the amount of interest Christopher's investment will earn.

19. Find (a) the perimeter and (b) the area of the rectangle.

9

16

20. Find (a) the perimeter and (b) the area of the rectangle.

20

32

21. Find (a) the perimeter and (b) the area of the rectangle.

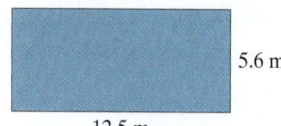

5.6 m

12.5 m

22. Find (a) the perimeter and (b) the area of the rectangle.

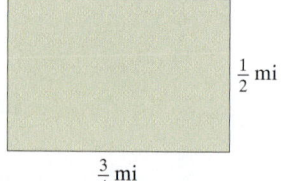

$\frac{1}{2}$ mi

$\frac{3}{4}$ mi

23. Find (a) the perimeter and (b) the area of the square.

9

24. Find (a) the perimeter and (b) the area of the square.

3.5

△ **25.** Find (a) the circumference and (b) the area of the circle with radius $r = 5$ cm. Use $\pi \approx 3.14$.

△ **26.** Find (a) the circumference and (b) the area of the circle with radius $r = 2.8$ yards. Use $\pi \approx 3.14$.

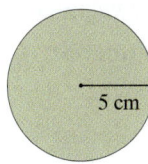

5 cm

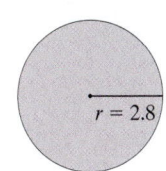

$r = 2.8$

△ **27. Area of a Circle** Find the area of a circle A when $\pi \approx \dfrac{22}{7}$ and $r = \dfrac{14}{3}$ inches.

△ **28. Area of a Circle** Find the area of a circle A when $\pi \approx 3.14$ and $r = 2.5$ km.

In Problems 29–42, solve each formula for the stated variable.

29. $d = rt$; solve for t

30. $A = lw$; solve for w

31. $C = \pi d$; solve for d

32. $F = mv^2$; solve for m

33. $I = Prt$; solve for r

34. $v = LWH$; solve for W

35. $A = \dfrac{1}{2}bh$; solve for h

36. $V = \dfrac{1}{3}Bh$; solve for B

37. $P = a + b + c$; solve for a

38. $S = a + b + c$; solve for b

39. $A = P + Prt$; solve for t

40. $P = 2l + 2w$; solve for w

41. $A = \dfrac{1}{2}h(B + b)$; solve for b

42. $S = 2\pi(h + r^2)$; solve for h

In Problems 43–50, solve for y.

43. $3x + y = 12$

44. $-2x + y = 18$

45. $10x - 5y = 25$

46. $12x - 6y = 18$

47. $4x + 3y = 13$

48. $5x + 6y = 18$

49. $\dfrac{1}{2}x - \dfrac{1}{6}y = 2$

50. $\dfrac{2}{3}x - \dfrac{5}{2}y = 5$

Applying the Concepts

In Problems 51–64, (a) solve for the indicated variable, and then (b) find the value of the unknown quantity. When given, label units in the answer.

51. Profit = Revenue − Cost: $P = R - C$

    (a) Solve for C.   **(b)** Find C when $P = \$1200$ and $R = \$1650$.

52. Profit = Revenue − Cost: $P = R - C$

    (a) Solve for R.   **(b)** Find R when $P = \$4525$ and $C = \$1475$.

53. Simple Interest: $I = Prt$

    (a) Solve for t.   **(b)** Find t when $I = \$42$, $P = \$525$, and $r = 4\%$.

54. Simple Interest: $I = Prt$

    (a) Solve for r.   **(b)** Find r when $I = \$225$, $P = \$5000$, and $t = 1.5$ years.

55. Physics Formula: $K = \dfrac{1}{2}mv^2$

    (a) Solve for m.   **(b)** Find m when $K = 8192$ and $v = 32$.

56. Statistics Formula: $Z = \dfrac{x - \mu}{\sigma}$

    (a) Solve for x.   **(b)** Find x when $Z = 2$, $\mu = 100$, and $\sigma = 15$.

57. Fahrenheit/Celsius Temperature Conversion: $F = \dfrac{9}{5}C + 32$

    (a) Solve for C.   **(b)** Find C when $F = 59°$.

58. **Algebra:** $y = mx + 5$
    (a) Solve for m.   (b) Find m when $x = 3$ and $y = -1$.

59. **New Amount = Principal + Interest:** $A = P + Prt$
    (a) Solve for r.   (b) Find r when $A = \$540$, $P = \$500$, and $t = 2$.

60. **New Amount = Principal + Interest:** $A = P + Prt$
    (a) Solve for t.   (b) Find t when $A = \$249$, $P = \$240$, and $r = 2.5\% = 0.025$.

△ 61. **Volume of a Right Circular Cylinder:** $V = \pi r^2 h$
    (a) Solve for h.   (b) Find h when $V = 320\pi$ mm^3 and $r = 8$ mm.

△ 62. **Volume of a Right Circular Cylinder:** $V = \pi r^2 h$
    (a) Solve for h.   (b) Find h when $V = 972\pi$ in.^{3} and $r = 9$ in.

△ 63. **Area of a Triangle:** $A = \dfrac{1}{2}bh$

    (a) Solve for b.   (b) Find b when $A = 45$ ft and $h = 5$ ft

△ 64. **Area of a Trapezoid:** $A = \dfrac{1}{2}h(b + B)$

    (a) Solve for h.   (b) Find h when $A = 99$ cm^2, $b = 19$ cm, and $B = 3$ cm.

65. **Energy Expenditure** Basal energy expenditure (E) is the amount of energy required to maintain the body's normal metabolic activity such as respiration, maintenance of body temperature, and so on. For males, the Basal energy expenditure is given by the formula

$$E = 66.67 + 13.75W + 5H - 6.76A$$

where W is the weight of the male (in kilograms), H is the height of the male (in centimeters), and A is the age of the male. Determine the Basal energy expenditure of a 37-year-old male who is 178 cm (5 feet, 10 inches) tall and weighs 82 kg (180 pounds).

66. **Energy Expenditure** See Problem 65. The Basal energy expenditure for females is given by

$$E = 665.1 + 9.56W + 1.85H - 4.68A$$

Compute the Basal energy expenditure of a 40-year-old female who is 168 cm tall (5 feet, 6 inches) and weighs 57 kg (125 pounds).

△ 67. **Soup Can** The formula $S = 2\pi rh + 2\pi r^2$ gives the surface area S of a right circular cylinder whose radius is r and height is h.
    (a) Solve the formula for h.
    (b) Find the height of a right circular cylinder whose surface area is 8.25π square inches and radius is 1.5 inches.

△ 68. **Cylinders** The volume V of a right circular cylinder is given by the formula $V = \pi r^2 h$, where r is the radius and h.
    (a) Solve the formula for h.
    (b) Find the height of a right circular cylinder whose volume is 90π cubic inches and whose radius is 3 inches.

△ 69. **Grocery Store** You are standing at the freezer case at your local grocery store trying to decide which is the "better" buy: a medium (12″) pizza for $9.99 or 2 small pizzas (8″) that are on special for $4.49 each. Which should you choose to get the best deal?

70. **Pizza for Dinner** Mama Mimi's Take and Bake Pizzeria is running a special on southwestern-style pizzas: a large 16″ pizza for $13.99 or two small 8″ pizzas for $12.99. Which should you choose to get the best deal?

71. **Taking a Trip** Jason drives a truck as an independent contractor. He bills himself out at $28 per hour. Suppose Jason has a contract that calls for him to leave a dock

at 9:00 A.M. and travel 600 miles to a warehouse. Jason has driven this route many times and figures that he can travel at an average speed of 50 miles per hour.

(a) Using the formula $d = rt$, where d is the distance traveled, r is the average speed, and t is the time spent traveling, determine how long Jason expects the trip to take.

(b) How much money does Jason expect to earn from this contract?

72. Taking a Trip Messai drives a truck as an independent contractor. He bills himself out at $32 per hour. Messai has a contract that calls for him to leave a dock at 8:00 A.M. and travel 145 miles to a warehouse. At the warehouse, he will wait while the truck is loaded (this takes 2 hours) and then return to his original dock. Messai has driven this route many times and figures that he can travel at an average speed of 58 miles per hour.

(a) Using the formula $d = rt$, where d is the distance traveled, r is the average speed, and t is the time spent traveling, determine how long Messai expects the roundtrip to take. Exclude the time Messia waits for the truck to be loaded.

(b) How much money does Messai expect to earn from this contract driving his truck?

△ **73. Area of a Region** Find the area of the figure below.

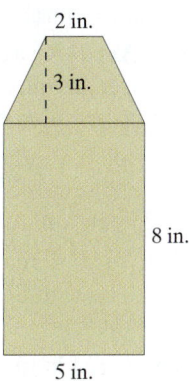

△ **74. Area of a Region** Find the area of the figure below.

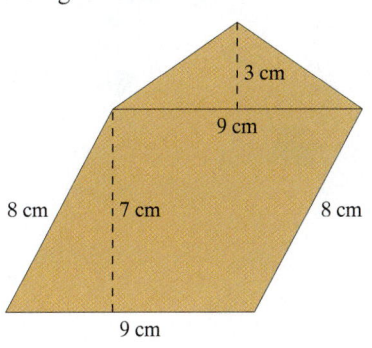

△ **75. Ice Cream Cone** Find the amount of ice cream in a cone if the radius of the cone is 4 cm and its height is 10 cm. The ice cream fully fills the cone and the hemisphere of ice cream on the top has a radius of 4 cm.

△ **76. Window** Find the area of the window given that the upper portion is a semicircle:

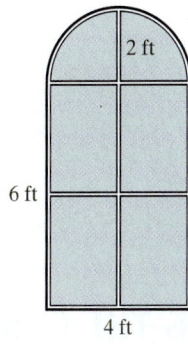

77. Federal Taxes According to the tax code in 2002, a married couple that earns over $137,300 per year filing a joint income tax return is subject to having their itemized deductions reduced. The formula $P = D - 0.03(I - 137{,}300)$ can be used to determine the permitted deductions P, where D represents the amount of deductions from Schedule A and I represents the couple's adjusted gross income.

(a) Solve the formula for I.

(b) Determine the adjusted gross income of a married couple filing a joint return whose allowed deductions were $15,200 and Schedule A deductions were $15,821.

78. Computing a Grade Jose's art history instructor uses the equation
$G = \dfrac{a + b + 2c + 2d}{6}$ to compute her students' semester grade. The variables a
and b represent the grades on two tests, c represents the grade on a research
paper, d represents the final exam grade, and G represents the student's average.

(a) Solve the equation for d, Jose's final exam grade.
(b) A final average of 84 will earn Jose a B in the course. Compute the grade Jose
must make on his final exam in order to earn a B for the semester, if he
scored 78 and 74 on his tests, and 84 on his research paper.

△ **79. Remodel a Bathroom** You plan to remodel your bathroom and you've chosen
1 foot-by-1 foot ceramic tiles for the floor. The bathroom is 7 feet 6 inches long
and 8 feet 2 inches wide.

(a) How many tiles do you need to cover the floor of your bathroom?
(b) Each tile costs $6. How much will it cost to tile your floor?
(c) The store from which you purchase the tile offers a discount of 10% on
orders over $350. Does your order qualify for the discount?

△ **80. Painting a Room** A gallon of paint can cover about 500 square feet. Find the
number of gallon containers of paint that must be purchased to paint two coats
on each wall of a rectangular room measuring 8 feet by 12 feet, with a 10-foot
ceiling. *Note:* You cannot purchase a partial can of paint!

△ **81. Landscaping a Back Yard** A circular swimming pool whose diameter is 24 feet
is to be installed in a rectangular yard that is 60 feet by 90 feet. Once the pool is
installed, grass is to be installed on the remaining land.

(a) Determine the area of land that is to receive grass. Round your answer to the
nearest foot. Use $\pi \approx 3.14159$.
(b) If sod costs $0.25 per square foot installed, what will be the cost of the lawn?

△ **82. Landscaping a Back Yard** A rectangular swimming pool whose dimensions are
12 feet by 24 feet is to be installed in a rectangular yard that is 80 feet by 40 feet.
Once the pool is installed, grass is to be installed on the remaining land.

(a) Determine the area of land that is to receive grass. Round your answer to the
nearest foot. Use $\pi \approx 3.14159$.
(b) A pallet of sod covers 500 square feet. How many pallets of sod are required?
(c) Each pallet of sod costs $96. What is the cost of the sod?

Extending the Concepts

△ **83.** A rectangle has length 5 feet and width 18 inches.

(a) What is the area in square inches?   (b) What is the area in square feet?

△ **84.** A rectangle has length 9 yards and width 8 feet.

(a) What is the area in square feet?   (b) What is the area in square yards?

△ **85.** Determine a formula for converting square inches to square feet.

△ **86.** Determine a formula for converting square yards to square feet.

PUTTING THE CONCEPTS TOGETHER (SECTIONS 2.1–2.4)

*These problems cover important concepts from Sections 2.1 to 2.4. We designed these problems
so that you can review the chapter so far and show your mastery of the concepts. Take time to work
these problems before proceeding with the next section. The answers to these problems are
located at the back of the text starting on page AN-3.*

1. Determine if the given value of the variable is a solution of the equation

$$4 - (6 - x) = 5x - 8$$

(a) $x = \dfrac{3}{2}$ (b) $x = -\dfrac{5}{2}$

2. Determine if the given value of the variable is a solution of the equation

$$\frac{1}{2}(x - 4) + 3x = x + \frac{1}{2}$$

(a) $x = -4$ **(b)** $x = 1$

In Problems 3–14, solve the equation and check the solution.

3. $x + \dfrac{1}{2} = -\dfrac{1}{6}$ **4.** $-0.4m = 16$ **5.** $14 = -\dfrac{7}{3}p$ **6.** $8n - 11 = 13$

7. $\dfrac{5}{2}n - 4 = -19$ **8.** $-(5 - x) = 2(5x + 8)$

9. $7(x + 6) = 2x + 3x - 15$ **10.** $-7a + 5 + 8a = 2a + 8 - 28$

11. $-\dfrac{1}{2}(x - 6) + \dfrac{1}{6}(x + 6) = 2$ **12.** $0.3x - 1.4 = -0.2x + 6$

13. $5 + 3(2x + 1) = 5x + x - 10$ **14.** $3 - 2(x + 5) = -2(x + 2) - 3$

15. Investment You have $7500 to invest and your financial advisor suggests that you put part of the money in a Certificate of Deposit that earns 2.4% simple interest and the remainder in bonds that earn 4% simple interest. To determine the amount you should invest in the CD to earn $220 interest at the end of one year, solve the equation $0.024x + 0.04(7500 - x) = 220$, where x represents the amount of money invested in CDs.

16. Area of a trapezoid: $A = \dfrac{1}{2}h(B + b)$

(a) Solve for b. **(b)** Find b when $A = 76$ in.^{2}, $h = 8$ in., and $B = 13$ in.

17. Volume of a right circular cylinder: $V = \pi r^2 h$

(a) Solve for h. **(b)** Find h when $V = 117\pi$ sq. in., and $r = 3$ in.

18. Solve the equation $3x + 2y = 14$ for y.

2.5 Introduction to Problem Solving: Direct Translation Problems

OBJECTIVES

1. Translate English Phrases to Algebraic Expressions
2. Translate English Sentences to Equations
3. Build Models for Solving Direct Translation Problems

Preparing for Introduction to Problem Solving: Direct Translation Problems
Before getting started, take this readiness quiz. If you get a problem wrong, go back to the section cited and review the material.

1. Solve the equation: $x + 34.95 = 60.03$ [Section 2.1, pp. 76–78]
2. Solve the equation: $x + 0.25x = 60$ [Section 2.2, p. 86]

1 Translate English Phrases to Algebraic Expressions

One of the neat features of mathematics is that the symbols we use allow us to express English phrases briefly and consistently. An algebraic expression is similar to an English phrase. For example, the English phrase "5 more than a number x" is represented algebraically as $x + 5$.

There are certain words or phrases in English that easily translate into mathematical symbols. Table 2 lists various English words or phrases and their corresponding math symbol.

Table 2 Math Symbols and the Words They Represent			
Add (+)	**Subtract (−)**	**Multiply (·)**	**Divide (/)**
sum	difference	product	quotient
plus	minus	times	divided by
greater than	subtracted from	of	per
more than	less	twice	ratio
exceeds by	less than	double	
in excess of	decreased by	half	
added to	fewer		
increased by			
combined			
altogether			

EXAMPLE 1   Writing English Phrases Using Math Symbols

Express each English phrase using mathematical symbols.

   (a) The sum of 2 and 5.

   (b) The difference of 12 and 7.

   (c) The product of −3 and 8.

   (d) The quotient of 10 and 2.

   (e) 9 less than 15.

   (f) A number z decreased by 11.

   (g) Three times the sum of a number x and 8.

Solution

   (a) Because we are talking about a sum, we know to use the + symbol, so "The sum of 2 and 5" is represented mathematically as $2 + 5$.

   (b) Because we are talking about a difference, we know to use the − symbol, so "The difference of 12 and 7" is represented mathematically as $12 - 7$.

   (c) "The product of −3 and 8" is represented as $-3 \cdot 8$.

   (d) "The quotient of 10 and 2" is represented mathematically as $\dfrac{10}{2}$.

   (e) "9 less than 15" is represented mathematically as $15 - 9$.

   (f) "A number z decreased by 11" is represented algebraically as $z - 11$.

   (g) "Three times the sum of a number x and 8" is represented as an algebraic expression as $3(x + 8)$.

Work Smart

When translating from English to math, try some specific examples. For example, to translate "a number z decreased by 11," pick specific values of z, as in "16 decreased by 11," which would be $16 - 11$ or 5. So "z decreased by 11" is $z - 11$.

In Example 1(g), we know the mathematical representation of the phrase is $3(x + 8)$ rather than $3x + 8$ because the phrase "three times the sum" means to multiply the sum of the two numbers by 3. The English phrase that would result in $3x + 8$ might be "the sum of three times a number and 8." Do you see the difference?

QUICK ✓ *Express each English phrase using mathematical symbols.*

1. The sum of 5 and 17.

2. The product of −2 and 6.

3. The quotient of 25 and 3.

4. The difference of 7 and 4.

5. Twice a less 2.

6. Three plus the quotient of z and 4.

EXAMPLE 2   Translate from an English Phrase to an Algebraic Expression

Write an algebraic expression for each problem.

(a) The Raiders scored p points in a football game. The Packers scored 12 more points than the Raiders. Write an algebraic expression for the number of points the Packers scored.

(b) A lumberman cuts a 50-foot log into 2 pieces. One piece is t feet long. Express the length of the second piece as an algebraic expression in t.

(c) The number of quarters in a drink machine is two fewer than the number of dimes, d, in the machine. Write an expression for the number of quarters as an algebraic expression in d.

Solution

(a) The phrase "more than" implies addition. The Packers scored $p + 12$ points in the football game.

(b) The log is 50 feet long. If the lumberman cuts one piece 20 feet long, then the other piece must be $50 - 20 = 30$ feet long. In general, if the lumberman cuts one piece that is t feet long, then the remaining piece must be $50 - t$ feet long.

(c) The phrase "less than" implies subtraction. Two less than the number of dimes is represented algebraically as $d - 2$.

QUICK  *Translate each phrase to an algebraic expression.*

7. Terry earned z dollars last week. Anne earned \$50 more than Terry last week. Write an algebraic expression for Anne's earnings in terms of z.

8. Melissa paid x dollars for her college math book and \$15 less than that for her college sociology book. Express the cost of her sociology book in terms of x.

9. Tim raided his piggy bank and found he had 75 dimes and quarters. Tim has d dimes. Express the number of quarters in his piggy bank as an algebraic expression in d.

EXAMPLE 3   Translate from an English Phrase to an Algebraic Expression

Write an algebraic expression for each problem.

(a) The number of student tickets sold to a play is five fewer than four times the number n of nonstudent tickets sold. Write an algebraic expression for the number of nonstudent tickets sold in terms of n.

(b) The height of a full-grown maple tree is ten feet more than three times the height h of a sapling. Write an algebraic expression for the height of the full-grown tree in terms of h.

(c) A pedestrian bridge over a river is f feet long. Further downstream, another pedestrian bridge over the same river is three feet less than twice the length of the first bridge. Write an algebraic expression for the length of the second bridge in terms of f.

Solution

(a) The phrase "fewer than" implies subtraction. The number of student tickets sold is five fewer than four times the number of nonstudent tickets sold. So the algebraic expression is 4 times the number of nonstudent tickets sold minus 5, or $4n - 5$.

(b) The full-grown tree is ten feet more than three times the height of the sapling, so we use the phrase "height of sapling times three plus ten" to obtain $3h + 10$ feet as the height of the full-grown tree.

(c) The second bridge is three feet less than twice the length of the first bridge, so we say "twice the length of the first bridge minus 3 feet" to obtain $2f - 3$ feet as the length of the second bridge.

QUICK ✔️ *Translate each phrase to an algebraic expression.*

10. The width of a platform is 2 feet less than three times the length, l. Express the width of the platform as an algebraic expression in terms of l.

11. T.J. has quarters and dimes in his piggy bank. The number of quarters is three more than twice the number of dimes. T.J. has q quarters. Express the number of dimes as an algebraic expression in terms of q.

12. The number of blue M&Ms in a bowl is five less than three times the number of brown M&Ms, b, in the bowl. Express the number of blue M&Ms in the bowl as an expression in terms of b.

② Translate English Sentences to Equations

In Words
English phrase is to algebraic expression as English sentence is to algebraic equation.

Work Smart
We learned in Section 2.1 that an equation is a statement in which two algebraic expressions are equal.

Nearly every word problem that we do in algebra requires some type of translation. Learning to speak the language of math is the same as learning to speak any language. Now that we have the ability to translate English phrases to algebraic expressions, we will extend the idea to translating English sentences to algebraic equations.

In English, a complete sentence must contain a subject and a verb, so expressions or "phrases" are not complete sentences. For example "Beats me!" is an expression, but it is not a complete sentence because it does not contain a subject. The expression "5 more than a number x" does not contain a verb and therefore is not a complete sentence either. The statement "5 more than a number x is 18" is a complete sentence because it contains a subject and a verb. Because this is a complete sentence, we can translate it into a mathematical statement. In mathematics, statements can be represented symbolically through equations.

In English, statements can be true or false. For example, "The moon is made of green cheese" is a false statement, while "The sky is blue" is a true statement. Mathematical statements can be true or false as well—we called them conditional equations.

Table 3 provides a summary of words that typically translate into an equal sign.

Table 3 Words That Translate into an Equal Sign			
is	yields	are	equals
was	gives	results in	is equal to
is equivalent to			

Notice that the words that translate into an equal sign are all verbs. So, the equal sign in an equation acts like a verb in a sentence.

Let's look at some examples where we translate English sentences into equations.

EXAMPLE 4 Translating English Sentences into Equations

Translate each of the following sentences into an equation. Do not solve the equation.

(a) Five more than a number x is 20.

(b) Four times the sum of a number z and 3 is 15.

(c) The difference of x and 5 equals the quotient of x and 2.

Solution

(a) 5 more than a number x is 20

$$x + 5 = 20$$

(b) Because the expression reads "Four times the sum," we first need to determine the sum and then multiply this result by 4.

Four times the sum of a number z and 3 is 15

$$4(z + 3) = 15$$

(c) The difference of x and 5 equals the quotient of x and 2

$$x - 5 = \frac{x}{2}$$

Work Smart

The English sentence "The sum of four times a number z and 3 is 15" would be expressed mathematically as $4z + 3 = 15$. Do you see how this differs from Example 4(b)?

QUICK ✓ *Translate each English statement into an equation. Do not solve the equation.*

13. The product of 3 and y is equal to 21.

14. The sum of 3 and x is equivalent to the product of 5 and x.

15. The difference of x and 10 equals the quotient of x and 2.

16. Three less than a number y is five times y.

An Introduction to Problem Solving and Mathematical Models

Every day we encounter various types of problems that must be solved. **Problem solving** is the ability to use information, tools, and our own skills to achieve a goal. For example, suppose 4-year-old Kevin wants a glass of water, but he is too short to reach the sink. Kevin has a problem. To solve the problem, he finds a step stool and pulls it over to the sink. He uses the step stool to climb on the counter, opens the kitchen cabinet and pulls out a cup. He then crawls along the counter top, turns on the faucet, fills the cup, and proceeds to drink the water. Problem solved!

Of course, this is not the only way that Kevin could solve the problem. Can you think of any other solutions? Just as there are various approaches to solving life's everyday problems, there are many ways to solve problems using mathematics. However, regardless of the approach, there are always some common aspects in solving any problem. For example, regardless of how Kevin ultimately ends up with his cup of water, someone must get a cup from the cabinet and someone must turn on the faucet.

One of the purposes of learning algebra is to be able to solve certain types of problems. To solve these problems, we will need techniques that can help us translate the verbal description of the problem into an equation that can be solved. The process of taking a verbal description of a problem and developing a mathematical equation that can be used to solve the problem is **mathematical modeling.**

Mathematical modeling begins with a problem. The problem is summarized as a verbal description. The verbal description is then translated into the language of mathematics. This translation results in an equation that can be solved (the mathematical problem). The solution must be checked against the mathematical problem (the equation) the verbal description. This entire process is called the **modeling process.** We call the equation that is developed the **mathematical model.** See Figure 3.

Not all models are mathematical. In general, a **model** is a way of using graphs, pictures, small-scale reproductions, equations, or even verbal descriptions to represent a real-life situation. Because the world is an extremely complex place, we often need to simplify information when we develop a model. For example, a map is a model of our road system. Maps don't show all the details of the system such as trees, buildings, or potholes, but they do a good job of describing how to get from point A to point B.

Figure 3

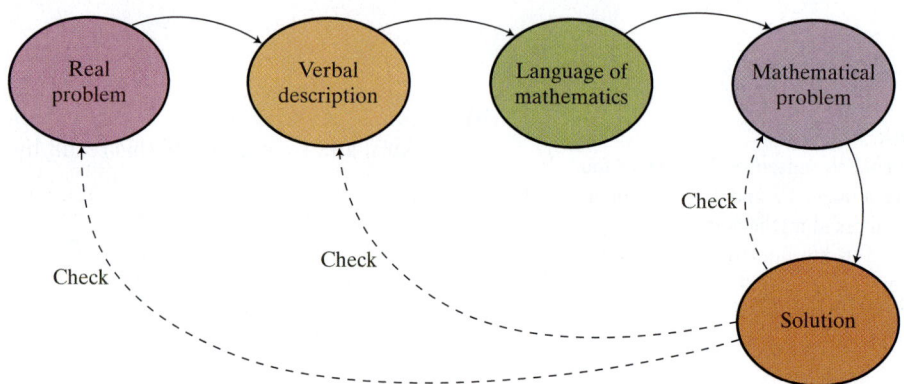

Mathematical models are similar in that we often make assumptions regarding our world in order to make the mathematics more manageable.

It is difficult to give a step-by-step approach for solving problems because each problem is unique in some way. However, because there are common links to many types of problems, we can categorize problems. In this text, we will present five categories of problems.

Five Categories of Problems

1. **Direct Translation**—problems in which we must translate from English into the language of mathematics by using key words in the verbal description.

2. **Geometry**—problems in which the unknown quantities are related through geometric formulas.

3. **Mixtures**—problems in which two or more quantities are combined in some fashion.

4. **Uniform Motion**—problems in which an object travels at a constant speed.

5. **Work Problems**—problems in which two or more entities join forces to complete a job.

We will present strategies for solving categories of problems throughout the text. In this section, we will concentrate on direct translation problems.

Regardless of the type of problem we solve, there are certain steps that should always be followed to assist in solving the problem. Below we provide you with a series of steps that should be followed when developing any mathematical model. As we proceed through this course, and in future courses, you will use the techniques that you have studied in this course to solve more complicated problems, but the approach remains the same.

Steps for Solving Problems with Mathematical Models

Step 1: Identify What You Are Looking For Read the problem very carefully, perhaps two or three times. Identify the type of problem. Identify the information that is given, and the information that we wish to learn from the problem. It is fairly typical that the last sentence in the problem indicates what it is we wish to solve for in the problem.

Step 2: Give Names to the Unknowns Assign variables to the unknown quantities in the problem. Choose a variable that is representative of the unknown quantity it represents. For example, use t for time.

Step 3: Translate the Problem into the Language of Mathematics Read the problem again. This time, after each sentence is read, determine if the sentence can be translated into a mathematical statement or expression in terms of the variables identified in Step 2. It is often helpful to create a table, chart, or figure. When you have finished reading the problem, if necessary, combine the mathematical statements or expressions into an equation that can be solved.

Step 4: Solve the Equation(s) Found in Step 3 Solve the equation for the variable and then answer the question posed by the original problem.

Step 5: Check the Reasonableness of Your Answer Check your answer to be sure that it makes sense. If it does not, go back and try again.

Step 6: Answer the Question Write your answer in a complete sentence.

Let's review each of these steps, one at a time.

- **Identify** Carefully read the problem. Reading a verbal description of a problem is not like reading a spy novel. You may need to read the problem three or four times. You may not know how to solve the problem while reading it, but you should get a sense of which of the five categories the problem falls into, what information you are given, and what you are being asked to do.

- **Name** Reread the problem and assign variables to the unknowns. You should write the name of each variable and what it represents. You will use this to check your final answer.

- **Translate** In this step, you develop a model (equation) that mathematically describes the problem. Be sure to use the guidelines presented in each category of problem. We will present these guidelines shortly.

- **Solve** the equation. This is generally the easy part. Most students say, "I could solve the problem, if I could find the right equation."

- **Check** Checking your answer can be difficult because you can make two types of errors while setting up or solving the problem. One type of error occurs if you correctly translate the problem into a model but then make an error solving the equation. A second type of error occurs if you misinterpret the problem and develop an incorrect model. The solution you obtain may still satisfy your model, but it probably will not be the solution to the original problem. We can check for this type of error by determining whether the solution is reasonable. Does your answer make sense? Always be sure that you are answering the question that is being asked.

- **Answer the question** identified in the problem in words.

Work Smart

Remember that you can use any letter to represent the unknown(s) when you make your model. Choose a letter that reminds you what it represents. For example, use t for time.

③ Build Models for Solving Direct Translation Problems

Let's look at a "direct translation" problem. Remember, these are problems that can be set up by reading the problem and using everyday language to translate the verbal description into a mathematical equation.

EXAMPLE 5 **Solving a Direct Translation Problem**

For the 2001–2002 season, the price of a ticket to a New York Knicks NBA game was $11 more than twice the price of a ticket to a Minnesota Timberwolves NBA game. If you buy a ticket to a Knicks game and a ticket to a Timberwolves game, and the total cost of the two tickets is $128, what does each of the tickets cost?

Solution

Step 1: Identify This is a direct translation problem. We can obtain an equation from the words of the problem. We want to know the price of a ticket to a NY Knicks basketball game and the price of a ticket to a Minnesota Timberwolves game.

Work Smart

It is helpful to assign the variable to the quantity that you know **least** about.

Step 2: Name We know that the price of a NY Knicks ticket was $11 more than twice the price of a ticket to a Minnesota Timberwolves game. We will let t represent the price of a ticket to a Timberwolves game. Then $2t + 11$ represents the price of a NY Knicks ticket.

Step 3: Translate Since we know that the total cost for a ticket to both games is $128, we use the equation

price of ticket to Timberwolves game plus price of ticket to Knicks game equals total cost

$$t \qquad + \qquad 2t + 11 \qquad = \qquad 128$$

Step 4: Solve We now solve the equation

$$t + (2t + 11) = 128$$

Combine like terms: $\qquad 3t + 11 = 128$

Subtract 11 from each side of the equation: $\quad 3t + 11 - 11 = 128 - 11$

$$3t = 117$$

Divide each side by 3: $\qquad \dfrac{3t}{3} = \dfrac{117}{3}$

$$t = 39$$

We have found $t = 39$. Recall that t represents the price of a ticket to a Timberwolves game. The price of a ticket to a Knicks game is $11 more than twice the price of a ticket to a Timberwolves game. So $2t + 11 = 2(39) + 11 = \$89$. It costs $89 for a ticket to a NY Knicks game.

Step 5: Check Is the total cost of a ticket to a Timberwolves game and a Knicks game $128? $39 + $89 = $128, so our answers are correct.

Step 6: Answer The price of a ticket to a Minnesota Timberwolves game is $39 and the price of a ticket to a NY Knicks game is $89. ∎

QUICK ✓ *Translate the problem to an algebraic equation and solve the equation for the unknowns.*

17. Sean and Connor decide to buy a pizza. The pizza costs $15 and they decide to split the cost based upon how much pizza each eats. Connor eats two-thirds of the amount that Sean eats, so Connor pays two-thirds of the amount that Sean pays. How much does each pay?

Consecutive Integer Problems

Work Smart

Examples of consecutive even integers are

16, 18, 20

78, 80, 82

752, 754, 756

Examples of consecutive odd integers are

9, 11, 13

31, 33, 35

623, 625, 627

Recall that an *integer* is a member of the set $\{\ldots, -3, -2, -1, 0, 1, 2, 3, \ldots\}$. An *even integer* is an integer that is divisible by two. For example, 16 and 124 are even integers. An *odd integer* is an integer that is not even. For example, 9 and 17 are odd integers. Consecutive integers differ by one, so if n represents the first integer, $n + 1$ represents the second integer, $n + 2$ represents the third integer, and so on. Consecutive *even* integers, such as 14 and 16, differ by two so if n represents the first even integer, $n + 2$ represents the second even integer and $n + 4$ represents the third even integer. Consecutive *odd* integers also differ by 2 (41 and 43, for example), so if n represents the first odd integer, $n + 2$ represents the second odd integer, and $n + 4$ represents the third odd integer. See Figure 4.

Figure 4

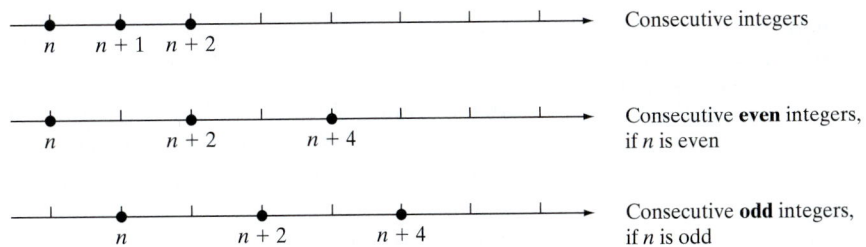

Consecutive integers

Consecutive even integers, if n is even

Consecutive odd integers, if n is odd

EXAMPLE 6 **Solve a Direct Translation Problem: Consecutive Integers**

The sum of three consecutive even integers is 324. Find the integers.

Solution

Step 1: Identify This is a direct translation problem. We are looking for three consecutive even integers, and we know that their sum is 324. Examples of consecutive even integers are $6, 8, 10, \ldots$.

Step 2: Name We will let n represent the first even integer, so $n + 2$ is the next even integer, and $n + 4$ is the third even integer.

Step 3: Translate We know that the sum of the three consecutive even integers is 324. We also know that the word "sum" translates to "addition" and the word "is" translates to "equals." So our equation is

$$\underbrace{n}_{\text{first even integer}} + \underbrace{n + 2}_{\text{second even integer}} + \underbrace{n + 4}_{\text{third even integer}} = \underbrace{324}_{\text{sum}}$$

Step 4: Solve We solve the equation.

$$n + (n + 2) + (n + 4) = 324$$

Combine like terms: $\quad 3n + 6 = 324$

Subtract 6 from each side of the equation: $\quad 3n + 6 - 6 = 324 - 6$

$$3n = 318$$

Divide each side by 3: $\quad \dfrac{3n}{3} = \dfrac{318}{3}$

$$n = 106$$

Since $n = 106$, is the first even integer, the remaining even integers are 108 and 110.

Step 5: Check These numbers are all even integers, so we know that they are possible solutions. Does $106 + 108 + 110 = 324$? Yes! We know we have the correct answer.

Step 6: Answer The three consecutive even integers are 106, 108, and 110. ▬

QUICK ✔️ *Translate the problem to an algebraic equation and solve the equation for the unknowns.*

18. The sum of three consecutive even integers is 270. Find the integers.

EXAMPLE 7 **Solve a Direct Translation Problem: Piece Lengths**

A carpenter is building a shelving system and cuts a 14-foot length of cherry shelving into three pieces. The second piece is twice as long as the first, and the third piece is 2 feet longer than the first. Find the length of each piece of cherry shelving.

Work Smart

Sometimes these problems are called "the whole equals the sum of the parts" problems because the values of the "parts" must sum to the value of the "whole."

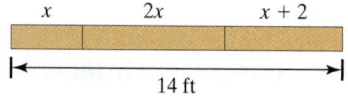

Solution

Step 1: Identify This is a direct translation problem. We are looking for the length of each piece of shelving. We know that the board is 14 feet long.

Step 2: Name Because the length of the second and third pieces are described in terms of the length of the first piece, we will let x represent the length of the first piece of shelving. The second piece is twice as long as the first piece, so we let $2x$ represent the length of the second piece. Because the third piece is 2 feet longer than the first piece, its length is $x + 2$.

Step 3: Translate We know that the lengths of the three pieces must total 14 feet, so our equation is

$$\underbrace{x}_{\substack{\text{length of} \\ \text{first piece}}} + \underbrace{2x}_{\substack{\text{length of} \\ \text{second piece}}} + \underbrace{x + 2}_{\substack{\text{length of} \\ \text{third piece}}} = \underbrace{14}_{\substack{\text{total} \\ \text{length}}}$$

Step 4: Solve We solve the equation.

$$x + 2x + (x + 2) = 14$$

Combine like terms: $\quad 4x + 2 = 14$

Subtract 2 from each side of the equation: $\quad 4x + 2 - 2 = 14 - 2$

$$4x = 12$$

Divide each side by 4: $\quad \dfrac{4x}{4} = \dfrac{12}{4}$

$$x = 3$$

Since $x = 3$ feet is the length of the first shelf, the second piece of shelving is $2x = 2 \cdot 3 = 6$ feet and the third piece is $x + 2 = 5$ feet long.

Step 5: Check Is the sum of the three pieces of cherry shelving equal to 14 feet? Does $3 + 6 + 5 = 14$? Yes! We know we have the correct answer.

Step 6: Answer The three pieces of shelving are 3 feet, 6 feet, and 5 feet long. ▬

QUICK ✓ *Translate the problem to an algebraic equation and solve.*

19. A 76-inch length of ribbon is to be cut into three pieces. The longest piece is to be 24 inches longer than the shortest piece, and the third piece is to be half the length of the longest piece. Find the length of each piece of ribbon.

EXAMPLE 8 Investment Decisions

A total of $25,000 is to be invested, some in bonds and some in certificates of deposit (CDs). The amount invested in CDs is to be $8000 less than the amount invested in bonds. How much is to be invested in each type of investment?

Solution

Step 1: Identify We want to know the amount invested in each type of investment.

Step 2: Name Let b represent the amount invested in bonds.

Step 3: Translate The equation that we will use is

Amount invested in bonds + Amount invested in CDs = Total investment

Suppose we invested $18,000 in bonds; then the amount in CDs will be $8000 less than this amount, or $10,000. In general, b is the amount invested in bonds, so $b - 8000$ represents the amount invested in CDs. Our total investment is $25,000. Substituting into the above equation, we have

Amount invested in bonds   Amount invested in CDs   Total investment

$$b \qquad + \qquad b - 8000 \qquad = \qquad 25{,}000$$

Step 4: Solve We now solve the equation

$$b + (b - 8000) = 25{,}000$$

Combine like terms: $2b - 8000 = 25{,}000$

Add 8000 to each side of the equation: $2b = 33{,}000$

Divide both sides by 2: $b = 16{,}500$

Step 5: Check If we invest \$16,500 in bonds, then the amount invested in CDs should be \$8000 less than this amount or \$8500. The total investment should then be \$25,000. Since the amount invested in bonds plus the amount invested in CDs is \$16,500 + \$8500 = \$25,000, the answer checks.

Step 6: Answer Invest \$16,500 in bonds and \$8500 in CDs.

QUICK ✓ *Translate the problem to an algebraic equation and solve.*

20. A total of \$18,000 is to be invested, some in stocks and some in bonds. If the amount invested in bonds is twice that invested in stocks, how much is invested in each category?

EXAMPLE 9 **Choosing a Long-Distance Carrier**

MCI has a long-distance phone plan that charges \$2.00 a month plus \$0.09 per minute of usage. Sprint has a long-distance phone plan that charges \$3.50 a month plus \$0.07 per minute of usage. For how many minutes of long distance calls will the costs for the two plans be the same? (*Source:* MCI and Sprint)

Solution

Step 1: Identify This is a direct translation problem. We are looking for the number of minutes for which the two plans cost the same.

Step 2: Name Let m represent the number of long distance minutes used in the month.

Step 3: Translate The monthly fee for MCI is \$2.00 plus \$0.09 for each minute used. So, if one minute is used, the fee is $2.00 + 0.09(1) = 2.09$ dollars. If two minutes are used, the fee is $2.00 + 0.09(2) = 2.18$ dollars. In general, if m minutes are used, the monthly fee is $2.00 + 0.09m$ dollars. Similar logic results in the monthly fee for Sprint being $3.50 + 0.07m$ dollars. We want to know the number of minutes for which the cost for the two plans will be the same, which means we need to solve

$$\text{Cost for MCI} = \text{Cost for Sprint}$$
$$2.00 + 0.09m = 3.50 + 0.07m$$

Step 4: Solve

Subtract 2.00 from both sides: $0.09m = 1.50 + 0.07m$

Subtract 0.07m from both sides: $0.02m = 1.50$

Divide both sides by 0.02: $m = 75$

Step 5: Check We believe the cost of the two plans will be the same if 75 minutes are used. The cost of MCI's plan will be $2.00 + 0.09(75) = \$8.75$. The cost of Sprint's plan will be $3.50 + 0.07(75) = \$8.75$. They are the same!

Step 6: Answer the Question The cost of the two plans will be the same if 75 minutes are used.

QUICK ✓

21. Truck Rentals You need to rent a moving truck. You have identified two companies that rent trucks. EZ-Rental charges $30 per day plus $0.15 per mile. Do It Yourself Rental charges $15 per day plus $0.25 per mile. For how many miles will the cost of renting be the same?

2.5 Exercises

For Extra Help:

Student Solutions Manual   CD Video   PH Math/Tutor Center   MathXL Tutorials on CD   MathXL®   MyMathLab

Concepts and Vocabulary

In Problems 1 and 2, fill in the blanks.

1. Letting variables represent unknown quantities and then expressing relationships among the variables in the form of equations is called _____ _____.

2. In English we use sentences. To translate a sentence into a mathematical statement we use _____.

In Problems 3 and 4, answer True or False to each statement.

3. If the first of three consecutive odd integers is represented by the variable n, the second consecutive odd integer would be $n + 1$ and the third odd integer would be $n + 3$.

4. Suppose a total of $10,000 is to be invested in stocks and bonds. If we let b represent the amount invested in bonds, then $b - 10,000$ represents the amount invested in stocks.

5. How is mathematical modeling related to problem solving? Why do we make assumptions when creating mathematical models?

6. What is the difference between an algebraic expression and an equation? How is each related to phrases and English statements?

7. Two students write an equation to solve a word problem with consecutive odd integers. One student assigns the variables as $n - 1, n + 1, n + 3$, where n is an even integer. A second student uses $n, n + 2, n + 4$, where n is an odd integer. Which student is correct? Will the value for n be the same for both? Make up a problem that can be solved in more than one way and explain how the variables were assigned.

8. Using the algebraic expression $3x + 5$, make up a problem that uses the direction evaluate. Using the same algebraic expression, make up a problem that uses the direction solve.

Building Skills

In Problems 9–26, translate each phrase to an algebraic expression. Let x represent the unknown number.

9. the sum of -5 and a number

10. a number increased by 32.3

11. the product of a number and $\dfrac{2}{3}$

12. the product of -2 and number

13. half of a number

14. double a number

15. a number less -25

16. 8 less than a number

17. the quotient of a number and 3

18. the quotient of -14 and a number

19. $\frac{1}{2}$ more than a number

20. $\frac{4}{5}$ of a number

21. 9 more than 6 times a number

22. 21 more than 4 times a number

23. twice the sum of 13.7 and a number

24. 50 less than half of a number

25. the sum of twice a number and 31

26. the sum of twice a number and 45

In Problems 27–34, translate each statement into an equation. Let x represent the unknown number. DO NOT SOLVE.

27. The sum of a number and 15 is -34.

28. The sum of 43 and a number is -72.

29. 35 is 7 less than triple a number.

30. 49 is 3 less than twice a number.

31. The quotient of a number and -4, increased by 5, is 36.

32. The quotient of a number and -6, decreased by 15, is 30.

33. Twice the sum of a number and 6 is the same as 3 more than the number.

34. Twice the sum of a number and 5 is the same as 7 more than the number.

In Problems 35–42, choose a variable to represent one quantity. State what that quantity represents and then express the second quantity in terms of the first.

35. The Columbus Clippers scored 5 more runs than the Richmond Braves.

36. The Toronto Blue Jays scored 3 fewer runs than the Cleveland Indians.

37. Jan has $0.55 more in her piggy bank than Bill.

38. Beryl has $0.25 more than 3 times the amount Ralph has.

39. Janet and Kathy will share the $200 grant.

40. Juan and Emilio will share the $1500 lottery winnings.

41. There were 1433 visitors to the Arts Center Spring show. Some were adults and some were children.

42. There were 12,765 fans at a recent NBA game. Some held paid admission tickets and some held special promotion tickets.

Applying the Concepts

43. Number Sense The sum of a number and -12 is 71. Find the number.

44. Number Sense The difference between a number and 13 is -29. Find the number.

45. Number Sense 25 less than twice a number is -53. Find the number.

46. Number Sense The sum of 13 and twice a number is -19. Find the number.

47. Consecutive Integers The sum of three consecutive integers is 165. Find the numbers.

48. Consecutive Integers The sum of three consecutive odd integers is 81. Find the numbers.

49. Bridges The longest bridge in the United States is the Verrazano-Narrows Bridge. The second-longest bridge in the United States is the Golden Gate Bridge, which is 60 feet shorter than the Verrazano-Narrows Bridge. The combined length of the two bridges is 8460 feet. Find the length of each bridge.

50. Towers The tallest towers in the world (those having the most stories) are the Sears Tower in Chicago, IL and the Taipei 101 Tower in Taipei, Taiwan. The Taipei Tower has 4 fewer stories than the Sears Tower. The two buildings together have 216 stories. Find the number of stories in each tower.

51. Buying a Motorcycle The total price for a new motorcycle is $11,894.79. The tax, title, and dealer preparation charges amount to $679.79. Find the price of the motorcycle before the extra charges.

52. Buying a Desk The total price for a new desk is $285.14, including sales tax of $16.14. Find the original cost of the desk.

53. Finance A total of $20,000 is to be invested, some in bonds and some in certificates of deposit (CDs). The amount invested in bonds is to be $3000 greater than the amount invested in CDs. How much is to be invested in each type of investment?

54. Finance A total of $10,000 is to be divided between Sean and George. George is to receive $3000 less than Sean. How much will each receive?

55. Investments Suppose that your Aunt May has left you an unexpected inheritance of $32,000. You have decided to invest the money rather than blow it on frivolous purchases. Your financial advisor has recommended that you diversify by placing some of the money in stocks and some in bonds. Based upon current market conditions, she has recommended that the amount in bonds should equal three-fifths of the amount invested in stocks. How much should be invested in stocks? How much should be invested in bonds?

56. Investments Jack and Diane have $40,000 to invest. Their financial advisor has recommended that they diversify by placing some of the money in stocks and some in bonds. Based upon current market conditions, he has recommended that the amount in bonds should equal two-thirds of the amount invested in stocks. How much should be invested in stocks? How much should be invested in bonds?

57. Cereal A serving of Kashi Go Lean Crunch cereal contains 4 times the amount of fiber as a serving of Kellogg's Smart Start Cereal. If you eat a serving of each cereal, you will consume 10 g of dietary fiber. Find the amount of dietary fiber in each cereal.

58. Books A paperback edition of a book costs $12.50 less than the hardback edition of the book. If you purchase one of each book, you will pay $37.40. Find the cost of the paperback edition of the book.

59. Income On a joint income tax return, Elizabeth Morrell's adjusted gross income was $2549 more than her husband Dan's adjusted gross income. Their combined adjusted gross income was $55,731. Find Elizabeth Morrell's adjusted gross income.

60. Spring Break Allison went shopping to prepare for her Spring Break trip. Her bathing suit cost $8 more than a pair of shorts, and a T-shirt cost $2 less than the shorts. Find the cost of the bathing suit if Allison spent $60 on the items, before sales tax.

61. Truck Rentals You need to rent a moving truck. You have identified two companies that rent trucks. EZ-Rental charges $35 per day plus $0.15 per mile. Do It Yourself Rental charges $20 per day plus $0.25 per mile. For how many miles will the cost of renting be the same?

62. Cellular Telephones You need a new cell phone for emergencies only. Company A charges $12 per month plus $0.10 per minute, while Company B charges $0.15 per minute with no monthly service charge. For how many minutes will the monthly cost be the same?

63. Comparing Printers Samuel is trying to decide between two laser printers, one manufactured by Hewlett-Packard, the other by Brother. Both have similar features and warranties, so price is the determining factor. The Hewlett-Packard costs $200 and printing costs are approximately $0.03 per page. The Brother costs $240 and printing costs are approximately $0.01 per page. How many pages need to be printed for the cost of the two printers to be the same?

64. Comparing Job Offers Hans has just been offered two sales jobs selling vacuums. The first job offer is a base monthly salary of $2000 plus a commission of $50 for each vacuum sold. The second job offer is a base monthly salary of $1200 plus a commission of $60 for each vacuum sold. How many vacuums must be sold for the two jobs to pay the same salary?

65. Adjusted Gross Income On a joint income tax return, Jensen Beck's adjusted gross income was $249 more than his wife Maureen's adjusted gross income. Their combined adjusted gross income was $72,193. Find Jensen and Maureen Beck's adjusted gross income.

66. **Camping Trip** Jaime Juarez purchased some new camping equipment. He spent $199 on a cookware set, a lantern, and a cook stove. The cookware set cost $30 more than the lantern, and the cook stove cost $34 more than the lantern. Find the cost of each item.

67. **Baseball Games** The Columbus Comets played a double-header and won both games. The scores of each of the two games were consecutive integers, and a total of 26 runs were scored. Find the number of runs scored in each of the two baseball games.

68. **Computing Grades** Going into the final exam, which will count as two tests, Brooke has test scores of 80, 83, 71, 61, and 95. What score does Brooke need on the final exam in order to have an average score of 80?

Extending the Concepts

In Problems 69–72, write a problem that would translate into the given equation.

69. $10x = 370$ **70.** $5n + 10 = 170$ **71.** $\dfrac{x + 74}{2} = 80$ **72.** $n + n + 2 = 98$

△ **73. Angles** The sum of the measures of the three angles in a triangle is 180 degrees. The measure of the smallest angle of a triangle is half the measure of the second angle. The measure of the third angle is 40° more than 4 times the measure of the first. Find the measure of each angle.

△ **74. Angles** The sum of the measures of the three angles in a triangle is 180 degrees. The measure of one angle of a triangle is one degree more than three times the measure of the smallest angle. The measure of the third angle is 13° less than twice the measure of the second angle. Find the measure of each angle.

2.6 Problem Solving: Direct Translation Problems Involving Percent

OBJECTIVES

1. Solve Direct Translation Problems Involving Percent

2. Model and Solve Direct Translation Problems from Business Involving Percent

Preparing for Problem Solving: Direct Translation Problems Involving Percent

Before getting started, take the following readiness quiz. If you get a problem wrong, go back to the section cited and review the material.

1. Write 45% as a decimal.   [Appendix, Section A.2, p. A11]

2. Write 0.2875 as a percent.   [Appendix, Section A.2, pp. A11–A12]

1 Solve Direct Translation Problems Involving Percent

Percent means "divided by 100" or "per hundred." We use the symbol % to denote percent, so 45% means 45 out of 100 or $\dfrac{45}{100}$ or 0.45. In applications involving percents, we often encounter the word "of," as in 20% of 100. The word "of" translates into "multiplication" in mathematics, so 20% of 100 means $0.20 \cdot 100 = 20$.

EXAMPLE 1   **Solving an Equation Involving Percent**

A number is 35% of 40. Find the number.

Solution

Step 1: Identify We want to know the unknown number.

Step 2: Name Let n represent the number.

Step 3: Translate We translate the words of the problem:

$$\underbrace{\text{a number}}_{n} \quad \underbrace{\text{is}}_{=} \quad \underbrace{35\%}_{0.35} \quad \underbrace{\text{of}}_{\cdot} \quad \underbrace{40}_{40}$$

The equation we want to solve is $n = 0.35(40)$.

Step 4: Solve We now solve the equation.

$$n = 0.35(40)$$

Multiply: $n = 14$

Step 5: Check Check the multiplication: $0.35(40) = 14$.

Step 6: Answer 14 is 35% of 40.

QUICK ✓ *Find the number.*

1. A number is 89% of 900. Find the number.

2. A number is 3.5% of 72. Find the number.

3. A number is 150% of 24. Find the number.

4. A number is $8\frac{3}{4}\%$ of 40. Find the number.

EXAMPLE 2 **Solving an Equation Involving Percent**

The number 240 is what percent of 800?

Solution

Step 1: Identify We want to know the percentage.

Step 2: Name Let x represent the percent.

Step 3: Translate We translate the words of the problem:

$$\underbrace{240}_{240} \quad \underbrace{\text{is}}_{=} \quad \underbrace{\text{what percent}}_{x} \quad \underbrace{\text{of}}_{\cdot} \quad \underbrace{800?}_{800}$$

The equation we will solve is $240 = 800x$.

Step 4: Solve We now solve the equation.

$$240 = 800x$$

Divide each side by 800: $$\frac{240}{800} = \frac{800x}{800}$$

$$0.3 = x$$

Since we are finding a percent and our answer is a decimal, we must change 0.3 to a percent by moving the decimal point two places to the right: $0.30 = 30\%$.

Step 5: Check Is 240 equal to 30% of 800? Does $(0.30)(800) = 240$? Yes!

Step 6: Answer The number 240 is 30% of 800.

QUICK ✓ *Find the percent.*

5. The number 8 is what percent of 20?

6. The number 15 is what percent of 40?

7. The number 12.3 is what percent of 60?

8. The number 44 is what percent of 40?

EXAMPLE 3   Solving an Equation Involving Percent

42 is 35% of what number?

Solution

Step 1: Identify We want to know a number.

Step 2: Name Let x represent the number.

Step 3: Translate We translate the words of the problem:

$$\underbrace{42}\; \underbrace{is}\; \underbrace{35\%}\; \underbrace{of}\; \underbrace{what\ number?}$$
$$42\; =\; 0.35\; \cdot\; x$$

The equation we will solve is $42 = 0.35x$.

Step 4: Solve We now solve the equation.

$$42 = 0.35x$$

Divide each side by 0.35: $\quad \dfrac{42}{0.35} = \dfrac{0.35x}{0.35}$

$$120 = x$$

Step 5: Check Is 35% of 120 equal to 42? Because $(0.35)(120) = 42$, the answer is correct.

Step 6: Answer 42 is 35% of 120. ■

QUICK ✓ *Find the number.*

 9. 14 is 28% of what number?   **10.** 111 is 74% of what number?

11. 14.8 is 18.5% of what number?   **12.** 102 is 136% of what number?

EXAMPLE 4   Educational Attainment of U.S. Residents

In 2003, the number of U.S. residents 25 years of age or older was approximately 185,000,000. If 30% of all U.S. residents 25 years of age or older are high school graduates, determine the number of U.S. residents 25 years of age or older in 2003 who were high school graduates. (*Source:* U.S. Census Bureau)

Solution

Step 1: Identify We want to know the number of U.S. residents 25 years of age or older who are high school graduates.

Step 2: Name Let x represent the number of high school graduates.

Step 3: Translate We know that 30% of U.S. residents 25 years of age or older are high school graduates. We also know that in the year 2003 there were approximately 185,000,000 U.S. residents 25 years of age or older. We translate the words of the problem:

$$\underbrace{30\%}\; \underbrace{of}\; \underbrace{U.S.\ residents}\; \underbrace{are}\; \underbrace{high\ school\ graduates}$$
$$0.30\; \cdot\; 185{,}000{,}000\; =\; x$$

Our equation is $(0.30)(185{,}000{,}000) = x$.

Step 4: Solve We solve the equation.

$$(0.30)(185{,}000{,}000) = x$$

Multiply: $\qquad 55{,}500{,}000 = x$

Step 5: Check We can recheck our arithmetic: Because $0.30 \cdot 185{,}000{,}000 = 55{,}500{,}000$, the answer is correct.

Step 6: Answer In the year 2003, the number of U.S. residents 25 years of age or older were high school graduates was 55,500,000.

QUICK ✓

13. In 2003, the number of U.S. residents 25 years of age or older was approximately 185,000,000. If 17% of all U.S. residents 25 years of age or older have bachelor's degrees, determine the number of U.S. residents 25 years of age or older in 2003 who have bachelor's degrees.

---

(2) Model and Solve Direct Translation Problems from Business Involving Percent

Now let's look at direct translation problems that involve percents. Typically, "percent problems" involve discounts or mark-ups that businesses use in determining their prices. They may also include finding the cost of an item including sales tax, as shown in Example 5.

EXAMPLE 5   Finding the Cost of an Item Excluding Sales Tax

You just purchased a new pair of jeans. The price of the jeans, including sales tax of 6% was $41.34. How much did the jeans cost excluding the sales tax?

Solution

Step 1: Identify We want to know the price of the jeans before sales tax.

Step 2: Name Let p represent the price of the jeans before sales tax.

Work Smart

To change a percent to a decimal, move the decimal point two places to the left.

Step 3: Translate The algebraic expression for new cost is computed by adding the price of the jeans before sales tax p and the amount of tax. The amount of tax is 6% of the price of jeans before sales tax.

$$\underbrace{\text{original cost}}_{p} + \underbrace{\text{amount of tax}}_{0.06p}$$

So the final cost of the jeans is represented by the algebraic expression $p + 0.06p$. We can now construct our equation.

$$\underbrace{\text{original cost}}_{p} + \underbrace{\text{amount of tax}}_{0.06p} = \underbrace{\text{total cost}}_{41.34}$$

Step 4: Solve We solve the equation.

$$p + 0.06p = 41.34$$

Combine like terms. Remember that the coefficient of p is 1. $1p + 0.06p = 41.34$

$$1.06p = 41.34$$

Divide each side of the equation by 1.06. $\dfrac{1.06p}{1.06} = \dfrac{41.34}{1.06}$

$$p = 39$$

The jeans cost $39.

Step 5: Check If the jeans cost $39, then the jeans plus the 6% sales tax on $39 amounts to $39 + (0.06)(39) = $39 + $2.34 = $41.34, so our answer is correct.

Step 6: Answer The jeans cost $39 before the sales tax.

QUICK ✓

14. As a reward for being named "Teacher of the Year," Janet received a 2.5% pay raise. If Janet's current salary is $39,000, determine Janet's new salary.

15. Suppose that you just purchased a used car. The price of the car including 7% sales tax was $7811. What was the price of the car excluding sales tax?

Another type of percent problem involves discounts or mark-ups that businesses use in determining their prices. When dealing with percents and the price of goods, it is helpful to remember the following:

$$\text{Original Price} - \text{Discount} = \text{Sale Price}$$
$$\text{Wholesale Price} + \text{Markup} = \text{Selling Price}$$

EXAMPLE 6   Markdown

Suppose you just learned that a local clothing store is going out of business and that all merchandise is marked down by 40%. The sale price of a jacket is $108. What was the original price?

Solution

Step 1: Identify This is a direct translation problem. We are looking for the original price of a jacket that was marked down by 40%, and we know that the sale price is $108.

Step 2: Name Let p represent the original price of the jacket.

Step 3: Translate We know that the original price minus the amount of discount will give us the sale price. We also know that the sale price was $108, so

$$p - \text{discount} = 108$$

"Marked down by 40%" means that the discount is 40% off of the original price, so discount is represented by the expression $0.40p$. Substituting into the equation $p - \text{discount} = 108$, we obtain the equation

$$p - 0.40p = 108$$

Step 4: Solve the equation.

$$p - 0.40p = 108$$

Combine like terms: $\quad 1p - 0.40p = 108$

$$0.60p = 108$$

Divide each side by 0.60: $\quad \dfrac{0.60p}{0.60} = \dfrac{108}{0.60}$

$$p = 180$$

Remember, p represents the original price, so the original price of the jacket was $180.

Step 5: Check If the original price of the jacket was $180, then the discount would be $0.40(180) = \$72$. Subtracting $72 from the original price of $180 results in a sale price of $108. This answer agrees with the information in the problem.

Step 6: Answer The original price of the jacket was $180. ■

QUICK ✓

16. Suppose that a gas station marks its gasoline up 80%. If the gas station charges $2.25 per gallon of 87 octane gasoline, what does it pay for the gasoline?

17. A furniture store marks recliners down by 25%. The sale price, excluding the sales tax, is $494.25. Find the original price of each recliner.

2.6 Exercises

For Extra Help:

Student Solutions Manual   CD Video   PH Math/Tutor Center   MathXL Tutorials on CD   MathXL®   MyMathLab

Concepts and Vocabulary

In Problems 1 and 2, fill in the blanks.

1. Expressed as a decimal, 3.2% = _____.

2. Wholesale Price + _____ = Selling Price

In Problems 3 and 4, answer True or False to each statement.

3. A computer is marked down 20% and sells for $1225. An equation to calculate the original price is $0.8x = 1225$.

4. Original Price − Discount = Sale Price.

5. The sales tax rate is 6%. Explain why $1.06x$ will correctly calculate the total purchase price of any item that sells for x dollars.

6. An item is reduced by 10% and then this is reduced by another 20%. Is this the same as reducing the item by 30%? Explain why or why not.

Building Skills

In Problems 7–24, find the unknown in each percent question.

7. What is 50% of 160?

8. What is 85% of 50?

9. 7% of 200 is what number?

10. 75% of 20 is what number?

11. What number is 16% of 30?

12. What number is 150% of 9?

13. 31.5 is 15% of what number?

14. 40% of what number is 122?

15. 60% of 120 is what number?

16. 45% of what number is 900?

17. 10 is 120% of what number?

18. 11 is 5.5% of what number?

19. What percent of 60 is 24?

20. 15 is what percent of 75?

21. 1.5 is what percent of 20?

22. 4 is what percent of 25?

23. What percent of 300 is 600?

24. What percent of 16 is 12?

Applying the Concepts

25. **Sales Tax** The sales tax in Delaware County, Ohio, is 6%. The total cost of purchasing a tennis racket, including tax, is $57.24. Find the cost of the tennis racket before sales tax.

26. **Sales Tax** The sales tax in Franklin County, Ohio, is 5.75%. The total cost of purchasing a used Honda Civic, including sales tax, is $8460. Find the cost of the car before sales tax.

27. **Pay Cut** Todd works for a computer firm, and in 2004 he earned a salary of $120,000. Recently, Todd was required to take a 15% pay cut. Find Todd's salary after the pay cut.

28. **Pay Raise** MaryBeth works from home as a graphic designer. Recently she raised her hourly rate by 5% to cover increased costs. Her new hourly rate is $23.70. Find MaryBeth's previous hourly rate.

29. **Bad Investment** After Mrs. Fisher lost 9% of her investment, she had $22,750. What was Mrs. Fisher's original investment?

30. **Good Investment** Perry just learned that his house increased in value by 4% over the past year. The value of the house is now $208,000. What was the value of the home one year ago?

31. **Bookstore Purchase** The bookstore at Marietta College had a one-day-only 25%-off sale on all merchandise, except for computer software and textbooks. You purchased pens, notebooks, and a book on study skills for $51. What was the price of the merchandise before the discount?

32. **Hailstorm** Toyota Town had a 15%-off sale on cars that had been damaged in a hailstorm. A new Toyota truck is on sale for $13,217.50. What was the price of the truck before the hailstorm discount?

33. **Business: Discount Pricing** A wool suit, discounted by 30% for a clearance sale, has a price tag of $399. What was the suit's original price?

34. **Business: Marking up the Price of Books** A college bookstore marks up the price that it pays the publisher for a book by 35%. If the selling price of a book is $56.00, how much did the bookstore pay for the book?

35. **Furniture Sale** A furniture store discounted a dining room table by 40%. The discounted price of the dining room table was $240. Determine the original price of the dining room table.

36. **Vacation Package** The Liberty Travel Agency advertised a 3-night vacation package in Jamaica for 30% off the regular price. The sale price of the package is $819. How much is the vacation package before the 30% off sale? (*Source: New York Times, 6/29/03*)

37. **Voting** In an election for school president, the loser received 60% of the winner's votes. If 848 votes were cast, how many did each receive?

38. **Voting** On a committee consisting of Republicans and Democrats, there are twice as many Republicans as Democrats. If 30% of the Republicans and 20% of the Democrats voted in favor of a bill and there were 160 yes votes, how many people are on the committee?

39. **Commission** Melanie receives a 3% commission on every house she sells. If she received a commission of $8571, what was the value of the house she sold?

40. **Commission** Mario collects a commission for bringing in advertisers to his magazine company. He receives 8% on $450 full-page ads and 5% for $300 half-page ads. If he sells twice as many half-page as full-page ads and his commission was $5610, how many of each type did he bring in?

41. **Voting** In the 2000 presidential election, George Bush received 48.85% of Florida's votes and Al Gore received 48.84% of the votes. If 5,963,070 votes were cast in Florida, how many votes separated the two candidates?

42. **Grades** If 15% of Grant's astronomy class received an A, how many students were in his astronomy class if 6 students earned A's this term?

43. **Bachelors** Based on data obtained from the U.S. Census Bureau, 30% of the 108 million males aged 15 years or older have never married. How many males aged 15 years or older have never married?

44. **Census Data** Based on data obtained from the U.S. Census Bureau, 25% of the 115 million females aged 15 years or older have never married. How many females aged 15 years or older have never married?

According to the U.S. Census Bureau, the types of households are changing. The table below shows the number of family households, single-occupant households, and other nonfamily households in 1990 and in 2000. In Problems 45–48, use the information found in the following table to answer each question.

	1990		2000	
	Number	**Percent**	**Number**	**Percent**
Households	91,947,410	100.0	105,480,101	100.0
Family households	64,517,947		71,787,347	
Single-occupant households	22,580,420		27,230,075	
Other nonfamily households	4,849,043		6,462,679	

SOURCE: *United States Census Bureau,* Census 2000.

45. Find, to the nearest tenth of a percent, the percent of family households in the United States in 1990.

46. Find, to the nearest tenth of a percent, the percent of family households in the United States in 2000.

47. Find, to the nearest tenth of a percent, the percent of single-occupant households in the United States in 1990.

48. Find, to the nearest tenth of a percent, the percent of nonfamily households in the United States in 2000.

Extending the Concepts

49. Discount Pricing Suppose that you are the manager of a clothing store and have just purchased 100 shirts for $12 each. After 1 month of selling the shirts at the regular price, you plan to have a sale giving 25% off the original selling price. However, you still want to make a profit of $6 on each shirt at the sale price. What should you price the shirts at initially to ensure this?

In Problems 50–55, find the percent increase or decrease. The percent increase or percent decrease is defined as $\dfrac{\text{amount of change}}{\text{original amount}} \times 100\%.$

50. Population Growth The population in a small fishing town grew from 2500 to 2825. Find the percent increase.

51. Gas Mileage The gas mileage on your VW van decreases due to the heavy weight of extra passengers and gear. With the extra weight your van gets 15 mpg and without the passengers and gear it gets 21 mpg. To the nearest tenth, what is your percent decrease in gas mileage with extra passengers and gear?

52. Teaching Salaries The highest average teaching salary in 2000–01 was $53,507 in Connecticut. You decided to move to Oregon where the average teaching salary was $44,988. After the move, to the nearest tenth, what percent decrease in salary do you expect to take?

53. Car Depreciation A new car decreases in value by 25% each year. At *www.bmwusa.com* Misha priced his convertible M3 at $55,670.

  (a) After two years, what will the car be worth?

  (b) To the nearest tenth of a percent, after 2 years, what is the overall decrease in the value of the car?

54. Gas Prices When gas prices went up from $1.69 to $2.09 per gallon, to the nearest hundredth, what was the percent increase?

55. Gas Prices Last week Danivan bought gas for $1.89 per gallon. This week gas is selling for $2.29 per gallon. To the nearest tenth, what is the percent increase in gas prices?

2.7 Problem Solving: Geometry and Uniform Motion

OBJECTIVES

1. Set Up and Solve Complementary and Supplementary Angle Problems
2. Set Up and Solve Angles of a Triangle Problems
3. Use Geometry Formulas to Solve Problems
4. Set Up and Solve Uniform Motion Problems

Preparing for Problem Solving: Geometry and Uniform Motion

Before getting started, take the following readiness quiz. If you get a problem wrong, go back to the section cited and review the material.

1. Solve: $q + 2q - 30 = 180$ [Section 2.2, p. 86]
2. Solve: $30w + 20(w + 5) = 300$ [Section 2.2, p. 87]

In this section we continue to solve problems using the six-step method introduced in Section 2.5.

1 Set Up and Solve Complementary and Supplementary Angle Problems

We begin by defining *complementary* and *supplementary angles*.

Figure 5

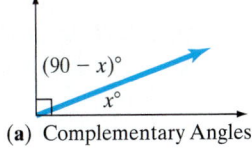

(**a**) Complementary Angles

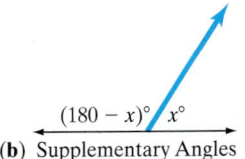

(**b**) Supplementary Angles

> **DEFINITION**
>
> Two angles whose measures sum to 90° are called **complementary angles.** Each angle is called the *complement* of the other.

For example, the angles shown in Figure 5(a) are complements because their sum is 90°. Notice the use of the symbol ⌐ to show the 90° angle.

> **DEFINITION**
>
> Two angles whose measures sum to 180° are called **supplementary angles.** Each angle is called the *supplement* of the other.

The angles shown in Figure 5(b) are supplementary. We use the notation $m\angle A$ to say "the measure of angle A."

EXAMPLE 1 Solving a Complementary Angle Problem

Find the measure of two complementary angles such that the measure of the larger angle is 6° greater than twice the measure of the smaller angle.

Solution

Step 1: Identify This is a complementary angle problem. We are looking for the measure of two angles whose sum is 90°.

Step 2: Name We know least about the measure of the smaller angle so we let x represent the measure of the smaller angle.

Step 3: Translate The measure of the larger angle is 6° more than twice the measure of the smaller angle, so $2x + 6$ represents the measure of the larger angle.

We use the formula

$$m\angle A + m\angle B = 90$$

$$\underbrace{x}_{\text{measure of angle A}} + \underbrace{(2x + 6)}_{\text{measure of angle B}} = 90$$

Step 4: Solve the equation.

$$x + (2x + 6) = 90$$

Combine like terms: $\quad 3x + 6 = 90$

Subtract 6 from each side of the equation: $\quad 3x = 84$

Divide each side by 3: $\quad x = 28$

Step 5: Check The measure of the smaller angle, $\angle A$, is $28°$. The measure of the larger angle, $\angle B$, is $(2x + 6)$ degrees: $2(28) + 6 = 56 + 6 = 62°$. Is the sum of the measures of these angles $90°$? $28° + 62° = 90°$. Our answer is correct.

Step 6: Answer The two complementary angles measure $28°$ and $62°$. ▬

QUICK ✓ *Solve each problem for the unknown angle measures.*

1. Find two complementary angles such that the measure of the larger angle is $12°$ more than the measure of the smaller angle.

2. Find two supplementary angles such that the measure of the larger angle is $30°$ less than twice the measure of the smaller angle.

Figure 6

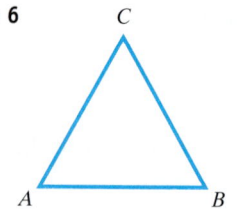

② Set Up and Solve Angles of a Triangle Problems

An important fact from geometry is that the sum of the measures of the interior angles of a triangle is $180°$. In the triangle in Figure 6, we have three angles, A, B, and C. Once again, we use the notation $m\angle A$ to represent the measure of an angle. In a triangle whose angles are A, B, and C, we have $m\angle A + m\angle B + m\angle C = 180°$.

EXAMPLE 2 **Solving Sum of Angles of a Triangle Problem**

The measure of the largest angle of a triangle is $20°$ more than twice the measure of the smallest angle, and the measure of the second angle is $10°$ more than twice the measure of the smallest angle. Find the measure of each angle of the triangle.

Solution

Step 1: Identify This is an "angles of a triangle" problem. We know that the sum of the measures of the interior angles of a triangle is $180°$.

Step 2: Name We know least about the measure of the smallest angle so we let x represent the measure of the smallest angle.

Step 3: Translate The measure of the largest angle is $20°$ more than twice the measure of the smallest angle, so $2x + 20$ represents the measure of the largest angle. The measure of the second angle is $10°$ more than twice the measure of the smallest angle so we let $2x + 10$ represent the measure of the second angle. Using the formula $m\angle A + m\angle B + m\angle C = 180°$, we have

measure of angle A $\quad$ measure of angle B $\quad$ measure of angle C

$$x \quad + \quad (2x + 10) \quad + \quad (2x + 20) \quad = 180°$$

Step 4: Solve the equation.

$$x + (2x + 10) + (2x + 20) = 180°$$

Combine like terms: $\quad 5x + 30 = 180°$

Subtract 30 from each side of the equation: $\quad 5x = 150°$

Divide each side by 5: $\quad x = 30°$

Step 5: Check 30 degrees is the measure of the smallest angle, $\angle A$. The largest angle, $\angle C$, has a measure of $(2x + 20)$ degrees: $2(30°) + 20 = 60 + 20 = 80°$. The second angle, $\angle B$, has a measure of $(2x + 10)$ degrees: $2(30°) + 10 = 60 + 10 = 70°$. Is the sum of the measures of these angles 180°? Because $30° + 70° + 80° = 180°$, our answer is correct.

Step 6: Answer The measures of the angles of the triangle are 30°, 70°, and 80°. ▬

QUICK ✓ *Solve for the unknown angle measures.*

3. The measure of the smallest angle of a triangle is one-third the measure of the largest angle. The measure of the second angle is 65° less than the measure of the largest angle. Find the measure of the angles of the triangle.

---

③ Use Geometry Formulas to Solve Problems

Recall from Section 2.4, that the perimeter of a figure is the sum of the lengths of its sides.

| EXAMPLE 3 | **Solving a Perimeter Problem**

The perimeter of the rectangular swimming pool shown in Figure 7 is 80 feet. If the length is 10 feet more than the width, find the length and the width of the pool.

Solution

Step 1: Identify This is a perimeter problem. We want to find the length and the width of the pool, given the perimeter. We know that the perimeter of a rectangle is the sum of the measures of the sides.

Figure 7

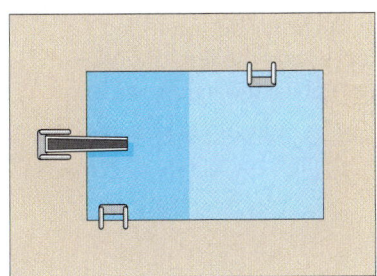

Step 2: Name Let w represent the width of the pool.

Step 3: Translate The length of the pool is 10 feet more than the width, so the length of the pool is $w + 10$. The formula for perimeter of a rectangle is $P = 2l + 2w$, where l represents the length of the rectangle and w represents the width. So we have

$$\underbrace{2l}_{2 \cdot \text{length}} + \underbrace{2w}_{2 \cdot \text{width}} = \underbrace{P}_{\text{perimeter}}$$
$$2(w + 10) + 2(w) = 80$$

Step 4: Solve the equation.

$$2(w + 10) + 2(w) = 80$$

$$\text{Use the Distributive Property:}\quad 2w + 20 + 2w = 80$$
$$\text{Combine like terms:}\quad 4w + 20 = 80$$
$$\text{Subtract 20 from each side of the equation:}\quad 4w = 60$$
$$\text{Divide each side by 4:}\quad w = 15$$

Step 5: Check The width of the pool is $w = 15$ feet, so the length is $w + 10 = 15 + 10 = 25$ feet. We need to see if the perimeter of the pool is 80 feet. Does $15 + 15 + 25 + 25 = 80$? Yes! Our answer is correct.

Work Smart

In solving problems dealing with geometric figures, it is helpful to draw a picture.

Step 6: Answer The length of the rectangular pool is 25 feet and the width of the pool is 15 feet. ▬

QUICK ✓ *Solve for the dimensions of the rectangle.*

4. The perimeter of a small rectangular garden is 9 feet. If the length is twice the width, find the width and length of the garden.

Recall, the area of a plane (two-dimensional) figure is the number of square units that the figure contains, such as square feet, square inches or square yards.

EXAMPLE 4 **Solving an Area Problem**

A garden in the shape of a trapezoid between a sidewalk and curb has an area of 18 square feet. The height is 3 feet and the shorter base is 2 feet less than the length of the longer base. Find the length of each base of the trapezoid. See Figure 8.

Figure 8

Solution

Step 1: Identify This problem is about the area of a trapezoid. The formula for the area of a trapezoid is $A = \frac{1}{2}h(B + b)$, where h is the height, B is the length of the longer base, and b is the length of the shorter base. We are given the area and the height of the trapezoid.

Step 2: Name We know that one base is 2 feet shorter than the other. Let B represent the length of the longer base.

Step 3: Translate Since one base is 2 feet shorter than the other base, and B represents the length of the longer base, then $B - 2$ represents the length of the shorter base. Using the area of a trapezoid formula, we replace the values we know for $A, h, B,$ and b.

$$A = \frac{1}{2}h(B + b)$$

$$\underbrace{18}_{\text{area}} = \frac{1}{2} \cdot \underbrace{3}_{\text{height}} \ (\underbrace{B}_{B} + \underbrace{B - 2}_{b})$$

Step 4: Solve the equation.

$$18 = \frac{1}{2} \cdot 3(B + B - 2)$$

Combine like terms in the parentheses: $\quad 18 = \frac{1}{2} \cdot 3(2B - 2)$

Multiply by 2 to clear fractions: $\quad 2[18] = 2\left[\frac{1}{2} \cdot 3(2B - 2)\right]$

$$36 = 3(2B - 2)$$

Use the Distributive Property: $\quad 36 = 6B - 6$

Add 6 to each side of the equation: $\quad 42 = 6B$

Divide each side by 6: $\quad 7 = B$

Work Smart

Instead of distributing the 3 in

$$36 = 3(2B - 2)$$

we could divide both sides by 3. Try it! Which approach do you prefer?

Step 5: Check The longer base is 7 feet long. The smaller base is $B - 2 = 7 - 2 = 5$ feet long. Is the area of the trapezoidal garden 18 square feet?

Because $\frac{1}{2} \cdot 3(7 + 5) = \frac{1}{2} \cdot 3(12) = 18$, the answers 7 feet and 5 feet are correct.

Step 6: Answer The lengths of the two bases are 5 feet and 7 feet.

QUICK ✅ *Solve for the unknown quantity.*

5. The surface area of a rectangular box is 62 square feet. If the length of the box is 3 feet and the width is 2 feet, find the height of the box.

④ Set Up and Solve Uniform Motion Problems

Objects that move at a constant velocity (speed) are said to be in **uniform motion.** When the average speed of an object is known, it can be interpreted as its constant velocity. For example, a car traveling at an average speed of 45 miles per hour is in uniform motion. An object traveling down an assembly line at a constant speed is also in uniform motion.

In Words
The uniform motion formula states that distance equals rate times time.

> **UNIFORM MOTION FORMULA**
>
> If an object moves at an average speed r, the distance d covered in time t is given by the formula
>
> $$d = rt$$

We will use a chart to set up uniform motion problems as shown in Table 4.

Table 4					
	Rate	•	Time	=	Distance
Object #1					distance 1
Object #2					distance 2

Rate, time, and distance must be expressed in corresponding units. For example, if rate (speed) is stated in miles per hour, then distance must be in miles and time must be in hours. If rate is measured in kilometers per minute, then distance is kilometers and time is minutes.

EXAMPLE 5   Solve a Uniform Motion Problem for Time

Bob and Karen drove from Atlanta, Georgia, to Durham, North Carolina, a distance of 390 miles, to attend a family reunion. Their average rate of speed for the first part of the trip was 60 miles per hour. Due to road construction, their average rate of speed for the remainder of the trip was 45 miles per hour. How long did they travel at 45 miles per hour if they drove 3 hours longer at 60 mph than at 45 mph?

Solution

Step 1: Identify This is a uniform motion problem. We wish to know the number of hours Bob and Karen drove at 45 mph.

Step 2: Name Let t represent the number of hours Bob and Karen drove at 45 miles per hour. Since they traveled 3 hours longer at 60 mph, $t + 3$ represents the number of hours driven at 60 mph.

Step 3: Translate We will set up Table 5, listing the information that we know.

Table 5					
	Rate (in mph)	•	Time	=	Distance
First part of trip	60		$t + 3$		$60(t + 3)$
Second part of trip	45		t		$45t$
Total					390

We summarize the information from Table 5. We know that the total distance that Bob and Karen traveled is 390 miles, so we state that

$$\underbrace{60(t + 3)}_{\substack{\text{distance traveled} \\ \text{at 60 mph}}} + \underbrace{45t}_{\substack{\text{distance traveled} \\ \text{at 45 mph}}} = \underbrace{390}_{\substack{\text{total} \\ \text{distance}}}$$

Step 4: Solve We wish to solve for t:

$$60(t + 3) + 45t = 390$$

Use the Distributive Property: $60t + 180 + 45t = 390$

Combine like terms: $105t + 180 = 390$

Subtract 180 from both sides: $105t + 180 - 180 = 390 - 180$

$$105t = 210$$

Divide both sides by 105: $\dfrac{105t}{105} = \dfrac{210}{105}$

$$t = 2$$

Work Smart

The variable time in the formula $d = rt$ stands for the number of hours traveled, not time of day.

Step 5: Check It appears that Bob and Karen drove for 2 hours at 45 mph. So they drove $t + 3 = 2 + 3 = 5$ hours at 60 miles per hour. Let's see if 2 hours driven at 45 mph plus 5 hours driven at 60 mph equals a distance of 390 miles. Because 2 hrs. · 45 mi./hr. + 5 hrs. · 60 mi./hr. = 90 miles + 300 miles = 390 miles, our answer checks!

Step 6: Answer the Question Bob and Karen drove for 2 hours at 45 mph.

EXAMPLE 6 Solve a Uniform Motion Problem for Rate

Two groups of friends took a canoe trip down Big Darby Creek. The first group left Dan's Canoe Livery at 12 noon. One-half hour later, the second group left Dan's Canoe Livery, traveling at an average speed that was 0.75 miles per hour faster than the first group. At 2:30 P.M. the second group caught up to the first group. How fast was each group paddling?

Solution

Step 1: Identify This is a uniform motion problem. We wish to know the rate of speed that each group is paddling.

Step 2: Name Let r represent the rate that the first group paddled. The second group's rate of paddling was 0.75 miles per hour greater than the first group, so we let $r + 0.75$ represent the rate of the second group.

Step 3: Translate Notice that we are given a specific time that the first group left, 12 noon. The second group left $\frac{1}{2}$ hour later than the first group. The variable time in our formula represents number of hours traveled. So the first group traveled for $2\frac{1}{2}$ (or 2.5) hours (12 noon until 2:30 P.M.), but the second group only traveled for 2 hours. We're now ready to fill in the chart as shown in Table 6.

Table 6				
	Rate (in mph)	**·**	**Time**	**=**	**Distance**
First group	r		2.5		$2.5r$
Second group	$r + 0.75$		2		$2(r + 0.75)$

We summarize the information from the chart. We know that the second group caught up to the first group, so the total distance the two groups traveled was the same.

$$\underbrace{2.5r}_{\text{distance traveled by first group}} = \underbrace{2(r + 0.75)}_{\text{distance traveled by second group}}$$

Step 4: Solve We wish to solve for r:

$$2.5r = 2(r + 0.75)$$

Use the Distributive Property: $2.5r = 2r + 1.5$

Subtract $2r$ from both sides: $2.5r - 2r = 2r - 2r + 1.5$

$$0.5r = 1.5$$

Divide both sides by 0.5: $\dfrac{0.5r}{0.5} = \dfrac{1.5}{0.5}$

$$r = 3$$

Step 5: Check It appears that the first group paddled at 3 miles per hour and the second group paddled at $r + 0.75 = 3 + 0.75 = 3.75$ miles per hour. Let's see if the distance traveled by each group is the same. Does 3 miles per hour · 2.5 hours = 2 hours · 3.75 miles per hour? $(3)(2.5) = 7.5$ miles and $(2)(3.75)$ also equals 7.5 miles, so our answers are correct.

Step 6: Answer the Question The first group paddled at 3 miles per hour and the second group paddled at 3.75 miles per hour. ▬

QUICK

6. Two bikers, José and Luis, start at the same point at the same time and travel in opposite directions. José's average speed is 5 miles per hour more than that of Luis, and after 3 hours the bikers are 63 miles apart. Find the average speed of each biker.

7. Tanya, a long-distance runner, runs at an average speed of 8 miles per hour. Two hours after Tanya leaves your house you leave in your car and follow the same route. If your average speed is 40 miles per hour, how long will it be before you catch up to Tanya? How far will each of you be from your home?

2.7 Exercises

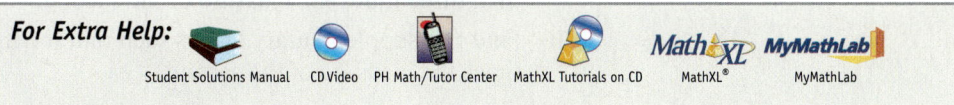

Concepts and Vocabulary

In Problems 1 and 2, fill in the blanks.

1. Complementary angles are angles whose measures sum to _____ degrees.

2. The sum of the measures of the angles of a triangle is _____ degrees.

In Problems 3–5, answer True or False to each statement.

3. The perimeter of a square is $4s$, where s is the length of a side of the square.

4. The perimeter of a rectangle can be found by multiplying the length of rectangle by the width of the rectangle.

5. When using $d = rt$ to calculate the distance traveled, it is not necessary to travel at a constant speed.

6. When setting up a uniform motion problem, you wrote $65t + 40t = 115$. Your classmate wrote $65t - 40t = 115$. Write a word problem for each of these equations, and explain the keys to recognizing the difference between the two types.

Building Skills

*In Problems 7–14, find the value of **x** and then identify the measure of each of the angles.*

△**7.**

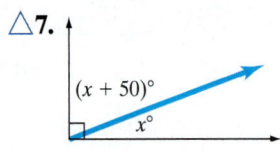

△**8.**

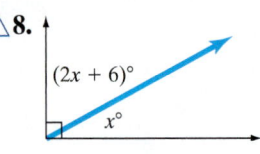

△**9.**

△**10.**

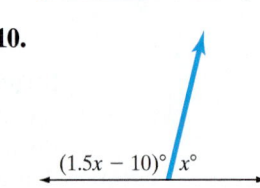

△**11.**

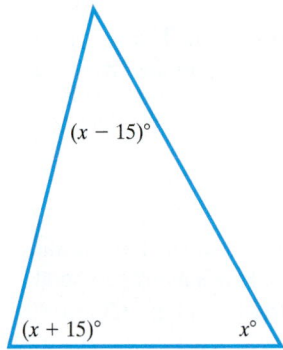

△**12.**

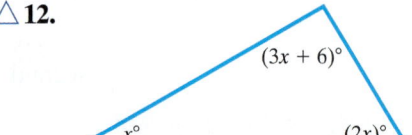

△**13.**

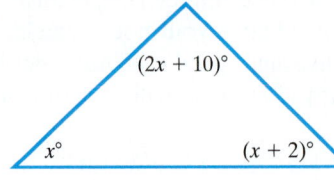

△**14.**

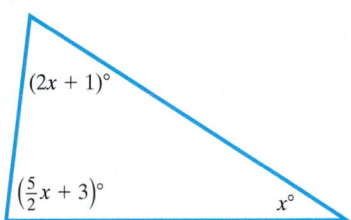

△ **15.** Find two supplementary angles such that the measure of the first angle is 10° less than three times the measure of the second.

△ **16.** Find two supplementary angles such that the measure of the first angle is four times the measure of the second.

△ **17.** Find two complementary angles such that the measure of the first angle is 15° more than the measure of the second.

△ **18.** Find two complementary angles such that the measure of the first angle is 25° less than the measure of the second.

△ **19.** The measures of two complementary angles are consecutive even integers. Find the measure of each angle.

△ **20.** The measures of the angles of a triangle are consecutive even integers. Find the measure of each angle.

△ **21.** In a triangle, the second angle measures four times the first. The measure of the third angle is 18° more than the second. Find the measures of the three angles.

△ **22.** In a triangle, the second angle measures 20° more than the first. The measure of the third angle is twice the second. Find the measures of the three angles.

△ **23.** The length of a rectangle is 8 ft longer than twice the width. If the perimeter is 88 ft, find the length and width of the rectangle.

△ **24.** The width of a rectangle is 10 m less than half of the length. If the perimeter is 52 meters, find the length of each side of the rectangle.

△ **25.** A rectangular field is divided into 2 squares of the same size and shape. If it takes 294 yards of fencing to enclose the field and divide the field into the two parcels, find the dimensions of the field. See the figure.

△ **26.** A rectangular field has been divided so that the length of one of the parcels is twice the other. The smaller parcel is a square and the larger parcel is a rectangle. If it takes 279 m of fencing to enclose the field and divide the two parcels, find the dimensions of the field.

△ **27.** An **isosceles triangle** has exactly two sides that are equal in length (*congruent*). If the base (the third side) measures 45 inches and the perimeter is 98 inches, find the length of the two congruent sides, called *legs*.

△ **28.** An isosceles triangle has a base of 17 cm. If the perimeter is 95 cm, find the length of each of the legs.

△ **29.** In an isosceles triangle, the base angles (angles opposite the two congruent legs) are equal in measure (*congruent*). Find the measures of the angles of an isosceles triangle in which the third angle (called the *vertex angle*) has a measure that is 16° less than twice the measures of the base angles.

△ **30.** In an isosceles triangle, the measure of the third angle is 4 degrees less than twice the measures of the base angles. Find the measure of each of the angles of the triangle.

31. Two cars leave Chicago, one traveling east and the other west. The car going east is traveling at 62 mph and the car going west is traveling at 68 mph. How long before they are 585 miles apart?

   (a) Write an algebraic expression for the distance traveled by the car going east.
   (b) Write an algebraic expression for the car going west.
   (c) Write an algebraic expression for the total distance traveled by the two cars.
   (d) Write an equation to answer the question.

32. Two trains leave Albuquerque, traveling the same direction on parallel tracks. One train is traveling at 72 mph and the other is traveling at 66 mph. How long before they are 45 miles apart?

   (a) Write an algebraic expression for the distance traveled by the faster train.
   (b) Write an algebraic expression for the slower train.
   (c) Write an algebraic expression for the difference in distance between the two.
   (d) Write an equation to answer the question.

In Problems 33 and 34, fill in the table from the information given. Then write the equation that will solve the problem. DO NOT SOLVE.

33. Martha is running in her first marathon. She can run at a rate of 528 ft per min. Ten minutes later her mom starts the same course, running at a rate of 880 ft per min. How long before Mom catches up to Martha?

	Rate	•	Time	=	Distance
Martha	?		?		?
Mom	?		?		?

34. A 580-mile trip in a small plane took a total of 5 hours. The first two hours were flown at one rate and then the plane encountered a head wind and was slowed by 10 mph. Find the rate for each portion of the trip.

	Rate	•	Time	=	Distance
Beginning of trip	?		?		?
Rest of the trip	?		?		?
Total			?		?

Applying the Concepts

△ **35.** **Rectangle** The width of a rectangle is 3 inches less than one-half the length. Find the length and the width of the rectangle if the perimeter of the rectangle is 36 inches.

△ **36.** **Garden** The length of a rectangular garden is 9 feet. If 26 feet of fencing are required to fence the garden, find the width of the garden.

△ **37.** **Billboard** A billboard along a highway has a perimeter of 110 feet. Find the length of the billboard if its height is 15 feet.

△ **38.** **Buying Wallpaper** Erika is buying wallpaper for her bedroom. She remembers that the perimeter of the room is 54 ft and that the room is twice as long as it is wide.

  (a) Find the dimensions of the room.
  (b) If the walls are 8 ft high, how many square feet of wallpaper does she need to buy?
  (c) Erika arrives at the decorating store and finds that wallpaper is sold by the square yard. How many square yards of wallpaper does Erika need to buy?

△ **39.** **Back Yard** Bob's back yard is in the shape of a trapezoid with height of 60 feet. The shorter base is 8 feet shorter than the longer base, and the area of the back yard is 2160 square feet. Find the length of each base of the trapezoidal yard.

△ **40.** **Buying Fertilizer** Melinda has to buy fertilizer for a flower garden in the shape of a right triangle. If the area of the garden is 54 square feet and the base of the garden measures 9 feet, find the height of the triangular garden.

△ **41.** **Garden** The perimeter of a rectangular garden is 60 yards. The width of the garden is three yards less than twice the length.

  (a) Find the length and width of the garden.
  (b) What is the area of the garden?

△ **42.** **Table** The Jacksons are having a custom rectangular table made for a small dining area. The length of the table is 18 inches more than the width, and the perimeter is 180 inches. Find the length and the width of the table.

43. **Boats** Two boats leave a port at the same time, one going north and the other traveling south. The northbound boat travels 16 mph faster than the southbound boat. If the southbound boat is traveling at 47 mph, how long will it be before they are 1430 miles apart?

44. **Cyclists** Two cyclists leave a city at the same time, one going east and the other going west. The west-bound cyclist bikes at 4 mph faster than the east-bound cyclist. After 5 hours they are 200 miles apart. How fast is the east-bound cyclist riding?

45. **Road Trip** Two cars leave a city on the same road, one driving 12 mph faster than the other. After 4 hours, the car traveling at the faster speed stops for lunch. After 4 hours and 30 minutes, the car traveling at the slower speed stops for lunch. Assuming that the person in the faster car is still eating lunch, the cars are now 24 miles apart. How fast is each car driving?

46. **Passenger and Freight Trains** Two trains leave a city on parallel tracks, traveling the same direction. The passenger train is going twice as fast as the freight train. After 45 minutes, the trains are 180 miles apart. Find the speed of each train.

47. **Down the Highway** A 360-mile trip began on a freeway in a car traveling at 62 mph. Once the road became a 2-lane highway, the car slowed to 54 mph. If the total trip took 6 hours, find the time spent on each type of road.

48. River Trip Max lives on a river, 30 miles from town. Max travels downstream (with the current) at 20 mph. Returning upstream (against the current) his progress is at 12 mph. If the total trip to town and back took 4 hours, how long did he spend returning from town?

49. Walking and Jogging Carol knows that when she jogs along her neighborhood greenway, she can complete the route in 10 minutes. It takes 30 minutes to cover the same distance when she walks. If her jogging rate is 4 mph faster than her walking rate, find the speed at which she jogs.

50. Trip to School Dien drives to school at 40 mph. Five minutes $\left(\frac{1}{12}\text{hour}\right)$ after he left home, his mother sees that he forgot his homework and leaves to take it to him, driving 48 mph. If they arrive at school at the same time, how far away is the school?

Extending the Concepts

Parallel lines are lines in the same plane that never intersect (think of railroad tracks going infinitely far out into space). A line that cuts two parallel lines is called a *transversal*. The transversal forms 8 different angles that are related in following ways:

> *Corresponding angles are equal in measure.*
>
> *Alternate interior angles are equal in measure.*
>
> *Interior angles on the same side of the transversal are supplementary.*

In the figure shown, lines l_1 and l_2 are parallel $(l_1 \| l_2)$ and the transversal is labeled t. In this figure, there are 4 pairs of corresponding angles:

$$\angle 1 \text{ and } \angle 5, \ \angle 2 \text{ and } \angle 6, \ \angle 3 \text{ and } \angle 7, \ \angle 4 \text{ and } \angle 8$$

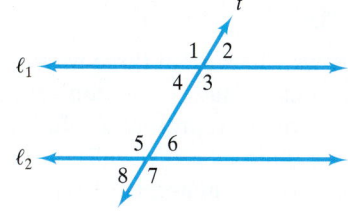

There are 2 pairs of alternate interior angles: $\angle 3$ and $\angle 5$, $\angle 4$ and $\angle 6$.

There are 2 pairs of interior angles on the same side of the transversal: $\angle 3$ and $\angle 6$, $\angle 4$ and $\angle 5$.

In Problems 51–56, given $l_1 \| l_2$, use the appropriate properties from geometry to solve for x.

△ **51.**   △ **52.**

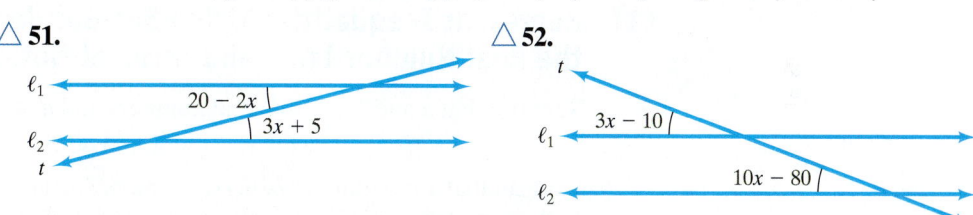

△ **53.**   △ **54.**   △ **55.**   △ **56.**

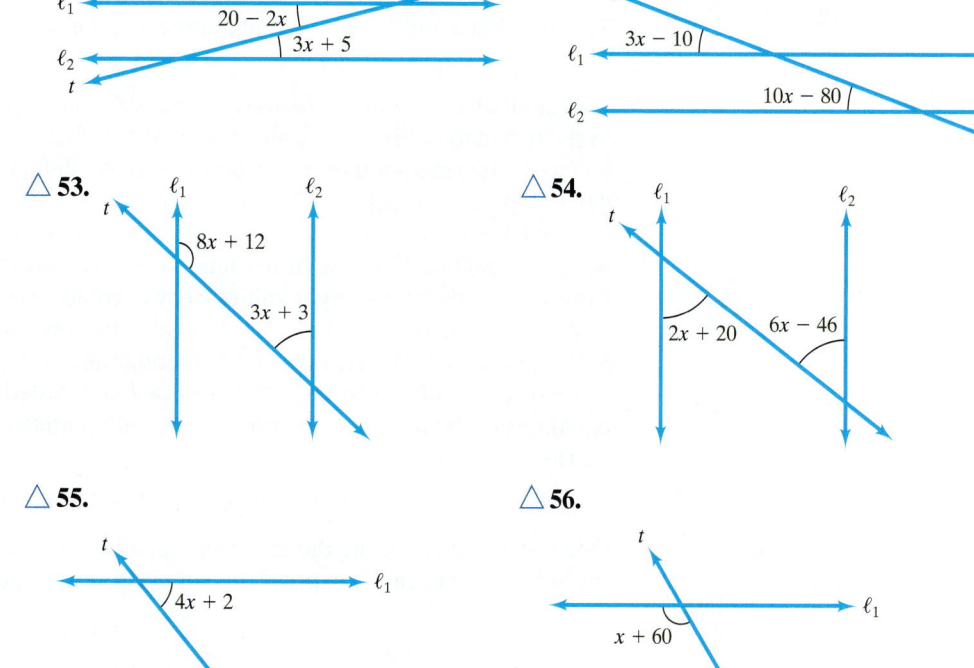

2.8 Solving Linear Inequalities in One Variable

OBJECTIVES

① Represent Inequalities Using Set-Builder Notation, the Real Number Line, and Interval Notation

② Solve Linear Inequalities

③ Solve Problems Involving Linear Inequalities

Preparing for Solving Linear Inequalities in One Variable
Before getting started, take the following readiness quiz. If you get a problem wrong, go back to the section cited and review the material.

In Problems 1–4, replace the question mark by $<$, $>$, or $=$ to make the statement true. *[Section 1.2, pp. 13–14]*

1. $3 \, ? \, 6$ **2.** $-3 \, ? \, -6$ **3.** $\dfrac{1}{2} \, ? \, 0.5$ **4.** $\dfrac{2}{3} \, ? \, \dfrac{3}{5}$

An **inequality in one variable** is a statement involving two expressions, at least one containing the variable, separated by one of the inequality symbols $<$, $\leq$, $>$, or $\geq$. To **solve an inequality** means to find all values of the variable for which the statement is true. These values are called **solutions** of the inequality. The set of all solutions is called the **solution set.**

> **DEFINITION**
>
> A **linear inequality in one variable** is an inequality that can be written in the form
> $$ax + b < c \quad \text{or} \quad ax + b \leq c \quad \text{or} \quad ax + b > c \quad \text{or} \quad ax + b \geq c$$
> where a, b, and c are real numbers and $a \neq 0$.

For example, the following are all linear inequalities involving one variable:

$$x - 4 > 9 \qquad 5x - 1 \leq 14 \qquad 8z < 0 \qquad 5x - 1 \geq 3x + 8$$

Before we discuss methods for solving linear inequalities, we will present three ways of representing the solution set. One of the methods for representing the solution set is through *set-builder notation.* A second, more streamlined, way to represent a solution set to an inequality is *interval notation.* Finally, because we often like to visualize solution sets, we present a method for graphing the solution set on a real number line.

① **Represent Inequalities Using Set-Builder Notation, the Real Number Line, and Interval Notation**

Suppose that a and b are two real numbers and $a < b$. We shall use the notation

$$a < x < b$$

to mean that x is a number *between* a and b. So, the expression $a < x < b$ is equivalent to the two inequalities $a < x$ and $x < b$. Similarly, the expression $a \leq x \leq b$ is equivalent to the two inequalities $a \leq x$ and $x \leq b$. We define $a \leq x < b$ and $a < x \leq b$ similarly. Expressions such as $-2 < x < 5$ or $x \geq 5$ are said to be in **inequality notation.**

While the expression $3 \geq x \geq 2$ is technically correct, it is not the preferred way to write inequalities. For ease of reading, we prefer that the numbers in the inequality go from smaller values to larger values. So, we would write $3 \geq x \geq 2$ as $2 \leq x \leq 3$.

A statement such as $3 \leq x \leq 1$ is false because there is no number x for which $3 \leq x$ and $x \leq 1$. We also never mix inequalities as in $2 \leq x \geq 3$.

Inequalities of the form $a < b$ or $a > b$ are called **strict inequalities,** whereas inequalities of the form $a \leq b$ or $a \geq b$ are called **nonstrict** (or **weak**) **inequalities.**

The inequality

$$x > 2 \quad \text{means} \quad \text{the set of all real numbers } x \text{ greater than 2}$$

One way of representing the set of all real numbers x that are greater than 2 is through **set-builder notation.** To express this set using set-builder notation, we write

$$\{x \quad | \quad x > 2\}$$

The set of all x such that x is greater than 2

In addition to representing inequalities using set-builder notation, we can use *interval notation.*

DEFINITION: INTERVAL NOTATION

Let a and b represent two real numbers with $a < b$.

A **closed interval,** denoted by $[a, b]$, consists of all real numbers x for which $a \le x \le b$.

An **open interval,** denoted by (a, b), consists of all real numbers x for which $a < x < b$.

The **half-open,** or **half-closed, intervals** are $(a, b]$, consisting of all real numbers x for which $a < x \le b$, and $[a, b)$, consisting of all real numbers x for which $a \le x < b$.

In each of these definitions, a is called the **left endpoint** and b is called the **right endpoint** of the interval.

The symbol ∞ (read as "infinity") is not a real number, but a notational device used to indicate unboundedness in the positive direction. In other words, the symbol ∞ means that there is no right endpoint on the inequality. The symbol $-\infty$ (read as "minus infinity" or "negative infinity") also is not a real number, but a notational device used to indicate unboundedness in the negative direction. The symbol $-\infty$ means that there is no left endpoint on the inequality. Using the symbols ∞ and $-\infty$, we can define five other kinds of intervals.

Work Smart

The symbols ∞ and $-\infty$ are never included as endpoints because they are not real numbers. So, we use parentheses when $-\infty$ or ∞ are endpoints.

INTERVALS INCLUDING ∞

$[a, \infty)$ consists of all real numbers x for which $x \ge a$

(a, ∞) consists of all real numbers x for which $x > a$

$(-\infty, a]$ consists of all real numbers x for which $x \le a$

$(-\infty, a)$ consists of all real numbers x for which $x < a$

$(-\infty, \infty)$ consists of all real numbers x (or $-\infty < x < \infty$)

We can also represent inequalities using a graph on the real number line. The inequality $x > 3$ or the interval $(3, \infty)$ consists of all numbers x that lie to the right of 3 on the real number line. We can represent these values by shading the real number line to the right of 3. To indicate that 3 is not included in the set, we will agree to use a parenthesis on the endpoint. See Figure 9.

To represent the inequality $x \ge 3$ or the interval $[3, \infty)$ graphically, we also shade to the right of 3, but this time we use a bracket on the endpoint to indicate that 3 is included in the set. See Figure 10.

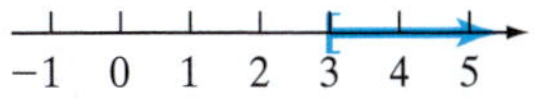

Table 7 summarizes interval notation, set-builder notation, and their graphs.

Figure 9

$x > 3$

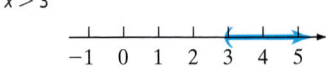

Figure 10

$x \ge 3$

Table 7		
Interval	**Set-Builder Notation**	**Graph**
The open interval (a, b)	$\{x \mid a < x < b\}$	$a \quad b$
The closed interval $[a, b]$	$\{x \mid a \le x \le b\}$	$a \quad b$
The half-open interval $[a, b)$	$\{x \mid a \le x < b\}$	$a \quad b$
The half-open interval $(a, b]$	$\{x \mid a < x \le b\}$	 $a \quad b$
The interval $[a, \infty)$	$\{x \mid x \ge a\}$	 a

(continued)

Table 7 (continued)		
Interval	**Set-Builder Notation**	**Graph**
The interval (a, ∞)	$\{x \mid x > a\}$	
The interval $(-\infty, a]$	$\{x \mid x \leq a\}$	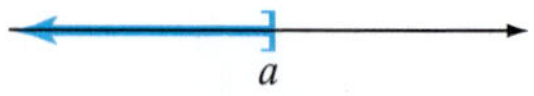
The interval $(-\infty, a)$	$\{x \mid x < a\}$	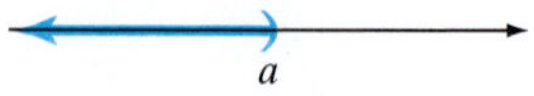
The interval $(-\infty, \infty)$	$\{x \mid x$ is a real number$\}$	

EXAMPLE 1 Using Set-Builder Notation, Interval Notation, and Graphing Inequalities

Write each inequality using set-builder notation and interval notation. Graph the inequality.

(a) $-2 \leq x \leq 4$

(b) $1 < x \leq 5$

Solution

(a) $-2 \leq x \leq 4$ represents all numbers x between -2 and 4, inclusive. Using set-builder notation, we write $\{x \mid -2 \leq x \leq 4\}$. In interval notation, we write $[-2, 4]$. To graph $-2 \leq x \leq 4$, we place brackets at -2 and 4 and shade in between. See Figure 11.

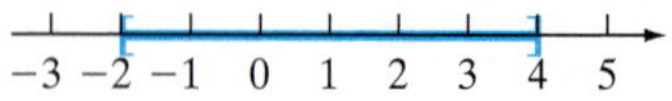

(b) $1 < x \leq 5$ represents all numbers x greater than 1 and less than or equal to 5. Using set-builder notation, we write $\{x \mid 1 < x \leq 5\}$. In interval notation, we write $(1, 5]$. To graph $1 < x \leq 5$, we place a parenthesis at 1 and a bracket at 5 and shade in between. See Figure 12.

Figure 11

$$\begin{array}{ccccccccc} \cdot & \cdot & \cdot & \cdot & \cdot & \cdot & \cdot & \cdot & \cdot \\ -3 & -2 & -1 & 0 & 1 & 2 & 3 & 4 & 5 \end{array}$$

Figure 12

$$\begin{array}{cccccccc} \cdot & \cdot & \cdot & \cdot & \cdot & \cdot & \cdot & \cdot \\ -1 & 0 & 1 & 2 & 3 & 4 & 5 & 6 \end{array}$$

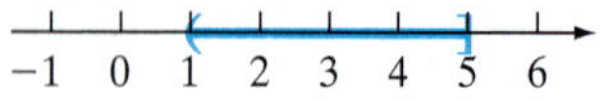

EXAMPLE 2 Using Interval Notation and Graphing Inequalities

Write each inequality using set-builder notation and interval notation. Graph the inequality.

(a) $x < 2$ (b) $x \geq -3$

Solution

(a) $x < 2$ represents all numbers x less than 2. Using set-builder notation, we write $\{x \mid x < 2\}$. In interval notation, we write $(-\infty, 2)$. To graph $x < 2$, we place a parenthesis at 2 and then shade to the left. See Figure 13.

Figure 13

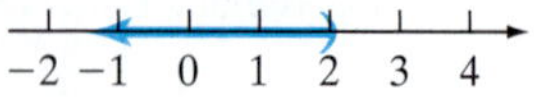

$$\begin{array}{ccccccc} \cdot & \cdot & \cdot & \cdot & \cdot & \cdot & \cdot \\ -2 & -1 & 0 & 1 & 2 & 3 & 4 \end{array}$$

(b) $x \geq -3$ represents all numbers x greater than or equal to -3. Using set-builder notation, we write $\{x \mid x \geq -3\}$. In interval notation, we write $[-3, \infty)$. To graph $x \geq -3$, we place a bracket at -3 and then shade to the right. See Figure 14.

Figure 14

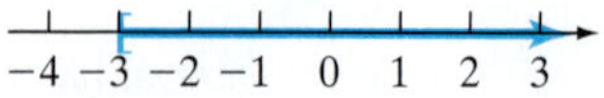

$$\begin{array}{cccccccc} \cdot & \cdot & \cdot & \cdot & \cdot & \cdot & \cdot & \cdot \\ -4 & -3 & -2 & -1 & 0 & 1 & 2 & 3 \end{array}$$

QUICK ✓ *Write each inequality in set-builder and interval notation. Graph the inequality.*

1. $-3 \leq x \leq 2$ **2.** $3 \leq x < 6$ **3.** $x \leq 3$ **4.** $\dfrac{1}{2} < x < \dfrac{7}{2}$

EXAMPLE 3   **Using Inequality Notation and Graphing Inequalities**

Write each interval in inequality notation involving x. Graph the inequality.

  (a) $[-2, 4)$ **(b)** $(1, 5)$

Solution

  (a) The interval $[-2, 4)$ consists of all numbers x for which $-2 \leq x < 4$. See Figure 15 for the graph.

  Figure 15
  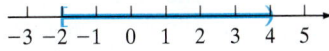

  (b) The interval $(1, 5)$ consists of all numbers x for which $1 < x < 5$. See Figure 16 for the graph.

  Figure 16
  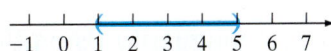

EXAMPLE 4   **Using Inequality Notation and Graphing Inequalities**

Write each interval in inequality notation involving x. Graph the inequality.

  (a) $\left[\dfrac{3}{2}, \infty\right)$ **(b)** $(-\infty, 1)$

Solution

  (a) The interval $\left[\dfrac{3}{2}, \infty\right)$ consists of all numbers x for which $x \geq \dfrac{3}{2}$. See Figure 17 for the graph.

  Figure 17
  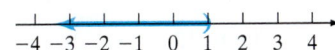

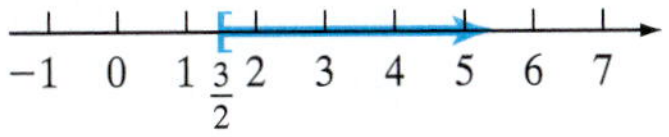

  (b) The interval $(-\infty, 1)$ consists of all numbers x for which $x < 1$. See Figure 18 for the graph.

  Figure 18

QUICK ✓ *Write each interval as an inequality. Graph the inequality.*

5. $(0, 5]$ **6.** $(-6, 0)$ **7.** $(5, \infty)$ **8.** $\left(-\infty, \dfrac{8}{3}\right]$

---

②   Solve Linear Inequalities

As with equations, one method for solving a linear inequality is to replace it by a series of equivalent inequalities until an inequality with an obvious solution, such as $x > 2$, is obtained.

In Words

The Addition Property of Inequality states that the direction of the inequality does not change when the same quantity is added to each side of the inequality.

Two inequalities that have exactly the same solution set are called **equivalent inequalities.** We obtain equivalent inequalities by applying some of the same operations as those used to find equivalent equations.

Consider the inequality $3 < 8$. If we add 2 to both sides of the inequality, the left side becomes 5 and the right side becomes 10. Since $5 < 10$, we see that adding the same quantity to both sides of an inequality does not change the sense, or direction, of the inequality. This result is called the **Addition Property of Inequality.**

ADDITION PROPERTY OF INEQUALITY

For real numbers a, b, and c

$$\text{If} \quad a < b, \quad \text{then} \quad a + c < b + c$$

$$\text{If} \quad a > b, \quad \text{then} \quad a + c > b + c$$

The Addition Property of Inequality also holds true for subtracting a real number from both sides of an inequality, since $a - b$ is equivalent to $a + (-b)$. In other words, subtracting a quantity from both sides of an inequality does not change the sense of the inequality.

EXAMPLE 5 **How to Solve an Inequality Using the Addition Property of Inequality**

Solve the linear inequality $3y + 5 \leq 2y + 8$ and express the solution set using set-builder notation and interval notation. Graph the solution set.

Step-by-Step Solution

Step 1: Get the expressions containing variables on the left side of the inequality.

Apply the Addition Property of Inequality and subtract 2y from both sides:

$$3y + 5 \leq 2y + 8$$
$$3y + 5 - 2y \leq 2y + 8 - 2y$$
$$y + 5 \leq 8$$

Step 2: Isolate the variable y on the left side.

Apply the Addition Property of Inequality and subtract 5 from both sides:

$$y + 5 - 5 \leq 8 - 5$$
$$y \leq 3$$

The solution set using set-builder notation is $\{y \,|\, y \leq 3\}$. The solution set using interval notation is $(-\infty, 3]$. The solution set is graphed in Figure 19.

Figure 19

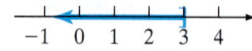

$-1 \quad 0 \quad 1 \quad 2 \quad 3 \quad 4$

QUICK ✓ *Find the solution of the linear inequality and express the solution set in set-builder notation or interval notation. Graph the solution set.*

9. $5n - 4 > 4n - 1$ **10.** $-2x + 3 < 7 - 3x$

11. $5n + 8 \leq 4n + 4$ **12.** $3(4x - 8) + 12 > 11x - 13$

We've seen what happens when we add a real number to both sides of an inequality. Let's look at two examples from arithmetic to see if we can figure out what happens when we multiply or divide both sides of an inequality by a non-zero constant.

EXAMPLE 6 **Multiplying or Dividing an Inequality by a Positive Number**

(a) Express the inequality that results by multiplying both sides of the inequality $-2 < 5$ by 3.

(b) Express the inequality that results by dividing both sides of the inequality $18 > 14$ by 2.

Solution

(a) We begin with

$$-2 < 5$$

Multiplying both sides by 3 results in the numbers -6 and 15 on each side of the inequality, so we have

$$-6 < 15$$

(b) We begin with

$$18 > 14$$

Dividing both sides by 2 results in the numbers 9 and 7 on each side of the inequality, so we have

$$9 > 7$$ ▬

Based on the results of Example 6, we see that multiplying (or dividing) both sides of an inequality by a positive real number does not affect the sense, or direction, of the inequality.

EXAMPLE 7 **Multiplying or Dividing an Inequality by a Negative Number**

(a) Express the inequality that results by multiplying both sides of the inequality $-2 < 5$ by -3.

(b) Express the inequality that results by dividing both sides of the inequality $18 > 14$ by -2.

Solution

(a) We begin with

$$-2 < 5$$

Multiplying both sides by -3 results in the numbers 6 and -15 on each side of the inequality, so we have

$$6 > -15$$

(b) We begin with

$$18 > 14$$

Dividing both sides by -2 results in the numbers -9 and -7 on each side of the inequality, so we have

$$-9 < -7$$ ▬

In Example 7, we see that multiplying (or dividing) both sides of an inequality by a negative real number produces an inequality that has the opposite sense, or direction, of the original inequality. The results of Examples 6 and 7 lead us to the **Multiplication Properties of Inequality.**

In Words
The Multiplication Property of Inequality states that if you multiply both sides of an inequality by a positive number, the inequality symbol remains the same, but if you multiply by a negative number, the inequality symbol reverses.

MULTIPLICATION PROPERTIES OF INEQUALITY

Let $a, b,$ and c be real numbers.

$$\text{If } a < b, \text{ and if } c > 0, \text{ then } ac < bc.$$
$$\text{If } a > b, \text{ and if } c > 0, \text{ then } ac > bc.$$

$$\text{If } a < b, \text{ and if } c < 0, \text{ then } ac > bc.$$
$$\text{If } a > b, \text{ and if } c < 0, \text{ then } ac < bc.$$

EXAMPLE 8 **Solving a Linear Inequality Using the Multiplication Properties of Inequality**

Solve each linear inequality and state the solution set using set-builder notation and interval notation. Graph the solution set.

(a) $7x > -35$ **(b)** $-4x \geq 24$

Solution

(a)
$$7x > -35$$

Divide both sides of the inequality by 7: $\dfrac{7x}{7} > \dfrac{-35}{7}$

$$x > -5$$

The solution set using set-builder notation is $\{x \mid x > -5\}$. The solution set using interval notation is $(-5, \infty)$. The graph of the solution set is shown in Figure 20.

Figure 20

(b)
$$-4x \geq 24$$

Divide both sides of the inequality by -4. Remember to reverse the inequality symbol! $\dfrac{-4x}{-4} \leq \dfrac{24}{-4}$

$$x \leq -6$$

The solution set using set-builder notation is $\{x \mid x \leq -6\}$. The solution set using interval notation is $(-\infty, -6]$. The graph of the solution set is shown in Figure 21.

Figure 21

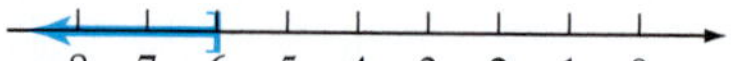

QUICK ✓ *Find the solution of the linear inequality and express the solution set in set-builder notation or interval notation. Graph the solution set.*

13. $6k < -36$ **14.** $2n \geq -5$ **15.** $-\dfrac{3}{2}k > 12$ **16.** $-\dfrac{4}{3}p \leq -\dfrac{4}{5}$

We are now ready to solve inequalities using both the Addition and Multiplication Properties of Inequalities. In each solution, we isolate the variable on the left side of the inequality, so the inequality is easier to read. Just remember that

$$a < x \text{ is equivalent to } x > a$$
$$\text{and} \quad a > x \text{ is equivalent to } x < a$$

In general, if the two sides of the inequality are interchanged, the direction of the inequality reverses.

<table>
<tr><td colspan="2">**EXAMPLE 9** **How to Solve an Inequality Using Both the Addition and Multiplication Properties of Inequalities**</td></tr>
</table>

Solve the inequality $4(x + 1) - 2x < 8x - 26$. State the solution set using set-builder notation and interval notation. Graph the solution set.

Step-by-Step Solution

Step 1: Remove parentheses.	$4(x + 1) - 2x < 8x - 26$
Use the Distributive Property:	$4x + 4 - 2x < 8x - 26$

Step 2: Combine like terms on each side of the inequality.	$2x + 4 < 8x - 26$

Step 3: Get the variable expressions on the left side of the inequality and the constants on the right side.	Subtract $8x$ from both sides: $2x + 4 - 8x < 8x - 26 - 8x$
	$-6x + 4 < -26$
	Subtract 4 from both sides: $-6x + 4 - 4 < -26 - 4$
	$-6x < -30$

Step 4: Get the coefficient of the variable term to be one.	Divide both sides by -6: Remember to reverse the inequality symbol! $\dfrac{-6x}{-6} > \dfrac{-30}{-6}$
	$x > 5$

The solution set using set-builder notation is $\{x \mid x > 5\}$, or using interval notation, $(5, \infty)$. The graph of the solution is given in Figure 22.

Figure 22

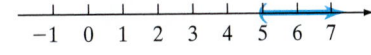

We can gather evidence to support that our solution to an inequality is correct in the same way we check the solution of an equation. We substitute a value for the variable that is in the solution set into the original inequality and see if we obtain a true statement. If we obtain a true statement, then we have evidence our solution is correct. Be warned, however, this does not prove your solution is correct. The check for Example 9 is shown below.

Check Since the solution of the inequality $4(x + 1) - 2x < 8x - 26$ is any real number greater than 5, let's replace x by 10.

$$4(x + 1) - 2x < 8x - 26$$

Replace x with 10: $\quad 4(10 + 1) - 2(10) \overset{?}{<} 8(10) - 26$

$$4(11) - 20 \overset{?}{<} 80 - 26$$

Perform the arithmetic: $\quad 24 < 54 \quad$ True

$24 < 54$ is a true statement, so $x = 10$ is in the solution set. We have some evidence that our solution set, $(5, \infty)$, is correct.

QUICK ✓ *Find the solution of the linear inequality and express the solution set using set-builder notation or interval notation. Graph the solution set.*

17. $3x - 7 > 14$

18. $-4n - 3 < 9$

19. $2x - 6 < 3(x + 1) - 5$

20. $-4(x + 6) + 18 \geq -2x + 6$

| **EXAMPLE 10** | **Solving a Linear Inequality Containing Fractions** |

Solve the inequality $\frac{1}{2}(x - 4) \geq \frac{3}{4}(2x + 1)$. Express the solution using set-builder notation and interval notation. Graph the solution set.

Solution

To clear the inequality of fractions, we multiply both sides of the inequality by 4, the least common denominator of $\frac{1}{2}$ and $\frac{3}{4}$.

$$\frac{1}{2}(x - 4) \geq \frac{3}{4}(2x + 1)$$

$$4 \cdot \frac{1}{2}(x - 4) \geq 4 \cdot \frac{3}{4}(2x + 1)$$

$$2(x - 4) \geq 3(2x + 1)$$

Use the Distributive Property: $\quad 2x - 8 \geq 6x + 3$

Subtract 6x from both sides: $\quad 2x - 8 - 6x \geq 6x + 3 - 6x$

$$-4x - 8 \geq 3$$

Add 8 to both sides: $\quad -4x - 8 + 8 \geq 3 + 8$

$$-4x \geq 11$$

Divide both sides by -4 and remember to reverse the inequality symbol! $\quad \dfrac{-4x}{-4} \leq \dfrac{11}{-4}$

$$x \leq -\frac{11}{4}$$

The solution of the inequality $\frac{1}{2}(x - 4) \geq \frac{3}{4}(2x + 1)$ is $\left\{ x \mid x \leq -\frac{11}{4} \right\}$, or using interval notation, $\left(-\infty, -\frac{11}{4} \right]$. The graph of the solution is given in Figure 23.

Figure 23

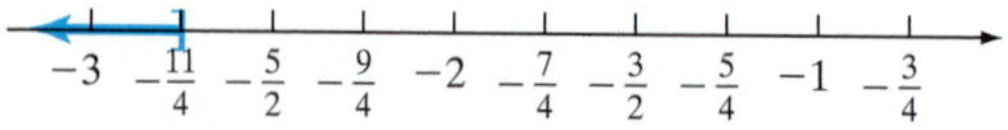

QUICK ✓ *Find the solution of the linear inequality and express the solution set using set-builder notation or interval notation. Graph the solution set.*

21. $\frac{1}{2}(x + 2) > \frac{1}{5}(x + 17)$

22. $\frac{4}{3}x - \frac{2}{3} \leq \frac{4}{5}x + \frac{3}{5}$

There are inequalities that are true for all values of the variable and inequalities that are false for all values of the variable. We present these special cases now.

EXAMPLE 11 Solving an Inequality for Which the Solution Set Is All Real Numbers

Solve the inequality $3(x + 4) - 5 > 7x - (4x + 2)$. Express the solution using set-builder notation and interval notation. Graph the solution set.

Solution

$$3(x + 4) - 5 > 7x - (4x + 2)$$

Use the Distributive Property: $3x + 12 - 5 > 7x - 4x - 2$

$$3x + 7 > 3x - 2$$

Subtract 3x from each side: $3x + 7 - 3x > 3x - 2 - 3x$

$$7 > -2$$

Figure 24

$-2\ -1\ \ 0\ \ 1\ \ 2$

Since 7 is greater than -2, the solution to this inequality is all real numbers. The solution set is $\{x | x$ is any real number$\}$. In interval notation, the solution is $(-\infty, \infty)$. Figure 24 shows the graph of the solution set.

EXAMPLE 12 Solving an Inequality for Which the Solution Set Is the Empty Set

Solve the inequality $8\left(\dfrac{1}{2}x - 1\right) + 2x \leq 6x - 10$. Express the solution using set-builder notation and interval notation, if possible. Graph the solution set.

Solution

$$8\left(\dfrac{1}{2}x - 1\right) + 2x \leq 6x - 10$$

Use the Distributive Property: $4x - 8 + 2x \leq 6x - 10$

$$6x - 8 \leq 6x - 10$$

Subtract 6x from each side: $6x - 8 - 6x \leq 6x - 10 - 6x$

$$-8 \leq -10$$

Figure 25

$-7\ -6\ -5\ -4\ -3\ -2\ -1\ \ \ 0$

The statement $-8 \leq -10$ is a false statement. Therefore, there is no solution to this inequality. The solution set is the empty set $\varnothing$ or $\{\ \}$. Figure 25 shows the graph of the solution set on a real number line.

QUICK ✓ *Find the solution of the linear inequality and express the solution set using set-builder notation and interval notation, if possible. Graph the solution set.*

23. $-2x + 7(x - 5) \leq 6x + 32$ **24.** $-x + 7 - 8x \geq 2(8 - 5x) + x$

25. $\dfrac{3}{2}x + 5 - \dfrac{5}{2}x < 4x - 3(x + 1)$ **26.** $0.8x + 3.2(x + 4) \geq 2x + 12.8 + 3x - x$

③ **Solve Problems Involving Linear Inequalities**

When you are confronted with a word problem, one of the first things that you need to do is look for key words that tip you off as to the type of word problem that it is. There are certain phrases that frequently occur in problems that lead to linear inequalities.

We list some of these phrases for you in Table 8.

Table 8	
Phrase	**Inequality**
At least	$\geq$
No less than	$\geq$
More than	$>$
Greater than	$>$
No more than	$\leq$
At most	$\leq$
Fewer than	$<$
Less than	$<$

When solving applications involving linear inequalities, we use the same steps for setting up applied problems that we introduced in Section 2.5 on pages 122–123.

EXAMPLE 13 Comparing Credit Cards

BankOne has offered you two different credit card options. The Southwest rewards card charges an annual fee of $39 plus 12.90% simple interest on all outstanding balances. The Marriott rewards card charges an annual fee of $30 plus 14.15% simple interest on all outstanding balances. What annual balance results in the Southwest card costing less than the Marriott card? (*Source:* BankOne.com)

Solution

Step 1: Identify We want to know the credit card balance for which the Southwest credit card costs less than the Marriott rewards card. The phrase "costs less" implies that this is an inequality problem.

Step 2: Name Let b represent the credit card balance.

Step 3: Translate Each card charges an annual fee plus simple interest. So, for each card the cost will be "annual fee + interest."

Recall from Section 2.4 that simple interest is found using the formula

$$I = Prt$$

where I is the interest charged, P is the balance on the credit card, r is the annual interest rate, and t is time.

In this problem, we let b represent the outstanding balance. The annual interest rate r will either be 0.129 (for Southwest) or 0.1415 for (Marriott). Because we are discussing annual cost, we have that $t = 1$.

Since we want to know what balance results in Southwest costing less than Marriott, we have the following inequality:

Southwest Credit Card Cost $<$ Marriott Credit Card Cost

Annual Fee for Southwest		Interest Charged by Southwest		Annual Fee for Marriott		Interest Charged by Marriott	
39	+	0.129b	$<$	30	+	0.1415b	The Model

Step 4: **Solve** Solve the inequality for b.

$$39 + 0.129b < 30 + 0.1415b$$

Subtract 39 from both sides: $$0.129b < -9 + 0.1415b$$

Subtract $0.1415b$ from both sides: $$-0.0125b < -9$$

Divide both sides by -0.0125.

Don't forget to reverse the
inequality symbol: $$b > 720$$

Step 5: **Check** If the balance is $720, then the annual cost for Southwest is $39 + 0.129(720) = \$131.88$. The annual cost for Marriott is $30 + 0.1415(720) = \$131.88$. For a balance greater than $720, say $750, the annual cost for Southwest is $39 + 0.129(750) = \$135.75$. The annual cost for Marriott is $30 + 0.1415(750) = \$136.13$.

Step 6: **Answer the Question** If the annual balance is greater than $720, then Southwest offers a better deal than Marriott.

QUICK ✔

27. You have just received two credit card applications in the mail. The card from Bank A has an annual fee of $25 and charges 9.9% simple interest. The card from Bank B has no annual fee, but charges 14.9% simple interest. For what annual balance will the card from Bank A cost less than the card from Bank B?

28. Suppose the daily revenue from selling x boxes of candy is given by the equation $R = 12x$. The daily cost of operating the store and making the candy is given by the equation $C = 8x + 96$. For how many boxes of candy will revenue exceed costs? That is, solve $R > C$.

2.8 Exercises

For Extra Help:

Student Solutions Manual   CD Video   PH Math/Tutor Center   MathXL Tutorials on CD   MathXL®   MyMathLab

Concepts and Vocabulary

In Problems 1–3, fill in the blanks.

1. A(n) _____ _____, denoted $[a, b]$, consists of all real numbers x for which $a \leq x \leq b$.

2. The _____ _____ states that the direction, or sense, of an inequality remains the same if each side is multiplied by a positive number, while the direction is reversed is each side is multiplied by a negative number.

3. In the interval $[a, b]$, a is called the _____ _____ and b is called the _____ _____ of the interval.

In Problems 4–6, answer True or False to each statement. In each statement, assume that $a < b$ and $c < 0$.

4. $a \pm c < b \pm c$ 5. $ac > bc$ 6. $\dfrac{a}{c} < \dfrac{b}{c}$

7. Explain why it is incorrect to write $x \geq 4$ in interval notation as $[4, \infty]$.

8. Explain the circumstances in which the direction of the inequality symbol is reversed when solving a linear inequality.

Building Skills

*In Problems 9–18, write the given statement using inequality symbols. Let **x** represent the unknown quantity.*

9. Karen's salary this year will be at least $16,000.

10. Bob's salary this year will be at most $120,000.

11. There will be at least 15,000 and at most 20,000 fans at the Cleveland Indians game today.

12. The cost of a new lawnmower is at least $250 and no more than $500.

13. The cost to remodel a kitchen is more than $12,000.

14. There are fewer than 25 students in your math class on any given day.

15. x is a positive number

16. x is a nonnegative number

17. x is a nonpositive number

18. x is a negative number

In Problems 19–32, graph each inequality on a number line, and write each inequality in interval notation.

19. $x > 2$

20. $n > 5$

21. $x \leq -1$

22. $x \leq 6$

23. $z \geq -3$

24. $x \geq -2$

25. $x < 4$

26. $y < -3$

27. $-2 \leq x \leq 5$

28. $-5 \leq x \leq 0$

29. $3 < x < 10$

30. $1 < z < 8$

31. $2 \leq y < 4$

32. $-8 \leq x \leq 3$

In Problems 33–42, use interval notation to express the inequality shown in each graph.

33.

34.

35.

36.

37.

38.

39.

40.

41.

42.

In Problems 43–60, solve the inequality and express the solution set in set-builder notation or interval notation. Graph the solution set on the real number line.

43. $x + 1 < 5$

44. $x + 4 \leq 3$

45. $3x \leq 15$

46. $4x > 12$

47. $-5x < 35$

48. $-7x \geq 28$

49. $3x - 7 > 2$

50. $2x + 5 > 1$

51. $3x - 1 \geq 3 + x$

52. $2x - 2 \geq 3 + x$

53. $1 - 2x \leq 3$

54. $2 - 3x \leq 5$

55. $-2(x + 3) < 8$

56. $-3(1 - x) > x + 8$

57. $4 - 3(1 - x) \leq 3$

58. $8 - 4(2 - x) \leq -2x$

59. $\frac{1}{2}(x - 4) > x + 8$

60. $3x + 4 > \frac{1}{3}(x - 2)$

In Problems 61–68, solve the inequality and express the solution set in set-builder notation or interval notation, if possible. Graph the solution set on the real number line.

61. $4(x - 1) > 3(x - 1) + x$

62. $2y - 5 + y < 3(y - 2)$

63. $5(n + 2) - 2n \leq 3(n + 4)$

64. $3(p + 1) - p \geq 2(p + 1)$

65. $2n - 3(n - 2) < n - 4$

66. $4x - 5(x + 1) \leq x - 3$

67. $3y - (5y + 2) > 4(y + 1) - 2y$

68. $8x - 3(x - 2) \geq x + 4(x + 1)$

Mixed Practice

In Problems 69–84, solve the inequality and express the solution set in set-builder notation or interval notation, if possible. Graph the solution set on the real number line.

69. $-1 < x - 5$

70. $6 \geq x + 15$

71. $-\frac{3}{4}x > -\frac{9}{16}$

72. $-\frac{5}{8}x > \frac{25}{48}$

73. $3(x + 1) > 2(x + 1) + x$

74. $5(x - 2) < 3(x + 1) + 2x$

75. $-4a + 1 > 9 + 3(2a + 1) + a$

76. $-5b + 2(b - 1) \leq 6 - 5b - 1$

77. $n + 3(2n + 3) > 7(n + 2)$

78. $2k - (k - 4) \geq 3k + 10 - 2k$

79. $\frac{x}{2} \geq 1 - \frac{x}{4}$

80. $\frac{x}{3} \geq 2 + \frac{x}{6}$

81. $\frac{x + 5}{2} + 4 > \frac{2x + 1}{3} + 2$

82. $\frac{3z - 1}{4} + 1 \leq \frac{6z + 5}{2} + 2$

83. $1.3x + 3.1 < 4.5x - 15.9$

84. $4.9 + 2.6x < 4.2x - 4.7$

Applying the Concepts

85. Auto Rental A car can be rented from Certified Auto Rental for $55 per week plus $0.18 per mile. How many miles can be driven if you have at most $280 to spend for weekly transportation?

86. Truck Rental A truck can be rented from Acme Truck Rental for $80 per week plus $0.28 per mile. How many miles can be driven if you have at most $100 to spend on truck rental?

87. Final Grade Yvette has scores of 72, 78, 66, and 81 on her algebra tests. Find the minimum score she can make on the final exam in order to pass the course with at least 360 points. The final exam counts as two test grades.

88. Final Grade To earn an A in Mrs. Smith's elementary statistics class, Elizabeth must earn at least 540 points. Thus far, Elizabeth has earned scores of 85, 83, 90, and 96. The final exam counts as two test grades. How many points does Elizabeth have to score on the final exam to earn an A?

89. Calling Plan Imperial Telephone has a long-distance calling plan that has a monthly fee of $10 and a charge of $0.03 per minute used. Mayflower Communications has a monthly fee of $6 and a fee of $0.04 per minute. For how many minutes is Imperial Telephone the cheaper plan?

90. **Commission** A recent college graduate had an offer of a sales position that pays $15,000 per year plus 1% of all sales.

    (a) Write an expression for the total annual salary based on sales of S dollars.
    (b) For what total sales amount will the college graduate earn in excess of $150,000 annually?

91. **Borrowing Money** The amount of money that a lending institution will allow you to borrow mainly depends on the interest rate and your annual income. The equation $L = 2.98I - 76.11$ describes the amount of money, L, that a bank will lend at an interest rate of 7.5% for 30 years, based upon annual income, I. For what annual income, I, will a bank lend at least $150,000? (*Source: Information Please Almanac*)

92. **Advertising** A marketing firm found that the equation $S = 2.1A + 224$ describes the amount of sales S of a product depends on A, the amount spent on advertising the product. Both S and A are measured in thousands of dollars. For what amount, A, is the sales of a product at least 350 thousand dollars?

Extending the Concepts

93. **Grades** In your Economics 101 class, you have scores of 68, 82, 87, and 89 on the first four of five tests. To earn a grade of B or higher, the average of the first five test scores must be greater than or equal to 80. Find the minimum score that you need on the last test to earn a B.

94. **Delivery Service** A messenger service charges $10 to make a delivery to an address. In addition, each letter delivered costs $3 and each package delivered costs $8. If there are 15 more letters than packages delivered to this address, what is the maximum number of items that can be delivered for $85?

CHAPTER 2 ACTIVITY: Pass to the Right

Focus: Solving linear equations and inequalities as a group.

Time: 20–30 minutes

Group size: 3–4

1. Each member of the group should choose one of the following four equations and write it down on a piece of paper. Do not begin to solve.

    (a) $\dfrac{x + 2}{2} - \dfrac{5x - 12}{6} = 1$ (b) $-\dfrac{1}{9}(x + 27) + \dfrac{1}{3}(x + 3) = x + 6$

    (c) $\dfrac{x + 3}{3} - \dfrac{2x - 12}{9} = 1$ (d) $-\dfrac{1}{2}(x + 6) + \dfrac{1}{7}(x + 7) = x + 3$

2. Each member of the group should pass their equation to the person on their right. This member should perform the first step in solving the equation. When you are done, pass your paper to the next group member on your right.

3. Upon receipt of the equation, each group member should check the previous member's work and then perform the next step. If an error is found, discuss and correct the error.

4. Continue passing the problems until all equations have been solved.

5. As a group, discuss the results.

6. If time permits, repeat this activity except choose one of the following inequalities. Express your final answer using interval notation.

    (a) $\dfrac{1}{4}(2x + 12) > \dfrac{3}{8}(x - 1)$ (b) $\dfrac{3x + 1}{10} - \dfrac{1 + 6x}{5} \le -\dfrac{1}{2}$

    (c) $\dfrac{1}{2}(2x + 14) > \dfrac{3}{4}(x - 1)$ (d) $\dfrac{3x + 1}{21} - \dfrac{1 + 4x}{7} \le -\dfrac{1}{3}$

CHAPTER 2 REVIEW

Section 2.1	Linear Equations: The Addition and Multiplication Properties of Equality

KEY CONCEPTS	KEY TERMS
• **Linear Equation in One Variable** An equation equivalent to one of the form $ax + b = c$, where a, b, and c are real numbers and $a \neq 0$. • **Addition Property of Equality** For real numbers a, b, and c, if $a = b$, then $a + c = b + c$. • **Multiplication Property of Equality** For real numbers a, b, and c where $c \neq 0$, if $a = b$, then $ac = bc$.	Linear equation Sides Left Side Right Side Solution Solution set Satisfies Equivalent equations

YOU SHOULD BE ABLE TO . . .	EXAMPLE	REVIEW EXERCISES
(1) Determine if a number is a solution of an equation (p. 75)	Example 1	1–4
(2) Use the Addition Property of Equality to solve linear equations (p. 76)	Examples 2 through 4	5–10, 15, 16, 19
(3) Use the Multiplication Property of Equality to solve linear equations (p. 78)	Examples 5 through 8	11–14, 17, 18, 20

In Problems 1–4, determine if the given value is the solution to the equation. Answer Yes or No.

1. $3x + 2 = 7$; $x = 5$

2. $5m - 1 = 17$; $m = 4$

3. $6x + 6 = 12$; $x = \dfrac{1}{2}$

4. $9k + 3 = 9$; $k = \dfrac{2}{3}$

In Problems 5–18, solve the equation. Check your solution.

5. $n - 6 = 10$

6. $n - 8 = 12$

7. $x + 6 = -10$

8. $x + 2 = -5$

9. $-100 = m - 5$

10. $-26 = m - 76$

11. $\dfrac{2}{3}y = 16$

12. $\dfrac{1}{4}x = 20$

13. $-6x = 36$

14. $-4x = -20$

15. $z + \dfrac{5}{6} = \dfrac{1}{2}$

16. $m - \dfrac{1}{8} = \dfrac{1}{4}$

17. $1.6x = 6.4$

18. $1.8m = 9$

19. Discount The cost of a new Honda Accord has been discounted by \$1200, so that the sale price is \$18,900. Find the original price of the Honda Accord, p, by solving the equation $p - 1200 = 18,900$.

20. Coffee While studying for an exam, Randi drank coffee at a local coffee house. She bought 3 cups of coffee for a total of \$7.65. Find the cost of each cup of coffee, c, by solving the equation $3c = 7.65$.

Section 2.2   Linear Equations: Using the Properties Together

KEY CONCEPT	KEY TERM
• Steps for solving an equation in one variable page 88.	Distributive Property

YOU SHOULD BE ABLE TO . . .	EXAMPLE	REVIEW EXERCISES
① Apply the Addition and Multiplication Properties of Equality to solve linear equations (p. 84)	Examples 1 and 2	21–24
② Combine like terms and apply the Distributive Property to solve linear equations (p. 86)	Examples 3 and 4	25–30
③ Solve a linear equation with the variable on both sides of the equation (p. 87)	Examples 5 and 6	31–34
④ Use linear equations to solve problems (p. 89)	Example 7	35, 36

In Problems 21–34, solve the equation. Check your solution.

21. $5x - 1 = -21$

22. $-3x + 7 = -5$

23. $\frac{2}{3}x + 5 = 11$

24. $\frac{5}{7}x - 2 = -17$

25. $-2x + 5 + 6x = -11$

26. $3x - 5x + 6 = 18$

27. $2m + 0.5m = 10$

28. $1.4m + m = -12$

29. $-2(x + 5) = -22$

30. $3(2x + 5) = -21$

31. $5x + 4 = -7x + 20$

32. $-3x + 5 = x - 15$

33. $4(x - 5) = -3x + 5x - 16$

34. $4(m + 1) = m + 5m - 10$

35. Ages Skye is 4 years older than Beth. The sum of their ages is 24. Find Skye's age by solving the equation $x + x + 4 = 24$, where x represents Beth's age and $x + 4$ represents Skye's age.

36. Parking Lot The length of a rectangular parking lot is 10 yards longer than the width, w, and the perimeter of the parking lot is 96 yards. Solve the equation $2w + 2(w + 10) = 96$ to find the width, w, of the parking lot. Then find the length of the parking lot, $w + 10$.

Section 2.3   Solving Linear Equations Involving Fractions and Decimals; Classifying Equations

KEY CONCEPTS	KEY TERMS
• A **conditional equation** is an equation that is true for some values of the variable and false for other values of the variable. • An equation that is false for every value of the variable is called a **contradiction.** • An equation that is satisfied for every choice of the variable for which both sides of the equation are defined is called an **identity.**	Least common denominator (LCD) Identity Conditional equation Contradiction

YOU SHOULD BE ABLE TO . . .	EXAMPLE	REVIEW EXERCISES
1 Use the least common denominator to solve a linear equation containing fractions (p. 93)	Examples 1 and 2	37–40, 45, 46
2 Solve a linear equation containing decimals (p. 95)	Examples 3 through 5	41–44, 47, 48
3 Classify linear equations as identity, conditional, or contradiction (p. 97)	Examples 6 through 8	49–54
4 Use linear equations to solve problems (p. 99)	Example 9	55, 56

In Problems 37–48, solve the equation. Check your solution.

37. $\dfrac{6}{7}x + 3 = \dfrac{1}{2}$

38. $\dfrac{1}{4}x + 6 = \dfrac{5}{6}$

39. $\dfrac{n}{2} + \dfrac{2}{3} = \dfrac{n}{6}$

40. $\dfrac{m}{8} + \dfrac{m}{2} = \dfrac{3}{4}$

41. $1.2r = -1 + 2.8$

42. $0.2x + 0.5x = 2.1$

43. $1.2m - 3.2 = 0.8m - 1.6$

44. $0.3m + 0.8 = 0.5m + 1$

45. $\dfrac{1}{2}(x + 5) = \dfrac{3}{4}$

46. $-\dfrac{1}{6}(x - 1) = \dfrac{2}{3}$

47. $0.1(x + 80) = -0.2 + 14$

48. $0.35(x + 6) = 0.45(x + 7)$

In Problems 49–54, determine if the equation is a contradiction, an identity, or a conditional equation. State the solution of the equation.

49. $4x + 2x - 10 = 6x + 5$

50. $-2(x + 5) = -5x + 3x + 2$

51. $-5(2n + 10) = 6n - 50$

52. $8m + 10 = -2(7m - 5)$

53. $10x - 2x + 18 = 2(4x + 9)$

54. $-3(2x - 8) = -3x - 3x + 24$

55. T-Shirt Al purchased a tie-dyed T-shirt at a "20% off" sale for $12.60. Solve the equation $p - 0.20p = 12.60$ to find the original price, p, of the shirt.

56. Couch Cushions Juanita was cleaning under her couch cushions and found some nickels and dimes. She found $0.55 in change, with the number of nickels one less than twice the number of dimes. Solve the equation $0.10x + 0.05(2x - 1) = 0.55$ to find x, the number of dimes Juanita found.

Section 2.4	**Evaluating Formulas and Solving Formulas for a Variable**

KEY CONCEPTS	KEY TERMS
• **Simple interest formula** $I = Prt$, I represents the amount of interest, P is the principal, r is the rate of interest, and t is time (expressed in years). • **Geometry Formulas (page 106)**	Formula Interest Principal Rate of interest

YOU SHOULD BE ABLE TO . . .	EXAMPLE	REVIEW EXERCISES
1 Evaluate a formula (p. 103)	Examples 1 through 6	57–60, 69, 70
2 Solve a formula for a variable (p. 108)	Examples 7 through 9	61–68

In Problems 57–60, substitute the given values into the formula and then simplify to find the unknown quantity. Label units in the answer.

57. area of a rectangle: $A = lw$; Find A when $l = 8$ inches and $w = 6$ inches.

58. perimeter of a square: $P = 4s$; Find P when $s = 16$ cm.

59. perimeter of a rectangle: $P = 2l + 2w$; Find w when $P = 16$ yards and $l = \dfrac{13}{2}$ yards.

60. circumference of a circle: $C = \pi d$; Find C when $d = \dfrac{15}{\pi}$ mm.

In Problems 61–66, solve each formula for the stated variable.

61. $V = LWH$; solve for H.

62. $I = Prt$; solve for P.

63. $S = 2LW + 2LH + 2WH$; solve for W.

64. $\rho = mv + MV$; solve for M.

65. $2x + 3y = 10$; solve for y.

66. $6x - 7y = 14$; solve for x.

67. Finance The formula $A = P(1 + r)^t$ can be used to find the future value A of a deposit of P dollars in an account that earns an annual interest rate r (expressed as a decimal) after t years.

   (a) Solve the formula for P.

   (b) How much would you have to deposit today in order to have $3000 in 6 years in a bank account that pays 5% annual interest? Round your answer to the nearest penny.

68. Cylinders The surface area A of a right circular cylinder is given by the formula $A = 2\pi rh + 2\pi r^2$, where r is the radius and h is the height.

   (a) Solve the formula for h.

   (b) Determine the height of a right circular cylinder whose surface area is 72π square centimeters and whose radius is 4 centimeters.

69. Christmas Bonus Samuel invested his $500 Christmas bonus in a 9-month Certificate of Deposit that earns 3% simple interest. Use the formula $I = Prt$ to find the amount of interest Samuel's investment will earn.

70. Coffee Table Find the area of a circular coffee table top with a diameter of 3 feet.

Section 2.5	Introduction to Problem Solving: Direct Translation Problems

KEY CONCEPTS	KEY TERMS
• **Six steps for solving problems with mathematical models** **Identify** what you are looking for **Name** the unknown(s) **Translate** to a mathematical equation **Solve** the equation **Check** the answer **Answer** the question in a complete sentence	Problem solving Mathematical modeling Modeling process Mathematical model Model Direct translation

YOU SHOULD BE ABLE TO...	EXAMPLE	REVIEW EXERCISES
① Translate English phrases to algebraic expressions (p. 117)	Examples 1 through 3	71–76; 83–86
② Translate English sentences to equations (p. 120)	Example 4	77–82
③ Build models for solving direct translation problems (p. 123)	Examples 5 through 9	87–90

In Problems 71–76, translate each phrase to an algebraic expression. Let x represent the unknown number.

71. the difference between a number and 6

72. eight subtracted from a number

73. the product of −8 and a number

74. the quotient of a number and 10

75. twice the sum of 6 and a number

76. four times the difference of 5 and a number

In Problems 77–82, translate each statement into an equation. Let x represent the unknown number. DO NOT SOLVE.

77. The sum of 6 and a number is equal to twice the number increased by 5.

78. The product of 6 and a number decreased by 10 is one more than double the number.

79. Eight less than a number is the same as half of the number.

80. The ratio of 6 to a number is the same as the number added to 10.

81. Four times the sum of twice a number and 8 is 16.

82. Five times the difference between double a number and 8 is −24.

In Problems 83–86, choose a variable to represent one quantity. State what the quantity represents, and then express the second quantity in terms of the first.

83. Jacob is seven years older than Sarah.

84. José runs twice as fast as Consuelo.

85. Irene has $6 less than Max.

86. Victor and Larry will share $350.

87. **Losing Weight** Over the past year Lee Lai lost 28 pounds. Find Lee Lai's weight one year ago if her current weight is 125 pounds.

88. **Consecutive Integers** The sum of three consecutive integers is 39. Find the integers.

89. **Finance** A total of $20,000 is to be divided between Roberto and Juan, with Roberto to receive $2000 less than Juan. How much will each receive?

90. **Truck Rentals** You need to rent a moving truck. You identified two companies that rent trucks. ABC-Rental charges $30 per day plus $0.15 per mile. U-Do-It Rental charges $15 per day plus $0.30 per mile. For how many miles will the cost of renting be the same?

Section 2.6	Problem Solving: Direct Translation Problems Involving Percent
KEY CONCEPTS	

- Percent means divided by 100 or per hundred.
- Total Cost = Original Cost + Sales Tax
- Original Price − Discount = Sale Price
- Wholesale Price + Markup = Selling Price

YOU SHOULD BE ABLE TO . . .	EXAMPLE	REVIEW EXERCISES
1 Solve direct translation problems involving percent (p. 131)	Examples 1 through 4	91–94
2 Model and solve direct translation problems from business involving percent (p. 134)	Examples 5 and 6	95–100

In Problems 91–94, find the unknown in each percent question.

91. What is 6.5% of 80?

92. 18 is 30% of what number?

93. 16 is what percent of 120?

94. 110% of what number is 55?

95. Sales Tax The sales tax in Florida is 6%. The total cost of purchasing a leotard, including tax, was $19.61. Find the cost of the leotard before sales tax.

96. Tutoring Mei Ling is a private tutor. She raised her hourly fee by 8.5% to cover her traveling expenses. Her new hourly rate is $32.55. Find Mei Ling's previous hourly fee.

97. Business: Discount Pricing A sweater, discounted by 70% for an end of the year clearance sale, has a price tag of $12. What was the sweater's original price?

98. Business: Mark Up A clothing store marks up the price that it pays for a suit 80%. If the selling price of a suit is $360, how much did the store pay for the suit?

99. Salary Tanya earns $500 each week plus 2% of the value of the computers she sells each week. If Tanya wishes to earn $3000 this week, what must be the total value of the computers she sells?

100. Voting In an election for school president, the loser received 80% of the winner's votes. If 900 votes were cast, how many did each receive?

Section 2.7	Problem Solving: Geometry and Uniform Motion	
KEY CONCEPTS		**KEY TERMS**
• Complementary angles are two angles whose measures sum to 90°. • Supplementary angles are two angles whose measures sum to 180°. • The sum of the measures of the angles of a triangle is 180°. • **Uniform Motion** If an object moves at an average speed r, the distance d covered in time t is given by the formula $d = rt$.		Complementary angles Supplementary angles Uniform motion

YOU SHOULD BE ABLE TO . . .	EXAMPLE	REVIEW EXERCISES
① Set up and solve complementary and supplementary angle problems (p. 139)	Example 1	101, 102
② Set up and solve angles of a triangle problems (p. 140)	Example 2	103, 104
③ Use geometry formulas to solve problems (p. 141)	Examples 3 and 4	105–108
④ Set up and solve uniform motion problems (p. 143)	Examples 5 and 6	109, 110

△ **101. Complementary Angles** Find the measure of two complementary angles such that the measure of the first angle is 20° more than six times the measure of the second.

△ **102. Supplementary Angles** Find the measure of two supplementary angles such that the measure of the first angle is 60° less than twice the measure of the second.

△ **103. Triangles** In a triangle, the measure of the second angle is twice the first. The measure of the third angle is 30° more than the second. Find the measure of each angle.

△ **104. Triangles** In a triangle, the measure of the second angle is 5° less than the first. The measure of the third angle is 5° less than twice the second. Find the measure of each angle.

△ **105. Rectangle** The length of a rectangle is 15 inches longer than twice its width. If the perimeter is 78 inches, find the length and width of the rectangle.

△ **106. Rectangle** The length of a rectangle is four times its width. If the perimeter is 70 cm, find the length and width of the rectangle.

△ **107. Garden** The perimeter of a rectangular garden is 120 feet. The width of the garden is twice the length.
  (a) Find the length and width of the garden.
  (b) What is the area of the garden?

△ **108. Back Yard** Yvonne's back yard is in the shape of a trapezoid with height of 80 feet. The shorter base is 10 feet shorter than the longer base, and the area of the back yard is 3600 square feet. Find the length of each base of the trapezoidal yard.

109. Boating Two motorboats leave the same dock at the same time traveling in the same direction. One boat travels at 18 miles per hour and the other travels at 25 miles per hour. In how many hours will the motorboats be 35 miles apart?

110. Train Station Two trains leave a station at the same time. One train is traveling east at 10 miles per hour faster than the other train, which is traveling west. After 6 hours, the two trains are 720 miles apart. At what speed did the faster train travel?

Section 2.8   Linear Inequalities

KEY CONCEPTS	KEY TERMS
	Solve an inequality

- **Linear Inequality in One Variable**
  An inequality of the form $ax + b < c, ax + b \le c, ax + b > c$, or $ax + b \ge c$, where a, b, and c are real numbers with $a \ne 0$.

- **Interval Notation and Set-Builder Notation**

Interval	Set-Builder Notation	Graph
The open interval (a, b)	$\{x \mid a < x < b\}$	
The closed interval $[a, b]$	$\{x \mid a \le x \le b\}$	
The half-open interval $[a, b)$	$\{x \mid a \le x < b\}$	
The half-open interval $(a, b]$	$\{x \mid a < x \le b\}$	
The interval $[a, \infty)$	$\{x \mid x \ge a\}$	
The interval (a, ∞)	$\{x \mid x > a\}$	
The interval $(-\infty, a]$	$\{x \mid x \le a\}$	
The interval $(-\infty, a)$	$\{x \mid x < a\}$	
The interval $(-\infty, \infty)$	$\{x \mid x \text{ is a real number}\}$	

KEY TERMS (continued):
Solutions
Solution set
Interval Notation
Closed interval
Open interval
Half-open or half-closed interval
Left endpoint
Right endpoint
Equivalent inequalities

- **Addition Property of Inequalities**
  For real numbers a, b, and c
  If $a < b$, then $a + c < b + c$.
  If $a > b$, then $a + b > b + c$.

(continued)

- **Multiplication Properties of Inequalities**

  For real numbers a, b, and c

  If $a < b$ and if $c > 0$, then $ac < bc$.

  If $a > b$ and if $c > 0$, then $ac > bc$.

  If $a < b$ and if $c < 0$, then $ac > bc$.

  If $a > b$ and if $c < 0$, then $ac < bc$.

YOU SHOULD BE ABLE TO . . .	EXAMPLE	REVIEW EXERCISES
1 Represent inequalities using set-builder notation, the real number line and interval notation (p. 150)	Examples 1 through 4	111–116
2 Solve linear inequalities (p. 153)	Examples 5 through 12	117–124
3 Solve problems involving linear inequalities (p. 159)	Example 13	125, 126

In Problems 111–116, write each inequality using interval notation and graph each inequality on a number line.

111. $x \le -3$ **112.** $x > 4$ **113.** $2 \le m \le 7$

114. $-1 < x < 5$ **115.** $0 < n \le 3$ **116.** $-3 \le n < 1$

In Problems 117–124, solve the inequality and express the solution in set-builder notation or interval notation if possible. Graph the solution set on the real number line.

117. $4x + 3 < 2x - 10$ **118.** $3x - 5 \ge -12$

119. $-4(x - 1) \le x + 8$ **120.** $6x - 10 < 7x + 2$

121. $-3(x + 7) > -x - 2x$ **122.** $4x + 10 \le 2(2x + 7)$

123. $\dfrac{1}{2}(3x - 1) > \dfrac{2}{3}(x + 3)$ **124.** $\dfrac{5}{4}x + 2 < \dfrac{5}{6}x - \dfrac{7}{6}$

125. Moving The Rent-A-Moving-Van Company charges a flat rate of $19.95 per day plus $0.20 for each mile driven. How many miles can be driven if a customer can afford to spend at most $32.95?

126. Bowling Travis is bowling three games in a tournament. In the first game, his score was 148. In the second game, his score was 155. What score must Travis get in the third game for his tournament average to be greater than 151?

CHAPTER 2 TEST

 Remember to use your Chapter Test Prep Video CD to see fully worked-out solutions to any of these problems you would like to review.

In Problems 1–8, solve the equation. Check your solution.

1. $x + 3 = -14$ **2.** $-\dfrac{2}{3}m = \dfrac{8}{27}$

3. $5(2x - 4) = 5x$ **4.** $-2(x - 5) = 5(-3x + 4)$

5. $-\dfrac{2}{3}x + \dfrac{3}{4} = \dfrac{1}{3}$ **6.** $-0.6 + 0.4y = 1.4$

7. $8x + 3(2 - x) = 5(x + 2)$ **8.** $2(x + 7) = 2x - 2 + 16$

In Problems 9 and 10, (a) solve for the indicated variable, and then (b) find the value of the unknown quantity. Label units in the answer.

9. Volume of a rectangular solid: $V = lwh$

   (a) Solve for l.
   (b) Find l when $V = 540$ in.^{3}, $w = 6$ in., and $h = 10$ in.

10. Equation of a line: $2x + 3y = 12$

   (a) Solve for y.
   (b) Find y when $x = 8$.

11. Translate the following statement into an equation: Six times the difference between a number and 8 is equal to 5 less than twice the number. DO NOT SOLVE.

12. 18 is 30% of a number. Find the number.

13. Consecutive Integers The sum of three consecutive integers is 48. Find the integers.

14. Trading Spaces On the show *Trading Spaces,* designer Vern Yip constructs a triangular art piece for a bedroom. The length of the longest side is two inches longer than the length of the middle side. The shortest side is 14 inches shorter than the middle side. If the perimeter of the art piece is 60 inches, what is the length of each side?

15. Buses Kimberly and Clay leave a concert hall at the same time traveling in buses going in opposite directions. Kimberly's bus travels at 40 mph and Clay's bus travels at 60 mph. In how many hours will Kimberly and Clay be 350 miles apart?

16. Construction A carpenter cuts an oak board 21 feet long into two pieces. The longer board is 1 foot longer than three times the shorter one. Find the length of each board.

17. New Backpack Sherry purchased a new backpack for her daughter. The backpack was on sale for 20% off the regular price. If Sherry paid $28.80 for the backpack without sales tax, what was the original price?

In Problems 18 and 19, solve the inequality and express the solution in set-builder notation or interval notation. Graph the solution set on the real number line.

18. $3(2x - 5) \leq x + 15$ **19.** $-6x - 4 < 2(x - 7)$

20. Cell Phone Danielle's cell phone plan has a $30 monthly fee and an extra charge of $0.35 a minute. For how many minutes can Danielle use her cell phone so that the monthly bill is at most $100?

Introduction to Graphing and Equations of Lines

Your electric bill arrives. How do electric companies compute the amount owed? They use linear equations! See Example 7 in Section 3.1.

OUTLINE

3.1 The Rectangular Coordinate System and Equations in Two Variables

3.2 Graphing Equations in Two Variables

3.3 Slope

3.4 Slope-Intercept Form of a Line

3.5 Point-Slope Form of a Line

3.6 Parallel and Perpendicular Lines

Putting the Concepts Together (Sections 3.1–3.6)

3.7 Variation

3.8 Linear Inequalities in Two Variables

Chapter 3 Activity: Graphing Practice
Chapter 3 Review
Chapter 3 Test
Cumulative Review Chapters 1–3

The Big Picture: Putting It Together

It is now time to switch gears. In Chapter 2, we solved linear equations involving one unknown. We are now going to focus our attention on linear equations and inequalities involving two unknowns.

When we dealt with a single unknown, we could represent solutions to equations or inequalities graphically using a real number line. In that case, we only needed one dimension to represent the solution. When we have two unknowns, we need to work in two dimensions. This is accomplished through the *rectangular coordinate system*.

3.1 The Rectangular Coordinate System and Equations in Two Variables

OBJECTIVES

1. Plot Points in the Rectangular Coordinate System
2. Determine If an Ordered Pair Satisfies an Equation
3. Create a Table of Values That Satisfies an Equation

Preparing for the Rectangular Coordinate System and Equations in Two Variables

Before getting started, take the following readiness quiz. If you get a problem wrong, go back to the section cited and review the material.

1. Plot the following points on the real number line: [Section 1.2, pp. 12–13]

$$4, -3, \frac{1}{2}, 5.5$$

2. Evaluate $3x + 5$ for (a) $x = 4$ (b) $x = -1$ [Section 1.7, pp. 57–58]
3. Evaluate $2x - 5y$ for (a) $x = 3, y = 2$ [Section 1.7, pp. 57–58]
   (b) $x = 1, y = -4$
4. Solve: $3x + 5 = 14$ [Section 2.2, pp. 84–86]
5. Solve: $5(x - 3) - 2x = 3x + 12$ [Section 2.2, pp. 86–89]

---

1 ## Plot Points in the Rectangular Coordinate System

Figure 1

We have all heard the saying, "A picture is worth a thousand words." Because pictures allow individuals to visualize ideas, they are typically more powerful than any other form of printed communication. Consider the picture shown in Figure 1, which shows the results of the Manhattan Project from the test conducted July 16, 1945. It illustrates the power of the atom in a way that words never could.

While the pictures that we use in mathematics might not deliver as powerful a message as the picture in Figure 1, they are powerful nonetheless. To draw pictures of mathematical relationships, we need a "canvas." The "canvas" that we use in this chapter is the *rectangular coordinate system.*

In Section 1.2, we learned how to plot points on the real number line. We locate a point on the real number line by assigning it a single real number, called the *coordinate of the point.* We can think of this as plotting in one dimension. In this chapter, we use the *rectangular coordinate system,* a system that allows us to plot points in two dimensions.

We begin by drawing two real number lines that intersect at right (90°) angles. One of the real number lines is drawn horizontally, while the other is drawn vertically. We call the horizontal real number line the **x-axis,** and the vertical real number line is called the **y-axis.** The point where the x-axis and y-axis intersect is called the **origin, O.** See Figure 2.

Figure 2
The rectangular coordinate system

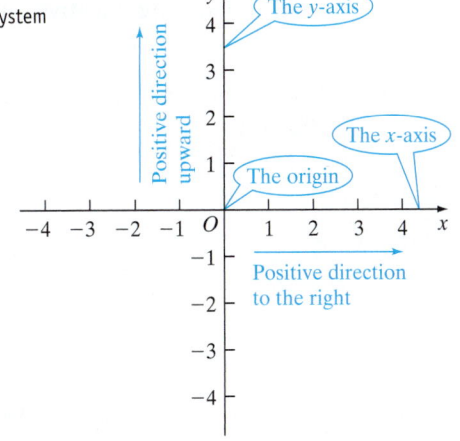

The origin O has a value of 0 on the x-axis and on the y-axis. Points on the x-axis to the right of O represent positive real numbers; points on the x-axis to the left of O represent negative real numbers. Points on the y-axis that are above O represent positive real numbers. Points on the y-axis that are below O represent negative real numbers.

Notice in Figure 2 that we label the *x*-axis "*x*" and the *y*-axis "*y*." An arrow is used at the end of each axis to denote the positive direction.

The coordinate system presented in Figure 2 is called a **rectangular** or **Cartesian coordinate system,** named after René Descartes (1596–1650), a French mathematician, philosopher, and theologian. The plane formed by the *x*-axis and *y*-axis is often referred to as the ***xy*-plane,** and the *x*-axis and *y*-axis are called the **coordinate axes.**

We can represent any point P in the rectangular coordinate system by using an **ordered pair (*x*, *y*)** of real numbers. We say that x represents the distance that P is from the *y*-axis. If $x > 0$ (that is, if x is positive), we travel x units to the right of the *y*-axis. If $x < 0$, we travel $|x|$ units to the left of the *y*-axis. We say that y represents the distance that P is from the *x*-axis. If $y > 0$, we travel y units above the *x*-axis. If $y < 0$, we travel $|y|$ units below the *x*-axis. The ordered pair (x, y) is also called the **coordinates** of P. For example, to plot the point $(2, 5)$, from the origin we would travel 2 units to the right along the *x*-axis and then 5 units up as shown in Figure 3(a).

Figure 3

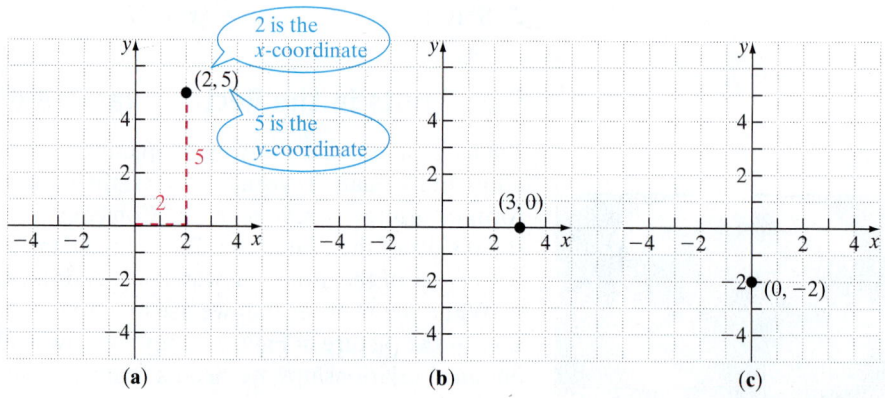

(a)   (b)   (c)

The origin O has coordinates $(0, 0)$. Any point on the *x*-axis has coordinates of the form $(x, 0)$, and any point on the *y*-axis has coordinates of the form $(0, y)$. For example, the point whose coordinates are $(3, 0)$ is 3 units to the right of the *y*-axis on the *x*-axis. See Figure 3(b). The point whose coordinates are $(0, -2)$ is 2 units below the *x*-axis on the *y*-axis. See Figure 3(c).

If (x, y) are the coordinates of a point P, then x is called the **x-coordinate** of P and y is called the **y-coordinate** of P.

If you look back at Figure 2, you should notice that the *x*- and *y*-axes divide the plane into four separate regions or **quadrants.** In quadrant I, both the *x*-coordinate and *y*-coordinate are positive. In quadrant II, *x* is negative and *y* is positive. In quadrant III, both *x* and *y* are negative. In quadrant IV, *x* is positive and *y* is negative. Points on the coordinate axes do not belong to a quadrant. See Figure 4.

Work Smart

Be careful: The order in which numbers appear in the ordered pairs matters. For example, (3, 2) represents a different point from (2, 3).

Figure 4

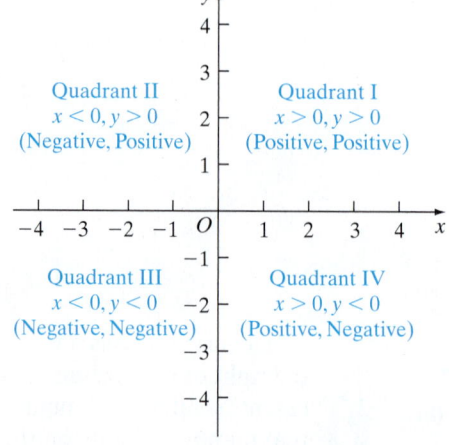

Let's summarize what we've learned so far.

SUMMARY: The Rectangular Coordinate System

- Composed of two real number lines—one horizontal (the x-axis) and one vertical (the y-axis). The x- and y-axes intersect at the origin.
- Also called the Cartesian coordinate system or xy-plane.
- Points in the rectangular coordinate system are denoted (x, y) and are called the coordinates of the point. We call x the x-coordinate and y the y-coordinate.
- If both x and y are positive, the point lies in quadrant I; if x is negative, but y is positive, the point lies in quadrant II; if x is negative and y is negative, the point lies in quadrant III; if x is positive and y is negative, the point lies in quadrant IV.
- Points on the x-axis have a y-coordinate of 0; points on the y-axis have an x-coordinate of 0.

EXAMPLE 1 Plotting Points in the Rectangular Coordinate System

Plot the following ordered pairs in the rectangular coordinate system. Tell which quadrant each point lies in or state that the point lies on the x- or y-axis.

(a) $A(3, 1)$ (b) $B(-4, 2)$ (c) $C(3, -5)$

(d) $D(4, 0)$ (e) $E(0, -3)$ (f) $F\left(\dfrac{5}{2}, -\dfrac{1}{2}\right)$

Solution

Before we can plot the points, we draw a rectangular or Cartesian coordinate system. See Figure 5(a). We now plot the points.

Figure 5

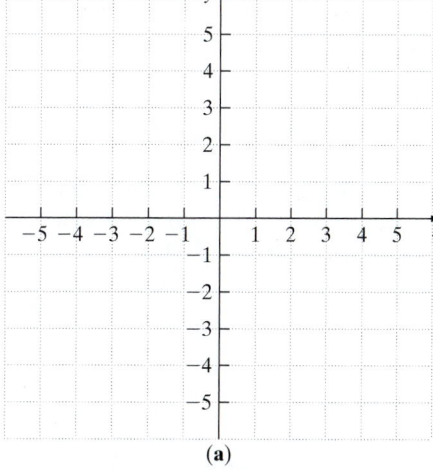

(a)

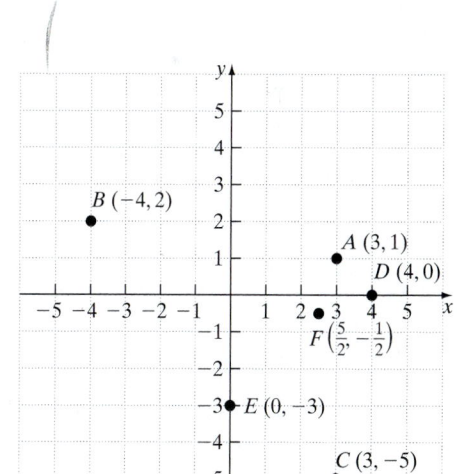

(b)

(a) To plot $A(3, 1)$, from the origin O, we travel 3 units to the right and then 1 unit up. Label the point A. Point A lies in quadrant I because both x and y are positive. See Figure 5(b).

(b) To plot $B(-4, 2)$, from the origin O, we travel 4 units to the left and then 2 units up. Label the point B. See Figure 5(b). Point B lies in quadrant II.

(c) See Figure 5(b). Point C lies in quadrant IV.

(d) See Figure 5(b). Point D does not lie in a quadrant because it lies on the x-axis.

(e) See Figure 5(b). Point E does not lie in a quadrant because it lies on the y-axis.

(f) It is helpful to convert the fractions to decimals, so $F\left(\frac{5}{2}, -\frac{1}{2}\right) = F(2.5, -0.5)$.

We see that the x-coordinate of the point is halfway between 2 and 3; the y-coordinate of the point is halfway between -1 and 0. See Figure 5(b). Point F lies in quadrant IV.

QUICK ✔ *Plot the ordered pairs in the rectangular coordinate system. Tell which quadrant each point lies in or state that the point lies on the x-axis or y-axis.*

1. (a) $(5, 2)$ **(b)** $(-4, -3)$ **(c)** $(1, -3)$ **(d)** $(-2, 0)$ **(e)** $(0, 6)$ **(f)** $\left(-\frac{3}{2}, \frac{5}{2}\right)$

2. (a) $(-6, 2)$ **(b)** $(1, 7)$ **(c)** $(-3, -2)$ **(d)** $(4, 0)$ **(e)** $(0, -1)$ **(f)** $\left(\frac{3}{2}, -\frac{7}{2}\right)$

EXAMPLE 2 **Identifying Points in the Rectangular Coordinate System**

Identify the coordinates of each point labeled in Figure 6.

Figure 6

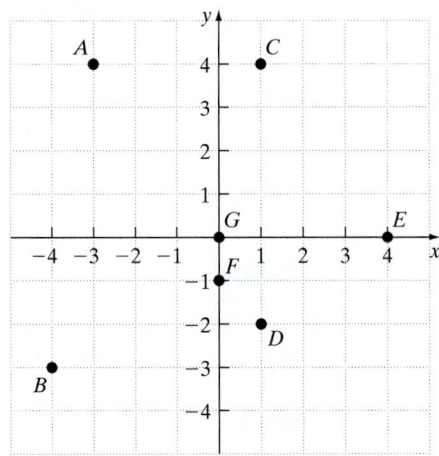

Solution

In an ordered pair (x, y), remember that x represents the position of the point left or right of the y-axis, while y represents the position of the point above or below the x-axis.

Since the point A is 3 units left of the y-axis, it has an x-coordinate of -3; since the point A is 4 units above the x-axis, it has a y-coordinate of 4. The ordered pair corresponding to the point A is $(-3, 4)$. We find the remaining coordinates in a similar fashion.

Point	Position	Ordered Pair
A	3 units left of the y-axis, 4 units above the x-axis	$(-3, 4)$
B	4 units left of the y-axis, 3 units below the x-axis	$(-4, -3)$
C	1 unit right of the y-axis, 4 units above the x-axis	$(1, 4)$
D	1 unit right of the y-axis, 2 units below the x-axis	$(1, -2)$
E	4 units right of the y-axis, on the x-axis	$(4, 0)$
F	On the y-axis, 1 unit below the x-axis	$(0, -1)$
G	On the x-axis; on the y-axis	$(0, 0)$

QUICK ✓ *Identify the coordinates of each point labeled in the figure below.*

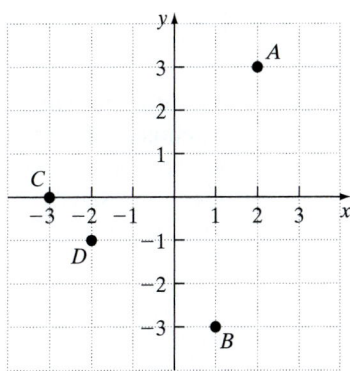

3. *A*   **4.** *B*   **5.** *C*   **6.** *D*

② Determine If an Ordered Pair Satisfies an Equation

In Sections 2.1–2.3, we solved linear equations in one variable. Recall that the solution of a linear equation is either a single value of the variable (conditional equation), the empty set (contradiction), or all real numbers (identity). Table 1 illustrates the three categories of linear equations in one variable.

Table 1		
Conditional Equation	**Contradiction**	**Identity**
$3x + 2 = 11$	$4(x - 2) - x = 2(x + 1) + x$	$-2(x + 3) + 4x = 2(x - 3)$
$3x = 9$	$4x - 8 - x = 2x + 2 + x$	$-2x - 6 + 4x = 2x - 6$
$x = 3$	$3x - 8 = 3x + 2$	$2x - 6 = 2x - 6$
	$-8 = 2$	$0 = 0$
	no solution	solution is all real numbers

We will now look at equations in two variables.

> **DEFINITION**
>
> An **equation in two variables,** x and y, is a statement in which the algebraic expressions involving x and y are equal. The expressions are called **sides** of the equation.

For example, the following are all equations in two variables.

$$x + 2y = 6 \qquad y = -3x + 7 \qquad y = x^2 + 3$$

Since an equation is a statement, it may be true or false, depending upon the values of the variables. Any values of the variable that make the equation a true statement are said to **satisfy** the equation.

The first equation $x + 2y = 6$ is satisfied when $x = 4$ and $y = 1$. Why? If we substitute 4 for x and 1 for y into the equation $x + 2y = 6$, we obtain a true statement.

$$x + 2y = 6$$

$$x = 4, y = 1: \quad 4 + 2(1) \overset{?}{=} 6$$

$$6 = 6 \quad \text{True}$$

Rather than saying the equation is satisfied when $x = 4$ and $y = 1$, we can say that the ordered pair $(4, 1)$ satisfies the equation. Does $x = 8$ and $y = -1$ satisfy the equation $x + 2y = 6$?

$$x + 2y = 6$$

$$x = 8, y = -1: \quad 8 + 2(-1) \stackrel{?}{=} 6$$

$$6 = 6 \quad \text{True}$$

So the ordered pair $(8, -1)$ satisfies the equation as well. In fact, there are infinitely many choices of x and y that satisfy the equation $x + 2y = 6$. However, there are some choices of x and y that do not satisfy the equation $x + 2y = 6$. For example, $x = 3$ and $y = 4$ do not satisfy the equation $x + 2y = 6$.

$$x + 2y = 6$$

$$x = 3, y = 4: \quad 3 + 2(4) \stackrel{?}{=} 6$$

$$11 = 6 \quad \text{False}$$

EXAMPLE 3 **Determining Whether an Ordered Pair Satisfies an Equation**

Determine if the following ordered pairs satisfy the equation $x + 2y = 8$.

(a) $(2, 3)$ **(b)** $(6, -1)$ **(c)** $(-2, 5)$

Solution

(a) For the ordered pair $(2, 3)$, we check to see if $x = 2$, $y = 3$ satisfies the equation $x + 2y = 8$.

$$x + 2y = 8$$

$$\text{Let } x = 2, y = 3: \quad 2 + 2(3) \stackrel{?}{=} 8$$

$$2 + 6 \stackrel{?}{=} 8$$

$$8 = 8 \quad \text{True}$$

The statement is true, so $(2, 3)$ satisfies the equation $x + 2y = 8$.

(b) For the ordered pair $(6, -1)$, we have

$$x + 2y = 8$$

$$\text{Let } x = 6, y = -1: \quad 6 + 2(-1) \stackrel{?}{=} 8$$

$$6 - 2 \stackrel{?}{=} 8$$

$$4 = 8 \quad \text{False}$$

The statement $4 = 8$ is false, so $(6, -1)$ does not satisfy the equation $x + 2y = 8$.

(c) For the ordered pair $(-2, 5)$, we have

$$x + 2y = 8$$

$$\text{Let } x = -2, y = 5: \quad -2 + 2(5) \stackrel{?}{=} 8$$

$$-2 + 10 \stackrel{?}{=} 8$$

$$8 = 8 \quad \text{True}$$

The statement is true, so $(-2, 5)$ satisfies the equation $x + 2y = 8$.

QUICK ✓

7. Determine if the following ordered pairs satisfy the equation $x + 4y = 12$.

(a) $(4, 2)$ **(b)** $(-2, 4)$ **(c)** $(1, 8)$

8. Determine if the following ordered pairs satisfy the equation $y = 4x + 3$.

(a) $(1, 3)$ **(b)** $(-2, -5)$ **(c)** $\left(-\dfrac{3}{2}, -3\right)$

(3) Create a Table of Values That Satisfies an Equation

In Example 3 we learned how to determine whether a given ordered pair satisfies an equation. However, we do not yet know how to find an ordered pair that satisfies an equation.

EXAMPLE 4   **How to Determine an Ordered Pair That Satisfies an Equation**

Find an ordered pair that satisfies the equation $3x + y = 5$.

Step-by-Step Solution

Step 1: Choose any value for one of the variables in the equation.	You may choose any value of x or y that you wish. In this example, we will let $x = 2$.

Step 2: Substitute the value of the variable chosen in Step 1 into the equation and then use the techniques learned in Chapter 2 to solve for the remaining variable.	Substitute 2 for x in the equation $3x + y = 5$ and then solve for y.

$$3x + y = 5$$

$$\text{Let } x = 2: \quad 3(2) + y = 5$$

$$\text{Simplify:} \quad 6 + y = 5$$

$$\text{Subtract 6 from both sides:} \quad 6 - 6 + y = 5 - 6$$

$$y = -1$$

The ordered pair that satisfies the equation is $(2, -1)$.

QUICK ✓ *Determine an ordered pair that satisfies the given equation by substituting the given value of the variable into the equation.*

9. $2x + y = 10; x = 3$ **10.** $-3x + 2y = 11; y = 1$

Look again at Example 3. Did you notice that two different ordered pairs satisfy the equation? In fact, there are an infinite number of ordered pairs that satisfy the equation $x + 2y = 8$ because for any real number y we can find a value of x that makes the equation a true statement. One approach to find some of the solutions of an equation in two variables is to create a table of values that satisfy the equation. The table is created by choosing values of x and using the equation to find the value of y (as we did in Example 4) or choosing a value of y and using the equation to find the value of x.

EXAMPLE 5   **Creating a Table of Values That Satisfies an Equation**

Use the equation $y = -2x + 5$ to complete Table 2 and list the ordered pairs that satisfy the equation.

Table 2		
x	**y**	**(x, y)**
−2		
0		
1		

Solution

The first entry in the table is $x = -2$. We substitute -2 for x and use the equation $y = -2x + 5$ to find y.

$$y = -2x + 5$$
$$x = -2: \quad y = -2(-2) + 5$$
$$y = 4 + 5$$
$$y = 9$$

Now substitute 0 for x in the equation $y = -2x + 5$.

$$y = -2x + 5$$
$$x = 0: \quad y = -2(0) + 5$$
$$y = 0 + 5$$
$$y = 5$$

Finally, substitute 1 for x in the equation $y = -2x + 5$.

$$y = -2x + 5$$
$$x = 1: \quad y = -2(1) + 5$$
$$y = -2 + 5$$
$$y = 3$$

The completed table is shown as Table 3.

Table 3		
x	**y**	**(x, y)**
-2	9	$(-2, 9)$
0	5	$(0, 5)$
1	3	$(1, 3)$

The ordered pairs that satisfy the equation are $(-2, 9)$, $(0, 5)$, and $(1, 3)$.

QUICK ✔ *Use the equation to complete the table. Use the table to list some of the ordered pairs that satisfy the equation.*

11. $y = 5x - 2$

x	**y**	**(x, y)**
-2		
0		
1		

12. $y = -3x + 4$

x	**y**	**(x, y)**
-1		
2		
5		

EXAMPLE 6 **Creating a Table of Values That Satisfies an Equation**

Use the equation $2x - 3y = 12$ to complete Table 4 and list the ordered pairs that satisfy the equation.

Table 4		
x	**y**	**(x, y)**
-3		
	-4	
6		

Solution

The first entry in the table is $x = -3$. We substitute -3 for x and use the equation $2x - 3y = 12$ to find y.

$$2x - 3y = 12$$

$x = -3:$ $\quad 2(-3) - 3y = 12$

$$-6 - 3y = 12$$

Add 6 to both sides: $\quad -3y = 18$

Divide both sides by -3: $\quad y = -6$

Substitute -4 for y in the equation $2x - 3y = 12$.

$$2x - 3y = 12$$

$y = -4:$ $\quad 2x - 3(-4) = 12$

$$2x + 12 = 12$$

Subtract 12 from both sides: $\quad 2x = 0$

Divide both sides by 2: $\quad x = 0$

Substitute 6 for x in the equation $2x - 3y = 12$.

$$2x - 3y = 12$$

$x = 6:$ $\quad 2(6) - 3y = 12$

$$12 - 3y = 12$$

Subtract 12 from both sides: $\quad -3y = 0$

Divide both sides by -3: $\quad y = 0$

The completed table is shown in Table 5.

Table 5		
x	**y**	**(x, y)**
-3	-6	$(-3, -6)$
0	-4	$(0, -4)$
6	0	$(6, 0)$

The ordered pairs that satisfy the equation are $(-3, -6)$, $(0, -4)$, and $(6, 0)$. ■

QUICK ✓ *Use the equation to complete the table. Use the table to list some of the ordered pairs that satisfy the equation.*

13. $2x + y = -8$

x	y	(x, y)
-5		
	-4	
2		

14. $2x - 5y = 18$

x	y	(x, y)
-6		
	-4	
2		

When we are working with equations that model a situation, we often do not use the variables x and y. Instead, we use variables that remind us of what they represent. For example, we might use the equation $C = 1.20m + 3$ to represent the cost of taking a taxi. Here C represents the cost (in dollars) and m represents the number of miles driven.

Recall from Section 1.2 that the scale of a number line refers to the distance between tick marks on the number line. Up to this point we have been using a scale of 1 on both the x-axis and y-axis. In applications, a different scale is often used to accomodate the ordered pairs that need to be plotted.

EXAMPLE 7 **An Electric Bill**

In North Carolina, Duke Power determines that the monthly electric bill for a household will be C dollars for using x kilowatt-hours (kWh) of electricity using the formula

$$C = 7.87 + 0.066508x$$

where $x \leq 300$ kWh. (*Source:* Duke Power)

(a) Complete Table 6 and use the table to list ordered pairs that satisfy the equation. Express answers rounded to the nearest penny.

Table 6			
x (kWh)	50 kWh	100 kWh	250 kWh
C ($)			

(b) Plot the ordered pairs (x, C) found in part (a) in a rectangular coordinate system.

Solution

(a) The first entry in the table is $x = 50$. We substitute 50 for x and use the equation $C = 7.87 + 0.066508x$ to find C. When we find C we will round our answer to two decimal places because C represents the cost, so our answer is correct to the nearest penny.

$$C = 7.87 + 0.066508x$$
$$x = 50: \quad C = 7.87 + 0.066508(50)$$
$$\text{Use a calculator:} \quad C = 11.20$$

Now substitute 100 for x and use the equation $C = 7.87 + 0.066508x$ to find C.

$$C = 7.87 + 0.066508x$$
$$x = 100: \quad C = 7.87 + 0.066508(100)$$
$$\text{Use a calculator:} \quad C = 14.52$$

Now substitute 250 for x and use the equation $C = 7.87 + 0.066508x$ to find C.

$$C = 7.87 + 0.066508x$$
$$x = 250: \quad C = 7.87 + 0.066508(250)$$
$$\text{Use a calculator:} \quad C = 24.50$$

Figure 7

Table 7 shows the completed table.

Table 7			
x (kWh)	50 kWh	100 kWh	250 kWh
C ($)	$11.20	$14.52	$24.50

The ordered pairs that satisfy the equation are $(50, 11.20)$, $(100, 14.52)$, and $(250, 24.50)$.

(b) Remember that x represents the number of kilowatt-hours used, so we label the horizontal axis x. Also, recall that C represents the bill, so we label the vertical axis C. Figure 7 shows a rectangular coordinate system with the ordered pairs found in part (a) plotted. Notice we use a different scale on the horizontal and vertical axis.

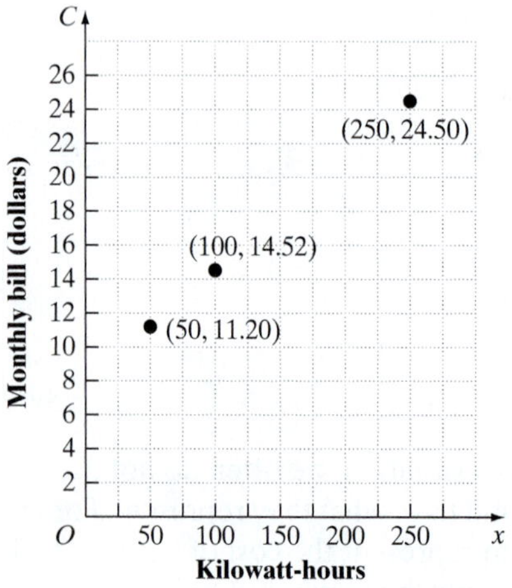

Notice in Figure 7 that we labeled the horizontal and vertical axis so that it is clear what they represent. Labeling the axes is a good practice to follow whenever you are drawing a graph.

QUICK ✓

15. Cinergy Corporation charges its customers in Cincinnati, Ohio, a monthly fee C dollars for using x cubic feet of natural gas using the formula

$$C = 5.50 + 0.666652x$$

(*Source*: Cinergy Corp.)

(a) Complete the table and use the results to list ordered pairs (x, C) that satisfy the equation. Express answers rounded to the nearest penny.

x (ft³)	50 ft³	100 ft³	150 ft³
C ($)			

(b) Plot the ordered pairs found in part (a) in a rectangular coordinate system.

3.1 Exercises

For Extra Help: Student Solutions Manual   CD Video   PH Math/Tutor Center   MathXL Tutorials on CD   MathXL®   MyMathLab

Concepts and Vocabulary

In Problems 1–3, fill in the blanks.

1. In the rectangular coordinate system, we call the horizontal real number line the _____ and we call the vertical real number line the _____. The point where these two axes intersect is called the _____.

2. If (x, y) are the coordinates of a point P, then x is called the _____ of P and y is called the _____ of P.

3. A(n) _____ to an equation in two variables x and y is an ordered pair, (x, y), that satisfies the equation.

In Problems 4–6, answer True or False to each statement.

4. The point whose ordered pair is $(-2, 4)$ is located in quadrant IV.

5. An equation in two variables can have more than one solution.

6. The ordered pairs $(3, 2)$ and $(2, 3)$ represent the same point in the Cartesian plane.

7. Describe how the quadrants in the rectangular coordinate system are labeled and how you can determine the quadrant in which a point lies. Describe the characteristics of a point that lies on either the x- or y-axis.

8. Describe how to plot the ordered pair $(3, -5)$.

Building Skills

In Problems 9–12, plot the following ordered pairs in the rectangular coordinate system. Tell which quadrant each point lies in or state that the point lies on the x-axis or y-axis.

9. $A(-3, 2)$; $B(4, 1)$; $C(-2, -4)$; $D(5, -4)$; $E(-1, 3)$; $F(2, -4)$

10. $P(-3, -2)$; $Q(2, -4)$; $R(4, 3)$; $S(-1, 4)$; $T(-2, -4)$; $U(3, -3)$

11. $A\left(\dfrac{1}{2}, 0\right)$; $B\left(\dfrac{3}{2}, -\dfrac{1}{2}\right)$; $C\left(4, \dfrac{7}{2}\right)$; $D\left(0, -\dfrac{5}{2}\right)$; $E\left(\dfrac{9}{2}, 2\right)$; $F\left(-\dfrac{5}{2}, -\dfrac{3}{2}\right)$; $G(0, 0)$

12. $P\left(\dfrac{3}{2}, -2\right)$; $Q\left(0, \dfrac{5}{2}\right)$; $R\left(-\dfrac{9}{2}, 0\right)$; $S(0, 0)$; $T\left(-\dfrac{3}{2}, -\dfrac{9}{2}\right)$; $U\left(3, \dfrac{1}{2}\right)$; $V\left(\dfrac{5}{2}, -\dfrac{7}{2}\right)$

In Problems 13 and 14, plot the following ordered pairs in the rectangular coordinate system. Tell the location of each point: positive x-axis, negative x-axis, positive y-axis, or negative y-axis.

13. $A(3, 0)$; $B(0, -1)$; $C(0, 3)$; $D(-4, 0)$

14. $P(0, -1)$; $Q(-2, 0)$; $R(0, 3)$; $S(1, 0)$

In Problems 15 and 16, identify the coordinates of each point labeled in the figure. Name the quadrant in which each point lies or state that the point lies on the x- or y-axis.

15.

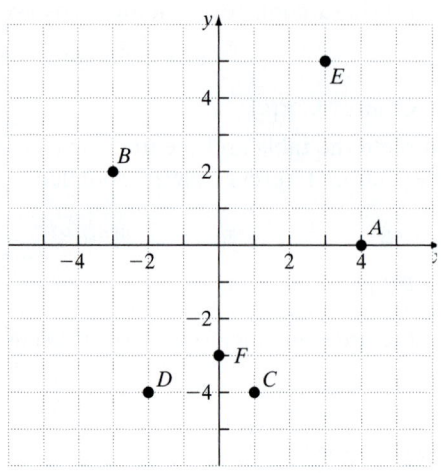

16.

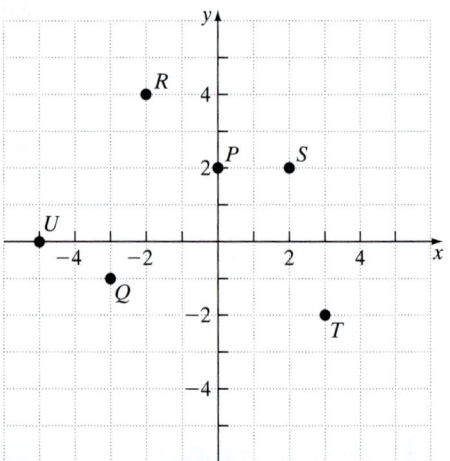

In Problems 17–22, determine whether or not the ordered pair is a solution to the equation.

17. $y = -3x + 5$

$A(-2, -1)$

$B(2, -1)$

$C\left(\frac{1}{3}, 4\right)$

18. $y = 2x - 3$

$A(-1, -5)$

$B(4, -5)$

$C(-2, -7)$

19. $3x + 2y = 4$

$A(0, 2)$

$B(1, 0)$

$C(4, -4)$

20. $5x - y = 12$

$A(2, 0)$

$B(0, 12)$

$C(-2, -22)$

21. $\frac{4}{3}x + y - 1 = 0$

$A(3, -3)$

$B(-6, -9)$

$C\left(\frac{3}{4}, 0\right)$

22. $\frac{3}{4}x + 2y = 0$

$A\left(-4, \frac{3}{2}\right)$

$B(0, 0)$

$C\left(1, -\frac{3}{2}\right)$

23. Find an ordered pair that satisfies the equation $x + y = 5$ by letting $x = 4$.

24. Find an ordered pair that satisfies the equation $x + y = 7$ by letting $x = 2$.

25. Find an ordered pair that satisfies the equation $2x + y = 9$ by letting $y = -1$.

26. Find an ordered pair that satisfies the equation $-4x - y = 5$ by letting $y = 7$.

27. Find an ordered pair that satisfies the equation $-3x + 2y = 15$ by letting $x = -3$.

28. Find an ordered pair that satisfies the equation $5x - 3y = 11$ by letting $y = 3$.

In Problems 29–42, use the equation to complete the table. Use the table to list some of the ordered pairs that satisfy the equation.

29. $y = -x$

x	y	(x, y)
-3		
0		
1		

30. $y = x$

x	y	(x, y)
-4		
0		
2		

31. $y = -3x + 1$

x	y	(x, y)
−2		
−1		
4		

32. $y = 4x - 5$

x	y	(x, y)
−3		
1		
2		

33. $2x + y = 6$

x	y	(x, y)
−1		
2		
3		

34. $3x + 4y = 2$

x	y	(x, y)
−2		
2		
4		

35. $y = 6$

x	y	(x, y)
−4		
1		
12		

36. $x = 2$

x	y	(x, y)
	−4	
	0	
	8	

37. $x - 2y + 6 = 0$

x	y	(x, y)
1		
	1	
−2		

38. $2x + y - 4 = 0$

x	y	(x, y)
−1		
	−4	
	2	

39. $y = 5 + \dfrac{1}{2}x$

x	y	(x, y)
	7	
−4		
	2	

40. $y = 8 - \dfrac{1}{3}x$

x	y	(x, y)
	10	
9		
	27	

41. $\dfrac{x}{2} + \dfrac{y}{3} = -1$

x	y	(x, y)
0		
	0	
	−6	

42. $\dfrac{x}{5} - \dfrac{y}{2} = 1$

x	y	(x, y)
	0	
0		
−5		

In Problems 43–54, for each equation find the missing value in the ordered pair.

43. $y = -3x - 10$

$A(\underline{\quad}, -16)$
$B(-3, \underline{\quad})$
$C(\underline{\quad}, -9)$

44. $y = 5x - 4$

$A(-1, \underline{\quad})$
$B(\underline{\quad}, 31)$
$C\left(-\dfrac{2}{5}, \underline{\quad}\right)$

45. $x = -\dfrac{1}{3}y$

$A(2, \underline{\quad})$
$B(\underline{\quad}, 0)$
$C\left(\underline{\quad}, -\dfrac{1}{2}\right)$

46. $y = \dfrac{2}{3}x$

$A\left(\underline{\quad}, -\dfrac{8}{3}\right)$
$B(0, \underline{\quad})$
$C\left(-\dfrac{5}{6}, \underline{\quad}\right)$

47. $x = 4$

$A(\underline{\quad}, -8)$

$B(\underline{\quad}, -19)$

$C(\underline{\quad}, 5)$

48. $y = -1$

$A(6, \underline{\quad})$

$B(-1, \underline{\quad})$

$C(0, \underline{\quad})$

49. $y = \dfrac{2}{3}x + 2$

$A(\underline{\quad}, 4)$

$B(-6, \underline{\quad})$

$C\left(\dfrac{1}{2}, \underline{\quad}\right)$

50. $y = -\dfrac{5}{4}x - 1$

$A(\underline{\quad}, -6)$

$B(-8, \underline{\quad})$

$C\left(\underline{\quad}, -\dfrac{11}{6}\right)$

51. $\dfrac{1}{2}x - 3y = 2$

$A(-4, \underline{\quad})$

$B(\underline{\quad}, -1)$

$C\left(-\dfrac{2}{3}, \underline{\quad}\right)$

52. $\dfrac{1}{3}x + 2y = -1$

$A(-4, \underline{\quad})$

$B\left(\underline{\quad}, -\dfrac{3}{4}\right)$

$C(0, \underline{\quad})$

53. $0.5x - 0.3y = 3.1$

$A(20, \underline{\quad})$

$B(\underline{\quad}, -17)$

$C(2.6, \underline{\quad})$

54. $-1.7x + 0.2y = -5$

$A(\underline{\quad}, -110)$

$B(40, \underline{\quad})$

$C(2.4, \underline{\quad})$

Applying the Concepts

55. Mail-Order CDs The cost of ordering CDs from a mail-order house is $9.95 per CD plus $4.95 for shipping and handling per order. The equation $C = 9.95n + 4.95$ represents the total cost, C, of ordering n CDs.

   (a) How much will it cost to order 2 CDs?
   (b) How much will it cost to order 5 CDs?
   (c) If you have $64.65 to spend, how many CDs can you order?
   (d) If (n, C) represents any ordered pair that satisfies $C = 9.95n + 4.95$, interpret the meaning of $(3, 34.8)$ in the context of this problem.

56. Taxi Ride The cost to take a taxi is $1.70 plus $2.00 per mile for each mile driven. The total cost, C, is given by the equation $C = 1.7 + 2m$, where m represents the total miles driven.

   (a) How much will it cost to take a taxi 5 miles?
   (b) How much will it cost to take a taxi 20 miles?
   (c) If you spent $32.70 on cab fare, how far was your trip?
   (d) If (m, C) represents any ordered pair that satisfies $C = 1.7 + 2m$, interpret the meaning of $(14, 29.7)$ in the context of this problem.

57. Cost of a Gallon of Gasoline An equation to approximate the future cost of a gallon of gasoline can be given by the model $C = \$0.0375n + \$1.15$, where n is the number of years after 1992.

   (a) According to the model, what was the cost of a gallon of gas in 1992 ($n = 0$)?
   (b) According to the model, what was the approximate cost of a gallon of gas in 1994 ($n = 2$)?
   (c) According to the model, what was the approximate cost of a gallon of gas in 2004 ($n = 12$)?
   (d) In what year was the cost of a gallon of gas $1.30?
   (e) The price will not increase at a constant rate forever. What factors do you think will affect the cost of gasoline in the future?

58. Life Expectancy The model $A = 0.183n + 67.895$ is used to estimate the life expectancy A of residents of the United States n years after 1950.

   (a) According to the model, what is the life expectancy for a person born in 1950?
   (b) According to the model, what is the life expectancy for a person born in 1980 ($n = 30$)?

(c) If the model holds true for future generations, what is the life expectancy for a person born in 2020?

(d) If a person has a life expectancy of 77 years, to the nearest year, when was the person born?

(e) Do you think life expectancy will continue to increase in the future? What could happen that would change this model?

In Problems 59–64, determine the value of k so that the given ordered pair is a solution to the equation.

59. Find the value of k for which $(1, 2)$ is a solution of $y = -2x + k$.

60. Find the value of k for which $(-1, 10)$ is a solution of $y = 3x + k$.

61. Find the value of k for which $(2, 9)$ is a solution of $7x - ky = -4$.

62. Find the value of k for which $(3, -1)$ is a solution of $4x + ky = 9$.

63. Find the value of k for which $\left(-8, -\dfrac{5}{2}\right)$ is a solution of $kx - 4y = 6$.

64. Find the value of k for which $\left(-9, \dfrac{1}{2}\right)$ is a solution of $kx + 2y = -2$.

In Problems 65–68, use the equation to complete the table. Use the table to list some of the ordered pairs that satisfy the equation.

65. $4a + 2b = -8$

a	b	(a, b)
2		
	-4	
	6	

66. $2r - 3s = 3$

r	s	(r, s)
	3	
	-1	
-3		

67. $\dfrac{2p}{5} + \dfrac{3q}{10} = 1$

p	q	(p, q)
0		
	0	
-10		

68. $\dfrac{4a}{3} + \dfrac{2b}{5} = -1$

a	b	(a, b)
	0	
0		
	10	

Extending the Concepts

In Problems 69 and 70, use the equation to complete the table. Choose any value for x and then solve the resulting equation to find the corresponding value for y. Then plot these ordered pairs in a rectangular coordinate system. Connect the points and describe the figure.

69. $3x - 2y = -6$

x	y	(x, y)

70. $-x + y = 4$

x	y	(x, y)

In Problems 71–74, use the equation to complete the table. Then plot the points in a rectangular coordinate system.

71. $y = x^2 - 4$

x	y	(x, y)
−2		
−1		
0		
1		
2		

72. $y = -x^2 + 3$

x	y	(x, y)
−2		
−1		
0		
1		
2		

73. $y = -x^3 + 2$

x	y	(x, y)
−2		
−1		
0		
1		
2		

74. $y = 2x^3 - 1$

x	y	(x, y)
−2		
−1		
0		
1		
2		

The Graphing Calculator

Graphing calculators can also create tables of values that satisfy an equation. To do this, we first solve the equation for y. For example, to obtain a table of values that satisfy the equation $2x - 3y = 12$ (Example 6), we solve for y as follows:

$$2x - 3y = 12$$

Subtract 2x from both sides: $\quad -3y = -2x + 12$

Divide both sides by −3: $\quad y = \dfrac{-2x + 12}{-3}$

Divide −3 into each term in the numerator: $\quad y = \dfrac{-2x}{-3} + \dfrac{12}{-3}$

Simplify: $\quad y = \dfrac{2}{3}x - 4$

We now enter the equation $y = \dfrac{2}{3}x - 4$ into the calculator and create the table shown below.

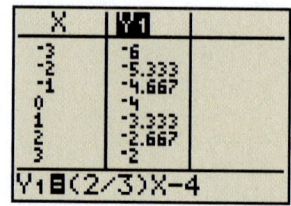

In Problems 75–82, use a graphing calculator to create a table of values that satisfy each equation. Have the table begin at −3 and increase by 1.

75. $y = 2x - 9$ **76.** $y = -3x + 8$ **77.** $y = -x + 8$ **78.** $y = 2x - 4$

79. $y + 2x = 13$ **80.** $y - x = -15$ **81.** $y = -6x^2 + 1$ **82.** $y = -x^2 + 3x$

3.2 Graphing Equations in Two Variables

OBJECTIVES

1 Graph a Line by Plotting Points

2 Graph a Line Using Intercepts

3 Graph Vertical and Horizontal Lines

Preparing for *Graphing Equations in Two Variables*

Before getting started, take the following readiness quiz. If you get a problem wrong, go back to the section cited and review the material.

1. Solve: $4x = 24$ [Section 2.1, pp. 78–80]

2. Solve: $-3y = 18$ [Section 2.1, pp. 78–80]

3. Solve: $2x + 5 = 13$ [Section 2.2, pp. 84–86]

1 **Graph a Line by Plotting Points**

In the previous section, we found values of x and y that satisfy an equation. What does this mean? Well, it means that the ordered pair (x, y) is a point on the graph of the equation.

In Words

The graph of an equation is a geometric way of representing the set of all ordered pairs that make the equation a true statement. Think of the graph as a picture of the solution set.

> **DEFINITION**
>
> The **graph of an equation in two variables** x and y is the set of points in the xy-plane whose coordinates (x, y) satisfy the equation.

But how do we obtain the graph of an equation? One method for graphing an equation is the **point-plotting method.**

EXAMPLE 1 **How to Graph an Equation Using the Point-Plotting Method**

Graph the equation $y = 2x - 3$ using the point-plotting method.

Step-by-Step Solution

Step 1: We find the ordered pairs that satisfy the equation by choosing some values of x and using the equation to find the corresponding values of y. See Table 8.

Table 8		
x	**y**	**(x, y)**
−2	$y = 2(-2) - 3$ $= -4 - 3$ $= -7$	$(-2, -7)$
−1	$y = 2(-1) - 3$ $= -5$	$(-1, -5)$
0	$y = 2(0) - 3$ $= -3$	$(0, -3)$
1	$y = 2(1) - 3$ $= -1$	$(1, -1)$
2	$y = 2(2) - 3$ $= 1$	$(2, 1)$

(continued)

Preparing for...Answers

1. {6} **2.** {−6} **3.** {4}

Step 2: Plot the points found in Step 1 in a rectangular coordinate system. See Figure 8.

Figure 8

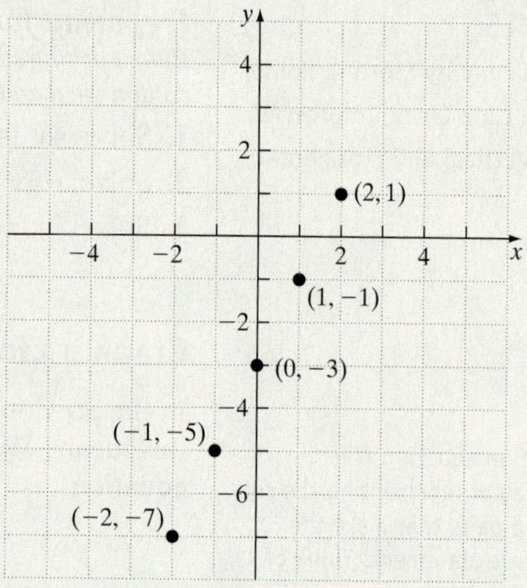

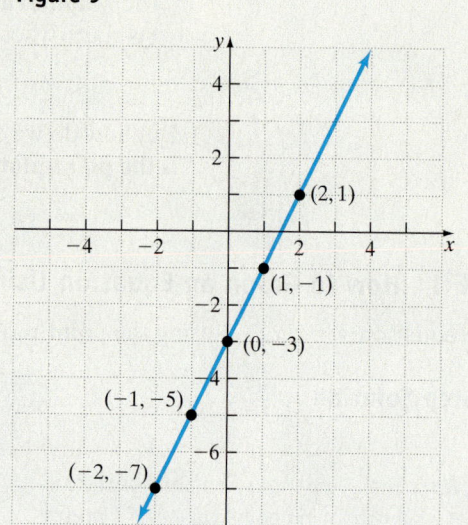

Step 3: Connect the points in a straight line. See Figure 9.

Figure 9

Work Smart

Remember, all coordinates of the points on the graph shown in Figure 9 satisfy the equation $y = 2x - 3$.

The graph of the equation shown in Figure 9 does not show all the points that satisfy the equation. For example, in Figure 9 the point (5, 7) is a part of the graph of $y = 2x - 3$, but it is not shown. Since the graph of $y = 2x - 3$ could be extended as far as we please, we use arrows to indicate that the pattern shown continues. It is important to show enough of the graph so that anyone who is looking at it will "see" the rest of it as an obvious continuation of what is there. This is called a **complete graph.**

We summarize the steps for graphing an equation using the point-plotting method below.

Graphing an Equation Using the Point-Plotting Method

Step 1: Find several ordered pairs that satisfy the equation.

Step 2: Plot the points found in Step 1 in a rectangular coordinate system.

Step 3: Connect the points in a smooth curve or line.

QUICK ✓ *Draw a complete graph of each equation using point plotting.*

1. $y = 3x - 2$ **2.** $y = -4x + 8$

A question you may be asking yourself is, "How many points do I need to find before I can be sure that I have a complete graph?" The answer is that it depends on the type of equation you are graphing. In mathematics, we classify equations as different types. For example, the equation that we graphed in Example 1 is called a *linear equation*.

> **DEFINITION**
>
> A **linear equation in two variables** is an equation of the form
>
> $$Ax + By = C$$
>
> where A, B, and C are real numbers. A and B cannot both be 0. When a linear equation is written in the form $Ax + By = C$, we say that the linear equation is in **standard form.**

EXAMPLE 2   Identifying Linear Equations in Two Variables

Determine whether or not the equation is a linear equation in two variables.

  (a) $3x - 4y = 9$ **(b)** $\dfrac{1}{2}x + \dfrac{2}{3}y = 4$ **(c)** $x^2 + 5y = 10$ **(d)** $-2y = 5$

Solution

  (a) The equation $3x - 4y = 9$ is a linear equation in two variables because it is written in the form $Ax + By = C$ with $A = 3$, $B = -4$, and $C = 9$.

  (b) The equation $\dfrac{1}{2}x + \dfrac{2}{3}y = 4$ is a linear equation in two variables because it is written in the form $Ax + By = C$ with $A = \dfrac{1}{2}$, $B = \dfrac{2}{3}$, and $C = 4$.

  (c) The equation $x^2 + 5y = 10$ is not a linear equation because x is squared.

  (d) The equation $-2y = 5$ is a linear equation in two variables because it is written in the form $Ax + By = C$ with $A = 0$, $B = -2$, and $C = 5$.   ▬

QUICK ✓ *Determine whether or not the equation is a linear equation in two variables.*

3. $4x - y = 12$ **4.** $5x - y^2 = 10$ **5.** $\dfrac{1}{2}x + \dfrac{2}{3}y = 4$ **6.** $5x = 20$

Work Smart

When graphing a line, be sure to find three points—just to be safe!

For the remainder of the text, we will refer to linear equations in two variables as **linear equations.** The graph of a linear equation is a **line.** To graph a linear equation requires only two points; however, we recommend that you find a third point as a check.

EXAMPLE 3   Graphing a Linear Equation Using the Point-Plotting Method

Graph the linear equation $2x + y = 4$.

Work Smart
Choose values of x (or y) that make the algebra easy.

Solution

We need to find ordered pairs that satisfy the equation. Because the coefficient of y is 1, it is easier to choose values of x and find the corresponding values of y. We will determine the value of y for $x = -2, 0$, and 2. There is nothing magical about these choices. Any three different values of x will give us the results we want.

$x = -2$:

Let $x = -2$:
$$2x + y = 4$$
$$2(-2) + y = 4$$
$$-4 + y = 4$$

Add 4 to both sides:
$$y = 8$$

$x = 0$:

Let $x = 0$:
$$2x + y = 4$$
$$2(0) + y = 4$$
$$y = 4$$

$x = 2$:

Let $x = 2$:
$$2x + y = 4$$
$$2(2) + y = 4$$
$$4 + y = 4$$

Subtract 4 from both sides:
$$y = 0$$

Table 9 summarizes the results. The ordered pairs $(-2, 8)$, $(0, 4)$, and $(2, 0)$ represent points that are on the graph of the equation. We plot these points in Figure 10(a). After connecting the points in a straight line we obtain the graph in Figure 10(b).

Table 9		
x	y	(x, y)
-2	8	$(-2, 8)$
0	4	$(0, 4)$
2	0	$(2, 0)$

Figure 10

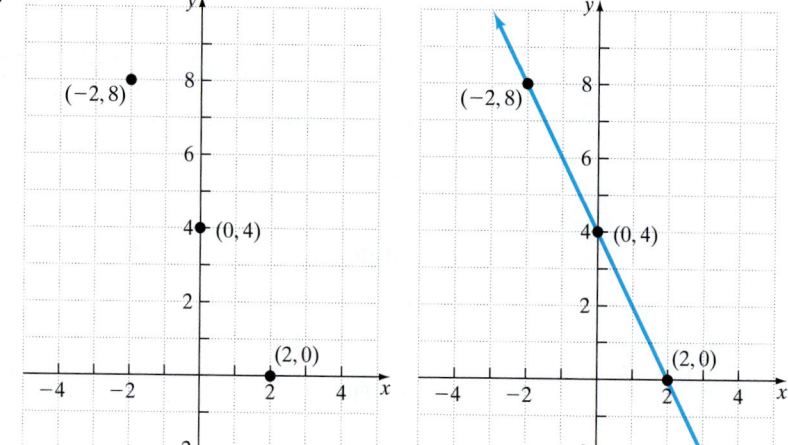

(a)　　　(b)

QUICK ✓ *Graph each linear equation using point plotting.*

7. $-3x + y = -6$　　　　**8.** $2x + 3y = 12$

EXAMPLE 4　Cost of Renting a Car

Your favorite car-rental agency quotes you the cost of renting a car in Washington, D.C., as \$30 per day plus \$0.20 per mile. The linear equation $C = 0.20m + 30$ models the cost, where C represents total cost and m represents the number of miles that were traveled.

(a) Complete Table 10 and use the results to list ordered pairs that satisfy the equation. Express answers rounded to the nearest penny.

(b) Graph the linear equation $C = 0.20m + 30$ using the points obtained in part (a).

Table 10		
m	C	(m, C)
0		
50		
100		

Solution

(a) The first entry in the table is $m = 0$. We substitute 0 for m and use the equation $C = 0.20m + 30$ to find C.

$$C = 0.20m + 30$$
$$m = 0: \quad C = 0.20(0) + 30$$
$$C = 30$$

Now substitute 50 for m and use the equation $C = 0.20m + 30$ to find C.

$$C = 0.20m + 30$$
$$m = 50: \quad C = 0.20(50) + 30$$
$$C = 40$$

Now substitute 100 for m and use the equation $C = 0.20m + 30$ to find C.

$$C = 0.20m + 30$$
$$m = 100: \quad C = 0.20(100) + 30$$
$$C = 50$$

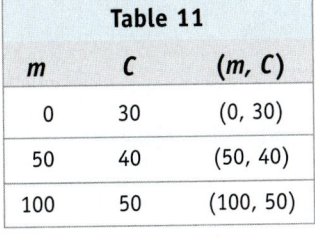

Table 11

m	C	(m, C)
0	30	$(0, 30)$
50	40	$(50, 40)$
100	50	$(100, 50)$

Table 11 shows the completed table.

The ordered pairs that satisfy the equation are $(0, 30)$, $(50, 40)$, and $(100, 50)$. The ordered pair $(50, 40)$ means if a car is driven 50 miles, the cost of renting the car will be \$40.

(b) Remember that m represents the number miles driven, so we label the horizontal axis m. Also, recall that C represents the cost of renting the car, so we label the vertical axis C. When drawing the horizontal axis, we set the scale to 10, which means each tick mark represents 10 miles. We scale the vertical axis to 5, which means each tick mark represents 5 dollars. Scaling in this way makes it easier to plot the ordered pairs. We plot the points found in part (a) in the rectangular coordinate system and then draw the line. See Figure 11.

Figure 11

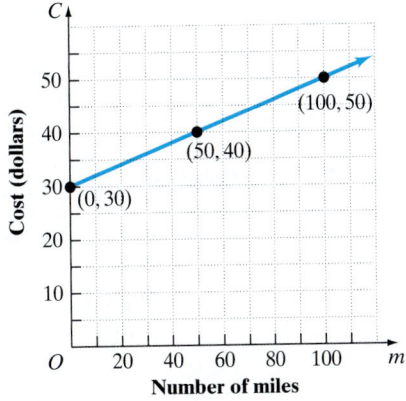

QUICK ✓

9. Michelle sells computers. Her monthly salary is \$3000 plus 8% of total sales. The linear equation $S = 0.08x + 3000$ models Michelle's monthly salary, S, where x represents her total sales in the month.

(a) Complete the table and use the results to list ordered pairs that satisfy the equation. Express answers rounded to the nearest penny.

x	S	(x, S)
0		
10,000		
25,000		

(b) Graph the linear equation $S = 0.08x + 3000$ using the points obtained in part (a).

② Graph a Line Using Intercepts

Intercepts should always be displayed in a complete graph.

Figure 12

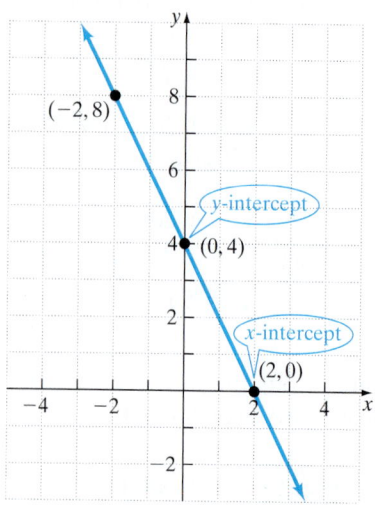

> **DEFINITIONS**
>
> The **intercepts** are the points, if any, where a graph crosses or touches the coordinate axes. The x-coordinate of a point at which the graph crosses or touches the x-axis is an ***x*-intercept,** and the y-coordinate of a point at which the graph crosses or touches the y-axis is a ***y*-intercept.**

See Figure 12 for an illustration. The graph in Figure 12 is the graph obtained in Example 3.

EXAMPLE 5 **Finding Intercepts from a Graph**

Find the intercepts of the graphs shown in Figures 13(a) and 13(b). What are the x-intercepts? What are the y-intercepts?

Figure 13

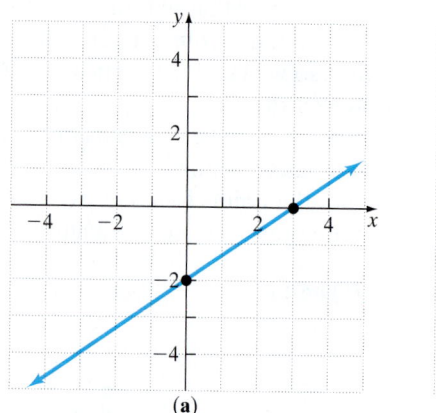
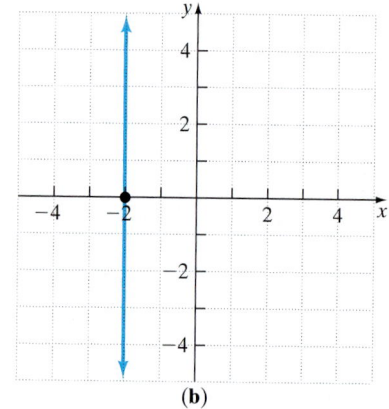

      (a)   (b)

In Words

An x-intercept exists when $y = 0$.
A y-intercept exists when $x = 0$.

Solution

(a) The intercepts of the graph in Figure 13(a) are the points $(0, -2)$ and $(3, 0)$. The x-intercept is 3. The y-intercept is -2.

(b) The intercept of the graph in Figure 13(b) is the point $(-2, 0)$. The x-intercept is -2. There are no y-intercepts.

In Example 5, you should notice the following: If we do not specify the type of intercept (x-intercept versus y-intercept), we report the intercept as an ordered pair. However, if we specify the type of intercept, then we only need report the coordinate of the intercept. For example, we would say $(4, 0)$ is an intercept, while we would say 4 is an x-intercept.

QUICK ✓ *Find the intercepts of the graph shown in the figure. What are the x-intercepts? What are the y-intercepts?*

10.

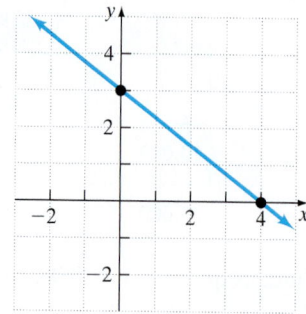

11.

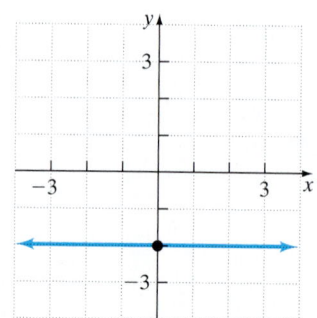

Now we will explain how to find the intercepts algebraically. From Figure 12 it should be apparent that an *x*-intercept exists when the value of *y* is 0 and that a *y*-intercept exists when the value of *x* is 0. This leads to the following procedure for finding intercepts.

Work Smart

Every point on the *x*-axis has a *y*-coordinate of 0. That's why we set $y = 0$ to find the *x*-intercept. Likewise, every point on the *y*-axis has an *x*-coordinate of 0. That's why we set $x = 0$ to find the *y*-intercept.

Procedure for Finding Intercepts

1. To find the *x*-intercept(s), if any, of the graph of an equation, let $y = 0$ in the equation and solve for *x*.
2. To find the *y*-intercept(s), if any, of the graph of an equation, let $x = 0$ in the equation and solve for *y*.

Because this method for graphing a linear equation results in only two points, we find a third point so that we can check our work.

EXAMPLE 6   **How to Graph a Linear Equation by Finding Its Intercepts**

Graph the linear equation $4x - 3y = 24$ by finding its intercepts.

Step-by-Step Solution

Step 1: Find the *y*-intercept by letting $x = 0$ and solving the equation for *y*.

$$4x - 3y = 24$$
$$\text{Let } x = 0: \quad 4(0) - 3y = 24$$
$$0 - 3y = 24$$
$$-3y = 24$$
$$\text{Divide both sides by } -3: \quad y = -8$$

The *y*-intercept is -8, so the point $(0, -8)$ is on the graph of the equation.

Step 2: Find the *x*-intercept by letting $y = 0$ and solving the equation for *x*.

$$4x - 3y = 24$$
$$\text{Let } y = 0: \quad 4x - 3(0) = 24$$
$$4x - 0 = 24$$
$$4x = 24$$
$$\text{Divide both sides by } 4: \quad x = 6$$

The *x*-intercept is 6, so the point $(6, 0)$ is on the graph of the equation.

Step 3: Find one additional point on the graph by choosing any value of *x* that is convenient and solving the equation for *y*.

We will let $x = 3$ and solve the equation $4x - 3y = 24$ for *y*.

$$\text{Let } x = 3: \quad 4(3) - 3y = 24$$
$$12 - 3y = 24$$
$$\text{Subtract 12 from both sides:} \quad -3y = 12$$
$$\text{Divide both sides by } -3: \quad y = -4$$

The point $(3, -4)$ is on the graph of the equation.

(continued)

Step 4: Plot the points found in Steps 1–3 and draw in the line.

We plot the points $(0, -8)$, $(6, 0)$, and $(3, -4)$. Connect the points in a straight line and obtain the graph in Figure 14.

Figure 14

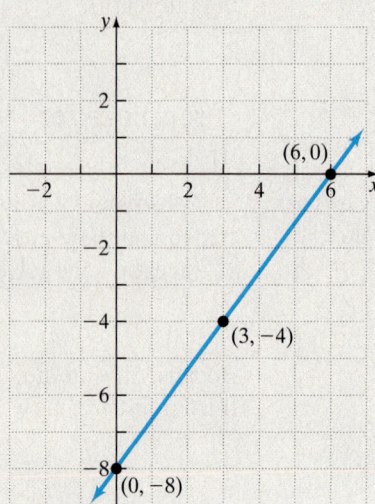

EXAMPLE 7   Graphing a Linear Equation by Finding Its Intercepts

Graph the linear equation $\frac{1}{2}x - 2y = 3$ by finding its intercepts.

Solution

x-intercept:

$$\frac{1}{2}x - 2y = 3$$

Let $y = 0$: $\frac{1}{2}x - 2(0) = 3$

$$\frac{1}{2}x = 3$$

Multiply both sides by 2: $x = 6$

y-intercept:

$$\frac{1}{2}x - 2y = 3$$

Let $x = 0$: $\frac{1}{2}(0) - 2y = 3$

$$-2y = 3$$

Divide both sides by -2: $y = -\frac{3}{2}$

Additional point (choose $x = 2$):

$$\frac{1}{2}x - 2y = 3$$

Let $x = 2$: $\frac{1}{2}(2) - 2y = 3$

$$1 - 2y = 3$$

Subtract 1 from both sides: $-2y = 2$

Divide both sides by -2: $y = -1$

Figure 15

Plot the points $(6, 0)$, $\left(0, -\frac{3}{2}\right)$, and $(2, -1)$. Connect the points in a straight line. See Figure 15.

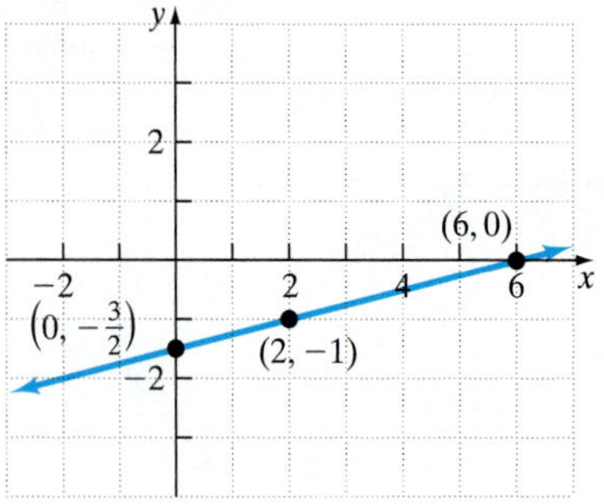

QUICK ✓ *Graph each linear equation by finding its intercepts.*

12. $x + y = 3$ **13.** $2x - 5y = 20$ **14.** $-3x + 4y = 9$ **15.** $\frac{3}{2}x - 2y = 9$

EXAMPLE 8   Graphing a Linear Equation of the Form $Ax + By = 0$

Graph the linear equation $2x + 3y = 0$ by finding its intercepts.

Solution

x-intercept:

$$2x + 3y = 0$$
Let $y = 0$: $\quad 2x + 3(0) = 0$
$$2x + 0 = 0$$
Divide both sides by 2: $\quad x = 0$

The x-intercept is 0, so the point $(0, 0)$ is on the graph of the equation.

y-intercept:

$$2x + 3y = 0$$
Let $x = 0$: $\quad 2(0) + 3y = 0$
$$3y = 0$$
Divide both sides by 3: $\quad y = 0$

The y-intercept is 0, so the point $(0, 0)$ is on the graph of the equation.

Additional point (choose $x = 3$):

$$2x + 3y = 0$$
Let $x = 3$: $\quad 2(3) + 3y = 0$
$$6 + 3y = 0$$
Subtract 6 from both sides: $\quad 3y = -6$
Divide both sides by 3: $\quad y = -2$

The point $(3, -2)$ is on the graph of the equation. Because both the x- and y-intercepts are 0, we find *two* additional points on the graph of the equation. We already have one additional point, $(3, -2)$, so we need one more. By letting $x = -3$, we find that $y = 2$. We plot the points $(0, 0)$, $(-3, 2)$, and $(3, -2)$. Connect the points in a straight line and obtain the graph in Figure 16.   ∎

Work Smart

Linear equations of the form $Ax + By = 0$, where $A \neq 0$ and $B \neq 0$ have only one intercept at $(0, 0)$, so two additional points should be plotted to obtain the graph.

Figure 16

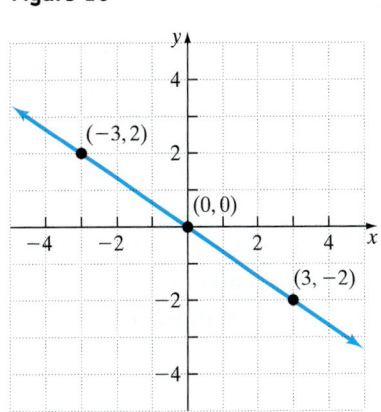

QUICK ✓ *Graph the equation by finding its intercepts.*

16. $y = \dfrac{1}{2}x$

17. $4x + y = 0$

---

③ Graph Vertical and Horizontal Lines

In the equation of a line, $Ax + By = C$, we said that A and B cannot both be zero. But what if $A = 0$ or $B = 0$? We find that this leads to special types of lines called *vertical lines* (when $B = 0$) and *horizontal lines* (when $A = 0$).

EXAMPLE 9   Graphing a Vertical Line

Graph the equation $x = 3$ using the point-plotting method.

Solution

Because the equation $x = 3$ can be written as $1x + 0y = 3$, we know that the graph is a line. When you look at the equation $x = 3$, notice that no matter what value of y we choose, the corresponding value of x is going to be 3. For example, if $y = -1$, then

$$1x + 0(-1) = 3$$
$$x = 3$$

See Table 12 for other choices for y. We see that the points $(3, -2)$, $(3, -1)$, $(3, 0)$, $(3, 1)$, and $(3, 2)$ are all points on the line. See Figure 17.

Table 12		
x	**y**	**(x, y)**
3	−2	(3, −2)
3	−1	(3, −1)
3	0	(3, 0)
3	1	(3, 1)
3	2	(3, 2)

Figure 17

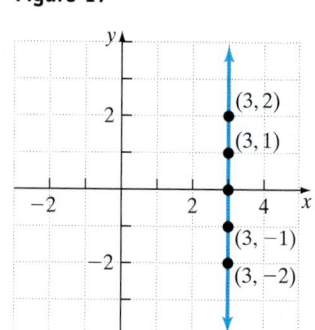

Based on the results of Example 9, we can write a definition of a vertical line:

> **DEFINITION: EQUATION OF A VERTICAL LINE**
> A vertical line is given by an equation of the form
> $$x = a$$
> where a is the x-intercept.

Now let's look at equations that lead to graphs that are horizontal lines.

EXAMPLE 10 Graphing a Horizontal Line

Graph the equation $y = -2$ using the point-plotting method.

Solution

Because the equation $y = -2$ can be written as $0x + 1y = -2$, we know that the graph is a line. In looking at the equation $y = -2$, notice that no matter what value of x we choose, the corresponding value of y is going to be -2. For example, if $x = -2$, then

$$0(-2) + 1y = -2$$
$$y = -2$$

See Table 13 for other choices of x. Therefore, the points $(-2, -2)$, $(-1, -2)$, $(0, -2)$, $(1, -2)$, and $(2, -2)$ are all points on the line. See Figure 18.

Table 13		
x	**y**	**(x, y)**
−2	−2	(−2, −2)
−1	−2	(−1, −2)
0	−2	(0, −2)
1	−2	(1, −2)
2	−2	(2, −2)

Figure 18

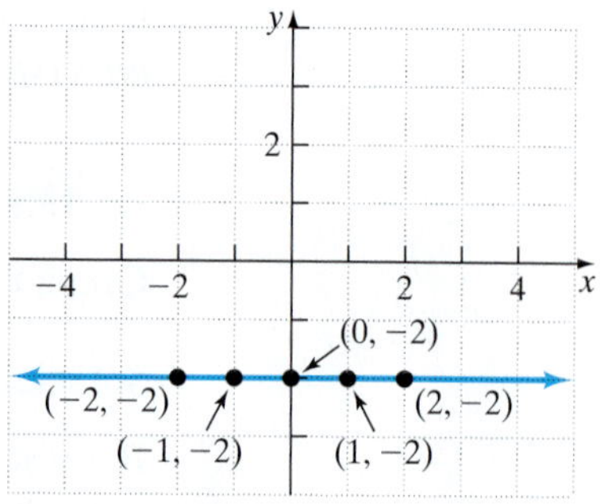

Based on the results of Example 10, we can generalize:

> **DEFINITION: EQUATION OF A HORIZONTAL LINE**
> A horizontal line is given by an equation of the form
> $$y = b$$
> where b is the y-intercept.

QUICK ✓ *Graph each equation.*

18. $x = -5$ **19.** $y = -4$ **20.** $x - 4 = 0$

We covered a lot of material in this section. We present a summary below to help you organize the information presented.

SUMMARY: Intercepts and Equations of Lines

Topic	Comments
Intercepts: Points where the graph crosses or touches a coordinate axis.	Intercepts need to be shown for a graph to be complete.
x-intercept: Point where the graph crosses or touches the x-axis. Found by letting $y = 0$ in the equation.	
y-intercept: Point where the graph crosses or touches the y-axis. Found by letting $x = 0$ in the equation.	
Standard Form of an Equation of a Line: $Ax + By = C$, where A and B are not both zero	Can be graphed using point-plotting or intercepts.
Equation of a Vertical Line: $x = a$	Graph is a vertical line whose x-intercept is a.
Equation of a Horizontal Line: $y = b$	Graph is a horizontal line whose y-intercept is b.

3.2 Exercises

For Extra Help: Student Solutions Manual   CD Video   PH Math/Tutor Center   MathXL Tutorials on CD   MathXL®   MyMathLab

Concepts and Vocabulary

In Problems 1–3, fill in the blanks.

1. A(n) _____ equation is an equation of the form $Ax + By = C$, where A, B, and C are real numbers, and A and B are not both zero. Equations written in this form are said to be in _____ _____.

2. The graph of a linear equation is a(n) _____.

3. The _____ are the points, if any, where a graph crosses or touches the coordinate axes.

In Problems 4–6, answer True or False to each statement.

4. To find the y-intercept(s), if any, of the graph of an equation, let $y = 0$ in the equation and solve for x.

5. All linear equations have exactly one x-intercept and one y-intercept.

6. A horizontal line can be represented by the equation $y = b$, where b is the y-intercept of the graph of the equation.

7. How many points are required to graph a line? Explain your reasoning and why you might include additional point(s) when graphing a line.

8. Explain how to use the intercepts to graph the equation $Ax + By = C$, where A, B, and C are not equal to zero. Explain how to graph the same equation when C is equal to zero. Can you use the same techniques for both equations? Why or why not?

Building Skills

In Problems 9–16, determine whether or not the equation is a linear equation in two variables.

9. $2x - 5y = 10$ **10.** $y^2 = 2x + 3$ **11.** $\sqrt{x} + y = 1$ **12.** $y - 2x = 9$

13. $y = \dfrac{4}{x}$ **14.** $x - 8 = 0$ **15.** $y - 1 = 0$ **16.** $y = \dfrac{-2}{x}$

In Problems 17–36, graph each linear equation using the point-plotting method.

17. $y = 2x$ **18.** $y = 3x$ **19.** $y = -5x$ **20.** $y = -4x$

21. $y = 4x - 2$ **22.** $y = -3x - 1$ **23.** $y = -2x + 5$ **24.** $y = x - 6$

25. $x + y = 5$ **26.** $x - y = 6$ **27.** $-2x + y = 6$ **28.** $5x - 2y = -10$

29. $4x - 2y = -8$ **30.** $x + 3y = 6$ **31.** $x = -4y$ **32.** $x = \dfrac{1}{2}y$

33. $y + 7 = 0$ **34.** $x - 6 = 0$

35. $y - 2 = 3(x + 1)$ **36.** $y + 3 = -2(x - 2)$

In Problems 37–44, find the intercepts of each graph.

37.

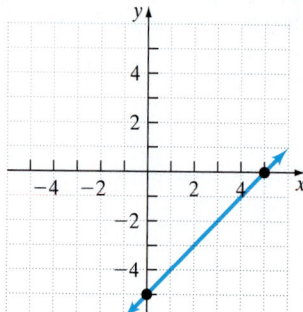

38.

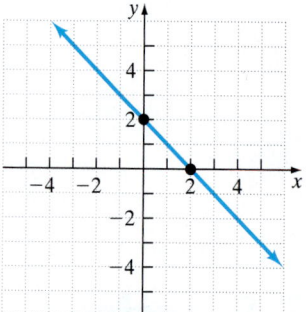

39.

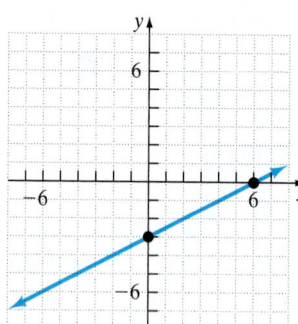

40.

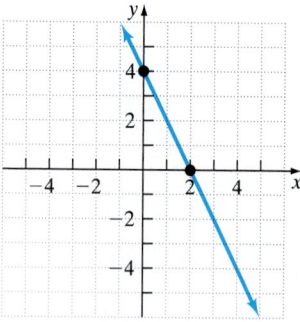

41.

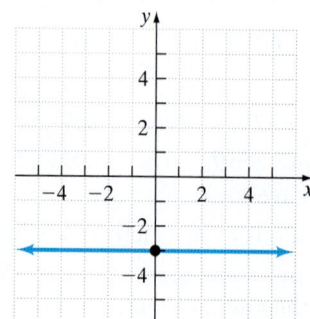

42.

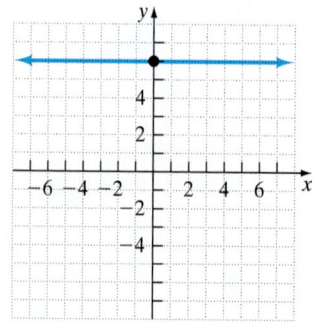

43.

44.

In Problems 45–56, find the intercepts of each equation.

45. $2x + 3y = -12$ **46.** $3x - 5y = 30$ **47.** $x = -6y$ **48.** $y = 10x$

49. $y = x - 5$ **50.** $y = -x + 7$ **51.** $\dfrac{x}{6} + \dfrac{y}{8} = 1$ **52.** $\dfrac{x}{2} - \dfrac{y}{8} = 1$

53. $x = 4$ **54.** $y = 6$ **55.** $y = -2$ **56.** $x = -8$

In Problems 57–72, graph each linear equation by finding its intercepts.

57. $3x + 6y = 18$ **58.** $4x - 2y = -8$ **59.** $-x + 5y = 15$ **60.** $-2x + y = 14$

61. $\dfrac{1}{2}x = y + 3$ **62.** $\dfrac{4}{3}x = -y + 1$ **63.** $9x - 2y = 0$ **64.** $\dfrac{1}{3}x - y = 0$

65. $y = -\dfrac{1}{2}x + 3$ **66.** $y = \dfrac{2}{3}x - 3$ **67.** $\dfrac{1}{3}y + 2 = 2x$ **68.** $\dfrac{1}{2}x - 3 = 3y$

69. $\dfrac{x}{2} + \dfrac{y}{3} = 1$ **70.** $\dfrac{y}{4} - \dfrac{x}{3} = 1$

71. $4y - 2x + 1 = 0$ **72.** $2y - 3x + 2 = 0$

In Problems 73–80, graph each horizontal or vertical line.

73. $x = 5$ **74.** $x = -7$ **75.** $y = -6$ **76.** $y = 2$

77. $y - 12 = 0$ **78.** $y + 3 = 0$ **79.** $3x - 5 = 0$ **80.** $2x - 7 = 0$

Mixed Practice

In Problems 81–92, graph each linear equation by any method.

81. $y = 2x - 5$ **82.** $y = -3x + 2$ **83.** $y = -5$ **84.** $x = 2$

85. $2x + 5y = -20$ **86.** $3x - 4y = 12$ **87.** $2x = -6y + 4$ **88.** $5x = 3y - 10$

89. $x - 3 = 0$ **90.** $y + 4 = 0$ **91.** $3y - 12 = 0$ **92.** $-4x + 8 = 0$

Applying the Concepts

93. Plot the points $(3, 5)$ and $(-2, 5)$ and draw a line through the points. What is the equation of this line?

94. Plot the points $(-1, 2)$ and $(5, 2)$ and draw a line through the points. What is the equation of this line?

95. Plot the points $(-2, -4)$ and $(-2, 1)$ and draw a line through the points. What is the equation of this line?

96. Plot the points $(3, -1)$ and $(3, 2)$ and draw a line through the points. What is the equation of this line?

In Problems 97–100, find the equation of each line.

97.

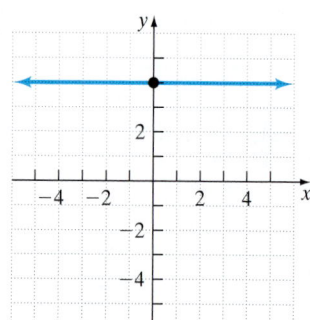

98.

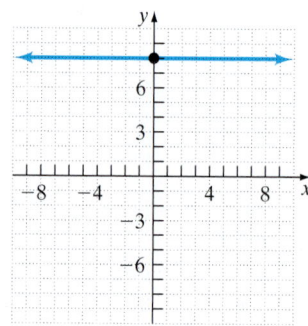

99.

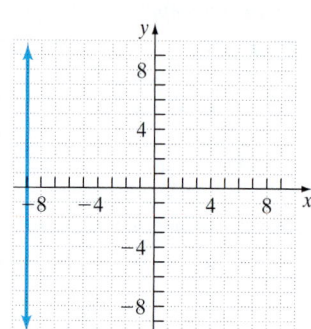

100.

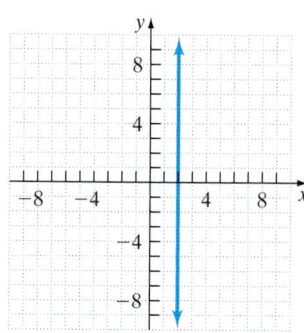

101. Create a set of ordered pairs in which the *x*-coordinate is twice the *y*-coordinate. What is the equation of this line?

102. Create a set of ordered pairs in which the *y*-coordinate is twice the *x*-coordinate. What is the equation of this line?

103. Create a set of ordered pairs in which the *y*-coordinate is 2 more than the *x*-coordinate. What is the equation of this line?

104. Create a set of ordered pairs in which the *x*-coordinate is 3 less than the *y*-coordinate. What is the equation of this line?

105. If $(3, y)$ is a point on the graph of $4x + 3y = 18$, find y.

106. If $(-4, y)$ is a point on the graph of $3x - 2y = 10$, find y.

107. If $(x, -2)$ is a point on the graph of $3x + 5y = 11$, find x.

108. If $(x, -3)$ is a point on the graph of $4x - 7y = 19$, find x.

109. Calculating Wages Marta earns $500 per week plus $100 in commission for every car she sells. The linear equation that calculates her weekly earnings is $E = 100n + 500$, where E represents her weekly earnings in dollars and n represents the number of cars she sold during the week.

   (a) Create a set of ordered pairs (n, E) if, in three consecutive weeks, she sold 0 cars, 4 cars, and 10 cars.

   (b) Graph the linear equation $E = 100n + 500$ using the ordered pairs obtained in part (a). Be sure to label the axes appropriately.

   (c) Explain the meaning of the E-intercept.

110. Carpet Cleaning Harry's Carpet Cleaning charges a $50 service charge plus $0.10 for each square foot of carpeting to be cleaned. The linear equation that calculates the total cost to clean a carpet is $C = 0.1f + 50$, where C is the total cost in dollars and f is the number of square feet of carpet.

   (a) Create a set of ordered pairs (f, C) for the following number of square feet to be cleaned: 1000 sq ft, 2000 sq ft, 2500 sq ft.

   (b) Graph the linear equation $C = 0.1f + 50$ using the ordered pairs obtained in part (a). Be sure to label the axes appropriately.

   (c) Explain the meaning of the C-intercept.

Extending the Concepts

111. Graph each of the following linear equations in the same xy-plane. What can you infer from these graphs?

$$y = 2x - 1 \qquad y = 2x + 3 \qquad 2x - y = 5$$

112. Graph each of the following linear equations in the same xy-plane. What can you infer from these graphs?

$$y = 3x + 2 \qquad 6x - 2y = -4 \qquad x = \frac{1}{3}y - \frac{2}{3}$$

113. Graph each of the following linear equations in the same xy-plane. What statement can you make about the steepness of the line as the coefficient of x gets larger?

$$y = x \qquad y = 2x \qquad y = 10x$$

114. Graph each of the following linear equations in the same xy-plane. What statement can you make about the steepness of the line as the coefficient of x gets smaller?

$$y = x + 2 \qquad y = \frac{1}{2}x + 2 \qquad y = \frac{1}{8}x + 2$$

In Problems 115–118, find the intercepts of each graph.

115.

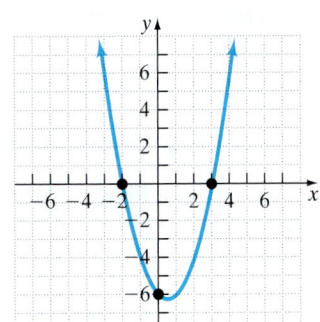

116.

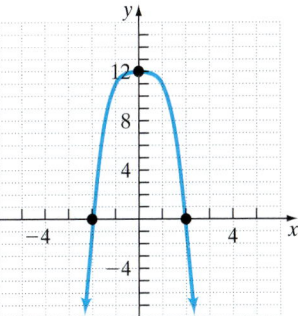

117.

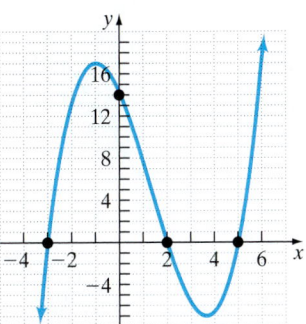

118.

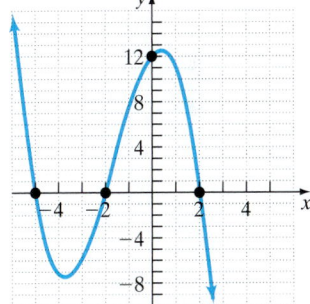

The Graphing Calculator

Graphing calculators can graph equations. In fact, graphing calculators also use the point-plotting method to obtain the graph by choosing 95 values of x and using the equation to find the corresponding value of y. As with creating tables, we first solve the equation for y. For example, to obtain a table of values that satisfy the equation $2x - 3y = 12$ (Example 6 from Section 3.1), we must solve for y

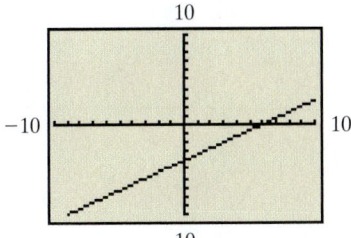

and we obtain $y = \frac{2}{3}x - 4$. We enter the equation $y = \frac{2}{3}x - 4$ into the calculator and create the graph shown in the right.

In Problems 119–126, use a graphing calculator to graph each equation.

119. $y = 2x - 9$ **120.** $y = -3x + 8$ **121.** $y = -x + 8$ **122.** $y = 2x - 4$

123. $y + 2x = 13$ **124.** $y - x = -15$ **125.** $y = -6x^2 + 1$ **126.** $y = -x^2 + 3x$

3.3 Slope

OBJECTIVES

1. Find the Slope of a Line Given Two Points
2. Find the Slope of Vertical and Horizontal Lines
3. Graph a Line Using Its Slope and a Point on the Line
4. Work with Applications of Slope

Preparing for Slope

Before getting started, take the following readiness quiz. If you get a problem wrong, go back to the section cited and review the material.

1. Evaluate: $\dfrac{5-2}{8-7}$ [Section 1.6, p. 52]

2. Evaluate: $\dfrac{3-7}{9-3}$ [Section 1.6, p. 52]

3. Evaluate: $\dfrac{-3-4}{6-(-1)}$ [Section 1.6, p. 52]

Figure 19

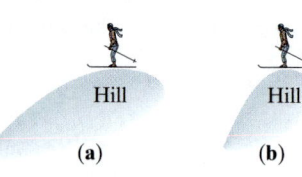

(a)   (b)

Pretend you are on snow skis for the first time in your life. The ski resort that you are visiting has two hills available to beginning skiers. The profile of each hill is shown in Figure 19. Which hill would you prefer to go down? Why?

It is clear from the figure that the hill in Figure 19(a) is not as steep as the hill in Figure 19(b). One of the things that mathematicians like to do is give numerical descriptions to situations such as the steepness of a hill. Measuring the steepness of each hill allows for them to be compared more easily. The numerical measure that we could use to describe the steepness of a hill is its *slope*.

1 Find the Slope of a Line Given Two Points

Consider the staircase drawn in Figure 20(a). If we draw a line through the top of each riser on the staircase (in blue), we can see that each step contains exactly the same horizontal change (or **run**) and the same vertical change (or **rise**).

Figure 20

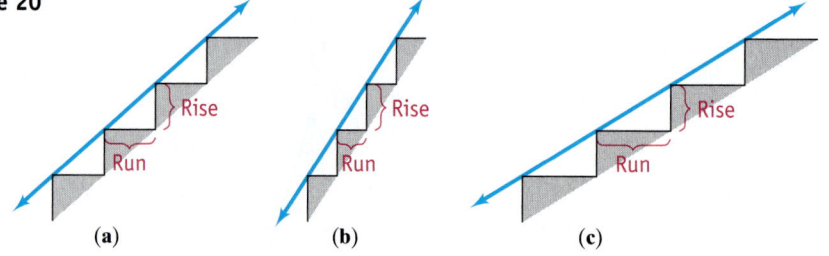

(a)   (b)   (c)

DEFINITION

The **slope** of a line, denoted by the letter m, is the ratio of the rise to the run. That is,

$$\text{Slope} = m = \frac{\text{Rise}}{\text{Run}}$$

Slope is a numerical measure of the steepness of the line. For example, if the run is decreased and the rise remains the same, then the staircase becomes steeper. See Figure 20(b). If the run is increased and the rise remains the same, then the staircase becomes less steep. See Figure 20(c).

Suppose that the staircase in Figure 20(a) has a rise of 6 inches and a run of 7 inches. Then the slope of the line is

$$m = \frac{\text{rise}}{\text{run}}$$

$$= \frac{6 \text{ inches}}{7 \text{ inches}}$$

If the rise of the stair is increased to 9 inches, then the slope of the line is

$$m = \frac{\text{rise}}{\text{run}}$$

$$= \frac{9 \text{ inches}}{7 \text{ inches}}$$

The main idea is the steeper the line, the larger the slope. We can define the slope of a line using rectangular coordinates.

Work Smart

The subscripts 1 and 2, on x_1, x_2, y_1, and y_2 do not represent a computation (as superscripts do in x^2). Instead, they are used to indicate that the values of the variable x_1 may be different from x_2, and y_1 may be different from y_2.

DEFINITION

If $x_1 \neq x_2$, the **slope m** of the line containing the points (x_1, y_1) and (x_2, y_2) is defined by the formula

$$m = \frac{y_2 - y_1}{x_2 - x_1}, \qquad x_1 \neq x_2$$

Figure 21 provides an illustration of the slope of a line.

In Words

The accepted symbol for the slope of a line is m. It comes from the French word *monter*, which means "to go up, ascend, or climb."

Figure 21

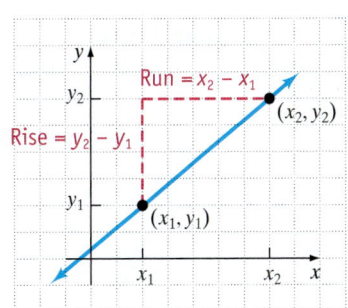

From Figure 21 we can see that the slope m of a line may be viewed as

$$m = \frac{\text{rise}}{\text{run}} = \frac{y_2 - y_1}{x_2 - x_1}$$

In Words

Slope is rise over run, or the change in y divided by the change in x.

We can also write the slope m of a line as

$$m = \frac{y_2 - y_1}{x_2 - x_1} = \frac{\text{change in } y}{\text{change in } x} = \frac{\Delta y}{\Delta x}$$

Work Smart

The symbol Δ comes from the first letter of the Greek word *dunamis*, which means "change."

The symbol Δ is the Greek letter "delta." In mathematics, we read the symbol Δ as "change in." So the notation $\dfrac{\Delta y}{\Delta x}$ is read "change in y divided by change in x."

EXAMPLE 1   **How to Find the Slope of a Line**

Find the slope of the line drawn in Figure 22.

Figure 22

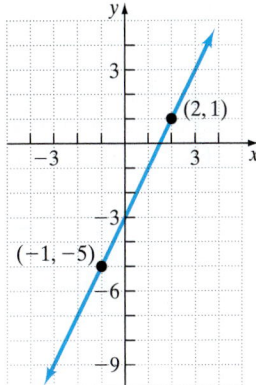

(continued)

Step-by-Step Solution

Step 1: Let one of the points be (x_1, y_1) and the other point be (x_2, y_2).

Let's say that $(x_1, y_1) = (-1, -5)$ and $(x_2, y_2) = (2, 1)$.

Step 2: Find the slope by evaluating

$$m = \frac{y_2 - y_1}{x_2 - x_1} = \frac{\text{Change in } y}{\text{Change in } x} = \frac{\Delta y}{\Delta x}$$

$$m = \frac{y_2 - y_1}{x_2 - x_1}$$

$$x_1 = -1, y_1 = -5; \quad = \frac{1 - (-5)}{2 - (-1)}$$
$$x_2 = 2, y_2 = 1:$$

$$= \frac{6}{3}$$

$$= 2$$

> **Work Smart**
>
> It doesn't matter which point is called (x_1, y_1) and which is called (x_2, y_2). The answer will be the same. In Example 1, if we let $(x_1, y_1) = (2, 1)$ and $(x_2, y_2) = (-1, -5)$, we obtain
>
> $$m = \frac{y_2 - y_1}{x_2 - x_1}$$
>
> $$= \frac{-5 - 1}{-1 - 2}$$
>
> $$= \frac{-6}{-3}$$
>
> $$= 2$$

Figure 23

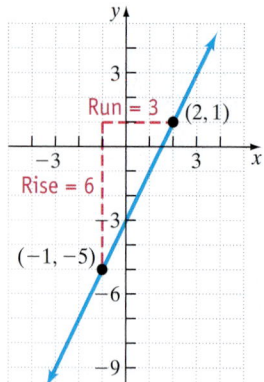

Remember—we said that the slope of the line can be thought of as "rise divided by run." This description of the slope of a line is illustrated in Figure 23.

We interpret the slope of the line drawn in Figure 23 as follows: "For every 6-unit increase in y, the values of x increase by 3 units." Or because $\frac{6}{3} = 2 = \frac{2}{1}$ "for every 2-unit increase in y, the value of x will increase by 1." Both interpretations are acceptable.

EXAMPLE 2 Finding and Interpreting the Slope of a Line

Plot the points $(-1, 3)$ and $(2, -2)$ in a rectangular coordinate system. Then draw a line through the two points. Find and interpret the slope of the line.

Solution

We plot the points $(x_1, y_1) = (-1, 3)$ and $(x_2, y_2) = (2, -2)$ in the rectangular coordinate system and draw a line through the two points. See Figure 24. The slope of the line drawn in Figure 24 is

$$m = \frac{y_2 - y_1}{x_2 - x_1} = \frac{-2 - 3}{2 - (-1)}$$

$$= \frac{-5}{3}$$

$$= -\frac{5}{3}$$

Figure 24

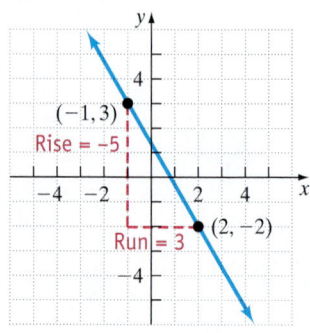

You can interpret a slope of $-\dfrac{5}{3} = \dfrac{-5}{3}$ this way: The value of y will go down 5 units whenever x increases by 3 units. Because $-\dfrac{5}{3} = \dfrac{5}{-3} = \dfrac{\text{rise}}{\text{run}}$, a second interpretation is as follows: The value of y will increase by 5 units whenever x decreases by 3 units. ■

Notice that the line drawn in Figure 23 goes up and to the right and the slope is positive, while the line drawn in Figure 24 goes down and to the right and slope is negative. In general, a line that goes down and to the right will have negative slope and a line that goes up and to the right will have positive slope. We illustrate this idea in Figure 25.

Figure 25

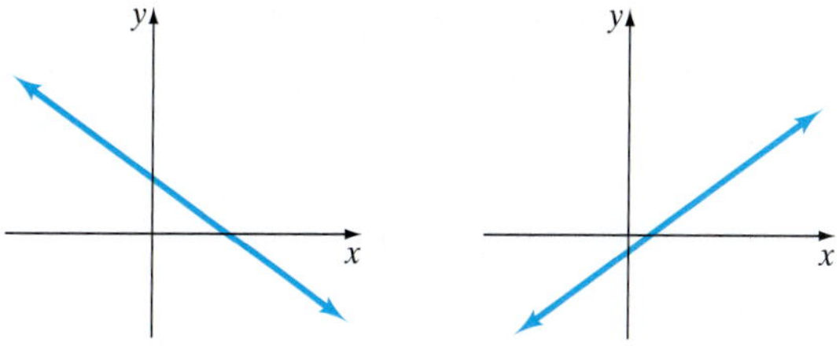

Line goes down and to the right: negative slope

Line goes up and to the right: positive slope

Work Smart

We read a graph from left to right, just like we read a book.

QUICK ✓ *Find and interpret the slope of the line containing the points.*

1. $(0, 4); (2, 10)$

2. $(-2, 3); (2, -7)$

---

(2) **Find the Slope of Vertical and Horizontal Lines**

Did you notice in the definition of slope, $m = \dfrac{y_2 - y_1}{x_2 - x_1}$, we have the restriction that $x_1 \neq x_2$? This means that the formula does not apply if the x-coordinates of the two points are the same. Why? Let's look at the following example.

EXAMPLE 3   **The Slope of a Vertical Line**

Plot the points $(2, -1)$ and $(2, 3)$ in a rectangular coordinate system. Then draw a line through the two points. Find and interpret the slope of the line.

Solution

We plot the points $(x_1, y_1) = (2, -1)$ and $(x_2, y_2) = (2, 3)$ in the rectangular coordinate system and draw a line through the two points. See Figure 26. The slope of the line drawn in Figure 26 is

Figure 26

$$m = \frac{y_2 - y_1}{x_2 - x_1} = \frac{3 - (-1)}{2 - 2}$$

$$= \frac{4}{0}$$

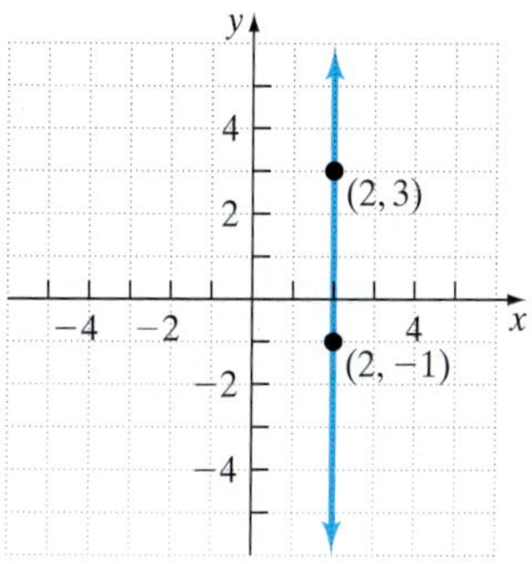

Because division by 0 is undefined, we say that the slope of the line is undefined. When y increases by 1, there is no change in x.

Let's generalize the results of Example 3. Let (x_1, y_1) and (x_2, y_2) be two distinct points. If $x_1 = x_2$, then we have a **vertical line** whose slope m is **undefined** (since this results in division by 0). Figure 27 illustrates a vertical line.

Figure 27
Slope of a vertical line.

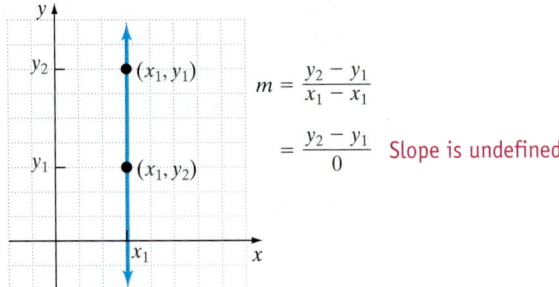

Okay, but what if $y_1 = y_2$?

EXAMPLE 4 The Slope of a Horizontal Line

Plot the points $(-2, 4)$ and $(3, 4)$ in a rectangular coordinate system. Then draw a line through the two points. Finally, find and interpret the slope of the line.

Solution

We plot the points $(x_1, y_1) = (-2, 4)$ and $(x_2, y_2) = (3, 4)$ in the rectangular coordinate system and draw a line through the two points. See Figure 28. The slope of the line drawn in Figure 28 is

$$m = \frac{y_2 - y_1}{x_2 - x_1} = \frac{4 - 4}{3 - (-2)}$$

$$= \frac{0}{5}$$

$$= 0$$

Figure 28

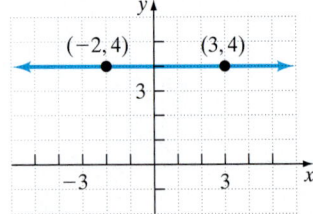

The slope of the line is 0. A slope of 0 can be interpreted as: There is no change in y when x increases by 1 unit.

Let's generalize the results of Example 4. Let $P = (x_1, y_1)$ and $Q = (x_2, y_2)$ be two distinct points. If $y_1 = y_2$, then we have a **horizontal line** whose slope m is 0. Figure 29 illustrates a horizontal line.

Figure 29

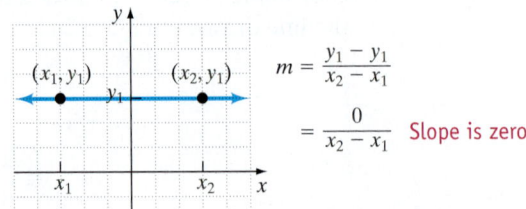

QUICK ✔ *Plot the given points in a rectangular coordinate system. Then draw a line through the two points. Finally, find and interpret the slope of the line.*

3. $(2, 5), (2, -1)$ **4.** $(2, 5), (6, 5)$ **5.** $(-3, -2), (1, -2)$ **6.** $\left(\frac{3}{2}, \frac{4}{3}\right), \left(\frac{3}{2}, \frac{8}{3}\right)$

SUMMARY: The Slope of a Line

Figure 30 illustrates the four possibilities for the slope of a line. Remember, just as we read a text from left to right, we also read graphs from left to right.

Figure 30

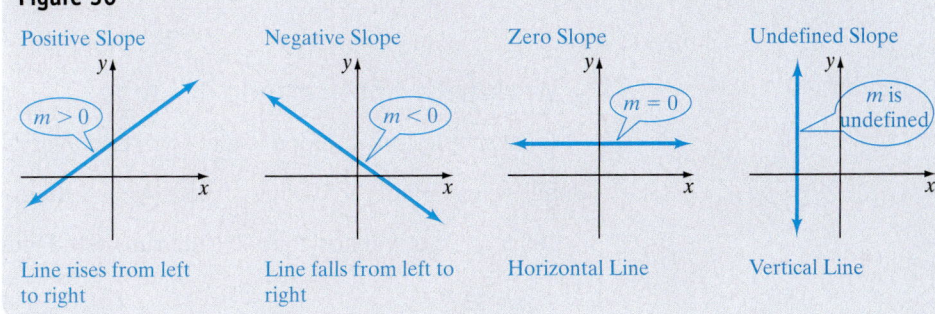

Positive Slope
$m > 0$
Line rises from left to right

Negative Slope
$m < 0$
Line falls from left to right

Zero Slope
$m = 0$
Horizontal Line

Undefined Slope
m is undefined
Vertical Line

(3) Graph a Line Using Its Slope and a Point on the Line

We now illustrate how to use the slope of a line to graph lines.

EXAMPLE 5 Graphing a Line Given a Point and Its Slope

Draw a graph of the line that contains the point $(1, 3)$ and has a slope of 2.

Solution

In Words
You can graph a line if you have a single point on the line and the slope of the line.

Because slope $= \dfrac{\text{rise}}{\text{run}}$, we have that $2 = \dfrac{2}{1} = \dfrac{\text{rise}}{\text{run}}$. This means that y will increase by 2 units (the rise), when x increases by 1 unit (the run). So if we start at $(1, 3)$ and move 2 units up and then 1 unit to the right, we end up at the point $(2, 5)$. We then draw a line through the points $(1, 3)$ and $(2, 5)$ to obtain the graph of the line. See Figure 31.

Figure 31

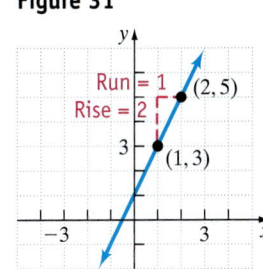

Run = 1
Rise = 2
$(2, 5)$
$(1, 3)$

EXAMPLE 6 Graphing a Line Given a Point and Its Slope

Draw a graph of the line that contains the point $(1, 3)$ and has a slope of $-\dfrac{2}{3}$.

Solution

Because slope $= \dfrac{\text{rise}}{\text{run}}$, we have $-\dfrac{2}{3} = \dfrac{-2}{3} = \dfrac{\text{rise}}{\text{run}}$. This means that y will decrease by 2 units when x increases by 3 units. If we start at $(1, 3)$ and move 2 units down and then 3 units to the right, we end up at the point $(4, 1)$. We then draw a line through the points $(1, 3)$ and $(4, 1)$ to obtain the graph of the line. See Figure 32.

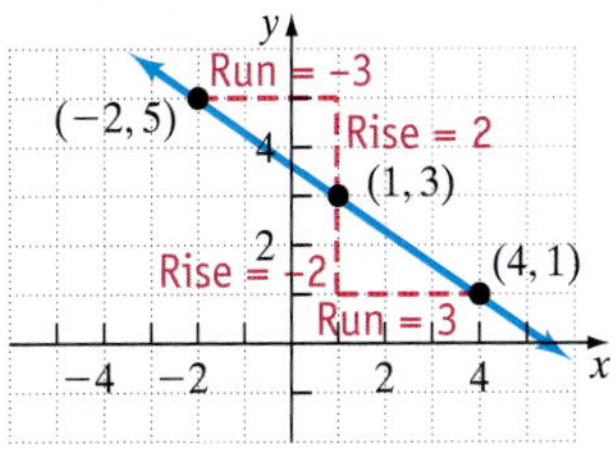

Work Smart
If the "rise" is positive, we go up. If the "rise" is negative, we go down. Similarly, if the "run" is positive, then we go to the right. If the "run" is negative, then we go to the left.

It is perfectly acceptable to set $\dfrac{\text{rise}}{\text{run}} = -\dfrac{2}{3} = \dfrac{2}{-3}$ so that we move 2 units up from $(1, 3)$ and then 3 units to the left. We would then end up at $(-2, 5)$, which is also on the graph of the line as indicated in Figure 32.

Figure 32

Run = −3
$(-2, 5)$
Rise = 2
$(1, 3)$
Rise = −2
$(4, 1)$
Run = 3

QUICK ✓

7. Draw a graph of the line that contains the point $(1, 2)$ and has a slope of

(a) $\dfrac{1}{2}$ **(b)** -3 **(c)** 0

---

④ Work with Applications of Slope

In its simplest form, slope is a ratio of rise over run. For example, if we are climbing a hill whose grade is 5% $\left(= 0.05 = \dfrac{5}{100} \right)$, then we go up 5 feet (the rise) for every 100 feet we travel horizontally (the run). See Figure 33.

Figure 33

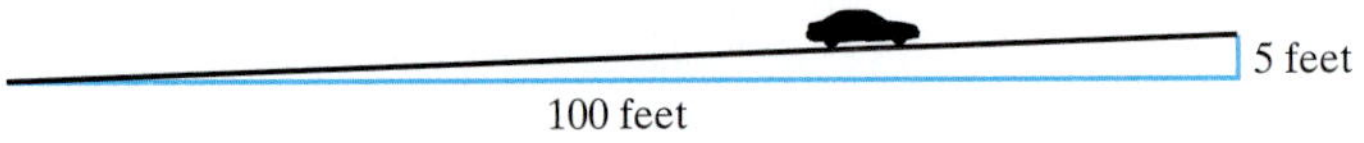

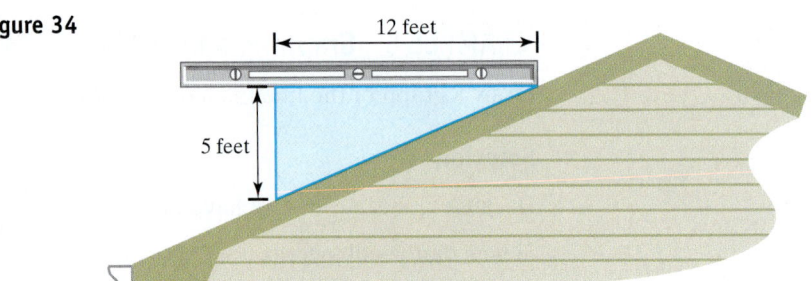

100 feet   5 feet

Or consider the pitch of a roof. If a roof's pitch is $\dfrac{5}{12}$, then every 5-foot measurement downward will result in a horizontal measurement of 12 feet. See Figure 34.

Work Smart

The pitch of a roof or grade of a road is always represented as a positive number.

Figure 34

12 feet

5 feet

EXAMPLE 7   **Finding the Grade of a Road**

In Heckman Pass, British Columbia, there is a road that rises 9 feet for every 50 feet of horizontal distance covered. What is the grade of the road?

Solution

The grade of the road is given by $\dfrac{\text{rise}}{\text{run}}$. Since a rise of 9 feet is accompanied by a run of 50 feet, the grade of the road is $\dfrac{9 \text{ feet}}{50 \text{ feet}} = 0.18 = 18\%$.

The slope m of a line measures the amount that y changes as x changes from x_1 to x_2. The slope of a line is also called the **average rate of change** of y with respect to x.

In applications, we are often interested in knowing how the change in one variable might impact some other variable. For example, if your income increases by \$1,000, how much will your spending (on average) change? Or, if the speed of your car increases by 10 miles per hour, how much (on average) will your car's gas mileage change?

EXAMPLE 8   **Slope as an Average Rate of Change**

In Naples, Florida, the price of a new three-bedroom house that is 1828 square feet costs \$280,000. The price of a new three-bedroom home that is 1987 square feet costs \$296,000. Find and interpret the slope of the line joining the points $(1828, 280000)$ and $(1987, 296000)$.

Solution

Let x represent the square footage of the house and y represent the price. Let $(x_1, y_1) = (1828, 280000)$ and $(x_2, y_2) = (1987, 296000)$ and compute the slope as

$$m = \frac{y_2 - y_1}{x_2 - x_1} = \frac{296{,}000 - 280{,}000}{1987 - 1828}$$

$$= \frac{16{,}000}{159}$$

$$= 100.63$$

The unit of measure of y is dollars while the unit of measure for x is square feet. So, the slope can be interpreted as follows: Between 1828 and 1987 square feet, the price increases by \$100.63 per square foot, on average. ■

QUICK ✓

8. A road rises 4 feet for every 50 feet of horizontal distance covered. What is the grade of the road?

9. A roof is pitched so that the vertical drop is 2 inches for every horizontal measurement of 12 inches. What is the pitch of the roof?

10. The annual cost of gasoline and maintenance on a Chevy Cobalt is \$1370 when it is driven 10,000 miles. The annual cost of gasoline and maintenance on a Chevy Cobalt is \$1850 when it is driven 14,000 miles. Find and interpret the slope of the line joining (10000, 1370) and (14000, 1850).

3.3 Exercises

For Extra Help:

Student Solutions Manual   CD Video   PH Math/Tutor Center   MathXL Tutorials on CD   MathXL®   MyMathLab

Concepts and Vocabulary

In Problems 1–3, fill in the blanks.

1. If the run of a line is 10 and its rise is 6, then its slope is _____.

2. The slope of a horizontal line is _____, while the slope of a vertical line is _____.

3. If the graph of a line goes up as you move to the right, then the slope of this line must be _____.

In Problems 4–6, answer True or False to each statement.

4. If $P = (x_1, y_1)$ and $Q = (x_2, y_2)$ are two distinct points with $y_1 \neq y_2$, the slope m of the line that contains points P and Q is defined by the formula $m = \dfrac{x_2 - x_1}{y_2 - y_1}$.

5. The slope of a line is also called the average rate of change of y with respect to x.

6. If the slope of a line is $\dfrac{3}{2}$, then y will increase by 3 units when x increases by 2 units.

7. Describe a line that has one x-intercept but no y-intercept. Give two ordered pairs that could lie on this line and then describe how to find its slope.

8. Describe a line that has one y-intercept but no x-intercept. Give two ordered pairs that could lie on this line and then describe how to find its slope.

Building Skills

In Problems 9–16, find the slope of the line whose graph is given.

9.

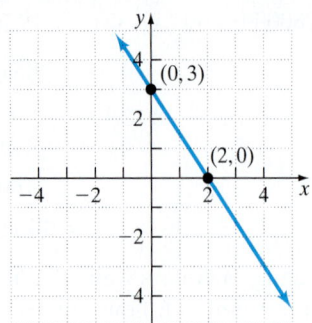

10.

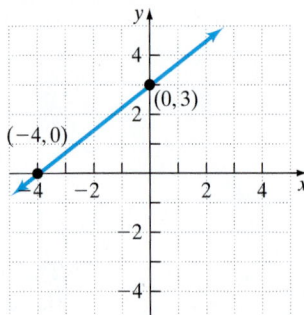

11.

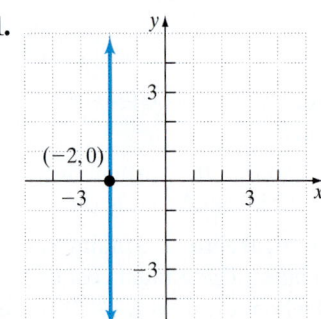

12.

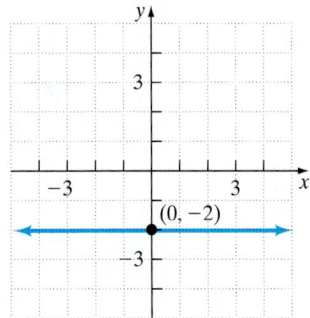

13.

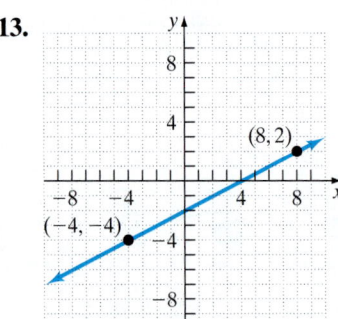

14.

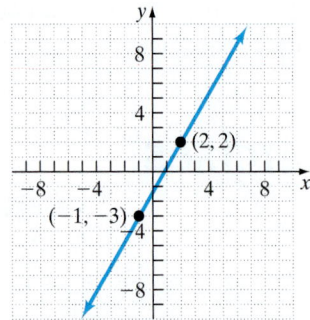

15.

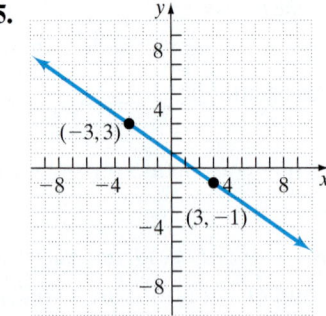

16.

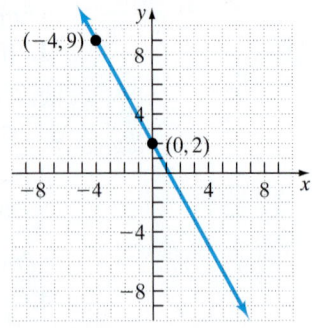

In Problems 17–20, (a) plot the points in a rectangular coordinate system, (b) draw a line through the points, (c) find and interpret the slope of the line.

17. $(-3, 2)$ and $(3, 5)$

18. $(-2, 6)$ and $(-2, -4)$

19. $(3, -1)$ and $(-2, -1)$

20. $(4, -5)$ and $(-2, -4)$

In Problems 21–40, find and interpret the slope of the line containing the given points.

21. $(10, 4)$ and $(6, 12)$

22. $(7, 3)$ and $(0, -11)$

23. $(4, -4)$ and $(12, -12)$

24. $(-3, 2)$ and $(2, -3)$

25. $(7, -2)$ and $(4, 3)$

26. $(-8, -1)$ and $(2, 3)$

27. $(0, 6)$ and $(-4, 0)$ **28.** $(-5, 0)$ and $(0, 3)$ ⊙ **29.** $(4, -6)$ and $(-1, -6)$

30. $(-1, -3)$ and $(-1, 2)$ **31.** $(-4, -1)$ and $(2, 3)$ **32.** $(5, 1)$ and $(-1, -1)$

⊙ **33.** $(3, 9)$ and $(3, -2)$ **34.** $(5, 1)$ and $(-2, 1)$ **35.** $\left(\frac{1}{2}, \frac{3}{4}\right)$ and $\left(-\frac{5}{2}, -\frac{1}{4}\right)$

36. $\left(-\frac{1}{3}, \frac{2}{5}\right)$ and $\left(\frac{2}{3}, -\frac{3}{5}\right)$ **37.** $\left(\frac{1}{3}, \frac{4}{9}\right)$ and $\left(-\frac{1}{3}, \frac{2}{9}\right)$ **38.** $\left(\frac{1}{4}, -\frac{4}{3}\right)$ and $\left(-\frac{5}{4}, \frac{1}{3}\right)$

39. $\left(\frac{1}{2}, \frac{1}{3}\right)$ and $\left(\frac{3}{4}, \frac{5}{6}\right)$ **40.** $\left(\frac{1}{3}, \frac{3}{4}\right)$ and $\left(-\frac{2}{5}, \frac{1}{2}\right)$

In Problems 41–58, draw a graph of the line that contains the given point and has the given slope.

41. $(4, 2)$; $m = 1$ **42.** $(3, -1)$; $m = -1$ **43.** $(0, 6)$; $m = -2$

44. $(-1, 3)$; $m = 3$ ⊙ **45.** $(-1, 0)$; $m = \frac{1}{4}$ **46.** $(5, 2)$; $m = -\frac{1}{2}$

47. $(2, -3)$; $m = 0$ **48.** $(1, 4)$; $m = $ undefined   **49.** $(2, 1)$; $m = \frac{2}{3}$

50. $(-2, -3)$; $m = \frac{5}{2}$ **51.** $(-1, -4)$; $m = \frac{5}{3}$ **52.** $(0, -2)$; $m = -\frac{3}{2}$

53. $(0, 0)$; $m = $ undefined   **54.** $(3, -1)$; $m = 0$ ⊙ **55.** $(0, 2)$; $m = -4$

56. $(0, 0)$; $m = \frac{1}{5}$ **57.** $(2, -3)$; $m = \frac{3}{4}$ **58.** $(-3, 0)$; $m = -3$

Applying the Concepts

In Problems 59–70, determine the missing value so that the line containing the two points will have the required slope.

59. $(3, 7)$ and $(x, 2)$; $m = 5$ **60.** $(-2, 6)$ and $(x, 4)$; $m = -2$

61. $(-1, 3)$ and $(9, y)$; $m = -\frac{1}{2}$ **62.** $(-1, -6)$ and $(4, y)$; $m = 2$

63. $(-2, 4)$ and $(x, 5)$; $m = $ undefined   **64.** $(0, 4)$ and $(5, y)$; $m = 0$

65. $(x, -8)$ and $(-2, -3)$; $m = -1$ **66.** $(-12, y)$ and $(3, -1)$; $m = -\frac{1}{3}$

67. $(x, -3)$ and $(-6, 0)$; $m = -\frac{1}{2}$ **68.** $(4, y)$ and $(-3, -1)$; $m = \frac{1}{7}$

69. $\left(\frac{1}{4}, \frac{1}{3}\right)$ and $\left(\frac{3}{4}, y\right)$; $m = 2$ **70.** $\left(\frac{1}{2}, \frac{3}{5}\right)$ and $\left(x, \frac{1}{5}\right)$; $m = \frac{2}{15}$

In Problems 71–74, draw the graph of the two lines with the given properties on the same rectangular coordinate system.

71. Both lines pass through the point $(2, -1)$. One has slope of 2 and the other has slope of $-\frac{1}{2}$.

72. Both lines pass through the point $(3, 0)$. One has slope of $\frac{2}{3}$ and the other has slope of $-\frac{3}{2}$.

73. Both lines have a slope of $\frac{3}{4}$. One passes through the point $(-1, -2)$, and the other passes through the point $(2, 1)$.

74. Both lines have a slope of -1. One passes through the point $(0, -3)$, and the other passes through the point $(2, -1)$.

75. Roof Pitch A carpenter who was installing a new roof on a garage noticed that for every one-foot horizontal run, the roof was elevated by 4 inches. What is the pitch of this roof?

76. Roof Pitch A canopy is set up on the football field. On the 45-yard line, the height of the canopy is 68 inches. The peak of the canopy is at the 50-yard marker where the height is 84 inches. What is the pitch of the roof of the canopy?

77. Building a Roof To build a shed in his back yard, Moises has decided to use a pitch of $\frac{2}{5}$ for his roof. The shed measures 30" from the side to the center. How much height should he add to the roof to get the desired pitch?

78. Building a Roof The design for the bedroom of a house requires a roof pitch of $\frac{7}{20}$. If the room measures 5 feet from the wall to the center, how high above the ceiling is the peak of the roof?

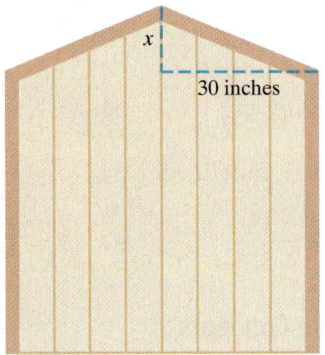

30 inches

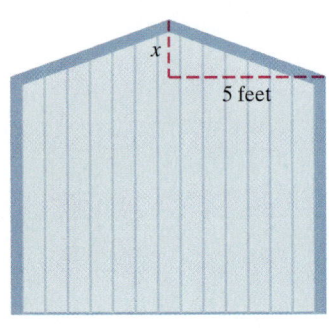

5 feet

79. Road Grade Fall River Road was completed in 1920 and was the first road built through the Rocky Mountains in Colorado. It was so steep that sometimes the early model cars had to drive up the hill in reverse to maximize their weak engines and fuel systems. If the road rises 200 feet for every 1250 feet of horizontal change, in percent, what is the grade of this road?

80. Road Grade Barbara decided to take a bicycle trip up to the observatory on Mauna Kea on the island of Hawaii. The road has a vertical rise of 120 feet for every 800 feet of horizontal change. In percent, what is the grade of this road?

81. Population Growth The population of the United States was 123,202,624 in 1930 and 281,421,906 in 2000. Use the ordered pairs $(0, 123 \text{ million})$ and $(70, 281 \text{ million})$ to find and interpret the slope of line representing the average rate of change in the population of the United States.

82. Earning Potential On average, a person who graduates from high school can expect to have lifetime earnings of 1.2 million dollars. It takes four years to earn a bachelor's degree, but the lifetime earnings will increase to 2.1 million dollars. Use the ordered pairs $(0, 1.2 \text{ million})$ and $(4, 2.1 \text{ million})$ to find and interpret the slope of the line representing the increase in earnings due to finishing college.

Extending the Concepts

In Problems 83–88, find any two ordered pairs that lie on the given line. Graph the line and then determine the slope of the line.

83. $3x + y = -5$

84. $2x + 5y = 12$

85. $y = -10$

86. $x = 7$

87. $y = 3x + 4$

88. $y = -x - 6$

In Problems 89–94, find the slope of the line containing the given points.

89. $(2a, a)$ and $(3a, -a)$

90. $(4p, 2p)$ and $(-2p, 5p)$

91. $(2p + 1, q - 4)$ and $(3p + 1, 2q - 4)$

92. $(3p + 1, 4q - 7)$ and $(5p + 1, 2q - 7)$

93. $(a + 1, b - 1)$ and $(2a - 5, b + 5)$

94. $(2a - 3, b + 4)$ and $(4a + 7, 5b - 1)$

*In economics, **marginal revenue** is a rate of change defined as the change in total revenue divided by the change in output. If Q_1 represents the number of units sold, then the total revenue from selling these goods is represented by R_1. If Q_2 represents a different number of units sold, then the total revenue from this sale is represented by R_2. We compute marginal revenue as*

$$MR = \frac{R_2 - R_1}{Q_2 - Q_1}.$$

So marginal revenue is a rate of change or slope. Marginal revenue is important in economics because it is used to determine the level of output that maximizes profits for a company. Use the marginal revenue formula to solve Problems 95 and 96.

95. Determine and interpret marginal revenue if total revenue is \$1000 when 400 hot dogs are sold at a baseball game and total revenue is \$1200 when 500 hot dogs are sold.

96. Determine and interpret marginal revenue if total revenue is \$300 when 30 compact disks are sold and total revenue is \$400 when 50 compact disks are sold.

3.4 Slope-Intercept Form of a Line

OBJECTIVES

1. Use the Slope-Intercept Form to Identify the Slope and y-Intercept of a Line
2. Graph a Line Whose Equation Is in Slope-Intercept Form
3. Graph a Line Whose Equation Is in the Form $Ax + By = C$
4. Find the Equation of a Line Given Its Slope and y-Intercept
5. Work with Linear Models in Slope-Intercept Form

Preparing for Slope-Intercept Form of a Line

Before getting started, take the following readiness quiz. If you get a problem wrong, go back to the section cited and review the material.

1. Solve $4x + 2y = 10$ for y. [Section 2.4, pp. 108–111]

2. Solve: $10 = 2x - 8$ [Section 2.2, pp. 84–86]

1 Use the Slope-Intercept Form to Identify the Slope and y-Intercept of a Line

We have defined a linear equation as an equation of the form $Ax + By = C$, where A and B are not both zero. So far, we have graphed equations of this form using point plotting or by finding its intercepts. From the previous section, we know the slope can be used to help us graph a line.

In this section, we use the slope and y-intercept to graph a line. This method for graphing will be more efficient than plotting points. Why? Well, suppose we wish to graph the equation $-2x + y = 5$ by plotting points. To do this, we will first solve the equation for y (get y by itself), then choose values of x and use the equation to find the corresponding value of y. The reason for doing this will become clear soon. We solve the equation $-2x + y = 5$ for y by adding $2x$ to both sides of the equation.

$$-2x + y = 5$$

Add 2x to both sides: $y = 2x + 5$

We now create Table 14, which gives us points on the graph of the equation. Figure 35 shows the graph of the line.

Notice two things about the line in Figure 35. First, the slope is $m = 2$. Second, the y-intercept is 5. If you look back at the form of the equation $-2x + y = 5$ after we solved for y, namely, $y = 2x + 5$, you should notice that the coefficient of the variable x is 2 and the constant is 5. This is no coincidence!

Table 14

x	$y = 2x + 5$	(x, y)
-2	$2(-2) + 5 = 1$	$(-2, 1)$
-1	$2(-1) + 5 = 3$	$(-1, 3)$
0	$2(0) + 5 = 5$	$(0, 5)$

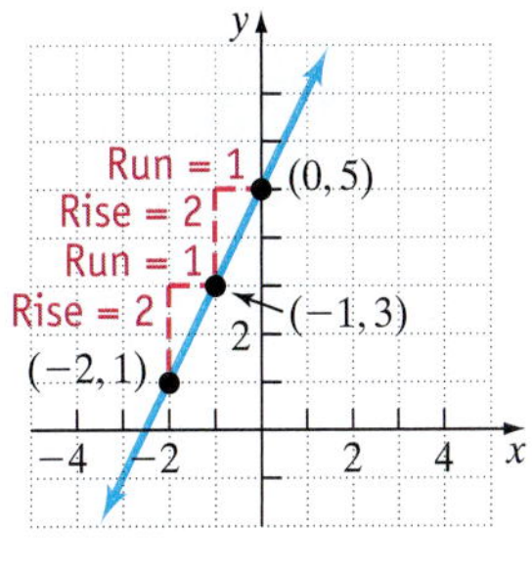

Figure 35

> **SLOPE-INTERCEPT FORM OF AN EQUATION OF A LINE**
> An equation of a line with slope m and y-intercept b is
> $$y = mx + b$$

EXAMPLE 1 | Finding the Slope and *y*-Intercept of a Line

Find the slope and y-intercept of the line whose equation is $y = -3x + 1$.

Solution

We compare the equation $y = -3x + 1$ to the slope-intercept form of a line $y = mx + b$ and find that the coefficient of x, -3, is the slope and the constant, 1, is the y-intercept.

EXAMPLE 2 | Finding the Slope and *y*-Intercept of a Line Whose Equation Is in Standard Form

Find the slope and y-intercept of the line whose equation is $3x + 2y = 6$.

Solution

First, we must rewrite the equation $3x + 2y = 6$ so that it is of the form $y = mx + b$. That is, we want to solve the equation for y.

$$3x + 2y = 6$$

Subtract 3x from both sides: $\quad 2y = -3x + 6$

Divide both sides by 2: $\quad y = \dfrac{-3x + 6}{2}$

$\dfrac{a + b}{c} = \dfrac{a}{c} + \dfrac{b}{c}:\quad y = \dfrac{-3}{2}x + \dfrac{6}{2}$

Simplify: $\quad y = -\dfrac{3}{2}x + 3$

Work Smart

Notice that after we subtracted $3x$ from both sides, we wrote the equation as $2y = -3x + 6$ rather than $2y = 6 - 3x$. This is because we want to get the equation in the form $y = mx + b$, so the term involving x should be first.

Now we compare the equation $y = -\dfrac{3}{2}x + 3$ to the slope-intercept form of a line $y = mx + b$ and find that the coefficient of x, $-\dfrac{3}{2}$, is the slope and the constant, 3, is the y-intercept. So for the line $3x + 2y = 6$, $m = -\dfrac{3}{2}$, and the y-intercept is $b = 3$.

QUICK ✔ *Find the slope and y-intercept of the line whose equation is given.*

1. $y = 4x - 3$ **2.** $3x + y = 7$ **3.** $2x + 5y = 15$

---

② Graph a Line Whose Equation Is in Slope-Intercept Form

In the previous section, we graphed an equation of a line using a point on the line and its slope. If an equation is in slope-intercept form, we can graph the line by plotting the y-intercept and using the slope to find another point on the line.

| EXAMPLE 3 | How to Graph a Line Whose Equation Is in Slope-Intercept Form |

Graph the line $y = 3x - 1$ using the slope and y-intercept.

Step-by-Step Solution

Step 1: Identify the slope and y-intercept of the line.

$$y = 3x - 1$$
$$y = 3x + (-1)$$

$m = 3$ $b = -1$

The slope is $m = 3$ and the y-intercept is $b = -1$.

Step 2: Plot the y-intercept and then use the slope to find a second point on the graph. Draw a straight line through the points.

Plot the y-intercept at $(0, -1)$. Use the slope
$$m = \frac{3}{1} = \frac{\text{rise}}{\text{run}}$$
to find a second point on the graph.
See Figure 36.

Figure 36

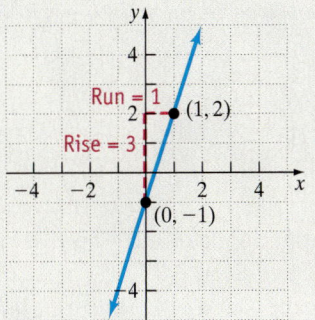

QUICK ✔ *Graph the line using the slope and y-intercept.*

4. $y = 2x - 5$

5. $y = \dfrac{1}{2}x - 5$

| EXAMPLE 4 | Graphing a Linear Equation Whose Equation Is in Slope-Intercept Form |

Graph the line $y = -\dfrac{4}{3}x + 2$ using the slope and y-intercept.

Solution

First, we determine the slope and y-intercept.

$$y = -\frac{4}{3}x + 2$$

$b = 2$

$m = -\dfrac{4}{3}$

Figure 37

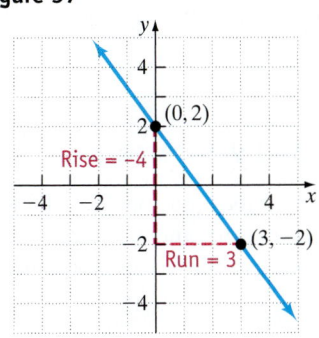

The slope is $m = -\dfrac{4}{3}$ and the y-intercept is $b = 2$. We plot the point $(0, 2)$. Now use

the slope $m = -\dfrac{4}{3} = \dfrac{-4}{3} = \dfrac{\text{rise}}{\text{run}}$ to find a second point on the graph. We then draw a

line through these two points. See Figure 37.

QUICK *Graph the line using the slope and y-intercept.*

6. $y = -3x + 1$

7. $y = -\dfrac{3}{2}x + 4$

---

③ Graph a Line Whose Equation Is in the Form $Ax + By = C$

If a linear equation is written in standard form $Ax + By = C$, we can still use the slope and y-intercept to obtain the graph of the equation. Let's see how.

EXAMPLE 5 **How to Graph a Line Whose Equation Is in the Form $Ax + By = C$**

Graph the line $8x + 2y = 10$ using the slope and y-intercept.

Step-by-Step Solution

Step 1: Solve the equation for y to put it in the form $y = mx + b$.

$$8x + 2y = 10$$

Subtract 8x from both sides: $2y = -8x + 10$

Divide both sides by 2: $y = \dfrac{-8x + 10}{2}$

Simplify: $y = -4x + 5$

Step 2: Identify the slope and y-intercept of the line.

The slope is -4 and the y-intercept is 5.

Step 3: Plot the y-intercept and then use the slope to find a second point on the graph. Draw a straight line through the points.

Plot the point $(0, 5)$ and use the slope $m = -4 = \dfrac{-4}{1} = \dfrac{\text{rise}}{\text{run}}$ to find a second point on the graph. See Figure 38.

Figure 38

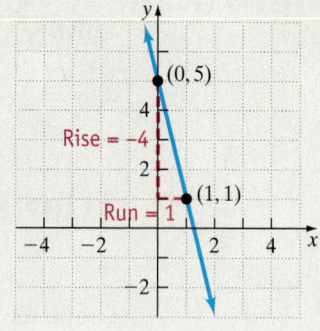

Work Smart

An alternative to graphing the equation in Example 5 using the slope and y-intercept would be to graph the line using intercepts.

QUICK *Graph each line using the slope and y-intercept.*

8. $-2x + y = -3$ **9.** $6x - 2y = 2$ **10.** $3x + 5y = 0$

---

④ Find the Equation of a Line Given Its Slope and y-Intercept

Up to now, we have identified the slope and y-intercept from its equation. We will now reverse the process and find the equation of a line whose slope and y-intercept are given. This is a fairly straightforward process—replace m with the given slope and b with the y-intercept.

EXAMPLE 6 **Finding the Equation of a Line Given Its Slope and *y*-Intercept**

Find the equation of a line whose slope is $\frac{3}{8}$ and whose *y*-intercept is -4. Graph the line.

Solution

The slope is $m = \frac{3}{8}$ and the *y*-intercept is $b = -4$. Substitute $\frac{3}{8}$ for m and -4 for b in the slope-intercept form of a line $y = mx + b$ to obtain

$$y = \frac{3}{8}x - 4$$

Figure 39 shows the graph of the equation.

Figure 39

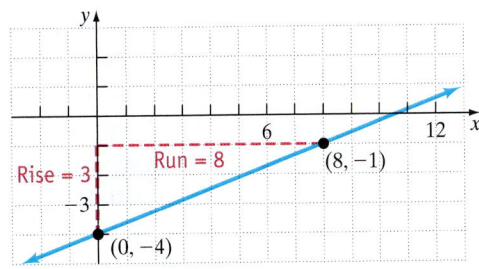

QUICK ✓ *Find the equation of the line whose slope and y-intercept are given. Graph the line.*

11. $m = 3, b = -2$ **12.** $m = -\frac{1}{4}, b = 3$ **13.** $m = 0, b = -1$

---

⑤ **Work with Linear Models in Slope-Intercept Form**

There are many situations where we can use a linear equation to describe the relationship that exists between two variables. For example, your long-distance phone bill depends linearly upon the number of minutes used or the cost of renting a moving truck depends linearly on the number of miles driven. Let's look at a couple of examples.

EXAMPLE 7 **A Model for Total Cholesterol**

When you have a physical exam, your doctor draws blood for your "cholesterol test." Your total cholesterol count is measured in milligrams per deciliter (mg/dL). It is the sum of low-density lipoprotein cholesterol (LDL)—sometimes called "bad cholesterol"—and high-density lipoprotein cholesterol (HDL)—sometimes called your "good cholesterol." Based on data from the National Center for Health Statistics, a woman's total cholesterol y is related to her age x by the following linear equation:

$$y = 1.1x + 157$$

(a) Use the equation to predict the total cholesterol of a woman who is 40 years old.

(b) Determine and interpret the slope of the equation.

(c) Determine and interpret the *y*-intercept of the equation.

(d) Graph the equation in a rectangular coordinate system.

Solution

(a) Because x represents the woman's age, we substitute 40 for x in the equation $y = 1.1x + 157$ to find the total cholesterol y.

$$y = 1.1x + 157$$
$$x = 40: \quad y = 1.1(40) + 157$$
$$= 201$$

We predict that a 40-year-old woman will have a total cholesterol of 201 mg/dL.

(b) The slope of the equation $y = 1.1x + 157$ is 1.1. Because slope equals $\dfrac{\text{rise}}{\text{run}} = \dfrac{1.1 \text{ mg/dL}}{1 \text{ year}}$, we interpret the slope as follows: The total cholesterol of a female increases by 1.1 mg/dL as age increases by 1 year."

(c) The y-intercept of the equation $y = 1.1x + 157$ is 157. The y-intercept is the value of total cholesterol, y, when $x = 0$. Since x represents age, we interpret the y-intercept as follows: "The total cholesterol of a newborn girl is 157 mg/dL."

(d) Figure 40 shows the graph of the equation. Because it does not make sense for x to be less than 0, we only graph the equation in quadrant I.

Figure 40

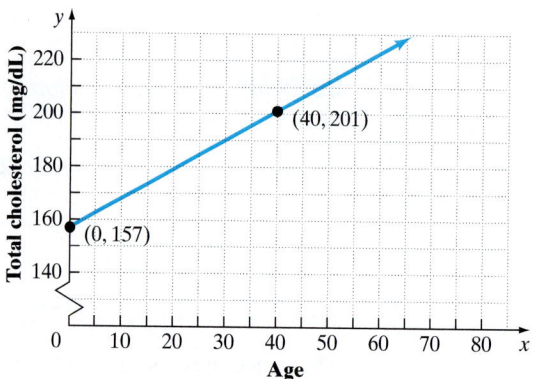

In Figure 40, notice the "broken line" ($\prec$) on the y-axis near the origin. We include this in the graph to indicate that a portion of the graph has been removed. This is done so that we do not have to start the y-axis at 0 and work our way higher. This avoids a lot of white space in the graph. Whenever you are reading a graph, always look carefully at how the axes are labeled and the units on each axis.

QUICK ✓

14. Based on data obtained from the National Center for Health Statistics, the birth weight y of a baby, measured in grams, is linearly related to gestation period x (in weeks) according to the equation

$$y = 143x - 2215$$

(a) Use the equation to predict the birth weight of a baby if the gestation period is 30 weeks.

(b) Use the equation to predict the birth weight of a baby if the gestation period is 36 weeks.

(c) Determine and interpret the slope of the equation.

(d) Explain why it does not make sense to interpret the y-intercept of the equation.

(e) Graph the equation in a rectangular coordinate system for $28 \leq x \leq 43$.

We know that the slope can be interpreted as a rate of change. For this reason, when information in a problem is given as a rate of change as in miles per gallon or dollars per pound, the rate of change will represent the slope in a linear model.

EXAMPLE 8 Cost of Owning and Operating a Car

There are many costs that factor into owning a car including gas, maintenance, and insurance. Some of these costs are affected by the number of miles that are driven (gas and maintenance), while others are not (comprehensive insurance, license plates, depreciation). Suppose the annual cost of operating a Chevy Cobalt is $0.25 per mile plus $3000.

(a) Write a linear equation that relates the annual cost of operating the car y to the number of miles driven in a year x.

(b) What is the annual cost of driving 11,000 miles?

(c) Graph the equation in a rectangular coordinate system.

Solution

(a) The rate of change in the problem is $0.25 per mile. We can express this as $\dfrac{\$0.25}{1 \text{ mile}}$, which is the slope m of the linear equation. The cost of $3000 is a cost that does not change with the number of miles driven. Put another way, if we drive 0 miles, the cost will be $3000, so this value represents the y-intercept, b. The linear equation that relates cost y to the number of miles driven x is

$$y = 0.25x + 3000$$

(b) Let $x = 11{,}000$ in the equation $y = 0.25x + 3000$.

$$y = 0.25(11{,}000) + 3000$$
$$= 2750 + 3000$$
$$= \$5750$$

The cost of driving 11,000 miles in a year is $5750. Remember, this cost includes gas, insurance, maintenance, and depreciation in the value of the vehicle!

(c) See Figure 41.

Figure 41

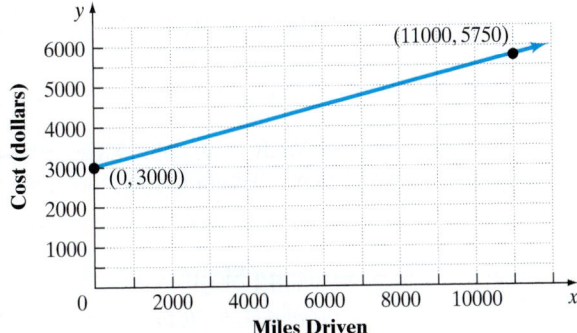

QUICK ✓

15. The daily cost, y, of renting a 16-foot moving truck for a day is $50 plus $0.38 per mile driven, x.

(a) Write a linear equation relating the daily cost y to the number of miles driven, x.

(b) Determine the cost of renting the truck if the truck is driven 75 miles.

(c) If the cost of renting the truck is $84.20, how many miles were driven?

(d) Graph the linear equation.

3.4 Exercises

Concepts and Vocabulary

In Problems 1–3, fill in the blanks.

1. The graph of the line whose equation is $y = 3x + 7$ has slope of _____ and y-intercept at _____.

2. To identify the slope and y-intercept in the equation $4x - 2y = 3$, you must first solve the equation for _____.

3. List three techniques that can be used to graph a line: _____, _____, _____.

In Problems 4–6, answer True or False to each statement.

4. The slope of the line $2y = 3x - 4$ is 3.

5. Every linear equation can be written in both standard form and slope-intercept form.

6. The graph below shows the graph of the linear equation $y = x + 2$.

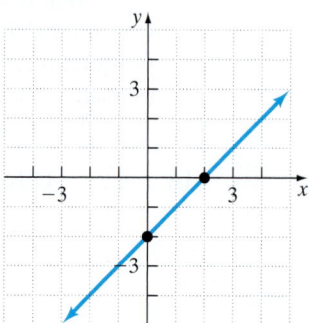

7. Which of the following equations could have the graph that is shown?

    (a) $y = 3x - 2$ **(b)** $y = -2x + 5$

    (c) $y = 3$ **(d)** $2x + 3y = 6$

    (e) $3x - 2y = 8$ **(f)** $4x - y = -4$

    (g) $-5x + 2y = 12$ **(h)** $x - y = -3$

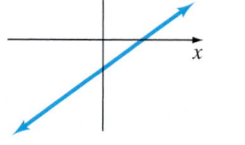

8. Without graphing, describe the orientation of each line (rises to the right, and so on). Explain how you came to this conclusion.

    (a) $y = 4x - 3$ **(b)** $y = -2x + 5$ **(c)** $y = x$ **(d)** $y = 4$

Building Skills

In Problems 9–30, find the slope and y-intercept of the line whose equation is given.

9. $y = 5x + 2$ **10.** $y = 7x + 1$ **11.** $y = 2x - 9$

12. $y = 3x - 7$ **13.** $y = -10x + 7$ **14.** $y = -6x + 2$

15. $y = -x - 9$ **16.** $y = -x - 12$ **17.** $y = -5$

18. $y = \dfrac{5}{4}x + 2$ **19.** $y = \dfrac{2}{3}x - 4$ **20.** $y = 3$

21. $2x - y = 4$ **22.** $3x - y = 9$ **23.** $2x + 3y = 24$

24. $6x - 8y = -24$ **25.** $5x - 3y = 1$ **26.** $2x + 6y = 8$

27. $x - 2y = 5$ **28.** $-x - 5y = 3$

29. $x = 6$ **30.** $x = -2$

In Problems 31–60, use the slope and y-intercept, if possible, to graph each line whose equation is given.

31. $y = x + 3$ **32.** $y = x + 4$ **33.** $y = -2x - 3$ **34.** $y = -4x - 1$

35. $y = -\dfrac{2}{3}x + 2$ **36.** $y = \dfrac{4}{3}x - 3$ **37.** $y = 0$ **38.** $x = 5$

39. $x = -6$ **40.** $y = 4$ **41.** $y = -\dfrac{5}{2}x - 2$ **42.** $y = -\dfrac{2}{5}x + 3$

43. $x + 2y = -6$ **44.** $x - 2y = -4$ **45.** $3x - 2y = 10$ **46.** $4x + 3y = -6$

47. $6x + 3y = -15$ **48.** $5x - 2y = 6$ **49.** $y = \dfrac{x}{3}$ **50.** $x = \dfrac{5}{2}$

51. $2x = -8y$ **52.** $y = -\dfrac{x}{4}$ **53.** $y = -\dfrac{4}{3}$ **54.** $-3x = 5y$

55. $y = -\dfrac{2x}{3} + 1$ **56.** $y = \dfrac{3x}{2} - 4$ **57.** $x + 2 = -7$ **58.** $y - 4 = -1$

59. $5x + y + 1 = 0$ **60.** $2x - y + 4 = 0$

In Problems 61–72, find the equation of the line with the given slope and intercept.

61. slope is -1; y-intercept is 8   **62.** slope is 1; y-intercept is 10

63. slope is $\dfrac{6}{7}$; y-intercept is -6 **64.** slope is $\dfrac{4}{7}$; y-intercept is -9

65. slope is $-\dfrac{1}{3}$; y-intercept is $\dfrac{2}{3}$ **66.** slope is $\dfrac{1}{4}$; y-intercept is $\dfrac{3}{8}$

67. slope is undefined; x-intercept is -5 **68.** slope is 0; y-intercept is -2

69. slope is 0; y-intercept is 3   **70.** slope is undefined; x-intercept is 4

71. slope is 5; y-intercept is 0   **72.** slope is -3; y-intercept is 0

Mixed Practice

In Problems 73–84, graph each equation using any method you wish.

73. $y = 2x - 7$ **74.** $y = -4x + 1$ **75.** $3x - 2y = 24$ **76.** $2x + 5y = 30$

77. $y = -5$ **78.** $x = -3$ **79.** $6x - 4y = 0$ **80.** $3x + 8y = 0$

81. $y = -\dfrac{5}{3}x + 6$ **82.** $y = -\dfrac{3}{5}x + 4$ **83.** $2y = x + 4$ **84.** $3y = x - 9$

Applying the Concepts

85. Weekly Salary Dien is paid a salary of \$400 per week plus an 8% commission on all sales he makes during the week.

    (a) Write a linear equation that calculates his weekly income, where y represents his income and x represents the amount of sales.

    (b) What is Dien's weekly income if he sold \$1200 worth of merchandise?

    (c) Graph the equation in a rectangular coordinate system. Label the axes appropriately.

86. Car Rental To rent a car for a day, Gloria pays \$75 plus \$0.10 per mile.

    (a) Write a linear equation that calculates the daily cost, y, to rent a car which will be driven x miles.

    (b) What is the cost to drive this car for 200 miles?

    (c) If Gloria paid \$87.50, how many miles did she drive?

    (d) Graph the equation in a rectangular coordinate system. Label the axes appropriately.

87. Cell Phone Costs The cost per minute for cell phone users has gone down over the years. In 1995, cell phone users paid, on the average, 56¢ per minute. In 2003, they paid 13¢ per minute. Assuming that the rate of decline of the cost per minute was constant, the cost per minute can be calculated by the equation $y = -5.375x + 56$, where x represents the number of years after 1995 and y represents the cost per minute of cell phone usage in cents.

(a) What was the cost per minute for a cell phone user in 1999?
(b) In which year did a cell phone user pay 23.75¢ per minute?
(c) Interpret the slope of $y = -5.375x + 56$.
(d) Can this trend continue indefinitely?
(e) Graph the equation in a rectangular coordinate system. Label the axes appropriately.

88. Counting Calories According to a 1989 National Academy of Sciences Report, the recommended daily intake of calories for males between the ages of 7 and 15 can be calculated by the equation $y = 125x + 1125$, where x represents the boy's age and y represents the recommended calorie intake.

(a) What is the recommended caloric intake for a 12-year-old boy?
(b) What is the age of a boy whose recommended caloric intake is 2250 calories?
(c) Interpret the slope of $y = 125x + 1125$.
(d) Why would this equation not be accurate for a 3-year-old male?
(e) Graph the equation in a rectangular coordinate system. Label the axes appropriately.

Extending the Concepts

In Problems 89–94, find the value of the missing coefficient so that the line will have the given property.

89. $2x + By = 12$; slope is $\dfrac{1}{2}$

90. $Ax + 2y = 5$; slope is $\dfrac{3}{2}$

91. $Ax - 2y = 10$; slope is -2

92. $12x + By = -1$; slope is -4

93. $x + By = \dfrac{1}{2}$; y-intercept is $-\dfrac{1}{6}$

94. $4x + By = \dfrac{4}{3}$; y-intercept is $\dfrac{2}{3}$

In Problems 95 and 96, use the following information. In business, a cost equation relates the total cost of producing a product or good such as a refrigerator, rug, or blender to the number of goods produced. The simplest cost model is the linear cost model. In the linear cost model, the slope of the linear equation represents the variable cost of producing a good—variable costs are costs that change with the level of output. Variable cost is reported as a rate of change, such as $40 per calculator. Examples of variable costs would be labor costs and materials. The y-intercept of the linear equation represents the fixed cost of production—these are costs that exist regardless of the level of production. Fixed costs would be cost of the manufacturing facility and insurance.

95. Cost Equations Suppose the variable cost of manufacturing a graphing calculator is $40 per calculator while the daily fixed cost is $4000.

(a) Write a linear equation that relates cost y to the number of calculators manufactured x.
(b) What is the daily cost of manufacturing 500 calculators?
(c) One day, the total cost was $19,000. How many calculators were manufactured?
(d) Graph the equation relating cost and number of calculators manufactured.

96. Cost Equations Suppose the variable cost of manufacturing a cellular telephone is $35 per phone, and the daily fixed cost is $3600.

(a) Write a linear equation that relates the daily cost y to the number of cellular telephones manufactured x.
(b) What is the daily cost of manufacturing 400 cellular phones?
(c) One day, the total cost was $13,225. How many cellular phones were manufactured?
(d) Graph the equation relating cost and number of cellular telephones manufactured.

3.5 Point-Slope Form of a Line

OBJECTIVES

1. Find the Equation of a Line Given a Point and a Slope
2. Find the Equation of a Line Given Two Points
3. Build Linear Models Using the Point-Slope Form of a Line

Preparing for Point-Slope Form of a Line
Before getting started, take the following readiness quiz. If you get a problem wrong, go back to the section cited and review the material.

1. Solve $y - 3 = 2(x + 1)$ for y.   [Section 2.4, pp. 108–111]

2. Evaluate $\dfrac{7 - 3}{4 - 2}$.   [Section 1.6, p. 52]

---

1 ## Find the Equation of a Line Given a Point and a Slope

We now have two forms for the equation of a line. We have the standard equation of a line $Ax + By = C$, where A and B are not both zero, and the slope-intercept form of a line $y = mx + b$, where m is the slope and b is the y-intercept. We now introduce another form for the equation of a line.

Suppose that we have a nonvertical line with slope m containing the point (x_1, y_1). For any other point (x, y) on the line, we know from the formula for the slope of a line that

$$m = \frac{y - y_1}{x - x_1}$$

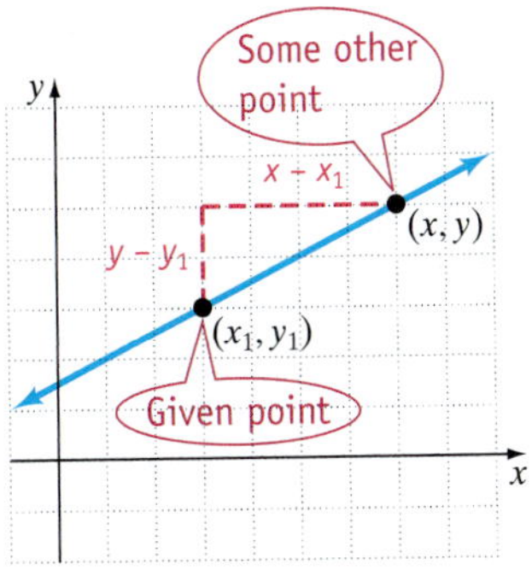

See Figure 42. Multiplying both sides by $x - x_1$, we can rewrite this expression as

$$m(x - x_1) = y - y_1 \quad \text{or} \quad y - y_1 = m(x - x_1)$$

Figure 42

> **POINT-SLOPE FORM OF AN EQUATION OF A LINE**
> An equation of a nonvertical line of slope m that contains the point (x_1, y_1) is
>
> $$\underset{\uparrow \text{ Given point } \uparrow}{y - y_1 = m\overset{\text{Slope}}{(x} - x_1)}$$

The point-slope form of a line can be used to write an equation in either slope-intercept form $(y = mx + b)$ or standard form $(Ax + By = C)$.

EXAMPLE 1 ### Using the Point-Slope Form of an Equation of a Line—Positive Slope

Find the equation of a line whose slope is 3 and contains the point $(-1, 3)$. Write the equation in slope-intercept form. Graph the line.

Figure 43

Solution

Because we are given the slope and a point on the line, we use the point-slope form of a line with $m = 3$ and $(x_1, y_1) = (-1, 3)$.

$$y - y_1 = m(x - x_1)$$

$m = 3, x_1 = -1, y_1 = 3$: $\quad y - \quad = (x - (\quad))$

$$y - 3 = 3(x + 1)$$

To put the equation in slope-intercept form, $y = mx + b$, we solve the equation for y.

Distribute: $\quad y - 3 = 3x + 3$

Add 3 to both sides: $\quad y = 3x + 6$

See Figure 43 for a graph of the line.

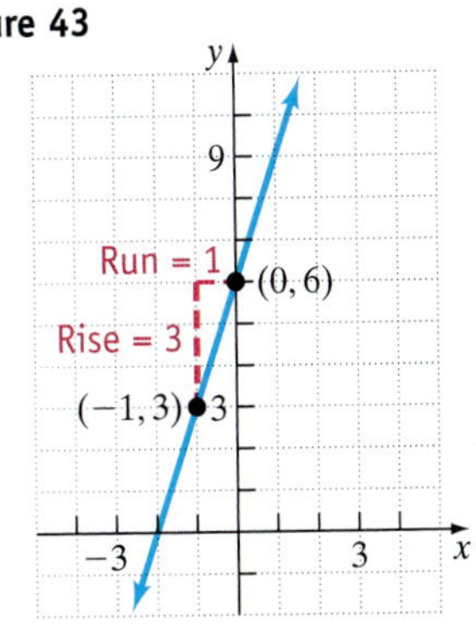

Preparing for...Answers **1.** $y = 2x + 5$
2. 2

QUICK ✓   *Find an equation of the line with the given properties. Write the equation in slope-intercept form. Graph the line.*

1. $m = 3$ containing $(x_1, y_1) = (2, 1)$ **2.** $m = \dfrac{1}{3}$ containing $(x_1, y_1) = (3, -4)$

EXAMPLE 2   **Using the Point-Slope Form of an Equation of a Line—Negative Slope**

Find the equation of a line whose slope is $-\dfrac{3}{4}$ and contains the point $(-4, 3)$. Write the equation in slope-intercept form. Graph the line.

Solution

Because we are given the slope and a point on the line, we use the point-slope form of a line with $m = -\dfrac{3}{4}$ and $(x_1, y_1) = (-4, 3)$.

Figure 44

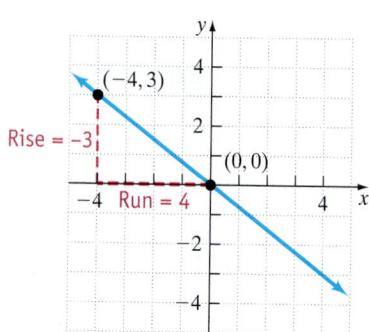

$$y - y_1 = m(x - x_1)$$

$m = -\frac{3}{4}, x_1 = -4, y_1 = 3{:}\quad y - 3 = -\dfrac{3}{4}(x - (-4))$

Simplify:$\quad y - 3 = -\dfrac{3}{4}(x + 4)$

Distribute:$\quad y - 3 = -\dfrac{3}{4}x - 3$

Add 3 to both sides:$\quad y = -\dfrac{3}{4}x$

See Figure 44 for a graph of the line.

QUICK ✓   *Find an equation of the line with the given properties. Write the equation in slope-intercept form. Graph the line.*

3. $m = -4$ containing $(x_1, y_1) = (-2, 5)$ **4.** $m = -\dfrac{5}{2}$ and $(x_1, y_1) = (-4, 5)$

EXAMPLE 3   **Finding the Equation of a Horizontal Line**

Find the equation of a horizontal line that contains the point $(-4, 2)$. Write the equation of the line in slope-intercept form. Graph the line.

Solution

The line is a horizontal line, so the slope of the line is 0. Because we know the slope and a point on the line, we use the point-slope form of a line with $m = 0$, $x_1 = -4$, and $y_1 = 2$.

$$y - y_1 = m(x - x_1)$$

$m = 0, x_1 = -4, y_1 = 2{:}\quad y - 2 = 0(x - (-4))$

$$y - 2 = 0$$

To put the equation of the line in slope-intercept form, we add 2 to both sides.

Add 2 to both sides:$\quad y = 2$ Slope-intercept form

See Figure 45 for a graph of the line.

Work Smart

When the slope of a line is 0, the equation of the line will always be in the form "$y =$ some number."

Figure 45

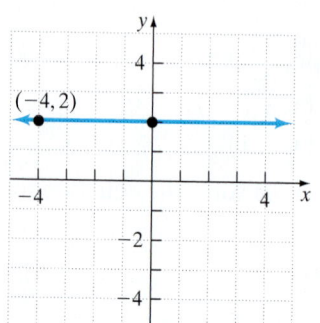

QUICK

5. Find the equation of a horizontal line that contains the point $(-2, 3)$. Write the equation of the line in slope-intercept form. Graph the line.

② Find the Equation of a Line Given Two Points

From Section 3.2, we know that two points are all that is needed to graph a line. If we are given two points, we can find an equation of the line through the points by first finding the slope of the line and then using the point-slope form of a line.

EXAMPLE 4   **How to Find the Equation of a Line from Two Points**

Find the equation of a line through the points $(1, 3)$ and $(4, 9)$. Write the equation in slope-intercept form. Graph the line.

Step-by-Step Solution

Step 1: Find the slope of the line containing the points.

Let $(x_1, y_1) = (1, 3)$ and $(x_2, y_2) = (4, 9)$. Substitute these values into the formula for the slope of a line.

$$m = \frac{y_2 - y_1}{x_2 - x_1} = \frac{9 - 3}{4 - 1} = \frac{6}{3} = 2$$

Step 2: Substitute the slope found in Step 1 and either point into the point-slope form of a line to find the equation.

With $m = 2$, $x_1 = 1$, and $y_1 = 3$, we have

$$y - y_1 = m(x - x_1)$$
$$m = 2, x_1 = 1, y_1 = 3: \quad y - 3 = 2(x - 1)$$

Step 3: Solve the equation for y.

Distribute the 2: $y - 3 = 2x - 2$

Add 3 to both sides: $y = 2x + 1$

The slope-intercept form of the equation is $y = 2x + 1$. The slope of the line is 2 and the y-intercept is 1. See Figure 46 for the graph.

Figure 46

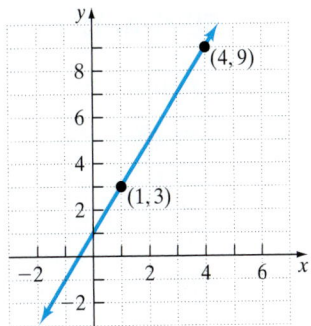

QUICK *Find the equation of the line containing the given points. Write the equation in slope-intercept form. Graph the line.*

6. $(0, 2); (3, 5)$

7. $(-1, 4); (1, -2)$

> **Work Smart: Study Skills**
>
> To write the equation of a nonvertical line, we must know either the slope of the line along with a point on the line or two points on the line.
>
> - If the **slope** and the **y-intercept** are known, use the slope-intercept form, $y = mx + b$.
> - If the **slope** and a **point** which is not the y-intercept are known, use the **point-slope** form, $y - y_1 = m(x - x_1)$.
> - If **two points** are known, first find the **slope,** then use that slope, along with one of the **points** in the **point-slope** formula, $y - y_1 = m(x - x_1)$.

EXAMPLE 5 **Finding an Equation of a Vertical Line from Two Points**

Find the equation of a line through the points $(-3, 2)$ and $(-3, -4)$. Write the equation in slope-intercept form. Graph the line.

Solution

Work Smart

The equation of a vertical line cannot be written in slope-intercept form.

Let $(x_1, y_1) = (-3, 2)$ and $(x_2, y_2) = (-3, -4)$. Substitute these values into the formula for the slope of a line.

$$m = \frac{y_2 - y_1}{x_2 - x_1} = \frac{-4 - 2}{-3 - (-3)} = \frac{-6}{0}$$

The slope is undefined, so the line is vertical. No matter what value of y we choose, the x-coordinate of the point on the line will be -3. For this reason, the equation of the line is $x = -3$. See Figure 47 for the graph.

Figure 47

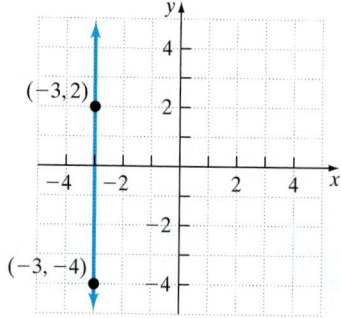

QUICK ✓

8. Find an equation of the line containing the points $(3, 2)$ and $(3, -4)$. If possible, write the answer in slope-intercept form. Graph the line.

SUMMARY: Equations of Lines

Form of Line	Formula	Comments
Horizontal line	$y = b$	Graph is a horizontal line (slope is 0) with y-intercept b.
Vertical line	$x = a$	Graph is a vertical line (undefined slope) with x-intercept a.
Point-slope	$y - y_1 = m(x - x_1)$	Useful for finding the equation of a line, given a point and a slope, or two points.
Slope-intercept	$y = mx + b$	Useful for finding the equation of a line, given the slope and y-intercept, or for quickly determining the slope and y-intercept of the line, given the equation of the line.
Standard form	$Ax + By = C$	Straightforward to find the x- and y-intercepts.

3 Build Linear Models Using the Point-slope Form of a Line

We can use the point-slope form of a line to build linear models from data.

| EXAMPLE 6 | Building a Linear Model from Data |

Healthcare costs are skyrocketing. For individuals 20 years of age or older, the percentage of total income y that an individual spends on healthcare increases linearly with age x. According to data obtained from the Bureau of Labor Statistics, a 35-year-old spends about 4.0% of income on healthcare, while a 65-year-old spends about 11.2% of income on healthcare.

(a) Plot the points $(35, 4.0)$ and $(65, 11.2)$ in a rectangular coordinate system and graph the line. Find the linear equation in slope-intercept form that relates the percent of income spent on healthcare y to the age x.

(b) Use the equation found in part (a) to predict the percentage of income that a 50-year-old spends on healthcare.

(c) Interpret the slope.

Solution

(a) We plot the ordered pairs $(35, 4.0)$ and $(65, 11.2)$ and draw a line through the points. See Figure 48.

Figure 48

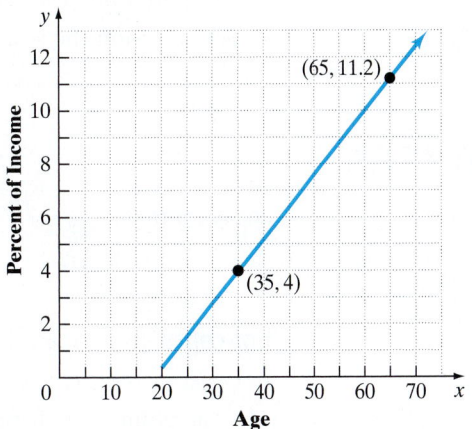

Because we know two points on the line, we will use the point-slope form of a line to find the equation of the line.

First, we must find the slope of the line:

$$m = \frac{y_2 - y_1}{x_2 - x_1} = \frac{11.2 - 4.0}{65 - 35}$$

$$= \frac{7.2}{30}$$

$$= 0.24$$

We use the point-slope form of a line with $m = 0.24$, $x_1 = 35$, and $y_1 = 4.0$:

$$y - y_1 = m(x - x_1)$$

$m = 0.24$, $x_1 = 35$, and $y_1 = 4.0$: $y - 4.0 = 0.24(x - 35)$

$$y - 4.0 = 0.24x - 8.4$$

Add 4.0 to both sides of the equation: $y = 0.24x - 4.4$

The equation $y = 0.24x - 4.4$ relates the percent of income spent on healthcare y to the age x.

(b) We substitute 50 for x in the equation found in part (a), so the predicted percentage of income spent on health care for a 50-year old is

$$y = 0.24x - 4.4$$
Let $x = 50$: $= 0.24(50) - 4.4$
$$= 7.6$$

We predict that 7.6% of a 50-year-old's income is spent on healthcare.

(c) The slope is 0.24. The percentage of income spent on healthcare for the individual increases by 0.24% as an individual ages by one year.

QUICK ✓

9. Armando owns a gas station. He has found that when the price of regular unleaded gasoline is \$2.20, he sells 400 gallons of gasoline between the hours of 7:00 A.M. and 8:00 A.M. When the price of regular unleaded gasoline is \$2.40, he sells 380 gallons of gasoline between the hours of 7:00 A.M. and 8:00 A.M. Suppose that the relation between the quantity of gasoline sold and price is linear.

(a) Plot the points in a rectangular coordinate system and graph the line. Find the linear equation in slope-intercept form that relates quantity of gasoline sold y to the price x.

(b) Use the equation found in part (a) to predict the number of gallons of gasoline sold if the price is \$2.30.

(c) Interpret the slope.

3.5 Exercises

For Extra Help:

Student Solutions Manual   CD Video   PH Math/Tutor Center   MathXL Tutorials on CD   MathXL®   MyMathLab

Concepts and Vocabulary

In Problems 1–3, fill in the blanks.

1. The point-slope form of a non-vertical line whose slope is m that contains the point (x_1, y_1) is _____.

2. The slope-intercept form of a non-vertical line whose slope is m and y-intercept is b is _____.

3. List the five forms that are used when writing the equation of a line: _____, _____, _____, _____, _____.

In Problems 4–6, answer True or False to each statement.

4. The line $y = 3$ is a vertical line.

5. The slope of the line $y - 3 = 4(x - 1)$ is 4.

6. The y-intercept of the line $y - 7 = 4(x - 1)$ is -1.

7. You are asked to write the equation of the line through the points $(3, 1)$ and $(4, 7)$ in slope-intercept form. After calculating the slope of the line, you choose the point-slope form and assign $x_1 = 3$ and $y_1 = 1$. Your friend lets $x_1 = 4$ and $y_1 = 7$. Will you and your friend obtain the same answer? Explain why or why not.

8. You are asked to write the equation of the line through $(-1, 3)$ and $(0, 4)$. Which form of a line would you choose to find the equation? Explain why you chose this form. Could you also use one of the other forms?

Building Skills

In Problems 9–26, find the equation of the line that contains the given point and slope. Write the equation in slope-intercept form and graph the line.

9. $(2, 5)$; slope $= 3$

10. $(4, 1)$; slope $= 6$

11. $(-1, 2)$; slope $= -2$

12. $(6, -3)$; slope $= -5$

13. $(8, -1)$; slope $= \dfrac{1}{4}$

14. $(-8, 2)$; slope $= -\dfrac{1}{2}$

15. $(0, 13)$; slope $= -6$

16. $(0, -4)$; slope $= 9$

17. $(5, -7)$; slope $= 0$

18. $(3, 12)$; undefined slope

19. $(-4, 5)$; undefined slope

20. $(-7, -1)$; slope $= 0$

21. $(-3, 0)$; slope $= \dfrac{2}{3}$

22. $(-10, 0)$; slope $= -\dfrac{4}{5}$

23. $(-5, -6)$; slope $= -\dfrac{3}{4}$

24. $(-5, -5)$; slope $= \dfrac{3}{2}$

25. $(4, -3)$; slope $= \dfrac{1}{2}$

26. $(2, -5)$; slope $= \dfrac{3}{2}$

In Problems 27–34, find the equation of the line that contains the given point and satisfies the given information. Write the equation in slope-intercept form, if possible.

27. Vertical line that contains $(-3, 10)$

28. Horizontal line that contains $(-6, -1)$

29. Horizontal line that contains $(-1, -5)$

30. Vertical line that contains $(4, -3)$

31. Horizontal line that contains $(0.2, -4.3)$

32. Vertical line that contains $(3.5, 2.4)$

33. Vertical line that contains $\left(\dfrac{1}{2}, \dfrac{7}{4}\right)$

34. Horizontal line that contains $\left(\dfrac{3}{2}, \dfrac{9}{4}\right)$

In Problems 35–50, find the equation of the line that contains the given points. Write the equation in slope-intercept form, if possible.

35. $(0, 4)$ and $(-2, 0)$

36. $(0, 3)$ and $(6, 0)$

37. $(1, 2)$ and $(0, 6)$

38. $(2, 4)$ and $(0, 8)$

39. $(-3, 2)$ and $(1, -4)$

40. $(-2, 4)$ and $(2, -2)$

41. $(-3, -11)$ and $(2, -1)$

42. $(4, 18)$ and $(-1, 3)$

43. $(4, -3)$ and $(-3, -3)$

44. $(-6, 5)$ and $(7, 5)$

45. $(2, -1)$ and $(2, -9)$

46. $(-3, 8)$ and $(-3, 1)$

47. $(0.1, 0.6)$ and $(0.5, 0.7)$

48. $(0.7, 0.8)$ and $(0.2, 0.4)$

49. $\left(\dfrac{1}{2}, -\dfrac{9}{4}\right)$ and $\left(\dfrac{5}{2}, -\dfrac{1}{4}\right)$

50. $\left(\dfrac{1}{3}, \dfrac{12}{5}\right)$ and $\left(\dfrac{4}{3}, \dfrac{2}{5}\right)$

Mixed Practice

In Problems 51–70, find the equation of the line described. Write the equation in slope-intercept form. Graph the line.

51. Contains $(4, -2)$ with slope $= 5$

52. Contains $(3, 2)$ with slope $= 4$

53. Horizontal line that contains $(-3, 5)$

54. Vertical line that contains $(-4, 2)$

55. Contains $(1, 3)$ and $(-4, -2)$

56. Contains $(-2, -8)$ and $(2, -6)$

57. Contains $(-2, 3)$ with slope $= \dfrac{1}{2}$

58. Contains $(-8, 3)$ with slope $= \dfrac{1}{4}$

59. Vertical line that contains $(5, 2)$

60. Horizontal line that contains $(-2, -6)$

61. Contains $(3, -19)$ and $(-1, 9)$

62. Contains $(-3, 13)$ and $(4, -22)$

63. Contains $(6, 3)$ with slope $= -\dfrac{2}{3}$

64. Contains $(-6, 3)$ with slope $= -\dfrac{2}{3}$

65. Contains $(-2, 3)$ and $(4, -6)$

66. Contains $(5, -3)$ and $(-3, 3)$

67. Contains $(-3, 0)$ with slope $= \dfrac{4}{5}$

68. Contains $(-10, 0)$ with slope $= -\dfrac{5}{4}$

69. Contains $(-3, 4)$ and $(6, 0)$

70. Contains $(8, -11)$ and $(-4, 4)$

Applying the Concepts

71. Shipping Packages The shipping department for a warehouse has noted that if 60 packages are shipped during a month, the total expenses for the department are \$1635. If 120 packages are shipped during a month, the total expenses for the shipping department are \$1770. Let x represent the number of packages and y represent the total expenses for the shipping department.

 (a) Interpret the meaning of the point $(60, 1635)$ in the context of this problem.
 (b) Plot the ordered pairs $(60, 1635)$ and $(120, 1770)$ in a rectangular coordinate system and graph the line through the points.
 (c) Find the linear equation in slope-intercept form that relates the total expenses for the shipping department, y, to the number of packages sent, x.
 (d) Use the equation found in part (c) to find the total expenses during a month when 200 packages were sent.
 (e) Interpret the slope.

72. Retirement Plans Based on the retirement plan available by his employer, Kei knows that if he retires after 20 years, his monthly retirement income will be \$3150. If he retires after 30 years, his monthly income increases to \$3600. Let x represent the number of years of service and y represent the monthly retirement income.

 (a) Interpret the meaning of the point $(30, 3600)$ in the context of this problem.
 (b) Plot the ordered pairs $(20, 3150)$ and $(30, 3600)$ in a rectangular coordinate system and graph the line through the points.
 (c) Find the linear equation in slope-intercept form that relates the monthly retirement income, y, to the number of years of service, x.
 (d) Use the equation found in part (c) to find the monthly income for 15 years of service.
 (e) Interpret the slope.

73. Traffic Fatalities In 1975, one region of the country had 181 traffic fatalities. In 2002, the same region had 276 fatalities. Let x represent the number of years after 1975 and let y represent the number of traffic fatalities.

 (a) Fill in the ordered pairs: (_____, 181); (_____, 276).
 (b) Plot the ordered pairs from part (a) in a rectangular coordinate system and graph the line through the points.
 (c) Find the linear equation in slope-intercept form that relates the number of traffic fatalities, y, to the number of years after 1975, x.
 (d) Use the equation found in part (c) to predict the number of traffic fatalities in 1990.
 (e) Interpret the slope.

74. U.S. Traffic Fatalities Nationwide, the statistics for traffic fatalities show a decline. In 1975, the United States had 39,161 fatal crashes while in 2002 the number dropped to 38,309. Let y represent the number of traffic fatalities, and x represent the number of years after 1975.

 (a) Fill in the ordered pairs: (_____, 39,161); (_____, 38,309).
 (b) Plot the ordered pairs from part (a) in a rectangular coordinate system and graph the line through the points.

(c) Find the linear equation in slope-intercept form that relates the number of traffic fatalities, y, to the number of years after 1975, x.

(d) Use the equation found in part (c) to find the number of traffic fatalities in 1987.

(e) Interpret the slope.

(f) Do you think that the slope is misleading in terms of describing the decline in traffic fatalities? Why?

Extending the Concepts

Up to this point, when we knew the slope of a line and a point on the line we found the equation of the line using point-slope form. We could also use the slope-intercept form to find this equation.

For example, suppose that we were asked to find the equation of the line whose slope is 6 and that passes through the point $(2, -5)$. Let's use the slope-intercept form, $y = mx + b$, to write the equation of this line. We know $m = 6$, so we have $y = 6x + b$. We also know that $y = -5$ when $x = 2$. So we substitute 2 for x and -5 for y into the equation and solve for b.

$$y = 6x + b$$

$$\text{Let } x = 2 \text{ and } y = -5: \qquad -5 = 6(2) + b$$

$$\text{Multiply:} \qquad -5 = 12 + b$$

$$\text{Subtract 12 from each side:} \qquad -5 - 12 = 12 - 12 + b$$

$$-17 = b$$

We now know that $b = -17$ and because we also know that $m = 6$, the equation of the line is $y = 6x - 17$. Use this technique to find the slope-intercept form of the line for Problems 75–86.

75. $(-4, 2)$; slope $= 3$ **76.** $(5, -2)$; slope $= 4$ **77.** $(3, -8)$; slope $= -2$

78. $(-1, 7)$; slope $= -5$ **79.** $\left(\dfrac{2}{3}, \dfrac{1}{2}\right)$; slope $= 6$ **80.** $\left(\dfrac{4}{3}, -\dfrac{3}{2}\right)$; slope $= -9$

81. $(6, -13)$ and $(-2, -5)$ **82.** $(-10, -5)$ and $(2, 7)$ **83.** $(5, -1)$ and $(-10, -4)$

84. $(-6, -3)$ and $(9, 2)$ **85.** $(-4, 8)$ and $(2, -1)$ **86.** $(4, -9)$ and $(8, -19)$

3.6 Parallel and Perpendicular Lines

OBJECTIVES

1. Determine Whether Two Lines Are Parallel
2. Find the Equation of a Line Parallel to a Given Line
3. Determine Whether Two Lines Are Perpendicular
4. Find the Equation of a Line Perpendicular to a Given Line

Preparing for Parallel and Perpendicular Lines

Before getting started, take the following readiness quiz. If you get a problem wrong, go back to the section cited and review the material.

1. Determine the reciprocal of 3.   [Section 1.4, pp. 24–25]

2. Determine the reciprocal of $-\dfrac{3}{5}$.   [Section 1.4, pp. 24–25]

① Determine Whether Two Lines Are Parallel

When two lines in the rectangular coordinate system do not intersect (that is, they have no points in common), they are said to be *parallel*. However, rather than looking at graphs of linear equations to determine whether they are parallel or not, we can look at the equations themselves to determine whether two lines might be parallel.

> **DEFINITION**
>
> Two nonvertical lines are **parallel** if and only if their slopes are equal and they have different y-intercepts. Vertical lines are parallel if they have different x-intercepts.

Work Smart

The use of the words "if and only if" given in the definition of parallel lines means that there are two statements being made:

If two nonvertical lines are parallel, then their slopes are equal and they have different y-intercepts.

If two nonvertical lines have equal slopes and different y-intercepts, then they are parallel.

Figure 49(a) shows nonvertical parallel lines. Figure 49(b) shows vertical parallel lines.

Figure 49
Parallel lines.

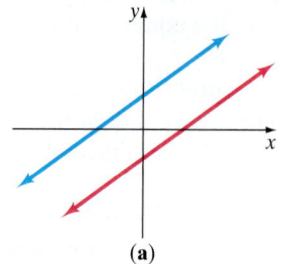 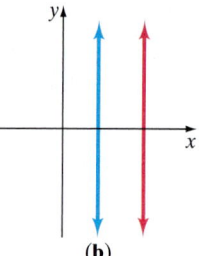

(a)   (b)

To determine whether two lines are parallel, we find the slope and y-intercept of each line by putting the equation of the line in slope-intercept form. If the slopes are the same, but the y-intercepts are different, then the lines are parallel.

EXAMPLE 1 Determining Whether Two Lines Are Parallel

Determine whether the line $y = 4x - 5$ is parallel to $y = 3x - 2$. Graph the lines to confirm your results.

Figure 50

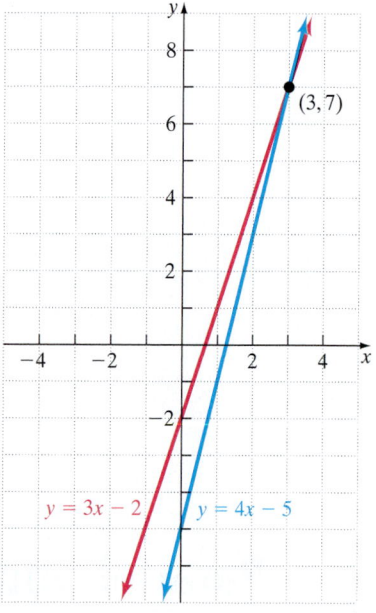

Solution

For the line $y = 4x - 5$, the slope is 4 and the y-intercept is -5. For the line $y = 3x - 2$, the slope is 3 and the y-intercept is -2. Because the lines have different slopes, they are not parallel. Figure 50 shows the graph of the two lines.

We can see from the graph in Figure 50 that the lines intersect at $(3, 7)$. Therefore, the lines are not parallel.

EXAMPLE 2 Determining Whether Two Lines Are Parallel

Determine whether the line $-3x + y = 5$ is parallel to $6x - 2y = -2$. Graph the lines to confirm your results.

Solution

To find the slope and y-intercept, we solve each equation for y so that each is in slope-intercept form.

$$-3x + y = 5$$

Add 3x to both sides: $y = 3x + 5$

The slope of the line $-3x + y = 5$ is 3 and the y-intercept is 5.

$$6x - 2y = -2$$

Subtract 6x from both sides: $-2y = -6x - 2$

Divide both sides by -2: $y = \dfrac{-6x - 2}{-2}$

Divide each term in the numerator by -2: $y = 3x + 1$

The slope of $6x - 2y = -2$ is 3 and the y-intercept is 1.

Because the lines have the same slope, 3, but different y-intercepts, the lines are parallel. Figure 51 shows a graph of the two lines.

Work Smart

Make sure both criteria for parallel lines are satisfied.

1. Same slope
2. Different y-intercepts

Lines with the same slope and same y-intercept are called *coincident lines*.

Figure 51

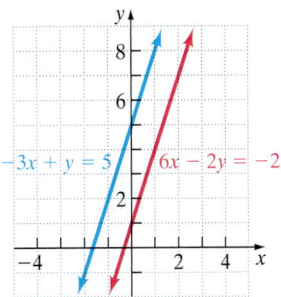

QUICK ✓ *Determine whether the two lines are parallel. Graph the lines to confirm your results.*

1. $y = 2x + 1$
$y = -2x - 3$

2. $6x + 3y = 3$
$10x + 5y = 10$

3. $4x + 5y = 10$
$8x + 10y = 20$

---

② **Find the Equation of a Line Parallel to a Given Line**

Now that we know how to identify parallel lines, we can find the equation of a line that is parallel to a given line.

EXAMPLE 3 **How to Find an Equation of a Line That Is Parallel to a Given Line**

Find an equation for the line that is parallel to $2x + y = 5$ and contains the point $(-1, 3)$. Write the equation of the line in slope-intercept form. Graph the lines.

Step-by-Step Solution

Step 1: Find the slope of the given line.

$$2x + y = 5$$

Subtract 2x from both sides: $y = -2x + 5$

The slope of the line is -2, so the slope of the parallel line is also -2.

Step 2: Use the point-slope form of a line with the given point and the slope found in Step 1 to find the equation of the parallel line.

$$y - y_1 = m(x - x_1)$$

$m = -2, x_1 = -1, y_1 = 3$: $y - 3 = -2(x - (-1))$

Step 3: Put the equation in slope-intercept form by solving for y.

$$y - 3 = -2(x + 1)$$

Distribute the -2: $y - 3 = -2x - 2$

Add 3 to both sides: $y = -2x + 1$

The equation of the line parallel to $2x + y = 5$ is $y = -2x + 1$. Figure 52 shows the graph of the parallel lines.

Figure 52

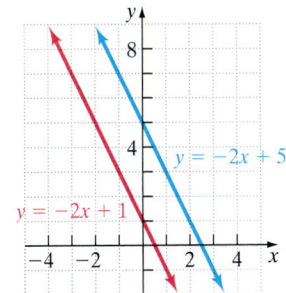

QUICK ✓ *Find the equation of the line that contains the given point and is parallel to the given line. Write the line in slope-intercept form. Graph the lines.*

4. $y = 2x + 1$ containing $(2, 3)$ **5.** $3x + 2y = 4$ containing $(-2, 3)$

EXAMPLE 4 **Finding an Equation of a Line That Is Parallel to a Given Line**

Find an equation for the line that is parallel to $x = 3$ and contains the point $(1, 5)$. Graph the lines.

Solution

The equation of the given line is $x = 3$. Because this is the equation of a vertical line, the line parallel to it will also be vertical. Vertical lines have equations of the form $x = a$. Since the line parallel to $x = 3$ contains the point $(1, 5)$, the equation of the parallel line is $x = 1$. Figure 53 shows the graphs of the lines $x = 3$ and $x = 1$.

Figure 53

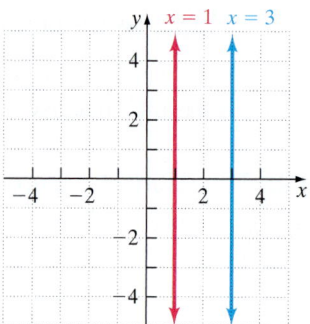

QUICK ✓ *Find the equation of the line that contains the given point and is parallel to the given line. Write the line in slope-intercept form, if possible. Graph the lines.*

6. $x = -2$ containing $(3, 1)$ **7.** $y + 3 = 0$ containing $(-2, 5)$

---

(3) Determine Whether Two Lines Are Perpendicular

When two lines intersect at a right (90°) angle, they are said to be **perpendicular.** See Figure 54.

Just as we use the slopes of lines to determine whether two lines are parallel, we also use slopes of lines to determine whether two lines are perpendicular.

Figure 54
Perpendicular lines.

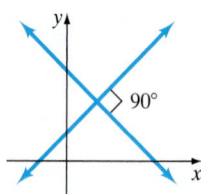

Work Smart

If m_1 and m_2 are negative reciprocals of each other, then $m_1 = \dfrac{-1}{m_2}$. For example, the numbers 4 and $-\dfrac{1}{4}$ are negative reciprocals. Watch out though, because the use of the word "*negative*" does not mean that the slope of the perpendicular line must be negative, it means that the nonvertical lines have slopes that are opposite in sign. One is positive and one is negative.

DEFINITION

Two nonvertical lines are **perpendicular** if and only if the product of their slopes is -1. Put another way, two nonvertical lines are perpendicular if their slopes are negative reciprocals of each other. Any vertical line is perpendicular to any horizontal line.

EXAMPLE 5 **Finding the Slope of a Line Perpendicular to a Given Line**

Find the slope of a line perpendicular to a line whose slope is (a) 5   (b) $-\dfrac{2}{3}$.

Solution

(a) To find the slope of a line perpendicular to a given line, we determine the negative reciprocal of the slope of the given line. The negative reciprocal of 5 is $\dfrac{-1}{5} = -\dfrac{1}{5}$. Any line whose slope is $-\dfrac{1}{5}$ will be perpendicular to the line whose slope is 5.

(b) The negative reciprocal of $-\dfrac{2}{3}$ is $\dfrac{-1}{-\dfrac{2}{3}} = \dfrac{3}{2}$. Any line whose slope is $\dfrac{3}{2}$ will be perpendicular to the line whose slope is $-\dfrac{2}{3}$. ▄

QUICK ✓ *Find the slope of a line perpendicular to the line whose slope is given.*

8. -4 **9.** $\dfrac{5}{4}$ **10.** $-\dfrac{1}{5}$

EXAMPLE 6 **Determining Whether Two Lines Are Perpendicular**

Determine whether the line $y = 3x - 2$ is perpendicular to $y = \dfrac{1}{3}x + 1$. Graph the lines to confirm your results.

Solution

To determine whether the lines are perpendicular, we first need to find the slope of each line. If the product of the slopes is -1 (the slopes are negative reciprocals of each other), then the lines are perpendicular. The slope of $y = 3x - 2$ is $m_1 = 3$. The slope of $y = \dfrac{1}{3}x + 1$ is $m_2 = \dfrac{1}{3}$. Because the product of the slopes,

$$m_1 \cdot m_2 = 3 \cdot \left(\dfrac{1}{3}\right) = 1 \neq -1,$$ the lines are not

Figure 55

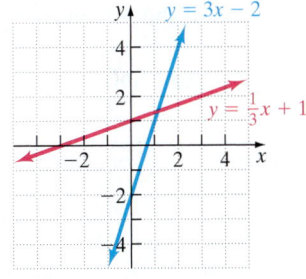

perpendicular. Notice that the slopes are reciprocals of each other, but are not *negative* reciprocals of each other. Figure 55 shows the graph of the two lines. ▄

EXAMPLE 7 **Determining Whether Two Lines Are Perpendicular**

Determine whether the line $2x + 3y = -6$ is perpendicular to $3x - 2y = 2$. Graph the lines to confirm your results.

Solution

To find the slopes of the two lines, we write the equations of the lines in slope-intercept form.

$$2x + 3y = -6$$

Subtract 2x from both sides: $$3y = -2x - 6$$

Divide both sides by 3: $$y = \dfrac{-2x - 6}{3}$$

Divide 3 into each term in the numerator: $$y = -\dfrac{2}{3}x - 2$$

The slope of $2x + 3y = -6$ is $m_1 = -\dfrac{2}{3}$.

$$3x - 2y = 2$$

Subtract $3x$ from both sides: $-2y = -3x + 2$

Divide both sides by -2: $y = \dfrac{-3x + 2}{-2}$

Divide -2 into each term in the numerator: $y = \dfrac{3}{2}x - 1$

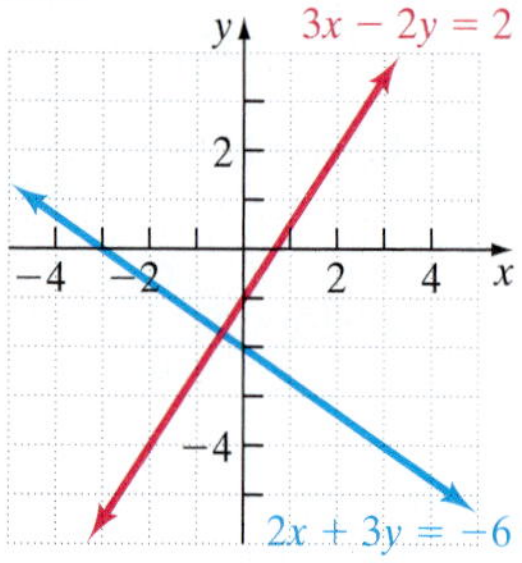

The slope of $3x - 2y = 2$ is $m_2 = \dfrac{3}{2}$.

The product of the slopes is $m_1 \cdot m_2 = -\dfrac{2}{3} \cdot \dfrac{3}{2} = -1$, so the lines are perpendicular.

Put another way, because the slopes are negative reciprocals of each other, the lines are perpendicular. See Figure 56 for the graph of the two lines.

Figure 56

QUICK ✓ *Determine whether the given lines are perpendicular. Graph the lines.*

11. $y = 4x - 3$
$\quad\ y = -\dfrac{1}{4}x - 4$

12. $2x - y = 3$
$\quad\ \ x - 2y = 2$

13. $5x + 2y = 8$
$\quad\ \ 2x - 5y = 10$

---

④ Find the Equation of a Line Perpendicular to a Given Line

Now that we know how to find the slope of a line perpendicular to a second line, we can find the equation of a line that is perpendicular to a given line.

EXAMPLE 8   **How to Find the Equation of a Line Perpendicular to a Given Line**

Find an equation of the line that is perpendicular to the line $y = 3x - 2$ and contains the point $(3, -1)$. Write the equation of the line in slope-intercept form. Graph the two lines.

Step-by-Step Solution

Step 1: Find the slope of the given line.

$\qquad\qquad\qquad\qquad\qquad\qquad y = 3x - 2$

The slope of the line is 3.

Step 2: Find the slope of the perpendicular line.

The slope of the perpendicular line is the negative reciprocal of 3, which is $-\dfrac{1}{3}$.

Step 3: Use the point-slope form of a line with the given point and the slope found in Step 2 to find the equation of the perpendicular line.

$\qquad\qquad\qquad\qquad\qquad\qquad y - y_1 = m(x - x_1)$

$m = -\dfrac{1}{3}, x_1 = 3, y_1 = -1$: $y - (-1) = -\dfrac{1}{3}(x - 3)$

Step 4: Put the equation in slope-intercept form by solving for y.

$\qquad\qquad\qquad\qquad\qquad\qquad y + 1 = -\dfrac{1}{3}(x - 3)$

Distribute the $-\dfrac{1}{3}$: $y + 1 = -\dfrac{1}{3}x + 1$

Subtract 1 from both sides: $y = -\dfrac{1}{3}x$

The equation of the line perpendicular to $y = 3x - 2$ is $y = -\dfrac{1}{3}x$.

Figure 57

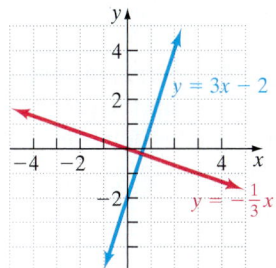

Figure 57 shows the graphs of the two lines.

QUICK ✓ *Find the equation of the line that contains the given point and is perpendicular to the given line. Write the line in slope-intercept form. Graph the lines.*

14. $(-4, 2)$; $y = 2x + 1$

15. $(-2, -1)$; $y = -\dfrac{2}{3}x + 1$

EXAMPLE 9 **Finding the Equation of a Line Perpendicular to a Given Line**

Find an equation of the line that is perpendicular to the line $x = 2$ and contains the point $(-4, 3)$. Write the equation of the line in slope-intercept form, if possible. Graph the two lines.

Solution

The line $x = 2$ is the equation of a vertical line. Therefore, the line perpendicular will be horizontal. Horizontal lines have a slope equal to 0. To find the equation of the perpendicular line we use the point-slope formula with $m = 0$, $x_1 = -4$, and $y_1 = 3$.

Work Smart

The line perpendicular to a vertical line is horizontal and vice versa.

$$y - y_1 = m(x - x_1)$$

$m = 0, x_1 = -4, y_1 = 3\text{:}\quad y - 3 = 0(x - (-4))$

$$y - 3 = 0$$

Add 3 to both sides: $y = 3$

Figure 58

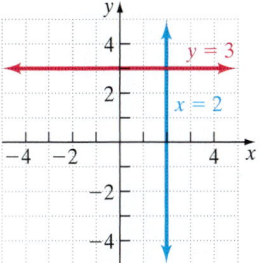

The equation of the line perpendicular to $x = 2$ through the point $(-4, 3)$ is $y = 3$. See Figure 58 for the graph of the two lines.

QUICK ✓ *Find the equation of the line that contains the given point and is perpendicular to the given line. Write the line in slope-intercept form, if possible. Graph the lines.*

16. $x = -4$ containing $(-1, -5)$

17. $y + 2 = 0$ containing $(3, -2)$

3.6 Exercises

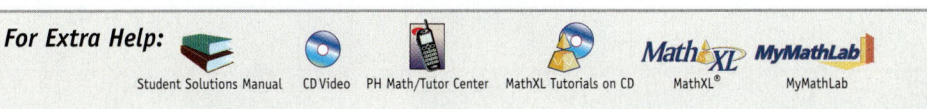

For Extra Help: Student Solutions Manual   CD Video   PH Math/Tutor Center   MathXL Tutorials on CD   MathXL®   MyMathLab

Concepts and Vocabulary

In Problems 1–3, fill in the blanks.

1. Given any two nonvertical lines, if the product of their slopes is -1, then the lines are _____.

2. A line has a slope of 4. The slope of the line parallel to this line is _____, while the slope of the line perpendicular to this line is _____.

3. If m_1 and m_2 are the slopes of two non-vertical lines with $m_1 = \dfrac{-1}{m_2}$, then the lines are _____.

In Problems 4–6, answer True or False to each statement.

4. Two different lines L_1 and L_2 have slopes $m_1 = 4$ and $m_2 = -4$, so L_1 is perpendicular to L_2.

5. Given any two lines, L_1 and L_2, if the slopes are equal (that is $m_1 = m_2$), then the lines must be parallel.

6. Lines in the same plane that never intersect are called parallel.

7. Describe the four possible relationships of the graphs of two lines in a rectangular coordinate system.

8. You are asked to determine if the following lines are parallel, perpendicular or neither: L_1 contains the points $(1, 1)$ and $(3, 5)$ and L_2 contains the points $(-1, -1)$ and $(4, 9)$. List the steps you would follow to make this determination, and describe the criteria you would use to answer the question.

Building Your Skills

In Problems 9–16, fill in the chart with the missing slopes.

	Slope of the Given Line	Slope of a Line Parallel to the Given Line	Slope of a Line Perpendicular to the Given Line
9.	$m = -3$		
10.	$m = 4$		
11.	$m = \dfrac{1}{2}$		
12.	$m = -\dfrac{1}{8}$		
13.	$m = -\dfrac{4}{9}$		
14.	$m = \dfrac{5}{2}$		
15.	$m = 0$		
16.	$m = \text{undefined}$		

In Problems 17–30, determine if the lines are parallel, perpendicular, or neither.

17. $L_1: y = x - 3$
$L_2: y = 1 - x$

18. $L_1: y = -4x + 3$
$L_2: y = 4x - 1$

19. $L_1: y = \dfrac{3}{4}x + 2$
$L_2: y = 0.75x - 1$

20. $L_1: y = 0.8x + 6$
$L_2: y = \dfrac{4}{5}x + \dfrac{19}{3}$

21. $L_1: y = -\dfrac{5}{3}x - 6$
$L_2: y = \dfrac{3}{5}x - 1$

22. $L_1: y = 3x - 1$
$L_2: y = 6 - \dfrac{x}{3}$

23. $L_1: x + y = -3$
$L_2: y - x = 1$

24. $L_1: x - 4y = 24$
$L_2: 2x - 8y = -8$

25. $L_1: 2x - 5y = 5$
$L_2: 5x + 2y = 4$

26. $L_1: x - 2y = -8$
$L_2: x + 2y = 2$

27. $L_1: 4x - 5y - 15 = 0$
$L_2: 8x - 10y + 5 = 0$

28. $L_1: x + y = 6$
$L_2: x - y = -2$

29. $L_1: 4x = 3y + 3$
$L_2: 6y = 8x + 36$

30. $L_1: 2x - 5y - 45 = 0$
$L_2: 5x + 2y - 8 = 0$

In Problems 31–38, each line contains the given points. (a) Find the slope of each line. (b) Determine if the lines are parallel, perpendicular, or neither.

31. $L_1: (0, -1)$ and $(-2, -7)$
$L_2: (-1, 5)$ and $(2, -4)$

32. $L_1: (-3, -14)$ and $(1, 2)$
$L_2: (0, 2)$ and $(-3, -10)$

33. $L_1: (2, 8)$ and $(7, 18)$
$L_2: (-2, -3)$ and $(6, 13)$

34. $L_1: (6, 0)$ and $(-2, 8)$
$L_2: (4, 1)$ and $(-6, -9)$

35. L_1: $(-2, -5)$ and $(4, -2)$
L_2: $(-8, -5)$ and $(0, -1)$

36. L_1: $(1, 6)$ and $(-1, -10)$
L_2: $(0, 1)$ and $(-2, 17)$

37. L_1: $(-6, -9)$ and $(3, 6)$
L_2: $(10, -8)$ and $(-5, 1)$

38. L_1: $(-8, -8)$ and $(4, 1)$
L_2: $(12, 8)$ and $(-4, -4)$

In Problems 39–50, find the equation of the line that contains the given point and is parallel to the given line. Write the equation in slope-intercept form, if possible.

39. $(4, -2)$; $y = 3x - 1$ 　 **40.** $(7, -5)$; $y = 2x + 6$ 　 **41.** $(-3, 8)$; $y = -4x + 5$

42. $(-2, 6)$; $y = -5x - 2$ 　 **43.** $(3, -7)$; $y = 4$ 　 **44.** $(-4, 5)$; $x = -3$

45. $(-1, 10)$; $x = 10$ 　 **46.** $(4, -8)$; $y = -1$ 　 **47.** $(10, 2)$; $3x - 2y = 5$

48. $(6, 7)$; $2x + 3y = 9$ 　 **49.** $(-1, -10)$; $x + 2y = 4$ 　 **50.** $(-3, -5)$; $2x - 5y = 6$

In Problems 51–62, find the equation of the line that contains the given point and is perpendicular to the given line. Write the equation in slope-intercept form, if possible.

51. $(3, 5)$; $y = \dfrac{1}{2}x - 2$

52. $(4, 7)$; $y = \dfrac{1}{3}x - 3$

53. $(-4, -1)$; $y = -4x + 1$

54. $(-2, -5)$; $y = -2x + 5$

55. $(-2, 1)$; x-axis 　 **56.** $(3, -6)$; y-axis 　 **57.** $(7, 5)$; y-axis

58. $(11, -6)$; x-axis 　 **59.** $(0, 0)$; $2x + 5y = 7$ 　 **60.** $(0, 0)$; $6x + 4y = 3$

61. $(-10, -3)$; $5x - 3y = 4$

62. $(-6, 10)$; $3x - 5y = 2$

Mixed Practice

In Problems 63–78, find the equation of the line that has the given properties. Write the equation in slope-intercept form, if possible. Graph each line.

63. Contains $(3, -5)$; slope $= 7$

64. Contains $(-2, 10)$; slope $= 3$

65. Contains $(2, 9)$; perpendicular to the line $y = -5x + 3$

66. Contains $(8, 3)$; parallel to the line $y = -4x + 2$

67. Contains $(6, -1)$; parallel to the line $y = -7x + 2$

68. Contains $(5, -2)$; perpendicular to the line $y = 4x + 3$

69. Contains $(-6, 2)$ and $(-1, -8)$

70. Contains $(-4, -1)$ and $(-3, -5)$

71. Slope $= 3$, y-intercept $= -2$

72. Slope $= -2$, y-intercept $= 7$

73. Contains $(5, 1)$; parallel to the line $x = -6$

74. Contains $(2, 7)$; perpendicular to the line $x = -8$

75. Contains $(3, -2)$; parallel to the line $4x + 3y = 9$

76. Contains $(-3, -2)$; perpendicular to the line $6x - 2y = 1$

77. Contains $(-1, -3)$; perpendicular to the line $x - 2y = -10$

78. Contains $(-7, 2)$; parallel to the line $x + 4y = 2$

Applying the Concepts

A parallelogram is a quadrilateral in which both pairs of opposite sides are parallel. In Problems 79 and 80, plot the following points, draw the figure, and then use slope to determine if the figure is a parallelogram.

79. $A(-1, 1)$; $B(3, 5)$; $C(6, 4)$; $D(2, 0)$

80. $A(-1, -3)$; $B(1, -1)$; $C(5, 1)$; $D(3, -2)$

A rectangle is a parallelogram that contains one right angle. That is, one pair of sides is perpendicular. In Problems 81 and 82, plot the following points, draw the figure, and then use slope to determine if the figure is a rectangle.

81. $A(6, -1)$; $B(-3, -2)$; $C(1, -6)$; $D(2, 3)$

82. $A(1, 1)$; $B(-1, 5)$; $C(5, 8)$; $D(7, 4)$

A right triangle is a triangle that contains one right angle. In Problems 83–86, plot each point and form the triangle ABC. Verify using slope that the triangle is a right triangle.

△ **83.** $A(-2, 5); B(1, 3); C(3, 6)$ △ **84.** $A(-5, 3); B(6, 0); C(5, 5)$

△ **85.** $A(4, -3); B(0, -3); C(4, 2)$ △ **86.** $A(-2, 5); B(12, 3); C(10, -11)$

Extending the Concepts

In Problems 87 and 88, find the missing coefficient so that the lines are parallel.

87. $-3y = 6x - 12$ and $4x + By = -2$

88. $Ax + 2y = 4$ and $15x = 5y + 20$

In Problems 89 and 90, find the missing coefficient so that the lines are perpendicular.

89. $Ax + 6y = -6$ and $12 - 6y = -9x$

90. $x - By = 10$ and $3y = -6x + 9$

△ **91.** The altitude of a triangle is a line segment drawn from a vertex of the triangle perpendicular to the opposite side. Plot the following points, draw triangle ABC and the segment joining points B and D, and then determine if $\overline{BD}$ is an altitude of the triangle. $A(-6, -3); B(-4, 7); C(-1, 2); D(-3, 0)$

△ **92.** The coordinates of the vertices of a quadrilateral are $A(2, 1)$, $B(4, 6)$, $C(6, 6)$, and $D(9, 2)$. Use slopes to show that the diagonals of the quadrilateral, $\overline{AC}$ and $\overline{BD}$, are perpendicular to each other.

PUTTING THE CONCEPTS TOGETHER (SECTIONS 3.1–3.6)

These problems cover important concepts from Sections 3.1 to 3.6. We designed these problems so that you can review the chapter so far and show your mastery of the concepts. Take time to work these problems before proceeding with the next section. The answers to these problems are located at the back of the text starting on page AN-10.

1. Given the linear equation $4x - 3y = 10$, determine whether the ordered pair $(1, -2)$ is a solution to the equation.

In Problems 2 and 3, graph each equation using the point-plotting method.

2. $y = \dfrac{2}{3}x - 1$ 3. $-5x + 2y = 10$

4. Given the equation $-8x + 2y = 6$, determine

   (a) the x-intercept   (b) the y-intercept

5. Graph the equation $4x + 3y = 6$ by finding the x-intercept and the y-intercept.

6. Given the equation $6x + 9y = -12$, determine

   (a) the slope   (b) the y-intercept

7. Find the slope of the line that contains the points $(3, -5)$ and $(-6, -2)$.

8. Given the linear equation $2y = -5x - 4$, find

   (a) the slope of the line perpendicular to the given line
   (b) the slope of the line parallel to the given line

9. Determine whether the lines are parallel, perpendicular, or neither. Explain how you came to your conclusion.

   $L_1: 10x + 5y = 2$ $L_2: y = -2x + 3$

In Problems 10–16, write the equation of the line that satisfies the given conditions. Write the equation in slope-intercept form, if possible.

10. slope $= 3$ and y-intercept is 1

11. slope $= -6$ and passes through $(-1, 4)$

12. through $(4, -1)$ and $(-2, 11)$

13. through $(-8, 0)$ and perpendicular to $y = \dfrac{2}{5}x - 5$

14. through $(-8, 3)$ and parallel to $-8y + 2x = -1$

15. horizontal line through $(-6, -8)$

16. through $(2, 6)$ with undefined slope

17. Shipping Expenses The shipping department records indicate that during a week when 80 packages were shipped, the total expenses recorded for the shipping department were \$1180. During a different week 50 packages were shipped and the expenses recorded were \$850. Use the ordered pairs $(80, 1180)$ and $(50, 850)$ to determine the average rate to ship an additional package.

18. Diamonds The relation between the cost of a diamond and its weight is linear. In looking at two diamonds, we find that one of the diamonds weighs 0.7 carats and costs \$3543, while the other diamond weighs 0.8 carats and costs \$4378. (*Source:* diamonds.com)

  (a) Use the ordered pairs $(0.7, 3543)$ and $(0.8, 4378)$ to find a linear equation that relates the price of a diamond to its weight.

  (b) Interpret the slope.

  (c) Predict the price of a diamond that weighs 0.76 carats.

3.7 Variation

OBJECTIVES

1. Model and Solve Direct Variation Problems
2. Model and Solve Inverse Variation Problems

Preparing for Variation
Before getting started, take the following readiness quiz. If you get a problem wrong, go back to the section cited and review the material.

1. Solve $10 = 2k$ for k.   [Section 2.1, pp. 78–81]

2. Solve $3 = \dfrac{k}{5}$ for k.   [Section 2.1, pp. 78–81]

1   Model and Solve Direct Variation Problems

Often two variables are related as proportions. For example, we say, "Revenue is proportional to sales" or "Force is proportional to acceleration." When we say that one variable is proportional to another variable, we are talking about *variation*. **Variation** describes how one quantity changes in relation to another quantity. In this text, we will discuss two types of variation: *direct* and *inverse*. We will discuss direct variation first.

In Words
When y is directly proportional to x, then y and x are related through a linear equation whose y-intercept is 0 and slope is k, the constant of proportionality.

DEFINITION

If x and y represent two quantities, we say that y **varies directly** with x, or y is **directly proportional to** x, if there is a nonzero number k such that

$$y = kx$$

The number k is called the **constant of proportionality,** or the **constant of variation.**

Preparing for...Answers   **1.** $\{5\}$ **2.** $\{15\}$

Figure 59
$y = kx, k > 0, x \geq 0$

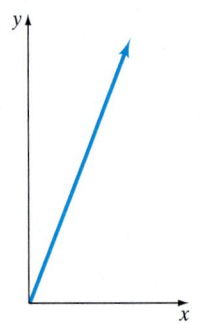

The graph in Figure 59 illustrates the relationship between y and x if y varies directly with x and $k > 0$, $x \geq 0$. Notice that the constant of proportionality k is the slope of the line.

If we know that two quantities vary directly, then knowing the value of each quantity in one instance allows us to write a formula that is true in all cases. Although the definition for direct variation uses y and x as the variables, any variables can be used.

EXAMPLE 1 Direct Variation

Suppose that y varies directly with x for $x \geq 0$. Find an equation that relates y and x if it is known that $y = 20$ when $x = 4$. Graph the equation.

Solution

Because y varies directly with x, we know that $y = kx$. Now we can use the given information to find k, the constant of proportionality.

$$y = kx$$
$$y = 20, x = 4: \quad 20 = k(4)$$
$$\text{Divide both sides by 4:} \quad 5 = k$$
$$\text{Symmetric Property:} \quad k = 5$$

With $k = 5$, we have that $y = 5x$. Because $x \geq 0$, we only need to graph the equation in quadrant I. See Figure 60.

Work Smart

Recall that the Symmetric Property states that if $a = b$, then $b = a$.

Figure 60

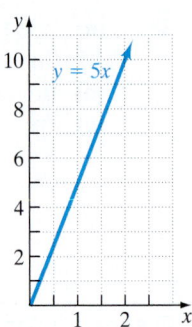

Work Smart

The constant of variation, k, is the slope of the line.

QUICK ✓

1. Suppose that y varies directly with x for $x \geq 0$. Find an equation that relates y and x if it is known that $y = 15$ when $x = 5$. Graph the equation.

2. Suppose that y varies directly with x for $x \geq 0$. Find an equation that relates y and x if it is known that $y = 6$ when $x = 18$. Graph the equation.

3. Suppose that q varies directly with w for $w \geq 0$.

   (a) Find an equation that relates q and w if it is known that $q = 10$ when $w = 40$.

   (b) Use the equation found in part (a) to determine q when $w = 60$.

   (c) Graph the equation found in part (a).

EXAMPLE 2 Car Payments

Brandon just bought a used car for $12,000. He decides to put $2000 down on the car and borrow the remaining $10,000. The bank lends Brandon $10,000 at 5.9% interest for 48 months. His payments are $234.39. The monthly payment p on a car varies directly with the amount borrowed b. If Brandon puts $3000 down on the car instead, what would his monthly payment be?

Solution

Step 1: Identify We want to know the monthly payment if Brandon puts $3000 down. The model will involve variables that are directly related.

Step 2: Name The variables have been named already: p is the monthly payment and b is the amount borrowed.

Step 3: Translate The monthly payment p varies directly with the amount borrowed b. So

$$p = kb$$

Since $p = 234.39$ when $b = 10{,}000$, we have that

$$234.39 = k(10{,}000)$$

Work Smart

If possible, find the exact value of k. If you cannot avoid approximate values, do not round the value of k to fewer than 5 decimal places.

Divide both sides by 10,000 to obtain $k = 0.023439$ and we have our model:

$$p = 0.023439b$$

Step 4: Solve If Brandon puts $3000 down, he will need to borrow $9000. We let $b = 9000$ in the model to determine the monthly payment.

$$p = 0.023439b$$
$$\text{Let } b = 9000: \quad p = 0.023439(9000)$$
$$\text{Evaluate:} \quad p \approx 210.95$$

Step 5: Check We can check the reasonableness of our answer. Brandon's payments are $234.39 when he borrows $10,000. It seems reasonable that his payments will decrease a little when he puts more money down. With $3000 down, we determine his payment will be $210.95. This seems reasonable.

Step 6: Answer When $3000 is put down, Brandon's monthly payment will be $210.95. ■

QUICK

4. The cost of gas C varies directly with the number of gallons pumped, g. Suppose that the cost of pumping 8 gallons of gas is $20.80. If 6.8 gallons are pumped into your car, what would the cost be?

---

(2)   Model and Solve Inverse Variation Problems

We now turn our attention to another kind of variation. This second type of variation is used to describe relations between two variables when an increase in one variable results in a decrease of a second variable. For example, as the price for a product increases, the quantity demanded of the product decreases.

Figure 61

$y = \dfrac{k}{x}, k > 0, x > 0$

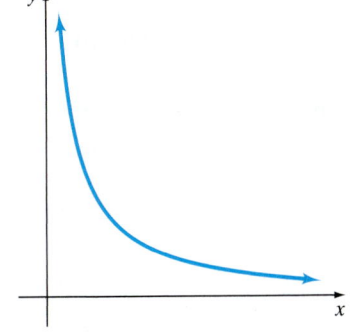

> **DEFINITION**
>
> If x and y represent two quantities, we say that y **varies inversely** with x, or y is **inversely proportional** to x, if there is a nonzero number k such that
>
> $$y = \frac{k}{x}$$
>
> where k is the **constant of variation.**

The graph in Figure 61 illustrates the relationship between y and x if y varies inversely with x with $k > 0$ and $x > 0$. Notice from the graph that as x increases the values of y decrease.

EXAMPLE 3 Inverse Variation

Suppose that y varies inversely with x for $x > 0$.

(a) Find an equation that relates y and x if it is known that $y = 20$ when $x = 3$.

(b) Use the equation found in part (a) to determine y when $x = 5$.

Solution

(a) Because y varies inversely with x, we know that $y = \dfrac{k}{x}$. Now we can use the given information to find k.

$$y = \frac{k}{x}$$

$$y = 20, x = 3: \quad 20 = \frac{k}{3}$$

$$\text{Multiply both sides by 3:} \quad 60 = k$$

$$\text{Symmetric Property:} \quad k = 60$$

With $k = 60$, we have that $y = \dfrac{60}{x}$.

(b) We can let $x = 5$ in the equation $y = \dfrac{60}{x}$ to find y. We obtain

$$y = \frac{60}{5}$$
$$= 12$$

When $x = 5$, $y = 12$.

QUICK ✔

5. Suppose that y varies inversely with x for $x > 0$.

(a) Find an equation that relates y and x if it is known that $y = 6$ when $x = 2$.
(b) Use the equation found in part (a) to determine y when $x = 4$.

6. Suppose that p varies inversely with q for $q > 0$.

(a) Find an equation that relates p and q if it is known that $p = 10$ when $q = 4$.
(b) Use the equation found in part (a) to determine p when $q = 8$.

EXAMPLE 4 Cleaning a Stadium

The amount of time T that it takes to clean a stadium after a game varies inversely with the number of employees cleaning, n. Suppose that it takes 5 hours to clean a stadium using 20 employees. How long will it take to clean the stadium using 16 employees?

Solution

Step 1: Identify We want to know the amount of time it will take to clean the stadium using 16 employees. In addition, we know that the model will involve variables that are inversely related.

Step 2: Name The variables have been named already: T is the time it takes to clean the stadium and n is the number of employees.

Step 3: Translate The time to clean T varies inversely with the number of employees n. So

$$T = \frac{k}{n}$$

Since $T = 5$ when $n = 20$, we have that

$$5 = \frac{k}{20}$$

Multiply both sides by 20 to obtain $k = 100$.

A model for the time T to clean the stadium with n employees is given by

$$T = \frac{100}{n}$$

Step 4: Solve We let $n = 16$ in the model to predict the time to clean the stadium.

$$T = \frac{100}{n}$$

Let $n = 16$: $T = \dfrac{100}{16}$

Evaluate: $T = 6.25$

Step 5: Check We can check the reasonableness of our answer. Because it takes 5 hours to clean the stadium when 20 employees are used, it seems reasonable that it will take more time with 16 employees (but not that much more). With 16 employees, we are predicting it will take 6.25 hours. This seems reasonable.

Step 6: Answer When 16 employees are used, we predict it will take 6.25 hours to clean the stadium. ▪

QUICK ✓

7. The rate of vibration (in oscillations per second) V of a string under constant tension varies inversely with the length l. If a string is 30 inches long and vibrates 20 times per second, determine the rate of vibration of a string that is 24 inches long.

3.7 Exercises

For Extra Help:

 Student Solutions Manual   CD Video   PH Math/Tutor Center   MathXL Tutorials on CD   MathXL®    MyMathLab

Concepts and Vocabulary

In Problems 1–3, fill in the blanks.

1. The statement t varies directly as s, is given by the equation _____ .

2. The statement f varies inversely as d, is given by the equation _____ .

3. In the equation $y = kx$, k is called the _____ _____ _____ and represents the rate of change between the variables x and y.

In Problems 4–6, answer True or False to each statement.

4. It is possible to solve a variation problem without knowing the constant of proportionality.

5. The constant of proportionality k in the direct variation model $y = kx$, is the slope.

6. If your employer reimburses you 36¢ per mile for driving your car on business trips, the amount you are reimbursed is inversely proportional to the miles you drive on business.

7. List 3 everyday occurrences of direct variation. Make up an example for each, and explain how to use this equation to acquire information about the situation.

8. Describe the conditions that are necessary for two variables to be directly proportional. Also describe the conditions necessary for two variables to be inversely proportional.

Building Skills

In Problems 9–14, suppose y varies directly with x. Find an equation that relates x and y given the following.

9. $y = 3$ when $x = 6$ **10.** $y = 15$ when $x = 45$ **11.** $x = 12$ when $y = 24$

12. $x = 2$ when $y = 8$ **13.** $y = -6$ when $x = 8$ **14.** $y = -10$ when $x = 24$

In Problems 15–20, suppose y varies inversely with x. Find an equation that relates x and y given the following.

15. $x = 2$ when $y = 3$ **16.** $x = 7$ when $y = 10$ **17.** $y = -3$ when $x = -4$

18. $y = -2$ when $x = -8$ 🔵 **19.** $x = -\dfrac{1}{2}$ when $y = -\dfrac{4}{7}$ **20.** $x = -\dfrac{3}{4}$ when $y = \dfrac{2}{9}$

In Problems 21–28, indicate whether the equation represents direct variation, inverse variation, or neither. If it is a variation equation, identify the constant of proportionality.

21. $y = \dfrac{2x}{3}$ **22.** $x = 4$ **23.** $y = \dfrac{1}{2}$

24. $4x = y$ **25.** $xy = 9$ **26.** $5y = 2x$

27. $6x = 3y$ **28.** $xy = 4$

In Problems 29–32, use the direct variation model to find each of the following.

29. Suppose p varies directly with g.

    (a) Find an equation that relates p and g if it is known that $p = 12$ when $g = 36$.
    (b) Use this equation to determine p if g is 9.

30. Suppose d varies directly with t.

    (a) Find an equation that relates d and t if it is known that $d = 320$ when $t = 8$.
    (b) Use this equation to determine d if t is 5.

31. Suppose y varies directly with x.

    (a) Find an equation that relates y and x if it is known that $x = 12$ when $y = -9$.
    (b) Use this equation to determine x if y is $\dfrac{5}{4}$.

32. Suppose n varies directly with m.

    (a) Find an equation that relates n and m if it is known that $m = 12$ when $n = 9$.
    (b) Use this equation to determine m if n is $-\dfrac{8}{5}$.

In Problems 33–36, use the inverse variation model to find each of the following.

33. Suppose e varies inversely with n.

    (a) Find an equation that relates e and n if it is known that $e = 4$ when $n = 2$.
    (b) Use this equation to determine e if n is 16.

34. Suppose y varies inversely with x.

    (a) Find an equation that relates y and x if it is known that $y = 4$ when $x = 3$.
    (b) Use this equation to determine y if x is 18.

35. Suppose b varies inversely with a.

    (a) Find an equation that relates b and a if it is known that $b = \dfrac{3}{2}$ when $a = \dfrac{1}{2}$.
    (b) Use this equation to determine a if b is $\dfrac{9}{10}$.

36. Suppose f varies inversely with d.

    (a) Find an equation that relates f and d if it is known that $f = \dfrac{2}{9}$ when $d = \dfrac{15}{2}$.
    (b) Use this equation to determine d if f is $\dfrac{15}{4}$.

Mixed Practice

In Problems 37–44, find the quantity indicated.

37. x varies inversely with y. If $x = 6$ when $y = 3$, find x when $y = 4$.

38. p varies inversely with v. If $p = 50$ when $v = 24$, find p when $v = 75$.

39. J is directly proportional to P. If $J = 80$ when $P = 120$, find P when $J = 50$.

40. A is directly proportional to B. If $A = 360$ when $B = 72$, find B when $A = 400$.

41. s varies directly with t. If $t = 18$ when $s = 21$, find t when $s = 49$.

42. m varies directly with r. If $r = 24$ when $m = 9$, find r when $m = 24$.

43. b is inversely proportional to a. If $a = 4$ when $b = 6$, find b when $a = \dfrac{4}{3}$.

44. x is inversely proportional to y. If $y = 14$ when $x = 4$, find x when $y = \dfrac{2}{3}$.

Applying the Concepts

45. **House of Representatives Apportionment** The number of representatives that each state has in the U.S. House of Representatives is directly proportional to the state's population. According to the 2000 Census, Ohio's 18 delegates represent a state whose population is 11.375 million. Find the number of representatives from Massachusetts if its population is 6.356 million.

46. **House of Representatives Apportionment** The number of representatives that each state has in the U.S. House of Representatives is directly proportional to the state's population. According to the 2000 Census, California's 53 delegates represent a state whose population is 33.931 million. Find the number of representatives from Florida if its population is 16.029 million.

47. **Buying Lumber** The cost of a certain type of lumber at Beechwold Lumber varies directly with the number of board feet purchased. If 70 board feet of this type of lumber cost Christine $717.50, find the number of board feet in Elizabeth's order of $1230.

48. **Buying Gasoline** The cost to purchase a tank of gasoline varies directly with the number of gallons purchased. You notice that the person in front of you spent $34.50 on 15 gallons of gas. If your SUV needs 35 gallons of gas, how much will you spend?

49. **Measuring Pressure** A fixed amount of gas is placed in a collapsible cylinder. The pressure in the cylinder, P, varies inversely with the volume V. If the pressure is measured at 5 atmospheres when the volume is 2.5 liters, what will the pressure be when the cylinder is collapsed to 2 liters?

50. **Measuring Sound** The frequency of sound varies inversely with the wavelength. If a radio station broadcasts at a frequency of 90 megahertz, the wave length of the sound is approximately 3 meters. Find the wave length of a radio broadcast at 20 megahertz.

51. **Demand** Suppose that the demand D for candy at the movie theater is inversely related to the price p. When the price of candy is $2.50 per bag, the theater sells 180 bags of candy. Determine the number of bags of candy that will be sold if the price is raised to $3 a bag.

52. **Driving to School** The time t that it takes to drive to school varies inversely with your average speed s. It takes you 24 minutes to drive to school when your average speed is 35 miles per hour. Suppose that your average speed to school yesterday was 30 miles per hour. How long did it take you to get to school?

Extending the Concepts

Joint variation is a model in which a variable is proportional to the product of two or more variables. For example, we may say that y varies jointly as x and z, or $y = kxz$. Use joint variation to solve Problems 53–56.

53. m varies jointly as r and s. If $m = 12$ when $r = 0.5$ and $s = 4$, find m when $s = 9$ and $r = \dfrac{5}{3}$.

54. m varies jointly as r and s. If $m = 9$ when $r = 6$ and $s = 2$, find m when $s = 3$ and $r = 16$.

55. p varies jointly as q and the square of r. If $p = 162$ when $q = 9$ and $r = 6$, find r when $p = 300$ and $q = 24$.

56. p varies jointly as q and the square of r. If $p = 6$ when $q = \dfrac{1}{8}$ and $r = 4$, find r when $p = 40$ and $q = 30$.

The energy produced by a photon of light, e, (in joules) is inversely proportional to the wavelength of the light, λ (in meters). The constant of variation is the product of Planck's Constant (6.63×10^{-34}) and the speed of light (3×10^{8}).

57. **(a)** Write an equation to calculate the energy produced by a photon with a wavelength, λ.

(b) Calculate the energy produced when the wavelength is 500×10^{-9} meters.

58. Use the equation from Problem 57(a) to calculate the energy produced when the wavelength is 600 nanometers. (1 nanometer $= 10^{-9}$ meters)

3.8 Linear Inequalities in Two Variables

OBJECTIVES

1. Determine Whether an Ordered Pair Is a Solution to a Linear Inequality
2. Graph Linear Inequalities
3. Solve Problems Involving Linear Inequalities

Preparing for Linear Inequalities in Two Variables

Before getting started, take the following readiness quiz. If you get a problem wrong, go back to the section cited and review the material.

1. Solve: $x - 4 > 5$ [Section 2.8, pp. 153–158]
2. Solve: $3x + 1 \le 10$ [Section 2.8, pp. 155–156]
3. Solve: $2(x + 1) - 6x > 18$ [Section 2.8, pp. 155–157]

1 **Determine Whether an Ordered Pair Is a Solution to a Linear Inequality**

In Chapter 2, we solved inequalities in one variable. In this section, we discuss linear inequalities in two variables.

> **DEFINITION**
>
> **Linear inequalities in two variables** are inequalities in one of the forms
>
> $$Ax + By < C \qquad Ax + By > C \qquad Ax + By \le C \qquad Ax + By \ge C$$
>
> where A and B are not both zero. A linear inequality in two variables x and y is **satisfied** by an ordered pair (a, b) if a true statement results when x is replaced by a and y is replaced by b.

EXAMPLE 1 **Determining Whether an Ordered Pair Is a Solution to a Linear Inequality in Two Variables**

Determine which of the following points are solutions to the linear inequality $2x + y \le 9$.

    (a) $(3, 5)$ **(b)** $(1, 3)$ **(c)** $(3, -2)$

Solution

(a) Let $x = 3$ and $y = 5$ in the inequality. If a true statement results, then $(3, 5)$ is a solution to the inequality.

$$2x + y \le 9$$
$$x = 3, y = 5: \quad 2(3) + 5 \overset{?}{\le} 9$$
$$6 + 5 \overset{?}{\le} 9$$
$$11 \le 9 \quad \text{False}$$

The statement $11 \le 9$ is false, so $(3, 5)$ is not a solution to the inequality.

(b) Let $x = 1$ and $y = 3$ in the inequality. If a true statement results, then $(1, 3)$ is a solution to the inequality.

$$2x + y \le 9$$
$$x = 1, y = 3: \quad 2(1) + 3 \overset{?}{\le} 9$$
$$2 + 3 \overset{?}{\le} 9$$
$$5 \le 9 \quad \text{True}$$

The statement $5 \le 9$ is true, so $(1, 3)$ is a solution to the inequality.

Preparing for...Answers **1.** $\{x | x > 9\}$
2. $\{x | x \le 3\}$ **3.** $\{x | x < -4\}$

(c) Let $x = 3$ and $y = -2$ in the inequality. If a true statement results, then $(3, -2)$ is a solution to the inequality.

$$2x + y \leq 9$$

$$x = 3, y = -2: \quad 2(3) + (-2) \overset{?}{\leq} 9$$

$$6 - 2 \overset{?}{\leq} 9$$

$$4 \leq 9 \quad \text{True}$$

The statement $4 \leq 9$ is true, so $(3, -2)$ is a solution to the inequality.   ▬

QUICK ✔ *Determine which of the following points are solutions to the given linear inequality.*

(a) $(2, 1)$ **(b)** $(3, 4)$ **(c)** $(-1, 10)$

1. $2x + y > 7$ **2.** $-3x + 2y \leq 8$

---

② Graph Linear Inequalities

Now that we know how to determine whether a point is a solution to a linear inequality in two variables, we are prepared to graph linear inequalities in two variables. A **graph of a linear inequality in two variables** x and y consists of all points (x, y) whose coordinates satisfy the inequality.

If we replace the inequality symbol with an equal sign in a linear inequality of the form

$$Ax + By < C \qquad Ax + By > C \qquad Ax + By \leq C \qquad Ax + By \geq C$$

we obtain the equation of a line, $Ax + By = C$. This **boundary line** separates the xy-plane into two regions, called **half-planes.** See Figure 62.

Let's consider the linear inequality $2x + y \leq 9$ in Example 1. Figure 63 shows the graph of the equation $2x + y = 9$ and the three points $(3, 5), (1, 3),$ and $(3, -2)$ examined in Example 1.

Work Smart

We graph the solutions of a linear inequality in one variable on the real number line. We graph the solutions of a linear inequality in two variables in the xy-plane.

Figure 62

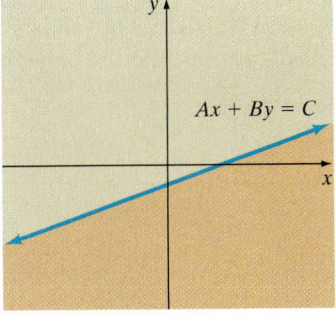

Figure 63

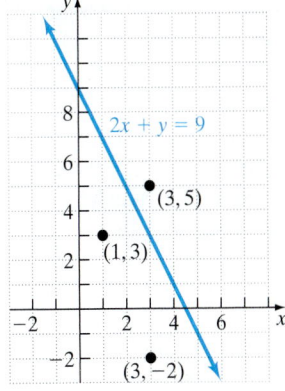

From the results of Example 1, we know that the two points below the line, $(1, 3)$ and $(3, -2)$, satisfy the inequality $2x + y \leq 9$ while $(3, 5)$ does not. In fact, all points below the line will satisfy the inequality and all points above the inequality will not satisfy the inequality. What can we learn from this? If you find a point that satisfies a linear inequality in two variables, then all points in the half-plane containing the point will satisfy the inequality. If you find a point that does not satisfy a linear inequality in two variables, then all points in the opposite half-plane will satisfy the inequality. For this reason, the use of a single **test point** is all that is required to obtain the graph of a linear inequality in two variables.

EXAMPLE 2 **How to Graph a Linear Inequality in Two Variables**

Graph the linear inequality $y < 2x - 5$.

Step-by-Step Solution

Step 1: Replace the inequality symbol with an equal sign and graph the resulting equation. If the inequality is strict ($<$ or $>$), use dashes to graph the line; if the inequality is non-strict ($\leq$ or $\geq$) use a solid line.

$$y < 2x - 5$$

We graph the line $y = 2x - 5$ whose slope is 2 and y-intercept is -5 using a dashed line because the inequality is strict ($<$). See Figure 64(a).

Figure 64 (a)

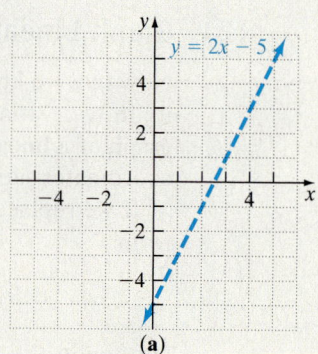

(a)

Work Smart

A strict inequality uses the symbol $<$ or $>$; a nonstrict inequality uses $\leq$ or $\geq$.

Work Smart

When the line does not contain the origin, it is usually easiest to choose the origin, (0, 0), as the test point.

Step 2: We select any test point that is not on the line and determine whether the test point satisfies the inequality. We will use (0, 0) as the test point. If (0, 0) satisfies the inequality, we shade the half-plane containing (0, 0); otherwise shade the half-plane opposite (0, 0).

$$y < 2x - 5$$

$$x = 0, y = 0: \quad 0 \overset{?}{<} 2(0) - 5$$

$$0 \overset{?}{<} 0 - 5$$

$$0 < -5 \quad \text{False}$$

Because (0, 0) does not satisfy the inequality $y < 2x - 5$, we shade the half-plane opposite the half-plane containing (0, 0). See Figure 64(b).

Work Smart

To double check your results, choose a point, such as (4, 0), in the shaded region. Does (4, 0) satisfy $y < 2x - 5$?

Figure 64 (b)

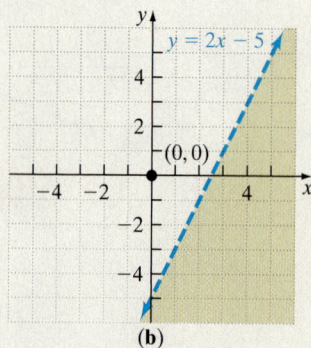

(b)

The shaded region represents the solution to the linear inequality. Because the inequality is strict, points on the line $y = 2x - 5$ do not satisfy the inequality.

We summarize the steps for graphing a linear inequality in two variables.

Steps for Graphing a Linear Inequality in Two Variables

Step 1: Replace the inequality symbol with an equal sign and graph the resulting equation. If the inequality is strict ($<$ or $>$), use dashes to graph the line; if the inequality is nonstrict ($\leq$ or $\geq$), use a solid line. The graph separates the xy-plane into two half-planes.

Step 2: Select a test point P that is not on the line (that is, select a test point in one of the half-planes).

(a) If the coordinates of P satisfy the inequality, then shade the half-plane containing P.

(b) If the coordinates of P do not satisfy the inequality, then shade the half-plane that does not contain P.

Note: If the inequality is nonstrict, then all points on the line also satisfy the inequality. If the inequality is strict, then all points on the line *do not* satisfy the inequality.

QUICK ✓ *Graph each linear inequality.*

3. $y < -2x + 1$ 　　　　　　　　　 **4.** $y \geq 3x + 2$

EXAMPLE 3   Graphing a Linear Inequality in Two Variables

Graph the linear inequality $5x + 2y \leq 10$.

Solution

First, we need to graph the equation $5x + 2y = 10$. We will do this by finding the intercepts of the equation. To find the x-intercept, let $y = 0$; to find the y-intercept, let $x = 0$.

$$5x + 2y = 10 \qquad\qquad\qquad 5x + 2y = 10$$

Let $x = 0$:　$5(0) + 2y = 10$ 　　　　　 Let $y = 0$:　$5x + 2(0) = 10$

$$2y = 10 \qquad\qquad\qquad\qquad 5x = 10$$

Divide both sides by 2:　$y = 5$ 　　 Divide both sides by 5:　$x = 2$

Plot the points $(0, 5)$ and $(2, 0)$ and draw the equation of the line as a solid line because the inequality is nonstrict ($\leq$). See Figure 65(a).

Because the graph of the equation does not contain the origin, we will select the origin $(0, 0)$ as our test point.

$$5x + 2y \leq 10$$

Let $x = 0$, let $y = 0$:　$5(\ ) + 2(\ ) \overset{?}{\leq} 10$

$$0 \leq 10 \quad \text{True}$$

Work Smart

The test point $(0, 0)$ is a solution of the linear inequality, so all other points in that half-plane are also solutions. That's why we shade the half-plane that contains the origin.

Because the test point $(0, 0)$ results in a true statement, we shade the half-plane that contains $(0, 0)$. See Figure 65(b). The shaded region and all points on the line $5x + 2y = 10$ represents the solution set.

Figure 65

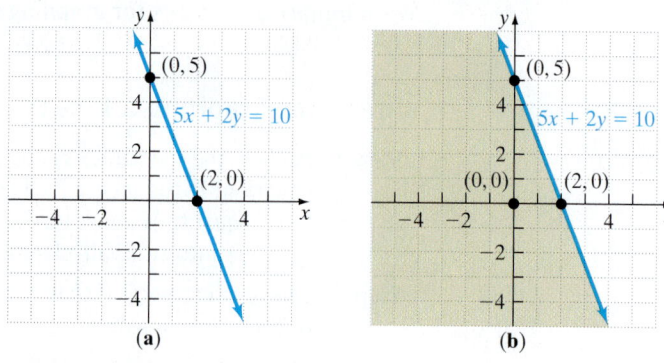

(a)   (b)

QUICK ✓ *Graph each linear inequality.*

5. $2x + 3y \leq 6$ **6.** $4x - 6y > 12$

EXAMPLE 4 **Graphing a Linear Inequality where the Line Goes through the Origin**

Graph the linear inequality $-4x + 3y > 0$.

Solution

We graph the equation $-4x + 3y = 0$ using a dashed line. Because this is an equation of the form $Ax + By = 0$, we know that the graph will pass through the origin. For this reason, we write the equation in slope-intercept form.

$$-4x + 3y = 0$$

Add 4x to both sides: $3y = 4x$

Divide both sides by 3: $y = \dfrac{4}{3}x$

The line has slope $\dfrac{4}{3}$ and y-intercept 0. See Figure 66(a).

The graph of the equation contains the origin, so we will use $(1, 3)$ as our test point.

$$-4x + 3y > 0$$

Let x = 1, let y = 3: $-4(1) + 3(3) \overset{?}{>} 0$

$$-4 + 9 \overset{?}{>} 0$$

$$5 > 0 \quad \text{True}$$

Because the test point $(1, 3)$ results in a true statement, we shade the half-plane that contains $(1, 3)$. See Figure 66(b). The shaded region represents the solution set.

Work Smart

Because the boundary line contains the origin, the test point cannot be the origin.

Figure 66

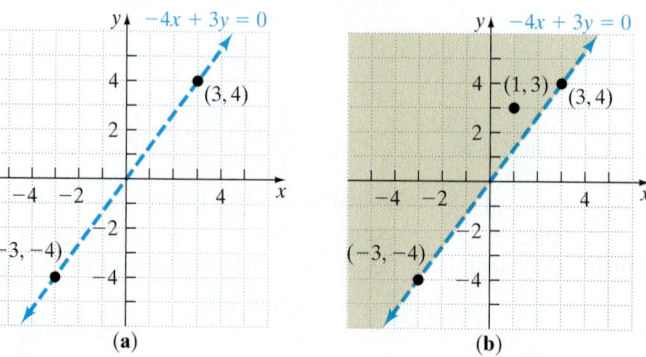

(a)   (b)

QUICK ✓ *Graph each linear inequality.*

7. $3x + y < 0$ **8.** $2x - 5y \leq 0$

EXAMPLE 5 Graphing a Linear Inequality Involving a Horizontal Line

Graph the linear inequality $y < 2$.

Solution

We begin by graphing the line horizontal line $y = 2$ using a dashed line (because the inequality is strict). See Figure 67(a). Next we select a test point. We can select the origin $(0, 0)$ as our test point because the line does not contain $(0, 0)$.

$$y < 2$$

Test point $(0, 0)$: $0 < 2$ True

Because the test point satisfies the inequality, we shade the half-plane that contains $(0, 0)$ as shown in Figure 67(b). The shaded region represents the solution to the linear inequality. Doesn't it seem intuitive that we shade below the line since these are the y-values that are less than 2?

Work Smart: Study Skills

Have you noticed the patterns?
If $y > mx + b$, shade above.
If $y < mx + b$, shade below.
If $y > b$, shade above.
If $y < b$, shade below.
If $x > a$, shade to the right.
If $x < a$, shade to the left.

Figure 67

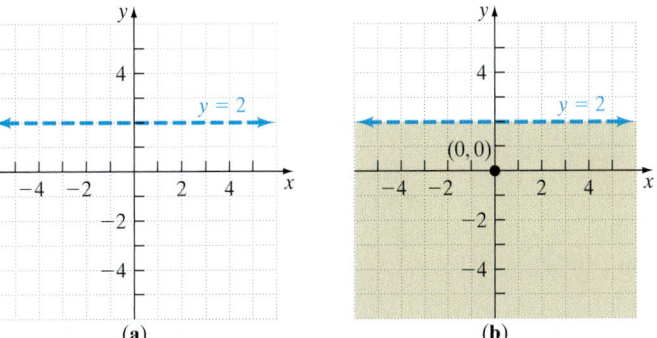

(a)   (b)

The process for graphing a linear inequality involving a vertical line parallels the process for graphing a linear inequality involving a horizontal line. That is, graph the vertical boundary line, choose a test point, and shade the half-plane containing the points that satisfy the inequality.

QUICK ✓ *Graph each linear inequality.*

9. $y < -1$ **10.** $3y - 9 \geq 0$ **11.** $x > 6$

---

③ Solve Problems Involving Linear Inequalities

Many applications of linear inequalities that involve two variables can be used to solve problems in areas such as nutrition, manufacturing, or sales. Linear inequalities can also be used to describe weight loads in elevators! Let's see how.

EXAMPLE 6 Elevator Capacity

An elevator designed and manufactured by Otis Elevator Company has a capacity of 3000 pounds (*Source: otis.com*). According to the National Center for Health Statistics, the average man 20 years or older weighs 180 pounds and the average woman 20 years or older weighs 150 pounds.

(a) Write a linear inequality that describes the various combinations of men and women who can ride this elevator.

(b) Can 10 men and 9 women ride in this elevator safely?

(c) Can 7 men and 11 women ride in this elevator safely?

Solution

(a) We are going to use the first three steps in problem solving strategy given in Chapter 2 on page 122 to help us develop the linear inequality.

Step 1: Identify We want to determine the number of men and women who can ride in the elevator without going over 3000 pounds.

Step 2: Name Unknowns Let m represent the number of males and f represent the number of females on the elevator.

Step 3: Translate For the sake of simplicity, assume that all the people who get on the elevator are "average." If there is one male on the elevator, the weight on the elevator will be 180 pounds. If there are two males on the elevator, the weight will be $2(180) = 360$ pounds. In general, if there are m males on the elevator, the weight will be $180m$. Similar logic for the females tells us that if there are f females on the elevator, the weight will be $150f$. Because the capacity of the elevator is 3000 pounds, we use a less than or equal ($\leq$) inequality. A linear inequality that describes the weight limitations on the elevator is

$$180m + 150f \leq 3000 \quad \text{The Model}$$

(b) Letting $m = 10$ and $f = 9$, we obtain

$$180(10) + 150(9) \overset{?}{\leq} 3000$$
$$3150 \leq 3000 \quad \text{False}$$

Because the inequality is false, 10 men and 9 women cannot ride the elevator safely.

(c) Letting $m = 7$ and $f = 11$, we obtain

$$180(7) + 150(11) \overset{?}{\leq} 3000$$
$$2910 \leq 3000 \quad \text{True}$$

Because the inequality is true, 7 men and 11 women can ride the elevator safely.

QUICK ✓

12. Kevin just received $2 from his grandma. He goes to a candy store where each sucker sells for $0.20 and each taffy stick sells for $0.25.

(a) Write a linear inequality that describes the various combinations of suckers, s, and taffy sticks, t, Kevin can buy.

(b) Can Kevin buy 6 suckers and 3 taffy sticks?

(c) Can Kevin buy 5 suckers and 5 taffy sticks?

3.8 Exercises

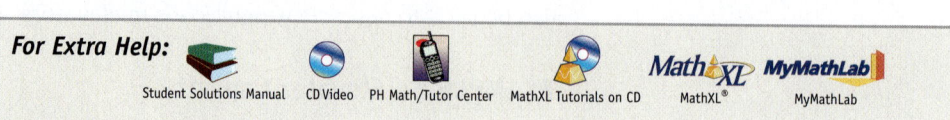

For Extra Help: Student Solutions Manual · CD Video · PH Math/Tutor Center · MathXL Tutorials on CD · Math XL · MathXL® · MyMathLab · MyMathLab

Concepts and Vocabulary

In Problems 1–3, fill in the blanks.

1. When drawing the boundary line for the graph of $Ax + By \geq C$, we use a _____ line.

2. When drawing the boundary line for the graph of $Ax + By < C$, we use a _____ line.

3. The boundary line separates the xy-plane into two regions, called _____.

In Problems 4–6, answer True or False to each statement.

4. After plotting the boundary line for the graph of $3x \le 2y$, we can use $(0,0)$ as a test point to decide which half-plane contains the solutions.

5. After drawing the graph of a linear inequality in two variables, it is a good idea to pick a test point in the shaded region to verify that the correct half-plane has been shaded.

6. The boundary line for the solution of the inequality $3x - 2y > -2$ is solid.

7. Describe how you can tell if a point (a, b) is a solution to a linear inequality in two variables. How can you decide if (a, b) lies on the boundary line?

8. Explain why $(0,0)$ is generally a good test point. Describe how you can decide when this is a good test point to use and when you should choose a different test point.

Building Skills

In Problems 9–20, determine which of the following points, if any, are solutions to the given linear inequality in two variables.

9. $y > -x + 2$
$A(2,4) \qquad B(3,-6) \quad C(0,0)$

10. $y \le x - 5$
$A(-4,-6) \quad B(0,0) \qquad C(3,10)$

11. $y \le 3x - 1$
$A(-6,-15) \quad B(0,0) \quad C(-1,-6)$

12. $y > -2x + 1$
$A(0,0) \qquad B(-2,3) \quad C(2,-1)$

13. $3x \ge 2y$
$A(-8,-12) \quad B(3,5) \qquad C(-5,-8)$

14. $4y < 5x$
$A(-4,-6) \quad B(5,6) \qquad C(-12,-15)$

15. $2x - 3y < -6$
$A(2,-1) \qquad B(4,8) \qquad C(-3,0)$

16. $3x + 5y \ge 4$
$A(2,0) \qquad B(-3,4) \quad C(6,-2)$

17. $x \le 2$
$A(7,2) \qquad B(2,5) \qquad C(4,2)$

18. $y \le -3$
$A(-3,-3) \quad B(-1,-4) \quad C(-6,-1)$

19. $y > -1$
$A(-1,1) \qquad B(3,-1) \quad C(4,-2)$

20. $x > 10$
$A(3,12) \qquad B(11,-6) \quad C(10,16)$

In Problems 21–50, graph each linear inequality.

21. $y > 3x - 2$

22. $y \le 4x + 2$

23. $y \le -x + 1$

24. $y > x - 3$

25. $y < \dfrac{x}{2}$

26. $y > \dfrac{x}{3}$

27. $y > 5$

28. $y < 4$

29. $y \le \dfrac{2}{5}x + 3$

30. $y \ge \dfrac{3}{2}x - 2$

31. $y \ge -\dfrac{4}{3}x + 2$

32. $y \le -\dfrac{3}{4}x + 1$

33. $x < 2$

34. $x > -1$

35. $3x - 4y < 12$

36. $2x + 6y < -6$

37. $2x + y \ge -4$

38. $6x - 8y \ge 24$

39. $x + y > 0$

40. $x - y > 0$

41. $5x - 2y < -8$

42. $3x + 2y \le -9$

43. $x > -1$

44. $y < -3$

45. $y \le 4$

46. $x \ge 2$

47. $\dfrac{x}{3} - \dfrac{y}{5} \ge 1$

48. $\dfrac{x}{4} + \dfrac{y}{2} < 1$

49. $-3 \ge x - y$

50. $-2 > x + y$

Applying the Concepts

In Problems 51–62, translate each statement into a linear inequality. Then graph the inequality.

51. The sum of two numbers, x and y, is at least 26.

52. The ratio of a number, x, and 3 is less than -2.

53. The ratio of a number, y, and -2 is at most 4.

54. One number, x, is at least 12 more than a second number, y.

55. One number, x, is no more than 3 less than a second number, y.

56. The difference of a number, x, and half a second number, y, is positive.

57. The sum of a number, x, and 3 times a second number, y, is negative.

58. The sum of two numbers, x and y, is at least -3.

59. The difference of twice a number, x, and half a second number, y, is at least 5.

60. The product of a number, y, and 2 is at most -10.

61. The product of a number, x, and -2 is more than -1.

62. The sum of half a number, x, and twice a second number, y, is at least 12.

63. Trip to the Aquarium A kindergarten class has a maximum of $120 to spend on a trip to the aquarium. The cost of admission for students is $3, while adults must pay $5.

   (a) Write a linear inequality that describes the various combinations of the number of students s and adults a that can go on the field trip.

   (b) Is there enough money to pay for 32 students and 6 adults?

   (c) Is there enough money to pay for 29 students and 4 adults?

64. Fishing Trip Patrick's row boat can hold a maximum of 500 pounds. In Patrick's circle of friends, the average adult weighs 160 pounds and the average child weighs 75 pounds.

   (a) Write a linear inequality that describes the various combinations of the number of adults a and children c that can go on a fishing trip in the row boat.

   (b) Will the boat sink with 2 adults and 4 children?

   (c) Will the boat sink with 1 adult and 5 children?

65. Backpacking Trip Scott's dad decided he can carry no more than 22 pounds in his backpack on the next group trip. He carries only camping stoves, which weigh 3.1 pounds, and sleeping bags, which weigh 5.7 pounds.

   (a) Write a linear inequality that describes the various combinations of the number of stoves s and sleeping bags b that he will carry.

   (b) Will his pack be too heavy if he carries 3 stoves and 2 sleeping bags?

   (c) Will his pack be too heavy if he carries 2 stoves and 3 sleeping bags?

66. School Fundraiser For the Clinton Elementary School fundraiser, Guillermo earns 5 points for each magazine subscription he sells and 2 points for each novelty. It takes at least 40 points to earn the portable CD player that Guillermo wants.

   (a) Write a linear inequality that describes the various combinations of subscriptions m and novelties n that Guillermo can sell to earn enough points for the portable CD player.

   (b) Will he earn the CD player if he sells 8 novelties and 5 subscriptions?

   (c) Will he earn the CD player if he sells 20 novelties and 2 subscriptions?

Extending the Concepts

Two inequalities joined by the word "and" or "or" form a **compound inequality.** *The directions "Solve $3x - 2y > 6$ and $x + y < 2$" mean to find the ordered pairs that satisfy both inequalities at the same time. The solution of a compound inequality of this type is found by graphing the two linear inequalities in the same Cartesian coordinate plane and then shading the area where the graphs overlap. Solve the compound inequalities in Problems 67–72.*

67. $3x - 2y > 6$ and $x + y < 2$ **68.** $2x - 4y \leq 4$ and $3x + 2y \geq 6$

69. $y > \dfrac{3}{4}x - 1$ and $x \geq 0$ **70.** $y > \dfrac{2}{3}x + 3$ and $y \geq 0$

71. $x < -3$ and $y \leq 4$ **72.** $y < -1$ and $x \geq -2$

The Graphing Calculator

Graphing calculators can be used to graph linear inequalities in two variables. The figure to the right shows the graph of $3x + y < 7$ using a TI-84 Plus graphing calculator. To obtain the graph, we solve the inequality for y and obtain $y < -3x + 7$. Graph the line $y = -3x + 7$ and shade below. Consult your owner's manual for specific keystrokes.

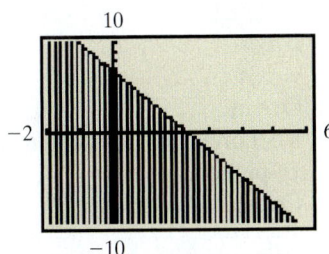

In Problems 73–84, graph the inequalities using a graphing calculator.

73. $y > 3$ **74.** $y < -2$ **75.** $y < 5x$ **76.** $y \geq \dfrac{2}{3}x$

77. $y > 2x + 3$ **78.** $y < -3x + 1$ **79.** $y \leq \dfrac{1}{2}x - 5$ **80.** $y \geq -\dfrac{4}{3}x + 5$

81. $3x + y \leq 4$ **82.** $-4x + y \geq -5$ **83.** $2x + 5y \leq -10$ **84.** $3x + 4y \geq 12$

CHAPTER 3 ACTIVITY: GRAPHING PRACTICE

Focus: Graphing and identifying the graphs of linear equations and inequalities.

Time: 15–20 minutes

Group size: 3–4

Materials needed: Blank piece of paper and graph paper for each group member.

1. On each of four pieces of paper, write one of the following equations or inequalities.

     (a) $x + 3 \geq -4(y - 1)$ **(b)** $4(y - 2) + 5 = 5(x + 1)$

     (c) $3(y - 7) + 4 = -2(x - 3) - 2$ **(d)** $\dfrac{2}{3}(x + 2y) + \dfrac{8}{3} \geq 0$

2. Place the four papers face down on the desk and mix them up. One by one, each group member should choose a piece of paper. Do not show your choice to the other members of the group.

3. Carefully graph that equation/inequality on your graph paper. Do not label your graphing.

4. When finished, place the unlabeled graphs in a pile.

5. As a group, with no help from the member who drew the graph, write the equation or inequality that is drawn.

6. As a group, discuss the outcome.

CHAPTER 3 REVIEW

Section 3.1	The Rectangular Coordinate System and Equations in Two Variables

KEY TERMS

x-axis	Ordered pair (x, y)	Quadrants
y-axis	Coordinates	Equation in two variables
Origin	x-coordinate	Sides
Rectangular or Cartesian coordinate system	y-coordinate	Satisfy
xy-plane		
Coordinate axes		

YOU SHOULD BE ABLE TO . . .	EXAMPLE	REVIEW EXERCISES
1 Plot points in the rectangular coordinate system (p. 175)	Examples 1 and 2	1–6
2 Determine if an ordered pair satisfies an equation (p. 179)	Example 3	7, 8
3 Create a table of values that satisfies an equation (p. 181)	Examples 4 through 7	9–16

In Problems 1–4, plot the following points in the rectangular coordinate system. Tell which quadrant each point belongs to or on which axis the point lies.

1. $A(3, -2)$ **2.** $B(-1, -3)$ **3.** $C(-4, 0)$ **4.** $D(0, 2)$

In Problems 5 and 6, identify the coordinates of each point labeled in the figure. Tell which quadrant each point belongs to (or on which axis the point lies).

5.
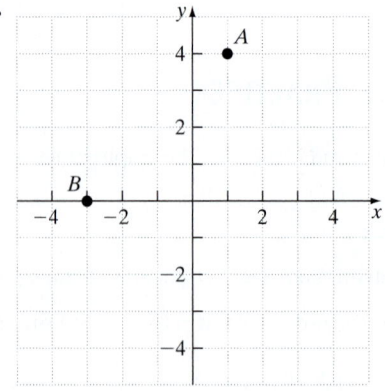

6.
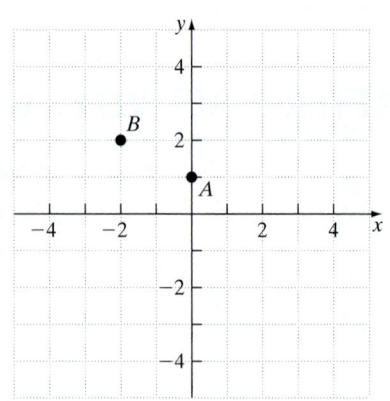

In Problems 7 and 8, determine whether the ordered pairs satisfy the given equation.

7. $y = 3x - 7$
$A(-1, -10)$
$B(-7, 0)$

8. $4x - 3y = 2$
$A(2, 2)$
$B(6, 5)$

9. Find an ordered pair that satisfies the equation $-x = 3y$ when
   (a) $x = 4$
   (b) $y = 2$

10. Find an ordered pair that satisfies the equation $x - y = 0$ when
   (a) $x = 4$
   (b) $y = -2$

In Problems 11–14, use the equation to complete the table. Use the table to list some of the ordered pairs that satisfy the equation.

11. $3x - 2y = 10$

x	y	(x, y)
−2		
0		
4		

12. $y = -x + 2$

x	y	(x, y)
−3		
2		
4		

13. $y = -\dfrac{1}{3}x - 4$

x	y	(x, y)
	2	
−6		
	−12	

14. $3x - 2y = 7$

x	y	(x, y)
	−8	
3		
	4	

15. Mail Order Shipping Martha purchases some clothes from a mail order catalogue. The shipping costs added to her order will be $5.00 plus $2.00 per item. The equation that calculates her total shipping cost, C, is given by the equation $C = 5 + 2x$, where x is the number of items purchased. Complete the table below and then graph the ordered pairs in a rectangular coordinate system.

x	C	(x, C)
1		
2		
3		

16. Department Store Wages Isabel works in a department store where her monthly earnings are $1000 plus 10% commission on her sales. The equation that calculates her earnings, E, is given by the equation: $E = 1000 + 0.10x$, where x is Isabel's sales for the month. Complete the table below and then graph the ordered pairs in a rectangular coordinate system.

x	E	(x, E)
500		
1000		
2000		

Section 3.2 Graphing Equations in Two Variables

KEY CONCEPTS	KEY TERMS

KEY CONCEPTS

- **Linear Equation**
  - A linear equation in two variables is an equation of the form $Ax + By = C$, where A, B, and C are real numbers. A and B cannot both be zero.
- **Procedure for Finding Intercepts**
  1. To find the x-intercept(s), if any, of the graph of an equation, let $y = 0$ in the equation and solve for x.
  2. To find the y-intercept(s), if any, of the graph of an equation, let $x = 0$ in the equation and solve for y.
- **Vertical Line**
  - A vertical line is given by the equation $x = a$, where a is the x-intercept.
- **Horizontal Line**
  - A horizontal line is given by the equation $y = b$ where b is the y-intercept.

KEY TERMS

Graph of an equation in two variables
Point-plotting method
Complete graph
Linear equation in two variables
Standard form of an equation of a line
Line
Intercepts
x-intercept
y-intercept

YOU SHOULD BE ABLE TO . . .	EXAMPLE	REVIEW EXERCISES
① Graph a line by plotting points (p. 191)	Examples 1, 3, and 4	17–22
② Graph a line using intercepts (p. 196)	Examples 5 through 8	23–32
③ Graph vertical and horizontal lines (p. 199)	Examples 9 and 10	33–36

In Problems 17–20, graph each linear equation using the point-plotting method.

17. $y = -2x$ **18.** $y = x$ **19.** $4x + y = -2$ **20.** $3x - y = -1$

21. Printing Cost The cost to print pamphlets for a new healthcare clinic is a $40 set-up fee plus $2.00 per pamphlet that will be printed. The equation that calculates the total cost for printing, C, is $C = 40 + 2p$, where p is the number of pamphlets to be printed. Complete the table below and then graph the equation.

p	C	(p, C)
20		
50		
80		

22. Performance Fees The Crickets are the newest band in town and have a gig performing at a local concert. They agree that their total fee will be $500 plus an additional $3.00 per person who attends the concert. The equation that calculates the total fee for performing, F, is $F = 500 + 3p$, where p is the number of people in attendance. Complete the table below and then graph the equation.

p	F	(p, F)
100		
200		
500		

In Problems 23 and 24, find the intercepts of each graph.

23.

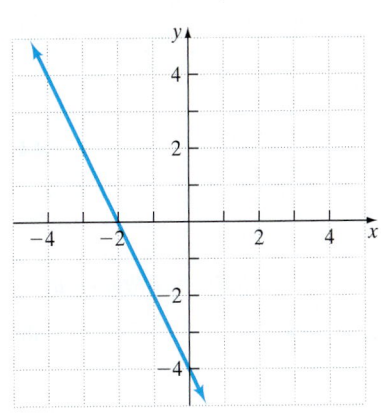

24.

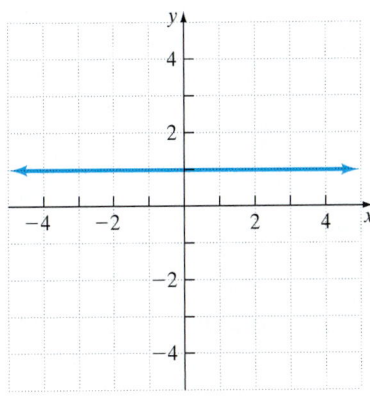

In Problems 25–28, find the x-intercept and y-intercept of each equation.

25. $-3x + y = 9$ **26.** $y = 2x - 6$ **27.** $x = 3$ **28.** $2x - 5y = 2$

In Problems 29–32, graph each linear equation by finding its intercepts.

29. $y - 3x = 3$ **30.** $2x + 5y = 0$ **31.** $\dfrac{x}{3} + \dfrac{y}{2} = 1$ **32.** $y = -\dfrac{3}{4}x + 3$

In Problems 33–36, graph each vertical or horizontal line.

33. $x = -2$ **34.** $y = 3$ **35.** $y = -4$ **36.** $x = 1$

Section 3.3	Slope

KEY CONCEPT	KEY TERMS
• **Slope** If $x_1 \neq x_2$, the **slope** *m* of the line containing (x_1, y_1) and (x_2, y_2) is defined by the formula $m = \dfrac{y_2 - y_1}{x_2 - x_1}$. The slope of vertical line is undefined. The slope of a horizontal line is 0.	Run Rise Slope Average rate of change

YOU SHOULD BE ABLE TO . . .	EXAMPLE	REVIEW EXERCISES
① Find the slope of a line given two points (p. 206)	Examples 1 and 2	37–42
② Find the slope of vertical and horizontal lines (p. 209)	Examples 3 and 4	43–46
③ Graph a line using its slope and a point on the line (p. 211)	Examples 5 and 6	47–50
④ Work with applications of slope (p. 212)	Examples 7 and 8	51, 52

In Problems 37 and 38, find the slope of the line whose graph is shown.

37.

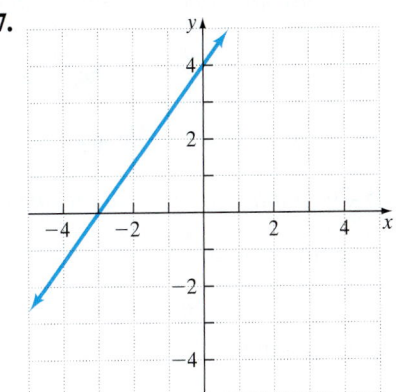

38.

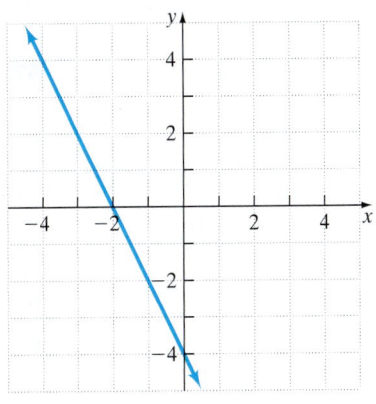

In Problems 39–46, find the slope of the line containing the given points.

39. $(-4, 6)$ and $(-3, -2)$

40. $(4, 1)$ and $(0, -7)$

41. $\left(\dfrac{1}{2}, -\dfrac{3}{4}\right)$ and $\left(\dfrac{5}{2}, -\dfrac{1}{4}\right)$

42. $\left(-\dfrac{1}{2}, \dfrac{2}{3}\right)$ and $\left(\dfrac{3}{2}, \dfrac{1}{3}\right)$

43. $(-3, -6)$ and $(-3, -10)$

44. $(-5, -1)$ and $(-1, -1)$

45. $\left(\dfrac{3}{4}, \dfrac{1}{2}\right)$ and $\left(-\dfrac{1}{4}, \dfrac{1}{2}\right)$

46. $\left(\dfrac{1}{3}, -\dfrac{3}{5}\right)$ and $\left(\dfrac{3}{9}, -\dfrac{1}{5}\right)$

In Problems 47–50, graph the line that contains the given point and has the given slope.

47. $(-2, -3); m = 4$

48. $(1, -3); m = -2$

49. $(0, 1); m = -\dfrac{2}{3}$

50. $(2, 3); m = 0$

51. Production Cost The total cost to produce 20 bicycles is \$1400 and the total cost to produce 50 bicycles is \$2750. Find and interpret the slope of the line containing the points $(20, 1400)$ and $(50, 2750)$.

52. Road Grade Scott is driving to the river on a road which falls 5 feet for every 100 feet of horizontal distance. What is the grade of this road? Express your answer as a percent.

Section 3.4 Slope-Intercept Form of a Line

KEY CONCEPT

- **Slope-intercept form of an equation of a line**
  - An equation of a line with slope m and y-intercept b is $y = mx + b$.

YOU SHOULD BE ABLE TO . . .	EXAMPLE	REVIEW EXERCISES
1 Use the slope-intercept form to identify the slope and y-intercept of a line (p. 217)	Examples 1 and 2	53–56
2 Graph a line whose equation is in slope-intercept form (p. 218)	Examples 3 and 4	57–62
3 Graph a line whose equation is in the form $Ax + By = C$ (p. 220)	Example 5	63, 64
4 Find the equation of a line given its slope and y-intercept (p. 220)	Example 6	65–70
5 Work with linear models in slope-intercept form. (p. 221)	Examples 7 and 8	71, 72

In Problems 53–56, find the slope and the y-intercept of the line whose equation is given.

53. $y = -x + \dfrac{1}{2}$
54. $y = x - \dfrac{3}{2}$
55. $3x - 4y = -4$
56. $2x + 5y = 8$

In Problems 57–64, use the slope and y-intercept to graph each line whose equation is given.

57. $y = \dfrac{1}{3}x + 1$
58. $y = -\dfrac{x}{2} - 1$
59. $y = -\dfrac{2}{3}x - 2$
60. $y = \dfrac{3x}{4} + 3$

61. $y = x$
62. $y = -2x$
63. $2x - y = -4$
64. $-4x + 2y = 2$

In Problems 65–70, find the equation of the line with the given slope and intercept.

65. slope is $-\dfrac{3}{4}$; y-intercept is $\dfrac{2}{3}$

66. slope is $\dfrac{1}{5}$; y-intercept is 10

67. slope is undefined; x-intercept is -12

68. slope is 0; y-intercept is -4

69. slope is 1; y-intercept is -20

70. slope is -1; y-intercept is -8

71. Car Rentals Hot Rod Car Rentals rents sports cars for $120 plus $80 per day. The equation that calculates the total cost C to rent a sports car is $C = 120 + 80d$, where d is the number of days that the car is used.
   (a) How much will is cost to rent the car for 3 days?
   (b) If the bill came to $680, how many days was the sports car rented out?
   (c) Graph the equation in a rectangular coordinate system.

72. Computer Rentals Anthony plans to rent a computer to finish his Master's thesis. There is no fixed fee, only a charge for each day the computer is checked out. If he keeps the computer for 22 days, he will pay $418. If his thesis advisor asks him to redo a section and he finds that he must keep the computer for 35 days, it will cost him $665.
   (a) Write two ordered pairs (d, C), where d represents the days the computer is rented and C represents the cost. Calculate the slope of the line that contains these two points.
   (b) Interpret the meaning of the slope of this line.
   (c) Write an equation that represents the total cost for Anthony to rent a computer.
   (d) Calculate the cost to rent a computer for 8 days.

Section 3.5	Point-Slope Form of a Line

KEY CONCEPT

- **Point-Slope Form of an Equation of a Line**
  An equation of a nonvertical line of slope m that contains the point (x_1, y_1) is $y - y_1 = m(x - x_1)$.

YOU SHOULD BE ABLE TO . . .	EXAMPLE	REVIEW EXERCISES
① Find the equation of a line given a point and a slope (p. 227)	Examples 1 through 3	73–80
② Find the equation of a line given two points (p. 229)	Examples 4 and 5	81–84
③ Build linear models using the point-slope form of a line (p. 231)	Example 6	85, 86

In Problems 73–80, find the equation of the line that contains the given point and the given slope. Write the equation in slope-intercept form, if possible.

73. $(0, -3)$; slope $= 6$

74. $(4, 0)$; slope $= -2$

75. $(3, -1)$; slope $= -\dfrac{1}{2}$

76. $(-1, -3)$; slope $= \dfrac{2}{3}$

77. $\left(-\dfrac{4}{3}, -\dfrac{1}{2}\right)$; horizontal

78. $\left(-\dfrac{4}{7}, \dfrac{8}{5}\right)$; vertical

79. $(-5, 2)$; slope is undefined

80. $(6, 0)$; slope $= 0$

In Problems 81–84, find the equation of the line that contains the given points. Write the equation of the line in slope-intercept form, if possible.

81. $(-7, 0)$ and $(0, 8)$

82. $(0, -6)$ and $(4, 0)$

83. $(3, 5)$ and $(-2, -10)$

84. $(-15, 1)$ and $(-5, -3)$

85. Harvesting Hay Farmer Myers noted that at the end of 3 days of harvesting 12 acres of hay remained in his field and at the end of 5 days, 4 acres remained. Use the ordered pairs $(3, 12)$ and $(5, 4)$ to write an equation that calculates how many acres of hay, A, are left to be harvested after d days.

86. Fish Population Pat works at Scripps Institute of Oceanography and is responsible for monitoring the fish population in various waters around the world. In one bay, the sunfish population was declining at a constant rate. His initial sample netted 15 sunfish and six months later, the same location netted 13 sunfish. Use the ordered pairs $(0, 15)$ and $(6, 13)$ to write an equation that will predict how many sunfish, F, the sample will yield after m months.

Section 3.6	Parallel and Perpendicular Lines	
KEY TERMS		
Parallel Perpendicular		
YOU SHOULD BE ABLE TO . . .	**EXAMPLE**	**REVIEW EXERCISES**
1 Determine whether two lines are parallel (p. 235)	Examples 1 and 2	87, 88
2 Find the equation of a line parallel to a given line (p. 237)	Examples 3 and 4	89–94
3 Determine whether two lines are perpendicular (p. 238)	Examples 5 through 7	95–98
4 Find the equation of a line perpendicular to a given line (p. 240)	Examples 8 and 9	99–102

In Problems 87 and 88, determine if the two lines are parallel.

87. $y = -\dfrac{1}{3}x + 2$ $\qquad$ **88.** $y = \dfrac{1}{2}x - 4$

$\qquad\quad x - 3y = 3$ $\qquad\qquad\qquad x - 2y = 6$

In Problems 89–94, find the equation of the line that contains the given point and is parallel to the given line. Write the equation in slope-intercept form, if possible.

89. $(3, -1)$; $y = -x + 5$ $\qquad$ **90.** $(-2, 4)$; $y = 2x - 1$

91. $(-1, 10)$; $3x + y = -7$ $\qquad$ **92.** $(4, -5)$; $6x + 2y = 5$

93. $(5, 19)$; y-axis $\qquad$ **94.** $(-1, -12)$; x-axis

In Problems 95 and 96, determine the slope of the line perpendicular to the given line.

95. $3x - 2y = 5$ $\qquad$ **96.** $4x - 9y = 1$

In Problems 97 and 98, determine if the two lines are perpendicular.

97. $x + 3y = 3$ $\qquad$ **98.** $5x - 2y = 2$

$\qquad\quad y = 3x + 1$ $\qquad\qquad\qquad y = \dfrac{2}{5}x + 12$

In Problems 99–102, write the equation of the line that contains the given point and is perpendicular to the given line. Write the equation in slope-intercept form, if possible.

99. $(-3, 4)$; $y = -3x + 1$ $\qquad$ **100.** $(4, -1)$; $y = 2x - 1$

101. $(1, -3)$; $2x - 3y = 6$ $\qquad$ **102.** $\left(-\dfrac{3}{5}, \dfrac{2}{5}\right)$; $x + y = -7$

Section 3.7	Variation	
KEY TERMS		
Variation Varies directly Directly proportional to Constant of proportionality	Constant of variation Varies inversely Inversely proportional	
YOU SHOULD BE ABLE TO . . .	**EXAMPLE**	**REVIEW EXERCISES**
1 Model and solve direct variation problems (p. 245)	Examples 1 and 2	103–108
2 Model and solve inverse variation problems (p. 247)	Examples 3 and 4	109–114

In Problems 103 and 104, assume that y varies directly with x. Find an equation that relates x and y given the following:

103. $x = 12$ when $y = 3$ **104.** $x = 3$ when $y = 18$

105. Suppose that f varies directly with g. If $f = 12$ when $g = 18$, find f when $g = 24$.

106. Suppose that p varies directly with q. If $p = 25$ when $q = 20$, find p when $q = 88$.

107. CD Sales The income from CD sales varies directly with the number of customers that enter the store. If 20 customers produce CD sales of $290, how much in sales will 35 customers produce?

108. Map Distances The distance between two points on a map varies directly with the actual distance between the cities. If two cities are 3 cm apart on a map and their actual distance is 60 miles, find the actual distance for two cities which measure 8 cm apart on the same map.

In Problems 109 and 110, suppose that y varies inversely with x. Find an equation that relates x and y given the following:

109. $x = 8$ when $y = 6$ **110.** $x = 6$ when $y = 9$

111. Suppose that r varies inversely with s. If $r = 18$ when $s = 2$, find r when $s = 4$.

112. Suppose s varies inversely with u. If $s = 3$ when $u = 2$, find s when $u = 24$.

113. Road Trip The time that it takes to drive from Los Angeles to San Francisco is inversely proportional to the speed the car is driven. If the trip takes 6.5 hours at 60 mph, what speed is necessary to make the trip in 5 hours?

114. Measuring Pressure The volume of a gas varies inversely with the pressure. If the volume is 8 liters when the pressure is 100 grams per liter, what is the volume of this gas if the pressure is 32 grams per liter?

Section 3.8	Linear Inequalities in Two Variables

KEY TERMS	
Satisfied Graph of a linear inequality in two variables	Half-planes Test point

YOU SHOULD BE ABLE TO . . .	EXAMPLE	REVIEW EXERCISES
1 Determine whether an ordered pair is a solution of a linear inequality (p. 252)	Example 1	115, 116
2 Graph linear inequalities (p. 253)	Examples 2 through 5	117–124
3 Solve problems involving linear inequalities (p. 257)	Example 6	125, 126

In Problems 115 and 116, determine which of the following points, if any, are solutions to the given linear inequality in two variables.

115. $y \le 3x + 4$

$A(2, 0)$ $B(-4, -8)$ $C(7, 26)$

116. $y > \dfrac{1}{3}x + 4$

$A(6, -2)$ $B(0, 4)$ $C(-18, -1)$

In Problems 117–124, graph each linear inequality.

117. $y < -\dfrac{1}{4}x + 2$ **118.** $y > 2x - 1$ **119.** $3x + 2y \ge -6$

120. $-2x + y \ge 4$ **121.** $x - 3y \le 0$ **122.** $x - 4y \ge 4$

123. $x < -3$ **124.** $y > 2$

125. Coin Jar A jar holding only quarters and dimes contains at least $12. If there are x quarters and y dimes in the jar, write a linear inequality in two variables that describes how many of each type of coin are in the jar.

126. Number The difference between twice a number, x, and half of a second number, y, is at most 10. Write a linear inequality in two variables that describes these two numbers.

CHAPTER 3 TEST

 Remember to use your Chapter Test Prep Video CD to see fully worked-out solutions to any of these problems you would like to review.

1. Determine whether the ordered pair $(-3, -2)$ is a solution to the equation $3x - 4y = -17$.

2. Given the equation $3x - 9y = 12$, determine

   (a) the x-intercept   **(b)** the y-intercept

3. Given the equation $4x - 3y = -24$, determine

   (a) the slope   **(b)** the y-intercept

In Problems 4 and 5, graph each linear equation.

4. $y = -\dfrac{3}{4}x + 2$ **5.** $3x - 6y = -12$

6. Find the slope of the line that contains the points $(2, -2)$ and $(-4, -1)$.

7. Given the linear equation $3y = 2x - 1$, find

   (a) the slope of a line perpendicular to the given line.

   (b) the slope of a line parallel to the given line.

8. Determine whether the lines are parallel, perpendicular, or neither. Explain how you came to your conclusion.

$$L_1: 3x - 7y = 2$$
$$L_2: y = \frac{7}{3}x + 4$$

In Problems 9–15, write the equation of the line that satisfies the given conditions. Write the equation in slope-intercept form, if possible.

9. slope $= -4$ and y-intercept is -15

10. slope $= 2$ and contains $(-3, 8)$

11. contains $(-3, -2)$ and $(-4, 1)$

12. contains $(4, 0)$ and parallel to $y = \dfrac{1}{2}x + 2$

13. contains $(4, 2)$ and perpendicular to $4x - 6y = 5$

14. horizontal line through $(3, 5)$

15. contains $(-2, -1)$ with undefined slope

16. m varies directly with n. If $m = 12$ when $n = 8$, find m when $n = 20$.

17. Shipping Packages The shipping department records indicate that during a week when 20 packages were shipped, the total expenses recorded for the shipping department were $560. During a different week when 30 packages were shipped, the expenses recorded were $640. Use the ordered pairs $(20, 560)$ and $(30, 640)$ to determine the average rate to ship a package.

In Problems 18–20, graph each inequality.

18. $y \geq x - 3$ **19.** $-2x - 4y < 8$ **20.** $x \leq -4$

CUMULATIVE REVIEW Chapters 1–3

In Problems 1–3, evaluate each expression.

1. $200 \div 25 \cdot (-2)$

2. $\dfrac{3}{4} + \dfrac{1}{6} - \dfrac{2}{3}$

3. $\dfrac{8 - 3(5 - 3^2)}{7 - 2 \cdot 6}$

4. Evaluate $x^3 + 3x^2 - 5x - 7$ for $x = -3$.

5. Simplify: $8m - 5m^2 - 3 + 9m^2 - 3m - 6$

In Problems 6 and 7, solve each equation.

6. $8(n + 2) - 7 = 6n - 5$

7. $\dfrac{2}{5}x + \dfrac{1}{6} = -\dfrac{2}{3}$

8. Solve $A = \dfrac{1}{2}h(b + B)$ for B.

In Problems 9 and 10, solve each linear inequality. Express the solution using set-builder notation and interval notation. Graph the solution set.

9. $6x - 7 > -31$

10. $5(x - 3) \geq 7(x - 4) + 3$

11. Plot the following ordered pairs in the same Cartesian plane.

$$A(-3, 0), \quad B(4, -2), \quad C(1, 5), \quad D(0, 3), \quad E(-4, -5), \quad F(-5, 2)$$

In Problems 12 and 13, graph the linear equation using the method you prefer.

12. $y = -\dfrac{1}{2}x + 4$

13. $4x - 5y = 15$

In Problems 14 and 15, find the equation of the line with the given properties. Express your answer in either slope-intercept or standard form, whichever you prefer.

14. Through the points $(3, -2)$ and $(-6, 10)$

15. Parallel to $y = -3x + 10$ and through the point $(-5, 7)$

16. Graph $x - 3y > 12$.

17. Computing Grades Shawn really wants an A in his geometry class. His four exam scores are $94, 95, 90$, and 97. The final exam is worth two exam scores. To have an A, his average must be at least 93. What scores can Sean score on the final exam to earn an A in the course?

18. Body Mass Index The body mass index (BMI) of a person 62 inches tall and weighing x pounds is given by $0.2x - 2$. A BMI of 30 or more is considered to be obese. For what weights would a person 62 inches tall be considered obese?

19. Supplementary Angles Two angles are supplementary. The measure of the larger angle is 15 degrees more than twice the measure of the smaller angle. Find the angle measures.

20. Cylinders Max has 100 square inches of aluminum with which to make a closed cylinder. If the radius of the cylinder must be 2 inches, how tall will the cylinder be? (Round to the nearest hundredth of an inch.)

21. Consecutive Integers Find three consecutive even integers such that the sum of the first two is 22 more than the third.

4 Systems of Linear Equations and Inequalities

It is well documented that American women can expect to live longer than American men. But is there ever going to be a time when men can expect to live as long as women? See Example 6 in Section 4.2.

OUTLINE

4.1 Solving Systems of Linear Equations by Graphing

4.2 Solving Systems of Linear Equations Using Substitution

4.3 Solving Systems of Linear Equations Using Elimination

   Putting the Concepts Together (Sections 4.1–4.3)

4.4 Solving Direct Translation, Geometry, and Uniform Motion Problems Using Systems of Linear Equations

4.5 Solving Mixture Problems Using Systems of Linear Equations

4.6 Systems of Linear Inequalities

   Chapter 4 Activity: Find the Numbers
   Chapter 4 Review
   Chapter 4 Test

The Big Picture: Putting It Together

In Chapter 2, we solved linear equations and inequalities in one variable. Recall that linear equations in one variable can have no solution (a contradiction), one solution, or infinitely many solutions (an identity). In Chapter 3, we learned how to graph both linear equations and linear inequalities in two variables. Remember, the graph of a linear equation in two variables represents the set of all points whose ordered pairs satisfy the equation. The graph of a linear equation will be used in this chapter to help us visualize results.

In this chapter, we will discuss methods for finding solutions that simultaneously satisfy two linear equations involving two variables. We are going to learn a graphical method for finding the solution and two algebraic methods. These systems can have no solution, one solution, or infinitely many solutions, just like linear equations in one variable. We conclude the chapter by looking at systems of linear inequalities. These systems require us to determine the region that satisfies two or more linear inequalities simultaneously.

4.1 Solving Systems of Linear Equations by Graphing

OBJECTIVES

1. Determine If an Ordered Pair Is a Solution of a System of Linear Equations
2. Solve a System of Linear Equations by Graphing
3. Classify Systems of Linear Equations as Consistent or Inconsistent
4. Solve Applied Problems Involving Systems of Linear Equations

Preparing for Solving Systems of Linear Equations by Graphing

Before getting started, take the following readiness quiz. If you get a problem wrong, go back to the section cited and review the material.

1. Graph: $y = 2x - 3$ [Section 3.4, pp. 218–220]
2. Graph: $3x + 4y = 12$ [Section 3.2, pp. 196–198]
3. Determine whether $2x + 6y = 12$ is parallel to $-3x - 9y = 18$.   [Section 3.6, pp. 235–237]

In Section 3.2, we learned that an equation in two variables is linear provided that it can be written in the form $Ax + By = C$, where A and B cannot both be zero. We now learn methods for finding ordered pairs that satisfy two linear equations at the same time.

> **DEFINITION**
>
> A **system of linear equations** is a grouping of two or more linear equations where each equation contains one or more variables.

In this chapter, we will discuss only systems of linear equations containing two variables and two equations. Example 1 gives some examples of systems of linear equations.

EXAMPLE 1 **Examples of Systems of Linear Equations**

(a) $\begin{cases} 2x + y = 5 \\ x - 5y = -10 \end{cases}$ Two equations containing two variables, x and y

(b) $\begin{cases} a + b = 5 \\ 2a - 3b = -3 \end{cases}$ Two equations containing two variables, a and b

We use a brace, as shown in the systems in Example 1, to remind us that we are dealing with a system of equations.

1 Determine If an Ordered Pair Is a Solution of a System of Linear Equations

> **DEFINITION**
>
> A **solution** of a system of equations consists of values for the variables that satisfy each equation of the system. When we are solving systems of two linear equations containing two unknowns, we represent the solution as an ordered pair, (x, y).

So to determine whether an ordered pair is a solution of a system of equations, we replace each variable with its value in each equation. If *both* the equations in the system are satisfied, then the ordered pair is a solution.

EXAMPLE 2 **Determining Whether an Ordered Pair Is a Solution of a System of Linear Equations**

Determine whether the given ordered pairs are solutions of the system of equations.

$$\begin{cases} x + 3y = 9 \\ 4x - 2y = 8 \end{cases}$$

(a) $(-3, 4)$ (b) $(3, 2)$

Preparing for...Answers

1.

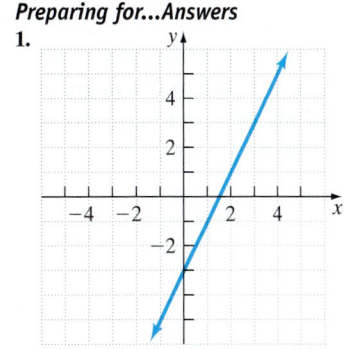

2.

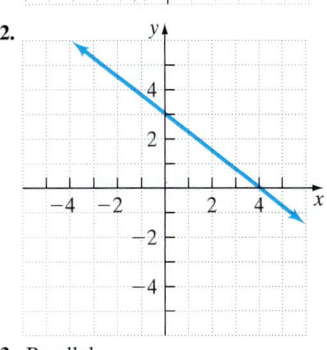

3. Parallel

Solution

(a) Let $x = -3$ and $y = 4$ in both equations. If both equations are true, then $(-3, 4)$ is a solution.

First equation: $x + 3y = 9$ Second equation: $4x - 2y = 8$

$x = -3, y = 4$: $-3 + 3(4) \stackrel{?}{=} 9$ $4(-3) - 2(4) \stackrel{?}{=} 8$

$-3 + 12 \stackrel{?}{=} 9$ $-12 - 8 \stackrel{?}{=} 8$

$9 = 9$ True $-20 = 8$ False

Although $(-3, 4)$ satisfies the first equation, it does not satisfy the second equation. Because the values $x = -3$ and $y = 4$ do not satisfy both equations, $(-3, 4)$ is not a solution of the system.

(b) Let $x = 3$ and $y = 2$ in both equations. If both equations are true, then $(3, 2)$ is a solution.

First equation: $x + 3y = 9$ Second equation: $4x - 2y = 8$

$x = 3, y = 2$: $3 + 3(2) \stackrel{?}{=} 9$ $4(3) - 2(2) \stackrel{?}{=} 8$

$3 + 6 \stackrel{?}{=} 9$ $12 - 4 \stackrel{?}{=} 8$

$9 = 9$ True $8 = 8$ True

Because the values $x = 3$ and $y = 2$ satisfy both equations, the ordered pair $(3, 2)$ is a solution of the system. ■

QUICK ✔️ *Determine whether the given set of ordered pairs is a solution of the system of equations.*

1. $\begin{cases} 2x + 3y = 7 \\ 3x + y = -7 \end{cases}$ **2.** $\begin{cases} 3x - 6y = 6 \\ -2x + 4y = -4 \end{cases}$

    (a) $(3, 1)$ **(b)** $(-4, 5)$ **(c)** $(-2, -1)$ **(a)** $(2, 0)$ **(b)** $(0, -1)$ **(c)** $(4, 1)$

---

② ## Solve a System of Linear Equations by Graphing

It is useful to visualize the problem of solving a system of two linear equations containing two variables as a geometry problem. The graph of each equation in the system is a line. So a system of two equations containing two variables represents a pair of lines. We know from Chapter 3 that the graph of an equation represents the set of all ordered pairs that make the equation a true statement. Therefore, **the point, or points, at which two lines intersect, if any, represents the solution of the system of equations.**

    Let's look at an example to learn how to solve a system of linear equations by graphing.

EXAMPLE 3   **How to Solve a System of Linear Equations by Graphing**

Solve the following system by graphing: $\begin{cases} 3x + y = 7 \\ -2x + 3y = -12 \end{cases}$

Step-by-Step Solution

Step 1: Graph the first equation in the system.   We graph the equation $3x + y = 7$ by putting the equation in slope-intercept form, $y = mx + b$.

$$3x + y = 7$$
$$y = -3x + 7$$

We graph the line whose slope is -3 and y-intercept is 7. See Figure 1(a).

Figure 1

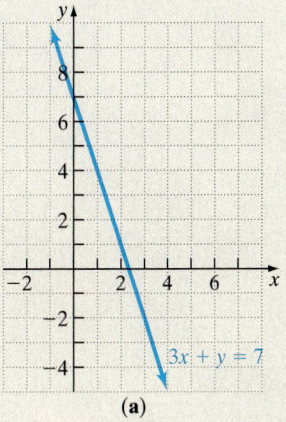

(a)

Step 2: Graph the second equation in the system.

Work Smart

We graph the line $-2x + 3y = -12$ using intercepts because the equation is in standard form and we can find the intercepts easily.

Graph the equation $-2x + 3y = -12$ using intercepts.

x-intercept: Let $y = 0$ and solve for x.

$$-2x + 3y = -12$$
$$-2x + 3(0) = -12$$
$$-2x = -12$$
$$x = 6$$

y-intercept: Let $x = 0$ and solve for y.

$$-2x + 3y = -12$$
$$-2(0) + 3y = -12$$
$$3y = -12$$
$$y = -4$$

Plot the points $(6, 0)$ and $(0, -4)$ and draw a line through these points on the same rectangular coordinate system that we graphed the line $3x + y = 7$. See Figure 1(b).

Figure 1

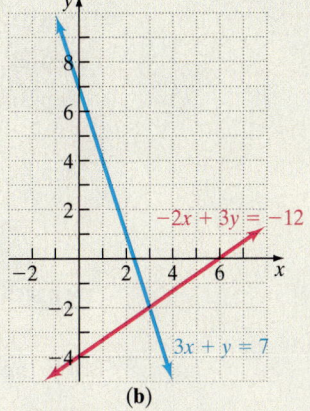

(b)

Step 3: Determine the point of intersection of the two lines, if any.

From the graphs in Figure 1(b), we can see that the lines appear to intersect at $(3, -2)$. We label this point in Figure 1(c).

(continued)

Figure 1 (Continued)

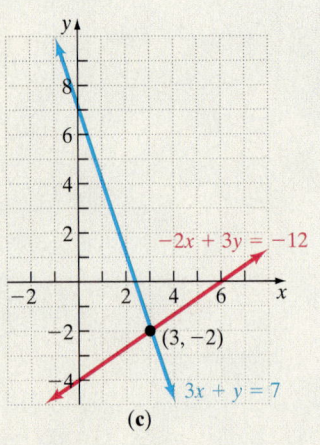

(c)

Step 4: Verify that the point of intersection determined in Step 3 is a solution of the system.

Let $x = 3$ and $y = -2$ in both equations.

First equation:
$$3x + y = 7$$
$$3(3) + (-2) \overset{?}{=} 7$$
$$9 - 2 \overset{?}{=} 7$$
$$7 = 7 \quad \text{True}$$

Second equation:
$$-2x + 3y = -12$$
$$-2(3) + 3(-2) \overset{?}{=} -12$$
$$-6 - 6 = -12$$
$$-12 = -12 \quad \text{True}$$

The solution of the system $\begin{cases} 3x + y = 7 \\ -2x + 3y = -12 \end{cases}$ is $(3, -2)$.

Below, we summarize the steps for obtaining the solution of a system of linear equations by graphing.

> ## Steps for obtaining the solution of a system of linear equations by graphing
>
> **Step 1:** Graph the first equation in the system.
>
> **Step 2:** Graph the second equation in the system.
>
> **Step 3:** Determine the point of intersection, if any. The point of intersection of the lines is the solution of the system.
>
> **Step 4:** Verify that the point of intersection determined in Step 3 is a solution of the system.

EXAMPLE 4 **Solving a System of Linear Equations Graphically**

Solve the following system by graphing: $\begin{cases} 3x - 2y = -16 \\ 5x + 2y = 0 \end{cases}$

Solution

We graph both equations in the system by putting each equation in slope-intercept form.

First equation: $\quad 3x - 2y = -16$

Subtract $3x$ from both sides: $\quad -2y = -3x - 16$

Divide both sides by -2: $\quad y = \dfrac{3}{2}x + 8$

Figure 2

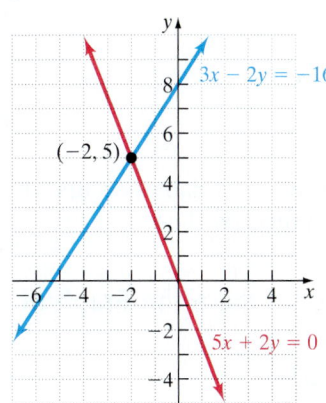

The slope of the first equation is $\dfrac{3}{2}$ and the y-intercept is 8.

Second equation:	$5x + 2y = 0$	
Subtract 5x from both sides:	$2y = -5x$	
Divide both sides by 2:	$y = -\dfrac{5}{2}x$	

The slope of the second equation is $-\dfrac{5}{2}$ and the y-intercept is 0. Figure 2 shows the graph of each equation. The point of intersection is $(-2, 5)$. Now, we check to see if the solution is $x = -2, y = 5$.

Check

First equation:	$3x - 2y = -16$		Second equation:	$5x + 2y = 0$
$x = -2, y = 5$:	$3(-2) - 2(5) \overset{?}{=} -16$		$x = -2, y = 5$:	$5(-2) + 2(5) \overset{?}{=} 0$
	$-6 - 10 \overset{?}{=} -16$			$-10 + 10 \overset{?}{=} 0$
	$-16 = -16$ True			$0 = 0$ True

Both equations check, so the solution is $(-2, 5)$.　■

QUICK ✓ *Solve each system by graphing.*

3. $\begin{cases} y = -2x + 9 \\ y = 3x - 11 \end{cases}$　　**4.** $\begin{cases} 4x + y = -3 \\ 3x - y = -11 \end{cases}$　　**5.** $\begin{cases} 2x + 3y = 1 \\ -4x + 2y = -26 \end{cases}$

Sometimes the lines that make up a system of linear equations in two variables do not intersect at all, as shown in Example 5.

EXAMPLE 5　**Solving a System of Linear Equations That Have No Point of Intersection**

Solve the following system by graphing: $\begin{cases} 3x - y = -4 \\ -6x + 2y = 2 \end{cases}$

Solution

We graph both equations in the system by putting each equation in slope-intercept form.

First equation:	$3x - y = -4$	
Subtract 3x from both sides:	$-y = -3x - 4$	
Divide both sides by -1:	$y = 3x + 4$	

The slope of the first equation is 3 and the y-intercept is 4.

Second equation:	$-6x + 2y = 2$	
Add 6x to both sides:	$2y = 6x + 2$	
Divide both sides by 2:	$y = 3x + 1$	

Work Smart

Notice that the slopes of the two lines in Example 5 are the same, and the y-intercepts are different.

Figure 3

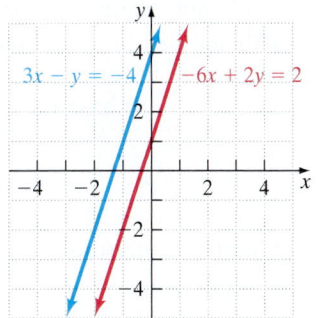

The slope of the second equation is 3 and the y-intercept is 1. Figure 3 shows the graph of each equation. The lines are parallel and therefore do not intersect anywhere. The system of equations has no solution. We could also say the solution set is the empty set, $\varnothing$.　■

QUICK ✔ *Solve each system by graphing.*

6. $\begin{cases} y = 2x + 4 \\ y = 2x - 1 \end{cases}$

7. $\begin{cases} 2x - 3y = -3 \\ -4x + 6y = -12 \end{cases}$

Another special case with systems of linear equations in two variables occurs when the graph of each of the equations in the system results in the same line.

EXAMPLE 6 **Solving a System of Linear Equations where the Graph of Each Equation Is the Same Line**

Solve the following system by graphing: $\begin{cases} 5x + 2y = -4 \\ 10x + 4y = -8 \end{cases}$

Solution

We graph both equations in the system by putting each equation in slope-intercept form.

$$\text{First equation:} \quad 5x + 2y = -4$$

$$\text{Subtract 5x from both sides:} \quad 2y = -5x - 4$$

$$\text{Divide both sides by 2:} \quad y = -\frac{5}{2}x - 2$$

Work Smart

Notice that the slopes and y-intercepts of the two lines in Example 6 are the same.

The slope of the first equation is $-\frac{5}{2}$ and the y-intercept is -2.

$$\text{Second equation:} \quad 10x + 4y = -8$$

$$\text{Subtract 10x from both sides:} \quad 4y = -10x - 8$$

$$\text{Divide both sides by 4:} \quad y = -\frac{5}{2}x - 2$$

Figure 4

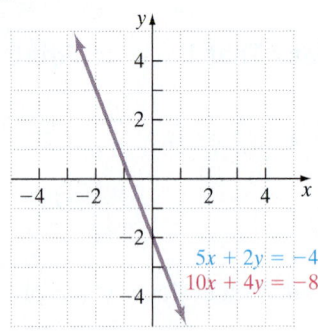

The slope of the second equation is $-\frac{5}{2}$ and the y-intercept is -2. Figure 4 shows the graph of each equation. The equations are the same so the lines coincide. The system of equations has infinitely many solutions. ■

QUICK ✔ *Solve each system by graphing.*

8. $\begin{cases} y = 3x + 2 \\ -6x + 2y = 4 \end{cases}$

9. $\begin{cases} 7x + 3y = -6 \\ -14x - 6y = 12 \end{cases}$

③ Classify Systems of Linear Equations as Consistent or Inconsistent

The results of Examples 3–6 tell us that there are three possibilities for the solution of a system of linear equations (one solution, no solution, or infinitely many solutions). In Examples 3 and 4, the system of linear equations had exactly one solution. In Example 5, the system of linear equations had no solution. In Example 6, the system of equations had infinitely many solutions. We classify systems of equations based on the number of solutions the system has.

> **DEFINITION**
>
> A **consistent system** is a system of equations that has at least one solution.
>
> An **inconsistent system** is a system of equations that has no solution.

We summarize these possibilities next.

CLASSIFYING A SYSTEM OF EQUATIONS GRAPHICALLY

Intersect: If the lines intersect, then the system of equations has one solution given by the point of intersection. We say that the system is **consistent** and the equations are **independent.** See Figure 5(a).

Parallel: If the lines are parallel, then the system of equations has no solution because the lines never intersect. We say that the system is **inconsistent.** See Figure 5(b).

Coincident: If the lines lie on top of each other (are coincident), then the system of equations has infinitely many solutions. The solution set is the set of all points on the line. The system is **consistent** and the equations are **dependent.** See Figure 5(c).

Figure 5

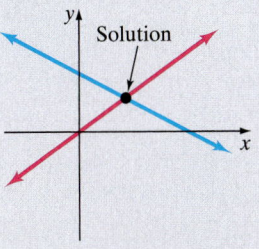
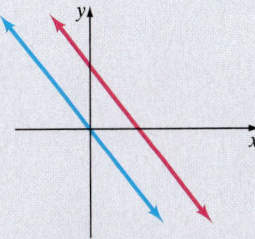
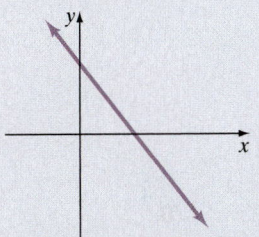

(a) Intersecting lines; system has exactly one solution

(b) Parallel lines; system has no solution

(c) Coincident lines; system has infinitely many solutions

Rather than graphing the lines in a system to classify the system, we can also classify a system algebraically.

Classifying a System of Equations Algebraically

Step 1: Write each equation in the system in slope-intercept form.

Step 2: **(a)** If the equations of the lines in the system have different slopes, then the lines will intersect. The point of intersection represents the solution. We say that the system is consistent and the equations are independent. See Examples 3 and 4.

  (b) If the equations of the lines have the same slope, but different y-intercepts, then the lines are parallel. We say that the system is inconsistent. See Example 5.

  (c) If the equations of the lines have the same slope and the same y-intercept, then the lines are coincident. We say that the system is consistent and the equations are dependent. See Example 6.

EXAMPLE 7 **Determining the Number of Solutions a System Has**

Without graphing, determine the number of solutions of the system:

$$\begin{cases} 12x + 4y = -16 \\ -9x - 3y = 3 \end{cases}$$

State whether the system is consistent or inconsistent. If the system is consistent, state whether the equations are dependent or independent.

Solution

We write the equations in slope-intercept form. We then determine the slope and y-intercept to determine the number of solutions.

First equation:	$12x + 4y = -16$
Subtract 12x from both sides:	$4y = -12x - 16$
Divide both sides by 4:	$y = -3x - 4$

Second equation:	$-9x - 3y = 3$
Add 9x to both sides:	$-3y = 9x + 3$
Divide both sides by -3:	$y = -3x - 1$

The slope of both equations is -3. The y-intercept of the first equation is -4, while the y-intercept of the second equation is -1. The equations have the same slope, but different y-intercepts, so the lines are parallel. Therefore, the system has no solution and is inconsistent. ▬

QUICK ✓ *Without graphing, determine the number of solutions of each system. State whether the system is consistent or inconsistent. For those systems that are consistent, state whether the equations are dependent or independent.*

10. $\begin{cases} 7x - 2y = 4 \\ 2x + 7y = 7 \end{cases}$ **11.** $\begin{cases} 6x + 4y = 4 \\ -12x - 8y = -8 \end{cases}$ **12.** $\begin{cases} 3x - 4y = 8 \\ -6x + 8y = 8 \end{cases}$

---

④ **Solve Applied Problems Involving Systems of Linear Equations**

We have seen in Chapter 3 that numerous situations, such as the cost of owning a car, can be modeled by linear equations. Many times we can use a system of two linear equations with two variables to answer interesting questions such as, "Which long-distance telephone plan should I choose?"

EXAMPLE 8 **SBC Long Distance**

SBC, a company that offers phone service in many states, has two domestic long-distance phone plans of interest. The "Just Call" plan charges $3.00 per month plus $0.03 per minute. The "Value Plus Flat Rate" plan has no monthly service fee and charges $0.05 per minute. Determine the number of minutes of long-distance phone calls that can be used each month in order for the two plans to charge the same fee.

Solution

Step 1: Identify We want to know the number of minutes of long-distance phone calls that can be used such that the two plans have the same fee.

Step 2: Name Let m represent the number of minutes used and let C represent the monthly cost.

Step 3: Translate For the "Just Call" plan, talking for 0 minutes results in a cost of $3.00; talking for 1 minute results in a cost of $3.00 plus $0.03; talking for two minutes results in a cost of $3.00 plus 2($0.03), or $3.06. To generalize, the monthly cost for the "Just Call" plan is $3.00 plus $0.03 times the number of minutes used. This leads to the following equation:

$$C = 0.03m + 3.00 \quad \text{Equation (1)}$$

For the "Value Plus Flat Rate" plan, talking for 0 minutes results in a cost of $0; talking for 1 minute results in a cost of $0.05; talking for two minutes results in a cost of $0.10. To generalize, the monthly cost for the "Value Plus Flat Rate" plan is $0.05 times the number of minutes used. This leads to the following equation:

$$C = 0.05m \quad \text{Equation (2)}$$

We use equations (1) and (2) to form the following system:

$$\begin{cases} C = 0.03m + 3.00 \\ C = 0.05m \end{cases} \quad \text{The Model}$$

Step 4: Solve To obtain the graph of each equation in the system, we create Table 1, which shows the monthly fee for various minutes used for each plan. Figure 6 shows the graph of the two equations in the system with the horizontal axis labeled m and the vertical axis labeled C.

Table 1

m	Cost for "Just Call" Plan	Cost for "Value Plus Flat Rate" Plan
0	$3.00	$ 0
100	$6.00	$ 5.00
200	$9.00	$10.00

Figure 6

From the graph, we see that the monthly charge will be the same if 150 minutes are used in a month.

Step 5: Check For the "Just Call" plan, the bill is $3.00 plus $0.03 times the number of minutes used. If 150 minutes are used, the bill is $3.00 plus $0.03 times 150 or $3.00 plus $4.50 for a total bill of $7.50. For the "Value Plus Flat Rate Plan," the bill is $0.05 times the number of minutes used. If 150 minutes are used, the bill is $0.05 times 150 or $7.50. Both bills are the same. Our answer checks.

Step 6: Answer The two plans result in the same monthly bill when 150 minutes are used. If you use the phone for fewer than 150 minutes, we can see that the "Value Plus Flat Rate" plan is the better deal; otherwise the "Just Call" plan is the better deal. ▬

QUICK ✓

13. John needs to rent a 12-foot moving truck for a day. He calls two rental companies to determine their fee structure. EZ-Rental charges $20 plus $0.40 per mile, while U-Move-It charges $35 plus $0.25 per mile. After how many miles will the cost of renting the truck be the same? What is the cost at this mileage?

Work Smart
Use the graphical method to solve a system of equations when it is helpful to visualize the solutions.

4.1 Exercises

For Extra Help: Student Solutions Manual   CD Video   PH Math/Tutor Center   MathXL Tutorials on CD   MathXL®   MyMathLab

Concepts and Vocabulary

In Problems 1–3, fill in the blanks.

1. In order for a system of linear equations to have exactly one solution, the graphs of the equations must _____.

2. If a system of linear equations has at least one solution, the system is called _____. In addition, we call this system _____ if it has exactly one solution and _____ if it has more than one solution.

3. A system with no solutions is called _____.

In Problems 4–6, answer True or False to each statement.

4. If a solution of a system of linear equations has more than one solution, it must have infinitely many solutions.

5. A solution of a system of two linear equations is an ordered pair that satisfies one of the equations in the system.

6. It is always possible to tell the exact solution of a system of linear equations by the graphing method.

7. The graphing method for solving a system of equations depends on your ability to graph a line. List all of the methods to graph a line discussed in Chapter 3 and then briefly describe the benefits and drawbacks to each method in the context of solving a system of linear equations. Can you think of any drawbacks to solving a system using the graphing method?

8. When solving a system of linear equations using the graphing method, you find that the solution appears to be $(3, 4)$. How do you verify that $(3, 4)$ is the solution? Is it possible that there are exactly two points that satisfy the given system? Why or why not? Make up a system of equations for which $(3, 4)$ is a solution.

Building Skills

In Problems 9–14, determine whether the ordered pair is a solution of the system of equations.

9. $\begin{cases} x - y = -4 \\ 3x + y = -4 \end{cases}$

    (a) $(2, 6)$
    (b) $(-2, 2)$
    (c) $(2, -2)$

10. $\begin{cases} 2x + 5y = 0 \\ x - 3y = 11 \end{cases}$

    (a) $(5, -2)$
    (b) $(-5, 2)$
    (c) $(2, -3)$

11. $\begin{cases} 3x - y = 2 \\ -15x + 5y = -10 \end{cases}$

    (a) $(1, -1)$
    (b) $(-2, -8)$
    (c) $(0, -2)$

12. $\begin{cases} x - 4y = 2 \\ 5x - 8y = -6 \end{cases}$

    (a) $(-8, -2)$
    (b) $(2, 2)$
    (c) $(-6, -3)$

13. $\begin{cases} 6x - 2y = 1 \\ y = 3x + 2 \end{cases}$

    (a) $\left(0, -\dfrac{1}{2}\right)$
    (b) $(-2, -4)$
    (c) $(0, 2)$

14. $\begin{cases} y = -4x - 1 \\ 2y + 8x = 2 \end{cases}$

    (a) $\left(-\dfrac{1}{2}, 0\right)$
    (b) $(0, 1)$
    (c) $(-1, 0)$

In Problems 15–30, solve each system of equations by graphing.

15. $\begin{cases} 2x - y = -1 \\ 3x + 2y = -5 \end{cases}$

16. $\begin{cases} x + 3y = 0 \\ x - y = 8 \end{cases}$

17. $\begin{cases} y = x + 5 \\ y = -\dfrac{1}{5}x - 1 \end{cases}$

18. $\begin{cases} y = -\dfrac{2}{3}x - 2 \\ y = \dfrac{1}{2}x + 5 \end{cases}$

19. $\begin{cases} y = \dfrac{3}{4}x - 4 \\ y = -\dfrac{1}{2}x + 1 \end{cases}$

20. $\begin{cases} y = -4x + 6 \\ y = -2x \end{cases}$

21. $\begin{cases} x - 4 = 0 \\ 3x - 5y = 22 \end{cases}$

22. $\begin{cases} y + 1 = 0 \\ x + 4y = -6 \end{cases}$

23. $\begin{cases} 3x - y = -1 \\ -6x + 2y = -4 \end{cases}$

24. $\begin{cases} -2x + 5y = -20 \\ 4x - 10y = 10 \end{cases}$

25. $\begin{cases} x + y = -2 \\ 3x - 4y = 8 \end{cases}$

26. $\begin{cases} 2x + 5y = 2 \\ -x + 2y = -1 \end{cases}$

27. $\begin{cases} 2y = 6 - 4x \\ 6x = 9 - 3y \end{cases}$

28. $\begin{cases} 3y = -4x - 4 \\ 8x + 2 = -6y - 6 \end{cases}$

29. $\begin{cases} y = -x + 3 \\ 3y = 2x + 9 \end{cases}$

30. $\begin{cases} 3y = 2x + 1 \\ y = 5x - 4 \end{cases}$

In Problems 31–38, determine the number of solutions of each system. State whether the system is consistent or inconsistent. For those systems that are consistent, state whether the equations are dependent or independent. State the solution of each system.

31.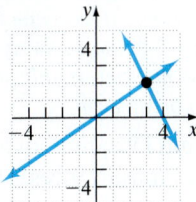

32.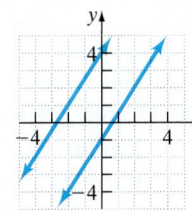

33.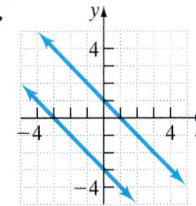

34.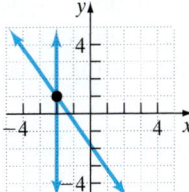

35.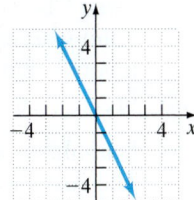

36.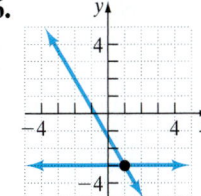

37.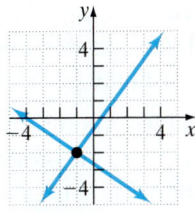

38.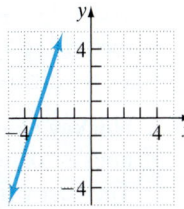

In Problems 39–50, without graphing, determine the number of solutions of each system of equations. State whether the system is consistent or inconsistent. For those systems that are consistent, state whether the equations are dependent or independent.

39. $\begin{cases} y = 2x + 3 \\ y = -2x + 3 \end{cases}$

40. $\begin{cases} y = -x + 5 \\ y = x - 5 \end{cases}$

41. $\begin{cases} 6x + 2y = 12 \\ 3x + y = 12 \end{cases}$

42. $\begin{cases} 2x + y = -3 \\ -2x - y = 6 \end{cases}$

43. $\begin{cases} x + 2y = 2 \\ 2x + 4y = 4 \end{cases}$

44. $\begin{cases} -x + y = 5 \\ x - y = -5 \end{cases}$

45. $\begin{cases} y = 4 \\ x = 4 \end{cases}$

46. $\begin{cases} y - 3 = 0 \\ x - 3 = 0 \end{cases}$

47. $\begin{cases} x - 2y = -4 \\ -x + 2y = -4 \end{cases}$

48. $\begin{cases} 5x + y = -1 \\ 10x + 2y = -2 \end{cases}$

49. $\begin{cases} x + 2y = 4 \\ x + 1 = 5 - 2y \end{cases}$

50. $\begin{cases} 2y + 6 = 2x \\ x + 4 = y - 7 \end{cases}$

Mixed Practice

In Problems 51–62, solve each system of equations by graphing. Based on the graph, state whether the system is consistent or inconsistent. For those systems that are consistent, state whether the equations are dependent or independent.

51. $\begin{cases} y = 2x + 9 \\ y = -3x - 6 \end{cases}$

52. $\begin{cases} y = 3x - 5 \\ y = -x + 11 \end{cases}$

53. $\begin{cases} x - 2y = 6 \\ 2x - 4y = 0 \end{cases}$

54. $\begin{cases} 2x + 3y = -3 \\ 4x + 6y = 6 \end{cases}$

55. $\begin{cases} 3x - 2y = -2 \\ 2x + y = 8 \end{cases}$

56. $\begin{cases} 3x - y = -11 \\ -2x + y = 9 \end{cases}$

57. $\begin{cases} y = x + 2 \\ 3x - 3y = -6 \end{cases}$

58. $\begin{cases} -2x + 6y = -6 \\ 3x - 9y = 9 \end{cases}$

59. $\begin{cases} -5x - 2y = -2 \\ 10x + 4y = 4 \end{cases}$

60. $\begin{cases} 3x + y = -2 \\ 6x + 2y = -4 \end{cases}$

61. $\begin{cases} 3x = 4y - 12 \\ 2x = -4y - 8 \end{cases}$

62. $\begin{cases} 2x = -5y - 25 \\ -3x = 10y + 50 \end{cases}$

Applying the Concepts

*In business, a company's **break-even point** is the number of units of a product that is produced so that the costs of production will be equal to the amount of revenue from the sale of the product. In Problems 63–66, solve the system of equations by graphing to find the number of units that need to be produced for the company to break even.*

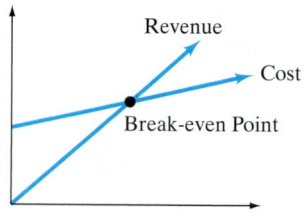

63. **Producing CDs** California Dreamin' produces CDs of local talent and sells the CDs to neighborhood music stores. The cost y to produce x CDs can be determined by the equation $y = 5x + 80$ and the revenue y from the sale of these CDs is given by the equation $y = 13x$. Find the break-even point: the point where the line that represents the cost, $y = 5x + 80$, intersects the revenue line, $y = 13x$. Interpret the x-coordinate and the y-coordinate of the break-even point.

64. **School Newspaper** School Printers, Inc. prints and sells the school newspaper. The cost y to produce x newspapers is determined by the equation $y = 0.05x + 8$ and the revenue y from the sale of these newspapers is given by the equation $y = 0.25x$. Find the break-even point: the point where the line that represents the cost, $y = 0.05x + 8$, intersects the revenue line, $y = 0.25x$. Interpret the x-coordinate and the y-coordinate of the break-even point.

65. **Hockey Skates** The Ice Blades Manufacturing Company produces and sells hockey skates to discount stores. The skates come in boxes of 10 pairs of skates. The cost y (in thousands of dollars) to produce x boxes of skates can be determined by the equation $y = 0.8x + 3.5$ and the revenue y (in thousands of dollars) from the sale of these skates is given by the equation $y = 1.5x$. Find the break-even point.

66. **Patio Chairs** Maumee Lumber Company produces wood patio chairs that they sell to local hardware and home improvement stores. The cost y (in hundreds of dollars) to produce x patio chairs can be determined by the equation $y = 0.5x + 6$. The revenue y (in hundreds of dollars) from the sale of x patio chairs is given by the equation $y = 1.5x$. Find the break-even point.

In Problems 67–70, set up a system of linear equations in two variables that models the problem. Then solve the system of linear equations.

67. **Car Rental** The Wheels-to-Go car rental agency rents cars for $50 daily plus $0.20 per mile. Acme Rental agency will rent the same car for $62 daily plus $0.12 per mile. On Liza's trip to Houston, she decides to rent a car but is not sure how far she will need to drive. Determine the number of miles for which the cost of the car rental will the same for both companies. If Liza plans to drive the car for 200 miles, which company should she use?

68. **Car Rental** The Speedy Car Rental agency charges $60 per day plus $0.14 per mile while the Slow-but-Cheap Car Rental agency charges $30 per day plus $0.20 per mile. Determine the number of miles for which the cost of the car rental will the same for both companies. If you plan to drive the car for 400 miles, which company should you use?

69. **Phone Charges** Sprint has two different long-distance phone plans. Plan A charges a monthly fee of $8.95 plus $0.05 per minute. Plan B charges a monthly fee of $5.95 plus $0.07 per minute. Determine the number of minutes for which the cost of each plan will be the same. If you typically use 100 long-distance minutes each month, which plan should you choose? (*Source:* Sprint.com)

70. **Political Flyers** A politician running for city council has found that PrintQuick can print pamphlets for her campaign for $0.04 per copy plus a one-time set-up fee of $10. She has also learned that Print-A-Lot will print the same pamphlet for a flat fee of $20. Determine the number of political pamphlets that can be printed for the cost at PrintQuick to be the same as at Print-A-Lot.

Extending the Concepts

In Problems 71 and 72, determine the value of the coefficient, c, so that the given system of equations is dependent.

71. $\begin{cases} 3x - y = -4 \\ \quad\quad y = cx + 4 \end{cases}$

72. $\begin{cases} 2(x + 4) = 4y + 4 \\ \quad\quad\quad y = cx + 1 \end{cases}$

In Problems 73 and 74, determine the value of c so that the given system of equations is dependent.

73. $\begin{cases} x + 3 = 3(x - y) \\ 2y + 3 = 2cx - y \end{cases}$

74. $\begin{cases} 5x + 2y = 3 \\ \quad 4y = -10x + c \end{cases}$

The Graphing Calculator

A graphing calculator can be used to approximate the point of intersection between two equations using its INTERSECT command. We illustrate this feature of the graphing calculator by solving the system $\begin{cases} x + y = -1 \\ -2x + y = -7 \end{cases}$. Start by graphing each equation in the system as shown in Figure 7(a). Then use the INTERSECT command and find that the lines intersect at $x = 2, y = -3$. See Figure 7(b) The solution is the ordered pair $(2, -3)$.

Figure 7

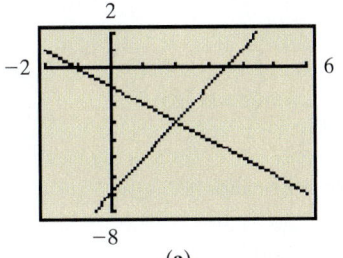
(a)

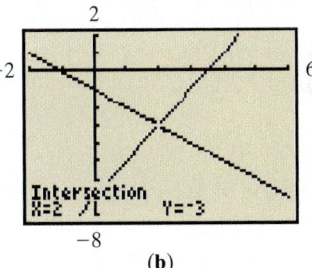
(b)

In Problems 75–80, use a graphing calculator to solve each system of equations. If necessary, express your solution rounded to two decimal places.

75. $\begin{cases} y = 3x - 1 \\ y = -2x + 5 \end{cases}$

76. $\begin{cases} y = \dfrac{3}{2}x - 4 \\ y = -\dfrac{1}{4}x + 3 \end{cases}$

77. $\begin{cases} 3x - y = -1 \\ -4x + y = -3 \end{cases}$

78. $\begin{cases} -6x - 2y = 4 \\ \quad 5x + 3y = -2 \end{cases}$

79. $\begin{cases} 4x - 3y = 1 \\ -8x + 6y = -2 \end{cases}$

80. $\begin{cases} -2x + 5y = -2 \\ \quad 4x - 10y = 1 \end{cases}$

4.2 Solving Systems of Linear Equations Using Substitution

OBJECTIVES

1. Solve a System of Linear Equations Using the Substitution Method

2. Solve Applied Problems Involving Systems of Linear Equations

Preparing for Solving Systems of Linear Equations Using Substitution

Before getting started, take the following readiness quiz. If you get a problem wrong, go back to the section cited and review the material.

1. Solve $3x - y = 2$ for y. [Section 2.4, pp. 108–111]
2. Solve $2x + 5y = 8$ for x. [Section 2.4, pp. 108–111]
3. Solve: $3x - 2(5x + 1) = 12$ [Section 2.2, pp. 88–89]

1 Solve a System of Linear Equations Using the Substitution Method

Preparing for...Answers

1. $y = 3x - 2$ **2.** $x = -\dfrac{5}{2}y + 4$
3. $\{-2\}$

If the x- and y-coordinates of the point of intersection of two lines are not integers, then obtaining an exact result using the graphing method can be difficult. For example,

Figure 8

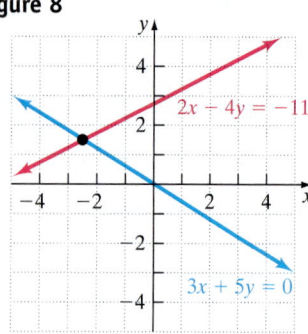

consider the following system of equations:

$$\begin{cases} 3x + 5y = 0 \\ 2x - 4y = -11 \end{cases}$$

Figure 8 shows the graph of the two equations in the system. The lines intersect at $\left(-\dfrac{5}{2}, \dfrac{3}{2}\right)$. Because the lines do not intersect at integer values, it is difficult to determine the solution of the system graphically. Therefore, rather than using graphical methods to obtain solutions of some systems of two linear equations, we prefer to use algebraic methods. The first algebraic method that we present is the *method of substitution*. Let's see how to solve a system of equations using this technique.

EXAMPLE 1 **How to Solve a System of Two Equations in Two Variables by Substitution**

Solve the following system by substitution: $\begin{cases} y = -2x + 10 \\ 3x + 5y = 8 \end{cases}$

Step-by-Step Solution

First, we will call the equation $y = -2x + 10$ equation (1) and $3x + 5y = 8$ equation (2), so that the system is displayed as follows:

$$\begin{cases} y = -2x + 10 & (1) \\ 3x + 5y = 8 & (2) \end{cases}$$

Step 1: Solve one of the equations for one of the unknowns.

Equation (1) is already solved for y: $\quad y = -2x + 10$

Step 2: Substitute $y = -2x + 10$ into equation (2).

$$3x + 5y = 8$$
Equation (2): $\quad 3x + 5(-2x + 10) = 8$

Step 3: Solve the equation for x.

Distribute the 5:	$3x - 10x + 50 = 8$
Combine like terms:	$-7x + 50 = 8$
Subtract 50 from both sides:	$-7x = -42$
Divide both sides by -7:	$x = 6$

Step 4: Let $x = 6$ in equation (1) to find the value of y.

Equation (1): $\quad y = -2x + 10$

$x = 6$: $\quad y = -2(6) + 10$

$y = -12 + 10$

$y = -2$

Step 5: Verify that $x = 6$ and $y = -2$ is the solution.

Equation (1): $\quad y = -2x + 10$

$-2 \stackrel{?}{=} -2(6) + 10$

$-2 \stackrel{?}{=} -12 + 10$

$-2 = -2$ True

Equation (2): $\quad 3x + 5y = 8$

$3(6) + 5(-2) \stackrel{?}{=} 8$

$18 - 10 \stackrel{?}{=} 8$

$8 = 8$ True

Both equations are satisfied so the solution of the system $\begin{cases} y = -2x + 10 \\ 3x + 5y = 8 \end{cases}$ is the ordered pair $(6, -2)$.

Figure 9 shows the graph of each equation in the system from Example 1. The point of intersection is $(6, -2)$, as we would expect.

Notice that we named each equation in the system in Example 1 before going through the steps for solving the system. For convenience, we shall name the equations in the problem for the remainder of the text.

Figure 9

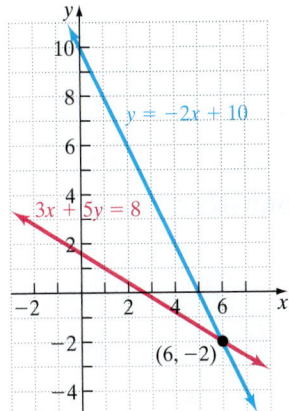

Below we summarize the steps for solving a system using substitution.

Steps for Solving a System of Two Linear Equations Containing Two Unknowns by Substitution

Step 1: Solve one of the equations for one of the unknowns.

Step 2: Substitute the expression solved for in Step 1 into the *other* equation. The result will be a single linear equation in one unknown.

Step 3: Solve the linear equation in one unknown found in Step 2.

Step 4: Substitute the value of the variable found in Step 3 into one of the *original* equations to find the value of the other variable.

Step 5: Check your answer.

EXAMPLE 2 **Solving a System of Equations by Substitution—First Solving for One Variable**

Solve the following system by substitution: $\begin{cases} 2x - y = -15 & (1) \\ 4x + 3y = 5 & (2) \end{cases}$

Solution

When using substitution, solve for the variable whose coefficient is 1 or -1, if possible, to simplify the algebra. We will solve equation (1) for y.

$$\text{Equation (1):} \qquad 2x - y = -15$$
$$\text{Subtract } 2x \text{ from both sides:} \qquad -y = -2x - 15$$
$$\text{Multiply both sides by } -1: \qquad y = 2x + 15$$

We substitute $2x + 15$ for y in equation (2).

$$\text{Equation (2):} \qquad 4x + 3y = 5$$
$$\text{Let } y = 2x + 15: \qquad 4x + 3(2x + 15) = 5$$

Work Smart

If you solve for a variable in equation (1), be sure to substitute into equation (2). If you solve for a variable in equation (2), be sure to substitute into equation (1).

Now we solve for x.

$$\text{Distribute to remove parentheses:} \qquad 4x + 6x + 45 = 5$$
$$\text{Combine like terms:} \qquad 10x + 45 = 5$$
$$\text{Subtract 45 from both sides:} \qquad 10x = -40$$
$$\text{Divide both sides by 10:} \qquad x = -4$$

Let $x = -4$ in equation (1) and solve for y.

$$\text{Equation (1):} \qquad 2x - y = -15$$
$$\text{Let } x = -4: \qquad 2(-4) - y = -15$$
$$-8 - y = -15$$
$$\text{Add 8 to both sides:} \qquad -y = -7$$
$$\text{Multiply both sides by } -1: \qquad y = 7$$

Check

$$\text{Equation (1):} \qquad 2x - y = -15 \qquad\qquad \text{Equation (2):} \qquad 4x + 3y = 5$$
$$\text{Let } x = -4, y = 7: \quad 2(-4) - 7 \overset{?}{=} -15 \qquad \text{Let } x = -4, y = 7: \quad 4(-4) + 3(7) \overset{?}{=} 5$$
$$-8 - 7 \overset{?}{=} -15 \qquad\qquad\qquad\qquad -16 + 21 \overset{?}{=} 5$$
$$-15 = -15 \quad \text{True} \qquad\qquad\qquad\qquad 5 = 5 \quad \text{True}$$

The solution of the system $\begin{cases} 2x - y = -15 \\ 4x + 3y = 5 \end{cases}$ is $(-4, 7)$.

QUICK ✓ *Solve the system using substitution.*

1. $\begin{cases} y = 3x - 2 \\ 2x - 3y = -8 \end{cases}$

2. $\begin{cases} 2x + y = -1 \\ 4x + 3y = 3 \end{cases}$

EXAMPLE 3 **Solving a System of Equations by Substitution—First Solving for One Variable**

Solve the following system by substitution:
$$\begin{cases} 2x + 3y = 9 & (1) \\ x + 2y = \dfrac{13}{2} & (2) \end{cases}$$

Solution

In this problem, we will solve equation (2) for x. Why did we choose this approach? Because the coefficient of x in equation (2) is one. That will make solving for x easier. If we try to solve for x or y in equation (1), or solve for y in equation (1), we will end up with an equation that contains a variable with a fractional coefficient. We'd like to avoid that situation, if possible.

Equation (2): $\qquad x + 2y = \dfrac{13}{2}$

Subtract 2y from both sides: $\qquad x = -2y + \dfrac{13}{2}$

We substitute $-2y + \dfrac{13}{2}$ for x in equation (1).

Equation (1): $\qquad 2x + 3y = 9$

Let $x = -2y + \dfrac{13}{2}$: $\qquad 2\left(-2y + \dfrac{13}{2}\right) + 3y = 9$

Now we solve this equation for y.

Distribute the 2: $\qquad 2(-2y) + 2\left(\dfrac{13}{2}\right) + 3y = 9$

$$-4y + 13 + 3y = 9$$

Combine like terms: $\qquad -y + 13 = 9$

Subtract 13 from each side: $\qquad -y = -4$

Multiply both sides by -1: $\qquad y = 4$

Let $y = 4$ in equation (1) and solve for x.

Equation (1): $\qquad 2x + 3y = 9$

Let $y = 4$: $\qquad 2x + 3(4) = 9$

$$2x + 12 = 9$$

Subtract 12 from both sides: $\qquad 2x = -3$

Divide both sides by 2: $\qquad \dfrac{2x}{2} = -\dfrac{3}{2}$

$$x = -\dfrac{3}{2}$$

Check

Equation (1): $\qquad 2x + 3y = 9$

Let $x = -\dfrac{3}{2}, y = 4$: $\quad 2\left(-\dfrac{3}{2}\right) + 3(4) \overset{?}{=} 9$

$$-3 + 12 \overset{?}{=} 9$$

$$9 = 9 \quad \text{True}$$

Equation (2): $\qquad x + 2y = \dfrac{13}{2}$

Let $x = -\dfrac{3}{2}, y = 4$: $\quad -\dfrac{3}{2} + 2(4) \overset{?}{=} \dfrac{13}{2}$

$$-\dfrac{3}{2} + 8 \overset{?}{=} \dfrac{13}{2}$$

$$\dfrac{13}{2} = \dfrac{13}{2} \quad \text{True}$$

The solution of the system $\begin{cases} 2x + 3y = 9 \\ x + 2y = \dfrac{13}{2} \end{cases}$ is $\left(-\dfrac{3}{2}, 4\right)$. ▬

Work Smart

Solving for x or for y in the first equation in Example 3 is possible, but will lead to fractions, which will make substitution more difficult. For example, if we solve for y in the first equation we obtain

$$2x + 3y = 9$$

$$3y = -2x + 9$$

$$y = -\dfrac{2}{3}x + 3$$

Then

$$x + 2\left(-\dfrac{2}{3}x + 3\right) = \dfrac{13}{2}$$

$$x - \dfrac{4}{3}x + 6 = \dfrac{13}{2}$$

$$-\dfrac{1}{3}x + 6 = \dfrac{13}{2}$$

$$-\dfrac{1}{3}x = \dfrac{1}{2}$$

$$x = -\dfrac{3}{2}$$

Whew!!!

QUICK ✔ *Solve the system using substitution.*

3. $\begin{cases} y = 4x + 1 \\ 8x - y = 5 \end{cases}$

4. $\begin{cases} 3x + 2y = -3 \\ -x + y = \dfrac{11}{6} \end{cases}$

Recall that a system of two linear equations containing two unknowns can be inconsistent. This means that the two lines in the system are parallel. What happens when we use the method of substitution on a system that is inconsistent? Let's find out!

EXAMPLE 4 **Solving an Inconsistent System Using Substitution**

Solve the following system by substitution: $\begin{cases} x - 3y = 5 \quad (1) \\ -2x + 6y = 3 \quad (2) \end{cases}$

Solution

Is the coefficient of one of the variables 1 or -1? Yes, the coefficient of x in equation (1) is 1, so we solve for x in that equation.

Equation (1):	$x - 3y = 5$
Add $3y$ to both sides:	$x = 3y + 5$
In equation (2), replace x by $3y + 5$:	$-2(3y + 5) + 6y = 3$
Distribute to remove parentheses:	$-6y - 10 + 6y = 3$
Combine like terms:	$-10 = 3$

Notice that the variable, y, was eliminated from the equation and that the last equation, $-10 = 3$, is a false statement. We conclude that the system $\begin{cases} x - 3y = 5 \\ -2x + 6y = 3 \end{cases}$ is inconsistent and has no solution.

So how do we determine if a system is inconsistent algebraically? If we end up with a false statement such as $-10 = 3$ or $-13 = 0$, we conclude the system is inconsistent.

How do we determine if a system is consistent, but dependent? That is, how do we determine if a system has infinitely many solutions? Example 5 supplies the answer.

EXAMPLE 5 **Solving a System with Infinitely Many Solutions Using Substitution**

Solve the following system by substitution: $\begin{cases} 6x - 2y = -4 & (1) \\ -3x + y = 2 & (2) \end{cases}$

Solution

It is easiest to solve equation (2) for y.

Equation (2):	$-3x + y = 2$
Add $3x$ to both sides:	$y = 3x + 2$
Let $y = 3x + 2$ in equation (1):	$6x - 2(3x + 2) = -4$
Distribute to remove parentheses:	$6x - 6x - 4 = -4$
Combine like terms:	$-4 = -4$

Notice that the variable x has been eliminated from the equation and the last equation, $-4 = -4$, is a true statement. We conclude that the system $\begin{cases} 6x - 2y = -4 \\ -3x + y = 2 \end{cases}$ is consistent, but dependent. The system has infinitely many solutions.

Notice that for both the system that has no solution and for the system that has infinitely many solutions, the variables dropped out. The difference between the systems is that the system with infinitely many solutions ends up with a statement that is true (such as $6 = 6$), while the system with no solution ends up with a statement that is false (such as $0 = 5$).

QUICK ✓ *Use the method of substitution to determine whether the system has infinitely many solutions or no solution.*

5. $\begin{cases} y = 5x + 2 \\ -10x + 2y = 4 \end{cases}$ **6.** $\begin{cases} 3x - 2y = 0 \\ -9x + 6y = 5 \end{cases}$ **7.** $\begin{cases} 2x - 6y = 2 \\ -3x + 9y = 4 \end{cases}$

② ## Solve Applied Problems Involving Systems of Linear Equations

Now let's look at a problem that can be solved using the method of substitution.

EXAMPLE 6 | Life Expectancy of Men versus Women

It is well documented that American women can expect to live longer than American men. But is there ever going to be a time when men can expect to live as long as women? The data given in Table 2 show the life expectancy at birth for both men and women from 1990 to 2001. The data are from the Centers for Disease Control and Prevention. Using techniques from statistics, we can determine an equation that models the life expectancy of Americans of each gender. These techniques indicate that the life expectancy E of women is given by $E = 0.09t - 96.63$, where t is the year. The life expectancy E of men is given by $E = 0.25t - 434.22$. The graphs in Figure 10 plot the data given in Table 2.

Table 2		
Year	Life Expectancy—Women	Life Expectancy—Men
1990	78.8	71.8
1991	78.9	72.0
1992	79.1	72.3
1993	78.8	72.2
1994	79.1	72.4
1995	78.8	72.5
1996	79.0	73.1
1997	78.9	73.6
1998	79.1	73.8
1999	79.4	73.9
2000	79.7	74.3
2001	79.8	74.4

Figure 10

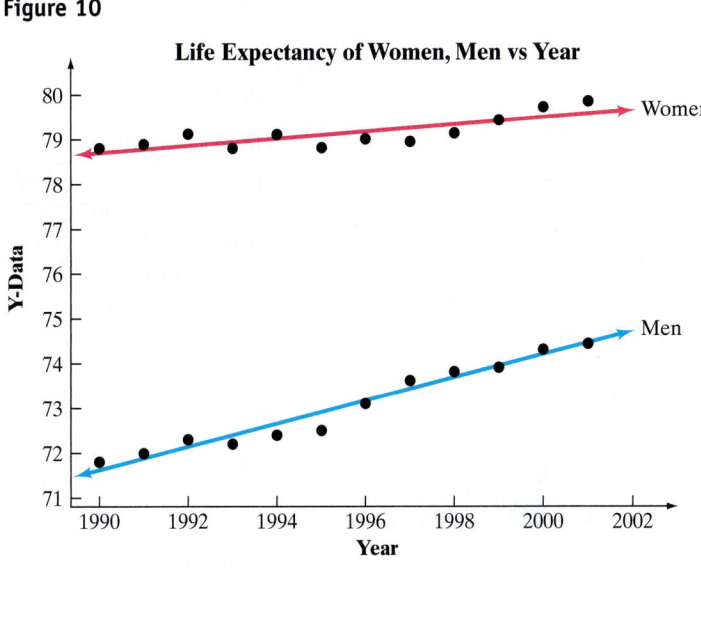

Life Expectancy of Women, Men vs Year

Projecting forward from these models, in what year, if ever, can men expect to live as long as women?

Solution

Step 1: Identify We want to know the year in which American men can expect to live as long as American women.

Step 2: Name The variables are E, which represents the life expectancy, and t, which represents the year.

Step 3: Translate The life expectancy for American women is given by the model $E = 0.09t - 96.63$. We call this equation (1). The life expectancy for American men is given by the model $E = 0.25t - 434.22$. We call this equation (2). We combine equations (1) and (2) to form the following system:

$$\begin{cases} E = 0.09t - 96.63 & (1) \\ E = 0.25t - 434.22 & (2) \end{cases} \quad \text{The Model}$$

Step 4: Solve We will solve this system of equations by substituting $0.09t - 96.63$ for E into equation (2). This leads to the following equation.

$$0.09t - 96.63 = 0.25t - 434.22$$

Add 96.63 to both sides: $\qquad 0.09t = 0.25t - 337.59$

Subtract $0.25t$ from both sides: $\qquad -0.16t = -337.59$

Divide both sides by -0.16: $\qquad t \approx 2109.9$

We round this result to the nearest year, 2110.

Step 5: Check For American women, the life expectancy in 2110 is predicted to be $E = 0.09(2110) - 96.63 = 93.27$ years. For American men, the life expectancy in 2110 is predicted to be $E = 0.25(2110) - 434.22 = 93.28$ years. The answers differ slightly because of rounding, but our answer checks.

Step 6: Answer According to these models, the life expectancy of American men will be the same as American women in 2110. One caveat to this prediction is that we cannot be sure that the trend that we see for the years 1990–2001 will continue for the next century.

In Words

Caveat (pronounced kä-vē-at) is Latin for "warning" or "beware!"

QUICK ✓

8. The participation rate in the workforce is defined by economists to be the number of individuals employed divided by the number of individuals 16 years of age or older. The participation rate P for men is given by $P = -0.08t + 243.96$, where t is the year. The participation rate P for women is given by $P = 0.24t - 415.10$, where t is the year. Based on these models, when, if ever, can we expect the participation rate of women in the workforce to equal the participation rate of men?

4.2 Exercises

For Extra Help:

Student Solutions Manual   CD Video   PH Math/Tutor Center   MathXL Tutorials on CD   MathXL®   MyMathLab

Concepts and Vocabulary

In Problems 1–3, fill in the blanks.

1. When solving a system of equations by substitution, whenever possible solve one of the equations for the variable that has a coefficient of _____ or _____.

2. While solving a system of equations by substitution, the variable expression has been eliminated and a *false* statement such as $-3 = 0$ results. This means that the solution of the system is _____.

3. While solving a system of equations by substitution, the variable expression has been eliminated and a *true* statement such as $0 = 0$ results. This means that the system has _____ _____ _____.

In Problems 4–6, answer True or False to each statement.

4. When solving a system of equations by substitution, you solve equation (1) for x and then you substitute this expression back into equation (1), resulting in the statement $5 = 5$. This means that there are infinitely many solutions of the system.

5. It is possible to solve a system of equations such as $\begin{cases} x = 10 \\ x - y = -2 \end{cases}$ by substitution.

6. A system of equations can have one solution or no solution, but not more than one solution.

7. A test question asks you to solve the following system of equations by substitution. What would be a reasonable first step? Explain which variable you would solve for and why.

$$\begin{cases} \dfrac{1}{3}x - \dfrac{1}{6}y = \dfrac{2}{3} \\ \dfrac{3}{2}x + \dfrac{1}{2}y = -\dfrac{5}{4} \end{cases}$$

8. List two advantages of solving a system of linear equations by substitution rather than by the graphing method.

Building Skills

In Problems 9–22, solve each system of equations using substitution.

9. $\begin{cases} x + 2y = 2 \\ y = 2x - 9 \end{cases}$

10. $\begin{cases} y = 3x - 11 \\ -x + 4y = -11 \end{cases}$

11. $\begin{cases} -2x + 5y = 7 \\ x = 3y - 4 \end{cases}$

12. $\begin{cases} -4x - y = -3 \\ x = y + 7 \end{cases}$

13. $\begin{cases} x + y = -7 \\ 2x - y = -2 \end{cases}$

14. $\begin{cases} 5x + 2y = -5 \\ 3x - y = -14 \end{cases}$

15. $\begin{cases} y = \dfrac{1}{2}x - 5 \\ y = -\dfrac{3}{4}x - 10 \end{cases}$

16. $\begin{cases} y = \dfrac{2}{3}x + 1 \\ y = -\dfrac{3}{2}x + 40 \end{cases}$

17. $\begin{cases} y = 3x + 4 \\ y = -\dfrac{1}{2}x + \dfrac{5}{3} \end{cases}$

18. $\begin{cases} y = 5x - 3 \\ y = 2x - \dfrac{21}{5} \end{cases}$

19. $\begin{cases} x = -6y \\ x - 3y = 3 \end{cases}$

20. $\begin{cases} y = 4x \\ 2x - 3y = 5 \end{cases}$

21. $\begin{cases} 2x - 3y = 0 \\ 8x + 6y = 3 \end{cases}$

22. $\begin{cases} 2x + 3y = -1 \\ 2x - 9y = -9 \end{cases}$

In Problems 23–30, determine if the given system is a dependent or inconsistent system. State the solution to each system.

23. $\begin{cases} x + 3y = -12 \\ x + 3y = 6 \end{cases}$

24. $\begin{cases} 2x - y = 3 \\ y - 2x = 3 \end{cases}$

25. $\begin{cases} y = 4x - 1 \\ 8x - 2y = 2 \end{cases}$

26. $\begin{cases} y = -x + 1 \\ x + y = 1 \end{cases}$

27. $\begin{cases} x + 2y = 6 \\ x = 3 - 2y \end{cases}$

28. $\begin{cases} 4 + y = 3x \\ 6x - 2y = -2 \end{cases}$

29. $\begin{cases} 5x + 6 = 2 - y \\ y = -5x - 4 \end{cases}$

30. $\begin{cases} x + y = 2 \\ 2x + 2y = 2 \end{cases}$

Mixed Practice

In Problems 31–50, solve each system of equations using substitution.

31. $\begin{cases} x - 3y = -2 \\ x = -2 \end{cases}$

32. $\begin{cases} y = -1 \\ -2x + 9y = 3 \end{cases}$

33. $\begin{cases} x = 2 \\ x - 6y = 4 \end{cases}$

34. $\begin{cases} 2x + y = 8 \\ y = 3 \end{cases}$

35. $\begin{cases} y = \dfrac{1}{2}x \\ x = 2(y + 1) \end{cases}$

36. $\begin{cases} y = \dfrac{3}{4}x \\ x = 4(y - 1) \end{cases}$

37. $\begin{cases} y = 2x - 8 \\ x - \dfrac{1}{2}y = 4 \end{cases}$

38. $\begin{cases} x = 3y + 6 \\ y - \dfrac{1}{3}x = -2 \end{cases}$

39. $\begin{cases} 3x + 2y = 4 \\ 3x + y = \dfrac{9}{2} \end{cases}$

40. $\begin{cases} 3x + 2y = 1 \\ x + 3y = -\dfrac{5}{3} \end{cases}$

41. $\begin{cases} x - 5y = 3 \\ -2x + 10y = 8 \end{cases}$

42. $\begin{cases} -4x + y = 3 \\ 8x - 2y = 1 \end{cases}$

43. $\begin{cases} 3x - 2y = 1 \\ -6x + 4y = -2 \end{cases}$

44. $\begin{cases} -x + 3y = 4 \\ 2x - 6y = -8 \end{cases}$

45. $\begin{cases} 4x + 8y = -9 \\ 8x + 4y = 3 \end{cases}$

46. $\begin{cases} 18x - 6y = -7 \\ x + 2y = 0 \end{cases}$

47. $\begin{cases} \dfrac{3}{2}x - y = 1 \\ 3x - 2y = 2 \end{cases}$

48. $\begin{cases} \dfrac{2}{3}x + \dfrac{1}{3}y = 1 \\ -4x - 2y = -6 \end{cases}$

49. $\begin{cases} \dfrac{x}{2} + \dfrac{y}{3} = \dfrac{1}{12} \\ \dfrac{5x}{4} + \dfrac{2y}{3} = -\dfrac{11}{24} \end{cases}$

50. $\begin{cases} \dfrac{x}{4} + \dfrac{y}{2} = \dfrac{3}{8} \\ \dfrac{x}{5} - \dfrac{3y}{4} = -\dfrac{11}{40} \end{cases}$

Applying the Concepts

Use substitution to solve each system of linear equations in two variables.

51. Fun with Numbers The sum of two numbers is 17, and their difference is 7. Determine the numbers by solving the following system of equations, using x and y to represent the unknown numbers.

$$\begin{cases} x + y = 17 & (1) \\ x - y = 7 & (2) \end{cases}$$

52. Fun with Numbers The sum of two numbers is 25, and their difference is 3. Determine the numbers by solving the following system of equations, using x and y to represent the unknown numbers.

$$\begin{cases} x + y = 25 & (1) \\ x - y = 3 & (2) \end{cases}$$

△ **53. Dimensions of a Garden** The perimeter of a rectangular garden is 34 feet. The length of the garden is 3 feet more than the width. Determine the dimensions of the garden by solving the following system of equations, where l and w represent the length and width of the garden.

$$\begin{cases} 2l + 2w = 34 & (1) \\ l = w + 3 & (2) \end{cases}$$

△ **54. Dimensions of a Rectangle** The perimeter of a rectangle is 52 inches. The length of the rectangle is 4 inches more than the width. Determine the dimensions of the rectangle by solving the following system of equations, where l and w represent the length and width of the rectangle.

$$\begin{cases} 2l + 2w = 52 & (1) \\ l \quad\quad = w + 4 & (2) \end{cases}$$

55. Investment Paul wants to invest part of his Ohio Lottery winnings in a safe money-market fund that earns 2.5% annual interest and the rest in a risky international fund that expects to yield of 9% annual interest. The amount of money invested in the money-market fund, x, is to be exactly twice the amount, y, invested in the international fund. Use the system of equations to determine the amount to be invested in each fund if a total of $560 is to be earned at the end of one year.

$$\begin{cases} x \quad\quad = 2y & (1) \\ 0.025x + 0.09y = 560 & (2) \end{cases}$$

56. Investment Elaine wants to invest part of her $24,000 Virginia Lottery winnings in an international stock fund that yields 4% annual interest and the remainder in a domestic growth fund that yields 6.5% annually. Use the system of equations to determine the amount Elaine should invest in each account to earn $1360 interest at the end of one year, where x represents the amount invested in the international stock fund and y represents the amount invested in the domestic growth fund.

$$\begin{cases} x + \quad y = 24{,}000 & (1) \\ 0.04x + 0.065y = 1360 & (2) \end{cases}$$

57. Sales of Music Videos According to the Recording Industry of America, the number of music videos sold in the U.S. annually since 1998 has been decreasing according to the equation $y = -2.17x + 24.94$, where y is the number of music videos sold (in millions) and x is the number of years since 1998. The sales of DVD videos have been increasing over this same period, according to the equation $y = 2.58x - 0.18$, where y is the number of DVD videos, in millions, and x is the number of years since 1998. Find, to the nearest year, when the sales of DVD videos will be equal to the sales of music videos. (*Source:* Recording Industry of America)

58. Sales of Music Videos According to the Recording Industry of America, the total value of DVD videos has increased since 1998, according to the equation $y = 57.2x + 2.7$, where y is the dollar value of U.S. sales and x is the number of years since 1998. During this same period, the dollar value of music videos has declined according to the equation $y = -48.8x + 454.4$, where y is the dollar value of U.S. music video sales and x is the number of years since 1998. Find, to the nearest year, when the dollar value of sales of music videos and DVD videos was the same. (*Source:* Recording Industry of America)

Extending the Concepts

59. For the system $\begin{cases} Ax + 3By = 2 \\ -3Ax + By = -11 \end{cases}$, find A and B such that $x = 3, y = 1$ is a solution.

60. Write a system of equations that has $(3, 5)$ as a solution.

61. Write a system of equations that has $(-1, 4)$ as a solution.

62. Write a system of equations that has infinitely many solutions.

63. Write a system of equations that has no solution.

4.3 Solving Systems of Linear Equations Using Elimination

OBJECTIVES

1. Solve a System of Linear Equations Using the Elimination Method
2. Solve Applied Problems Involving Systems of Linear Equations

Preparing for Solving Systems of Linear Equations Using Elimination

Before getting started, take the following readiness quiz. If you get a problem wrong, go back to the section cited and review the material.

1. What is the additive inverse of 5? [Section 1.3, p. 21]
2. What is the additive inverse of -8? [Section 1.3, p. 21]
3. Distribute: $\dfrac{2}{3}(3x - 9y)$ [Section 1.7, p. 61]
4. Solve: $2y - 5y = 12$ [Section 2.2, p. 86]

① Solve a System of Linear Equations Using the Elimination Method

Work Smart

Use the method of substitution if it is easy to solve for one of the variables in the system. Use elimination if substitution will lead to fractions or solving for a variable is not easy.

We have seen how to use the method of graphing and the method of substitution to solve a system of linear equations containing two unknowns. The method of graphing is nice because we can visualize the solutions, but it is not always easy to find the exact coordinates of the point of intersection. The method of substitution has the advantage of finding an exact solution because it is an algebraic method, but it may require that we introduce fractions to the solution process. We would rather avoid introducing fractions, if possible. For this reason, we introduce a second algebraic method that can be used to solve a system of equations called the *method of elimination*. This method is usually preferred over the method of substitution when substitution leads to fractions.

Remember that the Addition Property of Equality states that if we add the same quantity to both sides of an equation, we obtain an equivalent equation. Solving a system of linear equations using the elimination method takes this principle a step further. Adding equal quantities to both sides of an equation also results in a true statement.

EXAMPLE 1 Solving a System of Linear Equations Using Elimination

Solve the system $\begin{cases} x + y = 13 & (1) \\ 2x - y = -7 & (2) \end{cases}$

Solution

Equation (2) says that $2x - y$ equals -7. So, if we add $2x - y$ to the left side of equation (1) and -7 to the right side of equation (1), we are adding to the same value to each side of equation (1). We will perform this addition vertically.

Work Smart

This method is called "elimination" because one of the variables is eliminated through the process of addition. The elimination method is also sometimes referred to as the **addition method.**

$$\begin{array}{rl} x + y = 13 & (1) \\ 2x - y = -7 & (2) \\ \hline 3x + 0 = 6 \end{array}$$

Notice that the variable y has been eliminated, and we can solve for x.

$$3x + 0 = 6$$

Divide each side by 3: $\quad 3x = 6$

$$x = 2$$

Now that we have the value of x, we can substitute $x = 2$ into either equation (1) or equation (2) to find the value of y. Let's substitute $x = 2$ into equation (1).

$$x + y = 13$$

Substitute 2 for x: $\quad 2 + y = 13$

Subtract 2 from each side: $\quad\quad y = 11$

We have $x = 2$ and $y = 11$. Let's check this solution in both equations.

Check

Equation (1): $x + y = 13$

$x = 2, y = 11$: $2 + 11 \overset{?}{=} 13$

$13 = 13$ True

Equation (2): $2x - y = -7$

$x = 2, y = 11$: $2(2) - 11 \overset{?}{=} -7$

$4 - 11 \overset{?}{=} -7$

$-7 = -7$ True

The solution of the system $\begin{cases} x + y = 13 \\ 2x - y = -7 \end{cases}$ is $(2, 11)$.

The idea in using the elimination method is to get the coefficients of one of the variables to be additive inverses. In Example 1, the coefficients of the variable y were opposites: 1 and -1. Unfortunately, having linear equations in which one of the variables has opposite coefficients is uncommon, so we need to develop a strategy to solve systems in which the coefficients of one of the variables are not opposites.

EXAMPLE 2   **How to Solve a System of Linear Equations by Elimination—Multiply One Equation by a Number to Create Additive Inverses**

Solve the following system by elimination: $\begin{cases} 2x + 3y = -6 & (1) \\ -4x + 5y = -21 & (2) \end{cases}$

Step-by-Step Solution

Step 1: We need to get the coefficients of one of the variables to be additive inverses. In looking at the system, this can be accomplished by multiplying both sides of equation (1) by 2. Do you see why?

$$\begin{cases} 2x + 3y = -6 & (1) \\ -4x + 5y = -21 & (2) \end{cases}$$

Multiply (1) by 2: $\begin{cases} 2(2x + 3y) = 2(-6) & (1) \\ -4x + 5y = -21 & (2) \end{cases}$

Distribute the 2 in (1): $\begin{cases} 4x + 6y = -12 & (1) \\ -4x + 5y = -21 & (2) \end{cases}$

Step 2: Notice that the coefficients of the variable x are additive inverses. We now add equations (1) and (2) to eliminate the variable x and then solve for y.

$$\begin{cases} 4x + 6y = -12 & (1) \\ -4x + 5y = -21 & (2) \end{cases}$$

Add (1) and (2): $11y = -33$

Divide both sides by 11: $y = -3$

Step 3: We let $y = -3$ in either equation (1) or (2). Since equation (1) looks a little easier to work with, we will substitute $y = -3$ into equation (1) and solve for x.

Equation (1): $2x + 3y = -6$

Let $y = -3$: $2x + 3(-3) = -6$

$2x - 9 = -6$

Add 9 to both sides: $2x = 3$

Divide both sides by 2: $x = \dfrac{3}{2}$

We have that $x = \dfrac{3}{2}, y = -3$.

(continued)

Step 4: Check

| Equation (1): | $2x + 3y = -6$ | Equation (2): | $-4x + 5y = -21$ |

$x = \dfrac{3}{2}, y = -3$: $\quad 2\left(\dfrac{3}{2}\right) + 3(-3) \stackrel{?}{=} -6 \qquad x = \dfrac{3}{2}, y = -3$: $\quad -4\left(\dfrac{3}{2}\right) + 5(-3) \stackrel{?}{=} -21$

$$3 + (-9) \stackrel{?}{=} -6 \qquad\qquad\qquad -6 + (-15) \stackrel{?}{=} -21$$

$$-6 = -6 \quad \text{True} \qquad\qquad\qquad\qquad -21 = -21 \quad \text{True}$$

Both equations check, so the solution of the system $\begin{cases} 2x + 3y = -6 \\ -4x + 5y = -21 \end{cases}$ is $\left(\dfrac{3}{2}, -3\right)$.

The following summarizes the steps just used in Example 2.

Work Smart

The idea behind the elimination method is to get the coefficients of one of the variables to be additive inverses (opposites), so that one of the variables will be eliminated.

Steps for Solving a System of Linear Equations by Elimination

Step 1: Make sure that the coefficients on one of the variables are additive inverses by multiplying (or dividing) both sides of one (or both) equation(s) by a nonzero constant.

Step 2: Add the equations to eliminate the variable whose coefficients are now additive inverses. Solve the resulting equation for the remaining unknown.

Step 3: Substitute the value of the variable found in Step 2 into one of the *original* equations to find the value of the remaining variable.

Step 4: Check your answer.

QUICK ✓ *Solve the system using elimination.*

1. $\begin{cases} x - 3y = 2 \\ 2x + 3y = -14 \end{cases}$ **2.** $\begin{cases} x - 2y = 2 \\ -2x + 5y = -1 \end{cases}$

EXAMPLE 3 **Solve a System of Equations by Elimination—Multiply Both Equations by a Number to Create Additive Inverses**

Solve the following system by elimination: $\begin{cases} 2x + 3y = 3 & (1) \\ 3x + 5y = 7 & (2) \end{cases}$

Solution

We want the coefficients of one of the variables to be additive inverses. We cannot do this by multiplying a single equation by some nonzero constant, so we will need to multiply both equations by some nonzero constant. For example, we can multiply equation (1) by 3 and equation (2) by -2 so that the coefficients on x are additive inverses.

Work Smart

There is no single right way to multiply the equations in a system by a nonzero constant to get coefficients to be additive inverses. For example, we could also multiply equation (1) by 5 and equation (2) by -3. Do you see why this also works?

$$\begin{cases} 2x + 3y = 3 & (1) \\ 3x + 5y = 7 & (2) \end{cases}$$

Multiply equation (1) by 3:
Multiply equation (2) by -2:
$$\begin{cases} 3(2x + 3y) = 3 \cdot 3 & (1) \\ -2(3x + 5y) = -2 \cdot 7 & (2) \end{cases}$$

$$\text{Distribute:} \quad \begin{cases} 6x + 9y = 9 & (1) \\ -6x - 10y = -14 & (2) \end{cases}$$

$$\text{Add (1) and (2):} \qquad -y = -5$$
$$\text{Multiply both sides by } -1: \qquad y = 5$$

Now let $y = 5$ in equation (1) to determine the value of x.

$$\text{Equation (1):} \qquad 2x + 3y = 3$$
$$\text{Let } y = 5 \text{ in (1) and find } x: \quad 2x + 3(5) = 3$$
$$\text{Simplify:} \qquad 2x + 15 = 3$$
$$\text{Subtract 15 from both sides:} \qquad 2x = -12$$
$$\text{Divide both sides by 2:} \qquad x = -6$$

We have that $x = -6$ and $y = 5$. Now we check our answer.

Check

$$\text{Equation (1):} \qquad 2x + 3y = 3 \qquad\qquad \text{Equation (2):} \qquad 3x + 5y = 7$$
$$x = -6, y = 5: \quad 2(-6) + 3(5) \stackrel{?}{=} 3 \qquad x = -6, y = 5: \quad 3(-6) + 5(5) \stackrel{?}{=} 7$$
$$-12 + 15 \stackrel{?}{=} 3 \qquad\qquad -18 + 25 \stackrel{?}{=} 7$$
$$3 = 3 \quad \text{True} \qquad\qquad 7 = 7 \quad \text{True}$$

The solution to the system $\begin{cases} 2x + 3y = 3 \\ 3x + 5y = 7 \end{cases}$ is $(-6, 5)$.

QUICK ✓ *Solve the system using elimination.*

3. $\begin{cases} 5x + 4y = 10 \\ -2x + 3y = -27 \end{cases}$
4. $\begin{cases} 6x + 4y = 30 \\ 7x + 10y = 35 \end{cases}$

Now let's look at the results we obtain when we have systems that have no solution or infinitely many solutions.

EXAMPLE 4 **Solving a System of Equations with No Solution Using Elimination**

Solve the following system by elimination: $\begin{cases} 4x - 6y = -5 & (1) \\ 6x - 9y = 2 & (2) \end{cases}$

Solution

$$\begin{cases} 4x - 6y = -5 & (1) \\ 6x - 9y = 2 & (2) \end{cases}$$

$$\text{To eliminate } x, \text{ multiply equation (1) by } -3: \begin{cases} -3(4x - 6y) = -3 \cdot (-5) & (1) \\ 2(6x - 9y) = 2 \cdot 2 & (2) \end{cases}$$

$$\text{Distribute:} \begin{cases} -12x + 18y = 15 & (1) \\ 12x - 18y = 4 & (2) \end{cases}$$

$$\text{Add (1) and (2):} \qquad 0 = 19$$

The statement $0 = 19$ is false. Therefore, the system $\begin{cases} 4x - 6y = -5 \\ 6x - 9y = 2 \end{cases}$ is inconsistent and has no solution.

EXAMPLE 5 **Solving a System of Equations with Infinitely Many Solutions Using Elimination**

Solve the following system by elimination: $\begin{cases} 3x + 4y = -6 & (1) \\ \dfrac{3}{2}x + 2y = -3 & (2) \end{cases}$

Solution

$$\begin{cases} 3x + 4y = -6 & (1) \\ \dfrac{3}{2}x + 2y = -3 & (2) \end{cases}$$

To eliminate y, multiply equation (2) by -2:

$$\begin{cases} 3x + 4y = -6 & (1) \\ -2\left(\dfrac{3}{2}x + 2y\right) = -2(-3) & (2) \end{cases}$$

Distribute:

$$\begin{cases} 3x + 4y = -6 & (1) \\ -3x - 4y = 6 & (2) \end{cases}$$

Add (1) and (2):

$$0 = 0$$

Notice that both the variables were eliminated. However, the statement $0 = 0$ is true. Therefore, the system $\begin{cases} 3x + 4y = -6 \\ \dfrac{3}{2}x + 2y = -3 \end{cases}$ is consistent but dependent. The system has infinitely many solutions.

QUICK ✓ *Use elimination to determine whether the system has no solution or infinitely many solutions.*

5. $\begin{cases} 2x - 6y = 10 \\ 5x - 15y = 4 \end{cases}$ **6.** $\begin{cases} -x + 3y = 2 \\ 3x - 9y = -6 \end{cases}$ **7.** $\begin{cases} -4x + 8y = 4 \\ 3x - 6y = -3 \end{cases}$

We summarize the three methods that we have discussed for solving a system of two linear equations containing two unknowns.

SUMMARY: Which Method Should I Use?

We have presented three methods for solving systems of linear equations. Below, we present a summary of the questions you should ask before solving a system of linear equations. The answers to these questions tell us the most appropriate method. We also present an example of the situation and the advantages and disadvantages of each method.

Question	Method to Use	Example	Advantages/Disadvantages
Would it be beneficial to see the solutions visually?	Graphical	Find the break-even point of the cost equation $y = 20 + 4x$ and the revenue equation $y = 9x$, where x represents the number of calculators produced and sold.	Allows us to "see" the answer, but if the coordinates of the intersection point are not integers, it can be difficult to determine the answer.
Is one of the coefficients of the variables 1 or -1?	Substitution	$\begin{cases} -x + 2y = 7 \\ 2x - 3y = -15 \end{cases}$	Gives exact solutions. The algebra can be easy provided one of the variables has a coefficient of 1 or -1. If none of the coefficients are 1 or -1, the algebra can get messy.

Question	Method to Use	Example	Advantages/Disadvantages
Are the coefficients of one of the variables additive inverses? If the coefficients are not additive inverses, is it easy to get the coefficients to be additive inverses?	Elimination	$\begin{cases} 2x + 3y = -5 \\ -2x + y = -3 \end{cases}$	Gives exact solutions. It is easy to use when neither of the variables has a coefficient of 1 or -1.

② Solve Applied Problems Involving Systems of Linear Equations

We saw in the last section that applied problems are easily solved using substitution when at least one of the variables is isolated (that is, by itself). But many mathematical models that involve two equations containing two unknowns do not have one of the equations solved for one of the unknowns. If this occurs, it is better to use elimination to solve the problem.

EXAMPLE 6 Hot Dogs and Soda at the Game

Bill and Roger attend a baseball game with their kids. They get to the game early to watch batting practice and have dinner. Roger says that he will buy dinner. He gets 7 hot dogs and 5 Pepsis for $38.25. After the sixth inning everyone is hungry again, so Bill buys 5 hot dogs and 4 Pepsis for $28.50. We can determine the price of a hot dog and the price of a Pepsi by solving the system of equations

$$\begin{cases} 7h + 5p = 38.25 \\ 5h + 4p = 28.50 \end{cases}$$

where h represents the price of a hot dog and p represents the price of a Pepsi. How much does a hot dog cost at the ballpark? How much does a Pepsi cost?

Solution

We name the equation $7h + 5p = 38.25$ equation (1) and name $5h + 4p = 28.50$ equation (2). Now let's solve the system.

$$\begin{cases} 7h + 5p = 38.25 & (1) \\ 5h + 4p = 28.50 & (2) \end{cases}$$

Let's eliminate p. Multiply (1) by -4:
Multiply (2) by 5:
$$\begin{cases} -4(7h + 5p) = -4 \cdot 38.25 & (1) \\ 5(5h + 4p) = 5 \cdot 28.50 & (2) \end{cases}$$

Distribute:
$$\begin{cases} -28h - 20p = -153 & (1) \\ 25h + 20p = 142.50 & (2) \end{cases}$$

Add (1) and (2): $-3h = -10.50$

Divide both sides by -3: $h = 3.50$

Let $h = 3.50$ in (2): $5(3.50) + 4p = 28.50$

$17.50 + 4p = 28.50$

Subtract 17.50 from both sides: $4p = 11$

Divide both sides by 4: $p = 2.75$

We leave it to you to verify the answer. A hot dog costs $3.50 and a Pepsi costs $2.75 at the ballpark.

QUICK ✓

8. At a fast food joint, 5 cheeseburgers and 3 medium shakes cost $15.50. At the same fast food joint, 3 cheeseburgers and 2 medium shakes cost $9.75. We can determine the price of a cheeseburger and the price of a shake by solving the system of equations

$$\begin{cases} 5c + 3s = 15.50 \\ 3c + 2s = 9.75 \end{cases}$$

where c represents the price of a cheeseburger and s represents the price of a shake. How much does a cheeseburger cost at the fast food joint? How much does a shake cost?

4.3 Exercises

For Extra Help:
Student Solutions Manual   CD Video   PH Math/Tutor Center   MathXL Tutorials on CD   MathXL®   MyMathLab

Concepts and Vocabulary

In Problems 1–3, fill in the blanks.

1. List the three methods presented in this chapter for solving systems of equations: _____, _____, _____.

2. When using the elimination method to solve a system of equations, you add equation (1) and equation (2), resulting in the statement $-50 = -50$. This means that the system is _____ and has _____ many solutions.

3. The basic idea in using the elimination method is to get the coefficients of one of the variables to be _____ _____, such as 3 and -3.

In Problems 4–6, answer True or False to each statement.

4. It is possible to solve a system of equations using the elimination method without multiplying either equation by some nonzero constant.

5. Consider the following system of equations. This system could be solved by all of the following:

$$\begin{cases} 6x + 3y = 3 & (1) \\ -4x - y = 1 & (2) \end{cases}$$

(a) Multiply equation (1) by 4 and equation (2) by 6, then add.
(b) Multiply equation (1) by 2 and equation (2) by 3, then add.
(c) Multiply equation (2) by 3, then add.
(d) Multiply equation (1) one-third, then add.
(e) Solve equation (2) for y and substitute this expression into equation (1).

6. If an equation contains fractions, it is a good idea to multiply both sides of the equation by the LCD to clear the denominators before deciding which variable to eliminate.

7. Suppose you are given the system of equations $\begin{cases} x - 3y = 6 \\ 2x + y = 5 \end{cases}$. Which variable is easier to eliminate? Why? List the steps you would use to solve this system.

8. One of the methods for solving a system of equations that you have studied in this section has two commonly used titles: elimination method or addition method. Explain why either title would be appropriate for this process.

Building Skills

In Problems 9–16, solve each system of equations using elimination.

9. $\begin{cases} 2x + y = 3 \\ 5x - y = 11 \end{cases}$

10. $\begin{cases} x + y = -20 \\ x - y = 10 \end{cases}$

11. $\begin{cases} 3x - 2y = 10 \\ -3x + 12y = 30 \end{cases}$

12. $\begin{cases} 2x + 6y = 6 \\ -2x + y = 8 \end{cases}$

13. $\begin{cases} 2x + 3y = -4 \\ -2x + y = 6 \end{cases}$

14. $\begin{cases} 3x + 2y = 7 \\ -3x + 4y = 2 \end{cases}$

15. $\begin{cases} 6x - 2y = 0 \\ -9x - 4y = 21 \end{cases}$

16. $\begin{cases} 2x + 5y = -2 \\ -3x + 10y = -32 \end{cases}$

In Problems 17–24, determine if the system is dependent or inconsistent. State the number of solutions the system has.

17. $\begin{cases} 2x + 2y = 1 \\ -2x - 2y = 1 \end{cases}$

18. $\begin{cases} 3x - 2y = 4 \\ -3x + 2y = 4 \end{cases}$

19. $\begin{cases} 3x + y = -1 \\ 6x + 2y = -2 \end{cases}$

20. $\begin{cases} x - y = -4 \\ -2x + 2y = -8 \end{cases}$

21. $\begin{cases} 2x - 3y = 10 \\ -4x + 6y = -20 \end{cases}$

22. $\begin{cases} 2x + 5y = 15 \\ -6x - 15y = -45 \end{cases}$

23. $\begin{cases} -4x + 8y = 1 \\ 3x - 6y = 1 \end{cases}$

24. $\begin{cases} 4x + 6y = -10 \\ 9x + \dfrac{27}{2}y = 3 \end{cases}$

In Problems 25–44, solve each system of equations using elimination.

25. $\begin{cases} 2x + 3y = 14 \\ -3x + y = 23 \end{cases}$

26. $\begin{cases} 2x + y = -4 \\ 3x + 5y = 29 \end{cases}$

27. $\begin{cases} 2x + 4y = 0 \\ 5x + 2y = 6 \end{cases}$

28. $\begin{cases} -2x - 2y = 3 \\ x - 2y = 1 \end{cases}$

29. $\begin{cases} x - 3y = 4 \\ -2x + 6y = 3 \end{cases}$

30. $\begin{cases} 5x - y = 3 \\ -10x + 2y = 2 \end{cases}$

31. $\begin{cases} 2x + 3y = -3 \\ 3x + 5y = -9 \end{cases}$

32. $\begin{cases} 2x + 3y = 2 \\ 5x + 7y = 0 \end{cases}$

33. $\begin{cases} 5y = 2x - 5 \\ 2x - 5y = 1 \end{cases}$

34. $\begin{cases} x + 3y = 6 \\ 3y = -x + 12 \end{cases}$

35. $\begin{cases} 4x + 3y = 0 \\ 3x - 5y = 2 \end{cases}$

36. $\begin{cases} 5x + 7y = 6 \\ 2x - 3y = 11 \end{cases}$

37. $\begin{cases} 4x - 3y = -10 \\ -\dfrac{2}{3}x + y = \dfrac{11}{3} \end{cases}$

38. $\begin{cases} 12x + 15y = -65 \\ \dfrac{1}{2}y + 3x = -\dfrac{7}{2} \end{cases}$

39. $\begin{cases} 1.5x + 0.5y = -0.45 \\ -0.3x - 0.4y = -0.54 \end{cases}$

40. $\begin{cases} -2.4x - 0.4y = 0.32 \\ 4.2x + 0.6y = -0.54 \end{cases}$

41. $\begin{cases} \dfrac{1}{2}x + \dfrac{2}{3}y = -5 \\ \dfrac{5}{2}x + \dfrac{5}{6}y = -10 \end{cases}$

42. $\begin{cases} x + \dfrac{5}{3}y = -1 \\ \dfrac{1}{2}x + \dfrac{1}{4}y = \dfrac{1}{4} \end{cases}$

43. $\begin{cases} 0.05x + 0.10y = 5.50 \\ x + y = 80 \end{cases}$

44. $\begin{cases} -1.8x + 1.5y = 64.50 \\ 0.8x - 0.6y = -27 \end{cases}$

Mixed Practice

In Problems 45–68, solve by any method: graphing, substitution, or elimination.

45. $\begin{cases} x - y = -4 \\ 3x + y = 8 \end{cases}$

46. $\begin{cases} x - 2y = 0 \\ 3x + 5y = -11 \end{cases}$

47. $\begin{cases} 3x - 10y = -5 \\ 6x - 8y = 14 \end{cases}$

48. $\begin{cases} 3x - 5y = 14 \\ -2x + 6y = -16 \end{cases}$

49. $\begin{cases} x - 3y = -6 \\ 4x + 5y = 7 \end{cases}$

50. $\begin{cases} 2x - 3y = -10 \\ y - 3x = 1 \end{cases}$

51. $\begin{cases} 0.3x - 0.7y = 1.2 \\ 1.2x + 2.1y = 2 \end{cases}$

52. $\begin{cases} 0.25x + 0.10y = 3.70 \\ x + y = 25 \end{cases}$

53. $\begin{cases} x = 3y - 1 \\ 2x - 6y = -2 \end{cases}$

54. $\begin{cases} y = -4x + 3 \\ 8x + 2y = 6 \end{cases}$

55. $\begin{cases} 3x - 2y = 6 \\ \dfrac{3}{2}x - y = 3 \end{cases}$

56. $\begin{cases} 4x + 3y = 9 \\ \dfrac{4}{3}x + y = 3 \end{cases}$

57. $\begin{cases} y = -\dfrac{2}{3}x - \dfrac{7}{3} \\ y = \dfrac{3}{4}x - \dfrac{15}{4} \end{cases}$

58. $\begin{cases} y = \dfrac{1}{7}x - 4 \\ y = -\dfrac{3}{2}x + 19 \end{cases}$

59. $\begin{cases} \dfrac{x}{2} + \dfrac{y}{4} = -2 \\ \dfrac{3x}{2} + \dfrac{y}{5} = -6 \end{cases}$

60. $\begin{cases} \dfrac{x}{3} + \dfrac{y}{5} = 2 \\ \dfrac{x}{3} - \dfrac{2y}{5} = -1 \end{cases}$

61. $\begin{cases} x - 2y = -7 \\ 3x + 4y = 6 \end{cases}$

62. $\begin{cases} -5x + y = 3 \\ 10x + 3y = 14 \end{cases}$

63. $\begin{cases} 6x - 5y = 1 \\ 8x - 2y = -22 \end{cases}$

64. $\begin{cases} -3x - 2y = -19 \\ 4x + 5y = 30 \end{cases}$

65. $\begin{cases} y = 2x - 4y \\ 4x + 1 = 10y + 3 \end{cases}$

66. $\begin{cases} 12y = 8x + 3 \\ -2 + 4(3y - 2x) = 5 \end{cases}$

67. $\begin{cases} \dfrac{x}{2} - y = 1 \\ \dfrac{x}{5} + \dfrac{5y}{6} = \dfrac{14}{15} \end{cases}$

68. $\begin{cases} \dfrac{x}{2} + \dfrac{3y}{4} = \dfrac{1}{2} \\ -\dfrac{3x}{5} + \dfrac{3y}{4} = -\dfrac{1}{20} \end{cases}$

Applying the Concepts

69. Counting Calories Suppose that Kristin ate two McDonald's hamburgers and drank one medium Coke, for a total of 770 calories. Kristin's friend, Jack, ate three hamburgers and drank two medium Cokes (Jack takes advantage of free refills) for a total of 1260 calories. How many calories are in a McDonald's hamburger? How many calories are in a medium Coke? Solve the system

$$\begin{cases} 2h + c = 770 & (1) \\ 3h + 2c = 1260 & (2) \end{cases}$$

where h represents the number of calories in a hamburger and c represents the number of calories in a medium Coke to find the answers.

70. Carbs Yvette and José go to McDonald's for breakfast. Yvette orders two sausage biscuits and one 16-ounce orange juice. The entire meal had 98 grams of carbohydrates. José orders three sausage biscuits and two 16-ounce orange juices and his meal had 168 grams of carbohydrates. How many grams of carbohydrates are in a sausage biscuit? How many grams of carbohydrates are in a 16-ounce orange juice? Solve the system

$$\begin{cases} 2b + u = 98 & (1) \\ 3b + 2u = 168 & (2) \end{cases}$$

where b represents the number of grams of carbohydrates in a sausage biscuit and u represents the number of grams of carbohydrates in an orange juice to find the answers.

71. Planting Crops Farmer Green runs an organic farm and is planting his fields. He remembers from previous years that he planted 2 acres of tomatoes and 3 acres of zucchini in 65 hours. A different year he planted 3 acres of tomatoes and 4 acres of zucchini in 90 hours. If t represents the number of hours it takes to plant an acre of tomatoes, and z represents the number of hours it takes to plant one acre of zucchini, determine how long it takes Farmer Green to plant an acre of each crop by solving the following system of equations.

$$\begin{cases} 2t + 3z = 65 & (1) \\ 3t + 4z = 90 & (2) \end{cases}$$

If he wants to plant 5 acres of tomatoes and 2 acres of zucchini, will he get the crops in if the television meteorologist predicts rain in 72 hours?

72. Farmer's Daughter Farmer Green has talked his daughter, Haydee, into helping with some of the planting in the scenario in Problem 71. Haydee's specialty is planting corn and lettuce, which she has done for the past several years. One year it took her 48 hours to plant 3 acres of corn and 2 acres of lettuce. Another year it took her 68 hours to plant 4 acres of corn and 3 acres of lettuce. If c represents the number of hours to plant one acre of corn and l represents the number of hours required to plant one acre of lettuce, determine how long it takes Haydee to plant an acre of each crop by solving the following system of equations.

$$\begin{cases} 3c + 2l = 48 & (1) \\ 4c + 3l = 68 & (2) \end{cases}$$

If Farmer Green wants to have 4 acres of corn planted before the rains begin in 72 hours, how many acres of lettuce can Haydee expect to plant?

73. Making Coffee Suppose that you want to blend two coffees in order to obtain a new blend. The blend will be made with the best Arabica beans that sell for $9.00 per pound and select African Robustas that sell for $11.50 per pound to obtain 100 pounds of the new blend that will sell for $10.00 per pound. How many pounds of the Arabica and Robusta beans are required? Solve the system

$$\begin{cases} a + r = 100 & (1) \\ 9a + 11.50r = 1000 & (2) \end{cases}$$

where a represents the number of pounds of Arabica beans and r represents the number of pounds of African Robustas beans to determine the answer.

74. Candy A candy store sells chocolate-covered almonds for $6.50 per pound and chocolate-covered peanuts for $4.00 per pound. The manager decides to make a bridge mix that combines the almonds with the peanuts. She wants the bridge mix to sell for $6.00 per pound. How many pounds of chocolate-covered almonds and chocolate-covered peanuts are required to create 50 pounds of bridge mix? Solve the system

$$\begin{cases} a + p = 50 & (1) \\ 6.50a + 4.00p = 300 & (2) \end{cases}$$

where a represents the number of pounds of chocolate-covered almonds and p represents the number of pounds of chocolate-covered peanuts to determine the answer.

△ **75. Complementary Angles** Two angles are complementary if the sum of their measures is $90°$. The measure of one angle is $10°$ more than three times the measure of its complement. If A and B represent the measures of the two complementary angles, determine the measure of each of the angles by solving the following system of equations.

$$\begin{cases} A + B = 90 & (1) \\ A = 10 + 3B & (2) \end{cases}$$

△ **76. Supplementary Angles** Two angles are supplementary if the sum of their measures is $180°$. One-half the measure of one angle is $45°$ more than the measure of its supplement. If A and B represent the measures of the two supplementary angles, determine the measure of each of the angles by solving the following system of equations.

$$\begin{cases} A + B = 180 & (1) \\ \dfrac{A}{2} = 45 + B & (2) \end{cases}$$

Extending the Concepts

In Problems 77 and 78, solve the system of equations for x and y using the elimination method.

77. $\begin{cases} ax + 4y = 1 \\ -2ax - 3y = 3 \end{cases}$

78. $\begin{cases} 4x + 2by = 4 \\ 5x + 3by = 7 \end{cases}$

In Problems 79 and 80, solve the system of equations for x and y using the elimination method.

79. $\begin{cases} -3x + 2y = 6a \\ x - 2y = 2b \end{cases}$

80. $\begin{cases} -7x - 3y = 4b \\ 3x + y = -2a \end{cases}$

PUTTING THE CONCEPTS TOGETHER (Sections 4.1–4.3)

These problems cover important concepts from Sections 4.1 to 4.3. We designed these problems so that you can review the chapter so far and show your mastery of the concepts. Take time to work these problems before proceeding with the next section. The answers to these problems are located at the back of the text starting on page AN-14.

1. Determine whether the given ordered pairs are solutions of the given system of equations.

$$\begin{cases} 4x + y = -20 \\ y = -\dfrac{1}{6}x + 3 \end{cases}$$

(a) $\left(3, \dfrac{5}{2}\right)$ 　　　　**(b)** $(-6, 4)$ 　　　　**(c)** $(-4, -4)$

2. Suppose that you begin to solve a system of linear equations where each equation is written in slope-intercept form. You notice that the slopes and *y*-intercepts of the two lines are equal.

(a) How many solutions exist?
(b) State whether the system is consistent or inconsistent.
(c) If the system is consistent, state whether the system in dependent or independent.

3. Suppose that you begin to solve a system of linear equations where each equation is written in slope-intercept form. The slopes of the two lines are not equal, but the *y*-intercepts are the same.

(a) How many solutions exist?
(b) State whether the system is consistent or inconsistent.
(c) If the system is consistent, state whether the system in dependent or independent.

In Problems 4–13, solve each system of equations using any method that you wish.

4. $\begin{cases} 4x + y = 5 \\ -x + y = 0 \end{cases}$

5. $\begin{cases} y = -\dfrac{2}{5}x + 1 \\ y = -x + 4 \end{cases}$

6. $\begin{cases} x = 2y + 11 \\ 3x - y = 8 \end{cases}$

7. $\begin{cases} 4x + 3y = -4 \\ x + 5y = -1 \end{cases}$

8. $\begin{cases} y = -2x - 3 \\ y = \dfrac{1}{2}x + 7 \end{cases}$

9. $\begin{cases} -2x + 4y = 2 \\ 3x + 5y = -14 \end{cases}$

10. $\begin{cases} -\dfrac{3}{4}x + \dfrac{2}{3}y = \dfrac{9}{4} \\ 3x - \dfrac{1}{2}y = -\dfrac{5}{2} \end{cases}$

11. $\begin{cases} 3(y + 3) = 1 + 4(3x - 1) \\ x = \dfrac{y}{4} + 1 \end{cases}$

12. $\begin{cases} 0.4x - 2.5y = -6.5 \\ x + y = 5.5 \end{cases}$

13. $\begin{cases} 2(y + 1) = 3x + 4 \\ x = \dfrac{2}{3}y - 3 \end{cases}$

4.4 Solving Direct Translation, Geometry, and Uniform Motion Problems Using Systems of Linear Equations

OBJECTIVES

1. Model and Solve Direct Translation Problems
2. Model and Solve Geometry Problems
3. Model and Solve Uniform Motion Problems

Preparing for Solving Direct Translation, Geometry, and Uniform Motion Problems Using Systems of Linear Equations

Before getting started, take the following readiness quiz. If you get a problem wrong, go back to the section cited and review the material.

1. If you travel at an average speed of 45 miles per hour for 3 hours, how far will you travel? [Section 2.7, pp. 143–145]

2. Suppose that you have $3600 in a savings account. The bank pays 1.5% annual simple interest. What is the interest paid after 1 month? [Section 2.4, pp. 104–105]

Recall, in Sections 2.5–2.7 we modeled and solved problems using a linear equation with a single variable. In this section, we learn how to model problems using a system of equations. The problems that we present in this section will have two unknowns, and the unknowns will be related through a system of equations. This is different from the problems in Sections 2.5–2.7 because the problems in Sections 2.5–2.7 had a single unknown whose value was determined from a single equation. If you haven't done so already, go back to Section 2.5 and review the problem-solving strategy given on page 122.

1 Model and Solve Direct Translation Problems

One class of problems that we discussed in Chapter 2 was direct translation problems. Let's look at a couple of examples where direct translation results in a system of two linear equations containing two unknowns.

EXAMPLE 1 Fun with Numbers

The sum of two numbers is 45. Twice the first number minus the second number is 27. Find the numbers.

Solution

Step 1: Identify We are looking for two unknown numbers.

Step 2: Name Let x represent the first number and y represent the second number.

Step 3: Translate The sum of the two numbers is 45, so we know that

$$x + y = 45 \quad \text{Equation (1)}$$

Twice the first number minus the second number is 27. So

$$2x - y = 27 \quad \text{Equation (2)}$$

Equations (1) and (2) are combined to form the following system:

$$\begin{cases} x + y = 45 & (1) \\ 2x - y = 27 & (2) \end{cases} \quad \text{The Model}$$

Step 4: Solve We will use the method of elimination since the coefficients of y are opposites.

$$\begin{cases} x + y = 45 & (1) \\ 2x - y = 27 & (2) \end{cases}$$

Add equations (1) and (2): $3x \quad\quad = 72$

Divide both sides by 3: $x \quad\quad = 24$

Let $x = 24$ in equation (1) and solve for y.

Equation (1): $x + y = 45$

Let $x = 24$: $24 + y = 45$

Subtract 24 from both sides: $y = 21$

Step 5: Check The sum of 24 and 21 is 45. Twice 24 minus 21 is 48 minus 21, which equals 27.

Step 6: Answer The two numbers are 24 and 21.

QUICK ✓

1. The sum of two numbers is 104. The second number is 25 less than twice the first number. Find the numbers.

---

(2) Model and Solve Geometry Problems

Formulas from geometry are often needed to solve certain types of problems. The formula that you will need is determined by the problem. Remember that a formula is a model—and in geometry, formulas describe relationships and shapes.

EXAMPLE 2 Enclosing a Yard with a Fence

The Freese Family just bought a wooded lot that is on a lake. They want to enclose the lot with a fence but do not want to put up a fence along the lake. Dave Freese determined that he will need 240 feet of fence. He also knows that the width of the lot is 30 feet more than the length. See Figure 11. What are the dimensions of the lot?

Figure 11

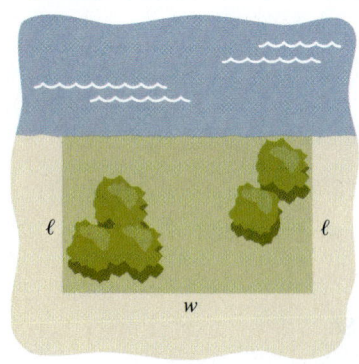

Solution

Step 1: Identify We are looking for the length and the width of the lot.

Step 2: Name Let l represent the length and w represent the width of the lot.

Step 3: Translate The perimeter P of a rectangle excluding the waterfront is $P = 2l + w$, where l is the length and w is the width. So we know that

$$2l + w = 240 \quad \text{Equation (1)}$$

In addition, the width is 30 feet more than the length.

$$w = l + 30 \quad \text{Equation (2)}$$

Equations (1) and (2) are combined to form the following system:

$$\begin{cases} 2l + w = 240 & (1) \\ w = l + 30 & (2) \end{cases} \quad \text{The Model}$$

Step 4: Solve We will use the method of substitution since equation (2) is already solved for w.

Equation (2): $w = l + 30$

Let $w = l + 30$ in equation (1): $2l + (l + 30) = 240$

Combine like terms: $3l + 30 = 240$

Subtract 30 from both sides: $3l = 210$

Divide both sides by 3: $l = 70$

Let $l = 70$ in equation (2) and solve for w.

$$\text{Equation (2):} \quad w = l + 30$$
$$\text{Let } l = 70: \quad w = 70 + 30$$
$$\text{Simplify:} \quad w = 100$$

Step 5: Check With $l = 70$ and $w = 100$, the perimeter (excluding the waterfront) would be $2(70) + 100 = 240$ feet. The width (100 feet) is 30 feet more than the length (70 feet).

Step 6: Answer The length of the backyard is 70 feet and the width is 100 feet. ▬

QUICK ✓

2. A rectangular field has a perimeter of 400 yards. The length of the field is three times the width. What is the length of the field? What is the width of the field?

Recall from Section 2.7 that complementary angles are two angles whose measures sum to 90°. Each angle is called the *complement* of the other. For example, the angles shown in Figure 12(a) below are complements because their measures sum to 90°. Supplementary angles are two angles whose measures sum to 180°. Each angle is called the *supplement* of the other. The angles shown in Figure 12(b) are supplementary.

Figure 12

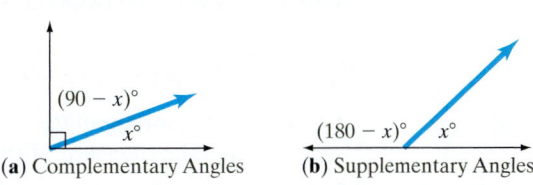

(**a**) Complementary Angles (**b**) Supplementary Angles

The next example was first presented in Section 2.7. In Section 2.7 we solved the problem by developing a model that involved only one unknown. We now show how the same problem can be solved by developing a model of two equations containing two unknowns.

EXAMPLE 3 Solve a Complementary Angle Problem

Find the measure of two complementary angles such that the measure of the larger angle is 6° greater than twice the measure of the smaller angle.

Solution

Step 1: Identify This is a complementary angle problem. We are looking for the measure of the two angles whose sum is 90°.

Step 2: Name Let x represent the measure of the smaller angle and y represent the measure of the other angle.

Step 3: Translate Because these are complementary angles, we know that the sum of the measures of the angles must be 90°. So we have

$$x + y = 90 \quad \text{Equation (1)}$$

The measure of the larger angle y is 6° more than twice the measure of the smaller angle, x. This leads to the following equation.

$$y = 2x + 6 \quad \text{Equation (2)}$$

Using equations (1) and (2), we obtain the following system.

$$\begin{cases} x + y = 90 & (1) \\ y = 2x + 6 & (2) \end{cases} \quad \text{The Model}$$

Step 4: Solve Since equation (2) is already solved for y, we let $y = 2x + 6$ in equation (1).

Equation (1):	$x + y = 90$
Let $y = 2x + 6$:	$x + (2x + 6) = 90$
Combine like terms:	$3x + 6 = 90$
Subtract 6 from each side of the equation:	$3x = 84$
Divide each side by 3:	$x = 28$

Now we let $x = 28$ in equation (2), $y = 2x + 6$, to find the measure of the larger angle, y.

$$y = 2x + 6$$
$$x = 28: \quad y = 2(28) + 6$$
$$y = 62$$

Step 5: Check The measure of the smaller angle, x, is $28°$. The measure of the larger angle, y, is $62°$. Is the sum of the measures of these angles $90°$? Yes, since $28° + 62° = 90°$. Is the measure of the larger angle $6°$ more than twice the measure of the smaller angle? Yes, since 2 times $28°$ is $56°$. Six more than this is $56° + 6° = 62°$. The answers check!

Step 6: Answer The two complementary angles measure $28°$ and $62°$. ▬

QUICK ✓ *Solve each problem for the unknown angle measures.*

3. Find two complementary angles such that the measure of the larger angle is $18°$ more than the measure of the smaller angle.

4. Find two supplementary angles such that the measure of the larger angle is $16°$ less than three times the measure of the smaller angle.

---

③ Model and Solve Uniform Motion Problems

Let's now look at a problem involving uniform motion. Remember, these problems use the fact that distance equals rate times time ($d = rt$).

EXAMPLE 4 **Uniform Motion—Flying a Piper Aircraft**

The airspeed of a plane is its speed through the air. This speed is different from the plane's groundspeed—its speed relative to the ground. The groundspeed of an airplane is affected by the speed of the wind. Suppose that a Piper aircraft flying west a distance of 500 miles takes 5 hours. The return trip takes 4 hours. Find the airspeed of the plane and the effect wind resistance has on the plane.

Solution

Step 1: Identify This is a uniform motion problem. We want to determine the airspeed of the plane and the effect of wind resistance.

Step 2: Name There are two unknowns in the problem—the airspeed of the plane and the impact of wind resistance on the plane. We will let a represent the airspeed of the plane and w represent the impact of wind resistance.

Step 3: Translate Going west, the plane is flying into the jet stream, so that the plane is slowed down by the wind. Therefore, the groundspeed of the plane will be $a - w$.

Going east, the wind is helping the plane. Therefore, the groundspeed of the plane will be $a + w$. We set up Table 3.

Table 3			
	Distance	**Rate**	**Time**
With Wind (East)	500	$a + w$	4
Against Wind (West)	500	$a - w$	5

Work Smart

Instead of distributing 4 into the equation $4(a + w) = 500$, we could divide both sides by 4. The same approach can be taken with the equation $5(a - w) = 500$—divide both sides of the equation by 5. Try solving the system this way. Is it easier?

Going with the wind, we have the equation

$$4(a + w) = 500 \quad \text{or} \quad 4a + 4w = 500 \quad (1)$$

Going against the wind, we have the equation

$$5(a - w) = 500 \quad \text{or} \quad 5a - 5w = 500 \quad (2)$$

Combining equations (1) and (2), we form a system of two linear equations containing two unknowns.

$$\begin{cases} 4a + 4w = 500 & (1) \\ 5a - 5w = 500 & (2) \end{cases} \quad \text{The Model}$$

Step 4: Solve We will use the elimination method by multiplying equation (1) by 5 and equation (2) by 4 and then adding equations (1) and (2).

$$\begin{cases} 5(4a + 4w) = 5(500) & (1) \\ 4(5a - 5w) = 4(500) & (2) \end{cases}$$

$$\text{Distribute:} \quad \begin{cases} 20a + 20w = 2500 & (1) \\ 20a - 20w = 2000 & (2) \end{cases}$$

$$\text{Add:} \quad 40a = 4500$$

$$\text{Divide both sides by 40:} \quad a = 112.5$$

We use equation (1) with $a = 112.5$ to find the effect of wind resistance.

$$\text{Equation (1):} \quad 4a + 4w = 500$$

$$a = 112.5: \quad 4(112.5) + 4w = 500$$

$$450 + 4w = 500$$

$$\text{Subtract 450 from both sides:} \quad 4w = 50$$

$$\text{Divide both sides by 4:} \quad w = 12.5$$

Step 5: Check Flying west, the groundspeed of the plane is $112.5 - 12.5 = 100$ miles per hour. Flying west, the plane flies 500 miles in 5 hours for an average speed of 100 miles per hour. Flying east, the groundspeed of the plane is $112.5 + 12.5 = 125$ miles per hour. Flying east, the plane flies 500 miles in 4 hours for an average speed of 125 miles per hour. Everything checks!

Step 6: Answer The airspeed of the plane is 100 miles per hour. The impact of wind resistance on the plane is 12.5 miles per hour. ▬

QUICK ✅

5. Suppose that a plane flying 1200 miles west requires 4 hours and flying 1200 miles east requires 3 hours. Find the airspeed of the plane and the effect wind resistance has on the plane.

4.4 Exercises

Concepts and Vocabulary

In Problems 1–3, fill in the blanks.

1. The distance around the outside of a rectangle is called the _____ and uses the formula _____.

2. If a number x is three more than twice a number y, then $x =$ _____.

3. Supplementary angles are two angles whose measures sum to _____, while the measures of two angles that are complementary sum to _____.

In Problems 4–6, answer True or False to each statement.

4. If the sum of a number x and three times a second number y is 45, then $x + 3y = 45$.

5. Complementary angles are angles whose measures sum to $90°$.

6. In uniform motion problems we use the equation $d = rt$, where d is distance, r is rate, and t is time.

7. While solving a supplementary angle problem, you find that one of the angles has a measure of $190°$. Is this possible? Why or why not?

8. While solving a uniform motion problem, you find that the average speed of a boat in still water is -12 miles per hour. Is this possible? Why or why not?

Building Skills

△ 9. The perimeter of a rectangle is 59 inches. The length is 5 inches less than twice the width. Let l represent the length of the rectangle and let w represent the width of the rectangle. Complete the system of equations.

$$\begin{cases} l = 2w - 5 \\ \underline{\hspace{2cm}} = 59 \end{cases}$$

△ 10. The perimeter of a rectangle is 212 centimeters. The length is 8 centimeters less than three times the width. Let l represent the length of the rectangle and let w represent the width of the rectangle. Complete the system of equations.

$$\begin{cases} 2w + 2l = 212 \\ \underline{\hspace{2cm}} \end{cases}$$

Applying the Concepts

11. **Fun with Numbers** Find two numbers whose sum is 82 and whose difference is 16.

12. **Fun with Numbers** Find two numbers whose sum is 55 and whose difference is 17.

13. **Fun with Numbers** Two numbers sum to 51. Twice the first subtracted from the second is 9. Find the numbers.

14. **Fun with Numbers** Two numbers sum to 32. Twice the first subtracted from the second is -22. Find the numbers.

15. **Oakland Baseball** The attendance at the games on two successive nights of Oakland A's baseball was 77,000. The attendance on Thursday's game was 7000 more than two-thirds of the attendance at Friday night's game. How many people attended the baseball game each night?

16. **Winning Baseball** The number of games the Oakland A's are expected to win this year is 8 less than two-thirds of the number that they are expected to lose. If there are 162 games in a season, how many games are the A's expected to win?

17. Investments Suppose that you received an unexpected inheritance of $21,000. You have decided to invest the money by placing some of the money in stocks and the remainder in bonds. To diversify, you decide that four times the amount invested in bonds should equal three times the amount invested in stocks. How much should be invested in stocks? How much should be invested in bonds?

18. Investments Marge and Homer have $40,000 to invest. Their financial advisor has recommended that they diversify by placing some of the money in stocks and the remainder in bonds. Based upon current market conditions, he has recommended that two times the amount in bonds should equal three times the amount invested in stocks. How much should be invested in stocks? How much should be invested in bonds?

△ **19. Fencing a Garden** Melody Jackson wishes to enclose a rectangular garden with fencing, using the side of her garage as one side of the rectangle. A neighbor gave her 30 feet of fencing, and Melody wants the length of the garden along the garage to be 3 feet more than the width. What are the dimensions of the garden?

△ **20. Perimeter of a Parking Lot** A rectangular parking lot has a perimeter of 125 feet. The length of the parking lot is 10 feet more than the width. What is the length of the parking lot? What is the width?

△ **21. Perimeter** The perimeter of a rectangle is 70 meters. If the width is 40% of the length, find the dimensions of the rectangle.

△ **22. Window Dimensions** The perimeter of a rectangular window is 162 inches. If the height of the window is 80% of the width, find the dimensions of the window.

△ **23. Working with Complements** The measure of one angle is 15° more than half the measure of its complement. Find the measures of the two angles.

△ **24. Working with Complements** The measure of one angle is 10° less than the measure of three times its complement. Find the measures of the two angles.

△ **25. Finding Supplements** The measure of one angle is 30° less than one-third the measure of its supplement. Find the measures of the two angles.

△ **26. Finding Supplements** The measure of one angle is 20° more than two-thirds the measure of its supplement. Find the measures of the two angles.

27. Kayaking Michael is kayaking in the Chesapeake Bay. He can paddle 3.5 mph against the current and 4.3 mph when he rows with the current. Find the speed of the current and the speed Michael can paddle in still water.

28. Southwest Airlines Plane A Southwest Airlines plane can fly 455 mph against the wind and 515 mph when it flies with the wind. Find the speed of the wind and the groundspeed of the airplane.

29. Biking Suppose that Jose bikes into the wind for 60 miles for 6 hours. After a long rest, he returns (with the wind at his back) in 5 hours. Determine the speed at which Jose can ride his bike in still air and determine the impact that the wind had on his speed.

30. Rowing On Monday afternoon, you rowed your boat with the current for 4.5 hours and covered 27 miles, stopping in the evening at a campground. On Tuesday morning you returned to your starting point against the current in 6.75 hours. Find the speed of the current and the rate at which you row in still water.

31. Outbound from Chicago Two trains leave Chicago going opposite directions, one going north and the other going south. The northbound train is traveling 12 mph slower than the southbound train. After 4 hours the trains are 528 miles apart. Find the speed of each train.

32. Horseback Riding Monica and Gabriella enjoy riding horses at a dude ranch in Colorado. They decide to go down different trails, agreeing to meet back at the ranch later in the day. Monica's horse is going 4 mph faster than the Gabriella's, and after 2.5 hours, they are 20 miles apart. Find the speed of each horse.

33. **Riding Bikes** Vanessa and Richie are riding their bikes down a trail to the next campground. Vanessa rides at 10 mph while Richie rides at 7 mph. Since Vanessa is a little speedier, she stays behind and cleans up camp for 30 minutes before leaving. How long has Richie been riding when Vanessa is 7 miles ahead of Richie?

34. **Running a Marathon** Rafael and Edith are running in a marathon to raise money for breast cancer. Rafael can run at 12 mph and Edith runs at 10 mph. Unfortunately, Rafael lost his car keys and went back to look for them while Edith started down the course. If Rafael was back at the starting line 15 minutes after Edith left, how long will it take him to catch up to her?

35. **Computing Wind Speed** With a tail wind, a small Piper aircraft can fly 600 miles in 3 hours. Against this same wind, the Piper can fly the same distance in 4 hours. Find the average wind speed and the average airspeed of the Piper.

36. **Computing Wind Speed** The average airspeed of a single engine aircraft is 150 miles per hour. If the aircraft flew the same distance in 2 hours with the wind as it flew in 3 hours against the wind, what was the wind speed?

Extending the Concepts

37. **Painting the House** Sam and Diane, working together, can paint a garage in 6 hours. When working alone, Diane paints a similar garage three times faster than her apprentice, Sam. Assuming no gain or loss of efficiency, how long should it take each person to complete such a job working alone.

4.5 Solving Mixture Problems Using Systems of Linear Equations

OBJECTIVES

1. Draw Up a Plan for Modeling Mixture Problems
2. Set Up and Solve Money Problems Using the Mixture Model
3. Set Up and Solve Dry Mixture and Percent Mixture Problems

Preparing for Solving Mixture Problems Using Systems of Linear Equations

Before getting started, take the following readiness quiz. If you get a problem wrong, go back to the section cited and review the material.

1. Suppose that Roberta has a credit card balance of $1200. Each month, the credit card charges 14% annual simple interest on any outstanding balances. What is the interest that Roberta will be charged on this loan after one month? What is Roberta's credit card balance after one month? [Section 2.4, pp. 104–105]

① Draw Up a Plan for Modeling Mixture Problems

Sometimes ingredients must be combined to form a new mixture. Problems that involve mixing two or more substances are called **mixture problems.** They can be set up using the model *number of units of the same kind · rate = amount*. The *rates* that we use in mixture problems are items such as cost per person, interest rates, or cost per pound. Using a chart such as the one below helps to organize the information given in the problem.

Work Smart
Use a chart to organize your thoughts and keep track of information.

	Number of Units ·	Rate =	Amount
Item 1			
Item 2			
Total			

Because each problem will be slightly different, we will adjust the titles of the categories, but the general equation *number of units of the same kind · rate = amount* will remain constant. Let's begin by practicing filling in the chart.

EXAMPLE 1 Set Up a Chart for Ticket Sales Problem Using the Mixture Model

A class of school children and their adult chaperones took a field trip to the San Diego Zoo. There were 20 more children than adults and total admission to the zoo was $230. If the children paid $5 each and the adults paid $8 each, how many adults and how many children went on the trip? Fill in a chart that summarizes the information in the problem. Do not solve the problem.

Solution

Both the number of children and the number of adults are unknown. We let *a* represent the number of adults on the field trip and *c* represent the number of children. We can now fill in Table 4 with the quantities that we know.

Table 4				
	Number	• Cost per Person	=	Amount
Adults	a	8		8a
Children	c	5		5c
Total				230

Work Smart
Not every entry in the chart must be filled.

QUICK ✓

1. A one-day admission ticket to Cedar Point Amusement Park costs $36.95 for adults and $21.95 for children. Two families purchased five more adult tickets than children's tickets and spent $302.55 for the tickets. How many adult tickets did the families purchase? Fill in a chart that summarizes the information in the problem. Do not solve the problem.

② Set Up and Solve Money Problems Using the Mixture Model

We are now ready to solve problems using the mixture model. We will set up a table to organize the given information, and then use the table to develop a model (equation). We continue to use the six-step procedure that was introduced in Section 2.5.

EXAMPLE 2 Solve a Coin Problem Using the Mixture Model

A third-grade class contributed $9.20 in dimes and quarters to the Red Cross. In all there were 56 coins. Find the number of dimes and the number of quarters that the children contributed to the Red Cross.

Solution

Step 1: Identify This is a money problem and we can use the mixture model to solve it. We want to know the number of dimes and the number of quarters that the children contributed, and we know that $9.20 was given.

Work Smart
It's always a good idea to name your variable so that it reminds you of what it represents, as in *q* for the number of quarters.

Step 2: Name Let *q* represent the number of quarters and let *d* represent the number of dimes.

Step 3: Translate We fill in Table 5 with the information that we know.

Table 5					
	Number of Coins	•	**Value per Coin in Dollars**	**=**	**Total Value**
Quarters	q		0.25		$0.25q$
Dimes	d		0.10		$0.10d$
Total	56				9.20

The total value of the quarters plus the total value of the dimes equals $9.20. Based on the information in Table 5, we have that

value of quarters   value of dimes   total value of coins

$$0.25q \quad + \quad 0.10d \quad = \quad 9.20 \qquad \text{Equation (1)}$$

There are a total of 56 coins, so

$$q + d = 56 \qquad \text{Equation (2)}$$

We use equations (1) and (2) to form a system of equations.

$$\begin{cases} 0.25q + 0.1d = 9.20 & (1) \\ q + d = 56 & (2) \end{cases} \quad \text{The Model}$$

Step 4: Solve We will use the method of elimination.

Multiply equation (2) by -0.1: $\begin{cases} 0.25q + 0.1d = 9.20 & (1) \\ -0.1(q + d) = -0.1(56) & (2) \end{cases}$

Distribute -0.1 in equation (2): $\begin{cases} 0.25q + 0.1d = 9.20 & (1) \\ -0.1q - 0.1d = -5.6 & (2) \end{cases}$

Add equations (1) and (2): $\quad 0.15q \qquad = 3.6$

Divide both sides by 0.15: $\qquad q \qquad = 24$

Now we need to determine the value of d. We let $q = 24$ in equation (2) and solve for d.

Equation (2): $\quad q + d = 56$

$q = 24$: $\quad 24 + d = 56$

Subtract 24 from both sides: $\quad d = 32$

Step 5: Check Because q represents the number of quarters, we know that the children collected 24 quarters. They collected 32 dimes. Is the total value of the dimes and quarters collected equal to $9.20? Because $24(\$0.25) + 32(\$0.10) = \$6.00 + \$3.20 = \$9.20$, our answer is correct.

Step 6: Answer The children collected 24 quarters and 32 dimes for the Red Cross.

QUICK ✓

2. You have a piggy bank containing a total of 85 coins in dimes and quarters. If the piggy bank contains $14.50, how many dimes are there in the piggy bank?

Recall, in Section 2.4 we introduced the simple interest formula $I = Prt$, where I is interest, P is principal (either an amount borrowed or deposited), r is the annual interest rate (expressed as a decimal), and t is time (measured in years). When solving interest problems using the mixture model in this section we will consider only the situation in which time is 1 year, so the simple interest formula reduces to $interest = principal \cdot rate \cdot 1$, or $interest = principal \cdot rate$.

EXAMPLE 3 Solve an Interest Problem Using the Mixture Model

You have $12,000 invested in a money market account that paid 6% annual interest and a stock fund that payed 8.5% annual interest. Suppose that you earned $807.50 in interest at the end of one year. How much was invested in each account?

Solution

Step 1: Identify This is a mixture problem involving simple interest. We need to know how much was invested in the money market account and how much was invested in the stock fund to earn $807.50 in interest.

Step 2: Name Let m represent the amount invested in the money market account and let s represent the amount invested in stocks.

Step 3: Translate We organize the information given in Table 6.

Table 6

	Principal $	•	Rate %	=	Interest $
Money Market	m		0.06		$0.06m$
Stock Fund	s		0.085		$0.085s$
Total	12,000				807.50

You earned interest of $807.50 on your principal of $12,000. The total interest is the sum of the interest from the money market account and the stock fund.

Interest from money market account + Interest from stock fund = $807.50

Work Smart

We use the simple interest formula $I = Prt$, where P is the principal, r is the interest rate (as a decimal), and t is time (in years).

Since $t = 1$ year, we have

$$\underset{\substack{\text{Interest earned} \\ \text{from money market}}}{0.06m} + \underset{\substack{\text{Interest earned} \\ \text{from stock fund}}}{0.085s} = \underset{\substack{\text{Total interest} \\ \text{earned}}}{807.50} \qquad \text{Equation (1)}$$

In addition, the total investment is to be $12,000 so that the amount invested in the money market account plus the amount invested in the stock fund equals $12,000:

$$m + s = 12,000 \qquad \text{Equation (2)}$$

We use equations (1) and (2) to form a system of equations.

$$\begin{cases} 0.06m + 0.085s = 807.50 & (1) \\ m + s = 12,000 & (2) \end{cases} \quad \text{The Model}$$

Work Smart

In Step 4, we could have multiplied both sides of equation (1) by 1000 to eliminate decimals and then solved the system

$$\begin{cases} 60m + 85s = 807,500 & (1) \\ m + s = 12,000 & (2) \end{cases}$$

Which method do you prefer?

Step 4: Solve We will use the method of elimination.

Multiply equation (2) by -0.06:
$$\begin{cases} 0.06m + 0.085s = 807.50 & (1) \\ -0.06(m + s) = -0.06(12,000) & (2) \end{cases}$$

Distribute -0.06 in equation (2):
$$\begin{cases} 0.06m + 0.085s = 807.50 & (1) \\ -0.06m - 0.06s = -720 & (2) \end{cases}$$

Add equations (1) and (2):
$$0.025s = 87.50$$

Divide both sides by 0.025:
$$s = 3500$$

Now we need to determine the value of m. We let $s = 3500$ in equation (2) and solve for m.

Equation (2): $m + s = 12,000$

$s = 3500$: $m + 3500 = 12,000$

Subtract 3500 from both sides: $m = 8500$

Step 5: Check The simple interest earned each year on the money market account is $(\$8500)(0.06)(1) = \$510$. The simple interest earned each year on the stock fund is $(\$3500)(0.085)(1) = \$297.50$. The total interest earned is $\$510 + \$297.50 = \$807.50$. You earned $\$807.50$ per year in interest, so this agrees with the information presented in the problem. In addition, the total amount invested is $\$8500 + \$3500 = \$12,000$.

Step 6: Answer You invested $\$8500$ in the money market account and $\$3500$ in the stock fund. ▪

QUICK ✓

3. Faye has recently retired and requires an extra $\$5500$ per year in income. She has $\$90,000$ to invest and can invest in either an Aa-bond that pays 5% per annum or a B-rated bond paying 7% annually. How much should be placed in each investment for Faye to achieve her goal?

---

③ Set Up and Solve Dry Mixture and Percent Mixture Problems

Mixture of Two Substances—Dry Mixture Problems

Often, new blends are created by mixing two quantities. For example, a chef might mix buckwheat flour with wheat flour to make buckwheat pancakes. Or a coffee shop might mix two different types of coffee to create a new coffee blend.

EXAMPLE 4 **Solve a Dry Mixture Problem Using the Mixture Model**

A store manager mixed together nuts and M&M's to make a trail mix. The manager wants to create 10 pounds of the mix and sell it for $\$5.10$ per pound. If the price of the nuts was $\$3$ per pound and the price of the M&M's was $\$6$ per pound, how many pounds of each type did he use so that the value of the mix is the same as the value of the individual items?

Solution

Step 1: Identify This is a mixture problem. We want to know the number of pounds of nuts and the number of pounds of M&M's that are required in the trail mix.

Step 2: Name Let n represent the number of pounds of nuts and m represent the number of pounds of M&M's that are required.

Step 3: Translate We are told that there is to be no difference in revenue between selling the nuts and M&M's separately versus the blend. This means that if the blend contains one pound of nuts and one pound of M&M's, we should collect $\$3(1) + \$6(1) = \$9$ because that is how much we would collect if we sold the nuts and M&M's separately.

We set up Table 7.

Table 7					
	Price \$/Pound	•	**Number of Pounds**	=	**Revenue**
Nuts	3		n		$3n$
M&M's	6		m		$6m$
Blend	5.10		10		$5.10(10) = 51$

In general, if the mixture contains n pounds of nuts, we should collect $\$3n$. If the mixture contains m pounds of M&M's, we should collect $\$6m$. If the mixture sells for $\$5.10$ per pound and we make 10 pounds of the blend, then we should collect $\$5.10(10) = \$51$ for the blend.

$$\left(\begin{array}{c}\text{Price per pound}\\\text{of nuts}\end{array}\right)\left(\begin{array}{c}\text{Pounds of}\\\text{nuts}\end{array}\right) + \left(\begin{array}{c}\text{Price per pound}\\\text{of M\&M's}\end{array}\right)\left(\begin{array}{c}\text{Pounds of}\\\text{M\&M's}\end{array}\right) = \left(\begin{array}{c}\text{Price per pound}\\\text{of blend}\end{array}\right)\left(\begin{array}{c}\text{Pounds of}\\\text{blend}\end{array}\right)$$

$$\$3 \quad \cdot \quad n \quad + \quad \$6 \quad \cdot \quad m \quad = \quad \$5.10 \quad \cdot \quad (10)$$

We have the equation

$$3n + 6m = 51 \quad \text{Equation (1)}$$

The number of pounds of nuts plus the number of M&M's should equal 10 pounds. We have the equation

$$n + m = 10 \quad \text{Equation (2)}$$

We use equations (1) and (2) to form a system of equations.

$$\begin{cases} 3n + 6m = 51 & (1) \\ n + m = 10 & (2) \end{cases} \quad \text{The Model}$$

Step 4: Solve We solve the system by substitution by solving equation (2) for n.

Equation (2):	$n + m = 10$
Subtract m from both sides:	$n = 10 - m$
Let $n = 10 - m$ in equation (1):	$3(10 - m) + 6m = 51$
Use the Distributive Property to remove the parentheses:	$30 - 3m + 6m = 51$
Combine like terms:	$30 + 3m = 51$
Subtract 30 from both sides:	$3m = 21$
Divide both sides by 3:	$m = 7$

With $m = 7$ pounds, we have that $n = 10 - m = 10 - 7 = 3$ pounds.

Step 5: Check It appears that we should mix 3 pounds of nuts with 7 pounds of M&M's. The 7 pounds of M&M's would sell for $6(7) = \$42$ and the 3 pounds of nuts would sell for $3(3) = \$9$; the total revenue would be $\$42 + \$9 = \$51$, which equals the revenue obtained from selling the blend. This checks with the information presented in the problem. In addition, the blend weighs 7 pounds + 3 pounds = 10 pounds, as required.

Step 6: Answer The manager should mix 3 pounds of nuts with 7 pounds of M&M's to make the blend. ▬

QUICK ✓

4. A coffee house has Brazilian coffee that sells for $6 per pound and Colombian coffee that sells for $10 per pound. How many pounds of each coffee should be mixed to obtain 20 pounds of a blend that costs $9 per pound?

Mixture of Two Substances—Percent Mixture Problems

The last example dealt with mixing two dry substances. We now turn our attention to mixing two liquids. These problems require using percents.

EXAMPLE 5 Solve a Percent Mixture Problem Using the Mixture Model

You work in the chemistry stock room and your instructor has asked you to prepare 4 liters of 15% hydrochloric acid (HCl). Looking through the supply room you see that there is a bottle of 12% HCl and another of 20% HCl. How much of each should you mix so that your instructor has the required solution?

Work Smart

Remember that mixtures can include interest (money), solids (nuts), liquids (chocolate milk), and even gases (Earth's atmosphere).

Work Smart

The system $\begin{cases} 3n + 6m = 51 \\ n + m = 10 \end{cases}$ could also have been solved by elimination. Which method do you prefer?

Solution

Step 1: Identify This is a percent mixture problem so we use the mixture model. We want to know the number of liters of 12% HCl that must be mixed with a 20% HCl solution to prepare 4 liters of 15% HCl. We also know that we need 4 liters of the solution.

Step 2: Name We let x represent the number of liters of the 12% HCl solution and let y represent the number of liters of 20% HCl.

Step 3: Translate We fill in Table 8 with the information that we know.

Table 8			
	Number of Liters •	Concentration (part of solution that is pure HCl per liter) =	Amount of Pure HCl
12% HCl Solution	x	0.12	$0.12x$
20% HCl Solution	y	0.20	$0.20y$
Total	4	0.15	$(0.15)(4)$

The amount of 12% HCl solution plus the amount of 20% HCl solution should yield 4 liters of 15% HCl solution. This leads to the following equation:

part that is 12% HCl   part that is 20% HCl   part of the total that is HCl

$$0.12x \quad + \quad 0.20y \quad = \quad (0.15)(4) \quad \text{Equation (1)}$$

The number of liters of HCl solution should equal 4 liters. We have the equation

$$x + y = 4 \quad \text{Equation (2)}$$

We use equations (1) and (2) to form a system of equations.

$$\begin{cases} 0.12x + 0.20y = 0.6 & (1) \\ x + \quad y = 4 & (2) \end{cases} \quad \text{The Model}$$

Step 4: Solve We solve the system by substitution by solving equation (2) for y.

Equation (2):	$x + y = 4$
Subtract x from both sides:	$y = 4 - x$
Substitute $y = 4 - x$ in equation (1):	$0.12x + 0.20(4 - x) = 0.6$
Use the Distributive Property:	$0.12x + 0.8 - 0.20x = 0.6$
Combine like terms:	$-0.08x + 0.8 = 0.6$
Subtract 0.8 from each side of the equation:	$-0.08x = -0.2$
Divide each side by -0.08:	$x = 2.5$

Because x represents the number of liters of 12% HCl solution, $x = 2.5$ liters of 12% HCl solution is required. We need $4 - x = 4 - 2.5 = 1.5$ liters of 20% HCl.

Step 5: Check The quantity of HCl in the 2.5 liters of 12% HCl is $0.12(2.5) = 0.3$ liters. The quantity of HCl in 1.5 liters of 20% HCl is $0.20(1.5) = 0.3$ liters. Combined, these solutions give us 0.6 liters of pure HCl. The total amount of HCl in 4 liters of 15% HCl is $0.15(4) = 0.6$ liters. Our answer is correct.

Step 6: Answer 2.5 liters of 12% HCl must be mixed with 1.5 liters of 20% HCl to form the required 4 liters of 15% HCl.

Quick

5. You are a wine maker and decide that wine with 9% alcohol is the best to sell. How many gallons of wine with 5% alcohol should be mixed with wine that is 15% alcohol to make 200 gallons of the desired blend?

4.5 Exercises

For Extra Help: Student Solutions Manual   CD Video   PH Math/Tutor Center   MathXL Tutorials on CD   MathXL®   MyMathLab

Concepts and Vocabulary

In Problems 1–3, fill in the blanks.

1. Problems that involve mixing two or more substances are called _____ _____.

2. The general equation we use to solve mixture problems is _____ · _____ = _____.

3. The simple interest formula states that interest = _____ · _____ · _____.

In Problems 4 and 5, answer True or False to each statement.

4. When we solve a simple interest problem using the mixture model, we assume that $t = 1$.

5. The system $\begin{cases} a + b = 700 \\ 0.05a + 0.1b = 10{,}000 \end{cases}$ correctly models the following problem.

  Molly invests $10,000 in two accounts, one paying 5% annual interest and the other paying 10% annual interest, and she earns $700 in interest at the end of the year on the two accounts. Let a represent the amount of money she invests in the account that pays 5% annual interest and b represent the amount of money she invests in the account that pays 10%. How much is invested in each account?

6. Without solving, explain what is wrong with the following mixture problem: How many liters of 25% ethanol should be added to 20 liters of 48% ethanol to obtain a solution of 58% ethanol?

Building Skills

In Problems 7–12, fill in the table from the information given. Then write the system that models the problem. DO NOT SOLVE.

7. The PTA had an ice cream social and sold adult tickets for $4 and student tickets for $1.50. The PTA treasurer found that 215 tickets had been sold and the receipts were $580.

	Number	·	Cost per Person	=	Total Value
Adults' Tickets	?		?		?
Students' Tickets	?		?		?
Total	?				?

8. John Murphy sells jewelry at art shows. He sells bracelets for $10 and necklaces for $15. At the end of one day John found that he had sold 69 pieces of jewelry and had receipts of $895.

	Number	•	Cost per Item	=	Total Value
Bracelets	?		?		?
Necklaces	?		?		?
Total	?				?

9. Maurice has a savings account that earns 5% simple interest per year and a money market account that earns 3% simple interest. At the end of one year, Maurice received $50 in interest on a total investment of $1600 in the two accounts.

	Principal	•	Rate	=	Interest
Savings Account	?		?		?
Money Market	?		?		?
Total	?				?

10. Sherry has a savings account that earns 2.75% simple interest per year and a certificate of deposit (CD) that earns 2% simple interest annually. At the end of one year, Sherry received $37.75 in interest on a total investment of $1700 in the two accounts.

	Principal	•	Rate	=	Interest
Savings Account	?		?		?
Certificate of Deposit	?		?		?
Total	?				?

11. A coffee shop wishes to blend two types of coffee to create a breakfast blend. They mix a mild coffee that sells for $7.50 per pound with a robust coffee that sells for $10.00 per pound. The owner wants to make 12 pounds of breakfast blend that will sell for $8.75 per pound.

	Number of Pounds	•	Price per Pound	=	Total Value
Mild Coffee	?		?		?
Robust Coffee	?		?		?
Total	?		?		?

12. A merchant wishes to mix peanuts worth $5 per pound and trail mix worth $2 per pound to yield 40 pounds of a nutty mixture that will sell for $3 per pound.

	Number of Pounds	•	Price per Pound	=	Total Value
Peanuts	?		?		?
Trail Mix	?		?		?
Total	?		?		?

In Problems 13–18, complete the system of linear equations to solve the problem. Do not solve.

13. A family of 11 decides to take a trip to Water World Water Park. The total cost of admission for the family is $296. If adult tickets cost $32 and children's tickets cost $24, how many adults and how many children went to Water World? Let a represent the number of adult tickets purchased and let c represent the number of children's tickets purchased.

$$\begin{cases} a + c = 11 \\ \underline{} + \underline{} = 296 \end{cases}$$

14. On a school field trip, 22 people attended a dress rehearsal of the Broadway play, "Avenue Q." They paid $274 for the tickets which cost $15 for each adult and $7 for each child. How many adults and how many children attended "Avenue Q"? Let a represent the number of adult tickets purchased and let c represent the number of children's tickets purchased.

$$\begin{cases} a + c = 22 \\ \underline{} + \underline{} = 274 \end{cases}$$

15. You have a total of $2250 to invest. Account A pays 10% annual interest and account B pays 7% annual interest. How much should you invest in each account if you would like the investment to earn $195 at the end of one year? Let A represent the amount of money invested in the account that earns 10% annual interest and let B represent the amount of money invested in the account that earns 7% annual interest.

$$\begin{cases} A + B = 2250 \\ \underline{} + \underline{} = 195 \end{cases}$$

16. You have a total of $2650 to invest. Account A pays 5% annual interest and account B pays 6.5% annual interest. How much should you invest in each account if you would like the investment to earn $155 at the end of one year? Let A represent the amount of money invested in the account that earns 5% annual interest and let B represent the amount of money invested in the account that earns 6.5% annual interest.

$$\begin{cases} A + B = \underline{} \\ \underline{} + \underline{} = 155 \end{cases}$$

17. Flowers on High sells flower bouquets at different prices depending on color: red for $5.85 per bouquet or yellow for $4.20 per bouquet. One day the number of red bouquets sold was 3 more than twice the number of yellow bouquets sold, and revenue from selling the bouquets was $128.85. How many of each color were sold? Let r represent the number of red bouquets sold and let y represent the number of yellow bouquets sold.

$$\begin{cases} r = 3 + 2y \\ \underline{} + \underline{} = \underline{} \end{cases}$$

18. The Latte Shoppe sells Bold Breakfast coffee for $8.60 per pound and Wake-Up coffee for $5.75 per pound. One day, the amount of Bold Breakfast coffee sold was 2 pounds less than twice the amount of Wake-Up coffee and the revenue received from selling both types of coffee was $143.45. How many pounds of each type of coffee were sold that day? Let B represent the number of pounds of Bold Breakfast coffee sold and let W represent the number of pounds of Wake-Up sold. Complete the system of equations:

$$\begin{cases} 8.60B + 5.75W = \underline{} \\ B = \underline{} - \underline{} \end{cases}$$

Applying the Concepts

19. Theater Tickets to a student theater production cost $8 for students and $10 for nonstudents. The receipts for opening night came to $3270 from selling 390 tickets. How many student tickets were sold?

20. Ticket Pricing A ticket on the roller coaster is priced differently for adults and children. One day there were 5 adults and 8 children in a group and the cost of their tickets was $48.50. Another group of 4 adults and 12 children paid $57. What is the price of each type of ticket?

21. Bedding Plants A girls' softball team is raising money for uniforms and travel expenses. They are selling flats of bedding plants for $13.00 and hanging baskets for $18.00. They hope to sell twice as many flats of bedding plants than hanging baskets. If they do, their total revenue will be $8800. How many flats of bedding plants do they hope to sell?

22. Amusement Park An adult discount ticket to King's Island Amusement Park costs $26, and a child's discount ticket to King's Island Amusement Park costs $24.50. A group of 13 friends purchased adult and children's tickets and paid $330.50. How many discount tickets for children were purchased?

23. Tip Jar The tip jar next to the cash register has 150 coins, all in nickels and dimes. If the total value of the tips is $12, how many of each coin are in the jar?

24. Bigger Tips Christa is a waitress and collects her tips at the table. At the end of the shift she has 68 bills in her tip wallet, all ones and fives. If the total value of her tips is $172, how many of each bill does she have?

25. Buying Stamps Rosemarie can buy stamps at two different prices. One day she bought 20 first-class stamps and 10 postcard stamps for $9.30. Another day she spent $39.10 on 80 first-class stamps and 50 postcard stamps. Find the cost of each type of stamp.

26. TV Commercials Advertising on television can be expensive depending on the expected number of viewers. The Fox network sells 30-second spots for $175,000 and one-minute spots for $250,000 during the basketball playoffs. If they sold 13 commercial spots and earned $2,650,000, how many of each type did they sell?

27. Investments Harry has $10,000 to invest. He invests in two different accounts, one expected to return 5% and the other expected to return 8%. If he wants to earn $575 for the year, how much should he invest at each rate?

28. Investments Ann has $5000 to invest. She invests in two different accounts, one expected to return 4.5% and the other expected to return 9%. In order to earn $382.50 for the year, how much should she invest at each rate?

29. Stock Return Esmeralda is investing $5000 in two stock plans. On the risky plan she hopes to earn 12% annual interest and on a safer plan she expects to earn 8% annual interest. If Esmeralda saw a return on her investment of $528 last year, how much did she invest in each of the stock plans?

30. Real Estate Return Victor, Esmeralda's wealthy brother, is also investing his money. He thinks he will do better if he invests his money in real estate partnerships. He has $12,000 to invest and has two groups in mind. One of the groups is promoting a 13% annual return on a downtown mall the other 10.5% annual return on a suburban office center. If Victor saw a profit of $1347.50 on his investment last year, how much did he invest with each group?

31. Olive Blend Juana works at a delicatessen, and it is her turn to make the zesty olive blend. If arbequina olives are $9 per pound and green olives are $4 per pound, how much of each kind of olive should she mix to get five pounds of zesty olive blend that will sell for $6 a pound?

32. Churrascaria Platter Churrascaria is a name that comes from the Brazilian gauchos of the 1800s. To celebrate, the cowboys would take a large variety of different meats and barbeque them. Today, some specialty restaurants still serve Brazilian barbeque in the Churrascaria style. Elis da Silva's Restaurant wants to offer a one-pound Churrascaria platter for $12. If it costs the restaurant $8 per pound to prepare grilled pork and $14 per pound to make barbecued flank steak, how many pounds of each meat can you serve on the platter so that the price is still $12?

33. Blending Coffee A coffee manufacturer wants to market a new blend of coffee that will cost $3.90 per pound by mixing two coffees that sell for $2.75 per pound and $5 per pound, respectively. What amounts of each coffee should be blended to obtain 100 pounds of the desired mixture?

34. Blending Nuts The Nutty Professor store wishes to mix peanuts that sell for $2 per pound with cashews that sell for $6 per pound. The store plans to use five pounds more of peanuts than cashews and sell the mixture for $26. How many pounds of each nut should be used in the mixture?

35. Grass Seed A nursery decides to make a grass-seed blend by mixing rye seed that sells for $4.20 per pound with blue grass seed that sells for $3.75 per pound. The final mix is 180 pounds and sells at $3.95. How much of each type was used?

36. Stock Account A stock portfolio is currently valued at $5872.44. IBM is currently selling at $82.01 per share and Microsoft is currently selling at $26.52 per share. If the number of IBM shares is 25 less than the number of Microsoft shares, find the number of shares of each.

37. Saline Solutions A lab technician needs 60 ml of a 50% saline solution. How many ml of 30% saline solution should she add to a 60% saline solution to obtain the required mixture?

38. Alcohol A laboratory assistant is asked to mix a 30% alcohol solution with 21 liters of an 80% alcohol solution to make a 60% alcohol solution. How many liters of the 30% alcohol solution should be used?

39. Silver Alloy How many liters of 10% silver must be added to 70 liters of 50% silver to make an alloy that is 30% silver?

40. Paint Four gallons of paint contain 2.5% pigment. How many gallons of paint that is 6% pigment must be added to make a paint that is 4% pigment?

Extending the Concepts

41. Antifreeze A radiator holds 3 gallons. How much of the 25% antifreeze solution should be drained and replaced with pure water to reduce the solution to 15% antifreeze?

42. Antifreeze A radiator holds 6 liters. How much of the 20% antifreeze solution should be drained and replaced with pure antifreeze to bring the solution up to 50% antifreeze?

43. Finance Jim Davidson invested $10,000 in two businesses. One business earned a profit of 5% for the year, while the other lost 7.5%. Find the amount he invested in each business if he earned $25 for the year.

44. Finance Marco invested money in two stocks. The stock that made a profit of 12% had $5000 less than the stock that lost 5.5%. Find the amount he invested in each if there was a net loss of $80 on his portfolio.

4.6 Systems of Linear Inequalities

OBJECTIVES

1. Determine Whether an Ordered Pair Is a Solution of a System of Linear Inequalities
2. Graph a System of Linear Inequalities
3. Solve Applied Problems Involving Systems of Linear Inequalities

Preparing for Systems of Linear Inequalities

Before getting started, take the following readiness quiz. If you get a problem wrong, go back to the section cited and review the material.

1. Solve: $3x - 2 \geq 7$ [Section 2.8, pp. 155–157]
2. Solve: $4(x - 1) < 6x + 4$ [Section 2.8, pp. 155–157]
3. Graph: $y > 2x - 5$ [Section 3.8, pp. 253–257]
4. Graph: $2x + 3y \leq 9$ [Section 3.8, pp. 253–257]

In Section 3.8, we graphed a single linear inequality in two variables. In this section, we discuss how to graph a system of linear inequalities in two variables.

1 Determine Whether an Ordered Pair Is a Solution of a System of Linear Inequalities

An ordered pair **satisfies** a system of linear inequalities if it makes each inequality in the system a true statement.

EXAMPLE 1   Determining Whether an Ordered Pair Is a Solution of a System of Linear Inequalities

Which of the following points, if any, satisfies the system of linear inequalities?

$$\begin{cases} 2x + y \leq 7 \\ 7x - 2y \geq 4 \end{cases}$$

(a) $(2, 1)$ **(b)** $(-2, 5)$

Solution

(a) Let $x = 2$ and $y = 1$ in each inequality in the system. If each statement is true, then $(2, 1)$ is a solution of the inequality.

$$2x + y \leq 7 \qquad\qquad 7x - 2y \geq 4$$

$x = 2, y = 1:$ $2(2) + 1 \overset{?}{\leq} 7 \qquad\qquad 7(2) - 2(1) \overset{?}{\geq} 4$

$$4 + 1 \overset{?}{\leq} 7 \qquad\qquad 14 - 2 \overset{?}{\geq} 4$$

$$5 \leq 7 \quad \text{True} \qquad\qquad 12 \geq 4 \quad \text{True}$$

Both inequalities are true when $x = 2$ and $y = 1$, so $(2, 1)$ is a solution of the system of inequalities.

(b) Let $x = -2$ and $y = 5$ in each inequality in the system. If each statement is true, then $(-2, 5)$ is a solution of the inequality.

$$2x + y \leq 7 \qquad\qquad 7x - 2y \geq 4$$

$x = -2, y = 5:$ $2(-2) + 5 \overset{?}{\leq} 7 \qquad 7(-2) - 2(5) \overset{?}{\geq} 4$

$$-4 + 5 \overset{?}{\leq} 7 \qquad\qquad -14 - 10 \overset{?}{\geq} 4$$

$$1 \leq 7 \quad \text{True} \qquad\qquad -24 \geq 4 \quad \text{False}$$

The inequality $7x - 2y \geq 4$ is not true when $x = -2$ and $y = 5$, so $(-2, 5)$ is not a solution of the system of inequalities.

Preparing for...Answers
1. $\{x | x \geq 3\}$ 2. $\{x | x > -4\}$
3.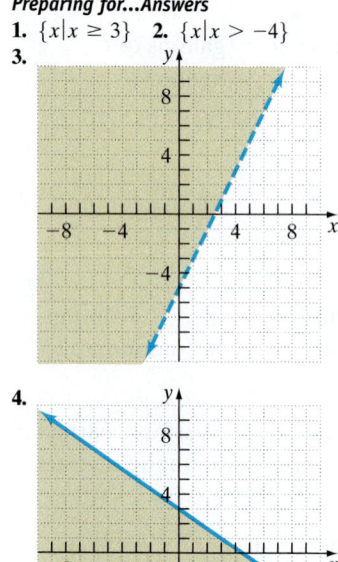
4.

QUICK

1. Determine which of the following points is a solution of the system of linear inequalities.

$$\begin{cases} 4x + y \le 6 \\ 2x - 5y < 10 \end{cases}$$

(a) $(1, 2)$ **(b)** $(-1, -3)$

② Graph a System of Linear Inequalities

Work Smart

Don't forget that we use a solid line when the inequality is nonstrict ($\le$ or $\ge$) and a dashed line when the inequality is strict ($<$ or $>$).

The graph of a system of inequalities in two variables x and y is the set of all points (x, y) that simultaneously satisfy *each* of the inequalities in the system. The graph of a system of linear inequalities can be obtained by graphing each linear inequality individually and then determining where, if at all, they intersect. The ONLY way we show the solution of a system of linear inequalities is by its graphical representation.

EXAMPLE 2 **How to Graph a System of Linear Inequalities**

Graph the system: $\begin{cases} y \ge 3x - 5 \\ y \le -2x + 5 \end{cases}$

Step-by-Step Solution

Step 1: We graph the first inequality in the system $y \ge 3x - 5$.

Graph $y \ge 3x - 5$ by graphing the line $y = 3x - 5$ (slope $= 3$, y-intercept $= -5$) with a solid line because the inequality is nonstrict ($\ge$). We choose to use the test point $(0, 0)$. Since $(0, 0)$ makes the inequality true ($0 \ge 3(0) - 5$), we shade the half-plane containing $(0, 0)$. See Figure 13.

Figure 13

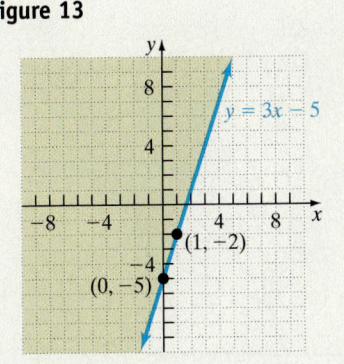

Step 2: Graph the second inequality in the system, $y \le -2x + 5$.

Work Smart

We can also graph inequalities by solving the inequality for y. Then, if the inequality is of the form $y > $ or $y \ge$, we shade above the line. If the inequality is of the form $y < $ or $y \le$, shade below the line.

Graph $y \le -2x + 5$ by graphing the line $y = -2x + 5$ (slope $= -2$, y-intercept $= 5$) with a solid line because the inequality is nonstrict ($\le$). Again, we choose to use the test point $(0, 0)$. Since $(0, 0)$ makes the inequality true ($0 \le -2(0) + 5$), we shade the half-plane containing $(0, 0)$. See Figure 14.

Figure 14

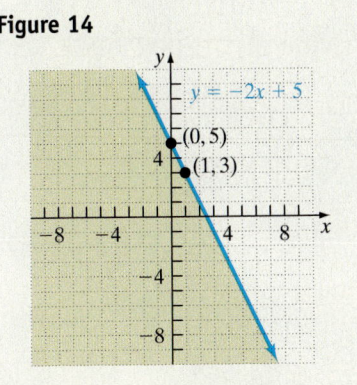

(continued)

Step 3: Combine the graphs in Steps 1 and 2 The overlapping shaded region is the solution of the system of linear inequalities.

See Figure 15. Note: The ordered pair $(2, 1)$ represents the solution to the system

$$\begin{cases} y = 3x - 5 \\ y = -2x + 5 \end{cases}$$

Figure 15

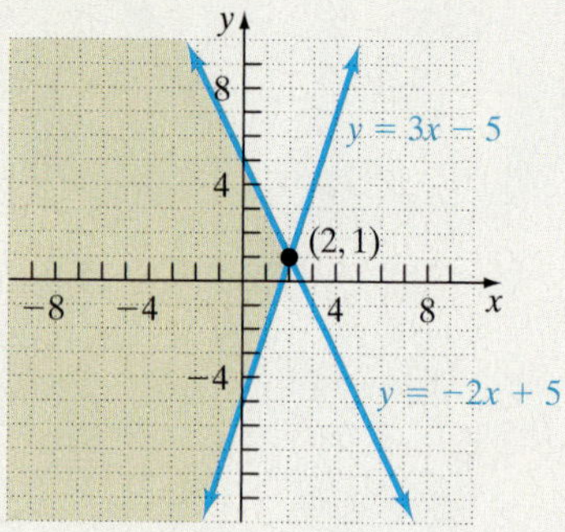

EXAMPLE 3 Graphing a System of Linear Inequalities

Graph the system: $\begin{cases} 2x + y < 7 \\ 7x - 2y > 4 \end{cases}$

Solution

We graph the inequality $2x + y < 7 \, (y < -2x + 7)$. See Figure 16(a). We then graph the inequality $7x - 2y > 4 \left(y < \dfrac{7}{2}x - 2 \right)$. See Figure 16(b). Don't forget to use dashed lines!

Now we combine the graphs in Figures 16(a) and (b). The overlapping shaded region is the solution to the system of linear inequalities. See Figure 16(c). The points on the two boundary lines are not solutions of the system. Do you know why?

Figure 16

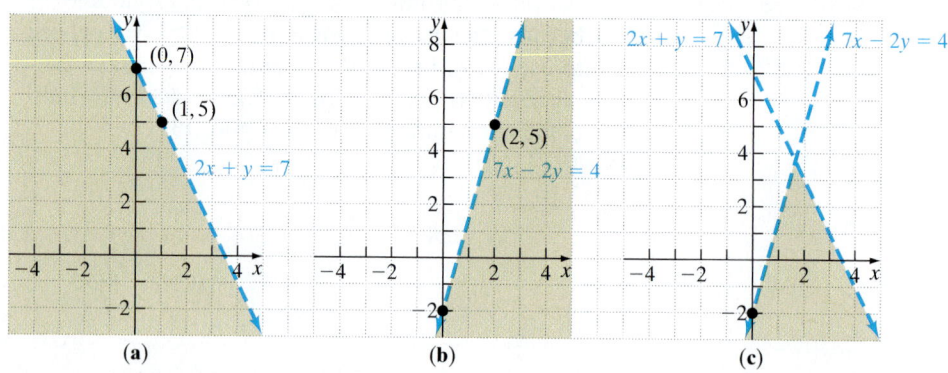

| (a) | (b) | (c) |

QUICK ✓ *Graph each system of linear inequalities.*

2. $\begin{cases} y \geq -3x + 8 \\ y \geq 2x - 7 \end{cases}$

3. $\begin{cases} 4x + 2y < -9 \\ x + 3y < -1 \end{cases}$

Rather than use multiple graphs to obtain the solution set to the system of linear inequalities, we can use a single graph to determine the overlapping region.

EXAMPLE 4 Graphing a System of Linear Inequalities

Graph the system: $\begin{cases} 2x + y \geq 3 \\ 3x - 2y < 8 \end{cases}$

Work Smart

$$3x - 2y < 8$$
$$-2y < -3x + 8$$
$$y > \frac{3}{2}x - 4$$

Remember to reverse the inequality symbol! Now graph $y = \frac{3}{2}x - 4$ using a dashed line and shade above.

Work Smart

The points on the solid boundary lines are also solutions of the system of linear inequalities.

Solution

We graph the inequality $2x + y \geq 3$ $(y \geq -2x + 3)$. On the same rectangular coordinate system, we graph the inequality $3x - 2y < 8$ $\left(y > \frac{3}{2}x - 4\right)$ using a different shading pattern.

The overlapping shaded region represents the solution. Figure 17 shows the solution. The points on the solid boundary line are solutions of the system, but the points on the dotted boundary line are not solutions.

Figure 17

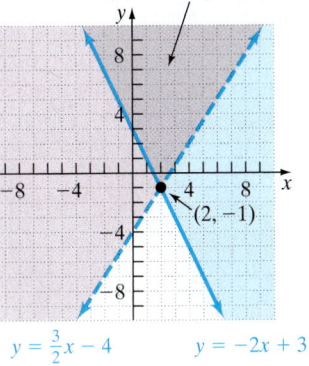

The points in this region satisfy both inequalities.

$$y = \tfrac{3}{2}x - 4 \qquad y = -2x + 3$$

QUICK ✓ *Graph the system of linear inequalities.*

4. $\begin{cases} x + y \leq 4 \\ -x + y \geq -4 \end{cases}$

5. $\begin{cases} 3x + y > -5 \\ x + 2y < 0 \end{cases}$

---

(3) ## Solve Applied Problems Involving Systems of Linear Inequalities

Now let's look at some problems that lead to systems of linear inequalities. As always, we use the problem-solving strategy presented in Section 2.5.

EXAMPLE 5 **Financial Planning**

Maurice recently retired and has up to $50,000 to invest. His financial advisor has recommended that he place at least $10,000 in Treasury notes and no more than $35,000 in corporate bonds. A system of linear inequalities that models this situation is given by

$$\begin{cases} c + t \leq 50{,}000 \\ c \leq 35{,}000 \\ t \geq 10{,}000 \end{cases}$$

where c represents the amount invested in corporate bonds and t represents the amount invested in Treasury notes.

(a) Graph the system.

(b) Can Maurice put $20,000 in corporate bonds and $30,000 in Treasury notes?

(c) Can Maurice put $40,000 in corporate bonds and $10,000 in Treasury notes?

Solution

(a) Draw a rectangular coordinate system with the horizontal axis labeled c (for corporate bonds) and the vertical axis labeled t (for Treasury notes). Each axis will be in thousands, so we draw the line $t + c = 50$ and shade below. We draw the line $t = 10$ and shade above. We draw the line $c = 35$ and shade to the left. Figure 18 shows the graph of the system of linear inequalities.

Figure 18

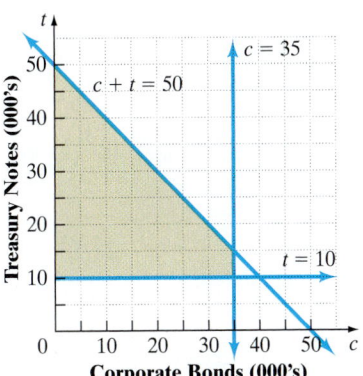

Corporate Bonds (000's)

(b) Yes, Maurice can put $20,000 in corporate bonds and $30,000 in Treasury notes because these values lie within the shaded region. Put another, way $c = 20$ and $t = 30$ satisfies all three inequalities.

(c) No, Maurice cannot put $40,000 in corporate bonds and $10,000 in Treasury notes because these values do not lie within the shaded region. Put another way, $c = 40$ and $t = 10$ does not satisfy the inequality $c \leq 35,000$.

QUICK ✓

6. Jack and Mary recently retired and have up to $75,000 to invest. Their financial advisor has recommended that they place no more than $50,000 in corporate bonds and at least $25,000 in Treasury notes. A system of linear inequalities that models this situation is given by

$$\begin{cases} c + t \leq 75,000 \\ c \quad\quad \leq 50,000 \\ \quad\quad t \geq 25,000 \end{cases}$$

where c represents the amount invested in corporate bonds and t represents the amount invested in Treasury notes.

(a) Graph the system.

(b) Can Jack and Mary invest $30,000 in corporate bonds and $35,000 in Treasury notes?

(c) Can Jack and Mary invest $60,000 in corporate bonds and $15,000 in Treasury notes?

4.6 Exercises

For Extra Help: Student Solutions Manual   CD Video   PH Math/Tutor Center   MathXL Tutorials on CD   MathXL®   MyMathLab

Concepts and Vocabulary

In Problems 1–3, fill in the blanks.

1. An ordered pair is a _____ of a system of linear inequalities if it makes each inequality in the system a true statement.

2. An ordered pair _____ a system of linear inequalities if it makes each inequality in the system a true statement.

3. When graphing linear inequalities, we use a _____ line when graphing strict inequalities ($>$ or $<$) and a _____ line when graphing nonstrict inequalities ($\geq$ or $\leq$).

In Problems 4–6, answer True or False to each statement.

4. We can list every ordered pair that satisfies a system of linear inequalities.

5. It is possible to have the system of linear inequalities graphed correctly and *not* have an overlapping shaded region to represent the solution.

6. Every point that is included in an overlapping shaded region on the graph of a system of linear inequalities satisfies all of the inequalities and therefore is a solution to the system.

7. Each of the graphs below uses two of the following lines:

$$l_1: y \geq -2x \qquad l_2: y \leq -2x \qquad l_3: y \geq x \qquad l_4: y \leq x$$

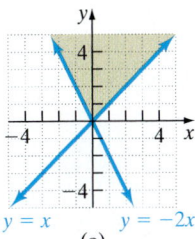
(a)

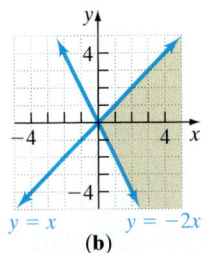
(b)

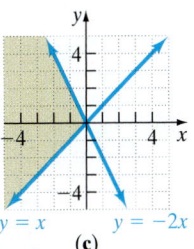
(c)

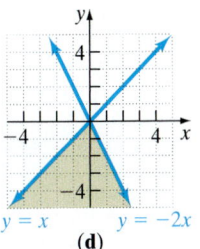
(d)

Write the system of linear inequalities for each graph and explain how you came to your conclusions.

8. When solving application problems which require solving a system of linear inequalities, typically the graph appears only in the first quadrant. Explain why this is so. Can you think of a situation when it would not be the case?

Building Skills

In Problems 9–12, determine which point(s), if any, is a solution of the system of linear inequalities.

9. $\begin{cases} x \geq 5 \\ y < -\dfrac{1}{2}x + 3 \end{cases}$

    (a) $(5, -2)$
    (b) $(10, -4)$
    (c) $(8, -3)$

10. $\begin{cases} 2x - 3y < 3 \\ 2x + y < -5 \end{cases}$

    (a) $(-4, 1)$
    (b) $\left(-\dfrac{3}{2}, -2\right)$
    (c) $(-1, -2)$

11. $\begin{cases} 2x + y > -4 \\ x - y \leq 1 \end{cases}$

    (a) $(-2, 1)$
    (b) $(-1, -2)$
    (c) $(2, -3)$

12. $\begin{cases} x - 2y > 2 \\ -3x - 2y \leq 6 \end{cases}$

    (a) $\left(-1, -\dfrac{3}{2}\right)$
    (b) $(0, -4)$
    (c) $(4, -1)$

In Problems 13–38, graph each system of linear inequalities.

13. $\begin{cases} x > 2 \\ y \leq -1 \end{cases}$

14. $\begin{cases} y > 4 \\ x < -3 \end{cases}$

15. $\begin{cases} y > -2 \\ x > -3 \end{cases}$

16. $\begin{cases} x \leq 3 \\ y < 1 \end{cases}$

17. $\begin{cases} x + y < 3 \\ x - y > 5 \end{cases}$

18. $\begin{cases} x - y > 2 \\ x + y \geq -2 \end{cases}$

19. $\begin{cases} x + y > 3 \\ 2x - y > 4 \end{cases}$

20. $\begin{cases} x - y \geq -1 \\ x + 2y > 4 \end{cases}$

21. $\begin{cases} x < 2 \\ y < \dfrac{1}{2}x + 3 \end{cases}$

22. $\begin{cases} y > 2 \\ y > \dfrac{2}{3}x - 1 \end{cases}$

23. $\begin{cases} x \geq -2 \\ y < 2x + 3 \end{cases}$

24. $\begin{cases} y > 2 \\ 2x - y \leq 1 \end{cases}$

25. $\begin{cases} x > 0 \\ y \leq \dfrac{2}{5}x - 1 \end{cases}$

26. $\begin{cases} y < -2 \\ y > 2x - 1 \end{cases}$

27. $\begin{cases} -y \leq x \\ 3x - y \geq -5 \end{cases}$

28. $\begin{cases} x + y < 2 \\ 3x - 5y \geq 0 \end{cases}$

29. $\begin{cases} x + y \leq -2 \\ y \geq x + 3 \end{cases}$

30. $\begin{cases} 3x + y > 3 \\ y > 4x - 2 \end{cases}$

31. $\begin{cases} x + 3y \geq 0 \\ 2y < x + 1 \end{cases}$

32. $\begin{cases} -y < \dfrac{2}{3}x + 1 \\ -3x + y \leq 2 \end{cases}$ **33.** $\begin{cases} x + y \geq 0 \\ x \quad\quad < 2y + 4 \end{cases}$ **34.** $\begin{cases} 2x - 3y \leq 9 \\ y < -2x + 3 \end{cases}$

35. $\begin{cases} x + 3y > 6 \\ 2x - y \leq 4 \end{cases}$ **36.** $\begin{cases} x - y \leq 3 \\ 2x + 3y \leq -9 \end{cases}$ **37.** $\begin{cases} -y \leq 3x - 4 \\ 2x + 3y \geq -3 \end{cases}$

38. $\begin{cases} x + 4y \leq 4 \\ 2x + 3y \geq 6 \end{cases}$

Applying the Concepts

39. House Blend The Coffee Cup coffee shop is experimenting with blending the "house" coffee. It will be made up of two varieties of coffee, French Roast and Hazelnut. The management has decided that they will make at most 30 pounds of "house" coffee in a day. The tasters have determined that the blend should be mixed so that the amount of French Roast coffee is at least twice the amount of Hazelnut coffee. A system of linear inequalities that models this situation is given by

$$\begin{cases} f + h \leq 30 \\ f \quad\quad \geq 2h \end{cases}$$

where f represents the number of pounds of French Roast coffee and h represents the number of pounds of Hazelnut coffee.

(a) Graph the system of linear inequalities.
(b) Is it possible to use 18 pounds of French Roast coffee and 11 pounds of Hazelnut coffee in the house blend?
(c) Is it possible to use 8 pounds of French Roast coffee and 4 pounds of Hazelnut coffee in the house blend?

40. Party Food Steven and Christopher are planning a party. They plan to buy bratwurst for $4.00 per pound and hamburger patties that cost $3.00 per pound. They can spend at most $70 and think they should have no more than 20 pounds of bratwurst and hamburger patties. A system of inequalities that models the situation is given by

$$\begin{cases} 4b + 3h \leq 70 \\ b + h \leq 20 \end{cases}$$

where b represents the number of pounds of bratwurst and h represents the number of pounds of hamburger patties.

(a) Graph the system of linear inequalities.
(b) Is it possible to purchase 10 pounds of bratwurst and 10 pounds of hamburger patties?
(c) Is it possible to purchase 15 pounds of bratwurst and 5 pounds of hamburger patties?

41. Auto Manufacturing A manufacturing plant has 360 hours of machine time budgeted for the production of cars and trucks on a given day. It takes 12 hours to build a car and 18 hours to build a truck. Based on experience, the plant manager knows that the number of trucks is fewer than 20 less than twice the number of cars produced. Due to limitations in the machinery, the maximum number of cars that can be produced in a day is 24. A system of linear inequalities that models this situation is given by

$$\begin{cases} 12c + 18t \leq 360 \\ t < 2c - 20 \\ c \quad\quad \leq 24 \end{cases}$$

where c represents the number of cars produced and t represents the number of trucks built.

(a) Graph the system of linear inequalities.

(b) Is it possible to build 17 cars and 9 trucks at this plant?

(c) Is it possible to build 12 cars and 11 trucks at this plant?

42. Breakfast at Burger King Aman decides to eat breakfast at Burger King. He is a big fan of their French toast sticks and he likes orange juice. He wants to eat no more than 500 calories and consume no more than 425 mg of sodium. Each French toast stick (with syrup) has 90 calories and 100 mg of sodium. Each small orange juice has 140 calories and 25 mg of sodium. The system of linear inequalities that represents the possible combination of French toast sticks and orange juice that Aman can consume is

$$\begin{cases} 90x + 140y \le 500 \\ 100x + 25y \le 425 \\ \quad x \quad\quad\quad \ge 0 \\ \quad\quad\quad\quad y \ge 0 \end{cases}$$

where x represents the number of Burger King French toast sticks and y represents the number of containers of orange juice.

(a) Graph the system of linear inequalities.

(b) Is it possible for Aman to consume 3 French toast sticks and 1 container of orange juice and stay within the allowance for calories and sodium?

(c) Is it possible for Aman to consume 2 French toast sticks and 2 containers of orange juice and stay within the allowance for calories and sodium?

Extending the Concepts

In Problems 43–46, the shaded region for the solution to each system of linear inequalities below differs from what we have seen previously. Graph and then state the solution of each system.

43. $\begin{cases} \dfrac{y}{2} - \dfrac{x}{6} \ge 1 \\ \dfrac{x}{3} - \dfrac{y}{1} \ge 1 \end{cases}$

44. $\begin{cases} \dfrac{x}{2} + \dfrac{y}{4} > 1 \\ -\dfrac{x}{4} - \dfrac{y}{8} > 1 \end{cases}$

45. $\begin{cases} x < \dfrac{3}{2}y + \dfrac{9}{2} \\ -2x < 3(y + 2) \end{cases}$

46. $\begin{cases} x > \dfrac{5}{4}y - \dfrac{1}{2} \\ 4x < 5(y + 3) \end{cases}$

In Problems 47 and 48, write a system of linear inequalities that will produce each shaded region as its solution.

47.

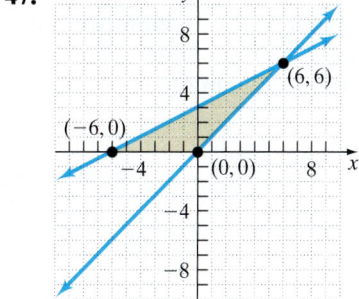

48.

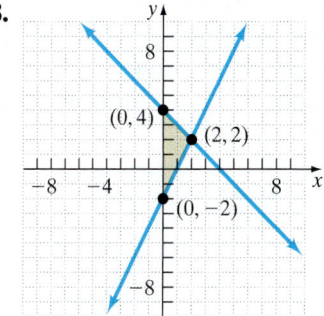

CHAPTER 4 ACTIVITY: FIND THE NUMBERS

Focus: Solving systems of equations

Time: 15 minutes

Group size: 2

Consider the following dialogue between two students:

> *Ryan:* Think of two numbers between 1 and 10, and don't tell me what they are.
>
> *Melissa:* Ok, I've thought of two numbers.
>
> *Ryan:* Now tell me the sum of the two numbers and the difference of the two numbers, and I'll tell you what your two numbers are.
>
> *Melissa:* Their sum is 14 and their difference is 6.
>
> *Ryan:* Your numbers are 10 and 4.
>
> *Melissa:* That's right! How did you do that?
>
> *Ryan:* I set up a system of equations using the sum and difference that you gave me, along with the variables x and y.

1. Each group member should set up and solve the system of equations described by Ryan. Discuss your results and be sure that you both arrive at the solutions 10 and 4.

2. Now each of you will think of two new numbers and the other will try to find the numbers by solving a system of equations. But this time the system will be a bit trickier! Give each other the following information about your two numbers, and then figure out each other's numbers:

   six more than three times the sum of the numbers

   two less than four times the difference of the numbers

3. Would the systems in this activity work if negative numbers were used? Try it and see!

CHAPTER 4 REVIEW

Section 4.1	Solve Systems of Linear Equations by Graphing

KEY CONCEPT	KEY TERMS

• **Recognizing Solutions of Systems of Two Linear Equations with Two Unknowns**
  • If the lines in a system of two linear equations containing two unknowns intersect, then the point of intersection is the solution and the system is consistent and independent.
  • If the lines in a system of two linear equations containing two unknowns are parallel, then the system has no solution and the system is inconsistent.
  • If the lines in a system of two linear equations containing two unknowns lie on top of each other, then the system has infinitely many solutions. The solution set is the set of all points on the line and the system is consistent, but dependent.
	System of linear equations
	Solution
	Consistent and independent
	Inconsistent
	Consistent and dependent

YOU SHOULD BE ABLE TO . . .	EXAMPLE	REVIEW EXERCISES
① Determine if an ordered pair is a solution of a system of linear equations (p. 273)	Example 2	1–4
② Solve a system of linear equations by graphing (p. 274)	Examples 3 through 6	5–12
③ Classify systems of linear equations as consistent or inconsistent (p. 278)	Example 7	13–18
④ Solve applied problems involving systems of linear equations (p. 280)	Example 8	19, 20

In Problems 1–4, determine whether the ordered pair is a solution to the system of equations.

1. $\begin{cases} x + 2y = 6 \\ 3x - y = -10 \end{cases}$

    (a) $(3, -1)$ **(b)** $(-2, 4)$ **(c)** $(4, 1)$

2. $\begin{cases} y = 3x - 5 \\ 3y = 6x - 5 \end{cases}$

    (a) $(2, 1)$ **(b)** $\left(0, -\dfrac{5}{3}\right)$ **(c)** $\left(\dfrac{10}{3}, 5\right)$

3. $\begin{cases} 3x - 4y = 2 \\ 20y = 15x - 10 \end{cases}$

    (a) $\left(\dfrac{1}{2}, -\dfrac{1}{8}\right)$ **(b)** $(6, 4)$

    (c) $(0.4, -0.2)$

4. $\begin{cases} x = -4y + 2 \\ 2x + 8y = 12 \end{cases}$

    (a) $(10, -1)$ **(b)** $\left(-\dfrac{1}{2}, \dfrac{1}{2}\right)$

    (c) $(2, 1)$

In Problems 5–12, solve each system of equations by graphing.

5. $\begin{cases} 2x - 4y = 8 \\ x + y = 7 \end{cases}$

6. $\begin{cases} x - y = -3 \\ 3x + 2y = 6 \end{cases}$

7. $\begin{cases} y = -\dfrac{x}{2} + 2 \\ y = x + 8 \end{cases}$

8. $\begin{cases} y = -x - 5 \\ y = \dfrac{3x}{4} + 2 \end{cases}$

9. $\begin{cases} 4x - 8 = 0 \\ 3y + 9 = 0 \end{cases}$

10. $\begin{cases} x = y \\ x + y = 0 \end{cases}$

11. $\begin{cases} 0.6x + 0.5y = 2 \\ 10y = -12x + 20 \end{cases}$

12. $\begin{cases} \dfrac{1}{4}x - \dfrac{1}{2}y = 1 \\ 3x - 6y = 12 \end{cases}$

In Problems 13–18, without graphing, determine the number of solutions to each system of equations. State whether the system is consistent or inconsistent. For those systems that are consistent, state whether the equations are dependent or independent.

13. $\begin{cases} 3x = y + 4 \\ 3x - y = -4 \end{cases}$

14. $\begin{cases} 4y = 2x - 8 \\ x - 2y = -4 \end{cases}$

15. $\begin{cases} -3x + 3y = -3 \\ \dfrac{1}{2}x - \dfrac{1}{2}y = 0.5 \end{cases}$

16. $\begin{cases} x - 2 = -\dfrac{2}{3}y - \dfrac{2}{3} \\ 3x = 4 - 2y \end{cases}$

17. $\begin{cases} 3 - 2x = y \\ \dfrac{y}{2} = x + 1.5 \end{cases}$

18. $\begin{cases} \dfrac{y}{2} = \dfrac{x}{4} + 2 \\ \dfrac{x}{8} + \dfrac{y}{4} = -1 \end{cases}$

19. Printing Costs Monique is creating a flyer to give to local businesses to advertise her new vintage clothing store. She is trying to decide between two quotes for the printing. Printer A has given her a quote of \$70 setup fee plus \$0.10 per flyer printed. Printer B has given her a quote of \$100 plus \$0.04 per flyer.

    (a) Write a system of linear equations that models the problem.
    (b) Graph the system of equations to determine how many fliers she needs to have printed in order for the cost to be the same at each printer.
    (c) If 400 fliers are printed, which printer should she choose to have the lower cost?

20. Flooring Installation To install carpeting in Juan's bedroom, it costs \$50 for installation plus \$40 per square yard of carpet. In the same room, Juan can install ceramic tile for a cost of \$350 for installation plus \$10 per square yard of tile.

    (a) Write a system of linear equations that models the problem.
    (b) Graph the system of equations to determine how many square yards of flooring he needs to have installed for the cost to be the same.
    (c) If the area of Juan's bedroom is 15 square yards, which material would be cheaper?

Section 4.2	Solving Systems of Linear Equations in Two Variables Using Substitution

KEY CONCEPT

- **Solving Systems of Linear Equations Using Substitution**

   Step 1: Solve one of the equations for one of the unknowns. For example, we might solve equation (1) for y in terms of x.

   Step 2: Substitute the expression solved for in Step 1 into the other equation. The result will be a single linear equation in one unknown. For example, if we solved equation (1) for y in terms of x in Step 1, then we would replace y in equation (2) with the expression in x.

   Step 3: Solve the linear equation in one unknown found in Step 2.

   Step 4: Substitute the value of the variable found in Step 3 into one of the original equations to find the value of the other variable.

   Step 5: Check your answer.

YOU SHOULD BE ABLE TO . . .	EXAMPLE	REVIEW EXERCISES
① Solve a system of linear equations using the substitution method (p. 285)	Examples 1 through 5	21–34
② Solve applied problems involving systems of linear equations (p. 291)	Example 6	35, 36

In Problems 21–34, solve each system of equations using substitution.

21. $\begin{cases} x + 4y = 6 \\ y = 2x - 3 \end{cases}$
22. $\begin{cases} 7x - 3y = 10 \\ y = 3x - 4 \end{cases}$
23. $\begin{cases} 2x + 5y = 4 \\ x = 3 - 2y \end{cases}$
24. $\begin{cases} 3x + y = 10 \\ x = 8 + 2y \end{cases}$

25. $\begin{cases} y = \dfrac{2}{3}x - 1 \\ y = \dfrac{1}{2}x + 2 \end{cases}$
26. $\begin{cases} y = -\dfrac{5}{6}x + 3 \\ y = -\dfrac{4}{3}x \end{cases}$
27. $\begin{cases} 2x - y = 6 \\ 4x + 3y = 2 \end{cases}$
28. $\begin{cases} 5x + 2y = 13 \\ x + 4y = -1 \end{cases}$

29. $\begin{cases} 6x + 3y = 12 \\ y = -2x + 4 \end{cases}$
30. $\begin{cases} x = 4y - 2 \\ 8y - 2x = 4 \end{cases}$
31. $\begin{cases} -6 - 2(3x - 6y) = 0 \\ 6 - 12(x - 2y) = 0 \end{cases}$

32. $\begin{cases} 6 - 2(3y + 4x) = 0 \\ 9(y - 1) + 12x = 0 \end{cases}$
33. $\begin{cases} \dfrac{1}{2}x - \dfrac{1}{4}y = \dfrac{1}{2} \\ \dfrac{1}{3}x - \dfrac{3}{4}y = -\dfrac{1}{4} \end{cases}$
34. $\begin{cases} -\dfrac{5x}{4} + \dfrac{y}{6} = \dfrac{7}{12} \\ \dfrac{3x}{2} - \dfrac{y}{10} = \dfrac{3}{5} \end{cases}$

△ **35. Walk around the Park** A rectangular park is surrounded by a walking path. The park is 75 meters longer than it is wide. If a person walks completely around the park and covers a distance of 650 meters, find the length and width of the park by solving the following system of equations where l is the length and w is the width.

$$\begin{cases} 2l + 2w = 650 & (1) \\ l = w + 75 & (2) \end{cases}$$

36. Fun with Numbers The sum of two numbers is 12. If twice the smaller number is subtracted from the larger number, the difference is 21. Determine the smaller of the two numbers by solving the following system of equations where x and y represent the unknown numbers.

$$\begin{cases} x + y = 12 & (1) \\ x - 2y = 21 & (2) \end{cases}$$

Section 4.3	Solving Systems of Linear Equations Using Elimination

KEY CONCEPT

- **Solving Systems of Linear Equations Using Elimination**

   Step 1: Make sure that the coefficients on one of the variables are additive inverses by multiplying (or dividing) both sides of one (or both) equation(s) by a nonzero constant.

   Step 2: Add the equations to eliminate the variable whose coefficients are now additive inverses. Solve the resulting equation for the remaining unknown.

Step 3: Substitute the value of the variable found in Step 2 into one of the original equations to find the value of the remaining variable.

Step 4: Check your answer.

YOU SHOULD BE ABLE TO . . .	EXAMPLE	REVIEW EXERCISES
1 Solve a system of linear equations using the elimination method (p. 296)	Examples 1 through 5	37–52
2 Solve applied problems involving systems of linear equations (p. 301)	Example 6	53, 54

In Problems 37–44, solve each system of equations using elimination.

37. $\begin{cases} 4x - y = 12 \\ 2x + y = -12 \end{cases}$

38. $\begin{cases} -2x + 3y = 27 \\ 2x - 5y = -41 \end{cases}$

39. $\begin{cases} -3x + 4y = 25 \\ x - 5y = -23 \end{cases}$

40. $\begin{cases} 5x + 8y = -15 \\ -2x + y = 6 \end{cases}$

41. $\begin{cases} 4x - 3y = -1 \\ 2x - 5y = 3 \end{cases}$

42. $\begin{cases} -2x + 5y = 0 \\ -3x - 2y = -19 \end{cases}$

43. $\begin{cases} 1.3x - 0.2y = -3 \\ -0.1x + 0.5y = 1.2 \end{cases}$

44. $\begin{cases} 2.5x + 0.5y = 6.5 \\ -0.5x - 1.2y = 4.6 \end{cases}$

In Problems 45–52, solve each system of equations by any method.

45. $\begin{cases} 2x + y = -1 \\ -6x - 8y = 13 \end{cases}$

46. $\begin{cases} \dfrac{x + 9}{6} = \dfrac{9 - 4y}{4} \\ \dfrac{2y - 8}{3} = 4x + 4 \end{cases}$

47. $\begin{cases} y + 5 = \dfrac{2}{3}x + 3 \\ \dfrac{2x + 3}{6} = \dfrac{y + 3}{2} \end{cases}$

48. $\begin{cases} \dfrac{1}{14} - \dfrac{x}{2} = -\dfrac{y}{7} \\ y = \dfrac{1}{2} + \dfrac{7x}{3} \end{cases}$

49. $\begin{cases} -x + y = 7 \\ -3x + 4y = 8 \end{cases}$

50. $\begin{cases} 4x + y = 2 \\ 9y - 3x = 5 \end{cases}$

51. $\begin{cases} 4x + 6 = 3y + 5 \\ 4(-2x - 4) = 6(-y - 3) \end{cases}$

52. $\begin{cases} 3y + 2x = 16x + 2 \\ x = -\dfrac{3}{14}y - \dfrac{1}{7} \end{cases}$

53. Valentine Gift The specialty bakery around the corner is putting together a Valentine's Day gift box to sell. The gift box is going to contain two items, heart-shaped cookies and chocolate candies. If the cookies sell for $1.50 per pound and the chocolates sell for $7.75 per pound, how many pounds of each type should be included in a four-pound box that sells for $4.00 per pound? Solve the system

$$\begin{cases} h + c = 4 \quad (1) \\ 1.50h + 7.75c = 16 \quad (2) \end{cases}$$

where h represents the number of pounds of heart-shaped cookies and c represents the number of pounds of candies to find the answers.

54. Raffle Tickets The parent booster club is selling 50–50 raffle tickets to raise money for the upcoming hockey banquet. Raffle tickets to win half the money collected are sold in two ways: Individual chances are sold for $1.00 each or a ticket with a block of 6 chances can be purchased for $5.00. On one particular night the booster club brought in $2025 from the sale of 725 raffle tickets. How many of each type of raffle ticket was sold? Solve the system

$$\begin{cases} i + b = 725 \quad (1) \\ 1i + 5b = 2025 \quad (2) \end{cases}$$

where i is the number of individual tickets and b is the number of block tickets to find the answers.

Section 4.4	Solving Direct Translation, Geometry, and Uniform Motion Problems Using Systems of Linear Equations		
YOU SHOULD BE ABLE TO . . .		**EXAMPLE**	**REVIEW EXERCISES**
① Model and solve direct translation problems (p. 307)		Example 1	55–58
② Model and solve geometry problems (p. 308)		Examples 2 and 3	59–62
③ Model and solve uniform motion problems (p. 310)		Example 4	63–66

55. **Fun with Numbers** Find two numbers whose difference is $\dfrac{1}{24}$ and whose sum is $\dfrac{17}{24}$.

56. **Fun with Numbers** The sum of two numbers is 5.8. If twice the smaller number is subtracted from the larger number, the difference is -2. Find the smaller number.

57. **Investments** Aaron has \$50,000 to invest. His financial advisor recommends he invest in stocks and bonds with the amount invested in stocks equaling \$4000 less than twice the amount invested in bonds. How much should be invested in stocks? How much should be invested in bonds?

58. **Bookstore Sales** A register at the bookstore sold only notebooks and scientific calculators. The cost of a notebook is $\dfrac{1}{3}$ of the cost of the calculator. If the calculator sells for \$7.50 and the register tape shows that 24 items resulted in revenue of \$110, how many of each was sold?

△ 59. **Geometry** The measure of one angle is 25° less than the measure of its supplement. Find the measures of the two angles.

△ 60. **Geometry** The measure of one angle is 15° more than half the measure of its complement. Find the measures of the two angles.

△ 61. **Horse Corral** The O'Connells are planning to build a rectangular corral for their horses using the river on their property as one of the sides. On the other three sides they are installing 52 meters of fencing. The longest side of the corral is opposite the water and is 8 meters less than twice the short side. Find the dimensions of the corral.

△ 62. **Geometry** In a triangle, the measure of the first angle is 10° less than three times the measure of the second. If the measure of the third angle is 30°, find the measure of the two unknown angles of the triangle.

63. **Wind Resistance** A plane can fly 2000 miles with a tailwind in 4 hours. It can make the return trip in 5 hours. Find the speed of the plane in still air and the speed of the wind.

64. **Fishing Spot** Dayle is paddling his canoe upstream, against the current, to a fishing spot 10 miles away. If he paddles upstream for 2.5 hours and his return trip takes 1.25 hours, find the speed of the current and his paddling speed in still water.

65. **Cycling in the Wind** A cyclist can go 36 miles with the wind blowing at her back in 3 hours. On the return trip, after 4 hours, the cyclist still has 4 miles remaining to return to the starting point. Find the speed of the cyclist and the speed of the wind.

66. **Out for a Run** One speedy jogger can run 2 mph faster than his running buddy. In the last race they entered, the faster runner covered 12 miles in the same time that it took for the slower jogger to cover 9 miles. Find the speed of each of the runners.

Section 4.5   Solving Mixture Problems Using Systems of Linear Equations in Two Variables

YOU SHOULD BE ABLE TO . . .	EXAMPLE	REVIEW EXERCISES
(1) Draw up a plan for modeling mixture problems (p. 314)	Example 1	67, 68
(2) Set up and solve money problems using the mixture model (p. 315)	Examples 2 and 3	69–72
(3) Set up and solve dry mixture and percent mixture problems (p. 318)	Examples 4 and 5	73–76

In Problems 67 and 68, fill in the table from the information given. DO NOT SOLVE.

67. Brendan has twice as many nickels as dimes. He has a total of $2.25 in change.

	Number of Coins	•	Value of Each Coin	=	Total Value
Dimes	?		?		?
Nickels	?		?		?
Total					?

68. Belinda invested part of her $15,000 inheritance in a savings account with an annual rate of 6.5% simple interest and the rest of her inheritance in a mutual fund with a rate of 8% simple interest. Her total yearly interest income from both accounts was $1050.

	Principal	Rate	Interest
Savings Account	?	?	?
Mutual Fund	?	?	?
Total	?		?

69. Movie Tickets A movie theater sells three types of tickets: children $4.00 each, adults $6.25 each, and seniors $5.00 each. At one particular show, the same number of children and adult tickets were sold. If a total of 300 tickets brought in $1531.25 in sales, how many of each type were sold?

70. Coins Sharona has $1.70 in change consisting of three more dimes than quarters. Find the number of quarters she has.

71. Investment Carlos Singer has some money invested at 5%, and $5000 more than that invested at 9%. His total annual interest income is $1430. Find the amount Carlos has invested at each rate.

72. Investment Hilda invested part of her $25,000 advance in savings bonds at 7% annual simple interest and the rest in a stock portfolio yielding 8% annual simple interest. If her total yearly interest income is $1900, find the amount Hilda has invested at each rate.

73. Sugar A baker wants to mix a 60% sugar solution with a 30% sugar solution to obtain 10 quarts of a 51% sugar solution. How much of the 30% sugar solution will the baker use?

74. Peroxide How many pints of a 25% peroxide solution should be added to 10 pints of a 60% peroxide solution to obtain a 30% peroxide solution?

75. Almond-Peanut Butter Joni works at a health food store and is planning to grind a special blend of peanut butter by mixing peanuts, which sell for $4.00 per pound, and almonds, which sell for $6.50 per pound. How many pounds of each type should she use if she needs 5 pounds of almond-peanut butter, which sells for $5.00 per pound?

76. Mixing Chemicals Harold needs 20 liters of 55% acid. If he has 35% acid and 60% acid that he can mix together, how many liters of each should he use to obtain the desired mixture?

Section 4.6 Solving Systems of Linear Inequalities

KEY TERM

Satisfies

YOU SHOULD BE ABLE TO ...	EXAMPLE	REVIEW EXERCISES
① Determine whether an ordered pair is a solution of a system of linear inequalities (p. 326)	Example 1	77–82
② Graph a system of linear inequalities (p. 327)	Examples 2 through 4	83–94
③ Solve applied problems involving systems of linear inequalities (p. 329)	Example 5	95

In Problems 77–82, determine which point(s), if any, is a solution to the system of linear inequalities.

77. $\begin{cases} x + y \le 2 \\ 3x - 2y > 6 \end{cases}$
   (a) $(-1, -5)$
   (b) $(3, -1)$
   (c) $(4, 1)$

78. $\begin{cases} y \ge 3x + 5 \\ y \ge -2x \end{cases}$
   (a) $(-1, 0)$
   (b) $(-2, 4)$
   (c) $(1, 8)$

79. $\begin{cases} x > 5 \\ y < -2 \end{cases}$
   (a) $(10, -10)$
   (b) $(5, -3)$
   (c) $(7, -2)$

80. $\begin{cases} y > x \\ 2x - y \le 3 \end{cases}$
   (a) $(4, 3)$
   (b) $(-3, -2)$
   (c) $(-1, 4)$

81. $\begin{cases} x + 2y < 6 \\ 4y - 2x > 16 \end{cases}$
   (a) $(-4, 2)$
   (b) $(4, 5)$
   (c) $(-2, -3)$

82. $\begin{cases} 2x - y \ge 3 \\ y \le 2x + 1 \end{cases}$
   (a) $(1, 2)$
   (b) $(3, -2)$
   (c) $(-1, 4)$

In Problems 83–94, graph the solution set of each system of linear inequalities.

83. $\begin{cases} x > -2 \\ y > 1 \end{cases}$

84. $\begin{cases} x \le 3 \\ y > -1 \end{cases}$

85. $\begin{cases} x + y \ge -2 \\ 2x - y \le -4 \end{cases}$

86. $\begin{cases} 3x + 2y < -6 \\ x - y < 2 \end{cases}$

87. $\begin{cases} x > 0 \\ y \le \dfrac{3}{4}x + 1 \end{cases}$

88. $\begin{cases} y \le 0 \\ y \le -\dfrac{1}{2}x - 3 \end{cases}$

89. $\begin{cases} -y \ge x \\ 4x - 3y \ge -12 \end{cases}$

90. $\begin{cases} -y \le x + 2 \\ 2x + 2y \ge -9 \end{cases}$

91. $\begin{cases} 2x + 3y > -3 \\ y > -\dfrac{2}{3}x + 2 \end{cases}$

92. $\begin{cases} x + 4y \le -4 \\ y \ge \dfrac{1}{4}x + 3 \end{cases}$

93. $\begin{cases} y > 2x - 5 \\ y - 2x \le 0 \end{cases}$

94. $\begin{cases} y > x + 2 \\ y < x - 4 \end{cases}$

95. Party Planning Alexis and Sarah are planning a barbeque for their friends. They plan to serve grilled fish and carne asada and want to spend at most $40 on the meat. The fish sells for $8 per pound and the carne asada is $5 per pound. Since most of their friends do not eat red meat, they plan to buy at least twice as much fish as carne asada. A system of linear inequalities that models this situation is given by

$$\begin{cases} 8x + 5y \le 40 \\ x \ge 2y \\ x \ge 0 \\ y \ge 0 \end{cases}$$

where x represents the number of pounds of fish and y represents the number of pounds of carne asada. Graph the system of linear inequalities that shows how many pounds of each they can buy and stay within their budget.

CHAPTER 4 TEST

Remember to use your Chapter Test Prep Video CD to see fully worked-out solutions to any of these problems you would like to review.

In Problems 1 and 2, determine which of the following is a solution to the system of linear equations or inequalities.

1. $\begin{cases} 3x - y = -5 \\ y = \dfrac{2}{3}x - 2 \end{cases}$

    (a) $(-1, 2)$
    (b) $(-9, -8)$
    (c) $(-3, -4)$

2. $\begin{cases} 2x + y \geq 10 \\ y < \dfrac{x}{2} + 1 \end{cases}$

    (a) $(5, 3)$
    (b) $(-2, 14)$
    (c) $(-3, -1)$

In Problems 3 and 4, you begin to solve a system of linear equations in two variables. Before you draw the graphs you note the given conditions.

3. The slopes of the two lines are negative reciprocals and the y-intercepts are the same.

    (a) Tell how many solutions exist.
    (b) State whether the system is consistent or inconsistent.
    (c) If the system is consistent, state whether the system is dependent or independent.

4. The slopes of the two lines are the same and the y-intercepts are different.

    (a) Tell how many solutions exist.
    (b) State whether the system is consistent or inconsistent.
    (c) If the system is consistent, state whether the system is dependent or independent.

In Problems 5 and 6, solve each system of equations by graphing.

5. $\begin{cases} 2x + 3y = 0 \\ x + 4y = 5 \end{cases}$

6. $\begin{cases} y = 2x - 6 \\ y = -\dfrac{1}{4}x + 3 \end{cases}$

In Problems 7 and 8, solve each system of equations using substitution.

7. $\begin{cases} 3x - y = 3 \\ 4x + 5y = -15 \end{cases}$

8. $\begin{cases} y = \dfrac{2}{3}x + 7 \\ y = \dfrac{1}{4}x + 2 \end{cases}$

In Problems 9 and 10, solve each system of equations using elimination.

9. $\begin{cases} 3x + 2y = -3 \\ 5x - y = -18 \end{cases}$

10. $\begin{cases} 4x - 5y = 12 \\ 3x + 4y = -22 \end{cases}$

In Problems 11–13, solve by any method.

11. $\begin{cases} \dfrac{2}{3}x - \dfrac{1}{6}y = \dfrac{25}{12} \\ -\dfrac{1}{4}x = \dfrac{3}{2}y \end{cases}$

12. $\begin{cases} 0.4x - 2.5y = -6.5 \\ x + y = 5.5 \end{cases}$

13. $\begin{cases} 4 + 3(y - 3) = -2(x + 1) \\ x + \dfrac{3y}{2} = \dfrac{3}{2} \end{cases}$

14. Plane Trip A small plane can fly 1575 miles in 7 hours with a tailwind. On the return trip, with a headwind, the same trip takes 9 hours. Find the speed of the plane in still air and the speed of the wind.

15. Ice Cream Manny is blending a peach swirl ice cream. He mixes together vanilla ice cream, which sells for $6 per container, and peach ice cream, which sells for $11.50 per container. How many containers of each should he use if he wants to have 11 containers of swirl ice cream, which he can sell for $8 each?

△ **16. Supplementary Angles** The measure of one angle is 30° less than three times the measure of its supplement. Find the measure of each angle.

17. Playground Equipment Moises is distributing new playground equipment to several elementary schools. Each school will receive some basketballs and some volleyballs. The cost of a basketball is $25 and the cost of a volleyball is $15. If each school needs 40 balls and he can spend $750 per school, how many of each type will a school receive?

In Problems 18–20, graph the solution set of each system of inequalities.

18. $\begin{cases} 4x - 2y < 8 \\ x + 3y < 6 \end{cases}$ **19.** $\begin{cases} x \le -2 \\ -2x - 4y \le 8 \end{cases}$ **20.** $\begin{cases} y \le \dfrac{2}{3}x + 4 \\ -2x + 3y > -3 \end{cases}$

CHAPTER
5 Exponents and Polynomials

Music producers follow trends in consumer taste. Did you know that the number of music videos sold in the United States reached its peak in 1998, and then began to decline? The polynomial $-0.425t^2 + 4.57t + 8.31$ describes the number of music videos sold, in millions, since 1993, where t represents the number of years since 1993. A recording company may want to use this model to decide how many music videos to produce in a given year. See Problem 108 in Section 5.1, p. 353.

OUTLINE

5.1 Adding and Subtracting Polynomials

5.2 Multiplying Monomials: The Product and Power Rules

5.3 Multiplying Polynomials

5.4 Dividing Monomials: The Quotient Rule and Integer Exponents

Putting the Concepts Together (Sections 5.1–5.4)

5.5 Dividing Polynomials

5.6 Applying Exponent Rules: Scientific Notation

Chapter 5 Activity: What Is the Question?
Chapter 5 Review
Chapter 5 Test
Cumulative Review Chapters 1–5

The Big Picture: Putting It Together

The first two chapters of the text were devoted to reviewing and developing fundamental skills with real numbers, algebraic expressions, and equations. We will use these skills often as we proceed through the course. In particular, we will describe how new topics and skills relate to those that we already know.

This chapter introduces polynomials. Polynomial expressions, such as the one given in the chapter opener, are used to describe many situations in business, consumer affairs, the physical and biological sciences, and even the entertainment industry. In this chapter, we will learn how to add, subtract, multiply, and divide polynomial expressions. You should pay attention to the techniques for performing these operations and how they are related to the techniques for adding, subtracting, multiplying, and dividing real numbers. Why? If we think of algebra as an extension of arithmetic skills, then comprehension and understanding of the material will come more quickly and easily.

5.1 Adding and Subtracting Polynomials

OBJECTIVES

① Define Monomial and Determine the Degree of a Monomial

② Define Polynomial and Determine the Degree of a Polynomial

③ Simplify Polynomials by Combining Like Terms

④ Evaluate Polynomials

Preparing for Adding and Subtracting Polynomials

Before getting started, take the following readiness quiz. If you get a problem wrong, go back to the section cited and review the material.

1. What is the coefficient of $-4x^5$? [Section 1.7, pp. 59–60]
2. Combine like terms: $-3x + 2 - 2x - 6x - 7$ [Section 1.7, pp. 61–62]
3. Use the Distributive Property to remove the parentheses: $-4(x - 3)$ [Section 1.7, p. 61]
4. Evaluate the expression $5 - 3x$ for $x = -2$. [Section 1.7, p. 58]

Recall from Section 1.7 that a term is a number or the product of a number and one or more variables raised to a power. The numerical factor of a term is the coefficient. A constant is a single number, such as 2 or $-\dfrac{3}{2}$. For example, consider Table 1, where some algebraic expressions are given. Note how the terms and coefficients of each algebraic expression are identified.

Table 1

Algebraic Expression	Terms	Coefficients
$3x + 2$	$3x, 2$	$3, 2$
$4x^2 - 7x + 5 = 4x^2 + (-7x) + 5$	$4x^2, -7x, 5$	$4, -7, 5$
$9x^2 + 7y^3$	$9x^2, 7y^3$	$9, 7$

Notice in the algebraic expression $4x^2 - 7x + 5$ given in Table 1 that we first rewrite it as a sum, $4x^2 + (-7x) + 5$, to identify the terms and coefficients.

① Define Monomial and Determine the Degree of a Monomial

In this chapter, we study *polynomials*. Polynomials are made up of terms that are **monomials.**

Work Smart

The prefix "mono" means "one." For example, a monorail has one rail. So a monomial in one variable is either a constant or a single term with a variable in it.

In Words

The degree of a monomial can be thought of as the number of times the variable occurs. For example, because $2x^4 = 2 \cdot x \cdot x \cdot x \cdot x$, the degree of $2x^4$ is 4.

> **DEFINITION**
>
> A **monomial** in one variable is the product of a constant and a variable raised to a whole number $(0, 1, 2, \dots)$ power. A monomial in one variable is of the form
>
> $$ax^k$$
>
> where a is a constant, x is a variable, and k is a whole number.

In a monomial of the form ax^k, where $a \neq 0$, we call k the **degree** of the monomial. The degree of a nonzero constant is zero. The number 0 has no degree.

EXAMPLE 1 Monomials

	MONOMIAL	COEFFICIENT	DEGREE
(a)	$2x^4$	2	4
(b)	$-\dfrac{7}{4}x^2$	$-\dfrac{7}{4}$	2
(c)	$-x = -1 \cdot x$	-1	1
(d)	$x^3 = 1 \cdot x^3$	1	3
(e)	8	8	0

Now let's look at some expressions that are not monomials.

EXAMPLE 2 **Examples of Expressions That Are Not Monomials**

   (a) $5x^{\frac{1}{2}}$ is not a monomial because the exponent of the variable x is $^1\!/_2$ and $^1\!/_2$
   is not a whole number.

   (b) $2x^{-4}$ is not a monomial because the exponent of the variable x is -4 and
   -4 is not a whole number.

QUICK ✔ *Determine whether the expression is a monomial. For those that are monomials, determine the coefficient and degree.*

1. $12x^6$ **2.** $3x^{-3}$ **3.** 10 **4.** $n^{\frac{1}{3}}$

So far, we have discussed only monomials in one variable. A monomial may contain more than one variable factor, such as $ax^m y^n$, where a is a constant (called the coefficient), x and y are variables, and m and n are whole numbers. An example of a monomial with more than one variable factor is $-4x^2 y^3$. The **degree of the monomial** $ax^m y^n$ is the sum of the exponents, $m + n$.

EXAMPLE 3 **Monomials in More than One Variable**

   (a) $-4x^3 y^4$ is a monomial in x and y of degree $3 + 4 = 7$. The coefficient is -4.

   (b) $10ab^5$ is a monomial in a and b of degree $1 + 5 = 6$. The coefficient is 10.

QUICK ✔ *Determine whether the expression is a monomial. For those that are monomials, determine the coefficient and degree.*

5. $3x^5 y^2$ **6.** $-2m^3 n$ **7.** $4ab^{1/2}$ **8.** $-xy$

(2) **Define Polynomial and Determine the Degree of a Polynomial**

We now turn our attention to polynomials.

> **DEFINITION**
>
> A **polynomial** is a monomial or the sum of monomials.

We give special names to certain polynomials. For example, a polynomial with exactly one term is a monomial; a polynomial that has two different monomials is called a **binomial;** and a polynomial that contains three different monomials is called a **trinomial.** So

Work Smart
The prefix "bi" means "two" as in bicycle. The prefix "tri" means "three" as in tricycle. The prefix "poly" means "many."

$-14x$ is a polynomial	but more specifically	$-14x$ is a monomial
$2x^3 - 5x$ is a polynomial	but more specifically	$2x^3 - 5x$ is a binomial
$-x^3 - 4x + 11$ is a polynomial	but more specifically	$-x^3 - 4x + 11$ is a trinomial
$3x^2 + 6xy - 2y^2$ is a polynomial	but more specifically	$3x^2 + 6xy - 2y^2$ is a trinomial

We use the term *monomial* to describe a polynomial with a *single* term, and the term *polynomial* to describe the sum of two or more monomials.

A polynomial is in **standard form** if it is written with the terms in descending order according to degree. The **degree of a polynomial** is the highest degree of all the terms of the polynomial. Remember, the degree of a nonzero constant is 0 and the number 0 has no degree.

EXAMPLE 4 Examples of Polynomials

POLYNOMIAL	DEGREE
(a) $7x^3 - 2x^2 + 6x + 4$	3
(b) $3 - 8x + x^2 = x^2 - 8x + 3$	2
(c) $-7x^4 + 24$	4
(d) $x^3y^4 - 3x^3y^2 + 2x^3y$	7
(e) $p^2q - 8p^3q^2 + 3 = -8p^3q^2 + p^2q + 3$	5
(f) 6	0
(g) 0	No Degree

Although we have been using x to represent the variable, other letters may also be used.

$$7t^4 + 14t^2 + 8 \text{ is a polynomial (in } t\text{) of degree 4}$$
$$-5z^2 + 3z - 6 \text{ is a polynomial (in } z\text{) of degree 2}$$
$$8y^3 - y^2 + 2y - 12 \text{ is a polynomial (in } y\text{) of degree 3}$$

Remember, polynomials consist of monomials. So, if any term in an algebraic expression is not a monomial, then the algebraic expression is not a polynomial.

EXAMPLE 5 Algebraic Expressions That Are Not Polynomials

(a) $4x^{-2} - 5x + 1$ is not a polynomial because the exponent on the first term, -2, is not a whole number.

(b) $\dfrac{4}{x^3}$ is not a polynomial because there is a variable expression in the denominator of the fraction.

(c) $4z^2 + 9z - 3z^{1/2}$ is not a polynomial because the exponent on the third term, $-3z^{1/2}$, is not a whole number.

QUICK ✓ *Determine whether the algebraic expression is a polynomial. For those that are polynomials, determine the degree.*

9. $-4x^3 + 2x^2 - 5x + 3$ **10.** $2m^{-1} + 7$ **11.** $\dfrac{-1}{x^2 + 1}$

12. $5n^3 - \dfrac{3}{2}n^2 + 7n^5 - 6$ **13.** $5p^3q - 8pq^2 + pq$

③ **Simplify Polynomials by Combining Like Terms**

In Section 1.7, we learned how to combine like terms. To simplify a polynomial means to perform all indicated operations and combine like terms. One operation we perform on polynomials is addition. To add polynomials, combine the like terms of the polynomials.

EXAMPLE 6 Simplifying Polynomials: Addition

Find the sum: $(-5x^3 + 6x^2 + 2x - 7) + (3x^3 + 4x + 1)$

Solution

We can find the sum using either horizontal addition or vertical addition.

Horizontal Addition:

The idea here is to combine like terms.

$$(-5x^3 + 6x^2 + 2x - 7) + (3x^3 + 4x + 1) = -5x^3 + 6x^2 + 2x - 7 + 3x^3 + 4x + 1$$
$$\text{Rearrange terms:} \quad = -5x^3 + 3x^3 + 6x^2 + 2x + 4x - 7 + 1$$
$$\text{Use the Distributive Property:} \quad = (-5 + 3)x^3 + 6x^2 + (2 + 4)x + (-7 + 1)$$
$$\text{Simplify:} \quad = -2x^3 + 6x^2 + 6x - 6$$

Work Smart

Remember, like terms have the same variable and the same exponent on the variable.

Vertical Addition:

The idea here is to line up like terms in each polynomial vertically and then add the coefficients.

$$
\begin{array}{r}
-5x^3 + 6x^2 + 2x - 7 \\
3x^3 \qquad\quad + 4x + 1 \\
\hline
-2x^3 + 6x^2 + 6x - 6
\end{array}
$$

QUICK ✓ *Add the polynomials using either horizontal or vertical addition.*

14. $(9x^2 - x + 5) + (3x^2 + 4x - 2)$

15. $(4z^4 - 2z^3 + z - 5) + (-2z^4 + 6z^3 - z^2 + 4)$

Adding polynomials in two variables is handled the same way as adding polynomials in a single variable, that is, by adding like terms.

EXAMPLE 7 Simplifying Polynomials in Two Variables: Addition

Find the sum: $(6a^2b - 4ab + 11ab^2) + (a^2b + 9ab - 3ab^2)$

Solution

To save space, we only present the horizontal format.

$$
\begin{aligned}
(6a^2b - 4ab + 11ab^2) + (a^2b + 9ab - 3ab^2) &= 6a^2b - 4ab + 11ab^2 + a^2b + 9ab - 3ab^2 \\
\text{Rearrange terms:}\quad &= 6a^2b + a^2b + 11ab^2 - 3ab^2 - 4ab + 9ab \\
\text{Use the Distributive Property:}\quad &= (6 + 1)a^2b + (11 - 3)ab^2 + (-4 + 9)ab \\
\text{Simplify:}\quad &= 7a^2b + 8ab^2 + 5ab
\end{aligned}
$$

QUICK ✓ *Simplify by adding the polynomials.*

16. $(7x^2y + x^2y^2 - 5xy^2) + (-2x^2y + 5x^2y^2 + 4xy^2)$

We can subtract polynomials using either the horizontal or vertical approach as well. However, the first step in subtracting polynomials requires that we remember the method for subtracting two real numbers. Remember,

$$
a - b = a + (-b)
$$

So, to subtract one polynomial from another, we add the opposite of each term in the polynomial following the subtraction sign and then combine like terms.

Work Smart

Taking the opposite of each term following the subtraction symbol is the same as multiplying each term of the polynomial by -1.

EXAMPLE 8 Simplifying Polynomials: Subtraction

Find the difference: $(6z^3 + 2z^2 - 5) - (-3z^3 + 9z^2 - z + 1)$

Solution

Horizontal Subtraction:

First, we write the problem as an addition problem and determine the opposite of each term following the subtraction sign.

$$
\begin{aligned}
(6z^3 + 2z^2 - 5) - (-3z^3 + 9z^2 - z + 1) &= (6z^3 + 2z^2 - 5) + (3z^3 - 9z^2 + z - 1) \\
\text{Remove parentheses:}\quad &= 6z^3 + 2z^2 - 5 + 3z^3 - 9z^2 + z - 1 \\
\text{Rearrange terms:}\quad &= 6z^3 + 3z^3 + 2z^2 - 9z^2 + z - 5 - 1 \\
\text{Combine like terms:}\quad &= 9z^3 - 7z^2 + z - 6
\end{aligned}
$$

Vertical Subtraction:

We line up like terms vertically,

$$6z^3 + 2z^2 \qquad - 5$$
$$\underline{-(-3z^3 + 9z^2 - z + 1)}$$

then change the sign of each coefficient of the second polynomial, and add.

$$6z^3 + 2z^2 \qquad - 5$$
$$\underline{+3z^3 - 9z^2 + z - 1}$$
$$9z^3 - 7z^2 + z - 6$$

QUICK ✓ *Simplify by subtracting the polynomials.*

17. $(7x^3 - 3x^2 + 2x + 8) - (2x^3 + 12x^2 - x + 1)$

18. $(6y^3 - 3y^2 + 2y + 4) - (-2y^3 + 5y + 10)$

EXAMPLE 9 **Simplifying Polynomials in Two Variables: Subtraction**

Find the difference: $(2p^2q + pq + 5pq^2) - (3p^2q - 6pq + 9pq^2)$

Solution

$$(2p^2q + pq + 5pq^2) - (3p^2q - 6pq + 9pq^2) = (2p^2q + pq + 5pq^2) + (-3p^2q + 6pq - 9pq^2)$$

Remove parentheses: $= 2p^2q + pq + 5pq^2 - 3p^2q + 6pq - 9pq^2$

Rearrange terms: $= 2p^2q - 3p^2q + 5pq^2 - 9pq^2 + pq + 6pq$

Simplify: $= -p^2q - 4pq^2 + 7pq$

QUICK ✓ *Subtract the polynomials.*

19. $(9x^2y + 6x^2y^2 - 3xy^2) - (-2x^2y + 4x^2y^2 + 6xy^2)$

EXAMPLE 10 **Simplifying Polynomials**

Perform the indicated operations: $(3x^2 - 2xy + y^2) - (7x^2 - y^2) + (3xy - 5y^2)$

Solution

$$(3x^2 - 2xy + y^2) - (7x^2 - y^2) + (3xy - 5y^2) = 3x^2 - 2xy + y^2 - 7x^2 + y^2 + 3xy - 5y^2$$

Rearrange terms: $= 3x^2 - 7x^2 - 2xy + 3xy + y^2 + y^2 - 5y^2$

Simplify: $= -4x^2 + xy - 3y^2$

QUICK ✓ *Perform the indicated operations.*

20. $(3a^2 - 5ab + 3b^2) + (4a^2 - 7b^2) - (8b^2 - ab)$

---

4 **Evaluate Polynomials**

To **evaluate** a polynomial, we substitute the given number for the value of the variable and simplify, just as we did in Section 1.7.

EXAMPLE 11 **Evaluating a Polynomial**

Evaluate the polynomial $2x^3 - 5x^2 + x - 3$ for

    (a) $x = 2$ **(b)** $x = -3$

Work Smart

In each example apply the exponent before you multiply.

Solution

(a) $2x^3 - 5x^2 + x - 3 = 2(2)^3 - 5(2)^2 + 2 - 3$
$$= 16 - 20 + 2 - 3$$
$$= -5$$

(b) $2x^3 - 5x^2 + x - 3 = 2(-3)^3 - 5(-3)^2 + (-3) - 3$
$$= -54 - 45 - 3 - 3$$
$$= -105$$

QUICK ✓

21. Evaluate the polynomial $-2x^3 + 7x + 1$ for

   (a) $x = 0$ (b) $x = 5$ (c) $x = -4$

EXAMPLE 12 Evaluating a Polynomial in Two Variables

Evaluate the polynomial $2a^2b + 3a^2b^2 - 4ab$ for $a = -3$ and $b = -1$.

Solution

Let $a = -3$ and $b = -1$ in $2a^2b + 3a^2b^2 - 4ab$.

$$2a^2b + 3a^2b^2 - 4ab = 2(-3)^2(-1) + 3(-3)^2(-1)^2 - 4(-3)(-1)$$
$$= 2(9)(-1) + 3(9)(1) - 4(-3)(-1)$$
$$= -18 + 27 - 12$$
$$= -3$$

QUICK ✓

22. Evaluate the polynomial $-2m^2n + 3mn - n^2$ for

   (a) $m = -2$ and $n = -4$ (b) $m = 1$ and $n = 5$

EXAMPLE 13 How Much Revenue?

The monthly revenue (in dollars) from selling x clocks is given by the polynomial $-0.3x^2 + 90x$. Evaluate the polynomial for $x = 200$ and describe the result in practical terms.

Solution

The variable x represents the number of clocks sold in a month. When we evaluate the polynomial $-0.3x^2 + 90x$ for $x = 200$, we are finding the revenue (in dollars) when 200 clocks are sold in a month. To find the revenue we replace x by 200 in the expression $-0.3x^2 + 90x$.

$$-0.3x^2 + 90x = -0.3(200)^2 + 90(200)$$
$$= -0.3(40,000) + 18,000$$
$$= -12,000 + 18,000$$
$$= 6000$$

According to the model, the revenue from selling 200 clocks in a month is $6000.

QUICK ✓ *Evaluate the polynomial for the given value.*

23. The polynomial $40x - 0.2x^2$ represents the monthly revenue (in dollars) from selling x wristwatches. Find the monthly revenue for selling 75 wristwatches by evaluating the polynomial $40x - 0.2x^2$ for $x = 75$.

5.1 Exercises

Concepts and Vocabulary

In Problems 1–3, fill in the blanks.

1. A _____ in one variable is the product of a number and a variable raised to a whole number power.

2. The implied coefficient of a term such as x^2 or z is _____.

3. The degree of the polynomial $3x^5 - 7x + 1$ is _____.

In Problems 4–6, answer True or False to each statement.

4. $3x + 4 + 6x$ is an example of a trinomial.

5. The degree of the polynomial $3ab^2 - 4a^2b^3 + \dfrac{6}{5}a^2b^2$ is 5.

6. The degree of the polynomial $-7y^3 + 2y^2 - \dfrac{1}{2}$ is 5.

7. When adding polynomials you may use either a horizontal or vertical format. Explain which format you prefer and what you view as its advantages.

8. Use the two polynomials $3x - 5$ and $2x + 3$ to make up three different problems. In the directions to the first problem use the word "evaluate"; in the second problem use the direction "solve"; and in the third use "simplify."

Building Skills

In Problems 9–20, determine whether the given expression is a monomial (Yes or No). For those that are monomials, state the coefficient and degree.

9. $\dfrac{1}{2}y^3$ 10. $-3x^2$ 11. $\dfrac{x^2}{7}$ 12. $4m^{101}$

13. z^{-6} 14. $\dfrac{1}{y^7}$ 15. $12mn^4$ 16. $-x^6y$

17. $\dfrac{3}{n^2}$ 18. y 19. 4 20. $\dfrac{2}{3}$

In Problems 21–24, write the polynomial in standard form.

21. $4x - 5 + x^3$ 22. $6 - 2m - m^2$

23. $-3y + 2 - y^4 + y^2$ 24. $2t^5 + 6t - t^3$

In Problems 25–44, determine whether the algebraic expression is a polynomial (Yes or No). If it is a polynomial, write the polynomial in standard form, determine its degree, and state if it is a monomial, binomial or trinomial. If it is a polynomial with more than 3 terms, identify the expression as polynomial.

25. $6x^2 - 10$ 26. $4x + 1$ 27. $\dfrac{-20}{n}$ 28. $\dfrac{1}{x}$

29. $3y^{1/3} + 2$ 30. $8m - 4m^{1/2}$ 31. $\dfrac{1}{8}$ 32. 32

33. $3t^2 - \dfrac{1}{2}t^4 + 6t$ 34. $4x^7 - 3x^3 - 1$ 35. $7x^{-1} + 4$ 36. $4y^{-2} + 6y - 1$

37. $5z^3 - 10z^2 + z + 12$ 38. $p^5 - 3p^4 + 7p + 8$

39. $3x^2y^2 + 2xy^4 + 4$ 40. $4mn^3 - 2m^2n^3 + mn^8$

41. $4x^3 - 3y^5$ 42. $-6s^{10} + 4t^3$

43. $4pqr + 2p^2q + 3pq^{1/4}$ 44. $-2xyz^2 + 7x^3z - 8y^{1/2}z$

In Problems 45–58, add the polynomials. Express your answer in standard form.

45. $(4x - 3) + (3x - 7)$ **46.** $(-13z + 4) + (9z - 10)$

47. $(-4m^2 + 2m - 1) + (2m^2 - 2m + 6)$ **48.** $(x^2 - 2) + (6x^2 - x - 1)$

49. $(p - p^3 + 2) + (6 - 2p^2 + p^3)$

50. $(4r^4 + 3r - 1) + (2r^4 - 7r^3 + r^2 - 10r)$

51. $(2y - 10) + (-3y^2 - 4y + 6)$ **52.** $(3 - 12w^2) + (2w^2 - 5 + 6w)$

53. $\left(\dfrac{1}{2}p^2 - \dfrac{2}{3}p + 2\right) + \left(\dfrac{3}{4}p^2 + \dfrac{5}{6}p - 5\right)$

54. $\left(\dfrac{3}{8}b^2 - \dfrac{3}{5}b + 1\right) + \left(\dfrac{5}{6}b^2 + \dfrac{2}{15}b - 1\right)$

55. $(5m^2 - 6mn + 2n^2) + (m^2 + 2mn - 3n^2)$

56. $(4a^2 + ab - 9b^2) + (-6a^2 - 4ab + b^2)$

57. $\begin{array}{r} 4n^2 - 2n + 1 \\ +(-6n + 4) \\ \hline \end{array}$ **58.** $\begin{array}{r} 8x^3 \qquad + 2x + 2 \\ +(-3x^2 - 4x - 2) \\ \hline \end{array}$

In Problems 59–72, subtract the polynomials. Express your answer in standard form.

59. $(3x - 10) - (4x + 6)$

60. $(-7t + 3) - (-2t - 4)$

61. $(12x^2 - 2x - 4) - (-2x^2 + x + 1)$

62. $(3x^2 + x - 3) - (x^2 - 2x + 4)$

63. $(y^3 - 2y + 1) - (-3y^3 + y + 5)$

64. $(m^4 - 3m^2 + 5) - (3m^4 - 5m^2 - 2)$

65. $(3y^3 - 2y) - (2y + y^2 + y^3)$

66. $(2x - 4x^3) - (-3 - 2x + x^3)$

67. $\left(\dfrac{5}{3}q^2 - \dfrac{5}{2}q + 4\right) - \left(\dfrac{1}{9}q^2 + \dfrac{3}{8}q + 2\right)$

68. $\left(\dfrac{7}{4}x^2 - \dfrac{5}{8}x - 1\right) - \left(\dfrac{7}{6}x^2 + \dfrac{5}{12}x + 5\right)$

69. $(-4m^2n^2 - 2mn + 3) - (4m^2n^2 + 2mn + 10)$

70. $(4m^2n - 2mn - 4) - (10m^2n - 6mn - 3)$

71. $\begin{array}{r} 6x - 3 \\ -(10x + 2) \\ \hline \end{array}$ **72.** $\begin{array}{r} 6n^2 - 2n - 3 \\ -(4n + 7) \\ \hline \end{array}$

In Problems 73–80, evaluate the polynomial for each of the given value(s).

73. $2x^2 - x + 3$ **74.** $-x^2 + 10$ **75.** $7 - x^2$ **76.** $2 + \dfrac{1}{2}n^2$

(a) $x = 0$ (a) $x = 0$ (a) $x = 3$

(b) $x = 5$ (b) $x = -1$ (b) $x = -\dfrac{5}{2}$ (a) $n = 4$

(c) $x = -2$ (c) $x = 1$ (b) $n = 0.5$

                                       (c) $x = -1.5$ (c) $n = -\dfrac{1}{4}$

77. $-x^2y + 2xy^2 - 3$
 for $x = 2$ and $y = -3$

78. $-2ab^2 - 2a^2b - b^3$
 for $a = 1$ and $b = -2$

79. $st + 2s^2t + 3st^2 - t^4$
 for $s = -2$ and $t = 4$

80. $m^2n^2 - mn^2 + 3m^2 - 2$
 for $m = \dfrac{1}{2}$ and $n = -1$

Mixed Practice

In Problems 81–96, perform the indicated operations. Express your answer in standard form.

81. $(7t - 3) - 4t$

82. $6x^2 - (18x^2 + 2)$

83. $(5x^2 + x - 4) + (-2x^2 - 4x + 1)$

84. $(4m^2 - m + 6) + (3m^2 - 4m - 10)$

85. $(2xy^2 - 3) + (7xy^2 + 4)$

86. $(-14xy + 3) - (-xy + 10)$

87. $(4 + 8y - 2y^2) - (3 - 7y - y^2)$

88. $(9 + 2z - 6z^2) - (-2 + z + 5z^2)$

89. $\left(\dfrac{5}{6}q^2 - \dfrac{1}{3}\right) + \left(\dfrac{3}{2}q^2 + 2\right)$

90. $\left(\dfrac{7}{10}t^2 - \dfrac{5}{12}t\right) + \left(\dfrac{3}{15}t^2 + \dfrac{3}{20}t - 3\right)$

91. $14d^2 - (2d - 10) - (d^2 - 3d)$

92. $3x - (5x + 1) - 4$

93. $(4a^2 - 1) + (a^2 + 5a + 2) - (-a^2 + 4)$

94. $(2b^2 + 3b - 5) - (b^2 - 4b + 1) + (b^2 + 1)$

95. $(x^2 - 2xy - y^2) - (3x^2 + xy - y^2) + (xy + y^2)$

96. $(2st^4 - 3s^2t^2 + t^4) + (7s^2t^2 - 3t^4 + 8st^4)$

97. Find the sum of $3x + 10$ and $-8x + 2$.

98. Find the sum of $4x^2 - 2x - 3$ and $-5x^2 - 2x + 7$.

99. Find the difference of $-x^2 + 2x + 3$ and $-4x^2 - 2x + 6$.

100. Find the difference of $4x - 3$ and $8x - 7$.

101. Subtract $14x^2 - 2x + 3$ from $2x - 10$.

102. Subtract $-4n - 8$ from $3n^2 + 8n - 9$.

103. What polynomial should be added to $3x - 5$ so that the sum is zero?

104. What polynomial should be added to $2a + 7b$ so that the sum is $a - b$?

Applying the Concepts

105. Height of a Ball The height above ground (in feet) of a ball dropped from the top of a 30-foot tall building after t seconds is given by the polynomial $-16t^2 + 30$. What is the height of the ball after $\dfrac{1}{2}$ second?

106. Height of a Ball The height above ground (in feet) of a ball tossed upward after t seconds is given by the polynomial $-16t^2 + 32t + 100$. What is the height of the ball after 2.5 seconds?

107. Income The bar graph shown represents the average per-capita income (in dollars) for residents of the United States by age in 2000. The polynomial $-50.27a^2 + 4426.16a - 40803.75$ can be used to approximate the average income I, where a is the age of the individual.

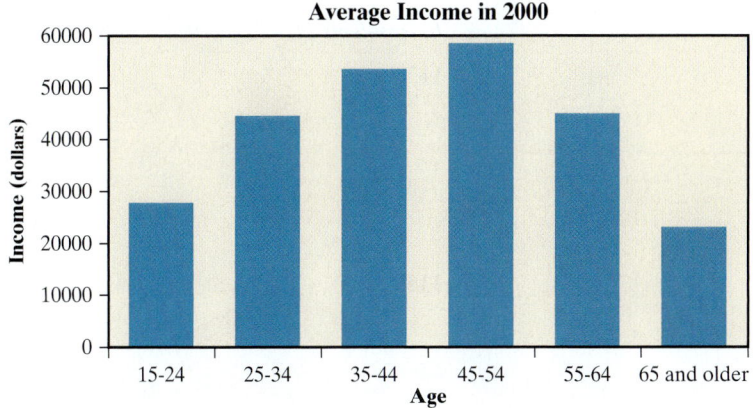

Average Income in 2000

(a) Use the polynomial to estimate the average income of an 18-year-old in 2000.
(b) Use the polynomial to estimate the average income of a 60-year-old in 2000.

108. Selling Music Videos The polynomial $-0.425t^2 + 4.57t + 8.31$ describes the number (in millions) of music videos sold since 1993, where t represents the number of years since 1993. Use the polynomial to estimate the number of music videos sold in 2005.

109. Manufacturing Calculators The revenue (in dollars) from manufacturing and selling x calculators in a day is given by the polynomial $-2x^2 + 120x$. The cost of manufacturing and selling x calculators in a day is given by the polynomial $0.125x^2 + 15x$.

(a) In business, profit is equal to revenue minus costs. Write a polynomial that represents the profit from manufacturing and selling x calculators in a day.
(b) Find the profit if 20 calculators are produced and sold each day.

110. Manufacturing DVDs The revenue (in dollars) from manufacturing and selling x DVDs each day is given by the polynomial $-0.001x^2 + 6x$. The cost of manufacturing and selling x DVDs each day is given by the polynomial $0.8x + 3000$.

(a) Write a polynomial that expresses the profit from manufacturing and selling x DVDs each day.
(b) Find the profit if 1600 DVDs are produced and sold in a day.

111. Lawn Service Marissa started a lawn-mowing business in Tampa, Florida, in which she has x lawns on her route. Her costs are $5 per lawn for fertilizer and $10 per lawn for labor.

(a) If she charges $25 per lawn, write a polynomial that represents her weekly profit.
(b) If Marissa mows 50 lawns in a week, what is her weekly profit?
(c) If Marissa mows 50 lawns per week, what is her annual profit?

112. Skateboard Business Kyle owns a skateboard shop. Kyle charges $40 for each skateboard he sells. The cost for manufacturing each skateboard is $12. In addition, Kyle has weekly costs of $504 regardless of how many skateboards he manufactures.

(a) If Kyle manufactures and sells x skateboards in a week, write a polynomial that represents his weekly profit.

(b) What is Kyle's profit if he sells 45 skateboards per week?

(c) How many skateboards does Kyle need to sell in order to break even (profit = 0)?

△ *In Problems 113–116, write the polynomial that represents the perimeter of each figure.*

113.

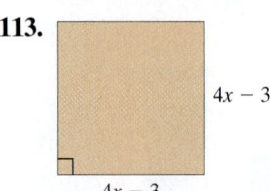

4x − 3

4x − 3

114.

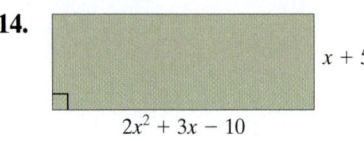

x + 5

2x² + 3x − 10

115.

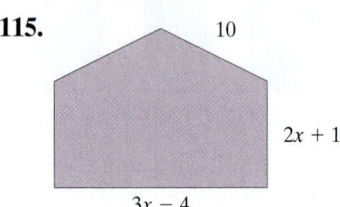

10

2x + 1

3x − 4

116.

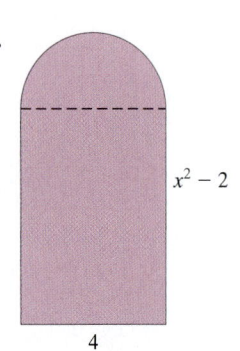

x² − 2

4

In Problems 117 and 118, write the polynomial that represents the unknown length.

117.

$$3x - 10$$

$$x - 5 \qquad ?$$

118.

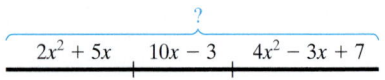

?

2x² + 5x   10x − 3   4x² − 3x + 7

Extending the Concepts

In Problems 119–124, simplify each of the following.

119. $(2x^2 - 3x + 1) + (x^2 + 9) - (4x^2 - 2x - 5)$

120. $(y^2 + 6y - 2) + (4y^2 - 9) - (2y^2 - 7y - 10)$

121. $(p^2 + 25) - (3p^2 - p + 4) - (-4p^2 - 9p + 5)$

122. $(n^2 + 4n - 5) - (2n^2 - 3n + 1) + (-n^2 - 7n + 6)$

123. $(4y - 3) + (y^3 - 8) - (2y^2 - 7y + 3)$

124. $(y^3 - 1) + (y^2 - 9) - (y^3 + 2y^2 - 7y + 3)$

5.2 Multiplying Monomials: The Product and Power Rules

OBJECTIVES

1. Simplify Exponential Expressions Using the Product Rule

2. Simplify Exponential Expressions Using the Power Rule

3. Simplify Exponential Expressions Containing Products

4. Multiply a Monomial by a Monomial

Preparing for...Answers **1.** 64
2. $12 - 15x$

Preparing for Multiplying Monomials: The Product and Power Rules

Before getting started, take the following readiness quiz. If you get a problem wrong, go back to the section cited and review the material.

1. Evaluate: 4^3 [Section 1.6, pp. 49–50]

2. Use the Distributive Property to simplify: $3(4 - 5x)$ [Section 1.7, p. 61]

In Section 1.6, we gave a definition for raising a real number to a natural number exponent. Recall that if a is a real number and n is a natural number, then the symbol a^n means that we should use a as a factor n times:

$$a^n = \underbrace{a \cdot a \cdot \ldots \cdot a}_{n \text{ factors}}$$

exponent

base

Work Smart

The natural numbers are 1, 2, 3, ...

For example,

$$4^3 = \underbrace{4 \cdot 4 \cdot 4}_{\text{3 factors}}$$

or

$$y^4 = \underbrace{y \cdot y \cdot y \cdot y}_{\text{4 factors}}$$

In the notation a^n we call a the base and n the power or exponent. We read a^n as "a raised to the power n" or "a raised to the nth power." We usually read a^2 as "a squared" and a^3 as "a cubed." The expression 4^3 is called **exponential form** and the expression $4 \cdot 4 \cdot 4$ is called **expanded form.**

① Simplify Exponential Expressions Using the Product Rule

We can discover several general rules for simplifying expressions with natural number exponents. The first rule that we introduce is used when multiplying two exponential expressions that have the same base. Consider the following:

$$\underset{\substack{\uparrow \ \uparrow \\ \text{Same base} \;\; \text{2 factors 4 factors}}}{x^2 \cdot x^4} = \underbrace{(x \cdot x)(x \cdot x \cdot x \cdot x)}_{} = \underbrace{x \cdot x \cdot x \cdot x \cdot x \cdot x}_{\text{6 factors}} = \underset{\substack{\uparrow \\ \text{Same base}}}{x^6}$$

Sum of powers 2 and 4

The following rule generalizes the result shown above.

In Words

The product of two exponential expressions with the same base is that base with the sum of the exponents as the power.

PRODUCT RULE FOR EXPONENTS

If a is a real number, and m and n are natural numbers, then

$$a^m \cdot a^n = a^{m+n}$$

EXAMPLE 1 Using the Product Rule to Evaluate Numerical Expressions with Exponents

Evaluate each expression:

(a) $2^2 \cdot 2^4$

(b) $(-3)^2(-3)$

Solution

(a) $2^2 \cdot 2^4 = 2^{2+4}$
$= 2^6$
$= 64$

(b) $(-3)^2(-3) = (-3)^{2+1}$
$= (-3)^3$
$= -27$

EXAMPLE 2 Using the Product Rule to Simplify Algebraic Expressions with Exponents

Simplify the following expressions:

(a) $t^4 \cdot t^7$

(b) $a^6 \cdot a \cdot a^4$

Solution

(a) $t^4 \cdot t^7 = t^{4+7}$
$= t^{11}$

(b) $a^6 \cdot a \cdot a^4 = a^6 \cdot a^1 \cdot a^4$
$= a^{6+1+4}$
$= a^{11}$

EXAMPLE 3 **Using the Product Rule to Simplify Algebraic Expressions with Exponents**

Simplify the expression: $m^3 \cdot m^5 \cdot n^9$

Solution

To use the Product Rule for Exponents, the bases must be the same.

$$m^3 \cdot m^5 \cdot n^9 = m^{3+5} \cdot n^9$$
$$= m^8 n^9$$

An expression such as $a^5 \cdot b^2$ is in simplest form because the bases are different.

QUICK ✓ *Evaluate or simplify each expression, if possible.*

1. $3^2 \cdot 3$ **2.** $(-5)^2(-5)^3$ **3.** $c^6 \cdot c^2$ **4.** $y^3 \cdot y \cdot y^5$ **5.** $a^4 \cdot a^5 \cdot b^6$

② Simplify Exponential Expressions Using the Power Rule

Another law of exponents applies when an exponential expression containing a power is itself raised to a power.

$$(3^2)^4 = 3^2 \cdot 3^2 \cdot 3^2 \cdot 3^2 = (3 \cdot 3) \cdot (3 \cdot 3) \cdot (3 \cdot 3) \cdot (3 \cdot 3) = 3^8$$

4 factors   2 factors   2 factors   2 factors   2 factors

$2 \cdot 4 = 8$ factors

We have the following result:

In Words

If an exponential expression contains a power raised to a power, keep the base and multiply the powers.

> **POWER RULE FOR EXPONENTS**
>
> If a is a real number and m, n are natural numbers, then
>
> $$(a^m)^n = a^{m \cdot n}$$

EXAMPLE 4 **Using the Power Rule to Simplify Exponential Expressions**

Simplify each expression. Write the answer in exponential form.

   (a) $(2^3)^2$ **(b)** $(z^4)^2$

Solution

   (a) $(2^3)^2 = 2^{3 \cdot 2}$ **(b)** $(z^4)^2 = z^{4 \cdot 2}$
                   $= 2^6$ $= z^8$

EXAMPLE 5 **Using the Power Rule to Simplify Exponential Expressions**

Simplify each expression. Write the answer in exponential form.

   (a) $[(-4)^2]^3$ **(b)** $[(-n)^3]^5$

Solution

   (a) $[(-4)^2]^3 = (-4)^{2 \cdot 3}$ **(b)** $[(-n)^3]^5 = (-n)^{3 \cdot 5}$
                    $= (-4)^6$ $= (-n)^{15}$

QUICK ✓ *Simplify each expression. Write the answer in exponential form.*

6. $(2^2)^4$ **7.** $[(-3)^3]^2$ **8.** $(b^2)^5$ **9.** $(z^4)^5$

③ Simplify Exponential Expressions Containing Products

The next law of exponents deals with raising a product to a power. Consider the following:

$$(x \cdot y)^3 = (x \cdot y) \cdot (x \cdot y) \cdot (x \cdot y) = (x \cdot x \cdot x) \cdot (y \cdot y \cdot y) = x^3 \cdot y^3$$

The following rule generalizes this result.

In Words
When we raise a product to a power, we raise each factor to the power. (Remember, each of the numbers or variables in a multiplication problem is a factor.)

> **PRODUCT TO A POWER RULE FOR EXPONENTS**
> If a, b are real numbers and n is a natural number, then
> $$(a \cdot b)^n = a^n \cdot b^n$$

> **EXAMPLE 6** **Using the Product to a Power Rule to Simplify Exponential Expressions**

Simplify each expression:

     (a) $(3b^2)^4$ **(b)** $(-2a^2b)^3$

Solution

Each expression contains the product of factors, so we use the Product to a Power Rule.

(a) $(3b^2)^4 = (3)^4(b^2)^4$
Evaluate 3^4: $= 81b^8$

(b) $(-2a^2b)^3 = (-2)^3(a^2)^3(b)^3$
Evaluate $(-2)^3$: $= -8a^{2 \cdot 3}b^3$
                     $= -8a^6b^3$

QUICK ✓ *Simplify each expression.*

10. $(2n)^3$ **11.** $(6y)^2$ **12.** $(-5x^4)^3$ **13.** $(-7a^3b)^2$

Let's summarize the three rules that we use when we find products of monomials.

Work Smart: Study Skills
Are you having trouble remembering the rules for multiplying exponents? Try making flash cards with the property on the front side, and an example or two on the back side. Also, write in words what each rule means and when it should be applied. This is a good study strategy that may apply to other topics.

> **RULES FOR MULTIPLYING MONOMIALS**
> - Product Rule for Exponents
>   If a is a real number and m and n are natural numbers, then $a^m \cdot a^n = a^{m+n}$.
> - Power Rule for Exponents
>   If a is a real number and m and n are natural numbers, then $(a^m)^n = a^{m \cdot n}$.
> - Product to a Power Rule for Exponents
>   If a and b are real numbers and n is a natural number, then $(a \cdot b)^n = a^n \cdot b^n$.

(4) Multiply a Monomial by a Monomial

To multiply two monomials, we multiply the coefficients of the monomials and use the Product Rule for Exponents to multiply the variable expressions.

EXAMPLE 7 **Multiplying a Monomial by a Monomial: One Variable**

Multiply and simplify:

(a) $(5x^2)(6x^4)$ 　　　　　　　　　　　**(b)** $(2p^3)(-5p^2)$

Solution

In each problem, we use the Commutative Property to rearrange factors

(a) $(5x^2)(6x^4) = 5 \cdot 6 \cdot x^2 \cdot x^4$ 　　　**(b)** $2p^3(-5p^2) = 2 \cdot (-5) \cdot p^3 \cdot p^2$
$$= 30x^{2+4}$$ 　　　　　　　　　　　$$= -10p^{3+2}$$
$$= 30x^6$$ 　　　　　　　　　　　　$$= -10p^5$$

EXAMPLE 8 **Multiplying a Monomial by a Monomial: Two Variables**

Multiply and simplify: $(-3x^2y)(-7x^5y^3)$

Solution

$$(-3x^2y)(-7x^5y^3) = -3 \cdot (-7) \cdot x^2 \cdot x^5 \cdot y \cdot y^3$$
$$= 21 \cdot x^{2+5} \cdot y^{1+3}$$
$$= 21x^7y^4$$

QUICK  *Multiply and simplify.*

14. $(2a^6)(4a^5)$ 　　　　　　　　　**15.** $(3p^2)(-4p^5)$

16. $(3m^2n^4)(-6mn^5)$ 　　　　　　**17.** $\left(\dfrac{4}{3}xy^2\right)\left(\dfrac{1}{2}x^2y\right)$

5.2 Exercises

For Extra Help:

Student Solutions Manual　CD Video　PH Math/Tutor Center　MathXL Tutorials on CD　MathXL®　MyMathLab

Concepts and Vocabulary

In Problems 1–3, fill in the blanks.

1. $(a^m)^n = $ _____.

2. In the exponential expression $10x^2$, x is called the _____ and 2 is called the _____ or _____.

3. The expression $y \cdot y \cdot y \cdot y$ is written in _____ _____.

In Problems 4 and 5, answer True or False to each statement.

4. $3^4 \cdot 3^2 = 9^6$

5. $(ab^3)^2 = ab^6$

6. Provide a justification for the product rule for exponential expressions.

7. Provide a justification for the power rule for exponential expressions.

8. Provide a justification for the product to a power rule for exponential expressions.

Building Skills

In Problems 9–44, simplify each expression.

9. $4^2 \cdot 4^3$

10. $3 \cdot 3^3$

11. $(-2)^3(-2)^4$

12. $(-0.3)^2(-0.3)^3$

13. $m^4 \cdot m^5$

14. $a^3 \cdot a^7$

15. $b^9 \cdot b^{11}$

16. $z^8 \cdot z^{23}$

17. $x^7 \cdot x$

18. $y^{13} \cdot y$

19. $p \cdot p^2 \cdot p^6$

20. $b^5 \cdot b \cdot b^7$

21. $(-n)^3(-n)^4$

22. $(-z)(-z)^5$

23. $a^5 \cdot a^{10} \cdot a^2$

24. $y \cdot y^2 \cdot y^5$

25. $(2^3)^2$

26. $(3^2)^2$

27. $(z^2)^5$

28. $(x^3)^4$

29. $[(-m)^9]^2$

30. $[(-n)^7]^3$

31. $(m^2)^7$

32. $(k^8)^3$

33. $[(-b)^4]^5$

34. $[(-a)^6]^3$

35. $(3 \cdot 5)^2$

36. $(8 \cdot 4)^2$

37. $(3x^2)^3$

38. $(4y^3)^2$

39. $(-3p^7q^2)^4$

40. $(-2m^2n^3)^4$

41. $(-2m^2n)^3$

42. $(-4ab^2)^3$

43. $(-5xy^2z^3)^2$

44. $(-3a^6bc^4)^4$

In Problems 45–56, multiply the monomials.

45. $(4x^2)(3x^3)$

46. $(5y^6)(3y^2)$

47. $(10a^3)(-4a^7)$

48. $(7b^5)(-2b^4)$

49. $(-m^3)(7m)$

50. $(-n^6)(5n)$

51. $\left(\frac{4}{5}x^4\right)\left(\frac{15}{2}x^3\right)$

52. $\left(\frac{3}{8}y^5\right)\left(\frac{4}{9}y^6\right)$

53. $(2x^2y^3)(3x^4y)$

54. $(6a^2b^3)(2a^5b)$

55. $\left(\frac{1}{4}mn^3\right)(-20mn)$

56. $\left(\frac{2}{3}s^2t^3\right)(-21st)$

Mixed Practice

In Problems 57–72, simplify each expression completely.

57. $x^2 \cdot 4x$

58. $z^3 \cdot 5z$

59. $(b^3)^2$

60. $(m^2)^4$

61. $(-3b)^4$

62. $(-4a)^2$

63. $(-6p)^2\left(\frac{1}{4}p^3\right)$

64. $(-8w)^2\left(\frac{3}{16}w^5\right)$

65. $(5x)^2(x^2)^3$

66. $(4y)^2(y^3)^5$

67. $(x^2y)^3(-2xy^3)$

68. $(s^2t^3)^2(3st)$

69. $(-3x)^2(2x^4)^3$

70. $(-2p^3)^2(3p)^3$

71. $\left(\frac{4}{5}q\right)^2(-5q)^2\left(\frac{1}{4}q^2\right)$

72. $\left(\frac{2}{3}m\right)^2(-3m)^3\left(\frac{3}{4}m^3\right)$

Applying the Concepts

△ **73. Cubes** Suppose the length of a side of a cube is x^2. Find the volume of the cube in terms of x.

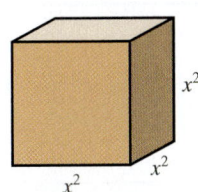

△ **74. Squares** Suppose that the length of a side of a square is $3s$. Find the area of the square in terms of s.

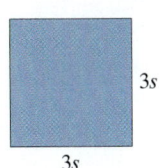

△ **75. Rectangle** Suppose that the width of a rectangle is $4x$ and its length is $12x$. Write an algebraic expression for the area of the rectangle in terms of x.

△ **76. Rectangle** Suppose that the width of a rectangle is $6a$ and its length is $8a$. Write an algebraic expression for the area of the rectangle in terms of a.

Extending the Concepts

In Problems 77–80, simplify each expression.

77. $(3x)^2(-2x)^4\left(\dfrac{1}{3}x^4\right)^2$

78. $(5y)^3(-3y)^2\left(\dfrac{1}{5}y^4\right)^2$

79. $-3(-5mn^3)^2\left(\dfrac{2}{5}m^5n\right)^3$

80. $-2(-4a^3b^2)^2\left(\dfrac{3}{2}ab^5\right)^3$

5.3 Multiplying Polynomials

OBJECTIVES

1. Multiply a Polynomial by a Monomial
2. Multiply Two Binomials Using the Distributive Property
3. Multiply Two Binomials Using the FOIL Method
4. Multiply the Sum and Difference of Two Terms
5. Square a Binomial
6. Multiply a Polynomial by a Polynomial

Preparing for Multiplying Polynomials

Before getting started, take the following readiness quiz. If you get a problem wrong, go back to the section cited and review the material.

1. Use the Distributive Property to simplify: $2(4x - 3)$ [Section 1.7, p. 61]

2. Find the product: $(3x^2)(-5x^3)$ [Section 5.2, p. 358]

3. Find the product: $(9a)^2$ [Section 5.2, p. 357]

1 Multiply a Polynomial by a Monomial

Now that we know how to multiply two monomials, we can extend our discussion to the product of a monomial and a polynomial. If we were asked to simplify $3(2x + 1)$, we would use the Distributive Property to obtain $3(2x) + 3(1)$. In general, when we multiply a polynomial by a monomial, we use the following property.

> **EXTENDED FORM OF THE DISTRIBUTIVE PROPERTY**
>
> $$a(b + c + \cdots + z) = a\cdot b + a\cdot c + \cdots + a\cdot z$$
>
> where $a, b, c, \ldots, z$ are real numbers.

In Words

The Extended Form of the Distributive Property says to multiply each term in parentheses by a.

EXAMPLE 1   **Multiplying a Monomial by a Trinomial Using the Extended Form of the Distributive Property**

Multiply and simplify: $2x^2(x^2 + 3x + 5)$

Solution

We are multiplying a monomial by a trinomial, so we use the Extended Form of the Distributive Property and multiply $2x^2$ by each term in parentheses.

$$2x^2(x^2 + 3x + 5) = 2x^2(x^2) + 2x^2(3x) + 2x^2(5)$$
$$= 2x^4 + 6x^3 + 10x^2$$

EXAMPLE 2   **Multiplying a Monomial by a Trinomial**

Multiply and simplify: $-2xy(3x^2 + 5xy + 2y^2)$

Solution

Use the Extended Form of the Distributive Property.

$$-2xy(3x^2 + 5xy + 2y^2) = -2xy(3x^2) + (-2xy)(5xy) + (-2xy)(2y^2)$$
$$= -6x^3y - 10x^2y^2 - 4xy^3$$

Multiplying a Trinomial by a Monomial

Multiply and simplify: $\left(\dfrac{4}{3}z^2 + 8z + \dfrac{1}{4}\right)\dfrac{1}{2}z^3$

Solution

$$\left(\frac{4}{3}z^2 + 8z + \frac{1}{4}\right)\frac{1}{2}z^3 = \frac{4}{3}z^2\left(\frac{1}{2}z^3\right) + 8z\left(\frac{1}{2}z^3\right) + \frac{1}{4}\left(\frac{1}{2}z^3\right)$$

$$= \frac{2}{3}z^5 + 4z^4 + \frac{1}{8}z^3$$

Work Smart

Example 3 is an algebraic **expression** so you do not clear fractions.

QUICK ✓ *Multiply and simplify each product.*

1. $3x(x^2 - 2x + 4)$ **2.** $-a^2b(2a^2b^2 - 4ab^2 + 3ab)$ **3.** $\left(5n^3 - \dfrac{15}{8}n^2 - \dfrac{10}{7}n\right)\dfrac{3}{5}n^2$

---

② Multiply Two Binomials Using the Distributive Property

We now discuss how to multiply two binomials. To help understand the process, we review multiplication of two-digit numbers. Suppose we want to find 32×14. We proceed as follows: Multiply 2 by 4 and obtain 8, multiply 3 by 4 and obtain 12, so that $32 \times 4 = 128$. Now, multiply 2 by 1 and obtain 2, multiply 3 by 1 and obtain 3, so that $32 \times 1 = 32$. We add vertically to obtain the product, 448.

$$
\begin{array}{r}
32 \\
\times 14 \\
\hline
4 \times 3 \rightarrow 128 \leftarrow 4 \times 2 \\
1 \times 3 \rightarrow \underline{32} \leftarrow 1 \times 2 \\
448
\end{array}
$$

Now suppose we want to multiply $(3x + 2)$ and $(x + 4)$. That is, we want to find

$$
\begin{array}{r}
3x + 2 \\
\times\ \ x + 4 \\
\hline
\end{array}
$$

We proceed in exactly the same way as we did when multiplying two-digit numbers. Multiply 2 by 4, multiply $3x$ by 4, so that $(3x + 2) \cdot 4 = 12x + 8$. Multiply 2 by x, multiply $3x$ by x, so that $(3x + 2) \cdot x = 3x^2 + 2x$. Now add vertically and obtain the product, $3x^2 + 14x + 8$.

$$
\begin{array}{r}
3x + 2 \\
\times\ \ x + 4 \\
\hline
4 \cdot 3x \longrightarrow 12x + 8 \leftarrow 4 \cdot 2 \\
x \cdot 3x \rightarrow \underline{3x^2 + 2x} \leftarrow x \cdot 2 \\
3x^2 + 14x + 8
\end{array}
$$

Multiplying two binomials uses exactly the same procedures that is used when multiplying two-digit numbers!

The multiplication that we just went through is known as *vertical multiplication*. We can also multiply using *horizontal multiplication*. Horizontal multiplication requires repeated use of the Distributive Property. For example, to find $(3x + 2)(x + 4)$, we distribute $3x + 2$ to x and then distribute $3x + 2$ to 4.

$$(3x + 2)(x + 4) = (3x + 2)x + (3x + 2) \cdot 4$$

Distribute x, distribute 4:
$$= 3x \cdot x + 2 \cdot x + 3x \cdot 4 + 2 \cdot 4$$
$$= 3x^2 + 2x + 12x + 8$$
$$= 3x^2 + 14x + 8$$

This is the same result that we obtained using vertical multiplication. Let's do a couple more examples using both horizontal and vertical multiplication.

EXAMPLE 4 Multiplying Two Binomials

Find the product: $(x + 3)(x - 8)$

Solution

Vertical Multiplication

$$
\begin{array}{r}
x + 3 \\
\times \ x - 8 \\
\hline
-8x - 24 \leftarrow -8(x + 3) \\
x^2 + 3x \longleftarrow x(x + 3) \\
\hline
x^2 - 5x - 24
\end{array}
$$

Horizontal Multiplication

We distribute the first binomial to each term in the second binomial.

$$(x + 3)(x - 8) = (x + 3)x + (x + 3)(-8)$$
$$= x^2 + 3x - 8x - 24$$

Combine like terms: $= x^2 - 5x - 24$

In either case, $(x + 3)(x - 8) = x^2 - 5x - 24$. ■

EXAMPLE 5 Multiplying Two Binomials

Find the product: $(3x + 4)(2x - 5)$

Solution

Vertical Multiplication

$$
\begin{array}{r}
3x + 4 \\
\times \ 2x - 5 \\
\hline
-15x - 20 \leftarrow -5(3x + 4) \\
6x^2 + 8x \longleftarrow 2x(3x + 4) \\
\hline
6x^2 - 7x - 20
\end{array}
$$

Horizontal Multiplication

We distribute the first binomial to each term in the second binomial.

$$(3x + 4)(2x - 5) = (3x + 4)(2x) + (3x + 4)(-5)$$
$$= 3x \cdot 2x + 4 \cdot 2x + 3x \cdot (-5) + 4 \cdot (-5)$$
$$= 6x^2 + 8x - 15x - 20$$

Combine like terms: $= 6x^2 - 7x - 20$

In either case, $(3x + 4)(2x - 5) = 6x^2 - 7x - 20$. ■

QUICK ✓ *Find the product.*

4. $(x + 7)(x + 2)$ **5.** $(2n + 1)(n - 3)$ **6.** $(3p - 2)(2p - 1)$

---

③ Multiply Two Binomials Using the FOIL Method

Work Smart

The FOIL method is a form of the Distributive Property, and FOIL can be used **only** when we multiply binomials.

When multiplying two binomials, there is a second method referred to as the **FOIL method**. FOIL stands for First, Outer, Inner, Last. "First" means that we are multiplying the *first* terms in each binomial, "Outer" means we are multiplying the *outside* terms of each binomial, "Inner" means we are multiplying the *innermost* terms of each binomial, and "Last" means we are multiplying the *last* terms of each binomial. The FOIL method is illustrated below. For comparison the horizontal method is illustrated as well.

The FOIL Method	Horizontal Multiplication Using the Distributive Property
First Last **F** **O** **I** **L** $(ax + b)(cx + d) = ax \cdot cx + ax \cdot d + b \cdot cx + b \cdot d$ Inner Outer	$(ax + b)(cx + d) = (ax + b)cx + (ax + b)d$ $= ax \cdot cx + b \cdot cx + ax \cdot d + b \cdot d$ Rearrange terms: $= ax \cdot cx + ax \cdot d + b \cdot cx + b \cdot d$

The FOIL method does not give different results than the Distributive Property. Instead, it is a memory device to help when multiplying two binomials.

EXAMPLE 6   Using the FOIL Method to Multiply Two Binomials

Find the product: $(y + 3)(y + 5)$

Solution

$$(y + 3)(y + 5) = (y)(y) + (y)(5) + (3)(y) + (3)(5)$$
$$= y^2 + 5y + 3y + 15$$
$$= y^2 + 8y + 15$$

EXAMPLE 7   Using the FOIL Method to Multiply Two Binomials

Find the product: $(a + 1)(a - 3)$

Solution

$$(a + 1)(a - 3) = (a)(a) + (a)(-3) + (1)(a) + (1)(-3)$$
$$= a^2 - 3a + a - 3$$
$$= a^2 - 2a - 3$$

QUICK ✓ *Find the product.*

7. $(x + 3)(x + 4)$ **8.** $(y + 5)(y - 3)$ **9.** $(a - 1)(a - 5)$

EXAMPLE 8   Using the FOIL Method to Multiply Two Binomials

Find the product: $(3a + 1)(a - 4)$

Solution

$$(3a + 1)(a - 4) = (3a)(a) + (3a)(-4) + (1)(a) + (1)(-4)$$
$$= 3a^2 - 12a + a - 4$$
$$= 3a^2 - 11a - 4$$

EXAMPLE 9   Using the FOIL Method to Multiply Two Binomials

Find the product: $(3a - 4b)(5a - 2b)$

Solution

$$(3a - 4b)(5a - 2b) = (3a)(5a) + (3a)(-2b) - (4b)(5a) - (4b)(-2b)$$
$$= 15a^2 - 6ab - 20ab + 8b^2$$
$$= 15a^2 - 26ab + 8b^2$$

QUICK ✓ *Find the product.*

10. $(2x + 3)(x + 4)$ **11.** $(3y + 5)(2y - 3)$

12. $(2a + 1)(3a - 4)$ **13.** $(5x + 2y)(3x - 4y)$

④ Multiply the Sum and Difference of Two Terms

Certain binomials lead to products that result in patterns. For this reason, we call these products **special products**. The first special product we discuss is a product of the form $(a - b)(a + b)$.

EXAMPLE 10 Finding a Product of the Form $(a - b)(a + b)$

Find the product: $(x - 5)(x + 5)$

Solution

$$
\begin{aligned}
(x - 5)(x + 5) &= \overset{F}{(x)(x)} + \overset{O}{(x)(5)} - \overset{I}{(5)(x)} - \overset{L}{(5)(5)} \\
&= x^2 + 5x - 5x - 25 \\
&= x^2 - 25
\end{aligned}
$$

We call products of the form $(a - b)(a + b)$ "the sum and difference of two terms." Do you see why?

In Example 10, did you notice that the outer product, $5x$, and the inner product, $-5x$, were opposites? This result is not a coincidence. The sum of the outer product and the inner product of two binomials in the form $(a - b)(a + b)$ is *always* zero, so the product $(a - b)(a + b)$ is the *difference* $a^2 - b^2$. We have the following formula based upon the results of Example 10.

> **PRODUCT OF THE SUM AND DIFFERENCE OF TWO TERMS**
>
> $$(a - b)(a + b) = a^2 - b^2$$

EXAMPLE 11 Finding the Product of the Sum and Difference of Two Terms

Find each product:

(a) $(2x + 5)(2x - 5)$ **(b)** $(4x - 3y)(4x + 3y)$

Solution

(a) $(2x + 5)(2x - 5) = (2x)^2 - 5^2$
$$= 4x^2 - 25$$

(b) $(4x - 3y)(4x + 3y) = (4x)^2 - (3y)^2$
$$= 16x^2 - 9y^2$$

QUICK ✓ *Find each product.*

14. $(a - 4)(a + 4)$ **15.** $(3w + 7)(3w - 7)$

16. $(6b - 2c)(6b + 2c)$ **17.** $(x - 2y^3)(x + 2y^3)$

⑤ Square a Binomial

Another special product involves squaring a binomial. For example, since $(x + 3)^2 = (x + 3)(x + 3)$, we use FOIL and obtain

$$(x + 3)^2 = (x + 3)(x + 3) = x^2 + 3x + 3x + 9 = x^2 + 6x + 9$$

Did you notice that the outer product and the inner product are the same, namely $3x$? Study Table 2 to discover the pattern.

Table 2

$(n + 5)^2 =$	$(n + 5)(n + 5) =$	$n^2 + 5n + 5n + 25 =$	$n^2 + 2(5n) + 25 =$	$n^2 + 10n + 25$
$(2a - 3)^2 =$	$(2a - 3)(2a - 3) =$	$4a^2 - 6a - 6a + 9 =$	$4a^2 - 2(6a) + 9 =$	$4a^2 - 12a + 9$
$(n + 5p)^2 =$		$n^2 + 5np + 5np + 25p^2 =$	$n^2 + 2(5np) + 25p^2 =$	$n^2 + 10np + 25p^2$
$(4b - 9)^2 =$			$16b^2 - 2(4b)(9) + 81 =$	$16b^2 - 72b + 81$

The results of Table 2 lead to the following.

$$(a + b)^2 = \underbrace{a^2}_{(\text{first term})^2} + \underbrace{2ab}_{2(\text{product of terms})} + \underbrace{b^2}_{(\text{second term})^2}$$

$$(a - b)^2 = \underbrace{a^2}_{(\text{first term})^2} - \underbrace{2ab}_{2(\text{product of terms})} + \underbrace{b^2}_{(\text{second term})^2}$$

Work Smart

$(x + y)^2 \neq x^2 + y^2$

$(x - y)^2 \neq x^2 - y^2$

Whenever you feel the urge to perform an operation that you're not quite sure about, try it with actual numbers. For example, does

$(3 + 2)^2 = 3^2 + 2^2?$

NO! So

$(x + y)^2 \neq x^2 + y^2$

SQUARES OF BINOMIALS

$$(a + b)^2 = a^2 + 2ab + b^2$$
$$(a - b)^2 = a^2 - 2ab + b^2$$

The trinomials $a^2 + 2ab + b^2$ and $a^2 - 2ab + b^2$ are called **perfect square trinomials**.

EXAMPLE 12   **How to Find a Product of the Form $(a + b)^2$**

Find the product: $(y + 7)^2$

Step-by-Step Solution

Step 1: Use the $(a + b)^2$ pattern.

$$\underbrace{(a + b)^2}_{} = \underbrace{a^2}_{} + \underbrace{2ab}_{} + \underbrace{b^2}_{}$$
$$(y + 7)^2 = (y)^2 + 2(y)(7) + 7^2$$

Step 2: Find each product.

$$= y^2 + 14y + 49$$

EXAMPLE 13   **Finding a Product of the Form $(a - b)^2$**

Find the product: $(3 - r)^2$

Solution

Use the $(a - b)^2$ pattern: $(3 - r)^2 = 3^2 - 2(3)(r) + r^2$
$$= 9 - 6r + r^2$$

QUICK ✓ *Find each product.*

18. $(z - 9)^2$ **19.** $(p + 1)^2$ **20.** $(4 - a)^2$ **21.** $(w + y)^2$

EXAMPLE 14   **Finding Products of the Form $(a + b)^2$ or $(a - b)^2$**

Find each product:

(a) $(9p - 4)^2$ **(b)** $(2x + 5y)^2$

Work Smart

If you can't remember the formulas for a perfect square, don't panic! Simply use the fact that $(x + a)^2 = (x + a)(x + a)$ and then FOIL. The same logic applies to perfect squares of the form $(x - a)^2$.

Solution

(a) $(9p - 4)^2 = (9p)^2 - 2(9p)4 + 4^2$
$$= 81p^2 - 72p + 16$$

(b) $(2x + 5y)^2 = (2x)^2 + 2(2x)(5y) + (5y)^2$
$$= 4x^2 + 20xy + 25y^2$$

QUICK ✓ *Find each product.*

22. $(3z - 4)^2$ **23.** $(5p + 1)^2$ **24.** $(4a - 5)^2$ **25.** $(2w + 7y)^2$

For convenience, we present a summary of binomial products.

SUMMARY OF BINOMIAL PRODUCTS

- When multiplying two binomials, we can use the Distributive Property.

$(3x + 5)(4x - 1) = (3x + 5)(4x) + (3x + 5)(-1)$
$$= 12x^2 + 20x - 3x - 5$$
$$= 12x^2 + 17x - 5$$

- When multiplying two binomials, we can use the FOIL pattern.

$(3x + 5)(4x - 1) = 3x(4x) + 3x(-1) + 5(4x) + 5(-1)$
$$= 12x^2 - 3x + 20x - 5$$
$$= 12x^2 + 17x - 5$$

- When finding a product in the form $(a - b)(a + b)$, we use the product of the sum and difference of two terms, $a^2 - b^2$.

$(7a + 2b)(7a - 2b) = (7a)^2 - (2b)^2$
$$= 49a^2 - 4b^2$$

- When squaring a binomial, $(a + b)^2 = a^2 + 2ab + b^2$ and $(a - b)^2 = a^2 - 2ab + b^2$. The trinomial product is called a perfect square trinomial.

$(3x + 5)^2 = (3x)^2 + 2(3x)(5) + (5)^2$
$$= 9x^2 + 30x + 25$$
$(4y - 3)^2 = (4y)^2 - 2(4y)(3) + (3)^2$
$$= 16y^2 - 24y + 9$$

⑥ Multiply a Polynomial by a Polynomial

When we find the product of two polynomials, we make repeated use of the Extended Form of the Distributive Property. We can use either a horizontal or vertical format. In addition, it is a good idea to write each polynomial in standard form.

EXAMPLE 15 **Multiplying Polynomials Using Horizontal Multiplication**

Find the product: $(2x + 3)(x^2 + 5x - 1)$

Solution

We distribute $2x + 3$ to each term in the trinomial.

$(2x + 3)(x^2 + 5x - 1) = (2x + 3)x^2 + (2x + 3) \cdot 5x + (2x + 3) \cdot (-1)$
$$= 2x(x^2) + 3(x^2) + 2x(5x) + 3(5x) + 2x(-1) + 3(-1)$$

Rearrange terms: $= 2x^3 + 3x^2 + 10x^2 + 15x - 2x - 3$

Combine like terms: $= 2x^3 + 13x^2 + 13x - 3$

QUICK ✓ *Find the product using horizontal multiplication.*

26. $(x - 2)(x^2 + 2x + 4)$ **27.** $(3y - 2)(y^2 + 2y + 4)$

The best way to use vertical multiplication to multiply two polynomials is to place the polynomial with more terms on top, align terms of the same degree, and then multiply. Make sure both polynomials are written in standard form.

EXAMPLE 16   **Multiplying a Binomial and a Trinomial Using Vertical Multiplication**

Find the product: $(2x + 3)(x^2 + 5x - 1)$

Solution

We arrange the polynomials vertically by placing the trinomial on top since it has more terms than the binomial. Now we multiply the trinomial by the 3 in the binomial and then multiply the trinomial by the $2x$ in the binomial.

$$
\begin{array}{r}
x^2 + 5x - 1 \\
\times \qquad 2x + 3 \\
\hline
3x^2 + 15x - 3 \\
\end{array}
$$

$3(x^2 + 5x - 1)$:

$2x(x^2 + 5x - 1)$: $2x^3 + 10x^2 - 2x$

Add vertically: $2x^3 + 13x^2 + 13x - 3$

QUICK ✔ *Find the product using vertical multiplication.*

28. $(x - 2)(x^2 + 2x + 4)$ **29.** $(3y - 2)(y^2 + 2y + 4)$

EXAMPLE 17   **Multiplying Three Polynomials**

Find the product: $2x(x - 4)(3x - 5)$

Solution

To find the product of three polynomials, multiply any two factors and then multiply that product by the remaining factor. We'll start by using the Distributive Property to multiply $2x$ and $(x - 4)$.

$$2x(x - 4)(3x - 5) = (2x^2 - 8x)(3x - 5)$$

FOIL: $= 2x^2 \cdot 3x - 2x^2 \cdot 5 - 8x \cdot 3x - 8x \cdot (-5)$

$= 6x^3 - 10x^2 - 24x^2 + 40x$

Combine like terms: $= 6x^3 - 34x^2 + 40x$

QUICK ✔ *Find the product.*

30. $-2a(4a - 1)(3a + 5)$ **31.** $\dfrac{3}{2}(5x - 2)(2x + 8)$

5.3 Exercises

For Extra Help: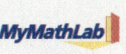

Student Solutions Manual   CD Video   PH Math/Tutor Center   MathXL Tutorials on CD   MathXL®   MyMathLab

Concepts and Vocabulary

In Problems 1–3, fill in the blanks.

1. The product $(a - b)(a + b)$ is equal to _____.

2. When using the FOIL method to multiply two binomials, the letter I indicates that the _____ terms should be multiplied.

3. $x^2 + 2xy + y^2$ is referred to as a _____ _____ _____.

In Problems 4–6, answer True or False to each statement.

4. The product $(x - y)(x^2 + 2xy + y^2)$ can be found using the FOIL method.

5. $(x - y)^2 = x^2 - y^2$

6. The product of a binomial and a binomial is always a trinomial.

7. Do you think there are values of x and y such that $(x + y)^2 = x^2 + y^2$? Use some numerical values for x and y to explain your reasoning.

8. Explain how the difference of squares can be used to calculate $19 \cdot 21$.

Building Skills

In Problems 9–20, use the Distributive Property to find each product.

9. $2x(3x - 5)$

10. $3m(2m - 7)$

11. $-7b(b - 5)$

12. $-5b(3b + 9)$

13. $\dfrac{1}{2}n(4n - 6)$

14. $\dfrac{3}{5}b(15b - 5)$

15. $3n^2(4n^2 + 2n - 5)$

16. $4w(2w^2 + 3w - 5)$

17. $2mn(6m^2 - 2mn + n^2)$

18. $6xy(9x^2 + 2xy - 3y^2)$

19. $(4x^2y - 3xy^2)x^2y$

20. $(7r + 3s^2)2r^2s$

In Problems 21–54, find each product.

21. $(x + 2)(x + 3)$

22. $(x + 3)(x + 7)$

23. $(q - 6)(q - 7)$

24. $(n - 4)(n - 5)$

25. $(y - 5)(y + 4)$

26. $(m - 3)(m + 6)$

27. $(x^2 + 3)(x^2 + 1)$

28. $(x^2 - 5)(x^2 - 2)$

29. $(7 - x)(6 - x)$

30. $(2 - y)(4 - y)$

31. $(1 - n)(n - 3)$

32. $(x - 5)(9 - x)$

33. $(3a + 2)(a + 1)$

34. $(2m - 1)(3m - 2)$

35. $(5u + 6v)(2u + v)$

36. $(3a - 2b)(a - 3b)$

37. $(2n - 3p)(2n + 3p)$

38. $(4x + 3y)(4x - 3y)$

39. $(x - 2)(x^2 + 3x + 1)$

40. $(3a - 1)(2a^2 - 5a - 3)$

41. $(2y^2 - 6y + 1)(y - 3)$

42. $(2m^2 - m + 2)(2m + 1)$

43. $(2x - 3)(x^2 - 2x - 1)$

44. $(4y + 1)(2y^2 - y - 3)$

45. $(3p^2 - 2p - 1)(2p - 3)$

46. $(m^2n - 2mn + 1)(mn - 3)$

47. $2b(b - 3)(b + 4)$

48. $5a(a + 6)(a - 1)$

49. $-\dfrac{1}{2}x(2x + 6)(x - 3)$

50. $-\dfrac{4}{3}k(k + 7)(3k - 9)$

51. $(x^2 - x + 4)(2x^2 - 3x - 1)$

52. $(2a^2 - a + 6)(a^2 + 3a + 3)$

53. $(5y^3 - y^2 + 2)(2y^2 + y + 1)$

54. $(2m^2 - m + 4)(-m^3 - 2m - 1)$

In Problems 55–66, use the FOIL method to find each product.

55. $(x - 7)(x + 3)$

56. $(p + 5)(p - 8)$

57. $(z - 10)(z + 3)$

58. $(p + 10)(p - 9)$

59. $(2x + 3)(3x - 1)$

60. $(3z - 2)(4z + 1)$

61. $(7k - 3)(k + 6)$

62. $(4r - 1)(3r - 5)$

63. $(7x + 3y)(2x + 5y)$

64. $(2m - 3n)(4m + n)$

65. $(2a - b)(5a + 2b)$

66. $(3r + 5s)(6r + 7s)$

In Problems 67–90, find the special products.

67. $(x - 3)(x + 3)$

68. $(y + 7)(y - 7)$

69. $(2z + 5)(2z - 5)$

70. $(6r - 1)(6r + 1)$

71. $(4x^2 + 1)(4x^2 - 1)$

72. $(3a^2 + 2)(3a^2 - 2)$

73. $(2x - 3y)(2x + 3y)$

74. $(8a - 5b)(8a + 5b)$

75. $\left(x - \dfrac{1}{3}\right)\left(x + \dfrac{1}{3}\right)$

76. $\left(y + \dfrac{2}{9}\right)\left(y - \dfrac{2}{9}\right)$ **77.** $(x^3 - y^2)(x^3 + y^2)$ **78.** $(2x^2 - 3z^3)(2x^2 + 3z^3)$

79. $(x - 2)^2$ **80.** $(x + 4)^2$ **81.** $(x + 2y)^2$

82. $(3x - 2y)^2$ **83.** $(2a - 3b)^2$ **84.** $(5x + 2y)^2$

85. $(4x + 3y)(4x + 3y)$ **86.** $(2p - 7q)(2p - 7q)$ **87.** $(5a - 2b^2)^2$

88. $(7m^2 + 2n^3)^2$ **89.** $\left(x + \dfrac{1}{2}\right)^2$ **90.** $\left(y - \dfrac{1}{3}\right)^2$

Mixed Practice

In Problems 91–110, perform the indicated operation.

91. $2x(x^2 - 3x + 2) + (x^3 - 4x + 1)$ **92.** $-4y^2(2y - 1) + (y^3 - 8y + 3)$

93. $-\dfrac{1}{2}x^2(10x^5 - 6x^4 + 12x^3) + (x^3)^2$ **94.** $-\dfrac{1}{3}(27y^2 - 9y + 6) - (3y)^2$

95. $7x^2(x + 3) - 2x(x^2 - 1)$ **96.** $2(3a^4 + 2b^4) - 3(b^4 - 2a^4)$

97. $-3w(w - 4)(w + 3)$ **98.** $5y(y + 5)(y - 3)$

99. $3a(a + 4)^2$ **100.** $2m(m - 3)^2$

101. $(n + 3)(n - 3) + (n + 3)^2$ **102.** $(s + 6)(s - 6) + (s - 6)^2$

103. $(a + 6b)^2 - (a - 6b)^2$ **104.** $(2a + 5b)^2 - (2a - 5b)^2$

105. $(x - 0.3)^2$ **106.** $(n + 0.2)^2$

107. $\left(\dfrac{2}{3}b - \dfrac{5}{9}\right)\left(\dfrac{2}{3}b + \dfrac{5}{9}\right)$ **108.** $\left(\dfrac{4}{5}b + \dfrac{2}{7}\right)\left(\dfrac{4}{5}b - \dfrac{2}{7}\right)$

109. $(x + 1)^2 - (2x + 1)(x - 1)$ **110.** $(a + 3)^2 - (a + 4)(3a - 1)$

111. Square $2x + 1$.   **112.** Square $3x - 2y$.

113. Find the cube of $x - 1$.   **114.** Find the cube of $2a + b$.

115. Subtract $x - 6$ from the product of $x + 3$ and $2x - 5$.

116. Add $2x + 3$ to the product of $x + 3$ and $2x - 3$.

Applying the Concepts

In Problems 117–124, find the area of the shaded region.

△ **117.**

$2x - 3$

$x + 5$

△ **118.**

$3x + 5$

$x + 9$

△ **119.**

$x + 5$

$x + 5$

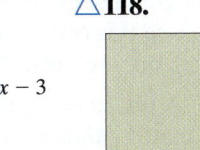

△ **120.**

$x - 3$

$x - 3$

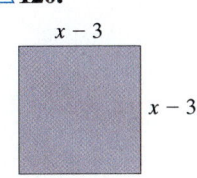

△ **121.**

$3x + 5$

$3x - 5$

△ **122.**

$2x - 3$

$2x + 3$

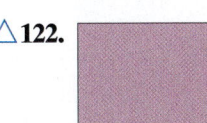

△ **123.**

$x - 1$

$x - 1$

x

$3x - 1$

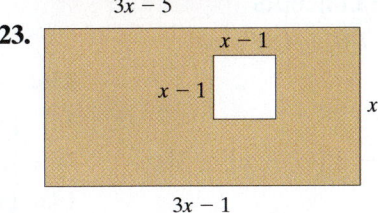

△ **124.**

x

$2x + 3$

$4x - 3$

$4x - 3$

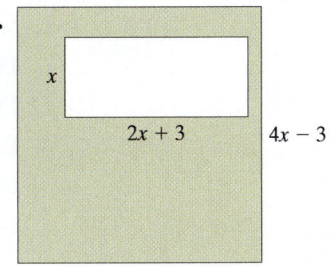

In Problems 125 and 126, find an algebraic expression for the area of the rectangle by finding the sum of the four interior rectangles. Then find the area of the rectangle by multiplying the width and height. Compare the two expressions. How is multiplying a binomial related to finding the area of the rectangle?

△ **125.** △ **126.**

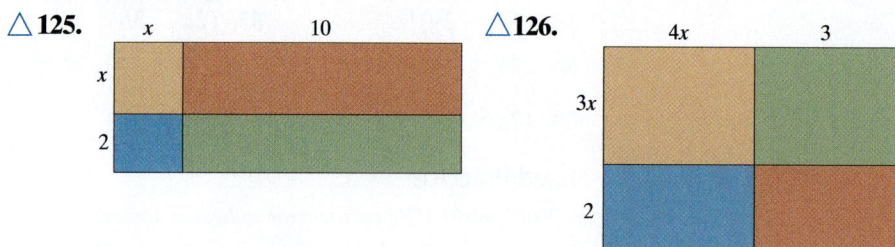

127. Consecutive Integers If x represents the first of three consecutive integers, write a polynomial that represents the product of the next two consecutive integers.

128. Consecutive Integers If x represents the first of three consecutive odd integers, write a polynomial that represents the product of the first and the third integer.

129. Sales Tax In a certain city, the sales tax rate has been increased by 2%. If the original sales tax rate was x%, write a polynomial that represents the total cost of a $40 purchase after the new sales tax has been imposed.

130. Property Tax In a certain city, the property tax rate has been decreased by 0.5%. If the original tax rate was x%, write a polynomial that represents the amount of property tax on a $120,000 home after the decrease in property tax rate.

△ **131. Area of a Circle** Write a polynomial for the area of a circle with radius $x + 2$ feet.

△ **132. Area of a Circle** Write a polynomial for the area of a circle with radius $2y - 3$ meters.

△ **133. Perfect Square** Why is the expression $(a + b)^2$ called a perfect square? Consider the figure to the right.

    (a) Find the area of each of the four quadrilaterals.
    (b) Use the result from part (a) to find the area of the entire region.
    (c) Find the length and width of the entire region in terms of a and b. Use this result to find the area of the entire region. What do you notice?

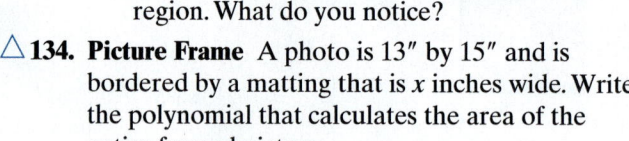

△ **134. Picture Frame** A photo is 13″ by 15″ and is bordered by a matting that is x inches wide. Write the polynomial that calculates the area of the entire framed picture.

Extending the Concepts

In Problems 135–144, find the product.

135. $(x - 2)(x + 2)^2$

136. $(x + 3)^3$

137. $[(x + 4) + 3]^2$

138. $[3 - (x + y)][3 + (x + y)]$

139. $[(x - y) + 3][(x - y) - 3]$

140. $[(a + 4) - 7][(a + 4) + 7]$

141. $(x + 2)^3$

142. $(y - 3)^3$

143. $(z + 3)^4$

144. $(m - 2)^4$

5.4 Dividing Monomials: The Quotient Rule and Integer Exponents

OBJECTIVES

1. Simplify Exponential Expressions Using the Quotient Rule
2. Simplify Exponential Expressions Using the Quotient to a Power Rule
3. Simplify Exponential Expressions Using Zero as an Exponent
4. Simplify Exponential Expressions Involving Negative Exponents
5. Simplify Exponential Expressions Using the Laws of Exponents

Preparing for Dividing Monomials: The Quotient Rule and Integer Exponents

Before getting started, take the following readiness quiz. If you get a problem wrong, go back to the section cited and review the material.

1. Find the product: $(3a^2)^3$ [Section 5.2, page 357]

2. Evaluate: $\left(\dfrac{2}{3}\right)^2$ [Section 1.6, page 49–50]

3. Find the reciprocal: (a) 5   (b) $-\dfrac{6}{7}$ [Section 1.4, page 30]

Before we can turn our attention to dividing polynomials, we must first discuss dividing monomials.

① Simplify Exponential Expressions Using the Quotient Rule

To find a general rule for the quotient of two exponential expressions, consider the following:

$$\frac{y^6}{y^2} = \frac{\overbrace{y \cdot y \cdot y \cdot y \cdot y \cdot y}^{6 \text{ factors}}}{\underbrace{y \cdot y}_{2 \text{ factors}}} = \underbrace{y \cdot y \cdot y \cdot y}_{4 \text{ factors}} = y^4$$

We might conclude from this result that

$$\frac{y^6}{y^2} = y^{6-2} = y^4$$

This result is true in general.

In Words

When dividing two exponential expressions with a common base, subtract the exponent in the denominator from the exponent in the numerator. Then write the base to that power.

QUOTIENT RULE FOR EXPONENTS

If a is a real number and if m and n are natural numbers, then

$$\frac{a^m}{a^n} = a^{m-n} \qquad \text{if } a \neq 0$$

EXAMPLE 1   **Using the Quotient Rule to Simplify Expressions**

Simplify each expression:

(a) $\dfrac{6^5}{6^3}$ (b) $\dfrac{n^6}{n^2}$

Solution

(a)
$$\frac{6^5}{6^3} = 6^{5-3}$$
$$= 6^2$$
Evaluate 6^2: $= 36$

(b) $\dfrac{n^6}{n^2} = n^{6-2}$
$$= n^4$$

EXAMPLE 2 Using the Quotient Rule to Simplify Expressions

Simplify each expression:

(a) $\dfrac{25b^9}{15b^6}$ **(b)** $\dfrac{-24x^6y^4}{10x^4y}$

Solution

(a)
$$\frac{25b^9}{15b^6} = \frac{25}{15} \cdot \frac{b^9}{b^6}$$

Divide out like factors; apply the Quotient Rule:
$$= \frac{\cancel{5} \cdot 5}{\cancel{5} \cdot 3} \cdot b^{9-6}$$

$$= \frac{5}{3}b^3$$

(b)
$$\frac{-24x^6y^4}{10x^4y} = \frac{-24}{10} \cdot \frac{x^6}{x^4} \cdot \frac{y^4}{y}$$

Divide out like factors; apply the Quotient Rule:
$$= \frac{-12 \cdot \cancel{2}}{5 \cdot \cancel{2}} \cdot x^{6-4} \cdot y^{4-1}$$

$$= -\frac{12}{5}x^2y^3$$

QUICK ☑ *Simplify each expression.*

1. $\dfrac{3^7}{3^5}$ **2.** $\dfrac{y^9}{y^3}$ **3.** $\dfrac{14c^6}{10c^5}$ **4.** $\dfrac{-21w^4z^8}{14w^3z}$

---

(2) ## Simplify Exponential Expressions Using the Quotient to a Power Rule

Now let's look at a quotient raised to a power:

$$\left(\frac{3}{2}\right)^4 = \left(\frac{3}{2}\right) \cdot \left(\frac{3}{2}\right) \cdot \left(\frac{3}{2}\right) \cdot \left(\frac{3}{2}\right) = \overbrace{\frac{3 \cdot 3 \cdot 3 \cdot 3}{2 \cdot 2 \cdot 2 \cdot 2}}^{4 \text{ factors}} = \frac{3^4}{2^4}$$

$\underbrace{}_{4 \text{ factors}}$

We are led to the following result:

In Words
When a quotient is raised to a power, both the numerator and the denominator are raised to the indicated power.

QUOTIENT TO A POWER RULE FOR EXPONENTS

If a, b are real numbers and n is a natural number, then

$$\left(\frac{a}{b}\right)^n = \frac{a^n}{b^n} \qquad \text{if } b \neq 0$$

EXAMPLE 3 Using the Quotient to a Power Rule to Simplify an Expression

Simplify the expression $\left(\dfrac{z}{3}\right)^2$.

Solution

This expression is a quotient, so we apply the Quotient to a Power Rule, raising both the numerator and the denominator to the indicated power.

$$\left(\frac{z}{3}\right)^2 = \frac{z^2}{3^2}$$

Evaluate 3^2: $= \dfrac{z^2}{9}$

EXAMPLE 4   **Using the Quotient to a Power Rule to Simplify Expressions**

Simplify the expressions:

(a) $\left(-\dfrac{2y}{z}\right)^3$

(b) $\left(\dfrac{3a^2}{b^3}\right)^4$

Solution

(a)
$$\left(-\frac{2y}{z}\right)^3 = \frac{(-2y)^3}{z^3}$$

Use the Power Rule, $(ab)^n = a^n b^n$: $= \dfrac{(-2)^3(y)^3}{z^3}$

Evaluate $(-2)^3$: $= \dfrac{-8y^3}{z^3}$

(b)
$$\left(\frac{3a^2}{b^3}\right)^4 = \frac{(3a^2)^4}{(b^3)^4}$$

Use the Power Rule, $(ab)^n = a^n b^n$: $= \dfrac{3^4(a^2)^4}{(b^3)^4}$

Evaluate $(3)^4$: $= \dfrac{81a^{2\cdot4}}{b^{3\cdot4}}$

$= \dfrac{81a^8}{b^{12}}$

QUICK ✓ *Simplify each expression.*

5. $\left(\dfrac{p}{2}\right)^4$ **6.** $\left(\dfrac{b}{3}\right)^3$ **7.** $\left(\dfrac{a^3}{b^4}\right)^2$ **8.** $\left(-\dfrac{2a^2}{b^4}\right)^3$

---

(3) ## Simplify Exponential Expressions Using Zero as an Exponent

Now that we have fully discussed exponential expressions with natural number (or positive integer) exponents, we will extend the definition of exponential expressions to integer exponents. We begin with raising a real number to the 0 power.

DEFINITION OF ZERO EXPONENT

If a is a nonzero real number (that is, $a \neq 0$), we define

$$a^0 = 1$$

The reason that $a^0 = 1$ is based upon the Quotient Rule for Exponents. That is,

$\dfrac{a^n}{a^n} = a^{n-n} = a^0$. In addition, $\dfrac{a^n}{a^n} = \dfrac{\overbrace{a \cdot a \cdot \ldots \cdot a}^{n \text{ factors}}}{\underbrace{a \cdot a \cdot \ldots \cdot a}_{n \text{ factors}}} = 1$. Therefore, $a^0 = 1$.

EXAMPLE 5 **Using Zero as an Exponent**

(a) $3^0 = 1$ **(b)** $x^0 = 1$ **(c)** $(4y)^0 = 1$

(d) $8w^0 = 8 \cdot 1$ **(e)** $-5^0 = -1 \cdot 5^0$

$\quad\quad = 8$ $\quad\quad\quad\quad = -1$

Work Smart
In Example 5(d), the exponent 0 applies *only* to the factor *w*.

QUICK ✓ *Simplify each expression. When they occur, assume variables are nonzero.*

9. 10^0 **10.** $4z^0$ **11.** $-p^0$

---

④ Simplify Exponential Expressions Involving Negative Exponents

Now let's look at exponents that are negative integers. Suppose that we wanted to simplify $\dfrac{z^3}{z^5}$. If we use the Quotient Rule for Exponents, we obtain

$$\frac{z^3}{z^5} = z^{3-5} = z^{-2}$$

We could also simplify this expression directly by dividing like factors:

$$\frac{z^3}{z^5} = \frac{\cancel{z} \cdot \cancel{z} \cdot \cancel{z}}{\cancel{z} \cdot \cancel{z} \cdot \cancel{z} \cdot z \cdot z} = \frac{1}{z^2}$$

This implies that $z^{-2} = \dfrac{1}{z^2}$. Considering this result, we define a raised to a negative integer power as follows:

DEFINITION OF NEGATIVE EXPONENT
If n is a positive integer and if a is a nonzero real number (that is, $a \neq 0$), then we define

$$a^{-n} = \frac{1}{a^n}$$

Remember, the reciprocal of a nonzero number a is $\dfrac{1}{a}$. For example, the reciprocal of 7 is $\dfrac{1}{7}$ and the reciprocal of $-\dfrac{5}{8}$ is $-\dfrac{8}{5}$. Whenever we encounter a negative exponent, we should think, "take the reciprocal of the base."

EXAMPLE 6 **Simplifying Exponential Expressions Containing Negative Integer Exponents**

(a) $x^{-4} = \dfrac{1}{x^4}$ **(b)** $2^{-3} = \dfrac{1}{2^3}$

$\quad\quad\quad\quad\quad\quad\quad\quad = \dfrac{1}{8}$

EXAMPLE 7 **Simplifying Exponential Expressions Containing Negative Integer Exponents**

Simplify each expression:

(a) $(-5)^{-2}$ **(b)** $(-n)^{-9}$

Solution

(a) $(-5)^{-2} = \dfrac{1}{(-5)^2}$

Evaluate: $= \dfrac{1}{25}$

(b) $(-n)^{-9} = \dfrac{1}{(-n)^9}$

$= \dfrac{1}{(-1)^9 \cdot n^9}$

Evaluate: $= -\dfrac{1}{n^9}$

EXAMPLE 8   **Simplifying the Sum of Exponential Expressions Containing Negative Integer Exponents**

Simplify: $2^{-2} + 4^{-1}$

Solution

$$2^{-2} + 4^{-1} = \frac{1}{2^2} + \frac{1}{4^1}$$

Evaluate: $= \dfrac{1}{4} + \dfrac{1}{4}$

Find the sum: $= \dfrac{2}{4}$

Write in lowest terms: $= \dfrac{1}{2}$

QUICK ✓   *Simplify each expression. Write answers with only positive exponents.*

12. z^{-5} **13.** $(-p)^{-4}$ **14.** $-x^{-3}$ **15.** $(-2)^{-4}$

16. -7^{-2} **17.** $(-7)^{-2}$ **18.** $4^{-1} - 2^{-3}$

How might we simplify $\dfrac{1}{a^{-n}}$? Because

$$\frac{1}{a^{-n}} = \frac{1}{\dfrac{1}{a^n}} = 1 \cdot \frac{a^n}{1} = a^n$$

we have the following result.

> If a is a real number and n is an integer, then
>
> $$\frac{1}{a^{-n}} = a^n \qquad \text{if } a \neq 0$$

EXAMPLE 9   **Simplifying Quotients Containing Negative Integer Exponents Using $\dfrac{1}{a^{-n}} = a^n$**

Simplify:

(a) $\dfrac{6}{2^{-3}}$ (b) $\dfrac{7}{n^{-2}}$

Work Smart

$\dfrac{6}{2^{-3}} = \dfrac{6}{\dfrac{1}{8}}$

$= 6 \div \dfrac{1}{8}$

$= 6 \cdot 8$

Solution

(a) $\dfrac{6}{2^{-3}} = 6 \cdot 2^3 = 6 \cdot 8 = 48$ (b) $\dfrac{7}{n^{-2}} = 7 \cdot n^2 = 7n^2$

QUICK ✓ *Simplify each numerical expression.*

19. $\dfrac{1}{3^{-2}}$ **20.** $\dfrac{1}{v^{-1}}$ **21.** $\dfrac{1}{-10^{-2}}$ **22.** $\dfrac{5}{2z^{-2}}$

EXAMPLE 10 **Simplifying Quotients Containing Negative Integer Exponents**

Simplify: $\left(\dfrac{3}{2}\right)^{-3}$

Solution

$$\left(\dfrac{3}{2}\right)^{-3} = \dfrac{1}{\left(\dfrac{3}{2}\right)^{3}}$$

Use $\left(\dfrac{a}{b}\right)^{n} = \dfrac{a^{n}}{b^{n}}$: $= \dfrac{1}{\dfrac{3^{3}}{2^{3}}}$

$$= \dfrac{1}{\dfrac{27}{8}}$$

$$= 1 \cdot \dfrac{8}{27}$$

$$= \dfrac{8}{27}$$

Work Smart

This "shortcut" says to simplify a quotient to a negative exponent, first take the reciprocal, then raise that expression to the positive exponent power.

The following shortcut is based upon the results of Example 10:

> If a and b are real numbers and n is an integer, then
> $$\left(\dfrac{a}{b}\right)^{-n} = \left(\dfrac{b}{a}\right)^{n} \qquad \text{if } a \neq 0, b \neq 0$$

For example, $\left(\dfrac{3}{2}\right)^{-3} = \left(\dfrac{2}{3}\right)^{3} = \dfrac{8}{27}$.

EXAMPLE 11 **Simplifying Quotients Involving Negative Exponents**

Simplify: $\left(-\dfrac{x^{3}}{4}\right)^{-2}$

Solution

$$\left(-\dfrac{x^{3}}{4}\right)^{-2} = \left(-\dfrac{4}{x^{3}}\right)^{2}$$

Use $\left(\dfrac{a}{b}\right)^{n} = \dfrac{a^{n}}{b^{n}}$: $= \dfrac{(-4)^{2}}{(x^{3})^{2}}$

$$= \dfrac{16}{x^{6}}$$

QUICK ✓ *Simplify each expression.*

23. $\left(\dfrac{7}{8}\right)^{-1}$ **24.** $\left(-\dfrac{1}{4}\right)^{-3}$ **25.** $\left(\dfrac{3a}{5}\right)^{-2}$ **26.** $\left(-\dfrac{2}{3n^{4}}\right)^{-3}$

(5) Simplify Exponential Expressions Using the Laws of Exponents

We now summarize the Laws of Exponents where the exponents are integers.

THE LAWS OF EXPONENTS

If a and b are real numbers and if m and n are integers, then, assuming the expression is defined,

Rule		Examples
Product Rule	$a^m \cdot a^n = a^{m+n}$	$7^3 \cdot 7 = 7^4; \ x^2 \cdot x^5 = x^7$
Power Rule	$(a^m)^n = a^{m \cdot n}$	$(3^4)^2 = 3^8; \ (x^3)^2 = x^6$
Product to Power Rule	$(a \cdot b)^n = a^n \cdot b^n$	$(x^3 y)^4 = x^{12} y^4$
Quotient Rule	$\dfrac{a^m}{a^n} = a^{m-n}$, if $a \neq 0$	$\dfrac{9^5}{9^3} = 9^2; \ \dfrac{y^{11}}{y^8} = y^3$
Quotient to Power Rule	$\left(\dfrac{a}{b}\right)^n = \dfrac{a^n}{b^n}$, if $b \neq 0$	$\left(\dfrac{3}{4}\right)^2 = \dfrac{9}{16}; \ \left(\dfrac{2x}{y^2}\right)^3 = \dfrac{8x^3}{y^6}$
Zero Exponent Rule	$a^0 = 1$, if $a \neq 0$	$3^0 = 1; \ (5m)^0 = 1$ $5m^0 = 5 \cdot 1 = 5$
Negative Exponent Rules	$a^{-n} = \dfrac{1}{a^n}$, if $a \neq 0$ $\dfrac{1}{a^{-n}} = a^n$, if $a \neq 0$	$2^{-3} = \dfrac{1}{8}; \ b^{-4} = \dfrac{1}{b^4}$ $\dfrac{1}{t^{-5}} = t^5$
Quotient to a Negative Power Rule	$\left(\dfrac{a}{b}\right)^{-n} = \left(\dfrac{b}{a}\right)^n$, if $a \neq 0, b \neq 0$	$\left(\dfrac{5}{k}\right)^{-2} = \left(\dfrac{k}{5}\right)^2 = \dfrac{k^2}{25}$

Let's do some examples where we use one or more of the rules listed above. To evaluate or simplify these exponential expressions, ask yourself the questions in the following display.

USING THE RULES OF EXPONENTS

- Does the exponential expression contain numerical expressions? If so, *evaluate* them. ("Evaluate" means "find the value.")
- Is the expression a product of monomials? If so, use the Product Rule:
$$a^m \cdot a^n = a^{m+n}$$
- Is the expression a quotient (division problem)? If so, use the Quotient Rule:
$$\frac{a^m}{a^n} = a^{m-n}$$
- Do you see a quantity raised to a power? If so, use one of the Power Rules:
$$(a^m)^n = a^{m \cdot n}, (a \cdot b)^n = a^n \cdot b^n, \text{ or } \left(\frac{a}{b}\right)^n = \frac{a^n}{b^n}.$$
- Is there a negative exponent in the expression? Remember, a negative in an exponent means "take the reciprocal of the base." Plus, an expression is not simplified if it contains negative exponents.
- Are you required to raise a quantity to the zero power? Remember, $a^0 = 1$ for $a \neq 0$.

EXAMPLE 12 How to Simplify Expressions Using the Product Rule and the Negative Exponent Rule

Simplify $\left(\dfrac{3}{2}a^3b^{-2}\right)(-12a^{-4}b^5)$. Write the answer with only positive exponents.

Step-by-Step Solution

Step 1: Rearrange factors.

$$\left(\frac{3}{2}a^3b^{-2}\right)(-12a^{-4}b^5) = \left(\frac{3}{2}\cdot(-12)\right)(a^3\cdot a^{-4})(b^{-2}\cdot b^5)$$

Step 2: Find each product.

Evaluate $\left(\dfrac{3}{2}\right)\cdot(-12)$. Apply the Product Rule to variable factors:

$$= -18a^{3+(-4)}b^{-2+5}$$
$$= -18a^{-1}b^3$$

Step 3: Simplify. Write the product so that the exponents are positive.

Apply the Negative Exponent Rule:

$$= -18\cdot\frac{1}{a}\cdot b^3$$
$$= \frac{-18b^3}{a}$$

Work Smart

$$\frac{-18b^3}{a} \neq \frac{b^3}{18a}$$

Do you know why?

QUICK ✓ *Simplify each expression. Write the answers with only positive exponents.*

27. $(-4a^{-3})(5a)$

28. $\left(-\dfrac{2}{5}m^{-2}n^{-1}\right)\left(-\dfrac{15}{3}mn^0\right)$

EXAMPLE 13 How to Simplify Expressions Using the Quotient Rule and the Negative Exponent Rule

Simplify $-\dfrac{27ab^4}{18a^3b}$. Write the answer with only positive exponents.

Step-by-Step Solution

Step 1: Write the quotient as the product of factors.

Use $-\dfrac{a}{b} = \dfrac{-a}{b}$: $\quad -\dfrac{27ab^4}{18a^3b} = \dfrac{-27}{18}\cdot\dfrac{a}{a^3}\cdot\dfrac{b^4}{b}$

Step 2: Find each quotient.

Simplify $\dfrac{-27}{18}$. Use the Quotient Rule.

$$= \frac{-3}{2}\cdot a^{1-3}\cdot b^{4-1}$$
$$= -\frac{3}{2}\cdot a^{-2}\cdot b^3$$

Step 3: Simplify. Write the quotient so that the exponents are positive.

Apply the Negative Exponent Rule:

$$= -\frac{3}{2}\cdot\frac{1}{a^2}\cdot b^3$$
$$= -\frac{3b^3}{2a^2}$$

QUICK ✓   *Simplify. Write answers with only positive exponents.*

29. $\dfrac{16m^2}{2m^7}$

30. $-\dfrac{16a^4b^{-1}}{12ab^{-4}}$

31. $\dfrac{45x^{-2}y^{-2}}{35x^{-4}y}$

EXAMPLE 14   **Simplifying Expressions Using the Power Rule and the Negative Exponent Rule**

Simplify $\left(\dfrac{25a^{-2}b}{10ab^{-1}}\right)^{-2}$. Write the answer with only positive exponents.

Solution

Because there are like factors inside the parentheses, we first perform operations within parentheses before we apply the Power Rule.

$$\left(\frac{25a^{-2}b}{10ab^{-1}}\right)^{-2} = \left(\frac{25 \cdot b \cdot b}{10 \cdot a^2 \cdot a}\right)^{-2}$$

Simplify $\dfrac{25}{10}$; use $a^m a^n = a^{m+n}$: $= \left(\dfrac{5b^2}{2a^3}\right)^{-2}$

Use $\left(\dfrac{a}{b}\right)^{-n} = \left(\dfrac{b}{a}\right)^{n}$: $= \left(\dfrac{2a^3}{5b^2}\right)^{2}$

Use $\left(\dfrac{a}{b}\right)^{n} = \dfrac{a^n}{b^n}$: $= \dfrac{(2a^3)^2}{(5b^2)^2}$

Use $(ab)^n = a^n b^n$ in the numerator and denominator: $= \dfrac{4a^6}{25b^4}$

QUICK ✓   *Simplify. Write the answers with only positive exponents.*

32. $(3y^2z^{-3})^{-2}$

33. $\left(\dfrac{2wz^{-3}}{7w^{-1}}\right)^{2}$

EXAMPLE 15   **Simplifying Exponential Expressions Using Exponent Rules**

Simplify: $(6c^{-2}d)\left(\dfrac{3c^{-3}d^4}{2d}\right)^{2}$

Write the answer with only positive exponents.

Solution

Since the quotient in the parentheses is messy, we will simplify it first.

$$(6c^{-2}d)\left(\frac{3c^{-3}d^4}{2d}\right)^{2} = \frac{6c^{-2}d}{1}\left(\frac{3c^{-3}d^{4-1}}{2}\right)^{2}$$

Simplify; use $a^{-n} = \dfrac{1}{a^n}$: $= \dfrac{6d}{c^2}\left(\dfrac{3d^3}{2c^3}\right)^{2}$

$$\text{Use } \left(\frac{a}{b}\right)^n = \frac{a^n}{b^n} \text{ and } (ab)^n = a^n b^n: \quad = \frac{6d}{c^2}\left(\frac{3^2(d^3)^2}{2^2(c^3)^2}\right)$$

$$\text{Use } (a^m)^n = a^{m \cdot n}: \quad = \frac{6d}{c^2}\left(\frac{9d^6}{4c^6}\right)$$

$$\text{Simplify; rearrange factors:} \quad = 6 \cdot \frac{9}{4} \cdot \frac{d \cdot d^6}{c^2 \cdot c^6}$$

$$\text{Evaluate the product } 6 \cdot \frac{9}{4}; \quad = \frac{27}{2} \cdot \frac{d^{1+6}}{c^{2+6}}$$

$$\text{Use } a^m a^n = a^{m+n}: \quad = \frac{27d^7}{2c^8}$$

QUICK ✓ *Simplify the expression. Write answers with only positive exponents.*

34. $\left(\dfrac{-6p^{-2}}{p}\right)(3p^8)^{-1}$

35. $(-25k^5r^{-2})\left(\dfrac{2}{5}k^{-3}r\right)^2$

5.4 Exercises

For Extra Help:

Student Solutions Manual · CD Video · PH Math/Tutor Center · MathXL Tutorials on CD · MathXL® · MyMathLab

Concepts and Vocabulary

In Problems 1–3, fill in the blanks.

1. $\dfrac{a^m}{a^n} = $ _____ provided that $a \neq 0$.

2. Any expression with a base (other than zero) raised to the zero power has the value of _____.

3. $a^{-n} = $ _____ provided that $a \neq 0$.

In Problems 4–6, answer True or False to each statement.

4. When simplifying an exponential expression, it is possible to apply different exponent laws in different orders and still obtain the correct result.

5. $(-a)^n \cdot (-a)^n$ will always represent a positive number.

6. $\dfrac{6^{10}}{6^4} = 1^6$

7. Explain two different approaches to simplify $\left(\dfrac{x^8}{x^2}\right)^3$. Which do you prefer?

8. Use the Quotient Rule and the expression $\dfrac{a^n}{a^n}, a \neq 0$, to explain why $a^0 = 1, a \neq 0$.

Building Skills

In Problems 9–24, use the Quotient Rule to simplify.

9. $\dfrac{3^5}{3^2}$

10. $\dfrac{4^{10}}{4^6}$

11. $\dfrac{2^{23}}{2^{19}}$

12. $\dfrac{10^5}{10^2}$

13. $\dfrac{x^{15}}{x^6}$

14. $\dfrac{x^{20}}{x^{14}}$

15. $\dfrac{a^{13}}{a}$

16. $\dfrac{a^{22}}{a}$

17. $\dfrac{16y^4}{4y}$

18. $\dfrac{9a^4}{27a}$

19. $\dfrac{-16m^{10}}{24m^3}$

20. $\dfrac{-36x^2y^5}{24xy^4}$

21. $\dfrac{-12m^9n^3}{-6mn}$ **22.** $\dfrac{-15r^9s^2}{-5r^8s}$ **23.** $\dfrac{22b^3c^7}{16bc^7}$ **24.** $\dfrac{39ab^6}{52ab}$

In Problems 25–32, use the Quotient to a Power Rule to simplify.

25. $\left(\dfrac{3}{2}\right)^3$ **26.** $\left(\dfrac{4}{9}\right)^2$ **27.** $\left(\dfrac{x^5}{3}\right)^3$ **28.** $\left(\dfrac{7}{y^2}\right)^2$

29. $\left(-\dfrac{x^5}{y^7}\right)^4$ **30.** $\left(-\dfrac{a^3}{b^{10}}\right)^5$ **31.** $\left(\dfrac{7a^2b}{c^3}\right)^2$ **32.** $\left(\dfrac{2mn^2}{q^3}\right)^4$

In Problems 33–38, use the Zero Exponent Rule to simplify.

33. 3^0 **34.** -100^0 **35.** $-\left(\dfrac{1}{2}\right)^0$ **36.** $\left(\dfrac{2}{5}\right)^0$

37. $18 \cdot 2^0$ **38.** $14^0 \cdot 3^0 \cdot 10$

In Problems 39–62, use the Negative Exponent Rules to simplify. Write answers with only positive exponents.

39. 3^{-1} **40.** 4^{-2} **41.** 10^{-3} **42.** 5^{-1}

43. m^{-2} **44.** k^{-2} **45.** $4a^{-2}$ **46.** $5b^{-3}$

47. $-4y^{-3}$ **48.** $-7z^{-5}$ **49.** $2^{-1} + 3^{-2}$ **50.** $4^{-2} - 2^{-3}$

51. $\left(\dfrac{1}{2}\right)^{-1}$ **52.** $\left(\dfrac{1}{3}\right)^{-2}$ **53.** $\left(\dfrac{2}{5}\right)^{-2}$ **54.** $\left(\dfrac{3}{2}\right)^{-3}$

55. $\left(\dfrac{3}{z^2}\right)^{-1}$ **56.** $\left(\dfrac{4}{p^2}\right)^{-2}$ **57.** $\left(-\dfrac{2n}{m^2}\right)^{-3}$ **58.** $\left(-\dfrac{5}{3b^2}\right)^{-3}$

59. $\dfrac{1}{4^{-2}}$ **60.** $\dfrac{1}{6^{-2}}$ **61.** $\dfrac{6}{x^{-4}}$ **62.** $\dfrac{4}{b^{-3}}$

Mixed Practice

In Problems 63–102, simplify. Write answers with only positive exponents.

63. $2^5 \cdot 2^{-3}$ **64.** $3^8 \cdot 3^{-6}$ **65.** $2^{-7} \cdot 2^4$ **66.** $10^{-4} \cdot 10^3$

67. $\dfrac{3}{3^{-3}}$ **68.** $\dfrac{5^4}{5^{-1}}$ **69.** $\dfrac{x^6}{x^{15}}$ **70.** $\dfrac{b^{24}}{b^{42}}$

71. $\dfrac{x^{10}}{x^{-3}}$ **72.** $\dfrac{a^{-4}}{a^6}$ **73.** $\dfrac{8x^2}{2x^5}$ **74.** $\dfrac{24m^4}{6m^8}$

75. $\dfrac{-27xy^3z^4}{18x^4y^3z}$ **76.** $\dfrac{8a^{15}b^2}{-18b^{20}}$ **77.** $\dfrac{-12a^2b^3c^6}{-15ab^4c^6}$

78. $(4x^{-6}y^3)(4^{-1}x^{10}y^{-7})$ **79.** $(3x^2y^{-3})(12^{-1}x^{-5}y^{-6})$ **80.** $(-14c^2d^4)(3c^{-6}d^{-4})$

81. $(-a^4)^{-3}$ **82.** $(-x^{-2})^5$ **83.** $(3m^{-2})^3$

84. $(2z^{-1})^3$ **85.** $(-3x^{-2}y^{-3})^{-2}$ **86.** $(-4a^{-3}b^{-2}c^{-1})^{-1}$

87. $2p^{-4} \cdot p^{-3} \cdot p^0$ **88.** $7a^{-2} \cdot a^0 \cdot a^{-3}$

89. $(-16a^3)(-3a^4)\left(\dfrac{1}{4}a^{-7}\right)$ **90.** $(-3x^{-4}y)(2x^{-3}y^{-6})\left(\dfrac{1}{36}x^{10}\right)$

91. $\dfrac{8x^2 \cdot x^7}{12x^{-3} \cdot x^4}$ **92.** $\dfrac{32x^{-3} \cdot x^2}{24x^2 \cdot x^5}$ **93.** $(2x^{-3}y^{-2})^4(3x^2y^{-3})^{-3}$

94. $(4a^{-2}b)^3(2a^2b^{-4})^{-2}$ **95.** $\left(\dfrac{y}{2z}\right)^{-3} \cdot \left(\dfrac{3y^2}{4z^3}\right)^2$ **96.** $\left(\dfrac{s}{4t^2}\right)^3 \cdot \left(\dfrac{3s^2}{2t^4}\right)^{-4}$

97. $(4x^2y)^3 \cdot \left(\dfrac{2x}{3y}\right)^{-3}$ **98.** $(2a^5b^2)^3 \cdot \left(\dfrac{4a^{-2}b^3}{3a}\right)^{-2}$ **99.** $\dfrac{(5a^{-3}b^2)^2}{a^{-4}b^{-4}}(15a^{-3}b)^{-1}$

100. $\left(\dfrac{2x^{-4}y^{-3}}{4xy^3}\right)^{-2}(4x^2y^{-1})^{-2}$ **101.** $\dfrac{4x^{-3}(2x^3)}{20(x^{-3})^2}$ **102.** $\dfrac{(-2a^{-3}b^2)^{-4}}{5a^2b^2c^0}$

Applying the Concepts

△ **103. Volume of a Cube** Write the polynomial in simplified form that represents the volume of a cube if each side has a length of $x + 2$ ft, where *volume* $=$ *side*^{3}.

△ **104. Volume of a Cube** Write the polynomial in simplified form that represents the volume of a cube if each side has a length of $2x - 1$ m, where *volume* $=$ *side*^{3}.

△ **105. Volume of a Box** The volume of a rectangular solid is given by the equation $V = l \cdot w \cdot h$. Find the volume of a shipping box whose dimensions are $l = x$ m, $w = x$ m, and $h = 3x$ m.

△ **106. Volume of a Box** The volume of a rectangular solid is given by the equation $V = l \cdot w \cdot h$. Find the volume of a shoe box whose dimensions are $l = 3n$ yards, $w = 2n$ yards, and $h = 6n$ yards.

△ **107. Volume of a Cylinder** The volume of a right circular cylinder is given by the equation $V = \pi r^2 h$, where r is the radius of the circular base of the cylinder and h is its height. Rewrite this equation in terms of the diameter, d, of the base of the cylinder.

△ **108. Volume of a Cylinder** The volume of a right circular cylinder is given by the equation $V = \pi r^2 h$, where r is the radius of the circular base of the cylinder and h is its height. Write an algebraic expression for the volume of a cylinder in which the radius of the circular base is equal to the height of the cylinder.

△ **109. Volume of a Trough** A trough is in the shape of a triangular prism. The volume of a prism is found by multiplying the area of the triangle by the height of prism (the distance between the two triangular bases). Use this fact to find the polynomial which represents the volume of the trough whose dimensions (in centimeters) are shown to the right.

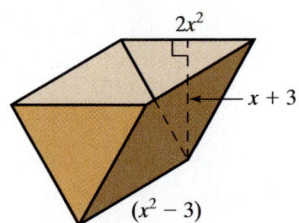

△ **110. Volume of a Trough** The triangular prism shown to the right has the following dimensions: $x + 2$ ft is the length of the base of the triangle and $4x^2$ ft is the height of the triangle. The length of the prism is $2x - 1$ ft. Write the polynomial which represents the volume of the trough.

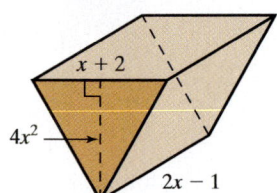

△ **111. Making Buttons** How much fabric is required to cover x buttons in the shape of a circle whose radius is $3x$?

△ **112. Making a Tablecloth** How much fabric is required to make a round tablecloth if the diameter of the tablecloth is $\left(\dfrac{1}{2}x\right)$ meters?

Extending the Concepts

In Problems 113–120, simplify. Write answers with only positive exponents.

113. $\dfrac{x^{2n}}{x^{3n}}$

114. $\dfrac{(x^{a-2})^3}{(x^{2a})^5}$

115. $\left(\dfrac{x^n y^m}{x^{4n-1} y^{m+1}}\right)^{-2}$

116. $\left(\dfrac{a^n b^{2m}}{ab^2}\right)^{-3}$

117. $(x^{2a} y^b z^{-c})^{3a}$

118. $(x^{-a} y^{-2b} z^{4c})^{2a}$

119. $\dfrac{(3a^n)^2}{(2a^{4n})^3}$

120. $\dfrac{y^a}{y^{4a}}$

PUTTING THE CONCEPTS TOGETHER (Sections 5.1–5.4)

These problems cover important concepts from Sections 5.1 to 5.4. We designed these problems so that you can review the chapter so far and show your mastery of the concepts. Take time to work these problems before proceeding with the next section. The answers to these problems are located at the back of the text starting on page AN-17.

1. Determine whether the following algebraic expression is a polynomial (Yes or No). If it is a polynomial, state the degree and then state if it is a monomial, binomial, or trinomial: $6x^2y^4 - 8x^5 + 3$.

2. Evaluate the polynomial $-x^2 + 3x$ for the given values: (a) 0   (b) −1   (c) 2.

In Problems 3–11, perform the indicated operation.

3. $(6x^4 - 2x^2 + 7) - (-2x^4 - 7 + 2x^2)$

4. $(2x^2y - xy + 3y^2) + (4xy - y^2 + 3x^2y)$

5. $-2mn(3m^2n - mn^3)$ 6. $(5x + 3)(x - 4)$

7. $(2x - 3y)(4x - 7y)$ 8. $(5x + 8)(5x - 8)$

9. $(2x + 3y)^2$ 10. $2a(3a - 4)(a + 5)$

11. $(4m + 3)(4m^3 - 2m^2 + 4m - 8)$

In Problems 12–20, simplify each expression. Write answers with only positive exponents.

12. $(-2x^7y^0z)(4xz^8)$ 13. $(5m^3n^{-2})(-3m^{-4}n)$ 14. $\dfrac{-18a^8b^3}{6a^5b}$

15. $\dfrac{16x^7y}{32x^9y^3}$ 16. $(4y^{-2}z^3)^{-2}$ 17. $\left(\dfrac{3}{2}r^2\right)^{-3}$

18. $\left(\dfrac{q^{-6}rt^5}{qr^{-4}t^{-3}}\right)^{-8}$ 19. $(4x^{-3}y^4)^{-2} \cdot (2x^4y^{-2})^{-3}$

20. $\left(\dfrac{2x^3y^{-3}}{4x^{-1}y^0}\right)^{-4} \cdot (-3x^4y^{-6})$

5.5 Dividing Polynomials

OBJECTIVES

1. Divide a Polynomial by a Monomial
2. Divide a Polynomial by a Binomial

Work Smart
A polynomial is a monomial or the sum of monomials.

Preparing for...Answers 1. $4a^2$ 2. $-\dfrac{7}{3x}$
3. $21x^2 - 6x$ 4. 2

Preparing for Dividing Polynomials
Before getting started, take the following readiness quiz. If you get a problem wrong, go back to the section cited and review the material.

1. Find the quotient: $\dfrac{24a^3}{6a}$ [Section 5.4, pp. 371–372]

2. Find the quotient: $\dfrac{-49x^3}{21x^4}$ [Section 5.4, p. 378]

3. Find the product: $3x(7x - 2)$ [Section 5.3, pp. 360–361]

4. Tell the degree of $4x^2 - 2x + 1$.   [Section 5.1, pp. 345–346]

We have now presented a discussion of adding, subtracting, and multiplying polynomials. All that's left to discuss is polynomial division! We begin with dividing a polynomial by a monomial. Recall, when we compute a quotient using long division, such as $15\overline{)345}$ with 23 above, the number 15 is called the *divisor*, the number 345 is called the *dividend*, and the number 23 is called the *quotient*. We use the same terminology with polynomial division.

① Divide a Polynomial by a Monomial

Dividing a polynomial by a monomial requires the Quotient Rule for Exponents. For convenience, we repeat the rule below.

> **QUOTIENT RULE FOR EXPONENTS**
> If a is a real number and m and n are integers, then
> $$\frac{a^m}{a^n} = a^{m-n}, \quad a \neq 0$$

In Words

To divide a polynomial by a monomial, divide each of the terms of the polynomial numerator (dividend) by the monomial denominator (divisor).

Remember, to add two rational numbers, the denominators must be the same. When we have the same denominator, we add the numerators and write the result over the common denominator. For example, $\frac{3}{11} + \frac{4}{11} = \frac{3+4}{11} = \frac{7}{11}$. When dividing a polynomial by a monomial, we reverse this process. So, if A, B, and C are monomials, we can write $\frac{A+B}{C}$ as $\frac{A}{C} + \frac{B}{C}$. We can extend this result to polynomials with three or more terms quite easily.

EXAMPLE 1 **Dividing a Binomial by a Monomial**

Divide and simplify: $\dfrac{9x^3 - 21x^2}{3x}$

Solution

We divide each term in the numerator by the denominator.

$$\frac{9x^3 - 21x^2}{3x} = \frac{9x^3}{3x} - \frac{21x^2}{3x}$$

Use the Quotient Rule, $\frac{a^m}{a^n} = a^{m-n}$: $= \frac{9}{3}x^{3-1} - \frac{21}{3}x^{2-1}$

Simplify: $= 3x^2 - 7x$ ∎

EXAMPLE 2 **Dividing a Trinomial by a Monomial**

Divide and simplify: $\dfrac{12p^4 + 24p^3 + 4p^2}{4p^2}$

Solution

Work Smart

$$\frac{12p^4 + 24p^3 + 4p^2}{4p^2} \neq$$

$$\frac{12p^4 + 24p^3 + 4p^2}{4p^2} \neq 12p^4 + 24p^3$$

You must divide *each* term in the numerator by the monomial in the denominator.

We divide each term in the numerator by the denominator.

$$\frac{12p^4 + 24p^3 + 4p^2}{4p^2} = \frac{12p^4}{4p^2} + \frac{24p^3}{4p^2} + \frac{4p^2}{4p^2}$$

Use the Quotient Rule, $\frac{a^m}{a^n} = a^{m-n}$: $= \frac{12}{4}p^{4-2} + \frac{24}{4}p^{3-2} + \frac{4}{4} \cdot \frac{p^2}{p^2}$

Simplify: $= 3p^2 + 6p + 1$ ∎

EXAMPLE 3   **Dividing a Trinomial by a Monomial**

Divide and simplify: $\dfrac{8a^2b^2 - 6a^2b + 5ab^2}{2a^2b^2}$

Solution

$$\frac{8a^2b^2 - 6a^2b + 5ab^2}{2a^2b^2} = \frac{8a^2b^2}{2a^2b^2} - \frac{6a^2b}{2a^2b^2} + \frac{5ab^2}{2a^2b^2}$$

Use the Quotient Rule, $\dfrac{a^m}{a^n} = a^{m-n}$: $= \dfrac{8}{2}a^{2-2}b^{2-2} - \dfrac{6}{2}a^{2-2}b^{1-2} + \dfrac{5}{2}a^{1-2}b^{2-2}$

Simplify: $= 4a^0b^0 - 3a^0b^{-1} + \dfrac{5}{2}a^{-1}b^0$

Use the Negative Exponent Rule, $a^{-n} = \dfrac{1}{a^n}$: $= 4 \cdot 1 \cdot 1 - 3 \cdot 1 \cdot \dfrac{1}{b} + \dfrac{5}{2} \cdot \dfrac{1}{a} \cdot 1$

$= 4 - \dfrac{3}{b} + \dfrac{5}{2a}$ ∎

QUICK ✓   *Find the quotient.*

1. $\dfrac{10n^4 - 20n^3 + 5n^2}{5n^2}$ **2.** $\dfrac{12k^4 - 18k^2 + 5}{2k^2}$

3. $\dfrac{x^4y^4 + 8x^2y^2 - 4xy}{4x^3y}$

---

(2)   **Divide a Polynomial by a Binomial**

The procedure for dividing a polynomial by a binomial is similar to the procedure for dividing two integers. Although this procedure should be familiar to you, we review it in the following example.

EXAMPLE 4   **Dividing an Integer by an Integer Using Long Division**

Divide 579 by 16.

Solution

$$
\begin{array}{r}
36 \leftarrow \text{quotient} \\
\text{divisor} \longrightarrow 16\overline{)579} \leftarrow \text{dividend} \\
\underline{48} \quad \leftarrow 3 \cdot 16 = 48 \text{ (subtract)} \\
99 \quad \text{bring down the 9} \\
\underline{96} \leftarrow 6 \cdot 16 = 96 \text{ (subtract)} \\
3 \leftarrow \text{remainder}
\end{array}
$$

So $\dfrac{579}{16} = 36\dfrac{3}{16}$.   ∎

We can always check our work after completing a long division problem by multiplying the quotient by the divisor and adding this product to the remainder. The result should be the dividend. That is,

$$(\text{Quotient})(\text{Divisor}) + \text{Remainder} = \text{Dividend}$$

For example, we can check the results of Example 4 as follows:

$$(36)(16) + 3 = 576 + 3 = 579$$

One other comment regarding Example 4: we wrote the solution as $\frac{579}{16} = 36\frac{3}{16}$. Remember, the mixed number $36\frac{3}{16}$ means $36 + \frac{3}{16}$. So, the answer is written in the form

$$\text{Quotient} + \frac{\text{Remainder}}{\text{Divisor}}$$

Now let's go over an example that introduces "how to" divide a polynomial by a binomial using long division.

EXAMPLE 5   How to Divide a Polynomial by a Binomial Using Long Division

Find the quotient when $x^2 + 10x + 21$ is divided by $x + 7$.

Step-by-Step Solution

To divide two polynomials, we first write each polynomial in standard form (descending order of degree). The dividend is $x^2 + 10x + 21$ and the divisor is $x + 7$.

Step 1: Divide the highest degree term of the dividend, x^2, by the highest degree term of the divisor, x. Enter the result over the term x^2.	$\begin{array}{r} x \\ x + 7 \overline{)\, x^2 + 10x + 21\,} \end{array}$ $\quad \frac{x^2}{x} = x$
Step 2: Multiply x by $x + 7$. Be sure to vertically align like terms.	$\begin{array}{r} x \\ x + 7 \overline{)\, x^2 + 10x + 21\,} \\ \underline{x^2 + 7x} \end{array}$ $\quad x(x + 7) = x^2 + 7x$
Step 3: Subtract $x^2 + 7x$ from $x^2 + 10x + 21$.	$\begin{array}{r} x \\ x + 7 \overline{)\, x^2 + 10x + 21\,} \\ \underline{-(x^2 + 7x)} \\ 3x + 21 \end{array}$ $\quad (x^2 + 10x + 21) - (x^2 + 7x) = 3x + 21$
Step 4: Repeat Steps 1–3 treating $3x + 21$ as the dividend.	$\begin{array}{r} x + 3 \\ x + 7 \overline{)\, x^2 + 10x + 21\,} \\ \underline{-(x^2 + 7x)} \\ 3x + 21 \\ \underline{-(3x + 21)} \\ 0 \end{array}$ $\quad \frac{3x}{x} = 3$ $\quad (3x + 21) - (3x + 21) = 0$ The quotient is $x + 3$ and the remainder is 0.
Step 5: Verify by showing that (Quotient)(Divisor) + Remainder = Dividend.	$(x + 3)(x + 7) + 0 = x^2 + 10x + 21$

The product checks, so $x^2 + 10x + 21$ divided by $x + 7$ is $x + 3$. We can express this division using fractions, as follows:

$$\frac{x^2 + 10x + 21}{x + 7} = x + 3$$

EXAMPLE 6   **How to Divide a Polynomial by a Binomial Using Long Division**

Find the quotient when $6x^2 + 9x - 10$ is divided by $2x - 1$.

Step-by-Step Solution

Each polynomial is in standard form. The dividend is $6x^2 + 9x - 10$ and the divisor is $2x - 1$.

Step 1: Divide the highest degree term of the dividend, $6x^2$, by the highest degree term of the divisor, $2x$. Enter the result over the term $6x^2$.

$$\frac{6x^2}{2x} = 3x$$

$$2x - 1 \overline{)6x^2 + 9x - 10} \quad\quad 3x$$

Step 2: Multiply $3x$ by $2x - 1$. Be sure to vertically align like terms.

$$
\begin{array}{r}
3x \\
2x - 1 \overline{)6x^2 + 9x - 10} \\
6x^2 - 3x
\end{array}
$$

$3x(2x - 1) = 6x^2 - 3x$

Step 3: Subtract $6x^2 - 3x$ from $6x^2 + 9x - 10$.

$$
\begin{array}{r}
3x \\
2x - 1 \overline{)6x^2 + 9x - 10} \\
-(6x^2 - 3x) \\
\hline
12x - 10
\end{array}
$$

$6x^2 + 9x - 10 - (6x^2 - 3x) = 12x - 10$

Step 4: Repeat Steps 1–3 treating $12x - 10$ as the dividend.

$$\frac{12x}{2x} = 6$$

$$
\begin{array}{r}
3x + 6 \\
2x - 1 \overline{)4x^2 + 9x - 10} \\
-(6x^2 - 3x) \\
\hline
12x - 10 \\
-(12x - 6) \\
\hline
-4
\end{array}
$$

$6(2x - 1) = 12x - 6$

$(12x - 10) - (12x - 6)$
$= 12x - 12x - 10 + 6 = -4$

Because -4 is a lower degree than the divisor, $2x - 1$, the process ends. The quotient is $3x + 6$ and the remainder is -4.

Step 5: Check We verify that (Quotient)(Divisor) + Remainder = Dividend.

$(3x + 6)(2x - 1) + (-4) = 6x^2 - 3x + 12x - 6 + (-4)$

Combine like terms: $= 6x^2 + 9x - 10$

The product checks, so our answer is correct. So

$$\frac{6x^2 + 9x - 10}{2x - 1} = 3x + 6 + \frac{-4}{2x - 1}.$$

Work Smart
You know that you're finished dividing when the degree of the remainder is less than the degree of the divisor.

QUICK ✓ *Find the quotient by performing long division.*

4. $\dfrac{x^2 - 3x - 40}{x + 5}$ **5.** $\dfrac{2x^2 - 5x - 12}{2x + 3}$ **6.** $\dfrac{4x^2 + 17x + 21}{x + 3}$

EXAMPLE 7   **Dividing Two Polynomials Using Long Division**

Simplify by performing long division: $\dfrac{8 - 9x + 2x^2 + 12x^3 + 5x^5}{x^2 + 3}$

Solution

Do you notice that the dividend is not written in standard form (that is, the dividend is not written in descending order of degree)? Also, when we write the dividend in standard form, do you notice that there is no x^4 term? When a term is missing, its coefficient is 0, so we rewrite the division problem as follows:

$$\frac{5x^5 + 0 \cdot x^4 + 12x^3 + 2x^2 - 9x + 8}{x^2 + 3}.$$

$$
\begin{array}{r}
5x^3 \qquad\quad - 3x\ + 2 \\
x^2 + 3\overline{)5x^5 + 0x^4 + 12x^3 + 2x^2 - 9x + 8} \\
-(5x^5 \qquad\quad + 15x^3) \qquad\qquad\quad 5x^3(x^2 + 3) \\
-3x^3 + 2x^2 - 9x + 8 \qquad 5x^5 + 12x^3 + 2x^2 - 9x + 8 - (5x^5 + 15x^3) = -3x^3 + 2x^2 - 9x + 8 \\
-(-3x^3 \qquad\quad - 9x) \qquad\quad -3x(x^2 + 3) \\
2x^2 \qquad\quad + 8 \qquad (-3x^3 + 2x^2 - 9x + 8) - (-3x^3 - 9x) = 2x^2 + 8 \\
-(2x^2 \qquad\quad + 6) \qquad 2(x^2 + 3) \\
2 \quad \leftarrow \text{Remainder}
\end{array}
$$

The quotient is $5x^3 - 3x + 2$ and the remainder is 2. We now check our work.

Check (Quotient)(Divisor) + Remainder = Dividend

$$
\begin{aligned}
(5x^3 - 3x + 2)(x^2 + 3) + 2 &= 5x^5 + 15x^3 - 3x^3 - 9x + 2x^2 + 6 + 2 \\
&= 5x^5 + 12x^3 + 2x^2 - 9x + 8
\end{aligned}
$$

Our answer checks, so $\dfrac{8 - 9x + 2x^2 + 12x^3 + 5x^5}{x^2 + 3} = 5x^3 - 3x + 2 + \dfrac{2}{x^2 + 3}.$ ∎

QUICK ✓ *Simplify by performing long division.*

7. $\dfrac{x + 1 - 3x^2 + 4x^3}{x + 2}$ **8.** $\dfrac{2x^3 + 3x^2 + 10}{2x - 5}$ **9.** $\dfrac{4x^3 - 3x^2 + x + 1}{x^2 + 2}$

5.5 Exercises

For Extra Help:

Student Solutions Manual   CD Video   PH Math/Tutor Center   MathXL Tutorials on CD   MathXL®   MyMathLab

Concepts and Vocabulary

In Problems 1–3, fill in the blanks.

1. To begin a polynomial division problem, write the divisor and the dividend in _____ form.

2. The first step to simplify $\dfrac{4x^4 + 8x^2}{2x}$ would be to rewrite $\dfrac{4x^4 + 8x^2}{2x}$ as _____ + _____.

3. To check the result of long division, multiply the _____ and the divisor and add this result to the _____. If correct, this result will be equal to the _____.

In Problems 4–6, answer True or False to each statement.

4. The division problem $\dfrac{3x^2 - 4}{x}$ can be written as $\dfrac{3x^2}{x} - \dfrac{4}{x}$.

5. $\dfrac{1}{x - 3} = \dfrac{1}{x} - \dfrac{1}{3}$

6. $\dfrac{x^2 - 2}{x + 2} = x - 1$

7. Explain how to divide polynomials when the divisor is a monomial and then when the divisor is a binomial. Which procedures are the same and which are different?

8. The first steps of a division problem are written below. Describe what has occurred. Are there any potential errors with this presentation? What would you recommend this student do to improve his or her chances of obtaining the correct answer?

$$
\begin{array}{r}
2x^2 \\
x^2 - 3 \overline{) \, 2x^4 - 3x^3 \; + 2x + 1} \\
-(2x^4 - 6x^2)
\end{array}
$$

Building Skills

In Problems 9–28, divide and simplify.

9. $\dfrac{4x^2 - 2x}{2x}$

10. $\dfrac{3x^3 - 6x^2}{3x^2}$

11. $\dfrac{9a^3 + 27a^2 - 3}{3a^2}$

12. $\dfrac{16m^3 + 8m^2 - 4}{8m^2}$

13. $\dfrac{5n^5 - 10n^3 - 25n}{25n}$

14. $\dfrac{5x^3 - 15x^2 + 10x}{5x^2}$

15. $\dfrac{15r^5 - 27r^3}{9r^3}$

16. $\dfrac{16a^5 - 12a^4 + 8a^3}{20a^4}$

17. $\dfrac{3x^7 - 9x^6 + 27x^3}{-3x^5}$

18. $\dfrac{7p^4 + 21p^3 - 3p^2}{-6p^4}$

19. $\dfrac{3z + 4z^3 - 2z^2}{8z}$

20. $\dfrac{-5y^2 + 15y^4 - 16y^5}{5y^2}$

21. $\dfrac{6x^2 + 9x^3 + 4}{6x}$

22. $\dfrac{4 + 7x^2 - 3x^4 + 6x^3}{2x^2}$

23. $\dfrac{14x - 10y}{-2}$

24. $\dfrac{35x + 20y}{-5}$

25. $\dfrac{12y - 30x}{-2x}$

26. $\dfrac{21y^2 + 35x^2}{-7x^2}$

27. $\dfrac{25a^3b^2c + 10a^2bc^3}{-5a^4b^2c}$

28. $\dfrac{16m^2n^3 - 24m^4n^3}{-8m^3n^4}$

In Problems 29–56, find the quotient using long division.

29. $\dfrac{x^2 - 4x - 21}{x + 3}$

30. $\dfrac{x^2 + 18x + 72}{x + 6}$

31. $\dfrac{x^2 - 9x + 20}{x - 4}$

32. $\dfrac{x^2 + 4x - 32}{x - 4}$

33. $\dfrac{x^3 + 4x^2 - 15x + 6}{x - 2}$

34. $\dfrac{x^3 - x^2 - 40x + 12}{x + 6}$

35. $\dfrac{x^4 - x^3 + 10x - 4}{x + 2}$

36. $\dfrac{x^4 - 2x^3 + x^2 + x - 1}{x - 1}$

37. $\dfrac{x^3 - x^2 + x + 8}{x + 1}$

38. $\dfrac{x^3 - 7x^2 + 15x - 11}{x - 3}$

39. $\dfrac{2x^2 - 7x - 15}{x - 5}$

40. $\dfrac{2x^2 + 5x - 42}{x + 6}$

41. $\dfrac{x^3 + 4x^2 - 5x + 2}{x - 2}$

42. $\dfrac{x^3 - 2x^2 + x + 6}{x + 1}$

43. $\dfrac{2x^4 - 3x^3 - 11x^2 - 40x - 1}{x - 4}$

44. $\dfrac{3x^4 + 7x^3 - 5x^2 + 8x + 12}{x + 3}$

45. $\dfrac{6x^2 - 28x + 30}{3x - 5}$

46. $\dfrac{12x^2 - 25x - 50}{3x - 10}$

47. $\dfrac{2x^3 + 7x^2 - 10x + 5}{2x - 1}$

48. $\dfrac{12a^3 + 11a^2 + 18a + 9}{4a + 1}$

49. $\dfrac{-24 + x^2 + x}{5 + x}$

50. $\dfrac{-16x + 70 + x^2}{-9 + x}$

51. $\dfrac{4x^2 + 5}{1 + 2x}$

52. $\dfrac{9x^2 - 14}{2 + 3x}$

53. $\dfrac{x^3 + 2x^2 - 8}{x^2 - 2}$

54. $\dfrac{64x^6 - 27}{4x^2 - 3}$

55. $\dfrac{18 + x^4 - 9x^2 + 3x^3 - 9x}{x^2 - 3}$

56. $\dfrac{x^2 - 1 + 6x^4 - x - 2x^3}{2x^2 + 1}$

Mixed Practice

In Problems 57–86, perform the indicated operation.

57. $(a - 5)(a + 6)$

58. $(7n + 3m^2 + 4m) + (-6m^2 - 7n + 4m)$

59. $(2x - 8) - (3x + x^2 - 2)$

60. $3x^2(7x^2 + 2x - 3)$

61. $\dfrac{6x^2 - 1 - x}{2x - 1}$

62. $\dfrac{2x^4 - 6x^3 - 5x + 1}{-2x^2}$

63. $(2ab + b^2 - a^2) + (b^2 - 4ab + a^2)$

64. $\dfrac{3a^4 b^{-2}}{6a^{-4} b^{-3}}$

65. $\dfrac{x^2 - 2x + 3}{x}$

66. $(b - 3)(b + 2)$

67. $\dfrac{18x^3 y^{-4} z^6}{27x^{-4} y^{-12} z^{-6}}$

68. $(4ab + 6ab^2) - (12a^2 b - 2ab - 3ab^2)$

69. $(n - 3)^2$

70. $\dfrac{3 + 6x^2 - 11x}{2x - 3}$

71. $(x^3 + x - 4x^4)(-10x^2)$

72. $(2x^3 + 6x) + (12x - x^2 - x^3)$

73. $(2pq - q^2) + (4pq - p^2 - q^2)$

74. $(4x^2 - 6x + 3) - (x - 7)$

75. $(4n - 2m)(n + 3m)$

76. $(2x - 3y)(3x - y)$

77. $\dfrac{x^2 - 8x + 5}{x - 5}$

78. $\dfrac{x^2 - 4x - 16}{x + 2}$

79. $(x^2 + x - 1)(x + 5)$

80. $(x^2 - 2x + 3)(x - 4)$

81. $(7rs^2 - 2r^2 s) - (2r^2 s - 8rs^2)$

82. $(x + 6)^2$

83. $\dfrac{-9mn^2 - 8m^2 n + 12mn}{3mn}$

84. $(x^4 - 2x^2 + x)(-3x)$

85. $2x^4(x^2 - 2x + 3)$

86. $\dfrac{2n^3 - 4n + 7n^2 - 1}{2n}$

Applying the Concepts

87. Find the quotient of $(3x^4 - 6x + 12x^2)$ and $-3x^3$.

88. Divide $x - 3$ by $x + 2$.

89. Divide the sum of $x^2 + 3x - 1$ and $x - 1$ by $-2x^3$.

90. Divide x^2 into the square of the difference of $x - 9$ and $x + 3$.

△ **91. Volume of a Box** The volume of a rectangular solid is $x^3 - 5x^2 + 6x$. One side measures $x - 2$ and the other measures $x - 3$. What is the measure of the third side?

△ **92. Area of a Rectangle** A rectangle has area $x^2 + 2x - 48$ square inches. If one side measures $x + 8$ inches, what is the measurement of the other side?

△ **93. Area of a Rectangle** If the area of a rectangle is $z^2 + 6z + 9$ and the length is $z + 3$, what is the width?

△ **94. Area of a Triangle** If the area of a triangle is $6a^2 - 5a - 6$ and the length of the base is $2a - 3$, what is the height?

△ **95. Area of a Triangle** If the area of a triangle is $6x^3 - 2x^2 - 8x$ and the height is $3x^2 - 4x$, what is the length of the base?

△ **96. Volume of a Box** The volume of a rectangular solid is $x^3 + 2x^2 - x - 2$ square feet. One side measures $x + 2$ feet and the other measures $x - 1$ feet. What is the measure of the third side?

97. Average Cost The average cost of manufacturing x computers per day is given by $\dfrac{0.004x^3 - 0.8x^2 + 180x + 5000}{x}$. Rewrite this quotient as the sum of four expressions by dividing x into each term in the numerator. Use this result to determine the average cost of manufacturing $x = 140$ computers in a day.

98. Average Cost The average cost of manufacturing x digital cameras per day is given by $\dfrac{0.0024x^3 - 0.4x^2 + 46x + 4000}{x}$. Rewrite this quotient as the sum of four expressions by dividing x into each term in the numerator. Use this result to determine the average cost of manufacturing $x = 90$ cameras in a day.

Extending the Concepts

In Problems 99 and 100, determine the value of the missing term so that the remainder is zero.

99. $\dfrac{(6x^2 - 13x + ?)}{2x - 3}$

100. $\dfrac{(3x^2 + ? - 5)}{x - 1}$

5.6 Applying Exponent Rules: Scientific Notation

OBJECTIVES

1. Convert Decimal Notation to Scientific Notation
2. Convert Scientific Notation to Decimal Notation
3. Use Scientific Notation to Multiply and Divide

Preparing for Applying Exponent Rules: Scientific Notation

Before getting started, take the following readiness quiz. If you get a problem wrong, go back to the section cited and review the material.

1. Find the product: $(3a^6)(4.5a^4)$ — [Section 5.2, p. 358]

2. Find the product: $(7n^3)(2n^{-2})$ — [Section 5.4, pp. 377–378]

3. Find the quotient: $\dfrac{3.6b^9}{0.9b^{-2}}$ — [Section 5.4, pp. 377–379]

Did you know that *Star Wars: Episode 1—The Phantom Menace* had box office revenues of $922,800,000? Did you know that the mass of a dust particle is 0.000000000753 kg? These numbers are difficult to write and difficult to read, so we use exponents to rewrite them.

1. ## Convert Decimal Notation to Scientific Notation

We use two types of notation to write numbers: decimal notation and scientific notation. Decimal notation is the notation we commonly see when we read the newspaper or a magazine. The numbers $922,800,000 and 0.000000000753 kg are written in decimal notation. When we write a number in scientific notation we express that number as the product of two factors: One factor is a number between 1 and 10, including 1 but not including 10, and the other factor is an integer power of 10.

> **DEFINITION**
>
> When a number has been written as the product of a number x, where $1 \le x < 10$, and a power of 10, it is said to be written in **scientific notation.** That is, a number is written in scientific notation when it is of the form
>
> $$x \times 10^N$$
>
> where
>
> $1 \le x < 10$ and N is an integer

Notice in the definition, $x < 10$. That's because when $x = 10$, we have 10^1, a power of ten.

For example, in scientific notation,

$$\text{Box Office Revenue for } \textit{Star Wars: The Phantom Menace} = \$9.228 \times 10^8 \text{ dollars}$$

$$\text{Mass of a dust particle} = 7.53 \times 10^{-10} \text{ kilograms}$$

In Words

When you have a number greater than or equal to 1, use $10^{\text{positive exponent}}$. When you have a number between 0 and 1, use $10^{\text{negative exponent}}$.

> ## Steps to Convert from Decimal Notation to Scientific Notation
>
> To change a positive number into scientific notation:
>
> **Step 1:** Count the number N of decimal places that the decimal point must be moved in order to arrive at a number x, where $1 \leq x < 10$.
>
> **Step 2:** If the original number is greater than or equal to 1, the scientific notation is $x \times 10^N$. If the original number is between 0 and 1, the scientific notation is $x \times 10^{-N}$.

EXAMPLE 1   How to Convert from Decimal Notation to Scientific Notation

Write 5283 in scientific notation.

Step-by-Step Solution

For a number to be in scientific notation, the decimal must be moved so that there is a single nonzero digit to the left of the decimal point. All remaining digits must appear to the right of the decimal point.

Step 1: The "understood" decimal point in 5283 follows the 3. Therefore, we will move the decimal to the left until it is between the 5 and the 2. Do you see why? This requires that we move the decimal $N = 3$ places.	5.283.
Step 2: The original number is greater than 1, so we write 5283 in scientific notation as	5.283×10^3

EXAMPLE 2   How to Convert from Decimal Notation to Scientific Notation

Write 0.054 in scientific notation.

Step-by-Step Solution

Step 1: Because 0.054 is less than 1, we shall move the decimal point to the right until it is between the 5 and the 4. This requires that we move the decimal $N = 2$ places.	0.054
Step 2: The original number is between 0 and 1, so we write 0.054 in scientific notation as	5.4×10^{-2}

QUICK ✓ *Write each number in scientific notation.*

1. 432   **2.** 10,302   **3.** 5,432,000   **4.** 0.093   **5.** 0.0000459   **6.** 0.00000008

② **Convert Scientific Notation to Decimal Notation**

Now we are going to convert a number from scientific notation to decimal notation. Study Table 3 to discover the pattern.

Table 3			
Scientific Notation	**Product**	**Decimal Notation**	**Location of Decimal Point**
3.69×10^2	3.69×100	369	moved 2 places to the right
3.69×10^1	3.69×10	36.9	moved 1 place to the right
3.69×10^0	3.69×1	3.69	didn't move
3.69×10^{-1}	3.69×0.1	0.369	moved 1 place to the left
3.69×10^{-2}	3.69×0.01	0.0369	moved 2 places to the left

We present the following steps for converting a number from scientific notation to decimal notation.

> **Steps to Convert a Number from Scientific Notation to Decimal Notation**
>
> **Step 1:** Determine the exponent on the number 10.
>
> **Step 2:** If the exponent is positive, then move the decimal N decimal places to the right. If the exponent is negative, then move the decimal $|N|$ decimal places to the left.

EXAMPLE 3 **How to Convert from Scientific Notation to Decimal Notation**

Write 2.3×10^3 in decimal notation.

Step-by-Step Solution

Step 1: Determine the exponent on the number 10.	The exponent on the 10 is 3.
Step 2: Since the exponent is positive, we move the decimal point 3 places to the right. Notice we add zeros to the right of 3, as needed.	2.300

So, $2.3 \times 10^3 = 2300$.

EXAMPLE 4 **How to Convert from Scientific Notation to Decimal Notation**

Write 4.57×10^{-5} in decimal notation.

Step-by-Step Solution

Step 1: Determine the exponent on the number 10.		The exponent on the 10 is -5.
Step 2: Since the exponent is negative, we move the decimal 5 places to the left.	Add zeros to the left of the original decimal point.	0.0000457

So, $4.57 \times 10^{-5} = 0.0000457$.

Work Smart

In Example 3, notice that

$$2.3 \times 10^3 = 2.3 \times 1000$$
$$= 2300$$

In Example 4, notice that

$$4.57 \times 10^{-5}$$
$$= 4.57 \times 0.00001$$
$$= 0.0000457$$

QUICK ✓ *Write each number in decimal notation.*

7. 3.1×10^2 **8.** 9.01×10^{-1} **9.** 1.7×10^5 **10.** 7×10^0 **11.** 8.9×10^{-4}

---

(3) ## Use Scientific Notation to Multiply and Divide

The Laws of Exponents make it relatively straightforward to multiply and divide numbers that are written in scientific notation. The two Laws of Exponents that we make use of are the Product Rule, $a^m \cdot a^n = a^{m+n}$, and the Quotient Rule, $\dfrac{a^m}{a^n} = a^{m-n}$. We will use these laws where the base is 10 as follows:

$$10^m \cdot 10^n = 10^{m+n} \quad \text{and} \quad \frac{10^m}{10^n} = 10^{m-n}$$

EXAMPLE 5 **Multiplying Using Scientific Notation**

Perform the indicated operation. Express the answer in scientific notation.

$$(3 \times 10^2) \cdot (2.5 \times 10^5)$$

Solution

$$(3 \times 10^2) \cdot (2.5 \times 10^5) = (3 \cdot 2.5) \times (10^2 \cdot 10^5)$$

Multiply $3 \cdot 2.5$; use the Product Rule for
Exponents on the base 10 factors: $= 7.5 \times 10^7$

EXAMPLE 6 **Multiplying Using Scientific Notation**

Perform the indicated operation. Express the answer in scientific notation.

(a) $(4 \times 10^{-2}) \cdot (6 \times 10^8)$ **(b)** $(3.2 \times 10^{-3}) \cdot (4.8 \times 10^{-4})$

Solution

(a) $(4 \times 10^{-2}) \cdot (6 \times 10^8) = (4 \cdot 6) \times (10^{-2} \cdot 10^8)$

Multiply $4 \cdot 6$; use the Product Rule: $= 24 \times 10^6$

Convert 24 to scientific notation: $= (2.4 \times 10^1) \times 10^6$

Apply the Product Rule for Exponents: $= 2.4 \times 10^7$

(b) $(3.2 \times 10^{-3}) \cdot (4.8 \times 10^{-4}) = (3.2 \cdot 4.8) \times (10^{-3} \cdot 10^{-4})$

Multiply $3.2 \cdot 4.8$; use the Product Rule: $= 15.36 \times 10^{-7}$

Convert 15.36 to scientific notation: $= (1.536 \times 10^1) \times 10^{-7}$

Apply the Product Rule for Exponents: $= 1.536 \times 10^{-6}$

QUICK ✓ *Perform the indicated operation. Express the answer in scientific notation.*

12. $(3 \times 10^4) \cdot (2 \times 10^3)$ **13.** $(2 \times 10^{-2}) \cdot (4 \times 10^{-1})$

14. $(5 \times 10^{-4}) \cdot (3 \times 10^7)$ **15.** $(8 \times 10^{-4}) \cdot (3.5 \times 10^{-2})$

EXAMPLE 7 Dividing Using Scientific Notation

Perform the indicated operation. Express the answer in scientific notation.

(a) $\dfrac{6 \times 10^6}{2 \times 10^2}$

(b) $\dfrac{2.4 \times 10^4}{3 \times 10^{-2}}$

Solution

(a) Divide $\dfrac{6}{2}$; use the Quotient Rule for Exponents:

$$\frac{6 \times 10^6}{2 \times 10^2} = \frac{6}{2} \times \frac{10^6}{10^2}$$
$$= 3 \times 10^4$$

(b) Divide $\dfrac{2.4}{3}$; use the Quotient Rule for Exponents:

$$\frac{2.4 \times 10^4}{3 \times 10^{-2}} = \frac{2.4}{3} \times \frac{10^4}{10^{-2}}$$
$$= 0.8 \times 10^{4-(-2)}$$

Convert 0.8 to scientific notation:

$$= (8 \times 10^{-1}) \times 10^6$$
$$= 8 \times 10^5 \quad \blacksquare$$

QUICK ✔ *Perform the indicated operation. Express the answer in scientific notation.*

16. $\dfrac{8 \times 10^6}{2 \times 10^1}$

17. $\dfrac{2.8 \times 10^{-7}}{1.4 \times 10^{-3}}$

18. $\dfrac{3.6 \times 10^3}{7.2 \times 10^{-1}}$

19. $\dfrac{5 \times 10^{-2}}{8 \times 10^2}$

EXAMPLE 8 Finding Newspaper Circulation

In 2003, *USA Today* had a daily circulation of 2.2×10^6 newspapers. What was the circulation of *USA Today* in the month of April when *USA Today* was published 21 times? (*Note*: *USA Today* does not publish on weekends.) Express the answer in scientific notation and in decimal notation.(*Source: Information Please* Almanac)

Solution

To find the total monthly circulation for the newspaper in April, we multiply the number of daily copies by the number of publication days. There are 21 publication days in April.

Let's first change 21 to scientific notation: $21 = 2.1 \times 10^1$. So we have

$$(2.2 \times 10^6) \cdot (2.1 \times 10^1) = (2.2 \cdot 2.1) \times (10^6 \cdot 10^1)$$
$$= 4.62 \times 10^7$$
$$= 46{,}200{,}000 \text{ newspapers.}$$

A total of 46,200,000 copies of *USA Today* were circulated in April 2003. $\quad \blacksquare$

QUICK ✔

20. Saudi Arabia produced nearly 8.4×10^6 barrels of oil per day in August 2003. How many barrels of oil did Saudi Arabia produce in August 2003? Express the solution in scientific notation and decimal notation.

5.6 Exercises

For Extra Help:

Student Solutions Manual   CD Video   PH Math/Tutor Center   MathXL Tutorials on CD   MathXL®   MyMathLab

Concepts and Vocabulary

In Problems 1–3, fill in the blanks.

1. A number written as 3.2×10^{-6} is said to be written in _____ _____.

2. To write 3.2×10^{-6} in decimal notation, move the decimal point in 3.2 six places to the _____.

3. When writing 47,000,000 in scientific notation, the power of 10 will be _____. (positive or negative).

In Problems 4–6, answer True or False to each statement.

4. To convert 2.4×10^3 to decimal notation, move the decimal three places to the right.

5. When a number is expressed in scientific notation, it is expressed as the product of a number $x, 0 \le x < 1$, and a power of 10.

6. It is useful to do calculations with scientific notation when using very large or very small numbers.

7. Explain the advantages of using scientific notation. Are there any disadvantages?

8. Explain why scientific notation is used to perform calculations that involve multiplying and dividing but not adding and subtracting.

Building Skills

In Problems 9–24, write each number in scientific notation.

9. 300,000	10. 421,000,000	11. 64,000,000	12. 8,000,000,000
13. 0.00051	14. 0.0000001	15. 0.000000001	16. 0.0000283
17. 8,007,000,000	18. 401,000,000	19. 0.0000309	20. 0.000201
21. 620	22. 8	23. 4	24. 120

In Problems 25–40, write each number in decimal notation.

25. 4.2×10^5	26. 3.75×10^2	27. 1×10^8	28. 6×10^6
29. 3.9×10^{-3}	30. 6.1×10^{-6}	31. 4×10^{-1}	32. 5×10^{-4}
33. 3.76×10^3	34. 4.9×10^{-1}	35. 8.2×10^{-3}	36. 5.4×10^5
37. 6×10^{-5}	38. 5.123×10^{-3}	39. 7.05×10^6	40. 7×10^8

In Problems 41–48, if a number is given in scientific notation, write it in decimal notation. If it is written in decimal notation, write it in scientific notation.

41. 0.0005	42. 1×10^9	43. 1.42×10^5	44. 6,000,000,000
45. 8×10^{-3}	46. 0.0035	47. 69,000	48. 4×10^{-1}

In Problems 49–54, a number is given in decimal notation. Write the number in scientific notation.

49. **World Population**   At the beginning of 2005, the population of the world was approximately 6,415,000,000 persons.

50. **United States Population**   At the beginning of 2005, the population of the United States was approximately 294,975,000 persons.

51. **Stock Exchange Daily Volume**   The average daily volume of the New York Stock Exchange in the year 2004 was 1.46 billion shares. (*Source:* New York Stock Exchange)

52. **Stock Exchange Total Volume**   The total volume of shares traded on the New York Stock Exchange in 2004 was approximately 367 billion shares. (*Source:* New York Stock Exchange)

53. Smallpox Virus The diameter of a smallpox virus is 0.00003 mm.

54. Human Blood Cell The diameter of a human blood cell is 0.0075 mm.

In Problems 55–60, a number is given in scientific notation. Write the number in decimal notation.

55. Farms In the year 2002, there were 2.158×10^6 farms in the United States.

56. Crude Oil Imports In the year 2002, the United States imported 3.302×10^9 barrels of crude oil.

57. Time A femtosecond is equal to 1×10^{-15} second.

58. Dust Particle The mass of a dust particle is 7.53×10^{-10} kg.

59. Vitamin A One-A-Day vitamin pill contains 2.25×10^{-3} grams of zinc.

60. Water Molecule The diameter of a water molecule is 3.85×10^{-7} m.

Mixed Practice

In Problems 61–78, perform the indicated operations. Express your answer in scientific notation.

61. $(2 \times 10^6)(1.5 \times 10^3)$ **62.** $(3 \times 10^{-4})(8 \times 10^{-5})$

63. $(1.2 \times 10^0)(7 \times 10^{-3})$ **64.** $(4 \times 10^7)(2.5 \times 10^{-4})$

65. $(7 \times 10^{-6})(1.5 \times 10^2)$ **66.** $(9 \times 10^{-1})(2 \times 10^6)$

67. $\dfrac{9 \times 10^4}{3 \times 10^{-4}}$ **68.** $\dfrac{6 \times 10^3}{1.2 \times 10^5}$ **69.** $\dfrac{2 \times 10^{-3}}{8 \times 10^{-5}}$ **70.** $\dfrac{4.8 \times 10^7}{1.2 \times 10^2}$

71. $\dfrac{5.4 \times 10^6}{3 \times 10^{-2}}$ **72.** $\dfrac{1.44 \times 10^{-3}}{1.2 \times 10^5}$ **73.** $\dfrac{56,000}{0.00007}$ **74.** $\dfrac{0.000275}{2,500}$

 **75.** $\dfrac{300,000 \times 15,000,000}{0.0005}$ **76.** $\dfrac{24,000,000,000}{0.00006 \times 2,000}$

77. $\dfrac{0.00072}{3,000 \times 8,000,000}$ **78.** $\dfrac{0.00075 \times 0.0000003}{500,000}$

Applying the Concepts

In 2003 and 2004, two NASA rovers landed on Mars. Landing a rover such as NASA's Opportunity or Spirit on the Martian surface required several course corrections as it traveled from Earth to Mars.

79. The rovers were launched from unmanned *Boeing Delta II* launch vehicles once out of our atmosphere and out of the grasp of Earth's gravity. The trajectory the rovers were originally launched on was designed to miss Mars by 211,000 miles. Write 211,000 in scientific notation.

80. The rovers traveled in an indirect path to Mars. The distance from Earth to Mars on the day the *Spirit* rover landed was 170,200,000 kilometers. Write 170,200,000 in scientific notation.

81. However, after traveling around Earth to gather speed and including the course corrections *Spirit* had to make to ensure a successful landing, the total distance traveled by the rover was 487,000,000 kilometers. Write 487,000,000 in scientific notation.

82. When the second NASA rover, *Opportunity,* landed on Mars in January 2004, the distance between the Earth and Mars was 198,700,000 miles. Write the distance between Earth and Mars in January 2004 in scientific notation.

83. Factoring in all of *Opportunity*'s course corrections and extra distance traveled, that craft traveled 456,000,000 kilometers. Write the total distance that *Opportunity* traveled in scientific notation.

84. The two-lander *NASA* project cost $820,000,000. Write $820,000,000 in scientific notation.

Once on Mars, the rovers looked for geological evidence that life may have existed at some point. NASA is planning future missions to Mars to look for a biological footprint of life.

85. Future missions are searching only for certain molecules present in Martian soil or ice. The tests will include the use of molecules engineered here on Earth that will give us a visible response if the molecules being sought after are found. These engineered molecules are expensive and difficult to create, however, so very dilute solutions will be used. Scientists express the concentration of solutions in molarity, abbreviated M. The solutions of engineered molecules in Martian experiments will be on the order of 0.000000025 M. Write 0.000000025 M in scientific notation.

86. This dilute solution will be so sensitive that it is able to detect these molecules of interest down to a concentration of less than 0.000000000214 M. Write 0.000000000214 M in scientific notation.

87. Garbage In 2000, the total waste generated in the United States was 4.638×10^{11} pounds. Also in 2000, the United States population was 2.81×10^8 people. Determine the garbage per capita (per person) in the United States in the year 2000.

88. Fossil Fuels In 2002, the United States imported 3.3×10^9 barrels of oil. The United States population in 2002 was 2.89×10^8 people. Determine the per capita number of barrels of oil imported into the United States in 2002.

89. Speed of Light Light travels at the rate of 1.86×10^5 miles per second. How far does light travel in one minute (6.0×10^1 seconds)?

90. Speed of Sound Sound travels at the rate of 1.127×10^3 feet per second . How far does sound travel in one minute (6.0×10^1 seconds)?

Extending the Concepts

Scientists often need to measure very small things, such as cells. They use the following units of measure:

millimeter (mm) $= 1 \times 10^{-3}$ *meter*   *micron (μm)* $= 1 \times 10^{-6}$ *meter*

nanometer (nm) $= 1 \times 10^{-9}$ *meter*   *picometer (pm)* $= 1 \times 10^{-12}$ *meter*

Write the following measurements in meters using scientific notation:

91. 250 μm **92.** 60.4 nm   **93.** 800 pm

94. 40 mm   **95.** 71.5 nm   **96.** 200 μm

Assume a cell is in the shape of a sphere. Given that the volume of a sphere is $V = \dfrac{4}{3}\pi r^3$, find the volume of a cell whose radius is given. Express the answer as a multiple of π in cubic meters.

97. 21 μm **98.** 0.75 nm   **99.** 6 nm   **100.** 108 μm

CHAPTER 5 ACTIVITY: WHAT IS THE QUESTION?

Focus: Using exponent rules, using scientific notation, and performing operations with polynomials

Time: 15–20 minutes

Group size: 2 or 4

In this activity you will work as a team to solve eight multiple-choice questions. However these questions are different from most multiple-choice questions. You are given the answer to a problem and must determine which of the multiple-choice options has the correct question for the given answer.

Before beginning the activity, decide how you will approach this task as a team. For example:

- If there are 2 members on your team, one member will always examine choices (a) and (b) and the other will always examine choices (c) and (d).

- If there are 4 members on your team, one member will always examine choice (a), another member will always examine choice (b), and so on.

1. The answer is $-3x^2 - 10x$. What is the question?

   (a) Simplify: $2x - 3x(x^2 + 4)$
   (b) Find the quotient: $(6x^3 - 20x^2) \div (-2x)$
   (c) Simplify: $-3x(x^2 + 3) + 1$
   (d) Find the quotient: $(-6x^4 - 20x^3) \div (2x^2)$

2. The answer is $-10x^4y^8$. What is the question?

   (a) Simplify: $(-5x^2y^4)^2$ **(b)** Simplify: $(5x^3y^3)(-2xy^5)$

   (c) Simplify: $\dfrac{-30x^{-2}y^6}{3x^2y^{-2}}$ **(d)** Simplify: $(5x^2y^4)(-2x^2y^2)$

3. The answer is $12x^2 - 16x - 3$. What is the question?

   (a) Multiply: $(6x - 1)(2x + 3)$
   (b) Divide: $(24x^3 - 32x^2 + 6x) \div (2x)$
   (c) Multiply: $(6x + 1)(2x - 3)$
   (d) Simplify: $(14x^2 - x + 2) - (2x^2 + 16x + 5)$

4. The answer is $x^2 + 5x + 6$. What is the question?

   (a) Find the product: $(x + 6)(x - 1)$
   (b) Simplify: $2x^2 + 7x + 9 - (x^2 - 2x - 3)$
   (c) Find the product: $(x + 2)(x + 3)$
   (d) Simplify: $(x + 6)^2$

5. The answer is 3. What is the question?

   (a) What is the name of the variable in $16z^2 + 3z - 5$?
   (b) What is the degree of the polynomial $2mn + 6m - 3$?
   (c) How many terms are in the polynomial $2mn + 6m - 3$?
   (d) What is the coefficient of b in the polynomial $3a^2b - 9a + 5b$?

6. The answer is 43. What is the question?

   (a) Evaluate: $4x^2 - 2x + 1$ for $x = -3$
   (b) Evaluate: $2x^2 + 3x - 4$ for $x = -2$
   (c) Evaluate: $-x^2 - 5x + 1$ for $x = -1$
   (d) Evaluate: $x^2 + 4x - 8$ for $x = -5$

7. The answer is $\dfrac{2x^5}{y^5}$. What is the question?

   (a) Simplify: $(6x^{-4}y)^0\left(\dfrac{12x^{-1}y^{-2}}{x^{-4}y^3}\right)$ **(b)** Simplify: $(6x^{-3}y^0)^{-1}\left(\dfrac{12xy^{-3}}{x^{-2}y^2}\right)$

   (c) Simplify: $(6x^2y^0)^{-1}\left(\dfrac{12x^{-2}y^3}{xy^{-4}}\right)$ **(d)** Simplify: $(6x^{-2}y^0)^{-1}\left(\dfrac{12x^0y^{-2}}{x^{-3}y^3}\right)$

8. The answer is 2.5×10^{-8}. What is the question?

   (a) Find the quotient: $\dfrac{2 \times 10^3}{8 \times 10^{-4}}$

   (b) Find the product: $(5 \times 10^{-4}) \cdot (5 \times 10^{-4})$

   (c) Find the quotient: $\dfrac{2 \times 10^{-3}}{8 \times 10^4}$

   (d) Find the product: $(5 \times 10^{12}) \cdot (5 \times 10^{-4})$

CHAPTER 5 REVIEW

Section 5.1	Adding and Subtracting Polynomials	

KEY CONCEPTS	KEY TERMS
• In a monomial in the form ax^k, k is the degree of the monomial. • The degree of a polynomial is the highest degree of all the terms of the polynomial.	Monomial Degree of a monomial Polynomial Binomial Trinomial Standard form Degree of a polynomial

YOU SHOULD BE ABLE TO . . .	EXAMPLE	REVIEW EXERCISES
① Define monomial and determine the degree of a monomial (p. 344)	Examples 1 through 3	1–4
② Define polynomial and determine the degree of a polynomial (p.345)	Examples 4 and 5	5–10
③ Simplify polynomials by combining like terms (p. 346)	Examples 6 through 10	11–16
④ Evaluate polynomials (p. 348)	Examples 11 through 13	17–20

In Problems 1–4, determine whether the given expression is a monomial (Yes or No). For those that are monomials, state the degree and the coefficient.

1. $4x^3$ **2.** $6x^{-3}$ **3.** $m^{1/2}$ **4.** mn^2

In Problems 5–10, determine whether the algebraic expression is a polynomial (Yes or No). If it is a polynomial, state the degree and then state if it is a monomial, binomial, or trinomial.

5. $4x^6 - 4x^{1/2}$ **6.** $\dfrac{3}{x} - \dfrac{1}{x^2}$ **7.** 6 **8.** $3x^3 - 4xy^4$

9. $-2x^5y - 7x^4y + 7$ **10.** $\dfrac{1}{2}x^3 + 2x^{10} - 5$

In Problems 11–14, perform the indicated operation.

11. $(6x^2 - 2x + 1) + (3x^2 + 10x - 3)$ **12.** $(-7m^3 - 2mn) + (8m^3 - 5m + 3mn)$

13. $(4x^2y + 10x) - (5x^2y - 2x)$

14. $(3y^2 - yz + 3z^2) - (10y^2 + 5yz - 6z^2)$

15. Find the sum of $-6x^2 + 5$ and $4x^2 - 7$.

16. Subtract $20y^2 - 10y + 5$ from $-18y + 10$.

In Problems 17–20, evaluate the polynomial for the given value(s).

17. $3x^2 - 5x$
    (a) $x = 0$
    (b) $x = -1$
    (c) $x = 2$

18. $-x^2 + 3$
    (a) $x = 0$
    (b) $x = -1$
    (c) $x = \dfrac{1}{2}$

19. $x^2y + 2xy^2$ for $x = -2$ and $y = 1$

20. $4a^2b^2 - 3ab + 2$ for $a = -1$ and $b = -3$

Section 5.2   Multiplying Monomials: The Product and Power Rules

KEY CONCEPTS	KEY TERMS
• **Product Rule for Exponents** If a is a real number and m, n are natural numbers, then $a^m \cdot a^n = a^{m+n}$ • **Power Rule for Exponents** If a is a real number and m, n are whole numbers greater than 0, then $(a^m)^n = a^{m \cdot n}$ • **Product to a Power Rule for Exponents** If a, b are real numbers and n is a whole number, then $(a \cdot b)^n = a^n \cdot b^n$	Base Power Exponent

YOU SHOULD BE ABLE TO . . .	EXAMPLE	REVIEW EXERCISES
1 Simplify exponential expressions using the Product Rule (p. 355)	Examples 1 through 3	21, 22, 25, 26
2 Simplify exponential expressions using the Power Rule (p. 356)	Examples 4 and 5	23, 24, 27, 28
3 Simplify exponential expressions containing products (p. 357)	Example 6	29–32
4 Multiply a monomial by a monomial (p. 358)	Examples 7 and 8	33–40

In Problems 21–32, simplify each expression.

21. $6^2 \cdot 6^5$ **22.** $\left(-\dfrac{1}{3}\right)^2 \left(-\dfrac{1}{3}\right)^3$ **23.** $(4^2)^6$ **24.** $[(-1)^4]^3$

25. $x^4 \cdot x^8 \cdot x$ **26.** $m^4 \cdot m^2$ **27.** $(r^3)^4$ **28.** $(m^8)^3$

29. $(4x)^3(4x)^2$ **30.** $(-2n)^3(-2n)^3$ **31.** $(-3x^2y)^4$ **32.** $(2x^3y^4)^2$

In Problems 33–40, multiply.

33. $3x^2 \cdot 5x^4$ **34.** $-4a \cdot 9a^3$ **35.** $-8y^4 \cdot (-2y)$ **36.** $12p \cdot (-p^5)$

37. $\dfrac{8}{3} w^3 \cdot \dfrac{9}{2} w$ **38.** $\dfrac{1}{3} z^2 \cdot \left(-\dfrac{9}{4} z\right)$ **39.** $(3x^2)^3 \cdot (2x)^2$ **40.** $(-4a)^2 \cdot (5a^4)$

Section 5.3   Multiplying Polynomials

KEY CONCEPTS	KEY TERMS
• **Extended form of the Distributive Property** $a(b + c + \cdots + z) = a \cdot b + a \cdot c + \cdots + a \cdot z$ where $a, b, c, \ldots, z$ are real numbers. • **FOIL Method for multiplying two binomials** $\quad$ **F** $\qquad$ **O** $\qquad$ **I** $\qquad$ **L** $(ax + b)(cx + d) = ax \cdot cx + ax \cdot d + b \cdot cx + b \cdot d$ • **Product of the Sum and Difference of Two Terms** $(a - b)(a + b) = a^2 - b^2$ • **Squares of Binomials** $(a + b)^2 = a^2 + 2ab + b^2,\ (a - b)^2 = a^2 - 2ab + b^2$	FOIL method Special products Sum and difference of two $\quad$ terms Difference of two squares Squares of binomials Perfect square trinomial

YOU SHOULD BE ABLE TO . . .	EXAMPLE	REVIEW EXERCISES
1 Multiply a polynomial by a monomial (p. 360)	Examples 1 through 3	41, 42
2 Multiply two binomials using the Distributive Property (p. 361)	Examples 4 and 5	43–46
3 Multiply two binomials using the FOIL method (p. 362)	Examples 6 through 9	49–54
4 Multiply the sum and difference of two terms (p. 364)	Examples 10 and 11	55, 56, 59, 60, 63, 64
5 Square a binomial (p. 364)	Examples 12 through 14	57, 58, 61, 62, 65, 66
6 Multiply a polynomial by a polynomial (p. 366)	Examples 15 through 17	47, 48

In Problems 41–48, multiply.

41. $-2x^3(4x^2 - 3x + 1)$ **42.** $\frac{1}{2}x^4(4x^3 + 8x^2 - 2)$ **43.** $(3x - 5)(2x + 1)$

44. $(4x + 3)(x - 2)$ **45.** $(x + 5)(x - 8)$ **46.** $(w - 1)(w + 10)$

47. $(4m - 3)(6m^2 - m + 1)$ **48.** $(2y + 3)(4y^4 + 2y^2 - 3)$

In Problems 49–54, use the FOIL method to find each product.

49. $(x + 5)(x + 3)$ **50.** $(2x - 1)(x - 8)$ **51.** $(2m + 7)(3m - 2)$

52. $(6m - 4)(8m + 1)$ **53.** $(3x + 2y)(7x - 3y)$ **54.** $(4x - y)(5x + 3y)$

In Problems 55–66, find the special products.

55. $(x - 4)(x + 4)$ **56.** $(2x + 5)(2x - 5)$ **57.** $(2x + 3)^2$

58. $(7x - 2)^2$ **59.** $(3x + 4y)(3x - 4y)$ **60.** $(8m - 6n)(8m + 6n)$

61. $(5x - 2y)^2$ **62.** $(2a + 3b)^2$ **63.** $(x - 0.5)(x + 0.5)$

64. $(r + 0.25)(r - 0.25)$ **65.** $\left(y + \dfrac{2}{3}\right)^2$ **66.** $\left(y - \dfrac{1}{2}\right)^2$

Section 5.4   Dividing Monomials: The Quotient Rule and Integer Exponents

KEY CONCEPTS

- **Quotient Rule for Exponents**
  If a is a nonzero real number and if m and n are integers, then $\dfrac{a^m}{a^n} = a^{m-n}$.

- **Definition of Zero as an Exponent**
  If a is a nonzero real number (that is, $a \neq 0$), we define $a^0 = 1$.

- **Quotient to a Power Rule for Exponents**
  If a, b are real numbers and n is an integer, then $\left(\dfrac{a}{b}\right)^n = \dfrac{a^n}{b^n}$ if $b \neq 0$.

  If n is negative or 0, then a cannot be 0.

- **Definition of a Negative Exponent**
  If n is a positive integer and if a is a nonzero real number (that is, $a \neq 0$), then we define
  $a^{-n} = \dfrac{1}{a^n}$ and $\dfrac{1}{a^{-n}} = a^n$ if $a \neq 0$.

- **Quotient to a Negative Power**
  If a and b are real numbers and n is an integer, then $\left(\dfrac{a}{b}\right)^{-n} = \left(\dfrac{b}{a}\right)^n$ if $a \neq 0, b \neq 0$.

YOU SHOULD BE ABLE TO . . .	EXAMPLE	REVIEW EXERCISES
① Simplify exponential expressions using the Quotient Rule (p. 371)	Examples 1 and 2	67–70, 75, 76
② Simplify exponential expressions using the Quotient to a Power Rule (p. 372)	Examples 3 and 4	77–80
③ Simplify exponential expressions using zero as an exponent (p. 373)	Example 5	71–74
④ Simplify exponential expressions involving negative exponents (p. 374)	Examples 6 through 11	81–86
⑤ Simplify exponential expressions using the Laws of Exponents (p. 377)	Examples 12 through 15	87–92

In Problems 67–92, simplify. Write answers with only positive exponents.

67. $\dfrac{6^5}{6^3}$ **68.** $\dfrac{7}{7^4}$ **69.** $\dfrac{x^{16}}{x^{12}}$ **70.** $\dfrac{x^3}{x^{11}}$

71. 5^0 **72.** -5^0 **73.** $m^0, m \neq 0$ **74.** $-m^0, m \neq 0$

75. $\dfrac{25x^3y^7}{10xy^{10}}$ **76.** $\dfrac{3x^4y^2}{9x^2y^{10}}$ **77.** $\left(\dfrac{x^3}{y^2}\right)^5$ **78.** $\left(\dfrac{7}{x^2}\right)^3$

79. $\left(\dfrac{2m^2n}{p^4}\right)^3$ **80.** $\left(\dfrac{3mn^2}{p^5}\right)^4$ **81.** -5^{-2} **82.** $\dfrac{1}{4^{-3}}$

83. $\left(\dfrac{2}{3}\right)^{-4}$ **84.** $\left(\dfrac{1}{3}\right)^{-3}$ **85.** $2^{-2} + 3^{-1}$ **86.** $4^{-1} - 2^{-3}$

87. $\dfrac{16x^{-3}y^4}{24x^{-6}y^{-1}}$ **88.** $\dfrac{15x^0y^{-6}}{35xy^4}$ **89.** $(2m^{-3}n)^{-4}(3m^{-4}n^2)^2$

90. $(4m^{-6}n^0)^3(3m^{-6}n^3)^{-2}$ **91.** $\left(\dfrac{3rs^{-1}}{4s^2}\right)^{-2} \cdot (2r^{-6}t^0)^{-1}$ **92.** $(6r^4s^{-3})^2 \cdot \left(\dfrac{3r^4s}{2r^{-2}s^{-2}}\right)^{-3}$

Section 5.5	Dividing Polynomials	
KEY CONCEPTS		**KEY TERMS**
• If A, B, and C are monomials, $\dfrac{A + B}{C} = \dfrac{A}{C} + \dfrac{B}{C}$. • (Quotient)(Divisor) + Remainder = Dividend		Divisor Dividend Quotient Remainder
YOU SHOULD BE ABLE TO . . .	**EXAMPLE**	**REVIEW EXERCISES**
① Divide a polynomial by a monomial (p. 384)	Examples 1 through 3	93–98
② Divide a polynomial by a binomial (p. 385)	Examples 4 through 7	99–104

In Problems 93–104, divide each of the following.

93. $\dfrac{36x^7 - 24x^6 + 30x^2}{6x^2}$ **94.** $\dfrac{15x^5 + 25x^3 - 30x^2}{5x}$ **95.** $\dfrac{16n^8 + 4n^5 - 10n}{4n^5}$

96. $\dfrac{30n^6 - 20n^5 - 16n^3}{5n^5}$ **97.** $\dfrac{2p^8 + 4p^5 - 8p^3}{-16p^5}$ **98.** $\dfrac{3p^4 - 6p^2 + 9}{-6p^2}$

99. $\dfrac{8x^2 - 2x - 21}{2x + 3}$ **100.** $\dfrac{3x^2 + 17x - 6}{3x - 1}$ **101.** $\dfrac{6x^2 + x^3 - 2x + 1}{x - 1}$

102. $\dfrac{-6x + 2x^3 - 7x^2 + 8}{x - 2}$ **103.** $\dfrac{x^3 + 8}{x + 2}$ **104.** $\dfrac{3x^3 + 2x - 7}{x - 5}$

Section 5.6   Applying Exponent Rules: Scientific Notation

KEY CONCEPTS	KEY TERMS
• **Definition of Scientific Notation** A number is written in scientific notation when it is of the form $x \times 10^N$, where $1 \leq x < 10$ and N is an integer. • **Convert from Decimal Notation to Scientific Notation** To change a positive number into scientific notation: **Step 1:** Count the number N of decimal places that the decimal point must be moved to arrive at a number x, where $1 \leq x < 10$. **Step 2:** If the original number is greater than or equal to 1, the scientific notation is $x \times 10^N$. If the original number is between 0 and 1, the scientific notation is $x \times 10^{-N}$. • **Convert from Scientific Notation to Decimal Notation** **Step 1:** Determine the exponent on the number 10. **Step 2:** If the exponent is positive, then move the decimal N decimal places to the right. If the exponent is negative, then move the decimal $|N|$ decimal places to the left.	Decimal notation Scientific notation

YOU SHOULD BE ABLE TO . . .	EXAMPLE	REVIEW EXERCISES
① Convert decimal notation to scientific notation (p. 391)	Examples 1 and 2	105–110
② Convert scientific notation to decimal notation (p. 393)	Examples 3 and 4	111–116
③ Use scientific notation to multiply and divide (p. 394)	Examples 5 through 8	117–122

In Problems 105–110, write in scientific notation.

105. 27,000,000   **106.** 1,230,000,000   **107.** 0.00006

108. 0.00000305   **109.** 3   **110.** 8

In Problems 111–116, write in decimal notation.

111. 6×10^{-4} **112.** 1.25×10^{-3} **113.** 6.13×10^5

114. 8×10^4 **115.** 3.7×10^{-1} **116.** 5.4×10^7

In Problems 117–122, perform the indicated operations. Express your answer in scientific notation.

117. $(1.2 \times 10^{-5})(5 \times 10^8)$ **118.** $(1.4 \times 10^{-10})(3 \times 10^2)$ **119.** $\dfrac{2.4 \times 10^{-6}}{1.2 \times 10^{-8}}$

120. $\dfrac{5 \times 10^6}{25 \times 10^{-3}}$ **121.** $\dfrac{200{,}000 \times 4{,}000{,}000}{0.0002}$ **122.** $\dfrac{1{,}200{,}000}{0.003 \times 2{,}000{,}000}$

CHAPTER 5 TEST

 Remember to use your Chapter Test Prep Video CD to see fully worked-out solutions to any of these problems you would like to review.

1. Determine whether the algebraic expression $6x^5 - 2x^4$ is a polynomial (Yes or No). If it is a polynomial, state the degree and then state if it is a monomial, binomial, or trinomial.

2. Evaluate the polynomial $3x^2 - 2x + 5$ for the given values:

    (a) $x = 0$ **(b)** $x = -2$ **(c)** $x = 3$

In Problems 3–11, perform the indicated operation.

    3. $(3x^2y^2 - 2x + 3y) + (-4x - 6y + 4x^2y^2)$

4. $(8m^3 + 6m^2 - 4) - (5m^2 - 2m^3 + 2)$

5. $-3x^3(2x^2 - 6x + 5)$

6. $(x - 5)(2x + 7)$

7. $(2x - 7)^2$

8. $(4x - 3y)(4x + 3y)$

9. $(3x - 1)(2x^2 + x - 8)$

10. $\dfrac{6x^4 - 8x^3 + 9}{3x^3}$

11. $\dfrac{3x^3 - 2x^2 + 5}{x + 3}$

In Problems 12–16, simplify each expression. Write answers with only positive exponents.

12. $(4x^3y^2)(-3xy^4)$

13. $\dfrac{18m^5n}{27m^2n^6}$

14. $\left(\dfrac{m^{-2}n^0}{m^{-7}n^4}\right)^{-6}$

15. $(4x^{-3}y)^{-2}(2x^4y^{-3})^4$

16. $(2m^{-4}n^2)^{-1} \cdot \left(\dfrac{16m^0n^{-3}}{m^{-3}n^2}\right)$

17. Write 0.000012 in scientific notation.

18. Write 2.101×10^5 in decimal notation.

In Problems 19 and 20, perform the indicated operation. Express your answer in scientific notation.

19. $(2.1 \times 10^{-6}) \cdot (1.7 \times 10^{10})$

20. $\dfrac{3 \times 10^{-4}}{15 \times 10^2}$

CUMULATIVE REVIEW CHAPTERS 1–5

1. Use the set $\left\{-6, -\dfrac{4}{2}, 0, 1.4, \sqrt{7}, \sqrt{25}\right\}$. List all of the elements that are

   (a) natural   **(b)** whole   **(c)** integers
   (d) rational   **(e)** irrational   **(f)** real

In Problems 2 and 3, evaluate each expression.

2. $-\dfrac{1}{2} + \dfrac{2}{3} \div 4 \cdot \dfrac{1}{3}$

3. $2 + 3[3 + 10(-1)]$

In Problems 4 and 5, simplify each algebraic expression.

4. $6x^3 - (-2x^2 + 3x) + 3x^2$

5. $-4(6x - 1) + 2(3x + 2)$

6. Solve: $-2(3x - 4) + 6 = 4x - 6x + 10$

7. Translate the following statement into an equation. DO NOT SOLVE.
   Four times the difference of a number and 5 is equal to 10 more than twice the number.

8. **Paycheck** Kathy's monthly paycheck from her part-time job working at an electronics store totaled $659.20. This amount included a 3% raise over her previous month's earnings. What were Kathy's monthly earnings before the 3% raise?

9. **Driving** Cheyenne and Amber live 306 miles apart. They start driving toward each other and meet in three hours. If Cheyenne drives 12 miles per hour faster than Amber, find Amber's driving speed.

10. Solve and graph the following inequality: $-5x + 2 > 17$

11. Evaluate $\dfrac{x^2 - y^2}{z}$ when $x = 3$, $y = -2$, and $z = -10$.

12. Find the slope of the line through $(-3, 8)$ and $(1, 2)$.

13. Graph the line $2x + 3y = 24$ by finding the intercepts.

14. Graph the line $y = -3x + 8$.

15. Solve the system of linear equations: $\begin{cases} 2x - 3y = 27 \\ -4x + 2y = -26 \end{cases}$

16. Solve the system of linear equations: $\begin{cases} 3x - 2y = 8 \\ -6x + 4y = 6 \end{cases}$

In Problems 17–23, perform the indicated operation.

17. $(4x^2 + 6x) - (-x + 5x^2) + (6x^3 - 2x^2)$

18. $(4m - 3)(7m + 2)$

19. $(3m - 2n)(3m + 2n)$

20. $(7x + y)^2$

21. $(2m + 5)(2m^2 - 5m + 3)$

22. $\dfrac{14xy^2 + 7x^2y}{7x^2y^2}$

23. $\dfrac{x^3 + 27}{x + 3}$

In Problems 24–27, simplify the expression. Write your answers with only positive exponents.

24. $(4m^0n^3)(-6n)$

25. $\dfrac{25m^{-6}n^{-2}}{-10m^{-4}n^{-10}}$

26. $\left(\dfrac{2xy^4}{z^{-2}}\right)^{-6}$

27. $(x^4y^{-2})^{-4} \cdot \left(\dfrac{6x^{-4}y^3}{3y^{-8}}\right)^{-1}$

28. Write 0.0000605 in scientific notation.

29. Write 2.175×10^6 in decimal notation.

30. Perform the indicated operation. Write your answer in scientific notation.

$$(3.4 \times 10^8)(2.1 \times 10^{-3})$$

6 Factoring Polynomials

The size of a television set is described by its diagonal measure. For example, suppose you want to purchase a 50-inch big-screen TV before the World Series. The TV measures 40 inches across the bottom, but how do you know if your TV cabinet will be big enough for this large TV? To find the height of a 50-inch television, we use the Pythagorean Theorem. See Problem 33 in Section 6.7 on page 467.

OUTLINE

6.1 Greatest Common Factor and Factoring by Grouping

6.2 Factoring Trinomials of the Form $x^2 + bx + c$

6.3 Factoring Trinomials of the Form $ax^2 + bx + c, a \neq 1$

6.4 Factoring Special Products

6.5 Summary of Factoring Techniques

Putting the Concepts Together (Sections 6.1–6.5)

6.6 Solving Polynomial Equations by Factoring

6.7 Modeling and Solving Problems with Quadratic Equations

Chapter 6 Activity: Which One Does Not Belong?

Chapter 6 Review

Chapter 6 Test

The Big Picture: Putting It Together

In Chapter 5, we learned how to multiply polynomials. We began by multiplying a monomial and a trinomial using the Distributive Property. We then learned how to multiply two binomials, and, in general, how to multiply two polynomials. In this chapter, we reverse the process. That is, we want to write a polynomial as a product. This process is called *factoring*.

In Chapter 2, we solved linear (first-degree) equations such as $2x + 5 = 8$. In this chapter, we discuss how factoring can be used to solve equations such as $2x^2 + 7x + 3 = 0$. The approach requires that we rewrite $2x^2 + 7x + 3$ as the product of two polynomials of degree 1. This ultimately leads us to solving two linear equations—something we already know how to do! This is one of the goals of algebra: Simplify a problem until it becomes a problem you already know how to solve!

Factoring is important for solving equations, but also will play a major role in Chapters 7 and 10, so be sure to work hard to learn the factoring techniques presented in this chapter.

6.1 Greatest Common Factor and Factoring by Grouping

OBJECTIVES

1. Find the Greatest Common Factor of Two or More Expressions
2. Factor Out the Greatest Common Factor in Polynomials
3. Factor Polynomials by Grouping

Preparing for Greatest Common Factor and Factoring by Grouping
Before getting started, take the following readiness quiz. If you get a problem wrong, go back to the section cited and review the material.

1. Write 48 as the product of prime numbers.   [Section A.1, p. A1]
2. Distribute: $2(5x - 3)$ [Section 1.7, p. 61]
3. Find the product: $(2x + 5)(x - 3)$ [Section 5.3, pp. 361–363]

Consider the following products:

$$5 \cdot 3 = 15$$
$$5(y + 5) = 5y + 25$$
$$(3x - 1)(x + 5) = 3x^2 + 14x - 5$$

The expressions on the left side are called **factors** of the expression on the right side. For example, $3x - 1$ and $x + 5$ are factors of $3x^2 + 14x - 5$. In the last chapter, we learned how to multiply factors to obtain a product. For example, we learned how to multiply expressions such as $3x - 1$ and $x + 5$ to obtain $3x^2 + 14x - 5$.

In this chapter, we learn how to obtain the factors of a polynomial such as $3x^2 + 14x - 5$. That is, we learn how to write $3x^2 + 14x - 5$ as $(3x - 1)(x + 5)$.

In Words
Factoring is "undoing" multiplication.

DEFINITION

To **factor** a polynomial means to write the polynomial as a product.

The process of factoring reverses the process of multiplying, as shown below.

Multiplication
$$\longrightarrow$$
Factored form $\longrightarrow$ $3(x - 7) = 3x - 21$ $\leftarrow$ Product
$$\longleftarrow$$
Factoring

1 Find the Greatest Common Factor of Two or More Expressions

If you look carefully at the illustration above, you will notice that 3 is the largest number that divides evenly into both $3x$ and 21 in the expression $3x - 21$. For this reason, 3 is the greatest common factor of $3x - 21$.

DEFINITION

The **greatest common factor (GCF)** of a list of polynomials is the largest expression that divides evenly into all the polynomials.

Writing $3x - 21$ as $3(x - 7)$ is referred to as factoring out the greatest common factor. But how can we find the GCF? The following example shows us how.

Preparing for...Answers **1.** $3 \cdot 2^4$
2. $10x - 6$ **3.** $2x^2 - x - 15$

EXAMPLE 1   How to Find the GCF of a List of Numbers

Find the GCF of 12 and 18.

Step-by-Step Solution

Step 1: Write each number as the product of prime factors.	$12 = 2 \cdot 2 \cdot 3$ $18 = 2 \quad \cdot 3 \cdot 3$
Step 2: Determine the common prime factors.	The common factors are 2 and 3.
Step 3: Find the product of the common factors found in Step 2. This number is the GCF.	The GCF is $2 \cdot 3 = 6$.

Work Smart

Remember, a prime number is a number greater than 1 that has no factors other than itself and 1. For example, 3, 7, and 13 are prime numbers while $4 (= 2 \cdot 2)$, $12 (= 2 \cdot 2 \cdot 3)$, and $35 (= 5 \cdot 7)$ are not prime.

Also, it is helpful to align the common factors vertically.

EXAMPLE 2   Finding the GCF of a List of Numbers

Find the GCF of 24, 40, and 72.

Solution

We write each number as the product of prime factors.

$$24 = 4 \cdot \ 6 = 2 \cdot 2 \cdot 2 \cdot 3$$
$$40 = 4 \cdot 10 = 2 \cdot 2 \cdot 2 \cdot \quad 5$$
$$72 = 8 \cdot \ 9 = 2 \cdot 2 \cdot 2 \cdot 3 \cdot 3$$

Because all three numbers contain three factors of 2, the GCF is $2 \cdot 2 \cdot 2 = 8$.

Notice the greatest common factor in Example 2 could be written as 2^3. The exponent 3 represents the number of times the factor 2 appears in the factorization of each number.

QUICK ✔ *Find the GCF of each list of numbers.*

1. 32, 40   **2.** 15, 25   **3.** 12, 45   **4.** 21, 35, 84

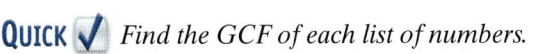

What if we want to find the greatest common factor among two or more expressions that contain variables? The approach to finding the GCF is the same as it is for numbers. Consider the terms x^3, x^5, and x^6. We can write each of these terms as the product of factors of x as follows:

$$x^3 = x \cdot x \cdot x$$
$$x^5 = x \cdot x \cdot x \cdot x \cdot x$$
$$x^6 = x \cdot x \cdot x \cdot x \cdot x \cdot x$$

Work Smart

The GCF of a variable factor is the lowest power of that variable.

Each of the terms contains three factors of x, so the greatest common factor is x^3. It is no coincidence that the exponent of the GCF is 3, the smallest exponent of the terms x^3, x^5, and x^6. This approach to finding the GCF for variable expressions will work in general. We illustrate how to find the GCF of two expressions in the next example.

EXAMPLE 3 **Finding the Greatest Common Factor of Two Expressions**

Find the greatest common factor (GCF) of $8y^4$, $12y^2$.

Solution

Step 1: The coefficients are 8 and 12. We need to determine the GCF of 8 and 12.

$$8 = 2 \cdot 2 \cdot 2$$
$$12 = 2 \cdot 2 \quad \cdot 3$$

The GCF of the coefficients is $2 \cdot 2 = 4$.

Step 2: The variable expressions are y^4 and y^2. For each variable, determine the smallest exponent that each variable is raised to.
    The GCF of y^4 and y^2 is y^2.

Step 3: Find the product of the common factors found in Steps 1 and 2. This expression is the GCF.
    The GCF is $4y^2$.

EXAMPLE 4 **Finding the Greatest Common Factor of Three Expressions**

Find the GCF of the expressions:

   (a) $3x^3, 9x^2, 21x$ **(b)** $10x^5y^4, 15x^2y^3, 25x^3y^5$

Solution

We determine the GCF of the coefficients and then determine the variable expression with the smallest exponent. The product of these two factors will be the GCF.

   (a) The coefficients, 3, 9, and 21, written as the product of prime numbers are

Factor the coefficients as a product of primes: $3 = 3$
$$9 = 3 \cdot 3$$
$$21 = 3 \cdot \quad 7$$

The GCF of the coefficients is 3.
    The variable expressions are x^3, x^2, and x. The smallest exponent is 1, so the GCF of the variable expressions is x. Therefore, the GCF of $3x^3$, $9x^2, 21x$ is $3x$.

   (b) The coefficients are 10, 15, and 25. We write these coefficients as the product of prime numbers.

Factor the coefficients as a product of primes: $10 = 5 \cdot 2$
$$15 = 5 \cdot \quad 3$$
$$25 = 5 \cdot \qquad 5$$

The GCF of the coefficients is 5.
    The GCF of x^5, x^2, and x^3 is x^2. The GCF of y^4, y^3, and y^5 is y^3. Therefore, the GCF of $10x^5y^4$, $15x^2y^3, 25x^3y^5$ is $5x^2y^3$.

QUICK ✓ *Find the greatest common factor (GCF) of the terms.*

5. $14y^3, 35y^2$ **6.** $6z^3, 8z^2, 12z$ **7.** $4x^3y^5, 8x^2y^3, 24xy^4$

The greatest common factor can be a binomial, as illustrated by the following example.

EXAMPLE 5 The GCF as a Binomial

Find the greatest common factor of each pair of expressions.

(a) $3(x - 1)$ and $8(x - 1)$ **(b)** $2(z + 3)(z + 5)$ and $4(z + 5)^2$

Solution

(a) There is no common factor between the coefficients, 3 and 8. However, each expression has $x - 1$ as a factor, so the GCF of $3(x - 1)$ and $8(x - 1)$ is $x - 1$.

(b) The GCF between 2 and 4 is 2. The GCF between $(z + 3)(z + 5)$ and $(z + 5)^2$ is $z + 5$. The GCF of the expressions $2(z + 3)(z + 5)$ and $4(z + 5)^2$ is $2(z + 5)$.

QUICK ✓ *Find the greatest common factor (GCF) of each pair of expressions.*

8. $7(2x + 3)$ and $-4(2x + 3)$ **9.** $9(k + 8)(3k - 2)$ and $12(k - 1)(k + 8)^2$

> ### SUMMARY: Steps to Find the Greatest Common Factor of Two or More Expressions
>
> **Step 1:** Find the GCF of the coefficients of each variable expression.
>
> **Step 2:** For each variable expression common to all the terms, determine the smallest exponent that the variable expression is raised to.
>
> **Step 3:** Find the product of the common factors found in Steps 1 and 2. This expression is the GCF.

② Factor Out the Greatest Common Factor in Polynomials

The first step in factoring any polynomial is to look for the greatest common factor. Once the GCF is identified, we use the Distributive Property "in reverse" to factor the polynomial as shown below.

$$ab + ac = a(b + c) \quad \text{or} \quad ab - ac = a(b - c)$$

When we use this method, we say that we "factor out" the greatest common factor. Example 6 illustrates the method.

EXAMPLE 6 How to Factor Out the Greatest Common Factor in a Polynomial

Factor $2x - 10$ by factoring out the greatest common factor.

Step-by-Step Solution

Step 1: Find the GCF.	GCF = 2
Step 2: Rewrite each term as the product of the GCF and remaining factor.	$2x - 10 = 2(x) - 2(5)$

(continued)

Step 3: Factor out the GCF.	$= 2(x - 5)$
Step 4: Check	$2(x - 5) = 2(x) - 2(5)$
	$= 2x - 10$

Work Smart

The Distributive Property comes in handy again for factoring out the GCF. Note how it is used in the check step, too.

So $2x - 10 = 2(x - 5)$.

Below we summarize the steps to follow when factoring out the greatest common factor.

Steps to Factor a Polynomial Using the Greatest Common Factor

Step 1: Identify the greatest common factor (GCF) of the terms that make up the polynomial.

Step 2: Rewrite each term as the product of the GCF and the remaining factor.

Step 3: Use the Distributive Property "in reverse" to factor out the GCF.

Step 4: Use the Distributive Property to verify that the factorization is correct.

EXAMPLE 7 Factoring Out the Greatest Common Factor in a Binomial

Factor the binomial $9z^3 + 36z^2$ by factoring out the greatest common factor.

Solution

The greatest common factor between 9 and 36 is 9. The greatest common factor between z^3 and z^2 is z^2. Therefore, the GCF is $9z^2$.

$$\begin{aligned} \text{Rewrite each term as the product} \\ \text{of the GCF and remaining factor:} \quad 9z^3 + 36z^2 &= 9z^2(z) + 9z^2(4) \\ \text{Factor out the GCF:} \quad &= 9z^2(z + 4) \\ \textbf{Check} \qquad 9z^2(z + 4) &= 9z^2(z) + 9z^2(4) \\ &= 9z^3 + 36z^2 \end{aligned}$$

So $9z^3 + 36z^2 = 9z^2(z + 4)$.

EXAMPLE 8 Factoring Out the Greatest Common Factor in a Trinomial

Factor the trinomial $6a^2b^2 - 8ab^3 + 18a^3b^4$ by factoring out the greatest common factor.

Solution

The GCF of $6a^2b^2 - 8ab^3 + 18a^3b^4$ is $2ab^2$. We now rewrite each term as the product of the GCF and the remaining factor.

$$\begin{aligned} 6a^2b^2 - 8ab^3 + 18a^3b^4 &= 2ab^2(3a) - 2ab^2(4b) + 2ab^2(9a^2b^2) \\ \text{Factor out the GCF:} \quad &= 2ab^2(3a - 4b + 9a^2b^2) \\ \textbf{Check} \qquad 2ab^2(3a - 4b + 9a^2b^2) &= 2ab^2(3a) - 2ab^2(4b) + 2ab^2(9a^2b^2) \\ &= 6a^2b^2 - 8ab^3 + 18a^3b^4 \end{aligned}$$

So $6a^2b^2 - 8ab^3 + 18a^3b^4 = 2ab^2(3a - 4b + 9a^2b^2)$.

QUICK ✔ *Factor each polynomial by factoring out the greatest common factor.*

10. $5z^2 - 30z$

11. $12p^2 + 30p^4$

12. $16y^3 - 12y^2 + 4y$

13. $6m^4n^2 + 18m^3n^4 - 22m^2n^5$

When the coefficient of the term of highest degree is negative, we factor the negative out of the polynomial as part of the GCF.

EXAMPLE 9 Factoring Out a Negative as Part of the GCF

Factor $-7a^3 + 14a$ by factoring out the greatest common factor.

Solution

This binomial is written in standard form. Since the coefficient on the highest-degree term, $-7a^3$, is negative, we factor the negative out as part of the GCF. So we use $-7a$ as the greatest common factor.

$$-7a^3 + 14a = -7a(a^2) + (-7a)(-2)$$

Factor out GCF: $= -7a(a^2 - 2)$

Check
$$-7a(a^2 - 2) = -7a(a^2) + (-7a)(-2)$$
$$= -7a^3 + 14a$$

So $-7a^3 + 14a = -7a(a^2 - 2)$.

QUICK ✔ *Factor out the greatest common factor.*

14. $-4y^2 + 8y$

15. $-6a^3 + 12a^2 - 3a$

Sometimes the greatest common factor is a binomial.

EXAMPLE 10 Factoring Out a Binomial as the Greatest Common Factor

Factor out the greatest common binomial factor: $5x(x - 2) + 3(x - 2)$

Solution

Do you see that $x - 2$ is common to both terms? The GCF is the binomial $x - 2$.

$$5x(x - 2) + 3(x - 2) = 5x(x - 2) + 3(x - 2)$$

Factor out $x - 2$: $= (x - 2)(5x + 3)$

Check
$$(x - 2)(5x + 3) = (x - 2)5x + (x - 2)3$$
$$= 5x(x - 2) + 3(x - 2)$$

So $5x(x - 2) + 3(x - 2) = (x - 2)(5x + 3)$.

QUICK ✔ *Factor out the greatest common factor.*

16. $2a(a - 5) + 3(a - 5)$

17. $7z(z + 5) - 4(z + 5)$

③ Factor Polynomials by Grouping

Sometimes a common factor does not occur in every term of the polynomial. If a polynomial contains four terms, it may be possible to find a GCF of the first two terms and a different GCF of the second two terms. When this happens, the common factor can be

Work Smart
Try factoring by grouping when a polynomial contains four terms.

factored out of each group of terms. This technique is called **factoring by grouping,** as is illustrated in the next example.

EXAMPLE 11 How to Factor by Grouping

Factor by grouping: $3x - 3y + ax - ay$

Step-by-Step Solution

Step 1: Group terms with common factors.	$3x - 3y + ax - ay = (3x - 3y) + (ax - ay)$
Step 2: In each grouping, factor out the GCF.	$= 3(x - y) + a(x - y)$
Step 3: Factor out the common factor, $x - y$, that remains.	$= (x - y)(3 + a)$
Step 4: Check	FOIL: $(x - y)(3 + a) = 3x + ax - 3y - ay$
	Rearrange terms: $= 3x - 3y + ax - ay$

So $3x - 3y + ax - ay = (x - y)(3 + a)$.

Based upon Example 11, we have the following steps for factoring by grouping.

Work Smart
We could have written the answer to Example 11 as $(3 + a)(x - y)$. Do you know why?

Steps to Factor a Polynomial by Grouping

Step 1: Group the terms with common factors.
Step 2: In each grouping, factor out the greatest common factor (GCF).
Step 3: Factor out the common factor that remains.
Step 4: Check your work by finding the product of the factors.

QUICK ✅ *Factor by grouping.*

18. $4x + 4y + bx + by$

EXAMPLE 12 Factoring by Grouping

Factor by grouping: $5x - 5y - 4bx + 4by$

Solution

First, we group terms with common factors.

$$5x - 5y - 4bx + 4by = (5x - 5y) + (-4bx + 4by)$$

Factor out the common factor in each group: $= 5(x - y) + (-4b)(x - y)$

Work Smart
Be careful when working with signs:
$-4b(x - y) = -4bx + 4by$

Factor out the common factor that remains: $= (x - y)(5 - 4b)$

Check   Use FOIL: $(x - y)(5 - 4b) = 5x - 4bx - 5y + 4by$

Rearrange terms: $= 5x - 5y - 4bx + 4by$

So $5x - 5y - 4bx + 4by = (x - y)(5 - 4b)$.

In Example 12, we could have rearranged terms using the Commutative Property of Addition to write the problem $5x - 5y - 4bx + 4by$ as $5x - 4bx - 5y + 4by$ and factored as follows.

$$5x - 4bx - 5y + 4by = x(5 - 4b) - y(5 - 4b)$$
$$= (5 - 4b)(x - y)$$

Notice that the answer contains the same factors, just written in reverse order. But $(5 - 4b)(x - y) = (x - y)(5 - 4b)$ by the Commutative Property of Multiplication, so we know that the answer is correct.

QUICK ✔ *Factor by grouping.*

Work Smart

Sometimes it will be necessary to rearrange terms before factoring by grouping.

19. $6az - 2a - 9bz + 3b$

20. $8b + 4 - 10ab - 5a$

Whenever we encounter a factoring problem, the first thing we should always do is look for a common factor. The next example illustrates this idea.

EXAMPLE 13   **Factoring by Grouping**

Factor: $3x^3 + 12x^2 - 6x - 24$

Solution

Work Smart

Whenever factoring, always look for a GCF first.

Do all four terms contain a common factor? Yes!! There is a GCF of 3, so we factor it out.

$$3x^3 + 12x^2 - 6x - 24 = 3(x^3 + 4x^2 - 2x - 8)$$

Now we group terms with common factors.

$$3(x^3 + 4x^2 - 2x - 8) = 3[(x^3 + 4x^2) + (-2x - 8)]$$

Factor out the common factor in each group: $= 3[(x^2(x + 4) + (-2)(x + 4)]$

Factor out the common factor that remains: $= 3(x + 4)(x^2 - 2)$

Check   Use FOIL: $3(x + 4)(x^2 - 2) = 3(x^3 - 2x + 4x^2 - 8)$

Distribute the 3: $= 3x^3 - 6x + 12x^2 - 24$

Rearrange terms: $= 3x^3 + 12x^2 - 6x - 24$

So $3x^3 + 12x^2 - 6x - 24 = 3(x + 4)(x^2 - 2)$. ∎

QUICK ✔ *Factor by grouping.*

21. $3z^3 + 12z^2 + 6z + 24$

22. $2n^4 + 2n^3 - 4n^2 - 4n$

6.1 Exercises

For Extra Help:

Student Solutions Manual   CD Video   PH Math/Tutor Center   MathXL Tutorials on CD   MathXL®   MyMathLab

Concepts and Vocabulary

In Problems 1–4, fill in the blanks.

1. The largest expression that divides evenly into a set of numbers is called the

_____ _____ _____ .

2. To _____ a polynomial means to write the polynomial as a product.

3. When we factor a polynomial using the GCF, we use the _____ Property in reverse.

4. When a polynomial contains four terms try factoring by _____.

In Problems 5–8, answer True or False to each statement.

5. The first step in any factoring problem is to look for the GCF of the polynomial.

6. There is more than one correct factorization of any given polynomial.

7. Every polynomial containing four terms can be factored by grouping.

8. We factor $(2x + 1)(x - 3) + (2x + 1)(2x + 7)$ as $(2x + 1)^2(3x + 4)$.

9. In your own words, write a list of steps for finding the greatest common factor and then write a second list of steps for factoring the GCF from a polynomial.

10. Describe how to factor a negative from a polynomial. What types of errors might happen during this process?

Building Skills

In Problems 11–38, find the greatest common factor, GCF, of each group of expressions.

11. $8, 6$ **12.** $49, 35$ **13.** $15, 14$ **14.** $6, 55$

15. $15, 20$ **16.** $24, 36$ **17.** $78, 104$ **18.** $65, 91$

19. $12, 28, 48$ **20.** $35, 42, 63$ **21.** $36, 54, 72$ **22.** $60, 75, 135$

23. x^{10}, x^2, x^8 **24.** y^3, y^5, y **25.** $7x, 14x^3$ **26.** $8a^4, 20a^2$

27. m^3n^2, m^2n^2 **28.** xy^2, x^4y **29.** $45a^2b^3, 75ab^2c$ **30.** $26xy^2, 39x^2y$

31. $4a^2bc^3, 6ab^2c^2, 8a^2b^2c^4$ **32.** $2x^2yz, xyz^2, 5x^3yz^2$

33. $3(x - 1)$ and $6(x + 1)$ **34.** $8(x + y)$ and $9(x + y)$

35. $2(x - 4)^2$ and $4(x - 4)^3$ **36.** $6(a - b)$ and $15(a - b)^3$

37. $12(x + 2)(x - 3)^2$ and $18(x - 3)^2(x - 2)$

38. $15(2a - 1)^2(2a + 1)$ and $18(2a - 1)^3(2a + 3)^2$

In Problems 39–44, identify the missing factor to make the statement true.

39. $12x^6 = 3x^3 \cdot ?$ **40.** $40x^5 = 10x^3 \cdot ?$ **41.** $9s^2t^2 = -3st^2 \cdot ?$

42. $-5y^4z^2 = yz \cdot ?$ **43.** $15m^3n^5 = 15m^2 \cdot ?$ **44.** $-120z^6y = -40z^5 \cdot ?$

In Problems 45–60, factor the GCF from the polynomial.

45. $12x - 18$ **46.** $3a + 6$

47. $5x^2y - 15x^3y^2$ **48.** $8a^3b^2 + 12a^5b^2$

49. $3x^3 + 6x^2 - 3x$ **50.** $5x^4 + 10x^3 - 25x^2$

51. $9m^5 - 18m^3 - 12m^2 + 81$ **52.** $5z^2 + 10z^4 - 15z^3 - 45z^5$

53. $15a^2b^4 - 60ab^3 + 45a^3b^2$ **54.** $12r^3s^2 + 3rs - 6rs^4$

55. $(x - 3)x - (x - 3)5$ **56.** $a(a - 5) + 6(a - 5)$

57. $x^2(x - 1) + y^2(x - 1)$ **58.** $(b + 2)a^2 - (b + 2)b$

59. $x^2(4x + 1) + 2x(4x + 1) + 5(4x + 1)$ **60.** $s^2(s^2 - 1) + 4s(s^2 - 1) + 7(s^2 - 1)$

In Problems 61–68, factor out the GCF using a negative sign.

61. $-3x^2 + 4x$ **62.** $-9b^2 - 6ab$ **63.** $-5x^3 + 10x^2 - 15x$

64. $-2y^2 + 10y - 14$ **65.** $-12z^3 + 16z^2 - 8z$ **66.** $-22n^4 + 18n^2 + 14n$

67. $10 - 5b - 15b^3$ **68.** $14m^2 - 16m^3 - 24m^4$

In Problems 69–78, factor by grouping.

69. $xy + 3y + 4x + 12$ **70.** $x^2 + ax + 2a + 2x$ **71.** $yz + z - y - 1$

72. $mn - 3n + 2n - 6$ ◉ **73.** $x^3 - x^2 + 2x - 2$ **74.** $z^3 + 4z^2 + 3z + 12$

75. $2t^3 - t^2 - 2t + 1$ **76.** $x^3 - x^2 - x + 1$

77. $2t^4 - t^3 - 2t + 1$ **78.** $6yz - 8y - 9z + 12$

Mixed Practice

In Problems 79–100, factor each polynomial.

79. $4y - 20$ **80.** $3z - 21$

81. $28m^3 + 7m^2 + 63m$ **82.** $10x^3 - 15x^2 + 5x$

83. $12m^3n^2p - 18m^2n$ **84.** $4s^2t^3 - 24st$

85. $(2p - 1)(p + 3) + (7p + 4)(p + 3)$ **86.** $(3x - 2)(4x + 1) + (3x - 2)(x - 10)$

87. $18ax - 9ay - 12bx + 6by$ **88.** $6xm + 3xn - 4ym - 2yn$

89. $(x - 2)(x - 3) + (x - 2)$ **90.** $(a + 2)(2b - 1) - (2b - 1)$

91. $15x^4 - 6x^3 + 30x^2 - 12x$ **92.** $2a^3 - 4a^2 + 8a - 16$

93. $-3x^3 + 6x^2 - 9x$ **94.** $-8z^4 - 12z^3 + 28z^2$

95. $-12b + 16b^2$ **96.** $-8a + 20a^2$

◉ **97.** $12xy + 9x - 8y - 6$ **98.** $-12mxz - 3xz + 24mx + 6x$

99. $\dfrac{1}{3}x^3 - \dfrac{2}{9}x^2$ **100.** $\dfrac{3}{4}p^4 - \dfrac{1}{4}p^3$

Applying the Concepts

101. Write any trinomial that has $-4xy$ as a GCF.

102. Write any binomial that has $3x^3$ as a GCF.

103. Write any binomial that has $(x - 1)$ as a GCF.

104. Write any trinomial that has $(y + 2)$ as a GCF.

105. Height of a Toy Rocket The height of a toy rocket after t seconds when fired straight up with an initial speed of 150 feet per second from the ground is given by the polynomial $-16t^2 + 150t$. Write the polynomial $-16t^2 + 150t$ in factored form.

106. Height of a Ball The height of a ball after t seconds when thrown straight up with an initial speed of 80 feet per second from a height of 48 feet is given by the polynomial $-16t^2 + 80t + 48$. Write the polynomial $-16t^2 + 80t + 48$ in factored form.

107. Selling Calculators A manufacturer of calculators found that the number of calculators sold at a price of p dollars is given by the polynomial $21{,}000 - 150p$. Write $21{,}000 - 150p$ in factored form.

108. Revenue A manufacturer of a gas clothes dryer has found that the revenue (in dollars) from selling clothes dryers at a price of p dollars is given by the expression $-4p^2 + 4000p$. Write $-4p^2 + 4000p$ in factored form.

△ **109. Area of a Rectangle** A rectangle has an area of $8x^5 - 28x^3$ square feet. Write $8x^5 - 28x^3$ in factored form.

△ **110. Area of a Parallelogram** A parallelogram has an area of $18n^4 - 15n^3 + 6n$ square cm. Write $18n^4 - 15n^3 + 6n$ in factored form.

△ **111. Surface Area** The surface area of a cylindrical can whose radius is r inches and height is 4 inches is given by $S = 2\pi r^2 + 8\pi r$ square inches. Express the surface area in factored form.

112. Surface Area The surface area of a cylindrical can whose radius is r inches and height is 8 inches is given by $S = 2\pi r^2 + 16\pi r$ square inches. Express the surface area in factored form.

In Problems 113 and 114, the area of the polygon is given. Write a polynomial that represents the missing length.

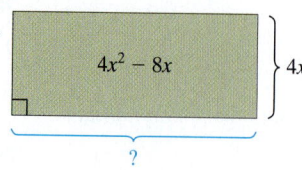 **113.**

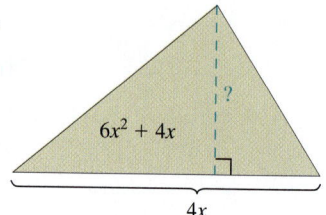 **114.**

Extending the Concepts

In Problems 115–122, find the missing factor.

115. $8x^{3n} + 10x^n = 2x^n \cdot ?$

116. $3x^{2n+2} + 6x^{6n} = 3x^{2n} \cdot ?$

117. $3 - 4x^{-1} + 2x^{-3} = x^{-3} \cdot ?$

118. $1 - 3x^{-1} + 2x^{-2} = x^{-2} \cdot ?$

119. $x^2 - 3x^{-1} + 2x^{-2} = x^{-2} \cdot ?$

120. $2x^2 + x^{-3} - x^{-4} = x^{-4} \cdot ?$

121. $\dfrac{6}{35}x^4 - \dfrac{1}{7}x^2 + \dfrac{2}{7}x = \dfrac{2}{7}x \cdot ?$

122. $\dfrac{2}{15}x^3 - \dfrac{1}{9}x^2 + \dfrac{1}{3}x = \dfrac{1}{3}x \cdot ?$

6.2 Factoring Trinomials of the Form $x^2 + bx + c$

OBJECTIVES

1. Factor Trinomials of the Form $x^2 + bx + c$
2. Factor Out the GCF, Then Factor $x^2 + bx + c$

Preparing for Factoring Trinomials of the Form $x^2 + bx + c$

Before getting started, take the following readiness quiz. If you get a problem wrong, go back to the section cited and review the material.

1. Find a pair of factors of 18 whose sum is 11.   [Section A.1, p. A1]
2. Find a pair of factors of -24 whose sum is -2.   [Section A.1, p. A1]
3. Determine the coefficients of $3x^2 - x - 4$.   [Section 1.7, pp. 59–60]
4. Find the product: $-5p(p + 4)$ [Section 5.3, pp. 360–361]
5. Find the product: $(z - 1)(z + 4)$ [Section 5.3, pp. 361–363]

In this section, we are going to factor trinomials that are of degree 2. Because the word **quadratic** means "relating to a square," these trinomials are called *quadratic trinomials*.

> **DEFINITION**
>
> A **quadratic trinomial** is a polynomial of the form $ax^2 + bx + c, a \neq 0$ where a represents the coefficient of the squared (second-degree) term, b represents the coefficient of the linear (first-degree) term, and c represents the constant.

When the trinomial is written in standard form (descending order of degree) the coefficient of the squared term is also called the **leading coefficient.** We begin by looking at quadratic trinomials where the leading coefficient, a, is 1. Examples of quadratic trinomials whose leading coefficient is 1 are

$$x^2 + 4x + 3 \qquad a^2 + 4a - 21 \qquad p^2 - 11p + 18$$

1 Factor Trinomials of the Form $x^2 + bx + c$

The idea behind factoring a second-degree trinomial of the form $x^2 + bx + c$ is to see whether it can be written as the product of two first-degree polynomials.

For example,

$$\text{Multiplication} \rightarrow$$

Factored form $\rightarrow$ $(x + 3)(x - 7) = x^2 - 4x - 21$ $\leftarrow$ Product

$\leftarrow$ Factoring

The factors of $x^2 - 4x - 21$ are $x + 3$ and $x - 7$. Notice the following:

$$x^2 - 4x - 21 = (x + 3)(x - 7)$$

The sum of -7 and 3 is -4.

The product of -7 and 3 is -21.

In general, if $x^2 + bx + c = (x + m)(x + n)$, then $mn = c$ and $m + n = b$. We illustrate how to factor a trinomial of the form $x^2 + bx + c$ in the following example.

EXAMPLE 1 **How to Factor a Trinomial of the Form $x^2 + bx + c$, where c Is Positive**

Factor: $x^2 + 7x + 12$

Step-by-Step Solution

Step 1: When we compare $x^2 + 7x + 12$ with $x^2 + bx + c$, we see that $b = 7$ and $c = 12$. We are looking for factors of $c = 12$ whose sum is $b = 7$. We begin by listing all factors of 12 and computing the sum of these factors.

Integers Whose Product Is 12	1, 12	2, 6	3, 4	−1, −12	−2, −6	−3, −4
Sum	13	8	7	−13	−8	−7

We can see that $3 \cdot 4 = 12$ and $3 + 4 = 7$, so $m = 3$ and $n = 4$.

Step 2: We write the trinomial in the form $(x + m)(x + n)$.

$$x^2 + 7x + 12 = (x + 3)(x + 4)$$

Step 3: Check

Use FOIL: $(x + 3)(x + 4) = x^2 + 4x + 3x + 3(4) = x^2 + 7x + 12$

So $x^2 + 7x + 12 = (x + 3)(x + 4)$.

We summarize the steps for factoring a trinomial of the form $x^2 + bx + c$ below.

Steps to Factor a Trinomial of the Form $x^2 + bx + c$

Step 1: Find the pair of integers whose product is c and whose sum is b. That is, determine m and n such that $mn = c$ and $m + n = b$.

Step 2: Write $x^2 + bx + c = (x + m)(x + n)$.

Step 3: Check your work by multiplying the binomials using the FOIL method.

In Example 1, notice that the coefficient of the middle term is positive and the constant is positive. **If the coefficient of the middle term and the constant are both positive, then m and n must both be positive.**

EXAMPLE 2 **Factoring a Trinomial of the Form $x^2 + bx + c$, where c Is Positive**

Factor: $p^2 - 11p + 24$

Solution

We first identify $b = -11$ and $c = 24$. We are looking for factors of $c = 24$ whose sum is $b = -11$. We begin by listing all factors of 24 and computing the sum of these factors.

Integers Whose Product Is 24	1, 24	2, 12	3, 8	4, 6	−1, −24	−2, −12	−3, −8	−4, −6
Sum	25	14	11	10	−25	−14	−11	−10

We can see that $-3 \cdot (-8) = 24$ and $-3 + (-8) = -11$, so $m = -3$ and $n = -8$.

Write the trinomial in the form $(p + m)(p + n)$:
$$p^2 - 11p + 24 = (p + (-3))(p + (-8))$$
$$= (p - 3)(p - 8)$$

Check

Use the FOIL pattern:
$$(p - 3)(p - 8) = p^2 - 8p - 3p + (-3)(-8)$$
$$= p^2 - 11p + 24$$

So $p^2 - 11p + 24 = (p - 3)(p - 8)$. ∎

In Example 2, notice that the coefficient of the middle term is negative and the constant is positive. **If the coefficient of the middle term is negative and the constant is positive, then m and n must both be negative.**

Work Smart

The sum of two even numbers is even, the sum of two odd numbers is even, the sum of an odd and even number is odd. In Example 2, the coefficient of the middle term is odd (-11). So one of the factors of 24 will be odd, and the other will be even.

Use this result on the remaining examples to reduce the list of possible factors.

QUICK ✔️ *Factor each trinomial.*

1. $y^2 + 9y + 20$ **2.** $w^2 + 10w + 24$ **3.** $q^2 - 5q + 6$ **4.** $z^2 - 9z + 14$

EXAMPLE 3   **Factoring a Trinomial of the Form $x^2 + bx + c$, where c Is Negative**

Factor: $z^2 + 3z - 28$

Solution

We see that $b = 3$ and $c = -28$. We are looking for factors of $c = -28$ whose sum is $b = 3$. We begin by listing all factors of -28 and computing the sum of these factors. Because c is negative, we know that one of the factors must be positive and the other negative.

Integers Whose Product Is −28	−1, 28	−2, 14	−4, 7	1, −28	2, −14	4, −7
Sum	27	12	3	−27	−12	−3

We can see that $-4 \cdot 7 = -28$ and $-4 + 7 = 3$, so $m = -4$ and $n = 7$.

Write the trinomial in the form $(z + m)(z + n)$:
$$z^2 + 3z - 28 = (z + (-4))(z + 7)$$
$$= (z - 4)(z + 7)$$

Check

Use the FOIL pattern:
$$(z - 4)(z + 7) = z^2 + 7z - 4z + (-4)(7)$$
$$= z^2 + 3z - 28$$

Thus, $z^2 + 3z - 28 = (z - 4)(z + 7)$. ∎

In Example 3, notice that the coefficient of the middle term is positive and the constant is negative. **If the constant is negative, then m and n must have opposite signs.** In addition, **if the coefficient of the middle term is positive, then the factor with the larger absolute value must be positive.**

EXAMPLE 4 **Factoring a Trinomial of the Form $x^2 + bx + c$, where c Is Negative**

Factor: $z^2 - 7z - 18$

Solution

To factor the expression, what do we want to find? We need factors of $c = -18$ whose sum is $b = -7$. We begin by listing all factors of -18 and computing the sum of these factors.

Integers Whose Product Is -18	$-1, 18$	$-2, 9$	$-3, 6$	$1, -18$	$2, -9$	$3, -6$
Sum	17	7	3	-17	-7	-3

We can see that $2 \cdot (-9) = -18$ and $2 + (-9) = -7$, so $m = 2$ and $n = -9$.

Write the trinomial in the form $(z + m)(z + n)$:
$$z^2 - 7z - 18 = (z + 2)(z + (-9))$$
$$= (z + 2)(z - 9)$$

Check

Use FOIL: $(z + 2)(z - 9) = z^2 - 9z + 2z + 2(-9)$
$$= z^2 - 7z - 18$$

So $z^2 - 7z - 18 = (z + 2)(z - 9)$. ■

In Example 4, notice that the coefficient of the middle term is negative and the constant is also negative. **If the coefficient of the middle term is negative and the constant is negative, then m and n must have opposite signs and the factor with the larger absolute value must be negative.**

QUICK ✔ *Factor each trinomial.*

5. $y^2 - 2y - 15$ **6.** $w^2 + w - 12$ **7.** $q^2 - 4q - 12$ **8.** $n^2 + 5n - 6$

Table 1 summarizes the four possibilities for factoring a quadratic trinomial in the form $x^2 + bx + c$.

Table 1		
Form	**Signs of m and n**	**Example**
$x^2 + bx + c$, where b and c are both positive	m and n are both positive.	$x^2 + 3x + 2 = (x + 2)(x + 1)$
$x^2 + bx + c$, where b is negative and c is positive	m and n are both negative.	$a^2 - 7a + 12 = (a - 4)(a - 3)$
$x^2 + bx + c$, where b is positive and c is negative	m and n are opposite in sign and the factor with the larger absolute value is positive.	$y^2 + 2y - 24 = (y + 6)(y - 4)$
$x^2 + bx + c$, where b is negative and c is negative	m and n are opposite in sign and the factor with the larger absolute value is negative.	$b^2 - 4b - 21 = (b - 7)(b + 3)$

DEFINITION
A polynomial that cannot be written as the product of two other polynomials (other than 1 or -1) is said to be a **prime polynomial.**

EXAMPLE 5 **Identifying a Prime Trinomial**

Show that $y^2 + 10y + 12$ is prime.

Solution

We are looking for factors of $c = 12$ whose sum is $b = 10$. Because both b and c are positive, we know that m and n must both be positive, so we only list positive factors of 12 and compute the sum of these factors.

Integers Whose Product Is 12	1, 12	2, 6	3, 4
Sum	13	8	7

There are no factors of 12 whose sum is 10. Therefore, $y^2 + 10y + 12$ is prime. ■

QUICK ✓ *Factor the trinomial. If the trinomial cannot be factored, state that it is prime.*

9. $c^2 - 5c + 8$ **10.** $q^2 + 4q - 45$

If a trinomial has more than one variable, we take the same approach as that used for trinomials with one variable. So trinomials of the form

$$x^2 + bxy + cy^2$$

will factor as

$$(x + my)(x + ny)$$

where

$$mn = c \quad \text{and} \quad m + n = b$$

EXAMPLE 6 **Factoring Trinomials with Two Variables**

Factor: $p^2 + 4pq - 21q^2$

Solution

The trinomial $p^2 + 4pq - 21q^2$ will factor as $(p + mq)(p + nq)$, where $mn = -21$ and $m + n = 4$. We are looking for factors of $c = -21$ whose sum is $b = 4$. We list factors of -21, but because c is negative and b is positive, we know the factor of -21 with the larger absolute value will be positive.

Integers Whose Product Is -21	$-1, 21$	$-3, 7$
Sum	20	4

We can see that $-3 \cdot 7 = -21$ and $-3 + 7 = 4$, so $m = -3$ and $n = 7$.

Write the trinomial in the
form $(p + mq)(p + nq)$: $\quad p^2 + 4pq - 21q^2 = (p + (-3)q)(p + 7q)$
$$= (p - 3q)(p + 7q)$$

Check

Use the FOIL pattern: $(p - 3q)(p + 7q) = p^2 + 7pq - 3pq - 21q^2$
$$= p^2 + 4pq - 21q^2$$

We conclude that $p^2 + 4pq - 21q^2 = (p - 3q)(p + 7q)$.   ▬

QUICK ✔ *Factor each trinomial.*

11. $x^2 + 9xy + 20y^2$ **12.** $m^2 + mn - 42n^2$

EXAMPLE 7   **Factoring a Trinomial Not Written in Standard Form**

Factor: $2w + w^2 - 8$

Solution

Do you notice that the trinomial is not in standard form $ax^2 + bx + c$? We will first write it in standard form (descending order of degree) and then factor.

Rearrange terms: $2w + w^2 - 8 = w^2 + 2w - 8$

We now look for factors of $c = -8$ whose sum is $b = 2$. Since the coefficient of w is positive and the constant term is negative, we know that the factor with the larger absolute value will be positive.

Integers Whose Product Is -8	$-1, 8$	$-2, 4$
Sum	7	2

We see that $-2 \cdot 4 = -8$ and $-2 + 4 = 2$, so $m = -2$ and $n = 4$. The factors are $(w - 2)(w + 4)$.

Check

Use the FOIL pattern: $(w - 2)(w + 4) = w^2 + 4w - 2w + (-2)(4)$
$$= w^2 + 2w - 8$$

So $w^2 + 2w - 8 = (w - 2)(w + 4)$.   ▬

QUICK ✔ *Completely factor each trinomial.*

13. $-56 + n^2 + n$ **14.** $y^2 + 35 - 12y$

---

② **Factor Out the GCF, Then Factor $x^2 + bx + c$**

Some algebraic expressions can be factored as trinomials in the form $x^2 + bx + c$ after we factor out a greatest common factor.

EXAMPLE 8   **Factoring Trinomials with a Common Factor**

Factor: $3u^3 - 21u^2 - 90u$

Solution

Did you notice that there is a GCF of $3u$ in the trinomial? We factor the $3u$ out:

$$3u^3 - 21u^2 - 90u = 3u(u^2 - 7u - 30)$$

We now concentrate on factoring the trinomial in parentheses, $u^2 - 7u - 30$. We are looking for factors of $c = -30$ whose sum is $b = -7$. Because c is negative and b is negative, the factor of -30 with the larger absolute value will be negative.

Integers Whose Product Is −30	1, −30	2, −15	3, −10	5, −6
Sum	−29	−13	−7	−1

We can see that $3(-10) = -30$ and $3 + (-10) = -7$, so $m = 3$ and $n = -10$.

Write the trinomial in the form $(u + m)(u + n)$:

$$u^2 - 7u - 30 = (u + 3)(u + (-10))$$
$$= (u + 3)(u - 10)$$

So we have

$$3u^3 - 9u^2 - 90u = 3u(u^2 - 7u - 30)$$
$$= 3u(u + 3)(u - 10)$$

Check

Use FOIL: $3u(u + 3)(u - 10) = 3u(u^2 - 10u + 3u - 30)$

Combine like terms: $= 3u(u^2 - 7u - 30)$

Distribute: $= 3u^3 - 21u^2 - 90u$

So $3u^3 - 21u^2 - 90u = 3u(u + 3)(u - 10)$.

Work Smart

We could have checked the result of Example 8 as follows:

$3u(u + 3)(u - 10)$
$= (3u^2 + 9u)(u - 10)$
$= 3u^3 - 30u^2 + 9u^2 - 90u$
$= 3u^3 - 21u^2 - 90u$

We say that a polynomial is **factored completely** if each factor in the final factorization is prime. For example, $3x^2 - 6x - 45 = (x - 5)(3x + 9)$ is not factored completely because the binomial $3x + 9$ has a common factor of 3. However, $3x^2 - 6x - 45 = 3(x - 5)(x + 3)$ is factored completely.

QUICK ✓ *Factor each trinomial completely.*

15. $4m^2 - 16m - 84$

16. $3z^3 + 12z^2 - 15z$

As we stated in the last section, sometimes the leading coefficient may be negative. If this is the case, we factor out the negative to make our factoring lives easier.

EXAMPLE 9 **Factoring Trinomials with a Leading Coefficient That Is Negative**

Factor completely: $-w^2 - 5w + 24$

Solution

Work Smart

It's easier to factor a trinomial in standard form when the leading coefficient is positive.

We notice that the leading coefficient is -1. It is easier to factor a trinomial when the leading coefficient is positive, so we will use -1 as the GCF and rewrite $-w^2 - 5w + 24$ as

$$-w^2 - 5w + 24 = -1(w^2 + 5w - 24)$$

To factor $w^2 + 5w - 24$, we look for two integers whose product is -24 and whose sum is 5. Since we have a negative product, but a positive sum, we know that one factor will be positive, and the other negative. Further, we know that the factor of -24 with the larger absolute value will be positive.

Integers Whose Product Is −24	−1, 24	−2, 12	−3, 8	−4, 6
Sum	23	10	5	2

We factor $w^2 + 5w - 24$ as $(w - 3)(w + 8)$. But remember that we already factored out the greatest common factor, -1, so

$$-w^2 - 5w + 24 = -1(w^2 + 5w - 24)$$
$$= -1(w - 3)(w + 8)$$

QUICK ✓ *Factor each trinomial completely.*

17. $-w^2 - 3w + 10$ **18.** $-2a^2 - 8a + 24$

6.2 Exercises

For Extra Help:

Student Solutions Manual   CD Video   PH Math/Tutor Center   MathXL Tutorials on CD   MathXL®   MyMathLab

Concepts and Vocabulary

In Problems 1–3, fill in the blanks.

1. A _____ _____ is a polynomial of the form $ax^2 + bx + c, a \neq 0$.

2. When factoring $x^2 - 10x + 24$, the sign of the factors of 24 must both be _____.

3. When factoring $x^2 - 10x - 24$, the factors of 24 must have _____ signs.

In Problems 4–6, answer True or False to each statement.

4. $(x - 3)(3x + 6)$ is factored completely.

5. A trinomial will always factor as a product of two binomials.

6. $4 + 4x + x^2$ has a leading coefficient of 4.

7. The answer key to your algebra exam said the factored form of a trinomial is $(x - 1)(x - 2)$. You have $(1 - x)(2 - x)$ on your paper. Is your answer marked correct or incorrect? Explain your reasoning.

8. In the trinomial $x^2 + bx + c$, both b and c are negative. Explain how to factor the trinomial. Make up trinomial and then use your rules to factor it.

Building Skills

In Problems 9–32, factor each trinomial completely. If the trinomial cannot be factored, say it is prime.

9. $x^2 + 5x + 6$ **10.** $p^2 + 7p + 6$ **11.** $m^2 + 9m + 18$

12. $n^2 + 12n + 20$ **13.** $x^2 - 15x + 36$ **14.** $z^2 - 13z + 36$

15. $a^2 - 8a + 12$ **16.** $b^2 - 7b + 12$ **17.** $x^2 - x - 12$

18. $y^2 - 8y - 9$ **19.** $x^2 + x - 20$ **20.** $a^2 + 6a - 40$

21. $z^2 + 12z - 45$ **22.** $t^2 + 2t - 38$ **23.** $x^2 - x - 32$

24. $x^2 - 11x - 12$ **25.** $x^2 - 5xy + 6y^2$ **26.** $x^2 - 14xy + 24y^2$

27. $r^2 + rs - 6s^2$ **28.** $p^2 + 5pq - 14q^2$ **29.** $x^2 - 3xy - 4y^2$

30. $x^2 + 9xy - 36y^2$ **31.** $z^2 + 7zy + y^2$ **32.** $m^2 + 16mn + 48n^2$

In Problems 33–40, factor each trinomial completely by factoring out the GCF first and then factoring the resulting trinomial.

33. $3x^2 + 3x - 6$ **34.** $5x^2 + 30x + 40$ **35.** $3a^3 - 24a^2 + 45a$

36. $4p^4 - 4p^3 - 8p^2$ **37.** $5x^2z - 20xz - 160z$ **38.** $8x^2z^2 - 56xz^2 + 80z^2$

39. $-2y^2 + 8y - 8$ **40.** $-3x^2 - 18x - 15$

In Problems 41–46, write the trinomial in standard form (descending order of degree) and then completely factor.

41. $-3x + x^2 - 18$ **42.** $50 - 15x + x^2$ **43.** $4x - 32 + x^2$

44. $-75 + x^2 + 10x$ **45.** $2x^2 + x^3 - 15x$ **46.** $x^3 - 20x - 8x^2$

Mixed Practice

In Problems 47–88, factor each polynomial completely. If the polynomial cannot be factored, say that it is prime.

47. $a^2 - 3ab - 28b^2$ **48.** $x^2 - 9xy - 36y^2$ **49.** $x^2 + x + 6$

50. $t^2 - 8t - 7$ **51.** $k^2 - k - 20$ **52.** $x^2 - 2x - 35$

53. $x^2 + xy - 30y^2$ **54.** $p^2 - 6pq + 5q^2$ **55.** $s^2t^2 - 8st + 15$

56. $x^2y^2 + 3xy + 2$ **57.** $-3c^3 + 3c^2 + 6c$ **58.** $-2z^3 - 2z^2 + 24z$

59. $-x^2 - x + 6$ **60.** $-r^2 - 12r - 36$ **61.** $g^2 - 4g + 21$

62. $x^2 - x + 6$ **63.** $n^4 - 30n^2 - n^3$ **64.** $-16x + x^3 - 6x^2$

65. $m^2 + 2mn - 15n^2$ **66.** $x^2 + 7xy + 12y^2$ **67.** $ab + 2b - 4a - 8$

68. $10ax - 5ay - 2x + y$ **69.** $35 + 12s + s^2$ **70.** $25 + 10x + x^2$

71. $3x^3y - 6x^2y - 6xy$ **72.** $3p^3q^2 - 6p^2q^2 + 15pq^2$ **73.** $n^2 - 9n - 45$

74. $x^2 - 6x - 42$ **75.** $a^2b^2 - 8ab + 12$ **76.** $x^2y^2 - 3xy - 18$

77. $-x^3 + 12x^2 + 28x$ **78.** $-3r^3 + 6r^2 - 3r$ **79.** $y^2 - 12y + 36$

80. $x^2 - 15x + 54$ **81.** $6rs - 9s - 4r + 6$ **82.** $12ab - 8b + 3a - 2$

83. $-36x + 18x^2 - 2x^3$ **84.** $-20mn^2 + 20m^2n - 5m^3$

85. $-21x^3y - 14xy^2$ **86.** $-12x^2y + 8xy^3$

87. $x^2 - ax + bx - ab$ **88.** $4mp - 6mq + 6np - 9nq$

Applying the Concepts

89. Punkin Chunkin In a recent Punkin Chunkin contest, the height of a pumpkin after t seconds was given by the trinomial $-16t^2 + 64t + 80$ feet. Write this polynomial in factored form.

90. Punkin Chunkin In a recent Punkin Chunkin contest, the height of a pumpkin after t seconds was given by the trinomial $-16t^2 + 16t + 32$ feet. Write this polynomial in factored form.

△ **91. A Rectangular Field** The trinomial $x^2 + 9x + 18$ square meters represents the area of a rectangular field. Find two binomials that represent the length and width of the field.

△ **92. Another Rectangular Field** The trinomial $x^2 + 6x + 8$ square yards represents the area of a rectangular field. Find two binomials that represent the length and width of the field.

△ **93. Area as a Polynomial** The area of a triangle is given by the trinomial $\frac{1}{2}x^2 + x - \frac{15}{2}$. Find algebraic expressions for the base and height of the triangle. (**Hint:** Factor out $\frac{1}{2}$ as a common factor.)

△ **94. Triangle Trinomial** The area of a triangle is given by the trinomial $\frac{1}{2}x^2 + 5x + 12$. Find algebraic expressions for the base and height of the triangle. (**Hint:** Factor out $\frac{1}{2}$ as a common factor.)

Extending the Concepts

In Problems 95–98, find all possible values of b so that the trinomial is factorable.

95. $x^2 + bx + 6$ **96.** $x^2 + bx - 10$ **97.** $x^2 + bx - 21$ **98.** $x^2 + bx + 12$

In Problems 99–102, find all possible positive values of c so that the trinomial is factorable.

99. $x^2 - 2x + c$ **100.** $x^2 - 3x + c$ **101.** $x^2 - 7x + c$ **102.** $x^2 + 6x + c$

6.3 Factoring Trinomials of the Form $ax^2 + bx + c, a \neq 1$

OBJECTIVES

1. Factor $ax^2 + bx + c, a \neq 1$ Using Trial and Error
2. Factor $ax^2 + bx + c, a \neq 1$ Using Grouping

Preparing for Factoring Trinomials of the Form $ax^2 + bx + c, a \neq 1$

Before getting started, take the following readiness quiz. If you get a problem wrong, go back to the section cited and review the material.

1. List the prime factorization of 24.   [Section A.1, p. A1]
2. Determine the coefficients of $5x^2 - 3x + 7$.   [Section 1.7, pp. 59–60]
3. Find the product: $(2x + 7)(3x - 1)$ [Section 5.3. pp. 361–363]

In this section, we factor quadratic trinomials of the form $ax^2 + bx + c$ in which the leading coefficient, a, is not 1. Examples of quadratic trinomials of the form $ax^2 + bx + c, a \neq 1$ are

$$2x^2 + 3x + 1 \qquad 5y^2 - y - 4 \qquad 10z^2 - 7z + 6$$

Before we begin to factor these quadratic trinomials, let's review multiplication of binomials using FOIL.

Factored Form	F	O	I	L	Polynomial
$(2x + 3)(x + 4) =$	$2x^2 +$	$8x +$	$3x +$	$12 =$	$2x^2 + 11x + 12$
$(3x - 4)(x + 7) =$	$3x^2 +$	$21x -$	$4x -$	$28 =$	$3x^2 + 17x - 28$
$(5a - 2b)(3a - b) =$	$15a^2 -$	$5ab -$	$6ab +$	$2b^2 =$	$15a^2 - 11ab + 2b^2$

To factor a trinomial in the form $ax^2 + bx + c, a \neq 1$, we reverse the FOIL multiplication process. That is, we will be given a trinomial such as $2x^2 + 11x + 12$, and we will write it as the product $(2x + 3)(x + 4)$. To factor trinomials in this form, we have two methods that can be used.

1. Trial and error
2. Factoring by grouping

There are pros and cons to both methods. We will point out these pros and cons as we proceed. We start with factoring by trial and error.

1   Factor $ax^2 + bx + c, a \neq 1$ Using Trial and Error

The idea behind using trial and error is to list various binomials and use FOIL to find their product until the combination of binomials is found that results in the trinomial to be factored. While this method may sound haphazard, experience and logic will play a role in minimizing the number of possibilities that you must try before a factored form is found. The following example illustrates the method.

Preparing for...Answers   **1.** $2^3 \cdot 3$
2. $5, -3, 7$ **3.** $6x^2 + 19x - 7$

EXAMPLE 1   **How to Factor $ax^2 + bx + c, a \neq 1$ Using Trial and Error**

Factor: $2x^2 + 7x + 5$

Step-by-Step Solution

Step 1: List the possibilities for the first terms of each binomial whose product is ax^2.	We list possible ways of representing the first term, $2x^2$. Since 2 is a prime number, we have one possibility: $$(2x + \underline{\quad})(x + \underline{\quad})$$

(continued)

Step 2: List the possibilities for the last terms of each binomial whose product is c.

The last term, 5, is also prime. It has the factors $(1)(5)$ and $(-1)(-5)$. However, did you notice that the coefficient of x, 7, is positive? To produce a positive sum, $7x$, we must have two positive factors. So we do not include the factors $(-1)(-5)$ in our list.

Step 3: Write out all the combinations of factors found in Steps 1 and 2. Multiply the binomials until a product is found that equals the trinomial.

Possible Factorization of $2x^2 + 7x + 5$	Product
$(2x + 1)(x + 5)$	$2x^2 + 11x + 5$
$(2x + 5)(x + 1)$	$2x^2 + 7x + 5$

The second row is the factorization that works, so $2x^2 + 7x + 5 = (2x + 5)(x + 1)$.

We summarize the steps used in Example 1 below.

Steps to Factor $ax^2 + bx + c$, $a \neq 1$ Using Trial and Error: a, b, and c Have No Common Factors

Step 1: List the possibilities for the first terms of each binomial whose product is ax^2.

$$(\_x + \quad)(\_x + \quad) = ax^2 + bx + c$$

Step 2: List the possibilities for the last terms of each binomial whose product is c.

$$(\_x + \square)(\_x + \square) = ax^2 + bx + c$$

Step 3: Write out all the combinations of factors found in Steps 1 and 2. Multiply the binomials until a product is found that equals the trinomial.

QUICK ✓ *Factor each trinomial completely.*

1. $3x^2 + 5x + 2$ **2.** $7y^2 + 22y + 3$

EXAMPLE 2 **Factoring $ax^2 + bx + c$, $a \neq 1$, Using Trial and Error**

Factor completely: $7x^2 - 18x + 8$

Solution

We list possible ways of representing the first term, $7x^2$. Since 7 is a prime number, we have

$$(7x + \_\_)(x + \_\_)$$

Let's look at the last term, 8, and list its factors:

Factors of 8	
$1 \cdot 8$	$-1 \cdot -8$
$2 \cdot 4$	$-2 \cdot -4$

Now let's concentrate on the middle term, $-18x$. To produce a negative sum, $-18x$, from a positive product, 8, we must have two *negative* factors. So we do not include the factors $1 \cdot 8$ or $2 \cdot 4$ in our list. Now we list the possible combinations of factors.

Work Smart

When factoring using trial and error it is only necessary to find the sum of the outer and inner products to see which factorization works.

Possible Factorization	Product
$(7x - 1)(x - 8)$	$7x^2 - 57x + 8$
$(7x - 8)(x - 1)$	$7x^2 - 15x + 8$
$(7x - 2)(x - 4)$	$7x^2 - 30x + 8$
$(7x - 4)(x - 2)$	$7x^2 - 18x + 8$

We see that $7x^2 - 18x + 8 = (7x - 4)(x - 2)$.

QUICK ✓ *Factor each trinomial completely.*

3. $3x^2 - 13x + 12$ **4.** $5p^2 - 21p + 4$

EXAMPLE 3   **Factoring $ax^2 + bx + c$, $a \neq 1$ Using Trial and Error**

Factor completely: $10x^2 - 13x - 3$

Solution

We list possible ways of representing the first term, $10x^2$.

$$(10x + \underline{\quad})(x + \underline{\quad})$$
$$(5x + \underline{\quad})(2x + \underline{\quad})$$

The last term, -3, has factors $-1 \cdot 3$ or $1 \cdot -3$.

We list the possible combinations of factors. We also highlight the sum of the "outer" and "inner" products.

Possible Factorization	Product
$(10x - 1)(x + 3)$	$10x^2 + 29x - 3$
$(10x - 3)(x + 1)$	$10x^2 + 7x - 3$
$(10x + 1)(x - 3)$	$10x^2 - 29x - 3$
$(10x + 3)(x - 1)$	$10x^2 - 7x - 3$
$(5x - 1)(2x + 3)$	$10x^2 + 13x - 3$
$(5x - 3)(2x + 1)$	$10x^2 - x - 3$
$(5x + 1)(2x - 3)$	$10x^2 - 13x - 3$
$(5x + 3)(2x - 1)$	$10x^2 + x - 3$

The row with the correct factorization is highlighted. So $10x^2 - 13x - 3 = (5x + 1)(2x - 3)$.

QUICK ✓ *Factor each trinomial completely.*

5. $2n^2 - 17n - 9$ **6.** $4b^2 - 5b - 6$

Factoring trinomials of the form $ax^2 + bx + c$, $a \neq 1$ can at first seem overwhelming. There are so many possibilities! Plus, the technique may at first seem haphazard. In fact, however, calling the technique "trial and error" is not quite truth in advertising because some thought is necessary to reduce the list of possible factors. To keep the list of possibilities small, here are some questions you should ask yourself before you begin to factor.

> **HINTS FOR USING TRIAL AND ERROR TO FACTOR $ax^2 + bx + c, a \neq 1$**
>
> - Is the constant c positive? If so, then the factors of the constant, c, must be the same sign as the coefficient of the middle term, b. For example,
> $$2a^2 + 11a + 5 = (2a + 1)(a + 5)$$
> $$2a^2 - 11a + 5 = (2a - 1)(a - 5)$$
> - Is the constant c negative? If so, then the factors of c must have opposite signs. For example,
> $$10b^2 + 19b - 15 = (5b - 3)(2b + 5)$$
> $$10b^2 - 19b - 15 = (5b + 3)(2b - 5)$$
> - If $ax^2 + bx + c$ has no common factor, then the binomials in the factored form cannot have common factors either.
> - Is the value of b small? If so, then choose factors of a and factors of c that are close to each other. If the value of b is large, then choose factors of a and factors of c that are far from each other.
> - Is the value of b correct, but has the wrong sign? Then interchange the signs in the binomial factors.

EXAMPLE 4 Factoring $ax^2 + bx + c, a \neq 1$ Using Trial and Error

Factor completely: $18x^2 + 3x - 10$

Solution

Remember, the first step in any factoring problem is to look for common factors. There are no common factors in $18x^2 + 3x - 10$.

List all possible ways of representing $18x^2$: $(18x + \_\_)(x + \_\_)$
$(9x + \_\_)(2x + \_\_)$
$(6x + \_\_)(3x + \_\_)$

Let's look at the last term, -10, and list its factors:

Factors of -10	
$-10 \cdot 1$	$10 \cdot -1$
$-5 \cdot 2$	$5 \cdot -2$

Before we list the possible combinations of factors, we ask some questions. Is the coefficient of the middle term of the polynomial $18x^2 + 3x - 10$ small? Yes, it is $+3$. Because this middle term is positive and small, the binomial factors we list should have outer and inner products that sum to a positive, small number. Therefore, we will start with $(6x + \_\_)(3x + \_\_)$ and the factors $(2)(-5)$ and $(-2)(5)$. We do not use $6x + 2$ or $6x - 2$ as possible factors because there is a common factor of 2 in these binomials and the trinomial to be factored has no common factors.

Let's try $(6x - 5)(3x + 2)$.

$$(6x - 5)(3x + 2) = 18x^2 + 12x - 15x - 10$$
$$= 18x^2 - 3x - 10$$

Close! The only problem is that the middle term has the opposite sign of the one that we want. Therefore, let's flip-flop the signs of -5 and 2 in the binomial and try $(6x + 5)(3x - 2)$.

$$(6x + 5)(3x - 2) = 18x^2 - 12x + 15x - 10$$
$$= 18x^2 + 3x - 10$$

So $18x^2 + 3x - 10 = (6x + 5)(3x - 2)$.

The moral of the story in Example 4 is that the name *trial and error* is a bit misleading. With some thought, you won't have to choose binomial factors haphazardly "until the cows come home" provided you use the helpful hints given and some care.

QUICK ✓ *Factor each trinomial completely.*

7. $12x^2 + 17x + 6$ **8.** $12y^2 + 32y - 35$

EXAMPLE 5   **Factoring Trinomials with Two Variables**

Factor completely: $48x^2 + 4xy - 30y^2$

Solution

Work Smart
Don't divide by the GCF because we have an algebraic expression, not an equation.

Did you notice that there is a greatest common factor of 2? We first factor out this GCF and obtain the polynomial $2(24x^2 + 2xy - 15y^2)$. The trinomial in parentheses will factor in the form $24x^2 + 2xy - 15y^2 = (\_\_x + \_\_y)(\_\_x + \_\_y)$.

List all possible ways of representing $24x^2$: $(24x + \_\_y)(x + \_\_y)$
$(12x + \_\_y)(2x + \_\_y)$
$(8x + \_\_y)(3x + \_\_y)$
$(6x + \_\_y)(4x + \_\_y)$

We list the factors of the last term, -15:

Factors of -15	
$-15 \cdot 1$	$15 \cdot -1$
$-5 \cdot 3$	$5 \cdot -3$

We do not use $24x + 3y$, $12x + 3y$, $3x + 3y$, or $6x + 3y$ as possible factors because there is a common factor in these binomials, and a common factor does not exist in the trinomial $24x^2 + 2xy - 15y$. Also, since the middle term has a small coefficient, we will start with $(6x + \_\_y)(4x + \_\_y)$ and the factors $(3)(-5)$ and $(-3)(5)$.
    Let's try $(6x - 5y)(4x + 3y)$.

$$(6x - 5y)(4x + 3y) = 24x^2 - 2xy - 15y^2$$

Close! The only problem is that the middle term is the opposite sign that we want. Therefore, let's flip-flop the signs of -5 and 3 in the binomial and try $(6x + 5y)(4x - 3y)$.

$$(6x + 5y)(4x - 3y) = 24x^2 - 18xy + 20xy - 15y^2$$
$$= 24x^2 + 2xy - 15y^2$$

So $48x^2 + 4xy - 30y^2 = 2(6x + 5y)(4x - 3y)$.   ■

QUICK ✓ *Factor the trinomial completely.*

9. $90x^2 + 21xy - 6y^2$

EXAMPLE 6   **Factoring Trinomials with a Negative Leading Coefficient**

Factor: $-14x^2 + 29x + 15$

Solution

Work Smart
It's easier to factor a trinomial in standard form when the leading coefficient is positive.

While there are no common factors in $-14x^2 + 29x + 15$, notice that the coefficient of the squared term is negative. Factor -1 out of the trinomial to obtain

$$-14x^2 + 29x + 15 = -1(14x^2 - 29x - 15)$$

Now factor the expression in parentheses and obtain

$$-14x^2 + 29x + 15 = -1(14x^2 - 29x - 15)$$
$$= -1(7x + 3)(2x - 5)$$

So $-14x^2 + 29x + 15 = -1(7x + 3)(2x - 5)$.

QUICK ✓ *Factor each trinomial completely.*

10. $-6y^2 + 23y + 4$ **11.** $-9x^2 - 21xy - 10y^2$

12. $-12x^2 - 10x + 8$ **13.** $-6x^2 - 3x + 45$

② Factor $ax^2 + bx + c$, $a \neq 1$ Using Grouping

We now introduce a second method for factoring trinomials of the form $ax^2 + bx + c$, $a \neq 1$. The second method uses a factoring technique introduced in the last section, factoring by grouping. The next example illustrates the procedure.

EXAMPLE 7 **How to Factor $ax^2 + bx + c$, $a \neq 1$ by Grouping**

Factor: $3x^2 + 14x + 15$

Step-by-Step Solution

First, we notice that $3x^2 + 14x + 15$ has no common factors. Comparing $3x^2 + 14x + 15$ to $ax^2 + bx + c$, we find that $a = 3$, $b = 14$, and $c = 15$.

Step 1: Find the value of *ac*.	The value of $a \cdot c = 3 \cdot 15 = 45$.

Step 2: Find the pair of integers, *m* and *n*, whose product is *ac* and whose sum is *b*.

We want to determine the integers whose product is 45 and whose sum is 14. Because both 14 and 45 are positive, we only list the positive factors of 45.

Integers Whose Product Is 45	1, 45	3, 15	5, 9
Sum	46	18	14

Step 3: Write $ax^2 + bx + c$ as $ax^2 + mx + nx + c$.

Write $3x^2 + 14x + 15$ as $3x^2 + 9x + 5x + 15$

$$14x = 9x + 5x$$

Step 4: Factor the expression in Step 3 by grouping.

$$3x^2 + 9x + 5x + 15 = (3x^2 + 9x) + (5x + 15)$$

Common factor in 1st group: $3x$;
common factor in 2nd group: 5: $\quad = 3x(x + 3) + 5(x + 3)$
Factor out $x + 3$: $\quad = (x + 3)(3x + 5)$

Step 5: Check Multiply out the factored form.

$$(x + 3)(3x + 5) = 3x^2 + 5x + 9x + 15$$
$$= 3x^2 + 14x + 15$$

So $3x^2 + 14x + 15 = (x + 3)(3x + 5)$.

We summarize below the steps used in Example 7.

Steps to Factor $ax^2 + bx + c$, $a \neq 1$ by Grouping: a, b, and c Have No Common Factors

Step 1: Find the value of ac.

Step 2: Find the pair of integers, m and n, whose product is ac and whose sum is b.

Step 3: Write $ax^2 + bx + c = ax^2 + mx + nx + c$.

Step 4: Factor the expression in Step 3 by grouping.

Step 5: Multiply out the factored form to verify your answer.

EXAMPLE 8   **Factoring $ax^2 + bx + c$, $a \neq 1$ by Grouping**

Factor: $12x^2 - x - 1$

Solution

First, we see that there is no greatest common factor in the expression $12x^2 - x - 1$. Comparing $12x^2 - x - 1$ to $ax^2 + bx + c$, we see that $a = 12$, $b = -1$, and $c = -1$.

The value of $a \cdot c = 12(-1) = -12$. We want to determine the integers whose product is -12 and whose sum is -1. Because the product $a \cdot c$ is negative, -12, we know that one integer will be positive and the other negative. Since the value of b is negative ($b = -1$), we know the factor of -12 with the larger absolute value will be negative.

Integers Whose Product Is -12	1, −12	2, −6	3, −4
Sum		−11	−4	−1

The integers whose product is -12 and whose sum is -1 are 3 and -4.

Work Smart

In Example 8, we could have written $12x^2 - x - 1$ as $12x^2 - 4x + 3x - 1$ and obtained the same result.

Write $12x^2 - x - 1$ as $12x^2 + 3x - 4x - 1$, and factor by grouping.

$$-x = 3x - 4x$$

$$12x^2 + 3x - 4x - 1 = (12x^2 + 3x) + (-4x - 1)$$

Common factor in 1st group: $3x$;
common factor in 2nd group: -1: $= 3x(4x + 1) - 1(4x + 1)$

Factor out $4x + 1$: $= (4x + 1)(3x - 1)$

Check

$$(4x + 1)(3x - 1) = 12x^2 - 4x + 3x - 1$$
$$= 12x^2 - x - 1$$

So $12x^2 - x - 1 = (4x + 1)(3x - 1)$.

QUICK ✓ *Factor each trinomial completely.*

14. $3x^2 - 2x - 8$ **15.** $10z^2 + 21z + 9$

The advantage of factoring quadratic trinomials of the form $ax^2 + bx + c$, $a \neq 1$ by grouping is that it is algorithmic (that is, step by step). However, if the product $a \cdot c$ gets large, then there are a lot of factors of ac whose sum must be determined. This can get overwhelming. Under these circumstances, it may be better to try trial and error.

EXAMPLE 9   **Factoring $ax^2 + bx + c$, $a \neq 1$ by Grouping**

Factor: $18x^2 - 33x - 30$

Solution

The first question we should ask is, "Is there is a greatest common factor in the expression $18x^2 - 33x - 30$?" Yes, there is a GCF of 3, so we first factor out the 3.

$$18x^2 - 33x - 30 = 3(6x^2 - 11x - 10)$$

Now we factor the remaining trinomial, $6x^2 - 11x - 10$, by grouping. We see that $a = 6$, $b = -11$, and $c = -10$. The value of $a \cdot c = 6(-10) = -60$.

Now, we want to determine the integers whose product is -60 and whose sum is -11.

Integers Whose Product Is -60	$1, -60$	$2, -30$	$3, -20$	$4, -15$	$5, -12$	$6, -10$
Sum	-59	-28	-17	-11	-7	-4

The integers whose product is -60 and whose sum is -11 are 4 and -15.
Write $6x^2 - 11x - 10$ as $6x^2 + 4x - 15x - 10$.

$$-11x = 4x - 15x$$

Now, factor by grouping.

$$6x^2 + 4x - 15x - 10 = (6x^2 + 4x) + (-15x - 10)$$

Common factor in 1st group: $2x$;
common factor in 2nd group: -5: $= 2x(3x + 2) - 5(3x + 2)$
Factor out $3x + 2$: $= (3x + 2)(2x - 5)$

Check

$$3(3x + 2)(2x - 5) = 3(6x^2 - 15x + 4x - 10)$$
$$= 3(6x^2 - 11x - 10)$$
$$= 18x^2 - 33x - 30$$

So $18x^2 - 33x - 30 = 3(3x + 2)(2x - 5)$.

QUICK ✓ *Factor the trinomial completely.*

16. $24x^2 + 6x - 9$ **17.** $-10n^2 + 17n - 3$

Work Smart

Let's compare the two methods presented in this section, trial and error and factoring by grouping, by factoring $3x^2 + 10x + 8$. The first question to ask is: Is there a GCF? No, there's not, so let's continue.

Trial and Error	Grouping
Step 1: The coefficient of x^2, 3, is prime, so we list the possibilities for the binomial factors: $(3x + \underline{\ \ })(x + \underline{\ \ })$.	**Step 1:** For the polynomial $3x^2 + 10x + 8$, $a = 3$ and $c = 8$; the value of ac is $3 \cdot 8 = 24$.
Step 2: The last term, 8, is not prime. Its factors are $(1)(8)$ or $(2)(4)$ or $(-1)(-8)$ or $(-2)(-4)$. Since $c = 8$ is positive and the coefficient of the middle term is also positive, we'll consider only $(1)(8)$ and $(2)(4)$.	**Step 2:** The two integers whose product is $ac = 24$ whose sum is $b = 10$ are 6 and 4.
	Step 3: Rewrite $3x^2 + 10x + 8$ as $3x^2 + 10x + 8 = 3x^2 + 6x + 4x + 8$.

(continued)

Trial and Error	Grouping
Step 3: The coefficient of the middle term is not large, so let's start by trying $(3x + 4)(x + 2)$ and $(3x + 2)(x + 4)$. We will multiply out $(3x + 4)(x + 2)$ first.	**Step 4:** Factor the expression $3x^2 + 6x + 4x + 8$ by grouping.

Trial and Error:

$(3x + 4)(x + 2) = 3x^2 + 10x + 8$

It works! Therefore,
$3x^2 + 10x + 8 = (3x + 4)(x + 2)$.

Grouping:

$3x^2 + 6x + 4x + 8$
$= 3x(x + 2) + 4(x + 2)$
$= (x + 2)(3x + 4)$

So $3x^2 + 10x + 8 = (x + 2)(3x + 4)$.

Because $(3x + 4)(x + 2)$ is equivalent to $(x + 2)(3x + 4)$, we see that both methods give the same result. Which method do you prefer?

6.3 Exercises

For Extra Help:

Student Solutions Manual CD Video PH Math/Tutor Center MathXL Tutorials on CD MathXL® MyMathLab

Concepts and Vocabulary

In Problems 1–3, fill in the blanks.

1. When factoring any polynomial, begin by determining whether each term contains a _____ _____ other than one.

2. When factoring $6x^2 + x - 1$ using grouping, $ac =$ _____ and $b =$ _____.

3. When factoring $ax^2 + bx + c$ and b is a small number, choose factors of the product ac that are _____ to each other.

In Problems 4–6, answer True or False to each statement.

4. The trinomial $12x^2 + 22x + 6$ is completely factored as $(4x + 6)(3x + 1)$.

5. To factor $2x^2 - 13x + 6$ using grouping, find factors whose product is 6 and sum is -13.

6. To check if a trinomial has been factored correctly, use FOIL to multiply the binomials and verify that the product is the original trinomial.

7. Describe when you would use the trial and error method and when you would use the grouping method to factor a trinomial. Make up two examples that demonstrate your reasoning.

8. How can you tell if a trinomial is not factorable? Make up an example to demonstrate your reasoning.

Building Skills

In Problems 9–36, factor completely. If a trinomial cannot be factored, say it is prime.

9. $6n^2 - 11n + 3$
10. $5w^2 + 13w - 6$
11. $4w^2 - 8w - 5$
12. $4x^2 - 9x + 4$
13. $5x^2 - 13x + 2$
14. $3x^2 + 16x - 12$
15. $4x^2 + 4x - 3$
16. $7a^2 + 37a + 10$
17. $27z^2 + 3z - 2$
18. $25t^2 + 5t - 2$
19. $9x^2 - 6x + 1$
20. $3x^2 - 17x + 10$
21. $2x^2 + 5x + 3$
22. $2x^2 - 7x + 3$
23. $6n^2 - 11n + 5$
24. $6y^2 - 5y - 6$
25. $12x^2 + 2xy - 4y^2$
26. $20t^2 + 23t + 8$
27. $4m^2 + 9m - 5$
28. $6x^2 - 14xy - 12y^2$
29. $6m^2 - 5m - 4$
30. $6m^2 - 23m - 4$
31. $-18n^2 + 39n - 18$
32. $-3x - 4 + 10x^2$
33. $12a^2 + 19a + 5$
34. $-5 - 9x + 18x^2$
35. $2x^2 - 5x + 3$
36. $2x^2 + 7x + 3$

Mixed Practice

In Problems 37–68, factor completely using any method you wish. If a polynomial cannot be factored, say it is prime.

37. $15x^2 - 23x + 4$

38. $12x^3 - 11x^2 - 15x$

39. $-13a + 12 - 4a^2$

40. $2x + 12x^2 - 24$

41. $10x^2 - 8xy - 24y^2$

42. $12n^2 + 7n - 10$

43. $8 - 18x + 9x^2$

44. $18x^2 + 88x - 10$

45. $-6x + 9 - 24x^2$

46. $-24z^2 + 18z + 2$

47. $4x^3y^2 - 8x^2y^3 - 4x^2y^2$

48. $9x^3y + 6x^2y + 3xy$

49. $x^2 - 13xy + 42y^2$

50. $k^2 + 2kp - 35p^2$

51. $4m^2 + 13mn + 3n^2$

52. $4m^2 - 19mn + 12n^2$

53. $10x - x^2 + 24$

54. $28 - x^2 + 3x$

55. $n^3 - n^2 + n - 1$

56. $ab - bx - ay + xy$

57. $6x^2 - 17x - 12$

58. $8x^2 + 14x - 15$

59. $48xy + 24x^2 - 30y^2$

60. $-20b^2 + 6a^2 + 7ab$

61. $4a(a + 2) - 8(a + 2)$

62. $2x(x + 1) - 2(x + 1)$

63. $18a^2 + 39ab - 24b^2$

64. $63y^3 + 60xy^3 + 12x^2y^3$

65. $2a^2b^2 - ac + 2ab^3 - bc$

66. $6xy^2 + 3x^3y^2 - 2y - x^2y$

67. $-6x^3 + 10x^2 - 4x^4$

68. $21n^2 - 18n^3 + 9n$

Applying the Concepts

△ **69. Area of a Triangle** A triangle has area described by the polynomial

$$3x^2 + \frac{13}{2}x - 14 \text{ square meters. Find the base and height of the triangle.}$$

(**Hint:** The area of a triangle is $A = \frac{1}{2}bh$.)

△ **70. Area of a Rectangle** A rectangle has area described by the polynomial $6x^2 + x - 1$ square centimeters. Find the length and width of the rectangle. (**Hint:** The area of a rectangle is $A = lw$.)

71. Suppose that we know one factor of $6x^2 - 11x - 10$ is $3x + 2$. What is the other factor?

72. Suppose that we know one factor of $8x^2 + 22x - 21$ is $2x + 7$. What is the other factor?

Extending the Concepts

In Problems 73–80, factor completely.

73. $30x + 22x^2 - 24x^3$

74. $48x - 74x^2 + 28x^3$

75. $27z^4 + 42z^2 + 16$

76. $15n^6 + 7n^3 - 2$

77. $3x^{2n} + 19x^n + 6$

78. $2x^{2n} - 3x^n - 5$

79. $6x^2(x^2 + 1) - 25x(x^2 + 1) + 14(x^2 + 1)$

80. $10x^2(x - 1) - x(x - 1) - 2(x - 1)$

In Problems 81 and 82, find all possible integer values of b so that the polynomial is factorable.

81. $3x^2 + bx - 5$

82. $6x^2 + bx + 7$

6.4 Factoring Special Products

OBJECTIVES

① Factor Perfect Square Trinomials

② Factor the Difference of Two Squares

③ Factor the Sum or Difference of Two Cubes

Preparing for Factoring Special Products

Before getting started, take the following readiness quiz. If you get a problem wrong, go back to the section cited and review the material.

1. Evaluate: 5^2 [Section 1.6, pp. 49–50]

2. Evaluate: $(-2)^3$ [Section 1.6, pp. 49–50]

3. Find the product: $(5p^2)^3$ [Section 5.2, p. 357]

4. Find the product: $(3z + 2)^2$ [Section 5.3, pp. 364–366]

5. Find the product: $(4m + 5)(4m - 5)$ [Section 5.3, p. 364]

Preparing for...Answers **1.** 25 **2.** −8
3. $125p^6$ **4.** $9z^2 + 12z + 4$
5. $16m^2 - 25$

In Section 5.3, we presented three binomial products that were "special products," that is, special cases of the FOIL pattern. We briefly review two of them here.

Pattern	Product	Example
Squaring a binomial	$(a + b)^2 = a^2 + 2ab + b^2$ $(a - b)^2 = a^2 - 2ab + b^2$	$(3a + 5)^2 = (3a)^2 + 2(3a)(5) + 5^2$ $= 9a^2 + 30a + 25$
Product of the sum and difference of two terms	$(a + b)(a - b) = a^2 - b^2$	$(4z + 7)(4z - 7) = (4z)^2 - 7^2$ $= 16z^2 - 49$

Recall that $a^2 + 2ab + b^2$ and $a^2 - 2ab + b^2$ are called perfect square trinomials. We call $a^2 - b^2$ is called the difference of two squares.

In this section, we look at polynomials that can be categorized as having "special formulas" for factoring. Why would we want to learn a method for factoring using "special formulas"? If you recognize a polynomial as a perfect square trinomial or the difference of two squares, you'll be able to factor it quickly without using trial and error or the grouping method.

We begin with factoring perfect square trinomials.

① Factor Perfect Square Trinomials

Reversing the formulas $(a + b)^2 = a^2 + 2ab + b^2$ and $(a - b)^2 = a^2 - 2ab + b^2$, we obtain a method for factoring perfect square trinomials.

In Words
A perfect square trinomial is a trinomial in which the first term and the third term are perfect squares and the second term is either 2 times or −2 times the product of the expressions being squared in the first and third terms.

> **PERFECT SQUARE TRINOMIALS**
> $$a^2 + 2ab + b^2 = (a + b)^2$$
> $$a^2 - 2ab + b^2 = (a - b)^2$$

Perfect square trinomials can be factored quickly when they are recognized. For a polynomial to be a perfect square trinomial, two conditions must be satisfied.

1. The first and last terms must be perfect squares. The perfect squares are $1^2 = 1, 2^2 = 4, 3^2 = 9$, and so on. Any variable raised to an even exponent is a perfect square. So, $x^2, x^4 = (x^2)^2, x^6 = (x^3)^2$ are all perfect squares. Examples of perfect squares are

$$
\begin{array}{lll}
49 & \text{because} & 7^2 = 49 \\
121 & \text{because} & 11^2 = 121 \\
9x^2 & \text{because} & (3x)^2 = 9x^2 \\
25a^4 & \text{because} & (5a^2)^2 = 25a^4
\end{array}
$$

2. The "middle term" must equal 2 times or −2 times the product of the expressions being squared in the first and last term.

Let's look at an example.

EXAMPLE 1   **How to Factor Perfect Square Trinomials**

Factor completely: $z^2 + 6z + 9$

Step-by-Step Solution

Step 1: Determine whether the first term and the third term are perfect squares.	The first term, z^2, is the square of z and the third term, 9, is the square of 3.

(continued)

Step 2: Determine whether the middle term is 2 times or -2 times the product of the expressions being squared in the first and last term.

The middle term, $6z$, is 2 times the product of z and 3.

Step 3: Use $a^2 + 2ab + b^2 = (a + b)^2$ to factor the expression.

$$\overset{a^2\quad +\ 2\cdot a\cdot b\ +\ b^2}{\underset{\downarrow\qquad\ \downarrow\ \downarrow\quad\ \downarrow}{}}$$
$$z^2 + 6z + 9 = z^2 + 2\cdot z\cdot 3 + 3^2$$

Factor as $(a + b)^2$ with $a = z$ and $b = 3$:　$= (z + 3)^2$

So $z^2 + 6z + 9 = (z + 3)^2$.

EXAMPLE 2　Factoring Perfect Square Trinomials

Factor completely:

    (a) $4x^2 - 20x + 25$　　　　**(b)** $9x^2 + 42xy + 49y^2$

Solution

  (a) The first term, $4x^2$, is the square of $2x$, and the third term, 25, is the square of 5. The middle term, $-20x$, is -2 times the product of $2x$ and 5. We use $a^2 - 2ab + b^2 = (a - b)^2$ to factor the expression.

$$\overset{a^2\quad\ -\ 2\cdot a\ \cdot\ b\ +\ b^2}{\underset{\downarrow\qquad\ \downarrow\ \downarrow\quad\downarrow}{}}$$
$$4x^2 - 20x + 25 = (2x)^2 - 2\cdot 2x\cdot 5 + (5)^2$$
$$= (2x - 5)^2$$

Therefore, $4x^2 - 20x + 25 = (2x - 5)^2$.

  (b) The first term, $9x^2$, is the square of $3x$, and the third term, $49y^2$, is the square of $7y$. The middle term, $42xy$, is 2 times the product of $3x$ and $7y$. We factor as $a^2 + 2ab + b^2 = (a + b)^2$.

$$\overset{a^2\quad\ +\ 2\cdot a\ \cdot\ b\ +\ b^2}{\underset{\downarrow\qquad\ \downarrow\ \downarrow\quad\downarrow}{}}$$
$$9x^2 + 42xy + 49y^2 = (3x)^2 + 2\cdot 3x\cdot 7y + (7y)^2$$
$$= (3x + 7y)^2$$

We see that $9x^2 + 42xy + 49y^2 = (3x + 7y)^2$.

QUICK ✓　*Factor each trinomial completely.*

1. $x^2 - 12x + 36$　　　**2.** $16x^2 + 40x + 25$　　　**3.** $9a^2 - 60ab + 100b^2$

EXAMPLE 3　Factoring a Trinomial

Factor completely, if possible: $b^2 - 10b + 36$

Solution

The first term, b^2, is the square of b. The third term, 36, is the square of 6. Is the middle term, $-10b$, -2 times the product of the expressions being squared in the first and last term?

$$-2(b)(6) = -12b \neq -10b.$$

So $b^2 - 10b + 36$ is not a perfect square trinomial. Can we factor $b^2 - 10b + 36$ using another strategy? The trinomial $b^2 - 10b + 36$ is in the form $x^2 + bx + c$, so we need two integers whose product is 36, whose sum is -10. We choose negative factors of 36, since the coefficient of the middle term is negative: The possibilities are $(-1)(-36), (-2)(-18), (-3)(-12),$ or $(-9)(-4)$. None of these factors sum to -10. We conclude that $b^2 - 10b + 36$ is prime.

QUICK ✓ *Factor each trinomial completely, if possible.*

4. $z^2 - 8z + 16$ **5.** $4n^2 + 12n + 9$ **6.** $y^2 + 13y + 35$

EXAMPLE 4   **Factoring a Perfect Square Trinomial Having a GCF**

Factor completely: $32m^4 - 48m^2 + 18$

Solution

Do you remember the first step in any factoring problem? It is to look for the GCF. Notice there is a common factor of 2 in the trinomial, so we factor it out.

$$32m^4 - 48m^2 + 18 = 2(16m^4 - 24m^2 + 9)$$

> **Work Smart**
> m^4 is a square because $m^4 = (m^2)^2$.

We now examine the trinomial $16m^4 - 24m^2 + 9$. The first term, $16m^4$, is the square of $4m^2$. The third term, 9, is the square of 3. The middle term, $-24m^2$, is -2 times the product of $4m^2$ and 3.

$$16m^4 - 24m^2 + 9 = \overset{a^2}{(4m^2)^2} - \overset{2\cdot a \cdot b}{2\cdot 4m^2 \cdot 3} + \overset{b^2}{3^2}$$
$$\text{Factor as } (a-b)^2: \quad = (4m^2 - 3)^2$$

> **Work Smart**
> Don't forget the GCF that was factored out in the first step.

Therefore,

$$32m^4 - 48m^2 + 18 = 2(16m^4 - 24m^2 + 9)$$
$$= 2(4m^2 - 3)^2$$

QUICK ✓ *Factor each trinomial completely.*

7. $4z^2 + 24z + 36$ **8.** $50a^3 + 80a^2 + 32a$ **9.** $50a^4 + 40a^2b + 8b^2$

Perfect square trinomials can also be factored using the methods introduced in the last section, but if you recognize the perfect square pattern, you won't have to use either trial and error or the grouping method.

(2)   **Factor the Difference of Two Squares**

> **In Words**
> The difference of two squares is just that! A perfect square minus a perfect square. This pattern is easy to recognize: Just look for two squares that have been subtracted.

Do you recall finding products such as $(x + 2y)(x - 2y)$ in Chapter 3? We found that $(x + 2y)(x - 2y) = x^2 - (2y)^2 = x^2 - 4y^2$. In general,

$$(a - b)(a + b) = a^2 - b^2$$

We will now reverse the multiplication process and find the factors of the difference of two squares.

> **DIFFERENCE OF TWO SQUARES**
> $$a^2 - b^2 = (a - b)(a + b)$$

EXAMPLE 5 **Factoring the Difference of Two Squares**

Factor completely:

(a) $n^2 - 81$ **(b)** $16x^2 - 9y^2$

Solution

(a) We notice that $n^2 - 81$ is the difference of two squares, n^2 and $81 = 9^2$. So

$$\overset{a^2 \quad - \quad b^2}{\downarrow \qquad \downarrow}$$
$$n^2 - 81 = (n)^2 - (9)^2$$
$$a^2 - b^2 = (a - b)(a + b): \quad = (n - 9)(n + 9)$$

Work Smart

Remember that the Commutative Property of Multiplication tells us that the order of the factors in the answer doesn't matter:

$(a - b)(a + b) = (a + b)(a - b)$

(b) We notice that $16x^2 - 9y^2$ is the difference of two squares, $16x^2 = (4x)^2$ and $9y^2 = (3y)^2$. So

$$\overset{a^2 \quad - \quad b^2}{\downarrow \qquad \downarrow}$$
$$16x^2 - 9y^2 = (4x)^2 - (3y)^2$$
$$a^2 - b^2 = (a - b)(a + b): \quad = (4x - 3y)(4x + 3y)$$

You can check the answer to any of these differences of two squares by using FOIL. To check Example 5(b),

Check $(4x - 3y)(4x + 3y) = 16x^2 + 12xy - 12xy - 9y^2$
$$= 16x^2 - 9y^2$$

Our answer checks, so $16x^2 - 9y^2 = (4x - 3y)(4x + 3y)$.

QUICK ✓ *Factor completely.*

10. $z^2 - 25$ **11.** $81m^2 - 16n^2$ **12.** $16a^2 - \dfrac{4}{9}b^2$

EXAMPLE 6 **Factoring the Difference of Two Squares**

Factor completely:

(a) $49k^4 - 100$ **(b)** $p^4 - 1$

Solution

(a) Because $49k^4 = (7k^2)^2$ and $100 = 10^2$, we have the difference of two squares. So

$$\overset{a^2 \quad - \quad b^2}{\downarrow \qquad \downarrow}$$
$$49k^4 - 100 = (7k^2)^2 - 10^2$$
$$a^2 - b^2 = (a - b)(a + b): \quad = (7k^2 - 10)(7k^2 + 10)$$

So $49k^4 - 100 = (7k^2 - 10)(7k^2 + 10)$.

(b) The expressions p^4 and 1 are both perfect squares. $p^4 = (p^2)^2$ and $1 = 1^2$. So

$$\overset{a^2 \quad - \quad b^2}{\downarrow \qquad \downarrow}$$
$$p^4 - 1 = (p^2)^2 - 1^2$$
$$a^2 - b^2 = (a - b)(a + b): \quad = (p^2 - 1)(p^2 + 1)$$

But $p^2 - 1$ is a difference of two squares! So we factor again:

$$= (p + 1)(p - 1)(p^2 + 1)$$

So $p^4 - 1 = (p + 1)(p - 1)(p^2 + 1)$.

You may be asking yourself, "What about the sum of two squares—how does it factor?" The answer is that if a and b are real numbers, **the sum of two squares, $a^2 + b^2$, is prime and does not factor.** So, if a variable represents a real number, binomials such as $a^2 + 1$ or $4y^2 + 81$ are prime.

QUICK ✔ *Factor each polynomial completely.*

13. $36b^4 - c^2$ **14.** $100k^4 - 81w^2$ **15.** $x^4 - 16$

EXAMPLE 7   **Factoring the Difference of Two Squares Containing a GCF**

Factor completely: $50x^2 - 72y^2$

Solution

What's the first thing we look for when we factor? A greatest common factor! The GCF of 50 and 72 is 2, so we proceed as follows:

$$\text{Factor out GCF} = 2: \quad 50x^2 - 72y^2 = 2(25x^2 - 36y^2)$$
$$\text{Difference of two squares with } a = 5x \text{ and } b = 6y: \quad = 2[(5x)^2 - (6y)^2]$$
$$\text{Factor as } (a - b)(a + b): \quad = 2(5x - 6y)(5x + 6y)$$

So $50x^2 - 72y^2 = 2(5x - 6y)(5x + 6y)$. ∎

QUICK ✔ *Factor each polynomial completely.*

16. $147x^2 - 48$ **17.** $-27a^3b + 75ab^3$

---

③ Factor the Sum or Difference of Two Cubes

Consider the following products:

$$(a + b)(a^2 - ab + b^2) = a^3 - a^2b + ab^2 + a^2b - ab^2 + b^3$$
$$= a^3 + b^3$$
$$(a - b)(a^2 + ab + b^2) = a^3 + a^2b + ab^2 - a^2b - ab^2 - b^3$$
$$= a^3 - b^3$$

These products show us that we can factor the sum or difference of two cubes as follows:

THE SUM OF TWO CUBES

$$a^3 + b^3 = (a + b)(a^2 - ab + b^2)$$

THE DIFFERENCE OF TWO CUBES

$$a^3 - b^3 = (a - b)(a^2 + ab + b^2)$$

Notice in the formulas that the sign between the cubes matches the sign in the first set of parentheses and that the sign in the middle of the trinomial factor must be opposite the sign between the cubes. Also remember that the perfect cubes are $1^3 = 1, 2^3 = 8, 3^3 = 27$, and so on. Any variable raised to a multiple of 3 is a perfect cube. So $x^3, x^6 = (x^2)^3, x^9 = (x^3)^3$, and so on are all perfect cubes.

EXAMPLE 8 **Factoring the Sum or Difference of Two Cubes**

Factor completely: $x^3 - 8$

Solution

We notice that we have the difference of two cubes, x^3 and $8 = 2^3$. We let $a = x$ and $b = 2$ in the factoring formula for the difference of two cubes.

$$a^3 - b^3 = (a - b)(a^2 + a\ b + b^2)$$

$$x^3 - 8 = x^3 - 2^3 = (x - 2)(x^2 + x(2) + 2^2)$$
$$= (x - 2)(x^2 + 2x + 4)$$

So $x^3 - 8 = (x - 2)(x^2 + 2x + 4)$. ▄

Work Smart

The trinomial in the factored form of the sum or difference of two cubes will always be prime.

EXAMPLE 9 **Factoring the Sum or Difference of Two Cubes**

Factor completely: $8m^3 + 125n^6$

Solution

Because $8m^3 = (2m)^3$ and $125n^6 = (5n^2)^3$, the expression $8m^3 + 125n^6$ is the sum of two cubes. We let $a = 2m$ and $b = 5n^2$ in the factoring formula for the sum of two cubes.

$$8m^3 + 125n^6 = (2m)^3 + (5n^2)^3$$
$$= (2m + 5n^2)[(2m)^2 - (2m)(5n^2) + (5n^2)^2]$$
$$= (2m + 5n^2)(4m^2 - 10mn^2 + 25n^4)$$

So $8m^3 + 125n^6 = (2m + 5n^2)(4m^2 - 10mn^2 + 25n^4)$. ▄

QUICK ✓ *Factor each polynomial completely.*

18. $z^3 + 125$ **19.** $8p^3 - 27q^6$

EXAMPLE 10 **Factoring the Sum or Difference of Two Cubes Having a GCF**

Factor completely: $250x^4 + 54x$

Solution

Remember, the first step in any factoring problem is to look for a common factor. Here, we have a common factor of $2x$.

$$250x^4 + 54x = 2x(125x^3 + 27)$$
$$(5x)^3 = 125x^3 \text{ and } 3^3 = 27: \quad = 2x[(5x)^3 + 3^3]$$
$$= 2x(5x + 3)(25x^2 - 15x + 9)$$

So $250x^4 + 54x = 2x(5x + 3)(25x^2 - 15x + 9)$. ▄

QUICK ✓ *Factor each polynomial completely.*

20. $54a - 16a^4$ **21.** $-375b^3 + 3$

6.4 Exercises

For Extra Help:

Student Solutions Manual   CD Video   PH Math/Tutor Center   MathXL Tutorials on CD   MathXL®   MyMathLab

Concepts and Vocabulary

In Problems 1–3, fill in the blanks.

1. The binomial $27x^3 + 64y^3$ is called the _____ of _____ _____.

2. $a^2 - 2ab + b^2 = $ _____.

3. $4m^2 - 81n^2$ is called the _____ of _____ _____ and factors into two binomials.

In Problems 4–6, answer True or False to each statement.

4. $x^2 + 9$ is prime.

5. $(x - 1)(x^2 - x + 1)$ can be factored further.

6. $4x^2 - 16y^2$ factors completely to $(2x - 4y)(2x + 4y)$.

7. In your own words, create a list of steps for factoring the sum or difference of two cubes.

8. You are asked to factor the polynomial $x^2 + 4$. What is your answer and what rationale would you give for this result?

Building Skills

In Problems 9–20, factor each perfect square trinomial completely.

9. $x^2 - 6x + 9$
10. $n^2 - 2n + 1$
11. $x^2 + 10x + 25$
12. $m^2 + 12m + 36$
13. $4p^2 - 4p + 1$
14. $9a^2 - 12a + 4$
15. $16x^2 + 24x + 9$
16. $16y^2 - 72y + 81$
17. $x^2 - 4xy + 4y^2$
18. $4a^2 + 20ab + 25b^2$
19. $4z^2 - 12z + 9$
20. $25k^2 - 70k + 49$

In Problems 21–32, factor each difference of two squares completely.

21. $x^2 - 121$
22. $x^2 - 9$
23. $16x^2 - 49y^2$
24. $36m^2 - 25n^2$
25. $4x^2 - 25$
26. $25m^2 - 9$
27. $100n^8 - 81p^4$
28. $36s^6 - 49t^4$
29. $k^8 - 256$
30. $a^4 - 16$
31. $25p^4 - 49q^2$
32. $36b^4 - 121a^2$

In Problems 33–42, factor each sum or difference of two cubes completely.

33. $1 + x^3$
34. $8v^3 + 1$
35. $8x^3 - 27y^3$
36. $64r^3 - 125s^3$
37. $x^6 - 8y^3$
38. $m^9 - 27n^6$
39. $27c^3 + 64d^9$
40. $125y^3 + 27z^6$
41. $16x^3 + 125y^3$
42. $25a^3 + 81b^6$

Mixed Practice

In Problems 43–84, factor completely. If the polynomial is prime, state so.

43. $x^2 - \dfrac{2}{3}x + \dfrac{1}{9}$
44. $x^2 + \dfrac{1}{2}x + \dfrac{1}{16}$
45. $16m^2 + 40mn + 25n^2$
46. $25p^2q^2 - 80pq + 64$
47. $18 - 12x + 2x^2$
48. $50 + 20x + 2x^2$
49. $48a^3 + 72a^2 + 27a$
50. $12n^3 - 36n^2 + 27n$
51. $81p^2 - \dfrac{1}{4}$
52. $\dfrac{x^2}{144} - \dfrac{1}{25}$
53. $x^4y^2 - x^2y^4$
54. $x^6y^2 - 16x^2y^2$

55. $2t^4 - 54t$ **56.** $16x^5 - 54x^2$ **57.** $3s^7 + 24s$

58. $32a^3 + 4b^6$ **59.** $4x - 4x^3$ **60.** $2x^5 - 162x$

61. $x^3y - xy^3$ **62.** $48z^4 - 3$ **63.** $x^3y^3 + 1$

64. $8r^3s^3 + t^3$ **65.** $x^8 - 25y^{10}$ **66.** $x^4 - 225x^2$

67. $x^3 - \dfrac{y^6}{64}$ **68.** $\dfrac{x^3}{125} + 8$ **69.** $2x^2 - 8x + 8$

70. $3x^2 + 18x + 27$ **71.** $16n^3 - 32n^5 - 2n$ **72.** $36x^3 - 12x^5 - 27x$

73. $-2x - 8x^2 - 8x^3$ **74.** $18x^2 - 24x^3 + 8x^4$ **75.** $x^6 - y^9$

76. $x^4y^3 + 216xy^3$ **77.** $9x^2 + y^2$ **78.** $16a^2 + 49b^2$

79. $2x(x^2 - 4) + 5(x^2 - 4)$ **80.** $4z(z^2 - 9) + 3(z^2 - 9)$

81. $2y^3 + 5y^2 - 32y - 80$ **82.** $m^3 + 2m^2 - 25m - 50$

83. $x^4 - 2x^3 + 8x - 16$ **84.** $b^4 - 4b^3 - 27b + 108$

Applying the Concepts

△ **85. Area of a Square** The area of a square is given by the polynomial $4x^2 + 20x + 25$. What is an algebraic expression for the length of one side of the square? (**Hint:** The area of a square is $A = s^2$.)

△ **86. Area of a Square** The area of a square is given by the polynomial $9x^2 - 6x + 1$. What is an algebraic expression for the length of one side of the square?

In Problems 87 and 88, calculate the area of the shaded region and then write this polynomial in factored form.

△ **87.**

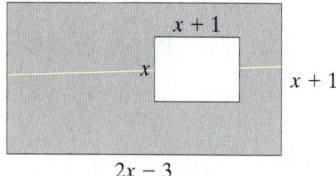

△ **88.**

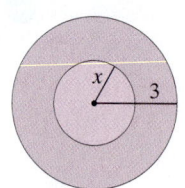

Extending the Concepts

In Problems 89–98, factor completely and then simplify, if possible.

89. $(x - 2)^2 - (x + 1)^2$ **90.** $(m - p)^2 - (m + p)^2$ **91.** $(x - y)^3 + y^3$

92. $(z + 1)^3 - z^6$ **93.** $(x + 1)^2 - 9$ **94.** $(x + 3)^2 - 25$

95. $2a^2(x + 1) - 17a(x + 1) + 30(x + 1)$

96. $25a^2(x - 1)^2 - 5a(x - 1)^2 - 2(x - 1)^2$

97. $5(x + 2)^2 - 7(x + 2) - 6$ **98.** $9(2a + b)^2 + 6(2a + b) - 8$

6.5 Summary of Factoring Techniques

OBJECTIVE

1 Factor Polynomials Completely

1 Factor Polynomials Completely

We have one objective in this section—to put together all the various factoring techniques we have discussed in Sections 6.1–6.4. Recall that we say that a polynomial is factored completely if each factor in the final factorization is prime. Here we present guidelines for factoring any polynomial.

Steps for Factoring

Step 1: Is there a greatest common factor? If so, factor out the greatest common factor (GCF).

Step 2: Count the number of terms.

Step 3: (a) 2 terms (Binomials)

- Is it the difference of two squares? If so,
$$a^2 - b^2 = (a - b)(a + b)$$
- Is it the difference of two cubes? If so,
$$a^3 - b^3 = (a - b)(a^2 + ab + b^2)$$
- Is it the sum of two cubes? If so,
$$a^3 + b^3 = (a + b)(a^2 - ab + b^2)$$

(b) 3 terms (Trinomials)

- Is it a perfect square trinomial? If so,
$$a^2 + 2ab + b^2 = (a + b)^2 \quad \text{or} \quad a^2 - 2ab + b^2 = (a - b)^2$$
- Is the coefficient of the square term 1? If so,
$$x^2 + bx + c = (x + m)(x + n) \text{ where } mn = c \text{ and } m + n = b$$
- Is the coefficient of the square term different from 1? If so,
  a. Use factoring by grouping.
  b. Use trial and error.

(c) 4 terms

- Use factoring by grouping.

Step 4: Check your work by multiplying out the factors.

Work Smart

Remember, if a and b are real numbers, the sum of two squares, $a^2 + b^2$ is prime.

| EXAMPLE 1 | How to Factor Completely |

Factor completely: $2x^2 - 4x - 48$

Step-by-Step Solution

Step 1: Is there a GCF? Yes. We factor out the GCF, 2.	$2x^2 - 4x - 48 = 2(x^2 - 2x - 24)$
Step 2: Identify the number of terms in the polynomial in parentheses.	This polynomial has three terms in parentheses.
Step 3: We concentrate on the trinomial in parentheses, $x^2 - 2x - 24$. It is not a perfect square trinomial. The leading coefficient is 1, so we try $(x + m)(x + n)$, where $mn = c$ and $m + n = b$.	We need to find two factors of -24 whose sum is -2. Since $-6(4) = -24$ and $-6 + 4 = -2$, we have that $m = -6$ and $n = 4$. $$2(x^2 - 2x - 24) = 2(x - 6)(x + 4)$$
Step 4: Check	FOIL: $2(x - 6)(x + 4) = 2(x^2 + 4x - 6x - 24)$ $\qquad\qquad\qquad\qquad = 2(x^2 - 2x - 24)$ Distribute: $\qquad\qquad\qquad\qquad = 2x^2 - 4x - 48$

So $2x^2 - 4x - 48 = 2(x - 6)(x + 4)$.

QUICK ✓ *Factor each polynomial completely.*

1. $2p^2 + 8p - 90$ **2.** $-45x^2 + 3xy + 6y^2$

EXAMPLE 2 **How to Factor Completely**

Factor completely: $25p^2 - 9q^2$

Step-by-Step Solution

Step 1: Is there a GCF?	There is no GCF.
Step 2: Identify the number of terms in the polynomial.	There are two terms.
Step 3: Because the first term, $25p^2 = (5p)^2$, and the second term, $9q^2 = (3q)^2$, are both perfect squares, we have the difference of two squares.	$$25p^2 - 9q^2 = ((5p)^2 - (3q)^2)$$ $$a^2 - b^2 = (a - b)(a + b): \quad = (5p - 3q)(5p + 3q)$$
Step 4: Check	FOIL: $(5p - 3q)(5p + 3q) = 25p^2 + 15pq - 15pq - 9q^2$ Combine like terms: $= 25p^2 - 9q^2$

So $25p^2 - 9q^2 = (5p - 3q)(5p + 3q)$.

QUICK ✓ *Factor each polynomial completely.*

3. $100x^2 - 81y^2$ **4.** $2ab^2 - 242a$

EXAMPLE 3 **How to Factor Completely**

Factor completely: $12x^2 + 36xy + 27y^2$

Step-by-Step Solution

Step 1: We notice that the coefficients 12, 36, and 27 are all multiples of 3, so we factor out the GCF of 3.	$12x^2 + 36xy + 27y^2 = 3(4x^2 + 12xy + 9y^2)$
Step 2: Identify the number of terms in the polynomial in parentheses.	There are three terms in the polynomial in parentheses.
Step 3: We concentrate on the polynomial in parentheses. Is it a perfect square trinomial? The first term is a perfect square, $4x^2 = (2x)^2$. The third term is also a perfect square, $9y^2 = (3y)^2$. The middle term is 2 times the product of 2x and 3y. The polynomial in parentheses is a perfect square trinomial.	$$3(4x^2 + 12xy + 9y^2) = 3[(2x)^2 + 2(2x)(3y) + (3y)^2]$$ $$a^2 + 2ab + b^2 = (a + b)^2: \quad = 3(2x + 3y)^2$$

Step 4: Check

$$3(2x + 3y)^2 = 3(2x + 3y)(2x + 3y)$$
$$= 3(4x^2 + 6xy + 6xy + 9y^2)$$
$$= 3(4x^2 + 12xy + 9y^2)$$
$$= 12x^2 + 36xy + 27y^2$$

So $12x^2 + 36xy + 9y^2 = 3(2x + 3y)^2$.

QUICK ✓ *Factor each polynomial completely.*

5. $p^2 - 12pq + 36q^2$ **6.** $75x^2 + 90x + 27$

EXAMPLE 4 **Factoring Completely**

Factor completely: $16m^3 + 54$

Solution

Notice that there is a greatest common factor of 2. Let's factor it out. The polynomial in the parentheses has two terms.

$$16m^3 + 54 = 2(8m^3 + 27)$$

Sum of two cubes with $a = 2m$ and $b = 3$: $= 2[(2m)^3 + 3^3]$

$a^3 + b^3 = (a + b)(a^2 - ab + b^2)$: $= 2[(2m + 3)((2m)^2 - (2m)(3) + (3)^2)]$

$$= 2[(2m + 3)(4m^2 - 6m + 9)]$$

Check $2[(2m + 3)(4m^2 - 6m + 9)] = 2[8m^3 - 12m^2 + 18m + 12m^2 - 18m + 27]$
$$= 2[8m^3 + 27]$$
$$= 16m^3 + 54$$

So $16m^3 + 54 = 2[(2m + 3)(4m^2 - 6m + 9)]$.

QUICK ✓ *Factor each polynomial completely.*

7. $125y^3 - 64$ **8.** $-24a^3 + 3b^3$

EXAMPLE 5 **Factoring Completely**

Factor completely: $-5x^5 + 80x$

Solution

We know that the first thing to look for is a greatest common factor. Because the leading coefficient is negative we factor out a negative as part of the GCF. The GCF is $-5x$.

$$-5x^5 + 80x = -5x(x^4 - 16)$$

There are two terms in the factor in parentheses. The terms x^4 and 16 are perfect squares since $x^4 = (x^2)^2$ and $16 = 4^2$. So $x^4 - 16$ is a difference of two squares:

Work Smart

Check each factor to determine if any factor can be factored again.

Difference of two squares with
$a = x^2$ and $b = 4$: $-5x(x^4 - 16) = -5x(x^2 - 4)(x^2 + 4)$.

Difference of two squares with $a = x$ and $b = 2$: $= -5x(x - 2)(x + 2)(x^2 + 4)$

Check $-5x(x - 2)(x + 2)(x^2 + 4) = -5x(x^2 - 4)(x^2 + 4)$
$$= -5x(x^4 - 16)$$
$$= -5x^5 + 80x$$

So $-5x^5 + 80x = -5x(x - 2)(x + 2)(x^2 + 4)$.

QUICK ✓ *Factor each polynomial completely.*

9. $x^4 - 1$ **10.** $-36x^2y + 16y$

EXAMPLE 6 **Factoring Completely**

Factor completely: $4x^3 - 6x^2 - 36x + 54$

Solution

Is there a greatest common factor among these four terms? Yes! The GCF is 2. Also notice that there are four terms, so we attempt to factor by grouping.

Factor out the GCF of 2: $4x^3 - 6x^2 - 36x + 54 = 2(2x^3 - 3x^2 - 18x + 27)$

Factor by grouping: $= 2[(2x^3 - 3x^2) + (-18x + 27)]$

Factor out common factor in each group: $= 2[x^2(2x - 3) - 9(2x - 3)]$

Factor out $2x - 3$: $= 2(2x - 3)(x^2 - 9)$

$x^2 - 9$ is the difference of two squares: $= 2(2x - 3)(x - 3)(x + 3)$

Check $2(2x - 3)(x - 3)(x + 3) = 2(2x - 3)(x^2 - 9)$

$= 2(2x^3 - 18x - 3x^2 + 27)$

$= 4x^3 - 6x^2 - 36x + 54$

So $4x^3 - 6x^2 - 36x + 54 = 2(2x - 3)(x - 3)(x + 3)$.   ∎

Work Smart: Study Skills

You know how to begin factoring a polynomial—look for the GCF. But do you know when a polynomial is factored completely? That is, do you know when to stop? Which of the following is not completely factored?

(a) $(x + 5)(2x + 6)$

(b) $(z - 3)(z + 3)(z^2 + 9)$

(c) $(2n - 5)(5n + 7)$

(d) $(4y - 3yz)(2y + 7yz)$

Did you recognize that (a) and (d) are not completely factored because each has a common factor in one of the binomials?

QUICK ✓ *Factor each polynomial completely.*

11. $2x^3 + 3x^2 + 4x + 6$ **12.** $6x^3 + 9x^2 - 6x - 9$

EXAMPLE 7 **Factoring Completely**

Factor completely: $-4xy^2 + 4xy + 12x$

Solution

Notice that each term has a common factor of $-4x$. Factor out the GCF.

GCF $= -4x$: $-4xy^2 + 4xy + 12x = -4x(y^2 - y - 3)$

There are three terms in the polynomial in parentheses, so we concentrate on the polynomial in parentheses, $y^2 - y - 3$. It is not a perfect square trinomial, so we need to find two factors of -3 whose sum is -1. There are no such factors. Therefore, $y^2 - y - 3$ is prime.

Check $-4x(y^2 - y - 3) = -4xy^2 + 4xy + 12x$

So $-4xy^2 + 4xy + 12x = -4x(y^2 - y - 3)$.   ∎

QUICK ✓ *Factor each polynomial completely.*

13. $-3z^2 + 9z - 21$ **14.** $6xy^2 + 15x^3$

6.5 Exercises

For Extra Help:

Student Solutions Manual CD Video PH Math/Tutor Center MathXL Tutorials on CD MathXL® MyMathLab

Concepts and Vocabulary

In Problems 1–3, fill in the blanks.

1. $x^2 + 2xy + y^2$ is called a _____ _____ _____ because x^2 and y^2 are _____ _____ and the middle term is twice xy.

2. When factoring a polynomial with four terms, try factoring by _____.

3. The first step in any factoring problem is to look for the _____ _____ _____.

In Problems 4–6, answer True or False to each statement.

4. The polynomial $x^4 - 16 = (x^2 + 4)(x^2 - 4)$ is factored completely.

5. There is only one technique that will factor a given polynomial correctly.

6. $a(x - y) + x - y$ can be factored by grouping.

7. Describe the method of factoring you find most difficult. Write a polynomial and factor it using this method.

8. You are grading the paper of a student who factors the polynomial $x^2 + 4x - 3x - 12$ as follows:

$$(x^2 + 4x)(-3x - 12) = x(x + 4) - 3(x - 4)$$
$$= (x - 3)(x + 4)$$

Did the student correctly factor the polynomial? What comments would you write on the student's paper?

Mixed Practice

In Problems 9–80, factor completely. If a polynomial cannot be factored, say it is prime.

9. $x^2 - 100$
10. $x^2 - 256$
11. $t^2 + t - 6$

12. $n^2 + 5n + 6$
13. $x + y + 2ax + 2ay$
14. $xy - 2ay + 3bx - 6ab$

15. $a^3 - 8$
16. $1 - y^9$
17. $a^2 - ab - 6b^2$

18. $x^2 - xy - 30y^2$
19. $2x^2 - 5x - 7$
20. $12n^2 - 23n + 5$

21. $2x^2 - 6xy - 20y^2$
22. $2x^2 + 4xy - 6y^2$
23. $9 - a^2$

24. $25 - m^2$
25. $u^2 - 14u + 33$
26. $x^2 + 5x - 6$

27. $xy - ay - bx + ab$
28. $2x^3 - x^2 - 18x + 9$
29. $w^2 + 6w + 8$

30. $x^2 - 8x + 15$
31. $36a^2 - 49b^4$
32. $100x^2 - 25y^2$

33. $x^2 + 2xm - 8m^2$
34. $s^2 - 11st + 24t^2$
35. $6x^2y^2 - 13xy + 6$

36. $6s^2t^2 + st - 1$
37. $x^3 + x^2 + x + 1$
38. $x^3 - 3x^2 + 2x - 6$

39. $12z^2 + 12z + 18$
40. $3y^2 + 6y + 3$
41. $14c^2 + 19c - 3$

42. $24x^2 + 66x + 45$
43. $27m^3 + 64n^6$
44. $8x^6 + 125y^3$

45. $2j^6 - 2j^2$
46. $48m - 3m^9$
47. $8a^2 + 18ab - 5b^2$

48. $10p^2 - 15q^2 + 19pq$
49. $2a^3 + 6a$
50. $4t^4 + 16t^2$

51. $12z^2 - 3$
52. $8n^3 - 18n$
53. $x^2 - x + 6$

54. $n^2 + 2n + 8$
55. $16a^4 + 2ab^3$
56. $24p^3q + 81q^4$

57. $p^2q^2 + 6pq - 7$
58. $x^2y^2 + 20xy + 96$
59. $s^2(s + 2) - 4(s + 2)$

60. $x^2(x + y) - 16(x + y)$
61. $-12x^3 + 2x^2 + 2x$
62. $-3x^3 - 13x^2 + 10x$

63. $10v^2 - 2 - v$
64. $27h - 5 + 18h^2$
65. $4n^2 - n^4 + 3n^3$

66. $4x - 2x^2 - 2x^3$
67. $-4a^3b + 2a^2b - 2ab$
68. $-4x^3y + 4x^2y - 4xy$

69. $12p - p^3 + p^2$ **70.** $-9x - 3x^3 - 12x^2$ **71.** $-32x^3 + 72xy^2$

72. $-48p^4 + 75p^2$ **73.** $2n^3 - 10n^2 - 6n + 30$

74. $6x^4 + 3x^3 - 24x^2 - 12x$ **75.** $16x^2 + 4x - 12$ **76.** $36x^2 + 30x - 24$

77. $14x^2 + 3x^4 + 8$ **78.** $14 + 53r^3 + 14r^6$ **79.** $-2x^3y + x^2y^2 + 3xy^3$

80. $6a^3 - 5a^2b + ab^2$

Applying the Concepts

81. Profit on Newspapers The revenue for selling x newspapers is given by the expression $x^3 + 8x$. The cost to produce x newspapers is $8x^2 - 7x$. If profit is calculated as revenue minus costs, write an expression, in factored form, that calculates the profit on x newspapers.

82. Profit on T-shirts Candy and her boyfriend produce silk-screened T-shirts. The cost to produce n T-shirts is given by the expression $12n - 2n^2$ and the revenue for selling the same number of T-shirts is $n^3 + 2n^2$. Write an expression, in factored form, that calculates the profit on n T-shirts.

△ **83. Volume of a Box** The volume of a box of nails is given by the formula $V = lwh$. The volume of the box is represented by the expression $4x^3 - 10x^2 - 6x$. Factor this expression to determine algebraic expressions for the dimensions of the box.

△ **84. Volume of a Box** The volume of a shoe box is given by the formula $V = lwh$. The volume of the box is represented by the expression $18x^3 + 3x^2 - 6x$. Factor this expression to determine algebraic expressions for the box.

Extending the Concepts

Although the most common pattern for factoring a polynomial with four terms is to group the first two terms and group the second two terms, it is not the only possibility. Here the polynomial has been written as three terms in the first group and a single term in the second group. Study the example and then try to find a grouping that will factor each polynomial.

$$x^2 + 2xy + y^2 - z^2 = (x^2 + 2xy + y^2) - z^2$$
$$= (x + y)^2 - z^2$$
$$= (x + y + z)(x + y - z)$$

In Problems 85–88, factor completely.

85. $4m^2 - 4mn + n^2 - p^2$ **86.** $b^2 + 2bc + c^2 - a^2$

87. $x^2 - y^2 + 2yz - z^2$ **88.** $16x^2 - y^2 - 8y - 16$

In Problems 89–92, see if you can extend this concept to find a creative way to group and then factor each of the following. You may need to rearrange the terms.

89. $x^2 - 2xy + y^2 - 6x + 6y + 8$ **90.** $a^2 + 2ab + b^2 - a - b - 6$

91. $x^2 + 2xy + y^2 - a^2 + 2ab - b^2$ **92.** $x^2 - 4mx + 4m^2 - y^2 - 2ny - 4n^2$

PUTTING THE CONCEPTS TOGETHER (Sections 6.1–6.5)

These problems cover important concepts from Sections 6.1 to 6.5. We designed these problems so that you can review the chapter so far and show your mastery of the concepts. Take time to work these problems before proceeding with the next section. The answers to these problems are located at the back of the text starting on page AN-19.

1. Find the GCF of $10x^3y^4z$, $15x^5y$, and $25x^2y^7z^3$.

In Problems 2–16, factor each polynomial completely. If the polynomial cannot be factored, say it is prime.

2. $x^2 - 3x - 4$ **3.** $x^6 - 27$

4. $6x(2x + 1) + 5z(2x + 1)$ **5.** $x^2 + 5xy - 6y^2$

6. $x^3 + 64$ **7.** $4x^2 + 49y^2$

8. $3x^2 + 12xy - 36y^2$

9. $12z^5 - 44z^3 - 24z^2$

10. $x^2 + 6x - 5$

11. $4m^4 + 5m^3 - 6m^2$

12. $5p^2 - 17p + 6$

13. $10m^2 + 25m - 6m - 15$

14. $36m^2 + 6m - 6$

15. $4m^2 - 20m + 25$

16. $5x^2 - xy - 4y^2$

17. Surface Area The surface area of a right circular cylinder is given by the formula $S = 2\pi rh + 2\pi r^2$. Factor the right side of this formula.

18. Rocket Height A toy rocket is launched upward from the ground with an initial velocity of 48 feet per second. Its height, h, in feet, after t seconds, is given by the equation $h = 48t - 16t^2$. Factor the right side of this equation.

6.6 Solving Polynomial Equations by Factoring

OBJECTIVES

1. Solve Quadratic Equations Using the Zero-Product Property

2. Solve Polynomial Equations of Degree Three or Higher Using the Zero-Product Property

Preparing for Solving Polynomial Equations by Factoring

Before getting started, take the following readiness quiz. If you get a problem wrong, go back to the cited section and review the material.

1. Solve: $x + 5 = 0$ [Section 2.1, pp. 76–78]

2. Solve: $2(x - 4) - 10 = 0$ [Section 2.2, pp. 86–87]

3. Evaluate $2x^2 + 3x - 4$ when (a) $x = 2$ (b) $x = -1$.   [Section 1.7, pp. 57–58]

In Sections 6.1–6.5, we learned how to factor polynomial expressions. A question that you may have been asking yourself is, "Why do I care about factoring? What good is it?" It turns out that there are many uses of factoring, but one use that we can present now is that factoring is essential for solving polynomial equations.

Work Smart

Remember, the degree of a polynomial is the value of the **largest exponent** on the variable. For example, the degree of $4x^3 - 9x^2 + 1$ is 3.

DEFINITIONS

A **polynomial equation** is any equation that contains only polynomial expressions. The **degree of a polynomial equation** is the degree of the polynomial expression in the equation.

Some examples of polynomial equations are

$$4x + 5 = 17 \qquad\qquad 2x^2 - 5x - 3 = 0 \qquad\qquad y^3 + 4y^2 = 3y + 18$$

Polynomial equation of degree 1   Polynomial equation of degree 2   Polynomial equation of degree 3

We learned how to solve polynomial equations of degree 1 back in Chapter 2. It seems logical that the next step is to learn how to solve polynomial equations of degree 2.

1   Solve Quadratic Equations Using the Zero-Product Property

If a polynomial equation can be solved using factoring, we make use of the following property.

THE ZERO-PRODUCT PROPERTY

If the product of two factors is zero, then at least one of the factors is 0. That is,

if $ab = 0$, then $a = 0$ or $b = 0$ or both a and b are 0.

Preparing for...Answers **1.** $\{-5\}$
2. $\{9\}$ **3. (a)** 10   **(b)** −5

For example, if $2x = 0$, then either $2 = 0$ or $x = 0$. Since $2 \neq 0$, it must be that $x = 0$.

EXAMPLE 1 **Using the Zero-Product Property**

Solve: $(x + 4)(2x - 5) = 0$

Solution

We have the product of two factors, $x + 4$ and $2x - 5$, set equal to 0. According to the Zero-Product Property, the value of at least one of the factors must equal 0. Therefore, we set each of the factors equal to 0 and solve each equation separately.

$$x + 4 = 0 \qquad \text{or} \qquad 2x - 5 = 0$$

Subtract 4 from both sides: $x = -4$ Add 5 to both sides: $2x = 5$

Divide both sides by 2: $\dfrac{2x}{2} = \dfrac{5}{2}$

$$x = \dfrac{5}{2}$$

Check

$$x = -4$$

$$(x + 4)(2x - 5) = 0$$

$$(-4 + 4)(2(-4) - 5) \stackrel{?}{=} 0$$

$$0(-13) \stackrel{?}{=} 0$$

$$0 = 0 \quad \text{True}$$

$$x = \dfrac{5}{2}$$

$$(x + 4)(2x - 5) = 0$$

$$\left(\dfrac{5}{2} + 4\right)\left(2 \cdot \left(\dfrac{5}{2}\right) - 5\right) \stackrel{?}{=} 0$$

$$\left(\dfrac{13}{2}\right)(5 - 5) \stackrel{?}{=} 0$$

$$\dfrac{13}{2} \cdot 0 \stackrel{?}{=} 0$$

$$0 = 0 \quad \text{True}$$

The solution set is $\left\{-4, \dfrac{5}{2}\right\}$.

QUICK ✓ *Use the Zero-Product Property to solve the equation.*

1. $x(x + 3) = 0$ **2.** $(x - 2)(4x + 5) = 0$

The Zero-Product Property comes in handy when we need to solve quadratic equations.

Work Smart

In a quadratic equation, why can't *a* equal 0? Because if *a* were equal to zero, the equation would be a linear equation.

> **DEFINITION**
>
> A **quadratic equation** is an equation equivalent to one of the form
>
> $$ax^2 + bx + c = 0$$
>
> where $a, b,$ and c are real numbers and $a \neq 0$.

The following are examples of quadratic equations.

$$2x^2 + 5x - 3 = 0 \qquad -6z^2 + 12z = 0 \qquad y^2 - 25 = 0 \qquad p^2 + 12p = -36$$

Notice that the equation $y^2 - 25 = 0$ is a quadratic equation even though it is missing the "y" term. The equation $-6z^2 + 12z = 0$ is a quadratic equation even though it is missing a constant term.

A quadratic equation is a specific type of a polynomial equation. The term "quadratic" means "of, or relating to, a square." There are many real-world situations that are modeled by these second-degree equations, such as problems involving revenue for selling x units of a good. Additionally, quadratic equations can be used to describe the height of a projectile over time.

Sometimes a quadratic equation is called a **second-degree equation** because the highest power of the variable in a quadratic equation is two.

In Words

A quadratic equation is in standard form if it is written in descending order of exponents, and is set equal to zero.

> **DEFINITION**
>
> A quadratic equation is said to be in **standard form** if it is written in the form $ax^2 + bx + c = 0$.

For example, the equation $2x^2 + x - 5 = 0$ is in standard form, while the equation $p^2 + 12p = -36$ is not in standard form. To place $p^2 + 12p = -36$ in standard form, we add 36 to both sides of the equation to obtain $p^2 + 12p + 36 = 0$.

When a quadratic equation is written in standard form, $ax^2 + bx + c = 0$, it may be possible to factor the expression $ax^2 + bx + c$ as the product of two first-degree polynomials. If it is possible to factor the trinomial, we can then use the Zero-Product Property to solve the quadratic equation.

EXAMPLE 2 **How to Solve a Quadratic Equation by Factoring**

Solve: $x^2 - 4x - 21 = 0$

Step-by-Step Solution

Step 1: Is the equation in standard form? Yes, it is written in the form $ax^2 + bx + c = 0$.	$x^2 - 4x - 21 = 0$
Step 2: Factor the expression on the left side of the equation.	Two integers whose product is -21 and whose sum is -4 are -7 and 3. $(x + 3)(x - 7) = 0$
Step 3: Set each factor to 0.	$x + 3 = 0$ or $x - 7 = 0$
Step 4: Solve each first-degree equation.	$x = -3$ or $x = 7$
Step 5: Check	$x^2 - 4x - 21 = 0$ $x = -3$: $(-3)^2 - 4(-3) - 21 \stackrel{?}{=} 0$ $9 + 12 - 21 \stackrel{?}{=} 0$ $0 = 0$ True $\qquad$ $x^2 - 4x - 21 = 0$ $x = 7$: $7^2 - 4(7) - 21 \stackrel{?}{=} 0$ $49 - 28 - 21 \stackrel{?}{=} 0$ $0 = 0$ True

The solution set is $\{-3, 7\}$.

Look at Step 3 in the solution to Example 2. To solve a quadratic equation, we wish to put the equation into a form that we already know how to solve. That is, we "transform" the quadratic equation $x^2 - 4x - 21 = 0$ into two linear equations, $x + 3 = 0$ or $x - 7 = 0$. We will present methods for solving $ax^2 + bx + c = 0$ when we cannot factor the expression $ax^2 + bx + c$ in Chapter 10.

Steps to Solve a Quadratic Equation by Factoring

Step 1: Write the quadratic equation in standard form, $ax^2 + bx + c = 0$.

Step 2: Factor the expression on the left side of the equation.

Step 3: Set each factor found in Step 2 equal to zero using the Zero-Product Property.

Step 4: Solve each first-degree equation for the variable.

Step 5: Be sure to check your answers by substituting into the *original* equation.

EXAMPLE 3　Solving a Quadratic Equation Not in Standard Form

Solve: $3x^2 + 5x = 14x$

Solution

We first write the equation in standard form, $ax^2 + bx + c = 0$.

$$3x^2 + 5x = 14x$$

Subtract 14x from both sides of the equation: $\quad 3x^2 - 9x = 0$

Factor: $\quad 3x(x - 3) = 0$

Set each factor to 0: $\quad 3x = 0 \quad \text{or} \quad x - 3 = 0$

Solve each first-degree equation: $\quad x = 0 \quad \text{or} \quad x = 3$

Check　Substitute $x = 0$ and $x = 3$ into the original equation.

$$3x^2 + 5x = 14x \qquad\qquad 3x^2 + 5x = 14x$$
$$x = 0: \quad 3(0)^2 + 5(0) \stackrel{?}{=} 14(0) \qquad x = 3: \quad 3(3)^2 + 5(3) \stackrel{?}{=} 14(3)$$
$$0 = 0 \quad \text{True} \qquad\qquad 27 + 15 \stackrel{?}{=} 42$$
$$42 = 42 \quad \text{True}$$

The solution set is $\{0, 3\}$.

QUICK ✔️　*Solve each quadratic equation by factoring.*

3. $p^2 - 6p + 8 = 0$ 　　　　　　**4.** $2t^2 - 5t = 3$

5. $2x^2 + 3x = 5$ 　　　　　　　**6.** $z^2 + 20 = -9z$

EXAMPLE 4　Solving a Quadratic Equation Not in Standard Form

Solve: $3m^2 - 3m = 2 - 2m$

Solution

Once again, we first write the quadratic equation in standard form, $ax^2 + bx + c = 0$.

$$3m^2 - 3m = 2 - 2m$$

Add 2m to both sides and subtract 2 from both sides to obtain standard form: $\quad 3m^2 - 3m + 2m - 2 = 2 - 2 - 2m + 2m$

$$3m^2 - m - 2 = 0$$

Factor $3m^2 - m - 2$: $\quad (3m + 2)(m - 1) = 0$

Set each factor to 0: $\quad 3m + 2 = 0 \quad \text{or} \quad m - 1 = 0$

Solve each first-degree equation: $\quad 3m = -2 \quad \text{or} \quad m = 1$

$$m = -\frac{2}{3}$$

Check   Check the two solutions by substituting $m = -\dfrac{2}{3}$ and $m = 1$ in the original equation. We leave this to you.

The solution set is $\left\{-\dfrac{2}{3}, 1\right\}$. ▬

QUICK ✓ *Solve each quadratic equation by factoring.*

7. $2k^2 - k - 4 = 1 + 2k$

8. $3x^2 + 9x = 4 - 2x$

EXAMPLE 5   **Solving a Quadratic Equation Not in Standard Form**

Solve: $(2x + 5)(x - 3) = 6x$

Solution

Work Smart

Do not attempt to solve $(2x + 5)(x - 3) = 6x$ by setting each factor to $6x$ as in $2x + 5 = 6x$ and $x - 3 = 6x$. The Zero-Product Property can only be applied when the product equals zero.

$$(2x + 5)(x - 3) = 6x$$

Multiply using FOIL: $2x^2 - x - 15 = 6x$

Write in standard form: $2x^2 - 7x - 15 = 0$

Factor the polynomial: $(2x + 3)(x - 5) = 0$

Set each factor to 0: $2x + 3 = 0 \quad \text{or} \quad x - 5 = 0$

Solve each first-degree equation: $2x = -3 \quad \text{or} \quad x = 5$

$$x = -\dfrac{3}{2}$$

Check   We leave it to you to substitute $x = -\dfrac{3}{2}$ and $x = 5$ into the original equation to verify the answer.

The solution set is $\left\{-\dfrac{3}{2}, 5\right\}$. ▬

QUICK ✓ *Solve each quadratic equation by factoring.*

9. $(x - 3)(x + 5) = 9$

10. $(x + 3)(2x - 1) = 7x - 3x^2$

EXAMPLE 6   **Solving a Quadratic Equation Not in Standard Form**

Solve: $4k^2 + 9 = -12k$

Solution

$$4k^2 + 9 = -12k$$

Write in standard form: $4k^2 + 12k + 9 = 0$

Factor the trinomial
$a^2 + 2ab + b^2 = (a + b)^2$: $(2k + 3)^2 = 0$

Set each factor to 0: $2k + 3 = 0 \quad \text{or} \quad 2k + 3 = 0$

Solve each first-degree equation: $2k = -3 \qquad\qquad 2k = -3$

$$k = -\dfrac{3}{2} \qquad\qquad k = -\dfrac{3}{2}$$

Check   Substitute $k = -\dfrac{3}{2}$ into the original equation to confirm that the solution is correct.

The solution set is $\left\{-\dfrac{3}{2}\right\}$. ▬

Notice in Example 6 that the solution $k = -\dfrac{3}{2}$ occurred twice. When this occurs, the solution is called a **double root.**

QUICK ✓ *Solve the quadratic equation by factoring.*

11. $9p^2 + 16 = 24p$

EXAMPLE 7   **Solving a Quadratic Equation Containing a GCF**

Solve: $-3x^2 + 6x + 72 = 0$

Solution

$$-3x^2 + 6x + 72 = 0$$

Factor the polynomial, GCF $= -3$: $-3(x^2 - 2x - 24) = 0$

$$-3(x - 6)(x + 4) = 0$$

Set each factor to 0: $-3 = 0$ or $x - 6 = 0$ or $x + 4 = 0$

Solve each first-degree equation: $x = 6$ $x = -4$

The statement $-3 = 0$ is false. So the solutions are 6 and -4.

Check   Substitute $x = 6$ and $x = -4$ into the original equation to check.

The solution set is $\{-4, 6\}$.

Another technique that we could have used in Example 7 is to use the Multiplication Property of Equality and first divide each term of the equation by -3. Then we would solve the quadratic equation $x^2 - 2x - 24 = 0$ and also obtain the solutions $x = -4$ or $x = 6$.

QUICK ✓ *Solve the quadratic equation by factoring.*

12. $4x^2 + 12x - 72 = 0$ **13.** $-2x^2 + 2x = -12$

EXAMPLE 8   **Throwing a Ball from the Top of a Building**

A ball is thrown vertically upward from the top of a building 96 feet tall with an initial velocity of 80 feet per second. Solve the equation $-16t^2 + 80t + 96 = 192$ to find the time t (in seconds) at which the ball is 192 feet from the ground. See Figure 1.

Figure 1

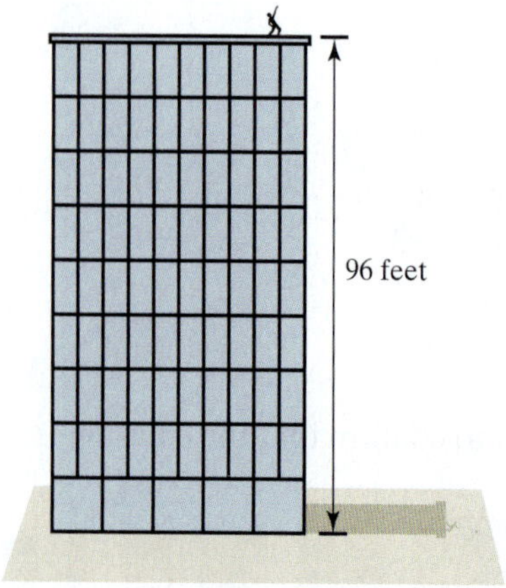

96 feet

Solution

$$-16t^2 + 80t + 96 = 192$$

Write the quadratic equation in standard form: $-16t^2 + 80t + 96 - 192 = 192 - 192$

$$-16t^2 + 80t - 96 = 0$$

Factor the polynomial: $-16(t^2 - 5t + 6) = 0$

$$-16(t - 3)(t - 2) = 0$$

Set each factor to 0: $-16 = 0$ or $t - 3 = 0$ or $t - 2 = 0$

Solve each first-degree equation: $t = 3$ or $t = 2$

The statement $-16 = 0$ is false. The solutions are $t = 3$ or $t = 2$. After both 2 seconds and 3 seconds the ball will be 192 feet from the ground. Do you see why?

QUICK ✓

14. A toy rocket is shot directly up from the ground with an initial velocity of 80 feet per second. Solve the equation $-16t^2 + 80t = 64$ to find the time (in seconds) at which the toy rocket is 64 feet from the ground.

(2) Solve Polynomial Equations of Degree Three or Higher Using the Zero-Product Property

The Zero-Product Property is also used to solve higher-degree polynomial equations.

EXAMPLE 9 **How to Solve a Polynomial Equation of Degree Three or Higher**

Solve: $18x^3 + 3x^2 - 6x = 0$

Step-by-Step Solution

Step 1: Write the equation in standard form.	$18x^3 + 3x^2 - 6x = 0$
Step 2: Factor the expression on the left side of the equation. Begin with the GCF, 3x.	$3x(6x^2 + x - 2) = 0$ $3x(3x + 2)(2x - 1) = 0$

Step 3: Set each factor to 0.	$3x = 0$	or	$3x + 2 = 0$	or	$2x - 1 = 0$
Step 4: Solve each first-degree equation.	$x = 0$		$3x = -2$ $x = -\dfrac{2}{3}$		$2x = 1$ $x = \dfrac{1}{2}$

Step 5: Check each solution by substituting $x = 0, x = -\dfrac{2}{3}$, and $x = \dfrac{1}{2}$ into the original equation.	We leave the check to you.

Work Smart
Do not divide both sides of the equation by the factor 3x. We can divide both sides by a constant, but not by a variable because the variable may have a value of zero.

The solution set is $\left\{ 0, -\dfrac{2}{3}, \dfrac{1}{2} \right\}$.

QUICK ✓ *Solve each polynomial equation and state the solutions.*

15. $(4x - 5)(x^2 - 9) = 0$ **16.** $3x^3 + 9x^2 + 6x = 0$

6.6 Exercises

For Extra Help:

Student Solutions Manual   CD Video   PH Math/Tutor Center   MathXL Tutorials on CD   MathXL®   MyMathLab

Concepts and Vocabulary

In Problems 1–3, fill in the blanks.

1. A _____ equation is an equation that can be written in the form $ax^2 + bx + c = 0$, where a, b, and c are real numbers and $a \neq 0$.

2. The Zero-Product Property states that if $ab = 0$, then either _____ or _____.

3. Quadratic equations are also known as _____-degree equations.

In Problems 4–6, answer True or False to each statement.

4. $x(x - 3) = 4$ means that $x = 4$ or $x - 3 = 4$.

5. $3x - 6 + x^2 = 0$ is written in standard form.

6. $x^3 - 4x^2 - 12x = 0$ can be solved using the Zero-Product Property.

7. When solving polynomial equations, we always begin by writing the equation in standard form. Explain why this important.

8. Explain the difference between a quadratic polynomial and a quadratic equation.

Building Skills

In Problems 9–12, identify each as a linear equation or a quadratic equation.

9. $3(x + 4) - 1 = 5x + 2$

10. $2x + 1 - (x + 7) = 3x + 1$

11. $x^2 - 2x = 8$

12. $(x + 2)(x - 2) = 14$

In Problems 13–20, solve each equation using the Zero-Product Property.

13. $2x(x + 4) = 0$

14. $3x(x + 9) = 0$

15. $(n + 3)(n - 9) = 0$

16. $(a + 8)(a - 4) = 0$

17. $(3p + 1)(p - 5) = 0$

18. $(4z - 3)(z + 4) = 0$

19. $(5y + 3)(2 - 7y) = 0$

20. $(6m - 5)(3 - 8m) = 0$

In Problems 21–44, solve each quadratic equation by factoring.

21. $x^2 - 3x - 4 = 0$

22. $x^2 + 2x - 63 = 0$

23. $n^2 + 9n + 14 = 0$

24. $p^2 - 5p - 24 = 0$

25. $4x^2 + 2x = 0$

26. $14x - 49x^2 = 0$

27. $2x^2 - 3x - 2 = 0$

28. $3x^2 + x - 14 = 0$

29. $a^2 - 6a + 9 = 0$

30. $k^2 + 12k + 36 = 0$

31. $6x^2 = 36x$

32. $2x^2 = 5x$

33. $n^2 - n = 6$

34. $a^2 - 6a = 16$

35. $b^2 + 18 = 11b$

36. $m^2 - 30 = 7m$

37. $1 - 5m = -4m^2$

38. $4p - 3 = -4p^2$

39. $n(n - 2) = 24$

40. $p(p + 1) = 2$

41. $(x - 2)(x - 3) = 56$

42. $(x + 5)(x - 3) = 9$

43. $(c + 2)^2 = 9$

44. $(2a - 1)^2 = 16$

In Problems 45–50, solve each polynomial equation by factoring.

45. $2x^3 + 2x^2 - 12x = 0$

46. $3x^3 + x^2 - 14x = 0$

47. $y^3 + 3y^2 - 4y - 12 = 0$

48. $m^3 + 2m^2 - 9m - 18 = 0$

49. $2x^3 + 3x^2 = 8x + 12$

50. $-2x + 3 = 3x^2 - 2x^3$

Mixed Practice

In Problems 51–76, solve each equation. Be careful; the problems represent a mix of linear, quadratic, and third-degree polynomial equations.

51. $(5x + 3)(x - 4) = 0$

52. $(3x - 2)(x + 5) = 0$

53. $p^2 - p - 20 = 0$

54. $z^2 - 13z + 40 = 0$

55. $4w + 3 = 2w - 7$

56. $7y + 3 = 2y - 12$

57. $4a^2 - 25a = 21$

58. $5m^2 = 18m + 8$

59. $2a(a + 1) = a^2 + 8$

60. $2y(y + 5) = y^2 + 11$

61. $2x^3 + x^2 = 32x + 16$

62. $2n^3 + 4 = n^2 + 8n$

63. $4(b - 3) - 3b = 8$

64. $2(p - 3) = p + 1$

65. $y^2 + 5y = 5(y + 20)$

66. $3z^2 + 7z = 7(z + 21)$

67. $(a + 3)(a - 5)(3a + 2) = 0$

68. $(w - 6)(w + 5)(2w - 3) = 0$

69. $(2k - 3)(2k^2 - 9k - 5) = 0$

70. $(7m - 11)(3m^2 - m - 2) = 0$

71. $(w - 3)^2 = 9 + 2w$

72. $3(k + 2)^2 = 5k + 8$

73. $\dfrac{1}{2}x^2 + \dfrac{5}{4}x = 3$

74. $z^2 + \dfrac{29}{4}z = 6$

75. $8x^2 + 44x = 24$

76. $9q^2 = 3q + 6$

Applying the Concepts

77. Tossing a Ball A ball is thrown vertically upward from the top of a building 80 feet tall with an initial velocity of 64 feet per second. Solve the equation $-16t^2 + 64t + 80 = 128$ to find the time t (in seconds) at which the ball is 128 feet from the ground.

78. Tossing a Ball A ball is thrown vertically upward from the ground with an initial velocity of 64 feet per second. Solve the equation $-16t^2 + 64t = 48$ to find the time t (in seconds) at which the ball is 48 feet from the ground.

79. Water Balloon A water balloon is dropped from a height of 40 feet. Solve the equation $-16t^2 + 40 = 24$ to find the time t (in seconds) at which the water balloon is 24 feet from the ground.

80. Flying Money A stunt-man dropped a bag containing fake money from a hot-air balloon at a height of 240 feet. Solve the equation $-16t^2 + 240 = 96$ to find the time t (in seconds) at which the bag was 96 feet from the ground.

81. Consecutive Integers The product of two consecutive integers is 12. Find the integers.

82. Consecutive Odd Integers The product of two consecutive odd integers is 143. Find the integers.

83. Consecutive Even Integers Find three consecutive even integers such that the product of the first and the third is 96.

84. Consecutive Odd Integers Find three consecutive odd integers such that the product of the second and the third is 99.

△ **85. Rectangle** The length and width of two sides of a rectangle are consecutive odd integers. The area of the rectangle is 255. Find the dimensions of the rectangle.

△ **86. Garden Area** The State University Landscape Club wants to establish a horticultural garden near the administration building. The length and width of the space that is available are consecutive even integers, and the area of the garden is 440 square feet. Find the dimensions of the garden.

The equation $N = \dfrac{t^2 - t}{2}$ models the number of soccer games that must be scheduled in a league with t teams, when each team plays every other team exactly once. Use this equation to solve Problems 87 and 88.

87. If a league has 28 games scheduled, how many teams are in the league?

88. If a league has 36 games scheduled, how many teams are in the league?

Extending the Concepts

89. Write a polynomial equation with integer coefficients that has $x = 0$, $x = 3$, and $x = -5$ as solutions.

90. Write a polynomial equation with integer coefficients that has $n = 0$, $n = 8$, and $n = -2$ as solutions.

91. Write a polynomial equation with integer coefficients that has $z = 6$ as a double root.

92. Write a polynomial equation with integer coefficients that has $x = -2$ as a double root.

93. Write a polynomial equation with integer coefficients that has $a = 4$, $a = -\dfrac{1}{2}$, and $a = \dfrac{2}{3}$ as solutions.

94. Write a polynomial equation with integer coefficients that has $a = 3$, $a = \dfrac{1}{3}$, and $a = \dfrac{4}{5}$ as solutions.

95. A student solved a quadratic equation using the following procedure. Explain the student's error and then work the problem correctly.

$$15x^2 = 5x$$
$$\frac{15x^2}{x} = \frac{5x}{x}$$
$$15x = 5$$
$$x = \frac{5}{15} = \frac{1}{3}$$

96. A student solved a quadratic equation using the following procedure. Explain the student's error and then work the problem correctly.

$$(x - 4)(x + 3) = -6$$
$$x - 4 = -6 \quad \text{or} \quad x + 3 = -6$$
$$x = -2 \qquad\qquad x = -9$$

In Problems 97–100, solve for x in each equation.

97. $x^2 - ax + bx - ab = 0$

98. $x^2 + ax - 6a^2 = 0$

99. $2x^3 - 4ax^2 = 0$

100. $4x^2 - 6ax + 10bx - 15ab = 0$

6.7 Modeling and Solving Problems with Quadratic Equations

OBJECTIVES

1. Model and Solve Problems Involving Quadratic Equations
2. Model and Solve Problems Using the Pythagorean Theorem

Preparing for Modeling and Solving Problems with Quadratic Equations

Before getting started, take the following readiness quiz. If you get a problem wrong, go back to the section cited and review the material.

1. Evaluate: 15^2. [Section 1.6, pp. 49–50]

2. Solve: $x^2 - 5x - 14 = 0$ [Section 6.6, pp. 451–454]

The solutions to many applied problems require solving polynomial equations by factoring. For example, the height of a projectile over time, the area of a triangular sail, and the dimensions of a big-screen TV may be found by solving a polynomial equation.

1 Model and Solve Problems Involving Quadratic Equations

Let's begin by solving a quadratic equation to determine the time at which a projectile is at a certain height.

EXAMPLE 1 Projectile Motion

A ball is thrown off a cliff from a height of 240 feet above sea level. The height h of the ball above the water (in feet) at any time t (in seconds) can be modeled by the equation

$$h = -16t^2 + 32t + 240$$

See Figure 2.

Figure 2

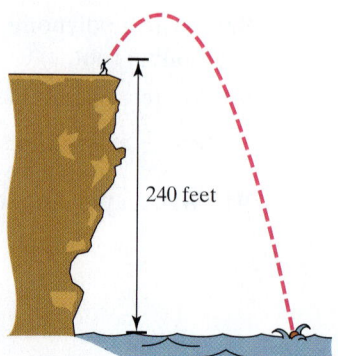

240 feet

(a) When will the height of the ball be 240 feet above sea level?

(b) When will the ball strike the water?

Solution

(a) To determine when the height of the ball will be 240 feet above sea level, we let $h = 240$ and solve the resulting equation.

$$-16t^2 + 32t + 240 = 240$$

Subtract 240 from both sides:	$-16t^2 + 32t = 0$
Factor out $-16t$:	$-16t(t - 2) = 0$
Set each factor to 0:	$-16t = 0$ or $t - 2 = 0$
Solve each equation:	$t = 0$ or $t = 2$

The ball will be at a height of 240 feet the instant the ball leaves the child's hand and after 2 seconds of flight.

(b) The ball will strike the water at the instant its height is 0. So we let $h = 0$ and solve the resulting equation.

$$-16t^2 + 32t + 240 = 0$$

Factor out -16:	$-16(t^2 - 2t - 15) = 0$
Factor:	$-16(t - 5)(t + 3) = 0$
Set each factor to 0:	$-16 = 0$ or $t - 5 = 0$ or $t + 3 = 0$
Solve each equation:	$t = 5$ or $t = -3$

The equation $-16 = 0$ is false, and since t represents time, we discard the solution $t = -3$. Therefore, the ball will strike the water after 5 seconds.   ▬

QUICK ✓

1. A model rocket is fired straight up from the ground. The height h of the rocket (in feet) at any time t (in seconds) can be modeled by the equation $h = -16t^2 + 160t$.

(a) When will the height of the rocket be 384 feet from the ground?

(b) When will the rocket strike the ground?

In the next two examples, we will employ the problem-solving strategy first presented in Section 2.5.

EXAMPLE 2   Geometry: Area of a Rectangle

A carpet installer finds that the length of a rectangular hallway is 3 feet more than twice the width. If the area of the hallway is 44 square feet, what are the dimensions of the hallway? See Figure 3.

Solution

Figure 3

Step 1: Identify This is a geometry problem involving the area of a rectangle.

Step 2: Name Because we know less about the width of the hallway, we let w represent the width. The length of the hallway is 3 feet more than twice the width, so we let $2w + 3$ represent the length.

Step 3: Translate We are given that the area of the hallway is 44 square feet. We know that the area of a rectangle = (length)(width), so we have

$$\text{area} = (\text{length})(\text{width})$$
$$44 = (2w + 3)(w)   \quad \text{The Model}$$

Step 4: Solve We now proceed to solve the equation. Do you recognize this equation as being a quadratic equation? The first step is to put the equation in standard form, $ax^2 + bx + c = 0$.

$$w(2w + 3) = 44$$

Distribute: $\qquad 2w^2 + 3w = 44$

Subtract 44 from both sides: $\qquad 2w^2 + 3w - 44 = 0$

Factor: $\qquad (2w + 11)(w - 4) = 0$

Set each factor to 0: $\qquad 2w + 11 = 0 \quad \text{or} \quad w - 4 = 0$

Solve: $\qquad 2w = -11 \quad \text{or} \quad w = 4$

$$\frac{2w}{2} = -\frac{11}{2}$$

$$w = -\frac{11}{2}$$

Step 5: Check Since w represents the width of the rectangular hallway, we discard the solution $w = -\dfrac{11}{2}$. If the width of the hallway is 4 feet, then the length would be $2w + 3 = 2(4) + 3 = 11$ feet. The area of a hallway that is 4 feet by 11 feet would be $4(11) = 44$ square feet. We have the right answer!

Step 6: Answer The dimensions of the hallway are 4 feet by 11 feet. ▬

QUICK ✓

2. A rectangular plot of land has length that is 3 kilometers less than twice its width. If the area of the land is 104 square kilometers, what are the dimensions of the land?

EXAMPLE 3 Geometry: Area of a Triangle

The height of a triangle is 5 inches less than the length of the base, and the area of the triangle is 42 square inches. Find the height of the triangle.

Solution

Step 1: Identify This is a geometry problem involving the area of a triangle.

Step 2: Name The height of the triangle is 5 inches less than the base. We will let b represent the length of the base and $b - 5$ represent the height of the triangle. See Figure 4.

Figure 4

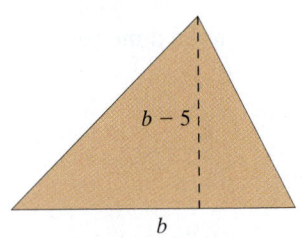

Step 3: Translate We know that the area of a triangle is given by the formula $\text{Area} = \dfrac{1}{2}(\text{base})(\text{height})$. In addition, we are given the area of the triangle to be 42 square feet, so

$$\text{Area} = \frac{1}{2}(\text{base})(\text{height})$$

$$42 = \frac{1}{2}(b)(b - 5) \qquad \text{The Model}$$

Work Smart

Multiply only the $\dfrac{1}{2}$ by 2—don't multiply the factors b and $(b - 5)$ by 2 also.

Step 4: Solve Our model is a quadratic equation, so we first put the equation in standard form, $ax^2 + bx + c = 0$.

$$42 = \frac{1}{2}(b)(b - 5)$$

Multiply by 2 to clear fractions: $\qquad 2(42) = 2\left[\dfrac{1}{2}(b)(b - 5)\right]$

$$84 = b(b - 5)$$

Distribute: $\qquad 84 = b^2 - 5b$

Subtract 84 from both sides: $\qquad 0 = b^2 - 5b - 84$

Factor: $\qquad 0 = (b - 12)(b + 7)$

Apply Zero-Product Property: $\qquad b - 12 = 0 \quad \text{or} \quad b + 7 = 0$

Solve: $\qquad b = 12 \quad \text{or} \qquad b = -7$

Step 5: Check Since b represents the base of the triangle, we discard the solution $b = -7$. Do you see why? The base of the triangle is 12 inches, so the height is $b - 5 = 12 - 5 = 7$ inches. The area of a triangle that has a base of 12 inches and a height of 7 inches is $\frac{1}{2} \cdot 12 \cdot 7 = 42$ square inches. We have the right answer!

Step 6: Answer The height of the triangle is 7 inches.

Quick ✓

3. The base of a triangular garden is 4 yards longer than the height, and the area of the garden is 48 square yards. Find the dimensions of the triangle.

---

② Model and Solve Problems Using the Pythagorean Theorem

The Pythagorean Theorem is a statement about right triangles.

> **DEFINITIONS**
>
> A **right triangle** is one that contains a **right angle,** that is, an angle of 90°. The side of the triangle opposite the 90° angle is called the **hypotenuse;** the remaining two sides are called **legs.**

In Figure 5 we use c to represent the length of the hypotenuse and a and b to represent the lengths of the legs. Notice the use of the symbol $\llcorner$ to show the 90° angle.

We now state the Pythagorean Theorem.

Figure 5

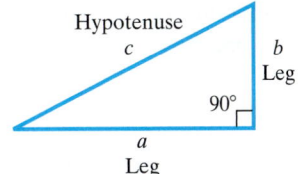

> **THE PYTHAGOREAN THEOREM**
>
> In a right triangle, the square of the length of the hypotenuse is equal to the sum of the squares of the lengths of the legs. That is, in the right triangle shown in Figure 5,
>
> $$a^2 + b^2 = c^2 \quad \text{or} \quad \text{leg}^2 + \text{leg}^2 = \text{hypotenuse}^2$$

EXAMPLE 4   Using the Pythagorean Theorem

Find the lengths of the sides of the right triangle in Figure 6 on page 464.

Solution

Figure 6 shows a right triangle, so we use the Pythagorean Theorem, $a^2 + b^2 = c^2$. The legs have lengths x and $x + 7$, and the hypotenuse has length 13. We substitute into the equation $a^2 + b^2 = c^2$.

Figure 6

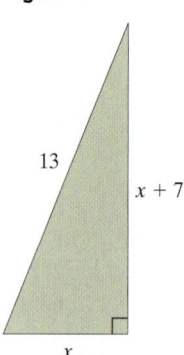

$$a^2 + b^2 = c^2$$

Substitute $a = x, b = x + 7, c = 13$: $\quad x^2 + (x + 7)^2 = 13^2$

$(a + b)^2 = a^2 + 2ab + b^2$: $\quad x^2 + x^2 + 14x + 49 = 169$

Combine like terms: $\quad 2x^2 + 14x + 49 = 169$

Subtract 169 from both sides: $\quad 2x^2 + 14x - 120 = 0$

Factor: $\quad 2(x^2 + 7x - 60) = 0$

$$2(x + 12)(x - 5) = 0$$

Set each factor $= 0$: $\quad 2 = 0 \quad$ or $\quad x + 12 = 0 \quad$ or $\quad x - 5 = 0$

Solve each equation: $\quad x = -12 \quad$ or $\quad x = 5$

The equation $2 = 0$ is false, and the solution $x = -12$ makes no sense because x represents the length of a leg of a right triangle. So $x = 5$ is the length of one leg of the triangle. The other leg is $x + 7 = 5 + 7 = 12$.

To check our solutions, let's replace a by 5 and b by 12 and see if our solutions satisfy the Pythagorean Theorem. Does $5^2 + 12^2 = 13^2$? Is $25 + 144 = 169$? Yes! Our answers are correct. ∎

QUICK ✓

4. Find the length of each leg of the right triangle pictured below.

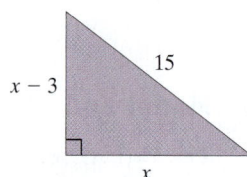

| **EXAMPLE 5** | **Will the Television Fit in the Media Cabinet?** |

A rectangular 20-inch television screen (measured diagonally) is 4 inches wider than it is tall. Will this TV fit in your new media cabinet that is 20 inches wide?

Solution

Step 1: Identify This is a geometry problem involving the lengths of the sides of a right triangle.

Step 2: Name We know that the diagonal of the television screen (hypotenuse) is 20 inches long. Let x represent the height of the screen. The screen is 4 inches wider than it is tall, so $x + 4$ represents the width.

Step 3: Translate We know that the Pythagorean Theorem tells us the relationship between the lengths of the sides of a right triangle, so we use $a^2 + b^2 = c^2$.

$$a^2 + b^2 = c^2$$
$$x^2 + (x + 4)^2 = 20^2$$

Work Smart

Remember

$$(x + 4)^2 \neq x^2 + 16$$
$$(x + 4)^2 = x^2 + 8x + 16$$

Step 4: Solve We now proceed to solve the equation. The first step is to put the quadratic equation in standard form, $ax^2 + bx + c = 0$.

$$x^2 + (x + 4)^2 = 20^2$$
$$x^2 + x^2 + 8x + 16 = 400$$
$$2x^2 + 8x + 16 = 400$$

Subtract 400 from both sides: $\quad 2x^2 + 8x - 384 = 0$

Factor: $\quad 2(x^2 + 4x - 192) = 0$

Factor: $\quad 2(x + 16)(x - 12) = 0$

Set each factor to 0: $\quad 2 = 0 \quad$ or $\quad x + 16 = 0 \quad$ or $\quad x - 12 = 0$

Solve: $\quad x = -16 \quad$ or $\quad x = 12$

Step 5: Check The statement $2 = 0$ is false. Since x represents the length of the shorter leg of the triangle, we discard the solution $x = -16$. If the shorter leg of the triangle (height of the TV) is 12 inches, then the longer leg (width of the television) would be $x + 4 = 12 + 4 = 16$ inches. Does $12^2 + 16^2 = 20^2$? Since $144 + 256 = 400$, our answer is correct!

Step 6: Answer The television screen has dimensions 12 inches by 16 inches, so it will fit in the new cabinet. ▬

QUICK ✓

5. A rectangular 10-inch television screen (measured diagonally) is 2 inches wider than it is tall. What are the dimensions of the TV screen?

6.7 Exercises

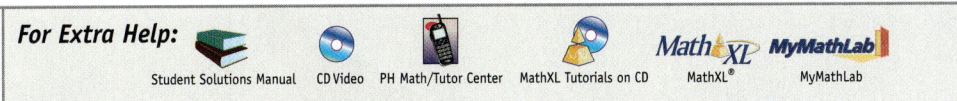

For Extra Help: Student Solutions Manual | CD Video | PH Math/Tutor Center | MathXL Tutorials on CD | MathXL® | MyMathLab

Concepts and Vocabulary

In Problems 1–3, fill in the blanks.

1. The _____ of a right triangle is always the side opposite the 90° angle.

2. The sides of a triangle other than the hypotenuse are called _____ and are labeled as _____ and _____.

3. The Pythagorean Theorem states that in a right triangle, if c is the hypotenuse and a, b are the legs, then _____.

In Problems 4–6, answer True or False to each statement.

4. If the length of the side of rectangle is 5 less than twice the width and the width is labeled x, the length should be labeled $5 - 2x$.

5. The difference of 25 and the square of a number is the same as the difference of 25 and a number squared.

6. The Pythagorean Theorem can be used to find the length of the side of any triangle.

✎ 7. In a projectile motion problem, you have two positive numbers that satisfy the word problem. Explain the meaning of each of these numbers and then give a case where you would have only one positive solution.

✎ 8. You have to give the class some steps to follow when solving a word problem. What steps would you include and what would you say about the numbers that might appear in the solution to your equation?

Building Skills

In Problems 9–12, use the given area to find the missing sides of the rectangle.

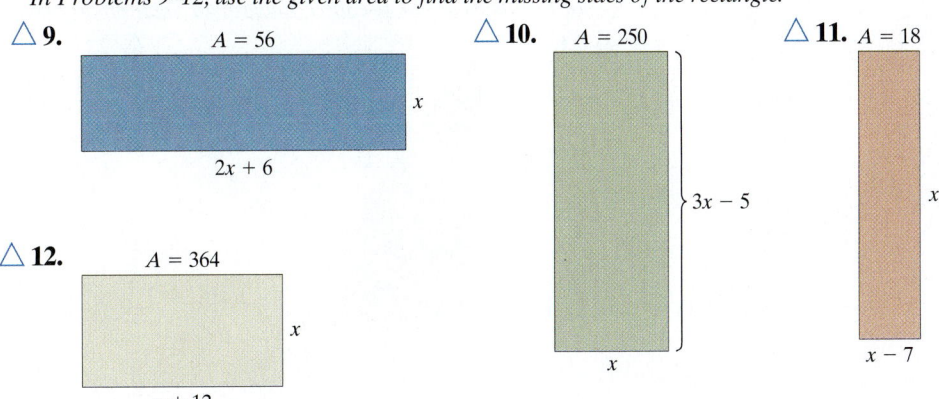

△ 9. $A = 56$, $2x + 6$

△ 10. $A = 250$, x, $3x - 5$

△ 11. $A = 18$, x, $x - 7$

△ 12. $A = 364$, x, $x + 12$

In Problems 13–16, use the given area to find the height and base of the triangle.

△ **13.**

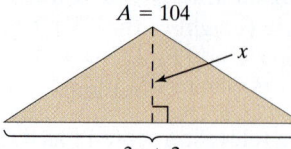

$A = 104$

x

$3x + 2$

△ **14.** $A = 42$

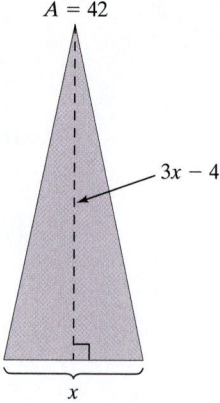

$3x - 4$

x

△ **15.** $A = 144$

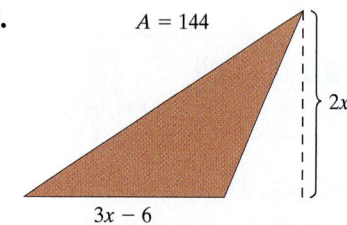

$2x$

$3x - 6$

△ **16.** $A = 84$

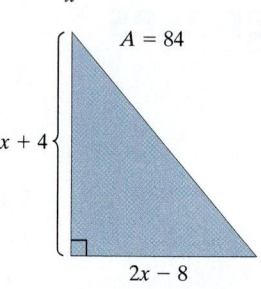

$x + 4$

$2x - 8$

In Problems 17–20, use the given area to find the dimensions of the quadrilateral.

△ **17.** $A = 143$

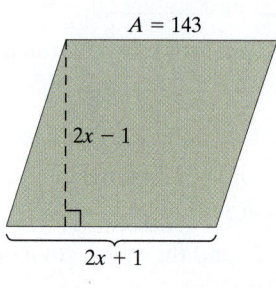

$2x - 1$

$2x + 1$

△ **18.** $A = 77$

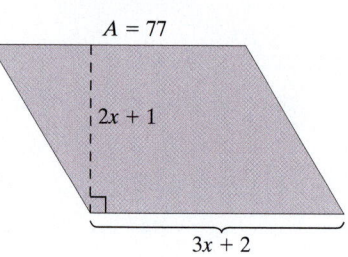

$2x + 1$

$3x + 2$

△ **19.**

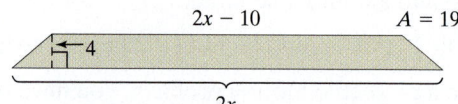

$2x - 10$

$A = 192$

4

$2x$

△ **20.**

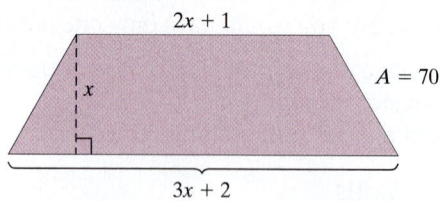

$2x + 1$

$A = 70$

x

$3x + 2$

In Problems 21–24, use the Pythagorean Theorem to find the lengths of the sides of the triangle.

△ **21.**

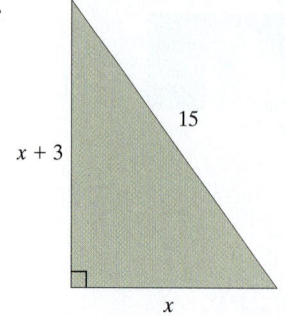

15

$x + 3$

x

△ **22.** x

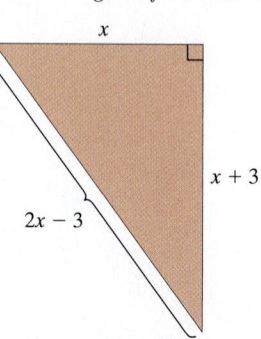

$x + 3$

$2x - 3$

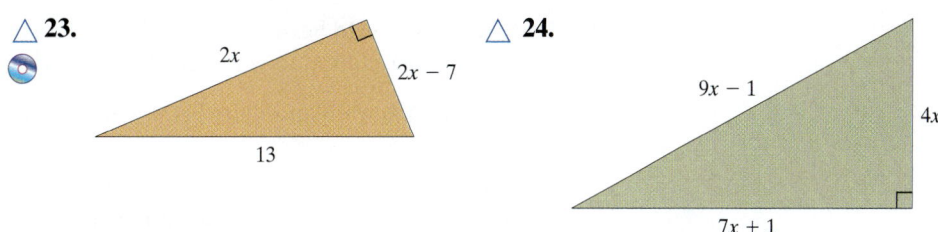

△ **23.**

$2x$

$2x - 7$

13

△ **24.**

$9x - 1$

$4x$

$7x + 1$

Applying the Concepts

25. **Projectile Motion** The height, h, of an object t seconds after it is dropped from a cliff 256 feet tall is given by the equation $h = -16t^2 + 256$. Suppose you dropped your glasses off the cliff. Fill in the table below to find the height of your glasses at each time, t.

Time, in Seconds	0	0.5	1	1.5	2	2.5	3	3.5	4
Height, in Feet									

26. **Projectile Motion** The height, h, of an object t seconds after it is propelled from ground level is given by the equation $h = -16t^2 + 64t$. Fill in the table below to find the height of the object at each time, t.

Time, in Seconds	0	0.5	1	1.5	2	2.5	3	3.5	4
Height, in Feet									

27. **Projectile Motion** If $h = -16t^2 + 96t$ represents the height of a rocket, in feet, t seconds after it was fired, when will the rocket hit the ground? (**Hint:** The rocket is on the ground when $h = 0$.)

28. **Projectile Motion** If $h = -16t^2 + 96t$ represents the height of a rocket, in feet, t seconds after it was fired, when will the rocket be 144 feet high?

△ 29. **Rectanglular Room** The length of a rectangular room is 8 meters more than the width. If the area of the room is 48 square meters, find the dimensions of the room.

△ 30. **Rectangular Room** The width of a rectangular room is 4 feet less than the length. If the area of the room is 21 square feet, find the dimensions of the room.

△ 31. **Sail Boat** The sail on a sail boat is in the shape of a triangle. If the height of the sail is 3 times the length of base and the area is 54 square feet, find the dimensions of the sail.

△ 32. **Triangle** The base of a triangle is 2 meters more than the height. If the area of the triangle is 24 square meters, find the base and height of the triangle.

33. **Big-Screen TV** Your big-screen TV measures 50 inches on the diagonal. If the front of the TV measures 40 inches across the bottom, find the height of the TV.

34. **Big-Screen TV** Hannah owns a 29-inch TV (that is, it measures 29 inches on the diagonal). If the television is 20 inches high, find the distance across the bottom.

△ 35. **Dimensions of a Rectangle** The length of a rectangle is 1 mm more than twice the width. If the area is 300 square mm, find the dimensions of the rectangle.

△ 36. **Playing Field Dimensions** The width of a rectangular playing field is 3 yards less than twice the length. If the area of the field is 104 square yards, what are its dimensions?

△ 37. **Rectangle and Square** A rectangle and a square have the same area. The width of the rectangle is 6 cm less than the side of the square and the length of the rectangle is 5 cm more than twice the side of the square. What are the dimensions of the rectangle?

△ 38. **Square and Rectangle** A rectangle and a square have the same area. The width of the rectangle is 2 in. less than the side of the square and the length of the rectangle is 3 in. less than twice the side of the square. What are the dimensions of the rectangle?

△ **39. Watching the World Series** David is an avid baseball fan and has purchased a plasma TV just in time to watch the World Series. The TV screen is 17 inches longer than it is wide and there is a $1\frac{1}{2}$ inch wide casing that surrounds the TV screen.

(a) David begins to install the TV and remembers that he was told that it measures 53 inches on the diagonal, including the casing. What are the dimensions of the TV screen?

(b) What size opening is required to fit the TV into David's entertainment center?

△ **40. Jasper's Big-Screen TV** Jasper purchased a new big-screen TV. The TV screen is 10 inches longer than it is wide and is surrounded by a casing that is 2 inches wide.

(a) Jasper lost his tape measure but sees on the box that the TV measures 50 inches on the diagonal, including the casing. What are the dimensions of the TV screen?

(b) Jasper knows that the size of the opening where he wants the TV installed is 29 by 42 inches. Will this TV fit into his space?

△ **41. How Tall Is the Pole?** A pole is supported by a 10-foot guy-wire that is attached to the ground. The distance from the pole to the point that wire attaches to the ground is 2 feet greater than the height of the pole. Find the height of the pole.

△ **42. Sailing on Lake Erie** A sail on a sailboat is in the shape of a right triangle. The longest side of the sail is 13 feet long and one side of the sail is 7 feet longer than the other. Find the dimensions of the sail.

Extending the Concepts

△ **43. Gardening** Beth has 28 feet of fence to enclose a small garden. One side of the garden lies along her house, so only three sides require fencing. Find the dimensions of the garden if she encloses 98 square feet of the garden with the fence.

△ **44. Volume of a Box** A rectangular solid has a square base and is 8 meters high. What are the dimensions of the base if the volume of the solid is 128 cubic meters? The volume of a rectangular solid is (length)(width)(height).

CHAPTER 6 ACTIVITY: WHICH ONE DOES NOT BELONG?

Focus: Factoring polynomials.

Time: 20–30 minutes

Group size: 3–4

Each group member should decide which polynomial from each row does not belong. Answers will vary. As a group, discuss each member's results. Each group member should be prepared to explain WHY that particular polynomial does not belong with the other three. Be creative!

	A	B	C	D
1	$15x^2 - 10ax - 15xy + 10ay$	$xr - 3xs - ry + 3sy$	$2ax - 14bx - 2ay + 14by$	$3ab - 3ay - 3bx + 3xy$
2	$-x^2 - x + 6$	$3x^2 + 39x + 120$	$2x^2 - 12xy - 54y^2$	$7x^2 + 7x - 140$
3	$3x^2 - 16x + 21$	$6x^2 - 3x + 15$	$3x^2 - 13x - 56$	$3x^2 + x + 1$
4	$9a^2 + 30a + 25$	$16x^2 - 24x + 9$	$4x^2 - 12xy + 9y^2$	$18s^2 - 60s + 9$
5	$4x^2 + 13x + 10$	$16x^2 + 40xy + 25y^2$	$16x^2 - 25y^2$	$8x^2 - 2xy - 15y^2$

CHAPTER 6 REVIEW

Section 6.1	Greatest Common Factor and Factoring by Grouping

KEY CONCEPTS	KEY TERMS
• **To find the greatest common factor of two or more expressions,** **Step 1:** Find the GCF of the coefficients of each variable expression. **Step 2:** For each variable expression, determine the smallest exponent that the variable expression is raised to. **Step 3:** Find the product of the common factors found in Steps 1 and 2. This expression is the GCF. • **To factor a polynomial using the greatest common factor,** **Step 1:** Identify the greatest common factor (GCF) of the terms that make up the polynomial. **Step 2:** Rewrite each term as the product of the GCF and remaining factor. **Step 3:** Use the Distributive Property "in reverse" to factor out the GCF. **Step 4:** Use the Distributive Property to verify that the factorization is correct.	Factors Greatest common factor Factoring by grouping

YOU SHOULD BE ABLE TO . . .	EXAMPLE	REVIEW EXERCISES
① Find the greatest common factor (GCF) of two or more expressions (p. 408)	Examples 1 through 5	1–10
② Factor out the greatest common factor in polynomials (p. 411)	Examples 6 through 10	11–16
③ Factor polynomials by grouping (p. 413)	Examples 11 through 13	17–20

In Problems 1–10, find the greatest common factor, GCF, of each group of expressions.

1. $24, 36$ **2.** $27, 54$ **3.** $10, 20, 30$ **4.** $8, 16, 28$

5. x^4, x^2, x^8 **6.** m^3, m, m^5 **7.** $30a^2b^4, 45ab^2$ **8.** $18x^4y^2z^3, 24x^3y^5z$

9. $4(2a + 1)^2$ and $6(2a + 1)^3$ **10.** $9(x - y)$ and $18(x - y)$

In Problems 11–16, factor the GCF from the polynomial.

11. $-18a^3 - 24a^2$ **12.** $-9x^2 + 12x$

13. $15y^2z + 5y^7z + 20y^3z$ **14.** $7x^3y - 21x^2y^2 + 14xy^3$

15. $x(5 - y) + 2(5 - y)$ **16.** $z(a + b) + y(a + b)$

In Problems 17–20, factor by grouping.

17. $5m^2 + 2mn + 15mn + 6n^2$ **18.** $2xy + y^2 + 2x^2 + xy$

19. $8x + 16 - xy - 2y$ **20.** $xy^2 + x - 3y^2 - 3$

Section 6.2	Factoring Trinomials of the Form $x^2 + bx + c$

KEY CONCEPT	KEY TERMS
• **To factor a trinomial of the form $x^2 + bx + c$, we use the following steps.** **Step 1:** Find the pair of integers whose product is c and whose sum is b. That is, determine m and n such that $mn = c$ and $m + n = b$. **Step 2:** Write $x^2 + bx + c = (x + m)(x + n)$. **Step 3:** Check your work by multiplying the binomials using the FOIL pattern.	Quadratic trinomial Leading coefficient Prime polynomial

YOU SHOULD BE ABLE TO...	EXAMPLE	REVIEW EXERCISES
① Factor trinomials of the form $x^2 + bx + c$ (p. 418)	Examples 1 through 7	21–28
② Factor out the GCF, then factor $x^2 + bx + c$ (p. 423)	Examples 8 and 9	29–34

In Problems 21–34, factor completely. If the polynomial cannot be factored, say it is prime.

21. $x^2 + 5x + 6$ **22.** $x^2 + 6x + 8$ **23.** $x^2 - 21 - 4x$

24. $3x + x^2 - 10$ **25.** $m^2 + m + 20$ **26.** $m^2 - 6m - 5$

27. $x^2 - 8xy + 15y^2$ **28.** $m^2 + 4mn - 5n^2$ **29.** $-p^2 - 11p - 30$

30. $-y^2 + 2y + 15$ **31.** $3x^3 + 33x^2 + 36x$ **32.** $4x^2 + 36x + 32$

33. $2x^2 - 2xy - 84y^2$ **34.** $4y^3 + 12y^2 - 40y$

Section 6.3   Factoring Trinomials of the Form $ax^2 + bx + c$, $a \neq 1$
KEY CONCEPTS

- **Factoring $ax^2 + bx + c$, $a \neq 1$ using trial and error: a, b, and c have no common factors**

  Step 1: List the possibilities for the first terms of each binomial whose product is ax^2.

  $$(\_x + \_)(\_x + \_) = ax^2 + bx + c$$

  Step 2: List the possibilities for the last terms of each binomial whose product is c.

  $$(\_x + \square)(\_x + \square) = ax^2 + bx + c$$

  Step 3: Write out all the combinations of factors found in Steps 1 and 2. Multiply the binomials out until a product is found that equals the trinomial.

- **Factoring $ax^2 + bx + c$, $a \neq 1$ by grouping: a, b, and c have no common factors**

  Step 1: Find the value of ac.

  Step 2: Find the pair of integers, m and n, whose product is ac and whose sum is b.

  Step 3: Write $ax^2 + bx + c = ax^2 + mx + nx + c$.

  Step 4: Factor the expression in Step 3 by grouping.

  Step 5: Multiply out the factored form to verify your answer.

YOU SHOULD BE ABLE TO...	EXAMPLE	REVIEW EXERCISES
① Factor $ax^2 + bx + c$, $a \neq 1$ using trial and error (p. 427)	Examples 1 through 6	35–44
② Factor $ax^2 + bx + c$, $a \neq 1$ using grouping (p. 432)	Examples 7 through 9	35–44

In Problems 35–44, factor completely using any method you wish. If a polynomial cannot be factored, say it is prime.

35. $5y^2 + 14y - 24$ **36.** $6y^2 - 41y - 7$ **37.** $-5x + 2x^2 + 3$

38. $23x + 6x^2 + 7$ **39.** $2x^2 - 7x - 6$ **40.** $8m^2 + 18m + 9$

41. $9m^3 + 30m^2n + 21mn^2$ **42.** $14m^2 + 16mn + 2n^2$ **43.** $15x^3 + x^2 - 2x$

44. $6p^4 + p^3 - p^2$

Section 6.4   Factoring Special Products

KEY CONCEPTS

- **Perfect Square Trinomials**

  $a^2 + 2ab + b^2 = (a + b)^2$

  $a^2 - 2ab + b^2 = (a - b)^2$

- **Difference of Two Squares**

  $a^2 - b^2 = (a - b)(a + b)$

- **Sum or Difference of Two Cubes**

  $a^3 + b^3 = (a + b)(a^2 - ab + b^2)$

  $a^3 - b^3 = (a - b)(a^2 + ab + b^2)$

KEY TERMS

Perfect square trinomial
Difference of two squares
Sum of two cubes
Difference of two cubes

YOU SHOULD BE ABLE TO . . .	EXAMPLE	REVIEW EXERCISES
1 Factor perfect square trinomials (p. 437)	Examples 1 through 4	45–50
2 Factor the difference of two squares (p. 439)	Examples 5 through 7	51–56
3 Factor the sum or difference of two cubes (p. 441)	Examples 8 through 10	57–62

In Problems 45–62, factor completely. If a polynomial cannot be factored, say it is prime.

45. $4x^2 - 12x + 9$ **46.** $x^2 - 10x + 25$ **47.** $x^2 + 6xy + 9y^2$

48. $9x^2 + 24xy + 4y^2$ **49.** $8m^2 + 8m + 2$ **50.** $2m^2 - 24m + 72$

51. $4x^2 - 25y^2$ **52.** $49x^2 - 36y^2$ **53.** $x^2 + 25$

54. $x^2 + 100$ **55.** $x^4 - 81$ **56.** $x^4 - 625$

57. $m^3 + 27$ **58.** $m^3 + 125$ **59.** $27p^3 - 8$

60. $64p^3 - 1$ **61.** $y^9 + 64z^6$ **62.** $8y^3 + 27z^6$

Section 6.5   Summary of Factoring Techniques

KEY CONCEPT

Steps for Factoring

Step 1: Is there a greatest common factor? If so, factor out the greatest common factor (GCF).

Step 2: Count the number of terms.

Step 3: (a) 2 terms

- Is it the difference of two squares? If so,

  $a^2 - b^2 = (a - b)(a + b)$

- Is it the difference of two cubes? If so,

  $a^3 - b^3 = (a - b)(a^2 + ab + b^2)$

- Is it the sum of two cubes? If so,

  $a^3 + b^3 = (a + b)(a^2 - ab + b^2)$

(b) 3 terms

- Is it a perfect square trinomial? If so,

  $a^2 + 2ab + b^2 = (a + b)^2$ or $a^2 - 2ab + b^2 = (a - b)^2$

- Is the leading coefficient 1? If so,

  $x^2 + bx + c = (x + m)(x + n)$ where $mn = c$ and $m + n = b$

- Is the coefficient of the square term different from 1? If so,
    - **a.** Use factoring by grouping
    - **b.** Use trial and error
- **(c)** 4 terms
    - Use factoring by grouping

Step 4: Check your work by multiplying out the factors.

YOU SHOULD BE ABLE TO . . .	EXAMPLE	REVIEW EXERCISES
1 Factor polynomials completely (p. 444)	Examples 1 through 7	63–78

In Problems 63–78, factor completely. If a polynomial cannot be factored, say it is prime.

63. $15a^3 - 6a^2b - 25ab^2 + 10b^3$ **64.** $12a^2 - 9ab + 4ab - 3b^2$

65. $x^2 - xy - 48y^2$ **66.** $x^2 - 10xy - 24y^2$

67. $x^3 - x^2 - 42x$ **68.** $3x^6 - 30x^5 + 63x^4$

69. $6x^2 + 11x + 3$ **70.** $10z^2 + 9z - 9$

71. $27x^3 + 8$ **72.** $8z^3 - 1$

73. $4y^2 + 18y - 10$ **74.** $5x^3y^2 - 8x^2y^2 + 3xy^2$

75. $25k^2 - 81m^2$ **76.** $x^4 - 9$

77. $16m^2 + 1$ **78.** $m^4 + 25$

Section 6.6   Solving Polynomial Equations by Factoring

KEY CONCEPT	KEY TERMS
• **The Zero-Product Property** If the product of two factors is zero, then at least one of the factors is 0. That is, if $ab = 0$, then $a = 0$ or $b = 0$ or both a and b are 0.	Polynomial equation Degree of a polynomial equation Quadratic equation Second-degree equation Standard form

YOU SHOULD BE ABLE TO . . .	EXAMPLE	REVIEW EXERCISES
1 Solve quadratic equations using the Zero-Product Property (p. 451)	Examples 1 through 8	79–88
2 Solve polynomial equations of degree three or higher using the Zero-Product Property (p. 457)	Example 9	89–92

In Problems 79–92, solve each equation by factoring.

79. $(x - 4)(2x - 3) = 0$ **80.** $(2x + 1)(x + 7) = 0$

81. $x^2 - 12x - 45 = 0$ **82.** $x^2 - 7x + 10 = 0$

83. $3x^2 + 6x = 0$ **84.** $4x^2 + 18x = 0$

85. $3x(x + 1) = 2x^2 + 5x + 3$ **86.** $2x^2 + 6x = (3x + 1)(x + 3)$

87. $5x^2 + 5 = -20x - 10$ **88.** $8x^2 - 10x = -2x + 6$

89. $x^3 = -11x^2 + 42x$ **90.** $-3x^2 = -x^3 + 18x$

91. $(3x - 4)(x^2 - 9) = 0$ **92.** $(2x + 5)(x^2 + 4x + 4) = 0$

Section 6.7	Modeling and Solving Problems with Quadratic Equations	
KEY CONCEPT		**KEY TERMS**
• **The Pythagorean Theorem** In a right triangle, the square of the length of the hypotenuse is equal to the sum of the squares of the lengths of the legs. That is, if a and b are the lengths of the legs and c is the length of the hypotenuse, then $\text{leg}^2 + \text{leg}^2 = \text{hypotenuse}^2$ or $c^2 = a^2 + b^2$.		Right triangle Right angle Hypotenuse Legs

YOU SHOULD BE ABLE TO . . .	EXAMPLE	REVIEW EXERCISES
1 Model and solve problems involving quadratic equations (p. 460)	Examples 1 through 3	93–96
2 Model and solve problems using the Pythagorean Theorem (p. 463)	Example 4 and 5	97, 98

93. If $h = -16t^2 + 80t$ represents the height of a jet of water from a geyser t seconds after the geyser erupts, when will the water hit the ground?

94. If $h = -16t^2 + 80t$ represents the height of the jet of water from a geyser, when will the jet of water be 96 feet high?

△**95. Tabletop** The length of a rectangular tabletop is 3 feet shorter than twice its width. If the area of the tabletop is 54 square feet, what are the dimensions of the tabletop?

△**96. Tarp** The length of a rectangular tarp is 1 yard shorter than twice its width. If the area of the tarp is 15 square yards, what are the dimensions of the tarp?

△**97. Right Triangle** The shorter leg of a right triangle is 2 feet shorter than the longer leg. The hypotenuse is 10 feet. How long is each leg?

△**98. Right Triangle** The shorter leg of a right triangle is 14 feet shorter than the longer leg. The hypotenuse is 26 feet. How long is each leg?

CHAPTER 6 TEST

 Remember to use your Chapter Test Prep Video CD to see fully worked-out solutions to any of these problems you would like to review.

1. Find the GCF of $16x^5y^2$, $20x^4y^6$, and $24x^6y^8$.

In Problems 2–16, factor each polynomial completely. If the polynomial cannot be factored, say it is prime.

2. $x^4 - 81$

3. $18x^5 - 9x^3 - 27x$

4. $xy - 7y - 4x + 28$

5. $27x^3 + 125$

6. $y^2 - 8y - 48$

7. $6m^2 - m - 5$

8. $4x^2 + 25$

9. $4(x - 5) + y(x - 5)$

10. $3x^2y - 15xy - 42y$

11. $x^2 + 4x + 12$

12. $3x^6 - 3$

13. $9x^3 + 39x^2 + 12x$

14. $6m^2 + 7m + 2$

15. $4m^2 - 6mn + 4$

16. $25x^2 + 70xy + 49y^2$

In Problems 17 and 18, solve the equation by factoring.

17. $5x^2 = -16x - 3$

18. $5x^3 - 20x^2 + 20x = 0$

19. The length of a rectangle is 8 inches shorter than three times its width. If the area of the rectangle is 35 square inches, what is the length of the rectangle?

20. The hypotenuse of a right triangle is one inch longer than the longer leg. The shorter leg is 7 inches shorter than the longer leg. Find the length of all three sides of the triangle.

Getting Ready for Intermediate Algebra: A Review of Chapters 1–6

The following problems cover important concepts from Chapters 1–6. We designed these problems so that you can review material that will be needed for the rest of the course. Take time to work these problems before proceeding to the next chapter. The answers to these problems are located at the back of the text starting on page AN-20. If you get any problems wrong, go to the section cited and review the material.

1. Using the set $\left\{ 0, 1, -6, \dfrac{2}{5}, -0.83, 0.5454\ldots, 1.010010001\ldots \right\}$,

   list all the numbers that are . . .   Section 1.2, pp. 9–12
   (a) Natural numbers, (b) Integers,
   (c) Rational numbers, (d) Irrational numbers,
   (e) Real numbers

In Problems 2–8, evaluate each expression.

2. $4(-5)$ Section 1.3, pp. 23–24

3. $\dfrac{-72}{60}$ Section 1.3, pp. 24–26

4. $\dfrac{12}{5} \cdot \left(-\dfrac{35}{8} \right)$ Section 1.4, p. 29

5. $\dfrac{5}{12} + \dfrac{11}{12}$ Section 1.4, p. 31

6. $\dfrac{5}{6} - \dfrac{7}{30}$ Section 1.4, pp. 32–34

7. $5 - 2(1 - 4)^3 + 5 \cdot 3$ Section 1.6, pp. 50–51

8. $\dfrac{|3 - 3 \cdot 5|}{-2^3}$ Section 1.6, p. 52

9. Solve: $5(x + 1) - (2x + 1) = x + 10$ Section 2.2, pp. 87–89

10. Solve: $\dfrac{1}{6}(x + 1) = 3 - \dfrac{1}{2}(x + 4)$ Section 2.3, pp. 93–95

11. Two angles are supplementary. The measure of the   Section 2.7, pp. 139–140
    larger angle is 15 degrees more than twice the
    measure of the smaller angle. Find the measure of
    each angle.

12. Solve $2(x + 3) \le 3x - 1$. Express your answer using   Section 2.8, pp. 153–159
    set-builder notation and interval notation. Graph
    the inequality.

13. The body mass index (BMI) of a person 62 inches   Section 2.8, pp. 159–161
    tall and weighing x pounds is given by $0.2x - 2$.
    A BMI of 30 or more is considered to be obese. For
    what weights would a person 62 inches tall be
    considered obese?

14. What is the degree of $3x^5 - 4x^3 + 2x^2 - 6x - 1$?   Section 5.1, pp. 345–346

15. Evaluate $3x^2 + 2x - 7$ when $x = -2$.   Section 5.1, pp. 348–349

16. Add: $(4y^3 + 8y^2 - y + 3) + (y^3 - 3y^2 - 9)$ Section 5.1, pp. 346–347

17. Simplify: $(-3x^4)^2$ Section 5.2, p. 357

In Problems 18–21, perform the indicated operation. Express your answer as a polynomial in standard form.

18. $-4x(x^2 - 5x + 3)$ Section 5.3, pp. 360–361

19. $(4x + 1)(3x - 5)$ Section 5.3, pp. 361–363

20. $(2x - 7)^2$ Section 5.3, pp. 364–366

21. $(5x + 3y)(5x - 3y)$ Section 5.3, p. 364

22. Simplify: $\dfrac{(4y^2z)^{-1}}{y^{-3}z^2}$ Section 5.4, pp. 377–380

In Problems 23–24, divide.

23. $\dfrac{4a^3 - 12a^2 + 2a}{2a}$ Section 5.5, pp. 384–385

24. $\dfrac{3x^3 + 10x^2 - 23x + 1}{x + 5}$ Section 5.5, pp. 385–388

In Problems 25–32, factor each polynomial completely.

25. $-4x^4 + 16x^2 - 20x$ Section 6.1, pp. 411–413

26. $8z^3 - 4z^2 + 6z - 3$ Section 6.1, pp. 413–415

27. $w^2 - 2w - 35$ Section 6.2, pp. 418–423

28. $-3c^3 + 15c^2 + 72c$ Section 6.2, pp. 423–425

29. $m^2 - 4m - 28$ Section 6.2, pp. 418–423

30. $3y^2 + 8y - 16$ Section 6.3, pp. 427–435

31. $p^4 - 16$ Section 6.4, pp. 439–441

32. $4a^2 + 20ab + 25b^2$ Section 6.4, pp. 437–439

33. Solve: $(b - 3)(2b + 5) = 0$ Section 6.6, pp. 451–452

34. Solve: $x^2 + 20 = 9x$ Section 6.6, pp. 452–456

35. Solve: $6k^2 + 17k - 14 = 0$ Section 6.6, pp. 452–456

To plan effectively for purchasing materials and hiring part-time college students, the president of a bicycle manufacturing company wants to know the average daily cost of manufacturing bicycles. A rational equation is used to model his average daily costs. See Example 7 in Section 7.7, page 536.

OUTLINE

7.1 Simplifying Rational Expressions

7.2 Multiplying and Dividing Rational Expressions

7.3 Adding and Subtracting Rational Expressions with a Common Denominator

7.4 Finding the Least Common Denominator and Forming Equivalent Rational Expressions

7.5 Adding and Subtracting Rational Expressions with Unlike Denominators

7.6 Complex Rational Expressions

   Putting the Concepts Together (Sections 7.1–7.6)

7.7 Rational Equations

7.8 Models Involving Rational Equations

   Chapter 7 Activity: Correct the Quiz
   Chapter 7 Review
   Chapter 7 Test

The Big Picture: Putting It Together

In this chapter, we discuss rational expressions. A rational expression is the ratio of two polynomials (that is, a polynomial divided by another polynomial). The techniques that we will learn in this chapter are similar to the techniques we learned when dealing with fractions. In the Appendix, Section A.1, we learned how to write a fraction in lowest terms by factoring the numerator and denominator and dividing out like factors—this same approach is used on rational expressions. The ability to factor a polynomial (discussed in Chapter 6) plays a *huge* role in being able to simplify a rational expression, so make sure you are good at factoring.

In Section 1.4 we discussed how to add, subtract, multiply, and divide rational numbers expressed as fractions. These same skills apply to adding, subtracting, multiplying, and dividing rational expressions. However, the skills learned in Chapter 5 plus the factoring skills acquired in Chapter 6 will also be needed to perform these operations. Just remember, the methods that we use to perform operations on rational expressions are identical to those that we use on rational numbers. So we really aren't learning any new methods here—just new ways to apply the skills that we already have!

7.1 Simplifying Rational Expressions

OBJECTIVES

1. Evaluate a Rational Expression
2. Determine Undefined Values of a Rational Expression
3. Simplify Rational Expressions

Preparing for Simplifying Rational Expressions

1. Evaluate $\dfrac{3x + y}{2}$ for $x = -1$ and $y = 7$. [Section 1.7, p. 58]

2. Factor: $2x^2 + x - 3$ [Section 6.3, pp. 427–435]

3. Solve: $3x^2 - 5x - 2 = 0$ [Section 6.6, pp. 451–457]

4. Write $\dfrac{21}{70}$ in lowest terms. [Appendix, Section A.1, p. A5]

5. Divide: $\dfrac{x^3 y^4}{x y^2}$ [Section 5.4, pp. 371–372]

Work Smart

The numbers $\dfrac{4}{3}, \dfrac{-5}{2}$, and 18 are examples of rational numbers.

Remember, a rational number is the quotient of two integers, where the denominator is not zero.

> **DEFINITION**
>
> A **rational expression** is the quotient of two polynomials. That is, a rational expression is written in the form $\dfrac{p}{q}$, where p and q are polynomials and $q \neq 0$.

Some examples of rational expressions are

(a) $\dfrac{x - 3}{4x + 1}$ (b) $\dfrac{x^2 - 4x - 12}{x^2 - 5}$ (c) $\dfrac{3a^2 + 7b + 2b^2}{a^2 - 2ab + 8b^2}$

Expressions (a) and (b) are rational expressions in one variable, x, while expression (c) is a rational expression in two variables, a and b.

1 Evaluate a Rational Expression

To **evaluate** a rational expression, we replace the variable with its assigned numerical value and perform the arithmetic, using the order of operations.

> **EXAMPLE 1 Evaluating a Rational Expression**
>
> Evaluate $\dfrac{-2}{x + 5}$ for (a) $x = -3$ (b) $x = 11$.

Solution

(a) Substitute -3 for x: $\dfrac{-2}{x + 5} = \dfrac{-2}{-3 + 5}$

$$= \dfrac{-2}{2} = -1$$

(b) Substitute 11 for x: $\dfrac{-2}{x + 5} = \dfrac{-2}{11 + 5}$

$$= \dfrac{-2}{16}$$

Divide out common factors: $= \dfrac{-1 \cdot 2}{8 \cdot 2}$

$$= -\dfrac{1}{8}$$

Preparing for...Answers **1.** 2

2. $(2x + 3)(x - 1)$ **3.** $\left\{ -\dfrac{1}{3}, 2 \right\}$

4. $\dfrac{3}{10}$ **5.** $x^2 y^2$

EXAMPLE 2 Evaluating a Rational Expression of a Higher Degree

Evaluate $\dfrac{p^2 - 9}{2p^2 + p - 10}$ for (a) $p = 1$ (b) $p = -2$.

Solution

(a) Substitute 1 for p:

$$\frac{p^2 - 9}{2p^2 + p - 10} = \frac{(1)^2 - 9}{2(1)^2 + 1 - 10}$$

$$= \frac{1 - 9}{2(1) + 1 - 10}$$

$$= \frac{-8}{2 + 1 - 10}$$

$$= \frac{-8}{-7}$$

$$= \frac{8}{7}$$

(b) Substitute -2 for p:

$$\frac{p^2 - 9}{2p^2 + p - 10} = \frac{(-2)^2 - 9}{2(-2)^2 + (-2) - 10}$$

$$= \frac{4 - 9}{2(4) - 2 - 10}$$

$$= \frac{-5}{8 - 2 - 10}$$

$$= \frac{-5}{-4}$$

$$= \frac{5}{4}$$

QUICK ✓ *Evaluate the rational expression for (a) $x = -5$ (b) $x = 3$.*

1. $\dfrac{3}{5x + 1}$ **2.** $\dfrac{x^2 + 6x + 9}{x + 1}$

EXAMPLE 3 Evaluating a Rational Expression with More than One Variable

Evaluate:

(a) $\dfrac{2a + 4b}{3c - 1}$ for $a = 1, b = -2, c = 3$ (b) $\dfrac{w - 3v}{v + 3w}$ for $w = 2$ and $v = -6$

Solution

(a) We substitute 1 for a, -2 for b, and 3 for c.

$$\frac{2a + 4b}{3c - 1} = \frac{2(1) + 4(-2)}{3(3) - 1}$$

$$= \frac{2 + (-8)}{9 - 1}$$

$$= \frac{-6}{8}$$

Divide out common factors: $\quad = \dfrac{\cancel{2} \cdot -3}{\cancel{2} \cdot 4}$

$$= -\frac{3}{4}$$

(b) We substitute 2 for w and -6 for v.

$$\frac{w - 3v}{v + 3w} = \frac{2 - 3(-6)}{-6 + 3(2)}$$

$$= \frac{2 + 18}{-6 + 6}$$

$$= \frac{20}{0}$$

Division by 0 is undefined, so the expression $\dfrac{w - 3v}{v + 3w}$ is not defined for $w = 2$ and $v = -6$. ∎

QUICK ✓ *Evaluate the rational expression for the given values of the variable.*

3. $\dfrac{3m - 5n}{p - 4}$ for $m = 2, n = -1$, and $p = 6$ **4.** $\dfrac{5y + 2}{2y - z}$ for $y = 3, z = 6$

---

(2) ## Determine Undefined Values of a Rational Expression

Work Smart

A rational expression is undefined if the *denominator* equals 0. It is okay for the *numerator* to equal 0.

A rational expression is **undefined** for those values of the variable(s) that make the denominator zero. **We find the values for which a rational expression is undefined by setting the denominator equal to zero and solving for the variable.**

| **EXAMPLE 4** | **Determining the Values for Which a Rational Expression Is Undefined** |

Find the value of x for which the expression $\dfrac{2}{x + 3}$ is undefined.

Solution

We want to find all values of x in the rational expression $\dfrac{2}{x + 3}$ that cause $x + 3$ to equal 0.

           Set the denominator equal to 0: $x + 3 = 0$

           Subtract 3 from both sides: $x = -3$

So -3 causes the denominator, $x + 3$, to equal 0. Therefore, the expression $\dfrac{2}{x + 3}$ is undefined for $x = -3$. ∎

| **EXAMPLE 5** | **Determining the Values for Which a Rational Expression Is Undefined** |

Find the values of x for which the expression $\dfrac{8x}{x^2 - 2x - 3}$ is undefined.

Solution

We want to find all values of x in the rational expression $\dfrac{8x}{x^2 - 2x - 3}$ that cause the denominator, $x^2 - 2x - 3$, to equal 0.

Work Smart

The equation $x^2 - 2x - 3 = 0$ is quadratic because the highest exponent on the variable term is 2.

         Set the denominator equal to 0: $x^2 - 2x - 3 = 0$

         This is a quadratic equation,
         so we factor the polynomial: $(x - 3)(x + 1) = 0$

         Set each factor equal to 0: $x - 3 = 0$ or $x + 1 = 0$

         Solve each equation for x: $x = 3$ or $x = -1$

Since 3 or -1 cause the denominator to equal zero, the rational expression $\dfrac{8x}{x^2 - 2x - 3}$ is undefined for $x = 3$ or $x = -1$.

Work Smart
In Example 5, the expression
$\frac{8x}{x^2 - 2x - 3}$ is *not* undefined at $x = 0$
because 0 causes the numerator to
equal 0, which is okay.

Check Let's verify our answer by evaluating $\frac{8x}{x^2 - 2x - 3}$ for each of these values.

If $x = 3$: $\frac{8(3)}{(3)^2 - 2(3) - 3} = \frac{24}{9 - 6 - 3} = \frac{24}{0}$ undefined

If $x = -1$: $\frac{8(-1)}{(-1)^2 - 2(-1) - 3} = \frac{-8}{1 + 2 - 3} = \frac{-8}{0}$ undefined

QUICK ✓ *Find the value(s) for which the rational expression is undefined.*

5. $\frac{3}{x + 7}$ **6.** $\frac{-4}{3n + 5}$ **7.** $\frac{8x}{x^2 + 2x - 3}$ **8.** $\frac{2k}{k^2 - 9}$

③ Simplify Rational Expressions

Remember that we write fractions in lowest terms by using the fact that $\frac{a \cdot c}{b \cdot c} = \frac{a}{b}$. For example, to write the fraction $\frac{18}{45}$ in lowest terms, we factor both 18 and 45 and divide out common factors.

$$\frac{18}{45} = \frac{2 \cdot 3 \cdot 3}{3 \cdot 3 \cdot 5} = \frac{2}{5}$$

To **simplify** a rational expression means to write the rational expression in the form $\frac{p}{q}$ where p and q are polynomials that have no common factors. We use the same ideas to simplify rational expressions as we do to write fractions in lowest terms.

> **SIMPLIFYING RATIONAL EXPRESSIONS**
> If $p, q,$ and r are polynomials, then
> $$\frac{p \cdot r}{q \cdot r} = \frac{p}{q} \quad \text{if } q \neq 0 \text{ and } r \neq 0$$

Work Smart
A rational expression is simplified if
the numerator and the denominator
share no common factor other than 1.

So to simplify a rational expression, factor the numerator, factor the denominator, and divide out common the factors.

EXAMPLE 6 **How to Simplify a Rational Expression**

Simplify: $\frac{7x + 14}{x^2 - 4}, x \neq -2, x \neq 2$

Step-by-Step Solution

We include the restrictions $x \neq -2, x \neq 2$ because these values of x cause division by 0.

Step 1: Completely factor the numerator and the denominator.	$\frac{7x + 14}{x^2 - 4} = \frac{7(x + 2)}{(x + 2)(x - 2)}$
Step 2: Divide out common factors.	$= \frac{7(x + 2)}{(x + 2)(x - 2)} = \frac{7}{x - 2}$

So $\frac{7x + 14}{x^2 - 4} = \frac{7}{x - 2}$.

When we simplify a rational expression by dividing out common factors, we are changing the values of the variable that are not allowed as replacement values. In Example 6 we include the restriction $x \neq -2$, $x \neq 2$ because the expression $\dfrac{7x + 14}{x^2 - 4}$ is equal to $\dfrac{7}{x - 2}$ for all values except $x = -2$ and $x = 2$. When $x = 2$, both expressions are undefined and when $x = -2$, the original expression is also undefined. By not allowing x to equal -2 in both instances, the expressions remain equal. In general, we must restrict all values of the variable that are allowed in the *original* rational expression. For the remainder of the text, we shall not include the restrictions on the variable, but you should be aware that the restrictions are necessary to maintain equality.

We summarize the steps that you can use to simplify a rational expression.

Steps to Simplify a Rational Expression

Step 1: Completely factor the numerator and denominator of the rational expression.

Step 2: Divide out common factors.

EXAMPLE 7 **Simplifying a Rational Expression**

Simplify:

(a) $\dfrac{2x - 10}{4x^2 - 20x}$

(b) $\dfrac{x^2 - 9}{2x^2 - 3x - 9}$

Solution

Work Smart

Notice in Example 7(a) that when all the factors in the numerator divide out, we are left with a factor of 1, not 0!

(a)   Factor the numerator and denominator: $\dfrac{2x - 10}{4x^2 - 20x} = \dfrac{2(x - 5)}{4x(x - 5)}$

Divide out common factors: $= \dfrac{2\,\cancel{(x - 5)}}{2 \cdot 2x\cancel{(x - 5)}}$

$= \dfrac{1}{2x}$

(b)   Factor the numerator and denominator: $\dfrac{x^2 - 9}{2x^2 - 3x - 9} = \dfrac{(x + 3)(x - 3)}{(2x + 3)(x - 3)}$

Divide out common factors: $= \dfrac{(x + 3)\cancel{(x - 3)}}{(2x + 3)\cancel{(x - 3)}}$

$= \dfrac{x + 3}{2x + 3}$

Work Smart

A common error students make is to divide out terms rather than to divide out factors. **When simplifying, we can only divide out common factors, not common terms.**

WRONG! $\dfrac{x + 3}{x} = \dfrac{\cancel{x} + 3}{\cancel{x}} = 3$ **WRONG!** $\dfrac{x^2 - 4x + 3}{x + 3} = \dfrac{x^2 - 4x + \cancel{3}}{\cancel{x} + \cancel{3}} = x^2 - 4$

If you aren't sure whether you can divide out, try the computation with numbers and see if it works.

For example, does $\dfrac{2 + 3}{2} = \dfrac{\cancel{2} + 3}{\cancel{2}} = 3$? No!

QUICK ✔ *Simplify the rational expression.*

9. $\dfrac{3n + 12}{6n + 24}$ **10.** $\dfrac{10p + 5}{4p^2 - 1}$ **11.** $\dfrac{2z^2 + 6z + 4}{-4z - 8}$ **12.** $\dfrac{a^2 + 3a - 28}{2a^2 - a - 28}$

EXAMPLE 8   Simplifying a Rational Expression

Simplify: $\dfrac{ab + 3b - ac - 3c}{a^2 + 6a + 9}$

Solution

Work Smart

Remember: you can simplify only after the numerator and denominator have been completely factored!

Factor the numerator and the denominator:
$$\dfrac{ab + 3b - ac - 3c}{a^2 + 6a + 9} = \dfrac{b(a + 3) - c(a + 3)}{(a + 3)^2}$$

$$= \dfrac{(a + 3)(b - c)}{(a + 3)^2}$$

Divide out common factors:
$$= \dfrac{\cancel{(a + 3)}(b - c)}{\cancel{(a + 3)}(a + 3)}$$

$$= \dfrac{b - c}{a + 3}$$

QUICK ✔ *Simplify the rational expression.*

13. $\dfrac{xz + x - yz - y}{z^2 - z - 2}$ **14.** $\dfrac{4k^2 + 4k + 1}{4k^2 - 1}$

In a rational expression, when a factor in the numerator and a factor in the denominator are identical, they form a quotient of 1. When a factor in the numerator is the *opposite* of the factor in the denominator, they form a quotient of -1.

Work Smart

$\dfrac{x - 7}{x + 7} \ne -1$

and, because addition is commutative,

$\dfrac{x + 7}{7 + x} = 1$

$\dfrac{3}{-3} = -1$ 3 and -3 are opposites, so $\dfrac{3}{-3} = -1$

$\dfrac{-42}{42} = -1$ -42 and 42 are opposites, so $\dfrac{-42}{42} = -1$

$\dfrac{5 - x}{x - 5} = -1$ $\dfrac{5 - x}{x - 5} = \dfrac{-1(-5 + x)}{x - 5} = \dfrac{-1(x - 5)}{x - 5} = -1$

$\dfrac{4x - 7}{7 - 4x} = -1$ $\dfrac{4x - 7}{7 - 4x} = \dfrac{4x - 7}{-1(-7 + 4x)} = \dfrac{4x - 7}{-1(4x - 7)} = -1$

EXAMPLE 9   Simplifying a Rational Expression Containing Opposite Factors

Simplify: $\dfrac{4 - x^2}{2x^2 - x - 6}$

Solution

Work Smart

In Example 9, we could also write

$\dfrac{4 - x^2}{2x^2 - x - 6} = \dfrac{-(2 + x)}{2x + 3}$

$= -\dfrac{2 + x}{2x + 3}$

$= -\dfrac{x + 2}{2x + 3}$

Factor the numerator and denominator:
$$\dfrac{4 - x^2}{2x^2 - x - 6} = \dfrac{(2 + x)(2 - x)}{(2x + 3)(x - 2)}$$

Factor -1 from $2 - x$:
$$= \dfrac{(2 + x)(-1)(-2 + x)}{(2x + 3)(x - 2)}$$

Divide out common factors:
$$= \dfrac{(2 + x)(-1)\cancel{(x - 2)}}{(2x + 3)\cancel{(x - 2)}}$$

$$= \dfrac{-(2 + x)}{2x + 3}$$

QUICK ✓ *Simplify the rational expression.*

15. $\dfrac{7a - 7b}{b - a}$

16. $\dfrac{12 - 4x}{4x^2 - 13x + 3}$

17. $\dfrac{25z^2 - 1}{3 - 15z}$

7.1 Exercises

For Extra Help:

Student Solutions Manual   CD Video   PH Math/Tutor Center   MathXL Tutorials on CD   MathXL®   MyMathLab

Concepts and Vocabulary

In Problems 1–3, fill in the blanks.

1. The quotient of two polynomials is called a _____ _____.

2. When the denominator of any rational expression has a value of zero, the expression is said to be _____.

3. To _____ a rational expression means to write the rational expression in the form $\dfrac{p}{q}$ where p and q are polynomials that have no common factors.

In Problems 4–6, answer True or False to each statement.

4. $\dfrac{a + b}{a - b} = -1$

5. $\dfrac{x + y}{y + x} = 1$

6. $\dfrac{2n + 3}{2n + 6} = \dfrac{3}{6} = \dfrac{1}{2}, n \neq -3$

7. Is $x^2 - 3x + 1$ a rational expression? Explain your response.

8. Rational expressions that represent the same quantity are said to be *equivalent*. Are the rational expressions $\dfrac{x - y}{y - x}$ and $\dfrac{y - x}{x - y}$ equivalent? Explain how to determine whether two rational expressions are equivalent.

Building Skills

In Problems 9–18, evaluate each expression for the given values.

9. $\dfrac{x}{x - 5}$
   (a) $x = 10$
   (b) $x = -5$
   (c) $x = 5$

10. $\dfrac{x}{x + 4}$
   (a) $x = 8$
   (b) $x = -4$
   (c) $x = -6$

11. $\dfrac{2a - 3}{a}$
   (a) $a = 0$
   (b) $a = -3$
   (c) $a = 9$

12. $\dfrac{2m - 1}{m}$
   (a) $m = 0$
   (b) $m = 1$
   (c) $m = -1$

13. $\dfrac{x + 2}{x - 2}$
   (a) $x = 4$
   (b) $x = 2$
   (c) $x = -2$

14. $\dfrac{a - 3}{a + 3}$
   (a) $a = 5$
   (b) $a = -3$
   (c) $a = 3$

15. $\dfrac{x^2 - 2x}{x - 4}$
   (a) $x = 3$
   (b) $x = 2$
   (c) $x = -3$

16. $\dfrac{a^2 - 2a}{a - 4}$
   (a) $a = 5$
   (b) $a = -1$
   (c) $a = -4$

17. $\dfrac{x^2 - y^2}{2x - y}$
   (a) $x = 2, y = 2$
   (b) $x = 2, y = 4$
   (c) $x = 1, y = 2$

18. $\dfrac{b^2 - a^2}{(a - b)^2}$
   (a) $a = 3, b = 2$
   (b) $a = 5, b = 4$
   (c) $a = -2, b = -2$

In Problems 19–30, find the value(s) of the variable for which the rational expression is undefined.

19. $\dfrac{2 - 4x}{3x}$

20. $\dfrac{-x + 1}{5x}$

21. $\dfrac{3p}{p - 5}$

22. $\dfrac{5m^3}{m + 8}$

23. $\dfrac{8}{3 - 2x}$

24. $\dfrac{12}{4a - 3}$

25. $\dfrac{6z}{z^2 - 36}$

26. $\dfrac{5x}{25 - x^2}$

27. $\dfrac{x}{x^2 - 7x + 10}$

28. $\dfrac{2x^2}{x^2 + x - 2}$

29. $\dfrac{12x + 5}{x^3 - x^2 - 6x}$

30. $\dfrac{3h + 2}{h^3 + 5h^2 + 4h}$

In Problems 31–42, simplify each rational expression. Assume that no variable has a value which results in a denominator with a value of zero.

31. $\dfrac{15}{5x - 10}$

32. $\dfrac{3}{3n + 9}$

33. $\dfrac{z - z^3}{3z}$

34. $\dfrac{6p^2 + 3p}{6p}$

35. $\dfrac{p - 3}{p^2 - p - 6}$

36. $\dfrac{x - 3}{x^2 - 4x + 3}$

37. $\dfrac{2 - x}{x - 2}$

38. $\dfrac{4 - z}{z - 4}$

39. $\dfrac{2k^2 - 14k}{7 - k}$

40. $\dfrac{2v^2 - 6v}{3 - v}$

41. $\dfrac{x^2 - 1}{x^2 + 5x + 4}$

42. $\dfrac{x^2 - 9}{x^2 + 5x + 6}$

Mixed Practice

In Problems 43–74, simplify each rational expression. Assume that no variable has a value which results in a denominator with a value of zero.

43. $\dfrac{20}{36}$

44. $\dfrac{49}{63}$

45. $\dfrac{150}{165}$

46. $\dfrac{456}{420}$

47. $\dfrac{3x^2}{6x^5}$

48. $\dfrac{3m}{9m^4}$

49. $\dfrac{24a^5b}{2a^7b}$

50. $\dfrac{45a^2b^3}{15ab}$

51. $\dfrac{-3x - 3y}{x + y}$

52. $\dfrac{-2a - 2b}{a + b}$

53. $\dfrac{b^2 - 25}{4b + 20}$

54. $\dfrac{3p - 12}{p^2 - 16}$

55. $\dfrac{x^2 - 2x - 15}{x^2 - 8x + 15}$

56. $\dfrac{x^2 + x - 2}{x^2 + 5x + 6}$

57. $\dfrac{x^3 + x^2 - 12x}{x^3 - x^2 - 20x}$

58. $\dfrac{x^3 + 3x^2 + 2x}{x^3 - 2x^2 - 3x}$

59. $\dfrac{x^2 - y^2}{y - x}$

60. $\dfrac{a - b}{b^2 - a^2}$

61. $\dfrac{4a - 4b}{a^2 - b^2}$

62. $\dfrac{5m + 5n}{m^2 + 2mn + n^2}$

63. $\dfrac{x^2}{x^2 - 4x}$

64. $\dfrac{9m}{9m^2 - 9m}$

65. $\dfrac{16 - c^2}{(c - 4)^2}$

66. $\dfrac{9 - x^2}{(x - 3)^2}$

67. $\dfrac{4x^2 - 20x + 24}{6x^2 - 48x + 90}$

68. $\dfrac{2x^2 + 5x - 3}{4x^2 - 8x + 3}$

69. $\dfrac{6 + x - x^2}{x^2 - 4}$

70. $\dfrac{5 + 4x - x^2}{x^2 - 25}$

71. $\dfrac{2t^2 - 18}{t^4 - 81}$

72. $\dfrac{x^3 + 4x}{x^4 - 16}$

73. $\dfrac{12w - 3w^2}{w^3 - 5w^2 + 4w}$

74. $\dfrac{7a - a^2}{a^3 - 5a^2 - 14a}$

Applying the Concepts

75. Drug Concentration The concentration, C, in mg/mL, of a drug in a patient's bloodstream t minutes after an injection is given by the formula $C = \dfrac{50t}{t^2 + 25}$.

  (a) Find the concentration in a patient 5 minutes after receiving an injection.
  (b) Find the concentration in a patient 10 minutes after receiving an injection.

76. Drug Concentration The concentration, D, in mg/mL, of a drug in a patient's bloodstream t minutes after an injection is given by the formula $D = \dfrac{t}{2t^2 + 1}$.

  (a) Find the concentration in a patient 30 minutes after receiving an injection.
  (b) Find the concentration in a patient 60 minutes after receiving an injection.

77. **Body Mass Index** BMI, or Body Mass Index, is used to determine if a person's weight is in a healthy range. The formula used to determine Body Mass Index is given by BMI $= \dfrac{k}{m^2}$, where k is the person's weight in kilograms and m is the height in meters. If a BMI range of 19 to 24.9 is considered healthy and 25 to 29.9 is considered overweight, is a person 2 meters tall weighing 110 kilograms considered overweight?

78. **Body Mass Index** A formula to calculate the BMI when the weight w, is in pounds, and the height h, is in inches, is given by BMI $= \dfrac{705w}{h^2}$. What is the BMI of a person weighing 120 pounds who is 5 feet tall?

79. **Cost of a Car** The average cost, in thousands of dollars, to produce Chevy Cobalts is given by the rational expression $\dfrac{0.2x^3 - 2.3x^2 + 14.3x + 10.2}{x}$, where x is the number of cars produced. If two Cobalts are produced, what is the average cost per car?

80. **Cost of a Car** Use the rational expression from Problem 79 to find the average cost of producing 10 Cobalts. Do you believe that the cost will continue to decrease as more cars are made?

Extending the Concepts

In Problems 81–86, simplify each rational expression.

81. $\dfrac{c^{12} - 1}{(c^4 - 1)(c^6 + 1)}$

82. $\dfrac{(1 - x)(x^6 + 1)}{x^4 - 1}$

83. $\dfrac{x^4 + x^2 - 12}{x^4 + 2x^3 - 9x - 18}$

84. $\dfrac{n^3 + 3n^2 - 8n - 24}{n^3 - 4n^2 + 3n - 12}$

85. $\dfrac{(t + 2)^3(t^4 - 16)}{(t^3 + 8)(t + 2)(t^2 - 4)}$

86. $\dfrac{24x^4 - 16x^3 + 27x - 18}{18 - 27x + 16x^2 - 24x^3}$

7.2 Multiplying and Dividing Rational Expressions

OBJECTIVES

1. Multiply Rational Expressions
2. Divide Rational Expressions

Preparing for Multiplying and Dividing Rational Expressions

Before getting started, take the following readiness quiz. If you get a problem wrong, go back to the section cited and review the material.

1. Find the product: $\dfrac{3}{14} \cdot \dfrac{28}{9}$ [Section 1.4, p. 29]

2. Find the reciprocal of $\dfrac{5}{8}$.   [Section 1.4, p. 30]

3. Find the quotient: $\dfrac{12}{25} \div \dfrac{12}{5}$ [Section 1.4, p. 30]

4. Factor: $3x^3 - 27x$ [Section 6.4, pp. 439–441]

5. Simplify the rational expression: $\dfrac{3x + 12}{5x^2 + 20x}$ [Section 7.1, pp. 480–483]

1   Multiply Rational Expressions

The steps for multiplying rational expressions follow the same logic as the steps for multiplying rational numbers. For example,

Preparing for...Answers 1. $\dfrac{2}{3}$ 2. $\dfrac{8}{5}$
3. $\dfrac{1}{5}$ 4. $3x(x + 3)(x - 3)$ 5. $\dfrac{3}{5x}$

$$\frac{2}{3} \cdot \frac{12}{5} = \frac{2 \cdot 12}{3 \cdot 5} = \frac{2 \cdot 3 \cdot 4}{3 \cdot 5} = \frac{2 \cdot 4}{5} = \frac{8}{5}$$

| EXAMPLE 1 | How to Multiply Rational Expressions |

Multiply $\dfrac{x-3}{5} \cdot \dfrac{5x+35}{x^2-9}$. Simplify the result, if possible.

Step-by-Step Solution

Step 1: Completely factor the polynomials in each numerator and denominator.	$\dfrac{x-3}{5} \cdot \dfrac{5x+35}{x^2-9} = \dfrac{x-3}{5} \cdot \dfrac{5(x+7)}{(x+3)(x-3)}$
Step 2: Multiply.	$= \dfrac{5(x-3)(x+7)}{5(x+3)(x-3)}$
Step 3: Divide out common factors in the numerator and denominator. The common factors are $x-3$ and 5.	$= \dfrac{\cancel{5}\,(\cancel{x-3})(x+7)}{\cancel{5}(x+3)(\cancel{x-3})}$ $= \dfrac{x+7}{x+3}$

Did you remember that $\dfrac{x+7}{x+3}$ cannot be simplified further? We divide *factors*, not terms, so $\dfrac{x+7}{x+3} \neq \dfrac{\cancel{x}+7}{\cancel{x}+3} \neq \dfrac{7}{3}$.

Steps to Multiply Rational Expressions

Step 1: Factor the polynomials in each numerator and denominator.

Step 2: Use the fact that if $\dfrac{a}{b}$ and $\dfrac{c}{d}$, $b \neq 0$, $d \neq 0$, are two rational expressions, then $\dfrac{a}{b} \cdot \dfrac{c}{d} = \dfrac{ac}{bd}$ to multiply the rational expressions.

Step 3: Divide out common factors in the numerator and denominator. Leave your answer in factored form.

In Words
To multiply two rational expressions, factor each polynomial, write the expression as a single fraction in factored form and then divide out common factors.

| EXAMPLE 2 | Multiplying Rational Expressions |

Multiply: $\dfrac{7y+28}{4y} \cdot \dfrac{8}{y^2+2y-8}$. Simplify the result, if possible.

Solution

Factor each polynomial: $\quad \dfrac{7y+28}{4y} \cdot \dfrac{8}{y^2+2y-8} = \dfrac{7(y+4)}{4y} \cdot \dfrac{4 \cdot 2}{(y+4)(y-2)}$

Multiply: $\quad = \dfrac{7(y+4) \cdot 4 \cdot 2}{4y(y+2)(y-2)}$

Divide out common factors, 4 and $y+4$: $\quad = \dfrac{7(\cancel{y+4}) \cdot \cancel{4} \cdot 2}{\cancel{4}y(\cancel{y+4})(y-2)}$

$= \dfrac{7 \cdot 2}{y(y-2)}$

Multiply: $\quad = \dfrac{14}{y(y-2)}$

QUICK ✓ *Multiply the rational expressions. Simplify the result, if possible.*

1. $\dfrac{p^2-9}{5} \cdot \dfrac{25p}{2p-6}$

2. $\dfrac{2x+8}{x^2+3x-4} \cdot \dfrac{7x-7}{6x+30}$

EXAMPLE 3 **Multiplying Rational Expressions**

Find the product of $\dfrac{p^2 - 9}{p^2 + 5p + 6}$ and $\dfrac{p + 2}{6 - 2p}$. Simplify, if possible.

Solution

Factor each polynomial: $\dfrac{p^2 - 9}{p^2 + 5p + 6} \cdot \dfrac{p + 2}{6 - 2p} = \dfrac{(p + 3)(p - 3)}{(p + 3)(p + 2)} \cdot \dfrac{p + 2}{2 \cdot (-1)(p - 3)}$

Work Smart
$6 - 2p = 2(3 - p) = -2(p - 3)$

Multiply: $= \dfrac{(p + 3)(p - 3)(p + 2)}{(p + 3)(p + 2) \cdot 2 \cdot (-1)(p - 3)}$

Work Smart
When all the factors in the numerator divide out. We are left with a factor of 1, *not* 0.

Divide out common factors: $p + 3$, $p - 3$, and $p + 2$: $= \dfrac{\cancel{(p + 3)}\,\cancel{(p - 3)}\,\cancel{(p + 2)}}{\cancel{(p + 3)}\,\cancel{(p + 2)} \cdot 2 \cdot (-1)\cancel{(p - 3)}}$

Multiply: $= -\dfrac{1}{2}$ ■

QUICK ✔ *Multiply the rational expressions. Simplify the result, if possible.*

3. $\dfrac{x^2 - 9}{x^2 - 25} \cdot \dfrac{2x - 10}{x - 3}$

4. $\dfrac{15a - 3a^2}{7} \cdot \dfrac{3 + 2a}{2a^2 - 7a - 15}$

EXAMPLE 4 **Multiplying Rational Expressions with Two Variables**

Find the product and simplify: $\dfrac{m^2 - n^2}{10m^2 - 10mn} \cdot \dfrac{10m + 5n}{2m^2 + 3mn + n^2}$

Solution

Factor each polynomial: $\dfrac{m^2 - n^2}{10m^2 - 10mn} \cdot \dfrac{10m + 5n}{2m^2 + 3mn + n^2} = \dfrac{(m + n)(m - n)}{10m(m - n)} \cdot \dfrac{5(2m + n)}{(2m + n)(m + n)}$

Multiply: $= \dfrac{(m + n)(m - n) \cdot 5(2m + n)}{5 \cdot 2m(m - n)(2m + n)(m + n)}$

Divide out common factors: $m + n$, $m - n$, $2m + n$, and 5: $= \dfrac{\cancel{(m + n)}\,\cancel{(m - n)} \cdot \cancel{5}\,\cancel{(2m + n)}}{\cancel{5} \cdot 2m\cancel{(m - n)}\,\cancel{(2m + n)}\,\cancel{(m + n)}}$

Multiply: $= \dfrac{1}{2m}$ ■

QUICK ✔ *Multiply the rational expressions. Simplify the result, if possible.*

5. $\dfrac{a^2 + 2ab + b^2}{3a + 3b} \cdot \dfrac{a - b}{a^2 - b^2}$

---

② **Divide Rational Expressions**

Let's review the language we use in division. In the expression $\dfrac{20x^5}{3y} \div \dfrac{4x^2}{15y^5}$, $\dfrac{20x^5}{3y}$ is

called the *dividend* and $\dfrac{4x^2}{15y^5}$ is called the *divisor*. The answer is called the *quotient*. We

could also write $\dfrac{20x^5}{3y} \div \dfrac{4x^2}{15y^5}$ as $\dfrac{\dfrac{20x^5}{3y}}{\dfrac{4x^2}{15y^5}}$.

The steps for finding the quotient of rational expressions are the same as the steps for finding the quotient of rational numbers. For example, to evaluate $\frac{2}{3} \div \frac{7}{9}$, we take the reciprocal of the divisor, $\frac{7}{9}$, and multiply.

$$\frac{2}{3} \div \frac{7}{9} = \frac{2}{3} \cdot \frac{9}{7} = \frac{2 \cdot \cancel{3} \cdot 3}{\cancel{3} \cdot 7} = \frac{2 \cdot 3}{1 \cdot 7} = \frac{6}{7}$$

reciprocal of divisor

dividend

EXAMPLE 5 **How to Divide Rational Expressions**

Find the quotient and simplify: $\dfrac{x + 3}{2x - 8} \div \dfrac{9}{4x}$

Step-by-Step Solution

Step 1: Multiply the dividend by the reciprocal of the divisor.	The reciprocal of $\frac{9}{4x}$ is $\frac{4x}{9}$: $\dfrac{x + 3}{2x - 8} \div \dfrac{9}{4x} = \dfrac{x + 3}{2x - 8} \cdot \dfrac{4x}{9}$
Step 2: Completely factor the polynomials in each numerator and denominator.	$= \dfrac{x + 3}{2(x - 4)} \cdot \dfrac{2 \cdot 2 \cdot x}{9}$
Step 3: Multiply.	$= \dfrac{(x + 3) \cdot 2 \cdot 2 \cdot x}{2(x - 4) \cdot 9}$
Step 4: Divide out common factors in the numerator and denominator. The common factor is 2.	Simplify: $= \dfrac{(x + 3) \cdot \cancel{2} \cdot 2 \cdot x}{\cancel{2}(x - 4) \cdot 9}$ $= \dfrac{2x(x + 3)}{9(x - 4)}$

So $\dfrac{x + 3}{2x - 8} \div \dfrac{9}{4x} = \dfrac{2x(x + 3)}{9(x - 4)}$.

Next, we summarize the procedure for dividing rational expressions.

Steps to Divide Rational Expressions

Step 1: Multiply the dividend by the reciprocal of the divisor.
Step 2: Factor each polynomial in the numerator and denominator.
Step 3: Multiply.
Step 4: Divide out common factors in the numerator and denominator. Leave the remaining factors in factored form.

EXAMPLE 6 **Dividing Rational Expressions**

Find the quotient and simplify: $\dfrac{y^2 - 9}{2y^2 - y - 15} \div \dfrac{3y^2 + 10y + 3}{2y^2 + y - 10}$

Solution

Multiply the dividend by the reciprocal of the divisor:
$$\frac{y^2 - 9}{2y^2 - y - 15} \div \frac{3y^2 + 10y + 3}{2y^2 + y - 10} = \frac{y^2 - 9}{2y^2 - y - 15} \cdot \frac{2y^2 + y - 10}{3y^2 + 10y + 3}$$

Factor each polynomial in the numerator and denominator:
$$= \frac{(y - 3)(y + 3)}{(2y + 5)(y - 3)} \cdot \frac{(2y + 5)(y - 2)}{(3y + 1)(y + 3)}$$

Multiply:
$$= \frac{(y - 3)(y + 3)(2y + 5)(y - 2)}{(2y + 5)(y - 3)(3y + 1)(y + 3)}$$

Divide out common factors: $y - 3$, $y + 3$, and $2y + 5$:
$$= \frac{\cancel{(y - 3)}\,\cancel{(y + 3)}\,\cancel{(2y + 5)}(y - 2)}{\cancel{(2y + 5)}\,\cancel{(y - 3)}(3y + 1)\cancel{(y + 3)}}$$

$$= \frac{y - 2}{3y + 1}$$

QUICK ✓ *Find the quotient and simplify, if possible.*

6. $\dfrac{12}{x^2 - x} \div \dfrac{4x - 2}{x^2 - 1}$

7. $\dfrac{x^2 - 9}{x^2 - 16} \div \dfrac{x^2 - x - 12}{x^2 + x - 12}$

EXAMPLE 7   **Dividing Rational Expressions**

Find the quotient and simplify: $\dfrac{a^2 + 3a + 2}{a^2 - 9} \div (a + 2)$

Solution

We write the divisor $a + 2$ as $\dfrac{a + 2}{1}$ and proceed as we did in Example 6.

$$\frac{a^2 + 3a + 2}{a^2 - 9} \div (a + 2) = \frac{a^2 + 3a + 2}{a^2 - 9} \div \frac{a + 2}{1}$$

$$= \frac{a^2 + 3a + 2}{a^2 - 9} \cdot \frac{1}{a + 2}$$

Factor:
$$= \frac{(a + 2)(a + 1)}{(a + 3)(a - 3)} \cdot \frac{1}{a + 2}$$

Multiply:
$$= \frac{(a + 2)(a + 1)}{(a + 3)(a - 3)(a + 2)}$$

Divide out common factors:
$$= \frac{\cancel{(a + 2)}(a + 1)}{(a + 3)(a - 3)\cancel{(a + 2)}}$$

$$= \frac{a + 1}{(a + 3)(a - 3)}$$

QUICK ✓ *Find the quotient and simplify.*

8. $\dfrac{q^2 - 6q - 7}{q^2 - 25} \div (q - 7)$

As we stated earlier, we could write an expression such as $\dfrac{20x^5}{3y} \div \dfrac{4x^2}{15y^5}$ vertically as
$\dfrac{\dfrac{20x^5}{3y}}{\dfrac{4x^2}{15y^5}}$. We use this form of division in the next example.

EXAMPLE 8 **Dividing Rational Expressions Written Vertically**

Find the quotient and simplify: $\dfrac{\dfrac{12x}{5x + 20}}{\dfrac{4x^2}{x^2 - 16}}$

Solution

Multiply by the reciprocal of the divisor: $\dfrac{\dfrac{12x}{5x + 20}}{\dfrac{4x^2}{x^2 - 16}} = \dfrac{12x}{5x + 20} \cdot \dfrac{x^2 - 16}{4x^2}$

Factor each polynomial in the numerator and denominator: $= \dfrac{4 \cdot 3 \cdot x}{5(x + 4)} \cdot \dfrac{(x - 4)(x + 4)}{4 \cdot x \cdot x}$

Multiply: $= \dfrac{4 \cdot 3 \cdot x(x - 4)(x + 4)}{5(x + 4) \cdot 4 \cdot x \cdot x}$

Divide out common factors 4, x, and $x + 4$: $= \dfrac{\cancel{4} \cdot 3 \cdot \cancel{x}(x - 4)\cancel{(x + 4)}}{5\cancel{(x + 4)} \cdot \cancel{4} \cdot \cancel{x} \cdot x}$

$= \dfrac{3(x - 4)}{5x}$

QUICK ✓ *Find the quotient and simplify.*

9. $\dfrac{\dfrac{x + 3}{x^2 - 4}}{\dfrac{4x + 12}{7x^2 + 14x}}$

EXAMPLE 9 **Dividing Rational Expressions with More Than One Variable**

Find the quotient and simplify: $\dfrac{5a - 5b}{7c^2} \div \dfrac{10b - 10a}{21c}$

Solution

Multiply by the reciprocal of the divisor: $\dfrac{5a - 5b}{7c^2} \div \dfrac{10b - 10a}{21c} = \dfrac{5a - 5b}{7c^2} \cdot \dfrac{21c}{10b - 10a}$

Factor each polynomial in the numerator and denominator: $= \dfrac{5(a - b)}{7 \cdot c \cdot c} \cdot \dfrac{7 \cdot 3 \cdot c}{5 \cdot 2 \cdot (-1)(a - b)}$

Multiply: $= \dfrac{5(a - b) \cdot 7 \cdot 3 \cdot c}{7 \cdot c \cdot c \cdot 5 \cdot 2 \cdot (-1)(a - b)}$

Divide out common factors, 5, 7, c, and $(a - b)$: $= \dfrac{\cancel{5}\cancel{(a - b)} \cdot \cancel{7} \cdot 3 \cdot \cancel{c}}{\cancel{7} \cdot c \cdot \cancel{c} \cdot \cancel{5} \cdot 2 \cdot (-1)\cancel{(a - b)}}$

$= \dfrac{3}{c \cdot 2 \cdot (-1)}$

Multiply: $= -\dfrac{3}{2c}$

So $\dfrac{5a - 5b}{7c^2} \div \dfrac{10b - 10a}{21c} = -\dfrac{3}{2c}$.

QUICK ✓ *Find the quotient and simplify.*

10. $\dfrac{3m - 6n}{5n} \div \dfrac{m^2 - 4n^2}{10mn}$

7.2 Exercises

Concepts and Vocabulary

In Problems 1–3, fill in the blanks.

1. The answer to a division problem is called the _____ and the answer to a multiplication problem is called the _____.

2. To divide two rational expressions, rewrite the problem as an equivalent multiplication problem by finding the _____ of the divisor.

3. To multiply two rational expressions, first _____ each numerator and denominator and then divide out any _____ _____.

In Problems 4–6, answer True or False to each statement.

4. $\dfrac{3}{x} \cdot \dfrac{x}{9} = 3$

5. $\dfrac{8}{x-7} \cdot \dfrac{7-x}{16x} = \dfrac{1}{2x}$

6. $\dfrac{3}{x^2} \div \dfrac{x}{9} = \dfrac{x^2}{3} \cdot \dfrac{x}{9}$

7. To multiply two fractions, we can multiply and then simplify or multiply, factor, and then divide out common factors. Use $\dfrac{4}{15} \cdot \dfrac{25}{28}$ to demonstrate these two techniques. When multiplying rational expressions, why would it not be a good idea to first multiply the numerators and denominators and then simplify the resulting rational expression?

8. A unit fraction is any representation of one. For instance, both $\dfrac{x}{x}$ and $\dfrac{1 \text{ foot}}{12 \text{ inches}}$ are unit fractions. Since multiplying by one does not change the value of any real number, use this concept to explain how to convert 56 inches into feet.

Building Skills

In Problems 9–16, multiply and simplify the result, if possible.

9. $\dfrac{4x^2}{9x^5} \cdot \dfrac{12}{21x^3}$

10. $-\dfrac{8y}{12y^5} \cdot \left(-\dfrac{9y^6}{4y^2}\right)$

11. $-\dfrac{15a^2b^2}{18ab} \cdot \dfrac{12a^3}{16b}$

12. $\dfrac{5x^2y}{26y^3} \cdot \dfrac{13xy^2}{20x^5}$

13. $\dfrac{x^2-x}{x^2-x-2} \cdot \dfrac{x-2}{x^2-1}$

14. $\dfrac{8n-8}{n^2-3n+2} \cdot \dfrac{n+2}{12}$

15. $\dfrac{p^2-1}{2p-3} \cdot \dfrac{2p^2+p-6}{p^2+3p+2}$

16. $\dfrac{z^2-4}{3z-2} \cdot \dfrac{3z^2+7z-6}{z^2+z-6}$

In Problems 17–24, divide and simplify the result, if possible.

17. $\dfrac{12z^2}{35} \div \dfrac{20z^6}{42z^3}$

18. $\dfrac{18a^2b}{5b^3} \div \dfrac{14a^3b^3}{15ab}$

19. $\dfrac{-\dfrac{3m^5}{8m^2}}{\dfrac{15m}{12m^3}}$

20. $\dfrac{-\dfrac{27xy^2}{60x^2y^4}}{-\dfrac{36x}{24y^5}}$

21. $\dfrac{x^2-x}{x^2-1} \div \dfrac{x+2}{x^2+3x+2}$

22. $\dfrac{3y^2+6y}{8} \div \dfrac{y+2}{12y-12}$

23. $\dfrac{(x+2)^2}{x^2-4} \div \dfrac{x^2-x-6}{x^2-5x+6}$

24. $\dfrac{4z^2+12z+9}{8z+16} \div \dfrac{(2z+3)^3}{12z+24}$

Mixed Practice

In Problems 25–60, perform the indicated operation and simplify.

25. $\dfrac{14}{9} \cdot \dfrac{15}{7}$

26. $\dfrac{7}{52} \cdot \dfrac{77}{13}$

27. $-\dfrac{20}{16} \div \left(-\dfrac{30}{24}\right)$

28. $-\dfrac{60}{35} \div \dfrac{15}{77}$

29. $\dfrac{8}{11} \div (-2)$

30. $\dfrac{16}{3} \div (-4)$

31. $\dfrac{3y}{y^2 - y - 6} \cdot \dfrac{4y + 8}{9y^2}$

32. $\dfrac{x - 4}{12x - 18} \cdot \dfrac{6}{x^2 - 16}$

33. $\dfrac{4a + 8b}{a^2 + 2ab} \cdot \dfrac{a^2}{12}$

34. $\dfrac{m^2 - n^2}{3m - 3n} \cdot \dfrac{6}{2m + 2n}$

35. $\dfrac{3x^2 - 6x}{x^2 - 2x - 8} \div \dfrac{x - 2}{x + 2}$

36. $\dfrac{x + 5}{3} \div \dfrac{30x}{4x + 20}$

37. $\dfrac{(w - 4)^2}{4 - w^2} \div \dfrac{w^2 - 16}{w - 2}$

38. $\dfrac{a^2 - b^2}{b - a} \cdot \dfrac{2a + 2b}{a^2 + 2ab + b^2}$

39. $\dfrac{3xy - 2y^2 - x^2}{x + y} \cdot \dfrac{x^2 - y^2}{x^2 - 2xy}$

40. $\dfrac{2r^2 + rs - 3s^2}{r^2 - s^2} \cdot \dfrac{r^2 - 2rs - 3s^2}{2r + 3s}$

41. $\dfrac{\dfrac{2c - 4}{8}}{\dfrac{2 - c}{2}}$

42. $\dfrac{\dfrac{2x - 3}{3}}{6x - 9}$

43. $(x + 1) \cdot \dfrac{x - 6}{x^2 - 5x - 6}$

44. $(x - 3) \cdot \dfrac{x + 2}{x^2 - 5x + 6}$

45. $\dfrac{4n^2 - 9}{6n + 18} \cdot \dfrac{9n^2 - 81}{2n^2 + 5n - 12}$

46. $\dfrac{2x^2 - 5x + 3}{x^2 - 1} \cdot \dfrac{x^2 + 1}{2x^2 - x - 3}$

47. $\dfrac{x^2 y}{2x^2 - 5xy + 2y^2} \div \dfrac{(2xy^2)^2}{2x^2 y - xy^2}$

48. $\dfrac{(x - 3)^2}{8xy^2} \div \dfrac{x^2 - 9}{(4x^2 y)^2}$

49. $\dfrac{2a^2 + 3ab - 2b^2}{a^2 - b^2} \cdot \dfrac{a^2 - ab}{2a^3 + 4a^2 b}$

50. $\dfrac{3y^2 - 3x^2}{2x^2 + xy - y^2} \cdot \dfrac{3x - 6y}{6x - 6y}$

51. $\dfrac{3t^2 - 27}{t + 2} \cdot \dfrac{t^2 - 4}{9t - 27}$

52. $\dfrac{p^2 - 49}{p^2 - 5p - 14} \cdot \dfrac{p - 7}{14 - 5p - p^2}$

53. $\dfrac{(x + 2)^2}{x^2 - 4} \div \dfrac{-x^2 + x + 6}{x^2 - 5x + 6}$

54. $\dfrac{4z^2 + 12z + 9}{8z + 16} \div \dfrac{(2z + 3)^3}{12z + 24}$

55. $\dfrac{9 - x^2}{x^2 + 5x + 4} \div \dfrac{x^2 - 2x - 3}{x^2 + 4x}$

56. $\dfrac{1}{b^2 + b - 12} \div \dfrac{1}{b^2 - 5b - 36}$

57. $\dfrac{\dfrac{a^2 - b^2}{a^2 + b^2}}{\dfrac{4a - 4b}{2a^2 + 2b^2}}$

58. $\dfrac{\dfrac{t^2}{t^2 - 16}}{\dfrac{t^2 - 3t}{t^2 - t - 12}}$

59. $\dfrac{x^3 - 1}{x^4 - 1} \div \dfrac{3x^2 + 3x + 3}{x^3 + x^2 + x + 1}$

60. $\dfrac{p^3 - 8q^3}{p^2 - 4q^2} \div \dfrac{p^2 + 4pq + 4q^2}{(p + 2q)^2}$

Applying the Concepts

61. Find the product of $\dfrac{x^2 + 3xy + 2y^2}{x^2 - y^2}$ and $\dfrac{3x - 3y}{9x^2 + 9xy - 18y^2}$.

62. Find the product of $\dfrac{x}{6y^2}$ and $\dfrac{21x^2 y}{(7x)^2}$.

63. Find the quotient of $\dfrac{x}{2y}$ and $\dfrac{(2xy)^2}{9xy^3}$.

64. Find the quotient of $\dfrac{x - 3}{2x + 6}$ and $\dfrac{x - 9}{4x^2 - 36}$.

65. Find $\dfrac{x - y}{x + y}$ squared divided by $x^2 - y^2$.

66. Find $(a - b)$ squared divided by $a^2 - b^2$.

67. What is $\dfrac{3x - 9}{2x + 4}$ divided into $\dfrac{6x}{x^2 - 4}$?

68. What is $\dfrac{2x^2}{3}$ divided into $\dfrac{x^3 - 3x}{6x}$?

△ **69. Area of a Rectangle** Write an algebraic expression for the area of the rectangle.

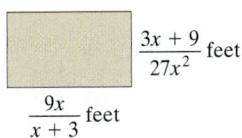

$\dfrac{3x + 9}{27x^2}$ feet

$\dfrac{9x}{x + 3}$ feet

△ **70. Area of a Rectangle** Write an algebraic expression for the area of the rectangle.

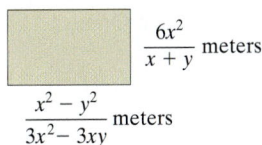

$\dfrac{6x^2}{x + y}$ meters

$\dfrac{x^2 - y^2}{3x^2 - 3xy}$ meters

△ **71. Area of a Triangle** Write an algebraic expression for the area of the triangle.

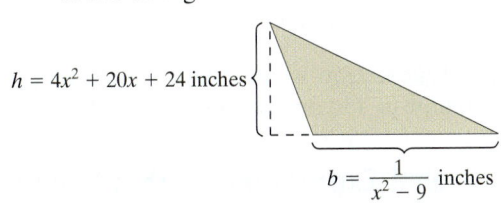

$h = 4x^2 + 20x + 24$ inches

$b = \dfrac{1}{x^2 - 9}$ inches

△ **72. Area of a Triangle** Write an algebraic expression for the area of the triangle.

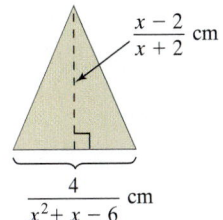

$\dfrac{x - 2}{x + 2}$ cm

$\dfrac{4}{x^2 + x - 6}$ cm

Extending the Concepts

In Problems 73–82, perform the indicated operation and simplify the result.

73. $\dfrac{xy - ay + xb - ab}{xy + ay - xb - ab} \cdot \dfrac{2xy - 2ay - 2xb + 2ab}{4b + 4y}$

74. $\dfrac{x^3 - x^2 + x - 1}{x + 1} \cdot \dfrac{x^2 + 2x + 1}{1 - x^2}$

75. $\dfrac{x^2 + x - 12}{x^2 - 2x - 35} \div \dfrac{x + 4}{x^2 + 4x - 5} \div \dfrac{12 - 4x}{x - 7}$

76. $\dfrac{x^2 - 6xy + 9y^2}{x^2 - 4y^2} \div \dfrac{x - 3y}{x^2 - 5xy + 6y^2} \div \dfrac{x^2 - 9y^2}{x^2 - xy - 6y^2}$

77. $\dfrac{a^2 - 2ab}{2b - 3a} \div \dfrac{3a^2 - 4ab - 4b^2}{16a^2b^2 - 36a^4} \div (6a)$

78. $\dfrac{(2v - 1)^6}{(1 - 2v)^5} \cdot \dfrac{1 - 8v^3}{4v^2 - 4v + 1} \div (2v - 1)$

79. $\dfrac{x^2 + xy - 3x - 3y}{x^3 + y^3} \cdot \dfrac{x^2 + 2xy + y^2}{x^2 - x - 6}$

80. $\dfrac{a^2 + 4a - 12}{a^2 + 6a - 27} \cdot \dfrac{a^2 - 9}{2a^2 + 12a} \div (4a^2)$

81. $\dfrac{p^3 - 27q^3}{9pq} \cdot \dfrac{(3p^2q)^3}{p^2 - 9q^2} \div \dfrac{1}{p^2 + 2pq - 3q^2}$

82. $\dfrac{x^2 - 1}{x^4 - 81} \div \dfrac{x^2 + 1}{(x - 3)^2} \cdot \dfrac{x^3 + x}{x^2 - 2x - 3}$

In Problems 83 and 84, find the missing expression.

83. $\dfrac{2x}{x^3 - 3x^2} \cdot \dfrac{x^2 - x - 6}{?} = \dfrac{1}{3x}$

84. $\dfrac{x^2 + 12x + 36}{?} \div \dfrac{x^2 + 11x + 30}{x^2 + 3x - 10} = \dfrac{x + 6}{x - 2}$

7.3 Adding and Subtracting Rational Expressions with a Common Denominator

OBJECTIVES

1. Add Rational Expressions with a Common Denominator
2. Subtract Rational Expressions with a Common Denominator
3. Add or Subtract Rational Expressions with Opposite Denominators

Preparing for Adding and Subtracting Rational Expressions with a Common Denominator

Before getting started, take the following readiness quiz. If you get a problem wrong, go back to the section cited and review the material.

1. Write $\dfrac{12}{15}$ in lowest terms.   [Appendix, Section A.1, p. A5]

2. Find each sum and write in lowest terms:

    (a) $\dfrac{7}{5} + \dfrac{2}{5}$ **(b)** $\dfrac{5}{6} + \dfrac{11}{6}$ [Section 1.4, p. 31]

3. Find each difference and write in lowest terms:

    (a) $\dfrac{7}{9} - \dfrac{5}{9}$ **(b)** $\dfrac{7}{8} - \dfrac{5}{8}$ [Section 1.4, p. 32]

4. Determine the additive inverse of 5.   [Section 1.3, p. 21]

5. Simplify: $-(x - 2)$ [Section 1.7, p. 61]

In the previous section, we learned how to multiply and divide rational expressions. We now learn how to add and subtract rational expressions.

1 Add Rational Expressions with a Common Denominator

The rules for adding rational expressions are the same as the rules for adding rational numbers. If the denominators of two rational expressions are the same, then we add the numerators, write the result over the common denominator, and simplify if possible. Two examples are worked below. In the left column, rational numbers are added; in the right column, rational algebraic expressions are added. Do you see that the steps are the same?

ADDING RATIONAL NUMBERS

$$\frac{3}{5} + \frac{8}{5} = \frac{3+8}{5} = \frac{11}{5}$$

$$\frac{7}{12} + \frac{11}{12} = \frac{18}{12} = \frac{\cancel{6} \cdot 3}{\cancel{6} \cdot 2} = \frac{3}{2}$$

ADDING RATIONAL EXPRESSIONS

$$\frac{3}{5a} + \frac{8}{5a} = \frac{3+8}{5a} = \frac{11}{5a}, a \neq 0$$

$$\frac{7}{12z} + \frac{11}{12z} = \frac{18}{12z} = \frac{\cancel{6} \cdot 3}{\cancel{6} \cdot 2 \cdot z} = \frac{3}{2z}, z \neq 0$$

Preparing for...Answers 1. $\dfrac{4}{5}$ **2. (a)** $\dfrac{9}{5}$ **(b)** $\dfrac{8}{3}$ **3. (a)** $\dfrac{2}{9}$ **(b)** $\dfrac{1}{4}$ **4.** -5 **5.** $2 - x$

Throughout the section, we assume that a variable cannot take on any values that result in division by zero.

EXAMPLE 1   How to Add Rational Expressions with a Common Denominator

Find the sum and simplify, if possible: $\dfrac{5}{x + 1} + \dfrac{3}{x + 1}$

Step-by-Step Solution

The rational expressions in the sum $\dfrac{5}{x + 1} + \dfrac{3}{x + 1}$ have a common denominator, $x + 1$.

Step 1: Add the numerators and write the result over the common denominator.

Use $\dfrac{a}{c} + \dfrac{b}{c} = \dfrac{a + b}{c}$: $\dfrac{5}{x + 1} + \dfrac{3}{x + 1} = \dfrac{5 + 3}{x + 1}$

Combine like terms in the numerator: $= \dfrac{8}{x + 1}$

Step 2: Simplify the rational expression by writing the rational expression in lowest terms.

The sum is already simplified.

$$\text{So } \frac{5}{x+1} + \frac{3}{x+1} = \frac{8}{x+1}.$$

Here are the steps we use to add rational expressions with a common denominator.

In Words
To add rational expressions with a common denominator, add the numerators and write the result over the common denominator. Then simplify, if necessary.

Steps to Add Rational Expressions

Step 1: Use the fact that if $\frac{a}{c}$ and $\frac{b}{c}$, $c \neq 0$, are two rational expressions, then

$$\frac{a}{c} + \frac{b}{c} = \frac{a+b}{c}$$

to add the rational expressions.

Step 2: Simplify the sum by writing the rational expression in lowest terms. This step will not always be necessary.

EXAMPLE 2 **Adding Rational Expressions with a Common Denominator**

Find the sum $\frac{x^2}{x+7} + \frac{7x}{x+7}$ and simplify the result, if possible.

Solution

$$\text{Use } \frac{a}{c} + \frac{b}{c} = \frac{a+b}{c}: \quad \frac{x^2}{x+7} + \frac{7x}{x+7} = \frac{x^2+7x}{x+7}$$

$$\text{Factor the numerator:} \quad = \frac{x(x+7)}{x+7}$$

$$\text{Divide out like factors:} \quad = \frac{x}{1} = x$$

QUICK ✓ *Find the sum and simplify the result, if possible.*

1. $\dfrac{1}{x-2} + \dfrac{3}{x-2}$

2. $\dfrac{2x+1}{x+1} + \dfrac{x^2}{x+1}$

EXAMPLE 3 **Adding Rational Expressions in Which a Numerator Contains More Than One Term**

Find the sum: $\dfrac{2x^2+x}{x^2-4} + \dfrac{x-x^2}{x^2-4}$. Simplify the result, if possible.

Solution

$$\text{Use } \frac{a}{c} + \frac{b}{c} = \frac{a+b}{c}: \quad \frac{2x^2+x}{x^2-4} + \frac{x-x^2}{x^2-4} = \frac{2x^2+x+x-x^2}{x^2-4}$$

$$\text{Combine like terms in the numerator:} \quad = \frac{x^2+2x}{x^2-4}$$

$$\text{Factor the numerator and denominator:} \quad = \frac{x(x+2)}{(x+2)(x-2)}$$

$$\text{Divide out like factors:} \quad = \frac{x}{x-2}$$

QUICK ✔ *Find the sum and simplify the result, if possible.*

3. $\dfrac{9x}{6x - 5} + \dfrac{2x - 3}{6x - 5}$

4. $\dfrac{2x - 2}{2x^2 - 7x - 15} + \dfrac{5}{2x^2 - 7x - 15}$

---

② **Subtract Rational Expressions with a Common Denominator**

The steps for subtracting rational expressions are the same as the steps for subtracting rational numbers. If the denominators of two rational expressions are the same, then we subtract the numerators and write the result over the common denominator.

SUBTRACTING RATIONAL NUMBERS

$$\frac{7}{3} - \frac{2}{3} = \frac{7 - 2}{3} = \frac{5}{3}$$

$$\frac{3}{10} - \frac{7}{10} = \frac{3 - 7}{10} = \frac{-4}{10} = \frac{-2 \cdot 2}{5 \cdot 2} = \frac{-2}{5}$$

SUBTRACTING RATIONAL EXPRESSIONS

$$\frac{7}{3x} - \frac{2}{3x} = \frac{7 - 2}{3x} = \frac{5}{3x}$$

$$\frac{3}{10b} - \frac{7}{10b} = \frac{3 - 7}{10b} = \frac{-4}{10b} = \frac{-2 \cdot 2}{5 \cdot 2 \cdot b} = \frac{-2}{5b}$$

Once again, we exclude values of the variable that result in division by zero.

EXAMPLE 4 **How to Subtract Rational Expressions with a Common Denominator**

Find the difference: $\dfrac{2n^2}{n^2 - 1} - \dfrac{2n}{n^2 - 1}$. Simplify the result, if possible.

Step-by-Step Solution

Step 1: Subtract the numerators and write the result over the common denominator.

Use $\dfrac{a}{c} - \dfrac{b}{c} = \dfrac{a - b}{c}$: $\quad \dfrac{2n^2}{n^2 - 1} - \dfrac{2n}{n^2 - 1} = \dfrac{2n^2 - 2n}{n^2 - 1}$

Step 2: Simplify the rational expression by writing the rational expression in lowest terms.

Factor the numerator and denominator: $\quad = \dfrac{2n(n - 1)}{(n - 1)(n + 1)}$

Divide out like factors: $\quad = \dfrac{2n}{n + 1}$

Therefore, $\dfrac{2n^2}{n^2 - 1} - \dfrac{2n}{n^2 - 1} = \dfrac{2n}{n + 1}$.

Here are the steps we use to subtract rational expressions with a common denominator.

Steps to Subtract Rational Expressions with a Common Denominator

In Words
To subtract rational expressions with a common denominator, subtract the numerators and write the result over the common denominator. Then simplify, if possible.

Step 1: Use the fact that if $\dfrac{a}{c}$ and $\dfrac{b}{c}$, $c \neq 0$, are two rational expressions, then

$$\frac{a}{c} - \frac{b}{c} = \frac{a - b}{c}$$

to subtract the rational expressions.

Step 2: Simplify the difference by writing the rational expression in lowest terms. This step will not always be necessary.

QUICK ✓ *Find the difference and simplify the result, if possible.*

5. $\dfrac{8y}{2y - 5} - \dfrac{6}{2y - 5}$

6. $\dfrac{10 + 3z}{6z} - \dfrac{7}{6z}$

When finding the difference between two rational expressions in which the numerator of the rational expression being subtracted contains more than one term, enclose the terms in the numerator that follow the subtraction sign in parentheses. This will remind you to distribute the minus sign across all the terms. The next example illustrates this point.

EXAMPLE 5   **Subtracting Rational Expressions in Which a Numerator Contains More than One Term**

Find the difference: $\dfrac{9x + 1}{x + 1} - \dfrac{6x - 2}{x + 1}$. Simplify the result.

Solution

First, we note that the rational expressions have the same denominator, $x + 1$.

Work Smart

Enclose the terms in the numerator that follow the subtraction sign in parentheses because this will remind you to distribute the minus sign across all the terms.

$$
\begin{aligned}
\text{Notice the use of parentheses:} \quad & \frac{9x + 1}{x + 1} - \frac{6x - 2}{x + 1} = \frac{9x + 1 - (6x - 2)}{x + 1} \\[2mm]
\text{Distribute the minus sign into the parentheses:} \quad & = \frac{9x + 1 - 6x + 2}{x + 1} \\[2mm]
\text{Combine like terms:} \quad & = \frac{3x + 3}{x + 1} \\[2mm]
\text{Factor the numerator:} \quad & = \frac{3(x + 1)}{x + 1} \\[2mm]
\text{Divide out like factors:} \quad & = \frac{3}{1} = 3
\end{aligned}
$$

QUICK ✓ *Find the difference and simplify the result.*

7. $\dfrac{2x^2 - 5x}{3x} - \dfrac{x^2 - 13x}{3x}$

8. $\dfrac{3x^2 + 8x - 1}{x^2 - 3x - 28} - \dfrac{2x^2 + 2x - 9}{x^2 - 3x - 28}$

③ Add or Subtract Rational Expressions with Opposite Denominators

Suppose you were asked to find the sum $\dfrac{w^2 - 3w}{w - 4} + \dfrac{4}{4 - w}$. Although the denominators of these rational expressions are different, you might have noticed that they are additive inverses (opposites) of each other. Recall that $4 - w = -w + 4 = -1(w - 4)$. We use this idea to rewrite the sum so that each rational expression has the same denominator.

EXAMPLE 6   **Adding Rational Expressions with Opposite Denominators**

Find the sum: $\dfrac{w^2 - 3w}{w - 4} + \dfrac{4}{4 - w}$. Simplify the result.

Solution

We rewrite the denominators so they are the same by factoring -1 from $4 - w$ in the second denominator.

Factor -1 from $4 - w$:
$$\frac{w^2 - 3w}{w - 4} + \frac{4}{4 - w} = \frac{w^2 - 3w}{w - 4} + \frac{4}{-1(w - 4)}$$

Use $\dfrac{a}{-b} = \dfrac{-a}{b}$:
$$= \frac{w^2 - 3w}{w - 4} + \frac{-4}{w - 4}$$

$$= \frac{w^2 - 3w - 4}{w - 4}$$

Factor:
$$= \frac{(w - 4)(w + 1)}{w - 4}$$

Divide out like factors:
$$= \frac{w + 1}{1}$$

$$= w + 1$$

So $\dfrac{w^2 - 3w}{w - 4} + \dfrac{4}{4 - w} = w + 1$.

QUICK ☑ *Find the sum and simplify the result.*

9. $\dfrac{3x}{x - 5} + \dfrac{1}{5 - x}$

10. $\dfrac{a^2 + 2a}{a - 7} + \dfrac{a^2 + 14}{7 - a}$

EXAMPLE 7 **Subtracting Rational Expressions with Opposite Denominators**

Find the difference: $\dfrac{b^2 - 11}{b^2 - 25} - \dfrac{3b + 1}{25 - b^2}$. Simplify the result.

Solution

We must first rewrite the denominators so that they are the same. This is done by factoring -1 from $25 - b^2$.

$$\frac{b^2 - 11}{b^2 - 25} - \frac{3b + 1}{25 - b^2} = \frac{b^2 - 11}{b^2 - 25} - \frac{3b + 1}{-(b^2 - 25)}$$

$-\dfrac{a}{-b} = +\dfrac{a}{b}$:
$$= \frac{b^2 - 11}{b^2 - 25} + \frac{3b + 1}{b^2 - 25}$$

Add numerators:
$$= \frac{b^2 - 11 + 3b + 1}{b^2 - 25}$$

Combine like terms:
$$= \frac{b^2 + 3b - 10}{b^2 - 25}$$

Factor:
$$= \frac{(b + 5)(b - 2)}{(b + 5)(b - 5)}$$

Divide out like factors:
$$= \frac{b - 2}{b - 5}$$

QUICK ☑ *Find the difference and simplify the result.*

11. $\dfrac{2n}{n^2 - 9} - \dfrac{6}{9 - n^2}$

12. $\dfrac{2k}{6k - 6} - \dfrac{9 + 4k}{6 - 6k}$

7.3 Exercises

Concepts and Vocabulary

In Problems 1–3, fill in the blanks.

1. When adding or subtracting rational expressions written with a common denominator, add or subtract the _____ only.

2. $\dfrac{3x + 1}{x - 4} - \dfrac{x + 3}{x - 4} = \dfrac{3x + 1 - (\quad)}{x - 4}$.

3. If $\dfrac{a}{c}$ and $\dfrac{b}{c}$, $c \neq 0$, are two rational expressions, then $\dfrac{a}{c} + \dfrac{b}{c} = \underline{\underline{\quad\quad}}$.

In Problems 4–6, answer True or False to each statement.

4. $\dfrac{2x}{a} + \dfrac{4x}{a} = \dfrac{6x}{2a} = \dfrac{3x}{a}$

5. When adding or subtracting rational expressions the result should always be simplified.

6. $8 - x = -1(x - 8)$

7. Is the following correct or incorrect? Explain your reasoning.

$$\frac{x - 2}{x} - \frac{x + 4}{x} = \frac{x - 2 - x + 4}{x} = \frac{2}{x}$$

8. The denominators in the expression $\dfrac{4x}{x - 2} - \dfrac{2x}{2 - x}$ are opposites. Explain how to rewrite this difference so that the terms have a common denominator and then describe the steps to simplify the result.

Building Skills

In Problems 9–22, add the rational expressions and simplify the result, if possible.

9. $\dfrac{3p}{8} + \dfrac{11p}{8}$

10. $\dfrac{2m}{9} + \dfrac{4m}{9}$

11. $\dfrac{n}{2} + \dfrac{3n}{2}$

12. $\dfrac{3}{x} + \dfrac{y}{x}$

13. $\dfrac{4a - 1}{3a} + \dfrac{2a - 2}{3a}$

14. $\dfrac{4n + 1}{8} + \dfrac{12n - 1}{8}$

15. $\dfrac{8c - 3}{c - 1} + \dfrac{2c - 1}{c - 1}$

16. $\dfrac{4n}{6n + 27} + \dfrac{18}{6n + 27}$

17. $\dfrac{2x}{x + y} + \dfrac{2y}{x + y}$

18. $\dfrac{4}{2x + 1} + \dfrac{8x}{2x + 1}$

19. $\dfrac{7x}{x + 2} + \dfrac{7x^2}{x + 2}$

20. $\dfrac{4p - 1}{p - 1} + \dfrac{p + 1}{p - 1}$

21. $\dfrac{12a - 1}{2a + 6} + \dfrac{13 - 8a}{2a + 6}$

22. $\dfrac{3x^2 + 4}{3x - 6} + \dfrac{3x^2 - 28}{3x - 6}$

In Problems 23–36, subtract the rational expressions and simplify the result, if possible.

23. $\dfrac{11x}{3} - \dfrac{5x}{3}$

24. $\dfrac{5n^2}{6} - \dfrac{n^2}{6}$

25. $\dfrac{4x - 3}{3} - \dfrac{x + 6}{3}$

26. $\dfrac{x + 2}{8} - \dfrac{2x - 5}{8}$

27. $\dfrac{x^2 - x}{2x} - \dfrac{2x^2 - x}{2x}$

28. $\dfrac{3x + 3x^2}{x} - \dfrac{x^2 - x}{x}$

29. $\dfrac{2c - 3}{4c} - \dfrac{6c + 9}{4c}$

30. $\dfrac{7x}{3} - \dfrac{x + 6}{3}$

31. $\dfrac{x^2 - 6}{x - 2} - \dfrac{x^2 - 3x}{x - 2}$

32. $\dfrac{2x - 3}{x - 1} - \dfrac{x + 1}{x - 1}$　　**33.** $\dfrac{2}{c^2 - 4} - \dfrac{c^2 - 2}{c^2 - 4}$　　**34.** $\dfrac{2z^2 + 7z}{z^2 - 1} - \dfrac{z^2 - 6}{z^2 - 1}$

35. $\dfrac{2x^2 + 5x}{2x + 3} - \dfrac{4x + 3}{2x + 3}$　　**36.** $\dfrac{2n^2 + n}{n^2 - n - 2} - \dfrac{n^2 + 6}{n^2 - n - 2}$

Mixed Practice

In Problems 37–72, perform the indicated operation and simplify the result, if possible.

37. $\dfrac{4}{7x} - \dfrac{11}{7x}$　　**38.** $\dfrac{2x}{5} - \left(-\dfrac{3x}{5}\right)$　　**39.** $\dfrac{5}{2x - 5} - \dfrac{2x}{2x - 5}$

40. $\dfrac{2b + 1}{b + 1} - \dfrac{b}{b + 1}$　　**41.** $\dfrac{2n^3}{n - 1} - \dfrac{2n^3}{n - 1}$　　**42.** $\dfrac{3n^2}{2n - 1} - \dfrac{6n}{2n - 1}$

43. $\dfrac{n^2 - 3}{2n + 3} + \dfrac{n^2 + n}{2n + 3}$　　**44.** $\dfrac{5x}{x^2 + 1} + \dfrac{x^2 - 3x}{x^2 + 1}$　　**45.** $\dfrac{3v - 1}{v^2 - 9} - \dfrac{8}{v^2 - 9}$

46. $\dfrac{x^2 - x}{x^2 - 1} - \dfrac{2}{x^2 - 1}$　　**47.** $\dfrac{x^2}{x^2 - 1} + \dfrac{2x + 1}{x^2 - 1}$　　**48.** $\dfrac{x^2}{x^2 - 9} + \dfrac{6x + 9}{x^2 - 9}$

49. $\dfrac{2x^2 + 3}{x^2 - 3x} - \dfrac{x^2}{x^2 - 3x}$　　　**50.** $\dfrac{4x}{x^2 - 4} + \dfrac{8}{x^2 - 4}$

51. $\dfrac{a}{a^2 - 3a - 10} + \dfrac{2}{a^2 - 3a - 10}$　　**52.** $\dfrac{2x}{2x^2 - x - 15} + \dfrac{5}{2x^2 - x - 15}$

53. $\dfrac{n}{n - 3} + \dfrac{3}{3 - n}$　　**54.** $\dfrac{4}{x - 1} + \dfrac{8x}{1 - x}$　　**55.** $\dfrac{12q}{2p - 2q} - \dfrac{8p}{2q - 2p}$

56. $\dfrac{2a}{a - b} - \dfrac{4b}{b - a}$　　**57.** $\dfrac{2p^2 - 1}{p - 1} - \dfrac{2p + 2}{1 - p}$　　**58.** $\dfrac{x^2 + 3}{x - 1} - \dfrac{x + 3}{1 - x}$

59. $\dfrac{2a}{a - b} + \dfrac{2a - 4b}{b - a}$　　**60.** $\dfrac{4k^2}{2k - 1} + \dfrac{2k}{1 - 2k}$　　**61.** $\dfrac{3}{p^2 - 3} + \dfrac{p^2}{3 - p^2}$

62. $\dfrac{x^2 + 6x}{x^2 - 1} + \dfrac{4x + 3}{1 - x^2}$　　**63.** $\dfrac{2x}{x^2 - y^2} - \dfrac{2y}{y^2 - x^2}$　　**64.** $\dfrac{m^2 + 2mn}{n^2 - m^2} - \dfrac{n^2}{m^2 - n^2}$

65. $\dfrac{3x^2}{x + 1} - \dfrac{x^2 + 2}{x + 1}$　　**66.** $\dfrac{x^2}{x^2 - 1} - \dfrac{3x + 2}{x^2 - 1}$　　**67.** $\dfrac{3x - 1}{x - y} + \dfrac{1 - 3y}{y - x}$

68. $\dfrac{2s - 3t}{s - t} + \dfrac{6s - 4t}{t - s}$　　**69.** $\dfrac{2p - 3q}{3p^2 - 18q^2} - \dfrac{9q + p}{18q^2 - 3p^2}$

70. $\dfrac{n + 3}{2n^2 - n - 3} - \dfrac{-n^2 - 3n}{2n^2 - n - 3}$　　**71.** $\dfrac{2g}{g - 3} - \left(\dfrac{g + 3}{g - 3} - \dfrac{g - 4}{g - 3}\right)$

72. $\dfrac{7x - 3y}{x^2 - y^2} - \left(\dfrac{2x - 3y}{x^2 - y^2} - \dfrac{x + 7y}{x^2 - y^2}\right)$

Applying the Concepts

73. Find the sum of $\dfrac{7a^2}{a}$ and $\dfrac{5a^2}{a}$.

74. Find the sum of $\dfrac{3n}{3n + 2}$ and $\dfrac{n^2 + 2}{3n + 2}$.

75. Find the difference of $\dfrac{2n}{n + 1}$ and $\dfrac{n + 3}{n + 1}$.

76. Find the difference of $\dfrac{p^2 - 4}{p - 4}$ and $\dfrac{4p}{p - 4}$.

77. Subtract $\dfrac{3}{2n}$ from $\dfrac{11}{2n}$.

78. Subtract $\dfrac{3x + 4}{4}$ from $\dfrac{x - 4}{4}$.

79. Find a rational expression which, when subtracted from $\dfrac{x^2}{x^2 - 9}$, gives a difference of $\dfrac{3x}{x^2 - 9}$.

80. Find a rational expression which, when subtracted from $\dfrac{6x}{x - 3}$, gives a difference of one.

81. Find a rational expression which, when added to $\dfrac{-3x + 4}{x + 2}$, gives a sum of one.

82. Find a rational expression which, when added to $\dfrac{x - 3}{x + 1}$, gives a sum of $\dfrac{2x - 5}{x + 1}$.

△ **83. Perimeter of a Rectangle** Find the perimeter of the rectangle.

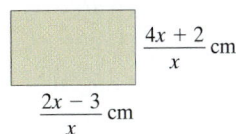

$\dfrac{4x + 2}{x}$ cm

$\dfrac{2x - 3}{x}$ cm

△ **84. Perimeter of a Rectangle** Find the perimeter of the rectangle.

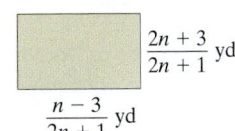

$\dfrac{2n + 3}{2n + 1}$ yd

$\dfrac{n - 3}{2n + 1}$ yd

Extending the Concepts

In Problems 85–88, perform the indicated operations and simplify the result, if possible.

85. $\dfrac{x^2}{3x^2 + 5x - 2} - \dfrac{x}{3x - 1} \cdot \dfrac{2x - 1}{x + 2}$

86. $\dfrac{4}{x - 2} \cdot \dfrac{x - 2}{x + 2} + \dfrac{3x - 1}{x^2 - 4}$

87. $\dfrac{5x}{x - 2} + \dfrac{2(x + 3)}{x - 2} - \dfrac{6(2x - 1)}{x - 2}$

88. $\dfrac{2a - b}{a - b} - \dfrac{6(a - 2b)}{a - b} + \dfrac{3(2b + a)}{b - a}$

In Problems 89 and 90, find the missing expression.

89. $\dfrac{2n + 1}{n - 3} - \dfrac{?}{n - 3} = \dfrac{6n + 7}{n - 3}$

90. $\dfrac{?}{n^2 - 1} - \dfrac{n - 3}{n^2 - 1} = \dfrac{1}{n + 1}$

7.4 Finding the Least Common Denominator and Forming Equivalent Rational Expressions

OBJECTIVES

1. Find the Least Common Denominator of Two or More Rational Expressions
2. Write a Rational Expression That Is Equivalent to a Given Rational Expression
3. Use the LCD to Write Equivalent Rational Expressions

Preparing for *Finding the Least Common Denominator and Forming Equivalent Rational Expressions*

Before getting started, take the following readiness quiz. If you get a problem wrong, go back to the section cited and review the material.

1. Write $\dfrac{5}{12}$ as a fraction with 24 as its denominator.   [Section A.1, pp. A3–A4]

2. Find the least common denominator (LCD) of $\dfrac{4}{15}$ and $\dfrac{7}{25}$.   [Section A.1, pp. A4–A5]

1   ## Find the Least Common Denominator of Two or More Rational Expressions

Work Smart

Don't confuse the LCD (least common denominator) with the GCF (greatest common factor)!

To add or subtract rational expressions the denominators must be the same. What if the denominators are different? We use the same ideas that we did for adding or subtracting rational numbers with unlike denominators: we rewrite each rational expression over the least common denominator.

Before we find the least common denominator (LCD) of rational expressions, let's review how to find the LCD of rational numbers.

Preparing for...Answers   **1.** $\dfrac{10}{24}$ **2.** 75

EXAMPLE 1 **How to Find the Least Common Denominator of Rational Numbers**

Find the least common denominator (LCD) of $\dfrac{5}{6}$ and $\dfrac{11}{45}$.

Step-by-Step Solution

Step 1: Write each denominator as the product of prime factors.	$6 = 2 \cdot 3$ $45 = 3 \cdot 3 \cdot 5$
Step 2: Write down the factor(s) that the denominators share. Then copy the remaining factors the greatest number of times that the factor occurs in a denominator.	The shared factor is 3 (indicated in blue). The remaining factors are 2, 3, and 5.
Step 3: Multiply the factors listed in Step 2. The product is the least common denominator (LCD).	The LCD is $3 \cdot 2 \cdot 3 \cdot 5 = 90$.

In Example 1, we could have written the factorization as follows:

$$6 = 2 \cdot 3$$
$$45 = 3^2 \cdot 5$$

To find the LCD, we notice that the factor 3 is written to the second power in the factorization of 45. We write 3^2 as part of the LCD because it is the "common factor to the highest power." We then write the factors that are not common, 2 and 5, as part of the LCD. The product of these factors gives us the least common denominator (LCD). So, the LCD is $3^2 \cdot 2 \cdot 5 = 90$.

Work Smart

It may helpful to align the factors vertically when finding the LCD.

Keep in mind that arithmetic is a special case of algebra, so the approach we just followed for finding the least common denominator of rational numbers will be the same approach we follow for finding the least common denominator of rational expressions.

> **DEFINITION**
>
> The **least common denominator (LCD)** of two or more rational expressions is the polynomial of least degree that is a multiple of each denominator in the rational expressions to be added or subtracted.

Let's see how to find the least common denominator of rational expressions.

EXAMPLE 2 **How to Find the Least Common Denominator of Rational Expressions with Monomial Denominators**

Find the least common denominator of the rational expressions $\dfrac{3}{4x^3}$ and $\dfrac{5}{6x}$.

Step-by-Step Solution

Step 1: Express each denominator as the product of factors.	$4x^3 = 2^2 \cdot x^3$ $6x = 2 \cdot 3 \cdot x$
Step 2: List the common factors raised to the highest power. Then list the factors that are not common.	• We list 2^2 because 2 is the highest power on the factor 2. • We list x^3 because 3 is the highest power on the factor x. • The factor that is not common is 3.

Step 3: Find the product of the factors written
Step 2. This algebraic expression is the LCD.

$$LCD = 2^2 \cdot 3 \cdot x^3$$
$$= 12x^3$$

We summarize the steps to finding the least common denominator below.

Steps to Find the Least Common Denominator of Rational Expressions

Step 1: Factor each denominator completely. When factoring, write the factored form using powers. For example, write $x^2 + 4x + 4$ as $(x + 2)^2$.

Step 2: If factors are common except for their power, then list the factor with the highest power. That is, list each factor the greatest number of times that it appears. Then list the factors that are not common.

Step 3: The LCD is the product of the factors written in Step 2.

EXAMPLE 3   Finding the Least Common Denominator of Rational Expressions with Monomial Denominators

Find the least common denominator of the rational expressions $\dfrac{1}{15xy^2}$ and $\dfrac{7}{18x^3y}$.

Solution

Work Smart
List the factor with the highest power in the LCD.

Express each denominator as the product of factors:
$$15xy^2 = 3 \cdot 5 \cdot x \cdot y^2$$
$$18x^3y = 2 \cdot 3^2 \cdot x^3 \cdot y$$

The product of each different factor the greatest number of times it appears is the LCD. So

$$LCD = 2 \cdot 3^2 \cdot 5 \cdot x^3 \cdot y^2$$
$$= 90x^3y^2$$

QUICK ✓ *Find the least common denominator of the rational expressions.*

1. $\dfrac{5}{8x^2y}$ and $\dfrac{1}{12xy^3}$

2. $\dfrac{1}{6a^3b^2}$ and $\dfrac{5}{21ab^3}$

EXAMPLE 4   Finding the Least Common Denominator of Rational Expressions with Polynomial Denominators

Find the LCD of the rational expressions $\dfrac{7}{3a}$ and $\dfrac{5}{6a + 6}$.

Solution

Factor the denominators:
$$3a \qquad\qquad = 3 \cdot a$$
$$6a + 6 = 6(a + 1) = 2 \cdot 3 \cdot (a + 1)$$

Work Smart
The factor a in the denominator $3a$ is different from the a in the binomial factor $(a + 1)$. The a in $(a + 1)$ is a *term*, not a *factor*.

The common factor is 3. The factors that are not common are $2, a,$ and $a + 1$. The product of these factors gives us the least common denominator.

$$LCD = 2 \cdot 3 \cdot a \cdot (a + 1)$$
$$= 6a(a + 1)$$

EXAMPLE 5 **Finding the Least Common Denominator of Rational Expressions with Polynomial Denominators**

Find the LCD of the rational expressions $\dfrac{7}{x^2 - x - 2}$ and $\dfrac{3}{x^2 + 2x + 1}$.

Solution

Factor the denominators:
$$x^2 - x - 2 = (x - 2)(x + 1)$$
$$x^2 + 2x + 1 = \qquad (x + 1)^2$$

The factor $x + 1$ is common except that it is written to the second power in the factorization of $x^2 + 2x + 1$. Therefore, $(x + 1)^2$ is part of the LCD. The factor that is not common is $x - 2$. The product of these factors gives us the least common denominator.

$$\text{LCD} = (x + 1)^2(x - 2)$$

QUICK ✓ *Find the least common denominator of the rational expressions.*

3. $\dfrac{2}{15z}$ and $\dfrac{7}{5z^2 + 5z}$

4. $\dfrac{3}{x^2 + 4x - 5}$ and $\dfrac{1}{x^2 + 10x + 25}$

EXAMPLE 6 **Finding the Least Common Denominator with Opposite Factors**

Find the LCD of the rational expressions $\dfrac{1}{x^2 - 25}$ and $\dfrac{9}{10 - 2x}$.

Solution

Factor the denominators:
$$x^2 - 25 = (x + 5)(x - 5)$$
$$10 - 2x = 2(5 - x)$$

It looks as though there are no common factors. But you should notice that the factors $x - 5$ and $5 - x$ are opposites. So we can write the factorization of $10 - 2x$ as follows:

$$10 - 2x = 2(5 - x)$$
Factor out -1: $\quad = 2(-1)(x - 5)$
$$= -2(x - 5)$$

Now, the common factor in each factorization is $x - 5$. The factors that are not common are -2 and $x + 5$. So the least common denominator is

$$\text{LCD} = -2(x + 5)(x - 5)$$

QUICK ✓ *Find the least common denominator of the rational expressions.*

5. $\dfrac{3}{21 - 3x}$ and $\dfrac{8}{x^2 - 49}$

---

② **Write a Rational Expression That Is Equivalent to a Given Rational Expression**

Rational expressions which represent the same quantity are said to be *equivalent*. Now that we know how to find the LCD, we learn to write a rational expression as an equivalent rational expression with the LCD as its denominator. We can obtain equivalent rational expressions by multiplying the numerator and denominator of a rational expression by the same quantity as follows:

$$\frac{p}{q} = \frac{p}{q} \cdot 1 = \frac{p}{q} \cdot \frac{r}{r} = \frac{pr}{qr}$$

In other words we obtain equivalent rational expressions by multiplying the rational expression by 1, the multiplicative identity. Let's review how to do this using rational numbers.

EXAMPLE 7   Forming Equivalent Rational Numbers

Write $\frac{2}{5}$ as an equivalent fraction with a denominator of 20.

Solution

We want to change from a denominator of 5 to a denominator of 20. In other words,

$$\frac{2}{5} \cdot \frac{?}{?} = \frac{\square}{20}$$

We know that $5 \cdot 4 = 20$, so we form the factor of $1 = \frac{4}{4}$.

$$\frac{2}{5} \cdot \frac{4}{4} = \frac{8}{20}$$

The equivalent fraction is $\frac{8}{20}$. ∎

The approach that we use to form equivalent fractions is the same as the approach we use to form equivalent rational expressions.

EXAMPLE 8   How to Form an Equivalent Rational Expression

Write the rational expression $\dfrac{4}{5x^2 + x}$ as an equivalent rational expression with denominator $10x^2 + 2x$.

Step-by-Step Solution

Step 1: Write each denominator in factored form.	$5x^2 + x = \quad x(5x + 1)$ $10x^2 + 2x = 2 \cdot x(5x + 1)$
Step 2: Determine the "missing factor(s)."	The new denominator has a factor of 2 that the denominator of the original rational expression is missing.
Step 3: Multiply the original rational expression by 1.	We multiply $\dfrac{4}{5x^2 + x}$ by $1 = \dfrac{2}{2}$. $\dfrac{4}{5x^2 + x} = \dfrac{4}{x(5x + 1)} \cdot$
Step 4: Find the product. Leave the denominator in factored form.	$= \dfrac{8}{2x(5x + 1)}$

So $\dfrac{4}{5x^2 + x} = \dfrac{8}{2x(5x + 1)}$. Notice that $2x(5x + 1) = 10x^2 + 2x$, as required. ∎

> ### SUMMARY: Steps to Form Equivalent Rational Expressions
>
> **Step 1:** Write each denominator in factored form.
>
> **Step 2:** Determine the "missing factor(s)." That is, what factor(s) does the new denominator have that is missing from the original denominator?
>
> **Step 3:** Multiply the original rational expression by $1 = \dfrac{\text{missing factor(s)}}{\text{missing factor(s)}}$.
>
> **Step 4:** Find the product. Leave the denominator in factored form.

QUICK ✓ *Write each rational expression as an equivalent rational expression with the given denominator.*

6. $\dfrac{3}{4p^2 - 8p}$ with denominator $16p^3(p - 2)$.

7. $\dfrac{2}{x^2 + 5x + 6}$ with denominator $(x + 2)(x - 4)(x + 3)$.

③ Use the LCD to Write Equivalent Rational Expressions

The last skill that we need before we add or subtract rational expressions with unlike denominators is to find the LCD of two or more rational expressions and then write them as equivalent rational expressions.

EXAMPLE 9 Using the LCD to Write Equivalent Rational Expressions

Find the LCD of the rational expressions $\dfrac{4}{x^2 + 3x + 2}$ and $\dfrac{9}{x^2 - 4}$. Then rewrite each rational expression with the LCD.

Solution

Factor each denominator:
$$x^2 + 3x + 2 = (x + 1)(x + 2)$$
$$x^2 - 4 = (x + 2)(x - 2)$$

Find the product of the factors written in the previous step to obtain the LCD:
$$\text{LCD} = (x + 2)(x + 1)(x - 2)$$

Rewrite each rational expression with a denominator of $(x + 2)(x + 1)(x - 2)$. Multiply out the numerators, but leave the denominator in factored form.
$$\frac{4}{x^2 + 3x + 2} = \frac{4}{(x + 2)(x + 1)} \cdot \frac{(x - 2)}{(x - 2)}$$
$$= \frac{4x - 8}{(x + 2)(x + 1)(x - 2)}$$
$$\frac{9}{x^2 - 4} = \frac{9}{(x - 2)(x + 2)} \cdot \frac{(x + 1)}{(x + 1)}$$
$$= \frac{9x + 9}{(x - 2)(x + 2)(x + 1)}$$

QUICK ✓ *Find the least common denominator of each rational expression. Then rewrite each rational expression with the LCD.*

8. $\dfrac{5}{x^2 - 4x - 5}$ and $\dfrac{-3}{x^2 - 7x + 10}$

7.4 Exercises

For Extra Help:

Student Solutions Manual   CD Video   PH Math/Tutor Center   MathXL Tutorials on CD   MathXL®   MyMathLab

Concepts and Vocabulary

In Problems 1–3, fill in the blanks.

1. The _____ _____ _____ of two or more rational expressions is the smallest polynomial that is a multiple of each denominator in the rational expressions to be added or subtracted.

2. If we wanted to rewrite the rational expression $\dfrac{2x + 1}{x - 1}$ with a denominator of $(x - 1)(x + 3)$, we would multiply the numerator and denominator of $\dfrac{2x + 1}{x - 1}$ by _____.

3. Any time we multiply a rational expression by _____, we form an equivalent rational expression.

In Problems 4–6, answer True or False to each statement.

4. To find the sum $\dfrac{3}{4a^2b} + \dfrac{5}{4ab}$, we begin by adding $3 + 5$.

5. In the expression $\dfrac{3}{4a^2b} + \dfrac{5}{4ab}$, the LCD is $16a^3b^2$.

6. The least common denominator of two rational expressions is unique.

7. If you were teaching the class how to write equivalent rational expressions with a common denominator, what steps would you list on the board for the class?

8. A member of your class says that $\dfrac{1}{(x + 3)(x + 2)}$ and $\dfrac{4}{(x + 2)}$ have an LCD of $(x + 3)(x + 2)^2$. This class member then rewrites the rational expressions as $\dfrac{x + 2}{(x + 3)(x + 2)^2}$ and $\dfrac{4(x + 3)(x + 2)}{(x + 3)(x + 2)^2}$. Use this result to explain why $(x + 3)(x + 2)^2$ is not the least common denominator.

Building Skills

In Problems 9–34, identify the LCD of the given rational expressions.

9. $\dfrac{5}{12}; \dfrac{4}{9}$

10. $\dfrac{7}{12}; \dfrac{3}{8}$

11. $\dfrac{4}{15}; \dfrac{3}{5}; \dfrac{1}{20}$

12. $\dfrac{1}{9}; \dfrac{3}{8}; \dfrac{7}{36}$

13. $\dfrac{14}{5x^2}; \dfrac{4}{5x}$

14. $\dfrac{3}{7y^2}; \dfrac{4}{49y}$

15. $\dfrac{7}{12xy^2}; \dfrac{4}{15x^3y}$

16. $\dfrac{5}{36xy^2}; \dfrac{1}{24x^2y}$

17. $\dfrac{3}{x}; \dfrac{4}{x + 1}$

18. $\dfrac{2x - 1}{2x + 1}; \dfrac{3x}{2}$

19. $\dfrac{4}{2x + 1}; 2$

20. $\dfrac{7}{y - 1}; 3$

21. $\dfrac{7}{2b - 6}; \dfrac{3b}{4b - 12}$

22. $\dfrac{2}{6x - 18}; \dfrac{3}{8x - 24}$

23. $\dfrac{6}{p^2 + p}; \dfrac{7}{p^2 - p - 2}$

24. $\dfrac{5}{2x^2 - 12x + 18}; \dfrac{4}{4x^2 - 36}$

25. $\dfrac{3}{r^2 + 4r + 4}; \dfrac{4}{r^2 - r - 2}$

26. $\dfrac{8}{x^2 - 1}; \dfrac{3}{x^2 - 2x + 1}$

27. $\dfrac{2}{x - 4}; \dfrac{3}{4 - x}$

28. $\dfrac{7}{3 - a}; \dfrac{1}{a - 3}$

29. $\dfrac{8}{x^2 - 9}; \dfrac{2}{6 - 2x}$

30. $\dfrac{-1}{c^2 - 49}; \dfrac{5}{21 - 3c}$

31. $\dfrac{1}{(x - 1)(x + 2)}; \dfrac{11x}{(1 - x)(2 + x)}$

32. $\dfrac{3z}{(z + 5)(z - 4)}; \dfrac{-z}{(4 - z)(5 + z)}$

33. $\dfrac{11}{p^2 - p}; \dfrac{-2}{p^3 - p^2}; \dfrac{8}{p^2 - 4p + 3}$

34. $\dfrac{7}{4n^3 - 2n^2}; \dfrac{9}{8n^2 - 4n}; \dfrac{6}{4n^2 - 1}$

In Problems 35–44, write an equivalent rational expression with the given denominator.

35. $\dfrac{4}{x}$ with denominator $3x^3$

36. $\dfrac{7}{x}$ with denominator $2x^2$

37. $\dfrac{3 + c}{a^2 b^2 c}$ with denominator $a^2 b^2 c^2$

38. $\dfrac{7a + 1}{abc}$ with denominator $a^2 b^2 c$

39. $\dfrac{x - 4}{x + 4}$ with denominator $x^2 - 16$

40. $\dfrac{3}{x + 2}$ with denominator $x^2 + 5x + 6$

41. $\dfrac{3n}{2n + 2}$ with denominator $6n^2 - 6$

42. $\dfrac{7a}{3a^2 + 9a + 6}$ with denominator $6a^2 + 18a + 12$

43. $4t$ with denominator $t - 1$

44. 7 with denominator $t^2 + 1$

Mixed Practice

In Problems 45–66, identify the LCD and then write each as an equivalent rational expression with that denominator.

45. $\dfrac{7}{12}; \dfrac{-8}{9}$

46. $\dfrac{-4}{9}; \dfrac{-3}{8}$

47. $\dfrac{7}{15}; 2$

48. $\dfrac{5}{6}; 4$

49. $\dfrac{2x}{3y}; \dfrac{4}{9y^2}$

50. $\dfrac{4}{5n^2}; \dfrac{3}{7n}$

51. $\dfrac{3a + 1}{2a^3}; \dfrac{4a - 1}{4a}$

52. $\dfrac{2p^2 - 1}{6p}; \dfrac{3p^3 + 2}{8p^2}$

53. $\dfrac{2}{m}; \dfrac{3}{m + 1}$

54. $\dfrac{x + 2}{x}; \dfrac{x}{x + 2}$

55. $\dfrac{y - 2}{4y}; \dfrac{y}{8y - 4}$

56. $\dfrac{3b}{7a}; \dfrac{2a}{7a + 14b}$

57. $\dfrac{1}{x - 1}; \dfrac{2x}{1 - x}$

58. $\dfrac{2a}{a - 2}; \dfrac{a}{2 - a}$

59. $\dfrac{3x}{x - 7}; \dfrac{-5}{7 - x}$

60. $\dfrac{5}{m - 6}; \dfrac{2m}{6 - m}$

61. $\dfrac{4x}{x^2 - 4}; \dfrac{2}{x + 2}$

62. $\dfrac{4a}{a^2 - 1}; \dfrac{7}{a + 1}$

63. $\dfrac{x + 1}{x^2 - 9}; \dfrac{x + 2}{x^2 - 2x - 3}$

64. $\dfrac{4n}{n^2 - n - 6}; \dfrac{2}{n^2 + 4n + 4}$

65. $\dfrac{3}{x + 4}; \dfrac{2x - 1}{2x^2 + 7x - 4}$

66. $\dfrac{3}{2n^2 - 7n + 3}; \dfrac{2n}{n - 3}$

Applying the Concepts

67. Painting a Room It takes an experienced painter x hours to paint a room alone; assuming that he works at a constant rate, he completes $\dfrac{1}{x}$ of the job per hour. It takes an apprentice 3 times as long to paint the same room as it takes the experienced painter, so the apprentice completes $\dfrac{1}{3x}$ of the job per hour. Identify the LCD of the two rational expressions that represent the rate of work of the experienced painter and the apprentice, and then write the equivalent fractions using this denominator.

68. Mowing the Lawn It takes Mario 2 hours longer to mow the lawn than it takes it Marco. If Marco takes x hours, he completes $\dfrac{1}{x}$ of the lawn per hour. Since Mario requires 2 more hours, he completes $\dfrac{1}{x+2}$ of the job per hour. Identify the LCD of the two rational expressions that represent the rate of work for Mario and Marco, and then write the equivalent fractions using this denominator.

69. Traveling on a River The time to complete a journey can be calculated by the formula $t = \dfrac{d}{r}$, where t is the time, d is the distance traveled, and r is rate. Suppose the time for a boat to travel a distance of 12 miles upstream on a river whose current is 4 miles per hour is given by $\dfrac{12}{r-4}$. The time for the boat to travel 12 miles downstream on the same river is $\dfrac{12}{r+4}$. Find the LCD of the two rational expressions and then write each as an equivalent rational expression using this denominator.

70. Traveling by Plane The time to complete a journey can be calculated by the formula $t = \dfrac{d}{r}$, where t is the time, d is the distance traveled, and r is rate. Suppose a plane flies 500 miles west into a headwind of 50 miles per hour in a time of $\dfrac{500}{r-50}$ hours, where r is the speed of plane in still air. The plane flies 500 miles east with a tailwind of 50 miles per hour in a time of $\dfrac{500}{r+50}$ hours. Find the LCD of the two rational expressions and then write each as an equivalent rational expression using this denominator.

Extending the Concepts

In Problems 71–74, identify the LCD of the rational expressions.

71. $\dfrac{4}{x^3+8}; \dfrac{1}{x^2-4}; \dfrac{5}{x^3-8}$

72. $\dfrac{x}{2x^3-2}; \dfrac{1}{3x^2-3}; \dfrac{2x+1}{4x-4}$

73. $\dfrac{a}{2a^4-2a^2b^2}; \dfrac{b}{4ab^2+4b^3}; \dfrac{ab}{a^3b-b^4}$

74. $\dfrac{5}{3x^2+xy-2y^2}; \dfrac{2}{x^2-xy-2y^2}$

7.5 Adding and Subtracting Rational Expressions with Unlike Denominators

OBJECTIVE

① Add and Subtract Rational Expressions with Unlike Denominators

Preparing for...Answers **1.** 120

2. $(2x+1)(x+1)$ **3.** $\dfrac{x+3}{2x}$

4. $\dfrac{-(1+x)}{2}$ **5.** $\dfrac{6}{x}$

Preparing for Adding and Subtracting Rational Expressions with Unlike Denominators

1. Find the least common denominator (LCD) of 15 and 24. [Section A.1, pp. A2–A5]

2. Factor: $2x^2+3x+1$ [Section 6.3, pp. 427–435]

3. Simplify: $\dfrac{3x+9}{6x}$ [Section 7.1, pp. 480–483]

4. Simplify: $\dfrac{1-x^2}{2x-2}$ [Section 7.1, pp. 480–483]

5. Find the sum: $\dfrac{5}{2x} + \dfrac{7}{2x}$ [Section 7.3, pp. 494–496]

(1)   Add and Subtract Rational Expressions with Unlike Denominators

We are now ready to add and subtract rational expressions with unlike denominators! Keep in mind that the steps for adding and subtracting rational expressions with unlike denominators parallel the steps for adding and subtracting rational numbers with unlike denominators.

Before we add and subtract rational expressions with unlike denominators, we provide an example to remind you how to add rational numbers with unlike denominators.

EXAMPLE 1   How to Add Rational Numbers with Unlike Denominators

Evaluate $\dfrac{1}{6} + \dfrac{9}{14}$. Write the sum in lowest terms.

Step-by-Step Solution

Step 1: Find the least common denominator of 14 and 6.	$6 = 2 \cdot 3$ $14 = 2 \cdot \quad 7$ $\text{LCD} = 2 \cdot 3 \cdot 7 = 42$
Step 2: Write each fraction as an equivalent fraction, with denominator of 42.	Because $6 \cdot 7 = 42$, we use $\dfrac{7}{7} = 1$ to rewrite the denominator 6 as 42. Since $14 \cdot 3 = 42$, we use $\dfrac{3}{3} = 1$ to rewrite the denominator 14 as 42. $\dfrac{1}{6} \cdot \dfrac{7}{7} + \dfrac{9}{14} \cdot \dfrac{3}{3} = \dfrac{7}{42} + \dfrac{27}{42}$
Step 3: Find the sum of the numerators, written over the common denominator.	$= \dfrac{7 + 27}{42}$ $= \dfrac{34}{42}$
Step 4: Simplify. Factor 34 and 42:	$= \dfrac{2 \cdot 17}{2 \cdot 21}$ $= \dfrac{\cancel{2} \cdot 17}{\cancel{2} \cdot 21}$ Divide out common factors: $= \dfrac{17}{21}$

Therefore, $\dfrac{1}{6} + \dfrac{9}{14} = \dfrac{17}{21}$.

QUICK ✓   *Find each sum or difference. Write the sum or difference in lowest terms.*

1. $\dfrac{5}{12} + \dfrac{5}{18}$ **2.** $\dfrac{5}{6} + \dfrac{8}{3}$ **3.** $\dfrac{19}{6} - \dfrac{5}{12}$ **4.** $\dfrac{8}{15} - \dfrac{3}{10}$

Now that you've practiced adding and subtracting rational numbers, let's learn how to add and subtract rational expressions.

| EXAMPLE 2 | **How to Add Rational Expressions with Unlike Monomial Denominators** |

Find the sum and simplify the result, if possible: $\dfrac{5}{6x^2} + \dfrac{4}{15x}$

Step-by-Step Solution

Step 1: Find the least common denominator.

$$6x^2 = 2 \cdot 3 \cdot x^2$$
$$15x = 3 \cdot 5 \cdot x$$
$$\text{The LCD is } 2 \cdot 3 \cdot 5 \cdot x^2 = 30x^2$$

Step 2: Rewrite each rational expression with the common denominator.

We multiply $\dfrac{5}{6x^2}$ by $\dfrac{5}{5}$ to get $30x^2$ in the denominator;

we multiply $\dfrac{4}{15x}$ by $\dfrac{2x}{2x}$ to get $30x^2$ in the denominator:

$$\dfrac{5}{6x^2} + \dfrac{4}{15x} = \dfrac{5}{6x^2} \cdot \dfrac{5}{5} + \dfrac{4}{15x} \cdot \dfrac{2x}{2x}$$
$$= \dfrac{25}{30x^2} + \dfrac{8x}{30x^2}$$

Step 3: Add the rational expressions found in Step 2.

Add using $\dfrac{a}{c} + \dfrac{b}{c} = \dfrac{a+b}{c}$:

$$= \dfrac{25 + 8x}{24x^2}$$

Step 4: Simplify.

The rational expression is already simplified.

Work Smart

Notice in Step 2, when we rewrite the rational expression with the common denominator, we are multiplying by 1.

So $\dfrac{5}{6x^2} + \dfrac{4}{15x} = \dfrac{25 + 8x}{24x^2}$.

We summarize the steps for adding or subtracting rational expressions with unlike denominators.

> ### Steps for Adding or Subtracting Rational Expressions with Unlike Denominators
>
> **Step 1:** Find the least common denominator.
> **Step 2:** Rewrite each rational expression with the common denominator.
> **Step 3:** Add or subtract the rational expressions found in Step 2.
> **Step 4:** Simplify the result.

QUICK ✓ *Find each sum and simplify the result, if possible.*

5. $\dfrac{1}{8a^3b} + \dfrac{5}{12ab^2}$ **6.** $\dfrac{1}{15xy^2} + \dfrac{7}{18x^3y}$

| EXAMPLE 3 | **Adding Rational Expressions with Unlike Polynomial Denominators** |

Find the sum and simplify the result, if possible: $\dfrac{-1}{x+3} + \dfrac{3}{x+2}$

Solution

First, we find that the least common denominator is $(x+3)(x+2)$ because the two denominators have no common factors. Now we need to rewrite each rational expression using the least common denominator. To do this, we multiply $\dfrac{-1}{x+3}$ by

$\dfrac{x + 2}{x + 2}$ to get $(x + 3)(x + 2)$ in the denominator and we multiply $\dfrac{3}{x + 2}$ by $\dfrac{x + 3}{x + 3}$ to get $(x + 3)(x + 2)$ in the denominator.

$$\frac{-1}{x + 3} + \frac{3}{x + 2} = \frac{-1}{x + 3} \cdot \frac{x + 2}{x + 2} + \frac{3}{x + 2} \cdot \frac{x + 3}{x + 3}$$

Multiply out the numerators; leave the denominators in factored form:
$$= \frac{-x - 2}{(x + 3)(x + 2)} + \frac{3x + 9}{(x + 2)(x + 3)}$$

Add using $\dfrac{a}{c} + \dfrac{b}{c} = \dfrac{a + b}{c}$:
$$= \frac{-x - 2 + 3x + 9}{(x + 3)(x + 2)}$$

Combine like terms:
$$= \frac{2x + 7}{(x + 3)(x + 2)}$$

There are no common factors, so $\dfrac{-1}{x + 3} + \dfrac{3}{x + 2} = \dfrac{2x + 7}{(x + 3)(x + 2)}$.

QUICK ✓ *Find each sum and simplify the result, if possible.*

7. $\dfrac{5}{x - 4} + \dfrac{3}{x + 2}$

8. $\dfrac{-1}{n - 3} + \dfrac{4}{n + 1}$

EXAMPLE 4 **Adding Rational Expressions with Unlike Polynomial Denominators**

Find the sum and simplify the result: $\dfrac{4}{x + 1} + \dfrac{16}{x^2 - 2x - 3}$

Solution

Notice that the denominators are not the same. So, what's our first step? We need to find the least common denominator. We begin by factoring each denominator.

$$x + 1 = \qquad x + 1$$
$$x^2 - 2x - 3 = (x - 3)(x + 1)$$

The LCD is $(x - 3)(x + 1)$. Now we rewrite each rational expression with the least common denominator.

$$\frac{4}{x + 1} + \frac{16}{x^2 - 2x - 3} = \frac{4}{x + 1} + \frac{16}{(x - 3)(x + 1)}$$

Rewrite each rational expression with the common denominator:
$$= \frac{4}{x + 1} \cdot \frac{x - 3}{x - 3} + \frac{16}{(x + 1)(x - 3)}$$

Don't reduce $\dfrac{4(x - 3)}{(x + 1)(x - 3)}$ or you'll be right back where you started!
$$= \frac{4(x - 3)}{(x + 1)(x - 3)} + \frac{16}{(x + 1)(x - 3)}$$

Multiply out the numerator:
$$= \frac{4x - 12}{(x + 1)(x - 3)} + \frac{16}{(x + 1)(x - 3)}$$

Add using $\dfrac{a}{c} + \dfrac{b}{c} = \dfrac{a + b}{c}$:
$$= \frac{4x - 12 + 16}{(x + 1)(x - 3)}$$

Add:
$$= \frac{4x + 4}{(x + 1)(x - 3)}$$

Factor the numerator:
$$= \frac{4(x + 1)}{(x + 1)(x - 3)}$$

Divide out like factors:
$$= \frac{4}{x - 3}$$

QUICK ✓ *Find each sum and simplify the result, if possible.*

9. $\dfrac{2}{x-3} + \dfrac{-2}{x^2 - 5x + 6}$ **10.** $\dfrac{-z+1}{z^2+7z+10} + \dfrac{2}{z+5}$ **11.** $\dfrac{1}{x^2+5x} + \dfrac{1}{x^2-5x}$

So far we have concentrated on adding rational expressions with unlike denominators. Now let's go over how to subtract rational expressions with unlike denominators.

EXAMPLE 5 **How to Subtract Rational Expressions with Unlike Denominators**

Find the difference and simplify the result, if possible: $\dfrac{8}{x} - \dfrac{2}{x+1}$

Step-by-Step Solution

Step 1: Find the least common denominator. The LCD is $x(x+1)$

Step 2: Rewrite each rational expression with the common denominator. Multiply out the numerators, but leave the denominators in factored form.

Multiply $\dfrac{8}{x}$ by $\dfrac{x+1}{x+1}$ to get $x(x+1)$ in the denominator; multiply $\dfrac{2}{x+1}$ by $\dfrac{x}{x}$ to get $x(x+1)$ in the denominator.

$$\dfrac{8}{x} - \dfrac{2}{x+1} = \dfrac{8}{x} \cdot \dfrac{x+1}{x+1} - \dfrac{2}{x+1} \cdot \dfrac{x}{x}$$

$$= \dfrac{8(x+1)}{x(x+1)} - \dfrac{2 \cdot x}{x(x+1)}$$

$$= \dfrac{8x+8}{x(x+1)} - \dfrac{2x}{x(x+1)}$$

Step 3: Subtract the rational expressions found in Step 2.

Subtract using $\dfrac{a}{c} - \dfrac{b}{c} = \dfrac{a-b}{c}$:

$$= \dfrac{6x+8}{x(x+1)}$$

Step 4: Simplify the result.

Factor the numerator:

$$= \dfrac{2(3x+4)}{x(x+1)}$$

Since there are no common factors, the expression is written in lowest terms. So $\dfrac{8}{x} - \dfrac{2}{x+1} = \dfrac{2(3x+4)}{x(x+1)}$.

QUICK ✓ *Find each difference and simplify the result, if possible.*

12. $\dfrac{-4}{5ab^2} - \dfrac{3}{4a^2b^3}$ **13.** $\dfrac{5}{x} - \dfrac{3}{x-4}$

EXAMPLE 6 **Adding Rational Expressions with Unlike Denominators— Both Denominators Factor**

Perform the indicated operation and simplify, if possible: $\dfrac{-2}{a^2-4a} + \dfrac{2}{4a-16}$

Solution

First, we factor each denominator to find the LCD.

$$a^2 - 4a = a \cdot (a-4)$$
$$4a - 16 = 4 \cdot (a-4)$$

The LCD is $4a(a - 4)$, so we must rewrite each denominator with a new denominator of $4a(a - 4)$. We multiply $\dfrac{-2}{a(a - 4)}$ by $\dfrac{4}{4}$ and we multiply $\dfrac{2}{4(a - 4)}$ by $\dfrac{a}{a}$.

$$\frac{-2}{a(a - 4)} + \frac{2}{4(a - 4)} = \frac{-2}{a(a - 4)} \cdot \frac{4}{4} + \frac{2}{4(a - 4)} \cdot \frac{a}{a}$$

Multiply numerators: $= \dfrac{-8}{4a(a - 4)} + \dfrac{2a}{4a(a - 4)}$

Add numerators using $\dfrac{a}{c} + \dfrac{b}{c} = \dfrac{a + b}{c}$: $= \dfrac{2a - 8}{4a(a - 4)}$

Factor the numerator and denominator: $= \dfrac{2(a - 4)}{2 \cdot 2a(a - 4)}$

Divide out like factors: $= \dfrac{1}{2a}$

Work Smart

When all the factors in the numerator divide out, the remaining factor is 1, not 0.

So $\dfrac{-2}{a^2 - 4a} + \dfrac{2}{4a - 16} = \dfrac{1}{2a}$. ■

EXAMPLE 7 Subtracting Rational Expressions with Unlike Denominators—Both Denominators Factor

Perform the indicated operation and simplify, if possible: $\dfrac{a - 2}{a^2 + 5a + 6} - \dfrac{2a - 3}{3a^2 + 9a}$

Solution

First, we factor each denominator.

$$a^2 + 5a + 6 = (a + 3)(a + 2)$$
$$3a^2 + 9a = 3a(a + 3)$$

The LCD is $3a(a + 3)(a + 2)$. Now we rewrite each expression with the common denominator.

$$\frac{a - 2}{a^2 + 5a + 6} - \frac{2a - 3}{3a^2 + 9a} = \frac{a - 2}{(a + 3)(a + 2)} \cdot \frac{3a}{3a} - \frac{2a - 3}{3a(a + 3)} \cdot \frac{a + 2}{a + 2}$$

Multiply the numerators: $= \dfrac{(a - 2) \cdot 3a}{3a(a + 3)(a + 2)} - \dfrac{(2a - 3)(a + 2)}{3a(a + 3)(a + 2)}$

$= \dfrac{3a^2 - 6a}{3a(a + 3)(a + 2)} - \dfrac{2a^2 + a - 6}{3a(a + 3)(a + 2)}$

Write numerators over the common denominator; don't forget the parentheses: $= \dfrac{3a^2 - 6a - (2a^2 + a - 6)}{3a(a + 3)(a + 2)}$

Distribute the minus sign: $= \dfrac{3a^2 - 6a - 2a^2 - a + 6}{3a(a + 3)(a + 2)}$

Combine like terms: $= \dfrac{a^2 - 7a + 6}{3a(a + 3)(a + 2)}$

Factor the numerator: $= \dfrac{(a - 6)(a - 1)}{3a(a + 3)(a + 2)}$

There are no common factors, so $\dfrac{a - 2}{a^2 + 5a + 6} - \dfrac{2a - 3}{3a^2 + 9a} = \dfrac{(a - 6)(a - 1)}{3a(a + 3)(a + 2)}$. ■

QUICK ✓ *Perform the indicated operation and simplify, if possible.*

14. $\dfrac{3}{(x-5)(x+4)} - \dfrac{2}{(x-5)(x-1)}$

15. $\dfrac{x-2}{x^2-3x} + \dfrac{x+3}{4x-12}$

EXAMPLE 8 **Adding Rational Expressions Containing Opposite Factors**

Find the sum and simplify, if possible: $\dfrac{1}{1-n} + \dfrac{2n}{n^2-1}$

Solution

First, we factor each denominator:

$$1-n = 1-n$$
$$n^2-1 = (n+1)(n-1)$$

Do you see that the factors $n-1$ and $1-n$ are opposites? We factor -1 from $1-n$ so that $1-n = -1(-1+n) = -1(n-1)$. We rewrite the sum as

$$\frac{1}{1-n} + \frac{2n}{n^2-1} = \frac{1}{-1(n-1)} + \frac{2n}{(n+1)(n-1)}$$

Use $\dfrac{a}{-b} = \dfrac{-a}{b}$: $= \dfrac{-1}{(n-1)} + \dfrac{2n}{(n+1)(n-1)}$

In this form, we can see that the LCD is $(n-1)(n+1)$. Now we rewrite each expression with the common denominator.

Multiply $\dfrac{-1}{n-1}$ by $\dfrac{n+1}{n+1}$: $= \dfrac{-1}{n-1} \cdot \dfrac{n+1}{n+1} + \dfrac{2n}{(n+1)(n-1)}$

Multiply the numerators: $= \dfrac{-1(n+1)}{(n+1)(n-1)} + \dfrac{2n}{(n+1)(n-1)}$

Distribute the -1: $= \dfrac{-n-1}{(n+1)(n-1)} + \dfrac{2n}{(n+1)(n-1)}$

Write numerators over the common denominator: $= \dfrac{-n-1+2n}{(n+1)(n-1)}$

Combine like terms: $= \dfrac{n-1}{(n+1)(n-1)}$

Divide out like factors: $= \dfrac{1}{n+1}$

Work Smart

Remember that when like factors divide, they form a quotient of 1. That's why

$$\frac{n-1}{(n+1)(n-1)}$$

simplifies to be

$$\frac{\overset{1}{\cancel{n-1}}}{(n+1)\underset{1}{\cancel{(n-1)}}} = \frac{1}{n+1}$$

So $\dfrac{1}{1-n} + \dfrac{2n}{n^2-1} = \dfrac{1}{n+1}$. ■

QUICK ✓ *Find the sum and simplify, if possible.*

16. $\dfrac{7}{3p^2-3p} + \dfrac{5}{6-6p}$

EXAMPLE 9 **Adding an Integer and a Rational Expression**

Find the sum and simplify, if possible: $5 + \dfrac{2}{3x-4}$

Solution

First, note that $5 = \dfrac{5}{1}$. Let's rewrite the expression as $\dfrac{5}{1} + \dfrac{2}{3x - 4}$ and proceed.

The LCD is $3x - 4$, so we rewrite the expression with the LCD by multiplying $\dfrac{5}{1}$ by $\dfrac{3x - 4}{3x - 4}$.

$$\frac{5}{1} + \frac{2}{3x - 4} = \frac{5}{1} \cdot \frac{3x - 4}{3x - 4} + \frac{2}{3x - 4}$$

Multiply the numerators:
$$= \frac{5(3x - 4)}{3x - 4} + \frac{2}{3x - 4}$$

$$= \frac{15x - 20}{3x - 4} + \frac{2}{3x - 4}$$

Write numerators over the common denominator:
$$= \frac{15x - 20 + 2}{3x - 4}$$

Combine like terms:
$$= \frac{15x - 18}{3x - 4}$$

Factor the numerator:
$$= \frac{3(5x - 6)}{3x - 4}$$

We see that $5 + \dfrac{2}{3x - 4} = \dfrac{3(5x - 6)}{3x - 4}$.

QUICK ✔ *Perform the indicated operation and simplify, if possible.*

17. $1 + \dfrac{3}{z - 5}$

18. $2 - \dfrac{3}{x - 1}$

EXAMPLE 10 **Adding and Subtracting Rational Expressions**

Perform the indicated operations and simplify, if possible.

$$\frac{5}{n - 2} + \frac{5}{n + 2} - \frac{6}{n^2 - 4}$$

Solution

To find the LCD, each denominator must be factored completely. We see that $n^2 - 4 = (n - 2)(n + 2)$, so the LCD is $(n - 2)(n + 2)$.

$$\frac{5}{n - 2} + \frac{5}{n + 2} - \frac{6}{n^2 - 4} = \frac{5}{n - 2} + \frac{5}{n + 2} - \frac{6}{(n - 2)(n + 2)}$$

Rewrite each expression with the common denominator:
$$= \frac{5}{n - 2} \cdot \frac{n + 2}{n + 2} + \frac{5}{n + 2} \cdot \frac{n - 2}{n - 2} - \frac{6}{(n - 2)(n + 2)}$$

Multiply the numerators:
$$= \frac{5n + 10}{(n - 2)(n + 2)} + \frac{5n - 10}{(n - 2)(n + 2)} - \frac{6}{(n - 2)(n + 2)}$$

Write numerators over the common denominator:
$$= \frac{5n + 10 + 5n - 10 - 6}{(n - 2)(n + 2)}$$

Combine like terms:
$$= \frac{10n - 6}{(n - 2)(n + 2)}$$

Factor the numerator:
$$= \frac{2(5n - 3)}{(n - 2)(n + 2)}$$

Thus, $\dfrac{5}{n - 2} + \dfrac{5}{n + 2} - \dfrac{6}{n^2 - 4} = \dfrac{2(5n - 3)}{(n - 2)(n + 2)}$.

QUICK ✓ *Perform the indicated operation and simplify, if possible.*

19. $\dfrac{1}{x+1} - \dfrac{2}{x^2-1} + \dfrac{3}{x-1}$

20. $\dfrac{3}{x} - \left(\dfrac{1}{x-2} - \dfrac{6}{x^2-2x} \right)$

7.5 Exercises

For Extra Help:

Student Solutions Manual CD Video PH Math/Tutor Center MathXL Tutorials on CD MathXL® MyMathLab

Concepts and Vocabulary

In Problems 1–3, fill in the blanks.

1. The first step in adding or subtracting rational expressions with unlike denominators is to determine the _____ _____ _____.

2. After adding or subtracting rational expressions, factor the numerator of the result to verify that the fraction is written in _____ _____.

3. To rewrite a rational expression with a least common denominator, multiply the numerator and _____ by the factor(s) missing from the LCD.

In Problems 4–6, answer True or False to each statement.

4. $\dfrac{3}{x} + \dfrac{6}{7x^2} = \dfrac{3+6}{7x^2}$

5. To add or subtract rational expressions, we write the rational expressions over a common denominator and then add or subtract the numerators.

6. $\dfrac{3}{x} + \dfrac{6}{-x} = \dfrac{3-6}{x}$

✎ 7. In your own words, explain how to add or subtract rational expressions with unlike denominators.

✎ 8. To find the product $\dfrac{1}{x-1} \cdot \dfrac{1}{x+1}$, your classmate decided that each fraction should be written with a common denominator and wrote the problem as $\dfrac{(x+1)}{(x-1)(x+1)} \cdot \dfrac{(x-1)}{(x+1)(x-1)}$. Is this a correct approach? Explain your response.

Building Skills

In Problems 9–20, find each sum and simplify, if possible.

9. $\dfrac{-4}{3} + \dfrac{1}{2}$

10. $\dfrac{7}{12} + \dfrac{3}{4}$

11. $\dfrac{2}{3x} + \dfrac{1}{x}$

12. $\dfrac{5}{2x} + \dfrac{6}{5}$

13. $\dfrac{a}{2a-1} + \dfrac{3}{2a+1}$

14. $\dfrac{2}{x-1} + \dfrac{x-1}{x+1}$

15. $\dfrac{3}{x-4} + \dfrac{5}{4-x}$

16. $\dfrac{7}{n-4} + \dfrac{8}{4-n}$

17. $\dfrac{a+3}{5a-a^2} + \dfrac{2a+1}{4a-20}$

18. $\dfrac{2x-6}{x^2-x-6} + \dfrac{x+4}{x+2}$

19. $\dfrac{2x+4}{x^2+2x} + \dfrac{3}{x}$

20. $\dfrac{3}{x-4} + \dfrac{x+4}{x^2-16}$

In Problems 21–32, find each difference and simplify, if possible.

21. $\dfrac{7}{15} - \dfrac{9}{25}$

22. $\dfrac{8}{21} - \dfrac{6}{35}$

23. $m - \dfrac{16}{m}$

24. $\dfrac{9}{x} - x$

25. $\dfrac{x}{x-3} - \dfrac{x-2}{x+3}$

26. $\dfrac{x-2}{x+2} - \dfrac{x+2}{x-2}$

27. $\dfrac{x+2}{x+3} - \dfrac{x^2-x}{x^2+6x+9}$

28. $\dfrac{x}{x+4} - \dfrac{-4}{x^2+8x+16}$

29. $\dfrac{6}{2a+6} - \dfrac{4a-1}{a+3}$

30. $\dfrac{3x+7}{2x-6} - \dfrac{x+2}{x-3}$

31. $\dfrac{-3x-9}{x^2+x-6} - \dfrac{x+3}{2-x}$

32. $\dfrac{p+1}{p^2+2p} - \dfrac{p^2-p}{p^2+p-2}$

Mixed Practice

In Problems 33–72, perform the indicated operation and simplify, if possible.

33. $\dfrac{-3}{4} + 2$

34. $4 - \dfrac{34}{3}$

35. $\dfrac{5}{3y^2} - \dfrac{3}{4y}$

36. $\dfrac{5}{4n} - \dfrac{3}{n^2}$

37. $\dfrac{9}{5x} - \dfrac{6}{10x}$

38. $\dfrac{7}{4b^2} - \dfrac{3b^2}{2b}$

39. $\dfrac{7}{2x^2y} + \dfrac{8}{4xy^2}$

40. $\dfrac{-4x}{y} + \dfrac{3x}{x^2y}$

41. $\dfrac{2x}{2x+3} - 1$

42. $a + \dfrac{a}{a-2}$

43. $\dfrac{n}{n-2} + \dfrac{n+2}{n}$

44. $\dfrac{6}{x+2} + \dfrac{4}{x-3}$

45. $\dfrac{2x}{x-3} - \dfrac{5}{x}$

46. $\dfrac{x}{x-2} - \dfrac{2}{x-1}$

47. $\dfrac{a}{a^2} - \dfrac{1}{a-1}$

48. $\dfrac{4}{5y} - \dfrac{2}{y+2}$

49. $\dfrac{4}{2n+1} + 1$

50. $\dfrac{x-2}{x+3} + 2$

51. $\dfrac{-12}{3x-6} + \dfrac{4x-1}{x-2}$

52. $\dfrac{2}{4x+4} + \dfrac{8}{3x+3}$

53. $\dfrac{3x-1}{x} - \dfrac{9}{x^2-9x}$

54. $\dfrac{2x}{8x-12} - \dfrac{3}{2x+2}$

55. $\dfrac{4}{n^2-3n} - \dfrac{3}{n^3-n^2}$

56. $\dfrac{x-1}{x^2-36} - \dfrac{x}{x^2-12x+36}$

57. $\dfrac{5}{2a-a^2} - \dfrac{3}{2a^2-4a}$

58. $\dfrac{-1}{3n-n^2} + \dfrac{1}{3n^2-9n}$

59. $\dfrac{2n+1}{n^2-4} + \dfrac{3n}{6-n^2-n}$

60. $\dfrac{a-5}{2a^2-6a} - \dfrac{6}{12a^2-4a^3}$

61. $\dfrac{x}{x^2-1} + \dfrac{x+1}{x^2-2x+1}$

62. $\dfrac{x+1}{x^2+4x+4} + \dfrac{3}{x^2+x-2}$

63. $\dfrac{2n+1}{n+3} + \dfrac{7-2n^2}{n^2+n-6}$

64. $\dfrac{4}{x^2+2x-15} + \dfrac{3}{x^2-x-6}$

65. $\dfrac{-3x-9}{x^2+x-6} - \dfrac{x+3}{2-x}$

66. $\dfrac{2x-1}{1-x} - \dfrac{-x^2-2x}{x^2+x-2}$

67. $\dfrac{4}{x+2} + \dfrac{-5x-2}{x^2+2x} - \dfrac{3-x}{x}$

68. $\dfrac{x+1}{x-3} + \dfrac{x+2}{x-2} - \dfrac{x^2+3}{x^2-x-6}$

69. $\dfrac{2}{m+2} - \dfrac{3}{m} + \dfrac{m+10}{m^2-4}$

70. $\dfrac{7}{w-3} - \dfrac{5}{w} - \dfrac{2w+6}{w^2-9}$

71. $\dfrac{2}{a} - \left(\dfrac{2}{a-1} - \dfrac{3}{(a-1)^2} \right)$

72. $\dfrac{2}{b} - \left(\dfrac{2}{b+2} - \dfrac{2}{(b+2)^2} \right)$

Applying the Concepts

In Problems 73–80, perform the indicated operation, if possible. Addition, subtraction, multiplication, and division of rational expressions are included.

73. Find the quotient of $\dfrac{x-3}{x+2}$ and $\dfrac{x^2-9}{x^2+4}$.

74. Find the quotient of $\dfrac{x}{x+1}$ and $\dfrac{5}{3x+3}$.

75. Find the difference of $\dfrac{4x+6}{2x^2+x-3}$ and $\dfrac{x-1}{x^2-1}$.

76. Find the difference of $\dfrac{x+4}{2x^2-8}$ and $\dfrac{3}{4x-8}$.

77. Find the sum of $\dfrac{x+2}{x^2+x-6}$ and $\dfrac{x-3}{x^2+5x+6}$.

78. Find the sum of $\dfrac{x^2}{x-4}$ and $\dfrac{16}{4-x}$.

79. Find the product of $\dfrac{2x-3}{x+6}$ and $\dfrac{x-1}{x-7}$.

80. Find the product of $\dfrac{x^2-3x+2}{x^2+2x-3}$ and $\dfrac{x^2+x-6}{x^2-4}$.

△ **81. Area of a Trapezoid** Use the formula $A = \dfrac{1}{2}h(B+b)$ to find the area of the trapezoid shown:

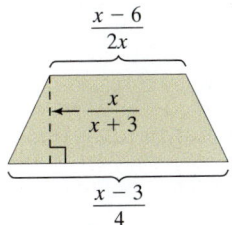

△ **82. Perimeter of a Rectangle** Find the perimeter of the rectangle:

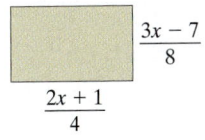

△ **83. Area of a Rectangle** Find the area of the rectangle:

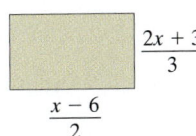

△ **84. Area of a Trapezoid** Find the area of the trapezoid:

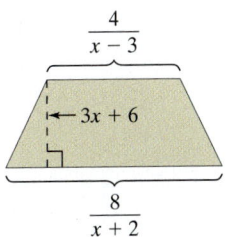

Extending the Concepts

In Problems 85–92, perform the indicated operations and simplify, if possible.

85. $\dfrac{1}{x} - \dfrac{2}{x^2+x} + \dfrac{3}{x^3-x^2}$

86. $\dfrac{x}{(x-1)^2} + \dfrac{2}{x} - \dfrac{x+1}{x^3-x^2}$

87. $\dfrac{2a+b}{a-b} \cdot \dfrac{2}{a+b} - \dfrac{3a+3b}{a^2-b^2}$

88. $\dfrac{2a+b}{a-b} \div \dfrac{a+b}{2} + \dfrac{3}{a+b}$

89. $\dfrac{x-3}{x-4} + \dfrac{x+2}{x-4} \cdot \dfrac{4}{x+1}$

90. $\dfrac{3}{m+n} - \dfrac{m-2n}{m-n} \cdot \dfrac{2}{m+n}$

91. $\dfrac{x+2}{x-2} - \dfrac{x-2}{x+2} \div \dfrac{1}{x^2-4}$

92. $\dfrac{a+3}{a-3} - \dfrac{a+3}{a-3} \cdot \dfrac{a^2-4a+3}{a^2+5a+6}$

7.6 Complex Rational Expressions

OBJECTIVES

① Simplify a Complex Rational Expression by Simplifying the Numerator and Denominator Separately

② Simplify a Complex Rational Expression Using the Least Common Denominator

Preparing for Complex Rational Expressions

Before getting started, take the following readiness quiz. If you get a problem wrong, go back to the section cited and review the material.

1. Factor: $6y^2 - 5y - 6$ [Section 6.3, pp. 427–435]

2. Find the quotient: $\dfrac{x+3}{12} \div \dfrac{x^2-9}{15}$ [Section 7.2, pp. 487–490]

DEFINITION

A **complex rational expression** is a fraction in which the numerator and/or the denominator contains the sum or difference of rational expressions.

The following are examples of complex rational expressions.

$$\dfrac{3}{\dfrac{4}{x}-\dfrac{x}{3}} \qquad \dfrac{\dfrac{x+1}{x+4}-1}{16} \qquad \dfrac{3-\dfrac{1}{x}}{1+\dfrac{1}{x}}$$

To **simplify** a complex rational expression means to write the rational expression in the form $\dfrac{p}{q}$, where p and q are polynomials that have no common factors. We can simplify complex rational expressions using one of two methods:

1. By simplifying the numerator and the denominator separately, or
2. By using the least common denominator.

The goal in each of these methods is to write the complex rational expression in the form $\dfrac{\text{rational expression}}{\text{rational expression}}$. Then, we can simplify the expression using the division techniques introduced in Section 7.2. To help refresh your memory regarding this technique, we present Example 1.

EXAMPLE 1 **Dividing Rational Expressions**

Find the quotient: $\dfrac{\dfrac{3}{x+4}}{\dfrac{9}{x^2-16}}$

Solution

Notice the rational expression is in the form $\dfrac{\text{rational expression}}{\text{rational expression}}$, where the rational expression in the numerator is $\dfrac{3}{x+4}$ and the rational expression in the denominator is $\dfrac{9}{x^2-16}$. To find the quotient, we multiply the rational expression in the numerator

---

Preparing for...Answers

1. $(3y+2)(2y-3)$ **2.** $\dfrac{5}{4(x-3)}$

by the reciprocal of the rational expression in the denominator.

$$\frac{\dfrac{3}{x+4}}{\dfrac{9}{x^2-16}} = \frac{3}{x+4} \cdot \frac{x^2-16}{9}$$

Factor the numerator and denominator:

$$= \frac{3}{x+4} \cdot \frac{(x+4)(x-4)}{3 \cdot 3}$$

Multiply:

$$= \frac{3(x+4)(x-4)}{(x+4) \cdot 3 \cdot 3}$$

Divide out common factors:

$$= \frac{3\,\cancel{(x+4)}(x-4)}{\cancel{(x+4)} \cdot 3 \cdot 3}$$

$$= \frac{x-4}{3}$$

QUICK ✓ *Simplify each expression.*

1. $\dfrac{\dfrac{2}{k+3}}{\dfrac{4}{k^2+4k+3}}$

2. $\dfrac{\dfrac{1}{n+3}}{\dfrac{8n}{2n+6}}$

---

① **Simplify a Complex Rational Expression by Simplifying the Numerator and Denominator Separately**

As mentioned, there are two methods that can be used to simplify a complex rational expression. We'll start by simplifying the numerator and denominator separately. We call this Method 1.

EXAMPLE 2 **How to Simplify a Complex Rational Expression (Method 1)**

Simplify: $\dfrac{\dfrac{1}{5}+\dfrac{1}{x}}{\dfrac{x+5}{2}}$

Step-by-Step Solution

Step 1: Write the numerator of the complex fraction as a single rational expression.

The numerator of the complex rational expression is $\dfrac{1}{5}+\dfrac{1}{x}$. Its least common denominator is $5x$.

LCD = $5x$:
$$\frac{1}{5}+\frac{1}{x} = \frac{1}{5}\cdot\frac{x}{x} + \frac{1}{x}\cdot\frac{5}{5}$$

$$= \frac{x}{5x}+\frac{5}{5x}$$

Use $\dfrac{a}{c}+\dfrac{b}{c}=\dfrac{a+b}{c}$:
$$= \frac{x+5}{5x}$$

Step 2: Write the denominator of the complex rational expression as a single rational expression.

The denominator of the complex rational expression is $\dfrac{x+5}{2}$. It is already written as a single rational expression.

(continued)

Step 3: Rewrite the complex rational expression using the rational expressions determined in Steps 1 and 2.

$$\frac{\dfrac{1}{5} + \dfrac{1}{x}}{\dfrac{x+5}{2}} = \frac{\dfrac{x+5}{5x}}{\dfrac{x+5}{2}}$$

Step 4: Simplify the rational expression using the techniques for dividing rational expressions from Section 7.2.

Rewrite the division problem as a multiplication problem: $= \dfrac{x+5}{5x} \cdot \dfrac{2}{x+5}$

Multiply: $= \dfrac{(x+5) \cdot 2}{5x(x+5)}$

Divide out like factors: $= \dfrac{2}{5x}$

So $\dfrac{\dfrac{1}{5} + \dfrac{1}{x}}{\dfrac{x+5}{2}} = \dfrac{2}{5x}$.

We summarize the steps that you can use to simplify a complex rational expression.

Work Smart

First simplify (add or subtract) both the numerator and the denominator into a single rational expression before you take the reciprocal and multiply.

> ### Steps to Simplify a Complex Rational Expression by Simplifying the Numerator and Denominator Separately (Method 1)
>
> **Step 1:** Write the numerator of the complex rational expression as a single rational expression.
>
> **Step 2:** Write the denominator of the complex rational expression as a single rational expression.
>
> **Step 3:** Rewrite the complex rational expression using the rational expressions determined in Steps 1 and 2.
>
> **Step 4:** Simplify the rational expression using the techniques for dividing rational expressions from Section 7.2.

EXAMPLE 3 Simplifying a Complex Rational Expression (Method 1)

Simplify: $\dfrac{\dfrac{2}{y} - \dfrac{8}{y^3}}{\dfrac{2}{y^2} + \dfrac{1}{y}}$

Solution

We write the numerator and denominator of the complex rational expression as a single rational expression.

NUMERATOR:

LCD is y^3: $\quad \dfrac{2}{y} - \dfrac{8}{y^3} = \dfrac{2}{y} \cdot \dfrac{y^2}{y^2} - \dfrac{8}{y^3}$

$= \dfrac{2y^2 - 8}{y^3}$

$= \dfrac{2(y^2 - 4)}{y^3}$

$= \dfrac{2(y+2)(y-2)}{y^3}$

DENOMINATOR:

LCD is y^2: $\quad \dfrac{2}{y^2} + \dfrac{1}{y} = \dfrac{2}{y^2} + \dfrac{1}{y} \cdot \dfrac{y}{y}$

$= \dfrac{2}{y^2} + \dfrac{y}{y^2}$

$= \dfrac{2 + y}{y^2}$

Now we rewrite the complex rational expression with the new numerator and denominator.

$$\dfrac{\dfrac{2}{y} - \dfrac{8}{y^3}}{\dfrac{2}{y^2} + \dfrac{1}{y}} = \dfrac{\dfrac{2(y+2)(y-2)}{y^3}}{\dfrac{2+y}{y^2}}$$

Multiply the numerator by the reciprocal of the denominator:
$$= \dfrac{2(y+2)(y-2)}{y^3} \cdot \dfrac{y^2}{2+y}$$

Simplify by dividing out common factors:
$$= \dfrac{2(y+2)(y-2)}{y^2 \cdot y} \cdot \dfrac{y^2}{y+2}$$

$$= \dfrac{2(y-2)}{y}$$

QUICK ✔ *Simplify each expression.*

3. $\dfrac{\dfrac{1}{3} + \dfrac{1}{x}}{\dfrac{x+3}{2}}$

4. $\dfrac{\dfrac{3}{y} - 1}{\dfrac{9}{y} - y}$

EXAMPLE 4   **Dividing Rational Expressions (Method 1)**

Simplify: $\dfrac{\dfrac{1}{x+2} + 1}{x - \dfrac{3}{x+2}}$

Solution

Write the numerator of the complex rational expression, $\dfrac{1}{x+2} + 1$, as a single rational expression. The LCD is $(x+2)$.

$$\dfrac{1}{x+2} + 1 = \dfrac{1}{x+2} + 1 \cdot \dfrac{x+2}{x+2}$$

Multiply:
$$= \dfrac{1}{x+2} + \dfrac{x+2}{x+2}$$

Write the numerators over a common denominator:
$$= \dfrac{1 + x + 2}{x+2}$$

Combine like terms:
$$= \dfrac{x+3}{x+2}$$

Write the denominator of the complex rational expression, $x - \dfrac{3}{x+2}$, as a single rational expression. The LCD of the expression is $x + 2$.

$$x - \dfrac{3}{x+2} = \dfrac{x}{1} \cdot \dfrac{x+2}{x+2} - \dfrac{3}{x+2}$$

Multiply:
$$= \dfrac{x^2 + 2x}{x+2} - \dfrac{3}{x+2}$$

Write the numerators over a common denominator:
$$= \dfrac{x^2 + 2x - 3}{x+2}$$

Factor:
$$= \dfrac{(x+3)(x-1)}{x+2}$$

Rewrite the complex rational expression using the numerator and denominator just found, and then simplify.

$$\frac{\dfrac{1}{x+2}+1}{x-\dfrac{3}{x+2}}=\frac{\dfrac{x+3}{x+2}}{\dfrac{(x+3)(x-1)}{x+2}}$$

Multiply the reciprocal of the denominator by the numerator:

$$=\frac{x+3}{x+2}\cdot\frac{x+2}{(x+3)(x-1)}$$

Multiply:

$$=\frac{\cancel{(x+3)}\,\cancel{(x+2)}}{\cancel{(x+2)}\,\cancel{(x+3)}\,(x-1)}$$

Divide like factors:

$$=\frac{1}{x-1}$$

QUICK ✔ *Simplify each expression.*

5. $\dfrac{\dfrac{1}{3}+\dfrac{1}{x+5}}{\dfrac{x+8}{9}}$

6. $\dfrac{\dfrac{8}{y+3}-2}{y-\dfrac{4}{y+3}}$

---

(2) Simplify a Complex Rational Expression Using the Least Common Denominator

We now introduce a second method for simplifying complex rational expressions.

EXAMPLE 5 **How to Simplify a Complex Rational Expression Using the Least Common Denominator (Method 2)**

Simplify: $\dfrac{\dfrac{2}{y}-\dfrac{8}{y^3}}{\dfrac{2}{y^2}+\dfrac{1}{y}}$

Step-by-Step Solution

Step 1: Find the least common denominator among each denominator in the complex rational expression.

The denominators of the complex rational expression are y, y^3, and y^2. The least common denominator is y^3.

Step 2: Multiply both the numerator and denominator of the complex rational expression by the least common denominator found in Step 1.

Multiply the numerator and denominator by y^3:

$$\left(\frac{\dfrac{2}{y}-\dfrac{8}{y^3}}{\dfrac{2}{y^2}+\dfrac{1}{y}}\right)\cdot\left(\frac{y^3}{y^3}\right)=\frac{\dfrac{2}{y}\cdot y^3-\dfrac{8}{y^3}\cdot y^3}{\dfrac{2}{y^2}\cdot y^3+\dfrac{1}{y}\cdot y^3}$$

Divide out common factors:

$$=\frac{\dfrac{2}{\cancel{y}}\cdot\cancel{y^3}y^2-\dfrac{8}{\cancel{y^3}}\cdot\cancel{y^3}1}{\dfrac{2}{\cancel{y^2}}\cdot\cancel{y^3}y+\dfrac{1}{\cancel{y}}\cdot\cancel{y^3}y^2}$$

Step 3: Simplify.

$$\begin{aligned} \text{Multiply:} \quad &= \frac{2y^2 - 8}{2y + y^2} \\[2mm] \text{Factor:} \quad &= \frac{2(y^2 - 4)}{y(2 + y)} \\[2mm] &= \frac{2(y + 2)(y - 2)}{y(y + 2)} \\[2mm] \text{Divide out like factors:} \quad &= \frac{2(y - 2)}{y} \end{aligned}$$

So $\dfrac{\dfrac{2}{y} - \dfrac{8}{y^3}}{\dfrac{2}{y^2} + \dfrac{1}{y}} = \dfrac{2(y - 2)}{y}$. This is the same result we obtained in Example 3.

Let's redo Example 4 using Method 2 so that we can compare the results.

EXAMPLE 6 **Simplifying a Complex Rational Expression Using the Least Common Denominator**

Simplify: $\dfrac{\dfrac{1}{x + 2} + 1}{x - \dfrac{3}{x + 2}}$

Solution

The least common denominator is $x + 2$. So, we multiply the numerator and denominator by $(x + 2)(x + 1)$.

$$\frac{\dfrac{1}{x + 2} + 1}{x - \dfrac{3}{x + 2}} = \frac{\dfrac{1}{x + 2} + 1}{\dfrac{x}{1} - \dfrac{3}{x + 2}} \cdot \frac{x + 2}{x + 2}$$

Distribute the LCD to each term: $\quad = \dfrac{\dfrac{1}{x + 2} \cdot (x + 2) + 1 \cdot (x + 2)}{\dfrac{x}{1} \cdot (x + 2) - \dfrac{3}{x + 2} \cdot (x + 2)}$

Divide out like factors: $\quad = \dfrac{\dfrac{1}{\cancel{x + 2}} \cdot \cancel{(x + 2)} + x + 2}{\dfrac{x}{1} \cdot (x + 2) - \dfrac{3}{\cancel{x + 2}} \cdot \cancel{(x + 2)}}$

Distribute: $\quad = \dfrac{1 + x + 2}{x^2 + 2x - 3}$

Combine like terms; factor: $\quad = \dfrac{x + 3}{(x + 3)(x - 1)}$

Divide out like factors: $\quad = \dfrac{1}{x - 1}$

This is the same result we obtained in Example 4.

Next, we summarize the steps you can use to simplify a complex rational expression using the least common denominator method.

> **Steps to Simplify a Complex Rational Expressions Using the Least Common Denominator (Method 2)**
>
> **Step 1:** Find the least common denominator among each denominator in the complex rational expression.
>
> **Step 2:** Multiply both the numerator and denominator of the complex rational expression by the least common denominator found in Step 1.
>
> **Step 3:** Simplify the rational expression, if possible.

QUICK ✓ *Simplify the complex rational expression using Method 2.*

7. $\dfrac{\dfrac{1}{x} + \dfrac{2}{y}}{\dfrac{2}{x} - \dfrac{1}{y}}$

7.6 Exercises

For Extra Help: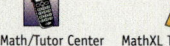

Student Solutions Manual CD Video PH Math/Tutor Center MathXL Tutorials on CD MathXL® MyMathLab

Concepts and Vocabulary

In Problems 1–3, fill in the blanks.

1. A rational expression such as $\dfrac{\dfrac{x}{2} + \dfrac{5}{x}}{\dfrac{2x - 1}{3}}$ is called a _____ _____ _____.

2. To _____ a complex rational expression means to write the rational expression in the form $\dfrac{p}{q}$, where p and q are polynomials that have no common factors.

3. To simplify $\dfrac{\dfrac{x + 1}{6}}{\dfrac{x^2 + 3x + 2}{3}}$ take the _____ of $\dfrac{x^2 + 3x + 2}{3}$ and _____ it by $\dfrac{x + 1}{6}$.

In Problems 4–6, answer True or False to each statement.

4. A complex rational expression is considered simplified if there is only one fraction bar and there are no factors common to both the numerator and denominator.

5. To simplify using Method 1, we rewrite $\dfrac{x - y}{\dfrac{1}{x} + \dfrac{2}{y}}$ as $\dfrac{x - y}{1} \cdot \left(\dfrac{x}{1} + \dfrac{y}{2}\right)$.

6. The expression $\dfrac{\dfrac{2}{x + 1}}{\dfrac{1}{x - 5}}$ is simplified.

7. State which method you prefer to use when simplifying complex rational expressions. State your reasons for choosing this method and then explain how to use this method to simplify $\dfrac{\dfrac{1}{x} + \dfrac{1}{y}}{\dfrac{1}{x^2} - \dfrac{1}{y^2}}$.

8. Explain the property of real numbers that is being applied in Method 2. Make up a complex rational expression and explain how to use Method 2 to simplify the expression.

Building Skills

In Problems 9–20, simplify the complex rational expressions using both Method 1 and Method 2.
State which method you prefer for the problem.

9. $\dfrac{\dfrac{4}{3}+1}{\dfrac{8}{15}-1}$

10. $\dfrac{\dfrac{5}{9}+1}{\dfrac{7}{3}-1}$

11. $\dfrac{1-\dfrac{3}{4}}{\dfrac{1}{8}+2}$

12. $\dfrac{\dfrac{2}{3}-\dfrac{3}{4}}{\dfrac{1}{6}+\dfrac{1}{2}}$

13. $\dfrac{1}{\dfrac{5}{6}-\dfrac{7}{8}}$

14. $\dfrac{\dfrac{3}{7}-\dfrac{1}{5}}{2}$

15. $\dfrac{\dfrac{x^2}{12}-\dfrac{1}{3}}{\dfrac{x+2}{18}}$

16. $\dfrac{\dfrac{4}{t^2}-1}{\dfrac{t+2}{t^3}}$

17. $\dfrac{\dfrac{x+3}{x^2}}{\dfrac{x^2}{9}-1}$

18. $\dfrac{\dfrac{n+2}{4}}{\dfrac{n^2}{n^2}-1}$

19. $\dfrac{\dfrac{m}{2}+n}{\dfrac{m}{n}}$

20. $\dfrac{\dfrac{1}{x}-\dfrac{1}{y}}{\dfrac{2}{xy}}$

Mixed Practice

In Problems 21–36, simplify the complex rational expression using either Method 1 or Method 2.

21. $\dfrac{\dfrac{1}{x}-\dfrac{3}{y}}{\dfrac{4}{x}+\dfrac{1}{y}}$

22. $\dfrac{\dfrac{a}{b}+\dfrac{b}{a}}{\dfrac{a}{b}-\dfrac{b}{a}}$

23. $\dfrac{\dfrac{3}{m}+\dfrac{2}{m^2}}{\dfrac{6}{m}+\dfrac{4}{m^2}}$

24. $\dfrac{\dfrac{3c}{4}+\dfrac{3d}{10}}{\dfrac{3c}{2}-\dfrac{6d}{5}}$

25. $\dfrac{1-\dfrac{49}{b^2}}{1+\dfrac{7}{b}}$

26. $\dfrac{\dfrac{a}{2}+4}{2-\dfrac{a}{2}}$

27. $\dfrac{1+\dfrac{5}{x}}{1+\dfrac{1}{x+4}}$

28. $\dfrac{\dfrac{x}{x+1}}{1+\dfrac{1}{x-1}}$

29. $\dfrac{\dfrac{1}{x^2}-\dfrac{1}{y^2}}{x-y}$

30. $\dfrac{6x+\dfrac{3}{y}}{\dfrac{9x+3}{y}}$

31. $\dfrac{\dfrac{x}{x-y}+\dfrac{y}{x+y}}{\dfrac{xy}{x^2-y^2}}$

32. $\dfrac{\dfrac{b}{b+1}-1}{\dfrac{b+3}{b}-2}$

33. $\dfrac{\dfrac{2}{x+4}}{\dfrac{2}{x+4}-4}$

34. $\dfrac{1-\dfrac{4}{x^2}}{1-\dfrac{1}{x}-\dfrac{6}{x^2}}$

35. $\dfrac{\dfrac{b^2}{b^2-16}-\dfrac{b}{b+4}}{\dfrac{b}{b^2-16}-\dfrac{1}{b-4}}$

36. $\dfrac{\dfrac{-6}{y^2+5y+6}}{\dfrac{2}{y+3}-\dfrac{3}{y+2}}$

Applying the Concepts

37. Finding the Mean The arithmetic mean of a set of numbers is found by adding the numbers and then dividing by the number of entries on the list. Write a complex rational expression to find the arithmetic mean of the expressions $\dfrac{n}{6}, \dfrac{n+3}{2}$, and $\dfrac{2n-1}{8}$ and then simplify the complex rational expression.

38. Finding the Mean See Problem 37. Write a complex rational expression to find the arithmetic mean of the expressions $\dfrac{z}{2}, \dfrac{z-3}{4}$, and $\dfrac{2z-1}{6}$ and then simplify the complex rational expression.

39. Area of a Rectangle In a rectangle, the length can be found by dividing the area by the width. If the area of a rectangle is $\dfrac{2x+3}{x^2}-\dfrac{3}{x}$ and the width is $\dfrac{x^2-9}{x^5}$, write a complex rational expression to find the length and then simplify the complex rational expression.

△ **40. Area of a Rectangle** In a rectangle, the length can be found by dividing the area by the width. If the area of a rectangle is $\dfrac{x-4}{x^3} - \dfrac{2}{x^2}$ and the width is $\dfrac{x^2-16}{x}$, write a complex rational expression to find the length and then simplify the complex rational expression.

41. Electric Circuits An electrical circuit contains two resistors connected in parallel, as shown in the figure. If the resistance of each is R_1 and R_2 ohms, respectively, then their combined resistance R is given by the formula

$$R = \dfrac{1}{\dfrac{1}{R_1} + \dfrac{1}{R_2}}$$

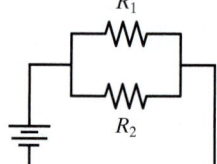

(a) Express R as a simplified rational expression.
(b) Evaluate the rational expression if $R_1 = 6$ ohms and $R_2 = 10$ ohms.

42. Electric Circuits An electrical circuit contains three resistors connected in parallel. If the resistance of each is R_1, R_2, and R_3 ohms, respectively, then their combined resistance is given by the formula

$$R = \dfrac{1}{\dfrac{1}{R_1} + \dfrac{1}{R_2} + \dfrac{1}{R_3}}$$

(a) Express R as a simplified rational expression.
(b) Evaluate the rational expression if $R_1 = 4$ ohms, $R_2 = 6$ ohms, and $R_3 = 10$ ohms.

Extending the Concepts

In Problems 43–46, write each expression in lowest terms.

43. $1 + \dfrac{1}{1 + \dfrac{1}{x}}$ **44.** $1 - \dfrac{1}{1 - \dfrac{1}{x-2}}$ **45.** $\dfrac{1}{1 - \dfrac{1}{2 - \dfrac{1}{3-x}}}$ **46.** $1 + \dfrac{2}{1 + \dfrac{2}{x + \dfrac{2}{x}}}$

PUTTING THE CONCEPTS TOGETHER (Sections 7.1–7.6)

These problems cover important concepts from Sections 7.1 to 7.6. We designed these problems so that you can review the chapter so far and show your mastery of the concepts. Take time to work these problems before proceeding with the next section. The answers to these problems are located at the back of the text starting on page AN-22.

1. Evaluate $\dfrac{a^3 - b^3}{a + b}$ when $a = 2$ and $b = -3$.

2. Find the values for which the following rational expressions are undefined:

(a) $\dfrac{-3a}{a - 6}$ (b) $\dfrac{y + 2}{y^2 + 4y}$

3. Simplify:

(a) $\dfrac{ax + ay - 4bx - 4by}{2x + 2y}$ (b) $\dfrac{x^2 + x - 2}{1 - x^2}$

4. Find the least common denominator (LCD) of the rational expressions:

$$\dfrac{5}{2a + 4b}, \dfrac{-1}{4a + 8b}, \dfrac{3ab}{8a - 32b}$$

5. Write $\dfrac{7}{3x^2 - x}$ as an equivalent rational expression with denominator $5x^2(3x - 1)$.

In Problems 6–13, perform the indicated operations and simplify, if possible.

6. $\dfrac{y^2 - y}{3y} \cdot \dfrac{6y^2}{1 - y^2}$

7. $\dfrac{m^2 + m - 2}{m^3 - 6m^2} \cdot \dfrac{2m^2 - 14m + 12}{m + 2}$

8. $\dfrac{4y + 12}{5y - 5} \div \dfrac{2y^2 - 18}{y^2 - 2y + 1}$

9. $\dfrac{4x^2 + x}{x + 3} + \dfrac{12x + 3}{x + 3}$

10. $\dfrac{x^2}{x^3 + 1} - \dfrac{x - 1}{x^3 + 1}$

11. $\dfrac{-8}{2x - 1} - \dfrac{9}{1 - 2x}$

12. $\dfrac{4}{m + 3} + \dfrac{3}{3m + 2}$

13. $\dfrac{3m}{m^2 + 7m + 10} - \dfrac{2m}{m^2 + 6m + 8}$

14. Simplify the complex rational expression: $\dfrac{\dfrac{-1}{m + 1} - 1}{m - \dfrac{2}{m + 1}}$

15. Simplify the complex rational expression: $\dfrac{\dfrac{4}{a} + \dfrac{1}{6}}{\dfrac{3}{a^2} + \dfrac{1}{2}}$

7.7 Rational Equations

OBJECTIVES

1 Solve Equations Containing Rational Expressions

2 Solve for a Variable in a Rational Equation

Preparing for Rational Equations

Before getting started, take the following readiness quiz. If you get a problem wrong, go back to the section cited and review the material.

1. Solve: $3k - 2(k + 1) = 6$ — [Section 2.2, pp. 86–87]

2. Factor: $3p^2 - 7p - 6$ — [Section 6.3, pp. 427–435]

3. Solve: $8z^2 - 10z - 3 = 0$ — [Section 6.6, pp. 451–457]

4. Find the values for which the expression $\dfrac{x + 4}{x^2 - 2x - 24}$ is undefined. — [Section 7.1, pp. 479–480]

5. Solve for y: $4x - 2y = 10$ — [Section 2.4, pp. 108–111]

1 Solve Equations Containing Rational Expressions

Up to this point, we have learned how to solve linear equations (Sections 2.1–2.3), quadratic equations (Section 6.6), and equations that contain polynomial expressions that can be factored (Section 6.6). We now introduce another type of equation.

> **DEFINITION**
>
> A **rational equation** is an equation that contains a rational expression.

Preparing for...Answers **1.** $\{8\}$

2. $(3p + 2)(p - 3)$ **3.** $\left\{-\dfrac{1}{4}, \dfrac{3}{2}\right\}$

4. $x = -4$ or $x = 6$ **5.** $y = 2x - 5$

Examples of rational equations are

$$\dfrac{3}{x + 4} = \dfrac{1}{x + 2} \quad \text{and} \quad \dfrac{4}{3} + \dfrac{7}{x - 4} = \dfrac{x - 1}{3x - 12}$$

The goal in solving these equations is to use algebraic techniques to rewrite the equation until it becomes one that you already know how to solve, such as a linear or quadratic equation. In Chapter 2, you learned how to solve equations such as $\frac{3x}{4} - \frac{x}{2} = \frac{1}{2}$. Recall, one approach to solving an equation whose coefficients are fractions is to multiply each side of the equation by the least common denominator of the fractions, which will eliminate the fractions.

To solve an equation such as $\frac{8}{p} + \frac{1}{4p} = \frac{11}{8}$ we use the same process. Let's compare the approaches.

EXAMPLE 1 Solving a Rational Equation

Solve: **(a)** $\frac{3x}{4} - \frac{x}{2} = \frac{1}{2}$ **(b)** $\frac{8}{p} + \frac{1}{4p} = \frac{11}{8}, \quad p \neq 0$

Solution

Notice Example 1(a) is a linear equation with fractional coefficients. Example 1(b) is a rational equation because the equation contains a rational expression. As we solve these two equations, notice that the steps for solving the two equations are identical. They may also look familiar because they are the same steps that we used back in Chapter 2.

(a)
$$\frac{3x}{4} - \frac{x}{2} = \frac{1}{2}$$

Find the LCD: The LCD is 4.

Multiply each side of the equation by the LCD:
$$4\left(\frac{3x}{4} - \frac{x}{2}\right) = 4\left(\frac{1}{2}\right)$$

Distribute and divide out common factors:
$$4\left(\frac{3x}{4}\right) - 4\left(\frac{x}{2}\right) = 4\left(\frac{1}{2}\right)$$
$$3x - 2x = 2$$
$$x = 2$$

(b)
$$\frac{8}{p} + \frac{1}{4p} = \frac{11}{8}, \quad p \neq 0$$

Find the LCD: The LCD is $8p$.

Multiply each side of the equation by the LCD:
$$8p\left(\frac{8}{p} + \frac{1}{4p}\right) = 8p\left(\frac{11}{8}\right)$$

Distribute and divide out common factors:
$$8p\left(\frac{8}{p}\right) + 8p\left(\frac{1}{4p}\right) = 8p\left(\frac{11}{8}\right)$$
$$64 \quad + \quad 2 \quad = \quad 11p$$

Solve the linear equation:
$$66 = 11p$$
$$6 = p$$

We leave the check up to you. The solution set of $\frac{3x}{4} - \frac{x}{2} = \frac{1}{2}$ is $\{2\}$. The solution set of $\frac{8}{p} + \frac{1}{4p} = \frac{11}{8}$ is $\{6\}$. ∎

Did you notice the restriction that $p \neq 0$ for the equation $\frac{8}{p} + \frac{1}{4p} = \frac{11}{8}$? We include that condition because if p is replaced by the number 0, both $\frac{8}{p}$ and $\frac{1}{4p}$ become undefined.

When solving a rational equation, always determine the value(s) of the variable that will cause the expression to be undefined.

The other interesting "tidbit" you should have noticed is that by multiplying both sides of the equation in Example 1(b) by the LCD, we obtain a linear equation—an equation that we already know how to solve. Again, this is the key to mathematics—solve a problem by rewriting it as a problem you already know how to solve.

EXAMPLE 2 **How to Solve a Rational Equation**

Solve: $\dfrac{3}{2b - 1} = \dfrac{4}{3b}$

Step-by-Step Solution

Step 1: Determine the value(s) of the variable that result in an undefined rational expression in the rational equation.

To find the undefined values, we determine the values of the variable that cause either denominator to equal 0. By solving $2b - 1 = 0$, we find $b = \dfrac{1}{2}$; by solving $3b = 0$, we find $b = 0$. The undefined values for the variable are $b = \dfrac{1}{2}$ and $b = 0$, so $b \neq \dfrac{1}{2}, 0$.

Step 2: Determine the least common denominator (LCD) of all the denominators.

The LCD is $3b(2b - 1)$.

Step 3: Multiply both sides of the equation by the LCD and simplify the expression on each side of the equation.

$$\frac{3}{2b - 1} = \frac{4}{3b}$$

Multiply both sides by $3b(2b - 1)$: $3b(2b - 1)\left(\dfrac{3}{2b - 1}\right) = 3b(2b - 1)\left(\dfrac{4}{3b}\right)$

Divide out common factors: $3b\cancel{(2b - 1)} \cdot \dfrac{3}{\cancel{2b - 1}} = \cancel{3b}(2b - 1) \cdot \dfrac{4}{\cancel{3b}}$

Step 4: Solve the resulting equation.

$$3b(3) = (2b - 1)4$$

Multiply and distribute: $9b = 8b - 4$

Subtract $8b$ from both sides: $b = -4$

Step 5: Verify your solution using the original equation.

Let $b = -4$: $\dfrac{3}{2(-4) - 1} \stackrel{?}{=} \dfrac{4}{3(-4)}$

$$\frac{3}{-8 - 1} \stackrel{?}{=} \frac{4}{-12}$$

$$\frac{3}{-9} \stackrel{?}{=} \frac{4}{-12}$$

$$-\frac{1}{3} = -\frac{1}{3} \quad \text{True}$$

The solution checks, so the solution set is $\{-4\}$.

The following steps can be used to solve any rational equation.

Steps to Solve a Rational Equation

Step 1: Determine the value(s) of the variable that result in an undefined rational expression in the rational equation.

Step 2: Determine the least common denominator (LCD) of all the denominators.

(continued)

Step 3: Multiply both sides of the equation by the LCD, and simplify the expression on each side of the equation.

Step 4: Solve the resulting equation.

Step 5: Verify your solution using the original equation.

QUICK ✓ *Solve the rational equation.*

1. $\dfrac{5}{2} + \dfrac{1}{z} = 4$ **2.** $\dfrac{8}{x+4} = \dfrac{12}{x-3}$ **3.** $\dfrac{3}{z} - \dfrac{1}{2z} = -\dfrac{5}{8}$ **4.** $\dfrac{4}{3b} + \dfrac{1}{6b} = \dfrac{7}{2b} + \dfrac{1}{3}$

EXAMPLE 3 Solving a Rational Equation

Solve: $\dfrac{4}{3} + \dfrac{7}{x-4} = \dfrac{x-1}{3x-12}$

Solution

We first factor the denominator $3x - 12$. Having this denominator written in factored form will help us to find the undefined values of the variable x and to find the LCD. Because $3x - 12 = 3(x - 4)$, we write the equation as

Work Smart

Writing the denominator in factored form makes finding the undefined values of the variable and the LCD easier.

$$\frac{4}{3} + \frac{7}{x-4} = \frac{x-1}{3(x-4)}$$

Now we set $x - 4 = 0$ and solve for x to find the undefined values of the rational equation.

$$x - 4 = 0$$
$$x = 4$$

The value of the variable for which the rational expression is undefined is $x = 4$, so $x \neq 4$. We also see from the factored form of the denominators that the LCD is $3(x - 4)$.

$$\frac{4}{3} + \frac{7}{x-4} = \frac{x-1}{3(x-4)}$$

Multiply both sides of the equation by the LCD, $3(x - 4)$:
$$3(x-4)\left(\frac{4}{3} + \frac{7}{x-4}\right) = 3(x-4) \cdot \frac{x-1}{3(x-4)}$$

Distribute:
$$3(x-4)\left(\frac{4}{3}\right) + 3(x-4)\left(\frac{7}{x-4}\right) = 3(x-4)\left(\frac{x-1}{3(x-4)}\right)$$

Divide common factors:
$$\cancel{3}(x-4) \cdot \frac{4}{\cancel{3}} + (3)\cancel{(x-4)} \cdot \frac{7}{\cancel{x-4}} = \cancel{3(x-4)} \cdot \frac{x-1}{\cancel{3(x-4)}}$$

$$4(x-4) + 3(7) = x - 1$$

Distribute the 4; multiply: $4x - 16 + 21 = x - 1$

Combine like terms: $4x + 5 = x - 1$

Subtract x from both sides: $3x + 5 = -1$

Subtract 5 from both sides: $3x = -6$

Divide both sides by 3: $x = -2$

Check Let $x = -2$ in the *original* equation.

$$\frac{4}{3} + \frac{7}{x-4} = \frac{x-1}{3x-12}$$

Let $x = -2$: $\dfrac{4}{3} + \dfrac{7}{-2-4} \stackrel{?}{=} \dfrac{-2-1}{3(-2)-12}$

$$\frac{4}{3} + \frac{7}{-6} \stackrel{?}{=} \frac{-3}{-6-12}$$

$$\frac{4}{3} - \frac{7}{6} \overset{?}{=} \frac{-3}{-18}$$

$$\frac{8}{6} - \frac{7}{6} \overset{?}{=} \frac{1}{6}$$

$$\frac{1}{6} = \frac{1}{6} \quad \text{True}$$

The solution checks, so the solution set is $\{-2\}$.

QUICK ✓ *Solve the rational equation.*

5. $\dfrac{3}{2} + \dfrac{5}{x - 3} = \dfrac{x + 9}{2x - 6}$

EXAMPLE 4   **Solving a Rational Equation in Which the LCD Is the Product of the Denominators**

Solve: $\dfrac{5}{x + 2} + \dfrac{3}{x - 2} = 1$

Solution

The rational expression $\dfrac{5}{x + 2} + \dfrac{3}{x - 2}$ is undefined at $x = -2$ and $x = 2$. The LCD is $(x + 2)(x - 2)$.

$$\frac{5}{x + 2} + \frac{3}{x - 2} = 1$$

Multiply both sides of the equation by the LCD: $(x + 2)(x - 2)\left[\dfrac{5}{x + 2} + \dfrac{3}{x - 2}\right] = 1(x + 2)(x - 2)$

Distribute: $(x + 2)(x - 2)\left(\dfrac{5}{x + 2}\right) + (x + 2)(x - 2)\left(\dfrac{3}{x - 2}\right) = (x + 2)(x - 2) \cdot 1$

Divide common factors: $\cancel{(x + 2)}(x - 2)\left(\dfrac{5}{\cancel{x + 2}}\right) + (x + 2)\cancel{(x - 2)}\left(\dfrac{3}{\cancel{x - 2}}\right) = (x + 2)(x - 2) \cdot 1$

$$(x - 2)5 + (x + 2)(3) = (x + 2)(x - 2)$$

Multiply and distribute: $5x - 10 + 3x + 6 = x^2 - 4$

Combine like terms: $8x - 4 = x^2 - 4$

Write the quadratic equation in standard form: $0 = x^2 - 8x$

Factor the polynomial: $0 = x(x - 8)$

Set each factor equal to zero and solve for x: $x = 0 \quad$ or $\quad x - 8 = 0$

$$x = 8$$

Work Smart

Recall that a quadratic equation is a second-degree equation. It is solved by writing the equation in standard form, $ax^2 + bx + c = 0$, factoring, and then using the Zero-Product Property.

Check

$$\frac{5}{x + 2} + \frac{3}{x - 2} = 1$$

$x = 0$:
$$\frac{5}{0 + 2} + \frac{3}{0 - 2} \overset{?}{=} 1$$
$$\frac{5}{2} + \frac{3}{-2} \overset{?}{=} 1$$
$$\frac{5}{2} - \frac{3}{2} \overset{?}{=} 1$$
$$\frac{2}{2} = 1$$
$$1 = 1 \quad \text{True}$$

$$\frac{5}{x + 2} + \frac{3}{x - 2} = 1$$

$x = 8$:
$$\frac{5}{8 + 2} + \frac{3}{8 - 2} \overset{?}{=} 1$$
$$\frac{5}{10} + \frac{3}{6} \overset{?}{=} 1$$
$$\frac{1}{2} + \frac{1}{2} \overset{?}{=} 1$$
$$1 = 1 \quad \text{True}$$

The solutions check, so the solution set is $\{0, 8\}$.

QUICK ✓ *Solve the rational equation.*

6. $\dfrac{4}{x-3} - \dfrac{3}{x+3} = 1$

Did you wonder why we determine the values of the variable that make a rational expression in the rational equation undefined? Well, in some instances the solution process gives results that do not satisfy the original equation.

> **DEFINITION**
>
> An **extraneous solution** is a solution that is obtained through the solving process that does not satisfy the original equation.

EXAMPLE 5 Solving a Rational Equation with an Extraneous Solution

Solve: $\dfrac{x^2 + 8x + 6}{x^2 + 3x - 4} = \dfrac{3}{x-1} - \dfrac{2}{x+4}$

Solution

Let's rewrite the equation so that the denominator $x^2 + 3x - 4$ is in factored form. This will allow us to find the values of x for which the rational expressions are undefined and determine the LCD. Because $x^2 + 3x - 4 = (x-1)(x+4)$, we write the equation as

$$\frac{x^2 + 8x + 6}{(x-1)(x+4)} = \frac{3}{x-1} - \frac{2}{x+4}$$

We can see that $x \neq 1$ and $x \neq -4$ since these values of x cause the denominator to equal 0. We also see that the LCD is $(x-1)(x+4)$.

Now we multiply both sides of the equation by the LCD and divide out like factors.

$$\frac{x^2 + 8x + 6}{(x-1)(x+4)} = \frac{3}{x-1} - \frac{2}{x+4}$$

$$(x-1)(x+4) \cdot \frac{x^2 + 8x + 6}{(x-1)(x+4)} = (x-1)(x+4)\left(\frac{3}{x-1}\right) - (x-1)(x+4)\left(\frac{2}{x+4}\right)$$

$$\cancel{(x-1)(x+4)}\left(\frac{x^2 + 8x + 6}{\cancel{(x-1)(x+4)}}\right) = \cancel{(x-1)}(x+4)\left(\frac{3}{\cancel{x-1}}\right) - (x-1)\cancel{(x+4)}\left(\frac{2}{\cancel{x+4}}\right)$$

$$x^2 + 8x + 6 = 3(x+4) - 2(x-1)$$

Multiply and distribute: $\quad x^2 + 8x + 6 = 3x + 12 - 2x + 2$

Combine like terms: $\quad x^2 + 8x + 6 = x + 14$

Write the quadratic equation in standard form: $\quad x^2 + 7x - 8 = 0$

Solve the quadratic equation: $\quad (x+8)(x-1) = 0$

$$x + 8 = 0 \quad \text{or} \quad x - 1 = 0$$

$$x = -8 \quad \text{or} \quad x = 1$$

Since $x = 1$ is one of the values of x that results in division by zero, we reject $x = 1$ as a solution. We leave it to you to verify that $x = -8$ is a solution.

The solution set of $\dfrac{x^2 + 8x + 6}{x^2 + 3x - 4} = \dfrac{3}{x-1} - \dfrac{2}{x+4}$ is $\{-8\}$.

QUICK ✓ *Solve the rational equation.*

7. $\dfrac{5}{z-4} + \dfrac{3}{z-2} = \dfrac{z^2 - z - 2}{z^2 - 6z + 8}$

You may recall from Chapter 2 that some equations in one variable have no solution. That's also true with rational equations.

EXAMPLE 6   Solving a Rational Equation with No Solution

Solve: $\dfrac{6}{z^2 - 1} = \dfrac{5}{z - 1} - \dfrac{3}{z + 1}$

Solution

$$\frac{6}{z^2 - 1} = \frac{5}{z - 1} - \frac{3}{z + 1}$$

Factor $z^2 - 1$: $\dfrac{6}{(z - 1)(z + 1)} = \dfrac{5}{z - 1} - \dfrac{3}{z + 1}$

Work Smart: Study Skills
Showing your work makes it easier for you to see how you get the extraneous solution. Don't skip steps to get the answer. Double check every step as you proceed through the problem to catch possible errors.

From the factored form of the denominator, we can see that $z \neq 1$ and $z \neq -1$. We also see that the LCD is $(z - 1)(z + 1)$. So we multiply both sides of the equation by the LCD, $(z - 1)(z + 1)$, and divide out common factors.

$$(z - 1)(z + 1) \cdot \frac{6}{(z - 1)(z + 1)} = (z - 1)(z + 1)\left(\frac{5}{z - 1}\right) - (z - 1)(z + 1)\left(\frac{3}{(z + 1)}\right)$$

$$\cancel{(z - 1)}(z + 1)\left(\frac{6}{\cancel{(z - 1)}(z + 1)}\right) = \cancel{(z - 1)}(z + 1)\left(\frac{5}{\cancel{z - 1}}\right) - (z - 1)\cancel{(z + 1)}\left(\frac{3}{\cancel{(z + 1)}}\right)$$

$$6 = 5(z + 1) - 3(z - 1)$$

Distribute: $6 = 5z + 5 - 3z + 3$

Combine like terms: $6 = 2z + 8$

Subtract 8 from both sides: $-2 = 2z$

Divide both sides by 2: $-1 = z$

Since z cannot equal -1 (because it results in division by 0), we reject $z = -1$ as a solution. Therefore, the equation $\dfrac{6}{z^2 - 1} = \dfrac{5}{z - 1} - \dfrac{3}{z + 1}$ has no solution. The solution set is $\{\ \}$ or $\varnothing$.

Work Smart
The solution set $\{\ \}$ is not the same as the solution set $\{0\}$.

QUICK ✓   *Solve the rational equation.*

8. $\dfrac{4}{y + 1} = \dfrac{7}{y - 1} - \dfrac{8}{y^2 - 1}$

Work Smart
When we solve a rational equation we multiply both sides of the equation by the LCD, but when we add or subtract rational expressions, we retain the LCD. Notice the difference in the directions: we *simplify* expressions and *solve* equations.

Simplify: $\dfrac{2}{x} - \dfrac{1}{6} + \dfrac{5}{2x} - \dfrac{1}{3}$

$\dfrac{2}{x} - \dfrac{1}{6} + \dfrac{5}{2x} - \dfrac{1}{3} = \dfrac{2}{x} \cdot \dfrac{6}{6} - \dfrac{1}{6} \cdot \dfrac{x}{x} + \dfrac{5}{2x} \cdot \dfrac{3}{3} - \dfrac{1}{3} \cdot \dfrac{2x}{2x}$

$= \dfrac{12}{6x} - \dfrac{x}{6x} + \dfrac{15}{6x} - \dfrac{2x}{6x}$

$= \dfrac{12 - x + 15 - 2x}{6x}$

$= \dfrac{17 - 3x}{6x}$

Solve: $\dfrac{2}{x} - \dfrac{1}{6} = \dfrac{5}{2x} - \dfrac{1}{3}$

$6x\left(\dfrac{2}{x} - \dfrac{1}{6}\right) = 6x\left(\dfrac{5}{2x} - \dfrac{1}{3}\right)$

$6x\left(\dfrac{2}{x}\right) - 6x\left(\dfrac{1}{6}\right) = 6x\left(\dfrac{5}{2x}\right) - 6x\left(\dfrac{1}{3}\right)$

$6(2) - x = 3(5) - 2x$

$12 - x = 15 - 2x$

$12 - x + 2x = 15 - 2x + 2x$

$12 + x = 15$

$x = 3$

EXAMPLE 7   An Application of Rational Equations: Average Daily Cost

Suppose that the average daily cost $\overline{C}$ of manufacturing x bicycles is given by the equation

$$\overline{C} = \frac{x^2 + 75x + 5000}{x}$$

Determine the number of bicycles for which the average daily cost will be $225.

Solution

First, notice that x must be greater than 0. Do you know why? Since we want to know the level of production to obtain an average daily cost of $225, we wish to solve the equation $\overline{C} = 225$.

$$\overline{C} = \frac{x^2 + 75x + 5000}{x}$$

$$225 = \frac{x^2 + 75x + 5000}{x}$$

Multiply both sides by x: $x \cdot 225 = \dfrac{x^2 + 75x + 5000}{x} \cdot x$

$$225x = x^2 + 75x + 5000$$

Subtract $225x$ from both sides: $0 = x^2 - 150x + 5000$

Factor: $0 = (x - 50)(x - 100)$

Zero-Product Property: $x - 50 = 0$ or $x - 100 = 0$

$$x = 50 \quad \text{or} \quad x = 100$$

The level of production for which the average daily cost will be $225 is at 50 bicycles or 100 bicycles.

QUICK ✓

9. The concentration C of a certain drug in a patient's bloodstream t hours after injection is given by $C = \dfrac{50t}{t^2 + 25}$. When will the concentration of the drug be 4 milligrams per liter?

② Solve for a Variable in a Rational Equation

Work Smart

The steps that we follow when solving formulas for a certain variable are identical to those that we followed when solving rational equations.

Recall from Section 2.4 that the expression "solve for the variable" means to get the variable by itself on one side of the equation with all other variables and constants, if any, on the other side. The steps that we follow when solving formulas for a certain variable are identical to those that we followed when solving rational equations. We compare solving the rational equation $5 = \dfrac{3}{1 + x}$ for x with solving the rational equation $P = \dfrac{A}{1 + r}$ for r.

$$5 = \frac{3}{1 + x} \qquad\qquad\qquad P = \frac{A}{1 + r}$$

Multiply by the LCD: $(1 + x)5 = (1 + x)\dfrac{3}{1 + x} \qquad (1 + r)P = (1 + r)\dfrac{A}{1 + r}$

Distribute: $5 + 5x = 3 \qquad\qquad\qquad P + Pr = A$

Isolate the unknown: $5 - 5 + 5x = 3 - 5 \qquad\qquad P - P + Pr = A - P$

$$5x = -2 \qquad\qquad\qquad Pr = A - P$$

$$\frac{5x}{5} = \frac{-2}{5} \qquad\qquad\qquad \frac{Pr}{P} = \frac{A - P}{P}$$

$$x = -\frac{2}{5} \qquad\qquad\qquad r = \frac{A - P}{P}$$

Do you see that the steps for solving for the variable r in $P = \dfrac{A}{1 + r}$ are identical to the steps in solving for x in $5 = \dfrac{3}{1 + x}$?

EXAMPLE 8   **How to Solve for a Variable in a Rational Equation**

Solve $\dfrac{x}{1 + y} = z$ for y.

Step-by-Step Solution

We need to isolate the variable y. That is, we need to get y by itself on one side of the equation and the other variables on the other side of the equation.

Step 1: Determine the value(s) of the variable that results in any undefined rational expression in the rational equation.

$$y \neq -1$$

Step 2: Determine the least common denominator (LCD) of all the denominators.

The LCD is $1 + y$.

Step 3: Multiply both sides of the equation by the LCD and simplify the expression on each side of the equation.

$$\frac{x}{1 + y} = z$$

$$(1 + y)\left(\frac{x}{1 + y}\right) = z(1 + y)$$

$$x = z(1 + y)$$

Step 4: Solve the resulting equation for y.

Divide both sides by z: $\dfrac{x}{z} = \dfrac{z(1 + y)}{z}$

$$\frac{x}{z} = 1 + y$$

Subtract 1 from both sides: $\dfrac{x}{z} - 1 = 1 - 1 + y$

$$\frac{x}{z} - 1 = y$$

Look back at Step 4, in which we divided both sides of the equation by z. Did you think about using the Distributive Property to simplify $x = z(1 + y)$? If you did, you are correct! Another way to complete solving $\dfrac{x}{1 + y} = z$ for y is as follows:

$$x = z(1 + y)$$

Use the Distributive Property: $x = z + zy$

Isolate the expression containing the variable y: $x - z = z - z + zy$

$$x - z = zy$$

Divide both sides by z: $\dfrac{x - z}{z} = \dfrac{zy}{z}$

$$\frac{x - z}{z} = y$$

The result that we get using this method produces the equation $y = \dfrac{x - z}{z}$, which is equivalent to the equation $y = \dfrac{x}{z} - 1$. Do you see why?

QUICK ✓ *Solve the equation for the indicated variable.*

10. Solve $R = \dfrac{4g}{x}$ for x **11.** $S = \dfrac{a}{1-r}$ for r

---

EXAMPLE 9 **Solving for a Variable in a Rational Equation**

Solve $\dfrac{1}{a} + \dfrac{1}{b} = \dfrac{1}{c}$ for b.

We can see that neither a, b, nor c can be equal to zero. Further, the LCD is abc.

$$\frac{1}{a} + \frac{1}{b} = \frac{1}{c}$$

Multiply both sides by the LCD: $(abc)\left(\dfrac{1}{a} + \dfrac{1}{b}\right) = (abc)\left(\dfrac{1}{c}\right)$

Distribute: $(abc)\left(\dfrac{1}{a}\right) + (abc)\left(\dfrac{1}{b}\right) = (abc)\left(\dfrac{1}{c}\right)$

Divide out common factors: $(abc)\left(\dfrac{1}{a}\right) + (abc)\left(\dfrac{1}{b}\right) = (abc)\left(\dfrac{1}{c}\right)$

$$bc + ac = ab$$

Work Smart
To solve for a variable that occurs on both sides of the equation, get all terms containing that variable on the same side of the equation. Then factor out the variable you wish to solve for.

To solve for b, we need to get both terms that contain the variable b on the same side of the equation. We'll subtract bc from both sides of the equation and continue.

Subtract bc from both sides: $ac = ab - bc$

Factor out b as a greatest common factor: $ac = b(a - c)$

Divide each side by $a - c$: $\dfrac{ac}{a-c} = \dfrac{b(a-c)}{a-c}$

$$\frac{ac}{a-c} = b$$

When we solve $\dfrac{1}{a} + \dfrac{1}{b} = \dfrac{1}{c}$ for b, we obtain the equation $b = \dfrac{ac}{a-c}$.

QUICK ✓ *Solve the equation for the indicated variable.*

12. $\dfrac{1}{f} = \dfrac{1}{p} + \dfrac{1}{q}$ for p

7.7 Exercises

For Extra Help: Student Solutions Manual CD Video PH Math/Tutor Center MathXL Tutorials on CD Math XL MathXL® MyMathLab MyMathLab

Concepts and Vocabulary

In Problems 1–3, fill in the blanks.

1. To solve an equation that contains one or more rational expressions, multiply both sides of the equation by the _____ _____ _____ to form an equivalent equation without any rational expressions.

2. Apparent solutions that do not satisfy the original equation are called _____ _____.

3. When solving an equation that contains rational expressions, extraneous solutions are values for which a rational expression in the equation is _____.

In Problems 4–6, answer True or False to each statement.

4. There are no values of x for which the rational expressions in the equation $\dfrac{3x + 1}{5} = \dfrac{10}{x + 2}$ are undefined.

5. Equivalent equations are different equations which have the same solution set.

6. A rational equation can have no solution.

7. Explain the difference between the direction *simplify* and *solve*. Make up a problem using each of these directions with the rational expressions $\dfrac{5}{x + 1}$ and $\dfrac{6}{x - 1}$.

8. Explain how using the LCD to solve an equation that contains rational expressions is different from using the LCD in a problem requiring adding and subtracting rational expressions with unlike denominators.

Building Skills

In Problems 9–36, solve each equation and state the solution set. Remember to state the values of the variable for which the expressions in each rational equation are undefined, if necessary.

9. $\dfrac{4}{x} = \dfrac{2}{9}$

10. $\dfrac{6}{x} = \dfrac{2}{9}$

11. $\dfrac{6}{x} + 5 = 3$

12. $\dfrac{12}{x} - 5 = -8$

13. $\dfrac{5}{3y} - \dfrac{1}{2} = \dfrac{5}{6y} - \dfrac{1}{12}$

14. $\dfrac{3}{z} - \dfrac{3}{2} = \dfrac{6}{z}$

15. $\dfrac{1}{3} + \dfrac{1}{x} = \dfrac{3}{x}$

16. $\dfrac{5}{6} + \dfrac{3}{x} = \dfrac{3}{4}$

17. $\dfrac{4}{x} - \dfrac{11}{5} = \dfrac{3}{2x} + \dfrac{6}{5}$

18. $\dfrac{6}{x} + \dfrac{2}{3} = \dfrac{4}{2x} - \dfrac{14}{3}$

19. $\dfrac{3}{2}m + 6 = 2m - \dfrac{7}{4}$

20. $\dfrac{4}{p} - \dfrac{5}{4} = \dfrac{5}{2p} + \dfrac{3}{8}$

21. $\dfrac{4}{x - 1} = \dfrac{3}{x + 4}$

22. $\dfrac{4}{x - 4} = \dfrac{5}{x + 4}$

23. $\dfrac{2}{x + 2} + 2 = \dfrac{7}{x + 2}$

24. $\dfrac{4}{x + 1} + 2 = \dfrac{3}{x + 1}$

25. $\dfrac{r - 4}{3r} + \dfrac{2}{5r} = \dfrac{1}{5}$

26. $\dfrac{x - 2}{4x} - \dfrac{x + 2}{3x} = \dfrac{1}{6x} + \dfrac{2}{3}$

27. $\dfrac{2}{a - 1} + \dfrac{3}{a + 1} = \dfrac{-6}{a^2 - 1}$

28. $\dfrac{6}{t} - \dfrac{2}{t - 1} = \dfrac{2 - 4t}{t^2 - t}$

29. $\dfrac{1}{4 - x} + \dfrac{2}{x^2 - 16} = \dfrac{1}{x - 4}$

30. $\dfrac{4}{x - 3} - \dfrac{3}{x - 2} = \dfrac{4x + 9}{x^2 - 5x + 6}$

31. $\dfrac{3}{2t - 2} - \dfrac{2t}{3t - 3} = -4$

32. $\dfrac{2x + 3}{x - 1} - 2 = \dfrac{3x - 1}{4x - 4}$

33. $\dfrac{2}{5} + \dfrac{3 - 2a}{10a - 20} = \dfrac{2a + 1}{a - 2}$

34. $\dfrac{1}{x - 2} + \dfrac{2}{2 - x} = \dfrac{5}{x + 1}$

35. $\dfrac{6}{j^2 - 1} - \dfrac{4j}{j^2 - 5j + 4} = -\dfrac{4}{j - 1}$

36. $\dfrac{3}{2x + 2} - \dfrac{5}{4x - 4} = \dfrac{2x}{x^2 - 1}$

Mixed Practice

In Problems 37–56, solve each equation and state the solution set. Be sure to check for extraneous solutions.

37. $\dfrac{x}{x-2} = \dfrac{3}{x+8}$

38. $\dfrac{x+5}{x-7} = \dfrac{x-3}{x+7}$

39. $\dfrac{x}{x+3} = \dfrac{6}{x-3} + 1$

40. $\dfrac{3x^2}{x+1} = 2 + \dfrac{3x}{x+1}$

41. $\dfrac{-p}{p-6} = \dfrac{p+6}{5}$

42. $\dfrac{1}{n-3} = \dfrac{3n-1}{9-n^2}$

43. $x = \dfrac{2-x}{6x}$

44. $x = \dfrac{6-5x}{6x}$

45. $\dfrac{2x+3}{x-1} = \dfrac{x-2}{x+1} + \dfrac{6x}{x^2-1}$

46. $\dfrac{2x}{x+4} = \dfrac{x+1}{x+2} - \dfrac{7x+12}{x^2+6x+8}$

47. $\dfrac{2x}{x+3} - \dfrac{2x^2+2}{x^2-9} = \dfrac{-6}{x-3} + 1$

48. $\dfrac{2}{b+2} - \dfrac{5b+6}{b^2-b-6} = \dfrac{-b}{b-3}$

49. $\dfrac{5x}{2x-3} = \dfrac{3x}{x-1} - \dfrac{5}{2x^2-5x+3}$

50. $\dfrac{2x+1}{x^2+2x-3} = \dfrac{x-1}{x^2+5x+6} + \dfrac{x+1}{x^2+x-2}$

51. $\dfrac{x}{x^2-1} - \dfrac{x+3}{x^2-x} = \dfrac{-3}{x^2+x}$

52. $\dfrac{2t}{t^2+2t+1} + \dfrac{t-1}{t^2+t} = \dfrac{6t+8}{t^3+2t^2+t}$

53. $\dfrac{5}{x-2} - \dfrac{2}{2-x} = \dfrac{4}{x+1}$

54. $\dfrac{x}{12} + \dfrac{1}{2} = \dfrac{1}{3x} + \dfrac{2}{x^2}$

55. $\dfrac{a}{2a-2} - \dfrac{2}{3a+3} = \dfrac{5a^2-2a+9}{12a^2-12}$

56. $\dfrac{2x+1}{x^2+2x-3} = \dfrac{x-1}{x^2+5x+6} + \dfrac{x+1}{x^2+x-2}$

In Problems 57–74, solve the equation for the indicated variable.

57. $x = \dfrac{2}{y}$ for y

58. $4 = \dfrac{k}{m}$ for m

59. $I = \dfrac{E}{R}$ for R

60. $T = \dfrac{D}{R}$ for R

61. $h = \dfrac{2A}{B+b}$ for b

62. $m = \dfrac{y-y_1}{x-x_1}$ for y

63. $\dfrac{x}{3+y} = z$ for y

64. $\dfrac{a}{b+2} = c$ for b

65. $\dfrac{1}{R} = \dfrac{1}{S} + \dfrac{1}{T}$ for S

66. $\dfrac{3}{i} - \dfrac{4}{j} = \dfrac{8}{k}$ for j

67. $m = \dfrac{n}{y} - \dfrac{p}{ay}$ for y

68. $\dfrac{1}{a} = \dfrac{1}{b} + \dfrac{1}{c}$ for c

69. $A = \dfrac{xy}{x+y}$ for x

70. $B = \dfrac{k}{x} - \dfrac{m}{cx}$ for x

71. $\dfrac{2}{x} - \dfrac{1}{y} = \dfrac{6}{z}$ for y

72. $y = \dfrac{a}{a+b}$ for b

73. $y = \dfrac{x}{x-c}$ for x

74. $X = \dfrac{ab}{a-b}$ for a

In Problems 75–84, simplify the expression or solve the equation.

75. $\dfrac{1}{x} + \dfrac{3}{x+5}$

76. $\dfrac{2}{x-5} + \dfrac{5}{x-5}$

77. $x - \dfrac{6}{x} = 1$

78. $z + \dfrac{3}{z} = 4$

79. $\dfrac{3}{x-1} \cdot \dfrac{x^2-1}{6} + 3$

80. $5 - \dfrac{n^2-3n-4}{n^2-4}$

81. $2b - \dfrac{5}{3} = \dfrac{1}{3b}$

82. $3a + \dfrac{7}{3} = \dfrac{2}{3a}$

83. $\dfrac{x^2}{x^2-4} - \dfrac{1}{x}$

84. $\dfrac{x}{x+1} + \dfrac{2x-3}{x-1}$

Applying the Concepts

85. Drug Concentration The concentration C of a drug in a patient's bloodstream in milligrams per liter t hours after ingestion is modeled by $C = \dfrac{40t}{t^2 + 9}$. When will the concentration of the drug be 4 milligrams per liter?

86. Drug Concentration The concentration C of a drug in a patient's bloodstream in milligrams per liter t hours after ingestion is modeled by $C = \dfrac{40t}{t^2 + 3}$. When will the concentration of the drug be 10 milligrams per liter?

87. Average Cost Suppose that the average daily cost $\overline{C}$ of manufacturing x bicycles is given by the equation $\overline{C} = \dfrac{x^2 + 75x + 5000}{x}$. Determine the level of production for which the average daily cost will be \$240.

88. Cost-Benefit Model Environmental scientists often use cost-benefit models to estimate the cost of removing a pollutant from the environment related to the percentage of pollutant removed. Suppose a cost-benefit model for the cost C (in millions of dollars) of removing x percent of the pollutants from Maple Lake is given by $C = \dfrac{25x}{100 - x}$. If the federal government budgets \$100 million to clean up the lake, what percent of the pollutants can be removed?

Extending the Concepts

89. For what value of k will the solution set of $\dfrac{4x + 3}{k} = \dfrac{x - 1}{3}$ be $\{2\}$?

90. For what value of k will the solution set of $\dfrac{1}{2} + \dfrac{3x}{k} = 1 + \dfrac{x}{3}$ be $\{1\}$?

7.8 Models Involving Rational Equations

OBJECTIVES

1 Model and Solve Ratio and Proportion Problems

2 Model and Solve Problems with Similar Figures

3 Model and Solve Work Problems

4 Model and Solve Uniform Motion Problems

Preparing for *Models Involving Rational Equations*
Before getting started, take the following readiness quiz. If you get a problem wrong, go back to the section cited and review the material.

1. Solve: $\dfrac{150}{r} = \dfrac{250}{r + 20}$ [Section 7.7, pp. 529–532]

1 Model and Solve Ratio and Proportion Problems

We begin with a definition.

> **DEFINITION**
> A **ratio** is the quotient of two numbers or two quantities. The ratio of two numbers a and b can be written as
> $$a \text{ to } b \qquad \text{or} \qquad a{:}b \qquad \text{or} \qquad \dfrac{a}{b}$$

When solving algebraic problems, we write ratios as $\dfrac{a}{b}$. For example, the odds of winning the "Pick Three" Instant Ohio Lottery are 1 in 1000, so we write that ratio as $\dfrac{1}{1000}$. Generally, a ratio is written as a fraction reduced to lowest terms. If 8 oz of a cleaner are required for every 4 gallons of water, the ratio is $\dfrac{8 \text{ oz}}{4 \text{ gal}} = \dfrac{2 \text{ oz}}{1 \text{ gal}}$.

A rational equation which involves two ratios is called a *proportion*.

> **DEFINITION**
> A **proportion** is an equation of the form $\dfrac{a}{b} = \dfrac{c}{d}$, where $b \neq 0$ and $d \neq 0$. We call a, b, c, and d the **terms** of the proportion. The terms b and c are called the **means** of the proportion and the terms a and d are called the **extremes** of the proportion.

We may solve a proportion using the same method that we used to solve a rational equation.

EXAMPLE 1 Solving a Proportion

Solve the proportion: $\dfrac{x}{5} = \dfrac{63}{105}$

Solution

Because $105 = 21 \cdot 5$, the LCD is 105. So we multiply both sides of the equation by 105.

$$\frac{x}{5} = \frac{63}{105}$$

$$105\left(\frac{x}{5}\right) = 105\left(\frac{63}{105}\right)$$

Divide out common factors: $\quad 21x = 63$

Divide both sides by 21: $\quad x = 3$

The solution set is $\{3\}$.

Work Smart

The equation $\dfrac{3}{8} = \dfrac{7}{x} + 1$ is NOT a proportion. Do you see why?

Work Smart

Another method frequently used to solve a proportion is the method of cross multiplication. When this process is used, the same result is obtained as multiplying both sides of the equation by the LCD.

> **CROSS-MULTIPLICATION PROPERTY**
>
> If $\dfrac{a}{b} = \dfrac{c}{d}$, where $b \neq 0$ and $d \neq 0$, then $a \cdot d = b \cdot c$.

We can use this method to solve a proportion because for the proportion $\dfrac{a}{b} = \dfrac{c}{d}$, the LCD is bd. If we multiply both sides of the equation $\dfrac{a}{b} = \dfrac{c}{d}$ by the LCD, we obtain the result given by the Cross-Multiplication Property:

$$bd\left(\frac{a}{b}\right) = \left(\frac{c}{d}\right)bd$$
$$ad = bc$$

If we solve the proportion in Example 1, $\dfrac{x}{5} = \dfrac{63}{105}$, using cross multiplication, we obtain

$$\frac{x}{5} = \frac{63}{105}$$

Cross multiply: $\quad 105x = 5 \cdot 63$

$$105x = 315$$

Divide both sides by 105: $\quad x = 3$

QUICK ✔ *Solve the proportion for the indicated variable.*

1. $\dfrac{x}{9} = \dfrac{-4}{3}$ **2.** $\dfrac{5}{y+4} = \dfrac{2}{y-1}$ **3.** $\dfrac{2p+1}{4} = \dfrac{p}{8}$ **4.** $\dfrac{6}{x^2} = \dfrac{2}{x}$

EXAMPLE 2   **Exchange Rates**

Last summer Laura took a Study Abroad trip to Italy. When she returned, she examined her credit card bill and found that she spent 199 euros (€) on the digital camera she bought in Venice. The bill stated that $1 US was equal to 0.796509 euros (€). How much did the camera cost in U.S. dollars?

Solution

Step 1: Identify   We want to know the cost of the camera in U.S. dollars.

Step 2: Name   We let c represent the cost of the camera in U.S. dollars.

Step 3: Translate   We want to know how many dollars are equivalent to 199 euros (€). Since c represents the cost of the camera (in dollars) and we know 1 dollar equals 0.796509 euros, we set up the proportion using $\dfrac{\text{dollars}}{\text{euros}}$.

$$\frac{\$1}{0.796509\,€} = \frac{c}{199\,€} \qquad \text{The Model}$$

Notice we include the units of measure in the model. This is done to keep track of the units. Our answer should be in dollars.

Step 4: Solve   We'll now solve the equation.

Cross multiply: $\$1 \cdot 199\,€ = [0.796509\,€]c$

Divide both sides by 0.796509: $\dfrac{\$1 \cdot 199\,€}{0.796509\,€} = \dfrac{[0.796509\,€]c}{0.796509\,€}$

Simplify using a calculator: $\$249.84 \approx c$

Step 5: Check   Is this answer reasonable? First, we notice our answer is in dollars. This is encouraging. Second, the ratio 1 dollar to 0.796509 euros should be approximately equal to the ratio of $249.84 to 199 euros. That is, $\dfrac{1}{0.796509}$ should equal $\dfrac{249.84}{199}$.

Because $\dfrac{1}{0.796509} \approx 1.2555$ and $\dfrac{249.84}{199} \approx 1.2555$, our answer checks.

Step 6: Answer   The camera that Laura purchased in Italy cost about $249.84 U.S.

Work Smart

You could also set up the model using $\dfrac{\text{euros}}{\text{dollars}}$.

$$\frac{0.796509\,€}{\$1} = \frac{199\,€}{c}$$

QUICK ✔

5. Taryn is reading the newest *Harry Potter* book. If she can read 54 pages in 60 minutes, how long will it take her to read the last 100 pages of the book?

6. An automobile manufacturer is advertising special financing on its autos. A buyer will have a monthly payment of $16.67 for every $1000 borrowed. If Clem borrows $14,000, find his monthly payment.

Sometimes proportions can be used as a general model for economic behavior.

EXAMPLE 3   **Model and Solve a Problem from Business**

A real estate agent knows that a 1800-square-foot house in a particular neighborhood sold for $150,000. How much should the real estate agent appraise a 2100-square-foot house in the same neighborhood?

Solution

Step 1: Identify   We want to know the price of a 2100-square-foot home in the same neighborhood as an 1800-square-foot house that sold for $150,000.

Step 2: Name   Let p represent the price of the 2100-square-foot house.

Step 3: Translate The selling price of an 1800-square-foot house is \$150,000. We wish to know the selling price p of a 2100-square foot house. We use the ratio $\dfrac{\text{square feet}}{\text{selling price}}$ to set up the proportion

$$\frac{1800 \text{ ft}^2}{\$150{,}000} = \frac{2100 \text{ ft}^2}{p}$$

Step 4: Solve

$$\text{Multiply by the LCD 150,000}p\colon \quad \$150{,}000p\!\left(\frac{1800 \text{ ft}^2}{\$150{,}000}\right) = \$150{,}000p\!\left(\frac{2100 \text{ ft}^2}{p}\right)$$

$$\text{Divide like factors:} \quad (1800 \text{ ft}^2)p = (\$150{,}000)(2100 \text{ ft}^2)$$

$$\text{Divide both sides by 1800:} \quad p = \frac{(\$150{,}000)(2100 \text{ ft}^2)}{1800 \text{ ft}^2}$$

$$p = \$175{,}000$$

Step 5: Check The answer is in dollars. Plus, the ratio $\dfrac{1800}{150{,}000} = 0.012$ equals the ratio $\dfrac{2100}{175{,}000} = 0.012$, so our answer checks.

Step 6: Answer We expect that the selling price of a 2100-square-foot house in this neighborhood will be about \$175,000.

QUICK ✓

7. On a map, $\dfrac{1}{4}$ inch represents a distance of 15 miles. According to the map, Springfield and Brookhaven are $3\dfrac{1}{2} = \dfrac{7}{2}$ inches apart. Find the number of miles between the two cities.

② Model and Solve Problems with Similar Figures

You may be familiar with using proportions to solve problems involving *similar figures* from geometry.

Figure 1

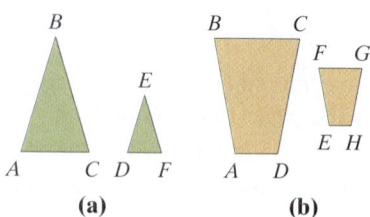

(a)   **(b)**

> **DEFINITION**
>
> Two figures are **similar** if their corresponding angle measures are equal and their corresponding sides are proportional.

Figure 1 shows examples of similar figures.

In Figure 1(a), $\triangle ABC$ is similar to $\triangle DEF$ and in Figure 1(b), quadrilateral $ABCD$ is similar to quadrilateral $EFGH$. Because $\triangle ABC$ is similar to $\triangle DEF$, we know that the ratio of AB to AC equals the ratio of DE to DF. That is,

$$\frac{AB}{AC} = \frac{DE}{DF}$$

The principle that the ratios of corresponding sides are equal can be used to find unknown lengths in similar figures.

EXAMPLE 4   Solving a Problem with Similar Triangles

Find the length of side DE in triangle $\triangle DEF$, given that $\triangle ABC$ is similar to $\triangle DEF$ based on the triangles shown in Figure 2. All measurements are in inches.

Solution

We know that $\triangle ABC$ is similar to $\triangle DEF$. We also know that the corresponding sides are proportional, so $\dfrac{AB}{AC} = \dfrac{DE}{DF}$. We substitute the known values and solve for the unknown value, letting $x = DE$.

Figure 2

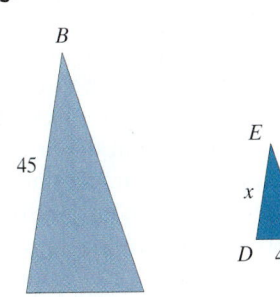

$$\frac{AB}{AC} = \frac{DE}{DF}$$

$AB = 45, AC = 10, DF = 4:$ $\dfrac{45}{10} = \dfrac{x}{4}$

Cross multiply: $10x = 180$

Divide both sides by 10: $x = 18$

The length of side DE is 18 inches.

In Example 4, we could also have used the proportion $\dfrac{AB}{DE} = \dfrac{AC}{DF}$. Then we would have solved the proportion $\dfrac{45}{x} = \dfrac{10}{4}$.

QUICK ✔ *Find the length of side XY given that $\triangle MNP$ is similar to $\triangle XYZ$.*

8.

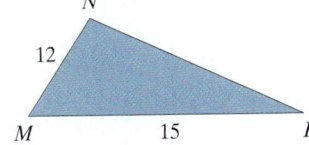   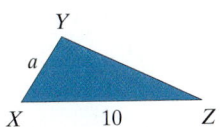

EXAMPLE 5   Finding the Height of a Tree

A fifth-grade student is conducting an experiment to find the height of a tree in the schoolyard. The student measures the length of the tree's shadow and then immediately measures the length of the shadow that a yardstick forms. The tree's shadow measures 24 feet, and the yardstick's shadow measures 4 feet. Find the height of the tree.

Solution

Figure 3

The ratio of the length of the tree's shadow to the height of the tree equals the ratio of the length of the yardstick's shadow to the height of the yardstick. This is because the tree and its shadow form a triangle that is similar to the yardstick and its shadow. See Figure 3.

Let h represent the height of the tree and use 3 feet for the length of the yardstick. We set up the proportion problem as

$$\frac{24}{h} = \frac{4}{3}\quad \text{The Model}$$

Now we solve the proportion.

$$\frac{24}{h} = \frac{4}{3}$$

Cross multiply: $4h = 72$

Divide both sides by 4: $h = 18$

Work Smart

We could have also solved the proportion $\dfrac{h}{3} = \dfrac{24}{4}$. Do you see why?

The height of the tree is 18 feet.

QUICK ✔

9. A 60-foot-tall tree casts a shadow of 25 feet. At the same time of day, a man casts a shadow of 2.5 feet. How tall is the man?

(3) Model and Solve Work Problems

We are now going to solve work problems. These problems assume that jobs are performed at a **constant rate,** which means an individual works at the same pace throughout the entire job. While this assumption is reasonable for machines, it is not likely to be true for people simply because of the old phrase "too many chefs spoil the broth." Think of it this way—if you continually add more people to paint a room, the time to complete the job may decrease initially, but eventually the painters get in each other's way and the time to completion actually increases. While we could take this into account when modeling situations such as this, we will make the "constant rate" assumption for humans as well to keep the mathematics manageable.

The constant-rate assumption states that if it takes t units of time to complete a job, then $\frac{1}{t}$ of the job is done in 1 unit of time. For example, if it takes 5 hours to paint a room, then $\frac{1}{5}$ of the room should be painted in 1 hour.

To solve these work problems, we'll use the six-step process that was introduced in Section 2.5.

EXAMPLE 6 **Planting Seedlings**

Dan, an experienced horticulturist, can plant 1000 seedlings in 2 hours. Hala, a student assistant, requires 5 hours to plant 1000 seedlings. How long would it take Dan and Hala working together to plant 1000 seedlings?

Solution

Step 1: Identify We want to know how long it will take for Dan and Hala to plant 1000 seedlings together.

Step 2: Name We let t represent the time (in hours) that it takes to plant the seedlings working together. Then in 1 hour they will complete $\frac{1}{t}$ of the job.

Step 3: Translate Since we know that Dan can finish the job in 2 hours, Dan will finish $\frac{1}{2}$ of the job in 1 hour. We know that Hala can finish the job in 5 hours, so Hala will finish $\frac{1}{5}$ of the job in 1 hour. Based on this information, we set up Table 1.

Table 1

	Number of Hours to Complete the Job	Part of the Job Completed per Hour
Dan	2	$\frac{1}{2}$
Hala	5	$\frac{1}{5}$
Together	t	$\frac{1}{t}$

We set up the model using the following logic:

$$\left(\begin{array}{c}\text{Part done by Dan}\\\text{in 1 hour}\end{array}\right) + \left(\begin{array}{c}\text{Part done by Hala}\\\text{in 1 hour}\end{array}\right) = \left(\begin{array}{c}\text{Part done together}\\\text{in 1 hour}\end{array}\right)$$

$$\frac{1}{2} \qquad + \qquad \frac{1}{5} \qquad = \qquad \frac{1}{t}$$

Step 4: Solve We now proceed to solve the equation.

$$\frac{1}{2} + \frac{1}{5} = \frac{1}{t}$$

Multiply both sides by the LCD, $10t$: $$10t \cdot \left(\frac{1}{2} + \frac{1}{5}\right) = 10t \cdot \frac{1}{t}$$

Distribute: $$10t \cdot \frac{1}{2} + 10t \cdot \frac{1}{5} = 10t \cdot \frac{1}{t}$$

Divide out common factors: $$5t + 2t = 10$$

Combine like terms: $$7t = 10$$

Divide both sides by 7: $$t = \frac{10}{7}$$

$$t \approx 1.428$$

Work Smart

We convert 0.428 hours to minutes by multiplying 0.428 by 60 minutes to obtain 24 minutes.

Step 5: Check It is always a good idea to make sure your answer is reasonable. We expect our answer to be greater than 0 but less than 2 (because it takes Dan 2 hours working by himself). Our answer of 1.428 hours or 1 hour, 24 minutes seems reasonable.

Step 6: Answer It will take Dan and Hala about 1 hour 24 minutes to plant 1000 seedlings.

QUICK ✓

10. It takes Molly 3 hours to shovel her driveway after a snowstorm working by herself. It takes her elderly neighbor 6 hours to shovel his driveway working by himself. How long would it take Molly and her neighbor to shovel the neighbor's driveway working together assuming each driveway is the same dimension?

EXAMPLE 7   **Mowing the Lawn**

Sara can cut the grass in 3 hours working by herself. When Sara cuts the grass with her younger brother Brian, it takes 2 hours. How long would it take Brian to cut the grass if he worked by himself?

Solution

Step 1: Identify We want to know how long it will take for Brian to cut the grass by himself.

Step 2: Name We let t represent the time (in hours) that it takes Brian to cut the grass by himself. Then, in 1 hour he will complete $\frac{1}{t}$ of the job.

Step 3: Translate Since we know that Sara can finish the job in 3 hours, Sara will finish $\frac{1}{3}$ of the job in 1 hour. We also know that together the job is completed in 2 hours, so $\frac{1}{2}$ of the job is completed in 1 hour. We set up the model using the following logic:

$$\begin{pmatrix} \text{Part done by} \\ \text{Sara in 1 hour} \end{pmatrix} + \begin{pmatrix} \text{Part done by} \\ \text{Brian in 1 hour} \end{pmatrix} = \begin{pmatrix} \text{Part done} \\ \text{together in 1 hour} \end{pmatrix}$$

$$\frac{1}{3} \qquad + \qquad \frac{1}{t} \qquad = \qquad \frac{1}{2}$$

Step 4: Solve We now proceed to solve the equation.

$$\frac{1}{3} + \frac{1}{t} = \frac{1}{2}$$

Multiply both sides by the LCD, $6t$: $6t \cdot \left(\frac{1}{3} + \frac{1}{t}\right) = 6t \cdot \frac{1}{2}$

Distribute: $6t \cdot \frac{1}{3} + 6t \cdot \frac{1}{t} = 6t \cdot \frac{1}{2}$

Divide out common factors: $2t + 6 = 3t$

Subtract $2t$ from each side: $6 = t$

Step 5: Check Is this answer correct? Does Brian do $\frac{1}{6}$ of the job per hour? Is $\frac{1}{3} + \frac{1}{6} = \frac{1}{2}$? Since $\frac{2}{6} + \frac{1}{6} = \frac{3}{6} = \frac{1}{2}$, the answer is correct.

Step 6: Answer It will take Brian 6 hours to mow the grass, working by himself. ▬

QUICK ✔

11. It takes Leon 5 hours to seal his driveway working by himself. If Michael helps Leon, they can seal the driveway in 2 hours. How long would it take Michael to seal the driveway alone?

---

④ **Model and Solve Uniform Motion Problems**

We first introduced uniform motion problems back in Section 2.7. Recall that uniform motion problems use the fact that distance equals rate times time, that is, $d = rt$. When modeling uniform motion problems that lead to rational equations, we usually end up using a variation on this model, $t = \frac{d}{r}$. These problems use the idea of constant rate, just like work problems. For example, a bicyclist may not ride for several hours at exactly the same speed, but we can assume that over the time period the average speed is constant. Again, the assumption makes the mathematics easier to deal with.

EXAMPLE 8 **A Boat Trip on the Mighty Olentangy**

The Olentangy River has a current of 2 miles per hour. A motorboat takes the same amount of time to go 48 miles downstream as it takes to go 36 miles upstream. What is the speed of the boat in still water?

Solution

Step 1: Identify This is a uniform motion problem. We wish to know the speed of the boat in still water.

Step 2: Name Let r represent the speed of the boat in still water.

Step 3: Translate The current pushes the boat when it is going downstream, so the total speed of the boat will be the speed of the boat in still water plus the speed of the current. We let $r + 2$ represent the speed of the boat downstream. Going upstream, the current slows the boat, so the total speed of the boat upstream will be $r - 2$. We set up Table 2.

The amount of time traveling downstream is the same as the amount of time spent traveling upstream, so

$$\frac{48}{r + 2} = \frac{36}{r - 2} \qquad \text{The Model}$$

Work Smart

Remember, $d = r \cdot t$, so $t = \dfrac{d}{r}$.

Table 2			
	Distance (miles)	Rate (miles per hour)	Time = $\dfrac{\text{Distance}}{\text{Rate}}$ (hours)
Downstream	48	$r + 2$	$\dfrac{48}{r + 2}$
Upstream	36	$r - 2$	$\dfrac{36}{r - 2}$

Step 4: Solve We wish to solve for r:

$$\frac{48}{r + 2} = \frac{36}{r - 2}$$

Multiply both sides by the LCD, $(r + 2)(r - 2)$: $\quad 48(r - 2) = 36(r + 2)$

Distribute: $\quad 48r - 96 = 36r + 72$

Subtract $36r$ from both sides: $\quad 12r - 96 = 72$

Add 96 to both sides: $\quad 12r = 168$

Divide each side by 12: $\quad \dfrac{12r}{12} = \dfrac{168}{12}$

$$r = 14$$

Step 5: Check Does the boat travel 48 miles downstream in the same time that it travels 36 miles upstream? Let's see. Going downstream, the boat travels at $14 + 2 = 16$ miles per hour, so it takes $\dfrac{48}{16} = 3$ hours to travel downstream. Going upstream, the boat travels at $14 - 2 = 12$ miles per hour, and it takes $\dfrac{36}{12} = 3$ hours to travel upstream. Our answer checks!

Step 6: Answer the Question The boat travels 14 miles per hour in still water. ▬

QUICK ✓

12. A small airplane can travel 120 mph in still air. The plane can fly 700 miles with the wind in the same time that it can travel 500 miles against the wind. Find the speed of the wind.

EXAMPLE 9   The Ride Back through the Forest

Every weekend, you ride your bicycle on a forest preserve path. The path is 12 miles long and ends at a waterfall, at which point you relax and then make the trip back to the starting point. One weekend, you notice that because the ride to the waterfall is partially uphill, your cycling speed returning is 2 miles per hour faster than the speed riding to the waterfall. If the round trip takes 5 hours (excluding resting time), find the average speed cycling back from the waterfall.

Solution

Step 1: Identify This is a uniform motion problem. We wish to know the average speed returning from the waterfall.

Step 2: Name Let r represent your cycling speed when you are cycling to the waterfall. Then $r + 2$ represents the speed cycling back from the waterfall to the starting point.

Step 3: Translate We know that the distance to the waterfall is 12 miles. We set up Table 3.

Table 3			
	Distance (miles)	Rate (miles per hour)	Time $=\dfrac{\text{Distance}}{\text{Rate}}$ (hours)
Going to the Waterfall	12	r	$\dfrac{12}{r}$
Returning Home	12	$r + 2$	$\dfrac{12}{r + 2}$

We know that the total time spent cycling was 5 hours, so we set up the equation

$$\left(\begin{array}{c}\text{time cycling to}\\\text{waterfall}\end{array}\right) + \left(\begin{array}{c}\text{time cycling to}\\\text{starting point}\end{array}\right) = 5 \text{ hours}$$

$$\frac{12}{r} + \frac{12}{r + 2} = 5 \quad \textcolor{red}{\text{The Model}}$$

Step 4: Solve We wish to solve for r:

$$\frac{12}{r} + \frac{12}{r + 2} = 5$$

$\textcolor{red}{\text{Multiply both sides by the LCD, } r(r + 2):}$ $\quad r(r + 2)\left(\dfrac{12}{r} + \dfrac{12}{r + 2}\right) = (5)r(r + 2)$

$\textcolor{red}{\text{Distribute:}}$ $\quad r(r + 2)\left(\dfrac{12}{r}\right) + r(r + 2)\left(\dfrac{12}{r + 2}\right) = (5)r(r + 2)$

$\textcolor{red}{\text{Divide out common factors:}}$ $\quad (r + 2)12 + r(12) = 5r(r + 2)$

$\textcolor{red}{\text{Multiply and distribute:}}$ $\quad 12r + 24 + 12r = 5r^2 + 10r$

$\textcolor{red}{\text{Combine like terms:}}$ $\quad 24r + 24 = 5r^2 + 10r$

$\textcolor{red}{\text{Write the equation in standard form } ax^2 + bx + c = 0:}$ $\quad 0 = 5r^2 - 14r - 24$

$\textcolor{red}{\text{Factor the polynomial:}}$ $\quad 0 = (5r + 6)(r - 4)$

$\textcolor{red}{\text{Set each factor equal to zero:}}$ $\quad 5r + 6 = 0 \quad \text{or} \quad r - 4 = 0$

$\textcolor{red}{\text{Solve each equation:}}$ $\quad 5r = -6 \qquad\qquad r = 4$

$$r = -\frac{6}{5}$$

We discard the solution $r = -\dfrac{6}{5}$ because r represents the cycling speed and speeds cannot be negative. So $r = 4$ miles per hour is the speed cycling to the waterfall. We are asked to find the speed cycling back to the starting point, $r + 2$, so $r + 2 = 4 + 2 = 6$ mph.

Step 5: Check Let's check our solution. We should find that the time cycling to the waterfall + the time returning to the starting point = 5 hours. Because

$$\frac{12 \text{ mi}}{4 \text{ mph}} + \frac{12 \text{ mi}}{6 \text{ mph}} = 3 \text{ hr} + 2 \text{ hr} = 5 \text{ hr, our answer is correct.}$$

Step 6: Answer Your average speed cycling from the waterfall to the starting point is 6 mph.

QUICK

13. To prepare for a half-marathon run, Sue ran 18 miles before she blistered her heel, and then she walked 1 additional mile. Her running speed was 12 times as fast as her walking speed. She was running and walking for 5 hours. Find Sue's running speed.

7.8 Exercises

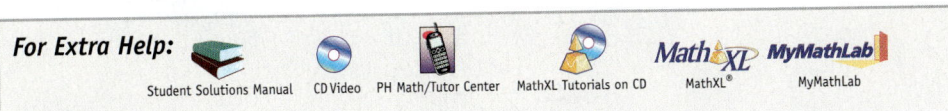

For Extra Help:
Student Solutions Manual   CD Video   PH Math/Tutor Center   MathXL Tutorials on CD   MathXL®   MyMathLab

Concepts and Vocabulary

In Problems 1–3, fill in the blanks.

1. In the proportion $\dfrac{x}{y} = \dfrac{3}{8}$, the terms x and 8 are called the _____ while y and 3 are called the _____.

2. In geometry, two figures are _____ if their corresponding angle measures are equal and their corresponding sides are proportional.

3. To solve uniform motion problems, we use the formula _____.

In Problems 4–6, answer True or False to each statement.

4. The following figures are similar.

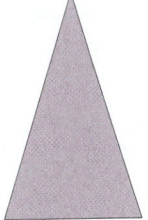

5. The following proportions all yield the same result.

(a) $\dfrac{x}{2} = \dfrac{4}{9}$ **(b)** $\dfrac{2}{x} = \dfrac{9}{4}$ **(c)** $\dfrac{x}{4} = \dfrac{2}{9}$ **(d)** $\dfrac{9}{2} = \dfrac{4}{x}$ **(e)** $\dfrac{4}{9} = \dfrac{x}{2}$

6. A student reads the following sentence from a word problem: *Barbara rows her boat in a stream that has a current of 1 mph. She travels 10 miles upstream and returns to her starting point 5 hours later.* She makes the following chart. The information is correctly placed for this part of the problem.

d	r	t
10	$r + 1$	5
10	$r - 1$	5

7. When modeling work problems, we assume that individuals work at a constant rate. When modeling uniform motion problems, we assume that individuals travel at a constant rate. Explain what these assumptions mean.

8. You are teaching the class how to solve the following problem: *Marcos can write a budget analysis in 8 hours, and Alonso can complete the same analysis in 6 hours. If they work together, how long will it take to finish the analysis?* Write an equation that could be used to find the answer to the question and then explain your justification for setting up the equation your way. If you can think of another way to do the problem, explain how to set it up differently and then explain why this logic will also solve the problem correctly.

Building Skills

In Problems 9–28, solve the proportion.

9. $\dfrac{9}{x} = \dfrac{3}{4}$

10. $\dfrac{x}{5} = \dfrac{12}{4}$

11. $\dfrac{4}{7} = \dfrac{2x}{9}$

12. $\dfrac{6}{5} = \dfrac{8}{3x}$

13. $\dfrac{6}{5} = \dfrac{x+2}{15}$

14. $\dfrac{5}{9} = \dfrac{2x+3}{6}$

15. $\dfrac{b}{b+6} = \dfrac{4}{9}$

16. $\dfrac{k}{k+3} = \dfrac{6}{15}$

17. $\dfrac{y}{y-10} = \dfrac{2}{27}$

18. $\dfrac{y}{y-3} = \dfrac{3}{10}$

19. $\dfrac{p+2}{4} = \dfrac{2p+4}{5}$

20. $\dfrac{n+6}{3} = \dfrac{n+4}{5}$

21. $\dfrac{2z-1}{z} = \dfrac{3}{5}$

22. $\dfrac{2k-3}{k} = \dfrac{4}{3}$

23. $\dfrac{2}{v^2-v} = \dfrac{1}{3-v}$

24. $\dfrac{n-2}{4n+7} = \dfrac{-2}{n+1}$

25. $\dfrac{10-x}{4x} = \dfrac{1}{x-1}$

26. $\dfrac{4}{x^2} = \dfrac{1}{x+3}$

27. $\dfrac{2p-3}{p^2+12p+6} = \dfrac{1}{p+4}$

28. $\dfrac{1}{2z-1} = \dfrac{z+4}{z^2+10z+14}$

In Problems 29–32, $\triangle ABC$ is similar to $\triangle XYZ$.

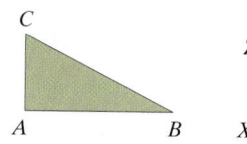

△ **29.** If $AB = 6$, $XY = 9$, $AC = 8$, find XZ.

△ **30.** If $XY = 8$, $YZ = 7$, $AB = 24$, find BC.

△ **31.** If $XY = n$, $ZY = 2n - 1$, $BC = 5$, and $AB = 3$, find n and ZY.

△ **32.** If $AB = n$, $BC = n + 2$, $XY = 5$, and $YZ = 15$, find n and BC.

In Problems 33 and 34, rectangle ABCD is similar to rectangle EFGH.

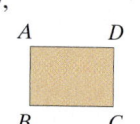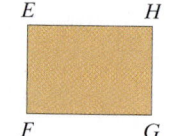

△ **33.** If $AB = x - 2$, $EF = 4$, $BC = 2x + 3$, and $FG = 9$, find x and AB.

△ **34.** If $DC = 6$, $HG = x - 1$, $DA = 8$, and $HE = 2x - 3$ find x and HE.

In Problems 35–40, write an algebraic expression that represents each phrase.

35. If Mariko worked x hours and Natalie worked twice as many hours as Mariko, write an algebraic expression that represents the number of hours Natalie worked.

36. If Brian worked for n days and Kellen worked for half as many days, write an algebraic expression that represents the number of days Kellen worked.

37. If David painted for 3 days less than Pierre and Pierre painted for t days, write an algebraic expression that represents the number of days David painted.

38. If Valerie danced for p minutes and Melody danced for 45 minutes less than Valerie, write an algebraic expression that represents the number of minutes Melody danced.

39. If the rate of the current of a stream is 2 mph and Joe can swim r mph in still water, write an algebraic expression that represents Joe's rate when he swims upstream.

40. If the wind is blowing at 25 mph and a plane flies at r mph in still air, write an algebraic expression that represents the rate the plane is traveling when it flies with the wind.

In Problems 41–46, write an equation that could be used to model each of the following. SET UP BUT DO NOT SOLVE THE EQUATION.

41. Painting Chairs Christina can paint a chair in 5 hours, and Victoria can paint a chair in 3 hours. How many hours will it take to paint the chair when the two girls work together?

42. **Weeding the Yard** Bill can weed the back yard in 6 hours. When he worked with Tamra, it only took 2 hours. How long would it take Tamra to weed the back yard if she worked alone?

43. **Filling a Tank** There are two different inlet pipes that can be used to fill a 3000-gallon tank. Pipe A takes 4 hours longer than Pipe B to fill the tank. With both pipes open, it takes 7 hours to fill the tank. How long would it take Pipe B alone to fill the tank?

44. **Canning Peaches** A factory that was producing canned peaches had an old canning machine. They decided to add a newer machine that could work twice as fast. With both machines on line, the plant could produce 10,000 cans in 6 hours. How long would it take to produce the same number of cans using only the newer machine?

45. **Bob's River Trip** Bob's boat travels at 14 mph in still water. Find the speed of the current if he can go 4 miles upstream in the same time that it takes to go 7 miles downstream.

46. **Marty's River Trip** A stream has a current of 3 mph. Find the speed of Marty's boat in still water if she can go 10 miles downstream in the same time it takes to go 4 miles upstream.

Applying the Concepts

47. **Yasmine's Map** On Yasmine's map, $\frac{1}{4}$ inch represents 10 miles. According to the map, Yellow Springs and Hillsboro are $3\frac{5}{8}$ inches apart. Find the number of miles between the two towns.

48. **Beth's Map** On Beth's map, 0.5 cm represents 60 miles. If it is 1840 miles from San Diego to New Orleans, how far is this distance on her map?

49. **Making Bread** If it takes 5 lb of flour to make 3 loaves of bread, how much flour is needed to make 5 loaves of bread?

50. **Buying Candy** If it costs $3.50 for 2 lb of candy, how much candy can be purchased for $8.75?

51. **Comparing Rubles and Dollars** $5 U.S. is worth approximately 143.25 Russian rubles. How many rubles can Hortencia purchase for $1300 U.S.?

52. **Comparing Pesos and Pounds** 50 Mexican pesos are worth approximately 2.5 British pounds. If Sean takes 80 pounds as spending money in Mexico City, how many pesos will he have?

53. **Tree Shadow** If a 6-ft man casts a shadow that is 2.5 ft long, how tall is a tree that has a shadow 10 ft long?

54. **Telephone Pole Shadow** A bush that is 4 m tall casts a shadow that is 1.75 m long. How tall is a telephone pole if its shadow is 8.75 m?

55. **Cleaning the Math Building** Josh can clean the math building on his campus in 3 hours. Ken takes 5 hours to complete the same building. If they work together, how long will it take them to finish cleaning this building?

56. **Pruning Trees** Martin can prune his fruit trees in 4 hours. His neighbor can prune the same trees for him in 7 hours. If they work together on this job, how long will take to prune the trees?

57. **Retrieving Volleyballs** After hitting practice for the Long Beach State volleyball team, Dyanne can retrieve all of the balls in the gym in 8 minutes. It takes Makini 6 minutes to retrieve all the balls. If they work together, how long will it take these two players to return the volleyballs and be ready to start the next round of hitting practice?

58. **Stuffing Envelopes** It took Kirsten 6 hours to stuff 3000 envelopes. When she worked with Brittany, it only took 4 hours to stuff the next 3000 envelopes. How long would it take Brittany working alone to stuff the 3000 envelopes?

59. **Painting City Hall** Three people were given the job of painting city hall. They divided the job into equal portions and decided they would each go home after completing their assigned portion. It took José 9 hours to paint his part and then he went home. It took Joaquín 12 hours to complete his part. The third person played around, never got started, and was promptly fired from the job. If José and Joaquín went back the second day and worked together to finish the undone portion, how long did it take them?

60. **Mowing the Grass** It takes 18 minutes to cut the grass in the outfield of the stadium. A new mower was purchased and with both mowers working, the same task takes 8 minutes. How long would it take the new mower to cut the outfield grass?

61. **Replacing Pipes** It takes an apprentice twice as long as the experienced plumber to replace the pipes under an old house. If it takes them 5 hours when they work together, how long would it take the apprentice alone?

62. **Making Care Packages** It takes Rocco 3 times as long as Traci to make 100 Red Cross care packages for refugees. If it takes them 10 hours when they work together to make 100 care packages, how long would it take Rocco working alone to make the same number?

63. **Filling the Pool** Using a single hose, Janet can fill a pool in 6 hours. The same pool can be drained in 8 hours by opening a drainpipe. If Janet forgets to close the drainpipe, how long would it take her to fill the pool?

64. **Jill's Bucket** Jill has a bucket with a hole in it. When the hole did not exist, Jill could fill the bucket in 30 seconds. With the hole, the bucket now empties in 210 seconds (3.5 minutes). How long will it take Jill to fill the bucket now that it has a hole in it?

65. **Travel by Boat** A boat can travel 10 km down the river in the same time it can go 4 km up the river. If the current in the river is 2 km per hour, how fast can the boat travel in still water?

66. **Travel by Plane** A small plane can travel 1000 miles with the wind in the same time it can go 600 miles against the wind. If the speed of the plane in still air is 180 mph, what is the speed of the wind?

67. **Iron Man Training** While training for an iron man competition, Tony bikes for 60 miles and runs for 15 miles. If his biking speed is 8 times his running speed and it takes 5 hours to complete the training, how long did he spend on his bike?

68. **Robin's Workout** Robin can run twice as fast as she walks. If she runs for 12 miles and walks for 8 miles, the total time to complete the trip is 5 hours. How many hours did she run?

69. **Driving during a Snowstorm** Claire and Chris were driving to their home in Milwaukee when they came upon a snowstorm. They drove at an average rate of 20 miles per hour slower for the last 60 miles during the snowstorm than they did for the first 90 miles. Assuming the time driven for the first 60 miles equals the time driven for the last 90 miles, find their average rate of speed during the last 60 miles.

70. **Driving during Road Construction** David and Lisa drove 150 miles through road construction at a certain average rate. By increasing their speed by 20 miles per hour, they traveled the next 250 miles without road construction in the same time they spent on the 150-mile leg of their trip. Find their average rate during the part of the trip that had road construction.

71. **Tough Commute** You have a 20-mile commute into work. Since you leave very early, the trip going to work is easier than the trip home. You can travel to work in the same time that it takes for you to make it 16 miles on the trip back home. Your average speed coming home is 7 miles per hour slower than your average speed going to work. What is your average speed going to work?

72. **A Bike Trip** A bicyclist rides his bicycle 12 miles up a hill and then 16 miles on level terrain. His speed on level ground is 3 miles per hour faster than his speed going uphill. The cyclist rides for the same amount of time going uphill and on level ground. Find his speed going uphill.

Extending the Concepts

In geometry, when a line is parallel to one side of a triangle, the intersection of the parallel line with the other two sides of the triangle will form a new triangle that is similar to the original triangle. In the figure below, $\overline{XY}$ is parallel to $\overline{BC}$, therefore $\triangle XAY$ is similar to $\triangle BAC$.

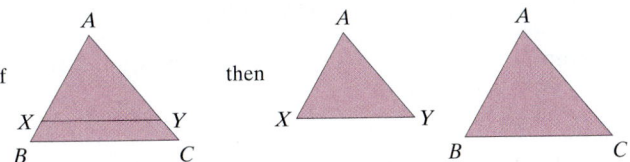

If   then

In Problems 73–78, solve for x.

△ **73.** $XA = 4$, $BA = 12$, $CA = 18$, and $YA = x$.

△ **74.** $XA = 4$, $BA = 10$, $CA = 30$, and $YA = x$.

△ **75.** $XA = 5$, $XB = 9$, $XY = 7$, and $BC = x$.

△ **76.** $XA = 3$, $XB = 15$, $XY = 4$, and $BC = x$.

△ **77.** $AY = 5$, $YC = 12$, $AX = 6$, and $XB = x$.

△ **78.** $AY = 4$, $YC = x$, $AX = x$, and $XB = 9$.

CHAPTER 7 ACTIVITY: Correct the Quiz

Focus: Performing operations with rational expressions and solving rational equations

Time: 15–20 minutes

Group size: 2

In this activity you will work as a team to grade the student quiz shown below. If an answer is correct, mark it correct. If an answer is wrong, mark it wrong and show the correct answer.

Once all of the quiz questions are graded, compute the final score for the quiz. Be prepared to discuss your results with the rest of the class.

Student Quiz	
Name: *Ima Student*	Quiz Score: _____
(1) Multiply: $\dfrac{6x - 12}{4x + 8} \cdot \dfrac{3x + 6}{2x - 4}$	Answer: $\dfrac{9}{4}$
(2) Subtract: $\dfrac{xy}{x^2 - y^2} - \dfrac{y}{x + y}$	Answer: $\dfrac{y^2}{x^2 - y^2}$
(3) Solve: $\dfrac{5x}{x + 1} = 2 + \dfrac{2x}{x + 1}$	Answer: $\{2\}$
(4) Divide: $\dfrac{x^2 + x - 6}{5x^2 - 7x - 6} \div \dfrac{3x^2 + 13x + 12}{6x^2 + 17x + 12}$	Answer: $\dfrac{2}{5}$
(5) Solve: Joe can mow his lawn in 2 hrs. Mike can mow the same lawn in 3 hrs. If they work together, how long will it take them to mow the lawn?	Answer: *1 hr, 12 min*
(6) Simplify: $\dfrac{\dfrac{1}{x + 5} - \dfrac{2}{x - 7}}{\dfrac{4}{x - 7} + \dfrac{1}{x + 5}}$	Answer: $\dfrac{x + 17}{5x + 13}$
(7) Add: $\dfrac{x}{x^2 + 10x + 25} + \dfrac{4}{x^2 + 6x + 5}$	Answer: $\dfrac{x^2 + 5x + 20}{(x + 5)(x + 1)}$

CHAPTER 7 REVIEW

Section 7.1	Simplifying Rational Expressions	

KEY CONCEPTS		KEY TERMS
• To determine undefined values of a rational expression, set the denominator equal to zero and solve for the variable. • **Simplifying a Rational Expression** First, completely factor the numerator and the denominator of the rational expression; then divide out common factors using the fact that if $p, q,$ and r are polynomials, then $\dfrac{p \cdot r}{q \cdot r} = \dfrac{p}{q}$ if $q \neq 0$ and $r \neq 0$.		Rational expression Evaluate Undefined Simplify

YOU SHOULD BE ABLE TO . . .	EXAMPLE	REVIEW EXERCISES
(1) Evaluate a rational expression (p. 477)	Examples 1 through 3	1–4
(2) Determine undefined values of a rational expression (p. 479)	Examples 4 and 5	5–10
(3) Simplify rational expressions (p. 480)	Examples 6–9	11–16

In Problems 1–4, evaluate each expression for the given values.

1. $\dfrac{2x}{x+3}$

    (a) $x = 1$ **(b)** $x = -3$
    (c) $x = 3$

2. $\dfrac{3x}{x-4}$

    (a) $x = 0$ **(b)** $x = 1$
    (c) $x = 4$

3. $\dfrac{x^2 + 2xy + y^2}{x - y}$

    (a) $x = 1$ **(b)** $x = 2$
         $y = 1$ $y = 1$
    (c) $x = -3$
         $y = -4$

4. $\dfrac{2x^2 - x - 3}{x + z}$

    (a) $x = 1$ **(b)** $x = 1$
         $z = 1$ $z = -1$
    (c) $x = 5$
         $z = -4$

In Problems 5–10, find the value(s) of the variable for which the rational expression is undefined.

5. $\dfrac{3x}{3x - 7}$ **6.** $\dfrac{x + 1}{4x - 2}$ **7.** $\dfrac{5}{x^2 + 25}$

8. $\dfrac{17}{4x^2 + 49}$ **9.** $\dfrac{5x + 2}{x^2 + 12x + 20}$ **10.** $\dfrac{3x - 1}{x^2 - 3x - 4}$

In Problems 11–16, simplify each rational expression, if possible. Assume that no variable has a value that results in a denominator with a value of zero.

11. $\dfrac{20x^5 y}{25xy^5}$ **12.** $\dfrac{45x^2 y^2}{60x^4 y^6}$ **13.** $\dfrac{3k - 21}{k^2 - 5k - 14}$

14. $\dfrac{-2x - 2}{x^2 - 2x - 3}$ **15.** $\dfrac{x^2 + 8x + 15}{2x^2 + 5x - 3}$ **16.** $\dfrac{3x^2 + 5x - 2}{2x^3 + 16}$

Section 7.2 Multiplying and Dividing Rational Expressions

KEY CONCEPTS

- **Multiplying Rational Expressions**

  If $\dfrac{a}{b}$ and $\dfrac{c}{d}$, $b \neq 0$, $d \neq 0$, are two rational expressions, then $\dfrac{a}{b} \cdot \dfrac{c}{d} = \dfrac{ac}{bd}$.

- **Dividing Rational Expressions**

  If $\dfrac{a}{b}$ and $\dfrac{c}{d}$, $b \neq 0$, $c \neq 0$, $d \neq 0$, are two rational expressions, then $\dfrac{\dfrac{a}{b}}{\dfrac{c}{d}} = \dfrac{a}{b} \div \dfrac{c}{d} = \dfrac{a}{b} \cdot \dfrac{d}{c} = \dfrac{ad}{bc}$.

YOU SHOULD BE ABLE TO . . .	EXAMPLE	REVIEW EXERCISES
(1) Multiply rational expressions (p. 485)	Examples 1 through 4	17, 18, 21, 22, 25, 26
(2) Divide rational expressions (p. 487)	Examples 5 through 9	19, 20, 23, 24, 27, 28

In Problems 17–28, perform the indicated operations and simplify, if possible.

17. $\dfrac{12m^4n^3}{7} \cdot \dfrac{21}{18m^2n^5}$

18. $\dfrac{3x^2y^4}{4} \cdot \dfrac{2}{9x^3y}$

19. $\dfrac{10m^2n^4}{9m^3n} \div \dfrac{15mn^6}{21m^2n}$

20. $\dfrac{5ab^3}{3b^4} \div \dfrac{10a^2b^8}{6b^2}$

21. $\dfrac{5x - 15}{x^2 - x - 12} \cdot \dfrac{x^2 - 6x + 8}{3 - x}$

22. $\dfrac{4x - 24}{x^2 - 18x + 81} \cdot \dfrac{x^2 - 9}{6 - x}$

23. $\dfrac{\dfrac{x^2 - 4}{x^2 - 8x + 15}}{\dfrac{12x + 24}{3x - 15}}$

24. $\dfrac{\dfrac{5x^3 + 10x^2}{3x}}{\dfrac{2x + 4}{18x^2}}$

25. $\dfrac{3x^2 + 14x - 5}{x^2 + x - 30} \cdot \dfrac{x^2 - 2x - 15}{3x^2 + 8x - 3}$

26. $\dfrac{y^2 - 5y - 14}{y^2 - 2y - 35} \cdot \dfrac{y^2 + 6y + 5}{y^2 - y - 6}$

27. $\dfrac{x^2 - 9x}{x^2 + 3x + 2} \div \dfrac{x^2 - 81}{x^2 + 2x}$

28. $\dfrac{y^2 - 9}{2y^2 - y - 15} \div \dfrac{3y^2 + 10y + 3}{2y^2 + y - 10}$

Section 7.3 Adding and Subtracting Rational Expressions with a Common Denominator

KEY CONCEPT

- **Adding/Subtracting Rational Expressions**

  To add or subtract rational expressions, if $\dfrac{a}{c}$ and $\dfrac{b}{c}$, $c \neq 0$, then $\dfrac{a}{c} + \dfrac{b}{c} = \dfrac{a + b}{c}$ and $\dfrac{a}{c} - \dfrac{b}{c} = \dfrac{a - b}{c}$.

YOU SHOULD BE ABLE TO . . .	EXAMPLE	REVIEW EXERCISES
(1) Add rational expressions with a common denominator (p. 494)	Examples 1 through 3	29–34
(2) Subtract rational expressions with a common denominator (p. 496)	Examples 4 and 5	35–38
(3) Add or subtract rational expressions with opposite denominators (p. 497)	Examples 6 and 7	39–42

In Problems 29–42, perform the indicated operations and simplify, if possible.

29. $\dfrac{4}{x-3} + \dfrac{5}{x-3}$

30. $\dfrac{3}{x+4} + \dfrac{7}{x+4}$

31. $\dfrac{m^2}{m+3} + \dfrac{3m}{m+3}$

32. $\dfrac{1}{6m} + \dfrac{5}{6m}$

33. $\dfrac{-m+1}{m^2-4} + \dfrac{2m+1}{m^2-4}$

34. $\dfrac{2m^2}{m+1} + \dfrac{5m+3}{m+1}$

35. $\dfrac{11}{15m} - \dfrac{6}{15m}$

36. $\dfrac{15b}{2b^2} - \dfrac{11b}{2b^2}$

37. $\dfrac{2y^2}{y-7} - \dfrac{y^2+49}{y-7}$

38. $\dfrac{6m+5}{m^2-36} - \dfrac{5m-1}{m^2-36}$

39. $\dfrac{3}{x-y} + \dfrac{10}{y-x}$

40. $\dfrac{7}{a-b} + \dfrac{3}{b-a}$

41. $\dfrac{2x}{x^2-25} - \dfrac{x-5}{25-x^2}$

42. $\dfrac{x+5}{2x-6} - \dfrac{x+3}{6-2x}$

Section 7.4   Finding the Least Common Denominator and Forming Equivalent Rational Expressions

KEY CONCEPTS	KEY TERMS
• The steps to find the LCD of rational expressions are given on page 503. • The steps to form equivalent rational expressions are given on page 506.	Least common denominator (LCD) Equivalent rational expressions

YOU SHOULD BE ABLE TO . . .	EXAMPLE	REVIEW EXERCISES
① Find the LCD of two or more rational expressions (p. 501)	Examples 1 through 6	43–48
② Write a rational expression that is equivalent to a given rational expression (p. 504)	Examples 7 and 8	49–52
③ Use the LCD to write equivalent rational expressions (p. 506)	Example 9	53–58

In Problems 43–48, identify the LCD of the given rational expressions.

43. $\dfrac{6}{4x^2y^7}; \dfrac{8}{6x^4y}$

44. $\dfrac{3}{20a^3bc^4}; \dfrac{7}{30ab^4c^7}$

45. $\dfrac{11}{4a}; \dfrac{7a}{8a+16}$

46. $\dfrac{6}{5a}; \dfrac{17}{a^2+6a}$

47. $\dfrac{x+1}{4x-12}; \dfrac{3x}{x^2-2x-3}$

48. $\dfrac{11}{x^2-7x}; \dfrac{x+2}{x^2-49}$

In Problems 49–52, write an equivalent rational expression with the given denominator.

49. $\dfrac{6}{x^3y}$ with denominator x^4y^7

50. $\dfrac{11}{a^3b^2}$ with denominator a^7b^5

51. $\dfrac{x-1}{x-2}$ with denominator x^2-4

52. $\dfrac{m+2}{m+7}$ with denominator $m^2+5m-14$

In Problems 53–58, identify the LCD and then write each as an equivalent rational expression with that denominator.

53. $\dfrac{5y}{6x^3}; \dfrac{7}{8x^5}$

54. $\dfrac{6}{5a^3}; \dfrac{11b}{10a}$

55. $\dfrac{4x}{x-2}; \dfrac{6}{2-x}$

56. $\dfrac{3}{m-5}; \dfrac{-2m}{5-m}$

57. $\dfrac{2}{m^2+5m-14}; \dfrac{m+1}{m^2+9m+14}$

58. $\dfrac{n-2}{n^2-5n}; \dfrac{n}{n^2-25}$

Section 7.5   Adding and Subtracting Rational Expressions with Unlike Denominators

KEY CONCEPT

- The steps for adding or subtracting rational expressions with unlike denominators are given on page 511.

YOU SHOULD BE ABLE TO . . .	EXAMPLE	REVIEW EXERCISES
① Add and subtract rational expressions with unlike denominators (p. 510)	Examples 1 through 10	59–72

In Problems 59–72, perform the indicated operations and simplify, if possible.

59. $\dfrac{4}{xy^3} + \dfrac{8y}{x^2z}$

60. $\dfrac{x}{2x^3y} + \dfrac{y}{10xy^3}$

61. $\dfrac{x}{x+7} + \dfrac{2}{x-7}$

62. $\dfrac{4}{2x+3} + \dfrac{x+1}{2x-3}$

63. $\dfrac{x+5}{x} - \dfrac{x+7}{x-2}$

64. $\dfrac{3x}{x+5} - \dfrac{x+1}{x}$

65. $\dfrac{3x-4}{4x+1} + \dfrac{3x+6}{4x^2+9x+2}$

66. $\dfrac{7}{3x^2+x-4} + \dfrac{9x+2}{3x^2-2x-8}$

67. $\dfrac{m}{m^2-9} - \dfrac{4m-12}{m+3}$

68. $\dfrac{2m+1}{m-5} - \dfrac{4}{m^2-3m-10}$

69. $\dfrac{3}{m-2} - \dfrac{1}{2-m}$

70. $\dfrac{m}{m-n} - \dfrac{n}{n-m}$

71. $4 + \dfrac{x}{x+3}$

72. $7 - \dfrac{1}{x+2}$

Section 7.6   Complex Rational Expressions

KEY CONCEPT	KEY TERMS
- There are two methods that can be used to simplify a complex rational expression. The steps for Method 1 are presented on page 522 while the steps for Method 2 are presented on page 526.	Complex rational expression Simplify

YOU SHOULD BE ABLE TO . . .	EXAMPLE	REVIEW EXERCISES
① Simplify a complex rational expression by simplifying the numerator and denominator separately (p. 521)	Examples 2 through 4	73–80
② Simplify a complex rational expression using the least common denominator (p. 524)	Examples 5 and 6	73–80

In Problems 73–80, simplify the complex rational expressions.

73. $\dfrac{\dfrac{1}{2} - \dfrac{2}{3}}{\dfrac{4}{9} + \dfrac{5}{6}}$

74. $\dfrac{\dfrac{1}{4} + \dfrac{1}{2}}{\dfrac{5}{8} - \dfrac{1}{6}}$

75. $\dfrac{\dfrac{1}{5} - \dfrac{1}{m}}{\dfrac{1}{10} + \dfrac{1}{m^2}}$

76. $\dfrac{\dfrac{1}{m^2} + \dfrac{2}{3}}{\dfrac{1}{m} - \dfrac{5}{6}}$

77. $\dfrac{\dfrac{x}{4} - \dfrac{1}{2}}{\dfrac{3x}{2} - 3}$

78. $\dfrac{\dfrac{7x}{3} + 7}{\dfrac{3x+9}{8}}$

79. $\dfrac{\dfrac{8}{y+4} + 2}{\dfrac{12}{y+4} - 2}$

80. $\dfrac{\dfrac{25}{y+5} + 5}{\dfrac{3}{y+5} - 5}$

Section 7.7   Rational Equations

KEY CONCEPTS	KEY TERMS
• The steps for solving any rational equation are given on pages 531–532. • To solve for a variable in a formula, get the variable by itself on one side of the equation.	Rational equation Extraneous solution

YOU SHOULD BE ABLE TO ...	EXAMPLE	REVIEW EXERCISES
(1) Solve equations containing rational expressions (p. 529)	Examples 1 through 7	81–92
(2) Solve for a variable in a rational equation (p. 536)	Examples 8 and 9	93–96

In Problems 81 and 82, state the values of the variable for which the expressions in the rational equation are undefined.

81. $\dfrac{4}{x + 5} + \dfrac{1}{x} = 3$

82. $\dfrac{2}{x^2 - 36} = \dfrac{1}{x + 6}$

In Problems 83–92, solve each equation and list the solution(s) in a solution set. Be sure to check for extraneous solutions.

83. $\dfrac{2}{x} - \dfrac{3}{4} = \dfrac{5}{x}$

84. $\dfrac{4}{x} + \dfrac{3}{4} = \dfrac{2}{3x} + \dfrac{23}{4}$

85. $\dfrac{4}{m} - \dfrac{3}{2m} = \dfrac{1}{2}$

86. $\dfrac{3}{m} + \dfrac{5}{3m} = 1$

87. $\dfrac{m + 4}{m - 3} = \dfrac{m + 10}{m + 2}$

88. $\dfrac{m - 6}{m + 5} = \dfrac{m - 3}{m + 1}$

89. $\dfrac{2x}{x - 1} - 5 = \dfrac{2}{x - 1}$

90. $\dfrac{2x}{x - 2} - 3 = \dfrac{4}{x - 2}$

91. $\dfrac{1}{x + 3} + \dfrac{1}{x - 3} = \dfrac{-5}{x^2 - 9}$

92. $\dfrac{3}{x - 5} - \dfrac{11}{x^2 - 25} = \dfrac{4}{x + 5}$

In Problems 93–96, solve the equation for the indicated variable.

93. $y = \dfrac{4}{k}$ for k **94.** $6 = \dfrac{x}{y}$ for y **95.** $\dfrac{1}{x} + \dfrac{1}{y} = \dfrac{1}{z}$ for y **96.** $\dfrac{1}{x} + \dfrac{1}{y} = \dfrac{1}{z}$ for z

Section 7.8   Models Involving Rational Equations

KEY CONCEPT	KEY TERMS
• **Cross-Multiplication Property** If $\dfrac{a}{b} = \dfrac{c}{d}$, then $a \cdot d = b \cdot c$, b and $d \neq 0$.	Ratio Proportion Terms Means Extremes Cross multiplication Similar figures Constant rate

YOU SHOULD BE ABLE TO ...	EXAMPLE	REVIEW EXERCISES
(1) Model and solve ratio and proportion problems (p. 541)	Examples 1 through 3	97–100, 103, 104
(2) Model and solve problems with similar figures (p. 544)	Examples 4 and 5	101, 102
(3) Model and solve work problems (p. 546)	Examples 6 and 7	105–108
(4) Model and solve uniform motion problems (p. 548)	Examples 8 and 9	109–112

In Problems 97–100, solve the proportion.

97. $\dfrac{6}{4y + 5} = \dfrac{2}{7}$ **98.** $\dfrac{2}{y - 3} = \dfrac{5}{y}$ **99.** $\dfrac{y + 1}{8} = \dfrac{1}{4}$ **100.** $\dfrac{6y + 7}{10} = \dfrac{2y + 9}{6}$

In Problems 101 and 102, use the similar triangles to solve for x.

101. 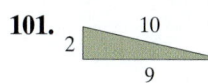 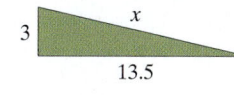   **102.**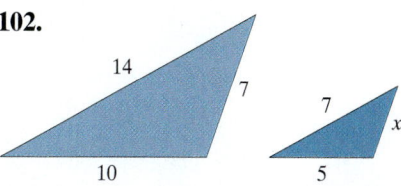

103. Cement A cement mixer uses 4 tanks of water to mix 15 bags of cement. How many tanks of water are needed to mix 30 bags of cement?

104. Pizza If 4 small pizzas cost $15.00, find the cost of 7 small pizzas.

105. Washing Lucille can wash the walls in 3 hours working alone, and Teresa can wash the walls in 2 hours. How long would it take them to wash the walls working together?

106. Carpet Fred can install the carpet in a room in 3 hours, but Barney needs 5 hours to install the carpet. How long would it take them to complete the carpet installation if they work together?

107. Dishes Working alone, it takes Jake 5 minutes longer to wash the dishes than it takes Adrienne when she washes the dishes alone. Washing the dishes together, Jake and Adrienne can finish the job in 6 minutes. How long does it take Jake to wash the dishes by himself?

108. Painting Working together, Donovan and Ben can paint a room in 4 hours. Working alone, it takes Donovan 7 hours to paint the room. How long would it take Ben to paint the room working alone?

109. Paddling Paul can paddle a kayak 15 miles upstream in the same amount of time it takes him to paddle a kayak 27 miles downstream. If the current is 2 mph, what is Paul's speed as he paddles downstream?

110. Cruising A cruise ship traveled for 275 miles with the current in the same amount of time it traveled 175 miles against the current. The speed of the current was 10 mph. What was the speed of the cruise ship as it traveled with the current?

111. Vacation On their vacation, a family traveled 135 miles by train and then traveled 855 miles by plane. The speed of the plane was three times the speed of the train. If the total time of the trip was 6 hours, what was the speed of the train?

112. Driving Tamika drove for 90 miles in the city. When she got on the highway, she increased her speed by 20 mph and drove for 130 miles. If Tamika drove a total of 4 hours, how fast did she drive in the city?

CHAPTER 7 TEST

 Remember to use your Chapter Test Prep Video CD to see fully worked-out solutions to any of these problems you would like to review.

1. Evaluate $\dfrac{3x - 2y^2}{6z}$ when $x = 2$, $y = -3$, and $z = -1$.

2. Find the values for which the following rational expression is undefined:
$\dfrac{x + 5}{x^2 - 3x - 10}$

3. Simplify: $\dfrac{x^2 - 4x - 21}{14 - 2x}$

In Problems 4–11, perform the indicated operations and simplify, if possible.

4. $\dfrac{\dfrac{35x^6}{9x^4}}{\dfrac{25x^5}{18x}}$

5. $\dfrac{5x - 15}{3x + 9} \cdot \dfrac{5x + 15}{3x - 9}$

6. $\dfrac{\dfrac{2x^2 - 5xy - 12y^2}{x^2 + xy - 20y^2}}{\dfrac{4x^2 - 9y^2}{x^2 + 4xy - 5y^2}}$

7. $\dfrac{y^2}{y + 3} + \dfrac{3y}{y + 3}$

8. $\dfrac{x^2}{x^2 - 9} - \dfrac{8x - 15}{x^2 - 9}$

9. $\dfrac{6}{y - z} + \dfrac{7}{z - y}$

10. $\dfrac{x}{x - 2} + \dfrac{3}{2x + 1}$

11. $\dfrac{2x}{x^2 + 5x + 6} - \dfrac{x + 1}{x^2 + 2x - 3}$

12. Simplify the following complex rational expression: $\dfrac{\dfrac{1}{9} - \dfrac{1}{y^2}}{\dfrac{1}{3} + \dfrac{1}{y}}$

In Problems 13 and 14, solve the rational equations. Check for extraneous solutions.

13. $\dfrac{m}{5} + \dfrac{5}{m} = \dfrac{m + 3}{4}$

14. $\dfrac{4}{x + 3} + \dfrac{5}{x - 6} = \dfrac{4x + 1}{x^2 - 3x - 18}$

15. Solve $\dfrac{1}{x} + \dfrac{1}{y} = \dfrac{1}{z}$ for x.

16. Solve the proportion: $\dfrac{2}{y + 1} = \dfrac{1}{y - 2}$

17. Use the similar triangles to solve for x.

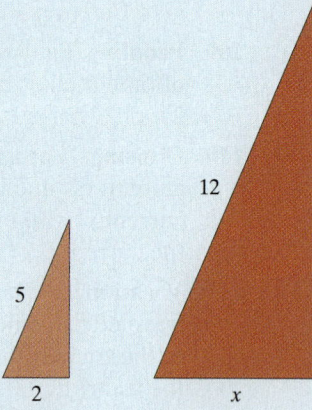

18. **Barrettes** If 8 hair barrettes cost $22.00, how much would 12 hair barrettes cost?

19. **Car Washing** It takes Frank 18 more minutes than Juan to wash a car. If they can wash the car together in 12 minutes, how long does it take Frank to wash the car by himself?

20. **Excursion** On a vacation excursion, some tourists walked 7 miles on a nature path and then hiked 12 miles up a mountainside. The tourists walked 3 mph faster than they hiked. The total time of the excursion was 4 hours. At what rate did the tourists hike?

Getting Ready for Intermediate Algebra: A Review of Chapters 1–7

The following problems cover important concepts from Chapters 1–7. We designed these problems so that you can review important concepts that will be needed for the rest of the course. Take time to work these problems before proceeding to the next chapter. The answers to these problems are located at the back of the text starting on page AN-23. If you get any problems wrong, go back to the section cited and review the material.

1. Evaluate: $\dfrac{4 - (-6)}{8 - 2}$ Section 1.6, p. 52

2. Evaluate $-\dfrac{2}{3}x + 5$ for $x = 6$.   Section 1.7, pp. 57–58

In Problems 3–6, solve each equation. State whether the equation is a contradiction, an identity, or a conditional equation.

3. $4x + 3 = 17$ Section 2.2, pp. 84–86

4. $5(x - 2) - 2x = 3x + 4$ Section 2.3, pp. 97–99

5. $\dfrac{1}{2}(x - 4) + \dfrac{2}{3}x = \dfrac{1}{6}(x - 4)$ Section 2.3, pp. 93–95

6. $0.3(x + 1) - 0.1(x - 7) = 0.4x - 0.2(x - 5)$ Section 2.3, pp. 97–99

In Problems 7 and 8, solve each linear inequality. Express the solution using set-builder notation and interval notation. Graph the solution set.

7. $6x - 7 > -31$ Section 2.8, pp. 153–158

8. $5(x - 3) \geq 7(x - 4) + 3$ Section 2.8, pp. 153–158

9. Use interval notation to express the inequality shown in the graph.   Section 2.8, pp. 150–153

10. Plot the following ordered pairs in the same Cartesian plane.   Section 3.1, pp. 175–179

$A(-3, 0)$, $B(4, -2)$, $C(1, 5)$, $D(0, 3)$, $E(-4, -5)$, $F(-5, 2)$

11. Graph $-4x + 3y = 24$ by finding its intercepts.   Section 3.2, pp. 196–199

12. Graph $x = 5$ Section 3.2, pp. 199–200

13. Find the slope of the line joining $(3, 6)$ and $(-1, -4)$ Section 3.3, pp. 206–209

14. Graph $y = 3x + 1$ Section 3.4, pp. 218–220

15. Find the equation of the line with slope $-\dfrac{4}{3}$ through $(-3, 1)$. Express your answer in slope-intercept form.   Section 3.5, pp. 227–229

16. Find an equation of the line through $(-2, 5)$ and $(2, 3)$. Express your answer in slope-intercept form.   Section 3.5, pp. 229–230

17. Find the equation of the line parallel to $y = -3x + 10$ through the point $(-5, 7)$. Express your answer in slope-intercept form.   Section 3.6, pp. 237–238

18. Graph: $x - 3y > 12$ Section 3.8, pp. 253–257

19. Solve: $\begin{cases} y = 4x - 3 \\ 4x - 3y = 5 \end{cases}$ Section 4.2, pp. 285–290

20. Solve: $\begin{cases} x + y = 3 \\ 3x + 2y = 2 \end{cases}$

Section 4.3, pp. 296–301

21. Graph the system: $\begin{cases} x + y \geq 2 \\ -3x + y \leq 10 \end{cases}$

Section 4.6, pp. 327–329

In Problem 22–24, perform the indicated operation.

22. $(12x^3 + 5x^2 - 3x + 1) - (2x^3 - 4x + 8)$

Section 5.1, pp. 347–348

23. $(2x - 5)(x + 3)$

Section 5.3, pp. 361–363

24. $\dfrac{x^3 - 2x^2 - 5x - 3}{x - 4}$

Section 5.5, pp. 385–388

In Problems 25–29, factor each polynomial completely.

25. $-2x^3 + 6x^2 - 8x + 24$

Section 6.1, pp. 413–415

26. $5p^3 + 50p^2 + 80p$

Section 6.2, pp. 423–425

27. $8y^2 - 2y - 3$

Section 6.3, pp. 427–435

28. $9a^2 + 24a + 16$

Section 6.4, pp. 436–439

29. $3k^4 - 27k^2$

Section 6.4, pp. 439–441

30. Solve: $8b^2 + 10b = 3$

Section 6.6, pp. 451–457

31. Solve: $(2x + 3)(x - 1) = 6x$

Section 6.6, pp. 451–457

32. Simplify: $\dfrac{2x^2 + x - 21}{x^2 + 6x - 27}$

Section 7.1, pp. 480–483

33. Simplify: $\dfrac{3x^2 + 14x - 5}{x^2 + x - 30} \cdot \dfrac{x^2 - 2x - 15}{3x^2 + 8x - 3}$

Section 7.2, pp. 485–487

34. Simplify: $\dfrac{\dfrac{y^2 - 9}{2y^2 - y - 15}}{\dfrac{3y^2 + 10y + 3}{2y^2 + y - 10}}$

Section 7.2, pp. 487–490

35. Subtract: $\dfrac{2x}{x - 3} - \dfrac{x + 1}{x + 2}$

Section 7.5, pp. 510–517

36. Solve: $\dfrac{9}{k - 2} = \dfrac{6}{k} + 3$

Section 7.7, pp. 529–536

37. Solve: $\dfrac{7}{y^2 + y - 12} - \dfrac{4y}{y^2 + 7y + 12} = \dfrac{6}{y^2 - 9}$

Section 7.7, pp. 529–536

8 Graphs, Relations, and Functions

Graphs come in many different forms. For example, the graph shown to the right was drawn by Florence Nightingale and is called a *polar area graph*. The area of each shaded region is proportional to the number of deaths represented by the shaded region. Nightingale used the graph to illustrate that more soldiers died because of unsanitary conditions in hospitals than on the battlefield during the Crimean War. This graph helped to increase the focus on the sanitary conditions in hospitals and shows how math can be used to save lives.

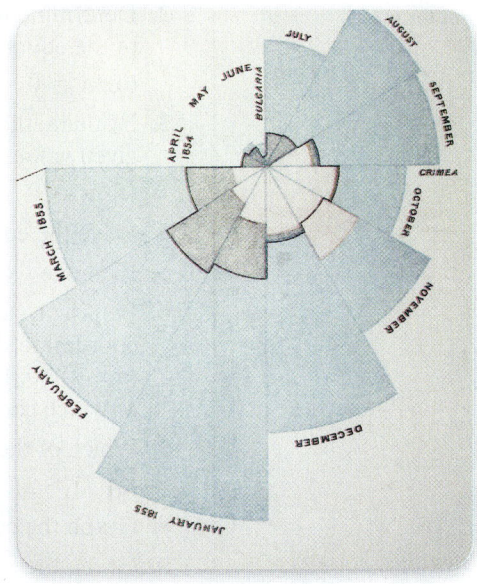

OUTLINE

8.1 Graphs of Equations

8.2 Relations

Putting the Concepts Together (Sections 8.1–8.2)

8.3 An Introduction to Functions

8.4 Functions and Their Graphs

8.5 Linear Functions

8.6 Compound Inequalities

8.7 Absolute Value Equations and Inequalities

Chapter 8 Activity: Shifting Discovery
Chapter 8 Review
Chapter 8 Test

The Big Picture: Putting It Together

In Chapter 3 we studied linear equations in two variables. We represented the solution of a linear equation in two variables using the rectangular coordinate system. The rectangular coordinate system provides the connection between algebra and geometry. Prior to the introduction of the rectangular coordinate system, algebra and geometry were thought to be separate subjects.

We begin this chapter with a brief review of the rectangular coordinate system and graphs of equations. We continue this chapter by introducing the concept of a function. The function is arguably the single most important concept of algebra.

8.1 Graphs of Equations

OBJECTIVES

1. Graph an Equation Using the Point-Plotting Method
2. Identify Intercepts from the Graph of an Equation
3. Interpret Graphs

Preparing for Graphs of Equations

Before getting started, take this readiness quiz. If you get a problem wrong, go back to the section cited and review the material.

1. Plot the following points on the real number line: $-2, 4, 0, \frac{1}{2}$. [Section 1.2, pp. 12–13]

2. Determine which of the following are solutions to the equation $3x - 5(x + 2) = 4$.
   (a) $x = 0$ **(b)** $x = -3$ **(c)** $x = -7$ [Section 2.1, pp. 75–76]

3. Evaluate the expression $2x^2 - 3x + 1$ for the given values of the variable.
   (a) $x = 0$ **(b)** $x = 2$ **(c)** $x = -3$ [Section 1.7, pp. 57–58]

4. Solve the equation $3x + 2y = 8$ for y. [Section 2.4, pp. 108–111]

5. Evaluate $|-4|$. [Section 1.2, pp. 14–15]

6. Plot the following ordered pairs in a rectangular coordinate system: $(3, 5), (-2, 3), (-1, -2),$ $(5, -3), (0, 4).$ Tell which quadrant each point lies in or which coordinate axis the point lies on. [Section 3.1, pp. 175–179]

7. Which of the following points satisfies $3x - 2y = 7$?
   (a) $(1, -2)$ **(b)** $(3, 1)$ **(c)** $(-2, -5)$ [Section 3.1, pp. 179–180]

8. Graph the equation: $y = -4x + 5$ [Section 3.4, pp. 218–220]

9. Graph the line $3x - 5y = 15$ using intercepts. [Section 3.2, pp. 196–199]

Remember from Section 1.2 that we locate a point on the real number line by assigning it a single real number, called the *coordinate of the point*. (See Problem 1 in *Preparing for . . .*) For work in a two-dimensional plane, we locate points by using two numbers, called an *ordered pair*. (See Problem 6 in *Preparing for . . .*) For convenience we provide a brief review of this two-dimensional plane before starting this section. A full discussion of this material can be found in Chapter 3, Section 1 on pages 175–185.

Draw two real number lines located in the same plane: one horizontal and the other vertical so that they intersect at a right (90°) angle. We call the horizontal real number line the **x-axis,** and the vertical real number line the **y-axis.** The point where the x-axis and y-axis intersect is called the **origin, O.** See Figure 1.

Preparing for...Answers

1.
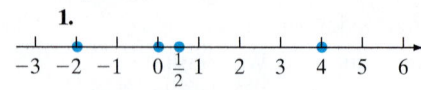

2. **(a)** No **(b)** No **(c)** Yes **3. (a)** 1
(b) 3 **(c)** 28 **4.** $y = -\frac{3}{2}x + 4$ **5.** 4

6.
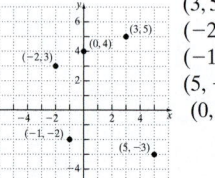

(3, 5): Quadrant I
(−2, 3): Quadrant II
(−1, −2): Quadrant III
(5, −3): Quadrant IV
(0, 4): y-axis

7. **(a)** Yes **(b)** Yes **(c)** No
8. 9.
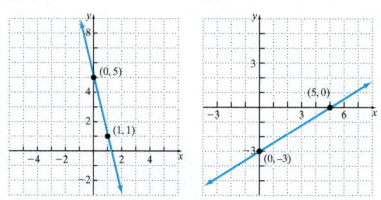

Figure 1

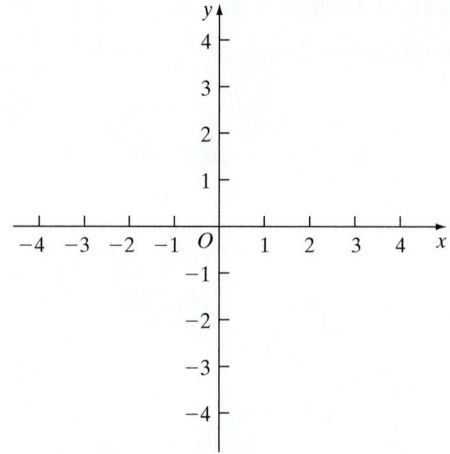

The origin O has a value of 0 on the x-axis and the y-axis. Points on the x-axis to the right of O are positive real numbers; points on the x-axis to the left of O are negative real numbers. Points on the y-axis that are above O are positive real numbers; points on the y-axis that are below O are negative real numbers. In Figure 1 we label the x-axis

"x" and the y-axis "y." Notice that an arrow is used at the end of each axis to denote the positive direction. We do not use an arrow to denote the negative direction.

The coordinate system presented in Figure 1 is called a **rectangular** or **Cartesian coordinate system,** named after René Descartes (1596–1650), a French mathematician, philosopher, and theologian. The plane formed by the x-axis and y-axis is often referred to as the **xy-plane,** and the x-axis and y-axis are called the **coordinate axes.**

We can represent any point P in the rectangular coordinate system by using an **ordered pair (x, y)** of real numbers. If $x > 0$, we travel x units to the right of the y-axis; if $x < 0$, we travel $|x|$ units to the left of the y-axis. If $y > 0$, we travel y units above the x-axis; if $y < 0$, we travel $|y|$ units below the x-axis. The ordered pair (x, y) is also called the **coordinates** of P.

The origin O has coordinates $(0, 0)$. Any point on the x-axis has coordinates of the form $(x, 0)$, and any point on the y-axis has coordinates of the form $(0, y)$.

If (x, y) are the coordinates of a point P, then x is called the **x-coordinate** or **abscissa,** of P and y is called the **y-coordinate** or **ordinate,** of P.

If you look back at Figure 1, you should notice that the x- and y-axes divide the plane into four separate regions or **quadrants.** In quadrant I, both the x- and y-coordinate are positive; in quadrant II, x is negative and y is positive; in quadrant III, both x and y are negative; and in quadrant IV, x is positive and y is negative. Points on the coordinate axes do not belong to a quadrant. See Figure 2.

Figure 2

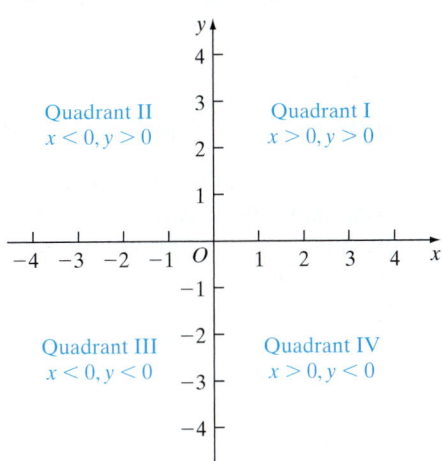

Figure 3

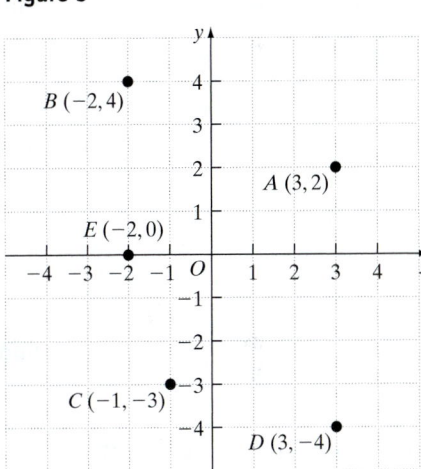

For example, Figure 3 shows the ordered pairs $A(3, 2)$, $B(-2, 4)$, $C(-1, -3)$, $D(3, -4)$, and $E(-2, 0)$. Point A lies in quadrant I, point B lies in quadrant II, point C lies in quadrant III, point D lies in quadrant IV, and point E lies on the negative x-axis.

QUICK ✓ *Plot each point in the xy-plane. Tell in which quadrant or on what coordinate axis each point lies. If you are struggling with these problems, go back and review pages 175–179 in Section 3.1.*

1. **(a)** $A(5, 2)$ **(b)** $B(4, -2)$
   (c) $C(0, -3)$ **(d)** $D(-4, -3)$

2. **(a)** $A(-3, 2)$ **(b)** $B(-4, 0)$
   (c) $C(3, -2)$ **(d)** $D(6, 1)$

Work Smart: Study Skills
The material in this chapter requires a working knowledge of the material in Chapter 3. If the terms and ideas that are being used in this section are unfamiliar, it would be a good idea to go back to Chapter 3 and review the material.

Recall from Chapter 3 that an **equation in two variables,** x and y, is a statement in which the algebraic expressions involving x and y are equal. The expressions are called **sides** of the equation. Since an equation is a statement, it may be true or false, depending upon the values of the variables.

For example,

$$x^2 = y + 2 \qquad 3x + 2y = 6 \qquad y = -4x + 5$$

are all equations in two variables. The equation $x^2 = y + 2$ is satisfied when $x = 3$ and $y = 7$ since $3^2 = 7 + 2$. It is also satisfied when $x = -2$ and $y = 2$. In fact, there are infinitely many choices of x and y that satisfy the equation $x^2 = y + 2$. Some choices of x and y do not satisfy the equation $x^2 = y + 2$. For example, $x = 3$ and $y = 4$ does not satisfy the equation because $3^2 \neq 4 + 2$ (that is, $9 \neq 6$).

QUICK ✓ *If you are struggling with these problems, go back and review pages 179–180 in Section 3.1.*

3. Determine if the following points are on the graph of $2x - 4y = 12$.

    (a) $(2, -3)$ **(b)** $(2, -2)$ **(c)** $\left(\dfrac{3}{2}, -\dfrac{9}{4}\right)$

4. Determine if the following points are on the graph of $y = x^2 + 3$.

    (a) $(1, 4)$ **(b)** $(-2, -1)$ **(c)** $(-3, 12)$

In Words
The graph of an equation is a geometric way of representing the set of all ordered pairs that make the equation a true statement.

The **graph of an equation in two variables** x and y is the set of all ordered pairs (x, y) in the xy-plane that satisfy the equation. Let's review the point-plotting method of obtaining the graph of an equation in two variables.

① Graph an Equation Using the Point-Plotting Method

Recall from Chapter 3, one of the most elementary methods for graphing an equation is the **point-plotting method.** With this method, we choose values for one of the variables and use the equation to determine the corresponding values of the remaining variable. If x and y are the variables in the equation, it does not matter whether we choose values of x and use the equation to find the corresponding y or choose y and find x. Convenience will determine which way we go.

EXAMPLE 1 **How to Graph an Equation by Plotting Points**

Graph the equation $y = -2x + 4$ by plotting points.

Step-by-Step Solution

Step 1: We want to find all points (x, y) that satisfy the equation. To determine these points we choose values of x (do you see why?) and use the equation to determine the corresponding values of y. See Table 1.

	Table 1	
x	**$y = -2x + 4$**	**(x, y)**
-3	$-2(-3) + 4 = 10$	$(-3, 10)$
-2	$-2(-2) + 4 = 8$	$(-2, 8)$
-1	$-2(-1) + 4 = 6$	$(-1, 6)$
0	$-2(0) + 4 = 4$	$(0, 4)$
1	$-2(1) + 4 = 2$	$(1, 2)$
2	$-2(2) + 4 = 0$	$(2, 0)$
3	$-2(3) + 4 = -2$	$(3, -2)$

Step 2: We plot the points listed in the third column of Table 1 as shown in Figure 4(a). Now connect the points to obtain the graph of the equation (*a line*) as shown in Figure 4(b).

Figure 4

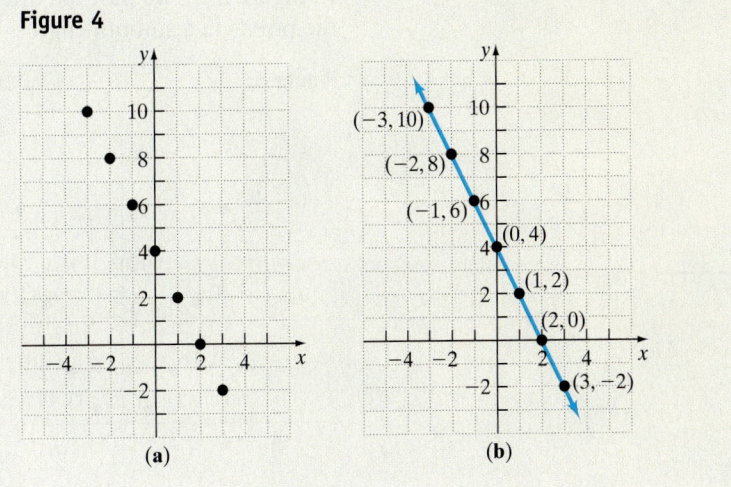

(a)   (b)

The equation $y = -2x + 4$ is a linear equation and its graph is a line. The graph of the equation shown in Figure 4(b) does not show all the points that satisfy $y = -2x + 4$. For example, in Figure 4(b) the point $(8, -12)$ is part of the graph of $y = -2x + 4$, but it is not shown. Since the graph of $y = -2x + 4$ can be extended as far as we please, we use arrows on the ends of the graph to indicate that the pattern shown continues. It is important to show enough of the graph so that anyone who is looking at it will "see" the rest of it as an obvious continuation of what is there. This is called a **complete graph.**

Now let's look at the graph of a **nonlinear equation.**

EXAMPLE 2   **Graphing an Equation by Plotting Points**

Graph the equation $y = x^2$ by plotting points.

Solution

Table 2 shows several points on the graph.

	Table 2	
x	**$y = x^2$**	**(x, y)**
−4	$y = (-4)^2 = 16$	(−4, 16)
−3	$y = (-3)^2 = 9$	(−3, 9)
−2	$y = (-2)^2 = 4$	(−2, 4)
−1	$y = (-1)^2 = 1$	(−1, 1)
0	$y = (0)^2 = 0$	(0, 0)
1	$y = (1)^2 = 1$	(1, 1)
2	$y = (2)^2 = 4$	(2, 4)
3	$y = (3)^2 = 9$	(3, 9)
4	$y = (4)^2 = 16$	(4, 16)

In Figure 5(a), we plot the ordered pairs listed in Table 2. In Figure 5(b), we connect the points in a smooth curve.

Figure 5

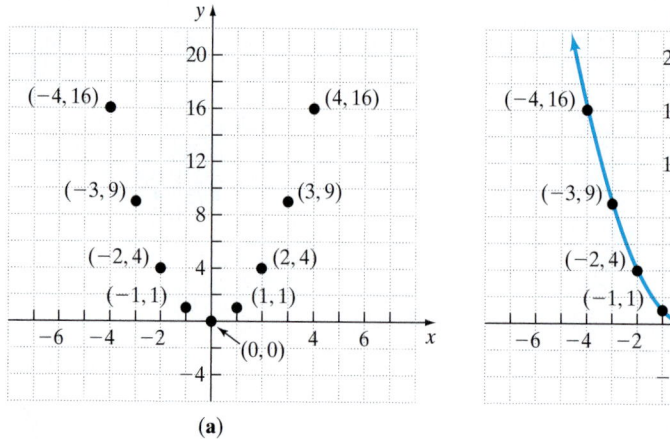

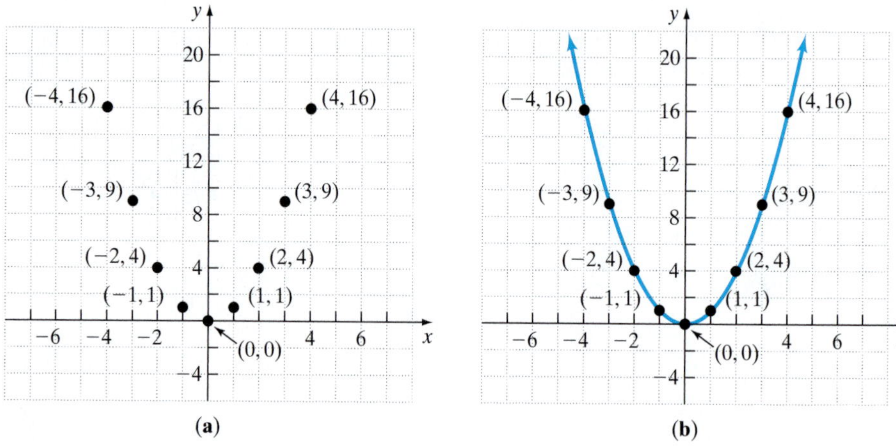

(a)   (b)

Work Smart
Notice we use a different scale on the *x*- and *y*-axis in Figure 5.

Work Smart
Experience will play a huge role in determining which *x*-values to choose in creating a table of values. For the time being, start by choosing values of *x* around $x = 0$ as in Table 2.

Two questions that you might be asking yourself right now are "How do I know how many points are sufficient?" and "How do I know which *x*-values (or *y*-values) I should choose in order to obtain points on the graph?" Often, the type of equation we wish to graph indicates the number of points that are necessary. For example, we know from Chapter 3 that if the equation is of the form $y = mx + b$, then its graph is a line and only two points are required to obtain the graph (as in Example 1). Other times, more points are required. At this stage in your math career, you will need to plot quite a few points to obtain a complete graph. However, as your experience and knowledge grows, you will learn to be more efficient in obtaining complete graphs.

QUICK ✓ *Graph each equation using the point-plotting method.*

5. $y = 3x + 1$ **6.** $2x + 3y = 8$ **7.** $y = x^2 + 3$

EXAMPLE 3 **Graphing the Equation $x = y^2$**

Graph the equation $x = y^2$ by plotting points.

Solution

Because the equation is solved for *x*, we will choose values of *y* and use the equation to find the corresponding values of *x*. See Table 3. We plot the ordered pairs listed in Table 3 and connect the points in a smooth curve. See Figure 6.

Table 3

y	$x = y^2$	(x, y)
-3	$(-3)^2 = 9$	$(9, -3)$
-2	$(-2)^2 = 4$	$(4, -2)$
-1	$(-1)^2 = 1$	$(1, -1)$
0	$0^2 = 0$	$(0, 0)$
1	$1^2 = 1$	$(1, 1)$
2	$2^2 = 4$	$(4, 2)$
3	$3^2 = 9$	$(9, 3)$

Figure 6

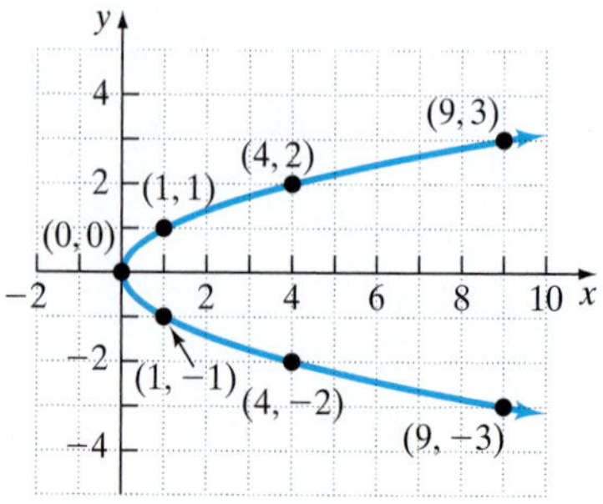

QUICK ✓ *Graph each equation using the point-plotting method.*

8. $x = y^2 + 2$ **9.** $x = (y - 1)^2$

Work Smart
In order for a graph to be complete, all of its intercepts must be displayed.

Figure 7

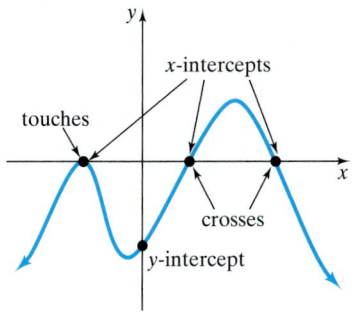

② Identify the Intercepts from the Graph of an Equation

One of the key components that should be displayed in a complete graph is the *intercepts* of the graph.

> **DEFINITION**
>
> The **intercepts** are the points, if any, where a graph crosses or touches the coordinate axes. The x-coordinate of a point at which the graph crosses or touches the x-axis is an **x-intercept,** and the y-coordinate of a point at which the graph crosses or touches the y-axis is a **y-intercept.**

See Figure 7 for an illustration. Notice that an x-intercept exists when $y = 0$ and a y-intercept exists when $x = 0$.

EXAMPLE 4 Finding Intercepts from a Graph

Find the intercepts of the graph shown in Figure 8. What are the x-intercepts? What are the y-intercepts?

Solution

The intercepts of the graph are the points

$$(-3, 0), \quad (0, 2), \quad (1, 0), \quad \text{and} \quad (3.8, 0)$$

The x-intercepts are $-3, 1,$ and 3.8. The y-intercept is 2. ∎

  In Example 4, you should notice the following: If we do not specify the type of intercept (x- versus y-), then we report the intercept as an ordered pair. However, if we specify the type of intercept, then we only need report the coordinate of the intercept. For x-intercepts, we report the x-coordinate of the intercept; for y-intercepts, we report the y-coordinate of the intercept.

Figure 8

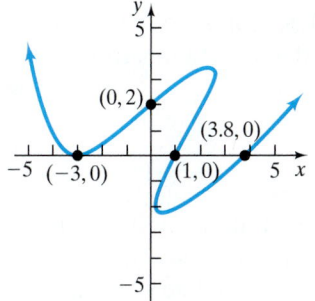

QUICK ✓

10. Find the intercepts of the graph shown in the figure. What are the x-intercepts? What are the y-intercepts?

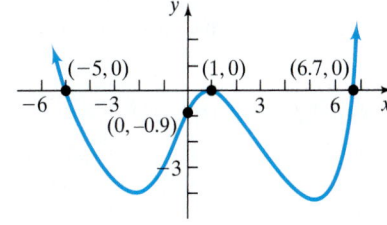

③ Interpret Graphs

Graphs play an important role in helping us to visualize relationships that exist between two variables or quantities. We have all heard the expression "A picture is worth a thousand words." A graph is a "picture" that illustrates the relationship between two variables. By visualizing this relationship, we are able to see important information and draw conclusions regarding the relationship between the two variables.

EXAMPLE 5 Interpret a Graph

The graph in Figure 9 shows the revenue R in dollars for selling x gallons of gasoline in an hour at a gas station. The vertical axis represents the revenue and the horizontal axis represents the number of gallons of gasoline sold.

   (a) What is the revenue if 150 gallons of gasoline are sold?

   (b) How many gallons of gasoline are sold when revenue is highest? What is the highest revenue?

   (c) Identify and interpret the intercepts.

Figure 9

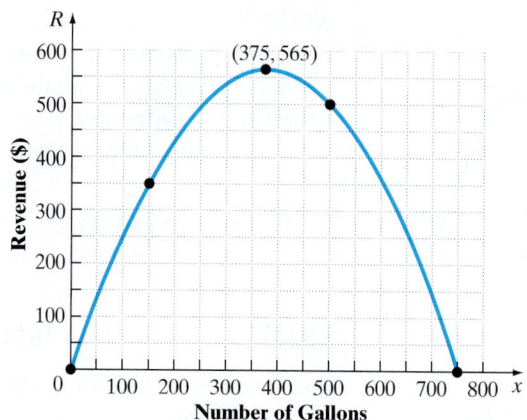

Solution

   (a) Draw a vertical line up from 150 on the horizontal axis until we reach the point on the graph. Then draw a horizontal line from this point to the vertical axis. The point where the horizontal line intersects the vertical axis is the revenue when 150 gallons of gasoline are sold. The revenue from selling 150 gallons of gasoline is \$350.

   (b) The revenue is highest when 375 gallons of gasoline are sold. The highest revenue is \$565.

   (c) The intercepts are $(0, 0)$ and $(750, 0)$. If the price of gasoline is too high, demand for gasoline (in theory) will be 0 gallons. This is the explanation for selling 0 gallons of gasoline. If the price of gasoline is \$0, then revenue will also be 0. The 750 gallons "sold" at a price of \$0 per gallon represents the maximum number of gallons that can be pumped at the station. In other words, it is the station's capacity.

QUICK ✓

11. The graph shown below represents the cost C (in thousands of dollars) of refining x gallons of gasoline per hour (in thousands). The vertical axis represents the cost and the horizontal axis represents the number of gallons of gasoline refined.

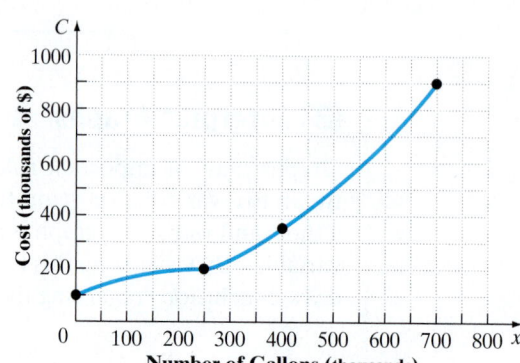

(a) What is the cost of refining 250 thousand gallons of gasoline per hour?

(b) What is the cost of refining 400 thousand gallons of gasoline per hour?

(c) In the context of the problem, explain the meaning of the graph ending at 700 thousand gallons of gasoline.

(d) Identify and interpret the intercept.

8.1 Exercises

For Extra Help:

Student Solutions Manual   CD Video   PH Math/Tutor Center   MathXL Tutorials on CD   MathXL®   MyMathLab

Concepts and Vocabulary

In Problems 1–3, fill in the blanks.

1. The point where the *x*-axis and *y*-axis intersect in the Cartesian coordinate system is called the _____.

2. If a point lies in quadrant II of the Cartesian coordinate system, then *x* _____ ($>, <$) 0 and *y* _____ ($>, <$) 0.

3. The points, if any, at which a graph crosses or touches a coordinate axis are called _____.

In Problems 4–6, answer True or False to each statement.

4. If a point lies in quadrant III of the Cartesian coordinate system, then both *x* and *y* are negative.

5. The graph of an equation must have at least one *x*-intercept.

6. The graph of an equation in two variables *x* and *y* is the set of all ordered pairs (x, y) in the *xy*-plane that satisfy the equation.

7. Explain what is meant by a complete graph.

8. Explain what the graph of an equation represents.

9. What is the point-plotting method for graphing an equation?

10. What is the *y*-coordinate of a point that is an *x*-intercept? What is the *x*-coordinate of a point that is a *y*-intercept?

Building Skills

11. Determine the coordinates of each of the points plotted. Tell in which quadrant or on what coordinate axis each point lies.

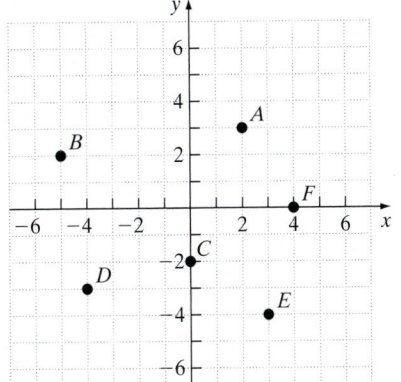

12. Determine the coordinates of each of the points plotted. Tell in which quadrant or on what coordinate axis each point lies.

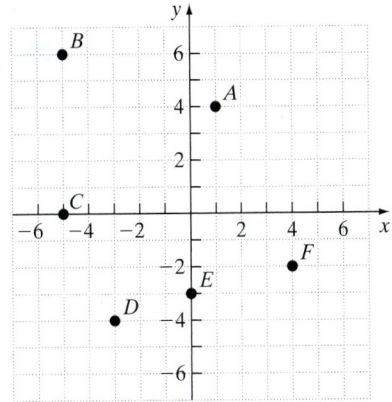

In Problems 13 and 14, plot each point in the xy-plane. Tell in which quadrant or on what coordinate axis each point lies.

13. $A(3, 5)$ 　　　　　　　　　　　　　　　　　**14.** $A(-3, 1)$
　　$B(-2, -6)$ 　　　　　　　　　　　　　　　　　$B(-6, 0)$
　　$C(5, 0)$ 　　　　　　　　　　　　　　　　　　$C(2, -5)$
　　$D(1, -6)$ 　　　　　　　　　　　　　　　　　$D(-6, -2)$
　　$E(0, 3)$ 　　　　　　　　　　　　　　　　　　$E(1, 2)$
　　$F(-4, 1)$ 　　　　　　　　　　　　　　　　　$F(0, -5)$

In Problems 15–20, determine whether the given points are on the graph of the equation.

15. $2x + 5y = 12$ 　　　　　　　　　　　　　**16.** $-4x + 3y = 18$
　　(a) $(1, 2)$ 　　　　　　　　　　　　　　　**(a)** $(1, 7)$
　　(b) $(-2, 3)$ 　　　　　　　　　　　　　　**(b)** $(0, 6)$
　　(c) $(-4, 4)$ 　　　　　　　　　　　　　　**(c)** $(-3, 10)$
　　(d) $\left(-\dfrac{3}{2}, 3\right)$ 　　　　　　　　　　　　　**(d)** $\left(\dfrac{3}{2}, 4\right)$

17. $y = -2x^2 + 3x - 1$ 　　　　　　　　　**18.** $y = x^3 - 3x$
　　(a) $(-2, -15)$ 　　　　　　　　　　　　　**(a)** $(2, 2)$
　　(b) $(3, 10)$ 　　　　　　　　　　　　　　**(b)** $(3, 8)$
　　(c) $(0, 1)$ 　　　　　　　　　　　　　　　**(c)** $(-3, -18)$
　　(d) $(2, -3)$ 　　　　　　　　　　　　　　**(d)** $(0, 0)$

19. $y = |x - 3|$ 　　　　　　　　　　　　　　**20.** $x^2 + y^2 = 1$
　　(a) $(1, 4)$ 　　　　　　　　　　　　　　　**(a)** $(0, 1)$
　　(b) $(4, 1)$ 　　　　　　　　　　　　　　　**(b)** $(1, 1)$
　　(c) $(-6, 9)$ 　　　　　　　　　　　　　　**(c)** $\left(\dfrac{1}{2}, \dfrac{1}{2}\right)$
　　(d) $(0, 3)$ 　　　　　　　　　　　　　　　**(d)** $\left(\dfrac{\sqrt{3}}{2}, \dfrac{1}{2}\right)$

In Problems 21–24, the graph of an equation is given. List the intercepts of the graph.

21. 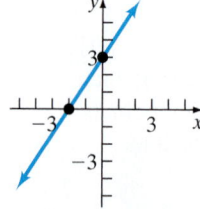 　　　　　　　　　　**22.**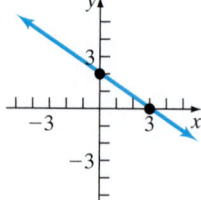

23. 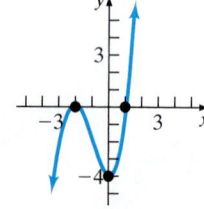 　　　　　　　　　　**24.**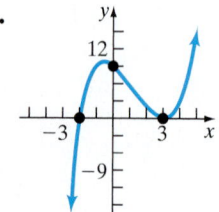

In Problems 25–52, graph each equation by plotting points.

25. $y = 4x$ 　　　　**26.** $y = 2x$ 　　　　**27.** $y = -\dfrac{1}{2}x$ 　　　　**28.** $y = -\dfrac{1}{3}x$

29. $y = x + 3$ **30.** $y = x - 2$ **31.** $y = -3x + 1$ **32.** $y = -4x + 2$

33. $y = \dfrac{1}{2}x - 4$ **34.** $y = -\dfrac{1}{2}x + 2$ **35.** $2x + y = 7$ **36.** $3x + y = 9$

37. $y = -x^2$ **38.** $y = x^2 - 2$ **39.** $y = 2x^2 - 8$ **40.** $y = -2x^2 + 8$

41. $y = |x|$ **42.** $y = |x| - 2$ **43.** $y = |x - 1|$ **44.** $y = -|x|$

45. $y = x^3$ **46.** $y = -x^3$ **47.** $y = x^3 + 1$ **48.** $y = x^3 - 2$

49. $x^2 - y = 4$ **50.** $x^2 + y = 5$ **51.** $x = y^2 - 1$ **52.** $x = y^2 + 2$

Applying the Concepts

53. If $(a, 4)$ is a point on the graph of $y = 4x - 3$, what is a?

54. If $(a, -2)$ is a point on the graph of $y = -3x + 5$, what is a?

55. If $(3, b)$ is a point on the graph of $y = x^2 - 2x + 1$, what is b?

56. If $(-2, b)$ is a point on the graph of $y = -2x^2 + 3x + 1$, what is b?

57. Area of a Window Bob Villa wishes to put a new window in his home. He wants the perimeter of the window to be 100 feet. The graph below shows the relation between the width, x, of the opening and the area of the opening.

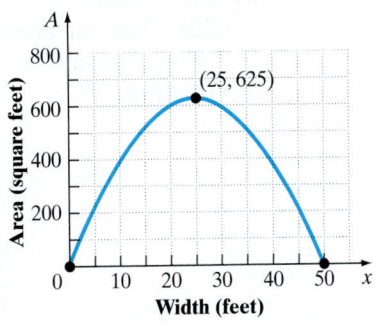

(a) What is the area of the opening if the width is 10 feet?
(b) What is the width of the opening in order for area to be a maximum? What is the maximum area of the opening?
(c) Identify and interpret the intercepts.

58. Projectile Motion The graph below shows the height, in feet, of a ball thrown straight up with an initial speed of 80 feet per second from an initial height of 96 feet after t seconds.

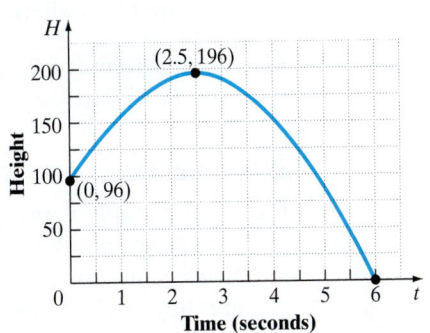

(a) What is the height of the object after 1.5 seconds?
(b) At what time is the height a maximum? What is the maximum height?
(c) Identify and interpret the intercepts.

59. Cell Phones We all struggle with selecting a cellular phone provider. The graph below shows the relation between the monthly cost (in dollars) of a cellular phone and the number of minutes used, m, when using the Sprint PCS 500-minute plan. (*Source: SprintPCS.com*)

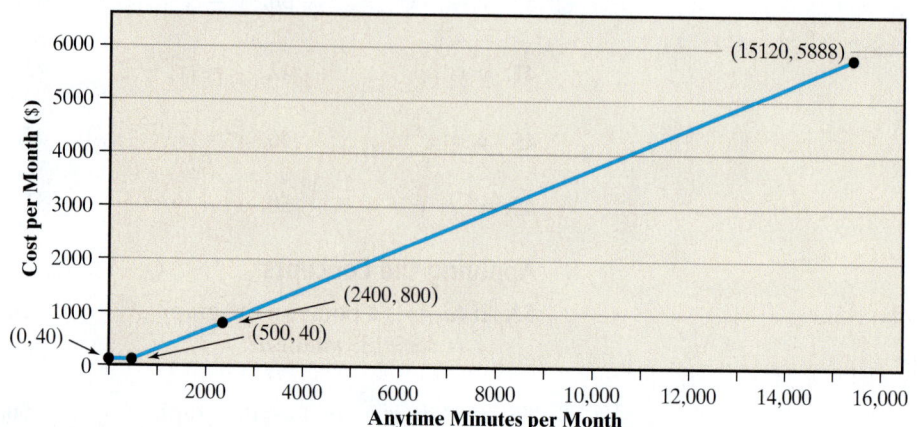

(a) What is the cost of talking for 200 minutes in a month? 500 minutes?
(b) What is the cost of talking 2400 minutes in a month?
(c) Identify and interpret the intercept.

60. Wind Chill It is $10°$ Celsius outside. The wind is calm but then gusts up to 20 meters per second. You feel the chill go right through your bones. The following graph shows the relation between the wind chill temperature (in degrees Celsius) and wind speed (in meters per second).

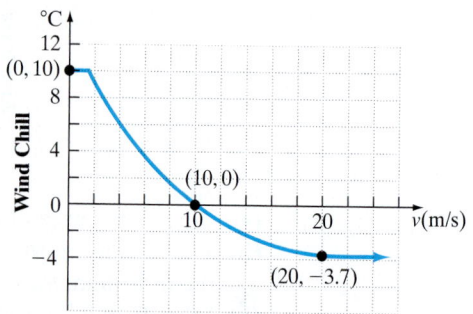

(a) What is the wind chill if the wind is blowing 4 meters per second?
(b) What is the wind chill if the wind is blowing 20 meters per second?
(c) Identify and interpret the intercepts.

61. Plot the points $(4, 0), (4, 2), (4, -3)$, and $(4, -6)$. Describe the set of all points of the form $(4, y)$ where y is a real number.

62. Plot the points $(4, 2), (1, 2), (0, 2)$, and $(-3, 2)$. Describe the set of all points of the form $(x, 2)$ where x is a real number.

Extending the Concepts

63. Draw a graph of an equation that contains two x-intercepts, -2 and 3. At the x-intercept -2, the graph crosses the x-axis; at the x-intercept 3, the graph touches the x-axis. Compare your graph with those of your classmates. How are they similar? How are they different?

64. Draw a graph that contains the points $(-3, -1), (-1, 1), (0, 3)$, and $(1, 5)$. Compare your graph with those of your classmates. How many of the graphs are straight lines? How many are "curved"?

65. Make up an equation that contains the points $(2, 0), (4, 0),$ and $(1, 0)$. Compare your equation with those of your classmates. How are they similar? How are they different?

66. Make up an equation that contains the points $(0, 3), (1, 3),$ and $(-4, 3)$. Compare your equation with those of your classmates. How many are the same?

The Graphing Calculator

Just as we have graphed equations using point-plotting, the graphing calculator also graphs equations by plotting points. Figure 10 shows the graph of $y = x^2$ and Table 4 shows points on the graph of $y = x^2$ using a TI-84 Plus graphing calculator.

Figure 10   **Table 4**

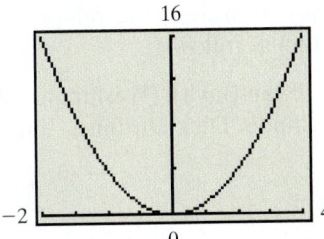

In Problems 67–74, use a graphing calculator to draw a complete graph of each equation. Use the TABLE feature to assist in selecting an appropriate viewing window.

67. $y = 3x - 9$ **68.** $y = -5x + 8$ **69.** $y = -x^2 + 8$ **70.** $y = 2x^2 - 4$

71. $y + 2x^2 = 13$ **72.** $y - x^2 = -15$ **73.** $y = x^3 - 6x + 1$ **74.** $y = -x^3 + 3x$

8.2 Relations

OBJECTIVES

1. Understand Relations
2. Find the Domain and the Range of a Relation
3. Graph a Relation Defined by an Equation

Preparing for Relations

Before getting started, take this readiness quiz. If you get a problem wrong, go back to the section cited and review the material.

1. Write the inequality $-4 \leq x \leq 4$ in interval notation.   [Section 2.8, pp. 150–153]

2. Write the interval $[2, \infty)$ using an inequality.   [Section 2.8, pp. 150–153]

① Understand Relations

We often see situations where one variable is somehow linked to the value of some other variable. For example, an individual's level of education is linked to annual income. Engine size is linked to gas mileage. When the value of one variable is related to the value of a second variable, we have a *relation*.

> **DEFINITION**
>
> When the elements in one set are linked to elements in a second set, we have a **relation.** If x and y are two elements in these sets and if a relation exists between x and y, then we say that x **corresponds** to y or that y **depends on** x, and we write $x \rightarrow y$. We may also write a relation where y depends on x as an ordered pair (x, y).

Preparing for...Answers **1.** $[-4, 4]$
2. $x \geq 2$

EXAMPLE 1 Illustrating a Relation

Consider the data presented in Figure 11, where a correspondence between states and senators in 2005 is shown for randomly selected senators. We might name the relation "is represented by." So, we would say "Indiana is represented by Evan Bayh."

Figure 11

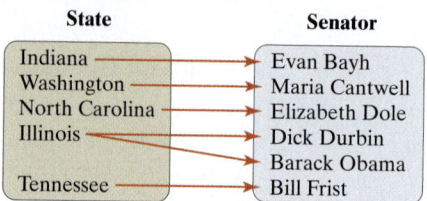

By representing the relation as in Figure 11, we are using **mapping,** in which we draw an arrow from an element from the set "state" to an element in the set "senator." We could also represent the relation in Figure 11 using ordered pairs in the form (state, senator) as follows:

{(Indiana, Evan Bayh), (Washington, Maria Cantwell), (North Carolina, Elizabeth Dole), (Illinois, Dick Durbin), (Illinois, Barack Obama), (Tennessee, Bill Frist)}

QUICK ✓

1. Use the map to represent the relation as a set of ordered pairs.

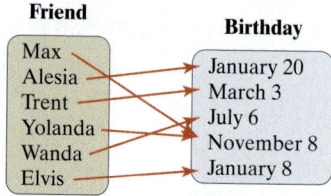

2. Use the set of ordered pairs to represent the relation as a map.

$$\{(1, 3), (5, 4), (8, 4), (10, 13)\}$$

---

2 Find the Domain and the Range of a Relation

In a relation we say that y depends on x and could write the relation as a set of ordered pairs (x, y). We can think of the set of all x as the **inputs** of the relation. The set of all y can be thought of as the **outputs** of the relation. We use this interpretation of a relation to define *domain* and *range*.

DEFINITION

The **domain** of a relation is the set of all inputs of the relation. The **range** is the set of all outputs of the relation.

EXAMPLE 2 Finding the Domain and the Range of a Relation

Find the domain and the range of the relation presented in Figure 11 from Example 1.

Solution

The domain is the set of all inputs and the range is the set of all outputs. The inputs, and therefore the domain, of the relation are

{Indiana, Washington, North Carolina, Illinois, Tennessee}

The outputs, and therefore the range, of the relation are

{Evan Bayh, Maria Cantwell, Elizabeth Dole, Dick Durbin, Barack Obama, Bill Frist}

The careful reader will notice that we did not list Illinois twice in the domain because the domain and the range are sets and we never list elements in a set more than once.

QUICK ✓

3. State the domain and the range of the relation.

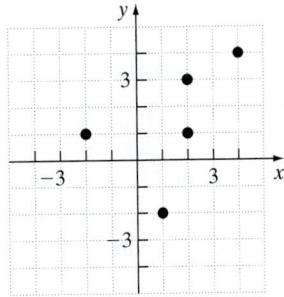

4. State the domain and the range of the relation.

$$\{(1, 3), (5, 4), (8, 4), (10, 13)\}$$

Relations can also be represented by plotting a set of ordered pairs. The set of all *x*-coordinates represents the domain of the relation and the set of all *y*-coordinates represents the range of the relation.

EXAMPLE 3 Finding the Domain and the Range of a Relation

Figure 12 shows the graph of a relation. Identify the domain and the range of the relation.

Figure 12

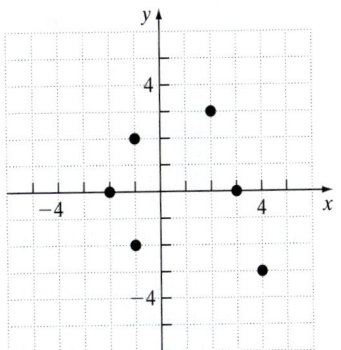

Solution

First, we notice that the ordered pairs in the graph are $(-2, 1), (1, -2), (2, 1), (2, 3),$ and $(4, 4)$. The domain is the set of all *x*-coordinates: $\{-2, 1, 2, 4\}$. The range is the set of all *y*-coordinates: $\{-2, 1, 3, 4\}$. ∎

QUICK ✓

5. Identify the domain and the range of the relation shown in the figure.

We have learned that a relation can be defined by a map or by a set of ordered pairs. A relation can also be defined by a graph. Remember that the graph of an equation is the set of all ordered pairs (x, y) such that the equation is a true statement. If a graph exists for some ordered pair (x, y), then the x-coordinate is in the domain and the y-coordinate is in the range. Think of it this way: When a graph of a relation is given, its domain may be viewed as the shadow created by the graph on the x-axis by vertical beams of light. Its range can be viewed as the shadow created by the graph on the y-axis by horizontal beams of light.

EXAMPLE 4 **Identifying the Domain and the Range of a Relation from Its Graph**

Figure 13 shows the graph of a relation. Determine the domain and the range of the relation.

Figure 13

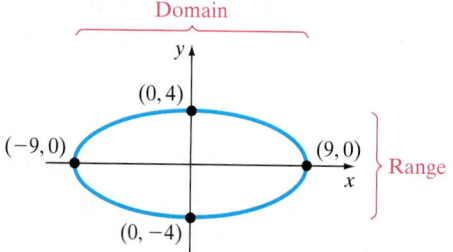

Solution

To find the domain of the relation, we determine the x-coordinates for which the graph exists. The graph exists for all x-values between -9 and 9, inclusive. Therefore, the domain is $\{x \mid -9 \le x \le 9\}$ or, using interval notation, $[-9, 9]$.

To find the range of the relation, we determine the y-coordinates for which the graph exists. The graph exists for all y-values between -4 and 4, inclusive. Therefore, the range is $\{y \mid -4 \le y \le 4\}$ or, using interval notation $[-4, 4]$. ■

QUICK *Identify the domain and the range of the relation from its graph.*

6.

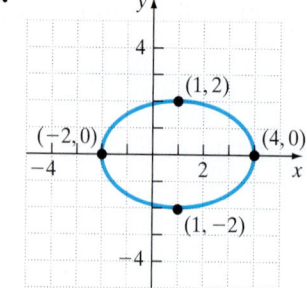

7.

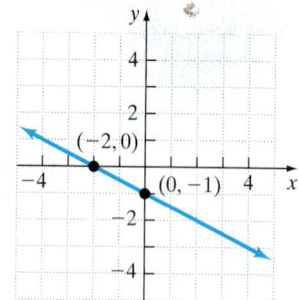

③ Graph a Relation Defined by an Equation

Another approach to defining a relation (instead of a map, a set of ordered pairs, or a graph) is to define relations through equations such as $x + y = 4$ or $x = y^2$. When relations are defined by equations, we typically graph the relation so that we can visualize how y depends upon x. As was seen in Example 4, the graph of the relation is also useful for helping us to identify the domain and the range of the relation.

EXAMPLE 5 **Relations Defined by Equations**

Graph the relation $y = -x^2 + 4$. Use the graph to determine the domain and the range of the relation.

Solution

The relation says to take the input x, square it, multiply this result by -1, and then add 4 to get the output y. We use the point-plotting method to graph the relation. Table 5 shows some points on the graph. Figure 14 shows a graph of the relation.

Figure 14

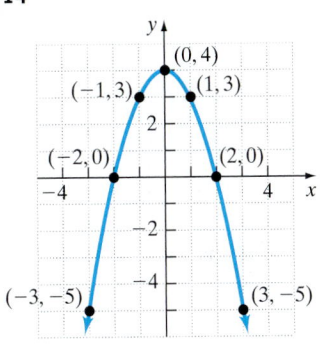

Table 5		
x	$y = -x^2 + 4$	(x, y)
-3	$-(-3)^2 + 4 = -5$	$(-3, -5)$
-2	$-(-2)^2 + 4 = 0$	$(-2, 0)$
-1	3	$(-1, 3)$
0	4	$(0, 4)$
1	3	$(1, 3)$
2	0	$(2, 0)$
3	-5	$(3, -5)$

From the graph, we can see that the graph extends indefinitely to the left and to the right (that is, the graph exists for all x-values). Therefore, the domain of the relation is the set of all real numbers, or $\{x \mid x \text{ is a real number}\}$, or using interval notation, $(-\infty, \infty)$. We also notice from the graph that there are no y-values greater than 4, but the graph exists everywhere for y-values less than or equal to 4. The range of the relation is $\{y \mid y \leq 4\}$, or using interval notation $(-\infty, 4]$. ∎

QUICK ✓ *Graph each relation. Use the graph to identify the domain and the range.*

8. $y = 3x - 8$ **9.** $y = x^2 - 8$ **10.** $x = y^2 + 1$

8.2 Exercises

For Extra Help: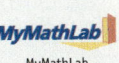

Student Solutions Manual   CD Video   PH Math/Tutor Center   MathXL Tutorials on CD   MathXL®   MyMathLab

Concepts and Vocabulary

In Problems 1 and 2, fill in the blanks.

1. If a relation exists between x and y, then we say that y _____ to x or that y _____ on x, and we write $x \rightarrow y$.

2. The _____ of a relation is the set of all inputs to the relation. The _____ is the set of all outputs of the relation.

In Problems 3 and 4, answer True or False to each statement.

3. If the graph of a relation does not exist at $x = 3$, then 3 is not in the domain of the relation.

4. The range of a relation is always the set of all real numbers.

5. In your own words, explain what a relation is. Be sure to include an explanation of domain and range.

✎ **6.** State the four methods for describing a relation presented in this section. When is using ordered pairs most appropriate? When is using a graph most appropriate? Support your opinion.

Building Skills

In Problems 7–10, write each relation as a set of ordered pairs. Then identify the domain and the range of the relation.

7.

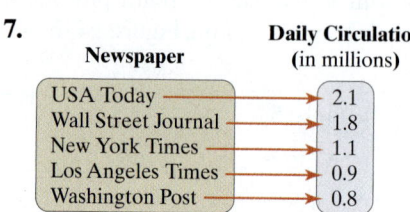

SOURCE: *Information Please Almanac*

8.

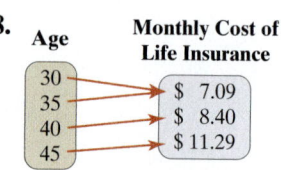

SOURCE: *eterm.com*

9.

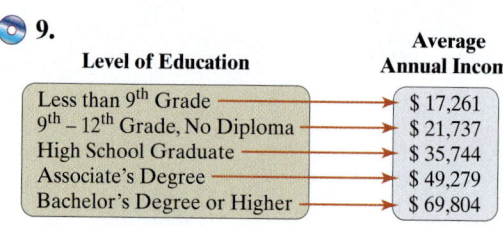

SOURCE: *United States Census Bureau*

10.

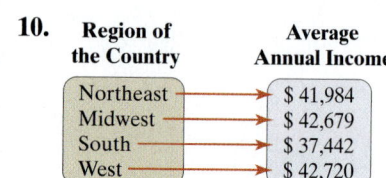

SOURCE: *United States Census Bureau*

In Problems 11–16, write each relation as a map. Then identify the domain and the range of the relation.

11. $\{(-3, 4), (-2, 6), (-1, 8), (0, 10), (1, 12)\}$

12. $\{(-2, 6), (-1, 3), (0, 0), (1, -3), (2, 6)\}$

13. $\{(-2, 4), (-1, 2), (0, 0), (1, 2), (2, 4)\}$

14. $\{(-2, -8), (-1, -1), (0, 0), (1, 1), (2, 8)\}$

15. $\{(0, -4), (-1, -1), (-2, 0), (-1, 1), (0, 4)\}$

16. $\{(-3, 0), (0, 3), (3, 0), (0, -3)\}$

In Problems 17–24, identify the domain and the range of the relation from the graph.

17.

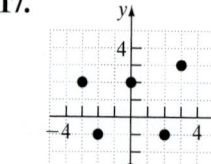

18.

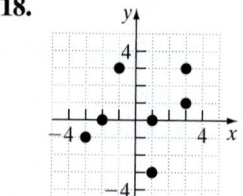

19.

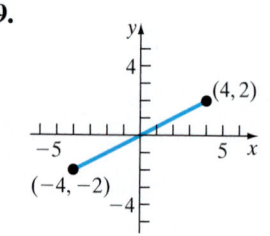

20.

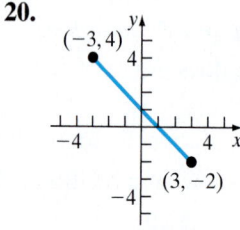

21.

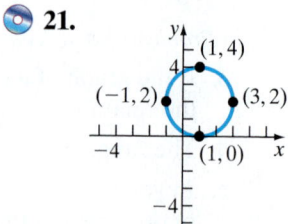

22.

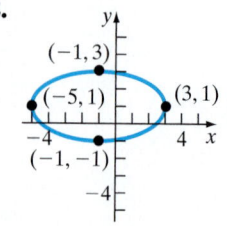

23.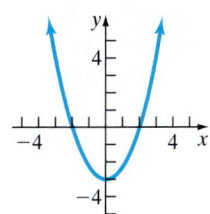

24.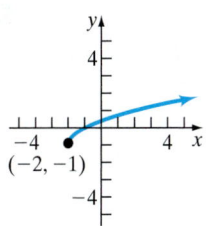

In Problems 25–52, use the graph the relation obtained in Problems 25–52 from Section 8.1 to identify the domain and the range of the relation.

25. $y = 4x$

26. $y = 2x$

27. $y = -\dfrac{1}{2}x$

28. $y = -\dfrac{1}{3}x$

29. $y = x + 3$

30. $y = x - 2$

31. $y = -3x + 1$

32. $y = -4x + 2$

33. $y = \dfrac{1}{2}x - 4$

34. $y = -\dfrac{1}{2}x + 2$

35. $2x + y = 7$

36. $3x + y = 9$

37. $y = -x^2$

38. $y = x^2 - 2$

39. $y = 2x^2 - 8$

40. $y = -2x^2 + 8$

41. $y = |x|$

42. $y = |x| - 2$

43. $y = |x - 1|$

44. $y = -|x|$

45. $y = x^3$

46. $y = -x^3$

47. $y = x^3 + 1$

48. $y = x^3 - 2$

49. $x^2 - y = 4$

50. $x^2 + y = 5$

51. $x = y^2 - 1$

52. $x = y^2 + 2$

Applying the Concepts

53. Area of a Window Bob Villa wishes to put a new window in his home. He wants the perimeter of the window to be 100 feet. The graph below shows the relation between the width, x, of the opening and the area of the opening.

   (a) Determine the domain and the range of the relation.

   (b) Provide an explanation as to why the domain obtained in part (a) is reasonable.

54. Projectile Motion The graph below shows the height, in feet, of a ball thrown straight up with an initial speed of 80 feet per second from an initial height of 96 feet after t seconds. Determine the domain and the range of the relation.

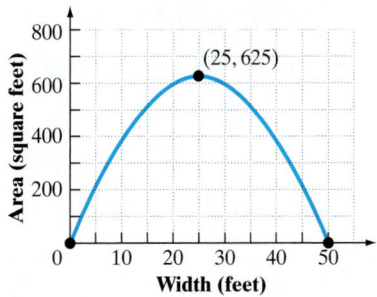

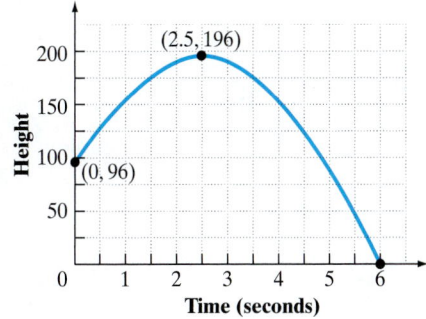

55. Cell Phones We all struggle with selecting a cellular phone provider. The graph to the right shows the relation between the monthly cost, C, of a cellular phone and the number of anytime minutes used, m, when using the Sprint PCS 500-minute plan. (*Source: SprintPCS.com*)

(a) Determine the domain and the range of the relation.

(b) If anytime minutes are from 7:00 A.M. to 7:00 P.M. Monday through Friday, provide an explanation as to why the domain obtained in part (a) is reasonable assuming there are 21 non-weekend days.

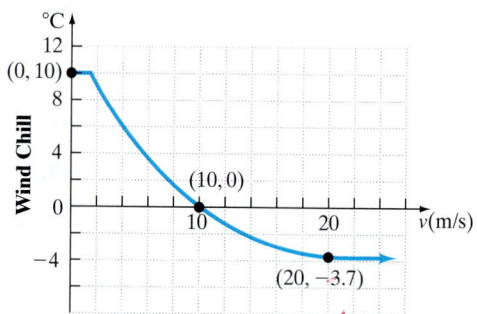

56. Wind Chill It is 10° Celsius outside. The wind is calm but then gusts up to 20 meters per second. You feel the chill go right through your bones. The graph to the right shows the relation between the wind chill temperature (in degrees Celsius) and wind speed (in meters per second). Determine the domain and the range of the relation.

Extending the Concepts

57. Draw the graph of a relation whose domain is all real numbers, but whose range is a single real number. Compare your graph with those of your classmates. How are they similar?

58. Draw the graph of a relation whose domain is a single real number, but whose range is all real numbers. Compare your graph with those of your classmates. How are they similar?

PUTTING THE CONCEPTS TOGETHER (Sections 8.1–8.2)

These problems cover important concepts from Sections 8.1 to 8.2. We designed these problems so that you can review the chapter so far and show your mastery of the concepts. Take time to work these problems before proceeding with the next section. The answers to these problems are located at the back of the text starting on page AN-25.

1. Plot the following ordered pairs in the same xy-plane. Tell in which quadrant or on what coordinate axis each point lies.

$$A(7, 0), B(-2, 6), C(8, 4), D(0, -9), E(-5, -10), F(6, -3)$$

2. Determine whether the ordered pair is a point on the graph of the equation $y = 4x - \dfrac{3}{2}$.

(a) $\left(1, \dfrac{5}{2}\right)$ (b) $\left(\dfrac{1}{2}, \dfrac{1}{2}\right)$ (c) $\left(\dfrac{1}{4}, \dfrac{1}{4}\right)$

In Problems 3 and 4, graph the equations by plotting points.

3. $y = |x| + 3$

4. $y = \dfrac{1}{2}x^2 - 1$

5. Identify the intercepts from the graph below.

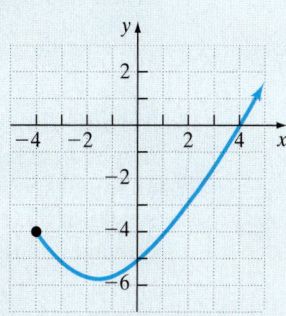

6. The graph below shows the average selling price of a new home from August 2002 to July 2003. The vertical axis represents the selling price, in thousands of dollars, and the horizontal axis represents the month.

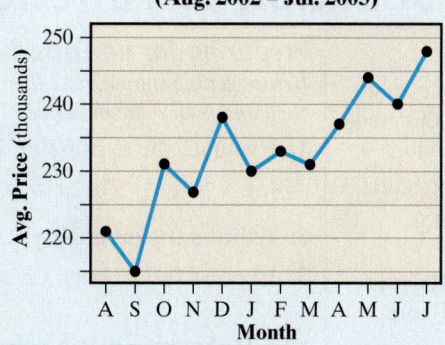

Source: U.S. Census Bureau

(a) What was the average selling price of a new home in January of 2003?

(b) In what month was the average selling price the highest? What was the approximate average selling price?

(c) Between what two consecutive months did the average selling price increase the most? What was the approximate increase?

7. Write the following relation as a set of ordered pairs.

Domain	Range
−2	−1
−1	0
0	1
1	2
2	3

8. Identify the domain and range of the relation from each graph.

(a)

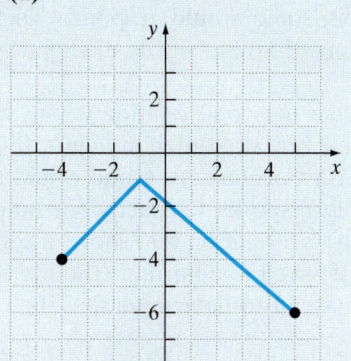

(b)

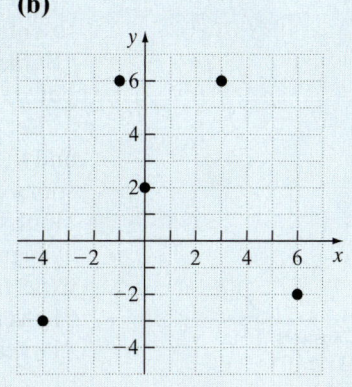

9. Graph each relation and use the graph to identify the domain and the range of the relation.

(a) $y = |x - 2| - 3$ (b) $y = \dfrac{1}{2}x^2 + 1$

10. Vertical Motion The graph to the right shows the height, h, in feet, of a ball thrown straight up with an initial speed of 40 feet per second from an initial height of 80 feet after t seconds. What are the domain and the range of the relation?

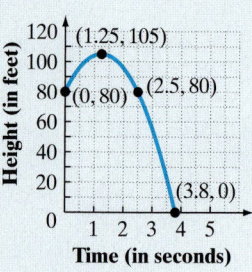

8.3 An Introduction to Functions

OBJECTIVES

1. Determine Whether a Relation Expressed as a Map or Ordered Pairs Represents a Function
2. Determine Whether a Relation Expressed as an Equation Represents a Function
3. Determine Whether a Relation Expressed as a Graph Represents a Function
4. Find the Value of a Function
5. Graph a Function
6. Work with Applications of Functions

Preparing for an Introduction to Functions

Before getting started, take this readiness quiz. If you get a problem wrong, go back to the section cited and review the material.

1. Evaluate the expression $2x^2 - 5x$ for

(a) $x = 1$ (b) $x = 4$ (c) $x = -3$ [Section 1.7, pp. 57–58]

2. Express the inequality $x \leq 5$ using interval notation.   [Section 2.8, pp. 150–153]

3. Express the interval $(2, \infty)$ using set-builder notation.   [Section 2.8, pp. 150–153]

1 Determine Whether a Relation Expressed as a Map or Ordered Pairs Represents a Function

We now present what is one of the most important concepts in algebra—the *function*. A function is a special type of relation. To understand the idea behind a function, let's revisit the relation presented in Example 1 from Section 8.2 shown again in Figure 15. Recall, this is a correspondence between states and their senators. We named the relation "is represented in the U.S. Senate by."

Figure 15

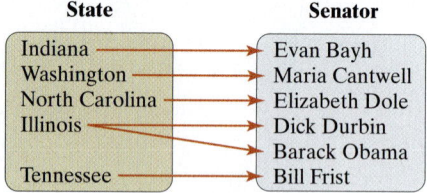

For the relation in Figure 15, if someone were asked to name the senator who represents Illinois, some would respond "Dick Durbin," while others would respond "Barack Obama." In other words, the input "state" does not correspond to a single output "senator."

Let's consider a second relation where we have a correspondence between states and their population presented in Figure 16(a). If asked for the population that corresponds to North Carolina, everyone would respond "8,320,146." In other words, each input "state" corresponds to exactly one output "population."

Figure 16(b) is a relation that shows a correspondence between "animals" and "life expectancy." If asked to determine the life expectancy of a dog, we would all respond "11 years." If asked to determine the life expectancy of a cat, we would all respond "11 years."

Preparing for...Answers **1.** (a) -3
(b) 12   (c) 33   **2.** $(-\infty, 5]$
3. $\{x \mid x > 2\}$

Figure 16

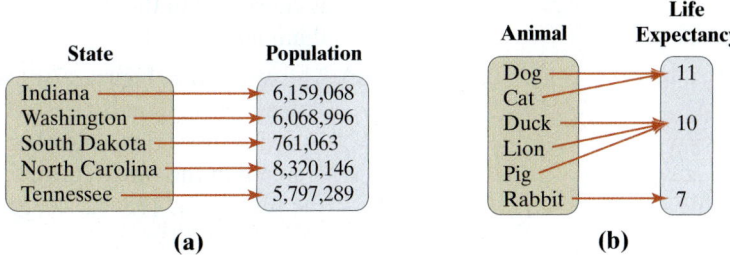

(a)   (b)

Looking carefully at the three relations, we should notice that the relations presented in Figures 16(a) and 16(b) have something in common, while the relation in Figure 15 is different. What is it? The common link between the relations in Figures 16(a) and (b) is that each input corresponds to only one output. However, in Figure 15 the input Illinois corresponds to two outputs—Dick Durbin and Barack Obama. This leads to the definition of a *function*.

In Words

For a relation to be classified as a function, each input may have only one output.

DEFINITION

A **function** is a relation in which each element in the domain (the inputs) of the relation corresponds to exactly one element in the range (the outputs) of the relation.

EXAMPLE 1   **Determining Whether a Relation Represents a Function**

Determine whether the following relations represent functions. If the relation is a function, then state its domain and range.

(a) See Figure 17(a). For this relation, the domain represents the length (mm) of the right humerus and the range represents the length (mm) of the right tibia for each of five rats sent to space. The lengths were measured once the rats returned from their trip.

(b) See Figure 17(b). For this relation, the domain represents the weight of pear-cut diamonds and the range represents their price.

(c) See Figure 17(c). For this relation, the domain represents the age of 5 males and the range represents their HDL (good) cholesterol (mg/dL).

Figure 17

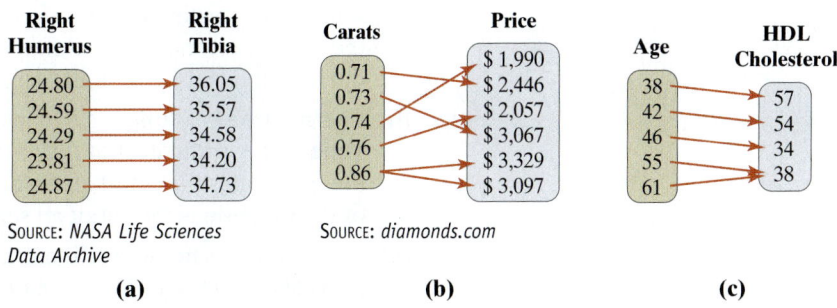

Source: *NASA Life Sciences Data Archive*

Source: *diamonds.com*

(a)   (b)   (c)

Solution

(a) The relation in Figure 17(a) is a function because each element in the domain corresponds to exactly one element in the range. The domain of the function is {24.80, 24.59, 24.29, 23.81, 24.87}. The range of the function is {36.05, 35.57, 34.58, 34.20, 34.73}.

(b) The relation in Figure 17(b) is not a function because there is an element in the domain, 0.86, that corresponds to two elements in the range. If 0.86

is chosen from the domain, a single price cannot be determined for the diamond.

(c) The relation in Figure 17(c) is a function because each element in the domain corresponds to exactly one element in the range. Notice that it is okay for more than one element in the domain to correspond to the same element in the range (both 55 and 61 correspond to 38). The domain of the function is $\{38, 42, 46, 55, 61\}$. The range of the function is $\{57, 54, 34, 38\}$.

QUICK ✓ *Determine whether the relation represents a function. If the relation is a function, state its domain and range.*

1.

Friend	Birthday
Max	January 20
Alesia	March 3
Trent	July 6
Yolanda	November 8
Wanda	January 8
Elvis	

2.

Calories per Serving of Cereal	Sugar Content (in grams)
170	16
200	18
210	23

We may also think of a function as a set of ordered pairs (x, y) in which no two ordered pairs have the same first coordinate, but different second coordinates.

EXAMPLE 2 **Determining Whether a Relation Represents a Function**

Determine whether each relation represents a function. If the relation is a function, then state its domain and range.

(a) $\{(1, 3), (-1, 4), (0, 6), (2, 8)\}$

(b) $\{(-2, 6), (-1, 3), (0, 2), (1, 3), (2, 6)\}$

(c) $\{(0, 3), (1, 4), (4, 5), (9, 5), (4, 1)\}$

Solution

(a) This relation is a function because there are no ordered pairs with the same first coordinate, but different second coordinates. The domain of the function is the set of all first coordinates, $\{-1, 0, 1, 2\}$. The range of the function is the set of all second coordinates, $\{3, 4, 6, 8\}$.

(b) This relation is a function because there are no ordered pairs with the same first coordinate, but different second coordinates. The domain of the function is the set of all first coordinates, $\{-2, -1, 0, 1, 2\}$. The range of the function is the set of all second coordinates, $\{2, 3, 6\}$.

(c) This relation is not a function because there are two ordered pairs, $(4, 5)$ and $(4, 1)$, with the same first coordinate, but different second coordinates.

In Example 2(b), notice that -2 and 2 in the domain each correspond to 6 in the range. This does not violate the definition of a function—two different first coordinates can have the same second coordinate. A violation of the definition occurs when two ordered pairs have the same first coordinate and different second coordinates as in Example 2(c).

QUICK ✔️ *Determine whether each relation represents a function. If the relation is a function, then state its domain and range.*

3. $\{(-3, 3), (-2, 2), (-1, 1), (0, 0), (1, 1)\}$ **4.** $\{(-3, 2), (-2, 5), (-1, 8), (-3, 6)\}$

---

②   Determine Whether a Relation Expressed as an Equation Represents a Function

At this point, we have shown how to identify when a relation defined by a map or ordered pairs is a function. In Section 8.2, we also learned how to express relations as equations and graphs. We will now address the circumstances under which equations are functions.

To determine whether an equation, where y depends upon x, is a function, it is often easiest to solve the equation for y. If a value of x corresponds to exactly one y, the equation defines a function; otherwise it does not define a function.

| EXAMPLE 3 | **Determining Whether an Equation Represents a Function** |

Determine whether the equation $y = 3x + 5$ shows y as a function of x.

Solution

The rule for getting from x to y is to multiply x by 3 and then add 5. Since there is only one output y that can result by performing these operations on any given input x, the equation is a function.   ▬

| EXAMPLE 4 | **Determining Whether an Equation Represents a Function** |

In Words
The symbol $\pm$ is a shorthand device and is read "plus or minus." For example, ± 4 means "negative four or positive four."

Determine whether the equation $y = \pm x^2$ shows y as a function of x.

Solution

Notice that for any single value of x (other than 0), two values of y result. For example, if $x = 2$, then $y = \pm 4$ (-4 or $+4$). Since a single x corresponds to more than one y, the equation is not that of a function.   ▬

QUICK ✔️ *Determine whether each equation shows y as a function of x.*

5. $y = -2x + 5$ **6.** $y = \pm 3x$ **7.** $y = x^2 + 5x$ **8.** $x + y^2 = 9$

---

③   Determine Whether a Relation Expressed as a Graph Represents a Function

Remember that the graph of an equation is the set of all ordered pairs (x, y) that satisfy the equation. For a relation to be a function, each number x in the domain can correspond to only one y in the range. This means that the graph of an equation will *not* represent a function if two points with the same x-coordinate have different y-coordinates.

> **VERTICAL LINE TEST**
>
> A set of points in the xy-plane is the graph of a function if and only if every vertical line intersects the graph in at most one point.

EXAMPLE 5 **Using the Vertical Line Test to Identify Graphs of Functions**

Which of the graphs in Figure 18 are graphs of functions?

Figure 18

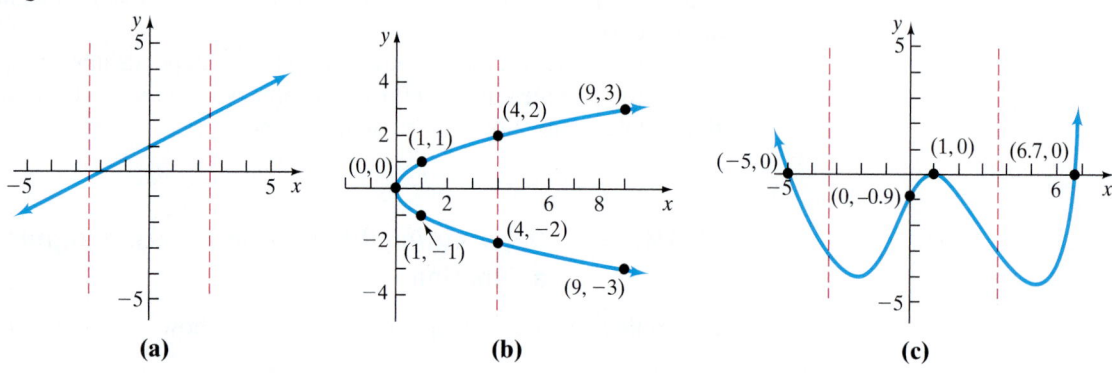

(a)   (b)   (c)

Solution

The graph in Figure 18(a) is a function, because every vertical line intersects the graph in at most one point. The graph in Figure 18(b) is not a function, because a vertical line intersects the graph in more than one point. The graph in Figure 18(c) is a function, because a vertical line intersects the graph in at most one point. ▬

Based on the results of Example 5, do you see why the vertical line test works? If a vertical line intersects the graph of an equation in two or more points, then the same x-coordinate corresponds to two or more different y-coordinates and we have violated the definition of a function.

QUICK *Use the vertical line test to determine whether the graph is that of a function.*

9.
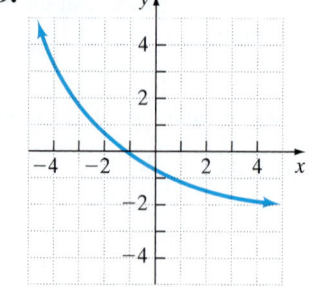

10.
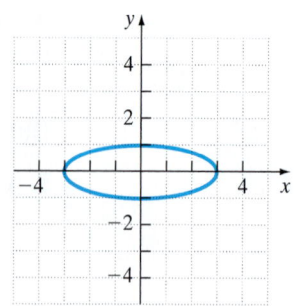

---

④ **Find the Value of a Function**

Functions are often denoted by letters such as f, F, g, G, and so on. If f is a function, then for each number x in its domain, the corresponding value in the range is denoted $f(x)$, read as "f of x" or as "f at x." We call $f(x)$ the **value of f at the number x**; $f(x)$ is the number that results when the function is applied to x; $f(x)$ does not mean "f times x." For example, the function $y = 3x + 5$ given in Example 3 may be written as $f(x) = 3x + 5$.

Work Smart

Be careful with function notation. In the expression $y = f(x)$, y is the dependent variable, x is the independent variable, and f is the name given to a rule that relates the input x to the output y.

For a function $y = f(x)$, the variable x is called the **independent variable,** because it can be assigned any of the numbers in the domain. The variable y is called the **dependent variable,** because its value depends on x.

Any symbol can be used to represent the independent variable. For example, if f is the *square function,* then f can be defined by $f(x) = x^2$, $f(t) = t^2$, or $f(z) = z^2$. All three functions are the same: Each tells us to square the independent variable.

In practice, the symbols used for the independent and dependent variables should remind us what they represent. For example, in economics, we use C for cost and q for quantity, so that $C(q)$ represents the cost of manufacturing q units of a good. Here, C is the dependent variable, q is the independent variable, and $C(q)$ is the rule that tells us how to get the output C from the input q.

The independent variable is also called the **argument** of the function. Thinking of the independent variable as an argument can sometimes make it easier to find the value of a function. For example, if f is the function defined by $f(x) = x^2$, then f tells us to square the argument. So, $f(2)$ means to square 2, $f(a)$ means to square a, and $f(x + h)$ means to square the quantity $x + h$.

EXAMPLE 6 Finding Values of a Function

For the function defined by $f(x) = x^2 + 6x$, evaluate:

(a) $f(3)$ **(b)** $f(-2)$

Solution

(a) Wherever we see an x in the equation defining the function f we substitute 3 to get

$$f(3) = 3^2 + 6(3)$$
$$= 9 + 18$$
$$= 27$$

(b) We substitute -2 for x in the expression $x^2 + 6x$ to get

$$f(-2) = (-2)^2 + 6(-2)$$
$$= 4 + (-12)$$
$$= -8$$

The notation $f(x)$ plays a dual role—it represents the rule for getting from the input to the output and its value represents the output y of the function. For example, in Example 6(a), the rule for getting from the input to the output is given by $f(x) = x^2 + 6x$. In words, the function says to "take some input x, square it, and add the result to six times the input x." If the input is 3, then $f(3)$ represents the output, 27.

EXAMPLE 7 Finding Values of a Function

For the function $F(z) = 4z + 7$, evaluate:

(a) $F(z + 3)$ **(b)** $F(z) + F(3)$

Solution

(a) Wherever we see a z in the equation defining F, we substitute $z + 3$ to get

$$F(z + 3) = 4(z + 3) + 7$$
$$= 4z + 12 + 7$$
$$= 4z + 19$$

(b) $F(z) + F(3) = \underbrace{4z + 7}_{F(z)} + \underbrace{4 \cdot 3 + 7}_{F(3)}$

$$= 4z + 7 + 12 + 7$$

$$= 4z + 26$$

QUICK ✓ *Let* $f(x) = 3x + 2$ *and* $g(x) = -2x^2 + x - 3$ *to evaluate each function.*

11. $f(4)$ **12.** $g(-2)$ **13.** $f(x - 2)$ **14.** $f(x) - f(2)$

SUMMARY: Important Facts about Functions

1. For each x in the domain there corresponds exactly one y in the range.

2. f is a symbol that we use to denote the function. It represents the equation that we use to get from an x in the domain to $f(x)$ in the range.

3. If $y = f(x)$, then x is called the independent variable or argument of f, and y is called the dependent variable or the value of f at x.

Figure 19

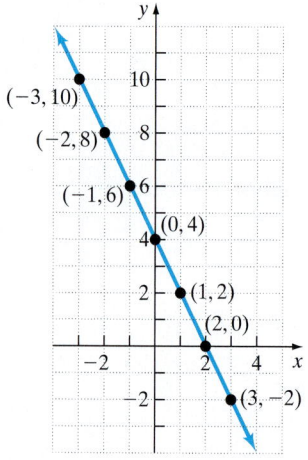

5 Graph a Function

In Example 1 from Section 8.1, we graphed the equation $y = -2x + 4$. We reproduce its graph in Figure 19. The graph passes the vertical line test so the equation $y = -2x + 4$ is that of a function. Therefore, we can express the relation using function notation as $f(x) = -2x + 4$. The graph of an equation is the same as the graph of the function where the horizontal axis represents the independent variable and the vertical axis represents the dependent variable. When we graph functions, we label the vertical axis either by y or by the name of the function.

DEFINITION

When a function is defined by an equation in x and y, the **graph of the function** is the set of *all* ordered pairs (x, y) such that $y = f(x)$.

EXAMPLE 8 Graphing a Function

Graph the function $f(x) = |x|$.

Solution

To graph the function $f(x) = |x|$, we first determine some ordered pairs $(x, f(x)) = (x, y)$ such that $y = |x|$. See Table 6. We now plot the ordered pairs (x, y) in Table 6 in an xy-plane and connect the points as shown in Figure 20.

Table 6		
x	$f(x)$	$(x, f(x))$
-3	$|-3| = 3$	$(-3, 3)$
-2	$|-2| = 2$	$(-2, 2)$
-1	$|-1| = 1$	$(-1, 1)$
0	$|0| = 0$	$(0, 0)$
1	$|1| = 1$	$(1, 1)$
2	$|2| = 2$	$(2, 2)$
3	$|3| = 3$	$(3, 3)$

Figure 20

$f(x) = |x|$

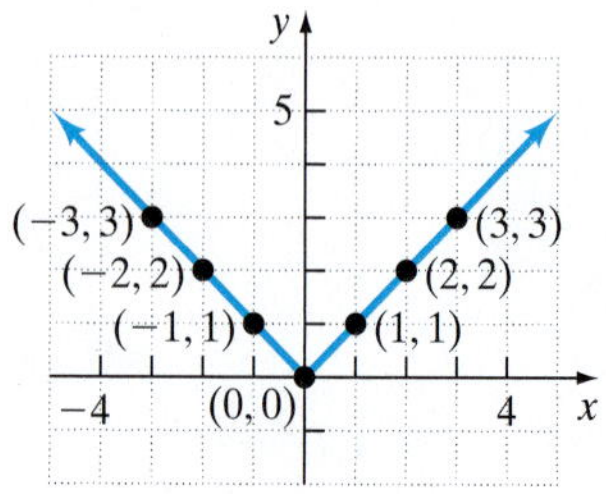

QUICK ✔ *Graph each function.*

15. $f(x) = -2x + 9$ **16.** $f(x) = x^2 + 2$ **17.** $f(x) = |x - 2|$

---

⑥   Work with Applications of Functions

EXAMPLE 9   Life Cycle Hypothesis

The Life Cycle Hypothesis from Economics was presented by Franco Modigliani in 1954. It states that income is a function of age. The function $I(a) = -55a^2 + 5119a - 54{,}448$ represents the relation between average annual income I and age a.

   (a) Identify the dependent and independent variables.

   (b) Evaluate $I(20)$. Provide a verbal explanation of the meaning of $I(20)$.

Solution

   (a) Because income depends upon age, we have that the dependent variable is income, I, and the independent variable is age, a.

   (b) We let $a = 20$ in the function.

$$I(20) = -55(20)^2 + 5119(20) - 54{,}448$$
$$= 25{,}932$$

   The average annual income of an individual who is 20 years of age is $25,932.

QUICK ✔

18. In 2002, the Prestige oil tanker sank and started leaking oil off the coast of Spain. The oil slick takes the shape of a circle. Suppose that the area A (in square miles) of the circle contaminated with oil can be determined using the function $A(t) = 0.25\pi t^2$, where t represents the number of days since the tanker sprung a leak.

   (a) Identify the dependent and independent variable.

   (b) Evaluate $A(30)$. Provide a verbal explanation of the meaning of $A(30)$.

8.3 Exercises

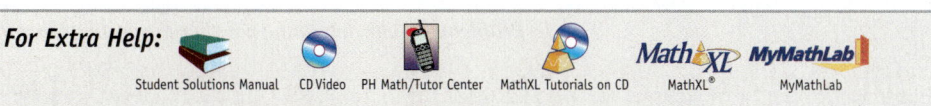

For Extra Help: Student Solutions Manual   CD Video   PH Math/Tutor Center   MathXL Tutorials on CD   MathXL®   MyMathLab

Concepts and Vocabulary

In Problems 1–3, fill in the blanks.

   1. A _____ is a relation in which each element in the domain of the relation corresponds to exactly one element in the range of the relation.

   2. In the function $H(q) = 2q^2 - 5q + 1$, H is called the _____ variable and q is called the _____ variable.

   3. The independent variable is also called the _____ of the function.

In Problems 4 and 5, answer True or False to each statement.

   4. Every relation is a function.

   5. In order for a graph to be that of a function, any vertical line can intersect the graph in at most one point.

✎ **6.** In your own words, explain why the vertical line test can be used to identify the graph of a function.

✎ **7.** What are the four forms of a function presented in this section?

✎ **8.** Explain why the term independent variable for x and dependent variable for y make sense in the function $y = f(x)$.

Building Skills

In Problems 9–18, determine whether each relation represents a function. State the domain and the range of each relation.

9.
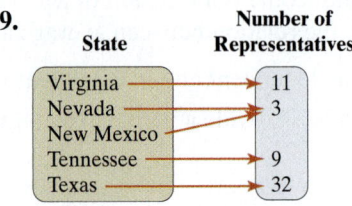

10.
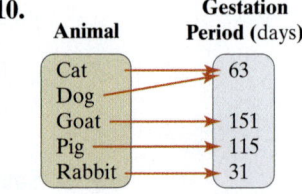

11.
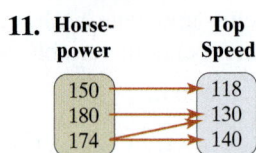

12.
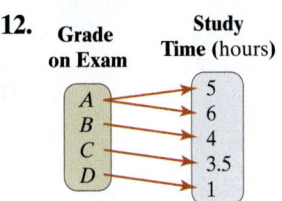

13. $\{(0, 3), (1, 4), (2, 5), (3, 6)\}$

14. $\{(-1, 4), (0, 1), (1, -2), (2, -5)\}$

15. $\{(-3, 5), (1, 5), (4, 5), (7, 5)\}$

16. $\{(-2, 3), (-2, 1), (-2, -3), (-2, 9)\}$

17. $\{(-10, 1), (-5, 4), (0, 3), (-5, 2)\}$

18. $\{(-5, 3), (-2, 1), (5, 1), (7, -3)\}$

In Problems 19–28, determine whether each equation shows y as a function of x.

19. $y = 2x + 9$ **20.** $y = -6x + 3$ **21.** $2x + y = 10$ **22.** $6x - 3y = 12$

23. $y = \pm 5x$ **24.** $y = \pm 2x^2$ **25.** $y = x^2 + 2$ **26.** $y = x^3 - 3$

27. $x + y^2 = 10$ **28.** $y^2 = x$

In Problems 29–36, determine whether the graph is that of a function.

29.

30.
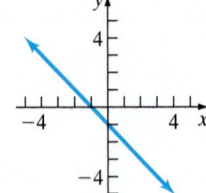

31.

32.

33.
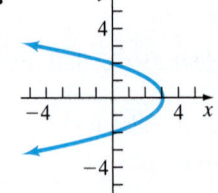

34.
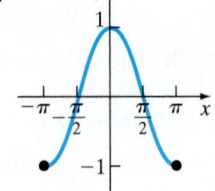

35.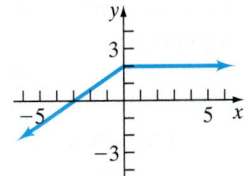

36.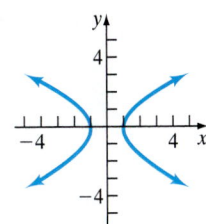

In Problems 37–40, find the following values for each function:

    (a) $f(0)$ **(b)** $f(3)$ **(c)** $f(-2)$ **(d)** $f(-x)$

    (e) $-f(x)$ **(f)** $f(x+2)$ **(g)** $f(2x)$ **(h)** $f(x+h)$

37. $f(x) = 2x + 3$ **38.** $f(x) = 3x + 1$

39. $f(x) = -5x + 2$ **40.** $f(x) = -2x - 3$

In Problems 41–48, find the value of each function.

41. $f(x) = x^2 + 3; f(2)$ **42.** $f(x) = -2x^2 + x + 1; f(-3)$

43. $s(t) = -t^3 - 4t; s(-2)$ **44.** $g(h) = -h^2 + 5h - 1; g(4)$

45. $F(x) = |x - 2|; F(-3)$ **46.** $G(z) = 2|z + 5|; G(-6)$

47. $F(z) = \dfrac{z + 2}{z - 5}; F(4)$ **48.** $h(q) = \dfrac{3q^2}{q + 2}; h(2)$

In Problems 49–56, graph each function.

49. $f(x) = 4x - 6$ **50.** $g(x) = -3x + 5$

51. $h(x) = x^2 - 2$ **52.** $F(x) = x^2 + 1$

53. $G(x) = |x - 1|$ **54.** $H(x) = |x + 1|$

55. $g(x) = x^3$ **56.** $h(x) = x^3 - 3$

Applying the Concepts

57. If $f(x) = 3x^2 - x + C$ and $f(3) = 18$, what is the value of C?

58. If $f(x) = -2x^2 + 5x + C$ and $f(-2) = -15$, what is the value of C?

59. If $f(x) = \dfrac{2x + 5}{x - A}$ and $f(0) = -1$, what is the value of A?

60. If $f(x) = \dfrac{-x + B}{x - 5}$ and $f(3) = -1$, what is the value of B?

△ **61. Geometry** Express the area A of a circle as a function of its radius, r. Determine the area of circle whose radius is 4 inches. That is, find $A(4)$.

△ **62. Geometry** Express the area A of a triangle as a function of its height h assuming that the length of the base is 8 centimeters. Determine the area of this triangle if its height is 5 centimeters. That is, find $A(5)$.

63. Salary Express the gross salary G of Jackie, who earns \$15 per hour as a function of the number of hours worked, h. Determine the gross salary of Jackie if she works 25 hours. That is, find $G(25)$.

64. Commissions Roberta is a commissioned salesperson. She earns a base weekly salary of \$250 per week plus 15% of the sales price of items sold. Express her gross salary G as a function of the price p of items sold. Determine the weekly gross salary of Roberta if the value of items sold is \$10,000. That is, find $G(10,000)$.

65. Population as a Function of Age The function $P(a) = 0.025a^2 - 5.633a + 300.517$ represents the population (in millions) of Americans in 2005, P, that are a years of age or older. (*Source:* United States Census Bureau)

   (a) Identify the dependent and independent variable.
   (b) Evaluate $P(20)$. Provide a verbal explanation of the meaning of $P(20)$.
   (c) Evaluate $P(0)$. Provide a verbal explanation of the meaning of $P(0)$.

66. Number of Rooms The function $N(r) = -1.33r^2 + 14.68r - 17.09$ represents the number of housing units (in millions), N, in 2005 that have r rooms, where $1 \le r \le 9$. (*Source*: United States Census Bureau)

   (a) Identify the dependent and independent variable.
   (b) Evaluate $N(3)$. Provide a verbal explanation of the meaning of $N(3)$.
   (c) Why is it unreasonable to evaluate $N(0)$?

67. Revenue Function The function $R(p) = -p^2 + 200p$ represents the daily revenue R earned from selling personal digital assistants (PDAs) at p dollars for $0 \le p \le 200$.

   (a) Identify the dependent and independent variable.
   (b) Evaluate $R(50)$. Provide a verbal explanation of the meaning of $R(50)$.
   (c) Evaluate $R(120)$. Provide a verbal explanation of the meaning of $R(120)$.

68. Average Trip Length The function $T(x) = 0.01x^2 - 0.12x + 8.89$ represents the average vehicle trip length T (in miles) x years since 1969.

   (a) Identify the dependent and independent variable.
   (b) Evaluate $T(35)$. Provide a verbal explanation of the meaning of $T(35)$.
   (c) Evaluate $T(0)$. Provide a verbal explanation of the meaning of $T(0)$.

Extending the Concepts

69. Investigate when the use of function notation $y = f(x)$ first appeared. Start by researching Lejeune Dirichlet.

70. Are all relations functions? Are all functions relations? Explain your answers.

The Graphing Calculator

Graphing calculators have the ability to evaluate any function you wish. Figure 21 shows the results obtained in Example 6 using a TI-84 Plus graphing calculator.

Figure 21

 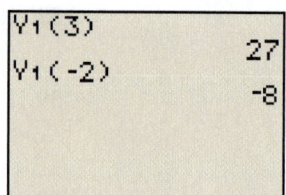

In Problems 71–78, use a graphing calculator to find the value of each function.

71. $f(x) = x^2 + 3; f(2)$

72. $f(x) = -2x^2 + x + 1; f(-3)$

73. $F(x) = |x - 2|; F(-3)$

74. $g(h) = \sqrt{2h + 1}; g(4)$

75. $H(x) = \sqrt{4x - 3}; H(7)$

76. $G(z) = 2|z + 5|; G(-6)$

77. $F(z) = \dfrac{z + 2}{z - 5}; F(4)$

78. $h(q) = \dfrac{3q^2}{q + 2}; h(2)$

8.4 **Functions and Their Graphs**

OBJECTIVES

1. Find the Domain of a Function
2. Obtain Information from the Graph of a Function
3. Interpret Graphs of Functions

Preparing for Functions and Their Graphs

Before getting started, take this readiness quiz. If you get a problem wrong, go back to the section cited, and review the material.

1. Solve: $3x - 12 = 0$ [Section 2.2, pp. 84–86]

2. Solve: $x^2 - 3x - 18 = 0$ [Section 6.6, pp. 451–457]

---

1 **Find the Domain of a Function**

When working with functions, we need to determine the set of inputs for which a function makes sense. Often the set of inputs for which the function makes sense is not specified; instead, only the equation defining the function is given.

In Words

The domain of a function is the set of all inputs for which the function gives an output that is a real number or makes sense.

DEFINITION

When only the equation of a function is given, we agree that the **domain of f** is the largest set of real numbers for which $f(x)$ is a real number.

When identifying the domain of a function don't forget that division by zero is undefined, so exclude values of the variable that cause division by zero.

EXAMPLE 1 **Finding the Domain of a Function**

Find the domain of each of the following functions:

(a) $G(x) = x^2 + 1$ **(b)** $g(z) = \dfrac{z - 3}{z + 1}$ **(c)** $P(t) = \dfrac{t - 5}{t^2 - 11t + 30}$

Solution

(a) The function G tells us to square a number x and then add 1 to that number. These operations can be performed on any real number, so the domain of G is the set of all real numbers. We can express the domain as $\{x \mid x \text{ is a real number}\}$ or, using interval notation, $(-\infty, \infty)$.

(b) The function g tells us to divide $z - 3$ by $z + 1$. Since division by 0 is not defined, the denominator $z + 1$ can never be 0. Therefore, z can never equal -1. The domain of g is $\{z \mid z \neq -1\}$.

(c) The function P tells us to divide $t - 5$ by $t^2 - 11t + 30$. The denominator cannot equal 0. To find the values of t to exclude from the domain, we solve

$$t^2 - 11t + 30 = 0$$
$$(t - 5)(t - 6) = 0$$
$$t = 5 \quad \text{or} \quad t = 6$$

The domain of P is $\{t \mid t \neq 5, t \neq 6\}$.   ▬

QUICK ✔ *Find the domain of each function.*

1. $f(x) = 3x^2 + 2$ **2.** $h(x) = \dfrac{x + 1}{x - 3}$ **3.** $F(x) = \dfrac{x + 2}{x^2 - 5x - 14}$

When we use functions in applications, the domain may be restricted by physical or geometric considerations, rather than by pure mathematical restrictions. For example, the domain of the function defined by $f(x) = x^2$ is the set of all real numbers.

However, if f is used to obtain the area of a square when the length x of a side is known, then we must restrict the domain of f to the positive real numbers, since the length of a side can never be 0 or negative.

EXAMPLE 2   Finding the Domain of a Function

The number N of computers produced at one of Dell Computers manufacturing facilities in one day after t hours is given by the function, $N(t) = 336t - 7t^2$. What is the domain of this function?

Solution

The independent variable in this function is t, where t represents the number of hours in the day. Therefore, the domain of the function is $\{t \mid 0 \leq t \leq 24\}$, or the interval $[0, 24]$.

QUICK ✓

4. The function $A(r) = \pi r^2$ gives the area of a circle A as a function of the radius r. What is the domain of the function?

---

② Obtain Information from the Graph of a Function

We can find the domain and the range of a function from its graph. The approach to finding the domain and the range of a function from its graph is identical to the approach taken to find the domain and the range of a relation from its graph.

EXAMPLE 3   Determining the Domain and the Range of a Function from Its Graph

Figure 22 shows the graph of a function.

(a) Determine the domain and the range of the function.

(b) Identify the intercepts.

Figure 22

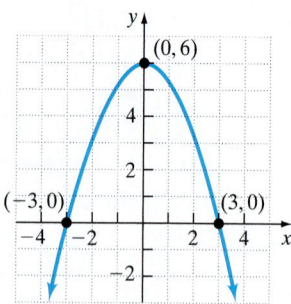

In Words
When the graph of a function is given, its domain may be viewed as the shadow created by the graph on the x-axis by vertical beams of light. Its range can be viewed as the shadow created by the graph on the y-axis by horizontal beams of light.

Solution

(a) To find the domain of the function, we determine the x-coordinates for which the graph of the function exists. Because the graph exists for all real numbers x, the domain is $\{x \mid x \text{ is a real number}\}$, or using interval notation, $(-\infty, \infty)$.

 To find the range of the function, we determine the y-coordinates for which the graph of the function exists. Because the graph exists for all real numbers y less than or equal to 6, the range is $\{y \mid y \leq 6\}$, or using interval notation, $(-\infty, 6]$.

(b) The intercepts are the points $(-3, 0)$, $(0, 6)$, and $(3, 0)$. The x-intercepts are -3 and 3. The y-intercept is 6.

QUICK ✓

5. Use the graph of the function to answer parts (a) and (b).

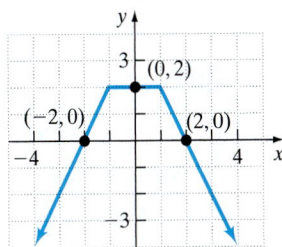

(a) Determine the domain and the range of the function.

(b) Identify the intercepts.

Remember, if (x, y) is a point on the graph of a function f, then y is the value of f at x, that is, $y = f(x)$. So, if $(1, 5)$ is a point on a function f, then $f(1) = 5$. The next example illustrates how to obtain information about a function if its graph is known.

EXAMPLE 4 | Obtaining Information from the Graph of a Function

The Wonder Wheel is a Ferris wheel located in Coney Island. See Figure 23. Let f be the distance above the ground of a person riding on the Wonder Wheel as a function of time x (in minutes). Figure 24 represents the graph of the function f. Use the graph to answer the following questions.

Figure 23

Figure 24

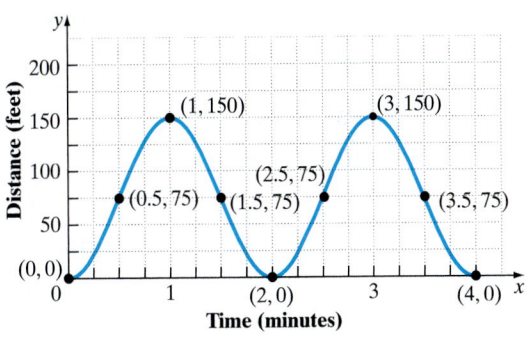

(a) What are $f(1.5)$ and $f(3)$? Interpret these values.

(b) What is the domain of f?

(c) What is the range of f?

(d) List the intercepts.

(e) For what values of x does $f(x) = 75$? That is, solve $f(x) = 75$.

Solution

(a) Since $(1.5, 75)$ is on the graph of f, then $f(1.5) = 75$. After 1.5 minutes, an individual on the Wonder Wheel is 75 feet in the air. Similarly, we find that since $(3, 150)$ is on the graph we have that $f(3) = 150$. After 3 minutes, an individual on the Wonder Wheel is 150 feet in the air.

(b) To determine the domain of f, we notice that for each number x between 0 and 4 inclusive, there are points $(x, f(x))$ on the graph of f. Therefore, the domain of f is $\{x \mid 0 \le x \le 4\}$, or the interval $[0, 4]$.

(c) The points on the graph have y-coordinates between 0 and 150, inclusive. Therefore, the range of f is $\{y \mid 0 \le y \le 150\}$, or the interval $[0, 150]$.

(d) The intercepts are $(0, 0), (2, 0),$ and $(4, 0)$.

(e) Since $(0.5, 75), (1.5, 75), (2.5, 75),$ and $(3.5, 75)$ are the only points on the graph for which $y = f(x) = 75$, the solution set to the equation $f(x) = 75$ is $\{0.5, 1.5, 2.5, 3.5\}$.

QUICK ✓

6. Use the graph to answer the following questions.

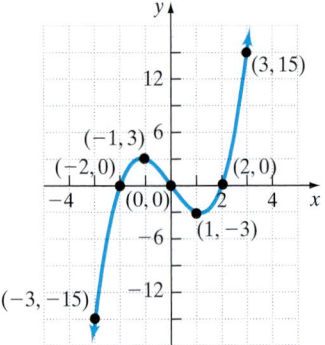

(a) What are $f(-3)$ and $f(1)$?

(b) What is the domain of f?

(c) What is the range of f?

(d) List the intercepts.

(e) For what value of x does $f(x) = 15$? That is, solve $f(x) = 15$.

EXAMPLE 5 Obtaining Information about the Graph of a Function

Consider the function $f(x) = 2x - 5$.

(a) Is the point $(3, -1)$ on the graph of the function?

(b) If $x = 1$, what is $f(x)$? What point is on the graph of the function?

(c) If $f(x) = 3$, what is x? What point is on the graph of f?

Solution

(a) When $x = 3$, then

$$f(x) = 2x - 5$$
$$f(3) = 2(3) - 5 = 6 - 5 = 1$$

Since $f(3) = 1$, the point $(3, 1)$ is on the graph; the point $(3, -1)$ is not on the graph.

(b) If $x = 1$, then

$$f(x) = 2x - 5$$
$$f(1) = 2(1) - 5 = 2 - 5 = -3$$

The point $(1, -3)$ is on the graph of f.

(c) If $f(x) = 3$, then

$$f(x) = 3$$
$$2x - 5 = 3$$

Add 5 to both sides: $\qquad 2x = 8$

Divide both sides by 2: $\qquad x = 4$

If $f(x) = 3$, then $x = 4$. The point $(4, 3)$ is on the graph of f.

Work Smart: Study Skills
Do not confuse the directions "Find $f(3)$" with "If $f(x) = 3$, what is x?" Write down and study errors that you commonly make so that you can avoid them.

QUICK ✓

7. Consider the function $f(x) = -3x + 7$.

(a) Is the point $(-2, 1)$ on the graph of the function?

(b) If $x = 3$, what is $f(x)$? What point is on the graph of the function?

(c) If $f(x) = -8$, what is x? What point is on the graph of f?

(3) **Interpret Graphs of Functions**

We can use the graph of a function to give a visual description of many different scenarios. Consider the following example.

EXAMPLE 6 **Graphing a Verbal Description**

Maria decides to take a walk. She leaves her house and walks 3 blocks in 2 minutes at a constant speed. She realizes that she left her front door unlocked, so she runs home in 1 minute. It takes Maria 1 minute to find her keys and lock the door. She then decides to run 10 blocks in 3 minutes. She is a little tired now, so she rests for 1 minute and then walks an additional 4 blocks in 10 minutes. She hitches a ride home with her neighbor who happens to drive by and gets home in 2 minutes. Draw a graph of Maria's distance from home (in blocks) as a function of time.

Solution

First, we recognize that distance from home is a function of time. Therefore, we draw a Cartesian Plane with the horizontal axis representing the independent variable, time, and the vertical axis representing the dependent variable, distance from home.

The ordered pair $(2, 3)$ corresponds to being 3 blocks from home after 2 minutes. We start the graph at the origin and then draw a straight line from $(0, 0)$ to $(2, 3)$. From the point $(2, 3)$, we draw a straight line to $(3, 0)$, which represents the return trip home to lock the door. Draw a line segment from $(3, 0)$ to $(4, 0)$ to represent the time it takes to lock the door. Draw a line segment from $(4, 0)$ to $(7, 10)$, which represents the 10 block run in 3 minutes. Now we draw a horizontal line from $(7, 10)$ to $(8, 10)$. This represents the resting period. Draw a line from $(8, 10)$ to $(18, 14)$ to represent the 4 block walk in 10 minutes. Finally, draw a line segment from $(18, 14)$ to $(20, 0)$ to represent the ride home. See Figure 25.

Figure 25

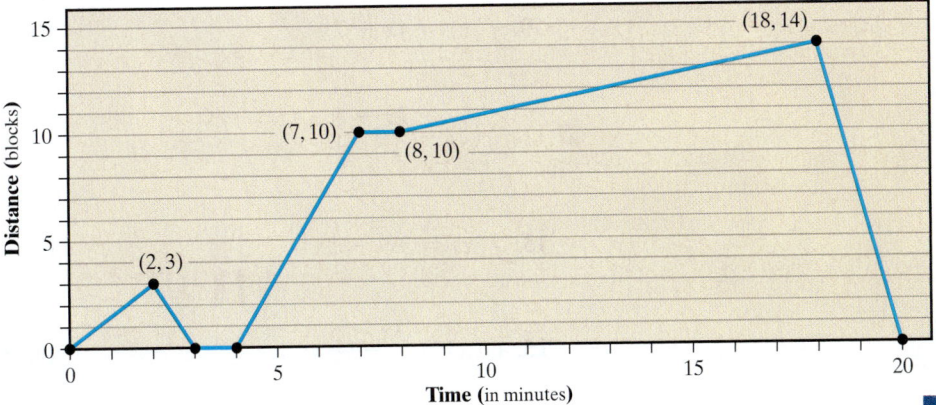

QUICK ✔

8. Maria decides to take a walk. She leaves her house and walks 5 blocks in 5 minutes at a constant speed. She realizes that she left her front door unlocked, so she runs home in 2 minutes. It takes Maria 1 minute to find her keys and lock the door. She then decides to jog 8 blocks in 5 minutes. She then runs 3 blocks in 1 minute. She is a little tired now, so she rests for 2 minutes and then walks home in 10 minutes. Draw a graph of Maria's distance from home (in blocks) as a function of time.

8.4 Exercises

For Extra Help:

Student Solutions Manual   CD Video   PH Math/Tutor Center   MathXL Tutorials on CD   MathXL®   MyMathLab

Concepts and Vocabulary

In Problems 1–3, fill in the blanks.

1. If the point $(3, 8)$ is on the graph of a function f, then $f(\_\_\_\_) = \_\_\_\_$.

2. If $g(-2) = 4$, then $(\_\_\_\_, \_\_\_\_)$ is a point on the graph of g.

3. When only the equation of a function is given, we agree that the ____ of f is the largest set of real numbers for which $f(x)$ is a real number.

In Problems 4 and 5, answer True or False to each statement.

4. A function can have more than one y-intercept.

5. A function can have more than one x-intercept.

6. Using the definition of a function, explain why the graph of a function can have at most one y-intercept.

7. In your own words, explain what the domain of a function is. In your explanation, provide a discussion as to how domains are determined in applications.

8. In your own words, explain what the range of a function is.

Building Skills

In Problems 9–18, find the domain of each function.

9. $f(x) = 4x + 7$

10. $G(x) = -8x + 3$

11. $F(z) = \dfrac{2z + 1}{z - 5}$

12. $H(x) = \dfrac{x + 5}{2x + 1}$

13. $f(x) = 3x^4 - 2x^2$

14. $s(t) = 2t^2 - 5t + 1$

15. $G(x) = \dfrac{3x - 5}{3x + 1}$

16. $H(q) = \dfrac{1}{6q + 5}$

17. $f(x) = \dfrac{10x + 7}{x^2 + 4x - 32}$

18. $f(x) = \dfrac{4x - 9}{2x^2 - 9x - 5}$

In Problems 19–28, for each graph of a function, find (a) the domain and the range, and (b) the intercepts, if any.

19.

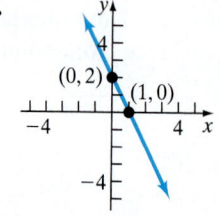

20.

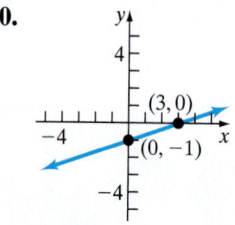

21.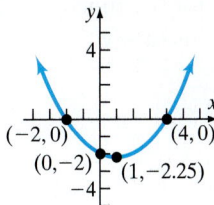

22.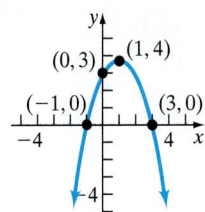

23.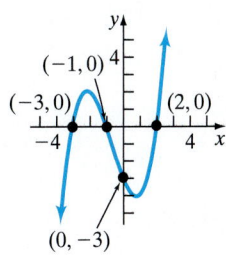

24.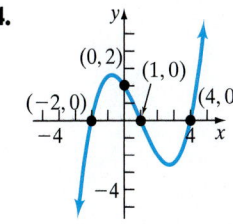

25.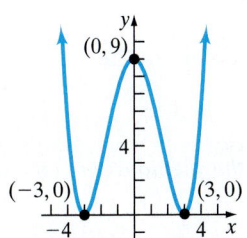

26.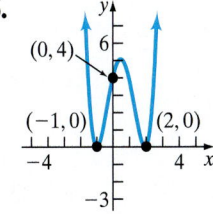

27.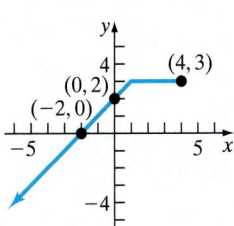

28.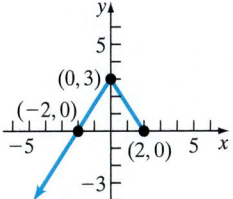

29. Use the graph of the function f shown to answer parts (a)–(k).

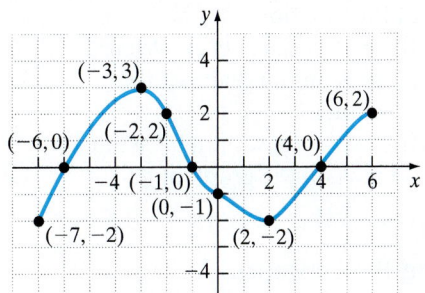

(a) Find $f(-7)$.
(b) Find $f(-3)$.
(c) Find $f(6)$.
(d) Is $f(2)$ positive or negative?
(e) For what numbers x is $f(x) = 0$?
(f) What is the domain of f?
(g) What is the range of f?
(h) What are the x-intercepts?
(i) What is the y-intercept?
(j) For what numbers x is $f(x) = -2$?
(k) For what number x is $f(x) = 3$?

30. Use the graph of the function g shown to answer parts (a)–(k).

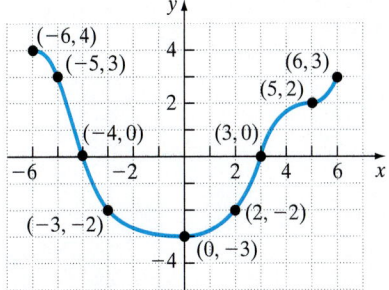

(a) Find $g(-3)$.
(b) Find $g(5)$.
(c) Find $g(6)$.
(d) Is $g(-5)$ positive or negative?
(e) For what numbers x is $g(x) = 0$?
(f) What is the domain of g?
(g) What is the range of g?
(h) What are the x-intercepts?
(i) What is the y-intercept?
(j) For what numbers x is $g(x) = -2$?
(k) For what number x is $g(x) = 3$?

31. Use the table of values for the function F to answer questions (a)–(e).

x	F(x)
−4	0
−2	3
−1	5
0	2
3	−6

(a) What is $F(-2)$?
(b) What is $F(3)$?
(c) For what number(s) x is $F(x) = 5$?
(d) What is the x-intercept of the graph of F?
(e) What is the y-intercept of the graph of F?

32. Use the table of values for the function G to answer questions (a)–(e).

x	G(x)
−5	−3
−4	0
0	5
3	8
7	5

(a) What is $G(3)$?
(b) What is $G(7)$?
(c) For what number(s) x is $G(x) = 5$?
(d) What is the x-intercept of the graph of G?
(e) What is the y-intercept of the graph of G?

In Problems 33–36, answer the questions about the given function.

33. $f(x) = 4x - 9$

(a) Is the point $(2, 1)$ on the graph of the function?
(b) If $x = 3$, what is $f(x)$? What point is on the graph of the function?
(c) If $f(x) = 7$, what is x? What point is on the graph of f?

34. $f(x) = 3x + 5$

(a) Is the point $(-2, 1)$ on the graph of the function?
(b) If $x = 4$, what is $f(x)$? What point is on the graph of the function?
(c) If $f(x) = -4$, what is x? What point is on the graph of f?

35. $g(x) = -\dfrac{1}{2}x + 4$

(a) Is the point $(4, 2)$ on the graph of the function?
(b) If $x = 6$, what is $g(x)$? What point is on the graph of the function?
(c) If $g(x) = 10$, what is x? What point is on the graph of g?

36. $H(x) = \dfrac{2}{3}x - 4$

(a) Is the point $(3, -2)$ on the graph of the function?
(b) If $x = 6$, what is $H(x)$? What point is on the graph of the function?
(c) If $H(x) = -4$, what is x? What point is on the graph of f?

Applying the Concepts

△ **37. Geometry** The volume V of a sphere as a function of its radius r is given by $V(r) = \dfrac{4}{3}\pi r^3$. What is the domain of this function?

△ **38. Geometry** The area A of a triangle as a function of its height h assuming that the length of the base is 5 centimeters is $A = \dfrac{5}{2}h$. What is the domain of the function?

39. Salary The gross salary G of Jackie as a function of the number of hours worked, h is given by $G(h) = 22.5h$. What is the domain of the function if she can work up to 60 hours per week?

40. Commissions Roberta is a commissioned salesperson. She earns a base weekly salary of $350 per week plus 12% of the sales price of items sold. Her gross salary G as a function of the price p of items sold is given by $G(p) = 350 + 0.12p$. What is the domain of the function?

41. Demand for Hot Dogs Suppose the function $D(p) = 1200 - 10p$ represents the demand for hot dogs, whose price is p, at a baseball game. Find the domain of the function.

42. Revenue Function The function $R(p) = -p^2 + 200p$ represents the daily revenue earned from selling personal digital assistants (PDAs) at p dollars for $0 \le p \le 200$. Explain why any p greater than $200 is not in the domain of the function.

43. Match each of the following functions with the graph that best describes the situation.

(a) The distance from ground level of a person who is jumping on a trampoline as a function of time

(b) The cost of a telephone call as a function of time

(c) The height of a human as a function of time

(d) The revenue earned from selling cars as a function of price

(e) The book value of a machine that is depreciated by equal amounts each year as a function of the year

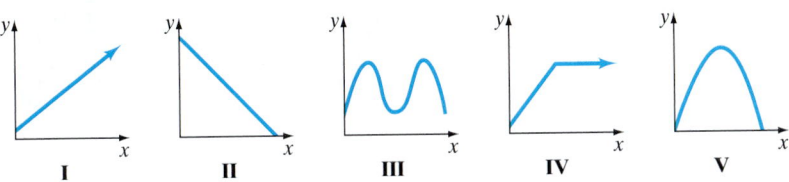

44. Match each of the following functions with the graph that best describes the situation.

(a) The average high temperature each day as a function of the day of the year

(b) The number of bacteria in a Petri dish as a function of time

(c) The distance that a person rides her bicycle at a constant speed as a function of time

(d) The temperature of a pizza after it is removed from the oven as a function of time

(e) The value of car as a function of time

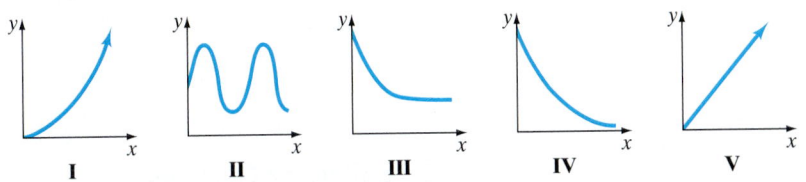

45. Pulse Rate Consider the following scenario: Zach starts jogging on a treadmill. His resting pulse rate is 70. As he continues to jog on the treadmill, his pulse increases at a constant rate until, after 10 minutes, his pulse is 120. He then starts jogging faster and his pulse increases at a constant rate for 2 minutes, at which time his pulse is up to 150. He then begins a cooling off period for 7 minutes until his pulse backs down to 110. He then gets off the treadmill and his pulse returns to 70 after 12 minutes. Draw a graph of Zach's pulse as a function of time.

46. Altitude of an Airplane Suppose that a plane is flying from Chicago to New Orleans. The plane leaves the gate and taxis for 5 minutes. The plane takes off and gets up to 10,000 feet after 5 minutes. The plane continues to ascend at constant rate until it reaches its cruising altitude of 35,000 feet after another 25 minutes. For the next 80 minutes, the plane maintains a constant height of 35,000 feet. The plane then descends at a constant rate until it lands after 20 minutes. It requires 5 minutes to taxi to the gate. Draw a graph of the height of the plane as a function of time.

47. Height of a Swing An 8-year-old girl gets on a swing and starts swinging for 10 minutes. Draw a graph that represents the height of the child from the ground as a function of time.

48. Temperature of Pizza Marissa is hungry and would like a pizza. Her mother pulls a frozen pizza out of the freezer and puts it in the oven. After 12 minutes the pizza is done, but Mom lets the pizza cool for 5 minutes before serving it to Marissa. Draw a graph that represents the temperature of the pizza as a function of time.

49. The graph below shows the weight of a person as a function of his age. Describe the weight of the individual over the course of his life.

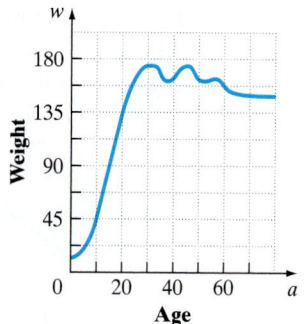

50. The graph below shows the depth of a lake (in feet) as a function of time (in days). Describe the depth of the lake over the course of the year.

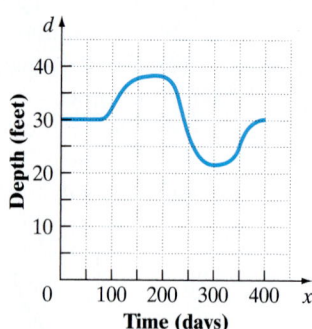

Extending the Concepts

51. Draw a graph of a function f with the following characteristics:
x-intercepts: $-4, -1, 2$; y-intercept: -2; $f(-3) = 7$ and $f(3) = 8$.

52. Draw a graph of a function f with the following characteristics:
x-intercepts: $-3, 2,$ and 5; y-intercept: 3; $f(3) = -2$.

8.5 Linear Functions

OBJECTIVES

1. Graph Linear Functions
2. Build Linear Models from Verbal Descriptions
3. Build Linear Models from Data

Preparing for Linear Functions

Before getting started, take this readiness quiz. If you get a problem wrong, go back to the section cited and review the material.

1. Graph: $y = 2x - 3$ [Section 3.4, pp. 218–220]
2. Graph: $\dfrac{1}{2}x + y = 2$ [Section 3.4, p. 220]
3. Graph: $y = -4$ [Section 3.2, pp. 200–201]
4. Graph: $x = 5$ [Section 3.2, pp. 199–200]
5. Find and interpret the slope of the line through $(-1, 3)$ and $(3, -4)$ [Section 3.3, pp. 206–209]
6. Find the equation of the line through $(1, 3)$ and $(4, 9)$ [Section 3.5, pp. 229–230]

① Graph Linear Functions

In Chapter 3, we discussed linear equations in two variables. Recall, a linear equation is an equation of the form $Ax + By = C$ where A, B, and C are real numbers. In addition, A and B cannot both be zero.

Remember, lines can have four basic shapes: (1) they can rise from left to right; (2) they can fall from left to right; (3) they can be horizontal; or (4) they can be vertical. Figure 26 shows these four basic shapes. Notice that lines that rise from left to right have a positive slope, lines that fall from left to right have a negative slope, horizontal lines have a 0 slope, and vertical lines have undefined slope.

Figure 26

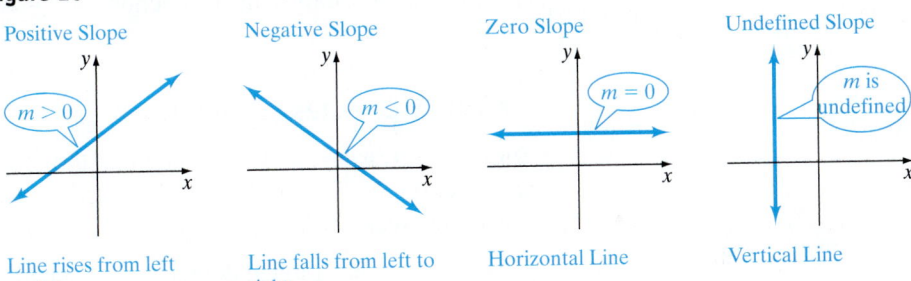

Positive Slope	Negative Slope	Zero Slope	Undefined Slope
$m > 0$	$m < 0$	$m = 0$	m is undefined
Line rises from left to right	Line falls from left to right	Horizontal Line	Vertical Line

If you look at the graphs shown in Figure 26, you should notice that all except one passes the vertical line test for identifying graphs of functions. Which one does not pass? We conclude the following: **All linear equations except equations of the form $x = a$, vertical lines, are functions.**

Because all linear equations except those of the form $x = a$ are functions, we can write any linear equation that is in the form $Ax + By = C$ using function notation provided that $B \neq 0$ as follows:

$$Ax + By = C \qquad B \neq 0$$

Subtract Ax from both sides: $By = -Ax + C$

Divide both sides by B: $\dfrac{By}{B} = \dfrac{-Ax + C}{B}$

Simplify: $y = -\dfrac{A}{B}x + \dfrac{C}{B}$

$$\underset{f(x)}{\updownarrow} \quad = \quad \underset{m}{\updownarrow} x \quad + \quad \underset{b}{\updownarrow}$$

This leads to the following definition:

DEFINITION

A **linear function** is a function of the form

$$f(x) = mx + b$$

where m and b are real numbers. The graph of a linear function is called a **line.**

Preparing for...Answers

1.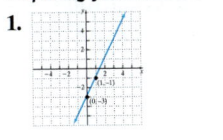
2.
3.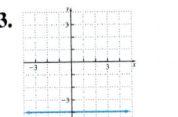
4.
5. $-\dfrac{7}{4}$; y decreases by 7 when x increases by 4.
6. $y = 2x + 1$

We can graph linear functions using the same techniques we used to graph linear equations that are written in slope-intercept form, $y = mx + b$ (See Section 3.4).

EXAMPLE 1 Graphing a Linear Function

Graph the linear function: $f(x) = 3x - 5$

Figure 27

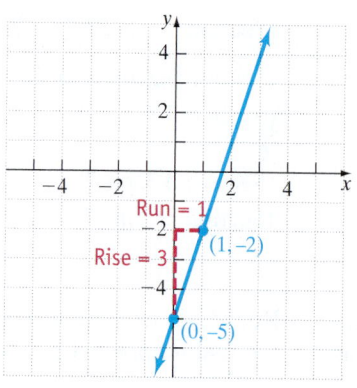

Solution

Comparing $f(x) = 3x - 5$ to $f(x) = mx + b$, we see that the slope m is 3 and the y-intercept b is -5. We begin by plotting the point $(0, -5)$. Because $m = 3 = \dfrac{3}{1} = \dfrac{\Delta y}{\Delta x} = \dfrac{\text{Rise}}{\text{Run}}$, from the point $(0, -5)$ we go up 3 units and to the right 1 unit and end up at $(1, -2)$. We draw a line through these two points and obtain the graph of $f(x) = 3x - 5$ shown in Figure 27. ∎

QUICK ✔ *Graph each linear function.*

1. $f(x) = 2x - 3$ **2.** $G(x) = -5x + 4$ **3.** $h(x) = \dfrac{3}{2}x + 1$ **4.** $f(x) = 4$

There are many applications of linear functions. For example, the cost of cab fare, sales commissions, or the cost of breakfast as a function of the number of eggs ordered can all be modeled by a linear function.

EXAMPLE 2 Sales Commissions

Tony's weekly salary at Apple Chevrolet is 0.75% of his weekly sales plus $450. The linear function $S(x) = 0.0075x + 450$ describes Tony's weekly salary S as a linear function of his weekly sales x.

(a) What is the implied domain of the function?

(b) Draw a graph of the function.

(c) If Tony sells cars worth a total of $55,000 one week, what is his salary?

(d) If Tony earned $723.75 one week, what was the value of the cars that he sold?

Solution

(a) The independent variable is weekly sales, x. Because it does not make sense to talk about negative weekly sales, we have that the domain of the function is $\{x \mid x \geq 0\}$ or, using interval notation, $[0, \infty)$.

(b) We plot the independent variable, *weekly sales,* on the horizontal axis and the dependent variable, *salary,* on the vertical axis. We graph the equation by plotting points. To obtain points on the graph of the function, evaluate the function for $x = \$0, \$10,000,$ and $\$20,000$.

$$S(0) = 0.0075(0) + 450$$
$$= 450$$
$(0, 450)$ is on the graph

$$S(10,000) = 0.0075(10,000) + 450$$
$$= \$525$$
$(10000, 525)$ is on the graph

$$S(20,000) = 0.0075(20,000) + 450$$
$$= \$600$$
$(20000, 600)$ is on the graph

We plot these points and obtain the graph of the linear function shown in Figure 28.

Figure 28

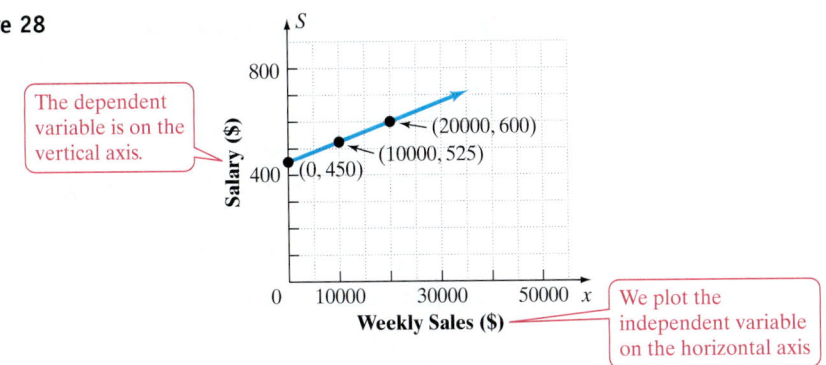

(c) We evaluate the function at $x = \$55{,}000$ to obtain

$$S(55000) = 0.0075(55000) + 450$$
$$= \$862.50$$

Tony will earn $862.50 for the week if he sells $55,000 worth of cars.

(d) Here, we need to solve the equation $S(x) = 723.75$.

$$0.0075x + 450 = 723.75$$

Subtract 450 from both sides: $\qquad 0.0075x = 273.75$

Divide both sides by 0.0075: $\qquad x = \$36{,}500$

If Tony sells $36,500 worth of cars in a week, he will earn $723.75. ∎

Notice a few details regarding the graph of the linear function in Figure 28. First, we only graph the function over its domain, $[0, \infty)$, so that in this case we only graph in quadrant I. Also notice that we labeled the horizontal axis x for the independent variable, *weekly sales,* and we labeled the vertical axis S for the dependent variable, *salary.* We also indicated what the independent and dependent variable represent on each coordinate axis. It is always a good practice to label your axes.

Quick ✓

5. The cost, C, of renting a 12-foot moving truck for a day is $40 plus $0.35 times the number of miles driven. The linear function $C(x) = 0.35x + 40$ describes the cost C of driving the truck x miles.

  (a) What is the implied domain of this linear function?

  (b) Determine the y-intercept of the graph of the linear function.

  (c) What is the rental cost if the truck is driven 80 miles?

  (d) Graph the linear function.

  (e) How many miles was the truck driven if the rental cost is $85.50?

  (**Hint:** Solve the equation $C(x) = 85.5$.)

② Build Linear Models from Verbal Descriptions

A linear function is a function of the form $f(x) = mx + b$, where m is the slope of the linear function and b is its y-intercept. In Section 3.3, we said that the slope m can be thought of as an average rate of change. The slope describes by how much a dependent variable changes for a given change in the independent variable. For example, in the linear function $f(x) = 4x + 3$, the slope is $4 = \dfrac{4}{1} = \dfrac{\Delta y}{\Delta x}$ so that the dependent variable y will increase by 4 units for every 1-unit increase in x (the independent variable). When the average rate of change of a function is constant, then we can use linear functions to model the situation. For example, if your phone company charges you $0.05 per minute to talk regardless of the number of minutes on the phone, then we can use a linear function to model the cost of talking with slope $m = \dfrac{0.05 \text{ dollars}}{1 \text{ minute}}$.

EXAMPLE 3 Cost Function

The simplest cost function is the linear cost function $C(x) = ax + b$, where b represents the fixed costs of operating a business and a represents the variable costs (the cost of manufacturing one additional item). Suppose that a small bicycle manufacturer has daily fixed costs of $2000 and each bicycle costs $80 to manufacture.

(a) Write a linear function that expresses the cost of manufacturing x bicycles in a day.

(b) Graph the linear function.

(c) What is the cost of manufacturing 18 bicycles in a day?

(d) How many bicycles can be manufactured for $4080? (**Hint:** Solve the equation $C(x) = 4080$.)

Figure 29

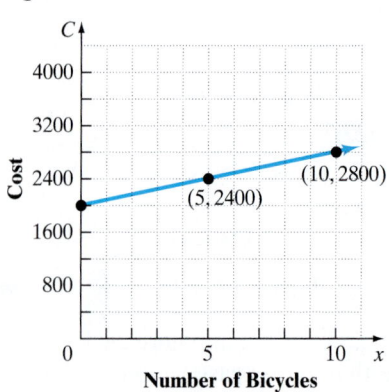

Number of Bicycles

Solution

(a) Because the variable costs are $80, we have that $a = 80$. The fixed costs are $2000 so that $b = 2000$. Therefore, the cost function is

$$C(x) = 80x + 2000$$

(b) Label the horizontal axis x and the vertical axis C. Figure 29 shows the graph of the cost function.

(c) We evaluate the function for $x = 18$ and obtain

$$C(18) = 80(18) + 2000$$
$$= \$3440$$

It will cost $3440 to manufacture 18 bicycles.

(d) We solve $C(x) = 4080$.

$$C(x) = 4080$$
$$80x + 2000 = 4080$$

Subtract 2000 from both sides: $\qquad 80x = 2080$

Divide both sides by 80: $\qquad x = 26$

So 26 bicycles can be manufactured for a cost of $4080.

QUICK ✓

6. Suppose that the government imposes a tax of $1 per bicycle manufactured for the business presented in Example 3.

(a) Write a linear function that expresses the cost C of manufacturing x bicycles in a day.

(b) Graph the linear function.

(c) What is the cost of manufacturing 18 bicycles in a day?

(d) How many bicycles can be manufactured for $4025?

EXAMPLE 4 Straight-Line Depreciation

Book value is the value of an asset such as a building or piece of machinery that a company uses to create its balance sheet. Some companies will use straight-line depreciation to depreciate their assets so that the value of the asset declines by a constant amount each year. The amount of the decline depends upon the useful life that the company places on the asset. Suppose that Prentice Hall Publishing Company just purchased a new fleet of cars for its sales force at a cost of $29,400 per car. The company will depreciate the cars using the straight-line method over 7 years, so that each car depreciates by $\dfrac{\$29,400}{7} = \$4200$ per year.

(a) Write a linear function that expresses the book value V of each car as a function of its age, x.

(b) What is the implied domain of this linear function?

(c) What is the book value of each car after 3 years?

(d) When will the book value of each car be \$12,600? (**Hint:** Solve the equation $V(x) = 12{,}600$.)

(e) Graph the linear function.

Solution

(a) We let $V(x)$ represent the book value of each car after x years, so $V(x) = mx + b$. The original value of the car is \$29,400, so $V(0) = 29{,}400$. The V-intercept of the linear function is \$29,400. Because each car depreciates by \$4200 per year, the slope of the linear function is -4200. The linear function that represents the book value of the car after x years is given by

$$V(x) = -4200x + 29{,}400$$

(b) Because the car cannot have a negative age, we know that the age, x, must be greater than or equal to zero. In addition, the car is depreciated over 7 years. After 7 years the book value of the car is $V(7) = 0$. Therefore, the implied domain of the function is $\{x \mid 0 \le x \le 7\}$, or using interval notation $[0, 7]$.

(c) The book value of the car after $x = 3$ years is given by $V(3)$.

$$\begin{aligned} V(3) &= -4200(3) + 29{,}400 \\ &= \$16{,}800 \end{aligned}$$

(d) To find when the book value is \$12,600, we solve the equation

$$V(x) = 12{,}600$$

$$-4200x + 29{,}400 = 12{,}600$$

Subtract 29,400 from both sides: $-4200x = -16{,}800$

Divide both sides by 4: $x = 4$

Each car will have a book value of \$12,600 after 4 years.

(e) Label the horizontal axis x and the vertical axis V. Since $V(0) = 29{,}400$, we know that $(0, 29400)$ is on the graph. Since $V(7) = 0$, we know that $(7, 0)$ is on the graph. To graph the function, we use these points (the intercepts), along with points $(3, 16800)$ and $(4, 12600)$ found in parts (c) and (d). See Figure 30.

Figure 30

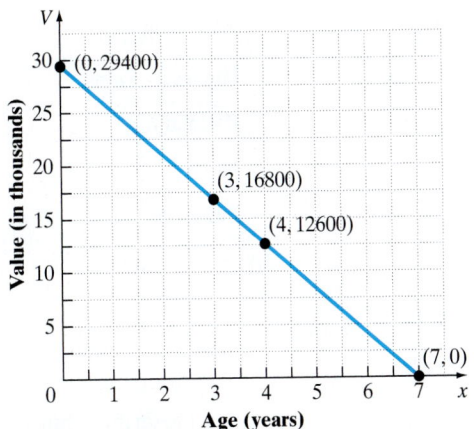

QUICK ✓

7. Roberta just purchased a new car. Her monthly payments are $250 per month. She estimates that maintenance and gas cost her $0.18 per mile.

(a) Write a linear function that relates the monthly cost C of operating the car as a function of miles driven, x.

(b) What is the implied domain of this linear function?

(c) What is the monthly cost of driving 320 miles?

(d) Graph the linear function.

(e) How many miles can Roberta drive each month if she can afford the monthly cost to be $282.40?

---

3 ## Build Linear Models from Data

We know from Section 3.5 that only two points are required to find the equation of a line. But what if we have a set of data with more than two points? How can we tell if the data (the two variables) are related linearly? While there are some rather sophisticated methods for determining whether two variables are linearly related (and beyond the scope of this course), we can draw a picture of the data and learn whether the variables might be linearly related. This picture is called a *scatter diagram.*

Scatter Diagrams

Often, we are interested in finding an equation that can describe the relation between two variables. The first step in determining the type of equation that should be used to describe the relation is to plot the ordered pairs that make up the relation in the Cartesian plane. The graph that results is called a **scatter diagram.**

EXAMPLE 5 Drawing a Scatter Diagram

In baseball, the on-base percentage for a team represents the percentage of time that the team safely reaches base. The data given in Table 7 represent the number of runs scored and the on-base percentage for various teams during the 2002 baseball season.

Table 7			
Team	**On-Base Percentage, x**	**Runs Scored, y**	**(x, y)**
NY Yankees	35.4	897	(35.4, 897)
Anaheim Angels	34.1	851	(34.1, 851)
Texas Rangers	33.8	843	(33.8, 843)
Toronto Blue Jays	32.7	813	(32.7, 813)
Minnesota Twins	33.2	768	(33.2, 768)
Oakland A's	33.9	800	(33.9, 800)
Kansas City Royals	32.3	737	(32.3, 737)
Baltimore Orioles	30.9	667	(30.9, 667)

SOURCE: *espn.com*

(a) Draw a scatter diagram of the data treating on-base percentage as the independent variable.

(b) Describe what happens as the on-base percentage increases.

Figure 31

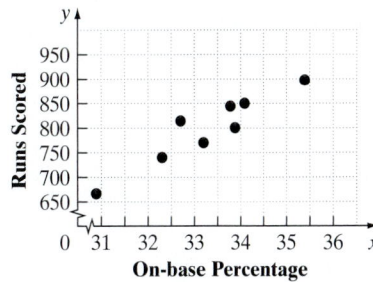

Solution

(a) To draw a scatter diagram we plot the ordered pairs listed in Table 7. See Figure 31.

(b) From the scatter diagram, we can see that as the on-base percentage increases, the number of runs scored also increases. While the relation between on-base percentage and number of runs scored does not follow a perfect linear relation (because the points don't all fall on a straight line), we can agree that the pattern of the data is linear.

QUICK ✓

8. The data listed below represent the total cholesterol (in mg/dL) and age of males.

Age	Total Cholesterol	Age	Total Cholesterol
25	180	38	239
25	195	48	204
28	186	51	243
32	180	62	228
32	197	65	269

(a) Draw a scatter diagram treating age as the independent variable.

(b) Describe the relation between age and total cholesterol.

Figure 32

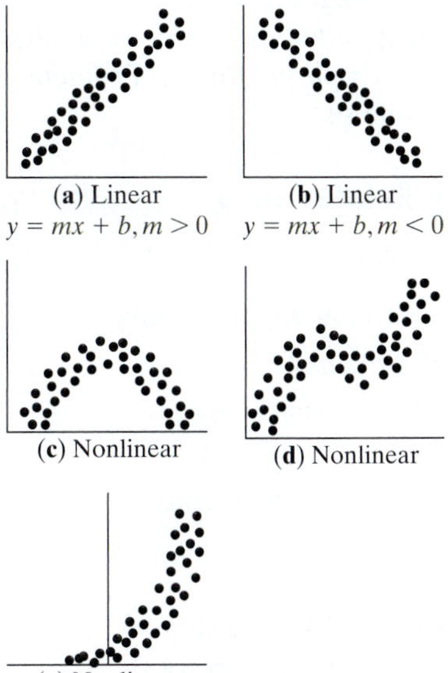

(a) Linear
$y = mx + b, m > 0$

(b) Linear
$y = mx + b, m < 0$

(c) Nonlinear

(d) Nonlinear

(e) Nonlinear

Recognizing the Type of Relation That Appears to Exist between Two Variables

We use scatter diagrams to help us see the type of relation that exists between two variables. In this text, we will look at a few different types of relations between two variables. For now, however, our only goal is to distinguish between linear and nonlinear relations. See Figure 32.

EXAMPLE 6   **Distinguishing between Linear and Nonlinear Relations**

Determine whether the relation between the two variables in Figure 33 is linear or nonlinear. If the relation is linear, indicate whether the slope is positive or negative.

Figure 33

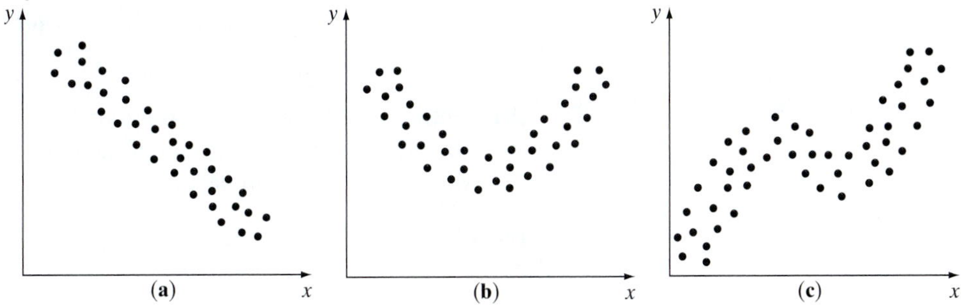

(a)   **(b)**   **(c)**

Solution

(a) Linear with negative slope   **(b)** Nonlinear   **(c)** Nonlinear

QUICK ✓ *Determine whether the relation between the two variables is linear or nonlinear. If it is linear, determine whether the slope is positive or negative.*

9.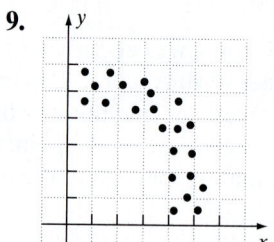

10.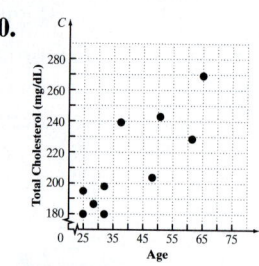

Fitting a Line to Data

Suppose that the scatter diagram of a set of data appears to be linearly related as in Figure 32(a) or (b). We might wish to find an equation of a line that relates the two variables. One way to obtain an equation for data that appears to follow a linear pattern is to draw a line through two points on the scatter diagram and determine the equation of the line through these points using the point-slope form of a line, $y - y_1 = m(x - x_1)$. To review using this formula work "*Preparing for* . . ." Problem 6 at the beginning of this section, if you haven't already done so.

EXAMPLE 7   Finding an Equation for Linearly Related Data

Using the data in Table 7 from Example 5,

(a) Select two points and find an equation of the line containing the points.

(b) Graph the line on the scatter diagram obtained in Example 5(a).

(c) Use the line found in part (a) to predict the number of runs scored by a team whose on-base percentage is 34.6%.

(d) Interpret the slope. Does it make sense to interpret the y-intercept?

Solution

(a) Select two points, for example, $(30.9, 667)$ and $(35.4, 897)$. (You should select your own two points and complete the solution.) The slope of the line joining the points $(30.9, 667)$ and $(35.4, 897)$ is

$$m = \frac{897 - 667}{35.4 - 30.9} = \frac{230}{4.5} = 51.1$$

The equation of the line with slope 51.1 and passing through $(30.9, 667)$ is found using the point-slope form with $m = 51.1$, $x_1 = 30.9$, and $y_1 = 667$.

$$\text{Point-slope form:} \quad y - y_1 = m(x - x_1)$$
$$m = 51.1;\ x_1 = 30.9,\ y_1 = 667: \quad y - 667 = 51.1(x - 30.9)$$
$$\text{Distribute 51.1:} \quad y - 667 = 51.1x - 1578.99$$
$$\text{Add 667 to both sides:} \quad y = f(x) = 51.1x - 911.99$$

(b) Figure 34 shows the scatter diagram with the graph of the line found in part (a). We obtain the graph of the line by drawing the line through the two points selected in part (a).

(c) We evaluate $f(x) = 51.1x - 911.99$ at $x = 34.6$.

$$f(34.6) = 51.1(34.6) - 911.99$$
$$= 856.07$$

We round this to the nearest whole number. We predict that a team whose on-base percentage is 34.6% will score 856 runs.

Figure 34

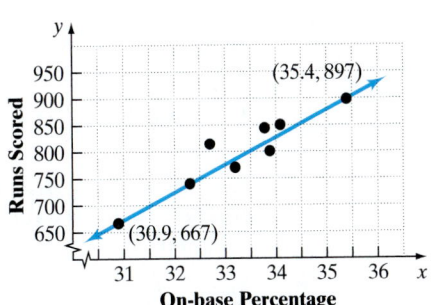

(d) The slope of the linear function is 51.1. This means that if the on-base percentage increases by 1%, then the number of runs scored will increase by about 51 runs. The y-intercept, −911.99, represents the runs scored when on-base percentage is 0. Since negative runs scored does not make sense and we do not have any observations near zero, it does not make sense to interpret the y-intercept.

QUICK ✓

11. Using the data from Quick Check Problem 8 on page 613,

   (a) Select two points and find an equation of the line containing the points.

   (b) Graph the line on the scatter diagram obtained in Quick Check Problem 8 (page 613).

   (c) Predict the total cholesterol of a 39-year-old male.

   (d) Interpret the slope. Does it make sense to interpret the y-intercept?

8.5 Exercises

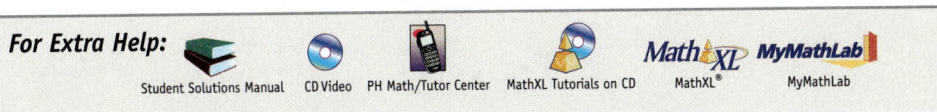

For Extra Help: Student Solutions Manual   CD Video   PH Math/Tutor Center   MathXL Tutorials on CD   MathXL®   MyMathLab

Concepts and Vocabulary

In Problems 1–3, fill in the blanks.

1. For the graph of a linear function $f(x) = mx + b$, m is the _____ and b is the _____.

2. The graph of a linear function is called a _____.

3. A _____ _____ is used to help us see the type of relation, if any, that may exist between two variables.

In Problems 4–6, answer True or False to the statement.

4. All linear equations are functions.

5. For the linear function $f(x) = 4x − 3$, the slope is 4 and the y-intercept is −3.

6. The graph of the linear function $G(x) = −2x + 3$ is a line that falls from left to right.

7. Under what circumstances are the x- and y-intercepts of a linear function the same?

8. Can there be a linear function that has an x-intercept, but no y-intercept?

Building Skills

For Problems 9–20, graph each linear function.

9. $F(x) = 5x − 2$

10. $F(x) = 4x + 1$

11. $G(x) = −3x + 7$

12. $G(x) = −2x + 5$

13. $H(x) = −2$

14. $P(x) = 5$

15. $f(x) = \dfrac{1}{2}x - 4$ **16.** $f(x) = \dfrac{1}{3}x - 3$ **17.** $F(x) = -\dfrac{5}{2}x + 5$

18. $P(x) = -\dfrac{3}{5}x - 1$ **19.** $G(x) = -\dfrac{3}{2}x$ **20.** $f(x) = \dfrac{4}{5}x$

In Problems 21–24, determine whether the scatter diagram indicates that a linear relation may exist between the two variables. If a linear relation does exist, indicate whether the slope is positive or negative.

21.

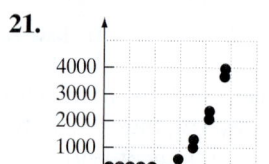

22.

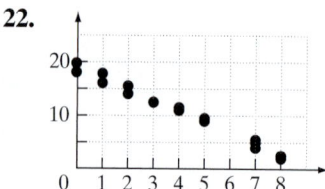

23.

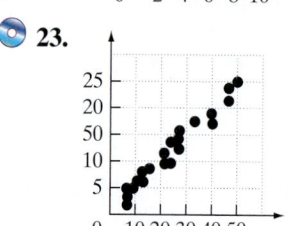

24.

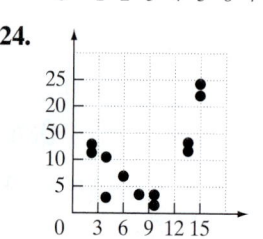

In Problems 25–28,

   (a) *Draw a scatter diagram of the data.*

   (b) *Select two points from the scatter diagram and find the equation of the line containing the points selected.* *

   (c) *Graph the line found in part (b) on the scatter diagram.*

25.

x	2	4	5	8	9
y	1.4	1.8	2.1	2.3	2.6

26.

x	2	3	5	6	7
y	5.7	5.2	2.8	1.9	1.8

27.

x	1.2	1.8	2.3	3.5	4.1
y	8.4	7.0	7.3	4.5	2.4

28.

x	0	0.5	1.4	2.1	3.9
y	0.8	1.3	1.9	2.5	5.0

Applying the Concepts

29. Taxes The function $T(x) = 0.15(x - 7300) + 730$ represents the tax bill T of a single person whose adjusted gross income in 2005 is x dollars for income between \$7300 and \$29,700, inclusive. (*Source:* Internal Revenue Service)

   (a) What is the implied domain of this linear function?
   (b) What is a single filer's tax bill if adjusted gross income is \$20,000?
   (c) Which variable is independent and which is dependent?
   (d) Graph the linear function over the domain specified in part (a).
   (e) What is a single filer's adjusted gross income if their tax bill is \$3385?

     (**Hint:** Solve the equation $T(x) = 3385$.)

30. Sales Commissions Tanya works for Prentice Hall as a book representative. The linear function $I(s) = 0.01s + 20{,}000$ describes the annual income I of Tanya when she has total sales s.

   (a) What is the implied domain of this linear function?
   (b) What is $I(0)$? Explain what this result means.
   (c) What is Tanya's salary if she sells \$500,000 in books for the year?
   (d) Graph the linear function.
   (e) At what level of sales will Tanya's income be \$45,000?

     (**Hint:** Solve the equation $I(s) = 45{,}000$.)

*Answers will vary.

31. Cab Fare The linear function $C(m) = 1.5m + 2$ describes the cab fare C for a ride of m miles.

(a) What is the implied domain of this linear function?
(b) What is $C(0)$? Explain what this result means.
(c) What is cab fare for a 5-mile ride?
(d) Graph the linear function.
(e) How many miles can you ride in a cab if you have $13.25?

32. Luxury Tax In 2002, Major League Baseball signed a labor agreement with the players. In this agreement, any team whose payroll exceeds $128 million in 2005 will have to pay a luxury tax of 22.5% (for first time offenses). The linear function $T(p) = 0.225(p - 128)$ describes the luxury tax T of a team whose payroll is p (in millions).

(a) What is the implied domain of this linear function?
(b) What is the luxury tax for a team whose payroll is $160 million?
(c) Graph the linear function.
(d) What is the payroll of a team that pays luxury tax of $11.7 million?

33. Health Costs The annual cost of health insurance H as a function of age a is given by the function $H(a) = 22.8a - 117.5$ for $15 \le a \le 90$. (*Source: Statistical Abstract*, 2005)

(a) What are the independent and dependent variables?
(b) What is the domain of this linear function?
(c) What is the health insurance premium of a 30 year old?
(d) Graph the linear function over its domain.
(e) What is the age of an individual whose health insurance premium is $976.90?

34. Birth Rate A multiple birth is any birth with 2 or more children born. The birth rate is the number of births per 1000 women. The birth rate B of multiple births as a function of age a is given by the function $B(a) = 1.73a - 14.56$ for $15 \le a \le 44$. (*Source:* Centers for Disease Control)

(a) What are the independent and dependent variables?
(b) What is the domain of this linear function?
(c) What is the multiple birth rate of women who are 22 years of age according to the model?
(d) Graph the linear function over its domain.
(e) What is the age of women whose multiple birth rate is 49.45?

35. Phone Charges Sprint has a long-distance phone plan that charges a monthly fee of $8.95 plus $0.05 per minute. (*Source: Sprint.com*)

(a) Find a linear function that expresses the monthly bill B as a function of minutes used m.
(b) What are the independent and dependent variables?
(c) What is the implied domain of this linear function?
(d) What is the monthly bill if 300 minutes are used for long-distance phone calls?
(e) How many minutes were used for long distance if the long-distance phone bill was $20.95?
(f) Graph the linear function.

36. RV Rental The rental cost R of a class C 20-foot recreational vehicle is $129.50 plus $0.15 per mile. (*Source: westernrv.com*)

(a) Find a linear function that expresses the cost R as a function of miles driven m.
(b) What are the independent and dependent variables?
(c) What is the implied domain of this linear function?
(d) What is the rental cost if 860 miles are driven?
(e) How many miles were driven if the rental cost is $213.80?
(f) Graph the linear function.

37. Depreciation Suppose that a company has just purchased a new computer for $2700. The company chooses to depreciate the computer using the straight-line method over 3 years.

(a) Find a linear function that expresses the book value V of the computer as a function of its age x.

(b) What is the implied domain of this linear function?

(c) What is the book value of the computer after the first year?

(d) What are the intercepts of the graph of the linear function?

(e) When will the book value of the computer be $900?

(f) Graph the linear function.

38. Depreciation Suppose that a company just purchased a new machine for its manufacturing facility for $1,200,000. The company chooses to depreciate the machine using the straight-line method over 20 years.

(a) Find a linear function that expresses the book value V of the machine as a function of its age x.

(b) What is the implied domain of this linear function?

(c) What is the book value of the machine after three years?

(d) What are the intercepts of the graph of the linear function?

(e) When will the book value of the machine be $480,000?

(f) Graph the linear function.

39. Concrete As concrete cures, it gains strength. The following data represent the 7-day and 28-day strength (in pounds per square inch) of a certain type of concrete.

7-day Strength, x	28-day Strength, y	7-day Strength, x	28-day Strength, y
2300	4070	2480	4120
3390	5220	3380	5020
2430	4640	2660	4890
2890	4620	2620	4190
3330	4850	3340	4630

(a) Draw a scatter diagram of the data treating 7-day strength as the independent variable.

(b) What type of relation appears to exist between 7-day strength and 28-day strength?

(c) Select two points and find an equation of the line containing the points.

(d) Graph the line on the scatter diagram drawn in part (a).

(e) Predict the 28-day strength of a slab of concrete if its 7-day strength is 3000 psi.

(f) Interpret the slope of the line found in part (c).

40. Candy The following data represent the weight (in grams) of various candy bars and the corresponding number of calories.

Candy Bar	Weight, x	Calories, y
Hershey's Milk Chocolate	44.28	230
Nestle Crunch	44.84	230
Butterfinger	61.30	270
Baby Ruth	66.45	280
Almond Joy	47.33	220
Twix (with Caramel)	58.00	280
Snickers	61.12	280
Heath	39.52	210

SOURCE: *Megan Pocius, student at Joliet Junior College*

(a) Draw a scatter diagram of the data treating weight as the independent variable.
(b) What type of relation appears to exist between the weight of a candy bar and the number of calories?
(c) Select two points and find an equation of the line containing the points.
(d) Graph the line on the scatter diagram drawn in part (a).
(e) Predict the number of calories in a candy bar that weighs 62.3 grams.
(f) Interpret the slope of the line found in part (c).

41. Raisins The following data represent the weight (in grams) of a box of raisins and the number of raisins in the box.

Weight, w	Number of Raisins, N	Weight, w	Number of Raisins, N
42.3	87	42.4	90
42.7	91	42.3	82
42.8	93	42.5	86
42.4	87	42.7	86
42.6	89	42.5	86

SOURCE: Jennifer Maxwell, student at Joliet Junior College

(a) Does the relation defined by the set of ordered pairs (w, N) represent a function?
(b) Draw a scatter diagram of the data treating weight as the independent variable.
(c) Select two points and find the equation of the line containing the points.
(d) Graph the line on the scatter diagram drawn in part (b).
(e) Express the relationship found in part (c) using function notation.
(f) Predict the number of raisins in a box that weighs 42.5 grams.
(g) Interpret the slope of the line found in part (c).

42. Height versus Head Circumference The following data represent the height (in inches) and head circumference (in inches) of 9 randomly selected children.

(a) Does the relation defined by the set of ordered pairs (h, C) represent a function?
(b) Draw a scatter diagram of the data treating height as the independent variable.
(c) Select two points and find the equation of the line containing the points.
(d) Graph the line on the scatter diagram drawn in part (b).
(e) Express the relationship found in part (c) using function notation.
(f) Predict the head circumference of a child who is 26.5 inches tall.
(g) Interpret the slope of the line found in part (c).

Height, h	Head Circumference, C
25.25	16.4
25.75	16.9
25	16.9
27.75	17.6
26.50	17.3
27.00	17.5
26.75	17.3
26.75	17.5
27.5	17.5

SOURCE: Denise Slucki, student at Joliet Junior College

Extending the Concepts

43. In parts (a)–(e), use the figure shown to the right.
(a) Solve $f(x) = 1$.
(b) Solve $f(x) = -3$.
(c) Solve $f(x) = 2$.
(d) What are the intercepts of the function $y = f(x)$?

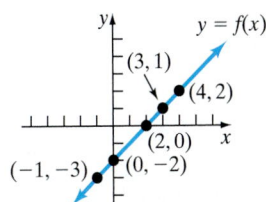

(e) The **zero** of a function is any value of the independent variable that causes the value of the function to be zero. What is the zero of the function?

44. In parts (a)–(e), use the figure shown to the right.
(a) Solve $g(x) = 1$.
(b) Solve $g(x) = -1$.
(c) Solve $g(x) = 4$.
(d) What are the intercepts of the function $y = g(x)$?
(e) See Problem 43(e). What is the zero of the function?

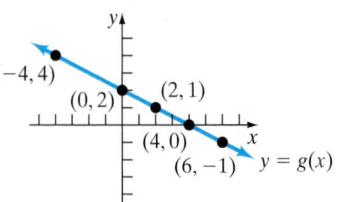

45. A strain of *E. coli* Beu 397-recA441 is placed into a Petri dish at 30° Celsius and allowed to grow. The population is estimated by means of an optical device in which the amount of light that passes through the Petri dish is measured. The data to the right are collected. Do you think that a linear function could be used to describe the relation between the two variables? Why or why not?

Time, x	Population, y
0	0.09
2.5	0.18
3.5	0.26
4.5	0.35
6	0.50

Source: *Dr. Polly Lavery, Joliet Junior College*

The Graphing Calculator

The equation of the line obtained in Example 7 depends on the points selected, which will vary from person to person. So the line we found might be different from the line that you found. Although the line that we found in Example 7 fits the data well, there may be a line that "fits it better." Do you think that your line fits the data better? Is there a line of *best fit*? As it turns out, there is a method for finding the line that best fits linearly related data (called the *line of best fit*).*

Graphing utilities can be used to draw scatter diagrams and find the line of best fit. Figure 35(a) shows a scatter diagram of the data presented in Table 7 from Example 5 drawn on a TI-84 Plus graphing calculator. Figure 35(b) shows the line of best fit from a TI-84 Plus graphing calculator.

The line of best fit is $y = 50.3x - 877.4$.

Figure 35

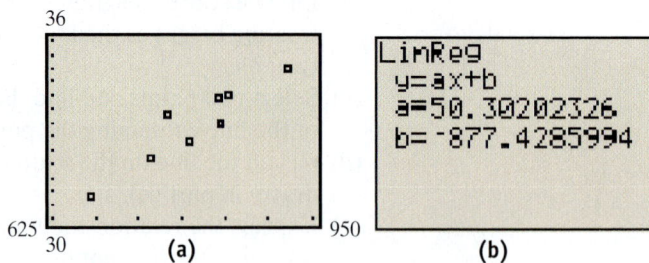

In Problems 46–49,

(a) *Draw a scatter diagram using a graphing calculator*

(b) *Find the line of best fit using a graphing calculator*

for the data in the problem specified.

46. Problem 39   **47.** Problem 40

48. Problem 41   **49.** Problem 42

---

*We shall not discuss in this book the underlying mathematics of lines of best fit. Books in elementary statistics discuss this topic.

8.6 Compound Inequalities

OBJECTIVES

1. Determine the Intersection or Union of Two Sets.
2. Solve Compound Inequalities Involving "and"
3. Solve Compound Inequalities Involving "or"
4. Solve Problems Using Compound Inequalities

Preparing for Compound Inequalities

Before getting started, take the following readiness quiz. If you get a problem wrong, go back to the section cited and review the material.

1. Use set-builder notation and interval notation to list the set of all real numbers x such that $-2 \le x \le 5$. [Section 2.8, pp. 150–153]
2. Graph the inequality $x \ge 4$. [Section 2.8, pp. 150–153]
3. Use interval notation to express the inequality shown in the graph. [Section 2.8, pp. 150–153]

$$\xleftarrow{\hspace{1em}} \begin{array}{ccccccc} & | & | & | & | & | \\ -2 & -1 & 0 & 1 & 2 & 3 & 4 \end{array} \xrightarrow{\hspace{1em}}$$

4. Solve: $2(x + 3) - 5x = 15$ [Section 2.2, pp. 86–89]
5. Solve: $2x + 3 > 11$ [Section 2.8, pp. 153–159]
6. Solve: $x + 8 \ge 4(x - 1) - x$ [Section 2.8, pp. 153–159]

---

1 ## Determine the Intersection or Union of Two Sets

Consider the information presented in Table 8 regarding students enrolled in an Intermediate Algebra course. We can classify the people in the course in a set. For example, suppose we define set A as the set of all students whose age is less than 25. Then

$$A = \{\text{Grace, Sophia, Kevin, Jack, George, Teresa}\}$$

Suppose we define set B as the set of all students who are female. Then

$$B = \{\text{Grace, Sophia, Mary, Nancy, Teresa}\}$$

Now list all the students that are in set A and set B. That is, list all the students who are less than 25 years of age and female.

$$A \text{ and } B = \{\text{Grace, Sophia, Teresa}\}$$

Now list all the students that are either in set A or set B or both.

$$A \text{ or } B = \{\text{Grace, Sophia, Kevin, Jack, George, Teresa, Mary, Nancy}\}$$

Figure 35 shows a Venn diagram illustrating the relation among A, B, A and B, and A or B. Notice that Grace, Sophia, and Teresa are in both A and B, while Robert is neither in A nor B.

Table 8

Student	Age	Gender
Grace	19	Female
Sophia	23	Female
Kevin	20	Male
Robert	32	Male
Jack	19	Male
Mary	35	Female
Nancy	40	Female
George	22	Male
Teresa	20	Female

Figure 35

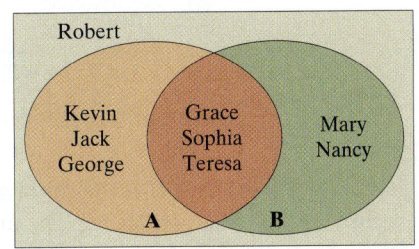

When we used the word *and* to obtain the set, we listed elements that were common to both set A and set B. When we used the word *or* to obtain the set, we listed elements that were in either set A or set B or both. These results lead us to the following definitions.

DEFINITIONS:

- The **intersection** of two sets A and B, denoted $A \cap B$, is the set of all elements that belong to both set A and set B.
- The **union** of two sets A and B, denoted $A \cup B$, is the set of all elements that are in the set A or in the set B or in both A and B.
- The word **and** implies intersection, while the word **or** implies union.

EXAMPLE 1 Finding the Intersection and Union of Sets

Let $A = \{1, 3, 5, 7, 9\}$ and let $B = \{1, 2, 3, 4, 5\}$. Find

(a) $A \cap B$ (b) $A \cup B$

Solution

(a) $A \cap B$ is the set of all elements that are in both A and B. So, $A \cap B = \{1, 3, 5\}$.

(b) $A \cup B$ is the set of all elements that are in A or B, or both. So,
$A \cup B = \{1, 2, 3, 4, 5, 7, 9\}$.

> **Work Smart**
> When finding the union of two sets, we list each element only once, even if it occurs in both sets.

QUICK ✓ Let $A = \{1, 2, 3, 4, 5, 6\}$, $B = \{1, 3, 5, 7\}$, and $C = \{2, 4, 6, 8\}$.

1. Find $A \cap B$. **2.** Find $A \cap C$. **3.** Find $A \cup B$.

4. Find $A \cup C$. **5.** Find $B \cap C$. **6.** Find $B \cup C$.

Let's look at the intersection and union of two sets from the point of view of inequalities.

EXAMPLE 2 Finding the Intersection and Union of Two Sets

Suppose $A = \{x \mid x \le 5\}$, $B = \{x \mid x \ge 1\}$, and $C = \{x \mid x < -2\}$.

(a) Determine $A \cap B$. Graph the set on a real number line. Write the set $A \cap B$ using both set-builder notation and interval notation.

(b) Determine $B \cup C$. Graph the set on a real number line. Write the set $B \cup C$ using both set-builder notation and interval notation.

Solution

(a) $A \cap B$ is the set of all real numbers that are less than or equal to 5 and greater than or equal to 1. We can identify this set by determining where the graphs of the inequalities overlap. See Figure 36.

Figure 36

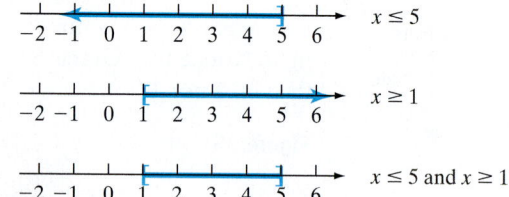

We can represent the set $A \cap B$ using set-builder notation as $\{x \mid 1 \le x \le 5\}$ or interval notation as $[1, 5]$.

> **Work Smart**
> Throughout the text, we will use the word "or" when using set-builder notation and we will use the union symbol, $\cup$, when using interval notation.

(b) $B \cup C$ is the set of all real numbers that are greater than or equal to 1 or less than -2. The union of these two sets would be all real numbers less than -2 or greater than or equal to 1. See Figure 37.

Figure 37

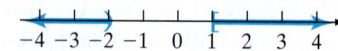

We can represent the set $B \cup C$ using set-builder notation as $\{x \mid x < -2 \text{ or } x \ge 1\}$ or interval notation as $(-\infty, -2) \cup [1, \infty)$.

QUICK ✓ Let $A = \{x \mid x > 2\}$, $B = \{x \mid x < 7\}$, and $C = \{x \mid x \le -3\}$.

7. Determine $A \cap B$. Graph the set on a real number line. Write the set $A \cap B$ using both set-builder notation and interval notation.

8. Determine $A \cup C$. Graph the set on a real number line. Write the set $A \cup C$ using both set-builder notation and interval notation.

② Solve Compound Inequalities Involving "and"

A **compound inequality** is formed by joining two inequalities with the word "and" or "or." For example,

$$3x + 1 > 4 \text{ and } 2x - 3 < 7$$
$$5x - 2 \le 13 \text{ or } 2x - 5 > 3$$

are examples of compound inequalities. To **solve a compound inequality** means to find all possible values of the variable such that the compound inequality results in a true statement. For example, the compound inequality

$$3x + 1 > 4 \text{ and } 2x - 3 < 7$$

is true for $x = 2$, but false for $x = 0$.

Let's look at an example that illustrates how to solve compound inequalities involving the word "and."

EXAMPLE 3 **How to Solve a Compound Inequality Involving "and"**

Solve $3x + 2 > -7$ and $4x + 1 \le 9$. Graph the solution set.

Step-by-Step Solution

Step 1: Solve each inequality separately.		

$$3x + 2 > -7 \qquad\qquad\qquad 4x + 1 \le 9$$

Subtract 2 from both sides: $\quad 3x > -9$ Subtract 1 from both sides: $\quad 4x \le 8$

Divide both sides by 3: $\quad x > -3$ Divide both sides by 4: $\quad x \le 2$

Step 2: Find the intersection of the solution sets, which will represent the solution set to the compound inequality.

To find the intersection of the two solution sets, we graph each inequality separately. See Figure 38.

Figure 38

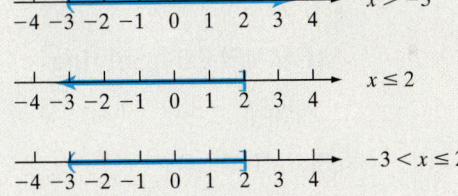

The intersection of $x > -3$ and $x \le 2$ is $-3 < x \le 2$.

The solution set is $\{x \mid -3 < x \le 2\}$ or, using interval notation, $(-3, 2]$.

The steps below summarize the procedure for solving compound inequalities involving "and."

> ### Steps for Solving Compound Inequalities Involving "and"
>
> **Step 1:** Solve each inequality separately.
>
> **Step 2:** Find the INTERSECTION of the solution sets of each inequality.

EXAMPLE 4 **Solving a Compound Inequality with "and"**

Solve $-2x + 5 > -1$ and $5x + 6 \le -4$. Graph the solution set.

Solution

We solve each inequality separately

$$-2x + 5 > -1$$
Subtract 5 from both sides: $\quad -2x > -6$
Divide both sides by -2; $\quad x < 3$
don't forget to reverse the direction of the inequality!

$$5x + 6 \le -4$$
Subtract 6 from both sides: $\quad 5x \le -10$
Divide both sides by 5: $\quad x \le -2$

Find the intersection of the solution sets, which will represent the solution set to the compound inequality. To find the intersection of the two solution sets, we graph each inequality separately. See Figure 39.

Figure 39

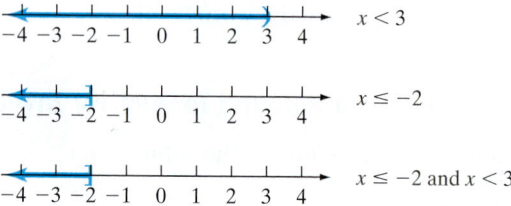

$x < 3$

$x \le -2$

$x \le -2$ and $x < 3$

The intersection of $x < 3$ and $x \le -2$ is $x \le -2$. The solution set is $\{x \mid x \le -2\}$ or, using interval notation, $(-\infty, -2]$. ■

QUICK ✓ *Solve each compound inequality. Express your solution using set-builder notation and interval notation. Graph the solution set.*

9. $2x + 1 \ge 5$ and $-3x + 2 < 5$ **10.** $4x - 5 < 7$ and $3x - 1 > -10$

11. $-8x + 3 < -5$ and $\dfrac{2}{3}x + 1 < 3$

EXAMPLE 5 **Solving a Compound Inequality with "and"**

Solve $x - 5 > -1$ and $2x - 3 \le -5$. Graph the solution set.

Solution

We solve each inequality separately

$$x - 5 > -1$$
Add 5 to both sides: $\quad x > 4$

$$2x - 3 \le -5$$
Add 3 to both sides: $\quad 2x \le -2$
Divide both sides by 2: $\quad x \le -1$

Find the intersection of the solution sets, which will represent the solution set to the compound inequality. To find the intersection of the two solution sets, we graph each inequality separately. See Figure 40.

Figure 40

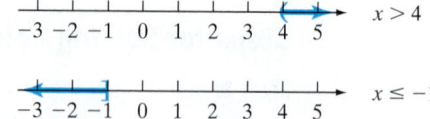

$x > 4$

$x \le -1$

Work Smart
The braces are empty in an empty set.

The intersection of $x > 4$ and $x \le -1$ is the empty set. The solution set is $\{\ \}$ or $\varnothing$. ■

QUICK ✓ *Solve each compound inequality. Express your solution using set-builder notation and interval notation. Graph the solution set.*

12. $3x - 5 < -8$ and $2x + 1 > 5$ **13.** $5x + 1 \le 6$ and $3x + 2 \ge 5$

Sometimes, we can combine "and" inequalities into a more streamlined notation.

> **WRITING INEQUALITIES INVOLVING "AND" COMPACTLY**
>
> If $a < b$, then we can write
>
> $$a < x \quad \text{and} \quad x < b$$
>
> more compactly as
>
> $$a < x < b$$

For example, we can write

$$-3 < -4x + 1 \quad \text{and} \quad -4x + 1 < 13$$

as

$$-3 < -4x + 1 < 13$$

When compound inequalities come in this form, we solve the inequality by getting the variable by itself in the "middle" with a coefficient of 1.

EXAMPLE 6 Solving a Compound Inequality

Solve $-3 < -4x + 1 < 13$ and graph the solution set.

Solution

Our goal is to get the variable by itself in the "middle" with a coefficient of 1.

$$-3 < -4x + 1 < 13$$

Subtract 1 from all three parts (Addition Property): $-3 - 1 < -4x + 1 - 1 < 13 - 1$

$$-4 < -4x < 12$$

Divide all three parts by -4. Don't forget to reverse the direction of the inequalities.

$$\frac{-4}{-4} > \frac{-4x}{-4} > \frac{12}{-4}$$

$$1 > x > -3$$

If $b > x > a$, then $a < x < b$: $-3 < x < 1$

Figure 41

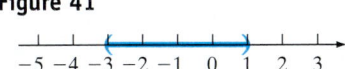

The solution using set-builder notation is $\{x \mid -3 < x < 1\}$. The solution using interval notation is $(-3, 1)$. Figure 41 shows the graph of the solution set. ∎

To visualize the results of Example 6, look at Figure 42, which shows the graph of $f(x) = -3$, $g(x) = -4x + 1$, and $h(x) = 13$. Notice the graph of $g(x) = -4x + 1$ is between the graphs of $f(x) = -3$ and $h(x) = 13$ for $-3 < x < 1$. So, the solution set of $-3 < -4x + 1 < 13$ is $\{x \mid -3 < x < 1\}$.

Figure 42

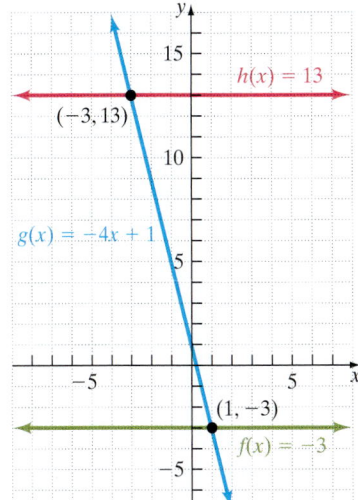

QUICK ✓ *Solve each compound inequality. Express your solution using set-builder notation and interval notation. Graph the solution set.*

14. $-2 < 3x + 1 < 10$ **15.** $0 < 4x - 5 \le 3$

16. $3 \le -2x - 1 \le 11$

③ Solve Compound Inequalities Involving "or"

We now address compound inequalities involving the word "or." The solution to these types of inequalities is the union of the solutions to each inequality.

EXAMPLE 7 **How to Solve a Compound Inequality Involving "or"**

Solve $3x - 5 < -2$ or $4 - 5x \le -16$. Graph the solution set.

Step-by-Step Solution

Step 1: Solve each inequality separately.		$3x - 5 < -2$		$4 - 5x \le -16$
	Add 5 to each side:	$3x < 3$	Subtract 4 from both sides:	$-5x \le -20$
	Divide both sides by 3:	$x < 1$	Divide both sides by -5: Don't forget to reverse the direction of the inequality.	$x \ge 4$

Step 2: Find the union of the solution sets, which will represent the solution set to the compound inequality.

The union of the two solution sets is $x < 1$ or $x \ge 4$. The solution set using set-builder notation is $\{x \mid x < 1 \text{ or } x \ge 4\}$. The solution set using interval notation is $(-\infty, 1) \cup [4, \infty)$. Figure 43 shows the graph of the solution set.

Figure 43

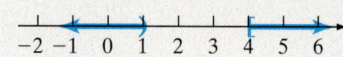

$$\begin{array}{cccccccccc} & & & & & & & & & \\ \hline -2 & -1 & 0 & 1 & 2 & 3 & 4 & 5 & 6 \end{array}$$

Below is a summary of the steps for solving compound inequalities involving "or."

> **Steps for Solving Compound Inequalities Involving "or"**
>
> **Step 1:** Solve each inequality separately.
>
> **Step 2:** Find the UNION of the solution sets of each inequality.

Work Smart

Remember, the union of sets A and B is the set of all elements that are in the set A or in the set B.

Work Smart

A common error to avoid is to write the solution $x < 1$ or $x > 4$ as $1 > x > 4$, which is incorrect. There are no real numbers that are less than 1 *and* greater than 4. Another common error is to "mix" symbols as in $1 < x > 4$. This makes no sense!

QUICK ✓ *Solve each compound inequality. Express your solution using set-builder notation and interval notation. Graph the solution set.*

17. $x + 3 < 1$ or $x - 2 > 3$

18. $3x + 1 \le 7$ or $2x - 3 > 9$

19. $2x - 3 \ge 1$ or $6x - 5 \ge 1$

20. $\dfrac{3}{4}(x + 4) < 6$ or $\dfrac{3}{2}(x + 1) > 15$

EXAMPLE 8 **Solving Compound Inequalities Involving "or"**

Solve $\dfrac{1}{2}x - 1 < 1$ or $\dfrac{2x - 1}{3} \ge -1$. Graph the solution set.

Solution

First, we solve each inequality separately.

	$\dfrac{1}{2}x - 1 < 1$		$\dfrac{2x - 1}{3} \ge -1$
Add 1 to each side:	$\dfrac{1}{2}x < 2$	Multiply both sides by 3:	$2x - 1 \ge -3$
Multiply both sides by 2:	$x < 4$	Add 1 to both sides:	$2x \ge -2$
		Divide both sides by 2:	$x \ge -1$

Find the union of the solution sets of each inequality. If we graph the solution set of each inequality separately, we notice that the union of the two solutions sets is the set of all real numbers. See Figure 44.

Figure 44

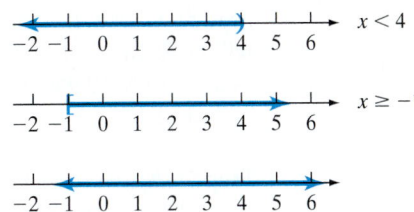

$x < 4$

$x \geq -1$

The solution set using set-builder notation is $\{x \mid x \text{ is any real number}\}$. The solution set using interval notation is $(-\infty, \infty)$.

QUICK ✔ *Solve each compound inequality. Express your solution using set-builder notation and interval notation. Graph the solution set.*

21. $3x - 2 > -5$ or $2x - 5 \leq 1$ **22.** $-5x - 2 \leq 3$ or $7x - 9 > 5$

---

④ Solve Problems Using Compound Inequalities

We now look at an application involving compound inequalities.

EXAMPLE 9   Federal Income Taxes

In 2005, a married couple filing a joint federal tax return whose income places them in the 25% tax bracket will pay federal income taxes between $8180 and $23,317.50, inclusive. The couple must pay federal income taxes equal to $8180 plus 25% of the amount over $59,400. Find the range of taxable income in order for a married couple to be in the 25% tax bracket. (*Source:* Internal Revenue Service)

Solution

Step 1: Identify We want to find the range of the taxable income for a married couple in the 25% tax bracket. This is a direct translation problem involving an inequality.

Work Smart
The word "range" tells us that an inequality is to be solved.

Step 2: Name We let t represent the taxable income.

Step 3: Translate The federal tax bill equals $8180 plus 25% of the taxable income over $59,400. If the couple has taxable income equal to $60,400, their tax bill will be $8180 plus 25% of $1000 ($1000 is the amount over $59,400). In general, if the couple has taxable income t, then their tax bill will be

$$\underbrace{8180}_{\$8180} \quad \underbrace{\text{plus}}_{+} \quad \underbrace{25\%}_{0.25} \quad \cdot \quad \underbrace{\text{of the amount over } \$59{,}400}_{(t - \$59{,}400)}$$

Because the tax bill is between $8180 and $23,317.50, we have

$$8180 \leq 8180 + 0.25(t - 59{,}400) \leq 23{,}317.50 \quad \text{The Model}$$

Step 4: Solve

$$8180 \leq 8180 + 0.25(t - 59{,}400) \leq 23{,}317.50$$

Remove the parentheses by distributing 0.25: $8180 \leq 8180 + 0.25t - 14{,}850 \leq 23{,}317.50$

Combine like terms: $8180 \leq -6670 + 0.25t \leq 23{,}317.50$

Add 6670 to all three parts: $14{,}850 \leq 0.25t \leq 29{,}987.50$

Divide all three parts by 0.25: $59{,}400 \leq t \leq 119{,}950$

Step 5: Check If a married couple has taxable income of $59,400, then their tax bill will be $8180 + 0.25($59,400 − $59,400) = $8180. If a married couple has taxable income of $119,950, then their tax bill will be $8180 + 0.25($119,950 − $59,400) = $23,317.50.

Step 6: Answer the Question A married couple who files a joint tax return with a tax bill between $8180 and $23,317.50 has taxable income between $59,400 and $119,950.

Quick ✓

23. In 2005, an individual filing a federal tax return whose income places them in the 25% tax bracket will pay federal income taxes between $4090 and $14,652.50. The individual must pay federal income taxes equal to $4090 plus 25% of the amount over $29,700. Find the range of taxable income in order for an individual to be in the 25% tax bracket. (*Source:* Internal Revenue Service)

24. AT&T offers a long-distance phone plan that charges $4.95 per month plus $0.07 per minute. During the course of a year, Sophia's long-distance phone bill ranges from $13.00 to $22.80. What was the range of monthly minutes?

8.6 Exercises

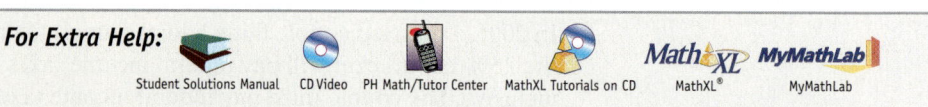

For Extra Help: Student Solutions Manual · CD Video · PH Math/Tutor Center · MathXL Tutorials on CD · MathXL® · MyMathLab

Concepts and Vocabulary

In Problems 1–3, fill in the blanks.

1. The _____ of two sets A and B, denoted $A \cap B$, is the set of all elements that belong to both set A and set B.

2. The word _____ implies intersection. The word _____ implies union.

3. A _____ _____ is formed by joining two inequalities with the word "and" or "or."

In Problems 4–6, answer True or False to each statement.

4. The intersection of two sets can be the empty set.

5. The symbol for the union of two sets is $\cap$.

6. The inequalities $5 > x > -2$ and $-2 < x < 5$ are equivalent.

7. Explain why the inequality $4 < x < 2$ makes no sense.

8. Explain why it is incorrect to write $-3 < x > 2$.

9. Is $x = 3$ a solution of $3x + 1 > 4$ and $2x - 3 < 7$?

10. Is $x = -1$ a solution of $5x - 2 \le 13$ or $2x - 5 > 3$?

Building Skills

In Problems 11–16, use $A = \{4, 5, 6, 7, 8, 9\}$, $B = \{1, 5, 7, 9\}$, and $C = \{2, 3, 4, 6\}$ to find each set.

11. $A \cup B$

12. $A \cup C$

13. $A \cap B$

14. $A \cap C$

15. $B \cap C$

16. $B \cup C$

In Problems 17–20, use the graph of the inequality to find each set.

17. $A = \{x \mid x \le 5\}; B = \{x \mid x > -2\}$.
Find (a) $A \cap B$ and (b) $A \cup B$.

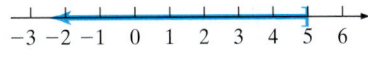

18. $A = \{x \mid x \ge 4\}; B = \{x \mid x < 1\}$.
Find (a) $A \cap B$ and (b) $A \cup B$.

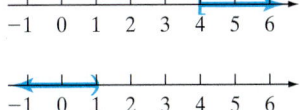

19. $E = \{x \mid x > 3\}; F = \{x \mid x < -1\}$.
Find (a) $E \cap F$ and (b) $E \cup F$.

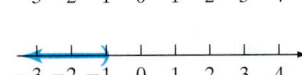

20. $E = \{x \mid x \le 2\}; F = \{x \mid x \ge -2\}$.
Find (a) $E \cap F$ and (b) $E \cup F$.

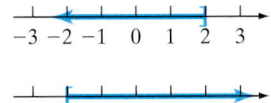

In Problems 21–24, use the graph to solve the compound inequality. Graph the solution set.

21. $-5 \le 2x - 1 \le 3$

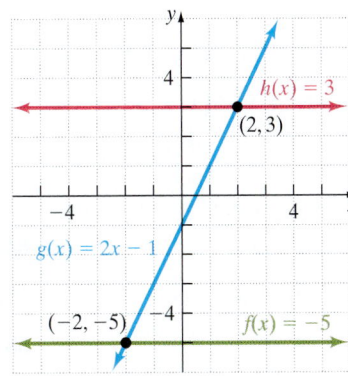

22. $-1 \le \dfrac{1}{2}x + 1 \le 3$

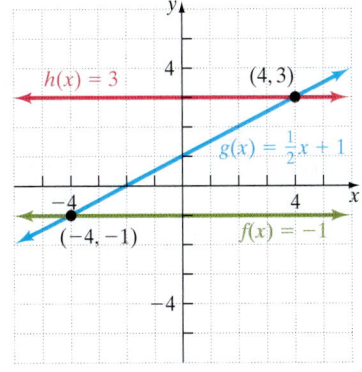

23. $-4 < -\dfrac{5}{3}x + 1 < 6$

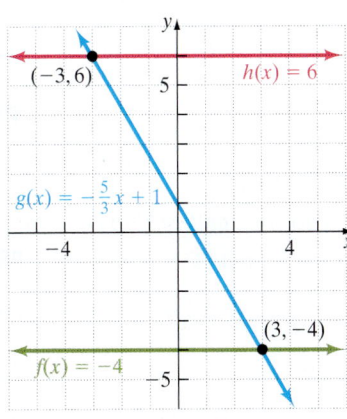

24. $-3 < \dfrac{5}{4}x + 2 < 7$

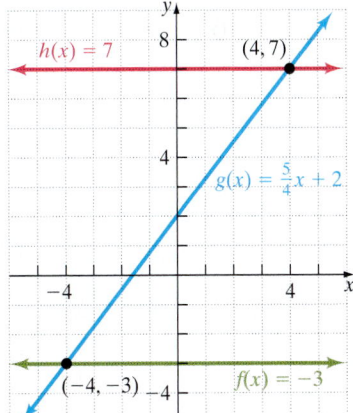

In Problems 25–62, solve each compound inequality. Graph the solution set.

25. $x < 3$ and $x \ge -2$

26. $x \le 5$ and $x > 0$

27. $x < -2$ or $x > 3$

28. $x < 0$ or $x \ge 6$

29. $4x - 4 < 0$ and $-5x + 1 \le -9$

30. $6x - 2 \le 10$ and $10x > -20$

31. $4x - 3 < 5$ and $-5x + 3 > 13$

32. $x - 3 \le 2$ and $6x + 5 \ge -1$

33. $-4x - 1 < 3$ and $-x - 2 > 3$

34. $7x + 2 \ge 9$ and $4x + 3 \le 7$

35. $x - 2 < -4$ or $x + 3 > 8$

36. $x + 3 \le 5$ or $x - 2 \ge 3$

37. $6(x - 2) < 12$ or $4(x + 3) > 12$

38. $4x + 3 > -5$ or $8x - 5 < 3$

39. $-8x + 6x - 2 > 0$ or $5x > 3x + 8$

40. $3x \ge 7x + 8$ or $x < 4x - 9$

41. $-3 \le 5x + 2 < 17$

42. $-10 < 6x + 8 \le -4$

43. $-3 \le 6x + 1 \le 10$

44. $-12 < 7x + 2 \le 6$

45. $2x + 5 \le -1$ or $\dfrac{4}{3}x - 3 > 5$

46. $-\dfrac{4}{5}x - 5 > 3$ or $7x - 3 > 4$

47. $3 \le -5x + 7 < 12$

48. $-6 < -3x + 6 \le 4$

49. $-1 \le \dfrac{1}{2}x - 1 \le 3$

50. $0 < \dfrac{3}{2}x - 3 \le 3$

51. $3 \le -2x - 1 \le 11$

52. $-3 < -4x + 1 < 17$

53. $\dfrac{1}{2}x < 3$ or $\dfrac{3x - 1}{2} > 4$

54. $\dfrac{2}{3}x + 2 \le 4$ or $\dfrac{5x - 3}{3} \ge 4$

55. $\dfrac{2}{3}x + \dfrac{1}{2} < \dfrac{5}{6}$ and $-\dfrac{1}{5}x + 1 < \dfrac{3}{10}$

56. $x - \dfrac{3}{2} \le \dfrac{5}{4}$ and $-\dfrac{2}{3}x - \dfrac{2}{9} < \dfrac{8}{9}$

57. $-2 < \dfrac{3x + 1}{2} \le 8$

58. $-4 \le \dfrac{4x - 3}{3} < 3$

59. $-8 \le -2(x + 1) < 6$

60. $-6 < -3(x - 2) < 15$

61. $3(x - 1) + 5 < 2$ or $-2(x - 3) < 1$

62. $2(x + 1) - 5 \le 4$ or $-(x + 3) \le -2$

Mixed Practice

In Problems 63–76, solve each compound inequality. Graph the solution set.

63. $3a + 5 < 5$ and $-2a + 1 \le 7$

64. $5x - 1 < 9$ and $5x > -20$

65. $5(x + 2) < 20$ or $4(x - 4) > -20$

66. $3(x + 7) < 24$ or $6(x - 4) > -30$

67. $-4 \le 3x + 2 \le 10$

68. $-8 \le 5x - 3 \le 4$

69. $2x + 7 < -13$ or $5x - 3 > 7$

70. $3x - 8 < -14$ or $4x - 5 > 7$

71. $5 < 3x - 1 < 14$

72. $-5 < 2x + 7 \le 5$

73. $\dfrac{x}{3} \le -1$ or $\dfrac{4x - 1}{2} > 7$

74. $\dfrac{x}{2} \le -4$ or $\dfrac{2x - 1}{3} \ge 2$

75. $-3 \le -2(x + 1) < 8$

76. $-15 < -3(x + 2) \le 1$

Applying the Concepts

In Problems 77–82, use the Addition Property and/or Multiplication Properties to find a and b.

77. If $-3 < x < 4$, then $a < x + 4 < b$.

78. If $-2 < x < 3$, then $a < x - 3 < b$.

79. If $4 < x < 10$, then $a < 3x < b$.

80. If $2 < x < 12$, then $a < \dfrac{1}{2}x < b$.

81. If $-2 < x < 6$, then $a < 3x + 5 < b$.

82. If $-4 < x < 3$, then $a < 2x - 7 < b$.

83. Systolic Blood Pressure Blood pressure is measured using two numbers. One of the numbers measures systolic blood pressure. The systolic blood pressure represents the pressure while the heart is beating. In a healthy person, the systolic blood pressure should be greater than 90 and less than 140. If we let the variable x represent a person's systolic blood pressure, express the systolic blood pressure of a healthy person using a compound inequality.

84. Diastolic Blood Pressure Blood pressure is measured using two numbers. One of the numbers measures diastolic blood pressure. The diastolic blood pressure represents the pressure while the heart is resting between beats. In a healthy person, the diastolic blood pressure should be greater than 60 and less than 90. If we let the variable x represent a person's diastolic blood pressure, express the diastolic blood pressure of a healthy person using a compound inequality.

85. Computing Grades Joanna desperately wants to earn a B in her history class. Her current test scores are 74, 86, 77, and 89. Her final exam is worth 2 test scores. In order to earn a B, Joanna's average must lie between 80 and 89, inclusive. What range of scores can Joanna receive on the final and earn a B in the course?

86. Computing Grades Jack needs to earn a C in his sociology class. His current test scores are 67, 72, 81, and 75. His final exam is worth 3 test scores. In order to earn a C, Jack's average must lie between 70 and 79, inclusive. What range of scores can Jack receive on the final exam and earn a C in the course?

87. Federal Tax Withholding The percentage method of withholding for federal income tax (2002) states that a single person whose weekly wages, after subtracting withholding allowances, are over $517, but not over $1105, shall have $69.60 plus 28% of the excess over $517 withheld. Over what range does the amount withheld vary if the weekly wages vary from $600 to $700, inclusive? (*Source:* Internal Revenue Service)

88. Federal Tax Withholding Rework Problem 87 if the weekly wages vary from $800 to $900, inclusive.

89. Gas Bill Pacific Gas and Electric Company charges $65.05 plus $1.15855 per therm for gas usage in excess of 70 therms. In the winter of 2004/2005, one homeowner's bill ranged from a low of $157.73 to a high of $175.11. Over what range did gas usage vary (in therms)? (*Source:* Pacific Gas and Electric Company)

90. Electric Bills In North Carolina, Duke Energy charges $31.52 plus $0.075895 for each additional kilowatt hour (kwh) used during the months from November through June for usage in excess of 350 kwh. Suppose one homeowner's electric bill ranged from a high of $69.47 to a low of $39.11 during this time period. Over what range did the usage vary (in kwh)? (*Source:* Duke Energy)

91. The Arithmetic Mean If $a < b$, show that $a < \dfrac{a+b}{2} < b$. We call $\dfrac{a+b}{2}$ the **arithmetic mean** of a and b.

92. Identifying Triangles A triangle is one such that the length of the longest side is greater than the difference of the other sides and the length of the longest side is less than the sum of the other sides. That is, if a, b, and c are sides such that $a \le b \le c$, then $b - a < c < b + a$. Determine which of the following could be lengths of the sides of a triangle.

(a) $3, 4, 5$ (b) $4, 7, 12$ (c) $3, 3, 5$ (d) $1, 9, 10$

Extending the Concepts

93. Solve $2x + 1 \le 5x + 7 \le x - 5$.

94. Solve $x - 3 \le 3x + 1 \le x + 11$.

95. Solve $4x + 1 > 2(2x + 1)$. Provide an explanation that generalizes the result.

96. Solve $4x + 1 > 2(2x - 1)$. Provide an explanation that generalizes the result.

97. Consider the following analysis assuming that $x < 2$.

$$5 > 2$$
$$5(x - 2) > 2(x - 2)$$
$$5x - 10 > 2x - 4$$
$$3x > 6$$
$$x > 2$$

How can it be that the final line in the analysis states that $x > 2$, when the original assumption stated that $x < 2$?

8.7 Absolute Value Equations and Inequalities

OBJECTIVES

1. Solve Absolute Value Equations
2. Solve Absolute Value Inequalities Involving $<$ or $\leq$
3. Solve Absolute Value Inequalities Involving $>$ or $\geq$
4. Solve Applied Problems Involving Absolute Value Inequalities

Preparing for Absolute Value Equations and Inequalities

Before getting started, take the following readiness quiz. If you get a problem wrong, go back to the section cited and review the material.

In Problems 1–4, evaluate each expression. [Section 1.2, pp. 14–15]

1. $|3|$ 2. $|-4|$ 3. $|-1.6|$ 4. $|0|$
5. Express the distance between the origin, 0, and 5 as an absolute value. [Section 1.2, pp. 14–15]
6. Express the distance between the origin, 0, and -8 as an absolute value. [Section 1.2, pp. 14–15]
7. Solve: $4x + 5 = -9$ [Section 2.2 pp. 84–86]
8. Solve: $-2x + 1 > 5$ [Section 2.8, pp. 153–158]

Figure 45

$|-5| = 5$

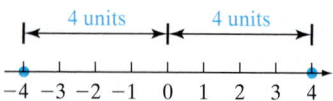

Recall from Section 1.2 that we defined the absolute value of a number as the distance between the number and the origin on the real number line. For example, $|-5| = 5$ because the distance on the real number from 0 to -5 is 5 units. See Figure 45 for a geometric interpretation of absolute value.

This interpretation of absolute value forms the basis for solving absolute value equations.

1 Solve Absolute Value Equations

We begin with an example.

EXAMPLE 1 Solving an Absolute Value Equation

Solve the equation $|x| = 4$.

Figure 46

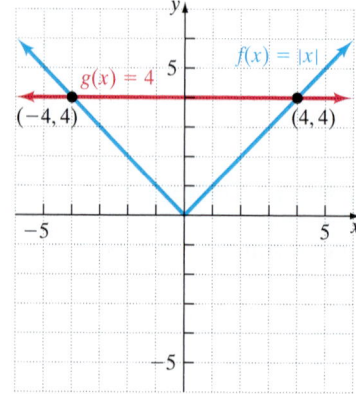

Solution

The equation $|x| = 4$ is asking, "Tell me all real numbers x such that the distance from the origin to x on the real number line is 4 units." There are two such numbers as indicated in Figure 46, -4 and 4. The solution set is $\{-4, 4\}$.

Figure 47

We can visualize the solution to $|x| = 4$ by graphing $f(x) = |x|$ and $g(x) = 4$ on the same xy-plane. Figure 47 shows the graphs of $f(x) = |x|$ and $g(x) = 4$. The x-coordinate of the points of intersection are -4 and 4, which represent the solutions to the equation $f(x) = g(x)$.

QUICK ✓ *Solve the equation.*

1. $|x| = 7$ 2. $|z| = 1$

The results of Example 1 lead us to the following result.

Preparing for...Answers 1. 3 2. 4
3. 1.6 4. 0 5. $|5|$ 6. $|-8|$ 7. $\left\{-\dfrac{7}{2}\right\}$
8. $\{x | x < -2\}; (-\infty, -2)$

EQUATIONS INVOLVING ABSOLUTE VALUE

If a is a positive real number and if u is any algebraic expression, then

$$|u| = a \quad \text{is equivalent to} \quad u = a \ \text{ or } \ u = -a$$

Note: If $a = 0$, the equation $|u| = 0$ is equivalent to $u = 0$. If $a < 0$, the equation $|u| = a$ has no real solution.

In the equation $|u| = a$, we require that a be a positive number. If a is negative, the equation has no real solution. To see why, consider the equation $|x| = -2$. Figure 48 shows the graph of $f(x) = |x|$ and $g(x) = -2$. Notice that the graphs do not intersect, which implies that the equation $|x| = -2$ has no real solution. The solution set is the empty set, $\varnothing$ or { }.

Figure 48

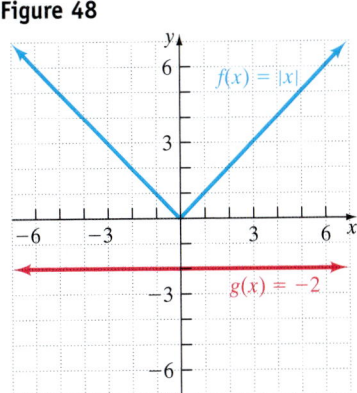

Example 2 illustrates how to solve an equation involving absolute value.

EXAMPLE 2   **How to Solve an Equation Involving Absolute Value**

Solve the equation $|2x - 1| + 3 = 12$.

Step-by-Step Solution

Step 1: Isolate the expression containing the absolute value.

$$|2x - 1| + 3 = 12$$
Subtract 3 from both sides: $|2x - 1| = 9$

Step 2: Rewrite the absolute value equation as two equations: $u = a$ and $u = -a$, where u is the algebraic expression in the absolute value symbol. Here $u = 2x - 1$ and $a = 9$.

$$2x - 1 = 9 \qquad \text{or} \qquad 2x - 1 = -9$$

Step 3: Solve each equation.

$2x - 1 = 9$

Add 1 to each side: $2x = 10$

Divide both sides by 2: $x = 5$

$2x - 1 = -9$

Add 1 to each side: $2x = -8$

Divide both sides by 2: $x = -4$

Step 4: Check Verify each solution.

Let $x = 5$:
$$|2x - 1| + 3 = 12$$
$$|2(5) - 1| + 3 \stackrel{?}{=} 12$$
$$|10 - 1| + 3 \stackrel{?}{=} 12$$
$$9 + 3 \stackrel{?}{=} 12$$
$$12 = 12 \quad \text{True}$$

Let $x = -4$:
$$|2x - 1| + 3 = 12$$
$$|2(-4) - 1| + 3 \stackrel{?}{=} 12$$
$$|-8 - 1| + 3 \stackrel{?}{=} 12$$
$$9 + 3 \stackrel{?}{=} 12$$
$$12 = 12 \quad \text{True}$$

Both solutions check, so the solution set is $\{-4, 5\}$.

The following steps can be used to solve an absolute value equation.

Steps for Solving Absolute Value Equations with One Absolute Value

Step 1: Isolate the expression containing the absolute value.

Step 2: Rewrite the absolute value equation as two equations: $u = a$ and $u = -a$, where u is the algebraic expression in the absolute value symbol.

Step 3: Solve each equation.

Step 4: Verify your solution.

QUICK ✓ *Solve each equation.*

3. $|2x - 3| = 7$ **4.** $|3x - 2| + 3 = 10$ **5.** $|-5x + 2| - 2 = 5$ **6.** $3|x + 2| - 4 = 5$

EXAMPLE 3 **Solving an Equation Involving Absolute Value with No Solution**

Solve the equation $|-x + 5| + 7 = 5$.

Solution

$$|-x + 5| + 7 = 5$$

Subtract 7 from both sides: $\qquad |-x + 5| = -2$

Work Smart

The equation $|u| = a$, where a is a negative real number has no real solution. See Figure 48 to see why.

Since the absolute value of any real number is always nonnegative (greater than or equal to zero), the equation has no real solution. The solution set is $\{\ \}$ or $\varnothing$.

QUICK ✓ *Solve each equation.*

7. $|5x + 3| = -2$ **8.** $|2x + 5| + 7 = 3$ **9.** $|x + 1| + 3 = 3$

What if an absolute value equation has two absolute values as in $|3x - 1| = |x + 5|$? How do we handle this situation? Well, there are four possibilities for the algebraic expressions in the absolute value symbol:

1. both algebraic expressions are positive,
2. both are negative,
3. the left is positive, and the right is negative, or
4. the left is negative and the right is positive.

To see how the solution works, we need to consider the algebraic definition for absolute value. This definition states that $|a| = a$, if $a \geq 0$ and $|a| = -a$ if $a < 0$.

So, if $3x - 1 \geq 0$, then $|3x - 1| = 3x - 1$. However, if $3x - 1 < 0$, then $|3x - 1| = -(3x - 1)$. This leads us to a method for solving absolute value equations with two absolute values.

Case 1: Both Algebraic Expressions Are Positive	Case 2: Both Algebraic Expressions Are Negative	Case 3: The Algebraic Expression on the Right Is Positive, and the Left Is Negative	Case 4: The Algebraic Expression on the Left Is Negative, and the Right Is Positive
$|3x - 1| = |x + 5|$	$|3x - 1| = |x + 5|$	$|3x - 1| = |x + 5|$	$|3x - 1| = |x + 5|$
$3x - 1 = x + 5$	$-(3x - 1) = -(x + 5)$	$3x - 1 = -(x + 5)$	$-(3x - 1) = x + 5$
	$3x - 1 = x + 5$		

Whether both algebraic expressions are positive, or both negative, we end up with equivalent equations. Also, if one side is positive and the other is negative, we end up with equivalent equations. So, the four possibilities reduce to two possibilities.

EQUATIONS INVOLVING TWO ABSOLUTE VALUES

If u and v are any algebraic expression, then

$$|u| = |v| \qquad \text{is equivalent to} \qquad u = v \quad \text{or} \quad u = -v$$

EXAMPLE 4 **Solving an Absolute Value Equation Involving Two Absolute Values**

Solve the equation $|2x - 3| = |x + 6|$.

Solution

The equation is in the form $|u| = |v|$, where $u = 2x - 3$ and $v = x + 6$. We rewrite the equation as two equations that do not involve absolute value:

$$2x - 3 = x + 6 \quad \text{or} \quad 2x - 3 = -(x + 6)$$

Now, we solve each equation.

	$2x - 3 = x + 6$			$2x - 3 = -(x + 6)$
			Distribute the -1:	$2x - 3 = -x - 6$
Add 3 to each side:	$2x = x + 9$		Add 3 to both sides:	$2x = -x - 3$
			Add x to both sides:	$3x = -3$
Subtract x from both sides:	$x = 9$		Divide both sides by 3:	$x = -1$

Check

$x = 9$:
$$|2(9) - 3| \stackrel{?}{=} |9 + 6|$$
$$|18 - 3| \stackrel{?}{=} |15|$$
$$|15| \stackrel{?}{=} 15$$
$$15 = 15 \quad \text{True}$$

$x = -1$:
$$|2(-1) - 3| \stackrel{?}{=} |-1 + 6|$$
$$|-2 - 3| \stackrel{?}{=} |5|$$
$$|-5| \stackrel{?}{=} 5$$
$$5 = 5 \quad \text{True}$$

Both solutions check, so the solution set is $\{-1, 9\}$.

QUICK ✓ *Solve each equation.*

10. $|x - 3| = |2x + 5|$ **11.** $|8z + 11| = |6z + 17|$
12. $|3 - 2y| = |4y + 3|$ **13.** $|2x - 3| = |5 - 2x|$

② Solve Absolute Value Inequalities Involving < or ≤

The method for solving absolute value equations relies on the geometric interpretation of absolute value. Namely, the absolute value of a real number x is the distance from the origin to x on the real number line. We use this same interpretation to solve absolute value inequalities.

EXAMPLE 5 **Solving an Absolute Value Inequality**

Solve the inequality $|x| < 4$. Graph the solution set.

Solution

The inequality $|x| < 4$ is asking, "Tell me all real numbers x such that the distance from the origin to x on the real number line is less than 4." Figure 49 illustrates the situation. We can see from the figure that any number between -4 and 4 satisfies the inequality. The solution set consists of all real numbers x for which $-4 < x < 4$ or, using interval notation, $(-4, 4)$.

Figure 49

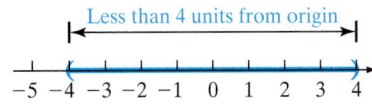

Less than 4 units from origin

Figure 50

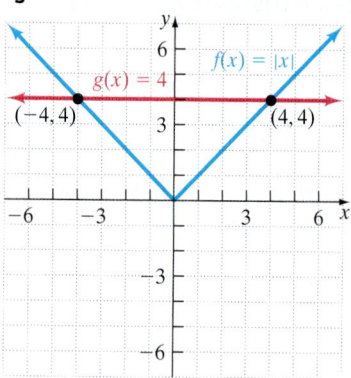

To visualize these results, we graph $f(x) = |x|$ and $g(x) = 4$. See Figure 50. Because we are solving $f(x) < g(x)$, we look for all x-coordinates such that the graph of $f(x)$ is below the graph of $g(x)$. From the graph, we can see that the graph of $f(x) = |x|$ is below the graph of $g(x) = 4$ for all x between -4 and 4. So the solution set consists of all x for which $-4 < x < 4$, or all x in the interval $(-4, 4)$.

The results of Example 5 lead to the following results.

INEQUALITIES OF THE FORM < OR ≤ INVOLVING ABSOLUTE VALUE

If a is a positive real number and if u is an algebraic expression, then

$$|u| < a \quad \text{is equivalent to} \quad -a < u < a$$
$$|u| \leq a \quad \text{is equivalent to} \quad -a \leq u \leq a$$

Note: If $a = 0$, $|u| < 0$ has no real solution, $|u| \leq 0$ is equivalent to $u = 0$. If $a < 0$, the inequality has no real solution.

QUICK ✓　*Solve each inequality. Graph the solution set.*

14. $|x| \leq 5$

15. $|x| < \dfrac{3}{2}$

EXAMPLE 6　**How to Solve an Absolute Value Inequality Involving ≤**

Solve the inequality $|2x + 3| \leq 5$. Graph the solution set.

Step-by-Step Solution

Step 1: The inequality is in the form $|u| \leq a$, where $u = 2x + 3$ and $a = 5$. We rewrite the inequality as a compound inequality that does not involve absolute value.

Use the fact that $|u| \leq a$ means $-a \leq u \leq a$:

$$|2x + 3| \leq 5$$
$$-5 \leq 2x + 3 \leq 5$$

Step 2: Solve the resulting compound inequality.

Subtract 3 from all three parts:　$-5 - 3 \leq 2x + 3 - 3 \leq 5 - 3$

$$-8 \leq 2x \leq 2$$

Divide all three parts of the inequality by 2:　$\dfrac{-8}{2} \leq \dfrac{2x}{2} \leq \dfrac{2}{2}$

$$-4 \leq x \leq 1$$

The solution using set-builder notation is $\{x \,|\, -4 \leq x \leq 1\}$. The solution using interval notation is $[-4, 1]$. Figure 51 shows the graph of the solution set.

Figure 51

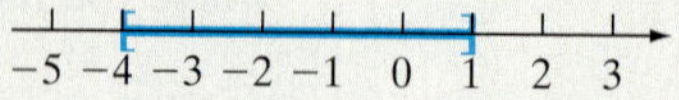

$$\overset{-5\ -4\ -3\ -2\ -1\ \ 0\ \ 1\ \ 2\ \ 3}{\longleftarrow\!|\!-\!\rule{0.6em}{0.8pt}\!-\!-\!-\!-\!|\!-\!-\!-\!|\!\longrightarrow}$$

Work Smart

Although not a complete check of the solution of Example 6, we can choose a number in the interval and see if it works. Let's try $x = -3$.

$$|2(-3) + 3| \overset{?}{\leq} 5$$
$$|-6 + 3| \overset{?}{\leq} 5$$
$$|-3| \overset{?}{\leq} 5$$
$$3 \leq 5$$

QUICK ✓　*Solve each inequality. Graph the solution set.*

16. $|x + 3| < 5$

17. $|2x - 3| \leq 7$

18. $|7x + 2| < -3$

EXAMPLE 7 Solving an Absolute Value Inequality Involving <

Solve the inequality $|-3x + 2| + 4 < 14$. Graph the solution set.

Solution

First, we want to isolate the absolute value by subtracting 4 from both sides of the inequality.

$$|-3x + 2| + 4 < 14$$

Subtract 4 from both sides:
$$|-3x + 2| < 10$$

Use $|u| < a$ means $-a < u < a$:
$$-10 < -3x + 2 < 10$$

Subtract 2 from all three parts:
$$-10 - 2 < -3x + 2 - 2 < 10 - 2$$
$$-12 < -3x < 8$$

Divide all three parts by -3. Be sure to reverse the direction of the inequalities.
$$\frac{-12}{-3} > \frac{-3x}{-3} > \frac{8}{-3}$$
$$4 > x > -\frac{8}{3}$$

Use $b > x > a$ is equivalent to $a < x < b$:
$$-\frac{8}{3} < x < 4$$

Figure 52

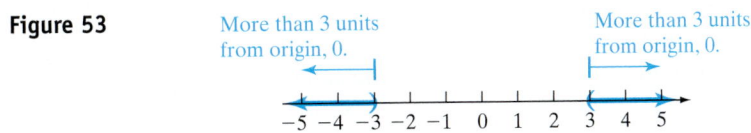

$$\xleftarrow{\;\;\;}\!\!\!\underset{-3\;-2\;-1\;\;0\;\;1\;\;2\;\;3\;\;4\;\;5}{|\;\;|\;\;|\;\;|\;\;|\;\;|\;\;|\;\;|\;\;|}\!\!\!\xrightarrow{\;\;\;}$$

The solution using set-builder notation is $\left\{ x \,\middle|\, -\dfrac{8}{3} < x < 4 \right\}$. The solution using interval notation is $\left(-\dfrac{8}{3}, 4 \right)$. Figure 52 shows the graph of the solution set. ▬

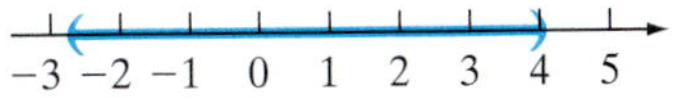

QUICK ✔ *Solve each inequality. Graph the solution set.*

19. $|x| + 4 < 6$ **20.** $|x - 3| + 4 \le 8$ **21.** $3|2x + 1| \le 9$ **22.** $|-3x + 1| - 5 < 3$

---

(3) ## Solve Absolute Value Inequalities Involving > or ≥

Now let's look at absolute value inequalities involving $>$ or $\ge$.

EXAMPLE 8 Solving an Absolute Value Inequality Involving >

Solve the inequality $|x| > 3$. Graph the solution set.

Solution

The inequality $|x| > 3$ is asking, "Tell me all real numbers x such that the distance from the origin to x on the real number line is more than 3 units." Figure 53 illustrates the situation.

Figure 53

More than 3 units from origin, 0.   More than 3 units from origin, 0.

$$\xleftarrow{\;\;\;}\!\!\!\underset{-5\;-4\;-3\;-2\;-1\;\;0\;\;1\;\;2\;\;3\;\;4\;\;5}{|\;\;|\;\;|\;\;|\;\;|\;\;|\;\;|\;\;|\;\;|\;\;|\;\;|}\!\!\!\xrightarrow{\;\;\;}$$

We can see from the figure that any number less than -3 or greater than 3 satisfies the inequality. The solution set consists of all real numbers x for which $x < -3$ or $x > 3$ or, using interval notation, $(-\infty, -3) \cup (3, \infty)$. ▬

To visualize these results, we graph $f(x) = |x|$ and $g(x) = 3$. See Figure 54. Because we are solving $f(x) > g(x)$, we look for all x-coordinates such that the graph of $f(x)$ is above the graph of $g(x)$. From the graph, we can see that the graph of $f(x) = |x|$ is above the graph of $g(x) = 3$ for all x less than -3 or all x greater than 3. So the solution set consists of all x for which $x < -3$ or $x > 3$, that is, all x in the interval $(-\infty, -3) \cup (3, \infty)$.

Figure 54

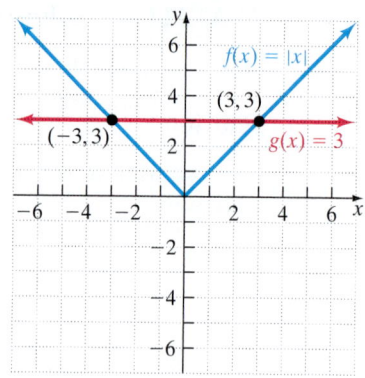

Based upon Example 8, we are led to the following results.

INEQUALITIES OF THE FORM > OR ≥ INVOLVING ABSOLUTE VALUE

If a is a positive real number and u is an algebraic expression, then

$$|u| > a \quad \text{is equivalent to} \quad u < -a \quad \text{or} \quad u > a$$
$$|u| \geq a \quad \text{is equivalent to} \quad u \leq -a \quad \text{or} \quad u \geq a$$

QUICK ✓ *Solve each inequality. Graph the solution set.*

23. $|x| \geq 6$　　　　**24.** $|x| > \dfrac{5}{2}$

EXAMPLE 9 **How to Solve an Inequality Involving >**

Solve the inequality $|2x - 5| > 3$. Graph the solution set.

Step-by-Step Solution

Step 1: The inequality is in the form $|u| > a$, where $u = 2x - 5$ and $a = 3$. We rewrite the inequality as a compound inequality that does not involve absolute value:	$|2x - 5| > 3$ $2x - 5 < -3 \quad \text{or} \quad 2x - 5 > 3$

Step 2: Solve each inequality separately.

	$2x - 5 < -3$		$2x - 5 > 3$
Add 5 to both sides:	$2x < 2$	Add 5 to both sides:	$2x > 8$
Divide both sides by 2:	$x < 1$	Divide both sides by 2:	$x > 4$

Step 3: Find the union of the solution sets of each inequality.

The solution set is $\{x \mid x < 1 \text{ or } x > 4\}$, or using interval notation, $(-\infty, 1) \cup (4, \infty)$. See Figure 55 for the graph of the solution set.

Figure 55

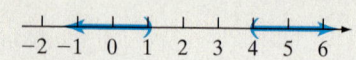

$$-2 \;-1 \;\; 0 \;\; 1 \;\; 2 \;\; 3 \;\; 4 \;\; 5 \;\; 6$$

Work Smart

$$|u| > a$$

CANNOT be written as

$$-a > u > a$$

QUICK ✓ *Solve each inequality. Graph the solution set.*

25. $|x + 3| > 4$　　　　**26.** $|4x - 3| \geq 5$　　　　**27.** $|-3x + 2| > 7$

28. $|2x + 5| - 2 > -2$　　**29.** $|6x - 5| \geq 0$　　　　**30.** $|2x + 1| > -3$

Below, we summarize the techniques for solving absolute value equations and inequalities.

SUMMARY: Solving Absolute Value Equations and Inequalities

Absolute Value Form	Equation/Inequality Form	Example
$\lvert u \rvert = a$	$u = -a$ or $u = a$	$\lvert 2x - 3 \rvert = 4$ $2x - 3 = -4$ or $2x - 3 = 4$ $2x = -1$ or $2x = 7$ $x = -\dfrac{1}{2}$ or $x = \dfrac{7}{2}$ The solution set is $\left\{ -\dfrac{1}{2}, \dfrac{7}{2} \right\}$.
$\lvert u \rvert = \lvert v \rvert$	$u = v$ or $u = -v$	$\lvert x + 6 \rvert = \lvert 3x - 4 \rvert$ $x + 6 = 3x - 4$ or $x + 6 = -(3x - 4)$ $x = 3x - 10$ or $x + 6 = -3x + 4$ $-2x = -10$ or $4x = -2$ $x = 5$ or $x = -\dfrac{1}{2}$ The solution set is $\left\{ -\dfrac{1}{2}, 5 \right\}$.
$\lvert u \rvert < a$ $\lvert u \rvert \leq a$	$-a < u < a$ $-a \leq u \leq a$	$2\lvert x - 1 \rvert + 3 \leq 11$ $2\lvert x - 1 \rvert \leq 8$ $\lvert x - 1 \rvert \leq 4$ $-4 \leq x - 1 \leq 4$ $-3 \leq x \leq 5$ The solution set is $\{x \mid -3 \leq x \leq 5\}$ or $[-3, 5]$.
$\lvert u \rvert > a$ $\lvert u \rvert \geq a$	$u < -a$ or $u > a$ $u \leq -a$ or $u \geq a$	$\lvert 3x + 2 \rvert > 8$ $3x + 2 < -8$ or $3x + 2 > 8$ $3x < -10$ or $3x > 6$ $x < -\dfrac{10}{3}$ or $x > 2$ The solution set is $\left\{ x \mid x < -\dfrac{10}{3} \text{ or } x > 2 \right\}$ or $\left(-\infty, -\dfrac{10}{3} \right) \cup (2, \infty)$.

(4) Solve Applied Problems Involving Absolute Value Inequalities

You may frequently read phrases such as "margin of error" and "tolerance" in the newspaper or on the Internet. For example, according to a Gallup poll conducted February 6, 2003, 57% of Americans felt that the United States rates favorably internationally. The poll had a margin of error of 3%. The 57% reported is an estimate of the true percentage of U.S. residents who believe that the United States rates favorably internationally. If we let p represent the true percentage of U.S. residents that believe that the United States rates favorably internationally, then we can represent the poll's margin of error mathematically as

$$\lvert p - 57 \rvert \leq 3$$

As another example, the tolerance of a belt whose width is 6 inches is $\dfrac{1}{16}$ inch. If x represents the actual width of the belt, then we can represent the acceptable belt widths as

$$|x - 6| \le \dfrac{1}{16}$$

EXAMPLE 10 **Analyzing the Margin of Error in a Poll**

The inequality

$$|p - 57| \le 3$$

represents the percentage of Americans who feel that the United States rates favorably internationally. Solve the inequality and interpret the results.

Solution

$$|p - 57| \le 3$$

Use $|u| \le a$ means $-a \le u \le a$: $\quad -3 \le p - 57 \le 3$

Add 57 to all three parts of the inequality: $\quad 54 \le p \le 60$

The percentage of Americans who feel that the United States rates favorably internationally is between 54% and 60%, inclusive.

QUICK ✓

31. The inequality $|x - 4| \le \dfrac{1}{32}$ represents the acceptable belt widths x (in inches) for a belt that is manufactured for a pulley system. Determine the acceptable belt widths.

32. In a recent poll conducted by ABC News, 9% of respondents stated that they have been shot at. The margin of error in the poll was 1.7%. If we let p represent the true percentage of people who have been shot at, we can represent the margin of error as

$$|p - 9| \le 1.7$$

Solve the inequality and interpret the results.

8.7 Exercises

For Extra Help:

Student Solutions Manual   CD Video   PH Math/Tutor Center   MathXL Tutorials on CD   MathXL®   MyMathLab

Concepts and Vocabulary

In Problems 1–3, fill in the blanks.

1. $|u| = a$ is equivalent to $u = $ _____ or $u = $ _____.

2. $|u| < a$ is equivalent to _____.

3. $|u| < a$ will have no solution if a _____ 0.

In Problems 4–6, answer True or False to each statement.

4. $|x| = -4$ has no real solution.

5. $|x| > -2$ has no real solution.

6. $|u| > a$ is equivalent to $-a < u < a$.

In Problems 7–10, use the graphs of the functions given to solve each problem.

7. $f(x) = |x|, g(x) = 5$
 (a) $f(x) = g(x)$
 (b) $f(x) \leq g(x)$
 (c) $f(x) > g(x)$

8. $f(x) = |x|, g(x) = 6$
 (a) $f(x) = g(x)$
 (b) $f(x) \leq g(x)$
 (c) $f(x) > g(x)$

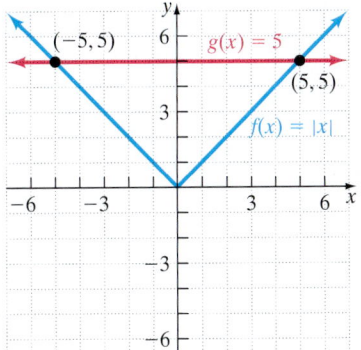
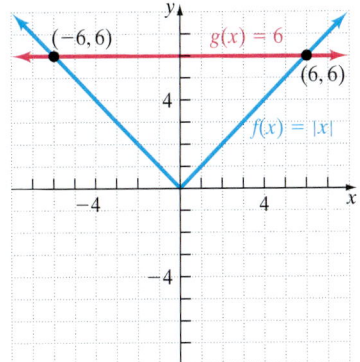

9. $f(x) = |x + 2|, g(x) = 3$
 (a) $f(x) = g(x)$
 (b) $f(x) < g(x)$
 (c) $f(x) \geq g(x)$

10. $f(x) = |2x|, g(x) = 10$
 (a) $f(x) = g(x)$
 (b) $f(x) < g(x)$
 (c) $f(x) \geq g(x)$

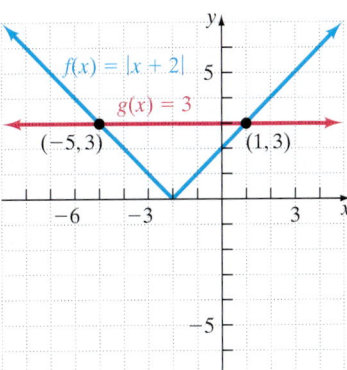
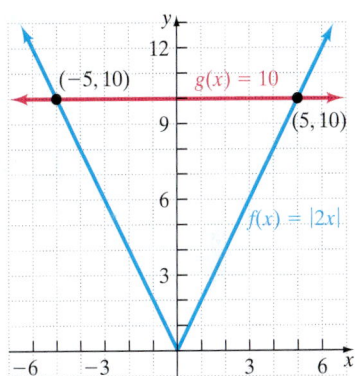

Building Skills

In Problems 11–32, solve each absolute value equation.

11. $|x| = 10$

12. $|z| = 9$

13. $|y - 3| = 4$

14. $|x + 3| = 5$

15. $|-3x + 5| = 8$

16. $|-4y + 3| = 9$

17. $|y| - 7 = -2$

18. $|x| + 3 = 5$

19. $|2x + 3| - 5 = 3$

20. $|3y + 1| - 5 = -3$

21. $-2|x - 3| + 10 = -4$

22. $3|y - 4| + 4 = 16$

23. $|-3x| - 5 = -5$

24. $|-2x| + 9 = 9$

25. $\left|\dfrac{3x - 1}{4}\right| = 2$

26. $\left|\dfrac{2x - 3}{5}\right| = 2$

27. $|3x + 2| = |2x - 5|$

28. $|5y - 2| = |4y + 7|$

29. $|8 - 3x| = |2x - 7|$

30. $|5x + 3| = |12 - 4x|$

31. $|4y - 7| = |9 - 4y|$

32. $|5x - 1| = |9 - 5x|$

In Problems 33–58, solve each absolute value inequality. Graph the solution set on a real number line.

33. $|x| < 9$

34. $|x| \leq \dfrac{5}{4}$

35. $|x - 4| \leq 7$

36. $|y + 4| < 6$

37. $|3x + 1| < 8$

38. $|4x - 3| \leq 9$

39. $|6x + 5| < -1$

40. $|4x + 3| \leq 0$

41. $|y - 5| > 2$

42. $|x + 4| \geq 7$

43. $|-4x - 3| \geq 5$

44. $|-5y + 3| > 7$

45. $2|y| + 3 > 1$

46. $3|z| + 8 > 2$

47. $2|x - 3| + 3 < 9$

48. $3|y + 2| - 2 > 7$

49. $|-5x - 3| > -3$

50. $|-9x + 2| \geq -1$

51. $|2 - 5x| + 3 < 10$

52. $|-3x + 2| - 7 \leq -2$

53. $|-2x + 1| > 1$

54. $|8x + 3| \geq 3$

55. $|1 - 2x| \geq |-5|$

56. $|3 - 5x| < |-7|$

57. $|(2x - 3) - 1| < 0.01$

58. $|(3x + 2) - 8| < 0.01$

Mixed Practice

In Problems 59–78, solve each absolute value equation or inequality. For absolute value inequalities, graph the solution set on a real number line.

59. $|x| > 5$

60. $|x| \geq \dfrac{8}{3}$

61. $|2x + 5| = 3$

62. $|4x + 3| = 1$

63. $7|x| = 35$

64. $8|y| = 32$

65. $|5x + 2| \leq 8$

66. $|7y - 3| < 11$

67. $|-2x + 3| = -4$

68. $|3x - 4| = -9$

69. $|3x + 2| \geq 5$

70. $|5y + 3| > 2$

71. $|3x - 2| + 7 > 9$

72. $|4y + 3| - 8 \geq -3$

73. $|5x + 3| = |3x + 5|$

74. $|3z - 2| = |z + 6|$

75. $|4x + 7| + 6 < 5$

76. $|4x + 1| > 0$

77. $\left|\dfrac{x - 2}{4}\right| = \left|\dfrac{2x + 1}{6}\right|$

78. $\left|\dfrac{1}{2}x - 3\right| = \left|\dfrac{2}{3}x + 1\right|$

Applying the Concepts

79. Express the fact that x differs from 5 by less than 3 as an inequality involving absolute value. Solve for x.

80. Express the fact that x differs from -4 by less than 2 as an inequality involving absolute value. Solve for x.

81. Express the fact that twice x differs from -6 by more than 3 as an inequality involving absolute value. Solve for x.

82. Express the fact that twice x differs from 7 by more than 3 as an inequality involving absolute value. Solve for x.

83. Tolerance A certain rod in an internal combustion engine is supposed to be 5.7 inches. The tolerance on the rod is 0.0005 inches. If x represents the length of a rod, the acceptable lengths of a rod can be expressed as $|x - 5.7| \le 0.0005$. Determine the acceptable lengths of the rod. (*Source:* WiseCo Piston)

84. Tolerance A certain rod in an internal combustion engine is supposed to be 6.125 inches. The tolerance on the rod is 0.0005 inches. If x represents the length of a rod, the acceptable lengths of a rod can be expressed as $|x - 6.125| \le 0.0005$. Determine the acceptable lengths of the rod.

85. IQ Scores According to the Stanford-Binet IQ test, a normal IQ score is 100. It can be shown that anyone with an IQ x that satisfies the inequality $\left|\dfrac{x - 100}{15}\right| > 1.96$ has an unusual IQ score. Determine the IQ scores that would be considered unusual.

86. Gestation Period The length of human pregnancy is about 266 days. It can be shown that a mother whose gestation period x satisfies the inequality $\left|\dfrac{x - 266}{16}\right| > 1.96$ has an unusual length of pregnancy. Determine the length of pregnancy that would be considered unusual.

Extending the Concepts

87. Explain why $|2x - 3| + 1 = 0$ has no solution.

88. Explain why the solution set of $|5x - 3| > -5$ is the set of all real numbers.

89. Explain why $|4x + 3| + 3 < 0$ has the empty set as the solution set.

90. Solve $|x - 5| = |5 - x|$. Explain why the result is reasonable. What do we call this type of equation?

In Problems 91–98, solve each equation.

91. $|x| - x = 5$

92. $|y| + y = 3$

93. $z + |-z| = 4$

94. $y - |-y| = 12$

95. $|4x + 1| = x - 2$

96. $|2x + 1| = x - 3$

97. $|x + 5| = -(x + 5)$

98. $|y - 4| = y - 4$

CHAPTER 8 ACTIVITY: SHIFTING DISCOVERY

Focus: Using graphing skills, discover the possible "rules" for graphing functions.

Time: 30–35 minutes

Group size: 4

Materials Needed: Graph paper (2–3 pieces)

Each member of the group needs to

1. Draw a coordinate plane and label the *x*-axis and *y*-axis.

2. By plotting points, graph the primary function: $f(x) = x^2$.

3. On the same coordinate plane, each group member graphs *one* of the following functions by plotting points. Be sure your graphs of the primary function and one of the functions (a)–(d) are on the same coordinate plane.

   (a) $f(x) = x^2 + 3$ **(b)** $f(x) = (x - 3)^2$

   (c) $f(x) = x^2 - 3$ **(d)** $f(x) = (x + 3)^2$

As a group, discuss the following:

4. What shape are the graphs?

5. Each member of the group, share the difference between the graph of your primary function and the other function you chose.

6. As a group, can you develop a possible rule for these differences?

With this possible rule in mind, each member of the group needs to

7. Draw a coordinate plane and label *x*-axis, *y*-axis, and -10 to 10 on each axis.

8. Graph the primary function by plotting points: $f(x) = |x|$.

9. On the same coordinate plane, each group member graphs *one* of the following functions by plotting points. Be sure your graphs of the primary function and one of the functions (a)–(d) are on the same coordinate plane.

   (a) $f(x) = |x| + 4$ **(b)** $f(x) = |x - 4|$

   (c) $f(x) = |x| - 4$ **(d)** $f(x) = |x + 4|$

10. Did your possible rules developed in Problem 6 hold true? Discuss.

CHAPTER 8 REVIEW

Section 8.1	Graphs of Equations

KEY CONCEPTS	KEY TERMS	

• **Graph of an Equation in Two Variables**
 The set of all ordered pairs (x, y) in the *xy*-plane that satisfy the equation
 	
 • **Intercepts**
 The points, if any, where a graph crosses or touches the coordinate axes
	x-axis
	 y-axis
	 Origin
	 Rectangular or
	     Cartesian coordinate
	     system
	 xy-plane
	 Coordinate axes
	 Ordered pair
	 Coordinates
	 x-coordinate
	 y-coordinate
	 Abscissa
		Ordinate
		 Quadrants
		 Equation in two variables
		 Sides
		 Satisfy
		 Graph of an equation in
		     two variables
		 Point-plotting method
		 Complete graph
		 Intercept
		 x-intercept
		 y-intercept

YOU SHOULD BE ABLE TO . . .	EXAMPLE	REVIEW EXERCISES
① Graph an equation using the point-plotting method (p. 568)	Examples 1 through 3	5–14
② Identify the intercepts from the graph of an equation (p. 571)	Example 4	15–16
③ Interpret graphs (p. 571)	Example 5	17–18

In Problems 1 and 2, plot each point in the same xy-plane. Tell in which quadrant or on what coordinate axis each point lies.

1. $A(2, -4)$
  $B(-1, -3)$
  $C(0, 4)$
  $D(-5, 1)$
  $E(1, 0)$
  $F(4, 3)$

2. $A(3, 0)$
  $B(1, 5)$
  $C(-3, -5)$
  $D(-1, 4)$
  $E(5, -2)$
  $F(0, -5)$

In Problems 3 and 4, determine whether the given points are on the graph of the equation.

3. $3x - 2y = 7$
  (a) $(3, 1)$
  (b) $(2, -1)$
  (c) $(4, 0)$
  (d) $\left(\dfrac{1}{3}, -3\right)$

4. $y = 2x^2 - 3x + 2$
  (a) $(-1, 3)$
  (b) $(1, 1)$
  (c) $(-2, 16)$
  (d) $\left(\dfrac{1}{2}, \dfrac{3}{2}\right)$

In Problems 5–14, graph each equation by plotting points.

5. $y = x + 2$ **6.** $2x + y = 3$ **7.** $y = 2x^2 - 3$ **8.** $y = -x^2 + 4$

9. $y = -|x| - 2$ **10.** $y = |x + 2| - 1$ **11.** $y = x^3 + 2$ **12.** $y = -x^3 + 1$

13. $x = y^2 + 1$ **14.** $y = \dfrac{1}{x - 2}$

In Problems 15 and 16, the graph of an equation is given. List the intercepts of the graph.

15.

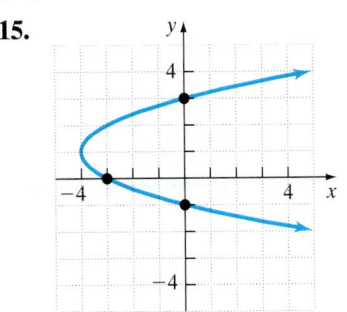

16.

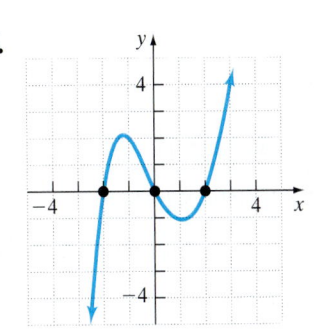

17. Cell Phones A cellular phone company offers a plan for $40 per month for 3000 minutes with additional minutes costing $0.05 per minute. The graph to the right shows the monthly cost, in dollars, when x minutes are used.

(a) If you talk for 2250 minutes in a month, how much is your monthly bill?

(b) Use the graph to estimate your monthly bill if you talk for 12 thousand minutes.

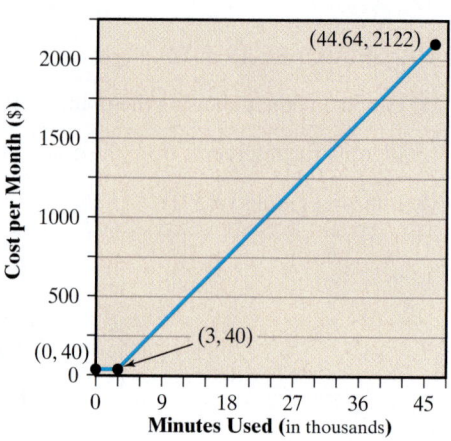

18. Kentucky Derby The graph to the right shows the winning times (to the nearest second) in the Kentucky Derby for the years 1995–2004. The vertical axis represents the winning time in seconds over 2 minutes and the horizontal axis represents the year. (*Source:* Churchill Downs Simulcast Network)

(a) Use the graph to determine the winning time of the Kentucky Derby in 1999.

(b) Use the graph to determine which year between 1995 and 2004 had the fastest winning time for the Kentucky Derby.

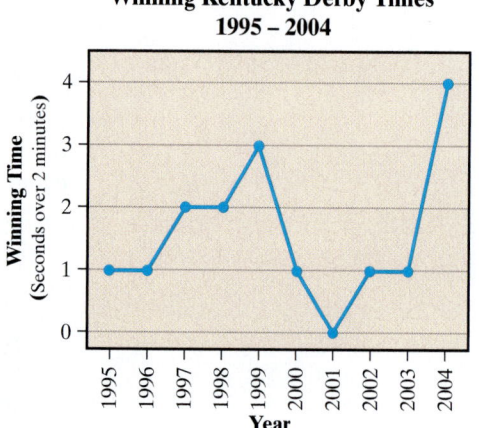

Winning Kentucky Derby Times 1995 – 2004

Section 8.2   Relations

KEY CONCEPT	KEY TERMS
• **Relation** A correspondence between two variables x and y where y depends on x. Relations can be represented through maps, sets of ordered pairs, equations, or graphs.	Relation Corresponds Depends on Mapping Inputs Outputs Domain Range

YOU SHOULD BE ABLE TO...	EXAMPLE	REVIEW EXERCISES
① Understand relations (p. 577)	Example 1	19–22
② Find the domain and the range of a relation (p. 578)	Examples 2 through 4	19–36, 37, 38
③ Graph a relation defined by an equation (p. 580)	Example 5	27–36

In Problems 19 and 20, write each relation as a set of ordered pairs. Then identify the domain and range of the relation.

19.

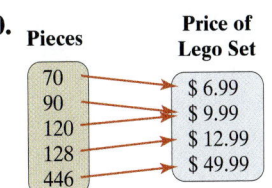

U.S. Coin	Weight (g)
Cent	2.500
Nickel	5.000
Dime	2.268
Quarter	5.670
Half Dollar	11.340
Dollar	8.100

SOURCE: *U.S. Mint Web site*

20.

Pieces	Price of Lego Set
70	$ 6.99
90	$ 9.99
120	$ 12.99
128	$ 49.99
446	

SOURCE: *Lego Web site*

In Problems 21 and 22, write each relation as a map. Then identify the domain and the range of the relation.

21. $\{(2, 7), (-4, 8), (3, 5), (6, -1), (-2, -9)\}$

22. $\{(3, 1), (3, 7), (5, 1), (-2, 8), (1, 4)\}$

In Problems 23–26, identify the domain and range of the relation from the graph.

23.

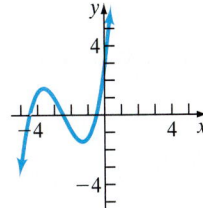

24.

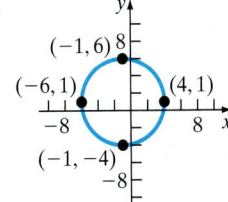

25.

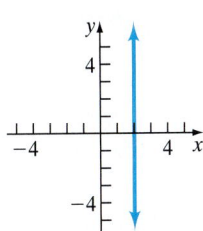

26.

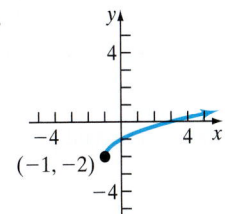

In Problems 27–36, graph the relation. Use the graph of the relation to identify the domain and the range of the relation. (See Problems 5–14.)

27. $y = x + 2$ **28.** $2x + y = 3$ **29.** $y = 2x^2 - 3$ **30.** $y = -x^2 + 4$

31. $y = -|x| - 2$ **32.** $y = |x + 2| - 1$ **33.** $y = x^3 + 2$ **34.** $y = -x^3 + 1$

35. $x = y^2 + 1$ **36.** $y = \dfrac{1}{x - 2}$

37. Cell Phones A cellular phone company offers a plan for $40 per month for 3000 minutes with additional minutes costing $0.05 per minute. The graph to the right shows the monthly cost, in dollars, when x minutes are used.

(a) What is the domain and the range of the relation?

(b) Explain why the domain obtained in part (a) is reasonable.

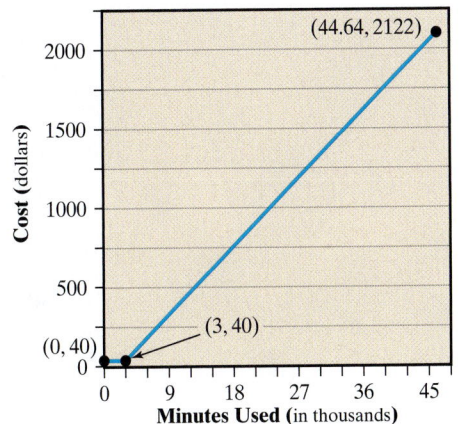

38. Vertical Motion The graph to the right shows the height, in feet, of a ball thrown straight up with an initial speed of 40 feet per second from an initial height of 96 feet after t seconds. What is the domain and the range of the relation?

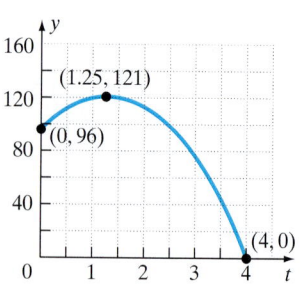

Section 8.3	An Introduction to Functions

KEY CONCEPTS	KEY TERMS
• **Functions** A special type of relation where any given input, x, corresponds to only one output y. Functions can be represented through maps, sets of ordered pairs, equations, or graphs. • **Vertical Line Test** A set of points in the xy-plane is the graph of a function if and only if every vertical line intersects the graph in at most one point. • **Graph of a Function** The graph of a function, f, is the set of all ordered pairs $(x, f(x))$.	Function Vertical Line Test Value of f at the number x Independent variable Dependent variable Argument Graph of the function

YOU SHOULD BE ABLE TO . . .	EXAMPLE	REVIEW EXERCISES
1 Determine whether a relation expressed as a map or ordered pairs represents a function (p. 586)	Examples 1 and 2	39, 40
2 Determine whether a relation expressed as an equation represents a function (p. 589)	Examples 3 and 4	41–44
3 Determine whether a relation expressed as a graph represents a function (p. 589)	Example 5	45–48
4 Find the value of a function (p. 590)	Examples 6 and 7	49–52
5 Graph a function (p. 592)	Example 8	53–56
6 Work with applications of functions (p. 593)	Example 9	57–58

In Problems 39 and 40, determine whether the given relation represents a function. State the domain and the range of each relation.

39. (a) $\{(-1, -2), (-1, 3), (5, 0), (7, 2), (9, 4)\}$

(b)

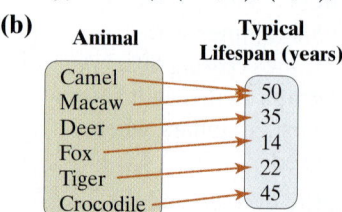

40. (a) $\{(-2, 4), (2, 3), (-3, -1), (5, 7), (4, 7)\}$

(b)

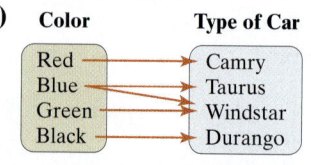

In Problems 41–44, determine whether each relation shows y as a function of x.

41. $3x - 5y = 18$

42. $x^2 + y^2 = 81$

43. $y = \pm 10x$

44. $y = x^2 - 14$

In Problems 45–48, determine whether the graph is that of a function.

45.

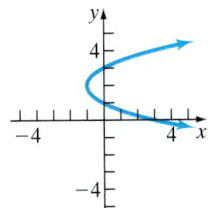

46.

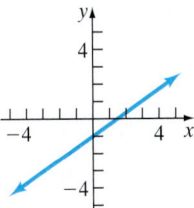

47.

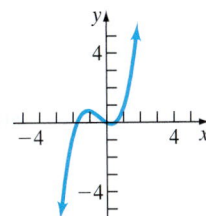

48.

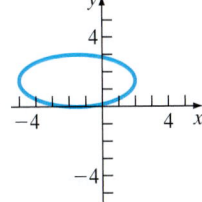

In Problems 49–52, find the indicated values for the given functions.

49. $f(x) = x^2 + 2x - 5$

    (a) $f(-2)$ **(b)** $f(3)$

50. $g(z) = \dfrac{2z + 1}{z - 3}$

    (a) $g(0)$ **(b)** $g(2)$

51. $F(x) = -2x + 7$

    (a) $F(5)$ **(b)** $F(-x)$

52. $G(x) = 2x + 1$

    (a) $G(7)$ **(b)** $G(x + h)$

In Problems 53–56, graph each function.

53. $f(x) = 2x - 5$

54. $g(x) = x^2 - 3x + 2$

55. $h(x) = (x - 1)^3 - 3$

56. $f(x) = |x + 1| - 4$

57. Population Using census data from 1900 to 2000, the function $P(t) = 0.144t^2 - 6.613t + 104.448$ represents the population, P, of Orange County in Florida (in thousands) t years after 1900.

    (a) Identify the dependent and independent variables.

    (b) Evaluate $P(110)$ and explain what it represents.

    (c) Evaluate $P(-70)$ and explain what it represents. Is the result reasonable? Explain.

58. Wages The function $W(a) = -0.058a^2 + 5.410a - 73.839$ represents the 2000 average annual wage, W, (in thousands) of a Wyoming resident in the mining industry who is a years old.

    (a) Identify the dependent and independent variables.

    (b) Evaluate $W(30)$ and explain what it represents.

    (c) Evaluate $W(16)$ and explain what it represents. Is this result reasonable? Explain.

Section 8.4	Functions and Their Graphs	
KEY CONCEPT		**KEY TERM**
• **Domain of a Function** When only an equation of a function is given, the domain of the function is the largest set of real numbers for which $f(x)$ is a real number. However, in applications, the domain of a function is the largest set of real numbers for which the output of the function is reasonable.		Domain of f

YOU SHOULD BE ABLE TO . . .	EXAMPLE	REVIEW EXERCISES
1 Find the domain of a function (p. 597)	Examples 1 and 2	59–70
2 Obtain information from the graph of a function (p. 598)	Examples 3 through 5	65–72
3 Interpret graphs of functions (p. 601)	Example 6	73, 74

For Problems 59–64, find the domain of each function.

59. $f(x) = -\dfrac{3}{2}x + 5$

60. $g(w) = \dfrac{w - 9}{2w + 5}$

61. $h(t) = \dfrac{t + 2}{t - 5}$

62. $F(x) = \dfrac{3}{x - 2}$

63. $G(t) = 3t^2 + 4t - 9$

64. $H(x) = x^4 - 2x^3 + 7$

For Problems 65–70, find (a) the domain and the range, and (b) the intercepts, if any.

65.

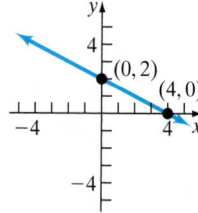

66.

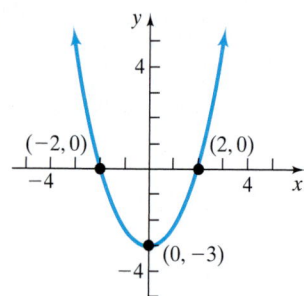

67.

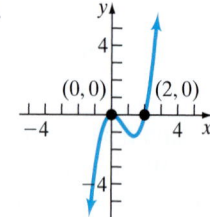

68.

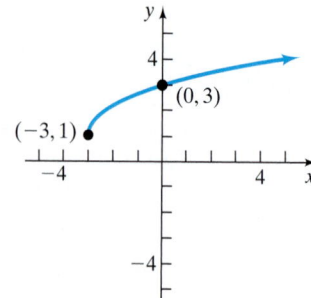

69.

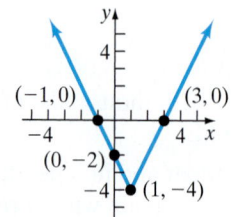

70.

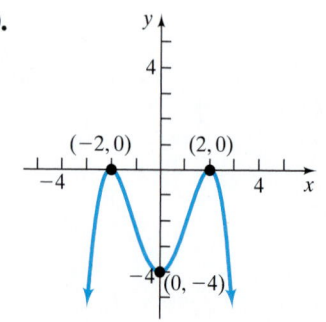

In Problems 71 and 72, answer the questions about the given function.

71. $h(x) = 2x - 7$

    (a) Is the point $(3, -1)$ on the graph of the function?

    (b) If $x = -2$, what is $h(x)$? What point is on the graph of the function?

    (c) If $h(x) = 4$, what is x? What point is on the graph of h?

72. $g(x) = \dfrac{3}{5}x + 4$

    (a) Is the point $(-5, 2)$ on the graph of the function?

    (b) If $x = 3$, what is $g(x)$? What point is on the graph of the function?

    (c) If $g(x) = -2$, what is x? What point is on the graph of g?

73. Travel by Train A Metrolink train leaves L.A. Union Station and travels 6 miles at a constant speed for 10 minutes arriving at the Glendale station where it waits 1 minute for passengers to board and depart. The train continues traveling at the same speed for 5 more minutes to reach downtown Burbank, which is 3 miles from Glendale. Sketch a graph that represents the distance of the train as a function of time until it reaches downtown Burbank.

74. Filling a Tub With the faucet running at a constant rate, it takes Angie 7 minutes to fill her bathtub. She turns off the faucet when the tub is full and realizes the water is too hot. She then opens the drain letting water out at a constant rate that is half the rate of the faucet. After draining for 2 minutes, she stops the drain and turns on the faucet at the same rate as before (but at a cooler temperature) until the tub is full. Sketch a graph that represents the amount of water in the tub as a function of time.

Section 8.5	Linear Functions		
KEY CONCEPT			**KEY TERMS**
• **Linear Function** A linear function is a function of the form $f(x) = mx + b$ where m is the slope and b is the y-intercept of the graph.			Linear function Line Scatter diagram

YOU SHOULD BE ABLE TO . . .	EXAMPLE	REVIEW EXERCISES
① Graph linear functions (p. 607)	Examples 1 and 2	75–80
② Build linear models from verbal descriptions (p. 609)	Examples 3 and 4	81–84
③ Build linear models from data (p. 612)	Examples 5 through 7	85–88

In Problems 75–78, graph each linear function.

75. $g(x) = 2x - 6$

76. $H(x) = -\dfrac{4}{3}x + 5$

77. $F(x) = -x - 3$

78. $f(x) = \dfrac{3}{4}x - 3$

79. Long Distance A phone company offers a plan for long-distance calls that charges \$5.00 per month plus 7¢ per minute. The monthly long-distance cost C for talking x minutes is given by the linear function $C(x) = 0.07x + 5$.

    (a) What is the implied domain of this linear function?

    (b) What is the cost if a person made 235 minutes worth of long-distance calls during one month?

    (c) Graph the linear function.

    (d) In one month, how many minutes of long-distance can be purchased for \$75?

80. Straight-Line Depreciation Using straight-line depreciation, the value V of a particular computer x years after purchase is given by the linear function $V(x) = 1800 - 360x$ for $0 \leq x \leq 5$.

(a) What are the independent and dependent variables?
(b) What is the domain of this linear function?
(c) What is the initial value of the computer?
(d) What is the value of the computer 2 years after purchase?
(e) Graph the linear function over its domain.
(f) After how long will the value of the computer be $0?

81. Federal Tax Returns In 1996, approximately 12.6% of U.S. Federal Tax Returns were filed electronically. In 2001, approximately 30.7% were filed electronically. (*Source:* Internal Revenue Service)

(a) Assuming a linear relation between year and percentage of electronic filings, find a linear function that relates the percentage of electronic filings to year, treating the number of years since 1996 as the independent variable, x.
(b) Predict the percentage of electronic filings in 2004 if the linear trend continues.
(c) Interpret the slope.
(d) Assuming the linear trend continues, in what year will the percentage of electronic returns be 48.8%?

82. Heart Rates According to the American Geriatric Society, the maximum recommended heart rate for a 20-year-old man under stress is 200 beats per minute. The maximum recommended heart rate for a 60-year-old man under stress is 160 beats per minute.

(a) Find a linear function that relates the maximum recommended heart rate for men to age.
(b) Predict the maximum recommended heart rate for a 45-year-old man under stress.
(c) Interpret the slope.
(d) For what age would the maximum recommended heart rate under stress be 168 beats per minute?

83. Car Rental The daily rental charge for a particular car is $35 plus 12¢ per mile.

(a) Find a linear function that expresses the rental cost C as a function of the miles driven m.
(b) What are the independent and dependent variables?
(c) What is the implied domain of this linear function?
(d) For a one-day rental, what is the rental cost if 124 miles are driven?
(e) For a one-day rental, how many miles were driven if the rental cost was $67.16?
(f) Graph the linear function.

84. Satellite Television Bill A satellite television company charges $33.99 per month for a 100-channel package, plus $3.50 for each pay-per-view movie watched that month.

(a) Find a linear function that expresses the monthly bill B as a function of x, the number of pay-per-view movies watched that month.
(b) What are the independent and dependent variables?
(c) What is the implied domain of this linear function?
(d) What is the monthly bill if 5 pay-per-view movies are watched that month?
(e) For one month, how many pay-per-view movies were watched if the bill was $58.49?
(f) Graph the linear function.

In Problems 85 and 86,

(a) *Draw a scatter diagram of the data.*
(b) *Select two points from the scatter diagram and find the equation of the line containing the points selected.**
(c) *Graph the line found in part* (b) *on the scatter diagram.*

85.

x	2	5	8	11	14
y	13.3	11.6	8.4	7.2	4.6

*Answers will vary.

86.

x	0	0.4	1.5	2.3	4.2
y	0.6	1.1	1.3	1.8	3.0

87. The table below gives the number of calories and the total carbohydrates (in grams) for a one-cup serving of seven name-brand cereals (not including milk).

Cereal	Calories, x	Total Carbohydrates (in grams), y
Rice Krispies®	96	23.2
Life®	160	33.3
Lucky Charms®	120	25.0
Kellogg's Complete®	120	30.7
Wheaties®	110	24.0
Cheerios®	110	22.0
Honey Nut Chex®	160	34.7

SOURCE: *Quaker Oats, General Mills, and Kellogg*

(a) Draw a scatter diagram of the data treating calories as the independent variable.
(b) What type of relation appears to exist between calories and total carbohydrates in a one-cup serving of cereal?
(c) Select two points and find an equation of the line containing the points.*
(d) Graph the line on the scatter diagram drawn in part (a).
(e) Predict the total carbohydrates in a one-cup serving of cereal that has 140 calories.
(f) Interpret the slope of the line found in part (c).

88. Second-Day Delivery Costs The table below lists some selected prices charged by Federal Express for FedEx 2Day delivery, depending on the weight of the package.

Weight (in pounds), x	FedEx 2Day® Delivery Charge, y
1	$9.25
3	$12.00
6	$17.25
8	$21.25
9	$23.00
11	$26.50

SOURCE: *Federal Express Corporation*

(a) Draw a scatter diagram of the data treating weight as the independent variable.
(b) What type of relation appears to exist between the weight of the package and the FedEx 2Day delivery charge?
(c) Select two points and find an equation of the line containing the points.*
(d) Graph the line on the scatter diagram drawn in part (a).
(e) Predict the FedEx 2Day delivery charge for shipping a 5-pound package.
(f) Interpret the slope of the line found in part (c).

---

*Answers will vary.

Section 8.6	Compound Inequalities
KEY CONCEPT	**KEY TERMS**
• If $a < b$, then we can write $a < x$ and $x < b$ as $a < x < b$.	Intersection Union Compound inequality Solve a compound inequality

YOU SHOULD BE ABLE TO . . .	EXAMPLE	REVIEW EXERCISES
① Determine the intersection or union of two sets (p. 621)	Examples 1 and 2	89–94
② Solve compound inequalities involving "and" (p. 623)	Examples 3 through 6	95, 96, 99, 100, 104
③ Solve compound inequalities involving "or" (p. 625)	Examples 7 and 8	97, 98, 101–103
④ Solve problems using compound inequalities (p. 627)	Example 9	105, 106

In Problems 89–92, use $A = \{2, 4, 6, 8\}$, $B = \{-1, 0, 1, 2, 3, 4\}$, and $C = \{1, 2, 3, 4\}$ to find each set.

89. $A \cup B$ **90.** $A \cap C$ **91.** $B \cap C$ **92.** $A \cup C$

In Problems 93 and 94, use the graph of the inequality to find each set.

93. $A = \{x \mid x \le 4\}$; $B = \{x \mid x > 2\}$.
Find (a) $A \cap B$ and (b) $A \cup B$.

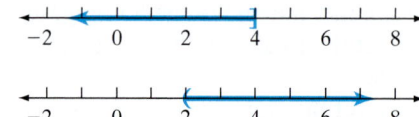

94. $E = \{x \mid x \ge 3\}$; $F = \{x \mid x < -2\}$.
Find (a) $E \cap F$ and (b) $E \cup F$.

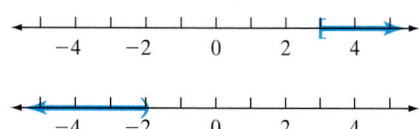

In Problems 95–104, solve each compound inequality. Graph the solution set.

95. $x < 4$ and $x + 3 > 2$

96. $3 < 2 - x < 7$

97. $x + 3 < 1$ or $x > 2$

98. $x + 6 \ge 10$ or $x \le 0$

99. $3x + 2 \le 5$ and $-4x + 2 \le -10$

100. $1 \le 2x + 5 < 13$

101. $x - 3 \le -5$ or $2x + 1 > 7$

102. $3x + 4 > -2$ or $4 - 2x \ge -6$

103. $\dfrac{1}{3}x > 2$ or $\dfrac{2}{5}x < -4$

104. $x + \dfrac{3}{2} \ge 0$ and $-2x + \dfrac{3}{2} > \dfrac{1}{4}$

105. Heart Rates The normal heart rate for healthy adults between the ages of 21 and 60 should be between 70 and 75 beats per minute (inclusive). If we let x represent the heart rate of an adult between the ages of 21 and 60, express the normal range of values using a compound inequality.

106. Heating Bills For usage above 300 kilowatt hours, the non–space heat winter energy charge for Illinois Power residential service was $23.12 plus $0.05947 per kilowatt hour over 300. During one winter, a customer's charge ranged from a low of $50.28 to a high of $121.43. Over what range of values did electric usage vary (in kilowatt hours)?

Section 8.7   Absolute Value Equations and Inequalities

KEY CONCEPTS

- **Equations Involving Absolute Value**
  If a is a positive real number and if u is any algebraic expression, then $|u| = a$ is equivalent to $u = a$ or $u = -a$.
- **Equations Involving Two Absolute Values**
  If u and v are any algebraic expression, then $|u| = |v|$ is equivalent to $u = v$ or $u = -v$.
- **Inequalities of the Form $<$ or $\le$ Involving Absolute Value**
  If a is a positive real number and if u is any algebraic expression, then $|u| < a$ is equivalent to $-a < u < a$
  and $|u| \le a$ is equivalent to $-a \le u \le a$.
- **Inequalities of the Form $>$ or $\ge$ Involving Absolute Value**
  If a is a positive real number and if u is any algebraic expression, then $|u| > a$ is equivalent to $u < -a$ or
  $u > a$ and $|u| \ge a$ is equivalent to $u \le -a$ or $u \ge a$.

YOU SHOULD BE ABLE TO . . .	EXAMPLE	REVIEW EXERCISES
(1) Solve absolute value equations (p. 632)	Examples 1 through 4	107–112
(2) Solve absolute value inequalities involving $<$ or $\le$ (p. 635)	Examples 5 through 7	113, 115, 118, 119
(3) Solve absolute value inequalities involving $>$ or $\ge$ (p. 637)	Examples 8 and 9	114, 116, 117, 120
(4) Solve applied problems involving absolute value inequalities (p. 639)	Example 10	121, 122

In Problems 107–112, solve the absolute value equation.

107. $|x| = 4$

108. $|3x - 5| = 4$

109. $|-y + 4| = 9$

110. $-3|x + 2| - 5 = -8$

111. $|2w - 7| = -3$

112. $|x + 3| = |3x - 1|$

In Problems 113–120, solve each absolute value inequality. Graph the solution set on a real number line.

113. $|x| < 2$

114. $|x| \ge \dfrac{7}{2}$

115. $|x + 2| \le 3$

116. $|4x - 3| \ge 1$

117. $3|x| + 6 \ge 1$

118. $|7x + 5| + 4 < 3$

119. $|(x - 3) - 2| \le 0.01$

120. $\left| \dfrac{2x - 3}{4} \right| > 1$

121. Tolerance The diameter of a certain ball bearing is required to be 0.503 inches. The tolerance on the bearing is 0.001 inches. If x represents the diameter of a bearing, the acceptable diameters of the bearing can be expressed as $|x - 0.503| \le 0.001$. Determine the acceptable diameters of the bearing.

122. Tensile Strength The tensile strength of paper used to make grocery bags is about 40 lb/in.^{2} A paper grocery bag whose tensile strength satisfies the inequality $\left| \dfrac{x - 40}{2} \right| > 1.96$ has an unusual tensile strength. Determine the tensile strengths that would be considered unusual.

CHAPTER 8 TEST

Remember to use your Chapter Test Prep Video CD to see fully worked-out solutions to any of these problems you would like to review.

1. Plot the following ordered pairs in the same xy-plane. Tell in which quadrant or on what coordinate axis each point lies.

$$A(3, -4), B(0, 2), C(3, 0), D(2, 1), E(-1, -4), F(-3, 5)$$

2. Determine whether the ordered pair is a point on the graph of the equation $y = 3x^2 + x - 5$.

(a) $(-2, 4)$ **(b)** $(-1, -3)$ **(c)** $(2, 9)$

In Problems 3 and 4, graph the equations by plotting points.

3. $y = 4x - 1$ **4.** $y = 4x^2$

5. Identify the intercepts from the graph below.

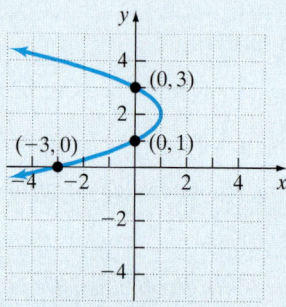

6. The graph below shows the monthly unemployment rate for South Bend, Indiana during 2002. The vertical axis represents the unemployment rate (as a percent) and the horizontal axis represents the month. (*Source:* U.S. Bureau of Labor Statistics)

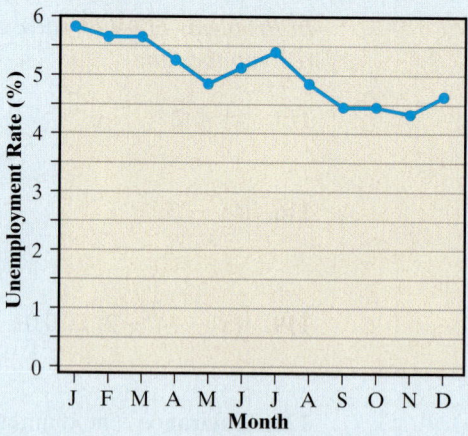

SOURCE: *U.S. Bureau of Labor Statistics*

(a) What was the approximate unemployment rate in South Bend for the month of May?

(b) In what month was the unemployment rate the highest? What was the approximate rate?

(c) In what month was the unemployment rate the lowest? What was the approximate rate?

(d) Describe the unemployment rate trend for South Bend during 2002.

7. Write the relation as a map. Then identify the domain and the range of the relation.

$$\{(2, 8), (5, -2), (7, 12), (-4, -7), (7, 3), (5, -1)\}$$

8. Identify the domain and range of the relation from the graph.

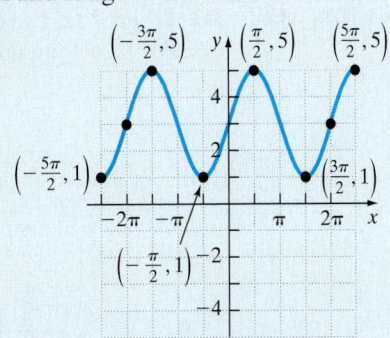

9. Graph the relation $y = x^2 - 3$ by plotting points. Use the graph of the relation to identify the domain and range.

In Problems 10 and 11, determine whether the relations represent functions. Identify the domain and the range of each relation.

10. Domain

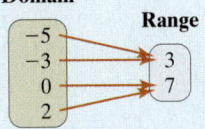

11.

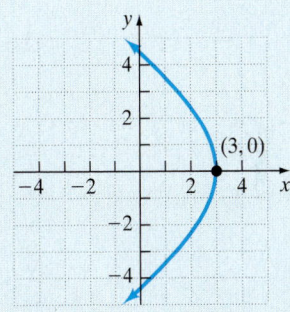

12. Does the equation $y = \pm 5x$ represent a function? Why or why not?

13. For $f(x) = -3x + 11$, find $f(x + h)$.

14. For $g(x) = 2x^2 + x - 1$, find the indicated values.

   (a) $g(-2)$ **(b)** $g(0)$ **(c)** $g(3)$

15. Sketch the graph of $f(x) = x^2 + 3$.

16. Using data from 1989 to 2002, the function $P(x) = 0.13x + 3.76$ approximates the average movie ticket price (in dollars) x years after 1989. (*Source:* National Association of Theater Owners)

   (a) Identify the dependent and independent variables.
   (b) Evaluate $P(15)$ and explain what it represents.

17. Using data from 1960 to 2000, the function $N(x) = 271.40x + 836.83$ represents the approximate number of registered climbers at Mt. Rainier x years after 1960. (*Source:* National Parks Service, U.S. Dept. of the Interior)

   (a) Identify the dependent and independent variables.
   (b) Evaluate $N(43)$ and explain what it represents.
   (c) There were 9714 registered climbers on Mt. Rainier in 2003. Compare this value to your result in part (b) and comment on any differences.

18. Find the domain of $f(x) = \dfrac{-15}{x + 2}$.

19. $h(x) = -5x + 12$

   (a) Is the point $(2, 2)$ on the graph of the function?
   (b) If $x = 3$, what is $h(x)$? What point is on the graph of the function?
   (c) If $h(x) = 0$, what is x? What point is on the graph of h?

20. The following graph represents the speed of a car as a function of time.

(a) When does the car stop accelerating?

(b) For how long does the car maintain a constant speed?

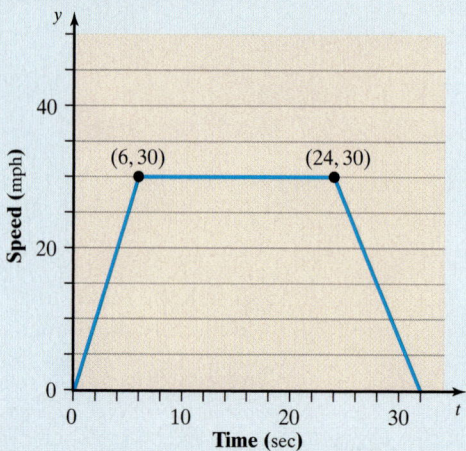

21. Crafts Fair Sales Henry plans to sell small wooden shelves at a crafts fair for $30 each. A booth at the fair costs $100 to rent. Henry estimates his expenses for producing the shelves to be $12 each, so his profit will be $18 per shelf.

(a) Write a function that expresses Henry's profit P as a function of the number of shelves x sold.

(b) What is the implied domain of this linear function?

(c) What is the profit if Henry sells 34 shelves?

(d) Graph the linear function.

(e) If Henry's profit is $764, how many shelves did he sell?

22. Shetland Pony Weights The table below lists the average weight of a Shetland pony, depending on the age of the pony.

(a) Draw a scatter diagram of the data treating age as the independent variable.

(b) What type of relation appears to exist between the age and the weight of the Shetland pony?

(c) Select two points and find an equation of the line containing the points.*

(d) Graph the line on the scatter diagram drawn in part (a).

(e) Predict the weight of a 9 month old Shetland pony.

(f) Interpret the slope of the line found in part (c).

Age (months), x	Average Weight (kilograms), y
3	60
6	95
12	140
18	170
24	185

Source: *The Merck Veterinary Manual*

23. Solve: $|2x + 5| - 3 = 0$

In Problems 24 and 25, solve each inequality and graph the solution set on a real number line.

24. $x + 2 < 8$ and $2x + 5 \geq 1$

25. $x > 4$ or $2(x - 1) + 3 < -2$

---

*Answers will vary.

9 Radicals and Rational Exponents

Where would we be without electricity? In 1752, when Ben Franklin first sent his kite into the clouds, the idea of a number whose square is -1 was still being developed by mathematicians. Yet it turns out that this concept is extremely useful in describing alternating electric currents. See Problems 113 and 114 from Section 9.9.

OUTLINE

9.1 Square Roots

9.2 nth Roots and Rational Exponents

9.3 Simplify Expressions Using the Laws of Exponents

9.4 Simplifying Radical Expressions

9.5 Adding, Subtracting, and Multiplying Radical Expressions

9.6 Rationalizing Radical Expressions

Putting the Concepts Together (Sections 9.1–9.6)

9.7 Functions Involving Radicals

9.8 Radical Equations and Their Applications

9.9 The Complex Number System

Chapter Activity: Which One Does Not Belong?
Chapter 9 Review
Chapter 9 Test
Cumulative Review Chapters 1–9

The Big Picture: Putting It Together

In Chapter 5 we simplified polynomial expressions by adding, subtracting, multiplying, and dividing. In Chapter 6, we factored polynomials. In Chapter 7 we used the skills learned in Chapters 5 and 6 to simplify rational expressions and perform operations on rational expressions.

We now present a similar discussion with radical expressions. We will learn how to add, subtract, multiply, and divide radical expressions. In addition, we will use factoring to simplify radical expressions. Throughout this discussion, keep in mind that radicals perform the "inverse" operation from raising a real number to an integer exponent. For example, a square root undoes the "squaring" operation.

9.1 Square Roots

OBJECTIVES

1. Evaluate Square Roots of Perfect Squares
2. Determine Whether a Square Root Is Rational, Irrational, or Not a Real Number
3. Find Square Roots of Variable Expressions

Preparing for Introduction to Square Roots

Before getting started, take the following readiness quiz. If you get a problem wrong, go to the section cited and review the material.

In Problems 1–3, use the set $\left\{-4, \dfrac{5}{3}, 0, \sqrt{2}, 6.95, 13, \pi\right\}$.

1. Which of the numbers are integers? [Section 1.2, pp. 9–12]
2. Which of the numbers are rational numbers? [Section 1.2, pp. 9–12]
3. Which of the numbers are irrational numbers? [Section 1.2, pp. 9–12]
4. Evaluate: (a) $\left(\dfrac{3}{2}\right)^2$ (b) $(0.4)^2$ [Section 1.6, pp. 49–50]

In Section 1.6, we introduced the concept of exponents. Exponents are used to indicate repeated multiplication. For example, 4^2 means $4 \cdot 4$, so $4^2 = 16$; $(-6)^2$ means $(-6) \cdot (-6)$, so $(-6)^2 = 36$. Now, we will reverse the process of raising a number to the second power and ask questions such as, "What number, or numbers, when squared, give me 16?"

1 ## Evaluate Square Roots of Perfect Squares

In Words

Taking the square root of a number is the "inverse" of squaring a number.

A real number is squared when it is raised to the power 2. The inverse of squaring a number is finding the **square root.** For example, since $5^2 = 25$ and $(-5)^2 = 25$, the square roots of 25 are -5 and 5. The square roots of $\dfrac{16}{49}$ are $-\dfrac{4}{7}$ and $\dfrac{4}{7}$. If we want only the positive square root of a number, we use the symbol $\sqrt{}$, called a **radical sign,** to denote the **principal square root,** or nonnegative (zero or positive) square root.

In Words

Nonnegative real numbers are positive real numbers or 0.

DEFINITION

If a is a nonnegative real number, the nonnegative real number b such that $b^2 = a$, is the **principal square root** of a and is denoted by $b = \sqrt{a}$.

For example, if we want the positive square root of 25, we would write $\sqrt{25} = 5$. We read $\sqrt{25} = 5$ as "the positive square root of 25 is 5." But what if we want the negative square root of a real number? In that case, we use the expression $-\sqrt{25} = -5$ to obtain the negative square root of 25.

PROPERTIES OF SQUARE ROOTS

- Every positive real number has two square roots, one positive and one negative.
- The square root of 0 is 0. That is, $\sqrt{0} = 0$.
- We use the symbol $\sqrt{}$, called a radical, to denote the nonnegative square root of a real number. The nonnegative square root is called the principal square root.
- The number under the radical is called the **radicand.** For example, the radicand in $\sqrt{25}$ is 25.
- For any real number c, such that $c \geq 0$, $\left(\sqrt{c}\right)^2 = c$. For example, $\left(\sqrt{4}\right)^2 = 4$ and $\left(\sqrt{8.3}\right)^2 = 8.3$.

To **evaluate** a square root, we ask ourselves, "What is the nonnegative number whose square is equal to the radicand?"

EXAMPLE 1 | Evaluating Square Roots

Evaluate each square root.

(a) $\sqrt{36}$ **(b)** $\sqrt{\dfrac{1}{9}}$ **(c)** $\sqrt{0.01}$ **(d)** $\left(\sqrt{2.3}\right)^2$

Solution

(a) Is there a positive number whose square is 36? Because $6^2 = 36$,
$$\sqrt{36} = 6.$$

(b) $\sqrt{\dfrac{1}{9}} = \dfrac{1}{3}$ because $\left(\dfrac{1}{3}\right)^2 = \dfrac{1}{9}$.

(c) $\sqrt{0.01} = 0.1$ because $0.1^2 = 0.01$.

(d) $\left(\sqrt{2.3}\right)^2 = 2.3$ because $\left(\sqrt{c}\right)^2 = c$ when $c \geq 0$. ▬

Figure 1

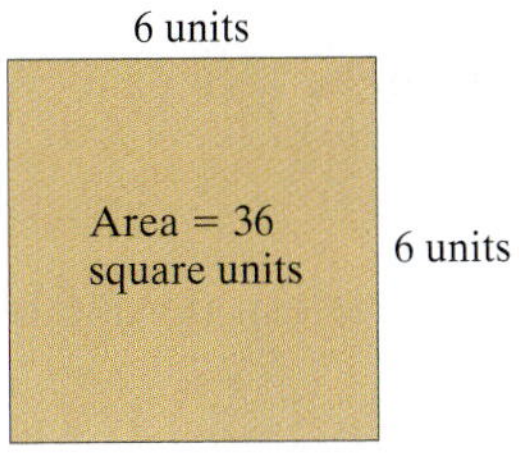

6 units

Area = 36
square units

6 units

A rational number is a **perfect square** if it is the square of a rational number. Examples 1(a), (b), and (c) are square roots of perfect squares since $6^2 = 36$, $\left(\dfrac{1}{3}\right)^2 = \dfrac{1}{9}$, and $0.1^2 = 0.01$. We can think of perfect squares geometrically as shown in Figure 1, where we have a square whose area is 36. The square root of the area, $\sqrt{36}$, gives us the length of each side of the square, 6 units.

QUICK ✓ *Evaluate each square root.*

1. $\sqrt{81}$ **2.** $\sqrt{900}$ **3.** $\sqrt{\dfrac{1}{4}}$ **4.** $\sqrt{0.16}$ **5.** $\left(\sqrt{13}\right)^2$

EXAMPLE 2 **Evaluating an Expression Containing Square Roots**

Evaluate each expression:

(a) $-4\sqrt{36}$ (b) $\sqrt{9} + \sqrt{16}$ (c) $\sqrt{9 + 16}$ (d) $\sqrt{64 - 4 \cdot 7 \cdot 1}$

Solution

(a) The expression $-4\sqrt{36}$ is asking us to find -4 times the positive square root of 36. So we first find the positive square root of 36 and then multiply this result by -4.
$$-4\sqrt{36} = -4 \cdot 6$$
$$= -24$$

Work Smart

In Examples 2(b) and (c), notice that
$$\sqrt{9} + \sqrt{16} \neq \sqrt{9 + 16}$$
In general,
$$\sqrt{a} + \sqrt{b} \neq \sqrt{a + b}$$
The radical acts like a grouping symbol, so always simplify the radicand before taking the square root.

(b) $\sqrt{9} + \sqrt{16} = 3 + 4$
$$= 7$$

(c) $\sqrt{9 + 16} = \sqrt{25}$
$$= 5$$

(d) $\sqrt{64 - 4 \cdot 7 \cdot 1} = \sqrt{64 - 28}$
$$= \sqrt{36}$$
$$= 6$$ ▬

QUICK ✓ *Evaluate each expression.*

6. $5\sqrt{9}$ **7.** $\sqrt{36 + 64}$ **8.** $\sqrt{36} + \sqrt{64}$ **9.** $\sqrt{25 - 4 \cdot 3 \cdot (-2)}$

---

② ## Determine Whether a Square Root Is Rational, Irrational, or Not a Real Number

Not all radical expressions will simplify to a rational number. For example, because there is no rational number whose square is 5, $\sqrt{5}$ is not a rational number. In fact, $\sqrt{5}$ is an *irrational* number. Remember, an irrational number is a number that cannot be written as the quotient of two integers.

Work Smart
The square roots of negative real numbers are not real.

What if we wanted to evaluate $\sqrt{-16}$? Because any positive real number squared is positive, any negative real number squared is also positive, and 0 squared is 0, there is no real number whose square is -16. We conclude: **Negative real numbers do not have square roots that are real numbers!**

The following comments regarding square roots are important.

> **MORE PROPERTIES OF SQUARE ROOTS**
> - The square root of a perfect square is a rational number.
> - The square root of a positive rational number that is not a perfect square is an irrational number. For example, $\sqrt{20}$ is an irrational number because 20 is not a perfect square.
> - The square root of a negative real number is not a real number. For example, $\sqrt{-2}$ is not a real number.

When a radical has a radicand that is not a perfect square, we can do one of two things:

1. Write a decimal approximation of the radical.
2. Simplify the radical using properties of radicals, if possible (Section 9.4).

EXAMPLE 3 **Writing a Radical as a Decimal Using a Calculator**

Write $\sqrt{5}$ as a decimal rounded to two decimal places.

Solution

We know that $\sqrt{4} = 2$ and $\sqrt{9} = 3$, so it seems reasonable to expect $\sqrt{5}$ to be between 2 and 3 since 5 is between 4 and 9. We can use a calculator or Appendix F to approximate $\sqrt{5}$. Figure 2 shows the results from a TI-84 Plus graphing calculator. From the display, we see that $\sqrt{5} \approx 2.24$.

Figure 2

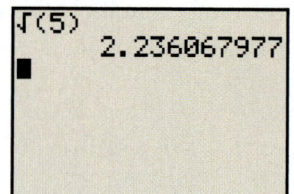

```
√(5)
            2.236067977
■
```

EXAMPLE 4 **Determining Whether a Square Root of an Integer Is Rational, Irrational, or Not a Real Number**

Determine if each square root is rational, irrational, or not a real number. Then evaluate each real square root. For each square root that is irrational, express the square root as a decimal rounded to two decimal places using either a calculator or Appendix F.

$$\textbf{(a)}\ \sqrt{51} \qquad \textbf{(b)}\ \sqrt{169} \qquad \textbf{(c)}\ \sqrt{-81}$$

Work Smart
$\sqrt{-81}$ is not a real number, but $-\sqrt{81}$ is a real number because $-\sqrt{81} = -9$. Note the placement of the negative sign!

Solution

(a) $\sqrt{51}$ is irrational because 51 is not a perfect square. That is, there is no rational number whose square is 51. Using a calculator, we find $\sqrt{51} \approx 7.14$.

(b) $\sqrt{169}$ is a rational number because $13^2 = 169$. So, $\sqrt{169} = 13$.

(c) $\sqrt{-81}$ is not a real number. There is no real number whose square is -81.

QUICK ✔ *Determine whether each square root is rational, irrational, or not a real number. Evaluate each square root that is rational. For each square root that is irrational, approximate the square root rounded to two decimal places using a calculator or Appendix F.*

10. $\sqrt{400}$ **11.** $\sqrt{40}$ **12.** $\sqrt{-25}$ **13.** $-\sqrt{196}$

③ Find Square Roots of Variable Expressions

What is $\sqrt{4^2}$? Because $4^2 = 16$, we have that $\sqrt{4^2} = \sqrt{16} = 4$. Based on this result, we might conclude that $\sqrt{a^2} = a$ for any real number a. Before we jump to this conclusion, let's consider $\sqrt{(-4)^2}$. Our "formula" says that $\sqrt{a^2} = a$, so we would think that $\sqrt{(-4)^2} = -4$, right? Wrong! $\sqrt{(-4)^2} = \sqrt{16} = 4$. So $\sqrt{4^2} = 4$ and $\sqrt{(-4)^2} = 4$. Regardless of whether the "a" in $\sqrt{a^2}$ is positive or negative, the result ends up being positive. So to say that $\sqrt{a^2} = a$ would not quite be correct. How can we fix our "formula"? In Section 1.2, we learned that $|a|$ will be a positive number if a is nonzero. From this, we have the following result:

In Words
The square root of a nonzero number squared will always be positive. The absolute value ensures this.

For any **real number** a,

$$\sqrt{a^2} = |a|$$

The bottom line is this—if you are taking the square root of some variable expression raised to the second power, the result will be the absolute value of the variable expression.

EXAMPLE 5   **Evaluating Square Roots**

Evaluate each square root.

     (a) $\sqrt{7^2}$ **(b)** $\sqrt{(-15)^2}$ **(c)** $\sqrt{x^2}$

     (d) $\sqrt{(3x-1)^2}$ **(e)** $\sqrt{x^2 + 6x + 9}$

Solution

     (a) $\sqrt{7^2} = 7$

     (b) $\sqrt{(-15)^2} = |-15| = 15$

     (c) We don't know whether the real number x is positive, negative, or zero. To ensure that the result is positive or zero, we write $\sqrt{x^2} = |x|$.

     (d) $\sqrt{(3x-1)^2} = |3x - 1|$

     (e) Notice that the radicand is a perfect square trinomial, so that $x^2 + 6x + 9$ factors to $(x + 3)^2$. Therefore,

Work Smart
Note how we use absolute values to get nonnegative answers in these examples.

$$\sqrt{x^2 + 6x + 9} = \sqrt{(x + 3)^2}$$
$$= |x + 3|$$

QUICK ✔ *Evaluate each square root.*

14. $\sqrt{(-14)^2}$ **15.** $\sqrt{z^2}$ **16.** $\sqrt{(2x + 3)^2}$ **17.** $\sqrt{p^2 - 12p + 36}$

9.1 Exercises

For Extra Help:

Student Solutions Manual   CD Video   PH Math/Tutor Center   MathXL Tutorials on CD   MathXL®   MyMathLab

Concepts and Vocabulary

In Problems 1–3, fill in the blanks.

1. The symbol $\sqrt{}$ is called a _____ _____.

2. If a is a nonnegative real number, the nonnegative number b such that $b^2 = a$ is the _____ _____ _____ of a and is denoted by $b = \sqrt{a}$.

3. For any real number a, $\sqrt{a^2} =$ _____.

In Problems 4–6, answer True or False to each statement.

4. The square root of 36 is 6.

5. Negative numbers do not have square roots in the real number system.

6. The square root of a negative real number is a negative real number.

7. Give four examples of perfect squares.

8. Explain why $\sqrt{a^2} = |a|$. Provide examples to support your explanation.

Building Skills

In Problems 9–18, evaluate each square root.

9. $\sqrt{1}$ **10.** $\sqrt{9}$ **11.** $-\sqrt{100}$ **12.** $-\sqrt{144}$ **13.** $\sqrt{\dfrac{1}{4}}$

14. $\sqrt{\dfrac{4}{81}}$ **15.** $\sqrt{0.36}$ **16.** $\sqrt{0.16}$ **17.** $\left(\sqrt{1.6}\right)^2$ **18.** $\left(\sqrt{3.7}\right)^2$

In Problems 19–30, tell if the square root is rational, irrational, or not a real number. If the square root is rational, find the exact value; if the square root is irrational, write the approximate value rounded to two decimal places using a calculator or Appendix F.

19. $\sqrt{-14}$ **20.** $\sqrt{-50}$ **21.** $\sqrt{64}$ **22.** $\sqrt{121}$

23. $\sqrt{\dfrac{1}{16}}$ **24.** $\sqrt{\dfrac{49}{100}}$ **25.** $\sqrt{44}$ **26.** $\sqrt{24}$

27. $\sqrt{50}$ **28.** $\sqrt{12}$ **29.** $\sqrt{-16}$ **30.** $\sqrt{-64}$

In Problems 31–42, simplify each square root.

31. $\sqrt{8^2}$ **32.** $\sqrt{5^2}$ **33.** $\sqrt{(-19)^2}$ **34.** $\sqrt{(-13)^2}$

35. $\sqrt{r^2}$ **36.** $\sqrt{w^2}$ **37.** $\sqrt{(x+4)^2}$ **38.** $\sqrt{(x-8)^2}$

39. $\sqrt{(4x-3)^2}$ **40.** $\sqrt{(5x+2)^2}$

41. $\sqrt{4y^2 + 12y + 9}$ **42.** $\sqrt{9z^2 - 24z + 16}$

Mixed Practice

In Problems 43–60, simplify each expression. If necessary, express results that are not rational numbers as a decimal rounded to two decimal places.

43. $\sqrt{25 + 144}$ **44.** $\sqrt{9 + 16}$ **45.** $\sqrt{25} + \sqrt{144}$ **46.** $\sqrt{9} + \sqrt{16}$

47. $\sqrt{-144}$ **48.** $\sqrt{-36}$ **49.** $3\sqrt{25}$ **50.** $-10\sqrt{16}$

51. $5\sqrt{\dfrac{16}{25}} - \sqrt{144}$ **52.** $2\sqrt{\dfrac{9}{4}} - \sqrt{4}$ **53.** $\sqrt{8^2 - 4\cdot 1\cdot 7}$

54. $\sqrt{9^2 - 4\cdot 1\cdot 20}$ **55.** $\sqrt{(-5)^2 - 4\cdot 2\cdot 5}$ **56.** $\sqrt{(-3)^2 - 4\cdot 3\cdot 2}$

57. $\dfrac{-(-1) + \sqrt{(-1)^2 - 4\cdot 6\cdot(-2)}}{2\cdot(-1)}$ **58.** $\dfrac{-7 + \sqrt{7^2 - 4\cdot 2\cdot 6}}{2\cdot 2}$

59. $\sqrt{(6-1)^2 + (15-3)^2}$ **60.** $\sqrt{(2-(-1))^2 + (6-2)^2}$

61. What are the square roots of 36? What is $\sqrt{36}$?

62. What are the square roots of 64? What is $\sqrt{64}$?

Applying the Concepts

For Problems 63 and 64, use the formula $Z = \dfrac{X - \mu}{\dfrac{\sigma}{\sqrt{n}}}$ from statistics (a formula used to determine the relative value of one observation to another) to evaluate each expression for the given values. Write the exact value and then write your answer rounded to two decimal places.

63. $X = 120, \mu = 100, \sigma = 15, n = 13$ **64.** $X = 40, \mu = 50, \sigma = 10, n = 5$

The area, A, of a square whose side has length s is given by $A = s^2$. We can calculate the length, s, of the side of a square as the positive square root of the area, A, using $s = \sqrt{A}$. In Problems 65–68, find the length of the side of the square whose area is given.

△ **65.** 625 square feet △ **66.** 169 square meters

△ **67.** 256 square kilometers △ **68.** 400 square inches

The area, A, of a circle whose radius is r is given by the formula $A = \pi r^2$. We can calculate the radius when the area is given using $r = \sqrt{\dfrac{A}{\pi}}$. In Problems 69–72, find the radius of the circle with the following areas.

△ **69.** 49π square meters △ **70.** 25π square feet

△ **71.** 196π square inches △ **72.** 289π square centimeters

△ **73.** **Great Pyramid at Giza** The volume V of a pyramid with a square base s and height h is $V = \dfrac{1}{3}s^2 h$. If we solve this formula for s, we obtain $s = \sqrt{\dfrac{3V}{h}}$. The Great Pyramid at Giza, built around 2500 B.C., has a volume of approximately 7,700,000 cubic meters. Find the length, to the nearest meter, of the side of the Great Pyramid if you know that the height is approximately 146 meters.

△ **74.** **Sun Pyramid** Use the formula given in Problem 73 to the find the length, to the nearest foot, of the side of the Sun Pyramid if the volume of the pyramid is approximately 114,000,000 cubic feet and the height is 210 feet. The Sun Pyramid was built in the ancient city of Teotihuacán, around 200 A.D.

Extending the Concepts

In Problems 75–78, simplify each expression.

75. $\sqrt{\sqrt{16}}$ **76.** $\sqrt{\sqrt{81}}$ **77.** $\sqrt{5\cdot\sqrt{25}}$ **78.** $\sqrt{2\cdot\sqrt{4}}$

9.2 *nth* Roots and Rational Exponents

OBJECTIVES

① Evaluate *n*th Roots
② Simplify Expressions of the Form $\sqrt[n]{a^n}$
③ Evaluate Expressions of the Form $a^{1/n}$
④ Evaluate Expressions of the Form $a^{m/n}$

Preparing for nth Root and Rational Exponents

Before getting started, take the following readiness quiz. If you get a problem wrong, go to the section cited and review the material.

1. Simplify: $\left(\dfrac{x^2 y}{xy^{-2}}\right)^{-3}$ [Section 5.4, pp. 377–380]

2. Simplify: $\left(\sqrt{7}\right)^2$ [Section 9.1, pp. 660–661]

3. Evaluate: $\sqrt{64}$ [Section 9.1, pp. 660–661]

In the last section, we learned skills for evaluating square roots. We now extend this skill to other types of roots.

① **Evaluate *n*th Roots**

A real number is cubed when it is raised to the power 3. The inverse of cubing a number is finding the *cube root*. For example, since $2^3 = 8$, the cube root of 8 is 2 and since $(-2)^3 = -8$, the cube root of -8 is -2. In general, we can find *n*th roots of numbers.

In Words
When you see the notation $\sqrt[n]{a} = b$, think to yourself, "Find a number b such that raising that number to the *n*th power gives me a."

> **DEFINITION**
>
> The **principal *n*th root of a number *a*,** symbolized by $\sqrt[n]{a}$, where $n \geq 2$ is an integer, is defined as follows:
>
> $$\sqrt[n]{a} = b \qquad \text{means} \qquad a = b^n$$
>
> - If $n \geq 2$ and even, then a and b must be greater than or equal to 0.
> - If $n \geq 3$ and odd, then a and b can be any real number.

Work Smart
If the index is even, then the radicand must be greater than or equal to zero in order for a radical to simplify to a real number. If the index is odd, the radicand can be any real number.

In the notation $\sqrt[n]{a}$, the integer n, $n \geq 2$, is called the **index.** If a radical is written without the index, it is understood that we mean the square root, so $\sqrt{a}$ represents the square root of a. If the index is 3, we call $\sqrt[3]{a}$ the **cube root** of a. If the index is even, then the radicand must be greater than or equal to 0. If the index is odd, then the radicand can be any real number. Do you know why? Since $\sqrt[n]{a} = b$ means $b^n = a$, if the index n is even, then $b^n \geq 0$ so $a \geq 0$. If we have an odd index n, then b^n can be any real number, so a can be any real number.

Before we evaluate *n*th roots, we list some "perfect" powers of 2, 3, 4, and 5. Having this list will be a great help in finding roots in the examples that follow. Some perfect cubes are $1^3 = 1$, $(-2)^3 = -8$, $3^3 = 27$, $4^3 = 64$ and $(-5)^3 = -125$, and so on. Some perfect fourths are $1^4 = 1$, $2^4 = 16$, $3^4 = 81$, $4^4 = 256$, $5^4 = 625$, and so on. Notice that perfect cubes can be negative, but perfect fourths cannot. Do you know why?

Perfect Squares	Perfect Cubes	Perfect Fourths	Perfect Fifths
$1^2 = 1$	$(-2)^3 = -8$	$1^4 = 1$	$(-2)^5 = -32$
$2^2 = 4$	$(-1)^3 = -1$	$2^4 = 16$	$(-1)^5 = -1$
$3^2 = 9$	$1^3 = 1$	$3^4 = 81$	$1^5 = 1$
$4^2 = 16$ and so on	$2^3 = 8$ and so on	$4^4 = 256$ and so on	$2^5 = 32$ and so on

EXAMPLE 1 **Evaluating *n*th Roots of Real Numbers**

Evaluate:

   (a) $\sqrt[3]{64}$ **(b)** $\sqrt[4]{16}$ **(c)** $\sqrt[3]{-8}$ **(d)** $\sqrt[4]{-81}$

Preparing for...Answers

1. $\dfrac{1}{x^3 y^9}$ **2.** 7   **3.** 8

Solution

(a) $\sqrt[3]{64} = 4$ since $4^3 = 64$.

(b) $\sqrt[4]{16} = 2$ since $2^4 = 16$.

(c) $\sqrt[3]{-8} = -2$ since $(-2)^3 = -8$.

(d) Because there is no real number b such that $b^4 = -81$, $\sqrt[4]{-81}$ is not a real number. ■

QUICK ✓ *Evaluate each root.*

1. $\sqrt[3]{125}$ **2.** $\sqrt[4]{81}$ **3.** $\sqrt[3]{-216}$ **4.** $\sqrt[4]{-32}$ **5.** $\sqrt[5]{\dfrac{1}{32}}$

The *n*th roots in Examples 1(a)–(c) were all rational numbers. This is not always the case. Just as we can approximate square roots using a calculator, we can also approximate *n*th roots.

EXAMPLE 2   Approximating an *n*th Root Using a Calculator

(a) Write $\sqrt[3]{25}$ as a decimal rounded to two decimal places.

(b) Write $\sqrt[4]{18}$ as a decimal rounded to two decimal places.

Solution

(a) Because $\sqrt[3]{8} = 2$ and $\sqrt[3]{27} = 3$, we expect $\sqrt[3]{25}$ is between 2 and 3 (closer to 3). Figure 1(a) shows the approximate value of $\sqrt[3]{25}$ obtained from a TI-84 Plus graphing calculator. So $\sqrt[3]{25} \approx 2.92$.

(b) Because $\sqrt[4]{16} = 2$ and $\sqrt[4]{81} = 3$, we expect $\sqrt[4]{18}$ is between 2 and 3 (closer to 2). Figure 1(b) shows the approximate value of $\sqrt[4]{18}$ obtained from a TI-84 Plus graphing calculator. So $\sqrt[4]{18} \approx 2.06$.

Figure 1

(a)   (b) ■

QUICK ✓ *Use a calculator to write the approximate value of each radical rounded to two decimal places.*

6. $\sqrt[3]{50}$ **7.** $\sqrt[4]{80}$ **8.** $\sqrt[5]{40}$

---

② Simplify Expressions of the Form $\sqrt[n]{a^n}$

We have already seen that $\sqrt{a^2} = |a|$. But what about $\sqrt[3]{a^3}$ or $\sqrt[4]{a^4}$? Recall that the definition of the principal *n*th root, $\sqrt[n]{a}$, requires that $a \geq 0$ when n is even and a can be any real number when n is odd.

> **SIMPLIFYING $\sqrt[n]{a^n}$**
>
> If $n \geq 2$ is a positive integer and a is a real number, then
>
> $$\sqrt[n]{a^n} = a \quad \text{if } n \geq 3 \text{ is odd}$$
> $$\sqrt[n]{a^n} = |a| \quad \text{if } n \geq 2 \text{ is even}$$

EXAMPLE 3 **Simplifying Radicals**

Simplify:

(a) $\sqrt[3]{x^3}$ (b) $\sqrt[4]{(x-7)^4}$ (c) $-\sqrt[6]{(-3)^6}$ (d) $\sqrt[3]{-\dfrac{8}{125}}$

Solution

(a) Because the index, 3, is odd, we have that $\sqrt[3]{x^3} = x$.

(b) Because the index, 4, is even, we have that $\sqrt[4]{(x-7)^4} = |x-7|$.

(c) $-\sqrt[6]{(-3)^6} = -|-3| = -3$

(d) $\sqrt[3]{-\dfrac{8}{125}} = \sqrt[3]{\dfrac{-8}{125}} = \sqrt[3]{\dfrac{(-2)^3}{5^3}} = \sqrt[3]{\left(\dfrac{-2}{5}\right)^3} = -\dfrac{2}{5}$

QUICK ✓ *Simplify each radical.*

9. $\sqrt[4]{5^4}$ **10.** $\sqrt[6]{z^6}$ **11.** $\sqrt[7]{(3x-2)^7}$ **12.** $\sqrt[8]{(-2)^8}$ **13.** $\sqrt[5]{\dfrac{-32}{243}}$

---

③ **Evaluate Expressions of the Form $a^{1/n}$**

In Sections 5.2 through 5.4, we carefully developed methods for simplifying algebraic expressions that contained integer exponents. This development started with methods for simplifying algebraic expressions containing only positive integer exponents. We then presented a definition for raising a nonzero real number to the power of 0. With this material in hand, we were able to develop rules for simplifying algebraic expressions involving *all* integer exponents. Of course, the world cannot easily be described using only integers, so it is logical that we would want to extend the rules of exponents to rational exponents.

Work Smart
Remember, a rational number is a number of the form $\dfrac{p}{q}$, where p and q are integers and $q \neq 0$.

We start by providing a definition for "a raised to the power $\dfrac{1}{n}$," where a is a real number and n is a positive integer. This definition needs to be written so that the laws of exponents apply. For example, we know that $a^2 = a \cdot a$, so

$$\left(5^{\frac{1}{2}}\right)^2 = 5^{\frac{1}{2}} \cdot 5^{\frac{1}{2}}$$

$a^m \cdot a^n = a^{m+n}:$ $= 5^{\frac{1}{2}+\frac{1}{2}}$

$$= 5^1$$

$$= 5$$

We also know that $\left(\sqrt{5}\right)^2 = 5$, so it is reasonable to conclude that

$$5^{\frac{1}{2}} = \sqrt{5}$$

This suggests the following definition:

> **DEFINITION OF $a^{1/n}$**
>
> If a is a real number and n is an integer with $n \geq 2$, then
>
> $$a^{\frac{1}{n}} = \sqrt[n]{a}$$
>
> provided that $\sqrt[n]{a}$ exists.

EXAMPLE 4 **Evaluating Expressions Containing Exponents of the Form 1/n**

Write each of the following expressions as a radical and simplify, if possible.

(a) $9^{\frac{1}{2}}$ (b) $(-64)^{\frac{1}{3}}$ (c) $-100^{\frac{1}{2}}$ (d) $(-100)^{\frac{1}{2}}$ (e) $z^{\frac{1}{2}}$

Work Smart

Notice the use of parentheses in Examples 4(c) and (d). In Example 4(c), $-100^{1/2}$ means that we should evaluate $100^{1/2}$ first and then multiply the result by -1. Remember, exponents before multiplication.

Solution

(a) $9^{\frac{1}{2}} = \sqrt{9} = 3$

(b) $(-64)^{\frac{1}{3}} = \sqrt[3]{-64} = -4$

(c) $-100^{\frac{1}{2}} = -1 \cdot 100^{\frac{1}{2}} = -\sqrt{100} = -10$

(d) $(-100)^{\frac{1}{2}} = \sqrt{-100}$ is not a real number because there is no real number whose square is -100.

(e) $z^{\frac{1}{2}} = \sqrt{z}$

QUICK ✔ *Write each of the following expressions and simplify, if possible.*

14. $25^{\frac{1}{2}}$ **15.** $(-27)^{\frac{1}{3}}$ **16.** $-64^{\frac{1}{2}}$ **17.** $(-64)^{\frac{1}{2}}$ **18.** $b^{\frac{1}{2}}$

EXAMPLE 5 Writing Radicals with Rational Exponents

Rewrite each of the following radicals with a rational exponent.

(a) $\sqrt[4]{7a}$ (b) $\sqrt[5]{\dfrac{xy^3}{4}}$

Solution

(a) The index on the radical is 4, so this becomes the denominator of the rational exponent. The parentheses are necessary because the radicand is $7a$ and the exponent $\frac{1}{4}$ is applied to both the 7 and the *a*.

$$\sqrt[4]{7a} = (7a)^{\frac{1}{4}}$$

(b) The index on the radical is 5, so this becomes the denominator of the rational exponent.

$$\sqrt[5]{\frac{xy^3}{4}} = \left(\frac{xy^3}{4}\right)^{\frac{1}{5}}$$

QUICK ✔ *Rewrite each of the following radicals with a rational exponent.*

19. $\sqrt[5]{8b}$ **20.** $\sqrt[8]{\dfrac{mn^5}{3}}$

---

(4) ## Evaluate Expressions of the Form $a^{\frac{m}{n}}$

We now look for a definition for $a^{\frac{m}{n}}$, where *m* and *n* are integers, $\dfrac{m}{n}$ is reduced to lowest terms, and $n \geq 2$. The definition we provide should obey all the laws of exponents presented earlier. For example,

$$a^{\frac{m}{n}} = a^{m \cdot \frac{1}{n}} = (a^m)^{\frac{1}{n}} = \sqrt[n]{a^m}$$

and

$$a^{\frac{m}{n}} = a^{\frac{1}{n} \cdot m} = \left(a^{\frac{1}{n}}\right)^m = \left(\sqrt[n]{a}\right)^m$$

This suggests the following definition:

In Words

The expression $a^{\frac{m}{n}} = \sqrt[n]{a^m}$ means that we will raise *a* to the *m*th power first, and then take the *n*th root. The expression $a^{\frac{m}{n}} = \left(\sqrt[n]{a}\right)^m$ means that we will take the *n*th root of *a* first, and then raise to the power of *m*.

> **DEFINITION OF $a^{\frac{m}{n}}$**
>
> If *a* is a real number, *m/n* is a rational number in lowest terms with $n \geq 2$, then
>
> $$a^{\frac{m}{n}} = \sqrt[n]{a^m} = \left(\sqrt[n]{a}\right)^m$$
>
> provided that $\sqrt[n]{a}$ exists.

When simplifying $a^{\frac{m}{n}}$ either $\sqrt[n]{a^m}$ or $\left(\sqrt[n]{a}\right)^m$ may be used. Use the one that makes simplifying the expression easier. Generally, taking the root first, as in $\left(\sqrt[n]{a}\right)^m$, is easier.

EXAMPLE 6 Evaluating Expressions of the Form $a^{m/n}$

Evaluate each of the following expressions, if possible.

(a) $25^{\frac{3}{2}}$ (b) $64^{\frac{2}{3}}$ (c) $-9^{\frac{5}{2}}$ (d) $(-8)^{\frac{4}{3}}$ (e) $(-81)^{\frac{7}{2}}$

Solution

(a) $25^{\frac{3}{2}} = \left(\sqrt{25}\right)^3 = 5^3 = 125$

(b) $64^{\frac{2}{3}} = \left(\sqrt[3]{64}\right)^2 = 4^2 = 16$

(c) $-9^{\frac{5}{2}} = -1 \cdot 9^{\frac{5}{2}} = -1 \cdot \left(\sqrt{9}\right)^5 = -1 \cdot 3^5 = -1 \cdot 243 = -243$

(d) $(-8)^{\frac{4}{3}} = \left(\sqrt[3]{-8}\right)^4 = (-2)^4 = 16$

(e) $(-81)^{\frac{7}{2}}$ is not a real number because $(-81)^{\frac{7}{2}} = \left(\sqrt{-81}\right)^7$ and $\sqrt{-81}$ is not a real number.

Work Smart

When simplifying expressions of the form $a^{\frac{m}{n}}$ it is typically easier to evaluate the radical first.

QUICK ✓ *Evaluate each expression, if possible.*

21. $16^{\frac{3}{2}}$ **22.** $27^{\frac{2}{3}}$ **23.** $-16^{\frac{3}{4}}$ **24.** $(-64)^{\frac{2}{3}}$ **25.** $(-25)^{\frac{5}{2}}$

The expressions in Examples 6(a)–(d) were all rational numbers. Not all expressions involving rational exponents will simplify to rational numbers.

EXAMPLE 7 Approximating Expressions Involving Rational Exponents

Write $35^{\frac{3}{4}}$ as a decimal rounded to two decimal places.

Solution

Figure 2 shows the results obtained from a TI-84 Plus graphing calculator. So $35^{\frac{3}{4}} \approx 14.39$.

Figure 2

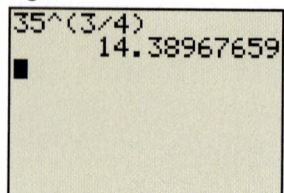

```
35^(3/4)
        14.38967659
```

QUICK ✓ *Approximate the expression rounded to two decimal places.*

26. $50^{\frac{2}{3}}$ **27.** $40^{0.15}$

EXAMPLE 8 Writing Radicals with Rational Exponents

Rewrite each of the following radicals with a rational exponent.

(a) $\sqrt[3]{x^2}$ (b) $\left(\sqrt[5]{10a^2b}\right)^4$

Solution

(a) The index, 3, is the denominator of the rational exponent and the power on the radicand, 2, is the numerator of the rational exponent.

$$\sqrt[3]{x^2} = x^{\frac{2}{3}}$$

(b) The index, 5, is the denominator of the rational exponent and the power, 4, is the numerator of the rational exponent.

$$\left(\sqrt[5]{10a^2b}\right)^4 = (10a^2b)^{\frac{4}{5}}$$

QUICK ✓ *Write each radical with a rational exponent.*

28. $\sqrt[8]{a^3}$ **29.** $\left(\sqrt[4]{12ab^3}\right)^9$

If a rational exponent is negative, then we can use the rule for negative rational exponents given below.

DEFINITION: NEGATIVE-EXPONENT RULE

If $\dfrac{m}{n}$ is a rational number, and if a is a nonzero real number (that is, if $a \neq 0$), then we define

$$a^{-\frac{m}{n}} = \frac{1}{a^{\frac{m}{n}}} \quad \text{and} \quad \frac{1}{a^{-\frac{m}{n}}} = a^{\frac{m}{n}} \quad \text{if } a \neq 0$$

EXAMPLE 9   **Evaluating Expressions with Negative Rational Exponents**

Rewrite each of the following with positive exponents, and completely simplify, if possible.

(a) $36^{-\frac{1}{2}}$ (b) $\dfrac{1}{27^{-\frac{2}{3}}}$ (c) $(6a)^{-\frac{5}{4}}$

Solution

(a) $36^{-\frac{1}{2}} = \dfrac{1}{36^{\frac{1}{2}}} = \dfrac{1}{\sqrt{36}} = \dfrac{1}{6}$

(b) Because the negative exponent is in the exponential expression in the denominator, we use $\dfrac{1}{a^{-\frac{m}{n}}} = a^{\frac{m}{n}}$ to simplify:

$$\frac{1}{27^{-\frac{2}{3}}} = 27^{\frac{2}{3}} = \left(\sqrt[3]{27}\right)^2 = 3^2 = 9$$

(c) $(6a)^{-\frac{5}{4}} = \dfrac{1}{(6a)^{\frac{5}{4}}}$

QUICK ✅ *Rewrite each of the following with positive exponents, and completely simplify, if possible.*

30. $81^{-\frac{1}{2}}$ **31.** $\dfrac{1}{8^{-\frac{2}{3}}}$ **32.** $(13x)^{-\frac{3}{2}}$

9.2 Exercises

For Extra Help: Student Solutions Manual   CD Video   PH Math/Tutor Center   MathXL Tutorials on CD   *Math* XL   *MyMathLab*   MathXL®   MyMathLab

Concepts and Vocabulary

In Problems 1–3, fill in the blanks.

1. If a is a nonnegative real number and $n \geq 2$ is an integer, then $a^{\frac{1}{n}} = $ _____.

2. If $\dfrac{m}{n}$ is a rational number and if a is a nonzero real number (that is, if $a \neq 0$), then we define $a^{-\frac{m}{n}} = $ _____.

3. In the notation $\sqrt[n]{a}$, the integer n, $n \geq 2$, is called the _____.

In Problems 4 and 5, answer True or False to each statement.

4. If a is a real number, and m/n is a rational number in lowest terms with $n \geq 2$, then $a^{\frac{m}{n}} = \sqrt[n]{a^m} = \left(\sqrt[n]{a}\right)^m$.

5. If $a < 0$ is a real number and $n \geq 2$ is an integer, then $a^{\frac{1}{n}}$ exists.

6. Explain why $(-9)^{\frac{1}{2}}$ is not a real number, but $-9^{\frac{1}{2}}$ is a real number.

7. In your own words, provide a justification for why $a^{\frac{1}{n}} = \sqrt[n]{a}$.

8. Under what conditions is $a^{\frac{m}{n}}$ a real number?

Building Skills

In Problems 9–26, simplify each radical.

9. $\sqrt[3]{125}$ **10.** $\sqrt[3]{216}$ **11.** $\sqrt[3]{-27}$ **12.** $\sqrt[3]{-64}$

13. $-\sqrt[4]{625}$ **14.** $-\sqrt[4]{256}$ **15.** $\sqrt[3]{-\dfrac{1}{8}}$ **16.** $\sqrt[3]{\dfrac{8}{125}}$

17. $-\sqrt[5]{-243}$ **18.** $-\sqrt[5]{-1024}$ **19.** $\sqrt[3]{5^3}$ **20.** $\sqrt[4]{6^4}$

21. $\sqrt[4]{m^4}$ **22.** $\sqrt[5]{n^5}$ **23.** $\sqrt[9]{(x-3)^9}$ **24.** $\sqrt[6]{(2x-3)^6}$

25. $-\sqrt[4]{(3p+1)^4}$ **26.** $-\sqrt[3]{(6z-5)^3}$

In Problems 27–56, evaluate each expression, if possible.

27. $4^{\frac{1}{2}}$ **28.** $16^{\frac{1}{2}}$ **29.** $-36^{\frac{1}{2}}$ **30.** $-25^{\frac{1}{2}}$ **31.** $8^{\frac{1}{3}}$

32. $27^{\frac{1}{3}}$ **33.** $-16^{\frac{1}{4}}$ **34.** $-81^{\frac{1}{4}}$ **35.** $\left(\dfrac{4}{25}\right)^{\frac{1}{2}}$ **36.** $\left(\dfrac{8}{27}\right)^{\frac{1}{3}}$

37. $(-125)^{\frac{1}{3}}$ **38.** $(-216)^{\frac{1}{3}}$ **39.** $(-4)^{\frac{1}{2}}$ **40.** $(-81)^{\frac{1}{2}}$ **41.** $4^{\frac{5}{2}}$

42. $25^{\frac{3}{2}}$ **43.** $-16^{\frac{3}{2}}$ **44.** $-100^{\frac{5}{2}}$ **45.** $8^{\frac{4}{3}}$ **46.** $27^{\frac{4}{3}}$

47. $(-64)^{\frac{2}{3}}$ **48.** $(-125)^{\frac{2}{3}}$ **49.** $-(-32)^{\frac{3}{5}}$ **50.** $-(-216)^{\frac{2}{3}}$ **51.** $144^{-\frac{1}{2}}$

52. $121^{-\frac{1}{2}}$ **53.** $\dfrac{1}{25^{-\frac{3}{2}}}$ **54.** $\dfrac{1}{49^{-\frac{3}{2}}}$ **55.** $\dfrac{1}{8^{-\frac{5}{3}}}$ **56.** $27^{-\frac{4}{3}}$

In Problems 57–68, rewrite each of the following radicals with a rational exponent.

57. $\sqrt[3]{3x}$ **58.** $\sqrt[5]{2y}$ **59.** $\sqrt[4]{\dfrac{x}{3}}$ **60.** $\sqrt{\dfrac{w}{2}}$ **61.** $\sqrt[4]{x^3}$

62. $\sqrt[3]{p^5}$ **63.** $\left(\sqrt[5]{3x}\right)^2$ **64.** $\left(\sqrt[4]{6z}\right)^3$ **65.** $\sqrt{\left(\dfrac{5x}{y}\right)^3}$ **66.** $\sqrt[6]{\left(\dfrac{2a}{b}\right)^5}$

67. $\sqrt[3]{(9ab)^4}$ **68.** $\sqrt[4]{(3pq)^7}$

In Problems 69–78, use a calculator to write each expression as a decimal rounded to two decimal places.

69. $\sqrt[3]{25}$ **70.** $\sqrt[3]{85}$ **71.** $\sqrt[4]{12}$ **72.** $\sqrt[4]{2}$ **73.** $20^{\frac{1}{2}}$

74. $5^{\frac{1}{2}}$ **75.** $4^{\frac{5}{3}}$ **76.** $100^{\frac{3}{4}}$ **77.** $10^{0.1}$ **78.** $100^{0.25}$

Mixed Practice

In Problems 79–92, evaluate each expression, if possible.

79. $\sqrt[3]{512}$ **80.** $\sqrt[3]{-125}$ **81.** $9^{\frac{5}{2}}$ **82.** $100^{\frac{3}{2}}$ **83.** $\sqrt[4]{-16}$

84. $\sqrt[4]{-1}$ **85.** $144^{-\frac{1}{2}}$ **86.** $125^{-\frac{1}{3}}$ **87.** $\sqrt[3]{0.008}$ **88.** $\sqrt[4]{0.0081}$

89. $4^{\frac{1}{2}} + 25^{\frac{3}{2}}$ **90.** $100^{\frac{1}{2}} - 4^{\frac{3}{2}}$ **91.** $(-25)^{\frac{5}{2}}$ **92.** $(-125)^{-\frac{1}{3}}$

Applying the Concepts

93. What are the cube roots of 1,000? What is $\sqrt[3]{1000}$?

94. What are the cube roots of 729? What is $\sqrt[3]{729}$?

95. **Wind Chill** According to the National Weather Service, the wind chill temperature is how cold people and animals feel when outside. Wind chill is based on the rate of heat loss from exposed skin caused by wind and cold. The formula for computing wind chill W is

$$W = 35.74 + 0.6215T - 35.75v^{0.16} + 0.4275Tv^{0.16}$$

where T is the air temperature in degrees Fahrenheit and v is the wind speed in miles per hour.

  (a) What is the wind chill if it is 30°F and the wind speed is 10 miles per hour?
  (b) What is the wind chill if it is 30°F and the wind speed is 20 miles per hour?
  (c) What is the wind chill if it is 0°F and the wind speed is 10 miles per hour?

96. **Money** The annual rate of interest r (expressed as a decimal) required to have A dollars after t years from an initial deposit of P dollars is given by

$$r = \left(\frac{A}{P}\right)^{\frac{1}{t}} - 1$$

  (a) If you deposit \$100 in a mutual fund today and have \$144 in the account in 2 years, what was your annual rate of interest earned?
  (b) If you deposit \$100 in a mutual fund today and have \$337.50 in 3 years, what was your annual rate of interest earned?
  (c) The Rule of 72 states that your money will double in $\dfrac{72}{100r}$ years where r is the rate of interest earned (expressed as a decimal). Suppose that you deposit \$1000 in a mutual fund today and have \$2000 in 8 years. What rate of interest did you earn? Compute $\dfrac{72}{100r}$ for this rate of interest. Is it close?

97. **Terminal Velocity** Terminal speed is the maximum speed that a body falling through air can reach. The speed is limited by air resistance. Terminal velocity is given by the formula $v_t = \sqrt{\dfrac{2mg}{C\rho A}}$, where m is the mass of the falling object, g is acceleration due to gravity (≈ 9.81 meters per second^2), C is a drag coefficient with $0.5 \le C \le 1.0$, ρ is the density of air (≈ 1.2 kg/m^3), and A is the cross-sectional area of the object. Suppose that a raindrop whose radius is 1.5 mm falls from the sky. The mass of the raindrop is given by $m = \dfrac{4}{3}\pi r^3 \rho_w$ where r is its radius and $\rho_w = 1000$ kg/m^3. The cross-sectional area of the raindrop is $A = \pi r^2$.

  (a) Substitute the formulas for the mass and area of a raindrop into the formula for terminal speed and simplify the expression.
  (b) Determine the terminal velocity of a raindrop whose radius is 0.0015 m with $C = 0.6$.

98. **Kepler's Law** Early in the seventeenth century, Johannes Kepler (1571–1630) discovered that the square of the period T of a planet varies directly with the cube of its mean distance r from the Sun. The period of a planet is the amount of time (in years) for the planet to complete one orbit around the Sun. Kepler's Law can be expressed using rational exponents as $T = kr^{\frac{3}{2}}$, where k is the constant of proportionality.

  (a) The period of Mercury is 0.241 years and its mean distance from the Sun is 5.79×10^{10} meters. Use this information to state Kepler's Law (find the value of k).
  (b) The mean distance of Mars to the Sun is 2.28×10^{11} m. Use this information along with the result of part (a) to find the amount of time it takes Mars to complete one orbit around the Sun.

Extending the Concepts

In Problems 99–102, evaluate each function.

99. $f(x) = x^{\frac{3}{2}}$; find $f(4)$

100. $g(x) = x^{-\frac{3}{2}}$; find $g(16)$

101. $F(z) = z^{\frac{4}{3}}$; find $F(-8)$

102. $G(a) = a^{\frac{5}{3}}$; find $G(-8)$

Synthesis Review

In Problems 103–107, completely simplify each expression.

103. $\left(\dfrac{x^2 y}{y^{-2}} \right)^3$

104. $\dfrac{(x+2)^2(x-1)^4}{(x+2)(x-1)}$

105. $\dfrac{(x-1)^2(x^2+5x+6)}{(x+2)\sqrt{x-1}}$

106. $\dfrac{(3a^2+5a-3)-(a^2-2a-9)}{4a^2+12a+9}$

107. $\dfrac{(4z^2-7z+3)+(-3z^2-z+9)}{(4z^2-2z-7)+(-3z^2-z+9)}$

9.3 Simplify Expressions Using the Laws of Exponents

OBJECTIVES

1 Use the Laws of Exponents to Simplify Expressions Involving Rational Exponents

2 Use the Laws of Exponents to Simplify Radical Expressions

3 Factor Expressions Containing Rational Exponents

Preparing for *Simplifying Expressions Using the Laws of Exponents*

Before getting started, take the following readiness quiz. If you get a problem wrong, go to the section cited and review the material.

1. Simplify: z^{-3} [Section 5.4, pp. 374–376]

2. Simplify: $x^{-2} \cdot x^5$ [Section 5.4, pp. 377–380]

3. Simplify: $\left(\dfrac{2a^2}{b^{-1}} \right)^3$ [Section 5.4, pp. 377–380]

4. Evaluate: $\sqrt{64}$ [Section 9.1, pp. 660–661]

1 **Use the Laws of Exponents to Simplify Expressions Involving Rational Exponents**

The Laws of Exponents that were presented in Chapter 5 on page 377 applied to integer exponents. These same laws apply to rational exponents as well.

THE LAWS OF EXPONENTS

If a and b are real numbers and if r and s are rational numbers, then assuming the expression is defined,

Zero-Exponent Rule:	$a^0 = 1$	if $a \neq 0$
Negative-Exponent Rule:	$a^{-r} = \dfrac{1}{a^r}$	if $a \neq 0$
Product Rule:	$a^r \cdot a^s = a^{r+s}$	
Quotient Rule:	$\dfrac{a^r}{a^s} = a^{r-s} = \dfrac{1}{a^{s-r}}$	if $a \neq 0$
Power Rule:	$(a^r)^s = a^{r \cdot s}$	
Product to Power Rule:	$(a \cdot b)^r = a^r \cdot b^r$	
Quotient to Power Rule:	$\left(\dfrac{a}{b} \right)^r = \dfrac{a^r}{b^r}$	if $b \neq 0$
Quotient to a Negative Power Rule:	$\left(\dfrac{a}{b} \right)^{-r} = \left(\dfrac{b}{a} \right)^r$	if $a \neq 0, b \neq 0$

Work Smart

We *simplify* expressions (no equal sign) and *solve* equations.

The direction **simplify** shall mean the following:

- All the exponents are positive.
- Each base only occurs once.
- There are no parentheses in the expression.
- There are no powers written to powers.

EXAMPLE 1   **Simplifying Expressions Involving Rational Exponents**

Simplify each of the following:

   (a) $27^{\frac{1}{2}} \cdot 27^{\frac{5}{6}}$

   (b) $\dfrac{8^{\frac{1}{3}}}{8^{\frac{5}{3}}}$

Solution

(a) $a^r \cdot a^s = a^{r+s}$

$$27^{\frac{1}{2}} \cdot 27^{\frac{5}{6}} = 27^{\frac{1}{2}+\frac{5}{6}}$$
$$= 27^{\frac{3}{6}+\frac{5}{6}}$$
$$= 27^{\frac{8}{6}}$$
$$= 27^{\frac{4}{3}}$$

$a^{\frac{m}{n}} = \left(\sqrt[n]{a}\right)^m$: $= \left(\sqrt[3]{27}\right)^4$
$$= 3^4$$
$$= 81$$

(b) $\dfrac{a^r}{a^s} = a^{r-s}$

$$\dfrac{8^{\frac{1}{3}}}{8^{\frac{5}{3}}} = 8^{\frac{1}{3}-\frac{5}{3}}$$
$$= 8^{-\frac{4}{3}}$$
$$= \dfrac{1}{8^{\frac{4}{3}}}$$

$a^{\frac{m}{n}} = \left(\sqrt[n]{a}\right)^m$: $= \dfrac{1}{\left(\sqrt[3]{8}\right)^4}$
$$= \dfrac{1}{2^4}$$
$$= \dfrac{1}{16}$$

EXAMPLE 2   **Simplifying Expressions Involving Rational Exponents**

Simplify each of the following:

   (a) $\left(36^{\frac{2}{5}}\right)^{\frac{5}{4}}$

   (b) $\left(x^{\frac{1}{2}} \cdot y^{\frac{2}{3}}\right)^{\frac{3}{2}}$

Solution

(a) $(a^r)^s = a^{r \cdot s}$

$$\left(36^{\frac{2}{5}}\right)^{\frac{5}{4}} = 36^{\frac{2}{5} \cdot \frac{5}{4}}$$
$$= 36^{\frac{10}{20}}$$
$$= 36^{\frac{1}{2}}$$
$$= 6$$

(b) $(ab)^r = a^r \cdot b^r$

$$\left(x^{\frac{1}{2}} \cdot y^{\frac{2}{3}}\right)^{\frac{3}{2}} = \left(x^{\frac{1}{2}}\right)^{\frac{3}{2}} \cdot \left(y^{\frac{2}{3}}\right)^{\frac{3}{2}}$$
$$= x^{\frac{1}{2} \cdot \frac{3}{2}} \cdot y^{\frac{2}{3} \cdot \frac{3}{2}}$$

$(a^r)^s = a^{r \cdot s}$: $= x^{\frac{3}{4}} y$

QUICK ✓ *Simplify each expression.*

1. $5^{\frac{3}{4}} \cdot 5^{\frac{1}{6}}$

2. $\dfrac{32^{\frac{6}{5}}}{32^{\frac{3}{5}}}$

3. $\left(100^{\frac{3}{8}}\right)^{\frac{4}{3}}$

4. $\left(a^{\frac{3}{2}} \cdot b^{\frac{5}{4}}\right)^{\frac{2}{3}}$

EXAMPLE 3 **Simplifying Expressions Involving Rational Exponents**

Simplify each of the following:

(a) $\left(x^{\frac{2}{3}}y^{-1}\right) \cdot \left(x^{-1}y^{\frac{1}{2}}\right)^{\frac{2}{3}}$

(b) $\left(\dfrac{9xy^{\frac{4}{3}}}{x^{\frac{5}{6}}y^{-\frac{2}{3}}}\right)^{\frac{1}{2}}$

Solution

(a)

Product to Power Rule: $(ab)^r = a^r b^r$
$$\left(x^{\frac{2}{3}}y^{-1}\right) \cdot \left(x^{-1}y^{\frac{1}{2}}\right)^{\frac{2}{3}} = x^{\frac{2}{3}}y^{-1}(x^{-1})^{\frac{2}{3}}\left(y^{\frac{1}{2}}\right)^{\frac{2}{3}}$$

Power Rule: $(a^r)^s = a^{rs}$: $\quad = x^{\frac{2}{3}}y^{-1}x^{-\frac{2}{3}}y^{\frac{1}{3}}$

Product Rule: $a^r \cdot a^s = a^{r+s}$: $\quad = x^{\frac{2}{3}+\left(-\frac{2}{3}\right)}y^{-1+\frac{1}{3}}$

$$= x^0 y^{-\frac{2}{3}}$$

$a^0 = 1$; Negative Exponent Rule: $a^{-r} = \dfrac{1}{a^r}$: $\quad = \dfrac{1}{y^{\frac{2}{3}}}$

(b)

Quotient Rule: $\dfrac{a^r}{a^s} = a^{r-s}$
$$\left(\dfrac{9xy^{\frac{4}{3}}}{x^{\frac{5}{6}}y^{-\frac{2}{3}}}\right)^{\frac{1}{2}} = \left(9x^{1-\frac{5}{6}}y^{\frac{4}{3}-\left(-\frac{2}{3}\right)}\right)^{\frac{1}{2}}$$

$x^{1-\frac{5}{6}} = x^{\frac{6}{6}-\frac{5}{6}} = x^{\frac{1}{6}}$;

$y^{\frac{4}{3}-\left(-\frac{2}{3}\right)} = y^{\frac{4}{3}+\frac{2}{3}} = y^{\frac{6}{3}} = y^2$: $\quad = \left(9x^{\frac{1}{6}}y^2\right)^{\frac{1}{2}}$

Power Rule: $(a^r)^s = a^{rs}$: $\quad = 9^{\frac{1}{2}} \cdot \left(x^{\frac{1}{6}}\right)^{\frac{1}{2}} \cdot (y^2)^{\frac{1}{2}}$

$9^{\frac{1}{2}} = \sqrt{9} = 3$; Power Rule: $(a^r)^s = a^{rs}$: $\quad = 3x^{\frac{1}{12}}y$

QUICK ✔ *Simplify each expression.*

5. $\left(8x^{\frac{3}{4}}y^{-1}\right)^{\frac{2}{3}}$ **6.** $\left(\dfrac{25x^{\frac{1}{2}}y^{\frac{3}{4}}}{x^{-\frac{3}{4}}y}\right)^{\frac{1}{2}}$ **7.** $8\left(125a^{\frac{3}{4}}b^{-1}\right)^{\frac{2}{3}}$

---

② Use the Laws of Exponents to Simplify Radical Expressions

Rational exponents can be used to simplify radicals.

EXAMPLE 4 **Simplifying Radicals Using Rational Exponents**

Use rational exponents to simplify the radicals.

(a) $\sqrt[8]{16^4}$ **(b)** $\sqrt[3]{64x^6y^3}$ **(c)** $\dfrac{\sqrt{x}}{\sqrt[3]{x^2}}$ **(d)** $\sqrt{\sqrt[3]{z}}$

Solution

The idea in all these problems is to rewrite the radical as an expression involving a rational exponent. Then use the Laws of Exponents to simplify the expression. Finally, write the simplified expression as a radical.

(a)

Write radical as a rational exponent using $\sqrt[n]{a^m} = a^{\frac{m}{n}}$.
$$\sqrt[8]{16^4} = 16^{\frac{4}{8}}$$

Simplify the exponent: $\quad = 16^{\frac{1}{2}}$

Write rational exponent as a radical: $\quad = \sqrt{16} = 4$

(b)

$$\sqrt[n]{a^m} = a^{\frac{m}{n}}$$
$$\downarrow$$
$$\sqrt[3]{64x^6y^3} = (64x^6y^3)^{\frac{1}{3}}$$

$(ab)^r = a^r \cdot b^r: \quad = 64^{\frac{1}{3}} \cdot (x^6)^{\frac{1}{3}} \cdot (y^3)^{\frac{1}{3}}$

$(a^r)^s = a^{rs}: \quad = 64^{\frac{1}{3}} \cdot x^{6 \cdot \frac{1}{3}} \cdot y^{3 \cdot \frac{1}{3}}$

$64^{\frac{1}{3}} = \sqrt[3]{64} = 4: \quad = 4x^2y$

(c)

$$\sqrt{a} = a^{\frac{1}{2}}; \ \sqrt[n]{a^m} = a^{\frac{m}{n}}$$
$$\downarrow$$
$$\frac{\sqrt{x}}{\sqrt[3]{x^2}} = \frac{x^{\frac{1}{2}}}{x^{\frac{2}{3}}}$$

$\dfrac{a^r}{a^s} = a^{r-s}: \quad = x^{\frac{1}{2} - \frac{2}{3}}$

$\text{LCD} = 6; \ \dfrac{1}{2} - \dfrac{2}{3} = \dfrac{3}{6} - \dfrac{4}{6} = -\dfrac{1}{6}: \quad = x^{-\frac{1}{6}}$

$a^{-r} = \dfrac{1}{a^r}: \quad = \dfrac{1}{x^{\frac{1}{6}}}$

Write rational exponent as a radical: $\quad = \dfrac{1}{\sqrt[6]{x}}$

(d)

Write radicand with a rational exponent.
$$\downarrow$$
$$\sqrt{\sqrt[3]{z}} = \sqrt{z^{\frac{1}{3}}}$$

$\sqrt{a} = a^{\frac{1}{2}}: \quad = \left(z^{\frac{1}{3}}\right)^{\frac{1}{2}}$

$(a^r)^s = a^{rs}: \quad = z^{\frac{1}{3} \cdot \frac{1}{2}}$

$\quad = z^{\frac{1}{6}}$

Write rational exponent as a radical: $\quad = \sqrt[6]{z}$

QUICK ✔ *Use rational exponents to simplify each radical.*

8. $\sqrt[10]{36^5}$ **9.** $\sqrt[4]{16a^8b^{12}}$ **10.** $\dfrac{\sqrt[3]{x^2}}{\sqrt[4]{x}}$ **11.** $\sqrt[4]{\sqrt[3]{a^2}}$

(3) ## Factor Expressions Containing Rational Exponents

Often, expressions involving rational exponents contain a common factor. When this occurs, we want to factor out the common factor to write the expression in simplified form. The goal of these types of problems is to write the expression as either a single product or a single quotient. We present two examples to illustrate the idea.

EXAMPLE 5 ### Writing an Expression Containing Rational Exponents as a Single Product

Simplify $9x^{\frac{4}{3}} + 4x^{\frac{1}{3}}(3x + 5)$ by factoring out $x^{\frac{1}{3}}$.

Solution

Clearly, $x^{\frac{1}{3}}$ is a factor of the second term, $4x^{\frac{1}{3}}(3x + 5)$. It is also a factor of the first term, $9x^{\frac{4}{3}}$. We can see this by rewriting $9x^{\frac{4}{3}}$ as $9x^{\frac{3}{3} + \frac{1}{3}} = 9x^{\frac{3}{3}} \cdot x^{\frac{1}{3}} = 9x \cdot x^{\frac{1}{3}}$. Now we

Work Smart

When factoring out the greatest common factor, factor out the variable expression raised to the smallest exponent that the expressions have in common. For example,

$$3x^5 + 12x^2 = 3x^2(x^3 + 4)$$

proceed to factor out $x^{\frac{1}{3}}$.

$$9x^{\frac{4}{3}} + 4x^{\frac{1}{3}}(3x + 5) = 9x \cdot x^{\frac{1}{3}} + 4x^{\frac{1}{3}}(3x + 5)$$

$$\text{Factor out } x^{1/3}: \quad = x^{\frac{1}{3}}(9x + 4(3x + 5))$$

$$\text{Distribute the 4:} \quad = x^{\frac{1}{3}}(9x + 12x + 20)$$

$$\text{Combine like terms:} \quad = x^{\frac{1}{3}}(21x + 20)$$

QUICK ✓

12. Simplify $8x^{\frac{3}{2}} + 3x^{\frac{1}{2}}(4x + 3)$ by factoring out $x^{\frac{1}{2}}$.

EXAMPLE 6 **Writing an Expression Containing Rational Exponents as a Single Quotient**

Simplify $4x^{\frac{1}{2}} + x^{-\frac{1}{2}}(2x + 1)$ by factoring out $x^{-\frac{1}{2}}$.

Solution

Clearly, $x^{-\frac{1}{2}}$ is a factor of the second term, $x^{-\frac{1}{2}}(2x + 1)$. It is also a factor of the first term, $4x^{\frac{1}{2}}$. We can see this by rewriting $4x^{\frac{1}{2}}$ as $4x^{\frac{2}{2} - \frac{1}{2}} = 4x^{\frac{2}{2}} \cdot x^{-\frac{1}{2}} = 4x \cdot x^{-\frac{1}{2}}$. Now we proceed to factor out $x^{-\frac{1}{2}}$.

$$4x^{\frac{1}{2}} + x^{-\frac{1}{2}}(2x + 1) = 4x \cdot x^{-\frac{1}{2}} + x^{-\frac{1}{2}}(2x + 1)$$

$$\text{Factor out } x^{-1/2}: \quad = x^{-\frac{1}{2}}(4x + (2x + 1))$$

$$\text{Combine like terms:} \quad = x^{-\frac{1}{2}}(6x + 1)$$

$$\text{Rewrite without negative exponents:} \quad = \frac{6x + 1}{x^{\frac{1}{2}}}$$

QUICK ✓

13. Simplify $9x^{\frac{1}{3}} + x^{-\frac{2}{3}}(3x + 1)$ by factoring out $x^{-\frac{2}{3}}$.

9.3 Exercises

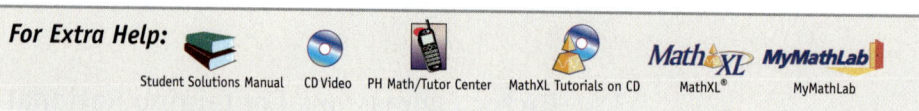

For Extra Help: Student Solutions Manual · CD Video · PH Math/Tutor Center · MathXL Tutorials on CD · MathXL® · MyMathLab

Concepts and Vocabulary

In Problems 1 and 2, fill in the blanks.

1. If a and b are real numbers and if r and s are rational numbers, then assuming the expression is defined, $(ab)^r = $ _____.

2. If a and b are real numbers and if r and s are rational numbers, then assuming the expression is defined, $a^r \cdot a^s = $ _____.

Building Skills

In Problems 3–24, simplify each of the following expressions.

3. $5^{\frac{1}{2}} \cdot 5^{\frac{3}{2}}$

4. $3^{\frac{1}{3}} \cdot 3^{\frac{5}{3}}$

5. $\dfrac{8^{\frac{5}{4}}}{8^{\frac{1}{4}}}$

6. $\dfrac{10^{\frac{7}{5}}}{10^{\frac{2}{5}}}$

7. $2^{\frac{1}{3}} \cdot 2^{-\frac{3}{2}}$

8. $9^{-\frac{5}{4}} \cdot 9^{\frac{1}{3}}$

9. $\dfrac{x^{\frac{1}{4}}}{x^{\frac{5}{6}}}$

10. $\dfrac{y^{\frac{1}{5}}}{y^{\frac{9}{10}}}$

11. $\left(4^{\frac{4}{3}}\right)^{\frac{3}{8}}$

12. $\left(9^{\frac{3}{5}}\right)^{\frac{5}{6}}$

13. $\left(25^{\frac{3}{4}} \cdot 4^{-\frac{3}{4}}\right)^2$

14. $\left(36^{-\frac{1}{4}} \cdot 9^{\frac{3}{4}}\right)^{-2}$

15. $\left(x^{\frac{3}{4}} \cdot y^{\frac{1}{3}}\right)^{\frac{2}{3}}$

16. $\left(a^{\frac{5}{4}} \cdot b^{\frac{3}{2}}\right)^{\frac{2}{5}}$

17. $\left(x^{-\frac{1}{3}} \cdot y\right)\left(x^{\frac{1}{2}} \cdot y^{-\frac{4}{3}}\right)$

18. $\left(a^{\frac{4}{3}} \cdot b^{-\frac{1}{2}}\right)\left(a^{-2} \cdot b^{\frac{5}{2}}\right)$

19. $\left(4a^2 b^{-\frac{3}{2}}\right)^{\frac{1}{2}}$

20. $\left(25 p^{\frac{2}{5}} q^{-1}\right)^{\frac{1}{2}}$

21. $\left(\dfrac{x^{\frac{2}{3}} y^{-\frac{1}{3}}}{8x^{\frac{1}{2}} y}\right)^{\frac{1}{3}}$

22. $\left(\dfrac{64 m^{\frac{1}{2}} n}{m^{-2} n^{\frac{4}{3}}}\right)^{\frac{1}{2}}$

23. $\left(\dfrac{50 x^{\frac{3}{4}} y}{2x^{\frac{1}{2}}}\right)^{\frac{1}{2}} + \left(\dfrac{x^{\frac{1}{2}} y^{\frac{1}{2}}}{9x^{\frac{3}{4}} y^{\frac{3}{2}}}\right)^{-\frac{1}{2}}$

24. $\left(\dfrac{27 x^{\frac{1}{2}} y^{-1}}{y^{-\frac{2}{3}} x^{-\frac{1}{2}}}\right)^{1/3} - \left(\dfrac{4x^{\frac{1}{3}} y^{\frac{4}{9}}}{x^{-\frac{1}{3}} y^{\frac{2}{3}}}\right)^{\frac{1}{2}}$

In Problems 25–30, distribute and simplify.

25. $x^{\frac{1}{2}}\left(x^{\frac{5}{2}} - 2\right)$

26. $x^{\frac{1}{3}}\left(x^{\frac{5}{3}} + 4\right)$

27. $2y^{-\frac{1}{3}}(1 + 3y)$

28. $3a^{-\frac{1}{2}}(2 - a)$

29. $4z^{\frac{3}{2}}\left(z^{\frac{5}{2}} - 8z^{-\frac{3}{2}}\right)$

30. $8p^{\frac{2}{3}}\left(p^{\frac{4}{3}} - 4p^{-\frac{2}{3}}\right)$

In Problems 31–46, use rational exponents to simplify each radical. Assume all variables are positive.

31. $\sqrt{x^8}$

32. $\sqrt[3]{x^6}$

33. $\sqrt[12]{8^4}$

34. $\sqrt[9]{125^6}$

35. $\sqrt[3]{8a^3 b^{12}}$

36. $\sqrt{25x^4 y^6}$

37. $\dfrac{\sqrt{x}}{\sqrt[4]{x}}$

38. $\dfrac{\sqrt[3]{y^2}}{\sqrt{y}}$

39. $\sqrt{x} \cdot \sqrt[3]{x}$

40. $\sqrt[4]{p^3} \cdot \sqrt[3]{p}$

41. $\sqrt{\sqrt[4]{x^3}}$

42. $\sqrt[3]{\sqrt{x^3}}$

43. $\sqrt{3} \cdot \sqrt[3]{9}$

44. $\sqrt{5} \cdot \sqrt[3]{25}$

45. $\dfrac{\sqrt{6}}{\sqrt[4]{36}}$

46. $\dfrac{\sqrt[4]{49}}{\sqrt{7}}$

47. Simplify $2x^{\frac{3}{2}} + 3x^{\frac{1}{2}}(x + 5)$ by factoring out $x^{\frac{1}{2}}$.

48. Simplify $6x^{\frac{4}{3}} + 4x^{\frac{1}{3}}(2x - 3)$ by factoring out $x^{\frac{1}{3}}$.

49. Simplify $5(x + 2)^{\frac{2}{3}}(3x - 2) + 9(x + 2)^{\frac{5}{3}}$ by factoring out $(x + 2)^{\frac{2}{3}}$.

50. Simplify $3(x - 5)^{\frac{1}{2}}(3x + 1) + 6(x - 5)^{\frac{3}{2}}$ by factoring out $(x - 5)^{\frac{1}{2}}$.

51. Simplify $x^{-\frac{1}{2}}(2x + 5) + 4x^{\frac{1}{2}}$ by factoring out $x^{-\frac{1}{2}}$.

52. Simplify $x^{-\frac{2}{3}}(3x + 2) + 9x^{\frac{1}{3}}$ by factoring out $x^{-\frac{2}{3}}$.

53. Simplify $2(x - 4)^{-\frac{1}{3}}(4x - 3) + 12(x - 4)^{\frac{2}{3}}$ by factoring out $2(x - 4)^{-\frac{1}{3}}$.

54. Simplify $4(x + 3)^{\frac{1}{2}} + (x + 3)^{-\frac{1}{2}}(2x + 1)$ by factoring out $(x + 3)^{-\frac{1}{2}}$.

55. Simplify $15x(x^2 + 4)^{\frac{1}{2}} + 5(x^2 + 4)^{\frac{3}{2}}$.

56. Simplify $24x(x^2 - 1)^{\frac{1}{3}} + 9(x^2 - 1)^{\frac{4}{3}}$.

Mixed Practice

In Problems 57–68, simplify each expression.

57. $\sqrt[8]{4^4}$

58. $\sqrt[6]{27^2}$

59. $(-2)^{\frac{1}{2}} \cdot (-2)^{\frac{3}{2}}$

60. $25^{\frac{3}{4}} \cdot 25^{\frac{3}{4}}$

61. $\left(100^{\frac{1}{3}}\right)^{\frac{3}{2}}$

62. $(8^4)^{\frac{5}{12}}$

63. $\left(\sqrt[4]{25}\right)^2$

64. $\left(\sqrt[6]{27}\right)^2$

65. $\sqrt[4]{x^2} - \dfrac{\sqrt[4]{x^6}}{x}$

66. $\sqrt[9]{a^6} - \dfrac{\sqrt[6]{a^5}}{\sqrt[6]{a}}$

67. $\left(4 \cdot 9^{\frac{1}{4}}\right)^{-2}$

68. $\left(4^{-1} \cdot 81^{\frac{1}{2}}\right)^{\frac{1}{2}}$

Applying the Concepts

69. If $3^x = 25$, what does $3^{\frac{x}{2}}$ equal?

70. If $5^x = 64$, what does $5^{\frac{x}{3}}$ equal?

71. If $7^x = 9$, what does $\sqrt{7^x}$ equal?

72. If $5^x = 27$, what does $\sqrt[3]{5^x}$ equal?

Extending the Concepts

In Problems 73 and 74, simplify the expression using rational exponents.

73. $\sqrt[4]{\sqrt[3]{\sqrt{x}}}$

74. $\sqrt[5]{\sqrt[3]{\sqrt{x^2}}}$

75. Without using a calculator, determine the value of $\left(6^{\sqrt{2}}\right)^{\sqrt{2}}$.

76. Determine the domain of $g(x) = (x-3)^{\frac{1}{2}}(x-1)^{-\frac{1}{2}}$.

77. Determine the domain of $f(x) = (x+3)^{\frac{1}{2}}(x+1)^{-\frac{1}{2}}$.

Synthesis Review

In Problems 78–81, simplify each expression.

78. $(2x-1)(x+4) - (x+1)(x-1)$

79. $3a(a-3) + (a+3)(a-2)$

80. $\dfrac{\sqrt{x^2+4x+4}}{x+2}, x+2 > 0$

81. $\dfrac{x^2-4}{x+2} \cdot (x+5) - (x+4)(x-1)$

9.4 Simplifying Radical Expressions

OBJECTIVES

1 Use the Product Property to Multiply Radical Expressions

2 Use the Product Property to Simplify Radical Expressions

3 Use the Quotient Property to Simplify Radical Expressions

4 Multiply Radicals with Unlike Indices

Preparing for Simplifying Radical Expressions

Before getting started, take the following readiness quiz. If you get a problem wrong, go back to the section cited and review the material.

1. List the perfect squares that are less than 200.

2. List the perfect cubes that are less than 200.

3. Simplify: (a) $\sqrt{16}$ (b) $\sqrt{p^2}$

[Section 9.1, pp. 660–663]

1 Use the Product Property to Multiply Radical Expressions

Perhaps you are noticing a trend at this point. When we introduce a new algebraic expression, we then learn how to multiply, divide, add and subtract the algebraic expression. Well, here we go again! First, we are going to learn how to multiply radical expressions when they have the same index.
    Consider the following:

$$\sqrt{4 \cdot 25} = \sqrt{100} = 10 \quad \text{and} \quad \sqrt{4} \cdot \sqrt{25} = 2 \cdot 5 = 10$$

This suggests the following result:

In Words
$\sqrt[n]{a} \cdot \sqrt[n]{b} = \sqrt[n]{ab}$ means "the product of the roots equals the root of the product provided the index is the same."

> **PRODUCT PROPERTY OF RADICALS**
>
> If $\sqrt[n]{a}$ and $\sqrt[n]{b}$ are real numbers, and $n \geq 2$ is an integer, then
>
> $$\sqrt[n]{a} \cdot \sqrt[n]{b} = \sqrt[n]{ab}$$

We can justify this formula using rational exponents.

$$\sqrt[n]{a} \cdot \sqrt[n]{b} = a^{\frac{1}{n}} \cdot b^{\frac{1}{n}}$$

Product to a Power Rule: $= (a \cdot b)^{\frac{1}{n}}$

$a^{\frac{1}{n}} = \sqrt[n]{a}$: $= \sqrt[n]{a \cdot b}$

EXAMPLE 1   **Using the Product Property to Multiply Radicals**

Multiply.

   (a) $\sqrt{5} \cdot \sqrt{3}$
   (b) $\sqrt[3]{2} \cdot \sqrt[3]{13}$
   (c) $\sqrt{x - 3} \cdot \sqrt{x + 3}$
   (d) $\sqrt[5]{6c} \cdot \sqrt[5]{7c^2}$

Solution

   (a) $\sqrt{5} \cdot \sqrt{3} = \sqrt{5 \cdot 3} = \sqrt{15}$
   (b) $\sqrt[3]{2} \cdot \sqrt[3]{13} = \sqrt[3]{2 \cdot 13} = \sqrt[3]{26}$
   (c) $\sqrt{x - 3} \cdot \sqrt{x + 3} = \sqrt{(x - 3)(x + 3)} = \sqrt{x^2 - 9}$
   (d) $\sqrt[5]{6c} \cdot \sqrt[5]{7c^2} = \sqrt[5]{6c \cdot 7c^2} = \sqrt[5]{42c^3}$

Work Smart
In Example 1(c), notice that $\sqrt{x^2 - 9}$ does not equal $\sqrt{x^2} - \sqrt{9}$.

QUICK ✔ *Multiply each radical expression.*

1. $\sqrt{11} \cdot \sqrt{7}$ **2.** $\sqrt[4]{6} \cdot \sqrt[4]{7}$ **3.** $\sqrt{x - 5} \cdot \sqrt{x + 5}$ **4.** $\sqrt[5]{5p} \cdot \sqrt[5]{4p^3}$

(2) ## Use the Product Property to Simplify Radical Expressions

Up to now, we have simplified radicals only when the radicand simplified to a perfect power, such as $\sqrt{81} = 9$ or $\sqrt[3]{\dfrac{1}{8}} = \dfrac{1}{2}$. When a radical does not simplify to a rational number, we can do one of two things:

   1. Write a decimal approximation of the radical.
   2. Simplify the radical using properties of radicals, if possible.

We learned how to approximate radicals using a calculator in Sections 9.1 and 9.2. Now we are going to learn how to use properties of radicals to write the radical in simplified form.

   Recall that a number that is the square of a rational number is called a perfect square. So $1^2 = 1, 2^2 = 4, 3^2 = 9$, and so on are perfect squares. A number that is the cube of a rational number is called a perfect cube. So $1^3 = 1, 2^3 = 8, 3^3 = 27$, and so on are perfect cubes. In general, if n is the index of a radical, then a^n is a perfect power of index where a is a rational number.

Work Smart
index $\rightarrow \sqrt[n]{a} \leftarrow$ radicand

   We say that a radical expression is **simplified** provided that the radicand does not contain any factors that are perfect powers of the index. For example, $\sqrt{50}$ is not simplified because 25 is a factor of 50 and 25 is a perfect square or $\sqrt[3]{16}$ is not simplified because 8 is a factor of 16 and 8 is a perfect cube. When the radicand contains variables, the exponent on the variable must be less than the index in order for the radical to be simplified.

To simplify radicals that contain perfect square factors, we use the Product Property of Radicals "in reverse." That is, we use $\sqrt[n]{ab} = \sqrt[n]{a} \cdot \sqrt[n]{b}$.

EXAMPLE 2 **How to Use the Product Property to Simplify a Radical**

Simplify: $\sqrt{18}$

Step-by-Step Solution

Step 1: What is the index on the radical? Since the index is 2, we write each factor of the radicand as the product of two factors, one of which is a perfect square.	The perfect squares are 1, 4, 9, 16, 25, ... Because 9 is a factor of 18 and 9 is a perfect square, we write 18 as $9 \cdot 2$.	$\sqrt{18} = \sqrt{9 \cdot 2}$
Step 2: Write the radicand as the product of two radicals, one of which contains a perfect square.		$= \sqrt{9} \cdot \sqrt{2}$
Step 3: Take the square root of each perfect power.		$= 3\sqrt{2}$

We summarize the steps used in Example 2 below.

Simplifying a Radical Expression

Step 1: Write each factor of the radicand as the product of two factors, one of which is a perfect power of the index.

Step 2: Write the radicand as the product of two radicals, one of which contains perfect squares using the Product Property of Radicals.

Step 3: Take the nth root of each perfect power.

EXAMPLE 3 **Using the Product Property to Simplify a Radical**

Simplify each of the following:

(a) $5\sqrt[3]{24}$ **(b)** $\sqrt{128x^2}$ **(c)** $\sqrt[4]{20}$

Solution

(a) We are looking for the largest factor of 24 that is a perfect cube. The perfect cubes are $1, 8, 27, \ldots$. Because 8 is a factor of 24 and 8 is a perfect cube we write 24 as $8 \cdot 3$.

$$5\sqrt[3]{24} = 5 \cdot \sqrt[3]{8 \cdot 3}$$
$$\sqrt[n]{ab} = \sqrt[n]{a} \cdot \sqrt[n]{b}: \quad = 5 \cdot \sqrt[3]{8} \cdot \sqrt[3]{3}$$
$$= 5 \cdot 2 \cdot \sqrt[3]{3}$$
$$= 10\sqrt[3]{3}$$

(b) Because 64 is a factor of 128 and 64 is a perfect square, we write 128 as $64 \cdot 2$, x^2 is a perfect square.

$$\sqrt{128x^2} = \sqrt{64x^2 \cdot 2}$$
$$\sqrt[n]{ab} = \sqrt[n]{a} \cdot \sqrt[n]{b}: \quad = \sqrt{64x^2} \cdot \sqrt{2}$$
$$\sqrt[n]{ab} = \sqrt[n]{a} \cdot \sqrt[n]{b}: \quad = \sqrt{64} \cdot \sqrt{x^2} \cdot \sqrt{2}$$
$$\sqrt{64} = 8, \ \sqrt{x^2} = |x|: \quad = 8|x|\sqrt{2}$$

(c) In $\sqrt[4]{20}$, the index is 4. The fourth powers (or perfect fourths) are 1, 16, 81, ... There are no factors of 20 that are fourth powers, so the radical $\sqrt[4]{20}$ cannot be simplified any further.

QUICK ✓ *Simplify each of the radical expressions.*

5. $\sqrt{48}$ **6.** $4\sqrt[3]{54}$ **7.** $\sqrt{200a^2}$ **8.** $\sqrt[4]{40}$

EXAMPLE 4   **Simplifying an Expression Involving a Square Root**

Simplify: $\dfrac{4 - \sqrt{20}}{2}$

Solution

We can simplify this expression using two different approaches.

Method 1:   **Method 2:**

4 is the largest perfect square factor of 20 ↓   4 is the largest perfect square factor of 20 ↓

$$\dfrac{4 - \sqrt{20}}{2} = \dfrac{4 - \sqrt{4 \cdot 5}}{2}$$ $$\dfrac{4 - \sqrt{20}}{2} = \dfrac{4 - \sqrt{4 \cdot 5}}{2}$$

Use $\sqrt{a \cdot b} = \sqrt{a} \cdot \sqrt{b}$: $= \dfrac{4 - \sqrt{4} \cdot \sqrt{5}}{2}$ Use $\sqrt{a \cdot b} = \sqrt{a} \cdot \sqrt{b}$: $= \dfrac{4 - \sqrt{4} \cdot \sqrt{5}}{2}$

$$= \dfrac{4 - 2 \cdot \sqrt{5}}{2}$$ $$= \dfrac{4 - 2 \cdot \sqrt{5}}{2}$$

Factor out the 2 in the numerator: $= \dfrac{2(2 - \sqrt{5})}{2}$ Use $\dfrac{A + B}{C} = \dfrac{A}{C} + \dfrac{B}{C}$: $= \dfrac{4}{2} - \dfrac{2 \cdot \sqrt{5}}{2}$

Divide out common factor: $= 2 - \sqrt{5}$ Divide out common factor: $= 2 - \sqrt{5}$

QUICK ✓ *Simplify the expression.*

9. $\dfrac{6 + \sqrt{45}}{3}$ **10.** $\dfrac{-2 + \sqrt{32}}{4}$

Recall how to simplify $\sqrt[n]{a^n}$ from Section 9.2.

> **SIMPLIFYING $\sqrt[n]{a^n}$**
>
> If $n \geq 2$ is a positive integer and a is a real number, then
>
> $$\sqrt[n]{a^n} = a \quad \text{if } n \geq 3 \text{ is odd}$$
> $$\sqrt[n]{a^n} = |a| \quad \text{if } n \geq 2 \text{ is even}$$

This means that

$$\sqrt{a^2} = |a| \qquad \sqrt[3]{a^3} = a \qquad \sqrt[4]{a^4} = |a| \qquad \sqrt[5]{a^5} = a \qquad \text{and so on}$$

In order to make our mathematical lives a little easier, for the remainder of the text, we shall assume that all variables that appear in the radicand are greater than or equal to zero (nonnegative). So

$$\sqrt{a^2} = a \qquad \sqrt[3]{a^3} = a \qquad \sqrt[4]{a^4} = a \qquad \sqrt[5]{a^5} = a \qquad \text{and so on}$$

What if the exponent on the radicand is greater than the index as in $\sqrt{x^3}$ or $\sqrt[3]{x^6}$? We could use the Laws of Exponents along with the rule for simplifying $\sqrt[n]{a^n}$ or we could use rational exponents.

$$\sqrt{x^6} = \sqrt{(x^3)^2} = x^3 \qquad \text{or} \qquad \sqrt{x^6} = x^{\frac{6}{2}} = x^3$$
$$\sqrt[3]{x^{12}} = \sqrt[3]{(x^4)^3} = x^4 \qquad \text{or} \qquad \sqrt[3]{x^{12}} = x^{\frac{12}{3}} = x^4$$

EXAMPLE 5 **Simplifying a Radical with a Variable Radicand**

Simplify $\sqrt{20x^{10}}$. Assume $x \geq 0$.

Solution

Because 4 is a factor of 20 and 4 is a perfect square, we write 20 as $4 \cdot 5$.

$$\sqrt{20x^{10}} = \sqrt{4 \cdot 5 \cdot x^{10}}$$
$$= \sqrt{4x^{10}} \cdot \sqrt{5}$$

$$\sqrt{4} = 2, \ \sqrt{x^{10}} = \sqrt{(x^5)^2} = x^5$$
$$\text{or } \sqrt{x^{10}} = x^{\frac{10}{2}} = x^5\text{:} \quad = 2x^5\sqrt{5}$$

∎

QUICK ✓

11. Simplify $\sqrt{75a^6}$. Assume $a \geq 0$.

What if the index does not divide evenly into the exponent on the variable in the radicand as in $\sqrt[3]{x^8}$? Under these circumstances, we rewrite the variable expression as the product of two variable expressions where one of the factors has an exponent that is a multiple of the index. For example, we can write

$$\sqrt[3]{x^8} \qquad \text{as} \qquad \sqrt[3]{x^6 \cdot x^2}$$

so that

$$\sqrt[3]{x^8} = \sqrt[3]{x^6 \cdot x^2} = \sqrt[3]{x^6} \cdot \sqrt[3]{x^2} = x^2\sqrt[3]{x^2}$$

EXAMPLE 6 **Simplifying Radicals**

Simplify:

(a) $\sqrt{80a^3}$ (b) $\sqrt[3]{27m^4n^{14}}$

Assume all variables are greater than or equal to zero.

Solution

(a)

$$80 = 16 \cdot 5; \ a^3 = a^2 \cdot a$$

$$\sqrt{80a^3} = \sqrt{16 \cdot 5 \cdot a^2 \cdot a}$$
$$= \sqrt{16a^2 \cdot 5a}$$
$$\sqrt[n]{ab} = \sqrt[n]{a} \cdot \sqrt[n]{b}\text{:} \quad = \sqrt{16a^2} \cdot \sqrt{5a}$$
$$\sqrt{a^2} = a \text{ assuming } a \geq 0\text{:} \quad = 4a\sqrt{5a}$$

(b)

$$m^4 = m^3 \cdot m; \; n^{14} = n^{12} \cdot n^2$$

$$\sqrt[3]{27m^4n^{14}} = \sqrt[3]{27 \cdot m^3 \cdot m \cdot n^{12} \cdot n^2}$$
$$= \sqrt[3]{27 \cdot m^3 \cdot n^{12} \cdot m \cdot n^2}$$
$$\sqrt[n]{ab} = \sqrt[n]{a} \cdot \sqrt[n]{b}: \quad = \sqrt[3]{27m^3n^{12}} \cdot \sqrt[3]{mn^2}$$
$$\sqrt[3]{27} = 3; \; \sqrt[3]{m^3} = m; \; \sqrt[3]{n^{12}} = n^{\frac{12}{3}} = n^4: \quad = 3mn^4\sqrt[3]{mn^2}$$

QUICK ✓ *Simplify each radical. Assume all variables are greater than or equal to zero.*

12. $\sqrt{18a^5}$ **13.** $\sqrt[3]{128x^6y^{10}}$ **14.** $\sqrt[4]{16a^5b^{11}}$

In this next example, we first multiply radical expressions and then simplify the product.

EXAMPLE 7   **Multiplying and Simplifying Radicals**

Multiply and simplify:

   (a) $\sqrt{3} \cdot \sqrt{15}$ **(b)** $3\sqrt[3]{4x} \cdot \sqrt[3]{2x^4}$ **(c)** $\sqrt[4]{27a^2b^5} \cdot \sqrt[4]{6a^3b^6}$

Assume all variables are greater than or equal to zero.

Solution

Remember, to multiply two radicals the index must be the same. When you have the same index, we multiply the radicands and then we simplify the product.

  (a) We start by looking to see if the index is the same. The index on both radicals is 2, so we multiply the radicands.

$$\sqrt{3} \cdot \sqrt{15} = \sqrt{3 \cdot 15}$$
$$= \sqrt{45}$$
9 is the largest factor of $\quad = \sqrt{9 \cdot 5}$
45 that is a perfect square:
Product Property of Radicals: $\quad = \sqrt{9} \cdot \sqrt{5}$
$$= 3\sqrt{5}$$

  (b) The index on both radicals is 3, so we multiply the radicands.

$$3\sqrt[3]{4x} \cdot \sqrt[3]{2x^4} = 3\sqrt[3]{4x \cdot 2x^4}$$
$$= 3\sqrt[3]{8x^5}$$
x^3 is a perfect cube; 8 is a perfect cube: $\quad = 3\sqrt[3]{8x^3 \cdot x^2}$
Product Property of Radicals: $\quad = 3\sqrt[3]{8x^3} \cdot \sqrt[3]{x^2}$
$\sqrt[3]{8} = 2; \; \sqrt[3]{x^3} = x: \quad = 3 \cdot 2 \cdot x \cdot \sqrt[3]{x^2}$
$$= 6x\sqrt[3]{x^2}$$

Work Smart

Notice that $27 = 3^3$ and that $6 = 3 \cdot 2$, so that $27 \cdot 6 = 3^3 \cdot 3 \cdot 2 = 3^4 \cdot 2$. This makes finding the perfect power of 4 a lot easier!

  (c) The index on both radicals is 4, so we multiply the radicands.

$$\sqrt[4]{27a^2b^5} \cdot \sqrt[4]{6a^3b^6} = \sqrt[4]{3^3 \cdot a^2b^5 \cdot 3 \cdot 2 \cdot a^3b^6}$$
$$= \sqrt[4]{3^4 \cdot 2 \cdot a^5b^{11}}$$
$$= \sqrt[4]{3^4 \cdot 2 \cdot a^4 \cdot a \cdot b^8 \cdot b^3}$$
Product Property of Radicals: $\quad = \sqrt[4]{3^4a^4b^8} \cdot \sqrt[4]{2ab^3}$
$$= 3ab^2\sqrt[4]{2ab^3}$$

QUICK ✅ *Multiply and simplify the radicals. Assume all variables are greater than or equal to zero.*

15. $\sqrt{6} \cdot \sqrt{8}$ **16.** $\sqrt[3]{12a^2} \cdot \sqrt[3]{10a^4}$ **17.** $4\sqrt[3]{8a^2b^5} \cdot \sqrt[3]{6a^2b^4}$

---

③   Use the Quotient Property to Simplify Radical Expressions

Now consider the following:

$$\sqrt{\frac{64}{4}} = \sqrt{\frac{4 \cdot 16}{4}} = \sqrt{16} = 4 \quad \text{and} \quad \frac{\sqrt{64}}{\sqrt{4}} = \frac{8}{2} = 4$$

This suggests the following result:

In Words

$$\sqrt[n]{\frac{a}{b}} = \frac{\sqrt[n]{a}}{\sqrt[n]{b}}$$

means "the root of the quotient equals the quotient of the roots" provided that the radicals have the same index.

> **QUOTIENT PROPERTY OF RADICALS**
>
> If $\sqrt[n]{a}$ and $\sqrt[n]{b}$ are real numbers, $b \neq 0$, and $n \geq 2$ is an integer, then
>
> $$\frac{\sqrt[n]{a}}{\sqrt[n]{b}} = \sqrt[n]{\frac{a}{b}}$$

We can justify this formula using rational exponents.

$$\frac{\sqrt[n]{a}}{\sqrt[n]{b}} = \frac{a^{\frac{1}{n}}}{b^{\frac{1}{n}}} = \left(\frac{a}{b}\right)^{\frac{1}{n}} = \sqrt[n]{\frac{a}{b}}$$

EXAMPLE 8   Using the Quotient Property to Simplify Radicals

Simplify:

(a) $\sqrt{\frac{18}{25}}$ **(b)** $\sqrt[3]{\frac{6z^3}{125}}$ **(c)** $\sqrt[4]{\frac{10a^2}{81b^4}}, b \neq 0$

Assume all variables are greater than or equal to zero.

Solution

In all three of these problems, you should notice that the expression in the denominator is a perfect power of the index. Therefore, we are going to use the Quotient Rule "in reverse" to simplify the expressions. That is, we use $\sqrt[n]{\frac{a}{b}} = \frac{\sqrt[n]{a}}{\sqrt[n]{b}}$.

(a) $\sqrt{\frac{18}{25}} = \frac{\sqrt{18}}{\sqrt{25}}$ **(b)** $\sqrt[3]{\frac{6z^3}{125}} = \frac{\sqrt[3]{6z^3}}{\sqrt[3]{125}}$ **(c)** $\sqrt[4]{\frac{10a^2}{81b^4}} = \frac{\sqrt[4]{10a^2}}{\sqrt[4]{81b^4}}$

$ = \frac{3\sqrt{2}}{5}$ $ = \frac{z\sqrt[3]{6}}{5}$ $ = \frac{\sqrt[4]{10a^2}}{3b}$

QUICK ✅ *Simplify the radicals. Assume all variables are greater than or equal to zero.*

18. $\sqrt{\frac{13}{49}}$ **19.** $\sqrt[3]{\frac{27p^3}{8}}$ **20.** $\sqrt[4]{\frac{3q^4}{16}}$

EXAMPLE 9   **Using the Quotient Property to Simplify Radicals**

Simplify:

(a) $\dfrac{\sqrt{24a^3}}{\sqrt{6a}}$ (b) $\dfrac{-2\sqrt[3]{54a}}{\sqrt[3]{2a^4}}$ (c) $\dfrac{\sqrt[3]{-375x^2y}}{\sqrt[3]{3x^{-1}y^7}}$

Assume all variables are greater than zero.

Solution

In these problems, we notice that the radical expression in the denominator cannot be simplified. However, the index on the numerator and denominator of each expression is the same, so we can write each expression as a single radical.

(a) $\dfrac{\sqrt{24a^3}}{\sqrt{6a}} = \sqrt{\dfrac{24a^3}{6a}}$

$\qquad = \sqrt{4a^2}$

$\qquad = 2a$

(b) $\dfrac{-2\sqrt[3]{54a}}{\sqrt[3]{2a^4}} = -2 \cdot \sqrt[3]{\dfrac{54a}{2a^4}}$

$\qquad = -2 \cdot \sqrt[3]{\dfrac{27}{a^3}}$

$\qquad = -2 \cdot \dfrac{3}{a}$

$\qquad = -\dfrac{6}{a}$

(c) $\dfrac{\sqrt[3]{-375x^2y}}{\sqrt[3]{3x^{-1}y^7}} = \sqrt[3]{\dfrac{-375x^2y}{3x^{-1}y^7}}$

$\qquad = \sqrt[3]{-125x^{2-(-1)}y^{1-7}}$

$\qquad = \sqrt[3]{-125x^3y^{-6}}$

$\qquad = \sqrt[3]{\dfrac{-125x^3}{y^6}}$

$\qquad = \dfrac{-5x}{y^2}$

QUICK ✔ *Simplify the radicals. Assume all variables are greater than zero.*

21. $\dfrac{\sqrt{12a^5}}{\sqrt{3a}}$ **22.** $\dfrac{\sqrt[3]{-24x^2}}{\sqrt[3]{3x^{-1}}}$ **23.** $\dfrac{\sqrt[3]{250a^5b^{-2}}}{\sqrt[3]{2ab}}$

---

④ **Multiply Radicals with Unlike Indices**

To multiply radicals we use the fact that $\sqrt[n]{a} \cdot \sqrt[n]{b} = \sqrt[n]{ab}$. This rule only works when the index on each radical is the same. What if the index on each radical is different? Can we still simplify the product? The answer is yes! To perform the multiplication we use the fact that $\sqrt[n]{a} = a^{\frac{1}{n}}$. Let's go over an example.

EXAMPLE 10   **Multiplying Radicals with Unlike Indices**

Multiply and simplify:

$$\sqrt[4]{8} \cdot \sqrt[3]{5}$$

Solution

Notice that the index is not the same, so $\sqrt[n]{a} \cdot \sqrt[n]{b} = \sqrt[n]{ab}$ cannot be used to find the product. We will use rational exponents along with $\sqrt[n]{a} = a^{\frac{1}{n}}$ instead.

$$\sqrt[4]{8} \cdot \sqrt[3]{5} = 8^{\frac{1}{4}} \cdot 5^{\frac{1}{3}}$$
$$\text{LCD} = 12: \quad = 8^{\frac{3}{12}} \cdot 5^{\frac{4}{12}}$$
$$a^{\frac{r}{s}} = (a^r)^{\frac{1}{s}}: \quad = \left[(8^3)^{\frac{1}{12}} \cdot (5^4)^{\frac{1}{12}}\right]$$
$$a^r \cdot b^r = (ab)^r: \quad = [(8^3)(5^4)]^{\frac{1}{12}}$$
$$= (320{,}000)^{\frac{1}{12}}$$
$$a^{\frac{1}{n}} = \sqrt[n]{a}: \quad = \sqrt[12]{320{,}000}$$

QUICK ✔ *Multiply and simplify.*

24. $\sqrt[4]{5} \cdot \sqrt[3]{3}$ **25.** $\sqrt{10} \cdot \sqrt[3]{12}$

9.4 Exercises

For Extra Help: Student Solutions Manual   CD Video   PH Math/Tutor Center   MathXL Tutorials on CD   MathXL®   MyMathLab

Concepts and Vocabulary

In Problems 1 and 2, fill in the blanks.

1. If $\sqrt[n]{a}$ and $\sqrt[n]{b}$ are real numbers and $n \geq 2$ is an integer, then $\sqrt[n]{a} \cdot \sqrt[n]{b} = $ _____.

2. A number that is the square of a rational number is called a _____ _____. A number that is the cube of a rational number is called a _____ _____.

In Problems 3 and 4, answer True or False to each statement.

3. The radical expression $\sqrt{8x}$ is simplified.

4. If $\sqrt[n]{a}$ and $\sqrt[n]{b}$ are real numbers, $b \neq 0$, and $n \geq 2$ is an integer, then $\dfrac{\sqrt[n]{a}}{\sqrt[n]{b}} = \sqrt[n]{\dfrac{a}{b}}$.

5. List the first six integers that are perfect squares.

6. List the first six positive integers that are perfect cubes.

7. In your own words, explain how you would simplify $\sqrt[3]{16a^5}$.

8. In order to use the Product Property to multiply radicals, what must be true about the index in each radical?

Building Skills

In Problems 9–18, use the Product Property to multiply. Assume that all variables can be any real number.

9. $\sqrt{2} \cdot \sqrt{7}$ **10.** $\sqrt{3} \cdot \sqrt{10}$ **11.** $\sqrt[3]{6} \cdot \sqrt[3]{10}$

12. $\sqrt[3]{-5} \cdot \sqrt[3]{7}$ **13.** $\sqrt{3a} \cdot \sqrt{5b}$ **14.** $\sqrt[4]{6a^2} \cdot \sqrt[4]{7b^2}$

15. $\sqrt{x-7} \cdot \sqrt{x+7}$ **16.** $\sqrt{p-5} \cdot \sqrt{p+5}$ **17.** $\sqrt{\dfrac{5x}{3}} \cdot \sqrt{\dfrac{3}{x}}$

18. $\sqrt[3]{\dfrac{-9x^2}{4}} \cdot \sqrt[3]{\dfrac{4}{3x}}$

In Problems 19–52, simplify each radical using the Product Property. Assume that all variables can be any real number.

19. $\sqrt{50}$ **20.** $\sqrt{32}$ **21.** $\sqrt[3]{54}$

22. $\sqrt[4]{162}$ **23.** $\sqrt{48x^2}$ **24.** $\sqrt{20a^2}$

25. $\sqrt[3]{-27x^3}$ **26.** $\sqrt[3]{-64p^3}$ **27.** $\sqrt[4]{32m^4}$

28. $\sqrt[4]{48z^4}$ **29.** $\sqrt{12p^2q}$ **30.** $\sqrt{45m^2n}$

31. $3\sqrt{24a}$ **32.** $4\sqrt{27b}$ **33.** $\sqrt{162m^4}$

34. $\sqrt{98w^8}$ **35.** $\sqrt{y^{13}}$ **36.** $\sqrt{s^9}$

37. $\sqrt[3]{c^8}$ **38.** $\sqrt[5]{x^{12}}$ **39.** $\sqrt{m^5n^3}$

40. $\sqrt{x^7y^2}$ **41.** $\sqrt{125p^3q^4}$ **42.** $\sqrt{243ab^5}$

43. $\sqrt[3]{-16x^9}$ **44.** $\sqrt[3]{-54q^{12}}$ **45.** $\sqrt{p^3q^5r}$

46. $\sqrt[4]{x^6y^9z^4}$ **47.** $\sqrt[5]{-16m^8n^2}$ **48.** $\sqrt{75x^6y}$

49. $\sqrt[4]{(x-y)^5}, x > y$ **50.** $\sqrt[3]{(a+b)^5}$ **51.** $\sqrt[3]{8x^3 - 8y^3}$

52. $\sqrt[3]{8a^3 + 8b^3}$

In Problems 53–60, simplify each expression.

53. $\dfrac{4 + \sqrt{36}}{2}$ **54.** $\dfrac{5 - \sqrt{100}}{5}$ **55.** $\dfrac{9 + \sqrt{18}}{3}$ **56.** $\dfrac{10 - \sqrt{75}}{5}$

57. $\dfrac{-4 - \sqrt{162}}{6}$ **58.** $\dfrac{-6 + \sqrt{48}}{8}$ **59.** $\dfrac{7 - \sqrt{98}}{14}$ **60.** $\dfrac{-6 + \sqrt{108}}{6}$

In Problems 61–78, multiply and simplify. Assume that all variables are greater than or equal to zero.

61. $\sqrt{5} \cdot \sqrt{5}$ **62.** $\sqrt{6} \cdot \sqrt{6}$

63. $\sqrt{2} \cdot \sqrt{8}$ **64.** $\sqrt{3} \cdot \sqrt{12}$

65. $\sqrt[3]{4} \cdot \sqrt[3]{2}$ **66.** $\sqrt[3]{9} \cdot \sqrt[3]{3}$

67. $\sqrt{5x} \cdot \sqrt{15x}$ **68.** $\sqrt{6x} \cdot \sqrt{30x}$

69. $\sqrt[3]{4b^2} \cdot \sqrt[3]{6b^2}$ **70.** $\sqrt[3]{9a} \cdot \sqrt[3]{6a^2}$

71. $2\sqrt{6ab} \cdot 3\sqrt{15ab^3}$ **72.** $3\sqrt{14pq^3} \cdot 2\sqrt{7pq}$

73. $\sqrt[4]{27p^3q^2} \cdot \sqrt[4]{12p^2q^2}$ **74.** $\sqrt[3]{16m^2n} \cdot \sqrt[3]{27m^2n}$

75. $\sqrt[5]{-8a^3b^4} \cdot \sqrt[5]{12a^3b}$ **76.** $\sqrt[5]{-27x^4y^2} \cdot \sqrt[5]{18x^3y^4}$

77. $\sqrt[4]{8(x-y)^2} \cdot \sqrt[4]{6(x-y)^3}, x > y$ **78.** $\sqrt[3]{9(a+b)^2} \cdot \sqrt[3]{6(a+b)^5}$

In Problems 79–88, simplify. Assume that all variables are greater than zero.

79. $\sqrt{\dfrac{3}{16}}$ **80.** $\sqrt{\dfrac{5}{36}}$ **81.** $\sqrt{\dfrac{121}{100}}$ **82.** $\sqrt{\dfrac{81}{64}}$

83. $\sqrt[4]{\dfrac{5x^4}{16}}$ **84.** $\sqrt[4]{\dfrac{2a^8}{81}}$ **85.** $\sqrt{\dfrac{9y^2}{25x^2}}$ **86.** $\sqrt{\dfrac{4a^4}{81b^2}}$

87. $\sqrt[3]{\dfrac{-27x^9}{64y^{12}}}$ **88.** $\sqrt[5]{\dfrac{-32a^{15}}{243b^{10}}}$

In Problems 89–102, divide and simplify. Assume that all variables are greater than zero.

89. $\dfrac{\sqrt{8}}{\sqrt{2}}$

90. $\dfrac{\sqrt{27}}{\sqrt{3}}$

91. $\dfrac{\sqrt[3]{128}}{\sqrt[3]{2}}$

92. $\dfrac{\sqrt[4]{64}}{\sqrt[4]{4}}$

93. $\dfrac{\sqrt{48a^3}}{\sqrt{6a}}$

94. $\dfrac{\sqrt{54y^5}}{\sqrt{3y}}$

95. $\dfrac{\sqrt{24a^5b}}{\sqrt{3ab^3}}$

96. $\dfrac{\sqrt{360m^7n^3}}{\sqrt{5mn^5}}$

97. $\dfrac{\sqrt{512a^7b}}{3\sqrt{2ab^3}}$

98. $\dfrac{\sqrt{375x^2y^7}}{10\sqrt{3y}}$

99. $\dfrac{\sqrt[3]{104a^5}}{\sqrt[3]{4a^{-1}}}$

100. $\dfrac{\sqrt[3]{-128x^8}}{\sqrt[3]{2x^{-1}}}$

101. $\dfrac{\sqrt{90x^3y^{-1}}}{\sqrt{2x^{-3}y}}$

102. $\dfrac{\sqrt{96a^5b^{-3}}}{\sqrt{3a^{-5}b}}$

In Problems 103–110, multiply and simplify.

103. $\sqrt{3}\cdot\sqrt[3]{4}$

104. $\sqrt{2}\cdot\sqrt[3]{7}$

105. $\sqrt[3]{2}\cdot\sqrt[6]{3}$

106. $\sqrt[4]{3}\cdot\sqrt[8]{5}$

107. $\sqrt{3}\cdot\sqrt[3]{18}$

108. $\sqrt{6}\cdot\sqrt[3]{9}$

109. $\sqrt[4]{9}\cdot\sqrt[6]{12}$

110. $\sqrt[5]{8}\cdot\sqrt[10]{16}$

Mixed Practice

In Problems 111–126, perform any indicated operation and simplify. Assume all variables are greater than zero.

111. $\sqrt[3]{\dfrac{5x}{8}}$

112. $\sqrt[3]{\dfrac{7a^2}{64}}$

113. $\sqrt[3]{5a}\cdot\sqrt[3]{9a}$

114. $\sqrt[5]{8b^2}\cdot\sqrt[5]{3b}$

115. $\sqrt{72a^4}$

116. $\sqrt{24b^6}$

117. $\sqrt[3]{6a^2b}\cdot\sqrt[3]{9ab}$

118. $\sqrt[4]{8x^3y^2}\cdot\sqrt[4]{4x^2y^3}$

119. $\dfrac{\sqrt[3]{-32a}}{\sqrt[3]{2a^4}}$

120. $\dfrac{\sqrt[3]{-250p^2}}{\sqrt[3]{2p^5}}$

121. $-5\sqrt[3]{32m^3}$

122. $-7\sqrt[3]{250p^3}$

123. $\sqrt[3]{81a^4b^7}$

124. $\sqrt[5]{32p^7q^{11}}$

125. $\sqrt[3]{12}\cdot\sqrt[3]{18}$

126. $\sqrt[4]{8}\cdot\sqrt[4]{18}$

Applying the Concepts

△ **127. Length of a Line Segment** The length of the line segment joining the points $(2, 5)$ and $(-1, -1)$ is given by

$$\sqrt{(5-(-1))^2 + (2-(-1))^2}$$

(a) Plot the points in the Cartesian plane and draw a line segment connecting the points.

(b) Express the length of the line segment as a radical in simplified form.

△ **128. Length of a Line Segment** The length of the line segment joining the points $(4, 2)$ and $(-2, 4)$ is given by

$$\sqrt{(4-2)^2 + (-2-4)^2}$$

(a) Plot the points in the Cartesian plane and draw a line segment connecting the points.

(b) Express the length of the line segment as a radical in simplified form.

129. Revenue Growth Suppose that the annual revenue R (in millions of dollars) of a company after t years of operating is modeled by the function

$$R(t) = \sqrt[3]{\frac{t}{2}}$$

(a) Predict the revenue of the company after 8 years of operation.

(b) Predict the revenue of the company after 27 years of operation.

△ **130. Sphere** The radius r of a sphere whose volume is V is given by

$$r = \sqrt[3]{\frac{3V}{4\pi}}$$

(a) Write the radius of a sphere whose volume is 9 cubic centimeters as a radical in simplified form.

(b) Write the radius of a sphere whose volume is 32π cubic meters as a radical in simplified form.

Extending the Concepts

131. Suppose that $f(x) = \sqrt{2x}$ and $g(x) = \sqrt{8x^3}$.

(a) Find $(f \cdot g)(x)$.

(b) Evaluate $(f \cdot g)(3)$.

In Problems 132–135, evaluate the formula

$$x = \frac{-b \pm \sqrt{b^2 - 4ac}}{2a}$$

for the given values of a, b, and c. Note that the symbol $\pm$ is shorthand notation to indicate that there are two solutions. One solution is obtained when you add the quantity after the $\pm$ symbol, and another is obtained when you subtract. This formula can be used to solve any equation of the form $ax^2 + bx + c = 0$.

132. $a = 1, b = 4, c = 1$

133. $a = 1, b = 6, c = 3$

134. $a = 2, b = 1, c = -1$

135. $a = 3, b = 4, c = -1$

Synthesis Review

In Problems 136–139, solve the following.

136. $4x + 3 = 13$

137. $2|7x - 1| + 4 = 16$

138. $\dfrac{3}{5}x + 2 \le 28$

139. $\dfrac{5}{2}|x + 1| + 1 \le 11$

140. How are the solutions in Problems 136 and 138 similar? How are the solutions in Problems 137 and 139 similar?

The Graphing Calculator

141. Exploration To understand the circumstances under which absolute value symbols are required when simplifying radicals, do the following.

(a) Graph $Y_1 = \sqrt{x^2}$ and $Y_2 = x$. Do you think that $\sqrt{x^2} = x$? Now graph $Y_1 = \sqrt{x^2}$ and $Y_2 = |x|$. Do you think that $\sqrt{x^2} = |x|$?

(b) Graph $Y_1 = \sqrt[3]{x^3}$ and $Y_2 = x$. Do you think that $\sqrt[3]{x^3} = x$? Now graph $Y_1 = \sqrt[3]{x^3}$ and $Y_2 = |x|$. Do you think that $\sqrt[3]{x^3} = |x|$?

(c) Graph $Y_1 = \sqrt[4]{x^4}$ and $Y_2 = x$. Do you think that $\sqrt[4]{x^4} = x$? Now graph $Y_1 = \sqrt[4]{x^4}$ and $Y_2 = |x|$. Do you think that $\sqrt[4]{x^4} = |x|$?

(d) In your own words, make a generalization about $\sqrt[n]{x^n}$.

9.5 Adding, Subtracting, and Multiplying Radical Expressions

OBJECTIVES

1. Add or Subtract Radical Expressions
2. Multiply Radical Expressions

Preparing for Adding, Subtracting, and Multiplying Radical Expressions

Before getting started, take the following readiness quiz. If you get a problem wrong, go back to the section cited and review the material.

1. Add: $4y^3 - 2y^2 + 8y - 1 + (-2y^3 + 7y^2 - 3y + 9)$ [Section 5.1, pp. 346–347]
2. Subtract: $5z^2 + 6 - (3z^2 - 8z - 3)$ [Section 5.1, pp. 347–348]
3. Multiply: $(4x + 3)(x - 5)$ [Section 5.3, pp. 361–363]
4. Multiply: $(2y - 3)(2y + 3)$ [Section 5.3, p. 364]

1 Add or Subtract Radical Expressions

Recall, a radical expression is an algebraic expression that contains a radical. We say that two radicals are **like radicals** if each radical has the same index and the same radicand. For example,

$$4\sqrt[3]{x - 4} \quad \text{and} \quad 10\sqrt[3]{x - 4}$$

are like radicals because each has the same index, 3, and the same radicand, $x - 4$.

To add or subtract radical expressions we combine like radicals using the Distributive Property in reverse as follows:

Work Smart

You will be doing three of the four basic operations in this section. Division of radical expressions is just a bit more involved, so we discuss it in the next section by itself.

$$\text{Factor out } \sqrt[3]{x - 4}$$
$$\downarrow$$
$$4\sqrt[3]{x - 4} + 10\sqrt[3]{x - 4} = (4 + 10)\sqrt[3]{x - 4}$$
$$4 + 10 = 14: \quad = 14\sqrt[3]{x - 4}$$

EXAMPLE 1 **Adding and Subtracting Radical Expressions**

Add or subtract, as indicated.

(a) $5\sqrt{2x} + 9\sqrt{2x}$

(b) $3\sqrt[3]{10} + 7\sqrt[3]{10} - 5\sqrt[3]{10}$

Solution

Work Smart

Remember that to add or subtract radicals both the index and the radicand must be the same.

(a) Both radicals have the same index, 2, and the same radicand, $2x$.

$$5\sqrt{2x} + 9\sqrt{2x} = (5 + 9)\sqrt{2x}$$
$$5 + 9 = 14: \quad = 14\sqrt{2x}$$

(b) All three radicals have the same index, 3, and the same radicand, 10.

$$3\sqrt[3]{10} + 7\sqrt[3]{10} - 5\sqrt[3]{10} = (3 + 7 - 5)\sqrt[3]{10}$$
$$3 + 7 - 5 = 5: \quad = 5\sqrt[3]{10}$$

Preparing for...Answers
1. $2y^3 + 5y^2 + 5y + 8$ 2. $2z^2 + 8z + 9$
3. $4x^2 - 17x - 15$ 4. $4y^2 - 9$

QUICK ✓ *Add or subtract, as indicated.*

1. $9\sqrt{13y} + 4\sqrt{13y}$ **2.** $\sqrt[4]{5} + 9\sqrt[4]{5} - 3\sqrt[4]{5}$

Sometimes we have to simplify the radical so that the radicands are the same before adding or subtracting.

EXAMPLE 2   Adding and Subtracting Radical Expressions

Add or subtract, as indicated. Assume all variables are real numbers greater than or equal to zero.

(a) $3\sqrt{12} + 7\sqrt{3}$ **(b)** $3x\sqrt{20x} - 7\sqrt{5x^3}$ **(c)** $3\sqrt{5} + 7\sqrt{13}$

Solution

(a) The index on each radical is the same, but the radicands are different. However, we can simplify the radicals to make the radicands the same.

$$\sqrt{12} = \sqrt{4} \cdot \sqrt{3}$$

$$
\begin{aligned}
3\sqrt{12} + 7\sqrt{3} &= 3\sqrt{4}\cdot\sqrt{3} + 7\sqrt{3} \\
\sqrt{4} = 2\text{:} \quad &= 3\cdot 2\sqrt{3} + 7\sqrt{3} \\
&= 6\sqrt{3} + 7\sqrt{3} \\
\text{Factor out } \sqrt{3}\text{:} \quad &= (6 + 7)\sqrt{3} \\
&= 13\sqrt{3}
\end{aligned}
$$

(b) The index on each radical is the same, but the radicands are different. However, we can simplify the radicals to make the radicands the same.

$$\sqrt{20x} = \sqrt{4} \cdot \sqrt{5x}; \quad \sqrt{5x^3} = \sqrt{x^2} \cdot \sqrt{5x}$$

$$
\begin{aligned}
3x\sqrt{20x} - 7\sqrt{5x^3} &= 3x\cdot\sqrt{4}\cdot\sqrt{5x} - 7\cdot\sqrt{x^2}\cdot\sqrt{5x} \\
\text{Simplify radicals:} \quad &= 3x\cdot 2\cdot\sqrt{5x} - 7\cdot x\cdot\sqrt{5x} \\
\text{Multiply:} \quad &= 6x\sqrt{5x} - 7x\sqrt{5x} \\
\text{Factor out } \sqrt{5x}\text{:} \quad &= (6x - 7x)\sqrt{5x} \\
6x - 7x = -x\text{:} \quad &= -x\sqrt{5x}
\end{aligned}
$$

(c) For $3\sqrt{5} + 7\sqrt{13}$ the index on each radical is the same, but the radicands are different and we cannot simplify the radicals.

QUICK ✓ *Add or subtract, as indicated. Assume all variables are real numbers greater than or equal to zero.*

3. $4\sqrt{18} - 3\sqrt{8}$ **4.** $-5x\sqrt[3]{54x} + 7\sqrt[3]{2x^4}$ **5.** $7\sqrt{10} - 6\sqrt{3}$

EXAMPLE 3   Adding or Subtracting Radical Expressions

Add or subtract, as indicated. Assume that all variables are real numbers greater than or equal to zero.

(a) $\sqrt[3]{16x^4} - 7x\sqrt[3]{-2x} + \sqrt[3]{54x}$ **(b)** $3\sqrt[4]{m^4n} - 5m\sqrt[8]{n^2}$

Solution

(a) The index on each radical is the same, but the radicands are different. However, we can simplify the radicals to make the radicands the same.

$$\sqrt[3]{16x^4} = \sqrt[3]{8x^3} \cdot \sqrt[3]{2x}; \sqrt[3]{-2x} = \sqrt[3]{-1} \cdot \sqrt[3]{2x}; \sqrt[3]{54x} = \sqrt[3]{27} \cdot \sqrt[3]{2x}$$

$$\sqrt[3]{16x^4} - 7x\sqrt[3]{-2x} + \sqrt[3]{54x} = \sqrt[3]{8x^3} \cdot \sqrt[3]{2x} - 7x\sqrt[3]{-1} \cdot \sqrt[3]{2x} + \sqrt[3]{27} \cdot \sqrt[3]{2x}$$

$\sqrt[3]{8x^3} = 2x; \sqrt[3]{-1} = -1; \sqrt[3]{27} = 3$: $\quad = 2x\sqrt[3]{2x} - 7x(-1)\sqrt[3]{2x} + 3\sqrt[3]{2x}$

$-7x(-1) = 7x$: $\quad = 2x\sqrt[3]{2x} + 7x\sqrt[3]{2x} + 3\sqrt[3]{2x}$

Factor out $\sqrt[3]{2x}$: $\quad = (2x + 7x + 3)\sqrt[3]{2x}$

Simplify: $\quad = (9x + 3)\sqrt[3]{2x}$

Factor out 3: $\quad = 3(3x + 1)\sqrt[3]{2x}$

Work Smart: Study Skills

Contrast adding radicals with multiplying radicals:

Add: $3\sqrt{5} + 8\sqrt{5} = (3 + 8)\sqrt{5}$
$= 11\sqrt{5}$

Multiply:
$3\sqrt{5} \cdot 8\sqrt{5} = 3 \cdot 8 \cdot \sqrt{5} \cdot \sqrt{5}$
$= 24\sqrt{25}$
$= 24 \cdot 5$
$= 120$

Ask yourself these questions:

How must the radicals be "like" to be added?

How must the radicals be "like" to be multiplied?

(b) Here, the index on the radical and the radicand are different. We start by dealing with the index using rational exponents.

$$\sqrt[n]{a^m} = a^{\frac{m}{n}}$$

$$3\sqrt[4]{m^4 n} - 5m\sqrt[8]{n^2} = 3\sqrt[4]{m^4 n} - 5m \cdot n^{\frac{2}{8}}$$

Reduce rational exponent: $\quad = 3\sqrt[4]{m^4 n} - 5m \cdot n^{\frac{1}{4}}$

Rewrite as radical: $\quad = 3\sqrt[4]{m^4 n} - 5m \cdot \sqrt[4]{n}$

Now the index is the same, but the radicands are different. We can deal with this issue as well.

$\sqrt[4]{m^4 n} = \sqrt[4]{m^4} \cdot \sqrt[4]{n}$: $\quad = 3\sqrt[4]{m^4} \cdot \sqrt[4]{n} - 5m \cdot \sqrt[4]{n}$

Simplify: $\quad = 3m \cdot \sqrt[4]{n} - 5m \cdot \sqrt[4]{n}$

Factor out $\sqrt[4]{n}$: $\quad = (3m - 5m) \cdot \sqrt[4]{n}$

Simplify: $\quad = -2m \cdot \sqrt[4]{n}$

QUICK ✔ *Add or subtract, as indicated. Assume all variables are real numbers greater than or equal to zero.*

6. $\sqrt[3]{8z^4} - 2z\sqrt[3]{-27z} + \sqrt[3]{125z}$ **7.** $\sqrt{25m} - 3\sqrt[4]{m^2}$

---

② **Multiply Radical Expressions**

We have already multiplied radical expressions in which a single radical was multiplied by a second single radical. Now we concentrate on multiplying radical expressions involving more than one radical. These expressions are multiplied in the same way that we multiplied polynomials.

EXAMPLE 4 **Multiplying Radical Expressions**

Multiply and simplify:

(a) $\sqrt{5}(3 - 4\sqrt{5})$ **(b)** $\sqrt[3]{2}(3 + \sqrt[3]{4})$ **(c)** $(3 + 2\sqrt{7})(2 - 3\sqrt{7})$

Solution

(a) We use the Distributive Property and multiply $\sqrt{5}$ by each term in the parentheses.

$$\sqrt{5}\left(3 - 4\sqrt{5}\right) = \sqrt{5}\cdot 3 - \sqrt{5}\cdot 4\sqrt{5}$$

Multiply radicals: $= 3\sqrt{5} - 4\cdot\sqrt{25}$

$\sqrt{25} = 5$: $= 3\sqrt{5} - 4\cdot 5$

Simplify: $= 3\sqrt{5} - 20$

(b) We use the Distributive Property and multiply $\sqrt[3]{2}$ by each term in parentheses.

$$\sqrt[3]{2}\left(3 + \sqrt[3]{4}\right) = \sqrt[3]{2}\cdot 3 + \sqrt[3]{2}\cdot\sqrt[3]{4}$$

Multiply radicals: $= 3\sqrt[3]{2} + \sqrt[3]{8}$

$\sqrt[3]{8} = 2$: $= 3\sqrt[3]{2} + 2$

(c) We treat this just like the product of two binomials and use the FOIL method to multiply.

First   Last

$$\left(3 + 2\sqrt{7}\right)\left(2 - 3\sqrt{7}\right) = 3\cdot 2 - 3\cdot 3\sqrt{7} + 2\sqrt{7}\cdot 2 - 2\sqrt{7}\cdot 3\sqrt{7}$$

Inner

Outer

Multiply: $= 6 - 9\sqrt{7} + 4\sqrt{7} - 6\sqrt{49}$

$\sqrt{49} = 7$: $= 6 - 9\sqrt{7} + 4\sqrt{7} - 6\cdot 7$

$= 6 - 9\sqrt{7} + 4\sqrt{7} - 42$

Simplify: $= -36 - 5\sqrt{7}$

QUICK ✓ *Multiply and simplify.*

8. $\sqrt{6}\left(3 - 5\sqrt{6}\right)$ **9.** $\sqrt[3]{12}\left(3 - \sqrt[3]{2}\right)$ **10.** $\left(2 - 7\sqrt{3}\right)\left(5 + 4\sqrt{3}\right)$

We can use our special products formulas (Section 5.3) to multiply radicals as well. In particular, we are going to use the formulas for perfect squares, $(a + b)^2 = a^2 + 2ab + b^2$ and $(a - b)^2 = a^2 - 2ab + b^2$, as well as the formula for the difference of two squares, $(a + b)(a - b) = a^2 - b^2$.

EXAMPLE 5   Multiplying Radical Expressions Involving Special Products

Multiply and simplify.

(a) $\left(2\sqrt{3} + \sqrt{5}\right)^2$ **(b)** $\left(3 + \sqrt{7}\right)\left(3 - \sqrt{7}\right)$

Solution $(\ a\ +\ b\ )^2 =\ a^2\ + 2\ a\ b\ +\ b^2$

(a) $\left(2\sqrt{3} + \sqrt{5}\right)^2 = \left(2\sqrt{3}\right)^2 + 2\cdot 2\sqrt{3}\cdot\sqrt{5} + \left(\sqrt{5}\right)^2$

Multiply: $= 4\sqrt{9} + 4\sqrt{15} + \sqrt{25}$

Simplify: $= 4\cdot 3 + 4\sqrt{15} + 5$

Combine like terms: $= 17 + 4\sqrt{15}$

(b) We should notice that $\left(3 + \sqrt{7}\right)\left(3 - \sqrt{7}\right)$ is in the form $(a + b)(a - b)$, so that

$$(a\ +\ b)(a\ -\ b\ ) = a^2\ -\ b^2$$

$$\left(3 + \sqrt{7}\right)\left(3 - \sqrt{7}\right) = 3^2 - \left(\sqrt{7}\right)^2$$

$$= 9 - 7$$

$$= 2$$

Notice that the product found in Example 5(b) is an integer. That is, there are no radicals in the product. Radical expressions such as $3 + \sqrt{7}$ and $3 - \sqrt{7}$ are called **conjugates** of each other. When we multiply radical expressions involving square roots that are conjugates, the result will never contain a radical. This result plays a huge role in the next section.

QUICK ✓ *Multiply and simplify.*

11. $\left(5\sqrt{2} + \sqrt{3}\right)^2$ **12.** $\left(\sqrt{7} - 3\sqrt{2}\right)^2$ **13.** $\left(\sqrt{3} + \sqrt{2}\right)\left(\sqrt{3} - \sqrt{2}\right)$

9.5 Exercises

For Extra Help:

Student Solutions Manual   CD Video   PH Math/Tutor Center   MathXL Tutorials on CD   MathXL®   MyMathLab

Concepts and Vocabulary

In Problems 1 and 2, fill in the blanks.

1. Two radicals are _____ _____ if each radical has the same index and the same radicand.

2. The radical expressions $4 + \sqrt{5}$ and $4 - \sqrt{5}$ are examples of _____.

In Problems 3 and 4, answer True or False to each statement.

3. To add or subtract radicals, the index and radicand must be the same.

4. The conjugate of $-5 + \sqrt{2}$ is $5 - \sqrt{2}$.

5. In your own words, explain how to add or subtract radicals.

6. Multiply $\left(\sqrt{a} - \sqrt{b}\right)\left(\sqrt{a} + \sqrt{b}\right)$ and provide a general result regarding the product of conjugates involving square roots.

Building Skills

In Problems 7–14, add or subtract as indicated.

7. $3\sqrt{2} + 7\sqrt{2}$ **8.** $6\sqrt{3} + 8\sqrt{3}$

9. $5\sqrt[3]{x} - 3\sqrt[3]{x}$ **10.** $12\sqrt[4]{z} - 5\sqrt[4]{z}$

11. $8\sqrt{5x} - 3\sqrt{5x} + 9\sqrt{5x}$ **12.** $4\sqrt[3]{3y} + 8\sqrt[3]{3y} - 10\sqrt[3]{3y}$

13. $4\sqrt[3]{5} - 3\sqrt{5} + 7\sqrt[3]{5} - 8\sqrt{5}$ **14.** $12\sqrt{7} + 5\sqrt[4]{7} - 5\sqrt{7} + 6\sqrt[4]{7}$

In Problems 15–36, add or subtract as indicated. Assume all variables are positive or zero.

15. $\sqrt{8} + 6\sqrt{2}$ **16.** $6\sqrt{3} + \sqrt{12}$ **17.** $\sqrt[3]{24} - 4\sqrt[3]{3}$

18. $\sqrt[3]{32} - 5\sqrt[3]{4}$ **19.** $\sqrt[3]{54} - 7\sqrt[3]{128}$ **20.** $7\sqrt[4]{48} - 4\sqrt[4]{243}$

21. $5\sqrt{54x} - 3\sqrt{24x}$ **22.** $2\sqrt{48z} - \sqrt{75z}$ **23.** $2\sqrt{8} + 3\sqrt{10}$

24. $4\sqrt{12} + 2\sqrt{20}$ **25.** $\sqrt{12x^3} + 5x\sqrt{108x}$ **26.** $3\sqrt{63z^3} + 2z\sqrt{28z}$

27. $\sqrt{12x^2} + 3x\sqrt{2} - 2\sqrt{98x^2}$ **28.** $\sqrt{48y^2} - 4y\sqrt{12} + \sqrt{108y^2}$

29. $\sqrt[3]{-54x^3} + 3x\sqrt[3]{16} - 2\sqrt[3]{128}$ **30.** $2\sqrt[3]{-5x^3} + 4x\sqrt[3]{40} - \sqrt[3]{135}$

31. $\sqrt{9x - 9} + \sqrt{4x - 4}$ **32.** $\sqrt{4x + 12} - \sqrt{9x + 27}$

33. $\sqrt{16x} - \sqrt[6]{x^3}$ **34.** $\sqrt{25x} - \sqrt[4]{x^2}$

35. $\sqrt[3]{27x} + 2\sqrt[9]{x^3}$ **36.** $\sqrt[4]{16y} + \sqrt[8]{y^2}$

In Problems 37–70, multiply and simplify. Assume all variables are positive or zero.

37. $\sqrt{3}(2 - 3\sqrt{2})$

38. $\sqrt{5}(5 + 3\sqrt{3})$

39. $\sqrt{3}(\sqrt{2} + \sqrt{6})$

40. $\sqrt{2}(\sqrt{5} - 2\sqrt{10})$

41. $\sqrt[3]{4}(\sqrt[3]{3} - \sqrt[3]{6})$

42. $\sqrt[3]{6}(\sqrt[3]{2} + \sqrt[3]{12})$

43. $\sqrt{2x}(3 - \sqrt{10x})$

44. $\sqrt{5x}(6 + \sqrt{15x})$

45. $(3 + \sqrt{2})(4 + \sqrt{3})$

46. $(5 + \sqrt{5})(3 + \sqrt{6})$

47. $(6 + \sqrt{3})(2 - \sqrt{7})$

48. $(7 - \sqrt{3})(6 + \sqrt{5})$

49. $(4 - 2\sqrt{7})(3 + 3\sqrt{7})$

50. $(9 + 5\sqrt{10})(1 - 3\sqrt{10})$

51. $(\sqrt{2} + 3\sqrt{6})(\sqrt{3} - 2\sqrt{2})$

52. $(2\sqrt{3} + \sqrt{10})(\sqrt{5} - 2\sqrt{2})$

53. $(2\sqrt{5} + \sqrt{3})(4\sqrt{5} - 3\sqrt{3})$

54. $(\sqrt{6} - 2\sqrt{2})(2\sqrt{6} + 3\sqrt{2})$

55. $(1 + \sqrt{3})^2$

56. $(2 - \sqrt{3})^2$

57. $(\sqrt{2} - \sqrt{5})^2$

58. $(\sqrt{7} - \sqrt{3})^2$

59. $(\sqrt{x} - \sqrt{2})^2$

60. $(\sqrt{z} + \sqrt{5})^2$

61. $(\sqrt{2} - 1)(\sqrt{2} + 1)$

62. $(\sqrt{3} - 1)(\sqrt{3} + 1)$

63. $(3 - 2\sqrt{5})(3 + 2\sqrt{5})$

64. $(6 + 3\sqrt{2})(6 - 3\sqrt{2})$

65. $(\sqrt{2x} + \sqrt{3y})(\sqrt{2x} - \sqrt{3y})$

66. $(\sqrt{5a} + \sqrt{7b})(\sqrt{5a} - \sqrt{7b})$

67. $(\sqrt[3]{x} + 4)(\sqrt[3]{x} - 3)$

68. $(\sqrt[3]{y} - 6)(\sqrt[3]{y} + 3)$

69. $(\sqrt[3]{2a} - 5)(\sqrt[3]{2a} + 5)$

70. $(\sqrt[3]{4p} - 1)(\sqrt[3]{4p} + 3)$

Mixed Practice

In Problems 71–90, perform the indicated operation and simplify. Assume all variables are positive or zero.

71. $\sqrt{5}(\sqrt{3} + \sqrt{10})$

72. $\sqrt{7}(\sqrt{14} + \sqrt{3})$

73. $\sqrt{28x^5} - x\sqrt{7x^3} + 5\sqrt{175x^5}$

74. $\sqrt{180a^5} + a^2\sqrt{20} - a\sqrt{80a^3}$

75. $(2\sqrt{3} + 5)(2\sqrt{3} - 5)$

76. $(4\sqrt{2} - 2)(4\sqrt{2} + 2)$

77. $\sqrt[3]{7}(2 + \sqrt[3]{4})$

78. $\sqrt[3]{9}(5 + 2\sqrt[3]{2})$

79. $(2\sqrt{2} + 5)(4\sqrt{2} - 4)$

80. $(5\sqrt{5} - 3)(3\sqrt{5} - 4)$

81. $4\sqrt{18} + 2\sqrt{32}$

82. $5\sqrt{20} + 2\sqrt{80}$

83. $(\sqrt{5} - \sqrt{3})^2$

84. $(\sqrt{2} - \sqrt{7})^2$

85. $3\sqrt[3]{5x^3y} + \sqrt[3]{40y}$

86. $5\sqrt[3]{3m^3n} + \sqrt[3]{81n}$

87. $\left(\sqrt{2x} - \sqrt{7y}\right)\left(\sqrt{2x} + \sqrt{7y}\right)$ **88.** $\left(\sqrt{3a} - \sqrt{4b}\right)\left(\sqrt{3a} + \sqrt{4b}\right)$

89. $-\dfrac{3}{5} \cdot \left(-\dfrac{\sqrt{5}}{5}\right) - \dfrac{4}{5} \cdot \left(-\dfrac{2\sqrt{5}}{5}\right)$ **90.** $\dfrac{4}{5} \cdot \left(-\dfrac{\sqrt{5}}{5}\right) + \left(-\dfrac{3}{5}\right) \cdot \left(-\dfrac{2\sqrt{5}}{5}\right)$

Applying the Concepts

In Problems 91 and 92, find the perimeter and area of the figures shown. Express your answer as a radical in simplified form.

91.

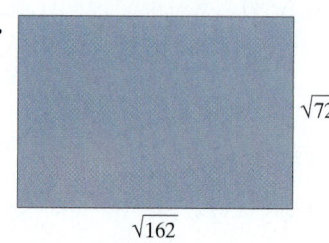

$\sqrt{72}$

$\sqrt{162}$

92.

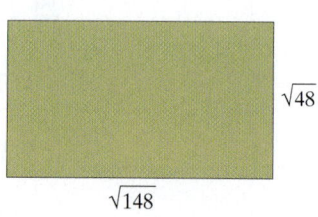

$\sqrt{48}$

$\sqrt{148}$

*Problems 93 and 94, use **Heron's Formula** for finding the area of a triangle whose sides are known. Heron's Formula states that the area A of a triangle with sides a, b, and c is*

$$A = \sqrt{s(s - a)(s - b)(s - c)}$$

where

$$s = \frac{1}{2}(a + b + c)$$

Find the area of the shaded region by computing the difference in the areas of each triangle. That is, compute "area of larger triangle minus area of smaller triangle." Write your answer as a radical in simplified form.

93.

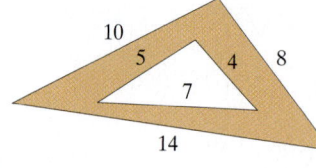

94.

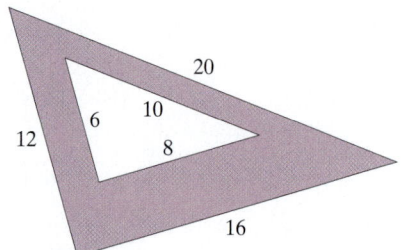

Extending the Concepts

95. Suppose that $f(x) = \sqrt{3x}$ and $g(x) = \sqrt{12x}$; find

    (a) $(f + g)(x)$ **(b)** $(f + g)(4)$ **(c)** $(f \cdot g)(x)$

96. Suppose that $f(x) = \sqrt{4x - 4}$ and $g(x) = \sqrt{25x - 25}$; find

    (a) $(f + g)(x)$ **(b)** $(f + g)(10)$ **(c)** $(f \cdot g)(x)$

97. Show that $-2 + \sqrt{5}$ is a solution to the equation $x^2 + 4x - 1 = 0$. Show that $-2 - \sqrt{5}$ is also a solution.

Synthesis Review

In Problems 98–103, multiply each of the following.

98. $(3a^3b)(4a^2b^4)$ **99.** $(3p - 1)(2p + 3)$ **100.** $(3y + 2)(2y - 1)$

101. $(m - 4)(m + 4)$ **102.** $(5w + 2)(5w - 2)$ **103.** $\left(\sqrt{x} + 2\right)\left(\sqrt{x} - 2\right)$

9.6 Rationalizing Radical Expressions

In Words
We call the process "rationalizing the denominator" because we are making the denominator a rational number (no radicals).

Preparing for Rationalizing Radical Expressions
Before getting started, take the following readiness quiz. If you get a problem wrong, go back to the section cited and review the material.

1. What would we need to multiply 12 by in order to make it the smallest perfect square that is a multiple of 12?

2. Simplify: $\sqrt{25x^2}$, $x > 0$ [Section 9.4, pp. 683–685]

When radical expressions appear in the denominator of a quotient, it is customary to rewrite the quotient so that the denominator does not contain any radicals. The process is referred to as **rationalizing the denominator.** In this section, we shall concentrate on rationalizing denominators that contain a single term and denominators that contain two terms.

① Rationalize a Denominator Containing One Term

To rationalize a denominator containing a single square root, we multiply the numerator and denominator of the quotient by a square root so that the radicand in the denominator becomes a perfect square. For example, if the denominator of a quotient contains $\sqrt{5}$, we would multiply the numerator and denominator by $\sqrt{5}$ because $5 \cdot 5 = 25$, which is a perfect square. If the denominator of a quotient contains $\sqrt{8}$, we would multiply the numerator and denominator by $\sqrt{2}$ because $8 \cdot 2 = 16$, which is a perfect square.

EXAMPLE 1   **Rationalizing a Denominator Containing a Square Root**

Rationalize the denominator of each expression:

(a) $\dfrac{1}{\sqrt{7}}$ **(b)** $\dfrac{\sqrt{5}}{\sqrt{12}}$ **(c)** $\dfrac{2}{3\sqrt{2x}}$

Solution

Work Smart
Remember that $a = 1 \cdot a$ and that 1 can take many forms. In Example 1(a),
$$1 = \frac{\sqrt{7}}{\sqrt{7}}.$$

(a) We have $\sqrt{7}$ in the denominator. So, we ask ourselves, "What can I multiply by 7 and obtain a perfect square?" Because $7 \cdot 7 = 49$ and 49 is a perfect square, we multiply the numerator and denominator by $\sqrt{7}$.

$$\frac{1}{\sqrt{7}} = \frac{1}{\sqrt{7}} \cdot \frac{\sqrt{7}}{\sqrt{7}}$$

Multiply numerators;
multiply denominators: $= \dfrac{\sqrt{7}}{\sqrt{49}}$

$\sqrt{49} = 7$: $= \dfrac{\sqrt{7}}{7}$

Work Smart
An alternative approach to Example 1(b) would be to simplify $\sqrt{12}$ first, as follows:

$$\frac{\sqrt{5}}{\sqrt{12}} = \frac{\sqrt{5}}{\sqrt{4 \cdot 3}}$$

$$= \frac{\sqrt{5}}{2\sqrt{3}}$$

$$= \frac{\sqrt{5}}{2\sqrt{3}} \cdot \frac{\sqrt{3}}{\sqrt{3}}$$

$$= \frac{\sqrt{15}}{6}$$

(b) We have $\sqrt{12}$ in the denominator. Again, we ask, "What can I multiply by 12 and obtain a perfect square?" Certainly, multiplying 12 by itself would result in a perfect square, but is this the best choice? No, you want to find the smallest integer that makes the radicand a perfect square. So, multiply 12 by 3 to obtain 36, a perfect square.

$$\frac{\sqrt{5}}{\sqrt{12}} = \frac{\sqrt{5}}{\sqrt{12}} \cdot \frac{\sqrt{3}}{\sqrt{3}}$$

Multiply numerators;
multiply denominators: $= \dfrac{\sqrt{15}}{\sqrt{36}}$

$\sqrt{36} = 6$: $= \dfrac{\sqrt{15}}{6}$

Preparing for...Answers **1.** 3 **2.** $5x$

(c)

$$\frac{2}{3\sqrt{2x}} = \frac{2}{3\sqrt{2x}} \cdot \frac{\sqrt{2x}}{\sqrt{2x}}$$

Multiply numerators;
multiply denominators:

$$= \frac{2\sqrt{2x}}{3\sqrt{4x^2}}$$

$\sqrt{4x^2} = 2x$:

$$= \frac{2\sqrt{2x}}{3 \cdot 2x}$$

Divide out common factor, 2:

$$= \frac{\sqrt{2x}}{3x}$$

QUICK ✓ *Rationalize each denominator.*

1. $\dfrac{1}{\sqrt{3}}$ **2.** $\dfrac{\sqrt{5}}{\sqrt{8}}$ **3.** $\dfrac{5}{\sqrt{10x}}$

In general, to rationalize a denominator when the denominator contains a single radical, we multiply the numerator and denominator of the quotient by a radical so that the product in the denominator has a radicand that is a perfect power of the index n. So, if the denominator contains a radical whose index is 3, we multiply the numerator and denominator by a cube root so that the radicand in the denominator becomes a perfect cube. For example, if the denominator of a quotient contains $\sqrt[3]{4}$, we would multiply the numerator and denominator of the quotient by $\sqrt[3]{2}$ because $4 \cdot 2 = 8$ and 8 is a perfect cube.

EXAMPLE 2 **Rationalizing a Denominator Containing Cube Roots and Fourth Roots**

Rationalize the denominator of each expression:

(a) $\dfrac{1}{\sqrt[3]{6}}$ **(b)** $\sqrt[3]{\dfrac{5}{18}}$ **(c)** $\dfrac{6}{\sqrt[4]{4z^3}}$

Assume all variables represent positive real numbers.

Solution

Work Smart

The radicand in $\sqrt[3]{6}$ is 6. Think of 6 as 6^1. To make it a perfect cube, 6^3, we need to multiply by 6^2 or 36. Then $6 \cdot 36 = 6^1 \cdot 6^2 = 6^3$. The cube root of 6^3 or 216 is 6.

(a) We have $\sqrt[3]{6}$ in the denominator. We want the radicand in the denominator to be a perfect cube (since the index is 3), so we multiply the numerator and denominator by $\sqrt[3]{6^2} = \sqrt[3]{36}$ since $6 \cdot 36 = 216$ and 216 is a perfect cube.

$$\frac{1}{\sqrt[3]{6}} = \frac{1}{\sqrt[3]{6}} \cdot \frac{\sqrt[3]{6^2}}{\sqrt[3]{6^2}}$$

Multiply numerators;
multiply denominators:

$$= \frac{\sqrt[3]{36}}{\sqrt[3]{6^3}}$$

$\sqrt[3]{6^3} = 6$:

$$= \frac{\sqrt[3]{36}}{6}$$

(b) First, we use the Quotient Property $\left(\sqrt[n]{\dfrac{a}{b}} = \dfrac{\sqrt[n]{a}}{\sqrt[n]{b}}\right)$ to rewrite the radical as the quotient of two radicals.

$$\sqrt[3]{\frac{5}{18}} = \frac{\sqrt[3]{5}}{\sqrt[3]{18}}$$

What do we need to multiply $\sqrt[3]{18}$ by in order to make it a perfect cube? We will write 18 as $9 \cdot 2 = 3^2 \cdot 2^1$. Therefore, if we multiply $18 = 3^2 \cdot 2^1$ by $3^1 \cdot 2^2 = 12$, we will have a perfect cube as the radicand in the denominator.

$$\sqrt[3]{\frac{5}{18}} = \frac{\sqrt[3]{5}}{\sqrt[3]{18}} = \frac{\sqrt[3]{5}}{\sqrt[3]{3^2 \cdot 2}} \cdot \frac{\sqrt[3]{3 \cdot 2^2}}{\sqrt[3]{3 \cdot 2^2}}$$

Multiply numerators; multiply denominators:
$$= \frac{\sqrt[3]{60}}{\sqrt[3]{3^3 \cdot 2^3}}$$

$$= \frac{\sqrt[3]{60}}{6}$$

(c) We rewrite the denominator as $\sqrt[4]{2^2 \cdot z^3}$. To make the radicand a perfect power of 4, we need to multiply $\sqrt[4]{2^2 \cdot z^3}$ by $\sqrt[4]{2^2 \cdot z} = \sqrt[4]{4z}$ to obtain $\sqrt[4]{2^4 z^4}$ in the denominator.

$$\frac{6}{\sqrt[4]{4z^3}} = \frac{6}{\sqrt[4]{2^2 \cdot z^3}} \cdot \frac{\sqrt[4]{2^2 \cdot z}}{\sqrt[4]{2^2 \cdot z}}$$

Multiply numerators; multiply denominators:
$$= \frac{6\sqrt[4]{4z}}{\sqrt[4]{2^4 \cdot z^4}}$$

$$= \frac{6\sqrt[4]{4z}}{2z}$$

Simplify:
$$= \frac{3\sqrt[4]{4z}}{z}$$

QUICK ✓   *Rationalize each denominator. Assume all variables are positive.*

4. $\dfrac{4}{\sqrt[3]{3}}$ **5.** $\sqrt[3]{\dfrac{3}{20}}$ **6.** $\dfrac{3}{\sqrt[4]{p}}$

---

②   Rationalize a Denominator Containing Two Terms

To rationalize a denominator containing two terms involving square roots, we use the fact that

$$(a + b)(a - b) = a^2 - b^2$$

and multiply both the numerator and denominator of the quotient by the conjugate of the denominator. For example, if the quotient is $\dfrac{3}{\sqrt{3} + 2}$, we would multiply both the numerator and the denominator by the conjugate of $\sqrt{3} + 2$, which is $\sqrt{3} - 2$. We know from the last section that the product $(\sqrt{3} + 2)(\sqrt{3} - 2)$ will not contain a radical.

EXAMPLE 3   **Rationalizing a Denominator Containing Two Terms**

Rationalize the denominator: $\dfrac{\sqrt{2}}{\sqrt{6} + 2}$

Solution

We see that $\sqrt{6} + 2$ is in the denominator of the quotient, so we multiply the numerator and denominator by the conjugate of $\sqrt{6} + 2$, $\sqrt{6} - 2$.

$$\frac{\sqrt{2}}{\sqrt{6} + 2} = \frac{\sqrt{2}}{\sqrt{6} + 2} \cdot \frac{\sqrt{6} - 2}{\sqrt{6} - 2}$$

Multiply the numerators
and denominators:
$$= \frac{\sqrt{2}(\sqrt{6} - 2)}{(\sqrt{6} + 2)(\sqrt{6} - 2)}$$

Distribute in the numerator;
$(a + b)(a - b) = a^2 - b^2$ in the denominator:
$$= \frac{\sqrt{12} - 2\sqrt{2}}{(\sqrt{6})^2 - 2^2}$$

$\sqrt{12} = 2\sqrt{3}$:
$$= \frac{2\sqrt{3} - 2\sqrt{2}}{6 - 4}$$

Factor out common factor of 2 in numerator:
$$= \frac{2(\sqrt{3} - \sqrt{2})}{2}$$

Divide out the 2s:
$$= \sqrt{3} - \sqrt{2}$$

QUICK ✓ *Rationalize the denominator.*

7. $\dfrac{4}{\sqrt{3} + 1}$

8. $\dfrac{\sqrt{2}}{\sqrt{6} - \sqrt{2}}$

EXAMPLE 4 **Rationalizing a Denominator Containing Two Terms**

Rationalize the denominator: $\dfrac{\sqrt{6} - 3}{\sqrt{10} - \sqrt{6}}$

Solution

We see that $\sqrt{10} - \sqrt{6}$ is in the denominator of the quotient, so we multiply the numerator and denominator by the conjugate of the denominator, $\sqrt{10} + \sqrt{6}$.

$$\frac{\sqrt{6} - 3}{\sqrt{10} - \sqrt{6}} = \frac{\sqrt{6} - 3}{\sqrt{10} - \sqrt{6}} \cdot \frac{\sqrt{10} + \sqrt{6}}{\sqrt{10} + \sqrt{6}}$$

Multiply the numerators
and denominators:
$$= \frac{(\sqrt{6} - 3)(\sqrt{10} + \sqrt{6})}{(\sqrt{10} - \sqrt{6})(\sqrt{10} + \sqrt{6})}$$

FOIL the numerator;
$(a + b)(a - b) = a^2 - b^2$ in the denominator:
$$= \frac{\sqrt{60} + \sqrt{36} - 3\sqrt{10} - 3\sqrt{6}}{(\sqrt{10})^2 - (\sqrt{6})^2}$$

Simplify radicals:
$$= \frac{2\sqrt{15} + 6 - 3\sqrt{10} - 3\sqrt{6}}{10 - 6}$$

$$= \frac{2\sqrt{15} + 6 - 3\sqrt{10} - 3\sqrt{6}}{4}$$

QUICK ✓ *Rationalize the denominator.*

9. $\dfrac{\sqrt{5} + 4}{\sqrt{5} - \sqrt{2}}$

9.6 Exercises

Concepts and Vocabulary

In Problems 1 and 2, fill in the blanks.

1. Rewriting a quotient to remove radicals from the denominator is called _____ _____ _____.

2. To rationalize the denominator of $\dfrac{\sqrt{5}}{\sqrt{7}}$, we would multiply the numerator and denominator by _____.

In Problem 3, answer True or False.

3. To rationalize the denominator of $\dfrac{4-\sqrt{3}}{-2+\sqrt{7}}$, we would multiply the numerator and denominator by $2-\sqrt{7}$.

4. Explain why it is necessary to multiply the numerator and denominator by the conjugate of the denominator when rationalizing a denominator containing two terms.

Building Skills

In Problems 5–48, rationalize each denominator. Assume all variables are positive.

5. $\dfrac{1}{\sqrt{2}}$

6. $\dfrac{2}{\sqrt{3}}$

7. $-\dfrac{6}{5\sqrt{3}}$

8. $-\dfrac{3}{2\sqrt{3}}$

9. $\dfrac{3}{\sqrt{12}}$

10. $\dfrac{5}{\sqrt{20}}$

11. $\dfrac{\sqrt{2}}{\sqrt{6}}$

12. $\dfrac{\sqrt{3}}{\sqrt{11}}$

13. $\sqrt{\dfrac{2}{p}}$

14. $\sqrt{\dfrac{5}{z}}$

15. $\dfrac{\sqrt{8}}{\sqrt{y^3}}$

16. $\dfrac{\sqrt{32}}{\sqrt{a^5}}$

17. $\dfrac{2}{\sqrt[3]{2}}$

18. $\dfrac{5}{\sqrt[3]{3}}$

19. $\sqrt[3]{\dfrac{7}{q}}$

20. $\sqrt[3]{\dfrac{-4}{p}}$

21. $\sqrt[3]{\dfrac{-3}{50}}$

22. $\sqrt[3]{\dfrac{-5}{72}}$

23. $\dfrac{2}{\sqrt[3]{20y}}$

24. $\dfrac{8}{\sqrt[3]{36z^2}}$

25. $\dfrac{-4}{\sqrt[4]{3x^3}}$

26. $\dfrac{6}{\sqrt[4]{9b^2}}$

27. $\dfrac{12}{\sqrt[5]{m^3n^2}}$

28. $\dfrac{-3}{\sqrt[5]{ab^3}}$

29. $\dfrac{4}{\sqrt{6}-2}$

30. $\dfrac{6}{\sqrt{7}-2}$

31. $\dfrac{5}{\sqrt{5}+2}$

32. $\dfrac{10}{\sqrt{10}+3}$

33. $\dfrac{8}{\sqrt{7}-\sqrt{3}}$

34. $\dfrac{12}{\sqrt{11}-\sqrt{7}}$

35. $\dfrac{\sqrt{2}}{\sqrt{10}-\sqrt{6}}$

36. $\dfrac{\sqrt{3}}{\sqrt{15}-\sqrt{6}}$

37. $\dfrac{\sqrt{p}}{\sqrt{p}+\sqrt{q}}$

38. $\dfrac{\sqrt{a}}{\sqrt{a}+\sqrt{b}}$

39. $\dfrac{18}{2\sqrt{3}+3\sqrt{2}}$

40. $\dfrac{15}{3\sqrt{5}+4\sqrt{3}}$

41. $\dfrac{\sqrt{7}+3}{\sqrt{7}-3}$

42. $\dfrac{\sqrt{5}+3}{\sqrt{5}-3}$

43. $\dfrac{\sqrt{3}-4\sqrt{2}}{2\sqrt{3}+5\sqrt{2}}$

44. $\dfrac{3\sqrt{6}+5\sqrt{7}}{2\sqrt{6}-3\sqrt{7}}$

45. $\dfrac{\sqrt{p}+2}{\sqrt{p}-2}$

46. $\dfrac{\sqrt{x}-4}{\sqrt{x}+4}$

47. $\dfrac{\sqrt{2}-3}{\sqrt{8}-\sqrt{2}}$

48. $\dfrac{2\sqrt{3}+3}{\sqrt{12}-\sqrt{3}}$

Mixed Practice

In Problems 49–56, perform the indicated operation and simplify.

49. $\sqrt{3} + \dfrac{1}{\sqrt{3}}$

50. $\sqrt{5} - \dfrac{1}{\sqrt{5}}$

51. $\dfrac{\sqrt{10}}{2} - \dfrac{1}{\sqrt{2}}$

52. $\dfrac{\sqrt{5}}{2} + \dfrac{3}{\sqrt{5}}$

53. $\sqrt{\dfrac{1}{3}} + \sqrt{12} + \sqrt{75}$

54. $\sqrt{\dfrac{2}{5}} + \sqrt{20} - \sqrt{45}$

55. $\dfrac{3}{\sqrt{18}} - \sqrt{\dfrac{1}{2}}$

56. $\sqrt{\dfrac{4}{3}} + \dfrac{4}{\sqrt{48}}$

In Problems 57–68, simplify each expression so that the denominator does not contain a radical. Work smart because in some of the problems, it will be easier if you divide the radicands before attempting to rationalize the denominator.

57. $\dfrac{\sqrt{3}}{\sqrt{12}}$

58. $\dfrac{\sqrt{2}}{\sqrt{18}}$

59. $\dfrac{3}{\sqrt{72}}$

60. $\dfrac{7}{\sqrt{98}}$

61. $\sqrt{\dfrac{4}{3}}$

62. $\sqrt{\dfrac{9}{5}}$

63. $\dfrac{\sqrt{3} - 3}{\sqrt{3} + 3}$

64. $\dfrac{\sqrt{2} - 5}{\sqrt{2} + 5}$

65. $\dfrac{2}{\sqrt{5} + 2}$

66. $\dfrac{5}{\sqrt{6} + 4}$

67. $\dfrac{\sqrt{8}}{\sqrt{2}}$

68. $\dfrac{\sqrt{75}}{\sqrt{3}}$

Applying the Concepts

In Problems 69–74, find the reciprocal of the given number. Be sure to rationalize the denominator.

69. $\sqrt{3}$

70. $\sqrt{7}$

71. $\sqrt[3]{12}$

72. $\sqrt[3]{18}$

73. $\sqrt{3} + 5$

74. $7 - \sqrt{2}$

Problems 75 and 76 contain expressions that are seen in a course in Trigonometry. Simplify each expression completely.

75. $\dfrac{1}{\sqrt{2}} \cdot \dfrac{\sqrt{3}}{2} - \dfrac{1}{\sqrt{2}} \cdot \dfrac{1}{2}$

76. $-\sqrt{\dfrac{2}{3}} \cdot \left(-\dfrac{2}{\sqrt{5}} \right) + \dfrac{1}{\sqrt{3}} \cdot \dfrac{1}{\sqrt{5}}$

Sometimes we are asked to rationalize a numerator. In Problems 77–80, rationalize each expression by multiplying the numerator and denominator by the conjugate of the numerator.

77. $\dfrac{\sqrt{2} + 1}{3}$

78. $\dfrac{\sqrt{3} + 2}{2}$

79. $\dfrac{\sqrt{x} - \sqrt{h}}{\sqrt{x}}$

80. $\dfrac{\sqrt{a} - \sqrt{b}}{\sqrt{2}}$

Extending the Concepts

81. When two quantities a and b are positive, we can verify that $a = b$, by showing that $a^2 = b^2$. Verify that $\dfrac{\sqrt{6} + \sqrt{2}}{4} = \dfrac{\sqrt{2} + \sqrt{3}}{2}$ by squaring each side.

82. Rationalize the denominator: $\dfrac{2}{\sqrt{2} + \sqrt{3} - \sqrt{9}}$

Synthesis Review

In Problems 83–86, graph each of the following functions using point plotting.

83. $f(x) = 5x - 3$

84. $g(x) = -3x + 9$

85. $G(x) = x^2$

86. $F(x) = x^3$

PUTTING THE CONCEPTS TOGETHER (Sections 9.1–9.6)

These problems cover important concepts from Sections 9.1–9.6. We designed these problems so that you can review the chapter so far and show your mastery of the concepts. Take time to work these problems before proceeding with the next section. The answers to these problems are located at the back of the text starting on page AN-35.

1. Evaluate: $-25^{\frac{1}{2}}$

2. Evaluate: $(-64)^{-2/3}$

3. Write the expression $\sqrt[4]{3x^3}$ with a rational exponent.

4. Write the expression $7z^{4/5}$ as a radical expression.

5. Evaluate using rational exponents: $\sqrt[3]{\sqrt{64x^3}}$

6. Distribute and simplify: $c^{1/2}(c^{3/2} + c^{5/2})$

In Problems 7–9, use Laws of Exponents to simplify each expression. Assume all variables in the radicand are greater than or equal to zero. Express answers with positive exponents.

7. $(a^{2/3}b^{-1/3})(a^{4/3}b^{-5/3})$

8. $\dfrac{x^{3/4}}{x^{1/8}}$

9. $(x^{3/4}y^{-1/8})^8$

In Problems 10–19, perform the indicated operation and simplify. Assume all variables in the radicand are greater than or equal to zero.

10. $\sqrt{15a} \cdot \sqrt{2b}$

11. $\sqrt{10m^3n^2} \cdot \sqrt{20mn}$

12. $\sqrt[3]{\dfrac{-32xy^4}{4x^{-2}y}}$

13. $2\sqrt{108} - 3\sqrt{75} + \sqrt{48}$

14. $-5b\sqrt{8b} + 7\sqrt{18b^3}$

15. $\sqrt[3]{16y^4} - y\sqrt[3]{2y}$

16. $(3\sqrt{x})(4\sqrt{x})$

17. $3\sqrt{x} + 4\sqrt{x}$

18. $(2 - 3\sqrt{2})(10 + \sqrt{2})$

19. $(4\sqrt{2} - 3)^2$

In Problems 20 and 21, rationalize the denominator.

20. $\dfrac{3}{2\sqrt{32}}$

21. $\dfrac{4}{\sqrt{3} - 8}$

9.7 Functions Involving Radicals

OBJECTIVES

1. Evaluate Functions Whose Rule Is a Radical Expression
2. Find the Domain of a Function Whose Rule Contains a Radical
3. Graph Functions Involving Square Roots
4. Graph Functions Involving Cube Roots

Preparing for Functions Involving Radicals

Before getting started, take the following readiness quiz. If you get a problem wrong, go back to the section cited and review the material.

1. Simplify: $\sqrt{121}$ [Section 9.1, pp. 660–661]

2. Simplify: $\sqrt{p^2}$ [Section 9.1, p. 663]

3. Given $f(x) = x^2 - 4$, find $f(3)$. [Section 8.3, pp. 590–592]

4. Solve: $-2x + 3 \ge 0$ [Section 2.8, pp. 153–158]

5. Graph $f(x) = x^2 + 1$ using point plotting. [Section 8.3, pp. 592–593]

Preparing for...Answers 1. 11 **2.** $|p|$

3. 5 **4.** $\left\{x \mid x \le \dfrac{3}{2}\right\}$ or $\left(-\infty, \dfrac{3}{2}\right]$

5.

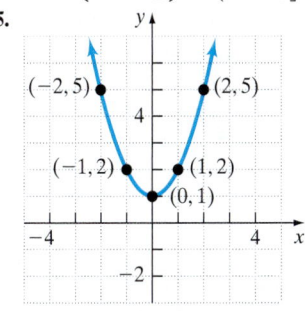

1. Evaluate Functions Whose Rule Is a Radical Expression

We can evaluate functions whose rule contains a radical by substituting the value of the independent variable into the rule, just as we did in Section 8.3.

EXAMPLE 1 | **Evaluating Functions for Which the Rule Is a Radical Expression**

For the functions $f(x) = \sqrt{x + 2}$ and $g(x) = \sqrt[3]{3x + 1}$, find

(a) $f(7)$

(b) $f(10)$

(c) $g(-3)$

Solution

(a) $f(x) = \sqrt{x + 2}$ (b) $f(x) = \sqrt{x + 2}$ (c) $g(x) = \sqrt[3]{3x + 1}$

$f(7) = \sqrt{7 + 2}$ $\qquad$ $f(10) = \sqrt{10 + 2}$ $\qquad$ $g(-3) = \sqrt[3]{3(-3) + 1}$

$\quad\ = \sqrt{9}$ $\qquad\qquad$ $\quad\ = \sqrt{12}$ $\qquad\qquad$ $\quad\ = \sqrt[3]{-8}$

$\quad\ = 3$ $\qquad\qquad\quad$ $\quad\ = 2\sqrt{3}$ $\qquad\qquad$ $\quad\ = -2$ ∎

QUICK ✔ *Find the following values for each function.*

1. $f(x) = \sqrt{3x + 7}$ $\qquad\qquad\qquad\qquad$ **2.** $g(x) = \sqrt[3]{2x + 7}$

$\quad$ **(a)** $f(3)$ $\qquad$ **(b)** $f(7)$ $\qquad\qquad\qquad\qquad$ **(a)** $g(-4)$ $\qquad$ **(b)** $g(10)$

---

② Find the Domain of a Function Whose Rule Contains a Radical

Recall the definition of the principal nth root of a number a, $\sqrt[n]{a}$. This definition states that $\sqrt[n]{a} = b$ means $a = b^n$. From this definition we learned that for $n \geq 2$ and even, the radicand, a, must be greater than or equal to 0. For $n \geq 3$ and odd, the radicand, a, can be any real number. This leads to a procedure for finding the domain of a function whose rule contains a radical.

> **FINDING THE DOMAIN OF A FUNCTION FOR WHICH THE RULE CONTAINS A RADICAL**
>
> • If the index on a radical is even, then the radicand must be greater than or equal to zero.
>
> • If the index on a radical is odd, then the radicand can be any real number.

EXAMPLE 2 Finding the Domain of a Radical Function

Find the domain of each of the following functions:

$\quad$ **(a)** $f(x) = \sqrt{x - 5}$ $\qquad$ **(b)** $G(x) = \sqrt[3]{2x + 1}$ $\qquad$ **(c)** $h(t) = \sqrt[4]{5 - 2t}$

Solution

(a) First, we take a look at the rule given in the function and interpret it. The function $f(x) = \sqrt{x - 5}$ tells us to take the square root of $x - 5$. We can only take square roots of numbers greater than or equal to zero, so the radicand, $x - 5$, must be greater than or equal to zero. This requires that

$$x - 5 \geq 0$$

Add 5 to both sides: $\quad x \geq 5$

The domain of f is $\{x \,|\, x \geq 5\}$ or the interval $[5, \infty)$.

(b) The function $G(x) = \sqrt[3]{2x + 1}$ tells us to take the cube root of $2x + 1$. We can take the cube root of any real number, so the domain of G is any real number.

(c) The function $h(t) = \sqrt[4]{5 - 2t}$ tells us to take the fourth root of $5 - 2t$. We can only take fourth roots of numbers greater than or equal to zero, so the radicand, $5 - 2t$, must be greater than or equal to zero. This requires that

$$5 - 2t \geq 0$$

Subtract 5 from both sides: $\quad -2t \geq -5$

Divide both sides by -2 (Don't forget to change the direction of the inequality!): $\quad t \leq \dfrac{5}{2}$

The domain of h is $\left\{ t \,\middle|\, t \leq \dfrac{5}{2} \right\}$ or the interval $\left(-\infty, \dfrac{5}{2} \right]$. ∎

QUICK ✓   *Find the domain of each function.*

3. $H(x) = \sqrt{x + 6}$ **4.** $g(t) = \sqrt[5]{3t - 1}$ **5.** $F(m) = \sqrt[4]{6 - 3m}$

③ Graph Functions Involving Square Roots

The **square root function** is given by $f(x) = \sqrt{x}$. The domain of the square root function is $\{x | x \geq 0\}$ or using interval notation $[0, \infty)$. We can obtain the graph of $f(x) = \sqrt{x}$ by determining some ordered pairs (x, y) such that $y = \sqrt{x}$. We then plot the ordered pairs in the xy-plane and connect the points. To make life easy, we choose values of x that are perfect squares ($0, 1, 4, 9$, and so on). Table 1 shows some points on the graph of $f(x) = \sqrt{x}$. Figure 3 shows the graph of $f(x) = \sqrt{x}$. From the graph of $f(x) = \sqrt{x}$ given in Figure 3, we can see that the range of $f(x) = \sqrt{x}$ is $[0, \infty)$.

Table 1		
x	**$f(x) = \sqrt{x}$**	**(x, y) or $(x, f(x))$**
0	$f(0) = 0$	$(0, 0)$
1	$f(1) = 1$	$(1, 1)$
4	$f(4) = 2$	$(4, 2)$
9	$f(9) = 3$	$(9, 3)$
16	$f(16) = 4$	$(16, 4)$

Figure 3
$f(x) = \sqrt{x}$

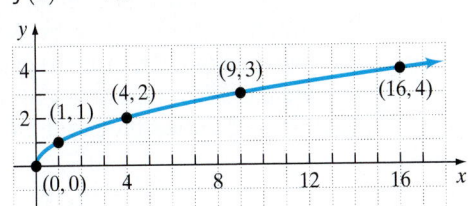

The point-plotting method can be used to graph a variety of functions involving square roots.

EXAMPLE 3   Graphing a Function Involving a Square Root

For the function $f(x) = \sqrt{x - 2}$,

(a) Find the domain.

(b) Graph the function using point plotting.

(c) Based on the graph, determine the range.

Solution

(a) The function $f(x) = \sqrt{x - 2}$ tells us to take the square root of $x - 2$. We can only take square roots of numbers greater than or equal to zero, so the radicand, $x - 2$, must be greater than or equal to zero. This requires that

$$x - 2 \geq 0$$

Add 2 to both sides: $x \geq 2$

The domain of f is $\{x | x \geq 2\}$ or the interval $[2, \infty)$.

(b) We choose values of x that are greater than or equal 2. Again, to make life easy, we choose values of x that will make the radicand a perfect square. See Table 2. Figure 4 shows the graph of $f(x) = \sqrt{x - 2}$.

Table 2

x	$f(x) = \sqrt{x - 2}$	(x, y) or $(x, f(x))$
2	$f(2) = 0$	$(2, 0)$
3	$f(3) = 1$	$(3, 1)$
6	$f(6) = 2$	$(6, 2)$
11	$f(11) = 3$	$(11, 3)$
18	$f(18) = 4$	$(18, 4)$

Figure 4

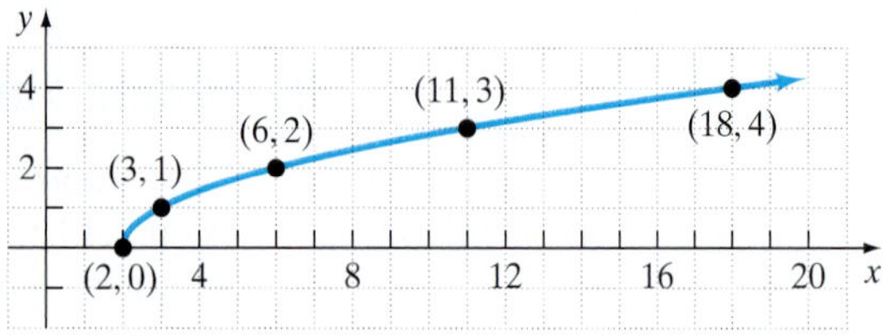

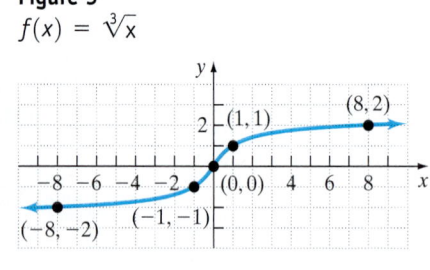

(c) From the graph of $f(x) = \sqrt{x - 2}$ given in Figure 4, we can see that the range of $f(x) = \sqrt{x - 2}$ is $[0, \infty)$.

QUICK ✓

6. For the function $f(x) = \sqrt{x + 3}$,

(a) Find the domain. **(b)** Graph the function using point plotting.

(c) Based on the graph, determine the range.

---

④ Graph Functions Involving Cube Roots

The **cube root function** is given by $f(x) = \sqrt[3]{x}$. The domain of the cube root function is $\{x \mid x \text{ is any real number}\}$ or using interval notation $(-\infty, \infty)$. We can obtain the graph of $f(x) = \sqrt[3]{x}$ by determining some ordered pairs (x, y) such that $y = \sqrt[3]{x}$. We then plot the ordered pairs in the xy-plane and connect the points. To make life easy, we choose values of x that are perfect cubes $(-8, -1, 0, 1, 8,$ and so on$)$. See Table 3. Figure 5 shows the graph of $f(x) = \sqrt[3]{x}$.

Table 3

x	$f(x) = \sqrt[3]{x}$	(x, y) or $(x, f(x))$
-8	$f(-8) = -2$	$(-8, -2)$
-1	$f(-1) = -1$	$(-1, -1)$
0	$f(0) = 0$	$(0, 0)$
1	$f(1) = 1$	$(1, 1)$
8	$f(8) = 2$	$(8, 2)$

Figure 5

$f(x) = \sqrt[3]{x}$

From the graph of $f(x) = \sqrt[3]{x}$ given in Figure 5, we can see that the range of $f(x) = \sqrt[3]{x}$ is $(-\infty, \infty)$.

The point-plotting method can be used to graph a variety of functions involving cube roots.

EXAMPLE 4 **Graphing a Function Involving a Cube Root**

For the function $g(x) = \sqrt[3]{x} + 2$,

(a) Find the domain. **(b)** Graph the function using point plotting.

(c) Based on the graph, determine the range.

Solution

(a) The function $g(x) = \sqrt[3]{x} + 2$ tells us to take the cube root of x and then add 2. We can take a cube root of any real number, so the domain of g is $\{x \mid x \text{ is any real number}\}$ or the interval $(-\infty, \infty)$.

(b) We choose values of x that make the radicand a perfect cube. See Table 4. Figure 6 shows the graph of $g(x) = \sqrt[3]{x} + 2$.

Table 4

x	$g(x) = \sqrt[3]{x} + 2$	(x, y) or $(x, g(x))$
-8	$g(-8) = \sqrt[3]{-8} + 2 = 0$	$(-8, 0)$
-1	$g(-1) = 1$	$(-1, 1)$
0	$g(0) = 2$	$(0, 2)$
1	$g(1) = 3$	$(1, 3)$
8	$g(8) = 4$	$(8, 4)$

Figure 6

(c) From the graph of $g(x) = \sqrt[3]{x} + 2$ given in Figure 6 we can see that the range of $g(x) = \sqrt[3]{x} + 2$ is $(-\infty, \infty)$.

Quick ✓

7. For the function $G(x) = \sqrt[3]{x} - 1$,

(a) Find the domain.

(b) Graph the function using point plotting.

(c) Based on the graph, determine the range.

9.7 Exercises

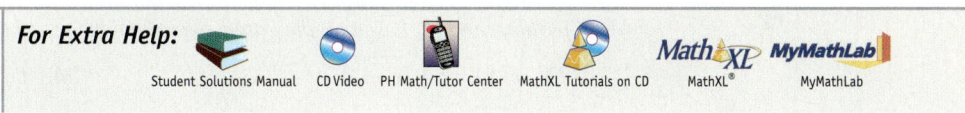

For Extra Help: Student Solutions Manual   CD Video   PH Math/Tutor Center   MathXL Tutorials on CD   Math XL   MathXL®   MyMathLab   MyMathLab

Concepts and Vocabulary

In Problems 1 and 2, fill in the blanks.

1. If the index on a radical expression is ____, then the radicand must be greater than or equal to zero.

2. If the index on a radical expression is ____, then the radicand can be any real number.

In Problem 3 and 4, answer True or False.

3. The domain and range of $f(x) = \sqrt{x}$ is $[0, \infty)$.

4. The domain and range of $f(x) = \sqrt[3]{x}$ is $(-\infty, \infty)$.

Building Skills

In Problems 5–16, evaluate each radical function at the indicated values.

5. $f(x) = \sqrt{x + 6}$

   (a) $f(3)$ (b) $f(8)$ (c) $f(-2)$

6. $f(x) = \sqrt{x + 10}$

   (a) $f(6)$ (b) $f(2)$ (c) $f(-6)$

7. $g(x) = -\sqrt{2x + 3}$

   (a) $g(11)$ (b) $g(-1)$ (c) $g\left(\dfrac{1}{8}\right)$

8. $g(x) = -\sqrt{4x + 5}$

   (a) $g(1)$ (b) $g(10)$ (c) $g\left(\dfrac{1}{8}\right)$

9. $G(m) = 2\sqrt{5m - 1}$

  (a) $G(1)$ (b) $G(5)$ (c) $G\left(\dfrac{1}{2}\right)$

10. $G(p) = 3\sqrt{4p + 1}$

  (a) $G(2)$ (b) $G(11)$ (c) $G\left(\dfrac{1}{8}\right)$

11. $H(z) = \sqrt[3]{z + 4}$

  (a) $H(4)$ (b) $H(-12)$ (c) $H(-20)$

12. $G(t) = \sqrt[3]{t - 6}$

  (a) $G(7)$ (b) $G(-21)$ (c) $G(22)$

13. $f(x) = \sqrt{\dfrac{x - 2}{x + 2}}$

  (a) $f(7)$ (b) $f(6)$ (c) $f(10)$

14. $f(x) = \sqrt{\dfrac{x - 4}{x + 4}}$

  (a) $f(5)$ (b) $f(8)$ (c) $f(12)$

15. $g(z) = \sqrt[3]{\dfrac{2z}{z - 4}}$

  (a) $g(-4)$ (b) $g(8)$ (c) $g(12)$

16. $H(z) = \sqrt[3]{\dfrac{3z}{z + 5}}$

  (a) $H(3)$ (b) $H(4)$ (c) $H(-1)$

In Problems 17–32, find the domain of the radical function.

17. $f(x) = \sqrt{x - 7}$

18. $f(x) = \sqrt{x + 4}$

19. $g(x) = \sqrt{2x + 7}$

20. $g(x) = \sqrt{3x + 7}$

21. $F(x) = \sqrt{4 - 3x}$

22. $G(x) = \sqrt{5 - 2x}$

23. $H(z) = \sqrt[3]{2z + 1}$

24. $G(z) = \sqrt[3]{5z - 3}$

25. $W(p) = \sqrt[4]{7p - 2}$

26. $C(y) = \sqrt[4]{3y - 2}$

27. $g(x) = \sqrt[5]{x - 3}$

28. $g(x) = \sqrt[5]{x + 9}$

29. $f(x) = \sqrt{\dfrac{3}{x + 5}}$

30. $f(x) = \sqrt{\dfrac{3}{x - 3}}$

31. $H(x) = \sqrt{\dfrac{x + 3}{x - 3}}$

32. $H(x) = \sqrt{\dfrac{x - 5}{x}}$

In Problems 33–54, (a) determine the domain of the function; (b) graph the function using point plotting; and (c) based on the graph, determine the range of the function.

33. $f(x) = \sqrt{x - 4}$

34. $f(x) = \sqrt{x - 1}$

35. $g(x) = \sqrt{x + 2}$

36. $g(x) = \sqrt{x + 5}$

37. $G(x) = \sqrt{2 - x}$

38. $F(x) = \sqrt{4 - x}$

39. $f(x) = \sqrt{x} + 3$

40. $f(x) = \sqrt{x} + 1$

41. $g(x) = \sqrt{x} - 4$

42. $g(x) = \sqrt{x} - 2$

43. $H(x) = 2\sqrt{x}$

44. $h(x) = 3\sqrt{x}$

45. $f(x) = \dfrac{1}{2}\sqrt{x}$

46. $g(x) = \dfrac{1}{4}\sqrt{x}$

47. $G(x) = -\sqrt{x}$

48. $F(x) = \sqrt{-x}$

49. $h(x) = \sqrt[3]{x + 2}$

50. $g(x) = \sqrt[3]{x - 4}$

51. $f(x) = \sqrt[3]{x} - 3$

52. $H(x) = \sqrt[3]{x} + 3$

53. $G(x) = 2\sqrt[3]{x}$

54. $F(x) = 3\sqrt[3]{x}$

Applying the Concepts

55. Distance to a Point on a Graph Suppose that $P = (x, y)$ is a point on the graph of $y = x^2 - 4$. The distance from P to $(0, 1)$ is given by the function

$$d(x) = \sqrt{x^4 - 9x^2 + 25}$$

See the figure.

(a) What is the distance from $P = (0, -4)$ to $(0, 1)$? That is, what is $d(0)$?

(b) What is the distance from $P = (1, -3)$ to $(0, 1)$? That is, what is $d(1)$?

(c) What is the distance from $P = (5, 21)$ to $(0, 1)$? That is, what is $d(5)$?

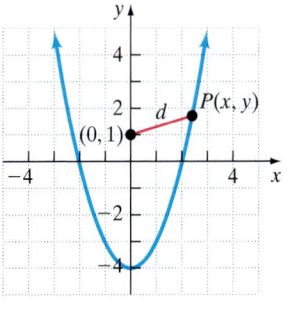

56. Distance to a Point on a Graph Suppose that $P = (x, y)$ is a point on the graph of $y = x^2 - 2$. The distance from P to $(0, 2)$ is given by the function

$$d(x) = \sqrt{x^4 - 7x^2 + 16}$$

See the figure.

(a) What is the distance from $P = (0, -2)$ to $(0, 2)$? That is, what is $d(0)$?

(b) What is the distance from $P = (1, -1)$ to $(0, 2)$? That is, what is $d(1)$?

(c) What is the distance from $P = (4, 14)$ to $(0, 2)$? That is, what is $d(4)$?

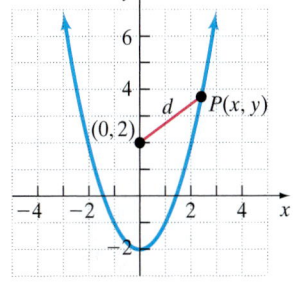

△ **57. Area** A rectangle is inscribed in a semicircle of radius 3 as shown in the figure. Let $P = (x, y)$ be the point in Quadrant I that is a vertex of the rectangle and is on the circle.

The area A of the rectangle as a function of x is given by

$$A(x) = 2x\sqrt{9 - x^2}$$

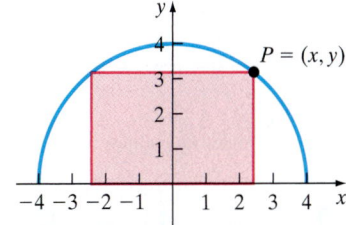

(a) What is the area of the rectangle whose vertex is at $\left(1, 2\sqrt{2}\right)$?

(b) What is the area of the rectangle whose vertex is at $\left(2, \sqrt{5}\right)$?

(c) What is the area of the rectangle whose vertex is at $\left(\sqrt{2}, \sqrt{7}\right)$?

△ **58. Area** A rectangle is inscribed in a semicircle of radius 4 as shown in the figure. Let $P = (x, y)$ be the point in Quadrant I that is a vertex of the rectangle and is on the circle.

The area A of the rectangle as a function of x is given by

$$A(x) = 2x\sqrt{16 - x^2}$$

(a) What is the area of the rectangle whose vertex is at $\left(1, \sqrt{15}\right)$?

(b) What is the area of the rectangle whose vertex is at $\left(2, 2\sqrt{3}\right)$?

(c) What is the area of the rectangle whose vertex is at $\left(2\sqrt{2}, 2\sqrt{2}\right)$?

Extending the Concepts

59. Use the results of Problems 33–36 to make a generalization about how to obtain the graph of $g(x) = \sqrt{x} + c$ from the graph of $f(x) = \sqrt{x}$.

60. Use the results of Problems 39–42 to make a generalization about how to obtain the graph of $g(x) = \sqrt{x + c}$ from the graph of $f(x) = \sqrt{x}$.

Synthesis Review

In Problems 61–65, add each of the following.

61. $\dfrac{1}{3} + \dfrac{1}{2}$

62. $\dfrac{1}{5} + \dfrac{3}{4}$

63. $\dfrac{1}{x} + \dfrac{3}{x+1}$

64. $\dfrac{5}{x-3} + \dfrac{2}{x+1}$

65. $\dfrac{4}{x-1} + \dfrac{3}{x+1}$

66. Explain how adding rational numbers that do not have denominators with any common factors is similar to adding rational expressions that do not have denominators with any common factors.

The Graphing Calculator

The graphing calculator can graph square root and cube root functions. The figure to the right shows the graph of $f(x) = \sqrt{x-2}$ using a TI-84 Plus graphing calculator. Note how the graph shown only exists for $x \geq 2$. This is useful in verifying the domain that we found algebraically.

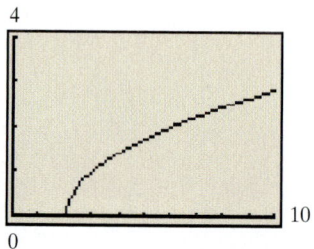

In Problems 67–88, graph the function using a graphing calculator. Compare the graphs obtained on the calculator to the hand drawn graphs in Problems 33–54.

67. $f(x) = \sqrt{x-4}$

68. $f(x) = \sqrt{x-1}$

69. $g(x) = \sqrt{x+2}$

70. $g(x) = \sqrt{x+5}$

71. $G(x) = \sqrt{2-x}$

72. $F(x) = \sqrt{4-x}$

73. $f(x) = \sqrt{x} + 3$

74. $f(x) = \sqrt{x} + 1$

75. $g(x) = \sqrt{x} - 4$

76. $g(x) = \sqrt{x} - 2$

77. $H(x) = 2\sqrt{x}$

78. $h(x) = 3\sqrt{x}$

79. $f(x) = \dfrac{1}{2}\sqrt{x}$

80. $g(x) = \dfrac{1}{4}\sqrt{x}$

81. $G(x) = -\sqrt{x}$

82. $F(x) = \sqrt{-x}$

83. $h(x) = \sqrt[3]{x+2}$

84. $g(x) = \sqrt[3]{x-4}$

85. $f(x) = \sqrt[3]{x} - 3$

86. $H(x) = \sqrt[3]{x} + 3$

87. $G(x) = 2\sqrt[3]{x}$

88. $F(x) = 3\sqrt[3]{x}$

9.8 Radical Equations and Their Applications

OBJECTIVES

1. Solve Radical Equations Containing One Radical
2. Solve Radical Equations Containing Two Radicals
3. Solve for a Variable in a Radical Equation

Preparing for Radical Equations and Their Applications

Before getting started, take the following readiness quiz. If you get a problem wrong, go back to the section cited and review the material.

1. Solve: $3x - 5 = 0$ [Section 2.2, pp. 84–86]

2. Solve: $2p^2 + 4p - 6 = 0$ [Section 6.6, pp. 451–457]

3. Simplify: $\left(\sqrt[3]{x} - 5\right)^3$ [Section 9.2, pp. 667–668]

Preparing for...Answers

1. $\left\{\dfrac{5}{3}\right\}$ **2.** $\{-3, 1\}$ **3.** $x - 5$

When the variable in an equation occurs in a radicand, the equation is called a **radical equation.** Examples of radical equations are

$$\sqrt{3x+1} = 5 \qquad \sqrt[3]{x-5} - 5 = 12 \qquad \sqrt{x-2} - \sqrt{2x+5} = 2$$

In this section, we are going to solve radical equations involving one or two radicals. We start with radical equations containing one radical.

① Solve Radical Equations Containing One Radical

Example 1 illustrates how to solve a radical equation.

EXAMPLE 1   **How to Solve a Radical Equation Containing One Radical**

Solve: $\sqrt{2x - 3} - 5 = 0$

Step-by-Step Solution

Step 1: Isolate the radical.	Add 5 to both sides:	$\sqrt{2x - 3} - 5 = 0$ $\sqrt{2x - 3} = 5$
Step 2: Raise both sides to the power of the index.	The index is 2, so we square both sides:	$\left(\sqrt{2x - 3}\right)^2 = 5^2$ $2x - 3 = 25$
Step 3: Solve the equation that results.	Add 3 to both sides: Divide both sides by 2:	$2x - 3 + 3 = 25 + 3$ $2x = 28$ $\dfrac{2x}{2} = \dfrac{28}{2}$ $x = 14$
Step 4: Check	Let $x = 14$ in the original equation:	$\sqrt{2x - 3} - 5 = 0$ $\sqrt{2 \cdot 14 - 3} - 5 \stackrel{?}{=} 0$ $\sqrt{28 - 3} - 5 \stackrel{?}{=} 0$ $5 - 5 = 0$ True

The solution set is $\{14\}$.

Below we summarize the steps to solve a radical equation that contains a single radical.

> ### Solving a Radical Equation Containing One Radical
>
> **Step 1:** Isolate the radical. That is, get the radical by itself on one side of the equation and everything else on the other side.
>
> **Step 2:** Raise both sides of the equation to the power of the index. This will eliminate the radical from the equation.
>
> **Step 3:** Solve the equation that results.
>
> **Step 4:** Check your answer. When solving radical equations containing an even index, apparent solutions that are not solutions to the original equation creep in. These solutions are called **extraneous solutions.**

QUICK

1. Solve: $\sqrt{3x + 1} - 4 = 0$

EXAMPLE 2 Solving a Radical Equation Containing One Radical

Solve:

(a) $\sqrt{4x + 1} - 2 = 1$ (b) $\sqrt{5x - 1} + 7 = 5$

Solution

(a)

$$\sqrt{4x + 1} - 2 = 1$$

Add 2 to both sides: $\sqrt{4x + 1} - 2 + 2 = 1 + 2$

$$\sqrt{4x + 1} = 3$$

The index is 2, so we square both sides: $\left(\sqrt{4x + 1}\right)^2 = 3^2$

$$4x + 1 = 9$$

Subtract 1 from both sides: $4x + 1 - 1 = 9 - 1$

$$4x = 8$$

Divide both sides by 4: $\dfrac{4x}{4} = \dfrac{8}{4}$

$$x = 2$$

Check

$$\sqrt{4x + 1} - 2 = 1$$

Let $x = 2$ in the original equation: $\sqrt{4 \cdot 2 + 1} - 2 \overset{?}{=} 1$

$$\sqrt{8 + 1} - 2 \overset{?}{=} 1$$

$$\sqrt{9} - 2 \overset{?}{=} 1$$

$$3 - 2 \overset{?}{=} 1$$

$$1 = 1 \qquad \text{True}$$

The solution set is $\{2\}$.

(b)

$$\sqrt{5x - 1} + 7 = 5$$

Subtract 7 from both sides: $\sqrt{5x - 1} + 7 - 7 = 5 - 7$

$$\sqrt{5x - 1} = -2$$

The equation has no real solution because the principal square root of a number cannot be less than 0. Put another way, there is no real number whose square root is -2, so the equation has no real solution. The solution set is $\varnothing$ or $\{\ \}$. ■

QUICK ✔

2. Solve: $\sqrt{6x + 4} - 5 = 3$ 3. Solve: $\sqrt{2x + 3} + 8 = 6$

EXAMPLE 3 Solving a Radical Equation Containing One Radical

Solve: $\sqrt{x + 5} = x - 1$

Solution

$$\sqrt{x + 5} = x - 1$$

The index is 2, so we square both sides: $\left(\sqrt{x + 5}\right)^2 = (x - 1)^2$

$$x + 5 = x^2 - 2x + 1$$

Subtract x and 5 from both sides: $x + 5 - x - 5 = x^2 - 2x + 1 - x - 5$

$$0 = x^2 - 3x - 4$$

Factor: $0 = (x - 4)(x + 1)$

Zero-Product Property: $x - 4 = 0 \quad \text{or} \quad x + 1 = 0$

$$x = 4 \quad \text{or} \qquad x = -1$$

Work Smart

$(x - 1)^2 = x^2 - 2x + 1$

Do not write

$(x - 1)^2 = x^2 + 1$

Check
$$\sqrt{x + 5} = x - 1$$

$x = 4$: $\sqrt{4 + 5} \overset{?}{=} 4 - 1$ $x = -1$: $\sqrt{-1 + 5} \overset{?}{=} -1 - 1$

$\qquad\qquad \sqrt{9} \overset{?}{=} 3$ $\qquad\qquad\qquad\qquad\qquad \sqrt{4} \overset{?}{=} -2$

$\qquad\qquad\quad 3 = 3$ True $\qquad\qquad\qquad\qquad\quad 2 = -2$ False

The apparent solution $x = -1$ does not check, so it is an extraneous solution. The solution set is $\{4\}$.

QUICK ✓

4. Solve: $\sqrt{2x + 1} = x - 1$

EXAMPLE 4   Solving a Radical Equation Containing One Radical

Solve: $\sqrt[3]{3x + 2} - 1 = 1$

Solution

$$\sqrt[3]{3x + 2} - 1 = 1$$

Add 1 to both sides: $\qquad \sqrt[3]{3x + 2} - 1 + 1 = 1 + 1$

$$\sqrt[3]{3x + 2} = 2$$

The index is 3, so we cube both sides: $\qquad \left(\sqrt[3]{3x + 2}\right)^3 = 2^3$

$$3x + 2 = 8$$

Subtract 2 from both sides: $\qquad 3x + 2 - 2 = 8 - 2$

$$3x = 6$$

Divide both sides by 3: $\qquad \dfrac{3x}{3} = \dfrac{6}{3}$

$$x = 2$$

Check
$$\sqrt[3]{3x + 2} - 1 = 1$$

Let $x = 2$ in the original equation: $\qquad \sqrt[3]{3 \cdot 2 + 2} - 1 \overset{?}{=} 1$

$$\sqrt[3]{6 + 2} - 1 \overset{?}{=} 1$$

$$\sqrt[3]{8} - 1 \overset{?}{=} 1$$

$$2 - 1 \overset{?}{=} 1$$

$$1 = 1 \quad \text{True}$$

The solution set is $\{2\}$.

QUICK ✓

5. Solve: $\sqrt[3]{6x + 3} - 4 = -1$

Sometimes, rather than an equation containing radicals, it will contain rational exponents. When solving these problems, we can rewrite the equation with a radical or we use the fact that $(a^r)^s = a^{r \cdot s}$.

EXAMPLE 5   Solving an Equation Containing a Rational Exponent

Solve: $(5x - 1)^{\frac{1}{2}} + 3 = 10$

Solution

$$(5x - 1)^{\frac{1}{2}} + 3 = 10$$

Subtract 3 from both sides: $(5x - 1)^{\frac{1}{2}} = 7$

Square both sides: $\left((5x - 1)^{\frac{1}{2}}\right)^2 = 7^2$

Use $(a^r)^s = a^{r \cdot s}$: $(5x - 1)^{\frac{1}{2} \cdot 2} = 49$

$$5x - 1 = 49$$

Add 1 to both sides: $5x = 50$

Divide both sides by 5: $x = 10$

Check

$$(5x - 1)^{\frac{1}{2}} + 3 = 10$$

Let $x = 10$ in the original equation: $(5 \cdot 10 - 1)^{\frac{1}{2}} + 3 \stackrel{?}{=} 10$

$$(49)^{\frac{1}{2}} + 3 \stackrel{?}{=} 10$$

$$7 + 3 = 10$$

$$10 = 10 \quad \text{True}$$

The solution set is $\{10\}$.

QUICK ✓

6. Solve: $(2x - 3)^{\frac{1}{3}} - 7 = -4$

---

② Solve Radical Equations Containing Two Radicals

Example 6 illustrates how to solve a radical equation containing two radicals.

EXAMPLE 6 How to Solve a Radical Equation Containing Two Radicals

Solve: $\sqrt[3]{p^2 - 4p - 4} = \sqrt[3]{-3p + 2}$

Step-by-Step-Solution

Step 1: Isolate one of the radicals.

The radical on the left side of the equation is isolated: $\sqrt[3]{p^2 - 4p - 4} = \sqrt[3]{-3p + 2}$

Step 2: Raise both sides to the power of the index.

The index is 3, so we cube both sides: $\left(\sqrt[3]{p^2 - 4p - 4}\right)^3 = \left(\sqrt[3]{-3p + 2}\right)^3$

$$p^2 - 4p - 4 = -3p + 2$$

Step 3: Because there is no radical, we solve the equation that results.

Add $3p$ to both sides; subtract 2 from both sides: $p^2 - 4p - 4 + 3p - 2 = -3p + 2 + 3p - 2$

Combine like terms: $p^2 - p - 6 = 0$

Factor: $(p - 3)(p + 2) = 0$

Zero-Product Property: $p - 3 = 0 \quad \text{or} \quad p + 2 = 0$

$$p = 3 \quad \text{or} \quad p = -2$$

Step 4: Check

$$\sqrt[3]{p^2 - 4p - 4} = \sqrt[3]{-3p + 2}$$

$p = -2$:

$$\sqrt[3]{(-2)^2 - 4(-2) - 4} \stackrel{?}{=} \sqrt[3]{-3(-2) + 2}$$

$$\sqrt[3]{4 + 8 - 4} \stackrel{?}{=} \sqrt[3]{6 + 2}$$

$$\sqrt[3]{8} = \sqrt[3]{8} \quad \text{True}$$

$p = 3$:

$$\sqrt[3]{(3)^2 - 4(3) - 4} \stackrel{?}{=} \sqrt[3]{-3(3) + 2}$$

$$\sqrt[3]{9 - 12 - 4} \stackrel{?}{=} \sqrt[3]{-9 + 2}$$

$$\sqrt[3]{-7} = \sqrt[3]{-7} \quad \text{True}$$

Both apparent solutions check. The solution set is $\{-2, 3\}$.

When a radical equation contains two radicals, the following steps should be used to solve the equation for the variable.

> ## Solving a Radical Equation Containing Two Radicals
>
> **Step 1:** Isolate one of the radicals. That is, get one of the radicals by itself on one side of the equation and everything else on the other side.
>
> **Step 2:** Raise both sides of the equation to the power of the index. This will eliminate one radical or both radicals from the equation.
>
> **Step 3:** If a radical remains in the equation, then follow the steps for solving a radical equation containing one radical. Otherwise, solve the equation that results.
>
> **Step 4:** Check your answer. When solving radical equations, apparent solutions that, in fact, are not solutions to the original equation may creep in. Remember, these solutions are called extraneous solutions.

QUICK ✓

7. Solve: $\sqrt[3]{m^2 + 4m + 4} = \sqrt[3]{2m + 7}$

EXAMPLE 7   **Solving a Radical Equation Containing Two Radicals**

Solve: $\sqrt{3x + 6} - \sqrt{x + 6} = 2$

Solution

Work Smart
When there is more than one radical, it is best to isolate the radical with the more complicated radicand.

$$\sqrt{3x + 6} - \sqrt{x + 6} = 2$$

Add $\sqrt{x + 6}$ to both sides: $\qquad \sqrt{3x + 6} = 2 + \sqrt{x + 6}$

Square both sides: $\qquad \left(\sqrt{3x + 6}\right)^2 = \left(2 + \sqrt{x + 6}\right)^2$

Use $(a + b)^2 = a^2 + 2ab + b^2$: $\qquad 3x + 6 = 4 + 4\sqrt{x + 6} + \left(\sqrt{x + 6}\right)^2$

$\left(\sqrt{x + 6}\right)^2 = x + 6$: $\qquad 3x + 6 = 4 + 4\sqrt{x + 6} + x + 6$

Isolate the radical: $\qquad 2x - 4 = 4\sqrt{x + 6}$

Factor out 2: $\qquad 2(x - 2) = 4\sqrt{x + 6}$

Divide both sides by 2: $\qquad x - 2 = 2\sqrt{x + 6}$

Square both sides: $\qquad (x - 2)^2 = \left(2\sqrt{x + 6}\right)^2$

$$x^2 - 4x + 4 = 4(x + 6)$$

Distribute: $\qquad x^2 - 4x + 4 = 4x + 24$

Subtract 4x and 24 from both sides: $\qquad x^2 - 8x - 20 = 0$

Factor: $\qquad (x - 10)(x + 2) = 0$

Zero-Product Property: $\qquad x - 10 = 0 \quad$ or $\quad x + 2 = 0$

$$x = 10 \quad \text{or} \quad x = -2$$

Check $x = -2$: $\sqrt{3 \cdot (-2) + 6} - \sqrt{-2 + 6} \overset{?}{=} 2$ $x = 10$: $\sqrt{3 \cdot 10 + 6} - \sqrt{10 + 6} \overset{?}{=} 2$

$$\sqrt{0} - \sqrt{4} \overset{?}{=} 2 \qquad\qquad \sqrt{36} - \sqrt{16} \overset{?}{=} 2$$

$$0 - 2 \overset{?}{=} 2 \qquad\qquad 6 - 4 \overset{?}{=} 2$$

$$-2 = 2 \quad \text{False} \qquad\qquad 2 = 2 \quad \text{True}$$

The apparent solution $x = -2$ does not check, so it is an extraneous solution. The solution set is $\{10\}$.

QUICK ✓

8. Solve: $\sqrt{2x + 1} - \sqrt{x + 4} = 1$

③ Solve for a Variable in a Radical Equation

In many situations, you will be required to solve for a variable in a formula. For instance, in Example 8, we are assessing how much error there is in estimates based upon a statistical study. This commonly used formula from statistics contains a radical.

EXAMPLE 8 Solving for a Variable

A formula from statistics for finding the margin of error in estimating a population mean is given by

$$E = z \cdot \frac{\sigma}{\sqrt{n}}$$

(a) Solve this equation for n.

(b) Find n when $\sigma = 12$, $z = 2$, and $E = 3$.

Solution

(a)
$$E = z \cdot \frac{\sigma}{\sqrt{n}}$$

Multiply both sides by $\sqrt{n}$: $\quad \sqrt{n} \cdot E = z\sigma$

Divide both sides by E: $\quad \sqrt{n} = \dfrac{z\sigma}{E}$

Square both sides: $\quad n = \left(\dfrac{z\sigma}{E}\right)^2$

In Words
The symbol σ is pronounced "sigma."

(b) $n = \left(\dfrac{z\sigma}{E}\right)^2 = \left(\dfrac{2 \cdot 12}{3}\right)^2$

$\qquad = 64$

QUICK ✓

9. The period of a pendulum is the time it takes to complete one trip back and forth. The period T, in seconds, of a pendulum of length L, in feet, may be approximated using the formula $T = 2\pi\sqrt{\dfrac{L}{32}}$.

(a) Solve the equation for L.

(b) Determine the length of a pendulum whose period is 2π seconds.

9.8 Exercises

For Extra Help: Student Solutions Manual   CD Video   PH Math/Tutor Center   MathXL Tutorials on CD   Math XL   MathXL®   MyMathLab   MyMathLab

Concepts and Vocabulary

In Problems 1 and 2, fill in the blanks.

1. When the variable in an equation occurs in a radical, the equation is called a

_____ _____.

2. When an apparent solution is not a solution of the original equation, we say the apparent solution is an _____ solution.

In Problems 3 and 4, answer True or False to each statement.

3. The first step in solving $x + \sqrt{x - 3} = 5$ is to square both sides of the equation.

4. When solving radical equations, extraneous solutions only occur when the index on the radical is even.

5. Why is it always necessary to check apparent solutions when solving radical equations?

6. How can you tell by inspection that the equation $\sqrt{x - 2} + 5 = 0$ will have no real solution?

Building Skills

In Problems 7–38, solve each equation.

7. $\sqrt{x} = 4$

8. $\sqrt{p} = 6$

9. $\sqrt{x - 3} = 2$

10. $\sqrt{y - 5} = 3$

11. $\sqrt{2t + 3} = 5$

12. $\sqrt{3w - 2} = 4$

13. $\sqrt{4x + 3} = -2$

14. $\sqrt{6p - 5} = -5$

15. $\sqrt[3]{4t} = 2$

16. $\sqrt[3]{9w} = 3$

17. $\sqrt[3]{5q + 4} = 4$

18. $\sqrt[3]{7m + 20} = 5$

19. $\sqrt{y} + 3 = 8$

20. $\sqrt{q} - 5 = 2$

21. $\sqrt{x + 5} - 3 = 1$

22. $\sqrt{x - 4} + 4 = 7$

23. $\sqrt{2x - 1} + 5 = 8$

24. $\sqrt{4x + 1} - 2 = 3$

25. $3\sqrt{x} + 5 = 8$

26. $4\sqrt{t} - 2 = 10$

27. $\sqrt{4 - x} - 3 = 0$

28. $\sqrt{6 - w} - 3 = 1$

29. $\sqrt{p} = 2p$

30. $\sqrt{q} = 3q$

31. $\sqrt{x + 6} = x$

32. $\sqrt{2p + 8} = p$

33. $\sqrt{w} = w - 6$

34. $\sqrt{m} = m - 12$

35. $\sqrt{17 - 2x} + 1 = x$

36. $\sqrt{1 - 4x} - 5 = x$

37. $\sqrt{w^2 - 11} + 5 = w + 4$

38. $\sqrt{z^2 - z - 7} + 3 = z + 2$

In Problems 39–52, solve each equation.

39. $\sqrt{x + 9} = \sqrt{2x + 5}$

40. $\sqrt{3x + 1} = \sqrt{2x + 7}$

41. $\sqrt[3]{4x - 3} = \sqrt[3]{2x - 9}$

42. $\sqrt[3]{3y - 2} = \sqrt[3]{5y + 8}$

43. $\sqrt{2w^2 - 3w - 4} = \sqrt{w^2 + 6w + 6}$ **44.** $\sqrt{2x^2 + 7x - 10} = \sqrt{x^2 + 4x + 8}$

45. $\sqrt{3w + 4} = 2 + \sqrt{w}$

46. $\sqrt{3y - 2} = 2 + \sqrt{y}$

47. $\sqrt{x + 1} - \sqrt{x - 2} = 1$

48. $\sqrt{2x - 1} - \sqrt{x - 1} = 1$

49. $\sqrt{2x + 6} - \sqrt{x - 1} = 2$

50. $\sqrt{2x + 6} - \sqrt{x - 6} = 3$

51. $\sqrt{2x + 5} - \sqrt{x - 1} = 2$

52. $\sqrt{4x + 1} - \sqrt{2x + 1} = 2$

In Problems 53–58, solve each equation.

53. $(2x + 3)^{\frac{1}{2}} = 3$

54. $(4x + 1)^{\frac{1}{2}} = 5$

55. $(6x - 1)^{\frac{1}{4}} = (2x + 15)^{\frac{1}{4}}$

56. $(6p + 3)^{\frac{1}{5}} = (4p - 9)^{\frac{1}{5}}$

57. $(x + 3)^{\frac{1}{2}} - (x - 5)^{\frac{1}{2}} = 2$

58. $(3x + 1)^{\frac{1}{2}} - (x - 1)^{\frac{1}{2}} = 2$

59. Finance Solve $A = P\sqrt{1 + r}$ for r.

60. Centripetal Acceleration Solve $v = \sqrt{ar}$ for a.

61. Volume of a Sphere Solve $r = \sqrt[3]{\dfrac{3V}{4\pi}}$ for V.

62. Surface Area of a Sphere Solve $r = \sqrt{\dfrac{S}{4\pi}}$ for S.

63. Coulomb's Law Solve $r = \sqrt{\dfrac{4F\pi\varepsilon_0}{q_1 q_2}}$ for F.

64. Potential Energy Solve $V = \sqrt{\dfrac{2U}{C}}$ for U.

Mixed Practice

In Problems 65–80, solve each equation.

65. $\sqrt{5p - 3} + 7 = 3$

66. $\sqrt{3b - 2} + 8 = 5$

67. $\sqrt{x + 12} = x$

68. $\sqrt{x + 20} = x$

69. $\sqrt{2p + 12} = 4$

70. $\sqrt{3a - 5} = 2$

71. $\sqrt[4]{x + 7} = 2$

72. $\sqrt[5]{x + 23} = 2$

73. $(3x + 1)^{\frac{1}{3}} + 2 = 0$

74. $(5x - 2)^{\frac{1}{3}} + 3 = 0$

75. $\sqrt{x} + 5 = 7$

76. $\sqrt{a} - 5 = -2$

77. $\sqrt{2x + 5} = \sqrt{3x - 4}$

78. $\sqrt{4c - 5} = \sqrt{3c + 1}$

79. $\sqrt{x - 1} + \sqrt{x + 4} = 5$

80. $\sqrt{x - 3} + \sqrt{x + 4} = 7$

Applying the Concepts

81. Suppose that $f(x) = \sqrt{x - 2}$.

    (a) Solve $f(x) = 0$. What point is on the graph of f?

    (b) Solve $f(x) = 1$. What point is on the graph of f?

    (c) Solve $f(x) = 2$. What point is on the graph of f?

    (d) Use the information obtained in parts (a)–(c) to graph $f(x) = \sqrt{x - 2}$.

    (e) Use the graph and the concept of the range of a function to explain why the equation $f(x) = -1$ has no solution.

82. Suppose that $g(x) = \sqrt{x} + 3$.

    (a) Solve $g(x) = 0$. What point is on the graph of g?

    (b) Solve $g(x) = 1$. What point is on the graph of g?

    (c) Solve $g(x) = 2$. What point is on the graph of g?

    (d) Use the information obtained in parts (a)–(c) to graph $g(x) = \sqrt{x} + 3$.

    (e) Use the graph and the concept of the range of a function to explain why the equation $g(x) = -1$ has no solution.

△ **83. Finding a *y*-Coordinate** The solutions to the equation

$$\sqrt{4^2 + (y - 2)^2} = 5$$

represent the *y*-coordinates such that the distance from the point $(3, 2)$ to $(-1, y)$ in the Cartesian plane is 5 units.

    (a) Solve the equation for y.

    (b) Plot the points in the Cartesian plane and label the lengths of the sides of the figure formed.

△ **84. Finding an *x*-Coordinate** The solutions to the equation

$$\sqrt{x^2 + 4^2} = 5$$

represent the *x*-coordinates such that the distance from the point $(0, 3)$ to $(x, -1)$ in the Cartesian plane is 5 units.

    (a) Solve the equation for x.

    (b) Plot the points in the Cartesian plane and label the lengths of the sides of the figure formed.

85. Revenue Growth Suppose that the annual revenue R (in millions of dollars) of a company after t years of operating is modeled by the function

$$R(t) = \sqrt[3]{\frac{t}{2}}$$

    (a) After how many years can the company expect to have annual revenue of $1 million?

    (b) After how many years can the company expect to have annual revenue of $2 million?

△ **86. Sphere** The radius r of a sphere whose volume is V is given by

$$r = \sqrt[3]{\frac{3V}{4\pi}}$$

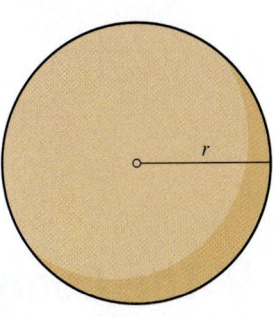

    (a) Find the volume of a sphere whose radius is 3 meters.

    (b) Find the volume of a sphere whose radius is 2 meters.

87. Birth Rates A plural birth is a live birth to twins, triplets, and so forth. The function $R(t) = 26 \cdot \sqrt[10]{t}$ models the plural birth rate R (live births per 1,000 live births), where t is the number of years since 1995.

    (a) Use the model to predict the year in which the plural birth rate will be 39.

    (b) Use the model to predict the year in which the plural birth rate will be 36.

88. Money The annual rate of interest r (expressed as a decimal) required to have A dollars after t years from an initial deposit of P dollars is given by

$$r = \sqrt[t]{\frac{A}{P}} - 1$$

    (a) Suppose that you deposit \$1,000 in an account that pays 5% annual interest so that $r = 0.05$. How much will you have after $t = 2$ years?

    (b) Suppose that you deposit \$1,000 in an account that pays 5% annual interest so that $r = 0.05$. How much will you have after $t = 3$ years?

Extending the Concepts

89. Solve: $\sqrt{3\sqrt{x+1}} = \sqrt{2x+3}$ **90.** Solve: $\sqrt[3]{2\sqrt{x-2}} = \sqrt[3]{x-1}$

91. Which step in the process of solving a radical equation leads to the possibility of extraneous solutions?

Synthesis Review

In Problems 92–95, identify which of the numbers in the set

$$\left\{0, -4, 12, \frac{2}{3}, 1.\overline{56}, \sqrt{2^3}, \pi, \sqrt{-5}, \sqrt[3]{-4}\right\}$$

are . . .

92. Integers   **93.** Rational numbers   **94.** Irrational numbers   **95.** Real numbers

96. State the difference between a rational number and an irrational number. Why is $\sqrt{-1}$ not real?

The Graphing Calculator

A graphing calculator can be used to verify solutions obtained algebraically. To solve $\sqrt{2x-3} = 5$ presented in Example 1, we graph $Y_1 = \sqrt{2x-3}$ and $Y_2 = 5$ and determine the x-coordinate of the point of intersection.

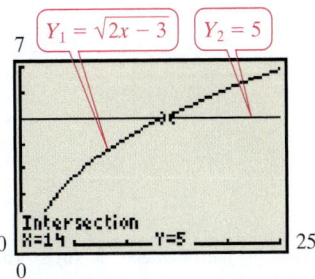

    The x-coordinate of the point of intersection is 14, so the solution set is $\{14\}$.

97. Verify your solution to Problem 11 by graphing $Y_1 = \sqrt{2x+3}$ and $Y_2 = 5$ and then finding the x-coordinate of their point of intersection.

98. Verify your solution to Problem 12 by graphing $Y_1 = \sqrt{3x-2}$ and $Y_2 = 4$ and then finding the x-coordinate of their point of intersection.

99. When you solved Problem 13 algebraically, you should have determined that the equation has no real solution. Verify this result by graphing $Y_1 = \sqrt{4x + 3}$ and $Y_2 = -2$. Explain what the algebraic solution means graphically.

100. When you solved Problem 14 algebraically, you should have determined that the equation has no real solution. Verify this result by graphing $Y_1 = \sqrt{6x - 5}$ and $Y_2 = -5$. Explain what the algebraic solution means graphically.

9.9 The Complex Number System

OBJECTIVES

1. Evaluate the Square Root of Negative Real Numbers
2. Add or Subtract Complex Numbers
3. Multiply Complex Numbers
4. Divide Complex Numbers
5. Evaluate the Powers of i

Preparing for *The Complex Number System*

Before getting started, take the following readiness quiz. If you get a problem wrong, go back to the section cited and review the material.

1. List the numbers in the set $\left\{ 8, -\frac{1}{3}, -23, 0, \sqrt{2}, 1.\overline{26}, -\frac{12}{3}, \sqrt{-5} \right\}$
   that are
   (a) Natural numbers   (b) Whole numbers
   (c) Integers   (d) Rational numbers
   (e) Irrational numbers   (f) Real numbers   [Section 1.2, pp. 9–12]

2. Distribute: $3x(4x - 3)$ [Section 5.3, pp. 360–361]

3. Multiply: $(z + 4)(3z - 2)$ [Section 5.3, pp. 361–363]

4. Multiply: $(2y + 5)(2y - 5)$ [Section 5.3, p. 364]

If you look back at Section 1.2, where we introduced the various number systems, you should notice that each time we encounter a situation where a number system can't handle a problem, we expand the number system. For example, if we only considered the whole numbers, we could not describe a negative balance in a checking account, so we introduced integers. If the world could only be described by integers, then we could not talk about parts of a whole as in $\frac{1}{2}$ a pizza or $\frac{3}{4}$ of a dollar, so we introduced rational numbers. If we only considered rational numbers, then we wouldn't be able to find a number whose square is 2, so we introduced the irrational numbers, so that $\left(\sqrt{2} \right)^2 = 2$. By combining the rational numbers with the irrational numbers, we created the real number system. The real number system is usually sufficient for solving most problems in mathematics, but not for all problems.

For example, suppose we wanted to determine a number whose square is -1. We know that when we square any real number, the result is never negative. We call this property of real numbers, the *Nonnegativity Property*.

NONNEGATIVITY PROPERTY OF REAL NUMBERS

For any real number $a, a^2 \geq 0$.

Because the square of any real number is never negative, there is no real number x for which

$$x^2 = -1$$

To remedy this situation, we introduce a new number.

DEFINITION

The **imaginary unit**, denoted by i, is the number whose square is -1. That is,

$$i^2 = -1$$

If we take the square root of both sides of $i^2 = -1$, we find that

$$i = \sqrt{-1}$$

Preparing for...Answers **1. (a)** 8   **(b)** 8, 0
(c) $8, -23, 0, -12/3$
(d) $8, -1/3, -23, 0, 1.\overline{26}, -12/3$
(e) $\sqrt{2}$
(f) $8, -\frac{1}{3}, -23, 0, \sqrt{2}, 1.\overline{26}, -\frac{12}{3}$
2. $12x^2 - 9x$ **3.** $3z^2 + 10z - 8$
4. $4y^2 - 25$

In looking at the development of the real number system, each new number system contained the earlier number system as a subset. By introducing the number i, we now have a new number system called the **complex number system.**

> **DEFINITION**
>
> **Complex numbers** are numbers of the form $a + bi$, where a and b are real numbers. The real number a is called the **real part** of the number $a + bi$; the real number b is called the **imaginary part** of $a + bi$.

In Words

The real number system is a subset of the complex number system. This means that all real numbers are, more generally, complex numbers.

For example, the complex number $6 + 2i$ has the real part 6 and the imaginary part 2. The complex number $4 - 3i = 4 + (-3)i$ has the real part 4 and the imaginary part -3.

When a complex number is written in the form $a + bi$, where a and b are real numbers, we say that it is in **standard form.** The complex number $a + 0i$ is typically written as a. This serves as a reminder that the real number system is a subset of the complex number system. The complex number $0 + bi$ is usually written as bi. Any number of the form bi is called a **pure imaginary number.** Figure 7 shows the relation between the number systems.

Figure 7
The Complex Number System

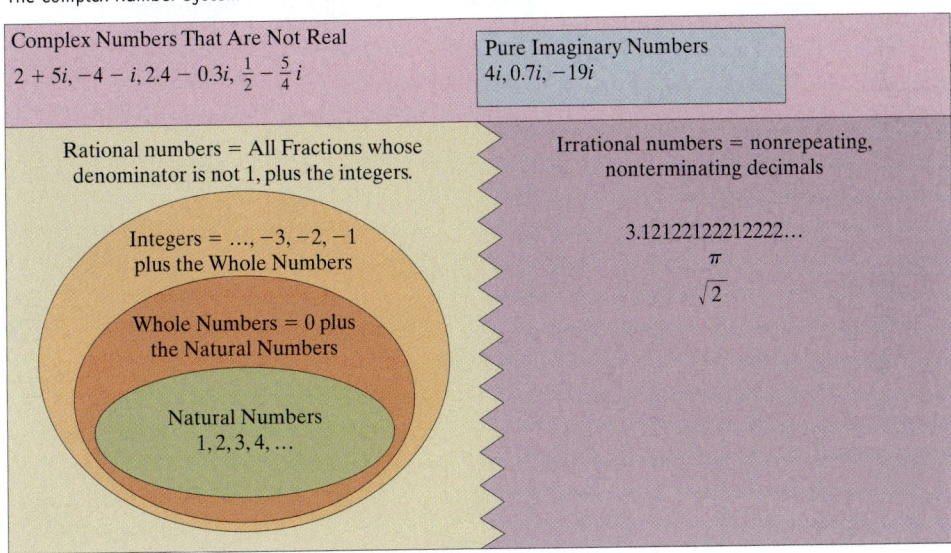

① Evaluate the Square Root of Negative Real Numbers

Using the definition of i along with the fact that $\sqrt{ab} = \sqrt{a} \cdot \sqrt{b}$ for real numbers a and b, we now have the ability to simplify square roots of negative numbers.

> **EVALUATING SQUARE ROOTS OF NEGATIVE NUMBERS**
>
> If N is a positive real number, we define the **principal square root of** $-N$, denoted by $\sqrt{-N}$, as
>
> $$\sqrt{-N} = \sqrt{N}\, i$$
>
> where $i = \sqrt{-1}$.

EXAMPLE 1 Evaluating the Square Root of a Negative Number

Write each of the following as a pure imaginary number.

    (a) $\sqrt{-16}$ **(b)** $\sqrt{-3}$ **(c)** $\sqrt{-18}$

Work Smart

When writing complex numbers whose imaginary part is a radical as in $\sqrt{3}i$ be sure that the "i" is not written under the radical.

Solution

(a) $\sqrt{-16} = \sqrt{16 \cdot (-1)}$ (b) $\sqrt{-3} = \sqrt{3 \cdot (-1)}$ (c) $\sqrt{-18} = \sqrt{18 \cdot (-1)}$

$\sqrt{ab} = \sqrt{a} \cdot \sqrt{b}:$ $= \sqrt{16} \cdot \sqrt{-1}$ $= \sqrt{3} \cdot \sqrt{-1}$ $= \sqrt{18} \cdot \sqrt{-1}$

$= 4i$ $= \sqrt{3}i$ $= 3\sqrt{2}i$

QUICK ✓ *Write each radical as a pure imaginary number.*

1. $\sqrt{-36}$ **2.** $\sqrt{-5}$ **3.** $\sqrt{-12}$

EXAMPLE 2 **Writing Complex Numbers in Standard Form**

Write each of the following in standard form.

    (a) $2 - \sqrt{-25}$ **(b)** $3 + \sqrt{-50}$ **(c)** $\dfrac{4 - \sqrt{-12}}{2}$

Solution

The standard form of a complex number is $a + bi$.

(a) $2 - \sqrt{-25} = 2 - \sqrt{25 \cdot (-1)}$

$\sqrt{ab} = \sqrt{a} \cdot \sqrt{b}:$ $= 2 - \sqrt{25} \cdot \sqrt{-1}$

$= 2 - 5i$

(b) $3 + \sqrt{-50} = 3 + \sqrt{50 \cdot (-1)}$

$\sqrt{ab} = \sqrt{a} \cdot \sqrt{b}:$ $= 3 + \sqrt{50} \cdot \sqrt{-1}$

$\sqrt{50} = \sqrt{25} \cdot \sqrt{2} = 5\sqrt{2}:$ $= 3 + 5\sqrt{2}i$

(c) $\dfrac{4 - \sqrt{-12}}{2} = \dfrac{4 - \sqrt{12 \cdot (-1)}}{2}$

$= \dfrac{4 - \sqrt{12} \cdot \sqrt{-1}}{2}$

$\sqrt{12} = \sqrt{4} \cdot \sqrt{3} = 2\sqrt{3}:$ $= \dfrac{4 - 2\sqrt{3}i}{2}$

Factor out 2: $= \dfrac{2(2 - \sqrt{3}i)}{2}$

Divide out the 2's: $= 2 - \sqrt{3}i$

QUICK ✓ *Write each expression in the standard form of a complex number, $a + bi$.*

4. $4 + \sqrt{-100}$ **5.** $-2 - \sqrt{-8}$ **6.** $\dfrac{6 - \sqrt{-72}}{3}$

---

② Add or Subtract Complex Numbers

Throughout your math career, whenever you were introduced to a new number system, you learned how to perform the four binary operations of addition, subtraction, multiplication, and division using this number system. Therefore, we will now show how to add and subtract complex numbers. Then we will discuss multiplying and dividing complex numbers.

Two complex numbers are added by adding the real parts and then adding the imaginary parts.

In Words
To add two complex numbers, add the real parts, then add the imaginary parts. To subtract two complex numbers, subtract the real parts, then subtract the imaginary parts.

> **SUM OF COMPLEX NUMBERS**
>
> $$(a + bi) + (c + di) = (a + c) + (b + d)i$$

To subtract two complex numbers, we use this rule:

> **DIFFERENCE OF COMPLEX NUMBERS**
>
> $$(a + bi) - (c + di) = (a - c) + (b - d)i$$

EXAMPLE 3   Adding Complex Numbers

Add:

(a) $(4 - 3i) + (-2 + 5i)$ (b) $\left(4 + \sqrt{-25}\right) + \left(6 - \sqrt{-16}\right)$.

Solution

(a) $(4 - 3i) + (-2 + 5i) = [4 + (-2)] + (-3 + 5)i$
$$= 2 + 2i$$

$\sqrt{-25} = 5i; \sqrt{-16} = 4i$

(b) $\left(4 + \sqrt{-25}\right) + \left(6 - \sqrt{-16}\right) = (4 + 5i) + (6 - 4i)$
$$= (4 + 6) + (5 - 4)i$$
$$= 10 + 1i$$
$$= 10 + i$$

Work Smart
Adding or subtracting complex numbers is just like combining like terms. For example,

$(4 - 3x) + (-2 + 5x)$
$= 4 + (-2) - 3x + 5x$
$= 2 + 2x$

so

$(4 - 3i) + (-2 + 5i)$
$= 4 + (-2) - 3i + 5i$
$= 2 + 2i$

EXAMPLE 4   Subtracting Complex Numbers

Subtract:

(a) $(-3 + 7i) - (5 - 4i)$ (b) $\left(3 + \sqrt{-12}\right) - \left(-2 - \sqrt{-27}\right)$.

Solution

(a) $(-3 + 7i) - (5 - 4i) = (-3 - 5) + (7 - (-4))i$
$$= -8 + 11i$$

$\sqrt{-12} = 2\sqrt{3}i; \sqrt{-27} = 3\sqrt{3}i$

(b) $\left(3 + \sqrt{-12}\right) - \left(-2 - \sqrt{-27}\right) = \left(3 + 2\sqrt{3}i\right) - \left(-2 - 3\sqrt{3}i\right)$

Distribute the minus: $= 3 + 2\sqrt{3}i + 2 + 3\sqrt{3}i$
$$= [3 + 2] + \left[2\sqrt{3} + 3\sqrt{3}\right]i$$
$$= 5 + 5\sqrt{3}i$$

QUICK ✓ *Add or subtract as indicated.*

7. $(4 + 6i) + (-3 + 5i)$ **8.** $(4 - 2i) - (-2 + 7i)$

9. $\left(4 - \sqrt{-4}\right) + \left(-7 + \sqrt{-9}\right)$

(3) ## Multiply Complex Numbers

We multiply complex numbers using the Distributive Property. The methods are almost the same methods that we used to multiply polynomials.

EXAMPLE 5 **Multiplying Complex Numbers**

Multiply:

    (a) $4i(3 - 6i)$ **(b)** $(-2 + 4i)(3 - i)$

Solution

(a) We distribute the $4i$ into each term in the parentheses.

$$4i(3 - 6i) = 4i \cdot 3 - 4i \cdot 6i$$
$$= 12i - 24i^2$$
$$i^2 = -1: \quad = 12i - 24 \cdot (-1)$$
$$= 24 + 12i$$

(b)
$$(-2 + 4i)(3 - i) = -2 \cdot 3 - 2 \cdot (-i) + 4i \cdot 3 + 4i \cdot (-i)$$
$$= -6 + 2i + 12i - 4i^2$$
$$\text{Combine like terms; } i^2 = -1: \quad = -6 + 14i - 4(-1)$$
$$= -6 + 14i + 4$$
$$= -2 + 14i$$

QUICK ✓ *Multiply.*

10. $3i(5 - 4i)$ **11.** $(-2 + 5i)(4 - 2i)$

Look back at the Product Property for Radicals on page 681. You should notice that the property only applies when $\sqrt[n]{a}$ and $\sqrt[n]{b}$ are real numbers. This means that

$$\sqrt{a} \cdot \sqrt{b} \neq \sqrt{ab} \quad \text{if } a < 0 \text{ or } b < 0$$

So how do we perform this multiplication? Well, we first need to write the radical as a complex number using the fact that $\sqrt{-N} = \sqrt{N}i$ and then perform the multiplication.

EXAMPLE 6 **Multiplying Square Roots of Negative Numbers**

Multiply:

    (a) $\sqrt{-25} \cdot \sqrt{-4}$ **(b)** $\left(2 + \sqrt{-16}\right)\left(1 - \sqrt{-4}\right)$

Solution

(a) We cannot use the Product Property of Radicals to multiply these radicals because neither $\sqrt{-25}$ nor $\sqrt{-4}$ are real numbers. Therefore, we express the radicals as pure imaginary numbers and then multiply.

$$\sqrt{-25} \cdot \sqrt{-4} = 5i \cdot 2i$$
$$= 10i^2$$
$$i^2 = -1: \quad = -10$$

Work Smart

$\sqrt{-25} \cdot \sqrt{-4} \neq \sqrt{(-25)(-4)}$
because the Product Property of
Radicals only applies when the radical
is a real number.

(b) First, we rewrite the expression as a complex number in standard form.

$$\left(2 + \sqrt{-16}\right)\left(1 - \sqrt{-4}\right) = (2 + 4i)(1 - 2i)$$

$$\text{FOIL:} \quad = 2 \cdot 1 + 2 \cdot (-2i) + 4i \cdot 1 + 4i \cdot (-2i)$$
$$= 2 - 4i + 4i - 8i^2$$
$$\text{Combine like terms; } i^2 = -1: \quad = 2 - 8(-1)$$
$$= 2 + 8$$
$$= 10$$

QUICK ✓ *Multiply.*

12. $\sqrt{-9} \cdot \sqrt{-36}$

13. $\left(2 + \sqrt{-36}\right)\left(4 - \sqrt{-25}\right)$

Complex Conjugates

We now introduce a special product that involves the *conjugate* of a complex number.

In Words
To find the complex conjugate of $a + bi$, simply change the sign from "+" to "−" or "−" to "+" between the "a" and "b" in the complex number.

> **COMPLEX CONJUGATE**
>
> If $a + bi$ is a complex number, then its **conjugate** is defined as $a - bi$.

For example,

Complex Number	Conjugate
$3 + 5i$	$3 - 5i$
$-10 - 3i$	$-10 + 3i$

Notice what happens when we multiply a complex number and its conjugate.

EXAMPLE 7 Multiplying a Complex Number by Its Conjugate

Find the product of $4 + 3i$ and its conjugate, $4 - 3i$.

Solution

$$(4 + 3i)(4 - 3i) = 4 \cdot 4 + 4 \cdot (-3i) + 3i \cdot 4 + 3i \cdot (-3i)$$
$$= 16 - 12i + 12i - 9i^2$$
$$= 16 - 9(-1)$$
$$= 16 + 9$$
$$= 25$$

Wow! The product of the complex number $4 + 3i$ and its conjugate $4 - 3i$ is 25—a real number! In fact, the results of Example 7 are true in general.

> **PRODUCT OF A COMPLEX NUMBER AND ITS CONJUGATE**
>
> The product of a complex number and its conjugate is a nonnegative real number. That is,
>
> $$(a + bi)(a - bi) = a^2 + b^2$$

Perhaps you noticed that multiplying a complex number and its conjugate is akin to multiplying $(a + b)(a - b) = a^2 - b^2$.

QUICK ✓ *Multiply.*

14. $(3 - 8i)(3 + 8i)$ **15.** $(-2 + 5i)(-2 - 5i)$

---

④ Divide Complex Numbers

Now that we understand the product of a complex number and its conjugate we can proceed to divide complex numbers.

EXAMPLE 8 How to Divide Complex Numbers

Divide: $\dfrac{-3 + i}{5 + 3i}$

Step-by-Step Solution

Step 1: Write the numerator and denominator in standard form, $a + bi$.

The numerator and denominator are already in standard form.

Step 2: Multiply the numerator and denominator by the complex conjugate of the denominator.

The complex conjugate of $5 + 3i$ is $5 - 3i$:

$$\frac{-3 + i}{5 + 3i} = \frac{-3 + i}{5 + 3i} \cdot \frac{5 - 3i}{5 - 3i}$$

$$= \frac{(-3 + i)(5 - 3i)}{(5 + 3i)(5 - 3i)}$$

Step 3: Simplify by writing the quotient in standard form, $a + bi$.

Multiply numerator and denominator; $(a + bi)(a - bi) = a^2 + b^2$:

$$= \frac{-3 \cdot 5 - 3 \cdot (-3i) + i \cdot 5 + i \cdot (-3i)}{5^2 + 3^2}$$

Combine like terms:

$$= \frac{-15 + 9i + 5i - 3i^2}{25 + 9}$$

$i^2 = -1$:

$$= \frac{-15 + 14i + 3}{34}$$

$$= \frac{-12 + 14i}{34}$$

Divide 34 into each term in the numerator to write in standard form:

$$= \frac{-12}{34} + \frac{14}{34}i$$

Write each fraction in lowest terms:

$$= -\frac{6}{17} + \frac{7}{17}i$$

We summarize the steps used in Example 8 below.

Dividing Complex Numbers

Step 1: Write the numerator and denominator in standard form, $a + bi$.

Step 2: Multiply the numerator and denominator by the complex conjugate of the denominator.

Step 3: Simplify by writing the quotient in standard form, $a + bi$.

EXAMPLE 9 **Dividing Complex Numbers**

Divide: $\dfrac{3 + 4i}{2i}$

Solution

$$\dfrac{3 + 4i}{2i} = \dfrac{3 + 4i}{0 + 2i}$$

The complex conjugate of $0 + 2i$ is $0 - 2i$: $= \dfrac{3 + 4i}{0 + 2i} \cdot \dfrac{0 - 2i}{0 - 2i}$

The 0s are not necessary: $= \dfrac{3 + 4i}{2i} \cdot \dfrac{-2i}{-2i}$

Multiply numerator; multiply denominator: $= \dfrac{(3 + 4i)(-2i)}{(2i)(-2i)}$

Distribute in numerator: $= \dfrac{-6i - 8i^2}{-4i^2}$

$i^2 = -1$: $= \dfrac{-6i + 8}{4}$

Divide 4 into each term in the numerator to write in standard form: $= \dfrac{8}{4} - \dfrac{6}{4}i$

Reduce each fraction to lowest terms: $= 2 - \dfrac{3}{2}i$

Work Smart

We could also have multiplied the numerator and denominator in Example 9 by i. Do you know why? Try it yourself! Which approach do you prefer?

Look back to Example 5(b), where we found $(-2 + 4i)(3 - i) = -2 + 14i$. Now find $\dfrac{-2 + 14i}{3 - i}$. What result do you expect? Verify that $\dfrac{-2 + 14i}{3 - i} = -2 + 4i$.

QUICK ✓ *Divide.*

16. $\dfrac{-4 + i}{3i}$

17. $\dfrac{4 + 3i}{1 - 3i}$

---

⑤ **Evaluate the Powers of i**

The **powers of i** follow a pattern.

$i^1 = i$ $\qquad i^2 = -1$ $\qquad i^3 = i^2 \cdot i^1 = -1 \cdot i = -i$ $\qquad i^4 = i^2 \cdot i^2 = (-1)(-1) = 1$

$i^5 = i^4 \cdot i = 1 \cdot i = i$ $\qquad i^6 = i^4 \cdot i^2 = 1 \cdot (-1) = -1$ $\qquad i^7 = i^4 \cdot i^3 = 1 \cdot (-i) = -i$ $\qquad i^8 = i^4 \cdot i^4 = 1 \cdot 1 = 1$

$i^9 = i^8 \cdot i = 1 \cdot i = i$ $\qquad i^{10} = i^8 \cdot i^2 = 1 \cdot (-1) = -1$ $\qquad i^{11} = i^8 \cdot i^3 = 1 \cdot (-i) = -i$ $\qquad i^{12} = (i^4)^3 = (1)^3 = 1$

Do you see the pattern? In the first column, the expressions all simplify to i; in the second column, the expressions all simplify to -1; in the third column, the expressions all simplify to $-i$; in the fourth column, the expressions all simplify to 1. That is, the powers of i repeat with every fourth power. Figure 8 shows the pattern.

Figure 8

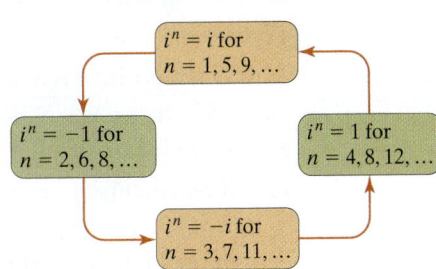

For this reason any power of i can be expressed in terms of 1, -1, i, or $-i$. The following steps can be used to simplify any power of i.

Simplifying the Powers of i

Step 1: Divide the exponent of i by 4. Rewrite i^n as $(i^4)^q \cdot i^r$, where q is the quotient and r is the remainder of the division.

Step 2: Simplify the product in Step 1 to i^r since $i^4 = 1$.

EXAMPLE 10　Simplifying Powers of i

Simplify:

(a) i^{34} 　　　　　　　　　　　**(b)** i^{101}

Solution

(a) We divide 34 by 4 and obtain of quotient of $q = 8$ and a remainder of $r = 2$. So

$$\begin{aligned} i^{34} &= (i^4)^8 \cdot i^2 \\ &= (1)^8 \cdot (-1) \\ &= -1 \end{aligned}$$

(b) We divide 101 by 4 and obtain a quotient of $q = 25$ and a remainder of $r = 1$. So

$$\begin{aligned} i^{101} &= (i^4)^{25} \cdot i^1 \\ &= (1)^{25} \cdot i \\ &= i \end{aligned}$$

QUICK ✓　*Simplify the power of i.*

18. i^{43} 　　　　　　　　　**19.** i^{98}

9.9 Exercises

For Extra Help:

Student Solutions Manual　CD Video　PH Math/Tutor Center　MathXL Tutorials on CD　MathXL®　MyMathLab

Concepts and Vocabulary

In Problems 1–3, fill in the blanks.

1. The _____ _____, denoted by i, is the number whose square is -1.

2. Any number of the form bi is called a _____ _____ _____.

3. If $a + bi$ is a complex number, then its _____ is defined as $a - bi$.

In Problems 4–6, answer True or False to each statement.

4. If N is a positive real number, we define the principal square root of $-N$, denoted by $\sqrt{-N}$, as $\sqrt{-N} = \sqrt{N}i$.

5. The conjugate of $-3 + 5i$ is $3 - 5i$.

6. All real numbers are complex numbers.

7. In your own words explain the relation between the natural numbers, whole numbers, integers, rational numbers, real numbers and complex number system.

8. In your own words, explain why the product of a complex number and its conjugate is a nonnegative real number. That is, explain why $(a + bi)(a - bi) = a^2 + b^2$.

9. How is multiplying two complex numbers related to multiplying two binomials?

10. How is the method used to rationalize denominators related to the method used to write the quotient of two complex numbers in standard form?

Building Skills

In Problems 11–20, write each expression as a pure imaginary number.

11. $\sqrt{-4}$ **12.** $\sqrt{-25}$ **13.** $-\sqrt{-81}$ **14.** $-\sqrt{-100}$

15. $\sqrt{-45}$ **16.** $\sqrt{-48}$ **17.** $\sqrt{-300}$ **18.** $\sqrt{-162}$

19. $\sqrt{-7}$ **20.** $\sqrt{-13}$

In Problems 21–28, write each expression as a complex number in standard form.

21. $5 + \sqrt{-49}$ **22.** $4 - \sqrt{-36}$ **23.** $-2 - \sqrt{-28}$ **24.** $10 + \sqrt{-32}$

25. $\dfrac{4 + \sqrt{-4}}{2}$ **26.** $\dfrac{10 - \sqrt{-25}}{5}$ **27.** $\dfrac{4 + \sqrt{-8}}{12}$ **28.** $\dfrac{15 - \sqrt{-50}}{5}$

In Problems 29–36, add or subtract as indicated.

29. $(4 + 5i) + (2 - 7i)$ **30.** $(-6 + 2i) + (3 + 12i)$

31. $(4 + i) - (8 - 5i)$ **32.** $(-7 + 3i) - (-3 + 2i)$

33. $\left(4 - \sqrt{-4}\right) - \left(2 + \sqrt{-9}\right)$ **34.** $\left(-4 + \sqrt{-25}\right) + \left(1 - \sqrt{-16}\right)$

35. $\left(-2 + \sqrt{-18}\right) + \left(5 - \sqrt{-50}\right)$ **36.** $\left(-10 + \sqrt{-20}\right) - \left(-6 + \sqrt{-45}\right)$

In Problems 37–60, multiply.

37. $6i(2 - 4i)$ **38.** $3i(-2 - 6i)$

39. $-\dfrac{1}{2}i(4 - 10i)$ **40.** $\dfrac{1}{3}i(12 + 15i)$

41. $(2 + i)(4 + 3i)$ **42.** $(3 - i)(1 + 2i)$

43. $(-3 - 5i)(2 + 4i)$ **44.** $(5 - 2i)(-1 + 2i)$

45. $(-3 + 5i)(5 - 3i)$ **46.** $(2 + 8i)(-3 - i)$

47. $\left(3 - \sqrt{2}i\right)\left(-2 + \sqrt{2}i\right)$ **48.** $\left(1 + \sqrt{3}i\right)\left(-4 - \sqrt{3}i\right)$

49. $\left(\dfrac{1}{2} - \dfrac{1}{4}i\right)\left(\dfrac{2}{3} + \dfrac{3}{4}i\right)$ **50.** $\left(-\dfrac{2}{3} + \dfrac{4}{3}i\right)\left(\dfrac{1}{2} - \dfrac{3}{2}i\right)$

51. $(3 + 2i)^2$ **52.** $(2 + 5i)^2$

53. $(-4 - 5i)^2$ **54.** $(2 - 7i)^2$

55. $\sqrt{-9} \cdot \sqrt{-4}$ **56.** $\sqrt{-36} \cdot \sqrt{-4}$

57. $\sqrt{-8} \cdot \sqrt{-10}$ **58.** $\sqrt{-12} \cdot \sqrt{-15}$

59. $\left(2 + \sqrt{-81}\right)\left(-3 - \sqrt{-100}\right)$ **60.** $\left(1 - \sqrt{-64}\right)\left(-2 + \sqrt{-49}\right)$

In Problems 61–66, (a) find the conjugate of the complex number, and (b) multiply the complex number by its conjugate.

61. $3 + 5i$ **62.** $5 + 2i$ **63.** $2 - 7i$

64. $9 - i$ **65.** $-7 + 2i$ **66.** $-1 - 4i$

In Problems 67–80, divide.

67. $\dfrac{1 + i}{3i}$ **68.** $\dfrac{2 - i}{2i}$ **69.** $\dfrac{-5 + 2i}{5i}$ **70.** $\dfrac{-4 + 5i}{6i}$

71. $\dfrac{3}{2 + i}$ **72.** $\dfrac{2}{4 + i}$ **73.** $\dfrac{-2}{-3 - 7i}$ **74.** $\dfrac{-4}{-5 - 3i}$

75. $\dfrac{2 + 3i}{3 - 2i}$ **76.** $\dfrac{2 + 5i}{5 - 2i}$ **77.** $\dfrac{4 + 2i}{1 - i}$ **78.** $\dfrac{-6 + 2i}{1 + i}$

79. $\dfrac{4 - 2i}{1 + 3i}$ **80.** $\dfrac{5 - 3i}{2 + 4i}$

In Problems 81–88, simplify.

81. i^{53} **82.** i^{72} **83.** i^{43} **84.** i^{110}

85. i^{153} **86.** i^{131} **87.** i^{-45} **88.** i^{-26}

Mixed Practice

In Problems 89–102, perform the indicated operation.

89. $(-4 - i)(4 + i)$ **90.** $(-5 + 2i)(5 - 2i)$ **91.** $(3 + 2i)^2$

92. $(-3 + 2i)^2$ **93.** $\dfrac{-3 + 2i}{3i}$ **94.** $\dfrac{5 - 3i}{4i}$

95. $\dfrac{-4 + i}{-5 - 3i}$ **96.** $\dfrac{-4 + 6i}{-5 - i}$ **97.** $(10 - 3i) + (2 + 3i)$

98. $(-4 + 5i) + (4 - 2i)$ **99.** $5i(-4 + 3i)$ **100.** $2i(3 - 4i)$

101. $\sqrt{-10} \cdot \sqrt{-15}$ **102.** $\sqrt{-8} \cdot \sqrt{-12}$

Applying the Concepts

In Problems 103–108, find the reciprocal of the complex number. Write each number in standard form.

103. $5i$ **104.** $7i$ **105.** $2 - i$

106. $3 - 5i$ **107.** $-4 + 5i$ **108.** $-6 + 2i$

109. Suppose that $f(x) = x^2$; find (a) $f(i)$ (b) $f(1 + i)$.

110. Suppose that $f(x) = x^2 + x$; find (a) $f(i)$ (b) $f(1 + i)$.

111. Suppose that $f(x) = x^2 + 2x + 2$; find (a) $f(3i)$ (b) $f(1 - i)$.

112. Suppose that $f(x) = x^2 + x - 1$; find (a) $f(2i)$ (b) $f(2 + i)$.

113. Impedance (Series Circuit) The total impedance, Z, of an ac circuit containing components in series is equivalent to the sum of the individual impedances. Impedance is measured in ohms (Ω) and is expressed as an imaginary number of the form $Z = R + i \cdot X$. Here, R represents resistance and X represents reactance.

    (a) If the impedance in one part of a series circuit is $7 + 3i$ ohms and the impedance of the remainder of the circuit is $3 - 4i$ ohms, find the total impedance of the circuit.

    (b) What is the total resistance of the circuit?

    (c) What is the total reactance of the circuit?

114. Impedance (Parallel Circuit) The total impedance, Z, of an ac circuit consisting of two parallel pathways is given by the formula $\dfrac{1}{Z} = \dfrac{1}{Z_1} + \dfrac{1}{Z_2}$, where Z_1 and Z_2 are the impedances of each pathway. If the impedances of the individual pathways are $Z_1 = 5$ ohms and $Z_2 = 1 - 2i$ ohms, find the total impedance of the circuit.

Extending the Concepts

115. For the function $f(x) = x^2 + 4x + 5$, find (a) $f(-2 + i)$ (b) $f(-2 - i)$.

116. For the function $f(x) = x^2 - 2x + 2$, find (a) $f(1 + i)$ (b) $f(1 - i)$.

117. For the function $f(x) = x^3 + 1$, find (a) $f(-1)$ (b) $f\left(\dfrac{1}{2} + \dfrac{\sqrt{3}}{2}i\right)$ (c) $f\left(\dfrac{1}{2} - \dfrac{\sqrt{3}}{2}i\right)$.

118. For the function $f(x) = x^3 - 1$, find (a) $f(1)$ (b) $f\left(-\dfrac{1}{2} + \dfrac{\sqrt{3}}{2}i\right)$

(c) $f\left(-\dfrac{1}{2} - \dfrac{\sqrt{3}}{2}i\right)$.

119. Any complex number z such that $f(z) = 0$ is called a **complex zero** of f. Look at the complex zeros in Problems 115–118. Conjecture a general result regarding the complex zeros of polynomials that have real coefficients.

Synthesis Review

120. Expand: $(x + 2)^3$

121. Expand: $(y - 4)^2$

122. Evaluate: $(3 + i)^3$

123. Evaluate: $(4 - 3i)^2$

124. How is raising a complex number to a positive integer power related to raising a binomial to a positive integer power?

The Graphing Calculator

Graphing calculators have the ability to add, subtract, multiply and divide complex numbers. First, put the calculator into complex mode as shown in Figure 9(a). Figure 9(b) shows the results of Examples 3(a) and 3(b). Figure 9(c) shows the results of Examples 5(b) and 8.

Figure 9

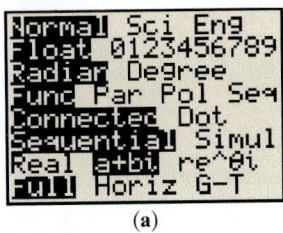
(a)

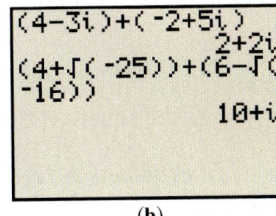
(b)

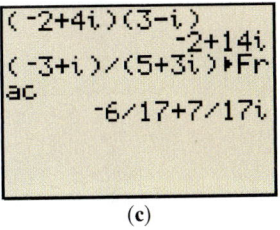
(c)

In Problems 125–132, use a graphing calculator to simplify each expression. Write your answers in standard form, $a + bi$.

125. $(1.3 - 4.3i) + (-5.3 + 0.7i)$

126. $(-3.4 + 1.9i) - (6.5 - 5.3i)$

127. $(0.3 - 5.2i)(1.2 + 3.9i)$

128. $(-4.3 + 0.2i)(7.2 - 0.5i)$

129. $\dfrac{4}{3 - 8i}$

130. $\dfrac{1 - 7i}{-4 + i}$

131. $3i^6 + (5 - 0.3i)(3 + 2i)$

132. $-6i^{14} + 5i^6 - (2 + i)(3 + 11i)$

CHAPTER 9 ACTIVITY: WHICH ONE DOES NOT BELONG?

Focus: Simplifying Radicals and Rational Exponents

Time: 15–25 minutes

Group Size: 2–4

5–10 Minutes: Individually, evaluate each row below. Decide which item does not belong and why. Be creative with your reasons!

10–15 Minutes: Discuss each row and make a list of each member's response. Is there any example that everyone agrees with?

	A	B	C	D
1	$16^{\frac{3}{2}}$	$-16^{\frac{3}{2}}$	$-16^{\frac{2}{3}}$	$-16^{-\frac{2}{3}}$
2	$\sqrt[2]{(a-b)^3},\, a > b$	$\sqrt[2]{a^3 - b^3}$	$\sqrt[2]{(a-b)^3},\, a < b$	$\sqrt[2]{a - b},\, a > b$
3	The range of x^2	The range of $\sqrt[3]{x}$	The range of $|x|$	The range of x^3
4	i^{212}	i^0	i^{-108}	i^1

CHAPTER 9 REVIEW

Section 9.1   Square Roots

KEY CONCEPTS

- The principal square root of any positive number is positive.
- The principal square root of 0 is 0 because $0^2 = 0$. That is, $\sqrt{0} = 0$.
- The square root of a perfect square is a rational number.
- The square root of a positive rational number that is not a perfect square is an irrational number.
- The square root of a negative real number is not a real number.
- For any real number a, $\sqrt{a^2} = |a|$. If $a \geq 0$, $\sqrt{a^2} = a$.

KEY TERMS

Square root
Radical
Principal square root
Radicand
Perfect square

YOU SHOULD BE ABLE TO . . .	EXAMPLE	REVIEW EXERCISES
① Evaluate square roots of perfect squares (p. 660)	Examples 1 and 2	1–12
② Determine whether a square root is rational, irrational, or not a real number (p. 661)	Examples 3 and 4	13–16
③ Find square roots of variable expressions (p. 663)	Example 5	17, 18

In Problems 1 and 2, find the value of each expression.

1. the square roots of 4

2. the square roots of 81

In Problems 3–12, find the exact value of each expression.

3. $-\sqrt{1}$

4. $-\sqrt{25}$

5. $\sqrt{0.16}$

6. $\sqrt{0.04}$

7. $\dfrac{3}{2}\sqrt{\dfrac{25}{36}}$

8. $\dfrac{4}{3}\sqrt{\dfrac{81}{4}}$

9. $\sqrt{25 - 9}$

10. $\sqrt{169 - 25}$

11. $\sqrt{9^2 - (4)(5)(-18)}$

12. $\sqrt{13^2 - (4)(-3)(-4)}$

In Problems 13–16, determine whether each square root is rational, irrational, or not a real number. Then evaluate the square root. For each square root that is irrational, use a calculator or Appendix F to round your answer to two decimal places.

13. $-\sqrt{9}$

14. $-\dfrac{1}{2}\sqrt{48}$

15. $\sqrt{14}$

16. $\sqrt{-2}$

In Problems 17 and 18, simplify each square root. Assume that the variable can be any real number.

17. $\sqrt{(4x - 9)^2}$

18. $\sqrt{(16m - 25)^2}$

Section 9.2   *n*th Roots and Rational Exponents

KEY CONCEPTS

- **Simplifying $\sqrt[n]{a^n}$**

  If   is a positive integer and a is a real number, then

  $\sqrt[n]{a^n} = a$ if $n \geq 3$ is odd

  $\sqrt[n]{a^n} = |a|$ if $n \geq 2$ is even

- If a is a real number and $n \geq 2$ is an integer, then $a^{\frac{1}{n}} = \sqrt[n]{a}$ provided that $\sqrt[n]{a}$ exists.

- If a is a real number, m/n is a rational number in lowest terms with $n \geq 2$, the $a^{\frac{m}{n}} = \sqrt[n]{a^m} = \left(\sqrt[n]{a}\right)^m$ provided that $\sqrt[n]{a}$ exists.

KEY TERMS

Principal *n*th root of a number a
Index
Cube root
$n \geq 2$

- If m/n is a rational number and if a is a nonzero real number, then $a^{-\frac{m}{n}} = \dfrac{1}{a^{\frac{m}{n}}}$ or $\dfrac{1}{a^{-\frac{m}{n}}} = a^{\frac{m}{n}}$.

YOU SHOULD BE ABLE TO . . .	EXAMPLE	REVIEW EXERCISES
① Evaluate nth roots (p. 666)	Examples 1 and 2	19–23, 37, 38
② Simplify expressions of the form $\sqrt[n]{a^n}$ (p. 667)	Example 3	24–26
③ Evaluate expressions of the form $a^{\frac{1}{n}}$ (p. 668)	Examples 4 and 5	27–30, 35, 39
④ Evaluate expressions of the form $a^{\frac{m}{n}}$ (p. 669)	Examples 6 through 9	31–34, 36, 40–42

In Problems 19–26, simplify each radical.

19. $\sqrt[3]{343}$ **20.** $\sqrt[3]{-125}$ **21.** $\sqrt[3]{\dfrac{8}{27}}$ **22.** $\sqrt[4]{81}$

23. $-\sqrt[5]{-243}$ **24.** $\sqrt[3]{10^3}$ **25.** $\sqrt[5]{z^5}$ **26.** $\sqrt[4]{(5p-3)^4}$

In Problems 27–34, evaluate each of the following expressions.

27. $81^{1/2}$ **28.** $(-256)^{1/4}$ **29.** $-4^{1/2}$ **30.** $729^{1/3}$

31. $16^{7/4}$ **32.** $-(-27)^{2/3}$ **33.** $-121^{3/2}$ **34.** $\dfrac{1}{36^{-1/2}}$

In Problems 35–38, use a calculator to write each expression rounded to two decimal places.

35. $(-65)^{1/3}$ **36.** $4^{3/5}$ **37.** $\sqrt[3]{100}$ **38.** $\sqrt[4]{10}$

In Problems 39–42, rewrite each of the following radicals with a rational exponent.

39. $\sqrt[3]{5a}$ **40.** $\sqrt[5]{p^7}$ **41.** $\left(\sqrt[4]{10z}\right)^3$ **42.** $\sqrt[6]{(2ab)^5}$

Section 9.3   Simplify Expressions Using the Laws of Exponents

KEY CONCEPTS		KEY TERM
• **If a and b are real numbers and if r and s are rational numbers, then assuming the expression is defined,**		Simplify
Zero-Exponent Rule:	$a^0 = 1$ if $a \neq 0$	
Negative-Exponent Rule:	$a^{-r} = \dfrac{1}{a^r}$ if $a \neq 0$	
Product Rule:	$a^r \cdot a^s = a^{r+s}$	
Quotient Rule:	$\dfrac{a^r}{a^s} = a^{r-s} = \dfrac{1}{a^{s-r}}$ if $a \neq 0$	
Power Rule:	$(a^r)^s = a^{r \cdot s}$	
Product to Power Rule:	$(a \cdot b)^r = a^r \cdot b^r$	
Quotient to Power Rule:	$\left(\dfrac{a}{b}\right)^r = \dfrac{a^r}{b^r}$ if $b \neq 0$	
Quotient to a Negative Power Rule:	$\left(\dfrac{a}{b}\right)^{-r} = \left(\dfrac{b}{a}\right)^r$ if $a \neq 0, b \neq 0$	

YOU SHOULD BE ABLE TO . . .	EXAMPLE	REVIEW EXERCISES
① Use the Laws of Exponents to simplify expressions involving rational exponents (p. 674)	Examples 1 through 3	43–48
② Use the Laws of Exponents to simplify radical expressions (p. 676)	Example 4	49–52
③ Factor expressions containing rational exponents (p. 677)	Examples 5 and 6	53, 54

In Problems 43–54, simplify the expression.

43. $4^{2/3} \cdot 4^{7/3}$

44. $\dfrac{k^{1/2}}{k^{3/4}}$

45. $(p^{4/3} \cdot q^4)^{3/2}$

46. $(32a^{-3/2} \cdot b^{1/4})^{1/5}$

47. $5m^{-2/3}(2m + m^{-1/3})$

48. $\left(\dfrac{16x^{1/3}}{x^{-1/3}}\right)^{-1/2} + \left(\dfrac{x^{-3/2}}{64x^{-1/2}}\right)^{1/3}$

49. $\sqrt[8]{x^6}$

50. $\sqrt{121x^4y^{10}}$

51. $\sqrt[3]{m^2} \cdot \sqrt{m^3}$

52. $\dfrac{\sqrt[3]{c}}{\sqrt[6]{c^4}}$

53. $2(3m - 1)^{1/4} + (m - 7)(3m - 1)^{5/4}$

54. $3(x^2 - 5)^{1/3} - 4x(x^2 - 5)^{-2/3}$

Section 9.4   Simplifying Radical Expressions

KEY CONCEPTS	KEY TERM
• **Product Property of Radicals** If $\sqrt[n]{a}$ and $\sqrt[n]{b}$ are real numbers and $n \geq 2$ is an integer, then $\sqrt[n]{a} \cdot \sqrt[n]{b} = \sqrt[n]{ab}$. • **Quotient Property of Radicals** If $\sqrt[n]{a}$ and $\sqrt[n]{b}$ are real numbers, $b \neq 0$ and $n \geq 2$ is an integer, then $\dfrac{\sqrt[n]{a}}{\sqrt[n]{b}} = \sqrt[n]{\dfrac{a}{b}}$.	Simplify

YOU SHOULD BE ABLE TO . . .	EXAMPLE	REVIEW EXERCISES
1 Use the Product Property to multiply radical expressions (p. 680)	Example 1	55, 56, 67–70
2 Use the Product Property to simplify radical expressions (p. 681)	Examples 2 through 7	57–70
3 Use the Quotient Property to simplify radical expressions (p. 686)	Examples 8 and 9	71–78
4 Multiply radicals with unlike indices (p. 687)	Example 10	79, 80

In Problems 55 and 56, use the Product Property to multiply. Assume that all variables can be any real number.

55. $\sqrt{15} \cdot \sqrt{7}$

56. $\sqrt[4]{2ab^2} \cdot \sqrt[4]{6a^2b}$

In Problems 57–62, simplify each radical using the Product Property. Assume that all variables can be any real number.

57. $\sqrt{80}$

58. $\sqrt[3]{-500}$

59. $\sqrt[3]{162m^6n^4}$

60. $\sqrt[4]{50p^8q^4}$

61. $2\sqrt{16x^6y}$

62. $\sqrt{(2x + 1)^3}$

In Problems 63–66, simplify each radical using the Product Property. Assume that all variables are greater than or equal to zero.

63. $\sqrt{w^3z^2}$

64. $\sqrt{45x^4yz^3}$

65. $\sqrt[3]{16a^{12}b^5}$

66. $\sqrt{4x^2 + 8x + 4}$

In Problems 67–70, multiply and simplify. Assume that all variables are greater than or equal to zero.

67. $\sqrt{15} \cdot \sqrt{18}$

68. $\sqrt[3]{20} \cdot \sqrt[3]{30}$

69. $\sqrt[3]{-3x^4y^7} \cdot \sqrt[3]{24x^3y^2}$

70. $3\sqrt{4xy^2} \cdot 5\sqrt{3x^2y}$

In Problems 71–78, simplify. Assume that all variables are greater than zero.

71. $\sqrt{\dfrac{121}{25}}$

72. $\sqrt{\dfrac{5a^4}{64b^2}}$

73. $\sqrt[3]{\dfrac{54k^2}{9k^5}}$

74. $\sqrt[3]{\dfrac{-160w^{11}}{343w^{-4}}}$

75. $\dfrac{\sqrt{12h^3}}{\sqrt{3h}}$

76. $\dfrac{\sqrt{50a^3b^3}}{\sqrt{8a^5b^{-3}}}$

77. $\dfrac{\sqrt[3]{-8x^7y}}{\sqrt[3]{27xy^4}}$

78. $\dfrac{\sqrt[4]{48m^2n^7}}{\sqrt[4]{3m^6n}}$

In Problems 79 and 80, multiply and simplify.

79. $\sqrt{5} \cdot \sqrt[3]{2}$

80. $\sqrt[4]{8} \cdot \sqrt[6]{4}$

Section 9.5	Adding, Subtracting, and Multiplying Radical Expressions	
KEY CONCEPTS		**KEY TERM**
• To add or subtract radicals, the index and the radicand must be the same. • When we multiply radicals, the index on each radical must be the same. Then we multiply the radicands.		Like radicals

YOU SHOULD BE ABLE TO . . .	EXAMPLE	REVIEW EXERCISES
1 Add or subtract radical expressions (p. 692)	Examples 1 through 3	81–90
2 Multiply radical expressions (p. 694)	Examples 4 and 5	91–100

In Problems 81–90, add or subtract as indicated. Assume all variables are positive or zero.

81. $2\sqrt[4]{x} + 6\sqrt[4]{x}$

82. $7\sqrt[3]{4y} + 2\sqrt[3]{4y} - 3\sqrt[3]{4y}$

83. $5\sqrt{2} - 2\sqrt{12}$

84. $\sqrt{18} + 2\sqrt{50}$

85. $\sqrt[3]{-16z} + \sqrt[3]{54z}$

86. $7\sqrt[3]{8x^2} - \sqrt[3]{-27x^2}$

87. $\sqrt{16a} + \sqrt[6]{729a^3}$

88. $\sqrt{27x^2} - x\sqrt{48} + 2\sqrt{75x^2}$

89. $5\sqrt[3]{4m^5y^2} - \sqrt[6]{16m^{10}y^4}$

90. $\sqrt{y^3 - 4y^2} - 2\sqrt{y - 4} + \sqrt[4]{y^2 - 8y + 16}$

In Problems 91–100, multiply and simplify.

91. $\sqrt{3}\left(\sqrt{5} - \sqrt{15}\right)$

92. $\sqrt[3]{5}\left(3 + \sqrt[3]{4}\right)$

93. $\left(3 + \sqrt{5}\right)\left(4 - \sqrt{5}\right)$

94. $\left(7 + \sqrt{3}\right)\left(6 + \sqrt{2}\right)$

95. $\left(1 - 3\sqrt{5}\right)\left(1 + 3\sqrt{5}\right)$

96. $\left(\sqrt[3]{x} + 1\right)\left(9\sqrt[3]{x} - 4\right)$

97. $\left(\sqrt{x} - \sqrt{5}\right)^2$

98. $\left(11\sqrt{2} + \sqrt{5}\right)^2$

99. $\left(\sqrt{2a} - b\right)\left(\sqrt{2a} + b\right)$

100. $\left(\sqrt[3]{6s} + 2\right)\left(\sqrt[3]{6s} - 7\right)$

Section 9.6   Rationalizing Radical Expressions

KEY CONCEPTS	KEY TERM
• To rationalize the denominator when the denominator contains a single radical, multiply the numerator and denominator by a radical such that the radicand in the denominator is a perfect power of the index, n. • To rationalize a denominator containing two terms, use the fact that $(a + b)(a - b) = a^2 - b^2$.	Rationalizing the denominator

YOU SHOULD BE ABLE TO . . .	EXAMPLE	REVIEW EXERCISES
1 Rationalize a denominator containing one term (p. 699)	Examples 1 and 2	101–108, 117
2 Rationalize a denominator containing two terms (p. 701)	Examples 3 and 4	109–116, 118

In Problems 101–118, rationalize the denominator.

101. $\dfrac{2}{\sqrt{6}}$ **102.** $\dfrac{6}{\sqrt{3}}$ **103.** $\dfrac{\sqrt{48}}{\sqrt{p^3}}$ **104.** $\dfrac{5}{\sqrt{2a}}$

105. $\dfrac{-2}{\sqrt{6y^3}}$ **106.** $\dfrac{3}{\sqrt[3]{5}}$ **107.** $\sqrt[3]{\dfrac{-4}{45}}$ **108.** $\dfrac{27}{\sqrt[5]{8p^3q^4}}$

109. $\dfrac{6}{7 - \sqrt{6}}$ **110.** $\dfrac{3}{\sqrt{3} - 9}$ **111.** $\dfrac{\sqrt{3}}{3 + \sqrt{2}}$ **112.** $\dfrac{\sqrt{k}}{\sqrt{k} - \sqrt{m}}$

113. $\dfrac{\sqrt{10} + 2}{\sqrt{10} - 2}$ **114.** $\dfrac{3 - \sqrt{y}}{3 + \sqrt{y}}$ **115.** $\dfrac{4}{2\sqrt{3} + 5\sqrt{2}}$ **116.** $\dfrac{\sqrt{5} - \sqrt{6}}{\sqrt{10} + \sqrt{3}}$

117. Simplify: $\dfrac{\sqrt{7}}{3} + \dfrac{6}{\sqrt{7}}$ **118.** Find the reciprocal: $4 - \sqrt{7}$

Section 9.7   Functions Involving Radicals

KEY CONCEPT	KEY TERMS

• **Finding the Domain of a Function Whose Rule Is a Radical Expression**
  1. If the index on a radical expression is even, then the radicand must be greater than or equal to zero.
  2. If the index on a radical expression is odd, then the radicand can be any real number.
	Square root function
	Cube root function

YOU SHOULD BE ABLE TO . . .	EXAMPLE	REVIEW EXERCISES
1 Evaluate functions whose rule is a radical expression (p. 705)	Example 1	119–122
2 Find the domain of a function whose rule is a radical (p. 706)	Example 2	123–128, 129(a), 130(a), 131(a), 132(a)
3 Graph functions involving square roots (p. 707)	Example 3	129(b)–131(b)
4 Graph functions involving cube roots (p. 708)	Example 4	132(b)

In Problems 119–122, evaluate each radical function at the indicated values.

119. $f(x) = \sqrt{x + 4}$

    (a) $f(-3)$ **(b)** $f(0)$ **(c)** $f(5)$

120. $g(x) = \sqrt{3x - 2}$

    (a) $g\left(\dfrac{2}{3}\right)$ **(b)** $g(2)$ **(c)** $g(6)$

121. $H(t) = \sqrt[3]{t + 3}$

    (a) $H(-2)$ **(b)** $H(-4)$ **(c)** $H(5)$

122. $G(z) = \sqrt{\dfrac{z - 1}{z + 2}}$

    (a) $G(1)$ **(b)** $G(-3)$ **(c)** $G(2)$

In Problems 123–128, find the domain of the radical function.

123. $f(x) = \sqrt{3x - 5}$ **124.** $g(x) = \sqrt[3]{2x - 7}$ **125.** $h(x) = \sqrt[4]{6x + 1}$

126. $F(x) = \sqrt[5]{2x - 9}$ **127.** $G(x) = \sqrt{\dfrac{4}{x - 2}}$ **128.** $H(x) = \sqrt{\dfrac{x - 3}{x}}$

In Problems 129–132, (a) determine the domain of the function; (b) graph the function using point-plotting; and (c) based on the graph, determine the range of the function.

129. $f(x) = \dfrac{1}{2}\sqrt{1 - x}$

130. $g(x) = \sqrt{x + 1} - 2$

131. $h(x) = -\sqrt{x + 3}$

132. $F(x) = \sqrt[3]{x + 1}$

Section 9.8	**Radical Equations and Their Applications**	
	KEY TERMS	
Radical equation	Extraneous solution	

YOU SHOULD BE ABLE TO . . .	**EXAMPLE**	**REVIEW EXERCISES**
① Solve radical equations containing one radical (p. 713)	Examples 1 through 5	133–142, 147, 148
② Solve radical equations containing two radicals (p. 716)	Examples 6 and 7	143–146
③ Solve for a variable in a radical equation (p. 718)	Example 8	149, 150

In Problems 133–148, solve each equation.

133. $\sqrt{m} = 13$

134. $\sqrt[3]{3t + 1} = -2$

135. $\sqrt[4]{3x - 8} = 3$

136. $\sqrt{2x + 5} + 4 = 2$

137. $\sqrt{k + 4} - 3 = -1$

138. $3\sqrt{t} - 4 = 11$

139. $2\sqrt[3]{m} + 5 = -11$

140. $\sqrt{q + 2} = q$

141. $\sqrt{w + 11} + 3 = w + 2$

142. $\sqrt{p^2 - 2p + 9} = p + 1$

143. $\sqrt{a + 10} = \sqrt{2a - 1}$

144. $\sqrt{5x + 9} = \sqrt{7x - 3}$

145. $\sqrt{c - 8} + \sqrt{c} = 4$

146. $\sqrt{x + 2} - \sqrt{x + 9} = 7$

147. $(4x - 3)^{1/3} - 3 = 0$

148. $(x^2 - 9)^{1/4} = 2$

149. Height of a Cone Solve $r = \sqrt{\dfrac{3V}{\pi h}}$ for h.

150. Ball Slide Speed Factor Solve $f_s = \sqrt[3]{\dfrac{30}{v}}$ for v.

Section 9.9	The Complex Number System

KEY CONCEPTS	KEY TERMS

KEY CONCEPTS

- **Nonnegativity Property of Real Numbers**
  For any real number a, $a^2 \geq 0$

- **Imaginary Unit**
  The imaginary unit, denoted by i, is the number whose square is -1.
  That is, $i^2 = -1$.

- **Complex numbers**
  Complex numbers are numbers of the form $a + bi$, where a and b are real numbers.
  The real number a is called the real part of the number $a + bi$; the real number b is
  called the imaginary part of $a + bi$.

- **Square Roots of Negative Numbers**
  If N is a positive real number, the principal square root of $-N$, denoted by $\sqrt{-N}$,
  is $\sqrt{-N} = \sqrt{N}\,i$, where $i = \sqrt{-1}$.

- **Sum of Complex Numbers**
  $(a + bi) + (c + di) = (a + c) + (b + d)i$

- **Difference of Complex Numbers**
  $(a + bi) - (c + di) = (a - c) + (b - d)i$

- **Complex Conjugate**
  If $a + bi$ is a complex number, then its conjugate is $a - bi$.

- **Product of a Complex Number and Its Conjugate**
  $(a + bi)(a - bi) = a^2 + b^2$

KEY TERMS

Imaginary unit
Complex number system
Complex number
Real part
Imaginary part
Standard form
Pure imaginary number
Principal square root
Conjugate
Powers of i

YOU SHOULD BE ABLE TO . . .	EXAMPLE	REVIEW EXERCISES
(1) Evaluate the square root of negative real numbers (p. 723)	Examples 1 and 2	151–154
(2) Add or subtract complex numbers (p. 724)	Examples 3 and 4	155–158
(3) Multiply complex numbers (p. 726)	Examples 5 through 7	159–164
(4) Divide complex numbers (p. 728)	Examples 8 and 9	165–168
(5) Evaluate the powers of i (p. 729)	Example 10	169, 170

In Problems 151 and 152, write each expression as a pure imaginary number.

151. $\sqrt{-29}$

152. $\sqrt{-54}$

In Problems 153 and 154, write each expression as a complex number in standard form.

153. $14 - \sqrt{-162}$

154. $\dfrac{6 + \sqrt{-45}}{3}$

In Problems 155–168, perform the indicated operation.

155. $(3 - 7i) + (-2 + 5i)$

156. $(4 + 2i) - (9 - 8i)$

157. $\left(8 - \sqrt{-45}\right) - \left(3 + \sqrt{-80}\right)$

158. $\left(1 + \sqrt{-9}\right) + \left(-6 + \sqrt{-16}\right)$

159. $(4 - 5i)(3 + 7i)$

160. $\left(\dfrac{1}{2} + \dfrac{2}{3}i\right)(4 - 9i)$

161. $\sqrt{-3} \cdot \sqrt{-27}$

162. $\left(1 + \sqrt{-36}\right)\left(-5 - \sqrt{-144}\right)$

163. $(1 + 12i)(1 - 12i)$

164. $(7 + 2i)(5 + 4i)$

165. $\dfrac{4}{3 + 5i}$

166. $\dfrac{-3}{7 - 2i}$

167. $\dfrac{2 - 3i}{5 + 2i}$

168. $\dfrac{4 + 3i}{1 - i}$

In Problems 169 and 170, simplify.

169. i^{59}

170. i^{173}

CHAPTER 9 TEST

 Remember to use your Chapter Test Prep Video CD to see fully worked-out solutions to any of these problems you would like to review.

1. Evaluate: $49^{-1/2}$

In Problems 2 and 3, simplify using rational exponents.

2. $\sqrt[3]{8x^{1/2}y^3} \cdot \sqrt{9xy^{1/2}}$

3. $\sqrt[5]{(2a^4b^3)^7}$

In Problems 4–9, perform the indicated operation and simplify. Assume all variables in the radicand are greater than or equal to zero.

4. $\sqrt{3m} \cdot \sqrt{13n}$

5. $\sqrt{32x^7y^4}$

6. $\dfrac{\sqrt{9a^3b^{-3}}}{\sqrt{4ab}}$

7. $\sqrt{5x^3} + 2\sqrt{45x}$

8. $\sqrt{9a^2b} - \sqrt[4]{16a^4b^2}$

9. $\left(11 + 2\sqrt{x}\right)\left(3 - \sqrt{x}\right)$

In Problems 10 and 11, rationalize the denominator.

10. $\dfrac{-2}{3\sqrt{72}}$

11. $\dfrac{\sqrt{5}}{\sqrt{5} + 2}$

12. For $f(x) = \sqrt{-2x + 3}$, find the following:

    (a) $f(1)$ **(b)** $f(-3)$

13. Determine the domain of the function $g(x) = \sqrt{-3x + 5}$.

14. For $f(x) = \sqrt{x} - 3$, do the following:

    (a) Determine the domain of the function.
    (b) Graph the function using point plotting.
    (c) From the graph, determine the range of the function.

In Problems 15–17, solve the given equations.

15. $\sqrt{x + 3} = 4$ **16.** $\sqrt{x + 13} - 4 = x - 3$ **17.** $\sqrt{x - 1} + \sqrt{x + 2} = 3$

In Problems 18–20, perform the indicated operation.

18. $(13 + 2i) + (4 - 15i)$ **19.** $(4 - 7i)(2 + 3i)$ **20.** $\dfrac{7 - i}{12 + 11i}$

CUMULATIVE REVIEW Chapters 1–9

1. Evaluate: $6 - 3^2 \div (9 - 3)$

2. Simplify: $(3x + 2y) - (2x - 5y + 3) + 9$

3. Solve: $(3x + 5) - 2 = 7x - 13$

4. Solve and write your answer in interval notation: $6x + \dfrac{1}{2}(4x - 2) \le 3x + 9$

5. For $f(x) = 3x^2 - x + 5$, find the following:

    (a) $f(-2)$ **(b)** $f(3)$

6. Find the domain of $g(x) = \dfrac{x^2 - 9}{x^2 - 2x - 8}$.

7. The annual sales for a certain computer game is given by $n(p) = -50p + 6000$, where n is the number of games sold and p is the price in dollars.

    (a) How many games will be sold in one year if the price were \$50?

    (b) At what price will there be no sales during a given year?

8. Find the equation of the line that passes through the points $(-1, 6)$ and $(3, -2)$.

9. Graph the inequality: $6x + 3y > 24$

10. Solve the system using substitution:

$$\begin{cases} 4x - y = 17 \\ 5x + 6y = 14 \end{cases}$$

11. Dried fruit is to be mixed with nuts to create a trail mix blend. The fruit costs \$3.45 per pound and the nuts cost \$2.10 per pound. How much of each should be used to make 10 pounds of trail mix that will sell for \$2.64 per pound if the total revenue should remain the same?

12. If $g(x) = x^3 + 4x - 16$, what is $g(-4)$?

13. Add: $(8x^3 - 4x^2 + 5x + 3) + (2x^2 - 8x + 7)$

14. Multiply: $(2x - 1)(4x^2 + 2x - 9)$

15. Divide: $\dfrac{6x^4 + 13x^3 - 21x^2 - 28x + 37}{2x^2 + 3x - 5}$

16. Factor completely: $8x^2 - 44x - 84$

17. Subtract: $\dfrac{2x}{x - 3} - \dfrac{x + 1}{x + 2}$

18. Solve: $\dfrac{9}{k - 2} = \dfrac{6}{k} + 3$

19. Solve and graph the solution set on a number line: $3x + 1 < 13$ or $-2x + 3 \le 23$

20. Shawn can paint a certain room in 4 hours. Payton can paint the same room in 6 hours. How long will it take them to paint the room if they work together?

21. Simplify: $\dfrac{\sqrt{50a^3b}}{\sqrt{2a^{-1}b^3}}$

22. Find the domain of the function $f(x) = \sqrt[4]{8 - 3x}$.

23. Solve: $\sqrt{x + 7} - 8 = x - 7$

24. Divide: $\dfrac{3i}{1 - 7i}$

25. Rationalize the denominator: $\dfrac{2}{4 - \sqrt{11}}$

10 Quadratic Equations and Functions

One of the more unusual sports is found in Millsboro, Delaware—Punkin Chunkin. Participants catapult or fire pumpkins to see who can toss them the farthest (and most accurately). Interestingly, this bizarre ritual is an application of a quadratic function at work. See Problems 73–76 in Section 10.5.

OUTLINE

10.1 Solving Quadratic Equations by Completing the Square

10.2 Solving Quadratic Equations by the Quadratic Formula

10.3 Solving Equations Quadratic in Form

  Putting the Concepts Together (Sections 10.1–10.3)

10.4 Graphing Quadratic Functions Using Transformations

10.5 Graphing Quadratic Functions Using Properties

10.6 Quadratic Inequalities

10.7 Rational Inequalities

  Chapter 10 Activity: Presidential Decision Making
  Chapter 10 Review
  Chapter 10 Test

The Big Picture: Putting It Together

In Chapter 2, we discussed solving linear equations and inequalities. In Chapter 3, we completed our discussion of "everything linear" after covering linear equations in two variables and their graphs.

This chapter is dedicated to completing our discussion of "everything quadratic" that began in Section 6.6, when we solved quadratic equations $ax^2 + bx + c = 0$, where the expression $ax^2 + bx + c$ was factorable. Remember, if the expression $ax^2 + bx + c$ was not factorable, we did not have a method for solving the equation. The missing piece for solving this type of quadratic equation was the idea of a radical. Having learned how to work with radicals in Chapter 9, we now have the tools necessary to expand our understanding of solving all quadratic equations $ax^2 + bx + c = 0$, where the expression $ax^2 + bx + c$ may or may not be factorable. The theme of the chapter will be to solve quadratic equations, graph quadratic functions, and solve quadratic inequalities.

10.1 Solving Quadratic Equations by Completing the Square

OBJECTIVES

1. Solve Quadratic Equations Using the Square Root Property
2. Complete the Square in One Variable
3. Solve Quadratic Equations by Completing the Square
4. Solve Problems Using the Pythagorean Theorem

Preparing for *Solving Quadratic Equations by Completing the Square*

Before getting started, take the following readiness quiz. If you get a problem wrong, go back to the section cited and review the material.

1. Multiply: $(2p + 3)^2$ [Section 5.3, pp. 364–366]

2. Factor: $y^2 - 8y + 16$ [Section 6.4, pp. 437–439]

3. Solve: $x^2 + 5x - 14 = 0$ [Section 6.6, pp. 451–457]

4. Solve: $x^2 - 16 = 0$ [Section 6.6, pp. 451–457]

5. Simplify: (a) $\sqrt{36}$ (b) $\sqrt{45}$ (c) $\sqrt{-12}$ [Section 9.1, pp. 660–661; Section 9.4, pp. 681–686; Section 9.9, pp. 723–724]

6. Find the complex conjugate of $-3 + 2i$. [Section 9.9, pp. 727–728]

1 Solve Quadratic Equations Using the Square Root Property

Suppose that we wanted to solve the quadratic equation

$$x^2 = p$$

where p is any real number. In words, this equation is saying, "give me all numbers whose square is p." So, if $p = 16$, then we would have the equation

$$x^2 = 16$$

which means we want "all numbers whose square is 16." There are two numbers whose square is 16, -4 and 4, so the solution set to the equation $x^2 = 16$ is $\{-4, 4\}$.

In general, we have the following method for solving equations of the form $x^2 = p$.

> **THE SQUARE ROOT PROPERTY**
>
> If $x^2 = p$, then $x = \sqrt{p}$ or $x = -\sqrt{p}$.

Work Smart

The Square Root Property is useful for solving equations of the form "some unknown squared equals a real number." To solve this equation, we take the square root of both sides of the equation, but don't forget the $\pm$ symbol to obtain the positive and negative square root.

When using the Square Root Property to solve an equation such as $x^2 = p$, we usually abbreviate the solutions as $x = \pm\sqrt{p}$, read "x equals plus or minus the square root of p." For example, the two solutions of the equation

$$x^2 = 16$$

are

$$x = \pm\sqrt{16}$$

and since $\sqrt{16} = 4$, we have

$$x = \pm 4$$

Let's look at an example that discusses how to solve a quadratic equation using the Square Root Property.

Preparing for...Answers

1. $4p^2 + 12p + 9$ **2.** $(y - 4)^2$
3. $\{-7, 2\}$ **4.** $\{-4, 4\}$ **5.** (a) 6
(b) $3\sqrt{5}$ (c) $2\sqrt{3}i$ **6.** $-3 - 2i$

EXAMPLE 1 How to Solve a Quadratic Equation Using the Square Root Property

Solve: $p^2 - 9 = 0$

Step-by-Step Solution

Step 1: Isolate the expression containing the square term.

$$p^2 - 9 = 0$$
Add 9 to both sides: $\quad p^2 = 9$

Step 2: Use the Square Root Property. Don't forget the $\pm$ symbol.

$$p = \pm\sqrt{9}$$
Simplify the radical: $\quad = \pm 3$

Step 3: Isolate the variable, if necessary.   The variable is already isolated.

Step 4: Verify your solution(s).

$$p = -3: \quad (-3)^2 - 9 \stackrel{?}{=} 0 \qquad p = 3: \quad 3^2 - 9 \stackrel{?}{=} 0$$
$$9 - 9 = 0 \quad \text{True} \qquad\qquad 9 - 9 = 0 \quad \text{True}$$

The solution set is $\{-3, 3\}$.

We summarize the steps used to solve quadratic equations using the Square Root Property.

> ## Solving Quadratic Equations Using the Square Root Property
>
> **Step 1:** Isolate the expression containing the square term.
>
> **Step 2:** Use the Square Root Property, which states if $x^2 = p$, then $x = \pm\sqrt{p}$. Don't forget the $\pm$ symbol.
>
> **Step 3:** Isolate the variable, if necessary.
>
> **Step 4:** Verify your solution(s).

You could also solve the equation in Example 1(a) by factoring the difference of two squares:

$$p^2 - 9 = 0$$
$$(p - 3)(p + 3) = 0$$

Zero-Product Property: $\quad p - 3 = 0 \quad \text{or} \quad p + 3 = 0$

$$p = 3 \quad \text{or} \qquad p = -3$$

Work Smart

As our mathematical knowledge develops, we will find there is more than one way to solve a problem.

So there is more than one way to obtain the solution! However, factoring only works nicely when solving equations of the form $x^2 = p$ when p is a perfect square. It doesn't work nicely when p is not a perfect square as Example 2 illustrates.

EXAMPLE 2 **Solving a Quadratic Equation Using the Square Root Property**

Solve: $3x^2 - 60 = 0$

Solution

$$3x^2 - 60 = 0$$

Add 60 to both sides of the equation: $\quad 3x^2 = 60$

Divide both sides by 3: $\quad x^2 = 20$

Use the Square Root Property: $\quad x = \pm\sqrt{20}$

Simplify the radical: $\quad = \pm 2\sqrt{5}$

Check

$$x = -2\sqrt{5}: \quad 3(-2\sqrt{5})^2 - 60 \stackrel{?}{=} 0 \qquad x = 2\sqrt{5}: \quad 3(2\sqrt{5})^2 - 60 \stackrel{?}{=} 0$$
$$(ab)^2 = a^2b^2: \quad 3(-2)^2(\sqrt{5})^2 - 60 \stackrel{?}{=} 0 \qquad\qquad 3 \cdot (2)^2(\sqrt{5})^2 - 60 \stackrel{?}{=} 0$$
$$3 \cdot 4 \cdot 5 - 60 \stackrel{?}{=} 0 \qquad\qquad\qquad 3 \cdot 4 \cdot 5 - 60 \stackrel{?}{=} 0$$
$$60 - 60 = 0 \quad \text{True} \qquad\qquad\qquad 60 - 60 = 0 \quad \text{True}$$

The solution set is $\{-2\sqrt{5}, 2\sqrt{5}\}$.

QUICK ✓ *Solve the quadratic equation using the Square Root Method.*

1. $p^2 = 48$ **2.** $3b^2 = 75$ **3.** $s^2 - 81 = 0$

There is no reason that the solution to a quadratic equation must be real. The next example illustrates a solution to a quadratic equation that is a non-real complex number.

EXAMPLE 3 Solving a Quadratic Equation Using the Square Root Property

Solve: $y^2 + 14 = 2$

Solution

$$y^2 + 14 = 2$$

Subtract 14 from both sides of the equation: $\quad y^2 = -12$

Use the Square Root Property: $\quad y = \pm\sqrt{-12}$

Simplify the radical: $\quad = \pm 2\sqrt{3}i$

Check

$y = -2\sqrt{3}i$: $\quad \left(-2\sqrt{3}i\right)^2 + 14 \overset{?}{=} 2$ $\qquad$ $y = 2\sqrt{3}i$: $\quad \left(2\sqrt{3}i\right)^2 + 14 \overset{?}{=} 2$

$\qquad\qquad\left(-2\sqrt{3}\right)^2 i^2 + 14 \overset{?}{=} 2$ $\qquad\qquad\qquad\left(2\sqrt{3}\right)^2 i^2 + 14 \overset{?}{=} 2$

$\qquad\qquad\quad 4\cdot 3\cdot(-1) + 14 \overset{?}{=} 2$ $\qquad\qquad\qquad\quad 4\cdot 3\cdot(-1) + 14 \overset{?}{=} 2$

$\qquad\qquad\qquad\quad -12 + 14 \overset{?}{=} 2$ $\qquad\qquad\qquad\qquad\quad -12 + 14 \overset{?}{=} 2$

$\qquad\qquad\qquad\qquad\qquad 2 = 2$ True $\qquad\qquad\qquad\qquad\qquad\qquad 2 = 2$ True

The solution set is $\left\{-2\sqrt{3}i, 2\sqrt{3}i\right\}$.

QUICK ✓ Solve the quadratic equation using the Square Root Property.

4. $d^2 = -72$ $\qquad\qquad$ **5.** $3q^2 + 27 = 0$

EXAMPLE 4 Solving Quadratic Equations Using the Square Root Property

Solve:
(a) $(x - 2)^2 = 25$ $\qquad$ **(b)** $(y + 5)^2 + 24 = 0$

Solution

(a) $\qquad\qquad\qquad\qquad\qquad\qquad (x - 2)^2 = 25$

Use the Square Root Property: $\qquad\qquad x - 2 = \pm\sqrt{25}$

Simplify the radical: $\qquad\qquad x - 2 = \pm 5$

Add 2 to each side: $\qquad\qquad\qquad x = 2 \pm 5$

2 ± 5 means $2 - 5$ or $2 + 5$: $\quad x = 2 - 5 \quad$ or $\quad x = 2 + 5$

$\qquad\qquad\qquad\qquad\qquad\qquad = -3 \qquad\qquad\qquad = 7$

Check

$x = -3$: $\quad (-3 - 2)^2 \overset{?}{=} 25$ $\qquad\qquad$ $x = 7$: $\quad (7 - 2)^2 \overset{?}{=} 25$

$\qquad\qquad\quad (-5)^2 \overset{?}{=} 25$ $\qquad\qquad\qquad\qquad\quad 5^2 \overset{?}{=} 25$

$\qquad\qquad\qquad 25 = 25$ True $\qquad\qquad\qquad\qquad 25 = 25$ True

The solution set is $\{-3, 7\}$.

(b)
$$(y + 5)^2 + 24 = 0$$

Subtract 24 from both sides: $\quad (y + 5)^2 = -24$

Use the Square Root Property: $\quad y + 5 = \pm\sqrt{-24}$

$\sqrt{-24} = \sqrt{24} \cdot \sqrt{-1} = \sqrt{4 \cdot 6}i = 2\sqrt{6}i$: $\quad y + 5 = \pm 2\sqrt{6}i$

Subtract 5 from each side: $\quad y = -5 \pm 2\sqrt{6}i$

$$y = -5 - 2\sqrt{6}i \quad \text{or} \quad y = -5 + 2\sqrt{6}i$$

Check

$y = -5 - 2\sqrt{6}i$:

$$\left(-5 - 2\sqrt{6}i + 5\right)^2 + 24 \overset{?}{=} 0$$
$$\left(-2\sqrt{6}i\right)^2 + 24 \overset{?}{=} 0$$
$$4 \cdot 6 \cdot i^2 + 24 \overset{?}{=} 0$$
$$-24 + 24 = 0 \quad \text{True}$$

$y = -5 + 2\sqrt{6}i$:

$$\left(-5 + 2\sqrt{6}i + 5\right)^2 + 24 \overset{?}{=} 0$$
$$\left(2\sqrt{6}i\right)^2 + 24 \overset{?}{=} 0$$
$$4 \cdot 6 \cdot i^2 + 24 \overset{?}{=} 0$$
$$-24 + 24 = 0 \quad \text{True}$$

The solution set is $\left\{-5 - 2\sqrt{6}i, -5 + 2\sqrt{6}i\right\}$.

QUICK ✓ *Solve the quadratic equation using the Square Root Property.*

6. $(y + 3)^2 = 100$ **7.** $(q - 5)^2 + 20 = 4$

---

② Complete the Square in One Variable

We now introduce the method of **completing the square.** The idea behind completing the square in one variable is to "adjust" the left side of a quadratic equation of the form $x^2 + bx + c$ in order to make it a perfect square trinomial. Recall that perfect square trinomials are trinomials of the form

$$A^2 + 2AB + B^2 = (A + B)^2$$

or

$$A^2 - 2AB + B^2 = (A - B)^2$$

For example, $x^2 + 6x + 9$ is a perfect square trinomial because $x^2 + 6x + 9 = (x + 3)^2$. Or $p^2 - 12p + 36$ is a perfect square trinomial because $p^2 - 12p + 36 = (p - 6)^2$.

We "adjust" the left side of $x^2 + bx + c$ by adding a number to make it a perfect square trinomial. For example, to make $x^2 + 6x$ a perfect square we would add 9. But where does this 9 come from? If we divide the coefficient on the first-degree term, 6, by 2, and then square the result, we obtain 9. This approach works in general.

Work Smart

To complete the square, the coefficient of x^2 must be 1.

OBTAINING A PERFECT SQUARE TRINOMIAL

Identify the coefficient of the first-degree term. Multiply this coefficient by $\dfrac{1}{2}$ and then square the result. That is, determine the value of b in $x^2 + bx + c$ and compute $\left(\dfrac{1}{2}b\right)^2$.

EXAMPLE 5   Obtaining a Perfect Square Trinomial

Determine the number that must be added to each expression in order to make it a perfect square trinomial. Then factor the expression.

Start	Add	Result	Factored Form
$y^2 + 8y$	$\left(\dfrac{1}{2} \cdot 8\right)^2 = 16$	$y^2 + 8y + 16$	$(y + 4)^2$
$x^2 + 12x$	$\left(\dfrac{1}{2} \cdot 12\right)^2 = 36$	$x^2 + 12x + 36$	$(x + 6)^2$
$a^2 - 20a$	$\left(\dfrac{1}{2} \cdot (-20)\right)^2 = 100$	$a^2 - 20a + 100$	$(a - 10)^2$
$p^2 - 5p$	$\left(\dfrac{1}{2} \cdot (-5)\right)^2 = \dfrac{25}{4}$	$p^2 - 5p + \dfrac{25}{4}$	$\left(p - \dfrac{5}{2}\right)^2$

Work Smart

It is common to write the value $\left(\dfrac{1}{2}b\right)^2$ as a fraction, not a decimal.

Did you notice in the factored form that the perfect square trinomial always factors so that

$$x^2 + bx + \left(\frac{b}{2}\right)^2 = \left(x + \frac{b}{2}\right)^2 \quad \text{or} \quad x^2 - bx + \left(\frac{b}{2}\right)^2 = \left(x - \frac{b}{2}\right)^2$$

That is, the perfect square trinomial will always factor as $\left(x \pm \dfrac{b}{2}\right)^2$, where we use the + if the coefficient of the first-degree term is positive and we use the − if the coefficient of the first-degree term is negative. The $\dfrac{b}{2}$ represents $\dfrac{1}{2}$ the value of the coefficient of the first-degree term.

QUICK ✓ *Determine the number that must be added to the expression to make it a perfect square trinomial. Then factor the expression.*

8. $p^2 + 14p$ **9.** $w^2 + 3w$

Figure 1

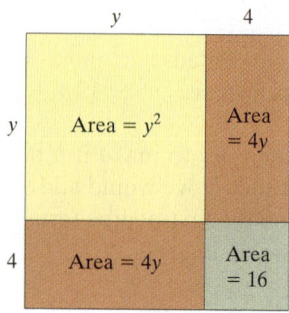

Are you wondering why we call making an expression a perfect square trinomial "completing the square"? Consider the expression $y^2 + 8y$ given in Example 5. We can geometrically represent this algebraic expression as shown in Figure 1. The yellow area is y^2 and each orange area is $4y$ (for a total area of $8y$). But what is the area of the green region in order to make the square complete? The dimensions of the green region must be 4 by 4, so the area of the green region is 16. The area of the entire square region, $(y + 4)^2$, equals the sum of the area of the regions that make up the square: $y^2 + 4y + 4y + 16 = y^2 + 8y + 16$.

③ **Solve Quadratic Equations by Completing the Square**

Up to this point, we have only been able to solve quadratic equations of the form $ax^2 + bx + c = 0$ when $ax^2 + bx + c$ was factorable. This raises the question, Is there a method to solve $ax^2 + bx + c = 0$ when $ax^2 + bx + c$ is not factorable? The answer is yes! We begin by presenting a method for solving quadratic equations of the form $x^2 + bx + c = 0$. That is, the coefficient of the square term is 1.

| **EXAMPLE 6** | **How to Solve a Quadratic Equation by Completing the Square** |

Solve: $x^2 + 6x + 1 = 0$

Step-by-Step Solution

Step 1: Rewrite $x^2 + bx + c = 0$ as $x^2 + bx = -c$ by subtracting the constant from both sides of the equation.	Subtract 1 from both sides:	$x^2 + 6x + 1 = 0$ $x^2 + 6x = -1$
Step 2: Complete the square in the expression $x^2 + bx$ by making it a perfect square trinomial.	$\left(\dfrac{1}{2} \cdot 6\right)^2 = 9$; Add 9 to both sides:	$x^2 + 6x + 9 = -1 + 9$ $x^2 + 6x + 9 = 8$
Step 3: Factor the perfect square trinomial on the left side of the equation.	$x^2 + 6x + 9 = (x + 3)^2$:	$(x + 3)^2 = 8$
Step 4: Solve the equation using the Square Root Property.	$\sqrt{8} = 2\sqrt{2}$: Subtract 3 from both sides: $a \pm b$ means $a - b$ or $a + b$:	$x + 3 = \pm\sqrt{8}$ $x + 3 = \pm 2\sqrt{2}$ $x = -3 \pm 2\sqrt{2}$ $x = -3 - 2\sqrt{2}$ or $x = -3 + 2\sqrt{2}$

Step 5: Verify your solution(s).

$$x^2 + 6x + 1 = 0$$

$x = -3 - 2\sqrt{2}$:

$$\left(-3 - 2\sqrt{2}\right)^2 + 6\left(-3 - 2\sqrt{2}\right) + 1 \stackrel{?}{=} 0$$
$$9 + 12\sqrt{2} + 8 - 18 - 12\sqrt{2} + 1 \stackrel{?}{=} 0$$
$$0 = 0 \quad \text{True}$$

$x = -3 + 2\sqrt{2}$:

$$\left(-3 + 2\sqrt{2}\right)^2 + 6\left(-3 + 2\sqrt{2}\right) + 1 \stackrel{?}{=} 0$$
$$9 - 12\sqrt{2} + 8 - 18 + 12\sqrt{2} + 1 \stackrel{?}{=} 0$$
$$0 = 0 \quad \text{True}$$

The solution set is $\left\{-3 - 2\sqrt{2}, -3 + 2\sqrt{2}\right\}$.

The following is a summary of the steps used to solve a quadratic equation by completing the square.

Solving a Quadratic Equation by Completing the Square

Step 1: Rewrite $x^2 + bx + c = 0$ as $x^2 + bx = -c$ by subtracting the constant from both sides of the equation.

Step 2: Complete the square in the expression $x^2 + bx$ by making it a perfect square trinomial. Don't forget, whatever you add to the left side of the equation must also be added to the right side.

Step 3: Factor the perfect square trinomial on the left side of the equation.

Step 4: Solve the equation using the Square Root Property.

Step 5: Verify your solutions.

QUICK ✓ *Solve the equation by completing the square.*

10. $b^2 + 2b - 8 = 0$

Work Smart

If the coefficient of the square term is not 1, we must divide each side of the equation by the coefficient of the square term so that it becomes 1 before using the method of completing the square.

Up to this point we have only looked at quadratic equations where the coefficient of the square term is 1. When the coefficient of the square term is not 1, we multiply or divide both sides of the equation by a nonzero constant so this coefficient becomes 1. The next example demonstrates this method.

EXAMPLE 7 **Solving a Quadratic Equation by Completing the Square When the Coefficient of the Square Term Is Not 1**

Solve: $2x^2 + 4x + 3 = 0$

Solution

First, we notice that the coefficient of the square term is 2, so we divide both sides of the equation by 2 in order to make the coefficient of the square term equal to 1.

$$2x^2 + 4x + 3 = 0$$

Divide both sides of the equation by 2: $\dfrac{2x^2 + 4x + 3}{2} = \dfrac{0}{2}$

Simplify: $x^2 + 2x + \dfrac{3}{2} = 0$

Now we are ready to solve the equation by completing the square.

Subtract $\dfrac{3}{2}$ from both sides: $\qquad x^2 + 2x = -\dfrac{3}{2}$

$\left(\dfrac{1}{2} \cdot 2\right)^2 = 1$; Add 1 to both sides: $\quad x^2 + 2x + 1 = -\dfrac{3}{2} + 1$

Simplify: $\quad x^2 + 2x + 1 = -\dfrac{1}{2}$

Factor expression on left: $\qquad (x + 1)^2 = -\dfrac{1}{2}$

Use Square Root Property: $\qquad x + 1 = \pm\sqrt{-\dfrac{1}{2}}$

$\sqrt{\dfrac{a}{b}} = \dfrac{\sqrt{a}}{\sqrt{b}}$: $\qquad x + 1 = \pm\dfrac{\sqrt{-1}}{\sqrt{2}}$

$\dfrac{\sqrt{-1}}{\sqrt{2}} = \dfrac{i}{\sqrt{2}} = \dfrac{\sqrt{2}}{2}i$: $\qquad x + 1 = \pm\dfrac{\sqrt{2}}{2}i$

Subtract 1 from both sides: $\qquad x = -1 \pm \dfrac{\sqrt{2}}{2}i$

$a \pm b$ means $a - b$ or $a + b$: $\qquad x = -1 - \dfrac{\sqrt{2}}{2}i \quad$ or $\quad x = -1 + \dfrac{\sqrt{2}}{2}i$

Work Smart

Notice the solutions in Example 7 are conjugates of each other.

We leave it to you to verify the solutions. The solution set is

$$\left\{-1 - \dfrac{\sqrt{2}}{2}i, \; -1 + \dfrac{\sqrt{2}}{2}i\right\}.$$

QUICK ✓ *Solve the quadratic equation by completing the square.*

11. $2q^2 + 6q - 1 = 0$

---

④ **Solve Problems Using the Pythagorean Theorem**

Figure 2

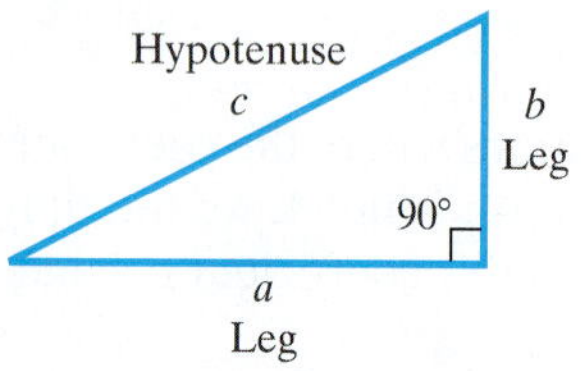

Hypotenuse
c
b
Leg
$90°$
a
Leg

The Pythagorean Theorem is a statement about *right triangles*. A **right triangle** is one that contains a **right angle,** that is, an angle of 90°. The side of the triangle opposite the 90° angle is called the **hypotenuse;** the remaining two sides are called **legs.** In Figure 2 we use c to represent the length of the hypotenuse and a and b to represent the lengths of the legs. Notice the use of the symbol ⌐ to show the 90° angle.

We now state the Pythagorean Theorem.

THE PYTHAGOREAN THEOREM

In a right triangle, the square of the length of the hypotenuse is equal to the sum of the squares of the lengths of the legs. That is, in the right triangle shown in Figure 2,

$$c^2 = a^2 + b^2$$

EXAMPLE 8 **Finding the Hypotenuse of a Right Triangle**

In a right triangle, one leg is of length 5 inches and the other is of length 12 inches. What is the length of the hypotenuse?

Solution

Since the triangle is a right triangle, we use the Pythagorean Theorem with $a = 5$ and $b = 12$ to find the length c of the hypotenuse.

$$c^2 = a^2 + b^2$$
$$c^2 = 5^2 + 12^2$$
$$= 25 + 144$$
$$= 169$$

We now use the Square Root Property to find c, the length of the hypotenuse.

$$c = \sqrt{169} = 13$$

The length of the hypotenuse is 13 inches. Notice that we only find the positive square root of 169 since c represents the length of a side of a triangle and a negative length does not make sense. ▬

QUICK ✓ *The lengths of the legs of a right triangle are given. Find the length of the hypotenuse.*

12. $a = 3, b = 4$ **13.** $a = 6, b = 6$

EXAMPLE 9 **How Far Can You See?**

The Currituck Lighthouse is located in Corolla, North Carolina. As part of North Carolina's Outer Banks, the lighthouse was completed in 1875. It stands 162 feet tall with the observation deck located 158 feet above the ground. See Figure 3.

Figure 3

The Web site for the Currituck Lighthouse states that if a person were standing on the observation deck, they could see approximately 18 miles. See Figure 4. Assuming that the radius of the Earth is 3960 miles, verify this claim.

Figure 4

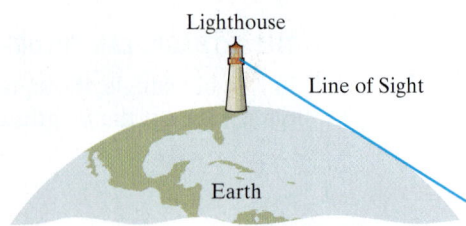

Solution

Step 1: Identify We want to know how far a person can see from the lighthouse.

Step 2: Name We will call this unknown distance, d.

Step 3: Translate To help with the translation, we draw a picture. From the center of Earth, draw two lines: one through the lighthouse and the other to the farthest point a person can see from the lighthouse. See Figure 5.

Figure 5

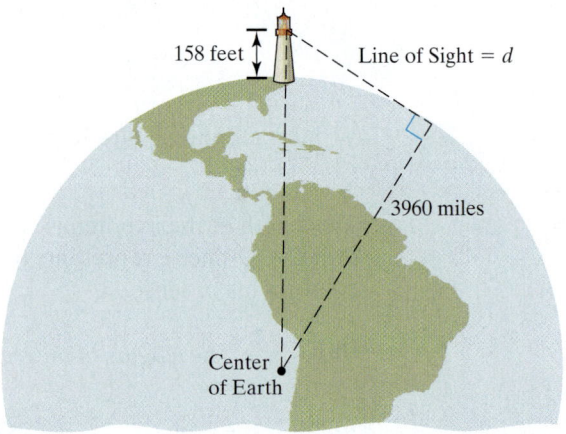

The line of sight and the two lines drawn from the center of Earth form a right triangle. So the angle where the line of sight touches the horizon measures 90°. From the Pythagorean Theorem, we know that

$$\text{Hypotenuse}^2 = \text{Leg}^2 + \text{Leg}^2$$

The length of the hypotenuse is 3960 miles plus 158 feet. We can't add 3960 miles to 158 feet without first converting the height of the tower to miles. Since 158 feet $=$ 158 feet $\cdot \dfrac{1 \text{ mile}}{5280 \text{ feet}} = \dfrac{158}{5280}$ mile, we have that the hypotenuse is $\left(3960 + \dfrac{158}{5280} \right)$ miles. One of the legs is 3960 miles. The length of the other leg is our unknown, d. So we have

$$3960^2 + d^2 = \left(3960 + \frac{158}{5280} \right)^2 \qquad \textcolor{red}{\text{The Model}}$$

Step 4: Solve

$$3960^2 + d^2 = \left(3960 + \frac{158}{5280} \right)^2$$

$$\textcolor{red}{\text{Subtract } 3960^2 \text{ from both sides:}} \quad d^2 = \left(3960 + \frac{158}{5280} \right)^2 - 3960^2$$

$$\textcolor{red}{\text{Use a calculator:}} \quad d^2 \approx 237.000895$$

$$\textcolor{red}{\text{Square Root Property:}} \quad d \approx \sqrt{237.000895}$$

$$\approx 15.39 \text{ miles}$$

Step 5: Check Our answer is less than the distance given on the Web site.

Step 6: Answer The distance given on the Currituck Lighthouse Web site appears to overstate the actual distance a person could see. Someone standing on the observation deck of the lighthouse could see about 15.39 miles.

QUICK ✓

14. The USS Constitution (aka *Old Ironsides*) is perhaps the most famous ship from United States Naval history. The mainmast of the Constitution is 220 feet high. Suppose that a sailor climbs the mainmast to a height of 200 feet in order to look for enemy vessels. How far could the sailor see? Assume the radius of the earth is 3960 miles.

10.1 Exercises

For Extra Help: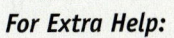

Student Solutions Manual CD Video PH Math/Tutor Center MathXL Tutorials on CD MathXL® MyMathLab

Concepts and Vocabulary

In Problems 1–3, fill in the blanks.

1. If $x^2 = p$, then $x =$ _____ or $x =$ _____.

2. The idea behind completing the square is to adjust the left side of a quadratic equation in order to make it a _____ _____ _____.

3. The side of a right triangle opposite the 90° angle is called the _____; the remaining two sides are called _____.

In Problems 4–6, answer True or False to each statement.

4. The solution set of the equation $(z - 2)^2 - 25 = 0$ is $\{-3, 7\}$.

5. The method of completing the square to solve a quadratic equation only works when the coefficient on the square term is 1.

6. The Pythagorean Theorem states that for any triangle, the length of the hypotenuse is equal to the sum of the squares of the lengths of the legs.

7. Explain in your own words what the expression "completing the square" means. You may want to use a figure similar to Figure 1 to assist in your explanation.

8. What would be your first step in solving the quadratic equation $3x^2 - 6x + 12 = 0$? Why?

Building Skills

In Problems 9–36, solve each equation using the Square Root Property.

9. $y^2 = 100$

10. $x^2 = 81$

11. $p^2 = 50$

12. $z^2 = 48$

13. $m^2 = -25$

14. $n^2 = -49$

15. $w^2 = \dfrac{5}{4}$

16. $z^2 = \dfrac{8}{9}$

17. $x^2 + 5 = 13$

18. $w^2 - 6 = 14$

19. $3z^2 = 48$

20. $4y^2 = 100$

21. $3x^2 = 8$

22. $5y^2 = 32$

23. $2p^2 + 23 = 15$

24. $-3x^2 - 5 = 22$

25. $(z + 3)^2 = 64$

26. $(y - 2)^2 = 9$

27. $(d - 1)^2 = -18$

28. $(z + 4)^2 = -24$

29. $3(q + 5)^2 - 1 = 8$

30. $5(x - 3)^2 + 2 = 27$

31. $(3q + 1)^2 = 9$

32. $(2p + 3)^2 = 16$

33. $\left(x - \dfrac{2}{3}\right)^2 = \dfrac{5}{9}$

34. $\left(y + \dfrac{3}{2}\right)^2 = \dfrac{3}{4}$

35. $x^2 + 8x + 16 = 81$

36. $q^2 - 6q + 9 = 16$

In Problems 37–46, complete the square in each expression. Then factor the perfect square trinomial.

37. $x^2 + 10x$ **38.** $y^2 + 16y$ 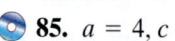 **39.** $z^2 - 18z$ **40.** $p^2 - 4p$

41. $y^2 + 7y$ **42.** $x^2 + x$ **43.** $w^2 + \dfrac{1}{2}w$ **44.** $z^2 - \dfrac{1}{3}z$

45. $q^2 - \dfrac{3}{5}q$ **46.** $m^2 + \dfrac{5}{2}m$

In Problems 47–74, solve each quadratic equation by completing the square.

47. $w^2 - 5w = 14$ **48.** $z^2 - 6z = 7$ **49.** $x^2 + 4x - 12 = 0$

50. $y^2 + 3y - 18 = 0$ **51.** $x^2 - 4x + 1 = 0$ **52.** $p^2 - 6p + 4 = 0$

53. $z^2 + 8z + 9 = 0$ **54.** $b^2 + 10b + 19 = 0$ **55.** $a^2 - 4a + 5 = 0$

56. $m^2 - 2m + 5 = 0$ **57.** $b^2 + 5b - 2 = 0$ **58.** $q^2 + 7q + 7 = 0$

59. $p^2 - 3p - 2 = 0$ **60.** $x^2 - 5x - 3 = 0$ **61.** $m^2 = 8m + 3$

62. $n^2 = 10n + 5$ **63.** $p^2 - p + 3 = 0$ **64.** $z^2 - 3z + 5 = 0$

65. $2y^2 - 5y - 12 = 0$ **66.** $3a^2 - 4a - 4 = 0$ **67.** $3y^2 - 6y + 2 = 0$

68. $2y^2 - 2y - 1 = 0$ **69.** $2z^2 - 5z + 1 = 0$ **70.** $2x^2 - 7x + 2 = 0$

71. $2x^2 + 4x + 5 = 0$ **72.** $2z^2 + 6z + 5 = 0$ **73.** $3b^2 + b - 5 = 0$

74. $3m^2 + 2m - 7 = 0$

In Problems 75–84, the lengths of the legs of a right triangle are given. Find the hypotenuse. Give exact answers and decimal approximations rounded to two decimal places.

75. $a = 6, b = 8$ **76.** $a = 7, b = 24$ **77.** $a = 12, b = 16$

78. $a = 15, b = 8$ **79.** $a = 5, b = 5$ **80.** $a = 3, b = 3$

81. $a = 1, b = \sqrt{3}$ **82.** $a = 2, b = \sqrt{5}$ **83.** $a = 6, b = 10$

84. $a = 8, b = 10$

In Problems 85–88, use the right triangle shown to the right and find the missing length. Give exact answers and decimal approximations rounded to two decimal places.

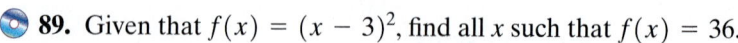

 85. $a = 4, c = 8$ **86.** $a = 4, c = 10$

87. $b = 8, c = 12$ **88.** $b = 2, c = 10$

 **89.** Given that $f(x) = (x - 3)^2$, find all x such that $f(x) = 36$.

90. Given that $f(x) = (x - 5)^2$, find all x such that $f(x) = 49$.

91. Given that $g(x) = (x + 2)^2$, find all x such that $g(x) = 18$.

92. Given that $h(x) = (x + 1)^2$, find all x such that $h(x) = 32$.

Applying the Concepts

In Problems 93 and 94, find the length of the diagonal in each figure.

93.

8

4

94.

6

9

95. Golf A golfer hits an errant tee shot that lands in the rough. The golfer finds that the ball is exactly 30 yards to the right of the 100 yard marker, which indicates the distance to the center of the green as shown in the figure. How far is the ball from the center of the green?

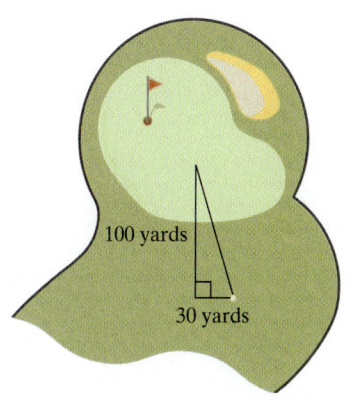

100 yards

30 yards

96. Baseball Jermaine Dye plays right field for the Chicago White Sox. He catches a fly ball 40 feet from the right field foul line, as indicated in the figure. How far is it to home plate?

97. Guy Wire A guy wire is a wire used to support telephone poles. Suppose that a guy wire is located 30 feet up a telephone pole and is anchored to the ground 10 feet from the base of the pole. How long is the guy wire?

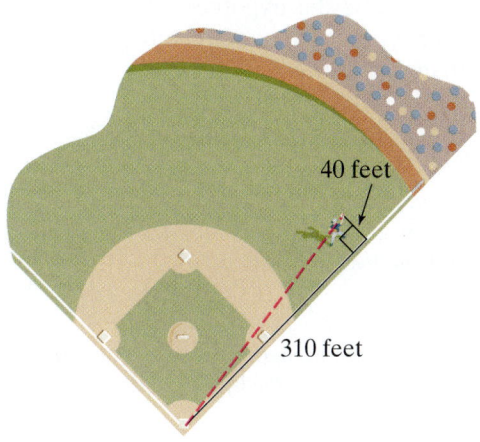

40 feet

310 feet

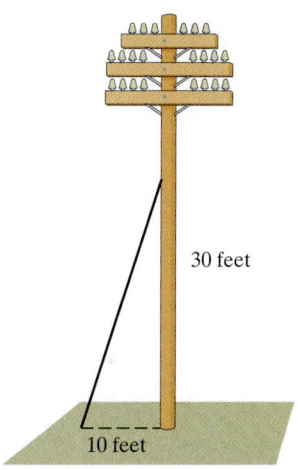

30 feet

10 feet

98. Guy Wire A guy wire is used to support an antenna on a roof top. The wire is located 40 feet up on the antenna and anchored to the roof 8 feet from the base of the antenna. What is the length of the guy wire?

99. Ladder Bob needs to wash the windows on his house. He has a 25-foot ladder and places the base of the ladder 10 feet from the wall on the house.

 (a) How far up the wall will the ladder reach?
 (b) If his windows are 20 feet above the ground, what is the farthest distance the base of the ladder can be from the wall?

100. Fire Truck Ladder A fire truck has a 75-foot ladder. If the truck can safely park 20 feet from a building, how far up the building can the ladder reach assuming that the top of the base of the ladder is resting on top of the truck and the truck is 10 feet tall?

101. Gravity The distance s that an object falls (in feet) after t seconds ignoring air resistance is given by the equation $s = 16t^2$.

 (a) How long does it take an object to fall 16 feet?
 (b) How long does it take an object to fall 48 feet?
 (c) How long does it take an object to fall 64 feet?

△ **102. Equilateral Triangles** An equilateral triangle is one whose sides are all the same length. The area A of an equilateral triangle whose sides are each length x is given by $A = \dfrac{\sqrt{3}}{4}x^2$.

(a) What is the length of each side of an equilateral triangle whose area is $\dfrac{8\sqrt{3}}{9}$ square feet?

(b) What is the length of each side of an equilateral triangle whose area is $\dfrac{25\sqrt{3}}{4}$ square meters?

Problems 103 and 104 are based on the following discussion. If P dollars are invested today at an annual interest rate r compounded once a year, then the value of the account A after 2 years is given by the formula $A = P(1 + r)^2$.

103. Value of Money Find the rate of interest required to turn an investment of $1000 into $1200 after 2 years. Express your answer as a percent rounded to two decimal places.

104. Value of Money Find the rate of interest required to turn an investment of $5000 into $6200 after 2 years. Express your answer as a percent rounded to two decimal places.

Extending the Concepts

If you look carefully at the Pythagorean Theorem, it states, "If we have a right triangle, then $c^2 = a^2 + b^2$, where c is the length of the hypotenuse." In this theorem "a right triangle" represents the hypothesis, while "$c^2 = a^2 + b^2$" represents the conclusion. The **converse** of a theorem interchanges the hypothesis and conclusion. The converse of the Pythagorean Theorem is true.

CONVERSE OF THE PYTHAGOREAN THEOREM

In a triangle, if the square of the length of one side equals the sum of the squares of the lengths of the other two sides, then the triangle is a right triangle. The 90° angle is opposite the longest side.

In Problems 105–108, the lengths of the sides of a triangle are given. Determine if the triangle is a right triangle. If it is, identify the hypotenuse.

105. 8, 15, 17

106. 4, 6, 8

107. 14, 18, 20

108. 20, 48, 52

109. Pythagorean Triples Suppose that m and n are positive integers with $m > n$. If $a = m^2 - n^2$, $b = 2mn$, and $c = m^2 + n^2$, show that a, b, and c are the lengths of the sides of a right triangle using the Converse of the Pythagorean Theorem. We call any numbers a, b, and c found from the above formulas **Pythagorean Triples.**

110. Solve $ax^2 + bx + c = 0$ for x by completing the square.

Synthesis Review

In Problems 111–114, solve each equation.

111. $a^2 - 5a - 36 = 0$

112. $p^2 + 4p = 32$

113. $|4q + 1| = 3$

114. $\left| \dfrac{3}{4}w - \dfrac{2}{3} \right| = \dfrac{5}{2}$

115. In Problems 111–114, you are asked to solve quadratic and absolute value equations. In solving both types of equations, we reduce the equation to a simpler equation. What is the simpler equation?

10.2 Solving Quadratic Equations by the Quadratic Formula

OBJECTIVES

1. Solve Quadratic Equations Using the Quadratic Formula
2. Use the Discriminant to Determine the Nature of Solutions in a Quadratic Equation
3. Model and Solve Problems Involving Quadratic Equations

Preparing for *Solving Quadratic Equations by the Quadratic Formula*

Before getting started, take the following readiness quiz. If you get a problem wrong, go back to the section cited and review the material.

1. Simplify: (a) $\sqrt{54}$ (b) $\sqrt{121}$ [Section 9.1, pp. 660–661; Section 9.4, pp. 681–686]

2. Simplify: (a) $\sqrt{-9}$ (b) $\sqrt{-72}$ [Section 9.9, pp. 723–724]

3. Simplify: $\dfrac{3 + \sqrt{18}}{6}$ [Section 9.4, p. 683]

At this stage of the course, we know three methods for solving quadratic equations: (1) factoring, (2) The Square Root Property, and (3) completing the square. Why do we need three methods? Well, each method provides the "quickest" route to the solution when used appropriately. For example, if the quadratic expression is easy to factor, the method of factoring will get you to the solution faster than the other two. If the equation is in the form $x^2 = p$, using the Square Root Property is fastest. If the quadratic expression is not factorable, we have no choice but to complete the square. But the method of completing the square is tedious. So you may be asking yourself if there is an alternative to this method. The answer is yes!

1   Solve Quadratic Equations Using the Quadratic Formula

We can use the method of completing the square to obtain a general formula for solving the quadratic equation

$$ax^2 + bx + c = 0, \qquad a \neq 0$$

To complete the square, we first must get the constant c on the right-hand side of the equation.

$$ax^2 + bx = -c, \qquad a \neq 0$$

Since $a \neq 0$, we can divide both sides of the equation by a to get

$$x^2 + \frac{b}{a}x = -\frac{c}{a}$$

Now that the coefficient of x^2 is 1, we complete the square on the left side by adding the square of $\frac{1}{2}$ of the coefficient of x to both sides of the equation. That is, we add

$$\left(\frac{1}{2} \cdot \frac{b}{a}\right)^2 = \frac{b^2}{4a^2}$$

to both sides of the equation. We now have

$$x^2 + \frac{b}{a}x + \frac{b^2}{4a^2} = -\frac{c}{a} + \frac{b^2}{4a^2}$$

or

$$x^2 + \frac{b}{a}x + \frac{b^2}{4a^2} = \frac{b^2}{4a^2} - \frac{c}{a}$$

We combine like terms on the right-hand side by writing the right-hand side over a common denominator. The least common denominator on the right-hand side is $4a^2$. So we multiply $-\dfrac{c}{a}$ by $\dfrac{4a}{4a}$:

$$x^2 + \frac{b}{a}x + \frac{b^2}{4a^2} = \frac{b^2}{4a^2} - \frac{c}{a}\cdot\frac{4a}{4a}$$

$$x^2 + \frac{b}{a}x + \frac{b^2}{4a^2} = \frac{b^2}{4a^2} - \frac{4ac}{4a^2}$$

$$x^2 + \frac{b}{a}x + \frac{b^2}{4a^2} = \frac{b^2 - 4ac}{4a^2}$$

Work Smart
To factor any perfect square trinomial of the form $x^2 + bx + c$, we write $\left(x + \dfrac{b}{2}\right)^2$.

The expression on the left-hand side is a perfect square trinomial. We factor the left-hand side and obtain

$$\left(x + \frac{b}{2a}\right)^2 = \frac{b^2 - 4ac}{4a^2}$$

At this point we will assume that $a > 0$ (you'll see why in a little while). This assumption does not compromise the results because if $a < 0$, we could multiply both sides of the equation $ax^2 + bx + c = 0$ by -1 to make it positive. With this assumption, we use the Square Root Property and get

$$x + \frac{b}{2a} = \pm\sqrt{\frac{b^2 - 4ac}{4a^2}}$$

$$\sqrt{\frac{a}{b}} = \frac{\sqrt{a}}{\sqrt{b}}: \quad x + \frac{b}{2a} = \pm\frac{\sqrt{b^2 - 4ac}}{\sqrt{4a^2}}$$

$$\sqrt{4a^2} = 2a \text{ since } a > 0: \quad x + \frac{b}{2a} = \pm\frac{\sqrt{b^2 - 4ac}}{2a}$$

$$\text{Subtract } \frac{b}{2a} \text{ from both sides:} \quad x = -\frac{b}{2a} \pm \frac{\sqrt{b^2 - 4ac}}{2a}$$

$$\text{Write over a common denominator:} \quad x = \frac{-b \pm \sqrt{b^2 - 4ac}}{2a}$$

In Words
The quadratic formula says that the solution(s) to the equation $ax^2 + bx + c = 0$ is (are) "the opposite of b plus or minus the square root of b squared minus $4ac$ all over $2a$."

This gives us the *quadratic formula*.

Work Smart: Study Skill
When solving homework problems always write the quadratic formula as part of the solution so you "accidentally" memorize the formula.

THE QUADRATIC FORMULA

The solution(s) to the quadratic equation $ax^2 + bx + c = 0$, $a \neq 0$, are given by the **quadratic formula**

$$x = \frac{-b \pm \sqrt{b^2 - 4ac}}{2a}$$

EXAMPLE 1 | **How to Solve a Quadratic Equation Using the Quadratic Formula**

Solve: $12x^2 + 5x - 3 = 0$

Step-by-Step Solution

Step 1: Write the equation in standard form $ax^2 + bx + c = 0$ and identify the values of a, b, and c.	$a = 12$ $b = 5$ $c = -3$ $12x^2 + 5x - 3 = 0$
Step 2: Substitute the values of a, b, and c into the quadratic formula.	$x = \dfrac{-b \pm \sqrt{b^2 - 4ac}}{2a}$ $x = \dfrac{-5 \pm \sqrt{5^2 - 4(12)(-3)}}{2(12)}$

Step 3: Simplify the expression found in Step 2.

$$= \frac{-5 \pm \sqrt{25 + 144}}{24}$$

$$= \frac{-5 \pm \sqrt{169}}{24}$$

$\sqrt{169} = 13$: $\quad = \dfrac{-5 \pm 13}{24}$

$a \pm b$ means $a - b$ or $a + b$: $\quad x = \dfrac{-5 - 13}{24} \quad \text{or} \quad x = \dfrac{-5 + 13}{24}$

$$= \frac{-18}{24} \quad \text{or} \quad = \frac{8}{24}$$

Reduce: $\quad = -\dfrac{3}{4} \quad \text{or} \quad = \dfrac{1}{3}$

Step 4: Check

$$12x^2 + 5x - 3 = 0$$

$x = -\dfrac{3}{4}$: $12\left(-\dfrac{3}{4}\right)^2 + 5\left(-\dfrac{3}{4}\right) - 3 \overset{?}{=} 0$ $\qquad$ $x = \dfrac{1}{3}$: $12\left(\dfrac{1}{3}\right)^2 + 5\left(\dfrac{1}{3}\right) - 3 \overset{?}{=} 0$

$12 \cdot \dfrac{9}{16} - \dfrac{15}{4} - 3 \overset{?}{=} 0$ $\qquad\qquad\qquad$ $12 \cdot \dfrac{1}{9} + \dfrac{5}{3} - 3 \overset{?}{=} 0$

$\dfrac{27}{4} - \dfrac{15}{4} - 3 \overset{?}{=} 0$ $\qquad\qquad\qquad\quad$ $\dfrac{4}{3} + \dfrac{5}{3} - 3 \overset{?}{=} 0$

$\dfrac{12}{4} - 3 \overset{?}{=} 0$ $\qquad\qquad\qquad\qquad\quad$ $\dfrac{9}{3} - 3 \overset{?}{=} 0$

$0 = 0$ True $\qquad\qquad\qquad\qquad\qquad$ $0 = 0$ True

The solution set is $\left\{ -\dfrac{3}{4}, \dfrac{1}{3} \right\}$.

Work Smart

If $b^2 - 4ac$ is a perfect square, then the quadratic equation can be solved by factoring.

Notice that the solutions to the equation in Example 1 are rational numbers and the expression $b^2 - 4ac$ under the radical in the quadratic formula, 169, is a perfect square. This leads to a generalization. Whenever the expression $b^2 - 4ac$ is a perfect square, then the quadratic equation will have rational solutions and the quadratic equation can be solved by factoring.

We summarize the steps used to solve a quadratic equation using the quadratic formula.

Solving a Quadratic Equation Using the Quadratic Formula

Step 1: Write the equation in standard form $ax^2 + bx + c = 0$ and identify the values of a, b, and c.

Step 2: Substitute the values of a, b, and c into the quadratic formula.

Step 3: Simplify the expression found in Step 2.

Step 4: Verify your solution(s).

QUICK ✓ *Solve each equation using the quadratic formula.*

1. $2x^2 - 3x - 9 = 0$ $\qquad\qquad$ **2.** $2x^2 + 7x = 4$

EXAMPLE 2 Solving a Quadratic Equation Using the Quadratic Formula

Solve: $3p^2 = 6p - 1$

Solution

First, we must write the equation in standard form to identify a, b, and c.

$$3p^2 = 6p - 1$$

Subtract $6p$ from both sides;
Add 1 to both sides: $3p^2 - 6p + 1 = 0$

The variable in the equation is p,
so write "$p = $": $p = \dfrac{-b \pm \sqrt{b^2 - 4ac}}{2a}$

$a = 3$, $b = -6$, $c = 1$: $p = \dfrac{-(-6) \pm \sqrt{(-6)^2 - 4(3)(1)}}{2(3)}$

$= \dfrac{6 \pm \sqrt{36 - 12}}{6}$

$= \dfrac{6 \pm \sqrt{24}}{6}$

$\sqrt{24} = \sqrt{4 \cdot 6} = 2\sqrt{6}$: $= \dfrac{6 \pm 2\sqrt{6}}{6}$

$\dfrac{a + b}{c} = \dfrac{a}{c} + \dfrac{b}{c}$: $= \dfrac{6}{6} \pm \dfrac{2\sqrt{6}}{6}$

Simplify: $= 1 \pm \dfrac{\sqrt{6}}{3}$

$a \pm b$ means $a - b$ or $a + b$: $p = 1 - \dfrac{\sqrt{6}}{3}$ or $p = 1 + \dfrac{\sqrt{6}}{3}$

Work Smart

We could also simplify $\dfrac{6 \pm 2\sqrt{6}}{6}$ by factoring:

$\dfrac{6 \pm 2\sqrt{6}}{6} = \dfrac{2(3 \pm \sqrt{2})}{6}$

$= \dfrac{3 \pm \sqrt{2}}{3}$

This is equivalent to $1 \pm \dfrac{\sqrt{6}}{3}$.
Ask your instructor which form of the solution is preferred, if any.

We leave it to you to verify the solutions. The solution set is $\left\{ 1 - \dfrac{\sqrt{6}}{3},\ 1 + \dfrac{\sqrt{6}}{3} \right\}$.

Notice in Example 2 that the value of $b^2 - 4ac$ is positive, but not a perfect square. There are two solutions to the quadratic equation and they are irrational.

QUICK ✓

3. Solve: $4z^2 + 1 = 8z$

EXAMPLE 3 Solving a Rational Equation That Leads to a Quadratic Equation

Solve: $9m + \dfrac{4}{m} = 12$

Solution

We first note that m cannot equal 0. To clear the equation of rational expressions, we multiply both sides of the equation by the LCD, m.

$$m\left(9m + \frac{4}{m}\right) = 12 \cdot m$$

Distribute the m:

$$9m^2 + 4 = 12m$$

Subtract $12m$ from both sides:

$$9m^2 - 12m + 4 = 0$$

$$m = \frac{-b \pm \sqrt{b^2 - 4ac}}{2a}$$

$a = 9, b = -12, c = 4:$

$$m = \frac{-(-12) \pm \sqrt{(-12)^2 - 4(9)(4)}}{2(9)}$$

$$= \frac{12 \pm \sqrt{144 - 144}}{18}$$

$$= \frac{12 \pm \sqrt{0}}{18}$$

$$= \frac{12}{18} = \frac{2}{3}$$

We leave it to you to verify the solution. The solution set is $\left\{\dfrac{2}{3}\right\}$.

In Example 3, we had one solution rather than two (as in Examples 1 and 2). In fact, the solution of $\dfrac{2}{3}$ is called a **repeated root** because it actually occurs twice! To see why, we solve the equation given in Example 3 by factoring.

$$9m^2 + 4 = 12m$$

Subtract $12m$ from both sides:

$$9m^2 - 12m + 4 = 0$$

$$(3m - 2)(3m - 2) = 0$$

Zero-Product Property:

$$3m - 2 = 0 \quad \text{or} \quad 3m - 2 = 0$$

$$m = \frac{2}{3} \quad \text{or} \quad m = \frac{2}{3}$$

So we obtain two solutions because the expression $9m^2 - 12m + 4$ is a perfect square trinomial. We know that the equation $9m^2 - 12m + 4 = 0$ has a repeated root because the value $b^2 - 4ac$ equals 0. We will have more to say about this soon.

QUICK ✔

4. Solve: $4w + \dfrac{25}{w} = 20$

EXAMPLE 4   **Solving a Quadratic Equation Using the Quadratic Formula**

Solve: $y^2 - 4y + 13 = 0$

Solution

$$y^2 - 4y + 13 = 0$$

$$1y^2 - 4y + 13 = 0$$

$$y = \frac{-b \pm \sqrt{b^2 - 4ac}}{2a}$$

$a = 1, b = -4, c = 13:$

$$y = \frac{-(-4) \pm \sqrt{(-4)^2 - 4(1)(13)}}{2(1)}$$

$$= \frac{4 \pm \sqrt{16 - 52}}{2}$$

$\sqrt{-36} = 6i$: $= \dfrac{4 \pm \sqrt{-36}}{2} = \dfrac{4 \pm 6i}{2}$

$\dfrac{a+b}{c} = \dfrac{a}{c} + \dfrac{b}{c}$: $= \dfrac{4}{2} \pm \dfrac{6}{2}i = 2 \pm 3i$

$a \pm b$ means $a - b$ or $a + b$: $x = 2 - 3i$ or $x = 2 + 3i$

We leave it to you to verify the solution. The solution set is $\{2 - 3i, 2 + 3i\}$.   ■

Notice in Example 4 that the value of $b^2 - 4ac$ is negative and the equation has two complex solutions that are not real.

QUICK ✓

5. Solve: $z^2 + 2z + 26 = 0$

---

(2) ## Use the Discriminant to Determine the Nature of Solutions in a Quadratic Equation

In the quadratic formula $x = \dfrac{-b \pm \sqrt{b^2 - 4ac}}{2a}$, the quantity $b^2 - 4ac$ is called the **discriminant** of the quadratic equation, because its value tells us the number of solutions and the type of solution to expect from the quadratic formula.

> **THE DISCRIMINANT AND THE NATURE OF THE SOLUTION OF A QUADRATIC EQUATION**
>
> For a quadratic equation $ax^2 + bx + c = 0$, the discriminant $b^2 - 4ac$ can be used to describe the nature of the solution as shown:
>
> | Discriminant | Number of Solutions | Type of Solution | Example |
> |---|---|---|---|
> | Positive and a perfect square | 2 | Rational | 1 |
> | Positive and not a perfect square | 2 | Irrational | 2 |
> | Zero | 1 (repeated root) | Rational | 3 |
> | Negative | 2 | Complex, nonreal | 4 |

If you look back at the results of Example 4, you should notice that the solutions are complex conjugates of each other. In general, for any quadratic equation of the form $ax^2 + bx + c = 0$, where a, b, and c are real numbers and $b^2 - 4ac < 0$, the equation will have two complex solutions that are not real and are complex conjugates of each other.

This result is a consequence of the quadratic formula. Suppose that $b^2 - 4ac = -N < 0$. Then, by the quadratic formula, the solutions are

$$x = \dfrac{-b + \sqrt{b^2 - 4ac}}{2a} = \dfrac{-b + \sqrt{-N}}{2a}$$

$$= \dfrac{-b + \sqrt{N}i}{2a} = \dfrac{-b}{2a} + \dfrac{\sqrt{N}}{2a}i$$

and

$$x = \dfrac{-b - \sqrt{b^2 - 4ac}}{2a} = \dfrac{-b - \sqrt{-N}}{2a}$$

$$= \dfrac{-b - \sqrt{N}i}{2a} = \dfrac{-b}{2a} - \dfrac{\sqrt{N}}{2a}i$$

which are conjugates of each other.

EXAMPLE 5 **Determining the Nature of the Solutions of a Quadratic Equation**

For each quadratic equation, determine the discriminant. Use the value of the discriminant to determine whether the quadratic equation has two unequal rational solutions, two irrational solutions, one repeated real solution, or two complex solutions that are not real.

(a) $x^2 - 5x + 2 = 0$ (b) $9y^2 + 6y + 1 = 0$ (c) $3p^2 - p = -5$

Solution

(a) We compare $x^2 - 5x + 2 = 0$ to the standard form $ax^2 + bx + c = 0$.

$$x^2 - 5x + 2 = 0$$
$$a = 1 \quad b = -5 \quad c = 2$$

We have that $a = 1$, $b = -5$, and $c = 2$. Substituting these values into the formula for the discriminant, $b^2 - 4ac$, we obtain

$$b^2 - 4ac = (-5)^2 - 4(1)(2) = 25 - 8 = 17$$

Because $b^2 - 4ac = 17$ and 17 is positive, but not a perfect square, the quadratic equation will have two irrational solutions.

(b) For the quadratic equation $9y^2 + 6y + 1 = 0$, we have that $a = 9$, $b = 6$, and $c = 1$. Substituting these values into the formula for the discriminant, $b^2 - 4ac$, we obtain

$$b^2 - 4ac = 6^2 - 4(9)(1) = 36 - 36 = 0$$

Because $b^2 - 4ac = 0$, the quadratic equation will have one repeated real solution.

(c) Is the quadratic equation $3p^2 - p = -5$ in standard form? No! We add 5 to both sides of the equation and write the equation as $3p^2 - p + 5 = 0$. So we have that $a = 3$, $b = -1$, and $c = 5$. Substituting these values into the formula for the discriminant, $b^2 - 4ac$, we obtain

$$b^2 - 4ac = (-1)^2 - 4(3)(5) = 1 - 60 = -59$$

Because $b^2 - 4ac = -59 < 0$, the quadratic equation will have two complex solutions that are not real. The solutions will be complex conjugates of each other.

QUICK ✔ *Use the value of the discriminant to determine whether the quadratic equation has two unequal real solutions, one repeated real solution, or two complex solutions that are not real.*

6. $2z^2 + 5z + 4 = 0$ **7.** $4y^2 + 12y = -9$ **8.** $2x^2 - 4x + 1 = 0$

Which Method Should I Use?

We have now introduced four methods for solving quadratic equations:

1. Factoring
2. Square Root Property
3. Completing the Square
4. The Quadratic Formula

You are probably asking yourself, "Which method should I use?" and "Does it matter which method I use?" The answer to the second question is that it does not matter which method you use, but one method may be more efficient than the others. Table 1 contains guidelines to help you solve any quadratic equation. Notice how the value of the discriminant can be used to guide us in choosing the most efficient method.

Table 1

Form of the Quadratic Equation	Method	Example
$x^2 = p$, where p is any real number	Square Root Property	$x^2 = 45$ Square Root Property: $x = \pm\sqrt{45}$ $= \pm 3\sqrt{5}$
$ax^2 + c = 0$	Square Root Property	$3p^2 + 12 = 0$ Subtract 12 from both sides: $3p^2 = -12$ Divide both sides by 3: $p^2 = -4$ Square Root Property: $p = \pm\sqrt{-4}$ $= \pm 2i$
$ax^2 + bx + c = 0$, where $b^2 - 4ac$ is a perfect square. That is, $b^2 - 4ac$ is 1, 4, 9, 16, 25, ...	Factoring or the Quadratic Formula	$a = 2, b = 1, c = -10$: $\quad 2m^2 + m - 10 = 0$ $b^2 - 4ac = 1^2 - 4(2)(-10) = 1 + 80 = 81$ 81 is a perfect square, so we can use factoring: $2m^2 + m - 10 = 0$ $(2m + 5)(m - 2) = 0$ $2m + 5 = 0 \quad$ or $\quad m - 2 = 0$ $m = -\dfrac{5}{2}$ or $\quad m = 2$
$ax^2 + bx + c = 0$, where $b^2 - 4ac$ is not a perfect square.	Quadratic Formula or Completing the Square	$a = 2, b = 4, c = -1$: $\quad 2x^2 + 4x - 1 = 0$ $b^2 - 4ac = 4^2 - 4(2)(-1) = 16 + 8 = 24$ 24 is not a perfect square, so we use the quadratic formula (since it's easier than completing the square): $x = \dfrac{-b \pm \sqrt{b^2 - 4ac}}{2a}$ $= \dfrac{-4 \pm \sqrt{24}}{2(2)}$ $= \dfrac{-4 \pm 2\sqrt{6}}{4}$ $= -1 \pm \dfrac{\sqrt{6}}{2}$ $x = -1 - \dfrac{\sqrt{6}}{2}$ or $x = -1 + \dfrac{\sqrt{6}}{2}$

Notice if the value of the discriminant is a perfect square, we can either factor or use the quadratic formula to solve the equation. We should factor if the quadratic expression is easy to factor, otherwise use the quadratic formula.

Also, you may have noticed that we did not recommend completing the square as one of the methods to use in solving a quadratic equation. This is because the quadratic formula was developed by completing the square of $ax^2 + bx + c = 0$. Besides, completing the square is a cumbersome task, whereas the quadratic formula is fairly straightforward to use. We did not waste your time by discussing completing the square, however, because it is needed to present a discussion of the quadratic formula. In addition, completing the square is a skill that you will need later in this course and in future math courses.

QUICK ✓ *Solve each quadratic equation using any method you wish.*

9. $5n^2 - 45 = 0$ **10.** $-2y^2 + 5y - 6 = 0$ **11.** $3w^2 + 2w = 5$

③ **Model and Solve Problems Involving Quadratic Equations**

Many applied problems require solving quadratic equations. In the example below, we use a quadratic equation to determine the number of units that a company must sell in order to earn a certain amount of revenue. As always, we shall employ the problem-solving strategy first presented in Section 2.5.

EXAMPLE 6   **Revenue**

The revenue R received by a company selling x specialty T-shirts per week is given by the function $R(x) = -0.005x^2 + 30x$.

(a) How many T-shirts must be sold in order for revenue to be \$25,000 per week?

(b) How many T-shirts must be sold in order for revenue to be \$45,000 per week?

Solution

(a) **Step 1: Identify** Here, we are looking to determine the number of T-shirts x required so that $R = \$25{,}000$.

Step 2: Name We know that x represents the number of T-shirts sold.

Step 3: Translate We need to solve the equation $R(x) = 25{,}000$.

$$R(x) = 25{,}000$$
$$-0.005x^2 + 30x = 25{,}000$$
$$-0.005x^2 + 30x - 25{,}000 = 0$$

Step 4: Solve $a = -0.005$, $b = 30$, $c = -25{,}000$

$$b^2 - 4ac = 30^2 - 4(-0.005)(-25{,}000)$$
$$= 400$$

Because 400 is a perfect square, we can solve the equation by factoring or using the quadratic formula. It is not obvious how to factor $-0.005x^2 + 30x - 25000$, so we will use the quadratic formula to solve the equation.

$$b^2 - 4ac = 400$$
$$\downarrow$$
$$x = \frac{-30 \pm \sqrt{400}}{2(-0.005)}$$
$$= \frac{-30 \pm 20}{-0.01}$$

$$x = \frac{-30 - 20}{-0.01} = 5000 \quad \text{or} \quad x = \frac{-30 + 20}{-0.01} = 1000$$

Step 5: Check If 1000 T-shirts are sold, then revenue is $R(1000) = -0.005(1000)^2 + 30(1000) = \$25{,}000$. If 5000 T-shirts are sold, then revenue is $R(5000) = -0.005(5000)^2 + 30(5000) = \$25{,}000$.

Step 6: Answer The company needs to sell either 1000 or 5000 T-shirts each week to earn \$25,000 in revenue.

(b) **Step 1: Identify** Here, we are looking to determine the number of T-shirts x required so that $R = \$45{,}000$. That is, we wish to solve the equation $R(x) = 45{,}000$.

Step 2: Name We know that x represents the number of T-shirts sold.

Step 3: Translate We need to solve the equation $R(x) = 45{,}000$.

$$R(x) = 45{,}000$$
$$-0.005x^2 + 30x = 45{,}000$$
$$-0.005x^2 + 30x - 45{,}000 = 0$$

Step 4: Solve $a = -0.005, b = 30, c = -45{,}000$

$$b^2 - 4ac = 30^2 - 4(-0.005)(-45{,}000) = 0$$

Because the discriminant is 0, the quadratic equation will have a single real solution. In addition, because the discriminant is 0, we can solve the equation by factoring or using the quadratic formula. It is not obvious how to factor $-0.005x^2 + 30x - 45{,}000$, so we will use the quadratic formula to solve the equation.

$$\begin{aligned} x &= \frac{-30 \pm \sqrt{0}}{2(-0.005)} \quad \color{red}{b^2 - 4ac = 0} \\ &= \frac{-30 \pm 0}{-0.01} \\ &= \frac{-30}{-0.01} = 3000 \end{aligned}$$

Step 5: Check If 3000 T-shirts are sold, then revenue is $R(3000) = -0.005(3000)^2 + 30(3000) = \$45{,}000$.

Step 6: Answer The company needs to sell 3,000 T-shirts each week to earn $45,000 in revenue.

QUICK ✓

12. The revenue R received by a video store renting x DVDs per day is given by the function $R(x) = -0.005x^2 + 4x$.

(a) How many DVDs must be rented in order for revenue to be $600 per day?

(b) How many DVDs must be rented in order for revenue to be $800 per day?

EXAMPLE 7 Designing a Window

A window designer wishes to design a window so that the diagonal is 20 feet. In addition, the length of the window needs to be 4 feet more than the height. What are the dimensions of the window?

Solution

Step 1: Identify We wish to know the dimensions of the window. That is, we want to know the length and height of the window.

Step 2: Name Let h represent the height of the window so that $h + 4$ is the length (since the length is 4 feet more than the height).

Figure 6

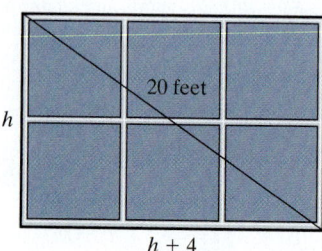

h

20 feet

$h + 4$

Step 3: Translate Figure 6 illustrates the situation. From the figure, we can see that the three sides form a right triangle. We can express the relation among the three sides using the Pythagorean Theorem.

$$\begin{aligned} \color{red}{\text{leg}^2 + \text{leg}^2 = \text{hypotenuse}^2:} & \quad h^2 + (h+4)^2 = 20^2 \\ \color{red}{\text{FOIL:}} & \quad h^2 + h^2 + 8h + 16 = 400 \\ \color{red}{\text{Combine like terms:}} & \quad 2h^2 + 8h + 16 = 400 \\ \color{red}{\text{Subtract 400 from both sides:}} & \quad 2h^2 + 8h - 384 = 0 \\ \color{red}{\text{Divide both sides by 2:}} & \quad h^2 + 4h - 192 = 0 \end{aligned}$$

Step 4: Solve In the model, we have that $a = 1, b = 4, c = -192$. The discriminant is $b^2 - 4ac = 4^2 - 4(1)(-192) = 784$ and $784 = 28^2$. So we can solve the equation by factoring.

$$h^2 + 4h - 192 = 0$$
$$(h + 16)(h - 12) = 0$$
$$h + 16 = 0 \quad \text{or} \quad h - 12 = 0$$
$$h = -16 \quad \text{or} \quad h = 12$$

Step 5: Check We disregard the solution $h = -16$ because h represents the height of the window. We see if a window whose dimensions are 12 feet by $12 + 4 = 16$ feet has a diagonal that is 20 feet by verifying that $12^2 + 16^2 = 20^2$.

$$12^2 + 16^2 \overset{?}{=} 20^2$$
$$144 + 256 \overset{?}{=} 400$$
$$400 = 400$$

Step 6: Answer The dimensions of the window are 12 feet by 16 feet.

QUICK

13. A rectangular plot of land is designed so that its length is 14 meters more than its width. The diagonal of the land is known to be 34 meters. What are the dimensions of the land?

10.2 Exercises

For Extra Help: Student Solutions Manual CD Video PH Math/Tutor Center MathXL Tutorials on CD MathXL® MyMathLab

Concepts and Vocabulary

In Problems 1–3, fill in the blanks.

1. The solution(s) to the quadratic equation $ax^2 + bx + c = 0, a \neq 0$, are given by the quadratic formula $x =$ _____.

2. In the quadratic formula, the quantity $b^2 - 4ac$ is called the _____ of the quadratic equation.

3. If the discriminant of a quadratic equation is _____, then the quadratic equation has two complex solutions that are not real.

In Problems 4–6, answer True or False to each statement.

4. If the discriminant of a quadratic equation is a perfect square, then the equation can be solved by factoring.

5. If the discriminant of a quadratic equation is zero, then the equation has no solution.

6. When solving a quadratic equation in which the solutions are complex numbers that are not real, the solutions will be complex conjugates of each other.

7. Explain the circumstances for which you would use factoring to solve a quadratic equation.

8. Explain the circumstances for which you would use the Square Root Property to solve a quadratic equation.

9. State the quadratic formula.

10. If you were to use the quadratic formula to solve $3x^2 - x = 5$, what would be the values of a, b, and c?

Building Skills

In Problems 11–28, solve each equation using the quadratic formula.

11. $x^2 - 4x - 12 = 0$ **12.** $p^2 - 4p - 32 = 0$ **13.** $6y^2 - y - 15 = 0$

14. $10x^2 + x - 2 = 0$ **15.** $4m^2 - 8m + 1 = 0$ **16.** $2q^2 - 4q + 1 = 0$

17. $3w - 6 = \dfrac{1}{w}$ **18.** $x + \dfrac{1}{x} = 3$ **19.** $3p^2 = -2p + 4$

20. $5w^2 = -3w + 1$ **21.** $x^2 - 2x + 7 = 0$ **22.** $y^2 - 4y + 5 = 0$

23. $2z^2 + 7 = 2z$ **24.** $2z^2 + 7 = 4z$ **25.** $4x^2 = 2x + 1$

26. $6p^2 = 4p + 1$ **27.** $1 = 3q^2 + 4q$ **28.** $1 = 5w^2 + 6w$

In Problems 29–38, determine the discriminant of each quadratic equation. Use the value of the discriminant to determine whether the quadratic equation has two rational solutions, two irrational solutions, one repeated real solution, or two complex solutions that are not real.

29. $x^2 - 5x + 1 = 0$ **30.** $p^2 + 4p - 2 = 0$ **31.** $3z^2 + 2z + 5 = 0$

32. $2y^2 - 3y + 5 = 0$ **33.** $9q^2 - 6q + 1 = 0$ **34.** $16x^2 + 24x + 9 = 0$

35. $3w^2 = 4w - 2$ **36.** $6x^2 - x = -4$ **37.** $6x = 2x^2 - 1$

38. $10w^2 = 3$

Mixed Practice

In Problems 39–64, solve each equation using any method you wish.

39. $w^2 - 5w + 5 = 0$ **40.** $q^2 - 7q + 7 = 0$ **41.** $3x^2 + 5x = 8$

42. $4p^2 + 5p = 9$ **43.** $2x^2 = 3x + 35$ **44.** $3x^2 + 5x = 2$

45. $q^2 + 2q + 8 = 0$ **46.** $w^2 + 4w + 9 = 0$ **47.** $5z^2 = 2z + 3$

48. $6x^2 = 2x + 4$ **49.** $7q - 2 = \dfrac{4}{q}$ **50.** $5m - 4 = \dfrac{5}{m}$

51. $5a^2 - 80 = 0$ **52.** $4p^2 - 100 = 0$ **53.** $8n^2 + 1 = 4n$

54. $4q^2 + 1 = 2q$ **55.** $27x^2 + 36x + 12 = 0$ **56.** $8p^2 - 40p + 50 = 0$

57. $\dfrac{1}{3}x^2 + \dfrac{2}{9}x - 1 = 0$ **58.** $\dfrac{1}{2}x^2 + \dfrac{3}{4}x - 1 = 0$ **59.** $(x - 5)(x + 1) = 4$

60. $(a - 3)(a + 1) = 2$ **61.** $\dfrac{x - 2}{x + 2} = x - 3$ **62.** $\dfrac{x - 5}{x + 3} = x - 3$

63. $\dfrac{x - 4}{x^2 + 2} = 2$ **64.** $\dfrac{x - 1}{x^2 + 4} = 1$

65. Suppose that $f(x) = x^2 + 4x - 21$.
 (a) Solve $f(x) = 0$ for x.
 (b) Solve $f(x) = -21$ for x.

66. Suppose that $f(x) = x^2 + 2x - 8$.
 (a) Solve $f(x) = 0$ for x.
 (b) Solve $f(x) = -8$ for x.

67. Suppose that $H(x) = -2x^2 - 4x + 1$.
 (a) Solve $H(x) = 0$ for x.
 (b) Solve $H(x) = 2$ for x.

68. Suppose that $g(x) = 3x^2 + x - 1$.
 (a) Solve $g(x) = 0$ for x.
 (b) Solve $g(x) = 4$ for x.

69. Suppose that $G(x) = 3x^2 + 2x + 2$.
 (a) Solve $G(x) = 0$ for x.
 (b) Solve $G(x) = 4$ for x.

70. Suppose that $F(x) = -x^2 + 3x - 3$.
 (a) Solve $F(x) = 0$ for x.
 (b) Solve $F(x) = -2$ for x.

Applying the Concepts

In Problems 71–74, use the Pythagorean Theorem to determine the value of x for the given measurements of each right triangle.

71.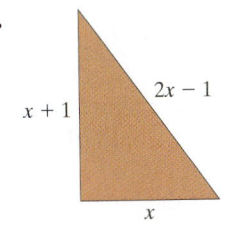

72.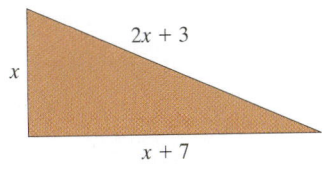

73.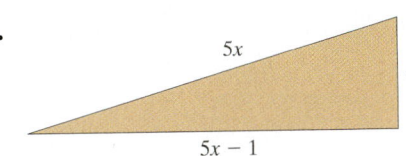

74.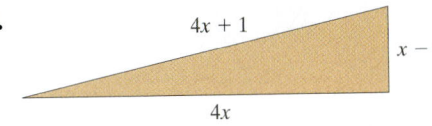

△ **75. Area** The area of a rectangle is 40 square inches. The width of the rectangle is 4 inches more than the length. What are the dimensions of the rectangle?

△ **76. Area** The area of a rectangle is 60 square inches. The width of the rectangle is 6 inches more than the length. What are the dimensions of the rectangle?

△ **77. Area** The area of a triangle is 25 square inches. The height of the triangle is 3 inches less than the base. What are the base and height of the triangle?

△ **78. Area** The area of a triangle is 35 square inches. The height of the triangle is 2 inches less than the base. What are the base and height of the triangle?

79. Revenue The revenue R received by a company selling x pairs of sunglasses per week is given by the function $R(x) = -0.1x^2 + 70x$.

(a) Find and interpret the values of $R(17)$ and $R(25)$.
(b) How many pairs of sunglasses must be sold in order for revenue to be $10,000 per week?
(c) How many pairs of sunglasses must be sold in order for revenue to be $12,250 per week?

80. Revenue The revenue R received by a company selling x "all day passes" to a small amusement park per day is given by the function $R(x) = -0.02x^2 + 24x$.

(a) Find and interpret the values of $R(300)$ and $R(800)$.
(b) How many tickets must be sold in order for revenue to be $4000 per day?
(c) How many tickets must be sold in order for revenue to be $7200 per day?

81. Projectile Motion The height s of a ball after t seconds when thrown straight up with an initial speed of 70 feet per second from an initial height of 5 feet can be modeled by the function

$$s(t) = -16t^2 + 70t + 5$$

(a) When will the height of the ball be 40 feet? Round your answer to the nearest tenth of a second.
(b) When will the height of the ball be 70 feet? Round your answer to the nearest tenth of a second.
(c) Will the ball ever reach a height of 150 feet? How does the result of the equation tell you this?

82. Projectile Motion The height s of a toy rocket after t seconds when fired straight up with an initial speed of 150 feet per second from an initial height of 2 feet can be modeled by the function

$$s(t) = -16t^2 + 150t + 2$$

(a) When will the height of the rocket be 200 feet? Round your answer to the nearest tenth of a second.

(b) When will the height of the rocket be 300 feet? Round your answer to the nearest tenth of a second.

(c) Will the rocket ever reach a height of 500 feet?

△ **83. Similar Triangles** Consult the figure. Suppose that $\triangle ABC \sim \triangle DEC$. The length of $\overline{BC}$ is 24 inches and the length of $\overline{DE}$ is 6 inches. If the length of $\overline{AB}$ equals the length of $\overline{CE}$, which we call x, find x.

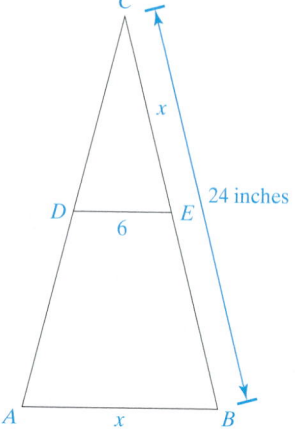

84. Number Sense Three times the square of a number equals the sum of two times the number and 5. Find the number(s).

85. Life Cycle Hypothesis The Life Cycle Hypothesis from Economics was presented by Franco Modigliani in 1954. One of its components states that income is a function of age. The function $I(a) = -55a^2 + 5119a - 54{,}448$ represents the relation between average annual income I and age a.

(a) For what age does average income I equal \$40,000? Round your answer to the nearest year.

(b) For what age does average income I equal \$50,000? Round your answer to the nearest year.

86. Population The function $P(a) = 0.015a^2 - 4.962a + 290.580$ represents the population (in millions) of Americans in 2001, P, that are a years of age or older. (*Source:* United States Census Bureau)

(a) For what age range is the population 200 million? Round your answer to the nearest year.

(b) For what age range is the population 50 million? Round your answer to the nearest year.

87. Upstream and Back Zene decides to canoe 4 miles upstream on a river to a waterfall and then canoe back. The total trip (excluding the time spent at the waterfall) takes 6 hours. Zene knows she can canoe at an average speed of 5 miles per hour in still water. What is the speed of the current?

88. Round Trip A Cessna aircraft flies 200 miles due west into the jet stream and flies back home on the same route. The total time of the trip (excluding the time on the ground) takes 4 hours. The Cessna aircraft can fly 120 miles per hour in still air. What is the net effect of the jet stream on the aircraft?

89. Work Robert and Susan have a newspaper route. When they work the route together, it takes 2 hours to deliver all the newspapers. One morning Robert told Susan he was too sick to deliver the papers. Susan doesn't remember how long it takes for her to deliver the newspapers working alone, but she does remember that Robert can finish the route one hour sooner than Susan can when working alone. How long will it take Susan to finish the route?

90. Work Demitrius needs to fill up his pool. When he rents a water tanker to fill the pool with the help of the hose from his house, it takes 5 hours to fill the pool. One year, money is tight and he can't afford to rent the water tanker to fill the pool. He doesn't remember how long it takes for his house hose to fill the pool, but does remember that the tanker hose filling the pool alone can finish the job in 8 fewer hours than using his house hose alone. How long will it take Demitrius to fill his pool using only his house hose?

Extending the Concepts

91. Show that the sum of the solutions to a quadratic equation is $-\dfrac{b}{a}$.

92. Show that the product of the solutions to a quadratic equation is $\dfrac{c}{a}$.

93. Show that the real solutions of the equation $ax^2 + bx + c = 0$ are the negatives of the real solutions of the equation $ax^2 - bx + c = 0$. Assume that $b^2 - 4ac \geq 0$.

94. Show that the real solutions of the equation $ax^2 + bx + c = 0$ are the reciprocals of the real solutions of the equation $cx^2 + bx + a = 0$. Assume that $b^2 - 4ac \geq 0$.

Synthesis Review

95. (a) Graph $f(x) = x^2 + 3x + 2$ by plotting points.
   (b) Solve the equation $x^2 + 3x + 2 = 0$.
   (c) Compare the solutions to the equation in part (b) to the x-intercepts of the graph drawn in part (a). What do you notice?

96. (a) Graph $f(x) = x^2 - x - 6$ by plotting points.
   (b) Solve the equation $x^2 - x - 6 = 0$.
   (c) Compare the solutions to the equation in part (b) to the x-intercepts of the graph drawn in part (a). What do you notice?

97. (a) Graph $g(x) = x^2 - 2x + 1$ by plotting points.
   (b) Solve the equation $x^2 - 2x + 1 = 0$.
   (c) Compare the solutions to the equation in part (b) to the x-intercepts of the graph drawn in part (a). What do you notice?

98. (a) Graph $g(x) = x^2 + 4x + 4$ by plotting points.
   (b) Solve the equation $x^2 + 4x + 4 = 0$.
   (c) Compare the solutions to the equation in part (b) to the x-intercepts of the graph drawn in part (a). What do you notice?

Graphing Calculator

In Problems 99–102, the graph of the quadratic function f is given. For each function determine the discriminant of the equation $f(x) = 0$ in order to determine the nature of the solutions the equation $f(x) = 0$ has. Compare the nature of solutions based on the discriminant to the graph of the function.

99. $f(x) = x^2 - 7x + 3$

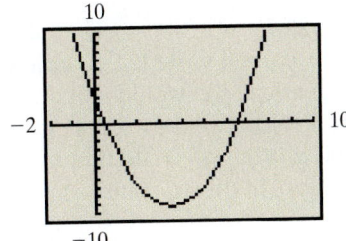

100. $f(x) = -x^2 - 5x + 1$

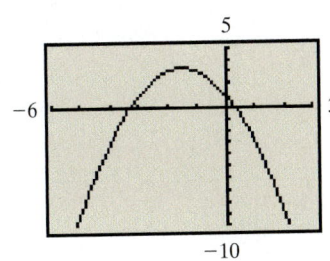

101. $f(x) = -x^2 - 3x - 4$

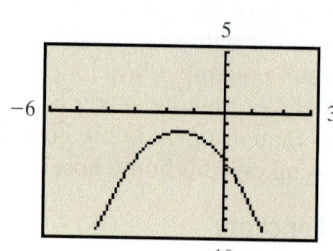

102. $f(x) = x^2 - 6x + 9$

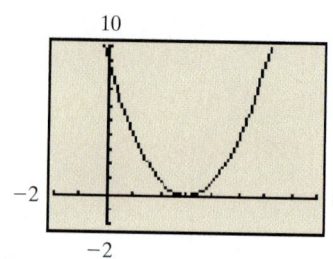

103. **(a)** Solve the equation $x^2 - 5x - 24 = 0$ algebraically.
(b) Graph $Y_1 = x^2 - 5x - 24$. Compare the x-intercepts of the graph to the solutions found in part (a).

104. **(a)** Solve the equation $x^2 - 4x - 45 = 0$ algebraically.
(b) Graph $Y_1 = x^2 - 4x - 45$. Compare the x-intercepts of the graph to the solutions found in part (a).

105. **(a)** Solve the equation $x^2 - 6x + 9 = 0$ algebraically.
(b) Graph $Y_1 = x^2 - 6x + 9$. Compare the x-intercepts of the graph to the solutions found in part (a).

106. **(a)** Solve the equation $x^2 + 10x + 25 = 0$ algebraically.
(b) Graph $Y_1 = x^2 + 10x + 25$. Compare the x-intercepts of the graph to the solutions found in part (a).

107. **(a)** Solve the equation $x^2 + 5x + 8 = 0$ algebraically.
(b) Graph $Y_1 = x^2 + 5x + 8$. How does the result of part (a) relate to the graph?

108. **(a)** Solve the equation $x^2 + 2x + 5 = 0$ algebraically.
(b) Graph $Y_1 = x^2 + 2x + 5$. How does the result of part (a) relate to the graph?

10.3 Solving Equations Quadratic in Form

OBJECTIVE

① Solve Equations That Are Quadratic in Form

Preparing for Solving Equations Quadratic in Form

Before getting started, take the following readiness quiz. If you get a problem wrong, go back to the section cited and review the material.

1. Factor: $x^2 - 5x - 6$ [Section 6.2, pp. 418–421]

2. Factor: $2u^2 + 3u - 5$ [Section 6.3, pp. 427–435]

3. Simplify: (a) $(x^2)^2$ (b) $(p^{-1})^2$ [Section 5.2, pp. 356–357]

① **Solve Equations That Are Quadratic in Form**

Consider the equation $x^4 - 4x^2 - 12 = 0$. While this equation is not of the form $ax^2 + bx + c = 0$, we can write the equation as $(x^2)^2 - 4x^2 - 12 = 0$. Then, if we let $u = x^2$ in the equation, we would obtain $u^2 - 4u - 12 = 0$, which is of the form $ax^2 + bx + c = 0$. Now we can solve $u^2 - 4u - 12 = 0$ for u by factoring. Then, using the fact that $u = x^2$, we could find x—which was our goal in the first place.

In general, if a substitution u transforms an equation into one of the form

$$au^2 + bu + c = 0$$

then the original equation is called an **equation quadratic in form.**

Preparing for...Answers

1. $(x - 6)(x + 1)$

2. $(2u + 5)(u - 1)$

3. (a) x^4 **(b)** $p^{-2} = \dfrac{1}{p^2}$

The difficulty in solving equations that are quadratic in form is that it is often hard to determine that the equation is, in fact, quadratic in form. Table 2 shows some equations that are quadratic in form and the appropriate substitution.

Table 2		
Original Equation	**Substitution**	**Equation with Substitution**
$2x^4 - 3x^2 + 5 = 0$ $2(x^2)^2 - 3x^2 + 5 = 0$	$u = x^2$	$2u^2 - 3u + 5 = 0$
$3(z - 5)^2 + 4(z - 5) + 1 = 0$	$u = z - 5$	$3u^2 + 4u + 1 = 0$
$-2y + 5\sqrt{y} - 2 = 0$ $-2(\sqrt{y})^2 + 5\sqrt{y} - 2 = 0$	$u = \sqrt{y}$	$-2u^2 + 5u - 2 = 0$

EXAMPLE 1   How to Solve Equations That Are Quadratic in Form

Solve: $x^4 + x^2 - 12 = 0$

Step-by-Step Solution

Step 1: Determine the appropriate substitution and write the equation in the form $au^2 + bu + c = 0$.

Let $u = x^2$:
$$x^4 + x^2 - 12 = 0$$
$$(x^2)^2 + x^2 - 12 = 0$$
$$u^2 + u - 12 = 0$$

Step 2: Solve the equation $au^2 + bu + c = 0$.

$$(u + 4)(u - 3) = 0$$
$$u + 4 = 0 \quad \text{or} \quad u - 3 = 0$$
$$u = -4 \quad \text{or} \quad u = 3$$

Step 3: Solve for the variable in the original equation using the value of u found in Step 2.

We want to know x, so replace u with x^2: $\quad x^2 = -4 \quad$ or $\quad x^2 = 3$

Square Root Property: $\quad x = \pm\sqrt{-4} \quad$ or $\quad x = \pm\sqrt{3}$
$$= \pm 2i$$

Step 4: Verify your solutions.

$x = 2i$:
$$(2i)^4 + (2i)^2 - 12 \stackrel{?}{=} 0$$
$$2^4 i^4 + 2^2 i^2 - 12 \stackrel{?}{=} 0$$
$$16(1) + 4(-1) - 12 \stackrel{?}{=} 0$$
$$16 - 4 - 12 \stackrel{?}{=} 0$$
$$0 = 0 \quad \text{True}$$

$x = -2i$:
$$(-2i)^4 + (-2i)^2 - 12 \stackrel{?}{=} 0$$
$$(-2)^4 i^4 + (-2)^2 i^2 - 12 \stackrel{?}{=} 0$$
$$16(1) + 4(-1) - 12 \stackrel{?}{=} 0$$
$$16 - 4 - 12 \stackrel{?}{=} 0$$
$$0 = 0 \quad \text{True}$$

$x = \sqrt{3}$:
$$(\sqrt{3})^4 + (\sqrt{3})^2 - 12 \stackrel{?}{=} 0$$
$$\sqrt{3^4} + \sqrt{3^2} - 12 \stackrel{?}{=} 0$$
$$\sqrt{81} + 3 - 12 \stackrel{?}{=} 0$$
$$9 + 3 - 12 \stackrel{?}{=} 0$$
$$0 = 0 \quad \text{True}$$

$x = -\sqrt{3}$:
$$(-\sqrt{3})^4 + (-\sqrt{3})^2 - 12 \stackrel{?}{=} 0$$
$$(-1)^4 \sqrt{3^4} + (-1)^2 \sqrt{3^2} - 12 \stackrel{?}{=} 0$$
$$\sqrt{81} + 3 - 12 \stackrel{?}{=} 0$$
$$9 + 3 - 12 \stackrel{?}{=} 0$$
$$0 = 0 \quad \text{True}$$

The solution set is $\left\{ 2i, -2i, \sqrt{3}, -\sqrt{3} \right\}$.

We summarize the steps can be used to solve an equation that is quadratic in form.

Solving Equations Quadratic in Form

Step 1: Determine the appropriate substitution and write the equation in the form $au^2 + bu + c = 0$.

Step 2: Solve the equation $au^2 + bu + c = 0$.

Step 3: Solve for the variable in the original equation using the value of u found in Step 2.

Step 4: Verify your solutions.

QUICK ✓ *Solve each equation.*

1. $x^4 - 13x^2 + 36 = 0$ **2.** $p^4 - 7p^2 = 18$

EXAMPLE 2 Solving Equations That Are Quadratic in Form

Solve: $(z^2 - 5)^2 - 3(z^2 - 5) - 4 = 0$

Solution

$$(z^2 - 5)^2 - 3(z^2 - 5) - 4 = 0$$

Let $u = z^2 - 5$: $u^2 - 3u - 4 = 0$

$$(u - 4)(u + 1) = 0$$

$$u - 4 = 0 \quad \text{or} \quad u + 1 = 0$$

$$u = 4 \quad \text{or} \quad u = -1$$

Replace u with $z^2 - 5$ and solve for z: $z^2 - 5 = 4 \quad \text{or} \quad z^2 - 5 = -1$

$$z^2 = 9 \quad \text{or} \quad z^2 = 4$$

Square Root Property: $z = \pm 3 \quad \text{or} \quad z = \pm 2$

Check

$z = -3$:

$$((-3)^2 - 5)^2 - 3((-3)^2 - 5) - 4 \overset{?}{=} 0$$
$$(9 - 5)^2 - 3(9 - 5) - 4 \overset{?}{=} 0$$
$$4^2 - 3(4) - 4 \overset{?}{=} 0$$
$$16 - 12 - 4 \overset{?}{=} 0$$
$$0 = 0 \quad \text{True}$$

$z = 3$:

$$((3)^2 - 5)^2 - 3((3)^2 - 5) - 4 \overset{?}{=} 0$$
$$(9 - 5)^2 - 3(9 - 5) - 4 \overset{?}{=} 0$$
$$4^2 - 3(4) - 4 \overset{?}{=} 0$$
$$16 - 12 - 4 \overset{?}{=} 0$$
$$0 = 0 \quad \text{True}$$

$z = -2$:

$$((-2)^2 - 5)^2 - 3((-2)^2 - 5) - 4 \overset{?}{=} 0$$
$$(4 - 5)^2 - 3(4 - 5) - 4 \overset{?}{=} 0$$
$$(-1)^2 - 3(-1) - 4 \overset{?}{=} 0$$
$$1 + 3 - 4 \overset{?}{=} 0$$
$$0 = 0 \quad \text{True}$$

$z = 2$:

$$((2)^2 - 5)^2 - 3((2)^2 - 5) - 4 \overset{?}{=} 0$$
$$(4 - 5)^2 - 3(4 - 5) - 4 \overset{?}{=} 0$$
$$(-1)^2 - 3(-1) - 4 \overset{?}{=} 0$$
$$1 + 3 - 4 \overset{?}{=} 0$$
$$0 = 0 \quad \text{True}$$

The solution set is $\{-3, -2, 2, 3\}$.

QUICK ✓   *Solve each equation.*

3. $(p^2 - 2)^2 - 9(p^2 - 2) + 14 = 0$ **4.** $2(2z^2 - 1)^2 + 5(2z^2 - 1) - 3 = 0$

When we have to raise both sides of an equation to an even power (such as squaring both sides of the equation), there is a possibility that we will introduce extraneous solutions to the equation. Under these circumstances, it is imperative that we verify our solutions.

EXAMPLE 3   **Solving Equations That Are Quadratic in Form**

Solve: $3x - 5\sqrt{x} - 2 = 0$

Solution

$$3x - 5\sqrt{x} - 2 = 0$$
$$3(\sqrt{x})^2 - 5\sqrt{x} - 2 = 0$$

Let $u = \sqrt{x}$:

$$3u^2 - 5u - 2 = 0$$
$$(3u + 1)(u - 2) = 0$$
$$3u + 1 = 0 \quad \text{or} \quad u - 2 = 0$$
$$3u = -1 \quad \text{or} \quad u = 2$$
$$u = \frac{-1}{3}$$

Replace u with $\sqrt{x}$ and solve for x:

$$\sqrt{x} = \frac{-1}{3} \quad \text{or} \quad \sqrt{x} = 2$$

Square both sides:

$$x = \frac{1}{9} \quad \text{or} \quad x = 4$$

Check

$x = \frac{1}{9}$: $3 \cdot \frac{1}{9} - 5\sqrt{\frac{1}{9}} - 2 = 0$ $x = 4$: $3 \cdot 4 - 5\sqrt{4} - 2 = 0$

$\qquad\qquad\qquad \frac{1}{3} - 5 \cdot \frac{1}{3} - 2 = 0$ $\qquad\qquad\qquad\qquad 12 - 5 \cdot 2 - 2 = 0$

$\qquad\qquad\qquad \frac{1}{3} - \frac{5}{3} + \frac{6}{3} = 0$ $\qquad\qquad\qquad\qquad\qquad 12 - 10 - 2 = 0$

$\qquad\qquad\qquad\qquad \frac{2}{3} = 0 \quad$ False $\qquad\qquad\qquad\qquad\qquad\qquad 0 = 0 \quad$ True

The apparent solution $x = \frac{1}{9}$ is extraneous. The solution set is $\{4\}$.

We could also have solved the equation in Example 3 using the methods introduced in Section 9.8 by isolating the radical and squaring both sides.

QUICK ✓   *Solve the equation.*

5. $3w - 14\sqrt{w} + 8 = 0$ $\qquad\qquad\qquad\qquad$ **6.** $2q - 9\sqrt{q} - 5 = 0$

EXAMPLE 4 Solving Equations That Are Quadratic in Form

Solve: $4x^{-2} + 13x^{-1} - 12 = 0$

Solution

$$4x^{-2} + 13x^{-1} - 12 = 0$$

$$4(x^{-1})^2 + 13x^{-1} - 12 = 0$$

Let $u = x^{-1}$:　　$4u^2 + 13u - 12 = 0$

Factor:　　$(4u - 3)(u + 4) = 0$

$$4u - 3 = 0 \quad \text{or} \quad u + 4 = 0$$

$$4u = 3 \quad \text{or} \quad u = -4$$

$$u = \frac{3}{4}$$

Replace u with x^{-1} and solve for x:

$$x^{-1} = \frac{3}{4} \quad \text{or} \quad x^{-1} = -4$$

$$x^{-1} = \frac{1}{x}: \qquad \frac{1}{x} = \frac{3}{4} \quad \text{or} \quad \frac{1}{x} = -4$$

Take the reciprocal of both sides of the equation:

$$x = \frac{4}{3} \quad \text{or} \quad x = \frac{1}{-4} = -\frac{1}{4}$$

Check　　$x = \frac{4}{3}$:

$$4\left(\frac{4}{3}\right)^{-2} + 13 \cdot \left(\frac{4}{3}\right)^{-1} - 12 \stackrel{?}{=} 0$$

$$4\left(\frac{3}{4}\right)^2 + 13 \cdot \frac{3}{4} - 12 \stackrel{?}{=} 0$$

$$4 \cdot \frac{9}{16} + \frac{39}{4} - 12 \stackrel{?}{=} 0$$

$$\frac{9}{4} + \frac{39}{4} - \frac{48}{4} \stackrel{?}{=} 0$$

$$0 = 0 \quad \text{True}$$

$x = -\frac{1}{4}$:

$$4\left(-\frac{1}{4}\right)^{-2} + 13 \cdot \left(-\frac{1}{4}\right)^{-1} - 12 \stackrel{?}{=} 0$$

$$4(-4)^2 + 13 \cdot (-4) - 12 \stackrel{?}{=} 0$$

$$4 \cdot 16 - 52 - 12 \stackrel{?}{=} 0$$

$$64 - 52 - 12 \stackrel{?}{=} 0$$

$$0 = 0 \quad \text{True}$$

The solution set is $\left\{-\frac{1}{4}, \frac{4}{3}\right\}$.

QUICK ✓ *Solve the equation.*

7. $5x^{-2} + 12x^{-1} + 4 = 0$

EXAMPLE 5 Solving Equations Quadratic in Form

Solve: $a^{2/3} + 3a^{1/3} - 28 = 0$

Solution

$$a^{2/3} + 3a^{1/3} - 28 = 0$$

$$\left(a^{1/3}\right)^2 + 3a^{1/3} - 28 = 0$$

Let $u = a^{1/3}$:　　$u^2 + 3u - 28 = 0$

$$(u + 7)(u - 4) = 0$$

$$u + 7 = 0 \quad \text{or} \quad u - 4 = 0$$

$$u = -7 \quad \text{or} \quad u = 4$$

$$a^{1/3} = -7 \quad \text{or} \quad a^{1/3} = 4$$

Replace u with $a^{1/3}$ and solve for a:　　$\left(a^{1/3}\right)^3 = (-7)^3 \quad \text{or} \quad \left(a^{1/3}\right)^3 = 4^3$

Cube both sides of the equation:　　$a = -343 \quad \text{or} \quad a = 64$

Check $\qquad a = -343:$ $\qquad\qquad\qquad\qquad a = 64:$

$$(-343)^{2/3} + 3(-343)^{1/3} - 28 = 0$$
$$\left(\sqrt[3]{-343}\right)^2 + 3 \cdot \sqrt[3]{-343} - 28 \stackrel{?}{=} 0$$
$$(-7)^2 + 3 \cdot (-7) - 28 \stackrel{?}{=} 0$$
$$49 - 21 - 28 \stackrel{?}{=} 0$$
$$0 = 0 \quad \text{True}$$

$$(64)^{2/3} + 3(64)^{1/3} - 28 = 0$$
$$\left(\sqrt[3]{64}\right)^2 + 3 \cdot \sqrt[3]{64} - 28 \stackrel{?}{=} 0$$
$$4^2 + 3 \cdot 4 - 28 \stackrel{?}{=} 0$$
$$16 + 12 - 28 \stackrel{?}{=} 0$$
$$0 = 0 \quad \text{True}$$

The solution set is $\{-343, 64\}$.

QUICK ✓ *Solve the equation.*

8. $p^{2/3} - 4p^{1/3} - 5 = 0$

10.3 Exercises

For Extra Help:

Student Solutions Manual CD Video PH Math/Tutor Center MathXL Tutorials on CD MathXL® MyMathLab

Concepts and Vocabulary

In Problems 1 and 2, fill in the blanks.

1. If a substitution u transforms an equation into one of the form $au^2 + bu + c = 0$, then the original equation is called a(n) ____ ____ ____ ____.

2. For the equation $2(3x + 1)^2 - 5(3x + 1) + 2 = 0$, an appropriate substitution would be $u =$ ____.

In Problems 3 and 4, answer True or False to each statement.

3. The equation $3\left(\dfrac{x}{x-2}\right)^2 - \dfrac{5x}{x-2} + 3 = 0$ is quadratic in form.

4. The equation $x - 5\sqrt{x} - 6 = 0$ can be solved either using the methods of this section or by isolating the radical and squaring both sides.

5. In your own words, explain the steps required to solve an equation quadratic in form. Be sure to include an explanation as to how to identify the appropriate substitution.

6. Under what circumstances might extraneous solutions occur when solving equations quadratic in form?

7. What is the appropriate choice for u when solving the equation $4(3x - 2)^2 + 7(3x - 2) + 2 = 0$?

8. What is the appropriate choice for u when solving the equation $2 \cdot \dfrac{1}{x^2} - 6 \cdot \dfrac{1}{x} + 3 = 0$?

Building Skills

In Problems 9–44, solve each equation.

9. $x^4 - 5x^2 + 4 = 0$

10. $x^4 - 10x^2 + 9 = 0$

11. $q^4 + 13q^2 + 36 = 0$

12. $z^4 + 10z^2 + 9 = 0$

13. $4a^4 - 17a^2 + 4 = 0$

14. $4b^4 - 5b^2 + 1 = 0$

15. $p^4 + 6 = 5p^2$

16. $q^4 + 15 = 8q^2$

17. $(x - 3)^2 - 6(x - 3) - 7 = 0$

18. $(x + 2)^2 - 3(x + 2) - 10 = 0$

19. $(x^2 - 1)^2 - 11(x^2 - 1) + 24 = 0$

20. $(p^2 - 2)^2 - 8(p^2 - 2) + 12 = 0$

21. $(y^2 + 2)^2 + 7(y^2 + 2) + 10 = 0$

22. $(q^2 + 4)^2 + 3(q^2 + 4) - 4 = 0$

23. $x - 3\sqrt{x} - 4 = 0$

24. $x - 5\sqrt{x} - 6 = 0$

25. $w + 5\sqrt{w} + 6 = 0$

26. $z + 7\sqrt{z} + 6 = 0$

27. $2x + 5\sqrt{x} = 3$

28. $3x = 11\sqrt{x} + 4$

29. $x^{-2} + 3x^{-1} = 28$

30. $q^{-2} + 2q^{-1} = 15$

31. $10z^{-2} + 11z^{-1} = 6$

32. $10a^{-2} + 23a^{-1} = 5$

33. $x^{2/3} + 3x^{1/3} - 4 = 0$

34. $y^{2/3} - 2y^{1/3} - 3 = 0$

35. $z^{2/3} - z^{1/3} = 2$

36. $w^{2/3} + 2w^{1/3} = 3$

37. $a + a^{1/2} = 30$

38. $b + 3b^{1/2} = 28$

39. $\dfrac{1}{x^2} - \dfrac{5}{x} + 6 = 0$

40. $\dfrac{1}{x^2} - \dfrac{7}{x} + 12 = 0$

41. $\left(\dfrac{1}{x + 2}\right)^2 + \dfrac{4}{x + 2} = 5$

42. $\left(\dfrac{1}{x + 2}\right)^2 + \dfrac{6}{x + 2} = 7$

43. $p^6 - 28p^3 + 27 = 0$

44. $y^6 - 7y^3 - 8 = 0$

Mixed Practice

In Problems 45–58, solve each equation.

45. $8a^{-2} + 2a^{-1} = 1$

46. $6b^{-2} - b^{-1} = 1$

47. $z^4 = 4z^2 + 32$

48. $x^4 + 3x^2 = 4$

49. $x^{1/2} + x^{1/4} - 6 = 0$

50. $c^{1/2} + c^{1/4} - 12 = 0$

51. $w^4 - 5w^2 - 36 = 0$

52. $p^4 - 15p^2 - 16 = 0$

53. $\left(\dfrac{1}{x + 3}\right)^2 + \dfrac{2}{x + 3} = 3$

54. $\left(\dfrac{1}{x - 1}\right)^2 + \dfrac{7}{x - 1} = 8$

55. $x - 7\sqrt{x} + 12 = 0$

56. $x - 8\sqrt{x} + 12 = 0$

57. $2(x - 1)^2 - 7(x - 1) = 4$

58. $3(y - 2)^2 - 4(y - 2) = 4$

59. Suppose that $f(x) = x^4 + 7x^2 + 12$. Find the values of x such that
(a) $f(x) = 12$ (b) $f(x) = 6$.

60. Suppose that $f(x) = x^4 + 5x^2 + 3$. Find the values of x such that
(a) $f(x) = 3$ (b) $f(x) = 17$.

61. Suppose that $g(x) = 2x^4 - 6x^2 - 5$. Find the values of x such that
(a) $g(x) = -5$ (b) $g(x) = 15$.

62. Suppose that $h(x) = 3x^4 - 9x^2 - 8$. Find the values of x such that
(a) $h(x) = -8$ (b) $h(x) = 22$.

63. Suppose that $F(x) = x^{-2} - 5x^{-1}$. Find the values of x such that
(a) $F(x) = 6$ (b) $F(x) = 14$.

64. Suppose that $f(x) = x^{-2} - 3x^{-1}$. Find the values of x such that
(a) $f(x) = 4$ (b) $f(x) = 18$.

In Problems 65–70, find the zeros of the function. [**Hint:** *Remember, r is a zero if* $f(r) = 0$.]

65. $f(x) = x^4 + 9x^2 + 14$

66. $f(x) = x^4 - 13x^2 + 42$

67. $g(t) = 6t - 25\sqrt{t} - 9$

68. $h(p) = 8p - 18\sqrt{p} - 35$

69. $s(d) = \dfrac{1}{(d + 3)^2} - \dfrac{4}{d + 3} + 3$

70. $f(a) = \dfrac{1}{(a - 2)^2} + \dfrac{3}{a - 2} - 4$

Applying the Concepts

71. (a) Solve $x^2 - 5x + 6 = 0$.
 (b) Solve $(x - 3)^2 - 5(x - 3) + 6 = 0$. Compare the solutions to part (a).
 (c) Solve $(x + 2)^2 - 5(x + 2) + 6 = 0$. Compare the solutions to part (a).
 (d) Solve $(x - 5)^2 - 5(x - 5) + 6 = 0$. Compare the solutions to part (a).
 (e) Conjecture a generalization for the solution of
 $(x - a)^2 - 5(x - a) + 6 = 0$.

72. (a) Solve $x^2 + 3x - 18 = 0$.
 (b) Solve $(x - 1)^2 + 3(x - 1) - 18 = 0$. Compare the solutions to part (a).
 (c) Solve $(x + 5)^2 + 3(x + 5) - 18 = 0$. Compare the solutions to part (a).
 (d) Solve $(x - 3)^2 + 3(x - 3) - 18 = 0$. Compare the solutions to part (a).
 (e) Conjecture a generalization for the solution of
 $(x - a)^2 + 3(x - a) - 18 = 0$.

73. For the function $f(x) = 2x^2 - 3x + 1$,
 (a) Solve $f(x) = 0$.
 (b) Solve $f(x - 2) = 0$. Compare the solutions to part (a).
 (c) Solve $f(x - 5) = 0$. Compare the solutions to part (a).
 (d) Conjecture a generalization for the zeros of $f(x - a)$.

74. For the function $f(x) = 3x^2 - 5x - 2$,
 (a) Solve $f(x) = 0$.
 (b) Solve $f(x - 1) = 0$. Compare the solutions to part (a).
 (c) Solve $f(x - 4) = 0$. Compare the solutions to part (a).
 (d) Conjecture a generalization for the zeros of $f(x - a)$.

75. Revenue The function $R(x) = \dfrac{(x - 1990)^2}{2} + \dfrac{3(x - 1990)}{2} + 3000$ models the
 revenue R (in thousands of dollars) of a start-up computer consulting firm in year
 x, where $x \geq 1990$.
 (a) Determine and interpret $R(1990)$.
 (b) Solve and interpret $R(x) = 3065$.
 (c) According to the model, in what year can the firm expect to receive $3350
 thousand in revenue?

76. Revenue The function $R(x) = \dfrac{(x - 2000)^2}{3} + \dfrac{5(x - 2000)}{3} + 2000$ models the
 revenue R (in thousands of dollars) of a start-up computer software firm in year
 x, where $x \geq 2000$.
 (a) Determine and interpret $R(2000)$.
 (b) Solve and interpret $R(x) = 2250$.
 (c) According to the model, in what year can the firm expect to receive $2350
 thousand in revenue?

Extending the Concepts

All of the problems given in this section resulted in equations quadratic in form that could be factored after the appropriate substitution. However, this is not a necessary requirement to solving equations quadratic in form. In Problems 77–80, determine the appropriate substitution, and then use the quadratic formula to find the value of u. Finally determine the value of the variable in the equation.

77. $x^4 + 5x^2 + 2 = 0$ **78.** $x^4 + 7x^2 + 4 = 0$

79. $2(x - 2)^2 + 8(x - 2) - 1 = 0$ **80.** $3(x + 1)^2 + 6(x + 1) - 1 = 0$

Synthesis Review

In Problems 81–85, add or subtract the expressions.

81. $(4x^2 - 3x - 1) + (-3x^2 + x + 5)$

82. $(5y^3 - 2y^2 + y + 4) - (2y^3 + 6y^2 - 3)$

83. $(3p^{-2} - 4p^{-1} + 8) - (2p^{-2} - 8p^{-1} - 1)$

84. $3\sqrt{2x} - \sqrt{8x} + \sqrt{50x}$

85. $\sqrt[3]{16a} + \sqrt[3]{54a} - \sqrt[3]{128a^4}$

86. Write a sentence or two that discusses how to add or subtract algebraic expressions, in general.

The Graphing Calculator

In Problems 87–92, use a graphing calculator to find the real solutions to the equations using either the ZERO or INTERSECT feature. Round your answers to two decimal places, if necessary.

87. $x^4 + 5x^2 - 14 = 0$

88. $x^4 - 4x^2 - 12 = 0$

89. $2(x - 2)^2 = 5(x - 2) + 1$

90. $3(x + 3)^2 = 2(x + 3) + 6$

91. $x - 5\sqrt{x} = -3$

92. $x + 4\sqrt{x} = 5$

93. **(a)** Graph $Y_1 = x^2 - 5x - 6$. Find the x-intercepts of the graph.
(b) Graph $Y_1 = (x + 2)^2 - 5(x + 2) - 6$. Find the x-intercepts of the graph.
(c) Graph $Y_1 = (x + 5)^2 - 5(x + 5) - 6$. Find the x-intercepts of the graph.
(d) Make a generalization based upon the results of parts (a), (b), and (c).

94. **(a)** Graph $Y_1 = x^2 + 4x + 3$. Find the x-intercepts of the graph.
(b) Graph $Y_1 = (x - 3)^2 + 4(x - 3) + 3$. Find the x-intercepts of the graph.
(c) Graph $Y_1 = (x - 6)^2 + 4(x - 6) + 3$. Find the x-intercepts of the graph.
(d) Make a generalization based upon the results of parts (a), (b), and (c).

PUTTING THE CONCEPTS TOGETHER (SECTIONS 10.1–10.3)

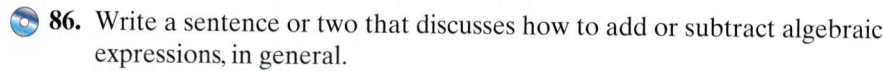

These problems cover important concepts from Sections 10.1 to 10.3. We designed these problems so that you can review the chapter so far and show your mastery of the concepts. Take time to work these problems before proceeding with the next section. The answers to these problems are located at the back of the text on page AN-41.

In Problems 1–3, complete the square in the given expression. Then factor the perfect square trinomial.

1. $z^2 + 10z$

2. $x^2 + 7x$

3. $n^2 - \dfrac{1}{4}n$

In Problems 4–6, solve each quadratic equation using the stated method.

4. $(2x - 3)^2 - 5 = -1$; square root method

5. $x^2 + 8x + 4 = 0$; completing the square

6. $x(x - 6) = -7$; quadratic formula

In Problems 7–10, solve each equation using the method you prefer.

7. $49x^2 - 80 = 0$

8. $p^2 - 8p + 6 = 0$

9. $3y^2 + 6y + 4 = 0$

10. $\dfrac{1}{4}n^2 + n = \dfrac{1}{6}$

In Problems 11–13, determine the discriminant of each quadratic equation. Use the value of the discriminant to determine whether the equation has two rational solutions, two irrational solutions, one repeated real solution, or two complex solutions that are not real.

11. $9x^2 + 12x + 4 = 0$ **12.** $3x^2 + 6x - 2 = 0$ **13.** $2x^2 + 6x + 5 = 0$

14. Find the missing length in the right triangle shown below.

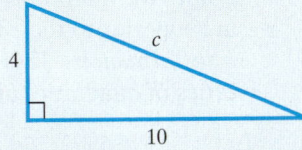

In Problems 15 and 16, solve each equation.

15. $2m + 7\sqrt{m} - 15 = 0$ **16.** $p^{-2} - 3p^{-1} - 18 = 0$

17. Revenue The revenue R received by a company selling x microwave ovens per day is given by the function $R(x) = -0.4x^2 + 140x$. How many microwave ovens must be sold in order for revenue to be \$12,000 per day?

18. Airplane Ride An airplane flies 300 miles into the wind and then flies home against the wind. The total time of the trip (excluding time on the ground) is 5 hours. If the plane can fly 140 miles per hour in still air, what was the speed of the wind? Round your answer to the nearest tenth.

10.4 Graphing Quadratic Functions Using Transformations

OBJECTIVES

1. Graph Quadratic Functions of the Form $f(x) = x^2 + k$
2. Graph Quadratic Functions of the Form $f(x) = (x - h)^2$
3. Graph Quadratic Functions of the Form $f(x) = ax^2$
4. Graph Quadratic Functions of the Form $f(x) = ax^2 + bx + c$
5. Find a Quadratic Function from Its Graph

Preparing for...Answers

1.

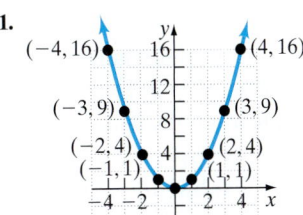

2.

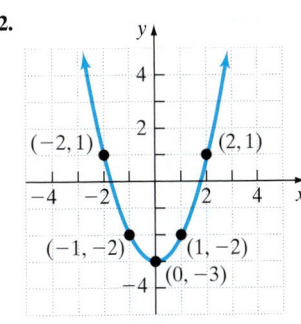

3. The set of all real numbers

Preparing for Graphing Quadratic Functions Using Transformations

Before getting started, take the following readiness quiz. If you get a problem wrong, go back to the section cited and review the material.

1. Graph $y = x^2$ using point plotting.   [Section 8.1, pp. 569–570]

2. Using the point-plotting method, graph $y = x^2 - 3$.   [Section 8.1, pp. 568–571]

3. What is the domain of $f(x) = 2x^2 + 5x + 1$?   [Section 8.4, pp. 597–598]

We begin with a definition.

DEFINITION

A **quadratic function** is a function of the form

$$f(x) = ax^2 + bx + c$$

where a, b, and c are real numbers and $a \neq 0$. The domain of a quadratic function consists of all real numbers.

Many situations can be modeled using quadratic functions. For example, we saw in Example 9 of Section 8.3 that Franco Modigliani used the quadratic function $I(a) = -55a^2 + 5119a - 54{,}448$ to model the relation between average annual income, I, and age, a.

A second situation in which a quadratic function appears involves the motion of a projectile. If we ignore the effect of air resistance on a projectile, the height H of the projectile as a function of horizontal distance traveled, x, can be modeled using a quadratic function. See Figure 7.

Figure 7

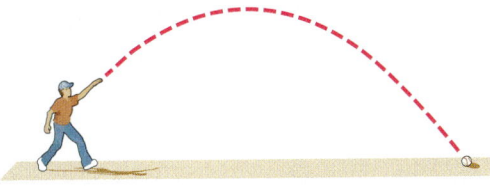

The major goal of this and the next section is to learn methods that will allow us to obtain the graph of a quadratic function. Back in Section 3.1 we learned how to graph virtually any type of equation using point plotting. However, we also discovered that this method is inefficient and could lead to incomplete graphs. Remember, a graph is complete if it shows all of the "interesting features" of the graph. Some of the interesting features that must be included are the intercepts and the high and low points of the graph. We will present two methods that can be used to graph quadratic functions that are superior to the point-plotting method. The first method utilizes a technique called *transformations*. This will be the subject of this section. The second method uses properties of quadratic functions, which is discussed in the next section.

Work Smart

Consider the quadratic function $f(x) = ax^2 + bx + c$, where $a = 1$, $b = 0$, and c is any real number. This is a function of the form $f(x) = x^2 + k$.

(1) Graph Quadratic Functions of the Form $f(x) = x^2 + k$

We begin by looking at the graph of any quadratic function of the form $f(x) = x^2 + k$ such as $f(x) = x^2 + 3$ or $f(x) = x^2 - 4$.

In Example 2 from Section 8.1, we graph the equation $y = x^2$. For convenience, we provide the graph of $y = f(x) = x^2$ in Figure 8.

Figure 8

What effect does adding a real number k to the function $f(x) = x^2$ have on its graph? Let's see!

EXAMPLE 1 Graphing a Quadratic Function of the Form $f(x) = x^2 + k$

On the same Cartesian plane, graph $g(x) = x^2$ and $f(x) = x^2 + 3$.

Solution

We begin by obtaining some points on the graphs of g and f. For example, when $x = 0$, then $y = g(0) = 0$ and $y = f(0) = 0^2 + 3 = 3$. When $x = 1$, then $y = g(1) = 1$ and $y = f(1) = 1^2 + 3 = 4$. Table 3 lists these points along with a few others. Notice that the y-coordinates on the graph of $f(x) = x^2 + 3$ are exactly 3 units larger than the corresponding y-coordinates on the graph of $g(x) = x^2$. Figure 9 shows the graphs of f and g.

Table 3

x	$g(x) = x^2$	$(x, g(x))$	$f(x) = x^2 + 3$	$(x, f(x))$
-2	$(-2)^2 = 4$	$(-2, 4)$	$(-2)^2 + 3 = 7$	$(-2, 7)$
-1	$(-1)^2 = 1$	$(-1, 1)$	$(-1)^2 + 3 = 4$	$(-1, 4)$
0	0	$(0, 0)$	3	$(0, 3)$
1	1	$(1, 1)$	4	$(1, 4)$
2	4	$(2, 4)$	7	$(2, 7)$

Figure 9

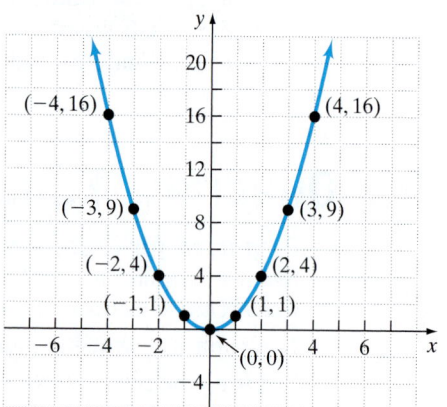

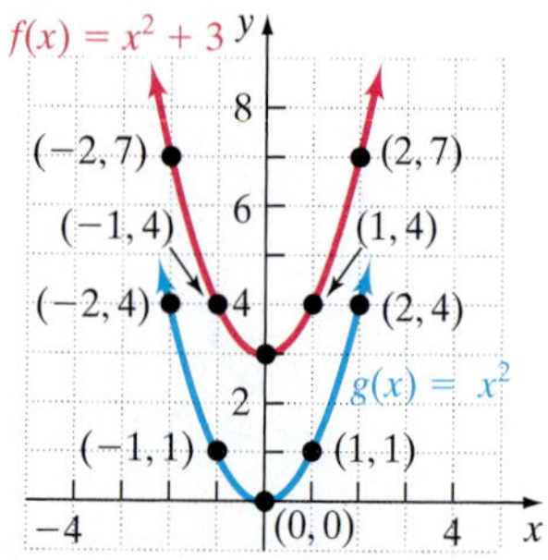

We conclude that the graph of f is identical to the graph of g except that it is shifted vertically up 3 units.

Let's look at another example.

EXAMPLE 2 **Graphing a Quadratic Function of the Form $f(x) = x^2 + k$**

On the same Cartesian plane, graph $g(x) = x^2$ and $f(x) = x^2 - 4$.

Solution

Table 4 lists some points on the graphs of g and f. Notice that the y-coordinates on the graph of $f(x) = x^2 - 4$ are exactly 4 units smaller than the corresponding y-coordinates on the graph of $g(x) = x^2$. Figure 10 shows the graphs of f and g.

Table 4

x	$g(x) = x^2$	$(x, g(x))$	$f(x) = x^2 - 4$	$(x, f(x))$
-2	$(-2)^2 = 4$	$(-2, 4)$	$(-2)^2 - 4 = 0$	$(-2, 0)$
-1	$(-1)^2 = 1$	$(-1, 1)$	$(-1)^2 - 4 = -3$	$(-1, -3)$
0	0	$(0, 0)$	-4	$(0, -4)$
1	1	$(1, 1)$	-3	$(1, -3)$
2	4	$(2, 4)$	0	$(2, 0)$

Figure 10

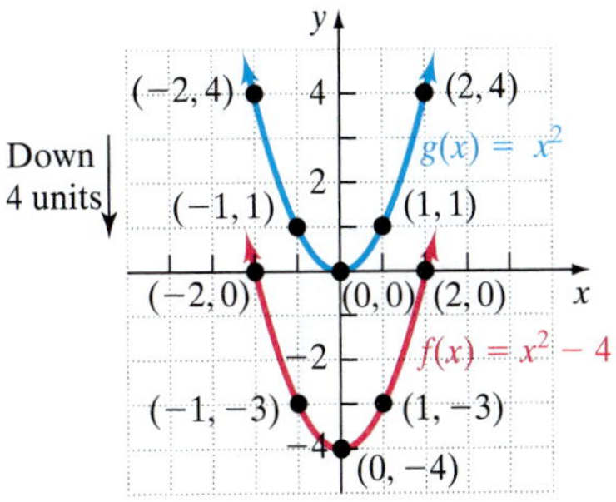

We conclude that the graph of g is identical to the graph of f except that it is shifted vertically down 4 units.

Based on the results of Examples 1 and 2, we are led to the following conclusion.

> **GRAPHING A FUNCTION OF THE FORM $f(x) = x^2 + k$**
> To obtain the graph of $f(x) = x^2 + k, k > 0$, from the graph of $y = x^2$, shift the graph of $y = x^2$ vertically up k units. To obtain the graph of $f(x) = x^2 - k, k > 0$, from the graph of $y = x^2$ shift the graph of $y = x^2$ vertically down k units.

QUICK ✓ *Use the graph of $g(x) = x^2$ to obtain graph the quadratic function.*

1. $f(x) = x^2 + 5$ **2.** $f(x) = x^2 - 2$

---

(2) **Graph Quadratic Functions of the Form $f(x) = (x - h)^2$**

We now look at the graph of any quadratic function of the form $f(x) = (x - h)^2$ such as $f(x) = (x + 3)^2$ or $f(x) = (x - 2)^2$. Our goal here is to determine the effect subtracting a real positive number h from x has on the graph of the function $f(x) = x^2$.

EXAMPLE 3 **Graphing a Quadratic Function of the Form $f(x) = (x - h)^2$**

On the same Cartesian plane, graph $g(x) = x^2$ and $f(x) = (x - 2)^2$.

Work Smart

Because we are subtracting 2 from each x-value in the function $f(x) = (x - 2)^2$, the x-values must be bigger by 2 to obtain the same y-value that was obtained in the graph of $g(x) = x^2$.

Solution

Again, we use the point-plotting method. Table 5 lists some points on the graphs of g and f. Notice when $g(x) = 0$, $x = 0$, and when $f(x) = 0$, $x = 2$. Also, when $g(x) = 4$, $x = -2$ or 2, and when $f(x) = 4$, $x = 0$ or 4. We conclude that the graph of f is identical to that of g, except that it is shifted 2 units to the right. See Figure 11.

			Table 5		
x	**$g(x) = x^2$**	**$(x, g(x))$**	**$f(x) = (x - 2)^2$**	**$(x, f(x))$**	
-2	$(-2)^2 = 4$	$(-2, 4)$	$(-2 - 2)^2 = 16$	$(-2, 16)$	
-1	$(-1)^2 = 1$	$(-1, 1)$	$(-1 - 2)^2 = 9$	$(-1, 9)$	
0	0	$(0, 0)$	4	$(0, 4)$	
1	1	$(1, 1)$	1	$(1, 1)$	
2	4	$(2, 4)$	0	$(2, 0)$	
3	9	$(3, 9)$	1	$(3, 1)$	
4	16	$(4, 16)$	4	$(4, 4)$	

Figure 11

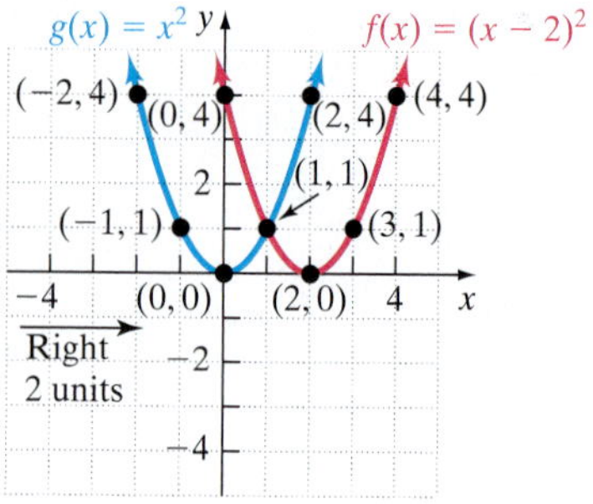

What if we add a positive number h to x?

EXAMPLE 4 **Graphing a Quadratic Function of the Form $f(x) = (x + h)^2$**

On the same Cartesian plane, graph $g(x) = x^2$ and $f(x) = (x + 3)^2$.

Solution

Table 6 lists some points on the graphs of g and f. Notice that when $g(x) = 0$, then $x = 0$, and when $f(x) = 0$, then $x = -3$. Also, when $g(x) = 4$, then $x = -2$ or 2, and when $f(x) = 4$, then $x = -5$ or -1. We conclude that the graph of f is identical to that of g, except that it is shifted 3 units to the left. See Figure 12.

			Table 6		
x	**$g(x) = x^2$**	**$(x, g(x))$**	**$f(x) = (x + 3)^2$**	**$(x, f(x))$**	
-5	$(-5)^2 = 25$	$(-5, 25)$	$(-5 + 3)^2 = 4$	$(-5, 4)$	
-4	$(-4)^2 = 16$	$(-4, 16)$	$(-4 + 3)^2 = 1$	$(-4, 1)$	
-3	$(-3)^2 = 9$	$(-3, 9)$	$(-3 + 3)^2 = 0$	$(-3, 0)$	
-2	4	$(-2, 4)$	1	$(-2, 1)$	
-1	1	$(-1, 1)$	4	$(-1, 4)$	
0	0	$(0, 0)$	9	$(0, 9)$	
1	1	$(1, 1)$	16	$(1, 16)$	
2	4	$(2, 4)$	25	$(2, 25)$	

Figure 12

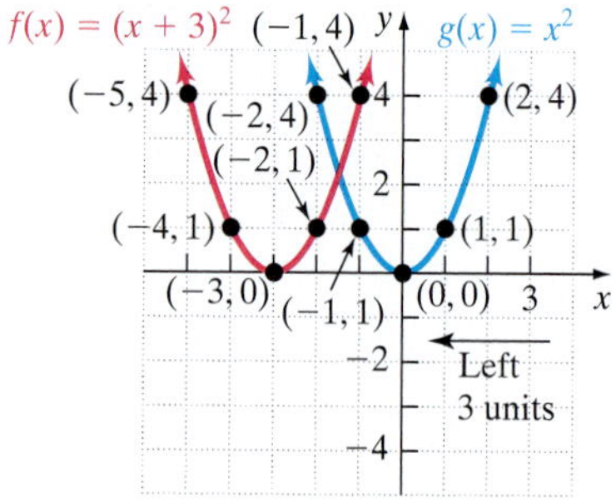

Work Smart

Because we are adding 3 to each x-value in the function $f(x) = (x + 3)^2$, the x-values must be smaller by 3 to obtain the same y-value that was obtained in the graph of $g(x) = x^2$.

Work Smart

If the function is of the form $f(x) = (x - h)^2$, shift the graph of $y = x^2$ right h units, if the function is of the form $f(x) = (x + h)^2$, shift the graph of $y = x^2$ left h units.

Based upon the results of Examples 3 and 4, we are led to the following conclusion.

GRAPHING A FUNCTION OF THE FORM $f(x) = (x - h)^2$

To obtain the graph of $f(x) = (x - h)^2$, $h > 0$, from the graph of $y = x^2$, shift the graph of $y = x^2$ horizontally to the right h units. To obtain the graph of $f(x) = (x + h)^2$ from the graph of $y = x^2$, $h > 0$, shift the graph of $y = x^2$ horizontally to the left h units.

QUICK ✓ *Use the graph of $y = x^2$ to obtain graph the quadratic function.*

3. $f(x) = (x + 5)^2$ **4.** $f(x) = (x - 1)^2$

Let's do an example where we combine a horizontal shift with a vertical shift.

EXAMPLE 5 **Combining Horizontal and Vertical Shifts**

Graph the function $f(x) = (x + 2)^2 - 3$.

Solution

We will graph f in steps. We begin with the graph of $y = x^2$ as shown in Figure 13(a). We shift the graph of $y = x^2$ horizontally 2 units to the left to get the graph of $y = (x + 2)^2$. See Figure 13(b). Then, we shift the graph of $y = (x + 2)^2$ vertically down 3 units to get the graph of $y = (x + 2)^2 - 3$. See Figure 13(c). Notice that we keep track of key points plotted on each graph.

Figure 13

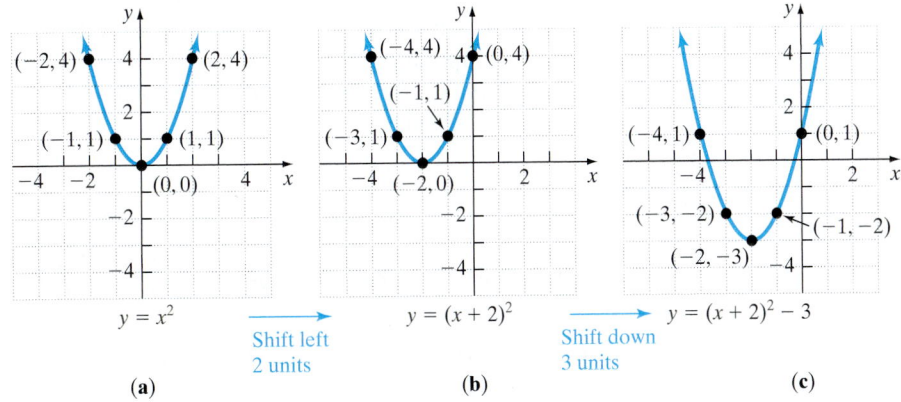

Note: The order in which the transformations take place does not matter. We could just as easily have shifted down 3 units first and then shifted left 2 units.

QUICK ✔ *Graph each quadratic function using horizontal and vertical shifts.*

5. $f(x) = (x - 3)^2 + 2$ **6.** $f(x) = (x + 1)^2 - 4$

---

③ **Graph Quadratic Functions of the Form $f(x) = ax^2$**

In the examples presented thus far, the coefficient of the square term has been equal to 1. We now discuss the impact that the value of a has on the graph of $f(x) = ax^2 + bx + c$. To make the discussion a little easier, we will only consider quadratic functions of the form $f(x) = ax^2, a \neq 0$.

First, let's consider situations in which the value of a is positive. Table 7 shows points on the graphs of $f(x) = x^2$, $g(x) = \frac{1}{2}x^2$ and $h(x) = 2x^2$. Figure 14 shows the graphs of

Table 7

x	$f(x) = x^2$	$g(x) = \dfrac{1}{2}x^2$	$h(x) = 2x^2$
-2	$(-2)^2 = 4$	$\dfrac{1}{2}(-2)^2 = 2$	$2(-2)^2 = 8$
-1	$(-1)^2 = 1$	$\dfrac{1}{2}(-1)^2 = \dfrac{1}{2}$	$2(-1)^2 = 2$
0	0	0	0
1	1	$\dfrac{1}{2}$	2
2	4	2	8

Figure 14
$f(x) = ax^2$. Since $a > 0$, the graphs open up.

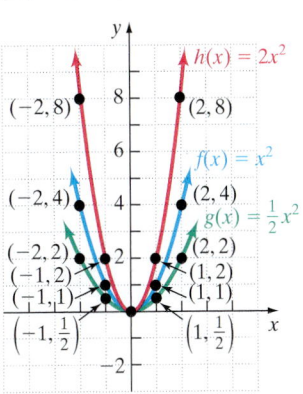

$f(x) = x^2$, $g(x) = \frac{1}{2}x^2$ and $h(x) = 2x^2$. Notice that the y-coordinates on the graph of g are exactly $\frac{1}{2}$ of the values of the y-coordinates on the graph of f. The y-coordinates on the graph of h are exactly 2 times the values of the y-coordinates on the graph of f. Put another way, the larger the value of a, the "taller" the graph is, and the smaller the value of a, the "shorter" the graph is. Also, notice that all three graphs open "up."

Now let's consider what happens when a is negative. Table 8 shows points on the graphs of $f(x) = -x^2$, $g(x) = -\frac{1}{2}x^2$ and $h(x) = -2x^2$. Figure 15 shows the graphs of $f(x) = -x^2$, $g(x) = -\frac{1}{2}x^2$ and $h(x) = -2x^2$. Notice that the y-coordinates on the graph of g are exactly $-\frac{1}{2}$ times the values of the y-coordinates on the graph of f. The y-coordinates on the graph of h are exactly -2 times the values of the y-coordinates on the graph of f. Put another way, the larger the value of $|a|$, the "taller" the graph is, and the smaller the value of $|a|$, the "shorter" the graph is. Also, notice that all three graphs open "down."

Table 8

x	$f(x) = -x^2$	$g(x) = -\frac{1}{2}x^2$	$h(x) = -2x^2$
-2	$-(-2)^2 = -4$	$-\frac{1}{2}(-2)^2 = -2$	$-2(-2)^2 = -8$
-1	$-(-1)^2 = -1$	$-\frac{1}{2}(-1)^2 = -\frac{1}{2}$	$-2(-1)^2 = -2$
0	0	0	0
1	-1	$-\frac{1}{2}$	-2
2	-4	-2	-8

Figure 15

$f(x) = ax^2$. Since $a < 0$, the graphs open down.

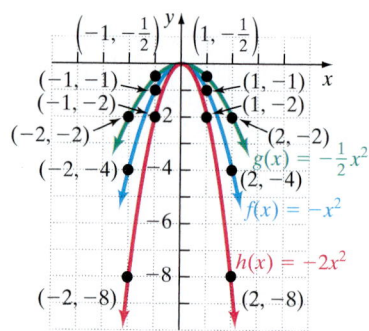

We summarize these conclusions below.

PROPERTIES OF THE GRAPH OF $f(x) = ax^2$

- If $a > 0$, the graph of $f(x) = ax^2$ will open upward. In addition, if $0 < a < 1$ (a is between 0 and 1), the opening in the graph will be "wider" than that of $y = x^2$. If $a > 1$, the opening in the graph will be "narrower" than that of $y = x^2$.

- If $a < 0$, the graph of $f(x) = ax^2$ will open downward. In addition, if $0 < |a| < 1$, the opening in the graph will be "wider" than that of $y = x^2$. If $|a| > 1$, the opening in the graph will be "narrower" than that of $y = x^2$.

- When $|a| > 1$, we say that the graph is **vertically stretched** by a factor of $|a|$. When $0 < |a| < 1$, we say that the graph is **vertically compressed** by a factor of $|a|$.

GRAPHING A FUNCTION OF THE FORM $f(x) = ax^2$

To obtain the graph of $f(x) = ax^2$ from the graph of $y = x^2$, multiply each y-coordinate on the graph of $y = x^2$ by a.

EXAMPLE 6 **Graphing a Quadratic Function of the Form $f(x) = ax^2$**

Use the graph of $y = x^2$ to obtain the graph of $f(x) = -2x^2$.

Solution

To obtain the graph of $f(x) = -2x^2$ from the graph of $y = x^2$, we multiply each y-coordinate on the graph of $y = x^2$ by -2 (the value of a). See Figure 16.

Figure 16

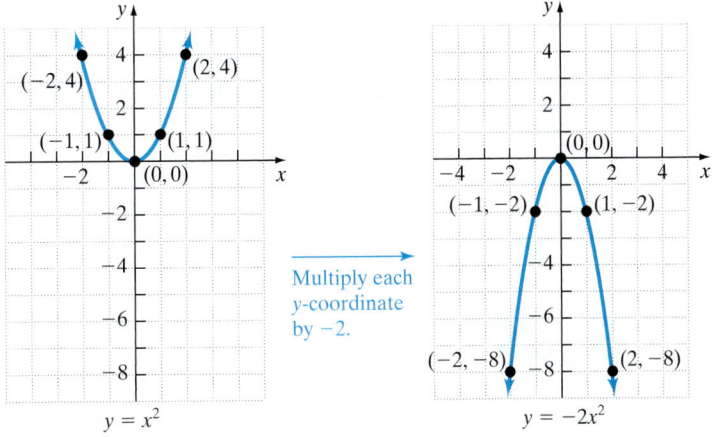

QUICK ✓ *Use the graph of $y = x^2$ to graph each quadratic function.*

7. $f(x) = 3x^2$

8. $f(x) = -\dfrac{1}{4}x^2$

---

④ **Graph Quadratic Functions of the Form $f(x) = ax^2 + bx + c$**

The graphs obtained in Examples 1–6 are typical of graphs of all quadratic functions. We call the graph of a quadratic function a **parabola** (pronounced puh-ráb-$\bar{o}$-luh). Refer to Figure 17, where two parabolas are shown.

Figure 17

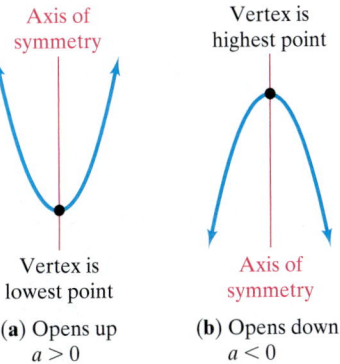

Axis of symmetry

Vertex is highest point

Vertex is lowest point

Axis of symmetry

(a) Opens up
$a > 0$

(b) Opens down
$a < 0$

Work Smart

The axis of symmetry is useful in graphing a quadratic function by hand.

The parabola in Figure 17(a) **opens up** (since $a > 0$) and has a lowest point; the parabola in Figure 17(b) **opens down** (since $a < 0$) and has a highest point. The lowest or highest point of a parabola is called the **vertex.** The vertical line passing through the vertex in each parabola in Figure 17 is called the **axis of symmetry** of the parabola. If we were to take the portion of the parabola to the right of the vertex and fold it over the axis of symmetry, it would lie directly on top of the portion of the parabola to the left of the vertex. Therefore, we say that the parabola is symmetric about its axis of symmetry. It is important to note that the axis of symmetry is not part of the graph of the quadratic function, but it will be useful when graphing quadratic functions using the methods we present in the next section.

Our goal right now is to combine the techniques learned from Examples 1–6 to graph any quadratic function. The techniques of shifting horizontally, shifting vertically, stretching, and compressing are collectively referred to as **transformations.** To graph any quadratic function of the form $f(x) = ax^2 + bx + c$ using transformations, we use the following steps.

Steps for Graphing Quadratic Functions Using Transformations

Step 1: Write the function $f(x) = ax^2 + bx + c$ as $f(x) = a(x - h)^2 + k$ by completing the square in x.

Step 2: Graph the function $f(x) = a(x - h)^2 + k$ using transformations.

Notice that we must first write the quadratic function $f(x) = ax^2 + bx + c$ as $f(x) = a(x - h)^2 + k$. This is necessary so we can determine the horizontal and vertical shifts.

EXAMPLE 7 **How to Graph a Quadratic Function of the Form $f(x) = ax^2 + bx + c$ Using Transformations**

Graph $f(x) = x^2 + 4x + 3$ using transformations. Identify the vertex and axis of symmetry of the parabola. Based on the graph, determine the domain and range of the quadratic function.

Step-by-Step Solution

Step 1: Write the function $f(x) = ax^2 + bx + c$ as $f(x) = a(x - h)^2 + k$ by completing the square in x.

$$f(x) = x^2 + 4x + 3$$

Group the terms involving x: $= (x^2 + 4x) + 3$

Complete the square in x by taking $\frac{1}{2}$ the coefficient on x and squaring the result: $\left(\frac{1}{2} \cdot 4\right)^2 = 4$.

Because we added 4, we must also subtract 4: $= (x^2 + 4x + 4) + 3 - 4$

$$= (x^2 + 4x + 4) - 1$$

Factor the perfect squared trinomial in parentheses: $= (x + 2)^2 - 1$

Step 2: Graph the function $f(x) = a(x - h)^2 + k$ using transformations.

The graph of $f(x) = (x + 2)^2 - 1$ is the graph of $y = x^2$ shifted 2 units left and 1 unit down. See Figure 18.

Figure 18

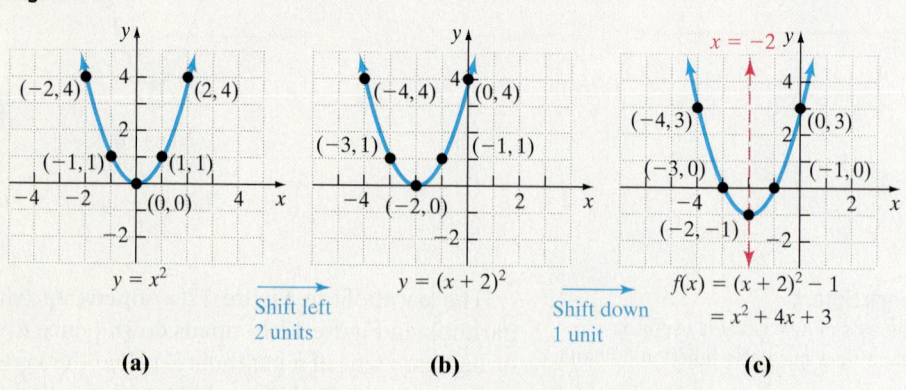

(a) $y = x^2$ — points $(-2, 4)$, $(2, 4)$, $(-1, 1)$, $(1, 1)$, $(0, 0)$

Shift left 2 units

(b) $y = (x + 2)^2$ — points $(-4, 4)$, $(0, 4)$, $(-3, 1)$, $(-1, 1)$, $(-2, 0)$

Shift down 1 unit

(c) $f(x) = (x + 2)^2 - 1 = x^2 + 4x + 3$ — axis $x = -2$; points $(-4, 3)$, $(0, 3)$, $(-3, 0)$, $(-1, 0)$, $(-2, -1)$

From the graph in Figure 18(c), we can see that the vertex of the parabola is $(-2, -1)$. In addition, because the parabola opens up (because $a = 1 > 0$), the vertex is the lowest point on the graph. The axis of symmetry is the line $x = -2$. The domain is the set of all real numbers, or using interval notation $(-\infty, \infty)$. The range is $\{y \mid y \geq -1\}$, or using interval notation $[-1, \infty)$.

EXAMPLE 8 **Graphing a Quadratic Function of the Form**
$f(x) = ax^2 + bx + c$ **Using Transformations**

Graph $f(x) = -2x^2 + 4x + 1$ using transformations. Identify the vertex and axis of symmetry of the parabola. Based on the graph, determine the domain and range of the quadratic function.

Solution

Write the function $f(x) = ax^2 + bx + c$ as $f(x) = a(x - h)^2 + k$ by completing the square in x.

$$f(x) = -2x^2 + 4x + 1$$

Group the terms involving x: $= (-2x^2 + 4x) + 1$

Factor out the coefficient of the square term, -2, from the parentheses: $= -2(x^2 - 2x) + 1$

Complete the square in x by taking $\frac{1}{2}$ the coefficient on x and squaring the result: $(\frac{1}{2} \cdot 2)^2 = 1$. We add 1 inside the parentheses. Because everything in the parentheses is multiplied by -2, so we really added -2, so we must add 2 to offset this: $= -2(x^2 - 2x + 1) + 1 + 2$

Factor the perfect square trinomial in parentheses: $= -2(x - 1)^2 + 3$

Now, graph the function $f(x) = a(x - h)^2 + k$ using transformations. Since $a = -2$, the parabola will open down and stretch by a factor of 2. Since $h = 1$ the parabola shifts 1 unit to the right; since $k = 3$, the parabola shifts 3 units up. See Figure 19.

Figure 19

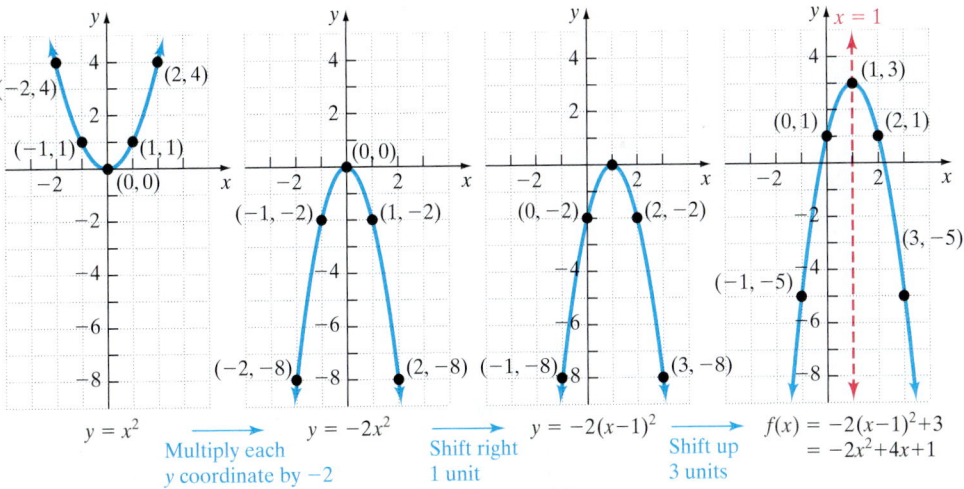

The vertex of the parabola is $(1, 3)$. In addition, because the parabola opens down (because $a = -2 < 0$), the vertex is the highest point on the graph. The axis of symmetry is the line $x = 1$. The domain is the set of all real numbers, or using interval notation $(-\infty, \infty)$. The range is $\{y \mid y \le 3\}$, or using interval notation $(-\infty, 3]$.

Work Smart: Study Skills

Be sure you know what $y = a(x - h)^2 + k$ represents in reference to the graph of $y = x^2$:

- $|a|$ represents the vertical stretch or compression factor: how wide or narrow the graph appears.
- The sign of a determines whether the parabola opens up or down.
- h represents the number of units the graph is shifted horizontally.
- k represents the number of units the graph is shifted vertically.

For example, $y = 4(x + 2)^2 - 3$ means that the graph of $y = x^2$ is stretched vertically by a factor of 4, is shifted 2 units horizontally to the left, and is shifted vertically down 3 units. The vertex of $y = 4(x + 2)^2 - 3$ is at $(-2, -3)$ and represents the low point of the graph since $a > 0$.

QUICK  *Graph each quadratic function using transformations. Based on the graph, determine the domain and range of each function.*

9. $f(x) = -3(x + 2)^2 + 1$ **10.** $f(x) = 2x^2 - 8x + 5$

---

⑤ **Find a Quadratic Function from Its Graph**

If we are given the vertex, (h, k), and one additional point on the graph of a quadratic function, we can find the quadratic function $f(x) = a(x - h)^2 + k$ that results in the given graph.

> **EXAMPLE 9** **Finding the Quadratic Function Given Its Vertex and One Other Point**

Determine the quadratic function whose graph is given in Figure 20. Write the function in the form $f(x) = a(x - h)^2 + k$.

Solution

The vertex is $(2, 3)$, so $h = 2$ and $k = 3$. Substitute these values into $f(x) = a(x - h)^2 + k$.

$$f(x) = a(x - h)^2 + k$$
$$h = 2, k = 3: \quad = a(x - 2)^2 + 3$$

To determine the value of a, we use the fact that $f(0) = -5$ (the y-intercept).

$$f(x) = a(x - 2)^2 + 3$$
$$x = 0, y = f(0) = -5: \quad -5 = a(0 - 2)^2 + 3$$
$$-5 = a(4) + 3$$
$$-5 = 4a + 3$$
$$\text{Subtract 3 from both sides:} \quad -8 = 4a$$
$$a = -2$$

The quadratic function whose graph is shown in Figure 20 is $f(x) = -2(x - 2)^2 + 3$. ∎

Figure 20

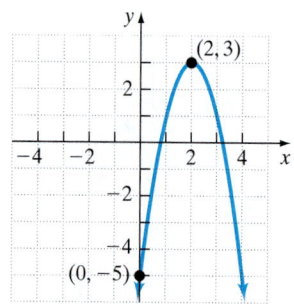

QUICK  *Find the quadratic function whose graph is given. Write the function in the form $f(x) = a(x - h)^2 + k$.*

11.

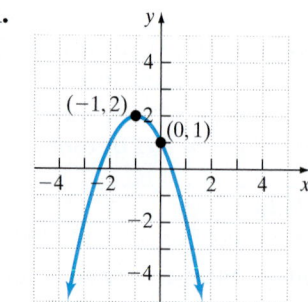

10.4 Exercises

Concepts and Vocabulary

In Problems 1–3, fill in the blanks.

1. A _____ _____ is a function of the form $f(x) = ax^2 + bx + c$ where a, b, and c are real numbers and $a \neq 0$.

2. To graph $f(x) = x^2 + k$ using the graph of $y = x^2$, shift the graph of $y = x^2$ vertically _____ k units if $k > 0$ and shift the graph of $y = x^2$ vertically _____ k units if $k < 0$.

3. When obtaining the graph of $f(x) = ax^2$ from the graph of $y = x^2$, we multiply each __-coordinate on the graph of $y = x^2$ by ___. If $|a| > 1$, we say that the graph is _____ _____ by a factor of $|a|$. If $0 < |a| < 1$, we say that the graph is _____ _____ by a factor of $|a|$.

In Problems 4–6, answer True or False to each statement.

4. To obtain the graph of $f(x) = x^2 + 5$ from the graph of $y = x^2$, shift the graph of $y = x^2$ vertically up 5 units.

5. To obtain the graph of $f(x) = (x + 5)^2$ from the graph of $y = x^2$, shift the graph of $y = x^2$ horizontally to the right 5 units.

6. The graph of $f(x) = -3x^2 + x + 6$ opens down.

7. What is the lowest or highest point on a parabola called? How do we know whether this point is a high point or a low point?

8. Why does the graph of a quadratic function open up if $a > 0$ and down if $a < 0$?

Building Skills

9. Match each quadratic function to its graph.

   (I) $f(x) = x^2 + 3$ **(II)** $f(x) = (x + 3)^2$

   (III) $f(x) = x^2 - 3$ **(IV)** $f(x) = (x - 3)^2$

(A)

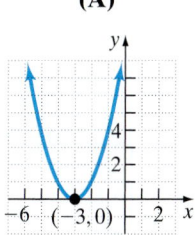

(B)

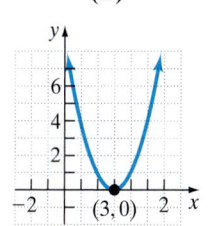

(C)

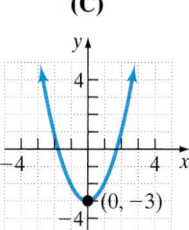

(D)

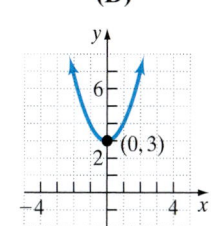

10. Match each quadratic function to its graph.

    (I) $f(x) = (x - 2)^2 - 4$ **(II)** $f(x) = -(x - 2)^2 + 4$

    (III) $f(x) = -(x + 2)^2 + 4$ **(IV)** $f(x) = 2(x - 2)^2 - 4$

                       (A)   **(B)**

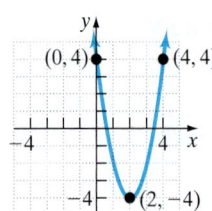 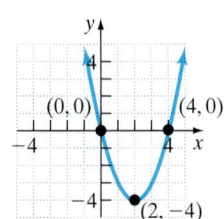

                       (C)   **(D)**

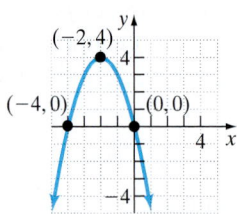 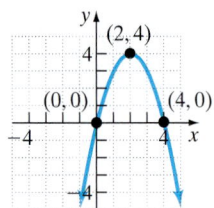

In Problems 11–18, verbally explain how to obtain the graph of the given quadratic function from the graph of $y = x^2$. For example, to obtain the graph of $f(x) = (x - 3)^2 - 6$ from the graph of $y = x^2$, we take the graph of $y = x^2$ and shift it 3 units to the right and 6 units down.

11. $f(x) = (x + 10)^2$ **12.** $G(x) = (x - 9)^2$

13. $F(x) = x^2 + 12$ **14.** $g(x) = x^2 - 8$

15. $H(x) = 2(x - 5)^2$ **16.** $h(x) = 4(x + 7)^2$

17. $f(x) = -3(x + 5)^2 + 8$ **18.** $F(x) = -\dfrac{1}{2}(x - 3)^2 - 5$

In Problems 19–42, use the graph of $y = x^2$ to graph the quadratic function.

19. $f(x) = x^2 + 1$ **20.** $f(x) = x^2 - 1$

21. $h(x) = x^2 + 6$ **22.** $g(x) = x^2 - 7$

23. $F(x) = (x - 3)^2$ **24.** $F(x) = (x - 2)^2$

25. $h(x) = (x + 2)^2$ **26.** $f(x) = (x + 4)^2$

27. $g(x) = 4x^2$ **28.** $G(x) = 5x^2$

29. $H(x) = \dfrac{1}{3}x^2$ **30.** $h(x) = \dfrac{3}{2}x^2$

31. $p(x) = -x^2$ **32.** $P(x) = -3x^2$

33. $f(x) = (x - 1)^2 - 3$ **34.** $g(x) = (x + 2)^2 - 1$

35. $F(x) = (x + 3)^2 + 1$ **36.** $G(x) = (x - 4)^2 + 2$

37. $h(x) = -(x + 3)^2 + 2$

38. $H(x) = -(x - 3)^2 + 5$

39. $G(x) = 2(x + 1)^2 - 2$

40. $F(x) = 3(x - 2)^2 - 1$

41. $H(x) = -\dfrac{1}{2}(x + 5)^2 + 3$

42. $f(x) = -\dfrac{1}{2}(x + 6)^2 + 2$

Mixed Practice

In Problems 43–64, write each function in the form $f(x) = a(x - h)^2 + k$. Then graph each quadratic function using transformations. Determine the vertex and axis of symmetry. Based on the graph, determine the domain and range of the quadratic function.

43. $f(x) = x^2 + 2x - 4$

44. $f(x) = x^2 + 4x - 1$

45. $g(x) = x^2 - 4x + 8$

46. $G(x) = x^2 - 2x + 7$

47. $f(x) = x^2 + 6x - 16$

48. $f(x) = x^2 + 4x + 5$

49. $F(x) = x^2 + x - 12$

50. $h(x) = x^2 - 7x + 10$

51. $H(x) = 2x^2 - 4x - 1$

52. $g(x) = 2x^2 + 4x - 3$

53. $P(x) = 3x^2 + 12x + 13$

54. $f(x) = 3x^2 + 18x + 25$

55. $F(x) = -x^2 - 10x - 21$

56. $g(x) = -x^2 - 8x - 14$

57. $g(x) = -x^2 + 6x - 1$

58. $f(x) = -x^2 + 10x - 17$

59. $H(x) = -2x^2 + 8x - 4$

60. $h(x) = -2x^2 + 12x - 17$

61. $f(x) = \dfrac{1}{3}x^2 - 2x + 4$

62. $f(x) = \dfrac{1}{2}x^2 + 2x - 1$

63. $G(x) = -12x^2 - 12x + 1$

64. $h(x) = -4x^2 + 4x$

Applying the Concepts

In Problems 65–74, write a quadratic function in the form $f(x) = a(x - h)^2 + k$ with the properties given.

65. Opens up; vertex at $(3, 0)$.

66. Opens up; vertex at $(0, 2)$.

67. Opens up; vertex at $(-3, 1)$

68. Opens up; vertex at $(4, -2)$

69. Opens down; vertex at $(5, -1)$

70. Opens down; vertex at $(-4, -7)$

71. Opens up; vertically stretched by a factor of 4; vertex at $(9, -6)$.

72. Opens up; vertically compressed by a factor of $\dfrac{1}{2}$; vertex at $(-5, 0)$.

73. Opens down; vertically compressed by a factor of $\dfrac{1}{3}$; vertex at $(0, 6)$.

74. Opens down; vertically stretched by a factor of 5; vertex at $(5, 8)$.

In Problems 75–80, determine the quadratic function whose graph is given.

75.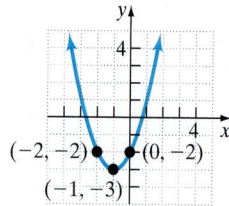

76.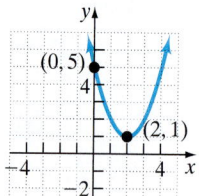

77.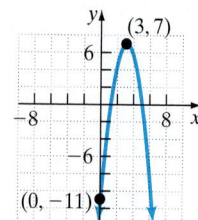

78.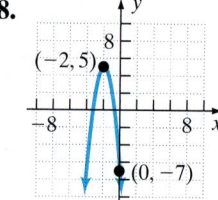

79.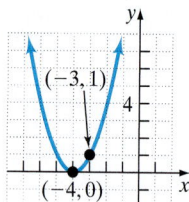

80.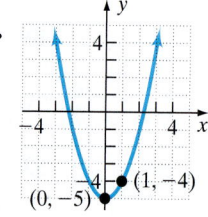

Extending the Concepts

81. Can a quadratic function have a range of $(-\infty, \infty)$? Justify your answer.

82. Can the graph of a quadratic function have more than one *y*-intercept? Justify your answer.

Synthesis Review

In Problems 83–85, divide.

83. $\dfrac{349}{12}$

84. $\dfrac{4x^2 + 19x - 1}{x + 5}$

85. $\dfrac{2x^4 - 11x^3 + 13x^2 - 8x}{2x - 1}$

86. Explain how division of real numbers is related to division of polynomials.

The Graphing Calculator

In Problems 87–94, graph each quadratic function. Determine the vertex and axis of symmetry. Based on the graph, determine the range of the function.

87. $f(x) = x^2 + 1.3$

88. $f(x) = x^2 - 3.5$

89. $g(x) = (x - 2.5)^2$

90. $G(x) = (x + 4.5)^2$

91. $h(x) = 2.3(x - 1.4)^2 + 0.5$

92. $H(x) = 1.2(x + 0.4)^2 - 1.3$

93. $F(x) = -3.4(x - 2.8)^2 + 5.9$

94. $f(x) = 0.3(x + 3.8)^2 - 8.9$

10.5 Graphing Quadratic Functions Using Properties

OBJECTIVES

1. Graph Quadratic Functions of the Form $f(x) = ax^2 + bx + c$
2. Find the Maximum or Minimum Value of a Quadratic Function
3. Model and Solve Optimization Problems Involving Quadratic Functions

Preparing for Graphing Quadratic Functions Using Properties

Before getting started, take the following readiness quiz. If you get a problem wrong, go back to the section cited and review the material.

1. Find the intercepts of the graph of $2x + 5y = 20$.   [Section 3.2, pp. 196–199]
2. Solve: $2x^2 - 3x - 20 = 0$ [Section 6.6, pp. 451–457]

In Section 10.4, we graphed quadratic functions using a method called transformations. We now introduce a second method for graphing quadratic equations that utilizes the properties of quadratic functions such as its intercepts, axis of symmetry, and vertex.

1 Graph Quadratic Functions of the Form $f(x) = ax^2 + bx + c$

We saw in Section 10.4 that a quadratic function $f(x) = ax^2 + bx + c$ can be written in the form $f(x) = a(x - h)^2 + k$ by completing the square in x. We also learned that the value of a determines whether the graph of the quadratic function (the parabola) opens up or down. In addition, we know that (h, k) is the vertex of the quadratic function.

We can obtain a formula for the vertex of a parabola by completing the square in $f(x) = ax^2 + bx + c, a \neq 0$, as follows:

$$f(x) = ax^2 + bx + c$$

Group terms involving x: $= (ax^2 + bx) + c$

Factor out a: $= a\left(x^2 + \dfrac{b}{a}x\right) + c$

Complete the square in x by taking $\frac{1}{2}$ the coefficient of x and squaring the result: $\left(\dfrac{1}{2} \cdot \dfrac{b}{a}\right)^2 = \dfrac{b^2}{4a^2}$. Because we added $\dfrac{b^2}{4a^2}$ inside the parentheses, we subtract

$a \cdot \dfrac{b^2}{4a^2} = \dfrac{b^2}{4a}$ outside the parentheses: $= a\left(x^2 + \dfrac{b}{a}x + \dfrac{b^2}{4a^2}\right) + c - \dfrac{b^2}{4a}$

Factor the perfect square trinomial; multiply c by $\dfrac{4a}{4a}$ to get a common denominator: $= a\left(x + \dfrac{b}{2a}\right)^2 + c \cdot \dfrac{4a}{4a} - \dfrac{b^2}{4a}$

Write expression in the form $f(x) = a(x - h)^2 + k$: $= a\left(x - \left(-\dfrac{b}{2a}\right)\right)^2 + \dfrac{4ac - b^2}{4a}$

If we compare $f(x) = a\left(x - \left(-\dfrac{b}{2a}\right)\right)^2 + \dfrac{4ac - b^2}{4a}$ to $f(x) = a(x - h)^2 + k$, we come to the following conclusion:

Work Smart

Another formula for the vertex is

$$\left(-\dfrac{b}{2a}, \dfrac{-D}{4a}\right)$$

where $D = b^2 - 4ac$, the discriminant.

THE VERTEX OF A PARABOLA

Any quadratic function $f(x) = ax^2 + bx + c, a \neq 0$, will have vertex

$$\left(-\dfrac{b}{2a}, \dfrac{4ac - b^2}{4a}\right)$$

Because the y-coordinate on the graph of any function can be found by evaluating the function at the corresponding x-coordinate, we can restate the coordinates of the vertex as

$$\left(-\dfrac{b}{2a}, f\left(-\dfrac{b}{2a}\right)\right)$$

Preparing for...Answers **1.** $(0, 4), (10, 0)$

2. $\left\{-\dfrac{5}{2}, 4\right\}$

Because the axis of symmetry intersects the vertex, we have that the axis of symmetry for any parabola is $x = -\dfrac{b}{2a}$. In addition, we know that the parabola will open up if $a > 0$ and down if $a < 0$. Using this information along with the intercepts of the graph of the quadratic function, we can obtain a complete graph.

The y-intercept is the value of the quadratic function $f(x) = ax^2 + bx + c$ at $x = 0$, that is, $f(0) = c$.

The x-intercepts, if there are any, are found by solving the quadratic equation

$$f(x) = ax^2 + bx + c = 0$$

As we learned in Section 10.2, this equation has two, one, or no real solutions, depending on the value of the discriminant $b^2 - 4ac$. We use the value of the discriminant to determine the number of x-intercepts the graph of the quadratic function will have.

> **THE x-INTERCEPTS OF THE GRAPH OF A QUADRATIC FUNCTION**
>
> 1. If the discriminant $b^2 - 4ac > 0$, the graph of $f(x) = ax^2 + bx + c$ has two different x-intercepts. The graph will cross the x-axis at the solutions to the equation $ax^2 + bx + c = 0$.
>
> 2. If the discriminant $b^2 - 4ac = 0$, the graph of $f(x) = ax^2 + bx + c$ has one x-intercept. The graph will touch the x-axis at the solution to the equation $ax^2 + bx + c = 0$.
>
> 3. If the discriminant $b^2 - 4ac < 0$, the graph of $f(x) = ax^2 + bx + c$ has no x-intercepts. The graph will not cross or touch the x-axis.

Figure 21 illustrates these possibilities for parabolas that open up.

Figure 21

$f(x) = ax^2 + bx + c,\ a > 0$

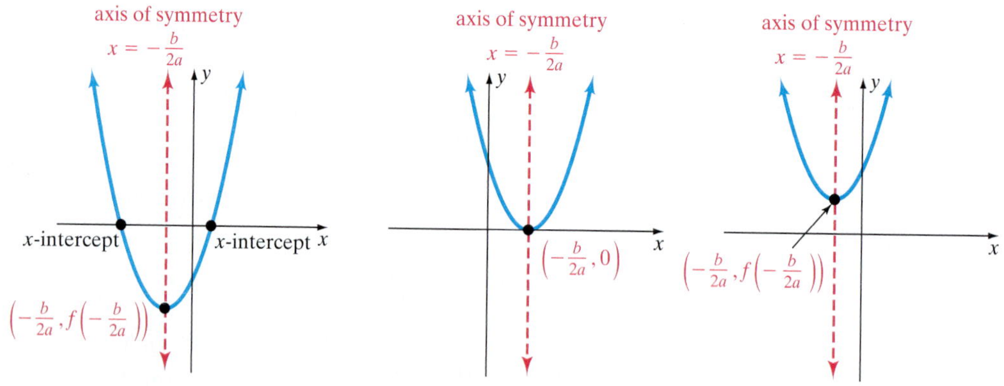

EXAMPLE 1 **How to Graph a Quadratic Function Using Its Properties**

Graph $f(x) = x^2 + 2x - 15$ using its properties.

Step-by-Step Solution

We compare $f(x) = x^2 + 2x - 15$ to $f(x) = ax^2 + bx + c$ and see that $a = 1, b = 2$, and $c = -15$.

Step 1: Determine whether the parabola opens up or down.   The parabola opens up because $a = 1 > 0$.

Step 2: Determine the vertex and axis of symmetry.

The x-coordinate of the vertex is

$$x = -\frac{b}{2a} = -\frac{2}{2(1)} = -1.$$

The y-coordinate of the vertex is

$$f\left(-\frac{b}{2a}\right) = f(-1)$$
$$= (-1)^2 + 2(-1) - 15$$
$$= 1 - 2 - 15$$
$$= -16$$

The vertex is $(-1, -16)$.
The axis of symmetry is the line

$$x = -\frac{b}{2a} = -1$$

Step 3: Determine the y-intercept, $f(0)$.

$$f(0) = 0^2 + 2(0) - 15$$
$$= -15$$

Step 4: Find the discriminant, $b^2 - 4ac$, to determine the number (and nature) of the x-intercepts. Then determine the x-intercepts, if any.

We have that $a = 1$, $b = 2$, and $c = -15$, so $b^2 - 4ac = (2)^2 - 4(1)(-15) = 64 > 0$. The parabola will have two different x-intercepts. We find the x-intercepts by solving

$$f(x) = 0$$
$$x^2 + 2x - 15 = 0$$

Factor: $(x + 5)(x - 3) = 0$

Zero-Product Property: $x + 5 = 0 \quad \text{or} \quad x - 3 = 0$
$$x = -5 \quad \text{or} \quad x = 3$$

Step 5: Plot the points. Use the axis of symmetry to find an additional point. Draw the graph of the quadratic function.

See Figure 22. Notice how we use the axis of symmetry to find the additional point $(-2, -15)$. The y-intercept, $(0, -15)$ is 1 unit to the right of the axis of symmetry, therefore, there must be a point 1 unit to the left of the axis of symmetry, $(-2, -15)$.

Figure 22

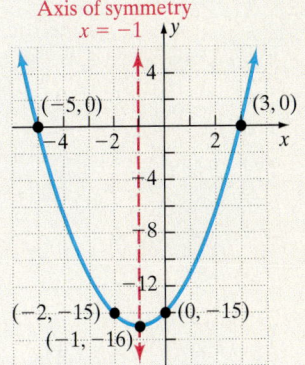

The following steps summarize how to graph any quadratic function using its properties.

Graphing a Quadratic Function Using Its Properties

To graph any quadratic function of the form $f(x) = ax^2 + bx + c$, $a \neq 0$, we use the following steps:

Step 1: Determine whether the parabola opens up or down.

Step 2: Determine the vertex and axis of symmetry.

Step 3: Determine the y-intercept, $f(0)$.

Step 4: Determine the discriminant, $b^2 - 4ac$.

- If $b^2 - 4ac > 0$, then the parabola has two x-intercepts, which are found by solving $f(x) = 0$ ($ax^2 + bx + c = 0$).
- If $b^2 - 4ac = 0$, the vertex is the x-intercept.
- If $b^2 - 4ac < 0$, there are no x-intercepts.

Step 5: Plot the points. Use the axis of symmetry to find an additional point. Draw the graph of the quadratic function.

QUICK ✓ *Graph the quadratic function using its properties.*

1. $f(x) = x^2 - 4x - 12$

In Example 1, the function was factorable, so the x-intercepts were rational numbers. When the quadratic function cannot be factored, we can use the quadratic formula to find the x-intercepts. For the purpose of graphing the quadratic function, we will approximate the x-intercepts rounded to two decimal places.

EXAMPLE 2 **Graphing a Quadratic Function Using Its Properties**

Graph $f(x) = -2x^2 + 12x - 5$ using its properties.

Solution

We compare $f(x) = -2x^2 + 12x - 5$ to $f(x) = ax^2 + bx + c$ and see that $a = -2$, $b = 12$, and $c = -5$. The parabola opens down because $a = -2 < 0$. The x-coordinate of the vertex is

$$x = -\frac{b}{2a} = -\frac{12}{2(-2)} = 3$$

The y-coordinate of the vertex is

$$f\left(-\frac{b}{2a}\right) = f(3)$$
$$= -2(3)^2 + 12(3) - 5$$
$$= -18 + 36 - 5$$
$$= 13$$

The vertex is $(3, 13)$. The axis of symmetry is the line

$$x = -\frac{b}{2a} = 3$$

The y-intercept is $f(0) = -2(0)^2 + 12(0) - 5 = -5$. Now we find the discriminant, $b^2 - 4ac$ to determine the number (and nature) of the x-intercepts. We have that

$a = -2$, $b = 12$, and $c = -5$, so $b^2 - 4ac = (12)^2 - 4(-2)(-5) = 104 > 0$. The parabola will have two different x-intercepts. We find the x-intercepts by solving

$$f(x) = 0$$
$$-2x^2 + 12x - 5 = 0$$

The equation cannot be solved by factoring (since $b^2 - 4ac$ is not a perfect square), so we use the quadratic formula:

$$x = \frac{-b \pm \sqrt{b^2 - 4ac}}{2a}$$

$a = -2$, $b = 12$, $b^2 - 4ac = 104$: $\quad = \dfrac{-12 \pm \sqrt{104}}{2(-2)}$

$\sqrt{104} = \sqrt{4 \cdot 26} = 2\sqrt{26}$: $\quad = \dfrac{-12 \pm 2\sqrt{26}}{-4}$

Divide -4 into each term in the numerator and simplify: $\quad x = 3 \pm \dfrac{\sqrt{26}}{-2}$ Exact solution

We evaluate the exact solution given by the quadratic formula and find the x-intercepts are approximately 0.45 and 5.55. Next, we plot the points and use the axis of symmetry to find an additional point by using the axis of symmetry. Finally, we draw the graph of the quadratic function. See Figure 23. Notice how we use the axis of symmetry to find the additional point $(6, -5)$. The y-intercept, $(0, -5)$ is 3 units to the left of the axis of symmetry, therefore, there must be a point 3 units to the right of the axis of symmetry, $(6, -5)$.

Figure 23

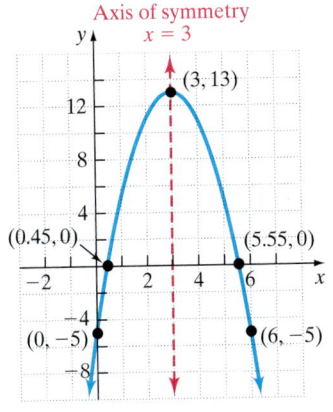

Work Smart

Notice that the vertex in Example 2 lies in quadrant I (above the x-axis) and the graph opens down. This tells us the graph must have two x-intercepts.

QUICK 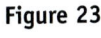 *Graph the quadratic function using its properties.*

2. $f(x) = -3x^2 + 12x - 7$

EXAMPLE 3 Graphing a Quadratic Function Using Its Properties

Graph $g(x) = x^2 - 8x + 16$ using its properties.

Solution

We compare $g(x) = x^2 - 8x + 16$ to $g(x) = ax^2 + bx + c$ and see that $a = 1$, $b = -8$, and $c = 16$. The parabola opens up because $a = 1 > 0$. The x-coordinate of the vertex is

$$x = -\frac{b}{2a} = -\frac{-8}{2(1)} = 4$$

The y-coordinate of the vertex is

$$f\left(-\frac{b}{2a}\right) = f(4)$$
$$= 4^2 - 8(4) + 16$$
$$= 16 - 32 + 16$$
$$= 0$$

The vertex is $(4, 0)$. The axis of symmetry is the line

$$x = -\frac{b}{2a} = 4$$

Figure 24

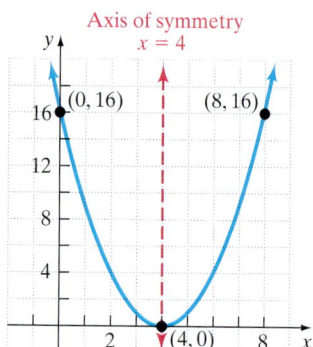

The y-intercept is $f(0) = 0^2 - 8(0) + 16 = 16$. We have that $a = 1, b = -8$, and $c = 16$, so $b^2 - 4ac = (-8)^2 - 4(1)(16) = 0$. The parabola will have one x-intercept, at the vertex. We verify this by solving

$$f(x) = 0$$
$$x^2 - 8x + 16 = 0$$
$$\text{Factor:} \quad (x - 4)^2 = 0$$
$$x = 4$$

Now we graph the parabola. See Figure 24. Notice how we use the axis of symmetry to find the additional point $(8, 16)$. The y-intercept, $(0, 16)$ is 4 units to the left of the axis of symmetry, so there must be a point 4 units to the right of the axis of symmetry, $(8, 16)$.

QUICK ✓ *Graph the quadratic function using its properties.*

3. $f(x) = x^2 + 6x + 9$

EXAMPLE 4   **Graphing a Quadratic Function Using Its Properties**

Graph $F(x) = 2x^2 + 6x + 5$ using its properties.

Solution

We compare $F(x) = 2x^2 + 6x + 5$ to $F(x) = ax^2 + bx + c$ and see that $a = 2$, $b = 6$, and $c = 5$. The parabola opens up because $a = 2 > 0$. The x-coordinate of the vertex is

$$x = -\frac{b}{2a} = -\frac{6}{2(2)} = -\frac{3}{2}$$

The y-coordinate of the vertex is

$$f\left(-\frac{b}{2a}\right) = f\left(-\frac{3}{2}\right)$$
$$= 2\left(-\frac{3}{2}\right)^2 + 6\left(-\frac{3}{2}\right) + 5$$
$$= 2\left(\frac{9}{4}\right) - 9 + 5$$
$$= \frac{1}{2}$$

Work Smart

Notice that the vertex lies in quadrant II (above the x-axis) and the graph opens up. This tells us the graph will have no x-intercepts.

The vertex is $\left(-\frac{3}{2}, \frac{1}{2}\right)$. The axis of symmetry is the line

$$x = -\frac{b}{2a} = -\frac{3}{2}$$

Figure 25

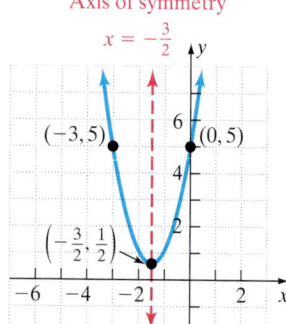

The y-intercept is $f(0) = 2(0)^2 + 6(0) + 5 = 5$. Because the parabola opens up and the vertex is in the second quadrant, we expect that the parabola will have no x-intercepts. We verify this by computing the discriminant with $a = 2, b = 6$, and $c = 5$:

$$b^2 - 4ac = 6^2 - 4(2)(5) = -4 < 0$$

The parabola will have no x-intercepts. Now we graph the parabola. See Figure 25. Notice how we use the axis of symmetry to find the additional point $(-3, 5)$. The y-intercept, $(0, 5)$ is $\frac{3}{2}(= 1.5)$ units to the right of the axis of symmetry, so there must be a point 1.5 units to the left of the axis of symmetry, $(-3, 5)$.

QUICK ✓ *Graph the quadratic function using its properties.*

4. $G(x) = -3x^2 + 9x - 8$

---

(2) Find the Maximum or Minimum Value of a Quadratic Function

Recall, the graph of a quadratic function $f(x) = ax^2 + bx + c$ is a parabola with vertex at $\left(-\dfrac{b}{2a}, f\left(-\dfrac{b}{2a}\right)\right)$. The vertex will be the highest point on the graph if $a < 0$ and the lowest point on the graph if $a > 0$. If the vertex is the highest point $(a < 0)$, then $f\left(-\dfrac{b}{2a}\right)$ is the **maximum value** of f. If the vertex is the lowest point $(a > 0)$, then $f\left(-\dfrac{b}{2a}\right)$ is the **minimum value** of f.

Work Smart
The vertex is not the maximum or minimum value—the y-coordinate of the vertex is the maximum or minimum value of the function.

This property of the graph of a quadratic function allows us to answer questions involving *optimization*. **Optimization** is the process whereby we find the maximum or minimum value(s) of a function. In the case of quadratic functions, the maximum $(a < 0)$ or minimum $(a > 0)$ is found through the vertex.

EXAMPLE 5 Finding the Maximum or Minimum Value of a Quadratic Function

Determine whether the quadratic function

$$f(x) = 3x^2 + 12x - 7$$

has a maximum or minimum value. Then find the maximum or minimum value of the function.

Solution

If we compare $f(x) = 3x^2 + 12x - 7$ to $f(x) = ax^2 + bx + c$, we find that $a = 3, b = 12$, and $c = -7$. Because $a = 3 > 0$, we know that the graph of the quadratic function will open up, so the function will have a minimum value. The minimum value of the function occurs at

$$x = -\frac{b}{2a} = -\frac{12}{2(3)} = -2$$

The minimum value of the function is

$$f\left(-\frac{b}{2a}\right) = f(-2) = 3(-2)^2 + 12(-2) - 7$$
$$= 3(4) - 24 - 7$$
$$= -19$$

So the minimum value of the function is -19 and occurs at $x = -2$. ∎

QUICK ✓ *Determine whether the quadratic function has a maximum or minimum value, then find the maximum or minimum value of the function.*

5. $f(x) = 2x^2 - 8x + 1$ **6.** $G(x) = -x^2 + 10x + 8$

As we stated at the beginning of Section 10.4, there are many applications that can be modeled using a quadratic function. Once a quadratic model has been determined, we can use properties of quadratic functions to answer interesting questions regarding the model.

EXAMPLE 6 Maximizing Revenue

Suppose that the marketing department of Dell Computer has found that, when a certain model of computer is sold at a price of p dollars, the daily revenue R (in dollars) as a function of the price p is

$$R(p) = -\frac{1}{4}p^2 + 400p$$

(a) For what price will the revenue be maximized?

(b) What is the maximum daily revenue?

Solution

(a) We notice that the revenue function is a quadratic function whose graph opens down since $a = -\frac{1}{4} < 0$. Therefore, the function will have a maximum at

$$p = -\frac{b}{2a} = -\frac{400}{2 \cdot \left(-\frac{1}{4}\right)} = -\frac{400}{-\frac{1}{2}} = 800$$

Revenue will be maximized when the price is $p = \$800$.

(b) The maximum daily revenue is found by letting $p = \$800$ in the revenue function.

$$R(800) = -\frac{1}{4} \cdot (800)^2 + 400 \cdot 800$$

$$= \$160{,}000$$

The maximum daily revenue is $\$160{,}000$. See Figure 26 for an illustration.

Figure 26

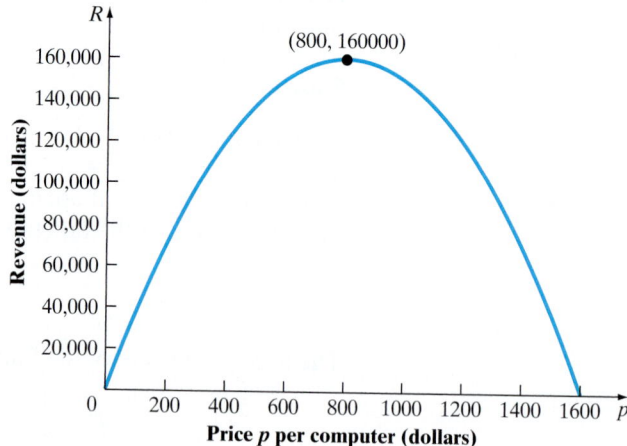

Quick ✔

7. Suppose that the marketing department of Texas Instruments has found that, when a certain model of calculator is sold at a price of p dollars, the daily revenue R (in dollars) as a function of the price p is $R(p) = -0.5p^2 + 75p$.

(a) For what price will the revenue be maximized?

(b) What is the maximum revenue?

③ Model and Solve Optimization Problems Involving Quadratic Functions

We now discuss models based on a verbal description that result in quadratic functions. As always, we shall use the problem solving strategy introduced in Section 2.5.

EXAMPLE 7 Maximizing the Area Enclosed by a Fence

A farmer has 3000 feet of fence to enclose a rectangular field. What is the maximum area that can be enclosed by the fence? What are the dimensions of the rectangle that encloses the most area?

Solution

Step 1: Identify We wish to determine the dimensions of a rectangle that maximize the area.

Step 2: Name We let w represent the width of the rectangle and l will represent the length.

Step 3: Translate Figure 27 illustrates the situation.

Figure 27

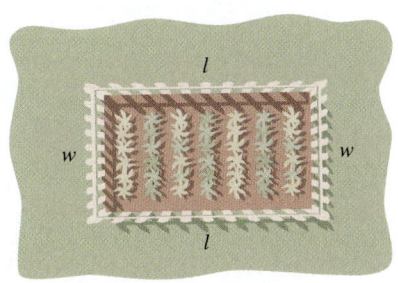

We know that the amount of fence available is 3000 feet. This means that the perimeter of the rectangle will be 3000 feet. Since the perimeter of a rectangle is $2l + 2w$ we have that

$$2l + 2w = 3000$$

The area A of the rectangle is

$$A = lw$$

To express A in terms of only one variable, we solve the equation $2l + 2w = 3000$ for l and then substitute for l in the area formula $A = lw$.

$$2l + 2w = 3000$$

Subtract $2w$ from both sides: $2l = 3000 - 2w$

Divide both sides by 2: $l = \dfrac{3000 - 2w}{2}$

Simplify: $l = 1500 - w$

Now let $l = 1500 - w$ in the formula $A = lw$.

$$A = (1500 - w)w$$
$$= -w^2 + 1500w$$

Now, A is a quadratic function of w.

$$A(w) = -w^2 + 1500w$$

Step 4: Solve We wish to find the dimensions that result in a maximum area enclosed by the fence. The model we developed in Step 3 is a quadratic function that opens down (because $a = -1 < 0$), so the vertex is a maximum point on the graph of A. The maximum value occurs at

$$w = -\frac{b}{2a} = -\frac{1500}{2(-1)} = 750$$

The maximum value of A is

$$A\left(-\frac{b}{2a}\right) = A(750) = -750^2 + 1500(750)$$
$$= 562{,}500 \text{ square feet}$$

We know that $l = 1500 - w$, so if the width is 750 feet, then the length will be $l = 1500 - 750 = 750$ feet.

Step 5: Check With a length and width of 750 feet, the perimeter is $2(750) + 2(750) = 3000$ feet. The area is $(750 \text{ feet})(750 \text{ feet}) = 562{,}500$ square feet. Everything checks!

Step 6: Answer The largest rectangle that can be enclosed by 3000 feet of fence has an area of 562,500 square feet. Its dimensions are 750 feet by 750 feet.

QUICK ✓

8. Roberta has 1000 yards of fence to enclose a rectangular field. What is the maximum area that can be enclosed by the fence? What are the dimensions of the rectangle that encloses the most area?

EXAMPLE 8 Pricing a Charter

Chicago Tours offers boat charters along the Chicago coastline on Lake Michigan. Normally, a ticket costs $20 per person, but for any group, Chicago Tours will lower the price of a ticket by $0.10 per person for each person in excess of 30. Determine the group size that will maximize revenue. What is the maximum revenue that can be earned from a group sale?

Solution

Step 1: Identify This is a direct translation problem involving revenue. Remember, revenue is price times quantity.

Step 2: Name We let x represent the number in the group in excess of 30.

Step 3: Translate Revenue is price times quantity. If 30 individuals make up a group, the revenue to Chicago Tours will be $20(30)$. If 31 individuals make up a group, revenue will be $19.90(31)$. If 32 individuals make up the group, revenue will be $19.80(32)$ In general, if x individuals make up a group in excess of 30, revenue will be $(20 - 0.1x)(x + 30)$. So the revenue R for a group that has $x + 30$ people in it is given by

$$R(x) = (20 - 0.1x)(x + 30)$$
$$\text{FOIL:} \quad = 20x + 600 - 0.1x^2 - 3x$$
$$\text{This is the model:} \quad = -0.1x^2 + 17x + 600$$

Step 4: Solve We wish to know the revenue maximizing number of individuals in a group. The function R is a quadratic function with a graph that opens down (because $a = -0.1 < 0$), so we know that the vertex is a maximum point. The value of x that results in a maximum is given by

$$x = -\frac{b}{2a} = -\frac{17}{2(-0.1)}$$
$$= 85$$

The maximum revenue is

$$R(85) = -0.1(85)^2 + 17(85) + 600$$
$$= \$1322.50$$

Step 5: Check Remember that x represents the number of passengers in excess of 30. Therefore, $30 + 85 = 115$ tickets should be sold to maximize revenue. The cost per ticket would be $\$20 - 0.1(85) = \$20 - \$8.50 = \$11.50$. Multiplying the cost per ticket by the number of passengers we obtain $\$11.50(115) = \$1322.50$. We have the right answer!

Step 6: Answer A group sale of 115 passengers will maximize revenue. The maximum revenue would be $1322.50.

Quick ✓

9. A compact disk manufacture charges $100 for each box of CDs ordered. However, it reduces the price by $1 for each box in excess of 30 boxes, but less than 90 boxes. Determine the number of boxes of CDs that should be sold to maximize revenue. What is the maximum revenue?

10.5 Exercises

For Extra Help:

Student Solutions Manual   CD Video   PH Math/Tutor Center   MathXL Tutorials on CD   MathXL®   MyMathLab

Concepts and Vocabulary

In Problems 1–3, fill in the blanks.

1. Any quadratic function $f(x) = ax^2 + bx + c, a \neq 0$, will have a vertex whose x-coordinate is $x =$ _____.

2. The graph of $f(x) = ax^2 + bx + c$ will have two different x-intercepts if $b^2 - 4ac$ _____ 0.

3. If the vertex of a quadratic function $f(x) = ax^2 + bx + c$ is the lowest point $(a > 0)$, then $f\left(-\dfrac{b}{2a}\right)$ is the _____ _____ of f.

In Problems 4–6, answer True or False to each statement.

4. The quadratic function $f(x) = -2x^2 - 3x + 6$ will have two x-intercepts.

5. The vertex of $f(x) = x^2 + 4x - 3$ is $(2, 29)$.

6. If the vertex of a quadratic function $f(x) = ax^2 + bx + c$ is the highest point, then $f\left(-\dfrac{b}{2a}\right)$ is the maximum value of f.

7. Explain how the discriminant is used to determine the number of x-intercepts the graph of a quadratic function will have.

8. Provide two methods for finding the vertex of any quadratic function $f(x) = ax^2 + bx + c$.

Building Skills

In Problems 9–16, use the discriminant to determine the number of x-intercepts the graph of each quadratic function will have. Then determine the x-intercepts.

9. $f(x) = x^2 - 6x - 16$ 10. $g(x) = 2x^2 - 7x - 4$ 11. $G(x) = -3x^2 + x - 1$

12. $H(x) = x^2 - 3x + 5$ 13. $h(x) = 4x^2 + 4x + 1$ 14. $f(x) = x^2 - 6x + 9$

15. $F(x) = 4x^2 - x - 1$ 16. $P(x) = -2x^2 + 3x + 1$

In Problems 17–56, graph each quadratic function using its properties by following Steps 1–5 on page 798. Based on the graph, determine the domain and range of the quadratic function.

17. $f(x) = x^2 - 4x - 5$ 18. $f(x) = x^2 - 2x - 8$

19. $G(x) = x^2 + 12x + 32$ 20. $g(x) = x^2 - 12x + 27$

21. $F(x) = -x^2 + 2x + 8$

22. $g(x) = -x^2 + 2x + 15$

23. $H(x) = x^2 - 4x + 4$

24. $h(x) = x^2 + 6x + 9$

25. $g(x) = x^2 + 2x + 5$

26. $f(x) = x^2 - 4x + 7$

27. $h(x) = -x^2 - 10x - 25$

28. $P(x) = -x^2 - 12x - 36$

29. $p(x) = -x^2 + 2x - 5$

30. $f(x) = -x^2 + 4x - 6$

31. $F(x) = 4x^2 - 4x - 3$

32. $f(x) = 4x^2 - 8x - 21$

33. $G(x) = -9x^2 + 18x + 7$

34. $g(x) = -9x^2 - 36x - 20$

35. $H(x) = 4x^2 - 4x + 1$

36. $h(x) = 9x^2 + 12x + 4$

37. $f(x) = -16x^2 - 24x - 9$

38. $F(x) = -4x^2 - 20x - 25$

39. $f(x) = 2x^2 + 8x + 11$

40. $F(x) = 3x^2 + 6x + 7$

41. $P(x) = -4x^2 + 6x - 3$

42. $p(x) = -2x^2 + 6x + 5$

43. $h(x) = x^2 + 5x + 3$

44. $H(x) = x^2 + 3x + 1$

45. $G(x) = -3x^2 + 8x + 2$

46. $F(x) = -2x^2 + 6x + 1$

47. $f(x) = 5x^2 - 5x + 2$

48. $F(x) = 4x^2 + 4x - 1$

49. $H(x) = -3x^2 + 6x$

50. $h(x) = -4x^2 + 8x$

51. $f(x) = x^2 - \dfrac{5}{2}x - \dfrac{3}{2}$

52. $g(x) = x^2 + \dfrac{5}{2}x - 6$

53. $G(x) = \dfrac{1}{2}x^2 + 2x - 6$

54. $H(x) = \dfrac{1}{4}x^2 + x - 8$

55. $F(x) = -\dfrac{1}{4}x^2 + x + 15$

56. $G(x) = -\dfrac{1}{2}x^2 - 8x - 24$

In Problems 57–68, determine whether the quadratic function has a maximum or minimum value. Then find the maximum or minimum value.

57. $f(x) = x^2 + 8x + 13$

58. $f(x) = x^2 - 6x + 3$

59. $G(x) = -x^2 - 10x + 3$

60. $g(x) = -x^2 + 4x + 12$

61. $F(x) = -2x^2 + 12x + 5$

62. $H(x) = -3x^2 + 12x - 1$

63. $h(x) = 4x^2 + 16x - 3$

64. $G(x) = 5x^2 + 10x - 1$

65. $f(x) = 2x^2 - 5x + 1$

66. $F(x) = 3x^2 + 4x - 3$

67. $H(x) = -3x^2 + 4x + 1$

68. $h(x) = -4x^2 - 6x + 1$

Applying the Concepts

69. Revenue Function Suppose that the marketing department of Panasonic has found that, when a certain model of DVD player is sold at a price of p dollars, the daily revenue R (in dollars) as a function of the price p is $R(p) = -2.5p^2 + 600p$.

(a) For what price will the daily revenue be maximized?

(b) What is the maximum daily revenue?

70. Revenue Function Suppose that the marketing department of Samsung has found that, when a certain model of cellular telephone is sold at a price of p dollars, the daily revenue R (in dollars) as a function of the price p is $R(p) = -5p^2 + 600p$.

(a) For what price will the daily revenue be maximized?

(b) What is the maximum daily revenue?

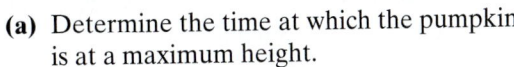 **71. Marginal Cost** The marginal cost of a product can be thought of as the cost of producing one additional unit of output. For example, if the marginal cost of producing the fiftieth product is \$6.30, then it costs \$6.30 to increase production from 49 to 50 units of output. Suppose that the marginal cost C (in dollars) to produce x digital cameras is given by $C(x) = 0.05x^2 - 6x + 215$. How many digital cameras should be produced to minimize marginal cost? What is the minimum marginal cost?

72. Marginal Cost (See Problem 71.) The marginal cost C (in dollars) of manufacturing x portable CD players is given by $C(x) = 0.05x^2 - 9x + 435$. How many portable CD players should be manufactured to minimize marginal cost? What is the minimum marginal cost?

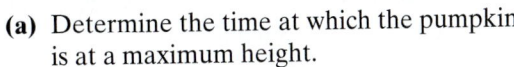 **73. Punkin Chunkin** Suppose that an air cannon in the Punkin Chunkin contest whose muzzle is 10 feet above the ground fires a pumpkin at an angle of 45° to the horizontal with a muzzle velocity of 335 feet per second. The model $s(t) = -16t^2 + 240t + 10$ can be used to estimate the height s of an object after t seconds.

(a) Determine the time at which the pumpkin is at a maximum height.

(b) Determine the maximum height of the pumpkin.

(c) After how long will the pumpkin strike the ground?

74. Punkin Chunkin Suppose that a catapult in the Punkin Chunkin contest releases a pumpkin 8 feet above the ground at an angle of 45° to the horizontal with an initial speed 220 feet per second. The model $s(t) = -16t^2 + 155t + 8$ can be used to estimate the height s of an object after t seconds.

(a) Determine the time at which the pumpkin is at a maximum height.

(b) Determine the maximum height of the pumpkin.

(c) After how long will the pumpkin strike the ground?

75. Punkin Chunkin Suppose that an air cannon in the Punkin Chunkin contest whose muzzle is 10 feet above the ground fires a pumpkin at an angle of 45° to the horizontal with a muzzle velocity of 335 feet per second. The model

$$h(x) = \frac{-32}{335^2}x^2 + x + 10$$ can be used to estimate the height h of an object after

the pumpkin has traveled x feet.

(a) How far from the cannon will the pumpkin reach a maximum height?

(b) What is the maximum height of the pumpkin?

(c) How far will the pumpkin travel before it strikes the ground?

(d) Compare your answer in part (b) of this problem with the answer found in part (b) of Problem 73. Why might the answers differ?

76. Punkin Chunkin Suppose that a catapult in the Punkin Chunkin contest releases a pumpkin 8 feet above the ground at an angle of 45° to the horizontal with an initial speed 220 feet per second. The model $h(x) = \dfrac{-32}{220^2}x^2 + x + 8$ can be used to estimate the height h of an object after the pumpkin has traveled x feet.

  (a) How far from the cannon will the pumpkin reach a maximum height?
  (b) What is the maximum height of the pumpkin?
  (c) How far will the pumpkin travel before it strikes the ground?
  (d) Compare your answer in part (b) of this problem with the answer found in part (b) of Problem 74. Why might the answers differ?

77. Life Cycle Hypothesis The Life Cycle Hypothesis from Economics was presented by Franco Modigliani in 1954. One of its components states that income is a function of age. The function $I(a) = -55a^2 + 5119a - 54{,}448$ represents the relation between average annual income I and age a.

  (a) According to the model, at what age will average income be a maximum?
  (b) According to the model, what is the maximum average income?

78. Advanced Degrees The function $P(x) = -0.008x^2 + 0.868x - 11.884$ models the percentage of the United States population whose age is given by x that have earned an advanced degree (more than a bachelor's degree) as of March, 2000. (*Source:* Based on data obtained from the U.S. Census Bureau)

  (a) What is the age for which the highest percentage of Americans have earned an advanced degree?
  (b) According to the model, what is the percentage of Americans that have earned an advanced degree at the age found in part (a)?

79. Fun with Numbers The sum of two numbers is 36. Find the numbers such that their product is a maximum.

80. Fun with Numbers The sum of two numbers is 50. Find the numbers such that their product is a maximum.

81. Fun with Numbers The difference of two numbers is 18. Find the numbers such that their product is a minimum.

82. Fun with Numbers The difference of two numbers is 10. Find the numbers such that their product is a minimum.

83. Enclosing a Rectangular Field Maurice has 500 yards of fencing and wishes to enclose a rectangular area. What is the maximum area that can be enclosed by the fence? What are the dimensions of the area enclosed?

84. Enclosing a Rectangular Field Maude has 800 yards of fencing and wishes to enclose a rectangular area. What is the maximum area that can be enclosed by the fence? What are the dimensions of the area enclosed?

85. Maximizing an Enclosed Area A farmer with 2000 meters of fencing wants to enclose a rectangular plot that borders on a river. If the farmer does not fence the side along the river, what is the largest area that can be enclosed? What is the length of the side of each enclosed area? See the figure.

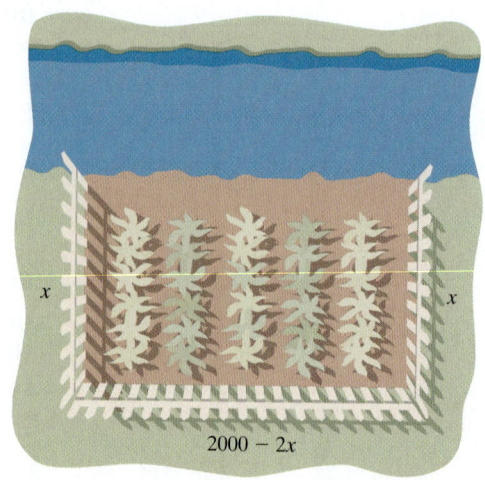

86. Maximizing an Enclosed Area A farmer with 8000 meters of fencing wants to enclose a rectangular plot and then divide it into two plots with a fence parallel to one of the sides. See the figure. What is the largest area that can be enclosed? What are the lengths of the sides of each part of the enclosed area?

87. Constructing Rain Gutters A rain gutter is to be made of aluminum sheets that are 20 inches wide by turning up the edges 90°. What depth will provide maximum cross-sectional area and hence allow the most water to flow?

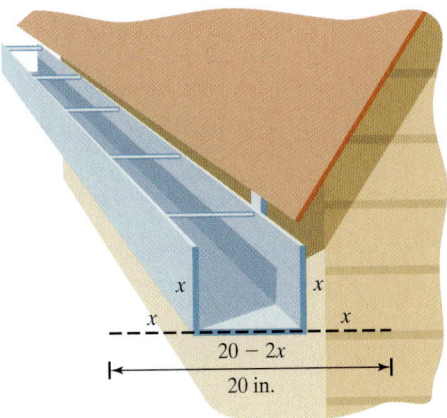

88. Maximizing the Volume of a Box A box with a rectangular base is to be constructed such that the perimeter of the base of the box is to be 40 inches. The height of the box must be 15 inches. Find the dimensions of the box such that the volume is maximized. What is the maximum volume?

89. Revenue Function Weekly demand for jeans at a department store obeys the demand equation

$$x = -p + 110$$

where x is the quantity demanded and p is the price (in dollars).

(a) Express the revenue R as a function of p. (**Hint:** $R = xp$)
(b) What price p maximizes revenue? What is the maximum revenue?
(c) How many pairs of jeans will be sold at the revenue maximizing price?

90. Revenue Function Demand for hot dogs at a baseball game obeys the demand equation

$$x = -800p + 8000$$

where x is the quantity demanded and p is the price (in dollars).

(a) Express the revenue R as a function of x. (**Hint:** $R = xp$)
(b) What price p maximizes revenue? What is the maximum revenue?
(c) How many hot dogs will be sold at the revenue maximizing price?

Extending the Concepts

Answer Problems 91 and 92 using the following information: A quadratic function of the form $f(x) = ax^2 + bx + c$ with $b^2 - 4ac > 0$ may also be written in the form $f(x) = a(x - r_1)(x - r_2)$, where r_1 and r_2 are the x-intercepts of the graph of the quadratic function.

91. (a) Find a quadratic function whose x-intercepts are 2 and 6 with $a = 1$; $a = 2$, and $a = -2$.
  (b) How does the value of a affect the intercepts?
  (c) How does the value of a affect the axis of symmetry?
  (d) How does the value of a affect the vertex?

92. (a) Find a quadratic function whose x-intercepts are -1 and 5 with $a = 1$; $a = 2$, and $a = -2$.
  (b) How does the value of a affect the intercepts?
  (c) How does the value of a affect the axis of symmetry?
  (d) How does the value of a affect the vertex?

93. Refer to Example 6 on page 802. Notice that if the price charged for the computer is $0 or $1,600 the revenue is $0. It is easy to explain why revenue would be $0 if the price charged is $0, but how can revenue be $0, if the price charged is $1600?

Synthesis Review

In Problems 94–97, graph each function using point-plotting.

94. $f(x) = -2x + 12$

95. $G(x) = \dfrac{1}{4}x - 2$

96. $f(x) = x^2 - 5$

97. $f(x) = (x + 2)^2 + 4$

98. For each function in Problems 94–97, explain an alternative method for graphing the function. Which method do you prefer? Why?

The Graphing Calculator: Finding the Vertex

Graphing calculators have a maximum and a minimum feature that allows us to determine the coordinates of the vertex of a parabola. For example, to find the vertex of $f(x) = x^2 + 2x - 15$ using a TI-84 graphing calculator, we use the MINIMUM feature. See Figure 28. The vertex is $(-1, -16)$.

Figure 28

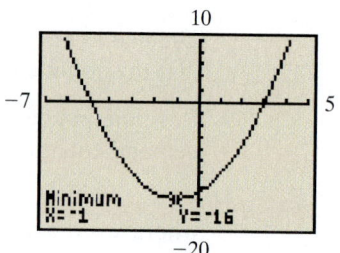

In Problems 99–106, use a graphing calculator to graph each quadratic function. Using the MAXIMUM or MINIMUM feature on the calculator, determine the vertex. If necessary, round your answers to two decimal places.

99. $f(x) = x^2 - 7x + 3$

100. $f(x) = x^2 + 3x + 8$

101. $G(x) = -2x^2 + 14x + 13$

102. $g(x) = -4x^2 - x + 11$

103. $F(x) = 5x^2 + 3x - 20$

104. $F(x) = 3x^2 + 2x - 21$

105. $H(x) = \dfrac{1}{2}x^2 - \dfrac{2}{3}x + 5$

106. $h(x) = \dfrac{3}{4}x^2 + \dfrac{4}{3}x - 1$

107. On the same screen, graph the family of parabolas $f(x) = x^2 + 2x + c$ for $c = -3$, $c = 0$, and $c = 1$. Describe the role that c plays in the graph for this family of functions.

108. On the same screen, graph the family of parabolas $f(x) = x^2 + bx + 1$ for $b = -4$, $b = 0$, and $b = 4$. Describe the role that b plays in the graph for this family of functions.

10.6 Quadratic Inequalities

(1) Solve Quadratic Inequalities

Preparing for Quadratic Inequalities

Before getting started, take the following readiness quiz. If you get a problem wrong, go back to the section cited and review the material.

1. Write $-4 \le x < 5$ in interval notation. [Section 2.8, pp. 150–153]

2. Solve: $3x + 5 > 5x - 3$ [Section 2.8, pp. 153–159]

In Section 2.8, we solved linear inequalities in one variable such as $2x - 3 > 4x + 5$. We were able to solve these inequalities using methods that were similar to solving linear equations. We also learned to represent the solution set to such an inequality using either set-builder notation or interval notation.

Unfortunately, the approach to solving inequalities involving quadratic expressions is not a simple extension of solving quadratic equations, but we will use the skills developed in solving quadratic equations to solve quadratic inequalities.

(1) Solve Quadratic Inequalities

We begin with a definition.

> **DEFINITION**
>
> A **quadratic inequality** is an inequality of the form
>
> $$ax^2 + bx + c > 0 \quad \text{or} \quad ax^2 + bx + c < 0 \quad \text{or}$$
> $$ax^2 + bx + c \ge 0 \quad \text{or} \quad ax^2 + bx + c \le 0$$
>
> where $a \ne 0$.

We will present two methods for solving quadratic inequalities. The first method is a graphical approach to the solution, while the second method is algebraic.

To help understand the logic behind the graphical approach, consider the following. Suppose we were asked to solve the inequality $ax^2 + bx + c > 0$. If we let $f(x) = ax^2 + bx + c$, then we are looking for all x-values such that $f(x) > 0$. Since f represents the y-values of the graph of the function $f(x) = ax^2 + bx + c$, we are basically looking for all x-values such that the graph of f is above the x-axis. This occurs when the graph lies in either quadrant I or II of the Cartesian plane. If we were asked to solve $f(x) < 0$, we would look for the x-values such that the graph is below the x-axis. That is, we would look for all x-values such that the graph lies in either quadrant III or IV of the Cartesian plane. See Figure 29 for an illustration of the idea.

Figure 29

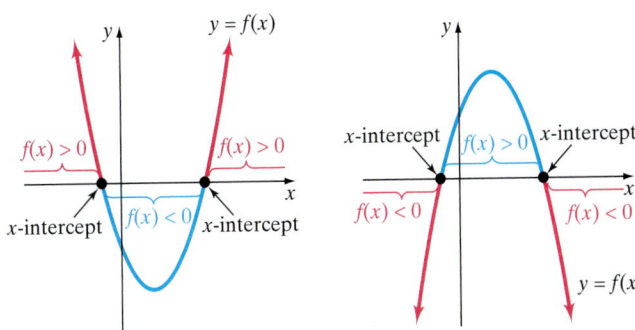

EXAMPLE 1   How to Solve a Quadratic Inequality Using the Graphical Method

Solve $x^2 - 4x - 5 \geq 0$ using the graphical method.

Step-by-Step Solution

Step 1: Write the inequality so that $ax^2 + bx + c$ is on one side of the inequality and 0 is on the other.

$$x^2 - 4x - 5 \geq 0$$

Step 2: Graph the function $f(x) = ax^2 + bx + c$. Be sure to label the x-intercepts of the graph.

We wish to graph $f(x) = x^2 - 4x - 5$.

x-intercepts: $\quad\quad\quad\quad f(x) = 0 \quad\quad$ y-intercept: $\quad f(0) = -5$

$$x^2 - 4x - 5 = 0$$
$$(x - 5)(x + 1) = 0$$
$$x = 5 \quad \text{or} \quad x = -1$$

Vertex: $\quad x = -\dfrac{b}{2a} = -\dfrac{-4}{2(1)} = 2$

$$f\left(-\dfrac{b}{2a}\right) = f(2) = -9$$

The vertex is at $(2, -9)$. Figure 30 shows the graph.

Figure 30

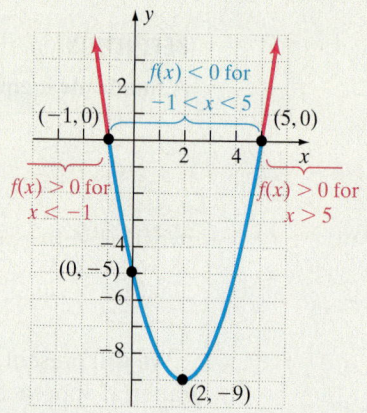

Step 3: From the graph, determine where the function is positive and determine where the function is negative. Use the graph to determine the solution set to the inequality.

From the graph shown in Figure 30, we can see that the graph of $f(x) = x^2 - 4x - 5$ is greater than 0 for $x < -1$ or $x > 5$. Because the inequality is nonstrict, we include the x-intercepts in the solution. So, the solution is $\{x | x \leq -1 \text{ or } x \geq 5\}$ using set-builder notation; the solution is $(-\infty, -1] \cup [5, \infty)$ using interval notation. See Figure 31 for a graph of the solution set.

Figure 31

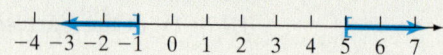

QUICK ✓ *Solve the quadratic inequality using the graphical method. Graph the solution set.*

1. $x^2 + 3x - 10 \geq 0$

The second method for solving inequalities is an algebraic method. This method is based upon the Rules of Signs when multiplying or dividing two (or more) real numbers. Recall that the product of two positive real numbers is positive, the product of a positive real number and a negative real number is negative, and the product of two negative real numbers is positive.

Let's solve the inequality from Example 1 using the algebraic method.

EXAMPLE 2   How to Solve a Quadratic Inequality Using the Algebraic Method

Solve $x^2 - 4x - 5 \geq 0$ using the algebraic method.

Step-by-Step Solution

Step 1: Write the inequality so that $ax^2 + bx + c$ is on one side of the inequality and 0 is on the other.

$$x^2 - 4x - 5 \geq 0$$

Step 2: Determine the solutions to the equation $ax^2 + bx + c = 0$.

$$x^2 - 4x - 5 = 0$$

Factor: $(x - 5)(x + 1) = 0$

Use Zero-Product Property: $x - 5 = 0$ or $x + 1 = 0$
$\qquad\qquad\qquad\qquad\qquad\qquad x = 5$ or $x = -1$

Step 3: Use the solutions to the equation solved in Step 2 to separate the real number line into intervals.

We separate the real number line into the following intervals:

$$(-\infty, -1) \qquad (-1, 5) \qquad (5, \infty)$$

Step 4: Write $x^2 - 4x - 5$ in factored form as $(x - 5)(x + 1)$. Within each interval formed in Step 3, choose a test point and determine the sign of each factor. Then determine the sign of the product. Also determine the value of $x^2 - 4x - 5$ at each solution found in Step 2.

In the interval $(-\infty, -1)$, we choose a test point of -2. The expression $x - 5$ equals -7 when $x = -2$. The expression $x + 1$ equals -1 when $x = -2$. Since the product of two negatives is positive, the expression $(x - 5)(x + 1)$ will be positive for all x in the interval $(-\infty, -1)$. In the interval $(-1, 5)$ we choose a test point of 0. For $x = 0$, $x - 5$ is negative, while $x + 1$ is positive, so $(x - 5)(x + 1)$ is negative. In the interval $(5, \infty)$ we choose a test point of 6. For $x = 6$, both $x - 5$ and $x + 1$ are positive, so $(x - 5)(x + 1)$ is positive. Table 9 shows these results and the sign of $(x - 5)(x + 1) = x^2 - 4x - 5$ in each interval. We also list the value of $(x - 5)(x + 1)$ at $x = -1$ and $x = 5$. We want to know where $x^2 - 4x - 5$ is greater than or equal to zero, so we include -1 and 5 in the solution, so the solution set is $\{x \mid x \leq -1 \text{ or } x \geq 5\}$ using set-builder notation; the solution is $(-\infty, -1] \cup [5, \infty)$ using interval notation.

Table 9

Interval	$(-\infty, -1)$		$(-1, 5)$		$(5, \infty)$
Test Point	-2	-1	0	5	6
Sign of $(x - 5)$	Negative	Negative	Negative	0	Positive
Sign of $(x + 1)$	Negative	0	Positive	Positive	Positive
Sign of $(x - 5)(x + 1)$	Positive	0	Negative	0	Positive
Conclusion	$(x - 5)(x + 1)$ is positive, so $(-\infty, -1)$ is part of the solution set	Because the inequality is nonstrict, -1 is part of the solution	$(x - 5)(x + 1)$ is negative, so $(-1, 5)$ is not part of the solution set	Because the inequality is nonstrict, 5 is part of the solution	$(x - 5)(x + 1)$ is positive, so $(5, \infty)$ is part of the solution set

QUICK ✓ *Solve the quadratic inequality using the algebraic method.*

2. $x^2 + 3x - 10 \geq 0$

SUMMARY: Solving Quadratic Inequalities

Graphical Method

Step 1: Write the inequality so that $ax^2 + bx + c$ is on one side of the inequality and 0 is on the other.

Step 2: Graph the function $f(x) = ax^2 + bx + c$. Be sure to label the x-intercepts of the graph.

Step 3: From the graph, determine where the function is positive and determine where the function is negative. Use the graph to determine the solution set to the inequality.

Algebraic Method

Step 1: Write the inequality so that $ax^2 + bx + c$ is on one side of the inequality and 0 is on the other.

Step 2: Determine the solutions to the equation $ax^2 + bx + c = 0$.

Step 3: Use the solutions to the equation solved in Step 2 to separate the real number line into intervals.

Step 4: Write $ax^2 + bx + c$ in factored form. Within each interval formed in Step 3, determine the sign of each factor. Then determine the sign of the product. Also determine the value of $ax^2 + bx + c$ at each solution found in Step 2.

(a) If the product of the factors is positive, then $ax^2 + bx + c > 0$ for all numbers x in the interval.

(b) If the product of the factors is negative, then $ax^2 + bx + c < 0$ for all numbers x in the interval.

If the inequality is not strict ($\leq$ or $\geq$), include the solutions of $ax^2 + bx + c = 0$ in the solution set.

Now let's do an example where we use both methods.

EXAMPLE 3 Solving a Quadratic Inequality

Solve: $-x^2 + 10 > 3x$

Solution

Graphical Method:

To make our lives a little easier, we will rearrange the inequality so that the coefficient on the square term is positive.

$$-x^2 + 10 > 3x$$

Add x^2 to both sides; subtract 10 from both sides: $\qquad 0 > x^2 + 3x - 10$

Use the fact that $0 > b$ is equivalent to $b < 0$: $\qquad x^2 + 3x - 10 < 0$

We wish to graph $f(x) = x^2 + 3x - 10$. We will graph the function using its properties by finding its intercepts and vertex.

x-intercepts: $\quad f(x) = 0$ $\qquad\qquad\qquad\qquad$ **y-intercept:** $\quad f(0) = -10$

$$x^2 + 3x - 10 = 0$$
$$(x + 5)(x - 2) = 0$$
$$x + 5 = 0 \quad \text{or} \quad x - 2 = 0$$
$$x = -5 \quad \text{or} \qquad x = 2$$

Figure 32

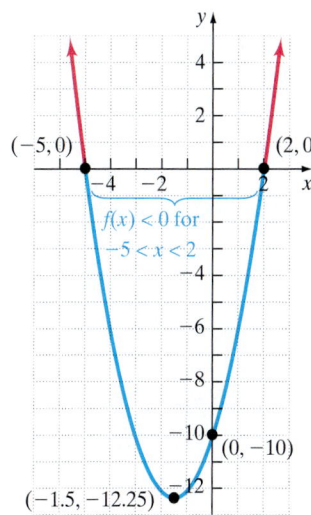

(−5, 0) (2, 0)

$f(x) < 0$ for $-5 < x < 2$

(0, −10)

(−1.5, −12.25)

Vertex:

$$x = -\frac{b}{2a} = -\frac{3}{2(1)} = -\frac{3}{2} = -1.5$$

$$f\left(-\frac{b}{2a}\right) = f\left(-\frac{3}{2}\right) = -\frac{49}{4} = -12.25$$

The vertex is at $(-1.5, -12.25)$. Figure 32 shows the graph.

From the graph shown in Figure 32, we can see that the graph of $f(x) = x^2 + 3x - 10$ is below the x-axis (and therefore $x^2 + 3x - 10 < 0$) for $-5 < x < 2$. Because the inequality is strict, we do not include the x-intercepts in the solution. So the solution is $\{x \mid -5 < x < 2\}$ using set-builder notation; the solution is $(-5, 2)$ using interval notation.

Algebraic Method:

Again, to make our lives a little easier, we will rearrange the inequality so that the coefficient on the square term is positive.

$$-x^2 + 10 > 3x$$

Add x^2 to both sides; subtract 10 from both sides: $0 > x^2 + 3x - 10$

Use the fact that $0 > b$ is equivalent to $b < 0$: $x^2 + 3x - 10 < 0$

Now, we solve the equation

$$x^2 + 3x - 10 = 0$$
$$(x + 5)(x - 2) = 0$$
$$x + 5 = 0 \quad \text{or} \quad x - 2 = 0$$
$$x = -5 \quad \text{or} \quad x = 2$$

We use the solutions to separate the real number line into the following intervals:

$$(-\infty, -5) \quad (-5, 2) \quad (2, \infty)$$

The factored form of $x^2 + 3x - 10$ is $(x + 5)(x - 2)$. Table 10 shows the sign of each factor and $(x + 5)(x - 2)$ in each interval. We also list the value of $(x + 5)(x - 2)$ at $x = -5$ and $x = 2$.

Table 10					
Interval	$(-\infty, -5)$		$(-5, 2)$		$(2, \infty)$
Test Point	−6	−5	0	2	3
Sign of $(x + 5)$	Negative	0	Positive	Positive	Positive
Sign of $(x - 2)$	Negative	Negative	Negative	0	Positive
Sign of $(x + 5)(x - 2)$	Positive	0	Negative	0	Positive
Conclusion	$(x + 5)(x - 2)$ is positive in the interval $(-\infty, -5)$, so it is not part of the solution set	The inequality is strict, so −5 is not part of the solution	$(x + 5)(x - 2)$ is negative in the interval $(-5, 2)$, so it is part of the solution set	The inequality is strict, so 2 is not part of the solution	$(x + 5)(x - 2)$ is positive in the interval $(2, \infty)$, so it is not part of the solution set

We want to know where $(x + 5)(x - 2) = x^2 + 3x - 10$ is negative, so we do not include −5 or 2 in the solution. The solution set is $\{x \mid -5 < x < 2\}$ using set-builder notation; the solution is $(-5, 2)$ using interval notation.

Figure 33 shows the graph of the solution set.

Figure 33

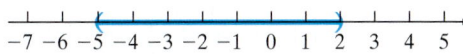

−7 −6 −5 −4 −3 −2 −1 0 1 2 3 4 5

QUICK ✔ *Solve the quadratic inequality. Graph the solution set.*

3. $-x^2 > 2x - 24$

EXAMPLE 4 **Solving a Quadratic Inequality**

Solve $2x^2 > 4x - 1$ using both the graphical and algebraic method. Graph the solution set.

Solution

Graphical Method:

$$2x^2 > 4x - 1$$

Subtract 4x from both sides; add 1 to both sides: $2x^2 - 4x + 1 > 0$

We wish to graph $f(x) = 2x^2 - 4x + 1$.

x-intercepts: $f(x) = 0$ **y-intercept:** $f(0) = 1$

$$2x^2 - 4x + 1 = 0$$

$a = 2; b = -4; c = 1$ $x = \dfrac{-(-4) \pm \sqrt{(-4)^2 - 4(2)(1)}}{2(2)}$

$$= \frac{4 \pm \sqrt{8}}{4}$$

$$= 1 \pm \frac{\sqrt{2}}{2}$$

$$\approx 0.29 \text{ or } 1.71$$

Vertex: $x = -\dfrac{b}{2a} = -\dfrac{-4}{2(2)} = 1$

$$f\left(-\frac{b}{2a}\right) = f(1) = -1$$

The vertex is at $(1, -1)$. Figure 34 shows the graph.

Figure 34

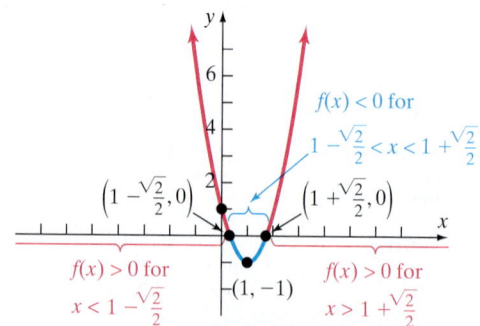

From the graph shown in Figure 34, we can see that the graph of $f(x) = 2x^2 - 4x + 1$ is greater than 0 for $x < 1 - \dfrac{\sqrt{2}}{2}$ or $x > 1 + \dfrac{\sqrt{2}}{2}$. Because the inequality is strict, we do not include the x-intercepts in the solution. So the solution is $\left\{ x \middle| x < 1 - \dfrac{\sqrt{2}}{2} \right.$ or $x > 1 + \dfrac{\sqrt{2}}{2} \right\}$ using set-builder notation; the solution is $\left(-\infty, 1 - \dfrac{\sqrt{2}}{2} \right) \cup \left(1 + \dfrac{\sqrt{2}}{2}, \infty \right)$ using interval notation.

Algebraic Method:

First, we put the inequality in the form $ax^2 + bx + c > 0$.

$$2x^2 > 4x - 1$$

Subtract $4x$ from both sides; add 1 to both sides: $2x^2 - 4x + 1 > 0$

Now we determine the solutions to the equation $2x^2 - 4x + 1 = 0$.

$$x = \frac{-(-4) \pm \sqrt{(-4)^2 - 4(2)(1)}}{2(2)}$$

$$= \frac{4 \pm \sqrt{8}}{4}$$

$$= 1 \pm \frac{\sqrt{2}}{2}$$

$$\approx 0.29 \text{ or } 1.71$$

We separate the real number line into the following intervals:

$$\left(-\infty, 1 - \frac{\sqrt{2}}{2}\right) \quad \left(1 - \frac{\sqrt{2}}{2}, 1 + \frac{\sqrt{2}}{2}\right) \quad \left(1 + \frac{\sqrt{2}}{2}, \infty\right)$$

Work Smart

The factor $\left(x - \left(1 - \frac{\sqrt{2}}{2}\right)\right)$ is close to the factor $x - 0.29$. So we might choose 0 as a test number. Now it's easy to see both factors are negative. Similarly, use $x - 1.71$ as an approximation of $\left(x - \left(1 + \frac{\sqrt{2}}{2}\right)\right)$.

The solutions to the equation $2x^2 - 4x + 1 = 0$ allow us to factor $2x^2 - 4x + 1$ as $\left(x - \left(1 - \frac{\sqrt{2}}{2}\right)\right)\left(x - \left(1 + \frac{\sqrt{2}}{2}\right)\right)$. We set up Table 11 using 0.29 as an approximation of $1 - \frac{\sqrt{2}}{2}$ and 1.71 as an approximation for $1 + \frac{\sqrt{2}}{2}$ as a guide to help us determine the sign of each factor.

Table 11

	$\left(-\infty, 1 - \frac{\sqrt{2}}{2}\right)$	$1 - \frac{\sqrt{2}}{2}$ ≈ 0.29	$\left(1 - \frac{\sqrt{2}}{2}, 1 + \frac{\sqrt{2}}{2}\right)$	$1 + \frac{\sqrt{2}}{2}$ ≈ 1.71	$\left(1 + \frac{\sqrt{2}}{2}, \infty\right)$
Interval	$\left(-\infty, 1 - \frac{\sqrt{2}}{2}\right)$		$\left(1 - \frac{\sqrt{2}}{2}, 1 + \frac{\sqrt{2}}{2}\right)$		$\left(1 + \frac{\sqrt{2}}{2}, \infty\right)$
Test Point	0	$1 - \frac{\sqrt{2}}{2}$	1	$1 + \frac{\sqrt{2}}{2}$	2
Sign of $\left(x - \left(1 - \frac{\sqrt{2}}{2}\right)\right)$	Negative	0	Positive	Positive	Positive
Sign of $\left(x - \left(1 + \frac{\sqrt{2}}{2}\right)\right)$	Negative	Negative	Negative	0	Positive
Sign of $\left(x - \left(1 - \frac{\sqrt{2}}{2}\right)\right)$ $\times \left(x - \left(1 + \frac{\sqrt{2}}{2}\right)\right)$	Positive	0	Negative	0	Positive
Conclusion	$2x^2 - 4x + 1$ is positive in the interval $\left(-\infty, 1 - \frac{\sqrt{2}}{2}\right)$, so it is part of the solution set	Since the inequality is strict, $1 - \frac{\sqrt{2}}{2}$ is not part of the solution	$2x^2 - 4x + 1$ is negative in the interval $\left(1 - \frac{\sqrt{2}}{2}, 1 + \frac{\sqrt{2}}{2}\right)$, so it is not part of the solution set	Since the inequality is strict, $1 + \frac{\sqrt{2}}{2}$ is not part of the solution	$2x^2 - 4x + 1$ is positive in the interval $\left(1 + \frac{\sqrt{2}}{2}, \infty\right)$, so it is part of the solution set

We want to know where $2x^2 - 4x + 1$ is positive, so we do not include $1 - \dfrac{\sqrt{2}}{2}$ or $1 + \dfrac{\sqrt{2}}{2}$ in the solution, so the solution set is $\left\{ x \mid x < 1 - \dfrac{\sqrt{2}}{2} \text{ or } x > 1 + \dfrac{\sqrt{2}}{2} \right\}$ using set-builder notation; the solution is $\left(-\infty, 1 - \dfrac{\sqrt{2}}{2} \right) \cup \left(1 + \dfrac{\sqrt{2}}{2}, \infty \right)$ using interval notation.

Figure 35 shows the graph of the solution set.

Figure 35

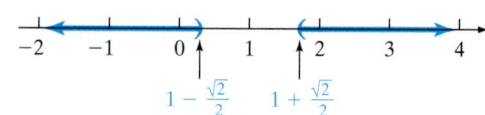

QUICK ✓

4. Solve $3x^2 > -x + 5$ using both the graphical and algebraic method. Graph the solution set.

10.6 Exercises

For Extra Help: Student Solutions Manual CD Video PH Math/Tutor Center MathXL Tutorials on CD MathXL® MyMathLab

Concepts and Vocabulary

In Problems 1 and 2, fill in the blanks.

1. The inequality $3x^2 - 7x + 2 < 0$ is an example of a(n) _____ inequality.

2. The sign of $(2x + 1)(x - 3)$ is _____ in the interval $\left(-\dfrac{1}{2}, 3 \right)$.

In Problems 3 and 4, answer True or False to each statement.

3. A test number for the interval $(-6, -2)$ could be -1.

4. The inequality $x^2 + 2 > 0$ is true for all real numbers.

5. The inequality $x^2 + 3 < -2$ has no solution. Explain why.

6. The inequality $x^2 - 1 \geq -1$ has the set of all real numbers as the solution. Explain why.

7. Explain when the endpoints of an interval are included in the solution set of a quadratic inequality.

8. Is the inequality $x^2 + 1 > 1$ true for all real numbers? Explain.

Building Skills

In Problems 9–12, use the graphs of the quadratic function f to determine the solution.

 9.

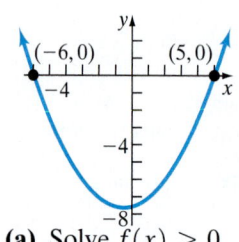

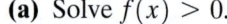

(a) Solve $f(x) > 0$.
(b) Solve $f(x) \leq 0$.

10.

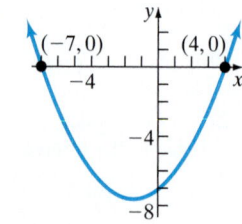

(a) Solve $f(x) > 0$.
(b) Solve $f(x) \leq 0$.

11.

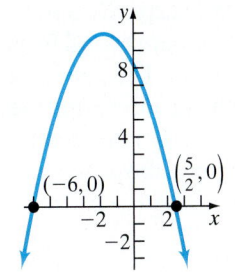

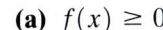

(a) $f(x) \geq 0$
(b) $f(x) < 0$

12.

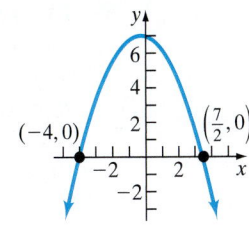

(a) $f(x) > 0$
(b) $f(x) \leq 0$

In Problems 13–40, solve each inequality. Graph the solution set.

13. $(x - 5)(x + 2) \geq 0$ **14.** $(x - 8)(x + 1) \leq 0$ **15.** $(x + 3)(x + 7) < 0$

16. $(x - 4)(x - 10) > 0$ **17.** $x^2 - 2x - 35 > 0$ **18.** $x^2 + 3x - 18 \geq 0$

19. $n^2 - 6n - 8 \leq 0$ **20.** $p^2 + 5p + 4 < 0$ **21.** $m^2 + 5m \geq 14$

22. $z^2 > 7z + 8$ **23.** $2q^2 \geq q + 15$ **24.** $2b^2 + 5b < 7$

25. $3x + 4 \geq x^2$ **26.** $x + 6 < x^2$ **27.** $-x^2 + 3x < -10$

28. $-x^2 > 4x - 21$ **29.** $-3x^2 \leq -10x - 8$ **30.** $-3m^2 \geq 16m + 5$

31. $x^2 + 4x + 1 < 0$ **32.** $x^2 - 3x - 5 \geq 0$ **33.** $-2a^2 + 7a \geq -4$

34. $-3p^2 < 3p - 5$ **35.** $z^2 + 2z + 3 > 0$ **36.** $y^2 + 3y + 5 \geq 0$

37. $2b^2 + 5b \leq -6$ **38.** $3w^2 + w < -2$ **39.** $x^2 + 6x + 9 \geq 0$

40. $p^2 - 8p + 16 \leq 0$

In Problems 41–46, for each function find the values of x that satisfy the given condition. Graph the solution set.

41. Solve $f(x) < 0$ if $f(x) = x^2 - 5x$.

42. Solve $f(x) > 0$ if $f(x) = x^2 + 4x$.

43. Solve $f(x) \geq 0$ if $f(x) = x^2 - 3x - 28$.

44. Solve $f(x) \leq 0$ if $f(x) = x^2 + 2x - 48$.

45. Solve $g(x) > 0$ if $g(x) = 2x^2 + x - 10$.

46. Solve $F(x) < 0$ if $F(x) = 2x^2 + 7x - 15$.

Applying the Concepts

In Problems 47–50, find the domain of the given function.

47. $f(x) = \sqrt{x^2 + 8x}$ **48.** $f(x) = \sqrt{x^2 - 5x}$

49. $g(x) = \sqrt{x^2 - x - 30}$ **50.** $G(x) = \sqrt{x^2 + 2x - 63}$

51. Physics A ball is thrown vertically upward with an initial speed of 80 feet per second from a cliff 500 feet above the sea level. The height s (in feet) of the ball from the ground after t seconds is $s(t) = -16t^2 + 80t + 500$. For what time t is the ball more than 596 feet above sea level?

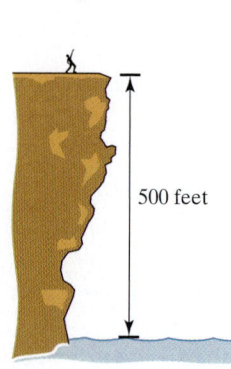

500 feet

52. Physics A water balloon is thrown vertically upward with an initial speed of 64 feet per second from the top of a building 200 feet above the ground. The height s (in feet) of the balloon from the ground after t seconds is $s(t) = -16t^2 + 64t + 200$. For what time t is the balloon more than 248 feet above the ground?

53. Revenue Function Suppose that the marketing department of Panasonic has found that, when a certain model of DVD player is sold a price of p dollars, the daily revenue R (in dollars) as a function of the price p is $R(p) = -2.5p^2 + 600p$. Determine the prices for which revenue will exceed \$35,750. That is, solve $R(p) > 35,750$.

54. Revenue Function Suppose that the marketing department of Samsung has found that, when a certain model of cellular telephone is sold a price of p dollars, the daily revenue R (in dollars) as a function of the price p is $R(p) = -5p^2 + 600p$. Determine the prices for which revenue will exceed \$17,500. That is, solve $R(p) > 17,500$.

Extending the Concepts

In Problems 55–58, solve each inequality algebraically by inspection. Then provide a verbal explanation of the solution.

55. $(x + 3)^2 \le 0$

56. $(x - 4)^2 > 0$

57. $(x + 8)^2 > -2$

58. $(3x + 1)^2 < -2$

59. Write a quadratic inequality that has $[-3, 2]$ as the solution set.

60. Write a quadratic inequality that has $(0, 5)$ as the solution set.

61. The inequalities $(3x + 2)^2 < 2$ and $(3x + 2)^{-2} > \frac{1}{2}$ have the same solution set. Why?

In Problems 62–67, use the techniques presented in this section to solve each inequality algebraically.

62. $(x + 1)(x - 2)(x - 5) > 0$

63. $(x + 3)(x - 1)(x - 3) < 0$

64. $(2x + 1)(x - 4)(x - 9) \le 0$

65. $(3x + 4)(x - 2)(x - 6) \ge 0$

66. $\dfrac{x^2 + 5x + 6}{x - 2} > 0$

67. $\dfrac{x^2 - 3x - 10}{x + 1} < 0$

Synthesis Review

In Problems 68–71, simplify each expression.

68. $\dfrac{3a^4 b}{12a^{-3}b^5}$

69. $(4mn^{-3})(-2m^4 n)$

70. $\left(\dfrac{3x^4 y}{6x^{-2}y^5}\right)^{1/2}$

71. $\left(\dfrac{9a^{2/3}b^{1/2}}{a^{-1/9}b^{3/4}}\right)^{-1}$

72. Do the Laws of Exponents presented in Section 5.4 also apply to the Laws of Exponents for rational exponents presented in Section 9.3?

The Graphing Calculator

A graphing calculator can be used to solve the quadratic inequality in Example 1 by graphing $Y_1 = x^2 - 4x - 5$. We use the ZERO feature of the calculator to find the x-intercepts of the graph. See Figures 36(a) and (b).

Figure 36

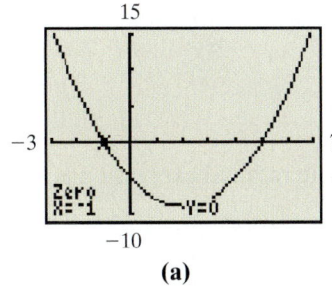
(a)

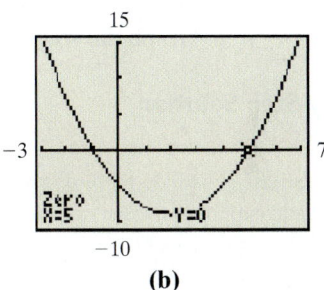
(b)

From Figures 36(a) and (b), we can see that $x^2 - 4x - 5 \geq 0$ for $x \leq -1$ or $x \geq 5$.

In Problems 73–76, solve each inequality using a graphing calculator.

73. $2x^2 + 7x - 49 > 0$

74. $2x^2 + 3x - 27 < 0$

75. $6x^2 + x \leq 40$

76. $8x^2 + 18x \geq 81$

10.7 Rational Inequalities

OBJECTIVE

① Solve a Rational Inequality

Preparing for Rational Inequalities

Before getting started, take the following readiness quiz. If you get a problem wrong, go back to the section cited and review the material.

1. Write $-1 < x \leq 8$ in interval notation.   [Section 2.8, pp. 150–153]

2. Solve: $2x + 3 > 4x - 9$ [Section 2.8, pp. 153–159]

3. Simplify: $\dfrac{x+2}{x-1} + 3$ [Section 7.5, pp. 510–517]

4. Solve: $\dfrac{x+1}{x-3} = \dfrac{5}{3}$ [Section 7.7, pp. 529–535]

Preparing for...Answers **1.** $(-1, 8]$
2. $\{x \mid x < 6\}$ or $(-\infty, 6)$
3. $\dfrac{4x-1}{x-1}$
4. $\{9\}$

Back in Section 2.8, we solved linear inequalities in one variable such as $2x - 3 > 4x + 5$. We were able to solve these inequalities using methods that were similar to solving linear equations. We also learned to represent the solution set to such an inequality using either set-builder notation or interval notation.

The approach to solving inequalities involving rational expressions is not a simple extension of solving rational equations. However, the approach to solving rational inequalities is similar to the algebraic approach to solving quadratic inequalities.

(1) Solve a Rational Inequality

A **rational inequality** is an inequality that contains a rational expression. Examples of rational inequalities include

$$\frac{1}{x} > 1 \qquad \frac{x-1}{x+5} \le 0 \qquad \frac{x^2+3x+2}{x-5} > 0 \qquad \frac{3}{x-5} < \frac{4x}{2x-1} + \frac{1}{x}$$

There are two keys to solving rational inequalities:

1. The quotient of two positive numbers is positive; the quotient of a positive and negative number is negative; and the quotient of two negative numbers is positive.

2. A rational expression may change signs (positive to negative or negative to positive) on either side of a value of the variable that makes the rational expression equal to 0 or for values for which the rational expression is undefined.

EXAMPLE 1 How to Solve a Rational Inequality

Solve $\dfrac{x+3}{x-4} \ge 0$. Graph the solution set.

Step-by-Step Solution

Step 1: Write the inequality so that a rational expression is on one side of the inequality and 0 is on the other. Be sure to write the rational expression as a single quotient.	The rational expression is already in the form that we need. $$\frac{x+3}{x-4} \ge 0$$
Step 2: Determine the numbers for which the rational expression equals 0 or is undefined.	The rational expression will equal 0 when $x = -3$. The rational expression is undefined when $x = 4$.
Step 3: Use the numbers found in Step 2 to separate the real number line into intervals.	We separate the real number line into the following intervals: $(-\infty, -3) \qquad (-3, 4) \qquad (4, \infty)$ Because the rational expression is undefined at $x = 4$, we plot an open circle at 4.
Step 4: Choose a test point within each interval formed in Step 3 to determine the sign of $x + 3$ and $x - 4$. Then determine the sign of the quotient.	In the interval $(-\infty, -3)$ we choose a test point of -4. The expression $x + 3$ equals -1 when $x = -4$. The expression $x - 4$ equals -8 when $x = -4$. Since the quotient of two negatives is positive, the expression $\frac{x+3}{x-4}$ will be positive when $x = -4$. So the expression $\frac{x+3}{x-4}$ will be positive for all x in the interval $(-\infty, -3)$. In the interval $(-3, 4)$ we choose a test point of 0. For $x = 0$, $x + 3$ is positive, while $x - 4$ is negative, so $\frac{x+3}{x-4}$ will be negative when $x = 0$. In the interval $(4, \infty)$ we choose a test point of 5. For $x = 5$, both $x + 3$ and $x - 4$ are positive, so $\frac{x+3}{x+4}$ will be positive. Table 12 shows these results, the sign of $\frac{x+3}{x-4}$ in each interval and the value of $\frac{x+3}{x-4}$ at $x = -3$ and $x = 4$. We want to know where $\frac{x+3}{x-4}$ is greater than or equal to zero, so we include the value of

x where $\dfrac{x+3}{x-4}$ is equal to zero. The solution set is $\{x \mid x \le -3 \text{ or } x > 4\}$ using set-builder notation; the solution is $(-\infty, -3] \cup (4, \infty)$ using interval notation. Notice that -3 is part of the solution set since $x = -3$ causes $\dfrac{x+3}{x-4}$ to equal zero, but 4 is not part of the solution set because it is not in the domain of $\dfrac{x+3}{x-4}$. Figure 37 shows the graph of the solution set.

Figure 37

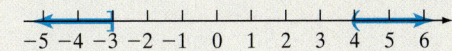

Table 12

	-3		0	4	
Interval	$(-\infty, -3)$		$(-3, 4)$		$(4, \infty)$
Test Point	-4	-3	0	4	5
Sign of $x + 3$	Negative	0	Positive	Positive	Positive
Sign of $x - 4$	Negative	Negative	Negative	0	Positive
Sign of $\dfrac{x+3}{x-4}$	Positive	0	Negative	Undefined	Positive
Conclusion	$\dfrac{x+3}{x-4}$ is positive, so $(-\infty, -3)$ is part of the solution set.	Because the inequality is nonstrict, -3 is part of the solution.	$\dfrac{x+3}{x-4}$ is negative, so $(-3, 4)$ is not part of the solution set.	4 cannot be part of the solution set because it causes division by 0.	$\dfrac{x+3}{x-4}$ is positive, so $(4, \infty)$ is part of the solution set.

Solving Rational Inequalities

Step 1: Write the inequality so that a rational expression is on one side of the inequality and 0 is on the other. Be sure to write the rational expression as a single quotient in factored form.

Step 2: Determine the numbers for which the rational expression equals 0 or is undefined.

Step 3: Use the numbers found in Step 2 to separate the real number line into intervals.

Step 4: Choose a test point within each interval formed in Step 3 to determine the sign of each factor in the numerator and denominator. Then determine the sign of the quotient.

- If the quotient is positive, then the rational expression is positive for all numbers x in the interval.
- If the quotient is negative, then the rational expression is negative for all numbers x in the interval.

Also determine the value of the rational expression at each value found in Step 2. If the inequality is not strict ($\le$ or $\ge$), include the values of the variable for which the rational expression equals 0 in the solution set, but do not include the values for which the rational expression is undefined!

QUICK ✓

1. Solve $\dfrac{x-7}{x+3} \geq 0$. Graph the solution set.

EXAMPLE 2 Solving a Rational Inequality

Solve $\dfrac{x+3}{x-1} > 2$. Graph the solution set.

Solution

First, we write the inequality so that a rational expression is on one side of the inequality and 0 is on the other.

$$\dfrac{x+3}{x-1} > 2$$

Subtract 2 from both sides: $\qquad \dfrac{x+3}{x-1} - 2 > 0$

LCD $= x - 1$; multiply -2 by $\dfrac{x-1}{x-1}$: $\qquad \dfrac{x+3}{x-1} - 2 \cdot \dfrac{x-1}{x-1} > 0$

Write rational expression over common denominator: $\qquad \dfrac{x+3-2(x-1)}{x-1} > 0$

Distribute -2: $\qquad \dfrac{x+3-2x+2}{x-1} > 0$

Combine like terms in numerator: $\qquad \dfrac{-x+5}{x-1} > 0$

We can see that the rational expression will equal 0 when $x = 5$. The rational expression is undefined when $x = 1$. We separate the real number line into the following intervals:

$$(-\infty, 1) \qquad (1, 5) \qquad (5, \infty)$$

$$-2 \;\; -1 \;\;\; 0 \;\;\; 1 \;\;\; 2 \;\;\; 3 \;\;\; 4 \;\;\; 5 \;\;\; 6 \;\;\; 7 \;\;\; 8 \;\;\; 9 \;\; 10$$

Table 13 shows the sign of $-x + 5$, $x - 1$, and $\dfrac{-x+5}{x-1}$ in each interval. In addition, it shows the value of $\dfrac{-x+5}{x-1}$ at $x = 1$ and $x = 5$.

Table 13					
		1		5	
Interval	$(-\infty, 1)$		$(1, 5)$		$(5, \infty)$
Test Point	0	1	3	5	6
Sign of $-x + 5$	Positive	Positive	Positive	0	Negative
Sign of $x - 1$	Negative	0	Positive	Positive	Positive
Sign of $\dfrac{-x+5}{x-1}$	Negative	Undefined	Positive	0	Negative
Conclusion	$\dfrac{-x+5}{x-1}$ is negative, so $(-\infty, 1)$ is not part of the solution set.	Because $\dfrac{-x+5}{x-1}$ is undefined at $x = 1$, it is not part of the solution set.	$\dfrac{-x+5}{x-1}$ is positive, so $(1, 5)$ is part of the solution set.	Because the inequality is strict, 5 is not part of the solution set.	$\dfrac{-x+5}{x-1}$ is negative, so $(5, \infty)$ is not part of the solution set.

We want to know the values of x such that $\dfrac{x + 3}{x - 1}$ is greater than 2. This is equivalent to determining where $\dfrac{-x + 5}{x - 1}$ is greater than zero. So the solution set is $\{x \mid 1 < x < 5\}$ using set-builder notation. The solution is $(1, 5)$ using interval notation. Notice that the endpoints of the interval are not part of the solution because the inequality in the original problem is strict. Figure 38 shows the graph of the solution set.

Figure 38

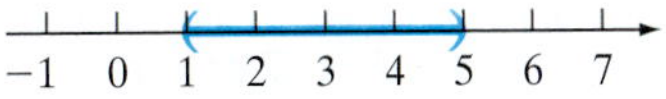

```
  +---+===+===+===+===+===+---+---+-->
 -1   0   1   2   3   4   5   6   7
```

QUICK ✓

2. Solve $\dfrac{4x + 5}{x + 2} < 3$. Graph the solution set.

10.7 Exercises

For Extra Help: Student Solutions Manual   CD Video   PH Math/Tutor Center   MathXL Tutorials on CD   MathXL®   MyMathLab

Concepts and Vocabulary

In Problem 1, fill in the blank.

1. The inequality $\dfrac{2x - 3}{x + 6} > 1$ is an example of a(n) _____ inequality.

In Problem 2, answer True or False to the statement.

2. A solution to the inequality $\dfrac{x + 1}{x - 1} > 0$ is 2.

3. In solving the rational inequality $\dfrac{x - 4}{x + 1} \leq 0$, a student determines that the only interval that makes the inequality true is $(-1, 4)$. He states that the solution set is $\{x \mid -1 \leq x \leq 4\}$. What is wrong with this solution?

4. In Step 2 of the steps for solving a rational inequality, we determine the numbers for which the rational expression equals 0 or is undefined. We then use these numbers to form intervals on the real number line. Explain why this guarantees that there is not a change in the sign of the rational expression within any given interval.

Building Skills

In Problems 5–26, solve each rational inequality. Graph the solution set.

5. $\dfrac{x - 4}{x + 1} > 0$

6. $\dfrac{x + 5}{x - 2} > 0$

7. $\dfrac{x + 9}{x - 3} < 0$

8. $\dfrac{x + 8}{x + 2} > 0$

9. $\dfrac{x + 10}{x - 4} \geq 0$

10. $\dfrac{x + 12}{x - 2} \geq 0$

11. $\dfrac{x + 7}{x - 8} \leq 0$

12. $\dfrac{x - 10}{x + 5} \leq 0$

13. $\dfrac{(2x - 1)(x + 3)}{x - 5} > 0$

14. $\dfrac{(5x - 2)(x + 4)}{x - 5} < 0$

15. $\dfrac{(3x + 5)(x + 8)}{x - 2} \leq 0$

16. $\dfrac{(3x - 2)(x - 6)}{x + 1} \geq 0$

17. $\dfrac{x - 5}{x + 1} < 1$

18. $\dfrac{x + 3}{x - 4} > 1$

19. $\dfrac{3x - 1}{x + 4} \geq 2$

20. $\dfrac{3x - 7}{x + 2} \leq 2$

21. $\dfrac{2x - 9}{x - 3} > 4$

22. $\dfrac{3x + 20}{x + 6} < 5$

23. $\dfrac{3}{x - 4} + \dfrac{1}{x} \geq 0$

24. $\dfrac{2}{x + 3} + \dfrac{2}{x} \leq 0$

25. $\dfrac{3}{x - 2} \leq \dfrac{4}{x + 5}$

26. $\dfrac{1}{x - 4} \geq \dfrac{3}{2x + 1}$

In Problems 27–30, for each function find the values of x that satisfy the given condition. Graph the solution set.

27. Solve $R(x) \leq 0$ if $R(x) = \dfrac{x - 6}{x + 1}$

28. Solve $R(x) \geq 0$ if $R(x) = \dfrac{x + 3}{x - 8}$

29. Solve $R(x) < 0$ if $R(x) = \dfrac{2x - 5}{x + 2}$

30. Solve $R(x) < 0$ if $R(x) = \dfrac{3x + 2}{x - 4}$

Applying the Concepts

31. Average Cost Suppose that the daily cost C of manufacturing x bicycles is given by $C(x) = 80x + 5000$. Then the average daily cost $\overline{C}$ is given by $\overline{C}(x) = \dfrac{80x + 5000}{x}$. How many bicycles must be produced each day in order for the average cost to be no more than \$130?

32. Average Cost See Problem 31. Suppose that the government imposes a \$10 tax on each bicycle manufactured so that the daily cost C of manufacturing x bicycles is now given by $C(x) = 90x + 5000$. Now the average daily cost $\overline{C}$ is given by $\overline{C}(x) = \dfrac{90x + 5000}{x}$. How many bicycles must be produced each day in order for the average cost to be no more than \$130?

Extending the Concepts

33. Write a rational inequality that has $(2, \infty)$ as the solution set.

34. Write a rational inequality that has $(-2, 5]$ as the solution set.

Synthesis Review

In Problems 35–40, find the x-intercepts of the graph of each function.

35. $F(x) = 6x - 12$

36. $G(x) = 5x + 30$

37. $f(x) = 2x^2 + 3x - 14$

38. $h(x) = -3x^2 - 7x + 20$

39. $R(x) = \dfrac{3x - 2}{x + 4}$

40. $R(x) = \dfrac{x^2 + 5x + 6}{x + 2}$

The Graphing Calculator

We can use a graphing calculator to approximate solutions to rational inequalities using the INTERSECT or ZERO (or ROOT) feature of the graphing calculator. We use the ZERO or ROOT feature of the graphing calculator when one side of the inequality is 0; we use the INTERSECT feature when neither side of the inequality is 0. For example, to solve $\dfrac{x-4}{x+1} \geq \dfrac{7}{2}$, we would graph $Y_1 = \dfrac{x-4}{x+1}$ and $Y_2 = \dfrac{7}{2}$. To determine the x-values such that the graph of Y_1 is above that of Y_2, we find x-coordinates of the point(s) of intersection. The graph of Y_1 is above that of Y_2 between $x = -3$ and $x = -1$. See Figure 39.

Figure 39

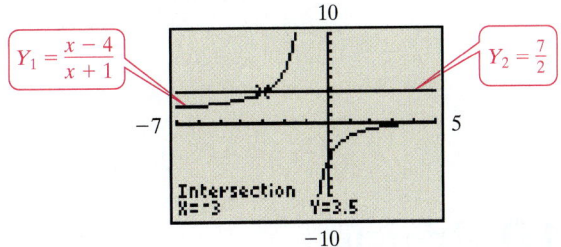

When using a graphing calculator to approximate solutions to inequalities, we typically express the solution as a decimal rounded to two decimal places, if exact answers cannot be found. The solution to the inequality $\dfrac{x-4}{x+1} \geq \dfrac{7}{2}$ is $\{x \mid -3 \leq x < -1\}$, or $[-3, -1)$ using interval notation.

In Problems 41–44, solve each inequality using a graphing calculator.

41. $\dfrac{x-5}{x+1} \leq 3$ **42.** $\dfrac{x+2}{x-5} > -2$ **43.** $\dfrac{2x+5}{x-7} > 3$ **44.** $\dfrac{2x-1}{x+5} \leq 4$

CHAPTER 10 ACTIVITY: PRESIDENTIAL DECISION MAKING

Focus: Developing quadratic equations.

Time: 30–35 minutes

Group size: 2–4

1. Your boss, Huntington Corporation's President, Gerald Cain, is very concerned about his approval rating with his employees. Last year, January 1, he made some policy changes and he saw his approval rating drop. In fact, his approval rating was 48% just before he made some policy changes. One month later, his rating was at 41%, two months later it was at 40%, and at three months it began to climb and was 45%. He discovered that the following function described his approval rating for that year:

$$R(x) = 3x^2 - 10x + 48$$

   (a) As a group, use the above function to find when President Cain's approval rating will return to the original rating.
   (b) If his approval rating continues to climb, when will he reach a 68% approval rating?

2. On January 1 of this year, President Cain surveyed his employees again and found that 68% of them approved of his leadership skills. At this time, President Cain decided to become very strict with his employees and began a series of new policies. He noticed that his approval rating began to steadily slip and reached an all-time low of 38% on March 30 (3 months later). President Cain was not worried because he knows from last year that this drop in popularity will bottom

out and eventually rise. He believes his popularity can be modeled by a quadratic function and needs your help. He has more bad news to deliver but does not want to begin the next round of policy changes until his approval rating is back to approximately 50%.

(a) As a group, write a quadratic function that would model President Cain's approval rating.

(b) As a group, develop different ways to advise President Cain what date to begin his policy changes. Use graphs and computations, to prove your point.

CHAPTER 10 REVIEW

Section 10.1   Solving Quadratic Equations by Completing the Square

KEY CONCEPTS

- **Square Root Property**
  If $x^2 = p$, then $x = \sqrt{p}$ or $x = -\sqrt{p}$.
- **Pythagorean Theorem**
  In a right triangle, the square of the length of the hypotenuse is equal to the sum of the squares of the lengths of the legs. That is, $\text{leg}^2 + \text{leg}^2 = \text{hypotenuse}^2$.

KEY TERMS

Completing the square
Right triangle
Right angle
Hypotenuse
Legs

YOU SHOULD BE ABLE TO . . .	EXAMPLE	REVIEW EXERCISES
① Solve quadratic equations using the square root property (p. 744)	Examples 1 through 4	1–10
② Complete the square in one variable (p. 747)	Example 5	11–16
③ Solve quadratic equations by completing the square (p. 748)	Examples 6 and 7	17–26
④ Solve problems using the Pythagorean Theorem (p. 750)	Examples 8 and 9	27–36

In Problems 1–10, solve each equation using the Square Root Property.

1. $m^2 = 169$ **2.** $n^2 = 75$ **3.** $a^2 = -16$

4. $b^2 = \dfrac{8}{9}$ **5.** $(x - 8)^2 = 81$ **6.** $(y - 2)^2 - 62 = 88$

7. $(3z + 5)^2 = 100$ **8.** $7p^2 = 18$ **9.** $3q^2 + 251 = 11$

10. $\left(x + \dfrac{3}{4}\right)^2 = \dfrac{13}{16}$

In Problems 11–16, complete the square in each expression. Then factor the perfect square trinomial.

11. $a^2 + 30a$ **12.** $b^2 - 14b$ **13.** $c^2 - 11c$

14. $d^2 + 9d$ **15.** $m^2 - \dfrac{1}{4}m$ **16.** $n^2 + \dfrac{6}{7}n$

In Problems 17–26, solve each quadratic equation by completing the square.

17. $x^2 - 10x + 16 = 0$ **18.** $y^2 - 3y - 28 = 0$ **19.** $z^2 - 6z - 3 = 0$

20. $a^2 - 5a - 7 = 0$ **21.** $b^2 + b + 7 = 0$ **22.** $c^2 - 6c + 17 = 0$

23. $2d^2 - 7d + 3 = 0$ **24.** $2w^2 + 2w + 5 = 0$ **25.** $3x^2 - 9x + 8 = 0$

26. $3x^2 + 4x - 2 = 0$

In Problems 27–32, the lengths of the legs of a right triangle are given. Find the hypotenuse.

27. $a = 9, b = 12$ **28.** $a = 8, b = 8$ **29.** $a = 3, b = 6$

30. $a = 10, b = 24$ **31.** $a = 5, b = \sqrt{11}$ **32.** $a = 6, b = \sqrt{13}$

In Problems 33–35, use the right triangle shown below and find the missing length.

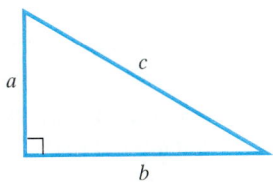

33. $a = 9, c = 12$ **34.** $b = 5, c = 10$ **35.** $b = 6, c = 17$

36. Baseball Diamond A baseball diamond is really a square that is 90 feet long on each side. (See the figure.) What is the distance between home plate and second base?

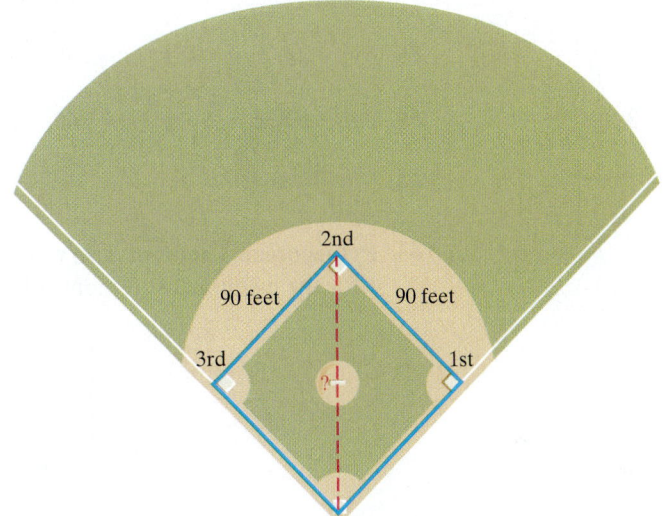

Section 10.2	Solving Quadratic Equations by the Quadratic Formula

KEY CONCEPTS	**KEY TERM**

• **The Quadratic Formula**
 	
 The solutions to the equation $ax^2 + bx + c = 0, a \neq 0$, are given by $x = \dfrac{-b \pm \sqrt{b^2 - 4ac}}{2a}$
 	
 • **Discriminant**
 For the quadratic equation $ax^2 + bx + c = 0, a \neq 0$:
   • If $b^2 - 4ac > 0$, the equation has two unequal real solutions.
     • If $b^2 - 4ac$ is a perfect square, the equation has two rational solutions.
     • If $b^2 - 4ac$ is not a perfect square, the equation has two irrational solutions.
   • If $b^2 - 4ac = 0$, the equation has a repeated real solution.
   • If $b^2 - 4ac < 0$, the equation has two complex solutions that are not real.
	Discriminant

YOU SHOULD BE ABLE TO . . .	EXAMPLE	REVIEW EXERCISES
① Solve quadratic equations using the quadratic formula (p. 757)	Examples 1 through 4	37–46
② Use the discriminant to determine the nature of solutions in a quadratic equation (p. 762)	Example 5	47–52
③ Model and solve problems involving quadratic equations (p. 765)	Examples 6 and 7	63–68

In Problems 37–46, solve each equation using the quadratic formula.

37. $x^2 - x - 20 = 0$ **38.** $4y^2 = 8y + 21$ **39.** $3p^2 + 8p = -3$

40. $2q^2 - 3 = 4q$ **41.** $3w^2 + w = -3$ **42.** $9z^2 + 16 = 24z$

43. $m^2 - 4m + 2 = 0$ **44.** $5n^2 + 4n + 1 = 0$ **45.** $5x + 13 = -x^2$

46. $-2y^2 = 6y + 7$

In Problems 47–52, determine the discriminant of each quadratic equation. Use the value of the discriminant to determine whether the quadratic equation has two rational solutions, two irrational solutions, one repeated real solution, or two complex solutions that are not real.

47. $p^2 - 5p - 8 = 0$ **48.** $m^2 + 8m + 16 = 0$ **49.** $3n^2 + n = -4$

50. $7w^2 + 3 = 8w$ **51.** $4x^2 + 49 = 28x$ **52.** $11z - 12 = 2z^2$

In Problems 53–62, solve each equation using any method you wish.

53. $x^2 + 8x - 9 = 0$ **54.** $6p^2 + 13p = 5$ **55.** $n^2 + 13 = -4n$

56. $5y^2 - 60 = 0$ **57.** $\dfrac{1}{4}q^2 - \dfrac{1}{2}q - \dfrac{3}{8} = 0$ **58.** $\dfrac{1}{8}m^2 + m + \dfrac{5}{2} = 0$

59. $(w - 8)(w + 6) = -33$ **60.** $(x - 3)(x + 1) = -2$ **61.** $9z^2 = 16$

62. $\dfrac{1 - 2x}{x^2 + 5} = 1$

63. Pythagorean Theorem Use the Pythagorean Theorem to determine the value of x for the given measurements of the right triangle shown below.

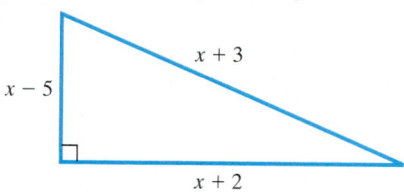

64. Area The area of a rectangle is 108 square centimeters. The width of the rectangle is 3 inches less than the length. What are the dimensions of the rectangle?

65. Revenue The revenue R received by a company selling x cellular phones per week is given by the function $R(x) = -0.2x^2 + 180x$.

  (a) How many cellular phones must be sold in order for revenue to be $36,000 per week?

  (b) How many cellular phones must be sold in order for revenue to be $40,500 per week?

66. Projectile Motion The height s of a ball after t seconds when thrown straight up with an initial speed of 50 feet per second from an initial height of 180 feet can be modeled by the function $s(t) = -16t^2 + 50t + 180$.

  (a) When will the height of the ball be 200 feet? Round your answer to the nearest tenth of a second.

  (b) When will the height of the ball be 100 feet? Round your answer to the nearest tenth of a second.

  (c) Will the ball ever reach a height of 300 feet? How does the result of the equation tell you this?

67. Pleasure Boat Ride A pleasure boat carries passengers 10 miles upstream and then returns to the starting point. The total time of the trip (excluding the time on the ground) takes 2 hours. If the speed of the current is 3 miles per hour, find the speed the boat in still water. Round your answer to the nearest tenth of an hour.

68. Work Together, Tom and Beth can wash their car in 30 minutes. By himself, Tom can wash the car in 14 minutes less time than Beth can by herself. How long will it take Beth to wash the car by herself? Round your answer to the nearest tenth of a minute.

Section 10.3	Solving Equations Quadratic in Form	
KEY TERM		
Equation quadratic in form		
YOU SHOULD BE ABLE TO . . .	**EXAMPLE**	**REVIEW EXERCISES**
① Solve equations that are quadratic in form (p. 772)	Examples 1 through 5	69–80

In Problems 69–78, solve each equation.

69. $x^4 + 7x^2 - 144 = 0$

70. $4w^4 + 5w^2 - 6 = 0$

71. $3(a + 4)^2 - 11(a + 4) + 6 = 0$

72. $(q^2 - 11)^2 - 2(q^2 - 11) - 15 = 0$

73. $y - 13\sqrt{y} + 36 = 0$

74. $5z + 2\sqrt{z} - 3 = 0$

75. $p^{-2} - 4p^{-1} - 21 = 0$

76. $2b^{2/3} + 13b^{1/3} - 7 = 0$

77. $m^{1/2} + 2m^{1/4} - 8 = 0$

78. $\left(\dfrac{1}{x + 5}\right)^2 + \dfrac{3}{x + 5} = 28$

In Problems 79 and 80, find the zeros of the function.

79. $f(x) = 4x - 20\sqrt{x} + 21$

80. $g(x) = x^4 - 17x^2 + 60$

Section 10.4	Graphing Quadratic Functions Using Transformations	
KEY CONCEPTS		**KEY TERMS**
• **Graphing a Function of the Form $f(x) = x^2 + k$** To obtain the graph of $f(x) = x^2 + k$ from the graph of $y = x^2$, shift the graph of $y = x^2$ vertically up k units if $k > 0$ and vertically down k units if $k < 0$. • **Graphing a Function of the Form $f(x) = (x - h)^2$** To obtain the graph of $f(x) = (x - h)^2$ from the graph of $y = x^2$, shift the graph of $y = x^2$ horizontally to the right h units if $h > 0$ and horizontally left h units if $h < 0$. • **Graphing a Function of the Form $f(x) = ax^2$** To obtain the graph of $f(x) = ax^2$ from the graph of $y = x^2$, multiply each y-coordinate on the graph of $y = x^2$ by a.		Quadratic function Vertically stretched Vertically compressed Parabola Opens up Opens down Vertex Axis of symmetry Transformations
YOU SHOULD BE ABLE TO . . .	**EXAMPLE**	**REVIEW EXERCISES**
① Graph quadratic functions of the form $f(x) = x^2 + k$ (p. 782)	Examples 1 and 2	81–82; 87–90
② Graph quadratic functions of the form $f(x) = (x - h)^2$ (p. 783)	Examples 3 through 5	83–84; 87–90
③ Graph quadratic functions of the form $f(x) = ax^2$ (p. 785)	Example 6	85–86; 89–90
④ Graph quadratic functions of the form $f(x) = ax^2 + bx + c$ (p. 787)	Examples 7 and 8	91–96
⑤ Find a quadratic function from its graph (p. 790)	Example 9	97–100

In Problems 81–90, use the graph of $y = x^2$ to graph the quadratic function.

81. $f(x) = x^2 + 4$ **82.** $g(x) = x^2 - 5$ **83.** $h(x) = (x + 1)^2$

84. $F(x) = (x - 4)^2$ **85.** $G(x) = -4x^2$ **86.** $H(x) = \dfrac{1}{5}x^2$

87. $p(x) = (x - 4)^2 - 3$ **88.** $P(x) = (x + 4)^2 + 2$ **89.** $f(x) = -(x - 1)^2 + 4$

90. $F(x) = \dfrac{1}{2}(x + 2)^2 - 1$

In Problems 91–96, graph each quadratic function using transformations. Determine the vertex and axis of symmetry. Based on the graph determine the domain and range of each function.

91. $g(x) = x^2 - 6x + 10$ **92.** $G(x) = x^2 + 8x + 11$

93. $h(x) = 2x^2 - 4x - 3$ **94.** $H(x) = -x^2 - 6x - 10$

95. $p(x) = -3x^2 + 12x - 8$ **96.** $P(x) = \dfrac{1}{2}x^2 - 2x + 5$

In Problems 97–100, determine the quadratic function whose graph is given.

97.

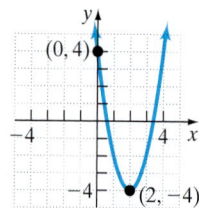

98.

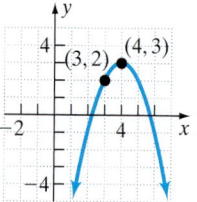

99.

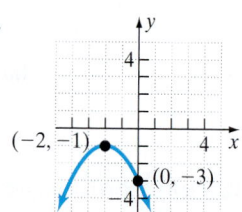

100.

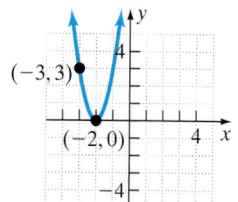

Section 10.5   Graphing Quadratic Functions Using Properties

KEY CONCEPTS

KEY TERMS

- **Vertex of a Parabola**

  Any quadratic function of the form $f(x) = ax^2 + bx + c, a \neq 0$, will have vertex

  $$\left(-\frac{b}{2a}, f\left(-\frac{b}{2a} \right) \right)$$

- **The x-Intercepts of the Graph of a Quadratic Function**

  1. If $b^2 - 4ac > 0$, the graph of $f(x) = ax^2 + bx + c$ has two different x-intercepts.

  2. If $b^2 - 4ac = 0$, the graph of $f(x) = ax^2 + bx + c$ has one x-intercept.

  3. If $b^2 - 4ac < 0$, the graph of $f(x) = ax^2 + bx + c$ has no x-intercepts.

Maximum value
Minimum value
Optimization

YOU SHOULD BE ABLE TO . . .	EXAMPLE	REVIEW EXERCISES
1 Graph quadratic functions of the form $f(x) = ax^2 + bx + c$ (p. 795)	Examples 1 through 4	101–108
2 Find the maximum or minimum value of a quadratic function (p. 801)	Examples 5 and 6	109–112
3 Model and solve optimization problems involving quadratic functions (p. 802)	Examples 7 and 8	113–118

In Problems 101–108, graph each quadratic function using its properties by following Steps 1–5 on page 798. Based on the graph determine the domain and range of each function.

101. $f(x) = x^2 + 2x - 8$

102. $F(x) = 2x^2 - 5x + 3$

103. $g(x) = -x^2 + 6x - 7$

104. $G(x) = -2x^2 + 4x + 3$

105. $h(x) = 4x^2 - 12x + 9$

106. $H(x) = \dfrac{1}{3}x^2 + 2x + 3$

107. $p(x) = \dfrac{1}{4}x^2 + 3x + 10$

108. $P(x) = -x^2 + 4x - 9$

In Problems 109–112, determine whether the quadratic function has a maximum or minimum value. Then find the maximum or minimum value.

109. $f(x) = -2x^2 + 16x - 10$

110. $g(x) = 6x^2 - 3x - 1$

111. $h(x) = -4x^2 + 8x + 3$

112. $F(x) = -\dfrac{1}{3}x^2 + 4x - 7$

113. Revenue Suppose that the marketing department of Zenith has found that, when a certain model of television is sold for a price of p dollars, the daily revenue R (in dollars) as a function of the price p is $R(p) = -\dfrac{1}{3}p^2 + 150p$.

(a) For what price will the daily revenue be maximized?

(b) What is this maximum daily revenue?

114. Electrical Power In a 120-volt electrical circuit having a resistance of 16 ohms, the available power P (in watts) is given by the function $P(I) = -16I^2 + 120I$, where I represents the current (in amperes).

(a) What current will produce the maximum power in the circuit?

(b) What is this maximum power?

115. Fun with Numbers The sum of two numbers is 24. Find the numbers such that their product is a maximum.

116. Maximizing an Enclosed Area Becky has 15 yards of fencing to make a rectangular kennel for her dog. She will build the kennel next to her garage, so she only needs to enclose three sides. (See the figure.)

(a) What dimensions maximize the area of the kennel?

(b) What is this maximum area?

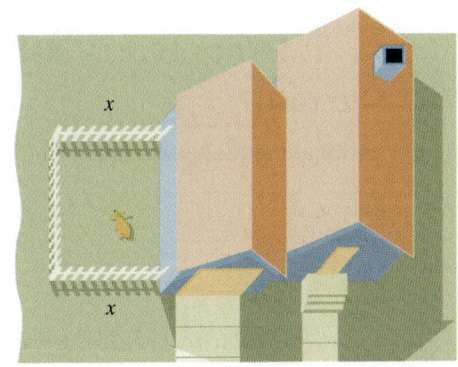

117. Kicking a Football Ted kicks a football at a 45° angle to the horizontal with an initial velocity of 80 feet per second. The model $h(x) = -0.005x^2 + x$ can be used to estimate the height h of the ball after it has traveled x feet.

(a) How far from Ted will the ball reach a maximum height?

(b) What is the maximum height of the ball?

(c) How far will the ball travel before it strikes the ground?

118. Revenue Monthly demand for automobiles at a certain dealership obeys the demand equation $x = -0.002p + 60$, where x is the quantity and p is the price (in dollars).

(a) Express the revenue R as a function of p. (**Hint:** $R = xp$)

(b) What price p maximizes revenue? What is the maximum revenue?

(c) How many automobiles will be sold at the revenue-maximizing price?

Section 10.6	Quadratic Inequalities	
KEY TERM		
Quadratic inequality		
YOU SHOULD BE ABLE TO . . .	**EXAMPLE**	**REVIEW EXERCISES**
① Solve quadratic inequalities (p. 811)	Examples 1 through 4	119–126

In Problems 119 and 120, use the graphs of the quadratic function f to determine the solution.

119.

(a) $f(x) > 0$ (b) $f(x) < 0$

120.

(a) $f(x) \geq 0$ (b) $f(x) \leq 0$

In Problems 120–126, solve each inequality. Graph the solution set.

121. $x^2 - 2x - 24 \leq 0$

122. $y^2 + 7y - 8 \geq 0$

123. $3z^2 - 19z + 20 > 0$

124. $p^2 + 4p - 2 < 0$

125. $4m^2 - 20m + 25 \geq 0$

126. $6w^2 - 19w - 7 \leq 0$

Section 10.7	Rational Inequalities	
KEY CONCEPT		**KEY TERM**
• The steps for solving any rational inequality are given on page 823.		Rational inequality
YOU SHOULD BE ABLE TO . . .	**EXAMPLE**	**REVIEW EXERCISES**
① Solve a rational inequality (p. 822)	Examples 1 and 2	127–136

In Problems 127–134, solve each rational inequality. Graph the solution set.

127. $\dfrac{x - 4}{x + 2} \geq 0$

128. $\dfrac{y - 5}{y + 4} < 0$

129. $\dfrac{4}{z^2 - 9} \leq 0$

130. $\dfrac{w^2 + 5w - 14}{w - 4} < 0$

131. $\dfrac{m - 5}{m^2 + 3m - 10} \geq 0$

132. $\dfrac{4}{n - 2} \leq -2$

133. $\dfrac{a + 1}{a - 2} > 3$

134. $\dfrac{4}{c - 2} - \dfrac{3}{c} < 0$

In Problems 135 and 136, for each function, find the values of x that satisfy the given condition. Graph the solution set.

135. Solve $Q(x) < 0$ if $Q(x) = \dfrac{2x + 3}{x - 4}$.　　**136.** Solve $R(x) \geq 0$ if $R(x) = \dfrac{x + 5}{x + 1}$.

CHAPTER 10 TEST

 Remember to use your Chapter Test Prep Video CD to see fully worked-out solutions to any of these problems you would like to review.

In Problems 1 and 2, complete the square in the given expression. Then factor the perfect square trinomial.

1. $x^2 - 3x$　　　　　　　　**2.** $m^2 + \dfrac{2}{5}m$

In Problems 3–6, solve each equation using the method you prefer.

3. $9\left(x + \dfrac{4}{3}\right)^2 = 1$　　　　**4.** $m^2 - 6m + 4 = 0$

5. $2w^2 - 4w + 3 = 0$　　　　**6.** $\dfrac{1}{2}z^2 - \dfrac{3}{2}z = -\dfrac{7}{6}$

7. Determine the discriminant of $2x^2 + 5x = 4$. Use the value of the discriminant to determine whether the quadratic equation has two rational solutions, two irrational solutions, one repeated real solution, or two complex solutions that are not real.

8. Find the missing length in the right triangle shown below.

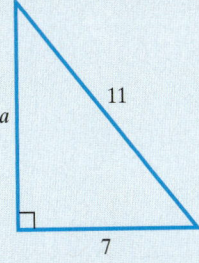

In Problems 9 and 10, solve each equation.

9. $x^4 - 5x^2 - 36 = 0$　　　　**10.** $6y^{1/2} + 13y^{1/4} - 5 = 0$

In Problems 11 and 12, graph each quadratic function. Determine the vertex and axis of symmetry. Based on the graph, determine the domain and range of the quadratic function.

11. $f(x) = (x + 2)^2 - 5$　　　　**12.** $g(x) = -2x^2 - 8x - 3$

13. Determine the quadratic function whose graph is given.

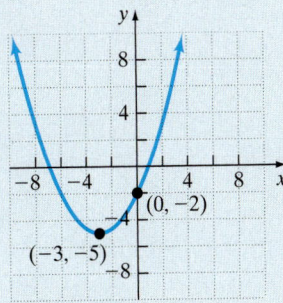

14. Determine whether the given quadratic function $h(x) = -\dfrac{1}{4}x^2 + x + 5$ has a maximum or minimum value. Then find the maximum or minimum value.

In Problems 15–17, solve each inequality. Graph the solution set.

15. $2m^2 + m - 15 > 0$ **16.** $z^2 + 6z - 1 \le 0$ **17.** $\dfrac{x + 5}{x - 2} \ge 3$

18. Projectile Motion The height s of a rock after t seconds when propelled straight up with an initial speed of 80 feet per second from an initial height of 20 feet can be modeled by the function $s(t) = -16t^2 + 80t + 20$. When will the height of the rock be 50 feet? Round your answer to the nearest tenth of a second.

19. Work Together, Lex and Rupert can roof a house in 16 hours. By himself, Rex can roof the house in 4 hours less time than Rupert can by himself. How long will it take Rupert to roof the house by himself? Round your answer to the nearest tenth of an hour.

20. Revenue A small company has found that, when their product is sold for a price of p dollars, the weekly revenue (in dollars) as a function of price p is $R(p) = -0.25p^2 + 170p$.

(a) For what price will the weekly revenue be maximized?

(b) What is this maximum weekly revenue?

21. Maximizing Volume A box with a rectangular base is to be constructed such that the perimeter of the base of the box is to be 50 inches. The height of the box must be 12 inches.

(a) Find the dimensions that maximize the volume of the box.

(b) What is this maximum volume?

11 Exponential and Logarithmic Functions

Earthquakes are one of the most powerful forces in nature. The powerful waves result from a movement of Earth's outer layer that move to the surface. One measure of the strength of an earthquake is the Richter scale. See Problems 111–114 in Section 11.3.

OUTLINE

11.1 Composite Functions and Inverse Functions

11.2 Exponential Functions

11.3 Logarithmic Functions

Putting the Concepts Together (Sections 11.1–11.3)

11.4 Properties of Logarithms

11.5 Exponential and Logarithmic Equations

Chapter 11 Activity: Correct the Quiz
Chapter 11 Review
Chapter 11 Test
Cumulative Review Chapters 1-11

The Big Picture: Putting It Together

Up to now, we have studied polynomial, rational, and radical expressions and functions. We learned how to evaluate, simplify, and solve equations involving these types of algebraic expressions. We also learned how to graph linear functions, quadratic functions and certain types of radical functions.

We now introduce two more types of functions, the *exponential* and *logarithmic functions*. The theme of the text will continue. We will evaluate these functions, graph them, and learn properties of the functions. We also will solve equations that involve the exponential or logarithmic expressions.

11.1 Composite Functions and Inverse Functions

OBJECTIVES

1. Form the Composite Function
2. Determine Whether or Not a Function Is One-to-One
3. Determine the Inverse of a Function Defined by a Map or Ordered Pair
4. Obtain the Graph of the Inverse Function from the Graph of the Function
5. Find the Inverse of a Function Defined by an Equation

Preparing for Composite Functions and Inverse Functions

Before getting started, take the following readiness quiz. If you get a problem wrong, go back to the section cited and review the material.

1. Determine the domain of $R(x) = \dfrac{x^2 - 9}{x^2 + 3x - 28}$. [Section 8.4, pp. 597–598]

2. If $f(x) = 2x^2 - x + 1$, find (a) $f(-2)$ (b) $f(a + 1)$. [Section 8.3, pp. 590–592]

3. The graph of a relation is given. Does the relation represent a function? [Section 8.3, pp. 589–590]

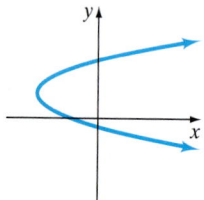

① Form the Composite Function

Consider the function $y = (x - 3)^4$. To evaluate this function, we first evaluate $x - 3$ and then raise the result to the fourth power. So, technically, two different functions form the function $y = (x - 3)^4$. Figure 1 illustrates the idea for $x = 5$.

Figure 1

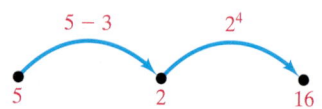

If we let $u = g(x) = x - 3$ and $y = f(u) = u^4$, then $u = g(5) = 5 - 3 = 2$ and $y = f(u) = f(2) = 2^4 = 16$. To shorten the notation, we can express $y = (x - 3)^4$ as $y = f(u) = f(g(x)) = (x - 3)^4$. This process is called **composition.**

In general, suppose that f and g are two functions and that x is a number in the domain of g. By evaluating g at x, we get $g(x)$. If $g(x)$ is in the domain of f, then we may evaluate f at $g(x)$ and obtain the expression $f(g(x))$. The correspondence from x to $f(g(x))$ is called a *composite function* $f \circ g$.

In Words

To find $f(g(x))$, first determine $g(x)$. This is the output of g. Then evaluate f at the output of g. The result is $f(g(x))$. The notation $f(g(x))$ is read "f of g of x."

DEFINITION

Given two functions f and g, the **composite function,** denoted by $f \circ g$ (read as "f composed with g"), is defined by

$$(f \circ g)(x) = f(g(x))$$

Figure 2 illustrates the definition. Notice that the "inside" function g in $f(g(x))$ is always evaluated first.

Figure 2

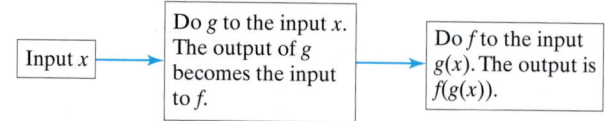

EXAMPLE 1 **Evaluating a Composite Function**

Suppose that $f(x) = x^2 - 3$ and $g(x) = 2x + 1$. Find

(a) $(f \circ g)(3)$ (b) $(g \circ f)(3)$ (c) $(f \circ f)(-2)$

Solution

(a) Using the flowchart in Figure 2, we evaluate $(f \circ g)(3)$ as follows:

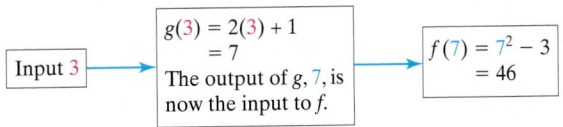

More directly,

$$(f \circ g)(3) = f(g(3)) = f(7) = 7^2 - 3 = 49 - 3 = 46$$

$$g(x) = 2x + 1 \qquad f(x) = x^2 - 3$$
$$g(3) = 2(3) + 1$$
$$= 7$$

(b) $(g \circ f)(3) = g(f(3)) = g(6) = 2(6) + 1 = 13$

$$f(x) = x^2 - 3 \quad g(x) = 2x + 1$$
$$f(3) = 3^2 - 3$$
$$= 6$$

(c) $(f \circ f)(-2) = f(f(-2)) = f(1) = 1^2 - 3 = -2$

$$f(x) = x^2 - 3 \qquad f(x) = x^2 - 3$$
$$f(-2) = (-2)^2 - 3$$
$$= 1$$

QUICK ✓

1. Suppose that $f(x) = 4x - 3$ and $g(x) = x^2 + 1$. Find

(a) $(f \circ g)(2)$ **(b)** $(g \circ f)(2)$ **(c)** $(f \circ f)(-3)$

Rather than evaluating a composite function at a specific value, we can also form a composite function that is dependent upon the independent variable, x.

EXAMPLE 2 Finding a Composite Function

Suppose that $f(x) = x^2 + 2x$ and $g(x) = 2x - 1$. Find

(a) $(f \circ g)(x)$ **(b)** $(g \circ f)(x)$ **(c)** $(f \circ g)(2)$

Solution

(a)
$$(f \circ g)(x) = f(g(x))$$
$$g(x) = 2x - 1: \quad = f(2x - 1)$$
$$f(x) = x^2 + 2x: \quad = (2x - 1)^2 + 2(2x - 1)$$
$$\text{FOIL; Distribute:} \quad = 4x^2 - 4x + 1 + 4x - 2$$
$$\text{Combine like terms:} \quad = 4x^2 - 1$$

Work Smart
$(f \circ g)(x)$ does not mean $(f \cdot g)(x)$.

(b)
$$(g \circ f)(x) = g(f(x))$$
$$f(x) = x^2 + 2x: \quad = g(x^2 + 2x)$$
$$g(x) = 2x - 1: \quad = 2(x^2 + 2x) - 1$$
$$\text{Distribute:} \quad = 2x^2 + 4x - 1$$

(c) Instead of finding $(f \circ g)(2)$ using the approach presented in Example 1, we will find $(f \circ g)(2)$ using the results from part (a).

$$(f \circ g)(2) = f(g(2)) = 4(2)^2 - 1 = 4(4) - 1 = 15$$

Notice that $(f \circ g)(x) \neq (g \circ f)(x)$ in Examples 2(a) and 2(b).

QUICK ✓

2. Suppose that $f(x) = x^2 - 3x + 1$ and $g(x) = 3x + 2$. Find

(a) $(f \circ g)(x)$ **(b)** $(g \circ f)(x)$ **(c)** $(f \circ g)(-2)$

② Determine Whether or Not a Function Is One-to-One

Figures 3 and 4 illustrate two different functions represented as mappings. The function in Figure 3 shows the correspondence between states and their population in millions. The function in Figure 4 shows a correspondence between "animals" and "life expectancy."

Figure 3

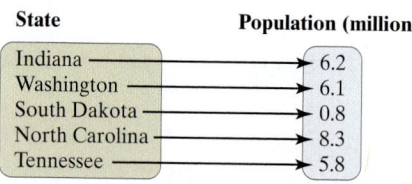

Figure 4

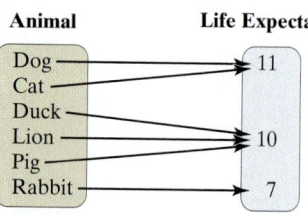

Suppose we asked a group of people to name the state which has a population of 0.8 million based on the function in Figure 3. Everyone in the group would respond "South Dakota." If we asked the same group of people to name the animal whose life expectancy is 11 years based on the function in Figure 4, some would respond "dog," while others would respond "cat." What is the difference between the functions in Figures 3 and 4? In Figure 3, each element in the domain corresponds to one (and only one) element in the range. In Figure 4, this is not the case—there is more than one element in the domain that corresponds to an element in the range. For example, "dog" corresponds to "11," but "cat" also corresponds to "11." We give functions such as the one in Figure 3 a special name.

In Words

A function is not one-to-one if two different inputs correspond to the same output.

DEFINITION

A function is **one-to-one** if any two different inputs in the domain correspond to two different outputs in the range. That is, if x_1 and x_2 are two different inputs of a function f, then $f(x_1) \neq f(x_2)$.

Put another way, a function is not one-to-one if two different elements in the domain correspond to the same element in the range. So, the function in Figure 4 is not one-to-one because two different elements in the domain, dog and cat, both correspond to 11.

EXAMPLE 3 **Determining Whether a Function Is One-to-One**

Determine which of the following functions is one-to-one.

(a) For the function to the right, the domain represents the age of 5 males and the range represents their HDL (good) cholesterol (mg/dL).

(b) $\{(-2, 6), (-1, 3), (0, 2), (1, 4), (2, 8)\}$

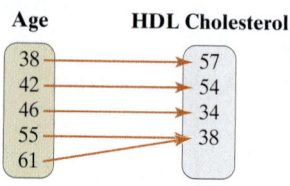

Solution

(a) The function is not one-to-one because there are two different inputs, 55 and 61, that correspond to the same output, 38.

(b) The function is one-to-one because there are no two distinct inputs that correspond to the same output.

QUICK ✔️ *Determine whether or not the function is one-to-one.*

3.

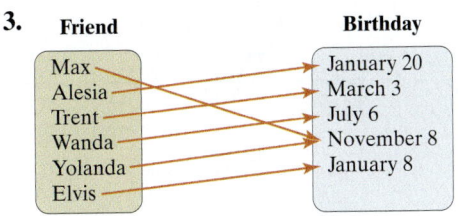

4. $\{(-3, 3), (-2, 2), (-1, 1), (0, 0), (1, -1)\}$

Consider the functions $y = 2x - 5$ and $y = x^2 - 4$ shown in Figure 5.

Figure 5

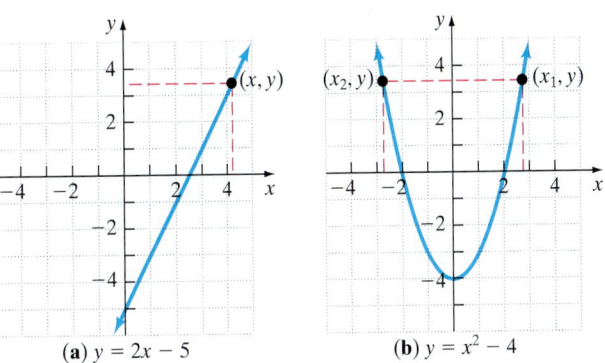

(a) $y = 2x - 5$ **(b)** $y = x^2 - 4$

Notice for the function $y = 2x - 5$ shown in Figure 5(a) that any output is the result of only one input. However, for the function $y = x^2 - 4$ shown in Figure 5(b) there are instances where a given output is the result of more than one input. For example, the output -3 is the image of the two inputs, -1 and 1.

If a horizontal line intersects the graph of a function at more than one point, the function cannot be one-to-one because this would mean that two different inputs give the same output. We state this result formally as the *horizontal line test*.

HORIZONTAL LINE TEST

If every horizontal line intersects the graph of a function f in at most one point, then f is one-to-one.

EXAMPLE 4 Using the Horizontal Line Test

For each function, use the graph to determine whether the function is one-to-one.

(a) $f(x) = -2x^3 + 1$ **(b)** $g(x) = x^3 - 4x$

Solution

(a) Figure 6(a) illustrates the horizontal line test for $f(x) = -2x^3 + 1$. Because every horizontal line will intersect the graph of f exactly once, it follows that f is one-to-one.

(b) Figure 6(b) illustrates the horizontal line test for $g(x) = x^3 - 4x$. The horizontal line $y = 1$ intersects the graph three times, so g is not one-to-one.

Figure 6

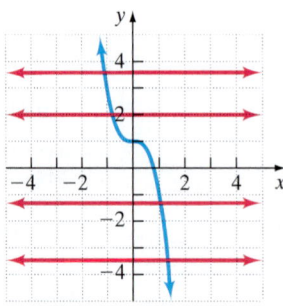

(a) Every horizontal line intersects the graph once; f is one-to-one.

(b) A horizontal line intersects the graph three times; g is not one-to-one.

QUICK ✓ *Use the graph to determine whether the given function is one-to-one.*

5. $f(x) = x^4 - 4x^2$

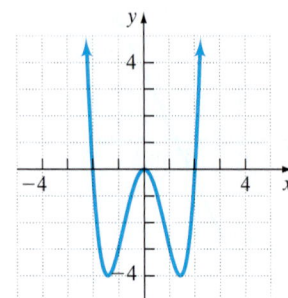

6. $f(x) = x^5 + 4x$

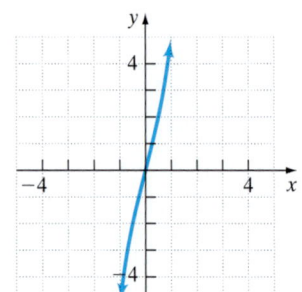

③ **Determine the Inverse of a Function Defined by a Map or Ordered Pair**

Now that we understand the concept of a one-to-one function, we can talk about *inverse functions*.

> **DEFINITION**
>
> If $f(x)$ is a one-to-one function with ordered pairs of the form (a, b), then the **inverse** function, denoted f^{-1}, is the set of ordered pairs of the form (b, a).

Work Smart

The symbol f^{-1} is used to represent the inverse function of f. The -1 used in f^{-1} is not an exponent. That is,

$$f^{-1}(x) \neq \frac{1}{f(x)}.$$

We begin by finding inverses of functions represented by maps and sets of ordered pairs. **In order for the inverse of a function to also be a function, the function must be one-to-one.** To find the inverse of a function defined by a map or a set of ordered pairs, we interchange the inputs and outputs, so for each ordered pair (a, b) that is defined in a function f, the ordered pair (b, a) is defined in the inverse function f^{-1}. The inverse undoes what the function does. Suppose a function takes the input 11 and gives the output 3. This can be represented as $(3, 11)$. The inverse would take as input 11 and give the output 3, which can be represented as $(11, 3)$.

EXAMPLE 5 **Finding the Inverse of a Function Defined by a Map**

Find the inverse of the following function. Let the domain of the function represent certain states and let the range represent the state's population in millions. State the domain and the range of the inverse function.

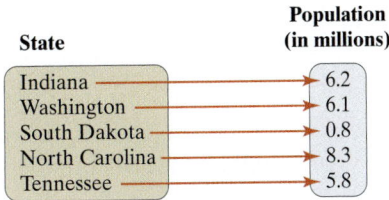

Solution

The elements in the domain represent the inputs to the function. The elements in the range represent the outputs to the function. The function is one-to-one, so the inverse will be a function. To find the inverse function, we interchange the elements in the domain with the elements in the range. For example, the function receives as input Indiana and outputs 6.2. So, the inverse receives as input 6.2 and outputs Indiana. The inverse function is shown below.

Population   **State**

6.2	→	Indiana
6.1	→	Washington
0.8	→	South Dakota
8.3	→	North Carolina
5.8	→	Tennessee

The domain of the inverse function is $\{6.2, 6.1, 0.8, 8.3, 5.8\}$. The range of the inverse function is $\{$Indiana, Washington, South Dakota, North Carolina, Tennessee$\}$. ■

QUICK ✓

7. Find the inverse of the one-to-one function shown to the right. Let the domain of the function represent the length of the right humerus and let the range represent the length of the right tibia of rats sent to space. State the domain and the range of the inverse function.

Right Humerus   **Right Tibia**

24.80	→	36.05
24.59	→	35.57
24.29	→	34.58
23.81	→	34.20
24.87	→	34.73

SOURCE: *NASA Life Sciences Data Archive*

EXAMPLE 6 **Finding the Inverse of a Function Defined by an Ordered Pair**

Find the inverse of the following function. State the domain and the range of the inverse function.

$$\{(-2, 6), (-1, 3), (0, 2), (1, 5), (2, 8)\}$$

Solution

The function is one-to-one, so the inverse will be a function. The inverse of a function defined by a set of ordered pairs is found by interchanging the entries in each ordered pair. So the inverse function is given by

$$\{(6, -2), (3, -1), (2, 0), (5, 1), (8, 2)\}$$

The domain of the inverse function is $\{6, 3, 2, 5, 8\}$. The range of the inverse function is $\{-2, -1, 0, 1, 2\}$. ■

QUICK ✓

8. Find the inverse of the following function. State the domain and the range of the inverse function.

$$\{(-3, 3), (-2, 2), (-1, 1), (0, 0), (1, -1)\}$$

Look back at Examples 5 and 6. Notice that the elements that are in the domain of f are the same elements that are in the range of its inverse. In addition, the elements that are in the range of f are the same elements that are in the domain of its inverse.

> **RELATION BETWEEN THE DOMAIN AND RANGE OF A FUNCTION AND ITS INVERSE**
>
> All elements in the domain of a function are also elements in the range of its inverse.
>
> All elements in the range of a function are also elements in the domain of its inverse.

Figure 7

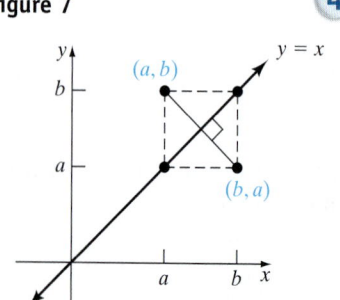

4 **Obtain the Graph of the Inverse Function from the Graph of the Function**

Look back at Example 6. In this example, we found that if a function is defined by a set of ordered pairs, we can find the inverse by interchanging the entries. So if (a, b) is a point on the graph of a function, then (b, a) is a point on the graph of the inverse function. See Figure 7. From the graph, it follows that the point (b, a) on the graph of the inverse function is the reflection about the line $y = x$ of the point (a, b).

 The graph of a function f and the graph of its inverse are symmetric with respect to the line $y = x$.

| **EXAMPLE 7** | **Graphing the Inverse Function** |

Figure 8(a) shows the graph of a one-to-one function $y = f(x)$. Draw the graph of its inverse.

Solution

We begin by adding the graph of $y = x$ to Figure 8(a). Since the points $(-4, -2)$, $(-3, -1)$, $(-1, 0)$, and $(4, 2)$ are on the graph of f, we know that the points $(-2, -4)$, $(-1, -3)$, $(0, -1)$, and $(2, 4)$ must be on the graph of the inverse of f. Using these points along with the fact that the graph of the inverse of f is a reflection about the line $y = x$ of the graph of f, we draw the graph of the inverse. See Figure 8(b).

Figure 8

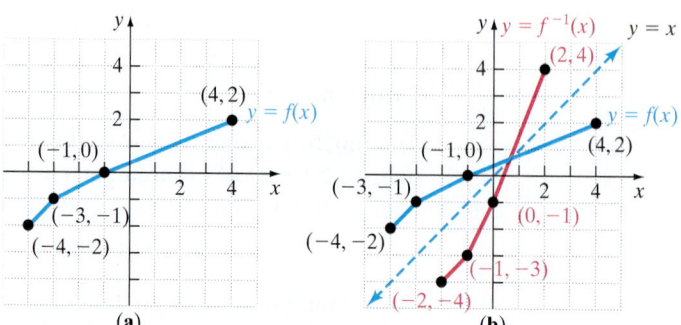

QUICK ✔

9. Below is the graph of a one-to-one function $y = f(x)$. Draw the graph of its inverse.

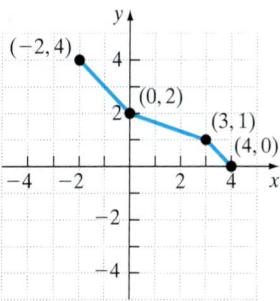

---

⑤ Find the Inverse of a Function Defined by an Equation

Work Smart

Suppose $f(x) = 2x$ so that each input gets doubled. The inverse function would be $f^{-1}(x) = x/2$ so that each input gets cut in half. So, if $x = 10$, then $f(10) = 20$. Now "plug" the output of f, 20, into f^{-1}, to get $f^{-1}(20) = 10$. Notice we are right back where we started!

We have learned how to find the inverse of a one-to-one function that is defined by a map, a set of ordered pairs, and a graph. We now discuss how to find the inverse of a function defined by an equation.

When a function is defined by an equation we use the notation $y = f(x)$. The notation $y = f^{-1}(x)$ is used to denote the equation whose rule is the inverse function of f.

Because the inverse of a function "undoes" what the original function does, we have the following relation between a function f and its inverse, f^{-1}.

$$f^{-1}(f(x)) = x \text{ for every } x \text{ in the domain of } f$$

$$f(f^{-1}(x)) = x \text{ for every } x \text{ in the domain of } f^{-1}$$

Let's go over an example to illustrate how to find the inverse of a one-to-one function defined by an equation.

EXAMPLE 8 **How to Find the Inverse of a One-to-One Function**

Find the inverse of $f(x) = 2x - 3$.

Step-by-Step Solution

Before we find the inverse function, we verify that the function is one-to-one. Because the graph of the function f is a line with y-intercept -3 and slope 2, we know by the horizontal line test that the function is one-to-one and therefore has an inverse function.

Step 1: Replace $f(x)$ with y in the equation for $f(x)$.	$f(x) = 2x - 3$ $y = 2x - 3$
Step 2: In $y = f(x)$, interchange the variables x and y to obtain $x = f(y)$.	$x = 2y - 3$
Step 3: Solve the equation found in Step 1 for y in terms of x.	Add 3 to both sides: $x + 3 = 2y$ Divide both sides by 2: $\dfrac{x + 3}{2} = y$

(Continued)

Step 4: Replace y with $f^{-1}(x)$.

$$f^{-1}(x) = \frac{x + 3}{2}$$

Step 5: Verify your result by showing that

$$f^{-1}(f(x)) = f^{-1}(2x - 3) \qquad f(f^{-1}(x)) = f\left(\frac{x + 3}{2}\right)$$

$$f^{-1}(f(x)) = x \text{ and}$$
$$f(f^{-1}(x)) = x.$$

$$= \frac{2x - 3 + 3}{2} \qquad = 2\left(\frac{x + 3}{2}\right) - 3$$

$$= \frac{2x}{2} \qquad\qquad = x + 3 - 3$$

$$= x \qquad\qquad\qquad = x$$

Everything checks, so $f^{-1}(x) = \dfrac{x + 3}{2}$.

Now, let's summarize the steps used in Example 8.

Steps for Finding the Inverse of a One-to-One Function Defined by an Equation

In Words

The function $f(x) = 2x - 3$ says we should take an input, double it, then subtract 3. The inverse function $f^{-1}(x) = \dfrac{x + 3}{2}$ says we should take an input add 3 and then cut it in half.

Step 1: Replace $f(x)$ with y in the equation for $f(x)$.

Step 2: In $y = f(x)$, interchange the variables x and y to obtain $x = f(y)$.

Step 3: Solve the equation found in Step 2 for y in terms of x.

Step 4: Replace y with $f^{-1}(x)$.

Step 5: Verify your result by showing that $f^{-1}(f(x)) = x$ and $f(f^{-1}(x)) = x$.

Consider the function $f(x) = 2x - 3$ and its inverse $f^{-1}(x) = \dfrac{x + 3}{2}$ from Example 8. Notice that $f(5) = 2(5) - 3 = 7$. What do you think $f^{-1}(7)$ will equal? Notice $f^{-1}(7) = \dfrac{7 + 3}{2} = \dfrac{10}{2} = 5$. So f^{-1} "undoes" what f did!

QUICK ✓

10. Find the inverse of $g(x) = 5x - 1$.

EXAMPLE 9 **Finding the Inverse of a One-to-One Function**

Find the inverse of $h(x) = x^3 + 4$.

Solution

$$h(x) = x^3 + 4$$

Replace $h(x)$ with y in the equation for $h(x)$: $\qquad y = x^3 + 4$

Interchange the variables x and y: $\qquad x = y^3 + 4$

Solve the equation for y in terms of x: $\qquad x - 4 = y^3$

Take the cube root of both sides: $\qquad \sqrt[3]{x - 4} = y$

Replace y with $h^{-1}(x)$: $\qquad h^{-1}(x) = \sqrt[3]{x - 4}$

Check Verify your result by showing that $h^{-1}(h(x)) = x$ and $h(h^{-1}(x)) = x$.

$$h^{-1}(h(x)) = h^{-1}(x^3 + 4) \qquad h(h^{-1}(x)) = h\left(\sqrt[3]{x - 4}\right)$$
$$= \sqrt[3]{x^3 + 4 - 4} \qquad\qquad = \left(\sqrt[3]{x - 4}\right)^3 + 4$$
$$= \sqrt[3]{x^3} \qquad\qquad\qquad = x - 4 + 4$$
$$= x \qquad\qquad\qquad\qquad = x$$

So $h^{-1}(x) = \sqrt[3]{x - 4}$. ▬

QUICK ✓

11. Find the inverse of $f(x) = x^5 + 3$.

11.1 Exercises

For Extra Help: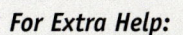

Student Solutions Manual   CD Video   PH Math/Tutor Center   MathXL Tutorials on CD   MathXL®   MyMathLab

Concepts and Vocabulary

In Problems 1–3, fill in the blanks.

1. Given two functions f and g, the _____ _____, denoted by $f \circ g$, is defined by $(f \circ g)(x) = f(g(x))$.

2. A function is _____ if any two different inputs in the domain correspond to two different outputs in the range. That is, if x_1 and x_2 are two different inputs of a function f, then $f(x_1) \neq f(x_2)$.

3. If f is some function, the _____ of f, denoted f^{-1}, receives as input $f(x)$, manipulates it and outputs the value of x.

In Problems 4–6, answer True or False to each statement.

4. If f is a one-to-one function so that its inverse is f^{-1}, then $f^{-1}(f(x)) = x$ for every x in the domain of f and $f(f^{-1}(x)) = x$ for every x in the domain of f^{-1}.

5. The notation $f^{-1}(x)$ is equivalent to $\dfrac{1}{f(x)}$.

6. $(f \circ g)(x) = f(x) \cdot g(x)$

7. In your own words, explain what it means for a function to be one-to-one. Why must a function be one-to-one in order for its inverse to be a function?

8. In your own words, explain why domain of f = range of f^{-1} and range of f = domain of f^{-1}.

9. State the horizontal line test. Why does it work?

10. If $f(g(x)) = (4x - 3)^5$ and $f(x) = x^5$, what is $g(x)$?

Building Skills

In Problems 11–18, for the given functions f and g, find

   (a) $(f \circ g)(3)$ **(b)** $(g \circ f)(-2)$ **(c)** $(f \circ f)(1)$ **(d)** $(g \circ g)(-4)$

11. $f(x) = 2x + 5; g(x) = x - 4$ **12.** $f(x) = 4x - 3; g(x) = x + 2$

13. $f(x) = x^2 + 4; g(x) = 2x + 3$ **14.** $f(x) = x^2 - 3; g(x) = 5x + 1$

15. $f(x) = 2x^3; g(x) = -2x^2 + 5$ **16.** $f(x) = -2x^3; g(x) = x^2 + 1$

17. $f(x) = |x - 10|; g(x) = \dfrac{12}{x + 3}$ **18.** $f(x) = \sqrt{x + 8}; g(x) = x^2 - 4$

In Problems 19–30, for the given functions f and g, find

(a) $(f \circ g)(x)$ **(b)** $(g \circ f)(x)$ **(c)** $(f \circ f)(x)$ **(d)** $(g \circ g)(x)$

19. $f(x) = x + 1; g(x) = 2x$

20. $f(x) = x - 3; g(x) = 4x$

21. $f(x) = 2x + 7; g(x) = -4x + 5$

22. $f(x) = 3x - 1; g(x) = -2x + 5$

23. $f(x) = x^2; g(x) = x - 3$

24. $f(x) = x^2 + 1; g(x) = x + 1$

25. $f(x) = \sqrt{x}; g(x) = x + 4$

26. $f(x) = \sqrt{x + 2}; g(x) = x - 2$

27. $f(x) = |x + 4|; g(x) = x^2 - 4$

28. $f(x) = |x - 3|; g(x) = x^3 + 3$

29. $f(x) = \dfrac{2}{x + 1}; g(x) = \dfrac{1}{x}$

30. $f(x) = \dfrac{2}{x - 1}; g(x) = \dfrac{4}{x}$

In Problems 31–40, determine which of the following functions is one-to-one.

31.

Level of Education	Average Annual Income
Less than 9th grade	$17,261
9th—12th grade, No diploma	$21,737
High School Graduate	$35,744
Associate's Degree	$49,279
Bachelor's Degree or Higher	$69,804

Source: *United States Census Bureau*

32.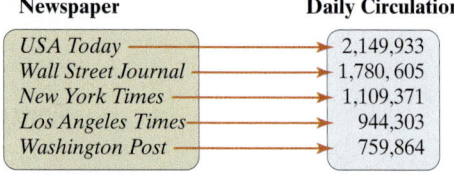

Newspaper	Daily Circulation
USA Today	2,149,933
Wall Street Journal	1,780,605
New York Times	1,109,371
Los Angeles Times	944,303
Washington Post	759,864

Source: *Information Please Almanac*, September 30, 2001.

33.

Age	Monthly Cost of Life Insurance
30	$7.09
35	$8.40
40	$11.29
45	

Source: *eterm.com*

34.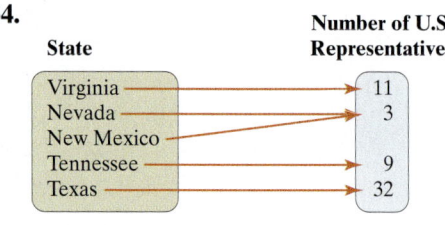

State	Number of U.S. Representatives
Virginia	11
Nevada	3
New Mexico	
Tennessee	9
Texas	32

35. $\{(-3, 4), (-2, 6), (-1, 8), (0, 10), (1, 12)\}$

36. $\{(-2, 6), (-1, 3), (0, 0), (1, -3), (2, 6)\}$

37. $\{(-2, 4), (-1, 2), (0, 0), (1, 2), (2, 4)\}$

38. $\{(-2, -8), (-1, -1), (0, 0), (1, 1), (2, 8)\}$

39. $\{(0, -4), (-1, -1), (-2, 0), (1, 1), (2, 4)\}$

40. $\{(-3, 0), (-2, 3), (-1, 0), (0, -3)\}$

In Problems 41–46, use the horizontal line test to determine whether the function whose graph is given is one-to-one.

41.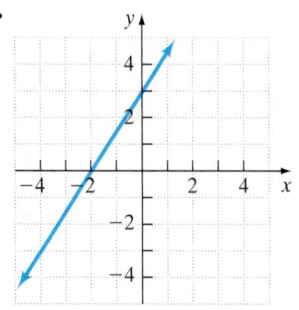

42.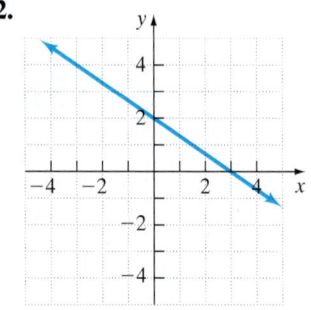

43.

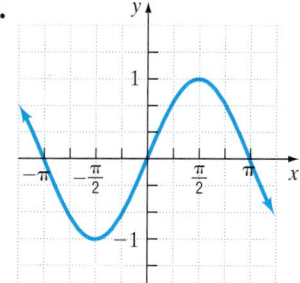

44.

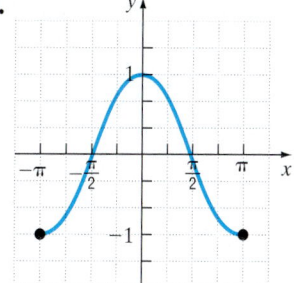

 45.

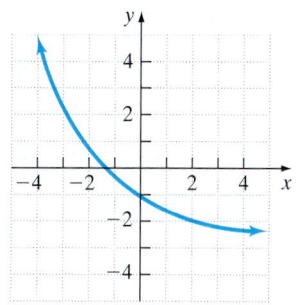

46.

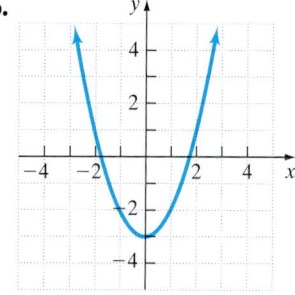

In Problems 47–52, find the inverse of the following one-to-one functions.

47.

U.S. Coin	Weight (g)
Cent	2.500
Nickel	5.000
Dime	2.268
Quarter	5.670
Half Dollar	11.340
Dollar	8.100

Source: *U.S. Mint Web site*

48.

Price ($)	Quantity Demanded
2300	152
2000	159
1700	164
1500	171
1300	176

49. $\{(0, 3), (1, 4), (2, 5), (3, 6)\}$

50. $\{(-1, 4), (0, 1), (1, -2), (2, -5)\}$

51. $\{(-2, 3), (-2, 1), (-2, -3), (-2, 9)\}$

52. $\{(-10, 1), (-5, 4), (0, 3), (-5, 2)\}$

In Problems 53–58, the graph of a one-to-one function f is given. Draw the graph of the inverse function f^{-1}.

53.

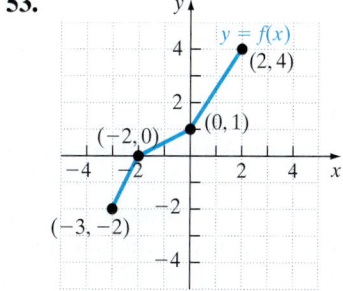

54.

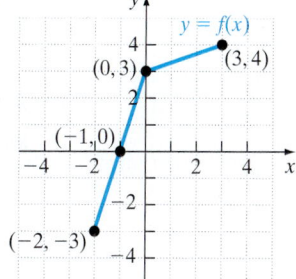

55.

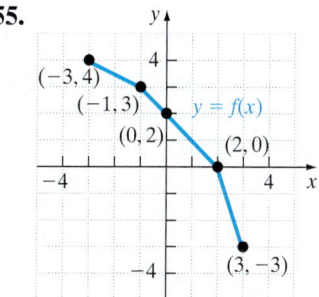

56.

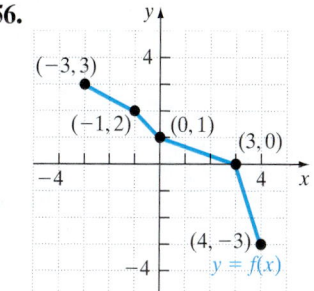

57.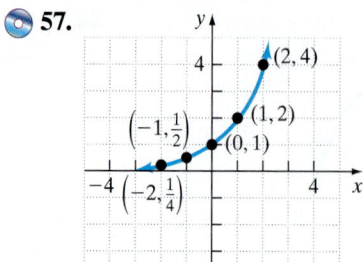

58.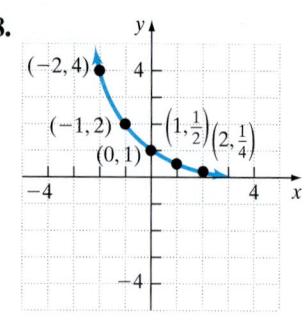

In Problems 59–68, verify that the functions f and g are inverses of each other.

59. $f(x) = x + 5; g(x) = x - 5$

60. $f(x) = 10x; g(x) = \dfrac{x}{10}$

61. $f(x) = 5x + 7; g(x) = \dfrac{x - 7}{5}$

62. $f(x) = 3x - 5; g(x) = \dfrac{x + 5}{3}$

63. $f(x) = 10x - 1; g(x) = \dfrac{x + 1}{10}$

64. $f(x) = 8x + 3; g(x) = \dfrac{x - 3}{8}$

65. $f(x) = \dfrac{3}{x - 1}; g(x) = \dfrac{3}{x} + 1$

66. $f(x) = \dfrac{2}{x + 4}; g(x) = \dfrac{2}{x} - 4$

67. $f(x) = \sqrt[3]{x + 4}; g(x) = x^3 - 4$

68. $f(x) = \sqrt[3]{2x + 1}; g(x) = \dfrac{x^3 - 1}{2}$

In Problems 69–88, find the inverse function of the given one-to-one function.

69. $f(x) = 6x$

70. $f(x) = 12x$

71. $f(x) = x + 4$

72. $g(x) = x + 6$

73. $h(x) = 2x - 7$

74. $H(x) = 3x + 8$

75. $G(x) = 2 - 5x$

76. $F(x) = 1 - 6x$

77. $g(x) = x^3 + 3$

78. $f(x) = x^3 - 2$

79. $p(x) = \dfrac{1}{x + 3}$

80. $P(x) = \dfrac{1}{x + 1}$

81. $F(x) = \dfrac{5}{2 - x}$

82. $G(x) = \dfrac{2}{3 - x}$

83. $f(x) = \sqrt[3]{x - 2}$

84. $f(x) = \sqrt[5]{x + 5}$

85. $R(x) = \dfrac{x}{x + 2}$

86. $R(x) = \dfrac{2x}{x + 4}$

87. $f(x) = \sqrt[3]{x - 1} + 4$

88. $g(x) = \sqrt[3]{x + 2} - 3$

Applying the Concepts

89. Environmental Disaster An oil tanker hits a rock that rips a hole in the hull of the ship. Oil leaking from the ship forms a circular region around the ship. If the radius r of the circle (in feet) as a function of time t (in hours) is $r(t) = 20t$, express the area A of the circular region contaminated with oil as a function of time. What will be the area of the circular region after 3 hours? (**Hint:** $A(r) = \pi r^2$)

90. Volume of a Balloon The volume V of a hot-air balloon (in cubic meters) as a function of its radius r is given by $V(r) = \dfrac{4}{3}\pi r^3$. If the radius r of the balloon is increasing as a function of time t (in minutes) according to $r(t) = 3\sqrt[3]{t}$, for $t \geq 0$, find the volume of the balloon as a function of time t. What will be the volume of the balloon after 30 minutes?

91. Buying Carpet You want to purchase new carpet for your family room. Carpet is sold by the square yard, but you have measured your family room in square feet. The function $A(x) = \dfrac{x}{9}$ converts the area of a room in square feet to an area in square yards. Suppose that the carpet you have selected is $18 per square yard installed. Then the function $C(A) = 18A$ represents the cost C of installing carpet in a room that is A square yards.

(a) Find the cost C as a function of the square footage of the room x.

(b) If your family room is 15 feet by 21 feet, what will it cost to install the carpet?

92. Tax Time You have a job that pays $20 per hour. Your gross salary G as a function of hours worked h is given by $G(h) = 20h$. Federal tax withholding T on your paycheck is equal to 18% of gross earnings G, so that federal tax withholding as a function of gross pay is given by the function $T(G) = 0.18G$.

(a) Find federal tax withholding T as a function of hours worked h.

(b) Suppose that you worked 28 hours last week. What will be the federal tax withholding on your paycheck?

93. If $f(4) = 12$, what is $f^{-1}(12)$?

94. If $g(-2) = 7$, what is $g^{-1}(7)$?

95. The domain of a function f is $[0, \infty)$, and its range is $[-5, \infty)$. State the domain and the range of f^{-1}.

96. The domain of a function f is $[5, \infty)$, and its range is $[0, \infty)$. State the domain and the range of f^{-1}.

97. The domain of a function g is $[-4, 10]$, and its range is $(-6, 12)$. State the domain and the range of g^{-1}.

98. The domain of a function g is $[0, 15]$, and its range is $(0, 8)$. State the domain and the range of g^{-1}.

99. Taxes The function $T(x) = 0.15(x - 6000) + 600$ represents the tax bill T of a single person whose adjusted gross income is x dollars for income between $6,000 and $28,400, inclusive. *(Source:* Internal Revenue Service) Find the inverse function that expresses adjusted gross income x as a function of taxes T. That is, find $x(T)$.

100. Health Costs The annual cost of health insurance H as a function of age a is given by the function $H(a) = 22.8a - 117.5$ for $15 \leq a \leq 90$. *(Source: Statistical Abstract, 2002)* Find the inverse function that expresses age a as a function of health insurance cost H. That is, find $a(H)$.

Extending the Concepts

101. If $f(x) = 2x^2 - x + 5$ and $g(x) = x + a$, find a so that the y-intercept of the graph of $(f \circ g)(x)$ is 20.

102. If $f(x) = x^2 - 3x + 1$ and $g(x) = x - a$, find a so that the y-intercept of the graph of $(f \circ g)(x)$ is -1.

The Graphing Calculator

Graphing calculators have the ability to evaluate composite functions. To obtain the results of Example 1, we would let $Y_1 = f(x) = x^2 - 3$ and $Y_2 = g(x) = 2x + 1$. Figure 9 shows the results of Example 1 using a TI-84 Plus graphing calculator.

Figure 9

```
Y₁(Y₂(3))
                46
Y₂(Y₁(3))
                13
Y₁(Y₁(-2))
                -2
```

In Problems 103–110, use a graphing calculator to evaluate the composite functions. Compare your answers with those found in Problems 11–18.

(a) $(f \circ g)(3)$ **(b)** $(g \circ f)(-2)$ **(c)** $(f \circ f)(1)$ **(d)** $(g \circ g)(-4)$

103. $f(x) = 2x + 5; g(x) = x - 4$ **104.** $f(x) = 4x - 3; g(x) = x + 2$

105. $f(x) = x^2 + 4; g(x) = 2x + 3$ **106.** $f(x) = x^2 - 3; g(x) = 5x + 1$

107. $f(x) = 2x^3; g(x) = -2x^2 + 5$ **108.** $f(x) = -2x^3; g(x) = x^2 + 1$

109. $f(x) = |x - 10|; g(x) = \dfrac{12}{x + 3}$ **110.** $f(x) = \sqrt{x + 8}; g(x) = x^2 - 4$

In Problems 111–116, the functions f and g are inverses. Graph both functions on the same screen along with the line $y = x$ to see the symmetry of the functions about the line $y = x$.

111. $f(x) = x + 5; g(x) = x - 5$ **112.** $f(x) = 10x; g(x) = \dfrac{x}{10}$

113. $f(x) = 5x + 7; g(x) = \dfrac{x - 7}{5}$ **114.** $f(x) = 3x - 5; g(x) = \dfrac{x + 5}{3}$

115. $f(x) = 10x - 1; g(x) = \dfrac{x + 1}{10}$ **116.** $f(x) = 8x + 3; g(x) = \dfrac{x - 3}{8}$

11.2 Exponential Functions

OBJECTIVES

1. Evaluate Exponential Expressions
2. Graph Exponential Functions
3. Define the Number e
4. Solve Exponential Equations
5. Study Exponential Models That Describe Our World

Preparing for Exponential Functions

Before getting started, take the following readiness quiz. If you get a problem wrong, go back to the section cited and review the material.

1. Evaluate: (a) 2^3 (b) 2^{-1} (c) 3^4 [Section 1.6, pp. 49–50; Section 5.4, pp. 374–376]

2. Graph $f(x) = x^2$. [Section 10.4, p. 782]

3. State the definition of a rational number. [Section 1.2, p. 10]

4. State the definition of an irrational number. [Section 1.2, p. 10]

5. Simplify: (a) $m^3 \cdot m^5$ (b) $\dfrac{a^7}{a^2}$ (c) $(z^3)^4$ [Section 5.4, pp. 371–380]

6. Solve: $x^2 - 5x = 14$ [Section 6.6, pp. 451–457]

Preparing for...Answers

1. (a) 8 **(b)** $\frac{1}{2}$ **(c)** 81

2.

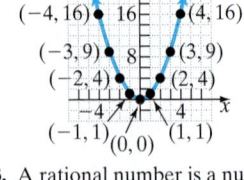

3. A rational number is a number that can be expressed as a quotient $\dfrac{p}{q}$ of two integers. The integer p is called the numerator, and the integer q, which cannot be 0, is called the denominator. The set of rational numbers is the numbers $\mathbb{Q} = \left\{ x \mid x = \dfrac{p}{q}, \text{where } p, q \right.$ are integers and $\left. q \neq 0 \right\}$.

4. An irrational number has a decimal representation that neither repeats nor terminates.

5. (a) m^8 **(b)** a^5 **(c)** z^{12}

6. $\{-2, 7\}$

Suppose that you have just been hired as a proofreader by Prentice Hall. Paul Murphy, Prentice Hall's editor, offers you two options for getting paid. Option A states that you will get paid $100 for each error you find in the final page proofs of a text. Option B states that you will get $2 for the first error you find in the final page proofs and your payment will double for each additional error you find. Based on your experience, you know there are typically about 15–20 errors in the final page proofs of a text. Which option will you go with?

If there is one error, Option A pays $100, while Option B pays $2. If there are two errors, Option A pays 2($100) = $200, while Option B pays $2^2 = $4. If there are three errors, Option A pays 3($100) = $300, while Option B pays $2^3 = $8. It's looking like Option A is the way to go. To complete the analysis, we set up Table 1, which lists the payment amount as a function of the number of errors in the page proof. Remember, in Option B, the payment amount doubles each time you find an error.

Holy cow! If you find 20 errors, you'll get paid over 1 million dollars! Paul Murphy better reconsider his offer! If we let x represent the number of errors, we can express the salary for Option A as a linear function, $f(x) = 100x$; we can express the salary for Option B as an *exponential function*, $g(x) = 2^x$.

Table 1					
Number of Errors	Option A Payment	Option B Payment	Number of Errors	Option A Payment	Option B Payment
0	$0	$0	11	$1,100	$2,048
1	$100	$2	12	$1,200	$4,096
2	$200	$4	13	$1,300	$8,192
3	$300	$8	14	$1,400	$16,384
4	$400	$16	15	$1,500	$32,768
5	$500	$32	16	$1,600	$65,536
6	$600	$64	17	$1,700	$131,072
7	$700	$128	18	$1,800	$262,144
8	$800	$256	19	$1,900	$524,288
9	$900	$512	20	$2,000	$1,048,576
10	$1,000	$1,024			

DEFINITION

An **exponential function** is a function of the form

$$f(x) = a^x$$

where a is a positive real number $(a > 0)$ and $a \neq 1$. The domain of the exponential function is the set of all real numbers.

We will address the restrictions on the base a shortly. The key point to understand with exponential functions is that the independent variable is in the exponent of the expression. Contrast this idea with polynomial functions (such as $f(x) = x^2 - 4x$ or $g(x) = 2x^3 + x^2 - 5$), where the independent variable is the base of the expression.

① Evaluate Exponential Expressions

In Section 9.2 we gave a definition for raising a real number a to a rational power. From that discussion, we gave meaning to expressions of the form

$$a^{m/n}$$

where the base a is a positive real number and the exponent m/n is a rational number.

However, our world does not only consist of rational numbers, so a logical question to ask is, "What if I want to raise the base a to any real number, rational or irrational?" Although the definition for raising a positive real number to any real number requires advanced mathematics, we can provide a discussion that is reasonable and intuitive.

Suppose that we wanted to determine the value of $3^{\sqrt{2}}$. Using our calculator, we know that $\sqrt{2} \approx 1.414213562$, so it should seem reasonable that we can approximate $3^{\sqrt{2}}$ as $3^{1.4}$, where the 1.4 comes from truncating the decimals to the right of the 4 in the tenths position. A better approximation of $3^{\sqrt{2}}$ would be $3^{1.4142}$, where the digits to the right of the ten-thousandths position have been truncated. The idea is this—the more decimals used in the approximation of an irrational number, the better the approximation of $3^{\sqrt{2}}$.

Fortunately, most calculators can easily evaluate expressions such as $3^{1.4142}$ using the $\boxed{x^y}$ key or a caret $\boxed{\wedge}$ key. To evaluate expressions of the form a^x using a scientific calculator, enter the base a, then press the $\boxed{x^y}$ key, enter the exponent x, and press $\boxed{=}$.

To evaluate expressions of the form a^x using a graphing calculator, enter the base a, then press the caret $\boxed{\wedge}$ key, enter the exponent x, and press $\boxed{\text{ENTER}}$.

EXAMPLE 1 **Evaluating Exponential Expressions**

Using a calculator, evaluate each of the following expressions. Write as many decimals as your calculator allows.

(a) $3^{1.4}$ (b) $3^{1.41}$ (c) $3^{1.414}$ (d) $3^{1.4142}$ (e) $3^{\sqrt{2}}$

Solution

(a) $3^{1.4} \approx 4.655536722$ (b) $3^{1.41} \approx 4.706965002$

(c) $3^{1.414} \approx 4.727695035$ (d) $3^{1.4142} \approx 4.72873393$

(e) $3^{\sqrt{2}} \approx 4.728804388$ ▬

QUICK ✔ *Using a calculator, evaluate each of the following expressions. Write as many decimals as your calculator allows.*

1. (a) $2^{1.7}$ **(b)** $2^{1.73}$ **(c)** $2^{1.732}$ **(d)** $2^{1.7321}$ **(e)** $2^{\sqrt{3}}$

Based on the results of Example 1, it should be clear that we can approximate the value of an exponential expression at any real number. This is why the domain of exponential functions is the set of all real numbers.

When we gave the definition of an exponential function, we excluded the possibility for the base a to equal 1 and we stated that the base a must be positive. We exclude the base $a = 1$ because this function is the constant function $f(x) = 1^x = 1$. We also exclude bases that are negative because we would run into problems for exponents such as $\frac{1}{2}$ or $\frac{3}{4}$. For example, suppose $f(x) = (-2)^x$. In the real number system, we could not evaluate $f\left(\frac{1}{2}\right)$ because $f\left(\frac{1}{2}\right) = (-2)^{1/2} = \sqrt{-2}$, which is not a real number. Finally, we exclude a base of 0, because this function is $f(x) = 0^x$ which equals zero for $x > 0$ and is undefined when $x < 0$. When $x = 0$, $f(x) = 0^x$ is *indeterminate* because its value is not precisely determined.

② Graph Exponential Functions

We use the point-plotting method to learn properties of exponential functions from their graphs.

EXAMPLE 2 **Graphing an Exponential Function**

Graph the exponential function $f(x) = 2^x$ using point plotting. From the graph, state the domain and the range of the function.

Solution

We begin by locating some points on the graph of $f(x) = 2^x$ as shown in Table 2. We plot the points in Table 2 and connect them in a smooth curve. Figure 10 shows the graph of $f(x) = 2^x$.

The domain of any exponential function is the set of all real numbers. Notice there is no x such that $2^x = 0$ and there is no x such that 2^x is negative. Based on this and the graph, we conclude that the range of $f(x) = 2^x$ is the set of all positive real numbers or $\{y | y > 0\}$ or $(0, \infty)$ using interval notation.

Figure 10

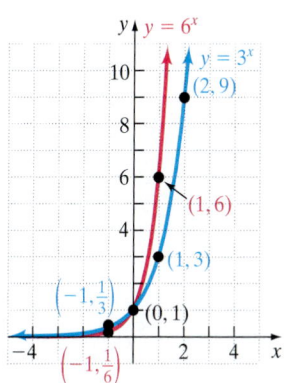

Table 2

x	$f(x) = 2^x$	$(x, f(x))$
-3	$f(-3) = 2^{-3} = \dfrac{1}{2^3} = \dfrac{1}{8}$	$\left(-3, \dfrac{1}{8}\right)$
-2	$f(-2) = 2^{-2} = \dfrac{1}{2^2} = \dfrac{1}{4}$	$\left(-2, \dfrac{1}{4}\right)$
-1	$f(-1) = 2^{-1} = \dfrac{1}{2^1} = \dfrac{1}{2}$	$\left(-1, \dfrac{1}{2}\right)$
0	$f(0) = 2^0 = 1$	$(0, 1)$
1	$f(1) = 2^1 = 2$	$(1, 2)$
2	$f(2) = 2^2 = 4$	$(2, 4)$
3	$f(3) = 2^3 = 8$	$(3, 8)$

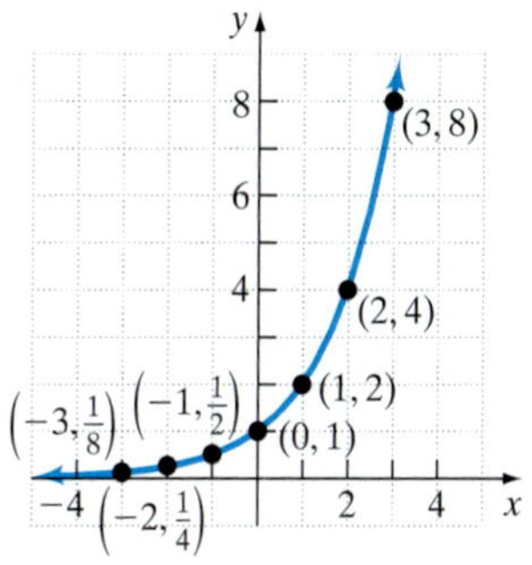

Figure 11

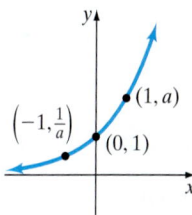

The graph of $f(x) = 2^x$ in Figure 10 is typical of all exponential functions that have a base larger than 1. Figure 11 shows the graph of two other exponential functions whose bases are larger than 1. Notice that the larger the base, the steeper the graph is for $x > 0$ and the closer the graph is to the x-axis for $x < 0$.

The following display summarizes the information that we have about $f(x) = a^x$ where the base is greater than $1 (a > 1)$.

Figure 12
$f(x) = a^x, a > 1$

> **PROPERTIES OF THE GRAPH OF AN EXPONENTIAL FUNCTION $f(x) = a^x$, $a > 1$**
>
> 1. The domain is the set of all real numbers. The range is the set of all positive real numbers.
> 2. There are no x-intercepts; the y-intercept is 1.
> 3. The graph of f contains the points $\left(-1, \dfrac{1}{a}\right)$, $(0, 1)$, and $(1, a)$.
>
> See Figure 12.

QUICK ✓

2. Graph the exponential function $f(x) = 4^x$ using point plotting. From the graph, state the domain and the range of the function.

Now we consider $f(x) = a^x, 0 < a < 1$.

EXAMPLE 3   Graphing an Exponential Function

Graph the exponential function $f(x) = \left(\dfrac{1}{2}\right)^x$ using point plotting. From the graph, state the domain and the range of the function.

Solution

We begin by locating some points on the graph of $f(x) = \left(\dfrac{1}{2}\right)^x$ as shown in Table 3 on page 856. We plot the points in Table 3 and connect them in a smooth curve. Figure 13 shows the graph of $f(x) = \left(\dfrac{1}{2}\right)^x$.

Table 3

x	$f(x) = \left(\dfrac{1}{2}\right)^x$	$(x, f(x))$
-3	$f(-3) = \left(\dfrac{1}{2}\right)^{-3} = 2^3 = 8$	$(-3, 8)$
-2	$f(-2) = \left(\dfrac{1}{2}\right)^{-2} = 2^2 = 4$	$(-2, 4)$
-1	$f(-1) = \left(\dfrac{1}{2}\right)^{-1} = 2^1 = 2$	$(-1, 2)$
0	$f(0) = \left(\dfrac{1}{2}\right)^{0} = 1$	$(0, 1)$
1	$f(1) = \left(\dfrac{1}{2}\right)^{1} = \dfrac{1}{2}$	$\left(1, \dfrac{1}{2}\right)$
2	$f(2) = \left(\dfrac{1}{2}\right)^{2} = \dfrac{1}{4}$	$\left(2, \dfrac{1}{4}\right)$
3	$f(3) = \left(\dfrac{1}{2}\right)^{3} = \dfrac{1}{8}$	$\left(3, \dfrac{1}{8}\right)$

Figure 13

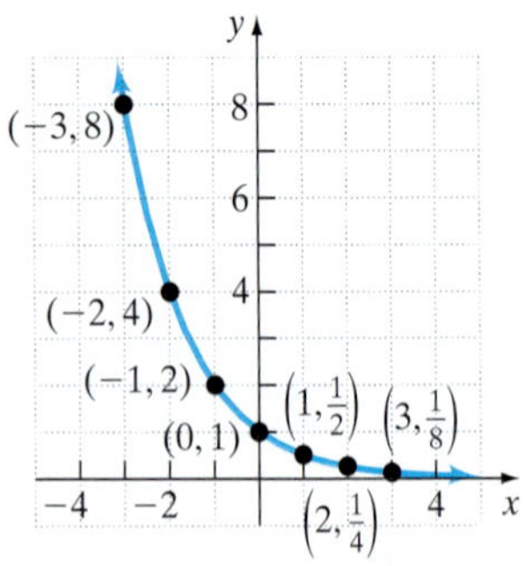

Figure 14

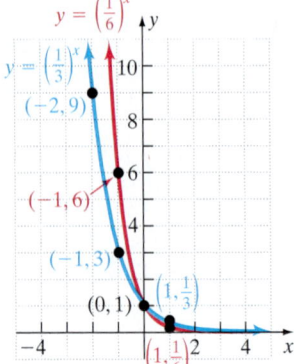

The domain of any exponential function is the set of all real numbers. From the graph, we conclude that the range of $f(x) = \left(\dfrac{1}{2}\right)^x$ is the set of all positive real numbers or $\{y \mid y > 0\}$ or $(0, \infty)$ using interval notation.

The graph of $f(x) = \left(\dfrac{1}{2}\right)^x$ in Figure 13 is typical of all exponential functions that have a base between 0 and 1. Figure 14 shows the graph of two additional exponential functions whose bases are between 0 and 1. Notice that the smaller the base, the closer the graph is to the x-axis for $x > 0$ and the steeper the graph is for $x < 0$.

The following display summarizes the information that we have about $f(x) = a^x$ where the base is between 0 and 1 ($0 < a < 1$).

Figure 15
$f(x) = a^x, 0 < a < 1$

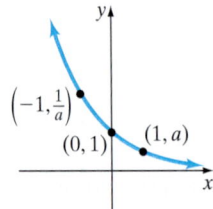

PROPERTIES OF THE GRAPH OF AN EXPONENTIAL FUNCTION $f(x) = a^x, 0 < a < 1$

1. The domain is the set of all real numbers. The range is the set of all positive real numbers.
2. There are no x-intercepts; the y-intercept is 1.
3. The graph of f contains the points $\left(-1, \dfrac{1}{a}\right)$, $(0, 1)$, and, $(1, a)$.

See Figure 15.

QUICK ✓

3. Graph the exponential function $f(x) = \left(\dfrac{1}{4}\right)^x$ using point plotting. From the graph, state the domain and the range of the function.

EXAMPLE 4   **Graphing an Exponential Function**

Use point plotting to graph $f(x) = 3^{x+1}$. From the graph, state the domain and the range of the function.

Solution

We choose values of x and use the equation that defines the function to find the value of the function. See Table 4. We then plot the ordered pairs and connect them in a smooth curve. See Figure 16.

Table 4

x	$f(x)$	$(x, f(x))$
-3	$f(-3) = 3^{-3+1} = 3^{-2} = \dfrac{1}{3^2} = \dfrac{1}{9}$	$\left(-3, \dfrac{1}{9}\right)$
-2	$f(-2) = 3^{-2+1} = 3^{-1} = \dfrac{1}{3^1} = \dfrac{1}{3}$	$\left(-2, \dfrac{1}{3}\right)$
-1	$f(-1) = 3^{-1+1} = 3^0 = 1$	$(-1, 1)$
0	$f(0) = 3^{0+1} = 3^1 = 3$	$(0, 3)$
1	$f(1) = 3^{1+1} = 3^2 = 9$	$(1, 9)$

Figure 16

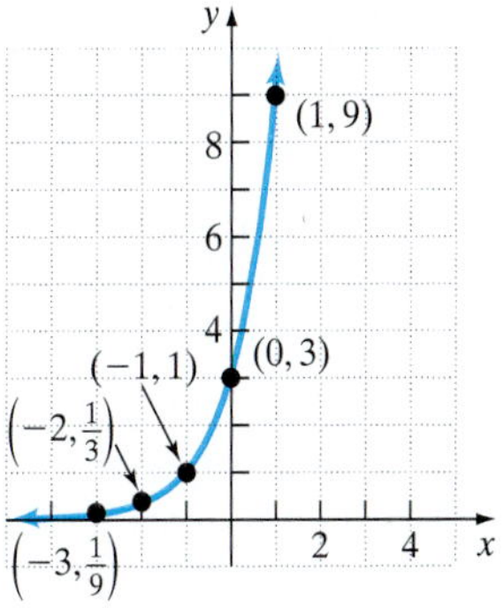

The domain is the set of all real numbers. The range is $\{y \mid y > 0\}$, or using interval notation, $(0, \infty)$.

QUICK ✔ *Graph each function using point plotting. From the graph, state the domain and the range of each function.*

4. $f(x) = 2^{x-1}$ **5.** $f(x) = 3^x + 1$

---

③ Define the Number e

Many problems that occur in nature require the use of an exponential function whose base is a certain irrational number, symbolized by the letter e. The number e can be used to model the growth of a stock's price or it can be used in a model to estimate the time of death of a carbon-based life form.

> **DEFINITION**
>
> The **number e** is defined as the number that the expression
> $$\left(1 + \frac{1}{n}\right)^n$$
> approaches as n becomes unbounded in the positive direction (that is, as n gets bigger).

Table 5 on page 858 illustrates what happens to the value of $\left(1 + \dfrac{1}{n}\right)^n$ as n takes on larger and larger values. The last number in the last column in Table 5 is the number e correct to nine decimal places.

Table 5

n	$\dfrac{1}{n}$	$1 + \dfrac{1}{n}$	$\left(1 + \dfrac{1}{n}\right)^{n}$
1	1	2	2
2	0.5	1.5	2.25
5	0.2	1.2	2.48832
10	0.1	1.1	2.59374246
100	0.01	1.01	2.704813829
1,000	0.001	1.001	2.716923932
10,000	0.0001	1.0001	2.718145927
100,000	0.00001	1.00001	2.718268237
1,000,000	0.000001	1.000001	2.718280469
1,000,000,000	10^{-9}	$1 + 10^{-9}$	2.718281827

The exponential function $f(x) = e^x$, whose base is the number e, occurs so often in applications that it is sometimes referred to as *the* exponential function. Most calculators have the key $\boxed{e^x}$ or $\boxed{\exp(x)}$, which may be used to evaluate the exponential function $f(x) = e^x$ for a given value of x.

Use your calculator to approximate the values of $f(x) = e^x$ for $x = -1, 0$, and 1 as we have done to create Table 6. The graph of the exponential function is shown in Figure 17(a). Since $2 < e < 3$, the graph of $f(x) = e^x$ lies between the graph of $y = 2^x$ and $y = 3^x$. See Figure 17(b).

Table 6

x	$f(x) = e^x$
-2	$e^{-2} \approx 0.135$
-1	$e^{-1} \approx 0.368$
0	$e^0 = 1$
1	$e^1 \approx 2.718$
2	$e^2 \approx 7.389$

Figure 17
$f(x) = e^x$

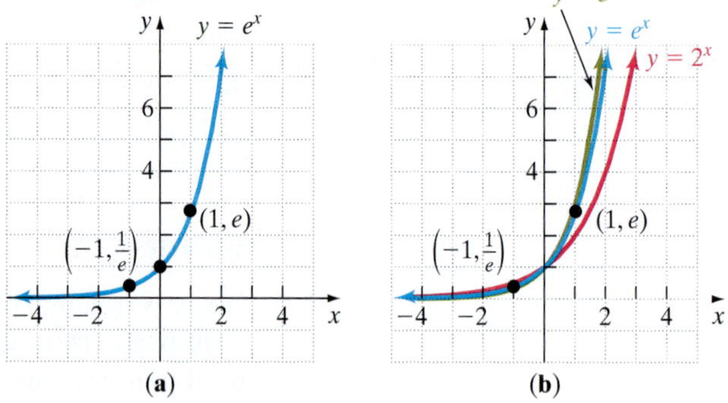

(a)

(b)

(4) ## Solve Exponential Equations

Equations that involve terms of the form a^x, $a > 0$, $a \neq 1$, are called **exponential equations.** Sometimes we can solve exponential equations using the Laws of Exponents and the following property.

Work Smart

To use the Property for Solving Exponential Equations, both sides of the equation must have the same base.

PROPERTY FOR SOLVING EXPONENTIAL EQUATIONS

$$\text{If} \quad a^u = a^v, \quad \text{then} \quad u = v.$$

This results from the fact that exponential functions are one-to-one. This property of exponential functions basically states that any output is the result of one (and only one) input. That is to say, two different inputs cannot yield the same output. For example, $y = x^2$ is not one-to-one because two different inputs, -2 and 2, correspond to the same output, 4.

To use the Property for Solving Exponential Equations, each side of the equality must be written with the same base.

EXAMPLE 5 **How to Solve an Exponential Equation**

Solve: $2^{x-3} = 32$

Step-by-Step Solution

Step 1: Use the Laws of Exponents to write both sides of the equation with the same base.	$32 = 2^5$:	$2^{x-3} = 32$ $2^{x-3} = 2^5$
Step 2: Set the exponents on each side of the equation equal to each other.		$x - 3 = 5$
Step 3: Solve the equation resulting from Step 2.	Add 3 to both sides:	$x - 3 + 3 = 5 + 3$ $x = 8$
Step 4: Verify your solution(s).	Let $x = 8$:	$2^{x-3} = 32$ $2^{8-3} \stackrel{?}{=} 32$ $2^5 \stackrel{?}{=} 32$ $32 = 32$ True

The solution set is $\{8\}$.

Below is a summary of the steps we used in Example 5.

Steps for Solving Exponential Equations of the Form $a^u = a^v$

Step 1: Use the Laws of Exponents to write both sides of the equation with the same base.

Step 2: Set the exponents on each side of the equation equal to each other.

Step 3: Solve the equation resulting from Step 2.

Step 4: Verify your solution(s).

QUICK ✓ *Solve each equation.*

6. $5^{x-4} = 5^{-1}$

7. $3^{x+2} = 81$

EXAMPLE 6 **Solving an Exponential Equation**

Solve the following equations:

(a) $4^{x^2} = 32$

(b) $\dfrac{e^{x^2}}{e^{2x}} = e^8$

Solution

(a)

$$4^{x^2} = 32$$

Rewrite exponential expressions with a common base:

$$(2^2)^{x^2} = 2^5$$

Use $(a^m)^n = a^{m \cdot n}$:

$$2^{2x^2} = 2^5$$

Set the exponents on each side of the equation equal to each other:

$$2x^2 = 5$$

Divide both sides by 5:

$$x^2 = \frac{5}{2}$$

Take the square root of both sides:

$$x = \pm\sqrt{\frac{5}{2}}$$

Rationalize the denominator:

$$x = \pm\frac{\sqrt{10}}{2}$$

Check

$x = \dfrac{-\sqrt{10}}{2}:$ $4^{\left(-\sqrt{10}/2\right)^2} \overset{?}{=} 32$

$$4^{10/4} \overset{?}{=} 32$$

$$4^{5/2} \overset{?}{=} 32$$

$a^{m/n} = \sqrt[n]{a^m}:$ $\sqrt{4^5} \overset{?}{=} 32$

$$2^5 \overset{?}{=} 32$$

$$32 = 32 \quad \text{True}$$

$x = \dfrac{\sqrt{10}}{2}:$ $4^{\left(\sqrt{10}/2\right)^2} \overset{?}{=} 32$

$$4^{10/4} \overset{?}{=} 32$$

$$4^{5/2} \overset{?}{=} 32$$

$$\sqrt{4^5} \overset{?}{=} 32$$

$$2^5 \overset{?}{=} 32$$

$$32 = 32 \quad \text{True}$$

The solution set is $\left\{ -\dfrac{\sqrt{10}}{2}, \dfrac{\sqrt{10}}{2} \right\}$.

(b)

$$\frac{e^{x^2}}{e^{2x}} = e^8$$

Use $\dfrac{a^m}{a^n} = a^{m-n}$:

$$e^{x^2 - 2x} = e^8$$

Set the exponents on each side of the equation equal to each other:

$$x^2 - 2x = 8$$

Subtract 8 from both sides:

$$x^2 - 2x - 8 = 0$$

Factor:

$$(x - 4)(x + 2) = 0$$

Zero-product property:

$$x = 4 \text{ or } x = -2$$

Check

$x = 4:$

$$\frac{e^{4^2}}{e^{2(4)}} \overset{?}{=} e^8$$

$$\frac{e^{16}}{e^8} \overset{?}{=} e^8$$

$$e^8 = e^8$$

$x = -2:$

$$\frac{e^{(-2)^2}}{e^{2(-2)}} \overset{?}{=} e^8$$

$$\frac{e^4}{e^{-4}} \overset{?}{=} e^8$$

$$e^8 = e^8$$

The solution set is $\{-2, 4\}$.

QUICK ✓ *Solve each equation.*

8. $e^{x^2} = e^x \cdot e^{4x}$

9. $\dfrac{2^{x^2}}{8} = 2^{2x}$

---

⑤ Study Exponential Models That Describe Our World

Exponential functions are used in many different disciplines such as biology (half-life), chemistry (carbon-dating), economics (time value of money), and psychology (learning curves). Exponential functions are also used in statistics, as shown in the next example.

EXAMPLE 7   **Exponential Probability**

From experience, the manager of a crisis helpline knows that between the hours of 3:00 A.M. and 5:00 A.M., calls occur at the rate of 3 calls per hour (0.05 calls per minute). The following formula from statistics can be used to determine the likelihood that a call will occur within t minutes of 3 A.M.

$$F(t) = 1 - e^{-0.05t}$$

(a) Determine the likelihood that a person will call within 5 minutes of 3:00 A.M.

(b) Determine the likelihood that a person will call within 20 minutes of 3:00 A.M.

Solution

(a) The likelihood that a call will occur within 5 minutes of 3:00 A.M. is found by evaluating the function $F(t) = 1 - e^{-0.05t}$ at $t = 5$.

$$F(5) = 1 - e^{-0.05(5)}$$
$$\approx 0.221$$

The likelihood that a call will occur within 5 minutes of 3:00 A.M. is $0.221 = 22.1\%$.

(b) The likelihood that a call will occur within 20 minutes of 3:00 A.M. is found by evaluating the function $F(t) = 1 - e^{-0.05t}$ at $t = 20$.

$$F(20) = 1 - e^{-0.05(20)}$$
$$\approx 0.632$$

The likelihood that a call will occur within 20 minutes of 3:00 A.M. is $0.632 = 63.2\%$. ■

QUICK ✓

10. From experience, the manager of a bank knows that between the hours of 3:00 P.M. and 5:00 P.M., people arrive at the rate of 15 people per hour (0.25 people per minute). The following formula from statistics can be used to determine the likelihood that a person will arrive within t minutes of 3:00 P.M.

$$F(t) = 1 - e^{-0.25t}$$

(a) Determine the likelihood that a person will arrive within 10 minutes of 3:00 P.M.

(b) Determine the likelihood that a person will arrive within 25 minutes of 3:00 P.M.

EXAMPLE 8 Radioactive Decay

The radioactive **half-life** for a given radioisotope of an element is the time for half the radioactive nuclei in any sample to decay to some other substance. For example, the half-life of plutonium-239 is 24,360 years. Plutonium-239 is particularly dangerous because it emits alpha particles that are absorbed into bone marrow. The maximum amount of plutonium-239 that an adult can handle without significant injury is 0.13 micrograms ($= 0.000000013$ grams). Suppose that a researcher possesses a 1-gram sample of Plutonium-239. The amount A (in grams) of Plutonium-239 after t years is given by

$$A(t) = 1 \cdot \left(\frac{1}{2}\right)^{t/24,360}$$

(a) How much plutonium-239 is left in the sample after 500 years?

(b) How much plutonium-239 is left in the sample after 24,360 years?

(c) How much plutonium-239 is left in the sample after 73,080 years?

Solution

(a) The amount of plutonium-239 left in the sample after 500 years is found by evaluating A at $t = 500$. That is, we determine $A(500)$.

$$A(500) = 1 \cdot \left(\frac{1}{2}\right)^{500/24,360}$$

Use a calculator: ≈ 0.986 gram

After 500 years, there will be approximately 0.986 gram of Plutonium-239 left in the sample.

(b) The amount of plutonium-239 left in the sample after 24,360 years is found by evaluating A at $t = 24,360$. That is, we determine $A(24,360)$.

$$A(24,360) = 1 \cdot \left(\frac{1}{2}\right)^{24,360/24,360}$$

$$= 1 \cdot \left(\frac{1}{2}\right)^{1}$$

$$= 0.5 \text{ gram}$$

After 24,360 years, there will be 0.5 gram of plutonium-239 left in the sample.

(c) The amount of plutonium-239 left in the sample after 73,080 years is found by evaluating A at $t = 73,080$. That is, we determine $A(73,080)$.

$$A(73,080) = 1 \cdot \left(\frac{1}{2}\right)^{73,080/24,360}$$

$$= 1 \cdot \left(\frac{1}{2}\right)^{3}$$

$$= \frac{1}{8} \text{ gram}$$

After 24,360 years, there will be $\frac{1}{8} = 0.125$ gram of plutonium-239 left in the sample.

QUICK ✓

11. The half-life of thorium-227 is 18.72 days. Suppose that a researcher possesses a 10-gram sample of thorium-227. The amount A (in grams) of thorium-227 after t days is given by

$$A(t) = 10 \cdot \left(\frac{1}{2}\right)^{t/18.72}$$

(a) How much thorium-227 is left in the sample after 10 days?

(b) How much thorium-227 is left in the sample after 18.72 days?

(c) How much thorium-227 is left in the sample after 74.88 days?

(d) How much thorium-227 is left in the sample after 100 days?

When we deposit money in a bank, the bank pays us interest on the balance in the account. When working with problems involving interest, we use the term **payment period** as shown in Table 7.

Table 7	
Payment Period	**Number of Times Interest Is Paid**
Annually	Once per year
Semiannually	Twice per year
Quarterly	4 times per year
Monthly	12 times per year
Daily	360 times per year

When the interest due at the end of a payment period is added to the principal so that the interest computed at the end of the next payment period is based on this new principal amount (old principal + interest), the interest is said to have been **compounded. Compound interest** is interest paid on previously earned interest.

The following formula can be used to determine the value of an account after a certain period of time.

Work Smart

When using the compound interest formula, be sure to express the interest as a decimal.

COMPOUND INTEREST FORMULA

The amount A after t years due to a principal P invested at an annual interest rate r compounded n times per year is

$$A = P\left(1 + \frac{r}{n}\right)^{nt}$$

For example, if you deposit $500 into an account paying 3% annual interest compounded monthly, then $P = \$500$, $r = 0.03$, and $n = 12$ (twelve compounding periods per year).

EXAMPLE 9 Future Value of Money

Suppose that you deposit $3000 into a Roth IRA today. Determine future value A of the deposit if it earns 8% interest compounded quarterly after

(a) 1 year   **(b)** 10 years   **(c)** 35 years, when you plan on retiring

Solution

We use the compound interest formula with $P = \$3000$, $r = 0.08$, and $n = 4$, so that

$$A = \$3000\left(1 + \frac{0.08}{4}\right)^{4t}$$

$$= \$3000(1 + 0.02)^{4t}$$

$$= \$3000(1.02)^{4t}$$

(a) The value of the account after $t = 1$ year is

$$A = \$3000(1.02)^{4(1)}$$

$$= \$3000(1.02)^4$$

Use a calculator: $= \$3000(1.08243216)$

$$= \$3247.30$$

(b) The value of the account after $t = 10$ years is

$$A = \$3000(1.02)^{4(10)}$$

$$= \$3000(1.02)^{40}$$

Use a calculator: $= \$3000(2.208039664)$

$$= \$6{,}624.12$$

(c) The value of the account after $t = 35$ years is

$$A = \$3000(1.02)^{4(35)}$$

$$= \$3000(1.02)^{140}$$

Use a calculator: $= \$3000(15.99646598)$

$$= \$47{,}989.40$$

QUICK ✓

12. Suppose that you deposit $2000 into a Roth IRA today. Determine the future value A of the deposit if it earns 12% interest compounded monthly (12 times per year) after

(a) 1 year   **(b)** 15 years   **(c)** 30 years, when you plan on retiring

11.2 Exercises

Concepts and Vocabulary

In Problems 1–3, fill in the blanks.

1. An exponential function is a function of the form $f(x) = a^x$ where a _____ 0 and a _____ 1.

2. The graph of every exponential function $f(x) = a^x$ passes through three points: _____, _____, and _____.

3. If $a^u = a^v$, then _____ = _____.

In Problems 4–6, answer True or False to each statement.

4. The domain of the exponential function $f(x) = a^x, a > 0, a \neq 1$, is the set of all real numbers.

5. The range of the exponential function $f(x) = a^x, a > 0, a \neq 1$, is the set of all real numbers.

6. The number e, rounded to five decimal places, is 2.71828.

7. As the base a of an exponential function $f(x) = a^x$ increases (for $a > 1$), what happens to the graph of the exponential function for $x > 0$? What happens to the behavior of the graph for $x < 0$?

8. The graphs of $f(x) = 2^{-x}$ and $g(x) = \left(\dfrac{1}{2}\right)^x$ are identical. Why?

9. Explain the difference between exponential functions and polynomial functions.

10. Can we solve the equation $2^x = 12$ using the fact that if $a^u = a^v$, then $u = v$. Why or why not?

Building Skills

In Problems 11–22, approximate each number using a calculator. Express your answer rounded to three decimal places.

11. (a) $3^{2.2}$ (b) $3^{2.23}$ (c) $3^{2.236}$ (d) $3^{2.2361}$ (e) $3^{\sqrt{5}}$

12. (a) $5^{1.4}$ (b) $5^{1.41}$ (c) $5^{1.414}$ (d) $5^{1.4142}$ (e) $5^{\sqrt{2}}$

13. (a) $4^{3.1}$ (b) $4^{3.14}$ (c) $4^{3.142}$ (d) $4^{3.1416}$ (e) 4^{π}

14. (a) $10^{2.7}$ (b) $10^{2.72}$ (c) $10^{2.718}$ (d) $10^{2.7183}$ (e) 10^{e}

15. (a) $3.1^{2.7}$ (b) $3.14^{2.72}$ (c) $3.142^{2.718}$ (d) $3.1416^{2.7183}$ (e) π^{e}

16. (a) $2.7^{3.1}$ (b) $2.72^{3.14}$ (c) $2.718^{3.142}$ (d) $2.7183^{3.1416}$ (e) e^{π}

17. e^2 **18.** e^3 **19.** e^{-2} **20.** e^{-3} **21.** $e^{2.3}$ **22.** $e^{1.5}$

In Problems 23–30, the graph of an exponential function is given. Match each graph to one of the following functions. It may prove useful to create a table of values for each function to assist in identifying the correct graph.

   (a) $f(x) = 2^x$ (b) $f(x) = 2^{-x}$ (c) $f(x) = 2^{x+1}$ (d) $f(x) = 2^{x-1}$

   (e) $f(x) = -2^x$ (f) $f(x) = 2^x + 1$ (g) $f(x) = 2^x - 1$ (h) $f(x) = -2^{-x}$

23.

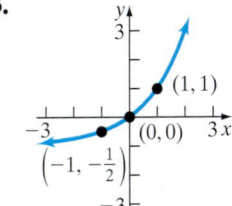

24.

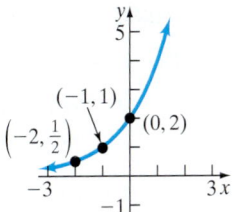

25.

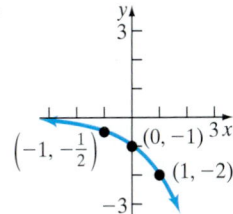

26.

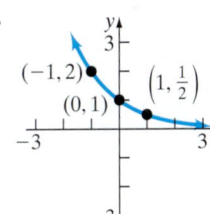

27.

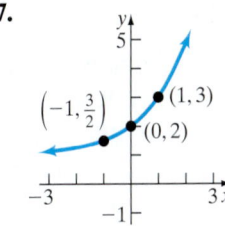

28.

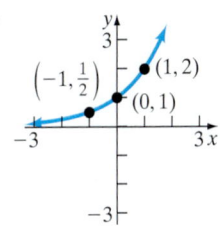

29.

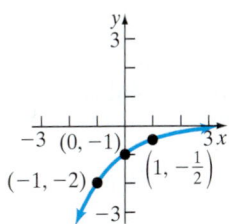

30.

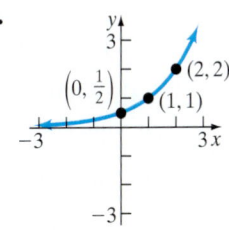

In Problems 31–48, graph each function. State the domain and the range of the function.

31. $f(x) = 5^x$

32. $f(x) = 7^x$

33. $g(x) = 10^x$

34. $G(x) = 8^x$

35. $F(x) = \left(\dfrac{1}{5}\right)^x$

36. $F(x) = \left(\dfrac{1}{7}\right)^x$

37. $G(x) = \left(\dfrac{1}{10}\right)^x$

38. $g(x) = \left(\dfrac{1}{8}\right)^x$

39. $h(x) = 2^{x+2}$

40. $H(x) = 2^{x-2}$

41. $f(x) = 2^x + 3$

42. $F(x) = 2^x - 3$

43. $F(x) = \left(\dfrac{1}{2}\right)^x - 1$

44. $G(x) = \left(\dfrac{1}{2}\right)^x + 2$

45. $P(x) = \left(\dfrac{1}{3}\right)^{x-2}$

46. $p(x) = \left(\dfrac{1}{3}\right)^{x+2}$

47. $g(x) = e^{x-1}$

48. $f(x) = e^x - 1$

In Problems 49–74, solve each equation.

49. $2^x = 2^5$

50. $3^x = 3^{-2}$

51. $3^{-x} = 81$

52. $4^{-x} = 64$

53. $\left(\dfrac{1}{2}\right)^x = \dfrac{1}{32}$

54. $\left(\dfrac{1}{3}\right)^x = \dfrac{1}{243}$

55. $5^{x-2} = 125$

56. $2^{x+3} = 128$

57. $4^x = 8$

58. $9^x = 27$

59. $2^{-x+5} = 16^x$

60. $3^{-x+4} = 27^x$

61. $3^{x^2-4} = 27^x$

62. $5^{x^2-10} = 125^x$

63. $4^x \cdot 2^{x^2} = 16^2$

64. $9^{2x} \cdot 27^{x^2} = 3^{-1}$

65. $2^x \cdot 8 = 4^{x-3}$

66. $3^x \cdot 9 = 27^x$

67. $\left(\dfrac{1}{5}\right)^x - 25 = 0$

68. $\left(\dfrac{1}{6}\right)^x - 36 = 0$

69. $(2^x)^x = 16$

70. $(3^x)^x = 81$

71. $e^x = e^{3x+4}$

72. $e^{3x} = e^2$

73. $(e^x)^2 = e^{3x-2}$

74. $(e^3)^x = e^2 \cdot e^x$

Mixed Practice

75. Suppose that $f(x) = 2^x$.

  (a) What is $f(3)$? What point is on the graph of f?

  (b) If $f(x) = \dfrac{1}{8}$, what is x? What point is on the graph of f?

76. Suppose that $f(x) = 3^x$.

  (a) What is $f(2)$? What point is on the graph of f?

  (b) If $f(x) = \dfrac{1}{81}$, what is x? What point is on the graph of f?

77. Suppose that $g(x) = 4^x - 1$.

  (a) What is $g(-1)$? What point is on the graph of g?

  (b) If $g(x) = 15$, what is x? What point is on the graph of g?

78. Suppose that $g(x) = 5^x + 1$.

  (a) What is $g(-1)$? What point is on the graph of g?

  (b) If $g(x) = 126$, what is x? What point is on the graph of g?

79. Suppose that $H(x) = 3 \cdot \left(\dfrac{1}{2}\right)^x$.

  (a) What is $H(-3)$? What point is on the graph of H?

  (b) If $H(x) = \dfrac{3}{4}$, what is x? What point is on the graph of H?

80. Suppose that $F(x) = -2 \cdot \left(\dfrac{1}{3}\right)^x$.

  (a) What is $F(-1)$? What point is on the graph of F?

  (b) If $F(x) = -18$, what is x? What point is on the graph of F?

Applying the Concepts

81. A Population Model According to the U.S. Census Bureau, the population of the United States in 2005 was 296 million people. In addition, the population of the United States was growing at a rate of 1.1% per year. Assuming that this growth rate continues, the model $P(t) = 296(1.011)^{t-2005}$ represents the population P (in millions of people) in year t.

(a) According to this model, what will be the population of the United States in 2008?

(b) According to this model, what will be the population of the United States in 2042?

(c) The United States Census Bureau predicts that the United States population will be 400 million in 2042. Compare this estimate to the one obtained in part (b). What might account for any differences?

82. A Population Model According to the *Statistical Abstract of the United States,* the population of the world in 2005 was 6,448 million people. In addition, the population of the world was growing at a rate of 1.26% per year. Assuming that this growth rate continues, the model $P(t) = 6,448(1.0126)^{t-2005}$ represents the population P (in millions of people) in year t.

(a) According to this model, what will be the population of the world in 2008?

(b) According to this model, what will be the population of the world in 2027?

(c) The United States Census Bureau predicts that the world population will be 8,000 million (8 billion) in 2027. Compare this estimate to the one obtained in part (b). What might account for any differences?

83. Time Is Money Suppose that you deposit $5000 into a Certificate of Deposit (CD) today. Determine the future value A of the deposit if it earns 6% interest compounded monthly after

(a) 1 year.   (b) 3 years.   (c) 5 years, when the CD comes due.

84. Time Is Money Suppose that you deposit $8000 into a Certificate of Deposit (CD) today. Determine the future value A of the deposit if it earns 4% interest compounded quarterly (4 times per year) after

(a) 1 year.   (b) 3 years.   (c) 5 years, when the CD comes due.

85. Do the Compounding Periods Matter? Suppose that you deposit $2000 into an account that pays 3% annual interest. How much will you have after 5 years if interest is compounded

(a) annually?   (b) quarterly?   (c) monthly?   (d) daily?

(e) Based on the results of parts (a)–(d), what impact does the number of compounding periods have on the future value, all other things equal?

86. Do the Compounding Periods Matter? Suppose that you deposit $1000 into an account that pays 6% annual interest. How much will you have after 3 years if interest is compounded

(a) annually?   (b) quarterly?   (c) monthly?   (d) daily?

(e) Based on the results of parts (a)–(d), what impact does the number of compounding periods have on the future value, all other things equal?

87. Depreciation Based on data obtained from the *Kelley Blue Book,* the value V of a Dodge Neon that is t years old can be modeled by $V(t) = 14,512(0.82)^t$.

(a) According to the model, what is the value of a brand-new Dodge Neon?

(b) According to the model, what is the value of a 2-year-old Dodge Neon?

(c) According to the model, what is the value of a 5-year-old Dodge Neon?

88. Depreciation Based on data obtained from the *Kelley Blue Book,* the value V of a Dodge Stratus that is t years old can be modeled by $V(t) = 19,282(0.84)^t$.

(a) According to the model, what is the value of a brand-new Dodge Stratus?

(b) According to the model, what is the value of a 2-year-old Dodge Stratus?

(c) According to the model, what is the value of a 5-year-old Dodge Stratus?

89. Radioactive Decay The half-life of beryllium-11 is 13.81 seconds. Suppose that a researcher possesses a 100-gram sample of beryllium-11. The amount A (in grams) of beryllium-11 after t seconds is given by

$$A(t) = 100 \cdot \left(\frac{1}{2}\right)^{t/13.81}$$

(a) How much beryllium-11 is left in the sample after 1 second?
(b) How much beryllium-11 is left in the sample after 13.81 seconds?
(c) How much beryllium-11 is left in the sample after 27.62 seconds?
(d) How much beryllium-11 is left in the sample after 100 seconds?

90. Radioactive Decay The half-life of carbon-10 is 19.255 seconds. Suppose that a researcher possesses a 200-gram sample of carbon-10. The amount A (in grams) of carbon-10 after t seconds is given by

$$A(t) = 100 \cdot \left(\frac{1}{2}\right)^{t/19.255}$$

(a) How much carbon-10 is left in the sample after 1 second?
(b) How much carbon-10 is left in the sample after 19.255 seconds?
(c) How much carbon-10 is left in the sample after 38.51 seconds?
(d) How much carbon-10 is left in the sample after 100 seconds?

91. Newton's Law of Cooling Newton's Law of Cooling states that the temperature of a heated object decreases exponentially over time toward the temperature of the surrounding medium. Suppose that a pizza is removed from a 400°F oven and placed in a room whose temperature is 70°F. The temperature u (in °F) of the pizza at time t (in minutes) can be modeled by $u(t) = 70 + 330e^{-0.072t}$.

(a) According to the model, what will be the temperature of the pizza after 5 minutes?
(b) According to the model, what will be the temperature of the pizza after 10 minutes?
(c) If the pizza can be safely consumed when its temperature is 200°F, will it be ready to eat after cooling for 13 minutes?

92. Newton's Law of Cooling Newton's Law of Cooling states that the temperature of a heated object decreases exponentially over time toward the temperature of the surrounding medium. Suppose that coffee that is 170°F is poured into a coffee mug and allowed to cool in a room whose temperature is 70°F. The temperature u (in °F) of the coffee at time t (in minutes) can be modeled by $u(t) = 70 + 100e^{-0.045t}$.

(a) According to the model, what will be the temperature of the coffee after 5 minutes?
(b) According to the model, what will be the temperature of the coffee after 10 minutes?
(c) If the coffee doesn't taste good once its temperature reaches 120°F, will it be bad after cooling for 20 minutes?

93. Learning Curve Suppose that a student has 200 vocabulary words to learn. If a student learns 20 words in 30 minutes, the function

$$L(t) = 200(1 - e^{-0.0035t})$$

models the number of words L that the student will learn after t minutes.

(a) How many words will the student learn after 45 minutes?
(b) How many words will the student learn after 60 minutes?

94. Learning Curve Suppose that a student has 50 biology terms to learn. If a student learns 10 terms in 30 minutes, the function

$$L(t) = 50(1 - e^{-0.0223t})$$

models the number of terms L that the student will learn after t minutes.

(a) How many words will the student learn after 45 minutes?

(b) How many words will the student learn after 60 minutes?

95. Current in an *RL* Circuit The equation governing the amount of current I (in amperes) after time t (in seconds) in a single *RL* circuit consisting of a resistance R (in ohms), an inductance L (in henrys), and an electromotive force E (in volts) is

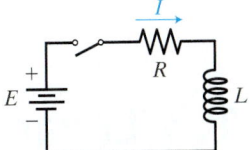

$$I = \frac{E}{R}[1 - e^{-(R/L)t}]$$

(a) If $E = 120$ volts, $R = 10$ ohms, and $L = 25$ henrys, how much current I is flowing after 0.05 second?

(b) If $E = 240$ volts, $R = 10$ ohms, and $L = 25$ henrys, how much current I is flowing after 0.05 second?

96. Current in an *RC* Circuit The equation governing the amount of current I (in amperes) after time t (in microseconds) in a single *RC* circuit consisting of a resistance R (in ohms), a capacitance C (in microfarads), and an electromotive force E (in volts) is

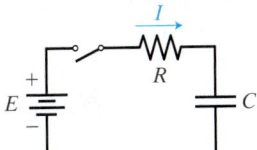

$$I = \frac{E}{R}e^{-t/(RC)}$$

(a) If $E = 120$ volts, $R = 2500$ ohms, and $C = 100$ microfarads, how much current I is flowing initially ($t = 0$)? After 50 microseconds?

(b) If $E = 240$ volts, $R = 2500$ ohms, and $C = 100$ microfarads, how much current I is flowing initially ($t = 0$)? After 50 microseconds?

Extending the Concepts

In Problems 97 and 98, find the exponential function whose graph is given.

97.

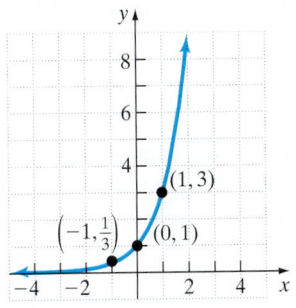

98.

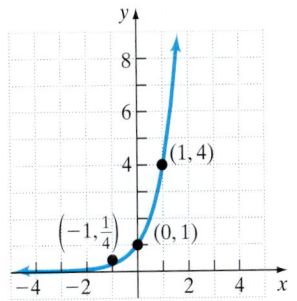

Synthesis Review

In Problems 99–104, evaluate each expression.

99. Evaluate $x^2 - 5x + 1$ at
(a) $x = -2$ and (b) $x = 3$.

100. Evaluate $x^3 - 5x + 2$ at
(a) $x = -4$ and (b) $x = 2$.

101. Evaluate $\dfrac{4}{x + 2}$ at (a) $x = -2$ and
(b) $x = 3$.

102. Evaluate $\dfrac{2x}{x + 1}$ at (a) $x = -4$ and
(b) $x = 5$.

103. Evaluate $\sqrt{2x + 5}$ at (a) $x = 2$ and
(b) $x = 11$.

104. Evaluate $\sqrt[3]{3x - 1}$ at (a) $x = 3$ and
(b) $x = 0$.

The Graphing Calculator

In Problems 105–112, graph each function using a graphing calculator. State the domain and range of each function.

105. $f(x) = 1.5^x$ **106.** $G(x) = 3.1^x$ **107.** $H(x) = 0.9^x$

108. $F(x) = 0.3^x$ **109.** $g(x) = 2.5^x + 3$ **110.** $f(x) = 1.7^x - 2$

111. $F(x) = 1.6^{x-3}$ **112.** $g(x) = 0.3^{x+2}$

11.3 Logarithmic Functions

OBJECTIVES

1. Change Exponential Expressions to Logarithmic Expressions
2. Change Logarithmic Expressions to Exponential Expressions
3. Evaluate Logarithmic Functions
4. Determine the Domain of a Logarithmic Function
5. Graph Logarithmic Functions
6. Work with Natural and Common Logarithms
7. Solve Logarithmic Equations
8. Study Logarithmic Models that Describe Our World

Preparing for Logarithmic Functions

Before getting started, take the following readiness quiz. If you get a problem wrong, go back to the section cited and review the material.

1. Solve: $3x + 2 > 0$ [Section 2.8, pp. 153–159]

2. Solve: $\sqrt{x + 2} = x$ [Section 9.8, pp. 713–716]

3. Solve: $x^2 = 6x + 7$ [Section 6.6, pp. 451–457]

We know about the "squaring function" $f(x) = x^2$ and we also know about the square root function, $g(x) = \sqrt{x}$. How are these two functions related? Well, if we evaluate $f(5)$, we obtain 25. If we evaluate $g(25)$, we obtain 5. That is, the input to f is the output of g and the output of f is the input to g. Basically, g "undoes" what f does.

In general, whenever we introduce a function in mathematics, we would also like to find a second function that "undoes" the function. For example, we have a square root function to undo the squaring function. We have a cube root function to undo the cubing function. A logical question is, "Do we have a function that "undoes" an exponential function?" The answer is yes. The function is called the *logarithmic function*.

Work Smart

We require that a is positive and not equal to 1 for the same reasons that we had these restrictions for the exponential function.

DEFINITION

The **logarithmic function to the base a**, where $a > 0$ and $a \neq 1$, is denoted by $y = \log_a x$ (read as "y is the logarithm to the base a of x") and is defined by

$$y = \log_a x \quad \text{is equivalent to} \quad x = a^y$$

In Words

The logarithm to the base a of x is the number y that we must raise a to in order to obtain x.

To evaluate logarithmic functions, we convert them to their equivalent exponential expression. Therefore, it is vital that we can easily go from logarithmic form to exponential form, and back. For example,

$$0 = \log_3 1 \quad \text{is equivalent to} \quad 3^0 = 1$$

$$2 = \log_5 25 \quad \text{is equivalent to} \quad 5^2 = 25$$

$$-2 = \log_4 \frac{1}{16} \quad \text{is equivalent to} \quad 4^{-2} = \frac{1}{16}$$

Notice that the base of the logarithm is the base of the exponential; the argument of the logarithm is what the exponential equals; and the value of the logarithm is the exponent of the exponential expression. A logarithm is just a fancy way of writing an exponential expression.

To help see how the logarithmic function "undoes" the exponential function, consider the function $y = 2^x$. If we input $x = 3$, then the output is $y = 2^3 = 8$. To undo this function would require that an input of 8 would give an output of 3. If we compare $2^3 = 8$ to $x = a^y$, we see that x is 8, a is 2, and y is 3, so that $3 = \log_2 8$ using the definition of a logarithm. The input of $\log_2 8$ is 8 and its output is 3.

Preparing for...Answers

1. $\{x \mid x > -2/3\}$ or $\left(-\frac{2}{3}, \infty\right)$

2. $\{2\}$ **3.** $\{-1, 7\}$

① Change Exponential Expressions to Logarithmic Expressions

We can use the definition of a logarithm to change from exponential expressions to logarithmic expressions.

EXAMPLE 1 **Changing Exponential Expressions to Logarithmic Expressions**

Rewrite each exponential expression to an equivalent expression involving a logarithm.

(a) $5^4 = b$ (b) $1.9^c = 12$ (c) $z^{1.2} = 7$

Solution

In each of these problems, we use the fact that $y = \log_a x$ is equivalent to $a^y = x$ provided that $a > 0$ and $a \neq 1$.

(a) If $5^4 = b$, then $4 = \log_5 b$.
(b) If $1.9^c = 12$, then $c = \log_{1.9} 12$.
(c) If $z^{1.2} = 7$, then $1.2 = \log_z 7$.

Work Smart: Study Skills

In doing problems similar to Examples 1 and 2, it is a good idea to say, "y equals the logarithm to the base a of x is equivalent to a to the y equals x." So that you memorize the definition of a logarithm.

QUICK ✓ *Rewrite each exponential expression to an equivalent expression involving a logarithm.*

1. $4^3 = w$ 2. $p^{-2} = 8$ 3. $5^b = 125$

---

② Change Logarithmic Expressions to Exponential Expressions

We can also use the definition of a logarithm to change from logarithmic expressions to exponential expressions.

EXAMPLE 2 **Changing Logarithmic Expressions to Exponential Expressions**

Change each logarithmic expression to an equivalent expression involving an exponent.

(a) $y = \log_3 81$ (b) $-3 = \log_a \dfrac{1}{27}$ (c) $2 = \log_4 x$

Solution

In each of these problems, we use the fact that $y = \log_a x$ is equivalent to $a^y = x$ provided that $a > 0$ and $a \neq 1$.

(a) If $y = \log_3 81$, then $3^y = 81$.
(b) If $-3 = \log_a \dfrac{1}{27}$, then $a^{-3} = \dfrac{1}{27}$.
(c) If $2 = \log_4 x$, then $4^2 = x$.

QUICK ✓ *Rewrite each logarithmic expression as an equivalent expression involving an exponent.*

4. $y = \log_2 16$ 5. $5 = \log_a 20$ 6. $-3 = \log_5 z$

(3) Evaluate Logarithmic Functions

To find the exact value of a logarithm, we write the logarithm in exponential notation and use the fact that if $a^u = a^v$, then $u = v$.

EXAMPLE 3 **Finding the Exact Value of a Logarithmic Expression**

Find the exact value of

(a) $\log_2 32$ 　　　　　　　　(b) $\log_4 \dfrac{1}{16}$

Solution

(a) We let $y = \log_2 32$ and convert this expression to an exponential expression.

$$y = \log_2 32$$

Write the logarithm as an exponent: $\quad 2^y = 32$

$32 = 2^5$: $\quad 2^y = 2^5$

Since we have the same base, we
set the exponents equal to each other: $\quad y = 5$

Therefore, $\log_2 32 = 5$.

(b) We let $y = \log_4 \dfrac{1}{16}$ and convert this expression to an exponential expression.

$$y = \log_4 \dfrac{1}{16}$$

Write the logarithm as an exponent: $\quad 4^y = \dfrac{1}{16}$

$\dfrac{1}{16} = 4^{-2}$: $\quad 4^y = 4^{-2}$

Since we have the same base, we
set the exponents equal to each other: $\quad y = -2$

Therefore, $\log_4 \dfrac{1}{16} = -2$. ▬

QUICK ✓ *Find the exact value of each logarithmic expression.*

7. $\log_5 25$ 　　　　　　　　**8.** $\log_2 \dfrac{1}{8}$

We could also write $y = \log_a x$ using function notation as $f(x) = \log_a x$. We use this notation in the next example to evaluate logarithmic functions.

EXAMPLE 4 **Evaluating Logarithmic Functions**

Find the value of each of the following given that $f(x) = \log_2 x$.

(a) $f(2)$ 　　　　　　　　(b) $f\left(\dfrac{1}{4}\right)$

Solution

(a) $f(2)$ means to evaluate $\log_2 x$ at $x = 2$. So we want to know the value of $\log_2 2$. To determine this value, we follow the approach of Example 3 by letting $y = \log_2 2$ and converting the expression to an exponential expression.

$$y = \log_2 2$$

Write the logarithm as an exponent: $2^y = 2$

$2 = 2^1$: $2^y = 2^1$

Since we have the same base, we set the exponents equal to each other: $y = 1$

Therefore, $f(2) = 1$.

(b) $f\left(\dfrac{1}{4}\right)$ means to evaluate $\log_2 x$ at $x = \dfrac{1}{4}$. So we want to know the value

of $\log_2\left(\dfrac{1}{4}\right)$. We let $y = \log_2\left(\dfrac{1}{4}\right)$ and convert this expression to an

exponential expression.

$$y = \log_2\left(\frac{1}{4}\right)$$

Write the logarithm as an exponent: $2^y = \dfrac{1}{4}$

$\dfrac{1}{4} = 2^{-2}$: $2^y = 2^{-2}$

Since we have the same base, we set the exponents equal to each other: $y = -2$

Therefore, $f\left(\dfrac{1}{4}\right) = -2$. ▬

QUICK ✔️ *Evaluate the function given that $g(x) = \log_5 x$.*

9. $g(25)$ **10.** $g\left(\dfrac{1}{5}\right)$

---

④ **Determine the Domain of a Logarithmic Function**

The domain of a function $y = f(x)$ is the set of all x such that the function makes sense and the range is the set of all images of x. To find the range of the logarithmic function, we recognize that $y = f(x) = \log_a x$ is equivalent to $x = a^y$. Because we can raise a to any real number (since $a > 0$ and $a \neq 1$), we conclude that y can be any real number in the function $y = f(x) = \log_a x$. In addition, because a^y is positive for any real number, we conclude that x must be positive. Since x represents the input of the logarithmic function, the domain of the logarithmic function is the set of all positive real numbers.

Work Smart

Notice that the domain of the logarithmic function is the same as the range of the exponential function. The range of the logarithmic function is the same as the domain of the exponential function.

> **DOMAIN AND RANGE OF THE LOGARITHMIC FUNCTION**
>
> Domain of the logarithmic function $= (0, \infty)$
>
> Range of the logarithmic function $= (-\infty, \infty)$

Because the domain of the logarithmic function is the set all positive real numbers, the argument of the logarithmic function must be greater than zero. For example, the logarithmic function $f(x) = \log_{10} x$ is defined for $x = 2$, but is not defined for $x = -1$ or $x = -8$ (or any other $x \leq 0$).

EXAMPLE 5 **Finding the Domain of a Logarithmic Function**

Find the domain of each logarithmic function.

(a) $f(x) = \log_6(x - 5)$ **(b)** $G(x) = \log_3(3x + 1)$

Solution

(a) The argument of the function $f(x) = \log_6(x - 5)$ is $x - 5$. The domain of f is the set of all real numbers x such that $x - 5 > 0$. We solve this inequality:

$$x - 5 > 0$$

Add 5 to both sides of the inequality: $x > 5$

The domain of f is $\{x \mid x > 5\}$, or using interval notation, $(5, \infty)$.

(b) The argument of the function $G(x) = \log_3(3x + 1)$ is $3x + 1$. The domain of G is the set of all real numbers x such that $3x + 1 > 0$. We solve this inequality:

$$3x + 1 > 0$$

Subtract 1 from both sides of the inequality: $3x > -1$

Divide both sides by 3: $x > -\dfrac{1}{3}$

The domain of G is $\left\{ x \mid x > -\dfrac{1}{3} \right\}$, or using interval notation, $\left(-\dfrac{1}{3}, \infty \right)$. ■

QUICK ✔ *Find the domain of each logarithmic function.*

11. $g(x) = \log_8(x + 3)$ **12.** $F(x) = \log_2(5 - 2x)$

---

⑤ Graph Logarithmic Functions

To graph a logarithmic function $y = \log_a x$, it is helpful to rewrite the function in exponential form as $x = a^y$. We would then choose "nice" values of y and use the expression $x = a^y$ to find the corresponding values of x.

EXAMPLE 6   Graphing a Logarithmic Function

Graph $f(x) = \log_2 x$ using point plotting. From the graph, state the domain and the range of the function.

Solution

We rewrite $y = f(x) = \log_2 x$, as $x = 2^y$. Table 7 shows various values of y, the corresponding values of x and points on the graph of $y = f(x) = \log_2 x$. We plot the ordered pairs in Table 7 and connect them in a smooth curve to obtain the graph of $f(x) = \log_2 x$. See Figure 18. The domain of f is $\{x \mid x > 0\}$ or, using interval notation, $(0, \infty)$. The range of f is the set of all real numbers or, using interval notation, $(-\infty, \infty)$.

Table 7

y	$x = 2^y$	(x, y)
-2	$\dfrac{1}{4}$	$\left(\dfrac{1}{4}, -2 \right)$
-1	$\dfrac{1}{2}$	$\left(\dfrac{1}{2}, -1 \right)$
0	1	$(1, 0)$
1	2	$(2, 1)$
2	4	$(4, 2)$

Figure 18

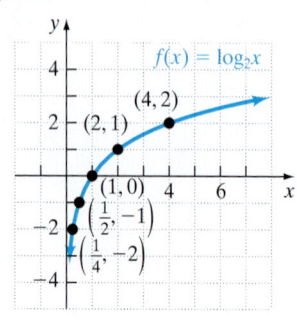

■

The graph of $f(x) = \log_2 x$ in Figure 18 is typical of all logarithmic functions that have a base larger than 1. Figure 19 shows the graph of two additional logarithmic functions whose bases are larger than 1, $y = \log_3 x$ and $y = \log_6 x$.

Figure 19

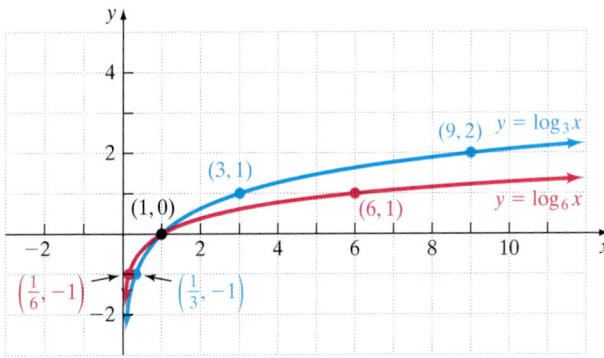

The following display summarizes the information that we have about $f(x) = \log_a x$, where the base is greater than $1(a > 1)$.

Figure 20
$f(x) = \log_a x, a > 1$

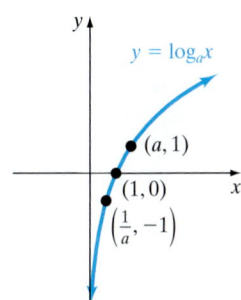

PROPERTIES OF THE GRAPH OF A LOGARITHMIC FUNCTION $f(x) = \log_a x, a > 1$

1. The domain is the set of all positive real numbers. The range is the set of all real numbers.

2. There are no y-intercepts; the x-intercept is 1.

3. The graph of f contains the points $\left(\dfrac{1}{a}, -1\right)$, $(1, 0)$, and $(a, 1)$.

See Figure 20.

QUICK ✓

13. Graph the logarithmic function $f(x) = \log_4 x$ using point plotting. From the graph, state the domain and the range of the function.

Now we consider $f(x) = \log_a x, 0 < a < 1$.

EXAMPLE 7 **Graphing a Logarithmic Function**

Graph $f(x) = \log_{1/2} x$ using point plotting. From the graph, state the domain and the range of the function.

Solution

We rewrite $y = f(x) = \log_{1/2} x$, as $x = \left(\dfrac{1}{2}\right)^y$. Table 8 on page 876 shows various values of y, the corresponding values of x and points on the graph of $y = f(x) = \log_{1/2} x$. We plot the ordered pairs in Table 8 and connect them in a smooth curve to obtain the graphs of $f(x) = \log_{1/2} x$. See Figure 21 on page 876. The domain of f is $\{x \mid x > 0\}$ or, using interval notation, $(0, \infty)$. The range of f is the set of all real numbers or, using interval notation, $(-\infty, \infty)$. ∎

The graph of $f(x) = \log_{1/2} x$ in Figure 21 is typical of all exponential functions that have a base between 0 and 1. Figure 22 shows the graph of two additional logarithmic functions whose bases are larger than 1, $y = \log_{1/3} x$ and $y = \log_{1/6} x$.

Table 8		
y	$x = \left(\dfrac{1}{2}\right)^y$	(x, y)
-2	4	$(4, -2)$
-1	2	$(2, -1)$
0	1	$(1, 0)$
1	$\dfrac{1}{2}$	$\left(\dfrac{1}{2}, 1\right)$
2	$\dfrac{1}{4}$	$\left(\dfrac{1}{4}, 2\right)$

Figure 21

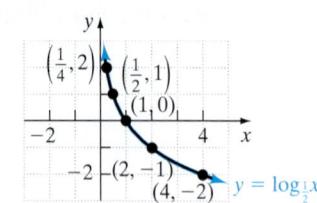

Figure 22

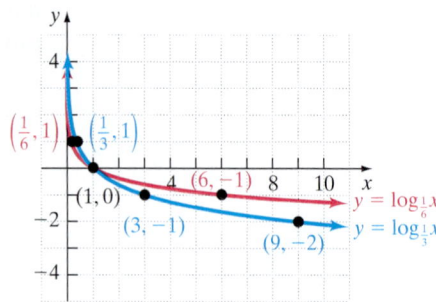

Figure 23

$f(x) = \log_a x,\ 0 < a < 1$

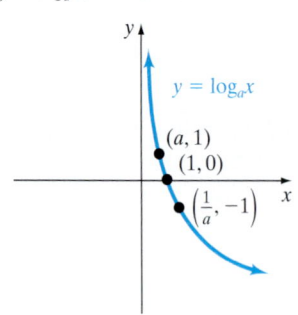

The following display summarizes the information that we have about $f(x) = \log_a x$, where the base a is between 0 and $1 (0 < a < 1)$.

> **PROPERTIES OF THE GRAPH OF AN EXPONENTIAL FUNCTION**
> $f(x) = \log_a x,\ 0 < a < 1$
>
> 1. The domain is the set of all positive real numbers. The range is the set of all real numbers.
> 2. There are no y-intercepts; the x-intercept is 1.
> 3. The graph of f contains the points $\left(\dfrac{1}{a}, -1\right)$, $(1, 0)$, and $(a, 1)$.
>
> See Figure 23.

QUICK ✓

14. Graph the logarithmic function $f(x) = \log_{\frac{1}{4}} x$ using point plotting. From the graph, state the domain and the range of the function.

---

(6) ## Work with Natural and Common Logarithms

In Words
We call a logarithm to the base e the natural logarithm because we can use this function to model many things in nature. In addition,

$\log_e x$ is written $\ln x$

If the base on a logarithmic function is the number e, then we have the **natural logarithm function.** This function occurs so frequently in applications, that it is given a special symbol, **ln** (from the Latin *logarithmus naturalis*). So we have the following definition.

> **DEFINITION**
>
> The natural logarithm: $y = \ln x$ if and only if $x = e^y$

Figure 24

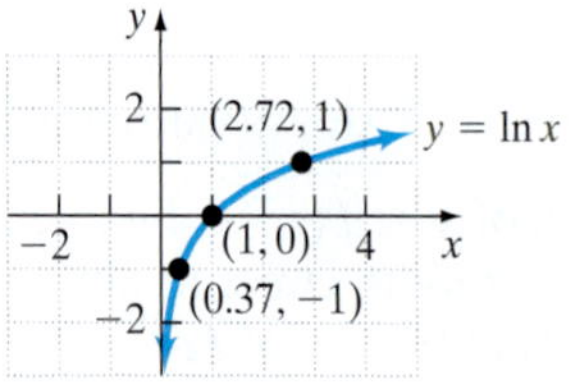

Figure 24 shows the graph of $y = \ln x$.

If the base of a logarithmic function is the number 10, then we have the **common logarithm function.** If the base a of the logarithmic function is not indicated, it is understood to be 10. So, we have the following definition.

In Words
When there is no base on a logarithm, then the base is understood to be 10. That is,

$\log x$ is written $\log_{10} x$

> **DEFINITION**
>
> The common logarithm: $y = \log x$ if and only if $x = 10^y$

Figure 25

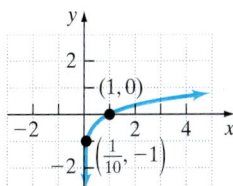

Figure 25 shows the graph of $y = \log x$.

Scientific and graphing calculators have both a natural logarithm button, $\boxed{\ln}$, and a common logarithm button, $\boxed{\log}$. This allows us to use the calculator to approximate the values of logarithms to the base e and base 10, when the results are not exact.

To evaluate logarithmic expressions using a scientific calculator, enter the argument of the logarithm and then press the $\boxed{\ln}$ or $\boxed{\log}$ button, depending on the base of the logarithm. For example, to evaluate $\log 80$, we would type in 80 and then press the $\boxed{\log}$ button. The display should show 1.90308999. Try it!

To evaluate logarithmic expressions using a graphing calculator, press the $\boxed{\ln}$ or $\boxed{\log}$ button, depending on the base of the logarithm, and then enter the argument of the logarithm. Finally, press $\boxed{\text{ENTER}}$. For example, to evaluate $\log 80$, we would press the $\boxed{\log}$ button, type in 80 and then press $\boxed{\text{ENTER}}$. The display should show 1.903089987. Try it!

By the way, it shouldn't be surprising that $\log 80$ is between 1 and 2 since $\log 10 = 1$ (because $10^1 = 10$) and $\log 100 = 2$ (because $10^2 = 100$). It is a good idea to get a sense of the value of the logarithm prior to using your calculator to approximate the value.

EXAMPLE 8   **Evaluating Natural and Common Logarithms on a Calculator**

Using a calculator, evaluate each of the following. Round your answers to 3 decimal places.

(a) $\ln 20$ (b) $\log 30$ (c) $\ln 0.5$

Solution

(a) $\ln 20 \approx 2.996$ (b) $\log 30 \approx 1.477$ (c) $\ln 0.5 \approx -0.693$

QUICK ✓ *Evaluate each logarithm using a calculator. Round your answers to 3 decimal places.*

15. $\log 1400$ **16.** $\ln 4.8$ **17.** $\log 0.3$

---

⑦ Solve Logarithmic Equations

Equations that contain logarithms are called **logarithmic equations.** Care must be taken when solving logarithmic equations because there is the possibility that extraneous solutions (an apparent solution that is not actually a solution to the original equation) creep in. This type of solution first appeared when we solved radical equations. To help locate extraneous solutions, we need to remember that in the expression $\log_a M$, both a and M must be positive with $a \neq 1$.

Remember how you learned to change from logarithmic form to exponential form? This skill is important for solving certain types of logarithmic equations!

Work Smart
Extraneous solutions can occur while solving logarithmic equations.

EXAMPLE 9   **Solving a Logarithmic Equation**

Solve:

(a) $\log_2(3x + 4) = 4$ (b) $\log_x 25 = 2$

Solution

(a) We obtain the solution by writing the logarithmic equation as an exponential equation using the fact that if $y = \log_a x$, then $a^y = x$.

$$\log_2(3x + 4) = 4$$

Write as an exponent; If $y = \log_a x$, then $a^y = x$: $\quad 2^4 = 3x + 4$

$2^4 = 16$: $\quad 16 = 3x + 4$

Subtract 4 from both sides: $\quad 12 = 3x$

Divide both sides by 3: $\quad 4 = x$

We want to verify our solution by letting $x = 4$ in the original equation:

$$\log_2(3 \cdot 4 + 4) \overset{?}{=} 4$$
$$\log_2(16) \overset{?}{=} 4$$

Since $2^4 = 16$, $\log_2 16 = 4$. Therefore, the solution set is $\{4\}$.

(b) Change the logarithmic equation to an exponential equation.

$$\log_x 25 = 2$$

Write as an exponent; If $y = \log_a x$, then $a^y = x$: $\quad x^2 = 25$

Use the Square Root Property: $\quad x = \pm\sqrt{25} = \pm 5$

Since the base of a logarithm must always be positive, we know that $x = -5$ is extraneous. We leave it to you to verify that $x = 5$ is a solution. The solution set is $\{5\}$.

QUICK ✓ *Solve each equation. Be sure to verify your solution.*

18. $\log_3(5x + 1) = 4$ **19.** $\log_x 16 = 2$

EXAMPLE 10 **Solving Logarithmic Equations**

Solve each equation and state the exact solution.

(a) $\ln x = 3$ **(b)** $\log(x + 1) = -2$

Solution

(a) We solve the equation by writing the logarithmic equation as an exponential equation.

$$\ln x = 3$$

Write as exponent; If $y = \ln x$, then $e^y = x$: $\quad e^3 = x$

We verify our solution by letting $x = e^3$ in the original equation:

$$\ln e^3 \overset{?}{=} 3$$

We know that $\ln e^3 = 3$ can be written as $\log_e e^3 = 3$, which is equivalent to $e^3 = e^3$, so we have a true statement. The solution set is $\{e^3\}$.

(b) Write the logarithmic equation as an exponential equation.

$$\log(x + 1) = -2$$

Write as exponent; If $y = \log x$, then $10^y = x$: $\quad 10^{-2} = x + 1$

$10^{-2} = 0.01$: $\quad x + 1 = 0.01$

Subtract 1 from both sides: $\quad x = -0.99$

We verify our solution by letting $x = -0.99$ in the original equation:

$$\log(-0.99 + 1) \overset{?}{=} -2$$
$$\log(0.01) \overset{?}{=} -2$$
$$10^{-2} = 0.01 \quad \text{True}$$

The solution set is $\{-0.99\}$.

QUICK ✔ *Solve each equation. Be sure to verify your solution.*

20. $\ln x = -2$ **21.** $\log(x - 20) = 4$

---

8 ## Study Logarithmic Models that Describe Our World

Common logarithms often are used when quantities vary from very large to very small numbers. The reason for this is that the common logarithm can "scale down" the measurement. For example, if a certain quantity can vary from $0.00000001 = 10^{-8}$ to $100,000,000 = 10^8$, the common logarithm of the same quantity would vary from $\log 10^{-8} = -8$ to $\log 10^8 = 8$.

Physicists define the **intensity of a sound wave** as the amount of energy that the sound wave transmits through a given area. For example, the least sound that a human ear can detect at a frequency of 100 hertz is about 10^{-12} watt per square meter. The *loudness L* (measured in **decibels** in honor of Alexander Graham Bell) of a sound of intensity x (measured in watts per square meter) is defined as follows.

> **DEFINITION**
>
> The **loudness** L, measured in decibels, of a sound of intensity x, measured in watts per square meter, is
>
> $$L(x) = 10 \log \frac{x}{10^{-12}}$$

The quantity 10^{-12} watt per square meter in the definition is the least intense sound that a human ear can detect. So, if we let $x = 10^{-12}$ watt per square meter, we obtain

$$L(10^{-12}) = 10 \log \frac{10^{-12}}{10^{-12}}$$
$$= 10 \log 1$$
$$= 10(0)$$
$$= 0$$

So at the least intense sound a human ear can detect, we measure sound at 0 decibels.

EXAMPLE 11 **Measuring the Loudness of a Sound**

Normal conversation has an intensity level of 10^{-6} watt per square meter. How many decibels is normal conversation?

Solution

We evaluate L at $x = 10^{-6}$.

$$L(10^{-6}) = 10 \log \frac{10^{-6}}{10^{-12}}$$

Laws of exponents; $\dfrac{a^m}{a^n} = a^{m-n}$: $= 10 \log 10^{-6-(-12)}$

Simplify: $= 10 \log 10^6$

$y = \log 10^6$ implies $10^y = 10^6$, so $y = 6$: $= 10(6)$

$= 60$ decibels

The loudness of normal conversation is 60 decibels.   ■

QUICK ✔

22. An MP3 player has an intensity level of 10^{-2} watt per square meter when set at its maximum level. How many decibels is the MP3 player on "full blast"?

11.3 Exercises

For Extra Help:

Student Solutions Manual CD Video PH Math/Tutor Center MathXL Tutorials on CD MathXL® MyMathLab

Concepts and Vocabulary

In Problems 1–3, fill in the blanks.

1. The logarithm to the base a of x, denoted $y = \log_a x$, can be expressed as an exponent as _____, where a _____ 0 and a _____ 1.

2. The domain of the logarithmic function $f(x) = \log_a x$ is _____.

3. The graph of every logarithmic function $f(x) = \log_a x$, $a > 0$, $a \neq 1$, passes through three points: _____, _____, and _____.

In Problems 4–6, answer True or False to each statement.

4. Logarithmic equations never have extraneous solutions.

5. If $y = \log_2 x$, then $y = 2^x$.

6. The base of the natural logarithmic function is e.

7. In the definition of the logarithmic function $f(x) = \log_a x$, the base a is not allowed to equal 1. Why?

8. The domain of $f(x) = \log_a(x^2 + 1)$ is the set of all real numbers. Explain why.

Building Skills

In Problems 9–20, change each exponential expression to an equivalent expression involving a logarithm.

9. $25 = 5^2$ 10. $64 = 8^2$ 11. $64 = 4^3$ 12. $16 = 2^4$

13. $\dfrac{1}{8} = 2^{-3}$ 14. $\dfrac{1}{9} = 3^{-2}$ 15. $e^x = 12$ 16. $e^{4.2} = M$

17. $a^3 = 19$ 18. $b^4 = 23$ 19. $5^{-6} = c$ 20. $10^{-3} = z$

In Problems 21–32, change each logarithmic expression to an equivalent expression involving an exponent.

21. $\log_2 16 = 4$ 22. $\log_3 81 = 4$ 23. $\log_3 \dfrac{1}{9} = -2$ 24. $\log_2 \dfrac{1}{32} = -5$

25. $\ln x = 4$ 26. $\ln(x - 1) = 3$ 27. $\log_5 a = -3$ 28. $\log_6 x = -4$

29. $\log_a 4 = 2$ 30. $\log_a 16 = 2$ 31. $\log_{1/2} 12 = y$ 32. $\log_{1/2} 18 = z$

In Problems 33–40, find the exact value of each logarithm without using a calculator.

33. $\log_3 1$ 34. $\log_5 5$ 35. $\log_2 8$ 36. $\log_4 16$

37. $\log_4\left(\dfrac{1}{16}\right)$ 38. $\log_5\left(\dfrac{1}{125}\right)$ 39. $\log_{\sqrt{2}} 4$ 40. $\log_{\sqrt{3}} 3$

In Problems 41–48, evaluate each function given that $f(x) = \log_3 x$, $g(x) = \log_5 x$, $H(x) = \log x$, and $P(x) = \ln x$.

41. $f(81)$ 42. $f(9)$ 43. $g\left(\sqrt{5}\right)$ 44. $g\left(\sqrt[3]{5}\right)$

45. $H(0.1)$ 46. $H(100{,}000)$ 47. $P(e^3)$ 48. $P(e^{-3})$

In Problems 49–60, find the domain of each function.

49. $f(x) = \log_2(x - 4)$ 50. $f(x) = \log_3(x - 2)$ 51. $G(x) = \log_5(x + 6)$

52. $g(x) = \log_4(x + 10)$ 53. $F(x) = \log_3(2x)$ 54. $h(x) = \log_4(5x)$

55. $f(x) = \log_8(3x - 2)$ 56. $F(x) = \log_2(4x - 3)$ 57. $H(x) = \log_7(2x + 1)$

58. $f(x) = \log_3(5x + 3)$ 59. $H(x) = \log_2(1 - 4x)$ 60. $G(x) = \log_4(3 - 5x)$

In Problems 61–70, graph each function. From the graph, state the domain and the range of each function.

61. $f(x) = \log_5 x$ **62.** $f(x) = \log_7 x$ **63.** $g(x) = \log_6 x$ **64.** $G(x) = \log_8 x$

65. $F(x) = \log_{1/5} x$ **66.** $F(x) = \log_{1/7} x$ **67.** $G(x) = \log_{1/6} x$ **68.** $g(x) = \log_{1/8} x$

69. $f(x) = \ln x$ **70.** $f(x) = \log x$

In Problems 71–82, use a calculator to evaluate each expression. Round your answers to three decimal places.

71. $\log 67$ **72.** $\log 106$ **73.** $\ln 5.4$ **74.** $\ln 10.4$

75. $\log 0.35$ **76.** $\log 0.78$ **77.** $\ln 0.2$ **78.** $\ln 0.4$

79. $\log \dfrac{5}{4}$ **80.** $\log \dfrac{10}{7}$ **81.** $\ln \dfrac{3}{8}$ **82.** $\ln \dfrac{1}{2}$

In Problems 83–102, solve each logarithmic equation.

83. $\log_3(2x + 1) = 2$ **84.** $\log_3(5x - 3) = 3$ **85.** $\log_5(20x - 5) = 3$

86. $\log_4(8x + 10) = 3$ **87.** $\log_a 36 = 2$ **88.** $\log_a 81 = 2$

89. $\log_a 18 = 2$ **90.** $\log_a 28 = 2$ **91.** $\log_a 1000 = 3$

92. $\log_a 243 = 5$ **93.** $\ln x = 5$ **94.** $\ln x = 10$

95. $\log(2x - 1) = -1$ **96.** $\log(2x + 3) = 1$ **97.** $\ln e^x = -3$

98. $\ln e^{2x} = 8$ **99.** $\log_3 81 = x$ **100.** $\log_4 16 = x + 1$

101. $\log_2(x^2 - 1) = 3$ **102.** $\log_3(x^2 + 1) = 2$

Mixed Practice

103. Suppose that $f(x) = \log_2 x$.

    (a) What is $f(16)$? What point is on the graph of f?

    (b) If $f(x) = -3$, what is x? What point is on the graph of f?

104. Suppose that $f(x) = \log_5 x$.

    (a) What is $f(5)$? What point is on the graph of f?

    (b) If $f(x) = -2$, what is x? What point is on the graph of f?

105. Suppose that $G(x) = \log_4(x + 1)$.

    (a) What is $G(7)$? What point is on the graph of G?

    (b) If $G(x) = 2$, what is x? What point is on the graph of G?

106. Suppose that $F(x) = \log_2 x - 3$.

    (a) What is $F(8)$? What point is on the graph of f?

    (b) If $F(x) = -1$, what is x? What point is on the graph of F?

Applying the Concepts

107. Loudness of a Whisper A whisper has an intensity level of 10^{-10} watt per square meter. How many decibels is a whisper?

108. Loudness of a Concert If you sit in the front row of a rock concert, you will experience an intensity level of 10^{-1} watt per square meter. How many decibels is a rock concert in the front row? If you move back to the 15th row, you will experience intensity of 10^{-2} watt per square meter. How many decibels is a rock concert in the 15th row?

109. Threshold of Pain The threshold of pain has an intensity level of 10^1 watt per square meter. How many decibels is the threshold of pain?

110. Exploding Eardrum Instant perforation of the eardrum occurs at an intensity level of 10^4 watt per square meter. At how many decibels will instant perforation of the eardrum occur?

*Problems 111–114 use the following discussion: The **Richter scale** is one way of converting seismographic readings into numbers that provide an easy reference for measuring the **magnitude M** of an earthquake. All earthquakes are compared to a **zero-level earthquake** whose* *seismographic reading measures 0.001 millimeter at a distance of 100 kilometers from the epicenter. An earthquake whose seismographic reading measures x millimeters has magnitude M given by*

$$M(x) = \log\left(\frac{x}{10^{-3}}\right)$$

where 10^{-3} is the reading of a zero-level earthquake 100 kilometers from its epicenter.

111. San Francisco, 1906 According to the United States Geological Survey, the San Francisco earthquake of 1906 resulted in a seismographic reading of 63,096 millimeters 100 kilometers from its epicenter. What was the magnitude of this earthquake?

112. Alaska, 1964 According to the United States Geological Survey, an earthquake on March 28, 1964 in Prince William Sound, Alaska resulted in a seismographic reading of 1,584,893 millimeters 100 kilometers from its epicenter. What was the magnitude of this earthquake? This earthquake was the second largest ever recorded, with the largest being the Great Chilean Earthquake of 1960, whose magnitude was 9.5 on the Richter scale.

113. Ecuador, 1906 According to the United States Geological Survey, an earthquake on January 31, 1906 off the coast of Ecuador had a magnitude of 8.8. What was the seismographic reading 100 kilometers from its epicenter?

114. South Carolina, 1886 According to the United States Geological Survey, an earthquake on September 1, 1886 in Charleston, South Carolina had a magnitude of 7.3. What was the seismographic reading 100 kilometers from its epicenter?

115. pH The pH of a chemical solution is given by the formula

$$pH = -\log[H^+]$$

where $[H^+]$ is the concentration of hydrogen ions in moles per liter. Values of pH range from 0 to 14. A solution whose pH is 7 is considered neutral. The pH of pure water at 25 degrees Celsius is 7. A solution whose pH is less than 7 is considered acidic, while a solution whose pH is greater than 7 is considered basic.

(a) What is the pH of household ammonia for which $[H^+]$ is 10^{-12}? Is ammonia basic or acidic?

(b) What is the pH of black coffee for which $[H^+]$ is 10^{-5}? Is black coffee basic or acidic?

(c) What is the pH of lemon juice for which $[H^+]$ is 10^{-2}? Is lemon juice basic or acidic?

(d) What is the concentration of hydrogen ions in human blood (pH = 7.4)?

116. Energy of an Earthquake The magnitude and the seismic moment are related to the amount of energy that is given off by an earthquake. The relationship between magnitude and energy is

$$\log E_S = 11.8 + 1.5M$$

where E_S is the energy (in ergs) for an earthquake whose magnitude is M. Note that E_S is the amount of energy given off from the earthquake as seismic waves.

(a) How much energy is given off by an earthquake that measures 5.8 on the Richter scale?

(b) The earthquake on the Rat Islands in Alaska on February 4, 1965 measured 8.7 on the Richter scale. How much energy was given off by this earthquake?

Extending the Concepts

In Problems 117–120, find the domain of each function.

117. $f(x) = \log_2(x^2 - 3x - 10)$

118. $f(x) = \log_5(x^2 + 2x - 24)$

119. $f(x) = \ln\left(\frac{x-3}{x+1}\right)$

120. $f(x) = \log\left(\frac{x+4}{x-3}\right)$

121. Find a so that the graph of $f(x) = \log_a x$ contains the point $(16, 2)$.

122. Find a so that the graph of $f(x) = \log_a x$ contains the point $\left(\dfrac{1}{4}, -2\right)$.

Synthesis Review

In Problems 123–128, add or subtract each expression.

123. $(2x^2 - 6x + 1) - (5x^2 + x - 9)$

124. $(x^3 - 2x^2 + 10x + 2) + (3x^3 - 4x - 5)$

125. $\dfrac{3x}{x^2 - 1} + \dfrac{x - 3}{x^2 + 3x + 2}$

126. $\dfrac{x + 5}{x^2 + 5x + 6} - \dfrac{2}{x^2 + 2x - 3}$

127. $\sqrt{8x^3} + x\sqrt{18x}$

128. $\sqrt[3]{27a} + 2\sqrt[3]{a} - \sqrt[3]{8a}$

The Graphing Calculator

In Problems 129–134, graph each logarithmic function using a graphing calculator. State the domain and the range of each function.

129. $f(x) = \log(x + 1)$

130. $g(x) = \log(x - 2)$

131. $G(x) = \ln(x) + 1$

132. $F(x) = \ln(x) - 4$

133. $f(x) = 2\log(x - 3) + 1$

134. $G(x) = -\log(x + 1) - 3$

PUTTING THE CONCEPTS TOGETHER (Sections 11.1–11.3)

These problems cover important concepts from Sections 11.1 to 11.3. We designed these problems so that you can review the chapter so far and show your mastery of the concepts. Take time to work these problems before proceeding with the next section. The answers to these problems are located at the back of the text starting on page AN–49.

1. Given the functions $f(x) = 2x + 3$ and $g(x) = 2x^2 - 4x$ find

    (a) $(f \circ g)(x)$ **(b)** $(g \circ f)(x)$ **(c)** $(f \circ g)(3)$

    (d) $(g \circ f)(-2)$ **(e)** $(f \circ f)(1)$

2. Find the inverse of the given one-to-one function.

    (a) $f(x) = 3x + 4$ **(b)** $g(x) = x^3 - 4$

3. Sketch the graph of the inverse of the given one-to-one function.

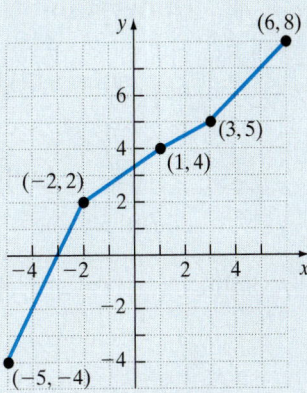

4. Approximate each number using a calculator. Express your answer rounded to three decimal places.

    (a) $2.7^{2.7}$ **(b)** $2.72^{2.72}$ **(c)** $2.718^{2.718}$

    (d) $2.7183^{2.7183}$ **(e)** e^e

5. Change each exponential expression to an equivalent expression involving a logarithm.

   (a) $a^4 = 6.4$ **(b)** $10^x = 278$

6. Change each logarithmic expression to an equivalent expression involving an exponent.

   (a) $\log_2 x = 7$ **(b)** $\ln 16 = M$

7. Find the exact value of each expression without using a calculator.

   (a) $\log_5 625$ **(b)** $\log_{\frac{2}{3}}\left(\dfrac{9}{4}\right)$

8. Determine the domain of the logarithmic function:
   $f(x) = \log_{13}(2x + 12)$.

In Problems 9 and 10, graph each function. State the domain and range of the function.

9. $f(x) = \left(\dfrac{1}{6}\right)^x$ 10. $g(x) = \log_{\frac{3}{2}} x$

In Problems 11–14, solve each equation. State the exact solution.

11. $3^{-x+2} = 27$ 12. $e^x = e^{2x+5}$ 13. $\log_2(2x + 5) = 4$ 14. $\ln x = 7$

15. Suppose that a student has 150 anatomy and physiology terms to learn. If the student learns 40 terms in 60 minutes, the function $L(t) = 150(1 - e^{-0.0052t})$ models the number of terms L that the student will learn after t minutes. According to the model, how many terms will the student learn after 90 minutes?

11.4 Properties of Logarithms

OBJECTIVES

1. Understand the Properties of Logarithms
2. Write a Logarithmic Expression as a Sum or Difference of Logarithms
3. Write a Logarithmic Expression as a Single Logarithm
4. Evaluate Logarithms Whose Base Is Neither 10 Nor e

Preparing for Properties of Logarithms

Before getting started, take the following readiness quiz. If you get a problem wrong, go back to the section cited.

1. Round 3.03468 to three decimal points.   [Section A.2, p. A9]
2. Evaluate a^0, $a \neq 0$.   [Section 5.4, pp. 373–374]

1   Understand the Properties of Logarithms

Logarithms have some very useful properties. These properties can be derived directly from the definition of a logarithm and the laws of exponents.

EXAMPLE 1   Establishing Properties of Logarithms

Determine the value of the following logarithmic expressions: (a) $\log_a 1$ (b) $\log_a a$.

Solution

(a) Remember, $y = \log_a x$ is equivalent to $x = a^y$.

$$y = \log_a 1$$

Change to an exponent: $a^y = 1$
Since $a \neq 0$, $a^0 = 1$: $a^y = a^0$
Since the bases are the same,
we equate the exponents: $y = 0$

So $\log_a 1 = 0$.

(b) Again, we need to write the logarithm as an exponent in order to evaluate the logarithm. Let $y = \log_a a$, so that

$$y = \log_a a$$

Change to an exponent: $a^y = a$

$a = a^1$: $a^y = a^1$

Since the bases are the same,
we equate the exponents: $y = 1$

So $\log_a a = 1$. ∎

We summarize the results of Example 1 in the box below.

$$\log_a 1 = 0 \qquad \log_a a = 1$$

QUICK ✓ *Evaluate each logarithm.*

1. $\log_5 1$ **2.** $\ln 1$ **3.** $\log_4 4$ **4.** $\log 10$

Suppose we were asked to evaluate $3^{\log_3 81}$. If we evaluate the exponent, $\log_3 81$, we obtain 4 (because $y = \log_3 81$ means $3^y = 81$ or $3^y = 3^4$, so $y = 4$). Now $3^{\log_3 81} = 3^4 = 81$. So $3^{\log_3 81} = 81$. This result is true in general.

In Words
If we raise the number a to $\log_a M$, we obtain M.

AN INVERSE PROPERTY OF LOGARITHMS

If a and M are positive real numbers, with $a \neq 1$, then

$$a^{\log_a M} = M$$

This is called an Inverse Property of Logarithms because if we compute the logarithm to the base a of a positive number M and then compute a raised to this power, we end up right back where we started, namely M.

EXAMPLE 2   **Using a Property of Logarithms**

    (a) $5^{\log_5 20} = 20$ **(b)** $0.8^{\log_{0.8} \sqrt{23}} = \sqrt{23}$ ∎

QUICK ✓ *Evaluate each logarithm.*

5. $12^{\log_{12} \sqrt{2}}$ **6.** $10^{\log 0.2}$

Suppose we were asked to evaluate $\log_5 5^6$. If we let $y = \log_5 5^6$, then $5^y = 5^6$ (since $y = \log_a x$ means $a^y = x$), so that $y = 6$. This result implies that the exponent of the argument is the value of the logarithm if the base of the logarithm and the base of the argument are the same.

In Words
The logarithm to the base a of a raised to the power of r is r.

AN INVERSE PROPERTY OF LOGARITHMS

If a is a positive real numbers, with $a \neq 1$, and r is any real number, then

$$\log_a a^r = r$$

This is also called an Inverse Property of Logarithms because if we compute a raised to some power r and then compute the logarithm to the base a of a^r, we end up right back where we started, namely r.

EXAMPLE 3   Using a Property of Logarithms

(a) $\log_4 4^3 = 3$ (b) $\ln e^{-0.5} = -0.5$ ■

QUICK ✓ *Evaluate each logarithm.*

7. $\log_8 8^{1.2}$ **8.** $\log 10^{-4}$

(2) Write a Logarithmic Expression as a Sum or Difference of Logarithms

The next two properties deal with working with logarithms where the argument is a product or quotient.

If we evaluate $\log_2 8$, we obtain 3. If we evaluate $\log_2 2 + \log_2 4$, we obtain $1 + 2 = 3$. Notice that $\log_2 8 = \log_2 2 + \log_2 4$. This suggests the following result.

In Words
The product rule of logarithms states that the log of a product equals the sum of the logs.

> ### THE PRODUCT RULE OF LOGARITHMS
> If M, N, and a are positive real numbers, with $a \neq 1$, then
> $$\log_a(MN) = \log_a M + \log_a N$$

EXAMPLE 4   Using the Product Rule of Logarithms

Write each of the following logarithms as the sum of logarithms.

(a) $\log_2(5 \cdot 3)$ (b) $\ln(6z)$

Solution

Do you notice that each argument contains a product? To simplify, we use the Product Rule of Logarithms.

(a) $\log_2(5 \cdot 3) = \log_2 5 + \log_2 3$
(b) $\ln(6z) = \ln 6 + \ln z$ ■

Work Smart
$\log_a(M + N)$ does not equal $\log_a M + \log_a N$.

QUICK ✓ *Write each logarithm as the sum of logarithms.*

9. $\log_4(9 \cdot 5)$ **10.** $\log(5w)$

If we evaluate $\log_2 8$, we obtain 3. If we evaluate $\log_2 16 - \log_2 2$, we obtain $4 - 1 = 3$. Notice that $\log_2 8 = \log_2 16 - \log_2 2$ and that $8 = \dfrac{16}{2}$. This suggests the following result.

In Words
The quotient rule states that the log of a quotient equals the difference of the logs.

> ### THE QUOTIENT RULE OF LOGARITHMS
> If M, N, and a are positive real numbers, with $a \neq 1$, then
> $$\log_a\left(\frac{M}{N}\right) = \log_a M - \log_a N$$

The results of the product and quotient rules for logarithms are not coincidental. After all, when we multiply exponential expressions with the same base, we add the exponents; when we divide exponential expressions with the same base, we subtract the exponents. The same thing is going on here!

EXAMPLE 5 **Using the Quotient Rule of Logarithms**

Write each of the following logarithms as the difference of logarithms.

(a) $\log_2\left(\dfrac{5}{3}\right)$ (b) $\log\left(\dfrac{y}{5}\right)$

Solution

Do you notice that each argument contains a quotient? To simplify, we use the Quotient Rule of Logarithms.

(a) $\log_2\left(\dfrac{5}{3}\right) = \log_2 5 - \log_2 3$ (b) $\log\left(\dfrac{y}{5}\right) = \log y - \log 5$ ▬

QUICK ✔ *Write each logarithm as the quotient of logarithms.*

Work Smart

$\log_a(M - N)$ does not equal $\log_a M - \log_a N$.

11. $\log_7\left(\dfrac{9}{5}\right)$ **12.** $\ln\left(\dfrac{p}{3}\right)$

We can write a single logarithm as the sum or difference of logs when the argument of the logarithm contains both products and quotients.

EXAMPLE 6 **Writing a Single Logarithm as the Sum or Difference of Logs**

Write $\log_3\left(\dfrac{4x}{y}\right)$ as the sum or difference of logarithms.

Solution

Notice that the argument of the logarithm contains a product and a quotient. So we will use both the Product and Quotient Rule of Logarithms. When using both rules, it is typically easiest to use the Quotient Rule first.

The Quotient Rule of Logarithms

↓

$\log_3\left(\dfrac{4x}{y}\right) = \log_3(4x) - \log_3 y$

Work Smart

When you need to use both the product and quotient rule, use the quotient rule first.

The Product Rule of Logarithms: $= \log_3 4 + \log_3 x - \log_3 y$ ▬

QUICK ✔ *Write each logarithm as the sum or difference of logarithms.*

13. $\log_2\left(\dfrac{3m}{n}\right)$ **14.** $\ln\left(\dfrac{q}{3p}\right)$

Another useful property of logarithms allows us to express powers on the argument of a logarithm as factors.

In Words

If an exponent exists on the quantity we are taking the log of it can be "brought down in front" of the log.

THE POWER RULE OF LOGARITHMS

If M and a are positive real numbers, with $a \neq 1$, and r is any real number, then

$$\log_a M^r = r \log_a M$$

EXAMPLE 7   **Using the Power Rule of Logarithms**

Use the Power Rule of Logarithms to express all powers as factors.

(a) $\log_8 3^5$ (b) $\ln x^{\sqrt{3}}$

Solution

(a) $\log_8 3^5 = 5 \log_8 3$ (b) $\ln x^{\sqrt{3}} = \sqrt{3} \ln x$

QUICK ✓ *Write each logarithm so that all powers are factors.*

15. $\log_2 5^{1.6}$ **16.** $\log b^5$

Work Smart

Whenever you see the word *factor*, you should think product (or multiplication).

We will use the direction *expand the logarithm* to mean to write a logarithm as a sum or difference with exponents written as factors.

EXAMPLE 8   **Expanding a Logarithm**

Expand the logarithm. That is, write each logarithm as the sum or difference of logarithms with all exponents written as factors.

(a) $\log_2(x^2 y^3)$ (b) $\log\left(\dfrac{100x}{\sqrt{y}}\right)$

Solution

(a) Do you see that the argument of the logarithm contains a product? For this reason, we use the Product Rule of Logarithms to write the single log as the sum of two logs.

$$\overset{\text{Product Rule}}{\underset{\downarrow}{}}$$

$$\log_2(x^2 y^3) = \log_2 x^2 + \log_2 y^3$$

Write exponents as factors: $= 2 \log_2 x + 3 \log_2 y$

Work Smart

If the argument of the logarithm contains both a quotient and a product, it is easiest to write the quotient as the difference of two logs first.

(b) The argument of the logarithm contains a quotient and a product.

$$\overset{\text{Quotient Rule}}{\underset{\downarrow}{}}$$

$$\log\left(\frac{100x}{\sqrt{y}}\right) = \log(100x) - \log\sqrt{y}$$

$\sqrt{y} = y^{1/2}:$ $= \log(100x) - \log y^{1/2}$

Product Rule: $= \log 100 + \log x - \log y^{1/2}$

Write exponents as factors: $= \log 100 + \log x - \frac{1}{2}\log y$

$\log 100 = 2:$ $= 2 + \log x - \frac{1}{2}\log y$

QUICK ✓ *Expand each logarithm.*

17. $\log_4(a^2 b)$ **18.** $\log_3\left(\dfrac{9m^4}{\sqrt[3]{n}}\right)$

③ **Write a Logarithmic Expression as a Single Logarithm**

Work Smart
To write two logs as a single log, the bases on the logs must be the same.

We can also use the Product Rule, Quotient Rule, and Power Rule of Logarithms to write sums and/or differences of logarithms that have the same base as a single logarithm. This skill is particularly useful when solving certain types of logarithmic equations (Section 11.5).

EXAMPLE 9 **Writing Expressions as a Single Logarithm**

Write each of the following as a single logarithm.

(a) $\log_6 3 + \log_6 12$ (b) $\log(x - 2) - \log x$

Solution

(a) To write the expression as a single logarithm, the base of each log must be the same. Both logarithms are base 6. Since we are adding the logs, we use the Product Rule to write the logs as a single log.

$$\log_6 3 + \log_6 12 = \log_6(3 \cdot 12) \quad \text{Product Rule}$$
$$3 \cdot 12 = 36: \quad = \log_6 36$$
$$6^2 = 36, \text{ so } \log_6 36 = 2: \quad = 2$$

(b) The base of each logarithm is 10. Since we are subtracting the logs, we use the Quotient Rule to write the logs as a single log.

$$\log(x - 2) - \log x = \log\left(\frac{x - 2}{x}\right)$$

QUICK ✓ *Write each expression as a single logarithm.*

19. $\log_8 4 + \log_8 16$ **20.** $\log_3(x + 4) - \log_3(x - 1)$

Work Smart
When coefficients appear on logarithms, write them as exponents before using the Product or Quotient Rule.

To use the Product or Quotient Rules to write the sum or difference of logs as a single log, the coefficients on the logarithms must be 1. Therefore, when coefficients appear on logarithms, we first use the Power Rule to write the coefficient as a power. For example, we would write $2 \log x$ as $\log x^2$.

EXAMPLE 10 **Writing Expressions as Single Logarithms**

Write each of the following as a single logarithm.

(a) $2 \log_2(x - 1) + \dfrac{1}{2}\log_2 x$ (b) $\log(x - 1) + \log(x + 1) - 3 \log x$

Solution

(a) The bases are the same for each logarithm. Because the logarithms have coefficients, we use the Power Rule to write the coefficients as exponents.

$$2 \log_2(x - 1) + \frac{1}{2}\log_2 x = \log_2(x - 1)^2 + \log_2 x^{1/2}$$
$$x^{1/2} = \sqrt{x}: \quad = \log_2(x - 1)^2 + \log_2\sqrt{x}$$
$$\text{Product Rule:} \quad = \log_2\left[(x - 1)^2\sqrt{x}\right]$$

(b) We will work from left to right in order to write the logarithm as a single logarithm. Since the first two logs are being added, we use the Product Rule to write these logs as a single log.

$$\begin{aligned}
\text{Product Rule} & \\
\downarrow & \\
\log(x - 1) + \log(x + 1) - 3 \log x &= \log[(x - 1)(x + 1)] - 3 \log x \\
(x - 1)(x + 1) = x^2 - 1: \quad &= \log(x^2 - 1) - 3 \log x \\
\text{Write the coefficient as an exponent:} \quad &= \log(x^2 - 1) - \log x^3 \\
\text{Quotient Rule:} \quad &= \log \frac{(x^2 - 1)}{x^3}
\end{aligned}$$

QUICK ✓ *Write each expression as a single logarithm.*

21. $\log_5 x - 3 \log_5 2$

22. $\log_2(x + 1) + \log_2(x + 2) - 2 \log_2 x$

---

4 **Evaluate Logarithms Whose Base Is Neither 10 Nor *e***

In Section 11.3, we learned how to use a calculator to approximate logarithms whose base was either 10 (the common logarithm) or *e* (the natural logarithm). But what if the base of the logarithm is neither 10 nor *e*? To approximate these types of logarithms, we present the following example.

EXAMPLE 11 **Approximating Logarithms Whose Base Is Neither 10 Nor *e***

Approximate $\log_2 5$. Round the answer to three decimal places.

Solution

We let $y = \log_2 5$ and then convert the logarithmic expression to an equivalent exponential expression using the fact that if $y = \log_a x$, then $x = a^y$.

$$y = \log_2 5$$
$$2^y = 5$$

In Words
"Taking the logarithm" of both sides of an equation is the same type of approach as squaring both sides of an equation.

Now, we will take the logarithm of both sides. Because we can use a calculator, we will take either the common log or natural log of both sides (it doesn't matter which). Let's take the common log of both sides.

$$\log 2^y = \log 5$$

Use the Power Rule to write the exponent, *y*, as a factor.

$$y \log 2 = \log 5$$

Now divide both sides by $\log 2$ to solve for *y*.

$$y = \frac{\log 5}{\log 2}$$

We approximate $\dfrac{\log 5}{\log 2}$ using a calculator and obtain $\dfrac{\log 5}{\log 2} \approx 2.322$. So $\log_2 5 \approx 2.322$.

Example 11 shows how to approximate a logarithm whose base is 2, but it certainly was a lot of work! The question you may be asking yourself is, "Is there an easier way to do this?" The answer, you'll be happy to hear, is yes!—it is called the **Change-of-Base Formula.**

CHANGE-OF-BASE FORMULA

If $a \neq 1$, $b \neq 1$, and M are positive real numbers, then

$$\log_a M = \frac{\log_b M}{\log_b a}$$

Because calculators only have keys for the common logarithm, $\boxed{\log}$, and natural logarithm, $\boxed{\ln}$, in practice, the Change-of-Base Formula uses either $b = 10$ or $b = e$, so that

$$\log_a M = \frac{\log M}{\log a} \quad \text{or} \quad \log_a M = \frac{\ln M}{\ln a}$$

EXAMPLE 12 Using the Change-of-Base Formula

Approximate $\log_4 45$. Round your answer to three decimal places.

Solution

We use the Change-of-Base Formula.

$$\text{Using Common Logarithms:} \quad \log_4 45 = \frac{\log 45}{\log 4}$$

$$\approx 2.746$$

$$\text{Using Natural Logarithms:} \quad \log_4 45 = \frac{\ln 45}{\ln 4}$$

$$\approx 2.746$$

So $\log_4 45 \approx 2.746$.

QUICK ✔ *Approximate each logarithm. Round your answers to three decimal places.*

23. $\log_3 32$ **24.** $\log_{\sqrt{2}} \sqrt{7}$

We have presented quite a few properties of logarithms. As a review and for convenience, we provide the following summary of the properties of logarithms.

SUMMARY: Properties of Logarithms

In the following properties, M, N, a, and b are positive real numbers, with $a \neq 1$, $b \neq 1$, and r is any real number.

- **Inverse Properties of Logarithms**
  $a^{\log_a M} = M$ and $\log_a a^r = r$
- **The Product Rule of Logarithms**
  $\log_a(MN) = \log_a M + \log_a N$
- **The Quotient Rule of Logarithms**
  $\log_a\left(\dfrac{M}{N}\right) = \log_a M - \log_a N$

- **The Power Rule of Logarithms**
  $\log_a M^r = r \log_a M$
- **Change-of-Base Formula**
  $\log_a M = \dfrac{\log_b M}{\log_b a} = \dfrac{\log M}{\log a} = \dfrac{\ln M}{\ln a}$

11.4 Exercises

For Extra Help:

Student Solutions Manual   CD Video   PH Math/Tutor Center   MathXL Tutorials on CD   MathXL®   MyMathLab

Concepts and Vocabulary

In Problems 1 – 3, fill in the blanks.

1. $\log_a 1 = $ _____ ; $\log_a a = $ _____.

2. $\log_a(xy) = $ _____.

3. $\log_3 10 = \dfrac{\log\_\_\_}{\log\_\_\_} = \dfrac{\ln\_\_}{\ln\_\_}$.

In Problems 4–6, answer True or False to each statement.

4. $\log(x + 4) = \log x + \log 4$.

5. $\log_2(x + 1) + \log_5(x - 2)$ can be written as a single logarithm using the Product Rule.

6. $\log_4(4x^2) = 1 + 2 \log_4 x$

7. State the Product Rule for Logarithms in your own words.

8. State the Quotient Rule for Logarithms in your own words.

9. Write an example that illustrates why $\log_2(x + y) \neq \log_2 x + \log_2 y$.

10. Write an example that illustrates why $(\log_a x)^r \neq r \log_a x$.

Building Skills

In Problems 11–24, use properties of logarithms to find the exact value of each expression. Do not use a calculator.

11. $\log_2 2^3$

12. $\log_5 5^{-3}$

13. $\ln e^{-7}$

14. $\ln e^9$

15. $3^{\log_3 5}$

16. $5^{\log_5 \sqrt{2}}$

17. $e^{\ln 2}$

18. $e^{\ln 10}$

19. $\log 2 + \log 5$

20. $\log_6 2 + \log_6 3$

21. $\log_3 12 - \log_3 4$

22. $\log_4 20 - \log_4 5$

23. $10^{\log 8 - \log 2}$

24. $e^{\ln 24 - \ln 3}$

In Problems 25–32, suppose that $\ln 2 = a$ and $\ln 3 = b$. Use properties of logarithms to write each logarithm in terms of a and b.

25. $\ln 6$

26. $\ln \dfrac{3}{2}$

27. $\ln 9$

28. $\ln 4$

29. $\ln 12$

30. $\ln 18$

31. $\ln \sqrt{2}$

32. $\ln \sqrt[4]{3}$

In Problems 33–54, write each expression as a sum and/or difference of logarithms. Express exponents as factors.

33. $\log(ab)$

34. $\log_4\left(\dfrac{a}{b}\right)$

35. $\log_5 x^4$

36. $\log_3 z^{-2}$

37. $\log_2(xy^2)$

38. $\log_3(a^3 b)$

39. $\log_5(25x)$

40. $\log_2(8z)$

41. $\log_7\left(\dfrac{49}{y}\right)$

42. $\log_2\left(\dfrac{16}{p}\right)$

43. $\ln(e^2 x)$

44. $\ln\left(\dfrac{x}{e^3}\right)$

45. $\log_3(27\sqrt{x})$

46. $\log_2\left(32\sqrt[4]{z}\right)$

47. $\log_5\left(x^2\sqrt{x^2 + 1}\right)$

48. $\log_3\left(x^3\sqrt{x^2 - 1}\right)$

49. $\log\left(\dfrac{x^4}{\sqrt[3]{(x - 1)}}\right)$

50. $\ln\left(\dfrac{\sqrt[5]{x}}{(x + 2)^2}\right)$

51. $\log_7\sqrt{\dfrac{x + 1}{x}}$

52. $\log_6\sqrt[3]{\dfrac{x - 2}{x + 1}}$

53. $\log_2\left[\dfrac{x(x - 1)^2}{\sqrt{x + 1}}\right]$

54. $\log_4\left[\dfrac{x^3(x - 3)}{\sqrt[3]{x + 1}}\right]$

In Problems 55–76, write each expression as a single logarithm.

55. $\log 25 + \log 4$

56. $\log_4 32 + \log_4 2$

57. $\log x + \log 3$

58. $\log_2 6 + \log_2 z$

59. $\log_3 36 - \log_3 4$

60. $\log_2 48 - \log_2 3$

61. $3 \log_3 x$

62. $8 \log_2 z$

63. $\log_4(x + 1) - \log_4 x$

64. $\log_5(2y - 1) - \log_5 y$

65. $2 \ln x + 3 \ln y$

66. $4 \log_2 a + 2 \log_2 b$

67. $\dfrac{1}{2}\log_3 x + 3 \log_3(x - 1)$

68. $\dfrac{1}{3}\log_4 z + 2 \log_4(2z + 1)$

69. $\log x^5 - 3 \log x$

70. $\log_7 x^4 - 2 \log_7 x$

71. $\dfrac{1}{2}[3 \log x + \log y]$

72. $\dfrac{1}{3}[\ln(x - 1) + \ln(x + 1)]$

73. $\log_8(x^2 - 1) - \log_8(x + 1)$

74. $\log_5(x^2 + 3x + 2) - \log_5(x + 2)$

75. $18 \log \sqrt{x} + 9 \log \sqrt[3]{x} - \log 10$

76. $10 \log_4 \sqrt[5]{x} + 4 \log_4 \sqrt{x} - \log_4 16$

In Problems 77–84, use the Change-of-Base Formula and a calculator to evaluate each logarithm. Round your answer to three decimal places.

77. $\log_2 10$

78. $\log_3 18$

79. $\log_8 3$

80. $\log_7 5$

81. $\log_{1/3} 19$

82. $\log_{1/4} 3$

83. $\log_{\sqrt{2}} 5$

84. $\log_{\sqrt{3}} \sqrt{6}$

Applying the Concepts

85. Find the value of $\log_2 3 \cdot \log_3 4 \cdot \log_4 5 \cdot \log_5 6 \cdot \log_6 7 \cdot \log_7 8$.

86. Find the value of $\log_2 4 \cdot \log_4 6 \cdot \log_6 8$.

87. Find the value of $\log_2 3 \cdot \log_3 4 \cdot \cdots \cdot \log_n(n + 1) \cdot \log_{n+1} 2$.

88. Find the value of $\log_3 3 \cdot \log_3 9 \cdot \log_3 27 \cdot \cdots \cdot \log_3 3^n$.

Extending the Concepts

89. Show that $\log_a\!\left(x + \sqrt{x^2 - 1}\right) + \log_a\!\left(x - \sqrt{x^2 - 1}\right) = 0$.

90. Show that $\log_a\!\left(\sqrt{x} + \sqrt{x - 1}\right) + \log_a\!\left(\sqrt{x} - \sqrt{x - 1}\right) = 0$.

91. If $f(x) = \log_a x$, show that $f(AB) = f(A) + f(B)$.

92. Find the domain of $f(x) = \log_a x^2$ and the domain of $g(x) = 2 \log_a x$. Since $\log_a x^2 = 2 \log_a x$, how can it be that the domains are not equal? Write a brief explanation.

Synthesis Review

In Problems 93–98, solve each equation.

93. $4x + 3 = 13$

94. $-3x + 10 = 4$

95. $x^2 + 4x + 2 = 0$

96. $3x^2 = 2x + 1$

97. $\sqrt{x + 2} - 3 = 4$

98. $\sqrt[3]{2x} - 2 = -5$

The Graphing Calculator

We can use the Change-of-Base Formula to graph any logarithmic function on a graphing calculator. For example, to graph $f(x) = \log_2 x$ we would graph $Y_1 = \dfrac{\log x}{\log 2}$ or $Y_1 = \dfrac{\ln x}{\ln 2}$.

In Problems 99–102, graph each logarithmic function using a graphing calculator. State the domain and the range of each function.

99. $f(x) = \log_3 x$

100. $f(x) = \log_5 x$

101. $F(x) = \log_{1/2} x$

102. $G(x) = \log_{1/3} x$

11.5 Exponential and Logarithmic Equations

OBJECTIVES

1 Solve Logarithmic Equations Using the Properties of Logarithms

2 Solve Exponential Equations

3 Solve Equations Involving Exponential Models

Preparing for Exponential and Logarithmic Equations

Before getting started, take the following readiness quiz. If you get a problem wrong, go back to the section cited and review the material.

1. Solve: $2x + 5 = 13$ [Section 2.2, pp. 84–86]
2. Solve: $x^2 - 4x = -3$ [Section 6.6, pp. 451–457]
3. Solve: $3a^2 = a + 5$ [Section 10.2, pp. 757–762]
4. Solve: $(x + 3)^2 + 2(x + 3) - 8 = 0$ [Section 10.3, pp. 772–777]

1 Solve Logarithmic Equations Using the Properties of Logarithms

In Section 11.3, we solved logarithmic equations of the form $\log_a x = y$. These equations are solved by changing the logarithmic equation to an equivalent exponential equation. Often, however, logarithmic equations have more than one logarithm in them. In this case, we need to use properties of logarithms to solve the logarithmic equation.

In the last section we learned that if $M = N$, then $\log_a M = \log_a N$. It turns out that the converse of this property is true as well.

ONE-TO-ONE PROPERTY OF LOGARITHMS

In the following property, M, N, and a are positive real numbers, with $a \neq 1$.

$$\text{If } \log_a M = \log_a N, \text{ then } M = N.$$

This property is useful for solving logarithmic equations that have the same base by setting the arguments equal to each other.

EXAMPLE 1   Solving a Logarithmic Equation

Solve: $2 \log_3 x = \log_3 25$

Solution

We notice that both logarithms are to the same base, 3. So if we can write the equation in the form $\log_a M = \log_a N$, we can use the One-to-One Property and set the arguments equal to each other.

$$2 \log_3 x = \log_3 25$$

$r \log_a M = \log_a M^r$: $\log_3 x^2 = \log_3 25$

Set the arguments equal to each other: $x^2 = 25$

Square Root Method: $x = -5 \quad \text{or} \quad x = 5$

The apparent solution $x = -5$ is extraneous because the argument of a logarithm must be greater than zero and -5 causes the argument to be negative. We now check the other apparent solution.

Check $x = 5$:

$$2 \log_3 5 \overset{?}{=} \log_3 25$$
$$2 \log_3 5 \overset{?}{=} \log_3 5^2$$
$$2 \log_3 5 = 2 \log_3 5 \quad \text{True}$$

The solution set is $\{5\}$.

QUICK ✓

1. Solve: $2 \log_4 x = \log_4 9$

___

If a logarithmic equation contains more than one logarithm on one side of the equation, then we can use properties of logarithms to rewrite the equation as a single logarithm. Once again, we use properties to reduce an equation into a form that is familiar. In this case, we use properties of logarithms to express the sum or difference of logarithms as a single logarithm. We then express the logarithmic equation as an exponential equation and solve for the unknown.

EXAMPLE 2 **Solving a Logarithmic Equation**

Solve: $\log_2(x - 2) + \log_2 x = 3$

Solution

We use the fact that the sum of two logarithms can be written as the logarithm of the product to write the equation with a single logarithm.

$$\log_2(x - 2) + \log_2 x = 3$$

$\log_a M + \log_a N = \log_a(MN)$: $\log_2[x(x - 2)] = 3$

If $y = \log_a M$, then $a^y = M$: $2^3 = x(x - 2)$

Distribute: $8 = x^2 - 2x$

Write in standard form: $x^2 - 2x - 8 = 0$

Factor: $(x - 4)(x + 2) = 0$

Zero-Product Property: $x - 4 = 0$ or $x + 2 = 0$

$x = 4$ or $x = -2$

Check $x = 4$: $\log_2(4 - 2) + \log_2 4 \stackrel{?}{=} 3$ $x = -2$: $\log_2(-2 - 2) + \log_2(-2) \stackrel{?}{=} 3$

$\log_2(2) + \log_2 4 \stackrel{?}{=} 3$ $x = -2$ is extraneous because it causes the argument to be negative

$1 + 2 = 3$ True

Work Smart

The apparent solution $x = -2$ in Example 2 is extraneous because it results in us attempting to find the log of a negative number, not because -2 is negative.

The solution set is $\{4\}$.   ▬

QUICK ✓

2. Solve: $\log_4(x - 6) + \log_4 x = 2$

___

② Solve Exponential Equations

In Section 11.2, we solved exponential equations by using the fact that if $a^u = a^v$, then $u = v$. However, in many situations, it is difficult (if not impossible) to write each side of the equation with the same base. For example, to solve the equation $3^x = 5$, using the method presented in Section 11.2, would require that we determine the quantity that we would raise 3 to in order to obtain 5 (but this is the original problem!).

EXAMPLE 3 **Using Logarithms to Solve Exponential Equations**

Solve: $3^x = 5$

Solution

We cannot write 5 so that it is 3 raised to some integer power. Therefore, we write the equation $3^x = 5$ as a logarithm using the fact that $a^y = x$ is equivalent to $y = \log_a x$.

$$3^x = 5$$

Write as a logarithm; If $a^y = x$, then $y = \log_a x$: $x = \log_3 5$ Exact solution

If we want a decimal approximation to the solution, we use the Change-of-Base Formula.

$$x = \log_3 5 = \frac{\log 5}{\log 3}$$

$$\approx 1.465 \quad \text{Approximate solution}$$

An alternative approach to solving the equation would be to take the logarithm of both sides of the equation. Typically, we take either the natural logarithm or common logarithm of both sides of the equation. If we take the natural logarithm of both sides of the equation, we obtain the following:

$$3^x = 5$$

$$\ln 3^x = \ln 5$$

$\log_a M^r = r \log_a M$: $x \ln 3 = \ln 5$

Divide both sides by $\ln 3$: $x = \dfrac{\ln 5}{\ln 3}$ Exact solution

$$\approx 1.465 \quad \text{Approximate solution}$$

The solution set is $\left\{\dfrac{\ln 5}{\ln 3}\right\}$. If we had taken the common logarithm of both sides, the solution set would be $\left\{\dfrac{\log 5}{\log 3}\right\}$.

QUICK ✓ *Solve each equation.*

3. $2^x = 11$ **4.** $5^{2x} = 3$

EXAMPLE 4   **Using Logarithms to Solve Exponential Equations**

Solve: $4e^{3x} = 10$

Solution

We first need to isolate the exponential expression by dividing both sides of the equation by 4.

$$4e^{3x} = 10$$

$$e^{3x} = \frac{5}{2}$$

We cannot express $\dfrac{5}{2}$ as e raised to an integer power, so we cannot solve the equation using the fact that if $a^u = a^v$, then $u = v$. However, we can solve the equation by writing the exponential equation as an equivalent logarithmic equation.

$$e^{3x} = \frac{5}{2}$$

Write as a logarithm; If $a^y = x$, then $y = \log_a x$: $\ln\dfrac{5}{2} = 3x$

Divide both sides by 3: $x = \dfrac{\ln\left(5/2\right)}{3}$ Exact solution

$x \approx 0.305$ Approximate solution

We leave it to you to verify the solution. The solution set is $\left\{\dfrac{\ln\left(5/2\right)}{3}\right\}$.

QUICK ✓ *Solve each equation.*

5. $e^{2x} = 5$ **6.** $3e^{-4x} = 20$

---

③ Solve Equations Involving Exponential Models

In Section 11.2, we looked at a variety of models from areas such as statistics, biology, and finance. Now that we have the ability to solve exponential equations, we can look at these models again. This time, rather than evaluating the models at certain values of the independent variable, we will solve equations involving the models.

EXAMPLE 5   Radioactive Decay

The half-life of plutonium-239 is 24,360 years. The maximum amount of plutonium-239 that an adult can handle without significant injury is 0.13 micrograms ($= 0.000000013$ grams). Suppose that a researcher possesses a 1-gram sample of plutonium-239. The amount A (in grams) of plutonium-239 after t years is given by

$$A(t) = 1 \cdot \left(\frac{1}{2}\right)^{t/24,360}$$

(a) How long will it take before 0.9 gram of plutonium-239 is left in the sample?

(b) How long will it take before the 1-gram sample is safe? That is, how long will it take before 0.000000013 gram is left?

Solution

(a) We need to determine the time until $A = 0.9$ gram. So we need to solve the equation

$$0.9 = 1 \cdot \left(\frac{1}{2}\right)^{t/24,360}$$

for t. But how can we get the t out of the exponent? Remember, one of the properties of logarithms that states $\log_a M^r = r \log_a M$. So if we take the logarithm of both sides of the equation, we can "get the variable down in front."

$$\log 0.9 = \log\left(\frac{1}{2}\right)^{t/24,360}$$

$$\log 0.9 = \frac{t}{24,360}\log\left(\frac{1}{2}\right)$$

Multiply both sides by 24,360: $24{,}360 \log 0.9 = t \log\left(\dfrac{1}{2}\right)$

Divide both sides by $\log\left(\dfrac{1}{2}\right)$: $\dfrac{24{,}360 \log 0.9}{\log\left(\dfrac{1}{2}\right)} = t$

So $t = \dfrac{24{,}360 \log 0.9}{\log\left(\dfrac{1}{2}\right)} \approx 3{,}702.8$. After approximately 3703 years, there

will be 0.9 gram of plutonium-239.

(b) We need to determine the time until $A = 0.000000013$ gram. So we solve the equation

$$0.000000013 = 1 \cdot \left(\frac{1}{2}\right)^{t/24{,}360}$$

Take the logarithm of both sides:

$$\log(0.000000013) = \log\left(\frac{1}{2}\right)^{t/24{,}360}$$

$\log_a M^r = r \log_a M$:

$$\log(0.000000013) = \frac{t}{24{,}360} \log\left(\frac{1}{2}\right)$$

Multiply both sides by 24,360:

$$24{,}360 \log(0.000000013) = t \log\left(\frac{1}{2}\right)$$

Divide both sides by $\log\left(\dfrac{1}{2}\right)$:

$$\frac{24{,}360 \log(0.000000013)}{\log\left(\dfrac{1}{2}\right)} = t$$

So $t = \dfrac{24{,}360 \log(0.000000013)}{\log\left(\dfrac{1}{2}\right)} \approx 638{,}156.8$ years. After approximately

638,157 years, the 1-gram sample will be safe! ■

QUICK ✓

7. The half-life of thorium-227 is 18.72 days. Suppose that a researcher possesses a 10-gram sample of thorium-227. The amount A (in grams) of thorium-227 after t days is given by

$$A(t) = 10 \cdot \left(\frac{1}{2}\right)^{t/18.72}$$

(a) How long will it take before 9 grams of thorium-227 is left in the sample?

(b) How long will it take before 3 grams of thorium-227 is left in the sample?

Now let's look at an example involving compound interest. Remember, the compound interest formula states that the future value of P dollars invested in an account paying an annual interest rate r, compounded n times per year for t years, is given by

$$A = P\left(1 + \frac{r}{n}\right)^{nt}.$$

EXAMPLE 6 Future Value of Money

Suppose that you deposit $3000 into a Roth IRA today. If the deposit earns 8% interest compounded quarterly, how long will it be before the account is worth

(a) $4,500?

(b) $6,000? That is, how long will you have to wait until your money doubles?

Solution

First, we write the model with the values of the variables entered. With $P = 3000$, $r = 0.08$ and $n = 4$ (compounded quarterly), we have

$$A = 3000\left(1 + \frac{0.08}{4}\right)^{4t} \quad \text{or} \quad A = 3000(1.02)^{4t}$$

(a) We want to know the time t until $A = 4500$. That is, we want to solve

$$4500 = 3000(1.02)^{4t}$$

Divide both sides by 3000: $\qquad 1.5 = (1.02)^{4t}$

Take the logarithm of both sides: $\qquad \log 1.5 = \log(1.02)^{4t}$

$\log_a M^r = r \log_a M$: $\qquad \log 1.5 = 4t \log(1.02)$

Divide both sides by 4 log (1.02): $\qquad \dfrac{\log 1.5}{4 \log(1.02)} = t$

So $t = \dfrac{\log 1.5}{4 \log(1.02)} \approx 5.12$. After approximately 5.12 years

(5 years, 1.5 months), the account will be worth \$4500.

(b) We want to know the time t until $A = 6000$. That is, we want to solve

$$6000 = 3000(1.02)^{4t}$$

Divide both sides by 3000: $\qquad 2 = (1.02)^{4t}$

Take the logarithm of both sides: $\qquad \log 2 = \log(1.02)^{4t}$

$\log_a M^r = r \log_a M$: $\qquad \log 2 = 4t \log(1.02)$

Divide both sides by 4 log(1.02): $\qquad \dfrac{\log 2}{4 \log(1.02)} = t$

So $t = \dfrac{\log 2}{4 \log(1.02)} \approx 8.75$. After approximately 8.75 years

(8 years, 9 months), the account will be worth \$6000. ▬

QUICK ✅

8. Suppose that you deposit \$2000 into a Roth IRA today. If the deposit earns 6% interest compounded monthly, how long will it be before the account is worth

(a) \$3000?

(b) \$4000? That is, how long will you have to wait until your money doubles?

11.5 Exercises

For Extra Help:

Student Solutions Manual CD Video PH Math/Tutor Center MathXL Tutorials on CD MathXL® MyMathLab

Concepts and Vocabulary

In Problems 1 and 2, fill in the blanks.

1. If $\log_a M = \log_a N$, then _____.

2. To solve $5^x = 12$, we would first take the _____ _____ or _____ _____ of both sides of the equation.

In Problems 3 and 4, answer True or False to each statement.

3. If a logarithmic equation contains more than one logarithm, we can use the properties of logarithms to write the equation with a single logarithm.

4. Logarithmic equations never have extraneous solutions.

5. Explain why we can't use the fact that if $a^u = a^v$, then $u = v$ to solve the equation $2^x = 7$.

6. Suppose you were solving a logarithmic equation that contains the term $\log_3(x + 3)$ and $x = -2$ was an apparent solution to the equation. Do you think that the solution is extraneous? Why or why not?

Building Skills

In Problems 7–44, solve each equation. Express irrational solutions in exact form and as a decimal rounded to three decimal places.

7. $\log_2 x = \log_2 7$

8. $\log_5 x = \log_5 13$

9. $2 \log_3 x = \log_3 81$

10. $2 \log_3 x = \log_3 4$

11. $\log_6(3x + 1) = \log_6 10$

12. $\log(2x - 3) = \log 11$

13. $\dfrac{1}{2}\ln x = 2 \ln 3$

14. $\dfrac{1}{2}\log_2 x = 2 \log_2 2$

15. $\log_2(x + 3) + \log_2 x = 2$

16. $\log_2(x - 7) + \log_2 x = 3$

17. $\log_2(x - 1) + \log_2(x + 5) = 4$

18. $\log_2(x - 2) + \log_2(x + 4) = 4$

19. $\log(x + 3) - \log x = 1$

20. $\log_3(x + 5) - \log_3 x = 2$

21. $\log_4(x + 5) - \log_4(x - 1) = 2$

22. $\log_3(x + 2) - \log_3(x - 2) = 4$

23. $\log_4(x + 3) + \log_4(x - 6) = \log_4 3$

24. $\log_5(x + 3) + \log_5(x - 4) = \log_5 8$

25. $2^x = 10$

26. $3^x = 8$

27. $5^x = 20$

28. $4^x = 20$

29. $\left(\dfrac{1}{2}\right)^x = 7$

30. $\left(\dfrac{1}{2}\right)^x = 10$

31. $e^x = 5$

32. $e^x = 3$

33. $10^x = 5$

34. $10^x = 0.2$

35. $3^{2x} = 13$

36. $2^{2x} = 5$

37. $\left(\dfrac{1}{2}\right)^{4x} = 3$

38. $\left(\dfrac{1}{3}\right)^{2x} = 4$

39. $4 \cdot 2^x + 3 = 8$

40. $3 \cdot 4^x - 5 = 10$

41. $-3e^x = -18$

42. $\dfrac{1}{2}e^x = 4$

43. $0.2^{x+1} = 3^x$

44. $0.4^x = 2^{x-3}$

Mixed Practice

In Problems 45–58, solve each equation. Express irrational solutions in exact form and as a decimal rounded to three decimal places.

45. $\log_4 x + \log_4(x - 6) = 2$

46. $\log_6 x + \log_6(x + 5) = 2$

47. $5^{3x} = 7$

48. $3^{2x} = 4$

49. $3 \log_2 x = \log_2 8$

50. $5 \log_4 x = \log_4 32$

51. $\dfrac{1}{3}e^x = 5$

52. $-4e^x = -16$

53. $\left(\dfrac{1}{4}\right)^{x+1} = 8^x$

54. $9^x = 27^{x-4}$

55. $\log_3 x = \log_3 16$

56. $\log_7 x = \log_7 8$

57. $\log_2(x + 4) + \log_2(x - 1) = \log_2 6$

58. $\log_3(x - 5) + \log_3(x + 1) = \log_3 7$

Applying the Concepts

59. A Population Model According to the U.S. Census Bureau, the population of the United States in 2005 was 296 million people. In addition, the population of the United States was growing at a rate of 1.1% per year. Assuming that this growth rate continues, the model $P(t) = 296(1.011)^{t-2005}$ represents the population P (in millions of people) in year t.

  (a) According to this model, when will the population of the United States be 351 million people?

  (b) According to this model, when will the population of the United States be 482 million people?

60. A Population Model According to the *Statistical Abstract of the United States,* the population of the world in 2005 was 6448 million people. In addition, the population of the world was growing at a rate of 1.26% per year. Assuming that this growth rate continues, the model $P(t) = 6448(1.0126)^{t-2005}$ represents the population P (in millions of people) in year t.

(a) According to this model, when will the population of the world be 9.65 billion people?

(b) According to this model, when will the population of the world be 11.55 billion people?

61. Time Is Money Suppose that you deposit $5000 into a Certificate of Deposit (CD) today. If the deposit earns 6% interest compounded monthly, how long will it be before the account is worth

(a) $7,000?

(b) $10,000? That is, how long will you have to wait until your money doubles?

62. Time Is Money Suppose that you deposit $8000 into a Certificate of Deposit (CD) today. If the deposit earns 4% interest compounded quarterly, how long will it be before the account is worth

(a) $10,000?

(b) $24,000? That is, how long will you have to wait until your money triples?

63. Depreciation Based on data obtained from the *Kelley Blue Book,* the value V of a Dodge Neon that is t years old can be modeled by $V(t) = 14{,}512(0.82)^t$.

(a) According to the model, when will the car be worth $10,000?

(b) According to the model, when will the car be worth $5000?

(c) According to the model, when will the car be worth $1000?

64. Depreciation Based on data obtained from the *Kelley Blue Book,* the value V of a Dodge Stratus that is t years old can be modeled by $V(t) = 19{,}282(0.84)^t$.

(a) According to the model, when will the car be worth $10,000?

(b) According to the model, when will the car be worth $5000?

(c) According to the model, when will the car be worth $1000?

65. Radioactive Decay The half-life of beryllium-11 is 13.81 seconds. Suppose that a researcher possesses a 100-gram sample of beryllium-11. The amount A (in grams) of beryllium-11 after t seconds is given by

$$A(t) = 100 \cdot \left(\frac{1}{2}\right)^{t/13.81}$$

(a) When will there be 90 grams of beryllium-11 left in the sample?

(b) When will there be 25 grams of beryllium-11 left in the sample?

(c) When will there be 10 grams of beryllium-11 left in the sample?

66. Radioactive Decay The half-life of carbon-10 is 19.255 seconds. Suppose that a researcher possesses a 100-gram sample of carbon-10. The amount A (in grams) of carbon-10 after t seconds is given by

$$A(t) = 100 \cdot \left(\frac{1}{2}\right)^{t/19.255}$$

(a) When will there be 90 grams of carbon-10 left in the sample?

(b) When will there be 25 grams of carbon-10 left in the sample?

(c) When will there be 10 grams of carbon-10 left in the sample?

67. Newton's Law of Cooling Newton's Law of Cooling states that the temperature of a heated object decreases exponentially over time toward the temperature of the surrounding medium. Suppose that a pizza is removed from a 400°F oven and placed in a room whose temperature is 70°F. The temperature u (in °F) of the pizza at time t (in minutes) can be modeled by $u(t) = 70 + 330e^{-0.072t}$.

(a) According to the model, when will be the temperature of the pizza be 300°F?

(b) According to the model, when will be the temperature of the pizza be 220°F?

68. Newton's Law of Cooling Newton's Law of Cooling states that the temperature of a heated object decreases exponentially over time toward the temperature of the surrounding medium. Suppose that coffee that is 170°F is poured into a coffee mug and allowed to cool in a room whose temperature is 70°F. The temperature u (in °F) of the coffee at time t (in minutes) can be modeled by $u(t) = 70 + 100e^{-0.045t}$.

(a) According to the model, when will be the temperature of the coffee be 120°F?

(b) According to the model, when will be the temperature of the coffee be 100°F?

69. Learning Curve Suppose that a student has 200 vocabulary words to learn. If a student learns 20 words in 30 minutes, the function

$$L(t) = 200(1 - e^{-0.0035t})$$

models the number of words L that the student will learn after t minutes.

(a) After how long will the student learn 50 words?

(b) After how long will the student learn 150 words?

70. Learning Curve Suppose that a student has 50 biology terms to learn. If a student learns 10 terms in 30 minutes, the function

$$L(t) = 50(1 - e^{-0.0223t})$$

models the number of terms L that the student will learn after t minutes.

(a) After how long will the student learn 10 words?

(b) After how long will the student learn 40 words?

Extending the Concepts

71. The Rule of 72 The Rule of 72 states that the time for an investment to double in value is approximately given by 72 divided by the annual interest rate. For example, an investment earning 10% annual interest will double in approximately $\dfrac{72}{10} = 7.2$ years.

(a) According to the Rule of 72, approximately how long will it take an investment to double if it earns 8% annual interest?

(b) Derive a formula that can be used to find the number of years required for an investment to double. (*Hint:* Let $A = 2P$ in the formula $A = P\left(1 + \dfrac{r}{n}\right)^{nt}$ and solve for t.)

(c) Use the formula derived in part (b) to determine the exact amount of time it takes an investment to double that earns 8% interest compounded monthly. Compare the result to the results given by the Rule of 72.

72. Critical Thinking Suppose you need to open up a savings account. Bank A offers 4% interest compounded daily, while Bank B offers 4.1% interest compounded quarterly. Which bank offers the better deal? Why?

73. Critical Thinking The bacteria in a 2-liter container double every minute. After 30 minutes the container is full. How long did it take to fill half the container?

Synthesis Review

In Problems 74–79, find the following values for each function.

(a) $f(3)$ (b) $f(-2)$ (c) $f(0)$

74. $f(x) = 5x + 2$ **75.** $f(x) = -2x + 7$ **76.** $f(x) = \dfrac{x + 3}{x - 2}$

77. $f(x) = \dfrac{x}{x - 5}$ **78.** $f(x) = \sqrt{x + 5}$ **79.** $f(x) = 2^x$

The Graphing Calculator

The techniques for solving equations introduced in this chapter apply only to certain types of exponential or logarithmic equations. Solutions for other types of equations are usually studied in calculus using numerical methods. However, a graphing calculator can be used to approximate solutions using the INTERSECT feature. For example, to solve

$e^x = 3x + 2$, we would graph $Y_1 = e^x$ and $Y_2 = 3x + 2$ on the same screen. We then use the INTERSECT feature to find the x-coordinate of the point of intersection. This x-coordinate represents the approximate solution as shown in Figure 26.

Figure 26

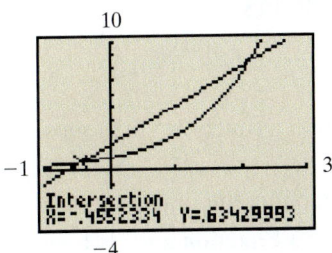
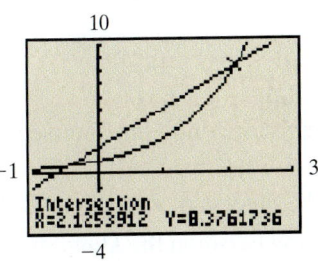

The solution set is $\{-0.46, 2.13\}$ rounded to two decimal places.

In Problems 80–87, solve each equation using a graphing calculator. Express your answer rounded to two decimal places.

80. $e^x = -3x + 2$ **81.** $e^x = -2x + 5$ **82.** $e^x = x + 2$

83. $e^x = x^2$ **84.** $e^x + \ln(x) = 2$ **85.** $e^x - \ln(x) = 4$

86. $\ln x = x^2 + 1$ **87.** $\ln x = x^2 - 1$

CHAPTER 11 ACTIVITY: CORRECT THE QUIZ

Focus: Solving exponential and logarithmic equations

Time: 10–15 minutes

Group size: 2

In this activity you will work as a team to grade the student quiz shown below. One of you will grade the odd questions, and the other will grade the even questions. If an answer is correct, mark it correct. If an answer is wrong, mark it wrong and show the correct answer.

Once all of the quiz questions are graded, explain your results to each other and compute the final score for the quiz. Be prepared to discuss your results with the rest of the class.

Student Quiz
Name: *Ima Student*   Quiz Score: _____
Solve the following equations. Express any irrational answers in exact form.
(1) $\log_2 x = 3$ Answer: $\{9\}$
(2) $\log_{16} x = \dfrac{3}{4}$ Answer: $\{8\}$
(3) $\log(2x) - \log 6 = \log(x - 8)$ Answer: $\{12\}$
(4) $3^x = 27$ Answer: $\{9\}$
(5) $6^{2x} = 18$ Answer: $\left\{ \dfrac{\log 18}{2 \log 6} \right\}$
(6) $4^{x+9} = 7$ Answer: $\left\{ \dfrac{\log 7}{\log 4} - 9 \right\}$
(7) $9^{3x-1} = 27^{4x}$ Answer: $\left\{ -\dfrac{1}{3} \right\}$
(8) $\log_3(2x - 3) = 2$ Answer: $\left\{ \dfrac{11}{2} \right\}$

CHAPTER 11 REVIEW

Section 11.1   Composite Functions and Inverse Functions

KEY CONCEPTS	KEY TERMS
• $(f \circ g)(x) = f(g(x))$ • **Horizontal Line Test** If every horizontal line intersects the graph of a function f in at most one point, then f is one-to-one. • For the inverse of a function to also be a function, the function must be one-to-one. • **Relation between the Domain and Range of a Function and Its Inverse** All elements in the domain of a function are also elements in the range of its inverse All elements in the range of a function are also elements in the domain of its inverse • The graph of a function f and the graph of its inverse are symmetric with respect to the line $y = x$. • $f^{-1}(f(x)) = x$ for every x in the domain of f and $f(f^{-1}(x)) = x$ for every x in the domain of f^{-1}	Composition Composite function One-to-one Inverse function

YOU SHOULD BE ABLE TO . . .	EXAMPLE	REVIEW EXERCISES
1 Form the composite function (p. 838)	Examples 1 and 2	1–8
2 Determine whether or not a function is one-to-one (p. 840)	Examples 3 and 4	9–12
3 Determine the inverse of a function defined by a map or ordered pair (p. 842)	Examples 5 and 6	13–16
4 Obtain the graph of the inverse function from the graph of a function (p. 844)	Example 7	17, 18
5 Find the inverse of a function defined by an equation (p. 845)	Examples 8 and 9	19–22

In Problems 1–4, for the given functions f and g, find:

    (a) $(f \circ g)(5)$ **(b)** $(g \circ f)(-3)$ **(c)** $(f \circ f)(-2)$ **(d)** $(g \circ g)(4)$

1. $f(x) = 3x + 5; g(x) = 2x - 1$ **2.** $f(x) = x - 3; g(x) = 5x + 2$

3. $f(x) = 2x^2 + 1; g(x) = x + 5$ **4.** $f(x) = x - 3; g(x) = x^2 + 1$

In Problems 5–8, for the given functions f and g, find:

    (a) $(f \circ g)(x)$ **(b)** $(g \circ f)(x)$ **(c)** $(f \circ f)(x)$ **(d)** $(g \circ g)(x)$

5. $f(x) = x + 1; g(x) = 5x$ **6.** $f(x) = 2x - 3; g(x) = x + 6$

7. $f(x) = x^2 + 1; g(x) = 2x + 1$ **8.** $f(x) = \dfrac{2}{x + 1}; g(x) = \dfrac{1}{x}$

In Problems 9–12, determine which of the following functions is one-to-one.

  9. $\{(-5, 8), (-3, 2), (-1, 8), (0, 12), (1, 15)\}$

  10. $\{(-4, 2), (-2, 1), (0, 0), (1, -1), (2, 8)\}$

  11.

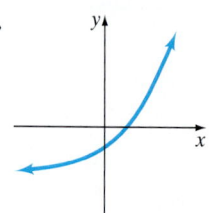

        12.

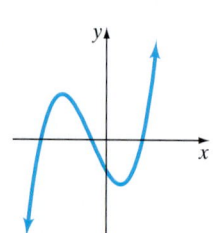

In Problems 13–16, find the inverse of the following one-to-one functions.

13.

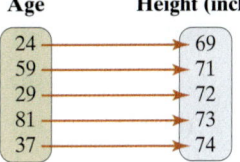

Age	Height (inches)
24	69
59	71
29	72
81	73
37	74

14.

Price ($)	Quantity Demanded
300	112
200	129
170	144
150	161
130	176

15. $\{(-5, 3), (-3, 1), (1, -3), (2, 9)\}$

16. $\{(-20, 1), (-15, 4), (5, 3), (25, 2)\}$

In Problems 17 and 18, the graph of a one-to-one function f is given. Draw the graph of the inverse function f^{-1}.

17.

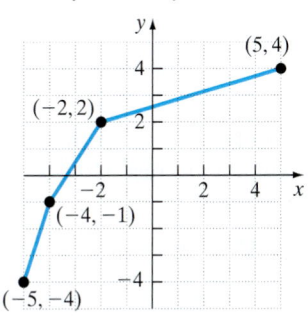

18.

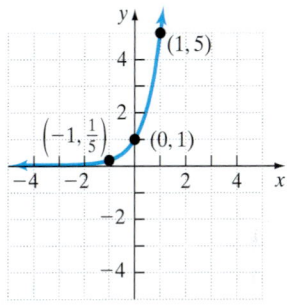

In Problems 19–22, find the inverse function of the given one-to-one function.

19. $f(x) = 5x$ **20.** $H(x) = 2x + 7$ **21.** $P(x) = \dfrac{4}{x + 2}$ **22.** $g(x) = 2x^3 - 1$

Section 11.2	**Exponential Functions**

KEY CONCEPTS	**KEY TERMS**

• **Properties of Exponential Functions**
  1. The domain is the set of all real numbers. The range is the set of all positive real numbers.
  2. There are no x-intercepts. The y-intercept is 1.
  3. The graph of an exponential function contains the points $\left(-1, \dfrac{1}{a}\right)$, $(0, 1)$, and $(1, a)$ if $a > 1$. The graph of an exponential function contains the points $(-1, a)$, $(0, 1)$, and $\left(1, \dfrac{1}{a}\right)$ if $0 < a < 1$.
• **Property for Solving Exponential Equations of the Form $a^u = a^v$**
  If $a^u = a^v$, then $u = v$.
	Exponential function
	The number e
	Exponential equation
	Half-life
	Payment period
	Compound interest

YOU SHOULD BE ABLE TO . . .	EXAMPLE	REVIEW EXERCISES
① Evaluate exponential expressions (p. 853)	Example 1	23–25
② Graph exponential functions (p. 854)	Examples 2 through 4	26–29
③ Define the number e (p. 857)		30
④ Solve exponential equations (p. 858)	Examples 5 and 6	31–36
⑤ Study exponential models that describe our world (p. 861)	Examples 7 through 9	37–40

In Problems 23–25, approximate each number using a calculator. Express your answer rounded to three decimal places.

23. (a) $7^{1.7}$ **(b)** $7^{1.73}$ **(c)** $7^{1.732}$ **(d)** $7^{1.7321}$ **(e)** $7^{\sqrt{3}}$

24. (a) $10^{3.1}$ **(b)** $10^{3.14}$ **(c)** $10^{3.142}$ **(d)** $10^{3.1416}$ **(e)** 10^{π}

25. (a) $e^{0.5}$ **(b)** e^{-1} **(c)** $e^{1.5}$ **(d)** $e^{-0.8}$ **(e)** $e^{\sqrt{\pi}}$

In Problems 26–29, graph each function. State the domain and range of the function.

26. $f(x) = 9^x$ **27.** $g(x) = \left(\dfrac{1}{9}\right)^x$ **28.** $H(x) = 4^{x-2}$ **29.** $h(x) = 4^x - 2$

30. State the definition of the number e.

In Problems 31–36, solve each equation.

31. $2^x = 64$ **32.** $25^{x-2} = 125$ **33.** $27^x \cdot 3^{x^2} = 9^2$

34. $\left(\dfrac{1}{4}\right)^x = 16$ **35.** $(e^2)^{x-1} = e^x \cdot e^7$ **36.** $(2^x)^x = 512$

37. Future Value of Money Suppose that you deposit $2500 into a traditional IRA that pays 4.5% annual interest. How much money will you have after 25 years if interest is compounded

(a) annually? **(b)** quarterly? **(c)** monthly? **(d)** daily?

38. Radioactive Decay The half-life of the radioactive gas radon is 3.5 days. Suppose a researcher possesses a 100-gram sample of radon gas. The amount A (in grams) of radon after t days is given by $A(t) = 100\left(\dfrac{1}{2}\right)^{t/3.5}$. How much radon gas is left in the sample after

(a) 1 day? **(b)** 3.5 days? **(c)** 7 days? **(d)** 30 days?

39. A Population Model According to the U.S. Census Bureau, the population of Nevada in 2000 was 1.998 million people. In addition, the population of Nevada was growing at a rate of 5.2% per year. Assuming that this growth rate continues, the model $P(t) = 1.998(1.052)^{t-2000}$ represents the population (in millions of people) in year t. According to this model, what will be the population of Nevada

(a) in 2006? **(b)** in 2010?

40. Newton's Law of Cooling A baker removes a cake from a 350°F oven and places it in a room whose temperature is 72°F. According to Newton's Law of Cooling, the temperature u (in °F) of the cake at time t (in minutes) can be modeled by $u(t) = 72 + 278e^{-0.0835t}$. According to this model, what will be the temperature of the cake

(a) after 15 minutes? **(b)** after 30 minutes?

Section 11.3	Logarithmic Functions

KEY CONCEPTS	KEY TERMS

• **The Logarithmic Function to the Base a**
$y = \log_a x$ is equivalent to $x = a^y$
• **Properties of the Logarithmic Function**
  1. The domain is the set of all positive real numbers. The range is the set of all real numbers.
  2. There are no y-intercepts. The x-intercept is 1.
  3. The graph of the logarithmic function contains the points $\left(\dfrac{1}{a}, -1\right)$, $(1, 0)$, and $(a, 1)$.
	Logarithmic function
	Natural logarithm function
	Logarithmic equation
	Intensity of a sound wave
	Decibels
	Loudness

YOU SHOULD BE ABLE TO . . .	EXAMPLE	REVIEW EXERCISES
① Change exponential expressions to logarithmic expressions (p. 871)	Example 1	41–44
② Change logarithmic expressions to exponential expressions (p. 871)	Example 2	45–48
③ Evaluate logarithmic functions (p. 872)	Examples 3 and 4	49–52
④ Determine the domain of a logarithmic function (p. 873)	Example 5	53–56
⑤ Graph logarithmic functions (p. 874)	Examples 6 and 7	57, 58
⑥ Work with natural and common logarithms (p. 876)	Example 8	59–62
⑦ Solve logarithmic equations (p. 877)	Examples 9 and 10	63–68
⑧ Study logarithmic models that describe our world (p. 879)	Example 11	69, 70

In Problems 41–44, change each exponential expression to an equivalent expression involving a logarithm.

41. $3^4 = 81$ **42.** $4^{-3} = \dfrac{1}{64}$ **43.** $b^3 = 5$ **44.** $10^{3.74} = x$

In Problems 45–48, change each logarithmic expression to an equivalent expression involving an exponent.

45. $\log_8 2 = \dfrac{1}{3}$ **46.** $\log_5 18 = r$ **47.** $\ln(x + 3) = 2$ **48.** $\log x = -4$

In Problems 49–52, find the exact value of each logarithm without using a calculator.

49. $\log_8 128$ **50.** $\log_6 1$ **51.** $\log \dfrac{1}{100}$ **52.** $\log_9 27$

In Problems 53–56, find the domain of each function.

53. $f(x) = \log_2(x + 5)$ **54.** $g(x) = \log_8(7 - 3x)$

55. $h(x) = \ln(3x)$ **56.** $F(x) = \log_{\frac{1}{3}}(4x + 10)$

In Problems 57 and 58, graph each function.

57. $f(x) = \log_{\frac{5}{2}} x$ **58.** $g(x) = \log_{\frac{2}{5}} x$

In Problems 59–62, use a calculator to evaluate each expression. Round your answers to three decimal places.

59. $\ln 24$ **60.** $\ln \dfrac{5}{6}$ **61.** $\log 257$ **62.** $\log 0.124$

In Problems 63–68, solve each logarithmic equation.

63. $\log_7(4x - 19) = 2$ **64.** $\log_{\frac{1}{3}}(x^2 + 8x) = -2$

65. $\log_a \dfrac{4}{9} = -2$ **66.** $\ln e^{5x} = 30$

67. $\log(6 - 7x) = 3$ **68.** $\log_b 75 = 2$

69. Loudness of a Vacuum Cleaner A vacuum cleaner has an intensity level of 10^{-4} watt per square meter. How many decibels is the vacuum cleaner?

70. The Great New Madrid Earthquake According to the United States Geological Survey, an earthquake on December 16, 1811 in New Madrid, Missouri had a magnitude of approximately 8.0. What would have been the seismographic reading 100 kilometers from its epicenter?

Section 11.4   Properties of Logarithms

KEY CONCEPTS

- **Properties of Logarithms**

  For the following properties, a, M, and N are positive real numbers with $a \neq 1$ and r is any real number.

  $$a^{\log_a M} = M \qquad \log_a(MN) = \log_a M + \log_a N \qquad \log_a M^r = r \log_a M$$

  $$\log_a a^r = r \qquad \log_a\left(\frac{M}{N}\right) = \log_a M - \log_a N$$

- **Change-of-Base Formula**

  If $a \neq 1$, $b \neq 1$, and M are positive real numbers, then

  $$\log_a M = \frac{\log_b M}{\log_b a} = \frac{\log M}{\log a} = \frac{\ln M}{\ln a}$$

YOU SHOULD BE ABLE TO . . .	EXAMPLE	REVIEW EXERCISES
1 Understand the properties of logarithms (p. 884)	Examples 1 through 3	71–76
2 Write a logarithmic expression as a sum or difference of logarithms (p. 886)	Examples 4 through 8	77–80
3 Write a logarithmic expression as a single logarithm (p. 889)	Examples 9 and 10	81–84
4 Evaluate logarithms whose base is neither 10 nor e (p. 890)	Examples 11 and 12	85–88

In Problems 71–76, use properties of logarithms to find the exact value of each expression. Do not use a calculator.

71. $\log_4 4^{21}$

72. $7^{\log_7 9.34}$

73. $\log_5 5$

74. $\log_9 1$

75. $\log_4 12 - \log_4 3$

76. $12^{\log_{12} 2 + \log_{12} 8}$

In Problems 77–80, write each expression as a sum and/or difference of logarithms. Write exponents as factors.

77. $\log_7\left(\dfrac{xy}{z}\right)$

78. $\log_3\left(\dfrac{81}{x^2}\right)$

79. $\log 1000r^4$

80. $\ln\sqrt{\dfrac{x-1}{x}}$

In Problems 81–84, write each expression as a single logarithm.

81. $4 \log_3 x + 2 \log_3 y$

82. $\dfrac{1}{4}\ln x + \ln 7 - 2 \ln 3$

83. $\log_2 3 - \log_2 6$

84. $\log_6(x^2 - 7x + 12) - \log_6(x - 3)$

In Problems 85–88, use the Change-of-Base Formula and a calculator to evaluate each logarithm. Round your answer to three decimal places.

85. $\log_6 50$

86. $\log_\pi 2$

87. $\log_{\frac{2}{3}} 6$

88. $\log_{\sqrt{5}} 20$

Section 11.5   Exponential and Logarithmic Equations

KEY CONCEPT

- In the following M, N, and a are positive real numbers with $a \neq 1$.

  If $\log_a M = \log_a N$, then $M = N$

YOU SHOULD BE ABLE TO . . .	EXAMPLE	REVIEW EXERCISES
1 Solve logarithmic equations using properties of logarithms (p. 894)	Examples 1 and 2	89–92
2 Solve exponential equations (p. 895)	Examples 3 and 4	93–96
3 Solve equations involving exponential models (p. 897)	Examples 5 and 6	97, 98

In Problems 89–96, solve each equation. Express irrational solutions in exact form and as a decimal rounded to three decimal places.

89. $3 \log_4 x = \log_4 1000$

90. $\log_3 x + \log_3(x + 6) = 3$

91. $\ln(x + 2) - \ln x = \ln(x + 1)$

92. $\frac{1}{3} \log_{12} x = 2 \log_{12} 2$

93. $2^x = 15$

94. $10^{3x} = 27$

95. $\frac{1}{3} e^{7x} = 13$

96. $3^x = 2^{x+1}$

97. Radioactive Decay The half-life of the radioactive gas radon is 3.5 days. Suppose a researcher possesses a 100-gram sample of radon gas. The amount A (in grams) of radon after t days is given by $A(t) = 100\left(\frac{1}{2}\right)^{t/3.5}$.

  (a) When will 75 grams of radon gas be left in the sample?
  (b) When will 1 gram of radon gas be left in the sample?

98. A Population Model According to the U.S. Census Bureau, the population of Nevada in 2000 was 1.998 million people. In addition, the population of Nevada was growing at a rate of 5.2% per year. Assuming that this growth rate continues, the model $P(t) = 1.998(1.052)^{t-2000}$ represents the population (in millions of people) in year t. According to this model, when will the population of Nevada be

  (a) 3.0 million people?
  (b) 4.5 million people?

CHAPTER 11 TEST

Remember to use your Chapter Test Prep Video CD to see fully worked-out solutions to any of these problems you would like to review.

1. Determine whether the following function is one-to-one:
$$\{(1, 4), (3, 2), (5, 8), (-1, 4)\}$$

2. Find the inverse of $f(x) = 4x - 3$.

3. Approximate each number using a calculator. Express your answer rounded to three decimal places.

  (a) $3.1^{3.1}$ **(b)** $3.14^{3.14}$ **(c)** $3.142^{3.142}$ **(d)** $3.1416^{3.1416}$ **(e)** p^p

4. Change $4^x = 19$ to an equivalent expression involving a logarithm.

5. Change $\log_b x = y$ to an equivalent expression involving an exponent.

6. Find the exact value of each expression without using a calculator.

  (a) $\log_3\left(\frac{1}{27}\right)$ **(b)** $\log 10,000$

7. Determine the domain of $f(x) = \log_5(7 - 4x)$.

In Problems 8 and 9, graph each function. State the domain and the range of the function.

8. $f(x) = 6^x$

9. $g(x) = \log_{\frac{1}{9}} x$

10. Use the properties of logarithms to find the exact value of each expression. Do not use a calculator.

  (a) $\log_7 7^{10}$ **(b)** $3^{\log_3 15}$

11. Write the expression $\log_4 \dfrac{\sqrt{x}}{y^3}$ as a sum and/or difference of logarithms. Express exponents as factors.

12. Write the expression $4 \log M + 3 \log N$ as a single logarithm.

13. Use the Change-of-Base Formula and a calculator to evaluate $\log_{\frac{3}{4}} 10$. Round to three decimal places.

In Problems 14–20, solve each equation. Express irrational solutions in exact form and as a decimal rounded to three places.

14. $4^{x+1} = 2^{3x+1}$

15. $5^{x^2} \cdot 125 = 25^{2x}$

16. $\log_a 64 = 3$

17. $\log_2(x^2 - 33) = 8$

18. $2 \log_7(x - 3) = \log_7 3 + \log_7 12$

19. $3^{x-1} = 17$

20. $\log(x - 2) + \log(x + 2) = 2$

21. According to the U.S. Bureau of Census, International Data Base, the population of Canada in 2002 was 31.9 million people. In addition, the population of Canada was growing at a rate of 0.8% annually. Assuming that this growth rate continues, the model $P(t) = 31.9(1.008)^{t-2002}$ represents the population (in millions of people) in year t.

   (a) According to the model, what will be the population of Canada in 2010?

   (b) According to the model, in what year will the population of Canada be 50 million?

22. Rustling leaves have an intensity of 10^{-11} watt per square meter. How many decibels are rustling leaves?

CUMULATIVE REVIEW Chapters 1–11

1. Solve: $3(5 - 2x) + 8 = 4(x - 7) + 1$

2. Solve: $5 - 3|x - 2| \geq -7$

3. Determine the domain of $f(x) = \dfrac{9 - x^2}{2x^2 - x - 21}$.

4. Graph the linear equation: $4x + 3y = 6$

5. Find the equation of the line that passes through the points $(-10, 17)$ and $(5, -4)$. Write your answer in either slope-intercept or standard form, whichever you prefer.

6. Graph the following system of linear inequalities.

$$\begin{cases} x + 2y \geq 8 \\ 2x - y < 1 \end{cases}$$

In Problems 7 and 8, add, subtract, multiply, or divide as indicated.

7. $(m^2 - 5m + 13) - (6 - 2m - 3m^2)$

8. $(2n + 3)(n^2 - 4n + 6)$

In Problems 9 and 10, factor completely.

9. $16a^2 + 8ab + b^2$

10. $6y^2 - 17y + 7$

In Problems 11 and 12, perform the indicated operations. Be sure to express the final answer in lowest terms.

11. $\dfrac{2x^2 - 9x - 5}{x^2 - 3x - 10} \cdot \dfrac{3x^2 + 2x - 8}{2x^2 - 13x - 7}$

12. $\dfrac{4}{p^2 - 6p + 5} + \dfrac{2}{p^2 - 3p - 10}$

13. Solve: $\dfrac{2}{x - 5} = \dfrac{x - 2}{x + 1} + \dfrac{6x - 12}{x^2 - 4x - 5}$

14. Simplify: $\sqrt{150} + 4\sqrt{6} - \sqrt{24}$

15. Rationalize the denominator: $\dfrac{1 + \sqrt{5}}{3 - \sqrt{5}}$

16. Solve: $\sqrt{x - 8} + \sqrt{x} = 4$

17. Solve: $3x^2 = 4x + 6$

18. Solve: $2a - 7\sqrt{a} + 6 = 0$

19. Graph: $f(x) = -x^2 + 6x - 4$

20. Solve: $3x^2 + 2x - 8 < 0$

21. Graph: $g(x) = 3^x - 4$

22. Evaluate: $\log_9\left(\dfrac{1}{27}\right)$

12 Conics

Earth is the center of the universe. While this statement seems ludicrous to us now, it was the common belief held up until the 1500s. A Polish astronomer named Nicolaus Copernicus published a book entitled *De Revolutionibus Orbium Coelestium* (On the Revolutions of the Celestial Spheres), which stated that the Earth (along with the other planets) orbited the Sun in a circular motion. Copernicus's ideas were not readily accepted by the geocentrists, who held on to the belief that Earth is at the center of the universe. Copernicus's model of planetary motion was later improved upon by the German astronomer Johannes Kepler. In 1609, Kepler published Astronomia nova (New Astronomy) in which he proved that the orbit of Mars is an ellipse, with the Sun occupying one of its two foci. See Problems 43–46 on page 947.

OUTLINE

12.1 Distance and Midpoint Formulas

12.2 Circles

12.3 Parabolas

12.4 Ellipses

12.5 Hyperbolas

Putting the Concepts Together (Sections 12.1–12.5)

12.6 Nonlinear Systems of Equations

Chapter 12 Activity: How Do You Know That...?
Chapter 12 Review
Chapter 12 Test

The Big Picture: Putting It Together

In Chapter 3, we introduced you to the Cartesian plane or rectangular coordinate system. In that chapter, we stated this system allows us to make connections between algebra and geometry. In this chapter, we develop this connection further by showing how geometric definitions of certain figures lead to algebraic equations.

We start the chapter by showing how to use algebra to find the distance between any two points in the Cartesian plane. The method used to find this distance algebraically is a direct consequence of the Pythagorean Theorem studied in Section 10.1. Knowing this formula allows us to present a complete discussion of the so-called *conic sections*.

If you were asked to tell someone what a circle is, what would you say? In all likelihood, you would draw a picture to illustrate your verbal description. By having the Cartesian plane and the distance formula, we can take the geometric definition of a circle and develop an algebraic equation whose graph would represent a circle. This powerful connection between geometry and algebra allows us to answer all types of interesting questions. The methods that we are about to present form the foundation of an area of mathematics called analytic geometry.

12.1 Distance and Midpoint Formulas

OBJECTIVES

1. Use the Distance Formula
2. Use the Midpoint Formula

Preparing for Distance and Midpoint Formulas

Before getting started, take the following readiness quiz. If you get a problem wrong, go back to the section cited and review the material.

1. Simplify: (a) $\sqrt{64}$ (b) $\sqrt{24}$

[Section 9.1, pp. 660–661; Section 9.4, pp. 681–683]

2. Find the length of the hypotenuse in a right triangle whose legs are 6 and 8.

[Section 10.1, pp. 750–753]

1 Use the Distance Formula

In Section 3.1, we learned how to plot points in the Cartesian plane. One thing we would like to be able to do is algebraically compute the distance between any two points plotted in the Cartesian plane because this allows us to see a connection between geometry (literally measuring the distance) and algebra. We can find the distance between two points in the Cartesian plane using the Pythagorean Theorem.

EXAMPLE 1 **Finding the Distance between Two Points**

Find the distance d between the points $(2, 4)$ and $(5, 8)$.

Solution

We first plot the points in the Cartesian plane and connect them with a straight line as shown in Figure 1(a). To find the length d, draw a horizontal line through the point $(2, 4)$ and a vertical line through the point $(5, 8)$ and form a right triangle. The right angle of this triangle is at the point $(5, 4)$. Do you see why? If we travel horizontally from the point $(2, 4)$, then there is no "up or down" movement. For this reason, the y-coordinate of the point at the right angle must by be 4. Similarly, if we travel vertically straight down from the point $(5, 8)$, we find that the x-coordinate of the point at the right angle must be 5. See Figure 1(b).

Figure 1

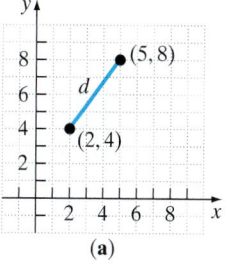
(a)

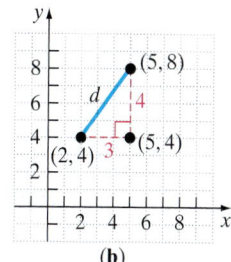
(b)

One leg of the triangle is of length 3 (since $5 - 2 = 3$). The other leg is of length 4 (because $8 - 4 = 4$). By the Pythagorean Theorem, we have that

$$d^2 = 3^2 + 4^2$$
$$= 9 + 16$$
$$= 25$$

Square Root Property: $d = \pm 5$

Because d represents the length of the hypotenuse, we discard the solution $d = -5$ and find that the hypotenuse is 5. Therefore, the distance between the points $(2, 4)$ and $(5, 8)$ is 5 units.

That was quite a bit of work to find the distance between $(2, 4)$ and $(5, 8)$. The *distance formula* can be used to find the distance between any two points in the Cartesian plane.

THE DISTANCE FORMULA

The distance between two points $P_1 = (x_1, y_1)$ and $P_2 = (x_2, y_2)$, denoted by $d(P_1, P_2)$, is

$$d(P_1, P_2) = \sqrt{(x_2 - x_1)^2 + (y_2 - y_1)^2}$$

Figure 2 illustrates the theorem.

We can provide a justification for the formula. Let (x_1, y_1) denote the coordinates of a point P_1 and let (x_2, y_2) denote the coordinates of point P_2. The line joining the points P_1 and P_2 is neither vertical nor horizontal. Form a right triangle so that the vertex of the right angle is at the point $P_3 = (x_2, y_1)$ as shown in Figure 3(a). The vertical distance from P_3 to P_2 is the absolute value of the difference of the y-coordinates, $|y_2 - y_1|$. The horizontal distance from P_1 to P_3 is the absolute value of the difference of the x-coordinates, $|x_2 - x_1|$. See Figure 3(b).

Figure 2

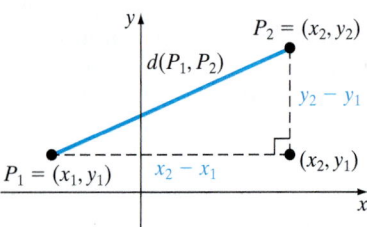

Figure 3

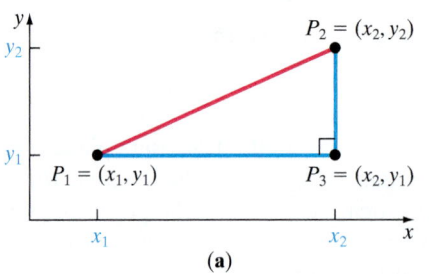 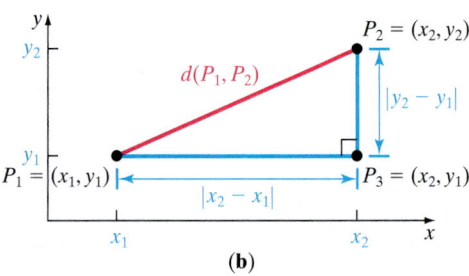

(a)   (b)

The distance $d(P_1, P_2)$ that we seek is the length of the hypotenuse of the right triangle, so by the Pythagorean Theorem, it follows that

$$[d(P_1, P_2)]^2 = |x_2 - x_1|^2 + |y_2 - y_1|^2$$
$$= (x_2 - x_1)^2 + (y_2 - y_1)^2$$

Take the square root of both sides: $d(P_1, P_2) = \sqrt{(x_2 - x_1)^2 + (y_2 - y_1)^2}$

If the line joining P_1 and P_2 is horizontal, then the y-coordinate of P_1 equals the y-coordinate of P_2; that is, $y_1 = y_2$. See Figure 4(a). In this case, the distance formula still works, because, for $y_1 = y_2$, it becomes

$$d(P_1, P_2) = \sqrt{(x_2 - x_1)^2 + 0^2}$$
$$= \sqrt{(x_2 - x_1)^2} = |x_2 - x_1|$$

A similar argument holds if the line joining P_1 and P_2 is vertical. See Figure 4(b). The distance formula works in all cases.

Figure 4

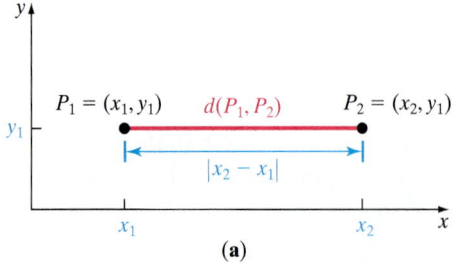 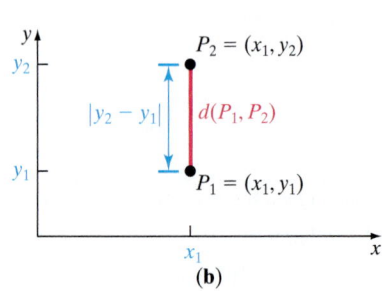

(a)   (b)

Figure 5

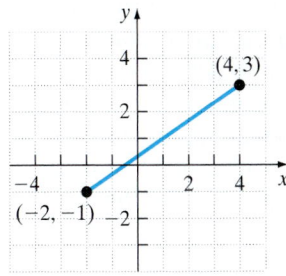

EXAMPLE 2   Finding the Length of a Line Segment

Find the length of the line segment shown in Figure 5. That is, find the distance between the points in Figure 5.

Solution

The length of the line segment is the distance between the points $(-2, -1)$ and $(4, 3)$. Using the distance formula with $P_1 = (x_1, y_1) = (-2, -1)$ and $P_2 = (x_2, y_2) = (4, 3)$, the length d is

$$d = \sqrt{(x_2 - x_1)^2 + (y_2 - y_1)^2}$$

$x_1 = -2, y_1 = -1, x_2 = 4, y_2 = 3: \quad = \sqrt{(4 - (-2))^2 + (3 - (-1))^2}$

$$= \sqrt{6^2 + 4^2}$$

$$= \sqrt{36 + 16}$$

$$= \sqrt{52}$$

$$= 2\sqrt{13} \approx 7.21$$

QUICK ✓ *Find the distance between the points.*

1. $(3, 8)$ and $(0, 4)$ **2.** $(-2, -5)$ and $(4, 7)$

The distance between two points $P_1 = (x_1, y_1)$ and $P_2 = (x_2, y_2)$ is never a negative number. In addition, the distance between two points is 0 only when the two points are identical, that is, when $x_1 = x_2$ and $y_1 = y_2$. Also, because $(x_2 - x_1)^2 = (x_1 - x_2)^2$ and $(y_2 - y_1)^2 = (y_1 - y_2)^2$, it does not matter whether we compute the distance from P_1 to P_2 or from P_2 to P_1. This should seem reasonable since the distance from P_1 to P_2 equals the distance from P_2 to P_1.

The next example shows how algebra (the distance formula) can be used to solve geometry problems.

EXAMPLE 3   Using Algebra to Solve Geometry Problems

Consider the three points $A = (-1, 1)$, $B = (2, -2)$, and $C = (3, 5)$.

(a) Plot each point in the Cartesian plane and form the triangle ABC.

(b) Find the length of each side of the triangle.

(c) Verify that the triangle is a right triangle.

(d) Find the area of the triangle.

Solution

(a) We plot points A, B, and C in Figure 6.

(b) We use the distance formula to find the length of each side of the triangle.

Figure 6

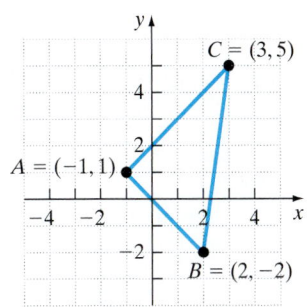

$d(A, B) = \sqrt{(2 - (-1))^2 + (-2 - 1)^2} = \sqrt{3^2 + (-3)^2}$

$\qquad = \sqrt{9 + 9} = \sqrt{18} = 3\sqrt{2}$

$d(A, C) = \sqrt{(3 - (-1))^2 + (5 - 1)^2} = \sqrt{4^2 + (-4)^2}$

$\qquad = \sqrt{16 + 16} = \sqrt{32} = 4\sqrt{2}$

$d(B, C) = \sqrt{(3 - 2)^2 + (5 - (-2))^2} = \sqrt{1^2 + 7^2} = \sqrt{1 + 49} = \sqrt{50} = 5\sqrt{2}$

(c) The Pythagorean Theorem says that if you have a right triangle, the sum of squares of the two legs will equal the square of the hypotenuse. The converse of this statement is true as well. That is, if the sum of squares of two sides of a triangle equals the square of the third side, then the triangle is a right triangle. So we need to show that the sum of squares of two sides of the triangle in Figure 6 equals the square of the third side. In looking at Figure 6, the right angle appears to be at vertex A, which means the side opposite vertex A should be the hypotenuse of the triangle. So we want to show that

$$[d(B, C)]^2 = [d(A, B)]^2 + [d(A, C)]^2$$

We know each of these distances from part (b), so

$$\left(5\sqrt{2}\right)^2 \overset{?}{=} \left(3\sqrt{2}\right)^2 + \left(4\sqrt{2}\right)^2$$

$(ab)^n = a^n \cdot b^n: \quad 25 \cdot 2 \overset{?}{=} 9 \cdot 2 + 16 \cdot 2$

$$50 = 18 + 32$$

$$50 = 50 \quad \text{True}$$

Since $[d(B, C)]^2 = [d(A, B)]^2 + [d(A, C)]^2$, we know that triangle ABC is a right triangle.

(d) The area of a triangle is $^1\!/_2$ times the product of the base and the height. From Figure 6, we will say that side AB forms the base and side AC forms the height. The length of side AB is $3\sqrt{2}$ and the length of side AC is $4\sqrt{2}$. So the area of triangle ABC is

$$\text{Area} = \frac{1}{2}(\text{Base})(\text{Height}) = \frac{1}{2}\left(3\sqrt{2}\right)\left(4\sqrt{2}\right) = 12 \text{ square units} \quad \blacksquare$$

QUICK ✓

3. Consider the three points $A = (-2, -1)$, $B = (4, 2)$, and $C = (0, 10)$.

(a) Plot each point in the Cartesian plane and form the triangle ABC.

(b) Find the length of each side of the triangle.

(c) Verify that the triangle is a right triangle.

(d) Find the area of the triangle.

---

② ## Use the Midpoint Formula

Suppose we had two points $P_1 = (x_1, y_1)$ and $P_2 = (x_2, y_2)$ in the Cartesian plane. Further suppose we wanted to find the point $M = (x, y)$ that is the same distance to each of these two points so that $d(P_1, M) = d(M, P_2)$. We can find this point M using the **midpoint formula.**

In Words
To find the midpoint of a line segment, average the x-coordinates and average the y-coordinates of the endpoints.

MIDPOINT FORMULA

The midpoint $M = (x, y)$ of the line segment from $P_1 = (x_1, y_1)$ to $P_2 = (x_2, y_2)$ is

$$M = \left(\frac{x_1 + x_2}{2}, \frac{y_1 + y_2}{2}\right)$$

EXAMPLE 4   **Finding the Midpoint of a Line Segment**

Find the midpoint of a line segment joining $P_1 = (-2, 3)$ and $P_2 = (4, 7)$. Plot the points P_1 and P_2 and their midpoint. Check your answer.

Solution

We substitute $x_1 = -2$, $y_1 = 3$, $x_2 = 4$, and $y_2 = 7$ into the midpoint formula. The coordinates (x, y) of the midpoint M are

$$x = \frac{x_1 + x_2}{2} = \frac{-2 + 4}{2} = \frac{2}{2} = 1$$

and

$$y = \frac{y_1 + y_2}{2} = \frac{3 + 7}{2} = \frac{10}{2} = 5$$

Figure 7

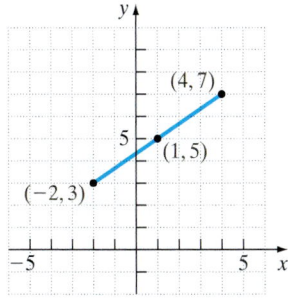

So the midpoint of the line segment joining $P_1 = (-2, 3)$ and $P_2 = (4, 7)$ is $M = (1, 5)$. See Figure 7.

Check We can check our solution by verifying that the distance from P_1 to M is equal to the distance from M to P_2.

$$d(P_1, M) = \sqrt{(1 - (-2))^2 + (5 - 3)^2} = \sqrt{3^2 + 2^2} = \sqrt{13}$$

$$d(M, P_2) = \sqrt{(4 - 1)^2 + (7 - 5)^2} = \sqrt{3^2 + 2^2} = \sqrt{13}$$

It checks!

QUICK ✓ *Find the midpoint of the line segment joining the points.*

4. $(3, 8)$ and $(0, 4)$

5. $(-2, -5)$ and $(4, 10)$

12.1 Exercises

For Extra Help:

Student Solutions Manual   CD Video   PH Math/Tutor Center   MathXL Tutorials on CD   MathXL®   MyMathLab

Concepts and Vocabulary

In Problems 1 and 2, fill in the blanks.

1. The distance between two points $P_1 = (x_1, y_1)$ and $P_2 = (x_2, y_2)$, denoted by $d(P_1, P_2)$, is _____.

2. The midpoint $M = (x, y)$ of the line segment from $P_1 = (x_1, y_1)$ to $P_2 = (x_2, y_2)$ is _____.

In Problems 3 and 4, answer True or False to each statement.

3. The distance between two points is sometimes a negative number.

4. The midpoint M of a line segment joining P_1 and P_2 is the point such that $d(P_1, M) = d(M, P_2)$.

5. How is the distance formula related to the Pythagorean Theorem?

6. How can the distance formula be used to verify that a point is the midpoint of a line segment?

Building Skills

In Problems 7–22, find the distance $d(P_1, P_2)$ between the points P_1 and P_2.

7.

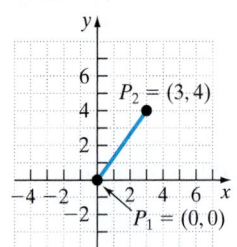

8.

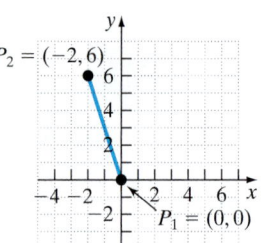

9.

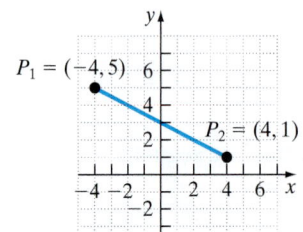

10.

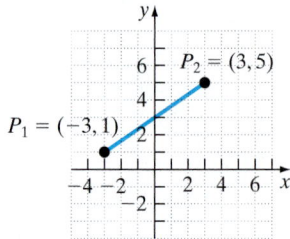

11. $P_1 = (2, 1)$; $P_2 = (6, 4)$ **12.** $P_1 = (1, 3)$; $P_2 = (4, 7)$

13. $P_1 = (-3, 2)$; $P_2 = (9, -3)$ **14.** $P_1 = (-10, -3)$; $P_2 = (14, 4)$

15. $P_1 = (-4, 2)$; $P_2 = (2, 2)$ **16.** $P_1 = (-1, 2)$; $P_2 = (-1, 0)$

17. $P_1 = (0, -3)$; $P_2 = (-3, 3)$ **18.** $P_1 = (5, 0)$; $P_2 = (-1, -4)$

19. $P_1 = \left(2\sqrt{2},\ \sqrt{5}\right)$; $P_2 = \left(5\sqrt{2}, 4\sqrt{5}\right)$

20. $P_1 = \left(\sqrt{6}, -2\sqrt{2}\right)$; $P_2 = \left(3\sqrt{6}, 10\sqrt{2}\right)$

21. $P_1 = (0.3, -3.3)$; $P_2 = (1.3, 0.1)$ **22.** $P_1 = (-1.7, 1.3)$; $P_2 = (0.3, 2.6)$

In Problems 23–34, find the midpoint of the line segment formed by joining the points P_1 and P_2.

23. $P_1 = (2, 2)$; $P_2 = (6, 4)$ **24.** $P_1 = (1, 3)$; $P_2 = (5, 7)$

25. $P_1 = (-3, 2)$; $P_2 = (9, -4)$ **26.** $P_1 = (-10, -3)$; $P_2 = (14, 7)$

27. $P_1 = (-4, 3)$; $P_2 = (2, 4)$ **28.** $P_1 = (-1, 2)$; $P_2 = (3, 9)$

29. $P_1 = (0, -3)$; $P_2 = (-3, 3)$ **30.** $P_1 = (5, 0)$; $P_2 = (-1, -4)$

31. $P_1 = \left(2\sqrt{2},\ \sqrt{5}\right)$; $P_2 = \left(5\sqrt{2}, 4\sqrt{5}\right)$

32. $P_1 = \left(\sqrt{6}, -2\sqrt{2}\right)$; $P_2 = \left(3\sqrt{6}, 10\sqrt{2}\right)$

33. $P_1 = (0.3, -3.3)$; $P_2 = (1.3, 0.1)$ **34.** $P_1 = (-1.7, 1.3)$; $P_2 = (0.3, 2.6)$

Applying the Concepts

35. Consider the three points $A = (0, 3)$, $B = (2, 1)$, $C = (6, 5)$.

    (a) Plot each point in the Cartesian plane and form the triangle ABC.
    (b) Find the length of each side of the triangle.
    (c) Verify that the triangle is a right triangle.
    (d) Find the area of the triangle.

36. Consider the three points $A = (0, 2)$, $B = (1, 4)$, $C = (4, 0)$.

    (a) Plot each point in the Cartesian plane and form the triangle ABC.
    (b) Find the length of each side of the triangle.
    (c) Verify that the triangle is a right triangle.
    (d) Find the area of the triangle.

37. Consider the three points $A = (-2, -4)$, $B = (3, 1)$, and $C = (15, -11)$.

    (a) Plot each point in the Cartesian plane and form the triangle ABC.
    (b) Find the length of each side of the triangle.
    (c) Verify that the triangle is a right triangle.
    (d) Find the area of the triangle.

38. Consider the three points $A = (-2, 3)$, $B = (2, 0)$, and $C = (5, 4)$.

   (a) Plot each point in the Cartesian plane and form the triangle ABC.

   (b) Find the length of each side of the triangle.

   (c) Verify that the triangle is a right triangle.

   (d) Find the area of the triangle.

39. Find all points having an x-coordinate of 2 whose distance from the point $(5, 1)$ is 5.

40. Find all points having an x-coordinate of 4 whose distance from the point $(0, 3)$ is 5.

41. Find all points having a y-coordinate of -3 whose distance from the point $(2, 3)$ is 10.

42. Find all points having a y-coordinate of -3 whose distance from the point $(-4, 2)$ is 13.

43. The City of Chicago The city of Chicago's road system is set up like a Cartesian plane, where streets are indicated by the number of blocks they are from Madison Street and State Street. For example, Wrigley Field in Chicago is located at 1060 West Addison, which is 10 blocks west of State Street and 36 blocks north of Madison Street.

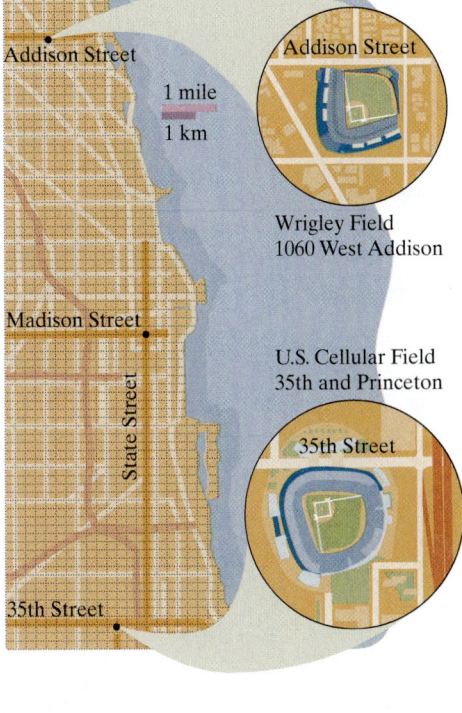

City of Chicago, Illinois

   (a) Find the distance "as the crow flies" from Madison and State Street to Wrigley Field. Use city blocks as the unit of measurement.

   (b) U.S. Cellular Field, home of the White Sox, is located at 35th and Princeton, which is 3 blocks west of State Street and 35 blocks south of Madison. Find the distance "as the crow flies" from Madison and State Street to U.S. Cellular Field.

   (c) Find the distance "as the crow flies" from Wrigley Field to U.S. Cellular Field.

44. Baseball A major league baseball "diamond" is actually a square, 90 feet on a side (see the figure). Overlay a Cartesian plane on a major league baseball diamond, so that the origin is at home plate, the positive x-axis lies in the direction from home plate to first base, and the positive y-axis lies in the direction from home plate to third base.

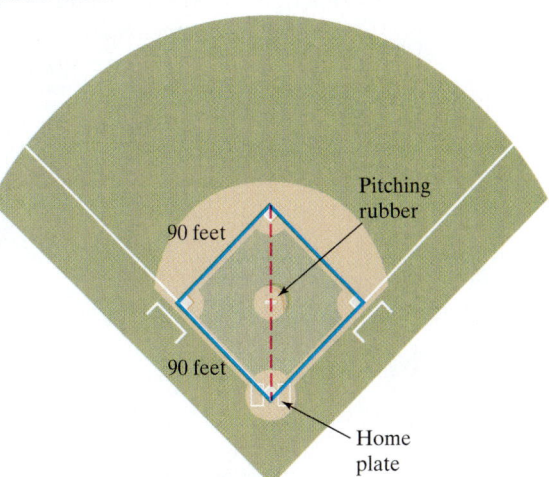

(a) What are the coordinates of home plate, first base, second base, and third base? Use feet as the unit of measurement.

(b) Suppose the center fielder is located at $(310, 260)$. How far is he from second base?

(c) Suppose the shortstop is located at $(60, 100)$. How far is he from second base?

Extending the Concepts

45. Baseball Refer to Problem 44.

(a) Suppose the right fielder catches a fly ball at $(320, 20)$. How many seconds will it take to throw the ball to second base if he can throw 130 feet per second? (**Hint:** time = distance divided by speed.)

(b) Suppose a runner "tagging up" from first base can run 27 feet per second. Would you "send the runner" as the first base coach if the right fielder requires 0.8 second to catch and throw? Why?

46. Let $P = (x, y)$ be a point on the graph of $y = x^2 - 4$.

(a) Express the distance d from P to the point $(1, 2)$ as a function of x using the distance formula. See the figure for an illustration of the problem.

(b) What is d if $x = 0$?

(c) What is d if $x = 3$?

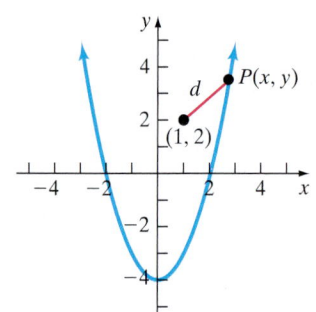

Synthesis Review

47. Evaluate 3^2. What is $\sqrt{9}$?

48. Evaluate 8^2. What is $\sqrt{64}$?

49. Evaluate $(-3)^4$. What is $\sqrt[4]{81}$?

50. Evaluate $(-3)^3$. What is $\sqrt[3]{-27}$?

51. Describe the relationship between raising a number to a positive integer power, n, and the nth root of a number.

12.2 Circles

OBJECTIVES

1. Write the Standard Form of the Equation of a Circle
2. Graph a Circle
3. Find the Center and Radius of a Circle From an Equation in General Form

Preparing for Circles

Before getting started, take the following readiness quiz. If you get the problem wrong, go back to the section cited and review the material.

1. Complete the square in x: $x^2 - 8x$ [Section 10.1, pp. 747–748]

Conics, an abbreviation for **conic sections,** are curves that result from the intersection of a right circular cone and a plane. The four conics that we study are shown in Figure 8. These conics are *circles* (Figure 8(a)); *ellipses* (Figure 8(b)); *parabolas* (Figure 8(c)); and *hyperbolas* (Figure 8(d)).

Figure 8

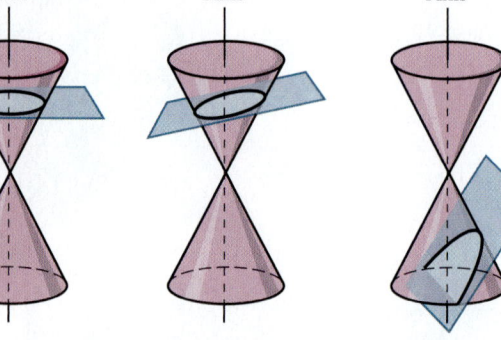

(a) Circle   (b) Ellipse   (c) Parabola   (d) Hyperbola

Preparing for...Answer

1. $x^2 - 8x + 16$

We study circles in this section, parabolas in Section 12.3, ellipses in Section 12.4, and hyperbolas in Section 12.5.

① Write the Standard Form of the Equation of a Circle

One advantage of a coordinate system (the Cartesian plane) is that it enables us to translate a geometric statement into an algebraic statement, and vice versa. Consider the following geometric statement that defines a circle.

Figure 9

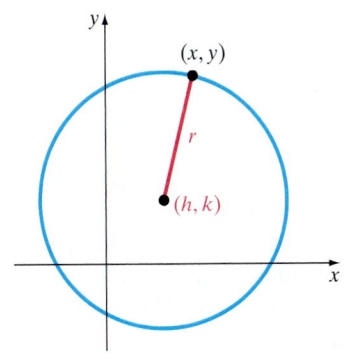

> **DEFINITION**
>
> A **circle** is the set of all points in the Cartesian plane that are a fixed distance r from a fixed point (h, k). The fixed distance r is called the **radius,** and the fixed point (h, k) is called the **center** of the circle. See Figure 9.

To find an equation that has this graph, we let (x, y) represent the coordinates of any point on a circle with radius r and center (h, k). Then the distance between the points (x, y) and (h, k) must always equal r. That is, by the distance formula,

$$\sqrt{(x - h)^2 + (y - k)^2} = r$$

If we square both sides of the equation, then

$$(x - h)^2 + (y - k)^2 = r^2$$

and we have the equation of a circle.

> **DEFINITION**
>
> The **standard form of an equation of a circle** with radius r and center (h, k) is
>
> $$(x - h)^2 + (y - k)^2 = r^2$$

EXAMPLE 1 Writing the Standard Form of the Equation of a Circle

Write the standard form of the equation of the circle with radius 4 and center $(2, -3)$.

Solution

We use the equation $(x - h)^2 + (y - k)^2 = r^2$ with $r = 4$, $h = 2$, $k = -3$ and obtain

$$(x - 2)^2 + (y - (-3))^2 = 4^2$$
$$(x - 2)^2 + (y + 3)^2 = 16$$

QUICK ✓ *Write the standard form of the equation of each circle whose radius is r and center is (h, k).*

1. $r = 5$; $(h, k) = (2, 4)$

2. $r = \sqrt{2}$; $(h, k) = (-2, 0)$

② Graph a Circle

The graph of any equation of the form $(x - h)^2 + (y - k)^2 = r^2$ is that of a circle with radius r and center (h, k).

EXAMPLE 2 **Graphing a Circle**

Graph the equation: $(x + 2)^2 + (y - 3)^2 = 9$

Solution

The graph of the equation is a circle because the equation is of the form $(x - h)^2 + (y - k)^2 = r^2$. To graph the equation, we first identify the center and radius of the circle by comparing the given equation to the standard form of the equation of a circle.

$$(x + 2)^2 + (y - 3)^2 = 9$$

$$(x - (-2))^2 + (y - 3)^2 = 3^2$$

$$(x - h)^2 + (y - k)^2 = r^2$$

We see that $h = -2$, $k = 3$, and $r = 3$. The circle has center $(-2, 3)$ and radius of 3 units. To graph this circle, we first plot the center $(-2, 3)$. Since the radius is 3 units, we can go 3 units in any direction from the center and find a point on the graph of the circle. It is easiest to find the four points left, right, up, and down from the center. These four points are $(-5, 3)$, $(1, 3)$, $(-2, 6)$, and $(-2, 0)$, respectively. We plot these points in Figure 10(a). We then use these points as guides to draw the graph of the circle shown in Figure 10(b).

Figure 10

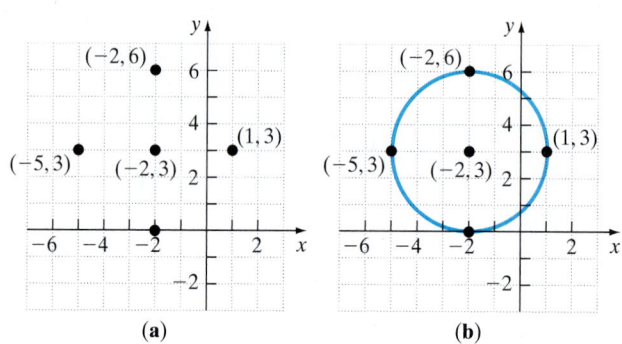

(a)   (b)

QUICK ✓ *Graph the circle.*

3. $(x - 3)^2 + (y - 1)^2 = 4$ **4.** $(x + 5)^2 + y^2 = 16$

---

(3) **Find the Center and Radius of a Circle from an Equation in General Form**

If we eliminate the parentheses from the standard form of the equation of the circle given in Example 2, we get

$$(x + 2)^2 + (y - 3)^2 = 9$$

FOIL: $x^2 + 4x + 4 + y^2 - 6y + 9 = 9$

Subtract 9 from both sides: $x^2 + y^2 + 4x - 6y + 4 = 0$

Any equation of the form

$$x^2 + y^2 + ax + by + c = 0$$

has a graph that is a circle, a point, or no graph at all. For example, the graph of the equation $x^2 + y^2 = 0$ is the single point $(0, 0)$. The equation $x^2 + y^2 + 4 = 0$, or $x^2 + y^2 = -4$ has no graph, because the sum of squares of real numbers are never negative.

In Words

The standard form of a circle is $(x - h)^2 + (y - k)^2 = r^2$. The general form is $x^2 + y^2 + ax + by + c = 0$.

DEFINITION

The **general form of the equation of a circle** is given by the equation

$$x^2 + y^2 + ax + by + c = 0$$

when the graph exists.

If an equation of a circle is in general form, we use the method of completing the square to put the equation in standard form, $(x - h)^2 + (y - k)^2 = r^2$, so that we can identify its center and radius.

EXAMPLE 3   **Graphing a Circle Whose Equation Is in General Form**

Graph the equation: $x^2 + y^2 + 8x - 2y - 8 = 0$

Solution

To determine the center and radius of the circle, we first need to put the equation in standard form by completing the square in both x and y (covered in Section 10.1). To do this, we group the terms involving x, group the terms involving y, and put the constant on the right side of the equation by adding 8 to both sides.

$$(x^2 + 8x) + (y^2 - 2y) = 8$$

Now we complete the square of each expression in parentheses. Remember, any number added to the left side of the equation must also be added to the right.

Figure 11

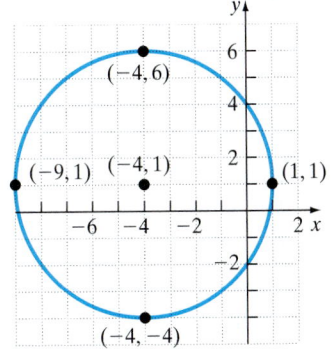

$$(x^2 + 8x + 16) + (y^2 - 2y + 1) = 8 + 16 + 1$$

$$\left(\frac{1}{2} \cdot 8\right)^2 = 16 \qquad \left(\frac{1}{2} \cdot (-2)\right)^2 = 1$$

Factor: $(x + 4)^2 + (y - 1)^2 = 25$

The equation is now in the standard form of a circle. The center of the circle is $(-4, 1)$ and its radius is 5. To graph the equation, we plot the center and the four points to the left, to the right, above, and below the center. Then trace in the circle. See Figure 11.

QUICK ✓   *Graph each circle.*

5. $x^2 + y^2 - 6x - 4y + 4 = 0$ **6.** $2x^2 + 2y^2 - 16x + 4y - 38 = 0$

12.2 Exercises

For Extra Help:

Student Solutions Manual   CD Video   PH Math/Tutor Center   MathXL Tutorials on CD   MathXL®   MyMathLab

Concepts and Vocabulary

In Problems 1 and 2, fill in the blanks.

1. A _____ is the set of all points in the Cartesian plane that are a fixed distance r from a fixed point (h, k).

2. For a circle, the _____ is the distance from the center to any point on the circle.

In Problems 3 and 4, answer True or False to each statement.

3. The center of the circle $(x + 1)^2 + (y - 3)^2 = 25$ is $(1, -3)$.

4. The center of the circle $x^2 + y^2 = 9$ is $(0, 0)$; its radius is 3.

5. How is the distance formula related to the definition of a circle?

6. Are circles functions? Why or why not?

7. Is $x^2 = 36 - y^2$ the equation of a circle? If so, what is the center and radius?

8. Is $3x^2 - 12x + 3y^2 - 15 = 0$ the equation of a circle? If so, what is the center and radius?

Building Skills

In Problems 9–12, find the center and radius of each circle. Write the standard form of the equation.

9.

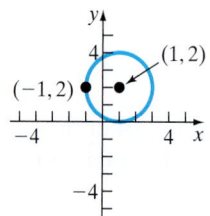

10.

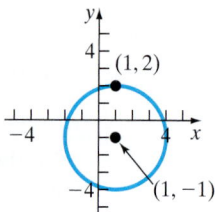

11.

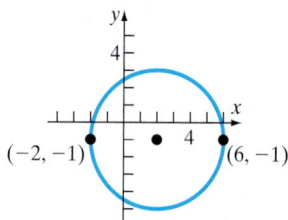

12.

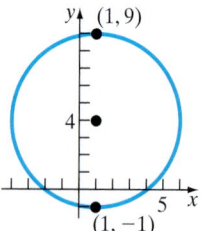

In Problems 13–24, write the standard form of the equation of each circle whose radius is r and center is (h, k). Graph each circle.

13. $r = 3$; $(h, k) = (0, 0)$

14. $r = 5$; $(h, k) = (0, 0)$

15. $r = 2$; $(h, k) = (1, 4)$

16. $r = 4$; $(h, k) = (3, 1)$

17. $r = 6$; $(h, k) = (-2, 4)$

18. $r = 3$; $(h, k) = (1, -4)$

19. $r = 4$; $(h, k) = (0, 3)$

20. $r = 2$; $(h, k) = (1, 0)$

21. $r = 5$; $(h, k) = (5, -5)$

22. $r = 4$; $(h, k) = (-4, 4)$

23. $r = \sqrt{5}$; $(h, k) = (1, 2)$

24. $r = \sqrt{7}$; $(h, k) = (5, 2)$

In Problems 25–34, find the center (h, k) and radius r of each circle. Graph each circle.

25. $x^2 + y^2 = 36$

26. $x^2 + y^2 = 144$

27. $(x - 4)^2 + (y - 1)^2 = 25$

28. $(x - 2)^2 + (y - 3)^2 = 9$

29. $(x + 3)^2 + (y - 2)^2 = 81$

30. $(x - 5)^2 + (y + 2)^2 = 49$

31. $x^2 + (y - 3)^2 = 64$

32. $(x - 6)^2 + y^2 = 36$

33. $(x - 1)^2 + (y + 1)^2 = \dfrac{1}{4}$

34. $(x - 2)^2 + (y + 2)^2 = \dfrac{1}{4}$

In Problems 35–40, find the center (h, k) and radius r of each circle. Graph each circle.

35. $x^2 + y^2 - 6x + 2y + 1 = 0$

36. $x^2 + y^2 + 2x - 8y + 8 = 0$

37. $x^2 + y^2 + 10x + 4y + 4 = 0$

38. $x^2 + y^2 + 4x - 12y + 36 = 0$

39. $2x^2 + 2y^2 - 12x + 24y - 72 = 0$

40. $2x^2 + 2y^2 - 28x + 20y + 20 = 0$

In Problems 41–46, find the standard form of the equation of each circle.

41. Center at the origin and containing the point $(4, -2)$.

42. Center at $(0, 3)$ and containing the point $(3, 7)$.

43. Center at $(-3, 2)$ and tangent to the y-axis.

44. Center at $(2, -3)$ and tangent to the x-axis.

45. With endpoints of a diameter at $(2, 3)$ and $(-4, -5)$.

46. With endpoints of a diameter at $(-5, -3)$ and $(7, 2)$.

Applying the Concepts

△ **47.** Find the area and circumference of the circle $(x - 3)^2 + (y - 8)^2 = 64$.

△ **48.** Find the area and circumference of the circle $(x - 1)^2 + (y - 4)^2 = 49$.

△ **49.** Find the area of the square in the figure.

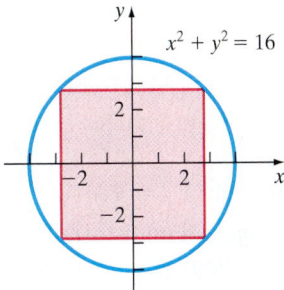

△ **50.** Find the area of the shaded region in the figure.

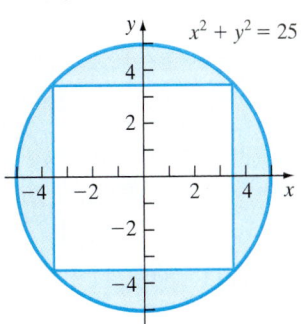

Extending the Concepts

51. Which of the following equations might have the graph shown? (More than one answer is possible.)

  (a) $(x - 2)^2 + y^2 = 1$

  (b) $x^2 + (y - 2)^2 = 1$

  (c) $(x + 4)^2 + y^2 = 9$

  (d) $(x - 5)^2 + y^2 = 25$

  (e) $x^2 + y^2 - 8x + 7 = 0$

  (f) $x^2 + y^2 + 10x + 18 = 0$

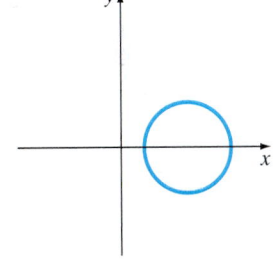

52. Which of the following equations might have the graph shown? (More than one answer is possible.)

  (a) $(x - 2)^2 + (y + 3)^2 = 4$

  (b) $(x - 3)^2 + (y - 4)^2 = 4$

  (c) $(x + 3)^2 + (y + 4)^2 = 9$

  (d) $(x - 5)^2 + (y - 5)^2 = 25$

  (e) $x^2 + y^2 + 8x + 10y + 32 = 0$

  (f) $x^2 + y^2 - 4x - 6y - 3 = 0$

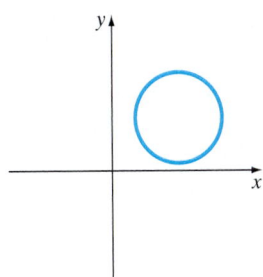

Synthesis Review

In Problems 53–57, graph each function using either point plotting or the properties of the function. For example, to graph a line, we would use the slope and y-intercept or the intercepts.

53. $f(x) = 4x - 3$

54. $2x + 5y = 20$

55. $g(x) = x^2 - 4x - 5$

56. $F(x) = -3x^2 + 12x - 12$

57. $G(x) = -2(x + 3)^2 - 5$

58. Present an argument that supports the approach you took to graphing each function in Problems 53–57. For example, in Problem 53 did you use point plotting? Intercepts? Slope? Which method did you use and why?

The Graphing Calculator

To graph a circle using a graphing calculator requires that we first solve the equation for y. For example, to graph $(x + 2)^2 + (y - 3)^2 = 9$, we solve for y as follows:

$$(x + 2)^2 + (y - 3)^2 = 9$$

Subtract $(x + 2)^2$ from both sides: $\quad (y - 3)^2 = 9 - (x + 2)^2$

Take the square root of both sides: $\quad y - 3 = \pm\sqrt{9 - (x + 2)^2}$

Add 3 to both sides: $\quad y = 3 \pm \sqrt{9 - (x + 2)^2}$

We graph the top half $Y_1 = 3 + \sqrt{9 - (x + 2)^2}$ and the bottom half $Y_2 = 3 - \sqrt{9 - (x + 2)^2}$. To get an undistorted view of the graph, be sure to use the ZOOM SQUARE feature on your graphing calculator. The figure below shows the graph of $(x + 2)^2 + (y - 3)^2 = 9$.

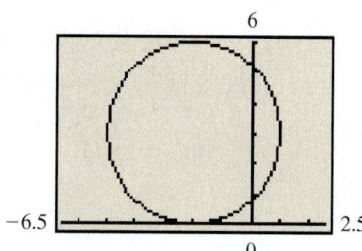

In Problems 59–68, graph each circle using a graphing calculator. Compare your graphs to the graphs drawn by hand in Problems 25–34.

59. $x^2 + y^2 = 36$

60. $x^2 + y^2 = 144$

61. $(x - 4)^2 + (y - 1)^2 = 25$

62. $(x - 2)^2 + (y - 3)^2 = 9$

63. $(x + 3)^2 + (y - 2)^2 = 81$

64. $(x - 5)^2 + (y + 2)^2 = 49$

65. $x^2 + (y - 3)^2 = 64$

66. $(x - 6)^2 + y^2 = 36$

67. $(x - 1)^2 + (y + 1)^2 = \dfrac{1}{4}$

68. $(x - 2)^2 + (y + 2)^2 = \dfrac{1}{4}$

12.3 Parabolas

OBJECTIVES

1. Graph Parabolas in Which the Vertex Is the Origin
2. Find the Equation of a Parabola
3. Graph Parabolas in Which the Vertex Is Not the Origin
4. Solve Applied Problems Involving Parabolas

Preparing for Parabolas

Before getting started, take the following readiness quiz. If you get a problem wrong, go back to the section cited and review the material.

1. Identify the vertex and axis of symmetry of $f(x) = -3(x + 4)^2 - 5$. Does the parabola open up or down? Why?   [Section 10.4, pp. 787–790]

2. Identify the vertex and axis of symmetry of the quadratic function $f(x) = 2x^2 - 8x + 1$. Does the parabola open up or down? Why?   [Section 10.5, pp. 795–801]

3. Complete the square of $x^2 - 12x$.   [Section 10.1, pp. 747–748]

4. Solve: $(x - 3)^2 = 25$ [Section 10.1, pp. 744–747]

We began a discussion of parabolas back in Section 10.4 when we studied quadratic functions. To refresh your memory, we include a summary of this information below.

SUMMARY: Parabolas That Open Up or Down

The graph of $y = a(x - h)^2 + k$ or $y = ax^2 + bx + c$ is a parabola that

1. opens up if $a > 0$ and opens down if $a < 0$.

2. has vertex (h, k) if the equation is of the form $y = a(x - h)^2 + k$.

3. has a vertex whose x-coordinate is $x = -\dfrac{b}{2a}$. The y-coordinate is found by evaluating the equation at the x-coordinate of the vertex.

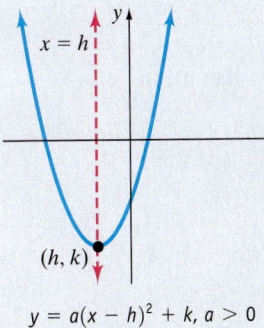
$$y = a(x - h)^2 + k, \; a > 0$$

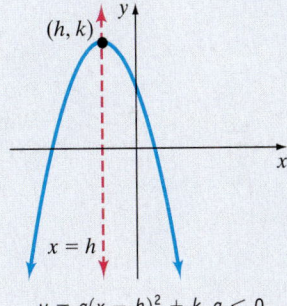
$$y = a(x - h)^2 + k, \; a < 0$$

The presentation of parabolas given in Sections 10.4 and 10.5 relied more on algebra. In this section, we are going to look at the parabola (and the other conic sections) from a geometric point of view.

Figure 12

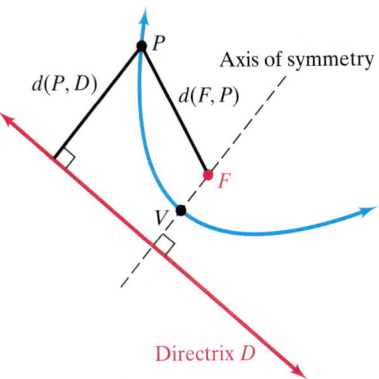

DEFINITION

A **parabola** is the collection of all points P in the plane that are the same distance from a fixed point F as they are from a fixed line D. The point F is called the **focus** of the parabola, and the line D is its **directrix**. In other words, a parabola is the set of points P for which

$$d(F, P) = d(P, D)$$

Figure 12 shows a parabola. The line through the focus F and perpendicular to the directrix D is called the **axis of symmetry** of the parabola. The point of intersection of the parabola with its axis of symmetry is called the **vertex** V.

Preparing for...Answers **1.** Vertex: $(-4, -5)$; axis of symmetry: $x = -4$; opens down since $a = -3 < 0$.
2. Vertex: $(2, -7)$; axis of symmetry: $x = 2$; opens up since $a = 2 > 0$.
3. $x^2 - 12x + 36$ **4.** $\{-2, 8\}$

1 Graph Parabolas in Which the Vertex Is the Origin

We want to develop an equation for the parabola based on the definition just given. For example, because the vertex is on the graph of the parabola, it must satisfy the definition that the distance from the focus to the vertex will equal the distance from the vertex to the directrix. Let a represent the distance from the focus to the vertex. To develop an equation for a parabola, we start by looking at parabolas whose vertex is at the origin. We consider four possibilities—parabolas that open left, parabolas that open right, parabolas that open up, and parabolas that open down.

Let's see how to obtain the equation of a parabola whose vertex is at the origin and opens right. To do this, set up a Cartesian plane and position the parabola so that its vertex is at the origin, $(0, 0)$, the focus is on the positive x-axis and the directrix is a vertical line in quadrants II and III. Because a represents the distance from the vertex to the focus, we have that the focus is located at $(a, 0)$. Also, because the distance from the vertex to the directrix is a units, the directrix is the line $x = -a$. See Figure 13. If $P = (x, y)$ is any point on the parabola, then the distance from P to the focus F, $(a, 0)$, must equal the distance from P to the directrix, $x = -a$. That is,

Figure 13

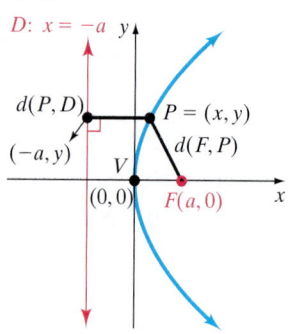

$$d(F, P) = d(P, D)$$

Use the distance formula: $\sqrt{(x - a)^2 + (y - 0)^2} = \sqrt{(x - (-a))^2 + (y - y)^2}$

Square both sides; simplify: $(x - a)^2 + y^2 = (x + a)^2$

Multiply out binomials: $x^2 - 2ax + a^2 + y^2 = x^2 + 2ax + a^2$

Combine like terms: $y^2 = 4ax$

So the equation of the parabola whose vertex is at the origin and opens to the right is $y^2 = 4ax$. We obtain the equations of the three other possibilities (opens left, opens up, or opens down) using logic similar to the logic we used to obtain the equation above. We summarize these possibilities in Table 1.

Table 1 Equations of a Parabola: Vertex at $(0, 0)$; Focus on an Axis; $a > 0$				
Vertex	**Focus**	**Directrix**	**Equation**	**Description**
$(0, 0)$	$(a, 0)$	$x = -a$	$y^2 = 4ax$	Parabola, axis of symmetry is the x-axis, opens to the right
$(0, 0)$	$(-a, 0)$	$x = a$	$y^2 = -4ax$	Parabola, axis of symmetry is the x-axis, opens to the left
$(0, 0)$	$(0, a)$	$y = -a$	$x^2 = 4ay$	Parabola, axis of symmetry is the y-axis, opens up
$(0, 0)$	$(0, -a)$	$y = a$	$x^2 = -4ay$	Parabola, axis of symmetry is the y-axis, opens down

The graphs of the four parabolas are given in Figure 14.

Figure 14

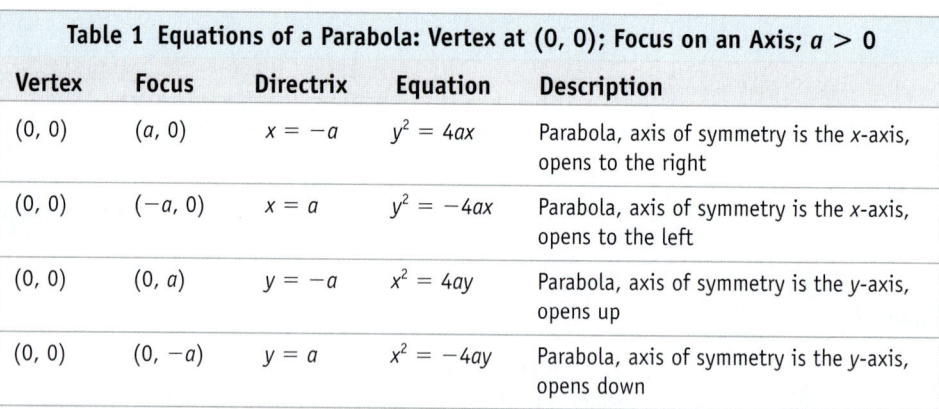
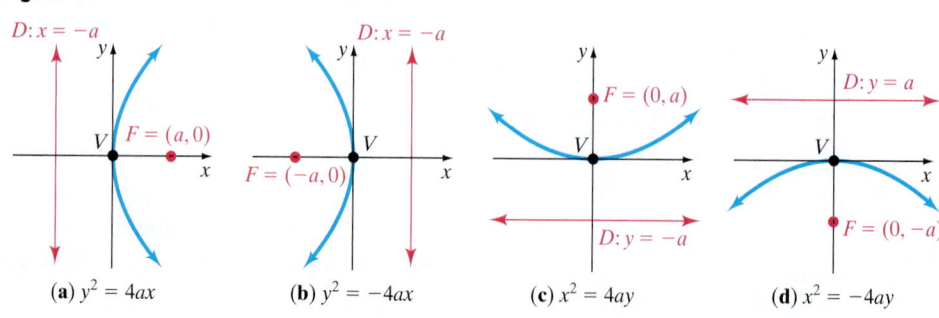

(a) $y^2 = 4ax$ **(b)** $y^2 = -4ax$ **(c)** $x^2 = 4ay$ **(d)** $x^2 = -4ay$

EXAMPLE 1 Graphing a Parabola That Opens Left

Graph the equation $y^2 = -12x$.

Solution

The equation $y^2 = -12x$ is of the form $y^2 = -4ax$, where $-4a = -12$, so that $a = 3$. The graph of the equation is a parabola with vertex at $(0, 0)$ and focus at $(-a, 0) = (-3, 0)$ so that the parabola opens to the left. The directrix is the line $x = 3$. To graph the parabola, it is helpful to plot the two points on the graph above and below the focus. Because the points are directly above and below the focus, we let $x = -3$ in the equation $y^2 = -12x$ and solve for y.

$$y^2 = -12x$$
$$\text{Let } x = -3: \quad = -12(-3)$$
$$= 36$$
$$\text{Take the square root of both sides:} \quad y = \pm 6$$

The points on the parabola above and below the focus are $(-3, 6)$ and $(-3, -6)$. These points help in graphing the parabola because they determine the "opening." See Figure 15.

Figure 15

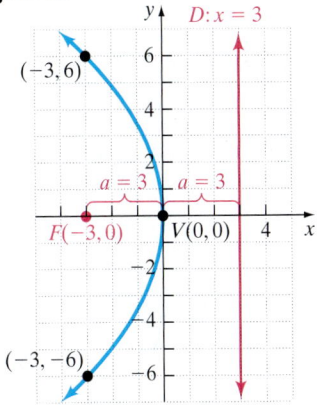

QUICK ✔ *Graph the equation.*

1. $y^2 = 8x$ **2.** $y^2 = -20x$

EXAMPLE 2 Graphing a Parabola That Opens Up

Graph the equation $x^2 = 8y$.

Solution

The equation $x^2 = 8y$ is of the form $x^2 = 4ay$, where $4a = 8$, so that $a = 2$. The graph of the equation is a parabola with vertex at $(0, 0)$ and focus at $(0, a) = (0, 2)$ so that the parabola opens up. The directrix is the line $y = -2$. To graph the parabola, it is helpful to plot the two points on the graph to the left and right of the focus. We let $y = 2$ in the equation $x^2 = 8y$ and solve for x.

$$x^2 = 8y = 8(2) = 16$$
$$\text{Take the square root of both sides:} \quad x = \pm 4$$

The points on the parabola to the left and right of the focus are $(-4, 2)$ and $(4, 2)$. See Figure 16.

Figure 16

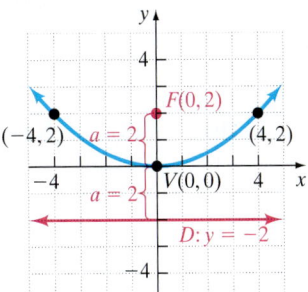

QUICK ✔ *Graph the equation.*

3. $x^2 = 4y$ **4.** $x^2 = -12y$

② Find the Equation of a Parabola

We are now going to change gears and use information regarding the equation of a parabola to obtain its equation.

EXAMPLE 3 | Finding an Equation of a Parabola

Find an equation of the parabola with vertex at $(0, 0)$ and focus at $(5, 0)$. Graph the equation.

Solution

Work Smart

It is helpful to plot the information given about the parabola before finding its equation. For example, if we plot the vertex and focus of the parabola in Example 3, we can see that it opens to the right.

The distance from the vertex $(0, 0)$ to the focus is $(5, 0)$ is $a = 5$. Because the focus lies on the positive x-axis, we know that the parabola will open to the right. This means the equation of the parabola is of the form $y^2 = 4ax$ with $a = 5$:

$$y^2 = 4(5)x = 20x$$

Figure 17 shows the graph of $y^2 = 20x$.

Figure 17

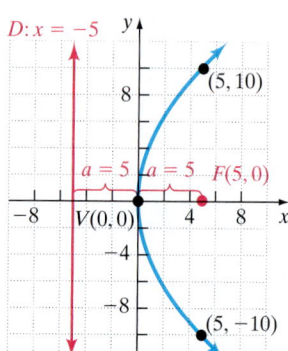

EXAMPLE 4 | Finding the Equation of a Parabola

Find the equation of a parabola with vertex at $(0, 0)$ if its axis of symmetry is the y-axis and its graph contains the point $(4, 3)$. Graph the equation.

Solution

Which of the four equations of parabolas listed in Table 1 should we use? The vertex is at the origin and the axis of symmetry is the y-axis, so the parabola either opens up or down. Because the graph contains the point $(4, 3)$, which is in quadrant I, the parabola must open up. Therefore, the equation of the parabola is of the form

$$x^2 = 4ay$$

Work Smart

Again, don't forget to draw a picture of the given information.

Because the point $(4, 3)$ is on the parabola, we let $x = 4$ and $y = 3$ in the equation $x^2 = 4ay$ to determine a.

$$x^2 = 4ay$$
$$x = 4, y = 3: \quad 4^2 = 4a(3)$$
$$16 = 12a$$
Divide both sides by 12: $\quad a = \dfrac{16}{12} = \dfrac{4}{3}$

The equation of the parabola is

$$x^2 = 4\left(\frac{4}{3}\right)y \quad \text{or} \quad x^2 = \frac{16}{3}y$$

Figure 18

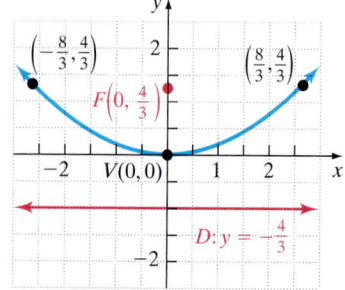

With $a = \dfrac{4}{3}$, we have that the focus is $\left(0, \dfrac{4}{3}\right)$ and the directrix is the line $y = -\dfrac{4}{3}$. We let $y = \dfrac{4}{3}$ to find points left and right of the focus to determine the "opening" and determine the points $\left(-\dfrac{8}{3}, \dfrac{4}{3}\right)$ and $\left(\dfrac{8}{3}, \dfrac{4}{3}\right)$ are on the graph. Figure 18 shows the graph of the parabola.

QUICK ✓ *Find the equation of the parabola described. Graph the equation.*

5. Vertex at $(0, 0)$; focus at $(0, -8)$

6. Vertex at $(0, 0)$; axis of symmetry the x-axis; contains the point $(3, 2)$

---

(3) ## Graph Parabolas in Which the Vertex Is Not the Origin

Work Smart

Think back to circles. The equation of a circle whose center is (h, k) is $(x - h)^2 + (y - k)^2 = r^2$. The center shifts horizontally h units and vertically k units.

If a parabola with vertex at the origin and axis of symmetry along a coordinate axis is shifted horizontally h units and then vertically k units, the result is a parabola with vertex at (h, k) and axis of symmetry parallel to either the x-axis or y-axis. The equations of these parabolas have the same form as those whose vertex is at the origin except that x is replaced with $x - h$ (the horizontal shift) and y is replaced with $y - k$ (the vertical shift). Table 2 givens the equations of the four parabolas. Figure 19(a)–(d) illustrates the graphs for $h > 0$ and $k > 0$.

Table 2 Parabolas with Vertex at (h, k); Axis of Symmetry Parallel to a Coordinate Axis, $a > 0$				
Vertex	**Focus**	**Directrix**	**Equation**	**Description**
(h, k)	$(h + a, k)$	$x = h - a$	$(y - k)^2 = 4a(x - h)$	Parabola, axis of symmetry parallel to x-axis, opens to the right
(h, k)	$(h - a, k)$	$x = h + a$	$(y - k)^2 = -4a(x - h)$	Parabola, axis of symmetry parallel to x-axis, opens to the left
(h, k)	$(h, k + a)$	$y = k - a$	$(x - h)^2 = 4a(y - k)$	Parabola, axis of symmetry parallel to y-axis, opens up
(h, k)	$(h, k - a)$	$y = k + a$	$(x - h)^2 = -4a(y - k)$	Parabola, axis of symmetry parallel to y-axis, opens down

Figure 19

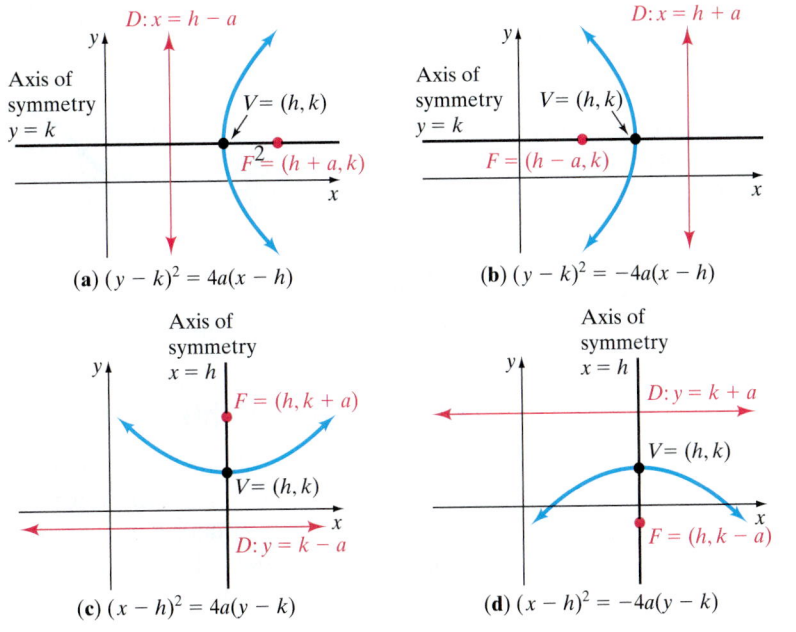

(a) $(y - k)^2 = 4a(x - h)$

(b) $(y - k)^2 = -4a(x - h)$

(c) $(x - h)^2 = 4a(y - k)$

(d) $(x - h)^2 = -4a(y - k)$

EXAMPLE 5 Graphing a Parabola Whose Vertex Is Not at the Origin

Graph the parabola $x^2 - 2x + 8y + 25 = 0$.

Solution

Notice the equation is not in any of the forms given in Table 2. We need to complete the square in x to write the equation in standard form.

$$x^2 - 2x + 8y + 25 = 0$$

$$\text{Isolate the terms involving } x: \quad x^2 - 2x = -8y - 25$$

$$\text{Complete the square:} \quad x^2 - 2x + 1 = -8y - 25 + 1$$

$$\text{Simplify:} \quad x^2 - 2x + 1 = -8y - 24$$

$$\text{Factor:} \quad (x - 1)^2 = -8(y + 3)$$

The equation is of the form $(x - h)^2 = -4a(y - k)$. This is a parabola that opens down with vertex $(h, k) = (1, -3)$. Since $-4a = -8$, we have that $a = 2$. Because the parabola opens down, the focus will be $a = 2$ units below the vertex at $(1, -5)$. We find two additional points on the graph to the left and right of the focus. To do this, we let $y = -5$ in the equation of the parabola.

$$\text{Let } y = -5: \quad (x - 1)^2 = -8(-5 + 3)$$

$$(x - 1)^2 = -8(-2)$$

$$(x - 1)^2 = 16$$

$$\text{Take the square root of both sides:} \quad x - 1 = \pm 4$$

$$\text{Add 1 to both sides:} \quad x = 1 \pm 4$$

$$x = 1 - 4 \quad \text{or} \quad x = 1 + 4$$

$$x = -3 \quad \text{or} \quad x = 5$$

The points $(-3, -5)$ and $(5, -5)$ are on the graph of the parabola. The directrix is $a = 2$ units above the vertex, so $y = -1$ is the directrix. Figure 20 shows the graph of the parabola.

Figure 20

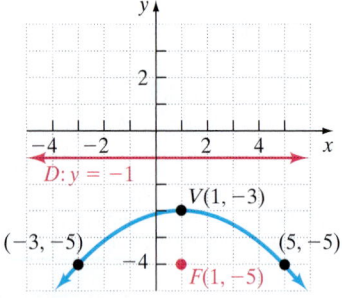

QUICK ✓

7. Graph the parabola $y^2 - 4y - 12x - 32 = 0$.

---

(4) Solve Applied Problems Involving Parabolas

The number of applications of parabolas is astounding. We have already seen how parabolas can be used to describe the shape of cables supporting a suspension bridge, but the uses of this equation do not stop there! For example, suppose that a mirror is shaped like a parabola. If a light bulb is placed at the focus of the parabola, then all the rays from the bulb will reflect off the mirror in lines parallel to the axis of symmetry. This concept is used in the design of a car's headlights, flashlights, and searchlights. See Figure 21.

As another example, suppose that rays of light are received by a parabola. When the rays strike the surface of a parabolic mirror whose axis of symmetry is parallel to these rays, they are all reflected to a single point—the focus. This idea is used in the design of some telescopes and satellite dishes. See Figure 22.

Figure 21
Searchlight

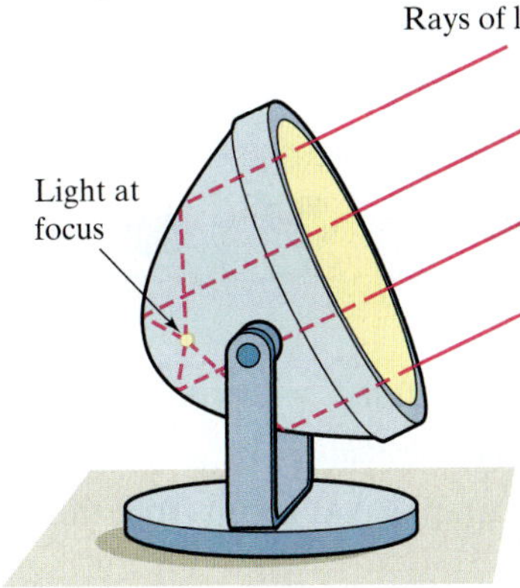

Rays of light

Light at focus

Figure 22
Telescope

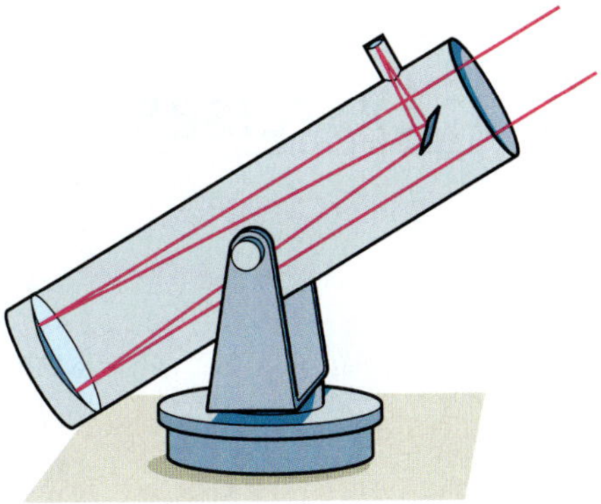

EXAMPLE 6   **A Satellite Dish**

A satellite dish is shaped like a parabola. See Figure 23(a). The signals that are received by the dish strike the surface of the dish and are reflected to a single point, where the receiver of the dish is located. If a satellite dish is 2 feet across at its opening and 6 inches deep (0.5 feet) at its center, at what position should the receiver be placed?

Solution

We want to know where to locate the receiver of the satellite dish. This means we want to know where the focus of the dish is. To solve this problem, we draw a parabola on a Cartesian plane so that the vertex of the parabola is the origin and the focus is on the positive y-axis. The width of the dish is 2 feet across and its height is 0.5 feet. So we know two points on the graph of the parabola indicated in Figure 23(b).

Figure 23

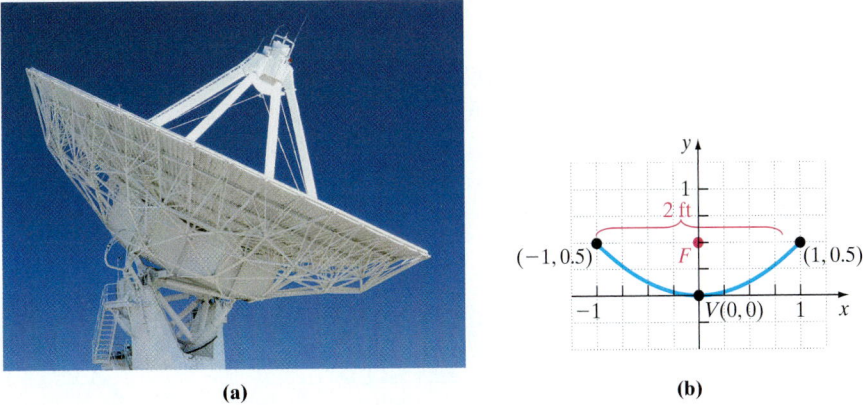

(a)   (b)

The parabola is an equation of the form $x^2 = 4ay$. Since $(1, 0.5)$ is a point on the graph, we have

$$x = 1, y = 0.5: \quad 1^2 = 4a(0.5)$$

$$1 = 2a$$

$$\text{Divide both sides by 2:} \quad a = \frac{1}{2}$$

The receiver should be located $\dfrac{1}{2}$ foot from the base of the dish along its axis of symmetry.

QUICK ✓

8. A satellite dish is shaped like a parabola. The signals that are received by the dish strike the surface of the dish and are reflected to a single point, where the receiver of the dish is located. If the dish is 4 feet across at its opening and 6 inches deep (0.5 feet) at its center, at what position should the receiver be placed?

12.3 Exercises

For Extra Help: | Student Solutions Manual | CD Video | PH Math/Tutor Center | MathXL Tutorials on CD | MathXL® | MyMathLab

Concepts and Vocabulary

In Problems 1–3, fill in the blanks.

1. _____ are graphs that result from the intersection of a right circular cone and a plane.

2. A _____ is the collection of all points P in the plane that are the same distance from a fixed point F as they are from a fixed line D.

3. The point of intersection of the parabola with its axis of symmetry is called the _____.

In Problems 4–6, answer True or False to each statement.

4. The line through the focus and perpendicular to the directrix is called the axis of symmetry.

5. The parabola $(x + 3)^2 = -14(y - 3)$ opens to the left.

6. The vertex of the parabola $(y + 2)^2 = 8(x - 3)$ is $(3, -2)$.

7. The distance from a point on a parabola to its focus is 8 units. What is the distance from the same point on the parabola to the directrix?

8. Write down the four equations that are parabolas with vertex at (h, k).

9. Draw a parabola and label the vertex, axis of symmetry, focus, and directrix.

10. Explain the difference between the discussion of parabolas presented in this section and the discussion presented in Sections 10.4 and 10.5.

Building Skills

In Problems 11–18, the graph of a parabola is given. Match each graph to its equation.

(a) $y^2 = 8x$ **(b)** $y^2 = -8x$ **(c)** $x^2 = 8y$
(d) $x^2 = -8y$ **(e)** $(y - 2)^2 = 8(x + 1)$ **(f)** $(y - 2)^2 = -8(x + 1)$
(g) $(x + 1)^2 = 8(y - 2)$ **(h)** $(x + 1)^2 = -8(y - 2)$

11.
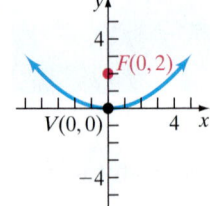

12.
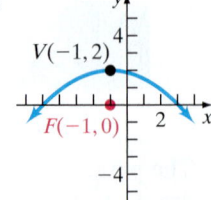

13.
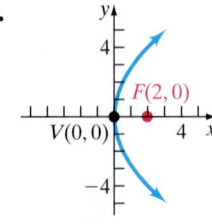

14.
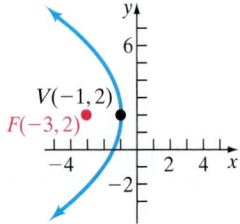

15.
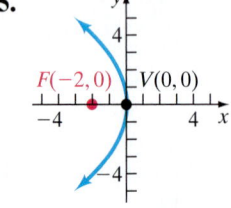

16.
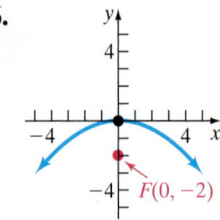

17.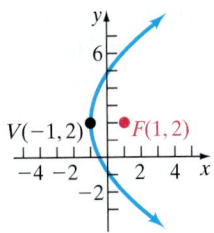

18.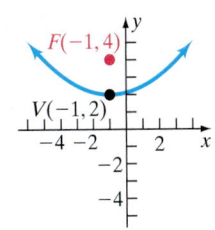

In Problems 19–28, find the equation of the parabola described. Graph the parabola.

19. Vertex at $(0,0)$; focus at $(5,0)$

20. Vertex at $(0,0)$; focus at $(0,5)$

21. Vertex at $(0,0)$; focus at $(0,-6)$

22. Vertex at $(0,0)$; focus at $(-8,0)$

23. Vertex at $(0,0)$; contains the point $(6,6)$; axis of symmetry the y-axis

24. Vertex at $(0,0)$; contains the point $(2,2)$; axis of symmetry the x-axis

25. Vertex at $(0,0)$; directrix the line $y=3$

26. Vertex at $(0,0)$; directrix the line $x=-4$

27. Focus at $(-3,0)$; directrix the line $x=3$

28. Focus at $(0,-2)$; directrix the line $y=2$.

In Problems 29–46, find the vertex, focus, and directrix of each parabola. Graph the parabola.

29. $x^2 = 24y$

30. $x^2 = 28y$

31. $y^2 = -6x$

32. $y^2 = 10x$

33. $x^2 = -8y$

34. $x^2 = -16y$

35. $(x-2)^2 = 4(y-4)$

36. $(x+4)^2 = -4(y-1)$

37. $(y+3)^2 = -8(x+2)$

38. $(y-2)^2 = 12(x+5)$

39. $(x+5)^2 = -20(y-1)$

40. $(x-6)^2 = 2(y-2)$

41. $x^2 + 4x + 12y + 16 = 0$

42. $x^2 + 2x - 8y + 25 = 0$

43. $y^2 - 8y - 4x + 20 = 0$

44. $y^2 - 8y + 16x - 16 = 0$

45. $x^2 + 10x + 6y + 13 = 0$

46. $x^2 - 4x + 10y + 4 = 0$

In Problems 47 and 48, write an equation for each parabola.

47.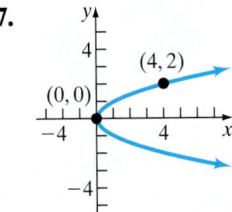

48.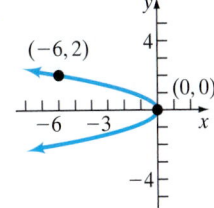

Applying the Concepts

49. A Headlight The headlight of a car is in the shape of a parabola. Its diameter is 4 inches and its depth is 1 inch. How far from the vertex should the light bulb be placed so that the rays will be reflected parallel to the axis?

50. A Headlight The headlight of a car is in the shape of a parabola. Suppose the engineers have designed the headlight to be 5 inches in diameter and wish the bulb to be placed at the focus 1 inch from the vertex. What is the depth of the headlight?

51. Suspension Bridge The cables of a suspension bridge are in the shape of a parabola, as shown in the figure. The towers supporting the cable are 500 feet apart and 60 feet high. If the cables touch the road surface midway between the towers, what is the height of the cable at a point 150 feet from the center of the bridge?

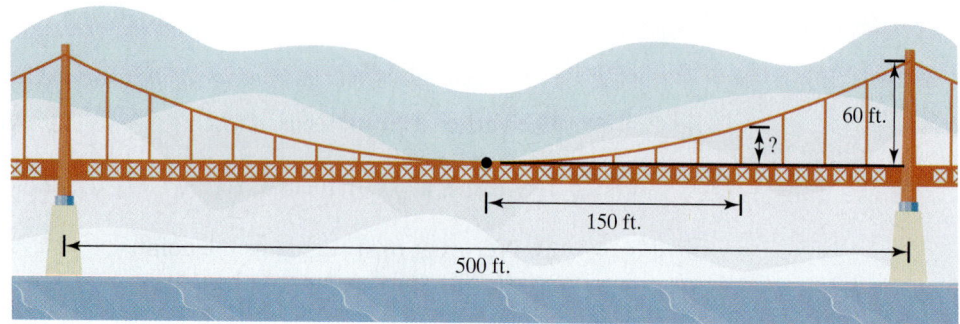

60 ft.
?
150 ft.
500 ft.

52. Suspension Bridge The cables of a suspension bridge are in the shape of a parabola. The towers supporting the cable are 400 feet apart and 80 feet high. If the cables touch the road surface midway between the towers, what is the height of the cable at a point 100 feet from the center of the bridge?

53. Parabolic Arch Bridge A bridge is built in the shape of a parabolic arch. The bridge has a span of 100 feet and a maximum height of 30 feet. See the illustration. Choose a suitable rectangular coordinate system and find the height of the arch at distances of 10, 30, and 50 feet from the center.

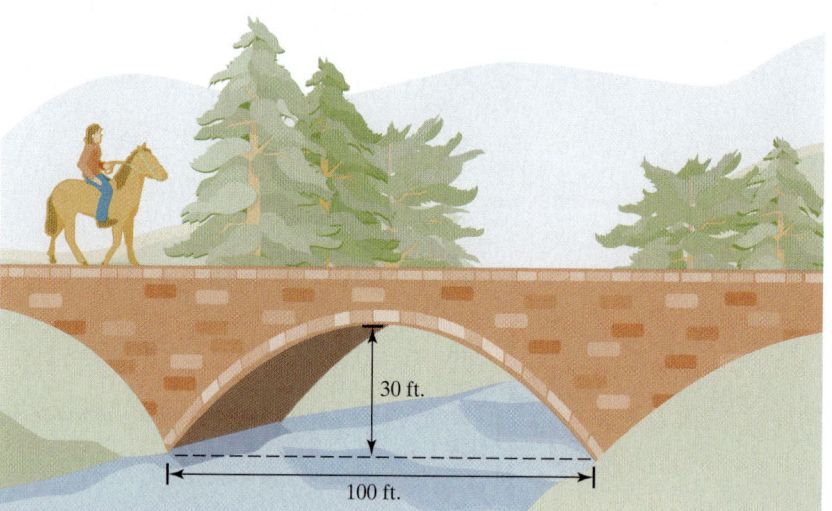

30 ft.
100 ft.

54. Parabolic Arch Bridge A bridge is to be built in the shape of a parabolic arch and is to have a span of 120 feet. The height of the arch a distance of 30 feet from the center is to be 15 feet. Find the height of the arch at its center.

Extending the Concepts

In Problems 55–58, write an equation for each parabola.

55.

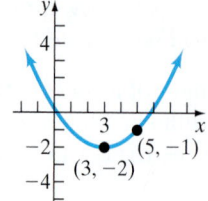

$(3, -2)$
$(5, -1)$

56.

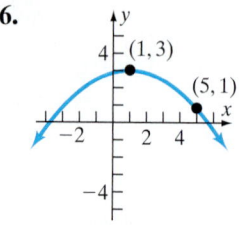

$(1, 3)$
$(5, 1)$

57.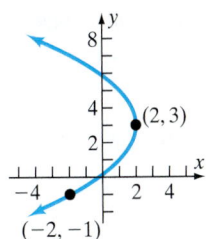

58.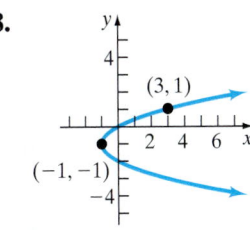

59. For the parabola $x^2 = 8y$, (a) verify that $(4, 2)$ is a point on the parabola, and (b) show that the distance from the focus to $(4, 2)$ equals the distance from $(4, 2)$ to the directrix.

60. For the parabola $(y - 3)^2 = 12(x + 1)$, (a) verify that $(2, 9)$ is a point on the parabola, and (b) show that the distance from the focus to $(2, 9)$ equals the distance from $(2, 9)$ to the directrix.

Synthesis Review

61. Graph $y = (x + 3)^2$ using the methods of Section 10.4.

62. Graph $y = (x + 3)^2 = x^2 + 6x + 9$ using the methods of Section 10.5.

63. Graph $y = (x + 3)^2$ using the methods of this section.

64. Graph $4(y + 2) = (x - 2)^2$ using the methods of Section 10.4. (**Hint:** Write the equation in the form $y = a(x - h)^2 + k$.)

65. Graph $4(y + 2) = (x - 2)^2$ using the methods of this section.

66. Compare and contrast the methods of graphing a parabola using the approach in Section 8.4 and the approach in this section. Which do you prefer? Why?

The Graphing Calculator

To graph a parabola using a graphing calculator, we must solve the equation for y, just as we did for circles. This is fairly straightforward if the equation is in the form given in either Table 1 or Table 2. If the equation is not in the form given in Table 1 or Table 2 and it is parabola that opens up or down, then solving for y is also straightforward. However, if it is a parabola that opens left or right, we will need to graph the parabola in two "pieces"—the top half and the bottom half. We can use the quadratic formula to accomplish this task. Consider the equation $y^2 + 8y + x - 4 = 0$. This equation is quadratic in y as shown:

$$y^2 + 8y + x - 4 = 0$$

$a = 1$ $b = 8$ $c = x - 4$

We use the quadratic formula and obtain

$$y = \frac{-8 \pm \sqrt{8^2 - 4(1)(x - 4)}}{2(1)}$$

So we graph

$$Y_1 = \frac{-8 - \sqrt{64 - 4(x - 4)}}{2} \quad \text{and} \quad Y_2 = \frac{-8 + \sqrt{64 - 4(x - 4)}}{2}$$

as shown in Figure 24.

Figure 24

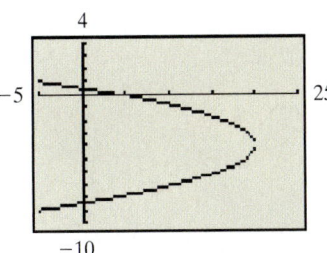

In Problems 67–78, graph each parabola using a graphing calculator.

67. $x^2 = 24y$ **68.** $x^2 = -8y$ **69.** $y^2 = -6x$

70. $y^2 = 10x$ **71.** $(x - 2)^2 = 4(y - 4)$ **72.** $(x + 4)^2 = -4(y - 1)$

73. $(y + 3)^2 = -8(x + 2)$ **74.** $(y - 2)^2 = 12(x + 5)$

75. $x^2 + 4x + 12y + 16 = 0$ **76.** $x^2 + 2x - 8y + 25 = 0$

77. $y^2 - 8y - 4x + 20 = 0$ **78.** $y^2 - 8y + 16x - 16 = 0$

12.4 Ellipses

OBJECTIVES

1. Graph Ellipses in Which the Center Is the Origin
2. Find the Equation of an Ellipse in Which the Center Is the Origin
3. Graph Ellipses in Which the Center Is Not the Origin
4. Solve Applied Problems Involving Ellipses

Preparing for Ellipses

Before getting started, take the following readiness quiz. If you get a problem wrong, go back to the section cited and review the material.

1. Complete the square of $x^2 + 10x$. [Section 10.1, pp. 747–748]
2. Graph $f(x) = (x + 2)^2 - 1$ using transformations. [Section 10.4, pp. 781–785]

1 Graph Ellipses in Which the Center Is the Origin

An ellipse is a conic section that is obtained through the intersection of a plane and a cone. See Figure 8(b) on page 920.

> **DEFINITION**
>
> An **ellipse** is the collection of points in the plane such that the sum of the distances from two fixed points, called the **foci,** is a constant.

The definition allows us to physically draw an ellipse. To do this, find a piece of string (the length of the string is the constant referred to in the definition). Now take two thumbtacks and stick them on a piece of cardboard so that the distance between them is less than the length of the string. The two thumbtacks represent the foci of the ellipse. Now attach the ends of the string to the thumbtacks and, using the point of a pencil, pull the string taut. Keeping the string taut, rotate the pencil around the two thumbtacks. The pencil traces out an ellipse as shown in Figure 25.

Figure 25

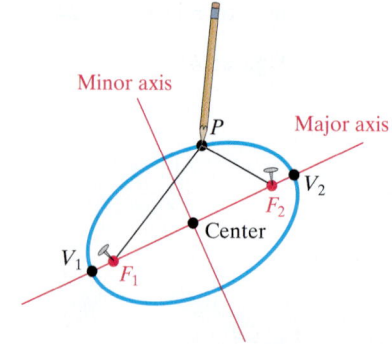

Preparing for...Answers
1. $x^2 + 10x + 25$
2.

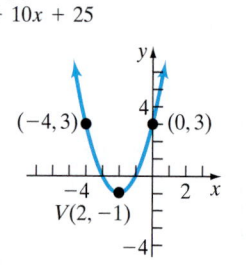

In Figure 25, the foci are labeled F_1 and F_2. The line containing the foci is called the **major axis.** The midpoint of the line segment joining the foci is called the **center** of the ellipse. The line through the center and perpendicular to the major axis is called the **minor axis.**

The two points of intersection of the ellipse and the major axis are the **vertices,** V_1 and V_2, of the ellipse. The distance from one vertex to the other is called the **length of the major axis.**

We are now ready to find the equation of an ellipse in a Cartesian plane. First, we place the center of the ellipse at the origin. Second, we position the ellipse so that its major axis coincides with a coordinate axis. Let's have the major axis coincide with the x-axis as shown in Figure 26 and call c the distance from the center of the ellipse to a focus, so that one focus is at $F_1 = (-c, 0)$ and the other focus is at $F_2 = (c, 0)$. Let $2a$ represent the constant distance referred to in the definition (the reason for this will be clear shortly).

Figure 26

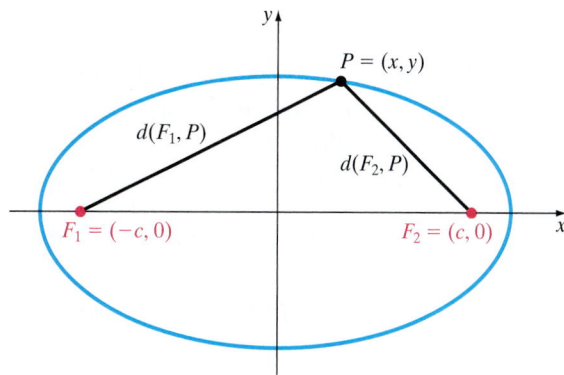

If $P = (x, y)$ is any point on the ellipse, we know from the definition of the ellipse that the sum of the distance from P to the foci equals a constant. Recall, we will let the constant equal $2a$, so that

$$d(F_1, P) + d(F_2, P) = 2a$$

By referring to Figure 26 and using the distance formula, we find that

$$\sqrt{(x - (-c))^2 + (y - 0)^2} + \sqrt{(x - c)^2 + (y - 0)^2} = 2a$$

This ultimately simplifies to

$$\frac{x^2}{a^2} + \frac{y^2}{b^2} = 1$$

So the equation of an ellipse whose center is at the origin and whose major axis is the x-axis is $\dfrac{x^2}{a^2} + \dfrac{y^2}{b^2} = 1$. Another possibility is that the major axis is the y-axis. We could obtain the equation for this ellipse using logic similar to the logic we used to obtain the equation above. We summarize the equations of ellipses with center at the origin in Table 3.

Table 3 Ellipses with Center at the Origin				
Center	**Major Axis**	**Foci**	**Vertices**	**Equation**
$(0, 0)$	x-axis	$(-c, 0)$ and $(c, 0)$	$(-a, 0)$ and $(a, 0)$	$\dfrac{x^2}{a^2} + \dfrac{y^2}{b^2} = 1$
$(0, 0)$	y-axis	$(0, -c)$ and $(0, c)$	$(0, -a)$ and $(0, a)$	$\dfrac{x^2}{b^2} + \dfrac{y^2}{a^2} = 1$

In both ellipses, $a > b$ and $b^2 = a^2 - c^2$. The graphs of the two ellipses are given in Figure 27.

Figure 27

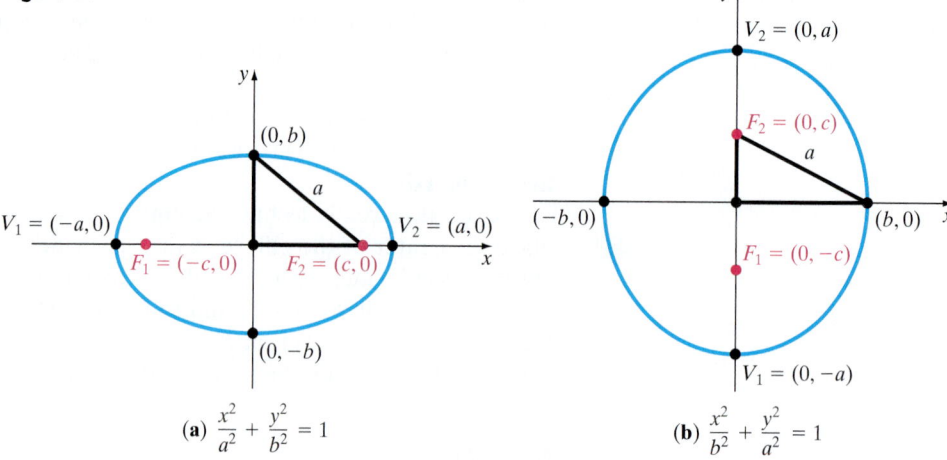

(a) $\dfrac{x^2}{a^2} + \dfrac{y^2}{b^2} = 1$

(b) $\dfrac{x^2}{b^2} + \dfrac{y^2}{a^2} = 1$

Work Smart

The term with the larger denominator tells us which axis is the major axis.

We use the fact that $a > b$ to determine whether the major axis is the x-axis or y-axis. If the larger denominator is associated with the x^2-term, then the major axis is the x-axis; if the larger denominator is associated with the y^2-term, then the major axis is the y-axis.

Graphing an ellipse whose center is at the origin is fairly straightforward because all we have to do is find the intercepts by letting $y = 0$ (x-intercepts) and letting $x = 0$ (y-intercepts). When you are asked to graph an ellipse, be sure to label the foci.

EXAMPLE 1 Graphing an Ellipse

Graph the ellipse: $\dfrac{x^2}{25} + \dfrac{y^2}{9} = 1$

Solution

First, we notice that the larger number, 25, is in the denominator of the x^2-term. This means that the major axis is the x-axis and the equation of the ellipse is of the form $\dfrac{x^2}{a^2} + \dfrac{y^2}{b^2} = 1$ so that $a^2 = 25$ and $b^2 = 9$. The center of the ellipse is the origin, $(0, 0)$. Because $b^2 = a^2 - c^2$, or $c^2 = a^2 - b^2$, we have that $c^2 = 25 - 9 = 16$, so that $c = \pm 4$. Since the major axis is the x-axis, the foci are $(-4, 0)$ and $(4, 0)$. We now find the intercepts:

x-intercepts: Let $y = 0$: $\dfrac{x^2}{25} + \dfrac{0^2}{9} = 1$ y-intercepts: Let $x = 0$: $\dfrac{0^2}{25} + \dfrac{y^2}{9} = 1$

$$\dfrac{x^2}{25} = 1 \qquad\qquad\qquad \dfrac{y^2}{9} = 1$$
$$x^2 = 25 \qquad\qquad\qquad\quad y^2 = 9$$
$$x = \pm 5 \qquad\qquad\qquad\quad y = \pm 3$$

The intercepts are $(-5, 0)$, $(5, 0)$, $(0, -3)$, and $(0, 3)$. Figure 28 shows the graph of the ellipse.

Figure 28

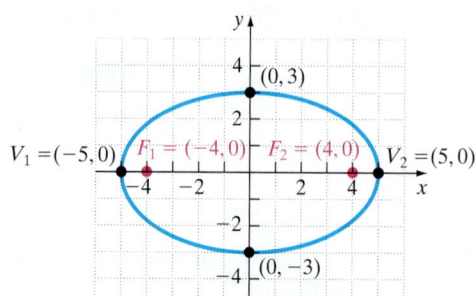

Let's try another one.

EXAMPLE 2 Graphing an Ellipse

Graph the ellipse: $\dfrac{x^2}{4} + \dfrac{y^2}{16} = 1$

Solution

First, we notice that the larger number, 16, is in the denominator of the y^2-term. This means that the major axis is the y-axis and the equation of the ellipse is of the form $\dfrac{x^2}{b^2} + \dfrac{y^2}{a^2} = 1$, so that $a^2 = 16$ and $b^2 = 4$. The center of the ellipse is the origin, $(0,0)$. Because $b^2 = a^2 - c^2$, or $c^2 = a^2 - b^2$, we have that $c^2 = 16 - 4 = 12$, so that $c = \pm\sqrt{12} = \pm 2\sqrt{3}$. Since the major axis is the y-axis, the foci are $\left(0, -2\sqrt{3}\right)$ and $\left(0, 2\sqrt{3}\right)$. We now find the intercepts:

x-intercepts: Let $y = 0$: $\quad \dfrac{x^2}{4} + \dfrac{0^2}{16} = 1 \qquad\qquad$ y-intercepts: Let $x = 0$: $\quad \dfrac{0^2}{4} + \dfrac{y^2}{16} = 1$

$$\dfrac{x^2}{4} = 1 \qquad\qquad\qquad\qquad \dfrac{y^2}{16} = 1$$
$$x^2 = 4 \qquad\qquad\qquad\qquad y^2 = 16$$
$$x = \pm 2 \qquad\qquad\qquad\qquad y = \pm 4$$

The intercepts are $(-2, 0)$, $(2, 0)$, $(0, -4)$, and $(0, 4)$. Figure 29 shows the graph of the ellipse. ∎

Figure 29

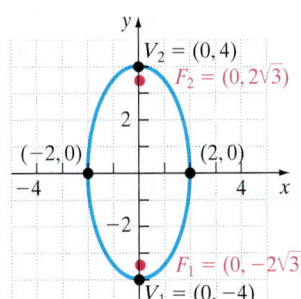

QUICK ✓ Graph each ellipse.

1. $\dfrac{x^2}{9} + \dfrac{y^2}{4} = 1$

2. $\dfrac{x^2}{16} + \dfrac{y^2}{36} = 1$

---

② Find the Equation of an Ellipse in Which the Center Is the Origin

Just as we did with parabolas, we are now going to use information about an ellipse to find its equation.

EXAMPLE 3 Finding the Equation of an Ellipse

Find the equation of the ellipse whose center is the origin with a focus at $(-2, 0)$ and vertex at $(-5, 0)$. Graph the ellipse.

Solution

By plotting the given focus and vertex, we find that the points lie on the x-axis. For this reason, the major axis is the x-axis. So the equation of the ellipse is of the form $\dfrac{x^2}{a^2} + \dfrac{y^2}{b^2} = 1$. The distance from the center of the ellipse to the vertex is $a = 5$ units.

The distance from the center of the ellipse to the focus is $c = 2$ units. Because $b^2 = a^2 - c^2$, we have that $b^2 = 5^2 - 2^2 = 25 - 4 = 21$. So the equation of the ellipse is

Work Smart

"a" is the distance from the center of an ellipse to one of its vertices. "c" is the distance from the center of an ellipse to one of its foci.

$$\dfrac{x^2}{25} + \dfrac{y^2}{21} = 1$$

Figure 30 shows the graph.

Figure 30

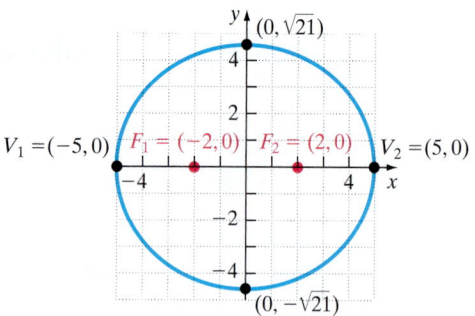

QUICK ✓

3. Find the equation of the ellipse whose center is the origin with a focus at $(0, 3)$ and vertex at $(0, 7)$. Graph the ellipse.

---

③ **Graph Ellipses in Which the Center Is Not the Origin**

If an ellipse with center at the origin and major axis coinciding with a coordinate axis is shifted horizontally h units and then vertically k units, the result is an ellipse with center at (h, k) and major axis parallel to a coordinate axis. The equations of these ellipses have the same forms as those given for ellipses whose center is the origin, except that x is replaced by $x - h$ (the horizontal shift) and y is replaced by $y - k$ (the vertical shift). Table 4 gives the forms of the equations for these ellipses. Figure 31 shows their graphs.

Table 4 Ellipses with Center at (h, k) and Major Axis Parallel to a Coordinate Axis

Center	Major Axis	Foci	Vertices	Equation
(h, k)	Parallel to x-axis	$(h + c, k)$	$(h + a, k)$	$\dfrac{(x - h)^2}{a^2} + \dfrac{(y - k)^2}{b^2} = 1$,
		$(h - c, k)$	$(h - a, k)$	$a > b$ and $b^2 = a^2 - c^2$
(h, k)	Parallel to y-axis	$(h, k + c)$	$(h, k + a)$	$\dfrac{(x - h)^2}{b^2} + \dfrac{(y - k)^2}{a^2} = 1$,
		$(h, k - c)$	$(h, k - a)$	$a > b$ and $b^2 = a^2 - c^2$

Work Smart

Do not attempt to memorize Table 4. Instead, understand the roles of a and c in an ellipse—a is the distance from the center to each vertex and c is the distance from the center to each focus.

Figure 31

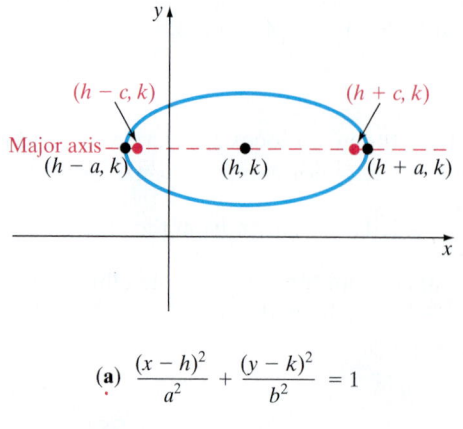

(a) $\dfrac{(x - h)^2}{a^2} + \dfrac{(y - k)^2}{b^2} = 1$

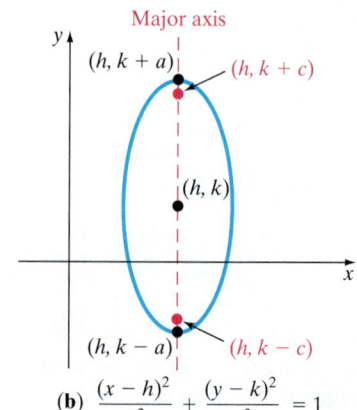

(b) $\dfrac{(x - h)^2}{b^2} + \dfrac{(y - k)^2}{a^2} = 1$

EXAMPLE 4 Graphing an Ellipse

Graph the equation: $x^2 + 16y^2 - 4x + 32y + 4 = 0$

Solution

We need to write the equation in one of the forms given in Table 4. To do this, we complete the squares in x and in y.

$$x^2 + 16y^2 - 4x + 32y + 4 = 0$$

Group like variables; place the constant on the right side:

$$(x^2 - 4x) + (16y^2 + 32y) = -4$$

Factor out 16 from the last two terms:

$$(x^2 - 4x) + 16(y^2 + 2y) = -4$$

Complete each square:

$$(x^2 - 4x + 4) + 16(y^2 + 2y + 1) = -4 + 4 + 16$$

Factor:

$$(x - 2)^2 + 16(y + 1)^2 = 16$$

Divide both sides by 16:

$$\frac{(x - 2)^2}{16} + \frac{(y + 1)^2}{1} = 1$$

Work Smart

When we completed the square in y, we added a 1, but because it was inside the parentheses with a factor of 16 in front, we must add $16 \cdot 1 = 16$ to the other side.

This is the equation of an ellipse with center at $(2, -1)$. Because 16 is the denominator of the x^2-term, we know the major axis is parallel to the x-axis. Because $a^2 = 16$ and $b^2 = 1$, we have that $c^2 = a^2 - b^2 = 16 - 1 = 15$. The vertices are $a = 4$ units to the left and right of center at $V_1 = (-2, -1)$ and $V_2 = (6, -1)$. The foci are $c = \sqrt{15}$ units to the left and right of center at $F_1 = \left(2 - \sqrt{15}, -1\right)$ and $F_2 = \left(2 + \sqrt{15}, -1\right)$. We then plot points $b = 1$ unit above and below the center at $(2, 0)$ and $(2, -2)$. Figure 32 shows the graph.

Figure 32

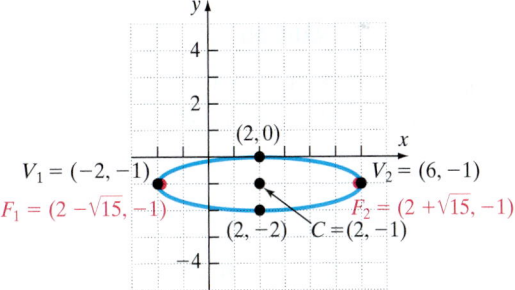

QUICK ✓

4. Graph the equation: $9x^2 + y^2 + 54x - 2y + 73 = 0$

---

④ ## Solve Applied Problems Involving Ellipses

Ellipses are found in many applications in science and engineering. For example, the orbits of the planets around the Sun are elliptical, with the Sun's position at a focus. See Figure 33.

Stone and concrete bridges are often the shape of semielliptical arches. Elliptical gears are used in machinery when a variable rate of motion is required.

Ellipses also have an interesting reflection property. If a source of light (or sound) is placed at one focus, the waves transmitted by the source will reflect off the ellipse and concentrate at the other focus. This is the principle behind whispering galleries, which

Figure 33

are rooms designed with elliptical ceilings. A person standing at one focus of the ellipse can whisper and be heard by a person standing at the other focus, because all the sound waves that reach the ceiling are reflected to the other person. National Statuary Hall in the United States Capitol used to be a whispering gallery. The acoustics in the room were such that the design of the room had to be changed.

EXAMPLE 5 **A Whispering Gallery**

The whispering gallery in the Museum of Science and Industry in Chicago is 47.3 feet long. The distance from the center of the room to the foci is 20.3 feet. Find an equation that describes the shape of the room. How high is the room at its center?

Solution

We set up a Cartesian plane so that the center of the ellipse is at the origin and the major axis is along the x-axis. The equation of the ellipse is

$$\frac{x^2}{a^2} + \frac{y^2}{b^2} = 1$$

Figure 34

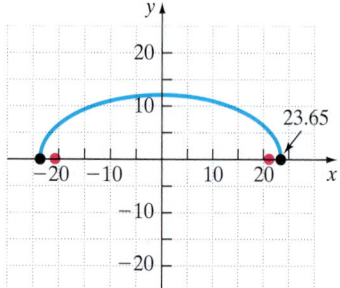

Since the length of the room is 47.3 feet, the distance from the center of the room to each vertex (the end of the room) will be $47.3/2 = 23.65$ feet, so $a = 23.65$ feet. The distance from the center of the room to each focus is $c = 20.3$ feet. See Figure 34. Since $b^2 = a^2 - c^2$, we have that $b^2 = 23.65^2 - 20.3^2 = 147.2325$. An equation that describes the shape of the room is given by

$$\frac{x^2}{23.65^2} + \frac{y^2}{147.2325} = 1$$

The height of the room at its center is $b = \sqrt{147.2325} \approx 12.1$ feet.

QUICK ✓

5. A hall 100 feet in length is to be designed as a whispering gallery. If the foci are to be located 30 feet from the center, determine an equation that describes the room. What is the height of the room at its center?

12.4 Exercises

Concepts and Vocabulary

In Problems 1–3, fill in the blanks.

1. An _____ is the collection of points in the plane such that the sum of the distances from two fixed points, called the _____, is a constant.

2. For an ellipse, the line containing the foci is called the _____ _____.

3. The two points of intersection of the ellipse and the major axis are the _____, V_1 and V_2, of the ellipse.

In Problems 4–6, answer True or False to each statement.

4. For any ellipse whose vertex is $(a, 0)$ and focus is $(c, 0)$, then $c^2 = a^2 + b^2$.

5. The minor axis of an ellipse is perpendicular to the major axis and contains the center of the ellipse.

6. The center of $\dfrac{(x-3)^2}{25} + \dfrac{(y+1)^2}{16} = 1$ is $(-3, 1)$.

7. In the ellipse drawn on the right the center is at the origin and the major axis is the x-axis. The point F is a focus. In the right triangle (drawn in red), label the lengths of the legs and the hypotenuse using the values of a, b, and c.

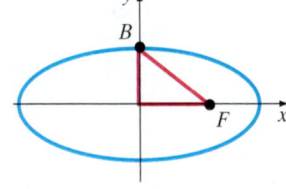

8. How does the ellipse given by the equation $\dfrac{x^2}{25} + \dfrac{y^2}{9} = 1$ differ from the ellipse given by the equation $\dfrac{x^2}{9} + \dfrac{y^2}{25} = 1$? How are they the same?

Building Skills

In Problems 9–12, the graph of an ellipse is given. Match each graph to its equation.

(a) $x^2 + \dfrac{y^2}{9} = 1$ **(b)** $\dfrac{x^2}{9} + y^2 = 1$ **(c)** $\dfrac{x^2}{16} + \dfrac{y^2}{9} = 1$ **(d)** $\dfrac{x^2}{9} + \dfrac{y^2}{16} = 1$

9.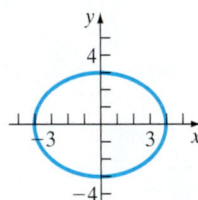

10.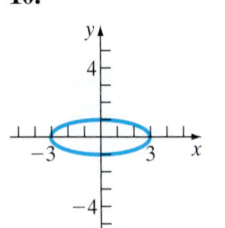

11.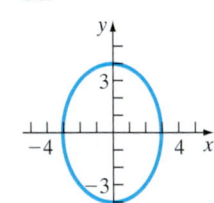

12.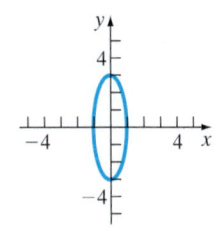

In Problems 13–22, find the vertices and foci of each ellipse. Graph each ellipse.

13. $\dfrac{x^2}{25} + \dfrac{y^2}{16} = 1$ **14.** $\dfrac{x^2}{25} + \dfrac{y^2}{4} = 1$ **15.** $\dfrac{x^2}{36} + \dfrac{y^2}{100} = 1$ **16.** $\dfrac{x^2}{16} + \dfrac{y^2}{36} = 1$

17. $\dfrac{x^2}{49} + \dfrac{y^2}{4} = 1$ **18.** $\dfrac{x^2}{121} + \dfrac{y^2}{100} = 1$ **19.** $x^2 + \dfrac{y^2}{49} = 1$ **20.** $\dfrac{x^2}{64} + y^2 = 1$

21. $4x^2 + y^2 = 16$ **22.** $9x^2 + y^2 = 81$

In Problems 23–30, find an equation for each ellipse. Graph each ellipse.

23. Center at $(0,0)$; focus at $(4,0)$; vertex at $(6,0)$

24. Center at $(0,0)$; focus at $(2,0)$; vertex at $(5,0)$

25. Center at $(0,0)$; focus at $(0,-4)$; vertex at $(0,7)$

26. Center at $(0,0)$; focus at $(0,-1)$; vertex at $(0,5)$

27. Foci at $(\pm 6, 0)$; Vertices at $(\pm 10, 0)$

28. Foci at $(0, \pm 2)$; Vertices at $(0, \pm 7)$

29. Foci at $(0, \pm 5)$; length of the major axis is 16

30. Foci at $(\pm 6, 0)$; length of the major axis is 20

In Problems 31–40, graph each ellipse.

31. $\dfrac{(x-3)^2}{9} + \dfrac{(y+2)^2}{25} = 1$ **32.** $\dfrac{(x-1)^2}{36} + \dfrac{(y+4)^2}{100} = 1$

33. $\dfrac{(x+2)^2}{16} + \dfrac{(y-5)^2}{4} = 1$ **34.** $\dfrac{(x+5)^2}{64} + \dfrac{(y+1)^2}{16} = 1$

35. $(x-5)^2 + \dfrac{(y+1)^2}{49} = 1$ **36.** $\dfrac{(x+8)^2}{81} + (y-3)^2 = 1$

37. $4(x+2)^2 + 16(y-1)^2 = 64$ **38.** $9(x-3)^2 + (y-4)^2 = 81$

39. $4x^2 + y^2 - 24x + 2y - 63 = 0$ **40.** $16x^2 + 9y^2 - 128x + 54y - 239 = 0$

Applying the Concepts

41. Semielliptical Arch Bridge An arch in the shape of the upper half of an ellipse is used to support a bridge that is to span a river 30 meters wide. The center of the arch is 10 meters above the center of the river (see the figure).

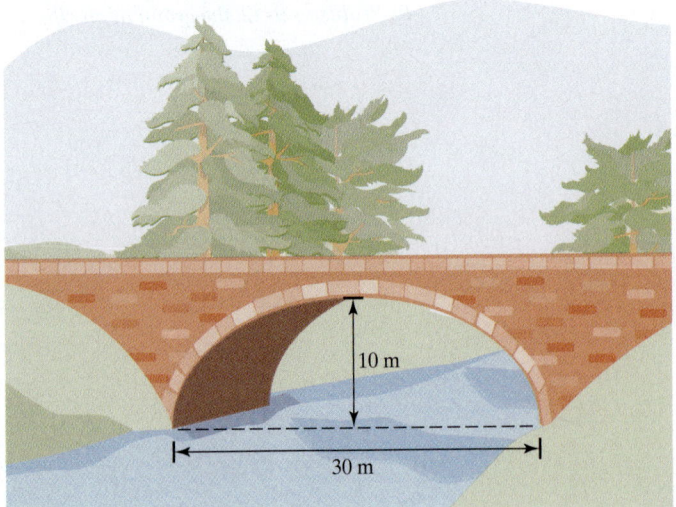

(a) Write the equation for the ellipse in which the *x*-axis coincides with the water and the *y*-axis passes through the center of the arch.

(b) Can a rectangular barge that is 18 meters wide and sits 7 meters above the surface of the water fit through the opening of the bridge?

(c) If heavy rains caused the river's level to increase 1.1 meters, will the barge make it through the opening?

42. London Bridge An arch in the shape of the upper half of an ellipse is used to support London Bridge. The main span is 45.6 meters wide. Suppose that the center of the arch is 15 meters above the center of the river.

(a) Write the equation for the ellipse in which the *x*-axis coincides with the water and the *y*-axis passes through the center of the arch.

(b) Can a rectangular barge that is 20 meters wide and sits 12 meters above the surface of the water fit through the opening of the bridge?

(c) If heavy rains caused the river's level to increase 1.5 meters, will the barge make it through the opening?

In Problems 43–46, use the fact that the orbit of a planet about the Sun is an ellipse, with the Sun at one focus. The **aphelion** *of a planet is its greatest distance from the Sun, and the* **perihelion** *is its shortest distance. The* **mean distance** *of a planet from the Sun is the length of the semimajor axis of the elliptical orbit. See the illustration.*

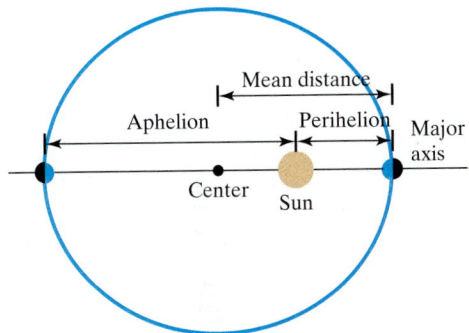

43. Earth The mean distance of Earth from the Sun is 93 million miles. If the aphelion of Earth is 94.5 million miles, what is the perihelion? Write an equation for the orbit of Earth around the Sun.

44. Mars The mean distance of Mars from the Sun is 142 million miles. If the perihelion of Mars is 128.5 million miles, what is the aphelion? Write an equation for the orbit of Mars about the Sun.

45. Jupiter The aphelion of Jupiter is 507 million miles. If the distance from the Sun to the center of its elliptical orbit is 23.2 million miles, what is the perihelion? What is the mean distance? Write an equation for the orbit of Jupiter around the Sun.

46. Pluto The perihelion of Pluto is 4551 million miles, and the distance of the Sun from the center of its elliptical orbit is 897.5 million miles. Find the aphelion of Pluto. What is the mean distance of Pluto from the Sun? Write an equation for the orbit of Pluto about the Sun.

Extending the Concepts

In Problems 47–50, write an equation for each ellipse.

47. **48.** **49.** **50.**

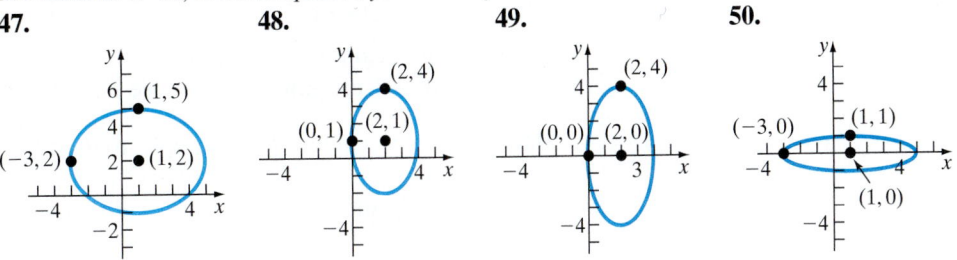

51. Show that a circle is a special kind of ellipse by letting $a = b$ in the equation of an ellipse centered at the origin. What is the value of c in a circle? What does this mean regarding the location of the foci?

52. The **eccentricity** e of an ellipse is defined as the number $\dfrac{c}{a}$, where a is the distance from the center of an ellipse to a vertex and c is the distance from the center of an ellipse to a focus. Because $a > c$, it follows that $e < 1$ for an ellipse. Write a paragraph about the general shape of each of the following ellipses. Be sure to justify your conclusion.

   (a) Eccentricity close to 0   **(b)** Eccentricity $= 0.5$ **(c)** Eccentricity close to 1

Synthesis Review

In Problems 53–56, fill in the following table for each function.

x	5	10	100	1000
$f(x)$				

53. $f(x) = \dfrac{5}{x + 2}$

54. $f(x) = \dfrac{x - 1}{x^2 + 4}$

55. $f(x) = \dfrac{2x + 1}{x - 3}$

56. $f(x) = \dfrac{3x^2 - x + 1}{x^2 + 1}$

In Problems 57 and 58, fill in the following table for each function.

x	5	10	100	1000
$f(x)$				
$g(x)$				

57. $f(x) = \dfrac{x^2 + 3x + 1}{x + 1}$; $g(x) = x + 2$ **58.** $f(x) = \dfrac{x^2 - 3x + 5}{x + 2}$; $g(x) = x - 5$

59. For the functions in Problems 53–56, compare the degree of the polynomial in the numerator to the degree of the polynomial in the denominator. Conjecture what happens to a rational function as x increases when the degree of the numerator is less than the degree of the denominator. Conjecture what happens to a rational function as x increases when the degree of the numerator equals the degree of the denominator.

60. For the functions in Problems 57 and 58, write f in the form quotient $+\ \dfrac{\text{remainder}}{\text{dividend}}$.

That is, perform the division indicated by the rational function. Now compare the values of f to those of g in the table. What does the function g represent?

Graphing Calculator

A graphing calculator can be used to graph ellipses by solving the equation for y. Because ellipses are not functions, we need to graph the upper half and lower half of

the ellipse in two pieces. For example, to graph $\dfrac{x^2}{4} + \dfrac{y^2}{9} = 1$, we would graph

$Y_1 = 3\sqrt{1 - \dfrac{x^2}{4}}$ and $Y_2 = -3\sqrt{1 - \dfrac{x^2}{4}}$. We obtain the graph shown in Figure 35.

Figure 35

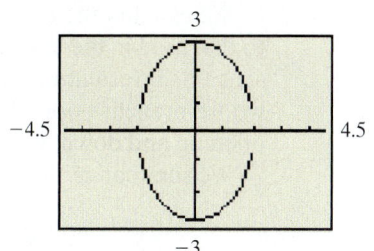

In Problems 61–68, graph each ellipse using a graphing calculator.

61. $\dfrac{x^2}{25} + \dfrac{y^2}{16} = 1$ **62.** $\dfrac{x^2}{25} + \dfrac{y^2}{4} = 1$ **63.** $4x^2 + y^2 = 16$

64. $9x^2 + y^2 = 81$ **65.** $\dfrac{(x-3)^2}{9} + \dfrac{(y+2)^2}{25} = 1$ **66.** $\dfrac{(x-1)^2}{36} + \dfrac{(y+4)^2}{100} = 1$

67. $\dfrac{(x+2)^2}{16} + \dfrac{(y-5)^2}{4} = 1$ **68.** $\dfrac{(x+5)^2}{64} + \dfrac{(y+1)^2}{16} = 1$

12.5 Hyperbolas

OBJECTIVES

1. Graph Hyperbolas in Which the Center Is the Origin
2. Find the Equation of a Hyperbola in Which the Center Is the Origin
3. Find the Asymptotes of a Hyperbola in Which the Center Is the Origin

Preparing for Hyperbolas

Before getting started, take the following readiness quiz. If you get a problem wrong, go back to the section cited and review the material.

1. Complete the square: $x^2 - 5x$ [Section 10.1, pp. 747–748]
2. Solve: $y^2 = 64$ [Section 10.1, pp. 744–747]

1 Graph Hyperbolas in Which the Center Is the Origin

Recall from Section 12.2 that a hyperbola is a conic section that is obtained through the intersection of a plane and two cones. See Figure 8(d) on page 920.

> A **hyperbola** is the collection of all points in the plane the difference of whose distances from two fixed points, called the **foci,** is a constant.

Figure 36 illustrates a hyperbola with foci F_1 and F_2. The line containing the foci is called the **transverse axis.** The midpoint of the line segment joining the foci is called the **center** of the hyperbola. The line through the center and perpendicular to the transverse axis is called the **conjugate axis.** The hyperbola consists of two separate curves called **branches.** The two points of intersection of the hyperbola and the transverse axis are the **vertices,** V_1 and V_2, of the hyperbola.

In Words

The distance from F_1 to P minus the distance from F_2 to P is a constant value for any point P on a hyperbola.

Figure 36

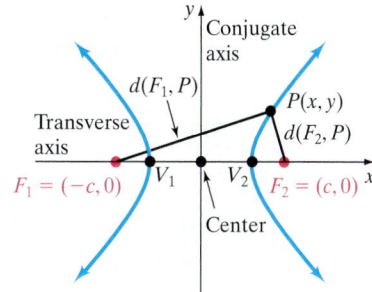

We can find the equation of a hyperbola using the distance formula in a way similar to that used in the last section to find the equation of an ellipse. However, we won't present this information here. Instead, we will give you the equations of the hyperbolas whose branches open left and right and the equations of hyperbolas whose branches open up and down with both hyperbolas centered at the origin.

We summarize the equations of hyperbolas in Table 5.

		Table 5 Hyperbolas with Center at the Origin		
Center	**Transverse Axis**	**Foci**	**Vertices**	**Equation**
$(0, 0)$	x-axis	$(-c, 0)$ and $(c, 0)$	$(-a, 0)$ and $(a, 0)$	$\dfrac{x^2}{a^2} - \dfrac{y^2}{b^2} = 1$ where $b^2 = c^2 - a^2$ or $c^2 = a^2 + b^2$
$(0, 0)$	y-axis	$(0, -c)$ and $(0, c)$	$(0, -a)$ and $(0, a)$	$\dfrac{y^2}{a^2} - \dfrac{x^2}{b^2} = 1$ where $b^2 = c^2 - a^2$ or $c^2 = a^2 + b^2$

The graphs of the two hyperbolas are given in Figure 37.

Figure 37

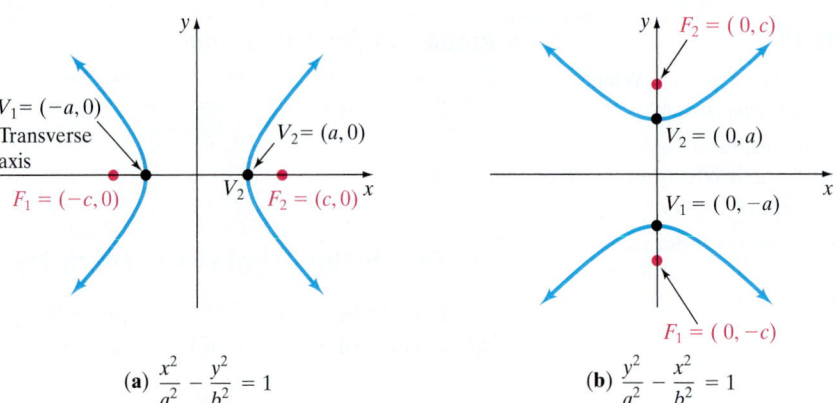

(a) $\dfrac{x^2}{a^2} - \dfrac{y^2}{b^2} = 1$ (b) $\dfrac{y^2}{a^2} - \dfrac{x^2}{b^2} = 1$

Notice the difference in the two equations given in Table 5. When the x^2-term is first, as in Figure 37(a), the transverse axis is the x-axis and the hyperbola opens left and right. When the y^2-term is first, as in Figure 37(b), the transverse axis is the y-axis and the hyperbola opens up and down. In both cases, the value of a^2 is in the denominator of the first term.

EXAMPLE 1 Graphing a Hyperbola

Graph the equation: $\dfrac{x^2}{16} - \dfrac{y^2}{4} = 1$

Solution

We notice the equation is of the form $\dfrac{x^2}{a^2} - \dfrac{y^2}{b^2} = 1$. Since the x^2-term is first, the graph of the hyperbola will open left and right. We have that $a^2 = 16$ and $b^2 = 4$. The center of the hyperbola is the origin, $(0, 0)$, and the transverse axis is the x-axis.

Because $b^2 = c^2 - a^2$ (or $c^2 = a^2 + b^2$), we find that $c^2 = 16 + 4 = 20$, so that $c = 2\sqrt{5}$. The vertices are at $(\pm a, 0) = (\pm 4, 0)$, and the foci are at $(\pm c, 0) = (\pm 2\sqrt{5}, 0)$.

To obtain the graph we plot the vertices and foci. Then, we locate points above and below the foci (just as we did when graphing parabolas). So we let $x = \pm 2\sqrt{5}$ in the equation $\dfrac{x^2}{16} - \dfrac{y^2}{4} = 1$.

$$\frac{\left(\pm 2\sqrt{5}\right)^2}{16} - \frac{y^2}{4} = 1$$

$$\frac{20}{16} - \frac{y^2}{4} = 1$$

Reduce the fraction: $\quad \dfrac{5}{4} - \dfrac{y^2}{4} = 1$

Subtract $\dfrac{5}{4}$ from both sides: $\quad -\dfrac{y^2}{4} = -\dfrac{1}{4}$

Multiply both sides by -4: $\quad y^2 = 1$

Take the square root of both sides: $\quad y = \pm 1$

The points above and below the foci are $\left(\pm 2\sqrt{5}, -1\right)$ and $\left(\pm 2\sqrt{5}, 1\right)$. See Figure 38 for the graph of the hyperbola.

Figure 38

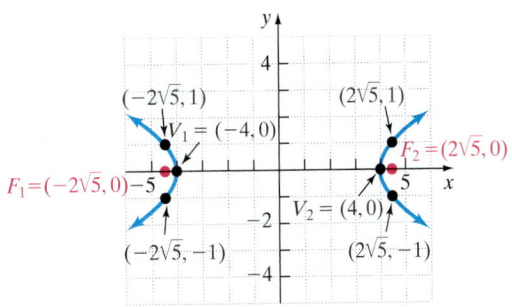

EXAMPLE 2   Graph a Hyperbola

Graph the equation: $4y^2 - 16x^2 = 16$

Solution

We wish to put the equation in one of the forms given in Table 5. To do this, we divide both sides of the equation by 16:

$$\frac{y^2}{4} - \frac{x^2}{1} = 1$$

Do you see that the y^2-term is first? This means that the hyperbola will open up and down. The center of this hyperbola is the origin, $(0, 0)$, and the transverse axis is the y-axis. Comparing the equation to $\dfrac{y^2}{a^2} - \dfrac{x^2}{b^2} = 1$, we find that $a^2 = 4$ and $b^2 = 1$. Because $b^2 = c^2 - a^2$ or $c^2 = a^2 + b^2$, we find that $c^2 = 4 + 1 = 5$. The vertices are at $(0, \pm a) = (0, \pm 2)$. The foci are at $(0, \pm c) = \left(0, \pm \sqrt{5}\right)$. Because the hyperbola opens up and down, we locate four additional points on the hyperbola by determining points left and right of each focus. To find these points, we let $y = \pm\sqrt{5}$ in the equation $\dfrac{y^2}{4} - \dfrac{x^2}{1} = 1$ and solve for x.

Figure 39

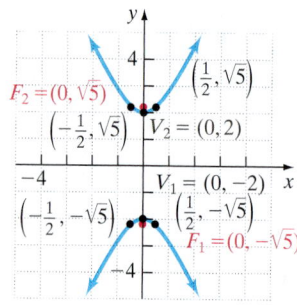

$$\frac{\left(\pm\sqrt{5}\right)^2}{4} - \frac{x^2}{1} = 1$$

$$\frac{5}{4} - \frac{x^2}{1} = 1$$

Subtract $\dfrac{5}{4}$ from both sides: $-x^2 = -\dfrac{1}{4}$

Multiply both sides by -1: $x^2 = \dfrac{1}{4}$

Take the square root of both sides: $x = \pm\dfrac{1}{2}$

Four additional points on the graph are $\left(-\dfrac{1}{2}, \pm\sqrt{5}\right)$ and $\left(\dfrac{1}{2}, \pm\sqrt{5}\right)$. See Figure 39 for the graph of the hyperbola.

QUICK ✔ *Graph each hyperbola.*

1. $\dfrac{x^2}{36} - \dfrac{y^2}{64} = 1$ **2.** $\dfrac{y^2}{9} - \dfrac{x^2}{16} = 1$

---

② Find the Equation of a Hyperbola in Which the Center Is the Origin

Just as we did with parabolas and ellipses, we are now going to use information about a hyperbola to find its equation.

EXAMPLE 3 Finding and Graphing the Equation of a Hyperbola

Find an equation of the hyperbola with center at the origin, one focus at $(0, -3)$, and vertices at $(0, -2)$ and $(0, 2)$. Graph the equation.

Solution

The center of a hyperbola is located at the midpoint of vertices (or foci). Because the vertices are at $(0, -2)$ and $(0, 2)$, the center must be at the origin, $(0, 0)$. If we were to plot the given focus and the vertices, we would notice that they all lie on the y-axis. Therefore, the transverse axis is the y-axis and the hyperbola opens up and down. Since there is a vertex at $(0, 2)$ and the center is the origin, we have that $a = 2$. Since there is a focus at $(0, -3)$ and the center is the origin, we have that $c = 3$. Since $b^2 = c^2 - a^2$, we have that $b^2 = 3^2 - 2^2 = 9 - 4 = 5$. Since the hyperbola opens up and down and the center is the origin, the equation of the hyperbola is of the form

$$\frac{y^2}{a^2} - \frac{x^2}{b^2} = 1$$

$a^2 = 4;\, b^2 = 5:$ $\dfrac{y^2}{4} - \dfrac{x^2}{5} = 1$

We wish to find points left and right of each focus, so we let $y = \pm3$ in the equation $\dfrac{y^2}{4} - \dfrac{x^2}{5} = 1$.

$$\frac{(\pm 3)^2}{4} - \frac{x^2}{5} = 1$$

$$\frac{9}{4} - \frac{x^2}{5} = 1$$

Subtract $\frac{9}{4}$ from both sides: $-\frac{x^2}{5} = -\frac{5}{4}$

Multiply both sides by -5: $x^2 = \frac{25}{4}$

Take the square root of both sides: $x = \pm\frac{5}{2}$

The points left and right of the foci are $\left(-\frac{5}{2}, \pm 3\right)$ and $\left(\frac{5}{2}, \pm 3\right)$. See Figure 40 for the graph of the hyperbola.

Figure 40

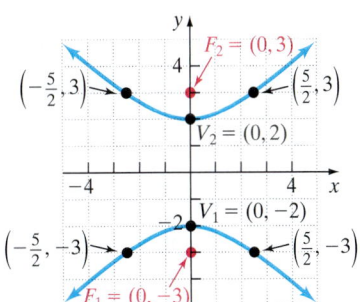

Look at the equations of the hyperbolas in Examples 1 and 3. For the hyperbola in Example 1, $a^2 = 16$ and $b^2 = 4$, so $a > b$; for the hyperbola in Example 3, $a^2 = 4$ and $b^2 = 5$, so $a < b$. We conclude that, for hyperbolas, there are no requirements involving the relative sizes for a and b. Contrast this situation to the case of an ellipse, in which the relative sizes of a and b dictate which axis is the major axis.

QUICK ✓

3. Find the equation of a hyperbola whose vertices are $(-4, 0)$ and $(4, 0)$ and has a focus at $(6, 0)$. Graph the hyperbola.

---

(3) **Find the Asymptotes of a Hyperbola in Which the Center Is the Origin**

Work Smart

Asymptotes provide an alternative method for determining the opening of each branch of the hyperbola, rather than finding and plotting four additional points.

As x and y get larger in both the positive and negative direction, the branches of the hyperbola approach two lines, called **asymptotes** of the hyperbola. The asymptotes provide guidance in graphing hyperbolas. Table 6 summarizes the asymptotes of the two hyperbolas discussed in this section.

Table 6 Asymptotes of a Hyperbola	
Hyperbola	**Asymptotes**
$\dfrac{x^2}{a^2} - \dfrac{y^2}{b^2} = 1$	$y = -\dfrac{b}{a}x$ and $y = \dfrac{b}{a}x$
$\dfrac{y^2}{a^2} - \dfrac{x^2}{b^2} = 1$	$y = -\dfrac{a}{b}x$ and $y = \dfrac{a}{b}x$

Figure 41 illustrates how the asymptotes can be used to help graph a hyperbola. It is important to remember that the asymptotes are not part of the hyperbola—they only serve as guides in graphing the hyperbola.

Figure 41

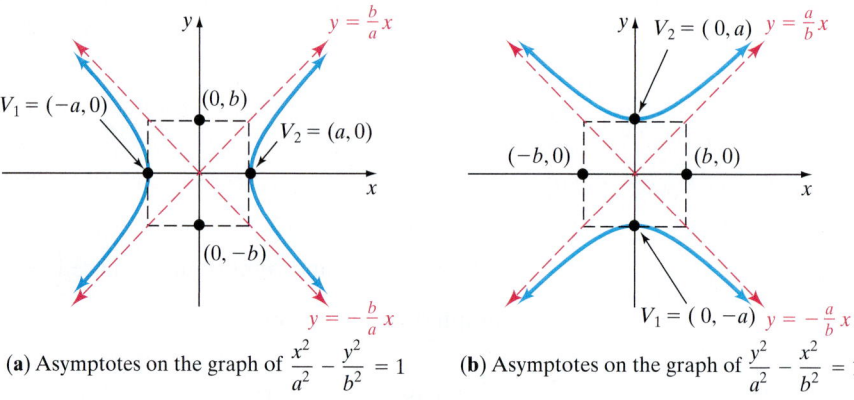

(a) Asymptotes on the graph of $\dfrac{x^2}{a^2} - \dfrac{y^2}{b^2} = 1$

(b) Asymptotes on the graph of $\dfrac{y^2}{a^2} - \dfrac{x^2}{b^2} = 1$

For example, suppose that we want to graph the equation

$$\frac{x^2}{a^2} - \frac{y^2}{b^2} = 1$$

We begin by plotting the vertices $(-a, 0)$ and $(a, 0)$. Then we plot the points $(0, -b)$ and $(0, b)$. We use these four points to construct a rectangle as shown in Figure 41(a). The diagonals of this rectangle have slopes $\dfrac{b}{a}$ and $-\dfrac{b}{a}$. If we draw a line through the corners of the rectangle, we have the asymptotes of the hyperbola. The equations of these asymptotes are $y = -\dfrac{b}{a}x$ and $y = \dfrac{b}{a}x$. Using this technique allows us to avoid plotting additional points, as was done earlier.

Figure 42

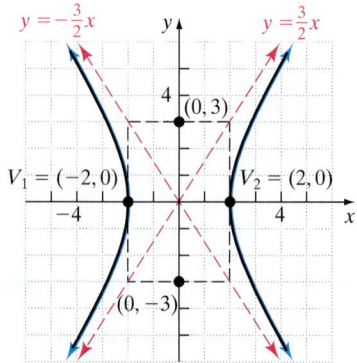

Work Smart: Study Skills

Which equation is that of an ellipse and which is the equation of a hyperbola?

$$\frac{x^2}{4} + y^2 = 1$$

$$\frac{x^2}{4} - y^2 = 1$$

Summarize how you can tell the difference between the equation of an ellipse and the equation of a hyperbola.

EXAMPLE 4 **Graphing a Hyperbola and Finding Its Asymptotes**

Graph the equation $9x^2 - 4y^2 = 36$ using the asymptotes as a guide.

Solution

We divide both sides of the equation by 36 to put the equation in the form $\dfrac{x^2}{a^2} - \dfrac{y^2}{b^2} = 1$.

$$\frac{x^2}{4} - \frac{y^2}{9} = 1$$

The center of the hyperbola is the origin, $(0, 0)$. Because the x^2-term is first, the hyperbola opens left and right. The transverse axis is along the x-axis. We can see that $a^2 = 4$ and $b^2 = 9$. Because $c^2 = a^2 + b^2$, we have that $c^2 = 4 + 9 = 13$, so that $c = \pm\sqrt{13}$. The vertices are at $(\pm a, 0) = (\pm 2, 0)$; the foci are at $(\pm c, 0) = \left(\pm\sqrt{13}, 0\right)$. Since $b^2 = 9$, we have that $b = \pm 3$. The equations of the asymptotes are

$$y = \frac{b}{a}x = \frac{3}{2}x \quad \text{and} \quad y = -\frac{b}{a}x = -\frac{3}{2}x$$

To graph the hyperbola, we form a rectangle using the points $(\pm a, 0) = (\pm 2, 0)$ and $(0, \pm b) = (0, \pm 3)$. The diagonals help us to draw the asymptotes. See Figure 42.

Quick ✓

4. Graph the equation $x^2 - 9y^2 = 9$ using the asymptotes as a guide.

5. Graph the equation $\dfrac{y^2}{16} - \dfrac{x^2}{9} = 1$ using the asymptotes as a guide.

12.5 Exercises

For Extra Help:

Student Solutions Manual   CD Video   PH Math/Tutor Center   MathXL Tutorials on CD   MathXL®   MyMathLab

Concepts and Vocabulary

In Problems 1–3, fill in the blanks.

1. A(n) _____ is the collection of points in the plane the difference of whose distances from two fixed points is a constant.

2. For a hyperbola, the foci lie on a line called the _____ _____.

3. The asymptotes of the hyperbola $\dfrac{x^2}{a^2} - \dfrac{y^2}{b^2} = 1$ are _____ and _____.

In Problems 4–6, answer True or False to each statement.

4. The line through the center of a hyperbola that is perpendicular to the transverse axis is called the conjugate axis.

5. Hyperbolas will always have asymptotes.

6. In any hyperbola, it must be the case that $a > b$.

7. Explain how the asymptotes of a hyperbola are helpful in obtaining its graph.

8. How can you tell the difference between the equation of a hyperbola and an ellipse just by looking at its equation?

Building Skills

In Problems 9–12, the graph of a hyperbola is given. Match each graph to its equation.

(a) $\dfrac{x^2}{4} - y^2 = 1$ **(b)** $x^2 - \dfrac{y^2}{4} = 1$ **(c)** $\dfrac{y^2}{4} - x^2 = 1$ **(d)** $y^2 - \dfrac{x^2}{4} = 1$

9. 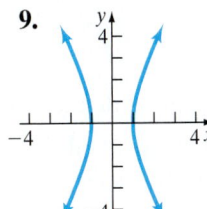   **10.** 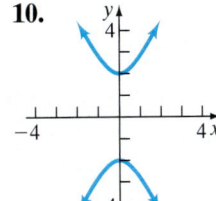   **11.** 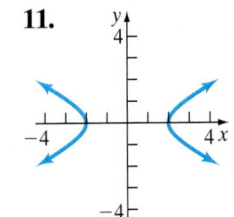   **12.**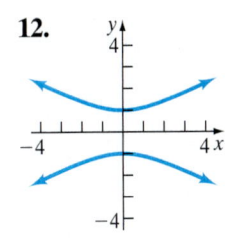

In Problems 13–20, graph each equation. You may either plot additional points or use the asymptotes to obtain the graph.

13. $\dfrac{x^2}{4} - \dfrac{y^2}{16} = 1$ **14.** $\dfrac{x^2}{9} - \dfrac{y^2}{16} = 1$ **15.** $\dfrac{y^2}{25} - \dfrac{x^2}{36} = 1$

16. $\dfrac{y^2}{81} - \dfrac{x^2}{9} = 1$ **17.** $4x^2 - y^2 = 36$ **18.** $x^2 - 9y^2 = 36$

19. $25y^2 - x^2 = 100$ **20.** $4y^2 - 9x^2 = 36$

In Problems 21–30, find the equation for the hyperbola described. Graph the equation.

21. Center at $(0, 0)$; focus at $(3, 0)$, vertex at $(2, 0)$

22. Center at $(0, 0)$; focus at $(-4, 0)$, vertex $(-1, 0)$

23. Vertices at $(0, 5)$ and $(0, -5)$; focus at $(0, 7)$

24. Vertices at $(0, 6)$ and $(0, -6)$; focus at $(0, 8)$

25. Foci at $(-10, 0)$ and $(10, 0)$; vertex at $(-7, 0)$

26. Foci at $(-5, 0)$ and $(5, 0)$; vertex at $(-3, 0)$

27. Vertices at $(0, -8)$ and $(0, 8)$; asymptote the line $y = 2x$

28. Vertices at $(0, -4)$ and $(0, 4)$; asymptote the line $y = 2x$

29. Foci at $(-3, 0)$ and $(3, 0)$; asymptote the line $y = x$

30. Foci at $(-9, 0)$ and $(9, 0)$; asymptote the line $y = -3x$

Applying the Concepts

In Problems 31–34, write the equation of the hyperbola.

31.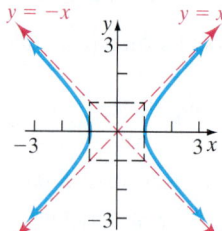

32.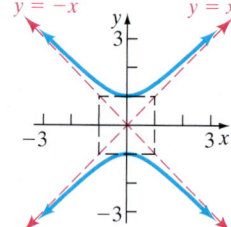

33.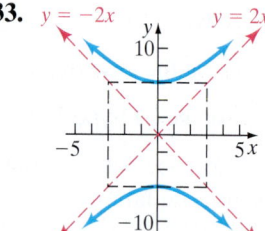

34.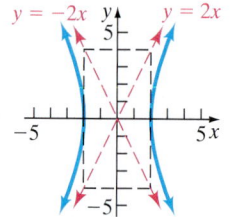

Extending the Concepts

35. Two hyperbolas that have the same set of asymptotes are called **conjugate.** Show that the hyperbolas

$$\frac{x^2}{4} - y^2 = 1 \qquad \text{and} \qquad y^2 - \frac{x^2}{4} = 1$$

are conjugate. Graph each hyperbola on the same Cartesian plane.

36. The **eccentricity** e of a hyperbola is defined as the number $\frac{c}{a}$. Because $c > a$, it follows that $e > 1$. Describe the general shape of a hyperbola whose eccentricity is close to 1. What is the shape if e is very large?

Synthesis Review

In Problems 37–41, solve each system of equations using either substitution or elimination.

37. $\begin{cases} 2x - 3y = -9 \\ -x + 5y = 8 \end{cases}$

38. $\begin{cases} 3x + 4y = 3 \\ -6x + 2y = -\dfrac{7}{2} \end{cases}$

39. $\begin{cases} 2x - 3y = 6 \\ -6x + 9y = -18 \end{cases}$

40. $\begin{cases} -2x + y = 8 \\ x - \dfrac{1}{2}y = -4 \end{cases}$

41. $\begin{cases} 6x + 3y = 4 \\ -2x - y = -\dfrac{4}{3} \end{cases}$

42. Which method did you use more often? Why? Do you think there are situations where substitution is superior to elimination? Are there situations where elimination is superior to substitution? Describe these circumstances.

The Graphing Calculator

A graphing calculator can be used to graph hyperbolas by solving the equation for y. Because hyperbolas are not functions, we need to graph the upper half and lower half of the hyperbola in two pieces. For example, to graph $\dfrac{x^2}{4} - \dfrac{y^2}{9} = 1$, we would graph $Y_1 = 3\sqrt{\dfrac{x^2}{4} - 1}$ and $Y_2 = -3\sqrt{\dfrac{x^2}{4} - 1}$. We obtain the graph shown in Figure 43.

Figure 43

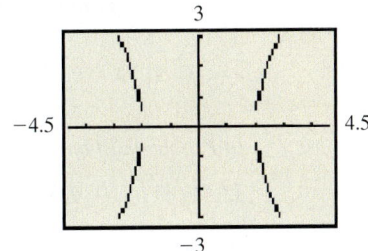

In Problems 43–50, graph each hyperbola using a graphing calculator.

43. $\dfrac{x^2}{4} - \dfrac{y^2}{16} = 1$

44. $\dfrac{x^2}{9} - \dfrac{y^2}{16} = 1$

45. $\dfrac{y^2}{25} - \dfrac{x^2}{36} = 1$

46. $\dfrac{y^2}{81} - \dfrac{x^2}{9} = 1$

47. $4x^2 - y^2 = 36$

48. $x^2 - 9y^2 = 36$

49. $25y^2 - x^2 = 100$

50. $4y^2 - 9x^2 = 36$

PUTTING THE CONCEPTS TOGETHER (Sections 12.1–12.5)

These problems cover important concepts from Sections 12.1 to 12.5. We designed these problems so that you can review the chapter so far and show your mastery of the concepts. Take time to work these problems before proceeding with the next section. The answers to these problems are located at the back of the text starting on page AN-55.

1. Find the exact distance $d(P_1, P_2)$ between points $P_1 = (-6, 4)$ and $P_2 = (3, -2)$.

2. Find the midpoint of the line segment formed by joining the points $P_1 = (-3, 1)$ and $P_2 = (5, -7)$.

In Problems 3 and 4, find the center (h, k) and the radius r of each circle. Graph each circle.

3. $(x + 2)^2 + (y - 8)^2 = 36$

4. $x^2 + y^2 + 6x - 4y - 3 = 0$

In Problems 5 and 6, find the standard form of the equation of each circle.

5. Center $(0, 0)$; contains the point $(-5, 12)$

6. With endpoints of a diameter at $(-1, 5)$ and $(5, -3)$.

In Problems 7 and 8, find the vertex, focus, and directrix of each parabola. Graph each parabola.

7. $(x + 2)^2 = -4(y - 4)$

8. $y^2 + 2y - 8x + 25 = 0$

In Problems 9 and 10, find an equation for each parabola described.

9. Vertex at $(-1, -2)$; focus at $(-1, -5)$

10. Vertex at $(-3, 3)$; contains the point $(-1, 7)$; axis of symmetry parallel to the x-axis

In Problems 11 and 12, find the vertices and foci of each ellipse. Graph each ellipse.

11. $x^2 + 9y^2 = 81$

12. $\dfrac{(x + 1)^2}{36} + \dfrac{(y - 2)^2}{49} = 1$

In Problems 13 and 14, find an equation for each ellipse described.

13. Foci at $(0, \pm 6)$; vertices at $(0, \pm 9)$

14. Center at $(3, -4)$; vertex at $(7, -4)$; focus at $(6, -4)$

In Problems 15 and 16, find the vertices, foci, and asymptotes for each hyperbola. Graph each hyperbola using the asymptotes as a guide.

15. $\dfrac{y^2}{81} - \dfrac{x^2}{9} = 1$

16. $25x^2 - y^2 = 25$

17. Find an equation for a hyperbola with center at $(0, 0)$, focus at $(0, -5)$ and vertex at $(0, -2)$.

18. A large flood light is in the shape of a parabola. Its diameter is 36 inches and its depth is 12 inches. How far from the vertex should the light bulb be placed so that the rays will be reflected parallel to the axis?

12.6 Nonlinear Systems of Equations

OBJECTIVES

1. Solve a System of Nonlinear Equations Using Substitution
2. Solve a System of Nonlinear Equations Using Elimination

Preparing for Nonlinear Systems of Equations

Before getting started, take the following readiness quiz. If you get a problem wrong, go back to the section cited and review the material.

1. Solve the system using substitution: $\begin{cases} y = 2x - 5 \\ 2x - 3y = 7 \end{cases}$ [Section 4.2, pp. 285–290]

2. Solve the system using elimination: $\begin{cases} 2x - 4y = -11 \\ -x + 5y = 13 \end{cases}$ [Section 4.3, pp. 296–300]

3. Solve the system: $\begin{cases} 3x - 5y = 4 \\ -6x + 10y = -8 \end{cases}$ [Section 4.3, pp. 300–301]

Recall from Section 4.1 that a system of equations is a collection of two or more equations, each containing one or more variables. We learned techniques for solving systems of linear equations back in Chapter 4. We now introduce techniques for solving systems of equations where the equations are not linear. We only deal with systems containing two equations with two unknowns. A **system of nonlinear equations** in two variables is a system of equations in which at least one of the equations is not linear. That is, at least

Preparing for...Answers **1.** $(2, -1)$
2. $(-1/2, 5/2)$
3. Infinitely many solutions

one of the equations cannot be written in the form $Ax + By = C$. The following are examples of nonlinear systems of equations containing two unknowns.

$$\begin{cases} x + y^2 = 5 & \text{(1) A parabola} \\ 2x + y = 4 & \text{(2) A line} \end{cases} \qquad \begin{cases} x^2 + y^2 = 9 & \text{(1) A circle} \\ -x^2 + y = 9 & \text{(2) A parabola} \end{cases}$$

In Section 4.1, we saw that the solution to a system of linear equations could be found geometrically by determining the point of intersection of the equations in the system. The same idea holds true for nonlinear systems—the point(s) of intersection represent the solution(s) to the system.

In looking back to the problems from Chapter 4, you probably started to get a sense as to when substitution was the best approach for solving a system and when elimination was the best approach. The same deal holds for nonlinear systems—sometimes substitution is best, sometimes elimination is best. Experience and a certain degree of imagination are your friends when solving these problems.

① Solve a System of Nonlinear Equations Using Substitution

The method of substitution for solving a system of nonlinear equations follows much the same approach as that for solving systems of linear equations using substitution.

EXAMPLE 1   **How to Solve a System of Nonlinear Equations Using Substitution**

Solve the following system of equations using substitution: $\begin{cases} 3x - y = -2 & \text{(1) A line} \\ 2x^2 - y = 0 & \text{(2) A parabola} \end{cases}$

Step-by-Step Solution

Step 1: Graph each equation in the system.

Figure 44 shows the graphs of $3x - y = -2 (y = 3x + 2)$ and $2x^2 - y = 0 (y = 2x^2)$. The system apparently has two solutions.

Figure 44

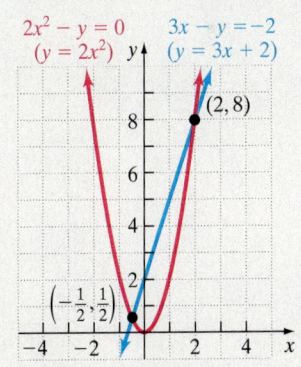

Step 2: Solve equation (1) for y.

Add y to both sides; $\quad 3x - y = -2$
add 2 to both sides: $\qquad\qquad y = 3x + 2$

Step 3: Substitute $3x + 2$ for y in equation (2).

Equation (2): $\qquad 2x^2 - y = 0$
$$2x^2 - (3x + 2) = 0$$
Distribute: $\quad 2x^2 - 3x - 2 = 0$

Step 4: Solve for x.

Factor: $\qquad\qquad (2x + 1)(x - 2) = 0$
Zero-Product Property: $\quad 2x + 1 = 0 \quad$ or $\quad x - 2 = 0$
$$x = -\frac{1}{2} \quad \text{or} \quad x = 2$$

(continued)

Step 5: Let $x = -\dfrac{1}{2}$ and $x = 2$ in equation (1) to determine y.

Equation (1): $\quad 3x - y = -2$

$x = -\dfrac{1}{2}: \quad 3\left(-\dfrac{1}{2}\right) - y = -2$

$$-\dfrac{3}{2} - y = -2$$

Add $\dfrac{3}{2}$ to both sides: $\qquad\qquad -y = -\dfrac{1}{2}$

Multiply both sides by -1: $\qquad\qquad y = \dfrac{1}{2}$

$x = 2: \quad 3(2) - y = -2$

$$6 - y = -2$$

Subtract 6 from both sides: $\qquad\qquad -y = -8$

Multiply both sides by -1: $\qquad\qquad y = 8$

The apparent solutions are $\left(-\dfrac{1}{2}, \dfrac{1}{2}\right)$ and $(2, 8)$.

Step 6: Check

$x = -\dfrac{1}{2}; y = \dfrac{1}{2}: \quad 3x - y = -2 \qquad\qquad\qquad 2x^2 - y = 0$

$3\left(-\dfrac{1}{2}\right) - \dfrac{1}{2} \overset{?}{=} -2 \qquad\qquad 2\left(-\dfrac{1}{2}\right)^2 - \dfrac{1}{2} \overset{?}{=} 0$

$-\dfrac{3}{2} - \dfrac{1}{2} \overset{?}{=} -2 \qquad\qquad\qquad 2\left(\dfrac{1}{4}\right) - \dfrac{1}{2} \overset{?}{=} 0$

$-\dfrac{4}{2} = -2 \quad \text{True} \qquad\qquad\qquad \dfrac{1}{2} - \dfrac{1}{2} = 0 \quad \text{True}$

$x = 2; y = 8: \quad 3x - y = -2 \qquad\qquad\qquad 2x^2 - y = 0$

$3(2) - 8 \overset{?}{=} -2 \qquad\qquad\qquad 2(2)^2 - 8 \overset{?}{=} 0$

$6 - 8 = -2 \quad \text{True} \qquad\qquad\qquad 2(4) - 8 = 0 \quad \text{True}$

Each solution checks. We now know that the graphs in Figure 44 intersect at $\left(-\dfrac{1}{2}, \dfrac{1}{2}\right)$ and $(2, 8)$.

Notice that the steps for solving a system of nonlinear equations using substitution are identical to the steps for solving a system of linear equations using substitution.

QUICK ✓

1. Solve the following system of equations using substitution: $\begin{cases} 2x + y = -1 \\ x^2 - y = 4 \end{cases}$

EXAMPLE 2 Solving a System of Nonlinear Equations Using Substitution

Solve the following system of equations using substitution:

$\begin{cases} x + y = 2 & \text{(1) A line} \\ (x + 2)^2 + (y - 1)^2 = 9 & \text{(2) A circle} \end{cases}$

Solution

Figure 45

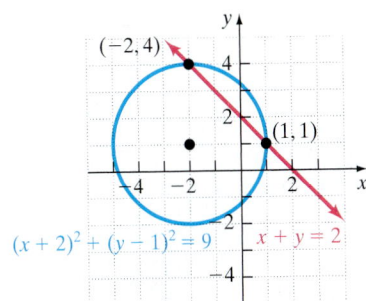

$(x+2)^2 + (y-1)^2 = 9$ $x + y = 2$

We start by graphing each equation in the system. Figure 45 shows the graphs of $x + y = 2 (y = -x + 2)$ and $(x + 2)^2 + (y - 1)^2 = 9$. The system apparently has two solutions.

Solve equation (1) for y.

$$x + y = 2$$

Add x to both sides: $y = -x + 2$

Substitute this expression for y into equation (2), $(x + 2)^2 + (y - 1)^2 = 9$, and then solve for x.

Substitute $-x + 2$ for y in equation (2):	$(x + 2)^2 + (\quad - 1)^2 = 9$
Combine like terms:	$(x + 2)^2 + (1 - x)^2 = 9$
FOIL	$x^2 + 4x + 4 + 1 - 2x + x^2 = 9$
Combine like terms:	$2x^2 + 2x + 5 = 9$
Put in standard form:	$2x^2 + 2x - 4 = 0$
Factor:	$2(x^2 + x - 2) = 0$
Divide both sides by 2:	$x^2 + x - 2 = 0$
Factor:	$(x + 2)(x - 1) = 0$
Zero-Product Property:	$x + 2 = 0$ or $x - 1 = 0$
Solve for x:	$x = -2$ or $x = 1$

Let $x = -2$ and $x = 1$ in equation (1) to determine y.

$$x + y = 2$$

$x = -2$: $-2 + y = 2$ $\qquad$ $x = 1$: $1 + y = 2$

Add 2 to both sides: $y = 4$ $\qquad$ Subtract 1 from both sides: $y = 1$

The apparent solutions are $(-2, 4)$ and $(1, 1)$.

Check

$x = -2$; $y = 4$: $x + y = 2$ $\qquad\qquad$ $(x + 2)^2 + (y - 1)^2 = 9$

$\qquad\qquad\qquad$ $-2 + 4 = 2$ True $\qquad$ $(-2 + 2)^2 + (4 - 1)^2 = 9$

$\qquad\qquad\qquad\qquad\qquad\qquad\qquad\qquad\qquad$ $3^2 = 9$ True

$x = 1$; $y = 1$: $x + y = 2$ $\qquad\qquad$ $(x + 2)^2 + (y - 1)^2 = 9$

$\qquad\qquad\qquad$ $1 + 1 = 2$ True $\qquad$ $(1 + 2)^2 + (1 - 1)^2 = 9$

$\qquad\qquad\qquad\qquad\qquad\qquad\qquad\qquad\qquad$ $3^2 = 9$

$\qquad\qquad\qquad\qquad\qquad\qquad\qquad\qquad\qquad$ $9 = 9$ True

Each solution checks. We now know that the graphs in Figure 45 intersect at $(-2, 4)$ and $(1, 1)$.

Quick ✔

2. Solve the following system of equations using substitution:

$$\begin{cases} 2x + y = 0 \\ (x - 4)^2 + (y + 2)^2 = 9 \end{cases}$$

---

② Solve a System of Nonlinear Equations Using Elimination

Now we discuss the method of elimination. The method of elimination for solving a system of nonlinear equations follows much the same approach as that for solving systems of linear equations using elimination. In fact, the steps are identical.

Recall, the basic idea in using the elimination is to get the coefficients of one of the variables to be additive inverses.

EXAMPLE 3 **How to Solve a System of Nonlinear Equations by Elimination**

Solve the following system of equations using elimination: $\begin{cases} x^2 + y^2 = 13 & \text{(1) A circle} \\ x^2 - y = 7 & \text{(2) A parabola} \end{cases}$

Step-by-Step Solution

Step 1: Graph each equation in the system.

Figure 46 shows the graphs of $x^2 + y^2 = 13$ and $x^2 - y = 7$ ($y = x^2 - 7$). The system apparently has four solutions.

Figure 46

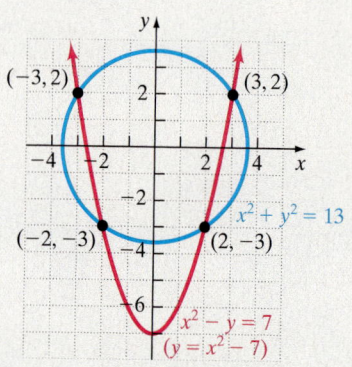

Step 2: Multiply equation (2) by -1 so that the coefficients on x^2 become additive inverses.

$$\begin{cases} x^2 + y^2 = 13 & \text{(1)} \\ -x^2 + y = -7 & \text{(2)} \end{cases}$$

Step 3: Add equations (1) and (2) to eliminate x^2. Solve the resulting equation for y.

Add: $\begin{cases} x^2 + y^2 = 13 & \text{(1)} \\ \underline{-x^2 + y = -7} & \text{(2)} \end{cases}$

$$y^2 + y = 6$$

Put in standard form: $\qquad y^2 + y - 6 = 0$

Factor: $\qquad (y + 3)(y - 2) = 0$

Zero-Product Property: $\quad y + 3 = 0 \quad \text{or} \quad y - 2 = 0$

$$y = -3 \quad \text{or} \qquad y = 2$$

Step 4: Solve for x using equation (2).

$$x^2 - y = 7 \quad \text{(2)}$$

Using $y = -3$: $\quad x^2 - (-3) = 7 \qquad$ Using $y = 2$: $\quad x^2 - 2 = 7$

$$x^2 + 3 = 7 \qquad\qquad\qquad\qquad x^2 = 9$$

$$x^2 = 4 \qquad\qquad\qquad\qquad\qquad x = \pm 3$$

Work Smart

We use equation (2) in Step 4 because it's easier to work with.

$$x = \pm 2$$

The apparent solutions are $(-2, -3)$, $(2, -3)$, $(-3, 2)$, and $(3, 2)$.

Step 5: Check

We leave it to you to verify that all four of the apparent solutions are solutions to the system. The four points $(-2, -3)$, $(2, -3)$, $(-3, 2)$, and $(3, 2)$ are the points of intersection of the graphs. Look again at Figure 46.

EXAMPLE 4 **Solving a System of Nonlinear Equations by Elimination**

Solve the following system of equations using elimination:

$$\begin{cases} x^2 - y^2 = 4 & \text{(1) A hyperbola} \\ x^2 - y = 0 & \text{(2) A parabola} \end{cases}$$

Solution

We graph each equation in the system. Figure 47 shows the graphs of $x^2 - y^2 = 4$ and $x^2 - y = 0 (y = x^2)$. The system apparently has no solution.

Now multiply equation (2) by -1 to get the coefficients on x^2 to be additive inverses.

$$\begin{cases} x^2 - y^2 = 4 & \text{(1)} \\ -x^2 + y = 0 & \text{(2)} \end{cases}$$

Figure 47

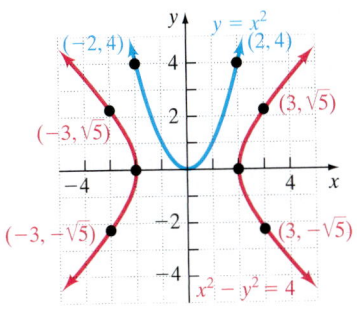

Add equations (1) and (2) to eliminate x^2. Solve the resulting equation for y.

$$\begin{cases} x^2 - y^2 = 4 & \text{(1)} \\ \underline{-x^2 + y = 0} & \text{(2)} \end{cases}$$
$$-y^2 + y = 4$$

Add y^2 to both sides; Add y to both sides: $\quad y^2 - y + 4 = 0$

This is a quadratic equation whose discriminant is $b^2 - 4ac = (-1)^2 - 4(1)(4) = 1 - 16 = -15$. The equation has no real solution. Therefore, the system of equations is inconsistent. The solution set is $\varnothing$ or $\{\ \}$. Figure 47 confirms this result. ▬

QUICK ✓ *Solve each system of nonlinear equations using elimination.*

3. $\begin{cases} x^2 + y^2 = 16 \\ x^2 - 2y = 8 \end{cases}$ **4.** $\begin{cases} x^2 - y = -4 \\ x^2 + y^2 = 9 \end{cases}$

12.6 Exercises

For Extra Help:    *Math*XL *MyMathLab*

Student Solutions Manual   CD Video   PH Math/Tutor Center   MathXL Tutorials on CD   MathXL®   MyMathLab

Concepts and Vocabulary

In Problem 1, fill in the blanks.

1. The two methods for solving a system of nonlinear equations in two unknowns discussed in this section are _____ and _____.

In Problem 2, answer True or False to the statement.

2. A system of nonlinear equations in two unknowns must have at least one solution.

3. How does a system of nonlinear equations differ from a system of linear equations?

4. Make up a system of nonlinear equations. Then solve the system.

Building Skills

In Problems 5–12, solve the system of nonlinear equations using the method of substitution.

5. $\begin{cases} y = x^2 + 4 \\ y = x + 4 \end{cases}$ **6.** $\begin{cases} y = x^3 + 2 \\ y = x + 2 \end{cases}$ **7.** $\begin{cases} y = \sqrt{25 - x^2} \\ x + y = 7 \end{cases}$

8. $\begin{cases} y = \sqrt{100 - x^2} \\ x + y = 14 \end{cases}$ **9.** $\begin{cases} x^2 + y^2 = 4 \\ y = x^2 - 2 \end{cases}$ **10.** $\begin{cases} x^2 + y^2 = 16 \\ y = x^2 - 4 \end{cases}$

11. $\begin{cases} xy = 4 \\ x^2 + y^2 = 8 \end{cases}$ **12.** $\begin{cases} xy = 1 \\ x^2 - y = 0 \end{cases}$

In Problems 13–20, solve the system of nonlinear equations using the method of elimination.

13. $\begin{cases} x^2 + y^2 = 4 \\ y^2 - x = 4 \end{cases}$

14. $\begin{cases} x^2 + y^2 = 8 \\ x^2 + y^2 + 4y = 0 \end{cases}$

15. $\begin{cases} x^2 + y^2 = 7 \\ x^2 - y^2 = 25 \end{cases}$

16. $\begin{cases} 4x^2 + 16y^2 = 16 \\ 2x^2 - 2y^2 = 8 \end{cases}$

17. $\begin{cases} x^2 + y^2 = 6y \\ x^2 = 3y \end{cases}$

18. $\begin{cases} 2x^2 + y^2 = 18 \\ x^2 - y^2 = 9 \end{cases}$

19. $\begin{cases} x^2 - 2x - y = 8 \\ 6x + 2y = -4 \end{cases}$

20. $\begin{cases} 2x^2 - 5x + y = 12 \\ 14x - 2y = -16 \end{cases}$

Mixed Practice

In Problems 21–36, solve the system of nonlinear equations using any method you wish.

21. $\begin{cases} y = x^2 - 6x + 4 \\ 5x + y = 6 \end{cases}$

22. $\begin{cases} y = x^2 + 4x + 5 \\ x - y = 9 \end{cases}$

23. $\begin{cases} x^2 + y^2 = 16 \\ x^2 - y^2 = 16 \end{cases}$

24. $\begin{cases} x^2 + y^2 = 25 \\ x^2 - y^2 = 25 \end{cases}$

25. $\begin{cases} (x - 4)^2 + y^2 = 25 \\ x - y = -3 \end{cases}$

26. $\begin{cases} (x + 5)^2 + (y - 2)^2 = 100 \\ 8x + y = 18 \end{cases}$

27. $\begin{cases} (x - 1)^2 + (y + 2)^2 = 4 \\ y^2 + 4y - x = -1 \end{cases}$

28. $\begin{cases} (x + 2)^2 + (y - 1)^2 = 4 \\ y^2 - 2y - x = 5 \end{cases}$

29. $\begin{cases} (x + 3)^2 + 4y^2 = 4 \\ x^2 + 6x - y = 13 \end{cases}$

30. $\begin{cases} 9x^2 + 4y^2 = 36 \\ x^2 + (y - 7)^2 = 4 \end{cases}$

31. $\begin{cases} x^2 - y^2 = 21 \\ x + y = 7 \end{cases}$

32. $\begin{cases} y - 2x = 1 \\ 2x^2 + y^2 = 1 \end{cases}$

33. $\begin{cases} x^2 + 2y^2 = 16 \\ 4x^2 - y^2 = 24 \end{cases}$

34. $\begin{cases} 4x^2 + 3y^2 = 4 \\ 6y^2 - 2x^2 = 3 \end{cases}$

35. $\begin{cases} x^2 + y^2 = 25 \\ y = -x^2 + 6x - 5 \end{cases}$

36. $\begin{cases} x^2 + y^2 = 65 \\ y = -x^2 + 9 \end{cases}$

Applying the Concepts

37. Fun with Numbers The difference of two numbers is 2. The sum of their squares is 34. Find the numbers.

38. Fun with Numbers The sum of two numbers is 8. The sum of their squares is 160. Find the numbers.

△ **39. Perimeter and Area of a Rectangle** The perimeter of a rectangle is 48 feet. The area of the rectangle is 140 square feet. Find the dimensions of the rectangle.

△ **40. Perimeter and Area of a Rectangle** The perimeter of a rectangle is 64 meters. The area of the rectangle is 240 square meters. Find the dimensions of the rectangle.

41. Constructing a Box A rectangular piece of cardboard, whose area is 190 square centimeters, is made into an open box by cutting a 2-centimeter square from each corner and turning up the sides. See the figure. If the box is to have a volume of 180 cubic centimeters, what size cardboard should you start with?

  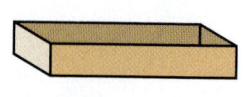

42. Fencing A farmer has 132 yards of fencing available to enclose a 900 square yard region in the shape of adjoining squares with sides of length x and y. See the figure. Find x and y.

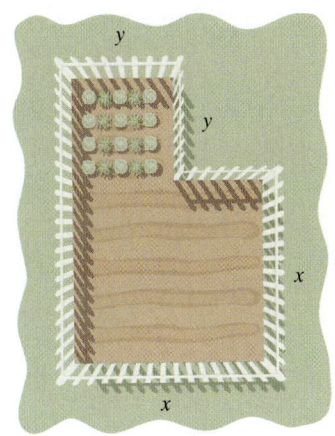

Extending the Concepts

In Problems 43–46, solve the system of nonlinear equations. Do not attempt to graph the equations in the system.

43. $\begin{cases} y^2 + y + x^2 - x - 2 = 0 \\ y + 1 + \dfrac{x - 2}{y} = 0 \end{cases}$

44. $\begin{cases} x^3 - 2x^2 + y^2 + 3y - 4 = 0 \\ x - 2 + \dfrac{y^2 - y}{x^2} = 0 \end{cases}$

45. $\begin{cases} \ln x = 4 \ln y \\ \log_3 x = 2 + 2 \log_3 y \end{cases}$

46. $\begin{cases} \ln x = 5 \ln y \\ \log_2 x = 3 + 2 \log_2 y \end{cases}$

47. If r_1 and r_2 are two solutions of a quadratic equation $ax^2 + bx + c = 0$, then it can be shown that

$$r_1 + r_2 = -\frac{b}{a} \quad \text{and} \quad r_1 r_2 = \frac{c}{a}$$

Solve this system of equations for r_1 and r_2.

48. A circle and a line intersect at most twice. A circle and a parabola intersect at most four times. How many times do you think a circle and the graph of a polynomial of degree 3 can intersect? What about a circle and the graph of a polynomial of degree 4? What about a circle and the graph of a polynomial of degree n? Explain your conclusions using an algebraic argument.

Synthesis Review

In Problems 49–53, evaluate the functions given that $f(x) = 3x + 4$ and $g(x) = 2^x$.

49. (a) $f(1)$
(b) $g(1)$

50. (a) $f(2)$
(b) $g(2)$

51. (a) $f(3)$
(b) $g(3)$

52. (a) $f(4)$
(b) $g(4)$

53. (a) $f(5)$
(b) $g(5)$

54. Use the results of Problems 49–53 to compute

(a) $f(2) - f(1)$ (b) $f(3) - f(2)$ (c) $f(4) - f(3)$ (d) $f(5) - f(4)$

(e) $\dfrac{g(2)}{g(1)}$ (f) $\dfrac{g(3)}{g(2)}$ (g) $\dfrac{g(4)}{g(3)}$ (h) $\dfrac{g(5)}{g(4)}$

(i) Make a generalization about $f(n + 1) - f(n)$ for $n \geq 1$ an integer. Make a generalization about $\dfrac{g(n + 1)}{g(n)}$ for $n \geq 1$ an integer.

The Graphing Calculator

In Problems 55–64, use a graphing calculator and the INTERSECT feature to solve the following systems of nonlinear equations.

55. $\begin{cases} y = x^2 - 6x + 4 \\ 5x + y = 6 \end{cases}$

56. $\begin{cases} y = x^2 + 4x + 5 \\ x - y = 9 \end{cases}$

57. $\begin{cases} x^2 + y^2 = 16 \\ x^2 - y^2 = 16 \end{cases}$

58. $\begin{cases} x^2 + y^2 = 25 \\ x^2 - y^2 = 25 \end{cases}$

59. $\begin{cases} (x - 4)^2 + y^2 = 25 \\ x - y = -3 \end{cases}$

60. $\begin{cases} (x + 5)^2 + (y - 2)^2 = 100 \\ 8x + y = 18 \end{cases}$

61. $\begin{cases} (x - 1)^2 + (y + 2)^2 = 4 \\ x^2 + 4y - x = -1 \end{cases}$

62. $\begin{cases} (x + 2)^2 + (y - 1)^2 = 4 \\ y^2 - 2y - x = 5 \end{cases}$

63. $\begin{cases} x^2 + 4y^2 = 4 \\ x^2 + 6x - y = -13 \end{cases}$

64. $\begin{cases} 9x^2 + 4y^2 = 36 \\ x^2 + (y - 7)^2 = 4 \end{cases}$

CHAPTER 12 ACTIVITY: How Do You Know That...?

Focus: A sharing of ideas to identify topics contained in the study of conics

Time: 30–35 minutes

Group size: 2–4

For each question, every member of the group should spend 1–2 minutes individually listing "How they know...." At the end of the allotted time, the group should convene and conduct a 2–3-minute discussion of the different responses.

How Do You Know...

...that a triangle with vertices at $(2, 6)$, $(0, -2)$ and $(5, 1)$ is an isosceles triangle?

...the coordinates for the midpoint between two given ordered pairs?

...that $x^2 + y^2 + 4x - 8y = 16$ is an equation of a circle and not a parabola?

...which way a parabola opens?

...whether an ellipse's center is the origin or another point?

...that a hyperbola will not intersect its asymptotes?

...that a circle and a parabola can have 0, 1, 2, 3, or 4 points of intersection? (If necessary, use a sketch to support your response.)

CHAPTER 12 REVIEW

Section 12.1	Distance and Midpoint Formulas

KEY CONCEPTS

- **The Distance Formula**
  The distance between two points $P_1 = (x_1, y_1)$ and $P_2 = (x_2, y_2)$, denoted by $d(P_1, P_2)$, is
  $d(P_1, P_2) = \sqrt{(x_2 - x_1)^2 + (y_2 - y_1)^2}$.

- **The Midpoint Formula**
  The midpoint $M = (x, y)$ of the line segment from $P_1 = (x_1, y_1)$ to $P_2 = (x_2, y_2)$ is $M = \left(\dfrac{x_1 + x_2}{2}, \dfrac{y_1 + y_2}{2} \right)$.

YOU SHOULD BE ABLE TO...	EXAMPLE	REVIEW EXERCISES
① Use the Distance Formula (p. 913)	Examples 1 through 3	1–6, 12
② Use the Midpoint Formula (p. 916)	Example 4	7–11

In Problems 1–6, find the distance $d(P_1, P_2)$ between points P_1 and P_2.

1. $P_1 = (0, 0)$ and $P_2 = (-4, -3)$

2. $P_1 = (-3, 2)$ and $P_2 = (5, -4)$

3. $P_1 = (-1, 1)$ and $P_2 = (5, 3)$

4. $P_1 = (6, -7)$ and $P_2 = (6, -1)$

5. $P_1 = \left(\sqrt{7}, -\sqrt{3}\right)$ and $P_2 = \left(4\sqrt{7}, 5\sqrt{3}\right)$

6. $P_1 = (-0.2, 1.7)$ and $P_2 = (1.3, 3.7)$

In Problems 7–11, find the midpoint of the line segment formed by joining the points P_1 and P_2.

7. $P_1 = (-1, 6)$ and $P_2 = (-3, 4)$

8. $P_1 = (7, 0)$ and $P_2 = (5, -4)$

9. $P_1 = \left(-\sqrt{3}, 2\sqrt{6}\right)$ and $P_2 = \left(-7\sqrt{3}, -8\sqrt{6}\right)$

10. $P_1 = (5, -2)$ and $P_2 = (0, 3)$

11. $P_1 = \left(\dfrac{1}{4}, \dfrac{2}{3}\right)$ and $P_2 = \left(\dfrac{5}{4}, \dfrac{1}{3}\right)$

12. Consider the three points $A = (-2, 2)$, $B = (1, -1)$, $C = (-1, -3)$.

    (a) Plot each point in the Cartesian plane and form the triangle ABC.

    (b) Find the length of each side of the triangle.

    (c) Verify that the triangle is a right triangle.

    (d) Find the area of the triangle.

Section 12.2 Circles	
KEY CONCEPTS	**KEY TERMS**
• **Standard Form of an Equation of a Circle** The standard form of a circle with radius r and center (h, k) is $(x - h)^2 + (y - k)^2 = r^2$ • **General Form of the Equation of a Circle** The general form of the equation of a circle is given by the equation $x^2 + y^2 + ax + by + c = 0$ when the graph exists.	Conic sections Circle Radius Center

YOU SHOULD BE ABLE TO . . .	EXAMPLE	REVIEW EXERCISES
① Write the standard form of the equation of a circle (p. 921)	Example 1	13–20
② Graph a circle (p. 921)	Example 2	15–18; 21–30
③ Find the center and radius of a circle from an equation in general form (p. 922)	Example 3	27–30

In Problems 13 and 14, find the center and radius of each circle. Write the standard form of the equation.

13.

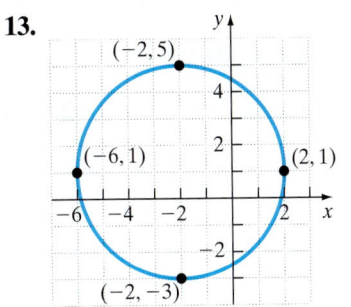

14.

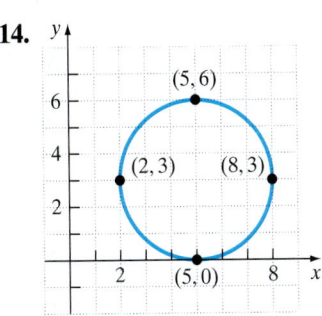

In Problems 15–18, write the standard form of the equation of each circle whose radius is r and center is (h, k). Graph each circle.

15. $r = 4$; $(h, k) = (0, 0)$

16. $r = 3$; $(h, k) = (-3, 1)$

17. $r = 1$; $(h, k) = (5, -2)$

18. $r = \sqrt{7}$; $(h, k) = (4, 0)$

In Problems 19 and 20, find the standard form of the equation of each circle.

19. Center at $(2, -1)$ and containing the point $(5, 3)$.

20. Endpoints of a diameter at $(1, 7)$ and $(-3, -1)$.

In Problems 21–26, find the center (h, k) and the radius r of each circle. Graph each circle.

21. $x^2 + y^2 = 25$

22. $(x - 1)^2 + (y - 2)^2 = 4$

23. $x^2 + (y - 4)^2 = 16$

24. $(x + 1)^2 + (y + 6)^2 = 49$

25. $(x + 2)^2 + \left(y - \dfrac{3}{2}\right)^2 = \dfrac{1}{4}$

26. $(x + 3)^2 + (y + 3)^2 = 4$

In Problems 27–30, find the center (h, k) and the radius r of each circle. Graph each circle.

27. $x^2 + y^2 + 6x + 10y - 2 = 0$

28. $x^2 + y^2 - 8x + 4y + 16 = 0$

29. $x^2 + y^2 + 2x - 4y - 4 = 0$

30. $x^2 + y^2 - 10x - 2y + 17 = 0$

Section 12.3	Parabolas

KEY CONCEPTS	**KEY TERMS**
• **Equations of a Parabola: Vertex at $(0, 0)$; $a > 0$** See Table 1 on page 928. • **Equations of a Parabola: Vertex at (h, k); $a > 0$** See Table 2 on page 931.	Parabola Focus Directrix Axis of symmetry Vertex

YOU SHOULD BE ABLE TO . . .	**EXAMPLE**	**REVIEW EXERCISES**
① Graph parabolas in which the vertex is the origin (p. 928)	Examples 1 and 2	31–36
② Find the equation of a parabola (p. 930)	Examples 3 and 4	31–34
③ Graph parabolas in which the vertex is not the origin (p. 931)	Example 5	37–39
④ Solve applied problems involving parabolas (p. 932)	Example 6	40

In Problems 31–34, find the equation of the parabola described. Graph each parabola.

31. Vertex at $(0, 0)$; focus at $(0, -3)$

32. Focus at $(-4, 0)$; directrix the line $x = 4$

33. Vertex at $(0, 0)$; contains the point $(8, -2)$; axis of symmetry the x-axis

34. Vertex at $(0, 0)$; directrix the line $y = -2$; axis of symmetry the y-axis

In Problems 35–39, find the vertex, focus, and directrix of each parabola. Graph each parabola.

35. $x^2 = 2y$

36. $y^2 = 16x$

37. $(x + 1)^2 = 8(y - 3)$

38. $(y - 4)^2 = -2(x + 3)$

39. $x^2 - 10x + 3y + 19 = 0$

40. **Radio Telescope** The U.S. Naval Research Laboratory has a giant radio telescope with a dish that is shaped like a parabola. The signals that are received by the dish strike the surface of the dish and are reflected to a single point, where the receiver is located. If the giant dish is 300 feet across and 44 feet deep at its center, at what position should the receiver be placed?

Section 12.4 Ellipses

KEY CONCEPTS	KEY TERMS

- **Ellipses with Center at (0, 0)**
  See Table 3 on page 939.
- **Ellipses with Center at (h, k)**
  See Table 4 on page 942.

Ellipse
Foci
Major axis
Center
Minor axis
Vertices

YOU SHOULD BE ABLE TO . . .	EXAMPLE	REVIEW EXERCISES
(1) Graph ellipses in which the center is the origin (p. 938)	Examples 1 and 2	41, 42
(2) Find the equation of an ellipse in which the center is the origin (p. 941)	Example 3	43–45
(3) Graph ellipses in which the center is not the origin (p. 942)	Example 4	46–47
(4) Solve applied problems involving ellipses (p. 943)	Example 5	48

In Problems 41 and 42, find the vertices and foci of each ellipse. Graph each ellipse.

41. $\dfrac{x^2}{9} + y^2 = 1$

42. $9x^2 + 4y^2 = 36$

In Problems 43–45, find an equation for each ellipse. Graph each ellipse.

43. Center at $(0, 0)$; focus at $(0, 3)$; vertex at $(0, 5)$
44. Center at $(0, 0)$; focus at $(-2, 0)$; vertex at $(-6, 0)$
45. Foci at $(\pm 8, 0)$; vertices at $(\pm 10, 0)$

In Problems 46 and 47, find the vertices and foci of each ellipse. Graph each ellipse.

46. $\dfrac{(x-1)^2}{49} + \dfrac{(y+2)^2}{25} = 1$

47. $25(x+3)^2 + 9(y-4)^2 = 225$

48. Semielliptical Arch Bridge An arch in the shape of the upper half of an ellipse is used to support a bridge that is to span a river 60 feet wide. The center of the arch is 16 feet above the center of the river.

(a) Write the equation for the ellipse in which the *x*-axis coincides with the water and the *y*-axis passes through the center of the arch.
(b) Can a rectangular barge that is 25 feet wide and sits 12 feet above the surface of the water fit through the opening of the bridge?

Section 12.5 Hyperbolas

KEY CONCEPTS	KEY TERMS

- **Hyperbolas with Center at the Origin**
  See Table 5 on page 950.
- **Asymptotes of a Hyperbola**
  See Table 6 on page 953.

Hyperbola
Foci
Transverse axis
Center
Conjugate axis
Branches
Vertices

YOU SHOULD BE ABLE TO . . .	EXAMPLE	REVIEW EXERCISES
(1) Graph hyperbolas in which the center is the origin (p. 949)	Examples 1 and 2	49–53
(2) Find the equation of a hyperbola in which the center is the origin (p. 952)	Example 3	54–56
(3) Find asymptotes of a hyperbola in which the center is the origin (p. 953)	Example 4	52, 53

In Problems 49–51, find the vertices and foci of each hyperbola. Graph each hyperbola.

49. $\dfrac{x^2}{4} - \dfrac{y^2}{9} = 1$ **50.** $\dfrac{y^2}{25} - \dfrac{x^2}{49} = 1$ **51.** $16y^2 - 25x^2 = 400$

In Problems 52 and 53, graph each hyperbola using the asymptotes as a guide.

52. $\dfrac{x^2}{36} - \dfrac{y^2}{36} = 1$ **53.** $\dfrac{y^2}{25} - \dfrac{x^2}{4} = 1$

In Problems 54–56, find an equation for each hyperbola described. Graph each hyperbola.

54. Center at $(0, 0)$; focus at $(-4, 0)$; vertex at $(-3, 0)$

55. Vertices at $(0, -3)$ and $(0, 3)$; focus at $(0, 5)$

56. Vertices at $(0, \pm 4)$; asymptote the line $y = \dfrac{4}{3}x$

Section 12.6	Nonlinear Systems of Equations	
KEY TERM		
System of nonlinear equations		
YOU SHOULD BE ABLE TO . . .	**EXAMPLE**	**REVIEW EXERCISES**
① Solve a system of nonlinear equations using substitution (p. 959)	Examples 1 and 2	57–60; 65–76
② Solve a system of nonlinear equations using elimination (p. 961)	Examples 3 and 4	61–64; 65–76

In Problems 57–60, solve the system of nonlinear equations using the method of substitution.

57. $\begin{cases} 4x^2 + y^2 = 10 \\ y = x \end{cases}$ **58.** $\begin{cases} y = 2x^2 + 1 \\ y = x + 2 \end{cases}$

59. $\begin{cases} 6x - y = 5 \\ xy = 1 \end{cases}$ **60.** $\begin{cases} x^2 + y^2 = 26 \\ x^2 - 2y^2 = 23 \end{cases}$

In Problems 61–64, solve the system of nonlinear equations using the method of elimination.

61. $\begin{cases} 4x - y^2 = 0 \\ 2x^2 + y^2 = 16 \end{cases}$ **62** $\begin{cases} x^2 - y = -2 \\ x^2 + y = 4 \end{cases}$

63. $\begin{cases} 4x^2 - 2y^2 = 2 \\ -x^2 + y^2 = 2 \end{cases}$ **64.** $\begin{cases} x^2 + y^2 = 8x \\ y^2 = 3x \end{cases}$

In Problems 65–72, solve the system of nonlinear equations using the method you prefer.

65. $\begin{cases} y = x + 2 \\ y = x^2 \end{cases}$ **66.** $\begin{cases} x^2 + 2y = 9 \\ 5x - 2y = 5 \end{cases}$

67. $\begin{cases} x^2 + y^2 = 36 \\ x - y = -6 \end{cases}$ **68.** $\begin{cases} y = 2x - 4 \\ y^2 = 4x \end{cases}$

69. $\begin{cases} x^2 + y^2 = 9 \\ x + y = 7 \end{cases}$ **70.** $\begin{cases} 2x^2 + 3y^2 = 14 \\ x^2 - y^2 = -3 \end{cases}$

71. $\begin{cases} x^2 + y^2 = 16 \\ x^2 + 4y = 16 \end{cases}$ **72.** $\begin{cases} x = 4 - y^2 \\ x = 2y + 4 \end{cases}$

73. **Fun with Numbers** The sum of two numbers is 12. The difference of their squares is 24. Find the two numbers.

△ 74. **Perimeter and Area of a Rectangle** The perimeter of a rectangle is 34 centimeters. The area of the rectangle is 60 square centimeters. Find the dimensions of the rectangle.

△ 75. **Dimensions of a Rectangle** The area of a rectangle is 2160 square inches. The diagonal of the rectangle is 78 inches. Find the dimensions of the rectangle.

△ 76. **Dimensions of a Triangle** A right triangle has a perimeter of 36 feet and a hypotenuse of 15 feet. Find the lengths of the legs of the right triangle.

CHAPTER 12 TEST

Remember to use your Chapter Test Prep Video CD to see fully worked-out solutions to any of these problems you would like to review.

1. Find the distance $d(P_1, P_2)$ between points $P_1 = (-1, 3)$ and $P_2 = (3, -5)$.

2. Find the midpoint of the line segment formed by joining the points $P_1 = (-7, 6)$ and $P_2 = (5, -2)$.

In Problems 3 and 4, find the center (h, k) and the radius r of each circle. Graph each circle.

3. $(x - 4)^2 + (y + 1)^2 = 9$

4. $x^2 + y^2 + 10x - 4y + 13 = 0$

In Problems 5 and 6, find the standard form of the equation of each circle.

5. Radius $r = 6$ and center $(h, k) = (-3, 7)$

6. Center at $(-5, 8)$ and containing the point $(3, 2)$.

In Problems 7 and 8, find the vertex, focus, and directrix of each parabola. Graph each parabola.

7. $(y + 2)^2 = 4(x - 1)$

8. $x^2 - 4x + 3y - 8 = 0$

In Problems 9 and 10, find an equation for each parabola described.

9. Vertex at $(0, 0)$; focus at $(0, -4)$

10. Focus at $(3, 4)$; directrix the line $x = -1$

In Problems 11 and 12, find the vertices and foci of each ellipse. Graph each ellipse.

11. $9x^2 + 25y^2 = 225$

12. $\dfrac{(x - 2)^2}{9} + \dfrac{(y + 4)^2}{16} = 1$

In Problems 13 and 14, find an equation for each ellipse described.

13. Center at $(0, 0)$; focus at $(0, -4)$; vertex at $(0, -5)$

14. Vertices at $(-1, 7)$ and $(-1, -3)$; focus at $(-1, -1)$

In Problems 15 and 16, find the vertices, foci, and asymptotes for each hyperbola. Graph each hyperbola using the asymptotes as a guide.

15. $x^2 - \dfrac{y^2}{4} = 1$

16. $16y^2 - 25x^2 = 1600$

17. Find an equation for a hyperbola with foci at $(\pm 8, 0)$ and vertex at $(-3, 0)$.

In Problems 18 and 19, solve the system of nonlinear equations using the method you prefer.

18. $\begin{cases} x^2 + y^2 = 17 \\ x + y = -3 \end{cases}$

19. $\begin{cases} x^2 + y^2 = 9 \\ 4x^2 - y^2 = 16 \end{cases}$

20. An arch in the shape of the upper half of an ellipse is used to support a bridge spanning a creek 30 feet wide. The center of the arch is 10 feet above the center of the creek.

   (a) Write the equation for the ellipse in which the x-axis coincides with the creek and the y-axis passes through the center of the arch.

   (b) What is the height of the arch at a distance 12 feet from the center of the creek?

13 Sequences, Series, and the Binomial Theorem

Population growth has been a topic of debate among scientists for a long time. In fact, over 200 years ago, the English economist and mathematician Thomas Robert Malthus anonymously published a paper predicting that the world's population would overwhelm the Earth's capacity to sustain it—he claimed that food supplies increase arithmetically while population grows geometrically. See Problems 53 and 54 in Section 13.1 and Problem 87 in Section 13.3.

OUTLINE

13.1 Sequences

13.2 Arithmetic Sequences

13.3 Geometric Sequences and Series

Putting the Concepts Together (Sections 13.1–13.3)

13.4 The Binomial Theorem

Chapter 13 Activity: Pass to the Right
Chapter 13 Review
Chapter 13 Test
Cumulative Review Chapters 1–13

The Big Picture: Putting It Together

In Chapter 8, we defined the domain of a function as the set of real numbers such that the function makes sense. In Sections 13.1–13.3, we introduce a special type of function called a *sequence*. The domain of a sequence is not the set of real numbers, but instead is the set of natural numbers (that is, the positive integers). Functions whose domain is the set all real numbers are useful for modeling situations where we wish to describe what happens to the value of some independent variable that is continuously changing, such as the population of bacteria. Sequences are powerful for modeling situations where we wish to describe what happens to the value of a dependent variable at discrete intervals of time, such as weekly changes in the value of a deposit at a bank.

In Chapter 5, we gave a formula for expanding $(x + a)^2$. A logical question to ask yourself is, "Does any formula exist for expanding $(x + a)^n$, where n is an integer greater than 2?" The answer is yes! We discuss this formula in Section 13.4.

13.1 Sequences

OBJECTIVES

1. Write the First Several Terms of a Sequence
2. Find a Formula for the nth Term of a Sequence
3. Use Summation Notation

Preparing for Sequences

Before getting started, take the following readiness quiz. If you get a problem wrong, go back to the section cited and review the material.

1. Evaluate $f(x) = x^2 - 4$ at (a) $x = 3$ (b) $x = -7$. [Section 8.3, pp. 590–592]
2. If $g(x) = 2x - 3$, find $g(1) + g(2) + g(3)$. [Section 8.3, pp. 590–592]
3. In the function $f(n) = n^2 - 4$, what is the independent variable? [Section 8.3, pp. 590–591]

We begin with a definition.

> **DEFINITION**
>
> A **sequence** is a function whose domain is the set of positive integers.

Because a sequence is a function, it will have a graph. In Figure 1(a), we have the graph of the function $f(x) = \dfrac{1}{x}$ for $x > 0$. If all the points on this graph were removed except those whose x-coordinates are positive integers—that is, if all points were removed except $(1, 1), \left(2, \dfrac{1}{2}\right), \left(3, \dfrac{1}{3}\right)$—and so on, the remaining points would be the graph of the sequence $f(n) = \dfrac{1}{n}$, as shown in Figure 1(b). Notice that we use n to represent the independent variable in a sequence. This serves to remind us that n is a positive integer, or natural number.

Figure 1

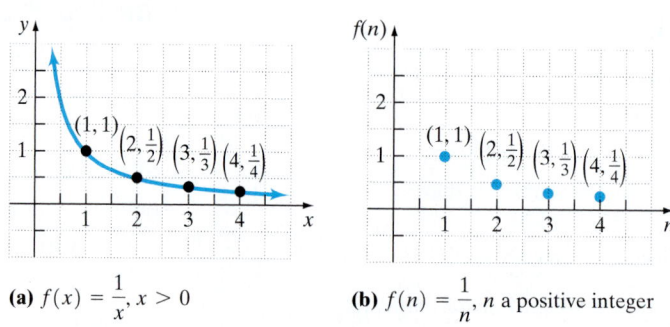

(a) $f(x) = \dfrac{1}{x}, x > 0$ **(b)** $f(n) = \dfrac{1}{n}, n$ a positive integer

A sequence may be represented by listing its values in order. For example, the sequence whose graph is given in Figure 1(b) might be represented as

$$f(1), f(2), f(3), f(4), \ldots \quad \text{or} \quad 1, \frac{1}{2}, \frac{1}{3}, \frac{1}{4}, \ldots$$

In Words
The word *infinite* means "without bound," so infinite sequences have no bound. The word *finite* means "bounded in number," so a finite sequence has a certain number of terms.

The numbers in the ordered list are called the **terms** of the sequence. Notice that the list never ends, as the **ellipsis** (the three dots) indicates. When a sequence does not end it is said to be an **infinite sequence.** We contrast this with **finite sequences** that have a domain that is the first n positive integers. For example,

$$40, 44, 48, 52, 56, 60, 64$$

is a finite sequence because it contains $n = 7$ terms.

1 Write the First Several Terms of a Sequence

In a sequence we do not use the traditional notation $f(n)$. This is done to distinguish sequences from functions whose domain is the set of all real numbers. Instead, we use the notation a_n to indicate that the function is a sequence. The value of the subscript represents the value of the independent variable. So a_4 means to evaluate the function defined by the sequence a_n at $n = 4$. For the sequence $f(n) = \dfrac{1}{n}$, we would write the function as $a_n = \dfrac{1}{n}$ and evaluate the function as follows:

$$a_1 = f(1) = 1, \quad a_2 = f(2) = \frac{1}{2}, \quad a_3 = f(3) = \frac{1}{3}, \quad a_4 = f(4) = \frac{1}{4}, \quad \ldots, \quad a_n = f(n) = \frac{1}{n}$$

In addition, just as we can name functions f, g, F, G and so on, we can name sequences. Typically, we use lower case letters from the beginning of the alphabet to name a sequence as in a_n, b_n, c_n, and so on (although this naming scheme is not necessary).

When a formula for the nth term (sometimes called the **general term**) of a sequence is known, rather than write out the terms of the sequence, we may represent the entire sequence by placing braces around the formula for the nth term. For example, the sequence whose nth term is $b_n = \left(\dfrac{1}{3}\right)^n$ can be written as

$$\{b_n\} = \left\{\left(\frac{1}{3}\right)^n\right\}$$

or by

$$b_1 = \frac{1}{3}, \quad b_2 = \frac{1}{9}, \quad b_3 = \frac{1}{27}, \quad \ldots, \quad b_n = \left(\frac{1}{3}\right)^n$$

EXAMPLE 1 Writing the First Five Terms of a Sequence

Write down the first five terms of the sequence $\{a_n\} = \left\{\dfrac{n}{n+1}\right\}$.

Solution

To find the first five terms of the sequence, we evaluate the function at $n = 1, 2, 3, 4,$ and 5.

$$a_1 = \frac{1}{1+1} = \frac{1}{2}, \quad a_2 = \frac{2}{2+1} = \frac{2}{3}, \quad a_3 = \frac{3}{3+1} = \frac{3}{4}, \quad a_4 = \frac{4}{4+1} = \frac{4}{5}, \quad a_5 = \frac{5}{5+1} = \frac{5}{6}$$

The first five terms of the sequence are $\dfrac{1}{2}, \dfrac{2}{3}, \dfrac{3}{4}, \dfrac{4}{5},$ and $\dfrac{5}{6}$. ■

EXAMPLE 2 Writing the First Five Terms of a Sequence

Write down the first five terms of the sequence $\{b_n\} = \{(-1)^n \cdot n^2\}$.

Solution

The first five terms of the sequence are

$$b_1 = (-1)^1 \cdot 1^2 = -1, \quad b_2 = (-1)^2 \cdot 2^2 = 4, \quad b_3 = (-1)^3 \cdot 3^2 = -9,$$
$$b_4 = (-1)^4 \cdot 4^2 = 16, \quad b_5 = (-1)^5 \cdot 5^2 = -25$$

The first five terms of the sequence are $-1, 4, -9, 16,$ and -25. ■

Notice in Example 2 that the signs of the terms in the sequence **alternate**. When this occurs, we use factors such as $(-1)^{n+1}$, which equals 1 if n is odd and -1 if n is even, or $(-1)^n$, which equals -1 if n is odd and 1 if n is even.

QUICK ✓ *Write the first five terms of the given sequence.*

1. $\{a_n\} = \{2n - 3\}$ **2.** $\{b_n\} = \{(-1)^n \cdot 4n\}$

② Find a Formula for the nth Term of a Sequence

Sometimes a sequence is indicated by an observed pattern in the first few terms that makes it possible to determine the formula for the nth term. In the example that follows, a sufficient number of terms of the sequence are given so that a natural choice for the nth term is suggested.

EXAMPLE 3 Determining a Sequence from a Pattern

Find a formula for the nth term of each sequence.

(a) $3, 7, 11, 15, \ldots$ (b) $2, 4, 8, 16, \ldots$ (c) $1, -8, 27, -64, 125, \ldots$

Solution

Our goal in all these problems is to find a formula in terms of n such that when $n = 1$ we get the first term, when $n = 2$ we get the second term, and so on.

(a) When $n = 1$, we have that $a_1 = 3$; when $n = 2$, we have that $a_2 = 7$. Notice that each subsequent term increases by 4. A formula for the nth term is given by $a_n = 4n - 1$.

(b) The terms of the sequence are all multiples of 2 with the first term equaling 2^1, the second term equaling 2^2, and so on. A formula for the nth term is given by $b_n = 2^n$.

(c) Notice that the terms alternate with the first term being positive. So $(-1)^{n+1}$ must be part of the formula. Ignoring the sign, we notice that the terms are all perfect cubes. A formula for the nth term is given by $c_n = (-1)^{n+1} \cdot n^3$. ▄

QUICK ✓ *Find a formula for the nth term of each sequence.*

3. $5, 7, 9, 11, \ldots$

4. $\dfrac{1}{2}, -\dfrac{1}{3}, \dfrac{1}{4}, -\dfrac{1}{5}, \ldots$

③ Use Summation Notation

In other mathematics courses, such as statistics or calculus, it is important to be able to find the sum of the first n terms of a sequence $\{a_n\}$, that is,

$$a_1 + a_2 + a_3 + \cdots + a_n$$

In Words
The symbol Σ is read "upper case sigma" or "sigma" for short.

Rather than write down all these terms, mathematicians use a more concise way to express the sum, called **summation notation.** Using summation notation, we would write $a_1 + a_2 + a_3 + \cdots + a_n$ as

$$\sum_{i=1}^{n} a_i = a_1 + a_2 + a_3 + \cdots + a_n$$

The symbol Σ is an instruction to sum, or add up, the terms. The integer i is called the **index** of the sum; it tells you where to start the sum and where to end it. The expression

$$\sum_{i=1}^{n} a_i$$

is an instruction to add the terms of the sequence $\{a_i\}$ from $i = 1$ through $i = n$. We read the expression $\sum_{i=1}^{n} a_i$ as "the sum of a sub i from $i = 1$ to $i = n$." When there are a finite number of terms to be added, the sum is called a **partial sum.**

EXAMPLE 4 **Finding a Partial Sum**

Write out the sum and determine its value.

(a) $\displaystyle\sum_{i=1}^{4}(2i + 5)$ 　　　　　　(b) $\displaystyle\sum_{i=1}^{5}(i^2 - 5)$

Solution

(a) $\displaystyle\sum_{i=1}^{4}(2i + 5) = \underbrace{(2\cdot1 + 5)}_{i\,=\,1} + \underbrace{(2\cdot2 + 5)}_{i\,=\,2} + \underbrace{(2\cdot3 + 5)}_{i\,=\,3} + \underbrace{(2\cdot4 + 5)}_{i\,=\,4}$

$$= 7 + 9 + 11 + 13$$
$$= 40$$

(b) $\displaystyle\sum_{i=1}^{5}(i^2 - 5) = \underbrace{(1^2 - 5)}_{i\,=\,1} + \underbrace{(2^2 - 5)}_{i\,=\,2} + \underbrace{(3^2 - 5)}_{i\,=\,3} + \underbrace{(4^2 - 5)}_{i\,=\,4} + \underbrace{(5^2 - 5)}_{i\,=\,5}$

$$= -4 + (-1) + 4 + 11 + 20$$
$$= 30$$

QUICK ✓ *Write out the sum and determine its value.*

5. $\displaystyle\sum_{i=1}^{3}(4i - 1)$ 　　　　　　6. $\displaystyle\sum_{i=1}^{5}(i^3 + 1)$

The index of summation does not have to begin with 1. In addition, the index of summation does not have to be i. For example, we might have

$$\sum_{k=3}^{6}(2k + 1) = \underbrace{(2\cdot3 + 1)}_{k\,=\,3} + \underbrace{(2\cdot4 + 1)}_{k\,=\,4} + \underbrace{(2\cdot5 + 1)}_{k\,=\,5} + \underbrace{(2\cdot6 + 1)}_{k\,=\,6}$$

$$= 7 + 9 + 11 + 13$$
$$= 40$$

Notice that the terms of this partial sum are identical to those given in Example 4(a). What is the moral of the story? We can make the same sum look entirely different by changing the starting point of the index of summation and changing the variable that represents the index.

Now let's reverse the process. Rather than writing out a sum, we will express a sum using summation notation.

EXAMPLE 5 **Writing a Sum in Summation Notation**

Express each sum using summation notation.

(a) $1^2 + 2^2 + 3^2 + \cdots + 10^2$ 　　　(b) $2 + 1 + \dfrac{2}{3} + \dfrac{1}{2} + \dfrac{2}{5} + \cdots + \dfrac{1}{6}$

Solution

(a) The sum $1^2 + 2^2 + 3^2 + \cdots + 10^2$ has 10 terms, each of the form i^2, and starts at $i = 1$ and ends at $i = 10$:

$$1^2 + 2^2 + 3^2 + \cdots + 10^2 = \sum_{i=1}^{10} i^2$$

(b) First, we need to figure out the pattern. With a little investigation, we discover that $2 + 1 + \dfrac{2}{3} + \dfrac{1}{2} + \dfrac{2}{5} + \cdots + \dfrac{1}{6}$ can be written as

$$\frac{2}{1} + \frac{2}{2} + \frac{2}{3} + \frac{2}{4} + \frac{2}{5} + \cdots + \frac{2}{12}, \text{ so that the } n\text{th term of the sum is } \frac{2}{n}.$$

There are $n = 12$ terms, so

$$2 + 1 + \frac{2}{3} + \frac{1}{2} + \frac{2}{5} + \cdots + \frac{1}{6} = \sum_{n=1}^{12} \frac{2}{n}$$

Work Smart

The index of summation can be any variable we desire and can start at any value we desire. Keep this in mind when checking your answers.

QUICK ✔ *Write each sum using summation notation.*

7. $1 + 4 + 9 + \cdots + 144$

8. $1 + \frac{1}{2} + \frac{1}{4} + \cdots + \frac{1}{32}$

13.1 Exercises

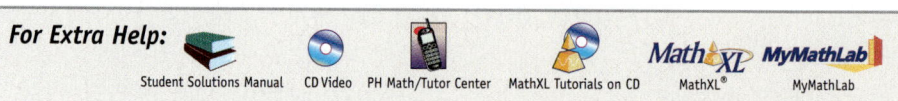

For Extra Help: Student Solutions Manual   CD Video   PH Math/Tutor Center   MathXL Tutorials on CD   MathXL®   MyMathLab

Concepts and Vocabulary

In Problems 1–3, fill in the blanks.

1. A(n) _____ is a function whose domain is the set of positive integers.

2. When a sequence does not end it is said to be a(n) _____ sequence. A(n) _____ sequence has a domain that is the first n positive integers.

3. When there is a finite number of terms to be added, the sum is called a _____ _____.

In Problems 4–6, answer True or False to each statement.

4. A sequence is a function.

5. In the expression $\sum_{k=1}^{6} 5k$, the expression $5k$ is called the index of summation.

6. We can evaluate the sequence $\{b_n\} = \{5n - 3\}$ at $n = 0$.

7. Explain how a sequence and a function are related.

8. What does the graph of a sequence look like when compared to the graph of a function? Use the function $f(x) = 3x + 1$ and the sequence $\{a_n\} = \{3n + 1\}$ when doing the comparison.

9. Write a sentence that explains the meaning of the symbol Σ.

10. What does it mean when a sequence alternates?

Building Skills

In Problems 11–22, write down the first five terms of each sequence.

11. $\{3n + 5\}$

12. $\{n - 4\}$

13. $\left\{\dfrac{n}{n + 2}\right\}$

14. $\left\{\dfrac{n + 4}{n}\right\}$

15. $\{(-1)^n n\}$

16. $\{(-1)^{n+1} n\}$

17. $\{2^n + 1\}$

18. $\{3^n - 1\}$

19. $\left\{\dfrac{2n}{2^n}\right\}$

20. $\left\{\dfrac{3n}{3^n}\right\}$

21. $\left\{\dfrac{n}{e^n}\right\}$

22. $\left\{\dfrac{n^2}{2}\right\}$

In Problems 23–30, the given pattern continues. Write down the nth term of each sequence suggested by the pattern.

23. $2, 4, 6, 8, \ldots$

24. $5, 10, 15, 20, \ldots$

25. $\dfrac{1}{2}, \dfrac{2}{3}, \dfrac{3}{4}, \dfrac{4}{5}, \ldots$

26. $\dfrac{1}{2}, 1, \dfrac{3}{2}, 2, \dfrac{5}{2}, \ldots$ **27.** $3, 6, 11, 18, \ldots$ **28.** $0, 7, 26, 63, \ldots$

29. $-1, 4, -9, 16, \ldots$ **30.** $1, -\dfrac{1}{2}, \dfrac{1}{4}, -\dfrac{1}{8}, \ldots$

In Problems 31–42, write out each sum and determine its value.

31. $\displaystyle\sum_{i=1}^{4} (5i + 1)$ **32.** $\displaystyle\sum_{i=1}^{5} (3i + 2)$ **33.** $\displaystyle\sum_{i=1}^{5} \dfrac{i^2}{2}$ **34.** $\displaystyle\sum_{i=1}^{4} \dfrac{i^3}{2}$

35. $\displaystyle\sum_{k=1}^{3} 2^k$ **36.** $\displaystyle\sum_{k=1}^{4} 3^k$ **37.** $\displaystyle\sum_{k=1}^{5} [(-1)^{k+1} \cdot 2k]$ **38.** $\displaystyle\sum_{k=1}^{8} [(-1)^k \cdot k]$

39. $\displaystyle\sum_{j=1}^{10} 5$ **40.** $\displaystyle\sum_{j=1}^{8} 2$ **41.** $\displaystyle\sum_{k=3}^{7} (2k - 1)$ **42.** $\displaystyle\sum_{j=5}^{10} (k + 4)$

In Problems 43–50, express each sum using summation notation.

43. $1 + 2 + 3 + \cdots + 15$ **44.** $1 + 3 + 5 + \cdots + 17$

45. $1 + \dfrac{1}{2} + \dfrac{1}{3} + \cdots + \dfrac{1}{12}$ **46.** $1 + \dfrac{1}{2} + \dfrac{1}{4} + \cdots + \dfrac{1}{2^{15}}$

47. $1 - \dfrac{1}{3} + \dfrac{1}{9} - \dfrac{1}{27} + \cdots + (-1)^{9+1}\left(\dfrac{1}{3^{9-1}}\right)$

48. $\dfrac{2}{3} - \dfrac{4}{9} + \dfrac{8}{27} + \cdots + (-1)^{15+1}\left(\dfrac{2}{3}\right)^{15}$

49. $5 + (5 + 2 \cdot 1) + (5 + 2 \cdot 2) + (5 + 2 \cdot 3) + \cdots + (5 + 2 \cdot 10)$

50. $3 + 3 \cdot \dfrac{1}{2} + 3 \cdot \dfrac{1}{4} + \cdots + 3 \cdot \left(\dfrac{1}{2}\right)^{11}$

Applying the Concepts

51. The Future Value of Money Suppose that you place \$12,000 in your company 401(k) plan that pays 6% interest compounded quarterly. The balance in the account after n quarters is given by

$$a_n = 12{,}000\left(1 + \dfrac{0.06}{4}\right)^n$$

  (a) Find the value in the account after 1 quarter.
  (b) Find the value in the account after 1 year.
  (c) Find the value in the account after 10 years.

52. The Future Value of Money Suppose that you place \$5000 into a company 401(k) plan that pays 8% interest compounded monthly. The balance in the account after n months is given by

$$a_n = 5000\left(1 + \dfrac{0.08}{12}\right)^n$$

  (a) Find the value in the account after 1 month.
  (b) Find the value in the account after 1 year.
  (c) Find the value in the account after 10 years.

53. Population Growth According to the U.S. Census Bureau, the population of the United States in 2005 was 296 million people. In addition, the population of the United States was growing at a rate of 1.1% per year. A model for the population of the United States is given by

$$p_n = 296(1.011)^n$$

where n is the number of years after 2005.

  (a) Use this model to predict the United States population in 2010 to the nearest million.
  (b) Use this model to predict the United States population in 2050 to the nearest million.

54. Population Growth According to the *Statistical Abstract of the United States,* the population of the world in 2005 was 6448 million people. In addition, the population of the world was growing at a rate of 1.26% per year. A model for the population of the world is given by

$$p_n = 6448(1.0126)^n$$

where n is the number of years after 2005.

(a) Use this model to predict the world population in 2010.

(b) Use this model to predict the world population in 2050.

55. Fibonacci Sequence Let

$$u_n = \frac{\left(1 + \sqrt{5}\right)^n - \left(1 - \sqrt{5}\right)^n}{2^n \cdot \sqrt{5}}$$

define the nth term of a sequence. Find the first 10 terms of the sequence. This sequence is called the **Fibonacci sequence.** The terms of the sequence are called **Fibonacci numbers.**

56. Pascal's Triangle The triangular array of numbers shown in the right is called Pascal's Triangle. Each number in the triangle is found by adding the entries directly above to the left and right of the number. For example, the 4 in the fourth row is found by adding the 1 and 3 above the 4.

Divide the triangular array using diagonal lines (as shown). Now find the sum of the numbers in each of the highlighted diagonal rows. Do you recognize the sequence? (**Hint:** See Problem 55.)

Extending the Concepts

A second way of defining a sequence is to assign a value to the first (or first few) term(s) and specify the nth term by a formula or equation that involves one or more of the terms preceding it. Sequences defined this way are said to be defined **recursively,** and the rule or formula is called a **recursive formula.** For example $s_1 = 3$ and $s_n = 2s_{n-1}$ is a recursively defined sequence where $s_1 = 3$, $s_2 = 2s_1 = 2(3) = 6$, $s_3 = 2s_2 = 2(6) = 12$, $s_4 = 2s_3 = 2(12) = 24$ and so on.

In Problems 57–60, a sequence is defined recursively. Write the first five terms of the sequence.

57. $a_1 = 10, a_n = 1.05a_{n-1}$

58. $b_1 = 20, b_n = 3b_{n-1}$

59. $b_1 = 8, b_n = n + b_{n-1}$

60. $c_1 = 1{,}000, c_n = 1.01c_{n-1} + 100$

61. Fibonacci Sequence Use the result of Problem 55 to do the following problems:

(a) Compute the ratio $\dfrac{u_{n+1}}{u_n}$ for the first 10 terms.

(b) As n gets large, what number does the ratio approach? This number is referred to as the **golden ratio.** Rectangles whose sides are in this ratio were considered pleasing to the eye by the Greeks. For example, the façade of the Parthenon was constructed using the golden ratio.

(c) Compute the ratio $\dfrac{u_n}{u_{n+1}}$ for the first 10 terms.

(d) As n gets large, what number does the ratio approach? This number is also referred to as the **golden ratio.** This ratio is believed to have been used in the construction of the Great Pyramid in Egypt. The ratio equals the sum of the areas of the four face triangles divided by the total surface area of the Great Pyramid.

62. Investigate various applications that lead to a Fibonacci sequence, such as art, architecture, or financial markets. Write an essay on these applications.

Synthesis Review

In Problems 63–65, (a) determine the slope of the linear function, (b) Compute $f(1), f(2), f(3)$, and $f(4)$.

63. $f(x) = 4x - 6$ **64.** $f(x) = 2x - 10$ **65.** $f(x) = -5x + 8$

66. Compute $f(2) - f(1), f(3) - f(2)$, and $f(4) - f(3)$ for each of the functions in Problems 63–65. How does the difference in the value of the function for consecutive values of the independent variable relate to the slope?

Graphing Calculator

Graphing calculators can be used to write the terms of a sequence. For example, Figure 2 shows the first five terms of the sequence $\{b_n\} = \{(-1)^n \cdot n^2\}$ that we studied in Example 2. Figure 3 shows how the terms of the sequence can be listed in a table using SEQuence mode.

Figure 2

Figure 3

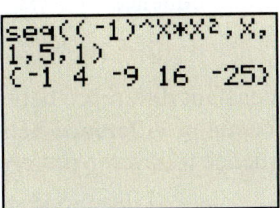

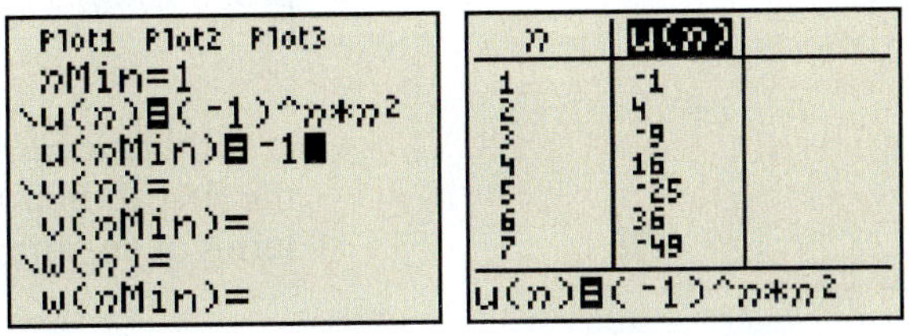

In Problems 67–74, use a graphing calculator to find the first five terms of the sequence.

67. $\{3n + 5\}$ **68.** $\{n - 4\}$ **69.** $\left\{\dfrac{n}{n + 2}\right\}$ **70.** $\left\{\dfrac{n + 4}{n}\right\}$

71. $\{(-1)^n n\}$ **72.** $\{(-1)^{n+1} n\}$ **73.** $\{2^n + 1\}$ **74.** $\{3^n - 1\}$

Graphing calculators can also be used to find the sum of a sequence. For example, Figure 4 shows the result of the sum $\displaystyle\sum_{i=1}^{5} (i^2 - 5)$ that we studied in Example 4(b).

Figure 4

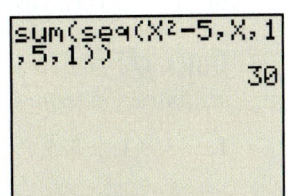

In Problems 75–82, use a graphing calculator to find the sum.

75. $\displaystyle\sum_{i=1}^{4} (5i + 1)$ **76.** $\displaystyle\sum_{i=1}^{5} (3i + 2)$ **77.** $\displaystyle\sum_{i=1}^{5} \dfrac{i^2}{2}$ **78.** $\displaystyle\sum_{i=1}^{4} \dfrac{i^3}{2}$

79. $\displaystyle\sum_{k=1}^{3} 2^k$ **80.** $\displaystyle\sum_{k=1}^{4} 3^k$ **81.** $\displaystyle\sum_{k=1}^{5} [(-1)^{k+1} \cdot 2k]$ **82.** $\displaystyle\sum_{k=1}^{8} [(-1)^k \cdot k]$

13.2 Arithmetic Sequences

OBJECTIVES

1. Determine If a Sequence Is Arithmetic
2. Find a Formula for the nth Term an Arithmetic Sequence
3. Find the Sum of an Arithmetic Sequence

Preparing for Arithmetic Sequences

Before getting started take the following readiness quiz. If you get a problem wrong, go back to the section cited and review the material.

1. Determine the slope of $y = -3x + 1$.   [Section 3.4, pp. 217–218]

2. If $g(x) = 5x + 2$, find $g(3)$.   [Section 8.3, pp. 590–592]

3. Solve: $\begin{cases} x - 3y = -17 \\ 2x + y = 1 \end{cases}$ [Section 4.2, pp. 285–290; Section 4.3, pp. 296–301]

In the last section, we looked at sequences in general. Now we look at a specific type of sequence, called an *arithmetic sequence*.

1 **Determine If a Sequence Is Arithmetic**

When the difference between successive terms of a sequence is always the same number, the sequence is called an **arithmetic sequence** (sometimes called an **arithmetic progression**). For example, the sequence

$$2, 6, 10, 14, \ldots$$

is arithmetic because the constant difference between consecutive terms is 4. If we call the first term a and the **common difference** between consecutive terms d, then the terms of an arithmetic sequence follow the pattern

$$a, a + d, a + 2d, a + 3d, \ldots$$

EXAMPLE 1 **Determining If a Sequence Is Arithmetic**

Determine if the sequence $3, 9, 15, 21, \ldots$ is arithmetic. If it is, determine the first term a and the common difference d.

Solution

To determine if the sequence is arithmetic, find the difference of consecutive terms. If this difference is constant, the sequence is arithmetic. The sequence $3, 9, 15, 21, \ldots$ is arithmetic because the difference between consecutive terms is 6 ($= 9 - 3$ or $15 - 9$ or $21 - 15$). The first term is $a = 3$ and the common difference is $d = 6$. ∎

QUICK ✓ *Determine which of the following sequences is arithmetic. If the sequence is arithmetic, determine the first term a and common difference d.*

1. $-3, -1, 1, 3, 5, \ldots$ **2.** $3, 9, 27, 81, \ldots$

EXAMPLE 2 **Determining If a Sequence Defined by a Function Is Arithmetic**

Show that the sequence $\{s_n\} = \{2n + 7\}$ is arithmetic. Find the first term and the common difference.

Solution

We could list out the first few terms and demonstrate that the difference between consecutive terms is the same, but this would be more of a demonstration, not a proof. To prove the sequence is arithmetic, we must show that for *any* consecutive terms, the difference is the same number. We do this by evaluating the sequence at the $(n - 1)$st term and nth term in the sequence and computing the difference. If it is a constant, we've shown the sequence is arithmetic.

$$s_{n-1} = 2(n-1) + 7 = 2n - 2 + 7 = 2n + 5 \quad \text{and} \quad s_n = 2n + 7$$

Now we compute $s_n - s_{n-1}$.

$$s_n - s_{n-1} = (2n + 7) - (2n + 5)$$
$$\text{Distribute:} \quad = 2n + 7 - 2n - 5$$
$$\text{Combine like terms:} \quad = 2$$

The difference between *any* consecutive terms is 2, so the sequence is arithmetic with common difference $d = 2$. To find the first term, we evaluate s_1 and find $s_1 = 2(1) + 7 = 9$, so that $a = 9$. ■

EXAMPLE 3 Determining Whether or Not a Sequence Defined by a Function Is Arithmetic

Show that the sequence $\{b_n\} = \{n^2\}$ is not arithmetic.

Solution

As in Example 2, we determine the $(n-1)$st term and the nth term. We then show that the difference between consecutive terms is not a constant.

$$b_{n-1} = (n-1)^2 = n^2 - 2n + 1 \quad \text{and} \quad b_n = n^2$$

Now we compute $b_n - b_{n-1}$.

$$b_n - b_{n-1} = n^2 - (n^2 - 2n + 1)$$
$$\text{Distribute:} \quad = n^2 - n^2 + 2n - 1$$
$$\text{Combine like terms:} \quad = 2n - 1$$

The difference between consecutive terms is not constant—its value depends upon n. Therefore, the sequence is not arithmetic. ■

QUICK ✔ *Determine if the sequence is arithmetic. If it is, state the first term a and the common difference d.*

3. $\{a_n\} = \{3n - 8\}$ **4.** $\{b_n\} = \{n^2 - 1\}$ **5.** $\{c_n\} = \{5 - 2n\}$

---

② Find a Formula for the nth Term of an Arithmetic Sequence

Suppose that a is the first term of an arithmetic sequence whose common difference is d. We want to find a formula for a_n, the nth term of the sequence. To do this, we write down the first few terms of the sequence.

$$a_1 = a$$
$$a_2 = a_1 + d = a + d$$
$$a_3 = a_2 + d = (a + d) + d = a + 2d$$
$$a_4 = a_3 + d = (a + 2d) + d = a + 3d$$
$$a_5 = a_4 + d = (a + 3d) + d = a + 4d$$
$$\vdots$$
$$a_n = a_{n-1} + d = [a + (n-2)d] + d = a + (n-1)d$$

We are led to the following result.

THE nTH TERM OF AN ARITHMETIC SEQUENCE

For an arithmetic sequence $\{a_n\}$ whose first term is a and whose common difference is d, the nth term is determined by the formula

$$a_n = a + (n-1)d$$

EXAMPLE 4 Finding a Formula for the nth Term of an Arithmetic Sequence

(a) Write a formula for the nth term of an arithmetic sequence whose fourth term is 8 and whose common difference is -3.
(b) Find the 14th term of the sequence.

Solution

(a) We wish to find a formula for the nth term of an arithmetic sequence. We know that $d = -3$ and that $a_4 = 8$.

$$a_n = a + (n - 1)d$$

$d = -3; a_4 = 8:$ $\quad a_4 = 8 = a + (4 - 1)(-3)$

We solve this equation for a, the first term of the sequence.

$$8 = a - 9$$

Add 9 to both sides: $\quad 17 = a$

So the formula for the nth term is

$a_n = a + (n - 1)d:$ $\quad a_n = 17 + (n - 1)(-3)$

Distribute: $\quad = 17 - 3n + 3$

Combine like terms: $\quad = -3n + 20$

(b) To find the fourteenth term, we let $n = 14$ in $a_n = -3n + 20$.

$$a_{14} = -3(14) + 20$$
$$= -42 + 20$$
$$= -22$$

The fourteenth term in the sequence is -22. ∎

QUICK ✓

6. (a) Write a formula for the nth term of an arithmetic sequence whose fifth term is 25 and whose common difference is 6.

(b) Find the 14th term of the sequence.

EXAMPLE 5 Finding a Formula for the nth Term of an Arithmetic Sequence

The fourth term of an arithmetic sequence is 7, and the tenth term is 31.
(a) Find the first term and the common difference.
(b) Give a formula for the nth term of the sequence.

Solution

(a) We know that the nth term of an arithmetic sequence is $a_n = a + (n - 1)d$, where a is the first term and d is the common difference. Since $a_4 = 7$ and $a_{10} = 31$, we have

$$\begin{cases} a_4 = a + (4 - 1)d \\ a_{10} = a + (10 - 1)d \end{cases} \quad \text{or} \quad \begin{cases} 7 = a + 3d \quad (1) \\ 31 = a + 9d \quad (2) \end{cases}$$

This is a system of two linear equations containing two variables, a and d. We can solve this system by elimination. If we subtract equation (2) from equation (1) we obtain

$$-24 = -6d$$

Divide both sides by -6: $\quad 4 = d$

Let $d = 4$ in equation (1) to find a.

$$7 = a + 3(4)$$
$$7 = a + 12$$

Subtract 12 from both sides: $-5 = a$

The first term is $a = -5$ and the common difference is $d = 4$.

(b) A formula for the nth term is

$$a_n = a + (n - 1)d$$

$a = -5; d = 4$: $= -5 + (n - 1)(4)$

Distribute: $= -5 + 4n - 4$

Combine like terms: $= 4n - 9$ ◼

QUICK

7. The fifth term of an arithmetic sequence is 7, and the thirteenth term is 31.

(a) Find the first term and the common difference.

(b) Give a formula for the nth term of the sequence.

---

③ Find the Sum of an Arithmetic Sequence

The next result gives a formula for finding the sum of the first n terms of an arithmetic sequence.

> **SUM OF n TERMS OF AN ARITHMETIC SEQUENCE**
>
> Let $\{a_n\}$ be an arithmetic sequence with first term a and common difference d. The sum S_n of the first n terms of $\{a_n\}$ is
>
> $$S_n = \frac{n}{2}[2a + (n - 1)d] \qquad \text{or} \qquad S_n = \frac{n}{2}(a + a_n)$$

We show where these results come from next.

$$S_n = a_1 + a_2 + a_3 + \cdots + a_n$$
$$= \underbrace{a}_{a_1} + \underbrace{(a + d)}_{a_2} + \underbrace{(a + 2d)}_{a_3} + \cdots + \underbrace{(a + (n - 1)d)}_{a_n}$$

We can also represent S_n by reversing the order in which we add the terms, so that

$$S_n = a_n + a_{n-1} + \cdots + a_1$$
$$= \underbrace{(a + (n - 1)d)}_{a_n} + \underbrace{(a + (n - 2)d)}_{a_{n-1}} + \cdots + \underbrace{a}_{a_1}$$

Add these two different representations of S_n as follows:

$$
\begin{aligned}
S_n &= a & + \ (a + d) & + (a + 2d) & + \cdots + (a + (n - 2)d) + (a + (n - 1)d)\\
S_n &= (a + (n - 1)d) + (a + (n - 2)d) + (a + (n - 3)d) + \cdots + & (a + d) & + \ a\\
\hline
2S_n &= 2a + (n - 1)d + 2a + (n - 1)d + 2a + (n - 1)d + \cdots + 2a + (n - 1)d + 2a + (n - 1)d
\end{aligned}
$$

On the right side of the equation, we are adding $2a + (n - 1)d$ to itself n times. This sum can be represented as $n[2a + (n - 1)d]$, so we obtain

$$2S_n = n[2a + (n - 1)d]$$

Divide both sides by 2: $S_n = \dfrac{n}{2}[2a + (n - 1)d]$ Formula (1)

This is one of the formulas listed in the box. If we rewrite the expression $2a + (n - 1)d$ as $a + a + (n - 1)d$, we notice that $a + (n - 1)d$ is a_n, so that

$$S_n = \frac{n}{2}[a + a_n] \quad \text{Formula (2)}$$

So we have two ways to find the sum of the first n terms of an arithmetic sequence. Notice that Formula (1) involves the first term a and common difference d, while Formula (2) involves the first term a and the last term a_n. You should use whichever is easier.

EXAMPLE 6 **Finding the Sum of *n* Terms of an Arithmetic Sequence**

Find the sum S_n of the first 50 terms of the arithmetic sequence $2, 5, 8, 11, \ldots$.

Solution

Because we know the first term is $a = 2$ and the common difference is $d = 5 - 2 = 3$, we use the formula $S_n = \frac{n}{2}[2a + (n - 1)d]$ to find the sum.

$$S_n = \frac{n}{2}[2a + (n - 1)d]$$

$n = 50,\, a = 2,\, d = 3: \quad S_{50} = \frac{50}{2}[2(2) + (50 - 1)(3)]$

$$= 25[4 + 49(3)]$$
$$= 25[151]$$
$$= 3{,}775$$

The sum of the first 50 terms, S_{50}, of the arithmetic sequence $2, 5, 8, 11, \ldots$ is 3,775.

> **Work Smart**
>
> There are two formulas for finding the sum of the first *n* terms of an arithmetic sequence. Use Formula (1) if you know *n*, the first term *a* and the common difference *d*; use Formula (2) if you know *n*, the first term *a*, and the last term a_n.

QUICK ✓

8. Find the sum S_n of the first 100 terms of the arithmetic sequence whose first term is 5 and whose common difference is 2.

9. Find the sum S_n of the first 70 terms of the arithmetic sequence $1, 5, 9, 13, \ldots$.

EXAMPLE 7 **Finding the Sum of *n* Terms of an Arithmetic Sequence**

Find the sum S_n of the first 40 terms of the arithmetic sequence $\{-2n + 50\}$.

Solution

We find that the first term is $a_1 = -2(1) + 50 = 48$ and the 40th term is $a_{40} = -2(40) + 50 = -30$. Since we know the first term and the last term of the sequence we use the formula $S_n = \frac{n}{2}[a + a_n]$ to find the sum.

$$S_n = \frac{n}{2}[a + a_n]$$

$n = 40,\, a = 48,\, a_{40} = -30: \quad S_{40} = \frac{40}{2}[48 + (-30)]$

$$= 360$$

The sum of the first 40 terms, S_{40}, of the arithmetic sequence $\{-2n + 50\}$ is 360.

QUICK ✓

10. Find the sum S_n of the first 50 terms of the arithmetic sequence whose first term is 4 and fiftieth term 298.

11. Find the sum S_n of the first 75 terms of the arithmetic sequence $\{-3n + 100\}$.

EXAMPLE 8   **Creating a Floor Design**

A ceramic tile floor is designed in the shape of a trapezoid 10 feet wide at the base and 5 feet wide at the top. See Figure 5. The tiles, 6 inches by 6 inches, are to be placed so that each successive row contains one fewer tile than the preceding row. How many tiles will be required?

Figure 5

Solution

Each tile is 6 inches (0.5 foot) and the bottom row is 10 feet wide, so the bottom row requires 20 tiles. Similar logic tells us the top row requires 10 tiles. Since each successive row has one fewer tile, the total number of tiles required is

$$S = 20 + 19 + 18 + \cdots + 11 + 10$$

This is the sum of an arithmetic sequence; the common difference is -1. The number of terms to be added is $n = 11$, with the first term $a = 20$ and the last terms $a_{11} = 10$. The sum S is

$$S_n = \frac{n}{2}[a + a_n] \qquad S_{11} = \frac{11}{2}(20 + 10) = 165$$

In all, 165 tiles will be required.

QUICK ✓

12. In the corner section of a theater, the first row has 20 seats. Each subsequent row has 2 more seats, and there are a total of 30 rows. How many seats are in this section?

13.2 Exercises

For Extra Help:

Student Solutions Manual   CD Video   PH Math/Tutor Center   MathXL Tutorials on CD   MathXL®   MyMathLab

Concepts and Vocabulary

In Problems 1 and 2, fill in the blanks.

1. In a(n) _____ sequence, the difference between consecutive terms is a constant.

2. For an arithmetic sequence $\{a_n\}$ whose first term is a and whose common difference is d, the nth term is determined by the formula _____.

In Problems 3 and 4, answer True or False to each statement.

3. In an arithmetic sequence, the sum of the first and last terms equals twice the sum of all the terms.

4. In an arithmetic sequence, the difference between the first and the last term is the common difference.

5. Explain how you can determine if a sequence is arithmetic.

6. Provide an explanation that justifies the formula for the nth term of an arithmetic sequence.

Building Skills

In Problems 7–14, an arithmetic sequence is given. Find the common difference, and write out the first four terms.

7. $\{n + 5\}$ 8. $\{n - 1\}$ 9. $\{7n + 2\}$ 10. $\{10n + 1\}$

11. $\{7 - 3n\}$ 12. $\{5 - 2n\}$ 13. $\left\{\dfrac{1}{2}n + 5\right\}$ 14. $\left\{\dfrac{1}{4}n + \dfrac{3}{4}\right\}$

In Problems 15–22, find a formula for the nth term of the arithmetic sequence whose first term a and common difference d are given. What is the fifth term?

15. $a = 4, d = 3$ 16. $a = 8, d = 3$ 17. $a = 10, d = -5$ 18. $a = 12, d = -3$

19. $a = 2; d = \dfrac{1}{3}$ 20. $a = -3; d = \dfrac{1}{2}$ 21. $a = 5; d = -\dfrac{1}{5}$ 22. $a = -\dfrac{4}{3}; d = -\dfrac{2}{3}$

In Problems 23–28, write a formula for the nth term of each arithmetic sequence. Use the formula to find the 20th term in each arithmetic sequence.

23. $2, 7, 12, 17, \ldots$ 24. $-5, -1, 3, 7, \ldots$ 25. $12, 9, 6, 3, \ldots$

26. $20, 14, 8, 2, \ldots$ 27. $1, \dfrac{5}{4}, \dfrac{3}{2}, \dfrac{7}{4}, \ldots$ 28. $10, \dfrac{19}{2}, 9, \dfrac{17}{2}, \ldots$

In Problems 29–36, find the first term and the common difference of the arithmetic sequence described. Give a formula for the nth term of the sequence.

29. 3rd term is 17; 7th term is 37   30. 5th term is 7; 9th term is 19

31. 4th term is -2; 8th term is 26   32. 2nd term is -9; 8th term is 15

33. 5th term is -1; 12th term is -22 34. 6th term is -8; 12th term is -38

35. 3rd term is 3; 9th term is 0   36. 5th term is 5; 13th term is 7

37. Find the sum of the first 30 terms of the sequence $2, 8, 14, 20, \ldots$.

38. Find the sum of the first 40 terms of the sequence $1, 8, 15, 22, \ldots$.

39. Find the sum of the first 25 terms of the sequence $-8, -5, -2, 1, \ldots$.

40. Find the sum of the first 75 terms of the sequence $-9, -5, -1, 3, \ldots$.

41. Find the sum of the first 40 terms of the sequence $10, 3, -4, -11, \ldots$.

42. Find the sum of the first 50 terms of the sequence $12, 4, -4, -12, \ldots$.

43. Find the sum of the first 40 terms of the arithmetic sequence $\{4n - 3\}$.

44. Find the sum of the first 80 terms of the arithmetic sequence $\{2n - 13\}$.

45. Find the sum of the first 75 terms of the arithmetic sequence $\{-5n + 70\}$.

46. Find the sum of the first 35 terms of the arithmetic sequence $\{-6n + 25\}$.

47. Find the sum of the first 30 terms of the arithmetic sequence $\left\{5 + \dfrac{2}{3}n\right\}$.

48. Find the sum of the first 28 terms of the arithmetic sequence $\left\{7 - \dfrac{3}{2}n\right\}$.

Applying the Concepts

49. Find x so that $x + 3, 2x + 1$, and $5x + 2$ are consecutive terms of an arithmetic sequence.

50. Find x so that $2x, 3x + 2$, and $5x + 3$ are consecutive terms of an arithmetic sequence.

51. A Stack of Cans Suppose that the bottom row in a stack of cans contains 35 cans. Each layer contains one fewer can than the layer below it. The top row has 1 can. How many cans are in the stack?

52. A Pile of Bricks Suppose that the bottom row in a pile of bricks contains 46 bricks. Each layer contains two fewer bricks than the layer below it. The top row has 2 bricks. How many bricks are in the stack?

53. The Theater An auditorium has 40 seats in the first row and 25 rows in all. Each successive row contains 2 additional seats. How many seats are in the auditorium?

54. Mosaic A mosaic is designed in the shape of an equilateral triangle, 20 feet on each side. Each tile in the mosaic is in the shape of an equilateral triangle, 12 inches to a side. The tiles are to alternate in color as shown in the illustration. How many tiles of each color will be required?

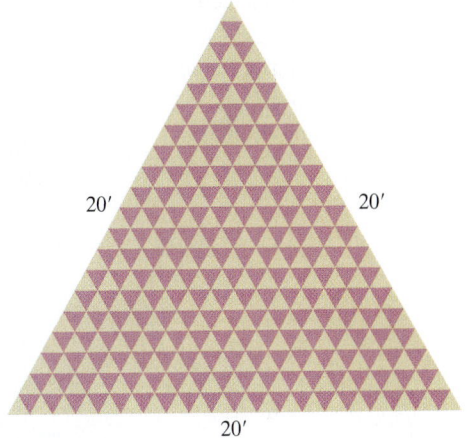

Extending the Concepts

In Problems 55–58, determine the number of terms that are in each arithmetic sequence.

55. $-5, -2, 1, \ldots, 244$

56. $-9, -5, -1, \ldots, 219$

57. $108, 101, 94, \ldots, -326$

58. $99, 93, 87, \ldots, -339$

59. Salary Suppose you have just been hired at the starting salary of \$32,000 per year. Your contract guarantees you a \$2500 raise each year. How many years will it take before your aggregate salary is \$757,500? (**Hint:** Your aggregate salary is \$32,000 + (32,000 + \$2500) + \ldots.)

60. Stadium Construction How many rows are in the corner section of a stadium containing 2030 seats if the first row has 14 seats and each successive row has 4 additional seats?

Synthesis Review

In Problems 61–63, (a) determine the base a of the exponential function, (b) Compute $f(1), f(2), f(3)$, and $f(4)$.

61. $f(x) = 3^x$

62. $f(x) = 4^x$

63. $f(x) = 10\left(\dfrac{1}{2}\right)^x$

64. Compute $\dfrac{f(2)}{f(1)}, \dfrac{f(3)}{f(2)},$ and $\dfrac{f(4)}{f(3)}$ for each of the functions in Problems 61–63. How does the ratio in the value of the function for consecutive values of the independent variable relate to the base?

The Graphing Calculator

In Problems 65–68, use a graphing calculator to find the sum of each sequence.

65. $\{3.45n + 4.12\}; n = 20$

66. $\{2.67n - 1.23\}; n = 25$

67. $85.9 + 83.5 + 81.1 + \cdots; n = 25$

68. $-11.8 + (-8.2) + (-4.6) + \cdots; n = 30$

13.3 Geometric Sequences and Series

OBJECTIVES

1. Determine If a Sequence Is Geometric
2. Find a Formula for the nth Term of a Geometric Sequence
3. Find the Sum of a Geometric Sequence
4. Find the Sum of a Geometric Series
5. Solve Annuity Problems

Preparing for Geometric Sequences and Series

Before getting started, take the following readiness quiz. If you get a problem wrong, go back to the section cited and review the material.

1. If $g(x) = 4^x$, evaluate $g(1), g(2),$ and $g(3)$. [Section 11.2, pp. 853–854]

2. Simplify $\dfrac{x^4}{x^3}$. [Section 5.4, pp. 371–372]

1 Determine If a Sequence Is Geometric

In the last section, we defined a sequence to be arithmetic if the difference between consecutive terms in the sequence was a constant. But what if the ratio of consecutive terms in the sequence is a constant? If the ratio of successive terms of a sequence is always the same number, the sequence is called a **geometric sequence.** For example, the sequence

$$2, 4, 8, 16, \ldots$$

is geometric because the constant ratio of consecutive terms is $2 \left(= \dfrac{4}{2} = \dfrac{8}{4} = \dfrac{16}{8} \right)$. If we call the first term a and the **common ratio** of consecutive terms r, then the terms of a geometric sequence follow the pattern

$$a, ar, ar^2, ar^3, \ldots$$

EXAMPLE 1   **Determining If a Sequence Is Geometric**

Determine if the sequence $2, 6, 18, 54, 162, \ldots$ is geometric. If it is geometric, state the first term and common ratio.

Solution

The sequence is geometric if the ratio of consecutive terms is a constant. Because the ratio of consecutive terms is $3 \left(= \dfrac{6}{2} = \dfrac{18}{6} = \dfrac{54}{18} = \dfrac{162}{54} \right)$, the sequence is geometric. The first term is $a = 2$ and the common ratio is $r = 3$. ∎

QUICK ✓ *Determine if the sequence is geometric. If it is, state the first term and common ratio.*

1. $4, 8, 16, 32, 64, \ldots$

2. $5, 10, 16, 23, 30, \ldots$

3. $9, 3, 1, \dfrac{1}{3}, \dfrac{1}{9}, \ldots$

Preparing for...Answers
1. $g(1) = 4, g(2) = 16, g(3) = 64$
2. x

EXAMPLE 2   **Determining If a Sequence Is Geometric**

Determine if the sequence $\{b_n\} = \{3^{-n}\}$ is geometric. If it is geometric, find the common ratio.

Solution

Work Smart
Notice the technique used in Example 2 is similar to the approach used in Section 13.2 to show a sequence in arithmetic.

We could list the first few terms and show that the ratio of consecutive terms is the same, but this would be more of a demonstration, not a proof. To prove the sequence is geometric, we must show that for *any* consecutive terms, the ratio is the same number. We do this by evaluating the sequence at the $(n - 1)$st term and nth term in the sequence and computing the ratio. If it is a constant, we've shown the sequence is geometric.

$$b_{n-1} = 3^{-(n-1)} = 3^{-n+1} \quad \text{and} \quad b_n = 3^{-n}$$

Now we compute $\dfrac{b_n}{b_{n-1}}$.

$$\frac{b_n}{b_{n-1}} = \frac{3^{-n}}{3^{-n+1}}$$

$$a^{m+n} = a^m \cdot a^n: \quad = \frac{3^{-n}}{3^{-n} \cdot 3^1}$$

$$\text{Divide out like factors:} \quad = \frac{1}{3}$$

The ratio of any two consecutive terms is $\dfrac{1}{3}$, so the common ratio is $\dfrac{1}{3}$. The sequence is geometric with $r = \dfrac{1}{3}$.   ▪

EXAMPLE 3   **Determining If a Sequence Is Geometric**

Determine if the sequence $\{a_n\} = \{n^2 + 1\}$ is geometric. If it is geometric, find the common ratio.

Solution

We proceed as we did in Example 2 and compute $\dfrac{a_n}{a_{n-1}}$. If this ratio is a constant, then the sequence is geometric.

$$\frac{a_n}{a_{n-1}} = \frac{n^2 + 1}{(n-1)^2 + 1}$$

$$\text{FOIL:} \quad = \frac{n^2 + 1}{n^2 - 2n + 1 + 1}$$

$$\text{Combine like terms:} \quad = \frac{n^2 + 1}{n^2 - 2n + 2}$$

We cannot simplify the rational expression any further. Because the ratio $\dfrac{a_n}{a_{n-1}}$ depends on the value of n, it is not a constant. Therefore, the sequence is not geometric.   ▪

QUICK ✓ *Determine if the sequence is geometric. If it is geometric, find the common ratio.*

4. $\{a_n\} = \{5^n\}$ **5.** $\{b_n\} = \{n^2\}$ **6.** $\{c_n\} = \left\{5\left(\dfrac{2}{3}\right)^n\right\}$

(2) ### Find a Formula for the nth Term of a Geometric Sequence

Suppose that a is the first term of a geometric sequence with common ratio $r \neq 0$. We seek a formula for the nth term of a_n. To see the pattern, we write down the first few terms.

$$a_1 = 1a = ar^0$$
$$a_2 = ra_1 = ar^1$$
$$a_3 = ra_2 = ar^2$$
$$a_4 = ra_3 = ar^3$$
$$a_5 = ra_4 = ar^4$$
$$\vdots$$
$$a_n = ra_{n-1} = r(ar^{n-2}) = ar^{n-1}$$

We are led to the following result.

THE nTH TERM OF A GEOMETRIC SEQUENCE

For a geometric sequence $\{a_n\}$ whose first term is a and whose common ratio is r, the nth term is determined by the formula

$$a_n = ar^{n-1}, \qquad r \neq 0$$

EXAMPLE 4 **Finding a Particular Term of a Geometric Sequence**

For the geometric sequence $8, 6, \dfrac{9}{2}, \dfrac{27}{8}, \ldots$

(a) Find a formula for the nth term.

(b) Find the eleventh term of the sequence.

Solution

(a) We are told that the sequence is geometric. The first term is $a = 8$. To find the common ratio, we compute the ratio of consecutive terms. So

$$r = \frac{6}{8} = \frac{9/2}{6} = \frac{27/8}{9/2} = \frac{3}{4}.$$ We now substitute these values into the formula for the nth term of a geometric sequence.

$$a_n = ar^{n-1}$$
$$= 8\left(\frac{3}{4}\right)^{n-1}$$

(b) The eleventh term is found by letting $n = 11$ in the formula found in part (a).

$$a_n = 8\left(\frac{3}{4}\right)^{n-1}$$
$$a_{11} = 8\left(\frac{3}{4}\right)^{11-1}$$
$$= 8\left(\frac{3}{4}\right)^{10}$$
$$\approx 0.45051$$

QUICK ✓ *Find a formula for the nth term of each geometric sequence. Use this result to find the ninth term of the sequence.*

7. $a = 5, r = 2$

8. $\{50, 25, 12.5, 6.25, \ldots\}$

③ Find the Sum of a Geometric Sequence

The next result gives us a formula for finding the sum of the first n terms of a geometric sequence.

SUM OF n TERMS OF A GEOMETRIC SEQUENCE

Let $\{a_n\}$ be a geometric sequence with first term a and common ratio r, where $r \neq 0, r \neq 1$. The sum S_n of the first n terms of $\{a_n\}$ is

$$S_n = a \cdot \frac{1 - r^n}{1 - r}, \qquad r \neq 0, r \neq 1$$

Where does this formula come from? Well, the sum S_n of the first n terms of $\{a_n\} = \{ar^{n-1}\}$ is

$$S_n = a + ar + ar^2 + \cdots + ar^{n-1}$$

Let's multiply both sides by r to obtain

$$rS_n = ar + ar^2 + ar^3 + \cdots + ar^n$$

Now subtract rS_n from S_n:

$$\begin{aligned} S_n &= a + ar + ar^2 + \cdots + ar^{n-1} \\ rS_n &= ar + ar^2 + ar^3 + \cdots + ar^n \\ \hline S_n - rS_n &= a - ar^n \end{aligned}$$

Factor S_n from the expression on the left and factor a from the expression on the right.

$$S_n(1 - r) = a(1 - r^n)$$

Divide both sides by $1 - r$ (since $r \neq 1$) and solve for S_n.

$$S_n = a \cdot \frac{1 - r^n}{1 - r}$$

EXAMPLE 5 **Finding the Sum of n Terms of a Geometric Sequence**

Find the sum of the first 10 terms of the sequence $2, 6, 18, 54, \ldots$.

Solution

This is a geometric sequence with $a = 2$ and common ratio $r = 3$ (Do you see why?). We wish to know the sum of the first 10 terms, S_{10}.

$$S_n = a \cdot \frac{1 - r^n}{1 - r}$$

$a = 2, r = 3, n = 10: \quad S_{10} = 2 \cdot \dfrac{1 - 3^{10}}{1 - 3}$

$$= 2 \cdot \frac{-59{,}048}{-2}$$

$$= 59{,}048 \qquad \blacksquare$$

EXAMPLE 6 **Finding the Sum of n Terms of a Geometric Sequence in Summation Notation**

Find the sum: $\displaystyle\sum_{n=1}^{8}\left[5 \cdot \left(\frac{1}{2}\right)^n\right]$

Express your answer to as many decimals as your calculator allows.

Solution

We can write this sum out as follows:

$$\sum_{n=1}^{8}\left[5\cdot\left(\frac{1}{2}\right)^n\right] = \frac{5}{2} + \frac{5}{4} + \cdots + \frac{5}{256}$$

We wish to find the sum of the first $n = 8$ terms of a geometric sequence with $a = \frac{5}{2}$ and common ratio $r = \frac{1}{2}$.

$$S_n = a\cdot\frac{1-r^n}{1-r}$$

$$a = \frac{5}{2},\ r = \frac{1}{2},\ n = 8:\quad S_8 = \frac{5}{2}\cdot\frac{1-\left(\frac{1}{2}\right)^8}{1-\frac{1}{2}}$$

$$= \frac{5}{2}\cdot\frac{0.99609375}{\frac{1}{2}}$$

$$= 4.98046875$$

QUICK ✓ *Find the sum.*

9. $3 + 6 + 12 + 24 + \cdots + 3\cdot 2^{12}$

10. $\displaystyle\sum_{n=1}^{10}\left[8\cdot\left(\frac{1}{2}\right)^n\right]$

---

④ ## Find the Sum of a Geometric Series

In Words
A sequence is a list of terms. A series is the sum of an infinite number of terms.

An infinite sum of the form

$$a + ar + ar^2 + \cdots + ar^{n-1} + \cdots$$

whose first term is a and common ratio is r, is called a **geometric series** and is denoted by

$$\sum_{n=1}^{\infty} ar^{n-1}$$

We know that the sum of the first n terms of a geometric sequence is given by the formula

$$S_n = a\cdot\frac{1-r^n}{1-r}$$

We can write this formula as $S_n = \dfrac{a - ar^n}{1 - r}$ by distributing the a. Now if we divide $1 - r$ into each term in the numerator, we obtain

$$S_n = \frac{a}{1-r} - \frac{ar^n}{1-r}$$

As n gets larger and larger, the expression S_n will approach the value $\dfrac{a}{1-r}$ because r^n approaches 0 provided that $-1 < r < 1$ and we have the following result.

> **SUM OF A GEOMETRIC SERIES**
>
> If $-1 < r < 1$, the sum of the geometric series $\displaystyle\sum_{n=1}^{\infty} ar^{n-1}$ is
>
> $$\sum_{n=1}^{\infty} ar^{n-1} = \frac{a}{1-r}$$

EXAMPLE 7 **Finding the Sum of an Geometric Series**

Find the sum of the geometric series: $4 + 2 + 1 + \dfrac{1}{2} + \cdots$.

Solution

This is an geometric series with $a = 4$ and common ratio $r = \dfrac{1}{2}\left(= \dfrac{2}{4} = \dfrac{1}{2} = \dfrac{1/2}{1}\right)$.

Since the common ratio r is between -1 and 1, we can use the formula for the sum of a geometric series to find that

$$4 + 2 + 1 + \frac{1}{2} + \cdots = \frac{4}{1 - \dfrac{1}{2}} = 8$$

QUICK ✓ *Find the sum of the geometric series.*

11. $10 + \dfrac{5}{2} + \dfrac{5}{8} + \dfrac{5}{32} + \cdots$ **12.** $\displaystyle\sum_{n=1}^{\infty} \left(\frac{1}{3}\right)^n$

Work Smart: Study Skills

Let's summarize the formulas for arithmetic and geometric sequences where a is the first term of the sequence, d is the common difference, and r is the common ratio.

	Arithmetic	**Geometric**
Find the nth term	$a_n = a + (n - 1)d$	$a_n = ar^{n-1}, r \neq 0$
Find the sum of the first n terms	$S_n = \dfrac{n}{2}[2a + (n - 1)d]$ $= \dfrac{n}{2}(a + a_n)$	$S_n = a \cdot \dfrac{1 - r^n}{1 - r}, r \neq 0, r \neq 1$

If you were asked to find the 5th term of the sequence $5, 20, 80, 320, \dots$, which formula would you use?

If you were asked to find the sum of the first 5 terms of the sequence $5, 9, 13, 17, \dots$, which formula would you use?

EXAMPLE 8 **Writing a Repeating Decimal as a Fraction**

Express $0.\overline{1}$ as a fraction in lowest terms.

Solution

The line over the 1 in the decimal indicates that the 1 repeats indefinitely. That is, we can write

$$0.\overline{1} = 0.11111\cdots$$
$$= 0.1 + 0.01 + 0.001 + \cdots$$

This is a geometric series with $a = 0.1$ and common ratio $r = 0.1 \left(= \dfrac{0.01}{0.1} = \dfrac{0.001}{0.01}\right)$.

Since the common ratio r is between -1 and 1, we can use the formula for the sum of a geometric series to find that

$$0.\overline{1} = 0.1 + 0.01 + 0.001 + \cdots$$
$$= \frac{0.1}{1 - 0.1}$$
$$= \frac{0.1}{0.9}$$

Multiply numerator and denominator by 10: $= \dfrac{1}{9}$

So $0.\overline{1} = \dfrac{1}{9}$.

QUICK ✓

13. Express $0.\overline{2}$ as a fraction in lowest terms.

EXAMPLE 9 **The Multiplier**

Suppose that, throughout the United States economy, individuals spend 90% of every additional dollar they earn. Economists would say that an individual's **marginal propensity to consume** is 0.90. For example, if Roberta earns an additional dollar, she will spend $0.9(\$1) = \$0.90$ of it and save $0.10. Whoever earns $0.90 (from Roberta) will spend 90% of it or $0.9(\$0.90) = \$0.81$. This process of spending continues and results in a geometric series as follows:

$$\$1 + \$0.90 + \$0.81 + \$0.729 + \cdots$$

The sum of this geometric series is called the **multiplier.** Suppose that the government gives a child-tax rebate of $500 to Roberta. What is the multiplier if individuals spend 90% of every additional dollar they earn?

Solution

The total impact of the $500 tax rebate on the U.S. economy is

$$\$500 + \$500(0.9) + \$500(0.9)^2 + \$500(0.9)^3 + \cdots$$

This is a geometric series with first term $a = 500$ and common ratio $r = 0.9$. The sum of this series is

$$\$500 + \$500(0.9) + \$500(0.9)^2 + \$500(0.9)^3 + \cdots = \frac{\$500}{1 - 0.9}$$

$$= \$5000$$

The United States economy will grow by $5000 because of the child-tax credit to Roberta.

QUICK ✓

14. Redo Example 9 if the marginal propensity to consume is 95%.

---

⑤ Solve Annuity Problems

In Section 11.2, we looked at the compound interest formula, which allows us to compute the future value of a lump sum of money that is deposited in an account that pays interest compounded periodically (say, monthly). Often, though, money is invested at periodic intervals of time. An **annuity** is a sequence of equal periodic deposits. The periodic deposits may be made annually, quarterly, monthly, or daily.

When deposits are made at the same time the interest is credited, the annuity is called **ordinary.** We will only deal with ordinary annuities here. The **amount of an annuity** is the sum of all deposits made plus all interest paid.

Suppose the interest an account earns is i percent per payment period (expressed as a decimal). For example, if an account pays 6% compounded monthly (12 times a year), then $i = \dfrac{0.06}{12} = 0.005$. To develop a formula for the amount of an annuity, suppose $\$P$ is deposited each payment period for n payment periods in an account that earns $i\%$ per payment period. When the last deposit is made at the nth payment period, the first deposit has earned interest compounded for $n - 1$ payment periods, the second deposit of $\$P$ has earned interest compounded for $n - 2$ payment periods, and so on. Table 1 shows the value of each deposit after n deposits have been made.

Table 1					
Deposit	1	2	3	$n-1$	n
Future Value of Deposit of $P	$P(1+i)^{n-1}$	$P(1+i)^{n-2}$	$P(1+i)^{n-3}$	$P(1+i)$	P

The amount A of the annuity is the sum of the amounts shown in Table 1, namely,

$$\begin{aligned} A &= P(1+i)^{n-1} + P(1+i)^{n-2} + P(1+i)^{n-3} + \cdots + P(1+i) + P \\ &= P[(1+i)^{n-1} + (1+i)^{n-2} + (1+i)^{n-3} + \cdots + (1+i) + 1] \\ &= P[1 + (1+i) + \cdots + (1+i)^{n-3} + (1+i)^{n-2} + (1+i)^{n-1}] \end{aligned}$$

The expression in brackets is the sum of a geometric sequence with n terms and a common ratio of $(1+i)$. As a result,

$$\begin{aligned} A &= P[1 + (1+i) + \cdots + (1+i)^{n-3} + (1+i)^{n-2} + (1+i)^{n-1}] \\ &= P \cdot \frac{1-(1+i)^n}{1-(1+i)} = P \cdot \frac{1-(1+i)^n}{-i} = P \cdot \frac{(1+i)^n - 1}{i} \end{aligned}$$

We have the following result.

AMOUNT OF AN ANNUITY

If P represents the deposit in dollars made at each payment period for an annuity at i percent interest per payment period, the amount A of the annuity after n payment periods is

$$A = P \cdot \frac{(1+i)^n - 1}{i}$$

EXAMPLE 10   Determining the Amount of an Annuity

To save for retirement, Alejandro decides to place $100 into a Roth Individual Retirement Account (IRA) every month for the next 30 years. What will be the value of the IRA after 30 years assuming that his account earns 6% interest compounded monthly?

Solution

This is an ordinary annuity with $n = 12 \cdot 30 = 360$ payments with deposits of $P = \$100$. The rate of interest per payment period is $i = \dfrac{0.06}{12} = 0.005$. The amount of 30 years (360 deposits) is

$$\begin{aligned} A &= \$100\left[\frac{(1+0.005)^{360} - 1}{0.005}\right] \\ &= \$100[1004.515042] \\ &= \$100{,}451.50 \end{aligned}$$

QUICK ✓

15. To save for retirement, Magglio decides to place $500 into a Roth Individual Retirement Account (IRA) every quarter (every three months) for the next 30 years. What will be the value of the IRA after 30 years assuming that his account earns 8% interest compounded quarterly?

13.3 Exercises

For Extra Help: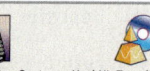
Student Solutions Manual   CD Video   PH Math/Tutor Center   MathXL Tutorials on CD   MathXL®   MyMathLab

Concepts and Vocabulary

In Problems 1 and 2, fill in the blanks.

1. In a(n) _____ sequence the ratio of successive terms is constant.

2. If $-1 < r < 1$, the sum of the geometric series $\sum\limits_{n=1}^{\infty} ar^{n-1} =$ _____.

In Problems 3 and 4, answer True or False to each statement.

3. In a geometric sequence, the common ratio is always a positive number.

4. For a geometric sequence with first term a and common ratio r, where $r \neq 0, r \neq 1$, the sum of the first n terms is $S_n = a \cdot \dfrac{1 - r^n}{1 - r}$.

5. How do you determine if a sequence is geometric?

6. How do you determine if a geometric series has a sum?

Building Skills

In Problems 7–14, a geometric sequence is given. Find the common ratio and write out the first four terms.

7. $\{4^n\}$ **8.** $\{(-2)^n\}$ **9.** $\left\{\left(\dfrac{2}{3}\right)^n\right\}$ **10.** $\left\{\dfrac{2^n}{3}\right\}$

11. $\{3 \cdot 2^{-n}\}$ **12.** $\left\{-10 \cdot \left(\dfrac{1}{2}\right)^n\right\}$ **13.** $\left\{\dfrac{5^{n-1}}{2^n}\right\}$ **14.** $\left\{\dfrac{3^{-n}}{2^{n-1}}\right\}$

In Problems 15–26, determine whether the given sequence is arithmetic, geometric, or neither. If the sequence is arithmetic, find the common difference; if it is geometric, find the common ratio.

15. $\{5n + 1\}$ **16.** $\{8 - 3n\}$ **17.** $\{2n^2\}$ **18.** $\{n^2 - 2\}$

19. $\left\{\dfrac{2^{-n}}{5}\right\}$ **20.** $\left\{\dfrac{2}{3^n}\right\}$ **21.** $54, 36, 24, 16, \ldots$ **22.** $100, 20, 4, \dfrac{4}{5}, \ldots$

23. $2, 6, 10, 14, \ldots$ **24.** $15, 12, 9, 6, \ldots$ **25.** $1, 2, 3, 5, 8, \ldots$ **26.** $5, -2, 3, -1, 2, \ldots$

In Problems 27–34, (a) find a formula for the nth term of the geometric sequence whose first term and common ratio are given, and (b) use the formula to find the eighth term.

27. $a = 10, r = 2$ **28.** $a = 2, r = 3$ **29.** $a = 100, r = \frac{1}{2}$

30. $a = 30, r = 1/3$ **31.** $a = 1, r = -3$ **32.** $a = 1, r = -4$

33. $a = 100, r = 1.05$ **34.** $a = 500, r = 1.04$

In Problems 35–40, find the indicated term of each geometric sequence.

35. 10th term of $3, 6, 12, 24, \ldots$ **36.** 12th term of $1, 3, 9, 27, \ldots$

37. 15th term of $4, -2, 1, -\frac{1}{2}, \ldots$ **38.** 8th term of $10, -20, 40, -80, \ldots$

39. 9th term of $0.5, 0.05, 0.005, 0.0005, \ldots$

40. 10th term of $0.4, 0.04, 0.004, 0.0004, \ldots$

In Problems 41–48, find the sum. If necessary, express your answer to as many decimals as your calculator allows.

41. $2 + 4 + 8 + \cdots + 2^{12}$ **42.** $3 + 9 + 27 + \cdots + 3^{10}$

43. $50 + 20 + 8 + \dfrac{16}{5} + \cdots + 50\left(\dfrac{2}{5}\right)^{10-1}$

44. $10 + 5 + \dfrac{5}{2} + \cdots + 10\left(\dfrac{1}{2}\right)^{12-1}$

45. $\displaystyle\sum_{n=1}^{10}[3 \cdot 2^n]$ **46.** $\displaystyle\sum_{n=1}^{12}[5 \cdot 2^n]$ **47.** $\displaystyle\sum_{n=1}^{8}\left[\dfrac{4}{2^{n-1}}\right]$ **48.** $\displaystyle\sum_{n=1}^{14}\left[10 \cdot \left(\dfrac{1}{2}\right)^{n-1}\right]$

In Problems 49–58, find the sum of each geometric series. If necessary, express your answer to as many decimals as your calculator allows.

49. $1 + \dfrac{1}{2} + \dfrac{1}{4} + \cdots$ **50.** $1 + \dfrac{1}{3} + \dfrac{1}{9} + \cdots$ **51.** $10 + \dfrac{10}{3} + \dfrac{10}{9} + \cdots$

52. $20 + 5 + \dfrac{5}{4} + \cdots$ **53.** $6 - 2 + \dfrac{2}{3} - \dfrac{2}{9} + \cdots$ **54.** $12 - 3 + \dfrac{3}{4} - \dfrac{3}{16} + \cdots$

55. $\displaystyle\sum_{n=1}^{\infty}\left(5 \cdot \left(\dfrac{1}{5}\right)^n\right)$ **56.** $\displaystyle\sum_{n=1}^{\infty}\left(10 \cdot \left(\dfrac{1}{3}\right)^n\right)$ **57.** $\displaystyle\sum_{n=1}^{\infty}\left(12 \cdot \left(-\dfrac{1}{3}\right)^{n-1}\right)$

58. $\displaystyle\sum_{n=1}^{\infty}\left(100 \cdot \left(-\dfrac{1}{2}\right)^{n-1}\right)$

In Problems 59–62, express each repeating decimal as a fraction in lowest terms.

59. $0.\overline{5}$ **60.** $0.\overline{3}$ **61.** $0.\overline{89}$ **62.** $0.\overline{45}$

Applying the Concepts

63. Find x so that x, $x + 2$, and $x + 3$ are consecutive terms of a geometric sequence.

64. Find x so that $x - 1$, x, and $x + 2$ are consecutive terms of a geometric sequence.

65. **Salary Increases** Suppose that you have been hired at an annual salary of $40,000 per year. You have been promised a raise of 5% for each of the next 10 years.

  (a) What will be your salary at the beginning of your second year?
  (b) What will be your salary at the beginning of your tenth year?
  (c) How much will you have earned cumulatively once you have finished your tenth year?

66. **Salary Increases** Suppose that you have been hired at an annual salary of $45,000 per year. Historically, the typical raise is 4% each year. You expect to be at the company for the next 10 years.

  (a) What will be your salary at the beginning of your second year?
  (b) What will be your salary at the beginning of your tenth year?
  (c) How much will you have earned cumulatively once you have finished your tenth year?

67. **Depreciation of a Car** Suppose that you have just purchased a Honda Accord for $20,000. Historically, the car depreciates by 8% each year, so that next year the car is worth $20,000(0.92). What will the value of the car be after you have owned it for five years?

68. **Depreciation of a Car** Suppose that you have just purchased a Chevy Cavalier for $16,000. Historically, the car depreciates by 10% each year, so that next year the car is worth $16,000(0.9). What will the value of the car be after you have owned it for four years?

69. **Pendulum Swings** Initially, a pendulum swings through an arc of 3 feet. On each successive swing, the length of the arc is 0.95 of the previous length.

  (a) What is the length of the arc after 10 swings?
  (b) On which swing is the length of the arc less than 1 foot for the first time?
  (c) After 10 swings, what total length will the pendulum have swung?
  (d) When it stops, what total length will the pendulum have swung?

70. Bouncing Balls A ball is dropped from a height of 30 feet. Each time it strikes the ground, it bounces up to 0.8 of the previous height.

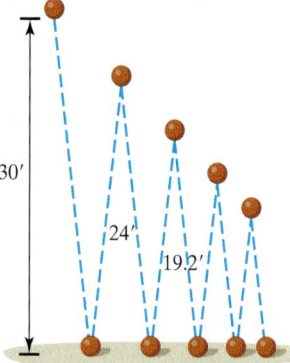

(a) What height will the ball bounce up to after it strikes the ground for the fourth time?

(b) How high will it bounce after it strikes the ground for the fifth time?

(c) How many times does the ball need to strike the ground before its bounce is less than 6 inches?

(d) What total distance does the ball travel before it stops bouncing?

71. A Job Offer You are interviewing for a job and receive two offers:

A: $30,000 to start, with guaranteed annual increases of 5% for the first 5 years
B: $31,000 to start with guaranteed annual increases of 4% for the first 5 years

Which offer is best if your goal is to be making as much money as possible after the fifth year? Which is best if your goal is to make as much money as possible over the entire 5 years of the contract?

72. Be an Agent Suppose that you are an agent for a professional baseball player. Management has just offered your client a 5-year contract with a first-year salary of $1,500,000. Beyond that, your client has three choices:

A. A bonus of $75,000 each year (including the first year)
B. An annual increase of 4% per year beginning after the first year
C. An annual increase of $80,000 per year beginning after the first year

Which option provides the most money over the 5-year period? Which the least? Which would you choose? Why?

73. The Multiplier Suppose the marginal propensity to consume throughout the U.S. economy is 0.98. What is the multiplier for the U.S. economy?

74. The Multiplier Suppose the marginal propensity to consume throughout the U.S. economy is 0.96. What is the multiplier for the U.S. economy?

75. Stock Price One method of pricing a stock is based upon the stream of future dividends of the stock. Suppose that a stock pays $P per year in dividends and, historically, the dividend has been increased by i% per year. If you desire an annual rate of return of r%, this method of pricing a stock states that the price you should pay is the present value of an infinite stream of payments:

$$\text{Price} = P + P \cdot \frac{1+i}{1+r} + P \cdot \left(\frac{1+i}{1+r}\right)^2 + P \cdot \left(\frac{1+i}{1+r}\right)^3 + \cdots$$

The price of the stock is the sum of a geometric series. Suppose that a stock pays an annual dividend of $2.00 and, historically, the dividend has been increased by 2% per year. You desire an annual rate of return of 9%. What is the most you should pay for the stock?

76. Stock Price Refer to Problem 75. Suppose that a stock pays an annual dividend of $3.00 and, historically, the dividend has been increased by 3% per year. You desire an annual rate of return of 10%. What is the most you should pay for the stock?

77. 401(k) Christine contributes $100 each month into her 401(k) retirement plan. What will be the value of Christine's 401(k) in 30 years if the per annum rate of return is assumed to be 8% compounded monthly?

78. Saving for a Home Jolene wants to purchase a new home. Suppose she invests $400 a month into a money market fund. If the per annum interest rate of return on the money market is 4% compounded monthly, how much will Jolene have for a down payment in 4 years?

79. **Roth IRA** Jackson contributes $500 each quarter into his Roth IRA. What will be the value of Jackson's IRA in 25 years if the per annum rate of return is assumed to be 6% compounded quarterly?

80. **Retirement** Raymont is planning on retiring in 15 years, so he contributes $1500 into his IRA every 6 months (semiannually). What will be the value of the IRA when Raymont retires if the per annum interest rate is 10% compounded semiannually?

81. **What's My Payment?** Suppose that Aaliyah wants to have $1,500,000 in her 401(k) retirement account in 35 years. How much does she need to contribute each month if the account earns 10% interest compounded monthly?

82. **What's My Payment?** Suppose that Sophia wants to have $2,000,000 in her 401(k) retirement account in 25 years. How much does she need to contribute each quarter if the account earns 12% interest compounded quarterly?

Extending the Concepts

83. Express $0.4\overline{9}$ as a fraction in lowest terms.

84. Express $0.85\overline{9}$ as a fraction in lowest terms.

85. Find the sum: $2 + 4 + 8 + \cdots + 1{,}073{,}741{,}824$

86. Can a sequence be both arithmetic and geometric? Give reasons for you answer.

87. Which yields faster growth in the terms of a sequence—an arithmetic sequence or a geometric sequence with $r > 1$? Why? Explain why Thomas Robert Malthus's conjecture that food supplies grow arithmetically while population grows geometrically provides a recipe for disaster unless population growth is curbed.

Synthesis Review

88. Express $\dfrac{1}{3}$ as a repeating decimal. Express $\dfrac{2}{3}$ as a repeating decimal.

89. What is $\dfrac{1}{3} + \dfrac{2}{3}$? Now add the repeating decimals found in Problem 88. Conjecture the value of $0.9999999\ldots$.

90. Prove that $0.9999\ldots = 0.\overline{9}$ equals 1.

The Graphing Calculator

In Problems 91–94, use a graphing calculator to find the sum of each sequence.

91. $4 + 4.8 + 5.76 + \cdots + 4(1.2)^{15-1}$

92. $3 + 4.8 + 7.68 + \cdots + 3(1.6)^{20-1}$

93. $\displaystyle\sum_{n=1}^{20}[1.2(1.05)^n]$

94. $\displaystyle\sum_{n=1}^{25}[1.3(0.55)^n]$

PUTTING THE CONCEPTS TOGETHER (Sections 13.1–13.3)

These problems cover important concepts from Sections 13.1 to 13.3. We designed these problems so that you can review the chapter so far and show your mastery of the concepts. Take time to work these problems before proceeding with the next section. The answers to these problems are located at the back of the text starting on page AN-59.

In Problems 1–6, determine if the sequence is arithmetic, geometric, or neither. If arithmetic or geometric, determine the first term and the common difference or common ratio.

1. $\dfrac{3}{4}, \dfrac{3}{16}, \dfrac{3}{64}, \dfrac{3}{256}, \ldots$

2. $\{2(n + 3)\}$

3. $\left\{\dfrac{7n + 2}{9}\right\}$

4. $1, -4, 9, -16, \ldots$ **5.** $\{3 \cdot 2^{n+1}\}$ **6.** $\{n^2 - 5\}$

7. Write out the sum and evaluate: $\displaystyle\sum_{k=1}^{6} [3k + 4]$

8. Express the sum using summation notation:

$$\frac{1}{2(6 + 1)} + \frac{1}{2(6 + 2)} + \frac{1}{2(6 + 3)} + \cdots + \frac{1}{2(6 + 12)}$$

In Problems 9–12, write a formula for the nth term of the indicated sequence. Write out the first five terms of each sequence.

9. arithmetic: $a = 25, d = -2$

10. arithmetic: $a_4 = 9, d = 11$

11. geometric: $a_4 = \dfrac{9}{25}, r = \dfrac{1}{5}$

12. geometric: $a = 150, r = 1.04$

In Problems 13–15, find the indicated sum.

13. $2 + 6 + 18 + \cdots + 2 \cdot (3)^{11-1}$

14. $2 + 7 + 12 + 17 + \cdots + [2 + (20 - 1) \cdot 5]$

15. $1000 + 100 + 10 + \cdots$

16. **Table Seating** A restaurant uses square tables in the main dining area that seat four people. For larger parties, tables can be placed together. Two tables will seat 6 people, three tables will seat 8 people, and so on (see diagram). How many tables could be needed to seat a party of 24 people?

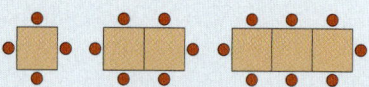

13.4 The Binomial Theorem

OBJECTIVES

1 Compute Factorials

2 Evaluate a Binomial Coefficient

3 Expand a Binomial

Preparing for the Binomial Theorem

Before getting started, take the following readiness quiz. If you get a problem wrong, go back to the section cited and review the material.

1. Multiply: $(x - 5)^2$ [Section 5.3, pp. 364–366]

2. Multiply: $(2x + 3)^2$ [Section 5.3, pp. 364–366]

1 Compute Factorials

Suppose we wanted to find the product of the first 12 integers. That is, suppose we wanted to compute

$$12 \cdot 11 \cdot 10 \cdot 9 \cdot 8 \cdot 7 \cdot 6 \cdot 5 \cdot 4 \cdot 3 \cdot 2 \cdot 1$$

This product is equal to 479,001,600. Not only is this a big number, writing out the product is long. A shorthand method for writing this product is *factorial notation*.

> **DEFINITION**
>
> If $n \geq 0$ is an integer, the **factorial symbol $n!$** (read "n factorial") is defined as follows:
>
> $$0! = 1 \qquad 1! = 1$$
> $$n! = n(n - 1)(n - 2) \cdot \cdots \cdot 3 \cdot 2 \cdot 1 \quad \text{if } n \geq 2$$

For example, $2! = 2 \cdot 1 = 2$, $3! = 3 \cdot 2 \cdot 1 = 6$, $4! = 4 \cdot 3 \cdot 2 \cdot 1 = 24$, and so on. Table 2 lists the values of $n!$ for $0 \le n \le 7$.

Table 2							
n	0	1	2	3	4	5	6	7
$n!$	1	1	2	6	24	120	720	5040

Because

$$\underbrace{n(n-1)(n-2) \cdot \cdots \cdot 3 \cdot 2 \cdot 1}_{(n-1)!}$$

we can use the formula

$$n! = n(n-1)!$$

to find successive factorials. For example, because $7! = 5040$, we have

$$8! = 8 \cdot 7! = 8(5040) = 40{,}320$$

Your calculator has a factorial key. Use it to see how fast factorials increase in value. Find the value of 69!. What happens when you try to find 70!? In fact, 70! is larger than 10^{100} (a **googol**), the largest number that most calculators can display.

EXAMPLE 1 Computing Factorials

Compute the value of $\dfrac{12!}{9!}$.

Solution

We could directly compute 12! and then 9!, but this would be inefficient. Rather, we shall use properties of factorials. Namely,

$$\frac{12!}{9!} = \frac{12 \cdot 11 \cdot 10 \cdot 9!}{9!}$$

Divide out 9!: $\quad = 12 \cdot 11 \cdot 10$

$$= 1320 \qquad \blacksquare$$

QUICK ✓ *Find the value of each factorial.*

1. 9! **2.** $\dfrac{7!}{3!}$

---

(2) ## Evaluate a Binomial Coefficient

A formula has been given for expanding $(x + a)^n$ for $n = 2$. The *Binomial Theorem* is a formula for the expansion of $(x + a)^n$ for any positive integer n. If $n = 1, 2, 3$, and 4, the expansion of $(x + a)^n$ is straightforward:

$(x + a)^1 = x + a$ — Two terms, beginning with x^1 and ending with a^1

$(x + a)^2 = x^2 + 2ax + a^2$ — Three terms, beginning with x^2 and ending with a^2

$(x + a)^3 = x^3 + 3ax^2 + 3a^2x + a^3$ — Four terms, beginning with x^3 and ending with a^3

$(x + a)^4 = x^4 + 4ax^3 + 6a^2x^2 + 4a^3x + a^4$ — Five terms, beginning with x^4 and ending with a^4

Notice that each expansion of $(x + a)^n$ begins with x^n and ends with a^n. As you read from left to right, the powers of x are decreasing by 1, while the powers of a are increasing by 1. Also, the number of terms that appears equals $n + 1$. Notice, too, that the degree of each monomial in the expansion equals n. For example, in the expansion of $(x + a)^3$, each monomial $(x^3, 3ax^2, 3a^2x, a^3)$ is of degree 3. As a result, we might conjecture that the expansion of $(x + a)^n$ would look like this:

$$(x + a)^n = x^n + \underline{}ax^{n-1} + \underline{}a^2x^{n-2} + \cdots + \underline{}a^{n-1}x + a^n$$

where the blanks are numbers to be found. This is, in fact, the case, as we shall see shortly.

First, we need to introduce a symbol. We define the symbol $\binom{n}{j}$, read "n taken j at a time" or "n choose j", as follows:

Work Smart

Do not write $\binom{n}{j}$ as $\left(\dfrac{n}{j}\right)$.

> **DEFINITION**
>
> If j and n are integers with $0 \le j \le n$, the symbol $\binom{n}{j}$ is defined as
>
> $$\binom{n}{j} = \frac{n!}{j!(n-j)!}$$

EXAMPLE 2 Evaluating $\binom{n}{j}$

Find:

(a) $\binom{4}{1}$ **(b)** $\binom{6}{2}$ **(c)** $\binom{5}{4}$

Solution

(a) Here, we have $n = 4$ and $j = 1$, so

$$\binom{4}{1} = \frac{4!}{1!(4-1)!} = \frac{4!}{1!3!} = \frac{4 \cdot 3 \cdot 2 \cdot 1}{1 \cdot 3 \cdot 2 \cdot 1} = \frac{4 \cdot \cancel{3} \cdot \cancel{2} \cdot 1}{1 \cdot \cancel{3} \cdot \cancel{2} \cdot 1} = 4$$

(b) $\binom{6}{2} = \frac{6!}{2!(6-2)!} = \frac{6!}{2! \cdot 4!} \underset{\underset{6! = 6 \cdot 5 \cdot 4!}{\uparrow}}{=} \frac{6 \cdot 5 \cdot 4!}{2 \cdot 1 \cdot 4!} = \frac{6 \cdot 5 \cdot \cancel{4!}}{2 \cdot 1 \cdot \cancel{4!}} = \frac{30}{2} = 15$

(c) $\binom{5}{4} = \frac{5!}{4!(5-4)!} = \frac{5!}{4! \cdot 1!} = \frac{5 \cdot 4!}{4! \cdot 1} = \frac{5 \cdot \cancel{4!}}{\cancel{4!} \cdot 1} = 5$

QUICK ✓ *Evaluate each expression.*

3. $\binom{7}{1}$ **4.** $\binom{6}{3}$

Four useful formulas involving the symbol $\binom{n}{j}$ are

$$\binom{n}{0} = 1 \qquad \binom{n}{1} = n \qquad \binom{n}{n-1} = n \qquad \binom{n}{n} = 1$$

Suppose that we arrange the various values of the symbol $\binom{n}{j}$ in a triangular display, as shown next and in Figure 6.

$$\binom{0}{0}$$

$$\binom{1}{0}\quad\binom{1}{1}$$

$$\binom{2}{0}\quad\binom{2}{1}\quad\binom{2}{2}$$

$$\binom{3}{0}\quad\binom{3}{1}\quad\binom{3}{2}\quad\binom{3}{3}$$

$$\binom{4}{0}\quad\binom{4}{1}\quad\binom{4}{2}\quad\binom{4}{3}\quad\binom{4}{4}$$

$$\binom{5}{0}\quad\binom{5}{1}\quad\binom{5}{2}\quad\binom{5}{3}\quad\binom{5}{4}\quad\binom{5}{5}$$

Figure 6
Pascal's Triangle

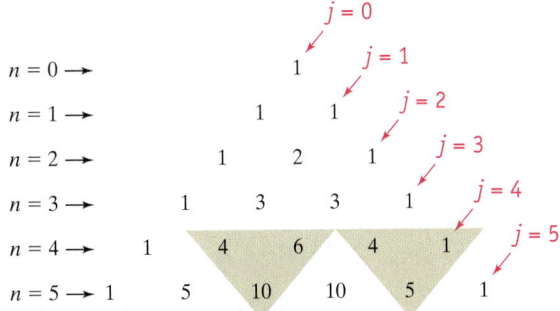

This display is called **Pascal's triangle,** named after Blaise Pascal (1623–1662), a French mathematician.

The Pascal's triangle has 1s down the sides. To get any other entry, add the two nearest entries in the row above it. The shaded triangles in Figure 6 illustrate this feature of the Pascal's triangle. Based on this feature, the row corresponding to $n = 6$ is found as follows:

$$n = 5 \rightarrow \quad 1 \quad 5 \quad 10 \quad 10 \quad 5 \quad 1$$
$$n = 6 \rightarrow \quad 1 \quad 6 \quad 15 \quad 20 \quad 15 \quad 6 \quad 1$$

Although the Pascal's triangle provides an interesting and organized display of the symbol $\binom{n}{j}$, in practice it is not all that helpful. For example, if you wanted to know the value of $\binom{12}{8}$, you would need to produce 12 rows of the triangle before seeing the answer. It is much faster to use the definition of $\binom{n}{j}$.

③ Expand a Binomial

Now we are ready to state the **Binomial Theorem.**

THE BINOMIAL THEOREM

Let x and a be real numbers. For any positive integer n, we have

$$(x + a)^n = \binom{n}{0}x^n + \binom{n}{1}ax^{n-1} + \binom{n}{2}a^2x^{n-2} + \cdots + \binom{n}{j}a^jx^{n-j} + \cdots + \binom{n}{n}a^n$$

You should now see why we needed to discuss the symbol $\binom{n}{j}$; these symbols are the numerical coefficients that appear in the expansion of $(x + a)^n$. Because of this, the symbol $\binom{n}{j}$ is called the **binomial coefficient.**

EXAMPLE 3 Expanding a Binomial

Use the Binomial Theorem to expand $(x + 3)^4$.

Work Smart

In the expansion, notice how the exponent on x **decreases** by one for each term as we move to the right, while the exponent on 3 **increases** by 1.

Solution

In the Binomial Theorem, let $a = 3$ and $n = 4$. Then

$$(x + 3)^4 = \binom{4}{0}x^4 + \binom{4}{1}3^1 \cdot x^{4-1} + \binom{4}{2}3^2 \cdot x^{4-2} + \binom{4}{3}3^3 \cdot x^{4-3} + \binom{4}{4}3^4$$

Use row 4 of the Pascal's triangle in Figure 6 or use $\binom{n}{j} = \dfrac{n!}{j!(n-j)!}$ to evaluate the binomial coefficients.

$$= 1 \cdot x^4 + 4 \cdot 3 \cdot x^3 + 6 \cdot 9 \cdot x^2 + 4 \cdot 27 \cdot x + 1 \cdot 81$$
$$= x^4 + 12x^3 + 54x^2 + 108x + 81$$

EXAMPLE 4 Expanding a Binomial

Expand $(2y - 3)^5$ using the Binomial Theorem.

Solution

First, we rewrite the expression $(2y - 3)^5$ as $[2y + (-3)]^5$. Now we use the Binomial Theorem with $n = 5$, $x = 2y$, and $a = -3$.

$$[2y + (-3)]^5 = \binom{5}{0}(2y)^5 + \binom{5}{1}(-3)^1(2y)^{5-1} + \binom{5}{2}(-3)^2(2y)^{5-2} + \binom{5}{3}(-3)^3(2y)^{5-3} + \binom{5}{4}(-3)^4(2y)^{5-4} + \binom{5}{5}(-3)^5$$

Use row 5 of Pascal's triangle or use the formula $\binom{n}{j} = \dfrac{n!}{j!(n-j)!}$ to evaluate binomial coefficients.

$$= (2y)^5 + 5(-3)^1(2y)^4 + 10(-3)^2(2y)^3 + 10(-3)^3(2y)^2 + 5(-3)^4(2y)^1 + (-3)^5$$
$$= 32y^5 + 5 \cdot -3 \cdot 16y^4 + 10 \cdot 9 \cdot 8y^3 + 10 \cdot -27 \cdot 4y^2 + 5 \cdot 81 \cdot 2y - 243$$
$$= 32y^5 - 240y^4 + 720y^3 - 1080y^2 + 810y - 243$$

QUICK ✓ *Expand each binomial using the Binomial Theorem.*

5. $(x + 2)^4$ **6.** $(2p - 1)^5$

13.4 Exercises

For Extra Help:

Student Solutions Manual   CD Video   PH Math/Tutor Center   MathXL Tutorials on CD   MathXL®   MyMathLab

Concepts and Vocabulary

In Problems 1–3, fill in the blanks.

1. If $n \geq 2$ is an integer, then $n! =$ _____.

2. $0! =$ _____; $1! =$ _____; $10! =$ _____.

3. _____ _____ is a triangular display of the binomial coefficients.

In Problems 4 and 5, answer True or False to each statement.

4. $\dbinom{n}{j} = \dfrac{j!}{n!(n-j)!}$

5. $\dbinom{n}{0} = 0$

6. Write down the first four rows of the Pascal's triangle.

7. Describe the pattern of exponents of x in the expansion of $(x + a)^n$.

8. Describe the pattern of exponents of a in the expansion of $(x + a)^n$.

9. What is true about the degree of each monomial in the expansion of $(x + a)^n$?

10. Explain how you might find a particular term in a binomial expansion. For example, how might you find the fifth term in the expansion of $(x + 3)^8$?

Building Skills

In Problems 11–26, expand each expression using the Binomial Theorem.

11. $(x + 1)^5$ **12.** $(x - 1)^4$ **13.** $(x - 4)^4$ **14.** $(x + 5)^5$

15. $(3p + 2)^4$ **16.** $(2q + 3)^4$ **17.** $(2z - 3)^5$ **18.** $(3w - 4)^4$

19. $(x^2 + 2)^4$ **20.** $(y^2 - 3)^4$ **21.** $(2p^3 + 1)^5$ **22.** $(3b^2 + 2)^5$

23. $(x + 2)^6$ **24.** $(p - 3)^6$ **25.** $(2p^2 - q^2)^4$ **26.** $(3x^2 + y^3)^4$

Applying the Concepts

27. Use the Binomial Theorem to find the numerical value of $(1.001)^4$ correct to five decimal places. [**Hint:** $(1.001)^4 = (1 + 10^{-3})^4$.]

28. Use the Binomial Theorem to find the numerical value of $(1.001)^5$ correct to five decimal places. [**Hint:** $(1.001)^5 = (1 + 10^{-3})^5$.]

29. Use the Binomial Theorem to find the numerical value of $(0.998)^5$ correct to five decimal places.

30. Use the Binomial Theorem to find the numerical value of $(0.997)^5$ correct to five decimal places.

Extending the Concepts

Notice in the formula for expanding a binomial $(x + a)^n$, the first term is $\dbinom{n}{0}a^0 \cdot x^n$, the second term is $\dbinom{n}{1}a^1 \cdot x^{n-1}$, the third term is $\dbinom{n}{2}a^2 \cdot x^{n-2}$, and so on. In general, the jth term in a binomial expansion of $(x + a)^n$ is $\dbinom{n}{j-1}a^{j-1}x^{n-j+1}$. Use this result to find the indicated term in Problems 31–34.

31. The third term in the expansion of $(x + 2)^7$

32. The fourth term in the expansion of $(x - 1)^{10}$

33. The sixth term in the expansion of $(2p - 3)^8$

34. The seventh term in the expansion of $(3p + 1)^9$

35. Show that $\dbinom{n}{n-1} = n$ and $\dbinom{n}{n} = 1$.

36. Show that if n and j are integers with $0 \le j \le n$, then $\dbinom{n}{j} = \dbinom{n}{n-j}$.

Synthesis Review

37. If $f(x) = x^4$, find $f(a - 2)$ with the aid of the Binomial Formula.

38. If $g(x) = x^5 + 3$, find $g(z + 1)$ with the aid of the Binomial Formula.

39. If $H(x) = x^5 - 4x^4$, find $H(p + 1)$ with the aid of the Binomial Formula.

40. If $h(x) = 2x^5 + 5x^4$, find $h(a + 3)$ with the aid of the Binomial Formula.

CHAPTER 13 ACTIVITY: PASS TO THE RIGHT

Focus: Review of objectives for arithmetic and geometric sequences and Binomial Theorem

Time: 30–35 minutes

Group size: 2–4

As a group, discover the relationship between Column A and Column B (i.e. $<$, $>$, or $=$). Be sure to discuss any differences in outcomes.

Column A	Column B
1 The sum of the first five terms of $$a_n = \frac{3}{5}n + 1$$	The sum of the first five terms of $$a_n = -\frac{1}{4}(n-1) + 4$$
2 $\displaystyle\sum_{i=1}^{4}(i^2 + 2i)$	$\displaystyle\sum_{i=1}^{4}i^2 + \sum_{i=1}^{4}2i$
3 The 8th term of an arithmetic sequence when $a_1 = 3$ and $d = \dfrac{3}{2}$	The 27th term of an arithmetic sequence when $a_1 = \dfrac{5}{3}$ and $d = \dfrac{1}{3}$
4 The 10th term of a geometric sequence with $a_1 = 3$ and $r = \sqrt{2}$	The 9th term of a geometric sequence with $a_1 = 5$ and $r = \sqrt{3}$
5 Find the coefficient of the 4th term of $(x + y)^{10}$	Find the coefficient of the 3rd term of $(8x - y)^4$

CHAPTER 13 REVIEW

Section 13.1	Sequences

KEY TERMS		
sequence	finite sequence	summation notation
terms	general term	index
ellipsis	alternate	partial sum
infinite sequence		

YOU SHOULD BE ABLE TO . . .	EXAMPLE	REVIEW EXERCISES
1 Write the first several terms of a sequence (p. 974)	Examples 1 and 2	1–6
2 Find a formula for the nth term of a sequence (p. 976)	Example 3	7–12
3 Use summation notation (p. 976)	Examples 4 and 5	13–20

In Problems 1–6, write down the first five terms of each sequence.

1. $\{-3n + 2\}$ **2.** $\left\{\dfrac{n-2}{n+4}\right\}$ **3.** $\{5^n + 1\}$

4. $\{(-1)^{n-1} \cdot 3n\}$ **5.** $\left\{\dfrac{n^2}{n+1}\right\}$ **6.** $\left\{\dfrac{\pi^n}{n}\right\}$

In Problems 7–12, the given pattern continues. Write down the nth term of each sequence suggested by the pattern.

7. $-3, -6, -9, -12, -15, \ldots$ **8.** $\dfrac{1}{3}, \dfrac{2}{3}, 1, \dfrac{4}{3}, \dfrac{5}{3}, \ldots$

9. $5, 10, 20, 40, 80, \ldots$

10. $-\dfrac{1}{2}, 1, -\dfrac{3}{2}, 2, \ldots$

11. $6, 9, 14, 21, 30, \ldots$

12. $0, \dfrac{1}{3}, \dfrac{1}{2}, \dfrac{3}{5}, \ldots$

In Problems 13–16, write out each sum and determine its value.

13. $\displaystyle\sum_{k=1}^{5}(5k - 2)$ **14.** $\displaystyle\sum_{k=1}^{6}\left(\dfrac{k + 2}{2}\right)$ **15.** $\displaystyle\sum_{i=1}^{5}(-2i)$ **16.** $\displaystyle\sum_{i=1}^{4}\dfrac{i^2 - 1}{3}$

In Problems 17–20, express each sum using summation notation.

17. $(4 + 3\cdot1) + (4 + 3\cdot2) + (4 + 3\cdot3) + \cdots + (4 + 3\cdot15)$

18. $\dfrac{1}{3^1} + \dfrac{1}{3^2} + \dfrac{1}{3^3} + \cdots + \dfrac{1}{3^8}$

19. $\dfrac{1^3 + 1}{1 + 1} + \dfrac{2^3 + 1}{2 + 1} + \dfrac{3^3 + 1}{3 + 1} + \cdots + \dfrac{10^3 + 1}{10 + 1}$

20. $(-1)^{1-1}\cdot1^2 + (-1)^{2-1}\cdot2^2 + (-1)^{3-1}\cdot3^2 + \cdots + (-1)^{7-1}\cdot7^2$

Section 13.2	**Arithmetic Sequences**

KEY CONCEPTS	**KEY TERMS**

• **The *n*th Term of an Arithmetic Sequence**
  For an arithmetic sequence whose first term is a and whose common difference is d,
  the nth term is $a_n = a + (n - 1)d$
 	
• **Sum of *n* Terms of an Arithmetic Sequence**
  For an arithmetic sequence whose first term is a and whose common difference is d,
  the sum of the first n terms is $S_n = \dfrac{n}{2}[2a + (n - 1)d]$ or $S_n = \dfrac{n}{2}(a + a_n)$.
	Arithmetic Sequence
	Common difference

YOU SHOULD BE ABLE TO . . .	EXAMPLE	REVIEW EXERCISES
1 Determine if a sequence is arithmetic (p. 982)	Examples 1 through 3	21–26
2 Find a formula for the nth term of an arithmetic sequence (p. 983)	Examples 4 and 5	27–32, 37
3 Find the sum of an arithmetic sequence (p. 985)	Examples 6 through 8	33–36, 38

In Problems 21–26, determine if the sequence is arithmetic. If so, find the common difference.

21. $4, 10, 16, 22, \ldots$

22. $-1, \dfrac{1}{2}, 2, \dfrac{7}{2}, \ldots$

23. $-2, -5, -9, -14, \ldots$

24. $-1, 3, -5, 7, \ldots$

25. $\{4n + 7\}$

26. $\left\{\dfrac{n + 1}{2n}\right\}$

In Problems 27–32, find a formula for the nth term of the arithmetic sequence. Use the formula to find the 25th term of the sequence.

27. $a = 3; d = 8$ **28.** $a = -4; d = -3$ **29.** $7, \dfrac{20}{3}, \dfrac{19}{3}, 6, \ldots$ **30.** $11, 17, 23, 29, \ldots$

31. 3rd term is 7 and the 8th term is 25.

32. 4th term is -20 and the 7th term is -32.

33. Find the sum of the first 30 terms of the sequence $-1, 9, 19, 29, \ldots$.

34. Find the sum of the first 40 terms of the sequence $5, 2, -1, -4, \ldots$.

35. Find the sum of the first 60 terms of the sequence $\{-2n - 7\}$.

36. Find the sum of the first 50 terms of the sequence $\left\{\frac{1}{4}n + 3\right\}$.

37. Cicadas Seventeen-year cicadas emerge every 17 years to mate, lay eggs, and start the next 17-year cycle. In 2004, the Brood X cicada (the largest brood of the 17-year cicada) emerged in Maryland, Kentucky, Tennessee, and parts of surrounding states. Determine when the Brood X cicada will first appear in the 22nd century.

38. Wind Sprints At a certain football practice, players would run wind sprints for exercise. Starting at the goal line, players would sprint to the 10-yard line and back to the goal line. The players would then sprint to the 20-yard line and back to the goal line. This continues for the 30-yard line, 40-yard line, and 50-yard line. Determine the total distance a player would run during the wind sprints.

Section 13.3 Geometric Sequences and Series

KEY CONCEPTS	KEY TERMS
• **The nth Term of a Geometric Sequence** For a geometric sequence whose first term is a and whose common ratio is r, the nth term is $a_n = ar^{n-1}$. • **Sum of n Terms of a Geometric Sequence** For a geometric sequence whose first term is a and whose common ratio is r, the sum of the first n terms is $S_n = a \cdot \dfrac{1 - r^n}{1 - r}$. • **Sum of a Geometric Series** If $-1 < r < 1$, the sum of the geometric series $\displaystyle\sum_{n=1}^{\infty} ar^{n-1} = \dfrac{a}{1 - r}$.	Geometric sequence Common ratio Geometric series Marginal propensity to consume Mulitplier

YOU SHOULD BE ABLE TO . . .	EXAMPLE	REVIEW EXERCISES
1 Determine if a sequence is geometric (p. 990)	Examples 1 through 3	39–44
2 Find a formula for the nth term of a geometric sequence (p. 992)	Example 4	45–48, 57
3 Find the sum of a geometric sequence (p. 993)	Examples 5 and 6	49–52, 58
4 Find the sum of a geometric series (p. 994)	Examples 7 through 9	53–56
5 Solve annuity problems (p. 996)	Example 10	59–62

In Problems 39–44, determine if the given sequence is geometric. If so, determine the common ratio.

39. $\dfrac{1}{3}, 2, 12, 72, \ldots$ **40.** $-1, 3, -9, 27, \ldots$ **41.** $1, 1, 2, 6, \ldots$

42. $6, 4, \dfrac{8}{3}, \dfrac{16}{9}, \ldots$ **43.** $\{5 \cdot (-2)^n\}$ **44.** $\{3n - 14\}$

In Problems 45–48, find a formula for the nth term of the geometric sequence. Use the formula to find the 10th term of the sequence.

45. $a = 4, r = 3$ **46.** $a = 8, r = \dfrac{1}{4}$ **47.** $a = 5, r = -2$ **48.** $a = 1000, r = 1.08$

In Problems 49–52, find the sum. If necessary, express your answer to as many decimal places as your calculator allows.

49. $2 + 4 + 8 + \cdots + 2^{15}$ **50.** $40 + 5 + \dfrac{5}{8} + \cdots + 40\left(\dfrac{1}{8}\right)^{13-1}$

51. $\sum_{n=1}^{12}\left[\frac{3}{4}\cdot(2)^{n-1}\right]$

52. $\sum_{n=1}^{16}\left[-4\cdot(3^n)\right]$

In Problems 53–56, find the sum of each geometric series. If necessary, express your answer to as many decimal places as your calculator allows.

53. $\sum_{n=1}^{\infty}\left[20\cdot\left(\frac{1}{4}\right)^n\right]$

54. $\sum_{n=1}^{\infty}\left[50\cdot\left(-\frac{1}{2}\right)^{n-1}\right]$

55. $1+\frac{1}{5}+\frac{1}{25}+\cdots$

56. $0.8+0.08+0.008+0.0008+\cdots$

57. Radioactive Decay The radioactive isotope Tritium has a half-life of about 12 years. If there were 200 grams of the isotope initially, use a geometric sequence to determine how much would remain after 72 years.

58. Computer Virus In January 2004, the Mydoom e-mail worm was declared the worst e-mail worm incident in virus history accounting for roughly 20–30% of worldwide e-mail traffic. Suppose the virus was initially sent to 5 e-mail addresses and that, upon receipt, sends itself out to 5 e-mail addresses from the address book of the infected computer. If each cycle of e-mails, including the initial sending, takes 1 minute to complete, how many total e-mails will have been sent after 15 minutes?

59. 403(b) Scott contributes $900 each quarter into a 403(b) plan at work. His employer agrees to match half of employee contributions up to $600 per quarter. What will be the value of Scott's 403(b) in 25 years if the per annum rate of return is assumed to be 7% compounded quarterly?

60. Lottery Payment The winner of a state lottery has the option of receiving about $2 million per year for 26 years (after taxes), or a lump sum payment of about $28 million (after taxes). Assuming all winnings will be invested at a per annum rate of return of 6.5% compounded annually, which option yields the most money after 26 years?

61. What's My Payment? Sheri starts her career when she is 22 years old and wants to have $2,500,000 in her 401(k) retirement account when she retires in 40 years. How much does she need to contribute each month if the account earns 9% interest compounded monthly?

62. College Savings Plan On Samantha's 8th birthday, her parents open a 529 college savings plan for her and plan to contribute $400 per month until she turns 18. The per annum rate of return is assumed to be 5.25% compounded monthly and the cost per credit hour at a private 4-year university is locked-in at a rate of $340 per hour. What is the value of the plan when Samantha turns 18, and how many credit hours will the plan cover?

Section 13.4 The Binomial Theorem

KEY CONCEPTS

- **Factorial symbol $n!$**
  If $n \geq 0$ is an integer, the factorial symbol $n!$ is defined as
  $$0! = 1 \quad 1! = 1 \quad n! = n(n-1)(n-2)\cdots\cdot3\cdot2\cdot1 \quad \text{if } n \geq 2$$

- **The symbol $\binom{n}{j}$**
  If j and n are integers with $0 \leq j \leq n$, then $\binom{n}{j} = \dfrac{n!}{j!(n-j)!}$

KEY TERMS

Factorial symbol
Googol
Pascal's triangle
Binomial coefficient

(continued)

- **The Binomial Theorem**

Let x and a be real numbers. For any positive integer n, we have

$$(x + a)^n = \binom{n}{0}x^n + \binom{n}{1}ax^{n-1} + \binom{n}{2}a^2x^{n-2} + \cdots + \binom{n}{j}a^jx^{n-j} + \cdots + \binom{n}{n}a^n$$

YOU SHOULD BE ABLE TO . . .	EXAMPLE	REVIEW EXERCISES
1 Compute factorials (p. 1002)	Example 1	63–66
2 Evaluate a binomial coefficient (p. 1003)	Example 2	67–70
3 Expand a binomial (p. 1005)	Examples 3 and 4	71–78

In Problems 63–66, evaluate the expression.

63. $5!$ **64.** $\dfrac{11!}{7!}$ **65.** $\dfrac{10!}{6!}$ **66.** $\dfrac{13!}{6!7!}$

In Problems 67–70, evaluate each binomial coefficient.

67. $\binom{7}{3}$ **68.** $\binom{10}{5}$ **69.** $\binom{8}{8}$ **70.** $\binom{6}{0}$

In Problems 71–76, expand each expression using the Binomial Theorem.

71. $(z + 1)^4$ **72.** $(y - 3)^5$ **73.** $(3y + 4)^6$

74. $(2x^2 - 3)^4$ **75.** $(3p - 2q)^4$ **76.** $(a^3 + 3b)^5$

77. Find the fourth term in the expansion of $(x - 2)^8$.

78. Find the seventh term in the expansion of $(2x + 1)^{11}$.

CHAPTER 13 TEST

 Remember to use your Chapter Test Prep Video CD to see fully worked-out solutions to any of these problems you would like to review.

In Problems 1–6, determine if the sequence is arithmetic, geometric, or neither. If arithmetic or geometric, determine the first term and the common difference or common ratio.

1. $-15, -7, 1, 9, \ldots$ **2.** $\{(-4)^n\}$ **3.** $\left\{\dfrac{4}{n!}\right\}$

4. $\left\{\dfrac{2n - 3}{5}\right\}$ **5.** $-3, 2, 0, 5, 3, \ldots$ **6.** $\{7 \cdot 3^n\}$

7. Write out the sum and evaluate: $\displaystyle\sum_{i=1}^{5}\left[\dfrac{3}{i^2} + 2\right]$

8. Express the sum using summation notation: $\dfrac{3}{5} + \dfrac{2}{3} + \dfrac{5}{7} + \dfrac{3}{4} + \cdots + \dfrac{5}{6}$

In Problems 9–12, write a formula for the nth term of the indicated sequence. Write out the first five terms of each sequence.

9. arithmetic: $a = 6, d = 10$ **10.** arithmetic: $a = 0, d = -4$

11. geometric: $a = 10, r = 2$ **12.** geometric: $a_3 = 9, r = -3$

In Problems 13–15, find the indicated sum.

13. $-2 + 2 + 6 + \cdots + [4 \cdot (20 - 1) - 2]$

14. $\dfrac{1}{9} - \dfrac{1}{3} + 1 - 3 + \cdots + \dfrac{1}{9} \cdot (-3)^{12-1}$

15. $216 + 72 + 24 + 8 + \cdots$ **16.** Evaluate $\dfrac{15!}{8!7!}$.   **17.** Evaluate $\dbinom{12}{5}$.

18. Expand $(5m - 2)^4$ using the Binomial Theorem.

19. Tuition Increase The average tuition and fees for in-state students at public four-year colleges and universities for the 2003–2004 academic year was $4694. This represented an increase of about 14% from the previous year. If this percent increase continues each year, what will the average tuition and fees for in-state students be in the 2023–2024 academic year?

20. Transit of Venus On June 8, 2004, the planet Venus passed between the sun and the Earth creating a rare type of solar eclipse. Such Venus transits continually recur at intervals of 8, 121.5, 8, and 105.5 years. Since the invention of the telescope, Venus transits have been recorded in the years 1631, 1639, 1761, 1769, 1874, 1882, and 2004. Use this information to determine when the next three Venus transits will occur.

CUMULATIVE REVIEW CHAPTERS 1–13

1. Solve $\dfrac{1}{2}(x + 2) = \dfrac{5}{4}(x - 3y)$ for y.

2. Evaluate $f(x) = x^2 - x + 7$ for $x = 2$ and $x = -3$.

In Problems 3–8, find all solutions to the indicated equation.

3. $\dfrac{1}{2}x - 2 = \dfrac{1}{3}(x + 1) + 3$ **4.** $5x^2 - 3x = 2$

5. $3x^2 + 7x - 2 = 0$ **6.** $\sqrt{2x + 1} - 3 = 8$

7. $4^{x+1} = 8^{2x-3}$ **8.** $x^2(2x + 1) + 40 = (x^2 - 8)(x - 5)$

In Problems 9 and 10, solve the indicated inequality. Write your answer in interval notation.

9. $\dfrac{2}{3}x + 1 > \dfrac{1}{4}x - \dfrac{3}{2}$ **10.** $3x^2 - 2x \le 3 - 10x$

In Problems 11 and 12, factor the expression completely.

11. $2x^2 - 5x - 18$ **12.** $6x^3 - 3x^2 + 4x - 2$

In Problems 13–15, perform the indicated operation and simplify. Write complex numbers in standard form.

13. $(5x - 3)(4x^2 - 2x + 1)$ **14.** $\dfrac{x}{x + 4} - \dfrac{3}{x - 1}$ **15.** $\dfrac{3 - i}{2 + i}$

16. Find the domain of the function $f(x) = \sqrt{x - 15} + \sqrt{2x - 5}$.

17. Find the equation of the line that passes through the points $(2, -3)$ and $(1, 4)$.

18. Solve the system: $\begin{cases} 2x + 3y = 5 \\ x - 2y = 6 \end{cases}$

19. Graph the quadratic function $f(x) = 2x^2 - 8x - 3$. Label the vertex and axis of symmetry.

20. Write the standard form of the equation of the circle whose center is $(4, -3)$ and whose radius is $r = 6$ units. Graph the circle.

21. Sketch the graph of the ellipse given by the equation $4x^2 + y^2 = 64$.

22. Find the sum of the first 20 terms of the arithmetic sequence whose first term is $a = -47$ and whose common difference is $d = 12$.

23. Find the sum of the infinite geometric series: $2, \dfrac{3}{2}, \dfrac{9}{8}, \dfrac{27}{32}, \ldots$.

24. Mowing Lawns The Robomower® automatic lawnmower can mow a 7500-square-foot lot in 5 hours. It takes the Mowbot® automatic lawnmower 6 hours to cut the same lot. How long would it take both machines to cut the lot if they work together?

25. Aluminum Alloy The most commonly used aluminum alloy is aluminum 3003 which is often used to make rain gutters. A manufacturer of rain gutters has 100 metric tons of an aluminum alloy that is 2.5% manganese, but this percent is too high. How much pure aluminum must be added to the 100 metric tons in order to obtain a desired alloy that is 1.2% manganese?

 Fractions, Decimals, and Percents

A.1 Fractions

OBJECTIVES

1. Factor a Number as a Product of Prime Factors
2. Find the Least Common Multiple of Two or More Numbers
3. Write Equivalent Fractions
4. Write a Fraction in Lowest Terms

We begin this section by reviewing two concepts from arithmetic: factoring a number as a product of primes and finding the least common multiple of a list of numbers. We base our discussion in this section on natural numbers.

1 Factor a Number as a Product of Prime Factors

When we multiply, the numbers that are multiplied together are called **factors** and the answer is called the **product.**

$$\underset{\text{factor}}{7} \cdot \underset{\text{factor}}{5} = \underset{\text{product}}{35}$$

When we write a number as a product, we say that we **factor** the number. For example, when we write 20 as the product $10 \cdot 2$, we say that we have factored 20.

Some natural numbers are prime numbers and others are composite.

> **DEFINITION**
>
> A natural number is **prime** if its factors are only one and itself. Natural numbers that are not prime are called **composite.** The number 1 is neither prime nor composite.

Work Smart
Remember, the natural numbers are 1, 2, 3, ...

Work Smart
The first six primes are 2, 3, 5, 7, 11, and 13.

Examples of prime numbers are 2, 3, 5, 7, 11, and 13. We use a *factor tree* to find the prime factorization of a number. The process begins with finding two factors of the given number. Continue to factor until all factors are primes.

When a composite number is written as the product of prime numbers, we say that we are writing the **prime factorization** of the number. One technique that may be used to write the prime factorization of a number is shown below.

EXAMPLE 1 Finding the Prime Factorization

Write the prime factorization of 24.

Solution

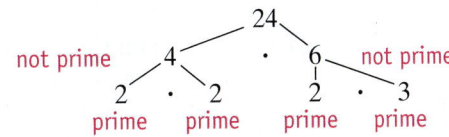

Work Smart
We could have begun the factorization of 24 with the factors 8 and 3 instead of 4 and 6. Try this for yourself.

All the numbers are prime, so we are done. The prime factorization of 24 is $2 \cdot 2 \cdot 2 \cdot 3$. Order is not important in multiplying factors. The product could also be written as $3 \cdot 2 \cdot 2 \cdot 2$ or $2 \cdot 3 \cdot 2 \cdot 2$. ∎

QUICK ✔ *Find the prime factorization of each of the following:*

1. 12   **2.** 18   **3.** 75   **4.** 120   **5.** 131   **6.** 459

② Find the Least Common Multiple of Two or More Numbers

A **multiple** of a number is the product of that number and any natural number. For example, the multiples of 2 are

$$2 \cdot 1 = 2, 2 \cdot 2 = 4, 2 \cdot 3 = 6, 2 \cdot 4 = 8, 2 \cdot 5 = 10, 2 \cdot 6 = 12, \text{ and so on.}$$

Multiples of 3 are

$$3 \cdot 1 = 3, 3 \cdot 2 = 6, 3 \cdot 3 = 9, 3 \cdot 4 = 12, 3 \cdot 5 = 15, 3 \cdot 6 = 18, \text{ and so on.}$$

Notice that the numbers 2 and 3 share two common multiples in this list: 6 and 12. The smallest common multiple, called the least common multiple, of 2 and 3 is 6.

> **DEFINITION**
>
> The **least common multiple (LCM)** of two or more natural numbers is the smallest number that is a multiple of each of the numbers.

EXAMPLE 2   **How to Find the Least Common Multiple**

Find the least common multiple of the numbers 6 and 15.

Step-by-Step Solution

Step 1: Write each number as the product of prime factors, aligning common factors vertically.	Arrange the common factor of 3 in its own column $\qquad \begin{aligned} 6 &= 2 \cdot 3 \\ 15 &= 3 \cdot 5 \end{aligned}$
Step 2: Write down the factor(s) that the numbers share, if any. Then write down the remaining factors the greatest number of times that the factors appear in any number.	The common factor is 3. The remaining factors are 2 and 5.
Step 3: Multiply the factors listed in Step 2. The product is the least common multiple (LCM).	The LCM is $2 \cdot 3 \cdot 5 = 30$

Work Smart

To see the individual factors more clearly, align the factors vertically.

The least common multiple of 6 and 15 is 30. We see that 30 is a multiple of 6 because $6 \cdot 5 = 30$, and 30 is a multiple of 15 because $15 \cdot 2 = 30$.

We could also find the least common multiple by listing the factors of each number until we find the smallest common multiple as follows:

Multiples of 6:	6, 12, 18, 24, 30, 36, 42, . . .
Multiples of 15:	15, 30, 45, 60, . . .

The least common multiple is 30 (indicated in red). This approach works just fine for numbers, but does not work for algebra. For this reason, it is recommended that you follow the steps used in Example 2 to find the least common multiple of two or more numbers so that you are better prepared when we discuss the least common multiple again later in the text.

EXAMPLE 3   Finding the Least Common Multiple

Find the LCM of the numbers 18 and 15.

Solution

We first write each number as the product of prime factors.

Write the factors in each column that the numbers share, if any. Then write down the remaining factors the greatest number of times that the factors appear in any number. Find the product of the factors.

$$18 = 2 \cdot 3 \cdot 3$$
$$15 = \quad 3 \cdot \quad 5$$
$$2 \cdot 3 \cdot 3 \cdot 5$$

The LCM is $2 \cdot 3 \cdot 3 \cdot 5 = 90$.

QUICK ✓ *Find the LCM of the numbers.*

7. 6 and 8   **8.** 5 and 10   **9.** 45 and 72   **10.** 7 and 3   **11.** 12, 18, and 30

---

(3) ## Write Equivalent Fractions

A fraction represents a part of a whole. For example, the fraction $\frac{5}{8}$ means "5 parts out of 8 parts." A fraction also indicates division: $\frac{5}{8}$ means "five divided by 8." Since $\frac{5}{8}$ indicates division, $\frac{5}{8}$ may be written as $8\overline{)5}$. Figure 1 shows the fraction $\frac{5}{8}$ visually.

In the fraction $\frac{5}{8}$, the number 5 is called the **numerator** and the number 8 is called the **denominator.** The denominator tells the number of equal parts that the whole is divided into, and the numerator tells the number of equal parts that are shaded. For example, in Figure 1 the box is divided into 8 equal parts, 5 of which are shaded.

We use whole numbers $(0, 1, 2, 3, \dots)$ for the numerators of fractions and natural numbers $(1, 2, 3, 4, \dots)$ for denominators.

Fractions without common denominators can be rewritten in an equivalent form so they have the same denominator.

Figure 1

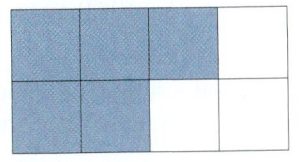

Work Smart

The denominator of a number such as 7 is 1 because $7 = \frac{7}{1}$.

> **DEFINITION**
>
> **Equivalent fractions** are fractions that represent the same part of a whole.

Figure 2

For example, $\frac{2}{3}$ and $\frac{8}{12}$ are equivalent fractions. To understand why, consider Figure 2. If we break the whole into 12 parts and shade 8 of these parts, the shaded region represents $\frac{8}{12}$ of the rectangle. If we only consider the 3 parts separated by the thick black lines, we can see that 2 parts are shaded for a fraction of $\frac{2}{3}$. In each case, the same portion of the rectangle is shaded, so $\frac{2}{3}$ and $\frac{8}{12}$ are equivalent fractions.

Now the question becomes, how do we obtain equivalent fractions? The answer lies in the following property.

In Words

We can obtain an equivalent fraction by multiplying the numerator and denominator of the fraction by the same nonzero number.

> If a, b, and c are whole numbers, then
>
> $$\frac{a}{b} = \frac{a \cdot c}{b \cdot c} \qquad \text{if } b \neq 0, c \neq 0$$

Let's see how this property works.

EXAMPLE 4 Writing an Equivalent Fraction

Write the fraction $\dfrac{3}{4}$ as an equivalent fraction with a denominator of 20.

Solution

We want to know $\dfrac{3}{4}$ equals "what" over 20, or $\dfrac{3}{4} = \dfrac{?}{20}$. To write the fraction $\dfrac{3}{4}$ with a denominator of 20, we multiply the numerator and denominator of $\dfrac{3}{4}$ by 5. Do you see why?

$$\frac{3}{4} = \frac{3 \cdot 5}{4 \cdot 5}$$
$$= \frac{15}{20}$$

QUICK ✓ *Rewrite each fraction with the denominator indicated.*

12. $\dfrac{1}{2}$; 10   **13.** $\dfrac{5}{8}$; 48

It is sometimes necessary to rewrite two or more fractions so that each has the same denominator. For example, we could rewrite the fractions $\dfrac{5}{6}$ and $\dfrac{3}{8}$ so that they have a common denominator of 24, 48, 96 and so on because these are common multiples of the denominators 6 and 8. Notice that 24 is the least common multiple of 6 and 8. When talking about the least common multiple as it applies to denominators of fractions, we use the phrase least common denominator.

> **DEFINITION**
>
> The **least common denominator (LCD)** is the least common multiple of the denominators of a group of fractions.

EXAMPLE 5 How to Write Two Fractions as Equivalent Fractions with the LCD

Write $\dfrac{5}{8}$ and $\dfrac{9}{20}$ as equivalent fractions with the least common denominator.

Step-by-Step Solution

Step 1: Find the least common denominator of the fractions.

The denominators of $\dfrac{5}{8}$ and $\dfrac{9}{20}$ are 8 and 20.

Write each denominator as the product of prime factors:
$$8 = 2 \cdot 2 \cdot 2$$
$$20 = 2 \cdot 2 \cdot \quad 5$$
$$\downarrow \ \downarrow \ \downarrow \ \downarrow$$
$$LCD = 2 \cdot 2 \cdot 2 \cdot 5$$
$$= 40$$

Step 2: Rewrite each fraction with the least common denominator.

Multiply the numerator and denominator of $\frac{5}{8}$ by 5:

$$\frac{5}{8} = \frac{5 \cdot 5}{8 \cdot 5}$$

$$= \frac{25}{40}$$

Multiply the numerator and denominator of $\frac{9}{20}$ by 2:

$$\frac{9}{20} = \frac{9 \cdot 2}{20 \cdot 2}$$

$$= \frac{18}{40}$$

So $\frac{5}{8} = \frac{25}{40}$ and $\frac{9}{20} = \frac{18}{40}$.

QUICK ✓ *Write the equivalent fractions with the least common denominator.*

14. $\frac{1}{4}$ and $\frac{5}{6}$

15. $\frac{5}{12}$ and $\frac{4}{15}$

16. $\frac{9}{20}$ and $\frac{11}{16}$

④ **Write a Fraction in Lowest Terms**

In Words
To write a fraction in lowest terms, find a common factor between the numerator and denominator and divide out the common factors.

DEFINITION
A fraction is written in lowest terms if the numerator and the denominator share no common factor other than 1.

We can write fractions in lowest terms using the fact that

$$\frac{a \cdot c}{b \cdot c} = \frac{a}{b}$$

So, to write a fraction in lowest terms, we write the numerator and the denominator as a product of primes, and then divide out common factors.

EXAMPLE 6 **Writing a Fraction in Lowest Terms**

Write $\frac{24}{40}$ in lowest terms.

Solution

Write the numerator and the denominator as the product of primes and divide out common factors.

$$\frac{24}{40} = \frac{2 \cdot 2 \cdot 2 \cdot 3}{2 \cdot 2 \cdot 2 \cdot 5}$$

Divide out common factors:

$$= \frac{\cancel{2} \cdot \cancel{2} \cdot \cancel{2} \cdot 3}{\cancel{2} \cdot \cancel{2} \cdot \cancel{2} \cdot 5}$$

$$= \frac{3}{5}$$

Work Smart
Use different slash marks to keep track of factors that have divided out. Also, you may wish to use nonprime factors when writing a fraction in lowest terms. In Example 6 we could have written $\frac{24}{40}$ in lowest terms as follows:

$$\frac{24}{40} = \frac{8 \cdot 3}{8 \cdot 5} = \frac{3}{5}$$

So $\frac{24}{40} = \frac{3}{5}$.

QUICK ✓ *Write each fraction in lowest terms if possible.*

17. $\dfrac{3}{6}$ **18.** $\dfrac{4}{9}$ **19.** $\dfrac{45}{80}$ **20.** $\dfrac{20}{50}$ **21.** $\dfrac{30}{105}$

A.1 Exercises

For Extra Help: 📕 💿 📱 🎯 Math XL MyMathLab
Student Solutions Manual   CD Video   PH Math/Tutor Center   MathXL Tutorials on CD   MathXL®   MyMathLab

Concepts and Vocabulary

In Problems 1–3, fill in the blanks.

1. In the statement $6 \cdot 8 = 48$, 6 and 8 are called _____, and 48 is called the _____.

2. The _____ _____ _____ of two natural numbers is the smallest number that is a multiple of the numbers.

3. Fractions which represent the same portion of a whole are called _____ _____.

In Problems 4–6, answer True or False to each statement.

4. The number 42 is a natural number.

5. The LCM of 5 and 15 is 75.

6. When finding the least common multiple of two numbers, list only the common or shared factors of the numbers.

✎ 7. You are explaining how to find the LCM to the members of your study group. In your own words, explain how you would find the LCM of 12, 18, and 45.

✎ 8. How can you tell if the number 91 is prime? In your own words, write a definition for *prime number*.

Building Skills

In Problems 9–32, find the prime factorization of each number.

9. 25	**10.** 9	**11.** 28	**12.** 100	**13.** 21
14. 35	**15.** 36	**16.** 54	**17.** 20	**18.** 63
19. 30	**20.** 45	**21.** 50	**22.** 70	**23.** 53
24. 79	**25.** 252	**26.** 315	**27.** 256	**28.** 243
29. 1300	**30.** 693	**31.** 2275	**32.** 6699	

In Problems 33–48, find the LCM of each set of numbers.

33. 6 and 15	**34.** 10 and 14	**35.** 12 and 10	**36.** 21 and 18
37. 15 and 14	**38.** 55 and 6	**39.** 30 and 45	**40.** 8 and 70
41. 42 and 14	**42.** 112 and 28	**43.** 5, 6 and 12	**44.** 9, 15 and 20
45. 3, 8 and 9	**46.** 4, 18 and 20	**47.** 4, 8, 12 and 18	**48.** 3, 6, 15 and 21

In Problems 49–52, write the fraction that is indicated by the shaded region.

49.
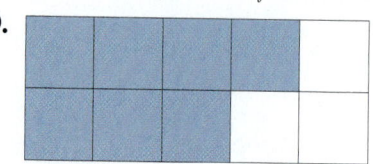

50.
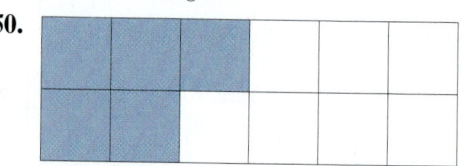

51.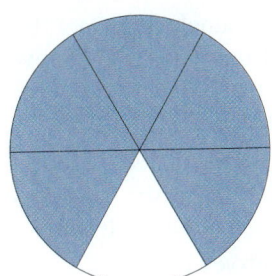

52.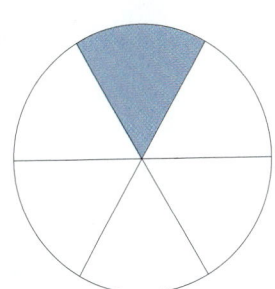

In Problems 53–58, write each fraction with the given denominator.

53. Write $\dfrac{2}{3}$ with denominator 12.

54. Write $\dfrac{4}{5}$ with denominator 15.

55. Write $\dfrac{3}{4}$ with denominator 24.

56. Write $\dfrac{5}{14}$ with denominator 28.

57. Write 7 with denominator 3.

58. Write 4 with denominator 10.

In Problems 59–72, write the equivalent fractions with the least common denominator.

59. $\dfrac{1}{2}$ and $\dfrac{3}{8}$

60. $\dfrac{3}{4}$ and $\dfrac{5}{12}$

61. $\dfrac{3}{5}$ and $\dfrac{2}{3}$

62. $\dfrac{1}{4}$ and $\dfrac{2}{9}$

63. $\dfrac{5}{6}$ and $\dfrac{5}{8}$

64. $\dfrac{1}{12}$ and $\dfrac{5}{18}$

65. $\dfrac{11}{15}$ and $\dfrac{7}{20}$

66. $\dfrac{5}{12}$ and $\dfrac{7}{15}$

67. $\dfrac{9}{16}$ and $\dfrac{7}{24}$

68. $\dfrac{23}{28}$ and $\dfrac{21}{24}$

69. $\dfrac{3}{24}$ and $\dfrac{7}{32}$

70. $\dfrac{17}{30}$ and $\dfrac{19}{45}$

71. $\dfrac{2}{9}$ and $\dfrac{7}{18}$ and $\dfrac{7}{30}$

72. $\dfrac{7}{10}$ and $\dfrac{1}{4}$ and $\dfrac{5}{6}$

In Problems 73–84, write each fraction in lowest terms if possible.

73. $\dfrac{14}{21}$

74. $\dfrac{9}{15}$

75. $\dfrac{38}{18}$

76. $\dfrac{81}{36}$

77. $\dfrac{22}{44}$

78. $\dfrac{24}{27}$

79. $\dfrac{32}{40}$

80. $\dfrac{49}{63}$

81. $\dfrac{15}{25}$

82. $\dfrac{34}{51}$

83. $\dfrac{150}{225}$

84. $\dfrac{144}{156}$

Applying the Concepts

85. Planets in Our Solar System At a certain point, Mercury, Venus, and Earth lie on a straight line. If it takes each planet 3, 7, and 12 months, respectively, to revolve around the sun, what is the fewest number of months until they align this way again?

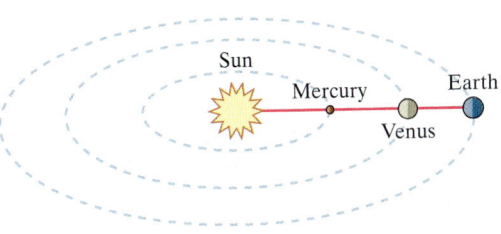

86. Talladega Raceway At Talladega one of the crew chiefs discovered that in a given time interval, Jeff Gordon completed 21 laps, Dale Earnhardt Jr. completed 18 laps, and Robby Gordon completed 15 laps. How many laps would a race have to have so that all three drivers were at the finish line at exactly the same time?

87. Sam's Medication Bob gives his dog Sam one type of medication every 4 days, and a second type of medication every 10 days. How often does Bob give Sam both types of medication on the same day?

88. Visiting Columbus Pamela and Geoff both visit Columbus on business. Pamela flies to Columbus from Atlanta about every 14 days, and Geoff takes the train to Columbus from Cincinnati every 20 days. How often are both Pamela and Geoff in Columbus on business?

89. Survey Data In a survey of 500 students, 325 stated that they work at least 25 hours per week. Express the fraction of students that work at least 25 hours per week as a fraction in lowest terms.

90. Survey Data In a survey of 750 students, 450 stated that they are enrolled in 15 or more semester hours. Express the fraction of students that are enrolled in 15 or more semester hours as a fraction in lowest terms.

91. Quiz Score A student earns 12 points out of a total of 20 points on a quiz. Express this score as a fraction in lowest terms.

92. Quiz Score A student earns 15 points out of a total of 25 points on a quiz. Express this score as a fraction in lowest terms.

93. Pizza Marko orders a pizza that is cut into 12 equal pieces. He eats 8 of the pieces. Write the amount of the pizza that Marko ate as a fraction in lowest terms.

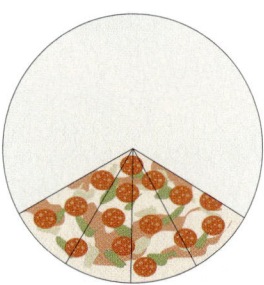

94. Sub Sandwich Aisha orders a 12-foot sub sandwich that is cut into 24 equal pieces for a party. After the party, she notices that 4 pieces remain. Write the amount of the sub sandwich that was eaten as a fraction in lowest terms.

95. The Sieve of Eratosthenes Eratosthenes (276 B.C.–194 B.C.) was born in Cyrene, which is now in Libya in North Africa. He devised an algorithm (a series of steps that are followed to solve a problem) for identifying prime numbers. The algorithm works as follows:

Step 1: List all the natural numbers that are greater than or equal to 2.

Step 2: The first number in the list, 2, is prime. Cross out all multiples of 2. For example, cross out 2, 4, 6, ...

Step 3: Identify the next number in the list after the most recently identified prime number. For example, we already know 2 is a prime number, so the next number in the list, 3, is also prime. Cross out all multiples of this number.

Step 4: Repeat Step 3.

Use the algorithm to find all the prime numbers less than 100.

A.2 Decimals and Percents

OBJECTIVES

1. Use Place Value
2. Round Decimals
3. Convert a Fraction to a Decimal and a Decimal to a Fraction
4. Convert a Percent to a Decimal and a Decimal to a Percent

Decimals and percentages commonly occur in everyday life. You received a 92% on your test, there is a 10% discount on jeans, we pay 7.75% in sales tax, 45% of the people polled support a proposition. Before we discuss decimals and percents, let us consider place value.

1 Use Place Value

Figure 3 shows how we interpret the place value of each digit in the number 9186.347. For example, the 7 is in the "thousandths" position, 3 is in the "tenths" position, and the 8 is in the "tens" position.

Figure 3

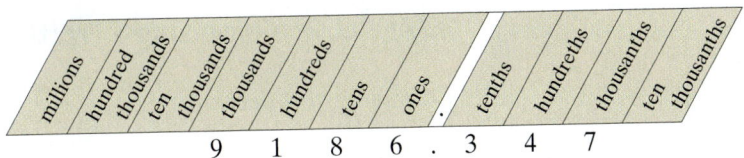

9 1 8 6 . 3 4 7

The number 9186.347 is read "nine thousand, one hundred eighty-six and three hundred forty-seven thousandths."

QUICK ✔ *Tell the place value of the digit in the given number.*

1. 235.71; the 1

2. 56,701.28; the 2

3. 278,403.95; the 8

4. 0.189; the 9

5. 3.590; the 3

6. 9,021,458.5; the 2

② Round Decimals

We round decimals in the same way we round whole numbers. First, identify the specified place value in the decimal. If the digit to the right is 5 or more, add 1 to the digit; if the next digit is 4 or less, leave the digit as it is. Then drop the digits to the right of the rounding number.

EXAMPLE 1   Rounding a Decimal Number

Round 8.726 to the nearest hundredth.

Solution

To round to the nearest hundredth, we determine that the number 2 is in the hundredths place: 8.726. The number to the right of 2 is 6. Since 6 is greater than 5, we round 8.726 to 8.73.

EXAMPLE 2   Rounding a Decimal Number

Round 0.9451 to the nearest thousandth.

Solution

To round 0.9451 to the nearest thousandth, we see that the number 5 is in the thousandths place: 0.9451. The number to the right of 5 is 1. Since 1 is less than 5, we round 0.9451 to 0.945.

QUICK ✔ *Round each number to the given decimal place.*

7. 0.17 to the nearest tenth

8. 0.932 to the nearest hundredth

9. 1.396 to the nearest hundredth

10. 14.3983 to the nearest thousandth

11. 690.004 to the nearest hundredth

12. 59.98 to the nearest tenth

③ Convert a Fraction to a Decimal and a Decimal to a Fraction

Convert a Fraction to a Decimal

To convert a fraction to a decimal, divide the numerator of the fraction by the denominator of the fraction until the remainder is 0 or the remainder repeats.

EXAMPLE 3 **Converting a Fraction to a Decimal**

Convert $\dfrac{9}{20}$ to a decimal.

Solution

$$\frac{9}{20} = 20\overline{\smash{\big)}\,9.00} \quad \begin{array}{r} 0.45 \\ \hline \end{array}$$

$$\begin{array}{r}
0.45 \\
20\overline{\smash{\big)}\,9.00} \\
\underline{8\ 0} \\
100 \\
\underline{100} \\
0
\end{array}$$

Therefore, $\dfrac{9}{20} = 0.45$.

EXAMPLE 4 **Converting a Fraction to a Decimal**

Convert $\dfrac{2}{3}$ to a decimal.

Solution

$$\begin{array}{r}
0.666 \\
3\overline{\smash{\big)}\,2.000} \\
\underline{1\ 8} \\
20 \\
\underline{18} \\
20 \\
\underline{18} \\
2
\end{array}$$

Notice that the remainder, 2, repeats. So $\dfrac{2}{3} = 0.666\ldots$.

In Example 3, the decimal 0.45 is called a **terminating decimal** because the decimal stops after the 5. In Example 4, the number $0.666\ldots$ is called a **repeating decimal** because the 6 continues repeating indefinitely. The decimal $0.666\ldots$ can also be written as $0.\overline{6}$. The bar over the 6 means the 6 repeats.

QUICK ✓ *Write the fraction as a decimal.*

13. $\dfrac{2}{5}$ **14.** $\dfrac{3}{7}$ **15.** $\dfrac{11}{8}$ **16.** $\dfrac{5}{6}$ **17.** $\dfrac{5}{9}$

Based on Examples 3 and 4 and Quick Check Problems 13–17, you should notice that **every fraction has a decimal representation that either terminates or repeats.**

Convert a Decimal to a Fraction

To convert a decimal to a fraction, identify the place value of the denominator, write the decimal as a fraction using the given denominator, and reduce.

EXAMPLE 5 **Writing a Decimal as a Fraction**

Convert each decimal to a fraction and write in lowest terms, if possible.

   (a) 0.8   **(b)** 0.77

Solution

(a) 0.8 is equivalent to 8 tenths, or $\dfrac{8}{10}$. Because, $\dfrac{8}{10} = \dfrac{4 \cdot 2}{5 \cdot 2} = \dfrac{4}{5}$, we write

$0.8 = \dfrac{4}{5}$.

(b) 0.77 is equivalent to 77 hundredths, or $\dfrac{77}{100}$.

QUICK ✓ *Write the decimal as a fraction and write in lowest terms, if possible.*

18. 0.65　　　　　　　**19.** 0.2　　　　　　　**20.** 0.625

---

④ Convert a Percent to a Decimal and a Decimal to a Percent

When computing with percents, it is convenient to write percents as decimals. How is a percent converted to a decimal? Let's see.

Convert a Percent to a Decimal

> **DEFINITION**
> The word **percent** means **parts per hundred** or **parts out of one hundred.**

So 25% means 25 parts out of 100 parts. Therefore, $25\% = \dfrac{25}{100} = \dfrac{1 \cdot 25}{4 \cdot 25} = \dfrac{1}{4}$.

　　Since the word percent means "parts per hundred," we have that 100% means 100 "parts per 100," so $100\% = 1$. Therefore, to convert from a percent to a decimal, multiply the percent by $\dfrac{1}{100\%}$.

EXAMPLE 6 Writing a Percent as a Decimal

Write the following percents as decimals:

(a) 17%　　　　　　　　　　　　**(b)** 150%

Solution

Work Smart
To convert from a percent to a decimal, move the decimal point two places to the left and drop the % symbol.

(a) $17\% = 17\% \cdot \dfrac{1}{100\%}$

$= \dfrac{17}{100}$

$= 0.17$

(b) $150\% = 150\% \cdot \dfrac{1}{100\%}$

$= \dfrac{150}{100}$

$= 1.5$

QUICK ✓ *Write the percent as a decimal.*

21. 23%　　**22.** 1%　　**23.** 72.4%　　**24.** 127%　　**25.** 89.26%

Convert a Decimal to a Percent

Because $100\% = 1$, to convert a decimal to a percent, multiply the decimal by $\dfrac{100\%}{1}$.

EXAMPLE 7   Writing a Decimal as a Percent

Write the following decimals in percent form:

(a) 0.445 **(b)** 1.42

Solution

(a) $0.445 = 0.445 \cdot \dfrac{100\%}{1}$

$= 44.5\%$

(b) $1.42 = 1.42 \cdot \dfrac{100\%}{1}$

$= 142\%$

Work Smart

To convert from a decimal to a percent, move the decimal point two places to the right and add the % symbol.

QUICK ✓ *Write the decimal as a percent.*

26. 0.15 **27.** 0.8 **28.** 1.3 **29.** 0.398 **30.** 0.004

A.2 Exercises

For Extra Help:

Student Solutions Manual   CD Video   PH Math/Tutor Center   MathXL Tutorials on CD   MathXL®   MyMathLab

Concepts and Vocabulary

In Problems 1–3, fill in the blanks.

1. The decimal 0.492 is called a _____ decimal.

2. The decimal $8.\overline{34}$ is called a _____ decimal.

3. In the number 465.39, the digit 5 is in the _____ place.

In Problems 4–6, answer True or False to each statement.

4. The word percent means parts per hundred.

5. When converting 9.2% to a decimal, the decimal point will shift 2 places to the left.

6. Every fraction is equivalent to either a terminating or a repeating decimal.

7. Your teacher called you to the board to solve the following problem: convert $\dfrac{7}{8}$ to a percent. As you walk by, your friend whispers, "make $\dfrac{7}{8}$ a decimal first." With a sigh of relief, you begin. Write out your explanation.

8. Explain why multiplying by the fraction $\dfrac{1}{100\%}$ correctly converts a percent to a decimal.

Building Skills

In Problems 9–14, tell the place value of the digit in the given number.

9. 3465.902; the 0 **10.** 549,813.0267; the 8 **11.** 357.469; the 5

12. 9124.786; the 7 **13.** 2018.3764; the 6 **14.** 539.016; the 9

In Problems 15–22, round each number to the given place.

15. 578.206 to the nearest tenth **16.** 7298.0845 to the nearest hundred

17. 354.678 to the nearest ten **18.** 543.06 to the nearest unit

19. 3682.0098 to the nearest thousandth **20.** 683.098 to the nearest hundredth

21. 29.96 to the nearest unit **22.** 37.439 to the nearest tenth

In Problems 23–32, write each fraction as a terminating or repeating decimal.

23. $\dfrac{5}{8}$ **24.** $\dfrac{3}{4}$ **25.** $\dfrac{2}{7}$ **26.** $\dfrac{5}{6}$ **27.** $\dfrac{5}{16}$

28. $\dfrac{11}{32}$ **29.** $\dfrac{3}{13}$ **30.** $\dfrac{6}{13}$ **31.** $\dfrac{29}{25}$ **32.** $\dfrac{57}{50}$

In Problems 33–40, write each fraction as a decimal, rounded to the indicated place.

33. $\dfrac{13}{6}$ to the nearest tenth   **34.** $\dfrac{15}{8}$ to the nearest tenth

35. $\dfrac{6}{21}$ to the nearest hundredth   **36.** $\dfrac{24}{28}$ to the nearest hundredth

37. $\dfrac{8}{3}$ to the nearest hundredth   **38.** $\dfrac{9}{7}$ to the nearest hundredth

39. $\dfrac{14}{27}$ to the nearest thousandth   **40.** $\dfrac{18}{31}$ to the nearest thousandth

In Problems 41–48, write each decimal as a fraction, written in lowest terms.

41. 0.75   **42.** 0.25   **43.** 0.9   **44.** 0.4

45. 0.982   **46.** 0.358   **47.** 0.2525   **48.** 0.3334

In Problems 49–54, write each percent as a decimal.

49. 37%   **50.** 59%   **51.** 6.02%   **52.** 8.25%   **53.** 0.1%   **54.** 0.5%

In Problems 55–60, write each decimal as a percent.

55. 0.2   **56.** 0.5   **57.** 0.275   **58.** 0.349   **59.** 2   **60.** 1

Applying the Concepts

61. Eating Healthy? In a poll conducted by Zogby International of 1200 adult Americans, 840 stated that they believe that they eat healthy. What percentage of adult Americans believe that they eat healthy?

62. Ghosts In a survey of 1100 adult women conducted by Harris Interactive, it was determined that 640 believe in ghosts. What percentage of adult women believe in ghosts?

63. Test Score A student earns 85 points out of a total of 110 points on an exam. Express this score as a percent.

64. Test Score A student earns 80 points out of a total of 115 points on an exam. Express this score as a percent.

65. Time Utilization In a 24-hour day, Jackson sleeps for 8 hours, works for 4 hours, and goes to school and studies for 6 hours.

   (a) What percent of the time does Jackson sleep?
   (b) What percent of the time does Jackson work?
   (c) What percent of the time does Jackson go to school and study?

66. United States Senate In 2004, the United States Senate was comprised of 48 Democrats, 51 Republicans, and 1 Independent.

   (a) What percent of the U.S. Senate was Democrat in 2004?
   (b) What percent of the U.S. Senate was Republican in 2004?
   (c) What percent of the U.S. Senate was Independent in 2004?

67. Cashews A single serving of cashews contains 14 grams of fat. Of this, 3 grams is saturated fat. What percentage of fat grams is saturated fat in a single serving of cashews?

68. Cheese Pizza A single serving of cheese pizza contains 11 grams of fat. Of this, 5 grams are saturated fat. What percentage of fat grams is saturated fat in a single serving of cheese pizza?

Synthetic Division

OBJECTIVES

1. Divide Polynomials Using Synthetic Division
2. Use the Remainder and Factor Theorems

Preparing for Dividing Polynomials and Synthetic Division

Before getting started, take the following readiness quiz. If you get a problem wrong, go back to the section cited and review the material.

1. Simplify: $\dfrac{15x^5}{12x^3}$ [Section 5.4, pp. 371–372]

2. Divide using long division: $\dfrac{2x^3 + 7x^2 - 17x - 10}{2x + 1}$ [Section 5.5, pp. 385–388]

When we added, subtracted, and multiplied polynomials, the result was also a polynomial. This is not true for polynomial division—a polynomial divided by a polynomial may not be a polynomial.

1   Divide Polynomials Using Synthetic Division

Work Smart

Synthetic division can be used only when the divisor is of the form $x - c$ or $x + c$.

To find the quotient and remainder when a polynomial of degree 1 or higher is divided by $x - c$, a shortened version of long division called **synthetic division** makes the task easier.

To see how synthetic division works, we will use long division to divide the polynomial $2x^3 - 5x^2 - 7x + 20$ by $x - 3$. Synthetic division comes from rewriting the long division in a more compact form. For example, in the long division below, the terms in red ink are not really necessary because they are identical to the terms directly above them. The subtraction signs are not necessary, because subtraction is understood. With these items removed, we have the division shown on the right.

$$
\begin{array}{r}
2x^2 + x - 4 \quad\longleftarrow \text{ Quotient}\\
x - 3{\overline{\smash{\big)}\,2x^3 - 5x^2 - 7x + 20}}\\
\underline{-(2x^3 - 6x^2)}\\
x^2 - 7x\\
\underline{-(x^2 - 3x)}\\
-4x + 20\\
\underline{-(-4x + 12)}\\
8 \quad\longleftarrow \text{ Remainder}
\end{array}
$$

$$
\begin{array}{r}
2x^2 + x - 4\\
x - 3{\overline{\smash{\big)}\,2x^3 - 5x^2 - 7x + 20}}\\
\underline{-6x^2}\\
x^2\\
\underline{-3x}\\
-4x\\
\underline{12}\\
8
\end{array}
$$

The x's that appear in the division on the right are not necessary if we are careful about positioning each coefficient. As long as the right-most number under the division symbol is the constant, the number to its left is the coefficient of x and so on, we can remove the x's. Now we have

$$
\begin{array}{r}
2x^2 + x - 4\\
x - 3{\overline{\smash{\big)}\,2 \;-5\;-7\quad 20}}\\
\underline{-6}\\
\boxed{1}\\
-3\\
\boxed{-4}\\
12\\
\boxed{8}
\end{array}
$$

Preparing for...Answers   **1.** $\dfrac{5}{4}x^2$

2. $x^2 + 3x - 10$

We can make this display more compact by moving the lines up until the "boxed" numbers align horizontally.

$$
\begin{array}{r}
2x^2 + x - 4 \\[-2pt]
x-3\overline{)2\;\;\;-5\;\;\;-7\;\;\;20} \\
\underline{-6\;\;\;-3\;\;\;12} \\
\square\;\;\;\;\;1\;\;\;-4\;\;\;\;8
\end{array}
$$

Because the leading coefficient of the divisor is always 1, we know that the leading coefficient of the dividend will always be the leading coefficient of the quotient. So, we place the leading coefficient of the quotient, 2, in the boxed position.

$$
\begin{array}{rl}
2x^2 + x - 4 & \text{Row 1} \\[-2pt]
x-3\overline{)2\;\;\;-5\;\;\;-7\;\;\;20} & \\
\underline{-6\;\;\;-3\;\;\;12} & \\
2\;\;\;\;\;1\;\;\;-4\;\;\;\;8 & \text{Row 4}
\end{array}
$$

The first three numbers in Row 4 are the coefficients of the quotient. The last number in row 4 is the remainder. Now, Row 1 above is not needed.

$$
\begin{array}{rl}
x-3\overline{)2\;\;\;-5\;\;\;-7\;\;\;20} & \text{Row 1} \\
\underline{-6\;\;\;-3\;\;\;12} & \text{Row 2} \\
2\;\;\;\;\;1\;\;\;-4\;\;\;\;8 & \text{Row 3}
\end{array}
$$

Remember, the entries in Row 3 are obtained by subtracting the entries in Row 2 from the entries in Row 1. Rather than subtracting the entries in Row 2, we can change the sign of each entry and then add. With this modification, our display becomes

$$
\begin{array}{rl}
x-3\overline{)2\;\;\;-5\;\;\;-7\;\;\;\;\;20} & \text{Row 1} \\
\underline{6\;\;\;\;\;3\;\;\;-12} & \text{Row 2 (add)} \\
2\;\;\;\;\;1\;\;\;-4\;\;\;\;\;\;8 & \text{Row 3}
\end{array}
$$

Work Smart

If there are any missing powers of x in the dividend, you must insert a coefficient of 0 for the missing term when doing synthetic division.

Notice that the entries in Row 2 are three times the entries one column to the left in Row 3 (for example, the 6 in Row 2 is 3 times 2; the 3 in Row 2 is 3 times 1, and so on). We remove the $x - 3$ and replace it with 3. The entries in Row 3 give us the quotient and remainder.

$$
\begin{array}{rl}
3\overline{)2\;\;\;-5\;\;\;-7\;\;\;\;\;20} & \text{Row 1} \\
\underline{6\;\;\;\;\;3\;\;\;-12} & \text{Row 2 (add)} \\
2\;\;\;\;\;1\;\;\;-4\;\;\;\;\;\;8 & \text{Row 3} \\
\underbrace{2x^2 + x - 4}\;\;\;8 &
\end{array}
$$

Quotient   Remainder

Let's go over an example step by step.

EXAMPLE 1   How to Use Synthetic Division to Divide Polynomials

Use synthetic division to find the quotient and remainder when $3x^3 + 11x^2 + 14$ is divided by $x + 4$.

Step-by-Step Solution

Step 1: Write the dividend in descending powers of x. Then copy the coefficients of the dividend. Remember to insert a 0 for any missing powers of x.	$3x^3 + 11x^2 + 14 = 3x^3 + 11x^2 + 0x + 14$ $\quad\quad 3 \quad\quad 11 \quad\quad 0 \quad\quad 14 \quad$ Row 1
Step 2: Insert the division symbol. Rewrite the divisor in the form $x - c$ and insert the value of c to the left of the division symbol.	$x + 4 = x - (-4)$ $-4\overline{)3 \quad\quad 11 \quad\quad 0 \quad\quad 14} \quad$ Row 1

(continued)

Step 3: Bring the 3 down two rows and enter it in Row 3.

$$-4\overline{)3 \qquad 11 \qquad 0 \qquad 14} \quad \text{Row 1}$$
$$\downarrow \qquad\qquad\qquad\qquad\qquad \text{Row 2}$$
$$3 \qquad\qquad\qquad\qquad\qquad \text{Row 3}$$

Step 4: Multiply the latest entry in Row 3 by -4 and place the result in Row 2, one column over to the right.

$$-4\overline{)3 \qquad 11 \qquad\quad 0 \qquad 14} \quad \text{Row 1}$$
$$\downarrow \qquad -12 \qquad\qquad\qquad \text{Row 2}$$
$$3^{\,-4(3)\nearrow} \qquad\qquad\qquad\qquad \text{Row 3}$$

Step 5: Add the entry in Row 2 to the entry above it in Row 1. Enter the sum in Row 3.

$$-4\overline{)3 \qquad\quad 11 \qquad\quad 0 \qquad 14} \quad \text{Row 1}$$
$$\downarrow \qquad -12 \qquad\qquad\qquad \text{Row 2}$$
$$3^{\,-4(3)\nearrow} \,-1 \qquad\qquad\qquad \text{Row 3}$$

Step 6: Repeat Steps 4 and 5 until no more entries are available in Row 1.

$$-4\overline{)3 \qquad\qquad 11 \qquad\qquad 0 \qquad\qquad 14} \quad \text{Row 1}$$
$$\downarrow \qquad -12 \qquad\quad 4 \qquad -16 \quad \text{Row 2}$$
$$3^{\,-4(3)\nearrow} -1^{\,-4(-1)\nearrow} 4^{\,-4(4)\nearrow} -2 \quad \text{Row 3}$$

Step 7: The final entry in Row 3, -2, is the remainder; the other entries in Row 3, 3, -1, and 4, are the coefficients of the quotient, in descending order. The quotient is a polynomial whose degree is one less than the degree of the dividend.

Quotient: $3x^2 - x + 4$

Remainder: -2

Step 8: Check

(Quotient)(Divisor) + Remainder = Dividend

$$(3x^2 - x + 4)(x + 4) - 2$$
$$= 3x^3 + 12x^2 - x^2 - 4x + 4x + 16 - 2$$
$$= 3x^3 + 11x^2 + 14$$

So $\dfrac{3x^3 + 11x^2 + 14}{x + 4} = 3x^2 - x + 4 - \dfrac{2}{x + 4}$.

Work Smart

The number opposite the number following the "x" in $x - c$ is "c." For example, in $x + 3$, $c = -3$. In $x - 5$, $c = 5$.

Let's do one more example where we consolidate all the steps given in Example 1.

EXAMPLE 2   **Dividing Two Polynomials Using Synthetic Division**

Use synthetic division to find the quotient and remainder when $x^4 - 5x^3 - 6x^2 + 33x - 15$ is divided by $x - 5$.

Solution

The divisor is $x - 5$ so that $c = 5$.

$$5\overline{)1 \quad -5 \quad -6 \quad\ 33 \quad -15}$$
$$\downarrow \quad\ \ 5 \quad\ \ 0 \quad -30 \quad\ \ 15$$
$$\overline{1 \quad\ \ 0 \quad -6 \quad\ \ 3 \quad\quad 0}$$

Work Smart: Study Skills

Knowing when a method **does not** apply is as essential as knowing when the method **does** apply. Identify when synthetic division can and cannot be used to divide polynomials.

The dividend is a fourth-degree polynomial, so the quotient is a third-degree polynomial. The quotient is $x^3 + 0x^2 - 6x + 3 = x^3 - 6x + 3$ and the remainder is 0. So

$$\frac{x^4 - 5x^3 - 6x^2 + 33x - 15}{x - 5} = x^3 - 6x + 3$$

In Example 2, because $\dfrac{x^4 - 5x^3 - 6x^2 + 33x - 15}{x - 5} = x^3 - 6x + 3$, we know that $x - 5$ and $x^3 - 6x + 3$ are factors of $x^4 - 5x^3 - 6x^2 + 33x - 15$. Therefore, we can write

$$x^4 - 5x^3 - 6x^2 + 33x - 15 = (x - 5)(x^3 - 6x + 3)$$

QUICK ✓ *Use synthetic division to find the quotient.*

1. $\dfrac{2x^3 + x^2 - 7x - 13}{x - 2}$

2. $\dfrac{x^4 + 8x^3 + 15x^2 - 2x - 6}{x + 3}$

2 **Use the Remainder and Factor Theorems**

Look back at Example 1 where we used synthetic division to find the quotient and remainder when $3x^3 + 11x^2 + 14$ is divided by $x + 4$. If we let $f(x) = 3x^3 + 11x^2 + 14$, we find that $f(-4) = -2$. Looking back at Example 1, we find that the remainder when $3x^3 + 11x^2 + 14$ is divided by $x + 4$ is -2. The value of the function f at $x = -4$ is the same as the remainder when f is divided by $x + 4 = x - (-4)$. This result is not a coincidence and is true in general! It is called the *Remainder Theorem*.

In Words
A theorem is a big idea that can be shown to be true in general. The word comes from a Greek verb meaning "to view."

THE REMAINDER THEOREM

Let f be a polynomial function. If $f(x)$ is divided by $x - c$, then the remainder is $f(c)$.

EXAMPLE 3 **Using the Remainder Theorem**

Use the Remainder Theorem to find the remainder if $f(x) = 2x^3 - 3x + 8$ is divided by $x + 3$.

Solution

The divisor is $x + 3 = x - (-3)$, so the Remainder Theorem says that the remainder is $f(-3)$.

$$f(x) = 2x^3 - 3x + 8$$
$$f(-3) = 2(-3)^3 - 3(-3) + 8$$
$$= 2(-27) + 9 + 8$$
$$= -54 + 9 + 8$$
$$= -37$$

When $f(x) = 2x^3 - 3x + 8$ is divided by $x + 3$, the remainder is -37.

Check Using synthetic division, we find that the remainder is, in fact, -37.

$$
\begin{array}{r|rrrr}
-3) & 2 & 0 & -3 & 8 \\
    &   & -6 & 18 & -45 \\
\hline
    & 2 & -6 & 15 & -37 \\
\end{array} \quad \leftarrow \text{Remainder}
$$

QUICK ✓

3. Use the Remainder Theorem to find the remainder if $f(x) = 3x^3 + 10x^2 - 9x - 4$ is divided by

(a) $x - 2$ **(b)** $x + 4$

We saw from Example 2 that when the remainder is 0, then the quotient and divisor are factors of the dividend. The Remainder Theorem can be used to determine whether an expression of the form $x - c$ is a factor of the dividend. This result is called the *Factor Theorem*.

Work Smart

"If and only if" statements are used to compress two statements into one. The Factor Theorem is two statements:

1. If $f(c) = 0$, then $x - c$ is a factor of f.
2. If $x - c$ is a factor of f, then $f(x) = 0$.

THE FACTOR THEOREM

Let f be a polynomial function. Then $x - c$ is a factor of $f(x)$ if and only if $f(c) = 0$.

We can use the Factor Theorem to determine whether a polynomial has a particular factor.

EXAMPLE 4 **Using the Factor Theorem**

Use the Factor Theorem to determine whether the function $f(x) = 2x^3 - 3x^2 - 18x - 8$ has the factor

(a) $x - 3$ **(b)** $x + 2$

Solution

The Factor Theorem states that if $f(c) = 0$, then $x - c$ is a factor of f.

(a) Because $x - 3$ is of the form $x - c$ with $c = 3$, we find the value of $f(3)$.

$$f(3) = 2(3)^3 - 3(3)^2 - 18(3) - 8 = -35 \neq 0$$

Since $f(3) \neq 0$, we know that $x - 3$ is not a factor of f.

(b) Because $x + 2 = x - (-2)$ is of the form $x - c$ with $c = -2$, we find the value of $f(-2)$. Rather than evaluating the function using substitution, we use synthetic division.

$$
\begin{array}{r|rrrr}
-2 & 2 & -3 & -18 & -8 \\
   &   & -4 & 14 & 8 \\
\hline
   & 2 & -7 & -4 & 0 \quad \leftarrow \text{Remainder}
\end{array}
$$

In Words

If $f(c) = 0$, then $f(x)$ can be written in factored form as

$$f(x) = (x - c)(\text{quotient})$$

The remainder is 0, so that $f(-2) = 0$. Because $f(-2) = 0$, we know that $x + 2$ is a factor of f. This means the dividend can be written as the product of the quotient and divisor. The quotient is $2x^2 - 7x - 4$ and the divisor is $x + 2$, so

$$2x^3 - 3x^2 - 18x - 8 = (x + 2)(2x^2 - 7x - 4)$$ ■

QUICK ✓

4. Use the Factor Theorem to determine whether $x - c$ is a factor of $f(x) = 2x^3 - 9x^2 - 6x + 5$ for the given values of c. If $x - c$ is a factor, then write f in factored form. That is, write $f(x) = (x - c)(\text{quotient})$.

   (a) $c = -2$ **(b)** $c = 5$

B Exercises

Concepts and Vocabulary

In Problems 1–2, fill in the blanks.

1. Because $\dfrac{3x^2 + 2x - 1}{x + 1} = 3x - 1$, the remainder when dividing $3x^2 + 2x - 1$ by $x + 1$ is _____ and $3x^2 + 2x - 1 =$ _____ · _____.

2. To find the quotient and remainder when a polynomial of degree 1 or higher is divided by $x - c$, a shortened version of long division, called _____ _____ can be used.

In Problems 3–5, answer True or False to each statement.

3. We can divide $-4x^3 + 5x^2 + 10x - 3$ by $x^2 - 2$ using synthetic division.

4. To check division, we compute $(\text{Quotient})(\text{Divisor}) + \text{Remainder}$ and determine whether it equals the Dividend.

5. If a polynomial function f is divided by $x - c$, then the remainder is $f(c)$.

6. If f is a polynomial of degree n and it is divided by $x + 4$, the quotient will be a polynomial of degree $n - 1$. Explain why.

7. Explain the Remainder Theorem in your own words. Explain the Factor Theorem in your own words.

8. Suppose that you were asked to divide $8x^3 - 3x + 1$ by $x + 3$. Would you use long division or synthetic division? Why?

Building Skills

In Problems 9–22, divide using synthetic division.

9. $\dfrac{x^2 - 3x - 10}{x - 5}$

10. $\dfrac{x^2 + 4x - 12}{x - 2}$

11. $\dfrac{2x^2 + 11x + 12}{x + 4}$

12. $\dfrac{3x^2 + 19x - 40}{x + 8}$

13. $\dfrac{x^2 - 3x - 14}{x - 6}$

14. $\dfrac{x^2 + 2x - 17}{x - 4}$

15. $\dfrac{x^3 - 19x - 15}{x - 5}$

16. $\dfrac{x^3 - 13x - 17}{x + 3}$

17. $\dfrac{3x^4 - 5x^3 - 21x^2 + 17x + 25}{x - 3}$

18. $\dfrac{2x^4 - x^3 - 38x^2 + 16x + 103}{x + 4}$

19. $\dfrac{x^4 - 40x^2 + 109}{x + 6}$

20. $\dfrac{a^4 - 65a^2 + 55}{a - 8}$

21. $\dfrac{2x^3 + 3x^2 - 14x - 15}{x - \dfrac{5}{2}}$

22. $\dfrac{3x^3 + 13x^2 + 8x - 12}{x - \dfrac{2}{3}}$

In Problems 23–30, use the Remainder Theorem to find the remainder.

23. $f(x) = x^2 - 5x + 1$ is divided by $x - 2$

24. $f(x) = x^2 + 4x - 5$ is divided by $x + 2$

25. $f(x) = x^3 - 2x^2 + 5x - 3$ is divided by $x + 4$

26. $f(x) = x^3 + 3x^2 - x + 1$ is divided by $x - 3$

27. $f(x) = 2x^3 - 4x + 1$ is divided by $x - 5$

28. $f(x) = 3x^3 + 2x^2 - 5$ is divided by $x + 3$

29. $f(x) = x^4 + 1$ is divided by $x - 1$

30. $f(x) = x^4 - 1$ is divided by $x - 1$

In Problems 31–38, use the Factor Theorem to determine whether $x - c$ is a factor of the given function for the given values of c. If $x - c$ is a factor, then write f in factored form. That is, write $f(x) = (x - c)(quotient)$.

31. $f(x) = x^2 - 3x + 2; c = 2$ **32.** $f(x) = x^2 + 5x + 6; c = 3$

33. $f(x) = 2x^2 + 5x + 2; c = -2$ **34.** $f(x) = 3x^2 + x - 2; c = 2$

35. $f(x) = 4x^3 - 9x^2 - 49x - 30; c = 3$ **36.** $f(x) = 2x^3 - 9x^2 - 2x + 24; c = 1$

37. $f(x) = 4x^3 - 7x^2 - 5x + 6; c = -1$ **38.** $f(x) = 5x^3 + 8x^2 - 7x - 6; c = -2$

Applying the Concepts

39. If $\dfrac{f(x)}{x - 5} = 3x + 5$, find $f(x)$. **40.** If $\dfrac{f(x)}{x + 3} = 2x + 7$, find $f(x)$.

41. If $\dfrac{f(x)}{x - 3} = x + 8 + \dfrac{4}{x - 3}$, find $f(x)$.

42. If $\dfrac{f(x)}{x - 3} = x^2 + 2 + \dfrac{7}{x - 3}$, find $f(x)$.

Extending the Concepts

43. Find the sum of a, b, c, and d if $\dfrac{2x^3 - 3x^2 - 26x - 37}{x + 2} = ax^2 + bx + c + \dfrac{d}{x + 2}$.

44. What is the remainder when $f(x) = 2x^{30} - 3x^{20} + 4x^{10} - 2$ is divided by $x - 1$?

Geometry Review

C.1 Lines and Angles

OBJECTIVES

1. Understand the Terms Point, Line, and Plane
2. Work with Angles
3. Find the Measures of Angles Formed by Parallel Lines

The word geometry comes from the Greek words "geo," meaning "earth" and "metra," meaning "measure." The Greek scholar Euclid collected and organized the geometry known in his day into a logical system more than two thousand years ago. Euclid's system forms the basis of the geometry we still study today.

(1) Understand the Terms Point, Line, and Plane

A **point** has no size, only position, and is usually designated by a capital letter as shown below.

$$P$$
•

A **line** is a set of points extending infinitely far in opposite directions. A line has no width or height, just length and is uniquely determined by two points. For example, the line in Figure 1 is passing through the points A and B. The notation for the line shown is $\overleftrightarrow{AB}$.

A **ray** is a half-line with one **endpoint,** which extends infinitely far in one direction. See Figure 2. The notation for the ray shown is $\overrightarrow{AB}$.

A **line segment** is a portion of a line that has a beginning and an end. See Figure 3. If two line segments have the same length, they are said to be **congruent.** Notation for the congruent line segments shown are $\overline{AB}$ and $\overline{CD}$.

A **plane** is the set of points that forms a flat surface that extends indefinitely. A plane has no thickness. See Figure 4. The arrows indicate that the plane extends indefinitely in each direction.

Figure 1

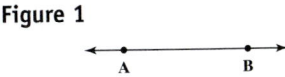

Figure 2

Figure 3

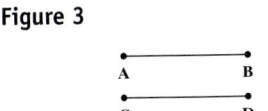

Figure 4

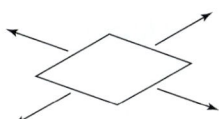

(2) Work with Angles

Figure 5

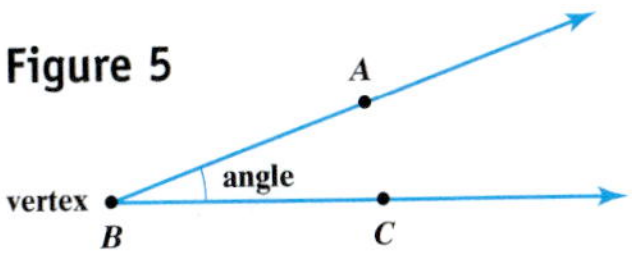

Suppose we draw two rays with a common endpoint as shown in Figure 5. The amount of rotation from one ray to the second ray is called the **angle** between the rays. The common endpoint is called the **vertex.** In Figure 5, the name of the angle is $\angle ABC$, $\angle CBA$ or $\angle B$. Angles are measured in **degrees,** which is symbolized °. One full rotation represents 360°. The notation $m\angle A = 60°$ means "the measure of angle A is 60 degrees." Because 60° is $\frac{1}{6}$ of 360°, an angle whose measure is 60° is $\frac{1}{6}$ of a full rotation. If two angles have the same measure, they are called **congruent.**

Work Smart
One full rotation is represented below.

360°

DEFINITION

An angle that measures 90° is called a **right angle.** The symbol ⌐ is used to denote a right angle. A right angle has $\frac{1}{4}$ of a full rotation.

An angle whose measure is between 0° and 90° is called an **acute angle.**

An angle whose measure is between 90° and 180° is called an **obtuse angle.**

An angle whose measure is 180° is called a **straight angle.** A straight angle has $\frac{1}{2}$ of a full rotation.

Figure 6

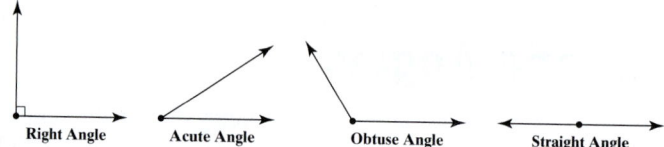

Right Angle   Acute Angle   Obtuse Angle   Straight Angle

QUICK ✓ *Classify each angle as right, acute, obtuse, or straight.*

1. 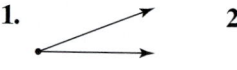   **2.**    **3.**    **4.**

DEFINITION

Two angles whose measures sum to 90° are called **complementary** angles. Each angle is the **complement** of the other. Two angles whose measures sum to 180° are called supplementary angles. Each angle is the **supplement** of the other. See Figure 7.

Figure 7

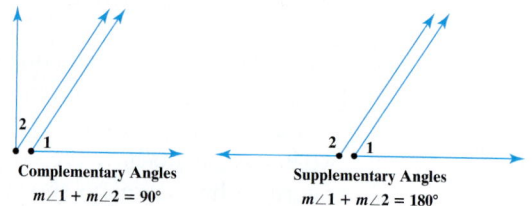

Complementary Angles   Supplementary Angles
$m\angle 1 + m\angle 2 = 90°$ $m\angle 1 + m\angle 2 = 180°$

EXAMPLE 1 **Finding the Complement of an Angle**

Find the complement of an angle whose measure is 18°.

Solution

Two angles are complementary if their sum is 90°. The measure of an angle that is complementary to an angle whose measure is 18° is $90° - 18° = 72°$.

EXAMPLE 2 **Finding the Supplement of an Angle**

Find the supplement of an angle whose measure is 97°.

Solution

Two angles are supplementary if their sum is 180°. The measure of an angle that is supplementary to an angle whose measure is 97° is $180° - 97° = 83°$.

QUICK ✓ *Find the complement and the supplement of each angle.*

5. 15°   **6.** 60°

③ **Find the Measures of Angles Formed by Parallel Lines**

Lines that lie in the same plane are called **coplanar.**

> **DEFINITION**
>
> **Parallel lines** are lines in the same plane that never meet. **Intersecting lines** meet or cross in one point. Two lines that intersect to form right (90°) angles are called **perpendicular lines.**

Figure 8

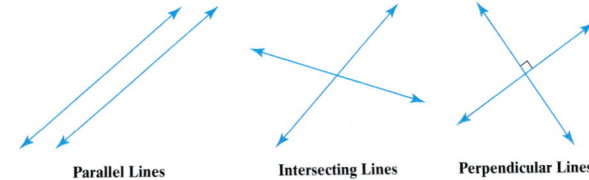

Parallel Lines   Intersecting Lines   Perpendicular Lines

Figure 9
$m\angle 1 = m\angle 3$
$m\angle 2 = m\angle 4$

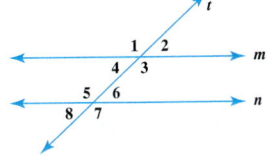

Figure 10

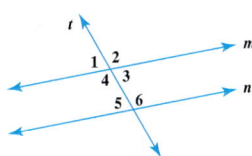

When two lines intersect they form four angles. Two angles that are opposite each other are called **vertical angles.** Vertical angles have equal measures. **Adjacent angles** have the same vertex and share a side. In Figure 9, angles 1 and 3 are vertical angles and angles 1 and 2 are adjacent angles. Angles 1 and 2 are also supplementary angles. Other pairs of adjacent angles are angles 2 and 3, angles 3 and 4, and angles 1 and 4. A line that cuts two parallel lines is called a **transversal.** In Figure 10, lines m and n are parallel and the transversal is labeled t. In this figure, there are certain angles with special names.

- There are 4 pairs of **corresponding angles:**

$$\angle 1 \text{ and } \angle 5, \quad \angle 2 \text{ and } \angle 6, \quad \angle 3 \text{ and } \angle 7, \quad \angle 4 \text{ and } \angle 8$$

- There are 2 pairs of **alternate interior angles:**

$$\angle 3 \text{ and } \angle 5, \quad \angle 4 \text{ and } \angle 6$$

Parallel lines and these angles are related in the following way.

> **PARALLEL LINES CUT BY A TRANSVERSAL**
>
> If two parallel lines are cut by a transversal, then
>
> - Corresponding angles are equal in measure.
> - Alternate interior angles are equal in measure.

EXAMPLE 3 **Finding the Measure of Corresponding and Alternate Interior Angles**

Given that lines m and n are parallel, t is a transversal, and the measure of angle 1 is 85°, find the measure of angles 2, 3, 4, 5 and 6.

Solution

$m\angle 2 = 180° - 85° = 95°$ because $\angle 1$ and $\angle 2$ are supplementary angles.

$m\angle 3 = 85°$ because $\angle 1$ and $\angle 3$ are vertical angles.

$m\angle 4 = 180° - 85° = 95°$ because $\angle 3$ and $\angle 4$ are supplementary angles.

$m\angle 5 = 85°$ because $\angle 1$ and $\angle 5$ are corresponding angles.

$m\angle 6 = 95°$ because $\angle 4$ and $\angle 6$ are alternate interior angles.

QUICK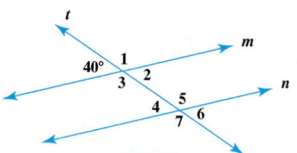

7. Find the measure of angles 1–7 given that lines *m* and *n* are parallel and *t* is a transversal.

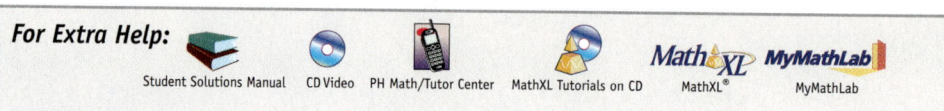

C.1 Exercises

For Extra Help: Student Solutions Manual   CD Video   PH Math/Tutor Center   MathXL Tutorials on CD   MathXL®   MyMathLab

Concepts and Vocabulary

In Problems 1–5, fill in the blanks.

1. If two line segments have the same length, they are _____.

2. Two rays that have a common endpoint form a(n) _____.

3. An angle that measures 90° is called a(n) _____ angle.

4. An angle whose measure is between 0° and 90° is called a(n) _____ angle.

5. Two angles whose measures sum to 180° are called _____ angles.

In Problems 6–8, answer True or False to each statement.

6. An angle whose measure is between 90° and 180° is called an obtuse angle.

7. Two intersecting lines form four angles. The two angles that are opposite each other are supplementary.

8. If two parallel lines are cut by a transversal, the corresponding angles are supplementary.

Building Skills

In Problems 9–16, classify each angle as right, acute, obtuse, or straight.

9.   **10.**   **11.**   **12.**

13.   **14.**   **15.**   **16.**

In Problems 17–20, find the complement of each angle.

17. 32°   **18.** 19°   **19.** 73°   **20.** 51°

In Problems 21–24, find the supplement of each angle.

21. 67°   **22.** 145°   **23.** 8°   **24.** 106°

25. Find the measure of angles 1–7 given that lines *m* and *n* are parallel and *t* is a transversal.

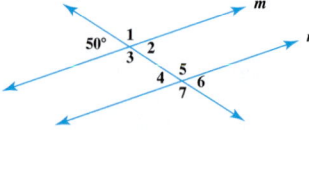

26. Find the measure of angles 1–7 given that lines *m* and *n* are parallel and *t* is a transversal.

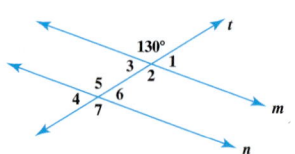

C.2 Polygons

OBJECTIVES

1. Define Polygon
2. Work with Triangles
3. Identify Quadrilaterals
4. Work with Circles

In Words

Vertices is the plural form of the word "vertex." The word polygon comes from the Greek words "poly," which means "many," and "gon," which means "angle."

(1) Define Polygon

A **polygon** is a closed figure in a plane consisting of line segments that meet at the **vertices.** A **regular polygon** is a polygon in which the sides are congruent and the angles are congruent. Figure 11 shows four regular polygons.

Figure 11
Regular polygons: All the sides are the same length; all the angles have the same measure.

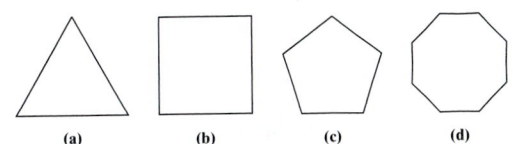

(a)   (b)   (c)   (d)

A polygon is named according to the number of sides. Table 1 summarizes the names of the polygons with 3 to 10 sides. A **triangle** is a polygon with three sides. Figure 11(a) is a regular triangle. A **quadrilateral** is a polygon with four sides. Figure 11(b) is a regular quadrilateral (also known as a *square*). A **pentagon** is a polygon with five sides. Figure 5(c) is a regular pentagon. An **octagon** is a polygon with eight sides. Figure 5(d) shows a regular octagon.

Table 1	
Polygons	
Number of Sides	**Name of Polygon**
3	Triangle
4	Quadrilateral
5	Pentagon
6	Hexagon
7	Heptagon
8	Octagon
9	Nonagon
10	Decagon

(2) Work with Triangles

Figure 12 shows triangle ABC. In triangle ABC angles A, B and C are called **interior angles.** The sum of the measures of the interior angles of a triangle is $180°$. If x, y, and z represent the measures of angles A, B, and C, respectively, then

$$x + y + z = 180°$$

Figure 12

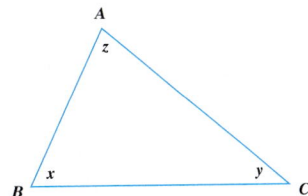

EXAMPLE 1 Finding the Measure of an Interior Angle of a Triangle

Find the measure of angle C in the triangle.

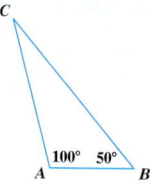

Solution

The measure of angle A is $100°$ and the measure of angle B is $50°$. We compute the measure of angle C as

$$m\angle C = 180° - 50° - 100° = 30°$$

We classify triangles by the lengths of their sides. A triangle in which all three sides are congruent is called an **equilateral** triangle. A triangle in which two sides are congruent is called an **isosceles** triangle, and a triangle in which none of the sides are congruent is called a **scalene** triangle. See Figure 13.

Figure 13

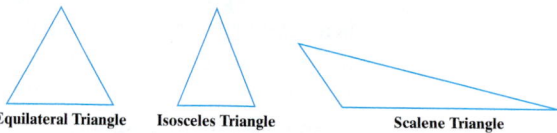

Equilateral Triangle   Isosceles Triangle   Scalene Triangle

A **right triangle** is a triangle that contains a right ($90°$) angle. In a right triangle, the longest side is called the hypotenuse, and the remaining two sides are called **legs.** See Figure 14.

Figure 14

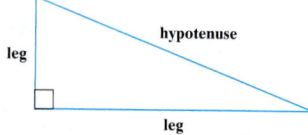

EXAMPLE 2 Finding the Measure of an Angle of a Right Triangle

Find the measure of angle B in the right triangle.

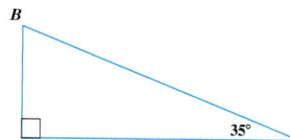

Solution

There are $180°$ degrees in a triangle and the right angle measures $90°$ so

$$m\angle B = 180° - 90° - 35° = 55°$$

QUICK ✓ *Find the measure of each angle B in the triangle.*

1.

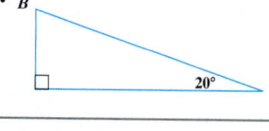

2.

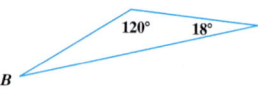

Figure 15

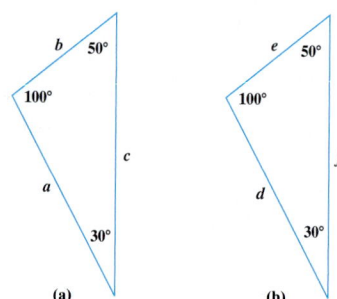

(a)   (b)

Congruent Triangles

Two triangles are congruent if the corresponding angles have the same measure and the corresponding sides have the same length. See Figures 15(a) and (b). We see that the corresponding angles in triangle (a) and (b) are equal. Also, the lengths of the corresponding sides are equal: $a = d, b = e$, and $c = f$.

It is not necessary to verify that all three angles and all three sides are the same measure to determine whether two triangles are congruent.

Determining Congruent Triangles

1. Two triangles are congruent if two of the angles are equal and the lengths of the corresponding sides between the two angles are equal. See Figure 16(a).

2. Two triangles are congruent if the lengths of the corresponding sides of the triangle are equal. See Figure 16(b).

3. Two triangles are congruent if the lengths of two corresponding sides are equal and the measures of the angles between the two sides are equal. See Figure 16(c).

Figure 16

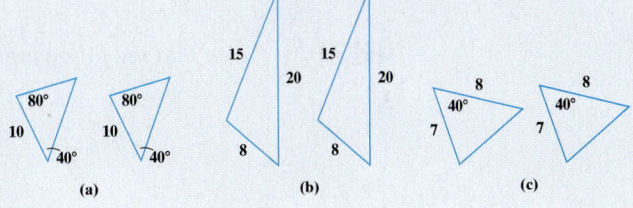

(a)   (b)   (c)

Similar Triangles

Figure 17

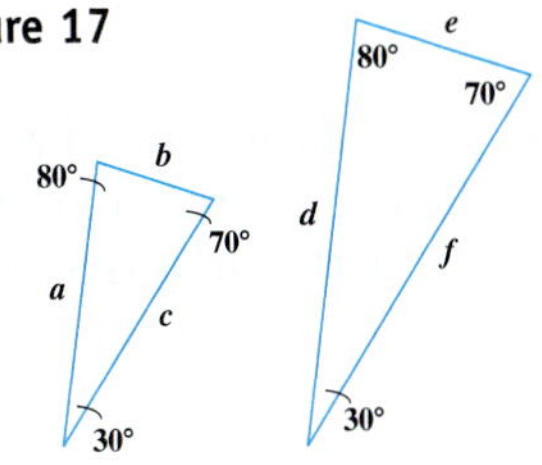

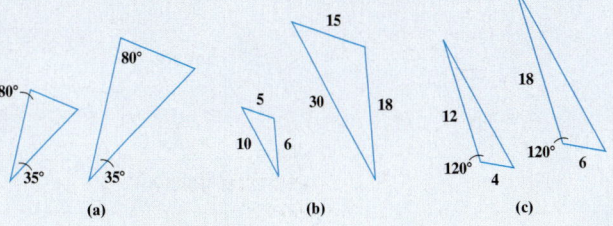

Two triangles are **similar** if they have the same shape. That is, the triangles are similar if the corresponding angles are equal and the lengths of the corresponding sides are proportional. In Figure 17 the triangles are similar because the corresponding angles are equal and the corresponding sides are proportional: $\dfrac{d}{a} = \dfrac{e}{b} = \dfrac{f}{c}$. It is not necessary to verify that all three angles are congruent and all three sides are proportional to determine whether two triangles are similar.

Determining Similar Triangles

1. Two triangles are similar if two of the corresponding angles are equal. See Figure 18(a).

2. Two triangles are similar if the lengths of the all three sides of each triangle are proportional. See Figure 18(b).

3. Two triangles are similar if two corresponding sides are proportional and the angles between the two sides are congruent. See Figure 18(c).

Figure 18

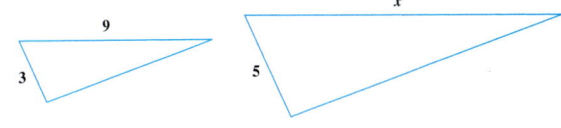

(a)   (b)   (c)

EXAMPLE 3

Given that the triangles in Figure 19 are similar, find the missing length.

Figure 19

Solution

Because the triangles are similar, the corresponding sides are proportional. That is, $\frac{3}{5} = \frac{9}{x}$. We solve this equation for x.

$$\frac{3}{5} = \frac{9}{x}$$

Multiply both sides by the LCD, $5x$: $\quad 5x \cdot \left(\frac{3}{5}\right) = 5x \cdot \left(\frac{9}{x}\right)$

Simplify: $\qquad 3x = 45$

Divide both sides by 3: $\qquad x = 15$

The missing length is 15 units.

QUICK ✓ *Given that the following triangles are similar, find the missing length.*

3.

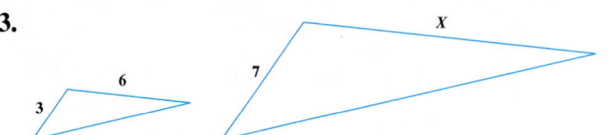

---

③ **Identify Quadrilaterals**

A **quadrilateral** is a polygon with four sides. A **parallelogram** is a quadrilateral in which both pairs of opposite sides are parallel. A **rectangle** is a parallelogram that contains a right angle. A **square** is a rectangle with all sides of equal length. A **rhombus** is a parallelogram that has all sides equal in length. A **trapezoid** is a quadrilateral with exactly one pair of opposite sides that are parallel.

Figure	Sketch
Parallelogram	
Rectangle	
Square	
Rhombus	
Trapezoid	

4 **Work with Circles**

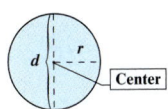

Figure 20

A *circle* is also a plane figure. A **circle** is a figure made up of all points in the plane that are a fixed distance from a point called the **center.** The **radius** of the circle is the line segment drawn from the center of the circle to any point on the circle. The **diameter** of the circle is the line segment which has endpoints on the circle. The diameter passes through the center of the circle. See Figure 20. Notice that the diameter d of a circle is twice the radius r. That is, $d = 2r$.

EXAMPLE 4 **Finding the Length of the Diameter of a Circle**

Find the length of the diameter of a circle with radius 4 cm.

Solution

The length of the diameter is twice the length of the radius.

$$d = 2 \cdot r$$
$$d = 2 \cdot 4 \text{ cm}$$
$$d = 8 \text{ cm}$$

The diameter of the circle is 8 cm.

EXAMPLE 5 **Finding the Length of the Radius of a Circle**

Find the length of the radius of the circle with diameter 18 yards.

Solution

The length of the radius is one-half the length of the diameter.

$$r = \frac{1}{2} \cdot d$$
$$r = \frac{1}{2} \cdot 18 \text{ yards}$$
$$r = 9 \text{ yards}$$

The radius of the circle is 9 yards.

QUICK ✓ *Find the length of the radius or diameter of each circle.*

4. $d = 15$ inches, find r.

5. $d = 24$ feet, find r.

6. $r = 3.6$ yards, find d.

7. $r = 9$ cm, find d.

C.2 Exercises

For Extra Help: Student Solutions Manual   CD Video   PH Math/Tutor Center   MathXL Tutorials on CD   MathXL®   MyMathLab

Concepts and Vocabulary

In Problems 1–4, fill in the blanks.

1. A _____ polygon is one in which the sides are congruent and the angles are congruent.

2. The sum of the measures of the interior angles of a triangle is _____ degrees.

3. A(n) _____ triangle is a triangle that has two congruent sides.

4. Two triangles are _____ if the measures of the angles of the triangle are equal and the sides are proportional.

In Problems 5–8, answer True or False to each statement.

5. A rhombus is a parallelogram that has all sides equal in length.

6. A heptagon is a polygon with six sides.

7. The diameter of a circle is exactly twice the length of the radius.

8. In triangle ABC, if $m\angle A = 83°$ and $m\angle B = 47°$, then $m\angle C = 50°$.

Building Skills

In Problems 9–12, find the measure of the missing angle of the triangle.

9.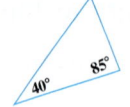

10.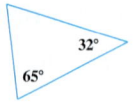

11.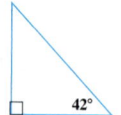

12.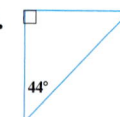

In Problems 13–16, determine the length of the missing side of the triangle. (These are similar triangles.)

13.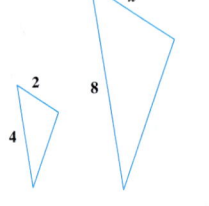

14.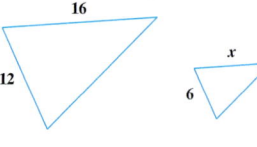

15.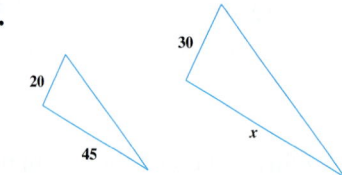

16.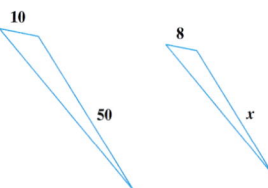

In Problems 17–20, find the length of the diameter of the circle.

17. $r = 5$ in

18. $r = 16$ feet

19. $r = 2.5$ cm

20. $r = 5.9$ in

In Problems 21–24, find the length of the radius of the circle.

21. $d = 14$ cm

22. $d = 58$ inches

23. $d = 11$ yards

24. $d = 27$ feet

C.3 Perimeter and Area of Polygons and Circles

OBJECTIVES

1. Find the Perimeter and Area of a Rectangle and a Square
2. Find the Perimeter and Area of a Parallelogram and a Trapezoid
3. Find the Perimeter and Area of a Triangle
4. Find the Circumference and Area of a Circle

The first concept we introduce in this section is the concept of *perimeter*. The **perimeter** of a polygon is the distance around the polygon. Put another way, the perimeter of a polygon is the sum of the lengths of the sides.

1 Find the Perimeter and Area of a Rectangle and a Square

A rectangle is a polygon, so the perimeter of a rectangle is the sum of the lengths of the sides.

EXAMPLE 1 Finding the Perimeter of a Rectangle

Find the perimeter of the rectangle in Figure 21.

Figure 21

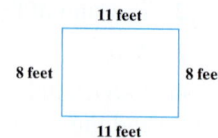

Solution

The perimeter of the rectangle is the sum of the lengths of the sides, so

$$\text{perimeter} = 11\text{ feet} + 8\text{ feet} + 11\text{ feet} + 8\text{ feet}$$
$$= 38\text{ feet}$$

Did you notice that the perimeter from Example 1 can also be written as

$$\text{perimeter} = 2 \cdot 11\text{ feet} + 2 \cdot 8\text{ feet}?$$

In general, the perimeter of a rectangle is written $P = 2l + 2w$.

A different measure of a polygon is its *area*. The **area** of a polygon is the amount of surface the polygon covers. Consider the rectangle shown in Figure 22. If we count the number of 1-unit by 1-unit squares within the rectangle, we see that the area of the rectangle is 6 square units. The area can also be found by multiplying the number of units of length by the number of units of width.

Figure 22

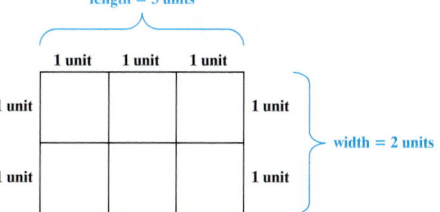

We can generalize this result to say that the area of a rectangle is the product of its length and width.

EXAMPLE 2   Finding the Area of a Rectangle

Find the area of the rectangle in Figure 23.

Solution

The area of the rectangle is the product of the length and the width, so

$$\text{area} = 6\text{ feet} \cdot 10\text{ feet}$$
$$= 60\text{ square feet}$$

We give a summary of how to find the perimeter and area of a rectangle.

Figure 23

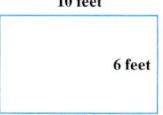

Figure	Sketch	Perimeter	Area
Rectangle		$P = 2l + 2w$	$A = lw$

EXAMPLE 3   Finding the Perimeter and Area of a Rectangle

Find (a) the perimeter and (b) the area of the rectangle shown in Figure 24.

Figure 24

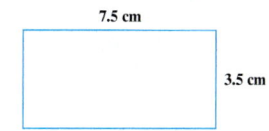

Solution

(a) The perimeter of the rectangle is

$$P = 2l + 2w$$
$$P = 2 \cdot 7.5 \text{ cm} + 2 \cdot 3.5 \text{ cm}$$
$$= 15 \text{ cm} + 7 \text{ cm}$$
$$= 22 \text{ cm}$$

The perimeter of the rectangle is 22 cm.

(b) The area of the rectangle is

$$A = lw$$
$$= 7.5 \text{ cm} \cdot 3.5 \text{ cm}$$
$$= 26.25 \text{ square cm}$$

The area of the given rectangle is 26.25 square cm.

QUICK ✓ *Find the perimeter and area of each rectangle.*

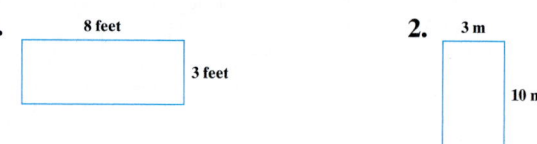

We can find the perimeter and area of a square using the same methods we use to find the perimeter and area of a rectangle. A square is a rectangle that has four congruent sides, so to find the perimeter of a square we use

$$\text{perimeter} = \text{side} + \text{side} + \text{side} + \text{side} = 4 \cdot \text{side} = 4s$$

where s is the length of a side.

EXAMPLE 4 Finding the Perimeter of a Square

Find the perimeter of the square in Figure 25.

Solution

Figure 25

The perimeter of the square is

$$\text{perimeter} = 4 \cdot s$$
$$= 4 \cdot 3 \text{ cm}$$
$$= 12 \text{ cm}$$

We know that the area of a rectangle is found by finding the product of the length and the width. In a square, the sides are congruent, so

$$\text{area} = \text{side} \cdot \text{side} = \text{side}^2$$

EXAMPLE 5 Finding the Area of a Square

Find the area of the square in Figure 26.

Solution

Figure 26

The area of the square is

$$\text{area} = \text{side}^2$$
$$= (7 \text{ inches})^2$$
$$= 49 \text{ square inches}$$

We summarize the formulas for the perimeter and area of a square in the following table.

Figure	Sketch	Perimeter	Area
Square		$P = 4s$	$A = s^2$

QUICK 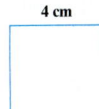 *Find the perimeter and area of each square.*

3. 4 cm

4. 1.5 yards

EXAMPLE 6 **Finding the Perimeter and Area of a Geometric Figure**

Find (a) the perimeter and (b) the area of the region shown in Figure 27.

Figure 27

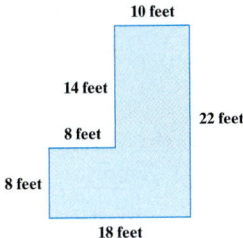

10 feet

14 feet

8 feet

22 feet

8 feet

18 feet

Solution

(a) The perimeter is the distance around the polygon. So,

Perimeter = 8 feet + 8 feet + 14 feet + 10 feet + 22 feet + 18 feet
= 80 feet

(b) Notice the region can be divided into an 8-foot by 8-foot square plus a 10-foot by 22-foot rectangle. To find the area, we find the area of the square and add it to the area of the rectangle.

$$\text{Area} = \text{Area of square} + \text{Area of rectangle}$$
$$= (8 \text{ feet})^2 + (10 \text{ feet})(22 \text{ feet})$$
$$= 64 \text{ feet}^2 + 220 \text{ feet}^2$$
$$= 284 \text{ feet}^2$$

QUICK  *Find the perimeter and area of the figure.*

5.

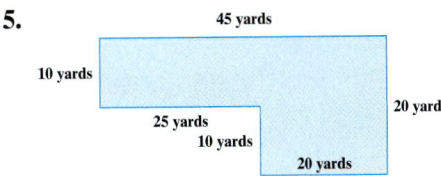

45 yards

10 yards

25 yards

10 yards

20 yards

20 yards

EXAMPLE 7 Painting a Room

You've decided to paint your rectangular bedroom. Two walls are 14 feet long and 7 feet high, and the other two walls are 10 feet long and 7 feet high.

(a) Ignoring the window and door openings in the bedroom, what is the area of the walls in your bedroom?

(b) You know that one gallon of paint will cover 300 square feet. A one-gallon can of paint costs $21.99. How many one-gallon cans of paint must you purchase to paint your bedroom?

Solution

(a) The bedroom consists of two walls that are 10 feet long and 7 feet high. The area of these two walls is

$$\text{area} = l \cdot w \cdot 2$$
$$\text{area} = 14 \text{ feet} \cdot 7 \text{ feet} \cdot 2$$
$$= 196 \text{ square feet}$$

The area of the other two walls is

$$\text{area} = l \cdot w \cdot 2$$
$$\text{area} = 10 \text{ feet} \cdot 7 \text{ feet} \cdot 2$$
$$= 140 \text{ square feet}$$

The total area to be painted is

$$\text{area} = 196 \text{ square feet} + 140 \text{ square feet}$$
$$= 336 \text{ square feet}$$

(b) A one-gallon can of paint covers 300 square feet. You have 336 square feet, so you will need to purchase 2 gallons of paint. ▬

---

② Find the Perimeter and Area of a Parallelogram and a Trapezoid

Recall that a parallelogram is a quadrilateral in which opposite sides are parallel. A trapezoid is a quadrilateral with exactly one pair of opposite sides that are parallel. The following table gives the formulas for the perimeter and area of parallelograms and trapezoids.

Figure	Sketch	Perimeter	Area
Parallelogram		$P = 2a + 2b$	$A = b \cdot h$
Trapezoid		$P = a + b + c + B$	$A = \dfrac{1}{2}h(b + B)$

EXAMPLE 8 Finding the Perimeter and Area of a Parallelogram

Find (a) the perimeter and (b) the area of parallelogram shown in Figure 28.

Figure 28

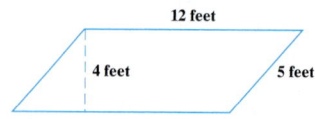

Solution

(a) The perimeter of the parallelogram is

$$P = 2l + 2w$$
$$= 2 \cdot 12 \text{ feet} + 2 \cdot 5 \text{ feet}$$
$$= 24 \text{ feet} + 10 \text{ feet}$$
$$= 34 \text{ feet}$$

The perimeter of the parallelogram is 34 feet.

(b) The area of the parallelogram is

$$A = b \cdot h$$
$$= 12 \text{ feet} \cdot 4 \text{ feet}$$
$$= 48 \text{ square feet}$$

The area of the parallelogram is 48 square feet.

EXAMPLE 9   Finding the Perimeter and Area of a Trapezoid

Find (a) the perimeter and (b) the area of trapezoid shown in Figure 29.

Figure 29

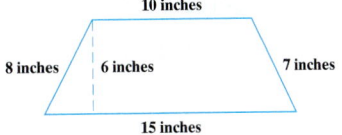

Solution

(a) The perimeter of the trapezoid is

$$\text{Perimeter} = 7 \text{ inches} + 8 \text{ inches} + 10 \text{ inches} + 15 \text{ inches}$$
$$= 40 \text{ inches}$$

The perimeter of the trapezoid is 40 inches.

(b) The area of the trapezoid is

$$A = \frac{1}{2}h(b + B)$$
$$= \frac{1}{2} \cdot 6 \text{ inches} \cdot (15 \text{ inches} + 10 \text{ inches})$$
$$= \frac{1}{2} \cdot 6 \text{ inches} \cdot 25 \text{ inches}$$
$$= 75 \text{ square inches}$$

The area of the trapezoid is 75 square inches.

QUICK ✓ *Find the perimeter and area of each figure.*

6.
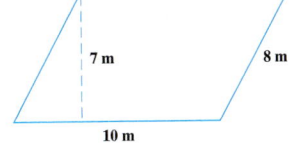

7.
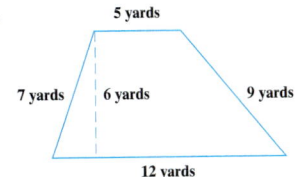

③ Find the Perimeter and Area of a Triangle

Recall that a triangle is a polygon with three sides. The following table gives the formulas for the perimeter and area of a triangle.

Figure	Sketch	Perimeter	Area
Triangle		$P = a + b + c$	$A = \dfrac{1}{2}bh$

EXAMPLE 10 Finding the Perimeter and Area of a Triangle

Find (a) the perimeter and (b) the area of the triangle shown in Figure 30.

Figure 30

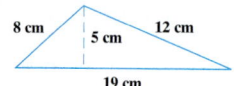

Solution

(a) To find the perimeter of the triangle, add the lengths of the three sides of the triangle.

$$\text{perimeter} = a + b + c$$
$$= 8 \text{ cm} + 12 \text{ cm} + 19 \text{ cm}$$
$$= 39 \text{ cm}$$

The perimeter of the triangle shown in Figure 30 is 39 cm.

(b) We use $A = \dfrac{1}{2}bh$ with base $= b = 19$ cm and height $= h = 5$ cm.

$$A = \frac{1}{2}bh = \frac{1}{2} \cdot 19 \text{ cm} \cdot 5 \text{ cm} = 47.5 \text{ square cm}$$

The area of the triangle in Figure 30 is 47.5 square cm.

QUICK ✓ *Find the perimeter and area of each triangle.*

8.

9.

④ Find the Circumference and Area of a Circle

The **circumference** of a circle is the distance around a circle. We use the diameter or the radius of the circle to find the circumference of a circle according to the formulas given below. We also give the formula for the area of a circle.

Figure	Sketch	Perimeter	Area
Circle	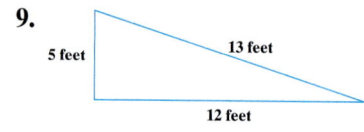	$C = \pi d$ where d is the diameter $C = 2\pi r$ where r is the radius	$A = \pi r^2$

EXAMPLE 11 Finding the Circumference of a Circle

Find the circumference of the circles in Figure 31.

Solution

(a) We know the length of the radius is 4 cm so we use the formula $C = 2\pi r$.

$$C = 2\pi r$$
$$r = 4\text{ cm:}\qquad = 2 \cdot \pi \cdot 4\text{ cm}$$
$$= 8 \cdot \pi\text{ cm}$$
$$\text{Use a calculator:}\qquad \approx 25.13\text{ cm}$$

The exact circumference of the circle is 8π cm, and 25.13 cm is the approximate circumference, to the nearest hundredth of a centimeter.

(b) The length of the diameter is given to be 12 inches, so we use $C = \pi d$.

$$C = \pi d$$
$$d = 12\text{ inches:}\qquad = \pi \cdot 12\text{ inches}$$
$$= 12\,\pi\text{ inches}$$
$$\text{Use a calculator:}\qquad \approx 37.70\text{ inches}$$

The exact circumference of the circle is 12π in., and 37.70 in. is the approximate circumference, to the nearest hundredth of an inch.

Figure 31

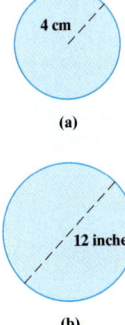

(a)

(b)

EXAMPLE 12 Finding the Area of a Circle

Find the area of the circle in Figure 32.

Solution

The circle in Figure 32 has radius 6 feet, so we substitute $r = 6$ in the equation $A = \pi r^2$.

$$A = \pi r^2$$
$$= \pi \cdot (6\text{ feet})^2$$
$$= \pi(36)\text{ square feet}$$
$$= 36\pi\text{ square feet}$$
$$\text{Use a calculator:}\qquad \approx 113.10\text{ square feet}$$

The area of the circle is exactly 36π square feet or approximately 113.10 square feet.

Figure 32

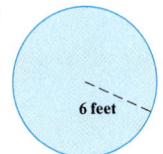

QUICK ✓ *Find the circumference and area of each circle.*

10.

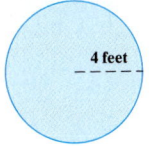

11.

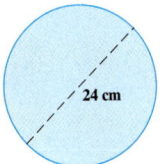

C.3 Exercises

Concepts and Vocabulary

In Problems 1–3, fill in the blanks.

1. The _____ of a polygon is the distance around the polygon.

2. The _____ of a circle is the distance around the circle.

3. To find the area of a trapezoid, we use the formula $A =$ _____, where _____ is the height of the trapezoid and the bases have lengths _____ and _____.

In Problems 4–6, answer True or False to each statement.

4. The area of a circle is $A = \dfrac{1}{2}\pi d^2$.

5. The length of the radius of a circle is twice the length of the diameter.

6. The area of a square is the sum of the lengths of the sides.

Building Skills

In Problems 7–10, find the perimeter and area of each rectangle.

7.

10 feet

4 feet

8.

12 miles

8 miles

9.

5 m

15 m

10.

2 cm

20 cm

In Problems 11 and 12, find the perimeter and area of each square.

11.

6 km

6 km

12.

13 yards

13 yards

In Problems 13–16, find the perimeter and area of each figure.

13.

7 feet

6 feet

15 feet

14 feet

8 feet

22 feet

14.

8 m

3 m

5 m

5 m

2 m

3 m

15.

13 m

2 m

8 m

2 m

8 m

2 m

16.

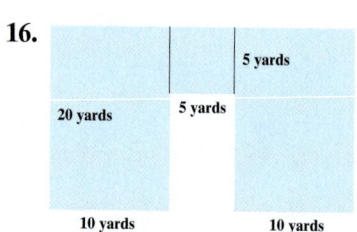

5 yards

20 yards

5 yards

10 yards

10 yards

In Problems 17–24, find the perimeter and area of each quadrilateral.

17.
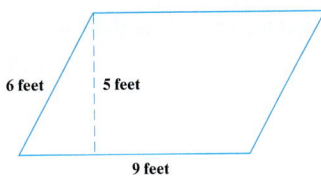

18.
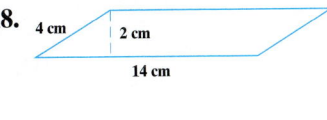

19.
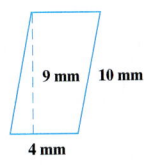

20.
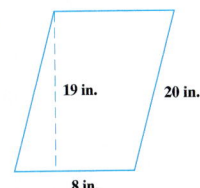

21.
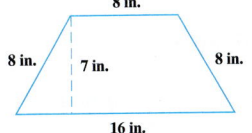

22.
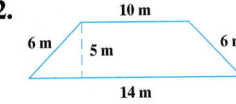

23.
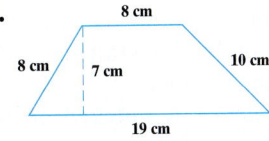

24.
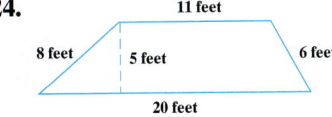

In Problems 25–28, find the perimeter and area of each triangle.

25.
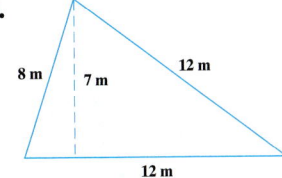

26.
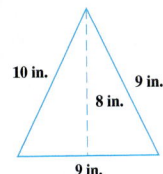

27.
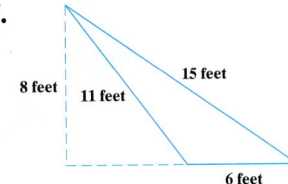

28.
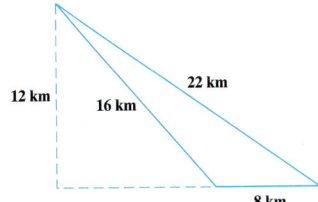

In Problems 29–32, find (a) the circumference and (b) the area of each circle. For both the circumference and area, provide exact answers and approximate answers rounded to the nearest hundredth.

29.
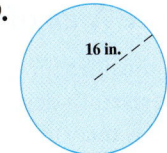

30.
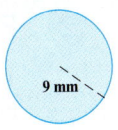

31.
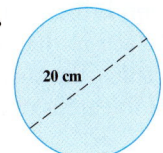

32.
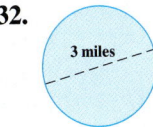

Applying the Concepts

In Problems 33–34, find the area of the shaded region.

33.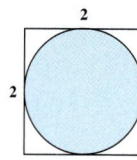

34.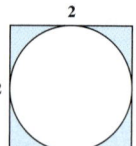

35. How many feet will a wheel with a diameter of 20 inches travel after 5 revolutions?

36. How many feet will a wheel with a diameter of 18 inches travel after 3 revolutions?

C.4 Volume and Surface Area

OBJECTIVES

1 Identify Solid Figures

2 Find the Volume and Surface Area of Solid Figures

In Words

Polyhedra is the plural form of the Greek word polyhedron.

1 Identify Solid Figures

A **geometric solid** is a three-dimensional region of space enclosed by planes and curved surfaces. Examples of geometric solids are cubes, pyramids, spheres, cylinders, and cones.

There are some geometric solids that you see and use every day. When you grab a box of cereal, you are holding a rectangular solid. When you reach for a can of soup, you are holding a circular cylinder.

We first discuss *polyhedra*. A **polyhedron** is a three-dimensional solid formed by connecting polygons. Figure 33 shows an example of a polyhedron called a hexagonal dipyramid. Notice that the top and bottom are formed by six connected triangles.

Figure 33
Hexagonal Dipyramid

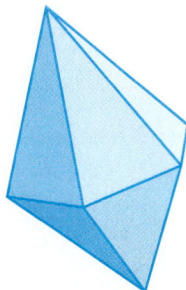

Each of the planes of the polyhedron is called a **face.** The line segment which is the intersection of any two faces of a polyhedron is called an **edge.** The point of intersection of three or more edges is called a **vertex.**

2 Find the Volume and Surface Area of Solid Figures

The **volume** of a polyhedron is the measure of the number of units of space contained in the solid. Volume can be used to describe the amount of soda in a can, or the amount of cereal in a box. Volume is measured in cubic units. For example, the cube in Figure 34 represents 1 cubic inch.

The **surface area** of a polyhedron is the sum of the areas of the faces of the polyhedron. For example, because each face in the cube shown in Figure 34 has an area of 1 square inch, and a cube has six sides, the surface area of the cube is 6 square inches. Because surface area is the sum of the areas of each polygon in the polyhedron, surface area is measured in square units.

Table 2 shows some common geometric solids, and formulas to find their volume and surface area.

Figure 34

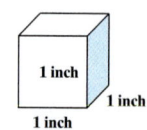

1 inch
1 inch
1 inch

Work Smart

Volume is measured in cubic units. Surface area is measured in square units.

Table 2		
Solids		**Formulas**
Cube		**Volume:** $V = s^3$ **Surface Area:** $S = 6s^2$
Rectangular Solid		**Volume:** $V = lwh$ **Surface Area:** $S = 2lw + 2lh + 2wh$
Sphere		**Volume:** $V = \dfrac{4}{3}\pi r^3$ **Surface Area:** $S = 4\pi r^2$
Right Circular Cylinder		**Volume:** $V = \pi r^2 h$ **Surface Area:** $S = 2\pi r^2 + 2\pi rh$
Cone		**Volume:** $V = \dfrac{1}{3}\pi r^2 h$
Square Pyramid		**Volume:** $V = \dfrac{1}{3}b^2 h$ **Surface Area:** $S = b^2 + 2bs$

EXAMPLE 1   Finding the Volume and Surface Area of a Rectangular Solid

Find the volume and surface area of the rectangular solid shown in Figure 35.

Solution

We have that $l = 10$ feet, $h = 5$ feet, and $w = 4$ feet. The volume of the rectangular solid is

$$V = lwh$$
$$= (10 \text{ feet})(4 \text{ feet})(5 \text{ feet})$$
$$= 200 \text{ cubic feet}$$

The volume of the rectangular solid is 200 cubic feet.

The surface area of the rectangular solid is

$$S = 2lw + 2lh + 2wh$$
$$= 2(10 \text{ feet})(4 \text{ feet}) + 2(10 \text{ feet})(5 \text{ feet}) + 2(4 \text{ feet})(5 \text{ feet})$$
$$= 220 \text{ square feet}$$

The surface area of the rectangular solid is 220 square feet.

Figure 35

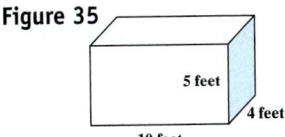

5 feet
4 feet
10 feet

EXAMPLE 2   Finding the Volume of a Right Circular Cylinder

Find the volume and surface area of the right circular cylinder shown in Figure 36.

Solution

We have that $h = 10$ inches and $r = 3$ inches. The volume of the right circular cylinder is

$$V = \pi r^2 h$$
$$= \pi(3 \text{ in.})^2(10 \text{ in.})$$
$$= 90\pi \text{ in.}^3$$
$$\approx 282.74 \text{ in.}^3$$

Figure 36

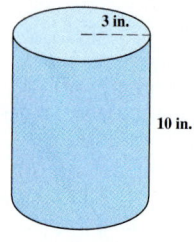

3 in.

10 in.

The volume of the right circular cylinder is 90π cubic inches exactly and 282.74 cubic inches approximately.

The surface area of the right circular cylinder is

$$V = 2\pi r^2 + 2\pi rh$$
$$= 2\pi(3 \text{ in.})^2 + 2\pi(3 \text{ in.})(10 \text{ in.})$$
$$= 18\pi \text{ in.}^2 + 60\pi \text{ in.}^2$$
$$= 78\pi \text{ in.}^2$$
$$\approx 245.04 \text{ in.}^2$$

The surface area of the right circular cylinder is 78π square inches exactly and 245.04 square inches approximately.

QUICK ✓ *Find the volume and surface area of the following polyhedra.*

1.

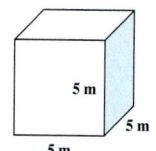

5 m

5 m

5 m

2.

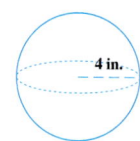

4 in.

C.4 Exercises

For Extra Help:

Student Solutions Manual   CD Video   PH Math/Tutor Center   MathXL Tutorials on CD   MathXL®   MyMathLab

Concepts and Vocabulary

In Problems 1–3, fill in the blanks.

1. A _____ is a three-dimensional solid formed by connecting polygons.

2. The _____ of a polyhedron is the measure of the number of units of space contained in the solid.

3. The _____ _____ of a polyhedron is the sum of the areas of the faces of the polyhedron.

In Problems 4–6, answer True or False to each statement.

4. Each of the planes of the polyhedron is called a face.

5. The volume of a polyhedron is measured in square units.

6. The volume of a rectangular solid is the product of the length, width, and height.

Building Skills

7. Find the volume V and surface area S of a rectangular box with length 10 feet, width 5 feet, and height 12 feet.

8. Find the volume V and surface area S of a rectangular box with length 2 meters, width 6 meters, and height 9 meters.

9. Find the volume V and surface area S of a sphere of radius 6 centimeters.

10. Find the volume V and surface area S of a sphere of radius 10 inches.

11. Find the volume V and surface area S of a right circular cylinder of radius 2 inches and height 8 inches.

12. Find the volume V and surface area S of a right circular cylinder of radius 3 inches and height 6 inches.

13. Find the volume V of a cone of radius 10 mm and height 8 mm.

14. Find the volume V of a cone of radius 20 feet and height 30 feet.

15. Find the volume V and surface area S of a square pyramid of height 10 feet, slant 12 feet, and base 8 feet.

16. Find the volume V and surface area S of a square pyramid of height 5 m, slant 8 m, and base 10 m.

Applying the Concepts

17. **Rain Gutter** A rain gutter is in the shape of a rectangular solid. How much water can the gutter hold if it is 4 inches in height, 3 inches wide and 12 feet long?

18. **Water for the Horses** A trough for horses is 10 feet long, 2 feet wide, and 3 feet deep. How much water can the trough hold?

19. **A Can of Peaches** A can of peaches is in the shape of a right circular cylinder. The can has a 4 inch diameter and is 6 inches tall. What is the volume of the can? What is the surface area of the can? Express your answer as a decimal rounded to the nearest hundredth.

20. **Coffee Can** A coffee can is in the shape of a right circular cylinder. The can has an 8-inch diameter and is 10 inches tall. What is the volume of the can? What is the surface area of the can? Express your answer as a decimal rounded to the nearest hundredth.

21. **Ice Cream Cone** A waffle cone for ice cream has a diameter of 8 cm and a height of 16 cm. How much ice cream can the cone hold if the ice cream is flush with the top of the cone? Express your answer as a decimal rounded to the nearest hundredth.

22. **Water Cooler** The cups at the water cooler are in the shape of a cone. How much water can a cup hold if it has a 5 inch diameter and is 8 inches in height? Express your answer as a decimal rounded to the nearest hundredth.

APPENDIX

D The Library of Functions

OBJECTIVE

1. Graph functions in the Library of Functions

Preparing for the Library of Functions

Before getting started, take the following readiness quiz. If you get a problem wrong, go back to the section cited and review the material.

1. Graph $y = x^2$ by point plotting.   [Section 8.1, pp. 569–570]
2. Graph $y = x^3$ by point plotting.   [Section 8.1, pp. 568–571]
3. Graph $x = y^2$ by point plotting.   [Section 8.1, pp. 570–571]

1   Graph Functions in the Library of Functions

In Table 1, we review a number of the functions that we have studied in the book. Pay attention to the properties listed and the shape of the graph.

Preparing for...Answers

1.

2.

3.

Table 1		
Function	**Properties**	**Graph**
Linear Function $f(x) = mx + b$ m and b are real numbers	• Domain and range are all real numbers. • Graph is nonvertical line with slope $= m$ y intercept $= b$.	
Identity Function (special type of linear function) $f(x) = x$	• Domain and range are all real numbers. • Graph is a line with slope of $m = 1$ y-intercept $= 0$. • The line consists of all points for which the x-coordinate equals the y-coordinate.	
Constant Function (special type of linear function) $f(x) = b$ b is a real number	• Domain is the set of all real numbers and range is the set consisting of a single number b. • Graph is a horizontal line with slope $m = 0$ y-intercept of b.	
Square Function $f(x) = x^2$	• Domain is the set of all real numbers; its range is the set of nonnegative real numbers. • The graph is a parabola whose intercept is (0, 0).	

Function	Properties	Graph
Cube Function $$f(x) = x^3$$	• Domain and the range are the set of all real numbers. • The intercept of the graph is at $(0, 0)$.	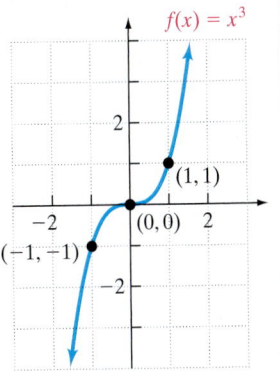
Square Root Function $$f(x) = \sqrt{x}$$	• Domain and the range are the set of nonnegative real numbers. • The intercept of the graph is at $(0, 0)$.	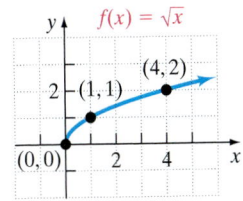
Cube Root Function $$f(x) = \sqrt[3]{x}$$	• Domain and the range are the set of real numbers. • The intercept of the graph is at $(0, 0)$.	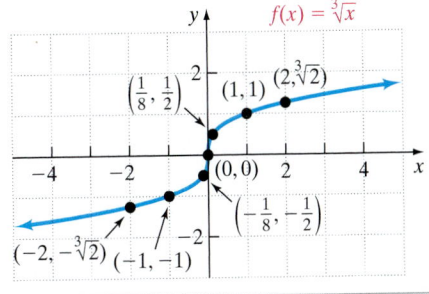
Reciprocal Function $$f(x) = \frac{1}{x}$$	• Domain and the range are the set of all nonzero real numbers. • The graph has no intercepts.	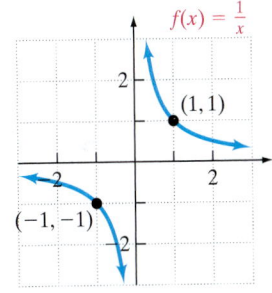
Absolute Value Function $$f(x) = |x|$$	• Domain of the is the set of all real numbers; its range is the set of nonnegative real numbers. • The intercept of the graph is $(0, 0)$. • If $x \geq 0$, then $f(x) = x$, and the graph of f is part of the line $y = x$; if $x < 0$, then $f(x) = -x$, and the graph of f is part of the line $y = -x$.	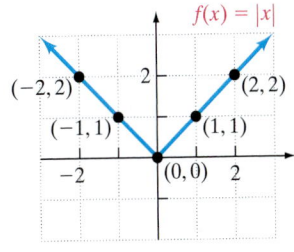

D Exercises

Concepts and Vocabulary

In Problems 1 and 2, fill in the blanks.

1. The range of the _____ _____ is the real number b.

2. The domain of the _____ _____ is the set of all nonzero real numbers.

In Problems 3 and 4, answer True or False to each statement.

3. The domain and the range of the square function is the set of all real numbers.

4. The domain and the range of the cube function is the set of all real numbers.

Building Skills

In Problems 5–12, match each graph to the function listed whose graph most resembles the one given.

    (a) Constant function   **(b)** Linear function   **(c)** Square function
    (d) Cube function   **(e)** Square root function   **(f)** Reciprocal function
    (g) Absolute value function   **(h)** Cube root function

5.  **6.** **7.**  **8.**

9.  **10.** **11.** 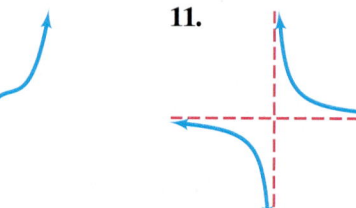 **12.**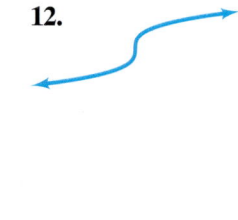

In Problems 13–18, sketch the graph of each function. Label at least three points.

13. $f(x) = x^2$ **14.** $f(x) = x^3$ **15.** $f(x) = \sqrt{x}$

16. $f(x) = \sqrt[3]{x}$ **17.** $f(x) = \dfrac{1}{x}$ **18.** $f(x) = |x|$

E More on Systems of Linear Equations

E.1 A Review of Systems of Linear Equations in Two Variables

OBJECTIVES

1. Determine Whether an Ordered Pair Is a Solution to a System of Linear Equations
2. Solve a System of Two Linear Equations Containing Two Unknowns by Graphing
3. Solve a System of Two Linear Equations Containing Two Unknowns by Substitution
4. Solve a System of Two Linear Equations Containing Two Unknowns by Elimination
5. Identify Inconsistent Systems
6. Express the Solution of a System of Dependent Equations

Preparing for Systems of Linear Equations in Two Variables

Before getting started, take the following readiness quiz. If you get a problem wrong, go back to the section cited and review the material.

1. Evaluate $2x - 3y$ for $x = 5$, $y = 4$. [Section 1.7, pp. 57–58]
2. Determine whether the point whose ordered pair is $(4, -1)$ is on the graph of the equation $2x - 3y = 11$. [Section 3.1, pp. 179–180]
3. Graph $y = 3x - 7$. [Section 3.4, pp. 218–220]
4. Find the equation of the line parallel to $y = -3x + 1$ containing the point $(2, 3)$. [Section 3.6, pp. 237–238]
5. Determine the slope and y-intercept of $4x - 3y = 15$. [Section 3.4, pp. 217–218]
6. What is the additive inverse of 4? [Section 1.3, p. 21]
7. Solve: $2x - 3(-3x + 1) = -36$ [Section 2.2, pp. 86–87]

Recall, from Section 3.2, that an equation in two variables is linear provided that it can be written in the form $Ax + By = C$, where A, B, and C are real numbers and A and B are not both zero. However, linear equations can have more than two variables.

Some examples of linear equations are

Linear equation in two variables, x and y	Linear equation in three variables, x, y, and z	Linear equation in four variables, w, x, y, and z
$4x - 3y = 9$	$-2x + y - 5z = -3$	$3w - x + 5y - 2z = 12$

A **system of linear equations** is a grouping of two or more linear equations, each of which contains one or more variables.

EXAMPLE 1 Examples of Systems of Linear Equations

(a) $\begin{cases} 2x + y = 5 \\ x - 5y = -10 \end{cases}$ Two equations containing two variables, x and y

(b) $\begin{cases} x + 3y + z = 8 \\ 3x - y + 6z = 12 \\ -4x - y + 2z = -1 \end{cases}$ Three equations containing three variables, x, y, and z

We use a brace, as shown in the systems in Example 1, to remind us that we are dealing with a system of equations. In this section, we concentrate on systems of two linear equations containing two variables such as the system in Example 1(a).

1 Determine Whether an Ordered Pair Is a Solution to a System of Linear Equations

A **solution** of a system of equations consists of values for the variables that are solutions of each equation of the system. When we are solving systems of two linear equations containing two unknowns, we represent the solution as an ordered pair, (x, y).

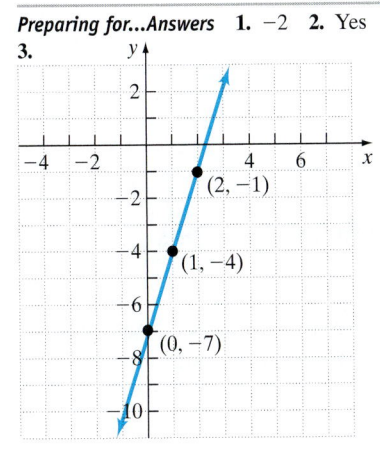

EXAMPLE 2   Determining Whether Values Are a Solution to a System of Linear Equations

Determine whether the given ordered pairs are solutions to the system of equations.

$$\begin{cases} 2x + 3y = 9 \\ -5x - 3y = 0 \end{cases}$$

(a) $(6, -1)$ **(b)** $(-3, 5)$

Solution

To help us organize our thoughts, we name $2x + 3y = 9$ equation (1) and $-5x - 3y = 0$ equation (2).

$$\begin{cases} 2x + 3y = 9 \quad (1) \\ -5x - 3y = 0 \quad (2) \end{cases}$$

(a) Let $x = 6$ and $y = -1$ in both equations (1) and (2). If both equations are true, then $(6, -1)$ is a solution.

Equation (1): $2x + 3y = 9$

$x = 6, y = -1$: $2(6) + 3(-1) \overset{?}{=} 9$

$12 - 3 \overset{?}{=} 9$

$9 = 9$ True

Equation (2): $-5x - 3y = 0$

$x = 6, y = -1$: $-5(6) - 3(-1) \overset{?}{=} 0$

$-30 + 3 \overset{?}{=} 0$

$-27 = 0$ False

Although $x = 6$, $y = -1$ satisfy equation (1), they do not satisfy equation (2); therefore, $(6, -1)$ is not a solution of the system of equations.

(b) Let $x = -3$ and $y = 5$ in both equations (1) and (2). If both equations are true, then $(-3, 5)$ is a solution.

Equation (1): $2x + 3y = 9$

$x = -3, y = 5$: $2(-3) + 3(5) \overset{?}{=} 9$

$-6 + 15 \overset{?}{=} 9$

$9 = 9$ True

Equation (2): $-5x - 3y = 0$

$x = -3, y = 5$: $-5(-3) - 3(5) \overset{?}{=} 0$

$15 - 15 \overset{?}{=} 0$

$0 = 0$ True

Because the values $x = -3$ and $y = 5$ satisfy both equations (1) and (2), the ordered pair $(-3, 5)$ is a solution of the system of equations.   ■

For the remainder of the chapter, we shall number each equation as we did in Example 1. When solving homework problems, you should do the same.

QUICK ✓

1. Which of the following points is a solution to the system of equations?

$$\begin{cases} 2x + 3y = 7 \\ 3x + y = -7 \end{cases}$$

(a) $(3, 1)$ **(b)** $(-4, 5)$ **(c)** $(-2, -1)$

Visualizing the Solutions in a System of Two Linear Equations Containing Two Unknowns

We can view the problem of solving a system of two linear equations containing two variables as a geometry problem. The graph of each equation in the system is a line. So, a system of two equations containing two variables represents a pair of lines. The graphs of the two lines can appear in one of three ways:

1. INTERSECT: If the lines intersect, then the system of equations has one solution given by the point of intersection. We say that the system is **consistent** and the equations are **independent.** See Figure 1(a).

2. **PARALLEL:** If the lines are parallel, then the system of equations has no solution because the lines never intersect. In this circumstance, we say that the system is **inconsistent.** See Figure 1(b).

3. **COINCIDENT:** If the lines lie on top of each other (are coincident), then the system of equations has infinitely many solutions. The solution set is the set of all points on the line. The system is **consistent** and the equations are **dependent.** See Figure 1(c).

Figure 1

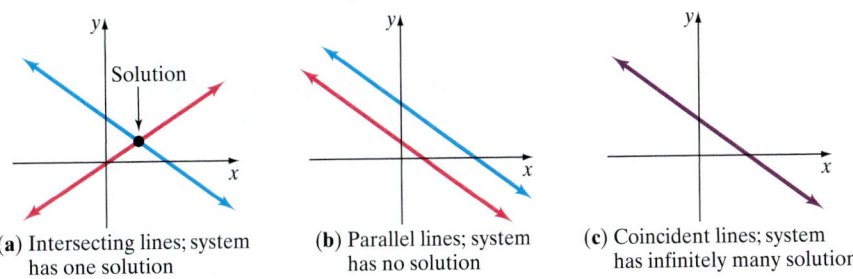

(**a**) Intersecting lines; system has one solution

(**b**) Parallel lines; system has no solution

(**c**) Coincident lines; system has infinitely many solutions

For now, we will concentrate on solving systems for which there is a single solution.

(2) ## Solve a System of Two Linear Equations Containing Two Unknowns by Graphing

Let's look at an example where we use graphing to solve a system.

EXAMPLE 3   **Solving a System of Two Linear Equations Using Graphing**

Solve the following system by graphing: $\begin{cases} x + y = -1 \\ -2x + y = -7 \end{cases}$

Solution

First, we name $x + y = -1$ equation (1) and $-2x + y = -7$ equation (2).

$$\begin{cases} x + y = -1 & (1) \\ -2x + y = -7 & (2) \end{cases}$$

In order to graph each equation, we put them in slope-intercept form. Equation (1) in slope-intercept form is $y = -x - 1$, which has slope -1 and y-intercept -1. Equation (2) in slope-intercept form is $y = 2x - 7$, which has slope 2 and y-intercept -7. Figure 2 shows their graphs (Note that we could also have graphed the lines using the intercepts). The lines appear to intersect at $(2, -3)$, so we believe that the ordered pair $(2, -3)$ is the solution to the system.

Figure 2

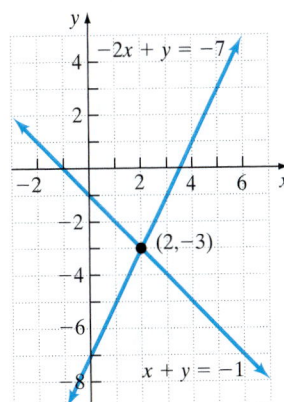

Check Let $x = 2$ and $y = -3$ in both equations in the system:

Equation (1):	$x + y = -1$	Equation (2):	$-2x + y = -7$
$x = 2, y = -3$:	$2 + (-3) \overset{?}{=} -1$	$x = 2, y = -3$:	$-2(2) + (-3) \overset{?}{=} -7$
	$2 - 3 \overset{?}{=} -1$		$-4 - 3 \overset{?}{=} -7$
	$-1 = -1$ True		$-7 = -7$ True

Both equations are true, so the solution is the ordered pair $(2, -3)$. ▬

QUICK ✔ *Solve the system by graphing.*

2. $\begin{cases} y = -3x + 10 \\ y = 2x - 5 \end{cases}$ 3. $\begin{cases} 2x + y = -1 \\ -2x + 2y = 10 \end{cases}$

Work Smart

Obtaining exact solutions using graphical methods can be difficult. Therefore, algebraic methods should be used.

3 Solve a System of Two Linear Equations Containing Two Unknowns by Substitution

If the x- and y-coordinates of the point of intersection between two lines are not integers, then obtaining an exact result graphically can be difficult. Therefore, rather than using graphical methods to obtain solutions to systems of two linear equations, we prefer to use algebraic methods. The first algebraic method that we present is the *method of substitution*. The goal of the method of substitution is to obtain a single linear equation involving a single unknown—something we already know how to solve.

Work Smart

When using substitution solve for the variable whose coefficient is 1 or -1 in order to simplify the algebra.

> **Steps for Solving a System of Two Linear Equations Containing Two Unknowns by Substitution**
>
> **Step 1:** Solve one of the equations for one of the unknowns. For example, we might solve equation (1) for y in terms of x. Choose the equation that is easiest to solve for a variable. Typically, this would be the equation that has a variable whose coefficient is 1 or -1.
>
> **Step 2:** Substitute the expression that equals the variable solved for in Step 1 into the other equation. The result will be a single linear equation in one unknown. For example, if we solved equation (1) for y in terms of x in Step 1, then we would replace y in equation (2) with the algebraic expression in x.
>
> **Step 3:** Solve the linear equation in one unknown found in Step 2.
>
> **Step 4:** Substitute the value of the variable into the expression found in Step 1 to find the value of the other variable.
>
> **Step 5:** Check your answer.

EXAMPLE 4 **Solving a System of Two Equations Containing Two Unknowns by Substitution**

Solve the following system by substitution: $\begin{cases} 2x - 3y = -6 & (1) \\ -8x + 3y = 3 & (2) \end{cases}$

Solution

We choose to solve equation (1) for x because it seems easiest.

$$\text{Equation (1):} \quad 2x - 3y = -6$$
$$\text{Add 3y to both sides:} \quad 2x = 3y - 6$$
$$\text{Divide both sides by 2:} \quad x = \frac{3y - 6}{2}$$
$$\text{Divide 2 into both terms in the numerator:} \quad x = \frac{3}{2}y - 3$$

Now substitute $\frac{3}{2}y - 3$ for x in equation (2) and then solve for y.

Work Smart

Notice how we use substitution to reduce a system of two linear equations involving two unknowns down to one linear equation involving one unknown. Again, we use algebraic techniques to reduce a problem down to one we already know how to solve!

$$\text{Equation (2):} \quad -8x + 3y = 3$$
$$-8\left(\frac{3}{2}y - 3\right) + 3y = 3$$
$$\text{Distribute the } -8: \quad -8 \cdot \frac{3}{2}y - (-8) \cdot 3 + 3y = 3$$
$$\text{Multiply:} \quad -12y + 24 + 3y = 3$$
$$\text{Combine like terms:} \quad -9y + 24 = 3$$
$$\text{Subtract 24 from both sides:} \quad -9y = -21$$
$$\text{Divide both sides by } -9: \quad y = \frac{-21}{-9}$$
$$y = \frac{7}{3}$$

Now we substitute $\dfrac{7}{3}$ for y into $x = \dfrac{3}{2}y - 3$ to find the value of x.

$$x = \frac{3}{2}\left(\frac{7}{3}\right) - 3$$

Multiply: $x = \dfrac{7}{2} - 3$

$$x = \frac{1}{2}$$

Check We check our answer that $x = \dfrac{1}{2}$ and $y = \dfrac{7}{3}$.

Equation (1): $2x - 3y = -6$ Equation (2): $-8x + 3y = 3$

$x = \dfrac{1}{2}, y = \dfrac{7}{3}$: $2\left(\dfrac{1}{2}\right) - 3\left(\dfrac{7}{3}\right) \overset{?}{=} -6$ $-8\left(\dfrac{1}{2}\right) + 3\left(\dfrac{7}{3}\right) \overset{?}{=} 3$

$1 - 7 \overset{?}{=} -6$ $-4 + 7 \overset{?}{=} 3$

$-6 = -6$ True $3 = 3$ True

Both equations are satisfied so the solution is the ordered pair $\left(\dfrac{1}{2}, \dfrac{7}{3}\right)$.

QUICK ✔ *Solve the system using substitution.*

4. $\begin{cases} y = -3x - 5 \\ 5x + 3y = 1 \end{cases}$ **5.** $\begin{cases} 2x + y = -2 \\ -3x - 2y = -2 \end{cases}$

Work Smart
Substitution is a method to use if one of the variables has a coefficient of 1 or if one of the variables is already solved for; otherwise use elimination.

(4) ## Solve a System of Two Linear Equations Containing Two Unknowns by Elimination

Using substitution to solve the system in Example 4 led to some rather complicated equations containing fractions. A second algebraic method for solving a system of linear equations is the *method of elimination*. This method is usually preferred over the method of substitution if substitution leads to fractions.

The basic idea in using elimination is to get the coefficients of one of the variables to be additive inverses, such as 5 and −5, so that we can add the equations together and get a single linear equation involving one unknown. Remember that this is the same goal we had when using the method of substitution.

In Words
"Back-substitute" means to "plug in" the known value of the variable into one of the equations in the system.

Steps for Solving a System of Linear Equations by Elimination

Step 1: Multiply both sides of one or both equations by a nonzero constant so that the coefficients of one of the variables are additive inverses.

Step 2: Add equations (1) and (2) to eliminate the variable whose coefficients are now additive inverses. Solve the resulting equation for the unknown.

Step 3: Back-substitute the value of the variable found in Step 2 into one of the original equations to find the value of the remaining variable.

Step 4: Check your answer.

What allows us to add two equations and use the result to replace an equation? Remember, an equation is a statement that the left side equals the right side. When we add equation (2) to equation (1), we are adding the same quantity to both sides of equation (1).

EXAMPLE 5 **Solving a System of Linear Equations by Elimination**

Solve: $\begin{cases} \dfrac{5}{2}x + 2y = 5 & (1) \\[2mm] \dfrac{3}{2}x + \dfrac{3}{2}y = \dfrac{9}{4} & (2) \end{cases}$

Solution

Because both equations (1) and (2) have fractions, our first goal is to get rid of the fractions by multiplying both sides of equation (1) by 2 and both sides of equation (2) by 4.

Work Smart

If you are not afraid of fractions, you could eliminate x by multiplying both sides of equation (1) by -3 and both sides of equation (2) by 5. Try it!

$$\begin{cases} \dfrac{5}{2}x + 2y = 5 & (1) \\[2mm] \dfrac{3}{2}x + \dfrac{3}{2}y = \dfrac{9}{4} & (2) \end{cases}$$

Multiply both sides of (1) by 2: $\quad \begin{cases} 2\left(\dfrac{5}{2}x + 2y\right) = 2 \cdot 5 & (1) \\[3mm] \end{cases}$

Multiply both sides of (2) by 4: $\quad \begin{cases} 4\left(\dfrac{3}{2}x + \dfrac{3}{2}y\right) = 4 \cdot \dfrac{9}{4} & (2) \end{cases}$

$$\begin{cases} 5x + 4y = 10 & (1) \\ 6x + 6y = 9 & (2) \end{cases}$$

Work Smart

Although it is not necessary to divide both sides of equation (2) by 3, it makes solving the problem easier. Do you see why?

Divide both sides of equation (2) by 3: $\quad \begin{cases} 5x + 4y = 10 & (1) \\ 2x + 2y = 3 & (2) \end{cases}$

Multiply both sides of equation (2) by -2: $\quad \begin{cases} 5x + 4y = 10 & (1) \\ -4x - 4y = -6 & (2) \end{cases}$

Add (1) and (2): $\quad \overline{\ x = 4}$

Back-substitute 4 for x into the original equation (1).

Equation (1): $\qquad \dfrac{5}{2}x + 2y = 5$

$x = 4$: $\qquad \dfrac{5}{2}(4) + 2y = 5$

$\qquad 10 + 2y = 5$

Subtract 10 from both sides: $\qquad 2y = -5$

Divide both sides by 2: $\qquad y = -\dfrac{5}{2}$

We have that $x = 4$ and $y = -\dfrac{5}{2}$.

We leave the check to you. Both equations are satisfied. The solution is the ordered pair $\left(4, -\dfrac{5}{2}\right)$.

QUICK ✔ *Solve the system using elimination.*

6. $\begin{cases} -2x + y = 4 \\ -5x + 3y = 7 \end{cases}$ **7.** $\begin{cases} -3x + 2y = 3 \\ 4x - 3y = -6 \end{cases}$

SUMMARY: Which Method Should I Use?

We have presented three methods for solving systems of two linear equations containing two unknowns. A question that remains unanswered is "When should I use each method?" Below, we present a summary of the methods, the advantages of each method, and when each method should be used.

Method	Advantages/Disadvantages	When Should I Use It?
Graphical	Allows us to "see" the answer but if the solutions are not integers, it can be difficult to determine the solution.	When a visual solution is required.
Substitution	Method gives exact solutions. The algebra can be easy provided one of the variables has a coefficient of 1. If none of the coefficients are one, the algebra can get messy.	If one of the coefficients of the variables is 1 or one of the variables is already solved for (as in $x =$ or $y =$).
Elimination	Method gives exact solutions. It is easy to use when none of the variables has a coefficient of 1.	If both equations are in standard form ($Ax + By = C$).

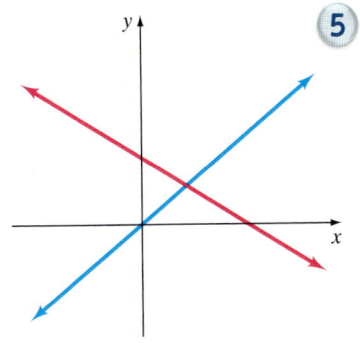

(5) ## Identify Inconsistent Systems

Examples 3–5 dealt only with consistent and independent systems of equations. That is, we only discussed systems of equations with a single solution. Remember that there are two other possibilities for the solution of a system of linear equations: (1) The system could be inconsistent, which means that the lines in the system are parallel, or (2) the system could be consistent, but dependent, which means that the lines in the system are coincident (the same line).

In this next example, we look at an inconsistent system of equations.

EXAMPLE 6   An Inconsistent System

Solve: $\begin{cases} 3x + 2y = 2 & (1) \\ -6x - 4y = 8 & (2) \end{cases}$

Solution

It seems easiest to use the method of elimination to solve this system because none of the variables have a coefficient of 1.

In looking at this system, we can make the coefficients on the variable x additive inverses by multiplying equation (1) by 2.

$$\begin{cases} 3x + 2y = 2 & (1) \\ -6x - 4y = 8 & (2) \end{cases}$$

Multiply (1) by 2: $\begin{cases} 2(3x + 2y) = 2(2) & (1) \\ -6x - 4y = 8 & (2) \end{cases}$

Use the Distributive Property: $\begin{cases} 6x + 4y = 4 & (1) \\ -6x - 4y = 8 & (2) \end{cases}$

Add (1) and (2): $\qquad 0 = 12$

The equation $0 = 12$ is false. We conclude that the system has no solution, so the solution set is $\emptyset$ or $\{\ \}$. The system is inconsistent.

Figure 3

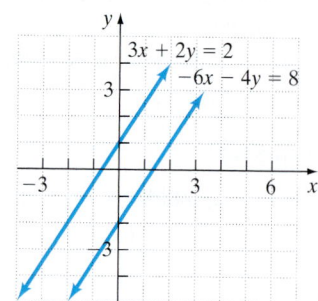

Figure 3 shows the pair of lines whose equations form the system in Example 6. Notice that the graphs of the two equations are lines, each with slope $-\frac{3}{2}$. Equation (1) has a y-intercept of 1, while equation (2) has a y-intercept of -2. Therefore, the lines are parallel and do not intersect. This geometric statement is equivalent to the algebraic statement that the system has no solution.

QUICK ✓ *Show that the system is inconsistent. Draw a graph to support your result.*

8. $\begin{cases} -3x + y = 2 \\ 6x - 2y = 1 \end{cases}$

---

6 **Express the Solution of a System of Dependent Equations**

The next example illustrates a system with infinitely many solutions.

EXAMPLE 7 **Solving a System of Dependent Equations**

Solve: $\begin{cases} 3x + y = 1 & (1) \\ -6x - 2y = -2 & (2) \end{cases}$

Solution

We choose to use the substitution method because solving equation (1) for y is straightforward.

$$\text{Equation (1):} \quad 3x + y = 1$$
$$\text{Subtract } 3x \text{ from both sides:} \quad y = -3x + 1$$

Substitute $-3x + 1$ for y in equation (2).

$$\text{Equation (2):} \quad -6x - 2y = -2$$
$$-6x - 2(-3x + 1) = -2$$
$$\text{Distribute the } -2\text{:} \quad -6x + 6x - 2 = -2$$
$$-2 = -2$$

The equation $-2 = -2$ is true. This means that as long as y is chosen so that it equals $-3x + 1$, we will have a solution to the system. For example, if $x = 0$, then $y = -3(0) + 1 = 1$; if $x = 1$, then $y = -3(1) + 1 = -2$; if $x = 2$, then $y = -3(2) + 1 = -5$. The system of equations is consistent, but dependent (the value of y that makes the equation true *depends* on the value of x) so that there are infinitely many solutions. We will write the solution as

$$\{(x, y) \mid 3x + y = 1\}$$

In Words

The solution to a dependent system is "the set of all ordered pairs such that one of the equations in the system is true."

Figure 4 illustrates the situation presented in Example 7. The graphs of the two equations are lines, each with slope -3 and each with y-intercept 1. The lines are coincident.

Look back at the equations in Example 7. Notice that the terms in equation (2) are two times the terms in equation (1). This is another way to identify dependent systems when you have two equations with two unknowns.

Figure 4

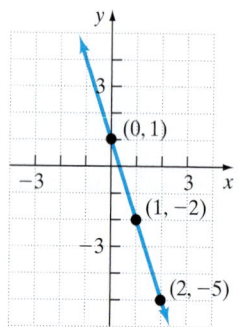

QUICK ✓ *Solve the system. Draw a graph to support your result.*

9. $\begin{cases} -3x + 2y = 8 \\ 6x - 4y = -16 \end{cases}$

E.1 Exercises

Concepts and Vocabulary

In Problems 1–3, fill in the blanks.

1. A _____ _____ _____ _____ is a grouping of two or more linear equations, each of which contains one or more variables.
2. If a system of equations has no solution, it is said to be _____.
3. If a system of equations has infinitely many solutions, the system is said to be _____ and the equations are _____.

In Problems 4–6, answer True or False to each statement.

4. A system of two linear equations containing two variables always has at least one solution.
5. When the lines in a system of equations are parallel, then the system is inconsistent and has no solution.
6. A method for obtaining an equivalent system of equations is to multiply both sides of an equation in the system by a nonzero constant.
7. In this section, we presented two algebraic methods for solving a system of linear equations. Are there any circumstances where one method is preferable to the other? What are these circumstances?
8. Describe geometrically the three possibilities for a solution to a system of two linear equations containing two variables.
9. The solution to a system of two linear equations in two unknowns is $x = 3$, $y = -2$. Where do the lines in the system intersect?
10. In the process of solving a system of linear equations, what tips you off that the system is consistent but dependent? What tips you off that the system is inconsistent?

Building Skills

In Problems 11–14, determine whether the given ordered pairs listed are solutions of the system of linear equations.

11. $\begin{cases} 2x + y = 13 \\ -5x + 3y = 6 \end{cases}$

    (a) $(5, 3)$ **(b)** $(3, 7)$

12. $\begin{cases} x - 2y = -11 \\ 3x + 2y = -1 \end{cases}$

    (a) $(-5, 3)$ **(b)** $(-3, 4)$

13. $\begin{cases} 5x + 2y = 9 \\ -10x - 4y = -18 \end{cases}$

    (a) $(1, 2)$ **(b)** $\left(2, -\dfrac{1}{2}\right)$

14. $\begin{cases} -3x + y = 5 \\ 6x - 2y = 6 \end{cases}$

    (a) $(-2, -1)$ **(b)** $(2, 0)$

In Problems 15–18, use the graph of the system to determine the solution.

15. $\begin{cases} x + y = 1 \\ x + 2y = 0 \end{cases}$

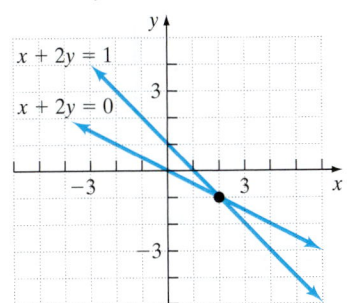

16. $\begin{cases} -2x + y = 4 \\ 2x + y = 0 \end{cases}$

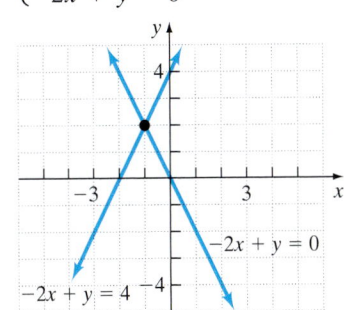

17. $\begin{cases} x - 2y = -2 \\ x - 2y = 2 \end{cases}$

18. $\begin{cases} 3x + y = 1 \\ -6x - 2y = -2 \end{cases}$

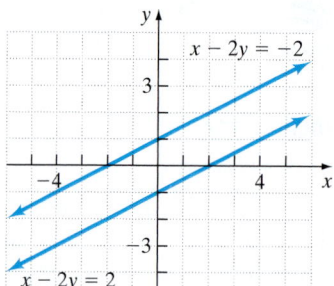

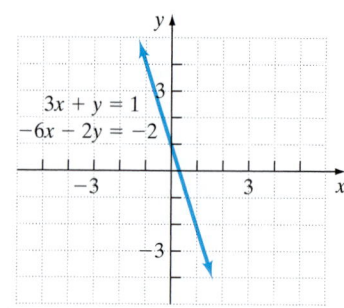

In Problems 19–24, solve the system of equations by graphing.

19. $\begin{cases} y = 3x \\ y = -2x + 5 \end{cases}$

20. $\begin{cases} y = -2x + 4 \\ y = 2x - 4 \end{cases}$

21. $\begin{cases} y = \dfrac{1}{2}x + 1 \\ 2x - 4y = -4 \end{cases}$

22. $\begin{cases} y = -\dfrac{2}{3}x + 3 \\ 2x + 3y = 9 \end{cases}$

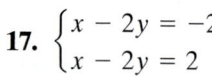 **23.** $\begin{cases} 2x + y = 2 \\ x + 3y = -9 \end{cases}$

24. $\begin{cases} -x + 2y = -9 \\ 2x + y = -2 \end{cases}$

In Problems 25–36, solve the system of equations using substitution.

25. $\begin{cases} y = -\dfrac{1}{2}x + 1 \\ y = -2x + 10 \end{cases}$

26. $\begin{cases} y = -3x - 4 \\ y = 4x + 17 \end{cases}$

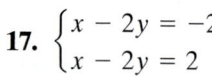 **27.** $\begin{cases} x = \dfrac{2}{3}y \\ 3x - y = -3 \end{cases}$

28. $\begin{cases} y = \dfrac{1}{2}x \\ x - 4y = -4 \end{cases}$

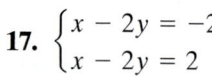 **29.** $\begin{cases} 3x + y = 1 \\ -6x - 2y = -4 \end{cases}$

30. $\begin{cases} -2x + 4y = 9 \\ x - 2y = -3 \end{cases}$

31. $\begin{cases} 2x - 4y = 2 \\ x + 2y = 0 \end{cases}$

32. $\begin{cases} 3x + 2y = 0 \\ 6x + 2y = 5 \end{cases}$

33. $\begin{cases} x + 3y = 6 \\ -\dfrac{x}{3} - y = -2 \end{cases}$

34. $\begin{cases} -4x + y = 8 \\ x - \dfrac{y}{4} = -2 \end{cases}$

35. $\begin{cases} x + y = 10{,}000 \\ 0.05x + 0.07y = 650 \end{cases}$

36. $\begin{cases} x + y = 5000 \\ 0.04x + 0.08y = 340 \end{cases}$

In Problems 37–48, solve the system of equations using elimination.

37. $\begin{cases} x + y = -5 \\ -x + 2y = 14 \end{cases}$

38. $\begin{cases} x + y = -6 \\ -2x - y = 0 \end{cases}$

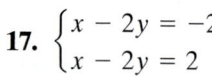 **39.** $\begin{cases} x + 2y = -5 \\ 3x + 3y = 9 \end{cases}$

40. $\begin{cases} -3x + 2y = -5 \\ 2x - y = 10 \end{cases}$

41. $\begin{cases} 5x - 2y = 2 \\ -10x + 4y = 3 \end{cases}$

42. $\begin{cases} 6x - 4y = 6 \\ -3x + 2y = 3 \end{cases}$

43. $\begin{cases} 2x + 5y = -3 \\ x + \dfrac{5}{4}y = -\dfrac{1}{2} \end{cases}$

44. $\begin{cases} x + 2y = -\dfrac{8}{3} \\ 3x - 3y = 5 \end{cases}$

45. $\begin{cases} \dfrac{1}{3}x - 2y = 6 \\ -\dfrac{1}{2}x + 3y = -9 \end{cases}$

46. $\begin{cases} \dfrac{5}{4}x - \dfrac{1}{2}y = 6 \\ -\dfrac{5}{3}x + \dfrac{2}{3}y = -8 \end{cases}$

47. $\begin{cases} 0.05x + 0.1y = 5.25 \\ 0.08x - 0.02y = 1.2 \end{cases}$

48. $\begin{cases} 0.04x + 0.06y = 2.1 \\ 0.06x - 0.03y = 0.15 \end{cases}$

Mixed Practice

In Problems 49–56, solve the system of equations using either substitution or elimination.

49. $\begin{cases} x + 3y = 0 \\ -2x + 4y = 30 \end{cases}$

50. $\begin{cases} 2x + y = -1 \\ -3x - 2y = 7 \end{cases}$

51. $\begin{cases} x = 5y - 3 \\ -3x + 15y = 9 \end{cases}$

52. $\begin{cases} y = \dfrac{1}{2}x + 2 \\ x - 2y = -4 \end{cases}$

53. $\begin{cases} 2x - 4y = 18 \\ 3x + 5y = -3 \end{cases}$

54. $\begin{cases} 12x + 45y = 0 \\ 8x + 6y = 24 \end{cases}$

55. $\begin{cases} \dfrac{5}{6}x - \dfrac{1}{3}y = -5 \\ -x + \dfrac{2}{5}y = 1 \end{cases}$

56. $\begin{cases} \dfrac{1}{3}x - \dfrac{1}{2}y = -5 \\ -\dfrac{4}{5}x + \dfrac{6}{5}y = 1 \end{cases}$

Applying the Concepts

In Problems 57–60, write each equation in the system of equations in slope-intercept form. Use the slope-intercept form to determine the number of solutions the system has.

57. $\begin{cases} 2x + y = -5 \\ 5x + 3y = 1 \end{cases}$

58. $\begin{cases} 4x - 2y = 8 \\ -10x + 5y = 5 \end{cases}$

59. $\begin{cases} 3x - 2y = -2 \\ -6x + 4y = 4 \end{cases}$

60. $\begin{cases} 2x - y = -5 \\ -4x + 3y = 9 \end{cases}$

△ **61. Parallelogram** Use the parallelogram shown to the right to answer parts (a) and (b).

    (a) Find the equation of the line for the diagonal through the points $(-1, 3)$ and $(3, 1)$. Find the equation for the line through the diagonal through $(-2, -1)$ and $(4, 5)$.

    (b) Find the point of intersection of the diagonals.

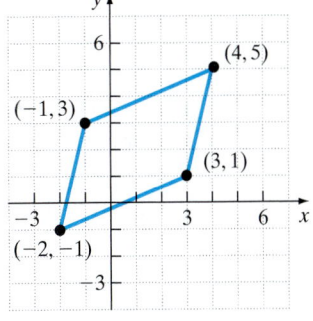

△ **62. Rhombus** A rhombus is a parallelogram whose adjacent sides are congruent. Use the rhombus to the right to answer parts (a), (b), and (c).

    (a) Find the equation of the line for the diagonal through the points $(-1, 3)$ and $(3, 1)$. Find the equation for the line through the diagonal through $(-1, -2)$ and $(3, 6)$.

    (b) Find the point of intersection of the diagonals.

    (c) Compare the slopes of the diagonals. What can be said about the diagonals?

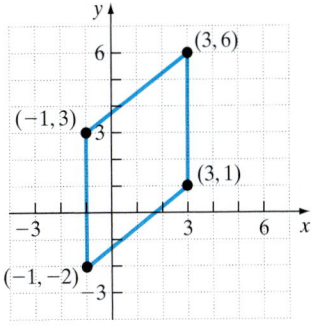

Extending the Concepts

63. Which of the following ordered pairs could be a solution to the system graphed to the right?

(a) $(2, 4)$
(b) $(-2, 0)$
(c) $(-3, 1)$
(d) $(5, -2)$
(e) $(-1, -3)$
(f) $(-1, 3)$

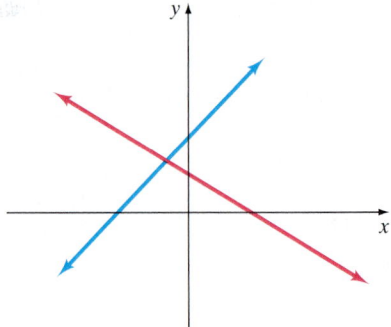

64. Which of the following systems of equations could have the graph to the right?

(a) $\begin{cases} 2x + 3y = 12 \\ 2x + y = -2 \end{cases}$

(b) $\begin{cases} 2x + 3y = 3 \\ -2x + y = 2 \end{cases}$

(c) $\begin{cases} 2x - 3y = 12 \\ x + 2y = 2 \end{cases}$

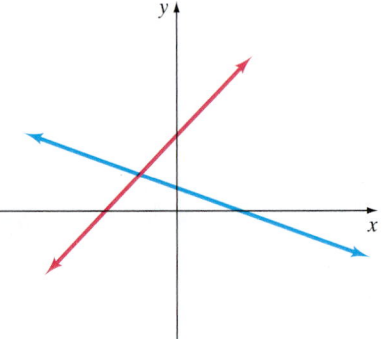

65. For the system $\begin{cases} Ax + 3By = 2 \\ -3Ax + By = -11 \end{cases}$, find A and B such that $x = 3$, $y = 1$ is a solution.

66. Write a system of equations that has $(3, 5)$ as a solution.

67. Write a system of equations that has $(-1, 4)$ as a solution.

△ **68. Centroid** The medians of a triangle are the line segments from each vertex to the midpoint of the opposite side. The centroid of a triangle is the point where the medians of the triangle intersect. Use the information given in the figure of the triangle to find its centroid.

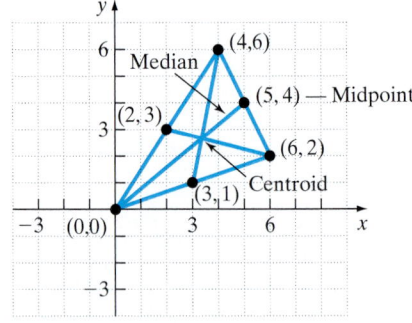

The Graphing Calculator

A graphing calculator can be used to approximate the point of intersection between two equations using its INTERSECT command. We illustrate this feature of the graphing calculator by doing Example 3. Start by graphing each equation in the system as shown in Figure 5(a). Then use the INTERSECT command and find that the lines intersect at $x = 2$, $y = -3$. See Figure 5(b). The solution is the ordered pair $(2, -3)$.

Figure 5

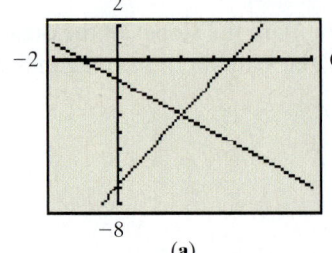

(a)

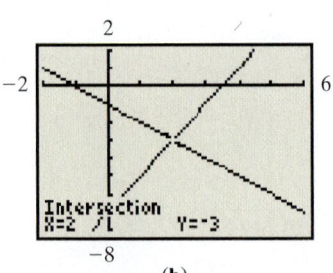

(b)

In Problems 69–76, use a graphing calculator to solve each system of equations. If necessary, express your solution rounded to two decimal places.

69. $\begin{cases} y = 3x - 1 \\ y = -2x + 5 \end{cases}$

70. $\begin{cases} y = \dfrac{3}{2}x - 4 \\ y = -\dfrac{1}{4}x + 3 \end{cases}$

71. $\begin{cases} 3x - y = -1 \\ -4x + y = -3 \end{cases}$

72. $\begin{cases} -6x - 2y = 4 \\ 5x + 3y = -2 \end{cases}$

73. $\begin{cases} 4x - 3y = 1 \\ -8x + 6y = -2 \end{cases}$

74. $\begin{cases} -2x + 5y = -2 \\ 4x - 10y = 1 \end{cases}$

75. $\begin{cases} 2x - 3y = 12 \\ 5x + y = -2 \end{cases}$

76. $\begin{cases} x - 3y = 21 \\ x + 6y = -2 \end{cases}$

E.2 Systems of Linear Equations in Three Variables

OBJECTIVES

1. Solve Systems of Three Linear Equations Containing Three Variables
2. Identify Inconsistent Systems
3. Express the Solution of a System of Dependent Equations
4. Model and Solve Problems Involving Three Linear Equations Containing Three Unknowns

Preparing for Systems of Linear Equations in Three Variables

1. Evaluate the expression $3x - 2y + 4z$ for
$x = 1$, $y = -2$, and $z = 3$ [Section 1.7, pp. 57–58]

1 Solve Systems of Three Linear Equations Containing Three Variables

An example of a linear equation in three variables is $2x - y + z = 8$. An example of a system of three linear equations containing three variables is

Three equations containing three variables, *x*, *y*, and *z* $\begin{cases} x + 3y + z = 8 \\ 3x - y + 6z = 12 \\ -4x - y + 2z = -1 \end{cases}$

Systems of three linear equations containing three variables have the same possible solutions as a system of two linear equations containing two variables:

1. **Exactly one solution**—A consistent system with independent equations
2. **No solution**—An inconsistent system
3. **Infinitely many solutions**—A consistent system with dependent equations

We can view the problem of solving a system of three linear equations containing three variables as a geometry problem. The graph of each equation in a system of linear equations containing three variables is a plane in space. A system of three linear equations containing three variables represents three planes in space. Figure 5 illustrates some of the possibilities.

Figure 5

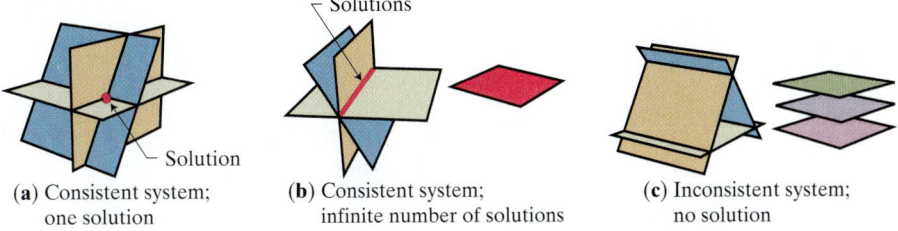

(**a**) Consistent system; one solution

(**b**) Consistent system; infinite number of solutions

(**c**) Inconsistent system; no solution

Recall that a **solution** to a system of equations consists of values for the variables that are solutions of each equation of the system. We write the solution to a system of three equations containing three unknowns as an **ordered triple** (x, y, z).

EXAMPLE 1 **Determining Whether Values Are a Solution to a System of Linear Equations**

Determine which of the following ordered triples are solutions to the system of equations.

$$\begin{cases} x + y + z = 0 \\ 2x - y + 3z = 17 \\ -3x + 2y - z = -21 \end{cases}$$

(a) $(1, 3, -4)$ **(b)** $(3, -5, 2)$

Solution

First, we name $x + y + z = 0$ equation (1), $2x - y + 3z = 17$ equation (2), and $-3x + 2y - z = -21$ equation (3) as follows:

$$\begin{cases} x + y + z = 0 & (1) \\ 2x - y + 3z = 17 & (2) \\ -3x + 2y - z = -21 & (3) \end{cases}$$

(a) Let $x = 1$, $y = 3$, and $z = -4$ in equations (1), (2), and (3). If all three equations are true, then $(1, 3, -4)$ is a solution.

Equation (1): $x + y + z = 0$ Equation (2): $2x - y + 3z = 17$

$1 + 3 + (-4) \stackrel{?}{=} 0$ $2(1) - 3 + 3(-4) \stackrel{?}{=} 17$

$0 = 0$ True $2 - 3 - 12 \stackrel{?}{=} 17$

$-13 = 17$ False

Equation (3): $-3x + 2y - z = -21$

$-3(1) + 2(3) - (-4) \stackrel{?}{=} -21$

$-3 + 6 + 4 \stackrel{?}{=} -21$

$7 = -21$ False

Although these values satisfy equation (1), they do not satisfy equations (2) or (3). Because the values $x = 1$, $y = 3$, and $z = -4$ do not satisfy equations (2) or (3), $(1, 3, -4)$ is not a solution.

(b) Let $x = 3$, $y = -5$, and $z = 2$ in equations (1), (2), and (3). If all three equations are true, then $(3, -5, 2)$ is a solution.

Equation (1): $x + y + z = 0$ Equation (2): $2x - y + 3z = 17$

$3 + (-5) + (2) \stackrel{?}{=} 0$ $2(3) - (-5) + 3(2) \stackrel{?}{=} 17$

$0 = 0$ True $6 + 5 + 6 \stackrel{?}{=} 17$

$17 = 17$ True

Equation (3): $-3x + 2y - z = -21$

$-3(3) + 2(-5) - (2) \stackrel{?}{=} -21$

$-9 - 10 - 2 \stackrel{?}{=} -21$

$-21 = -21$ True

Because the values $x = 3$, $y = -5$, and $z = 2$ satisfy all three equations, the ordered triple $(3, -5, 2)$ is a solution.

QUICK ✓

1. Determine which of the following ordered triples are solutions to the system of equations.

$$\begin{cases} x + y + z = 3 \\ 3x + y - 2z = -23 \\ -2x - 3y + 2z = 17 \end{cases}$$

(a) $(3, 2, -2)$ **(b)** $(-4, 1, 6)$

Typically, when solving a system of three linear equations containing three variables, we use the method of elimination. Recall that the idea behind the method of elimination is to eliminate variables from the system. We eliminate variables by multiplying equations by nonzero constants in order to get the coefficients of the variables to be additive inverses. We then add the equations to remove the variable that has coefficients that are additive inverses. Other methods that we can use are to interchange any two equations or multiply (or divide) each side of an equation by the same non-zero constant.

The process for solving a system of three linear equations containing three unknowns is an extension of the elimination method presented in Section 4.3. Let's look at an example to see how to solve a system of three linear equations with three variables.

EXAMPLE 2   How to Solve a System of Three Linear Equations with Three Variables

Use the method of elimination to solve the system: $\begin{cases} x + y - z = -1 & (1) \\ 2x - y + 2z = 8 & (2) \\ -3x + 2y + z = -9 & (3) \end{cases}$

Step-by-Step Solution

Step 1: Our goal is to eliminate the same variable from two of the equations. In looking at the system, we notice that we can use equation (1) to eliminate the variable x from equations (2) and (3). We can do this by multiplying equation (1) by -2 and adding the result to equation (2). The equation that results becomes equation (4). Why do we do this? Because the coefficients on x are now additive inverses and adding the equations eliminates the variable x. We also multiply equation (1) by 3 and add the result to equation (3). The equation that results becomes equation (5).

$\begin{array}{l} x + y - z = -1 \quad (1) \\ 2x - y + 2z = 8 \quad (2) \end{array}$

Multiply (1) by –2: $\begin{array}{r} -2x - 2y + 2z = 2 \quad (1) \\ 2x - y + 2z = 8 \quad (2) \end{array}$

Add: $\quad -3y + 4z = 10$

$\begin{array}{l} x + y - z = -1 \quad (1) \\ -3x + 2y + z = -9 \quad (3) \end{array}$

Multiply (1) by 3: $\begin{array}{r} 3x + 3y - 3z = -3 \quad (1) \\ -3x + 2y + z = -9 \quad (3) \end{array}$

Add: $\quad 5y - 2z = -12$

$\begin{cases} x + y - z = -1 & (1) \\ -3y + 4z = 10 & (4) \\ 5y - 2z = -12 & (5) \end{cases}$

Step 2: We now concentrate on equations (4) and (5), treating them as a system of two equations containing two variables. It is easiest to eliminate the variable z by multiplying equation (5) by 2 and then add equations (4) and (5). The result will be equation (6).

$\begin{array}{l} -3y + 4z = 10 \quad (4) \\ 5y - 2z = -12 \quad (5) \end{array}$

Multiply (5) by 2: $\begin{array}{r} -3y + 4z = 10 \quad (4) \\ 10y - 4z = -24 \quad (5) \end{array}$

Add: $\quad 7y = -14$

$\begin{cases} x + y - z = -1 & (1) \\ -3y + 4z = 10 & (4) \\ 7y = -14 & (6) \end{cases}$

(continued)

Step 3: We solve equation (6) for y by dividing both sides of the equation by 7.

$$\begin{cases} x + y - z = -1 & (1) \\ -3y + 4z = 10 & (4) \\ y = -2 & (6) \end{cases}$$

Step 4: Back-substitute -2 for y in equation (4) and solve for z.

$$-3y + 4z = 10$$
$$-3(-2) + 4z = 10$$
$$6 + 4z = 10$$
$$4z = 4$$
$$z = 1$$

Step 5: Back-substitute -2 for y and 1 for z into equation (1) and solve for x.

$$x + y - z = -1$$
$$x + (-2) - 1 = -1$$
$$x - 3 = -1$$
$$x = 2$$

The solution appears to be $x = 2$, $y = -2$, $z = 1$.

Step 6: Check Verify that $x = 2$, $y = -2$, and $z = 1$ is the solution.

Equation (1): $x = 2, y = -2, z = 1$:

$$x + y - z = -1$$
$$2 + (-2) - 1 \overset{?}{=} -1$$
$$-1 = -1 \quad \text{True}$$

Equation (2):

$$2x - y + 2z = 8$$
$$2(2) - (-2) + 2(1) \overset{?}{=} 8$$
$$4 + 2 + 2 \overset{?}{=} 8$$
$$8 = 8 \quad \text{True}$$

Equation (3):

$$-3x + 2y + z = -9$$
$$-3(2) + 2(-2) + 1 \overset{?}{=} -9$$
$$-6 - 4 + 1 \overset{?}{=} -9$$
$$-9 = -9 \quad \text{True}$$

The solution is the ordered triple $(2, -2, 1)$.

We now summarize the steps used in Example 1.

> ## Steps for Solving a System of Three Linear Equations Containing Three Unknowns by Elimination
>
> **Step 1:** Select two of the equations and eliminate one of the variables from one of the equations. Select any two other equations and eliminate the *same variable* from one of the equations.
>
> **Step 2:** You will have two equations that have only two unknowns. Eliminate a second variable from the two linear equations in two unknowns.
>
> **Step 3:** Solve for the remaining variable.
>
> **Step 4:** Use the value of the variable found in Step 3 to find the value of a second variable.
>
> **Step 5:** Use the two known values of the variables identified in Steps 3 and 4 to find the value of the third variable.
>
> **Step 6:** Check your answer.

Work Smart

By eliminating a variable in two of the equations in Step 1, we are creating a system of 2 equations with 2 unknowns that can be solved. Remember, this is the goal in mathematics—reduce the problem to one you already know how to solve.

QUICK ✓

2. Use the method of elimination to solve the system:
$$\begin{cases} x + y + z = -3 \\ 2x - 2y - z = -7 \\ -3x + y + 5z = 5 \end{cases}$$

EXAMPLE 3 Solving a System of Three Linear Equations with Three Variables

Use the method of elimination to solve the system:
$$\begin{cases} 4x \qquad + z = 4 & (1) \\ 2x + 3y \qquad = -4 & (2) \\ 2y - 4z = -15 & (3) \end{cases}$$

Solution

We eliminate z from equation (3) by multiplying equation (1) by 4 and adding the result to equation (3). The equation that results becomes equation (4).

$$\begin{array}{ll} 4x \quad + z = 4 & (1) \\ 2y - 4z = -15 & (3) \end{array}$$
Multiply (1) by 4:
$$\begin{array}{ll} 16x \quad + 4z = 16 & (1) \\ \underline{2y - 4z = -15} & (3) \\ 16x + 2y \quad = 1 \end{array} \longrightarrow$$
$$\begin{cases} 4x \qquad + z = 4 & (1) \\ 2x + 3y \qquad = -4 & (2) \\ 16x + 2y \qquad = 1 & (4) \end{cases}$$

We now concentrate on equations (2) and (4), treating them as a system of two equations containing two variables. It is easiest to eliminate the variable x by multiplying equation (2) by -8 and then add equations (2) and (4). The result will be equation (5).

$$\begin{array}{ll} 2x + 3y = -4 & (2) \\ 16x + 2y = 1 & (4) \end{array}$$
Multiply (2) by -8:
$$\begin{array}{ll} -16x - 24y = 32 & (2) \\ \underline{16x + 2y = 1} & (4) \\ -22y = 33 \end{array} \longrightarrow$$
$$\begin{cases} 4x \qquad + z = 4 & (1) \\ 2x + 3y \qquad = -4 & (2) \\ \qquad - 22y \qquad = 33 & (5) \end{cases}$$

Work Smart

We chose to eliminate z from equation (3) because the coefficient of z in equation (1) is 1. We could also have eliminated x from equation (2) by multiplying both sides of equation (2) by -2 and then added (1) and (2). There is more than one approach to solving the system!

We solve equation (5) for y by dividing both sides of the equation by -22.

Equation (5): $-22y = 33$
$$y = \frac{33}{-22} = -\frac{3}{2}$$

Now we back-substitute $-\dfrac{3}{2}$ for y into equation (2) and solve for x.

Equation (2): $2x + 3y = -4$
$$2x + 3\left(-\frac{3}{2}\right) = -4$$
$$2x - \frac{9}{2} = -4$$
$$2x = \frac{1}{2}$$
$$x = \frac{1}{4}$$

Now back-substitute $\dfrac{1}{4}$ for x into equation (1) and solve for z.

$$\text{Equation (1):} \qquad 4x + z = 4$$
$$4\left(\dfrac{1}{4}\right) + z = 4$$
$$1 + z = 4$$
$$z = 3$$

The solution is the ordered triple $\left(\dfrac{1}{4}, -\dfrac{3}{2}, 3\right)$. ▬

QUICK ✓

3. Use the method of elimination to solve the system: $\begin{cases} 2x \qquad -4z = -7 \\ x + 6y \qquad = 5 \\ \qquad 2y - z = 2 \end{cases}$

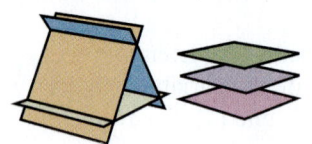

② Identify Inconsistent Systems

Examples 2 and 3 were consistent and independent systems resulting in a single solution. We now look at an inconsistent system.

EXAMPLE 4 An Inconsistent System of Linear Equations

Use the method of elimination to solve the system: $\begin{cases} x + 2y - z = 4 & (1) \\ -2x + 3y + z = -4 & (2) \\ x + 9y - 2z = 1 & (3) \end{cases}$

Solution

In looking at the system, notice that we can use equation (1) to eliminate the variable x from equations (2) and (3). We can do this by multiplying equation (1) by 2 and adding the result to equation (2). The equation that results becomes equation (4). We also multiply equation (1) by -1 and add the result to equation (3). The equation that results becomes equation (5).

$$\begin{array}{ll} x + 2y - z = 4 & (1) \\ -2x + 3y + z = -4 & (2) \end{array} \qquad \text{Multiply by 2:} \qquad \begin{array}{ll} 2x + 4y - 2z = 8 & (1) \\ \underline{-2x + 3y + z = -4} & (2) \\ 7y - z = 4 \end{array}$$

$$\begin{array}{ll} x + 2y - z = 4 & (1) \\ x + 9y - 2z = 1 & (3) \end{array} \qquad \text{Multiply by } -1: \qquad \begin{array}{ll} -x - 2y + z = -4 & (1) \\ \underline{x + 9y - 2z = 1} & (3) \\ 7y - z = -3 \end{array} \qquad \begin{cases} x + 2y - z = 4 & (1) \\ 7y - z = 4 & (4) \\ 7y - z = -3 & (5) \end{cases}$$

We now concentrate on equations (4) and (5), treating them as a system of two equations containing two variables. We multiply equation (4) by -1 and then add equations (4) and (5). The result will be equation (6).

$$\begin{array}{ll} 7y - z = 4 & (4) \\ 7y - z = -3 & (5) \end{array} \qquad \text{Multiply by } -1: \qquad \begin{array}{ll} -7y + z = -4 & (4) \\ \underline{7y - z = -3} & (5) \\ 0 = -7 \end{array} \qquad \begin{cases} x + 2y - z = 4 & (1) \\ 7y - z = 4 & (4) \\ 0 = -7 & (6) \quad \text{False} \end{cases}$$

Equation (6) now states that $0 = -7$, which is a false statement. Therefore, the system is inconsistent. The solution set is $\varnothing$ or $\{\ \}$. ▬

Quick ✓

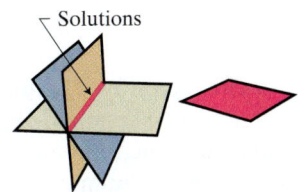
Solutions

4. Use the method of elimination to solve the system: $\begin{cases} x - y + 2z = -7 \\ -2x + y - 3z = 5 \\ x - 2y + 3z = 2 \end{cases}$

③ Express the Solution of a System of Dependent Equations

Now we look at a system of dependent equations.

EXAMPLE 5   Solving a System of Dependent Equations

Use the method of elimination to solve the system: $\begin{cases} x - 3y - z = 4 & (1) \\ x - 2y + 2z = 5 & (2) \\ 2x - 5y + z = 9 & (3) \end{cases}$

Solution

We can use equation (1) to eliminate the variable x from equations (2) and (3). We can do this by multiplying equation (1) by -1 and adding the result to equation (2). The equation that results becomes equation (4). We also multiply equation (1) by -2 and add the result to equation (3). The equation that results becomes equation (5).

$$\begin{array}{ll} x - 3y - z = 4 & (1) \\ x - 2y + 2z = 5 & (2) \end{array} \quad \text{Multiply by } -1: \quad \begin{array}{ll} -x + 3y + z = -4 & (1) \\ \underline{x - 2y + 2z = 5} & (2) \\ y + 3z = 1 \end{array}$$

$$\begin{array}{ll} x - 3y - z = 4 & (1) \\ 2x - 5y + z = 9 & (3) \end{array} \quad \text{Multiply by } -2: \quad \begin{array}{ll} -2x + 6y + 2z = -8 & (1) \\ \underline{2x - 5y + z = 9} & (3) \\ y + 3z = 1 \end{array} \quad \begin{cases} x - 3y - z = 4 & (1) \\ y + 3z = 1 & (4) \\ y + 3z = 1 & (5) \end{cases}$$

We now concentrate on equations (4) and (5), treating them as a system of two equations containing two variables. We multiply equation (4) by -1 and then add equations (4) and (5). The result will be equation (6).

$$\begin{array}{ll} y + 3z = 1 & (4) \\ y + 3z = 1 & (5) \end{array} \quad \text{Multiply by } -1: \quad \begin{array}{ll} -y - 3z = -1 & (4) \\ \underline{y + 3z = 1} & (5) \\ 0 = 0 \end{array} \quad \begin{cases} x - 3y - z = 4 & (1) \\ y + 3z = 1 & (4) \\ 0 = 0 & (6) \end{cases}$$

The statement $0 = 0$ in equation (6) indicates that we have a dependent system, so the system has infinitely many solutions. We can express how the values for x and y *depend* on the value of z by letting z represent any real number. Then, we can solve equation (4) for y and obtain y in terms of z.

$$\begin{array}{lr} \text{Equation (4):} & y + 3z = 1 \\ \text{Subtract } 3z \text{ from both sides:} & y = -3z + 1 \end{array}$$

Let $y = -3z + 1$ in equation (1) and solve for x in terms of z.

$$\begin{array}{lr} \text{Equation (1):} & x - 3y - z = 4 \\ \text{Let } y = -3z + 1 \text{ in (1):} & x - 3(-3z + 1) - z = 4 \\ \text{Distribute:} & x + 9z - 3 - z = 4 \\ \text{Combine like terms:} & x + 8z - 3 = 4 \\ \text{Subtract } 8z \text{ from both sides; Add 3 to both sides:} & x = -8z + 7 \end{array}$$

The solution to the system is $\{(x, y, z) \mid x = -8z + 7, y = -3z + 1, z \text{ is any real number}\}$. To find specific solutions to the system, choose any value of z and use the equations $x = -8z + 7$ and $y = -3z + 1$ to determine x and y. Some specific solutions to the system are $x = 7, y = 1, z = 0$; $x = -1, y = -2, z = 1$; or $x = 15, y = 4, z = -1$.

QUICK ✅

5. Use the method of elimination to solve the system: $\begin{cases} x - y + 3z = 2 \\ -x + 2y - 5z = -3 \\ 2x - y + 4z = 3 \end{cases}$

If you look back at Examples 2–5, you will notice that we did not always eliminate x first and y second. The order in which variables are eliminated from a system does not matter and different approaches to solving the problem will lead to the right answer if done correctly. For example, in Example 5 we could have chosen to eliminate z from equations (1) and (3). We would have ended up with the same solution.

④ **Model and Solve Problems Involving Three Linear Equations Containing Three Unknowns**

Now let's look at problems that can be solved using three equations containing three variables.

EXAMPLE 6 **Production**

A swing-set manufacturer has three different models of swing sets. The Monkey requires 2 hours to cut the wood, 2 hours to stain, and 3 hours to assemble. The Gorilla requires 3 hours to cut the wood, 4 hours to stain, and 4 hours to assemble. The King Kong requires 4 hours to cut the wood, 5 hours to stain, and 5 hours to assemble. The company has 61 hours available to cut the wood, 73 hours available to stain, and 83 hours available to assemble each day. How many of each type of swing set can be manufactured each day?

Solution

Step 1: Identify We want to determine the number of Monkey swing sets, Gorilla swing sets, and King Kong swing sets that can be manufactured each day.

Step 2: Name Let m represent the number of Monkey swing sets, g represent the number of Gorilla swing sets, and k represent the number of King Kong swing sets.

Step 3: Translate We organize the information given in Table 4.

Table 4				
	Monkey	**Gorilla**	**King Kong**	**Total Hours Available**
Cut Wood	2	3	4	61
Stain	2	4	5	73
Assemble	3	4	5	83

If we manufacture m Monkey swing sets, then we need $2m$ hours to cut wood. If we manufacture g Gorilla swing sets, then we need $3g$ hours to cut wood. If we manufacture k King Kong swing sets, then we need $4k$ hours to cut wood. There are a total of 61 hours available, so

$$2m + 3g + 4k = 61 \quad \text{Equation (1)}$$

If we manufacture m Monkey swing sets, then we need $2m$ hours to stain. If we manufacture g Gorilla swing sets, then we need $4g$ hours to stain. If we manufacture k King Kong swing sets, then we need $5k$ hours to stain. There are a total of 73 hours available, so

$$2m + 4g + 5k = 73 \quad \text{Equation (2)}$$

If we manufacture m Monkey swing sets, then we need $3m$ hours to assemble. If we manufacture g Gorilla swing sets, then we need $4g$ hours to assemble. If we manufacture k King Kong swing sets, then we need $5k$ hours to assemble. There are a total of 83 hours available, so

$$3m + 4g + 5k = 83 \quad \text{Equation (3)}$$

We combine equations (1), (2), and (3) to form the following system:

$$\begin{cases} 2m + 3g + 4k = 61 \quad (1) \\ 2m + 4g + 5k = 73 \quad (2) \quad \text{The Model} \\ 3m + 4g + 5k = 83 \quad (3) \end{cases}$$

Step 4: Solve If we solve the system of equations found in Step 3, we find that $m = 10$, $g = 7$, and $k = 5$.

Step 5: Check By manufacturing 10 Monkeys, 7 Gorillas, and 5 King Kongs, we need $2(10) + 3(7) + 4(5) = 61$ hours to cut wood; $2(10) + 4(7) + 5(5) = 73$ hours to stain; and $3(10) + 4(7) + 5(5) = 83$ hours to assemble.

Step 6: Answer The company should manufacture 10 Monkey swing sets, 7 Gorilla swing sets, and 5 King Kong swing sets. ▬

Quick

6. The Mowing 'Em Down lawn mower company manufactures three styles of lawn mower. The 21-inch model requires 2 hours to mold, 3 hours for engine manufacturing, and 1 hour to assemble. The 24-inch model requires 3 hours to mold, 3 hours for engine manufacturing, and 1 hour to assemble. The 40-inch riding mower requires 4 hours to mold, 4 hours for engine manufacturing, and 2 hours to assemble. The company has 81 hours available to mold, 95 hours available for engine manufacturing, and 35 hours available to assemble each day. How many of each type of mower can be manufactured each day?

E.2 Exercises

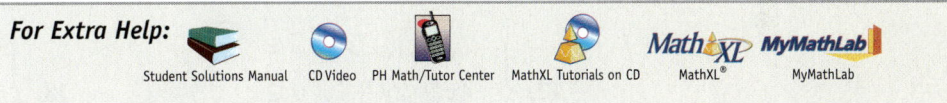

For Extra Help: Student Solutions Manual CD Video PH Math/Tutor Center MathXL Tutorials on CD MathXL® MyMathLab

Concepts and Vocabulary

In Problems 1–3, fill in the blanks.

1. If a system of equations has no solution, it is said to be _____.

2. If a system of equations has infinitely many solutions, the system is said to be _____ and the equations are _____.

3. A _____ to a system of equations consists of values for the variables that are solutions of each equation of the system.

In Problems 4–6, answer True or False to each statement.

4. A system of three linear equations containing three variables always has at least one solution.

5. When the planes in a system of equations are parallel, then the system is inconsistent and has no solution.

6. One of the rules for obtaining an equivalent system of equations allows us to replace any equation in the system by the sum (or difference) of that equation and a nonzero multiple of any other equation in the system.

7. Suppose that $(3, 2, 5)$ is the only solution to a system of three linear equations containing three variables. What does this mean geometrically?

8. Why is it necessary to eliminate the same variable in the first step of the "Steps for solving a system of three linear equations containing three unknowns by elimination" (p. E16)?

Building Skills

In Problems 9 and 10, determine whether the given ordered triples listed are solutions of the system of linear equations.

9. $\begin{cases} x + y + 2z = 6 \\ -2x - 3y + 5z = 1 \\ 2x + y + 3z = 5 \end{cases}$ **(a)** $(6, 2, -1)$ **(b)** $(-3, 5, 2)$

10. $\begin{cases} 2x + y - 2z = 6 \\ -2x + y + 5z = 1 \\ 2x + 3y + z = 13 \end{cases}$ **(a)** $(3, 2, 1)$ **(b)** $(10, -4, 5)$

In Problems 11–22, solve each system of three linear equations containing three unknowns.
(Hint: *Each system has exactly one solution.)*

11. $\begin{cases} x + y + z = 5 \\ -2x - 3y + 2z = 8 \\ 3x - y - 2z = 3 \end{cases}$ **12.** $\begin{cases} x + 2y - z = 4 \\ 2x - y + 3z = 8 \\ -2x + 3y - 2z = 10 \end{cases}$

13. $\begin{cases} x - 3y + z = 13 \\ 3x + y - 4z = 13 \\ -4x - 4y + 2z = 0 \end{cases}$ **14.** $\begin{cases} x + 2y - 3z = -19 \\ 3x + 2y - z = -9 \\ -2x - y + 3z = 26 \end{cases}$

15. $\begin{cases} 2x - y + 2z = 1 \\ -2x + 3y - 2z = 3 \\ 4x - y + 6z = 7 \end{cases}$ **16.** $\begin{cases} x - y + 3z = 2 \\ -2x + 3y - 8z = -1 \\ 2x - 2y + 4z = 7 \end{cases}$

17. $\begin{cases} x - 4y + z = 5 \\ 4x + 2y + z = 2 \\ -4x + y - 3z = -8 \end{cases}$ **18.** $\begin{cases} 2x + 2y - z = -7 \\ x + 2y - 3z = -8 \\ 4x - 2y + z = -11 \end{cases}$

19. $\begin{cases} x - 3y = 12 \\ 2y - 3z = -9 \\ 2x + z = 7 \end{cases}$ **20.** $\begin{cases} 2x + z = -7 \\ 3y - 2z = 17 \\ -4x - y = 7 \end{cases}$

21. $\begin{cases} 2y - z = -3 \\ -2x + 3y = 10 \\ 4x + 3z = -11 \end{cases}$ **22.** $\begin{cases} x - 3z = -3 \\ 3y + 4z = -5 \\ 3x - 2y = 6 \end{cases}$

Mixed Practice

In Problems 23–32, solve each system of equations.

23. $\begin{cases} x - y + z = 5 \\ -2x + y - z = 2 \\ x - 2y + 2z = 1 \end{cases}$ **24.** $\begin{cases} x - y + 2z = 3 \\ 2x + y - 2z = 1 \\ 4x - y + 2z = 0 \end{cases}$

25. $\begin{cases} x - 2y + z = 5 \\ -2x + y - z = 2 \\ x - 5y - 4z = 8 \end{cases}$

26. $\begin{cases} x + 2y - z = -4 \\ -2x + 4y - z = 6 \\ 2x + 2y + 3z = 1 \end{cases}$

27. $\begin{cases} x + 2y - z = 1 \\ 2x + 7y + 4z = 11 \\ x + 3y + z = 4 \end{cases}$

28. $\begin{cases} x + y - 2z = 3 \\ -2x - 3y + z = -7 \\ x + 2y + z = 4 \end{cases}$

29. $\begin{cases} x + y + z = 5 \\ 3x + 4y + z = 16 \\ -x - 4y + z = -6 \end{cases}$

30. $\begin{cases} x + y + z = 4 \\ 2x + 3y - z = 8 \\ x + y - z = 3 \end{cases}$

31. $\begin{cases} x + y + z = 3 \\ -x + \dfrac{1}{2}y + z = \dfrac{1}{2} \\ -x + 2y + 3z = 4 \end{cases}$

32. $\begin{cases} x + \dfrac{1}{2}y + \dfrac{1}{2}z = \dfrac{3}{2} \\ -x + 2y + 3z = 1 \\ 3x + 4y + 5z = 7 \end{cases}$

Applying the Concepts

33. Role Reversal Write a system of three linear equations containing three unknowns that has the solution $(2, -1, 3)$.

34. Role Reversal Write a system of three linear equations containing three unknowns that has the solution $(-4, 1, -3)$.

35. Curve Fitting The function $f(x) = ax^2 + bx + c$ is a quadratic function, where $a, b,$ and c are constants.

   (a) If $f(1) = 4$, then $4 = a(1)^2 + b(1) + c$ or $a + b + c = 4$. Find two additional linear equations if $f(-1) = -6$, and $f(2) = 3$.

   (b) Use the three linear equations found in part (a) to determine $a, b,$ and c. What is the quadratic function that contains the points $(-1, -6), (1, 4),$ and $(2, 3)$?

36. Curve Fitting The function $f(x) = ax^2 + bx + c$ is a quadratic function, where $a, b,$ and c are constants.

   (a) If $f(-1) = 6$, then $6 = a(-1)^2 + b(-1) + c$ or $a - b + c = 6$. Find two additional linear equations if $f(1) = 2$, and $f(2) = 9$.

   (b) Use the three linear equations found in part (a) to determine $a, b,$ and c. What is the quadratic function that contains the points $(-1, 6), (1, 2),$ and $(2, 9)$?

37. Electricity: Kirchhoff's Rules An application of Kirchhoff's Rule to the circuit shown results in the following system of equations:

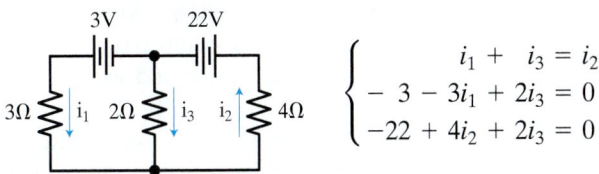

$$\begin{cases} i_1 + i_3 = i_2 \\ -3 - 3i_1 + 2i_3 = 0 \\ -22 + 4i_2 + 2i_3 = 0 \end{cases}$$

In the system circuit V is the voltage, Ω is resistance, and i is the current. Find the currents $i_1, i_2,$ and i_3.

38. Electricity: Kirchhoff's Rules An application of Kirchhoff's Rule to the circuit shown results in the following system of equations:

$$\begin{cases} i_1 + i_3 = i_2 \\ -8 - 5i_1 + 8i_3 = 0 \\ -48 + 6i_2 + 8i_3 = 0 \end{cases}$$

In the system circuit V is the voltage, Ω is resistance, and i is the current. Find the currents $i_1, i_2,$ and i_3.

39. Minor League Baseball In the Joliet Jackhammers baseball stadium, there are three types of seats available. Box seats are $9, reserved seats are $7, and lawn seats are $5. The stadium capacity is 4100. If all the seats are sold, the total revenue to the club is $28,400. If $\frac{1}{2}$ of the box seats are sold, $\frac{1}{2}$ of the reserved seats are sold, and all the lawn seats are sold, the total revenue is $18,300. How many are there of each kind of seat?

40. Theater Revenues A theater has 600 seats, divided into orchestra, main floor, and balcony seating. Orchestra seats sell for $80, main floor seats for $60, and balcony seats for $25. If all the seats are sold, the total revenue to the theater is $33,500. One evening, all the orchestra seats were sold, $\frac{3}{5}$ of the main seats were sold, and only $\frac{4}{5}$ of the balcony seats were sold. The total revenue collected was $24,640. How many are there of each kind of seat?

41. Nutrition Nancy's dietitian wants her to consume 470 mg of sodium, 89 g of carbohydrates, and 20 g of protein for breakfast. This morning, Nancy wants to have Chex® cereal, 2% milk, and orange juice for breakfast. Each serving of Chex® cereal contains 220 mg of sodium, 26 g of carbohydrates, and 1 g of protein. Each serving of 2% milk contains 125 mg of sodium, 12 g of carbohydrates, and 8 g of protein. Each serving of orange juice contains 0 mg of sodium, 26 mg of carbohydrates, and 2 g of protein. How many servings of each does Nancy need?

42. Nutrition Antonio is on a special diet that requires he consume 1325 calories, 172 grams of carbohydrates, and 63 grams of protein for lunch. He goes to Wendy's and wishes to have their Broccoli and Cheese Baked Potato, Chicken BLT Salad, and a medium Coke. Each Broccoli and Cheese Baked Potato has 480 calories, 80 g of carbohydrates, and 9 g of protein. Each Chicken BLT Salad has 310 calories, 10 g of carbohydrates, and 33 g of protein. Each Coke has 140 calories, 37 g of carbohydrates, and 0 g of protein. How many servings of each does Antonio need?

43. Finance Sachi has $25,000 to invest. Her financial planner suggests that she diversify her investment into three investment categories: Treasury bills that yield 3% simple interest annually, municipal bonds that yield 5% simple interest annually, and corporate bonds that yield 9% simple interest annually. Sachi would like to earn $1210 per year in income. In addition, Sachi wants her investment in Treasury bills to be $7000 more than her investment in corporate bonds. How much should Sachi invest in each investment category?

44. Finance Delu has $15,000 to invest. She decides to place some of the money into a savings account paying 2% annual interest, some in Treasury bonds paying 5% annual interest and some in a mutual fund paying 10% annual interest. Delu would like to earn $720 per year in income. In addition, Delu wants her investment in the savings account to be twice the amount in the mutual fund. How much should Delu invest in each investment category?

△ **45. Geometry** A circle is inscribed in $\triangle ABC$ as shown in the figure. Suppose that $AB = 6$, $AC = 14$, and $BC = 12$. Find the length of $\overline{AM}$, $\overline{BN}$, and $\overline{CO}$. (**Hint:** $\overline{AM} \cong \overline{AO}$; $\overline{BM} \cong \overline{BN}$; $\overline{NC} \cong \overline{OC}$)

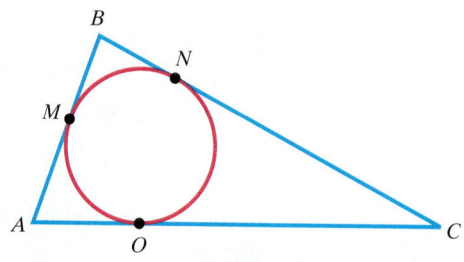

Extending the Concepts

In Problems 46–49, solve each system of equations.

46. $\begin{cases} \dfrac{1}{4}x + \dfrac{1}{4}y + \dfrac{1}{2}z = 6 \\ -\dfrac{1}{8}x + \dfrac{1}{2}y - \dfrac{1}{5}z = -5 \\ \dfrac{1}{2}x + \dfrac{1}{2}y - \dfrac{1}{2}z = -3 \end{cases}$

47. $\begin{cases} \dfrac{2}{5}x + \dfrac{1}{2}y - \dfrac{1}{3}z = 0 \\ \dfrac{3}{5}x - \dfrac{1}{4}y + \dfrac{1}{2}z = 10 \\ -\dfrac{1}{5}x + \dfrac{1}{4}y - \dfrac{1}{6}z = -4 \end{cases}$

48. $\begin{cases} x + y + z + w = 0 \\ 2x - 3y - z + w = -17 \\ 3x + y + 2z - w = 8 \\ -x + 2y - 3z + 2w = -7 \end{cases}$

49. $\begin{cases} x + y + z + w = 3 \\ -2x - y + 3z - w = -1 \\ 2x + 2y - 2z + w = 2 \\ -x + 2y - 3z + 2w = 12 \end{cases}$

E.3 Using Matrices to Solve Systems

OBJECTIVES

1. Write the Augmented Matrix of a System of Linear Equations
2. Write the System from the Augmented Matrix
3. Perform Row Operations on a Matrix
4. Solve Systems of Linear Equations Using Matrices
5. Solve Dependent and Inconsistent Systems

Preparing for Using Matrices to Solve Systems

Before getting started, take the following readiness quiz. If you get a problem wrong, go back to the section cited and review the material.

1. Determine the coefficients of the expression $4x - 2y + z$. [Section 1.7, pp. 59–60]
2. Solve $x - 4y = 3$ for x. [Section 2.4, pp. 108–111]
3. Evaluate $3x - 2y + z$ when $x = 1$, $y = -3$, and $z = 2$. [Section 1.7, pp. 57–58]

We now present an alternative approach to solving systems of linear equations. The benefit in this approach is that it streamlines the notation and makes working with the system more manageable.

Before we discuss this method for solving systems of linear equations, we need to introduce a new concept, called the *matrix*.

> **DEFINITION**
>
> A **matrix** is a rectangular array of numbers.

A matrix has rows and columns. The number of rows and columns are used to "name" the matrix. For example, a matrix with 2 rows and 3 columns is called a "2 by 3 matrix" and is denoted "2×3."

Below are some examples of matrices.

Work Smart

A spreadsheet such as Microsoft Excel is a matrix!

2 × 3 matrix
$$\begin{bmatrix} 3 & -1 & 4 \\ 8 & 0 & -5 \end{bmatrix}$$

3 × 3 matrix
$$\begin{bmatrix} 2 & -8 & 12 \\ 0 & 7 & -2 \\ 5 & -2 & 1 \end{bmatrix}$$

1 Write the Augmented Matrix of a System of Linear Equations

We will use matrix notation to represent a system of linear equations. The matrix used to represent a system of linear equations is called an **augmented matrix.** In order to write the augmented matrix of a system, the terms containing the variables of each equation must be on the left side of the equal sign in the same order (x, y, and z, for example) and the constants on the right side. A variable that does not appear in an equation has a coefficient of 0. For example, the system of two linear equations containing two unknowns,

$$\begin{cases} a_1 x + b_1 y = c_1 \\ a_2 x + b_2 y = c_2 \end{cases} \quad \text{is written as an augmented matrix as} \quad \left[\begin{array}{cc|c} a_1 & b_1 & c_1 \\ a_2 & b_2 & c_2 \end{array}\right]$$

Notice that the first column represents the coefficients on the variable x, the second column represents the coefficients on the variable y, the vertical bar represents the equal signs, and the constants are to the right of the vertical bar. The first row in the matrix is equation (1) and the second row in the matrix is equation (2).

Preparing for...Answers **1.** $4, -2, 1$ **2.** $x = 4y + 3$ **3.** 11

EXAMPLE 1 **Writing the Augmented Matrix of a System of Linear Equations**

Write each system of linear equations as an augmented matrix.

(a) $\begin{cases} x - 3y = 2 \\ -2x + 5y = 7 \end{cases}$ (b) $\begin{cases} 2x + 3y - z = 1 \\ x - 2z + 1 = 0 \\ -4x - y + 3z = 5 \end{cases}$

Solution

(a) In the augmented matrix, we let the first column represent the coefficients on the variable x. The second column represents the coefficients on the variable y. The vertical line signifies the equal signs. The third column represents the constants to the right of the equal sign.

$$\begin{array}{cc} x & y \\ \left[\begin{array}{cc|c} 1 & -3 & 2 \\ -2 & 5 & 7 \end{array}\right] \end{array} \quad \begin{array}{c} x - 3y = 2 \\ -2x + 5y = 7 \end{array}$$

(b) We must be careful to write the system of equations so that the variables are all on the left side of the equal sign and the constants are on the right. Also, if a variable is missing from an equation, then its coefficient is understood to be 0. The system

$$\begin{cases} 2x + 3y - z = 1 \\ x - 2z + 1 = 0 \\ -4x - y + 3z = 5 \end{cases} \quad \text{gets rearranged as} \quad \begin{cases} 2x + 3y - z = 1 \\ x + 0y - 2z = -1 \\ -4x - y + 3z = 5 \end{cases}$$

The augmented matrix is

$$\left[\begin{array}{ccc|c} 2 & 3 & -1 & 1 \\ 1 & 0 & -2 & -1 \\ -4 & -1 & 3 & 5 \end{array}\right]$$

QUICK ✓ *Write the augmented matrix of each system of equations.*

1. $\begin{cases} 3x - y = -10 \\ -5x + 2y = 0 \end{cases}$ 2. $\begin{cases} x + 2y - 2z = 11 \\ -x - 2z = 4 \\ 4x - y + z - 3 = 0 \end{cases}$

(2) **Write the System from the Augmented Matrix**

We are now going to write the system of linear equations that corresponds to a given augmented matrix.

EXAMPLE 2 **Writing the System of Linear Equations from the Augmented Matrix**

Write the system of linear equations corresponding to each augmented matrix.

(a) $\left[\begin{array}{cc|c} 2 & 1 & -5 \\ -1 & 3 & 2 \end{array}\right]$ (b) $\left[\begin{array}{ccc|c} 1 & -3 & 2 & 5 \\ -2 & 0 & 4 & -3 \\ -1 & 4 & 1 & 0 \end{array}\right]$

Solution

(a) The augmented matrix has two rows and so represents a system of two equations. Because there are two columns to the left of the vertical bar, the system has two variables. If we call these two variables x and y, the system of equations is

$$\begin{cases} 2x + y = -5 \\ -x + 3y = 2 \end{cases}$$

(b) Since the augmented matrix has three rows, it represents a system of three equations. Since there are three columns to the left of the vertical bar, the system contains three variables. If we call these three variables x, y, and z, the system of equations is

$$\begin{cases} x - 3y + 2z = 5 \\ -2x \quad\quad + 4z = -3 \\ -x + 4y + \ z = 0 \end{cases}$$

QUICK ✓ *Write the system of linear equations corresponding to the given augmented matrix.*

3. $\begin{bmatrix} 1 & -3 & | & 7 \\ -2 & 5 & | & -3 \end{bmatrix}$ **4.** $\begin{bmatrix} 1 & -3 & 2 & | & 4 \\ 3 & 0 & -1 & | & -1 \\ -1 & 4 & 0 & | & 0 \end{bmatrix}$

③ Perform Row Operations on a Matrix

Row operations are used on an augmented matrix to solve the corresponding system of equations. There are three basic row operations.

In Words

The first row operation is like "flip-flopping" two equations in a system. The second row operation is like multiplying both sides of the equation by a nonzero constant. The third row operation is like adding two equations and replacing an equation with the sum.

> **ROW OPERATIONS**
>
> **1.** Interchange any two rows.
>
> **2.** Replace a row by a nonzero multiple of that row.
>
> **3.** Replace a row by the sum of that row and a nonzero multiple of some other row.

If you look carefully, these are the same types of operations that we can perform on a system of equations from Section E.2. The main reason for using a matrix to solve a system of equations is that it is more efficient notation and helps us to organize the mathematics. To see how row operations work, consider the following augmented matrix.

$$\begin{bmatrix} 1 & 3 & | & -1 \\ -2 & 1 & | & 3 \end{bmatrix}$$

Suppose that we want to apply a row operation to this matrix that results in a matrix whose entry in row 2, column 1 is a 0. The current entry in row 2, column 1 is -2. The row operation to use is

Multiply each entry in row 1 by 2 and then add the result to the corresponding entry in row 2. Have this result replace the current row 2.

That's a whole lot of words! To streamline this a little, we introduce some notation. If we use R_2 to represent the new entries in row 2 and we use r_1 and r_2 to represent the original entries in rows 1 and 2 respectively, then we can represent the row operation given above by

> Multiply each entry in row 1 by 2 and add the result to the corresponding entry in row 2 . . .

$$R_2 = 2r_1 + r_2$$

> . . . to obtain the "new" row 2.

We demonstrate this row operation below.

$$\begin{bmatrix} 1 & 3 & | & -1 \\ -2 & 1 & | & 3 \end{bmatrix} \xrightarrow{R_2 = 2r_1 + r_2} \begin{bmatrix} 1 & 3 & | & -1 \\ 2(1) + (-2) & 2(3) + 1 & | & 2(-1) + 3 \end{bmatrix} = \begin{bmatrix} 1 & 3 & | & -1 \\ 0 & 7 & | & 1 \end{bmatrix}$$

Notice that we now have a 0 in row 2, column 1, as desired.

EXAMPLE 3 **Applying a Row Operation to an Augmented Matrix**

Apply the row operation $R_2 = -3r_1 + r_2$ to the augmented matrix

$$\begin{bmatrix} 1 & 2 & | & -3 \\ 3 & 4 & | & 5 \end{bmatrix}$$

Solution

The row operation $R_2 = -3r_1 + r_2$ tells us to multiply the entries in row 1 by -3 and then add the result to the entries in row 2. The result should replace the current row 2.

$$\begin{bmatrix} 1 & 2 & | & -3 \\ 3 & 4 & | & 5 \end{bmatrix} \xrightarrow{R_2 = -3r_1 + r_2} \begin{bmatrix} 1 & 2 & | & -3 \\ -3(1) + 3 & -3(2) + 4 & | & -3(-3) + 5 \end{bmatrix} = \begin{bmatrix} 1 & 2 & | & -3 \\ 0 & -2 & | & 14 \end{bmatrix}$$

Quick ✓

5. Apply the row operation $R_2 = 4r_1 + r_2$ to the augmented matrix
$$\begin{bmatrix} 1 & -2 & | & 5 \\ -4 & 5 & | & -11 \end{bmatrix}.$$

EXAMPLE 4 **Finding a Particular Row Operation**

For the augmented matrix

$$\begin{bmatrix} 1 & -3 & | & -12 \\ 0 & 1 & | & 5 \end{bmatrix}$$

find a row operation that will result in the entry in row 1, column 2 becoming a 0 and perform the row operation.

Solution

We want a 0 in row 1, column 2. We can accomplish this by multiplying row 2 by 3 and adding the result to row 1. That is, we apply the row operation $R_1 = 3r_2 + r_1$.

$$\begin{bmatrix} 1 & -3 & | & -12 \\ 0 & 1 & | & 5 \end{bmatrix} \xrightarrow{R_1 = 3r_2 + r_1} \begin{bmatrix} 3(0) + 1 & 3(1) + (-3) & | & 3(5) + (-12) \\ 0 & 1 & | & 5 \end{bmatrix} = \begin{bmatrix} 1 & 0 & | & 3 \\ 0 & 1 & | & 5 \end{bmatrix}$$

Work Smart
If you want to change the entries in row 1, then you should multiply the entries in some other row and then add this result to the entries in row 1.

Quick ✓

6. For the augmented matrix $\begin{bmatrix} 1 & 5 & | & 13 \\ 0 & 1 & | & 2 \end{bmatrix}$, find a row operation that will result in the entry in row 1, column 2 becoming a 0 and perform the row operation.

---

④ Solve Systems of Linear Equations Using Matrices

To solve a system of linear equations using matrices, we use row operations on the augmented matrix of the system to obtain a matrix that is in *row echelon form*.

> **DEFINITION**
>
> A matrix is in **row echelon form** when
>
> **1.** The entry in row 1, column 1 is a 1, and 0s appear below it.
> **2.** The first nonzero entry in each row after the first row is a 1, 0s appear below it, and it appears to the right of the first nonzero entry in any row above.
> **3.** Any rows that contain all 0s to the left of the vertical bar appear at the bottom.

For a system of two equations containing two variables with a single solution (that is, a system that is consistent and independent), the augmented matrix is in row echelon form if it is of the form

$$\begin{bmatrix} 1 & a & | & b \\ 0 & 1 & | & c \end{bmatrix}$$

where a, b, and c are real numbers. A system of three linear equations containing three variables that is consistent and independent is in row echelon form if it is of the form

$$\begin{bmatrix} 1 & a & b & | & d \\ 0 & 1 & c & | & e \\ 0 & 0 & 1 & | & f \end{bmatrix}$$

where a, b, c, d, e, and f are real numbers.

The augmented matrix $\begin{bmatrix} 1 & a & | & b \\ 0 & 1 & | & c \end{bmatrix}$, is equivalent to the system $\begin{cases} x + ay = b \ (1) \\ \qquad y = c \ (2) \end{cases}$.

In other words, because c is a known number, we know y. We can then back-substitute to find x. For example, the augmented matrix $\begin{bmatrix} 1 & 3 & | & -5 \\ 0 & 1 & | & -3 \end{bmatrix}$ is in row echelon form

and is equivalent to the system $\begin{cases} x + 3y = -5 \ (1) \\ \qquad y = -3 \ (2) \end{cases}$. From equation (2), we know that

$y = -3$. Using back-substitution, we find that $x = 4$. So, the solution is $(4, -3)$.

EXAMPLE 5 **How to Solve a System of Two Linear Equations Containing Two Variables Using Matrices**

Solve: $\begin{cases} 3x - 2y = -19 \\ x + 2y = 7 \end{cases}$

Step-by-Step Solution

Step 1: Write the augmented matrix of the system.	$\begin{bmatrix} 3 & -2 & | & -19 \\ 1 & 2 & | & 7 \end{bmatrix}$
Step 2: We want the entry in row 1, column 1 to be 1. We can interchange rows 1 and 2 in the augmented matrix.	$\begin{bmatrix} 1 & 2 & | & 7 \\ 3 & -2 & | & -19 \end{bmatrix}$
Step 3: We want the entry in row 2, column 1 to be 0. We use the row operation $R_2 = -3r_1 + r_2$ to accomplish this. The entries in row 1 remain unchanged.	$\begin{bmatrix} 1 & 2 & | & 7 \\ 0 & -8 & | & -40 \end{bmatrix}$
Step 4: Now we want the entry in row 2, column 2 to be 1. This is accomplished by multiplying row 2 by $-\dfrac{1}{8}$. We use the row operation $R_2 = -\dfrac{1}{8}r_2$.	$\begin{bmatrix} 1 & 2 & | & 7 \\ 0 & 1 & | & 5 \end{bmatrix}$
Step 5: From row 2 we have that $y = 5$. Row 1 represents the equation $x + 2y = 7$. Back-substitute 5 for y and solve for x.	$\begin{aligned} x + 2y &= 7 \\ x + 2(5) &= 7 \\ x + 10 &= 7 \\ x &= -3 \end{aligned}$

(continued)

Step 6: Check We let $x = -3$ and $y = 5$ in both equations (1) and (2) to verify our solution.

Equation (1): $3x - 2y = -19$

$x = -3, y = 5$: $3(-3) - 2(5) \overset{?}{=} -19$

$-9 - 10 \overset{?}{=} -19$

$-19 = -19$ True

Equation (2): $x + 2y = 7$

$x = -3, y = 5$: $-3 + 2(5) \overset{?}{=} 7$

$-3 + 10 \overset{?}{=} 7$

$7 = 7$ True

The solution is the ordered pair $(-3, 5)$.

QUICK ✓

7. Solve the following system using matrices. $\begin{cases} 2x - 4y = 20 \\ 3x + y = 16 \end{cases}$

EXAMPLE 6   **How to Solve a System of Three Linear Equations Containing Three Variables Using Matrices**

Solve: $\begin{cases} x + y + z = 3 \\ -2x - 3y + 2z = 13 \\ 4x + 5y + z = -3 \end{cases}$

Step-by-Step Solution

Step 1: Write the augmented matrix of the system.

$$\begin{bmatrix} 1 & 1 & 1 & | & 3 \\ -2 & -3 & 2 & | & 13 \\ 4 & 5 & 1 & | & -3 \end{bmatrix}$$

Step 2: We want the entry in row 1, column 1 to be 1. This is already done.

$$\begin{bmatrix} 1 & 1 & 1 & | & 3 \\ -2 & -3 & 2 & | & 13 \\ 4 & 5 & 1 & | & -3 \end{bmatrix}$$

Step 3: We want the entry in row 2, column 1 to be 0. We use the row operation $R_2 = 2r_1 + r_2$ to accomplish this. We also want the entry in row 3, column 1 to be 0. We use the row operation $R_3 = -4r_1 + r_3$ to accomplish this. The entries in row 1 remain unchanged.

$R_2 = 2r_1 + r_2$:
$R_3 = -4r_1 + r_3$:
$$\begin{bmatrix} 1 & 1 & 1 & | & 3 \\ 0 & -1 & 4 & | & 19 \\ 0 & 1 & -3 & | & -15 \end{bmatrix}$$

Step 4: Now we want the entry in row 2, column 2 to be 1. This is accomplished by interchanging rows 2 and 3.

$$\begin{bmatrix} 1 & 1 & 1 & | & 3 \\ 0 & 1 & -3 & | & -15 \\ 0 & -1 & 4 & | & 19 \end{bmatrix}$$

Step 5: We need the entry in row 3, column 2 to be 0. We use the row operation $R_3 = r_2 + r_3$.

$R_3 = r_2 + r_3$:
$$\begin{bmatrix} 1 & 1 & 1 & | & 3 \\ 0 & 1 & -3 & | & -15 \\ 0 & 0 & 1 & | & 4 \end{bmatrix}$$

Step 6: We want the entry in row 3, column 3 to be 1. This is already done.

$$\begin{bmatrix} 1 & 1 & 1 & | & 3 \\ 0 & 1 & -3 & | & -15 \\ 0 & 0 & 1 & | & 4 \end{bmatrix}$$

Step 7: The augmented matrix is in row echelon form. Write the system of equations corresponding to the augmented matrix and solve.

$$\begin{cases} x + y + z = 3 & (1) \\ \quad\;\; y - 3z = -15 & (2) \\ \quad\qquad\;\; z = 4 & (3) \end{cases}$$

From equation (3), we have that $z = 4$. Letting $z = 4$ in equation (2), we can determine y.

$$\begin{aligned} \text{Equation (2):} \quad & y - 3z = -15 \\ z = 4: \quad & y - 3(4) = -15 \\ & y - 12 = -15 \\ & y = -3 \end{aligned}$$

Let $y = -3$ and $z = 4$ in equation (1) to find that $x = 2$.

Step 8: Check We let $x = 2$, $y = -3$, and $z = 4$ in each equation in the original system to verify our solution.

We leave it to you to verify the solution.

The solution is the ordered triple $(2, -3, 4)$.

Work Smart

Look back at Step 4 in Example 6. There is more than one option for obtaining a 1 in row 2, column 2. We could also have used the row operation $R_2 = 2r_3 + r_2$ or $R_2 = -r_2$.

QUICK ✓

8. Solve: $\begin{cases} x - y + 2z = 7 \\ 2x - 2y + z = 11 \\ -3x + y - 3z = -14 \end{cases}$

SUMMARY: Steps for Solving a System of Linear Equations Using Matrices

Step 1: Write the augmented matrix of the system.

Step 2: Perform row operations so that the entry in row 1, column 1 is 1.

Step 3: Perform row operations so that all the entries below the 1 in row 1, column 1 are 0's.

Step 4: Perform row operations so that the entry in row 2, column 2 is 1. Make sure that the entries in column 1 remain unchanged. If it is impossible to place a 1 in row 2, column 2, then proceed to use operations to place a 1 in row 2, column 3. (*Note:* If a row with all 0's is obtained, then it should be placed in the last row of the matrix.)

Step 5: Once a 1 is in place, perform row operations to place 0s below it.

Step 6: Repeat Steps 4 and 5 until you have the augmented matrix in echelon form.

Step 7: With the augmented matrix in echelon form, write the corresponding system of equations. Back-substitute to find the values of the variables.

Step 8: Check your answer.

5 Solve Dependent and Inconsistent Systems

The matrix method for solving a system of linear equations also identifies systems that have infinitely many solutions (dependent systems) and systems with no solution (inconsistent).

EXAMPLE 7 Solving a Dependent System Using Matrices

Solve:
$$\begin{cases} x + y + 3z = 3 \\ -2x - y - 8z = -5 \\ 3x + 2y + 11z = 8 \end{cases}$$

Solution

We will write the augmented matrix of the system and perform row operations to get the matrix into row echelon form.

$$\begin{bmatrix} 1 & 1 & 3 & | & 3 \\ -2 & -1 & -8 & | & -5 \\ 3 & 2 & 11 & | & 8 \end{bmatrix} \xrightarrow[R_3 = -3r_1 + r_3]{R_2 = 2r_1 + r_2} \begin{bmatrix} 1 & 1 & 3 & | & 3 \\ 0 & 1 & -2 & | & 1 \\ 0 & -1 & 2 & | & -1 \end{bmatrix} \xrightarrow{R_3 = r_2 + r_3} \begin{bmatrix} 1 & 1 & 3 & | & 3 \\ 0 & 1 & -2 & | & 1 \\ 0 & 0 & 0 & | & 0 \end{bmatrix}$$

Notice that the last row is all 0s. The augmented matrix is in row echelon form. The system of equations corresponding to this matrix is

$$\begin{cases} x + y + 3z = 3 & (1) \\ y - 2z = 1 & (2) \\ 0 = 0 & (3) \end{cases}$$

The statement $0 = 0$ in equation (3) indicates that we have a dependent system. If we let z represent any real number, then we can solve equation (2) for y in terms of z.

$$\begin{aligned} \text{Equation (2):} \quad & y - 2z = 1 \\ \text{Add } 2z \text{ to both sides:} \quad & y = 2z + 1 \end{aligned}$$

Let $y = 2z + 1$ in equation (1) and solve for x in terms of z.

$$\begin{aligned} \text{Equation (1):} \quad & x + y + 3z = 3 \\ \text{Let } y = 2z + 1: \quad & x + (2z + 1) + 3z = 3 \\ \text{Combine like terms:} \quad & x + 5z + 1 = 3 \\ \text{Subtract } 5z + 1 \text{ from both sides:} \quad & x = -5z + 2 \end{aligned}$$

The solution to the system is $\{(x, y, z) \mid x = -5z + 2, y = 2z + 1, z \text{ is any real number}\}$. To find specific solutions to the system, choose any value of z and use the equations $x = -5z + 2$ and $y = 2z + 1$ to determine x and y. Some specific solutions to the system are $x = 2, y = 1, z = 0$; $x = -3, y = 3, z = 1$; or $x = 7, y = -1$, $z = -1$.

If we evaluate the system at some of the specific solutions, we see that all of them satisfy the original system. This leads us to believe that our solution is correct.

QUICK ✔

9. Solve:
$$\begin{cases} x + y - 3z = 8 \\ 2x + 3y - 10z = 19 \\ -x - 2y + 7z = -11 \end{cases}$$

EXAMPLE 8 Solving an Inconsistent System Using Matrices

Solve: $\begin{cases} 2x + y - z = 5 \\ -x + 2y + z = 1 \\ 3x + 4y - z = -2 \end{cases}$

Solution

We will write the augmented matrix of the system and perform row operations to get the matrix into row echelon form.

$$\begin{bmatrix} 2 & 1 & -1 & | & 5 \\ -1 & 2 & 1 & | & 1 \\ 3 & 4 & -1 & | & -2 \end{bmatrix} \xrightarrow{R_1 = r_2 + r_1} \begin{bmatrix} 1 & 3 & 0 & | & 6 \\ -1 & 2 & 1 & | & 1 \\ 3 & 4 & -1 & | & -2 \end{bmatrix} \xrightarrow[R_3 = -3r_1 + r_3]{R_2 = r_1 + r_2} \begin{bmatrix} 1 & 3 & 0 & | & 6 \\ 0 & 5 & 1 & | & 7 \\ 0 & -5 & -1 & | & -20 \end{bmatrix}$$

$$\xrightarrow{R_2 = \frac{1}{5}r_2} \begin{bmatrix} 1 & 3 & 0 & | & 6 \\ 0 & 1 & \frac{1}{5} & | & \frac{7}{5} \\ 0 & -5 & -1 & | & -20 \end{bmatrix} \xrightarrow{R_3 = 5r_2 + r_3} \begin{bmatrix} 1 & 3 & 0 & | & 6 \\ 0 & 1 & \frac{1}{5} & | & \frac{7}{5} \\ 0 & 0 & 0 & | & -13 \end{bmatrix}$$

We want the entry in row 3, column 3 to be 1. This cannot be done. The augmented matrix is in row echelon form. The system of equations corresponding to this matrix is

$$\begin{cases} x + 3y \quad\quad = 6 \quad (1) \\ \quad\quad y + \frac{1}{5}z = \frac{7}{5} \quad (2) \\ \quad\quad\quad\quad 0 = -13 \quad (3) \end{cases}$$

The statement $0 = -13$ in equation (3) indicates that we have an inconsistent system. The solution is $\varnothing$ or $\{\ \}$. ■

QUICK ✓

10. Solve: $\begin{cases} -x + 2y - z = 5 \\ 2x + y + 4z = 3 \\ 3x - y + 5z = 0 \end{cases}$

E.3 Exercises

For Extra Help: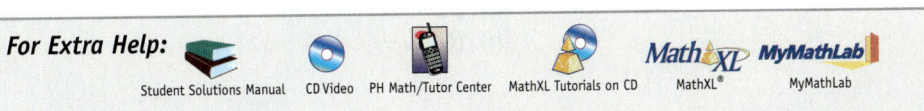

Student Solutions Manual　　CD Video　PH Math/Tutor Center　MathXL Tutorials on CD　MathXL®　MyMathLab

Concepts and Vocabulary

In Problems 1–3, fill in the blanks.

1. An m by n rectangular array of numbers is called a(n) _____.

2. The matrix used to represent a system of linear equations is called a(n) _____ matrix.

3. A 4×3 matrix has _____ rows and _____ columns.

In Problems 4–6, answer True or False to each statement.

4. The augmented matrix of a system of two equations containing 2 unknowns has 2 rows and 2 columns.

5. The matrix $\begin{bmatrix} 1 & 3 & | & 2 \\ 0 & 1 & | & -5 \end{bmatrix}$ is in row echelon form.

6. A 2×5 matrix has 2 rows and 5 columns.

7. Write a paragraph that outlines the strategy for putting an augmented matrix in row echelon form.

8. When solving a system of linear equations using matrices, how do you know that the system is inconsistent?

9. What would be the next row operation on the given augmented matrix?

$$\begin{bmatrix} 1 & 3 & | & 8 \\ 0 & 5 & | & 10 \end{bmatrix}$$

10. What would your recommend as the next row operation on the given augmented matrix? Why?

$$\begin{bmatrix} 1 & 3 & -2 & | & 4 \\ 0 & 5 & 3 & | & 2 \\ 0 & -4 & 6 & | & -5 \end{bmatrix}$$

Building Skills

In Problems 11–18, write the augmented matrix of the given system of equations.

11. $\begin{cases} x - 3y = 2 \\ 2x + 5y = 1 \end{cases}$

12. $\begin{cases} -x + y = 6 \\ 5x - y = -3 \end{cases}$

13. $\begin{cases} x + y + z = 3 \\ 2x - y + 3z = 1 \\ -4x + 2y - 5z = -3 \end{cases}$

14. $\begin{cases} x + y - z = 2 \\ -2x + y - 4z = 13 \\ 3x - y - 2z = -4 \end{cases}$

15. $\begin{cases} -x + y - 2 = 0 \\ 5x + y + 5 = 0 \end{cases}$

16. $\begin{cases} 6x + 4y + 2 = 0 \\ -x - y + 1 = 0 \end{cases}$

17. $\begin{cases} x + z = 2 \\ 2x + y = 13 \\ x - y + 4z + 4 = 0 \end{cases}$

18. $\begin{cases} 2x + 7 = 1 \\ -x - 6z = 5 \\ 5x + 2y - 4z + 1 = 0 \end{cases}$

In Problems 19–24, perform each row operation on the given augmented matrix.

19. $\begin{bmatrix} 1 & -3 & | & 2 \\ -2 & 5 & | & 1 \end{bmatrix}$

   (a) $R_2 = 2r_1 + r_2$ followed by
   (b) $R_2 = -r_2$

20. $\begin{bmatrix} 1 & 5 & | & 7 \\ 3 & 11 & | & 13 \end{bmatrix}$

   (a) $R_2 = -3r_1 + r_2$ followed by
   (b) $R_2 = -\dfrac{1}{4}r_2$

21. $\begin{bmatrix} 1 & 1 & -1 & | & 4 \\ 2 & 5 & 3 & | & -3 \\ -1 & -3 & 2 & | & 1 \end{bmatrix}$

   (a) $R_2 = -2r_1 + r_2$ followed by
   (b) $R_3 = r_1 + r_3$

22. $\begin{bmatrix} 1 & -1 & 1 & | & 6 \\ -2 & 1 & -3 & | & 3 \\ 3 & 2 & -2 & | & -5 \end{bmatrix}$

   (a) $R_2 = 2r_1 + r_2$ followed by
   (b) $R_3 = -3r_1 + r_3$

23. $\begin{bmatrix} 1 & 1 & 1 & | & 4 \\ 0 & 5 & 3 & | & -3 \\ 0 & -4 & 2 & | & 8 \end{bmatrix}$

   (a) $R_2 = r_3 + r_2$ followed by
   (b) $R_3 = \dfrac{1}{2}r_3$

24. $\begin{bmatrix} 1 & -3 & 4 & | & 11 \\ 0 & 3 & 6 & | & -12 \\ 0 & -2 & -3 & | & 8 \end{bmatrix}$

   (a) $R_2 = \dfrac{1}{3}r_2$ followed by
   (b) $R_3 = 2r_2 + r_3$

In Problems 25–30, the reduced row echelon form of a system of linear equations is given. Write the system of equations corresponding to the given matrix. Use x, y; or x, y, z as variables. Determine whether the system is consistent and independent, consistent and dependent, or inconsistent. If it is consistent and independent, give the solution.

25. $\begin{bmatrix} 1 & 4 & | & -5 \\ 0 & 1 & | & -2 \end{bmatrix}$

26. $\begin{bmatrix} 1 & -2 & | & 3 \\ 0 & 1 & | & -5 \end{bmatrix}$

27. $\begin{bmatrix} 1 & 3 & -2 & | & 6 \\ 0 & 1 & 5 & | & -2 \\ 0 & 0 & 0 & | & 4 \end{bmatrix}$

28. $\begin{bmatrix} 1 & -2 & -4 & | & 6 \\ 0 & 1 & -5 & | & -3 \\ 0 & 0 & 0 & | & 0 \end{bmatrix}$

29. $\begin{bmatrix} 1 & -2 & -1 & | & 3 \\ 0 & 1 & -2 & | & -8 \\ 0 & 0 & 1 & | & 5 \end{bmatrix}$

30. $\begin{bmatrix} 1 & 2 & -1 & | & -7 \\ 0 & 1 & 2 & | & -4 \\ 0 & 0 & 1 & | & -3 \end{bmatrix}$

In Problems 31–52, solve each system of equations using matrices. If the system has no solution, say that it is inconsistent.

31. $\begin{cases} x - 3y = 18 \\ 2x + y = 1 \end{cases}$

32. $\begin{cases} x + 5y = 2 \\ -2x + 3y = 9 \end{cases}$

33. $\begin{cases} 2x + 4y = 10 \\ x + 2y = 3 \end{cases}$

34. $\begin{cases} 5x - 2y = 3 \\ -15x + 6y = -9 \end{cases}$

35. $\begin{cases} x - 6y = 8 \\ 2x + 8y = -9 \end{cases}$

36. $\begin{cases} 3x + 3y = -1 \\ 2x + y = 1 \end{cases}$

37. $\begin{cases} 4x - y = 8 \\ 2x - \dfrac{1}{2}y = 4 \end{cases}$

38. $\begin{cases} 5x - 2y = 10 \\ 2x - \dfrac{4}{5}y = 4 \end{cases}$

39. $\begin{cases} x + y + z = 0 \\ 2x - 3y + z = 19 \\ -3x + y - 2z = -15 \end{cases}$

40. $\begin{cases} x + y + z = 5 \\ 2x - y + 3z = -3 \\ -x + 2y - z = 10 \end{cases}$

41. $\begin{cases} 2x + y - z = 13 \\ -x - 3y + 2z = -14 \\ -3x + 2y - 3z = 3 \end{cases}$

42. $\begin{cases} 2x - y + 2z = 13 \\ -x + 2y - z = -14 \\ 3x + y - 2z = -13 \end{cases}$

43. $\begin{cases} 2x - y + 3z = 1 \\ -x + 3y + z = -4 \\ 3x + y + 7z = -2 \end{cases}$

44. $\begin{cases} -x + 2y + z = 1 \\ 2x - y + 3z = -3 \\ -x + 5y + 6z = 2 \end{cases}$

45. $\begin{cases} 3x + y - 4z = 0 \\ -2x - 3y + z = 5 \\ -x - 5y - 2z = 3 \end{cases}$

46. $\begin{cases} -x + 4y - 3z = 1 \\ 3x + y - z = -3 \\ x + 9y - 7z = -1 \end{cases}$

47. $\begin{cases} 2x - y + 3z = -1 \\ 3x + y - 4z = 3 \\ x + 7y - 2z = 2 \end{cases}$

48. $\begin{cases} x + y + z = 8 \\ 2x + 3y + z = 19 \\ 2x + 2y + 4z = 21 \end{cases}$

49. $\begin{cases} 3x + 5y + 2z = 6 \\ 10y - 2z = 5 \\ 6x + 4z = 8 \end{cases}$

50. $\begin{cases} 2x + y + 3z = 3 \\ 2x - 3y = 7 \\ 4y + 6z = -2 \end{cases}$

51. $\begin{cases} x - z = 3 \\ 2x + y = -3 \\ 2y - z = 7 \end{cases}$

52. $\begin{cases} x + 2y - z = 3 \\ y + z = 1 \\ x - 3z = 2 \end{cases}$

Applying the Concepts

53. Curve Fitting The function $f(x) = ax^2 + bx + c$ is a quadratic function, where $a, b,$ and c are constants.

    (a) If $f(-1) = 6$, then $6 = a(-1)^2 + b(-1) + c$ or $a - b + c = 6$. Find two additional linear equations if $f(1) = 0$, and $f(2) = 3$.

    (b) Use the three linear equations found in part (a) to determine $a, b,$ and c. What is the quadratic function that contains the points $(-1, 6), (1, 0)$, and $(2, 3)$?

54. Curve Fitting The function $f(x) = ax^2 + bx + c$ is a quadratic function, where $a, b,$ and c are constants.

    (a) If $f(-1) = -6$, then $-6 = a(-1)^2 + b(-1) + c$ or $a - b + c = -6$. Find two additional linear equations if $f(1) = 0$, and $f(2) = -3$.

    (b) Use the three linear equations found in part (a) to determine $a, b,$ and c. What is the quadratic function that contains the points $(-1, -6), (1, 0)$, and $(2, -3)$?

55. Finance Carissa has \$20,000 to invest. Her financial planner suggests that she diversify her investment into three investment categories: Treasury bills that yield 4% simple interest, municipal bonds that yield 5% simple interest, and corporate bonds that yield 8% simple interest. Carissa would like to earn \$1070 per year in income. In addition, she wants her investment in Treasury bills to be \$3000 more than her investment in corporate bonds. How much should Carissa invest in each investment category?

56. Finance Marlon has \$12,000 to invest. He decides to place some of the money into a savings account paying 2% interest, some in Treasury bonds paying 4% interest, and some in a mutual fund paying 9% interest. Marlon would like to earn \$440 per year in income. In addition, Marlon wants his investment in the savings account to be \$4000 more than the amount in Treasury bonds. How much should Marlon invest in each investment category?

Extending the Concepts

Sometimes it is advantageous to write a matrix in **reduced row echelon form.** In this form, row operations are used to obtain entries that are 0 above and below the leading 1 in a row. Augmented matrices for systems in reduced row echelon form with 2 equations containing 2 variables and 3 equations containing 3 variables are shown below.

$$\begin{bmatrix} 1 & 0 & | & a \\ 0 & 1 & | & b \end{bmatrix} \qquad \begin{bmatrix} 1 & 0 & 0 & | & a \\ 0 & 1 & 0 & | & b \\ 0 & 0 & 1 & | & c \end{bmatrix}$$

The obvious advantage to writing an augmented matrix in reduced row echelon form is that the solution to the system is readily seen. In the system with 2 equations containing 2 variables, the solution is $x = a$ and $y = b$. In the system with 3 equations containing 3 variables, the solution is $x = a$, $y = b$, and $z = c$.

In Problems 57–60, solve the system of equations by writing the augmented matrix in reduced row echelon form.

57. $\begin{cases} 2x + y = 1 \\ -3x - 2y = -5 \end{cases}$

58. $\begin{cases} 2x + y = -1 \\ -3x - 2y = -3 \end{cases}$

59. $\begin{cases} x + y + z = 3 \\ 2x + y - 4z = 25 \\ -3x + 2y + z = 0 \end{cases}$

60. $\begin{cases} x + y + z = 3 \\ 3x - 2y + 2z = 38 \\ -2x - 3z = -19 \end{cases}$

The Graphing Calculator

Graphing calculators have the ability to solve systems of linear equations by writing the equation in echelon form. We enter the augmented matrix into the graphing calculator and name it A. See Figure 6(a). Figures 6(b) shows the results of Example 6 using a TI-84 Plus graphing calculator. Since the entire matrix does not fit on the screen, we need to scroll right to the see the rest of it. See Figure 6(c).

Figure 6

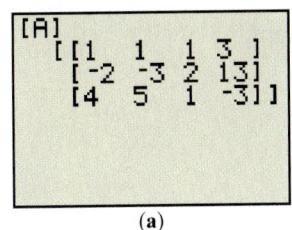   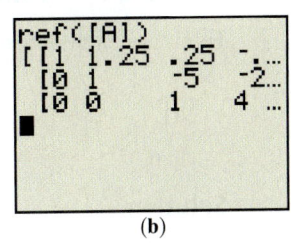   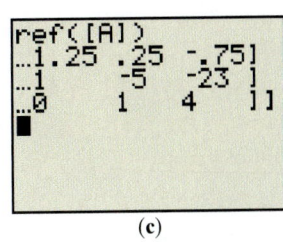

(a)   (b)   (c)

Notice that the row echelon form of the augmented matrix using the graphing calculator differs from the row echelon form in the algebraic solution presented in Example 6, yet both matrices provide the same solution! This is because the two solutions use different row operations to obtain the row echelon form.

In Problems 61–66, solve each system of equations using a graphing calculator.

61. $\begin{cases} 2x + 3y = 1 \\ -3x - 4y = -3 \end{cases}$

62. $\begin{cases} 3x + 2y = 4 \\ -5x - 3y = -4 \end{cases}$

63. $\begin{cases} 2x + 3y - 2z = -12 \\ -3x + y + 2z = 0 \\ 4x + 3y - z = 3 \end{cases}$

64. $\begin{cases} 2x - 3y - 4z = 16 \\ -3x + y + 2z = -23 \\ 4x + 3y - z = 13 \end{cases}$

E.4 Determinants and Cramer's Rule

OBJECTIVES

1. Evaluate the Determinant of a 2 × 2 Matrix
2. Use Cramer's Rule to Solve a System of Two Equations Containing Two Variables
3. Evaluate the Determinant of a 3 × 3 Matrix
4. Use Cramer's Rule to Solve a System of Three Equations Containing Three Variables

Preparing for Determinants and Cramer's Rule

Before getting started, take the following readiness quiz. If you get a problem wrong, go back to the section cited and review the material.

1. Evaluate $4 \cdot 2 - 3 \cdot (-3)$ [Section 1.6, pp. 50–51]

2. Simplify $\dfrac{18}{6}$ [Section 1.3, pp. 24–26]

Up to this point, we have learned how to solve systems of linear equations using substitution, elimination and row operations on augmented matrices. This section presents another method for solving a system of linear equations. However, to use this method, the number of equations must equal the number of variables. The method for solving these systems is called *Cramer's Rule* and is based on the concept of a *determinant*.

① Evaluate the Determinant of a 2 × 2 Matrix

A matrix is **square** provided that the number of rows equals the number of columns. For any square matrix, we can compute its *determinant*.

Work Smart

The methods for solving systems presented in this section only work when the number of equations equals the number of variables.

> **DEFINITION**
>
> Suppose that a, b, c, and d are four real numbers. The **determinant** of a 2 × 2
>
> matrix $\begin{bmatrix} a & b \\ c & d \end{bmatrix}$, denoted $\begin{vmatrix} a & b \\ c & d \end{vmatrix}$, is
>
> $$\begin{vmatrix} a & b \\ c & d \end{vmatrix} = ad - bc$$

Preparing for...Answers   **1.** 17   **2.** 3

EXAMPLE 1 **Evaluating a 2 × 2 Determinant**

Evaluate each determinant:

(a) $\begin{vmatrix} 5 & 2 \\ 3 & 4 \end{vmatrix}$

(b) $\begin{vmatrix} -1 & 3 \\ -4 & 5 \end{vmatrix}$

Solution

(a) $\begin{vmatrix} 5 & 2 \\ 3 & 4 \end{vmatrix} = 5(4) - 3(2)$

$= 20 - 6$

$= 14$

(b) $\begin{vmatrix} -1 & 3 \\ -4 & 5 \end{vmatrix} = -1(5) - (-4)(3)$

$= -5 - (-12)$

$= -5 + 12$

$= 7$

QUICK ✓ *Evaluate each determinant.*

1. $\begin{vmatrix} 5 & 3 \\ 4 & 6 \end{vmatrix}$

2. $\begin{vmatrix} -2 & -5 \\ 1 & 7 \end{vmatrix}$

② **Use Cramer's Rule to Solve a System of Two Equations Containing Two Variables**

We can use 2×2 determinants to solve a system of two equations containing two variables.

CRAMER'S RULE FOR TWO EQUATIONS CONTAINING TWO VARIABLES

The solution to the system of equations

$$\begin{cases} ax + by = s \quad (1) \\ cx + dy = t \quad (2) \end{cases}$$

is given by

$$x = \frac{\begin{vmatrix} s & b \\ t & d \end{vmatrix}}{\begin{vmatrix} a & b \\ c & d \end{vmatrix}} = \frac{D_x}{D}, \qquad y = \frac{\begin{vmatrix} a & s \\ c & t \end{vmatrix}}{\begin{vmatrix} a & b \\ c & d \end{vmatrix}} = \frac{D_y}{D}$$

provided that

$$D = \begin{vmatrix} a & b \\ c & d \end{vmatrix} = ad - bc \neq 0$$

Look very carefully at the pattern in Cramer's Rule. The denominator in the solution is the determinant of the coefficients of the variables.

$$\begin{cases} ax + by = s \\ cx + dy = t \end{cases}, \qquad D = \begin{vmatrix} a & b \\ c & d \end{vmatrix}$$

In the solution for x, the numerator is the determinant formed by replacing the entries in the first column (the coefficients of x) in D by the constants on the right side of the equal sign. That is,

$$D_x = \begin{vmatrix} s & b \\ t & d \end{vmatrix}$$

In the solution for y, the numerator is the determinant formed by replacing the entries in the second column (the coefficients of y) in D by the constants on the right side of the equal sign. That is,

$$D_y = \begin{vmatrix} a & s \\ c & t \end{vmatrix}$$

EXAMPLE 2 **How to Solve a System of Two Linear Equations Containing Two Variables Using Cramer's Rule**

Use Cramer's Rule to solve the system $\begin{cases} x + 2y = 0 \\ -2x - 8y = -9 \end{cases}$

Step-by-Step Solution

Step 1: Determine the determinant of the coefficients of the variables, D.

$$D = \begin{vmatrix} 1 & 2 \\ -2 & -8 \end{vmatrix} = 1(-8) - (-2)(2) = -8 - (-4) = -8 + 4 = -4$$

Step 2: Because $D \neq 0$, we continue by determining D_x by replacing the first column in D with the constants on the right side of the equal sign in the system. We determine D_y by replacing the second column in D with the constants on the right side of the equal sign in the system.

$$D_x = \begin{vmatrix} 0 & 2 \\ -9 & -8 \end{vmatrix} = 0(-8) - (-9)(2) = 0 - (-18) = 18$$

$$D_y = \begin{vmatrix} 1 & 0 \\ -2 & -9 \end{vmatrix} = 1(-9) - (-2)(0) = -9 - 0 = -9$$

Step 3: Find $x = \dfrac{D_x}{D}$ and $y = \dfrac{D_y}{D}$

$$x = \frac{D_x}{D} = \frac{18}{-4} = -\frac{9}{2} \qquad y = \frac{D_y}{D} = \frac{-9}{-4} = \frac{9}{4}$$

Step 4: Check $x = -\dfrac{9}{2}$ and $y = \dfrac{9}{4}$.

We leave the check to you.

The solution is the ordered pair $\left(-\dfrac{9}{2}, \dfrac{9}{4} \right)$.

If the determinant of the coefficients of the variables, D, is found to be 0 when using Cramer's Rule, then the system of equations is either consistent, but dependent, or inconsistent. We will learn how to deal with these possibilities shortly.

QUICK ✓ *Use Cramer's Rule to solve the system, if possible.*

3. $\begin{cases} 3x + 2y = 1 \\ -2x - y = 1 \end{cases}$ **4.** $\begin{cases} 4x - 2y = 8 \\ -6x + 3y = 3 \end{cases}$

---

③ Evaluate the Determinant of a 3 × 3 Matrix

To use Cramer's Rule to solve a system of three equations containing three variables, we need to define a 3 × 3 determinant.

The **determinant of a 3 × 3 matrix** is symbolized by

$$\begin{vmatrix} a_{1,1} & a_{1,2} & a_{1,3} \\ a_{2,1} & a_{2,2} & a_{2,3} \\ a_{3,1} & a_{3,2} & a_{3,3} \end{vmatrix}$$

where $a_{1,1}, a_{1,2}, \ldots, a_{3,3}$ are real numbers.

As with matrices, the subscript is used to identify the row and column of an entry. For example, $a_{2,3}$ is the entry in row 2, column 3.

DEFINITION

The **value of a determinant of a 3 × 3 matrix** may be defined in terms of 2 × 2 determinants as follows:

$$\begin{vmatrix} a_{1,1} & a_{1,2} & a_{1,3} \\ a_{2,1} & a_{2,2} & a_{2,3} \\ a_{3,1} & a_{3,2} & a_{3,3} \end{vmatrix} = a_{1,1} \begin{vmatrix} a_{2,2} & a_{2,3} \\ a_{3,2} & a_{3,3} \end{vmatrix} \overset{\text{Minus}}{-} a_{1,2} \begin{vmatrix} a_{2,1} & a_{2,3} \\ a_{3,1} & a_{3,3} \end{vmatrix} \overset{\text{Plus}}{+} a_{1,3} \begin{vmatrix} a_{2,1} & a_{2,2} \\ a_{3,1} & a_{3,2} \end{vmatrix}$$

2 × 2 determinant left after removing the row and column containing $a_{1,1}$.

2 × 2 determinant left after removing the row and column containing $a_{1,2}$.

2 × 2 determinant left after removing the row and column containing $a_{1,3}$.

The 2 × 2 determinants shown in the definition above are called **minors** of the 3 × 3 determinant. Notice that once again we have reduced a problem into something we already know how to do—here we reduced the 3 × 3 determinant into 3 different 2 × 2 determinants.

The formula given in the definition above is easiest to remember by noting that each entry in row 1 is multiplied by the 2 × 2 determinant that remains after the row and column containing the entry have been removed.

EXAMPLE 3   **Evaluating a 3 × 3 Determinant**

Evaluate: $\begin{vmatrix} 3 & 2 & -4 \\ 1 & 7 & -3 \\ 0 & 2 & -5 \end{vmatrix}$

Solution

$$\begin{vmatrix} 3 & 2 & -4 \\ 1 & 7 & -3 \\ 0 & 2 & -5 \end{vmatrix} = 3\begin{vmatrix} 7 & -3 \\ 2 & -5 \end{vmatrix} - 2\begin{vmatrix} 1 & -3 \\ 0 & -5 \end{vmatrix} + (-4)\begin{vmatrix} 1 & 7 \\ 0 & 2 \end{vmatrix}$$

$$= 3(-35 - (-6)) - 2(-5 - 0) - 4(2 - 0)$$
$$= 3(-29) - 2(-5) - 4(2)$$
$$= -87 + 10 - 8$$
$$= -85$$

The definition demonstrates one way to find the value of a 3 × 3 determinant—by expanding across row 1. In fact, the expansion can take place across any row or down any column. The terms to be added or subtracted consist of the row (or column) entry times the value of the 2×2 determinant that remains after removing the row and column containing the entry. There is only one glitch — the signs of the terms in the expansion change depending upon the row or column that is expanded on. The signs of the terms obey the following scheme:

$$\begin{array}{ccc} + & - & + \\ - & + & - \\ + & - & + \end{array}$$

For example, if we choose to expand down column 2, we obtain

Minus   Plus   Minus

$$\begin{vmatrix} a_{1,1} & a_{1,2} & a_{1,3} \\ a_{2,1} & a_{2,2} & a_{2,3} \\ a_{3,1} & a_{3,2} & a_{3,3} \end{vmatrix} = -a_{1,2}\begin{vmatrix} a_{2,1} & a_{2,3} \\ a_{3,1} & a_{3,3} \end{vmatrix} + a_{2,2}\begin{vmatrix} a_{1,1} & a_{1,3} \\ a_{3,1} & a_{3,3} \end{vmatrix} - a_{3,2}\begin{vmatrix} a_{1,1} & a_{1,3} \\ a_{2,1} & a_{2,3} \end{vmatrix}$$

EXAMPLE 4   Evaluating a 3 × 3 Determinant

Redo Example 3 by expanding down column 1. That is, evaluate $\begin{vmatrix} 3 & 2 & -4 \\ 1 & 7 & -3 \\ 0 & 2 & -5 \end{vmatrix}$ by expanding down column 1.

Solution

We choose to expand down column 1 because it has a 0 in it.

$$\begin{vmatrix} 3 & 2 & -4 \\ 1 & 7 & -3 \\ 0 & 2 & -5 \end{vmatrix} = 3\begin{vmatrix} 7 & -3 \\ 2 & -5 \end{vmatrix} - 1\begin{vmatrix} 2 & -4 \\ 2 & -5 \end{vmatrix} + 0\begin{vmatrix} 2 & -4 \\ 7 & -3 \end{vmatrix}$$

$$= 3(-35 - (-6)) - 1(-10 - (-8)) + 0(-6 - (-28))$$
$$= 3(-29) - 1(-2) + 0$$
$$= -87 + 2$$
$$= -85$$

Work Smart

Expand across the row or down the column with the most 0s.

Notice that the results of Example 3 and 4 are the same! However, the computation in Example 4 was a little easier because we expanded down the column that contains a 0. To make the computation easier, expand across the row or down the column that contains the most 0s.

QUICK ✓

5. Evaluate $\begin{vmatrix} 2 & -3 & 5 \\ 0 & 4 & -1 \\ 3 & 8 & -7 \end{vmatrix}$

④ Use Cramer's Rule to Solve a System of Three Equations Containing Three Variables

Cramer's Rule can also be applied to a system of three linear equations containing three unknowns.

> **CRAMER'S RULE FOR THREE EQUATIONS CONTAINING THREE VARIABLES**
>
> For the system of three equations containing three variables
>
> $$\begin{cases} a_1x + b_1y + c_1z = d_1 \\ a_2x + b_2y + c_2z = d_2 \\ a_3x + b_3y + c_3z = d_3 \end{cases}$$
>
> with
>
> $$D = \begin{vmatrix} a_1 & b_1 & c_1 \\ a_2 & b_2 & c_2 \\ a_3 & b_3 & c_3 \end{vmatrix} \neq 0 \quad D_x = \begin{vmatrix} d_1 & b_1 & c_1 \\ d_2 & b_2 & c_2 \\ d_3 & b_3 & c_3 \end{vmatrix} \quad D_y = \begin{vmatrix} a_1 & d_1 & c_1 \\ a_2 & d_2 & c_2 \\ a_3 & d_3 & c_3 \end{vmatrix} \quad D_z = \begin{vmatrix} a_1 & b_1 & d_1 \\ a_2 & b_2 & d_2 \\ a_3 & b_3 & d_3 \end{vmatrix}$$
>
> then
>
> $$x = \frac{D_x}{D} \qquad y = \frac{D_y}{D} \qquad z = \frac{D_z}{D}$$

EXAMPLE 5   How to Use Cramer's Rule

Use Cramer's Rule, if applicable, to solve the following system:

$$\begin{cases} x - 2y + z = -9 \\ -3x + y - 2z = 5 \\ 4x + 3z = 1 \end{cases}$$

Step-by-Step Solution

Step 1: Find the determinant of the coefficients of the variables, D. We choose to expand down column 2. Do you see why?

$$D = \begin{vmatrix} 1 & -2 & 1 \\ -3 & 1 & -2 \\ 4 & 0 & 3 \end{vmatrix} = -(-2)\begin{vmatrix} -3 & -2 \\ 4 & 3 \end{vmatrix} + 1\begin{vmatrix} 1 & 1 \\ 4 & 3 \end{vmatrix} - 0\begin{vmatrix} 1 & 1 \\ -3 & -2 \end{vmatrix}$$

$$= 2[-3(3) - 4(-2)] + 1[1(3) - 4(1)] - 0$$

$$= 2(-1) + 1(-1) - 0$$

$$= -2 - 1$$

$$= -3$$

Step 2: Because $D \neq 0$, we continue by determining D_x by replacing the first column in D with the constants on the right side of the equal sign in the system. We determine D_y by replacing the second column in D with the constants on the right side of the equal sign in the system. We determine D_z by replacing the third column in D with the constants on the right side of the equal sign in the system.

Expand down column 1

$$D_x = \begin{vmatrix} -9 & -2 & 1 \\ 5 & 1 & -2 \\ 1 & 0 & 3 \end{vmatrix} = -(-2)\begin{vmatrix} 5 & -2 \\ 1 & 3 \end{vmatrix} + 1\begin{vmatrix} -9 & 1 \\ 1 & 3 \end{vmatrix} - 0\begin{vmatrix} -9 & 1 \\ 5 & -2 \end{vmatrix}$$

$$= 2[15 - (-2)] + 1[-27 - 1] - 0$$

$$= 2(17) - 28$$

$$= 6$$

Expand down column 1

$$D_y = \begin{vmatrix} 1 & -9 & 1 \\ -3 & 5 & -2 \\ 4 & 1 & 3 \end{vmatrix} = 1\begin{vmatrix} 5 & -2 \\ 1 & 3 \end{vmatrix} - (-3)\begin{vmatrix} -9 & 1 \\ 1 & 3 \end{vmatrix} - 4\begin{vmatrix} -9 & 1 \\ 5 & -2 \end{vmatrix}$$

$$= -15$$

Expand down column 2

$$D_z = \begin{vmatrix} 1 & -2 & -9 \\ -3 & 1 & 5 \\ 4 & 0 & 1 \end{vmatrix} = -(-2)\begin{vmatrix} -3 & 5 \\ 4 & 1 \end{vmatrix} + 1\begin{vmatrix} 1 & -9 \\ 4 & 1 \end{vmatrix} - 0\begin{vmatrix} 1 & -9 \\ -3 & 5 \end{vmatrix}$$

$$= -9$$

Step 3: Find $x = \dfrac{D_x}{D}$, $y = \dfrac{D_y}{D}$, and $z = \dfrac{D_z}{D}$.

$$x = \frac{D_x}{D} = \frac{6}{-3} = -2$$

$$y = \frac{D_y}{D} = \frac{-15}{-3} = 5$$

$$z = \frac{D_z}{D} = \frac{-9}{-3} = 3$$

Step 4: Check your answer.

$$x - 2y + z = -9$$
$$-2 - 2(5) + 3 \overset{?}{=} -9$$
$$-2 - 10 + 3 \overset{?}{=} -9$$
$$-9 = -9 \quad \text{True}$$

$$-3x + y - 2z = 5$$
$$-3(-2) + 5 - 2(3) \overset{?}{=} 5$$
$$6 + 5 - 6 \overset{?}{=} 5$$
$$5 = 5 \quad \text{True}$$

$$4x + 3z = 1$$
$$4(-2) + 3(3) \overset{?}{=} 1$$
$$-8 + 9 \overset{?}{=} 1$$
$$1 = 1 \quad \text{True}$$

The solution is the ordered triple $(-2, 5, 3)$.

QUICK ✓

6. Use Cramer's Rule to solve the system $\begin{cases} x - y + 3z = -2 \\ 4x + 3y + z = 9 \\ -2x + 5z = 7 \end{cases}$

We already know that Cramer's Rule does not apply when the determinant of the coefficients on the variables, D, is 0. But can we learn anything about the system other than it is not a consistent and independent system if $D = 0$? The answer is yes!

Work Smart: Study Skills
At the end of Section E.1, we summarized the methods that can be used to solve a system of linear equations in two variables. Make up a summary table of your own for solving systems of linear equations in three variables. When is elimination appropriate? When would you use matrices? When is Cramer's Rule best?

> **CRAMER'S RULE WITH INCONSISTENT OR DEPENDENT SYSTEMS**
> - If $D = 0$ and at least one of the determinants D_x, D_y, or D_z is different from 0, then the system is inconsistent and the solution set is $\emptyset$ or $\{\ \}$.
> - If $D = 0$ and all the determinants D_x, D_y, or D_z equal 0, then the system is consistent and dependent so that there are infinitely many solutions.

E.4 Exercises

For Extra Help: Student Solutions Manual CD Video PH Math/Tutor Center MathXL Tutorials on CD MathXL® MyMathLab

Concepts and Vocabulary

In Problems 1–3, fill in the blanks.

1. Cramer's Rule uses _____ to solve a system of linear equations.

2. $D = \begin{vmatrix} a & b \\ c & d \end{vmatrix} = $ _____.

3. A matrix is _____ provided that the number of rows equals the number of columns.

In Problems 4 and 5, answer True or False to each statement.

4. A 3×3 determinant can never equal 0.

5. If $D = 0$, then the system of equations is inconsistent.

6. Why can't Cramer's Rule be used if the determinant of the coefficient matrix is zero? What are the possibilities for the solution to the system of equations if the coefficient matrix is zero?

7. Suppose that you wish to solve a system of equations using Cramer's Rule and find that $D = 4$, $D_x = -4$, $D_y = 8$, and $D_z = 0$. What is the solution to the system?

8. Suppose that you wish to solve a system of equations using Cramer's Rule and find that $D = 0$, $D_x = 3$, $D_y = 7$, and $D_z = -13$. What is the solution to the system?

Building Skills

In Problems 9–18, find the value of each determinant.

9. $\begin{vmatrix} 4 & 2 \\ 1 & 3 \end{vmatrix}$

10. $\begin{vmatrix} 5 & 3 \\ 2 & 4 \end{vmatrix}$

11. $\begin{vmatrix} -2 & -4 \\ 1 & 3 \end{vmatrix}$

12. $\begin{vmatrix} -8 & 5 \\ -4 & 3 \end{vmatrix}$

13. $\begin{vmatrix} 2 & 0 & -1 \\ 3 & 8 & -3 \\ 1 & 5 & -2 \end{vmatrix}$

14. $\begin{vmatrix} -2 & 1 & 6 \\ -3 & 2 & 5 \\ 1 & 0 & -2 \end{vmatrix}$

15. $\begin{vmatrix} -3 & 2 & 3 \\ 0 & 5 & -2 \\ 1 & 4 & 8 \end{vmatrix}$

16. $\begin{vmatrix} 8 & 4 & -1 \\ 2 & -7 & 1 \\ 0 & 5 & -3 \end{vmatrix}$

17. $\begin{vmatrix} 0 & 2 & 1 \\ 1 & -6 & -4 \\ -3 & 4 & 5 \end{vmatrix}$

18. $\begin{vmatrix} -3 & 4 & -2 \\ 1 & -2 & 0 \\ 0 & 6 & 6 \end{vmatrix}$

In Problems 19–26, solve each system of equations using Cramer's Rule, if applicable.

19. $\begin{cases} x + y = -4 \\ x - y = -12 \end{cases}$

20. $\begin{cases} x + y = 6 \\ x - y = 4 \end{cases}$

21. $\begin{cases} 2x + 3y = 3 \\ -3x + y = -10 \end{cases}$

22. $\begin{cases} 2x + 4y = -6 \\ 3x + 2y = 7 \end{cases}$

23. $\begin{cases} 3x + 4y = 1 \\ -6x + 8y = 4 \end{cases}$

24. $\begin{cases} 2x + 4y = 6 \\ 3x + 6y = 1 \end{cases}$

25. $\begin{cases} 2x - 6y - 12 = 0 \\ 3x - 5y - 11 = 0 \end{cases}$

26. $\begin{cases} 3x - 6y - 2 = 0 \\ x + 2y - 4 = 0 \end{cases}$

In Problems 27–38, solve each system of equations using Cramer's Rule, if applicable.

27. $\begin{cases} x - y + z = -4 \\ x + 2y - z = 1 \\ 2x + y + 2z = -5 \end{cases}$

28. $\begin{cases} x + y - z = 6 \\ x + 2y + z = 6 \\ -x - y + 2z = -7 \end{cases}$

29. $\begin{cases} x + y + z = 4 \\ 5x + 2y - 3z = 7 \\ 2x - y - z = 5 \end{cases}$

30. $\begin{cases} x + y + z = -3 \\ -2x - 3y - z = 1 \\ 2x - y - 3z = -5 \end{cases}$

31. $\begin{cases} 2x + y - z = 4 \\ -x + 2y + 2z = -6 \\ 5x + 5y - z = 6 \end{cases}$

32. $\begin{cases} -x + 2y - z = 2 \\ 2x + y + 2z = -6 \\ -x + 7y - z = 0 \end{cases}$

33. $\begin{cases} 3x + y + z = 5 \\ x + y - 3z = 9 \\ 4x + 3y + z = 11 \end{cases}$

34. $\begin{cases} x - 2y - z = 1 \\ 2x + 2y + z = 3 \\ x - 4y + 3z = 14 \end{cases}$

35. $\begin{cases} 2x + z = 27 \\ -x - 3y = 6 \\ x - 2y + z = 27 \end{cases}$

36. $\begin{cases} x + 2y + z = 0 \\ -x - 3y - 2z = 3 \\ 2x - 3z = 7 \end{cases}$

37. $\begin{cases} 5x + 3y & = 2 \\ -10x & + 3z = -3 \\ & y - 2z = -9 \end{cases}$

38. $\begin{cases} 2x + 4y & = 0 \\ -2x & + z = -5 \\ & -4y - 3z = 1 \end{cases}$

Applying the Concepts

In Problems 39–42, solve for x.

39. $\begin{vmatrix} x & 3 \\ 1 & 2 \end{vmatrix} = 7$

40. $\begin{vmatrix} -2 & x \\ 3 & 4 \end{vmatrix} = 1$

41. $\begin{vmatrix} x & -1 & -2 \\ 1 & 0 & 4 \\ 3 & 2 & 5 \end{vmatrix} = 5$

42. $\begin{vmatrix} 2 & x & -1 \\ 3 & 5 & 0 \\ -4 & 1 & 2 \end{vmatrix} = 0$

Problems 43–46 use the following result. Determinants can be used to find the area of a triangle. If (a_1, b_1), (a_2, b_2), and (a_3, b_3) are the vertices of a triangle, the area of the triangle is $|D|$, where

$$D = \frac{1}{2} \begin{vmatrix} a_1 & a_2 & a_3 \\ b_1 & b_2 & b_3 \\ 1 & 1 & 1 \end{vmatrix}$$

△ **43. Geometry: Area of a Triangle** Given the points $A = (1, 1)$, $B = (5, 1)$, and $C = (5, 6)$,

   (a) Plot the points in the Cartesian plane and form triangle ABC.
   (b) Find the area of the triangle ABC.

△ **44. Geometry: Area of a Triangle** Given the points $A = (-1, -1)$, $B = (3, 2)$, and $C = (0, 6)$,

   (a) Plot the points in the Cartesian plane and form triangle ABC.
   (b) Find the area of the triangle ABC.

△ **45. Geometry: Area of a Parallelogram** Find the area of a parallelogram by doing the following.

   (a) Plot the points $A = (2, 1)$, $B = (7, 2)$, $C = (8, 4)$, and $D = (3, 3)$ in the Cartesian plane.
   (b) Form triangle ABC and find the area of triangle ABC.
   (c) Find the area of the triangle ADC.
   (d) Conclude that the diagonal of the parallelogram forms two triangles of equal area. Use this result to find the area of the parallelogram.

△ **46. Geometry: Area of a Parallelogram** Find the area of a parallelogram by doing the following.

   (a) Plot the points $A = (-3, -2)$, $B = (3, 1)$, $C = (4, 4)$, and $D = (-2, 1)$ in the Cartesian plane.
   (b) Form triangle ABC and find the area of triangle ABC.
   (c) Find the area of the triangle ADC.
   (d) Conclude that the diagonal of the parallelogram forms two triangles of equal area. Use this result to find the area of the parallelogram.

△ **47. Geometry: Equation of a Line** An equation of the line containing the two points (x_1, y_1) and (x_2, y_2) may be expressed as the determinant

$$\begin{vmatrix} x & y & 1 \\ x_1 & y_1 & 1 \\ x_2 & y_2 & 1 \end{vmatrix} = 0$$

   (a) Find the equation of the line containing $(3, 2)$ and $(5, 1)$ using the determinant.
   (b) Verify your result by using the slope formula and the point-slope formula.

△ **48. Geometry: Collinear Points** The distinct points (x_1, y_1), (x_2, y_2), and (x_3, y_3) are collinear if and only if

$$\begin{vmatrix} x_1 & y_1 & 1 \\ x_2 & y_2 & 1 \\ x_3 & y_3 & 1 \end{vmatrix} = 0$$

(a) Plot the points $(-3, -2)$, $(1, 2)$, and $(7, 8)$ in the Cartesian plane.
(b) Show that the points are collinear using the determinant.
(c) Show that the points are collinear using the idea of slopes.

Extending the Concepts

49. Evaluate the determinant $\begin{vmatrix} 3 & -2 \\ 1 & 4 \end{vmatrix}$. Now interchange rows 1 and 2 and

recompute the determinant. What do you notice? Do you think that this result is true in general?

50. Evaluate the determinant $\begin{vmatrix} -3 & 1 \\ 6 & 5 \end{vmatrix}$. Multiply the entries in column 2 by 3 and

recompute the determinant. What do you notice? Do you think that this result is true in general?

The Graphing Calculator

Graphing calculators have the ability to evaluate 2×2 determinants. First, we enter the matrix into the graphing calculator and name it A. Then, we compute the determinant of matrix A. Figure 7 shows the results of Example 1(a) using a TI-84 Plus graphing calculator.

Because graphing calculators have the ability to compute determinants of matrices, they can be used to solve systems of equations using Cramer's Rule. Figure 8 shows the results of Example 2 using Cramer's Rule, where $A = D$, $B = D_x$, and $C = D_y$.

Figure 7

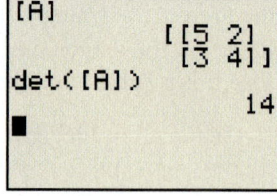

Figure 8

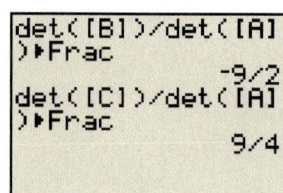

In Problems 51–56, use a graphing calculator to solve the system of equations.

51. $\begin{cases} x + y = -4 \\ x - y = -12 \end{cases}$

52. $\begin{cases} x + y = 6 \\ x - y = 4 \end{cases}$

53. $\begin{cases} 2x + 3y = 3 \\ -3x + y = -10 \end{cases}$

54. $\begin{cases} 2x + 4y = -6 \\ 3x + 2y = 7 \end{cases}$

55. $\begin{cases} x - y + z = -4 \\ x + 2y - z = 1 \\ 2x + y + 2z = -5 \end{cases}$

56. $\begin{cases} x + y - z = 6 \\ x + 2y + z = 6 \\ -x - y + 2z = -7 \end{cases}$

Table of Square Roots

n	$\sqrt{n}$	n	$\sqrt{n}$	n	$\sqrt{n}$	n	$\sqrt{n}$
1	1	26	5.09902	51	7.14143	76	8.71780
2	1.41421	27	5.19615	52	7.21110	77	8.77496
3	1.73205	28	5.29150	53	7.28011	78	8.83176
4	2	29	5.38516	54	7.34847	79	8.88819
5	2.23607	30	5.47723	55	7.41620	80	8.94427
6	2.44949	31	5.56776	56	7.48331	81	9
7	2.64575	32	5.65685	57	7.54983	82	9.05539
8	2.82843	33	5.74456	58	7.61577	83	9.11043
9	3	34	5.83095	59	7.68115	84	9.16515
10	3.16228	35	5.91608	60	7.74597	85	9.21954
11	3.31662	36	6	61	7.81025	86	9.27362
12	3.46410	37	6.08276	62	7.87401	87	9.32738
13	3.60555	38	6.16441	63	7.93725	88	9.38083
14	3.74166	39	6.24500	64	8	89	9.43398
15	3.87298	40	6.32456	65	8.06226	90	9.48683
16	4	41	6.40312	66	8.12404	91	9.53939
17	4.12311	42	6.48074	67	8.18535	92	9.59166
18	4.24264	43	6.55744	68	8.24621	93	9.64365
19	4.35890	44	6.63325	69	8.30662	94	9.69536
20	4.47214	45	6.70820	70	8.36660	95	9.74679
21	4.58258	46	6.78233	71	8.42615	96	9.79796
22	4.69042	47	6.85565	72	8.48528	97	9.84886
23	4.79583	48	6.92820	73	8.54400	98	9.89949
24	4.89898	49	7	74	8.60233	99	9.94987
25	5	50	7.07107	75	8.66025	100	10

Answers to Quick ✔ Exercises

Chapter 1

Section 1.2

1. $\{1, 3, 5, 9\}$ **2.** $\{$Alabama, Alaska, Arkansas, Arizona$\}$ **3.** $\emptyset$ or $\{\ \}$ **4.** 12 **5.** 12, 0 **6.** $-5, 12, 0$ **7.** $\frac{11}{5}, -5, 12, 2.\overline{76}, 0, \frac{18}{4}$

8. π **9.** All **10.** [number line from -4 to 4 with points at $\frac{1}{2}$, 1, and 3.5] **11.** $<$ **12.** $<$ **13.** $>$ **14.** $>$ **15.** $=$ **16.** 15 **17.** $\frac{3}{4}$

Section 1.3

1. 27 **2.** -8 **3.** 14 **4.** -7 **5.** -17 **6.** 3 **7.** -1 **8.** -2 **9.** -4 **10.** -20 **11.** -6 **12.** 10 **13.** -4 **14.** -3

15. -24 **16.** -56 **17.** -7 **18.** $-\frac{3}{7}$ **19.** 21 **20.** $\frac{8}{5}$ **21.** 5.75 **22.** 80 **23.** -178 **24.** -18 **25.** -2572 **26.** -21

27. -58 **28.** 12 **29.** 550 **30.** -21 **31.** -52 **32.** 80 **33.** 108 **34.** 325 **35.** 108 **36.** 360 **37.** $\frac{1}{6}$ **38.** $-\frac{1}{2}$ **39.** -5

40. -7 **41.** 9 **42.** $-\frac{27}{2}$

Section 1.4

1. $-\frac{2}{7}$ **2.** $-\frac{3}{5}$ **3.** -6 **4.** $\frac{27}{32}$ **5.** $-\frac{8}{3}$ **6.** $-\frac{6}{25}$ **7.** $\frac{3}{4}$ **8.** $-\frac{3}{22}$ **9.** $\frac{1}{12}$ **10.** $\frac{5}{7}$ **11.** -4 **12.** $-\frac{20}{31}$ **13.** $\frac{50}{49}$ **14.** $-\frac{3}{8}$

15. $-\frac{16}{7}$ **16.** $\frac{5}{27}$ **17.** $-\frac{6}{5}$ **18.** $\frac{10}{11}$ **19.** $-\frac{3}{7}$ **20.** $\frac{1}{7}$ **21.** $\frac{5}{36}$ **22.** $\frac{29}{42}$ **23.** $-\frac{13}{4}$ **24.** $\frac{13}{55}$ **25.** $-\frac{25}{16}$ **26.** $\frac{15}{4}$

27. 21.014 **28.** 64.57 **29.** 22.368 **30.** 337.5788 **31.** -71.412 **32.** -89.112 **33.** 4.78 **34.** 65.884 **35.** -0.1035 **36.** -899.5

37. 0.0135 **38.** 0.0198 **39.** 0.25 **40.** 8.36 **41.** -0.094 **42.** -0.03

Section 1.5

1. 8 feet **2.** 8 hours, 20 minutes **3.** 5 pounds, 8 ounces **4.** 22 **5.** $\frac{3}{20}$ **6.** 11.98 **7.** -78 **8.** $-\frac{36}{331}$ **9.** 349 **10.** 14

11. 14 **12.** -24.2 **13.** $\frac{50}{13}$ **14.** 0 **15.** undefined **16.** 0 **17.** undefined

Section 1.6

1. 11^5 **2.** 7^8 **3.** $(-2)^3$ **4.** 16 **5.** 64 **6.** $-\frac{1}{216}$ **7.** 0.81 **8.** -16 **9.** 16 **10.** 15 **11.** -31 **12.** 12 **13.** 31

14. -3 **15.** 19 **16.** -63 **17.** 4 **18.** $-\frac{8}{7}$ **19.** $\frac{4}{7}$ **20.** $\frac{5}{3}$ **21.** 20 **22.** 40 **23.** -9 **24.** 48 **25.** $-\frac{5}{11}$ **26.** 41

27. 12 **28.** -108 **29.** 10

Section 1.7

1. -7 **2.** 9 **3.** 2 **4.** \$220 **5.** $5x^2; 3xy$ **6.** $9ab; -3bc, 5ac; -ac^2$ **7.** $\frac{2mn}{5}; \frac{-3n}{7}$ **8.** $\frac{m^2}{3}; -8$ **9.** 2 **10.** 1 **11.** -1

12. 5 **13.** $-\frac{2}{3}$ **14.** $\frac{1}{6}$ **15.** like **16.** like **17.** unlike **18.** unlike **19.** $6x + 12$ **20.** $-5x - 10$ **21.** $-2k + 14$

22. $6x + 9$ **23.** $-5x$ **24.** $-4x^2$ **25.** $-8x + 3$ **26.** $-8x + 14$ **27.** $-2a + 9b - 4$ **28.** $12ac - 5a + b$ **29.** $8ab^2 - a^2b$

30. $2rs - \frac{3}{2}r^2 - 5$ **31.** $-2x - 1$ **32.** $-2m - n - 7$ **33.** $a - 11b$ **34.** $6x - 2$

Chapter 2

Section 2.1

1. yes **2.** no **3.** yes **4.** no **5.** $\{32\}$ **6.** $\{14\}$ **7.** $\{12\}$ **8.** $\{-15\}$ **9.** $\left\{\frac{7}{3}\right\}$ **10.** $\{-1\}$ **11.** $\left\{\frac{5}{8}\right\}$ **12.** $\left\{-\frac{13}{12}\right\}$

13. $p = \$12,455$ **14.** $\{2\}$ **15.** $\{-2\}$ **16.** $\left\{\frac{5}{2}\right\}$ **17.** $\left\{-\frac{7}{3}\right\}$ **18.** $\{9\}$ **19.** $\{-9\}$ **20.** $\left\{-\frac{10}{3}\right\}$ **21.** $\{6\}$ **22.** $\left\{\frac{8}{3}\right\}$

23. $\left\{-\frac{5}{14}\right\}$

Section 2.2

1. $\{3\}$ **2.** $\{2\}$ **3.** $\{18\}$ **4.** $\left\{\frac{3}{2}\right\}$ **5.** $\{2\}$ **6.** $\{-18\}$ **7.** $\left\{\frac{5}{2}\right\}$ **8.** $\{-8\}$ **9.** $\{2\}$ **10.** $\left\{\frac{4}{3}\right\}$ **11.** $\{13\}$ **12.** $\{6\}$

13. $\{6\}$ **14.** $\left\{-\frac{7}{2}\right\}$ **15.** $\{1\}$ **16.** $\left\{-\frac{5}{3}\right\}$ **17.** 52 hours

Section 2.3

1. $\{10\}$ **2.** $\left\{-\frac{5}{18}\right\}$ **3.** $\{-14\}$ **4.** $\left\{\frac{35}{3}\right\}$ **5.** $\{100\}$ **6.** $\{50\}$ **7.** $\{50\}$ **8.** $\{160\}$ **9.** $\{4\}$ **10.** $\{3000\}$ **11.** $\emptyset$

12. all real numbers **13.** all real numbers **14.** $\emptyset$ **15.** identity; all real numbers **16.** conditional; $\{4\}$ **17.** contradiction; $\emptyset$

18. contradiction; $\emptyset$ **19.** \$500

Section 2.4

1. $59°$ **2.** multiply; 4   **3.** \$312.50   **4.** 5936 persons   **5.** \$50 interest; \$2550 total   **6.** 36 square inches   **7. (a)** 187 square feet **(b)** \$46.75

8. 28.27 sq feet   **9.** The extra large pizza is the better buy.   **10.** $C = \dfrac{5}{9}(F - 32)$ **11.** $h = \dfrac{A - 2\pi r^2}{2\pi r}$ **12.** $y = \dfrac{7 - x}{2}$ **13.** $y = \dfrac{15 - 5x}{-3}$

14. $b = \dfrac{28 - 3a}{8}$ **15.** $t = 24 - 6rs$ **16.** $\dfrac{d}{r} = t; t = 9\dfrac{1}{6}$ hours, or 9 hours 10 min.   **17.** $\dfrac{I}{Pr} = t; t = 0.5$ year

Section 2.5

1. $5 + 17$ **2.** $-2 \cdot 6$ **3.** $\dfrac{25}{3}$ **4.** $7 - 4$ **5.** $2a - 2$ **6.** $3 + \dfrac{z}{4}$ **7.** $z + 50$ **8.** $x - 15$ **9.** $75 - d$ **10.** $3l - 2$ **11.** $2q + 3$

12. $3b - 5$ **13.** $3y = 21$ **14.** $3 + x = 5x$ **15.** $x - 10 = \dfrac{x}{2}$ **16.** $y - 3 = 5y$

17. $s + \dfrac{2}{3}s = 15$; Sean pays \$9 and Connor pays \$6   **18.** $n + (n + 2) + (n + 4) = 270; 88, 90, 92$

19. $76 = x + (x + 24) + \dfrac{1}{2}(x + 24)$, where x = length of smallest piece of ribbon; 16 inches, 40 inches, 20 inches

20. $x + 2x = 18{,}000$, where x = amount invested in stocks; \$6,000 in stocks; \$12,000 in bonds   **21.** 150 miles

Section 2.6

1. 801   **2.** 2.52   **3.** 36   **4.** 3.5   **5.** 40%   **6.** 37.5%   **7.** 20.5%   **8.** 110%   **9.** 50   **10.** 150   **11.** 80   **12.** 75
13. 31,450,000 residents   **14.** \$39,975   **15.** \$7300   **16.** \$1.25 per gallon.   **17.** \$659

Section 2.7

1. $39°$ and $51°$ **2.** $70°$ and $110°$ **3.** $35°, 40°$, and $105°$ **4.** width is $1\dfrac{1}{2}$ feet; length is 3 feet   **5.** 5 feet

6. José's speed is 13 mph and Luis' speed is 8 mph.   **7.** It takes $\dfrac{1}{2}$ hour to catch up to Tanya. Each of you has traveled 20 miles.

Section 2.8

1. $\{x | -3 \le x \le 2\}; [-3, 2]$

2. $\{x | 3 \le x < 6\}; [3, 6)$

3. $\{x | x \le 3\}; (-\infty, 3]$

4. $\left\{ x \middle| \dfrac{1}{2} < x < \dfrac{7}{2} \right\}; \left(\dfrac{1}{2}, \dfrac{7}{2} \right)$

5. $0 < x \le 5$

6. $-6 < x < 0$

7. $x > 5$

8. $x \le \dfrac{8}{3}$

9. $\{n | n > 3\}; (3, \infty)$

10. $\{x | x < 4\}; (-\infty, 4)$

11. $\{n | n \le -4\}; (-\infty, -4]$

12. $\{x | x > -1\}; (-1, \infty)$

13. $\{k | k < -6\}; (-\infty, -6)$

14. $\left\{ n \middle| n \ge -\dfrac{5}{2} \right\}; \left[-\dfrac{5}{2}, \infty \right)$

15. $\{k | k < -8\}; (-\infty, -8)$

16. $\left\{ p \middle| p \ge \dfrac{3}{5} \right\}; \left[\dfrac{3}{5}, \infty \right)$

17. $\{x | x > 7\}; (7, \infty)$

18. $\{n | n > -3\}; (-3, \infty)$

19. $\{x | x > -4\}; (-4, \infty)$

20. $\{x | x \le -6\}; (-\infty, -6]$

21. $\{x | x > 8\}; (8, \infty)$

22. $\left\{ x \middle| x \le \dfrac{19}{8} \right\}; \left(-\infty, \dfrac{19}{8} \right]$

23. $\{x | x \ge -67\}; [-67, \infty)$

24. $\varnothing$ or $\{\ \}$

25. $\{x | x > 4\}; (4, \infty)$

26. $\{x | x$ is any real number$\}; (-\infty, \infty)$

27. Any balance over \$500   **28.** For more than 24 boxes revenue exceeds cost.

Chapter 3

Section 3.1

1. $\left(-\dfrac{3}{2}, \dfrac{5}{2} \right)$ **(a)** I,   **(b)** III,   **2.**   **(a)** II,   **(b)** I,   **3.** $(2, 3)$ **4.** $(1, -3)$ **5.** $(-3, 0)$
(c) IV,   **(d)** x-axis,   **(c)** III,   **(d)** x-axis,
(e) y-axis,   **(f)** II   **(e)** y-axis,   **(f)** IV

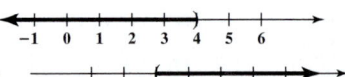

6. $(-2, -1)$ **7. (a)** Yes   **(b)** No   **(c)** No   **8. (a)** No   **(b)** Yes   **(c)** Yes   **9.** $(3, 4)$ **10.** $(-3, 1)$

11.

x	y	(x, y)
-2	-12	$(-2, -12)$
0	-2	$(0, -2)$
1	3	$(1, 3)$

12.

x	y	(x, y)
-1	7	$(-1, 7)$
2	-2	$(2, -2)$
5	-11	$(5, -11)$

13.

x	y	(x, y)
-5	2	$(-5, 2)$
-2	-4	$(-2, -4)$
2	-12	$(2, -12)$

14.

x	y	(x, y)
-6	-6	$(-6, -6)$
-1	-4	$(-1, -4)$
2	$-\frac{14}{5}$	$\left(2, -\frac{14}{5}\right)$

15. (a)

x (ft^3)	50 ft^3	100 ft^3	150 ft^3
C ($)	$38.83	$72.17	$105.50

$(50, 38.83), (100, 72.17), (150, 105.50)$

(b)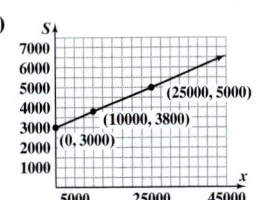

Section 3.2

1. 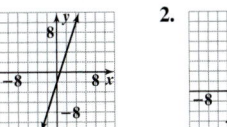   **2.**   **3.** Linear   **4.** Not linear   **5.** Linear   **6.** Linear

7. 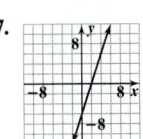   **8.** 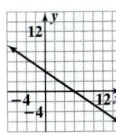   **9. (a)** $(0, 3000), (10000, 3800), (25000, 5000)$ **(b)**   **10.** Intercepts: $(0, 3)$, $(4, 0)$; x-intercept: 4; y-intercept: 3   **11.** Intercept: $(0, -2)$; y-intercept: -2

12. 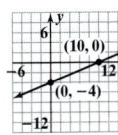   **13.** 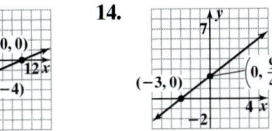   **14.** 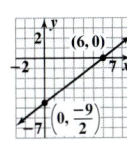   **15.**   **16.** 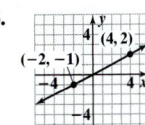   **17.**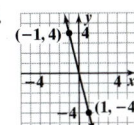

18. 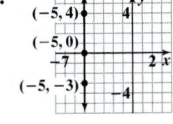   **19.** 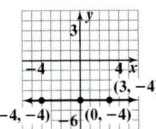   **20.**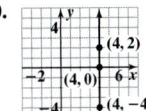

Section 3.3

1. 3; The value of y increases by 3 when x increases by 1.   **2.** $-\frac{5}{2}$; The value of y decreases by 5 when x increases by 2.

3. Slope undefined; when y increases by 1, there is no change in x.   **4.** $m = 0$; there is no change in y when x increases by 1 unit.   **5.** $m = 0$; there is no change in y when x increases by 1 unit.   **6.** Slope undefined; when y increases by 1, there is no change in x.

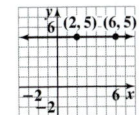   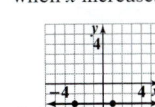   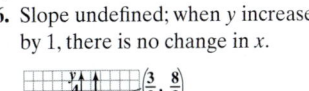

7. (a) 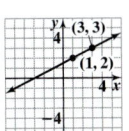   **(b)** 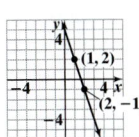   **(c)** 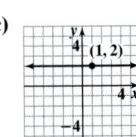   **8.** 8%   **9.** $\frac{1}{6}$ **10.** 0.12; between 10,000 and 14,000 miles driven, the annual cost of gasoline and maintenance on a Chevy Cavalier is $0.12 per mile.

Section 3.4

1. slope: 4, y-intercept: -3 **2.** slope: -3, y-intercept: 7   **3.** slope: $-\frac{2}{5}$, y-intercept: 3

4. 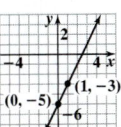   **5.** 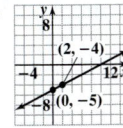   **6.** 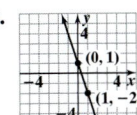   **7.** 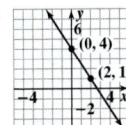   **8.** 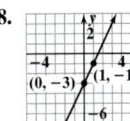   **9.** 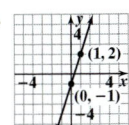   **10.**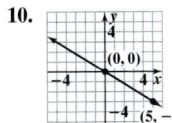

11. $y = 3x - 2$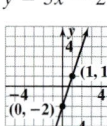

12. $y = -\frac{1}{4}x + 3$

13. $y = -1$

14. (a) 2075 grams **(b)** 2933 grams **(c)** slope = 143. Birth weight increases by 143 grams for each additional week of pregnancy.
(d) A gestation period of 0 weeks does not make sense.
(e)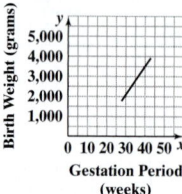

15. (a) $y = 0.38x + 50$ **(b)** $78.50 **(c)** 90 miles
(d)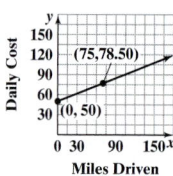

Section 3.5

1. $y = 3x - 5$ **2.** $y = \frac{1}{3}x - 5$ **3.** $y = -4x - 3$ **4.** $y = -\frac{5}{2}x - 5$ **5.** $y = 3$ **6.** $y = x + 2$ **7.** $y = -3x + 1$

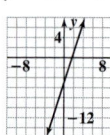

8. $x = 3$ **9. (a)** $y = -100x + 620$ **(b)** 390 gallons **(c)** The number of gallons sold will decrease by 100 if the price per gallon of gasoline increases by $1.

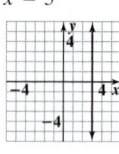

Section 3.6

1. Not parallel **2.** Parallel **3.** Not parallel **4.** $y = 2x - 1$ **5.** $y = -\frac{3}{2}x$ **6.** $x = 3$ **7.** $y = 5$

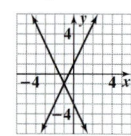

8. $\frac{1}{4}$ **9.** $-\frac{4}{5}$ **10.** 5 **11.** Perpendicular **12.** Not perpendicular **13.** Perpendicular **14.** $y = -\frac{1}{2}x$ **15.** $y = \frac{3}{2}x + 2$

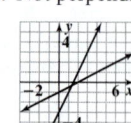

16. $y = -5$ **17.** $x = 3$

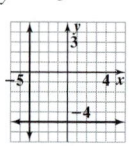

Section 3.7

1. $y = 3x$ **2.** $y = \frac{1}{3}x$ **3. (a)** $q = \frac{1}{4}w$ **(b)** 15 **(c)** **4.** $17.68 **5. (a)** $y = \frac{12}{x}$ **(b)** 3
6. (a) $p = \frac{40}{q}$ **(b)** 5 **7.** 25 times per second

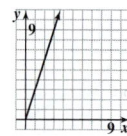

Section 3.8

1. (a) No **(b)** Yes **(c)** Yes **2. (a)** Yes **(b)** Yes **(c)** No

3. 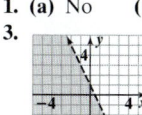 **4.** 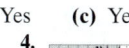 **5.** 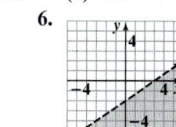 **6.**   **7.** 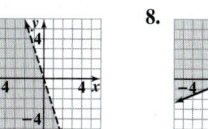 **8.** 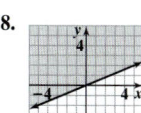 **9.**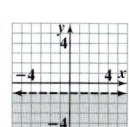

10. 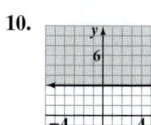 **11.** 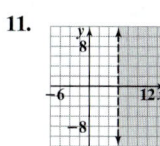 **12. (a)** $0.2s + 0.25t \le 2$ **(b)** Yes **(c)** No

Chapter 4

Section 4.1

1. (a) No **(b)** Yes **(c)** No **2. (a)** Yes **(b)** Yes **(c)** Yes **3.** $(4, 1)$ **4.** $(-2, 5)$ **5.** $(5, -3)$ **6.** No solution; $\{\ \}$ or $\varnothing$
7. No solution; $\{\ \}$ or $\varnothing$ **8.** Infinitely many solutions **9.** Infinitely many solutions **10.** One solution; consistent; independent
11. Infinitely many solutions; consistent; dependent **12.** No solution; inconsistent **13.** After 100 miles the cost is $60.

Section 4.2

1. $(2, 4)$ **2.** $(-3, 5)$ **3.** $\left(\dfrac{3}{2}, 7\right)$ **4.** $\left(-\dfrac{4}{3}, \dfrac{1}{2}\right)$ **5.** Infinitely many solutions **6.** No solution **7.** No solution **8.** During the year 2060

Section 4.3

1. $(-4, -2)$ **2.** $(8, 3)$ **3.** $(6, -5)$ **4.** $(5, 0)$ **5.** No solution **6.** Infinitely many solutions **7.** Infinitely many solutions
8. Cheeseburger: $1.75; shake: $2.25

Section 4.4

1. 43 and 61 **2.** Width: 50 yards; length: 150 yards **3.** $36°$ and $54°$ **4.** $49°$ and $131°$ **5.** Airspeed: 350 mph; wind resistance: 50 mph

Section 4.5

1.

	Number	·	Cost per Person	=	Amount
Adults	a		36.95		36.95a
Children	c		21.95		21.95c
Total					302.55

2. 45 dimes **3.** $40,000 in Aa-rated bonds: $50,000 in B-rated bond
4. 5 pounds of Brazilian coffee and 15 pounds of Colombian coffee
5. Mix 120 gallons of the wine with 5% alcohol and 80 gallons of the wine with 15% alcohol.

Section 4.6

1. (a) solution **(b)** not a solution **2.** 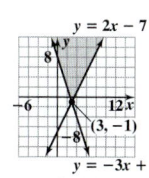 **3.** 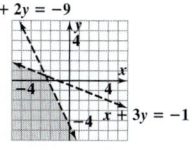 **4.** 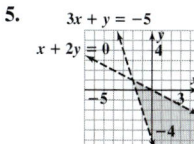 **5.**

6. (a) 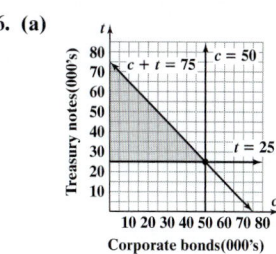 **(b)** Yes **(c)** No

Chapter 5

Section 5.1

1. Monomial, Coefficient $= 12$, Degree $= 6$ **2.** Not a monomial **3.** Monomial, Coefficient $= 10$, Degree $= 0$ **4.** Not a monomial
5. Coefficient: 3, Degree: 7 **6.** Coefficient: -2, Degree: 4 **7.** Not a monomial **8.** Coefficient: -1, Degree: 2 **9.** Polynomial; degree: 3
10. Not a polynomial **11.** Not a polynomial **12.** Polynomial; degree: 5 **13.** Polynomial; degree: 4 **14.** $12x^2 + 3x + 3$
15. $2z^4 + 4z^3 - z^2 + z - 1$ **16.** $5x^2y + 6x^2y^2 - xy^2$ **17.** $5x^3 - 15x^2 + 3x + 7$ **18.** $8y^3 - 3y^2 - 3y - 6$ **19.** $11x^2y + 2x^2y^2 - 9xy^2$
20. $7a^2 - 4ab - 12b^2$ **21. (a)** 1 **(b)** -214 **(c)** 101 **22. (a)** 40 **(b)** -20 **23.** $1875

Section 5.2
1. 27 **2.** -3125 **3.** c^8 **4.** y^9 **5.** a^9b^6 **6.** 2^8 **7.** $(-3)^6$ **8.** b^{10} **9.** z^{20} **10.** $8n^3$ **11.** $36y^2$ **12.** $-125x^{12}$

13. $49a^6b^2$ **14.** $8a^{11}$ **15.** $-12p^7$ **16.** $-18m^3n^9$ **17.** $\dfrac{2}{3}x^3y^3$

Section 5.3
1. $3x^3 - 6x^2 + 12x$ **2.** $-2a^4b^3 + 4a^3b^3 - 3a^3b^2$ **3.** $3n^5 - \dfrac{9}{8}n^4 - \dfrac{6}{7}n^3$ **4.** $x^2 + 9x + 14$ **5.** $2n^2 - 5n - 3$ **6.** $6p^2 - 7p + 2$
7. $x^2 + 7x + 12$ **8.** $y^2 + 2y - 15$ **9.** $a^2 - 6a + 5$ **10.** $2x^2 + 11x + 12$ **11.** $6y^2 + y - 15$ **12.** $6a^2 - 5a - 4$
13. $15x^2 - 14xy - 8y^2$ **14.** $a^2 - 16$ **15.** $9w^2 - 49$ **16.** $36b^2 - 4c^2$ **17.** $x^2 - 4y^6$ **18.** $z^2 - 18z + 81$ **19.** $p^2 + 2p + 1$
20. $16 - 8a + a^2$ **21.** $w^2 + 2wy + y^2$ **22.** $9z^2 - 24z + 16$ **23.** $25p^2 + 10p + 1$ **24.** $16a^2 - 40a + 25$ **25.** $4w^2 + 28wy + 49y^2$
26. $x^3 - 8$ **27.** $3y^3 + 4y^2 + 8y - 8$ **28.** $x^3 - 8$ **29.** $3y^3 + 4y^2 + 8y - 8$ **30.** $-24a^3 - 34a^2 + 10a$ **31.** $15x^2 + 54x - 24$

Section 5.4
1. 9 **2.** y^6 **3.** $\dfrac{7}{5}c$ **4.** $-\dfrac{3}{2}wz^7$ **5.** $\dfrac{p^4}{16}$ **6.** $\dfrac{b^3}{27}$ **7.** $\dfrac{a^6}{b^8}$ **8.** $-\dfrac{8a^6}{b^{12}}$ **9.** 1 **10.** 4 **11.** -1 **12.** $\dfrac{1}{z^5}$ **13.** $\dfrac{1}{p^4}$ **14.** $-\dfrac{1}{x^3}$

15. $\dfrac{1}{16}$ **16.** $-\dfrac{1}{49}$ **17.** $\dfrac{1}{49}$ **18.** $\dfrac{1}{8}$ **19.** 9 **20.** v **21.** -100 **22.** $\dfrac{5}{2}z^2$ **23.** $\dfrac{8}{7}$ **24.** -64 **25.** $\dfrac{25}{9a^2}$ **26.** $-\dfrac{27}{8}n^{12}$

27. $\dfrac{-20}{a^2}$ **28.** $\dfrac{2}{mn}$ **29.** $\dfrac{8}{m^5}$ **30.** $-\dfrac{4}{3}a^3b^3$ **31.** $\dfrac{9x^2}{7y^3}$ **32.** $\dfrac{z^6}{9y^4}$ **33.** $\dfrac{4w^4}{49z^6}$ **34.** $-\dfrac{2}{p^{11}}$ **35.** $\dfrac{-4}{k}$

Section 5.5
1. $2n^2 - 4n + 1$ **2.** $6k^2 - 9 + \dfrac{5}{2k^2}$ **3.** $\dfrac{xy^3}{4} + \dfrac{2y}{x} - \dfrac{1}{x^2}$ **4.** $x - 8$ **5.** $x - 4$

6. $4x + 5 + \dfrac{6}{x + 3}$ **7.** $4x^2 - 11x + 23 - \dfrac{45}{x + 2}$ **8.** $x^2 + 4x + 10 + \dfrac{60}{2x - 5}$ **9.** $4x - 3 + \dfrac{-7x + 7}{x^2 + 2}$

Section 5.6
1. 4.32×10^2 **2.** 1.0302×10^4 **3.** 5.432×10^6 **4.** 9.3×10^{-2} **5.** 4.59×10^{-5} **6.** 8×10^{-8} **7.** 310 **8.** 0.901 **9.** 170,000
10. 7 **11.** 0.00089 **12.** 6×10^7 **13.** 8×10^{-3} **14.** 1.5×10^4 **15.** 2.8×10^{-5} **16.** 4×10^5 **17.** 2×10^{-4} **18.** 5×10^3
19. 6.25×10^{-5} **20.** $2.604 \times 10^8 = 260{,}400{,}000$ barrels

Chapter 6

Section 6.1
1. 8 **2.** 5 **3.** 3 **4.** 7 **5.** $7y^2$ **6.** $2z$ **7.** $4xy^3$ **8.** $(2x + 3)$ **9.** $3(k + 8)$ **10.** $5z(z - 6)$ **11.** $6p^2(2 + 5p^2)$
12. $4y(4y^2 - 3y + 1)$ **13.** $2m^2n^2(3m^2 + 9mn^2 - 11n^3)$ **14.** $-4y(y - 2)$ **15.** $-3a(2a^2 - 4a + 1)$ **16.** $(a - 5)(2a + 3)$
17. $(z + 5)(7z - 4)$ **18.** $(x + y)(4 + b)$ **19.** $(2a - 3b)(3z - 1)$ **20.** $(2b + 1)(4 - 5a)$ **21.** $3(z + 4)(z^2 + 2)$
22. $2n(n + 1)(n^2 - 2)$

Section 6.2
1. $(y + 5)(y + 4)$ **2.** $(w + 6)(w + 4)$ **3.** $(q - 3)(q - 2)$ **4.** $(z - 7)(z - 2)$ **5.** $(y - 5)(y + 3)$ **6.** $(w - 3)(w + 4)$
7. $(q - 6)(q + 2)$ **8.** $(n + 6)(n - 1)$ **9.** Prime **10.** $(q + 9)(q - 5)$ **11.** $(x + 4y)(x + 5y)$ **12.** $(m + 7n)(m - 6n)$
13. $(n + 8)(n - 7)$ **14.** $(y - 7)(y - 5)$ **15.** $4(m + 3)(m - 7)$ **16.** $3z(z - 1)(z + 5)$ **17.** $-1(w + 5)(w - 2)$
18. $-2(a + 6)(a - 2)$

Section 6.3
1. $(3x + 2)(x + 1)$ **2.** $(7y + 1)(y + 3)$ **3.** $(3x - 4)(x - 3)$ **4.** $(5p - 1)(p - 4)$ **5.** $(2n + 1)(n - 9)$ **6.** $(4b + 3)(b - 2)$
7. $(3x + 2)(4x + 3)$ **8.** $(6y - 5)(2y + 7)$ **9.** $3(6x - y)(5x + 2y)$ **10.** $-1(6y + 1)(y - 4)$ **11.** $-1(3x + 2y)(3x + 5y)$
12. $-2(2x - 1)(3x + 4)$ **13.** $-3(2x - 5)(x + 3)$ **14.** $(3x + 4)(x - 2)$ **15.** $(5z + 3)(2z + 3)$ **16.** $3(4x + 3)(2x - 1)$
17. $-(5n - 1)(2n - 3)$

Section 6.4
1. $(x - 6)^2$ **2.** $(4x + 5)^2$ **3.** $(3a - 10b)^2$ **4.** $(z - 4)^2$ **5.** $(2n + 3)^2$ **6.** prime **7.** $4(z + 3)^2$ **8.** $2a(5a + 4)^2$
9. $2(5a^2 + 2b)^2$ **10.** $(z - 5)(z + 5)$ **11.** $(9m - 4n)(9m + 4n)$ **12.** $\left(4a - \dfrac{2}{3}b\right)\left(4a + \dfrac{2}{3}b\right)$ **13.** $(6b^2 - c)(6b^2 + c)$
14. $(10k^2 - 9w)(10k^2 + 9w)$ **15.** $(x^2 + 4)(x - 2)(x + 2)$ **16.** $3(7x - 4)(7x + 4)$ **17.** $-3ab(3a - 5b)(3a + 5b)$
18. $(z + 5)(z^2 - 5z + 25)$ **19.** $(2p - 3q^2)(4p^2 + 6pq^2 + 9q^4)$ **20.** $2a(3 - 2a)(9 + 6a + 4a^2)$ **21.** $-3(5b - 1)(25b^2 + 5b + 1)$

Section 6.5
1. $2(p - 5)(p + 9)$ **2.** $-3(3x + y)(5x - 2y)$ **3.** $(10x - 9y)(10x + 9y)$ **4.** $2a(b - 11)(b + 11)$ **5.** $(p - 6q)^2$
6. $3(5x + 3)^2$ **7.** $(5y - 4)(25y^2 + 20y + 16)$ **8.** $-3(2a - b)(4a^2 + 2ab + b^2)$ **9.** $(x^2 + 1)(x - 1)(x + 1)$
10. $-4y(3x - 2)(3x + 2)$ **11.** $(x^2 + 2)(2x + 3)$ **12.** $3(2x + 3)(x - 1)(x + 1)$ **13.** $-3(z^2 - 3z + 7)$ **14.** $3x(2y^2 + 5x^2)$

Section 6.6
1. $\{-3, 0\}$ **2.** $\left\{-\dfrac{5}{4}, 2\right\}$ **3.** $\{2, 4\}$ **4.** $\left\{-\dfrac{1}{2}, 3\right\}$ **5.** $\left\{-\dfrac{5}{2}, 1\right\}$ **6.** $\{-5, -4\}$ **7.** $\left\{-1, \dfrac{5}{2}\right\}$ **8.** $\left\{-4, \dfrac{1}{3}\right\}$ **9.** $\{-6, 4\}$

10. $\left\{-\dfrac{3}{5}, 1\right\}$ **11.** $\left\{\dfrac{4}{3}\right\}$ **12.** $\{-6, 3\}$ **13.** $\{-2, 3\}$ **14.** After 1 second or 4 seconds **15.** $\left\{-3, 3, \dfrac{5}{4}\right\}$ **16.** $\{-2, -1, 0\}$

Section 6.7

1. (a) After 4 seconds and after 6 seconds **(b)** After 10 seconds **2.** 8 km by 13 km **3.** base = 12 yards, height = 8 yards
4. $x = 12, x - 3 = 9$ **5.** 6 inches by 8 inches

Chapter 7

Section 7.1

1. (a) $-\dfrac{1}{8}$ **(b)** $\dfrac{3}{16}$ **2. (a)** -1 **(b)** 9 **3.** $\dfrac{11}{2}$ **4.** undefined **5.** $x = -7$ **6.** $n = -\dfrac{5}{3}$ **7.** $x = -3$ or $x = 1$ **8.** $k = -3$ or $k = 3$ **9.** $\dfrac{1}{2}$
10. $\dfrac{5}{2p - 1}$ **11.** $-\dfrac{z + 1}{2}$ **12.** $\dfrac{a + 7}{2a + 7}$ **13.** $\dfrac{x - y}{z - 2}$ **14.** $\dfrac{2k + 1}{2k - 1}$ **15.** -7 **16.** $\dfrac{-4}{4x - 1}$ **17.** $\dfrac{-(5z + 1)}{3}$

Section 7.2

1. $\dfrac{5p(p + 3)}{2}$ **2.** $\dfrac{7}{3(x + 5)}$ **3.** $\dfrac{2(x + 3)}{x + 5}$ **4.** $\dfrac{-3a}{7}$ **5.** $\dfrac{1}{3}$ **6.** $\dfrac{6(x + 1)}{x(2x - 1)}$ **7.** $\dfrac{(x - 3)^2}{(x - 4)^2}$ **8.** $\dfrac{q + 1}{(q - 5)(q + 5)}$ **9.** $\dfrac{7x}{4(x - 2)}$
10. $\dfrac{6m}{m + 2n}$

Section 7.3

1. $\dfrac{4}{x - 2}$ **2.** $x + 1$ **3.** $\dfrac{11x - 3}{6x - 5}$ **4.** $\dfrac{1}{x - 5}$ **5.** $\dfrac{2(4y - 3)}{2y - 5}$ **6.** $\dfrac{z + 1}{2z}$ **7.** $\dfrac{x + 8}{3}$ **8.** $\dfrac{x + 2}{x - 7}$ **9.** $\dfrac{3x - 1}{x - 5}$ **10.** 2 **11.** $\dfrac{2}{n - 3}$
12. $\dfrac{2k + 3}{2(k - 1)}$

Section 7.4

1. $24x^2y^3$ **2.** $42a^3b^3$ **3.** $15z(z + 1)$ **4.** $(x - 1)(x + 5)^2$ **5.** $-3(x + 7)(x - 7)$ **6.** $\dfrac{12p^2}{16p^3(p - 2)}$
7. $\dfrac{2x - 8}{(x + 2)(x - 4)(x + 3)}$ **8.** LCD $= (x - 5)(x - 2)(x + 1)$; $\dfrac{5}{x^2 - 4x - 5} = \dfrac{5x - 10}{(x - 5)(x - 2)(x + 1)}$; $\dfrac{-3}{x^2 - 7x + 10} = \dfrac{-3x - 3}{(x - 5)(x - 2)(x + 1)}$

Section 7.5

1. $\dfrac{25}{36}$ **2.** $\dfrac{7}{2}$ **3.** $\dfrac{11}{4}$ **4.** $\dfrac{7}{30}$ **5.** $\dfrac{3b + 10a^2}{24a^3b^2}$ **6.** $\dfrac{6x^2 + 35y}{90x^3y^2}$ **7.** $\dfrac{2(4x - 1)}{(x - 4)(x + 2)}$ **8.** $\dfrac{3n - 13}{(n - 3)(n + 1)}$ **9.** $\dfrac{2}{x - 2}$ **10.** $\dfrac{1}{z + 2}$
11. $\dfrac{2}{(x + 5)(x - 5)}$ **12.** $\dfrac{-16ab - 15}{20a^2b^3}$ **13.** $\dfrac{2(x - 10)}{x(x - 4)}$ **14.** $\dfrac{x - 11}{(x - 5)(x + 4)(x - 1)}$ **15.** $\dfrac{(x + 8)(x - 1)}{4x(x - 3)}$ **16.** $\dfrac{14 - 5p}{6p(p - 1)}$
17. $\dfrac{z - 2}{z - 5}$ **18.** $\dfrac{2x - 5}{x - 1}$ **19.** $\dfrac{4x}{(x + 1)(x - 1)}$ **20.** $\dfrac{2}{x - 2}$

Section 7.6

1. $\dfrac{k + 1}{2}$ **2.** $\dfrac{1}{4n}$ **3.** $\dfrac{2}{3x}$ **4.** $\dfrac{1}{3 + y}$ **5.** $\dfrac{3}{x + 5}$ **6.** $\dfrac{-2}{y + 4}$ **7.** $\dfrac{y + 2x}{2y - x}$

Section 7.7

1. $\left\{\dfrac{2}{3}\right\}$ **2.** $\{-18\}$ **3.** $\{-4\}$ **4.** $\{-6\}$ **5.** $\{4\}$ **6.** $\{-5, 6\}$ **7.** $\{5\}$ **8.** $\{\ \}$ or $\varnothing$ **9.** After $2\dfrac{1}{2}$ hours and 10 hours
10. $x = \dfrac{4g}{R}$ **11.** $r = 1 - \dfrac{a}{S}$ or $r = \dfrac{S - a}{S}$ **12.** $p = \dfrac{fq}{q - f}$

Section 7.8

1. $\{-12\}$ **2.** $\left\{\dfrac{13}{3}\right\}$ **3.** $\left\{-\dfrac{2}{3}\right\}$ **4.** $\{3\}$ **5.** about 111 minutes **6.** \$233.38 **7.** 210 miles **8.** $XY = 8$ **9.** 6 feet
10. 2 hours **11.** $3\dfrac{1}{3}$ hours **12.** 20 mph **13.** 6 mph

Chapter 8

Section 8.1
1. A: Quadrant I ** ** B: Quadrant IV **2.** A: Quadrant II ** ** B: x-axis **3. (a)** No **(b)** Yes **(c)** Yes
    C: y-axis **   ** D: Quadrant III ** ** C: Quadrant IV ** ** D: Quadrant I

4. (a) Yes   **(b)** No   **(c)** Yes   **5.** 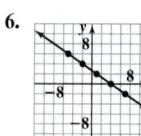   **6.** 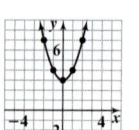   **7.** 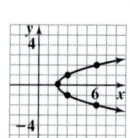   **8.**   **9.**

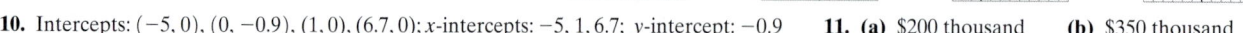

10. Intercepts: $(-5, 0), (0, -0.9), (1, 0), (6.7, 0)$; x-intercepts: $-5, 1, 6.7$; y-intercept: -0.9 **11. (a)** \$200 thousand   **(b)** \$350 thousand
(c) The capacity of the refinery is 700 thousand gallons of gasoline per hour.   **(d)** The intercept is $(0, 100)$. The cost of \$100 thousand for producing 0 gallons of gasoline can be thought of as fixed costs.

Section 8.2

1. {(Max, November 8), (Alesia, January 20), (Trent, March 3), (Yolanda, November 8), (Wanda, July 6), (Elvis, January 8)}
2.

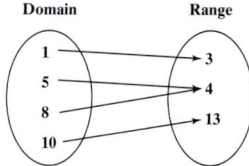

Domain   **Range**

3. Domain: {Max, Alesia, Trent, Yolanda, Wanda, Elvis}; Range: {January 20, March 3, July 6, November 8, January 8}
4. Domain: $\{1, 5, 8, 10\}$; Range: $\{3, 4, 13\}$ **5.** Domain: $\{-2, -1, 2, 3, 4\}$; Range: $\{-3, -2, 0, 2, 3\}$
6. Domain: $\{x | -2 \le x \le 4\}$ or $[-2, 4]$; Range: $\{y | -2 \le y \le 2\}$ or $[-2, 2]$
7. Domain: $\{x | x$ is a real number$\}$ or $(-\infty, \infty)$; Range: $\{y | y$ is a real number$\}$ or $(-\infty, \infty)$

8.   **9.**   **10.**

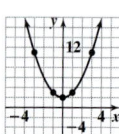

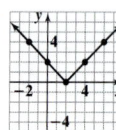

Domain: $\{x | x$ is a real number$\}$ or $(-\infty, \infty)$
Range: $\{y | y$ is a real number$\}$ or $(-\infty, \infty)$

Domain: $\{x | x$ is a real number$\}$ or $(-\infty, \infty)$
Range: $\{y | y \ge -8\}$ or $[-8, \infty)$

Domain: $\{x | x \ge 1\}$ or $[1, \infty)$
Range: $\{y | y$ is a real number$\}$ or $(-\infty, \infty)$

Section 8.3

1. Function; Domain: {Max, Alesia, Trent, Yolanda, Wanda, Elvis}; Range: {January 20, March 3, July 6, November 8, January 8}   **2.** Not a function
3. Function; Domain: $\{-3, -2, -1, 0, 1\}$, Range: $\{0, 1, 2, 3\}$ **4.** Not a function   **5.** Function   **6.** Not a function   **7.** Function
8. Not a function   **9.** Function   **10.** Not a function   **11.** 14   **12.** -13 **13.** $3x - 4$ **14.** $3x - 6$
15.   **16.**   **17.**   **18. (a)** Independent variable: t; dependent variable: A
(b) $A(30) \approx 706.86$ square miles. After 30 days, the area contaminated with oil will be a circle covering about 706.86 square miles.

Section 8.4

1. $\{x | x$ is a real number$\}$; $(-\infty, \infty)$ **2.** $\{x | x \ne 3\}$ **3.** $\{x | x \ne -2, 7\}$ **4.** $\{r | r > 0\}$; $(0, \infty)$
5. (a) Domain: $\{x | x$ is a real number$\}$; $(-\infty, \infty)$; Range: $\{y | y \le 2\}$; $(-\infty, 2]$
(b) $(-2, 0), (0, 2), (2, 0)$; x-intercepts: -2 and 2; y-intercept: 2
6. (a) $f(-3) = -15$; $f(1) = -3$ **(b)** Domain: $\{x | x$ is a real number$\}$ or $(-\infty, \infty)$
(c) Range: $\{y | y$ is a real number$\}$ or $(-\infty, \infty)$
(d) $(-2, 0), (0, 0), (2, 0)$; x-intercepts: $-2, 0$, and 2; y-intercept: 0   **(e)** $\{3\}$
7. (a) No   **(b)** $f(3) = -2$; $(3, -2)$ is on the graph
(c) $x = 5$; $(5, -8)$ is on the graph
8.

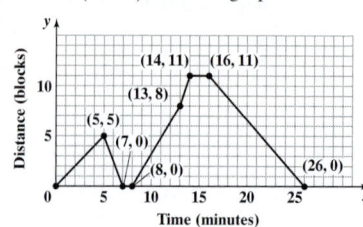

Section 8.5

1.   **2.**   **3.**   **4.**

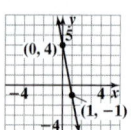

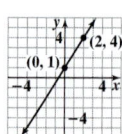

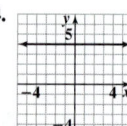

5. (a) $\{x | x \ge 0\}$ or $[0, \infty)$ **(b)** $(0, 40)$ **(c)** \$68   **(d)**   **(e)** 130 miles

6. (a) $C(x) = 81x + 2000$

(b)

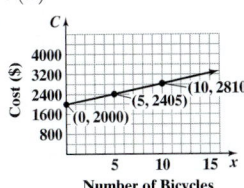

(c) $3458

(d) 25

7. (a) $C(x) = 0.18x + 250$

(b) $[0, \infty)$

(c) $307.60

(d)

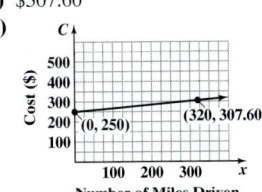

(e) 180 miles

8. (a)

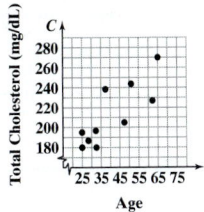

(b) As age increases, total cholesterol also increases.

9. Nonlinear   **10.** Linear with positive slope

11. (a) Answers will vary. Using $(25, 180)$ and $(65, 269)$: $y = f(x) = 2.225x + 124.375$

(b) Answers will vary.

(c) 211   **(d)** Each year, a male's total cholesterol increases by 2.225 mg/dL; No

Section 8.6

1. $\{1, 3, 5\}$ **2.** $\{2, 4, 6\}$ **3.** $\{1, 2, 3, 4, 5, 6, 7\}$ **4.** $\{1, 2, 3, 4, 5, 6, 8\}$ **5.** $\emptyset$ or $\{\ \}$ **6.** $\{1, 2, 3, 4, 5, 6, 7, 8\}$

7. $\{x | 2 < x < 7\}$; $(2, 7)$

8. $\{x | x \leq -3 \text{ or } x > 2\}$; $(-\infty, -3] \cup (2, \infty)$

9. $\{x | x \geq 2\}$; $[2, \infty)$

10. $\{x | -3 < x < 3\}$; $(-3, 3)$

11. $\{x | 1 < x < 3\}$; $(1, 3)$

12. $\{\ \}$ or $\emptyset$ **13.** $\{1\}$

14. $\{x | -1 < x < 3\}$; $(-1, 3)$

15. $\left\{x \left| \frac{5}{4} < x \leq 2\right.\right\}$; $\left(\frac{5}{4}, 2\right]$

16. $\{x | -6 \leq x \leq -2\}$; $[-6, -2]$

17. $\{x | x < -2 \text{ or } x > 5\}$; $(-\infty, -2) \cup (5, \infty)$

18. $\{x | x \leq 2 \text{ or } x > 6\}$; $(-\infty, 2] \cup (6, \infty)$

19. $\{x | x \geq 1\}$; $[1, \infty)$

20. $\{x | x < 4 \text{ or } x > 9\}$; $(-\infty, 4) \cup (9, \infty)$

21. $\{x | x \text{ is any real number}\}$; $(-\infty, \infty)$

22. $\{x | x \geq -1\}$; $[-1, \infty)$

23. To be in the 25% tax bracket, an individual must have income between $29,700 and $71,950.

24. Sophia's monthly minutes for the year were between 115 minutes and 255 minutes.

Section 8.7

1. $\{-7, 7\}$ **2.** $\{-1, 1\}$ **3.** $\{-2, 5\}$ **4.** $\left\{-\frac{5}{3}, 3\right\}$ **5.** $\left\{-1, \frac{9}{5}\right\}$ **6.** $\{-5, 1\}$ **7.** $\{\ \}$ or $\emptyset$ **8.** $\{\ \}$ or $\emptyset$ **9.** $\{-1\}$

10. $\left\{-8, -\frac{2}{3}\right\}$ **11.** $\{-2, 3\}$ **12.** $\{-3, 0\}$ **13.** $\{2\}$ **14.** $\{x | -5 \leq x \leq 5\}$; $[-5, 5]$

15. $\left\{x \left| -\frac{3}{2} < x < \frac{3}{2}\right.\right\}$; $\left(-\frac{3}{2}, \frac{3}{2}\right)$

16. $\{x | -8 < x < 2\}$; $(-8, 2)$

17. $\{x | -2 \leq x \leq 5\}$; $[-2, 5]$

18. $\{\ \}$ or $\emptyset$

19. $\{x | -2 < x < 2\}$; $(-2, 2)$

20. $\{x | -1 \leq x \leq 7\}$; $[-1, 7]$

21. $\{x | -2 \leq x \leq 1\}$; $[-2, 1]$

22. $\left\{x \left| -\frac{7}{3} < x < 3\right.\right\}$; $\left(-\frac{7}{3}, 3\right)$

23. $\{x | x \leq -6 \text{ or } x \geq 6\}$; $(-\infty, -6] \cup [6, \infty)$

24. $\left\{x \left| x < -\frac{5}{2} \text{ or } x > \frac{5}{2}\right.\right\}$; $\left(-\infty, -\frac{5}{2}\right) \cup \left(\frac{5}{2}, \infty\right)$

25. $\{x | x < -7 \text{ or } x > 1\}$; $(-\infty, -7) \cup (1, \infty)$

26. $\left\{x \mid x \le -\frac{1}{2} \text{ or } x \ge 2\right\}; \left(-\infty, -\frac{1}{2}\right] \cup [2, \infty)$

27. $\left\{x \mid x < -\frac{5}{3} \text{ or } x > 3\right\}; \left(-\infty, -\frac{5}{3}\right) \cup (3, \infty)$

28. $\left\{x \mid x \ne -\frac{5}{2}\right\}; \left(-\infty, -\frac{5}{2}\right) \cup \left(-\frac{5}{2}, \infty\right)$

29. $\{x \mid x \text{ is any real number}\}; (-\infty, \infty)$

30. $\{x \mid x \text{ is any real number}\}; (-\infty, \infty)$

31. The acceptable belt width is between 127/32 inches and 129/32 inches.

32. The percentage of people that have been shot at is between 7.3 percent and 10.7 percent.

Chapter 9

Chapter 9.1

1. 9   **2.** 30   **3.** $\frac{1}{2}$ **4.** 0.4   **5.** 13   **6.** 15   **7.** 10   **8.** 14   **9.** 7   **10.** rational; 20   **11.** irrational; ≈ 6.32

12. not a real number   **13.** rational; -14 **14.** 14   **15.** $|z|$ **16.** $|2x + 3|$ **17.** $|p - 6|$

Section 9.2

1. 5   **2.** 3   **3.** -6 **4.** not a real number   **5.** $\frac{1}{2}$ **6.** 3.68   **7.** 2.99   **8.** 2.09   **9.** 5   **10.** $|z|$ **11.** $3x - 2$ **12.** 2   **13.** $-\frac{2}{3}$

14. 5   **15.** -3 **16.** -8 **17.** not a real number   **18.** $\sqrt{b}$ **19.** $(8b)^{\frac{1}{5}}$ **20.** $\left(\frac{mn^5}{3}\right)^{\frac{1}{8}}$ **21.** 64   **22.** 9   **23.** -8 **24.** 16

25. not a real number   **26.** 13.57   **27.** 1.74   **28.** $a^{\frac{3}{8}}$ **29.** $(12ab^3)^{\frac{9}{4}}$ **30.** $\frac{1}{9}$ **31.** 4   **32.** $\frac{1}{(13x)^{3/2}}$

Section 9.3

1. $5^{11/12}$ **2.** 8   **3.** 10   **4.** $a \cdot b^{5/6}$ **5.** $\frac{4x^{1/2}}{y^{2/3}}$ **6.** $\frac{5x^{5/8}}{y^{1/8}}$ **7.** $\frac{200a^{\frac{1}{2}}}{b^{\frac{2}{3}}}$ **8.** 6   **9.** $2a^2b^3$ **10.** $\sqrt[12]{x^5}$ **11.** $\sqrt[6]{a}$ **12.** $x^{1/2}(20x + 9)$

13. $\frac{12x + 1}{x^{2/3}}$

Section 9.4

1. $\sqrt{77}$ **2.** $\sqrt[4]{42}$ **3.** $\sqrt{x^2 - 25}$ **4.** $\sqrt[5]{20p^4}$ **5.** $4\sqrt{3}$ **6.** $12\sqrt[3]{2}$ **7.** $10|a|\sqrt{2}$ **8.** Fully simplified   **9.** $2 + \sqrt{5}$

10. $\frac{-1 + 2\sqrt{2}}{2}$ or $-\frac{1}{2} + \sqrt{2}$ **11.** $5a^3\sqrt{3}$ **12.** $3a^2\sqrt{2a}$ **13.** $4x^2y^3\sqrt[3]{2y}$ **14.** $2ab^2\sqrt[4]{ab^3}$ **15.** $4\sqrt{3}$ **16.** $2a^2\sqrt[3]{15}$

17. $8ab^3\sqrt[3]{6a}$ **18.** $\frac{\sqrt{13}}{7}$ **19.** $\frac{3p}{2}$ **20.** $\frac{q\sqrt[4]{3}}{2}$ **21.** $2a^2$ **22.** $-2x$ **23.** $\frac{5a\sqrt[3]{a}}{b}$ **24.** $\sqrt[12]{10{,}125}$ **25.** $2\sqrt[6]{2250}$

Section 9.5

1. $13\sqrt{13y}$ **2.** $7\sqrt[4]{5}$ **3.** $6\sqrt{2}$ **4.** $-8x\sqrt[3]{2x}$ **5.** Cannot be simplified   **6.** $(8z + 5)\sqrt[3]{z}$ **7.** $2\sqrt{m}$ **8.** $3(\sqrt{6} - 10)$

9. $3\sqrt[3]{12} - 2\sqrt[3]{3}$ **10.** $-74 - 27\sqrt{3}$ **11.** $53 + 10\sqrt{6}$ **12.** $25 - 6\sqrt{14}$ **13.** 1

Section 9.6

1. $\frac{\sqrt{3}}{3}$ **2.** $\frac{\sqrt{10}}{4}$ **3.** $\frac{\sqrt{10x}}{2x}$ **4.** $\frac{4\sqrt[3]{9}}{3}$ **5.** $\frac{\sqrt[3]{150}}{10}$ **6.** $\frac{3\sqrt[4]{p^3}}{p}$ **7.** $2(\sqrt{3} - 1)$ **8.** $\frac{\sqrt{3} + 1}{2}$ **9.** $\frac{5 + \sqrt{10} + 4\sqrt{5} + 4\sqrt{2}}{3}$

Section 9.7

1. (a) 4   **(b)** $2\sqrt{7}$ **2. (a)** -1 **(b)** 3   **3.** $\{x \mid x \ge -6\}$ or $[-6, \infty)$ **4.** $\{t \mid t \text{ is any real number}\}$ or $(-\infty, \infty)$

5. $\{m \mid m \le 2\}$ or $(-\infty, 2]$ **6. (a)** $\{x \mid x \ge -3\}$ or $[-3, \infty)$ **7. (a)** $\{x \mid x \text{ is any real number}\}$ or $(-\infty, \infty)$

(b) 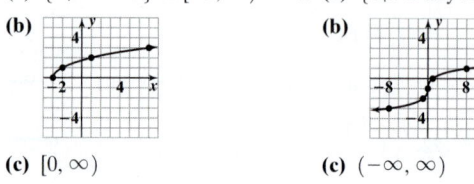   **(b)**

(c) $[0, \infty)$ **(c)** $(-\infty, \infty)$

Section 9.8

1. $\{5\}$ **2.** $\{10\}$ **3.** No real solution   **4.** $\{4\}$ **5.** $\{4\}$ **6.** $\{15\}$ **7.** $\{-3, 1\}$ **8.** $\{12\}$ **9. (a)** $L = \frac{8T^2}{\pi^2}$ **(b)** 32 feet

Section 9.9

1. $6i$ **2.** $\sqrt{5}i$ **3.** $2\sqrt{3}i$ **4.** $4 + 10i$ **5.** $-2 - 2\sqrt{2}i$ **6.** $2 - 2\sqrt{2}i$ **7.** $1 + 11i$ **8.** $6 - 9i$ **9.** $-3 + i$

10. $12 + 15i$ **11.** $2 + 24i$ **12.** -18 **13.** $38 + 14i$ **14.** 73 **15.** 29 **16.** $\dfrac{1}{3} + \dfrac{4}{3}i$ **17.** $-\dfrac{1}{2} + \dfrac{3}{2}i$ **18.** $-i$ **19.** -1

Chapter 10

Section 10.1

1. $\{-4\sqrt{3}, 4\sqrt{3}\}$ **2.** $\{-5, 5\}$ **3.** $\{-9, 9\}$ **4.** $\{-6\sqrt{2}i, 6\sqrt{2}i\}$ **5.** $\{-3i, 3i\}$ **6.** $\{-13, 7\}$ **7.** $\{5 - 4i, 5 + 4i\}$ **8.** $49; (p + 7)^2$

9. $\dfrac{9}{4}; \left(w + \dfrac{3}{2}\right)^2$ **10.** $\{-4, 2\}$ **11.** $\left\{\dfrac{-3 - \sqrt{11}}{2}, \dfrac{-3 + \sqrt{11}}{2}\right\}$ **12.** $c = 5$ **13.** $c = 6\sqrt{2}$ **14.** Approximately 17.32 miles.

Section 10.2

1. $\left\{-\dfrac{3}{2}, 3\right\}$ **2.** $\left\{-4, \dfrac{1}{2}\right\}$ **3.** $\left\{1 - \dfrac{\sqrt{3}}{2}, 1 + \dfrac{\sqrt{3}}{2}\right\}$ **4.** $\left\{\dfrac{5}{2}\right\}$ **5.** $\{-1 - 5i, -1 + 5i\}$ **6.** Two complex solutions that are not real

7. One repeated real solution **8.** Two irrational solutions **9.** $\{-3, 3\}$ **10.** $\left\{\dfrac{5 - \sqrt{23}i}{4}, \dfrac{5 + \sqrt{23}i}{4}\right\}$ **11.** $\left\{-\dfrac{5}{3}, 1\right\}$

12. (a) 200 or 600 DVDs **(b)** 400 DVDs **13.** 16 meters by 30 meters

Section 10.3

1. $\{-3, -2, 2, 3\}$ **2.** $\{-3, 3, \sqrt{2}i, -\sqrt{2}i\}$ **3.** $\{-3, -2, 2, 3\}$ **4.** $\left\{-\dfrac{\sqrt{3}}{2}, \dfrac{\sqrt{3}}{2}, -i, i\right\}$ **5.** $\left\{\dfrac{4}{9}, 16\right\}$ **6.** $\{25\}$ **7.** $\left\{-\dfrac{5}{2}, -\dfrac{1}{2}\right\}$
8. $\{-1, 125\}$

Section 10.4

1. 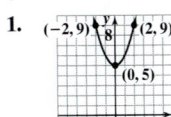 **2.** 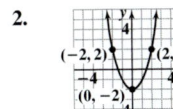 **3.** 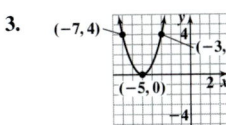 **4.** 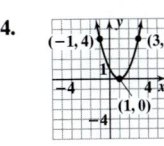 **5.**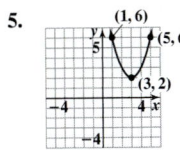

6. 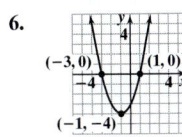 **7.** 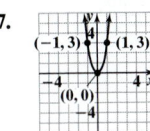 **8.** 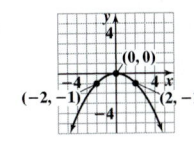 **9.** 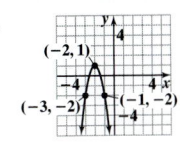 Domain: $\{x \,|\, x \text{ is any real number}\}$ or $(-\infty, \infty)$
Range: $\{y \,|\, y \le 1\}$ or $(-\infty, 1]$

10. 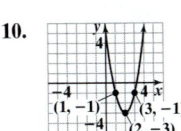 Domain: $\{x \,|\, x \text{ is any real number}\}$ or $(-\infty, \infty)$ **11.** $f(x) = -(x + 1)^2 + 2$
Range: $\{y \,|\, y \ge -3\}$ or $[-3, \infty)$

Section 10.5

1. 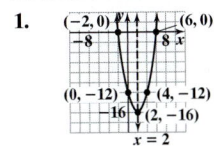 **2.** 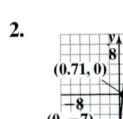 **3.** 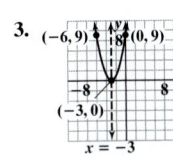 **4.**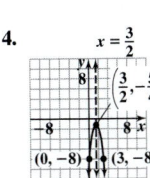

5. Minimum; -7 **6.** Maximum; 33 **7. (a)** Revenue will be maximized when the price is $75. **(b)** The maximum daily revenue is $2812.50.
8. The maximum area that can be enclosed is 62,500 feet. The dimensions of the enclosed area are 250 feet by 250 feet.
9. There should be 65 boxes of CDs sold to maximize revenue. The maximum revenue would be $4225.

Section 10.6

1. $\{x \,|\, x \le -5 \text{ or } x \ge 2\}; (-\infty, -5] \cup [2, \infty)$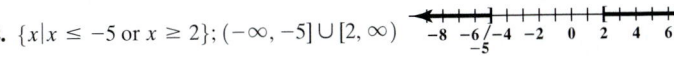

2. $\{x \,|\, x \le -5 \text{ or } x \ge 2\}; (-\infty, -5] \cup [2, \infty)$ **3.** $\{x \,|\, -6 < x < 4\}; (-6, 4)$

4. $\left\{x \,\middle|\, x < \dfrac{-1 - \sqrt{61}}{6} \text{ or } x > \dfrac{-1 + \sqrt{61}}{6}\right\}; \left(-\infty, \dfrac{-1 - \sqrt{61}}{6}\right) \cup \left(\dfrac{-1 + \sqrt{61}}{6}, \infty\right)$

Section 10.7

1. $\{x \,|\, x < -3 \text{ or } x \ge 7\}; (-\infty, -3) \cup [7, \infty)$ 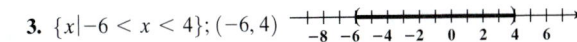 **2.** $\{x \,|\, -2 < x < 1\}; (-2, 1)$

Chapter 11

Section 11.1

1. (a) 17 **(b)** 26 **(c)** -63 **2. (a)** $9x^2 + 3x - 1$ **(b)** $3x^2 - 9x + 5$ **(c)** 29 **3.** not one-to-one **4.** one-to-one
5. not one-to-one **6.** one-to-one

7.
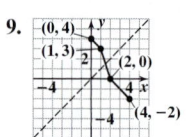

Right Tibia	Right Humerus
36.05	24.80
35.57	24.59
34.58	24.29
34.20	23.81
34.73	24.87

The domain of the inverse function is $\{36.05, 35.57, 34.58, 34.20, 34.73\}$. The range of the inverse function is $\{24.80, 24.59, 24.29, 23.81, 24.87\}$.

8. $\{(3, -3), (2, -2), (1, -1), (0, 0), (-1, 1)\}$. The domain of the inverse function is $\{3, 2, 1, 0, -1\}$. The range of the inverse function is $\{-3, -2, -1, 0, 1\}$.

9.
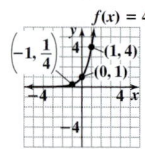

10. $g^{-1}(x) = \dfrac{x + 1}{5}$ **11.** $f^{-1}(x) = \sqrt[5]{x - 3}$

Section 11.2

1. (a) 3.249009585 **(b)** 3.317278183 **(c)** 3.321880096 **(d)** 3.32211036 **(e)** 3.321997085

2. The domain of f is all real numbers or, using interval notation, $(-\infty, \infty)$. The range of f is $\{y \mid y > 0\}$ or, using interval notation, $(0, \infty)$.

3. The domain of f is all real numbers or, using interval notation, $(-\infty, \infty)$. The range of f is $\{y \mid y > 0\}$ or, using interval notation, $(0, \infty)$.

4. The domain of f is all real numbers or, using interval notation, $(-\infty, \infty)$. The range of f is $\{y \mid y > 0\}$ or, using interval notation, $(0, \infty)$.

5. The domain of f is all real numbers or, using interval notation, $(-\infty, \infty)$. The range of f is $\{y \mid y > 1\}$ or, using interval notation, $(1, \infty)$.

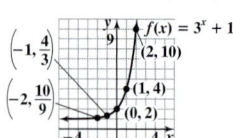

6. $\{3\}$ **7.** $\{2\}$ **8.** $\{0, 5\}$ **9.** $\{-1, 3\}$ **10. (a)** 0.918 or 91.8% **(b)** 0.998 or 99.8%
11. (a) approximately 6.91 grams **(b)** 5 grams **(c)** 0.625 gram **(d)** approximately 0.247 gram
12. (a) $2,253.65 **(b)** $11,991.60 **(c)** $71,899.28

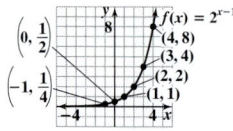

Section 11.3

1. $3 = \log_4 w$ **2.** $-2 = \log_p 8$ **3.** $b = \log_5 125$ **4.** $2^y = 16$ **5.** $a^5 = 20$ **6.** $5^{-3} = z$ **7.** 2 **8.** -3 **9.** 2 **10.** -1

11. $\{x \mid x > -3\}$ or $(-3, \infty)$ **12.** $\left\{x \mid x < \dfrac{5}{2}\right\}$ or $\left(-\infty, \dfrac{5}{2}\right)$

13. The domain of f is $\{x \mid x > 0\}$ or, using interval notation, $(0, \infty)$. The range of f is all real numbers or, using interval notation, $(-\infty, \infty)$.

14. The domain of f is $\{x \mid x > 0\}$ or, using interval notation, $(0, \infty)$. The range of f is all real numbers or, using interval notation, $(-\infty, \infty)$.

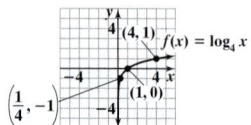

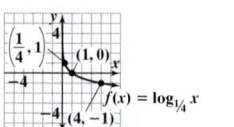

15. 3.146 **16.** 1.569 **17.** -0.523 **18.** $\{16\}$ **19.** $\{4\}$ **20.** $\{e^{-2}\}$ **21.** $\{10,020\}$ **22.** 100 decibels

Section 11.4

1. 0 **2.** 0 **3.** 1 **4.** 1 **5.** $\sqrt{2}$ **6.** 0.2 **7.** 1.2 **8.** -4 **9.** $\log_4 9 + \log_4 5$ **10.** $\log 5 + \log w$ **11.** $\log_7 9 - \log_7 5$
12. $\ln p - \ln 3$ **13.** $\log_2 3 + \log_2 m - \log_2 n$ **14.** $\ln q - \ln 3 - \ln p$ **15.** $1.6 \log_2 5$ **16.** $5 \log b$ **17.** $2 \log_4 a + \log_4 b$

18. $2 + 4 \log_3 m - \dfrac{1}{3} \log_3 n$ **19.** 2 **20.** $\log_3 \left(\dfrac{x + 4}{x - 1}\right)$ **21.** $\log_5 \dfrac{x}{8}$ **22.** $\log_2 \dfrac{x^2 + 3x + 2}{x^2}$ **23.** 3.155 **24.** 2.807

Section 11.5

1. $\{3\}$ **2.** $\{8\}$ **3.** $\left\{\dfrac{\ln 11}{\ln 2}\right\}$ or $\left\{\dfrac{\log 11}{\log 2}\right\}$; $\approx \{3.459\}$ **4.** $\left\{\dfrac{\ln 3}{2 \ln 5}\right\}$ or $\left\{\dfrac{\log 3}{2 \log 5}\right\}$; $\approx \{0.341\}$ **5.** $\left\{\dfrac{\ln 5}{2}\right\}$; $\approx \{0.805\}$

6. $\left\{-\dfrac{\ln\left(^{20}/_3\right)}{4}\right\}$; $\approx \{-0.474\}$ **7. (a)** approximately 2.85 days **(b)** approximately 32.52 days **8. (a)** approximately 6.77 years

(b) approximately 11.58 years

Chapter 12

Section 12.1

1. 5   **2.** $6\sqrt{5} \approx 13.42$ **3. (a)**

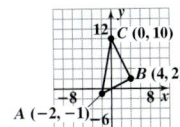

(b) $d(A, B) = 3\sqrt{5}; d(A, C) = 5\sqrt{5}; d(B, C) = 4\sqrt{5}$
(c) $[d(A, C)]^2 = [d(A, B)]^2 + [d(B, C)]^2$ **(d)** 30 square units

4. $\left(\frac{3}{2}, 6\right)$ **5.** $\left(1, \frac{5}{2}\right)$

Section 12.2

1. $(x - 2)^2 + (y - 4)^2 = 25$ **2.** $(x + 2)^2 + y^2 = 2$

3.

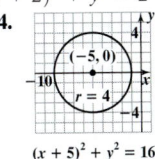

$(x - 3)^2 + (y - 1)^2 = 4$

4.

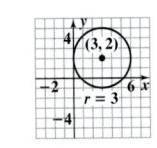

$(x + 5)^2 + y^2 = 16$

5.

$x^2 + y^2 - 6x - 4y + 4 = 0$

6.

$2x^2 + 2y^2 - 16x + 4y - 38 = 0$

Section 12.3

1.

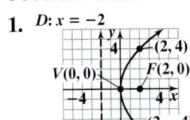

2.

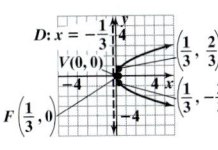

3.

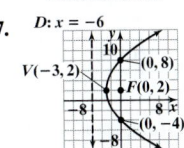

4.

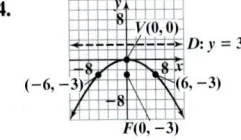

5.

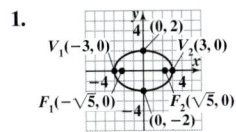

6.

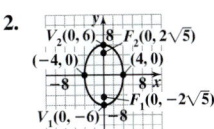

7.

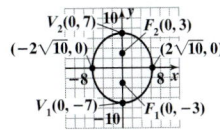

8. The receiver should be located 2 feet from the base of the dish along its axis of symmetry.

Section 12.4

1.

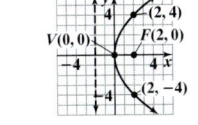

2.

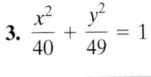

3. $\dfrac{x^2}{40} + \dfrac{y^2}{49} = 1$

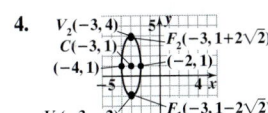

4.

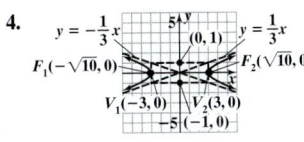

5. $\dfrac{x^2}{2500} + \dfrac{y^2}{1600} = 1$; 40 feet

Section 12.5

1.

2.

3. $\dfrac{x^2}{16} - \dfrac{y^2}{20} = 1$

4.

5.

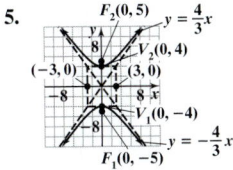

Section 12.6

1. $(-3, 5)$ and $(1, -3)$ **2.** $\left(\frac{11}{5}, -\frac{22}{5}\right)$ and $(1, -2)$ **3.** $(0, -4), \left(-2\sqrt{3}, 2\right), \left(2\sqrt{3}, 2\right)$ **4.** $\varnothing$ or $\{\ \}$

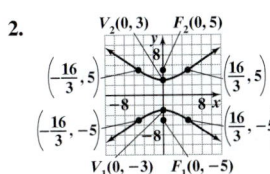
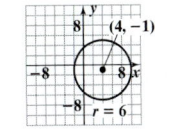

Chapter 13

Section 13.1
1. $-1, 1, 3, 5, 7$ **2.** $-4, 8, -12, 16, -20$ **3.** $a_n = 2n + 3$ **4.** $b_n = \dfrac{(-1)^{n+1}}{n+1}$ **5.** $3 + 7 + 11 = 21$ **6.** $2 + 9 + 28 + 65 + 126 = 230$
7. $\sum_{k=1}^{12} k^2$ **8.** $\sum_{n=1}^{6} \dfrac{1}{2^{n-1}}$

Section 13.2
1. Arithmetic; $a = -3, d = 2$ **2.** Not arithmetic **3.** Arithmetic; $a = -5, d = 3$ **4.** Not arithmetic **5.** Arithmetic; $a = 3, d = -2$
6. (a) $a_n = 6n - 5$ (b) 79 **7.** (a) $a = -5, d = 3$ (b) $a_n = 3n - 8$ **8.** 10,400 **9.** 9,730 **10.** 7,550 **11.** $-1,050$ **12.** 1,470 seats

Section 13.3
1. Geometric; $a = 4, r = 2$ **2.** Not geometric **3.** Geometric; $a = 9, r = \dfrac{1}{3}$ **4.** Geometric; $r = 5$ **5.** Not geometric
6. Geometric; $r = \dfrac{2}{3}$ **7.** $a_n = 5 \cdot 2^{n-1}; a_9 = 1280$ **8.** $a_n = 50 \cdot \left(\dfrac{1}{2}\right)^{n-1}; a_9 = 0.1953125$ or $a_9 = \dfrac{25}{128}$ **9.** 24,573 **10.** 7.9921875 **11.** $\dfrac{40}{3}$
12. $\dfrac{1}{2}$ **13.** $\dfrac{2}{9}$ **14.** The U.S. economy will increase by \$10,000. **15.** \$244,129.08

Section 13.4
1. 362,880 **2.** 840 **3.** 7 **4.** 20 **5.** $x^4 + 8x^3 + 24x^2 + 32x + 16$ **6.** $32p^5 - 80p^4 + 80p^3 - 40p^2 + 10p - 1$

Appendix A

Section A.1
1. $2 \cdot 2 \cdot 3$ **2.** $2 \cdot 3 \cdot 3$ **3.** $3 \cdot 5 \cdot 5$ **4.** $2 \cdot 2 \cdot 2 \cdot 3 \cdot 5$ **5.** prime **6.** $3 \cdot 3 \cdot 3 \cdot 17$ **7.** 24 **8.** 10 **9.** 360 **10.** 21 **11.** 180
12. $\dfrac{5}{10}$ **13.** $\dfrac{30}{48}$ **14.** $\dfrac{1}{4} = \dfrac{3}{12}; \dfrac{5}{6} = \dfrac{10}{12}$ **15.** $\dfrac{5}{12} = \dfrac{25}{60}; \dfrac{4}{15} = \dfrac{16}{60}$ **16.** $\dfrac{9}{20} = \dfrac{36}{80}; \dfrac{11}{16} = \dfrac{55}{80}$ **17.** $\dfrac{1}{2}$ **18.** $\dfrac{4}{9}$ **19.** $\dfrac{9}{16}$ **20.** $\dfrac{2}{5}$ **21.** $\dfrac{2}{7}$

Section A.2
1. hundredths **2.** tenths **3.** thousands **4.** thousandths **5.** ones **6.** ten-thousands **7.** 0.2 **8.** 0.93 **9.** 1.40 **10.** 14.398
11. 690.00 **12.** 60.0 **13.** 0.4 **14.** $0.\overline{428571}$ **15.** 1.375 **16.** $0.833\ldots$ **17.** $0.\overline{5}$ **18.** $\dfrac{13}{20}$ **19.** $\dfrac{1}{5}$ **20.** $\dfrac{5}{8}$ **21.** 0.23 **22.** 0.01
23. 0.724 **24.** 1.27 **25.** 0.8926 **26.** 15% **27.** 80% **28.** 130% **29.** 39.8% **30.** 0.4%

Appendix B

1. $2x^2 + 5x + 3 - \dfrac{7}{x-2}$ **2.** $x^3 + 5x^2 - 2$ **3.** (a) 42 (b) 0 **4.** (a) $f(-2) = -35; x + 2$ is not a factor
(b) $f(5) = 0; f(x) = (x - 5)(2x^2 + x - 1)$

Appendix C

Section C.1
1. Acute **2.** Obtuse **3.** Straight **4.** Right **5.** Complement: 75°; Supplement: 165° **6.** Complement: 30°; Supplement: 120°
7. $m\angle 1 = 140°; m\angle 2 = 40°; m\angle 3 = 140°; m\angle 4 = 40°; m\angle 5 = 140°; m\angle 6 = 40°; m\angle 7 = 140°$

Section C.2
1. 70° **2.** 42° **3.** 14 units **4.** $\dfrac{15}{2}$ inches or 7.5 inches **5.** 12 feet **6.** 7.2 yards **7.** 18 cm

Section C.3
1. Perimeter: 22 feet; Area: 24 square feet **2.** Perimeter: 26 m; Area: 30 m² **3.** Perimeter: 16 cm; Area 16 cm² **4.** Perimeter: 6 yards;
2.25 square yards **5.** Perimeter: 130 yards; 650 square yards **6.** Perimeter: 36 m; Area: 70 m² **7.** Perimeter: 33 yards; Area: 51 square yards
8. Perimeter: 19 mm; Area: 12 mm² **9.** Perimeter: 30 feet; Area: 30 square feet **10.** Circumference: 8π feet ≈ 25.13 feet;
Area: 16π square feet ≈ 50.27 square feet **11.** Circumference: 24π cm ≈ 75.40 cm; Area: 144π cm² ≈ 452.39 cm²

Section C.4
1. Volume: 125 m³; Surface area: 150 m² **2.** Volume: $\dfrac{256}{3}\pi$ in.³ ≈ 268.08 in.³; Surface area: 64π in.² ≈ 201.06 in.²

Appendix E

Section E.1
1. (a) No (b) Yes (c) No **2.** $(3, 1)$ **3.** $(-2, 3)$ **4.** $(-4, 7)$ **5.** $(-6, 10)$ **6.** $(-5, -6)$ **7.** $(3, 6)$ **8.** $\varnothing$ or { }; inconsistent
9. $\{(x, y) | -3x + 2y = 8\}$

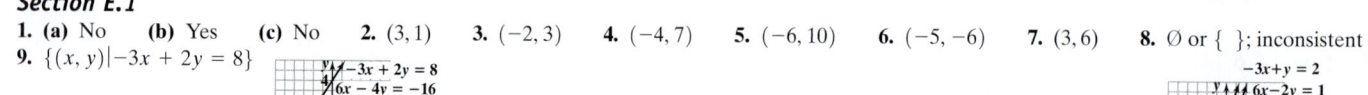
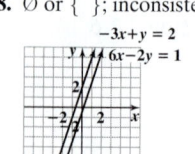

Section E.2

1. (a) No   **(b)** Yes   **2.** $(-3, 1, -1)$ **3.** $\left(-\dfrac{5}{2}, \dfrac{5}{4}, \dfrac{1}{2}\right)$ **4.** $\emptyset$ or $\{\ \}$; the system is inconsistent

5. $\{(x, y, z) \mid x = -z + 1, y = 2z - 1, z \text{ is any real number}\}$ **6.** Fourteen 21-inch, eleven 24-inch, and 5 riders

Section E.3

1. $\begin{bmatrix} 3 & -1 & -10 \\ -5 & 2 & 0 \end{bmatrix}$ **2.** $\begin{bmatrix} 1 & 2 & -2 & 11 \\ -1 & 0 & -2 & 4 \\ 4 & -1 & 1 & 3 \end{bmatrix}$ **3.** $\begin{cases} x - 3y = 7 \\ -2x + 5y = -3 \end{cases}$ **4.** $\begin{cases} x - 3y + 2z = 4 \\ 3x \quad\ - z = -1 \\ -x + 4y \quad = 0 \end{cases}$ **5.** $\begin{bmatrix} 1 & -2 & 5 \\ 0 & -3 & 9 \end{bmatrix}$

6. $R_1 = -5r_2 + r_1;\ \begin{bmatrix} 1 & 0 & 3 \\ 0 & 1 & 2 \end{bmatrix}$ **7.** $(6, -2)$ **8.** $(3, -2, 1)$ **9.** $\{(x, y, z) \mid x = -z + 5, y = 4z + 3, z \text{ is any real number}\}$.

10. $\emptyset$ or $\{\ \}$; inconsistent

Section E.4

1. 18   **2.** -9 **3.** $(-3, 5)$ **4.** Dependent or inconsistent system   **5.** -91 **6.** $(-1, 4, 1)$

Answers to Selected Exercises

Chapter 1

Section 1.1
Answers will vary.

Section 1.2
1. rational　**3.** absolute value　**5.** True　**7.** Answers may vary.　**9.** $A = \{0, 1, 2, 3, 4\}$　**11.** $D = \{1, 2, 3, 4\}$　**13.** $E = \{6, 8, 10, 12, 14\}$

15. 3　**17.** $-4, 3, 0$　**19.** $2.303003000\ldots$　**21.** All numbers listed　**23.** π　**25.** $\dfrac{5}{5} = 1$　**27.** (number line from -3 to 3, points at -1.5, $\frac{3}{3}$, $\frac{4}{3}$)

29. 12　**31.** 4　**33.** $\dfrac{3}{8}$　**35.** 2.1　**37.** True　**39.** True　**41.** True　**43.** False　**45.** <　**47.** >　**49.** >　**51.** =

53. **(a)** (number line with points at $-4.5, -4, -2, -\frac{1}{2}, 0, \frac{3}{5}, 2, 3.5, 4, 6, 8$)　**(b)** $-4.5, -1, -\dfrac{1}{2}, \dfrac{3}{5}, 1, 3.5, |-7| = 7$　**(c) (i)** $-1, 1, |-7| = 7$　**(ii)** All numbers listed

55. -100, Integers, Rational, Real　**57.** -10.5 Rational, Real　**59.** $\dfrac{75}{25}$, Natural, Whole, Integers, Rational, Real　**61.** $7.56556555\ldots$, Irrational, Real
63. True　**65.** False　**67.** True　**69.** True　**71.** True　**73.** Irrational numbers　**75.** Real numbers　**77.** 0　**79.** True　**81.** True
83. $\{7, 8, 9, 10, 11, 12, 13, 14, 15\}$　**85.** $\{10, 11, 12, 13, 14, 15\}$　**87.** $\{11, 12\}$　**89.** $\{2, 4, 6, 8, 10\}$　**91. (a)** $\{1\}, \{2\}, \{3\}, \{4\}, \{1, 2\}, \{1, 3\},$
$\{1, 4\}, \{2, 3\}, \{2, 4\}, \{3, 4\}, \{1, 2, 3\}, \{1, 2, 4\}, \{1, 3, 4\}, \{2, 3, 4\}, \{1, 2, 3, 4\}, \varnothing$　**(b)** 16

Section 1.3
1. sum　**3.** product　**5.** False　**7.** Answers may vary.　**9.** 15　**11.** 4　**13.** 4　**15.** -19　**17.** 21　**19.** -328　**21.** -11　**23.** 43
25. 325　**27.** -125　**29.** 11　**31.** -8　**33.** -28　**35.** 54　**37.** 0　**39.** -41　**41.** -31　**43.** 172　**45.** 40　**47.** -56　**49.** 0
51. 144　**53.** -126　**55.** -90　**57.** 210　**59.** 120　**61.** $\dfrac{1}{8}$　**63.** $-\dfrac{1}{4}$　**65.** 1　**67.** 5　**69.** 7　**71.** -15　**73.** $\dfrac{7}{2}$　**75.** $-\dfrac{10}{7}$
77. $\dfrac{35}{4}$　**79.** -72　**81.** 60　**83.** 171　**85.** -15　**87.** -42　**89.** $\dfrac{15}{4}$　**91.** 40　**93.** -238　**95.** $28 + (-21) = 7$
97. $-21 - 47 = -68$　**99.** $-12 \cdot 18 = -216$　**101.** $-36 \div (-108)$ or $\dfrac{-36}{-108} = \dfrac{1}{3}$　**103.** -3.25 points　**105.** -6 yards　**107.** $-\$48$
109. 14 miles　**111.** 655　**113.** No; -125 cases　**115.** 25,725 feet　**117.** $-3, -5$　**119.** $-12, 2$
121. (a) $1, 2, 1.5, 1.\overline{6}, 1.6, 1.625, 1.615, 1.619, 1.618, \ldots$　**(b)** 1.618　**(c)** Answers may vary.

Section 1.4
1. reciprocals, multiplicative inverses　**3.** $\dfrac{a - b}{c}$　**5.** False　**7.** Answers may vary.　**9.** $\dfrac{2}{3}$　**11.** $-\dfrac{19}{9}$　**13.** $-\dfrac{1}{2}$　**15.** $\dfrac{4}{5}$　**17.** $-\dfrac{3}{5}$
19. $\dfrac{2}{3}$　**21.** $\dfrac{12}{25}$　**23.** -25　**25.** $-\dfrac{2}{3}$　**27.** 8　**29.** $\dfrac{31}{3}$　**31.** $\dfrac{6}{11}$　**33.** $\dfrac{5}{3}$　**35.** $-\dfrac{1}{5}$　**37.** $\dfrac{5}{6}$　**39.** $-\dfrac{1}{9}$　**41.** $-\dfrac{2}{3}$　**43.** $\dfrac{9}{11}$　**45.** -12
47. 32　**49.** $\dfrac{3}{2}$　**51.** $\dfrac{1}{2}$　**53.** 2　**55.** $\dfrac{1}{3}$　**57.** $\dfrac{5}{2}$　**59.** $\dfrac{13}{12}$　**61.** $\dfrac{1}{4}$　**63.** $\dfrac{9}{5}$　**65.** $-\dfrac{1}{6}$　**67.** $\dfrac{33}{40}$　**69.** $\dfrac{139}{36}$　**71.** $-\dfrac{7}{18}$　**73.** -6.5
75. 8.4　**77.** 55.92　**79.** 1.49　**81.** 42.55　**83.** 24.94　**85.** 9　**87.** 24.3　**89.** 490　**91.** 0.95　**93.** $-\dfrac{11}{30}$　**95.** $-\dfrac{4}{3}$　**97.** $-\dfrac{2}{3}$
99. $-\dfrac{1}{4}$　**101.** $-\dfrac{129}{35}$　**103.** 1.6　**105.** $-\dfrac{2}{21}$　**107.** -50.526　**109.** -16　**111.** -15　**113.** -6.58　**115.** -7.9　**117.** $-\dfrac{25}{14}$
119. 58.39　**121.** 30.96　**123.** $\dfrac{1}{8}$　**125.** 21 hours　**127.** 18 students　**129.** $-\$68.79$　**131.** $\$2.40$　**133.** 13.2　**135.** $\dfrac{86}{15}$

Putting the Concepts Together
1. (a) $-12, -\dfrac{14}{7} = -2, 0, 3$　**(b)** $-12, -\dfrac{14}{7} = -2, -1.25, 0, 3, 11.2$　**(c)** $\sqrt{2}$　**(d)** All the numbers in the set are real numbers.　**2.** <　**3.** -11
4. -65　**5.** -27　**6.** 27　**7.** -5.5　**8.** -10　**9.** -100　**10.** 72　**11.** -5　**12.** 16　**13.** -9　**14.** -3　**15.** $\dfrac{31}{5}$　**16.** $\dfrac{31}{36}$
17. $-\dfrac{17}{36}$　**18.** $\dfrac{9}{5}$　**19.** $-\dfrac{1}{28}$　**20.** 0　**21.** 10.76　**22.** 7.646　**23.** 1.46　**24.** 22.232

Section 1.5
1. Identity Property; Addition　**3.** undefined　**5.** True　**7.** Answers may vary.　**9.** Additive Inverse Property　**11.** Multiplicative Identity
Property　**13.** Multiplicative Inverse Property　**15.** Additive Inverse Property　**17.** Commutative Property of Multiplication
19. Commutative Property of Addition　**21.** 156 inches　**23.** 45 meters　**25.** $10\dfrac{1}{2}$ gallons　**27.** $11\dfrac{1}{4}$ pounds　**29.** $4\dfrac{1}{2}$ hours = 4 hours,
30 minutes　**31.** 29　**33.** 18　**35.** -65　**37.** 347　**39.** -90　**41.** undefined　**43.** -34　**45.** 0　**47.** 0　**49.** $-\dfrac{20}{3}$
51. $\$203.16$　**53.** Answers may vary.　**55.** Answers may vary.　**57.** $-6 - (4 + 10)$　**59.** $25 - (6 - 10) - 1$　**61.** $58\dfrac{2}{3}$ feet per second

Section 1.6

1. base **3.** grouping symbols **5.** False **7.** Answers may vary. **9.** 5^2 **11.** $\left(\frac{3}{5}\right)^3$ **13.** 64 **15.** 64 **17.** 1000 **19.** $\frac{27}{64}$

21. 2.25 **23.** -9 **25.** -1 **27.** 0 **29.** $\frac{1}{64}$ **31.** $-\frac{1}{27}$ **33.** 14 **35.** -3 **37.** 2500 **39.** 160 **41.** 20 **43.** 4 **45.** $\frac{3}{5}$

47. -1 **49.** 42 **51.** 5 **53.** -4 **55.** 115 **57.** -5 **59.** $-\frac{65}{4}$ **61.** -24 **63.** $-\frac{13}{12}$ **65.** 0.5 **67.** 12 **69.** $-\frac{1}{2}$ **71.** $\frac{3}{4}$

73. 1 **75.** 24 **77.** $\frac{133}{8}$ **79.** $\frac{2}{5}$ **81.** 3 **83.** $\frac{3}{2}$ **85.** $\frac{64}{27}$ **87.** $-\frac{1}{6}$ **89.** $2^3 \cdot 3^2$ **91.** $2^4 \cdot 3$ **93.** $(4 \cdot 3 + 6) \cdot 2$

95. $(4 + 3) \cdot (4 + 2)$ **97.** $(6 - 4) + (3 - 1)$ **99.** \$514.93 **101.** 603.19 in.^{2} **103.** \$1060.90 in.^{2} **105.** 115.75°

Section 1.7

1. Distributive **3.** -1 **5.** False **7.** Answers may vary. **9.** $2x^3, \frac{x^2}{4}, -x, 6; 2, \frac{1}{4}, -1, 6$ **11.** $z^2, \frac{2y}{3}; 1, \frac{2}{3}$ **13.** 13 **15.** 17 **17.** -21

19. 104 **21.** $\frac{17}{5}$ **23.** 225 **25.** 153 **27.** unlike **29.** like **31.** like **33.** unlike **35.** $3m + 6n$ **37.** $18n^2 + 12n - 6$

39. $-x + y$ **41.** $-4x + 3y$ **43.** $3x$ **45.** $6z$ **47.** $10m + 10n$ **49.** $2.2x^7$ **51.** $10y^6$ **53.** $-6w - 12y + 13z$ **55.** $-3k + 15$

57. $4n - 8$ **59.** $-4n + 20$ **61.** $\frac{5}{6}x$ **63.** $-\frac{11}{2}$ **65.** $-3.5x - 6$ **67.** $6x - 0.06$ **69.** 32 **71.** 27 **73.** 0 **75.** -13

77. -7 **79.** -12 **81.** 44 **83.** $-\frac{3}{2}$ **85.** 36 **87.** \$78.70 **89.** \$4819 **91. (a)** $8w - 8$ **(b)** 32 yards **93.** \$228.88

95. Answers may vary; $-3x^2 + 7x - 3$

Chapter 1 Review

1. $A = \{0, 1, 2, 3, 4, 5, 6\}$ **2.** $B = \{1, 2, 3\}$ **3.** $C = \{-2, -1, 0, 1, 2, 3, 4, 5\}$ **4.** $D = \{-2, -1, 0, 1, 2, 3\}$ **5.** $\frac{9}{3} = 3, 11$ **6.** $0, \frac{9}{3} = 3, 11$

7. $-6, 0, \frac{9}{3} = 3, 11$ **8.** $-6, -3.25, 0, \frac{9}{3} = 3, 11, \frac{5}{7}$ **9.** $5.030030003\ldots$ **10.** All numbers listed

11. **12.** **13.** False **14.** True

15. True **16.** True **17.** $-\frac{1}{2}$ **18.** 7 **19.** -6 **20.** -8.2 **21.** $=$ **22.** $<$ **23.** $>$ **24.** $>$ **25.** $>$ **26.** $<$

27. Answers may vary. **28.** natural numbers **29.** 7 **30.** -4 **31.** -34 **32.** -95 **33.** -4 **34.** 47 **35.** -53 **36.** -115

37. -22 **38.** -7 **39.** 21 **40.** 67 **41.** 26 **42.** -36 **43.** 12 **44.** -40 **45.** -1118 **46.** -8037 **47.** -715 **48.** $-11,130$

49. 5 **50.** -12 **51.** 5 **52.** -25 **53.** -8 **54.** $-\frac{16}{5}$ **55.** $-\frac{10}{3}$ **56.** $-\frac{30}{7}$ **57.** -13 **58.** 45 **59.** $-43 + 101 = 58$

60. $45 + (-28) = 17$ **61.** $-10 - (-116) = 106$ **62.** $74 - 56 = 18$ **63.** $13 + (-8) = 5$ **64.** $-60 - (-10) = -50$

65. $-21 \cdot (-3) = 63$ **66.** $54 \cdot (-18) = -972$ **67.** $-34 \div (-2)$ or $\frac{-34}{-2} = 17$ **68.** $-49 \div 14$ or $\frac{-49}{14} = -\frac{7}{2}$ **69.** 26 yards **70.** $-3°F$

71. $24°F$ **72.** 87 points **73.** $\frac{1}{2}$ **74.** $-\frac{1}{3}$ **75.** $-\frac{2}{3}$ **76.** $-\frac{7}{5}$ **77.** $\frac{5}{4}$ **78.** $-\frac{5}{28}$ **79.** $-\frac{1}{20}$ **80.** $-\frac{3}{2}$ **81.** $\frac{4}{17}$

82. $-\frac{2}{3}$ **83.** $-\frac{3}{10}$ **84.** -32 **85.** $\frac{1}{3}$ **86.** $-\frac{2}{5}$ **87.** $\frac{3}{7}$ **88.** 3 **89.** $\frac{7}{20}$ **90.** $\frac{31}{36}$ **91.** $-\frac{59}{245}$ **92.** $\frac{13}{12}$ **93.** $-\frac{19}{12}$

94. $-\frac{11}{4}$ **95.** 0 **96.** $-\frac{23}{24}$ **97.** 48.5 **98.** -24.66 **99.** 82.98 **100.** -53.74 **101.** 0.0804 **102.** -260.154 **103.** 18.4

104. -25.79 **105.** 2.3 **106.** -22.9 **107.** -5.418 **108.** -0.732 **109.** $-\$98.93$ **110.** 24 friends **111.** $\frac{23}{2}$ or $11\frac{1}{2}$ inches

112. \$186.81 **113.** Associative Property of Multiplication **114.** Multiplicative Inverse Property **115.** Multiplicative Inverse Property
116. Commutative Property of Multiplication **117.** Commutative Property of Multiplication **118.** Additive Inverse Property **119.** Identity
Property of Addition **120.** Identity Property of Addition **121.** Commutative Property of Addition **122.** Multiplicative Identity Property
123. Multiplication Property of Zero **124.** Associative Property of Addition **125.** 29 **126.** 99 **127.** 18 **128.** 121 **129.** 3.4

130. 5.3 **131.** -33 **132.** 6 **133.** undefined **134.** 0 **135.** -334 **136.** 2 **137.** 0 **138.** 1 **139.** 0 **140.** 130 **141.** $-\frac{5}{3}$
142. $-\frac{150}{13}$ **143.** 3^4 **144.** $\left(\frac{2}{3}\right)^3$ **145.** $(-4)^2$ **146.** $(-3)^3$ **147.** 125 **148.** 32 **149.** 81 **150.** -64 **151.** -81

152. $\frac{1}{64}$ **153.** -4 **154.** -76 **155.** 206 **156.** 32 **157.** 2 **158.** $\frac{1}{2}$ **159.** $\frac{6}{5}$ **160.** $\frac{7}{5}$ **161.** 21 **162.** -18 **163.** -729

164. -3 **165.** $3x^2, -x, 6; 3, -1, 6$ **166.** $2x^2y^3, -\frac{y}{5}; 2, -\frac{1}{5}$ **167.** like **168.** unlike **169.** unlike **170.** like **171.** $-3x$

172. $-4x - 15$ **173.** $-4.1x^4 + 0.3x^3$ **174.** $-3x^4 + 6x^2 + 12$ **175.** $18 - x$ **176.** $4x - 18$ **177.** $9x - 4$ **178.** -1 **179.** \$98.70

Chapter 1 Test

1. $\frac{1}{3}$ **2.** $\frac{9}{4}$ **3.** $-\frac{320}{3}$ **4.** -102 **5.** -12.16 **6.** -20 **7.** undefined **8.** -14 **9.** 55 **10. (a)** 6 **(b)** 0, 6 **(c)** $-2, 0, 6$

(d) $-2, -\frac{1}{2}, 0, 2.5, 6$ **(e)** none **(f)** All those listed. **11.** $<$ **12.** $=$ **13.** -7 **14.** 15 **15.** -102 **16.** -343 **17.** $-16x - 28$

18. $-4x^2 + 5x + 4$ **19.** \$531.85 **20.** $4x + 10$

Chapter 2

Section 2.1

1. Addition Property of Equality   **3.** subtract 9   **5.** False   **7.** Answers may vary.   **9.** Answers may vary.   **11.** Yes   **13.** No   **15.** Yes

17. Yes   **19.** $\{20\}$ **21.** $\{-12\}$ **23.** $\{19\}$ **25.** $\{-13\}$ **27.** $\{2\}$ **29.** $\left\{\dfrac{1}{4}\right\}$ **31.** $\left\{\dfrac{19}{24}\right\}$ **33.** $\{-6.1\}$ **35.** $\{5\}$ **37.** $\{-4\}$

39. $\left\{\dfrac{7}{2}\right\}$ **41.** $\left\{-\dfrac{5}{2}\right\}$ **43.** $\{21\}$ **45.** $\{121\}$ **47.** $\left\{\dfrac{3}{5}\right\}$ **49.** $\left\{-\dfrac{1}{3}\right\}$ **51.** $\{9\}$ **53.** $\left\{-\dfrac{4}{9}\right\}$ **55.** $\left\{-\dfrac{5}{3}\right\}$ **57.** $\{2\}$ **59.** $\{-3\}$

61. $\left\{\dfrac{2}{3}\right\}$ **63.** $\{-6\}$ **65.** $\{19\}$ **67.** $\{283\}$ **69.** $\{-50\}$ **71.** $\{41.1\}$ **73.** $\left\{\dfrac{20}{3}\right\}$ **75.** $\{-4\}$ **77.** $\{-12\}$ **79.** $\left\{\dfrac{1}{2}\right\}$ **81.** $\left\{\dfrac{3}{16}\right\}$

83. $\left\{-\dfrac{5}{4}\right\}$ **85.** \$18,499.80   **87.** \$68   **89.** 12   **91.** 0.18 or 18%   **93.** $x = 48 - \lambda$ **95.** $x = \dfrac{14}{\theta}$ **97.** $\lambda = \dfrac{50}{9}$ **99.** $\theta = -\dfrac{6}{7}$

Section 2.2

1. add 8   **3.** add x **5.** True   **7.** Answers may vary.   **9.** Answers may vary.   **11.** $\{1\}$ **13.** $\{-2\}$ **15.** $\{-3\}$ **17.** $\left\{\dfrac{3}{2}\right\}$

19. $\{-3\}$ **21.** $\{12\}$ **23.** $\{4\}$ **25.** $\{-2\}$ **27.** $\{-5\}$ **29.** $\{-8\}$ **31.** $\{-5\}$ **33.** $\{-8\}$ **35.** $\{3\}$ **37.** $\left\{-\dfrac{7}{2}\right\}$ **39.** $\{7\}$

41. $\{-18\}$ **43.** $\left\{\dfrac{7}{2}\right\}$ **45.** $\left\{-\dfrac{27}{16}\right\}$ **47.** $\{2\}$ **49.** $\left\{-\dfrac{3}{4}\right\}$ **51.** $\left\{\dfrac{1}{3}\right\}$ **53.** $\left\{\dfrac{1}{16}\right\}$ **55.** $\left\{\dfrac{15}{2}\right\}$ **57.** $\left\{\dfrac{1}{2}\right\}$ **59.** $\{-7\}$

61. $\left\{\dfrac{51}{7}\right\}$ **63.** McDonald's: 23 g; Burger King: 27 g   **65.** width: $\dfrac{13}{3}$ or $4\dfrac{1}{3}$ feet; length: $\dfrac{32}{3}$ or $10\dfrac{2}{3}$ feet   **67.** \$8   **69.** Yes   **71.** $\{2.47\}$

73. $\{2.6\}$ **75.** $\dfrac{20}{3}$ **77.** $-\dfrac{5}{4}$

Section 2.3

1. conditional equation   **3.** 100   **5.** False   **7.** Answers may vary.   **9.** $\left\{\dfrac{9}{2}\right\}$ **11.** $\{-2\}$ **13.** $\{-6\}$ **15.** $\left\{\dfrac{2}{3}\right\}$ **17.** $\{30\}$

19. $\{-4\}$ **21.** $\{50\}$ **23.** $\{4.8\}$ **25.** $\{150\}$ **27.** $\{6\}$ **29.** $\left\{\dfrac{25}{8}\right\}$ **31.** $\{-20\}$ **33.** $\{1\}$ **35.** $\{-12\}$ **37.** $\{60\}$

39. $\{5\}$ **41.** $\{2\}$ **43.** $\{75\}$ **45.** contradiction; $\varnothing$ or $\{\ \}$ **47.** identity; all real numbers   **49.** conditional equation; $\left\{-\dfrac{1}{2}\right\}$

51. contradiction; $\varnothing$ or $\{\ \}$ **53.** contradiction; $\varnothing$ or $\{\ \}$ **55.** identity; all real numbers   **57.** $\left\{-\dfrac{3}{4}\right\}$ **59.** identity; all real numbers

61. $\left\{-\dfrac{1}{3}\right\}$ **63.** $\{-20\}$ **65.** $\left\{\dfrac{7}{2}\right\}$ **67.** contradiction; $\varnothing$ or $\{\ \}$ **69.** $\{-2\}$ **71.** $\{3\}$ **73.** contradiction; $\varnothing$ or $\{\ \}$ **75.** $\{-4\}$

77. $\{39\}$ **79.** $\{0\}$ **81.** $\left\{\dfrac{2}{3}\right\}$ **83.** -1.70 **85.** -13.2 **87.** \$50   **89.** \$18,000   **91.** \$8.50   **93.** \$95   **95.** 15 quarters

97. 6 units   **99.** \$12,000   **101.** Answers may vary.

Section 2.4

1. formula   **3.** subtract   **5.** False   **7.** Answers may vary.   **9.** Rogan 1099.14 feet; Nurek 984.3 feet   **11.** \$19.20   **13.** \$650

15. 20°C   **17.** \$3   **19. (a)** 50 units   **(b)** 144 square units   **21. (a)** 36.2 meters   **(b)** 70 square meters

23. (a) 36 units   **(b)** 81 square units   **25. (a)** 31.4 cm   **(b)** 78.5 cm^2 **27.** 68.44 square inches   **29.** $t = \dfrac{d}{r}$ **31.** $d = \dfrac{C}{\pi}$

33. $r = \dfrac{I}{Pt}$ **35.** $h = \dfrac{2A}{b}$ **37.** $a = P - b - c$ **39.** $t = \dfrac{A - P}{Pr}$ **41.** $b = \dfrac{2A}{h} - B$ **43.** $y = -3x + 12$ **45.** $y = 2x - 5$

47. $y = \dfrac{-4x + 13}{3}$ **49.** $y = 3x - 12$ **51. (a)** $C = R - P$ **(b)** \$450   **53. (a)** $t = \dfrac{I}{Pr}$ **(b)** 2 years   **55. (a)** $m = \dfrac{2K}{v^2}$ **(b)** 16

57. (a) $C = \dfrac{5}{9}(F - 32)$ **(b)** 15°C   **59. (a)** $r = \dfrac{A - P}{Pt}$ **(b)** 0.04 or 4%   **61. (a)** $h = \dfrac{V}{\pi r^2}$ **(b)** 5 mm

63. (a) $b = \dfrac{2A}{h}$ **(b)** 18 ft   **65.** 1834.05   **67. (a)** $h = \dfrac{S - 2\pi r^2}{2\pi r}$ **(b)** 1.25 inches   **69.** medium 12"

71. (a) 12 hours   **(b)** \$336   **73.** 50.5 square inches   **75.** 96π cm$^3 \approx 301.59$ cm^3 **77. (a)** $I = \dfrac{D - P}{0.03} + 137{,}300$ **(b)** \$158,000

79. (a) 62   **(b)** \$372   **(c)** Yes   **81. (a)** 4948 ft^2 **(b)** \$1237   **83. (a)** 1080 in.^{2} **(b)** 7.5 ft^2 **85.** Multiply by $\dfrac{1\ \text{ft}^2}{144\ \text{in}^2}$.

Putting the Concepts Together

1. (a) Yes   **(b)** No   **2. (a)** No   **(b)** Yes   **3.** $\left\{-\dfrac{2}{3}\right\}$ **4.** $\{-40\}$ **5.** $\{-6\}$ **6.** $\{3\}$ **7.** $\{-6\}$ **8.** $\left\{-\dfrac{7}{3}\right\}$

9. $\left\{-\dfrac{57}{2}\right\}$ **10.** $\{25\}$ **11.** $\{6\}$ **12.** $\left\{\dfrac{74}{5}\right\}$ or $\{14.8\}$ **13.** contradiction; $\varnothing$ or $\{\ \}$ **14.** identity; all real numbers   **15.** \$5000

16. (a) $b = \dfrac{2A}{h} - B$ **(b)** 6 in.   **17. (a)** $h = \dfrac{V}{\pi r^2}$ **(b)** 13 in.   **18.** $y = -\dfrac{3}{2}x + 7$

Section 2.5

1. mathematical modeling   **3.** False   **5.** Answers may vary.   **7.** Answers may vary.   **9.** $-5 + x$ **11.** $x\left(\dfrac{2}{3}\right)$ or $\dfrac{2}{3}x$ **13.** $\dfrac{1}{2}x$

15. $x - (-25)$ **17.** $\dfrac{x}{3}$ **19.** $x + \dfrac{1}{2}$ **21.** $6x + 9$ **23.** $2(13.7 + x)$ **25.** $2x + 31$ **27.** $x + 15 = -34$ **29.** $35 = 3x - 7$

31. $\dfrac{x}{-4} + 5 = 36$ **33.** $2[x + 6] = x + 3$ **35.** Braves: r; Clippers: $r + 5$ **37.** Bill's amount: b; Jan's amount: $b + 0.55$

39. Janet's share: j; Kathy's share: $200 - j$ **41.** number adults: a; number children: $1433 - a$ **43.** 83   **45.** -14 **47.** 54, 55, 56

49. Verrazano-Narrows Bridge: 4260 ft; Golden Gate Bridge: 4200 ft   **51.** \$11,215   **53.** CD: \$8500; Bonds: \$11,500

55. Stocks: \$20,000; Bonds: \$12,000   **57.** Smart Start: 2 g; Go Lean: 8 g   **59.** \$29,140   **61.** 150 miles   **63.** 2000 pages

65. Jensen: \$36,221; Maureen: \$35,972   **67.** 5, 6, 7, and 8   **69.** Answers may vary.   **71.** Answers may vary.   **73.** $20°, 40°, 120°$

Section 2.6

1. 0.032   **3.** True   **5.** Answers may vary.   **7.** 80   **9.** 14   **11.** 4.8   **13.** 210   **15.** 72   **17.** $8\dfrac{1}{3}$ **19.** 40%   **21.** 7.5%

23. 200%   **25.** \$54   **27.** \$102,000   **29.** \$25,000   **31.** \$68   **33.** \$570   **35.** \$400   **37.** winner: 530; loser: 318   **39.** \$285,700

41. 597   **43.** 32.4 million   **45.** 70.2%   **47.** 24.6%   **49.** \$24   **51.** 28.6%   **53.** (a) \$31,314.38   (b) 43.7%   **55.** 21.2%

Section 2.7

1. 90   **3.** True   **5.** False   **7.** $20°, 70°$ **9.** $50°, 130°$ **11.** $60°, 75°, 45°$ **13.** $42°, 44°, 94°$ **15.** $47.5°, 132.5°$ **17.** $52.5°, 37.5°$

19. $44°, 46°$ **21.** $18°, 72°, 90°$ **23.** $l = 32$ feet; $w = 12$ feet   **25.** 42 ft by 84 ft   **27.** 26.5 in.   **29.** $49°, 49°, 82°$

31. (a) $62t$ (b) $68t$ (c) $62t + 68t$ (d) $62t + 68t = 585$ **33.** $528(t + 10) = 880t$ **35.** length $= 14$ in.; width $= 4$ in.   **37.** 40 ft

39. 40 ft; 32 ft   **41.** (a) length $= 11$ ft; width $= 19$ ft   (b) 209 ft^2 **43.** 13 hours   **45.** fast car: 60 mph; slow car: 48 mph

47. freeway: 4.5 hours; 2-lane: 1.5 hours   **49.** 6 mph   **51.** $x = 3$ **53.** $x = 32.5$ **55.** $x = 18$

Section 2.8

1. closed interval   **3.** left endpoint; right endpoint   **5.** True   **7.** The symbol ∞ means unboundedness in the positive direction and is not a real number.   **9.** $x \geq 16,000$ **11.** $15,000 \leq x \leq 20,000$ **13.** $x > 12,000$ **15.** $x > 0$ **17.** $x \leq 0$ **19.** $(2, \infty)$

21. $(-\infty, -1]$ **23.** $[-3, \infty)$ **25.** $(-\infty, 4)$

27. $[-2, 5]$ **29.** $(3, 10)$ **31.** $[2, 4)$

33. $(-\infty, 2)$ **35.** $\varnothing$ or $\{\ \}$ **37.** $(-\infty, \infty)$ **39.** $[-2, 1]$ **41.** $(-3, 1]$ **43.** $\{x \mid x < 4\}; (-\infty, 4)$

45. $\{x \mid x \leq 5\}; (-\infty, 5]$ **47.** $\{x \mid x > -7\}; (-7, \infty)$ **49.** $\{x \mid x > 3\}; (3, \infty)$

51. $\{x \mid x \geq 2\}; [2, \infty)$ **53.** $\{x \mid x \geq -1\}; [-1, \infty)$ **55.** $\{x \mid x > -7\}; (-7, \infty)$

57. $\left\{x \mid x \leq \dfrac{2}{3}\right\}; \left(-\infty, \dfrac{2}{3}\right]$ **59.** $\{x \mid x < -20\}; (-\infty, -20)$

61. $\varnothing$ or $\{\ \}$ **63.** $\{n \mid n$ is any real number$\}; (-\infty, \infty)$

65. $\{n \mid n > 5\}; (5, \infty)$ **67.** $\left\{y \mid y < -\dfrac{3}{2}\right\}; \left(-\infty, -\dfrac{3}{2}\right)$

69. $\{x \mid x > 4\}; (4, \infty)$ **71.** $\left\{x \mid x < \dfrac{3}{4}\right\}; \left(-\infty, \dfrac{3}{4}\right)$

73. $\{x \mid x$ is any real number$\}; (-\infty, \infty)$ **75.** $\{a \mid a < -1\}; (-\infty, -1)$

77. $\varnothing$ or $\{\ \}$ **79.** $\left\{x \mid x \geq \dfrac{4}{3}\right\}; \left[\dfrac{4}{3}, \infty\right)$ **81.** $\{x \mid x < 25\}; (-\infty, 25)$

83. $\{x \mid x > 5.9375\}; (5.9375, \infty)$ **85.** at most 1250 miles   **87.** 32   **89.** more than 400 minutes

91. greater than \$50,361.11   **93.** at least 74

Chapter Review

1. No   **2.** No   **3.** No   **4.** Yes   **5.** $\{16\}$ **6.** $\{20\}$ **7.** $\{-16\}$ **8.** $\{-7\}$ **9.** $\{-95\}$ **10.** $\{50\}$ **11.** $\{24\}$ **12.** $\{80\}$

13. $\{-6\}$ **14.** $\{5\}$ **15.** $\left\{-\dfrac{1}{3}\right\}$ **16.** $\left\{\dfrac{3}{8}\right\}$ **17.** $\{4\}$ **18.** $\{5\}$ **19.** \$20,100   **20.** \$2.55   **21.** $\{-4\}$ **22.** $\{4\}$ **23.** $\{9\}$

24. $\{-21\}$ **25.** $\{-4\}$ **26.** $\{-6\}$ **27.** $\{4\}$ **28.** $\{-5\}$ **29.** $\{6\}$ **30.** $\{-6\}$ **31.** $\left\{\dfrac{4}{3}\right\}$ **32.** $\{5\}$ **33.** $\{2\}$ **34.** $\{7\}$

35. 14 years   **36.** width $= 19$ yards; length $= 29$ yards   **37.** $\left\{-\dfrac{35}{12}\right\}$ **38.** $\left\{-\dfrac{62}{3}\right\}$ **39.** $\{-2\}$ **40.** $\left\{\dfrac{6}{5}\right\}$ **41.** $\left\{\dfrac{3}{2}\right\}$ **42.** $\{3\}$

43. $\{4\}$ **44.** $\{-1\}$ **45.** $\left\{-\dfrac{7}{2}\right\}$ **46.** $\{-3\}$ **47.** $\{58\}$ **48.** $\{-10.5\}$ **49.** contradiction; $\emptyset$ or $\{\ \}$ **50.** contradiction; $\emptyset$ or $\{\ \}$

51. conditional equation; $\{0\}$ **52.** conditional equation; $\{0\}$ **53.** identity; all real numbers   **54.** identity; all real numbers

55. \$15.75   **56.** 3 dimes   **57.** 48 in.^{2} **58.** 64 cm   **59.** $\dfrac{3}{2}$ yards   **60.** 15 mm   **61.** $H = \dfrac{V}{LW}$ **62.** $P = \dfrac{I}{rt}$ **63.** $W = \dfrac{S - 2LH}{2L + 2H}$

64. $M = \dfrac{\rho - mv}{V}$ **65.** $y = \dfrac{-2x + 10}{3}$ **66.** $x = \dfrac{14 + 7y}{6}$ **67. (a)** $P = \dfrac{A}{(1 + r)^t}$ **(b)** \$2238.65   **68. (a)** $h = \dfrac{A - 2\pi r^2}{2\pi r}$ **(b)** 5 cm

69. \$11.25   **70.** $\dfrac{9}{4}\pi$ ft$^2 \approx 7.1$ ft^2 **71.** $x - 6$ **72.** $x - 8$ **73.** $-8x$ **74.** $\dfrac{x}{10}$ **75.** $2(6 + x)$ **76.** $4(5 - x)$ **77.** $6 + x = 2x + 5$

78. $6x - 10 = 2x + 1$ **79.** $x - 8 = \dfrac{1}{2}x$ **80.** $\dfrac{6}{x} = 10 + x$ **81.** $4(2x + 8) = 16$ **82.** $5(2x - 8) = -24$ **83.** Sarah's age: s; Jacob's age: $s + 7$

84. Consuelo's speed: c; Jose's speed: $2c$ **85.** Max's amount: m; Irene's amount: $m - 6$ **86.** Victor's amount: v; Larry's amount: $350 - v$

87. 153 pounds   **88.** 12, 13, 14   **89.** Juan: \$11,000; Roberto: \$9000   **90.** 100 miles   **91.** 5.2   **92.** 60   **93.** 13%   **94.** 50   **95.** \$18.50

96. \$30   **97.** \$40   **98.** \$200   **99.** \$125,000   **100.** winner: 500 votes; loser: 400 votes   **101.** $10°, 80°$ **102.** $80°, 100°$ **103.** $30°, 60°, 90°$

104. $50°, 45°, 85°$ **105.** length $= 31$ in.; width $= 8$ in.   **106.** length $= 28$ cm; width $= 7$ cm   **107. (a)** length $= 20$ ft; width $= 40$ ft   **(b)** 800 ft^2

108. 50 ft; 40 ft   **109.** 5 hours   **110.** 65 mph   **111.** $(-\infty, -3]$ **112.** $(4, \infty)$

113. $[2, 7]$ **114.** $(-1, 5)$ **115.** $(0, 3]$

116. $[-3, 1)$ **117.** $\left\{x \mid x < -\dfrac{13}{2}\right\}; \left(-\infty, -\dfrac{13}{2}\right)$

118. $\left\{x \mid x \geq -\dfrac{7}{3}\right\}; \left[-\dfrac{7}{3}, \infty\right)$ **119.** $\left\{x \mid x \geq -\dfrac{4}{5}\right\}; \left[-\dfrac{4}{5}, \infty\right)$

120. $\{x \mid x > -12\}; (-12, \infty)$ **121.** $\emptyset$ or $\{\ \}$

122. $\{x \mid x$ is any real number$\}, (-\infty, \infty)$ **123.** $\{x \mid x > 3\}; (3, \infty)$

124. $\left\{x \mid x < -\dfrac{38}{5}\right\}; \left(-\infty, -\dfrac{38}{5}\right)$ **125.** at most 65 miles   **126.** more than 150

Chapter 2 Test

1. $\{-17\}$ **2.** $\left\{-\dfrac{4}{9}\right\}$ **3.** $\{4\}$ **4.** $\left\{\dfrac{10}{13}\right\}$ **5.** $\left\{\dfrac{5}{8}\right\}$ **6.** $\{5\}$ **7.** $\emptyset$ or $\{\ \}$ **8.** all real numbers

9. (a) $l = \dfrac{V}{wh}$ **(b)** 9 in.   **10. (a)** $y = -\dfrac{2}{3}x + 4$ **(b)** $y = -\dfrac{4}{3}$ **11.** $6(x - 8) = 2x - 5$ **12.** 60

13. 15, 16, 17   **14.** 10 in., 24 in., 26 in.   **15.** 3.5 hours   **16.** shorter piece is 5 feet; longer is 16 feet   **17.** \$36

18. $\{x \mid x \leq 6\}; (-\infty, 6]$ **19.** $\left\{x \mid x > \dfrac{5}{4}\right\}; \left(\dfrac{5}{4}, \infty\right)$ **20.** at most 200 minutes

Chapter 3

Section 3.1

1. x-axis, y-axis, origin   **3.** solution   **5.** True   **7.** Answers may vary.

9.

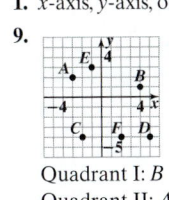

Quadrant I: B
Quadrant II: A, E
Quadrant III: C
Quadrant IV: D, F

11.

Quadrant I: C, E
Quadrant III: F
Quadrant IV: B
x-axis: A, G
y-axis: D, G

13.

Positive x-axis: A
Negative x-axis: D
Positive y-axis: C
Negative y-axis: B

15. $A(4, 0)$; positive x-axis $B(-3, 2)$; quadrant II $C(1, -4)$; quadrant IV $D(-2, -4)$; quadrant III $E(3, 5)$; quadrant I $F(0, -3)$; negative y-axis

17. A No   **19.** A Yes   **21.** A Yes
B Yes B No B No
C Yes C Yes C Yes

23. $(4, 1)$ **25.** $(5, -1)$ **27.** $(-3, 3)$

29.

x	y	(x, y)
−3	3	(−3, 3)
0	0	(0, 0)
1	−1	(1, −1)

31.

x	y	(x, y)
−2	7	(−2, 7)
−1	4	(−1, 4)
4	−11	(4, −11)

33.

x	y	(x, y)
−1	8	(−1, 8)
2	2	(2, 2)
3	0	(3, 0)

35.

x	y	(x, y)
−4	6	(−4, 6)
1	6	(1, 6)
12	6	(12, 6)

37.

x	y	(x, y)
1	$\frac{7}{2}$	$\left(1, \frac{7}{2}\right)$
−4	1	(−4, 1)
−2	2	(−2, 2)

39.

x	y	(x, y)
4	7	(4, 7)
−4	3	(−4, 3)
−6	2	(−6, 2)

41.

x	y	(x, y)
0	−3	(0, −3)
−2	0	(−2, 0)
2	−6	(2, −6)

43. $A(2, -16)$
$B(-3, -1)$
$C\left(-\frac{1}{3}, -9\right)$

45. $A(2, -6)$
$B(0, 0)$
$C\left(\frac{1}{6}, -\frac{1}{2}\right)$

47. $A(4, -8)$
$B(4, -19)$
$C(4, 5)$

49. $A(3, 4)$
$B(-6, -2)$
$C\left(\frac{1}{2}, \frac{7}{3}\right)$

51. $A\left(-4, -\frac{4}{3}\right)$
$B(-2, -1)$
$C\left(-\frac{2}{3}, -\frac{7}{9}\right)$

53. $A(20, 23)$
$B(-4, -17)$
$C(2.6, -6)$

55. (a) $24.85 **(b)** $54.70 **(c)** 6 CDs
(d) It costs $34.80 to order 3 CDs.
57. (a) $1.15 **(b)** $1.23 **(c)** $1.60 **(d)** 1996
(e) Answers may vary. **59.** $k = 4$ **61.** $k = 2$ **63.** $k = \frac{1}{2}$

65.

a	b	(a, b)
2	−8	(2, −8)
0	−4	(0, −4)
−5	6	(−5, 6)

67.

p	q	(p, q)
0	$\frac{10}{3}$	$\left(0, \frac{10}{3}\right)$
$\frac{5}{2}$	0	$\left(\frac{5}{2}, 0\right)$
−10	$\frac{50}{3}$	$\left(-10, \frac{50}{3}\right)$

69. Points may vary; line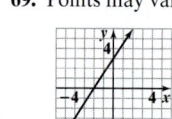

71.

x	y	(x, y)
−2	0	(−2, 0)
−1	−3	(−1, −3)
0	−4	(0, −4)
1	−3	(1, −3)
2	0	(2, 0)

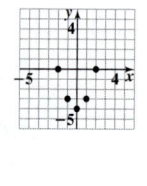

73.

x	y	(x, y)
−2	10	(−2, 10)
−1	3	(−1, 3)
0	2	(0, 2)
1	1	(1, 1)
2	−6	(2, −6)

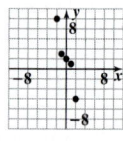

75.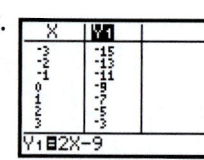

77.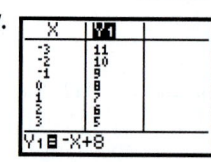

79.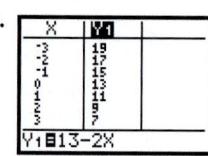

81.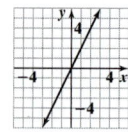

Section 3.2

1. linear, standard form **3.** intercepts **5.** False **7.** Two; three is recommended for a check. **9.** linear **11.** nonlinear **13.** nonlinear
15. linear **17.** $y = 2x$ **19.** $y = -5x$ **21.** $y = 4x - 2$ **23.** $y = -2x + 5$ **25.** $x + y = 5$ **27.** $-2x + y = 6$

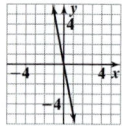 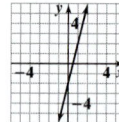 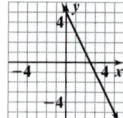 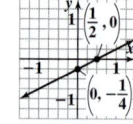 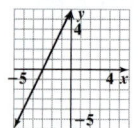 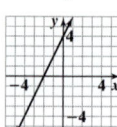

29. $4x - 2y = -8$ **31.** $x = -4y$ **33.** $y + 7 = 0$ **35.** $y - 2 = 3(x + 1)$ **37.** $(0, -5), (5, 0)$ **39.** $(0, -3), (6, 0)$
41. $(0, -3)$ **43.** $(-5, 0)$ **45.** $(0, -4), (-6, 0)$
47. $(0, 0)$ **49.** $(0, -5), (5, 0)$ **51.** $(0, 8), (6, 0)$
53. $(4, 0)$ **55.** $(0, -2)$
$3x + 6y = 18$

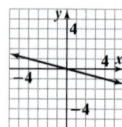 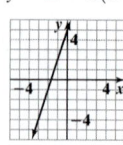 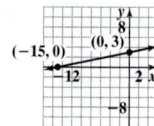

57. **59.** $-x + 5y = 15$ **61.** $\frac{1}{2}x = y + 3$ **63.** $9x - 2y = 0$ **65.** $y = -\frac{1}{2}x + 3$ **67.** $\frac{1}{3}y + 2 = 2x$

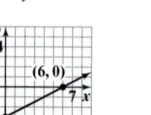 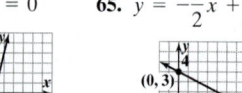 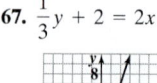 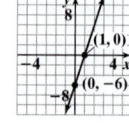 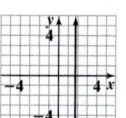

69. $\frac{x}{2} + \frac{y}{3} = 1$ **71.** $4y - 2x + 1 = 0$ **73.** $x = 5$ **75.** $y = -6$ **77.** $y - 12 = 0$ **79.** $3x - 5 = 0$

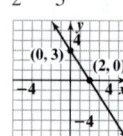 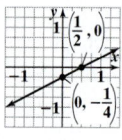 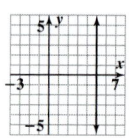 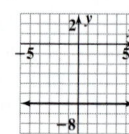 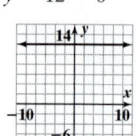

81. $y = 2x - 5$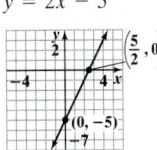

83. $y = -5$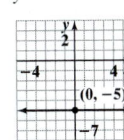

85. $2x + 5y = -20$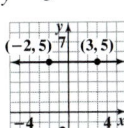

87. $2x = -6y + 4$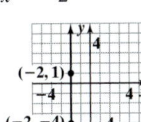

89. $x - 3 = 0$

91. $3y - 12 = 0$

93. $y = 5$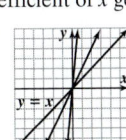

95. $x = -2$

97. $y = 4$
99. $x = -9$
101. $x = 2y$
103. $y = x + 2$
105. $y = 2$
107. $x = 7$

109. (a) $(0, 500), (4, 900), (10, 1500)$
(b)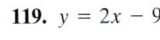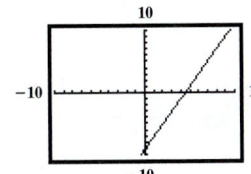
Number of cars sold

(c) If she sells 0 cars, her earnings are $500.

111. The "steepness" of the lines is the same.
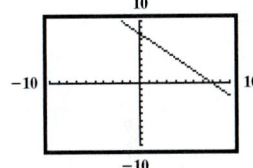

113. The lines get more steep as the coefficient of x gets larger.
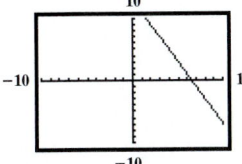

115. $(0, -6), (-2, 0), (3, 0)$
117. $(0, 14), (-3, 0), (2, 0), (5, 0)$

119. $y = 2x - 9$
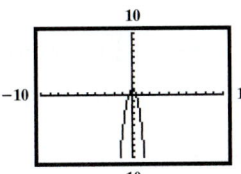

121. $y = -x + 8$

123. $y = 2x + 13$ or $y = -2x + 13$
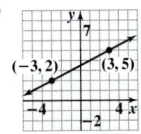

125. $y = -6x^2 + 1$
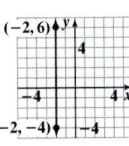

Section 3.3

1. $\frac{6}{10}$ or $\frac{3}{5}$ **3.** positive   **5.** True   **7.** vertical line; Answers may vary   **9.** $-\frac{3}{2}$ **11.** undefined   **13.** $\frac{1}{2}$ **15.** $-\frac{2}{3}$

17. (a), (b)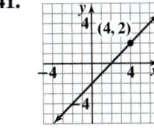

(c) $m = \frac{1}{2}$; for every 2-unit increase in x, there is a 1-unit increase in y.

19. (a), (b)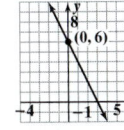

(c) $m = 0$; the line is horizontal

21. $m = -2$; for every 1-unit increase in x, there is a 2-unit decrease in y.
23. $m = -1$; for every 1-unit increase in x, there is a 1-unit decrease in y.
25. $m = -\frac{5}{3}$; for every 3-unit increase in x, there is a 5-unit decrease in y.
27. $m = \frac{3}{2}$; for every 2-unit increase in x, there is a 3-unit increase in y.

29. $m = 0$; there is no change in the y-values. The line is horizontal.   **31.** $m = \frac{2}{3}$; for every 3-unit increase in x, there is a 2-unit increase in y.

33. m is undefined; the line is vertical.   **35.** $m = \frac{1}{3}$; for every 3-unit increase in x, there is a 1-unit increase in y.

37. $m = \frac{1}{3}$; for every 3-unit increase in x, there is a 1-unit increase in y.   **39.** $m = 2$; for every 1-unit increase in x, there is a 2-unit increase in y.

41.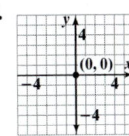
43.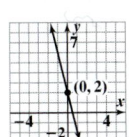
45.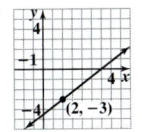
47.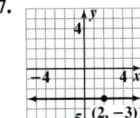
49.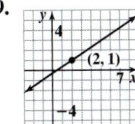
51.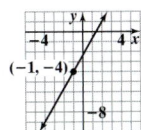

53.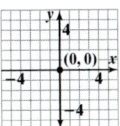
55.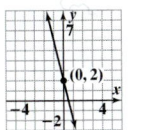
57.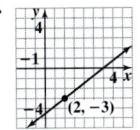
59. $x = 2$
61. $y = -2$
63. $x = -2$
65. $x = 3$
67. $x = 0$
69. $y = \frac{4}{3}$
71. 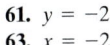 $m = -\frac{1}{2}$, $m = 2$
73. 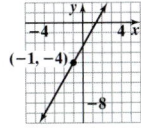 $(2, 1)$, $(-1, -2)$

75. $\frac{1}{3}$ **77.** 12 in. or 1 ft **79.** 16% **81.** $m = 2.26$ million; the population is increasing at an average rate of about 2.26 million people per year.

83. Points may vary.
$(-2, 1), (0, -5)$
$m = -3$
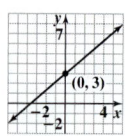

85. Points may vary.
$(4, -10), (-2, -10)$
$m = 0$
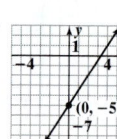

87. Points may vary. $(-2, -2), (0, 4)$
$m = 3$
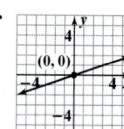

89. $m = -2$ **91.** $m = \dfrac{q}{p}$

93. $m = \dfrac{6}{a - 6}$

95. $MR = 2$; For every hot dog sold, revenue increases by \$2, on average.

Section 3.4

1. $3; 7$ **3.** plotting points, using intercepts, using the slope and a point **5.** False **7. (a)** (a), (e)

9. $m = 5; b = 2$ **11.** $m = 2; b = -9$ **13.** $m = -10; b = 7$ **15.** $m = -1; b = -9$ **17.** $m = 0, b = -5$ **19.** $m = \dfrac{2}{3}; b = -4$

21. $m = 2; b = -4$ **23.** $m = -\dfrac{2}{3}; b = 8$ **25.** $m = \dfrac{5}{3}; b = -\dfrac{1}{3}$ **27.** $m = \dfrac{1}{2}; b = -\dfrac{5}{2}$ **29.** m is undefined; no y-intercept

31. **33.** **35.** **37.** **39.** **41.**

43. **45.** **47.** **49.** **51.** **53.**

55. **57.** **59.** **61.** $y = -x + 8$ **63.** $y = \dfrac{6}{7}x - 6$ **65.** $y = -\dfrac{1}{3}x + \dfrac{2}{3}$

67. $x = -5$ **69.** $y = 3$ **71.** $y = 5x$

73. **75.** **77.** **79.** **81.**

83.
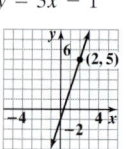
85. (a) $y = 0.08x + 400$
(b) \$496 **(c)**
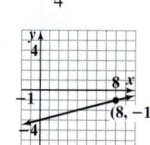

87. (a) 34.5¢ per year. **(b)** 2001 **(c)** The cost per minute is decreasing by 5.375¢ per year. **(d)** No, the cost will never be 0 or negative.
(e)
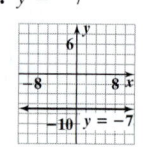

89. $B = -4$ **91.** $A = -4$ **93.** $B = -3$ **95. (a)** $y = 40x + 4000$ **(b)** \$24,000 **(c)** 375 calculators **(d)**

Section 3.5

1. $y - y_1 = m(x - x_1)$ **3.** standard form, slope-intercept form, point-slope form; $x = a$, $y = b$ **5.** True **7.** Yes; Answers may vary.

9. $y = 3x - 1$ **11.** $y = -2x$ **13.** $y = \dfrac{1}{4}x - 3$ **15.** $y = -6x + 13$ **17.** $y = -7$ **19.** $x = -4$

21. $y = \frac{2}{3}x + 2$ **23.** $y = -\frac{3}{4}x - \frac{39}{4}$ **25.** $y = \frac{1}{2}x - 5$ **27.** $x = -3$ **29.** $y = -5$ **31.** $y = -4.3$ **33.** $x = \frac{1}{2}$

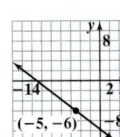   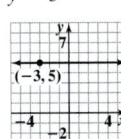   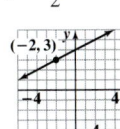

35. $y = 2x + 4$ **37.** $y = -4x + 6$ **39.** $y = -\frac{3}{2}x - \frac{5}{2}$

41. $y = 2x - 5$ **43.** $y = -3$ **45.** $x = 2$

47. $y = 0.25x + 0.575$ **49.** $y = x - \frac{11}{4}$

51. $y = 5x - 22$ **53.** $y = 5$ **55.** $y = x + 2$ **57.** $y = \frac{1}{2}x + 4$ **59.** $x = 5$ **61.** $y = -7x + 2$

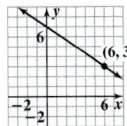   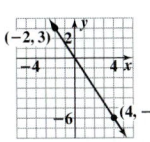   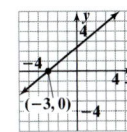   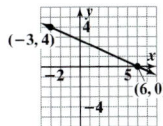   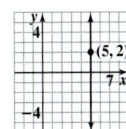   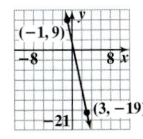

63. $y = -\frac{2}{3}x + 7$ **65.** $y = -\frac{3}{2}x$ **67.** $y = \frac{4}{5}x + \frac{12}{5}$ **69.** $y = -\frac{4}{9}x + \frac{8}{3}$

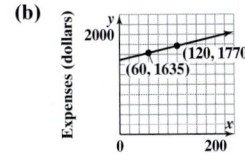   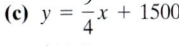   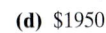

71. (a) When 60 packages are shipped, the expenses are \$1635.

(b) 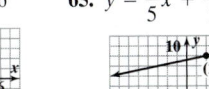   **(c)** $y = \frac{9}{4}x + 1500$ **(d)** \$1950

(e) Expenses increase by \$2.25 for each additional package.

73. (a) $(0, 181), (27, 276)$ **(b)** 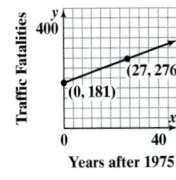   **(c)** $y = \frac{95}{27}x + 181$ **(d)** about 234

(e) The number of traffic fatalities in this region increases by about 3.5 each year.

75. $y = 3x + 14$ **77.** $y = -2x - 2$ **79.** $y = 6x - \frac{7}{2}$ **81.** $y = -x - 7$ **83.** $y = \frac{1}{5}x - 2$ **85.** $y = -\frac{3}{2}x + 2$

Section 3.6

1. perpendicular   **3.** perpendicular   **5.** False   **7.** L_1 could be parallel to L_2. L_1 could be perpendicular to L_2. L_1 and L_2 could intersect, but not at right angles, L_1 could be coincident with L_2.

	Slope of the Given Line	Slope of a Line Parallel to the Given Line	Slope of a Line Perpendicular to the Given Line
9.	$m = -3$	$m_1 = -3$	$m_2 = \frac{1}{3}$
11.	$m = \frac{1}{2}$	$m_1 = \frac{1}{2}$	$m_2 = -2$
13.	$m = -\frac{4}{9}$	$m_1 = -\frac{4}{9}$	$m_2 = \frac{9}{4}$
15.	$m = 0$	$m_1 = 0$	$m_2 =$ undefined

17. perpendicular   **19.** parallel   **21.** perpendicular   **23.** perpendicular   **25.** perpendicular   **27.** parallel   **29.** parallel

31. (a) $m_1 = 3, m_2 = -3$ **(b)** neither   **33. (a)** $m_1 = 2, m_2 = 2$ **(b)** parallel   **35. (a)** $m_1 = \frac{1}{2}, m_2 = \frac{1}{2}$ **(b)** parallel

37. (a) $m_1 = \frac{5}{3}, m_2 = -\frac{3}{5}$ **(b)** perpendicular   **39.** $y = 3x - 14$ **41.** $y = -4x - 4$ **43.** $y = -7$ **45.** $x = -1$ **47.** $y = \frac{3}{2}x - 13$

49. $y = -\frac{1}{2}x - \frac{21}{2}$ **51.** $y = -2x + 11$ **53.** $y = \frac{1}{4}x$ **55.** $x = -2$ **57.** $y = 5$ **59.** $y = \frac{5}{2}x$ **61.** $y = -\frac{3}{5}x - 9$

63. $y = 7x - 26$ **65.** $y = \frac{1}{5}x + \frac{43}{5}$ **67.** $y = -7x + 41$ **69.** $y = -2x - 10$ **71.** $y = 3x - 2$ **73.** $x = 5$

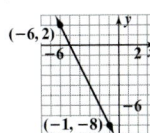   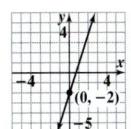   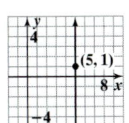

75. $y = -\dfrac{4}{3}x + 2$ **77.** $y = -2x - 5$ **79.** 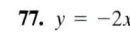   **81.** 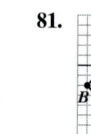   **83.** 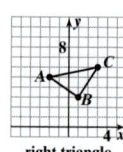   **85.**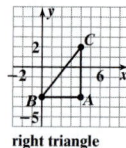

parallelogram   rectangle   right triangle   right triangle

87. $B = 2$ **89.** $A = 4$ **91.**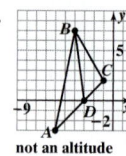

not an altitude

Putting the Concepts Together

1. yes, $(1, -2)$ is a solution   **2.** 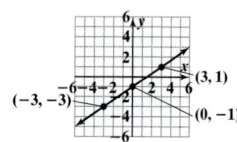   **3.** 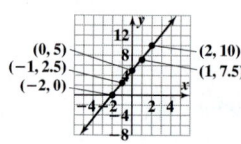   **4. (a)** x-intercept is $-\dfrac{3}{4}$ **(b)** y-intercept is 3

5. (a) x-intercept is $\dfrac{3}{2}$ **(b)** y-intercept is 2   **(c)** 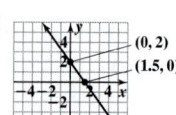   **6. (a)** slope $= -\dfrac{2}{3}$ **(b)** y-intercept $= -\dfrac{4}{3}$ **7.** slope $= -\dfrac{1}{3}$

8. (a) $m = \dfrac{2}{5}$ **(b)** $m = -\dfrac{5}{2}$ **9.** The lines are parallel. Answers may vary.   **10.** $y = 3x + 1$ **11.** $y = -6x - 2$ **12.** $y = -2x + 7$

13. $y = -\dfrac{5}{2}x - 20$ **14.** $y = \dfrac{1}{4}x + 5$ **15.** $y = -8$ **16.** $x = 2$ **17.** $m = 11$. Every package increases expenses by \$11.

18. (a) $y = 8350x - 2302$, where x is the weight, in carats, and y is the price.   **(b)** For every 1-carat increase in weight, the cost increases by \$8350.
(c) \$4044

Section 3.7

1. $t = ks$ **3.** constant of proportionality or constant of variation   **5.** True   **7.** Answers may vary.   **9.** $y = \dfrac{1}{2}x$ **11.** $y = 2x$

13. $y = -\dfrac{3}{4}x$ **15.** $y = \dfrac{6}{x}$ **17.** $y = \dfrac{12}{x}$ **19.** $y = \dfrac{2}{7x}$ **21.** direct variation; $\dfrac{2}{3}$ **23.** neither   **25.** inverse variation; 9

27. direct variation; 2   **29. (a)** $p = \dfrac{1}{3}g$ **(b)** $p = 3$ **31. (a)** $y = -\dfrac{3}{4}x$ **(b)** $x = -\dfrac{5}{3}$ **33. (a)** $e = \dfrac{8}{n}$ **(b)** $e = \dfrac{1}{2}$ **35. (a)** $b = \dfrac{3}{4a}$

(b) $a = \dfrac{5}{6}$ **37.** $\dfrac{9}{2}$ **39.** $P = 75$ **41.** $t = 42$ **43.** $b = 18$ **45.** 10 representatives   **47.** 120 board feet   **49.** 6.25 atmospheres

51. 150 bags   **53.** $m = 90$ **55.** $r = 5$ **57. (a)** $e = \dfrac{1.989 \times 10^{-25}}{\lambda}$ **(b)** 3.978×10^{-19} joules

Section 3.8

1. solid   **3.** half-planes   **5.** True   **7.** Answers may vary.   **9.** A is a solution.   **11.** C is a solution.   **13.** A and C are solutions.
15. B is a solution.   **17.** B is a solution.   **19.** A is a solution.

21. $y > 3x - 2$ **23.** $y \le -x + 1$ **25.** $y < \dfrac{x}{2}$ **27.** $y > 5$ **29.** $y \le \dfrac{2}{5}x + 3$ **31.** $y \ge -\dfrac{4}{3}x + 2$ **33.** $x < 2$

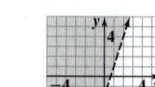 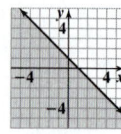 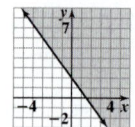 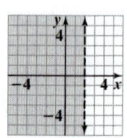

35. $3x - 4y < 12$ **37.** $2x + y \ge -4$ **39.** $x + y > 0$ **41.** $5x - 2y < -8$ **43.** $x > -1$ **45.** $y \le 4$ **47.** $\dfrac{x}{3} - \dfrac{y}{5} \ge 1$

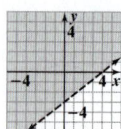 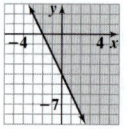 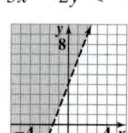 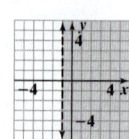 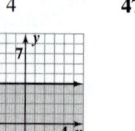 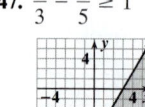

49. $-3 \geq x - y$ **51.** $x + y \geq 26$ **53.** $\dfrac{y}{-2} \leq 4$ **55.** $x \leq y - 3$ **57.** $x + 3y < 0$ **59.** $2x - \dfrac{1}{2}y \geq 5$ **61.** $-2x > -1$

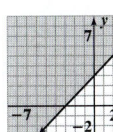 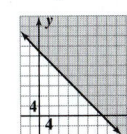 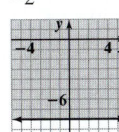 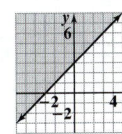 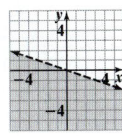 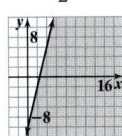 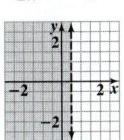

63. (a) $3s + 5a \leq 120$ **(b)** No   **(c)** Yes   **65. (a)** $3.1s + 5.7b \leq 22$ **(b)** No   **(c)** Yes

67. $3x - 2y > 6$ and $x + y < 2$ **69.** $y > \dfrac{3}{4}x - 1$ and $x \geq 0$ **71.** $x < -3$ and $y \leq 4$ **73.** $y > 3$ **75.** $y < 5x$

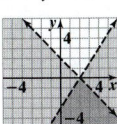 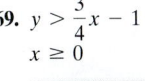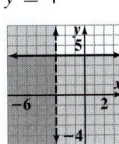 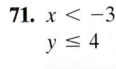 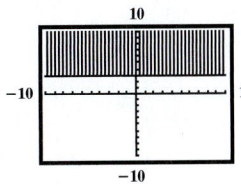 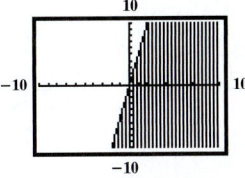

77. $y > 2x + 3$ **79.** $y \leq \dfrac{1}{2}x - 5$ **81.** $3x + y \leq 4$ **83.** $2x + 5y \leq -10$

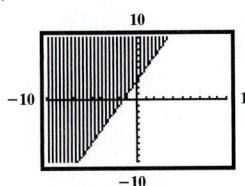 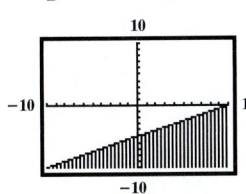 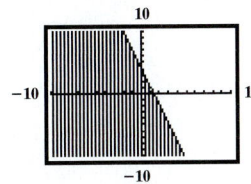 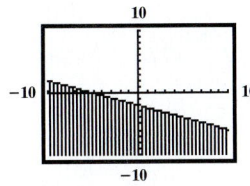

Chapter 3 Review

1–4.

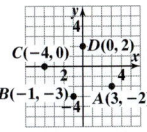

$A(3, -2)$ quadrant IV;
$B(-1, -3)$ quadrant III;
$C(-4, 0)$ negative x-axis;
$D(0, 2)$ positive y-axis

5. $A(1, 4)$; quadrant I
$B(-3, 0)$; x-axis

6. $A(0, 1)$; y-axis
$B(-2, 2)$; quadrant II

7. A. Yes
B. No

8. A. Yes
B. No

9. (a) $\left(4, -\dfrac{4}{3}\right)$
(b) $(-6, 2)$

10. (a) $(4, 4)$
(b) $(-2, -2)$

11.

x	y	(x, y)
-2	-8	$(-2, -8)$
0	-5	$(0, -5)$
4	1	$(4, 1)$

12.

x	y	(x, y)
-3	5	$(-3, 5)$
2	0	$(2, 0)$
4	-2	$(4, -2)$

13.

x	y	(x, y)
-18	2	$(-18, 2)$
-6	-2	$(-6, -2)$
24	-12	$(24, -12)$

14.

x	y	(x, y)
-3	-8	$(-3, -8)$
3	1	$(3, 1)$
5	4	$(5, 4)$

15.

x	y	(x, y)
1	7	$(1, 7)$
2	9	$(2, 9)$
3	11	$(3, 11)$

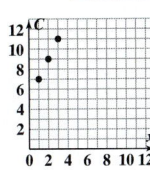

16.

x	E	(x, E)
500	1050	$(500, 1050)$
1000	1100	$(1000, 1100)$
2000	1200	$(2000, 1200)$

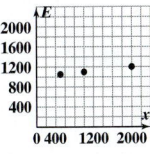

17. $y = -2x$

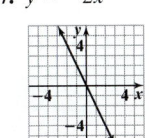

18. $y = x$

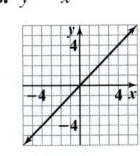

19. $4x + y = -2$

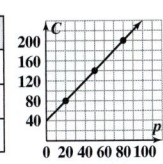

20. $3x - y = -1$

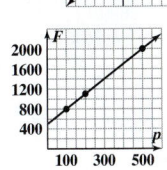

21.

p	C	(p, C)
20	80	$(20, 80)$
50	140	$(50, 140)$
80	200	$(80, 200)$

22.

p	F	(p, F)
100	800	$(100, 800)$
200	1100	$(200, 1100)$
500	2000	$(500, 2000)$

23. $(-2, 0), (0, -4)$ **24.** $(0, 1)$ **25.** $(0, 9), (-3, 0)$ **26.** $(0, -6), (3, 0)$ **27.** $(3, 0)$ **28.** $\left(0, -\dfrac{2}{5}\right), (1, 0)$

29. $y - 3x = 3$ **30.** $2x + 5y = 0$ **31.** $\dfrac{x}{3} + \dfrac{y}{2} = 1$ **32.** $y = -\dfrac{3}{4}x + 3$ **33.** $x = -2$ **34.** $y = 3$ **35.** $y = -4$

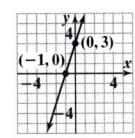 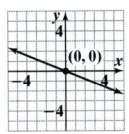 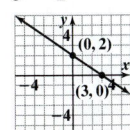 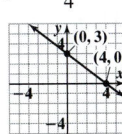 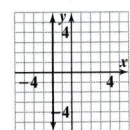 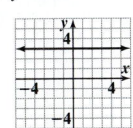 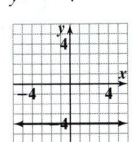

36. $x = 1$ **37.** $\dfrac{4}{3}$ **38.** -2 **39.** -8 **40.** 2 **41.** $\dfrac{1}{4}$ **42.** $-\dfrac{1}{6}$ **43.** undefined   **44.** 0 **45.** 0 **46.** undefined

47. 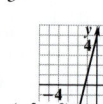   **48.** 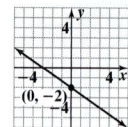   **49.** 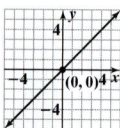   **50.** 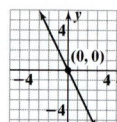   **51.** $m = 45$; the cost to produce 1 additional bicycle is $45.

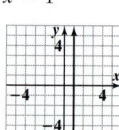

52. 5% **53.** $m = -1; b = \dfrac{1}{2}$ **54.** $m = 1; b = -\dfrac{3}{2}$ **55.** $m = \dfrac{3}{4}; b = 1$ **56.** $m = -\dfrac{2}{5}; b = \dfrac{8}{5}$

57. $y = \dfrac{1}{3}x + 1$ **58.** $y = -\dfrac{x}{2} - 1$ **59.** $y = -\dfrac{2}{3}x - 2$ **60.** $y = \dfrac{3x}{4} + 3$ **61.** $y = x$ **62.** $y = -2x$ **63.** $2x - y = -4$

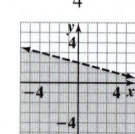

64. $-4x + 2y = 2$ **65.** $y = -\dfrac{3}{4}x + \dfrac{2}{3}$ **66.** $y = \dfrac{1}{5}x + 10$ **67.** $x = -12$ **68.** $y = -4$ **69.** $y = x - 20$ **70.** $y = -x - 8$

71. (a) $360   (b) 7 days   (c)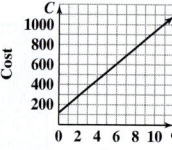

72. (a) $(22, 418), (35, 665), m = 19$
(b) It costs $19 more for each additional day.
(c) $C = 19d$
(d) $152

73. $y = 6x - 3$ **74.** $y = -2x + 8$ **75.** $y = -\dfrac{1}{2}x + \dfrac{1}{2}$ **76.** $y = \dfrac{2}{3}x - \dfrac{7}{3}$ **77.** $y = -\dfrac{1}{2}$ **78.** $x = -\dfrac{4}{7}$ **79.** $x = -5$ **80.** $y = 0$

81. $y = \dfrac{8}{7}x + 8$ **82.** $y = \dfrac{3}{2}x - 6$ **83.** $y = 3x - 4$ **84.** $y = -\dfrac{2}{5}x - 5$ **85.** $A = -4d + 24$ **86.** $F = -\dfrac{1}{3}m + 15$ **87.** not parallel

88. parallel   **89.** $y = -x + 2$ **90.** $y = 2x + 8$ **91.** $y = -3x + 7$ **92.** $y = -3x + 7$ **93.** $x = 5$ **94.** $y = -12$ **95.** $-\dfrac{2}{3}$

96. $-\dfrac{9}{4}$ **97.** perpendicular   **98.** not perpendicular   **99.** $y = \dfrac{1}{3}x + 5$ **100.** $y = -\dfrac{1}{2}x + 1$ **101.** $y = -\dfrac{3}{2}x - \dfrac{3}{2}$ **102.** $y = x + 1$

103. $y = \dfrac{1}{4}x$ **104.** $y = 6x$ **105.** $f = 16$ **106.** $p = 110$ **107.** $507.50   **108.** 160 mi   **109.** $y = \dfrac{48}{x}$ **110.** $y = \dfrac{54}{x}$ **111.** $r = 9$

112. $s = \dfrac{1}{4}$ **113.** 78 mph   **114.** 25 liters   **115.** A, B are solutions.   **116.** C is a solution.

117. $y < -\dfrac{1}{4}x + 2$ **118.** $y > 2x - 1$ **119.** $3x + 2y \geq -6$ **120.** $-2x + y \geq 4$ **121.** $x - 3y \leq 0$ **122.** $x - 4y \geq 4$

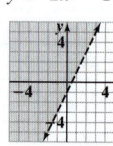   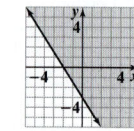   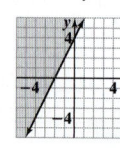   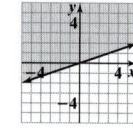   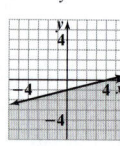

123. $x < -3$ **124.** $y > 2$ **125.** $0.25x + 0.1y \geq 12$ **126.** $2x - \dfrac{1}{2}y \leq 10$

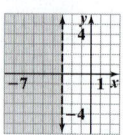   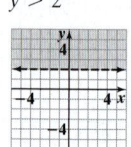

Chapter 3 Test

1. No   **2.** (a) $(4, 0)$ (b) $\left(0, -\dfrac{4}{3}\right)$ **3.** (a) $m = \dfrac{4}{3}$ (b) $b = 8$

4. $y = -\dfrac{3}{4}x + 2$ **5.** $3x - 6y = -12$ **6.** $-\dfrac{1}{6}$ **7.** (a) $-\dfrac{3}{2}$ (b) $\dfrac{2}{3}$ **8.** neither; Answers may vary.   **9.** $y = -4x - 15$ **10.** $y = 2x + 14$

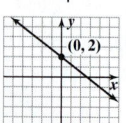   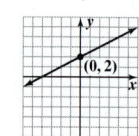

11. $y = -3x - 11$ **12.** $y = \dfrac{1}{2}x - 2$ **13.** $y = -\dfrac{3}{2}x + 8$ **14.** $y = 5$ **15.** $x = -2$

16. $m = 30$ **17.** $8 per package   **18.** $y \geq x - 3$ **19.** $-2x - 4y < 8$ **20.** $x \leq -4$

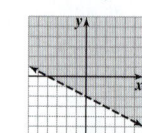   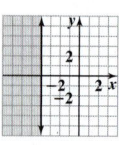

Cumulative Review Chapters 1–3

1. -16　**2.** $\dfrac{1}{4}$　**3.** -4　**4.** 8　**5.** $4m^2 + 5m - 9$　**6.** $\{-7\}$　**7.** $\left\{-\dfrac{25}{12}\right\}$　**8.** $B = \dfrac{2A - hb}{h}$ or $B = \dfrac{2A}{h} - b$

9. $\{x \mid x > -4\}$ or $(-4, \infty)$　　　　**10.** $\{x \mid x \le 5\}$ or $(-\infty, 5]$　　　**11.** 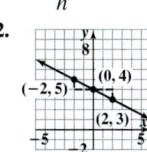　**12.**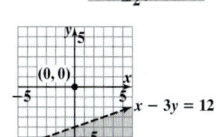

13. 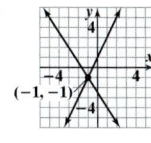　**14.** $y = -\dfrac{4}{3}x + 2$ or $4x + 3y = 6$　**15.** $y = -3x - 8$ or $3x + y = -8$　**16.**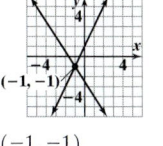

17. Shawn needs to score at least 91 on the final exam to earn an A.　**18.** A person 62 inches tall would be considered obese if they weighed 160 pounds or more.　**19.** The angles measure 55° and 125°.　**20.** The cylinder should be about 5.96 inches tall.　**21.** The three consecutive even integers are 24, 26, and 28.

Chapter 4

Section 4.1

1. intersect in one point　**3.** inconsistent　**5.** False　**7.** Answers may vary.　**9. (a)** No　**(b)** Yes　**(c)** Not

11. (a) No　**(b)** Yes　**(c)** Yes　**13. (a)** No　**(b)** No　**(c)** No

15. 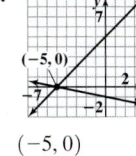　**17.** 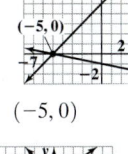　**19.** 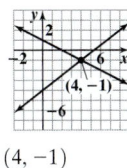　**21.** 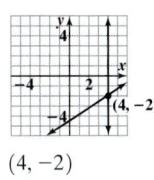　**23.** 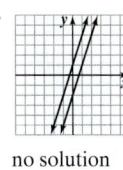　**25.**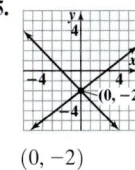

$(-1, -1)$　　　$(-5, 0)$　　　$(4, -1)$　　　$(4, -2)$　　　no solution　　　$(0, -2)$

27. 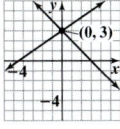　**29.**

infinitely many solutions　$(0, 3)$

31. one solution; consistent; independent; $(3, 2)$　**33.** no solution; inconsistent

35. infinitely many solutions; consistent; dependent　**37.** one solution; consistent; independent; $(-1, -2)$　**39.** one solution; consistent; independent　**41.** no solution; inconsistent

43. infinitely many solutions; consistent; dependent　**45.** one solution; consistent; independent

47. no solution; inconsistent　**49.** infinitely many solutions; consistent; dependent

51.

consistent; independent; $(-3, 3)$

53.

inconsistent; no solution

55.

consistent; independent; $(2, 4)$

57.

consistent; dependent; infinitely many solutions

59.

consistent; dependent; infinitely many solutions

61.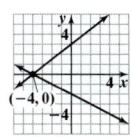

consistent; independent; $(-4, 0)$

63.

The break-even point is $(10, 130)$. The company needs to sell 10 CDs to break even at a cost/revenue of $130.

65.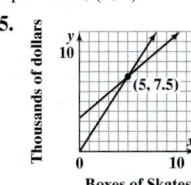

The break-even point is $(5, 7.5)$. The company needs to sell 5 boxes of skates to break even at a cost/revenue of $7500.

67.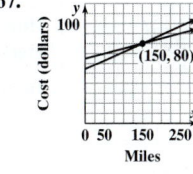

The cost for driving 150 miles is the same for both companies ($80). Choose Acme to drive 200 miles.

69.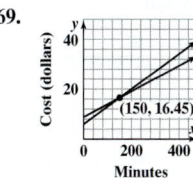

The cost is the same ($16.45) when 150 minutes are used. Choose Plan B for 100 long-distance minutes.

71. $c = 3$　**73.** $c = 1$　**75.**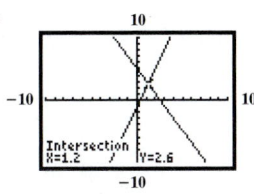

$(1.2, 2.6)$

77.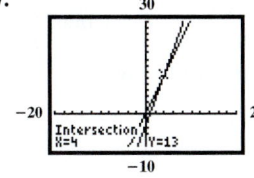

$(4, 13)$

79.

infinitely many solutions

Section 4.2

1. $1, -1$ **3.** infinitely many solutions **5.** True **7.** Answers may vary. **9.** $(4, -1)$ **11.** $(-1, 1)$ **13.** $(-3, -4)$ **15.** $(-4, -7)$

17. $\left(-\dfrac{2}{3}, 2\right)$ **19.** $\left(2, -\dfrac{1}{3}\right)$ **21.** $\left(\dfrac{1}{4}, \dfrac{1}{6}\right)$ **23.** inconsistent; no solution **25.** dependent; infinitely many solutions **27.** inconsistent; no solution

29. dependent; infinitely many solutions **31.** $(-2, 0)$ **33.** $\left(2, -\dfrac{1}{3}\right)$ **35.** no solution **37.** infinitely many solutions **39.** $\left(\dfrac{5}{3}, -\dfrac{1}{2}\right)$

41. no solution **43.** infinitely many solutions **45.** $\left(\dfrac{5}{4}, -\dfrac{7}{4}\right)$ **47.** infinitely many solutions **49.** $\left(-\dfrac{5}{2}, 4\right)$ **51.** 5 and 12

53. length 10 ft; width 7 ft **55.** \$8000 in money market; \$4000 in international fund **57.** 2003 **59.** $A = \dfrac{7}{6}; B = -\dfrac{1}{2}$ **61.** Answers may vary.
63. Answers may vary.

Section 4.3

1. graphing, substitution, elimination **3.** additive inverses **5.** True **7.** Answers may vary. **9.** $(2, -1)$ **11.** $(6, 4)$ **13.** $\left(-\dfrac{11}{4}, \dfrac{1}{2}\right)$
15. $(-1, -3)$ **17.** inconsistent; 0 **19.** dependent; infinitely many **21.** dependent; infinitely many **23.** inconsistent; 0 **25.** $(-5, 8)$

27. $\left(\dfrac{3}{2}, -\dfrac{3}{4}\right)$ **29.** no solution **31.** $(12, -9)$ **33.** no solution **35.** $\left(\dfrac{6}{29}, -\dfrac{8}{29}\right)$ **37.** $\left(\dfrac{1}{2}, 4\right)$ **39.** $(-1, 2.1)$ **41.** $(-2, -6)$

43. $(50, 30)$ **45.** $(1, 5)$ **47.** $(5, 2)$ **49.** $\left(-\dfrac{9}{17}, \dfrac{31}{17}\right)$ **51.** $\left(\dfrac{8}{3}, -\dfrac{4}{7}\right)$ **53.** infinitely many solutions **55.** infinitely many solutions

57. $(1, -3)$ **59.** $(-4, 0)$ **61.** $\left(-\dfrac{8}{5}, \dfrac{27}{10}\right)$ **63.** $(-4, -5)$ **65.** no solution **67.** $\left(\dfrac{106}{37}, \dfrac{16}{37}\right)$

69. hamburger 280 cal; Coke 210 cal **71.** tomatoes 10 hr; zucchini 15 hr; no **73.** 60 lb Arabica; 40 lb Robusta **75.** 70° and 20°
77. $\left(-\dfrac{3}{a}, 1\right)$ **79.** $\left(-3a - b, -\dfrac{3}{2}a - \dfrac{3}{2}b\right)$

Putting the Concepts Together

1. (a) No **(b)** Yes **(c)** No **2. (a)** infinitely many **(b)** consistent **(c)** dependent
3. (a) one **(b)** consistent **(c)** independent **4.** $(1, 1)$ **5.** $(5, -1)$ **6.** $(1, -5)$ **7.** $(-1, 0)$ **8.** $(-4, 5)$ **9.** $(-3, -1)$
10. $\left(-\dfrac{1}{3}, 3\right)$ **11.** infinitely many solutions **12.** $(2.5, 3)$ **13.** no solution

Section 4.4

1. perimeter; $P = 2l + 2w$ **3.** $180°; 90°$ **5.** True **7.** No; answers may vary. **9.** $2l + 2w$ **11.** $33, 49$ **13.** $14, 37$
15. Thursday 35,000; Friday 42,000 **17.** \$12,000 in stocks; \$9000 in bonds **19.** length 12 ft; width 9 ft **21.** length 25 m; width 10 m
23. $40°, 50°$ **25.** $22.5°, 157.5°$ **27.** current 0.4 mph; still-water 3.9 mph **29.** bike 11 mph; wind 1 mph
31. northbound 60 mph; southbound 72 mph **33.** 4 hr **35.** wind 25 mph; Piper 175 mph **37.** Sam 24 hr; Diane 8 hr

Section 4.5

1. mixture problems **3.** P, r, t **5.** False

7.

	Number	· Cost	= Total
Adult	a	4	$4a$
Student	s	1.5	$1.5s$
Total	215		580

$$\begin{cases} a + s = 215 \\ 4a + 1.5s = 580 \end{cases}$$

9.

	P	· r	= I
Savings	s	0.05	$0.05s$
Money Market	m	0.03	$0.03m$
Total	1600		50

$$\begin{cases} s + m = 1600 \\ 0.05s + 0.03m = 50 \end{cases}$$

11.

	lb	· Price	= Total
Mild	m	7.5	$7.5m$
Robust	r	10	$10r$
Total	12	8.75	$8.75(12)$

$$\begin{cases} m + r = 12 \\ 7.5m + 10r = 8.75(12) \end{cases}$$

13. $32a + 24c$ **15.** $0.1A + 0.07B$ **17.** $5.85r + 4.20y = 128.85$ **19.** 315 **21.** 400 **23.** 60 nickels, 90 dimes
25. first-class \$0.37; postcard \$0.19 **27.** \$7500 in 5% account; \$2500 in 8% account **29.** \$3200 in risky plan; \$1800 in safer plan
31. 2 lb arbequina; 3 lb green **33.** 48.9 lb of the \$2.75 per pound coffee; 51.1 lb of the \$5 per pound coffee **35.** 80 lb rye; 100 lb blue-grass
37. 20 ml 30% saline solution; 40 ml 60% saline solution **39.** 70 l **41.** 1.2 gal **43.** \$6200 at 5%; \$3800 at $7\dfrac{1}{2}$% loss

Section 4.6

1. solution **3.** dashed; solid **5.** True **7. (a)** $\begin{cases} y \geq x \\ y \geq -2x \end{cases}$ **(b)** $\begin{cases} y \leq x \\ y \geq -2x \end{cases}$ **(c)** $\begin{cases} y \geq x \\ y \leq -2x \end{cases}$ **(d)** $\begin{cases} y \leq x \\ y \leq -2x \end{cases}$

9. (a) Yes **(b)** Yes **(c)** Yes **11. (a)** Yes **(b)** No **(c)** No

13. $\begin{cases} x > 2 \\ y \le -1 \end{cases}$ **15.** $\begin{cases} y > -2 \\ x > -3 \end{cases}$ **17.** $\begin{cases} x + y < 3 \\ x - y > 5 \end{cases}$ **19.** $\begin{cases} x + y > 3 \\ 2x - y > 4 \end{cases}$ **21.** $\begin{cases} x < 2 \\ y < \frac{1}{2}x + 3 \end{cases}$ **23.** $\begin{cases} x \ge -2 \\ y < 2x + 3 \end{cases}$

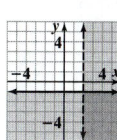 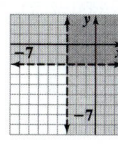 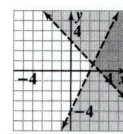 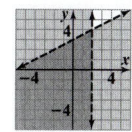 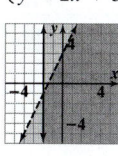

25. $\begin{cases} x > 0 \\ y \le \frac{2}{5}x - 1 \end{cases}$ **27.** $\begin{cases} -y \le x \\ 3x - y \ge -5 \end{cases}$ **29.** $\begin{cases} x + y \le -2 \\ y \ge x + 3 \end{cases}$ **31.** $\begin{cases} x + 3y \ge 0 \\ 2y < x + 1 \end{cases}$ **33.** $\begin{cases} x + y \ge 0 \\ x < 2y + 4 \end{cases}$ **35.** $\begin{cases} x + 3y > 6 \\ 2x - y \le 4 \end{cases}$

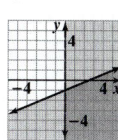 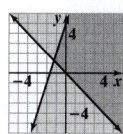 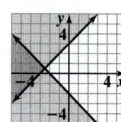 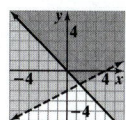 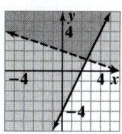

37. $\begin{cases} -y \le 3x - 4 \\ 2x + 3y \ge -3 \end{cases}$ **39. (a)** **41. (a)** **43.** $\begin{cases} \dfrac{y}{2} - \dfrac{x}{6} \ge 1 \\ \dfrac{x}{3} - \dfrac{y}{1} \ge 1 \end{cases}$ **45.** $\begin{cases} x < \dfrac{3}{2}y + \dfrac{9}{2} \\ -2x < 3(y + 2) \end{cases}$ **47.** $\begin{cases} y \ge 0 \\ y \ge x \\ y \le \dfrac{1}{2}(x + 6) \end{cases}$

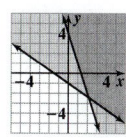 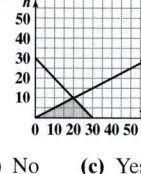 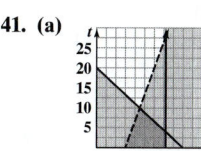

(b) No **(c)** Yes **(b)** No **(c)** No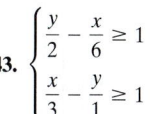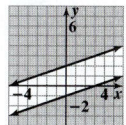

Chapter 4 Review

1. (a) No **(b)** Yes **(c)** No **2. (a)** No **(b)** No **(c)** Yes **3. (a)** Yes **(b)** Yes **(c)** Yes **4. (a)** No **(b)** No **(c)** No

5. **6.** **7.** **8.** **9.** **10.**

 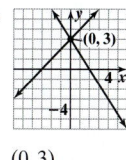 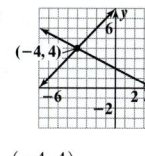 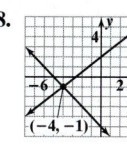 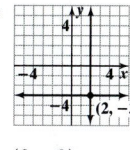 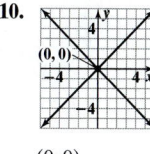

$(6, 1)$ $(0, 3)$ $(-4, 4)$ $(-4, -1)$ $(2, -3)$ $(0, 0)$

11. **12.**

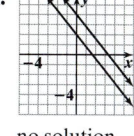 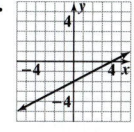

13. none; inconsistent **14.** none; inconsistent **15.** infinitely many; consistent; dependent
16. infinitely many; consistent; dependent **17.** one; consistent; independent
18. one; consistent; independent

no solution   infinitely many solutions

19. (a) $\begin{cases} y = 70 + 0.1x \\ y = 100 + 0.04x \end{cases}$ **20. (a)** $\begin{cases} y = 50 + 40x \\ y = 350 + 10x \end{cases}$ **21.** $(2, 1)$ **22.** $(1, -1)$ **23.** $(7, -2)$ **24.** $(4, -2)$ **25.** $(18, 11)$
26. $(-6, 8)$ **27.** $(2, -2)$ **28.** $(3, -1)$

(b) 500 fliers **(b)** 10 sq yd **29.** infinitely many solutions **30.** infinitely many solutions

31. no solution **32.** infinitely many solutions **33.** $\left(\dfrac{3}{2}, 1\right)$

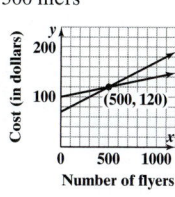

34. $\left(\dfrac{1}{3}, -1\right)$ **35.** width 125 m; length 200 m **36.** -3

(c) Printer A **(c)** tile

37. $(0, -12)$ **38.** $(-3, 7)$ **39.** $(-3, 4)$ **40.** $(-3, 0)$ **41.** $(-1, -1)$ **42.** $(5, 2)$ **43.** $(-2, 2)$ **44.** $\left(\dfrac{202}{55}, -\dfrac{59}{11}\right)$ **45.** $\left(\dfrac{1}{2}, -2\right)$

46. $\left(-\dfrac{3}{2}, 1\right)$ **47.** infinitely many solutions **48.** $\left(\dfrac{6}{7}, \dfrac{5}{2}\right)$ **49.** $(-20, -13)$ **50.** $\left(\dfrac{1}{3}, \dfrac{2}{3}\right)$ **51.** no solution **52.** $\left(-\dfrac{1}{7}, 0\right)$

53. 2.4 lb cookies; 1.6 lb chocolates **54. (a)** 400 individual tickets; 325 block tickets **55.** $\dfrac{3}{8}, \dfrac{1}{3}$ **56.** 2.6, 3.2

57. \$32,000 in stocks, \$18,000 in bonds **58.** 14 notebooks, 10 calculators **59.** $77.5°, 102.5°$ **60.** $40°, 50°$ **61.** 15 m by 22 m **62.** $110°, 40°$
63. plane 450 mph, wind 50 mph **64.** current 2 mph, paddling 6 mph **65.** cyclist 10 mph, wind 2 mph **66.** faster 8 mph; slower 6 mph

67.

	Number	·	Value	=	Total Value
Dimes	d		0.10		$0.10d$
Nickels	n		0.05		$0.05n$
Total					2.25

68.

		P	·	r	=	I
Savings		s		0.065		$0.065s$
Mutual Fund		m		0.08		$0.08m$
Total		15,000				2200

69. 125 children; 125 adult; 50 senior
70. 7 dimes; 4 quarters
71. $7000 at 5%, $12,000 at 9%
72. $10,000 bonds, $15,000 stocks
73. 7 quarts 60% sugar solution; 3 quarts 30% sugar solution

74. 60 pints   **75.** 3 lb peanuts; 2 lb almonds   **76.** 4 l 35% acid; 16 l 60% acid   **77. (a)** Yes   **(b)** Yes   **(c)** No
78. (a) No   **(b)** Yes   **(c)** Yes   **79. (a)** Yes   **(b)** No   **(c)** No   **80. (a)** No   **(b)** Yes   **(c)** Yes
81. (a) No   **(b)** No   **(c)** No   **82. (a)** No   **(b)** Yes   **(c)** No

83. $\begin{cases} x > -2 \\ y > 1 \end{cases}$ **84.** $\begin{cases} x \le 3 \\ y > -1 \end{cases}$ **85.** $\begin{cases} x + y \ge -2 \\ 2x - y \le -4 \end{cases}$ **86.** $\begin{cases} 3x + 2y < -6 \\ x - y < 2 \end{cases}$ **87.** $\begin{cases} x > 0 \\ y \le \frac{3}{4}x + 1 \end{cases}$ **88.** $\begin{cases} y \le 0 \\ y \le -\frac{1}{2}x - 3 \end{cases}$

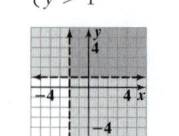 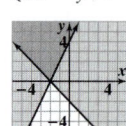 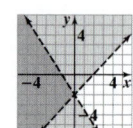 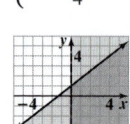 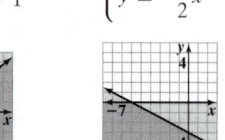

89. $\begin{cases} -y \ge x \\ 4x - 3y > -12 \end{cases}$ **90.** $\begin{cases} -y < x + 2 \\ 2x + 2y \ge -9 \end{cases}$ **91.** $\begin{cases} 2x + 3y > -3 \\ y > -\frac{2}{3}x + 2 \end{cases}$ **92.** $\begin{cases} x + 4y \le -4 \\ y \ge \frac{1}{4}x + 3 \end{cases}$ **93.** $\begin{cases} y > 2x - 5 \\ y - 2x \le 0 \end{cases}$

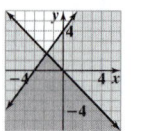 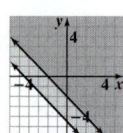 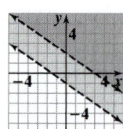 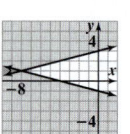 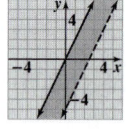

94. $\begin{cases} y > x + 2 \\ y < x - 4 \end{cases}$ **95.** $x = $ fish, $y = $ carne asada

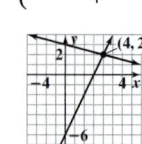

Chapter 4 Test

1. (a) No   **(b)** No   **(c)** Yes   **2. (a)** Yes   **(b)** No   **(c)** No   **3. (a)** one   **(b)** consistent   **(c)** independent
4. (a) none   **(b)** inconsistent   **(c)** not applicable

5. $\begin{cases} 2x + 3y = 0 \\ x + 4y = 5 \end{cases}$ **6.** $\begin{cases} y = 2x - 6 \\ y = -\frac{1}{4}x + 3 \end{cases}$ **7.** $(0, -3)$ **8.** $(-12, -1)$ **9.** $(-3, 3)$ **10.** $(-2, -4)$ **11.** $\left(3, -\frac{1}{2}\right)$ **12.** $(2.5, 3)$

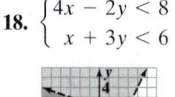 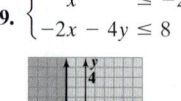

13. infinitely many solutions   **14.** airplane 200 mph; wind 25 mph

15. 7 containers vanilla; 4 containers peach   **16.** $52.5°, 127.5°$ **17.** 15 basketballs, 25 volleyballs

$(-3, 2)$ $(4, 2)$

18. $\begin{cases} 4x - 2y < 8 \\ x + 3y < 6 \end{cases}$ **19.** $\begin{cases} x \le -2 \\ -2x - 4y \le 8 \end{cases}$ **20.** $\begin{cases} y \le \frac{2}{3}x + 4 \\ -2x + 3y > -3 \end{cases}$

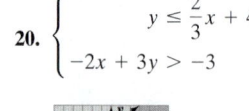 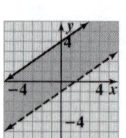

Chapter 5

Section 5.1
1. monomial   **3.** 5   **5.** True   **7.** Answers may vary.   **9.** Yes; coefficient $\frac{1}{2}$; degree 3   **11.** Yes; coefficient $\frac{1}{7}$; degree 2   **13.** No

15. Yes; coefficient 12; degree 5   **17.** No   **19.** Yes; coefficient 4; degree 0   **21.** $x^3 + 4x - 5$ **23.** $-y^4 + y^2 - 3y + 2$

25. Yes; $6x^2 - 10$; degree 2; binomial   **27.** No   **29.** No   **31.** Yes; $\frac{1}{8}$; degree 0; monomial   **33.** Yes; $-\frac{1}{2}t^4 + 3t^2 + 6t$; degree 4; trinomial

35. No   **37.** Yes; $5z^3 - 10z^2 + z + 12$; degree 3; polynomial   **39.** Yes; $2xy^4 + 3x^2y^2 + 4$; degree 5; trinomial   **41.** Yes; $-3y^5 + 4x^3$; degree 5;

binomial   **43.** No   **45.** $7x - 10$ **47.** $-2m^2 + 5$ **49.** $-2p^2 + p + 8$ **51.** $-3y^2 - 2y - 4$ **53.** $\frac{5}{4}p^2 + \frac{1}{6}p - 3$

55. $6m^2 - 4mn - n^2$ **57.** $4n^2 - 8n + 5$ **59.** $-x - 16$ **61.** $14x^2 - 3x - 5$ **63.** $4y^3 - 3y - 4$ **65.** $2y^3 - y^2 - 4y$

67. $\frac{14}{9}q^2 - \frac{23}{8}q + 2$ **69.** $-8m^2n^2 - 4mn - 7$ **71.** $-4x - 5$ **73. (a)** 3   **(b)** 48   **(c)** 13   **75. (a)** -2 **(b)** $\frac{3}{4}$ **(c)** 4.75

77. 45   **79.** -328 **81.** $3t - 3$ **83.** $3x^2 - 3x - 3$ **85.** $9xy^2 + 1$ **87.** $-y^2 + 15y + 1$ **89.** $\frac{7}{3}q^2 + \frac{5}{3}$ **91.** $13d^2 + d + 10$

93. $6a^2 + 5a - 3$ **95.** $-2x^2 - 2xy + y^2$ **97.** $-5x + 12$ **99.** $3x^2 + 4x - 3$ **101.** $-14x^2 + 4x - 13$ **103.** $-3x + 5$

105. 26 ft   **107. (a)** \$22,579.65   **(b)** \$43,793.85   **109. (a)** $-2.125x^2 + 105x$ **(b)** \$1250   **111. (a)** $10x$ **(b)** \$500   **(c)** \$26,000

113. $16x - 12$ **115.** $7x + 18$ **117.** $2x - 5$ **119.** $-x^2 - x + 15$ **121.** $2p^2 + 10p + 16$ **123.** $y^3 - 2y^2 + 11y - 14$

Section 5.2

1. a^{mn} **3.** expanded form   **5.** False   **7.** Answers may vary.   **9.** 1024   **11.** -128 **13.** m^9 **15.** b^{20} **17.** x^8 **19.** p^9 **21.** $(-n)^7$

23. a^{17} **25.** 64   **27.** z^{10} **29.** $(-m)^{18}$ **31.** m^{14} **33.** $(-b)^{20}$ **35.** 225   **37.** $27x^6$ **39.** $81p^{28}q^8$ **41.** $-8m^6n^3$ **43.** $25x^2y^4z^6$

45. $12x^5$ **47.** $-40a^{10}$ **49.** $-7m^4$ **51.** $6x^7$ **53.** $6x^6y^4$ **55.** $-5m^2n^4$ **57.** $4x^3$ **59.** b^6 **61.** $81b^4$ **63.** $9p^5$ **65.** $25x^8$

67. $-2x^7y^6$ **69.** $72x^{14}$ **71.** $4q^6$ **73.** x^6 **75.** $48x^2$ **77.** $16x^{14}$ **79.** $-\frac{24}{5}m^{17}n^9$

Section 5.3

1. $a^2 - b^2$ **3.** perfect square trinomial   **5.** False   **7.** Yes; $x = 0, y = 0$ **9.** $6x^2 - 10x$ **11.** $-7b^2 + 35b$ **13.** $2n^2 - 3n$

15. $12n^4 + 6n^3 - 15n^2$ **17.** $12m^3n - 4m^2n^2 + 2mn^3$ **19.** $4x^4y^2 - 3x^3y^3$ **21.** $x^2 + 5x + 6$ **23.** $q^2 - 13q + 42$

25. $y^2 - y - 20$ **27.** $x^4 + 4x^2 + 3$ **29.** $42 - 13x + x^2$ **31.** $-n^2 + 4n - 3$ **33.** $3a^2 + 5a + 2$ **35.** $10u^2 + 17uv + 6v^2$

37. $4n^2 - 9p^2$ **39.** $x^3 + x^2 - 5x - 2$ **41.** $2y^3 - 12y^2 + 19y - 3$ **43.** $2x^3 - 7x^2 + 4x + 3$ **45.** $6p^3 - 13p^2 + 4p + 3$

47. $2b^3 + 2b^2 - 24b$ **49.** $-x^3 + 9x$ **51.** $2x^4 - 5x^3 + 10x^2 - 11x - 4$ **53.** $10y^5 + 3y^4 + 4y^3 + 3y^2 + 2y + 2$ **55.** $x^2 - 4x - 21$

57. $z^2 - 7z - 30$ **59.** $6x^2 + 7x - 3$ **61.** $7k^2 + 39k - 18$ **63.** $14x^2 + 41xy + 15y^2$ **65.** $10a^2 - ab - 2b^2$ **67.** $x^2 - 9$

69. $4z^2 - 25$ **71.** $16x^4 - 1$ **73.** $4x^2 - 9y^2$ **75.** $x^2 - \frac{1}{9}$ **77.** $x^6 - y^4$ **79.** $x^2 - 4x + 4$ **81.** $x^2 + 4xy + 4y^2$

83. $4a^2 - 12ab + 9b^2$ **85.** $16x^2 + 24xy + 9y^2$ **87.** $25a^2 - 20ab^2 + 4b^4$ **89.** $x^2 + x + \frac{1}{4}$ **91.** $3x^3 - 6x^2 + 1$ **93.** $-5x^7 + 4x^6 - 6x^5$

95. $5x^3 + 21x^2 + 2x$ **97.** $-3w^3 + 3w^2 + 36w$ **99.** $3a^3 + 24a^2 + 48a$ **101.** $2n^2 + 6n$ **103.** $24ab$ **105.** $x^2 - 0.6x + 0.09$

107. $\frac{4}{9}b^2 - \frac{25}{81}$ **109.** $-x^2 + 3x + 2$ **111.** $4x^2 + 4x + 1$ **113.** $x^3 - 3x^2 + 3x - 1$ **115.** $2x^2 - 9$ **117.** $2x^2 + 7x - 15$

119. $x^2 + 10x + 25$ **121.** $9x^2 - 25$ **123.** $2x^2 + x - 1$ **125.** $x^2 + 12x + 20$ **127.** $x^2 + 3x + 2$ **129.** $40.8 + 0.4x$

131. $\pi x^2 + 4\pi x + 4\pi$ feet   **133. (a)** a^2, ab, ab, b^2 **(b)** $a^2 + 2ab + b^2$ **(c)** length: $a + b$, width: $a + b$, area: $(a + b)^2$

135. $x^3 + 2x^2 - 4x - 8$ **137.** $x^2 + 14x + 49$ **139.** $x^2 - 2xy + y^2 - 9$ **141.** $x^3 + 6x^2 + 12x + 8$ **143.** $z^4 + 12z^3 + 54z^2 + 108z + 81$

Section 5.4

1. a^{m-n} **3.** $\frac{1}{a^n}$ **5.** False   **7.** Answers may vary.   **9.** 27   **11.** 16   **13.** x^9 **15.** a^{12} **17.** $4y^3$ **19.** $-\frac{2}{3}m^7$ **21.** $2m^8n^2$

23. $\frac{11}{8}b^2$ **25.** $\frac{27}{8}$ **27.** $\frac{x^{15}}{27}$ **29.** $\frac{x^{20}}{y^{28}}$ **31.** $\frac{49a^4b^2}{c^6}$ **33.** 1   **35.** -1 **37.** 18   **39.** $\frac{1}{3}$ **41.** $\frac{1}{1000}$ **43.** $\frac{1}{m^2}$ **45.** $\frac{4}{a^2}$

47. $-\frac{4}{y^3}$ **49.** $\frac{11}{18}$ **51.** 2   **53.** $\frac{25}{4}$ **55.** $\frac{z^2}{3}$ **57.** $-\frac{m^6}{8n^3}$ **59.** 16   **61.** $6x^4$ **63.** 4   **65.** $\frac{1}{8}$ **67.** 81   **69.** $\frac{1}{x^9}$ **71.** x^{13}

73. $\frac{4}{x^3}$ **75.** $-\frac{3z^3}{2x^3}$ **77.** $\frac{4a}{5b}$ **79.** $\frac{1}{4x^3y^9}$ **81.** $-\frac{1}{a^{12}}$ **83.** $\frac{27}{m^6}$ **85.** $\frac{x^4y^6}{9}$ **87.** $\frac{2}{p^7}$ **89.** 12   **91.** $\frac{2x^8}{3}$ **93.** $\frac{16y}{27x^{18}}$ **95.** $\frac{9y}{2z^3}$

97. $216x^3y^6$ **99.** $\frac{5ab^7}{3}$ **101.** $\frac{2x^6}{5}$ **103.** $x^3 + 6x^2 + 12x + 8$ cubic feet   **105.** $3x^3$ cubic meters   **107.** $V = \frac{\pi d^2 h}{4}$

109. $x^5 + 3x^4 - 3x^3 - 9x^2$ cubic cm   **111.** $9\pi x^3$ square units   **113.** $\frac{1}{x^n}$ **115.** $x^{6n-2}y^2$ **117.** $\frac{x^{6a^2}y^{3ab}}{z^{3ac}}$ **119.** $\frac{9}{8a^{10n}}$

Putting the Concepts Together

1. Yes; degree 6; trinomial   **2. (a)** 0   **(b)** -4 **(c)** 2   **3.** $8x^4 - 4x^2 + 14$ **4.** $5x^2y + 3xy + 2y^2$ **5.** $-6m^3n^2 + 2m^2n^4$

6. $5x^2 - 17x - 12$ **7.** $8x^2 - 26xy + 21y^2$ **8.** $25x^2 - 64$ **9.** $4x^2 + 12xy + 9y^2$ **10.** $6a^3 + 22a^2 - 40a$

11. $16m^4 + 4m^3 + 10m^2 - 20m - 24$ **12.** $-8x^8z^9$ **13.** $-\frac{15}{mn}$ **14.** $-3a^3b^2$ **15.** $\frac{1}{2x^2y^2}$ **16.** $\frac{y^4}{16z^6}$ **17.** $\frac{8}{27r^6}$ **18.** $\frac{q^{56}}{r^{40}t^{64}}$

19. $\frac{1}{128x^6y^2}$ **20.** $-\frac{48y^6}{x^{12}}$

Section 5.5

1. standard   **3.** quotient; remainder; dividend   **5.** False   **7.** Answers may vary.   **9.** $2x - 1$ **11.** $3a + 9 - \frac{1}{a^2}$ **13.** $\frac{n^4}{5} - \frac{2n^2}{5} - 1$

15. $\frac{5r^2}{3} - 3$ **17.** $-x^2 + 3x - \frac{9}{x^2}$ **19.** $\frac{3}{8} + \frac{z^2}{2} - \frac{z}{4}$ **21.** $x + \frac{3x^2}{2} + \frac{2}{3x}$ **23.** $-7x + 5y$ **25.** $-\frac{6y}{x} + 15$ **27.** $-\frac{5}{a} - \frac{2c^2}{a^2b}$

29. $x - 7$ **31.** $x - 5$ **33.** $x^2 + 6x - 3$ **35.** $x^3 - 3x^2 + 6x - 2$ **37.** $x^2 - 2x + 3 + \frac{5}{x + 1}$ **39.** $2x + 3$ **41.** $x^2 + 6x + 7 + \frac{16}{x - 2}$

43. $2x^3 + 5x^2 + 9x - 4 + \frac{-17}{x - 4}$ **45.** $2x - 6$ **47.** $x^2 + 4x - 3 + \frac{2}{2x - 1}$ **49.** $x - 4 + \frac{-4}{5 + x}$ **51.** $2x - 1 + \frac{6}{1 + 2x}$

53. $x + 2 + \frac{2x - 4}{x^2 - 2}$ **55.** $x^2 + 3x - 6$ **57.** $a^2 + a - 30$ **59.** $-x^2 - x - 6$ **61.** $3x + 1$ **63.** $-2ab + 2b^2$ **65.** $x - 2 + \frac{3}{x}$

67. $\frac{2}{3}x^7y^8z^{12}$ **69.** $n^2 - 6n + 9$ **71.** $-10x^5 - 10x^3 + 40x^6$ **73.** $6pq - 2q^2 - p^2$ **75.** $4n^2 + 10nm - 6m^2$ **77.** $x - 3 + \frac{-10}{x-5}$

79. $x^3 + 6x^2 + 4x - 5$ **81.** $15rs^2 - 4r^2s$ **83.** $-3n - \frac{8}{3}m + 4$ **85.** $2x^6 - 4x^5 + 6x^4$ **87.** $-x + \frac{2}{x^2} - \frac{4}{x}$ **89.** $-\frac{1}{2x} - \frac{2}{x^2} + \frac{1}{x^3}$

91. x **93.** $z + 3$ **95.** $4x + 4$ **97.** \$182.11 per computer **99.** 6

Section 5.6

1. scientific notation **3.** positive **5.** False **7.** Answers may vary. **9.** 3×10^5 **11.** 6.4×10^7 **13.** 5.1×10^{-4} **15.** 1×10^{-9}
17. 8.007×10^9 **19.** 3.09×10^{-5} **21.** 6.2×10^2 **23.** 4×10^0 **25.** 420,000 **27.** 100,000,000 **29.** 0.0039 **31.** 0.4 **33.** 3760
35. 0.0082 **37.** 0.00006 **39.** 7,050,000 **41.** 5×10^{-4} **43.** 142,000 **45.** 0.008 **47.** 6.9×10^4 **49.** 6.415×10^9 **51.** 1.46×10^9
53. 3×10^{-5} **55.** 2,158,000 **57.** 0.000000000000001 **59.** 0.00225 **61.** 3×10^9 **63.** 8.4×10^{-3} **65.** 1.05×10^{-3} **67.** 3×10^8
69. 2.5×10^1 **71.** 1.8×10^8 **73.** 8×10^8 **75.** 9×10^{15} **77.** 3×10^{-14} **79.** 2.11×10^5 mi **81.** 4.87×10^8 km **83.** 4.56×10^8
85. 2.5×10^{-8} M **87.** $\approx 1.65 \times 10^3$ lb/person **89.** 1.116×10^7 miles **91.** 2.5×10^{-4} m **93.** 8×10^{-10} m **95.** 7.15×10^{-8} m
97. $1.2348\pi \times 10^{-14}$ m^3 **99.** $2.88\pi \times 10^{-25}$ m^3

Chapter 5 Review

1. Yes; degree 3; coefficient 4 **2.** No **3.** No **4.** Yes; degree 3; coefficient 1 **5.** No **6.** No **7.** Yes; degree 0; monomial
8. Yes; degree 5; binomial **9.** Yes; degree 6; trinomial **10.** Yes; 10; trinomial **11.** $9x^2 + 8x - 2$ **12.** $m^3 + mn - 5m$ **13.** $-x^2y + 12x$

14. $-7y^2 - 6yz + 9z^2$ **15.** $-2x^2 - 2$ **16.** $-20y^2 - 8y + 5$ **17. (a)** 0 **(b)** 8 **(c)** 2 **18. (a)** 3 **(b)** 2 **(c)** $\frac{11}{4}$ **19.** 0

20. 29 **21.** 279,936 **22.** $-\frac{1}{243}$ **23.** 16,777,216 **24.** 1 **25.** x^{13} **26.** m^6 **27.** r^{12} **28.** m^{24} **29.** $1024x^5$ **30.** $64n^6$

31. $81x^8y^4$ **32.** $4x^6y^8$ **33.** $15x^6$ **34.** $-36a^4$ **35.** $16y^5$ **36.** $-12p^6$ **37.** $12w^4$ **38.** $-\frac{3}{4}z^3$ **39.** $108x^8$ **40.** $80a^6$

41. $-8x^5 + 6x^4 - 2x^3$ **42.** $2x^7 + 4x^6 - x^4$ **43.** $6x^2 - 7x - 5$ **44.** $4x^2 - 5x - 6$ **45.** $x^2 - 3x - 40$ **46.** $w^2 + 9w - 10$
47. $24m^3 - 22m^2 + 7m - 3$ **48.** $8y^5 + 12y^4 + 4y^3 + 6y^2 - 6y - 9$ **49.** $x^2 + 8x + 15$ **50.** $2x^2 - 17x + 8$ **51.** $6m^2 + 17m - 14$
52. $48m^2 - 26m - 4$ **53.** $21x^2 + 5xy - 6y^2$ **54.** $20x^2 + 7xy - 3y^2$ **55.** $x^2 - 16$ **56.** $4x^2 - 25$ **57.** $4x^2 + 12x + 9$
58. $49x^2 - 28x + 4$ **59.** $9x^2 - 16y^2$ **60.** $64m^2 - 36n^2$ **61.** $25x^2 - 20xy + 4y^2$ **62.** $4a^2 + 12ab + 9b^2$ **63.** $x^2 - 0.25$

64. $r^2 - 0.0625$ **65.** $y^2 + \frac{4}{3}y + \frac{4}{9}$ **66.** $y^2 - y + \frac{1}{4}$ **67.** 36 **68.** $\frac{1}{343}$ **69.** x^4 **70.** $\frac{1}{x^8}$ **71.** 1 **72.** -1 **73.** 1 **74.** -1

75. $\frac{5x^2}{2y^3}$ **76.** $\frac{x^2}{3y^8}$ **77.** $\frac{x^{15}}{y^{10}}$ **78.** $\frac{343}{x^6}$ **79.** $\frac{8m^6n^3}{p^{12}}$ **80.** $\frac{81m^4n^8}{p^{20}}$ **81.** $-\frac{1}{25}$ **82.** 64 **83.** $\frac{81}{16}$ **84.** 27 **85.** $\frac{7}{12}$ **86.** $\frac{1}{8}$

87. $\frac{2x^3y^5}{3}$ **88.** $\frac{3}{7xy^{10}}$ **89.** $\frac{9m^4}{16}$ **90.** $\frac{64}{9m^6n^6}$ **91.** $\frac{8r^4s^6}{9}$ **92.** $\frac{32}{3r^{10}s^{15}}$ **93.** $6x^5 - 4x^4 + 5$ **94.** $3x^4 + 5x^2 - 6x$ **95.** $4n^3 + 1 - \frac{5}{2n^4}$

96. $6n - 4 - \frac{16}{5n^2}$ **97.** $-\frac{p^3}{8} - \frac{1}{4} + \frac{1}{2p^2}$ **98.** $-\frac{p^2}{2} + 1 - \frac{3}{2p^2}$ **99.** $4x - 7$ **100.** $x + 6$ **101.** $x^2 + 7x + 5 + \frac{6}{x-1}$

102. $2x^2 - 3x - 12 + \frac{-16}{x-2}$ **103.** $x^2 - 2x + 4$ **104.** $3x^2 + 15x + 77 + \frac{378}{x-5}$ **105.** 2.7×10^7 **106.** 1.23×10^9 **107.** 6×10^{-5}
108. 3.05×10^{-6} **109.** 3×10^0 **110.** 8×10^0 **111.** 0.0006 **112.** 0.00125 **113.** 613,000 **114.** 80,000 **115.** 0.37
116. 54,000,000 **117.** 6×10^3 **118.** 4.2×10^{-8} **119.** 2×10^2 **120.** 2×10^8 **121.** 4×10^{15} **122.** 2×10^2

Chapter 5 Test

1. Yes; degree 5; binomial **2. (a)** 5 **(b)** 21 **(c)** 26 **3.** $7x^2y^2 - 6x - 3y$ **4.** $10m^3 + m^2 - 6$ **5.** $-6x^5 + 18x^4 - 15x^3$
6. $2x^2 - 3x - 35$ **7.** $4x^2 - 28x + 49$ **8.** $16x^2 - 9y^2$ **9.** $6x^3 + x^2 - 25x + 8$ **10.** $2x - \frac{8}{3} + \frac{3}{x^3}$ **11.** $3x^2 - 11x + 33 + \frac{-94}{x+3}$
12. $-12x^4y^6$ **13.** $\frac{2m^3}{3n^5}$ **14.** $\frac{n^{24}}{m^{30}}$ **15.** $\frac{x^{22}}{y^{14}}$ **16.** $\frac{8m^7}{n^7}$ **17.** 1.2×10^{-5} **18.** 210,100 **19.** 3.57×10^4 **20.** 2×10^{-7}

Chapters 1–5 Cumulative Review

1. (a) $\left\{\sqrt{25}\right\}$ **(b)** $\left\{0, \sqrt{25}\right\}$ **(c)** $\left\{-6, -\frac{4}{2}, 0, \sqrt{25}\right\}$ **(d)** $\left\{-6, -\frac{4}{2}, 0, 1.4, \sqrt{25}\right\}$ **(e)** $\left\{\sqrt{7}\right\}$ **(f)** $\left\{-6, -\frac{4}{2}, 0, 1.4, \sqrt{7}, \sqrt{25}\right\}$

2. $-\frac{4}{9}$ **3.** -19 **4.** $6x^3 + 5x^2 - 3x$ **5.** $-18x + 8$ **6.** $\{1\}$ **7.** $4(x - 5) = 2x + 10$ **8.** \$640 **9.** 45 mi/h

10. $\{x | x < -3\}$ or $(-\infty, -3)$; **11.** $-\frac{1}{2}$ **12.** $-\frac{3}{2}$ **13.** **14.** **15.** $(3, -7)$

16. No solution **17.** $6x^3 - 3x^2 + 7x$ **18.** $28m^2 - 13m - 6$ **19.** $9m^2 - 4n^2$ **20.** $49x^2 + 14xy + y^2$ **21.** $4m^3 - 19m + 15$ **22.** $\frac{2}{x} + \frac{1}{y}$

23. $x^2 - 3x + 9$ **24.** $-24n^4$ **25.** $-\frac{5n^8}{2m^2}$ **26.** $\frac{1}{64x^6y^{24}z^{12}}$ **27.** $\frac{1}{2x^{12}y^3}$ **28.** 6.05×10^{-5} **29.** 2,175,000 **30.** 7.14×10^5

Chapter 6

Section 6.1

1. greatest common factor **3.** Distributive **5.** true **7.** false **9.** Answers may vary **11.** 2 **13.** 1 **15.** 5 **17.** 26 **19.** 4
21. 18 **23.** x^2 **25.** $7x$ **27.** m^2n^2 **29.** $15ab^2$ **31.** $2abc^2$ **33.** 3 **35.** $2(x - 4)^2$ **37.** $6(x - 3)^2$ **39.** $4x^3$ **41.** $-3s$

43. mn^5 **45.** $6(2x - 3)$ **47.** $5x^2y(1 - 3xy)$ **49.** $3x(x^2 + 2x - 1)$ **51.** $3(3m^5 - 6m^3 - 4m^2 + 27)$ **53.** $15ab^2(ab^2 - 4b + 3a^2)$

55. $(x - 3)(x - 5)$ **57.** $(x - 1)(x^2 + y^2)$ **59.** $(4x + 1)(x^2 + 2x + 5)$ **61.** $-x(3x - 4)$ **63.** $-5x(x^2 - 2x + 3)$

65. $-4z(3z^2 - 4z + 2)$ **67.** $-5(3b^3 + b - 2)$ **69.** $(x + 3)(y + 4)$ **71.** $(y + 1)(z - 1)$ **73.** $(x - 1)(x^2 + 2)$ **75.** $(2t - 1)(t^2 - 1)$

77. $(2t - 1)(t^3 - 1)$ **79.** $4(y - 5)$ **81.** $7m(4m^2 + m + 9)$ **83.** $6m^2n(2mnp - 3)$ **85.** $3(p + 3)(3p + 1)$

87. $3(2x - y)(3a - 2b)$ **89.** $(x - 2)(x - 2)$ or $(x - 2)^2$ **91.** $3x(5x - 2)(x^2 + 2)$ **93.** $-3x(x^2 - 2x + 3)$ **95.** $-4b(3 - 4b)$

97. $(4y + 3)(3x - 2)$ **99.** $\frac{1}{3}x^2\left(x - \frac{2}{3}\right)$ **101.** Answers may vary.   **103.** Answers may vary.   **105.** $-2t(8t - 75)$ **107.** $150(140 - p)$

109. $4x^3(2x^2 - 7)$ **111.** $2\pi r(r + 4)$ **113.** $x - 2$ **115.** $4x^{2n} + 5$ **117.** $3x^3 - 4x^2 + 2$ **119.** $x^4 - 3x + 2$ **121.** $\frac{3}{5}x^3 - \frac{1}{2}x + 1$

Section 6.2

1. quadratic trinomial   **3.** opposite   **5.** false   **7.** Answers may vary.   **9.** $(x + 2)(x + 3)$ **11.** $(m + 3)(m + 6)$ **13.** $(x - 3)(x - 12)$

15. $(a - 2)(a - 6)$ **17.** $(x - 4)(x + 3)$ **19.** $(x - 4)(x + 5)$ **21.** $(z - 3)(z + 15)$ **23.** prime   **25.** $(x - 2y)(x - 3y)$

27. $(r - 2s)(r + 3s)$ **29.** $(x - 4y)(x + y)$ **31.** prime   **33.** $3(x + 2)(x - 1)$ **35.** $3a(a - 3)(a - 5)$ **37.** $5z(x - 8)(x + 4)$

39. $-2(y - 2)^2$ **41.** $(x - 6)(x + 3)$ **43.** $(x + 8)(x - 4)$ **45.** $x(x + 5)(x - 3)$ **47.** $(a + 4b)(a - 7b)$ **49.** prime   **51.** $(k - 5)(k + 4)$

53. $(x - 5y)(x + 6y)$ **55.** $(st - 3)(st - 5)$ **57.** $-3c(c - 2)(c + 1)$ **59.** $-(x + 3)(x - 2)$ **61.** prime   **63.** $n^2(n - 6)(n + 5)$

65. $(m + 5n)(m - 3n)$ **67.** $(a + 2)(b - 4)$ **69.** $(s + 5)(s + 7)$ **71.** $3xy(x^2 - 2x - 2)$ **73.** prime   **75.** $(ab - 6)(ab - 2)$

77. $-x(x - 14)(x + 2)$ **79.** $(y - 6)^2$ **81.** $(2r - 3)(3s - 2)$ **83.** $-2x(x - 3)(x - 6)$ **85.** $-7xy(3x^2 + 2y)$ **87.** $(x - a)(x + b)$

89. $-16(t + 1)(t - 5)$ **91.** $(x + 6)$ and $(x + 3)$ **93.** $(x + 5)$ and $(x - 3)$ **95.** $-7, -5, 5, 7$ **97.** $-20, -4, 4, 20$ **99.** 1   **101.** $6, 10, 12$

Section 6.3

1. common factor   **3.** close   **5.** false   **7.** Answers may vary.   **9.** $(2n - 3)(3n - 1)$ **11.** $(2w - 5)(2w + 1)$ **13.** prime

15. $(2x - 1)(2x + 3)$ **17.** $(9z - 2)(3z + 1)$ **19.** $(3x - 1)^2$ **21.** $(2x + 3)(x + 1)$ **23.** $(6n - 5)(n - 1)$ **25.** $2(2x - y)(3x + 2y)$

27. prime   **29.** $(3m - 4)(2m + 1)$ **31.** $-3(3n - 2)(2n - 3)$ **33.** $(4a + 5)(3a + 1)$ **35.** $(2x - 3)(x - 1)$ **37.** $(3x - 4)(5x - 1)$

39. $-(a + 4)(4a - 3)$ **41.** $2(x - 2y)(5x + 6y)$ **43.** $(3x - 4)(3x - 2)$ **45.** $-3(4x + 3)(2x - 1)$ **47.** $4x^2y^2(x - 2y - 1)$

49. $(x - 6y)(x - 7y)$ **51.** $(m + 3n)(4m + n)$ **53.** $-(x - 12)(x + 2)$ **55.** $(n - 1)(n^2 + 1)$ **57.** prime   **59.** $6(2x + 5y)(2x - y)$

61. $4(a + 2)(a - 2)$ **63.** $3(3a + 8b)(2a - b)$ **65.** $(2ab^2 - c)(a + b)$ **67.** $-2x^2(2x + 5)(x - 1)$ **69.** $(2x + 7)$ m and $(3x - 4)$ m

71. $2x - 5$ **73.** $-2x(4x + 3)(3x - 5)$ **75.** $(9z^2 + 8)(3z^2 + 2)$ **77.** $(3x^n + 1)(x^n + 6)$ **79.** $(x^2 + 1)(2x - 7)(3x - 2)$ **81.** ±2 or ±14

Section 6.4

1. sum, two cubes   **3.** difference, two squares   **5.** false   **7.** Answers may vary.   **9.** $(x - 3)^2$ **11.** $(x + 5)^2$ **13.** $(2p - 1)^2$

15. $(4x + 3)^2$ **17.** $(x - 2y)^2$ **19.** $(2z - 3)^2$ **21.** $(x - 11)(x + 11)$ **23.** $(4x - 7y)(4x + 7y)$ **25.** $(2x - 5)(2x + 5)$

27. $(10n^4 - 9p^2)(10n^4 + 9p^2)$ **29.** $(k - 2)(k + 2)(k^2 + 4)(k^4 + 16)$ **31.** $(5p^2 - 7q)(5p^2 + 7q)$ **33.** $(1 + x)(1 - x + x^2)$

35. $(2x - 3y)(4x^2 + 6xy + 9y^2)$ **37.** $(x^2 - 2y)(x^4 + 2x^2y + 4y^2)$ **39.** $(3c + 4d^3)(9c^2 - 12cd^3 + 16d^6)$ **41.** prime   **43.** $\left(x - \frac{1}{3}\right)^2$

45. $(4m + 5n)^2$ **47.** $2(x - 3)^2$ **49.** $3a(4a + 3)^2$ **51.** $\left(9p - \frac{1}{2}\right)\left(9p + \frac{1}{2}\right)$ **53.** $x^2y^2(x - y)(x + y)$ **55.** $2t(t - 3)(t^2 + 3t + 9)$

57. $3s(s^2 + 2)(s^4 - 2s^2 + 4)$ **59.** $4x(1 - x)(1 + x)$ **61.** $xy(x - y)(x + y)$ **63.** $(xy + 1)(x^2y^2 - xy + 1)$ **65.** $(x^4 - 5y^5)(x^4 + 5y^5)$

67. $\left(x - \frac{y^2}{4}\right)\left(x^2 + \frac{xy^2}{4} + \frac{y^4}{16}\right)$ **69.** $2(x - 2)^2$ **71.** $-2n(2n - 1)^2(2n + 1)^2$ **73.** $-2x(1 + 2x)^2$ **75.** $(x^2 - y^3)(x^4 + x^2y^3 + y^6)$

77. prime   **79.** $(x - 2)(x + 2)(2x + 5)$ **81.** $(2y + 5)(y - 4)(y + 4)$ **83.** $(x - 2)(x + 2)(x^2 - 2x + 4)$ **85.** $2x + 5$

87. $(x - 3)(x + 1)$ **89.** $-3(2x - 1)$ **91.** $x(x^2 - 3xy + 3y^2)$ **93.** $(x - 2)(x + 4)$ **95.** $(x + 1)(2a - 5)(a - 6)$ **97.** $x(5x + 13)$

Section 6.5

1. perfect square trinomial, perfect squares   **3.** greatest common factor   **5.** false   **7.** Answers may vary.   **9.** $(x - 10)(x + 10)$

11. $(t + 3)(t - 2)$ **13.** $(x + y)(1 + 2a)$ **15.** $(a - 2)(a^2 + 2a + 4)$ **17.** $(a - 3b)(a + 2b)$ **19.** $(2x - 7)(x + 1)$

21. $2(x - 5y)(x + 2y)$ **23.** $(3 - a)(3 + a)$ **25.** $(u - 11)(u - 3)$ **27.** $(x - a)(y - b)$ **29.** $(w + 2)(w + 4)$

31. $(6a - 7b^2)(6a + 7b^2)$ **33.** $(x + 4m)(x - 2m)$ **35.** $(3xy - 2)(2xy - 3)$ **37.** $(x + 1)(x^2 + 1)$ **39.** $6(2z^2 + 2z + 3)$

41. $(7c - 1)(2c + 3)$ **43.** $(3m + 4n^2)(9m^2 - 12mn^2 + 16n^4)$ **45.** $2j^2(j - 1)(j + 1)(j^2 + 1)$ **47.** $(4a - b)(2a + 5b)$ **49.** $2a(a^2 + 3)$

51. $3(2z - 1)(2z + 1)$ **53.** prime   **55.** $2a(2a + b)(4a^2 - 2ab + b^2)$ **57.** $(pq + 7)(pq - 1)$ **59.** $(s + 2)^2(s - 2)$

61. $-2x(3x + 1)(2x - 1)$ **63.** $(5v + 2)(2v - 1)$ **65.** $-n^2(n - 4)(n + 1)$ **67.** $-2ab(2a^2 - a + 1)$ **69.** $-p(p - 4)(p + 3)$

71. $-8x(2x - 3y)(2x + 3y)$ **73.** $2(n - 5)(n^2 - 3)$ **75.** $4(4x - 3)(x + 1)$ **77.** $(3x^2 + 2)(x^2 + 4)$ **79.** $-xy(2x - 3y)(x + y)$

81. $x(x - 5)(x - 3)$ **83.** $2x, 2x + 1, x - 3$ **85.** $(2m - n + p)(2m - n - p)$ **87.** $(x + y - z)(x - y + z)$

89. $(x - y - 4)(x - y - 2)$ **91.** $(x + y - a + b)(x + y + a - b)$

Putting the Concepts Together (Sections 6.1–6.5)

1. $5x^2y$ **2.** $(x - 4)(x + 1)$ **3.** $(x^2 - 3)(x^4 + 3x^2 + 9)$ **4.** $(2x + 1)(6x + 5z)$ **5.** $(x + 6y)(x - y)$ **6.** $(x + 4)(x^2 - 4x + 16)$

7. prime   **8.** $3(x + 6y)(x - 2y)$ **9.** $4z^2(3z^3 - 11z - 6)$ **10.** prime   **11.** $m^2(m + 2)(4m - 3)$ **12.** $(5p - 2)(p - 3)$

13. $(2m + 5)(5m - 3)$ **14.** $6(3m - 1)(2m + 1)$ **15.** $(2m - 5)^2$ **16.** $(5x + 4y)(x - y)$ **17.** $S = 2\pi r(h + r)$ **18.** $h = 16t(3 - t)$

Section 6.6

1. quadratic **3.** second **5.** false **7.** Answers may vary. **9.** linear **11.** quadratic **13.** $\{-4, 0\}$ **15.** $\{-3, 9\}$ **17.** $\left\{-\frac{1}{3}, 5\right\}$

19. $\left\{-\frac{3}{5}, \frac{2}{7}\right\}$ **21.** $\{-1, 4\}$ **23.** $\{-7, -2\}$ **25.** $\left\{-\frac{1}{2}, 0\right\}$ **27.** $\left\{-\frac{1}{2}, 2\right\}$ **29.** $\{3\}$ **31.** $\{0, 6\}$ **33.** $\{-2, 3\}$ **35.** $\{2, 9\}$

37. $\left\{\frac{1}{4}, 1\right\}$ **39.** $\{-4, 6\}$ **41.** $\{-5, 10\}$ **43.** $\{-5, 1\}$ **45.** $\{-3, 0, 2\}$ **47.** $\{-3, -2, 2\}$ **49.** $\left\{-\frac{3}{2}, 2, -2\right\}$ **51.** $\left\{-\frac{3}{5}, 4\right\}$

53. $\{-4, 5\}$ **55.** $\{-5\}$ **57.** $\left\{-\frac{3}{4}, 7\right\}$ **59.** $\{-4, 2\}$ **61.** $\left\{-4, -\frac{1}{2}, 4\right\}$ **63.** $\{20\}$ **65.** $\{-10, 10\}$ **67.** $\left\{-3, -\frac{2}{3}, 5\right\}$

69. $\left\{-\frac{1}{2}, \frac{3}{2}, 5\right\}$ **71.** $\{0, 8\}$ **73.** $\left\{-4, \frac{3}{2}\right\}$ **75.** $\left\{-6, \frac{1}{2}\right\}$ **77.** 1 sec, 3 sec **79.** 1 sec **81.** -4 and -3 or 3 and 4

83. $-12, -10,$ and $-8,$ or 8, 10, and 12 **85.** 15 by 17 **87.** 8 teams **89.** $x(x - 3)(x + 5) = 0$ **91.** $(z - 6)(z - 6) = 0$ or $(z - 6)^2 = 0$

93. $(a - 4)(2a + 1)(3a - 2) = 0$ **95.** Student divided by a variable expression; $\left\{0, \frac{1}{3}\right\}$ **97.** $\{a, -b\}$ **99.** $\{0, 2a\}$

Section 6.7

1. hypotenuse **3.** $a^2 + b^2 = c^2$ **5.** false **7.** Answers may vary. **9.** 4, 14 **11.** 2, 9 **13.** base = 26; height = 8
15. base = 18; height = 16 **17.** base = 13; height = 11 **19.** $B = 53; b = 43$ **21.** 9, 12 **23.** 12, 5
25. 256, 252, 240, 220, 192, 156, 112, 60, 0 feet **27.** 6 sec **29.** width = 4 m; length = 12 m **31.** base = 6 ft; height = 18 ft **33.** 30 inches
35. width = 12 mm; length = 25 mm **37.** width = 4 cm; length = 25 cm **39. (a)** width = 25 in.; length = 42 in. **(b)** 28 in. by 45 in.
41. 6 ft **43.** width = 7 ft; length = 14 feet

Chapter 6 Review

1. 12 **2.** 27 **3.** 10 **4.** 4 **5.** x^2 **6.** m **7.** $15ab^2$ **8.** $6x^3y^2z$ **9.** $2(2a + 1)^2$ **10.** $9(x - y)$ **11.** $-6a^2(3a + 4)$
12. $-3x(3x - 4)$ **13.** $5y^2z(3 + y^5 + 4y)$ **14.** $7xy(x^2 - 3xy + 2y^2)$ **15.** $(5 - y)(x + 2)$ **16.** $(a + b)(z + y)$
17. $(5m + 2n)(m + 3n)$ **18.** $(2x + y)(y + x)$ **19.** $(x + 2)(8 - y)$ **20.** $(y^2 + 1)(x - 3)$ **21.** $(x + 3)(x + 2)$ **22.** $(x + 2)(x + 4)$
23. $(x - 7)(x + 3)$ **24.** $(x + 5)(x - 2)$ **25.** prime **26.** prime **27.** $(x - 3y)(x - 5y)$ **28.** $(m + 5n)(m - n)$
29. $-(p + 6)(p + 5)$ **30.** $-(y - 5)(y + 3)$ **31.** $3x(x^2 + 11x + 12)$ **32.** $4(x + 1)(x + 8)$ **33.** $2(x - 7y)(x + 6y)$
34. $4y(y + 5)(y - 2)$ **35.** $(5y - 6)(y + 4)$ **36.** $(y - 7)(6y + 1)$ **37.** $(2x - 3)(x - 1)$ **38.** $(2x + 7)(3x + 1)$ **39.** prime
40. $(4m + 3)(2m + 3)$ **41.** $3m(m + n)(3m + 7n)$ **42.** $2(7m + n)(m + n)$ **43.** $x(5x + 2)(3x - 1)$ **44.** $p^2(3p - 1)(2p + 1)$
45. $(2x - 3)^2$ **46.** $(x - 5)^2$ **47.** $(x + 3y)^2$ **48.** prime **49.** $2(2m + 1)^2$ **50.** $2(m - 6)^2$ **51.** $(2x - 5y)(2x + 5y)$
52. $(7x - 6y)(7x + 6y)$ **53.** prime **54.** prime **55.** $(x - 3)(x + 3)(x^2 + 9)$ **56.** $(x - 5)(x + 5)(x^2 + 25)$ **57.** $(m + 3)(m^2 - 3m + 9)$
58. $(m + 5)(m^2 - 5m + 25)$ **59.** $(3p - 2)(9p^2 + 6p + 4)$ **60.** $(4p - 1)(16p^2 + 4p + 1)$ **61.** $(y^3 + 4z^2)(y^6 - 4y^3z^2 + 16z^4)$
62. $(2y + 3z^2)(4y^2 - 6yz^2 + 9z^4)$ **63.** $(5a - 2b)(3a^2 - 5b^2)$ **64.** $(4a - 3b)(3a + b)$ **65.** prime **66.** $(x - 12y)(x + 2y)$
67. $x(x - 7)(x + 6)$ **68.** $3x^4(x - 7)(x - 3)$ **69.** $(3x + 1)(2x + 3)$ **70.** $(2z + 3)(5z - 3)$ **71.** $(3x + 2)(9x^2 - 6x + 4)$
72. $(2z - 1)(4z^2 + 2z + 1)$ **73.** $2(2y - 1)(y + 5)$ **74.** $xy^2(5x - 3)(x - 1)$ **75.** $(5k - 9m)(5k + 9m)$ **76.** $(x^2 - 3)(x^2 + 3)$

77. prime **78.** prime **79.** $\left\{\frac{3}{2}, 4\right\}$ **80.** $\left\{-7, -\frac{1}{2}\right\}$ **81.** $\{-3, 15\}$ **82.** $\{2, 5\}$ **83.** $\{0, -2\}$ **84.** $\left\{-\frac{9}{2}, 0\right\}$ **85.** $\{-1, 3\}$

86. $\{-3, -1\}$ **87.** $\{-3, -1\}$ **88.** $\left\{-\frac{1}{2}, \frac{3}{2}\right\}$ **89.** $\{-14, 0, 3\}$ **90.** $\{-3, 0, 6\}$ **91.** $\left\{-3, \frac{4}{3}, 3\right\}$ **92.** $\left\{-\frac{5}{2}, -2\right\}$ **93.** 5 sec

94. 2 sec or 3 sec **95.** 9 ft, 6 ft **96.** 3 yd, 5 yd **97.** 8 ft, 6 ft **98.** 24 ft, 10 ft

Chapter 6 Test

1. $4x^4y^2$ **2.** $(x + 3)(x - 3)(x^2 + 9)$ **3.** $9x(2x^2 - 3)(x^2 + 1)$ **4.** $(x - 7)(y - 4)$ **5.** $(3x + 5)(9x^2 - 15x + 25)$ **6.** $(y - 12)(y + 4)$
7. $(6m + 5)(m - 1)$ **8.** prime **9.** $(x - 5)(4 + y)$ **10.** $3y(x - 7)(x + 2)$ **11.** prime **12.** $3(x - 1)(x + 1)(x^2 + x + 1)(x^2 - x + 1)$
13. $3x(3x + 1)(x + 4)$ **14.** $(3m + 2)(2m + 1)$ **15.** $2(2m^2 - 3mn + 2)$ **16.** $(5x + 7y)^2$ **17.** $\left\{-\frac{1}{5}, -3\right\}$ **18.** $\{0, 2\}$
19. length = 7 in.; width = 5 in. **20.** legs: 12 in., 5 in., hypotenuse: 13 in.

Getting Ready: A Review of Chapters 1–6

1. (a) 1 **(b)** $0, 1, -6$ **(c)** $0, 1, -6, \frac{2}{5}, -0.83, 0.5454\ldots$ **(d)** $1.010010001\ldots$ **(e)** $0, 1, -6, \frac{2}{5}, -0.83, 0.5454\ldots, 1.010010001\ldots$ **2.** -20

3. $-\frac{6}{5}$ **4.** $-\frac{21}{2}$ **5.** $\frac{4}{3}$ **6.** $\frac{3}{5}$ **7.** 74 **8.** $-\frac{3}{2}$ **9.** $\{3\}$ **10.** $\left\{\frac{5}{4}\right\}$ **11.** The angles measure $55°$ and $125°$.

12. $\{x \mid x \geq 7\}; [7, \infty)$ ⊢—————→ **13.** A person 62 inches tall would be considered obese if they weighed 160 pounds or more.

14. 5 **15.** 1 **16.** $5y^3 + 5y^2 - y - 6$ **17.** $9x^8$ **18.** $-4x^3 + 20x^2 - 12x$ **19.** $12x^2 - 17x - 5$ **20.** $4x^2 - 28x + 49$ **21.** $25x^2 - 9y^2$

22. $\frac{y}{4z^3}$ **23.** $2a^2 - 6a + 1$ **24.** $3x^2 - 5x + 2 - \frac{9}{x + 5}$ **25.** $-4x(x^3 - 4x + 5)$ **26.** $(2z - 1)(4z^2 + 3)$ **27.** $(w + 5)(w - 7)$

28. $-3c(c - 8)(c + 3)$ **29.** prime **30.** $(3y - 4)(y + 4)$ **31.** $(p + 2)(p - 2)(p^2 + 4)$ **32.** $(2a + 5b)^2$ **33.** $\left\{-\frac{5}{2}, 3\right\}$ **34.** $\{4, 5\}$

35. $\left\{-\frac{7}{2}, \frac{2}{3}\right\}$

Chapter 7

Section 7.1

1. rational expression **3.** simplify **5.** True **7.** Answers may vary. **9. (a)** 2 **(b)** $\frac{1}{2}$ **(c)** undefined **11. (a)** undefined **(b)** 3

(c) $\frac{5}{3}$ **13. (a)** 3 **(b)** undefined **(c)** 0 **15. (a)** -3 **(b)** 0 **(c)** $-\frac{15}{7}$ **17. (a)** 0 **(b)** undefined **(c)** undefined **19.** 0

21. 5 **23.** $\frac{3}{2}$ **25.** $-6, 6$ **27.** $2, 5$ **29.** $-2, 0, 3$ **31.** $\frac{3}{x-2}$ **33.** $\frac{(1+z)(1-z)}{3}$ **35.** $\frac{1}{p+2}$ **37.** -1 **39.** $-2k$ **41.** $\frac{x-1}{x+4}$

43. $\frac{5}{9}$ **45.** $\frac{10}{11}$ **47.** $\frac{1}{2x^3}$ **49.** $\frac{12}{a^2}$ **51.** -3 **53.** $\frac{b-5}{4}$ **55.** $\frac{x+3}{x-3}$ **57.** $\frac{x-3}{x-5}$ **59.** $-(x+y)$ **61.** $\frac{4}{a+b}$

63. $\frac{x}{x-4}$ **65.** $\frac{-(4+c)}{c-4}$ **67.** $\frac{2(x-2)}{3(x-5)}$ **69.** $\frac{-(x-3)}{x-2}$ **71.** $\frac{2}{t^2+9}$ **73.** $\frac{-3}{w-1}$ **75. (a)** 5 mg/mL **(b)** 4 mg/mL

77. Yes; BMI $= 27.5$ **79.** \$15,600 **81.** $\frac{(c^2-c+1)(c^2+c+1)}{c^2+1}$ **83.** $\frac{(x^2+4)(x^2-3)}{(x+2)(x^3-9)}$ **85.** $\frac{(t^2+4)(t+2)}{t^2-2t+4}$

Section 7.2

1. quotient, product **3.** factor, common factors **5.** False **7.** Answers may vary. **9.** $\frac{16}{63x^6}$ **11.** $-\frac{5a^4}{8}$ **13.** $\frac{x}{(x+1)^2}$ **15.** $p-1$

17. $\frac{18}{25z}$ **19.** $-\frac{3m^5}{10}$ **21.** x **23.** 1 **25.** $\frac{10}{3}$ **27.** 1 **29.** $-\frac{4}{11}$ **31.** $\frac{4}{3y(y-3)}$ **33.** $\frac{a}{3}$ **35.** $\frac{3x}{x-4}$ **37.** $\frac{4-w}{(w+2)(w+4)}$

39. $-\frac{(x-y)^2}{x}$ **41.** $-\frac{1}{2}$ **43.** 1 **45.** $\frac{3(2n+3)(n-3)}{2(n+4)}$ **47.** $\frac{x}{4y^2(x-2y)}$ **49.** $\frac{2a-b}{2a(a+b)}$ **51.** $\frac{(t+3)(t-2)}{3}$ **53.** -1

55. $-\frac{x(x+3)}{(x+1)^2}$ **57.** $\frac{a+b}{2}$ **59.** $\frac{1}{3}$ **61.** $\frac{1}{3(x-y)}$ **63.** $\frac{9}{8}$ **65.** $\frac{x-y}{(x+y)^3}$ **67.** $\frac{4x}{(x-2)(x-3)}$ **69.** $\frac{1}{x}$ square feet

71. $\frac{2(x+2)}{x-3}$ square inches **73.** $\frac{(x-a)^2}{2(x+a)}$ **75.** $\frac{1-x}{4}$ **77.** $\frac{2a^2}{3}$ **79.** $\frac{(x+y)^2}{(x^2-xy+y^2)(x+2)}$ **81.** $3p^5q^2(p^2+3pq+9q^2)(p-q)$

83. $6x+12$

Section 7.3

1. numerators **3.** $a+b; c$ **5.** True **7.** Incorrect; Answers may vary. **9.** $\frac{7p}{4}$ **11.** $2n$ **13.** $\frac{2a-1}{a}$ **15.** $\frac{2(5c-2)}{c-1}$ **17.** 2

19. $\frac{7x(1+x)}{x+2}$ **21.** 2 **23.** $2x$ **25.** $x-3$ **27.** $-\frac{x}{2}$ **29.** $-\frac{c+3}{c}$ **31.** 3 **33.** -1 **35.** $x-1$ **37.** $-\frac{1}{x}$ **39.** -1 **41.** 0

43. $n-1$ **45.** $\frac{3}{v+3}$ **47.** $\frac{x+1}{x-1}$ **49.** $\frac{x^2+3}{x(x-3)}$ **51.** $\frac{1}{a-5}$ **53.** 1 **55.** $\frac{2(2p+3q)}{p-q}$ **57.** $\frac{2p^2+2p+1}{p-1}$ **59.** $\frac{4b}{a-b}$ **61.** -1

63. $\frac{2}{x-y}$ **65.** $2(x-1)$ **67.** $\frac{3x+3y-2}{x-y}$ **69.** $\frac{p+2q}{p^2-6q^2}$ **71.** $\frac{2g-7}{g-3}$ **73.** $12a$ **75.** $\frac{n-3}{n+1}$ **77.** $\frac{4}{n}$ **79.** $\frac{x}{x+3}$

81. $\frac{4x-2}{x+2}$ **83.** $\frac{2(6x-1)}{x}$ cm **85.** $\frac{-x(x-1)}{(3x-1)(x+2)}$ **87.** $\frac{-5x+12}{x-2}$ **89.** $-4n-6$

Section 7.4

1. least common denominator **3.** 1 **5.** False **7.** Answers may vary. **9.** 36 **11.** 60 **13.** $5x^2$ **15.** $60x^3y^2$ **17.** $x(x+1)$

19. $2x+1$ **21.** $4(b-3)$ **23.** $p(p+1)(p-2)$ **25.** $(r+2)^2(r+1)(r-2)$ **27.** $-(x-4)$ **29.** $-2(x+3)(x-3)$

31. $-(x-1)(x+2)$ **33.** $p^2(p-1)(p-3)$ **35.** $\frac{12x^2}{3x^3}$ **37.** $\frac{3c+c^2}{a^2b^2c^2}$ **39.** $\frac{(x-4)^2}{x^2-16}$ **41.** $\frac{9n^2-9n}{6n^2-6}$ **43.** $\frac{4t^2-4t}{t-1}$ **45.** $\frac{21}{36}; \frac{-32}{36}$

47. $\frac{7}{15}; \frac{30}{15}$ **49.** $\frac{6xy}{9y^2}; \frac{4}{9y^2}$ **51.** $\frac{6a+2}{4a^3}; \frac{4a^3-a^2}{4a^3}$ **53.** $\frac{2m+2}{m(m+1)}; \frac{3m}{m(m+1)}$ **55.** $\frac{2y^2-5y+2}{4y(2y-1)}; \frac{y^2}{4y(2y-1)}$

57. $\frac{-1}{-(x-1)}; \frac{2x}{-(x-1)}$ **59.** $\frac{-3x}{-(x-7)}; \frac{-5}{-(x-7)}$ **61.** $\frac{4x}{(x+2)(x-2)}; \frac{2x-4}{(x+2)(x-2)}$

63. $\frac{x^2+2x+1}{(x+3)(x-3)(x+1)}; \frac{x^2+5x+6}{(x+3)(x-3)(x+1)}$ **65.** $\frac{6x-3}{(x+4)(2x-1)}; \frac{2x-1}{(x+4)(2x-1)}$ **67.** $3x; \frac{3}{3x}; \frac{1}{3x}$

69. $(r-4)(r+4); \frac{12r+48}{(r-4)(r+4)}; \frac{12r-48}{(r-4)(r+4)}$ **71.** $(x+2)(x-2)(x^2-2x+4)(x^2+2x+4)$

73. $4a^2b^2(a+b)(a-b)(a^2+ab+b^2)$

Section 7.5

1. least common denominator **3.** denominator **5.** True **7.** Answers may vary. **9.** $-\frac{5}{6}$ **11.** $\frac{5}{3x}$ **13.** $\frac{2a^2+7a-3}{(2a-1)(2a+1)}$

15. $\frac{-2}{x-4}$ **17.** $\frac{2a^2-3a-12}{4a(a-5)}$ **19.** $\frac{5}{x}$ **21.** $\frac{8}{75}$ **23.** $\frac{(m-4)(m+4)}{m}$ **25.** $\frac{2(4x-3)}{(x-3)(x+3)}$ **27.** $\frac{6(x+1)}{(x+3)^2}$ **29.** $\frac{-4(a-1)}{a+3}$

31. $\dfrac{x}{x-2}$ **33.** $\dfrac{5}{4}$ **35.** $\dfrac{20-9y}{12y^2}$ **37.** $\dfrac{6}{5x}$ **39.** $\dfrac{7y+4x}{2x^2y^2}$ **41.** $\dfrac{-3}{2x+3}$ **43.** $\dfrac{2(n^2-2)}{n(n-2)}$ **45.** $\dfrac{2x^2-5x+15}{x(x-3)}$ **47.** $\dfrac{-1}{a(a-1)}$

49. $\dfrac{2n+5}{2n+1}$ **51.** $\dfrac{4x-5}{x-2}$ **53.** $\dfrac{3x-28}{x-9}$ **55.** $\dfrac{4n^2-7n+9}{n^2(n-3)(n-1)}$ **57.** $\dfrac{-13}{2a(a-2)}$ **59.** $\dfrac{n^2-n-3}{-(n+2)(n-2)(n+3)}$ **61.** $\dfrac{2x^2+x+1}{(x+1)(x-1)^2}$

63. $\dfrac{-3n+5}{(n+3)(n-2)}$ **65.** $\dfrac{x}{x-2}$ **67.** $\dfrac{x-4}{x}$ **69.** $\dfrac{6}{m(m-2)}$ **71.** $\dfrac{a+2}{a(a-1)^2}$ **73.** $\dfrac{x^2+4}{(x+2)(x+3)}$ **75.** $\dfrac{x+3}{(x+1)(x-1)}$

77. $\dfrac{2x^2-x+10}{(x+3)(x-2)(x+2)}$ **79.** $\dfrac{2x^2-5x+3}{(x+6)(x-7)}$ **81.** $\dfrac{x-4}{8}$ square units **83.** $\dfrac{2x^2-9x-18}{6}$ square units **85.** $\dfrac{x^3-2x^2+4x+3}{x^2(x+1)(x-1)}$

87. $\dfrac{1}{a+b}$ **89.** $\dfrac{x^2+2x+5}{(x-4)(x+1)}$ **91.** $\dfrac{-x^3+6x^2-11x+10}{x-2}$

Section 7.6

1. complex rational expression **3.** reciprocal, multiply **5.** False **7.** Answers may vary. **9.** -5 **11.** $\dfrac{2}{17}$ **13.** -24 **15.** $\dfrac{3(x-2)}{2}$

17. $\dfrac{9}{x-3}$ **19.** $\dfrac{n(m+2n)}{2m}$ **21.** $\dfrac{y-3x}{4y+x}$ **23.** $\dfrac{1}{2}$ **25.** $\dfrac{b-7}{b}$ **27.** $\dfrac{x+4}{x}$ **29.** $\dfrac{-(x+y)}{x^2y^2}$ **31.** $\dfrac{x^2+2xy-y^2}{xy}$ **33.** $\dfrac{-1}{2x+7}$

35. $-b$ **37.** $\dfrac{11(2n+3)}{72}$ **39.** $\dfrac{-x^3}{x+3}$ **41. (a)** $R=\dfrac{R_1R_2}{R_2+R_1}$ **(b)** $\dfrac{15}{4}$ ohms **43.** $\dfrac{2x+1}{x+1}$ **45.** $\dfrac{5-2x}{2-x}$

Putting the Concepts Together (Sections 7.1–7.6)

1. -35 **2. (a)** The expression is undefined for $a=6$. **(b)** The expression is undefined for $y=0$ or $y=-4$. **3. (a)** $\dfrac{a-4b}{2}$ **(b)** $\dfrac{-(x+2)}{x+1}$

4. The LCD is $8(a+2b)(a-4b)$. **5.** $\dfrac{7}{3x^2-x}=\dfrac{35x}{5x^2(3x-1)}$ **6.** $\dfrac{-2y^2}{y+1}$ **7.** $\dfrac{2(m-1)^2}{m^2}$ **8.** $\dfrac{2(y-1)}{5(y-3)}$ **9.** $4x+1$ **10.** $\dfrac{1}{x+1}$

11. $\dfrac{1}{2x-1}$ **12.** $\dfrac{15m+17}{(m+3)(3m+2)}$ **13.** $\dfrac{m}{(m+5)(m+4)}$ **14.** $-\dfrac{1}{m-1}$ **15.** $\dfrac{a(24+a)}{3(6+a^2)}$

Section 7.7

1. least common denominator **3.** undefined **5.** True **7.** Answers may vary. **9.** $\{18\}$ **11.** $\{-3\}$ **13.** $\{2\}$ **15.** $\{6\}$

17. $\left\{\dfrac{5}{2}\right\}$ **19.** $\left\{\dfrac{31}{2}\right\}$ **21.** $\{-7\}$ **23.** $\left\{\dfrac{1}{2}\right\}$ **25.** $\{7\}$ **27.** $\{\ \}$ or $\varnothing$ **29.** $\{-3\}$ **31.** $\left\{\dfrac{3}{4}\right\}$ **33.** $\left\{-\dfrac{5}{6}\right\}$ **35.** $\{-4\}$

37. $\{-3,-2\}$ **39.** $\{-1\}$ **41.** $\{-9,4\}$ **43.** $\left\{-\dfrac{2}{3},\dfrac{1}{2}\right\}$ **45.** $\{\ \}$ or $\varnothing$ **47.** $\{-5,5\}$ **49.** $\{-1,5\}$ **51.** $\{-6\}$ **53.** $\{-5\}$

55. $\{\ \}$ or $\varnothing$ **57.** $y=\dfrac{2}{x}$ **59.** $R=\dfrac{E}{I}$ **61.** $b=\dfrac{2A}{h}-B$ **63.** $y=\dfrac{x}{z}-3$ **65.** $S=\dfrac{RT}{T-R}$ **67.** $y=\dfrac{an-p}{am}$ **69.** $x=\dfrac{Ay}{y-A}$

71. $y=\dfrac{xz}{2z-6x}$ **73.** $x=\dfrac{cy}{y-1}$ **75.** $\dfrac{4x+5}{x(x+5)}$ **77.** $\{-2,3\}$ **79.** $\dfrac{x+7}{2}$ **81.** $\left\{-\dfrac{1}{6},1\right\}$ **83.** $\dfrac{x^3-x^2+4}{x(x+2)(x-2)}$

85. 1 hour and 9 hours **87.** 40 or 125 bicycles **89.** $k=33$

Section 7.8

1. extremes, means **3.** $d=rt$ **5.** True **7.** Answers may vary. **9.** $\{12\}$ **11.** $\left\{\dfrac{18}{7}\right\}$ **13.** $\{16\}$ **15.** $\left\{\dfrac{24}{5}\right\}$ **17.** $\left\{-\dfrac{4}{5}\right\}$

19. $\{-2\}$ **21.** $\left\{\dfrac{5}{7}\right\}$ **23.** $\{-3,2\}$ **25.** $\{2,5\}$ **27.** $\{-2,9\}$ **29.** $XZ=12$ **31.** $n=3; ZY=5$ **33.** $x=30; AB=28$

35. $2x$ **37.** $t-3$ **39.** $r-2$ **41.** $\dfrac{1}{5}+\dfrac{1}{3}=\dfrac{1}{t}$ **43.** $\dfrac{1}{b+4}+\dfrac{1}{b}=\dfrac{1}{7}$ **45.** $\dfrac{4}{14-c}=\dfrac{7}{14+c}$ **47.** 145 miles **49.** $8\dfrac{1}{3}$ lb

51. 37,245 rubles **53.** 24 ft **55.** $1\dfrac{7}{8}$ hr **57.** $3\dfrac{3}{7}$ minutes **59.** $5\dfrac{1}{7}$ hr **61.** 15 hr **63.** 24 hr **65.** $\dfrac{14}{3}$ km per hour

67. 1 hour, 40 minutes **69.** 40 mph **71.** 35 mph **73.** $x=6$ **75.** $x=\dfrac{98}{5}$ **77.** $x=\dfrac{72}{5}$

Chapter 7 Review

1. (a) $\dfrac{1}{2}$ **(b)** undefined **(c)** 1 **2. (a)** 0 **(b)** -1 **(c)** undefined **3. (a)** undefined **(b)** 9 **(c)** 49 **4. (a)** -1

(b) undefined **(c)** 42 **5.** $\dfrac{7}{3}$ **6.** $\dfrac{1}{2}$ **7.** none **8.** none **9.** -10 or -2 **10.** -1 or 4 **11.** $\dfrac{4x^4}{5y^4}$ **12.** $\dfrac{3}{4x^2y^4}$ **13.** $\dfrac{3}{k+2}$

14. $-\dfrac{2}{x-3}$ **15.** $\dfrac{x+5}{2x-1}$ **16.** $\dfrac{3x-1}{2(x^2-2x+4)}$ **17.** $\dfrac{2m^2}{n^2}$ **18.** $\dfrac{y^3}{6x}$ **19.** $\dfrac{14}{9n^2}$ **20.** $\dfrac{1}{ab^7}$ **21.** $-\dfrac{5(x-2)}{x+3}$ **22.** $\dfrac{4(x-3)(x+3)}{(x-9)(x-9)}$

23. $\dfrac{x-2}{4(x-3)}$ **24.** $15x^3$ **25.** $\dfrac{x+5}{x+6}$ **26.** $\dfrac{y+1}{y-3}$ **27.** $\dfrac{x^2}{(x+1)(x+9)}$ **28.** $\dfrac{y-2}{3y+1}$ **29.** $\dfrac{9}{x-3}$ **30.** $\dfrac{10}{x+4}$ **31.** m **32.** $\dfrac{1}{m}$

33. $\dfrac{1}{m-2}$ **34.** $2m+3$ **35.** $\dfrac{1}{3m}$ **36.** $\dfrac{2}{b}$ **37.** $y+7$ **38.** $\dfrac{1}{m-6}$ **39.** $-\dfrac{7}{x-y}$ **40.** $\dfrac{4}{a-b}$ **41.** $\dfrac{3x-5}{x^2-25}$ **42.** $\dfrac{x+4}{x-3}$

43. $12x^4y^7$ **44.** $60a^3b^4c^7$ **45.** $8a(a+2)$ **46.** $5a(a+6)$ **47.** $4(x-3)(x+1)$ **48.** $x(x-7)(x+7)$ **49.** $\dfrac{6xy^6}{x^4y^7}$ **50.** $\dfrac{11a^4b^3}{a^7b^5}$

51. $\dfrac{(x-1)(x+2)}{(x-2)(x+2)}$ **52.** $\dfrac{(m+2)(m-2)}{(m+7)(m-2)}$ **53.** $\dfrac{20x^2y}{24x^5}$; $\dfrac{21}{24x^5}$ **54.** $\dfrac{12}{10a^3}$; $\dfrac{11a^2b}{10a^3}$ **55.** $\dfrac{-4x}{-(x-2)}$; $\dfrac{6}{-(x-2)}$ **56.** $\dfrac{-3}{-(m-5)}$; $\dfrac{-2m}{-(m-5)}$

57. $\dfrac{2m+4}{(m+7)(m-2)(m+2)}$; $\dfrac{m^2-m-2}{(m+7)(m-2)(m+2)}$ **58.** $\dfrac{n^2+3n-10}{n(n-5)(n+5)}$; $\dfrac{n^2}{n(n-5)(n+5)}$ **59.** $\dfrac{4xz+8y^4}{x^2y^3z}$ **60.** $\dfrac{5xy^2+x^2y}{10x^3y^3}$

61. $\dfrac{x^2-5x+14}{(x+7)(x-7)}$ **62.** $\dfrac{2x^2+13x-9}{(2x+3)(2x-3)}$ **63.** $\dfrac{-4x-10}{x(x-2)}$ **64.** $\dfrac{2x^2-6x-5}{x(x+5)}$ **65.** $\dfrac{3x-1}{4x+1}$ **66.** $\dfrac{3x-4}{(x-1)(x-2)}$

67. $\dfrac{-(4m-9)(m-4)}{(m+3)(m-3)}$ **68.** $\dfrac{2m^2+5m-2}{(m-5)(m+2)}$ **69.** $\dfrac{4}{m-2}$ **70.** $\dfrac{m+n}{m-n}$ **71.** $\dfrac{5x+12}{x+3}$ **72.** $\dfrac{7x+13}{x+2}$ **73.** $-\dfrac{3}{23}$ **74.** $\dfrac{18}{11}$

75. $\dfrac{2m^2-10m}{m^2+10}$ **76.** $\dfrac{6+4m^2}{6m-5m^2}$ **77.** $\dfrac{1}{6}$ **78.** $\dfrac{56}{9}$ **79.** $\dfrac{y+8}{-y+2}$ **80.** $\dfrac{5y+50}{-5y-22}$ **81.** $x=-5$ or $x=0$ **82.** $x=-6$ or $x=6$

83. $\{-4\}$ **84.** $\left\{\dfrac{2}{3}\right\}$ **85.** $\{5\}$ **86.** $\left\{\dfrac{14}{3}\right\}$ **87.** $\{38\}$ **88.** $\left\{\dfrac{9}{7}\right\}$ **89.** $\{\ \}$ or $\varnothing$ **90.** $\{\ \}$ or $\varnothing$ **91.** $\left\{-\dfrac{5}{2}\right\}$ **92.** $\{24\}$

93. $k=\dfrac{4}{y}$ **94.** $y=\dfrac{x}{6}$ **95.** $y=\dfrac{xz}{x-z}$ **96.** $z=\dfrac{xy}{x+y}$ **97.** $\{4\}$ **98.** $\{5\}$ **99.** $\{1\}$ **100.** $\{3\}$ **101.** $x=15$ **102.** $x=3.5$

103. 8 tanks **104.** \$26.25 **105.** $1\dfrac{1}{5}$ hours **106.** $1\dfrac{7}{8}$ hours **107.** 15 min **108.** $9\dfrac{1}{3}$ hours **109.** 9 mph **110.** 55 mph

111. 70 mph **112.** 45 mph

Chapter 7 Test

1. 2 **2.** $x=5$ or $x=-2$ **3.** $-\dfrac{x+3}{2}$ **4.** $\dfrac{14}{5x^2}$ **5.** $\dfrac{25}{9}$ **6.** $\dfrac{x-y}{2x-3y}$ **7.** y **8.** $\dfrac{x-5}{x+3}$ **9.** $\dfrac{-1}{y-z}$ or $\dfrac{1}{z-y}$

10. $\dfrac{2(x+3)(x-1)}{(x-2)(2x+1)}$ **11.** $\dfrac{x^2-5x-2}{(x+2)(x+3)(x-1)}$ **12.** $\dfrac{y-3}{3y}$ **13.** $\{-20,5\}$ **14.** $\{2\}$ **15.** $x=\dfrac{yz}{y-z}$ **16.** $\{5\}$ **17.** $x=\dfrac{24}{5}$

18. \$33 **19.** 36 min **20.** 4 mph

Getting Ready: A Review of Chapters 1–7

1. $\dfrac{5}{3}$ **2.** 1 **3.** $\left\{\dfrac{7}{2}\right\}$; conditional **4.** $\varnothing$ or $\{\ \}$; contradiction **5.** $\left\{\dfrac{4}{3}\right\}$; conditional **6.** all real numbers; identity

7. $\{x\,|\,x>-4\}$; $(-4,\infty)$ 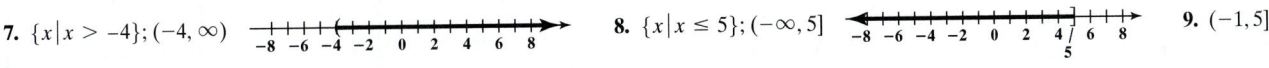 **8.** $\{x\,|\,x\le 5\}$; $(-\infty,5]$ 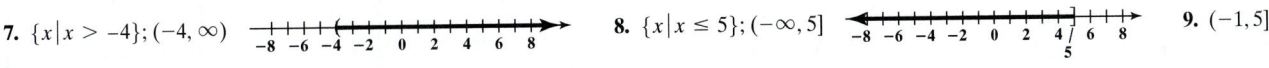 **9.** $(-1,5]$

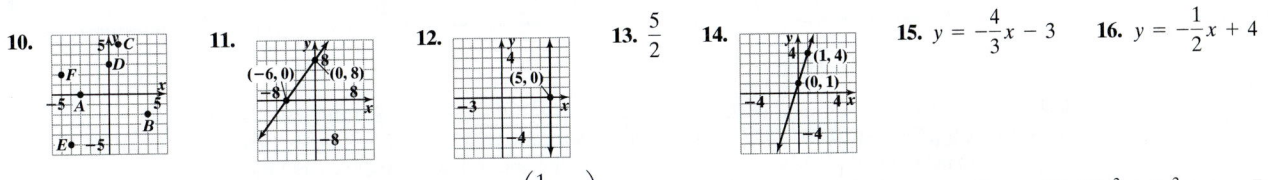

10. **11.** **12.** **13.** $\dfrac{5}{2}$ **14.** **15.** $y=-\dfrac{4}{3}x-3$ **16.** $y=-\dfrac{1}{2}x+4$

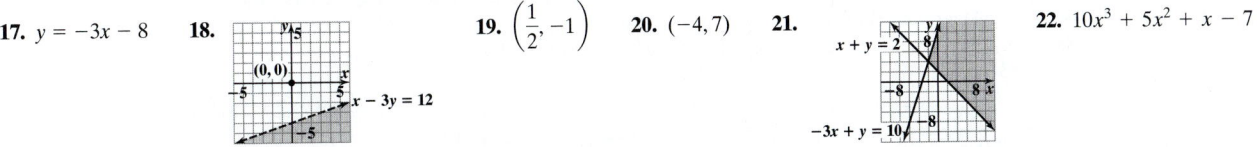

17. $y=-3x-8$ **18.** **19.** $\left(\dfrac{1}{2},-1\right)$ **20.** $(-4,7)$ **21.** **22.** $10x^3+5x^2+x-7$

23. $2x^2+x-15$ **24.** $x^2+2x+3+\dfrac{9}{x-4}$ **25.** $-2(x-3)(x^2+4)$ **26.** $5p(p+8)(p+2)$ **27.** $(4y-3)(2y+1)$ **28.** $(3a+4)^2$

29. $3k^2(k+3)(k-3)$ **30.** $\left\{-\dfrac{3}{2},\dfrac{1}{4}\right\}$ **31.** $\left\{-\dfrac{1}{2},3\right\}$ **32.** $\dfrac{2x+7}{x+9}$ **33.** $\dfrac{x+5}{x+6}$ **34.** $\dfrac{y-2}{3y+1}$ **35.** $\dfrac{x^2+6x+3}{(x-3)(x+2)}$ **36.** $\{-1,4\}$

37. $\left\{\dfrac{1}{4}\right\}$

Chapter 8

Section 8.1

1. origin **3.** intercepts **5.** False **7.** Answers will vary. **9.** Answers will vary.

11. A: $(2,3)$; I B: $(-5,2)$; II C: $(0,-2)$; y-axis D: $(-4,-3)$; III E: $(3,-4)$; IV F: $(4,0)$; x-axis

13. 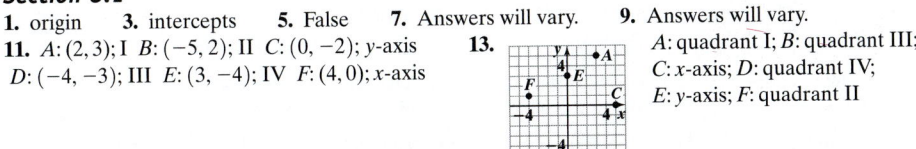 A: quadrant I; B: quadrant III; C: x-axis; D: quadrant IV; E: y-axis; F: quadrant II

15. (a) yes **(b)** no **(c)** yes **(d)** yes **17. (a)** yes **(b)** no **(c)** no **(d)** yes **19. (a)** no **(b)** yes **(c)** yes **(d)** yes

21. $(-2, 0)$ and $(0, 3)$. **23.** $(-2, 0)$, $(1, 0)$, and $(0, -4)$ **25.** $y = 4x$ **27.** $y = -\dfrac{1}{2}x$ **29.** $y = x + 3$

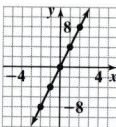 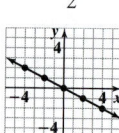 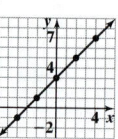

31. $y = -3x + 1$ **33.** $y = \dfrac{1}{2}x - 4$ **35.** $2x + y = 7$ **37.** $y = -x^2$ **39.** $y = 2x^2 - 8$ **41.** $y = |x|$

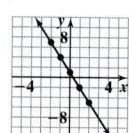 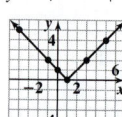 (35 graph) 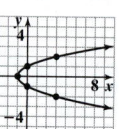 (39 graph)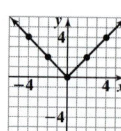

43. $y = |x - 1|$ **45.** $y = x^3$ **47.** $y = x^3 + 1$ **49.** $x^2 - y = 4$ **51.** $x = y^2 - 1$

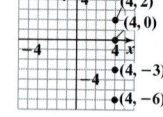 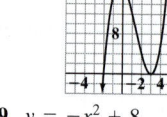 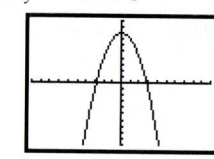 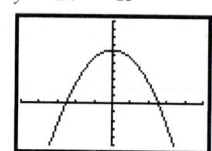 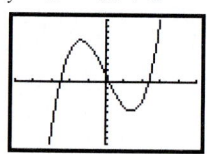

53. $a = \dfrac{7}{4}$ **55.** $b = 4$ **57. (a)** 400 ft^2 **(b)** 25 feet; 625 ft^2 **(c)** The x-intercepts are $x = 0$ and $x = 50$. These values form the bounds for the width of the opening. The y-intercept is $y = 0$. The area of the opening will be 0 ft^2 when the width is 0 feet. **59. (a)** \$40; \$40 **(b)** \$800
(c) 40; the monthly cost will be \$40 if no minutes are used.

61. Vertical line with an x-intercept of 4. **63.** Answers will vary. One possible graph is below. **65.** Answers will vary. One possibility: $y = 0$

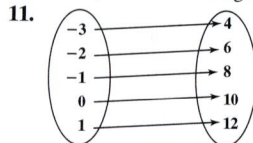
(4, 2)
(4, 0)
(4, −3)
(4, −6)

67. $y = 3x - 9$ **69.** $y = -x^2 + 8$ **71.** $y + 2x^2 = 13$ **73.** $y = x^3 - 6x + 1$

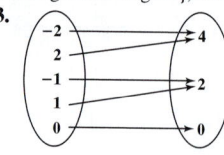 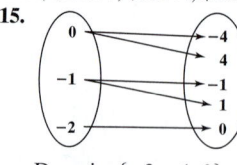

Section 8.2

1. corresponds; depends on **3.** True **5.** Answers will vary. **7.** {(USA Today, 2.1), (Wall Street Journal, 1.8), (New York Times, 1.1),
(Los Angeles Times, 0.9), (Washington Post, 0.8)}; Domain: {USA Today, Wall Street Journal, New York Times, Los Angeles Times, Washington Post};
Range: {0.8, 0.9, 1.1, 1.8, 2.1} **9.** {(Less than 9th Grade, \$17261), (9th–12th Grade no diploma, \$21737), (High School Graduate, \$35744),
(Associate's Degree, \$49279), (Bachelor's Degree or Higher, \$69804)}; Domain: {Less than 9th Grade, 9th–12th Grade no diploma, High School
Graduate, Associate's Degree, Bachelor's Degree or Higher}; Range: {\$17261, \$21737, \$35744, \$49279, \$69804}

11.

Domain: $\{-3, -2, -1, 0, 1\}$
Range: $\{4, 6, 8, 10, 12\}$

13.

Domain: $\{-2, -1, 0, 1, 2\}$
Range: $\{0, 2, 4\}$

15.

Domain: $\{-2, -1, 0\}$
Range: $\{-4, -1, 0, 1, 4\}$

17. Domain: $\{-3, -2, 0, 2, 3\}$
Range: $\{-3, -1, 2, 3\}$

19. Domain: $\{x \mid -4 \le x \le 4\}$ or $[-4, 4]$
Range: $\{y \mid -2 \le y \le 2\}$ or $[-2, 2]$

21. Domain: $\{x \mid -1 \le x \le 3\}$ or $[-1, 3]$
Range: $\{y \mid 0 \le y \le 4\}$ or $[0, 4]$

23. Domain: $\{x \mid x$ is a real number$\}$ or $(-\infty, \infty)$
Range: $\{y \mid y \ge -3\}$ or $[-3, \infty)$

25. Domain: $\{x \mid x$ is a real number$\}$ or $(-\infty, \infty)$
Range: $\{y \mid y$ is a real number$\}$ or $(-\infty, \infty)$

27. Domain: $\{x \mid x$ is a real number$\}$ or $(-\infty, \infty)$
Range: $\{y \mid y$ is a real number$\}$ or $(-\infty, \infty)$

29. Domain: $\{x \mid x$ is a real number$\}$ or $(-\infty, \infty)$
Range: $\{y \mid y$ is a real number$\}$ or $(-\infty, \infty)$

31. Domain: $\{x \mid x$ is a real number$\}$ or $(-\infty, \infty)$
Range: $\{y \mid y$ is a real number$\}$ or $(-\infty, \infty)$

33. Domain: $\{x \mid x$ is a real number$\}$ or $(-\infty, \infty)$
Range: $\{y \mid y$ is a real number$\}$ or $(-\infty, \infty)$

35. Domain: $\{x \mid x$ is a real number$\}$ or $(-\infty, \infty)$
Range: $\{y \mid y$ is a real number$\}$ or $(-\infty, \infty)$

37. Domain: $\{x \mid x$ is a real number$\}$ or $(-\infty, \infty)$
Range: $\{y \mid y \le 0\}$ or $(-\infty, 0]$

39. Domain: $\{x \mid x$ is a real number$\}$ or $(-\infty, \infty)$
Range: $\{y \mid y \ge -8\}$ or $[-8, \infty)$

41. Domain: $\{x \mid x$ is a real number$\}$ or $(-\infty, \infty)$
Range: $\{y \mid y \ge 0\}$ or $[0, \infty)$

43. Domain: $\{x \mid x$ is a real number$\}$ or $(-\infty, \infty)$
Range: $\{y \mid y \ge 0\}$ or $[0, \infty)$

45. Domain: $\{x \mid x$ is a real number$\}$ or $(-\infty, \infty)$
Range: $\{y \mid y$ is a real number$\}$ or $(-\infty, \infty)$

47. Domain: $\{x \mid x$ is a real number$\}$ or $(-\infty, \infty)$
Range: $\{y \mid y$ is a real number$\}$ or $(-\infty, \infty)$

49. Domain: $\{x \mid x$ is a real number$\}$ or $(-\infty, \infty)$
Range: $\{y \mid y \ge -4\}$ or $[-4, \infty)$

51. Domain: $\{x \mid x \ge -1\}$ or $[-1, \infty)$
Range: $\{y \mid y$ is a real number$\}$ or $(-\infty, \infty)$

53. (a) Domain: $\{x \mid 0 \le x \le 50\}$ or $[0, 50]$
Range: $\{y \mid 0 \le y \le 625\}$ or $[0, 625]$
(b) Answers may vary.

55. (a) Domain: $\{m|0 \le m \le 15{,}120\}$ or $[0, 15120]$
Range: $\{c|40 \le c \le 5888\}$ or $[40, 5888]$
(b) Answers will vary.

57. Actual graphs will vary but all should be horizontal lines.

Putting the Concepts Together (Sections 8.1–8.2)

1. A: x-axis; B: quadrant II;
C: quadrant I; D: y-axis;
E: quadrant III; F: quadrant IV.

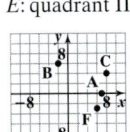

2. (a) yes   **(b)** yes   **(c)** no
3. $y = |x| + 3$

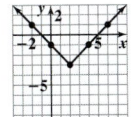

4. $y = \frac{1}{2}x^2 - 1$

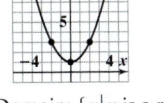

5. $(0, -5)$ and $(4, 0)$

6. (a) About $230,000.   **(b)** July of 2003; About $248,000.   **(c)** Between September 2002 and October 2002; About $16,000

7. $\{(-2, -1), (-1, 0), (0, 1),$
$(1, 2), (2, 3)\}$

8. (a) Domain: $\{x|-4 \le x \le 5\}$ or $[-4, 5]$
Range: $\{y|-6 \le y \le -1\}$ or $[-6, -1]$
(b) Domain: $\{-4, -1, 0, 3, 6\}$
Range: $\{-3, -2, 2, 6\}$

9. (a) $y = |x - 2| - 3$

Domain: $\{x|x \text{ is a real number}\}$
or $(-\infty, \infty)$
Range: $\{y|y \ge -3\}$ or $[-3, \infty)$

(b) $y = \frac{1}{2}x^2 + 1$

Domain: $\{x|x \text{ is a real number}\}$
or $(-\infty, \infty)$
Range: $\{y|y \ge 1\}$
or $[1, \infty)$

10. Domain: $\{t|0 \le t \le 3.8\}$ or $[0, 3.8]$
Range: $\{h|0 \le h \le 105\}$ or $[0, 105]$

Section 8.3

1. function   **3.** argument   **5.** True   **7.** map; ordered pairs; equation; graph
9. Function.
Domain: {Virginia, Nevada, New Mexico, Tennessee, Texas}
Range: $\{3, 9, 11, 32\}$

11. Not a function.
Domain: $\{150, 174, 180\}$
Range: $\{118, 130, 140\}$

13. Function.
Domain: $\{0, 1, 2, 3\}$
Range: $\{3, 4, 5, 6\}$

15. Function.
Domain: $\{-3, 1, 4, 7\}$
Range: $\{5\}$

17. Not a function.
Domain: $\{-10, -5, 0\}$
Range: $\{1, 2, 3, 4\}$

19. Function   **21.** Function   **23.** Not a function   **25.** Function

27. Not a function   **29.** Function   **31.** Not a function   **33.** Function   **35.** Function   **37. (a)** $f(0) = 3$ **(b)** $f(3) = 9$
(c) $f(-2) = -1$ **(d)** $f(-x) = -2x + 3$ **(e)** $-f(x) = -2x - 3$ **(f)** $f(x + 2) = 2x + 7$ **(g)** $f(2x) = 4x + 3$
(h) $f(x + h) = 2x + 2h + 3$ **39. (a)** $f(0) = 2$ **(b)** $f(3) = -13$ **(c)** $f(-2) = 12$ **(d)** $f(-x) = 5x + 2$
(e) $-f(x) = 5x - 2$ **(f)** $f(x + 2) = -5x - 8$ **(g)** $f(2x) = -10x + 2$ **(h)** $f(x + h) = -5x - 5h + 2$ **41.** $f(2) = 7$
43. $s(-2) = 16$ **45.** $F(-3) = 5$ **47.** $F(4) = -6$
49. $f(x) = 4x - 6$ **51.** $h(x) = x^2 - 2$ **53.** $G(x) = |x - 1|$ **55.** $g(x) = x^3$ **57.** $-6 = C$ **61.** $A(r) = \pi r^2$; 50.27 in.^{2}
59. $A = 5$ **63.** $G(h) = 15h$; $375

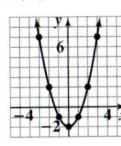
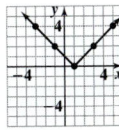
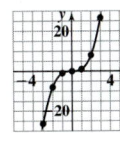

65. (a) The dependent variable is the population, P, and the independent variable is the age, a.
(b) $P(20) = 197.857$; The population of Americans that were 20 years of age or older in 2005 was roughly 198 million.
(c) $P(0) = 300.517$; $P(0)$ represents the entire population of the U.S. since every member of the population is at least 0 years of age. The population of
the U.S. in 2005 was roughly 301 million.
67. (a) The dependent variable is revenue, R, and the independent variable is price, p.
(b) $R(50) = 7500$; Selling PDAs for $50 will yield a daily revenue of $7500 for the company.
(c) $R(120) = 9600$; Selling PDAs for $120 will yield a daily revenue of $9600 for the company.
69. Answers will vary.   **71.** $f(2) = 7$ **73.** $F(-3) = 5$ **75.** $H(7) = 5$ **77.** $F(4) = -6$

Section 8.4

1. $f(3) = 8$ **3.** domain   **5.** True   **7.** Answers will vary.   **9.** $\{x|x \text{ is a real number}\}$ or $(-\infty, \infty)$ **11.** $\{z|z \ne 5\}$

13. $\{x|x \text{ is a real number}\}$ or $(-\infty, \infty)$ **15.** $\left\{x|x \ne -\frac{1}{3}\right\}$ **17.** $\{x|x \ne -8, x \ne 4\}$

19. (a) Domain: $\{x|x \text{ is a real number}\}$ or $(-\infty, \infty)$
Range: $\{y|y \text{ is a real number}\}$ or $(-\infty, \infty)$
(b) $(0, 2)$ and $(1, 0)$

21. (a) Domain: $\{x|x \text{ is a real number}\}$ or $(-\infty, \infty)$
Range: $\{y|y \ge -2.25\}$ or $[-2.25, \infty)$
(b) $(-2, 0), (4, 0)$, and $(0, -2)$

23. (a) Domain: $\{x \mid x \text{ is a real number}\}$ or $(-\infty, \infty)$ **25. (a)** Domain: $\{x \mid x \text{ is a real number}\}$ or $(-\infty, \infty)$
Range: $\{y \mid y \text{ is a real number}\}$ or $(-\infty, \infty)$ Range: $\{y \mid y \geq 0\}$ or $[0, \infty)$
(b) $(-3, 0), (-1, 0), (2, 0),$ and $(0, -3)$ **(b)** $(-3, 0), (3, 0),$ and $(0, 9)$

27. (a) Domain: $\{x \mid x \leq 4\}$ or $(-\infty, 4]$ **29. (a)** $f(-7) = -2$ **(b)** $f(-3) = 3$ **(c)** $f(6) = 2$ **(d)** negative
Range: $\{y \mid y \leq 3\}$ or $(-\infty, 3]$ **(e)** $\{-6, -1, 4\}$ **(f)** $\{x \mid -7 \leq x \leq 6\}$ or $[-7, 6]$ **(g)** $\{y \mid -2 \leq y \leq 3\}$ or $[-2, 3]$ **(h)** $-6, -1,$ and 4
(b) $(-2, 0)$ and $(0, 2)$ **(i)** -1 **(j)** $\{-7, 2\}$ **(k)** $x = -3$

31. (a) $F(-2) = 3$ **(b)** $F(3) = -6$ **(c)** $x = -1$ **(d)** $x = -4$ **(e)** $y = 2$ **33. (a)** no **(b)** $f(3) = 3; (3, 3)$ **(c)** $x = 4; (4, 7)$

35. (a) yes **(b)** $g(6) = 1; (6, 1)$ **(c)** $x = -12; (-12, 10)$ **37.** $\{r \mid r \geq 0\}$ or $[0, \infty)$ **39.** $\{h \mid 0 \leq h \leq 60\}$ or $[0, 60]$

41. $\{p \mid 0 \leq p \leq 120\}$ or $[0, 120]$ **43. (a)** III **(b)** I **(c)** IV **(d)** V **(e)** II

45.

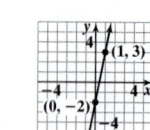

Pulse rate (beats per min.)

47. Answers will vary.

49. Answers will vary.

51. Answers will vary. One possibility.

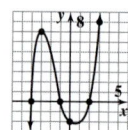

Section 8.5

1. slope; y-intercept **3.** scatter diagram **5.** True **7.** A non-horizontal line that passes through the origin.

9. $f(x) = 5x - 2$ **11.** $G(x) = -3x + 7$ **13.** $H(x) = -2$ **15.** $f(x) = \dfrac{1}{2}x - 4$ **17.** $F(x) = -\dfrac{5}{2}x + 5$ **19.** $G(x) = -\dfrac{3}{2}x$

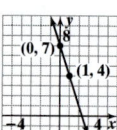 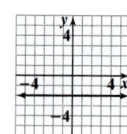 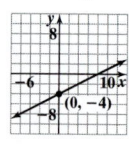 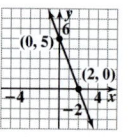 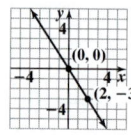

21. Nonlinear **23.** Linear with positive slope

25. (a)

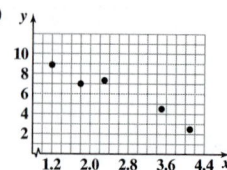

(b) Answers will vary. Using the points $(4, 1.8)$ and $(9, 2.6)$, the equation is $y = 0.16x + 1.16$.

(c)

27. (a)

(b) Answers will vary. Using the points $(1.2, 8.4)$ and $(4.1, 2.4)$, the equation is $y = -2.1x + 10.92$.

(c)

29. (a) $\{x \mid 7300 \leq x \leq 29{,}700\}$ or $[7300, 29700]$ **(b)** \$2635
(c) The independent variable is x; the dependent variable is T.
(d) **(e)** \$25,000

Tax Bill (\$)

$(29700, 4090)$
$(18000, 2335)$
$(7300, 730)$

Adjusted Gross Income (\$)

31. (a) $\{m \mid m \geq 0\}$ or $[0, \infty)$ **(b)** 2; The base fare is \$2.00 before any distance is driven. **(c)** \$9.50
(d) **(e)** A person can travel 7.5 miles in a cab for \$13.25.

Cab fare (\$)

$(15, 24.5)$
$(10, 17)$
$(0, 2)$

Distance (miles)

33. (a) The independent variable is a; the dependent variable is H.
(b) $\{a \mid 15 \leq a \leq 90\}$ or $[15, 90]$ **(c)** \$566.50
(d) **(e)** 48 years

Premium (\$)

$(90, 1934.5)$
$(50, 1022.5)$
$(15, 224.5)$

Age (years)

35. (a) $B(m) = 0.05m + 8.95$
(b) The independent variable is m; the dependent variable is B.
(c) $\{m \mid m \geq 0\}$ or $[0, \infty)$
(d) \$23.95
(e) 240 minutes
(f)

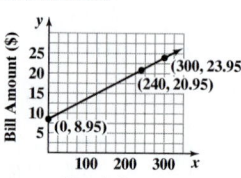

Bill Amount (\$)

$(300, 23.95)$
$(240, 20.95)$
$(0, 8.95)$

Time (minutes)

37. (a) $V(x) = -900x + 2700$
  (b) $\{x | 0 \le x \le 3\}$ or $[0, 3]$
  (c) \$1800
  (d) The y-intercept is 2700 and the x-intercept is 3.
  (e) After two years
  (f)

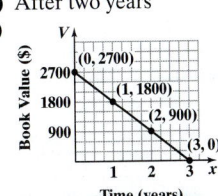

39. (a)

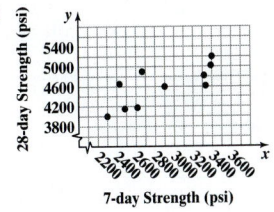

  (b) Linear
  (c) Answers will vary. Using the points (2300, 4070) and (3390, 5220), the equation is $y = 1.06x + 1632$.

(d)

  (e) 4812 psi
  (f) If the 7-day strength is increased by 1 psi, then the 28-day strength will increase by 1.06 psi.

41. (a) No
  (b)

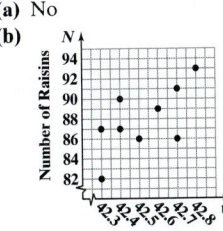

  (c) Answers will vary. Using the points (42.3, 82) and (42.8, 93), the equation is $N = 22w - 848.6$.
  (d)

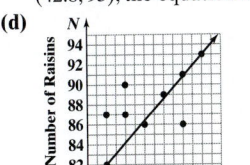

  (e) $N(w) = 22w - 848.6$
  (f) approximately 86 raisins
  (g) If the weight increases by one gram, then the number of raisins increases by 22 raisins.

43. (a) $x = 3$ **(b)** $x = -1$
  (c) $x = 4$
  (d) The y-intercept is -2 and the x-intercept is 2.
  (e) 2

45. Answers will vary.

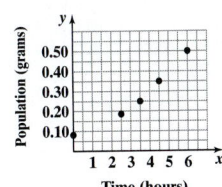

47. (a)

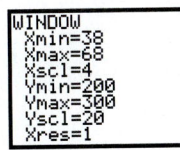

  (b) $y = 2.884x + 97.587$

49. (a)

  ```
  WINDOW
  Xmin=24.5
  Xmax=28
  Xscl=.5
  Ymin=16.2
  Ymax=17.8
  Yscl=.2
  Xres=1
  ```

  (b) $y = 0.373x + 7.327$

Section 8.6

1. intersection   **3.** compound inequality   **5.** False   **7.** There is no real number x such that $x > 4$ and $x < 2$.   **9.** Yes
11. $\{1, 4, 5, 6, 7, 8, 9\}$ **13.** $\{5, 7, 9\}$ **15.** $\emptyset$ or $\{\ \}$ **17. (a)** $A \cap B = \{x | -2 < x \le 5\}; (-2, 5]$

(b) $A \cup B = \{x | x \text{ is any real number}\}; (-\infty, \infty)$

19. (a) $E \cap F = \emptyset$

(b) $E \cup F = \{x | x < -1 \text{ or } x > 3\}; (-\infty, -1) \cup (3, \infty)$

21. $\{x | -2 \le x \le 2\}; [-2, 2]$

23. $\{x | -3 < x < 3\}; (-3, 3)$

25. $\{x | -2 \le x < 3\}; [-2, 3)$

27. $\{x | x < -2 \text{ or } x > 3\}; (-\infty, -2) \cup (3, \infty)$

29. $\emptyset$ or $\{\ \}$

31. $\{x | x < -2\}; (-\infty, -2)$

33. $\emptyset$ or $\{\ \}$

35. $\{x | x < -2 \text{ or } x > 5\}; (-\infty, -2) \cup (5, \infty)$

37. $\{x | x \text{ is any real number}\}; (-\infty, \infty)$

39. $\{x | x < -1 \text{ or } x > 4\}; (-\infty, -1) \cup (4, \infty)$

41. $\{x | -1 \le x < 3\}; [-1, 3)$

43. $\left\{ x | -\dfrac{2}{3} \le x \le \dfrac{3}{2} \right\}; \left[-\dfrac{2}{3}, \dfrac{3}{2} \right]$

45. $\{x | x \le -3 \text{ or } x > 6\}; (-\infty, -3] \cup (6, \infty)$

47. $\left\{x \mid -1 < x \le \dfrac{4}{5}\right\}; \left(-1, \dfrac{4}{5}\right]$ **49.** $\{x \mid 0 \le x \le 8\}; [0, 8]$

51. $\{x \mid -6 \le x \le -2\}; [-6, -2]$

53. $\{x \mid x \text{ is any real number}\};$ Interval: $(-\infty, \infty)$ **55.** $\emptyset$ or $\{\ \}$

57. $\left\{x \mid -\dfrac{5}{3} < x \le 5\right\}; \left(-\dfrac{5}{3}, 5\right]$ **59.** $\{x \mid -4 < x \le 3\}; (-4, 3]$

61. $\left\{x \mid x < 0 \text{ or } x > \dfrac{5}{2}\right\}; (-\infty, 0) \cup \left(\dfrac{5}{2}, \infty\right)$

63. $\{a \mid -3 \le a < 0\}; [-3, 0)$ **65.** $\{x \mid x \text{ is any real number}\}; (-\infty, \infty)$

67. $\left\{x \mid -2 \le x \le \dfrac{8}{3}\right\}; \left[-2, \dfrac{8}{3}\right]$

69. $\{x \mid x < -10 \text{ or } x > 2\}; (-\infty, -10) \cup (2, \infty)$ **71.** $\{x \mid 2 < x < 5\}; (2, 5)$

73. $\left\{x \mid x \le -3 \text{ or } x > \dfrac{15}{4}\right\}; (-\infty, -3] \cup \left(\dfrac{15}{4}, \infty\right)$

75. $\left\{x \mid -5 < x \le \dfrac{1}{2}\right\}; \left(-5, \dfrac{1}{2}\right]$ **77.** $a = 1$ and $b = 8$ **79.** $a = 12$ and $b = 30$ **81.** $a = -1$ and $b = 23$

83. $90 < x < 140$ **85.** Joanna needs to score at least a 77 on the final. That is, $77 \le x \le 100$ (assuming 100 is the max score, otherwise $77 \le x \le 104$). **87.** The amount withheld ranges between \$92.84 and \$120.84, inclusive. **89.** The gas usage ranged from 150 to 165 therms.

91. Step 1:
$$a < b$$
$$a + a < a + b$$
$$2a < a + b$$
$$\frac{2a}{2} < \frac{a+b}{2}$$
$$a < \frac{a+b}{2}$$

Step 2:
$$a < b$$
$$a + b < b + b$$
$$a + b < 2b$$
$$\frac{a+b}{2} < \frac{2b}{2}$$
$$\frac{a+b}{2} < b$$

Step 3: Since $a < \dfrac{a+b}{2}$ and $\dfrac{a+b}{2} < b$, it follows that $a < \dfrac{a+b}{2} < b$.

93. $\{\ \}$ or $\emptyset$ **95.** This is a contradiction. There is no solution. If, during simplification, the variable terms are all eliminated and a contradiction results, then there is no solution to the inequality. **97.** If $x < 2$ then $x - 2 < 2 - 2 \Rightarrow x - 2 < 0$. When multiplying both sides of the inequality by $x - 2$ in the second step, the direction of the inequality must switch.

Section 8.7

1. $u = a$ or $u = -a$ **3.** $a < 0$ **5.** False **7. (a)** $\{-5, 5\}$ **(b)** $\{x \mid -5 \le x \le 5\}; [-5, 5]$ **(c)** $\{x \mid x < -5 \text{ or } x > 5\}; (-\infty, -5) \cup (5, \infty)$
9. (a) $\{-5, 1\}$ **(b)** $\{x \mid -5 < x < 1\}; (-5, 1)$ **(c)** $\{x \mid x \le -5 \text{ or } x \ge 1\}; (-\infty, -5] \cup [1, \infty)$ **11.** $\{-10, 10\}$ **13.** $\{-1, 7\}$
15. $\left\{-1, \dfrac{13}{3}\right\}$ **17** $\{-5, 5\}$ **19.** $\left\{-\dfrac{11}{2}, \dfrac{5}{2}\right\}$ **21.** $\{-4, 10\}$ **23.** $\{0\}$ **25.** $\left\{-\dfrac{7}{3}, 3\right\}$ **27.** $\left\{-7, \dfrac{3}{5}\right\}$ **29.** $\{1, 3\}$ **31.** $\{2\}$
33. $\{x \mid -9 < x < 9\}; (-9, 9)$ **35.** $\{x \mid -3 \le x \le 11\}; [-3, 11]$

37. $\left\{x \mid -3 < x < \dfrac{7}{3}\right\}; \left(-3, \dfrac{7}{3}\right)$ **39.** $\emptyset$ or $\{\ \}$ **41.** $\{y \mid y < 3 \text{ or } y > 7\}; (-\infty, 3) \cup (7, \infty)$ **43.** $\left\{x \mid x \le -2 \text{ or } x \ge \dfrac{1}{2}\right\}; (-\infty, -2] \cup \left[\dfrac{1}{2}, \infty\right)$

45. $\{y \mid y \text{ is any real number}\}(-\infty, \infty)$ **47.** $\{x \mid 0 < x < 6\}; (0, 6)$

49. $\{x \mid x \text{ is any real number}\}; (-\infty, \infty)$ **51.** $\left\{x \mid -1 < x < \dfrac{9}{5}\right\}; \left(-1, \dfrac{9}{5}\right)$

53. $\{x \mid x < 0 \text{ or } x > 1\}; (-\infty, 0) \cup (1, \infty)$

55. $\{x \mid x \le -2 \text{ or } x \ge 3\}; (-\infty, -2] \cup [3, \infty)$

57. $\{x \mid 1.995 < x < 2.005\}; (1.995, 2.005)$

59. $\{x \mid x < -5 \text{ or } x > 5\}; (-\infty, -5) \cup (5, \infty)$ **61.** $\{-4, -1\}$ **63.** $\{-5, 5\}$

65. $\left\{x \mid -2 \le x \le \dfrac{6}{5}\right\}; \left[-2, \dfrac{6}{5}\right]$ **67.** $\emptyset$ or $\{\ \}$

69. $\left\{x \mid x \le -\dfrac{7}{3} \text{ or } x \ge 1\right\}; \left(-\infty, -\dfrac{7}{3}\right] \cup [1, \infty)$

71. $\left\{x \mid x < 0 \text{ or } x > \dfrac{4}{3}\right\}; (-\infty, 0) \cup \left(\dfrac{4}{3}, \infty\right)$ **73.** $\{-1, 1\}$ **75.** $\emptyset$ or $\{\ \}$ **77.** $\left\{-8, \dfrac{4}{7}\right\}$

79. $|5 - x| < 3$ $\{x|2 < x < 8\}; (2, 8)$ **81.** $|2x - (-6)| > 3$ $\left\{x|x < -\dfrac{9}{2} \text{ or } x > -\dfrac{3}{2}\right\}; \left(-\infty, -\dfrac{9}{2}\right) \cup \left(-\dfrac{3}{2}, \infty\right)$ **83.** The acceptable rod

lengths are between 5.6995 inches and 5.7005 inches, inclusive. **85.** An unusual IQ score would be less than 70.6 or greater than 129.4.

87. The absolute value, when isolated, is equal to a negative number which is not possible. **89.** The absolute value, when isolated, is less than -3.

Since absolute values are always nonnegative, this is not possible. **91.** $\left\{-\dfrac{5}{2}\right\}$ **93.** $\{2\}$ **95.** $\varnothing$ or $\{\ \}$ **97.** $\{x|x \le -5\}; (-\infty, -5]$

Chapter 8 Review

1. A: quadrant IV;
B: quadrant III;
C: y-axis; D: quadrant II;
E: x-axis; F: quadrant I.

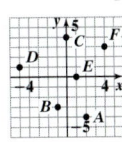

2. A: x-axis; B: quadrant I;
C: quadrant III; D: quadrant II;
E: quadrant IV; F: y-axis.

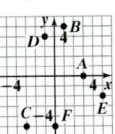

3. (a) yes **(b)** no **(c)** no **(d)** yes
4. (a) no **(b)** yes **(c)** yes **(d)** no

6. $2x + y = 3$

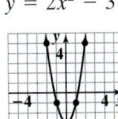

7. $y = 2x^2 - 3$

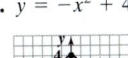

8. $y = -x^2 + 4$

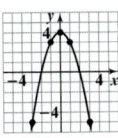

9. $y = -|x| - 2$

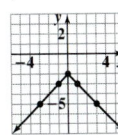

10. $y = |x + 2| - 1$

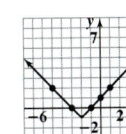

11. $y = x^3 + 2$

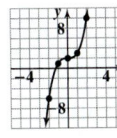

12. $y = -x^3 + 1$

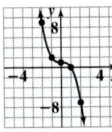

13. $x = y^2 + 1$

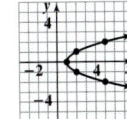

14. $y = \dfrac{1}{x - 2}$

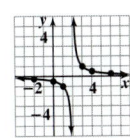

15. $(-3, 0), (0, -1), (0, 3)$ **16.** $(-2, 0), (0, 0), (2, 0)$ **17. (a)** \$40 **(b)** About \$500 **18. (a)** 2:03 or 123 seconds **(b)** 2001; 2:00 or 120 seconds
19. $\{(\text{Cent}, 2.500), (\text{Nickel}, 5.000), (\text{Dime}, 2.268), (\text{Quarter}, 5.670), (\text{Half Dollar}, 11.340), (\text{Dollar}, 8.100)\}$
Domain: {Cent, Nickel, Dime, Quarter, Half Dollar, Dollar}
Range: {2.268, 2.500, 5.000, 5.670, 8.100, 11.340}
20. $\{(70, \$6.99), (90, \$9.99), (120, \$9.99), (128, \$12.99), (446, \$49.99)\}$
Domain: {70, 90, 120, 128, 446}
Range: {\$6.99, \$9.99, \$12.99, \$49.99}

21. Domain: $\{-4, -2, 2, 3, 6\}$
Range: $\{-9, -1, 5, 7, 8\}$

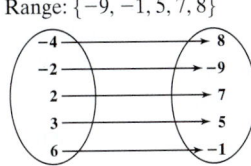

22. Domain: $\{-2, 1, 3, 5\}$
Range: $\{1, 4, 7, 8\}$

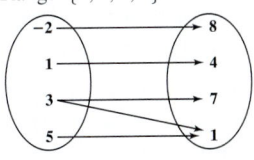

23. Domain: $\{x|x \text{ is a real number}\}$ or $(-\infty, \infty)$
Range: $\{y|y \text{ is a real number}\}$ or $(-\infty, \infty)$

24. Domain: $\{x|-6 \le x \le 4\}$ or $[-6, 4]$
Range: $\{y|-4 \le y \le 6\}$ or $[-4, 6]$

25. Domain: $\{2\}$
Range: $\{y|y \text{ is a real number}\}$ or $(-\infty, \infty)$

26. Domain: $\{x|x \ge -1\}$ or $[-1, \infty)$
Range: $\{y|y \ge -2\}$ or $[-2, \infty)$

27. $y = x + 2$

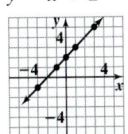

Domain: $\{x|x \text{ is a real number}\}$ or $(-\infty, \infty)$
Range: $\{y|y \text{ is a real number}\}$ or $(-\infty, \infty)$

28. $2x + y = 3$

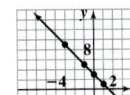

Domain: $\{x|x \text{ is a real number}\}$ or $(-\infty, \infty)$
Range: $\{y|y \text{ is a real number}\}$ or $(-\infty, \infty)$

29. $y = 2x^2 - 3$

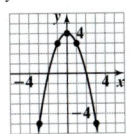

Domain: $\{x|x \text{ is a real number}\}$ or $(-\infty, \infty)$
Range: $\{y|y \ge -3\}$ or $[-3, \infty)$

30. $y = -x^2 + 4$

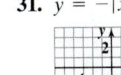

Domain: $\{x|x \text{ is a real number}\}$ or $(-\infty, \infty)$
Range: $\{y|y \le 4\}$ or $(-\infty, 4]$

31. $y = -|x| - 2$

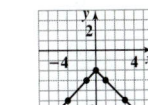

Domain: $\{x|x \text{ is a real number}\}$ or $(-\infty, \infty)$
Range: $\{y|y \le -2\}$ or $(-\infty, -2]$

32. $y = |x + 2| - 1$

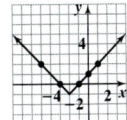

Domain: $\{x|x \text{ is a real number}\}$ or $(-\infty, \infty)$
Range: $\{y|y \ge -1\}$ or $[-1, \infty)$

33. $y = x^3 + 2$

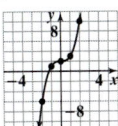

Domain: $\{x|x \text{ is a real number}\}$ or $(-\infty, \infty)$
Range: $\{y|y \text{ is a real number}\}$ or $(-\infty, \infty)$

34. $y = -x^3 + 1$

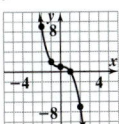

Domain: $\{x|x \text{ is a real number}\}$ or $(-\infty, \infty)$
Range: $\{y|y \text{ is a real number}\}$ or $(-\infty, \infty)$

35. $x = y^2 + 1$

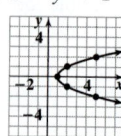

Domain: $\{x \mid x \geq 1\}$
or $[1, \infty)$
Range: $\{y \mid y \text{ is a real number}\}$
or $(-\infty, \infty)$

36. $y = \dfrac{1}{x-2}$

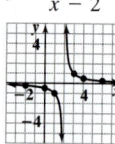

Domain: $\{x \mid x \neq 2\}$ or $(-\infty, 2) \cup (2, \infty)$
Range: $\{y \mid y \neq 0\}$ or $(-\infty, 0) \cup (0, \infty)$

37. (a) Domain: $\{x \mid 0 \leq x \leq 44.64\}$ or $[0, 44.64]$
Range: $\{y \mid 40 \leq y \leq 2122\}$ or $[40, 2122]$
(b) Answers may vary.
38. Domain: $\{t \mid 0 \leq t \leq 4\}$ or $[0, 4]$
Range: $\{y \mid 0 \leq y \leq 121\}$ or $[0, 121]$
39. (a) Not a function.
Domain: $\{-1, 5, 7, 9\}$
Range: $\{-2, 0, 2, 3, 4\}$
(b) Function.
Domain: {Camel, Macaw, Deer, Fox, Tiger, Crocodile}
Range: $\{14, 22, 35, 45, 50\}$

40. (a) Function
Domain: $\{-3, -2, 2, 4, 5\}$
Range: $\{-1, 3, 4, 7\}$
(b) Not a function
Domain: {Red, Blue, Green, Black}
Range: {Camry, Taurus, Windstar, Durango}
41. Function
42. Not a function
43. Not a function
44. Function
45. Not a function
46. Function

47. Function
48. Not a function
49. (a) $f(-2) = -5$
(b) $f(3) = 10$
50. (a) $g(0) = -\dfrac{1}{3}$
(b) $g(2) = -5$
51. (a) $F(5) = -3$
(b) $F(-x) = 2x + 7$
52. (a) $G(7) = 15$
(b) $G(x + h) = 2x + 2h + 1$
53. $f(x) = 2x - 5$
54. $g(x) = x^2 - 3x + 2$
55. $h(x) = (x - 1)^3 - 3$
56. $f(x) = |x + 1| - 4$

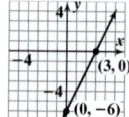 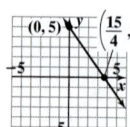 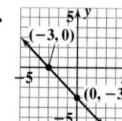 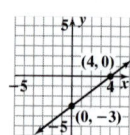

57. (a) The dependent variable is the population, P, and the independent variable is the number of years after 1900, t.
(b) $P(110) = 1119.418$; The population of Orange County will be roughly 1,119,418 in 2010.
(c) $P(-70) = 1272.958$; The population of Orange County was roughly 1,272,958 in 1830. This is not reasonable. (The population of the entire Florida territory was roughly 35,000 in 1830.)
58. (a) The dependent variable is the annual wage, W, and the independent variable is age, a.
(b) $W(30) = 36.261$; According to the model, a 30-year-old Wyoming resident working in the mining industry in 2000 made about $36,261 annually on average.
(c) $W(16) = -2.127$; $W(16)$ represents the average annual salary of a 16-year-old Wyoming resident working in the mining industry in 2000. This result is unreasonable since annual salaries should not be negative.

59. $\{x \mid x \text{ is a real number}\}$ or $(-\infty, \infty)$
60. $\{w \mid w \neq -\frac{5}{2}\}$
61. $\{t \mid t \neq 5\}$
62. $\{x \mid x \neq 2\}$
63. $\{t \mid t \text{ is a real number}\}$ or $(-\infty, \infty)$
64. $\{x \mid x \text{ is a real number}\}$ or $(-\infty, \infty)$
65. (a) Domain: $\{x \mid x \text{ is a real number}\}$ or $(-\infty, \infty)$; Range: $\{y \mid y \text{ is a real number}\}$ or $(-\infty, \infty)$
b. $(0, 2)$ and $(4, 0)$
66. (a) Domain: $\{x \mid x \text{ is a real number}\}$ or $(-\infty, \infty)$; Range: $\{y \mid y \geq -3\}$ or $[-3, \infty)$
(b) $(-2, 0), (2, 0), (0, -3)$
67. (a) Domain: $\{x \mid x \text{ is a real number}\}$ or $(-\infty, \infty)$; Range: $\{y \mid y \text{ is a real number}\}$ or $(-\infty, \infty)$
(b) $(0, 0)$ and $(2, 0)$
68. (a) Domain: $\{x \mid x \geq -3\}$ or $[-3, \infty)$; Range: $\{y \mid y \geq 1\}$ or $[1, \infty)$
(b) $(0, 3)$
69. (a) Domain: $\{x \mid x \text{ is a real number}\}$ or $(-\infty, \infty)$; Range: $\{y \mid y \geq -4\}$ or $[-4, \infty)$
(b) $(-1, 0), (3, 0), (0, -2)$
70. (a) Domain: $\{x \mid x \text{ is a real number}\}$ or $(-\infty, \infty)$; Range: $\{y \mid y \leq 0\}$ or $(-\infty, 0]$
(b) $(-2, 0), (2, 0), (0, -4)$
71. (a) yes
(b) $h(-2) = -11; (-2, -11)$
(c) $x = \dfrac{11}{2}; \left(\dfrac{11}{2}, 4\right)$
72. (a) no
(b) $g(3) = \dfrac{29}{5}; \left(3, \dfrac{29}{5}\right)$
(c) $x = -10; (-10, -2)$
73.

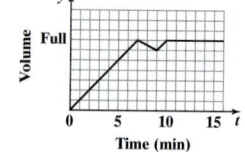

74.

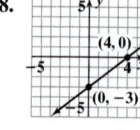

75.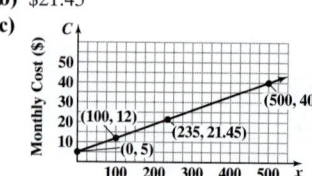
76.
77.
78.

79. (a) $\{x \mid x \geq 0\}$ or $[0, \infty)$
(b) \$21.45
(c)

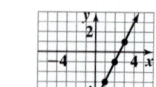

80. (a) The independent variable is x; the dependent variable is V.
(b) $\{x \mid 0 \leq x \leq 5\}$ or $[0, 5]$
(c) \$1800
(d) \$1080
(e)

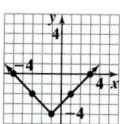

(f) After 5 years

(d) 1000 minutes

81. (a) $E(x) = 3.62x + 12.6$
(b) 41.56%
(c) Electronically filed tax returns are increasing at a rate of 3.62% per year.
(d) 2006
82. (a) $H(x) = -x + 220$
(b) 175 beats per minute
(c) The maximum recommended heart rate for men under stress decreases at a rate of 1 beat per minute per year.
(d) 52 years

83. (a) $C(m) = 0.12m + 35$
(b) The independent variable is m; the dependent variable is C.
(c) $\{m \mid m \geq 0\}$ or $[0, \infty)$
(d) $49.88   **(e)** 268 miles were driven

(f)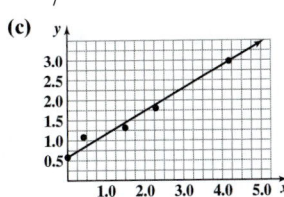

84. (a) $B(x) = 3.50x + 33.99$
(b) The independent variable is x; the dependent variable is B.
(c) $\{x \mid x \geq 0\}$ or $[0, \infty)$
(d) $51.49
(e) 7 pay-per-view movies

(f)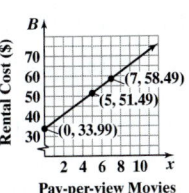

85. (a)

(b) Answers will vary. Using the points $(2, 13.3)$ and $(14, 4.6)$, the equation is $y = -0.725x + 14.75$.

(c)

86. (a)

(b) Answers will vary. Using the points $(0, 0.6)$ and $(4.2, 3.0)$, the equation is $y = \dfrac{4}{7}x + 0.6$.

(c)

87. (a)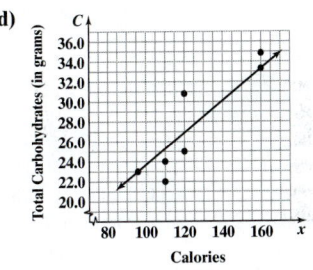

(b) Approximately linear.
(c) Answers will vary. Using the points $(96, 23.2)$ and $(160, 33.3)$, the equation is $y = 0.158x + 8.032$.

(d)

(e) 30.2 grams   **(f)** In a one-cup serving of cereal, total carbohydrates will increase by 0.158 grams for each one-calorie increase.

88. (a)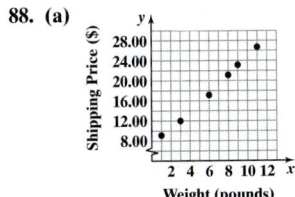

(b) Approximately linear.
(c) Answers will vary. Using the points $(1, 9.25)$ and $(11, 26.50)$, the equation is $y = 1.725x + 7.525$.

(d)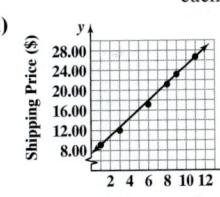

(e) $16.15
(f) The FedEx 2Day shipping price increases by $1.725 for each one-pound increase in the weight of a package.

89. $\{-1, 0, 1, 2, 3, 4, 6, 8\}$ **90.** $\{2, 4\}$ **91.** $\{1, 2, 3, 4\}$ **92.** $\{1, 2, 3, 4, 6, 8\}$ **93. (a)** $\{x \mid 2 < x \leq 4\}; (2, 4]$

(b) $\{x \mid x \text{ is any real number}\}; (-\infty, \infty)$

94. (a) $\{\ \}$ or $\varnothing$ **(b)** $\{x \mid x < -2 \text{ or } x \geq 3\}; (-\infty, -2) \cup [3, \infty)$

95. $\{x \mid -1 < x < 4\}; (-1, 4)$ Graph: **96.** $\{x \mid -5 < x < -1\}; (-5, -1)$

Graph: **97.** $\{x \mid x < -2 \text{ or } x > 2\}; (-\infty, -2) \cup (2, \infty)$ Graph:

98. $\{x \mid x \leq 0 \text{ or } x \geq 4\}; (-\infty, 0] \cup [4, \infty)$ Graph: **99.** $\{\ \}$ or $\varnothing$ **100.** $\{x \mid -2 \leq x < 4\}; [-2, 4)$

Graph: **101.** $\{x \mid x \leq -2 \text{ or } x > 3\}; (-\infty, -2] \cup (3, \infty)$ Graph:

102. $\{x \mid x \text{ is any real number}\}; (-\infty, \infty)$ Graph: **103.** $\{x \mid x < -10 \text{ or } x > 6\}; (-\infty, -10) \cup (6, \infty)$

Graph: **104.** $\left\{x \mid -\dfrac{3}{2} \leq x < \dfrac{5}{8}\right\}; \left[-\dfrac{3}{2}, \dfrac{5}{8}\right)$ Graph: **105.** $70 \leq x \leq 75$

106. The electric usage varied from roughly 756.7 kilowatt hours up to roughly 1953.1 kilowatt hours. (recall, x is the number *above* 300).

107. $\{-4, 4\}$ **108.** $\left\{\dfrac{1}{3}, 3\right\}$ **109.** $\{-5, 13\}$ **110.** $\{-3, -1\}$ **111.** $\{\ \}$ or $\varnothing$ **112.** $\left\{-\dfrac{1}{2}, 2\right\}$

113. $\{x \mid -2 < x < 2\}; (-2, 2)$ **114.** $\left\{x \mid x \leq -\dfrac{7}{2} \text{ or } x \geq \dfrac{7}{2}\right\}; \left(-\infty, -\dfrac{7}{2}\right] \cup \left[\dfrac{7}{2}, \infty\right)$

115. $\{x \mid -5 \leq x \leq 1\}; [-5, 1]$

116. $\left\{x \mid x \leq \dfrac{1}{2} \text{ or } x \geq 1\right\}; \left(-\infty, \dfrac{1}{2}\right] \cup [1, \infty)$

117. $\{x \mid x$ is a real number$\}$; $(-\infty, \infty)$ 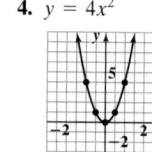 **118.** $\{\ \}$ or $\varnothing$

119. $\{x \mid 4.99 \le x \le 5.01\}$; $[4.99, 5.01]$

120. $\left\{x \mid x < -\frac{1}{2} \text{ or } x > \frac{7}{2}\right\}$; $\left(-\infty, -\frac{1}{2}\right) \cup \left(\frac{7}{2}, \infty\right)$

121. The acceptable diameters of the bearing are between 0.502 inches and 0.504 inches, inclusive. **122.** Tensile strengths below 36.08 lb/in.^{2} or above 43.92 lb/in.^{2} would be considered unusual.

Chapter 8 Test

1. A: quadrant IV; B: y-axis; C: x-axis; D: quadrant I; E: quadrant III; F: quadrant II.

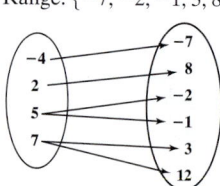

2. (a) no (b) yes (c) yes

3. $y = 4x - 1$

4. $y = 4x^2$

5. $(-3, 0), (0, 1), (0, 3)$

6. (a) $\approx 4.9\%$ (b) January; $\approx 5.9\%$ (c) November; $\approx 4.4\%$ (d) Answers will vary.

7. Domain: $\{-4, 2, 5, 7\}$ Range: $\{-7, -2, -1, 3, 8, 12\}$

8. Domain: $\left\{x \mid -\dfrac{5\pi}{2} \le x \le \dfrac{5\pi}{2}\right\}$ or $\left[-\dfrac{5\pi}{2}, \dfrac{5\pi}{2}\right]$ Range: $\{y \mid 1 \le y \le 5\}$ or $[1, 5]$

9. $y = x^2 - 3$

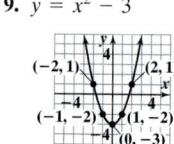

Domain: $\{x \mid x$ is a real number$\}$ or $(-\infty, \infty)$ Range: $\{y \mid y \ge -3\}$ or $[-3, \infty)$

10. Function. Domain: $\{-5, -3, 0, 2\}$ Range: $\{3, 7\}$

11. Not a function. Domain: $\{x \mid x \le 3\}$ or $(-\infty, 3]$ Range: $\{y \mid y$ is a real number$\}$ or $(-\infty, \infty)$

12. No **13.** $f(x + h) = -3x - 3h + 11$ **14.** (a) $g(-2) = 5$ (b) $g(0) = -1$ (c) $g(3) = 20$

15. $f(x) = x^2 + 3$

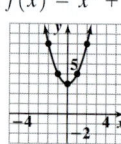

16. (a) The dependent variable is the ticket price, P, and the independent variable is the number of years after 1989, x. (b) $P(15) = 5.71$; According to the model, the average ticket price in 2004 ($x = 15$) was \$5.71.

17. (a) The dependent variable is the number of registered climbers, N, and the independent variable is the number of years after 1960, x. (b) $N(43) \approx 12,507$; According to the model, there were about 12,507 registered climbers on Mt. Rainier in 2003 ($x = 43$). (c) Answers may vary.

18. $\{x \mid x \ne -2\}$ or $(-\infty, -2) \cup (-2, \infty)$

19. (a) yes (b) $h(3) = -3$; $(3, -3)$ (c) $x = \dfrac{12}{5}$; $\left(\dfrac{12}{5}, 0\right)$

20. (a) The car stops accelerating when the speed stops increasing. Thus, the car stops accelerating after 6 seconds. (b) The car has a constant speed when the graph is horizontal. Thus, the car maintains a constant speed for 18 seconds.

21. (a) $P(x) = 18x - 100$ (b) $\{x \mid x \ge 0\}$ or $[0, \infty)$ (c) \$512 (d)

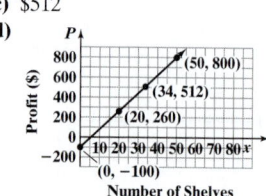

(e) 48 shelves

22. (a)

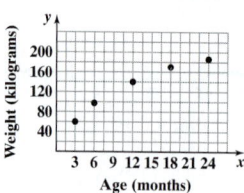

(b) Approximately linear. (c) Answers will vary. Using the points $(6, 95)$ and $(18, 170)$, the equation is $y = 6.25x + 57.5$.

(d)

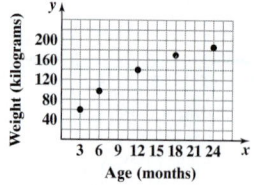

(e) 113.75 kilograms (f) A Shetland pony's weight will increase by 6.25 kilograms for each one-month increase in age.

23. $\{-4, -1\}$ **24.** $\{x \mid -2 \le x < 6\}$; $[-2, 6)$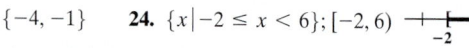

25. $\left\{x \mid x < -\dfrac{3}{2} \text{ or } x > 4\right\}$; $\left(-\infty, -\dfrac{3}{2}\right) \cup (4, \infty)$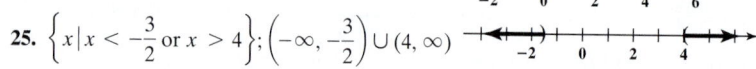

Chapter 9

Section 9.1

1. radical sign   **3.** $|a|$ **5.** True   **7.** Perfect squares: 4, 9, 16, 25   **9.** 1   **11.** -10 **13.** $\dfrac{1}{2}$ **15.** 0.6   **17.** 1.6   **19.** not a real number

21. rational; 8   **23.** rational; $\dfrac{1}{4}$ **25.** irrational; 6.63   **27.** irrational; 7.07   **29.** not a real number   **31.** 8   **33.** 19   **35.** $|r|$ **37.** $|x + 4|$

39. $|4x - 3|$ **41.** $|2y + 3|$ **43.** 13   **45.** 17   **47.** not a real number   **49.** 15   **51.** -8 **53.** 6   **55.** not a real number

53. 6   **55.** not a real number   **57.** -4 **59.** 13   **61.** The square roots of 36 are -6 and 6; $\sqrt{36} = 6$ **63.** $\dfrac{4\sqrt{13}}{3}$; 4.81

65. 25 ft   **67.** 16 km   **69.** 7 m   **71.** 14 in.   **73.** 398 m   **75.** 2   **77.** 5

Section 9.2

1. $\sqrt[n]{a}$ **3.** index   **5.** False   **7.** Answers will vary.   **9.** 5   **11.** -3 **13.** -5 **15.** $-\dfrac{1}{2}$ **17.** 3   **19.** 5   **21.** $|m|$

23. $x - 3$ **25.** $-|3p + 1|$ **27.** 2   **29.** -6 **31.** 2   **33.** -2 **35.** $\dfrac{2}{5}$ **37.** -5 **39.** not a real number   **41.** 32   **43.** -64

45. 16   **47.** 16   **49.** 8   **51.** $\dfrac{1}{12}$ **53.** 125   **55.** 32   **57.** $(3x)^{1/3}$ **59.** $\left(\dfrac{x}{3}\right)^{1/4}$ **61.** $x^{3/4}$ **63.** $(3x)^{2/5}$ **65.** $\left(\dfrac{5x}{y}\right)^{3/2}$

67. $(9ab)^{4/3}$ **69.** 2.92   **71.** 1.86   **73.** 4.47   **75.** 10.08   **77.** 1.26   **79.** 8   **81.** 243   **83.** not a real number   **85.** $\dfrac{1}{12}$ **87.** 0.2

89. 127   **91.** not a real number   **93.** 10; 10   **95. (a)** about $21.25°F$ **(b)** about $17.36°F$ **(c)** about $-15.93°F$

97. (a) $\sqrt{\dfrac{8r\rho_w g}{3C\rho}}$ m/s   **(b)** about 7.38 m/s   **99.** 8   **101.** 16   **103.** $x^6 y^9$ **105.** $(x - 1)^{3/2}(x + 3)$ **107.** $\dfrac{z - 6}{z - 1}$

Section 9.3

1. $a^r b^r$ **3.** 25   **5.** 8   **7.** $\dfrac{1}{2^{1/6}}$ **9.** $\dfrac{1}{x^{7/12}}$ **11.** 2   **13.** $\dfrac{125}{8}$ **15.** $x^{1/2} y^{2/9}$ **17.** $\dfrac{x^{1/6}}{y^{1/3}}$ **19.** $\dfrac{2a}{b^{3/4}}$ **21.** $\dfrac{x^{1/18}}{2y^{4/9}}$ **23.** $8x^{1/8} y^{1/2}$

25. $x^2 - 2x^{1/2}$ **27.** $\dfrac{2}{y^{1/3}} + 6y^{2/3}$ **29.** $4z^3 - 32 = 4(z - 2)(z^2 + 2z + 4)$ **31.** x^4 **33.** 2   **35.** $2ab^4$ **37.** $\sqrt[4]{x}$ **39.** $\sqrt[3]{x^5}$ **41.** $\sqrt[8]{x^3}$

43. $\sqrt[6]{3^7}$ **45.** 1   **47.** $5x^{1/2}(x + 3)$ **49.** $8(x + 2)^{2/3}(3x + 1)$ **51.** $\dfrac{6x + 5}{x^{1/2}}$ **53.** $\dfrac{2(10x - 27)}{(x - 4)^{1/3}}$ **55.** $5(x^2 + 4)^{1/2}(x^2 + 3x + 4)$ **57.** 2

59. 4   **61.** 10   **63.** 5   **65.** 0   **67.** $\dfrac{1}{48}$ **69.** 5   **71.** 3   **73.** $\sqrt[24]{x}$ **75.** 36   **77.** $\{x \mid x > -1\}$ or $(-1, \infty)$ **79.** $2(2a^2 - 4a - 3)$ **81.** -6

Section 9.4

1. $\sqrt[n]{a \cdot b}$ **3.** False   **5.** 0, 1, 4, 9, 16, 25   **7.** Answers may vary.   **9.** $\sqrt{14}$ **11.** $\sqrt[3]{60}$ **13.** $\sqrt{15ab}$ if $a,b \geq 0$ **15.** $\sqrt{x^2 - 49}$ if $|x| \geq 7$

17. $\sqrt{5}$ if $x > 0$ **19.** $5\sqrt{2}$ **21.** $3\sqrt[3]{2}$ **23.** $4|x|\sqrt{3}$ **25.** $-3x$ **27.** $2|m|\sqrt[4]{2}$ **29.** $2|p|\sqrt{3q}$ **31.** $6\sqrt{6a}$ **33.** $9m^2\sqrt{2}$

35. $y^6\sqrt{y}$ **37.** $c^2\sqrt[3]{c^2}$ **39.** $m^2 n\sqrt{mn}$ **41.** $5pq^2\sqrt{5p}$ **43.** $-2x^3\sqrt[3]{2}$ **45.** $pq^2\sqrt{pqr}$ **47.** $-m\sqrt[5]{16m^3 n^2}$ **49.** $(x - y)\sqrt[4]{x - y}$

51. $2\sqrt[3]{x^3 - y^3}$ **53.** 5   **55.** $3 + \sqrt{2}$ **57.** $\dfrac{-4 - 9\sqrt{2}}{6}$ **59.** $\dfrac{1 - \sqrt{2}}{2}$ **61.** 5   **63.** 4   **65.** 2   **67.** $5x\sqrt{3}$ **69.** $2b\sqrt[3]{3b}$

71. $18ab^2\sqrt{10}$ **73.** $3pq\sqrt[4]{4p}$ **75.** $-2ab\sqrt[5]{3a}$ **77.** $2(x - y)\sqrt[4]{3(x - y)}$ **79.** $\dfrac{\sqrt{3}}{4}$ **81.** $\dfrac{11}{10}$ **83.** $\dfrac{x\sqrt[4]{5}}{2}$ **85.** $\dfrac{3y}{5x}$ **87.** $-\dfrac{3x^3}{4y^4}$

89. 2   **91.** 4   **93.** $2a\sqrt{2}$ **95.** $\dfrac{2a^2\sqrt{2}}{b}$ **97.** $\dfrac{16a^3}{3b}$ **99.** $a^2\sqrt[3]{26}$ **101.** $\dfrac{3x^3\sqrt{5}}{y}$ **103.** $\sqrt[6]{432}$ **105.** $\sqrt[6]{12}$ **107.** $3\sqrt[6]{12}$

109. $\sqrt[3]{18}$ **111.** $\dfrac{\sqrt[3]{5x}}{2}$ **113.** $\sqrt[3]{45a^2}$ **115.** $6a^2\sqrt{2}$ **117.** $3a\sqrt[3]{2b^2}$ **119.** $\dfrac{-2\sqrt[3]{2}}{a}$ **121.** $-10m\sqrt[3]{4}$ **123.** $3ab^2\sqrt[3]{3ab}$ **125.** 6

127. (a)   **(b)** $3\sqrt{5}$ units   **129. (a)** roughly \$1,587,000   **(b)** roughly \$2,381,000   **131. (a)** $4x^2$ **(b)** 36

133. $x = -3 - \sqrt{6}$ or $x = -3 + \sqrt{6}$ **135.** $x = \dfrac{-2 - \sqrt{7}}{3}$ or $x = \dfrac{-2 + \sqrt{7}}{3}$ **137.** $x = -\dfrac{5}{7}$ or $x = 1$ **139.** $-5 \leq x \leq 3$

141. (a)

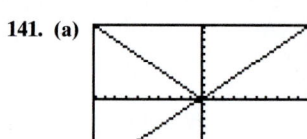

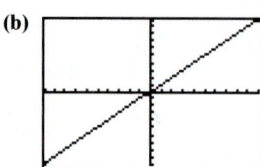

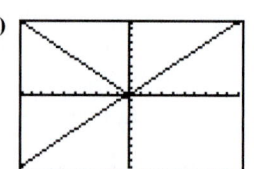

The graphs of $y = \sqrt{x^2}$ and $y = x$ are not the same so the expressions cannot be equal.

The graphs of $y = \sqrt[3]{x^3}$ and $y = x$ appear to be the same. $\sqrt[3]{x^3} = x$

The graphs of $y = \sqrt[4]{x^4}$ and $y = x$ are not the same so the expressions cannot be equal.

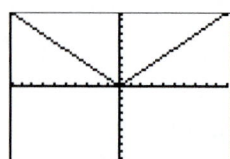

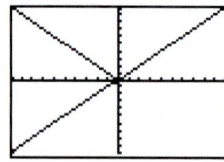

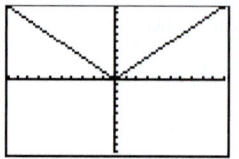

The graphs of $y = \sqrt{x^2}$ and $y = |x|$ appear to be the same. $\sqrt{x^2} = |x|$

The graph of $y = \sqrt[3]{x^3}$ and $y = |x|$ are not the same so the expressions cannot be equal.

The graphs of $y = \sqrt[4]{x^4}$ and $y = |x|$ appear to be the same. $\sqrt[4]{x^4} = |x|$

(d) Answers will vary.

Section 9.5

1. like radicals   **3.** True   **5.** Answers may vary.   **7.** $10\sqrt{2}$ **9.** $2\sqrt[3]{x}$ **11.** $14\sqrt{5x}$ **13.** $11\sqrt[3]{5} - 11\sqrt{5}$ **15.** $8\sqrt{2}$ **17.** $-2\sqrt[3]{3}$
19. $-25\sqrt[3]{2}$ **21.** $9\sqrt{6x}$ **23.** $4\sqrt{2} + 3\sqrt{10}$ **25.** $32x\sqrt{3x}$ **27.** $2x\sqrt{3} - 11x\sqrt{2}$ **29.** $(3x - 8)\sqrt[3]{2}$ **31.** $5\sqrt{x} - 1$ **33.** $3\sqrt{x}$
35. $5\sqrt[3]{x}$ **37.** $2\sqrt{3} - 3\sqrt{6}$ **39.** $\sqrt{6} + 3\sqrt{2}$ **41.** $\sqrt[3]{12} - 2\sqrt[3]{3}$ **43.** $3\sqrt{2x} - 2x\sqrt{5}$ **45.** $12 + 3\sqrt{3} + 4\sqrt{2} + \sqrt{6}$
47. $12 - 6\sqrt{7} + 2\sqrt{3} - \sqrt{21}$ **49.** $6\sqrt{7} - 30$ **51.** $\sqrt{6} - 12\sqrt{3} + 9\sqrt{2} - 4$ **53.** $31 - 2\sqrt{15}$ **55.** $4 + 2\sqrt{3}$ **57.** $7 - 2\sqrt{10}$
59. $x - 2\sqrt{2x} + 2$ **61.** 1 **63.** -11 **65.** $2x - 3y$ **67.** $\sqrt[3]{x^2} + \sqrt[3]{x} - 12$ **69.** $\sqrt[3]{4a^2} - 25$ **71.** $\sqrt{15} + 5\sqrt{2}$ **73.** $26x^2\sqrt{7x}$

75. -13 **77.** $2\sqrt[3]{7} + \sqrt[3]{28}$ **79.** $-4 + 12\sqrt{2}$ **81.** $20\sqrt{2}$ **83.** $8 - 2\sqrt{15}$ **85.** $(3x + 2)\sqrt[3]{5y}$ **87.** $2x - 7y$ **89.** $\dfrac{11\sqrt{5}}{25}$

91. The area is 108 square units. The perimeter is $30\sqrt{2}$ units.   **93.** $12\sqrt{6}$ square units   **95. (a)** $3\sqrt{3x}$ **(b)** $6\sqrt{3}$ **(c)** $6x$

97. Check $x = -2 + \sqrt{5}$:
$0 \overset{?}{=} x^2 + 4x - 1$
$0 \overset{?}{=} (-2 + \sqrt{5})^2 + 4(-2 + \sqrt{5}) - 1$
$0 \overset{?}{=} (-2)^2 + 2(-2)(\sqrt{5}) + (\sqrt{5})^2 - 8 + 4\sqrt{5} - 1$
$0 \overset{?}{=} 4 - 4\sqrt{5} + \sqrt{25} - 8 + 4\sqrt{5} - 1$
$0 \overset{?}{=} 9 - 9$
$0 = 0$ true
The value is a solution.

Check $x = -2 - \sqrt{5}$:
$0 \overset{?}{=} x^2 + 4x - 1$
$0 \overset{?}{=} (-2 - \sqrt{5})^2 + 4(-2 - \sqrt{5}) - 1$
$0 \overset{?}{=} (-2)^2 - 2(-2)(\sqrt{5}) + (\sqrt{5})^2 - 8 - 4\sqrt{5} - 1$
$0 \overset{?}{=} 4 + 4\sqrt{5} + \sqrt{25} - 8 - 4\sqrt{5} - 1$
$0 \overset{?}{=} 9 - 9$
$0 = 0$ true
The value is a solution.

99. $6p^2 + 7p - 3$ **101.** $m^2 - 16$ **103.** $x - 4$

Section 9.6

1. rationalizing the denominator   **3.** False   **5.** $\dfrac{\sqrt{2}}{2}$ **7.** $-\dfrac{2\sqrt{3}}{5}$ **9.** $\dfrac{\sqrt{3}}{2}$ **11.** $\dfrac{\sqrt{3}}{3}$ **13.** $\dfrac{\sqrt{2p}}{p}$ **15.** $\dfrac{2\sqrt{2y}}{y^2}$ **17.** $\sqrt[3]{4}$

19. $\dfrac{\sqrt[3]{7q^2}}{q}$ **21.** $-\dfrac{\sqrt[3]{60}}{10}$ **23.** $\dfrac{\sqrt[3]{50y^2}}{5y}$ **25.** $-\dfrac{4\sqrt[4]{27x}}{3x}$ **27.** $\dfrac{12\sqrt[5]{m^2n^3}}{mn}$ **29.** $2(\sqrt{6} + 2)$ **31.** $5(\sqrt{5} - 2)$ **33.** $2(\sqrt{7} + \sqrt{3})$

35. $\dfrac{\sqrt{5} + \sqrt{3}}{2}$ **37.** $\dfrac{p - \sqrt{pq}}{p - q}$ **39.** $-3(2\sqrt{3} - 3\sqrt{2})$ or $3(3\sqrt{2} - 2\sqrt{3})$ **41.** $-8 - 3\sqrt{7}$ or $-3\sqrt{7} - 8$ **43.** $\dfrac{13\sqrt{6} - 46}{38}$

45. $\dfrac{p + 4\sqrt{p} + 4}{p - 4}$ **47.** $\dfrac{2 - 3\sqrt{2}}{2}$ **49.** $\dfrac{4\sqrt{3}}{3}$ **51.** $\dfrac{\sqrt{10} - \sqrt{2}}{2}$ **53.** $\dfrac{22\sqrt{3}}{3}$ **55.** 0 **57.** $\dfrac{1}{2}$ **59.** $\dfrac{\sqrt{2}}{4}$ **61.** $\dfrac{2\sqrt{3}}{3}$

63. $\sqrt{3} - 2$ **65.** $2(\sqrt{5} - 2)$ **67.** 2 **69.** $\dfrac{\sqrt{3}}{3}$ **71.** $\dfrac{\sqrt[3]{18}}{6}$ **73.** $\dfrac{5 - \sqrt{3}}{22}$ **75.** $\dfrac{\sqrt{6} - \sqrt{2}}{4}$ **77.** $\dfrac{1}{3(\sqrt{2} - 1)}$

79. $\dfrac{x - h}{x + \sqrt{xh}}$ **81.** $\dfrac{(\sqrt{6})^2 + 2 \cdot \sqrt{6} \cdot \sqrt{2} + (\sqrt{2})^2}{4^2} \overset{?}{=} \left(\dfrac{\sqrt{2} + \sqrt{3}}{2}\right)^2$

$\dfrac{6 + 2\sqrt{12} + 2}{16} \overset{?}{=} \dfrac{2 + \sqrt{3}}{4}$

$\dfrac{8 + 2 \cdot 2\sqrt{3}}{16} \overset{?}{=} \dfrac{2 + \sqrt{3}}{4}$

$\dfrac{8 + 4\sqrt{3}}{16} \overset{?}{=} \dfrac{2 + \sqrt{3}}{4}$

$\dfrac{2 + \sqrt{3}}{4} = \dfrac{2 + \sqrt{3}}{4}$

83.

85.

Putting the Concepts Together

1. -5 **2.** $\dfrac{1}{16}$ **3.** $(3x^3)^{1/4}$ **4.** $7\sqrt[5]{z^4}$ **5.** $2x^{1/2}$ or $2\sqrt{x}$ **6.** $c^2 + c^3$ **7.** $\dfrac{a^2}{b^2}$ **8.** $x^{5/8}$ or $\sqrt[8]{x^5}$ **9.** $\dfrac{x^6}{y}$ **10.** $\sqrt{30ab}$ **11.** $10m^2n\sqrt{2n}$

12. $-2xy$ **13.** $\sqrt{3}$ **14.** $11b\sqrt{2b}$ **15.** $y\sqrt[3]{2y}$ **16.** $12x$ **17.** $7\sqrt{x}$ **18.** $14 - 28\sqrt{2} = 14(1 - 2\sqrt{2})$ **19.** $41 - 24\sqrt{2}$

20. $\dfrac{3\sqrt{2}}{16}$ **21.** $-\dfrac{4(\sqrt{3} + 8)}{61}$

Section 9.7

1. even   **3.** True   **5. (a)** 3 **(b)** $\sqrt{14}$ **(c)** 2 **7. (a)** -5 **(b)** -1 **(c)** $-\dfrac{\sqrt{13}}{2}$ **9. (a)** 4 **(b)** $4\sqrt{6}$ **(c)** $\sqrt{6}$

11. (a) 2 **(b)** -2 **(c)** $-2\sqrt[3]{2}$ **13. (a)** $\dfrac{\sqrt{5}}{3}$ **(b)** $\dfrac{\sqrt{2}}{2}$ **(c)** $\dfrac{\sqrt{6}}{3}$ **15. (a)** 1 **(b)** $\sqrt[3]{4}$ **(c)** $\sqrt[3]{3}$

17. $\{x|x \ge 7\}$ or $[7, \infty)$ **19.** $\left\{x\middle|x \ge -\dfrac{7}{2}\right\}$ or $\left[-\dfrac{7}{2}, \infty\right)$ **21.** $\left\{x\middle|x \le \dfrac{4}{3}\right\}$ or $\left(-\infty, \dfrac{4}{3}\right]$

23. $\{z|z \text{ is any real number}\}$ or $(-\infty, \infty)$ **25.** $\left\{p\middle|p \ge \dfrac{2}{7}\right\}$ or $\left[\dfrac{2}{7}, \infty\right)$ **27.** $\{x|x \text{ is any real number}\}$ or $(-\infty, \infty)$

29. $\{x|x > -5\}$ or $(-5, \infty)$ **31.** $\{x|x \le -3 \text{ or } x > 3\}$ or $(-\infty, -3] \cup (3, \infty)$

33. (a) $\{x|x \ge 4\}$ or $[4, \infty)$ **(b)**

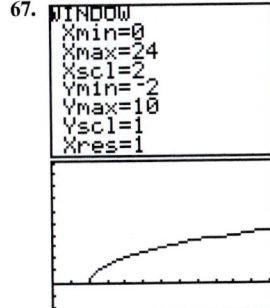

(c) $[0, \infty)$

35. (a) $\{x|x \ge -2\}$ or $[-2, \infty)$ **(b)**

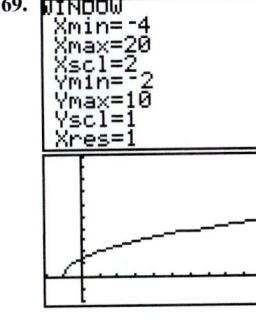

(c) $[0, \infty)$

37. (a) $\{x|x \le 2\}$ or $(-\infty, 2]$ **(b)**

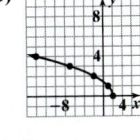

(c) $[0, \infty)$

39. (a) $\{x|x \ge 0\}$ or $[0, \infty)$ **(b)**

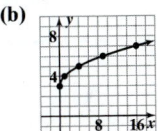

(c) $[3, \infty)$

41. (a) $\{x|x \ge 0\}$ or $[0, \infty)$ **(b)**

(c) $[-4, \infty)$

43. (a) $\{x|x \ge 0\}$ or $[0, \infty)$ **(b)**

(c) $[0, \infty)$

45. (a) $\{x|x \ge 0\}$ or $[0, \infty)$ **(b)**

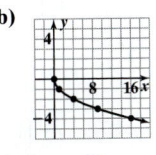

(c) $[0, \infty)$

47. (a) $\{x|x \ge 0\}$ or $[0, \infty)$ **(b)**

(c) $(-\infty, 0]$

49. (a) all real numbers or $(-\infty, \infty)$ **(b)**

(c) $(-\infty, \infty)$

51. (a) all real numbers or $(-\infty, \infty)$ **(b)**

(c) $(-\infty, \infty)$

53. (a) all real numbers or $(-\infty, \infty)$ **(b)**

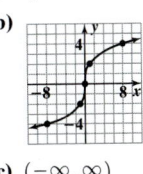

(c) $(-\infty, \infty)$

55. (a) 5 units   **(b)** $\sqrt{17} \approx 4.123$ units   **(c)** $5\sqrt{17} \approx 20.616$ units
57. (a) $4\sqrt{2} \approx 5.657$ square units   **(b)** $4\sqrt{5} \approx 8.944$ square units   **(c)** $2\sqrt{14} \approx 7.483$ square units

59. Shift the graph of $f(x)$ c units to the right (if $c < 0$) or left (if $c > 0$).   **61.** $\dfrac{5}{6}$ **63.** $\dfrac{4x + 1}{x(x + 1)}$ **65.** $\dfrac{7x + 1}{(x - 1)(x + 1)}$ or $\dfrac{7x + 1}{x^2 - 1}$

67.

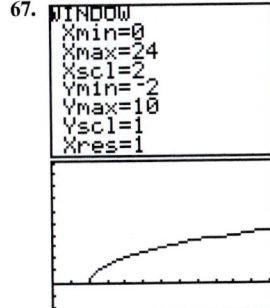

69.

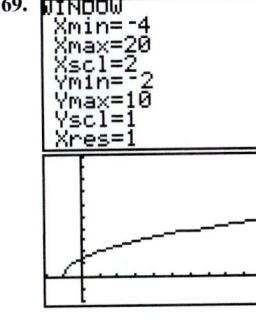

71.

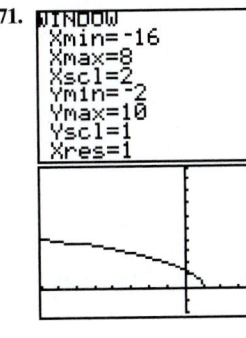

73.

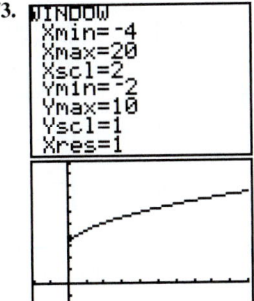

75. 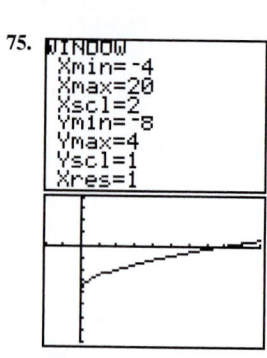 **77.** 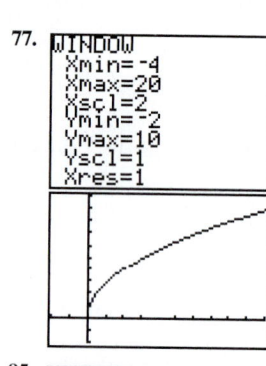 **79.** 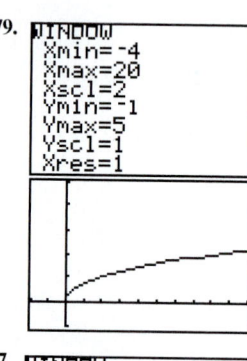 **81.**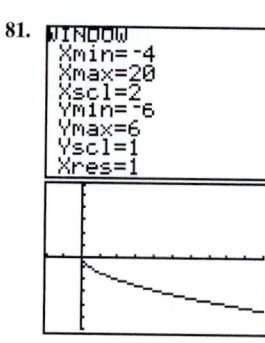

83. 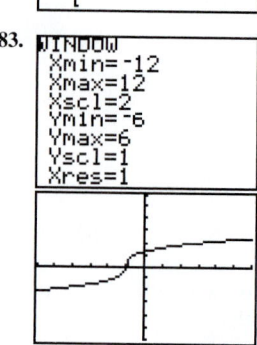 **85.** 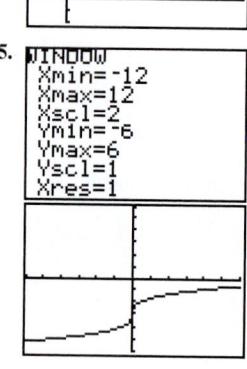 **87.**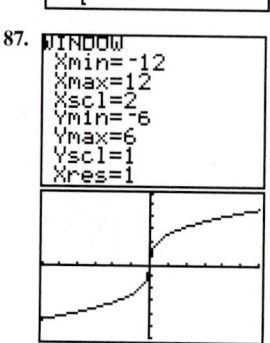

Section 9.8

1. radical equation **3.** False **5.** Answers will vary. **7.** $\{16\}$ **9.** $\{7\}$ **11.** $\{11\}$ **13.** no real solution; $\varnothing$ or $\{\ \}$ **15.** $\{2\}$

17. $\{12\}$ **19.** $\{25\}$ **21.** $\{11\}$ **23.** $\{5\}$ **25.** $\{1\}$ **27.** $\{-5\}$ **29.** $\left\{0, \dfrac{1}{4}\right\}$ **31.** $\{3\}$ **33.** $\{9\}$ **35.** $\{4\}$ **37.** $\{6\}$ **39.** $\{4\}$

41. $\{-3\}$ **43.** $\{-1, 10\}$ **45.** $\{0, 4\}$ **47.** $\{3\}$ **49.** $\{5\}$ **51.** $\{2, 10\}$ **53.** $\{3\}$ **55.** $\{4\}$ **57.** $\{6\}$ **59.** $r = \dfrac{A^2 - P^2}{P^2}$

61. $V = \dfrac{4}{3}\pi r^3$ **63.** $F = \dfrac{q_1 q_2 r^2}{4\pi\varepsilon_0}$ **65.** no real solution; $\varnothing$ or $\{\}$ **67.** $\{4\}$ **69.** $\{2\}$ **71.** $\{9\}$ **73.** $\{-3\}$ **75.** $\{4\}$ **77.** $\{9\}$ **79.** $\{5\}$

81. (a) $(2, 0)$ **(b)** $(3, 1)$ **(c)** $(6, 2)$ **(d)** [graph] **(e)** The equation $f(x) = -1$ has no solution because the graph of the function does not go below the x-axis.

83. (a) $y = 5$ or $y = -1$ **(b)** [graph] **85. (a)** after 2 years **(b)** after 16 years **87. (a)** in the year 2053 **(b)** in the year 2021 **89.** $\left\{-\dfrac{3}{4}, 0\right\}$

91. Raising both sides of the equation to an even power. **93.** $0, -4, 12, \dfrac{2}{3}$, and $1.\overline{56}$ **95.** $0, -4, 12, \dfrac{2}{3}, 1.\overline{56}, \sqrt{2}, \pi$, and $\sqrt[3]{-4}$

97. 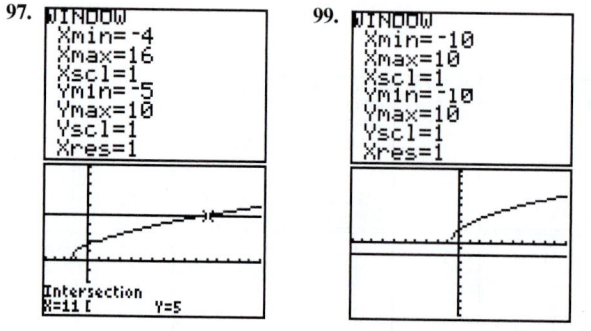 **99.** [graph]

The two graphs do not intersect. Therefore, the equation has no real solution.

Section 9.9

1. imaginary unit **3.** conjugate **5.** False **7.** Answers will vary. **9.** Answers will vary. **11.** $2i$ **13.** $-9i$ **15.** $3\sqrt{5}i$ **17.** $10\sqrt{3}i$

19. $\sqrt{7}i$ **21.** $5 + 7i$ **23.** $-2 - 2\sqrt{7}i$ **25.** $2 + i$ **27.** $\dfrac{1}{3} + \dfrac{\sqrt{2}}{6}i$ **29.** $6 - 2i$ **31.** $-4 + 6i$ **33.** $2 - 5i$ **35.** $3 - 2\sqrt{2}i$

37. $24 + 12i$ **39.** $-5 - 2i$ **41.** $5 + 10i$ **43.** $14 - 22i$ **45.** $34i$ **47.** $-4 + 5\sqrt{2}i$ **49.** $\dfrac{25}{48} + \dfrac{5}{24}i$ **51.** $5 + 12i$ **53.** $-9 + 40i$

55. -6 **57.** $-4\sqrt{5}$ **59.** $84 - 47i$ **61. (a)** $3 - 5i$ **(b)** 34 **63. (a)** $2 + 7i$ **(b)** 53 **65. (a)** $-7 - 2i$ **(b)** 53

67. $\dfrac{1}{3} - \dfrac{1}{3}i$ **69.** $\dfrac{2}{5} + i$ **71.** $\dfrac{6}{5} - \dfrac{3}{5}i$ **73.** $\dfrac{3}{29} - \dfrac{7}{29}i$ **75.** i **77.** $1 + 3i$ **79.** $-\dfrac{1}{5} - \dfrac{7}{5}i$ **81.** i **83.** $-i$ **85.** i **87.** $-i$

89. $-15 - 8i$ **91.** $5 + 12i$ **93.** $\dfrac{2}{3} + i$ **95.** $\dfrac{1}{2} - \dfrac{1}{2}i$ **97.** 12 **99.** $-15 - 20i$ **101.** $-5\sqrt{6}$ **103.** $-\dfrac{1}{5}i$ **105.** $\dfrac{2}{5} + \dfrac{1}{5}i$

107. $-\dfrac{4}{41} - \dfrac{5}{41}i$ **109. (a)** -1 **(b)** $2i$ **111. (a)** $-7 + 6i$ **(b)** $4 - 4i$ **113. (a)** $10 - i$ ohms **(b)** 10 ohms **(c)** -1 ohm

115. (a) 0 **(b)** 0 **117. (a)** 0 **(b)** 0 **(c)** 0

119. For a polynomial with real coefficients, the zeros will be real numbers or will occur in conjugate pairs. If the complex number $a + bi$ is a complex zero of the polynomial, then its conjugate $a - bi$ is also a complex conjugate. **121.** $y^2 - 8y + 16$ **123.** $7 - 24i$

125. 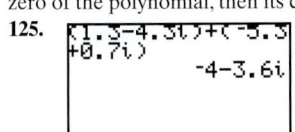 **127.** 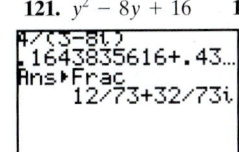 **129.** 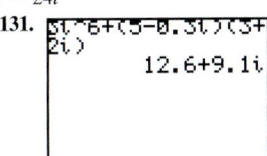 **131.**

Chapter 9 Review

1. $-2, 2$ **2.** $-9, 9$ **3.** -1 **4.** -5 **5.** 0.4 **6.** 0.2 **7.** $\dfrac{5}{4}$ **8.** 6 **9.** 4 **10.** 12 **11.** 21 **12.** 11 **13.** rational: -3

14. irrational; -3.46 **15.** irrational; 3.74 **16.** not a real number **17.** $|4x - 9|$ **18.** $|16m - 25|$ **19.** 7 **20.** -5 **21.** $\dfrac{2}{3}$

22. 3 **23.** 3 **24.** 10 **25.** z **26.** $|5p - 3|$ **27.** 9 **28.** not a real number **29.** -2 **30.** 9 **31.** 128 **32.** -9 **33.** -1331

34. 6 **35.** -4.02 **36.** 2.30 **37.** 4.64 **38.** 1.78 **39.** $(5a)^{1/3}$ **40.** $p^{7/5}$ **41.** $(10z)^{3/4}$ **42.** $(2ab)^{5/6}$ **43.** 64 **44.** $\dfrac{1}{k^{1/4}}$

45. $p^2 \cdot q^6$ or $(p \cdot q^3)^2$ **46.** $\dfrac{2b^{1/20}}{a^{3/10}}$ or $2\left(\dfrac{b}{a^6}\right)^{1/20}$ **47.** $10m^{1/3} + \dfrac{5}{m}$ **48.** $\dfrac{1}{2x^{1/3}}$ **49.** $\sqrt[4]{x^3}$ **50.** $11x^2y^5$ **51.** $m^2\sqrt[6]{m}$ **52.** $\dfrac{1}{\sqrt[3]{c}}$

53. $(3m - 1)^{1/4}(3m^2 - 22m + 9)$ **54.** $\dfrac{(3x + 5)(x - 3)}{(x^2 - 5)^{2/3}}$ **55.** $\sqrt{105}$ **56.** $\sqrt[4]{12a^3b^3}$ **57.** $4\sqrt{5}$ **58.** $-5\sqrt[3]{4}$ **59.** $3m^2n\sqrt[6]{6n}$

60. $p^2|q|\sqrt[4]{50}$ **61.** $8|x^3|\sqrt{y}$ as long as $y \geq 0$ **62.** $(2x + 1)\sqrt{2x + 1}$ as long as $2x + 1 \geq 0$ **63.** $wz\sqrt{w}$ **64.** $3x^2z\sqrt{5yz}$

65. $2a^4b\sqrt[3]{2b^2}$ **66.** $2(x + 1)$ **67.** $3\sqrt{30}$ **68.** $2\sqrt[3]{75}$ **69.** $-2x^2y^3\sqrt[3]{9x}$ **70.** $30xy\sqrt{3xy}$ **71.** $\dfrac{11}{5}$ **72.** $\dfrac{a^2\sqrt{5}}{8b}$ **73.** $\dfrac{\sqrt[3]{6}}{k}$

74. $\dfrac{-2w^5\sqrt[3]{20}}{7}$ **75.** $2h$ **76.** $\dfrac{5b^3}{2a}$ **77.** $\dfrac{-2x^2}{3y}$ **78.** $\dfrac{2n\sqrt{n}}{m}$ **79.** $\sqrt[6]{500}$ **80.** $2\sqrt[12]{2}$ **81.** $8\sqrt[4]{x}$ **82.** $6\sqrt[3]{4y}$ **83.** $5\sqrt{2} - 4\sqrt{3}$

84. $13\sqrt{2}$ **85.** $\sqrt[3]{2z}$ **86.** $17\sqrt[3]{x^2}$ **87.** $7\sqrt{a}$ **88.** $9x\sqrt{3}$ **89.** $4m\sqrt[3]{4m^2y^2}$ **90.** $(y - 1)\sqrt{y - 4}$ **91.** $\sqrt{15} - 3\sqrt{5}$

92. $3\sqrt[3]{5} + \sqrt[3]{20}$ **93.** $7 + \sqrt{5}$ **94.** $42 + 7\sqrt{2} + 6\sqrt{3} + \sqrt{6}$ **95.** -44 **96.** $9\sqrt[3]{x^2} + 5\sqrt[3]{x} - 4$ **97.** $x - 2\sqrt{5x} + 5$

98. $247 + 22\sqrt{10}$ **99.** $2a - b^2$ **100.** $\sqrt[3]{36s^2} - 5\sqrt[3]{6s} - 14$ **101.** $\dfrac{\sqrt{6}}{3}$ **102.** $2\sqrt{3}$ **103.** $\dfrac{4\sqrt{3p}}{p^2}$ **104.** $\dfrac{5\sqrt{2a}}{2a}$ **105.** $-\dfrac{\sqrt{6y}}{3y^2}$

106. $\dfrac{3\sqrt[3]{25}}{5}$ **107.** $-\dfrac{\sqrt[3]{300}}{15}$ **108.** $\dfrac{27\sqrt[5]{4p^2q}}{2pq}$ **109.** $\dfrac{42 + 6\sqrt{6}}{43}$ **110.** $-\dfrac{\sqrt{3} + 9}{26}$ **111.** $\dfrac{\sqrt{3}(3 - \sqrt{2})}{7}$ or $\dfrac{3\sqrt{3} - \sqrt{6}}{7}$ **112.** $\dfrac{k + \sqrt{km}}{k - m}$

113. $\dfrac{7 + 2\sqrt{10}}{3}$ **114.** $\dfrac{9 - 6\sqrt{y} + y}{9 - y}$ or $\dfrac{y - 6\sqrt{y} + 9}{9 - y}$ **115.** $\dfrac{10\sqrt{2} - 4\sqrt{3}}{19}$ **116.** $\dfrac{8\sqrt{2} - 3\sqrt{15}}{7}$ **117.** $\dfrac{25\sqrt{7}}{21}$ **118.** $\dfrac{4 + \sqrt{7}}{9}$

119. (a) 1 **(b)** 2 **(c)** 3 **120. (a)** 0 **(b)** 2 **(c)** 4 **121. (a)** 1 **(b)** -1 **(c)** 2 **122. (a)** 0 **(b)** 2 **(c)** $\dfrac{1}{2}$

123. $\left\{x \mid x \geq \dfrac{5}{3}\right\}$ or $\left[\dfrac{5}{3}, \infty\right)$ **124.** $\{x \mid x$ is any real number$\}$ or $(-\infty, \infty)$ **125.** $\left\{x \mid x \geq -\dfrac{1}{6}\right\}$ or $\left[-\dfrac{1}{6}, \infty\right)$

126. $\{x \mid x$ is any real number$\}$ or $(-\infty, \infty)$ **127.** $\{x \mid x > 2\}$ or $(2, \infty)$ **128.** $\{x \mid x < 0$ or $x \geq 3\}$ or $(-\infty, 0) \cup [3, \infty)$

129. (a) $\{x \mid x \leq 1\}$ or $(-\infty, 1]$ **130. (a)** $\{x \mid x \geq -1\}$ or $[-1, \infty)$ **131. (a)** $\{x \mid x \geq -3\}$ or $[-3, \infty)$ **132. (a)** $\{x \mid x$ is any real number$\}$ or $(-\infty, \infty)$

(b)

(c) $[0, \infty)$ **(c)** $[-2, \infty)$ **(c)** $(-\infty, 0]$ **(c)** $(-\infty, \infty)$

133. $\{169\}$ **134.** $\{-3\}$ **135.** $\left\{\dfrac{89}{3}\right\}$ **136.** no real solution; $\varnothing$ or $\{\ \}$ **137.** $\{0\}$ **138.** $\{25\}$ **139.** $\{-512\}$ **140.** $\{2\}$

141. $\{5\}$ **142.** $\{2\}$ **143.** $\{11\}$ **144.** $\{6\}$ **145.** $\{9\}$ **146.** no real solution; $\varnothing$ or $\{\ \}$ **147.** $\left\{\dfrac{15}{2}\right\}$ **148.** $\{-5, 5\}$

149. $h = \dfrac{3V}{\pi r^2}$ **150.** $v = \dfrac{30}{f_s^3}$ **151.** $\sqrt{29}i$ **152.** $3\sqrt{6}i$ **153.** $14 - 9\sqrt{2}i$ **154.** $2 + \sqrt{5}i$ **155.** $1 - 2i$ **156.** $-5 + 10i$

157. $5 - 7\sqrt{5}i$ **158.** $-5 + 7i$ **159.** $47 + 13i$ **160.** $8 - \dfrac{11}{6}i$ **161.** -9 **162.** $67 - 42i$ **163.** 145 **164.** $27 + 38i$

165. $\dfrac{6}{17} - \dfrac{10}{17}i$ **166.** $-\dfrac{21}{53} - \dfrac{6}{53}i$ **167.** $\dfrac{4}{29} - \dfrac{19}{29}i$ **168.** $\dfrac{1}{2} + \dfrac{7}{2}i$ **169.** $-i$ **170.** i

Chapter 9 Test

1. $\dfrac{1}{7}$ **2.** $6y\sqrt[12]{x^8 y^3}$ **3.** $2a^5 b^4 \sqrt[5]{4a^3 b}$ **4.** $\sqrt{39mn}$ **5.** $4x^3 y^2 \sqrt{2x}$ **6.** $\dfrac{3a}{2b^2}$ **7.** $(x + 6)\sqrt{5x}$ **8.** $a\sqrt{b}$ **9.** $33 - 5\sqrt{x} - 2x$

10. $\dfrac{-\sqrt{2}}{18}$ **11.** $5 - 2\sqrt{5}$ **12.** (a) 1 (b) 3 **13.** $\left\{x \mid x \le \dfrac{5}{3}\right\}$ or $\left(-\infty, \dfrac{5}{3}\right]$ **14.** (a) $\{x \mid x \ge 0\}$ or $[0, \infty)$ (b)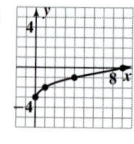

(c) $[-3, \infty)$ **15.** $\{13\}$ **16.** $\{3\}$ **17.** $\{2\}$ **18.** $17 - 13i$ **19.** $29 - 2i$ **20.** $\dfrac{73}{265} - \dfrac{89}{265}i$

Cumulative Review 1–9

1. $\dfrac{9}{2}$ **2.** $x + 7y + 6$ **3.** $\{4\}$ **4.** $(-\infty, 2]$ **5.** (a) 19 (b) 29 **6.** $\{x \mid x \ne 4, -2\}$ **7.** (a) 3500 games (b) $\$120$

8. $y = -2x + 4$ **9.** 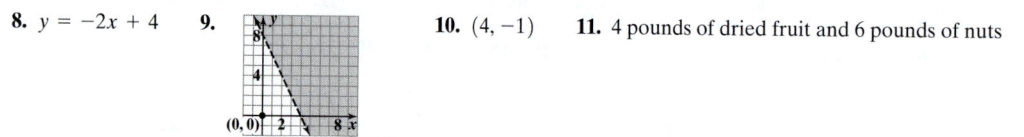 **10.** $(4, -1)$ **11.** 4 pounds of dried fruit and 6 pounds of nuts **12.** -96

13. $8x^3 - 2x^2 - 3x + 10$ **14.** $8x^3 - 20x + 9$ **15.** $3x^2 + 2x - 6 + \dfrac{7}{2x^2 + 3x - 5}$ **16.** $4(2x + 3)(x - 7)$

17. $\dfrac{x^2 + 6x + 3}{(x - 3)(x + 2)}$ **18.** $\{-1, 4\}$ **19.** $\{x \mid x < 4 \text{ or } x \ge 10\}$, or $(-\infty, 4) \cup [10, \infty)$

20. 2.4 hours **21.** $\dfrac{5a^2}{b}$ **22.** $\left\{x \mid x \le \dfrac{8}{3}\right\}$ or $\left(-\infty, \dfrac{8}{3}\right]$ **23.** $\{2\}$ **24.** $-\dfrac{21}{50} + \dfrac{3}{50}i$ **25.** $\dfrac{8 + 2\sqrt{11}}{5}$

Chapter 10

Section 10.1

1. $\sqrt{p}; -\sqrt{p}$ **3.** hypotenuse; legs **5.** False **7.** Answers will vary. **9.** $\{-10, 10\}$ **11.** $\{-5\sqrt{2}, 5\sqrt{2}\}$ **13.** $\{-5i, 5i\}$

15. $\left\{-\dfrac{\sqrt{5}}{2}, \dfrac{\sqrt{5}}{2}\right\}$ **17.** $\{-2\sqrt{2}, 2\sqrt{2}\}$ **19.** $\{-4, 4\}$ **21.** $\left\{-\dfrac{2\sqrt{6}}{3}, \dfrac{2\sqrt{6}}{3}\right\}$ **23.** $\{-2i, 2i\}$ **25.** $\{-11, 5\}$

27. $\{1 - 3\sqrt{2}i, 1 + 3\sqrt{2}i\}$ **29.** $\{-5 - \sqrt{3}, -5 + \sqrt{3}\}$ **31.** $\left\{-\dfrac{4}{3}, \dfrac{2}{3}\right\}$ **33.** $\left\{\dfrac{2}{3} - \dfrac{\sqrt{5}}{3}, \dfrac{2}{3} + \dfrac{\sqrt{5}}{3}\right\}$ **35.** $\{-13, 5\}$

37. $x^2 + 10x + 25; (x + 5)^2$ **39.** $z^2 - 18z + 81; (z - 9)^2$ **41.** $y^2 + 7y + \dfrac{49}{4}; \left(y + \dfrac{7}{2}\right)^2$ **43.** $w^2 + \dfrac{1}{2}w + \dfrac{1}{16}; \left(w + \dfrac{1}{4}\right)^2$

45. $q^2 - \dfrac{3}{5}q + \dfrac{9}{100}; \left(q - \dfrac{3}{10}\right)^2$ **47.** $\{-2, 7\}$ **49.** $\{-6, 2\}$ **51.** $\{2 - \sqrt{3}, 2 + \sqrt{3}\}$ **53.** $\{-4 - \sqrt{7}, -4 + \sqrt{7}\}$ **55.** $\{2 - i, 2 + i\}$

57. $\left\{-\dfrac{5}{2} - \dfrac{\sqrt{33}}{2}, -\dfrac{5}{2} + \dfrac{\sqrt{33}}{2}\right\}$ **59.** $\left\{\dfrac{3}{2} - \dfrac{\sqrt{17}}{2}, \dfrac{3}{2} + \dfrac{\sqrt{17}}{2}\right\}$ **61.** $\{4 - \sqrt{19}, 4 + \sqrt{19}\}$ **63.** $\left\{\dfrac{1}{2} - \dfrac{\sqrt{11}i}{2}, \dfrac{1}{2} + \dfrac{\sqrt{11}i}{2}\right\}$

65. $\left\{-\dfrac{3}{2}, 4\right\}$ **67.** $\left\{1 - \dfrac{\sqrt{3}}{3}, 1 + \dfrac{\sqrt{3}}{3}\right\}$ **69.** $\left\{\dfrac{5}{4} - \dfrac{\sqrt{17}}{4}, \dfrac{5}{4} + \dfrac{\sqrt{17}}{4}\right\}$ **71.** $\left\{-1 - \dfrac{\sqrt{6}i}{2}, -1 + \dfrac{\sqrt{6}i}{2}\right\}$

73. $\left\{-\dfrac{1}{6} - \dfrac{\sqrt{61}}{6}, -\dfrac{1}{6} + \dfrac{\sqrt{61}}{6}\right\}$ **75.** 10 **77.** 20 **79.** $5\sqrt{2}; 7.07$ **81.** 2 **83.** $2\sqrt{34}; 11.66$ **85.** $b = 4\sqrt{3} \approx 6.93$

87. $a = 4\sqrt{5} \approx 8.94$ **89.** $x = -3$ or $x = 9$ **91.** $x = -2 \pm 3\sqrt{2}$ **93.** approximately 8.944 units **95.** approximately 104.403 yards

97. approximately 31.623 feet **99.** (a) approximately 22.913 feet (b) 15 feet **101.** (a) 1 second (b) approximately 1.732 seconds (c) 2 seconds **103.** approximately 9.54% **105.** The triangle is a right triangle; the hypotenuse is 17. **107.** The triangle is not a right triangle.

109. $c^2 = (m^2 + n^2)^2 = m^4 + 2m^2 n^2 + n^4$

$a^2 + b^2 = (m^2 - n^2)^2 + (2mn)^2$

$= m^4 - 2m^2 n^2 + n^4 + 4m^2 n^2$

$= m^4 + 2m^2 n^2 + n^4$

Because c^2 and $a^2 + b^2$ result in the same expression, a, b, and c are the lengths of the sides of a right triangle.

111. $\{-4, 9\}$ **113.** $\left\{-1, \dfrac{1}{2}\right\}$

115. In both cases, the simpler equations are linear.

Section 10.2

1. $\dfrac{-b \pm \sqrt{b^2 - 4ac}}{2a}$ **3.** negative   **5.** False   **7.** Answers may vary.

9. The solutions to the equation $ax^2 + bx + c = 0, a \neq 0$, are given by $x = \dfrac{-b \pm \sqrt{b^2 - 4ac}}{2a}$.   **11.** $\{-2, 6\}$ **13.** $\left\{-\dfrac{3}{2}, \dfrac{5}{3}\right\}$

15. $\left\{1 - \dfrac{\sqrt{3}}{2}, 1 + \dfrac{\sqrt{3}}{2}\right\}$ **17.** $\left\{1 - \dfrac{2\sqrt{3}}{3}, 1 + \dfrac{2\sqrt{3}}{3}\right\}$ **19.** $\left\{-\dfrac{1}{3} - \dfrac{\sqrt{13}}{3}, -\dfrac{1}{3} + \dfrac{\sqrt{13}}{3}\right\}$ **21.** $\{1 - \sqrt{6}i, 1 + \sqrt{6}i\}$

23. $\left\{\dfrac{1}{2} - \dfrac{\sqrt{13}}{2}i, \dfrac{1}{2} + \dfrac{\sqrt{13}}{2}i\right\}$ **25.** $\left\{\dfrac{1}{4} - \dfrac{\sqrt{5}}{4}, \dfrac{1}{4} + \dfrac{\sqrt{5}}{4}\right\}$ **27.** $\left\{-\dfrac{2}{3} - \dfrac{\sqrt{7}}{3}, -\dfrac{2}{3} + \dfrac{\sqrt{7}}{3}\right\}$ **29.** 21; two irrational solutions

31. -56; two complex solutions that are not real   **33.** 0; one repeated real solution   **35.** -8; two complex solutions that are not real

37. 44; two irrational solutions   **39.** $\left\{\dfrac{5}{2} - \dfrac{\sqrt{5}}{2}, \dfrac{5}{2} + \dfrac{\sqrt{5}}{2}\right\}$ **41.** $\left\{-\dfrac{8}{3}, 1\right\}$ **43.** $\left\{-\dfrac{7}{2}, 5\right\}$ **45.** $\{-1 - \sqrt{7}i, -1 + \sqrt{7}i\}$

47. $\left\{-\dfrac{3}{5}, 1\right\}$ **49.** $\left\{\dfrac{1}{7} - \dfrac{\sqrt{29}}{7}, \dfrac{1}{7} + \dfrac{\sqrt{29}}{7}\right\}$ **51.** $\{-4, 4\}$ **53.** $\left\{\dfrac{1}{4} - \dfrac{1}{4}i, \dfrac{1}{4} + \dfrac{1}{4}i\right\}$ **55.** $\left\{-\dfrac{2}{3}\right\}$ **57.** $\left\{-\dfrac{1}{3} - \dfrac{2\sqrt{7}}{3}, -\dfrac{1}{3} + \dfrac{2\sqrt{7}}{3}\right\}$

59. $\{2 - \sqrt{13}, 2 + \sqrt{13}\}$ **61.** $\{1 - \sqrt{5}, 1 + \sqrt{5}\}$ **63.** $\left\{\dfrac{1}{4} - \dfrac{3\sqrt{7}}{4}i, \dfrac{1}{4} + \dfrac{3\sqrt{7}}{4}i\right\}$ **65. (a)** $x = -7$ or $x = 3$

(b) $x = -4$ or $x = 0$ **67. (a)** $n = -1 \pm \dfrac{\sqrt{6}}{2}$ **(b)** $x = -1 \pm \dfrac{\sqrt{2}}{2}$ **69. (a)** $x = -\dfrac{1}{3} \pm \dfrac{\sqrt{5}}{3}i$ **(b)** $x = -\dfrac{1}{3} \pm \dfrac{\sqrt{7}}{3}$

71. $x = 3$; the three sides measure 3, 4, and 5 units   **73.** Either $x = 1$ and the three sides measure 3, 4, and 5 units, or $x = 5$ and the three sides measure 7, 24, and 25 units.

75. $-2 + 2\sqrt{11}$ inches by $2 + 2\sqrt{11}$ inches, which is approximately 4.633 inches by 8.633 inches.   **77.** The base is $\dfrac{3}{2} + \dfrac{\sqrt{209}}{2}$ inches, which is approximately 8.728 inches; the height is $-\dfrac{3}{2} + \dfrac{\sqrt{209}}{2}$ inches, which is approximately 5.728 inches.

79. (a) $R(17) = 1161.1$; if 17 pairs of sunglasses are sold per week, then the company's revenue will be $1161.10. $R(25) = 1687.5$; if 25 pairs of sunglasses are sold per week, then the company's revenue will be $1687.50.
(b) either 200 or 500 pairs of sunglasses
(c) 350 pairs of sunglasses

81. (a) after approximately 0.6 seconds and after approximately 3.8 seconds
(b) after approximately 1.3 seconds and after approximately 3.0 seconds
(c) No; the solutions to the equation are complex solutions that are not real.

83. 12 inches   **85. (a)** ages 25 and 68   **(b)** ages 30 and 63   **87.** approximately 4.3 miles per hour   **89.** approximately 4.6 hours

91. By the quadratic formula, the solutions of the equation $ax^2 + bx + c = 0$ are $x = \dfrac{-b - \sqrt{b^2 - 4ac}}{2a}$ and $x = \dfrac{-b + \sqrt{b^2 - 4ac}}{2a}$.

The sum of these two solutions is

$$\dfrac{-b - \sqrt{b^2 - 4ac}}{2a} + \dfrac{-b + \sqrt{b^2 - 4ac}}{2a} = \dfrac{-2b}{2a} = \dfrac{b}{a}.$$

93. The solutions of $ax^2 + bx + c = 0$ are $x = \dfrac{-b \pm \sqrt{b^2 - 4ac}}{2a}$.

The solutions of $ax^2 - bx + c = 0$ are

$$x = \dfrac{-(-b) \pm \sqrt{(-b)^2 - 4ac}}{2a} = \dfrac{b \pm \sqrt{b^2 - 4ac}}{2a}.$$

Now, the negatives of the solutions to $ax^2 - bx + c = 0$

are $-\left(\dfrac{b \pm \sqrt{b^2 - 4ac}}{2a}\right) = \dfrac{-b \mp \sqrt{b^2 - 4ac}}{2a} = \dfrac{-b \pm \sqrt{b^2 - 4ac}}{2a}$

which are the solutions to $ax^2 + bx + c = 0$.

95. (a)

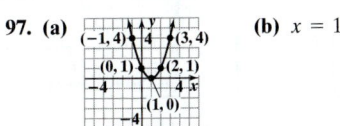

(b) $x = -1$ or $x = -2$

(c) The x-intercepts of the function $f(x) = x^2 + 3x + 2$ are -2 and -1, which are the same as the solutions of the equation $x^2 + 3x + 2 = 0$.

97. (a)

(b) $x = 1$

(c) The x-intercept of the function $g(x) = x^2 - 2x + 1$ is 1, which is the same as the solution of the equation $x^2 - 2x + 1 = 0$.

99. The discriminant is 37; the equation has two irrational solutions. This conclusion based on the discriminant is apparent in the graph because the graph has two x-intercepts.

101. The discriminant is -7; the equation has two complex solutions that are not real. This conclusion based on the discriminant is apparent in the graph because the graph has no x-intercept.

103. (a) $x = -3$ or $x = 8$
(b) The x-intercepts are -3 and 8.

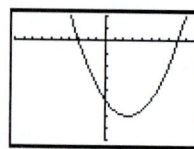

The x-intercepts of $y = x^2 - 5x - 24$ are the same as the solutions of $x^2 - 5x - 24 = 0$.

105. (a) $x = 3$
(b) The x-intercept is 3.

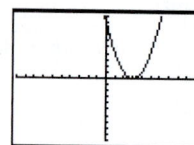

The x-intercept of $y = x^2 - 6x + 9$ is the same as the solution of $x^2 - 6x + 9 = 0$.

107. (a) $x = -\dfrac{5}{2} \pm \dfrac{\sqrt{7}}{2}i$

(b) The graph has no x-intercepts.

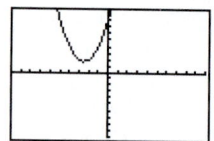

$y = x^2 + 5x + 8$ has no x-intercepts, and the solutions of $x^2 + 5x + 8 = 0$ are not real.

Section 10.3

1. equation quadratic in form **3.** True **5.** Answers will vary. **7.** $3x - 2$ **9.** $\{-2, -1, 1, 2\}$ **11.** $\{-3i, -2i, 2i, 3i\}$

13. $\left\{-\dfrac{1}{2}, -2, \dfrac{1}{2}, 2\right\}$ **15.** $\{-\sqrt{2}, -\sqrt{3}, \sqrt{2}, \sqrt{3}\}$ **17.** $\{10, 2\}$ **19.** $\{-2, -3, 2, 3\}$ **21.** $\{-2i, -\sqrt{7}i, 2i, \sqrt{7}i\}$

23. $\{16\}$ **25.** no solution; $\emptyset$ or $\{\ \}$ **27.** $\left\{\dfrac{1}{4}\right\}$ **29.** $\left\{-\dfrac{1}{7}, \dfrac{1}{4}\right\}$ **31.** $\left\{-\dfrac{2}{3}, \dfrac{5}{2}\right\}$ **33.** $\{-64, 1\}$ **35.** $\{-1, 8\}$ **37.** $\{25\}$ **39.** $\left\{\dfrac{1}{3}, \dfrac{1}{2}\right\}$

41. $\left\{-\dfrac{11}{5}, -1\right\}$ **43.** $\left\{1, 3, -\dfrac{1}{2} - \dfrac{\sqrt{3}}{2}i, -\dfrac{3}{2} - \dfrac{3\sqrt{3}}{2}i, -\dfrac{1}{2} + \dfrac{\sqrt{3}}{2}i, -\dfrac{3}{2} + \dfrac{3\sqrt{3}}{2}i\right\}$ **45.** $\{-2, 4\}$ **47.** $\{-2i, -2\sqrt{2}, 2i, 2\sqrt{2}\}$ **49.** $\{16\}$

51. $\{-3, -2i, 3, 2i\}$ **53.** $\left\{-\dfrac{10}{3}, -2\right\}$ **55.** $\{9, 16\}$ **57.** $\left\{\dfrac{1}{2}, 5\right\}$ **59. (a)** $0, -\sqrt{7}i, \sqrt{7}i$ **(b)** $-\sqrt{6}i, \sqrt{6}i, -i, i$ **61. (a)** $0, -\sqrt{3}, \sqrt{3}$

(b) $-\sqrt{2}i, \sqrt{2}i, -\sqrt{5}, \sqrt{5}$ **63. (a)** $-1, \dfrac{1}{6}$ **(b)** $-\dfrac{1}{2}, \dfrac{1}{7}$ **65.** $-\sqrt{7}i, \sqrt{7}i, -\sqrt{2}i$ and $\sqrt{2}i$ **67.** $\dfrac{81}{4}$ **69.** $-\dfrac{8}{3}, -2$

71. (a) $x = 2$ or $x = 3$ **(b)** $x = 5$ or $x = 6$; comparing these solutions to those in part (a), we note that $5 = 2 + 3$ and $6 = 3 + 3$.
(c) $x = 0$ or $x = 1$; comparing these solutions to those in part (a), we note that $0 = 2 - 2$ and $1 = 3 - 2$.
(d) $x = 7$ or $x = 8$; comparing these solutions to those in part (a), we note that $7 = 2 + 5$ and $8 = 3 + 5$.
(e) The solution set of the equation $(x - a)^2 - 5(x - a) + 6 = 0$ is $\{2 + a, 3 + a\}$.

73. (a) $x = \dfrac{1}{2}$ or $x = 1$ **(b)** $x = \dfrac{5}{2}$ or $x = 3$; comparing these solutions to those in part (a), we note that $\dfrac{5}{2} = \dfrac{1}{2} + 2$ and $3 = 1 + 2$.

(c) $x = \dfrac{11}{2}$ or $x = 6$; comparing these solutions to those in part (a), we note that $\dfrac{11}{2} = \dfrac{1}{2} + 5$ and $6 = 1 + 5$.

(d) For $f(x) = 2x^2 - 3x + 1$, the zeros of $f(x - a)$ are $\dfrac{1}{2} + a$ and $1 + a$

75. (a) $R(1990) = 3000$; the revenue in 1990 was \$3,000 thousand (or \$3,000,000)

(b) $x = 2000$; in the year 2000, revenue was \$3,065 thousand (or \$3,065,000) **(c)** 2015 **77.** $x = \pm\dfrac{\sqrt{10 + 2\sqrt{17}}}{2}i$ or $x = \pm\dfrac{\sqrt{10 - 2\sqrt{17}}}{2}i$

79. $x = \pm\dfrac{3\sqrt{2}}{2}$ **81.** $x^2 - 2x + 4$ **83.** $p^{-2} + 4p^{-1} + 9$ **85.** $5\sqrt[3]{2a} - 4a\sqrt[3]{2a} = (5 - 4a)\sqrt[3]{2a}$

87. Let $Y_1 = x^4 + 5x^2 - 14$.

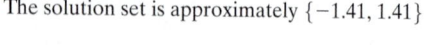

The solution set is approximately $\{-1.41, 1.41\}$.

89. Let $Y_1 = 2(x - 2)^2$ and $Y_2 = 5(x - 2) + 1$.

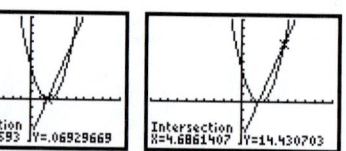

The solution set is approximately $\{1.81, 4.69\}$.

91. Let $Y_1 = x - 5\sqrt{x}$ and $Y_2 = -3$.

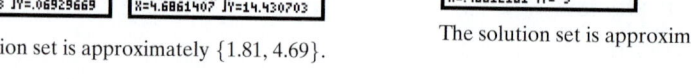

The solution set is approximately $\{0.49\}$.

93. (a) $Y_1 = x^2 - 5x - 6$

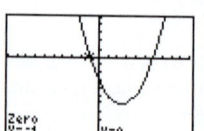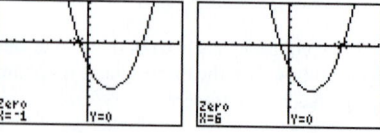

The x-intercepts are -1 and 6.

(b) $Y_1 = (x + 2)^2 - 5(x + 2) - 6$

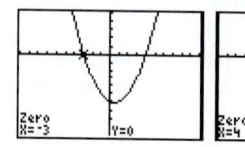

The x-intercepts are -3 and 4.

(c) $Y_1 = (x + 5)^2 - 5(x + 5) - 6$

(d) The x-intercepts of the graph of $y = f(x) = x^2 - 5x - 6$ are -1 and 6. The x-intercepts of the graph of $y = f(x + a) = (x + a)^2 - 5(x + a) - 6$ are $-1 - a$ and $6 - a$.

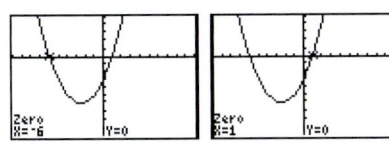

The x-intercepts are -6 and 1.

Putting the Concepts Together

1. $z^2 + 10z + 25 = (z + 5)^2$ **2.** $x^2 + 7x + \dfrac{49}{4} = \left(x + \dfrac{7}{2}\right)^2$ **3.** $n^2 - \dfrac{1}{4}n + \dfrac{1}{64} = \left(n - \dfrac{1}{8}\right)^2$ **4.** $\left\{\dfrac{1}{2}, \dfrac{5}{2}\right\}$ **5.** $\left\{4 - 2\sqrt{3}, 4 + 2\sqrt{3}\right\}$

6. $\left\{3 - \sqrt{2}, 3 + \sqrt{2}\right\}$ **7.** $\left\{-\dfrac{4\sqrt{5}}{7}, \dfrac{4\sqrt{5}}{7}\right\}$ **8.** $\left\{4 - \sqrt{10}, 4 + \sqrt{10}\right\}$ **9.** $\left\{-1 - \dfrac{\sqrt{3}}{3}i, -1 + \dfrac{\sqrt{3}}{3}i\right\}$ **10.** $\left\{-2 - \dfrac{\sqrt{42}}{3}, -2 + \dfrac{\sqrt{42}}{3}\right\}$

11. $b^2 - 4ac = 0$; the quadratic equation will have one repeated real solution.
12. $b^2 - 4ac = 60$; the quadratic equation will have two irrational solutions.
13. $b^2 - 4ac = -4$; the quadratic equation will have two complex solutions that are not real.

14. $c = \sqrt{116} = 2\sqrt{29}$ **15.** $\left\{\dfrac{9}{4}\right\}$ **16.** $\left\{\dfrac{1}{6}, -\dfrac{1}{3}\right\}$ **17.** Revenue will be \$12,000 when either 150 microwaves or 200 microwaves are sold.

18. The speed of the wind was approximately 52.9 miles per hour.

Section 10.4

1. quadratic function   **3.** y; a; vertically stretched; vertically compressed   **5.** False

7. the vertex; if $a > 0$, the graph opens up and the vertex is the low point, if $a < 0$, the graph opens down and the vertex is the high point.

9. (I) (D)   (II) (A)   (III) (C)   (IV) (B)   **11.** shift 10 units to the left   **13.** shift 12 units up

15. shift 5 units to the right and vertically stretch by a factor of 2 (multiply the y-coordinates by 2)

17. shift 5 units to the left, multiply the y-coordinates by -3 (which means it opens down and is stretched vertically by a factor of 3), and shift up 8 units

19. 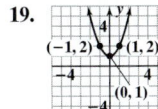   **21.** 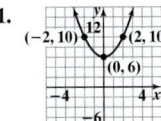   **23.** 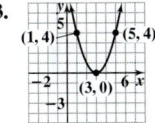   **25.** 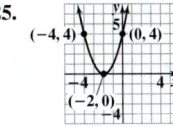   **27.** 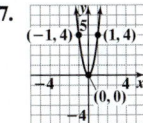   **29.**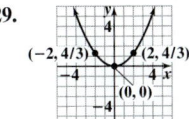

31. 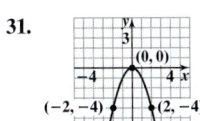   **33.** 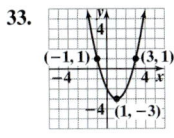   **35.** 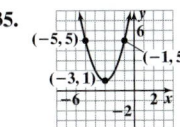   **37.** 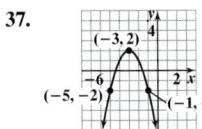   **39.** 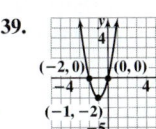   **41.**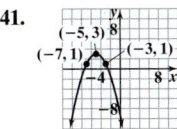

43. $f(x) = (x + 1)^2 - 5$

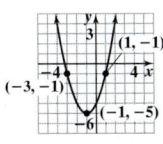
vertex is $(-1, -5)$; axis of symmetry is $x = -1$; domain is the set of all real numbers, $(-\infty, \infty)$; range is $\{y \,|\, y \geq -5\}$, or $[-5, \infty)$

45. $g(x) = (x - 2)^2 + 4$

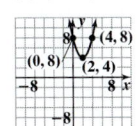
vertex is $(2, 4)$; axis of symmetry is $x = 2$; domain is the set of all real numbers, $(-\infty, \infty)$; range is $\{y \,|\, y \geq 4\}$, $[4, \infty)$

47. $f(x) = (x + 3)^2 - 25$

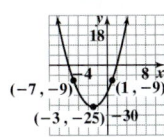
vertex is $(-3, -25)$; axis of symmetry is $x = -3$; domain is the set of all real numbers $(-\infty, \infty)$; range is $\{y \,|\, y \geq -25\}$, or $[-25, \infty)$

49. $F(x) = \left(x + \dfrac{1}{2}\right)^2 - \dfrac{49}{4}$

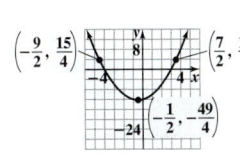
vertex is $\left(-\dfrac{1}{2}, -\dfrac{49}{4}\right)$; axis of symmetry is $x = -\dfrac{1}{2}$ domain is the set of all real numbers, or $(-\infty, \infty)$; range is $\left\{y \,|\, y \geq -\dfrac{49}{4}\right\}$, or $\left[-\dfrac{49}{4}, \infty\right)$

51. $H(x) = 2(x - 1)^2 - 3$

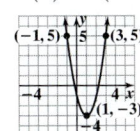
vertex is $(1, -3)$; axis of symmetry is $x = 1$; domain is the set of all real numbers, $(-\infty, \infty)$; range is $\{y \,|\, y \geq -3\}$, or $[-3, \infty)$

53. $P(x) = 3(x + 2)^2 + 1$

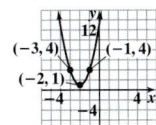
vertex is $(-2, 1)$; axis of symmetry is $x = -2$; domain is the set of all real numbers, or $(-\infty, \infty)$; range is $\{y \,|\, y \geq 1\}$, or $[1, \infty)$

55. $F(x) = -(x + 5)^2 + 4$

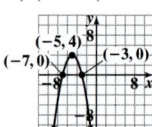 vertex is $(-5, 4)$; axis of symmetry is $x = -5$; domain is the set of all real numbers, or $(-\infty, \infty)$; range is $\{y \mid y \leq 4\}$, or $(-\infty, 4]$

57. $g(x) = -(x - 3)^2 + 8$

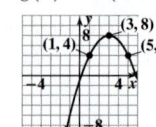 vertex is $(3, 8)$; axis of symmetry is $x = 3$; domain is the set of all real numbers, or $(-\infty, \infty)$; range is $\{y \mid y \leq 8\}$, or $(-\infty, 8]$

59. $H(x) = -2(x - 2)^2 + 4$

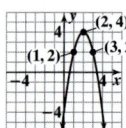 vertex is $(2, 4)$; axis of symmetry is $x = 2$; domain is the set of all real numbers, or $(-\infty, \infty)$; range is $\{y \mid y \leq 4\}$, or $(-\infty, 4]$

61. $f(x) = \dfrac{1}{3}(x - 3)^2 + 1$

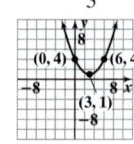 vertex is $(3, 1)$; axis of symmetry is $x = 1$; domain is the set of all real numbers, or $(-\infty, \infty)$; range is $\{y \mid y \geq 1\}$, or $[1, \infty)$

63. $G(x) = -12\left(x + \dfrac{1}{2}\right)^2 + 4$ 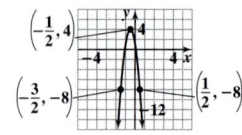 vertex is $\left(-\dfrac{1}{2}, 4\right)$; axis of symmetry is $x = -\dfrac{1}{2}$; domain is the set of all real numbers, or $(-\infty, \infty)$; range is $\{y \mid y \leq 4\}$, or $(-\infty, 4]$

65. Answers may vary. One possibility: $y = f(x) = (x - 3)^2$ **67.** Answers may vary. One possibility: $y = f(x) = (x + 3)^2 + 1$

69. Answers may vary. One possibility: $y = f(x) = -(x - 5)^2 - 1$ **71.** $y = f(x) = 4(x - 9)^2 - 6$

73. $y = f(x) = -\dfrac{1}{3}x^2 + 6$ **75.** $f(x) = (x + 1)^2 - 3$ **77.** $f(x) = -2(x - 3)^2 + 7$ **79.** $f(x) = (x + 4)^2$ **81.** No; explanations may vary.

83. $29 + \dfrac{1}{12}$ **85.** $x^3 - 5x^2 + 4x - 2 + \dfrac{-2}{2x - 1}$

87.

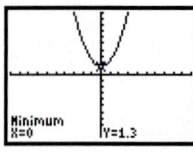

Vertex: $(0, 1.3)$
Axis of symmetry: $x = 0$
Range: $\{y \mid y \geq 1.3\} = [1.3, \infty)$

89.

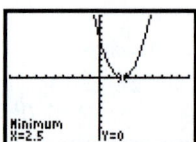

Vertex: $(2.5, 0)$
Axis of symmetry: $x = 2.5$
Range: $\{y \mid y \geq 0\} = [0, \infty)$

91.

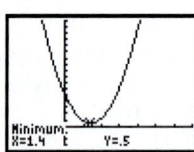

Vertex: $(1.4, 0.5)$
Axis of symmetry: $x = 1.4$
Range: $\{y \mid y \geq 0.5\} = [0.5, \infty)$

93.

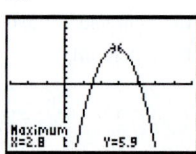

Vertex: $(2.8, 5.9)$
Axis of symmetry: $x = 2.8$
Range: $\{y \mid y \leq 5.9\} = (-\infty, 5.9]$

Section 10.5

1. $x = -\dfrac{b}{2a}$ **3.** minimum value **5.** False **7.** Answers may vary. **9.** the discriminant is positive; there are two distinct x-intercepts: 8 and -2

11. the discriminant is negative; there are no x-intercepts **13.** the discriminant is zero; there is one x-intercept: $-\dfrac{1}{2}$

15. the discriminant is positive; there are two distinct x-intercepts: approximately -0.39 and approximately 0.64

17.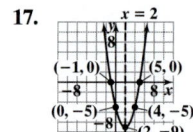

Domain: $\{x \mid x$ is any real number$\}$ or $(-\infty, \infty)$
Range: $\{y \mid y \geq -9\}$ or $[-9, \infty)$

19.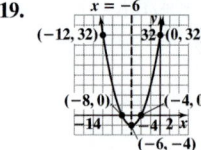

Domain: $\{x \mid x$ is any real number$\}$ or $(-\infty, \infty)$
Range: $\{y \mid y \geq -4\}$ or $[-4, \infty)$

21.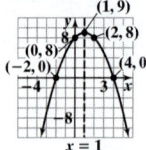

Domain: $\{x \mid x$ is any real number$\}$ or $(-\infty, \infty)$
Range: $\{y \mid y \leq 9\}$ or $(-\infty, 9]$

23.

Domain: $\{x \mid x$ is any real number$\}$ or $(-\infty, \infty)$
Range: $\{y \mid y \geq 0\}$ or $[0, \infty)$

25.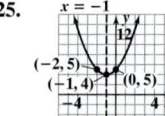

Domain: $\{x \mid x$ is any real number$\}$ or $(-\infty, \infty)$
Range: $\{y \mid y \geq 4\}$ or $[4, \infty)$

27.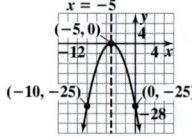

Domain: $\{x \mid x$ is any real number$\}$ or $(-\infty, \infty)$
Range: $\{y \mid y \leq 0\}$ or $(-\infty, 0]$

29.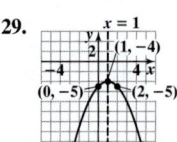

Domain: $\{x \mid x$ is any real number$\}$ or $(-\infty, \infty)$
Range: $\{y \mid y \leq -4\}$ or $(-\infty, -4]$

31.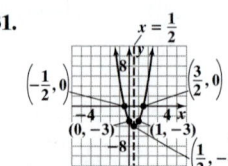

Domain: $\{x \mid x$ is any real number$\}$ or $(-\infty, \infty)$
Range: $\{y \mid y \geq -4\}$ or $[-4, \infty)$

33.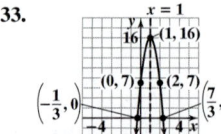

Domain: $\{x \mid x$ is any real number$\}$ or $(-\infty, \infty)$
Range: $\{y \mid y \leq 16\}$ or $(-\infty, 16]$

35.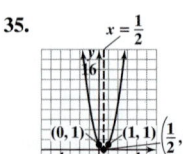

Domain: $\{x \mid x$ is any real number$\}$ or $(-\infty, \infty)$
Range: $\{y \mid y \geq 0\}$ or $[0, \infty)$

37.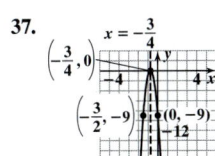

Domain: $\{x \mid x$ is any real number$\}$ or $(-\infty, \infty)$
Range: $\{y \mid y \leq 0\}$ or $(-\infty, 0]$

39.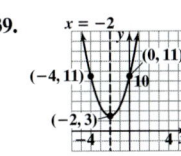

Domain: $\{x \mid x$ is any real number$\}$ or $(-\infty, \infty)$
Range: $\{y \mid y \geq 3\}$ or $[3, \infty)$

41.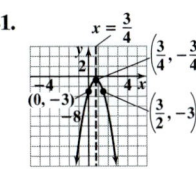

Domain: $\{x \mid x$ is any real number$\}$ or $(-\infty, \infty)$
Range: $\left\{ y \mid y \leq -\dfrac{3}{4} \right\}$ or $\left(-\infty, -\dfrac{3}{4} \right]$

43.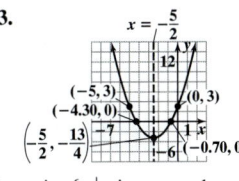

Domain: $\{x \mid x$ is any real number$\}$ or $(-\infty, \infty)$
Range: $\left\{ y \mid y \geq -\dfrac{13}{4} \right\}$ or $\left[-\dfrac{13}{4}, \infty \right)$

45.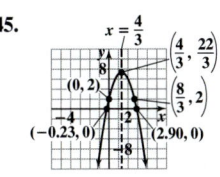

Domain: $\{x \mid x$ is any real number$\}$ or $(-\infty, \infty)$
Range: $\left\{ y \mid y \leq \dfrac{22}{3} \right\}$ or $\left(-\infty, \dfrac{22}{3} \right]$

47.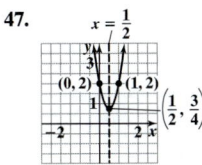

Domain: $\{x \mid x$ is any real number$\}$ or $(-\infty, \infty)$
Range: $\left\{ y \mid y \geq \dfrac{3}{4} \right\}$ or $\left[\dfrac{3}{4}, \infty \right)$

49.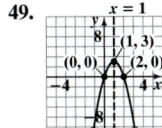

Domain: $\{x \mid x$ is any real number$\}$ or $(-\infty, \infty)$
Range: $\{y \mid y \leq 3\}$ or $(-\infty, 3]$

51.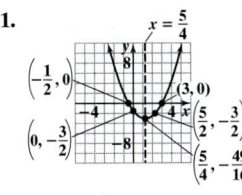

Domain: $\{x \mid x$ is any real number$\}$ or $(-\infty, \infty)$
Range: $\left\{ y \mid y \geq -\dfrac{49}{16} \right\}$ or $\left[-\dfrac{49}{16}, \infty \right)$

53.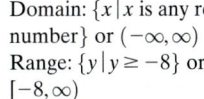

Domain: $\{x \mid x$ is any real number$\}$ or $(-\infty, \infty)$
Range: $\{y \mid y \geq -8\}$ or $[-8, \infty)$

55.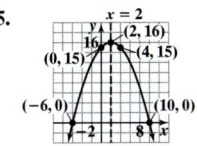

Domain: $\{x \mid x$ is any real number$\}$ or $(-\infty, \infty)$
Range: $\{y \mid y \leq 16\}$ or $(-\infty, 16]$

57. minimum; -3 **59.** maximum; 28 **61.** maximum; 23 **63.** minimum; -19 **65.** minimum; $-\frac{17}{8}$ **67.** maximum; $\frac{7}{3}$

69. (a) \$120   **(b)** \$36,000   **71.** $60; \$35$ **73. (a)** after 7.5 seconds   **(b)** 910 feet   **(c)** about 15.042 seconds
75. (a) about 1753.52 feet from the cannon   **(b)** about 886.76 feet   **(c)** about 3517 feet from the cannon   **(d)** The two answers are close.
Explanations may vary.   **77. (a)** about 46.5 years   **(b)** \$64,661.75   **79.** 18 and 18   **81.** -9 and 9
83. 15,625 square yards; 125 yards $\times$ 125 yards   **85.** 500,000 square meters; 500 m $\times$ 1000 m and the long side is parallel to the river
87. 5 inches   **89. (a)** $R = -p^2 + 110p$ **(b)** \$55; \$3025   **(c)** 55 pairs   **91. (a)** $f(x) = x^2 - 8x + 12$; $f(x) = 2x^2 - 16x + 24$;
$f(x) = -2x^2 + 16x - 24$ **(b)** The value of a has no effect on the x-intercepts.   **(c)** The value of a has no effect on the axis of symmetry.
(d) The x-coordinate of the vertex is 4, which does not depend on a. However, the y-coordinate is $-4a$, which does depend on a.

93. A revenue of \$0 when the price charged is some positive number is an indication that the price was too high to keep any consumer demand. Regardless of the price charged, if no items are sold, then no revenue can be generated.

95.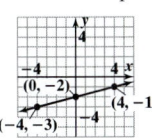

97.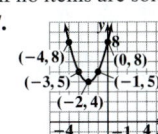

99. Vertex: $(3.5, -9.25)$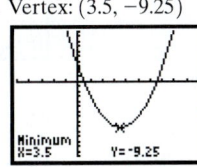

101. Vertex: $(3.5, 37.5)$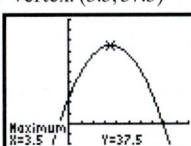

103. Vertex: $(-0.3, -20.45)$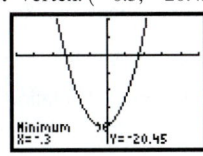

105. Vertex: $(0.67, 4.78)$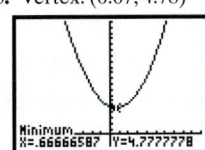

107. c is the y-intercept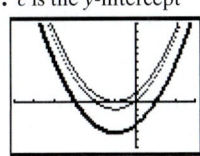

Section 10.6
1. quadratic   **3.** False   **5.** Answers may vary,   **7.** Answers may vary.   **9. (a)** $\{x \mid x < -6 \text{ or } x > 5\}$ or $(-\infty, -6) \cup (5, \infty)$

(b) $\{x \mid -6 \leq x \leq 5\}$ or $[-6, 5]$ **11. (a)** $\left\{ x \mid -6 \leq x \leq \dfrac{5}{2} \right\}$ or $\left[-6, \dfrac{5}{2} \right]$ **(b)** $\left\{ x \mid x < -6 \text{ or } x > \dfrac{5}{2} \right\}$ or $(-\infty, -6) \cup \left(\dfrac{5}{2}, \infty \right)$

13. $\{x \mid x \leq -2 \text{ or } x \geq 5\}$ or $(-\infty, -2] \cup [5, \infty)$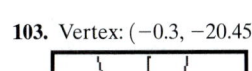

15. $\{x \mid -7 < x < -3\}$ or $(-7, -3)$

17. $\{x \mid x < -5 \text{ or } x > 7\}$ or $(-\infty, -5) \cup (7, \infty)$

19. $\left\{ n \mid 3 - \sqrt{17} \leq n \leq 3 + \sqrt{17} \right\}$ or $\left[3 - \sqrt{17}, 3 + \sqrt{17} \right]$

21. $\{m \mid m \leq -7 \text{ or } m \geq 2\}$ or $(-\infty, -7] \cup [2, \infty)$

23. $\left\{q \mid q \leq -\frac{5}{2} \text{ or } q \geq 3\right\}$ or $\left(-\infty, -\frac{5}{2}\right] \cup [3, \infty)$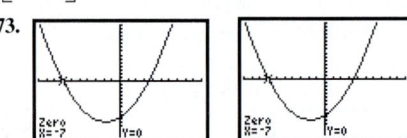

25. $\{x \mid -1 \leq x \leq 4\}$ or $[-1, 4]$

27. $\{x \mid x < -2 \text{ or } x > 5\}$ or $(-\infty, -2) \cup (5, \infty)$

29. $\left\{x \mid x \leq -\frac{2}{3} \text{ or } x \geq 4\right\}$ or $\left(-\infty, -\frac{2}{3}\right] \cup [4, \infty)$

31. $\left\{x \mid -2 - \sqrt{3} < x < -2 + \sqrt{3}\right\}$ or $\left(-2 - \sqrt{3}, -2 + \sqrt{3}\right)$

33. $\left\{a \mid -\frac{1}{2} \leq a \leq 4\right\}$ or $\left[-\frac{1}{2}, 4\right]$ **35.** $\{z \mid z \text{ is any real number}\}$ or $(-\infty, \infty)$

37. no solution   **39.** $\{x \mid x \text{ is any real number}\}$ or $(-\infty, \infty)$

41. $\{x \mid 0 < x < 5\}$ or $(0, 5)$ **43.** $\{x \mid x \leq -4 \text{ or } x \geq 7\}$ or $(-\infty, -4] \cup [7, \infty)$

45. $\left\{x \mid x < -\frac{5}{2} \text{ or } x > 2\right\}$ or $\left(-\infty, -\frac{5}{2}\right) \cup (2, \infty)$ **47.** $\{x \mid x \leq -8 \text{ or } x \geq 0\}$ or $(-\infty, -8] \cup [0, \infty)$

49. $\{x \mid x \leq -5 \text{ or } x \geq 6\}$ or $(-\infty, -5] \cup [6, \infty)$ **51.** between 2 and 3 seconds after the ball is thrown   **53.** between \$110 and \$130
55. $x = -3$; a perfect square cannot be negative. Therefore, the only solution will be where the perfect square expression equals zero, which is -3.
57. all real numbers; a perfect square must always be zero or greater. Therefore, it must always be larger than -2. Thus, all values of x will make the
inequality true.   **59.** Answers may vary. One possibility follows: $x^2 + x - 6 \leq 0$ **61.** Answer will vary. One possibility follows: The inequalities
have the same solution set because they are equivalent.   **63.** $\{x \mid x < -3 \text{ or } 1 < x < 3\}$ or $(-\infty, -3) \cup (1, 3)$ **65.** $\left\{x \mid -\frac{4}{3} \leq x \leq 2 \text{ or } x \geq 6\right\}$ or

$\left[-\frac{4}{3}, 2\right] \cup [6, \infty)$ **67.** $\{x \mid x < -2 \text{ or } -1 < x < 5\}$ or $(-\infty, -2) \cup (-1, 5)$ **69.** $\dfrac{-8m^5}{n^2}$ **71.** $\dfrac{b^{\frac{1}{4}}}{9a^{\frac{7}{9}}}$

73.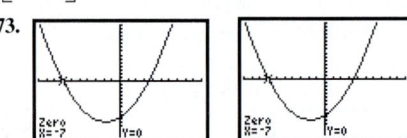

75.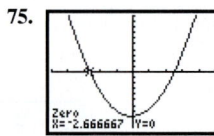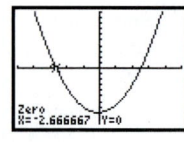

$\{x \mid x < -7 \text{ or } x > 3.5\}$ or $(-\infty, -7)$ or $(3.5, \infty)$ $\left\{x \mid -\frac{8}{3} \leq x \leq \frac{5}{2}\right\}$ or $\left[-\frac{8}{3}, \frac{5}{2}\right]$

Section 10.7

1. rational   **3.** The statement $(-1, 4)$ indicates that neither endpoint is included in the solution set. The statement $\{x \mid -1 \leq x \leq 4\}$ indicates that
both endpoints are included in the solution set. In fact, the solution of the inequality $\dfrac{x - 4}{x + 1} \leq 0$ should include the endpoint 4, but it should not include
the endpoint -1. Therefore, the statement of the solution set should be $\{x \mid -1 < x \leq 4\}$ using set-builder notation or $(-1, 4]$ using interval notation.
5. $\{x \mid x < -1 \text{ or } x > 4\}$ or $(-\infty, -1) \cup (4, \infty)$ **7.** $\{x \mid -9 < x < 3\}$ or $(-9, 3)$ **9.** $\{x \mid x \leq -10 \text{ or } x > 4\}$ or $(-\infty, -10] \cup (4, \infty)$

11. $\{x \mid -7 \leq x < 8\}$ or $[-7, 8)$ **13.** $\left\{x \mid -3 < x < \frac{1}{2} \text{ or } x > 5\right\}$ or $\left(-3, \frac{1}{2}\right) \cup (5, \infty)$ **15.** $\left\{x \mid x \leq -8 \text{ or } -\frac{5}{3} \leq x < 2\right\}$ or

$(-\infty, -8] \cup \left[-\frac{5}{3}, 2\right)$

17. $\{x \mid x > -1\}$ or $(-1, \infty)$ **19.** $\{x \mid x < -4 \text{ or } x \geq 9\}$ or $(-\infty, -4) \cup [9, \infty)$ **21.** $\left\{x \mid \frac{3}{2} < x < 3\right\}$ or $\left(\frac{3}{2}, 3\right)$

23. $\{x \mid 0 < x \leq 1 \text{ or } x > 4\}$ or $(0, 1] \cup (4, \infty)$ **25.** $\{x \mid -5 < x < 2 \text{ or } x \geq 23\}$ or $(-5, 2) \cup [23, \infty)$ **27.** $\{x \mid -1 < x \leq 6\}$ or $(-1, 6]$

29. $\left\{x \mid -2 < x < \dfrac{5}{2}\right\}$ or $\left(-2, \dfrac{5}{2}\right)$ **31.** 100 or more bicycles **33.** Answers may vary. One possibility follows: $\dfrac{10}{x-2} > 0$ **35.** 2

37. $-\dfrac{7}{2}$ and 2 **39.** $\dfrac{2}{3}$

41. $\{x \mid x \le -4 \text{ or } x > -1\}$ or $(-\infty, -4] \cup (-1, \infty)$ **43.** $\{x \mid 7 < x < 26\}$ or $(7, 26)$

Chapter 10 Review

1. $\{-13, 13\}$ **2.** $\{-5\sqrt{3}, 5\sqrt{3}\}$ **3.** $\{-4i, 4i\}$ **4.** $\left\{-\dfrac{2\sqrt{2}}{3}, \dfrac{2\sqrt{2}}{3}\right\}$ **5.** $\{-1, 17\}$ **6.** $\{2 - 5\sqrt{6}, 2 + 5\sqrt{6}\}$ **7.** $\left\{-5, \dfrac{5}{3}\right\}$

8. $\left\{-\dfrac{3\sqrt{14}}{7}, \dfrac{3\sqrt{14}}{7}\right\}$ **9.** $\{-4\sqrt{5}i, 4\sqrt{5}i\}$ **10.** $\left\{-\dfrac{3}{4} - \dfrac{\sqrt{13}}{4}, -\dfrac{3}{4} + \dfrac{\sqrt{13}}{4}\right\}$ **11.** $a^2 + 30a + 225; (a + 15)^2$ **12.** $b^2 - 14b + 49; (b - 7)^2$

13. $c^2 - 11c + \dfrac{121}{4}; \left(c - \dfrac{11}{2}\right)^2$ **14.** $d^2 + 9d + \dfrac{81}{4}; \left(d + \dfrac{9}{2}\right)^2$ **15.** $m^2 - \dfrac{1}{4}m + \dfrac{1}{64}; \left(m - \dfrac{1}{8}\right)^2$ **16.** $n^2 + \dfrac{6}{7}n + \dfrac{9}{49}; \left(n + \dfrac{3}{7}\right)^2$

17. $\{2, 8\}$ **18.** $\{-4, 7\}$ **19.** $\{3 - 2\sqrt{3}, 3 + 2\sqrt{3}\}$ **20.** $\left\{\dfrac{5}{2} - \dfrac{\sqrt{53}}{2}, \dfrac{5}{2} + \dfrac{\sqrt{53}}{2}\right\}$ **21.** $\left\{-\dfrac{1}{2} - \dfrac{3\sqrt{3}}{2}i, -\dfrac{1}{2} + \dfrac{3\sqrt{3}}{2}i\right\}$

22. $\{3 - 2\sqrt{2}i, 3 + 2\sqrt{2}i\}$ **23.** $\left\{\dfrac{1}{2}, 3\right\}$ **24.** $\left\{-\dfrac{1}{2} - \dfrac{3}{2}i, -\dfrac{1}{2} + \dfrac{3}{2}i\right\}$ **25.** $\left\{\dfrac{3}{2} - \dfrac{\sqrt{15}}{6}i, \dfrac{3}{2} + \dfrac{\sqrt{15}}{6}i\right\}$ **26.** $\left\{-\dfrac{2}{3} - \dfrac{\sqrt{10}}{3}, -\dfrac{2}{3} + \dfrac{\sqrt{10}}{3}\right\}$

27. $c = 15$ **28.** $c = 8\sqrt{2}$ **29.** $c = 3\sqrt{5}$ **30.** $c = 26$ **31.** $c = 6$ **32.** $c = 7$ **33.** $b = 3\sqrt{7}$ **34.** $a = 5\sqrt{3}$ **35.** $a = \sqrt{253}$

36. approximately 127.3 feet **37.** $\{-4, 5\}$ **38.** $\left\{-\dfrac{3}{2}, \dfrac{7}{2}\right\}$ **39.** $\left\{-\dfrac{4}{3} - \dfrac{\sqrt{7}}{3}, -\dfrac{4}{3} + \dfrac{\sqrt{7}}{3}\right\}$ **40.** $\left\{1 - \dfrac{\sqrt{10}}{2}, 1 + \dfrac{\sqrt{10}}{2}\right\}$

41. $\left\{-\dfrac{1}{6} - \dfrac{\sqrt{35}}{6}i, -\dfrac{1}{6} + \dfrac{\sqrt{35}}{6}i\right\}$ **42.** $\left\{\dfrac{4}{3}\right\}$ **43.** $\{2 - \sqrt{2}, 2 + \sqrt{2}\}$ **44.** $\left\{-\dfrac{2}{5} - \dfrac{1}{5}i, -\dfrac{2}{5} + \dfrac{1}{5}i\right\}$ **45.** $\left\{-\dfrac{5}{2} - \dfrac{3\sqrt{3}}{2}i, -\dfrac{5}{2} + \dfrac{3\sqrt{3}}{2}i\right\}$

46. $\left\{-\dfrac{3}{2} - \dfrac{\sqrt{5}}{2}i, -\dfrac{3}{2} + \dfrac{\sqrt{5}}{2}i\right\}$ **47.** 57; two irrational solutions **48.** 0; one repeated real solution

49. -47; two complex solutions that are not real **50.** -20; two complex solutions that are not real **51.** 0; one repeated real solution

52. 25; two rational solutions **53.** $\{-9, 1\}$ **54.** $\left\{-\dfrac{5}{2}, \dfrac{1}{3}\right\}$ **55.** $\{-2 - 3i, -2 + 3i\}$ **56.** $\{-2\sqrt{3}, 2\sqrt{3}\}$ **57.** $\left\{1 - \dfrac{\sqrt{10}}{2}, 1 + \dfrac{\sqrt{10}}{2}\right\}$

58. $\{-4 - 2i, -4 + 2i\}$ **59.** $\{-3, 5\}$ **60.** $\{1 - \sqrt{2}, 1 + \sqrt{2}\}$ **61.** $\left\{-\dfrac{4}{3}, \dfrac{4}{3}\right\}$ **62.** $\{-1 - \sqrt{3}i, -1 + \sqrt{3}i\}$

63. $x = 10$; the three sides measure 5, 12, and 13 **64.** 12 centimeters by 9 centimeters **65. (a)** either 300 or 600 cellular phones
(b) 450 cellular phones **66. (a)** after approximately 0.5 second and after approximately 2.7 seconds **(b)** after approximately 4.3 seconds
(c) No; the solutions to the equation are complex solutions that are not real **67.** approximately 10.8 miles per hour

68. approximately 67.8 minutes **69.** $\{-4i, 4i, -3, 3\}$ **70.** $\left\{-\dfrac{\sqrt{3}}{2}, \dfrac{\sqrt{3}}{2}, -\sqrt{2}i, \sqrt{2}i\right\}$ **71.** $\left\{-\dfrac{10}{3}, -1\right\}$

72. $\{-4, 4, -2\sqrt{2}, 2\sqrt{2}\}$ **73.** $\{16, 81\}$ **74.** $\left\{\dfrac{9}{25}\right\}$ **75.** $\left\{-\dfrac{1}{3}, \dfrac{1}{7}\right\}$ **76.** $\left\{-343, \dfrac{1}{8}\right\}$ **77.** $\{16\}$

78. $\left\{-\dfrac{36}{7}, -\dfrac{19}{4}\right\}$ **79.** $\left\{\dfrac{9}{4}, \dfrac{49}{4}\right\}$ **80.** $\{-2\sqrt{3}, -\sqrt{5}, \sqrt{5}, 2\sqrt{3}\}$

81. **82.** **83.** **84.** **85.**

86. **87.** **88.** **89.** **90.**

91. vertex is $(3, 1)$; axis of symmetry is $x = 3$
Domain: $\{x \mid x \text{ is any real number}\}$ or $(-\infty, \infty)$
Range: $\{y \mid y \ge 1\}$ or $[1, \infty)$

92. vertex is $(-4, -5)$; axis of symmetry is $x = -4$
Domain: $\{x \mid x \text{ is any real number}\}$ or $(-\infty, \infty)$
Range: $\{y \mid y \ge -5\}$ or $[-5, \infty)$

93. $H(x) = 2(x - 1)^2 - 5$ vertex is $(1, -5)$; axis of symmetry is $x = 1$
Domain: $\{x \mid x \text{ is any real number}\}$ or $(-\infty, \infty)$
Range: $\{y \mid y \ge -5\}$ or $[-5, \infty)$

94. 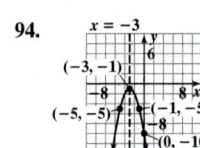 vertex is $(-3, -1)$; axis of symmetry is $x = -3$ Domain: $\{x \mid x$ is any real number$\}$ or $(-\infty, \infty)$ Range: $\{y \mid y \le -1\}$ or $(-\infty, -1]$

95. 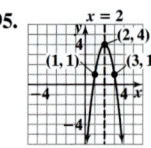 vertex is $(2, 4)$; axis of symmetry is $x = 2$ Domain: $\{x \mid x$ is any real number$\}$ or $(-\infty, \infty)$ Range: $\{y \mid y \le 4\}$ or $(-\infty, 4]$

96. 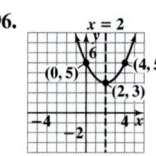 vertex is $(2, 3)$; axis of symmetry is $x = 2$ Domain: $\{x \mid x$ is any real number$\}$ or $(-\infty, \infty)$ Range: $\{y \mid y \ge 3\}$ or $[3, \infty)$

97. $f(x) = 2(x - 2)^2 - 4$ or $f(x) = 2x^2 - 8x + 4$ **98.** $f(x) = -(x - 4)^2 + 3$ or $f(x) = -x^2 + 8x - 13$

99. $f(x) = -\dfrac{1}{2}(x + 2)^2 - 1$ or $f(x) = -\dfrac{1}{2}x^2 - 2x - 3$ **100.** $f(x) = 3(x + 2)^2$ or $f(x) = 3x^2 + 12x + 12$

101.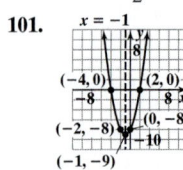
Domain: $\{x \mid x$ is any real number$\}$ or $(-\infty, \infty)$ Range: $\{y \mid y \ge -8\}$ or $[-8, \infty)$

102.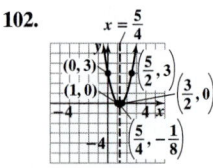
Domain: $\{x \mid x$ is any real number$\}$ or $(-\infty, \infty)$ Range: $\left\{y \mid y \ge -\dfrac{1}{8}\right\}$ or $\left[-\dfrac{1}{8}, \infty\right)$

103.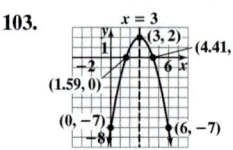
Domain: $\{x \mid x$ is any real number$\}$ or $(-\infty, \infty)$ Range: $\{y \mid y \le 2\}$ or $(-\infty, 2]$

104.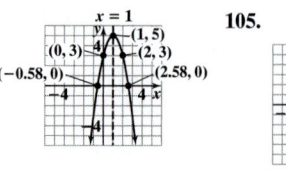
Domain: $\{x \mid x$ is any real number$\}$ or $(-\infty, \infty)$ Range: $\{y \mid y \le 5\}$ or $(-\infty, 5]$

105.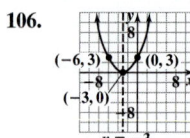
Domain: $\{x \mid x$ is any real number$\}$ or $(-\infty, \infty)$ Range: $\{y \mid y \ge 0\}$ or $[0, \infty)$

106.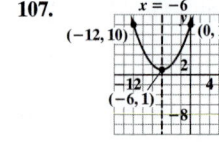
Domain: $\{x \mid x$ is any real number$\}$ or $(-\infty, \infty)$ Range: $\{y \mid y \ge 0\}$ or $[0, \infty)$

107.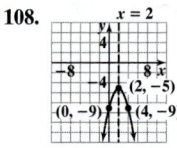
Domain: $\{x \mid x$ is any real number$\}$ or $(-\infty, \infty)$ Range: $\{y \mid y \ge 1\}$ or $[1, \infty)$

108.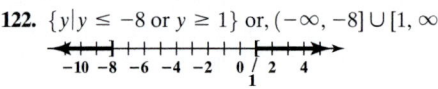
Domain: $\{x \mid x$ is any real number$\}$ or $(-\infty, \infty)$ Range: $\{y \mid y \le -5\}$ or $(-\infty, -5]$

109. maximum; 22   **110.** minimum; $-\dfrac{11}{8}$

111. maximum; 7   **112.** maximum; 5

113. (a) \$225   **(b)** \$16,875

114. (a) 3.75 amperes   **(b)** 225 watts

115. both numbers are 12

116. (a) 3.75 yards by 7.5 yards   **(b)** 28.125 square yards

117. (a) 100 feet   **(b)** 50 feet   **(c)** 200 feet

118. (a) $R = -0.002p^2 + 60p$ **(b)** \$15,000; \$450,000

(c) 30 automobiles per month

119. (a) $\{x \mid x < -2$ or $x > 3\}$ or $(-\infty, -2) \cup (3, \infty)$ **(b)** $\{x \mid -2 < x < 3\}$ or $(-2, 3)$

120. (a) $\left\{x \mid -\dfrac{7}{2} \le x \le 1\right\}$ or $\left[-\dfrac{7}{2}, 1\right]$ **(b)** $\left\{x \mid x \le -\dfrac{7}{2}$ or $x \ge 1\right\}$ or $\left(-\infty, -\dfrac{7}{2}\right] \cup [1, \infty)$

121. $\{x \mid -4 \le x \le 6\}$ or $[-4, 6]$
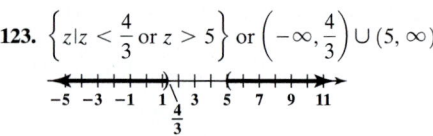

122. $\{y \mid y \le -8$ or $y \ge 1\}$ or, $(-\infty, -8] \cup [1, \infty)$
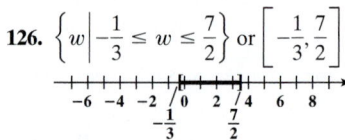

123. $\left\{z \mid z < \dfrac{4}{3}$ or $z > 5\right\}$ or $\left(-\infty, \dfrac{4}{3}\right) \cup (5, \infty)$

124. $\{p \mid -2 - \sqrt{6} < p < -2 + \sqrt{6}\}$ or $(-2 - \sqrt{6}, -2 + \sqrt{6})$

125. $\{m \mid m$ is any real number$\}$ or $(-\infty, \infty)$

126. $\left\{w \mid -\dfrac{1}{3} \le w \le \dfrac{7}{2}\right\}$ or $\left[-\dfrac{1}{3}, \dfrac{7}{2}\right]$

127. $\{x \mid x < -2$ or $x \ge 4\}$ or $(-\infty, -2) \cup [4, \infty)$ **128.** $\{y \mid -4 < y < 5\}$ or $(-4, 5)$ **129.** $\{z \mid -3 < z < 3\}$ or $(-3, 3)$

130. $\{w \mid w < -7$ or $2 < w < 4\}$ or $(-\infty, -7) \cup (2, 4)$ **131.** $\{m \mid -5 < m < 2$ or $m \ge 5\}$ or $(-5, 2) \cup [5, \infty)$

132. $\{n \mid 0 \le n < 2\}$ or $[0, 2)$ **133.** $\left\{a \mid 2 < a < \dfrac{7}{2}\right\}$ or $\left(2, \dfrac{7}{2}\right)$ **134.** $\{c \mid c < -6$ or $0 < c < 2\}$ or $(-\infty, -6) \cup (0, 2)$

135. $\left\{x \mid -\dfrac{3}{2} < x < 4\right\}$ or $\left(-\dfrac{3}{2}, 4\right)$ **136.** $\{x \mid x \le -5$ or $x > -1\}$ or $(-\infty, -5] \cup (-1, \infty)$

Chapter 10 Test

1. $x^2 - 3x + \dfrac{9}{4}; \left(x - \dfrac{3}{2}\right)^2$ **2.** $m^2 + \dfrac{2}{5}m + \dfrac{1}{25}; \left(m + \dfrac{1}{5}\right)^2$ **3.** $\left\{-\dfrac{5}{3}, -1\right\}$ **4.** $\{3 - \sqrt{5}, 3 + \sqrt{5}\}$ **5.** $\left\{1 - \dfrac{\sqrt{2}}{2}i, 1 + \dfrac{\sqrt{2}}{2}i\right\}$

6. $\left\{\dfrac{3}{2} - \dfrac{\sqrt{3}}{6}i, \dfrac{3}{2} + \dfrac{\sqrt{3}}{6}i\right\}$ **7.** 57; two irrational solutions   **8.** $a = 6\sqrt{2}$ **9.** $\{-3, 3, -2i, 2i\}$ **10.** $\left\{\dfrac{1}{81}\right\}$

11. 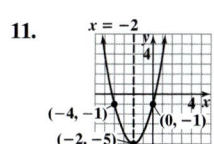 vertex is $(-2, -5)$; axis of symmetry
is $x = -2$
Domain: $\{x \mid x$ is any real number$\}$ or $(-\infty, \infty)$
Range: $\{y \mid y \geq -5\}$ or $[-5, \infty)$

12. 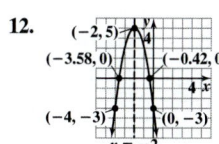 vertex is $(-2, 5)$; axis of symmetry
is $x = -2$
Domain: $\{x \mid x$ is any real number$\}$ or $(-\infty, \infty)$
Range: $\{y \mid y \leq 5\}$ or $(-\infty, 5]$

13. $f(x) = \frac{1}{3}(x + 3)^2 - 5$ or $f(x) = \frac{1}{3}x^2 + 2x - 2$ **14.** maximum; 6

15. $\left\{m \mid m < -3 \text{ or } m > \frac{5}{2}\right\}$ or $(-\infty, -3) \cup \left(\frac{5}{2}, \infty\right)$ **16.** $\left\{z \mid -3 - \sqrt{10} \leq z \leq -3 + \sqrt{10}\right\}$ or $\left[-3 - \sqrt{10}, -3 + \sqrt{10}\right]$

17. $\left\{x \mid 2 < x \leq \frac{11}{2}\right\}$ or $\left(2, \frac{11}{2}\right]$

18. 0.4 second and 4.6 seconds   **19.** 34.1 hours   **20. (a)** $340   **(b)** $28,900   **21. (a)** 12.5 in. by 12.5 in. by 12 in.   **(b)** 1875 cubic inches

Chapter 11

Section 11.1

1. composite function   **3.** inverse   **5.** false   **7.** Answers may vary.   **9.** Horizontal Line Test: If every horizontal line intersects the graph of a function f in at most one point, then f is one-to-one.   **11. (a)** 3   **(b)** -3 **(c)** 19   **(d)** -12 **13. (a)** 85   **(b)** 19   **(c)** 29   **(d)** -7

15. (a) -4394 **(b)** -507 **(c)** 16   **(d)** -1453 **17. (a)** 8   **(b)** $\frac{4}{5}$ **(c)** 1   **(d)** $-\frac{4}{3}$ **19. (a)** $(f \circ g)(x) = 2x + 1$

(b) $(g \circ f)(x) = 2x + 2$ **(c)** $(f \circ f)(x) = x + 2$ **(d)** $(g \circ g)(x) = 4x$ **21. (a)** $(f \circ g)(x) = -8x + 17$ **(b)** $(g \circ f)(x) = -8x - 23$
(c) $(f \circ f)(x) = 4x + 21$ **(d)** $(g \circ g)(x) = 16x - 15$ **23. (a)** $(f \circ g)(x) = x^2 - 6x + 9$ **(b)** $(g \circ f)(x) = x^2 - 3$ **(c)** $(f \circ f)(x) = x^4$
(d) $(g \circ g)(x) = x - 6$ **25. (a)** $(f \circ g)(x) = \sqrt{x + 4}$ **(b)** $(g \circ f)(x) = \sqrt{x} + 4$ **(c)** $(f \circ f)(x) = \sqrt[4]{x}$ **(d)** $(g \circ g)(x) = x + 8$
27. (a) $(f \circ g)(x) = x^2$ **(b)** $(g \circ f)(x) = x^2 + 8x + 12$ **(c)** $(f \circ f)(x) = \|x + 4\| + 4\|$ **(d)** $(g \circ g)(x) = x^4 - 8x^2 + 12$

29. (a) $(f \circ g)(x) = \dfrac{2x}{x + 1}$, where $x \neq -1, 0$ **(b)** $(g \circ f)(x) = \dfrac{x + 1}{2}$, where $x \neq -1$ **(c)** $(f \circ f)(x) = \dfrac{2(x + 1)}{x + 3}$, where $x \neq -1, -3$

(d) $(g \circ g)(x) = x$, where $x \neq 0$ **31.** one-to-one   **33.** not one-to-one   **35.** one-to-one   **37.** not one-to-one   **39.** one-to-one
41. one-to-one   **43.** not one-to-one   **45.** one-to-one
47. 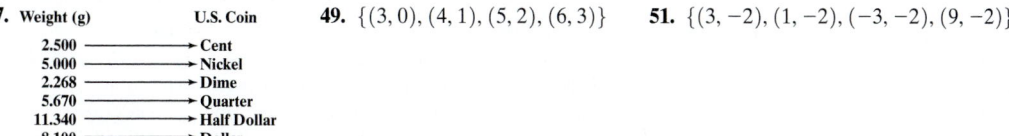   **49.** $\{(3, 0), (4, 1), (5, 2), (6, 3)\}$ **51.** $\{(3, -2), (1, -2), (-3, -2), (9, -2)\}$

53. 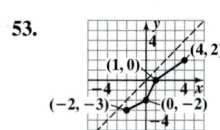   **55.** 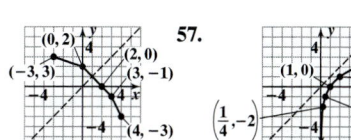   **57.** 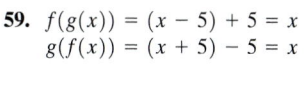   **59.** $f(g(x)) = (x - 5) + 5 = x$
$g(f(x)) = (x + 5) - 5 = x$

61. $f(g(x)) = 5\left(\dfrac{x - 7}{5}\right) + 7 = x - 7 + 7 = x$ **63.** $f(g(x)) = 10\left(\dfrac{x + 1}{10}\right) - 1 = x + 1 - 1 = x$

$g(f(x)) = \dfrac{(5x + 7) - 7}{5} = \dfrac{5x}{5} = x$ $g(f(x)) = \dfrac{(10x - 1) + 1}{10} = \dfrac{10x}{10} = x$

65. $f(g(x)) = \dfrac{3}{\left(\dfrac{3}{x} + 1\right) - 1} = \dfrac{3}{\dfrac{3}{x}} = 3 \cdot \dfrac{x}{3} = x$ **67.** $f(g(x)) = \sqrt[3]{(x^3 - 4) + 4} = \sqrt[3]{x^3} = x$

$g(f(x)) = \left(\sqrt[3]{x + 4}\right)^3 - 4 = x + 4 - 4 = x$

$g(f(x)) = \dfrac{3}{\dfrac{3}{x - 1}} + 1 = 3 \cdot \dfrac{x - 1}{3} + 1 = x - 1 + 1 = x$

69. $f^{-1}(x) = \dfrac{x}{6}$ **71.** $f^{-1}(x) = x - 4$ **73.** $h^{-1}(x) = \dfrac{x + 7}{2}$ **75.** $G^{-1}(x) = \dfrac{2 - x}{5}$ **77.** $g^{-1}(x) = \sqrt[3]{x} - 3$ **79.** $p^{-1}(x) = \dfrac{1}{x} - 3$

81. $F^{-1}(x) = 2 - \dfrac{5}{x}$ **83.** $f^{-1}(x) = x^3 + 2$ **85.** $R^{-1}(x) = \dfrac{2x}{1 - x}$ **87.** $f^{-1}(x) = (x - 4)^3 + 1$ **89.** $A(t) = 400\pi t^2; 3600\pi \approx 11{,}309.73$ sq ft

91. (a) $C(x) = 2x$ **(b)** $630   **93.** $f^{-1}(12) = 4$ **95.** Domain of f^{-1}: $[-5, \infty)$ Range of f^{-1}: $[0, \infty)$ **97.** Domain of g^{-1}: $(-6, 12)$

Range of g^{-1}: $[-4, 10]$ **99.** $x(T) = \dfrac{T + 300}{0.15}$ for $600 \leq T \leq 3960$ **101.** $\left\{-\dfrac{5}{2}, 3\right\}$ **103. (a)** 3   **(b)** -3 **(c)** 19   **(d)** -12

105. (a) 85   **(b)** 19   **(c)** 29   **(d)** -7 **107. (a)** -4394 **(b)** -507 **(c)** 16   **(d)** -1453 **109. (a)** 8   **(b)** $\dfrac{4}{5}$ **(c)** 1   **(d)** $-\dfrac{4}{3}$

111. $f(x) = x + 5; g(x) = x - 5$

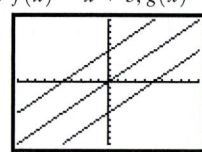

113. $f(x) = 5x + 7; g(x) = \dfrac{x - 7}{5}$

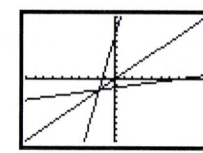

115. $f(x) = 10x - 1; g(x) = \dfrac{x + 1}{10}$

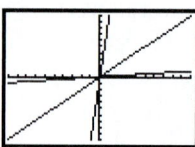

Section 11.2

1. $>; \neq$ **3.** $u = v$ **5.** false **7.** As the base a increases, the steeper the graph of $f(x) = a^x (a > 1)$ is for $x > 0$ and the closer the graph is to the x-axis for $x < 0$. **9.** Answers may vary. **11. (a)** 11.212 **(b)** 11.587 **(c)** 11.664 **(d)** 11.665 **(e)** 11.665 **13. (a)** 73.517 **(b)** 77.708 **(c)** 77.924 **(d)** 77.881 **(e)** 77.880 **15. (a)** 21.217 **(b)** 22.472 **(c)** 22.460 **(d)** 22.460 **(e)** 22.459 **17.** 7.389 **19.** 0.135 **21.** 9.974 **23.** g **25.** e **27.** f **29.** h

31. 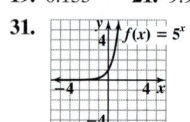 Domain: all real numbers or $(-\infty, \infty)$ Range: $\{y | y > 0\}$ or $(0, \infty)$

33. 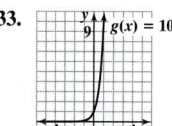 Domain: all real numbers or $(-\infty, \infty)$ Range: $\{y | y > 0\}$ or $(0, \infty)$

35. 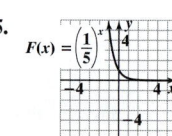 Domain: all real numbers or $(-\infty, \infty)$ Range: $\{y | y > 0\}$ or $(0, \infty)$

37. 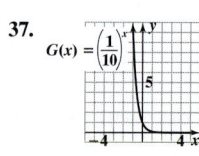 Domain: all real numbers or $(-\infty, \infty)$ Range: $\{y | y > 0\}$ or $(0, \infty)$

39. 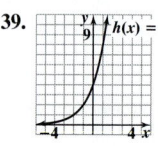 Domain: all real numbers or $(-\infty, \infty)$ Range: $\{y | y > 0\}$ or $(0, \infty)$

41. 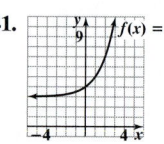 Domain: all real numbers or $(-\infty, \infty)$ Range: $\{y | y > 3\}$ or $(3, \infty)$

43. 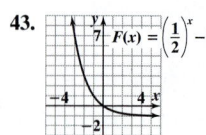 Domain: all real numbers or $(-\infty, \infty)$ Range: $\{y | y > -1\}$ or $(-1, \infty)$

45. 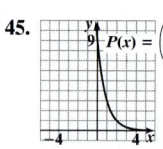 Domain: all real numbers or $(-\infty, \infty)$ Range: $\{y | y > 0\}$ or $(0, \infty)$

47. 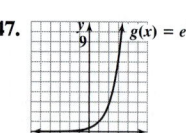 Domain: all real numbers or $(-\infty, \infty)$ Range: $\{y | y > 0\}$ or $(0, \infty)$

49. $\{5\}$ **51.** $\{-4\}$ **53.** $\{5\}$ **55.** $\{5\}$ **57.** $\left\{\dfrac{3}{2}\right\}$ **59.** $\{1\}$ **61.** $\{-1, 4\}$ **63.** $\{-4, 2\}$ **65.** $\{9\}$ **67.** $\{-2\}$ **69.** $\{-2, 2\}$

71. $\{-2\}$ **73.** $\{2\}$ **75. (a)** $f(3) = 8; (3, 8)$ **(b)** $x = -3; \left(-3, \dfrac{1}{8}\right)$ **77. (a)** $g(-1) = -\dfrac{3}{4}; \left(-1, -\dfrac{3}{4}\right)$ **(b)** $x = 2; (2, 15)$

79. (a) $H(-3) = 24; (-3, 24)$ **(b)** $x = 2; \left(2, \dfrac{3}{4}\right)$ **81. (a)** approximately 306 million people **(b)** approximately 444 million people

(c) The U.S. Census Bureau's prediction is 44 million people fewer than that of the model. Reasons given may vary. **83. (a)** $5308.39 **(b)** $5983.40 **(c)** $6744.25 **85. (a)** $2318.55 **(b)** $2322.37 **(c)** $2323.23 **(d)** $2323.65 **(e)** Answers may vary. **87. (a)** $14,512 **(b)** $9757.87 **(c)** $5380.18 **89. (a)** approximately 95.105 grams **(b)** 50 grams **(c)** 25 grams **(d)** approximately 0.661 gram **91. (a)** approximately 300.233°F **(b)** approximately 230.628°F **(c)** yes **93. (a)** approximately 29 words **(b)** approximately 38 words **95. (a)** approximately 0.238 ampere **(b)** approximately 0.475 ampere **97.** $y = 3^x$ **99. (a)** 15 **(b)** -5 **101. (a)** undefined **(b)** $\dfrac{4}{5}$ **103. (a)** 3 **(b)** $3\sqrt{3}$

105. $f(x) = 1.5^x$

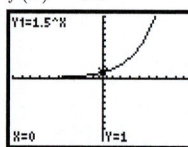 Domain: all real numbers or $(-\infty, \infty)$ Range: $\{y | y > 0\}$ or $(0, \infty)$

107. $H(x) = 0.9^x$

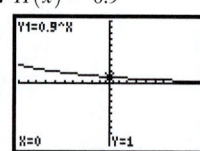 Domain: all real numbers or $(-\infty, \infty)$ Range: $\{y | y > 0\}$ or $(0, \infty)$

109. $g(x) = 2.5^x + 3$

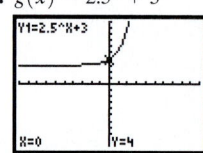 Domain: all real numbers or $(-\infty, \infty)$ Range: $\{y | y > 3\}$ or $(3, \infty)$

111. $F(x) = 1.6^{x-3}$

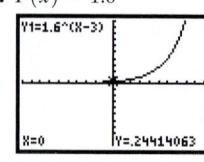 Domain: all real numbers or $(-\infty, \infty)$ Range: $\{y | y > 0\}$ or $(0, \infty)$

Section 11.3

1. $x = a^y; >; \neq$ **3.** $\left(\dfrac{1}{a}, -1\right), (1, 0), (a, 1)$ **5.** false **7.** Answers may vary. **9.** $2 = \log_5 25$ **11.** $3 = \log_4 64$ **13.** $-3 = \log_2 \left(\dfrac{1}{8}\right)$

15. $\ln 12 = x$ **17.** $\log_a 19 = 3$ **19.** $\log_5 c = -6$ **21.** $2^4 = 16$ **23.** $3^{-2} = \dfrac{1}{9}$ **25.** $e^4 = x$ **27.** $5^{-3} = a$ **29.** $a^2 = 4$

31. $\left(\dfrac{1}{2}\right)^y = 12$ **33.** 0 **35.** 3 **37.** -2 **39.** 4 **41.** 4 **43.** $\dfrac{1}{2}$ **45.** -1 **47.** 3 **49.** $\{x | x > 4\}$ or $(4, \infty)$

51. $\{x | x > -6\}$ or $(-6, \infty)$ **53.** $\{x | x > 0\}$ or $(0, \infty)$ **55.** $\left\{x | x > \dfrac{2}{3}\right\}$ or $\left(\dfrac{2}{3}, \infty\right)$ **57.** $\left\{x | x > -\dfrac{1}{2}\right\}$ or $\left(-\dfrac{1}{2}, \infty\right)$

59. $\left\{x | x < \dfrac{1}{4}\right\}$ or $\left(-\infty, \dfrac{1}{4}\right)$

61.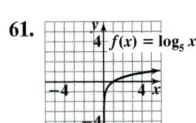
Domain: $\{x \mid x > 0\}$ or $(0, \infty)$
Range: all real numbers or $(-\infty, \infty)$

63.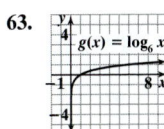
Domain: $\{x \mid x > 0\}$ or $(0, \infty)$
Range: all real numbers or $(-\infty, \infty)$

65.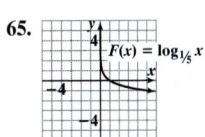
Domain: $\{x \mid x > 0\}$ or $(0, \infty)$
Range: all real numbers or $(-\infty, \infty)$

67.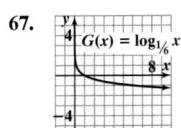
Domain: $\{x \mid x > 0\}$ or $(0, \infty)$
Range: all real numbers or $(-\infty, \infty)$

69.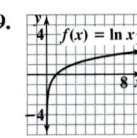
Domain: $\{x \mid x > 0\}$ or $(0, \infty)$
Range: all real numbers or $(-\infty, \infty)$

71. 1.826 **73.** 1.686 **75.** -0.456
77. -1.609 **79.** 0.097 **81.** -0.981
83. $\{4\}$ **85.** $\left\{\dfrac{13}{2}\right\}$ **87.** $\{6\}$
89. $\{3\sqrt{2}\}$ **91.** $\{10\}$ **93.** $\{e^5\}$
95. $\left\{\dfrac{11}{20}\right\}$ **97.** $\{-3\}$ **99.** $\{4\}$

101. $\{-3, 3\}$ **103. (a)** $f(16) = 4; (16, 4)$ **(b)** $x = \dfrac{1}{8}; \left(\dfrac{1}{8}, -3\right)$ **105. (a)** $G(7) = \dfrac{3}{2}; \left(7, \dfrac{3}{2}\right)$ **(b)** $x = 15; (15, 2)$ **107.** 20 decibels

109. 130 decibels **111.** approximately 7.8 on the Richter scale **113.** approximately 630,957 millimeters **115. (a)** 12; basic **(b)** 5; acidic
(c) 2; acidic **(d)** $10^{-7.4}$ moles per liter **117.** $\{x \mid x < -2 \text{ or } x > 5\}$ or $(-\infty, -2) \cup (5, \infty)$ **119.** $\{x \mid x < -1 \text{ or } x > 3\}$ or $(-\infty, -1) \cup (3, \infty)$

121. $a = 4$ **123.** $-3x^2 - 7x + 10$ **125.** $\dfrac{4x^2 + 2x + 3}{(x + 1)(x - 1)(x + 2)}$ **127.** $5x\sqrt{2x}$

129. $f(x) = \log(x + 1)$
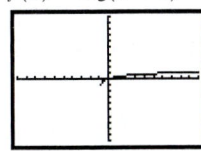
Domain: $\{x \mid x > -1\}$ or $(-1, \infty)$
Range: all real number or $(-\infty, \infty)$

131. $G(x) = \ln(x) + 1$
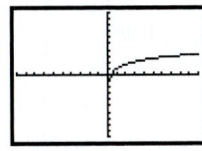
Domain: $\{x \mid x > 0\}$ or $(0, \infty)$
Range: all real number or $(-\infty, \infty)$

133. $f(x) = 2\log(x - 3) + 1$
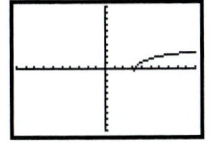
Domain: $\{x \mid x > 3\}$ or $(3, \infty)$
Range: all real number or $(-\infty, \infty)$

Putting the Concepts Together (Sections 1–3)
1. (a) $(f \circ g)(x) = 4x^2 - 8x + 3$ **(b)** $(g \circ f)(x) = 8x^2 + 16x + 6$ **(c)** $(f \circ g)(3) = 15$ **(d)** $(g \circ f)(-2) = 6$ **(e)** $(f \circ f)(1) = 13$

2. (a) $f^{-1}(x) = \dfrac{x - 4}{3}$ **(b)** $g^{-1}(x) = \sqrt[3]{x + 4}$

3. 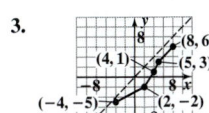 **4. (a)** 14.611 **(b)** 15.206 **(c)** 15.146 **(d)** 15.155 **(e)** 15.154 **5. (a)** $\log_a 6.4 = 4$ **(b)** $\log 278 = x$
6. (a) $2^7 = x$ **(b)** $e^M = 16$ **7. (a)** 4 **(b)** -2 **8.** $\{x \mid x > -6\}$ or $(-6, \infty)$

9.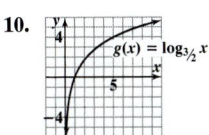
Domain: all real numbers or $(-\infty, \infty)$
Range: $\{y \mid y > 0\}$ or $(0, \infty)$

10.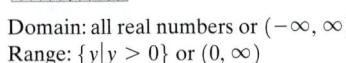
Domain: $\{x \mid x > 0\}$ or $(0, \infty)$
Range: all real numbers or $(-\infty, \infty)$

11. $\{-1\}$ **12.** $\{-5\}$ **13.** $\left\{\dfrac{11}{2}\right\}$
14. $\{e^7\}$ **15.** approximately 56 terms

Section 11.4
1. $0; 1$ **3.** $\dfrac{\log 10}{\log 3} = \dfrac{\ln 10}{\ln 3}$ **5.** false **7.** Answers may vary. **9.** Answers may vary. **11.** 3 **13.** -7 **15.** 5 **17.** 2 **19.** 1

21. 1 **23.** 4 **25.** $a + b$ **27.** $2b$ **29.** $2a + b$ **31.** $\dfrac{1}{2}a$ **33.** $\log a + \log b$ **35.** $4\log_5 x$ **37.** $\log_2 x + 2\log_2 y$ **39.** $2 + \log_5 x$

41. $2 - \log_7 y$ **43.** $2 + \ln x$ **45.** $3 + \dfrac{1}{2}\log_3 x$ **47.** $2\log_5 x + \dfrac{1}{2}\log_5(x^2 + 1)$ **49.** $4\log x - \dfrac{1}{3}\log(x - 1)$

51. $\dfrac{1}{2}\log_7(x + 1) - \dfrac{1}{2}\log_7 x$ **53.** $\log_2 x + 2\log_2(x - 1) - \dfrac{1}{2}\log_2(x + 1)$ **55.** 2 **57.** $\log(3x)$ **59.** 2 **61.** $\log_3 x^3$ **63.** $\log_4\left(\dfrac{x + 1}{x}\right)$

65. $\ln(x^2 y^3)$ **67.** $\log_3\left[\sqrt{x}(x - 1)^3\right]$ **69.** $\log(x^2)$ **71.** $\log\left(x\sqrt{xy}\right)$ **73.** $\log_8(x - 1)$ **75.** $\log\left(\dfrac{x^{12}}{10}\right)$ **77.** 3.322

79. 0.528 **81.** -2.680 **83.** 4.644 **85.** 3 **87.** 1

89. $\log_a(x + \sqrt{x^2 - 1}) + \log_a(x - \sqrt{x^2 - 1})$
$= \log_a[(x + \sqrt{x^2 - 1})(x - \sqrt{x^2 - 1})]$
$= \log_a[x^2 - x\sqrt{x^2 - 1} + x\sqrt{x^2 - 1} - (x^2 - 1)]$
$= \log_a(x^2 - x^2 + 1)$
$= \log_a 1$
$= 0$

91. If $f(x) = \log_a x$, then
$f(AB) = \log_a(AB)$
$= \log_a A + \log_a B$
$= f(A) + f(B)$

93. $\left\{\dfrac{5}{2}\right\}$ **95.** $\{-2 - \sqrt{2}, -2 + \sqrt{2}\}$

97. $\{47\}$

99. $f(x) = \log_3 x = \dfrac{\log x}{\log 3}$

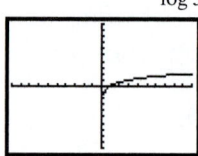

Domain: $\{x \mid x > 0\}$ or $(0, \infty)$
Range: all real numbers or $(-\infty, \infty)$

101. $F(x) = \log_{1/2} x = \dfrac{\log x}{\log\left(\frac{1}{2}\right)}$

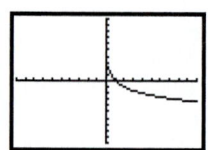

Domain: $\{x \mid x > 0\}$ or $(0, \infty)$
Range: all real numbers or $(-\infty, \infty)$

Section 11.5

1. $M = N$ **3.** True (so long as the logarithms have the same base)   **5.** Answers may vary.   **7.** $\{7\}$ **9.** $\{9\}$ **11.** $\{3\}$ **13.** $\{81\}$

15. $\{1\}$ **17.** $\{3\}$ **19.** $\left\{\dfrac{1}{3}\right\}$ **21.** $\left\{\dfrac{7}{5}\right\}$ **23.** $\left\{\dfrac{3 + \sqrt{93}}{2}\right\} \approx \{6.322\}$ **25.** $\left\{\dfrac{1}{\log 2}\right\} \approx \{3.322\}$ or $\left\{\dfrac{\ln 10}{\ln 2}\right\} \approx \{3.322\}$

27. $\left\{\dfrac{\log 20}{\log 5}\right\} \approx \{1.861\}$ or $\left\{\dfrac{\ln 20}{\ln 5}\right\} \approx \{1.861\}$ **29.** $\left\{\dfrac{\log 7}{\log\left(\frac{1}{2}\right)}\right\} \approx \{-2.807\}$ or $\left\{\dfrac{\ln 7}{\ln\left(\frac{1}{2}\right)}\right\} \approx \{-2.807\}$ **31.** $\{\ln 5\} \approx \{1.609\}$

33. $\{\log 5\} \approx \{0.699\}$ **35.** $\left\{\dfrac{\log 13}{2 \log 3}\right\} \approx \{1.167\}$ or $\left\{\dfrac{\ln 13}{2 \ln 3}\right\} \approx \{1.167\}$ **37.** $\left\{\dfrac{\log 3}{4 \log\left(\frac{1}{2}\right)}\right\} \approx \{-0.396\}$ or $\left\{\dfrac{\ln 3}{4 \ln\left(\frac{1}{2}\right)}\right\} \approx \{-0.396\}$

39. $\left\{\dfrac{\log\left(\frac{5}{4}\right)}{\log 2}\right\} \approx \{0.322\}$ or $\left\{\dfrac{\ln\left(\frac{5}{4}\right)}{\ln 2}\right\} \approx \{0.322\}$ **41.** $\{\ln 6\} \approx \{1.792\}$ **43.** $\left\{\dfrac{\log 0.2}{\log 3 - \log 0.2}\right\} \approx \{-0.594\}$ or $\left\{\dfrac{\ln 0.2}{\ln 3 - \ln 0.2}\right\} \approx \{-0.594\}$

45. $\{8\}$ **47.** $\left\{\dfrac{\log 7}{3 \log 5}\right\} \approx \{0.403\}$ or $\left\{\dfrac{\ln 7}{3 \ln 5}\right\} \approx \{0.403\}$ **49.** $\{2\}$ **51.** $\{\ln 15\} \approx \{2.708\}$ **53.** $\left\{-\dfrac{2}{5}\right\}$ **55.** $\{16\}$ **57.** $\{2\}$

59. (a) about the year 2021   (b) about the year 2050   **61.** (a) approximately 5.6 years   (b) approximately 11.6 years
63. (a) approximately 1.876 years   (b) approximately 5.369 years   (c) approximately 13.479 years   **65.** (a) approximately 2.099 seconds
(b) 27.62 seconds   (c) approximately 45.876 seconds   **67.** (a) approximately 5 minutes   (b) approximately 11 minutes
69. (a) approximately 82 minutes   (b) approximately 396 minutes (or 6.6 hours)   **71.** (a) approximately 9 years   (b) $t = \dfrac{\log 2}{n \log\left(1 + \dfrac{r}{n}\right)}$

(c) approximately 8.693 years, which is about the same as the result from the Rule of 72   **73.** 29 minutes   **75.** (a) 1   (b) 11   (c) 7
77. (a) $-\dfrac{3}{2}$ (b) $\dfrac{2}{7}$ (c) 0   **79.** (a) 8   (b) $\dfrac{1}{4}$ (c) 1   **81.** approximately $\{1.06\}$ **83.** approximately $\{-0.70\}$
85. approximately $\{0.05, 1.48\}$ **87.** approximately $\{0.45, 1\}$

Chapter 11 Review

1. (a) 32   (b) -9 (c) 2   (d) 13   **2.** (a) 24   (b) -28 (c) -8 (d) 112   **3.** (a) 201   (b) 24   (c) 163   (d) 14
4. (a) 23   (b) 37   (c) -8 (d) 290   **5.** (a) $(f \circ g)(x) = 5x + 1$ (b) $(g \circ f)(x) = 5x + 5$ (c) $(f \circ f)(x) = x + 2$
(d) $(g \circ g)(x) = 25x$ **6.** (a) $(f \circ g)(x) = 2x + 9$ (b) $(g \circ f)(x) = 2x + 3$ (c) $(f \circ f)(x) = 4x - 9$ (d) $(g \circ g)(x) = x + 12$
7. (a) $(f \circ g)(x) = 4x^2 + 4x + 2$ (b) $(g \circ f)(x) = 2x^2 + 3$ (c) $(f \circ f)(x) = x^4 + 2x^2 + 2$ (d) $(g \circ g)(x) = 4x + 3$
8. (a) $(f \circ g)(x) = \dfrac{2x}{x + 1}$, where $x \neq -1, 0$ (b) $(g \circ f)(x) = \dfrac{x + 1}{2}$, where $x \neq -1$ (c) $(f \circ f)(x) = \dfrac{2(x + 1)}{x + 3}$, where $x \neq -1, -3$
(d) $(g \circ g)(x) = x$, where $x \neq 0$ **9.** not one-to-one   **10.** one-to-one   **11.** one-to-one   **12.** not one-to-one

13.

Height (inches)	Age
69	24
71	59
72	29
73	81
74	37

14.

Quantity Demanded	Price ($)
112	300
129	200
144	170
161	150
176	130

15. $\{(3, -5), (1, -3), (-3, 1), (9, 2)\}$ **16.** $\{(1, -20), (4, -15), (3, 5), (2, 25)\}$

17.

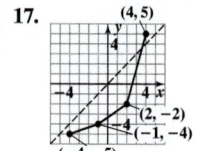

18.

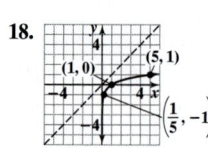

19. $f^{-1}(x) = \dfrac{x}{5}$ **20.** $H^{-1}(x) = \dfrac{x - 7}{2}$ **21.** $P^{-1}(x) = \dfrac{4}{x} - 2$ **22.** $g^{-1}(x) = \sqrt[3]{\dfrac{x + 1}{2}}$

23. (a) 27.332   (b) 28.975   (c) 29.088   (d) 29.093   (e) 29.091   **24.** (a) 1258.925
(b) 1380.384   (c) 1386.756   (d) 1385.479   (e) 1385.456   **25.** (a) 1.649   (b) 0.368
(c) 4.482   (d) 0.449   (e) 5.885

26. Domain: all real numbers or $(-\infty, \infty)$
Range: $\{y \mid y > 0\}$ or $(0, \infty)$

27. Domain: all real numbers or $(-\infty, \infty)$
Range: $\{y \mid y > 0\}$ or $(0, \infty)$

28. Domain: all real numbers or $(-\infty, \infty)$
Range: $\{y \mid y > 0\}$ or $(0, \infty)$

29. Domain: all real numbers or $(-\infty, \infty)$
Range: $\{y \mid y > -2\}$ or $(-2, \infty)$

30. The number e is defined as the number that the expression $\left(1 + \dfrac{1}{n}\right)^n$ approaches as n becomes unbounded in the positive direction.

31. $\{6\}$ **32.** $\left\{\dfrac{7}{2}\right\}$ **33.** $\{-4, 1\}$ **34.** $\{-2\}$ **35.** $\{9\}$ **36.** $\{-3, 3\}$ **37. (a)** \$7513.59 **(b)** \$7652.33 **(c)** \$7684.36

(d) \$7700.01 **38. (a)** approximately 82.034 grams **(b)** 50 grams **(c)** 25 grams **(d)** approximately 0.263 gram

39. (a) approximately 2.708 million people **(b)** approximately 3.317 million people **40. (a)** approximately 151.449°F **(b)** approximately 94.706°F

41. $\log_3 81 = 4$ **42.** $\log_4\left(\dfrac{1}{64}\right) = -3$ **43.** $\log_b 5 = 3$ **44.** $\log x = 3.74$ **45.** $8^{1/3} = 2$ **46.** $5^r = 18$ **47.** $e^2 = x + 3$ **48.** $10^{-4} = x$

49. $\dfrac{7}{3}$ **50.** 0 **51.** -2 **52.** $\dfrac{3}{2}$ **53.** $\{x \mid x > -5\}$ or $(-5, \infty)$ **54.** $\left\{x \mid x < \dfrac{7}{3}\right\}$ or $\left(-\infty, \dfrac{7}{3}\right)$ **55.** $\{x \mid x > 0\}$ or $(0, \infty)$

56. $\left\{x \mid x > -\dfrac{5}{2}\right\}$ or $\left(-\dfrac{5}{2}, \infty\right)$

57. **58.** **59.** 3.178 **60.** -0.182 **61.** 2.410 **62.** -0.907 **63.** $\{17\}$ **64.** $\{-9, 1\}$

65. $\left\{\dfrac{3}{2}\right\}$ **66.** $\{6\}$ **67.** $\{-142\}$ **68.** $\{5\sqrt{3}\}$ **69.** 80 decibels

70. 100,000 millimeters **71.** 21 **72.** 9.34 **73.** 1 **74.** 0 **75.** 1

76. 16 **77.** $\log_7 x + \log_7 y - \log_7 z$ **78.** $4 - 2\log_3 x$ **79.** $3 + 4\log r$

80. $\dfrac{1}{2}\ln(x - 1) - \dfrac{1}{2}\ln x$ **81.** $\log_3(x^4 y^2)$ **82.** $\ln\left(\dfrac{7\sqrt[4]{x}}{9}\right)$ **83.** -1 **84.** $\log_6(x - 4)$ **85.** 2.183 **86.** 0.606 **87.** -4.419

88. 3.723 **89.** $\{10\}$ **90.** $\{3\}$ **91.** $\{\sqrt{2}\} \approx \{1.414\}$ **92.** $\{64\}$ **93.** $\left\{\dfrac{\log 15}{\log 2}\right\} \approx \{3.907\}$ or $\left\{\dfrac{\ln 15}{\ln 2}\right\} \approx \{3.907\}$ **94.** $\left\{\dfrac{\log 27}{3}\right\} \approx \{0.477\}$

95. $\left\{\dfrac{\ln 39}{7}\right\} \approx \{0.523\}$ **96.** $\left\{\dfrac{\log 2}{\log 3 - \log 2}\right\} \approx \{1.710\}$ or $\left\{\dfrac{\ln 2}{\ln 3 - \ln 2}\right\} \approx \{1.710\}$ **97. (a)** approximately 1.453 days

(b) approximately 23.253 days **98. (a)** about 2008 **(b)** about 2016

Chapter 11 Test

1. not one-to-one **2.** $f^{-1}(x) = \dfrac{x + 3}{4}$ **3. (a)** 33.360 **(b)** 36.338 **(c)** 36.494 **(d)** 36.463 **(e)** 36.462

4. $\log_4 19 = x$ **5.** $b^y = x$ **6. (a)** -3 **(b)** 4 **7.** $\left\{x \mid x < \dfrac{7}{4}\right\}$ or $\left(-\infty, \dfrac{7}{4}\right)$

8. Domain: all real numbers or $(-\infty, \infty)$
Range: $\{y \mid y > 0\}$ or $(0, \infty)$

9. Domain: $\{x \mid x > 0\}$ or $(0, \infty)$
Range: all real numbers or $(-\infty, \infty)$

10. (a) 10 **(b)** 15 **11.** $\dfrac{1}{2}\log_4 x - 3\log_4 y$ **12.** $\log(M^4 N^3)$ **13.** -8.004 **14.** $\{1\}$ **15.** $\{1, 3\}$ **16.** $\{4\}$ **17.** $\{-17, 17\}$

18. $\{9\}$ **19.** $\left\{\dfrac{\log 17 + \log 3}{\log 3}\right\} \approx \{3.579\}$ or $\left\{\dfrac{\ln 17 + \ln 3}{\ln 3}\right\} \approx \{3.579\}$ **20.** $\{2\sqrt{26}\} \approx \{10.198\}$ **21. (a)** approximately 34 million people

(b) about 2058 **22.** 10 decibels

Cumulative Review 1–11

1. $\{5\}$ **2.** $\{x \mid -2 \le x \le 6\}$ or $[-2, 6]$ **3.** $\left\{x \mid x \ne -3 \text{ and } x \ne \dfrac{7}{2}\right\}$ **4.** **5.** $y = -\dfrac{7}{5}x + 3$ or $7x + 5y = 15$

6. 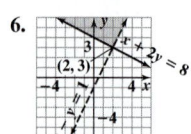   **7.** $4m^2 - 3m + 7$ **8.** $2n^3 - 5n^2 + 18$ **9.** $(4a + b)^2$ **10.** $(3y - 7)(2y - 1)$ **11.** $\dfrac{3x - 4}{x - 7}$

12. $\dfrac{6(p + 1)}{(p - 5)(p - 1)(p + 2)}$ **13.** $\{4\}$ **14.** $7\sqrt{6}$ **15.** $2 + \sqrt{5}$ **16.** $\{9\}$ **17.** $\left\{\dfrac{2 - \sqrt{22}}{3}, \dfrac{2 + \sqrt{22}}{3}\right\}$

18. $\left\{\dfrac{9}{4}, 4\right\}$ **19.** 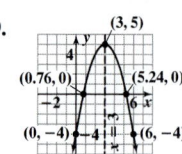   **20.** $\left\{x \mid -2 < x < \dfrac{4}{3}\right\}$ or $\left(-2, \dfrac{4}{3}\right)$ 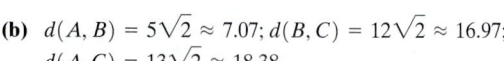   **21.** 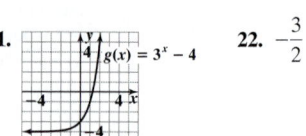   **22.** $-\dfrac{3}{2}$

Chapter 12

Section 12.1

1. $\sqrt{(x_2 - x_1)^2 + (y_2 - y_1)^2}$ **3.** false   **5.** Answers may vary.   **7.** 5   **9.** $4\sqrt{5} \approx 8.94$ **11.** 5   **13.** 13   **15.** 6   **17.** $3\sqrt{5} \approx 6.71$

19. $3\sqrt{7} \approx 7.94$ **21.** $\sqrt{12.56} \approx 3.54$ **23.** $(4, 3)$ **25.** $(3, -1)$ **27.** $\left(-1, \dfrac{7}{2}\right)$ **29.** $\left(-\dfrac{3}{2}, 0\right)$ **31.** $\left(\dfrac{7\sqrt{2}}{2}, \dfrac{5\sqrt{5}}{2}\right)$

33. $(0.8, -1.6)$ **35. (a)** 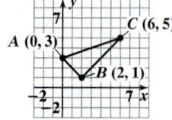   **37. (a)**

(b) $d(A, B) = 2\sqrt{2} \approx 2.83; d(B, C) = 4\sqrt{2} \approx 5.66;$
$d(A, C) = 2\sqrt{10} \approx 6.32$

(c) $[d(A, B)]^2 + [d(B, C)]^2 \overset{?}{=} [d(A, C)]^2$
$(2\sqrt{2})^2 + (4\sqrt{2})^2 \overset{?}{=} (2\sqrt{10})^2$
$4 \cdot 2 + 16 \cdot 2 \overset{?}{=} 4 \cdot 10$
$8 + 32 \overset{?}{=} 40$
$40 = 40 \leftarrow$ True
Therefore, triangle ABC is a right triangle.
(d) 8 square units

(b) $d(A, B) = 5\sqrt{2} \approx 7.07; d(B, C) = 12\sqrt{2} \approx 16.97;$
$d(A, C) = 13\sqrt{2} \approx 18.38$

(c) $[d(A, B)]^2 + [d(B, C)]^2 \overset{?}{=} [d(A, C)]^2$
$(5\sqrt{2})^2 + (12\sqrt{2})^2 \overset{?}{=} (13\sqrt{2})^2$
$25 \cdot 2 + 144 \cdot 2 \overset{?}{=} 169 \cdot 2$
$50 + 288 \overset{?}{=} 338$
$338 = 338 \leftarrow$ True
Therefore, triangle ABC is a right triangle.
(d) 60 square units

39. $(2, -3), (2, 5)$ **41.** $(-6, -3), (10, -3)$ **43. (a)** approximately 37.36 blocks   **(b)** approximately 35.13 blocks
(c) approximately 71.34 blocks   **45. (a)** approximately 1.85 seconds   **(b)** No. The ball will reach second base (2.65 seconds) before the runner
(3.33 seconds).   **47.** 9; 3   **49.** 81; 3   **51.** If n is a positive integer and $a^n = b$, then $\sqrt[n]{b} = \begin{cases} |a|, & \text{if } n \text{ is even} \\ a, & \text{if } n \text{ is odd} \end{cases}$

Section 12.2

1. circle   **3.** false   **5.** Answers may vary.   **7.** Yes; Center: $(0, 0); r = 6$ **9.** $(x - 1)^2 + (y - 2)^2 = 4$ **11.** $(x - 2)^2 + (y + 1)^2 = 16$
13. $x^2 + y^2 = 9$ **15.** $(x - 1)^2 + (y - 4)^2 = 4$ **17.** $(x + 2)^2 + (y - 4)^2 = 36$ **19.** $x^2 + (y - 3)^2 = 16$ **21.** $(x - 5)^2 + (y + 5)^2 = 25$

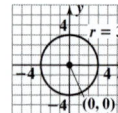   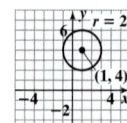   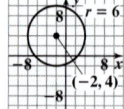   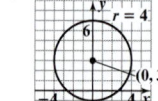   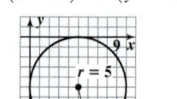

23. $(x - 1)^2 + (y - 2)^2 = 5$ **25.** $C = (0, 0), r = 6$ **27.** $C = (4, 1), r = 5$ **29.** $C = (-3, 2), r = 9$ **31.** $C = (0, 3), r = 8$

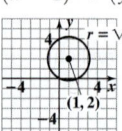   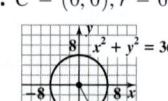   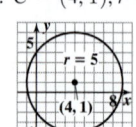   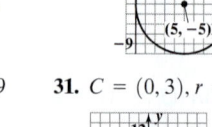   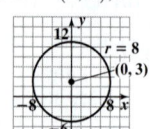

33. $C = (1, -1), r = \dfrac{1}{2}$ **35.** $C = (3, -1), r = 3$ **37.** $C = (-5, -2), r = 5$ **39.** $C = (3, -6), r = 9$

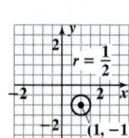   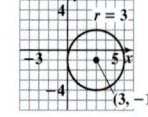   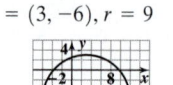   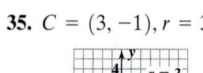

$x^2 + y^2 - 6x + 2y + 1 = 0$ $x^2 + y^2 + 10x + 4y + 4 = 0$ $2x^2 + 2y^2 - 12x + 24y - 72 = 0$

41. $x^2 + y^2 = 20$ **43.** $(x + 3)^2 + (y - 2)^2 = 9$ **45.** $(x + 1)^2 + (y + 1)^2 = 25$ **47.** $A = 64\pi$ square units; $C = 16\pi$ units
49. 32 square units   **51.** $(x - 2)^2 + y^2 = 1; x^2 + y^2 - 8x + 7 = 0$

53. $f(x) = 4x - 3$ **55.** $g(x) = x^2 - 4x - 5$ **57.** $G(x) = -2(x + 3)^2 - 5$ **59.**

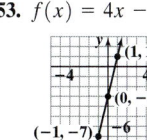 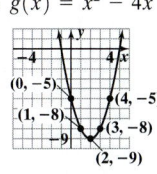 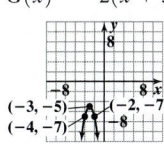 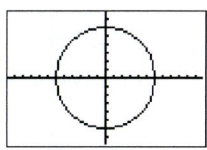

The graph here agrees with that in Problem 25.

61. 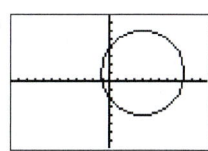   **63.** 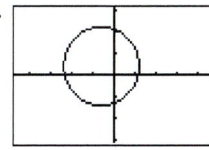   **65.** 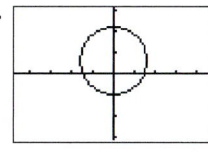   **67.**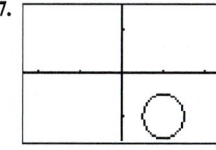

The graph here agrees with that in Problem 27.   The graph here agrees with that in Problem 29.   The graph here agrees with that in Problem 31.   The graph here agrees with that in Problem 33.

Section 12.3

1. Conics   **3.** vertex   **5.** false   **7.** 8 units   **9.** Answers may vary.   **11.** c   **13.** a   **15.** b   **17.** e
$y^2 = 20x$

19. 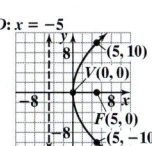   **21.** $x^2 = -24y$ 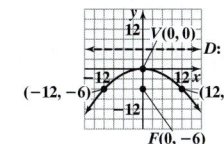   **23.** $x^2 = 6y$ 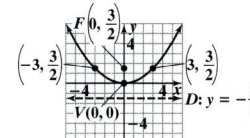   **25.** $x^2 = -12y$ 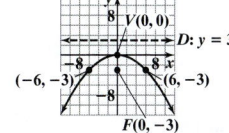   **27.** $y^2 = -12x$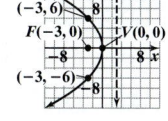

29. vertex $(0, 0)$, focus $(0, 6)$, directrix $y = -6$ 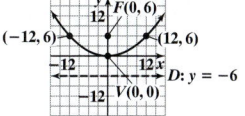   **31.** vertex $(0, 0)$, focus $\left(-\dfrac{3}{2}, 0\right)$, directrix $x = \dfrac{3}{2}$ 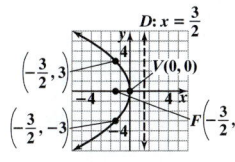   **33.** vertex $(0, 0)$, focus $(0, -2)$, directrix $y = 2$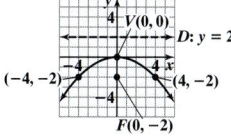

35. vertex $(2, 4)$, focus $(2, 5)$, directrix $y = 3$ 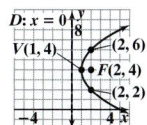   **37.** vertex $(-2, -3)$, focus $(-4, -3)$, directrix $x = 0$ **39.** vertex $(-5, 1)$, focus $(-5, -4)$, directrix $y = 6$ **41.** vertex $(-2, -1)$, focus $(-2, -4)$, directrix $y = 2$

43. vertex $(1, 4)$, focus $(2, 4)$, directrix $x = 0$ **45.** vertex $(-5, 2)$, focus $\left(-5, \dfrac{1}{2}\right)$, directrix $y = \dfrac{7}{2}$ 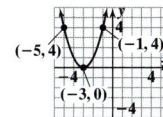   **47.** $y^2 = x$ **49.** 1 inch above the vertex along its axis of symmetry   **51.** 21.6 feet
53. The height of the bridge is 28.8 feet at a distance of 10 feet from the center, 19.2 feet at a distance of 30 feet from the center, and 0 feet (i.e., ground level) at a distance of 50 feet from the center.

55. $(x - 3)^2 = 4(y + 2)$ **57.** $(y - 3)^2 = -4(x - 2)$

59. (a) Let $x = 4$ and $y = 2$:
$4^2 \stackrel{?}{=} 8 \cdot 2$
$16 = 16 \leftarrow$ True
Thus, $(4, 2)$ in on the parabola.
(b) The focus of the parabola is $F(0, 2)$, and the directrix is $D: y = -2$.
$d(F, P) = \sqrt{(0 - 4)^2 + (2 - 2)^2} = \sqrt{16} = 4$,
$d(P, D) = 2 - (-2) = 4$.
Thus, $d(F, P) = d(P, D) = 4$.

61. $y = (x + 3)^2$ 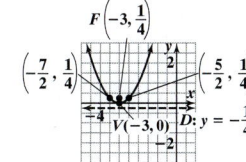   **63.** $y = (x + 3)^2$ 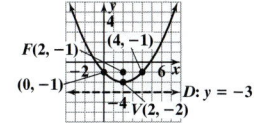   **65.** $4(y + 2) = (x - 2)^2$

67. 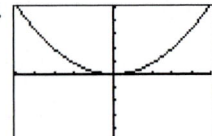 **69.** 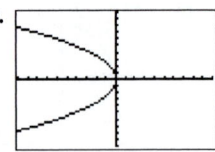 **71.** 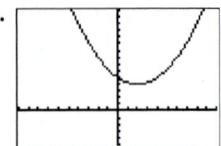 **73.**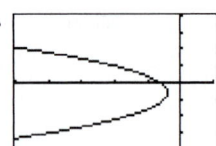

75. 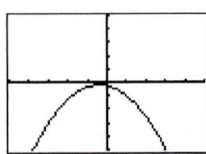 **77.**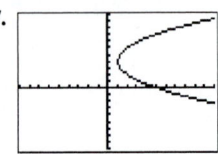

Section 12.4

1. ellipse, foci **3.** vertices **5.** true

7. 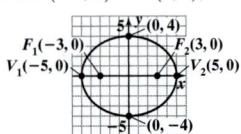 **9.** c **11.** d **13.** Foci: $(-3, 0)$ and $(3, 0)$; Vertices: $(-5, 0)$ and $(5, 0)$ **15.** Foci: $(0, -8)$ and $(0, 8)$; Vertices: $(0, -10)$, and $(0, 10)$

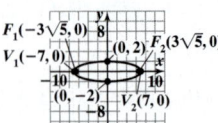 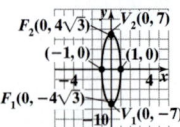 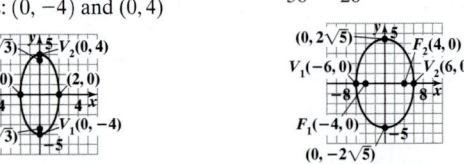

17. Foci: $\left(-3\sqrt{5}, 0\right)$ and $\left(3\sqrt{5}, 0\right)$; Vertices: $(-7, 0)$ and $(7, 0)$ **19.** Foci: $\left(0, -4\sqrt{3}\right)$ and $\left(0, 4\sqrt{3}\right)$; Vertices: $(0, -7)$ and $(0, 7)$ **21.** Foci: $\left(0, -2\sqrt{3}\right)$ and $\left(0, 2\sqrt{3}\right)$; Vertices: $(0, -4)$ and $(0, 4)$ **23.** $\dfrac{x^2}{36} + \dfrac{y^2}{20} = 1$

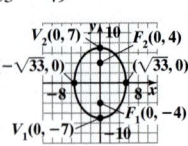 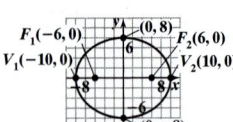 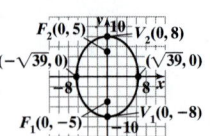 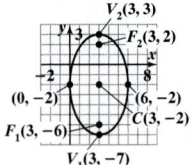

25. $\dfrac{x^2}{33} + \dfrac{y^2}{49} = 1$ **27.** $\dfrac{x^2}{100} + \dfrac{y^2}{64} = 1$ **29.** $\dfrac{x^2}{39} + \dfrac{y^2}{64} = 1$ **31.** $\dfrac{(x-3)^2}{9} + \dfrac{(y+2)^2}{25} = 1$

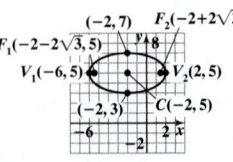 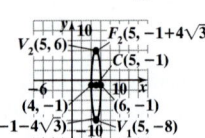 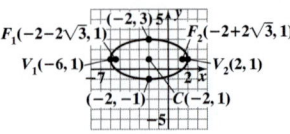 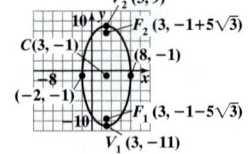

33. $\dfrac{(x+2)^2}{16} + \dfrac{(y-5)^2}{4} = 1$ **35.** $(x-5)^2 + \dfrac{(y+1)^2}{49} = 1$ **37.** $4(x+2)^2 + 16(y-1)^2 = 64$ **39.** $4x^2 + y^2 - 24x + 2y - 63 = 0$

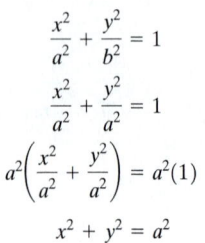

41. (a) $\dfrac{x^2}{225} + \dfrac{y^2}{100} = 1$ **(b)** Yes **(c)** No **43.** Perihelion = 91.5 million miles; $\dfrac{x^2}{8649} + \dfrac{y^2}{8646.75} = 1$ **45.** Perihelion = 460.6 million miles;

Mean distance = 483.8 million miles; $\dfrac{x^2}{234{,}062.44} + \dfrac{y^2}{233{,}524.2} = 1$ **47.** $\dfrac{(x-1)^2}{16} + \dfrac{(y-2)^2}{9} = 1$ **49.** $\dfrac{(x-2)^2}{4} + \dfrac{y^2}{16} = 1$

51. Let $a = b$, then

$$\frac{x^2}{a^2} + \frac{y^2}{b^2} = 1$$

$$\frac{x^2}{a^2} + \frac{y^2}{a^2} = 1$$

$$a^2\left(\frac{x^2}{a^2} + \frac{y^2}{a^2}\right) = a^2(1)$$

$$x^2 + y^2 = a^2$$

which is the equation of a circle with center $(0, 0)$ and radius a. $c = 0$; The foci are located at the center point.

53.

x	5	10	100	1000
$f(x)$	0.71429	0.41667	0.04902	0.00499

55.

x	5	10	100	1000
$f(x)$	5.5	3	2.07216	2.00702

57.

x	5	10	100	1000
$f(x)$	6.83333	11.90909	101.99010	1001.99900
$g(x)$	7	12	102	1002

59. In Problems 53 and 54, the degree of the numerator is less than the degree of the denominator. In Problems 55 and 56, the degree of the numerator and denominator are the same.

Conjecture 1: If the degree of the numerator of a rational function is less than the degree of the denominator, then as x increases, the value of the function will approach zero (0).

Conjecture 2: If the degree of the numerator of a rational function equals the degree of the denominator, then as x increases, the value of the function will approach the ratio of the leading coefficients of the numerator and denominator.

61. 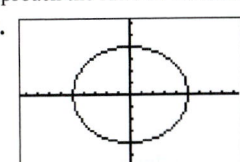 **63.** 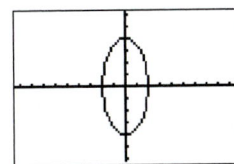 **65.** 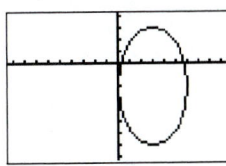 **67.**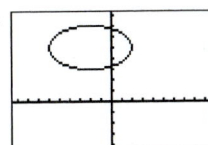

Section 12.5

1. hyperbola **3.** $y = -\dfrac{b}{a}x;\ y = \dfrac{b}{a}x$ **5.** True **7.** Answers may vary. **9.** b **11.** a

13. $\dfrac{x^2}{4} - \dfrac{y^2}{16} = 1$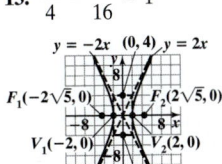

15. $\dfrac{y^2}{25} - \dfrac{x^2}{36} = 1$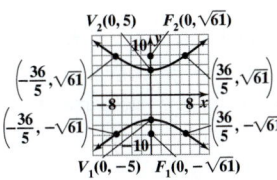

17. $4x^2 - y^2 = 36$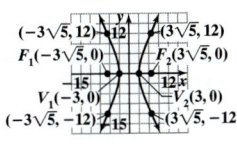

19. $25y^2 - x^2 = 100$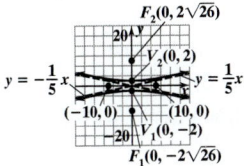

21. $\dfrac{x^2}{4} - \dfrac{y^2}{5} = 1$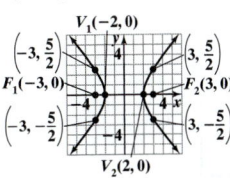

23. $\dfrac{y^2}{25} - \dfrac{x^2}{24} = 1$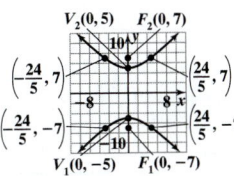

25. $\dfrac{x^2}{49} - \dfrac{y^2}{51} = 1$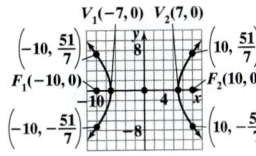

27. $\dfrac{y^2}{64} - \dfrac{x^2}{16} = 1$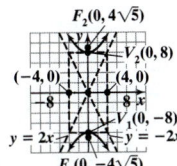

29. $\dfrac{x^2}{4.5} - \dfrac{y^2}{4.5} = 1$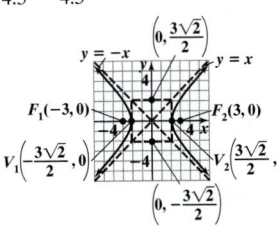

31. $x^2 - y^2 = 1$ **33.** $\dfrac{y^2}{36} - \dfrac{x^2}{9} = 1$

35. The asymptotes of both hyperbolas are $y = -\dfrac{1}{2}x$ and $y = \dfrac{1}{2}x$. Thus, they are conjugates.

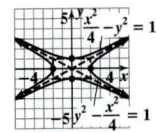

37. $(-3, 1)$
39. $\{(x, y) \mid 2x - 3y = 6\}$
41. $\{(x, y) \mid 6x + 3y = 4\}$
43.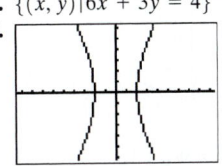

45. 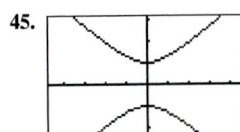 **47.** 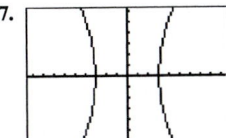 **49.**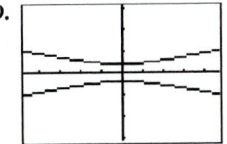

Putting the Concepts Together (Sections 1–5)

1. $3\sqrt{13}$ **2.** $(1, -3)$

3. $C = (-2, 8), r = 6$ **4.** $C = (-3, 2), r = 4$ **5.** $x^2 + y^2 = 169$ **6.** $(x - 2)^2 + (y - 1)^2 = 25$ **7.** Vertex: $(-2, 4)$; focus: $(-2, 3)$; directrix: $y = 5$

8. Vertex: $(3, -1)$; focus: $(5, -1)$; directrix: $x = 1$

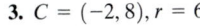 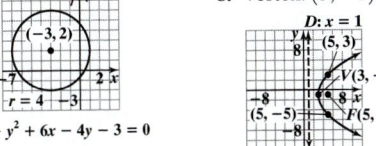 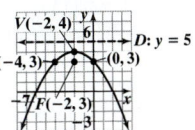

$(x + 2)^2 + (y - 8)^2 = 36$ $x^2 + y^2 + 6x - 4y - 3 = 0$

9. $(x + 1)^2 = -12(y + 2)$ **10.** $(y - 3)^2 = 8(x + 3)$

11. Foci: $\left(-6\sqrt{2}, 0\right)$ and $\left(6\sqrt{2}, 0\right)$; vertices: $(-9, 0)$ and $(9, 0)$

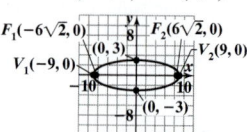

12. Center: $(-1, 2)$; vertices: $(-1, -5)$ and $(-1, 9)$; foci: $\left(-1, 2 - \sqrt{13}\right)$ and $\left(-1, 2 + \sqrt{13}\right)$

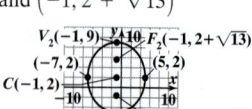

13. $\dfrac{x^2}{45} + \dfrac{y^2}{81} = 1$

14. $\dfrac{(x-3)^2}{16} + \dfrac{(y+4)^2}{7} = 1$

15. Vertices: $(0, -9)$ and $(0, 9)$; foci: $\left(0, -3\sqrt{10}\right)$ and $\left(0, 3\sqrt{10}\right)$; asymptotes: $y = 3x$ and $y = -3x$

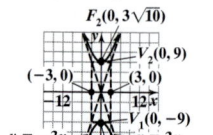

16. Vertices: $(-1, 0)$ and $(1, 0)$; foci: $\left(-\sqrt{26}, 0\right)$ and $\left(\sqrt{26}, 0\right)$; asymptotes: $y = -5x$ and $y = 5x$

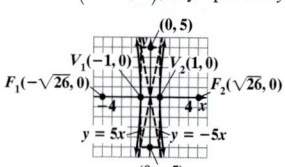

17. $\dfrac{y^2}{4} - \dfrac{x^2}{21} = 1$

18. 6.75 inches above the vertex along its axis of symmetry.

Section 12.6

1. substitution, elimination   **3.** In a system of nonlinear equations, at least one of its equations is not linear.   **5.** $(0, 4)$ and $(1, 5)$
7. $(3, 4)$ and $(4, 3)$ **9.** $(0, -2), \left(-\sqrt{3}, 1\right)$, and $\left(\sqrt{3}, 1\right)$ **11.** $(-2, -2)$ and $(2, 2)$ **13.** $(0, -2), (0, 2), \left(-1, -\sqrt{3}\right)$, and $\left(-1, \sqrt{3}\right)$
15. $\varnothing$ **17.** $(0, 0), (-3, 3)$, and $(3, 3)$ **19.** $(-3, 7)$ and $(2, -8)$ **21.** $(-1, 11)$ and $(2, -4)$ **23.** $(-4, 0)$ and $(4, 0)$ **25.** $(0, 3)$ and $(1, 4)$
27. $\left(0, -2 - \sqrt{3}\right), \left(0, -2 + \sqrt{3}\right), (1, -4)$, and $(1, 0)$ **29.** $\varnothing$ **31.** $(5, 2)$ **33.** $\left(-\dfrac{8}{3}, -\dfrac{2\sqrt{10}}{3}\right), \left(-\dfrac{8}{3}, \dfrac{2\sqrt{10}}{3}\right), \left(\dfrac{8}{3}, -\dfrac{2\sqrt{10}}{3}\right)$, and $\left(\dfrac{8}{3}, \dfrac{2\sqrt{10}}{3}\right)$
35. $(0, -5), (3, 4), (4, 3)$, and $(5, 0)$ **37.** Either -5 and -3, or 3 and 5   **39.** 14 feet by 10 feet   **41.** 19 cm by 10 cm   **43.** $(0, -2), (0, 1)$, and $(2, -1)$
45. $(81, 3)$
47. If $r_1 = \dfrac{-b + \sqrt{b^2 - 4ac}}{2a}$, then $r_2 = \dfrac{-b - \sqrt{b^2 - 4ac}}{2a}$; if $r_1 = \dfrac{-b - \sqrt{b^2 - 4ac}}{2a}$, then $r_2 = \dfrac{-b + \sqrt{b^2 - 4ac}}{2a}$.
49. (a) $f(1) = 7$ (b) $g(1) = 2$ **51.** (a) $f(3) = 13$ (b) $g(3) = 8$ **53.** (a) $f(5) = 19$ (b) $g(5) = 32$

55.

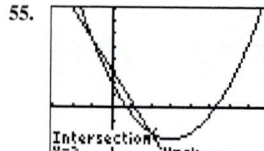 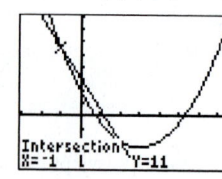

$(-1, 11)$ and $(2, -4)$

57.

 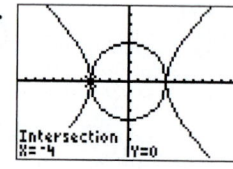 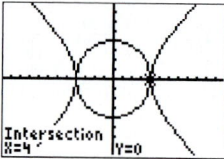

$(-4, 0)$ and $(4, 0)$

59.

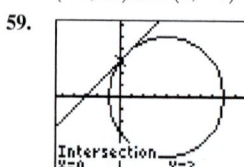 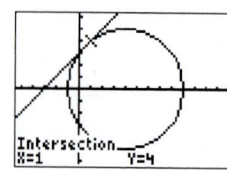

$(0, 3)$ and $(1, 4)$

61.

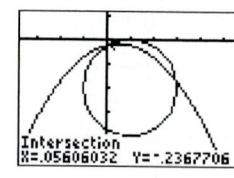 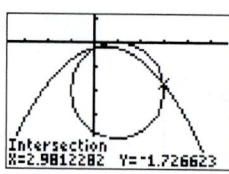

approximately $(0.056, -0.237)$ and $(2.981, -1.727)$

63.

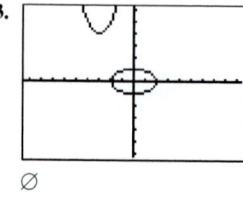

$\varnothing$

Chapter 12 Review

1. 5   **2.** 10   **3.** $2\sqrt{10} \approx 6.32$ **4.** 6   **5.** $3\sqrt{19} \approx 13.08$ **6.** 2.5   **7.** $(-2, 5)$ **8.** $(6, -2)$ **9.** $\left(-4\sqrt{3}, -3\sqrt{6}\right)$ **10.** $\left(\dfrac{5}{2}, \dfrac{1}{2}\right)$
11. $\left(\dfrac{3}{4}, \dfrac{1}{2}\right)$ **12.** (a)

$A(-2, 2)$, $C(-1, -3)$, $B(1, -1)$

(b) $d(A, B) = 3\sqrt{2} \approx 4.24$;
$d(B, C) = 2\sqrt{2} \approx 2.83$;
$d(A, C) = \sqrt{26} \approx 5.10$

(c) $[d(A, B)]^2 + [d(B, C)]^2 \stackrel{?}{=} [d(A, C)]^2$
$\left(3\sqrt{2}\right)^2 + \left(2\sqrt{2}\right)^2 \stackrel{?}{=} \left(\sqrt{26}\right)^2$
$9 \cdot 2 + 4 \cdot 2 \stackrel{?}{=} 26$
$18 + 8 \stackrel{?}{=} 26$
$26 = 26 \leftarrow$ True
Therefore, triangle ABC is a right triangle.
(d) 6 square units

13. $C = (-2, 1); r = 4; (x + 2)^2 + (y - 1)^2 = 16$
14. $C = (5, 3); r = 3; (x - 5)^2 + (y - 3)^2 = 9$
15. $x^2 + y^2 = 16$ **16.** $(x + 3)^2 + (y - 1)^2 = 9$

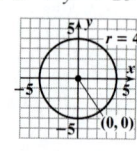 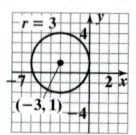

17. $(x - 5)^2 + (y + 2)^2 = 1$ **18.** $(x - 4)^2 + y^2 = 7$ **19.** $(x - 2)^2 + (y + 1)^2 = 25$ **21.** $C = (0, 0), r = 5$ **22.** $C = (1, 2), r = 2$

20. $(x + 1)^2 + (y - 3)^2 = 20$

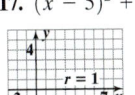
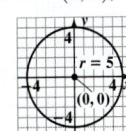

23. $C = (0, 4), r = 4$ **24.** $C = (-1, -6), r = 7$ **25.** $C = \left(-2, \dfrac{3}{2}\right), r = \dfrac{1}{2}$ **26.** $C = (-3, -3), r = 2$ **27.** $C = (-3, -5), r = 6$

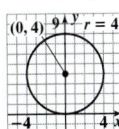
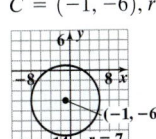
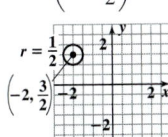
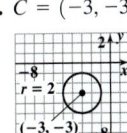
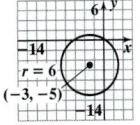

28. $C = (4, -2), r = 2$ **29.** $C = (-1, 2), r = 3$ **30.** $C = (5, 1), r = 3$ **31.** $x^2 = -12y$ **32.** $y^2 = -16x$

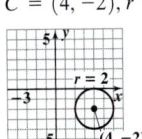
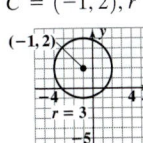
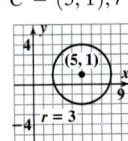
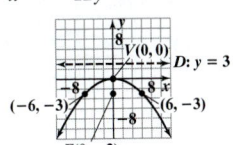
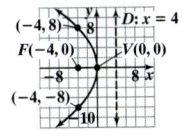

33. $y^2 = \dfrac{1}{2}x$ **34.** $x^2 = 8y$ **35.** Vertex $(0, 0)$; focus $\left(0, \dfrac{1}{2}\right)$; directrix $y = -\dfrac{1}{2}$.   **36.** Vertex $(0, 0)$; focus $(4, 0)$; directrix $x = -4$

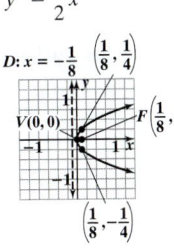
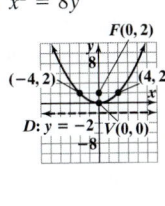
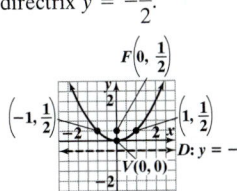
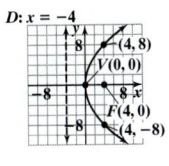

37. Vertex $(-1, 3)$; focus $(-1, 5)$; directrix $y = 1$ **38.** Vertex $(-3, 4)$; focus $\left(-\dfrac{7}{2}, 4\right)$; directrix $x = -\dfrac{5}{2}$ **39.** Vertex $(5, 2)$; focus $\left(5, \dfrac{5}{4}\right)$; directrix $y = \dfrac{11}{4}$

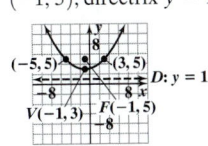
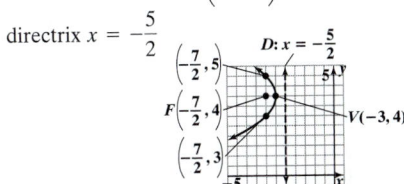
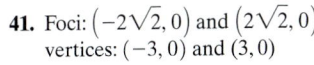
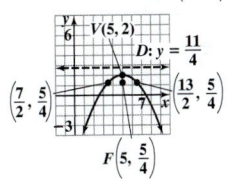

40. approximately 127.84 feet above the center of the dish along its axis of symmetry   **41.** Foci: $\left(-2\sqrt{2}, 0\right)$ and $\left(2\sqrt{2}, 0\right)$; vertices: $(-3, 0)$ and $(3, 0)$ **42.** Foci: $\left(0, -\sqrt{5}\right)$ and $\left(0, \sqrt{5}\right)$; vertices: $(0, -3)$, and $(0, 3)$

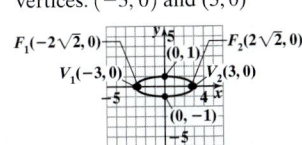
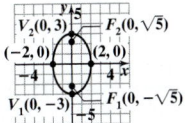

43. $\dfrac{x^2}{16} + \dfrac{y^2}{25} = 1$ **44.** $\dfrac{x^2}{36} + \dfrac{y^2}{32} = 1$ **45.** $\dfrac{x^2}{100} + \dfrac{y^2}{36} = 1$ **46.** Vertices: $(-6, -2)$ and $(8, -2)$; foci: $\left(1 - 2\sqrt{6}, -2\right)$ and $\left(1 + 2\sqrt{6}, -2\right)$

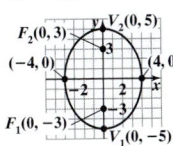
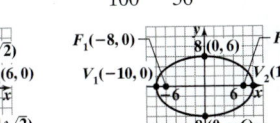
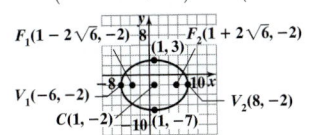

47. Vertices: $(-3, -1)$ and $(-3, 9)$; foci: $(-3, 0)$ and $(-3, 8)$ **48. (a)** $\dfrac{x^2}{900} + \dfrac{y^2}{256} = 1$ **(b)** Yes   **49.** Vertices: $(-2, 0)$ and $(2, 0)$; foci: $\left(-\sqrt{13}, 0\right)$ and $\left(\sqrt{13}, 0\right)$ **50.** Vertices: $(0, -5)$ and $(0, 5)$; foci: $\left(0, -\sqrt{74}\right)$ and $\left(0, \sqrt{74}\right)$

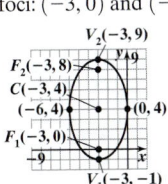
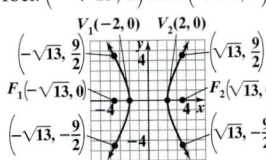
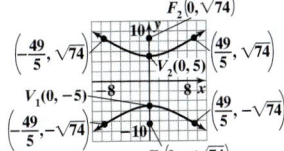

51. Vertices: $(0, -5)$ and $(0, 5)$; foci: $\left(0, -\sqrt{41}\right)$ and $\left(0, \sqrt{41}\right)$

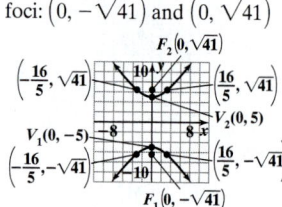

52. Asymptotes: $y = x$ and $y = -x$

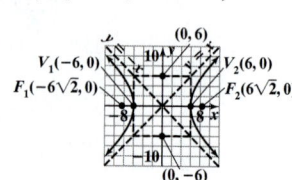

53. Asymptotes: $y = \dfrac{5}{2}x$ and $y = -\dfrac{5}{2}x$

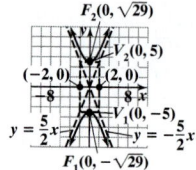

54. $\dfrac{x^2}{9} - \dfrac{y^2}{7} = 1$

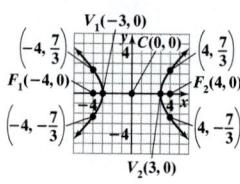

55. $\dfrac{y^2}{9} - \dfrac{x^2}{16} = 1$

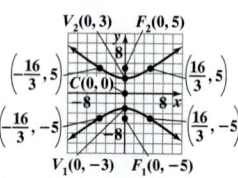

56. $\dfrac{y^2}{16} - \dfrac{x^2}{9} = 1$

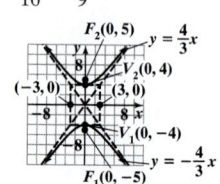

57. $\left(\sqrt{2}, \sqrt{2}\right)$ and $\left(-\sqrt{2}, -\sqrt{2}\right)$

58. $\left(-\dfrac{1}{2}, \dfrac{3}{2}\right)$ and $(1, 3)$ **59.** $\left(-\dfrac{1}{6}, -6\right)$ and $(1, 1)$

60. $(-5, -1)$, $(-5, 1)$, $(5, -1)$, and $(5, 1)$

61. $\left(2, -2\sqrt{2}\right)$ and $\left(2, 2\sqrt{2}\right)$

62. $(-1, 3)$ and $(1, 3)$

63. $\left(-\sqrt{3}, -\sqrt{5}\right)$, $\left(-\sqrt{3}, \sqrt{5}\right)$, $\left(\sqrt{3}, -\sqrt{5}\right)$, and $\left(\sqrt{3}, \sqrt{5}\right)$ **64.** $(0, 0)$, $\left(5, -\sqrt{15}\right)$, and $\left(5, \sqrt{15}\right)$ **65.** $(2, 4)$ and $(-1, 1)$

66. $(-7, -20)$ and $\left(2, \dfrac{5}{2}\right)$ **67.** $(0, 6)$ and $(-6, 0)$ **68.** $(4, 4)$ and $(1, -2)$ **69.** $\varnothing$ **70.** $(-1, -2)$, $(-1, 2)$, $(1, -2)$, and $(1, 2)$

71. $(-4, 0)$, $(4, 0)$, and $(0, 4)$ **72.** $(4, 0)$ and $(0, -2)$ **73.** 7 and 5 **74.** 12 cm by 5 cm **75.** 72 inches by 30 inches **76.** 12 inches and 9 inches

Chapter 12 Test

1. $4\sqrt{5}$ **2.** $(-1, 2)$

3. $C = (4, -1)$, $r = 3$

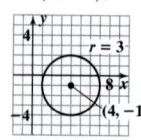

4. $C = (-5, 2)$, $r = 4$

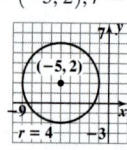

5. $(x + 3)^2 + (y - 7)^2 = 36$

6. $(x + 5)^2 + (y - 8)^2 = 100$

7. Vertex $(1, -2)$; focus $(2, -2)$; directrix $x = 0$

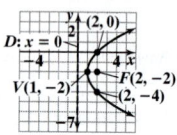

8. Vertex $(2, 4)$; focus $\left(2, \dfrac{13}{4}\right)$; directrix $y = \dfrac{19}{4}$

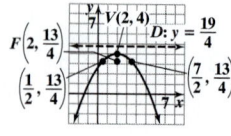

9. $x^2 = -16y$ **10.** $(y - 4)^2 = 8(x - 1)$

11. Foci: $(-4, 0)$ and $(4, 0)$; vertices: $(-5, 0)$ and $(5, 0)$

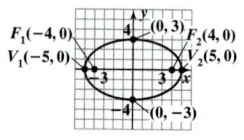

12. Vertices: $(2, -8)$ and $(2, 0)$; foci: $\left(2, -4 - \sqrt{7}\right)$ and $\left(2, -4 + \sqrt{7}\right)$

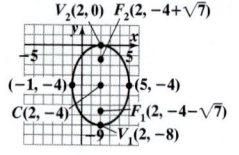

13. $\dfrac{x^2}{9} + \dfrac{y^2}{25} = 1$ **14.** $\dfrac{(x + 1)^2}{16} + \dfrac{(y - 2)^2}{25} = 1$ **15.** Vertices: $(-1, 0)$ and $(1, 0)$; foci: $\left(-\sqrt{5}, 0\right)$ and $\left(\sqrt{5}, 0\right)$; asymptotes: $y = 2x$ and $y = -2x$

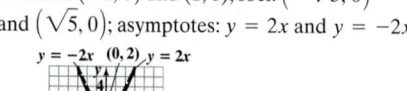
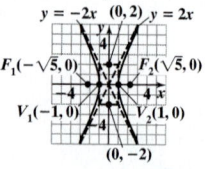

16. Vertices: $(0, -10)$ and $(0, 10)$; foci: $\left(0, -2\sqrt{41}\right)$ and $\left(0, 2\sqrt{41}\right)$; asymptotes: $y = \dfrac{5}{4}x$ and $y = -\dfrac{5}{4}x$

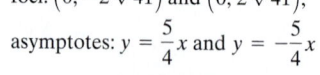
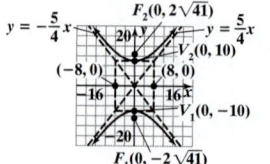

17. $\dfrac{x^2}{9} - \dfrac{y^2}{55} = 1$ **18.** $(-4, 1)$ and $(1, -4)$ **19.** $\left(-\sqrt{5}, -2\right)$, $\left(-\sqrt{5}, 2\right)$, $\left(\sqrt{5}, -2\right)$, and $\left(\sqrt{5}, 2\right)$ **20. (a)** $\dfrac{x^2}{225} + \dfrac{y^2}{100} = 1$ **(b)** 6 feet

Chapter 13

Section 13.1

1. sequence **3.** partial sum **5.** false **7.** Answers will vary. **9.** Answers will vary. **11.** 8, 11, 14, 17, and 20 **13.** $\dfrac{1}{3}, \dfrac{1}{2}, \dfrac{3}{5}, \dfrac{2}{3}$, and $\dfrac{5}{7}$

15. $-1, 2, -3, 4$, and -5 **17.** 3, 5, 9, 17, and 33 **19.** $1, 1, \dfrac{3}{4}, \dfrac{1}{2}$, and $\dfrac{5}{16}$ **21.** $\dfrac{1}{e}, \dfrac{2}{e^2}, \dfrac{3}{e^3}, \dfrac{4}{e^4}$, and $\dfrac{5}{e^5}$ **23.** $a_n = 2n$ **25.** $a_n = \dfrac{n}{n + 1}$

27. $a_n = n^2 + 2$ **29.** $a_n = (-1)^n n^2$ **31.** 54   **33.** $\dfrac{55}{2}$ **35.** 14   **37.** 6   **39.** 50   **41.** 45   **43.** $\displaystyle\sum_{k=1}^{15} k$ **45.** $\displaystyle\sum_{i=1}^{12} \dfrac{1}{i}$

47. $\displaystyle\sum_{i=1}^{9} (-1)^{i+1}\left(\dfrac{1}{3^{i-1}}\right)$ **49.** $\displaystyle\sum_{k=1}^{11} (2k + 3)$ **51. (a)** \$12,180   **(b)** \$12,736.36   **(c)** \$21,768.22   **53. (a)** 313 million   **(b)** 484 million

55. $1, 1, 2, 3, 5, 8, 13, 21, 34,$ and 55 **57.** $10, 10.5, 11.025, 11.57625,$ and 12.1550625 **59.** $8, 10, 13, 17,$ and 22

61. (a) $1; 2; 1.5; 1.\overline{6}; 1.6; 1.625; \dfrac{21}{13} \approx 1.615385; \dfrac{34}{21} \approx 1.619048; \dfrac{55}{34} \approx 1.617647; \dfrac{89}{55} \approx 1.618182$ **(b)** around 1.618

(c) $1; 0.5; \dfrac{2}{3} = 0.\overline{6}; 0.6; 0.625; \dfrac{8}{13} \approx 0.615385; \dfrac{13}{21} \approx 0.619048; \dfrac{21}{34} \approx 0.617647; \dfrac{34}{55} \approx 0.618182; \dfrac{55}{89} \approx 0.617978$ **(d)** around 0.618

63. (a) $m = 4$ **(b)** $f(1) = -2; f(2) = 2; f(3) = 6; f(4) = 10$ **65. (a)** $m = -5$ **(b)** $f(1) = 3; f(2) = -2; f(3) = -7; f(4) = -12$

67. $8, 11, 14, 17,$ and 20 **69.** $\dfrac{1}{3}, \dfrac{1}{2}, \dfrac{3}{5}, \dfrac{2}{3},$ and $\dfrac{5}{7}$ **71.** $-1, 2, -3, 4,$ and -5 **73.** $3, 5, 9, 17,$ and 33 **75.** 54   **77.** $\dfrac{55}{2}$ **79.** 14   **81.** 6

Section 13.2

1. arithmetic   **3.** false   **5.** Answers will vary.   **7.** $d = 1; 6, 7, 8,$ and 9 **9.** $d = 7; 9, 16, 23,$ and 30 **11.** $d = -3; 4, 1, -2,$ and -5

13. $d = \dfrac{1}{2}; \dfrac{11}{2}, 6, \dfrac{13}{2},$ and 7 **15.** $a_n = 3n + 1; a_5 = 16$ **17.** $a_n = -5n + 15; a_5 = -10$ **19.** $a_n = \dfrac{1}{3}n + \dfrac{5}{3}; a_5 = \dfrac{10}{3}$

21. $a_n = -\dfrac{1}{5}n + \dfrac{26}{5}; a_5 = \dfrac{21}{5}$ **23.** $a_n = 5n - 3; a_{20} = 97$ **25.** $a_n = -3n + 15; a_{20} = -45$ **27.** $a_n = \dfrac{1}{4}n + \dfrac{3}{4}; a_{20} = \dfrac{23}{4}$

29. $a = 7; d = 5; a_n = 5n + 2$ **31.** $a = -23; d = 7; a_n = 7n - 30$ **33.** $a = 11; d = -3; a_n = -3n + 14$ **35.** $a = 4; d = -\dfrac{1}{2}; a_n = -\dfrac{1}{2}n + \dfrac{9}{2}$

37. $S_{30} = 2670$ **39.** $S_{25} = 700$ **41.** $S_{40} = -5060$ **43.** $S_{40} = 3160$ **45.** $S_{75} = -9000$ **47.** $S_{30} = 460$ **49.** $x = -\dfrac{3}{2}$

51. There are 630 cans in the stack.   **53.** There are 1600 seats in the auditorium.   **55.** There are 84 terms in the sequence.

57. There are 63 terms in the sequence.   **59.** It will take about 15.22 years.   **61. (a)** 3   **(b)** $3; 9; 27; 81$ **63. (a)** $\dfrac{1}{2}$ **(b)** $5; \dfrac{5}{2}, \dfrac{5}{4}, \dfrac{5}{8}$

65. 806.9   **67.** 1427.5

Section 13.3

1. geometric   **3.** false   **5.** Answers will vary.   **7.** $r = 4; 4, 16, 64,$ and 256 **9.** $r = \dfrac{2}{3}; \dfrac{2}{3}, \dfrac{4}{9}, \dfrac{8}{27},$ and $\dfrac{16}{81}$ **11.** $r = \dfrac{1}{2}; \dfrac{3}{2}, \dfrac{3}{4}, \dfrac{3}{8},$ and $\dfrac{3}{16}$

13. $r = \dfrac{5}{2}; \dfrac{1}{2}, \dfrac{5}{4}, \dfrac{25}{8},$ and $\dfrac{125}{16}$ **15.** arithmetic; $d = 5$ **17.** Neither   **19.** geometric; $r = \dfrac{1}{2}$ **21.** geometric; $r = \dfrac{2}{3}$ **23.** arithmetic; $d = 4$

25. Neither   **27. (a)** $a_n = 10 \cdot 2^{n-1}$ **(b)** $a_8 = 1280$ **29. (a)** $a_n = 100 \cdot \left(\dfrac{1}{2}\right)^{n-1}$ **(b)** $a_8 = \dfrac{25}{32}$ **31. (a)** $a_n = (-3)^{n-1}$ **(b)** $a_8 = -2187$

33. (a) $a_n = 100 \cdot (1.05)^{n-1}$ **(b)** $a_8 = 100 \cdot (1.05)^7$ **35.** $a_{10} = 1536$ **37.** $a_{15} = \dfrac{1}{4096}$ **39.** $a_9 = 0.000000005$ **41.** 8190   **43.** 83.3245952

45. 6138   **47.** 7.96875   **49.** 2   **51.** 15   **53.** $\dfrac{9}{2}$ **55.** $\dfrac{5}{4}$ **57.** 9   **59.** $\dfrac{5}{9}$ **61.** $\dfrac{89}{99}$ **63.** $x = -4$ **65. (a)** \$42,000   **(b)** \$62,053

(c) \$503,116   **67.** \$13,182   **69. (a)** About 1.891 feet   **(b)** On the 23rd swing   **(c)** About 24.08 feet   **(d)** The pendulum will swing a total of 60 feet.   **71.** Option A will yield the larger annual salary in the final year of the contract and option B will yield the larger cumulative salary over the life of the contract.   **73.** The multiplier is 50.   **75.** \$31.14 per share   **77.** \$149,035.94   **79.** \$114,401.52   **81.** \$395.09, or about \$395

83. $0.4\overline{9} = \dfrac{1}{2}$ **85.** 2,147,483,646   **87.** Answers will vary.   **89.** 1   **91.** approximately 288.1404315   **93.** 41.66310217

Putting the Concepts Together (Sections 1–3)

1. Geometric with $a = \dfrac{3}{4}$ and common ratio $r = \dfrac{1}{4}$ **2.** Arithmetic with $a = 8$ and $d = 2$ **3.** Arithmetic with $a = 1$ and $d = \dfrac{7}{9}$

4. Neither arithmetic nor geometric   **5.** Geometric with $a = 12$ and $r = 2$ **6.** Neither arithmetic nor geometric   **7.** 87   **8.** $\displaystyle\sum_{i=1}^{12} \dfrac{1}{2(6 + i)}$

9. $a_n = 27 - 2n; 25, 23, 21, 19,$ and 17 **10.** $a_n = 11n - 35; -24, -13, -2, 9,$ and 20 **11.** $a_n = 45 \cdot \left(\dfrac{1}{5}\right)^{n-1}; 45, 9, \dfrac{9}{5}, \dfrac{9}{25},$ and $\dfrac{9}{125}$

12. $a_n = 150 \cdot (1.04)^{n-1}; 150, 156, 162.24, 168.7296,$ and 175.478784 **13.** $S_{11} = 177,146$ **14.** $S_{20} = 990$ **15.** $\dfrac{10,000}{9}$ **16.** A party of 24 people would require 11 tables.

Section 13.4

1. $n(n - 1)(n - 2) \cdots \cdots 3 \cdot 2 \cdot 1$ **3.** Pascal's triangle   **5.** false   **7.** Beginning with n, the exponents of x decrease by 1 in subsequent terms until the exponent is 0.   **9.** The degree of each monomial is equal to n.   **11.** $x^5 + 5x^4 + 10x^3 + 10x^2 + 5x + 1$

13. $x^4 - 16x^3 + 96x^2 - 256x + 256$ **15.** $81p^4 + 216p^3 + 216p^2 + 96p + 16$ **17.** $32z^5 - 240z^4 + 720z^3 - 1080z^2 + 810z - 243$

19. $x^8 + 8x^6 + 24x^4 + 32x^2 + 16$ **21.** $32p^{15} + 80p^{12} + 80p^9 + 40p^6 + 10p^3 + 1$ **23.** $x^6 + 12x^5 + 60x^4 + 160x^3 + 240x^2 + 192x + 64$

25. $16p^8 - 32p^6q^2 + 24p^4q^4 - 8p^2q^6 + q^8$ **27.** 1.00401   **29.** 0.99004   **31.** $84x^5$ **33.** $-108,864p^3$

35. $\dbinom{n}{n-1} = \dfrac{n!}{(n-1)!(n-(n-1))!} = \dfrac{n \cdot (n-1)!}{(n-1)!1!} = \dfrac{n \cdot (n-1)!}{(n-1)!} = n$ $\quad \dbinom{n}{n} = \dfrac{n!}{n!(n-n)!} = \dfrac{n!}{n!0!} = \dfrac{n!}{n!} = 1$

37. $a^4 - 8a^3 + 24a^2 - 32a + 16$ **39.** $p^5 + p^4 - 6p^3 - 14p^2 - 11p - 3$

Chapter 13 Review

1. $-1, -4, -7, -10,$ and -13 **2.** $-\dfrac{1}{5}, 0, \dfrac{1}{7}, \dfrac{1}{4},$ and $\dfrac{1}{3}$ **3.** $6, 26, 126, 626,$ and 3126 **4.** $3, -6, 9, -12,$ and 15 **5.** $\dfrac{1}{2}, \dfrac{4}{3}, \dfrac{9}{4}, \dfrac{16}{5},$ and $\dfrac{25}{6}$

6. $\pi, \dfrac{\pi^2}{2}, \dfrac{\pi^3}{3}, \dfrac{\pi^4}{4},$ and $\dfrac{\pi^5}{5}$ **7.** $a_n = -3n$ **8.** $a_n = \dfrac{n}{3}$ **9.** $a_n = 5 \cdot 2^{n-1}$ **10.** $a_n = (-1)^n \cdot \dfrac{n}{2}$ **11.** $a_n = n^2 + 5$ **12.** $a_n = \dfrac{n-1}{n+1}$ **13.** 65

14. $\dfrac{33}{2}$ **15.** -30 **16.** $\dfrac{26}{3}$ **17.** $\displaystyle\sum_{i=1}^{15}(4 + 3i)$ **18.** $\displaystyle\sum_{i=1}^{8}\dfrac{1}{3^i}$ **19.** $\displaystyle\sum_{i=1}^{10}\dfrac{i^3 + 1}{i + 1}$ **20.** $\displaystyle\sum_{i=1}^{7}[(-1)^{i-1} \cdot i^2]$ **21.** arithmetic with $d = 6$

22. arithmetic with $d = \dfrac{3}{2}$ **23.** not arithmetic   **24.** not arithmetic   **25.** arithmetic with $d = 4$ **26.** not arithmetic

27. $a_n = 8n - 5; a_{25} = 195$ **28.** $a_n = -3n - 1; a_{25} = -76$ **29.** $a_n = -\dfrac{1}{3}n + \dfrac{22}{3}; a_{25} = -1$ **30.** $a_n = 6n + 5; a_{25} = 155$

31. $a_n = \dfrac{18}{5}n - \dfrac{19}{5}; a_{25} = \dfrac{431}{5}$ **32.** $a_n = -4n - 4; a_{25} = -104$ **33.** 4320 **34.** -2140 **35.** -4080 **36.** $\dfrac{1875}{4}$ or 468.75 **37.** 2106

38. 300 yards   **39.** geometric with $r = 6$ **40.** geometric with $r = -3$ **41.** not geometric   **42.** geometric with $r = \dfrac{2}{3}$

43. geometric with $r = -2$ **44.** not geometric   **45.** $a_n = 4 \cdot 3^{n-1}; a_{10} = 78{,}732$ **46.** $a_n = 8 \cdot \left(\dfrac{1}{4}\right)^{n-1}; a_{10} = \dfrac{1}{32{,}768}$

47. $a_n = 5 \cdot (-2)^{n-1}; a_{10} = -2560$ **48.** $a_n = 1000 \cdot (1.08)^{n-1}; a_{10} \approx 1999.005$ **49.** $65{,}534$ **50.** ≈ 45.71428571 **51.** $\dfrac{12{,}285}{4}$ or 3071.25

52. $-258{,}280{,}320$ **53.** $\dfrac{20}{3}$ **54.** $\dfrac{100}{3}$ **55.** $\dfrac{5}{4}$ **56.** $\dfrac{8}{9}$ **57.** After 72 years, there will be 3.125 grams of the Tritium remaining.
58. After 15 minutes a total of about 38.15 billion e-mails will have been sent.   **59.** After 25 years, Scott's 403(b) will be worth \$360,114.89.
60. The lump sum option would yield more money after 26 years.   **61.** Sheri would need to contribute \$534.04, or about \$534, each month to reach her goal.   **62.** When Samantha turns 18, the plan will be worth \$62,950.79 and will cover about 185 credit hours.   **63.** 120   **64.** 7920
65. 5040   **66.** 1716   **67.** 35   **68.** 252   **69.** 1   **70.** 1   **71.** $z^4 + 4z^3 + 6z^2 + 4z + 1$ **72.** $y^5 - 15y^4 + 90y^3 - 270y^2 + 405y - 243$
73. $729y^6 + 5832y^5 + 19{,}440y^4 + 34{,}560y^3 + 34{,}560y^2 + 18{,}432y + 4096$ **74.** $16x^8 - 96x^6 + 216x^4 - 216x^2 + 81$
75. $81p^4 - 216p^3q + 216p^2q^2 - 96pq^3 + 16q^4$ **76.** $a^{15} + 15a^{12}b + 90a^9b^2 + 270a^6b^3 + 405a^3b^4 + 243b^5$ **77.** $-448x^5$ **78.** $14{,}784x^5$

Chapter 13 Test
1. arithmetic with $a = -15$ and $d = 8$ **2.** geometric with $a = -4$ and $r = -4$ **3.** neither arithmetic nor geometric
4. arithmetic with $a = -\dfrac{1}{5}$ and $d = \dfrac{2}{5}$ **5.** neither arithmetic nor geometric   **6.** geometric with $a = 21$ and $r = 3$ **7.** $\dfrac{17269}{1200}$ **8.** $\displaystyle\sum_{i=1}^{8}\dfrac{i + 2}{i + 4}$
9. $a_n = 10n - 4; 6, 16, 26, 36,$ and 46 **10.** $a_n = 4 - 4n; 0, -4, -8, -12,$ and -16 **11.** $a_n = 10 \cdot 2^{n-1}; 10, 20, 40, 80,$ and 160
12. $a_n = (-3)^{n-1}; 1, -3, 9, -27,$ and 81 **13.** 720   **14.** $-\dfrac{132{,}860}{9}$ **15.** 324   **16.** 6435   **17.** 792
18. $625m^4 - 1000m^3 + 600m^2 - 160m + 16$ **19.** about \$64,512   **20.** 2012, 2117, and 2125

Cumulative Review Chapters 1–13
1. $y = \dfrac{3x - 4}{15}$ or $y = \dfrac{1}{5}x - \dfrac{4}{15}$ **2.** $f(2) = 9; f(-3) = 19$ **3.** $\{32\}$ **4.** $\left\{-\dfrac{2}{5}, 1\right\}$ **5.** $\left\{\dfrac{-7 - \sqrt{73}}{6}, \dfrac{-7 + \sqrt{73}}{6}\right\}$ **6.** $\{60\}$
7. $\left\{\dfrac{11}{4}\right\}$ **8.** $\{-4, -2, 0\}$ **9.** $(-6, \infty)$ **10.** $\left[-3, \dfrac{1}{3}\right]$ **11.** $(x + 2)(2x - 9)$ **12.** $(2x - 1)(3x^2 + 2)$ **13.** $20x^3 - 22x^2 + 11x - 3$
14. $\dfrac{(x - 6)(x + 2)}{(x + 4)(x - 1)}$ **15.** $1 - i$ **16.** $\{x \mid x \geq 15\}$ or $[15, \infty)$ **17.** $y = -7x + 11$ **18.** $(4, -1)$
19.   **20.** $(x - 4)^2 + (y + 3)^2 = 36$ **21.**   **22.** 1340   **23.** 8

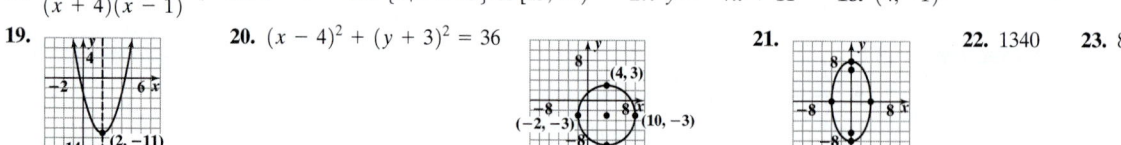

24. It would take about 2.73 hours to cut the lot if both machines worked together.   **25.** $\dfrac{325}{3}$ metric tons of pure aluminum must be added.

Appendix A

Section A.1
1. factors, product   **3.** equivalent fractions   **5.** False   **7.** Answers will vary.   **9.** $5 \cdot 5$ **11.** $2 \cdot 2 \cdot 7$ **13.** $3 \cdot 7$ **15.** $2 \cdot 2 \cdot 3 \cdot 3$
17. $2 \cdot 2 \cdot 5$ **19.** $2 \cdot 3 \cdot 5$ **21.** $2 \cdot 5 \cdot 5$ **23.** 53 is prime.   **25.** $2 \cdot 2 \cdot 3 \cdot 3 \cdot 7$ **27.** $2 \cdot 2 \cdot 2 \cdot 2 \cdot 2 \cdot 2 \cdot 2$ **29.** $2 \cdot 2 \cdot 5 \cdot 5 \cdot 13$
31. $5 \cdot 5 \cdot 7 \cdot 13$ **33.** 30   **35.** 60   **37.** 210   **39.** 90   **41.** 42   **43.** 60   **45.** 72   **47.** 72   **49.** $\dfrac{7}{10}$ **51.** $\dfrac{5}{6}$ **53.** $\dfrac{8}{12}$
55. $\dfrac{18}{24}$ **57.** $\dfrac{21}{3}$ **59.** $\dfrac{4}{8}$ and $\dfrac{3}{8}$ **61.** $\dfrac{9}{15}$ and $\dfrac{10}{15}$ **63.** $\dfrac{20}{24}$ and $\dfrac{15}{24}$ **65.** $\dfrac{44}{60}$ and $\dfrac{21}{60}$ **67.** $\dfrac{27}{48}$ and $\dfrac{14}{48}$ **69.** $\dfrac{12}{96}$ and $\dfrac{21}{96}$
71. $\dfrac{20}{90}$ and $\dfrac{35}{90}$ and $\dfrac{21}{90}$ **73.** $\dfrac{2}{3}$ **75.** $\dfrac{19}{9}$ **77.** $\dfrac{1}{2}$ **79.** $\dfrac{4}{5}$ **81.** $\dfrac{3}{5}$ **83.** $\dfrac{2}{3}$ **85.** 84 months   **87.** 20 days   **89.** $\dfrac{13}{20}$ **91.** $\dfrac{3}{5}$
93. $\dfrac{2}{3}$ **95.** $2, 3, 5, 7, 11, 13, 17, 19, 23, 29, 31, 37, 41, 43, 47, 53, 59, 61, 67, 71, 73, 79, 83, 89, 97$

Section A.2

1. terminating **3.** ones **5.** True **7.** Answers will vary. **9.** hundredths place **11.** tens place **13.** thousandths place **15.** 578.2
17. 350 **19.** 3682.010 **21.** 30 **23.** 0.625 **25.** $0.\overline{285714}$ **27.** 0.3125 **29.** $0.\overline{230769}$ **31.** 1.16 **33.** 2.2 **35.** 0.29 **37.** 2.67
39. 0.519 **41.** $\dfrac{3}{4}$ **43.** $\dfrac{9}{10}$ **45.** $\dfrac{491}{500}$ **47.** $\dfrac{101}{400}$ **49.** 0.37 **51.** 0.0602 **53.** 0.001 **55.** 20% **57.** 27.5% **59.** 200%
61. 70% **63.** 77.27% **65. (a)** 33.33% **(b)** 16.67% **(c)** 25% **67.** 21.43%

Appendix B

1. $0; (x + 1) \cdot (3x - 1)$ **3.** False **5.** True **7.** Answers will vary. **9.** $x + 2$ **11.** $2x + 3$ **13.** $x + 3 + \dfrac{4}{x - 6}$
15. $x^2 + 5x + 6 + \dfrac{15}{x - 5}$ **17.** $3x^3 + 4x^2 - 9x - 10 - \dfrac{5}{x - 3}$ **19.** $x^3 - 6x^2 - 4x + 24 - \dfrac{35}{x + 6}$ **21.** $2x^2 + 8x + 6$
23. -5 **25.** -119 **27.** 231 **29.** 2 **31.** $x - 2$ is a factor; $f(x) = (x - 2)(x - 1)$ **33.** $x + 2$ is a factor; $f(x) = (x + 2)(2x + 1)$
35. $x - 3$ is not a factor **37.** $x + 1$ is a factor; $f(x) = (x + 1)(4x^2 - 11x + 6)$ **39.** $f(x) = 3x^2 - 10x - 25$ **41.** $f(x) = x^2 + 5x - 20$
43. $a = 2, b = -7, c = -12$, and $d = -13$, thus, $a + b + c + d = -30$

Appendix C

Section C.1

1. congruent **3.** right **5.** supplementary **7.** False **9.** acute **11.** right **13.** straight **15.** obtuse **17.** 58° **19.** 17°
21. 113° **23.** 172° **25.** $m\angle 1 = 130°; m\angle 2 = 50°; m\angle 3 = 130°; m\angle 4 = 50°; m\angle 5 = 130°; m\angle 6 = 50°; m\angle 7 = 130°$

Section C.2

1. regular **3.** isosceles **5.** True **7.** True **9.** 55° **11.** 48° **13.** 4 units **15.** 67.5 units **17.** 10 inches **19.** 5 cm **21.** 7 cm
23. $\dfrac{11}{2}$ yards or 5.5 yards

Section C.3

1. perimeter **3.** $\dfrac{1}{2}h(b + B); h; b; B$ **5.** False **7.** Perimeter: 28 feet; Area: 40 square feet **9.** Perimeter: 40 m; Area: 75 m²
11. Perimeter: 24 km; Area: 36 km² **13.** Perimeter: 72 feet; Area: 218 square feet **15.** Perimeter: 54 m; Area: 62 m² **17.** Perimeter: 30 feet;
Area: 45 square feet **19.** Perimeter: 28 mm; Area: 36 mm² **21.** Perimeter: 40 in; Area: 84 in.² **23.** Perimeter: 45 cm; Area: 94.5 cm²
25. Perimeter: 32 m; Area: 42 m² **27.** Perimeter: 32 ft; Area: 24 ft² **29.** Circumference: 32π in. ≈ 100.53 in; Area: 256π in.² ≈ 804.25 in.²
31. Circumference: 20π cm ≈ 62.83 cm; Area: 100π cm² ≈ 314.16 cm² **33.** π square units **35.** about 26.18 feet

Section C.4

1. polyhedron **3.** surface area **5.** False **7.** Volume: 600 cubic feet; Surface Area: 460 square feet
9. Volume: 288π cubic centimeters ≈ 904.78 cubic centimeters; Surface Area: 144π square centimeters ≈ 452.39 square centimeters
11. Volume: 32π cubic inches ≈ 100.53 cubic inches; Surface Area: 40π square inches ≈ 125.66 square inches
13. Volume: $\dfrac{800}{3}\pi$ cubic millimeters ≈ 837.76 cubic millimeters **15.** Volume: $\dfrac{640}{3}$ cubic feet; Surface Area: 256 square feet
17. 1728 cubic inches **19.** 75.40 cubic inches; 100.53 square inches **21.** approximately 268.08 cubic centimeters

Appendix D

1. constant function **3.** False **5. (c)** **7. (e)** **9. (b)** **11. (f)**
13. 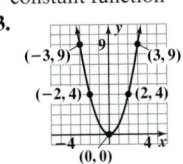 **15.** 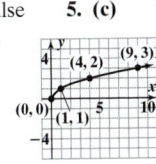 **17.**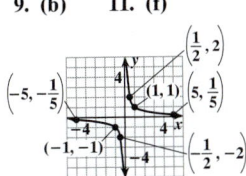

Appendix E

Section E.1

1. system of linear equations **3.** consistent; dependent **5.** True **7.** Yes; answers may vary **9.** at the point $(3, -2)$ **11. (a)** no **(b)** yes
13. (a) yes **(b)** yes **15.** $(2, -1)$ **17.** no solution **19.** $(1, 3)$ **21.** $\left\{ (x, y) \,\middle|\, y = \dfrac{1}{2}x + 1 \right\}$ **23.** $(3, -4)$

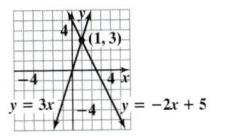 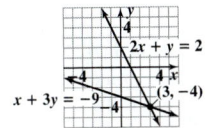

25. $(6, -2)$ **27.** $(-2, -3)$ **29.** no solution **31.** $\left(\dfrac{1}{2}, -\dfrac{1}{4}\right)$ **33.** $\{(x, y)\,|\,x + 3y = 6\}$ **35.** $(2500, 7500)$ **37.** $(-8, 3)$ **39.** $(11, -8)$

41. no solution **43.** $\left(\dfrac{1}{2}, -\dfrac{4}{5}\right)$ **45.** $\left\{(x,y)\,\Big|\,\dfrac{1}{3}x - 2y = 6\right\}$ **47.** $(25, 40)$ **49.** $(-9, 3)$ **51.** $\{(x,y)\,|\,x = 5y - 3\}$ **53.** $\left(\dfrac{39}{11}, -\dfrac{30}{11}\right)$

55. no solution **57.** $y = -2x - 5;\ y = -\dfrac{5}{3}x + \dfrac{1}{3};$ exactly one solution **59.** $y = \dfrac{3}{2}x + 1;\ y = \dfrac{3}{2}x + 1;$ infinite number of solutions

61. (a) $y = -\dfrac{1}{2}x + \dfrac{5}{2};\ y = x + 1$ **(b)** $(1, 2)$ **63.** (c) and (f) **65.** $A = \dfrac{7}{6}$ and $B = -\dfrac{1}{2}$

67. Answers will vary. One possibility follows: $\begin{cases} x + y = 3 \\ x - y = -5 \end{cases}$

69. $(1.2, 2.6)$

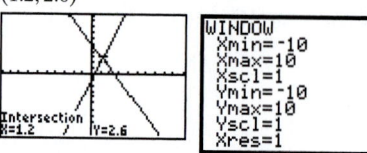

71. $(4, 13)$

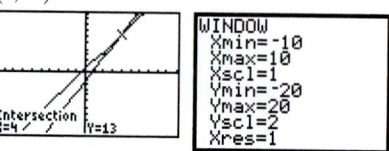

73. $\{(x, y)\,|\,4x - 3y = 1\}$

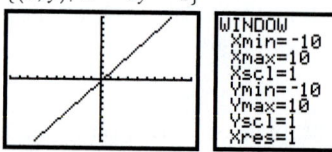

75. approximately $(0.35, -3.76)$

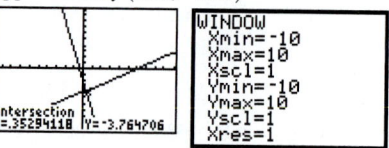

Section E.2

1. inconsistent **3.** solution **5.** True **7.** The three planes formed by the system will intersect at the point $(3, 2, 5)$. **9. (a)** no **(b)** yes

11. $(3, -2, 4)$ **13.** $(3, -4, -2)$ **15.** $\left(0, 2, \dfrac{3}{2}\right)$ **17.** $\left(\dfrac{1}{3}, -\dfrac{2}{3}, 2\right)$ **19.** $(3, -3, 1)$ **21.** $(-5, 0, 3)$ **23.** no solution **25.** $\left(-\dfrac{7}{2}, -\dfrac{7}{2}, \dfrac{3}{2}\right)$

27. $\{(x, y, z)\,|\,x = 5z - 5,\ y = -2z + 3,\ z \text{ is any real number}\}$ **29.** $\left(\dfrac{11}{2}, 0, -\dfrac{1}{2}\right)$ **31.** $\left\{(x, y, z)\,\Big|\,x = \dfrac{1}{3}z + \dfrac{2}{3},\ y = -\dfrac{4}{3}z + \dfrac{7}{3},\ z \text{ is any real number}\right\}$

33. Answers will vary. One possibility follows. $\begin{cases} x + y + z = 4 \\ x - y + z = 6 \\ x + y - z = -2 \end{cases}$ **35. (a)** $a - b + c = -6;\ 4a + 2b + c = 3$ **(b)** $a = -2, b = 5, c = 1;\ f(x) = -2x^2 + 5x + 1$ **37.** $i_1 = 1, i_2 = 4,$ and $i_3 = 3$ **39.** There are 1490 box seats, 970 reserve seats, and 1640 lawn seats in the stadium. **41.** Nancy needs 1 serving of Chex® cereal, 2 servings of 2% milk, and 1.5 servings of orange juice. **43.** \$12,000 in Treasury bills, \$8000 in municipal bonds, and \$5000 in corporate bonds.

45. $\overline{AM} = 4, \overline{BN} = 2,$ and $\overline{OC} = 10$ **47.** $(10, -4, 6)$ **49.** $(-2, 1, 0, 4)$

Section E.3

1. matrix **3.** $4; 3$ **5.** True **7.** Answers will vary. **9.** Multiply each entry in row 2 by $\dfrac{1}{5}$ (or divide each entry of row 2 by 5).

11. $\begin{bmatrix} 1 & -3 & | & 2 \\ 2 & 5 & | & 1 \end{bmatrix}$ **13.** $\begin{bmatrix} 1 & 1 & 1 & | & 3 \\ 2 & -1 & 3 & | & 1 \\ -4 & 2 & -5 & | & -3 \end{bmatrix}$ **15.** $\begin{bmatrix} -1 & 1 & | & 2 \\ 5 & 1 & | & -5 \end{bmatrix}$ **17.** $\begin{bmatrix} 1 & 0 & 1 & | & 2 \\ 2 & 1 & 0 & | & 13 \\ 1 & -1 & 4 & | & -4 \end{bmatrix}$ **19. (a)** $\begin{bmatrix} 1 & -3 & | & 2 \\ 0 & -1 & | & 5 \end{bmatrix}$

(b) $\begin{bmatrix} 1 & -3 & | & 2 \\ 0 & 1 & | & -5 \end{bmatrix}$ **21. (a)** $\begin{bmatrix} 1 & 1 & -1 & | & 4 \\ 0 & 3 & 5 & | & -11 \\ -1 & -3 & 2 & | & 1 \end{bmatrix}$ **(b)** $\begin{bmatrix} 1 & 1 & -1 & | & 4 \\ 0 & 3 & 5 & | & -11 \\ 0 & -2 & 1 & | & 5 \end{bmatrix}$ **23. (a)** $\begin{bmatrix} 1 & 1 & 1 & | & 4 \\ 0 & 1 & 5 & | & 5 \\ 0 & -4 & 2 & | & 8 \end{bmatrix}$ **(b)** $\begin{bmatrix} 1 & 1 & 1 & | & 4 \\ 0 & 1 & 5 & | & 5 \\ 0 & -2 & 1 & | & 4 \end{bmatrix}$

25. $\begin{cases} x + 4y = -5 & (1) \\ \quad\ y = -2 & (2) \end{cases}$ consistent and independent; $(3, -2)$ **27.** $\begin{cases} x + 3y - 2z = 6 & (1) \\ \quad\ y + 5z = -2 & (2) \\ \quad\quad 0 = 4 & (3) \end{cases}$ inconsistent; $\varnothing$ or $\{\ \}$ **29.** $\begin{cases} x - 2y - z = 3 & (1) \\ \quad\ y - 2z = -8 & (2) \\ \quad\quad z = 5 & (3) \end{cases}$ consistent and independent; $(12, 2, 5)$ **31.** $(3, -5)$ **33.** no solution

35. $\left(\dfrac{1}{2}, -\dfrac{5}{4}\right)$ **37.** $\{(x, y)\,|\,4x - y = 8\}$ **39.** $(3, -4, 1)$ **41.** $(4, 0, -5)$ **43.** $\{(x, y, z)\,|\,x = -2z - 0.2,\ y = -z - 1.4,\ z \text{ is any real number}\}$

45. no solution **47.** $\left(\dfrac{3}{10}, \dfrac{1}{10}, -\dfrac{1}{2}\right)$ **49.** $\left(\dfrac{5}{3}, \dfrac{2}{5}, -\dfrac{1}{2}\right)$ **51.** $(-2, 1, -5)$ **53. (a)** $a + b + c = 0;\ 4a + 2b + c = 3$

(b) $a = 2, b = -3, c = 1;\ f(x) = 2x^2 - 3x + 1$ **55.** \$8000 in Treasury bills, \$7000 in municipal bonds, and \$5000 in corporate bonds **57.** $(-3, 7)$ **59.** $(2, 5, -4)$ **61.** $(5, -3)$ **63.** $(4, -2, 7)$

Section E.4

1. determinants **3.** square **5.** False **7.** $(-1, 2, 0)$ **9.** 10 **11.** -2 **13.** -9 **15.** -163 **17.** 0 **19.** $(-8, 4)$ **21.** $(3, -1)$

23. $\left(-\dfrac{1}{6}, \dfrac{3}{8}\right)$ **25.** $\left(\dfrac{3}{4}, -\dfrac{7}{4}\right)$ **27.** $(-2, 1, -1)$ **29.** $(3, -1, 2)$ **31.** $\left\{(x, y, z)\,\Big|\,x = \dfrac{4}{5}z + \dfrac{14}{5},\ y = -\dfrac{3}{5}z - \dfrac{8}{5},\ z \text{ is any real number}\right\}$

33. $\left(\dfrac{3}{2}, \dfrac{9}{4}, -\dfrac{7}{4}\right)$ **35.** $\left\{(x, y, z) \,|\, x = -\dfrac{1}{2}z - 4,\ y = \dfrac{1}{4}z + \dfrac{3}{2},\ z \text{ is any real number}\right\}$ **37.** $\left(\dfrac{7}{5}, -\dfrac{5}{3}, \dfrac{11}{3}\right)$ **39.** $x = 5$ **41.** $x = -2$

43. (a)

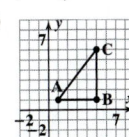

45. (a)

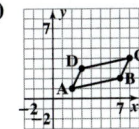

(b)

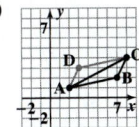

(c) The area of triangle ADC is 4.5.   **(d)** 9

  (b) The area of triangle ABC is 10.

The area of triangle ABC is 4.5.

47. (a) $x + 2y = 7$ **51.** $(-8, 4)$

  (b) $x + 2y = 7$ **53.** $(3, -1)$

49. 14; -14; answers may vary.   **55.** $(-2, 1, -1)$

Applications Index

Acoustics
loudness, 879
  in concerts, 881
    decibels for threshold of pain, 881
    of vacuum cleaner, 907
    of whisper, 881
measuring sound, 251
whispering gallery, 944

Agriculture, 397
harvesting hay, 267
planting crops, 304–305

Air travel
altitude of airplane, 606
jet stream effects on aircraft, 770
wind speed, 781

Archaeology
Great Pyramid of Giza, 665
Sun Pyramid, 665

Architecture
whispering gallery, 944
windows
  area, 115, 575, 583
  dimensions, 212

Art
for bedroom, 172
dancing, 552
picture frame, 370

Astronomy
distance of planet to Sun, 947
Kepler's Law, 673
Martian rovers, 397–398
orbit of planet around Sun, 947
radio telescope, 968
transit of Venus, A13

Biology
basal energy expenditure (E), 114
blood pressure, 630–631
Body Mass Index, 271, 485
cholesterol, 221–222
cicada cycle, 1010
E. coli population, 620
gestation period, 643
heart rate, 652
human blood cell, 397
pulse rate, 605
smallpox virus, 397

Business, 833
advertising, 162, 543
  on television, 324
billboard height, 148
break-even point, 284
carpet cleaning, 204
Christmas bonus, 167
commission, 137, 162, 596, 605, 608–609, 616
cost(s)
  average, 541
  of manufacturing, 56, 391, 826
  marginal, 807
  printing, 335
  production, 266
cost equation, 226
delivery service, 163
demand
  for candy, 251
  for hot dogs, 605
depreciation, 617–618, 652, 867–868
  straight-line, 610–611, 652
discount, 138, 169
  on backpack, 172

  on new car, 164
  price before, 137
earning potential, 216
earnings, 104
  of salesperson, 112
gas station, 232
income from CD sales, 269
job offers, 130, 1000
lawn mower company, E21
lawn service, 353
manufacturing
  of automobiles, 332–333
  of bicycles, 826
  of calculators, 353, 417
  cost of, 56, 391, 826
  of DVDs, 353
  of rain gutters, A14
  swing-set, E20–E21
markdown, 135
markups, 137, 169
music videos sale, 353
newspaper circulation, 395
planning a trip, 112
price(s)
  for charter, 804–805
  of movie tickets, 657
profit, 27, 110, 113
  on newspapers, 450
  on T-shirts, 450
real estate, 543–544
renting
  a car, 64
  car rental agencies, 284
  a truck, 64, 128
retirement plans, 234
revenue, 417, 596, 765–766, 769, 779, 781, 806–807, 809, 820, 830, 833, 834, 836
  growth of, 691, 720
  maximizing, 802
  monthly, 349
  theater, E24
revenue function, 605
sale(s)
  bookstore, 338
  of coffee, 323
  of computer game, 742
  of DVD videos, 295
  of music videos, 295
  tax on, 370
shipping expenses, 234, 245, 270
skateboard shop, 353–354
tiered prices, 323
truck driving, 114–115
wages and salaries, 169, 595, 604, 649, 989
  calculating, 204
  department store, 263
  hourly wage, 89, 102
  increases in, 999
  monthly salary, 195, 405
  overtime pay, 92
  pay cut, 136
  pay raise, 137
  weekly salary, 225
warehouse inventory, 28

Chemistry
aluminum alloy, A14
concentration of solutions, 398
pH, 882
radioactive decay, 862–863, 868, 897–898, 901, 906, 909, 1011
silver alloy, 325
water molecule, 397

Combinatorics
ceramic tile floor design, 987
mosaic design, 989
seats in auditorium, 989
stack of objects, 989
table seating, 1002

Communications
cell phones, 130, 576, 584, 646, 647
  costs of, 226
  plans, 172
long-distance calling plan, 162, 280–281
long-distance carrier, 127, 651
phone charges, 284, 617
satellite dish, 933–934

Computers
printers, 130
rentals, 267
viruses, 1011

Construction
of box, 964
  maximizing volume, 809
of buttons, 382
cement mixing, 561
of enclosures, 808
  fences, 803–804, 965
  maximizing area, 803–804, 808–809, 833
  maximizing volume, 836
foundation design, 89
length of oak board, 172
lumber purchase, 251
number of rooms in housing units, 596
of rain gutters, 809
of roof, 216, 836
roof pitch, 212, 216
of shelving system, 125–126
of stadium, 989
of table, 148
tablecloth, 382

Demographics
bachelors, 137
birth rate(s), 617, 721
census, 27
household types, 138
life expectancy, 188–189
  men versus women, 291–292
never-married females, 137
population
  by age cohort, 770
  bacterial, 902
  of Canada, 910
  fish, 267
  as function of age, 596
  growth of, 104, 138, 216, 867, 900–901
  of Nevada, 906, 909
  of Orange County, FL, 649
  of United States, 396, 867, 900, 979
  of world, 396, 901, 980

Design
ceramic tile floor, 987
of foundation, 89
of window, 766–767

Distance
to catch up, 145
depth of lake, 606
height
  of airplane, 28, 606
  of ball, 352, 417
  of dam, 111
  of Mt. Whitney, 28
  of rocket, 451, 457

Distance (continued)
of swing, 606
of telephone pole, 553
of toy rocket, 417
of tree, 545, 553
between home plate and second base, 829
ladder's reach, 755
length
of bridge, 111
of guy wire, 755
of pole, 468
of ribbon, 69
line of sight
from lighthouse, 751–752
from mainmast of the Constitution, 753
on map, 269
from school, 149
of walk, 601

Economics
demand, 251
Life Cycle Hypothesis, 593, 770, 808
marginal propensity to consume, 996, 1000
unemployment rate, 656

Education
advanced degrees, 808
biology class, 39
earning potential and, 216
educational attainment of U.S. residents, 133–134
field trip, 260, 323
grades, 137, 162, 163
computing, 116, 131, 271, 631
learning curve, 868–869, 902
school fundraiser, 260
school newspaper, 284
scores
multiple choice, 68
study-abroad trip, 112
teacher salary, 135, 138
tuition increase, A13
tutoring, 169

Electricity and electronics
circuits, 528
current
in an *RC* circuit, 869
in an *RL* circuit, 869
electric bill, 184
impedance
parallel circuit, 732
series circuit, 732
Kirchhoff's Rule, E23
power, 833

Energy
crude oil imports, 397
electric bill, 184
fossil fuels, 398
gasoline refining, 572–573
natural gas fee, 185
oil tanker leak, 593
photon, 251

Engineering
bridges, 129
London, 947
parabolic arch, 936
semielliptical arch, 946–947, 969
suspension, 936
Chicago's road system, 919
concrete strength, 618
elevator capacity, 257–258
grade of road, 212
parking lot dimensions, 165, 313
road grade, 216, 266
tensile strength, 655
tolerances, 643, 655
towers, 129

Entertainment
budget for, 652
movie tickets, 339, 657
party food, 332
party planning, 340
performance fees, 264
satellite television bill, 652
theater tickets, 324
TV, 56
big-screen, 467, 468
watching, 39

Environment
cost-benefit model of pollution, 541
oil tanker leak, 593, 850

Exercise
workout, 554

Finance, 114. *See also* **Investment(s); Money**
annuities, 996–997
bank balance, 27, 40, 69, 73
buying
clothes, 69
computer, 129
motorcycle, 129
used cars, 135
cab fare, 617
car payments, 246–247
carpet purchase, 850–851
checking account, 27, 69
balancing, 48
overdrawn, 39
Christmas bonus, 167
college savings plan, 1011
commission, 137
comparison shopping, 114
cost(s)
of barrettes, 562
of books, 84, 130
of car, 83, 102, 223, 485
of car rental, 194–195
of CD player, 111
of CDs, 78, 188
of cell phone, 226
of coffee, 164
comparison of, 284
of computer rentals, 267
excluding sales tax, 134, 135
of gasoline, 188, 251
of Happy Meal, 84
of healthcare, 231–232
of hot dogs and soda at baseball game, 301–302
of kayak, 83, 102
of moving, 171
of moving van rental, 72
of pizza, 108
of printing, 264, 335
before sales tax, 136
of second-day delivery, 653
of shipping, 263, 270
of sod, 107–108
of stamps, 324
of taxi ride, 188
of tickets, 323
of TV, 56
credit card comparison, 96–97
discounts, 83, 102
on computer, 112
on T-shirts, 166
division of funds, 130, 168
earnings, 263, 658
electric bill, 184, 631
entertainment budget, 652
exchange rates, 543
fees
natural gas, 185
performance, 264
phone, 617

future value of money, 167, 863–864, 898–899, 906, 979
income
adjusted gross, 130
average per-capita, 353
hourly wage, 102
monthly paycheck, 405
inheritance, 65
interest
amount earned, 673, 721
compounded monthly, 901, 979
compounded quarterly, 901, 979
compounding periods, 867, 902
on credit card debt, 84
rate of, 756
simple, 99–100, 113, 322
movie ticket price, 657
price
of charter, 804
of gas, 138
of three-bedroom house, 212–213
of tickets, 324
profit, 27
raffle tickets, 337
rentals
car, 64, 102, 162, 225, 267, 485, 652
rototiller, 102
RV, 617
truck, 64, 130, 162, 168
retirement plans, 234
revenue from school play, 58
salary, 595
sales, 104
of bedding plants, 324
crafts fair, 658
of jewelry, 322
of tickets, 65, 321
satellite television bill, 652
saving for home, 1000
school fundraiser, 260
Stock Exchange daily volume, 396
tax(es), 616, 617
adjusted gross income, 851
federal, 103, 115, 627–628
luxury, 617
property, 370
sales, 102, 136, 169, 370
withholding, 631, 851
tax returns, 652
teacher salary, 135
tip jar, 324
trip to the aquarium, 260
tuition increase, A13

Food and nutrition
bread baking, 553
breakfast at Burger King, 333
calories, 226
in fast food, 304
candy, 553, 618–619
carbohydrates in fast food, 304
cereal, 130
grams of fat, 91
Halloween candy, 39
hot dogs, 301–302, 605
ice cream cone, 115, C23
pizza, 39, 108, 114, 561
temperature of, 606
prescribed diet, E24
raisins, 619
soda, 301–302

Games
lottery payment, 1011

Gardening. *See* **Landscaping**

Gemology
diamonds, 245

Geometry

angles
  complementary, 169, 305, 309–310, 313
  measures of two, 338
  supplementary, 169, 305, 310,
    313, 342
area of region, 115
box, volume of, 382, 390, 450, 468,
    809, 964
centroid, E12
circle
  area of, 113, 370, 595
  circumference of, 113
collinear points, E46
cone
  height of, 739
  volume of, 57
cube, volume of, 359, 382
cylinders, 271
equation of line, E11
parallelogram, E11
  area of, 417, E45
quadrilateral, vertices of, 244
rectangle
  area of, 359, 390, 417, 436, 461–462,
    493, 519, 527–528, 769
  dimensions of, 170, 295, 313, 459, 467,
    769, 830, 964, 971
  length of, 147, 473
  perimeter of, 65, 73, 102–103, 148, 501, 519
  width of, 65
rhombus, E11
right circular cylinder, 114
  surface area of, 114, 167, 417–418, 451
  volume of, 114, 382
sphere
  radius of, 691, 721
  volume of, 604
square, area of, 359, 444
supplementary angles, 271
surface area, 56
trapezoid, area of, 114, 519
triangle, 631
  altitude of, 244
  angles of, 338
  area of, 114, 390, 426, 436, 462–463,
    493, 595, 604, 769, E45
  base and height of, 467, 769
  dimensions of, 971
  equilateral, 755–756
  isosceles, 147
  length of leg, 473
  measures of, 169–170
  Pascal's, 980
  perimeter of, 92
  sides of, 473
  similar, 544–545, 770

Health

calories, 226
cholesterol level, 221–222
eardrum perforation, 881
healthcare costs, 231–232
insurance cost, 617, 851
vitamins, 397
weight as function of age, 606
weight loss, 168

Home improvement

bathroom remodeling, 116
carpet installation, 561
carpet purchase, 850–851
flooring installation, 335
hanging wallpaper, 92
painting, 116, 314, 508, 552, 561, C14
replacing pipes, 554
wallpaper purchase, 148
washing walls, 561

Investment(s)

annual interest, 57
asset allocation, 126–127, 130, 295
in bonds, 313, 338, E24, E36
in Certificate of Deposit (CD), 99–100, 105, 112,
    117, 867
college savings plan, A11
diversification, 130, 313
401(k) plan, 1000, 1001, A11
in funds, 65
  domestic, 295
  international, 295
  money-market, 295
house price appreciation, 137
income from, E24, E36
of inheritance, 112, 130, 313, 339
interest on, 321, 323, 339
  in money market account, 317–318
IRA, 1001
  Roth, 1001
loss, 137
original, 137
in real estate, 324
for retirement, 329–330
return on, 324
Rule of 72, 902
stock account, 325
Stock Exchange daily volume, 396
stock prices, 27, 40, 48, 1000
in stocks, 313, 324, 325, 338
in two businesses, 325

Landscaping

back yard, 116
  dimensions of, 170
cost of sod, 107–108
garden
  area of, 142
  dimensions of, 92, 148, 294, 459
  fencing for, 107, 313, 468
  fertilizer for, 148
  length and width of, 170
  rototilling, 102
  width of, 148
grass seed, 325
lawn area, 107–108
lawn mowing, 509, 547–548, 554, A14
lawn service, 353
planting seedlings, 546–547
pruning trees, 553
swimming pool, 116
weeding the yard, 553
yard dimensions, 148
yard fencing, 308–309

Leisure and recreation

backpacking, 260
bike riding, 313, 314
camping trip, 131
canoeing, 144–145
fishing, 260
hiking, 27
horseback riding, 313
kayaking, 313
pleasure boat ride, 830
Punkin Chunkin contest, 426, 807–808
reading, 543
rowing, 313
Spring Break, 130
swimming pool, 141
swinging, 606
ticket pricing, 324
vacation, 137, 561, 562
walk around the park, 336
Wonder Wheel Ferris wheel, 599–600

Medicine

cholesterol, 221–222
drug concentration, 484, 541

medication, A7
smallpox virus, 397

Miscellaneous

age differences, 165
balloon volume, 850
coffee can volume, C23
coffee table area, 167
dog run, 91
dress size, 104
Fibonacci sequence, 980
field dimensions, 147, 309, 426, 462, 467
fitting television in media cabinet, 464–465
garbage, 398
horse corral, 338
maps, 553
peach can volume, C23
rain gutter capacity, C23
room dimensions, 467
sail dimensions, 467, 468
tabletop dimensions, 473
tarp dimensions, 473
trough volume, 382, C23
water cooler cup capacity, C23

Mixtures

almond-peanut butter, 339
of candies, 305
Churrascaria platter, 325
of coffee, 305, 322, 325, 332
of coins, 315–316, 339
gift box, 337
of grass seed, 325
ice cream, 342
of nuts, 325
of olives, 324
paint, 325
of party food, 332, 340
silver alloy, 325
solutions, 319–320
  acid, 339
  alcohol, 325
  antifreeze, 325
  peroxide, 339
  saline, 325
  sugar, 339
trail mix, 318–319, 322, 742
wine, 321

Money. *See also* **Finance; Investment(s)**

borrowing, 162
coin jar, 270
coin mixture, 315–316, 339
coins found, 102, 166
comparing pesos and pounds, 553
comparing rubles and dollars, 553
division of, 168
future value of, 167
interest on, 317–318
piggy bank contents, 102

Motion. *See also* **Physics**

pendulum, 999
projectile, 575, 583, 769–770, 830
  Punkin Chunkin, 807–808
vertical, 586, 648
  thrown object, 820

Motor vehicles

buying gasoline, 251
car costs, 83, 223, 485
car depreciation, 138
car manufacturing, 332–333
car payments, 246–247
car purchase, 102
car rental, 194–195, 267, 652
car washing, 562
depreciation of, 901, 999
discount on new car, 164
driving speed, 406, 561

Motor vehicles (continued)
gas capacity, 103–104
gas mileage, 138
gas prices, 138
headlight, 935
motorcycle purchase, 129
traffic fatalities, 234–235

Pediatrics
birth weight, 222
height versus head circumference, 619

Physics, 113
bouncing balls, 1000
dropped object, 459
energy of earthquake, 882
gravity, 755
height of ball, 352
jet of water, 473
measuring sound, 251
Newton's Law of Cooling, 868, 901–902, 906
pendulum motion, 999
photon energy, 251
pressure, 251, 269
projectile motion, 460–461, 467, 575, 583,
      769–770, 830, 836
   Punkin Chunkin, 807–808
speed
   of light, 398
   of sound, 398
terminal velocity, 673
uniform motion, 310–311, 548
vertically thrown ball, 456, 459
vertical motion, 586, 648
   thrown object, 820

Play
playground equipment, 342

Politics
committee membership, 137
election for school president, 137, 169
House of Representatives apportionment, 251
political flyers, 284
presidential election of 2000, 137

Probability
of crisis helpline call, 861
Psychometrics
IQ scores, 643

Rate. See also Speed
of canoe paddling, 144–145
of cyclists, 148
of flying with the wind, 552
of plane travel, 148
of swimming, 552
of two cars, 148
uniform motion problem for,
      144–145

Real estate
investment in, 324
price of new three-bedroom
      house, 212–213
saving for home, 1000

Recreation. See Leisure and recreation

Seismology
earthquake energy, 882
earthquake magnitude, 882
Great New Madrid earthquake, 907
seismographic reading some distance from
      earthquake epicenter, 882

Speed. See also Rate
average, 145
ball slide speed factor, 739
of bike, 313, 338
of boat, 313, 830
of current, 313, 338, 770
driving, 406, 561
of horses, 313
jogging, 149
of kayak, 561
of light, 398
of paddler, 338
of plane, 310–311, 313
during road trip, 269
of sound, 398
of train(s), 148, 170
   moving in opposite directions, 313
of wind, 314, 338, 341, 781

Sports
baseball, 131, 755
   agent for player, 1000
   attendance, 312
   diamond, 919, 920–921
   distance between home plate and second
      base, 829
   games won, 312
   hot dogs and soda at the
      game, 301–302
   minor league, E24
   throw from right field to second
      base, 920
   World Series, 468
basketball, ticket prices, 123–124
bowling, 171
car racing, A7
football, 27, 68, 833
   Super Bowl party, 69
   wind sprints, 1010
golf, 755
hockey, 284
iron man training, 554
jogging, 605
Kentucky Derby, 646
marathon, 147, 314, 551
volleyball, 553
Statistics, 113
mean, 527
Supernatural
belief in ghosts, A13
Surveys, A8

Temperature, 27, 40
conversions, 109, 113
in Paris, 104
of pizza, 606
wind chill, 576, 584, 673
winter, 68

Time
average trip length, 596
to can peaches, 553
to fill a tank, 553
to fill bucket, 554
to fill pool, 554
for object to fall a given distance, 459
of riding bikes, 314
for river trip, 149
to travel, 143–144
   a given distance apart, 147, 148, 170, 172
   on two types of road, 148
uniform motion problem for, 143–144

Transportation. See also Motor vehicles; Travel
bicycling through the forest, 549–550
boat trip, 548–549

Travel
to the aquarium, 260
bike trip, 554
by boat, 548–549, 554
commuting, 554
by cruise ship, 561
driving to school, 251
exchange rates, 543
by plane, 341, 509, 554
on a river, 509, 553
during road construction, 554
road trip, 269
during a snowstorm, 554
by train, 651
to Water World Water Park, 323

Volume. See also Geometry
of balloon, 850
of box, 382, 390, 450, 468, 809, 964
of can, C23
of trough, 382, C23
of water in tub, 651

Weather
wind chill, 576, 584, 673

Weight
at birth, 222
as function of age, 606
loss of, 168
of Shetland pony, 658

Work
car washing, 562, 831
cleaning
   the math building, 553
   a stadium, 248–249
earning potential, 216
to fill pool, 771
hours of, 552
job offers, 130, 1000
making care packages, 554
painting city hall, 554
planting seedlings, 546–547
roofing a house, 836
stuffing envelopes, 553
washing the dishes, 561
weekly salary, 225
working together on a job, 770

Subject Index

Abscissa (*x*-coordinate), 566
Absolute value
  adding integers using, 20–21
  computing, 15
  definition of, 14
  of real number, 14–15
Absolute value equations, 632–35
  solving, 632–34
Absolute value function, D2
Absolute value inequalities, 635–40
  involving < or =, 635–37, 639–40
  involving > or =, 637–39
  solving applied problems involving, 639–40
Acute angle, C2
Addition
  Associative Property of, 44–45, 46
  Commutative Property of, 43–44, 46, 62
  of complex numbers, 724–25
  of decimals, 34–35
  of fractions, 31, 32–33
  horizontal, 346
  Identity Property of, 41, 46
  in order of operations, 51, 54
  of polynomials, 346–47
  of radical expressions, 692–94
  of rational expressions, 494–96, 497–98
    with a common denominator, 494–96
    containing opposite factors, 515
    and integer, 515–16
    numerator containing more than one term, 495–96
    with opposite denominators, 497–98
    with unlike denominators, 509–19
  of rational numbers, 494
  vertical, 347
  words representing, 118
Addition Property of Equality, 76–78, 84–86, 296
Addition Property of Inequality, 153, 156
Additive inverse (opposite), 21, 46
Adjacent angles, C3
Algebra, definition of, 57
Algebraic expressions
  definition of, 58
  evaluating, 57–58
  for revenue, 58
  simplifying, 57–65
    by combining like terms, 60, 61–63
    distributive property for, 61
    like terms and unlike terms, 59–60
  translating English phrases to, 117–20
Al-jabr, 74
Alternate interior angles, C3
Angle(s), C1–C3
  acute, C2
  adjacent, C3
  complementary, 139–40, 309–10, C2
  complement of, C2
  congruent, C1
  corresponding, C3
  formed by parallel lines, C3–C4
  interior, C5–C6
    alternate, C3
  obtuse, C2
  right, 463, 750, C2, C6
  straight, C2
  supplementary, 139–40, 309
  supplement of, C2
  vertex, 147
  vertical, C3
Annotated examples, 5
Annuity, 996–97
  amount of, 996, 997
  ordinary, 996
Aphelion, 947
Area, C10–C20
  of circle, C16–C17

defined, 105
  formulas of, 106
  of parallelogram, C14–C15
  of polygon, C11
  problems of, 142
  of rectangle, 103, C11–C12
  of trapezoid, C15
  of triangle, C16
Argument of function, 591
Arithmetic sequences (arithmetic progression), 982–90, 995
  common difference between consecutive terms of, 982
  definition of, 982
  determining, 982–83
  formula for *n*th term of, 983–85
  sum of, 985–87, 990
Associative property of addition, 44–45, 46, 47
Associative property of multiplication, 44–45, 47
Astronomia nova (Kepler), 912
Asymptotes, 953–55
  conjugate, 956
Augmented matrix, E25–E27
  in reduced row echelon form, E36
  row operation on, E27–E28
Average rate of change, 212–13
Axis/axes
  coordinate, 176, 566
  of ellipse, 939
  of hyperbola, 949, 950
  of symmetry, 927
    of parabola, 787, 796

Basal energy expenditure (*E*), 114
Base, 49
Bell, Alexander Graham, 879
Best fit line, 620
Binomial(s), 345. *See also* Polynomial(s)
  division of polynomial by, 386–88
  expanding, A5–A6
  greatest common factor as, 411, 413
    factoring out, 412
  multiplication of, 362, 363, 427
  squares of (perfect squares), 364–66, 437, 660–61, 666
Binomial products, 364–66
  product of sum and difference of two terms, 437
  squaring binomials, 364–66, 437
  using the Distributive Property, 361, 362
  using the FOIL method, 362–63
Binomial Theorem, A2–A7
  binomial coefficient, A3–A5, A6
  to expand a binomial, A5–A6
  factorial and, A2–A3
  statement of, A5
Book value, 610
Boundary line, 253
Box, volume and surface area of, 106
Branches of hyperbola, 949
Break-even point, 284

Calculators. *See also* Graphing calculator
  *n*th root approximated using, 667
  writing radical as decimal using, 662
Cartesian (rectangular) coordinate system, 176, 566
Center
  of circle, 921, 922–23, C9
  of ellipse, 939
  of hyperbola, 949, 950
Change-of-Base Formula, 890–91
Chapter review, 6
Chapter test, 6
Circle, 920–26, C9
  area of, 106, C16–C17
  center of, 921, 922–23, C9
  circumference of, 105, C16–C17
  definition of, 921, C9

diameter of, C9
  equation of
    general form of, 923
    standard form of, 921, 923
  graph of, 921–22, 923, 926
  perimeter of, 106
  radius of, 921, 922–23, C9
Circumference, 105, C16–C17
Classes, 2–4
  activities before, during, and after, 3–4
Closed interval, 90
Coefficient, 59–60, 345
  binomial, A3–A5, A6
Collinear points, E46
Combining like terms, 60, 61–63
  simplifying polynomials by, 346–48
Common difference, 982
Common logarithm function, 876–77
Common ratio, 990
Commutative Property of Addition, 43–44, 46, 62
Commutative Property of Multiplication, 43–44, 47
Complementary angles, 139–40, 309–10, C2
  defined, 139
Complement of angle, C2
Complete graph, 192, 569
Completing the square, 744–56, 764
  in one variable, 747–48
  steps in, 749
  when coefficient of square term is not 1, 750
Complex conjugates, 727–28
Complex number(s), 722–33
  adding or subtracting, 724–25
  definition of, 723
  dividing, 728–29
  multiplying, 726–28
    complex conjugates, 727–28
    square roots of negative numbers, 726–27
  powers of *i*, 729–30
  square root of negative real numbers, 723–24
  in standard form, 723
Complex rational expressions, 520–29
  definition of, 520
  dividing, 520–21
  simplifying, 521–26
    by simplifying numerator and denominator separately, 521–23
    using least common denominator, 524–26
Complex zero, 733
Composite functions, 838–40
  definition of, 838
  evaluating, 838–39
  finding, 839–40
  graphing calculator to evaluate, 851
Composite number, 1001
Composition, 838
Compound inequalities, 261, 621–31
  defined, 623
  intersection or union of two sets, 621–22
  involving "and," 623–25
  involving "or," 625–27
  solving, 623
  solving problems using, 627–28
Compound interest, 863–64
Conditional equation, 97, 99
Cone
  surface area of, C21
  volume of, 106, C21
Congruent angles, C1
Congruent line segments, C1
Congruent sides, 147
Congruent triangles, C6–C7
Conics, 912–72
  circle, 920–26, C9
    area of, 106, C16–C17

center of, 921, 922–23, C9
  circumference of, 105, C16–C17
  definition of, 921, C9
  general form of equation, 923
  graph of, 921–22, 923, 926
  radius of, 921, 922–23, C9
  standard form of equation of, 921, 923
  definition of, 920
  ellipse, 920, 938–49
    applied problems involving, 943–44
    axes of, 939
    center of, 939
    definition of, 938
    equation of, 941–42
    foci of, 938, 942
    graph of, 938–41, 942–43, 948–49
    vertices of, 939, 942
  hyperbola, 920, 949–57
    asymptotes of, 953–55
    axes of, 949, 950
    branches of, 949
    center of, 949, 950
    definition of, 949
    equation of, 950, 952–53
    foci of, 949, 950
    graph of, 949–53, 957
    vertices of, 949, 950
  parabola, 787, 920, 927–38
    applied problems involving, 932–34
    axis of symmetry of, 787, 796, 927
    definition of, 927
    directrix of, 927, 928, 931
    equation of, 928, 930–31
    focus of, 927, 928, 931
    graph of, 928–29, 931–32, 937
    opening up or down, 787, 927
    vertex of, 787, 795, 927, 928, 931
Conjugate asymptotes, 956
Conjugate axis, 949
Conjugates, 696
  complex, 727–28
Consecutive integer problems, 124–25
Consistent system of equations, 278–80
Constant
  definition of, 57
  of proportionality (constant of variation), 245, 247
Constant function, D1
Constant rate, 546
Contradiction, 97–98, 99
Coordinate, 12, 176
  of point, 566
Coordinate axes, 176, 566
Copernicus, Nicolaus, 912
Coplanar lines, C3
Corollary, 90
Correspondence, 577, 586–87
Corresponding angles, C3
Cost equation, 226
Cost function, linear, 610
Counting numbers (natural numbers), 9, 11
Cramer's Rule, E37–E46
  with inconsistent or dependent systems, E43
  solving systems of linear equations with
    in three variables, E41–E43
    in two variables, E38–E39
Cross-Multiplication Property, 542
Cube(s)
  perfect, 666
  sum or difference of two, 441–42
  volume and surface area of, 106, C21
Cube function, D2
Cube root, 666, D2
  graphing functions involving, 708–9
  rationalizing denominator containing, 700–701
Cylinder, volume and surface area, C21

Decibels, 879
Decimal(s), 10, 1008–13
  addition of, 34–35
  converting between fractions and,
    1009–11
  converting between percent and,
    1011–12
  division of, 36–37
  linear equations with, 95–97
  multiplication of, 35–36
  place value, 1008–9
  radical written as, 662
  repeating, 995, 1010
  rounding, 1009
  subtraction of, 34–35
  terminating, 1010
Decimal notation, 391
  converting between scientific
    notation and, 391–94
Degree(s), C1
  of monomial, 344, 345
  of polynomial, 345, 451
Denominator, 1003
  least common (LCD), 93–95
  rationalizing the, 699–702
    containing a square root, 699–700
    containing cube roots and fourth
      roots, 700–701
    containing one term, 699–701
    containing two terms, 701–2
Dependence, 577
Dependent systems of equations, 279,
    E19–E20, E43
  linear
    Cramer's Rule with, E43
    matrices to solve, E32
    in three variables, E19–E20, E43
    in two variables, E8
Dependent variable, 591
Depreciation, straight-line, 610–12
De Revolutionibus Orbium Coelestium
    (Copernicus), 912
Descartes, René, 176, 566
Determinants, E37–E46
  of 2 ∞ 2 matrix, E37–E38, E46
  of 3 ∞ 3 matrix, E39–E41
  definition of, E37
  minors of, E40
Diameter, 105, C9
Difference, 18
  common, 982
  definition of, 22
  of two cubes, 441–42
  of two squares, 439–41
  formula for, 364
Directly proportional, 245
Directrix, 927, 928, 931
Direct translation problems, 117–38
  defined, 122
  involving percent, 131–38
  models for solving, 123–28
  systems of linear equations to
    solve, 307–8
  translating English phrases to
    algebraic expressions, 117–20
  translating English sentences to
    equations, 120–23
Direct variation, 245–47
  definition of, 245
Discriminant, 762–64
Distance
  mean, 947
  between two points, 40
Distance formula, 913–16
  statement of, 914
Distributive Property, 62
  binomial products using, 361, 362
  Extended Form of, 360, 366
  factoring polynomials using, 411–12
  simplifying algebraic expressions
    using, 61
  to solve linear equations, 86–87
Dividend, 24, 36, 487
Division
  of complex numbers, 728–29
  of decimals, 36–37
  of fractions, 30

of inequality, 154
of integer by integer, 385–86
long, 385–88
  dividing polynomial by binomial
    using, 386–88
  of monomials, 371–83
    involving negative exponents,
      374–76
    using the Laws of Exponents,
      377–80
    using the Quotient Rule for
      Exponents, 371–72
    using the Quotient to a Power
      Rule, 372–73
    using zero as an exponent, 373–74
  in order of operation, 51, 54
  of polynomials, 383–91
    by a binomial, 386–88
    by a monomial, 384–85
    using synthetic division, B1–B7
  of rational expressions, 487–88,
    520–21
    complex, 520–21
    with more than one variable, 490
    written vertically, 490
  using scientific notation, 395
  words representing, 118
Division bar, order of operations and, 52
Division properties of zero, 46, 47
Divisor, 24, 36, 487
Domain
  of function, 597–98
    inverse functions and, 844
    logarithmic, 873–74
    radical, 705–6
  of relation, 578–80
Double root (repeated solution), 456
Dry mixture problems, 318–19

e, 857–58
Earthquakes, 837, 881–82
Eccentricity, 948, 956
Edge of polyhedron, C20
Elements in set, 8
Elimination method
  nonlinear equations solved using,
    961–63
  systems of linear equations solved
    using, 296–306
    applied problems involving, 301–2
    with infinitely many solutions, 300
    by multiplying both equations by
      a number to create additive
      inverses, 297–99
    with no solution, 299
    steps for, 297–98
    in three variables, E15–E18
    in two variables, E5–E6, E7
Ellipse, 920, 938–49
  applied problems involving, 943–44
  axes of, 939
  center of, 939
  definition of, 938
  equation of, 941–42
  foci of, 938, 942
  graph of, 938–41, 942–43, 948–49
  vertices of, 939, 942
Ellipsis, 9, 974
Embedded grouping symbols, 52–54
Empty set, 8, 158
Endpoint, C1
  of interval, 90
Energy expenditure, basal (E), 114
Equality
  Addition Property of, 76–78, 84–86, 296
  Multiplication Property of, 78–82,
    84–86, 93
Equation(s)
  absolute value, 632–35
    solving, 632–34
  conditional, 97, 99
  cost, 226
  of ellipse, 941–42
  equivalent, 76
  exponential, 858–61, 895–97
    logarithms to solve, 895–97
    steps for solving, 859

of hyperbola, 950, 952–53
logarithmic, 877–79, 894–95
  properties of logarithms to solve,
    894–95
ordered pair satisfying, 179–81
of parabola, 928, 930–31
quadratic in form, 772–81
  steps for solving, 773–74
relation expressed as, 589
  graph of, 580–81
satisfying, 568
sides of, 179, 567
translating English sentences to,
  120–23
in two variables, graphing, 179,
  191–205, 567
  by plotting points, 191–95
  using intercepts, 196–99
  vertical and horizontal lines,
    199–201
Equilateral triangle, C6
Equivalent equations, 76
Equivalent fractions, 1003–5
Equivalent inequalities, 153
Equivalent rational expressions, 504–6
  forming, 505–6
  LCD to write, 506
Eratosthenes, 1008
Error, margin of, 639, 640
Evaluating, definition of, 22
Even integer, 124
Example formats, 5
Exams, preparing for, 6–7
Expanded form, 49
Exponent(s), 49–57
  defined, 49
  evaluating exponential expressions,
    49–50
  of form 1/n, 668–69
  of form $a^{m/n}$, 669–71
  Laws of, 377–80
  negative, 374–76
  order of operations on, 50–55
    division bar and, 52
    embedded grouping symbols
      and, 52–54
    evaluating expression using, 54–55
    parentheses and, 51
  Power Rule for, 356–57, 377
  Product Rule for, 355–56, 357, 377
  Quotient Rule for, 371–72, 377,
    384–85
  Quotient to a Power Rule for,
    372–73
  rational
    approximating expressions
      involving, 670
    negative, 671
    radicals with, 669–71
    simplifying, 674–76
    simplifying radicals using, 676–77
  zero, 373–74
Exponential equations, 858–61, 895–97
  logarithms to solve, 895–97
  steps for solving, 859
Exponential expressions
  changing to logarithmic
    expressions, 871
  simplifying, 356–57
Exponential form, 49
Exponential functions, 852–70
  definition of, 853
  e, 857–58
  evaluating, 853–54
  graph of, 854–57
  properties of, 855, 856
  indeterminate, 854
Exponential models, 861–64, 897–99
  exponential probability, 861
  future value of money, 863–64,
    898–99
  radioactive decay, 862–63, 897–98
Exponential notation, 49
Extended Form of Distributive
  Property, 360, 366
Extraneous solution, 534, 713, 877
Extremes of proportion, 542

Face of polyhedron, C20
Factorial, A2–A3
Factorial notation, A2
Factorial symbol, A2
Factoring
  expressions containing rational
    exponents, 677–78
  as a product of prime factors, 1001
  to simplify rational expressions,
    480–81
  to solve quadratic equations, 764
Factoring polynomials, 407–73
  completely, 424, 444–48
  of form $ax^2 + bx + c, a \neq 1$, 427–36
    with a negative leading coefficient,
      431–32
    steps for, 433
    with two variables, 431
    using grouping, 432–35
    using trial and error, 427–32,
      434–35
  greatest common factor and, 408–13
  by grouping, 413–15
  polynomial equations solved by,
    451–60
  quadratic equations, 451–57
  special products, 436–44
    difference of two squares, 439–41
    perfect square trinomials, 437–39
    sum or difference of two cubes,
      441–42
  trinomials of form $x^2 + bx + c$,
    418–26
    with a common factor, 423, 424
    with a negative leading coefficient,
      424–25
    not written in standard form, 423
    steps for, 419
    with two variables, 422–23
    where c is negative, 420–21
    where c is positive, 419–20, 421
Factors, 23
  definition of, 408, 1001
Factor Theorem, B4–B5
Factor tree, 1001
Fibonacci numbers, 980
Fibonacci sequence, 28, 980
Fifths, perfect, 666
Finite sequences, 974
Focus/foci
  of ellipse, 938, 942
  of hyperbola, 949, 950
  of parabola, 927, 928, 931
FOIL method, 362–63, 427
  product of two binomials using,
    362–63
Formula(s), 103–17
  containing a percent, 104
  definition of, 103
  difference of two squares, 364
  evaluating, 103–5, 110–11
  geometry, 105–8, 141–43
  recursive, 980
  simple interest, 104–5, 316–17
  solving for variable, 108–11
  uniform motion, 143
Fourths, perfect, 666
Fraction(s), 10, 1001–8
  addition of, 31, 32–33
  converting between decimals and,
    1009–11
  dividing, 30
  equivalent, 1003–5
  linear equations with, 80, 81,
    93–95
  least common denominator
    to solve, 93–95
  linear inequalities containing, 157
  in lowest terms, 1005–6
  multiplying, 29
  repeating decimal as, 995
  subtraction of, 32, 33
Franklin, Ben, 659
Function(s), 586–620
  absolute value, D2
  applications of, 593
  argument of, 591

composite, 838–40
  definition of, 838
  evaluating, 838–39, 851
  finding, 839–40
constant, D1
cube, D2
cube root, D2
  definition of, 587
  domain of, 597–98
  graph of, 592–93
    definition of, 592
    information from, 598–600
    interpreting, 601
    vertical line test of, 590
identity, D1
involving radicals, 705–12
  domain of, 706–7
  graph of, 707–9
  where rule is radical expression,
    705–6
library of, D1–D3
linear, D1
one-to-one, 840–42
  definition of, 840
  horizontal line test for, 841–42
reciprocal, D2
relation as, 586–90
square, 591, D1
square root, D2
value of, 590–92
Future value, 863–64, 898–99

General form of equation of
  circle, 923
General term of sequence, 975
Geometric figure, perimeter and
  area of, C13
Geometric sequences, 990–94, 995
  common ratio of consecutive terms
    of, 990
  definition of, 990
  determining, 990–91
  formula for *n*th term of, 992
  sum of, 993, 996
Geometric series, 994–96
  definition of, 994
  sum of, 994–95
Geometric solid, C20
Geometry, C1–C23
  angles, C1–C3
  lines, C1
  polygons, C5–C20
    circles, C9
    defined, C5
    perimeter and area of, C10–C20
    quadrilaterals, C8
    triangles, C5–C8
  volume and surface area, C20–C23
    of solid figures, C20–C22
Geometry formulas, 105–8, 141–43
Geometry problems, 139–43
  complementary and supplementary
    angle problems, 139–40
  defined, 122
  geometry formulas to solve, 141–43
  systems of linear equations to solve,
    308–10
  triangle problems, 140–41
Golden ratio, 28, 980
Googol, 1003
Graph(s), 565, 566–77
  of circle, 921–22, 923, 926
  complete, 569
  determining if point is on, 568
  of ellipse
    center at origin, 938–41
    center not at origin, 942–43
  of equation in two variables, 568
  of exponential functions, 854–57
    properties of, 855, 856
  of function, 592–93
    definition of, 592
    information from, 598–600
    interpreting, 601
    vertical line test of, 590
  of hyperbola, 949–53, 957
  of inequalities, 91–93

intercepts from, 571
interpreting, 571–73
  of inverse functions, 844–45
  of linear inequalities, 91–93
  of logarithmic functions, 874–76
    properties of, 875–76
  of parabolas
    vertex at origin, 928–29
    vertex not at origin, 931–32
  point-plotting method for, 568–71
  polar area, 565
  of quadratic functions, 781–810
    finding quadratic function
      from, 790
    of form $f(x) = ax^2$, 785–87
    of form $f(x) = ax^2 + bx + c$,
      787–90, 795–801
    of form $f(x) = x^2 + k$, 782–83
    of form $f(x) = (x - h)^2$, 783–85
    steps for, 788, 796–98
    using properties, 795–810
    using transformations, 781–94
    x-intercepts of, 796
  of radical functions, 707–9
    cube roots, 708–9
    square roots, 707–8
  of relation, 589–90
    defined by equation, 580–81
    domain and the range from, 580
  vertically compressed, 786
  vertically stretched, 786
  of $x = y^2$, 570
Graph(s)/graphing, 174–217
  complete, 192
  of equations in two variables, 191–205
    by plotting points, 191–95
    using intercepts, 196–99
    vertical and horizontal lines,
      199–201
  of line
    equation in form $Ax + By = C$, 220
    given a point and its slope, 211–12
    in slope-intercept form, 218–20
  of linear inequalities in two variables,
    253–57
    involving horizontal line, 257
    steps for, 255
    where line goes through origin,
      256–57
  of parallel lines, 236
  plotting points in rectangular
    coordinate system, 175–79
  slope, 206–17
    given two points, 206–9
    of horizontal line, 210–11
    of vertical line, 209–10
  systems of linear equations solved by,
    273–85, 300
    with no point of intersection,
      277–78
    ordered pair as solution of, 273–74
    steps for, 274–76
    in two variables, E3, E7
    where the graph of each equation
      is the same line, 278
  systems of linear inequalities, 327–29
Graphing calculator, 190, 205, 261,
  285, 577
  approximating solutions with
    to exponential equations, 902–3
    to rational inequalities, 826–27
  composite functions evaluated
    using, 851
  ellipses graphed on, 948–49
  to graph circle, 926
  hyperbolas graphed on, 957
  INTERSECT feature, 826,
    902–3, 966
  maximum and minimum feature, 810
  parabola graphed on, 937
  quadratic inequality solved using, 821
  scatter diagrams on, 620
  solution to radical equations verified
    using, 721–22
  sum of sequence using, 990
  systems of linear equations solved
    using, E36–E37

terms of sequence written with, 981
  $2 \infty 2$ determinants evaluated
    with, E46
  ZERO (or ROOT) feature, 826
Greatest common factor (GCF), 408–13
  as binomial, 411
  binomial as, 413
  factoring perfect square trinomial
    having, 439
  factoring the sum or difference of two
    cubes having, 442
  of list of numbers, 409
  in polynomials, factoring out, 411–13
  quadratic equations containing, 456
  of two or more expressions, 408–11
Grouping, factoring by, 413–15
  of quadratic trinomials, 432–35
Grouping symbols, embedded, 52–54

Half-life, 862–63
Half-open, or half-closed, intervals, 90
Half-planes, 253–54
Heron's Formula, 698
Hexagonal dipyramid, C20
Homework, 4
Horizontal addition, 346
Horizontal lines
  equation of, 200, 201, 230
    point-slope form of, 228–29
  graphing, 199–201
  slope of, 209–10
Horizontal line test, 841–42
Horizontal multiplication, 361, 362, 366
Hyperbola, 920, 949–57
  asymptotes of, 953–55
  axes of, 949, 950
  branches of, 949
  center of, 949, 950
  definition of, 949
  equation of, 950, 952–53
  foci of, 949, 950
  graph of, 949–53, 957
  vertices of, 949, 950
Hypotenuse, 463, 750, 751

i, 729–30
Identity, defined, 98, 99
Identity function, D1
Identity property of addition, 41, 46
Imaginary numbers, 8
Inconsistent system of equations,
  278–80, 289–90
  linear equations
    Cramer's Rule with, E43
    matrices to solve, E33
    in three variables, E18–E19, E43
    in two variables, E7–E8
Independent system of equations, 279
Independent variable, 591
Indeterminate functions, 854
Index, 666
  of sum, 976
Inequality(ies)
  absolute value, 635–40
    involving < or ≤, 635–37, 639–40
    involving > or ≥, 637–39
    solving applied problems
      involving, 639–40
  Addition Property of, 153, 156
  compound, 261, 621–31
    defined, 623
    intersection or union of two sets,
      621–22
    involving "and," 623–25
    involving "or," 625–27
    solving, 623
    solving problems using, 627–28
  equivalent, 153
  linear, 89–101
    graph of, 91–93
    in one variable, 89–90
    real number line and interval
      notation to represent, 90–93
    solving, 89
  linear, in two variables, 252–61
    definition of, 252
    graphing, 253–57

ordered pair as solution to,
    252–53
  satisfying, 252
  solving problems involving,
    257–58
  Multiplication Properties of, 154–57
  to order real numbers, 13–14
  rational, 821–27
    definition of, 822
    solving, 822–25, 826–27
  solutions of, 89
Inequality notation, 90
Inequality symbols, 13–14
Infinite sequences, 974
Infinity (8), 90–91
Inputs of relation, 578
Integer(s), 9, 11, 18–28
  adding, 18–21
    using absolute value, 20–21
    using number line, 18–20
  consecutive integer problems, 124–25
  definition of, 9
  dividing, 18, 24–26, 29
    by integer, 385–86
  even, 124
  multiplying, 18, 23–24
  odd, 124
  as rational numbers, 10
  square root of, 662–63
  subtracting, 18, 21–23
Intensity of sound wave, 879
Intercepts, 201
  definition of, 196, 571
  finding, 196–97
  from graph, 571
  graphing equations in two variables
    using, 196–99
Interest
  compound, 863–64
  simple, 104–5, 316–17
Interior angles, C5–C6
Intersecting lines, C3
Intersection of sets, 17, 621–22
Interval notation, 90
Inverse, multiplicative, 47
Inverse functions, 842–47
  defined by a map, 843
  defined by an equation, 845–46
  defined by an ordered pair, 843–44
  definition of, 842
  domain and range of function and its
    inverse, 844
  graph of, 844–45
Inversely proportional, 247
Inverse Property of Logarithms,
  885–86, 891
Inverse variation, 247–49
  definition of, 247
Irrational numbers, 10, 11, 661
  square root as, 661–63
Isolating variable, 77
Isosceles triangle, 147, C6

Joint variation, 251

Kepler, Johannes, 673, 912
Kepler's Law, 673
Kirchhoff's Rule, E23

Laws of Exponents, 377–80
  exponential equations solved
    using, 858
  Negative-Exponent Rule, 671, 674
  Power Rule, 674
  Product Rule, 674
  Product to a Power, 674
  Quotient Rule, 674
  Quotient to a Negative Power
    Rule, 674
  Quotient to a Power Rule, 674
  simplifying expressions using, 674–80
    factoring expressions containing
      rational exponents, 677–78
    involving rational exponents,
      674–76
    radical expressions, 676–77
  Zero-Exponent Rule, 674

Least common denominator (LCD), 93–95, 1004
of rational expressions, 501–4
definition of, 502
with monomial denominators, 502–3
with opposite factors, 504
with polynomial denominators, 503–4
to write equivalent rational expressions, 506
of rational numbers, 501–2
simplify complex rational expression using, 524–26
Least common multiple (LCM), 1002–3
Left endpoint, 90
Left side of equation, 75
Legs of triangle, 147, 463, 750, C6
Length of major axis, 939
Like radicals, 692
Like terms, combining, 60, 61–63
simplifying polynomials by, 346–48
using the commutative property, 62
using the distributive property, 62
Line(s), 607, C1
of best fit, 620
coplanar, C3
graph of
equation in form $Ax + By = C$, 220
given a point and its slope, 211–12
in slope-intercept form, 218–20
horizontal, 199–201, 210–11
intersecting, C3
parallel, 235–38
definition of, 235
determining, 235–37
equation of, 237–38
perpendicular, 238–41, C3
definition of, 238
determining, 239–40
equation of, 240–41
slope of, 239
slope of, 206–17, 239
given two points, 206–9
of horizontal line, 210–11
of vertical line, 209–10
vertical, 199–201, 209–10
Linear equation(s)
cost functions, 610
definition of, 607
of form $Ax + By = 0$, 199
of horizontal line, 200, 201, 228–30
of line parallel to a given line, 237–38
of line perpendicular to a given line, 240–41
point-slope form of, 227–35
given two points, 229–30
horizontal line, 228–29
linear models using, 231–32
with negative slope, 228
with positive slope, 227–28
statement of, 227
of sales commissions, 608–9
slope-intercept form of, 217–26, 230
in form $Ax + By = C$, 220
graph of line in, 218–20
linear models in, 221–23
slope and y-intercept of line using, 217–18
statement of, 218
in standard form, 230
straight-line depreciation, 610
of vertical line, 200, 201, 230
Linear equations in one variable, 74–103
addition property of equality to solve, 76–78, 84–86
classifying, 97–99
combining like terms to solve, 86
with decimals, 95–97
definition of, 75
distributive property to solve, 86–87
with fractions, 80, 81, 93–95
least common denominator (LCD) to solve, 93–95
multiplication property of equality to solve, 78–82, 84–86
sides of, 75

solution of, 75–76
to solve problems, 89, 99–100
with variable on both sides of, 87–89
Linear equations in two variables
definition of, 193
graph of, 193
using intercepts, 196–99
using point-plotting method, 193–94
identifying, 193
in standard form, 193, 201
Linear functions, 607–20, D1
Linear inequalities, 89–101
graph of, 91–93
in one variable, 89–90
real number line and interval notation to represent, 90–93
solving, 89
Linear inequalities in one variable
containing fractions, 157
solution sets for, 153, 158
solving, 153
Addition Property of Inequality for, 153, 156
Multiplication Properties of Inequality for, 154–57
Linear inequalities in two variables, 252–61
definition of, 252
graphing, 253–57
involving horizontal line, 257
steps for, 255
where line goes through origin, 256–57
ordered pair as solution to, 252–53
satisfying, 252
solving problems involving, 257–58
Linear models, 609–15
from data, 612–15
from verbal descriptions, 609–12
Linear relations, 613–14
Line segment, C1
length of, 915
midpoint of, 917
Logarithm(s), 884–93
Change-of-Base Formula, 890–91
common, 876–77
evaluating, for base not 10 or e, 890–91
expanding, 888
Inverse Property of, 885–86, 891
logarithmic expression written as single, 889–90
natural, 876–77
One-to-One Property of, 894
Power Rule of, 887–88, 891
Product Rule of, 886, 891
Quotient Rule of, 886–87, 891
to solve exponential equations, 895–97
Logarithmic equations, 877–79, 894–95
properties of logarithms to solve, 894–95
Logarithmic functions, 870–83
definition of, 870
domain and range of, 873–74
evaluating, 872–73
exponential expressions to logarithmic expressions, 871
graph of, 874–76
properties of, 875–76
logarithmic expressions to exponential expressions, 871
Logarithmic models, 879
Long division, 385–88
dividing polynomial by binomial using, 386–88
Loudness, 879
Lowest terms, 29, 1005–6

Magnitude, 881
Major axis of ellipse, 939, 942
Malthus, Thomas Robert, 973
Mapping, 578
Maps, 121
relation expressed as, 586–89
Marginal propensity to consume, 996

Marginal revenue, 217
Margin of error, 639, 640
Mathematical modeling. See Model(s)/modeling
Mathematics, in everyday language, 5–6
Mathematics symbols, 118
Matrix/matrices, E25–E37
augmented, E25–E28, E36
definition of, E25
determinant of, E37–E46
in reduced row echelon form, E36
in row echelon form, E28–E29
row operations on, E27–E28
solving systems of linear equations using, E28–E31
dependent and inconsistent systems, E32–E33
in three variables, E30
in two variables, E29–E30
square, E37
Mean distance, 947
Means of proportion, 542
Midpoint formula, 916–17
Minor axis of ellipse, 939
Minors of determinants, E40
Mixed number, 18
Mixture problems
defined, 122
systems of linear equations to solve, 314–21
dry mixtures, 318–19
money problems, 315–18
percent mixture, 319–21
plan for modeling, 314–15
Model(s)/modeling, 121–28
defined, 121
exponential, 861–64, 897–99
exponential probability, 861
future value of money, 863–64, 898–99
radioactive decay, 862–63, 897–98
involving rational equations, 541–55
containing opposite factors, 515
problems with similar figures, 544–45
ratio and proportion problems, 541–44
uniform motion problems, 548–51
work problems, 546–48
linear, 609–15
from data, 612–15
from verbal descriptions, 609–12
logarithmic, 879
mathematical, 121–23
of mixture problems, 314–15
of optimization problems, 802–5
process of, 121
with quadratic equations, 460–68, 765–67
Pythagorean Theorem and, 463–65
steps for using, 122–23
of systems of linear equations in three variables, E20–E21
Molarity, 398
Money
future value of, 863–64, 898–99
problems involving, 315–18
Monomial(s)
definition of, 344
degree of, 344, 345
dividing, 371–83
involving negative exponents, 374–76
using the Laws of Exponents, 377–80
using the Quotient Rule for Exponents, 371–72
using the Quotient to a Power Rule, 372–73
using zero as an exponent, 373–74
division of polynomial by, 384–85
in more than one variable, 345
multiplying, 354–60
by a monomial, 358
using the Power Rule for Exponents, 356–57

using the Product Rule for Exponents, 355–56, 357
using the Product to a Power Rule for Exponents, 357
multiplying polynomial by, 360–61
Motion, uniform, 548–51
Multiple, 1002–3
Multiplication
Associative Property of, 44–45, 47
of binomials, 427
Commutative Property of, 43–44, 47
of complex numbers, 726–28
complex conjugates, 727–28
square roots of negative numbers, 726–27
of decimals, 35–36
of fractions, 29
horizontal, 361, 362, 366
of inequality, 154–56
of integers, 18, 23–24
of monomials, 354–60
by a monomial, 358
using the Power Rule for Exponents, 356–57
using the Product Rule for Exponents, 355–56, 357
using the Product to a Power Rule for Exponents, 357
in order of operation, 51, 53, 54
of polynomials, 360–70
by a monomial, 360–61
by a polynomial, 366–67
squaring a binomial, 364–66
sum and difference of two terms, 364
two binomials using the Distributive Property, 361, 362
two binomials using the FOIL method, 362–63
of radical expressions, 694–96
involving special products, 695–96
Product Property for, 681
with unlike indices, 687–88
of rational expressions, 485–87
with two variables, 487
using scientific notation, 394
vertical, 361, 362, 367
words representing, 118
Multiplication Properties of Inequality, 154–57
Multiplication Property of Equality, 78–82, 84–86, 93, 456
Multiplication Property of Zero, 45–46, 47
Multiplicative identity, 41–42, 47
Multiplicative inverse (reciprocal) property, 24–25, 47
Multiplier, 996
Music producers, 343
Music videos, 343

Natural logarithm function, 876–77
Natural numbers (counting numbers), 9, 11
Negative exponent, 374–76
Negative-Exponent Rule, 377, 378, 379, 671, 674
Negative real numbers, 13, 662
Newton's Law of Cooling, 901–2
Nightingale, Florence, 565
Nonlinear relations, 613–14
Nonlinear systems of equations, 958–66
definition of, 958–59
elimination method for solving, 961–63
substitution method for solving, 959–61
nth roots, 666–67
Number(s). See also Rational numbers; Real numbers
classifying, 9–12
imaginary, 8
irrational, 10, 11–12
mixed, 18
whole, 9, 11
Number line, adding integers using, 18–20

Number line, real
to represent linear inequalities, 90–93
Number systems, 8–12
Numerator, 1003

Obtuse angle, C2
Octagon, C5
Odd integer, 124
One-to-one function, 840–42
definition of, 840
horizontal line test for, 841–42
One-to-One Property of
Logarithms, 894
Open interval, 90
Operations, 18
Opposite (additive inverse), 21, 46
Optimization, 801, 802–5
Ordered pair(s), 176, 566
inverse defined by, 843–44
as point on graph of equation, 566
relation expressed as, 586–89
satisfying an equation, 179–81
as solution of systems of linear
equations, 273–74
in two variables, E1–E3
as solution of systems of linear
inequalities, 326–27
Ordered triple, E13
Order of operations, 50–55
division bar and, 52
embedded grouping symbols
and, 52–54
evaluating expression using, 54–55
parentheses and, 51
Ordinary annuity, 996
Ordinate (y-coordinate), 566
Origin (O), 12, 175, 566
Outputs of relation, 578

Parabola, 787, 920, 927–38
applied problems involving, 932–34
axis of symmetry of, 787, 796, 927
definition of, 927
directrix of, 927, 928, 931
equation of, 928, 930–31
focus of, 927, 928, 931
graph of, 928–29, 931–32, 937
opening up and opening down,
787, 927
vertex of, 787, 795, 927, 928, 931
Parallel lines, 235–38
angles formed by, C3–C4
definition of, 235, C3
determining, 235–37
equation of, 237–38
Parallelogram, C8
perimeter and area of, 106, C14–C15
Parentheses, order of operations and, 51
Partial sum, 976–77
Pascal, Blaise, A5
Pascal's Triangle, 980, A5
Payment period, 863
Pentagon, C5
Percent, 104, 131–38
converting between decimal and,
1011–12
defined, 131, 1011
direct translation problems involving,
131–38
discounts or mark-ups, 134–35
formulas containing, 104
mixture problems involving, 319–21
steps in solving problems involving,
131–33
Perfect cubes, 666
Perfect fifths, 666
Perfect fourths, 666
Perfect squares (squares of binomials),
364–66, 437, 660–61, 666
Perfect square trinomials, 437–39,
748–49
factoring, 437–39
Perihelion, 947
Perimeter, C10–C20
defined, 105
formulas of, 106
of parallelogram, C14–C15

problems involving, 141
of rectangle, 107, C10–C12
of trapezoid, C15, C16
Perpendicular lines, 238–41, C3
definition of, 238
determining, 239–40
equation of, 240–41
slope of, 239
pH, 882
Photon, energy of, 251
Place value, 1008–9
Planck's Constant, 251
Plane, C1
Point(s)
collinear, E46
coordinate of, 566
distance between two, 40
plotting, graphing by
of equations in two variables,
191–95
of linear equation, 193–94
in rectangular coordinate system,
175–79
Point-plotting method, 568–71
Point-slope form of line, 227–35
given two points, 229–30
horizontal line, 228–29
linear models using, 231–32
with negative slope, 228
with positive slope, 227–28
statement of, 227
Polar area graph, 565
Polygons, C5–C20. See also Circle;
Triangle(s)
area of, C11
defined, C5
perimeter and area of, C10–C20
quadrilaterals, C8
regular, C5
Polyhedron, C20
Polynomial(s), 343, 344–54. See also
Binomial(s); Trinomial(s)
addition of, 346–47
definition of, 344, 345
degree of, 345
dividing, 383–91
by a binomial, 386–88
by a monomial, 384–85
using synthetic division, B1–B7
evaluating, 348–49
examples of, 346
factoring. See Factoring polynomials
multiplying, 360–70
by a monomial, 360–61
by a polynomial, 366–67
squaring a binomial, 364–66
sum and difference of two
terms, 364
two binomials using the
Distributive Property, 361, 362
two binomials using the FOIL
method, 362–63
prime, 422
simplifying
by combining like terms, 346–48
in two variables, 347, 348
in standard form, 345
subtracting, 347–48
Polynomial equation(s)
definition of, 451
degree of, 451
examples of, 451
factoring to solve, 451–60
quadratic equations, 451–57
Zero-Product Property to solve third-
or higher-degree, 457
Polynomial functions
Factor Theorem for, B4–B5
Remainder Theorem for, B4–B5
Population growth, 973
Positive real numbers, 13
Power(s), 49. See also Exponent(s)
of i, 729–30
perfect, 666
Power Rule
for Exponential Expressions, 674
for Exponents, 356–57, 377

of Logarithms, 887–88, 891
simplifying expressions using, 379
Prime factorization, 1001
Prime number, 1001
Prime polynomial, 422
Principal, defined, 105
Principal nth root, 666
Principal square root, 660
Probability, exponential, 861
Problems, categories of, 122
Problem solving, 117–49
definition of, 121
direct translation problems, 117–38
defined, 122
involving percent, 131–38
models for solving, 123–28
translating English phrases to
algebraic expressions, 117–20
translating English sentences to
equations, 120–23
using systems of linear equations,
307–8
geometry problems, 139–43
complementary and
supplementary angle problems,
139–40
defined, 122
geometry formulas to solve, 141–43
triangle problems, 140–41
using systems of linear equations,
308–10
uniform motion problems, 122,
143–45, 548–51
for rate, 144–45
for time, 143–44
using systems of linear equations,
310–11
Product(s), 18, 1001. See also
Multiplication
binomial, 364–66
squaring, 364–66
using the Distributive Property,
361, 362
using the FOIL method, 362–63
definition of, 23
special, 364
multiplying radical expressions
involving, 695–96
of sum and difference of two terms,
437
Product Property of Radicals, 680–86
to multiply radicals, 681
to simplify radical expressions, 681–86
Product Rule, 674
for Exponents, 355–56, 357, 377
of Logarithms, 886, 891
simplifying expressions using, 378
Product to a Power Rule, 377, 674
for Exponents, 357
Progressions. See Arithmetic sequences
(arithmetic progression)
Proportion
definition of, 542
extremes of, 542
means of, 542
problems involving, 541–43
terms of, 542
Proportionality, constant of (constant of
variation), 245, 247
Pyramid, volume and surface area, C21
Pythagorean Theorem, 463–65,
750–53, 916
converse of, 756
definition of, 751
distance between two points using,
913–14
Pythagorean Triples, 756

Quadrants, 176, 567
Quadratic equation(s), 451–57, 460–68,
744–81
completing the square to solve,
744–56, 764
in one variable, 747–48
steps in, 749
when coefficient of square term is
not 1, 750

containing a greatest common
factor, 456
definition of, 452
factoring, 453–54, 764
modeling and problem solving
involving, 460–68, 765–67
using Pythagorean Theorem,
463–65
not in standard form, 454–56
Pythagorean Theorem to solve,
750–53
quadratic formula to solve, 757–72
discriminant to determine nature
of solutions, 762–64
statement of, 758
steps in, 758–59
solving rational equation leading to,
760–61
Square Root Property to solve,
744–47, 764
in standard form, 453
Zero-Product Property to solve,
451–57
Quadratic functions
definition of, 781
given vertex and one other point, 790
graph of, 781–810
finding quadratic function from, 790
of form $f(x) = ax^2$, 785–87
of form $f(x) = ax^2 + bx + c$,
787–90, 795–801
of form $f(x) = x^2 + k$, 782–83
of form $f(x) = (x - h)^2$, 783–85
steps for, 788, 796–98
using properties, 795–810
using transformations, 781–94
x-intercepts of, 796
maximum or minimum value of,
801–2
modeling and solving optimization
problems involving, 802–5
Quadratic inequalities, 811–21
definition of, 811
solving
algebraic method of, 813–14, 815,
817–18
graphical method of, 812–13,
814–15, 816
with graphing calculator, 821
steps for, 812, 813, 814
Quadratic trinomials, 418–36
of form $ax^2 + bx + c, a \neq 1$, 427–36
grouping to solve, 432–35
with a negative leading coefficient,
431–32
steps for solving, 433
trial and error method of solving,
427–32, 434–35
with two variables, 431
of form $x^2 + bx + c$, 418–26
with a common factor, 423, 424
with a negative leading coefficient,
424–25
not written in standard form, 423
steps for solving, 419
with two variables, 422–23
where c is negative, 420–21
where c is positive, 419–20, 421
Quadrilaterals, C8
defined, C5
Questions, asking, 4
Quotient(s), 18, 36, 487. See also
Division
definition of, 24, 25
of integers, 29
simplifying, 376
Quotient Property of Radicals, 686–87
Quotient Rule
for Exponents, 371–72, 377,
384–85, 674
of Logarithms, 886–87, 891
for simplifying expressions, 371–72,
378–79
Quotient to a Negative Power Rule,
377, 674
Quotient to a Power Rule, 372–73,
377, 674

Radical(s)
functions involving, 705–12
domain of, 706–7
graph of, 707–9
where rule is radical expression, 705–6
like, 692
with rational exponents, 669–71
simplifying, 667–68
using rational exponents, 676–77
written as decimal, 662
Radical equations, 712–22
containing a rational exponent, 715–16
containing one radical, 713–15
containing two radicals, 716–18
definition of, 712
graphing calculator to verify solutions to, 721–22
solving for variable in, 718
Radical expressions
adding and subtracting, 692–94
multiplying, 694–96
involving special products, 695–96
Product Property for, 681
with unlike indices, 687–88
rationalizing, 699–705
rationalizing the denominator, 699–702
simplifying, 676–77, 680–91, 693
Product Property for, 681–86
Quotient Property for, 686–87
with variable radicand, 684
Radical sign, 660
Radicand, 660
simplifying radical with variable, 684
Radioactive decay, 862–63, 897–98
Radius, defined, 105
Radius of circle, 921, 922–23, C9
Range
inverse functions and, 844
of logarithmic functions, 873–74
of relation, 578–80
Rate(s)
of change, average, 212–13
constant, 546
in mixture problems, 314
in uniform motion problems, 144–45
Ratio, 25
common, 990
definition of, 541
golden, 28, 980
problems involving, 541–43
Rational equation(s), 529–55
applications of, 536
containing rational expressions, 529–36
LCD is product of denominators, 533–34
definition of, 529
with extraneous solution, 534
models involving, 541–55
containing opposite factors, 515
problems with similar figures, 544–45
ratio and proportion problems, 541–44
uniform motion problems, 548–51
work problems, 546–48
with no solution, 535
solution leading to quadratic equation, 760–61
solving for variable in, 536–38
Rational exponents
approximating expressions involving, 670
negative, 671
radicals with, 669–71
simplifying, 674–76
simplifying radicals using, 676–77
Rational expression(s), 476–529
adding, 494–96, 497–98
with a common denominator, 494–96
and integer, 515–16
numerator containing more than one term, 495–96

with opposite denominators, 497–98
with unlike denominators, 509–19
complex, 520–29
definition of, 520
simplifying, 521–26
defined, 476, 477
dividing, 487–90, 520–21
with more than one variable, 490
written vertically, 490
equivalent, 504–6
forming, 505–6
evaluating, 477–79
of higher degree, 478
with more than one variable, 478–79
examples of, 477
least common denominator of, 501–4
definition of, 502
with monomial denominators, 502–3
with opposite factors, 504
with polynomial denominators, 503–4
to write equivalent rational expressions, 506
multiplying, 485–87
with two variables, 487
rational equations containing, 529–36
LCD is product of denominators, 533–34
simplifying, 480–85
containing opposite factors, 482–83
subtracting, 496–98
with a common denominator, 496–97
numerator containing more than one term, 497
with opposite denominators, 498
with unlike denominators, 509–19
undefined values of, 479–80
Rational inequalities, 821–27
definition of, 822
solving, 822–25
graphing calculator for, 826–27
Rationalizing radical expressions, 699–705
Rationalizing the denominator, 699–702
containing a square root, 699–700
containing cube roots and fourth roots, 700–701
containing one term, 699–701
containing two terms, 701–2
Rational numbers, 10, 11, 28–41
adding, 494
adding and subtracting decimals, 34–35
adding fractions, 31, 32–33
dividing decimals, 36–37
dividing fractions, 30
evaluating expression containing rational, 34
least common denominator of, 501–2
multiplying decimals, 35–36
multiplying fractions, 29
square root as, 661–63
subtracting fractions, 32, 33
Ray, C1
Real number line
constructing, 12
origin of, 12
plot points on, 12–13
to represent linear inequalities, 90–93
scale of, 12
Real numbers, 8–73. See also Integer(s); Rational numbers
absolute value of, 14–15
definition of, 11
irrational numbers, 10, 11
natural numbers (counting numbers), 9, 11
negative, 13, 662
ordering
inequalities in, 13–14
positive, 13
properties of, 41–48

associative properties of addition and multiplication, 44–45, 46, 47
commutative properties of addition and multiplication, 43–44, 46, 47
identity property of addition, 41, 46
multiplication and division properties of zero, 45–46, 47
multiplicative identity, 41–42
set of, 11
whole numbers, 9, 11
Real number system, 8
Reciprocal function, D2
Reciprocal (multiplicative inverse) property, 24–25, 47
Rectangle, C8
area of, 103, 106, C11–C12
perimeter of, 106, 107, C10–C12
Rectangular box, volume and surface area of, 106
Rectangular (Cartesian) coordinate system, 175–79, 566
identifying points in, 178–79
plotting points in, 177–78, 566
Rectangular solid, volume and surface area, C21
Recursive formula, 980
Regular polygon, C5
Relation(s), 577–84
defined, 577
defined by equation, graph of, 580–81
domain and range of, 578–80
expressed as an equation, 589
as function, 586–90
graph of, 589–90
illustration of, 578
inputs of, 578
linear vs. nonlinear, 613–14
outputs of, 578
Remainder Theorem, B4–B5
Repeated root, 761
Repeated solution (double root), 456
Repeating decimal, 1010
Revenue, marginal, 217
Rhombus, C8
Richter scale, 837, 881
Right angle, 463, 750, C2
Right circular cylinder, volume and surface area of, 106, C21, C22
Right endpoint, 90
Right side of equation, 75
Right triangles, 463, 750, 751, C6
Roots, repeated, 761
Rounding, 1009
Row echelon form, E28–E29
reduced, E36
Row operations on matrices, E27–E28
Rule of 72, 902

Satisfying an equation, 179–85, 568
table of values for, 181–85
Scale of number line, 12
Scalene triangle, C6
Scatter diagrams, 612–13, 620
Scientific notation, 391–98
to decimal notation, 393–94
decimal notation to, 391–92
definition of, 391
to multiply and divide, 394–95
Second-degree equation, 453
Sequences, 974–1002
annuities, 996–97
arithmetic (arithmetic progression), 982–90, 995
common difference between consecutive terms of, 982
definition of, 982
determining, 982–83
formula for nth term of, 983–85
sum of, 985–87, 990
definition of, 974
Fibonacci, 980
finite, 974
geometric, 990–94, 995
common ratio of consecutive terms of, 990
definition of, 990

determining, 990–91
formula for nth term of, 992
sum of, 993, 996
infinite, 974
summation notation for, 976–78
terms of, 974
alternating signs of, 975
formula for nth term, 976
general, 975
graphing calculators to write, 981
writing first several, 974–75
Set(s), 8
defined, 8
elements in, 8
empty, 8, 158
intersection of, 17, 621–22
of real numbers, 11
solution, 75, 89–90
union of, 17, 621–22
writing, 9
Set notation, 75
Sides of equation, 179, 567
of linear equations, 75
Sieve of Eratosthenes, 1008
Sign(s)
dividing two integers and, 25
multiplying two integers and, 23
of number, 13
Similar figures
definition of, 544
problems with, 544–45
Similar triangles, C7–C8
Simple interest formula, 104–5, 316–17
Simplifying expressions
algebraic expressions, 57–65
by combining like terms, 60, 61–63
distributive property for, 61
with exponents, 355–56
like terms and unlike terms, 59–60
exponential expressions, 356–57, 377–80
involving negative exponents, 374–76
Laws of Exponents for, 674–80
using zero as an exponent, 373–74
polynomials, by combining like terms, 346–48
Power Rule and the Negative Exponent Rule for, 379
Product Rule and the Negative Exponent Rule for, 378
Quotient Rule and the Negative Exponent Rule for, 378–79
Quotient Rule for, 371–72
Quotient to a Power Rule for Exponents for, 372–73
radical expressions, 676–77, 680–91, 693
Product Property for, 681–86
Quotient Property for, 686–87
with variable radicand, 684
radicals, 667–68
using rational exponents, 676–77
rational expressions, 480–85
complex, 521–26
containing opposite factors, 482–83
Slope, 206–17
applications of, 212–13
as average rate of change, 212–13
definition of, 206–7
finding, 208–9
given two points, 206–9
of horizontal line, 210–11
interpreting, 208–9
negative, 211, 228
positive, 211, 227–28
undefined, 210, 211
of vertical line, 209–10
zero, 210, 211
Slope-intercept form of line, 217–26, 230
in form $Ax + By = C$, 220
graph of, 218–20
linear models in, 221–23
slope and y-intercept of line identified using, 217–18
statement of, 218
Solid, geometric, C20

Solution(s). *See also* Zero(s)
extraneous, 534, 713, 877
of inequality, 89
of linear equation, 75–76
to system of equations, E13
Solution set, 75, 89–90
for linear identities, 98
for linear inequalities, 153, 158
Sound wave, intensity of, 879
Special products, 364
factoring, 436–44
difference of two squares, 439–41
perfect square trinomials, 437–39
sum or difference of two cubes, 441–42
multiplying radical expressions involving, 695–96
Sphere, volume and surface area of, 106, C21
Square(s), C8
area and perimeter of, 106
difference of two, 439–41
perfect (squares of binomials), 660–61, 666
perimeter and area of, C12–C13
Square function, 591, D1
Square matrix, E37
Square pyramid, volume and surface area, C21
Square root(s), 660–65
definition of, 660
evaluating, 660–61, 663
graphing functions involving, 707–8
of negative numbers, 726–27
real numbers, 723–24
of perfect squares, 660–61
principal, 660
properties of, 660, 662
as rational, irrational, or not a real number, 661–63
rationalizing denominator containing, 699–700
simplifying expression involving, 683
of variable expressions, 663
Square root function, D2
Square Root Property, 744–47, 764
Squaring a binomial, 364–66, 437
Standard form
complex number in, 723
of equation of circle, 921, 923
linear equation in, 193, 201, 230
polynomial in, 345
quadratic equations in, 453
Step-by-step examples, 5
Straight angle, C2
Straight-line depreciation, 610–12
Study skills, 4, 6
Subset(s), 17
Substitution method
nonlinear systems of equations solved using, 959–61
systems of linear equations solved using, 285–95, 300
applied problems, 291–92
first solving for one variable, 287–89
inconsistent system, 289–90
with infinitely many solutions, 290

steps for, 286–87
in two variables, E4–E5, E7
Subtraction
of complex numbers, 724–25
of decimals, 34–35
of fractions, 32, 33
of integers, 18, 21–23
in order of operation, 51, 54
of polynomials, 347–48
of radical expressions, 692–94
of rational expressions, 496–98
with a common denominator, 496–97
numerator containing more than one term, 497
with opposite denominators, 498
with unlike denominators, 509–19
words representing, 118
Sum, 18. *See also* Addition
of geometric series, 994–95
partial, 976–77
of sequences
arithmetic, 985–87, 990
geometric, 993, 996
of two cubes, 441–42
Sum and difference of two terms, 364
Summation notation, 976–78
Supplementary angles, 139–40, 309
defined, 139
Supplement of angle, C2
Surface area
defined, 105
formulas of, 106
Symmetric Property, 109
Synthetic division of polynomial, B1–B7
System of equations, solution to, E13
Systems of linear equations, 272–325
applied problems involving, 280–81, 291–92, 301–2
augmented matrix of, E25–E26
classifying algebraically, 279
classifying graphically, 279
consistent, 278–80
definition of, 273, E1
dependent, 279, E32
examples of, 273, E1
inconsistent, 278–80, 289–90, E18–E19, E33, E43
independent, 279
matrices to solve, E28–E31
ordered pair as solution of, 273–74
problem solving using, 307–25
direct translation problems, 307–8
geometry problems, 308–10
mixture problems, 314–21
uniform motion problems, 310–11
solution of, 273
number of, 279–80
solving
by graphing, 273–85, 300
using elimination, 296–306
using substitution, 285–95, 300
in three variables, E13–E25
Cramer's rule to solve, E41–E43
dependent, E19–E20, E43
elimination method for solving, E15–E18
inconsistent, E18–E19, E43
matrices to solve, E30–E31

modeling, E20–E21
solving, E13–E18, E20–E21
in two variables, E1–E13
Cramer's rule to solve, E38–E39
dependent, E8
elimination method for solving, E5–E6, E7
graphing solution for, E3, E7
inconsistent systems of, E7–E8
matrices to solve, E29–E30
ordered pair as solution to, E1–E3
substitution method for solving, E4–E5, E7
visualizing solutions in, E2–E3
Systems of linear inequalities, 326–33
applied problems involving, 329–30
graphing, 327–29
ordered pair as solution of, 326–27
satisfying, 326
Systems of nonlinear equations, 958–66
definition of, 958–59
elimination method for solving, 961–63
substitution method for solving, 959–61

Terminating decimal, 1010
Terms, 344
coefficient of, 59–60
definition of, 59
like, 59–60
combining, 60, 61–63
of proportion, 542
sum and difference of two, 364
unlike, 59–60
Test, chapter, 6
Test point, 254
TI-84 Plus graphing calculator. *See also* Graphing calculator
Time, in uniform motion problems, 143–44
Tolerance, 639
Transformations
definition of, 782, 788
graphing quadratic functions using, 781–94
steps for, 788
Transversal, C3
Transverse axis, 949, 950
Trapezoid, C8
perimeter and area of, 106, C15
Triangle(s), C5–C8
area and perimeter of, 106, C16
congruent, C6–C7
defined, C5
equilateral, C6
interior angles of, C5–C6
isosceles, 147, C6
legs of, 147, 463, 750, C6
Pascal's, 980, A5
problems involving, 140–41
right, 463, 750, 751, C6
scalene, C6
similar, C7–C8
Trinomial(s), 345
dividing by a monomial, 385
factoring out greatest common factor in, 412
multiplying a monomial by, 360–61

multiplying by a monomial, 361
perfect square, 365, 437–39, 748–49
quadratic, 418

Uniform motion problems, 122, 143–45, 548–51
for rate, 144–45
systems of linear equations to solve, 310–11
for time, 143–44
Union of sets, 17, 621–22

Variable(s), 58
definition of, 57
dependent, 591
independent, 591
isolating, 77
Variation, 245–51
constant of, 245, 247
direct, 245–47
definition of, 245
inverse, 247–49
definition of, 247
joint, 251
Vertex angle, 147
Vertex/vertices, C1
of ellipse, 939, 942
of hyperbola, 949, 950
of parabola, 787, 795, 927, 928, 931
of polygon, C5
of polyhedron, C20
Vertical addition, 347
Vertical angles, C3
Vertical lines
equation of, 200, 201, 230
graphing, 199–201
slope of, 209–10
Vertical line test, 590
Vertical multiplication, 361, 362, 367
Videos, music, 343
Volume
defined, 105
formulas of, 106

Whole numbers, 9, 11
Work problems, 546–48
defined, 122

x-axis, 175, 566
x-coordinate (abscissa), 176, 566
x-intercepts, 196, 201, 571
of graph of quadratic function, 796
xy-plane, 176, 566

y-axis, 175, 566
y-coordinate (ordinate), 176, 566
y-intercept, 196, 201, 571

Zero(s), 9, 13. *See also* Solution(s)
complex, 733
division properties of, 46, 47
multiplication property of, 45–46, 47
Zero exponent, 373–74
Zero Exponent Rule, 377, 674
Zero-level earthquake, 881–82
Zero-Product Property
quadratic equations solved using, 451–57
to solve third- or higher-degree polynomial equations, 457

Photo Credits

Chapter 1 Pages 1, 28, Cosmo Condina/The Stock Connection

Chapter 2 Pages 74, 116, Randy M. Ury/CORBIS-NY

Chapter 3 Page 174, Getty Images; Page 175, U.S. Air Force

Chapter 4 Page 272, Patrik Giardino/CORBIS/Bettmann

Chapter 5 Pages 343, 353, Getty Images-Digital Vision;
Page 397, Karl Ronstrom/CORBIS/Reuters America LLC

Chapter 6 Page 407, Tim Boyle/Getty Images

Chapter 7 Pages 476, 536, Carlo Orlandi/AP Wide World Photos

Chapter 8 Page 565, Courtesy of Florence Nightengale Museum, London;
Page 599, Joel Rogers

Chapter 9 Page 659, The Isamu Noguchi Foundation, Inc.;
Page 665, Norbert Schiller/The Image Works

Chapter 10 Pages 743, 807, AP Wide World Photos; Page 751, Jill Davis

Chapter 11 Page 837, David Weintraub/Photo Researchers, Inc.

Chapter 12 Pages 912, 944 (left) Victor Habbick Visions/Photo Researchers, Inc.;
Page 933, Alan Becker/Getty Images, Inc.-Image Bank;
Page 944 (right) Scott McDonald/Hedrich Blessing

Chapter 13 Page 973, Ambient Images; Page 980, Getty Images, Inc.-Image Bank

Chapter Test Prep Video CD
ELEMENTARY & INTERMEDIATE ALGEBRA
Michael Sullivan, III, Katherine R. Struve & Janet Mazzarella
ISBN 0-13-188794-7
CD License Agreement
© 2007 Pearson Education, Inc.
Pearson Prentice Hall
Pearson Education, Inc.
Upper Saddle River, NJ 07458
All rights reserved.
Pearson Prentice Hall™ is a trademark of Pearson Education, Inc.

YOU SHOULD CAREFULLY READ THE TERMS AND CONDITIONS BEFORE USING THE CD-ROM PACKAGE. USING THIS CD-ROM PACKAGE INDICATES YOUR ACCEPTANCE OF THESE TERMS AND CONDITIONS.

Pearson Education, Inc. provides this program and licenses its use. You assume responsibility for the selection of the program to achieve your intended results, and for the installation, use, and results obtained from the program. This license extends only to use of the program in the United States or countries in which the program is marketed by authorized distributors.

LICENSE GRANT
You hereby accept a nonexclusive, nontransferable, permanent license to install and use the program ON A SINGLE COMPUTER at any given time. You may copy the program solely for backup or archival purposes in support of your use of the program on the single computer. You may not modify, translate, disassemble, decompile, or reverse engineer the program, in whole or in part.

TERM
The License is effective until terminated. Pearson Education, Inc. reserves the right to terminate this License automatically if any provision of the License is violated. You may terminate the License at any time. To terminate this License, you must return the program, including documentation, along with a written warranty stating that all copies in your possession have been returned or destroyed.

LIMITED WARRANTY
THE PROGRAM IS PROVIDED "AS IS" WITHOUT WARRANTY OF ANY KIND, EITHER EXPRESSED OR IMPLIED, INCLUDING, BUT NOT LIMITED TO, THE IMPLIED WARRANTIES OF MERCHANTABILITY AND FITNESS FOR A PARTICULAR PURPOSE. THE ENTIRE RISK AS TO THE QUALITY AND PERFORMANCE OF THE PROGRAM IS WITH YOU. SHOULD THE PROGRAM PROVE DEFECTIVE, YOU (AND NOT PEARSON EDUCATION, INC. OR ANY AUTHORIZED DEALER) ASSUME THE ENTIRE COST OF ALL NECESSARY SERVICING, REPAIR, OR CORRECTION. NO ORAL OR WRITTEN INFORMATION OR ADVICE GIVEN BY PEARSON EDUCATION, INC., ITS DEALERS, DISTRIBUTORS, OR AGENTS SHALL CREATE A WARRANTY OR INCREASE THE SCOPE OF THIS WARRANTY.

SOME STATES DO NOT ALLOW THE EXCLUSION OF IMPLIED WARRANTIES, SO THE ABOVE EXCLUSION MAY NOT APPLY TO YOU. THIS WARRANTY GIVES YOU SPECIFIC LEGAL RIGHTS AND YOU MAY ALSO HAVE OTHER LEGAL RIGHTS THAT VARY FROM STATE TO STATE.

Pearson Education, Inc. does not warrant that the functions contained in the program will meet your requirements or that the operation of the program will be uninterrupted or error-free. However, Pearson Education, Inc. warrants the CD-ROM(s) on which the program is furnished to be free from defects in material and workmanship under normal use for a period of ninety (90) days from the date of delivery to you as evidenced by a copy of your receipt. The program should not be relied on as the sole basis to solve a problem whose incorrect solution could result in injury to person or property. If the program is employed in such a manner, it is at the user's own risk and Pearson Education, Inc. explicitly disclaims all liability for such misuse.

LIMITATION OF REMEDIES
Pearson Education, Inc.'s entire liability and your exclusive remedy shall be: 1. the replacement of any CD-ROM not meeting Pearson Education, Inc.'s "LIMITED WARRANTY" and that is returned to Pearson Education, or 2. if Pearson Education is unable to deliver a replacement CD-ROM that is free of defects in materials or workmanship, you may terminate this agreement by returning the program.

IN NO EVENT WILL PEARSON EDUCATION, INC. BE LIABLE TO YOU FOR ANY DAMAGES, INCLUDING ANY LOST PROFITS, LOST SAVINGS, OR OTHER INCIDENTAL OR CONSEQUENTIAL DAMAGES ARISING OUT OF THE USE OR INABILITY TO USE SUCH PROGRAM EVEN IF PEARSON EDUCATION, INC. OR AN AUTHORIZED DISTRIBUTOR HAS BEEN ADVISED OF THE POSSIBILITY OF SUCH DAMAGES, OR FOR ANY CLAIM BY ANY OTHER PARTY.

SOME STATES DO NOT ALLOW FOR THE LIMITATION OR EXCLUSION OF LIABILITY FOR INCIDENTAL OR CONSEQUENTIAL DAMAGES, SO THE ABOVE LIMITATION OR EXCLUSION MAY NOT APPLY TO YOU.

GENERAL
You may not sublicense, assign, or transfer the license of the program. Any attempt to sublicense, assign or transfer any of the rights, duties, or obligations hereunder is void.

This Agreement will be governed by the laws of the State of New York.

Should you have any questions concerning this Agreement, you may contact Pearson Education, Inc. by writing to:
ESM Media Development
Higher Education Division
Pearson Education, Inc.
1 Lake Street
Upper Saddle River, NJ 07458

Should you have any questions concerning technical support, you may write to:
New Media Production
Higher Education Division
Pearson Education, Inc.
1 Lake Street
Upper Saddle River, NJ 07458

YOU ACKNOWLEDGE THAT YOU HAVE READ THIS AGREEMENT, UNDERSTAND IT, AND AGREE TO BE BOUND BY ITS TERMS AND CONDITIONS. YOU FURTHER AGREE THAT IT IS THE COMPLETE AND EXCLUSIVE STATEMENT OF THE AGREEMENT BETWEEN US THAT SUPERSEDES ANY PROPOSAL OR PRIOR AGREEMENT, ORAL OR WRITTEN, AND ANY OTHER COMMUNICATIONS BETWEEN US RELATING TO THE SUBJECT MATTER OF THIS AGREEMENT.

System Requirements
Windows
Pentium II 300 MHz processor
Windows NT, 2000, ME, or XP
64 MB RAM (128 MB RAM required for Windows XP)
4.3 MB available hard drive space (optional-for minimum QuickTime installation)
800 x 600 resolution
8x or faster CD-ROM drive
QuickTime 6.x
Sound card

Macintosh
PowerPC G3 233 MHz or better
Mac OS 9.x or 10.x
64 MB RAM
10 MB available hard drive space for Mac OS 9, 19 MB on OS X
(optional—if QuickTime installation is needed)
800 x 600 resolution
8x or faster CD-ROM drive
QuickTime 6.x

Support Information
If you are having problems with this software, call (800) 677-6337 between 8:00 a.m. and 8:00 p.m. EST, Monday through Friday, and 5:00 p.m. through 12:00 a.m. EST on Sundays. You can also get support by filling out the web form located at: http://247.prenhall.com/mediaform

Our technical staff will need to know certain things about your system in order to help us solve your problems more quickly and efficiently. If possible, please be at your computer when you call for support. You should have the following information ready:
• Textbook ISBN
• CD-ROM ISBN
• corresponding product and title
• computer make and model
• Operating System (Windows or Macintosh) and Version
• RAM available
• hard disk space available
• Sound card? Yes or No
• printer make and model
• network connection
• detailed description of the problem, including the exact wording of any error messages.

NOTE: Pearson does not support and/or assist with the following:
• third-party software (i.e. Microsoft including Microsoft Office suite, Apple, Borland, etc.)
• homework assistance
• Textbooks and CD-ROMs purchased used are not supported and are non-replaceable. To purchase a new CD-ROM, contact Pearson Individual Order Copies at 1-800-282-0693.

Working with Radicals (Chapter 9)

Simplifying $\sqrt[n]{a^n}$

- If $n \geq 2$ is a positive integer and a is a real number, then

  $\sqrt[n]{a^n} = a$ if $n \geq 3$ is odd

  $\sqrt[n]{a^n} = |a|$ if $n \geq 2$ is even

- If a is a real number and $n \geq 2$ is an integer, then $a^{1/n} = \sqrt[n]{a}$ provided that $\sqrt[n]{a}$ exists.

- If a is a real number, m/n is a rational number in lowest terms with $n \geq 2$, then $a^{m/n} = \sqrt[n]{a^m} = \left(\sqrt[n]{a}\right)^m$ provided that $\sqrt[n]{a}$ exists.

- If m/n is a rational number and if a is a nonzero real number, then $a^{-m/n} = \dfrac{1}{a^{m/n}}$ or $\dfrac{1}{a^{-m/n}} = a^{m/n}$

Product Property of Radicals

If $\sqrt[n]{a}$ and $\sqrt[n]{b}$ are real numbers and $n \geq 2$ is an integer, then $\sqrt[n]{a} \cdot \sqrt[n]{b} = \sqrt[n]{ab}$

Quotient Property of Radicals

If $\sqrt[n]{a}$ and $\sqrt[n]{b}$ are real numbers, $b \neq 0$ and $n \geq 2$ is an integer, then $\dfrac{\sqrt[n]{a}}{\sqrt[n]{b}} = \sqrt[n]{\dfrac{a}{b}}$

Quadratic Equations and Quadratic Functions (Chapter 10)

- **Square Root Property**

  If $x^2 = p$, then $x = \sqrt{p}$ or $x = -\sqrt{p}$

- **Pythagorean Theorem**

  In a right triangle, the square of the length of the hypotenuse is equal to the sum of the squares of the lengths of the legs. That is, $\text{leg}^2 + \text{leg}^2 = \text{hypotenuse}^2$.

- **The Quadratic Formula**

  The solutions to the equation $ax^2 + bx + c = 0$, $a \neq 0$, are given by $x = \dfrac{-b \pm \sqrt{b^2 - 4ac}}{2a}$

- **Discriminant**

  For the quadratic equation $ax^2 + bx + c = 0$, $a \neq 0$:

  - If $b^2 - 4ac > 0$, the equation has two unequal real solutions.
    - If $b^2 - 4ac$ is a perfect square, the equation has two rational solutions.
    - If $b^2 - 4ac$ is not a perfect square, the equation has two irrational solutions.
  - If $b^2 - 4ac = 0$, the equation has a repeated real solution.
  - If $b^2 - 4ac < 0$, the equation has two complex solutions that are not real.

- **Vertex of a Parabola**

  Any quadratic function of the form $f(x) = ax^2 + bx + c$, $a \neq 0$, will have vertex $\left(-\dfrac{b}{2a}, f\left(-\dfrac{b}{2a}\right)\right)$

- **The x-Intercepts of the Graph of a Quadratic Function**

  1. If $b^2 - 4ac > 0$, the graph of $f(x) = ax^2 + bx + c$ has two different x-intercepts.
  2. If $b^2 - 4ac = 0$, the graph of $f(x) = ax^2 + bx + c$ has one x-intercept.
  3. If $b^2 - 4ac < 0$, the graph of $f(x) = ax^2 + bx + c$ has no x-intercepts.

Properties of Logarithms (Chapter 11)

$a^{\log_a M} = M$ $\log_a a^r = r$

$\log_a(MN) = \log_a M + \log_a N$

$\log_a\left(\dfrac{M}{N}\right) = \log_a M - \log_a N$

$\log_a M^r = r \log_a M$

$\log_a M = \dfrac{\log_b M}{\log_b a} = \dfrac{\log M}{\log a} = \dfrac{\ln M}{\ln a}$

Formulas from Chapter 12

The Distance Formula

$d(P_1, P_2) = \sqrt{(x_2 - x_1)^2 + (y_2 - y_1)^2}$

The Midpoint Formula

$M = \left(\dfrac{x_1 + x_2}{2}, \dfrac{y_1 + y_2}{2}\right)$

Standard Form of an Equation of a Circle

$(x - h)^2 + (y - k)^2 = r^2$ with radius r and center (h, k)

General Form of the Equation of a Circle

$x^2 + y^2 + ax + by + c = 0$ when the graph exists

Formulas for Lines and Slope (Chapter 3)

Standard form of a line	$Ax + By = C$
Equation of a vertical line	$x = a$ where a is the x-intercept
Equation of a horizontal line	$y = b$ where b is the y-intercept
Slope of a line	$m = \dfrac{y_2 - y_1}{x_2 - x_1}$, $x_1 \neq x_2$ Slope undefined if $x_1 = x_2$
Point-slope form of a line	$y - y_1 = m(x - x_1)$
Slope-intercept form of a line	$y = mx + b$

The Rules of Exponents (Chapter 5, Chapter 9)

If a and b are real numbers and if r and s are rational numbers, then assuming the expression is defined,

Zero Exponent Rule:	$a^0 = 1$	if $a \neq 0$
Negative Exponent Rule:	$a^{-r} = \dfrac{1}{a^r}$	if $a \neq 0$
Product Rule:	$a^r \cdot a^s = a^{r+s}$	
Quotient Rule:	$\dfrac{a^r}{a^s} = a^{r-s} = \dfrac{1}{a^{s-r}}$	if $a \neq 0$
Power Rule:	$(a^r)^s = a^{r \cdot s}$	
Product to Power Rule:	$(a \cdot b)^r = a^r \cdot b^r$	
Quotient to Power Rule:	$\left(\dfrac{a}{b}\right)^r = \dfrac{a^r}{b^r}$	if $b \neq 0$
Quotient to a Negative Power Rule:	$\left(\dfrac{a}{b}\right)^{-r} = \left(\dfrac{b}{a}\right)^r$	if $a \neq 0$, $b \neq 0$

Working with Rational Expressions (Chapter 7)

Multiplying Rational Expressions	$\dfrac{a}{b} \cdot \dfrac{c}{d} = \dfrac{ac}{bd}$	$b \neq 0, d \neq 0$
Adding Rational Expressions	$\dfrac{a}{c} + \dfrac{b}{c} = \dfrac{a + b}{c}$	$c \neq 0$
Subtracting Rational Expressions	$\dfrac{a}{c} - \dfrac{b}{c} = \dfrac{a - b}{c}$	$c \neq 0$
Dividing Rational Expressions	$\dfrac{a}{b} \div \dfrac{c}{d} = \dfrac{\frac{a}{b}}{\frac{c}{d}} = \dfrac{a}{b} \cdot \dfrac{d}{c} = \dfrac{ad}{bc}$	$b \neq 0, c \neq 0,$ $d \neq 0$

Steps for Factoring (Chapter 6)

Step 1: Factor out the Greatest Common Factor (GCF), if any exists.

Step 2: Count the number of terms.

Step 3: **(a)** 2 terms

- Is it the difference of two squares? If so,
  $$A^2 - B^2 = (A - B)(A + B)$$
- Is it the difference of two cubes? If so,
  $$A^3 - B^3 = (A - B)(A^2 + AB + B^2)$$
- Is it the sum of two cubes? If so,
  $$A^3 + B^3 = (A + B)(A^2 - AB + B^2)$$

(b) 3 terms

- Is it a perfect square trinomial? If so,
  $$A^2 + 2AB + B^2 = (A + B)^2 \quad \text{or}$$
  $$A^2 - 2AB + B^2 = (A - B)^2$$
- Is the coefficient of the square term 1? If so, $x^2 + bx + c = (x + m)(x + n)$ where $mn = c$ and $m + n = b$
- Is the coefficient of the square term different from 1? If so,
  a. Use factoring by grouping
  b. Use trial and error

(c) 4 terms

- Use factoring by grouping

Step 4: Check your work by multiplying out the factored form.

Functions (Chapter 8)

- A **function** is a special type of relation where any given input, x, corresponds to only one output y. Functions can be represented through maps, sets of ordered pairs, equations, or graphs.
- **Vertical Line Test:** A set of points in the xy-plane is the graph of a function if and only if every vertical line intersects the graph in at most one point.
- The graph of a function, f, is the set of all ordered pairs $(x, f(x))$.
- When only an equation of a function is given, the **domain** of the function is the largest set of real numbers for which $f(x)$ is a real number.
- The **range** of a function is the set of all outputs of the function.